W9-AYY-018

Mass

$1\ g = 10^{-3}\ kg$
$1\ kg = 10^3\ g$
$1\ u = 1.66 \times 10^{-24}\ g = 1.66 \times 10^{-27}\ kg$
$1\ metric\ ton = 1000\ kg$

Length

$1\ nm = 10^{-9}\ m$
$1\ cm = 10^{-2}\ m = 0.394\ in.$
$1\ m = 10^{-3}\ km = 3.28\ ft = 39.4\ in.$
$1\ km = 10^3\ m = 0.621\ mi$
$1\ in. = 2.54\ cm = 2.54 \times 10^{-2}\ m$
$1\ ft = 0.305\ m = 30.5\ cm$
$1\ mi = 5280\ ft = 1609\ m = 1.609\ km$

Area

$1\ cm^2 = 10^{-4}\ m^2 = 0.1550\ in^2$
$\qquad = 1.08 \times 10^{-3}\ ft^2$
$1\ m^2 = 10^4\ cm^2 = 10.76\ ft^2 = 1550\ in^2$
$1\ in^2 = 6.94 \times 10^{-3}\ ft^2 = 6.45\ cm^2$
$\qquad = 6.45 \times 10^{-4}\ m^2$
$1\ ft^2 = 144\ in^2 = 9.29 \times 10^{-2}\ m^2 = 929\ cm^2$

Volume

$1\ cm^3 = 10^{-6}\ m^3 = 3.35 \times 10^{-5}\ ft^3$
$\qquad = 6.10 \times 10^{-2}\ in^3$
$1\ m^3 = 10^6\ cm^3 = 10^3\ L = 35.3\ ft^3$
$\qquad = 6.10 \times 10^4\ in^3 = 264\ gal$
$1\ liter = 10^3\ cm^3 = 10^{-3}\ m^3 = 1.056\ qt$
$\qquad = 0.264\ gal$
$1\ in^3 = 5.79 \times 10^{-4}\ ft^3 = 16.4\ cm^3$
$\qquad = 1.64 \times 10^{-5}\ m^3$
$1\ ft^3 = 1728\ in^3 = 7.48\ gal = 0.0283\ m^3$
$\qquad = 28.3\ L$
$1\ qt = 2\ pt = 946\ cm^3 = 0.946\ L$
$1\ gal = 4\ qt = 231\ in^3 = 0.134\ ft^3 = 3.785\ L$

Time

$1\ h = 60\ min = 3600\ s$
$1\ day = 24\ h = 1440\ min = 8.64 \times 10^4\ s$
$1\ y = 365\ days = 8.76 \times 10^3\ h$
$\qquad = 5.26 \times 10^5\ min = 3.16 \times 10^7\ s$

Angle

$1\ rad = 57.3°$
$\quad 1° = 0.0175\ rad \qquad 60° = \pi/3\ rad$
$\quad 15° = \pi/12\ rad \qquad 90° = \pi/2\ rad$
$\quad 30° = \pi/6\ rad \qquad 180° = \pi\ rad$
$\quad 45° = \pi/4\ rad \qquad 360° = 2\pi\ rad$
$1\ rev/min = \pi/30\ rad/s = 0.1047\ rad/s$

Speed

$1\ m/s = 3.60\ km/h = 3.28\ ft/s$
$\qquad = 2.24\ mi/h$
$1\ km/h = 0.278\ m/s = 0.621\ mi/h$
$\qquad = 0.911\ ft/s$
$1\ ft/s = 0.682\ mi/h = 0.305\ m/s$
$\qquad = 1.10\ km/h$
$1\ mi/h = 1.467\ ft/s = 1.609\ km/h$
$\qquad = 0.447\ m/s$
$60\ mi/h = 88\ ft/s$

Force

$1\ N = 0.225\ lb$
$1\ lb = 4.45\ N$
Equivalent weight of a mass of 1 kg
$\quad$ on Earth's surface $= 2.2\ lb = 9.8\ N$

Pressure

$1\ Pa\ (N/m^2) = 1.45 \times 10^{-4}\ lb/in^2$
$\qquad = 7.5 \times 10^{-3}\ torr\ (mm\ Hg)$
$1\ torr\ (mm\ Hg) = 133\ Pa\ (N/m^2)$
$\qquad = 0.02\ lb/in^2$
$1\ atm = 14.7\ lb/in^2 = 1.013 \times 10^5\ N/m^2$
$\qquad = 30\ in.\ Hg = 76\ cm\ Hg$
$1\ lb/in^2 = 6.90 \times 10^5\ Pa\ (N/m^2)$
$1\ bar = 10^5\ Pa$
$1\ millibar = 10^2\ Pa$

Energy

$1\ J = 0.738\ ft\cdot lb = 0.239\ cal$
$\qquad = 9.48 \times 10^{-4}\ Btu = 6.24 \times 10^{18}\ eV$
$1\ kcal = 4186\ J = 3.968\ Btu$
$1\ Btu = 1055\ J = 778\ ft\cdot lb = 0.252\ kcal$
$1\ cal = 4.186\ J = 3.97 \times 10^{-3}\ Btu$
$\qquad = 3.09\ ft\cdot lb$
$1\ ft\cdot lb = 1.36\ J = 1.29 \times 10^{-3}\ Btu$
$1\ eV = 1.60 \times 10^{-19}\ J$
$1\ kWh = 3.6 \times 10^6\ J$

Power

$1\ W = 0.738\ ft\cdot lb/s = 1.34 \times 10^{-3}\ hp$
$\qquad = 3.41\ Btu/h$
$1\ ft\cdot lb/s = 1.36\ W = 1.82 \times 10^{-3}\ hp$
$1\ hp = 550\ ft\cdot lb/s = 745.7\ W$
$\qquad = 2545\ Btu/h$

Mass–Energy Equivalents

$1\ u = 1.66 \times 10^{-27}\ kg \leftrightarrow 931.5\ MeV$
$1\ electron\ mass = 9.11 \times 10^{-31}\ kg$
$\qquad = 5.49 \times 10^{-4}\ u \leftrightarrow 0.511\ MeV$
$1\ proton\ mass = 1.673 \times 10^{-27}\ kg$
$\qquad = 1.007\ 267\ u \leftrightarrow 938.28\ MeV$
$1\ neutron\ mass = 1.675 \times 10^{-27}\ kg$
$\qquad = 1.008\ 665\ u \leftrightarrow 939.57\ MeV$

Temperature

$T_F = \frac{9}{5} T_C + 32$
$T_C = \frac{5}{9}(T_F - 32)$
$T_K = T_C + 273$

cgs Force

$1\ dyne = 10^{-5}\ N = 2.25 \times 10^{-6}\ lb$

cgs Energy

$1\ erg = 10^{-7}\ J = 7.38 \times 10^{-6}\ ft\cdot lb$

College Physics

College Physics

Fourth Edition

Jerry D. Wilson

Lander University

Anthony J. Buffa

California Polytechnic State University
San Luis Obispo

PRENTICE HALL
Upper Saddle River, NJ 07458

Library of Congress Cataloging-in-Publication Data

Wilson, Jerry D.
   College physics.— 4th ed. / Jerry D. Wilson, Anthony J. Buffa.
     p. cm.
   Includes bibliographical references and index.
   ISBN 0-13-082444-5 (Vol. I)
   1. Physics. I. Buffa, Anthony J. II. Title.
QC21.2.W548 2000
530—dc21
                                     99–12353
                                       CIP

Executive Editor: Alison Reeves
Senior Development Editor: Karen Karlin
Editor in Chief: Paul F. Corey
Assistant Vice President of Production and Manufacturing: David W. Riccardi
Executive Managing Editor: Kathleen Schiaparelli
Assistant Managing Editor: Lisa Kinne
Editor in Chief–Development: Carol Trueheart
Senior Marketing Manager: Danny Hoyt
Project Manager: Elizabeth Kell
Director of Marketing: John Tweeddale
Manufacturing Manager: Trudy Pisciotti
Director of Creative Services: Paula Maylahn
Associate Creative Director: Amy Rosen
Art Director: Amy Rosen/Joseph Sengotta
Art Manager: Gus Vibal
Art Editor: Karen Branson
Art Studio: Academy Artworks
Interior Design: Rosemarie Votta
Cover Design: Joseph Sengotta
Assistant to Art Director: John M. Christiana
Cover Photographs: U.S. skater Christine Witty from West Allis, Wisconsin, skates during the women's 1500-meter speed skating competition at the Winter Olympics in Nagano on Monday, February 16, 1998. Witty won the bronze medal. Lionel Cironneau, AP/Wide World Photos; Homer Levi Dodge (1887–1983), the first President of the American Association of Physics Teachers, and Mrs. Margaret Dodge skating on the Iowa River, ca. January 1921. American Institute of Physics/Niels Bohr Library; Bont Sonic Clap Skate. Bont Skates Pty Ltd.
Photo Editor: Melinda Reo
Photo Researcher: Beaura K. Ringrose
Editorial Assistant: Gillian Buonanno
Assistant Editor: Wendy Rivers/Amanda Griffith
Senior Media Editor: Laura Pople
Production Supervision/Composition: WestWords, Inc.

Printed in the United States of America
10 9 8 7 6 5 4 3 2 1

ISBN 0-13-082444-5

Prentice-Hall International (UK) Limited, *London*
Prentice-Hall of Australia Pty. Limited, *Sydney*
Prentice-Hall Canada Inc., *Toronto*
Prentice-Hall Hispanoamericana, S.A., *Mexico*
Prentice-Hall of India Private Limited, *New Delhi*
Prentice-Hall of Japan, Inc., *Tokyo*
Prentice-Hall (Singapore) Pte Ltd
Editora Prentice-Hall do Brasil, Ltda., *Rio de Janeiro*

About the Authors

Jerry D. Wilson   Jerry Wilson, a native of Ohio, is now Emeritus Professor of Physics and former Chair of the Division of Biological and Physical Sciences at Lander University in Greenwood, South Carolina. He received his B.S. degree from Ohio University, M.S. degree from Union College, and in 1970, a Ph.D. from Ohio University. He earned his M.S. degree while employed as a Materials Behavior Physicist by the General Electric Co.

As a doctoral graduate student, Professor Wilson held the faculty rank of Instructor and began teaching physical science courses. During this time, he co-authored a physical science text that is now in its eighth edition. In conjunction with his teaching career, Professor Wilson continued his writing and has authored or co-authored six titles. Having retired from full-time teaching, he continues to write, including *The Curiosity Corner*, a weekly column for local newspapers, which now can also be found on the Internet.

With several competitive books available, one may wonder why another algebra-based physics text was written. Having taught introductory physics many times, I was well aware of the needs of students and the difficulties they have in mastering the subject. I decided to write a text that presents the basic physics principles in a clear and concise manner, with illustrative examples that help resolve the major difficulty in learning physics: problem solving. Also, I wanted to write a text that is relevant so as to show students how physics applies in their everyday world—how things work and why things happen. Once the basics are learned, these follow naturally.

—*Jerry Wilson*

Anthony J. Buffa   Anthony Buffa received his B.S. degree in physics from Rensselaer Polytechnic Institute and both his M.S. and Ph.D. degrees in physics from the University of Illinois, Urbana–Champaign. In 1970, Professor Buffa joined the faculty at California Polytechnic State University, San Luis Obispo, where he is currently Professor of Physics, and has been a research associate with the Department of Physics Radioanalytical Facility since 1980.

Professor Buffa's main interest continues to be teaching. He has taught courses at Cal Poly ranging from introductory physical science to quantum mechanics, has developed and revised many laboratory experiments, and has taught elementary physics to local teachers in an NSF-sponsored workshop. Combining physics with his interests in art and architecture, Dr. Buffa develops his own artwork and sketches, which he uses to increase his effectiveness in teaching physics.

I try to teach my students the crucial role physics plays in understanding all aspects of the world around them—whether it be technology, biology, astronomy, or any other field. In that regard, I emphasize conceptual understanding before number crunching. To this end, I rely heavily on visual methods. I hope the artwork and other pedagogical features in this book assist you in achieving your own teaching goals for your students.

—*Tony Buffa*

Brief Contents

Contents

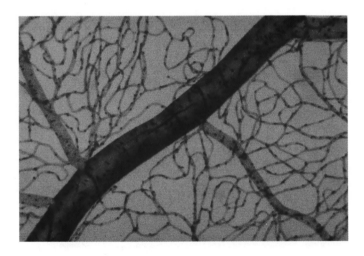

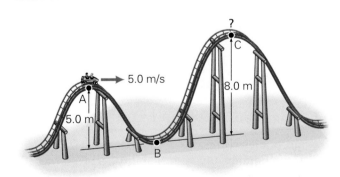

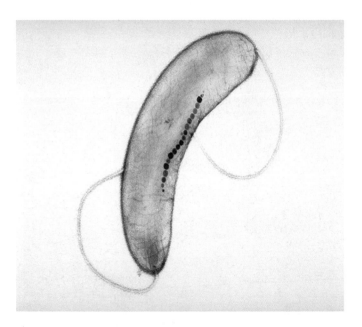

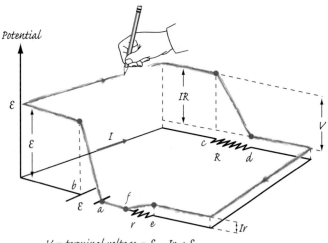

$V = terminal\ voltage = \mathcal{E} - Ir < \mathcal{E}$

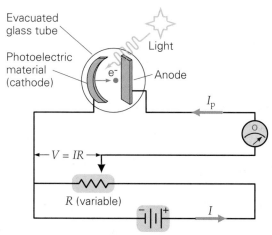

Learn by Drawing

Demonstrations

Applications (Insights in **boldface**)

Suggested Problem-Solving Procedure. An extensive section (Section 1.7) provides a framework for thinking about problem solving. This section includes:

- An overview of problem-solving strategies;
- A seven-step procedure that is general enough to apply to most problems in physics but is easily used in specific situations;
- Three Examples that illustrate the problem-solving process, showing how the general procedure is applied in practice.

Problem-Solving Strategies and Hints. The initial treatment of problem solving is followed up throughout *College Physics* with an abundance of suggestions, tips, cautions, shortcuts, and useful techniques for solving specific kinds of problems. These strategies and hints help students apply general principles to specific contexts as well as avoid common pitfalls and misunderstandings.

Conceptual Examples. *College Physics* was among the first physics text to include examples that are conceptual in nature in addition to quantitative ones. Our Conceptual Examples ask students to think about a physical situation and choose the correct prediction on the basis of an understanding of relevant principles. The discussion that follows (Reasoning and Answer) explains clearly how the correct answer can be identified as well as why the other answers are wrong.

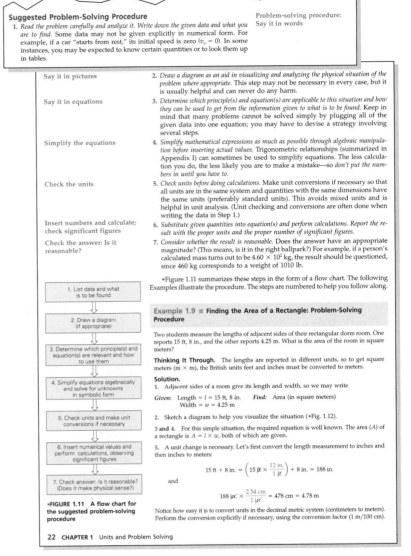

Suggested Problem-Solving Procedure

1. *Read the problem carefully and analyze it.* Write down the given data and what you are to find. Some data may not be given explicitly in numerical form. For example, if a car "starts from rest," its initial speed is zero ($v_o = 0$). In some instances, you may be expected to know certain quantities or to look them up in tables.

Problem-solving procedure:
Say it in words

Say it in pictures

2. *Draw a diagram as an aid in visualizing and analyzing the physical situation of the problem where appropriate.* This step may not be necessary in every case, but it is usually helpful and can never do any harm.

Say it in equations

3. *Determine which principle(s) and equation(s) are applicable to this situation and how they can be used to get from the information given to what is to be found.* Keep in mind that many problems cannot be solved simply by plugging all of the given data into one equation; you may have to devise a strategy involving several steps.

Simplify the equations

4. *Simplify mathematical expressions as much as possible through algebraic manipulation before inserting actual values.* Trigonometric relationships (summarized in Appendix I) can sometimes be used to simplify equations. The less calculation you do, the less likely you are to make a mistake—so *don't put the numbers in until you have to.*

Check the units

5. *Check units before doing calculations.* Make unit conversions if necessary so that all units are in the same system and quantities with the same dimensions have the same units (preferably standard units). This avoids mixed units and is helpful in unit analysis. (Unit checking and conversions are often done when writing the data in Step 1.)

Insert numbers and calculate; check significant figures

6. *Substitute given quantities into equation(s) and perform calculations. Report the result with the proper units and the proper number of significant figures.*

Check the answer: Is it reasonable?

7. *Consider whether the result is reasonable.* Does the answer have an appropriate magnitude? (This means, is it in the right ballpark?) For example, if a person's calculated mass turns out to be 4.60×10^2 kg, the result should be questioned, since 460 kg corresponds to a weight of 1010 lb.

•Figure 1.11 summarizes these steps in the form of a flow chart. The following Examples illustrate the procedure. The steps are numbered to help you follow along.

1. List data and what is to be found

2. Draw a diagram (if appropriate)

3. Determine which principle(s) and equation(s) are relevant and how to use them

4. Simplify equations algebraically and solve for unknowns in symbolic form

5. Check units and make unit conversions if necessary

6. Insert numerical values and perform calculations, observing significant figures

7. Check answer: Is it reasonable? (Does it make physical sense?)

•FIGURE 1.11 A flow chart for the suggested problem-solving procedure

Example 1.9 ■ **Finding the Area of a Rectangle: Problem-Solving Procedure**

Two students measure the lengths of adjacent sides of their rectangular dorm room. One reports 15 ft, 8 in., and the other reports 4.25 m. What is the area of the room in square meters?

Thinking It Through. The lengths are reported in different units, so to get square meters (m × m), the British units feet and inches must be converted to meters.

Solution.
1. Adjacent sides of a room give its length and width, so we may write

Given: Length $= l = 15$ ft, 8 in. *Find:* Area (in square meters)
Width $= w = 4.25$ m

2. Sketch a diagram to help you visualize the situation (•Fig. 1.12).

3 and 4. For this simple situation, the required equation is well known. The area (A) of a rectangle is $A = l \times w$, both of which are given.

5. A unit change is necessary. Let's first convert the length measurement to inches and then inches to meters:

$$15 \text{ ft} + 8 \text{ in.} = \left(15 \text{ ft} \times \frac{12 \text{ in.}}{1 \text{ ft}}\right) + 8 \text{ in.} = 188 \text{ in.}$$

and

$$188 \text{ in.} \times \frac{2.54 \text{ cm}}{1 \text{ in.}} = 478 \text{ cm} = 4.78 \text{ m}$$

Notice how easy it is to convert units in the decimal metric system (centimeters to meters). Perform the conversion explicitly if necessary, using the conversion factor (1 m/100 cm).

22 CHAPTER 1 Units and Problem Solving

More Explanation in Examples. Too many solutions to worked examples in other texts rely on formulas such as "From Eq. 6.7 we have. . . ." We have tried to make the solutions to in-text Examples as clear, patient, and detailed as possible. The aim is not merely to show students which equations to use but to explain the strategy being employed and the role of each step in the overall plan. Students are encouraged to learn the "why" of each step along with the "how." This technique will make it easier for students to apply the demonstrated techniques to other problems that are not identical in structure.

Thinking It Through. New to the Fourth Edition, every worked Example now includes a "Thinking It Through" section after the problem statement and before the Solution to focus students on the critical thinking and analysis they should do before beginning to use equations.

Example 4.5 ■ A Braking Car: Finding a Force From Motional Effects

A car traveling at 72.0 km/h along a straight, level road is brought uniformly to a stop in a distance of 40.0 m. If the car weighs 8.80×10^3 N, what is the braking force?

Thinking It Through. It is noted that the car's velocity changes, so there is an acceleration. Given a distance, we might surmise that first the acceleration is found by using a kinematic equation, and then the acceleration is used to compute the force.

Solution. In bringing the car to a stop, the braking force caused an acceleration (actually a deceleration), as illustrated in •Fig. 4.13. Listing what is given and what we must find, we have

Given: $v_o = 72.0$ km/h = 20.0 m/s $\quad$ *Find:* F_b (braking force)
$\qquad\quad v = 0$
$\qquad\quad x = 40.0$ m
$\qquad\quad w = 8.80 \times 10^3$ N

(F_b is net force)

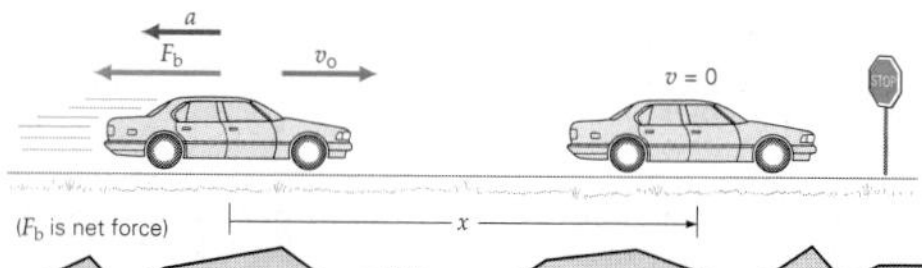

•**FIGURE 4.13 Finding force from motional effects** See Example 4.5.

Follow-up Exercises. Follow-up Exercises at the end of each Conceptual Example and each regular worked Example further reinforce the importance of conceptual understanding and offer additional practice. (Answers to Follow-up Exercises are given at the back of the book.)

Chapter Review. Each Chapter Review is made up of three parts:

1. *Important Terms:* A listing, with page references, of the key terms introduced in the chapter that students should be able to define and explain.
2. *Important Concepts:* A summary of the key principles of each chapter.
3. *Important Equations:* A listing, cross-referenced to the equations in the chapter, of the major laws and mathematical relationships introduced. Specific applicability and limiting conditions are clearly stated for each expression.

---

10. If your professor's car is traveling west with a constant velocity of 55 mi/h west on a straight highway, what is the net force acting on it?

11. ■ Which has more inertia, 20 cm³ of water or 10 cm³ aluminum, and how many times more? (See Table 9.2.)

12. ■■ ◯ You are told that an object has zero acceleration. Two forces on the object are $F_1 = 3.6$ N at 74° below the $+x$ axis and $F_2 = 3.6$ N at 34° above the $-x$ axis. Is there a third force on the object? If yes, what is it? Can you tell whether the object is at rest or in motion?

13. ■■ ◯ A 5.0-kilogram block at rest on a frictionless surface is acted on by forces $F_1 = 5.5$ N and $F_2 = 3.5$ N, as illustrated in •Fig. 4.30. What additional horizontal force will keep the block at rest?

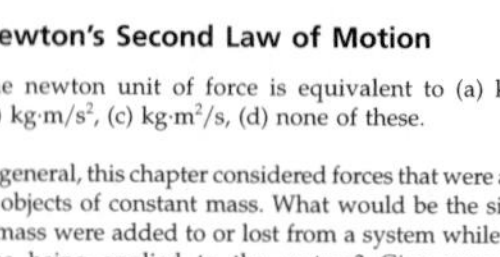

•**FIGURE 4.30 Two applied forces** See Exercises 13 and 86.

14. ■■■ ◯ A 1.5-kilogram object moves up the y axis with a constant speed. When it reaches the origin, the forces $F_1 = 5.0$ N at 37° above the $+x$ axis, $F_2 = 2.5$ N in the $+x$ direction, $F_3 = 3.5$ N at 45° below the $-x$ axis, and $F_4 = 1.5$ N in the $-y$ direction are applied to it. (a) Will the object continue to move along the y axis? (b) If not, what simultaneously applied force will keep it moving along the y axis with a constant speed?

4.3 Newton's Second Law of Motion

15. The newton unit of force is equivalent to (a) kg·m/s, (b) kg·m/s², (c) kg·m²/s, (d) none of these.

16. In general, this chapter considered forces that were applied to objects of constant mass. What would be the situation if mass were added to or lost from a system while a force was being applied to the system? Give examples of situations in which this might happen.

17. Good football wide receivers usually have "soft" hands for catching balls (•Fig. 4.31). How would you interpret this description on the basis of Newton's second law?

18. ■ A 3.0-newton net force is applied to a 1.5-kilogram mass. What is its acceleration?

19. ■ What is the mass of an object that accelerates at 3.0 m/s² under the influence of a 5.0-newton net force?

20. ■ A worker pushes on a crate, which experiences a net force of 75 N. If the crate also experiences an acceleration of 0.50 m/s², what is its weight?

•**FIGURE 4.31 Soft hands** See Exercise 17.

21. ■ An ocean tanker has a gross mass of 7.0×10^7 kg. What constant net force would give the tanker an acceleration of 0.10 m/s²?

22. ■ A 6.0-kilogram object is brought to the Moon, where the acceleration due to gravity is only one-sixth of that on the Earth. What is the mass of the object on the Moon? (a) 1.0 kg (b) 6.0 kg (c) 59 kg (d) 9.8 kg

23. ■ What is the mass of a person weighing 740 N?

24. ■ What is the net force acting on a 1.0-kilogram object in free fall?

25. ■ What is the weight of an 8.0-kilogram mass in newtons? How about in pounds?

26. ■■ What is the weight of a 150-pound person in newtons? What is his mass in kilograms?

27. ■■ (a) Is the product label shown in •Fig. 4.32 correct on the Earth? (b) Is it correct on the Moon, where the acceleration due to gravity is only about one-sixth of that on Earth? If not, what should the label be on the Moon?

San Giorgio
Rippled Edge
Lasagne®
AN ENRICHED MACARONI PRODUCT
NET WT. 16 OZ. (1 LB.) 454 g

•**FIGURE 4.32 Correct label?** See Exercise 27.

Exercises. Each chapter ends with a wealth of Exercises, organized by chapter section and ranked by general level of difficulty. In addition, the Exercises offer the following special features to help students refine both their conceptual understanding and their problem-solving skills:

• *Integration of Conceptual and Quantitative Exercises.* To help break down the artificial and ultimately counterproductive barrier between conceptual questions and quantitative problems, we do not distinguish between these categories in the end-of-chapter Exercises. Instead, each section begins with a series of multiple-choice and short-answer questions that provide content review, test conceptual understanding, and ask students to reason from principles. The aim is to show students that the same kind of conceptual insight is required regardless of whether the desired answer involves words, equations, or numbers. The conceptual or "thought" questions are marked by a bold **TQ** in the *Annotated Instructor's Edition* of the text for easy reference when assigning questions. Unlike most other texts, *College Physics* offers short answers to all odd-numbered conceptual questions (as well as to all odd-numbered quantitative problems) in the back of the text so that students can check their understanding. About 35% of all Thought Questions and Exercises in the Fourth Edition are new.

- **Interactive Exercises.** New to the Fourth Edition, many of the end-of-chapter Exercises are keyed to simulations on Prentice Hall's multimedia study guide, the *Interactive Journey through Physics*. Exercises that have a corresponding simulation are indicated with a CD-ROM icon. ● The *College Physics Media Pack* (ISBN 0-13-085346-1), a specially discounted package consisting of the text and *Interactive Journey through Physics* CD-ROM, includes a cross-reference/location guide to allow you to match Exercise numbers to corresponding simulations.

- **Paired and Trio Exercises.** Most numbered sections include at least one set of paired Exercises and, new to the Fourth Edition, one set of trio Exercises, that deal with similar situations. The first problem in a pair or trio is solved in the Study Guide; the second problem, exploring a similar situation, has only an answer at the back of the book, thereby encouraging students to work out the problem on their own. The third problem in a trio is answered in the *Student Study Guide and Solutions Manual*.

- **Additional Exercises.** Each chapter includes a supplemental section of Additional Exercises drawn from all sections of the chapter to ensure that students can synthesize concepts.

The Absolutely Zero Tolerance for Errors Club (The AZTECs). This team approach to accuracy checking worked quite well in the third edition, so we did it again. Bo Lou of Ferris State University, the author of our *Instructor's Solutions Manual*, headed the AZTEC team and was supported by the text authors and two additional accuracy checkers, Bill McCorkle of West Liberty State University and Dave Curott of the University of North Alabama. Each member of the team individually and independently worked all end-of-chapter Exercises. The results were then collected, and any discrepancies were resolved by a "team" discussion. All data in the chapters, as well as the answers at the back of the book, were checked and rechecked in first and second page proofs. In addition, two other physics teachers—J. Erik Hendrickson and K. W. Nicholson—read all first pages in detail, checking for errors in the chapter narrative and text art. Although it is probably not humanly possible to produce a physics text with absolutely no errors, that was our goal; we worked very hard to make the book as error-free as we could.

New Multimedia Explorations of Physics

New to the Fourth Edition are a state-of-the art Website and a CD-ROM media package.

Companion Website. Our Website (at http://www.prenhall.com/wilson), with contributions from leaders in physics education research, provides students with a variety of interactive explorations of each chapter's topics, easily accommodating differences in learning styles. Student tools include Warm-Ups, Puzzles, and "What Is Physics Good For?" applications by Gregor Novak and Andy Gavrin (Indiana University–Purdue University, Indianapolis); award-winning Java-based Physlet problems by Wolfgang Christian (Davidson College); algorithmically generated numerical Practice Problems, multiple-choice Practice Questions, on-line Destinations, and Net Search key words by Carl Adler (East Carolina University); Ranking Task Exercises edited by Tom O'Kuma (Lee College), David Maloney (Indiana University–Purdue University, Fort Wayne) and Curtis Hieggelke (Joliet Junior College); downloadable PDF files for a Mechanics Problem-Solving Workbook by Dan Smith (South Carolina State University); and MCAT Questions by Glen Terrell (University of Texas at Arlington) and from ARCO's MCAT Supercourse. Using the Preferences module at the opening of the site or the tool in the "Results reporter" part of each module, students can, at a professor's request, have the results of their work on the Companion Website e-mailed to the professor or teaching assistant. Instructor tools include on-line grading capabilities and a Syllabus Manager. See pp. xxviii–xxiv for further information about the modules in this site.

Media Pack for College Physics, Fourth Edition (0-13-085346-1). *College Physics, Fourth Edition* can be purchased in a specially discounted package called the Media Pack, which includes the student text, the dual-platform *Interactive Journey through Physics* (*IJTP*) CD-ROM by Cindy Schwarz (Vassar College) and Logal, Inc., *Science on the Internet: A Student's Guide, 1999* by Andrew Stull, and a cross-reference/location guide to correlate Exercises in the text marked with a CD-ROM icon and corresponding simulations on the *IJTP* CD-ROM. This CD-ROM is a multimedia study guide for physics, with simulations, animations, videos, hyperlinked topic reviews, MCAT review questions and problems (including Ranking Task Exercises, context-rich problems, and video problems) for mechanics, thermodynamics, electricity and magnetism, and light and optics. It also includes a built-in scientific calculator with a library of key physics constants, a glossary, and pertinent tables and equations. See pp. xxx–xxxi for a complete description of the types of materials on the *IJTP* CD-ROM.

Additional Supplements

The pedagogical value of *College Physics* is enhanced by a variety of supplements developed to address the needs of both students and instructors.

For the Instructor

Annotated Instructor's Edition (0-13-084167-6). The margins of the *Annotated Instructor's Edition* (*AIE*) contain an abundance of suggestions for classroom demonstrations and activities, along with teaching tips (points to emphasize, discussion suggestions, and common misunderstandings to avoid). In addition, the *AIE* contains:

- Icons that identify each illustration reproduced as a transparency in the *Transparency Pack*.
- Answers to end-of-chapter Exercises (following each Exercise).
- References to applicable video demonstrations from the *Physics You Can See* videotape.

Instructor's Solutions Manual (0-13-084168-4). Prepared by Bo Lou of Ferris State University, the *Instructor's Solutions Manual* supplies answers with complete, worked-out solutions to all end-of-chapter Exercises. Each solution has been checked for accuracy by a minimum of five instructors. This manual is also available electronically on both Windows and Macintosh platforms.

Test Item File (0-13-084160-9). Fully revised by Dave Curott of the University of North Alabama, the *Test Item File* now offers more than 2300 questions—approximately 30% of them new to this edition—and includes several new conceptual questions per chapter. The questions are now organized and referenced by type and by section.

Prentice Hall Custom Test (Windows: 0-13-084171-4; Macintosh: 0-13-084172-2). Based on the powerful testing technology developed by Engineering Software Associates, Inc. (ESA), the *Prentice Hall Custom Test* allows instructors to create and tailor exams to their own needs. With the On-line Testing Program, exams can also be administered on-line and data can then be automatically transferred for evaluation. A comprehensive desk reference guide is included, along with on-line assistance.

Transparency Pack (0-13-084175-7). The *Transparency Pack* contains more than 300 full-color acetates of text illustrations useful for class lectures. It is available upon adoption of the text.

Physics You Can See Video Demonstrations (0-205-12393-7). Each segment, 2–5 minutes long, demonstrates a classical physics experiment. Eleven segments are included, such as "Coin & Feather" (acceleration due to gravity), "Monkey & Gun" (rate of vertical free fall), "Swivel Hips" (force pairs), and "Collapse a Can" (atmospheric pressure).

Presentation Manager CD-ROM (0-13-084174-9). This new CD-ROM contains all the text art and videos from the *Physics You Can See* videotape as well as additional lab and demonstration videos and animations from the *Interactive Journey through Physics* CD-ROM, which is also available from Prentice Hall (see below).

Just-in-Time Teaching: Blending Active Learning with Web Technology (0-13-085034-9). Just-in-Time Teaching (JiTT) is an exciting new teaching and learning methodology designed to engage students. Using feedback from pre-class Web assignments, instructors can adjust classroom lessons so that students receive rapid response to the specific questions and problems they are having—instead of more generic lectures that may or may not address topics with which students actually need help. Many teachers have found that this process makes students become active and interested learners. In this resource book for educators, authors Gregor Novak (Indiana University–Purdue University, Indianapolis), Evelyn Patterson (United States Air Force Academy), Andrew Gavrin (Indiana University–Purdue University, Indianapolis), and Wolfgang Christian (Davidson College) more fully explain what Just-in-Time Teaching is, its underlying goals and philosophies, and how to implement it. They also provide an extensive section of tested resource materials that can be used in introductory physics courses with the JiTT approach.

For the Student

Student Study Guide and Solutions Manual (0-13-084365-2). Significantly revised by Bo Lou of Ferris State University, the *Student Study Guide and Solutions Manual* presents chapter-by chapter reviews, chapter summaries, key terms, additional worked problems, and solutions to selected problems.

The New York Times "Themes of the Times" Program. This innovative program, made possible through an exclusive partnership between Prentice Hall and *The New York Times*, brings current and relevant applications into the classroom. Through this program, adopters of *College Physics, Fourth Edition* are eligible to receive our free, unique "mini-newspapers," which bring together a collection of the latest and best physics articles from the highly respected pages of *The New York Times*. They are updated annually and are free to qualified adopters up to the quantity of texts purchased. Contact your local representative for ordering.

MCAT Physics Study Guide (0-13-627951-1). This study resource by Joseph Boone of California Polytechnic State University–San Luis Obispo references all of the physics topics on the MCAT to the appropriate sections in the text. Since most MCAT questions require more thought and reasoning than simply plugging numbers into an equation, this study guide is designed to refresh students' memory about the topics they've covered in class. Additional review, practice problems, and review questions are included.

Science on the Internet: A Student's Guide, 1999 (0-13-021308-X). This guide helps students gain a greater understanding of the Internet and the ways in which they can access information on the Web relating to their study of physics.

Other Related Multimedia Materials

Interactive Physics II Player Workbook (Windows: 0-13-667312-0; Macintosh: 0-13-477670-4). Written by Cindy Schwarz of Vassar College, this highly interactive workbook/software package contains simulation projects of varying difficulty. Each includes a physics review, simulation details, hints, an explanation of results, math help, and a self-test.

Interactive Journey through Physics CD-ROM (0-13-254103-3). This highly interactive CD-ROM can be used as a stand-alone supplement for any introductory physics course or as a general reference tool. Through simulation, animation, video, and interactive problem solving, students can visualize difficult physics concepts in ways not available through the traditional lecture, lab, and text. The innovative concept checks and extension exercises within the simulations facilitate the reinforcement of important physical concepts. The content of this CD-ROM is organized according to the main topics in physics—mechanics, electricity and magnetism, thermodynamics, and light and optics. The numerous analysis tools are easily navigated through a user-friendly interface. See the previous description under Media Pack and pp. xxx–xxxi for more detail about the *IJTP CD-ROM*, and see ordering information for the discounted price when purchased with *College Physics, Fourth Edition*, in the Media Pack.

Acknowledgments

We would like to acknowledge the generous assistance we received from many people during the preparation of the Fourth Edition. First, our sincere thanks go to Bo Lou of Ferris State University for his vital contributions and meticulous, conscientious help with checking problem solutions and answers, preparing the *Instructor's Solutions Manual* as well as the answer keys for the back of the book, and preparing a major revision of the *Student Study Guide and Solutions Manual*. We are similarly grateful to Dave Curott of the University of North Alabama for preparing the *Test Item File* as well as for his participation as an accuracy checker for all solutions to end-of-chapter Exercises.

Indeed, all the members of AZTEC—Bo Lou, Dave Curott, and Bill McCorkle (West Liberty State University)—as well as the reviewers of first and second page proofs—Xiaochun He (Georgia State University), J. Erik Hendrickson (University of Wisconsin–Eau Claire), K. W. Nicholson (Central Alabama Community College), Peter Shull (Oklahoma State University), and Lorin Vant-Hull (University of Houston)—deserve more than a special thanks for their tireless, timely, and extremely thorough review of all materials in the book for scientific accuracy.

Dozens of other colleagues, listed next helped us with reviews of the Third Edition to help us plan the Fourth Edition as well as with reviews of manuscript as it was developed. We are indebted to them, as their thoughtful and constructive suggestions benefited the book greatly.

At Prentice Hall, the editorial staff continued to be particularly helpful. We are especially grateful to Karen Karlin, Senior Development Editor, for everything— her cheerful competence, experienced hand, insight, creativity, extreme attention to detail, and even her queries, all of which have helped make this book one of the most carefully crafted introductory physics texts available. Pat McCutcheon, Project Manager for the book, and Lisa Kinne, Assistant Managing Editor, kept the whole complex endeavor moving forward, while designers Joseph Sengotta and Rosemarie Votta made sure the ultimate physical presentation would be both visually engaging and clean and easy to use. We also thank Danny Hoyt, Senior Marketing Manager, for his cheerful enthusiasm; Wendy Rivers for her work on the supplements; Alison Reeves, Executive Editor, and Gillian Buonanno, Editorial Assistant, for their help in coordinating all of these facets; and Paul Corey, Editor in Chief, for his support and encouragement.

In addition, I (Tony Buffa) once again extend my thanks to my coauthor, Jerry Wilson, for his helpfulness and professional approach to the work at hand during this revision. As always, several colleagues of mine at Cal Poly gave unselfishly of their time for fruitful discussions. Among them are Professors Joseph Boone, Ronald Brown, Theodore Foster, and Richard Frankel. My family—my wife, Connie, and daughters, Jeanne and Julie—was a continuous source of support and inspiration that helped me through the hectic parts of the schedule. I also acknowledge my recently deceased mother, Florence, for her help and advice over the years.

Finally, both of us would like to urge anyone using the book—student or instructor—to pass on to us any suggestions that you have for its improvement. We look forward to hearing from you.

—*Jerry D. Wilson*
—*Anthony J. Buffa*

Reviewers of the Fourth Edition

Anand Batra
Howard University

Jeffrey Braun
University of Evansville

Aaron Chesir
Lucent Technologies

Lattie Collins
East Tennessee State University

James Cook
Middle Tennessee State University

James Ellingson
College of DuPage

John Flaherty
Yuba College

Rex Gandy
Auburn University

Xiaochun He
Georgia State University

J. Erik Hendrickson
University of Wisconsin–Eau Claire

Jacob W. Huang
Towson University

John Kenny
Bradley University

Dana Klinck
Hillsborough Community College

Chantana Lane
University of Tennessee–Chattanooga

R. Gary Layton
Northern Arizona University

Mark Lindsay
University of Louisville

Dan MacIsaac
Northern Arizona University

Trecia Markes
University of Nebraska–Kearney

Aaron McAlexander
Central Piedmont Community College

William McCorkle
West Liberty State University

Michael McGie
California State University–Chico

Gerhard Muller
University of Rhode Island

K. W. Nicholson
Central Alabama Community College

William Pollard
Valdosta State University

David Rafaelle
Glendale Community College

Robert Ross
University of Detroit–Mercy

Gerald Royce
Mary Washington College

Peter Shull
Oklahoma State University

Larry Silva
Appalachian State University

Christopher Sirola
Tri-County Technical College

Soren P. Sorensen
University of Tennessee–Knoxville

Frederick Thomas
Sinclair Community College

Jacqueline Thornton
St. Petersburg Junior College

Anthony Trippe
ITT Technical Institute–San Diego

Lorin Vant-Hull
University of Houston

Kevin Williams
ITT Technical Institute–Earth City

Linda Winkler
Appalachian State University

John Zelinsky
Southern Illinois University

Reviewers of Previous Editions

William Achor
West Maryland College

Arthur Alt
College of Great Falls

Zaven Altounian
McGill University

Frederick Anderson
University of Vermont

Charles Bacon
Ferris State College

Ali Badakhshan
University of Northern Iowa

William Berres
Wayne State University

Hugo Borja
Macomb Community College

Bennet Brabson
Indiana University

Michael Browne
University of Idaho

David Bushnell
Northern Illinois University

Lyle Campbell
Oklahoma Christian University

Lowell Christensen
American River College

Philip A. Chute
University of Wisconsin–Eau Claire

Lawrence Coleman
University of California–Davis

Lattie F. Collins
East Tennessee State University

David M. Cordes
Belleville Area Community College

James R. Crawford
Southwest Texas State University

William Dabby
Edison Community College

J. P. Davidson
University of Kansas

Donald Day
Montgomery College

Richard Delaney
College of Aeronautics

Arnold Feldman
University of Hawaii

Rober J. Foley
University of Wisconsin–Stout

Donald Foster
Wichita State University

Donald R. Franceschetti
Memphis State University

Frank Gaev
ITT Technical Institute–Ft. Lauderdale

Simon George
California State–Long Beach

Barry Gilbert
Rhode Island College

Richard Grahm
Ricks College

Tom J. Gray
University of Nebraska

Douglas Al Harrington
Northeastern State University

Al Hilgendorf
University of Wisconsin–Stout

Joseph M. Hoffman
Frostburg State University

Omar Ahmad Karim
University of North Carolina–Wilmington

S.D. Kaviani
El Camino College

Victor Keh
ITT Technical Institute–Norwalk, California

James Kettler
Ohio University, Eastern Campus

Phillip Laroe
Carroll College

Rubin Laudan
Oregon State University

Bruce A. Layton
Mississippi Gulf Coast Community College

Federic Liebrand
Walla Walla College

Bryan Long
Columbia State Community College

Michael LoPresto
Henry Ford Community College

Robert March
University of Wisconsin

John D. McCullen
University of Arizona

Gary Motta
Lassen College

J. Ronald Mowrey
Harrisburg Area Community College

R. Daryl Pedigo
Austin Community College

New Multimedia Explorations of Physics

Companion Website www.prenhall.com/wilson

This exciting text-specific interactive Website for introductory physics provides students and instructors with a wealth of innovative on-line materials for use with *College Physics, Fourth Edition.*

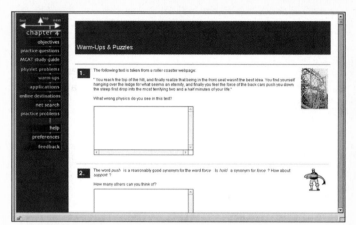

Warm-Ups & Puzzles by Gregor Novak and Andrew Gavrin (Indiana University–Purdue University, Indianapolis)

Warm-Up and Puzzle questions are real-world short-answer questions based on important concepts in the text chapters. Both types of questions get students' attention, often refer to current events, and are good discussion starters. The **Warm-Ups** are designed to help introduce a topic, whereas the **Puzzles** are more complex and often require the integration of more than one concept. Thus, professors can assign Warm-Up questions after students have read a chapter but before the class lecture on that topic, and Puzzle questions as follow-up assignments submitted after class.

Applications by Gregor Novak and Andrew Gavrin (Indiana University–Purdue University, Indianapolis)

The **Applications** modules answer the question "What is physics good for?" by connecting physics concepts to real-world phenomena and new developments in science and technology. These illustrated essays include embedded Web links to related sites, one for each chapter. Each essay is followed by short-answer/essay questions, which professors can assign for extra credit.

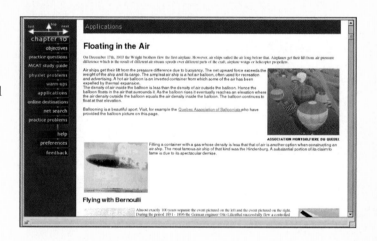

Physlet Problems by Wolfgang Christian (Davidson College)

Physlet Problems are multimedia-focused problems based on Wolfgang Christian's award-winning Java applets for physics, called *Physlets*. With these problems, students use multimedia elements to help solve a problem by observing, applying appropriate physics concepts, and making measurements of parameters they deem important. No numbers are given, so students are required to consider a problem qualitatively instead of plugging in formulas.

Practice Questions by Carl Adler
(East Carolina University)
Twenty-five to 30 multiple-choice **Practice Questions** are available for review and drill with each chapter.

Practice Problems by Carl Adler
(East Carolina University)
Ten algorithmically generated numerical **Practice Problems** per chapter allow students to get multiple iterations of each problem set for practice.

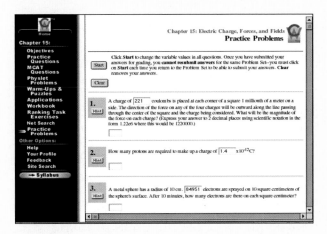

Mechanics Problem-Solving Workbook
by Dan Smith (South Carolina State University)
The **Mechanics Problem-Solving Workbook** is a self-contained workbook designed for student self-study and is available from the Wilson/Buffa Website in downloadable PDF files. It helps students explore issues of central importance in mechanics by beginning with conceptual solutions to problems and then building slowly to final results. Each topic has between 4 and 10 problems that students work out in successive stages, beginning with a conceptual description of the problem and ending with algebraic analyses. Students must complete each stage before proceeding to the next stage, thereby ensuring that they understand the physics before using the algebra.

Ranking Task Exercises
edited by Tom O'Kuma (Lee College), David Maloney (Indiana University–Purdue University, Fort Wayne), and Curtis Hieggelke (Joliet Junior College)
Available for most text chapters as PDF files downloadable from the Wilson/Buffa Website, **Ranking Tasks** are conceptual exercises that require students to rank a number of situations or variations of a situation. Engaging in this process of making comparative judgments helps students reason about physical situations and often gives them new insights into relationships among various concepts and principles.

MCAT Study Guide by Glen Terrell
(University of Texas at Arlington) and
ARCO's MCAT Supercourse
For all relevant chapters, the MCAT Study Guide module provides students with an average of 25 multiple-choice questions on topics and concepts covered on the MCAT exam. As with all multiple-choice modules, the computer automatically grades and scores student responses and provides cross-references to corresponding text sections.

Destinations
Destinations are links to relevant Websites for each chapter, either about the physics topic in the chapter or about related applications.

Net Search
The **Net Search** feature automatically configures key words according to the key word search conventions of the top three Internet search engines.

Syllabus Manager
Wilson/Buffa's **Syllabus Manager** provides instructors with an easy, step-by-step process for creating and revising a class syllabus with direct links to the text's Companion Website and other on-line content. Through this on-line syllabus, instructors can add assignments and send announcements to the class with the click of a button. The completed syllabus is hosted on Prentice Hall's servers, allowing the syllabus to be updated from any computer with Internet access.

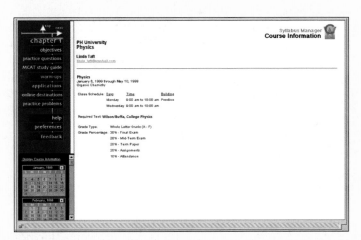

On-line Grading
Scoring for all objective questions and problems, as well as responses to essay questions, can be e-mailed back to the instructor.

Media Pack for *College Physics, Fourth Edition*

To help your students succeed in physics, the Fourth Edition includes end-of-chapter Exercises that have corresponding simulations on the *Interactive Journey through Physics* CD-ROM. These problems are marked with a CD-ROM icon, and a key in the front of the textbook gives the location of the relevant simulation on the CD-ROM. The **Media Pack** version of *College Physics, Fourth Edition* (ISBN 0-13-085346-1) includes the textbook, the *Interactive Journey through Physics* CD-ROM, and *Science on the Internet: A Student's Guide, 1999* by Andrew Stull. Encourage your students to explore the world of physics further with the interactive learning tools found on the CD-ROM:

Interactive Journey through Physics

Whether your students are exploring concepts, improving their grades, or reviewing for a test such as the MCAT, this interactive CD-ROM augments the traditional learning experiences of lecture, lab, and text.

Four dynamic components—simulation, animation, video, and interactive problem solving—enable students to visualize and interact with concepts in ways not possible in traditional learning programs. The content of the CD-ROM is organized according to the main topics in physics—**Mechanics, Electricity and Magnetism, Thermodynamics,** and **Light and Optics.** Although the CD is designed to be interactive, instructors can also use any of the simulations, animations, or videos to add multimedia to their lectures.

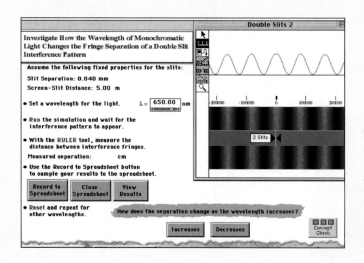

Topic Coverage

I. Mechanics. *Overview:* **1D-2D Motion**—Horizontal motion. Free fall. Projectile motion. Boats. Airplanes. Inclined planes. Frames of reference. **Forces**—Inclined planes. Collisions. Friction. Vector components. Kinds of forces. Elevators. **Energy**—Kinetic energy. Potential energy. Work. Work/energy connections. Friction. **Rotation**—Circular motion. Planets. Rotation and translation. Angular momentum. Circular motion II. Vibrations and Waves—Simple harmonic motion. Pendulums. Standing waves. Sound waves. Doppler effect. **II. Thermodynamics.** *Overview:* **Gas Laws**—*PV* plots. *PT* Plots. *TV* plots. $PV = nRT$. **Temperature Changes**—Temperature scales. Calorimetry. Effects on solids. Kinetic theory. **First Law**—First law. Special processes. Cycles. Carnot cycle. **III. Electricity & Magnetism.** *Overview:* **Circuits**—Ohm's law. Resistors. Capacitors. RC circuits. **Fixed charges**—Coloumb's law. Electric field. Electric potential. Moving Charges—Constant *E* field. Constant *B* field. Through a potential. *E* and *B* fields. Induced EMF. **IV. Light & Optics.** *Overview:* **Reflection**—Law of reflection. Concave mirror. Convex mirror. Mystery mirrors. Mirror equation. **Refraction**—Snell's law. Critical angle. Index of refraction. Converging lenses. Diverging lenses. Prisms. Lens equation. **Diffraction**—Principle. Single slit. Gratings. **Interference**—Double slit. Thin films.

SIMULATE

Students can ask questions and alter the parameters of experiments to discover the answers. More than 120 simulations of physical systems are provided, with Concept Check questions, built-in spreadsheets, and Extensions for further exploration.

SOLVE

A collection of 231 interactive problems (including "content-rich," ranking-task, and video problems) helps students build critical problem-solving skills. Also featured are 149 MCAT review questions that are based on ARCO's bestselling MCAT review guides and offer pre-meds review both during and after their physics course.

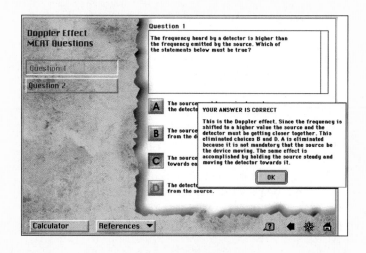

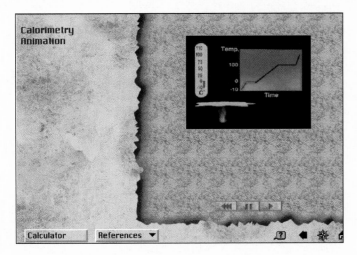

ANIMATE

Thirty-nine narrated animations illustrate physics principles at work, guiding students through explorations of different physical phenomena. This CD-ROM also includes animations that allow students to observe a simulation in an automated mode with audio descriptions to clarify the content.

VISUALIZE

Sixty-two narrated video segments allow students to view actual laboratory experiments and demonstrations. Students can use a special video analysis tool to conduct "video-based labs," a research-proven way to help students make mental links between a physical event and its graphical representation.

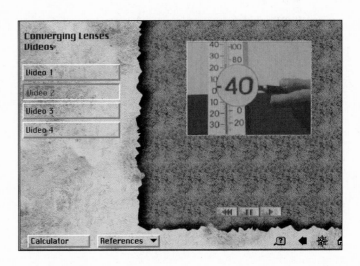

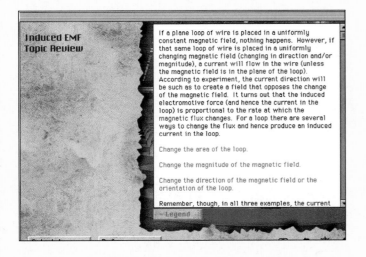

REVIEW

For a brief overview of concepts, students can access a written review of the subtopics covered on the CD-ROM. Hyperlinks let students check definitions, reference equations, or run a related animation or video in the context of the written review.

Units and Problem Solving

Not everyone has had the experience of laying out a course on a marine chart, as in the photo above. But most people have done something quite similar with a road map—plotting the shortest distance between two cities, measuring it, and estimating how long the trip will take. You might be planning a 500-mile trip; traveling at an average speed of 50 mi/h, the trip would take about 10 hours. Knowing your car's fuel economy, you might also estimate the amount of gas you expect to need and what it is likely to cost. (At 25 mpg and $1.25 per gallon, how much should you be prepared to spend?)

Measurement and problem solving are not confined to science—they are part of our lives. But they play a particularly central role in our attempts to describe and understand the physical world, as we shall see in this chapter.

1.1 Why and How We Measure

Objective: **To distinguish standard units and systems of units.**

Imagine that someone is giving you directions to her house. Would you find it helpful to be told, "Drive along Elm Street for a little while and turn right at one of the lights, then keep going for quite a long way"? Or would you want to deal with a bank that sent you a statement at the end of the month saying, "You still have some money left in your account. Not a great deal, though."?

Measurement is important to all of us. It is one of the concrete ways in which we deal with our world. This is particularly true in physics. *Physics is concerned with the description and understanding of nature,* and measurement is one of its most important tools.

There are certainly ways of describing the physical world that do not involve measurement. For instance, we might talk about the color of a flower or a dress. But color perception is subjective: It may vary from one person to another. Indeed, many people are color blind and cannot tell certain colors apart. Light can also be described in terms of wavelengths and frequencies. Different wavelengths are associated with different colors because of the physiological response of our eyes to light. But unlike the sensations or perceptions of color, the wavelengths can be measured. They are the same for everyone. In other words, they are *objective*. *Physics attempts to describe nature in an objective way through measurements.*

Standard Units

Definition of:
Standard unit

Systems of units

Measurements are expressed in unit values, or units. As you are probably aware, a large variety of units are used to express measured values. Some of the earliest units of measurement, such as the foot, were originally referenced to parts of the human body. (Even today, the hand is still used as a unit to measure the height of horses.) If a unit becomes officially accepted, it is called a **standard unit**. Traditionally, a government or international body establishes standard units.

A group of standard units and their combinations is called a **system of units**. Two major systems of units are in use today—the metric system and the British system. The latter is still widely used in the United States but has virtually disappeared in the rest of the world, having been replaced by the metric system.

Different units in the same system or units from different systems can be used to describe the same thing. For example, your height can be expressed in inches, feet, centimeters, meters—or even miles, for that matter (although this would not be a very *convenient* unit). It is always possible to convert from one unit to another, however, and this is sometimes necessary.

You may be surprised at how easy it is to determine very basic facts about our world through simple measurements and calculations that almost anyone can perform. Some 2200 years ago, a Greek scientist, using only some basic facts of geometry, was able to calculate the size of the Earth with considerable accuracy (see the Insight on p. 3).

Modern science may utilize more sophisticated mathematics and technology, but the approach is still the same. Much of our knowledge rests on a foundation of ingenious measurement and simple calculation. In a sense, mathematics is the language of physics. Principles can be expressed in words but are stated most clearly—and usually most concisely—in mathematical terms. One equation is worth a thousand words, so to speak. But, in order to use equations to do calculations, you must have the proper background and understanding. You will begin your study of physics with a look at some basic quantities and units and a general approach to problem solving. A firm understanding of these will benefit you throughout this entire course of study.

1.2   SI Units of Length, Mass, and Time

Objectives: To (a) describe the SI and (b) specify the references for the three main base quantities of this system.

Length, mass, and time are fundamental physical quantities that describe a great many objects and phenomena. In fact, the topics of mechanics (the study of motion and force) covered in the first part of this book require *only* these physical

A 2000-Year-Old Calculation: The Earth's Circumference

One of the first measurements of the Earth's circumference was done sometime before 200 B.C. by the Greek scientist Eratosthenes (276–196 B.C.). His method, although quite simple, shows a great deal of insight and ingenuity. The basis for Eratosthenes' technique lies in the understanding that because the distance between the Earth and the Sun is so great compared to our planet's size, the Sun's rays arriving at Earth are virtually parallel. The Sun would have to be at an infinite distance for the rays to be exactly parallel, but the divergence is so small that it is insignificant for practical purposes.

Eratosthenes lived in Alexandria, Egypt. He had learned that on the first day of summer at the town of Syene (now Aswan), some 5000 stadia to the south of Alexandria, the noon Sun was directly overhead. (A *stadium*—plural, *stadia*—was a Greek unit of length believed to be equal to about $\frac{1}{6}$ km.) The Sun's position overhead was apparent because deep wells at Syene were lighted all the way to the bottom, and a vertical stick would cast no shadow at noon on that day.

However, on the first day of summer at Alexandria, some 800 km to the north, Eratosthenes observed that a vertical stick *did* cast a noon-day shadow with a length $\frac{1}{8}$ that of the stick. This corresponds to an angle of 7.1° between the Sun's rays and the vertical (see Fig. 1 insert). Since lines extending through the well at Syene and through the vertical stick at Alexandria intersect at the Earth's center, the angular distance between these locations is also 7.1° (Fig. 1).

Then Eratosthenes reasoned that the 800-kilometer arc length was to the 7.1° angle as the Earth's total circumference was to 360°, the number of degrees in a complete circle. That is, in ratio form,

$$\frac{\text{circumference}}{360°} = \frac{800 \text{ km}}{7.1°}$$

and

$$\text{circumference} = \frac{360°}{7.1°} \times 800 \text{ km} = 40\,600 \text{ km}$$

The Earth's radius (R) can easily be calculated from the circumference (c), since $c = 2\pi R$:

$$R = \frac{c}{2\pi} = \frac{40\,600 \text{ km}}{2\pi} = 6460 \text{ km}$$

This is quite close to the modern value of 6378 km for Earth's equatorial radius (Appendix I). We cannot be certain of the accuracy of Eratosthenes' result because several different Greek stadia of various lengths were in use at that time. (You can see why it is important to have *standard* units.) Even so, his method was sound, and the circumference of the Earth was measured more than 2000 years ago.

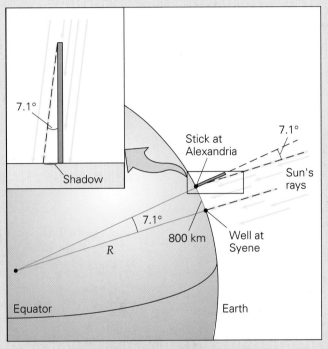

FIGURE 1   A 2000-year-old calculation   Observations on the incidence of sunlight allowed the Greek scientist Eratosthenes to calculate the circumference and radius of the Earth sometime before 200 B.C.

quantities. The system of units used by scientists to represent these and other quantities is based on the metric system.

Historically, the metric system was the outgrowth of proposals for a more uniform system of weights and measures in France during the seventeenth and eighteenth centuries. The modernized version of the metric system is called the **International System of Units**, officially abbreviated as **SI** (from the French *Système International des Unités*).

The SI

The SI includes *base quantities* and *derived quantities*, which are described by base units and derived units, respectively. **Base units,** such as the meter and the kilogram, are represented by standards. Other quantities that may be expressed in terms of combinations of base units are called **derived units**. (Think of how we commonly measure the length of a trip in miles and the time in hours. To express how fast we travel, we use the derived unit of miles per hour, which represents distance per unit of time, or length per time.)

One of the refinements of the SI was the adoption of new standard references for some base units, including those of length and of time.

Length

Length is the base quantity used to measure distances or dimensions in space. We commonly say that length is the distance between two points. But the distance between any two points depends on how the space between them is traversed, which may be in a straight or a curved line.

The SI unit of length is the **meter** (**m**). The meter was originally defined as $1/10\,000\,000$ of the distance from the North Pole to the Equator along a meridian running through Paris (•Fig. 1.1a).* A portion of this meridian between Dunkirk, France, and Barcelona, Spain, was surveyed to establish the standard length, which was assigned the name *metre* (from the Greek word *metron*, meaning "a measure"; the American spelling is *meter*). A meter is 39.37 inches—slightly longer than a yard.

The length of the meter was initially preserved in the form of a material standard: the distance between two marks on a metal bar (made of a platinum–iridium alloy) that was stored under controlled conditions and called the Meter of the Archives. However, it is not desirable to have a reference standard that changes with external conditions, such as temperature. Research provided more accurate standards. In 1983, the meter was redefined in terms of an unvarying property of light: the length of the path traveled by light in a vacuum during an interval of $1/299\,792\,458$ of a second (Fig. 1.1b). In other words, light travels $299\,792\,458$ meters in a second, and the speed of light in a vacuum is defined to be $299\,792\,458$ meters per second. Note that the length standard is referenced to time, which can be measured with great accuracy.

Mass

Mass is the base quantity used to describe amounts of matter. The more massive an object, the more matter it contains. (We will encounter more precise definitions of mass in Chapters 4 and 7.)

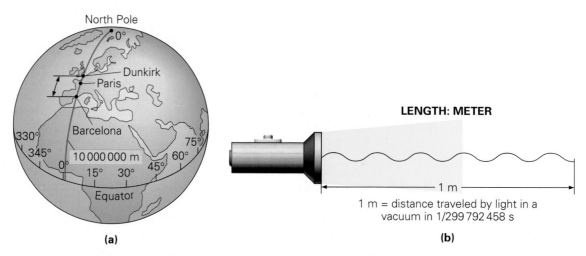

(a)   **(b)**

•**FIGURE 1.1 The SI length standard: the meter** **(a)** The meter was originally defined as $1/10\,000\,000$ (or 10^{-7}) of the distance from the North Pole to the Equator along a meridian running through Paris, of which a portion was measured between Dunkirk and Barcelona. A metal bar (called the Meter of the Archives) was constructed as a standard. **(b)** The meter is currently defined in terms of the speed of light.

*Note that this book and most physicists have adopted the practice of writing large numbers with a thin space for three-digit groups—for example, $10\,000\,000$ (not 10,000,000). The reason for this practice is to avoid confusion with the European practice of using a comma as a decimal point. For instance, 3.141 in the United States would be written as 3,141 in Europe. Large decimal numbers, such as $0.537\,84$, may also be separated for consistency. Spaces are generally used for numbers with more than four digits on either side of the decimal point.

The SI unit of mass is the **kilogram (kg)**. The kilogram was originally defined in terms of a specific volume of water but is now referenced to a specific material standard: the mass of a prototype platinum-iridium cylinder kept at the International Bureau of Weights and Measures in Sèvres, France. The United States has a duplicate of the prototype cylinder (•Fig. 1.2). (The kilogram may eventually be referenced to something less subject to time and chance than a piece of metal.)

You may have noticed that the phrase "weights and measures" is generally used instead of "masses and measures." In the SI, mass is a base quantity, but in the familiar British system, weight is used instead to describe amounts of mass—for example, weight in pounds instead of mass in kilograms. The weight of an object is the gravitational attraction that Earth exerts on an object. For example, when you weigh yourself on a scale, your weight is a measure of the downward gravitational force exerted on you by Earth. We can use weight in this way because near Earth's surface, mass and weight are directly proportional to each other: They differ only by a particular constant.

But treating weight as a base quantity creates some problems. A base quantity is naturally most useful if its value is the same everywhere. This is the case with mass—an object has the same mass, or amount of matter, regardless of its location. *But this is not true of its weight.* For example, the weight of an object on the Moon is less than its weight on Earth. This is because the Moon is less massive than Earth, and the gravitational attraction (weight) for an object on the Moon is less than that on Earth. That is, an object with a given amount of mass has a particular weight on Earth, but on the Moon, the same amount of mass will weigh only about one-sixth as much. Similarly, the weight of an object would vary on the surfaces of different planets.

For now, keep in mind that *weight is related to mass, but they are not the same*. Since the weight of an object of a certain mass can vary with location, it is much more useful to take mass as the base quantity, as the SI does. Base quantities should remain the same regardless of where they are measured, under normal or standard conditions. The distinction between mass and weight will be more fully explained in a later chapter. Our discussion until then will be chiefly concerned with mass.

Time

Time is a difficult concept to define. (Try it.) A common definition is that time is the forward flow of events. This is not so much a definition as an observation that time has never been known to run backward (as it might appear to do when you view a film run backward in a projector). Time is sometimes said to be a fourth dimension, accompanying the three dimensions of space. Thus, if something exists in space, it also exists in time. In any case, events can be used to mark time measurements. The events are analogous to the marks on a meterstick used for length measurements.

The SI unit of time is the **second (s)**. The solar "clock" was originally used to define the second. A solar day is the interval of time that elapses between two successive crossings of the same longitude line (meridian) by the Sun, at its highest point in the sky at that longitude (•Fig. 1.3a). A second was fixed as $1/86\,400$ of this apparent solar day (1 day = 24 h = 1440 min = 86 400 s). However, the elliptical path of the Earth's motion around the Sun causes apparent solar days to vary in length.

As a more precise standard, an average, or mean, solar day was computed from the lengths of the apparent solar days during a solar year. In 1956, the second was referenced to this mean solar day. But the mean solar day is not exactly the same for each yearly period because of minor variations in the Earth's motions and a steady slowing of its rate of rotation due to tidal friction. So scientists kept looking for something better.

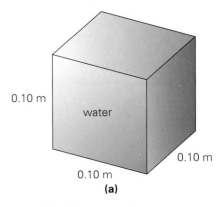

0.10 m

water

0.10 m

0.10 m

(a)

(b)

•**FIGURE 1.2 The SI mass standard: the kilogram (a)** The kilogram was originally defined in terms of a specific volume of water, that of a cube 0.10 m on a side, thereby associating the mass standard with the length standard. The standard kilogram is now defined by a metal cylinder. **(b)** Prototype Kilogram No. 20 is the national standard of mass for the United States. It is kept under a double bell jar in a vault at the National Institute of Standards and Technology (NIST, formerly the National Bureau of Standards) in Washington, D.C.

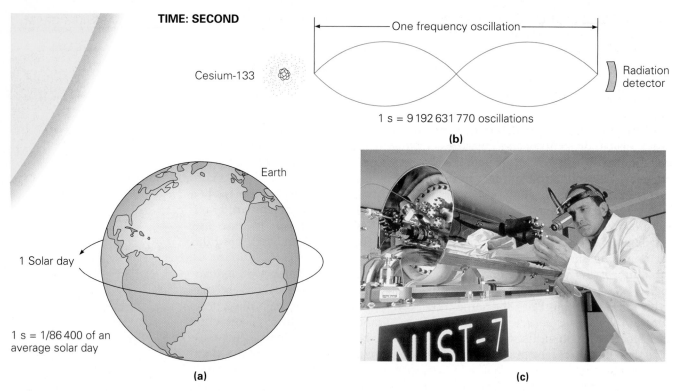

TIME: SECOND

One frequency oscillation

Cesium-133

Radiation detector

1 s = 9 192 631 770 oscillations

(b)

Earth

1 Solar day

1 s = 1/86 400 of an average solar day

(a)

NIST-7

(c)

•**FIGURE 1.3   The SI time standard: the second   (a)** The second was once defined in terms of the average solar day. **(b)** It is currently defined in terms of the frequency of the radiation associated with an atomic transition. **(c)** This atomic "clock" is the primary frequency standard at the National Institute of Standards and Technology. The device keeps time with an accuracy of about three millionths of a second per year.

In 1967, an atomic standard was adopted as a more precise reference. Currently, the second is defined by the radiation frequency of the cesium-133 atom. This particular cesium atom is used in an atomic "clock" that maintains our time standard. The atomic clock doesn't look much like an ordinary clock (Fig. 1.3b,c).

SI Base Units

The complete SI has seven base quantities and base units. In addition to the meter, kilogram, and second for (1) length, (2) mass, and (3) time, respectively, we measure (4) electric current (charge) in amperes (A), (5) temperature in kelvins (K), (6) amount of substance in moles (mol), and (7) luminous intensity in candelas (cd). See Table 1.1.

This is thought to be the smallest number of base quantities needed for a full description of everything observed or measured in nature. There are also two supplemental units for angular measure: the two-dimensional plane angle and the three-dimensional solid angle. There has not been general agreement about whether these geometric units are base or derived.

TABLE 1.1   The Seven Base Units of the SI

Name of Unit (abbreviation)	Property Measured
meter (m)	length
kilogram (kg)	mass
second (s)	time
ampere (A)	electric current
kelvin (K)	temperature
mole (mol)	amount of substance
candela (cd)	luminous intensity

1.3   More about the Metric System

Objectives: To use common **(a)** metric prefixes and **(b)** nonstandard metric units.

The metric system involving the standard units of length, mass, and time now incorporated in the SI was once called the **mks system** (for *m*eter-*k*ilogram-*s*econd). Another metric system that has been used in dealing with relatively small quantities is the **cgs system** (for *c*entimeter-*g*ram-*s*econd). In the United States, the system still generally in use is the British (or English) engineering system, in which the standard units of length, mass, and time are foot, slug, and second, respectively. You may not have heard of the slug, because, as we mentioned earlier, gravitational force (weight) is commonly used instead of mass—pounds instead of slugs—to describe quantities of matter. As a result, the British system is sometimes called the **fps system** (*foot-pound-second*).

The metric system is predominant throughout the world and is coming into increasing use in the United States. Primarily because it is simple mathematically, it is the preferred system of units for science and technology. SI units are used throughout most of this book. All quantities can be expressed in SI units. However, some units from other systems are accepted for limited use as a matter of practicality—for example, the time unit of hour and the temperature unit of degree Celsius. British units will sometimes be used in the early chapters for comparison purposes, since these units are still employed in everyday activities and for many practical applications.

The increasing worldwide use of the metric system means that you should be familiar with it. One of the greatest advantages of the metric system is that it is a decimal, or base-10, system. This means that larger or smaller units are obtained by multiplying or dividing a base unit by powers of 10. A list of the various multiples and corresponding prefixes is given in Table 1.2. In decimal measurements, the prefixes milli-, centi-, and kilo- are the ones most commonly used. The decimal characteristic makes it convenient to change measurements from one size metric unit to another. With the familiar British system, different factors must be used, such as 16 for converting pounds to ounces and 12 for converting feet to inches. The British system developed historically and not very scientifically (see the Insight on p. 8).

You are already familiar with one base-10 system—U.S. currency. Just as a meter can be divided into 10 decimeters, 100 centimeters, or 1000 millimeters, the "base unit" of the dollar can be broken down into 10 "decidollars" (dimes), 100 "centidollars" (cents), or 1000 "millidollars" (tenths of a cent, or mills, used in figuring property taxes and bond levies). Since all the metric prefixes are powers of 10, there are no metric analogues for quarters or nickels.

TABLE 1.2 Multiples and Prefixes for Metric Units*

Multiple[†]	Prefix (and Abbreviation)	Pronunciation	Multiple[†]	Prefix (and Abbreviation)	Pronunciation
10^{24}	yotta- (Y)	yot'ta (*a* as in *a*bout)	10^{-1}	deci- (d)	des'i (as in *deci*mal)
10^{21}	zetta- (Z)	zet'ta (*a* as in *a*bout)	10^{-2}	centi- (c)	sen'ti (as in *senti*mental)
10^{18}	exa- (E)	ex'a (*a* as in *a*bout)	10^{-3}	milli- (m)	mil'li (as in *mili*tary)
10^{15}	peta- (P)	pet'a (as in *petal*)	10^{-6}	micro- (μ)	mi'kro (as in *micro*phone)
10^{12}	tera- (T)	ter'a (as in *terr*ace)	10^{-9}	nano- (n)	nan'oh (*an* as in *ann*ual)
10^{9}	giga- (G)	ji'ga (*ji* as in *jiggle, a* as in *a*bout)	10^{-12}	pico- (p)	pe'ko (*peek-oh*)
10^{6}	mega- (M)	meg'a (as in *mega*phone)	10^{-15}	femto- (f)	fem'toe (*fem* as in *feminine*)
10^{3}	kilo- (k)	kil'o (as in *kilo*watt)	10^{-18}	atto- (a)	at'toe (as in *anatomy*)
10^{2}	hecto- (h)	hek'to (*heck-toe*)	10^{-21}	zepto- (z)	zep'toe (as in *zeppelin*)
10	deka- (da)	dek'a (*deck* plus *a* as in *a*bout)	10^{-24}	yocto- (y)	yock'toe (as in *sock*)

*For example, 1 gram (g) multiplied by 1000 (10^3) is 1 kilogram (kg); 1 gram multiplied by 1/1000 (10^{-3}) is 1 milligram (mg).

[†]The most commonly used prefixes are printed in color. Note that the abbreviations for the multiples 10^6 and greater are capitalized, whereas the abbreviations for the smaller multiples are lowercased.

A Bit of History: Why 5280 Feet in a Mile?

Have you ever wondered why there are 5280 feet in a mile? It is certainly a strange number, and 1000 meters in a kilometer is clearly more convenient. It all has to do with the historical development of British units.

The word "mile" come from the Latin *mille passus*, literally meaning one thousand (*mille*, as in millennium) paces. This unit was introduced to Britain during the Roman occupation, circa A.D. 50. Each *passus* or pace contained 5 *pes*, the Roman foot unit; thus a *mille passus* was 5000 *pes* or feet, even though the Roman foot was slightly shorter than our current foot. (The length of the foot has varied with different cultures, as you might expect when a length is referenced to an anatomical unit.)

A common length in early England was the furlong, which is said to have originated as the average length of a furrow a yoke of oxen could plow before needing a rest—a "furrow long." The furlong was used in measuring land acreage and is still used today in horse racing. About A.D. 1500, the "Old London" mile was defined to be 8 furlongs. At the

time, the furlong, in terms of a longer (German) foot, contained 625 feet. So, by this definition, the mile again equaled (8 × 625 ft =) 5000 ft.

The other 280 feet were added in 1593 when Queen Elizabeth I decreed (by statute) that the furlong would be defined in terms of the shorter British foot, which made the furlong 660 feet. Hence, the 8-furlong mile became 5280 ft (= 8 × 660 ft).

The 5280-foot mile is called a *statute* mile, because it was established by the queen's statute. We also have a *nautical* mile, which is generally used at sea (and in air transportation), as the name implies. How does it differ from a regular, or statute, mile? The nautical mile's definition is a bit more scientific and somewhat like that of the meter. In principle, the nautical mile is equal to the length of one minute of arc (latitude) on a meridian, or a great circle. An international committee in 1929 established this length to be 6076 ft, or 1.852 km. The knot, a unit of speed, is one nautical mile per hour.

With unusual numbers such as these, it is easy to see the advantages of the metric decimal system.

The official metric prefixes can help eliminate confusion. In the United States, a billion is a thousand million (10^9); in Great Britain, a billion is a million million (10^{12}). The use of metric prefixes eliminates any confusion, since giga- indicates 10^9 and tera- stands for 10^{12}.

Volume

In the SI, the standard unit of volume is the cubic meter (m^3)—the three-dimensional derived unit of the meter base unit. Since this is a rather large unit, it is often more convenient to use the nonstandard unit of volume (or capacity) of a cube 10 cm (centimeters) on a side (•Fig. 1.4a). This volume was given the

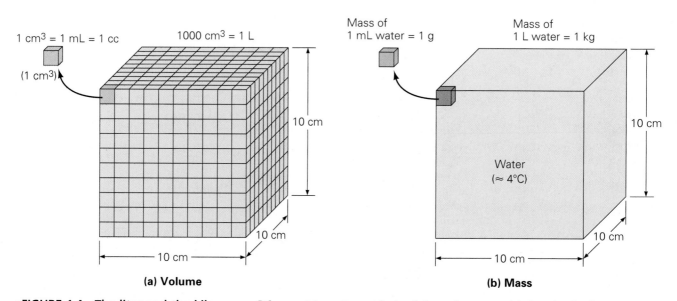

(a) Volume   **(b) Mass**

•**FIGURE 1.4   The liter and the kilogram**   Other metric units are derived from the meter. **(a)** A unit of volume (capacity) was taken to be the volume of a cube 1.0 dm (10 cm or 0.10 m) on a side and was given the name liter. **(b)** The mass of a liter of water (at its maximum density) was defined to be 1 kg. Note that the decimeter cube contains 1000 cm^3, or 1000 mL. Thus, 1 cm^3, or 1 mL, of water has a mass of 1 g.

name *litre*, which is spelled **liter** (**L**) in the United States. The volume of a liter is 1000 cm³ (10 cm × 10 cm × 10 cm). Since 1 L = 1000 mL (milliliters), it follows that 1 mL = 1 cm³. (The cubic centimeter is sometimes abbreviated cc, particularly in chemistry and biology.)

Recall that the standard unit of mass, the kilogram, was originally defined to be the mass of a cubic volume of water 10 cm or 0.10 m on a side, or the volume of one *liter* of water (at its maximum density, at about 4°C). The density of water does not vary greatly with temperature, so we can say with fairly good accuracy that *1 L of water has a mass of 1 kg*, an approximation that we will use throughout this book (Fig. 1.4b). Also, since 1 kg = 1000 g and 1 L = 1000 cm³, 1 cm³ (or 1 mL) of water has a mass of 1 g.

You are probably more familiar with the liter than you think. The use of the liter is becoming quite common in some instances in the United States, as •Fig. 1.5 indicates.

Note: Liter is sometimes abbreviated as a lowercase "ell" (l), but a capital "ell" (L) is preferred in the United States so that the abbreviation is less likely to be confused with the numeral one. (Isn't 1 L clearer than 1 l?)

•**FIGURE 1.5 Two, three, one, and one-half liters** The liter is now a common volume unit for soft drinks.

Example 1.1 ■ The Metric Ton (or Tonne): Another Unit of Mass

As we have seen, the metric unit of mass was originally related to the length standard, with a liter (1000 cm³) of water having a mass of 1 kg. The standard metric unit of volume is the cubic meter (m³), and this volume of water was used to define a larger mass unit called the *metric ton* (or *tonne*, as it is sometimes spelled). A metric ton is equivalent to how many kilograms?

Thinking It Through. A cubic meter is a relatively large volume and holds a large amount of water (more than a cubic yard; why?). The key is to find how many cubic volumes measuring 10 cm on a side (liters) are in a cubic meter. We expect, therefore, a large number.

Solution. Each liter of water has a mass of 1 kg, so we must find how many liters there are in 1 m³. Since there are 100 cm in a meter, we can visualize a cubic meter as simply a cube with sides 100 cm in length. Therefore, a cubic meter (1 m³) has a volume of 10² cm × 10² cm × 10² cm = 10⁶ cm³. Since 1 L = 10³ cm³, there must be 10⁶ cm³/(10³ cm³/L) = 1000 L in 1 m³. Thus, 1 metric ton is equivalent to 1000 kg.

Note that this entire line of reasoning can be expressed very concisely in a single calculation:

$$\frac{1 \text{ m}^3}{1 \text{ L}} = \frac{100 \text{ cm} \times 100 \text{ cm} \times 100 \text{ cm}}{10 \text{ cm} \times 10 \text{ cm} \times 10 \text{ cm}} = 1000$$

Follow-up Exercise. One kilogram has an equivalent weight of 2.2 lb. Using this relationship, how does the metric ton compare with the British ton? (*See the Answers to Follow-up Exercises section.*)*

Because the metric system is coming into increasing use in the United States, you may find it helpful to have an idea of how metric and British units compare. The relative sizes of some units are illustrated in •Fig. 1.6. The mathematical conversion from one unit to another will be discussed shortly.

1.4 Dimensional Analysis and Unit Analysis

Objectives: To explain the advantages of and apply (a) dimensional analysis and (b) unit analysis.

The fundamental, or base, quantities used in physical descriptions are called *dimensions*. For example, length, mass, and time are dimensions. You could measure the distance between two points and express it in units of meters, centimeters, or feet, but the quantity would still have the dimension of length.

---

*The Answers to Follow-up Exercises section after the appendices contains the answers—and, for Conceptual Exercises, the reasoning—for all Follow-up Exercises in this book. We'll remind you here in Chapter 1 to refer to that section, but then you're on your own.

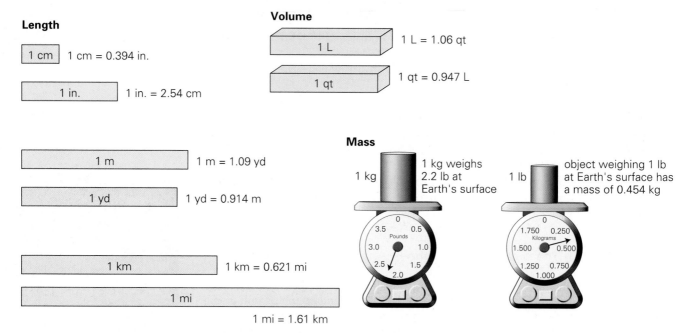

•FIGURE 1.6 Comparison of some SI and British units The bars illustrate the relative magnitudes of each pair of units. (The comparison scales are different in each case.)

Definition of:
Dimensional analysis

It is common to express dimensional quantities by bracketed symbols, such as [L], [M], and [T] for length, mass, and time, respectively. Derived quantities are combinations of dimensions; for example, velocity (*v*) has the dimensions [L]/[T] (think of miles/hour or meters/second), and volume (*V*) has the dimensions [L] × [L] × [L], or [L³]. Addition and subtraction can be done only with quantities that have the same dimensions, for example, 10 s + 20 s = 30 s, or [T] + [T] = [T].

Dimensional analysis is a procedure by which the dimensional consistency of any equation may be checked. You have used equations and know that an equation is a mathematical equality. Since physical quantities used in equations have dimensions, *the two sides of an equation must be equal not only in numerical value, but also in dimensions.* For example, suppose you had the length quantities *a* = 3.0 m and *b* = 4.0 m. Inserting them into the equation *a* × *b* = *c* gives 3.0 m × 4.0 m = 12 m². Both sides of the equation are numerically equal (3 × 4 = 12), and both sides are dimensionally equal ([L] × [L] = [L²]). Notice that we are not algebraically manipulating the dimension symbol [L] in the same way you normally would a variable, such as *L*. (Note that variables are italicized.) The expression [L] × [L] = [L²] is read, "when you take a length dimension and multiply it by a length dimension, the result has the dimension of length squared." A second example will help clarify the point: If *a* and *b* are inserted into the equation *a* + *b* = *c*, we get 3.0 m + 4.0 m = 7.0 m. Comparing both sides of the equation dimensionally, we would say [L] + [L] = [L], or a length plus a length gives another length, but *not* [L] + [L] = 2[L].

One of the major advantages of dimensional analysis is its usefulness for checking whether an equation that has been derived or is being used in solving a problem has the correct form. For example, suppose that you think you can solve a problem about the distance an object travels by using the equation *x* = *at*, where *x* represents distance, or length, and has the dimension [L]; *a* represents acceleration, which, as we will see in Chapter 2, has the dimensions [L]/[T²]; and *t* represents time and has the dimension [T]. (The dimensions and units of some common quantities are given in Table 1.3.) First, before trying to use numbers with the equation, you can check to see if it is dimensionally correct:

TABLE 1.3   Dimensions and Units

Quantity	Dimension	Unit
mass	[M]	kg
time	[T]	s
length	[L]	m
area	[L²]	m²
volume	[L³]	m³
velocity (*v*)	$\dfrac{[L]}{[T]}$	$\dfrac{m}{s}$
acceleration (*a* or *g*)	$\dfrac{[L]}{[T^2]}$	$\dfrac{m}{s^2}$

$$x = at$$

is expressed dimensionally as

$$[L] = \frac{[L]}{[T^2]} \times [T] \quad \text{or} \quad [L] = \frac{[L]}{[T]}$$

which is not true. Thus, $x = at$ *cannot* be a correct equation.

Dimensional analysis will tell you if an equation is incorrect, but a dimensionally consistent equation may *not* correctly express the real relationship of quantities. For example, in terms of dimensions,

$$x = at^2$$

is

$$[L] = \frac{[L]}{[T^2]} \times [T^2] \quad \text{or} \quad [L] = [L]$$

This equation is dimensionally correct. But, as you will see in Chapter 2, it is not physically correct. The correct form of the equation—both dimensionally and physically—is $x = \frac{1}{2}at^2$. The fraction $\frac{1}{2}$ has no dimensions; it is a dimensionless number. Thus, you must be sure you have the correct form of an equation before you use it to find solutions to Exercises.

Doing dimensional analysis with the symbols [L], [T], and [M] is fine, but in practice it is often more convenient to use the actual unit abbreviations, such as m, s, and kg. Units can also be treated as algebraic quantities, and like symbols can be canceled. Using units instead of symbols in dimensional analysis is called **unit analysis**. If an equation is correct by unit analysis, it must be dimensionally correct. The following Example demonstrates the use of unit analysis.

Definition of:
Unit analysis

Example 1.2 ■ Checking Dimensions: Unit Analysis

A professor puts two equations on the board: (a) $v = v_0 + at$ and (b) $x = v/2a$, where x is distance in meters (m), v and v_0 are velocities in meters/second (m/s), a is acceleration in (meters/second)/second or meters/second^2 (m/s^2), and t is time in seconds (s). Are the equations dimensionally correct? Use unit analysis to find out.

Thinking It Through. Simply insert the units for the quantities in each equation, cancel, and check the units on both sides.

Solution.
(a) The equation is

$$v = v_0 + at$$

Inserting units for the physical quantities gives

$$\frac{m}{s} = \frac{m}{s} + \left(\frac{m}{s^2} \times s\right) \quad \text{or} \quad \frac{m}{s} = \frac{m}{s} + \left(\frac{m}{s \times \cancel{s}} \times \cancel{s}\right)$$

Notice that units cancel like numbers in a fraction. Then, we have

$$\frac{m}{s} = \frac{m}{s} + \frac{m}{s} \quad \begin{bmatrix} \textit{Dimensionally} \\ \textit{correct} \end{bmatrix}$$

The equation is dimensionally correct, since the units on each side are meters per second. (The equation is also a correct relationship, as we shall see in Chapter 2.)

(b) By unit analysis, the equation

$$x = \frac{v}{2a}$$

is

$$m = \frac{\left(\dfrac{m}{s}\right)}{\left(\dfrac{m}{s^2}\right)} = \frac{\cancel{m}}{\cancel{s}} \times \frac{s^{\cancel{2}}}{\cancel{m}} \qquad \text{or} \qquad m = s \quad \left[\begin{array}{l} Not\ dimensionally \\ correct \end{array}\right]$$

Meters (m) cannot equal seconds (s), so in this case, the equation is dimensionally incorrect and therefore not physically correct.

Follow-up Exercise. Is the equation $x = v^2/a$ dimensionally correct? *(See the Answers to Follow-up Exercises section.)*

Mixed Units

Unit analysis also allows you to check for mixed units. In general, in working problems, you should always use the same unit for a given dimension throughout a problem. For example, suppose that you wanted to buy a new carpet to fit a rectangular room. To compute the floor area, you would measure the lengths of two adjacent sides of the room and multiply them. Would you measure the sides in different units such as 10 ft × 3.0 m and express the area as 30 ft·m? Probably not. Such mixed units are not usually very useful.

Let's look at mixed units in an equation. Suppose that you used centimeters as the unit for x in the equation

$$v^2 = v_o^2 + 2ax$$

and the units for the other quantities as in Example 1.2. In terms of units, this would give

$$\left(\frac{m}{s}\right)^2 = \left(\frac{m}{s}\right)^2 + \left(\frac{m}{s^2} \times cm\right)$$

or

$$\frac{m^2}{s^2} = \frac{m^2}{s^2} + \frac{m \times cm}{s^2}$$

which is dimensionally correct. That is, it is dimensionally equivalent to

$$\frac{[L^2]}{[T^2]} = \frac{[L^2]}{[T^2]} + \frac{[L^2]}{[T^2]}$$

But the units are mixed (m and cm). The terms on the right-hand side should not be added together without centimeters first being changed to meters.

Determining the Units of Quantities

Another aspect of unit analysis that is very important in physics is the determination of the units of quantities from defining equations. For example, **density** (the Greek letter rho, ρ) is defined by the equation

$$\rho = \frac{m}{V} \tag{1.1}$$

where m is mass and V is volume. (Density is the mass of an object or substance per unit volume and is a measure of the compactness of the mass of that object or substance.) What are the units of density? In SI units, mass is measured in kilograms and volume in cubic meters. Hence the defining equation

$$\rho = \frac{m}{V} \left(\frac{kg}{m^3} \right)$$

gives the derived SI unit for density as kilograms per cubic meter (kg/m^3).

What are the units of π? The relationship between the circumference (c) and the diameter (d) of a circle is given by the equation $c = \pi d$, so $\pi = c/d$. If length is measured in meters, then

$$\pi = \frac{c}{d} \left(\frac{\not{m}}{\not{m}} \right)$$

Thus, the constant π has no units, because they cancel out. It is unitless or a dimensionless constant.

1.5 Unit Conversions

Objectives: To (a) explain conversion factor relationships, and (b) apply them in converting units within a system or from one system of units to another.

Because units in different systems, or even different units in the same system, can express the same quantity, it is sometimes necessary to convert the units of a quantity from one unit to another. For example, we may need to convert feet to yards or inches to centimeters. You already know how to do many unit conversions. If a room is 12 ft long, what is its length in yards? Your immediate answer is 4 yd.

How did you do this? Well, you must have known a relationship between the units of foot and yard. That is, you know that 1 yd = 3 ft. This is what we call an *equivalence statement*. As we saw in Section 1.4, the numerical values and units on both sides of an equation must be the same. In equivalence statements, we commonly use an equal sign to indicate that 1 yd and 3 ft stand for the *same* or *equivalent length*. The numbers are different because they stand for different *units* of length.

Mathematically to change units, we use **conversion factors**, which are simply equivalence statements expressed in the form of ratios—for example, 1 yd/3 ft or 3 ft/1 yd. (The 1 is often omitted in the denominators of such ratios for convenience; for example, 3 ft/yd.) To understand why such ratios are useful, note that dividing the expression 1 yd = 3 ft by 3 ft (or 3 ft = 1 yd by 1 yd) on both sides gives

$$\frac{(1 \text{ yd} = 3 \text{ ft})}{3 \text{ ft}} = \frac{1 \text{ yd}}{3 \text{ ft}} = \frac{3 \text{ ft}}{3 \text{ ft}} = 1 \quad \text{or} \quad \frac{(3 \text{ ft} = 1 \text{ yd})}{1 \text{ yd}} = \frac{3 \text{ ft}}{1 \text{ yd}} = \frac{1 \text{ yd}}{1 \text{ yd}} = 1$$

As you can see from these examples, a conversion factor always has an actual value of 1—and you can multiply any quantity by 1 without changing its value, or size. Thus, a *conversion factor simply lets you express a quantity in terms of other units without changing its physical value or size.*

What is done in converting 12 feet to yards may be expressed mathematically as follows:

$$12 \not{ft} \times \frac{1 \text{ yd}}{3 \not{ft}} = 4 \text{ yd} \quad \text{(feet cancel)}$$

Using the appropriate conversion factor form, the feet cancel, as shown by the slash marks, giving the correct unit analysis, yd = yd.

Suppose you are asked to convert 12.0 inches to centimeters. You may not know the conversion factor in this case, but you can get it from a table (such as the one that appears inside the front cover of this book) that gives the needed relationships: 1 in. = 2.54 cm or 1 cm = 0.394 in. It makes no difference which of these equivalence statements is used. The question, once you have expressed the equivalence statement as a conversion factor, is whether to divide or multiply by that factor to make the conversion. *In doing unit conversions, take advantage of unit analysis—that is,* let the units determine the appropriate form of conversion factor, so to speak.

Note that the equivalence statement 1 in. = 2.54 cm can give rise to two forms of the conversion factor: 1 in./2.54 cm or 2.54 cm/1 in. Unit analysis tells you that the second form is the appropriate one to multiply by in this case:

$$12.0 \text{ in.} \times \frac{2.54 \text{ cm}}{1 \text{ in.}} = 30.5 \text{ cm}$$

What would be the result if you used the other form of the first conversion factor, 1 in./2.54 cm? Let's see:

$$12.0 \text{ in.} \times \frac{1 \text{ in.}}{2.54 \text{ cm}} = 4.72 \frac{\text{in}^2}{\text{cm}}$$

This result is dimensionally correct (why?), but it is not a conversion to centimeters. The units in²/cm for length are nonstandard and confusing. To use this ratio properly, you must *divide* a quantity in inches by the conversion factor:

$$\frac{12.0 \text{ in.}}{\left(\dfrac{1 \text{ in.}}{2.54 \text{ cm}}\right)} = 12.0 \left(\frac{2.54 \text{ cm}}{1}\right) = 30.5 \text{ cm}$$

(Remember to invert the fraction in the denominator.) Compare this procedure with the first conversion of 12.0 in. above. Can you see why they are equivalent?

Alternatively, you could multiply by a conversion factor derived from the equivalence statement 1 cm = 0.394 in.:

$$12.0 \text{ in.} \times \frac{1 \text{ cm}}{0.394 \text{ in.}} = 30.5 \text{ cm}$$

Or you could divide by the reciprocal form of this conversion factor:

$$\frac{12.0 \text{ in.}}{\left(\dfrac{0.394 \text{ in.}}{1 \text{ cm}}\right)} = 30.5 \text{ cm}$$

A few commonly used equivalence statements are not dimensionally correct; for example, consider 1 kg = 2.2 lb, which is used for conversions. The kilogram is a unit of mass, and the pound is a unit of weight. This means that 1 kilogram is *equivalent* to 2.2 pounds; that is, a 1-kilogram *mass* has a *weight* of 2.2 lb, but only on the Earth's surface. Since mass and weight are directly proportional, differing only by a particular constant, we can use the dimensionally incorrect conversion factor 1 kg/2.2 lb (near the Earth's surface).

Note: 1 kg of mass has an equivalent weight of 2.2 lb near the surface of the Earth.

Example 1.3 ■ Converting Units: Use of Conversion Factors

(a) A basketball player is 6.5 ft tall. What is the player's height in meters? (b) How many seconds are there in a 30-day month? (c) What is 50 mi/h in meters per second?

Thinking It Through. If we use the correct conversion factors, the rest is basic mathematics.

Solution.
(a) From the conversion table, 1 ft = 0.3048 m, so

$$6.5 \text{ ft} \times \frac{0.3048 \text{ m}}{1 \text{ ft}} = 2.0 \text{ m}$$

Another feet–meter conversion is shown in •Fig. 1.7 (top). Is it correct? You should be able to do this one in your head.

(b) The conversion factor for days and seconds is available from the table (1 day = 86 400 s), but you may not always have a table handy. You can always use several better-known conversion factors to get the result:

$$30 \text{ days} \times \frac{24 \text{ h}}{\text{day}} \times \frac{60 \text{ min}}{\text{h}} \times \frac{60 \text{ s}}{\text{min}} = 2.6 \times 10^6 \text{ s}$$

Note how unit analysis checks the conversion factors for you. The rest is simple arithmetic.

(c) In this case, from the conversion table, 1 mi = 1609 m and 1 h = 3600 s. (The latter is easily computed.) These ratios are used to cancel the units that are to be changed, leaving behind the ones that are wanted:

$$\frac{50 \text{ mi}}{1 \text{ h}} \times \frac{1609 \text{ m}}{1 \text{ mi}} \times \frac{1 \text{ h}}{3600 \text{ s}} = 22 \text{ m/s}$$

Follow-up Exercise. (a) Convert 50 mi/h directly to meters per second by using a single conversion factor, and (b) show that this single conversion factor can be derived from those in part c of this Example. *(See the Answers to Follow-up Exercises section.)*

Example 1.4 ■ More Conversions: A Really Long Capillary System

Capillaries, the smallest blood vessels of the body, connect the arterial system with the venous system and supply our tissues with oxygen and nutrients (•Fig. 1.8). It is estimated that if all of the capillaries of an average adult were unwound and spread out end to end, they would extend to a length of about 64 000 km. (a) How many miles is this? (b) Compare this length to the circumference of the Earth.

Thinking It Through. (a) This is a simple conversion—just use the right conversion factor. (b) How do we calculate the circumference of a circle or sphere? There is an equation, and the radius or diameter of the Earth must be known. (If you do not remember it, see the table inside the back cover.)

Solution.
(a) We see in the conversion table that 1 km = 0.621 mi, so

$$64\,000 \text{ km} \times \frac{0.621 \text{ mi}}{1 \text{ km}} = 40\,000 \text{ mi} \quad \text{(rounded off)}$$

•**FIGURE 1.7 Converting units** Road signs show the elevation in both feet and meters, and distances in both miles and kilometers.

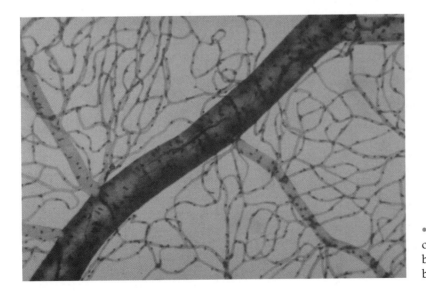

•**FIGURE 1.8 Capillary system** Capillaries connect the arterial and venous systems in our bodies. They are the smallest blood vessels, but their total combined length is impressive.

(b) A length of 40 000 mi is substantial. To see how this length compares to the circumference (*c*) of the Earth, recall that the radius of the Earth is about 4000 mi, or a diameter (*d*) of 8000 mi. The circumference of a circle is given by $c = \pi d$, as we saw in Section 1.4. Then,

$$c = \pi d \approx 3 \times 8000 \text{ mi} = 24\,000 \text{ mi}$$

[To make a general comparison, we round π (= 3.14 …) off to 3.] So,

$$\frac{\text{capillary length}}{\text{Earth's circumference}} = \frac{40\,000 \text{ mi}}{24\,000 \text{ mi}} = 1.7$$

The capillaries of your body have a total length that would extend 1.7 times around the world. Wow!

Follow-up Exercise.   Taking the average distance between the east and west coasts of the continental United States to be 4800 km, how many times would the capillary length of your body cross the country? *(See the Answers to Follow-up Exercises section.)*

Example 1.5 ■ Converting Units of Area: Choosing the Correct Conversion Factor

A hall bulletin board has an area of 2.5 m². What is this area in square centimeters (cm²)?

Thinking It Through.   This is a conversion of area units, and we know that 1 m = 100 cm. So, some squaring must be done to get square meters and square centimeters.

Solution.   A common error in such conversions is the use of incorrect conversion factors. Because 1 m = 100 cm, it is sometimes assumed that 1 m² = 100 cm², which is *wrong*. The correct area conversion factor may be obtained directly from the correct linear conversion factor, 100 cm/1 m or 10^2 cm/1 m, by squaring it:

$$\left(\frac{10^2 \text{ cm}}{1 \text{ m}}\right)^2 = \frac{10^4 \text{ cm}^2}{1 \text{ m}^2}$$

Hence, 1 m² = 10^4 cm² (= 10 000 cm²). We can therefore write:

$$2.5 \text{ m}^2 \times \left(\frac{10^2 \text{ cm}}{1 \text{ m}}\right)^2 = 2.5 \text{ m}^2 \times \frac{10^4 \text{ cm}^2}{1 \text{ m}^2} = 2.5 \times 10^4 \text{ cm}^2$$

Follow-up Exercise.   How many cubic centimeters are there in one cubic meter? *(See the Answers to Follow-up Exercises section.)*

Conceptual Example 1.6

Selecting Units: Units and Magnitudes

You are selling a car and want to express its fuel economy so that it will sound the most impressive. Which set of units would give the largest numerical value? (a) miles per gallon, (b) kilometers per gallon, (c) miles per liter, (d) kilometers per liter. (Note: 1 mi = 1.61 km.) *Clearly establish the reasoning used in determining your answer before you check it below. That is, **why** did you select your answer?*

Reasoning and Answer.   To obtain the largest numerical value, we require the larger numerical value distance unit and the smaller numerical value volume unit. Since 1 mi is equivalent to 1.61 km, 1.61 is the larger numerical value, (b) would have a larger numerical value than (a), and (d) would have a larger value than (c). That is, 1 mi/gal = 1.61 km/gal and 1 mi/L = 1.61 km/L. Hence, the length unit would be kilometers, and we must decide which volume unit would give the larger apparent value—kilometers per gallon or kilometers per liter. Since a liter is roughly a quart, a gallon is much larger than a liter (about 4 times larger). A gallon of gas would therefore run a car many more kilometers than would a liter of gas. That is, 1 gal ≈ 4 L; so, for example, 40 km/gal ≈ 10 km/L. Thus the answer is (b).

Follow-up Exercise. (a) Which set of units would give the *smallest* numerical value? (b) Given that 1 L = 0.264 gal, justify your answer mathematically. *(See the Answers to Follow-up Exercises section.)*

1.6 Significant Figures

Objectives: To (a) determine the number of significant figures in a numerical value and (b) report the proper number of significant figures after performing simple calculations.

Most of the time, you will be given numerical data when asked to solve a problem. In general, such data are either exact numbers or measured numbers (quantities). **Exact numbers** are those without any uncertainty or error. These include numbers such as the 100 used to calculate a percentage and the 2 in the equation $r = d/2$ relating the radius and diameter of a circle. **Measured numbers** are those obtained from measurement processes and so generally have some degree of uncertainty or error.

When calculations are done with measured numbers, the error of measurement is propagated, or carried along, by the mathematical operations. A question of how to report a result arises. For example, suppose that you are asked to find time (t) from the formula $x = vt$ and are given that $x = 5.3$ m and $v = 1.67$ m/s. Then

$$t = \frac{x}{v} = \frac{5.3 \text{ m}}{1.67 \text{ m/s}} = ?$$

Doing the division operation on a calculator yields a result such as 3.173 652 695 (•Fig. 1.9). How many figures, or digits, should you report in the answer?

The error or uncertainty of the result of a mathematical operation may be computed by statistical methods. A simpler, widely used procedure for estimating this uncertainty involves the use of **significant figures** (**sf**), sometimes called significant digits. The degree of accuracy of a measured quantity depends on how finely divided the measuring scale of the instrument is. For example, you might measure the length of an object as 2.5 cm with one instrument and 2.54 cm with another; the second instrument provides more significant figures and a greater degree of accuracy.

In general, *the number of significant figures of a numerical quantity is the number of reliably known digits it contains.* For a measured quantity, this is usually defined as all of the digits that can be read directly from the instrument used in making the measurement plus one uncertain digit that is obtained by estimating the fraction of the smallest division of the instrument's scale.

The quantities 2.5 cm and 2.54 cm have two and three significant figures, respectively. This is rather evident. However, some confusion may arise when a quantity contains one or more zeros. For example, how many significant figures does the quantity 0.0254 m have? What about 104.6 m? 2705.0 m? In such cases, we will use the following rules:

1. Zeros at the beginning of a number are not significant. They merely locate the decimal point.

    0.0254 m   three significant figures   (2, 5, 4)

•**FIGURE 1.9 Significant figures and insignificant figures** For the division operation of 5.3/1.67, a calculator with a floating decimal point gives many digits. A calculated quantity can be no more accurate than the least accurate quantity involved in the calculation, so this result should be rounded off to two significant figures—that is, 3.2.

2. Zeros within a number are significant.

104.6 m   four significant figures   (1, 0, 4, 6)

Rules for determining the number of significant figures in quantities with zeros

3. Zeros at the end of a number after the decimal point are significant.

2705.0 m   five significant figures   (2, 7, 0, 5, 0)

4. In whole numbers without a decimal point that end in one or more zeros (trailing zeros)—for example, 500 kg—the zeros may or may not be significant. In such cases, it is not clear which zeros serve only to locate the decimal point and which are actually part of the measurement. That is, if the first zero from the left (5<u>0</u>0 kg) was the estimated digit in the measurement, then only two digits are reliably known and there are only two significant figures. Similarly, if the last zero was the estimated digit (50<u>0</u> kg), then there are three significant figures. This ambiguity may be removed by using scientific (powers-of-ten) notation:

5.0×10^2 kg   two significant figures

5.00×10^2 kg   three significant figures

This notation is helpful in expressing the results of calculations with the proper numbers of significant figures, as we shall see shortly. (Appendix I includes a review of scientific notation.)

Note: To avoid confusion with numbers having trailing zeroes used as given quantities in text examples and exercises, we will consider the trailing zeroes to be significant. For example, assume that a time of 20 s has two significant figures, even if it is not written out as 2.0×10^1 s.

It is important to report the results of mathematical operations with the proper number of significant figures. The following general rule tells how to determine the number of significant figures in the result of a multiplication or division:

Multiplication and division with significant figures

> The final result of a multiplication or division operation should have the same number of significant figures as the quantity with the least number of significant figures that was used in the calculation.

What this means is that the result of a calculation can be no more accurate than the least accurate quantity used; that is, you cannot gain accuracy in performing mathematical operations. Thus, the result that should be reported for the division operation discussed at the beginning of this section is

$$\frac{\overset{(2\,sf)}{5.3 \text{ m}}}{\underset{(3\,sf)}{1.67 \text{ m/s}}} = 3.2 \text{ s } {\scriptstyle (2\,sf)}$$

The result is rounded off to two significant figures (see Fig. 1.9).

For multiple operations, rounding off to the proper number of significant figures should not be done at each step, because rounding errors may accumulate. It is usually suggested that one or two insignificant figures be carried along, or if a calculator is being used, rounding off may be done on the final result of the multiple calculations.

Rounding rules

The rules for rounding off numbers are as follows:

1. If the next digit after the last significant figure is 5 or greater, round up: Increase the last significant figure by 1. (For example, 2.136 becomes 2.14 rounded to 3 significant figures.)

2. If the next digit after the last significant figure is less than 5, round down: Do not change the last significant figure. (For example, 2.132 becomes 2.13 rounded to 3 significant figures.)

Example 1.7 ■ Using Significant Figures in Multiplication and Division: Rounding Applications

The following operations are performed and the results rounded off to the proper number of significant figures (sf).

Multiplication

$$2.4 \text{ m} \times 3.65 \text{ m} = 8.76 \text{ m}^2 = 8.8 \text{ m}^2 \quad (rounded \text{ to } 2 \text{ sf})$$
$$\underset{(2 \text{ sf})}{\phantom{2.4 \text{ m}}} \quad \underset{(3 \text{ sf})}{\phantom{3.65 \text{ m}}}$$

Division

$$\frac{\overset{(4 \text{ sf})}{725.0 \text{ m}}}{\underset{(3 \text{ sf})}{0.125 \text{ s}}} = 5800 \text{ m/s} = 5.80 \times 10^3 \text{ m/s} \quad (represented \text{ with } 3 \text{ sf; why?})$$

Follow-up Exercise. Perform the following operations with answers expressed in the standard powers-of-ten notation (one digit to the left of the decimal point) with the proper number of significant figures. (a) $(2.0 \times 10^5 \text{ kg})(0.035 \times 10^2 \text{ kg})$ and (b) $(148 \times 10^{-6} \text{ m})/(0.4906 \times 10^{-6} \text{ m})$ *(See the Answers to Follow-up Exercises section.)*

There is also a general rule for determining the number of significant figures for the results of addition and subtraction.

> The final result of the addition or subtraction of numbers should have the same number of decimal places as the quantity with the least number of decimal places that was used in the calculation.

Addition and subtraction with significant figures

This rule is applied by rounding off numbers to the least number of decimal places *before* adding or subtracting. The rounding off ensures that a result will have no more reliability than the quantity read from the measuring instrument with the least fine scale. Another way of looking at this is that the rounding off is determined by the quantity with the first doubtful or estimated figure from the left. (Why?)

Example 1.8 ■ Using Significant Figures in Addition and Subtraction: Application of Rules

The following operations are performed by finding the number that has the first doubtful figure in a column (counting from the left with the decimal points aligned) and rounding the other numbers to this column. (Units have been omitted for convenience.)

Addition
In the numbers to be added, note that 23.1 has the least number of decimal places, or has the first doubtful figure in a column counting from the left, the 1 (evaluating from the left with the decimal points of the numbers aligned). The other numbers are rounded to this column and the addition performed:

$$
\begin{array}{ll}
23.1 & 23.1 \\
0.546 \longrightarrow & 0.5 \\
\underline{1.\overset{\cdot}{4}5} \quad (rounding \text{ off}) & \underline{1.5} \\
& 25.1
\end{array}
$$

Subtraction

The same rounding procedure is used. Here, the 157 has the least number of decimal places (none). Rounding the 5.5 and subtracting, we get

$$
\begin{array}{r}
157 \\
-\ 5.5 \\
\hline
\end{array}
\quad \xrightarrow{\text{\textit{(rounding off)}}} \quad
\begin{array}{r}
157 \\
-\ \ 6 \\
\hline
151
\end{array}
$$

(The decimal point, although omitted, is understood in aligning the numbers.)

Follow-up Exercise.   Given the numbers 23.15, 0.546, and 1.058, (a) add the first two and (b) subtract the last number from the first. (*See the Answers to Follow-up Exercises section.*)

Suppose you must deal with mixed operations—multiplication and/or division *and* addition and/or subtraction. What do you do in this case? Just follow your regular rules for order of algebraic operations, and observe significant figures as you go.

The number of digits reported in a result depends on the number of digits in the given data. The rules for rounding will generally be observed for examples in this book. However, there will be exceptions that may make a difference, as explained in the following hint.

Problem-Solving Hint: The "Correct" Answer

When working problems, you naturally strive to get the correct answer and will probably want to check your answers against those listed in the Answers to Odd-numbered Exercises section in the back of the book. However, on occasion you may find that your answer differs slightly from that given even though you have solved the problem correctly. There are several reasons why this could happen.

As stated previously, it is best to round off only the final result of a multi-part calculation, but this is not always convenient in elaborate calculations. Sometimes the results of intermediate steps are important in themselves and need to be rounded off to the appropriate number of digits as if each were a final answer. Similarly, text Examples in this book are often worked in steps for pedagogical reasons, to show the stages in the *reasoning* of the solution. The results obtained when intermediate steps are rounded off may differ slightly from those obtained when only the final answer is rounded.

Rounding differences may also occur when using conversion factors. For example, in changing 5.0 mi to kilometers with the conversion factors listed in the front of the book,

$$
5.0 \ \text{mi} \left(\frac{1.609 \ \text{km}}{1 \ \text{mi}} \right) = (8.045 \ \text{km}) = 8.0 \ \text{km} \quad \textit{(2 significant figures)}
$$

and

$$
5.0 \ \text{mi} \left(\frac{1 \ \text{km}}{0.621 \ \text{mi}} \right) = (8.051 \ \text{km}) = 8.1 \ \text{km} \quad \textit{(2 significant figures)}
$$

The difference arises because of rounding of the conversion factors. Actually, 1 km = 0.6214 mi, so 1 mi = (1/0.6214) km = 1.609 269 km = 1.609 km. (Try repeating these conversions with the unrounded factors and see what you get.) To avoid rounding differences in conversions, we will generally use the multiplication form of a conversion factor, as in the first of the equations above, unless there is a convenient exact factor, such as 1 min/60 s.

Finally, slight differences in answers may occur when different methods are used to solve a problem, even though both may be legitimate. Keep in mind that when solving a problem (a general procedure for which is given in Section 1.7), *if your answer differs from that in the text in only the last digit, the disparity is most likely the result of rounding difference or of an alternative method of solution being used.*

1.7 Problem Solving

An important aspect of physics is problem solving. In general, problem solving involves the application of physical principles and equations to data from a particular situation in order to find some unknown or wanted quantity. There is no universal method for approaching a problem that will automatically produce a solution. (•Fig 1.10).

A few general points, though, are worth keeping in mind:

- *Make sure you understand the problem.* The foundation of successful problem solving is having a thorough grasp of the problem. All too often a student will start working on a problem without really understanding the situation described or knowing what is to be found. (If you don't know exactly what you want, it is very difficult to find it!) Before starting to work a problem, make it a habit to list everything that is given or known (as we will do for most of the Examples in this text). Otherwise, it's all too easy to overlook a critical piece of information.

- To proceed from the given data to the ultimate solution, you may have to devise a strategy or plan. Many problems cannot be solved merely by finding one equation and "plugging in" the given quantities to get the solution. Often you will need to perform intermediate steps, each of which will bring you closer to the final answer. The process is a lot like crossing a stream by stepping from stone to stone—you need to keep in mind where you are and where you're going, and as you plan each step you must think ahead to the next one.

- Remember that equations are expressions of physical principles. Solving problems involves translating principles into the "language" of equations, but you should also be able to look at an equation and say what physical relationship it embodies. If you use an equation without knowing what it means, you *may* be lucky and get the right answer, but you will not be learning much physics—and you probably won't be able to repeat your success when confronted with a slightly different problem.

- A common cause of failure in problem solving is the *misapplication* of physical principles or equations. Principles and equations are generally subject to certain conditions or are limited in their applicability to physical situations.

- Many problems can be solved by more than one method. As with many things in life, however, there is often a hard way and an easy way. If you understand the problem and the relevant principles well, you can probably figure out how to solve the problem in the fastest and easiest way.

Although there is no magic formula for problem solving, there are some sound practices that can be very useful. The steps in the following procedure are intended to provide you with a framework that can be applied to solving most of the problems that you will encounter in this text. We will generally use these steps in dealing with the Example problems throughout this text. Additional helpful problem-solving hints will be given where appropriate in subsequent chapters.

Let's see…how do I solve this?
What are they telling me?
What principles apply?
Oh! I see what to do.

•FIGURE 1.10 Problem solving
There is no set approach in working problems. However, you may find the suggested procedure given here to be helpful.

Suggested Problem-Solving Procedure

Problem-solving procedure:
Say it in words

1. *Read the problem carefully and analyze it. Write down the given data and what you are to find.* Some data may not be given explicitly in numerical form. For example, if a car "starts from rest," its initial speed is zero ($v_o = 0$). In some instances, you may be expected to know certain quantities or to look them up in tables.

Say it in pictures

Say it in equations

Simplify the equations

Check the units

Insert numbers and calculate; check significant figures

Check the answer: Is it reasonable?

2. *Draw a diagram as an aid in visualizing and analyzing the physical situation of the problem where appropriate.* This step may not be necessary in every case, but it is usually helpful and can never do any harm.

3. *Determine which principle(s) and equation(s) are applicable to this situation and how they can be used to get from the information given to what is to be found.* Keep in mind that many problems cannot be solved simply by plugging all of the given data into one equation; you may have to devise a strategy involving several steps.

4. *Simplify mathematical expressions as much as possible through algebraic manipulation before inserting actual values.* Trigonometric relationships (summarized in Appendix I) can sometimes be used to simplify equations. The less calculation you do, the less likely you are to make a mistake—so *don't put the numbers in until you have to.*

5. *Check units before doing calculations.* Make unit conversions if necessary so that all units are in the same system and quantities with the same dimensions have the same units (preferably standard units). This avoids mixed units and is helpful in unit analysis. (Unit checking and conversions are often done when writing the data in Step 1.)

6. *Substitute given quantities into equation(s) and perform calculations. Report the result with the proper units and the proper number of significant figures.*

7. *Consider whether the result is reasonable.* Does the answer have an appropriate magnitude? (This means, is it in the right ballpark?) For example, if a person's calculated mass turns out to be 4.60×10^2 kg, the result should be questioned, since 460 kg corresponds to a weight of 1010 lb.

•Figure 1.11 summarizes these steps in the form of a flow chart. The following Examples illustrate the procedure. The steps are numbered to help you follow along.

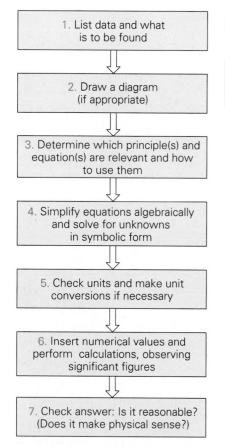

•FIGURE 1.11 A flow chart for the suggested problem-solving procedure

Example 1.9 ■ Finding the Area of a Rectangle: Problem-Solving Procedure

Two students measure the lengths of adjacent sides of their rectangular dorm room. One reports 15 ft, 8 in., and the other reports 4.25 m. What is the area of the room in square meters?

Thinking It Through. The lengths are reported in different units, so to get square meters (m × m), the British units feet and inches must be converted to meters.

Solution.

1. Adjacent sides of a room give its length and width, so we may write

Given: Length = l = 15 ft, 8 in.   *Find:* Area (in square meters)
          Width = w = 4.25 m

2. Sketch a diagram to help you visualize the situation (•Fig. 1.12).

3 **and** 4. For this simple situation, the required equation is well known. The area (A) of a rectangle is $A = l \times w$, both of which are given.

5. A unit change is necessary. Let's first convert the length measurement to inches and then inches to meters:

$$15 \text{ ft} + 8 \text{ in.} = \left(15 \text{ ft} \times \frac{12 \text{ in.}}{1 \text{ ft}} \right) + 8 \text{ in.} = 188 \text{ in.}$$

and

$$188 \text{ in.} \times \frac{2.54 \text{ cm}}{1 \text{ in.}} = 478 \text{ cm} = 4.78 \text{ m}$$

Notice how easy it is to convert units in the decimal metric system (centimeters to meters). Perform the conversion explicitly if necessary, using the conversion factor (1 m/100 cm).

6. Now perform the calculation:

$$A = l \times w = 4.78 \text{ m} \times 4.25 \text{ m}$$

$$= 20.315 \text{ m}^2 = 20.3 \text{ m}^2 \quad \textit{(computed value rounded to 3 sf—why?)}$$

7. The answer appears reasonable. Since 1 m ≈ 3 ft, the dorm room would be about 13 ft by 14 ft, which is about right (but, as always, too small). Suppose you had inadvertently input 47.8 instead of 4.78 on your calculator. The result would be $A = 47.8$ m $\times$ 4.25 m = 203 m². A room with an area of about 200 m² would have dimensions of about 10 m by 20 m, which is roughly 30 ft by 60 ft. Since this is not the size of a typical dorm room, the magnitude of the result should make you suspect there may be an error.

Follow-up Exercise. The dimensions of a textbook are 0.22 m × 0.26 m × 4.0 cm. What volume in a backpack would the book take up? Give the answer in both cubic meters and cubic centimeters. *(See the Answers to Follow-up Exercises section.)*

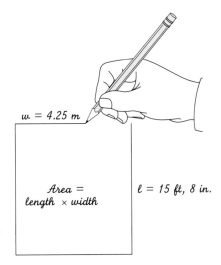

•**FIGURE 1.12 A helpful step in problem solving** Drawing a diagram helps you visualize and better understand the situation. See Example 1.9.

Many problems will involve basic trigonometric functions. The most common functions are given in the marginal note on this page; for other trigonometric relations, see Appendix I or the tables inside the back cover.

Example 1.10 ■ Finding the Length of One Side of a Triangle: Trigonometry Application

A flower bed is laid out in the form of a triangle as shown in •Fig. 1.13. What is the length of the side of the bed that runs along the flagstone walkway?

Thinking It Through. The length of one side of a right triangle and an angle are given. Basic trigonometry applies.

Solution.
1 and 2. In some problems we are given diagrams with data. Here we have

Given: $x = 5.4$ m **Find:** r (long side of triangle)
$\theta = 40°$

3 and 4. Noting in the figure that the flower bed is a right triangle (as indicated by the right angle symbol at the large angle), we can use a trigonometric function to find the hypotenuse r. Recalling that x and r are associated with the cosine for a standard right triangle, we can write

$$\cos \theta = \frac{x}{r} \quad \text{or} \quad r = \frac{x}{\cos \theta}$$

5 and 6. The units are OK. (Since x is in meters and $\cos \theta$ is dimensionless, r will come out in meters.) So put in the data:

$$r = \frac{x}{\cos \theta} = \frac{5.4 \text{ m}}{\cos 40°} = \frac{5.4 \text{ m}}{0.766} = 7.0 \text{ m}$$

7. The magnitude of r seems reasonable compared to the value of x.

Follow-up Exercise. To keep animals out of the flower bed in Fig. 1.13, the owner decides to fence in the perimeter of the bed. What will be the total length of the fence? *(See the Answers to Follow-up Exercises section.)*

Trigonometric functions:

$$\cos \theta = \frac{x}{r} \left(\frac{\text{side adjacent}}{\text{hypotenuse}} \right)$$

$$\sin \theta = \frac{y}{r} \left(\frac{\text{side opposite}}{\text{hypotenuse}} \right)$$

$$\tan \theta = \frac{\sin \theta}{\cos \theta} = \frac{y}{x} \left(\frac{\text{side opposite}}{\text{side adjacent}} \right)$$

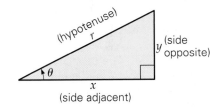

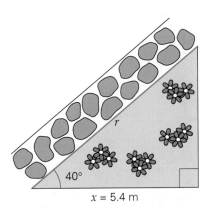

•**FIGURE 1.13 The length of a triangle's side** See Example 1.10.

It is understood that you do not need to write out the steps of problem-solving procedure each time. However, it is a good practice to run through them mentally when working a problem. In the examples in the following chapters, the problem-solving steps will not always be listed as was done here for illustration. Even so, you should be able to see the general pattern outlined above. In some instances, especially when introducing the application of a new principle or concept, problem-solving steps or hints will be listed to promote better understanding.

The main point of this section is that in solving problems, you should have some systematic procedure for analyzing the situation and extracting the information wanted. The suggested problem-solving procedure given here is one option. Let's review it again.

Example 1.11 ■ Finding the Capacity of a Cylinder: Problem Solving in Three Dimensions

A cylindrical container with a height of 28.5 cm and an inside diameter of 10.4 cm is filled with water. What is the mass of the water in kilograms?

Thinking It Through. The dimensions of the cylinder are given, so we can find its volume, or the volume of water it would hold. How is volume related to mass? By density ($\rho = m/V$), and the density of water is known.

Solution.

1. *Read the problem carefully and analyze it.*
   From the problem, we have

   Given: $h = 28.5$ cm (height)   *Find:* m (mass in kg)
   $\qquad\quad d = 10.4$ cm (diameter)

2. *Draw a diagram as an aid in visualizing and analyzing the situation.*
   Sketch the cylinder, showing its dimensions, and add any general information you might know about a cylinder. For example, the circular base has an area of $A = \pi r^2$, where r is the radius and $r = d/2$ (•Fig. 1.14).

3. *Determine which principle(s) and equation(s) are applicable and how to use them.*
   We are given the height and diameter of the cylinder, so we can find its volume, which is the volume of the water it contains. But how is volume related to mass, which is what we want? The link is density (ρ), which we saw earlier (Eq. 1.1) is mass per unit volume:

$$\rho = \frac{m}{V} \quad \text{or} \quad m = \rho V$$

The density of water is not given. However, you can look that up in a table, or you may remember from the definition of the kilogram that $\rho_{H_2O} = 1.00$ g/cm^3 (at maximum density, which will be assumed here). So, knowing the volume and the density of water, we can find the mass.

4. *Simplify expressions algebraically if possible.*
   Should you not remember the formula for the volume of a cylinder, your sketch can be helpful. The area of the circular end of the cylinder is $A = \pi r^2 = \pi d^2/4$ (since $r = d/2$), and the volume is this area times the height of the cylinder, or $V = Ah$. (Formulas for areas and volumes of common shapes are given in Table 5 of Appendix I.)
   We can therefore write an equation for the mass entirely in terms of known quantities:

$$m = \rho V = \rho Ah = \rho \left(\frac{\pi d^2}{4} \right) h$$

5. *Check units before doing calculations.*
   The cylinder dimensions are in centimeters and the density in grams per cubic centimeter. Hence, the mass will come out in grams, which we can convert to kilograms as requested in the problem. (Another option would be to convert to meter and kilogram units first, but this would require more steps.)
   Note: If the given data have mixed units, it is a good practice to convert to consistent units when writing down what is given. This avoids mistakes that might be made later if this step were overlooked.

6. *Substitute the given quantities into the equation and perform calculations, observing significant figures.*

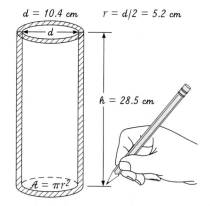

$d = 10.4$ cm $r = d/2 = 5.2$ cm

$h = 28.5$ cm

$A = \pi r^2$

•**FIGURE 1.14 The capacity of a cylinder** Drawing a diagram helps in problem solving. See Example 1.11.

Using the equation derived above and inserting the known quantities, we have

$$m = \rho V = \rho A h = \rho \left(\frac{\pi d^2}{4} \right) h$$

$$= \left(\frac{1.00 \text{ g}}{\text{cm}^3} \right) \left[\frac{\pi (10.4 \text{ cm})^2}{4} \right] (28.5 \text{ cm}) = 2.42 \times 10^3 \text{ g} = 2.42 \text{ kg}$$

7. *Consider whether the result is reasonable.*
It may be difficult to determine whether a result is reasonable because of unfamiliarity with units. Here, we might calculate that about 2.5 kg of water has a weight of about 5.0 lb (since 1 kg is equivalent to 2.2 lb or $\approx$ 2.0 lb). This seems to be a reasonable weight for this volume of water. (The dimensions are approximately those of a 2-liter soda bottle.)

Follow-up Exercise. If the outside diameter of the cylinder in Fig. 1.14 is 10.5 cm, what is the total outside surface area of the sides and base of the cylinder? *(See the Answers to Follow-up Exercises section.)*

Conceptual Example 1.12

Doubling: Big Time

A wise woman (early physicist?) did a favor for a king. The king wanted to repay her and asked what she would like. The wise woman noticed a chessboard in the throne room and told the king she had a humble request, only a single grain of wheat. But, for each of the next 64 days, this grain of wheat would be doubled on each square of the chessboard (64 total squares, 8 squares by 8 squares, as on a checkerboard). She would then take only what was on the last square. The king was delighted and granted the wish. However, the king and his kingdom went bankrupt. Why? *Clearly establish the reasoning used in determining your answer before you check it below. That is, **why** did you select your answer?*

Reasoning and Answer. To use something more familiar for our analysis, let's use a penny instead of a grain of wheat. The king would still think it's a bargain, right? (Be careful.)

This doubling builds faster than you think—something like compounding interest rates, but a bit faster. So, on the first day of doubling, there would be 2^1 (two) pennies on the first square; on the second day, 2^2 ($2 \times 2 = 4$) pennies on the next square; on the third day, 2^3 ($2 \times 2 \times 2 = 8$) pennies on the next; on the fourth day, 2^4 ($2 \times 2 \times 2 \times 2 = 16$) pennies on that square, and so on. The doubling would continue for exactly 64 days.

Rather than doing 2×2 this many times, we simply want to know the value of 2^{64}. On your scientific calculator, you should have a y^x function: Enter ($y =$) 2, push the y^x key, and enter ($x =$) 64. You should get 1.8×10^{19} pennies! Slipping a couple digits (dividing by 100) to convert to dollars, you have 1.8×10^{17} dollars.

A trillion is 10^{12}, so this is $18\,000$ trillion dollars—which beats the U.S. national debt of $5–6 trillion dollars. A very wise, and rich, woman.

Follow-up Exercise. Back to the king and the wheat. Let's say an average grain of wheat has mass 0.010 g. How many metric tons of wheat would the king owe the wise woman? (As a comparison, the annual output of wheat in the United States is on the order of 60 million metric tons.) *(See the Answers to Follow-up Exercises section.)*

Chapter Review

Important Terms

You should be able to define and explain these chapter terms clearly.

standard unit   *2*
system of units   *2*
International System
  of Units (SI)   *3*
SI base units   *3*
SI derived units   *3*
meter (m)   *4*

kilogram (kg)   *5*
second (s)   *5*
mks system   *6*
cgs system   *6*
fps system   *7*
liter (L)   *9*
dimensional analysis   *10*

unit analysis   *11*
density (ρ)   *12*
conversion factor   *13*
exact number   *17*
measured number   *17*
significant figures (sf)   *17*

Important Concepts

- The SI units of length, mass, and time are the meter (m), kilogram (kg), and second (s), respectively.
- To a good approximation, a liter (L) of water has a mass of 1 kg.
- Dimensional analysis and/or unit analysis can be used to determine whether an equation is correct, and unit analysis can be used to find the unit of a quantity.

- The number of significant figures of a numerical quantity is the number of reliably known digits it contains.
- Exercises should be worked with a consistent problem solving procedure.

Important Equations

This equation is a mathematical statement of a concept or principle presented in the chapter. It will be needed in working some Exercises, so a good understanding is important. You should be able to identify the symbols and to explain their relationships before proceeding. (The in-text equation reference number is given for convenience.) Most chapters will have more than one important equation.

Density: $\qquad \rho = \dfrac{m}{V} \quad \left(\dfrac{\text{mass}}{\text{volume}} \right)$ $\qquad$ (1.1)

Exercises

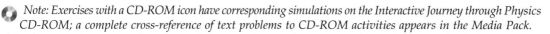 *Note: Exercises with a CD-ROM icon have corresponding simulations on the Interactive Journey through Physics CD-ROM; a complete cross-reference of text problems to CD-ROM activities appears in the Media Pack.*

Throughout the text, many exercise sections will include "paired" and "trio" exercises. These exercise pairs and trios, identified with colored numbers, are intended to assist you in problem solving and learning. In a pair, the first exercise (even numbered) is worked out in the Study Guide so that you can consult it should you need assistance in solving it. The second exercise (odd numbered) is similar in nature, and its answer is given at the back of the book. In a trio, the first exercise (even numbered) is worked out in the Study Guide. The second exercise (odd numbered) is similar in nature, and its answer is given at the back of the book. The third exercise (even numbered) is based on the same physical principle but is to be solved in a different manner. Its answer is given in the Study Guide.

1.2 Units

1. The only SI standard currently represented by an artifact is the (a) meter, (b) kilogram, (c) second, (d) electric charge.

2. Which of the following is *not* an SI base quantity? (a) length, (b) mass, (c) weight, (d) time.

3. Which of the following is the SI unit of mass? (a) pound, (b) gram, (c) kilogram, (d) ton.

4. Are the following reasonable statements? (Justify your answers.) (a) It took 300 L of gasoline to fill up the car's tank. (b) The center on the basketball team is 225 cm tall. (c) The area of a dorm room is 120 m².

5. ■ The metric system is a decimal (base-10) system, and the British system is in part a duodecimal (base-12) system. Discuss the ramifications if the monetary system of the United States used a duodecimal base. What would be the possible values of our coins if this were the case?

6. ■ (a) In the British system, 16 oz = 1 pt and 16 oz = 1 lb. Is there something wrong here? Explain. (b) Here's an old one: a pound of feathers weighs more than a pound of gold. How can that be? (Hint: Look up *ounce* in the dictionary.)

7. ■■ A sailor told you that if his ship is traveling at 25 knots (nautical miles per hour), it is faster than the 25 mi/h your car travels. How can that be?

1.4 Dimensional Analysis*

8. Both sides of an equation are equal in (a) numerical value, (b) units, (c) dimensions, (d) all of the preceding.

9. Unit analysis of an equation cannot tell you if (a) it is dimensionally correct, (b) the equation is correct, (c) the mathematics is correct, (d) both (b) and (c).

10. Can dimensional analysis tell you whether you have used the correct equation in solving a problem? Explain.

11. If an equation has the same dimensions or units on both sides, the equation is (a) dimensionally correct, (b) correct, (c) may be correct, (d) both (a) and (c).

12. ■ Show that the equation $x = x_0 + vt$, where v is velocity and x and x_0 are lengths, is dimensionally correct.

13. ■ If x refers to distance, v_0 and v to velocities, a to acceleration, and t to time, which of the following equations is dimensionally correct? (a) $x = v_0 t + at^3$, (b) $v^2 = v_0^2 + 2at$, (c) $x = at + vt^2$, (d) $v^2 = v_0^2 + 2ax$.

14. ■ Show that $v^2 = v_0^2 + 2ax$ is dimensionally correct. (Here a is acceleration, v and v_0 are velocities, and x is length.)

15. ■■ Use SI unit analysis to show that the equation $A = 4\pi r^2$, where A is the area and r is the radius of a sphere, is dimensionally correct.

16. ■■ You are told that the volume of a sphere is given by $V = \pi d^3/4$, where V is the volume and d is the diameter of a sphere. Is this equation dimensionally correct? (Use SI unit analysis to find out.)

17. ■■ The correct equation for the volume of a sphere is $V = 4\pi r^3/3$, where r is the radius of the sphere. Is the equation in Exercise 16 correct?

18. ■■ If $x = gt^2/2$, where x is length and t is time, is dimensionally correct, what are the SI units of the constant g?

*Dimensions and/or units of velocity and acceleration are given in the chapter.

19. ■■ Is the equation $v = v_0 \sin \theta - gt$ dimensionally correct? Show by using SI unit analysis. (Here v and v_0 are velocities, θ is an angle, t is time, and g is acceleration.)

20. ■■ Using unit analysis, determine whether the following equations are dimensionally correct (t is time, x is length, a is acceleration, and v is velocity): (a) $t = \sqrt{2x/a}$ and (b) $v = \frac{1}{2}(v_0 + a)t$. (See Section 1.4 for units.)

21. ■■ Is the equation for the area of a trapezoid, $A = \frac{1}{2}a(b_1 + b_2)$, where a is the height and b_2 and b_1 are the bases, dimensionally correct (•Fig. 1.15)? If not, how should it be changed to correct it?

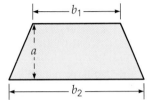

•**FIGURE 1.15**
The area of a trapezoid
See Exercise 21.

22. ■■ One student, using unit analysis, says that the equation $v = \sqrt{2ax}$ is dimensionally correct. Another says it isn't. With whom do you agree?

23. ■■ The equation for the frequency (f) of oscillation of a simple pendulum is $f = \dfrac{1}{2\pi}\sqrt{\dfrac{g}{L}}$, where L is the length of the pendulum string and g is the acceleration due to gravity. Frequency is commonly expressed in units of hertz (Hz). What is a hertz unit in terms of SI base units?

24. ■■■ Newton's second law of motion (Chapter 4) is expressed by the equation $F = ma$, where F represents force, m is mass, and a is acceleration. (a) The SI unit of force is, appropriately, called the newton (N). What are the units of the newton in terms of SI base quantities? (b) Another equation for force associated with uniform circular motion (Chapter 7) is $F = mv^2/r$, where v is speed and r is the radius of the circular path. Does this equation give the same units for the newton?

25. ■■■ Einstein's famous mass–energy equivalence is expressed by the equation $E = mc^2$, where E is energy, m is mass, and c is the speed of light. (a) What are the SI base units of energy? (b) Another equation for energy is $E = mgh$, where m is mass, g is the acceleration due to gravity, and h is height. Does this equation give the same units as in (a)?

1.5 Unit Conversions*

26. A good way to ensure proper unit conversion is to (a) use another measurement instrument, (b) always work in one system of units, (c) use unit analysis, (d) use dimensional analysis.

*Conversion factors are listed inside the front cover.

27. A conversion factor written as 1 mi = 1609 m means that (a) 1 mi is equivalent to 1609 m, (b) this is a true equation, (c) 1 m = 1609 mi, (d) none of the preceding.

28. ■ Figure 1.7 (top) shows the elevation of a location in both feet and meters. If a town is 130 ft above sea level, what is the elevation in meters?

29. ■ If you wanted to express your height as a larger number, which unit in each of the following pairs would you use: (a) meter or yard; (b) decimeter or foot; (c) centimeter or inch?

30. ■ What is the length in feet of (a) a 100-meter dash and (b) a 2.4-meter high jump?

31. ■ If the capillaries of an average adult were unwound and spread out end to end, they would extend to a length of more than 40 000 mi (Fig. 1.8). If you are 1.75 m tall, how many times would the capillary length equal your height?

32. ■ Standing at 1454 ft, the Sears Tower in Chicago is one of the tallest buildings in the world. What is its height in meters?

33. ■ A Boeing 777 jet has a length of 209 ft, 1 in.; a wingspan of 199 ft, 11 in.; and a fuselage diameter of 20 ft, 4 in. What are these dimensions in meters, and how big around is the fuselage?

34. ■ (a) Which holds more soda, and how much more: a half-gallon bottle or a 2-liter bottle? (b) Suppose that a 16-ounce soda and a 500-milliliter soda sell for the same price. Which would give you the most for your money, and how much more (in milliliters) would you get? [Hint: 1 pint = 16 oz.]

35. ■ Jeff regularly buys 18 gal of gas, but the gas station has installed new pumps that are measured in liters. How many liters of gas (rounded off to a whole number) should he ask for?

36. ■ (a) A football field is 300 ft long and 160 ft wide. What are the field's dimensions in meters? (b) A football is 11.0 to $11\frac{1}{4}$ in. long. What is its length in centimeters?

37. ■ Suppose that when the United States goes completely metric, the dimensions of a football field are established as 100 m by 54 m. Which would be larger, the metric football field or a current football field (see Exercise 36a), and what would be the difference in the areas?

38. ■■ A car travels at a constant speed of 15 m/s. How many miles does it travel in 1 h?

39. ■■ Royal Air Force pilot Andy Green broke the sound barrier on land for the first time and achieved a record land speed of more than 763 mi/h in Black Rock Desert, NV, Oct. 15, 1997 (•Fig. 1.16). (a) What is this speed expressed in meters per second? (b) How long would it take the jet-powered car to travel the length of a 300-foot football field at this speed?

•**Figure 1.16 Record run** See Exercise 39.

40. ■■ Which one of the following quantities represents the greatest speed? (a) 1 m/s, (b) 1 km/h, (c) 1 ft/s, (d) 1 mi/h.

41. ■■ An automobile speedometer is shown in •Fig. 1.17. (a) What would the equivalent scale readings (for each empty box) be in kilometers per hour? (b) What would the 65 mi/h speed limit be in kilometers per hour?

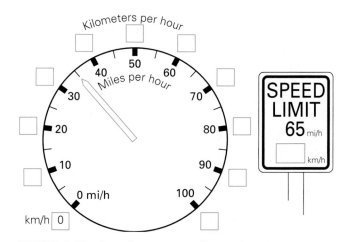
•**FIGURE 1.17 Speedometer readings** See Exercise 41.

42. ■■ A student has a car that gets on average 25.0 mi/gal of gasoline. She plans to spend a year in Europe and take her car with her. (a) What should she expect the car's average gas mileage to be in kilometers per liter? (b) During the year there, she drove 6000 km. Assuming gas costs $5.00/gal in Europe, how much did she spend on fuel? (Compute to the nearest dollar.)

43. ■■ Some common product labels are shown in •Fig. 1.18. From the units on the labels, find (a) the number of milliliters in 2.0 fl. oz and (b) the number of ounces in 100 g.

•**FIGURE 1.18   Conversion factors**   See Exercise 43.

44. ■■ Another set of product labels is shown in •Fig. 1.19, but there is an inconsistency. What is the problem, and which is correct? [Hint: 1 L = 1.057 618 qt.]

•**FIGURE 1.19   An extra milliliter?**   See Exercise 44.

45. ■■ A student was 18 in. long when she was born. She is now 5 ft 6 in. and 20 years old. On average, how many centimeters a year did she grow?

46. ■■ In a "19-inch TV" the measurement indicates the diagonal length of the TV tube. Assuming that the face of the tube is flat and rectangular and that the diagonal makes a 37° angle with the base of the tube, what is the area of the tube face in (a) square inches and (b) square centimeters? (Hint: A diagram as suggested in the problem-solving procedure is helpful here.)

47. ■■ The width and length of a room are 3.2 yd and 4.0 yd, respectively. If the height of the room is 8.0 ft, what is the volume of the room in (a) cubic meters and (b) cubic feet?

48. ■■■ The density of metal mercury is 13.6 g/cm^3. (a) What is this density expressed in kilograms per cubic meter? (b) How many kilograms of mercury would be required to fill a 0.250-liter container?

49. ■■■ (a) What is the density, in kilograms per cubic meter, of a sphere with radius 10 cm and a mass of 10 kg? (b) A second sphere of the same density has a radius of 20 cm. What is its mass?

50. ■■■ In the Bible, Noah is instructed to build an ark 300 cubits long, 50.0 cubits wide, and 30.0 cubits high (•Fig. 1.20).

Historical records indicate a cubit is equal to half of a yard. (a) What would the dimensions of the ark be in meters? (b) What would its volume be in cubic meters? To approximate, assume that the ark was to be rectangular.

•**FIGURE 1.20   Noah and his ark**   See Exercise 50.

1.6 Significant Figures

51. Which of the following has the greatest number of significant figures? (a) 103.07, (b) 124.5, (c) 0.099 16, (d) 5.408×10^5

52. In a multiplication and/or division operation involving the numbers 15 437; 201.08; and 408.0×10^5, the result should be rounded to how many significant figures? (a) 3, (b) 4, (c) 5, (d) any number.

53. ■ Express the length 50 500 μm (micrometers) in centimeters, decimeters, and meters to three significant figures.

54. ■ Using a meterstick, a student measures a length and reports it to be 0.8755 m. What is the smallest division on the meterstick scale?

55. ■ If a measured length is reported as 25.483 cm, could this length have been measured with an ordinary meterstick whose smallest division is millimeters? Discuss in terms of significant figures.

56. ■ Determine the number of significant figures in the following measured numbers: (a) 1.007 m, (b) 8.03 cm, (c) 16.272 kg, (d) 0.015 μs (microseconds).

57. ■ Express each of the numbers in Exercise 56 with two significant figures.

58. ■ Which of the following quantities has three significant figures: (a) 305.0 cm, (b) 0.0500 mm, (c) 1.000 81 kg, (d) 8.06×10^4 m^2?

59. ■ Express each of the following numbers to only three significant figures: (a) 10.072 m, (b) 775.4 km, (c) 0.002 549 kg, (d) 93 000 000 mi.

60. ■■ A rectangle is 3.7 m long and 2.37 m wide. What is its area?

61. ■■ What is the area of a circle of radius 6.37×10^6 m?

62. ■■ The side of a cube is measured, and its volume is reported to be 2.5×10^2 cm^3. What was the measured length of the side of the cube?

63. ■■ The outside dimensions of a cylindrical soda can are reported as 12.559 cm for the diameter and 5.62 cm for the height. What is the total outside area of the can?

64. ■■■ In doing a problem, a student adds 46.9 m and 5.72 m and then subtracts 38 m from the result. What is the final answer?

65. ■■■ Work the following Exercise by two procedures as directed, commenting on and explaining any difference in the answers. Use your calculator for the calculations. Compute $p = mv$, where $v = x/t$. (a) First compute v and then p. (b) Compute $p = mx/t$ without an intermediate step. Given: $x = 8.5$ m, $t = 2.7$ s, and $m = 0.66$ kg.

1.7 Problem Solving

66. An important step in problem solving before mathematically solving an equation is (a) checking units, (b) checking significant figures, (c) consulting with a friend, (d) checking to see if the result is reasonable.

67. An important final step in problem solving before reporting an answer is (a) reading the problem again, (b) saving your calculations, (c) seeing if the answer is reasonable, (d) checking your results with another student.

68. ■ Earth has a mass of 6.0×10^{24} kg and a volume of 1.1×10^{21} m^3. What is Earth's average density?

69. ■ The density of sea water is 1.03 g/cm^3. What is the mass of 1 L of sea water?

70. ■ Does each reported measurement have an appropriate magnitude? (a) A subcompact car has a mass of 1000 kg. (b) A person is 283 cm tall. (c) A runner runs a mile in 3.00×10^8 μs. (d) A normal highway driving speed is 25 m/s. (e) A dachshund stands 10 dm tall. (f) A professor has a waist of 40 cm.

71. ■■ Nutrition Facts labels now appear on most foods. An abbreviated label concerned with fat is shown in •Fig. 1.21. When burned in the body, each gram of fat supplies 9 Calories. (A food Calorie is really a kilocalorie, as we shall see in Chapter 11.) (a) What percentage of the Calories in one serving is supplied by fat? (b) You may notice that your answer doesn't agree with the listed Total Fat percentage in Fig. 1.21. This is because the given Percent Daily Values are the percentages of the maximum recommended amounts of nutrients (in grams) in a 2000-Calorie diet. What are the maximum recommended amounts of total fat and saturated fat for a 2000-Calorie diet?

Nutrition Facts
Serving Size: 1 can
Calories: 310

Amount Per Serving	**% Daily Value***
Total Fat 18 g	28%
Saturated Fat 7g	35%

* Percent Daily Values are based on a 2,000 Calorie diet.

•**FIGURE 1.21   Nutrition Facts**   See Exercise 71.

72. ■■ The thickness of the numbered pages of a textbook is measured to be 3.75 cm. If the last page of the book is numbered 860, what is the average thickness of a page?

73. ■■ A light-year is a unit of distance corresponding to the distance light travels in a vacuum in 1 y. If the speed of light is 3.00×10^8 m/s, what is the length of a light-year in meters?

74. ■■ To go to a football game, you first drive 1000 m north and then 500 m west. What is the straight-line distance from your home to the football field?

75. ■■ Two equally long chains of length 1.0 m are used to support a lamp as shown in •Fig. 1.22. The distance between the two chains is 1.0 m along the ceiling. What is the vertical distance from the lamp to the ceiling?

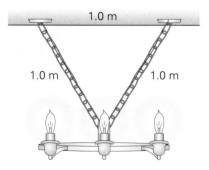

•**FIGURE 1.22   Support the lamp**   See Exercise 75.

76. ■■ Tony's Pizza Palace sells a 9.0-inch (diameter) pizza for $7.95 and a 12-inch pizza for $13.50. Which is the better buy?

77. ■■ In •Fig. 1.23, which black region has greater area, the center circle or the outer ring?

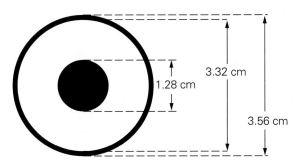

•**FIGURE 1.23 Which black area is greater?** See Exercise 77.

78. ■■ The distance between the bases in a baseball field is 90 ft. What is the straight-line distance in meters between first base and third base?

79. ■■ The Channel Tunnel, or "Chunnel," that runs under the English Channel between Great Britain and France is 31 mi long. (There are actually three separate tunnels.) A shuttle train that carries passengers through the tunnel travels with an average speed of 75 mi/h. On the average, how long in minutes does it take the shuttle to make a one-way trip through the Chunnel?

80. ■■■ When trains travel through the Chunnel (see Exercise 79), friction heats the air. Left unchecked, the air temperature in the tunnel would be unbearable, so a giant air-conditioning system was installed. In the cooling process, water is pumped through a 300-mile network of 24-inch-diameter pipes. How many metric tons of water does the network contain?

81. ■■■ A student wants to determine the distance of a small island from the lake shore (•Fig. 1.24). He first draws a 50-meter line parallel to the shore. Then, he goes to the ends of the line and measures the angles of the lines of sight from the island relative to the line he has drawn. The angles are 30° and 40°. How far is the island from the shore?

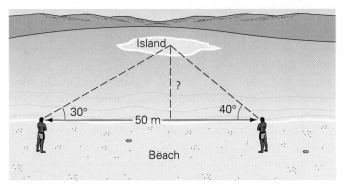

•**FIGURE 1.24 Measuring with lines of sight** See Exercise 81.

Additional Exercises

82. Express the following calculations to the proper number of significant figures: (a) 12.634 + 2.1, (b) 13.5 − 2.134, (c) $\pi(0.25 \text{ m})^2$, (d) $\sqrt{2.37}/3.5$

83. Suppose you are paying $1.20 for 1 gal of gas. Then the United States switches to SI units, and you find that gas costs $0.32/L. Are you paying more after the switch? If so, how much would you be paying?

84. On the average, the human heart beats 70 times a minute (called the pulse rate). On the average, how many times does the heart beat in a 70-year lifetime?

85. The base and the height of a right triangle are 11.2 cm long and 7.5 cm high, respectively. What is the area of the triangle?

86. The general equation for a parabola is $y = ax^2 + bx + c$, where a, b, and c are constants. What are the units of each constant if y and x are in meters?

87. A rectangular block has the dimensions 4.85 cm, 6.5 cm, and 15.51 cm. What is the volume of the block in cubic centimeters?

88. The radius of a solid sphere is 12 cm. What is the surface area of the sphere in (a) square centimeters and (b) square meters? (c) If the mass of the sphere is 9.0 kg, what is the sphere's density in kilograms per cubic meter?

89. A cylindrical drinking glass has an inside diameter of 8.0 cm and a depth of 12 cm. If a person drinks a completely full glass of water, how much (in liters) will be consumed?

90. The top of a rectangular table measures 1.245 m by 0.760 m. (a) What is the smallest division on the scale of the measurement instrument? (b) What is the area of the tabletop?

91. When computing the average speed of a cross-country runner, a student gets 25 m/s. Is this a reasonable result? Justify your answer.

92. The average density of the Moon is 3.36 g/cm³, and the Moon's diameter is 2160 mi. What is the total mass of the Moon in kilograms? (Compare your answer to the value given inside the back cover.)

93. The inner and outer diameters of a thick-walled spherical metal shell are 18.5 cm and 24.6 cm, respectively. What is the volume occupied by the shell itself?

94. An airplane flies 100 mi south from city A to city B, 200 mi east from city B to city C, and then 300 mi north from city C to city D. (a) What is the straight-line distance from A to D? (b) What is the direction of city D relative to city A?

Kinematics: Description of Motion

The cheetah is running at full stride. This fastest of all land animals is capable of attaining speeds up to 113 km/h, or 70 mi/h. The sense of motion in this photograph is so strong you can almost feel the air rushing by you. And yet, this is an illusion. Motion takes place in time, but the photo can only "freeze" a single instant. You'll find that, without the dimension of time, you can hardly describe motion at all.

The description of motion involves the representation of a restless world. Nothing is ever perfectly still. You may sit, apparently at rest, but your blood flows, and air moves into and out of your lungs. The air is composed of gas molecules moving at different speeds and in different directions. And, while you experience stillness, you, your chair, the building you are in, and the air you breathe are all revolving through space with the Earth, part of a solar system in a spiraling galaxy in an expanding universe.

The branch of physics concerned with the study of motion and what produces and affects motion is called **mechanics**. The roots of mechanics and of human interest in motion go back to early civilizations. The study of the motions of heavenly bodies, or celestial mechanics, grew out of the need to measure time and location. Several early Greek scientists, notably Aristotle, put forth theories of motion that were useful descriptions but were later proved to be incorrect. Our currently accepted concepts of motion were formulated in large part by Galileo (1564–1642) and Isaac Newton (1642–1727).

Mechanics is usually divided into two parts: (1) kinematics and (2) dynamics. **Kinematics** deals with the description of the motion of objects without consideration of what causes the motion. **Dynamics** analyzes the causes of motion. This chapter covers kinematics and reduces the description of motion to its simplest terms by considering the simple case of motion in a straight line, or linear motion, which is motion in one dimension (of space). We'll learn to analyze changes in motion—speeding up, slowing down, stopping. Along the way, we'll deal with a particularly interesting case of accelerated motion: free fall under the influence only of gravity. Chapter 3 focuses on motion in two dimensions (which can easily be extended to three dimensions). Chapter 4 investigates dynamics to show what causes changes in motion.

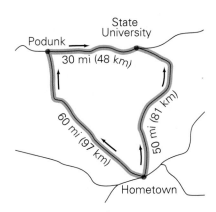

•FIGURE 2.1 **Distance—total path length** In driving to State University from Hometown, one student may take the shortest route and travel a distance of 50 mi (81 km). Another student takes a longer route in order to visit a friend in Podunk before returning to school. The longer trip is in two segments; the distance traveled is the total path length, 60 mi + 30 mi = 90 mi (145 km).

2.1 Distance and Speed: Scalar Quantities

Objectives: To (a) define distance and calculate speed and (b) explain what is meant by a scalar quantity.

Distance

We can observe motion in many instances around us. What is motion? This seems a simple question, but you might have some difficulty giving an immediate answer (and it's not fair to use the verb "moving" to describe motion). After a little thought, you should be able to conclude that **motion** (or moving) involves the changing of position. Motion can be described in part by specifying *how far* something travels in changing position—that is, the distance it travels. **Distance** is simply the *total path length* traversed in moving from one location to another. For example, you may drive to school from your hometown and express the distance traveled in miles or kilometers. In general, the distance between two points depends on the path traveled (•Fig. 2.1).

Along with many other quantities in physics, we say that distance is a scalar quantity. A **scalar quantity** is one with only magnitude, or size. That is, a *scalar* (short for scalar quantity) has only a numerical value, such as 160 km or 100 mi. (Note that the magnitude includes units.) Distance is a scalar quantity; it tells you the magnitude only—how far, but not how far in any physical direction. Other examples of scalars are quantities such as 10 s (time), 3.0 kg (mass), and −20°C (temperature). As we will learn in Section 2.2, there is another type of quantity (a *vector*) that has direction associated with it.

Distance: How far

A scalar quantity has magnitude but no direction.

Speed

When something is in motion, its position changes with time. That is, it moves a certain distance in a given time. Both length and time are therefore important quantities in describing motion. For example, imagine a car and a pedestrian moving down a street and traveling a distance (length) of one block. You would expect the car to travel faster, to cover the distance in a shorter time, than the person does. This relation can be expressed by using length and time to give the rate at which distance is traveled, or **speed**, for each.

Average speed is the distance d traveled, the actual path length, divided by the total time t elapsed in traveling that distance:

Speed: How fast

$$\text{average speed} = \frac{\text{distance traveled}}{\text{total time to travel that distance}}$$

$$avg. \; sp. = \frac{d}{t} \tag{2.1}$$

SI unit of speed: meters per second (m/s)

The SI standard unit of speed is meters per second (length/time), although kilometers per hour is used in many everyday applications. The British standard unit is feet per second, but we often use miles per hour.

Since distance is a scalar quantity (as is time), speed is also a scalar. The distance does not have to be in a straight line (see Fig. 2.1). For example, you probably have computed the average speed of an automobile trip by using the distance obtained from the starting and ending odometer readings. Suppose these readings were 17 455 km and 17 775 km, respectively, for a 4-hour trip. (We'll assume that you have a car with odometer readings in kilometers.) Subtracting the readings gives a traveled distance d of 320 km, so the average speed of the trip is $d/t = 320$ km/4.0 h = 80 km/h (or about 50 mi/h).

Average speed gives a general description of motion over a time interval t. In the case of the auto trip with an average speed of 80 km/h, the car's speed wasn't *always* 80 km/h. With various stops and starts on the trip, the car must have been moving more slowly than the average speed part of the time. It therefore had to be moving more rapidly than the average speed another part of the time. With an average speed, you really don't know how fast the car was moving at any particular time during the trip. Similarly, the average test score of a class doesn't tell you the score of any particular student.

If the time interval t considered becomes smaller and smaller and approaches zero, the speed calculation gives an **instantaneous speed**. This is how fast something is moving *at a particular instant of time*. The speedometer of a car gives an approximate instantaneous speed. For example, the speedometer shown in •Fig. 2.2 indicates a speed of about 44 mi/h, or 70 km/h. If the car travels with constant speed (so the speedometer reading does not change), then the average and instantaneous speeds will be equal. (Do you agree? Think of the average test-score analogy.)

•FIGURE 2.2 Instantaneous speed The speedometer of a car gives the speed over a very short interval of time, so its reading approaches the instantaneous speed. Add the direction of the car's motion, and you have a good approximation of the instantaneous velocity (covered in Section 2.2).

Example 2.1 ■ Slow Motion: *Sojourner* Moves Along

On July 4, 1997, the *Pathfinder Lander* touched down on the surface of Mars. Out rolled the rover named *Sojourner* (•Fig. 2.3)

Sojourner could move at a maximum average speed of about 0.600 m/min. At this speed, how long would it take the rover to travel 3.00 m to get to another rock to analyze?

Thinking It Through. Knowing the average speed and distance, we can compute the time to travel the distance from the defining equation of average speed.

Solution. We list the data and what is to be found in symbol form:

Given: $avg. \; sp. = 0.600 \; \dfrac{m}{min} \left(\dfrac{1 \; min}{60 \; s} \right)$ *Find:* t (time to travel distance d)

$\qquad\qquad = 0.0100 \; m/s$

$\qquad d = 3.0 \; m$

where the unit meters per minute was converted to the standard meters per second.
From Eq. 2.1, we have

•FIGURE 2.3 Away we go! *Sojourner* speeding at 0.600 m/min (0.0100 m/s) along the Martian surface. See Example 2.1.

$$avg. \; sp. = \frac{d}{t}$$

Rearranging, we get

$$t = \frac{d}{avg.\ sp.} = \frac{3.00\ m}{0.0100\ m/s} = 300\ s = 5.00\ min$$

Note that this value is at *Sojourner's maximum* average speed. The trip would probably take longer, as the rover usually had to maneuver around or over small rocks.

Follow-up Exercise. (a) Was it necessary to convert meters per minute to meters per second? Explain. (b) Suppose *Sojourner* took 15.0 min to travel the 3.00 m. What would be the rover's average speed in this case?

2.2 One-Dimensional Displacement and Velocity: Vector Quantities

Objectives: **To (a) define displacement and calculate velocity, and (b) explain the difference between scalar and vector quantities.**

Displacement

As we have seen, distance is a scalar quantity with only magnitude (and units). However, often when we describe motion, more information can be given by adding a *direction*. This is particularly convenient for a change of position in a straight line. We define **displacement** as the straight-line distance between two points, along with the *direction* from the starting point to the final position.

As such, displacement is a **vector quantity**. A vector (short for vector quantity) has both magnitude *and* direction. For example, when we describe the displacement of an airplane as 25 km north, we are giving a *vector* description (magnitude and direction). Other vector quantities include velocity and acceleration, as we will see shortly.

Graphically, vectors are represented by arrows, with the direction of the arrowhead giving the direction of the vector (•Fig. 2.4). The length of the arrow can be drawn to scale—that is, made proportional to the magnitude or numerical value of the vector. Note in the figure that one vector is twice as long as the other.

For straight-line, or linear, motion, it is convenient to specify position by using the familiar two-dimensional Cartesian coordinate system with x and y axes at right angles. A straight-line path can be in any direction, but for convenience we usually orient the coordinate axes so that the motion is along one of them. (See Learn by Drawing.)

Algebra applies to vectors, but we have to know how to specify and deal with the direction part of the vector. This process is relatively simple in one dimension. Vectors are discussed more fully in Chapter 3, but let's take a sneak preview here for displacements used in straight-line, or linear, motion.

To illustrate this, suppose a student moves in a straight line from the lockers toward the physics lab as shown in •Fig. 2.5. The scalar distance traveled (along the x axis) is determined to be 8.0 m (Fig. 2.5a). But, in using vectors to specify the student's displacement, we first use vectors x_1, and x_2 to locate the positions x_1 and x_2 relative to the origin (Fig. 2.5b). It is a common practice to use boldface symbols to represent vectors. So, when you see such symbols, keep in mind that they have both magnitude *and* direction.

To specify the student's displacement between x_1 and x_2, we subtract the position vectors,

$$\Delta \mathbf{x} = \mathbf{x}_2 - \mathbf{x}_1 \tag{2.2}$$

where the Greek letter Δ (delta) is used to represent a change or difference in a quantity. Here, it means the difference in *position* vectors along the x axis.

Equation 2.2 essentially means to subtract the length of $\mathbf{x}_1$ from that of $\mathbf{x}_2$. As a result, we have a new vector $\Delta \mathbf{x}$ that represents the displacement from position x_1 to x_2 (Fig. 2.5c). (Note that we have indicated a direction.)

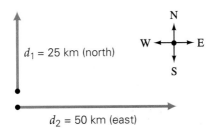

•**FIGURE 2.4 Vector representation** A vector quantity is represented graphically by an arrow. The vector direction is indicated by the arrow direction, and the length of the arrow is proportional to the magnitude of the quantity.

Displacement: How far and in what direction

LEARN BY DRAWING

Cartesian Coordinates and One-dimensional Displacement

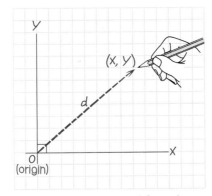

(a) A two-dimensional Cartesian coordinate system. A displacement vector **d** locates a point (x, y).

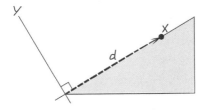

(b) For one-dimensional, or straight-line, motion, it is convenient to orient the motion along one of the coordinate axes.

•FIGURE 2.5 Distance (scalar) and displacement (vector) (a) The distance (straight-line path length) between the lockers and the physics lab is 8.0 m and is a scalar quantity. (b) To indicate a displacement, x_1 and x_2 locate the specific initial and final positions. (c) The displacement is then $\Delta x = x_2 - x_1 = 9.0\text{ m} - 1.0\text{ m} = +8.0\text{ m}$—that is, 8.0 m in the $+x$ direction.

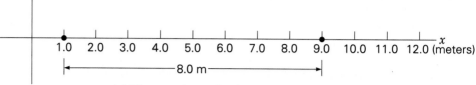

(a) Distance (magnitude or numerical value)

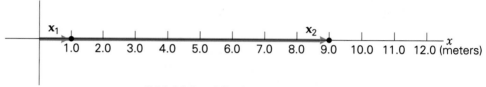

(b) Initial and final position vectors

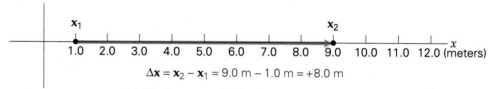

$\Delta \mathbf{x} = \mathbf{x}_2 - \mathbf{x}_1 = 9.0\text{ m} - 1.0\text{ m} = +8.0\text{ m}$

(c) Displacement (magnitude and direction)

Note: In the classroom, many instructors may use arrow notation for vectors: $\vec{A}$. This notation should not be confused with the short line that denotes an average value. Using precise notation is well worth the extra time it takes. Poor or confusing notation can cause problems.

In one dimension, or along a coordinate axis, it is convenient to indicate the direction of a displacement by a plus sign ($+$) or minus sign ($-$). The boldface vector notation has the direction implied. But by using magnitudes and directional signs explicitly for the student's displacement in Fig. 2.5, we have

$$\Delta \mathbf{x} = \mathbf{x}_2 - \mathbf{x}_1 = 9.0\text{ m} - 1.0\text{ m} = +8.0\text{ m}$$

Hence, his displacement (magnitude and direction) is 8.0 m in the positive x direction, as indicated by the positive ($+$) result. (As in "regular" mathematics, the plus sign is often omitted as being understood.)

Suppose the other student in Fig. 2.5 walks from the physics lab (initial, $x_1 = 9.0$ m) to the end of the lockers (final, $x_2 = 1.0$ m). Her displacement would be

$$\Delta \mathbf{x} = \mathbf{x}_2 - \mathbf{x}_1 = 1.0\text{ m} - 9.0\text{ m} = -8.0\text{ m}$$

The minus sign indicates that the direction of the displacement was in the negative x direction, and a representative vector arrow would point that way. We say that the two students' displacements are equal and opposite.

Velocity

As we have seen, speed, like the distance it incorporates, is a scalar quantity—it has magnitude only. Another quantity used to describe motion is *velocity*. Speed and velocity are often used synonomously in everyday conversation, but the terms have different meanings in physics. Speed is a scalar, and velocity is a vector—it has both magnitude and direction.

Velocity tells you how fast something is moving *and* in what direction. And just as we can speak of average and instantaneous speeds, we have average and instantaneous velocities, involving vector displacements. The **average velocity** is the displacement divided by the total travel time:

Velocity: How fast and in what direction

$$\text{average velocity} = \frac{\text{displacement}}{\text{total travel time}}$$

$$\overline{\mathbf{v}} = \frac{\Delta \mathbf{x}}{\Delta t} = \frac{\mathbf{x} - \mathbf{x}_o}{t - t_o} \qquad (2.3)$$

SI unit of velocity: meters per second (m/s)

Definition of:
Average velocity

where the boldfacing indicates vector quantities. (A bar above a symbol, such as $\overline{\mathbf{v}}$, is a common way of indicating an average value.)

Note the general notation that is more convenient with fewer subscripts. In the previous discussion on displacement, $\mathbf{x}_1$ and $\mathbf{x}_2$ were used specifically to refer to the initial and final displacements for clarity. But we can generalize. The subscripts on the $\mathbf{x}_o$ and the t_o refer to the *original*, or initial, conditions—where you start the measurements. The $\mathbf{x}$ and t are the displacement and time, respectively, for some later condition—after a change has occurred. These general variables fit whatever the described condition may be. No numerical subscript is needed.

It is common to take $\mathbf{x}_o = 0$ and $t_o = 0$, since we generally (but not always) start a measurement at the zero end of the measurement scale. Keeping this in mind, we have $\Delta \mathbf{x} = \mathbf{x}$ and $\Delta t = t$ in Eq. 2.3, which give the simpler form of

$$\overline{\mathbf{v}} = \frac{\mathbf{x}}{t} \quad \text{or} \quad \mathbf{x} = \overline{\mathbf{v}}t \qquad (2.4)$$

Note: In Eq. 2.4, t means Δt and x means Δx.

As this equation indicates, the standard units of velocity are the same as those for speed: meters per second (m/s) or feet per second (ft/s).

Also, as discussed for motion in one dimension, it is convenient to use plus and minus signs to indicate the directions of displacements and velocities along the positive and negative axes—for example, $+x$ and $+v$ (or simply x and v) and $-x$ and $-v$. In the case of more than one displacement, the average velocity is equal to the total or net displacement divided by the total time. The total displacement is found by adding the displacements algebraically according to the directional signs.

You might be wondering whether there is a relationship between average speed and average velocity. A quick look at Fig. 2.5 will show you that *if* all the motion is in one direction, the distance is equal to the magnitude of the displacement, and the average speed is the magnitude of the average velocity. However, be careful. This is not true if there is motion in both directions, as the following Example shows.

Example 2.2 ■ There and Back: Average Speed versus Average Velocity

A jogger jogs from one end to the other of a straight 300-meter track (from point A to point B in •Fig. 2.6) in 2.50 min and then turns around and jogs 100 m back toward the starting point (to point C) in another 1.00 min. What are the jogger's average speeds and average velocities in going (a) from A to B and (b) from A to C after turning around?

Thinking It Through. The average speed and average velocity are computed by using the defining equations. But keep in mind that velocity and displacement are vectors, so their directions must be specified.

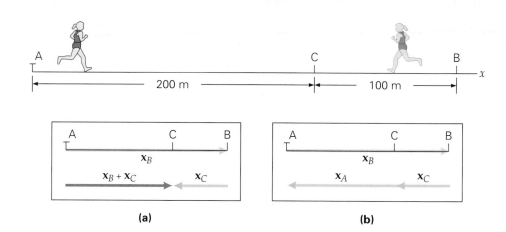

•**FIGURE 2.6 Average speed and velocity** (a) The jogger starts at A, jogs to B, then back to C. What are the average speeds and velocities of both segments of the lap? See Example 2.2. **(b)** If the jogger returns to the starting point (A), the total or net displacement, $\mathbf{x}_C + \mathbf{x}_B + \mathbf{x}_A$, is zero.

Solution. From the problem we have

Given: $\mathbf{x}_B = 300$ m $- 0 = +300$ m (from A to B)   *Find:* Average speeds and
$\qquad \mathbf{x}_C = 200$ m $- 300$ m $= -100$ m (from B to C)   average velocities

$\qquad t_B = (2.50 \text{ min})(60 \text{ s/min}) = 150 \text{ s}$ ⎱ (conversion to
$\qquad t_C = (1.00 \text{ min})(60 \text{ s/min}) = 60.0 \text{ s}$ ⎰ standard units)

(a) The jogger's average speed in going from A to B is computed as follows:

$$avg. \ sp._B = \frac{d}{t} = \frac{300 \text{ m}}{150 \text{ s}} = 2.00 \text{ m/s} \quad \text{(a scalar)}$$

The average velocity in going from A to B with $x_1 = 0$ and $t_1 = 0$ is also easily found, but direction must be indicated. Using Eq. 2.4 and indicating directions by signs,

$$\bar{\mathbf{v}} = \frac{\mathbf{x}_B}{t_B} = \frac{+300 \text{ m}}{150 \text{ s}} = +2.00 \text{ m/s} \quad \text{(a vector)}$$

The positive direction is to the right in Fig. 2.6. Note that the average speed is equal to the magnitude of the average velocity in this instance.

(b) The average speed in going from A to C involves the total *distance* traveled, so

$$avg. \ sp._C = \frac{d}{t} = \frac{300 \text{ m} + 100 \text{ m}}{150 \text{ s} + 60.0 \text{ s}} = 1.90 \text{ m/s}$$

where there are no directional signs. (Why?)

The average velocity, as noted above, involves the *sum* of the vector displacements (vector addition):

$$\bar{\mathbf{v}}_C = \frac{\mathbf{x}_B + \mathbf{x}_C}{t_B + t_C} = \frac{+300 \text{ m} + (-100 \text{ m})}{150 \text{ s} + 60.0 \text{ s}} = +0.952 \text{ m/s}$$

Note that direction makes a difference; the average speed is not equal to the magnitude of the average velocity in this case. This is because the displacements add vectorially. Notice in Fig. 2.6a, which illustrates this simple case of vector addition, that adding $\mathbf{x}_B$ and $\mathbf{x}_C$ gives an effective, or resultant, displacement from the initial starting point (A) to the stopping point (C), or $+200$ m.

Suppose that next the jogger sprints from point C back to the starting point (A) in 0.500 min. What are the average speed and average velocity for the whole lap? The additional data are $\mathbf{x}_A = 0 - 200$ m $= -200$ m (Fig. 2.6b) and $t_A = 30.0$ s. Then, for the round trip, the average speed is

$$avg. \ sp._A = \frac{d}{t} = \frac{300 \text{ m} + 100 \text{ m} + 200 \text{ m}}{150 \text{ s} + 60.0 \text{ s} + 30.0 \text{ s}} = 2.50 \text{ m/s}$$

The average velocity is

$$\overline{\mathbf{v}}_A = \frac{\mathbf{x}_B + \mathbf{x}_C + \mathbf{x}_A}{t_B + t_C + t_A} = \frac{+300 \text{ m} + (-100 \text{ m}) + (-200 \text{ m})}{240 \text{ s}} = 0$$

The average velocity in this case is zero!

The total displacement is measured from the initial starting point to the final stopping point. When the jogger comes back to the starting point, the displacement is zero and so, therefore, is the average velocity. This is true for any round trip (•Fig. 2.7). Notice in Fig. 2.6b that the (vector) addition of $\mathbf{x}_C + \mathbf{x}_A$ gives a combined, or resultant, vector that is equal and opposite (in direction) to $\mathbf{x}_B$. The sum of all three vectors is zero—they cancel each other, so to speak.

Follow-up Exercise. (a) Example 2.2 shows that it is possible to have a zero average velocity. Is it possible for the average speed to be zero? Explain. (b) It took 6 y and 2.3 billion mi for the *Galileo* spacecraft to reach Jupiter in late 1995, where it went into orbit and sent a probe into the Jovian atmosphere. What was *Galileo's* average speed in miles per hour for the trip to Jupiter?

As Example 2.2 shows, average velocity provides only a very general description of motion. One way to obtain a closer look at motion is to take smaller time segments, that is, to let the time of observation (Δt) become smaller. As with speed, when Δt approaches zero, we obtain the **instantaneous velocity,** which describes how fast something is moving and in what direction at a particular instant of time.

In one dimension, uniform motion means motion with a constant velocity (constant magnitude *and* constant direction). For example, the car in •Fig. 2.8 has a uniform velocity (as well as a uniform speed). It travels the same distance in equal time intervals (50 km each hour), and the direction of its motion does not change.

Graphical Analysis

Graphical analysis is often helpful in understanding motion and its related quantities. For example, the motion of the car in Fig. 2.8a may be represented on a plot of position versus time, or x versus t. As can be seen from Fig. 2.8b, a straight line is obtained for a uniform, or constant, velocity on such a graph.

Recall from Cartesian graphs of y versus x that the slope of a straight line is given by $\Delta y/\Delta x$. Here, with a plot of x versus t, the slope of the line, $\Delta x/\Delta t$, is equal to the average velocity $\overline{v} = \Delta x/\Delta t$). For uniform motion, this is equal to the instantaneous velocity. That is, $\overline{v} = v$. (Why?) The numerical value of the slope is the magnitude of the velocity and the sign of the slope gives direction. A positive slope indicates that x increases with time, so the motion is in the positive x direction. (The plus sign is often omitted as being understood.)

Suppose that a plot of position versus time for a car's motion was a straight line with a negative slope, as in •Fig. 2.9. What does this slope indicate? As the figure shows, the position (x) values get smaller with time at a constant rate, indicating that the car was traveling in uniform motion in the negative x direction.

In most instances the motion of an object is nonuniform, meaning that different distances are covered in equal intervals of time. An x versus t plot for such motion in one dimension is a curved line, as illustrated in •Fig. 2.10. The average velocity of a particular interval of time is the slope of a straight line between the two points on the curve that correspond to the starting and ending times of the interval. In the figure, the average velocity of the total trip is the slope of the straight line joining the beginning and ending points of the curve (t_1 and t_2).

•FIGURE 2.7 Back home again! Despite having covered nearly 110 m on the base paths at the moment the runner slides through the batter's box (his original position) into home plate, his displacement is zero—at least, if he is a right-handed hitter. No matter how fast he ran the bases, his average velocity for the round trip is also zero. (What would his displacement be if he bats lefty? his average velocity?)

Instantaneous velocity: How fast and in what direction right now

The word *uniform* means constant.

Distance

| | 50 km | 100 km | 150 km |
| | 1.0 h | 2.0 h | 3.0 h |

Time

Δx (km)	Δt (h)	$\Delta x/\Delta t$
50	1.0	50 km/1.0 h = 50 km/h
100	2.0	100 km/2.0 h = 50 km/h
150	3.0	150 km/3.0 h = 50 km/h

(a)

•**FIGURE 2.8 Uniform linear motion—constant velocity** In uniform linear motion, an object travels at a constant velocity, covering the same distance in equal time intervals. **(a)** Here, a car travels exactly 50 km each hour. **(b)** An x-versus-t plot is a straight line, since equal displacements are covered in equal times. The numerical value of the slope of the line is equal to the magnitude of the velocity, and the sign of the slope gives its direction. (Average velocity equals instantaneous velocity in this case. Why?)

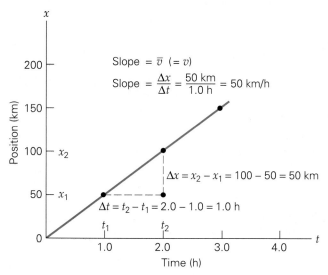

Slope $= \bar{v} \ (= v)$

Slope $= \dfrac{\Delta x}{\Delta t} = \dfrac{50 \text{ km}}{1.0 \text{ h}} = 50 \text{ km/h}$

$\Delta x = x_2 - x_1 = 100 - 50 = 50$ km

$\Delta t = t_2 - t_1 = 2.0 - 1.0 = 1.0$ h

Uniform velocity

(b)

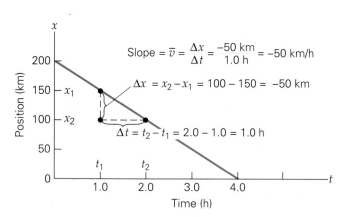

Slope $= \bar{v} = \dfrac{\Delta x}{\Delta t} = \dfrac{-50 \text{ km}}{1.0 \text{ h}} = -50 \text{ km/h}$

$\Delta x = x_2 - x_1 = 100 - 150 = -50$ km

$\Delta t = t_2 - t_1 = 2.0 - 1.0 = 1.0$ h

•**FIGURE 2.9 Position-versus-time graph for an object in uniform motion in the −x direction** A straight line on an x-versus-t plot with a negative slope indicates uniform motion in the −x direction. Note that the object's location changes at a constant rate. At $t = 4.0$ h, the object is at $x = 0$. How does the graph look if the motion continues for $t > 4.0$ h?

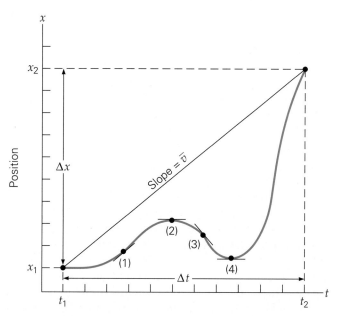

•**FIGURE 2.10 Position-versus-time graph for an object in nonuniform linear motion** For a nonuniform velocity, an x-versus-t plot is a curved line. The slope of the line between two positions is the average velocity between those positions, and the instantaneous velocity is the slope of a line tangent to the curve at any point. Four tangent lines are shown. Can you describe the object's motion in words?

The instantaneous velocity is equal to the slope of a straight line tangent to the curve at a specific point. Four tangent lines are shown in Fig. 2.10. At (1) the slope is positive, and the motion is in the positive x direction. At (2) the slope of a horizontal tangent line is zero, so there is no motion. That is, the object has stopped ($v = 0$). At (3) the slope is negative, so the object is moving in the negative x direction. Thus, it stopped and changed direction at point (2). What is happening at point (4)?

Drawing various tangent lines along the curve, we see that their slopes vary, indicating that the instantaneous velocity is changing with time. An object in nonuniform motion speeds up and/or slows down. How we describe motion with changing velocity is the topic of Section 2.3.

But, before we speed things up (or slow them down), consider the following Example.

Conceptual Example 2.3

A Round Trip: Using Definitions

In driving the usual route to school, a student computes the average speed for the trip to be 20 km/h. On the return trip along the same route, there is less traffic and the average speed is 30 km/h. What is the average speed for the total trip: (a) less than 25 km/h, (b) 25 km/h, (c) greater than 25 km/h? *Clearly establish the reasoning used in determining your answer before checking it below. That is, **why** did you select your answer?*

Reasoning and Answer. It is tempting to choose 25 km/h as the correct answer, as this is the average of the two average speeds. However, this is not the case. The average of the average speeds would be 25 km/h only if the student spent the *same amount of time* traveling at each speed. For example, suppose you spent an hour traveling at each of the given speeds. Then it is easy to see that the average speed for the total trip would be 25 km/h: *avg. sp.* = (20 km + 30 km)/2.0 h = 25 km/h.

But in this case, the *distances traveled* rather than the times traveled are the same. To travel the same distance, more time must be spent traveling at 20 km/h than at 30 km/h. Hence the total average is weighted toward the slower average speed, and the answer is (a).

You can confirm this by finding the actual average speed, using the definition of average speed: the total distance divided by the time to travel this distance. We are not given the distance, but let's say that it is d one way, for a total round-trip distance of $2d$. The times for each part of the trip are given by $t_1 = d/avg.\ sp._1$ and $t_2 = d/avg.\ sp._2$. Thus, the average speed of the total trip, by definition, is

$$avg.\ sp. = \frac{\text{total distance}}{\text{time to travel distance}}$$

$$= \frac{2d}{t_1 + t_2} = \frac{2d}{d/avg.\ sp._1 + d/avg.\ sp._2} = \frac{2}{1/avg.\ sp._1 + 1/avg.\ sp._2}$$

where the d in the denominator was factored out and canceled. So,

$$avg.\ sp. = \frac{2}{1/avg.\ sp._1 + 1/avg.\ sp._2}$$

$$= \frac{2}{1/20 + 1/30} = \frac{2}{3/60 + 2/60} = \frac{2}{5/60} = 120/5 = 24\ \text{km/h}$$

where the units were omitted in the calculation for convenience.

Follow-up Exercise. What average speed on the return trip would give an average of 25 km/h for the total trip?

2.3 Acceleration

Objectives: To (a) explain the relationship between velocity and acceleration and (b) perform graphical analyses of acceleration.

The basic description of motion involves the time rate of change of position, which we call velocity. Going one step further, we can consider how this *rate of change* changes. Suppose that something is moving at a constant velocity and then the velocity changes. Such a change in velocity is called an *acceleration*. The gas pedal on an automobile is commonly called the accelerator. When you press down on the accelerator, the car speeds up; when you let up on the accelerator, the car slows down. In either case, there is a change in velocity with time. We define **acceleration** as the time rate of change of velocity.

Analogous to average velocity is the **average acceleration**, or the change in velocity divided by the time taken to make the change:

$$\text{average acceleration} = \frac{\text{change in velocity}}{\text{time to make the change}}$$

$$\bar{\mathbf{a}} = \frac{\Delta \mathbf{v}}{\Delta t} = \frac{\mathbf{v} - \mathbf{v_o}}{t - t_o} \tag{2.5}$$

SI unit of acceleration: meters per second squared (m/s^2)

In Eq. 2.5, we use the boldface vector notation because, in general, the velocities may be in different directions. Since velocity is a vector quantity, so is acceleration. Analogous to instantaneous velocity is **instantaneous acceleration**, which is the acceleration at a particular instant of time.

The dimensions of acceleration are $([L]/[T])/[T]$, from $\Delta v/\Delta t$. The SI units of acceleration are therefore meters per second per second, that is, $(m/s)/s$ or $m/s \cdot s$, commonly expressed as meters per second squared (m/s^2). In the British system, the units are feet per second squared (ft/s^2).

Since velocity is a vector quantity, having both magnitude and direction, a change in velocity may involve either or both of these factors. An acceleration, therefore, may result from a change in *speed* (magnitude), a change in *direction*, or a change in *both*, as illustrated in •Fig. 2.11.

For the special case of straight-line, or linear, motion, plus and minus signs can be used to indicate velocity directions (so we can do away with the boldface), as was done for linear displacements. Then, Eq. 2.5 can be written as

$$\bar{a} = \frac{v - v_o}{t} \tag{2.6}$$

where t_o is taken to be zero (v_o may not be zero, so it cannot generally be omitted).

Margin notes:

Acceleration: The time rate of change of velocity

Definition of:
Average acceleration

Instantaneous acceleration: a change in velocity *right now*

Note: In compound units, multiplication is indicated by a dot.

LEARN BY DRAWING

Signs of Velocity and Acceleration

a positive
v positive
Result: Faster in +x direction
$-x$ .... $+x$

a negative
v positive
Result: Slower in +x direction
$-x$ .... $+x$

a positive
v negative
Result: Slower in −x direction
$-x$ .... $+x$

a negative
v negative
Result: Faster in −x direction
$-x$ .... $+x$

Example 2.4 ■ Slowing It Down: Average Acceleration

A couple in their sport utility vehicle (SUV) is traveling 90 km/h down a straight highway. They see an accident in the distance, so the driver slows down to 40 km/h in 5.0 s. What is the average acceleration of the SUV?

Thinking It Through. To find the average acceleration, the variables as defined in Eq. 2.6 must be given, and they are.

Solution. From the problem, we have the following data. [With the motion in a straight line, the instantaneous velocities are assumed to be in the positive direction, and conversions to standard units (kilometers per second to meters per second) are

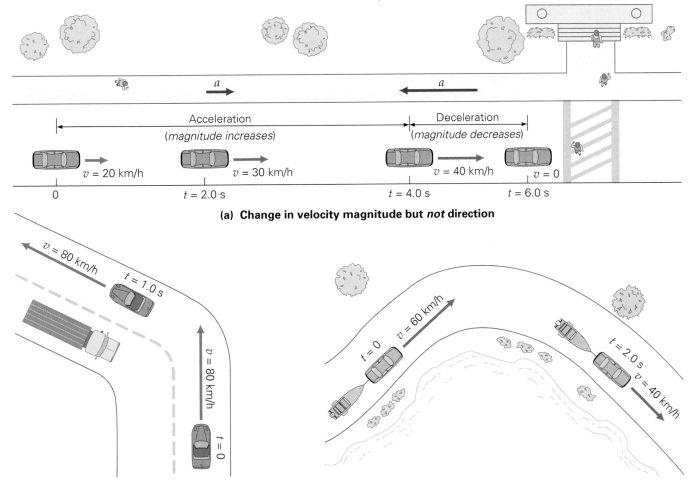

(a) Change in velocity magnitude but *not* direction

(b) Change in velocity direction but *not* magnitude

(c) Change in velocity magnitude *and* direction

•**FIGURE 2.11   Acceleration—the time rate of change of velocity**   Since velocity is a vector quantity, with magnitude and direction, an acceleration can occur when there is **(a)** a change in magnitude but not direction, **(b)** a change in direction but not magnitude, or **(c)** a change in both magnitude and direction.

made right away since it is noted that the time is given in seconds. In general, we always work with acceleration in standard units.]

Given: $v_0 = (90 \text{ km/h}) \left(\dfrac{0.278 \text{ m/s}}{1 \text{ km/h}} \right)$ **Find:** $\bar{a}$ (average acceleration)

$\qquad = 25 \text{ m/s}$

$\qquad v = (40 \text{ km/h}) \left(\dfrac{0.278 \text{ m/s}}{1 \text{ km/h}} \right)$

$\qquad = 11 \text{ m/s}$

$\qquad t = 5.0 \text{ s}$

Given the initial and final velocities and the time interval, the average acceleration can be found by using Eq. 2.6:

$$\bar{a} = \frac{v - v_0}{t} = \frac{11 \text{ m/s} - 25 \text{ m/s}}{5.0 \text{ s}} = -2.8 \text{ m/s}^2$$

The minus sign indicates the direction of the (vector) acceleration. In this case, the acceleration is opposite to the direction of the motion ($+v$), and the car slows. Such an acceleration is sometimes called a *deceleration*, since the car is slowing.

Follow-up Exercise.   Does a negative acceleration necessarily mean that a moving object is slowing down (decelerating) or its speed is decreasing?

Although acceleration can vary with time, our study of motion in this chapter will be restricted to constant accelerations for simplicity. (An important constant acceleration is the acceleration due to gravity near the Earth's surface, which will be considered in the next section.) Since for a constant acceleration the average is equal to the constant value ($\bar{a} = a$), the bar over the acceleration in Eq. 2.6 may be omitted. Thus, for a constant acceleration, the equation relating velocity, acceleration, and time is commonly written (rearranging Eq. 2.6) as follows:

$$v = v_\mathrm{o} + at \qquad \textit{(constant acceleration only)} \qquad (2.7)$$

(Note that the term at represents the *change* in velocity that occurs, since $at = v - v_\mathrm{o} = \Delta v$.)

Example 2.5 ■ Fast Start, Slow Stop: Motion with Constant Acceleration

A drag racer starting from rest accelerates in a straight line at a constant rate of 5.5 m/s² for 6.0 s. (a) What is the racer's velocity at the end of this time? (b) If a parachute deployed at this time causes the racer to slow down uniformly at a rate of 2.4 m/s², how long will it take the racer to come to a stop?

Thinking It Through. It is noted that the racer first speeds up and then slows down, so close attention must be given to the directional signs of the vector quantities. The answers can then be found by using the appropriate equations.

Solution. Taking the initial motion to be in the positive direction, we have

Given: (a) $v_\mathrm{o} = 0$ (at rest)   *Find:* v (final velocity)
         $a = 5.5$ m/s²
         $t = 6.0$ s
     (b) $v_\mathrm{o} = v$ [from part (a)]   *Find:* t (time)
         $v = 0$ (comes to stop)
         $a = -2.4$ m/s² (opposite direction of v_o)

The data have been listed in two parts. This helps avoid confusion with symbols. Note that the final velocity (v) that is to be found in part (a) becomes the initial velocity (v_o) for part (b).

(a) To find v, we use Eq. 2.7 directly:

$$v = v_\mathrm{o} + at = 0 + (5.5 \text{ m/s}^2)(6.0 \text{ s}) = 33 \text{ m/s}$$

(b) Here we want time, so solving Eq. 2.6 for t and using $v_\mathrm{o} = 33$ m/s from part (a), we have

$$t = \frac{v - v_\mathrm{o}}{a} = \frac{0 - 33 \text{ m/s}}{-2.4 \text{ m/s}^2} = 14 \text{ s}$$

Note that the time comes out positive, as it should. We start implicitly at zero time (taking $t_\mathrm{o} = 0$ when the parachute is deployed), and time goes forward or in a "positive direction."

Follow-up Exercise. What is the racer's instantaneous velocity 10 seconds after the parachute is deployed?

Motions with constant accelerations are easy to represent graphically by using instantaneous velocity versus time. A v versus t plot is a straight line whose slope is equal to the acceleration, as illustrated in •Fig. 2.12. Note that Eq. 2.7 can be written $v = at + v_\mathrm{o}$, which, as you may recognize, has the form of an equation of a straight line, $y = mx + b$ (slope m and intercept b). In Fig. 2.12a, the motion is in the positive direction and the acceleration adds to the velocity for a time t, as illustrated by the

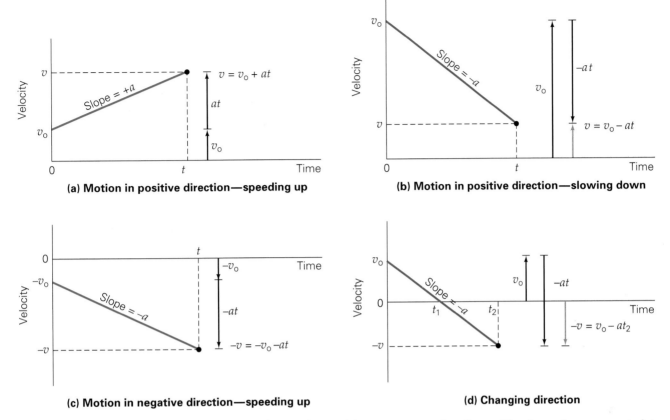

(a) Motion in positive direction—speeding up

$v = v_0 + at$

Slope = $+a$

(b) Motion in positive direction—slowing down

Slope = $-a$

$-at$

$v = v_0 - at$

(c) Motion in negative direction—speeding up

Slope = $-a$

$-v_0$

$-at$

$-v = -v_0 - at$

(d) Changing direction

Slope = $-a$

$-at$

$-v = v_0 - at_2$

•**FIGURE 2.12 Velocity-versus-time graphs for motions with constant accelerations** The slope of a v versus t plot is the acceleration. **(a)** A positive slope indicates an increase in the velocity in the positive direction. The vertical arrows to the right indicate how the acceleration adds velocity to the initial velocity v_0. **(b)** A negative slope indicates a decrease in the initial velocity v_0, or a deceleration. **(c)** Here a negative slope indicates a negative acceleration, but the initial velocity is in the negative direction,$-v_0$, so the speed of the object increases in that direction. **(d)** The situation here is initially similar to that in (b) but ends by resembling that in (c). Can you explain what happened at time t_1?

vertical arrows at the right of the graph. Here the slope is positive, $+a$. In Fig. 2.12b, the negative slope $(-a)$ indicates a negative acceleration that produces a slowing down, or deceleration. However, Fig. 2.12c illustrates how a negative acceleration can speed things up (for motion in the negative direction). The situation in Fig. 2.12d is slightly more complex. Can you explain what is happening there?

When an object moves at a constant acceleration, its velocity changes by the same amount in each time unit. For example, if the acceleration is 10 m/s² in the same direction as the initial velocity, the object's velocity increases by 10 m/s in each second. Suppose that the object has an initial velocity (v_0) of 20 m/s at $t_0 = 0$. Then, for $t = 0$, 1.0, 2.0, 3.0, and 4.0 s, the velocities are 20, 30, 40, 50, and 60 m/s, respectively. The average velocity over the 4-second interval is $\bar{v} = 40$ m/s.

This average velocity may be computed in the regular manner, or you may immediately recognize that the uniformly increasing series of numbers 20, 30, 40, 50, and 60 has an average value of 40 (the midway value of the series). Note that the average of the extreme (initial and final) values also gives the average of the series—that is, $(20 + 60)/2 = 40$. This is true in general: When the velocity changes at a uniform rate because of a constant acceleration, $\bar{v}$ is the average of the initial and final velocities:

$$\bar{v} = \frac{v + v_0}{2} \qquad \text{(constant acceleration only)} \qquad (2.8)$$

Example 2.6 ■ On the Water: Using Multiple Equations

A motorboat starting from rest on a lake accelerates in a straight line at a constant rate of 3.0 m/s² for 8.0 s. How far does the boat travel during this time?

Thinking It Through. Since we have only one equation for distance (Eq. 2.4, $x = \bar{v}t$), this equation cannot be used directly. The average velocity must first be found, so multiple equations are involved.

Solution. Reading the problem and summarizing the given data and what is to be found, we have

Given: $v_\text{o} = 0$ *Find:* x (distance)
         $a = 3.0$ m/s²
         $t = 8.0$ s

(It is noted that all of the units are standard.)

In analyzing the problem, we might reason as follows: To find x, we will need to use Eq. 2.4, $x = \bar{v}t$. (The average velocity $\bar{v}$ must be used because the velocity is changing and not constant.) With t given, the solution to the problem then involves finding $\bar{v}$. By Eq. 2.8, $\bar{v} = (v + v_\text{o})/2$, so with $v_\text{o} = 0$, we need find only the final velocity v to solve the problem. Equation 2.7, $v = v_\text{o} + at$, enables us to calculate v from the given data. So, we have the following:

The velocity of the boat at the end of 8.0 s is

$$v = v_\text{o} + at = 0 + (3.0 \text{ m/s}^2)(8.0 \text{ s}) = 24 \text{ m/s}$$

The average velocity over that time interval is

$$\bar{v} = \frac{v + v_\text{o}}{2} = \frac{24 \text{ m/s} + 0}{2} = 12 \text{ m/s}$$

Finally, the magnitude of the displacement, which in this case is the same as the distance traveled, is

$$x = \bar{v}t = (12 \text{ m/s})(8.0 \text{ s}) = 96 \text{ m}$$

Follow-up Exercise. (Sneak preview.) In Section 2.4, the following equation will be derived: $x = v_\text{o}t + \frac{1}{2}at^2$. Use the data in this Example to see if this equation gives the distance traveled.

2.4 Kinematic Equations (Constant Acceleration)

Objectives: **To (a) explain the constant acceleration kinematic equations and (b) apply them to physical situations.**

The description of motion in one dimension with constant acceleration requires only three basic equations. From previous sections, these are

$$x = \bar{v}t \tag{2.4}$$

$$\bar{v} = \frac{v + v_\text{o}}{2} \qquad \textit{(constant acceleration only)} \tag{2.8}$$

$$v = v_\text{o} + at \qquad \textit{(constant acceleration only)} \tag{2.7}$$

(Keep in mind that the first equation is general and is not limited to situations in which there is constant acceleration, as the latter two are.)

However, as Example 2.6 showed, the description of motion in some instances requires multiple applications of these equations, which may not be obvious at first. It would be helpful if there were a way to reduce the number of operations in solving kinematic problems, and there is—combining equations algebraically.

For instance, suppose we want an expression that gives displacement (x) in terms of time and acceleration rather than in terms of time and average velocity (as in Eq. 2.4). We can eliminate v from Eq. 2.4 by substituting for v from Eq. 2.8 into Eq. 2.4:

$$x = \bar{v}t = \left(\frac{v + v_0}{2}\right)t$$

Then substituting for v from Eq. 2.7 gives

$$x = \left(\frac{v + v_0}{2}\right)t = \left[\frac{(v_0 + at) + v_0}{2}\right]t$$

Simplifying gives

$$x = v_0 t + \tfrac{1}{2}at^2 \qquad \textit{(constant acceleration only)} \qquad (2.9)$$

Essentially, this series of steps was done in Example 2.6. This combined equation allows the distance traveled by the motorboat in that Example to be computed directly:

$$x = v_0 t + \tfrac{1}{2}at^2 = 0 + \tfrac{1}{2}(3.0 \text{ m/s}^2)(8.0 \text{ s})^2 = 96 \text{ m}$$

Much easier, isn't it?

Perhaps we want an expression that gives velocity as a function of position (x) rather than time (as in Eq. 2.7). We can eliminate t from Eq. 2.7 by using Eq. 2.4 in the form $t = x/\bar{v}$:

$$v = v_0 + a\left(\frac{x}{\bar{v}}\right)$$

Then we replace $\bar{v}$, using Eq. 2.8:

$$v = v_0 + \frac{ax}{\dfrac{v + v_0}{2}} \qquad \text{or} \qquad v - v_0 = \frac{2ax}{v + v_0}$$

Simplify by using the algebraic relationship $(v + v_0)(v - v_0) = v^2 - v_0^2$, we have

$$v^2 = v_0^2 + 2ax \qquad \textit{(constant acceleration only)} \qquad (2.10)$$

Problem-Solving Hint

Students in introductory physics courses are sometimes overwhelmed by the various kinematic equations—and this is only Chapter 2! Keep in mind that equations and mathematics are the "tools" of physics; as any mechanic or carpenter will tell you, tools make your work easier as long as you are familiar with them and know how to use them. The same is true with our physics "tools."

Basically, just this set of equations

$$v = v_0 + at \qquad\qquad (2.7)$$

$$x = v_0 t + \tfrac{1}{2}at^2 \qquad\qquad (2.9)$$

$$v^2 = v_0^2 + 2ax \qquad\qquad (2.10)$$

is used to solve the majority of kinematic problems. (Occasionally we are interested in average speed or velocity, but, as pointed out earlier, averages don't tell you a great deal.) Now, note that each of these three equations has four variables. Three of these must be known in a particular equation to solve for the fourth, unknown variable. Always try to understand and visualize a problem, but listing the data as in the Suggested Problem-Solving Procedure in Chapter 1 may help you decide which equation to use, since you generally need to know three variables. Remember this as you go over the remaining Examples in the chapter.

Also, don't overlook any *implied data*. For example, what does "dropped" in the following Example imply?

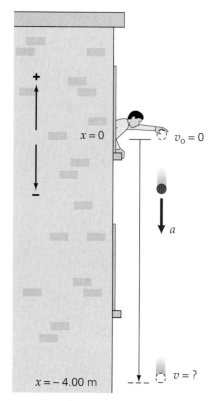

x = 0 $v_0 = 0$

a

x = −4.00 m   v = ?

•**FIGURE 2.13   A dropped ball**   A sketch to help visualize Example 2.7.

Example 2.7 ■ A Dropped Ball: Using the Kinematic Equations

A ball is dropped out of a window near the top of a building. If the ball accelerates toward the ground at a rate of 9.80 m/s², what is its velocity when it has fallen 4.00 m?

Thinking It Through.   Usually, one of the kinematic equations will provide a solution. A good way to determine which equation to use is to look at what is given and what is to be found.

Solution.   Here, we list the given data and make a sketch to help visualize the situation (•Fig. 2.13).

Given: $v_0 = 0$ (dropped)   *Find:* v (velocity)
         $a = -9.80$ m/s²
         $x = -4.00$ m

Note that three variables are given: The statement that the ball is dropped means that the initial velocity is zero, $v_0 = 0$. Upward is usually taken as the positive direction and downward as the negative direction. The magnitudes of **a** and **x** are then written with minus signs to indicate their directions.

No directional sign is explicitly assigned to the velocity we wish to find. In this case you should realize that the final velocity will be negative, or in the downward direction.

Since time is not directly involved, the velocity may be obtained directly from Eq. 2.10:

$$v^2 = v_0{}^2 + 2ax = 0 + 2(-9.80 \text{ m/s}^2)(-4.00 \text{ m}) = 78.4 \text{ m}^2/\text{s}^2$$

and

$$v = \pm\sqrt{78.4 \text{ m}^2/\text{s}^2} = -8.85 \text{ m/s}$$

where the negative root is chosen because the motion is downward (that is, in the negative direction) in this physical situation. (Keep in mind that when you take the square root of a quantity you always obtain both positive and negative roots.)

Using the combined forms of the equations can often save you a lot of steps and calculations.

Follow-up Exercise.   In this Example, how long does it take for the ball to fall the 4.00 m?

TABLE 2.1   Equations for Linear Motion with Constant Acceleration*

$x = \bar{v}t$	(2.4)†
$\bar{v} = \dfrac{v + v_0}{2}$	(2.8)
$v = v_0 + at$	(2.7)
$x = v_0 t + \frac{1}{2}at^2$	(2.9)
$v^2 = v_0{}^2 + 2ax$	(2.10)

*It is assumed that $x_0 = 0$ and that the velocity is v_0 at $t_0 = 0$. The initial position and time, x_0 and t_0, may be included for general cases, for example, $x - x_0 = \bar{v}(t - t_0)$.

†Note that Eq. 2.4 is not limited to constant acceleration but applies generally.

For your convenience, the equations for linear motion with constant acceleration are summarized in Table 2.1. Note that only the first three are basic equations; the last two are convenient combinations. Also, they are quite versatile.

Conceptual Example 2.8

Two Race Cars: The Effect of Squared Quantities

During some time trials, race car A, starting from rest, accelerates uniformly along a straight, level track for a particular time interval. Car B, also starting from rest, accelerates at the same rate but for twice the time. At the ends of their respective acceleration periods, which of these statement is true: (a) Car A has traveled a greater distance, (b) car B has traveled twice as far as car A, (c) car B has traveled four times as far as car A, (d) both cars have traveled the same distance? *Clearly establish the reasoning and physical principle(s) used in determining your answer before checking it below. That is, why did you select your answer?*

Reasoning and Answer. It is given that $v_o = 0$ and that the acceleration is the same for both cars. To find the distances traveled in a time t, you would use Eq. 2.9, which, with $v_o = 0$, becomes $x = \frac{1}{2}at^2$. The important thing to note here is that the distance increases as t^2. That is, if you double the time, the distance quadruples (increases by a factor of four).

In this example, car B accelerates twice as long as car A, or $t_B = 2t_A$, so car B travels four times as far as car A and the answer is (c). Expressed mathematically,

$$x_A = \tfrac{1}{2}at_A{}^2 \quad \text{and} \quad x_B = \tfrac{1}{2}at_B{}^2 = \tfrac{1}{2}a(2t_A)^2 = \tfrac{1}{2}a(4t_A{}^2) = 4(\tfrac{1}{2}at_A{}^2) = 4x_A$$

How do their distances compare if $t_B = 3t_A$?

Follow-up Exercise. In this Example, how do the speeds of the cars compare at the ends of the acceleration periods?

Example 2.9 ■ Putting on the Brakes: Vehicle Stopping Distance

The stopping distance of a vehicle after the brakes have been applied is an important factor in road safety. This distance depends on the initial speed $(+v_o)$ and the braking capacity, or deceleration, $-a$, which is assumed to be constant. (Recall that the minus sign indicates that the acceleration is in the negative direction. In this case the sign of acceleration is opposite that of the velocity, which is taken to be positive. Thus the car slows to a stop.) Express the stopping distance x in terms of these quantities.

Thinking It Through. Again, a kinematic equation is required, and the appropriate one is determined by what is given and what is to be found. Notice the distance x is wanted, and time is not involved.

Solution. Here, we are working with variables, so we can represent quantities only in symbolic form.

Given: v_o (positive direction)   *Find:* x (in terms of the given variables)
$\quad\quad\;\; -a$ (opposite direction of v_o)
$\quad\quad\;\; v = 0$ (car comes to stop)

Again, it is helpful to make a sketch of the situation, particularly when directional vector quantities are involved (•Fig. 2.14). Since Eq. 2.10 has the variables we want, it should allow us to find the stopping distance. Expressing the negative acceleration explicitly gives

$$v^2 = v_o{}^2 + 2(-a)x = v_o{}^2 - 2ax$$

Since the vehicle comes to a stop ($v = 0$), we can solve for x:

$$x = \frac{v_o{}^2}{2a}$$

This gives us x expressed in terms of the vehicle's initial speed and stopping acceleration.

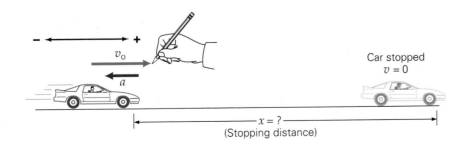

•FIGURE 2.14 Vehicle stopping distance A sketch to help visualize Example 2.9.

Notice that the stopping distance x is proportional to the *square* of the initial speed. Doubling the initial speed therefore increases the stopping distance by a factor of 4 (for the same deceleration). That is, if the stopping distance is x_1 for an initial speed of v_1, then for a two-fold initial speed ($v_2 = 2v_1$), the stopping distance would increase four-fold:

$$x_1 = \frac{v_1^2}{2a}$$

$$x_2 = \frac{v_2^2}{2a} = \frac{(2v_1)^2}{2a} = 4\left(\frac{v_1^2}{2a}\right) = 4x_1$$

We can get the same result by directly using ratios:

$$\frac{x_2}{x_1} = \frac{v_2^2}{v_1^2} = \left(\frac{v_2}{v_1}\right)^2 = 2^2 = 4.$$

Do you think this is an important consideration in setting speed limits, for example, in school zones? (The driver's reaction time should also be considered. A method for approximating a person's reaction time is given in Section 2.5.)

Follow-up Exercise. A driver traveling at 25 mi/h in a school zone can brake to an emergency stop in 8.0 ft. Under similar conditions, what would be the braking distance if the car were traveling at 40 mi/h?

2.5 Free Fall

Objective: **To use the kinematic equations to analyze free fall.**

One of the more common cases of constant acceleration is the acceleration due to gravity near the Earth's surface. When an object is dropped, its initial velocity is zero (at the instant it is released). At a later time while falling, it has a nonzero velocity. There has been a change in velocity and so, by definition, an acceleration. This **acceleration due to gravity** (g) has an approximate magnitude of

Definition of:
Acceleration due to gravity

$$g = 9.80 \text{ m/s}^2 \qquad \textit{acceleration due to gravity}$$

or 980 cm/s^2 and is directed downward (toward the center of the Earth). In British units, the value of g is about 32.2 ft/s^2.

The values given here for g are only approximate because the acceleration due to gravity varies slightly at different locations as a result of differences in elevation and regional average mass density of the Earth. These small variations will be ignored in this book unless otherwise noted. (Gravitation is studied in more detail in Chapter 7.) Air resistance is another factor that affects the acceleration of a falling object. But for relatively dense objects and over the short distances of fall commonly encountered, air resistance produces only a small effect, which will also be ignored here for simplicity. (The frictional effect of air resistance will be considered in Chapter 4.)

Definition of:
Free fall

*Objects in motion solely under the influence of gravity are said to be in **free fall**.* The words "free fall" bring to mind dropped objects that are moving downward under the influence of gravity ($g = 9.80$ m/s^2 in the absence of air resistance). However, the term can be applied in general to any vertical motion under the sole influence of gravity. Objects released from rest or thrown upward or downward are in free fall once they are released. That is, after $t = 0$ (the time of release), only gravity is acting and influencing the motion. (Even when an object projected upward is traveling upward, *it is still accelerating downward*.) Thus, the set of equations for motion in one dimension (in Table 2.1) can be used to describe generalized free fall.

The acceleration due to gravity g is the *constant* acceleration for all free-falling objects, regardless of their mass or weight. It was once thought that heavier bodies fall faster than lighter bodies. This was part of Aristotle's theory of motion. You can easily observe that a coin falls faster than a sheet of paper when dropped simultaneously from the same height. But in this case air resistance plays a noticeable role. If the paper is crumpled into a compact ball, it gives the coin a better race. Similarly, a feather "floats" down much more slowly than a coin falls. However, in a near-vacuum, where there is negligible air resistance, the feather and the coin fall with the same acceleration—the acceleration due to gravity (•Fig. 2.15).

Astronaut David Scott performed a similar experiment on the Moon in 1971 by simultaneously dropping a feather and a hammer from the same height. He did not need a vacuum pump: The Moon has no atmosphere and therefore no air resistance. The hammer and the feather reached the lunar surface together, but both fell at a slower rate than on Earth. The acceleration due to gravity near the Moon's surface is about one-sixth of that near the Earth's surface ($g_M \approx g/6$).

Currently accepted ideas about the motion of falling bodies are due in large part to Galileo. He challenged Aristotle's theory and experimentally investigated the motion of objects. Legend has it that Galileo studied the accelerations of falling bodies by dropping objects of different weights from the top of the Leaning Tower of Pisa (see the Insight on p. 52).

It is customary to use y to represent the vertical direction and to take upward as positive (as with the vertical y axis of Cartesian coordinates). Since the acceleration due to gravity is always downward, it is in the negative direction. This negative acceleration, $a = -g = -9.80 \text{ m/s}^2$, should be substituted into the equations of motion. However, the relationship $a = -g$ may be expressed explicitly in the equations for linear motion (see Table 2.1):

Note: While traveling upward ($+v$), a projected object is accelerating downward, or has an acceleration in the downward direction ($-g$).

$$y = \bar{v}t \tag{2.4'}$$

$$\bar{v} = \frac{v + v_o}{2} \tag{2.8'}$$

Free-fall equations with

$$v = v_o - gt \qquad \text{—g expressed explicitly} \tag{2.7'}$$

$$y = v_o t - \tfrac{1}{2}gt^2 \tag{2.9'}$$

$$v^2 = v_o{}^2 - 2gy \tag{2.10'}$$

Note: Here g refers to the magnitude of acceleration due to gravity.

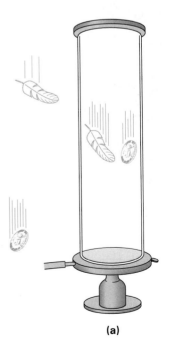

(a)

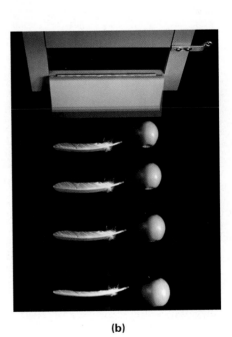

(b)

•**FIGURE 2.15 Free fall and air resistance** (a) When dropped simultaneously from the same height, a feather falls more slowly than a coin because of air resistance. But when both objects are dropped in an evacuated container with a good partial vacuum, where air resistance is negligible, the feather and the coin fall together with a constant acceleration. (b) An actual demonstration with multiflash photography. An apple and a feather are released simultaneously through a trap door into a large vacuum chamber, and they fall together—almost. Although the chamber has a partial vacuum, there is still some air resistance. (How can you tell?)

Galileo Galilei and the Leaning Tower of Pisa

Galileo Galilei (Fig. 1) was born in Pisa, Italy, in 1564 during the Renaissance. Today, he is known throughout the world by his first name and often referred to as the father of modern science or the father of modern mechanics and experimental physics, which attests to the magnitude of his scientific contributions.

One of Galileo's greatest contributions to science was the establishment of the scientific method—that is, investigation through experiment. In contrast, Aristotle's approach was based on logical deduction. By the scientific method, for a theory to be valid it must predict or agree with experimental results. If it doesn't, it is invalid or requires modification. Galileo said, "I think that in the discussion of natural problems we ought not to begin at the authority of places of Scripture, but at sensible experiments and necessary demonstrations."[*]

Probably the most popular and well-known legend about Galileo is that he performed experiments with falling bodies by dropping objects from the Leaning Tower of Pisa (Fig. 2). There is some doubt as to whether Galileo actually did this, but there is little doubt that he questioned Aristotle's view on the motion of falling objects. In 1638, Galileo wrote:

> Aristotle says that an iron ball of one hundred pounds falling from a height of one hundred cubits reaches the ground before a one-pound ball has fallen a single cubit. I say that they arrive at the same time. You find, on making the experiment, that the larger outstrips the smaller by two finger-breadths, that is, when the larger has reached the ground, the other is short of it by two finger-breadths; now you would not hide behind these two fingers the ninety-nine cubits of Aristotle.[†]

This and other writings show that Galileo was aware of the effect of air resistance.

The experiments at the Tower of Pisa were supposed to have taken place around 1590. In his writings of about that time, Galileo mentions dropping objects from a high tower, but never specifically names the Tower of Pisa. A letter written to Galileo from another scientist in 1641 describes the dropping of a cannon ball and a musket ball from the Tower of Pisa. The first account of Galileo doing a similar experiment was written a dozen years after his death by Vincenzo Viviani, his last pupil and first biographer. It is not known whether Galileo told this story to Viviani in his declining years or Viviani created this picture of his former teacher.

The important point is that Galileo recognized (and probably experimentally showed) that free-falling objects fall with the same acceleration regardless of their mass or weight (see Fig. 2.15). Galileo gave no reason why all objects in free fall have the same acceleration, but Newton did, as you will learn in a later chapter.

[*]From *Growth of Biological Thought: Diversity, Evolution & Inheritance*, by F. Meyr (Cambridge, MA: Harvard University Press, 1982).

[†]From *Aristotle, Galileo, and the Tower of Pisa*, by L. Cooper (Ithaca, NY: Cornell University Press, 1935).

FIGURE 2   The Leaning Tower of Pisa   The tower, constructed as a belfry for the nearby cathedral, was built on shifting subsoil. Construction began in 1173, and the tower started to shift one way and then the other, before inclining in its present direction. Today, the tower leans about 5 m (16 ft) from the vertical at the top. It has been closed since 1990, and efforts are being made to stabilize the leaning.

FIGURE 1   Galileo   Galileo is alleged to have performed free-fall experiments by dropping objects off the Leaning Tower of Pisa.

The origin ($y = 0$) of the reference frame is usually taken to be at the initial position of the object. Since upward is generally taken to be in the positive direction (+y axis on a graph), writing $-g$ explicitly in the equations reminds you of directional differences, and the value of g is inserted as 9.80 m/s². However, the choice is arbitrary. The equations can be written with $a = g$, for example, $v = v_o + gt$, with the directional minus sign associated directly with g. In this case, a value of -9.80 m/s² must be substituted for g each time.

Note that you have to be explicit about the directions of vector quantities. The displacement y and the velocities v and v_o may be positive or negative, depending on the direction of motion. The use of these equations and the sign convention are illustrated in the following Examples.

Example 2.10 ■ A Stone Thrown Downward: The Kinematic Equations Revisited

A boy on a bridge throws a stone vertically downward with an initial velocity of 14.7 m/s toward the river below. If the stone hits the water 2.00 s later, what is the height of the bridge above the water?

Thinking It Through. A free-fall problem, but note that the initial velocity is downward, or negative. It is important to express this explicitly.

Solution. As usual, we first write down what is given and what is to be found.

Given: $v_o = -14.7$ m/s (downward taken as   *Find:* y (height)
    $t = 2.00$ s   the negative direction)
    g (= 9.80 m/s²)

Notice that g is now a positive number, since the directional minus sign has already been put into the previous equations of motion. After a while, you will probably just write down the symbol g, since you will be familiar with its numerical value. This time, draw a sketch of your own to help you analyze the situation.

Which equation(s) will provide the solution using the given data? It should be evident that the distance the stone travels in a time t is given directly by Eq. 2.9':

$$y = v_o t - \tfrac{1}{2}gt^2 = (-14.7 \text{ m/s})(2.00 \text{ s}) - \tfrac{1}{2}(9.80 \text{ m/s}^2)(2.00 \text{ s})^2$$

$$= -29.4 \text{ m} - 19.6 \text{ m} = -49.0 \text{ m}$$

The minus sign indicates that the displacement is downward, which agrees with what you know from the statement of the problem.

Follow-up Exercise. How much longer would it take for the stone to reach the river if the boy in this Example had dropped the ball rather than thrown it?

Example 2.11 ■ Measuring Reaction Time: Free Fall

Reaction time is the time it takes a person to notice, think, and act in response to a situation—for example, the time between first observing and then responding to an obstruction on the road ahead while you are driving an automobile. Reaction time varies with the complexity of the situation (and with the individual). In general, the largest part of a person's reaction time is spent thinking, but practice in dealing with a given situation can reduce this time.

A person's reaction time can be measured by having another person drop a ruler (without warning) from above and through the thumb and forefinger as shown in •Fig. 2.16. The first person grasps the falling ruler as quickly as possible, and the length of the ruler below the top of the finger is noted. If on the average the ruler descends 18.0 cm before it is caught, what is the person's average reaction time?

•**FIGURE 2.16 Reaction time**
A person's reaction time can be measured by having the person grasp a dropped ruler. See Example 2.11.

Thinking It Through. Both distance and time are involved. This indicates which kinematic equation should be used.

Solution. Notice that only the distance of fall is given. However, we know a couple of other things, such as v_o and g, so

Given: $y = -18.0$ cm $= -0.180$ m   *Find:* t (reaction time)
$v_o = 0$
$g \; (= 9.80 \text{ m/s}^2)$

(Note that the distance y has been converted directly to meters.) We can see that Eq. 2.9' applies here:

$$y = v_o t - \tfrac{1}{2}gt^2$$

or

$$y = -\tfrac{1}{2}gt^2 \quad (\text{with } v_o = 0)$$

Solving for t gives

$$t = \sqrt{\frac{2y}{-g}} = \sqrt{\frac{2(-0.180 \text{ m})}{-9.80 \text{ m/s}^2}} = 0.192 \text{ s}$$

Try this with a fellow student and measure your reaction time. Why do you think another person should drop the ruler rather than yourself?

Follow-up Exercise. A popular trick is to substitute a crisp dollar bill lengthwise for the ruler in Fig. 2.16, telling the person that he or she can have the dollar if caught. Is this a good deal? (The length of a dollar is 15.7 cm.)

Example 2.12 ■ Free Fall Up and Down: Using Implicit Data

A worker on a scaffold on a billboard throws a ball straight up. It has an initial velocity of 11.2 m/s when it leaves his hand at the top of the billboard (•Fig. 2.17). (a) What is the maximum height the ball reaches relative to the top of the billboard? (b) How long does it take to reach this height? (c) What is the position of the ball at $t = 2.00$ s?

Thinking It Through. In (a), only the upward part of the motion has to be considered. Note that the ball stops (zero velocity) at the maximum height, which allows this height to be found. (b) Knowing the maximum height, we can determine the upward time of flight. In (c), the distance–time equation (Eq. 2.9') applies at any time and gives the position (y) of the ball relative to the launch point.

Solution. It might appear that all that is given in the general problem is the initial velocity v_o. However, a couple of other things are implied because they are understood. One is the acceleration g, and the other is the velocity at the maximum height where the ball stops. Here, in changing direction, the velocity of the ball is momentarily zero, so we have

Given: $v_o = 11.2$ m/s   *Find:* (a) y_{max} (maximum height)
$g \; (= 9.80 \text{ m/s}^2)$ (b) t_u (time upward)
$v = 0 \; (\text{at } y_{max})$ (c) y (at $t = 2.00$ s)
$t = 2.00$ s (for part c)

(a) Notice that we reference the height ($y = 0$) to the top of the billboard. For this part of the problem, we need be concerned with only the upward motion—a ball is thrown upward and stops at its maximum height y_{max}. With $v = 0$ at this height, y_{max} may be found directly from Eq. 2.10':

$$v^2 = 0 = v_o{}^2 - 2gy_{max}$$

and

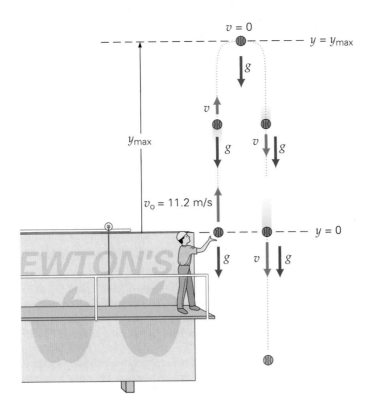

$v = 0$

$y = y_{max}$

g

v

g

v

g

y_{max}

$v_o = 11.2$ m/s

$y = 0$

g

v

g

$$y_{max} = \frac{v_o^2}{2g} = \frac{(11.2 \text{ m/s})^2}{2(9.80 \text{ m/s}^2)} = 6.40 \text{ m}$$

relative to the top of billboard ($y = 0$, see Fig. 2.17).

(b)   The time the ball travels upward is designated t_u. This is the time it takes to reach y_{max}, where $v = 0$. Then, we know v_o and v, so the time t_u can be found directly from Eq. 2.7':

$$v = 0 = v_o - gt_u$$

and

$$t_u = \frac{v_o}{g} = \frac{11.2 \text{ m/s}}{9.80 \text{ m/s}^2} = 1.14 \text{ s}$$

(c)   The height of the ball at $t = 2.00$ s is given directly by Eq. 2.9':

$$y = v_o t - \tfrac{1}{2}gt^2$$

$$= (11.2 \text{ m/s})(2.00 \text{ s}) - \tfrac{1}{2}(9.80 \text{ m/s}^2)(2.00 \text{ s})^2 = 22.4 \text{ m} - 19.6 \text{ m} = 2.8 \text{ m}$$

Note that this is 2.8 m above, or measured upward from, the reference point ($y = 0$). The ball has reached its maximum height and is on its way back down.

Considered from another reference point, this situation is like dropping a ball from a height of y_{max} above the top of the billboard with $v_o = 0$ and asking how far it falls in a time $t = 2.00$ s $- t_u = 2.00$ s $- 1.14$ s $= 0.86$ s. The answer is

$$y = v_o t - \tfrac{1}{2}gt^2 = 0 - \tfrac{1}{2}(9.80 \text{ m/s}^2)(0.86 \text{ s})^2 = -3.6 \text{ m}$$

This is the same as the position found above but is measured with respect to the maximum height as the reference point; that is,

$$y_{max} - 3.6 \text{ m} = 6.4 \text{ m} - 3.6 \text{ m} = 2.8 \text{ m}$$

Follow-up Exercise.   At what height does the ball in this Example have a speed of 5.00 m/s? (Hint: This occurs twice—once on the way up, and once on the way down.)

Here are a couple of interesting facts about the vertical projectile motion of an object thrown upward in the absence of significant air resistance. First, the times of flight upward and downward are the same. That is, the time it takes the object to reach its maximum height is the same as the time it takes the object to fall from the maximum height back to the initial starting point. Note that at the very top of the trajectory, the object's velocity is zero for an instant, but the acceleration (even at the top) remains a constant 9.8 m/s^2 downward. If the acceleration went to zero, it would remain there and gravity would be turned off!

Second, the object returns to the starting point with the same speed as it was launched. (The velocities have the same magnitude but are opposite in direction.)

These facts can be shown mathematically. See Exercises 68 and 81.

Problem-Solving Hint

When working vertical projection problems involving motions up and down, it is often convenient to divide the problem into two parts and consider each separately. As seen in Example 2.12, for the upward part of the motion, the velocity is zero at the maximum height. A zero quantity simplifies the calculations. Similarly, the downward part of the motion is analogous to that of an object dropped from a height where the initial velocity is zero.

However, as Example 2.12 shows, the appropriate equations may be used directly at any position or time of the motion. For instance, note in part (c) that the height was found directly for a time *after* the ball had reached the maximum height. The velocity of the ball at that time could also have been found directly from Eq. 2.7′, $v = v_o - gt$.

Chapter Review

Important Terms

You should be able to define and explain these chapter terms clearly.

mechanics   *32*
kinematics   *33*
dynamics   *33*
motion   *33*
distance   *33*
scalar (quantity)   *33*
speed   *33*

average speed   *33*
instantaneous speed   *34*
displacement   *35*
vector (quantity)   *35*
velocity   *37*
average velocity   *37*
instantaneous velocity   *39*

acceleration   *42*
average acceleration   *42*
instantaneous acceleration   *42*
acceleration due to gravity   *50*
free fall   *50*

Important Concepts

- Motion involves a change of position; it can be described in terms of the distance moved (a scalar) or the displacement (a vector).

- A scalar quantity has magnitude (value and units) only; a vector quantity has magnitude *and* direction.

- Speed (a scalar) is the time rate of change of position, and average velocity (a vector) is the time rate of change of displacement.

- Acceleration is the time rate of change of velocity, and hence it is a vector quantity.

- Most kinematic equations are limited to constant accelerations only.

- An object in free fall has a constant acceleration of magnitude $g = 9.80$ m/s^2 (acceleration due to gravity) near the surface of the Earth.

Important Equations

These equations are mathematical statements of the concepts and principles presented in the chapter. They will be needed in working problem Exercises, so a good understanding is important. You should be able to identify the symbols and to explain the relationships and limitations before proceeding. (In-text equation reference numbers are given for convenience.)

Average speed: $avg.\ sp. = \dfrac{d}{t}$ (2.2)

Kinematic Equations for Linear Motion with Constant Acceleration

(with $x_o = 0$): $x = \bar{v}t$ (2.4)

(general, not limited to constant acceleration):

$\bar{v} = \dfrac{v + v_o}{2}$ (2.8)

$v = v_o + at$ (2.7)

$x = v_o t + \frac{1}{2}at^2$ (2.9)

$v^2 = v_o^2 + 2ax$ (2.10)

constant acceleration only

Acceleration Due to Gravity:

$g = 9.80\ \text{m/s}^2 = 980\ \text{cm/s}^2\ (\approx 32.2\ \text{ft/s}^2)$

Kinematic Equations Applied to Free Fall
(with downward taken as the negative direction, $y_o = 0$, and $-g$ expressed explicitly):

$y = \bar{v}t$ (2.4')

$\bar{v} = \dfrac{v + v_o}{2}$ (2.8')

$v = v_o - gt$ (2.7')

$y = v_o t - \frac{1}{2}gt^2$ (2.9')

$v^2 = v_o^2 - 2gy$ (2.10')

g constant

Exercises

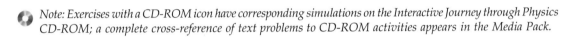 Note: Exercises with a CD-ROM icon have corresponding simulations on the Interactive Journey through Physics CD-ROM; a complete cross-reference of text problems to CD-ROM activities appears in the Media Pack.

2.1 Distance and Speed: Scalar Quantities
and
2.2 Displacment and Velocity: Vector Quantities

1. Distance is always (a) equal to the magnitude of the corresponding displacement, (b) less than or equal to the magnitude of the corresponding displacement, (c) greater than or equal to the magnitude of the corresponding displacement.

2. Two people choose different reference points to specify an object's position. Does this difference affect the coordinates of the object? How about its displacement? Explain.

3. A scalar quantity has (a) only magnitude, (b) only direction, (c) both direction and magnitude.

4. Can a displacement from one point to another be zero, yet the distance involved in moving between these points be nonzero? How about the reverse situation? Explain.

5. You are told that a person walks 500 m. What can you safely say about the person's final position relative to the starting point?

6. An object travels at a constant velocity. What is the relationship of the object's speed to its velocity?

7. Speed is the magnitude of velocity. Is average speed the magnitude of average velocity? Explain.

8. ■ What is the magnitude of the displacement of a car that travels half a lap along a circle of radius of 150 m?

9. ■ A student throws a rock straight upward from shoulder level, which is 1.65 m above the ground. What is the displacement of the rock when it hits the ground?

10. ■ A small airplane flies in a straight line at a speed of 150 km/h. How long does it take the plane to fly 250 km?

11. ■ A bus travels at an average speed of 90 km/h. On the average, how far does the bus travel in 20 min? Would this be the actual travel distance? Explain.

12. ■ A motorist drives 150 km from one city to another in 2.5 h but makes the return trip in only 2.0 h. What are the average speeds for (a) each half of the round trip and (b) the total trip?

13. ■ (a) A senior citizen walks 0.30 km in 10 min going around a shopping mall. What is her average speed in meters per second? (b) If she wants to increase her average speed by 20% in walking a second lap, what would her travel time in minutes have to be?

14. ■ A race car travels a complete lap on a circular track of radius 500 m in 50 s. (a) What is the average speed of the race car? (b) What is the average velocity of the race car?

15. ■■ A boy runs 30 m east, 40 m north, 50 m west. What is his net displacement?

16. ■■ A student throws a ball vertically upward such that it travels 10 m to its maximum height. If the ball is caught at the initial height 2.4 s after being thrown, (a) what is the ball's average speed? (b) Its average velocity?

17. ■■ An insect crawls along the edge of a rectangular swimming pool of length 27 m and width 21 m (•Fig. 2.18). If it crawls from corner A to corner B in 30 min, (a) what is its average speed in centimeters per second? (b) What is the magnitude of its average velocity?

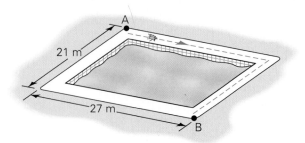

•FIGURE 2.18 Speed versus velocity See Exercise 17. (Not drawn to scale; insect is displaced for clarity.)

18. ■■ The distance of one lap around an oval dirt bike track is 1.50 km. If a rider going at a constant speed makes one lap in 1.10 min, what is the speed of the bike and rider in meters per second? Is the velocity of the bike also constant? Explain.

19. ■■ (a) Given that the speed of sound is 340 m/s and the speed of light is 3.00×10^8 m/s (186 000 mi/s), how much time will elapse between a lightning flash and the resulting thunder if the lightning strikes 2.50 km away from the observer? (b) Does your answer change if you assume light travels with an infinite speed (that is, the lightning flash is seen instantaneously)?

20. ■■ 💿 A plot of position versus time is in •Fig. 2.19 for an object in linear motion. (a) What are the average velocities

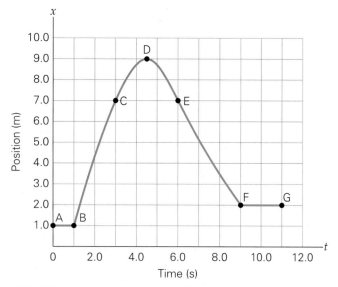

•FIGURE 2.19 Position versus time See Exercise 20.

for the segments AB, BC, CD, DE, EF, FG, and BG? (b) State whether the motion is uniform or nonuniform in each case. (c) What is the instantaneous velocity at point D?

21. ■■ 💿 In demonstrating a dance step, a person moves in one dimension as shown in •Fig. 2.20. What are the (a) average speed and (b) average velocity for each phase of the motion? (c) What are the instantaneous velocities at $t = 1.0$ s, 2.5 s, 4.5 s, and 6.0 s? (d) What is the average velocity for the interval between $t = 4.5$ s and $t = 9.0$ s? [Hint: Recall that the overall displacement is the displacement between the starting point and the ending point.]

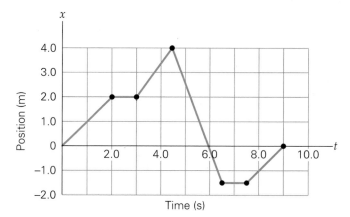

•FIGURE 2.20 Position versus time See Exercise 21.

22. ■■ You can determine the speed of a car by measuring the time it takes to travel between the mile markers. (a) How many seconds should elapse between mile markers if the car's average speed is 65 mi/h? (b) What is the average speed in miles per hour if it takes 65 s to travel between mile markers?

23. ■■ An earthquake releases two types of traveling seismic waves, called *transverse* and *longitudinal* waves. The average speeds of transverse and longitudinal waves in rock are 8.9 km/s and 5.1 km/s, respectively. A seismograph records the arrival of the transverse waves 73 s before that of the longitudinal waves. Assuming the waves travel in straight lines, how far away is the center of the earthquake?

24. ■■ The Indianapolis 500, a 500-mile auto race, was first run in 1911 in a time of 6 h, 42 min, and 8 s. In 1990, the race was run in a record time of 2 h, 41 min, and 18 s. (a) What were the average speeds for the Indy 500 for these years in miles per hour (to 0.1 mi/h)? (b) What was the percentage change in the average speed from 1911 to 1990?

25. ■■■ A student driving home for the holidays starts at 8:00 A.M. to make the 675-kilometer trip, practically all of which is on non-urban interstate highway. If she wants to arrive home no later than 3:00 P.M., what must be her minimum average speed? Will she have to exceed the 65 mi/h speed limit?

26. ■■■ Two runners approaching each other on a straight track have constant speeds of +4.50 m/s and −3.50 m/s, respectively, when they are 100 m apart (•Fig. 2.21). How long will it take for the runners to meet, and at what position will this occur if they maintain these speeds?

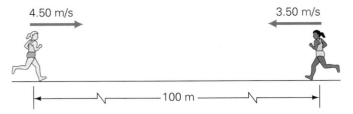

•**FIGURE 2.21** **When and where do they meet?** See Exercise 26.

27. ■■■ In driving the usual route to school, a student computes his average speed to be 30 km/h. In a hurry to get home that afternoon, he wants to average 60 km/h for the round trip. What would his average speed for the return trip over the same route have to be in order to do this? (Hint: See Example 2.3, and be ready for a surprise.)

2.3 Acceleration

28. The gas pedal of an automobile is commonly referred to as the accelerator. Which of the following might also be called an accelerator: (a) the brakes, (b) the steering wheel, (c) the gear shift, (d) all of the preceding?

29. A car is traveling at a constant speed of 55 mi/h on a circular track. Is the car accelerating? Explain.

30. 💿 On a velocity-versus-time plot for an initially moving object that has a constant acceleration, (a) the graph is a curved line, (b) the y-intercept is nonzero, (c) the slope of the line is necessarily positive, (d) the slope of the line is necessarily negative.

31. An object traveling at a constant velocity v_o experiences a constant acceleration in the same direction for a time t. Then, an acceleration of equal magnitude is experienced in the opposite direction of v_o for the same time t. What is the velocity after all of this?

32. Describe the motions of the two objects that have the velocity-versus-time plots shown in •Fig. 2.22.

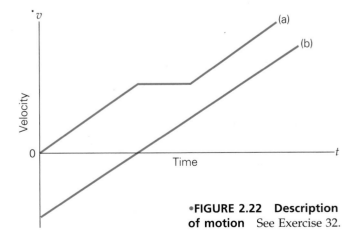

•**FIGURE 2.22** **Description of motion** See Exercise 32.

33. ■ An automobile traveling at 25.0 km/h along a straight road accelerates to 65.0 km/h in 6.00 s. What is the magnitude of the average acceleration?

34. ■ A sports car can accelerate from zero to 60 mi/h in 3.9 s. What is the magnitude of the average acceleration of the car in meters per second squared?

35. ■ If the sports car in Exercise 34 can accelerate at a rate of 7.2 m/s², how long does it take for the car to accelerate from zero to 60 mi/h?

36. ■ If an automobile moving with a velocity of 40 km/h along a straight road brakes uniformly to rest in 5.0 s, by how much must the automobile's velocity change each second?

37. ■ A skier with speed of 5.00 m/s crests a hill. On the downslope, she accelerates at 3.00 m/s² for 3.50 s. What is her final speed?

38. ■■ An object moves in the positive x direction at an initial speed of 3.5 m/s. If the object experiences an acceleration of 0.50 m/s² in the negative x direction, after how long a time will the object have a negative velocity?

39. ■■ What is the acceleration for each graph segment in •Fig. 2.23? Describe the motion of the object over the total time interval.

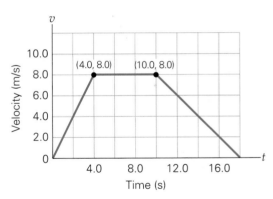

•**FIGURE 2.23** **Velocity versus time** See Exercises 39 and 61.

40. ■■ •Figure 2.24 shows a plot of velocity versus time for an object in linear motion. (a) Compute the acceleration for each phase of motion. (b) Describe how the object moves during the last time segment.

41. ■■■ A train normally travels at a uniform speed of 72 km/h on a long stretch of straight, level track. On a particular day, the train must make a 2.0-minute stop at a station along this track. If the train decelerates at a uniform rate of 1.0 m/s² and accelerates at a rate of 0.50 m/s², how much time is lost in stopping at the station?

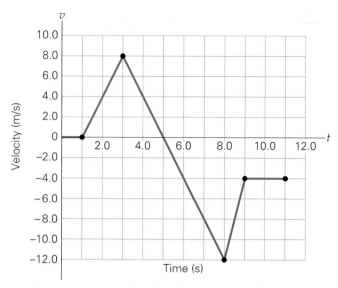

•FIGURE 2.24 **Velocity versus time** See Exercises 40 and 62.

2.4 Kinematic Equations (Constant Acceleration)

42. For a constant linear acceleration, the velocity-versus-time graph would be (a) a horizontal line, (b) a vertical line, (c) a non-horizontal and non-vertical straight line, (d) a curve.

43. An object accelerates uniformly from rest for t seconds. The average speed for this time interval is (a) $\frac{1}{2}at$, (b) $\frac{1}{2}at^2$ (c) $2at$, (d) $2at^2$.

44. A classmate states that a negative acceleration always means that a moving object is decelerating. Is this true? Explain.

45. ■ At a sports car rally, a car starting from rest accelerates uniformly at a rate of 9.0 m/s² over a straight-line distance of 100 m. The time to beat is 4.5 s. Does the driver do it? If not, what must the acceleration be to do so?

46. ■ A car accelerates from rest at a constant rate of 2.0 m/s² for 5.0 s. (a) What is the speed of the car at that time? (b) How far does the car travel in this time?

47. ■ A car traveling at 35 mi/h is to stop on a 35-meter-long shoulder of the road. What minimum deceleration is required?

48. ■ A motorboat traveling on a straight course slows down uniformly from 75 km/h to 40 km/h in a distance of 50 m. What is the required acceleration?

49. ■■ The driver of a pickup truck going 100 km/h applies the brakes, giving the truck a uniform deceleration of 6.50 m/s² while it travels 20.0 m. (a) What is the speed of the truck in kilometers per hour at the end of this distance? (b) How much time has elapsed?

50. ■■ An experimental rocket car starting from rest reaches a speed of 560 km/h after a straight 400-meter run on a level salt flat. Assuming the acceleration to be constant, what was the time of the run?

51. ■■ A rocket car is traveling at a constant speed of 250 km/h on a salt flat. The driver gives the car a reverse thrust, and it experiences a continuous and constant deceleration of 8.25 m/s². How much time elapses until the car is 175 m from the point where the reverse thrust is applied?

52. ■■ Two identical cars capable of accelerating at 3.00 m/s² are racing on a straight track and cross the starting line together with running starts. Car A has an initial speed of 2.50 m/s; car B starts with an initial speed of 5.00 m/s. (a) What is the separation of the two cars after 10 s? (b) Which car is moving faster after 10 s?

53. ■■ An object moves in the positive x direction at a speed of 40 m/s. As it passes through the origin, it starts to experience a constant acceleration of 3.5 m/s² in the negative x direction. How much time elapses before the object returns to the origin?

54. ■■ A rifle bullet with a muzzle speed of 330 m/s is fired directly into a special, dense material that stops the bullet in 30 cm. Assuming the deceleration to be constant, what is its magnitude?

55. ■■ A bullet traveling horizontally at a speed of 350 m/s hits a board perpendicular to the surface, passes through it, and emerges on the other side at a speed of 210 m/s. If the board is 4.00 cm thick, how long does the bullet take to pass through it?

56. ■■ A rock hits the ground at a speed of 10 m/s and leaves a hole 25 cm deep. What is the magnitude of the (assumed) uniform deceleration of the rock?

57. ■■ A jet aircraft being launched from an aircraft carrier is accelerated from rest along a 94-meter track for 2.5 s. What is the launch speed of the jet?

58. ■■ The speed limit in a school zone is 40 km/h (about 25 mi/h). A driver traveling at this speed sees a child run into the road 13 m ahead of his car. He applies the brakes, and the car decelerates at a uniform rate of 8.0 m/s². If the driver's reaction time is 0.25 s, will the car stop before hitting the child?

59. ■■ Assuming a reaction time of 0.50 s for the driver in Exercise 58, will the car stop before hitting the child?

60. ■■ Show that the area under the curve of a velocity-versus-time plot for a constant acceleration is equal to the displacement. Do this for the cases in which (a) $a = 0$, (b) $v_0 = 0$, and (c) $v_0 \neq 0$. [Hint: The area of a triangle is $ab/2$, or half the altitude times the base.]

61. ■■ Compute the distance traveled for the motion represented by Fig. 2.23. [Hint: See Exercise 60.]

62. ■■■ Figure 2.24 shows velocity versus time for an object in linear motion. (a) What are the instantaneous velocities at $t = 8.0$ s and $t = 11.0$ s? (b) Compute the total displacement of the object. (c) Compute the total distance the object travels.

63. ■■■ In our kinematic equations, it was assumed for simplicity and convenience that $x_o = 0$ and $t_o = 0$. Rewrite the equations with x_o and t_o expressed explicitly. What is the difference between the x in Eq. 2.9 and the x in your rewritten equation?

64. ■■■ An object moves in the positive x direction with a constant acceleration. At $x = 5.0$ m, its speed is 10 m/s; 2.5 s later, the object is at $x = 65$ m. What is its acceleration?

2.5 Free Fall
(Neglect air resistance throughout.)

65. 🌐 An object is thrown vertically upward. Which one of the following is true? (a) Its velocity changes nonuniformly, (b) the maximum height is independent of the initial velocity, (c) the travel time upward is slightly greater than its travel time downward, (d) the speed on returning to its starting point is the same as its initial speed.

66. A dropped object in free fall (a) falls 9.8 m each second, (b) falls 9.8 m during the first second, (c) has an increase in speed of 9.8 m/s each second, (d) has an increase in acceleration of 9.8 m/s each second.

67. An object is thrown straight upward. When it reaches the highest point in the air, what is its acceleration? Explain.

68. Given the data for an object projected vertically upward, you are asked to find the time the object is at height y. The equation $y = v_o t - \frac{1}{2}gt^2$ is solved by using the quadratic formula, and two roots are obtained. Does this mean there are two answers? Explain. [Hint: Do the roots of a quadratic equation necessarily have to be both positive and negative?] ?

69. ■ If a dropped object falls 19.6 m in 2.00 s, how far will it fall in 4.00 s?

70. ■ 🌐 For the motion of a dropped object in free fall, sketch the general forms of the graphs of (a) v versus t and (b) y versus t.

71. ■ 🌐 Sketch the general forms of the graphs of (a) v versus t and (b) y versus t for an object projected vertically upward.

72. ■ A student drops a ball from the top of a tall building; it takes 2.8 s to reach the ground. (a) What was its speed just before hitting the ground? (b) What was the height of the tall building?

73. ■ An object dropped from the top of a cliff takes 1.80 s to hit the water in the lake below. What is the cliff's height above the water?

74. ■ A boy throws a stone straight upward with an initial speed of 15 m/s on the surface of the Earth. What maximum height above the starting point will the stone reach before falling back down?

75. ■ In Exercise 74, what would be the maximum height if the boy and the stone were on the surface of the Moon, where the acceleration due to gravity is only 1.67 m/s²?

76. ■■ A spring-loaded gun shoots a 0.0050-kilogram bullet vertically upward with an initial velocity of 21 m/s. (a) What is the height of the bullet 3.0 s after firing? (b) At what times is the bullet 12 m above the muzzle of the gun?

77. ■■ The ceiling of a classroom is 3.75 m above the floor. A student tosses an apple vertically upward, releasing it 0.50 m above the floor. What is the maximum initial speed that can be given to the apple if it is not to touch the ceiling?

78. ■■ (a) The World Trade Center and the Empire State Building in New York City have heights of about 417 m and 381 m, respectively. If objects were dropped from the top of each, what would be the difference in time in their reaching the ground? (b) What would be the time of fall for Chicago's Sears Tower, at 443 m?

79. ■■ A stone is thrown vertically downward at an initial speed of 14 m/s from a height of 65 m above the ground. (a) How far does the stone travel in 2.0 s? (b) What is its velocity just before it hits the ground?

80. ■■ A ball is projected vertically downward at a speed of 4.00 m/s. (a) How far does the ball travel in 1.80 s? (b) What is the velocity of the ball at that time?

81. ■■ Referring to the thrown ball in Example 2.12 and Fig. 2.17, (a) compare the travel time upward (t_u) with the travel time downward (t_d) required for the ball to return to its starting point. (b) Compare the initial velocity of the ball with the velocity it has when back at the starting point.

82. ■■ 🌐 Draw graphs of y versus t and v versus t for the following: (a) an object dropped from rest from a height of 30 m above the ground and (b) an object that is projected vertically upward with an initial velocity of 34.3 m/s and that returns to the same point. (c) Draw plots of a versus t for these motions.

83. ■■ A superball is dropped from a height of 4.00 m. Assuming the ball rebounds with 95% of its impact speed, how high will the ball go? (Would it be 95.0% of the initial height?)

84. ■■ Two balls are thrown vertically, both at the initial speed of 10.0 m/s from a height of 60.0 m above the ground. Ball A is thrown upward, while ball B is projected downward. (a) Which ball will hit the ground first? (b) What is the time

lapse between the two balls hitting the ground? (c) Do the masses of the balls matter?

85. ■■■ In •Fig. 2.25, a student at a window on the top floor of a dorm sees his math professor walking beside the building. He drops a water balloon from 18.0 m above the ground when the prof is 1.00 m from the point directly beneath the window. If the prof is 170 cm tall and walks at a rate of 0.450 m/s, does the balloon hit her? If not, how close does it come?

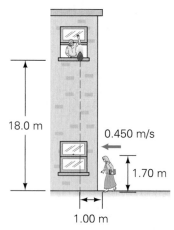

•FIGURE 2.25   **Hit the prof.**   See Exercise 85. (Not drawn to scale.)

86. ■■■ A photographer in a helicopter ascending vertically at a constant rate of 12.5 m/s accidentally drops a camera out the window when the helicopter is 60.0 m above the ground. (a) How long will it take the camera to reach the ground? (b) What will its speed be when it hits?

87. ■■■ The acceleration due to gravity on the Moon is about one-sixth of that on Earth. (a) If an object were dropped from the same height on the Moon and on Earth, how much longer (by what factor) would it take it to hit the surface of the Moon? (b) For a projectile with an initial velocity of 18.0 m/s upward, what would be the maximum height and total time of flight on the Moon and on Earth?

88. ■■■ It takes 0.210 s for a dropped object to pass a window that is 1.35 m tall (•Fig. 2.26). From what height above the top of the window was the object released?

Additional Exercises

89. In throwing an object vertically upward at a speed of 7.25 m/s from the top of a tall building, Jennifer leans over the edge so that the object will not strike the building on the return trip. (a) What is the velocity of the object when it has traveled a total distance of 25.0 m? (b) How long does it take to travel this distance?

90. On landing, a jet plane on a straight runway comes to rest at an average velocity of −35.0 km/h. If this takes 7.00 s, what is the plane's acceleration?

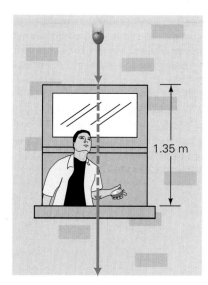

•FIGURE 2.26   **From where did it come?**   See Exercise 88.

91. A vertically moving projectile reaches a maximum height of 23 m above its starting position. (a) What was the projectile's initial speed? (b) What is its height above the starting point at $t = 1.3$ s?

92. A drag racer traveling at a speed of 200 km/h on a straight track ejects a parachute and slows uniformly to a speed of 20 km/h in 12 s. (a) What is the racer's acceleration? (b) How far does it travel in the 12-second interval?

93. On a cross-country trip, a couple drives 500 mi in 10 h on the first day, 380 mi in 8.0 h on the second day, and 600 mi in 15 h on the third day. What was their average speed for the whole trip?

94. A car traveling at 25 mi/h has 1.5 s to come to a stop approaching a red light. The constant deceleration of the car is 7.0 m/s². Will the car stop safely before it reaches the red light?

95. What is the magnitude of the displacement if a car travels three quarters of a lap on a circular track of radius 50 m?

96. A circular track has a diameter of 0.50 km. A go-cart with a constant speed of 10.0 m/s makes two complete laps around the track. How long does it take the cart to complete the two laps?

97. At what speed must an object be projected vertically upward for the object to reach a maximum height of 14.0 m above its starting point?

98. A train on a straight, level track has an initial speed of 45.0 km/h. A uniform acceleration of 1.50 m/s² is applied while the train travels 200 m. (a) What is the speed of the train at the end of this distance? (b) How long did it take for the train to travel the 200 m?

99. A car going 85 km/h on a straight road is brought uniformly to a stop in 10 s. How far does the car travel during that time?

100. A car and a motorcycle start from rest at the same time on a straight track, but the motorcycle is 25.0 m behind the car (•Fig. 2.27). The car accelerates at a uniform rate of 3.70 m/s^2, and the motorcycle at a uniform rate of 4.40 m/s^2. (a) How much time elapses before the motorcycle overtakes the car? (b) How far will each have traveled during that time? (c) How far ahead of the car will the motorcycle be 2.00 s later? (Both vehicles are still accelerating.)

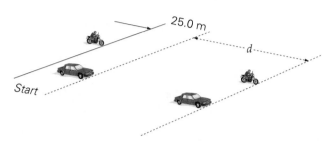

•**FIGURE 2.27   A tie race**   See Exercise 100. (Not drawn to scale.)

101. A person throws a stone straight upward at an initial speed of 15 m/s on a bridge that is 25 m above the surface of the water. If the stone just misses the bridge on the way down, (a) what is the speed of the stone just before it hits the water? (b) What is the total time the stone is in the air?

102. An object initially at rest experiences an acceleration of 1.5 m/s^2 for 6.0 s and then travels at that constant velocity for another 8.0 s. What is the magnitude of the object's average velocity over the 14 s interval?

103. To find the depth of the water surface in a well, a person drops a stone from the top of the well and simultaneously starts a stopwatch. The watch is stopped when the splash is heard, giving a reading of 3.65 s. The speed of sound is 340 m/s. Find the depth of the water surface below the top of the well. Take the person's reaction time for stopping the watch to be 0.250 s.

104. An arrow shot from a bow vertically upward has an initial speed of 50.0 m/s. (a) What is the maximum height of the arrow above its launch point? (b) How long does it take for the arrow to return to its launch point? (Assume that the arrow travels in a straight line.)

105. An object starting from rest and traveling in only one direction has velocities of 5.0 m/s, 10 m/s, 15 m/s, 20 m/s, and 25 m/s at the ends of the first, second, third, fourth, and fifth seconds, respectively. (a) What is the magnitude of the object's acceleration? (b) What is the object's average velocity for the 5-second interval? (c) Sketch a graph of v versus t. (d) Sketch a graph of x versus t.

106. On a walk in the country, a couple goes 1.80 km east along a straight road in 20.0 min and then 2.40 km directly north in 35.0 min. Answer the following in units of kilometers per hour. (a) What is the average velocity for each segment of their walk? (b) What is the average speed for the total trip? (c) If they take a straight-line path back to their original starting place in 25.0 min, what are the average speed and average velocity for the total trip?

Motion in Two Dimensions

You *can* get there from here! It's just a matter of knowing which way to head at the crossroads. But did you ever wonder why so many roads meet at right angles? There's a good reason. Living on Earth's surface, we are used to describing locations in two dimensions, and one of the easiest ways to do this is by referring to a pair of mutually perpendicular axes. When you want to tell someone how to get to a particular place in the city, you might say, "Go uptown 4 blocks and then crosstown for three more." In the country, it's "Go south for 5 miles and then another half mile east." Either way, though, you need to know how far to go in each of *two* directions that are 90° apart.

You can use the same approach to describe motion—and the motion doesn't have to be in a straight line. As you will shortly see, we can use vectors, introduced in Chapter 2, to describe motion in curved paths as well. Such analysis of *curvilinear* motion will eventually allow you to analyze the behavior of batted balls, planets circling the Sun, and even electrons in atoms.

Curvilinear motion can be analyzed by using rectangular components of motion. Essentially, you break down, or *resolve*, the curved motion into rectangular (*x* and *y*) components and look at the motion in both dimensions simultaneously. To those components you can apply the kinematic equations introduced in Chapter 2. For an object moving in a curved path, for example, the x and y coordinates of the motion at any time give the object's position as the point (x, y).

3.1 Components of Motion

Objectives: To (a) analyze motion in terms of its components and (b) apply the kinematic equations to components of motion.

An object moving in a straight line was considered in Chapter 2 to be moving along one of the Cartesian axes (x or y). But what if the motion is not along an axis? For example, consider the situation illustrated in •Fig. 3.1. Here, the balls are moving uniformly across a tabletop. The ball rolling in a straight line along the side of the table designated as the x direction is moving in one dimension. That is, its motion can be described with a single coordinate, x, as was done for motions in Chapter 2. Similarly, the motion of the ball rolling in the y direction can be described by a single y coordinate. However, both x and y coordinates are needed to describe the motion of the ball rolling diagonally across the table. We call this motion in two dimensions.

You might observe that if the diagonally moving ball were the only object you had to consider, the x axis could be chosen in that direction and the motion reduced to one dimension. This is true, but once the coordinate axes are fixed, motions not along the axes must be described with two coordinates (x, y) or in two

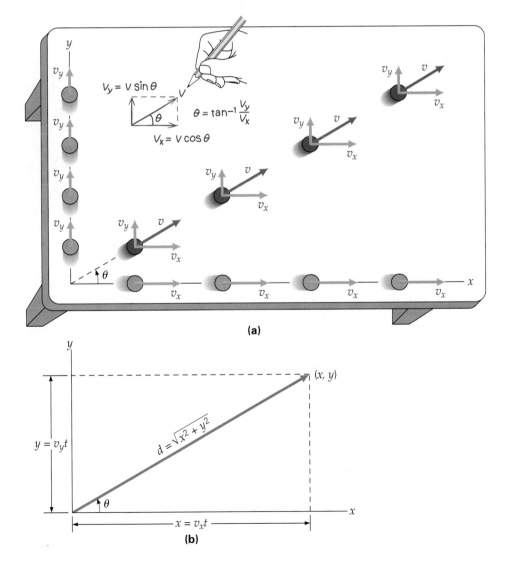

(a)

(b)

•FIGURE 3.1 **Components of motion** **(a)** The velocity (and displacement) for uniform straight-line motion—the dark purple ball—may have x and y components (v_x and v_y as shown in the pencil drawing) because of the chosen orientation of the coordinate axes. Note that the velocity and displacement of the ball in the x direction are exactly the same as those that a ball rolling along the x axis with a uniform velocity of v_x would have. A comparable relationship holds true for the ball's motion in the y direction. Since the motion is uniform, the ratio v_y/v_x (and therefore θ) is constant. **(b)** The coordinates (x, y) of the ball's position and the distance d it has traveled from the origin can be found for any time t.

dimensions. Also, keep in mind that not all motions in a plane (two dimensions) are in straight lines. Think about the path of a ball you toss to another person.

In considering the motion of the ball moving diagonally across the table in Fig. 3.1a, we can think of the ball moving in the x and y directions simultaneously. That is, it has a velocity in the x direction (v_x) *and* a velocity in the y direction (v_y) at the same time. The combined velocity components describe the actual motion of the ball. If the ball has a constant velocity (**v**) in a direction at an angle θ relative to the x axis, then the velocities in the x and y directions are obtained by resolving, or breaking down, the velocity vector into **components of motion** in these directions, as shown in the pencil drawing in Fig. 3.1a. As this drawing shows, the v_x and v_y components have magnitudes of

Magnitudes of velocity components in x and y directions

$$v_x = v \cos \theta \tag{3.1a}$$

$$v_y = v \sin \theta \tag{3.1b}$$

(Notice that $v = \sqrt{v_x{}^2 + v_y{}^2}$, so v is a combination of the velocities in the x and y directions.)

You are familiar with the use of two-dimensional length components in finding the x and y coordinates in a Cartesian system. For the ball rolling on the table, its position, (x, y), or the distance traveled from the origin in each of the component directions at time t, is given by

$$x = v_x t \qquad \text{\textit{Magnitude of displacement}} \tag{3.2a}$$
$$\text{\textit{components (constant velocity—}}$$
$$y = v_y t \qquad \text{\textit{zero acceleration—only)}} \tag{3.2b}$$

just as it is for motion in the x and y directions separately. (Here, as usual, we assume that $x_o = y_o = 0$ at $t = 0$.) The ball's straight-line distance from the origin is then $d = \sqrt{x^2 + y^2}$ (Fig. 3.1b).

Note that $\tan \theta = v_y/v_x$, so the direction of the motion relative to the x axis is given by $\theta = \tan^{-1}(v_y/v_x)$. (See the hand-drawn sketch in Fig. 3.1a.) Also, $\theta = \tan^{-1}(y/x)$. Why?

Example 3.1 ■ On a Roll: Using Components of Motion

If the diagonally moving ball in Fig. 3.1a has a velocity of 0.50 m/s at an angle of 37° relative to the x axis, find how far it travels in 3.0 s by using x and y components.

Thinking It Through. Given the magnitude and direction (angle) of the velocity of the ball, the x and y components of the velocity can be found. Then, the distance in each direction can be computed. Since x and y are at right angles, the Pythagorean theorem gives the straight-line path distance of the ball. (Note the procedure: Separate the motion into components, calculate what is needed in each direction, and recombine if necessary.)

Solution. Organizing the data, we have

Given: $v = 0.50$ m/s   *Find:* d (distance)
     $\theta = 37°$
     $t = 3.0$ s

The distance traveled by the ball in terms of its x and y components is given by $d = \sqrt{x^2 + y^2}$. To find x and y as given by Eq. 3.2, we must first compute the velocity components v_x and v_y (Eq. 3.1):

$$v_x = v \cos 37° = (0.50 \text{ m/s})(0.80) = 0.40 \text{ m/s}$$

$$v_y = v \sin 37° = (0.50 \text{ m/s})(0.60) = 0.30 \text{ m/s}$$

Then, the component distances are

$$x = v_x t = (0.40 \text{ m/s})(3.0 \text{ s}) = 1.2 \text{ m}$$

$$y = v_y t = (0.30 \text{ m/s})(3.0 \text{ s}) = 0.90 \text{ m}$$

and the actual path distance is

$$d = \sqrt{x^2 + y^2} = \sqrt{(1.2 \text{ m})^2 + (0.90 \text{ m})^2} = 1.5 \text{ m}$$

Follow-up Exercise. Suppose that a ball were rolling diagonally across a table with the same speed as in Example 3.1 *but* from the lower right corner, which is taken as the origin of the coordinate system, toward the upper left corner at an angle of 37° relative to the $-x$ axis. What would be the velocity components in this case?

Problem-Solving Hint

Note that for this simple case, the distance can also be obtained directly from $d = vt = (0.50 \text{ m/s})(3.0 \text{ s}) = 1.5 \text{ m}$. However, we have solved this Example in a more general way to illustrate the use of components of motion. The direct solution would have been evident if the equations had been combined algebraically before calculation, that is,

$$x = v_x t = (v \cos \theta)t$$

$$y = v_y t = (v \sin \theta)t$$

and

$$d = \sqrt{x^2 + y^2} = \sqrt{(v \cos \theta)^2 t^2 + (v \sin \theta)^2 t^2} = \sqrt{v^2 t^2 (\cos^2 \theta + \sin^2 \theta)} = vt$$

Before embarking on the first solution strategy that occurs to you, pause for a moment to see whether there might be an easier or more direct way of approaching the problem.

Kinematic Equations for Components of Motion

Example 3.1 involved two-dimensional motion in a plane. With a constant velocity (constant components v_x and v_y), the motion is in a straight line. The motion may also be accelerated. For motion in a plane *with a constant acceleration* having components a_x and a_y, the displacement (from the origin) and velocity components are given by the kinematic equations of Chapter 2 for the x and y directions:

$$x = v_{x_0} t + \tfrac{1}{2} a_x t^2 \qquad (3.3a)$$

$$y = v_{y_0} t + \tfrac{1}{2} a_y t^2 \qquad (3.3b)$$

(constant acceleration only)

$$v_x = v_{x_0} + a_x t \qquad (3.3c)$$

$$v_y = v_{y_0} + a_y t \qquad (3.3d)$$

Kinematic equations for displacement and velocity components

If an object is initially moving with a constant velocity and suddenly experiences an acceleration in the direction of the velocity (0°) or opposite to it (180°), it would continue in a straight-line path, either speeding up or slowing down, respectively.

If, however, the acceleration vector is at some angle other than 0° or 180° to the velocity vector, the motion is along a curved path. For the motion of an object to be *curvilinear*—that is, to vary from a straight-line path—an acceleration is required. For a curved path, the ratio of the velocity components varies with time. That is, the direction of the motion, $\theta = \tan^{-1}(v_y/v_x)$, varies with time because one or both of the velocity components do, so the motion is not in a straight line.

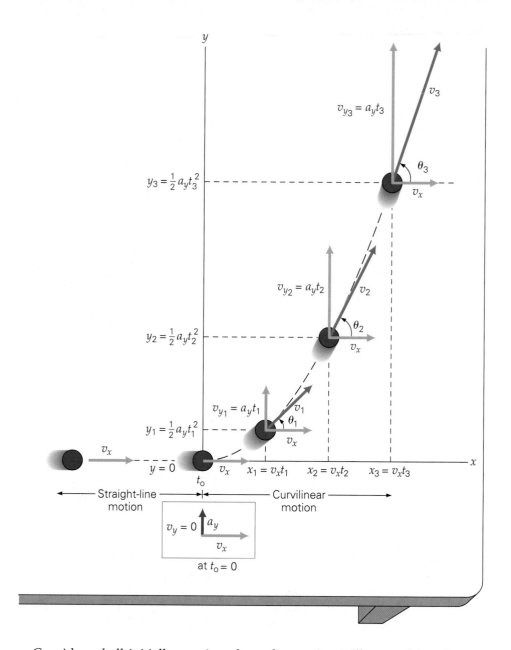

•FIGURE 3.2 Curvilinear motion
An acceleration not parallel to the instantaneous velocity produces a curved path. Here an acceleration $(\mathbf{a}_y)$ is applied at $t_o = 0$ to a ball initially moving with a constant velocity $(\mathbf{v}_x)$. The result is a curved path with the velocity components as shown. Notice how $\mathbf{v}_y$ increases with time.

Consider a ball initially moving along the x axis, as illustrated in •Fig. 3.2. Assume that starting at a time $t_o = 0$ it receives a constant acceleration a_y in the y direction. The magnitude of the x component of the ball's displacement is given by $x = v_x t$; the $\frac{1}{2}a_x t^2$ term of Eq. 3.3a drops out since there is no acceleration in the x direction. Prior to t_o, the motion is in a straight line along the x axis. But at any time after t_o, there is a y displacement with a magnitude of $y = \frac{1}{2}a_y t^2$ (as given by Eq. 3.3b with $v_{y_o} = 0$). The result is a curved path for the ball.

Note that the length (magnitude) of the velocity component v_y changes with time. The total velocity vector (**v**) *at any time* is tangent to the curved path of the ball. It is at an angle θ relative to the x axis, given by $\theta = \tan^{-1}(v_y/v_x)$, which now changes with time as we see in Fig. 3.2 and in the following Example.

Example 3.2 ■ A Curving Path: Vector Components

Suppose that the ball in Fig. 3.2 has an initial velocity of 1.50 m/s along the x axis and, starting at $t_o = 0$, receives an acceleration of 2.80 m/s² in the y direction. (a) What is the position of the ball 3.00 s after t_o? (b) What is the velocity of the ball at that time?

Thinking It Through. Keep in mind that the motions in the x and y directions can be analyzed independently—the connecting factor is time. That is, both component motions occur at the same time. In (a), simply compute the x and y positions at the given time, taking into account that there is an acceleration in the y direction. For (b), find the component velocities and combine to get the velocity.

Solution. Referring to Fig. 3.2, we have the following:

Given: $v_{x_0} = v_x = 1.50$ m/s   *Find:* (a) (x, y) (position coordinates)
$\quad\quad\; v_{y_0} = 0$ (b) $\mathbf{v}$ (velocity)
$\quad\quad\; a_x = 0$
$\quad\quad\; a_y = 2.80$ m/s^2
$\quad\quad\; t = 3.00$ s

(a) At 3.00 s after $t_0 = 0$, Eqs. 3.3a and 3.3b tell us that the ball has traveled the following distances from the origin in the x and y directions:

$$x = v_{x_0}t + \tfrac{1}{2}a_x t^2 = (1.50 \text{ m/s})(3.00 \text{ s}) + 0 = 4.50 \text{ m}$$

$$y = v_{y_0}t + \tfrac{1}{2}a_y t^2 = 0 + \tfrac{1}{2}(2.80 \text{ m/s}^2)(3.00 \text{ s})^2 = 12.6 \text{ m}$$

Thus, its position is $(x, y) = (4.50 \text{ m}, 12.6 \text{ m})$. If you computed the distance $d = \sqrt{x^2 + y^2}$, what would this be? (Note that it is not the actual distance the ball has traveled in 3.00 s but the magnitude of the *displacement*, or straight-line distance, from the origin at $t = 3.00$ s.)

(b) The x component of the velocity is given by Eq. 3.3c:

$$v_x = v_{x_0} + a_x t = 1.50 \text{ m/s} + 0 = 1.50 \text{ m/s}$$

(This is constant since there is no acceleration in the x direction.) Similarly, the y component of the velocity is given by Eq. 3.3d:

$$v_y = v_{y_0} + a_y t = 0 + (2.80 \text{ m/s}^2)(3.00 \text{ s}) = 8.40 \text{ m/s}$$

The velocity therefore has a magnitude of

$$v = \sqrt{v_x^2 + v_y^2} = \sqrt{(1.50 \text{ m/s})^2 + (8.40 \text{ m/s})^2} = 8.53 \text{ m/s}$$

and its direction relative to the x axis is

$$\theta = \tan^{-1}\left(\frac{v_y}{v_x}\right) = \tan^{-1}\left(\frac{8.40 \text{ m/s}}{1.50 \text{ m/s}}\right) = 79.9°$$

Note: Don't confuse the direction of the velocity with the direction of the displacement from the origin. The direction of the velocity is always tangent to the path.

Follow-up Exercise. Suppose that the ball in Example 3.2 also received an acceleration of 1.00 m/s^2 in the $+x$ direction at t_0. What would be the position of the ball 3.00 s after t_0 in this case?

3.2 Vector Addition and Subtraction

Objectives: **To add and subtract vectors (a) graphically and (b) analytically.**

Many physical quantities have a direction associated with them—that is, they are vectors. You have worked with a few such quantities related to motion (displacement, velocity, and acceleration) and will encounter more during this course of study. A very important technique in the analysis of many physical situations is the addition (and subtraction) of vectors. By adding or combining such quantities (**vector addition**), you can obtain the overall, or net, effect that occurs—the *resultant*, as we call the vector sum.

You have already been adding vectors. In Chapter 2, displacements were added to get the net displacement. In this chapter, vector components of motion were added to give net effects. Notice in the preceding Example 3.2 that the velocity components v_x and v_y were added to give the resultant velocity.

Note: One thing that scalar addition and vector addition have in common is that the quantities being added must have the same units. You cannot meaningfully add a displacement vector to a velocity vector, just as you cannot add the scalars of distance and speed and obtain a result with any physical meaning.

In this section, we will look at vector addition and subtraction in general, along with common vector notation. As you will learn, these operations are not the same as scalar or numerical addition and subtraction, with which you are already familiar. Vectors have magnitudes *and* directions, so different rules apply.

In general, there are geometric (graphical) methods and analytical (computational) methods of vector addition. The geometric methods are useful in helping you visualize the concepts of vector addition. Analytical methods are more commonly used, however, because they are faster and more precise.

Vector Addition: Geometric Methods

Triangle Method. To add two vectors—say, to add **B** to **A** (that is, to find **A** + **B**) by the **triangle method**—you first draw **A** on a sheet of graph paper to some scale (•Fig. 3.3a). For example, if **A** is a displacement in meters, a convenient scale is 1 cm:1 m, or 1 cm of vector length on the graph corresponds to 1 m of displacement. As shown in Fig. 3.3b, the direction of the **A** vector is specified as being at an angle (θ_A) relative to a coordinate axis, usually the *x* axis.

Next draw **B** with its tail starting at the tip of **A**. (Thus, this method is also called the *tip-to-tail method*.) The vector from the tail of **A** to the tip of **B** is then the vector sum, or the resultant of the two vectors: **R** = **A** + **B**.

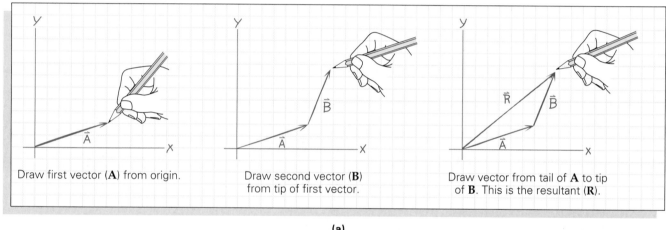

| Draw first vector (**A**) from origin. | Draw second vector (**B**) from tip of first vector. | Draw vector from tail of **A** to tip of **B**. This is the resultant (**R**). |

(a)

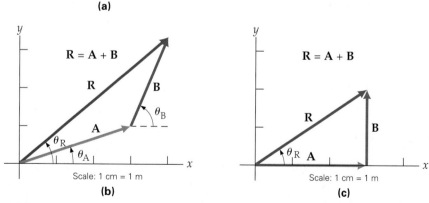

(b)　　　　**(c)**

•**FIGURE 3.3 Triangle method of vector addition**　(a) The vectors **A** and **B** are placed tip to tail. The vector that extends from the tail of **A** to the tip of **B**, forming the third side of the triangle, is the resultant **R** = **A** + **B**. (b) When drawn to scale, the magnitude of **R** can be found by measuring its length and using the scale conversion, and the direction angle θ_R can be measured with a protractor. Analytical methods can also be used. For a non-right triangle as in (b), the laws of sines and cosines would be used. (c) If the vector triangle is a right triangle, **R** is easily obtained via the Pythagorean theorem and the direction angle is given by an inverse trigonometric function.

If drawn to scale, the magnitude of **R** can be found by measuring its length and using the scale conversion. In such a graphical approach, the direction angle (θ_R) is measured with a protractor. Knowing the magnitudes and directions (angles θ) of **A** and **B**, the magnitude and direction of **R** can also be found analytically by using trigonometric methods. For the non-right triangle in Fig. 3.3b, the laws of sines and cosines would be used (see Appendix I).

The resultant of the vector right triangle in Fig. 3.3c would be much easier to find by using the Pythagorean theorem for the magnitude and an inverse trig function to find the direction angle. [Notice how **R** in this case is made up of x and y components (**A** and **B**).]

Parallelogram Method. Another graphical method of vector addition similar to the triangle method is the **parallelogram method.** In •Fig. 3.4, **A** and **B** are drawn tail to tail, and a parallelogram is formed as shown. The resultant **R** lies along the diagonal of the parallelogram. Drawing the diagram to scale with proper orientations, the magnitude and direction of **R** can be measured directly from the diagram as in the triangle method.

Notice that **B** could be moved to the other side of the parallelogram, forming the **A** + **B** triangle (and demonstrating why the triangle and parallelogram methods are equivalent). In general, a vector (arrow) can be moved around in vector addition methods. As long as you don't change its length (magnitude) or direction, you don't change the vector. In Fig. 3.4, this shifting of vector arrows shows that **A** + **B** = **B** + **A**—that is, the vectors can be added in either order.

Polygon Method. The triangle method can be extended to include the addition of any number of vectors. The method is then called the **polygon method** because the resulting graphical figure is a polygon. This is illustrated for four vectors in •Fig. 3.5, where **R** = **A** + **B** + **C** + **D**. Note that this addition is essentially three applications of the triangle method. The length and direction of the resultant could be found analytically by successive applications of the laws of sines and cosines, but an easier analytical method, the component method, will be described shortly. As in the parallelogram method, the four vectors (or any number of vectors) can be added in any order.

Vector Subtraction. Vector subtraction is a special case of vector addition:

$$\mathbf{A} - \mathbf{B} = \mathbf{A} + (-\mathbf{B})$$

That is, to subtract **B** from **A**, a *negative* **B** is added to **A**. In Chapter 2, you learned that a minus sign simply means that the direction of a vector is opposite that of one with a plus sign (for example, +x and −x). The same is true of a vector. The

Note: A common *error* is to draw **R** from the tip of **A** to the tip of **B**.

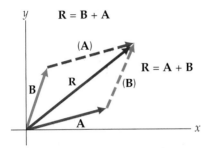

•**FIGURE 3.4 Parallelogram method of vector addition** The diagonal of the parallelogram formed after **A** and **B** are drawn with their tails at the origin is the resultant **R** = **A** + **B**. Drawn to scale, the magnitude and direction of **R** can be measured directly from the diagram as in the triangle method. The parallelogram method is equivalent to the triangle method. Shifting **B** to the right forms the same **A** + **B** vector triangle shown in Fig. 3.3b. Shifting **A** up has the same effect, with **R** = **B** + **A**.

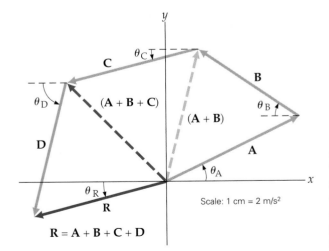

•**FIGURE 3.5 Polygon method of vector addition** The vectors to be added are placed tip to tail. The resultant **R** is the vector from the tail of the first vector (**A**) to the tip of the last vector (**D**), which completes the polygon. This method is essentially multiple applications of the triangle method: (**A** + **B**) + **C**, and so on.

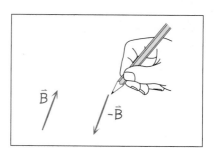

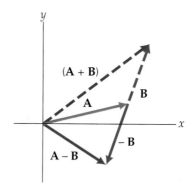

•**FIGURE 3.6 Vector subtraction**
Vector subtraction is a special case
of vector addition; that is,
A − **B** = **A** + (−**B**), where −**B** has
the same magnitude as **B** but is in
the opposite direction (see sketch).
Thus **A** + **B** is *not* the same as
A − **B**. Can you show that
B − **A** = −(**A** − **B**) geometrically?

vector −**B** has the same magnitude as the vector **B** but is in the opposite direction (•Fig. 3.6). The vector diagram in the figure provides a graphical representation of **A** − **B**.

Example 3.3 ■ Vector Subtraction: A Special Case of Vector Addition

Consider yourself in a car (A) traveling along a straight, level highway with a speed v_A = 75 km/h. Another car (B) travels at a speed v_B = 90 km/h. Find the differences in the velocities $\mathbf{v}_{BA} = \mathbf{v}_B - \mathbf{v}_A$ when (a) the other car travels in the same direction in front of you and (b) the other car is approaching you traveling in the opposite direction.

Thinking It Through.   This is a case of simple vector addition (or subtraction). The important things to note are the directions of the velocities so that you will use the appropriate signs.

Solution.   Taking the direction in which you are traveling as positive, we have, where the signs indicate vector directions,

Given: v_A = +75 km/h   *Find:* $\mathbf{v}_{BA} = \mathbf{v}_B - \mathbf{v}_A$
     (a) v_B = +90 km/h (same direction)
     (b) v_B = −90 km/h (opposite direction)

(a)   With both cars traveling in the same (+) direction,

$$\mathbf{v}_{BA} = \mathbf{v}_B - \mathbf{v}_A = 90 \text{ km/h} - (+75 \text{ km/h}) = +15 \text{ km/h}$$

(b)   When traveling in opposite directions,

$$\mathbf{v}_{BA} = \mathbf{v}_B - \mathbf{v}_A = -90 \text{ km/h} - (+75 \text{ km/h}) = -165 \text{ km/h}$$

Here, $\mathbf{v}_{BA}$ is what is called a *relative* velocity—the velocity of B relative to A. In part (a), relative to you in car A, car B appears to travel at a speed of 15 km/h. However, when approaching you, car B appears to travel at 165 km/h. It's as though you consider yourself or your car's reference frame to be at rest. (We will further explore the subject of relative velocity in Section 3.3.)

Follow-up Exercise.   What is the reference frame for the given velocities in Example 3.3?

Vector Components and the Analytical Component Method

Probably the most widely used analytical method for adding multiple vectors is the **component method**. It will be used again and again throughout the course of our study, so a basic understanding of the method is *essential*. Learn this section well.

Adding Rectangular Vector Components.   By *rectangular components*, we mean those at right (90°) angles to each other and usually taken in the rectangular coordinate *x* and *y* directions. You have already had an introduction to the addition of such components in the discussion of the velocity components of motion. For the general case, suppose that **A** and **B**, two vectors at right angles, are added, as illustrated in •Fig. 3.7a. The right angle makes the math easy. The magnitude of **C** is given by the Pythagorean theorem:

$$C = \sqrt{A^2 + B^2} \tag{3.4a}$$

Note: The notation $\tan^{-1} x$ stands for arctangent *x*, or the angle whose tangent is *x*.

The orientation of **C** relative to the *x* axis is given by the angle

$$\theta = \tan^{-1}\left(\frac{B}{A}\right) \tag{3.4b}$$

Magnitude-angle form of a vector

This is how a resultant is expressed in **magnitude-angle form**.

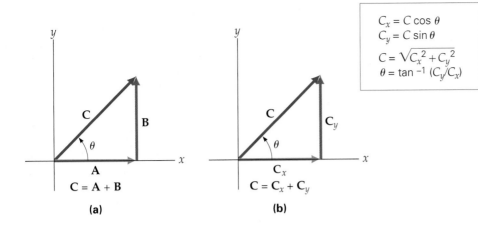

•FIGURE 3.7 Vector components
(a) The vectors **A** and **B** along the
x and y axes, respectively, add to
give **C**. (b) A vector **C** may be
resolved into rectangular
components $\mathbf{C}_x$ and $\mathbf{C}_y$.

$$C_x = C \cos \theta$$
$$C_y = C \sin \theta$$
$$C = \sqrt{C_x^2 + C_y^2}$$
$$\theta = \tan^{-1}(C_y/C_x)$$

(a)   (b)

Resolving a Vector into Rectangular Components; Unit Vectors. Resolving a vector into rectangular components is essentially the reverse of adding the components. Given a vector **C**, Fig. 3.7b illustrates how it may be resolved into x and y vector components, $\mathbf{C}_x$ and $\mathbf{C}_y$. You simply complete the vector triangle with x and y components. As the diagram shows, the magnitudes or vector lengths of these components are given by

$$C_x = C \cos \theta \qquad \text{(3.5a)}$$
$$C_y = C \sin \theta \qquad \text{(3.5b)}$$

magnitudes of components

(similar to $v_x = v \cos \theta$ and $v_y = v \sin \theta$ in Example 3.1).* The angle of direction of **C** can also be expressed in terms of the components, since $\tan \theta = C_y/C_x$, or

$$\theta = \tan^{-1}\left(\frac{C_y}{C_x}\right) \qquad \begin{array}{l}\textit{direction of vector from} \\ \textit{magnitudes of components}\end{array} \qquad \text{(3.6)}$$

A general notation for expressing the magnitude and direction of a vector involves the use of unit vectors. For example, as illustrated in •Fig. 3.8, a vector **A** can be written as $\mathbf{A} = A\hat{\mathbf{a}}$. The numerical magnitude is represented by A, and $\hat{\mathbf{a}}$ is called a **unit vector**. That is, it has a magnitude of unity, or one, but no units and so simply indicates the vector's direction. For example, a velocity along the x axis might be written $\mathbf{v} = (4.0 \text{ m/s}) \, \hat{\mathbf{x}}$ (that is, 4.0 m/s magnitude in the +x direction).

Note in Fig. 3.8 how −**A** would be represented in this notation. Although the minus sign is sometimes put in front of the numerical magnitude, this is an absolute number; the minus actually goes with the unit vector; $-\mathbf{A} = -A\hat{\mathbf{a}} = A(-\hat{\mathbf{a}})$.[†] That is, the unit vector is in the $-\hat{\mathbf{a}}$ direction (opposite $\hat{\mathbf{a}}$). A velocity of $\mathbf{v} = (-4.0 \text{ m/s}) \, \hat{\mathbf{x}}$ has a magnitude of 4.0 m/s in the −x direction; that is, $\mathbf{v} = (4.0 \text{ m/s})(-\hat{\mathbf{x}})$.

This notation can be used to express explicitly the rectangular components of a vector. For example, the ball's displacement from the origin in Example 3.2 could be written $\mathbf{d} = (4.50 \text{ m}) \, \hat{\mathbf{x}} + (12.6 \text{ m}) \, \hat{\mathbf{y}}$, where $\hat{\mathbf{x}}$ and $\hat{\mathbf{y}}$ are unit vectors in the x and y directions, respectively. In some instances it may be more convenient to express a general vector in this unit vector **component form**:

$$\mathbf{C} = C_x \hat{\mathbf{x}} + C_y \hat{\mathbf{y}} \qquad \text{(3.7)}$$

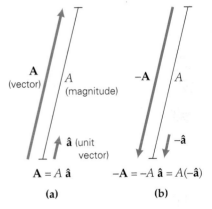

$$\mathbf{A} = A \, \hat{\mathbf{a}} \qquad -\mathbf{A} = -A \, \hat{\mathbf{a}} = A(-\hat{\mathbf{a}})$$
(a)   (b)

•FIGURE 3.8 Unit vectors (a) A unit vector $\hat{\mathbf{a}}$ has a magnitude of unity, or one, and so simply indicates a vector's direction. Written with the magnitude A, it represents the vector **A**, and $\mathbf{A} = A\hat{\mathbf{a}}$. (b) For the vector −**A**, the unit vector is $-\hat{\mathbf{a}}$, and $-\mathbf{A} = -A\hat{\mathbf{a}} = A(-\hat{\mathbf{a}})$. The magnitude A is a positive number— that is, $|A|$.

Component form of a vector

---

*Only a vector in the first quadrant is illustrated in Fig. 3.7b, but the equations hold for all quadrants when vectors are referenced to the nearest x axis. The directions of the components will be indicated by + and − signs, as will be shown shortly.

[†]The notation is sometimes written $\mathbf{A} = |A|\hat{\mathbf{a}}$ or $-\mathbf{A} = -|A|\hat{\mathbf{a}}$ so as to express the absolute value of the magnitude of A.

Vector Addition Using Components.   The **analytical component method** of vector addition involves resolving the vectors into rectangular components and adding the components for each axis independently. This method is illustrated graphically in •Fig. 3.9 for two vectors, $\mathbf{F}_1$ and $\mathbf{F}_2$. The sums of the x and y components of the vectors being added are then equal to the corresponding components of the resultant vector.

The same principle applies if you are given three (or more) vectors to add. You could find the resultant by applying the graphical tip-to-tail method as illustrated in •Fig. 3.10a. However, this technique involves drawing the vectors to scale and using a protractor to measure angles, which is time-consuming. Note that the magnitude of the directional angle θ is not obvious in the figure and would have to be measured (or computed trigonometrically).

However, if you use the component method, you do not have to draw the vectors tip to tail. In fact, it is usually more convenient to put all of the tails together at the origin, as shown in Fig. 3.10b. Also, the vectors do not have to be drawn to scale, since the sketch is just a visual aid in applying the analytical method.

In the component method, you simply resolve the vectors to be added into their x and y components, add the respective components, and recombine to find the resultant. This procedure is illustrated in Fig. 3.10c. Notice that it is just linear vector addition in the component directions and that components may add or subtract. A vector in the x or y direction has a "component" only in that direction (for example, $\mathbf{v}_2$ in the y direction in the figure).

Notice also in Fig. 3.10c that the directional angle θ of the resultant is referenced to the x axis, as are the individual vectors in Fig. 3.10b. *In adding vectors by the component method, we will reference all vectors to the nearest x axis—that is, the $+x$ axis or $-x$ axis.* This policy eliminates angles greater than 90° (as occurs when we customarily measure angles counterclockwise from the $+x$ axis) and the use of

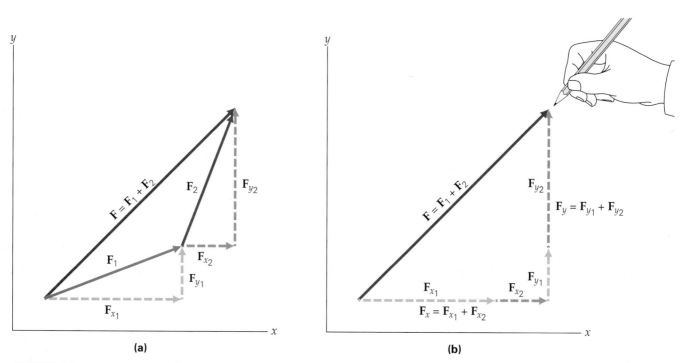

(a)   (b)

•**FIGURE 3.9 Component addition**   (a) In adding vectors by the component method, each vector is first resolved into its x and y components. (b) The sums of the x and y components of $\mathbf{F}_1$ and $\mathbf{F}_2$ are the x and y components of the resultant $\mathbf{F}$; that is, $\mathbf{F}_x = \mathbf{F}_{x_1} + \mathbf{F}_{x_2}$ and $\mathbf{F}_y = \mathbf{F}_{y_1} + \mathbf{F}_{y_2}$.

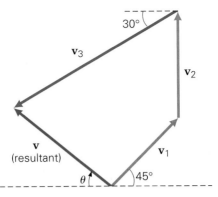

(a) Polygon method

•**FIGURE 3.10** **Component method of vector addition** **(a)** Several vectors may be added graphically to find the resultant **v**, but this technique is time-consuming. **(b)** In the analytical component method all the vectors to be added are first placed with their tails at the origin so that they may be easily resolved into rectangular components. **(c)** The respective summations of all the x components and all the y components are then added to give the components of the resultant **v**. Notice how the negative components subtract. See Example 3.4.

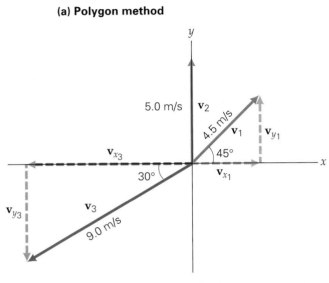

(b) Component method
(resolving components)

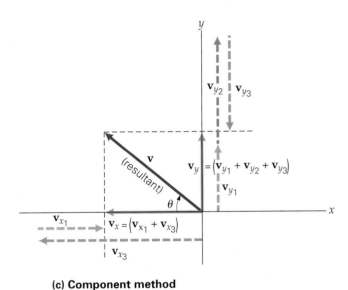

(c) Component method
(adding x and y components, shown as offset dashed arrows, and finding resultant)

double angle formulas, such as $\cos(\theta + 90°)$. This restriction greatly simplifies calculations. The recommended procedures for adding vectors analytically by the component method can be summarized as follows:

Procedures for Adding Vectors by the Component Method

1. Resolve the vectors to be added into their x and y components. Use the acute angles (those less than 90°) between the vectors and the x axis, and indicate the directions of the components by plus and minus signs (•Fig. 3.11).
2. Add all of the x components together and all of the y components together vectorially to obtain the x and y components of the resultant, or vector sum. (This is done algebraically by using plus and minus signs.)
3. Express the resultant vector, using:
   (a) the component form—for example, $\mathbf{C} = C_x\,\hat{\mathbf{x}} + C_y\,\hat{\mathbf{y}}$ —or
   (b) the magnitude-angle form.

For the latter, find the magnitude of the resultant by using the summed x and y components and the Pythagorean theorem:

$$C = \sqrt{C_x^2 + C_y^2}$$

Find the angle of direction (relative to the x axis) by taking the inverse tangent

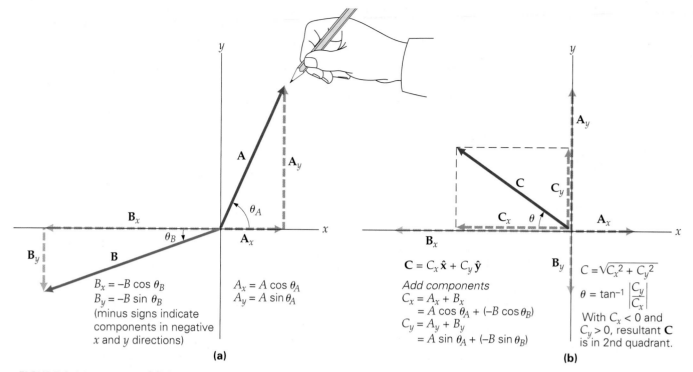

$B_x = -B \cos\theta_B$
$B_y = -B \sin\theta_B$
(minus signs indicate components in negative x and y directions)

$A_x = A \cos\theta_A$
$A_y = A \sin\theta_A$

$\mathbf{C} = C_x\,\hat{\mathbf{x}} + C_y\,\hat{\mathbf{y}}$
Add components
$C_x = A_x + B_x$
$\quad = A\cos\theta_A + (-B\cos\theta_B)$
$C_y = A_y + B_y$
$\quad = A\sin\theta_A + (-B\sin\theta_B)$

$C = \sqrt{C_x^2 + C_y^2}$
$\theta = \tan^{-1}\left|\dfrac{C_y}{C_x}\right|$
With $C_x < 0$ and $C_y > 0$, resultant $\mathbf{C}$ is in 2nd quadrant.

(a)   **(b)**

•**FIGURE 3.11   Vector addition by the analytical component method**   (a) Resolve the vectors into their x and y components. **(b)** Add all of the x components and all of the y components together vectorially to obtain the x and y components (C_x and C_y) of the resultant. Express the resultant in either component form or magnitude-angle form. All angles are referenced to the $+x$ or $-x$ axis to keep them less than 90°.

Note: The absolute value indicates that minus signs are ignored (for example, $|-3| = 3$). This is done to avoid negative values and angles greater than 90°.

($\tan^{-1}$) of the *absolute value* of the ratio of the y and x components (that is, the positive value, ignoring, any minus signs):

$$\theta = \tan^{-1}\left|\frac{C_y}{C_x}\right|$$

Designate the quadrant in which the resultant lies. This information is obtained from the signs of the summed components or from a sketch of their addition via the triangle (or parallelogram) method (see Fig. 3.11). The angle θ is the angle between the resultant and the x axis in that quadrant.

Example 3.4 ■ Applying the Analytical Component Method: Separating and Combining x and y Components

Let's apply the procedural steps of the component method to the addition of the vectors in Fig. 3.10b. The vectors with units of meters per second represent velocities.

Thinking It Through.   Follow and learn the steps given on page 75. Basically, you resolve the vectors into components and add the respective components to get the components of the resultant, which then may be expressed in component form or magnitude-angle form.

Solution.

1. The rectangular components of the vectors are shown in Fig. 3.10b.
2. Summing these components gives

$$\mathbf{v} = v_x\,\hat{\mathbf{x}} + v_y\,\hat{\mathbf{y}} = (v_{x_1} + v_{x_2} + v_{x_3})\,\hat{\mathbf{x}} + (v_{y_1} + v_{y_2} + v_{y_3})\,\hat{\mathbf{y}}$$

where

$$v_x = v_{x_1} + v_{x_2} + v_{x_3} = v_1\cos 45° + v_2\cos 90° - v_3\cos 30°$$

$$= (4.5\text{ m/s})(0.707) + (5.0\text{ m/s})(0) - (9.0\text{ m/s})(0.866) = -4.6\text{ m/s}$$

$$v_y = v_{y_1} + v_{y_2} + v_{y_3} = v_1 \sin 45° + v_2 \sin 90° - v_3 \sin 30°$$

$$= (4.5 \text{ m/s})(0.707) + (5.0 \text{ m/s})(1) - (9.0 \text{ m/s})(0.50) = 3.7 \text{ m/s}$$

In tabular form, the components are:

	x components		y components
v_{x_1}	$v_1 \cos 45° = +3.2$ m/s	v_{y_1}	$v_1 \sin 45° = +3.2$ m/s
v_{x_2}	$v_2 \cos 90° = 0$ m/s	v_{y_2}	$v_2 \sin 0° = +5.0$ m/s
v_{x_3}	$v_3 \cos 30° = -7.8$ m/s	v_{y_3}	$v_3 \sin 30° = -4.5$ m/s
Sums:	$v_x = -4.6$ m/s		$v_y = +3.7$ m/s

The directions of the components are indicated by signs (the + sign is usually omitted as being understood). In this case, v_2 has no x component. Note that in general for the analytical component method, the x components are cosine functions and the y components are sine functions, as long as we reference to the nearest x axis.

3. In component form, the resultant vector is

$$\mathbf{v} = (-4.6 \text{ m/s})\,\hat{\mathbf{x}} + (3.7 \text{ m/s})\,\hat{\mathbf{y}}$$

In magnitude-angle form, the resultant velocity has a magnitude of

$$v = \sqrt{v_x^2 + v_y^2} = \sqrt{(-4.6 \text{ m/s})^2 + (3.7 \text{ m/s})^2} = 5.9 \text{ m/s}$$

Since the x component is negative and the y component is positive, the resultant lies in the second quadrant at an angle of

$$\theta = \tan^{-1}\left|\frac{v_y}{v_x}\right| = \tan^{-1}\left(\frac{3.7 \text{ m/s}}{4.6 \text{ m/s}}\right) = 39°$$

above the negative x axis (see Fig. 3.10c).

Follow-up Exercise.   Suppose in Example 3.4 there were an additional $\mathbf{v_4} = (+4.6 \text{ m/s})\,\hat{\mathbf{x}}$ velocity vector. What would be the resultant of all four vectors in this case?

Although this discussion is limited to motion in two dimensions (in a plane), the component method is easily extended to three dimensions. For a velocity in three dimensions, the vector has x, y, and z components: $\mathbf{v} = v_x\,\hat{\mathbf{x}} + v_y\,\hat{\mathbf{y}} + v_z\,\hat{\mathbf{z}}$.

Conceptual Example 3.5

Vector Components in Action: Sailing "Into" the Wind

A sailboat (or wind surfer) on a lake travels in the direction the wind is blowing and then returns. Traveling "into" the wind, how does the boat get back home? *Clearly establish the reasoning and physical principle(s) used in determining your answer before checking it below. That is, **why** did you select your answer?*

Reasoning and Answer.   Sailing into the wind is called *tacking*. This is not a direct mode of sailing but a wise use of vector components. Here, we will consider components of force, which is a vector quantity. As you know from experience, a single force (or force component), such as a push, gives rise to motion in the direction in which the force is applied. (We shall learn much more about forces in Chapter 4.) For simplicity, let's assume that the wind fills the sail and exerts a force perpendicular to the sail (F_s), as illustrated in •Fig. 3.12a. (Ignore friction and water current effects.)

Note that this force is resolved into components. The force parallel ($F_{\parallel}$) to the boat is at an acute angle ($<90°$) to the direction of the wind and is an "upwind" force on the boat.

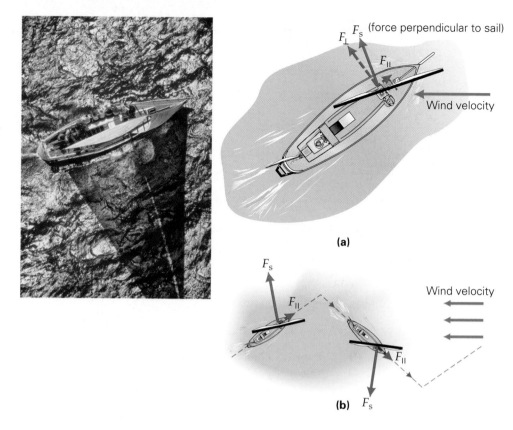

•**FIGURE 3.12 Let's go tacking.**
(a) The wind filling the sail exerts a force perpendicular to the sail (F_s). We can resolve this force vector into components. The one parallel to the motion of the boat ($F_\parallel$) has an upwind component, and there is motion in the upwind direction. **(b)** By changing the direction of the sail, the sailor can "tack" upwind. See Conceptual Example 3.5.

But, before heading too far to the northeast (as in Fig. 3.12a), the sail is turned so that the direction of $F_\parallel$ is changed by about 90° (Fig. 3.12b). The boat then comes back more in line with the desired upwind course. Using this zig-zag process, the boat sails "into" the wind and eventually gets home.

Follow-up Exercise.   The zig-zag, straight-line path in Fig. 3.12b was drawn for simplicity. Would this be the actual path of the boat? Explain. [Hint: Consider $F_\perp$.]

3.3   Relative Velocity

Objective: **To determine relative velocities through vector addition and subtraction.**

Measurements must be made with respect to some reference. This reference is usually taken to be the origin of a coordinate system. The point you designate as the origin of a set of coordinate axes is arbitrary and entirely a matter of choice. For example, you may "attach" the coordinate system to the road or the ground and then measure the displacement or velocity of a car relative to these axes.

We can analyze a situation from any frame of reference. For example, the origin of the coordinate axes may be attached to a car moving along a highway. In analyzing motion from another reference frame, you do not change the physical situation or what is taking place, only the point of view from which you describe it. Hence, we say that motion is *relative* (to some reference frame), and we refer to **relative velocity**. Since velocity is a vector, vector addition and subtraction are helpful in determining relative velocities.

Relative Velocities in One Dimension

When the velocities are linear (along a straight line) in the same or opposite directions and all have the same reference (such as the ground), we can find relative velocities by vector subtraction. As an illustration, consider cars moving with

•**FIGURE 3.13 Relative velocity** The observed velocity of a car depends on, or is relative to, the frame of reference. The velocities shown in **(a)** are relative to the ground or to the parked car A. In **(b)** the frame of reference is with respect to car B, and the velocities are those that the driver of car B would observe. **(c)** These aircraft, performing air-to-air refueling, are normally described as traveling at hundreds of kilometers per hour. To what frame of reference do their velocities refer? What is their velocity relative to each other?

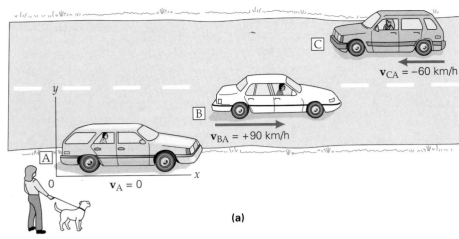

(a)

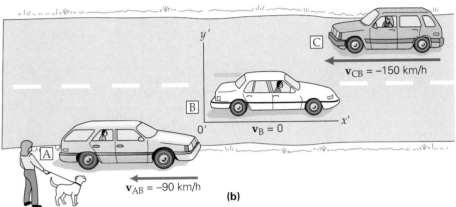

(b)

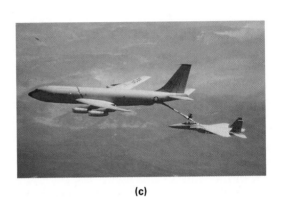

(c)

constant velocities along a straight, level highway, as in •Fig. 3.13. The velocities shown in the figure are *relative to the Earth or the ground*, as indicated by the reference set of coordinate axes in Fig. 3.13a. They are also relative to the stationary observers standing by the highway and sitting in the parked car A. That is, these observers see the cars as moving with velocities $v_B = +90$ km/h and $v_C = -60$ km/h. The relative velocity of two objects is given by the velocity (vector) difference between them. For example, the velocity of car B *relative to car A* is given by

$$\mathbf{v}_{BA} = \mathbf{v}_B - \mathbf{v}_A = +90 \text{ km/h} - 0 = +90 \text{ km/h}$$

Thus, a person sitting in car A would see car B move away (in the positive x direction) with a speed of 90 km/h. For this linear case, the directions of the velocities are indicated by plus and minus signs (in addition to the minus sign in the formula).

Similarly, the velocity of car C relative to an observer in car A is

$$\mathbf{v}_{CA} = \mathbf{v}_C - \mathbf{v}_A = -60 \text{ km/h} - 0 = -60 \text{ km/h}$$

The person in car A would see car C approaching (in the negative x direction) with a speed of 60 km/h.

But suppose that you want to know the velocities of the other cars *relative to car B* (that is, from the point of view of an observer in car B) or relative to a set of coordinate axes with the origin fixed to car B (Fig. 3.13b). Relative to those axes, car B is not moving and acts as the fixed reference point. The other cars are moving relative to car B. The velocity of car C relative to car B is

$$\mathbf{v}_{CB} = \mathbf{v}_C - \mathbf{v}_B = -60 \text{ km/h} - (+90 \text{ km/h}) = -150 \text{ km/h}$$

Note: Use subscripts carefully!
$v_{AB} = $ velocity *of* A *relative to* B.

Similarly, car A has a velocity relative to car B of

$$\mathbf{v}_{AB} = \mathbf{v}_A - \mathbf{v}_B = 0 - (+90 \text{ km/h}) = -90 \text{ km/h}$$

Notice that relative to B, the other cars are both moving in the negative x direction. That is, C is approaching B with a velocity of 150 km/h in the $-x$ direction and A appears to be receding from B with a velocity of 90 km/h in the $-x$ direction. (Imagine yourself in car B, and take that position as stationary. Car C would appear to be coming toward you at a high rate of speed, and car A would be getting farther and farther away, as though it were moving backward relative to you.) Note that, in general,

$$\mathbf{v}_{AB} = -\mathbf{v}_{BA}$$

What about the velocities relative to car C? From the point of view (or reference point) of car C, cars A and B would both appear to be approaching or moving in the positive x direction. For B relative to C,

$$\mathbf{v}_{BC} = \mathbf{v}_B - \mathbf{v}_C = 90 \text{ km/h} - (-60 \text{ km/h}) = +150 \text{ km/h}$$

Can you show that $\mathbf{v}_{AC} = +60$ km/h? Also note the situation in Fig. 3.13c.

In some instances, we may need to work with velocities that do not all have the same reference point. In such cases, relative velocities can be found by means of vector addition. To solve problems of this kind, *it is essential to identify the velocity references with care.*

Let's look first at a one-dimensional (linear) example. Suppose that a straight moving walkway in a major airport moves with a velocity of $\mathbf{v}_{wg} = +1.0$ m/s, where the subscripts indicate the velocity of the walkway (w) with respect to, or relative to, the ground (g). A passenger (p) on the walkway trying to make a flight connection walks with a velocity of $\mathbf{v}_{pw} = +2.0$ m/s relative to the walkway. What is the passenger's velocity relative to an observer standing next to the walkway (that is, relative to the ground)?

The velocity we are seeking, $\mathbf{v}_{pg}$, is given by

$$\mathbf{v}_{pg} = \mathbf{v}_{pw} + \mathbf{v}_{wg} = +2.0 \text{ m/s} + 1.0 \text{ m/s} = +3.0 \text{ m/s}$$

Thus the stationary observer sees the passenger as traveling with speed of 3.0 m/s down the walkway. (Make a sketch and show how the vectors add.)

Problem-Solving Hint

Notice the pattern of the subscripts in this example. On the right side of the equation, the two inner subscripts are the same (w). The outer subscripts (p and g) are sequentially the same as those for the relative velocity on the left side of the equation. When adding relative velocities, always check to make sure that the subscripts have this relationship—it indicates that you have set up the equation correctly.

What if a passenger got on the walkway going in the opposite direction and walked with the same speed as that of the walkway? Now it is essential to indicate the direction in which the passenger is walking by means of a minus sign: $\mathbf{v}_{pw} = -1.0$ m/s. In this case, relative to the stationary observer,

$$\mathbf{v}_{pg} = \mathbf{v}_{pw} + \mathbf{v}_{wg} = -1.0 \text{ m/s} + 1.0 \text{ m/s} = 0$$

so the passenger is stationary with respect to the ground and the walkway acts as a treadmill. (Excellent physical exercise!)

Relative Velocities in Two Dimensions

Of course, velocities are not always in the same or opposite directions. However, knowing how to use rectangular components to add or subtract vectors, we can solve problems involving relative velocities in two dimensions, as Examples 3.6 and 3.7 show.

Example 3.6 ■ Across and Down the River: Relative Velocity and Components of Motion

The current of a 500-meter-wide straight river has a flow rate of 2.55 km/h. A motorboat that travels with a constant speed of 8.00 km/h in still water crosses the river (•Fig. 3.14). (a) If the boat's bow points directly across the river toward the opposite shore, what is the velocity of the boat relative to the stationary observer sitting at the corner of the bridge? (b) How far downstream will the boat's landing point be from the point directly opposite its starting point? (c) What is the distance traveled by the boat in crossing the river? (Assume that the boat comes instantaneously to rest on a grassy shore.)

Thinking It Through. Careful designation of the given quantities is very important—the velocity of what, relative to what. Once this is done, part (a) should be straightforward. (See the previous Problem-Solving Hint.) For parts (b) and (c), we use kinematics, where the time it takes the boat to cross the river is the key.

Solution. As indicated in Fig. 3.14, we take the river's flow velocity ($\mathbf{v}_{rs}$, river to shore) to be in the x direction and the boat's velocity ($\mathbf{v}_{br}$, boat to river) to be in the y direction. Note that the river's flow velocity is *relative to the shore* and that the boat's velocity is *relative to the river*, as indicated by the subscripts. Listing the data, we have

Given: y_{max} = 500 m (river width)
$\mathbf{v}_{rs}$ = (2.55 km/h) $\hat{\mathbf{x}}$
$\qquad$ = (0.709 m/s) $\hat{\mathbf{x}}$
$\qquad$ (velocity of river
$\qquad$ relative to shore)
$\mathbf{v}_{br}$ = (8.00 km/h) $\hat{\mathbf{y}}$
$\qquad$ = (2.22 m/s) $\hat{\mathbf{y}}$
$\qquad$ (velocity of boat
$\qquad$ relative to river)

Find: (a) $\mathbf{v}_{bs}$ (velocity of boat
$\qquad$ relative to shore)
(b) x (distance downstream)
(c) d (distance traveled
$\qquad$ by boat)

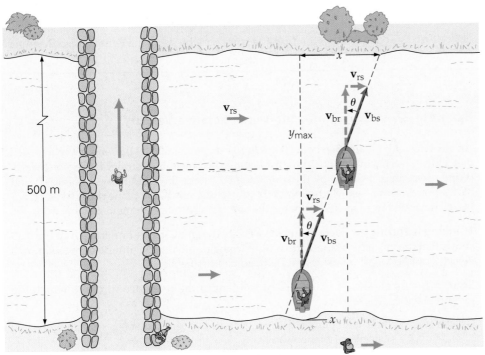

•**FIGURE 3.14 Relative velocity and components of motion** As the boat moves across the river, it is carried downstream by the current. See Example 3.6.

Notice that as the boat moves toward the opposite shore, it is also carried downstream by the current. These velocity components would be clearly apparent relative to the jogger crossing the bridge and to the person sauntering downstream in Fig. 3.14. If both observers stay even with the boat, the velocity of each will match one of the components of the boat's velocity. Since the velocity components are constant, the boat travels in a straight line diagonally across the river (much like the ball rolling across the table in Example 3.1).

(a) The velocity of the boat relative to the shore ($\mathbf{v}_{bs}$) is given by vector addition. In this case, we have

$$\mathbf{v}_{bs} = \mathbf{v}_{br} + \mathbf{v}_{rs}$$

Since the velocities are not along one axis, their magnitudes cannot be added directly. Notice in Fig. 3.14 that the vectors form a right triangle, so we can apply the Pythagorean theorem to find the magnitude of $\mathbf{v}_{bs}$:

$$v_{bs} = \sqrt{(v_{br})^2 + (v_{rs})^2} = \sqrt{(2.22 \text{ m/s})^2 + (0.709 \text{ m/s})^2}$$
$$= 2.33 \text{ m/s}$$

The direction of this velocity is defined by

$$\theta = \tan^{-1}\left(\frac{v_{rs}}{v_{br}}\right) = \tan^{-1}\left(\frac{0.709 \text{ m/s}}{2.22 \text{ m/s}}\right) = 17.7°$$

(b) To find the distance x that the current carries the boat downstream, we use components. Note that in the y direction, $y_{max} = v_{br}t$, and

$$t = \frac{y_{max}}{v_{br}} = \frac{500 \text{ m}}{2.22 \text{ m/s}} = 225 \text{ s}$$

which is the time it takes the boat to cross the river.

During this time, the boat is carried downstream by the current a distance of

$$x = v_{rs}t = (0.709 \text{ m/s})(225 \text{ s}) = 160 \text{ m}$$

(c) We could find the distance d the boat travels by using x and y_{max} with the Pythagorean theorem again, but let's use the magnitude of the relative velocity and time:

$$d = v_{bs}t = (2.33 \text{ m/s})(225 \text{ s}) = 524 \text{ m}$$

Follow-up Exercise. In part (c), find the distance d by using the Pythagorean theorem. (The answer may be slightly different. Why?)

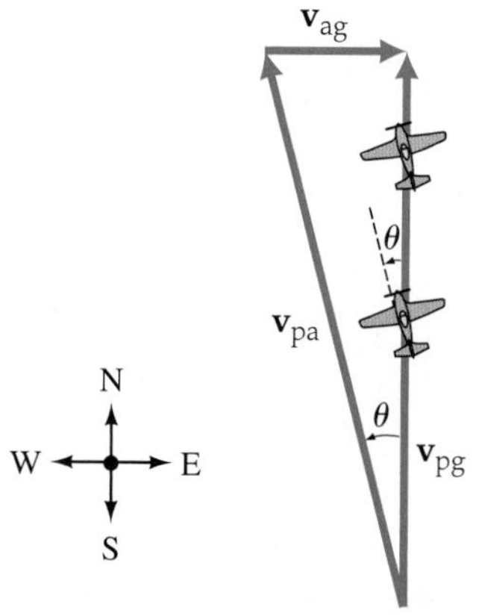

•FIGURE 3.15 Flying into the wind To fly directly north, the plane's heading (θ direction) must be west of north. See Example 3.7.

Example 3.7 ■ Flying Into the Wind: Relative Velocity

An airplane with an air speed of 200 km/h (its speed in still air) flies in a direction such that with a west wind of 50.0 km/h blowing, it travels in a straight line northward. (Wind direction is specified by the direction *from* which the wind blows, so a west wind blows from west to east.) To maintain its course due north, the plane must fly at an angle as illustrated in •Fig. 3.15. What is the speed of the plane along its northward path?

Thinking It Through. Here again, the velocity designations are important, but Fig. 3.15 shows that the velocity vectors form a right triangle, and the magnitude of the unknown velocity can be found by using the Pythagorean theorem.

Solution. As always, it is important to identify the reference frame to which the given velocities are relative.

Given: $\mathbf{v}_{pa} = 200$ km/h at angle θ *Find:* v_{pg} (ground speed of plane)
(velocity of plane with
respect to still air = air speed)

$$\mathbf{v}_{ag} = 50.0 \text{ km/h east (velocity of air with respect to Earth or ground} = \text{wind speed)}$$

plane flies due north with velocity $\mathbf{v}_{pg}$

The speed of the plane with respect to the Earth or the ground, v_{pg}, is called its ground speed; v_{pa} is its air speed. Vectorially, the respective velocities are related by

$$\mathbf{v}_{pg} = \mathbf{v}_{pa} + \mathbf{v}_{ag}$$

If there were no wind blowing ($v_{ag} = 0$), the ground speed and air speed would be equal. However, a head wind (one blowing directly toward the plane) would cause a slower ground speed; and a tail wind, a faster ground speed. The situation is analogous to that of a boat going upstream versus downstream.

Here $\mathbf{v}_{pg}$ is the resultant of the other two vectors, which can be added by the triangle method. We use the Pythagorean theorem to find v_{pg}, noting that v_{pa} is the hypotenuse of the triangle:

$$v_{pg} = \sqrt{v_{pa}{}^2 - v_{ag}{}^2} = \sqrt{(200 \text{ km/h})^2 - (50.0 \text{ km/h})^2} = 194 \text{ km/h}$$

(Note that it was convenient to use the units of kilometers per hour, since the calculation did not involve any other units.)

Follow-up Exercise. What must be the plane's heading (θ direction) in Example 3.7 for the plane to fly directly north?

3.4 Projectile Motion

Objectives: To analyze projectile motion to find (a) position, (b) time of flight, and (c) range.

A familiar example of two-dimensional, curvilinear motion is the motion of objects that are thrown or projected by some means. The motion of a stone thrown across a stream or a golf ball driven off a tee is **projectile motion.** A special case of projectile motion in one dimension occurs when an object is projected vertically upward. This case was treated in Chapter 2 in terms of free fall with air resistance neglected. We will generally neglect air resistance here also so that the only acceleration acting on a projectile is that due to gravity.

Note: Review Section 2.5 (free fall).

We can use vector components to analyze projectile motion. You simply break up the motion into its x and y components and treat them separately.

Horizontal Projections

It is worthwhile to analyze first the special case of the motion of an object projected horizontally, or parallel to a level surface. Suppose that you throw an object horizontally with an initial velocity v_{x_o} (●Fig. 3.16a). Projectile motion is analyzed beginning at the instant of release ($t = 0$). Once the object is released, there is no longer a horizontal acceleration ($a_x = 0$), so throughout the object's path the horizontal velocity remains constant: $v_x = v_{x_o}$

According to the equation $x = v_x t$ (Eq. 3.2a), the projected object would continue to travel in the horizontal direction indefinitely. However, you know that this is not what happens. As soon as the object is projected, it is in free fall in the vertical direction, with $v_{y_o} = 0$ (as though it were dropped) and $a_y = -g$. In other words, the projected object travels at a uniform velocity in the horizontal direction while *at the same time* undergoing acceleration in the downward direction under the influence of gravity. The result is a curved path, as illustrated in Fig. 3.16. (Compare the motions in Fig. 3.16 and Fig. 3.2. Do you see any similarities?) If there were no horizontal motion, the object would simply drop to the ground in

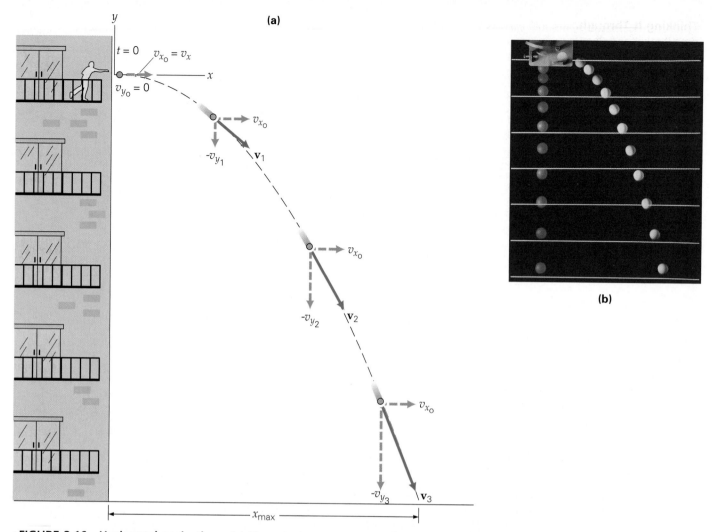

(a)

(b)

•**FIGURE 3.16 Horizontal projection** (a) The velocity components of a projectile launched horizontally show that it travels to the right as it falls downward. **(b)** A multiflash photograph shows the paths of two golf balls. One was projected horizontally at the same time that the other was dropped straight down. The horizontal lines are 15 cm apart, and the time interval between flashes was 1/30 s. The vertical motions of the balls are the same. Why? Can you describe the horizontal motion of the yellow ball?

a straight line. In fact, the time of flight of the projected object is *exactly the same as if it were falling vertically.*

Note the components of the velocity vector in Fig. 3.16a. The length of the horizontal component of the velocity vector remains the same, but the length of the vertical component increases with time. What is the instantaneous velocity at any point along the path? (Think in terms of vector addition, covered in Section 3.3.) The photo in Fig. 3.16b shows the actual motions of a horizontally projected golf ball and one that is simultaneously dropped from rest. The horizontal reference lines show that the balls fall vertically at the same rate. The only difference is that the horizontally projected ball also travels to the right as it falls.

Example 3.8 ■ Starting at the Top: Horizontal Projection

Suppose that the ball in Fig. 3.16a is projected from a height of 25.0 m above the ground and is thrown with an initial horizontal velocity of 8.25 m/s. (a) How long is the ball in flight before striking the ground? (b) How far from the building does the ball strike the ground?

Thinking It Through. In looking at the components of motion, part (a) involves the time it take the ball to fall vertically, analogous to a ball dropped from that height. This is also the time the ball travels in the horizontal direction. The horizontal speed is constant, so we can find the horizontal distance, part (b).

Solution. Writing the data with the origin chosen as the point from which the ball is thrown and downward taken as the negative direction, we have

Given: $y = -25.0$ m   *Find:* (a) t (time of flight)
$\qquad v_{x_0} = 8.25$ m/s $\qquad$ (b) x (horizontal distance)
$\qquad a_x = 0$
$\qquad v_{y_0} = 0$
$\qquad a_y = -g$

(a) As noted previously, the time of flight is the same as the time it takes for the ball to fall vertically to the ground. To find this time, we can use the equation $y = v_{y_0}t - \frac{1}{2}gt^2$, in which the negative direction of g is expressed explicitly as was done in Chapter 2. With $v_{y_0} = 0$,

$$y = -\tfrac{1}{2}gt^2$$

and

$$t = \sqrt{\frac{2y}{-g}} = \sqrt{\frac{2(-25.0 \text{ m})}{-9.80 \text{ m/s}^2}} = 2.26 \text{ s}$$

(b) The ball travels in the x direction for the same amount of time it travels in the y direction (that is, 2.26 s). Since there is no acceleration in the horizontal direction, it does so with a uniform velocity. Thus, with $a_x = 0$,

$$x = v_{x_0}t = (8.25 \text{ m/s})(2.26 \text{ s}) = 18.6 \text{ m}$$

Follow-up Exercise. What is the velocity (in component form) of the ball just before it strikes the ground?

Projections at Arbitrary Angles

The general case of projectile motion involves an object projected at an arbitrary angle θ relative to the horizontal; for example, a golf ball struck by a club (•Fig. 3.17). During projectile motion, the object travels up and down while traveling horizontally with a constant velocity.

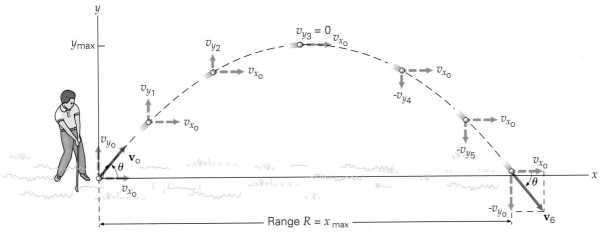

•**FIGURE 3.17   Projection at an angle**   The velocity components of the ball are shown for various times. Note that $v_y = 0$ at the top of the arc, or at y_{max}. The range (R) is the maximum horizontal distance, or x_{max}.

This motion is also analyzed by using its components. As before, upward is taken as the positive direction and downward as the negative direction. The initial velocity v_o is first resolved into rectangular components:

$$v_{x_o} = v_o \cos \theta \qquad \text{initial velocity} \qquad \text{(3.8a)}$$
$$v_{y_o} = v_o \sin \theta \qquad \text{components } (t_o = 0) \qquad \text{(3.8b)}$$

Since there is no horizontal acceleration and gravity acts in the negative y direction, the x component of the velocity is constant and the y component varies with time (see Eq. 3.3):

$$v_x = v_{x_o} = v_o \cos \theta \qquad \text{projectile motion} \qquad \text{(3.9a)}$$
$$v_y = v_{y_o} - gt = (v_o \sin \theta) - gt \qquad \text{velocity components} \qquad \text{(3.9b)}$$

The components of the instantaneous velocity at various times are illustrated in Fig. 3.17. The instantaneous velocity is the sum of these components and is tangent to the curved path of the ball at any point. Notice that the ball strikes the ground at the same speed as it was launched (but with $-v_{y_o}$) and at the same angle below the horizontal.

Similarly, the displacement components are given by

$$x = v_{x_o}t = (v_o \cos \theta)t \qquad \text{projectile motion} \qquad \text{(3.10a)}$$
$$y = v_{y_o}t - \tfrac{1}{2}gt^2 = (v_o \sin \theta)t - \tfrac{1}{2}gt^2 \qquad \begin{array}{l}\text{displacement}\\\text{components}\end{array} \qquad \text{(3.10b)}$$

The curve produced by these equations, or the path of motion of the projectile, is called a **parabola**. The path of projectile motion is often referred to as a parabolic arc. Such arcs are commonly observed (•Fig. 3.18).

Note as in the case of horizontal projection, *time is the common feature shared by the components of motion.* Aspects of projectile motion that may be of interest in various situations include the time of flight, the maximum height reached, and the **range** (R), which is the maximum horizontal distance traveled.

(a)

(b)

•**FIGURE 3.18 Parabolic arcs**
(a) Fireworks and (b) streams of water move in parabolic arcs, though air resistance may distort the trajectories to some extent.

Example 3.9 ■ Teeing Off: Projection at an Angle

Suppose a golf ball is hit off the tee with an initial velocity of 30.0 m/s at an angle of 35° to the horizontal as in Fig. 3.17. (a) What is the maximum height reached by the ball? (b) What is its range?

Thinking It Through. The maximum height involves the y component; finding it is like finding the maximum height of a ball projected vertically upward. The ball travels in the x direction for the same amount of time it would take the ball to go up and down.

Solution.

Given: $v_o = 30.0$ m/s   *Find:* (a) y_{max}
$\qquad\quad \theta = 35°$ $\qquad\qquad\qquad$ (b) $R = x_{max}$
$\qquad\quad a_y = -g$

Let us compute v_{x_o} and v_{y_o} explicitly to be able to use simplified kinematic equations:

$$v_{x_o} = v_o \cos 35° = (30.0 \text{ m/s})(0.819) = 24.6 \text{ m/s}$$
$$v_{y_o} = v_o \sin 35° = (30.0 \text{ m/s})(0.574) = 17.2 \text{ m/s}$$

(a) Just as for an object thrown vertically upward, $v_y = 0$ at the maximum height (y_{max}). Thus, we can find the time to reach the maximum height (t_{max}) by using Eq. 3.9b with v_y set equal to zero:

$$v_y = 0 = v_{y_o} - gt_{max}$$
$$t_{max} = \frac{v_{y_o}}{g} = \frac{17.2 \text{ m/s}}{9.80 \text{ m/s}^2} = 1.76 \text{ s}$$

As in the case of vertical projection, the time in going up is equal to the time in coming down, so the total time of flight is $t = 2t_{max}$ (to return to the elevation from which the object was projected).

The maximum height (y_{max}) is then obtained by substituting t_{max} into Eq. 3.10b:

$$y_{max} = v_{y_0}t_{max} - \tfrac{1}{2}gt_{max}^2 = (17.2 \text{ m/s})(1.76 \text{ s}) - \tfrac{1}{2}(9.80 \text{ m/s}^2)(1.76 \text{ s})^2 = 15.1 \text{ m}$$

The maximum height could also be obtained directly from Eq. 2.10', $v_y^2 = v_{y_0}^2 - 2gy$, with $y = y_{max}$ and $v_y = 0$. However, the method of solution used here illustrates how the time of flight is obtained.

(b) The range (R) is equal to the horizontal distance traveled (x_{max}), which is easily found by substituting the total time of flight $t = 2t_{max} = 2(1.76 \text{ s}) = 3.52$ s into Eq. 3.10a:

$$R = x_{max} = v_x t = v_{x_0}(2t_{max}) = (24.6 \text{ m/s})(3.52 \text{ s}) = 86.6 \text{ m}$$

Follow-up Exercise. What is the location of the golf ball in Example 3.9 at $t = 3.00$ s?

The range of a projectile is an important consideration in various applications. This is particularly true in those sports in which a maximum range is desired, such as in driving a golf ball or throwing a javelin.

Projectile range—the maximum horizontal distance traveled

In general, what is the range of a projectile launched with velocity v_0 at an angle θ? To answer this question requires us to consider the equation used in Example 3.9 to calculate the range, $R = v_x t$. First let's look at the expressions for v_x and t. Since there is no acceleration in the horizontal direction, we know that

$$v_x = v_{x_0} = v_0 \cos \theta$$

and $t = 2t_{max}$, where (as we found in Example 3.9)

$$t_{max} = \frac{v_{y_0}}{g} = \frac{v_0 \sin \theta}{g}$$

Then,

$$R = v_x t = (v_0 \cos \theta)(2t_{max}) = (v_0 \cos \theta)\left(\frac{2v_0 \sin \theta}{g}\right) = \frac{2v_0^2 \sin \theta \cos \theta}{g}$$

Using the trigonometric identity $\sin 2\theta = 2 \cos \theta \sin \theta$ (see Appendix I), we have

Projectile range—for $y_{initial} = y_{final}$ only

$$R = \frac{v_0^2 \sin 2\theta}{g} \qquad \begin{array}{l}\textit{projectile range, } x_{max} \\ (\textit{only for } y_{initial} = y_{final})\end{array} \qquad (3.11)$$

Note that the range depends on the magnitude of the initial velocity (or speed, v_0) and the angle of projection (θ); g is assumed to be constant. Keep in mind that this equation applies only to the *special*, but common, case of $y_{initial} = y_{final}$—that is, the landing point is at the same height as the launch point.

Example 3.10 ■ The Last Shot: A Win or a Loss?

The college basketball game is close. The home team trails by two points, and 1.00 s remains in the game. The hometown guard takes a jump shot (in other words, his feet leave the floor). The ball is launched at the angle of 60° relative to the horizontal, with a speed of 10.0 m/s, and an initial height of 3.05 m (10 ft) above the floor, the same height as the hoop (•Fig. 3.19). The ball goes in, and the crowd goes wild!

(a) Baskets made from outside a semicircle of radius 6.02 m (19 ft, 9 in.) from beneath the center of the hoop are given 3 points, whereas those made from inside the semicircle are worth 2 points (see Fig 3.19). Did the home team win, or did the game go into overtime in a tie?

(b) Suppose the home guard stops at a distance of 7.00 m from a point directly below the point nearest to him on the hoop and takes a set shot (that is, his feet are on the floor) with the same launch speed as in part (a) but at an angle of 30° and from an initial height of 1.80 m above the floor. Is there a possibility of him making the shot?

•FIGURE 3.19 A close game! A home player
takes a final shot, and the basket is good. But is
it a win or a tie? See Example 3.10.

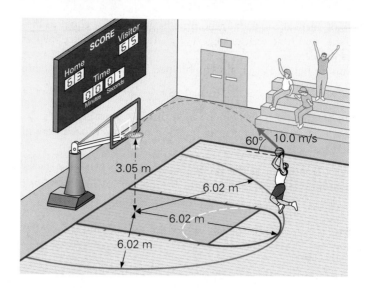

Thinking It Through. Part (a) involves the range of the ball. Note that the launch
point is the same height as the basket. In part (b), there is a question of the height of
the ball at the hoop. Was there enough arc for the ball to get there and go through? The
vertical height of the ball at the basket must be determined, but *note* that the range
equation (Eq. 3.11) *cannot* be used. Why?

Solution.

Given: (a) $v_0 = 10.0$ m/s   *Find:* (a) if $R > 6.02$ m
            $\theta = 60°$ (b) if $y_{final} > 3.05$ m
            $t = 1.00$ s   (note $y_{initial} = y_{final}$)
         (b) $x = 7.00$ m
            $\theta = 30°$
            $v_0 = 10.0$ m/s   (note $y_{initial} \neq y_{final}$)

(a) We must determine whether the ball was shot at a distance greater than 6.02 m—
that is, whether the range was greater than 6.02 m. Since the launch point and the basket are at the same height, Eq. 3.11 applies, and

$$R = \frac{v_0^2 \sin 2\theta}{g} = \frac{(10.0 \text{ m/s})^2 \sin 2(60°)}{9.80 \text{ m/s}^2} = 8.84 \text{ m}$$

The shot was made from outside the semicircle, and the home team wins!

(b) Here, we want to know where the ball is vertically when it gets to the basket—that
is, its y position. Knowing the horizontal distance and the initial velocity, we have

$$v_x = v_0 \cos 30° = (10.0 \text{ m/s})(0.866) = 8.66 \text{ m/s}$$

and

$$v_y = v_0 \sin 30° = (10.0 \text{ m/s})(0.500) = 5.00 \text{ m/s}$$

The time of flight is found from the x direction kinematic equation:

$$t = \frac{x}{v_x} = \frac{7.00 \text{ m}}{8.66 \text{ m/s}} = 0.808 \text{ s}$$

And at this time, the vertical height of the ball is

$$y = v_{y_0}t - \tfrac{1}{2}gt^2 = (5.00 \text{ m/s})(0.808 \text{ s}) - \tfrac{1}{2}(9.80 \text{ m/s}^2)(0.808 \text{ s})^2 = 0.84 \text{ m}$$

Note: This is the height of the ball *above* the height of the launch point. (Why?) Hence,
the ball is at a height of 1.80 m + 0.84 m = 2.64 m from the floor—well under the 3.02 m
of the basket height, so the shot would not be good.

Follow-up Exercise. In part (a) of Example 3.10, did time run out before the ball went
through the hoop? (Even if it did, the basket counts.)

Equation 3.11 allows us to compute the range for a particular projection angle and initial velocity. However, we are sometimes interested in the maximum range for a given initial velocity—for example, the maximum range of an artillery piece that fires a projectile with a particular muzzle velocity. Is there an optimum angle? Under ideal conditions, the answer is yes.

For a particular v_o, the range is a maximum (R_{max}) when $\sin 2\theta = 1$, since this is the maximum value of the sine function (which varies from 0 to 1). Thus

$$R_{max} = \frac{v_o^2}{g} \qquad (3.12)$$

Because this maximum range is obtained when $\sin 2\theta = 1$ and because $\sin 90° = 1$, we have

$$2\theta = 90° \qquad \text{or} \qquad \theta = 45°$$

for the maximum range for a given initial speed when the projectile returns to the elevation from which it was projected. At a greater or smaller angle, for a projectile with the same initial speed, the range will be less, as illustrated in •Fig. 3.20. Also, the range is the same for angles equally above and below 45°, such as 30° and 60°. (See Exercise 88.)

Thus, to get the maximum range, a projectile should *ideally* be projected at an angle of 45°. However, up to now we have neglected air resistance. In actual situations, such as when a ball or object is thrown or hit hard, this factor may have a significant effect. Air resistance reduces the speed of the projectile, thereby reducing the range. As a result, when air resistance is a factor the angle of projection for maximum range is less than 45°, which gives a greater initial horizontal velocity (•Fig. 3.21). Other factors, such as spin and wind, may affect the range of a projectile. For example, backspin on a driven golf ball provides lift, and the projection angle for the maximum range may be considerably less than 45°.

Keep in mind that the 45° maximum range requires that the initial velocity components be equal: $\tan^{-1}(v_{y_o}/v_{x_o}) = 45°$ and $\tan 45° = 1$, so $v_{y_o} = v_{x_o}$. However, this condition may not always be physically possible, as the following Example shows.

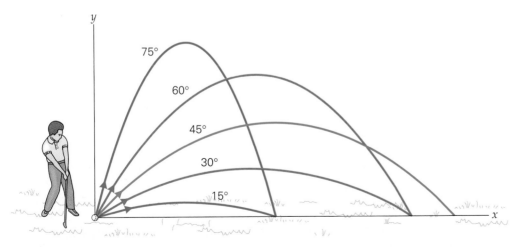

•**FIGURE 3.20 Range** For a projectile with a given initial speed, the maximum range is ideally attained with a projection angle of 45°. For projection angles above and below 45°, the range is shorter, and it is equal for angles equally different from 45° (for example, 30° and 60°).

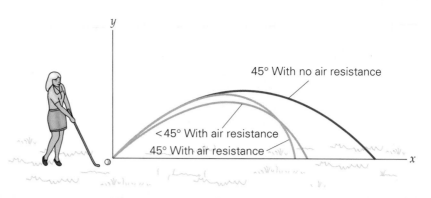

•**FIGURE 3.21 Air resistance and range** When air resistance is a factor, the angle of projection for maximum range is less than 45°.

45° With no air resistance

<45° With air resistance

45° With air resistance

•**FIGURE 3.22   Hanging in there** Basketball players seem to "hang" in the air at the peak of their jump. Why is this? See the Follow-up Exercise, Conceptual Example 3.11.

Conceptual Example 3.11

The Longest Jump: Theory and Practice

In a long jump event, the jumper normally has a launch angle of (a) less than 45°, (b) exactly 45°, (c) greater than 45°? *Clearly establish the reasoning and physical principle(s) used in determining your answer before checking it below. That is, **why** did you select your answer?*

Reasoning and Answer.   Air resistance is not a major factor here (although wind speed is taken into account for record settings in track and field events). Therefore, it would seem that, to achieve maximum range, the jumper would take off at an angle of 45°. But there is another physical consideration. Let's look more closely at the jumper's initial velocity components.

To maximize a long jump, the jumper runs as fast as possible and then pushes upward as strongly as possible to maximize the velocity components. The initial vertical velocity component (v_{y_o}) depends on the upward push of the jumper's legs, whereas the initial horizontal component (v_{x_o}) depends mostly on the running speed toward the jump point. In general, a greater velocity can be achieved by running than by jumping, so $v_{x_o} > v_{y_o}$. Then, since $\theta = \tan^{-1}(v_{y_o}/v_{x_o})$, we have $\theta < 45°$, where $v_{y_o}/v_{x_o} < 1$ in this case. Hence, the answer is (a) and certainly could not be (c). A typical launch angle for a long jump is 20° to 25°. (If a jumper increased her launch angle closer to the ideal 45°, her running speed would have to decrease, resulting in a decrease in range.)

Follow-up Exercise.   When jumping to score, basketball players seem to be suspended momentarily or to "hang" in the air (•Fig. 3.22). Explain the physics of this effect.

Chapter Review

Important Terms

components of motion   66
vector addition (subtraction)   69
triangle method   70
parallelogram method   71
polygon method   71

component method   72
magnitude-angle form   72
unit vector   73
component form   73
analytical component method   74

relative velocity   78
projectile motion   83
parabola   86
range   86

Important Concepts

• Motion in two dimensions is analyzed by considering linear component motion in each dimension individually.

• Vector subtraction is a special case of vector addition.

• Of the various methods of vector addition, the component method is the most powerful. A resultant vector can be expressed in the magnitude-angle form or the unit-vector component form.

- Relative velocity is a velocity expressed *relative* to a particular reference frame.
- Projectile motion is analyzed by considering horizontal and vertical components separately—constant velocity in the horizontal direction and an acceleration due to gravity, *g*, in the downward vertical direction.

- To get the maximum range, a projectile should ideally be projected at an angle of 45°. However, when air resistance is a factor, the angle of projection for the maximum range is less than 45°.

Important Equations

Components of Initial Velocity:

$$v_{x_0} = v_0 \cos \theta \qquad (3.1a)$$
$$v_{y_0} = v_0 \sin \theta \qquad (3.1b)$$

Components of Displacement
(constant acceleration only):

$$x = v_{x_0} t + \tfrac{1}{2} a_x t^2 \qquad (3.3a)$$
$$y = v_{y_0} t + \tfrac{1}{2} a_y t^2 \qquad (3.3b)$$

Component of Velocity
(constant acceleration only):

$$v_x = v_{x_0} + a_x t \qquad (3.3c)$$
$$v_y = v_{y_0} + a_y t \qquad (3.3d)$$

Vector Representation:

$$\left. \begin{array}{l} C = \sqrt{C_x{}^2 + C_y{}^2} \\[2mm] \theta = \tan^{-1} \left| \dfrac{C_y}{C_x} \right| \end{array} \right\} \quad \textit{magnitude-angle form}$$

(3.4a)

(3.4b)

$$\mathbf{C} = C_x \,\hat{\mathbf{x}} + C_y \,\hat{\mathbf{y}} \qquad \textit{component form} \qquad (3.7)$$

Exercises

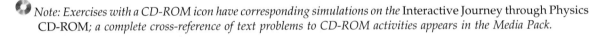

Note: Exercises with a CD-ROM icon have corresponding simulations on the Interactive Journey through Physics CD-ROM; a complete cross-reference of text problems to CD-ROM activities appears in the Media Pack.

3.1 Components of Motion

1. On Cartesian axes, the *x* component of a vector is generally associated with a (a) cosine, (b) sine, (c) tangent, (d) none of these.

2. For an object in curvilinear motion, (a) its velocity components are constant, (b) the *y*-velocity component is necessarily greater than the *x*-velocity component, (c) there is an acceleration nonparallel to the object's path, (d) the velocity and acceleration vectors must be at right angles (90°).

3. Describe the motion of an object that is initially traveling with a constant velocity and then receives an acceleration of constant magnitude (a) in a direction parallel to the initial velocity, (b) in a direction perpendicular to the initial velocity, (c) that is always perpendicular to the instantaneous velocity or direction of motion.

4. ■ A golf ball is hit with an initial speed of 35 m/s at an angle 37° above the horizontal. What are the initial horizontal and vertical velocity components?

5. ■ The *x* and *y* components of an acceleration vector are 3.0 m/s² and 4.0 m/s², respectively. What are the magnitude and direction of the acceleration vector?

6. ■ If the magnitude of a velocity vector is 7.0 m/s and the *x* component is +3.0 m/s, what are the possible values of its *y* component?

7. ■■ The *x* component of a velocity vector that has an angle of 37° to the +*x* axis has a magnitude of 4.8 m/s. (a) What is the magnitude of the velocity? (b) What is the magnitude of the *y* component of the velocity?

8. ■■ A student walks 100 m west and 50 m south. What displacement will bring him back to the starting point?

9. ■■ A student strolls diagonally across a level rectangular campus plaza, covering the 50-meter distance in 1.0 min (•Fig. 3.23). (a) If the diagonal route makes a 37° angle with the long side of the plaza, what would be the distance if she walked half way around the outside of the

•FIGURE 3.23 **Which way?** See Exercise 9.

plaza instead of along the diagonal route? (b) If she walked the outside route in 1.0 min at a constant speed, how much time in minutes would she spend on each leg?

10. ■■ The displacement vector of a moving object initially at the origin has a magnitude of 12.5 cm and is at an angle of 30° below the $-x$ axis at a particular instant. What are the coordinates of the object at that instant?

11. ■■ A ball rolls with a constant velocity of 1.50 m/s at an angle of 45° below the $+x$ axis in the 4th quadrant. Taking the ball to be at the origin at $t = 0$, what are its (x, y) coordinates 1.65 s later?

12. ■■ A ball rolling on a table has velocity with rectangular components $v_x = 0.60$ m/s and $v_y = 0.80$ m/s. What is the displacement of the ball in an interval of 2.5 s?

13. ■■ A ball has an initial velocity of 1.30 m/s along the $+y$ axis and, starting at t_o, receives an acceleration of 2.10 m/s^2 in the $+x$ direction. (a) What is the position of the ball 2.50 s after t_o? (b) What is the velocity of the ball at that time?

14. ■■ A small plane takes off with a constant velocity of 150 km/h at an angle of 37°. In 3.00 s, (a) how high is the plane above the ground, and (b) what horizontal distance has it traveled from the lift-off point?

15. ■■ A ball rolls diagonally across a table top from corner to corner with a constant speed of 0.75 m/s. The table top is 3.0 m wide and 4.0 m long. Another ball, starting at the same time as the first one, rolls along the long edge of the table. What constant speed must the second ball have in order to reach the same corner at the same time as the ball traveling diagonally?

16. ■■ A particle moves with a speed of 2.5 m/s in the $+x$ direction. Upon reaching the origin, it receives a continuous constant acceleration of 0.75 m/s^2 in the $-y$ direction. What is the position of the particle 4.0 s later?

17. ■■■ An automobile travels 800 m with a constant speed of 60 km/h along a straight highway that is inclined 15° to the horizontal. An observer notes only the vertical motion of the car. What is (a) the magnitude of the car's vertical velocity and (b) the vertical travel distance?

3.2 Vector Addition and Subtraction

18. Two vectors, of magnitudes 3 and 4, respectively, are added. The magnitude of the resultant vector is (a) 1, (b) 7, (c) between 1 and 7.

19. In Exercise 18, under what condition will the magnitude of the resultant equal 1? How about 7?

20. The resultant of $\mathbf{A} - \mathbf{B}$ is the same as that of (a) $\mathbf{B} - \mathbf{A}$, (b) $-\mathbf{A} + \mathbf{B}$, (c) $-(\mathbf{A} + \mathbf{B})$, (d) $-(\mathbf{B} - \mathbf{A})$.

21. ■ Find the rectangular components of a velocity vector that has a magnitude of 10.0 m/s and is oriented at an angle of 30° above the $+x$ axis.

22. ■ Via the triangle method, show graphically that (a) $\mathbf{A} + \mathbf{B} = \mathbf{B} + \mathbf{A}$ and that (b) if $\mathbf{A} - \mathbf{B} = \mathbf{C}$, then $\mathbf{A} = \mathbf{B} + \mathbf{C}$.

23. ■ Is vector addition associative; that is, does $(\mathbf{A} + \mathbf{B}) + \mathbf{C} = \mathbf{A} + (\mathbf{B} + \mathbf{C})$? Justify your answer.

24. ■ A vector has an x component of -2.5 m and a y component of 4.2 m. What is the vector expressed in magnitude-angle form?

25. ■ In Chapter 2, the displacement between two points on the x axis was written $\Delta x = x_2 - x_1$. Show what this means graphically in terms of vector subtraction.

26. ■ For the two vectors $\mathbf{x}_1 = (20\ \text{m})\ \hat{\mathbf{x}}$ and $\mathbf{x}_2 = (15\ \text{m})\ \hat{\mathbf{x}}$, compute and show graphically (a) $\mathbf{x}_1 + \mathbf{x}_2$, (b) $\mathbf{x}_1 - \mathbf{x}_2$, and (c) $\mathbf{x}_2 - \mathbf{x}_1$.

27. ■ During a takeoff with no wind, an airplane moves with a speed of 120 mi/h at an angle of 25° above the ground. What is the ground speed of the plane?

28. ■■ A dog walks northeast for 15 m and then east for 25 m. Find the resultant (or sum) displacement vector by (a) the graphical method and (b) the component method.

29. ■■ An airplane flies northwest for 250 mi and then west for 150 mi. Find the resultant (or sum) displacement vector by (a) the graphical method and (b) the component method.

30. ■■ Two boys are pulling a box across a horizontal floor as shown in •Fig 3.24. If $F_1 = 50.0$ N and $F_2 = 100$ N, find the resultant (or sum) force by (a) the graphical method and (b) the component method.

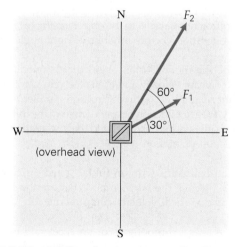

•**FIGURE 3.24   Adding force vectors**   See Exercises 30 and 47.

31. ■■ For each of the following vectors, give another vector that, when added to the first, yields a *null vector* (a vector with a magnitude of zero). Express the vector in the alternate component or magnitude-angle form: (a) **A** = 4.5 cm at an angle of 40° above the +x axis, (b) **B** = (2.0 cm) $\hat{\mathbf{x}}$ − (4.0 cm) $\hat{\mathbf{y}}$, (c) **C** = 8.0 cm at an angle of 60° above the −x axis.

32. ■■ Find the resultant (or sum) of the three vectors **A** = (−4.0 m) $\hat{\mathbf{x}}$ + (2.0 m) $\hat{\mathbf{y}}$, **B** = (6.0 m) $\hat{\mathbf{x}}$ + (3.5 m) $\hat{\mathbf{y}}$, and **C** = (−5.5 m) $\hat{\mathbf{y}}$.

33. ■■ (a) Find the resultant (or sum) of the force vectors in •Fig. 3.25. (b) If **F**$_1$ in the figure were at an angle of 27° instead of 37° with the +x axis, what would be the resultant (or sum) of **A** and **B**?

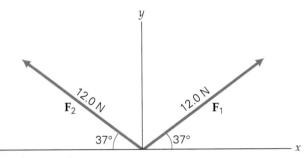

•**FIGURE 3.25   Vector addition**   See Exercises 33 and 34.

34. ■■ For the vectors in Fig. 3.25, determine (a) **F** = **F**$_1$ − **F**$_2$ and (b) **F** = **F**$_2$ − **F**$_1$. Do you think that the resultants will be the same? That is, does **F**$_1$ − **F**$_2$ = **F**$_2$ − **F**$_1$? (c) Is the following a true relationship: **F**$_1$ − **F**$_2$ = −(**F**$_2$ − **F**$_1$)?

35. ■■ For the vectors shown in •Fig. 3.26, determine **A** + **B** + **C**.

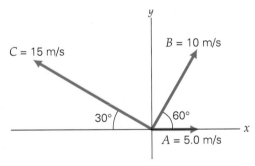

•**FIGURE 3.26   Adding velocity vectors**   See Exercises 35, 36, and 37.

36. ■■ For the vectors shown in Fig. 3.26, determine **A** − **B** − **C**.

37. ■■ For the vectors shown in Fig. 3.26, determine **A** + **B** − **C**.

38. ■■ Two force vectors, **F**$_1$ = (3.0 N) $\hat{\mathbf{x}}$ − (3.0 N) $\hat{\mathbf{y}}$ and **F**$_2$ = (−6.0 N) $\hat{\mathbf{x}}$ + (4.5 N) $\hat{\mathbf{y}}$, are applied to a particle. What third force **F**$_3$ would make the net or resultant force on the particle zero?

39. ■■ Two force vectors, **F**$_1$ = 8.0 N at an angle of 60° above the +x axis and **F**$_2$ = 5.5 N at an angle of 45° below the +x axis, are applied to a particle at the origin. What third force **F**$_3$ would make the net or resultant force on the particle zero?

40. ■■ A student works three problems involving the addition of different vectors **F**$_1$ and **F**$_2$. He states that the magnitudes of the three resultants are given by (a) $F_1 + F_2$, (b) $F_1 − F_2$, (c) $\sqrt{F_1^2 + F_2^2}$. Is this possible? If so, describe the vectors in each case.

41. ■■ A block weighing 50 N rests on an inclined plane. Its weight is a force directed vertically downward, as illustrated in •Fig. 3.27. Find the components of the force along the surface of the plane and perpendicular to it.

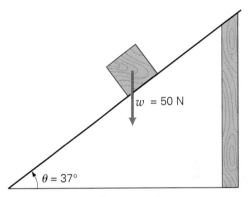

•**FIGURE 3.27   Block on an inclined plane**   See Exercise 41.

42. ■■■ An amusement park roller coaster starts out on a level track 50.0 m long and then goes up a 25.0-meter incline at an angle of 30° to the horizontal. It then goes down a 15.0-meter ramp with an incline of 40° to the horizontal. When the roller coaster has reached the bottom of the ramp, what is its displacement from its starting point?

43. ■■■ A person walks from point A to point B as shown in •Fig. 3.28. What is the person's displacement relative to A?

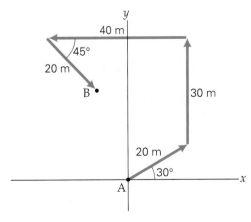

•**FIGURE 3.28   Adding displacement vectors**   See Exercise 43.

44. ■■■ A meteorologist tracks the movement of a cloud with Doppler radar. At 8:00 P.M., the cloud was 60 mi northeast of her station. At 10:00 P.M., the cloud is at 75 mi north. What is the average velocity of the cloud?

45. ■■■ A flight controller determines that an airplane is 20.0 mi south of him. Half an hour later the same plane is 35.0 mi northwest of him. If the plane is flying with a constant velocity, what is its velocity during this time?

46. ■■■ A ship was seen on a radar screen to be 10 km east of a particular location. Some time later, the ship is at 15 km north of west. What is the displacement of the ship?

47. ■■■ Two students pull on a box at rest as shown in Fig. 3.24. If $F_1 = 100$ N and $F_2 = 150$ N and a third student wants to stop the box from moving initially, what force should he apply?

3.3 Relative Velocity

48. A student walks on a treadmill moving at 4.0 m/s and remains at the same place in the gym. (a) What is the student's velocity relative to the gym floor? (b) What is the student's speed relative to the treadmill?

49. You are walking along a straight sidewalk to your dorm in the rain. (Your umbrella is in your room, of course.) A wind is blowing horizontally in the direction of your motion. Which part of your body (face or back) will stay dry if you walk (a) slower than the wind? (b) as fast as the wind? (c) faster than the wind?

50. ■ While you are traveling in a car on a straight, level interstate highway at 90 km/h, another car passes you in the same direction; its speedometer reads 120 km/h. (a) What is your velocity relative to the other driver? (b) What is the other car's velocity relative to you?

51. ■ A shopper is in a hurry to catch a bargain in a department store. She walks up the escalator, rather than letting it carry her, with a speed of 1.0 m/s relative to the escalator. If the escalator is 20 m long and moves with a speed of 0.50 m/s, how long does it take the shopper to get to the next floor?

52. ■ 🌐 A person riding in the back of a pickup truck traveling at 70 km/h on a straight, level road throws a ball with a speed of 15 km/h relative to the truck in the direction opposite its motion. What is the velocity of the ball (a) relative to a stationary observer by the side of the road? (b) relative to the driver of a car moving in the same direction as the truck at a speed of 90 km/h?

53. ■ In Exercise 52, what are the relative velocities if the ball is thrown in the direction of the truck?

54. ■ A boat can travel with a speed of 5.0 m/s relative to still water. If the boat is heading upstream in a river with a cur-

rent of 3.0 m/s, what distance does the boat travel in 30 s?

55. ■■ In a 500-meter stretch of a river, the speed of the current is a steady 5.0 m/s. How long does it take in minutes for a boat to finish one round trip (upstream and downstream) if the speed of the boat is 7.5 m/s relative to still water?

56. ■■ A passenger on a train traveling 90 km/h on a straight, level track notes that it takes another train (180 m long and moving with constant velocity) 4.0 s to pass by. What is the velocity of that passing train relative to the ground?

57. ■■ A moving walkway in an airport is 75 m long and moves with a speed of 0.30 m/s. A passenger, after traveling 25 m while standing on the walkway, starts to walk with a speed of 0.50 m/s relative to the walkway surface. How long does it take her to travel the total walkway distance?

58. ■■ A swimmer swims north across a river that flows at 0.20 m/s from west to east. If the speed of the swimmer is 0.15 m/s relative to still water, what is the swimmer's velocity relative to the river bank?

59. ■■ 🌐 A swimmer maintains a speed of 0.15 m/s relative to the water when he swims directly toward the opposite shore of a river. The river is straight and has a current that flows at 0.75 m/s. (a) How far downstream is he carried in 1.5 min? (b) What is his velocity relative to an observer on shore?

60. ■■ Swimming at 0.15 m/s relative to still water, a swimmer heads directly across a river of width 100 m. He arrives 50 m downstream from the point directly across the river from his starting point. (a) What is the speed of the current in the river? (b) What direction should he head so as to arrive at a point directly opposite his starting point?

61. ■■ It is raining and there is no wind. (a) When you are sitting in a stationary car, the rain falls straight down relative to the car and the ground. But when you're driving, the rain appears to hit the windshield at an angle, which increases as the velocity of the car increases. Explain this effect in terms of velocity components. [Hint: Consider the velocity of the rain relative to that of the car.] (b) If the raindrops fall straight down with a speed of 10 m/s but appear to make an angle of 25° to the vertical, what is the speed of the car?

62. ■■ A boat that travels with a speed of 6.75 m/s in still water is to go directly across a river and back (●Fig. 3.29). The current flows at 0.50 m/s. (a) At what angle(s) must the boat be steered? (b) How long does it take to make a round trip? (Assume that the boat's speed is constant at all times, and neglect turnaround time.)

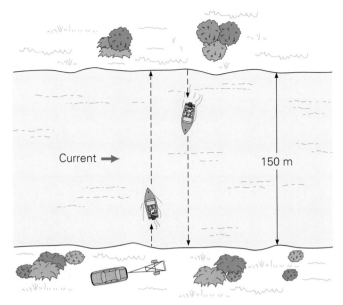

•FIGURE 3.29 **A boat race** See Exercise 62. (Not drawn to scale.)

63. ■■■ Two cars travel on level streets that intersect each other at right angles. Car A approaches the intersection from the east with a speed of 30 km/h. Car B has just passed through the intersection and is traveling 25 km/h northward. (a) What is the velocity of car B relative to car A? (b) the velocity of car A relative to car B?

3.4 Projectile Motion*

64. If air resistance is neglected, the motion of an object projected at an angle consists of a uniform downward acceleration combined with (a) an equal horizontal acceleration, (b) a uniform horizontal velocity, (c) a constant upward velocity, (d) an acceleration always perpendicular to the path of motion.

65. Using a sketch of components, show that the velocity of a projectile at the same elevation as the launch point ($y_{final} = y_{initial}$) has the same magnitude as the initial velocity and the same size directional angle but is directed below the horizontal.

66. Figure 3.16b shows the multiflash photographs of one ball dropping from rest and at the same time, another ball projected horizontally at the same height. The two balls hit the ground at the same time. Why? Explain.

67. In •Fig. 3.30, a spring-loaded "cannon" on a wheeled car can fire a metal ball vertically. The car is given a push and set in motion horizontally with constant velocity. A pin is pulled with a string to launch the ball, which travels upward and then falls back into the moving cannon every time. Why? Explain.

*Assume angles to be exact for significant figures.

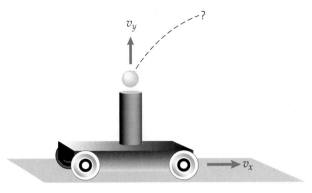

•FIGURE 3.30 **A ballistics car** See Exercises 67 and 77.

68. ■ A ball with a horizontal speed of 1.5 m/s rolls off a bench 2.0 m high. (a) How long will it take the ball to reach the floor? (b) How far from a point on the floor directly below the edge of the bench will the ball land?

69. ■ An electron is ejected horizontally, with a speed of 1.5×10^6 m/s, from the electron gun of a TV set. If the viewing screen is 35 cm away from the end of the gun, how far will the electron fall or travel in the vertical direction before hitting the screen?

70. ■ A ball is thrown horizontally, with a speed of 15 m/s, from the top of a 6.0-meter-tall hill. How far from the point on the ground directly below the launch point does the ball strike the ground?

71. ■ If Exercise 70 were taking place on the surface of the Moon, where the acceleration due to gravity is only 1.67 m/s², what would be the answer?

72. ■ A ball rolls horizontally off the edge of a 6.5-meter-tall platform. If the ball lands 8.7 m from the point on the ground directly below the edge of the platform, what is the speed of the ball when it rolls off the platform?

73. ■ A golf ball is hit with a speed of 30 m/s at an angle of 30° above the horizontal. What are the horizontal and vertical components of the ball's velocity?

74. ■■ A pitcher throws a fastball horizontally with a speed of 140 km/h toward home plate 18.4 m away. (a) If the batter's combined reaction and swing times total 0.350 s, how long before swinging can the batter watch the ball after it has left the pitcher's hand? (A major reason why some players illegally use hollowed-out bats with cork or superball filling is to make the bat lighter, which decreases the swing time and allows more time to follow the ball before deciding to swing.) (b) In traveling to the plate, how far does the ball drop from its original horizontal line?

75. ■■ One ball rolls with a constant speed of 0.25 m/s on a table 0.95 m above a floor, and another ball rolls on the floor directly under the first ball and with the same speed and direction. Will the balls collide after the first ball rolls off the table? If so, how far from the point directly below

the edge of the table will they be when they strike each other?

76. ■■ An airplane must drop a package of supplies so that it hits the ground at a designated spot near some campers. The airplane, moving horizontally with a constant velocity of 140 km/h, approaches the spot at an altitude of 0.500 km above level ground. Having the designated point in sight, the pilot prepares to drop the package. (a) What should the angle between the horizontal and the pilot's line of sight be when the package is released? (b) What is the location of the plane when the package hits the ground?

77. ■■ A wheeled car with a spring-loaded cannon fires a metal ball vertically (Fig. 3.30). If the initial vertical speed of the ball is 5.0 m/s as the cannon moves horizontally with a speed of 0.75 m/s, (a) how far from the launch point does the ball fall back into the cannon? (b) What would happen if the cannon were accelerating?

78. ■■ ⦿ A soccer player kicks a stationary ball, giving it a speed of 20.0 m/s at an angle of 15.0° to the horizontal. (a) What is the maximum height reached by the ball? (b) What is its range? (c) How could the range be increased?

79. ■■ ⦿ A rifle fires a bullet with a speed of 250 m/s at an angle of 37° above the horizontal. (a) Relative to the launch height, what height does the bullet reach? (b) How long is the bullet in the air? (c) What is its horizontal range?

80. ■■ A golf ball, hit off a tee, has an initial speed of 30.0 m/s at an angle of 37° to the horizontal. At what other time would the golf ball have the same vertical height as it has at $t = 3.00$ s? [Hint: See Example 3.9.]

81. ■■ An arrow has an initial launch speed of 18 m/s. If it must strike a target 31 m away at the same elevation, what should be the projection angle?

82. ■■ An astronaut on the Moon fires a projectile from a launcher on a level surface so as to get the maximum range. If the launcher gives the projectile a muzzle velocity of 25 m/s, what is the range of the projectile? [Hint: Don't forget where the launcher is and gravitational effects. The acceleration due to gravity on the Moon is only one-sixth of that on the Earth.]

83. ■■ A stone thrown off a bridge 20 m above a river has an initial velocity of 12 m/s at an angle of 45° above the horizontal (•Fig. 3.31). (a) What is the range of the stone? (b) At what velocity does the stone strike the water?

84. ■■ William Tell is said to have shot an apple off his son's head with an arrow. If the arrow was shot with an initial speed of 55 m/s and the boy was 15 m away, at what launch angle did Bill aim the arrow? (Assume that the arrow and apple are initially at the same height above the ground.)

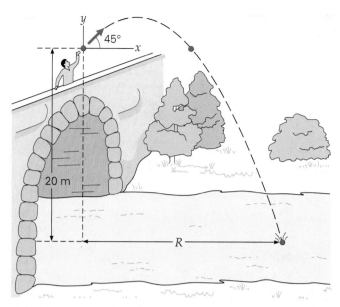

•FIGURE 3.31 A view from the bridge See Exercise 83.

85. ■■■ This time William Tell is shooting at an apple that hangs on a tree (•Fig. 3.32). The apple is a horizontal distance of 20.0 m away and at a height of 4.00 m above the ground. If the arrow is released from a height of 1.00 m above the ground and hits the apple 0.500 s later, what is its initial velocity?

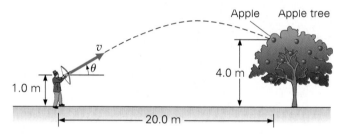

•FIGURE 3.32 Hit the apple See Exercise 85. (Not drawn to scale.)

86. ■■■ A ditch 2.5 m wide crosses a trailbike path (•Fig. 3.33). An upward incline of 15° has been built up on the approach so that the top of the incline is level with the top of the other side of the ditch. With what minimum speed must a trailbike be moving to clear the ditch? (Add 1.4 m to the range for the back of the bike to clear the ditch safely.)

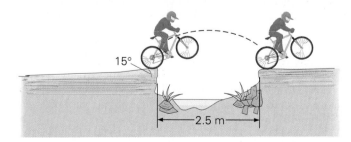

•FIGURE 3.33 Clear the ditch? See Exercise 86. (Not drawn to scale.)

87. ■■■ A quarterback passes a football with a speed of 50 ft/s at an angle of 40° to the horizontal toward an intended receiver 30 yd downfield. The pass is released 5.0 ft above the ground. Assume that the receiver is stationary and that he will catch the ball if it comes to him. Will the pass be completed? If not, will the throw be long or short?

88. ■■■ Three identical balls are thrown off a cliff at a height (h) above a horizontal plain. One ball is projected at an angle of 45° above the horizontal, the second is thrown horizontally, and the third is thrown at an angle of 30° below the horizontal (downward). If all of the balls have the same initial speed, which one will have the greatest speed when it hits the ground? (Justify your answer mathematically.)

89. ■■■ The hole on a level, elevated golf green is a horizontal distance of 150 m from the tee and at an elevation of 12.0 m above the tee. A golfer hits a ball at an angle 10.0° greater than that of the hole above the tee and makes a hole-in-one! (a) What was the initial speed of the ball? (b) Suppose the next golfer hits her ball toward the hole with the same speed but at an angle of 10.5° greater than that of the hole above the tee. Does the ball go in the hole, or is the shot too long or too short?

Additional Exercises

90. The apparatus for a popular lecture demonstration is shown in ●Fig. 3.34. The gun is aimed directly at the can, which is simultaneously released when the gun is fired. This gun won't miss as long as the initial speed of the bullet is sufficient to reach the falling target before the target hits the floor. Verify this, using the figure. [Hint: Note that $y_0 = x \tan \theta$.]

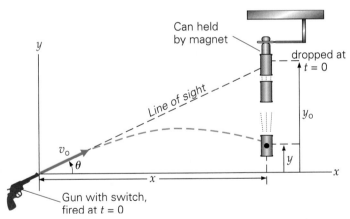

●FIGURE 3.34 **A sure shot** See Exercise 90. (Not drawn to scale.)

91. What is the resultant velocity vector of the following three velocity vectors: $\mathbf{v}_1 = (3.0 \text{ m/s})\,\hat{\mathbf{x}} + (4.0 \text{ m/s})\,\hat{\mathbf{y}}$; $\mathbf{v}_2 = (-4.0 \text{ m/s})\,\hat{\mathbf{x}} + (5.0 \text{ m/s})\,\hat{\mathbf{y}}$; $\mathbf{v}_3 = (5.0 \text{ m/s})\,\hat{\mathbf{x}} - (7.0 \text{ m/s})\,\hat{\mathbf{y}}$? Express your result in both component and magnitude-angle forms.

92. ● A motorboat travels with a speed of 40.0 km/h in a straight path on a still lake. Suddenly, a strong, steady wind pushes the boat with a speed of 15.0 km/h perpendicularly to its straight-line path for 5.00 s. Relative to its position just when the wind started to blow, where is the boat located at the end of this time?

93. A golf ball lies 2.00 m directly south of the hole on a level green. On the first putt, the ball travels 3.00 m along a straight-line path at an angle of 5° east of north; on the second putt, it travels a straight-line distance of 1.20 m at an angle of 6° south of west. What would be the displacement of a third putt that would put the ball in the hole?

94. If the flow rate of the current in a straight river is greater than the speed of a boat in the water, the boat cannot make a trip *directly across* the river. Prove this statement.

95. A student throws a softball horizontally from a dorm window 15.0 m above the ground. Another student standing 10.0 m away from the dorm catches the ball at a height of 1.50 m above the ground. What is the initial speed of the ball?

96. ● A javelin is thrown at angles of 35° and 60°, respectively, to the horizontal from the same height and with the same speed in each case. For which throw does the javelin go farther, and how many times farther? (Assume that the landing place is at the same height as the launching.)

97. A firefighter holds the nozzle of a hose a horizontal distance of 25.0 m from a flaming building. If the speed of the water coming from the nozzle is 20.0 m/s, (a) show that the stream will not reach a third-story window 11.0 m above the nozzle level for a projection angle of 45°. (b) A greater height can be achieved by directing the nozzle at a greater angle. Could the window be reached with a stream projection angle of 50°?

98. In a movie, a monster climbs to the top of a building 30 m above the ground and hurls a boulder downward with a speed of 25 m/s at an angle of 45° below the horizontal. How far from the base of the building does the boulder land?

99. The shells fired from an artillery piece have a muzzle speed of 150 m/s, and the target is at a horizontal distance of 2.00 km. (a) At what angle relative to the horizontal should the gun be aimed? (b) Could the gun hit a target 3.00 km away?

100. A ball moving with a speed of 2.5 m/s along the $+y$ axis experiences an acceleration of 0.45 m/s² in the $+x$ direction when the ball is at $(x, y) = (0, 1.0 \text{ m})$. What are (a) the velocity components and (b) the position of the ball 4.0 s after the acceleration is applied?

101. ⊕ A ball is thrown horizontally from the top of a building at a height of 32.5 m above the ground and hits the level ground 56.0 m from the base of the building. (a) What is the initial speed of the ball? (b) What is the velocity of the ball just before it hits the ground?

102. A field goal is attempted when the football is at the center of the field 40 yd from the goalposts. If the kicker gives the ball a velocity of 70 ft/s toward the goalposts at an angle of 45° to the horizontal, will the kick be good? (The crossbar of the goalposts is 10 ft above the ground, and the ball must be higher than this when it reaches the goalposts for the field goal to be good.)

103. (a) An object moving with a speed of 10 m/s in a straight line has a velocity with an x component of +6.0 m/s.

In what direction is the object moving? (b) What is the y component of the velocity?

104. At a track and field meet, the best long jump is measured as 8.20 m. The jumper took off at an angle of 37° to the horizontal. (a) What was the jumper's initial speed? (b) If there were another meet on the Moon and the same jumper could attain only half of the initial speed he had on the Earth, what would be the maximum jump there? (Air resistance does not have to be neglected in part (b). Why?)

105. ⊕ An airplane with a speed of 150 km/h heads directly north while a west wind blows with a constant speed of 30 km/h. How far does the plane travel during the 4.0 h of flight?

Force and Motion

Y ou don't have to understand any physics to know what's needed to get the car (or anything else) moving: a push or a pull. If the frustrated motorist (or the tow truck that he will soon call) can apply enough *force*, the car will move. But what's keeping the car stuck in the snow? A car's engine can generate plenty of force—so why doesn't he just put the car into reverse and back out? For a car to move, another force is needed besides that exerted by the engine: *friction*. Here, the problem is most likely that there is not enough friction between the tires and the snow.

In Chapters 2 and 3 we learned how to analyze motion in terms of kinematics. Now we need to know more about the *dynamics* of motion—that is, what causes motion and changes in motion. This inquiry leads us to the concept of force.

The study of force and motion occupied many early scientists. It was the English scientist Isaac Newton (1642–1727) (•Fig. 4.1) who summarized the various relationships and principles of those early scientists into three statements, or laws, which not surprisingly are known as Newton's laws of motion. These laws sum up the concepts of dynamics. In this chapter, you'll learn what Newton had to say about force and motion.

4.1 The Concepts of Force and Net Force

Objectives: To (a) relate force and motion and (b) explain what is meant by a net or unbalanced force.

Let's first take a closer look at the meaning of force. It is easy to give examples of forces, but how would you generally define this concept? An operational definition of force is based on observed effects. That is, a force is described in terms of what it does. From your own experience, you know that *forces can produce changes in motion*. A force can set a stationary object into motion. It can also speed up or slow down a moving object, or change the direction of its motion. In other words, a force can produce a *change* in velocity (speed and/or direction)—that is, an acceleration. Therefore, an observed change in motion, including motion starting from rest, is evidence of a force. This leads to a common definition of **force**:

A force is something capable of changing an object's state of motion (its velocity).

The word "capable" is very significant here. It takes into account the fact that a force may be acting on an object, but its capability to produce a change in motion may be balanced or canceled by one or more other forces. The net effect is then zero. Thus, a force may not *necessarily* produce a change in motion. However, it follows that if a force acts *alone*, the object on which it acts *will* accelerate.

Since a force can produce an acceleration—a vector quantity—force must itself be a vector quantity, with both magnitude and direction. When several forces act on an object, you will often be interested in their combined effect—the unbalanced or net force. The **unbalanced** or **net force** is the vector sum, $\Sigma_i \mathbf{F}_i$, or resultant, of all the forces acting on an object or system. Consider the opposite forces illustrated in •Fig. 4.2a. The net force is zero when forces of equal magnitude act in opposite directions (Fig. 4.2b). Such forces are said to be *balanced forces*. A nonzero net force is referred to as an *unbalanced force* (Fig. 4.2c). In this case, *the situation can be analyzed as though only one force equal to the net force were acting*. An unbalanced or net force produces an acceleration. In some instances, an applied

•**FIGURE 4.1 Isaac Newton** Newton (1642–1727), one of the greatest scientific minds of all time, made fundamental contributions to mathematics, astronomy, and several branches of physics, including optics and mechanics. He formulated the laws of motion and universal gravitation (Chapter 7) and was one of the inventors of calculus. He did some of his most profound work when he was in his middle twenties.

Note: In the notation $\Sigma_i \mathbf{F}_i$, the Greek letter sigma means the "sum of" the individual forces, as indicated by the *i* subscripts: $\Sigma_i \mathbf{F}_i = \mathbf{F}_1 + \mathbf{F}_2 + \mathbf{F}_3 + \cdots$ — a vector sum. The *i* subscripts are sometimes omitted as being understood, and we write $\Sigma \mathbf{F}$.

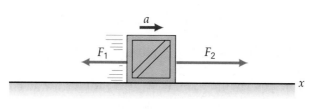

(a)

•**FIGURE 4.2 Net force**
(a) Opposite forces are applied to a crate. **(b)** If the forces are of equal magnitude, the vector resultant, or the net force acting on the crate in the *x* direction, is zero. The forces acting on the crate are said to be balanced. **(c)** If the forces are unequal in magnitude, the resultant is not zero. A nonzero net force, or an unbalanced force, then acts on the crate, producing an acceleration (for example, setting it into motion if it was initially at rest).

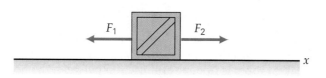

(b) Zero net force (balanced forces)

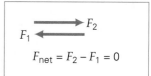

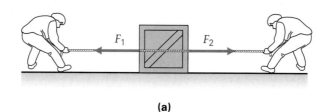

(c) Nonzero net force (unbalanced forces)

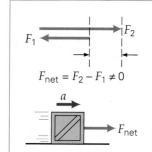

unbalanced force may also deform an object, that is, change its size and/or shape (as we shall see in Chapter 9). A deformation involves a change in motion for some part of an object, hence there is an acceleration.

Forces are sometimes divided into two types or classes. The more familiar of these are *contact forces*. Such forces arise because of physical contact between objects. For example, when you push on a door to open it or throw or kick a ball, you exert a contact force on the door or ball.

The other class of forces is called *action-at-a-distance forces*. Examples of these forces are gravity, the electrical force between two charges, and the magnetic force between two magnets. The Moon is attracted to Earth and maintained in orbit by a gravitational force, but there seems to be nothing physically transmitting that force. In Chapter 30, you will learn the modern view of how such action-at-a-distance forces are thought to be transmitted.

Now, with our better understanding of the concept of force, let's see how force and motion are related through Newton's laws.

4.2 Inertia and Newton's First Law of Motion

Objectives: To (a) state and explain Newton's first law of motion and (b) describe inertia and its relationship to mass.

The groundwork for Newton's first law of motion was laid by Galileo. In his experimental investigations, Galileo dropped objects to observe motion under the influence of gravity. However, the relatively large acceleration due to gravity causes dropped objects to move quite fast and quite far in a short time. From the kinematic equations in Chapter 2, you can see that 3.0 s after being dropped (if we neglect air resistance), an object in free fall has a speed of about 29 m/s (64 mi/h) and has fallen a distance of 44 m (about 48 yd, or almost half the length of a football field). Thus, experimental measurements of free-fall distance versus time were particularly difficult to make with the instrumentation available in Galileo's time.

To slow things down so that he could study motion, Galileo used balls rolling on inclined planes. He allowed a ball to roll down one inclined plane and then up another with a different degree of incline (•Fig. 4.3). Galileo noted that the ball rolled to approximately the same height in every case, but it rolled farther in the horizontal direction when the angle of this incline was smaller. When allowed to roll onto a horizontal surface, the ball traveled a considerable distance and went even farther when the surface was made smoother. Galileo wondered how far the ball would travel if the horizontal surface could be made perfectly smooth (frictionless). Although this situation was impossible to attain experimentally, Galileo reasoned that in this ideal case with an infinitely long surface, the ball would continue to travel (slide) indefinitely with straight-line, uniform motion, since there would be nothing (no force) to cause its motion to change.

According to Aristotle's theory of motion, which had been accepted for about 1500 years prior to Galileo's time, the normal state of a body was to be at rest (with the exception of celestial bodies, which were naturally in motion). Aristotle probably observed that objects moving on a surface tend to slow down and come to rest, so this conclusion would have seemed logical to him. However, from his experiments, Galileo concluded that bodies in motion exhibit the behavior of

•FIGURE 4.3 Galileo's experiment
A ball rolls farther along the upward incline as the angle of incline is decreased. On a smooth, horizontal surface, the ball rolls a greater distance before coming to rest. How far would the ball travel on an ideal, perfectly smooth surface?

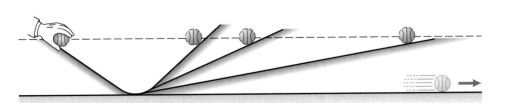

maintaining that motion, and that if an object is initially at rest, it will remain so, unless something causes it to move.

Galileo called this tendency of an object to maintain its initial state of motion **inertia**. That is,

Definition of: Inertia

> Inertia is the natural tendency of an object to maintain a state of rest or to remain in uniform motion in a straight line (constant velocity).

For example, if you've ever tried to stop a slowly rolling automobile by pushing on it, you felt its resistance to a change in motion, to slowing down. Physicists describe the property of inertia in terms of observed behavior, as they do for all physical phenomena. A comparative example of inertia is illustrated in •Fig. 4.4. If two punching bags have the same density (mass per unit volume; see Chapter 1), the large one will have more mass and more inertia, as you would quickly notice when you tried to punch them.

Newton related the concept of inertia to mass. Originally, he called mass a quantity of matter, but he later redefined it as follows:

Relationship of mass and inertia

> Mass is a measure of inertia.

That is, a massive object has more inertia, or more resistance to a change in motion, than a less massive object does. For example, a car has more inertia than a bicycle.

Newton's first law of motion, sometimes called the *law of inertia*, summarizes these observations:

Newton's first law—the law of inertia

> In the absence of an unbalanced applied force, a body at rest remains at rest and a body already in motion remains in motion with a constant velocity (constant speed and direction).

That is, if the net force acting on an object is zero, then its acceleration is zero. (See Demonstration 1.)

4.3 Newton's Second Law of Motion

Objectives: To (a) state and explain Newton's second law of motion, (b) apply it to physical situations, and (c) distinguish between weight and mass.

A change in motion, or an acceleration (change in speed and/or direction), is evidence of a force. All experiments indicate that an acceleration is directly proportional to, and in the direction of, the applied net force; that is,

$$\mathbf{a} \propto \mathbf{F}_{net}$$

where the boldface symbols indicate vector quantities. For example, if you hit a ball twice as hard (applied twice as much force), you would expect the acceleration of the ball to be twice as great (but still in the direction of the force).

However, as Newton recognized, the inertia or mass of the object also plays a role. For a given force, the more massive the object, the less its acceleration will be. That is, the acceleration and mass (m) of an object are inversely proportional:

$$a \propto \frac{1}{m}$$

For example, if you hit two different balls with the same force, the less massive ball would acquire a greater acceleration.

Then combining these relationships, we have,

$$\mathbf{a} \propto \frac{\mathbf{F}_{net}}{m}$$

or in words,

•**FIGURE 4.4   A difference in inertia**   The larger punching bag has more mass and hence more inertia or resistance to a change in motion.

Demonstration 1 ■ Newton's First Law and Inertia

According to Newton's first law, an object remains at rest or in motion with a constant velocity unless acted on by an unbalanced force.

(a) A pen is at rest on an embroidery hoop on top of a bottle.

(b) The hoop is struck sharply and accelerates horizontally. Because the friction between the pen and hoop is small and acts only for a short time, the pen does not move appreciably in the horizontal direction. However, there is now an unbalanced force acting vertically on it—gravity.

(c) The pen falls into the bottle.

The acceleration of an object is directly proportional to the *net* force acting on it and inversely proportional to its mass. The direction of the acceleration is in the direction of the applied net force.

Newton's second law—cause and effect

• Figure 4.5 presents some illustrations of this principle.

The SI unit of force is, appropriately, called the **newton** (N). With $\mathbf{F}_{net} \propto m\mathbf{a}$, in equation form **Newton's second law of motion** is commonly expressed as follows:

The newton (N), unit of force

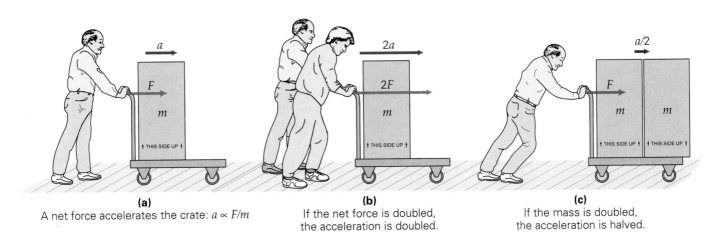

(a)
A net force accelerates the crate: $a \propto F/m$

(b)
If the net force is doubled, the acceleration is doubled.

(c)
If the mass is doubled, the acceleration is halved.

• **FIGURE 4.5 Newton's second law** The relationships among force, acceleration, and mass shown here are expressed by Newton's second law of motion (assuming no friction).

$$\mathbf{F} = m\mathbf{a} \qquad \textit{Newton's second law} \qquad (4.1)$$

SI unit of force: newton (N) or
kilogram-meter per second squared (kg · m/s^2)

*(Note: The net force is commonly written as just **F** for convenience, with the understanding that this is the vector sum $\mathbf{F} = \Sigma_i\mathbf{F}_i = \mathbf{F}_{net}$. This simpler notation will generally be used hereafter; however, you may wish to write $\mathbf{F}_{net}$, or $\Sigma_i\mathbf{F}_i$, explicitly as a reminder.)*

Equation 4.1 shows that (by unit analysis) a newton in base units is defined as 1 N ≡ 1 kg·m/s^2. That is, a net force of 1 N gives a mass of 1 kg an acceleration of 1 m/s^2 (•Fig. 4.6). The British-system unit of force is the pound (lb). One pound is equivalent to about 4.5 N (actually, 4.448 N). An average apple weighs about 1 N.

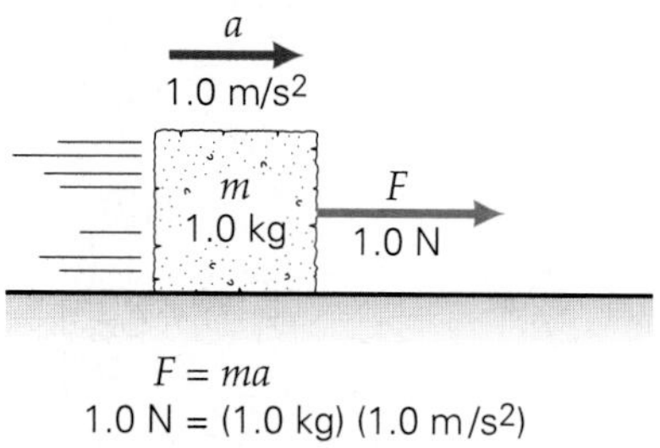

Thus, if the net force acting on an object is zero its acceleration is zero and it remains at rest or in uniform motion, which is consistent with the first law. For a nonzero net force (an unbalanced force), the resulting acceleration is in the same direction as the force.*

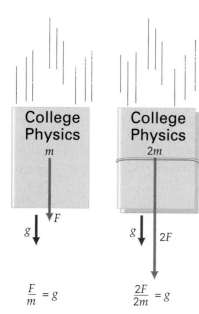

•**FIGURE 4.6   The newton (N)**   A net force of 1.0 N acting on a mass of 1.0 kg produces an acceleration of 1.0 m/s^2 (on a frictionless surface).

Weight

Equation 4.1 can be used to relate mass and weight. Recall from Chapter 1 that weight is the gravitational force of attraction that a celestial body exerts on an object. For us, this is the gravitational attraction of Earth. Its effects are easily demonstrated: When you drop an object, it falls (accelerates) toward Earth. Since there is only one force acting on the object, its **weight (w)** is the net force **F**, and the acceleration due to gravity (**g**) can be substituted for a in Eq. 4.1. We can therefore write, in terms of magnitude,

$$w = mg \qquad (4.2)$$

$$(F = ma)$$

The weight of 1.0 kg of mass is then $w = mg = (1.0 \text{ kg})(9.8 \text{ m/s}^2) = 9.8$ N.

Thus, 1.0 kg of mass has a weight of approximately 9.8 N, or 2.2 lb. But, although weight and mass are simply related through Eq. 4.2, keep in mind that *mass is the fundamental property*. Mass doesn't depend on the value of g, but weight does. As pointed out previously, the acceleration due to gravity on the Moon is about $\frac{1}{6}$ that on Earth. The weight of an object on the Moon would thus be one-sixth of its weight on Earth; but its mass, which reflects the quantity of matter it contains, would be the same in both places.

Newton's second law also shows why all objects in free fall have the same acceleration. Consider, for example, two falling objects, one with twice the mass of the other. The object with twice as much mass would have twice as much weight, or two times as much gravitational force acting on it. But the more massive object also has twice the inertia, so twice as much force is needed to give it the same acceleration. Expressing this relationship mathematically, for the smaller mass (m) we can write $F/m = g$, and for the larger mass ($2m$), we have the same acceleration: $2F/2m = g$ (•Fig. 4.7).

Newton's second law allows you to analyze dynamic situations. In using this equation, you should keep in mind that F is the *magnitude of the net force* and m

•**FIGURE 4.7   Newton's second law and free fall**   In free fall, all objects fall with the same constant acceleration g. An object with twice the mass of another has twice as much gravitational force acting on it. But with twice the mass, it also has twice as much inertia, so twice as much force is needed to give it the same acceleration.

*It may appear that Newton's first law is a special case of the second law, but that is not so. The first law *defines* what is called an *inertial reference system* (as we shall see in Chapter 26): one on which there is no net force or that is not accelerating, or one in which an isolated object is stationary or moves with a constant velocity. If Newton's first law holds, then the second law in the form $\mathbf{F} = m\mathbf{a}$ applies to the system.

is the *total mass of the system*. The boundaries defining a system may be real or imaginary. For example, a system might consist of all the gas molecules in a particular sealed vessel. But you might also define a system to be all the gas molecules in an arbitrary cubic meter of air. In studying dynamics we often have occasion to work with systems made up of one or more discrete masses: the Earth and Moon, for instance, or a series of blocks on a table top, or a tractor and wagon, as will be seen in the following Example.

In $F = ma$, m is the *total* mass of the system.

Example 4.1 ■ Newton's Second Law: Finding Acceleration

A tractor pulls a loaded wagon with a constant force of 440 N (•Fig. 4.8). If the total mass of the wagon and its contents is 275 kg, what is the wagon's acceleration? (Ignore any frictional forces.)

Thinking It Through. This is a direct application of Newton's second law. Note that the *total* mass is given; we treat the two separate masses as one.

Solution. Listing the data, we have

Given: $F = 440$ N   *Find:* a (acceleration)
       $m = 275$ kg

In this case, F is the net force and the acceleration is given by Eq. 4.1, $F = ma$. Solving for a,

$$a = \frac{F}{m} = \frac{440 \text{ N}}{275 \text{ kg}} = 1.60 \text{ m/s}^2$$

in the direction that the tractor is pulling.

Note that m was the *total* mass of the wagon and its contents. If the masses of wagon and contents had been given separately—say, $m_1 = 75$ kg and $m_2 = 200$ kg, respectively—they would be added together in Newton's law, $F = ma = (m_1 + m_2)a$.

In reality there would be an opposing force of friction. Suppose there were an effective frictional force of $f = 140$ N. In this case, the net force would be the vector sum of the force exerted by the tractor and the frictional force, and the acceleration would be

$$a = \frac{F - f}{m} = \frac{440 \text{ N} - 140 \text{ N}}{275 \text{ kg}} = 1.09 \text{ m/s}^2$$

in the direction that the tractor is pulling.

With a constant net force, the acceleration is also constant, so the kinematic equations of Chapter 2 can be applied. Suppose the wagon started from rest ($v_o = 0$). Could you find how far it traveled in 4.00 s? Using the appropriate kinematic equation,

$$x = v_o t + \tfrac{1}{2}at^2 = 0 + \tfrac{1}{2}(1.09 \text{ m/s}^2)(4.00 \text{ s})^2 = 8.72 \text{ m}$$

Follow-up Exercise. Suppose the applied force on the wagon is 550 N. With the same frictional force, what would be the wagon's velocity 4.0 s after starting from rest?

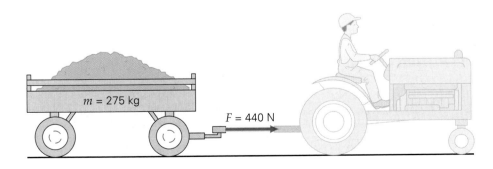

$m = 275$ kg

$F = 440$ N

•FIGURE 4.8 **Force and acceleration** See Example 4.1.

Example 4.2 ■ Newton's Second Law: Finding Mass

A student weighs 588 N. What is her mass?

Thinking It Through. Newton's second law allows us to determine a mass if we know the weight (force), since, g is known.

Solution.

Given: $w = 588$ N   *Find:* m (mass)

Recall that weight is a (gravitational) force and that Newton's second law, $F = ma$, can be written in the form $w = mg$ (Eq. 4.2), where g is the acceleration due to gravity (9.80 m/s²). Rearranging the equation, we have

$$m = \frac{w}{g} = \frac{588 \text{ N}}{9.80 \text{ m/s}^2} = 60.0 \text{ kg}$$

On the surface of Earth, this is equivalent to 60.0 kg (2.2 lb/kg) = 132 lb. In countries using the metric system, the kilogram unit of mass, rather than a force unit, is used to express "weight." It would be said that the student weighs 60.0 "kilos."

Follow-up Exercise. (a) A person in Europe is a bit overweight and would like to lose 5.0 "kilos." What would be the equivalent pound loss? (b) What is your "weight" in kilos?

A dynamic system may consist of more than one discrete mass. In applications of Newton's second law, it is often advantageous, and sometimes necessary, to isolate a given mass within a system. This is possible because any part of a system can be treated as a discrete system to which the second law can be applied, as the following Example shows.

Example 4.3 ■ Newton's Second Law: All or Part of the System

Two blocks with masses, $m_1 = 2.5$ kg and $m_2 = 3.5$ kg, rest on a frictionless surface and are connected by a light string (•Fig. 4.9). A horizontal force (F) of 12.0 N is applied to m_1 as shown in the figure. (a) What is the magnitude of the acceleration of the masses (that is, of the system)? (b) What is the magnitude of the force (T) in the string? [When a rope or string is stretched taut, it is said to be under *tension*, which is represented by

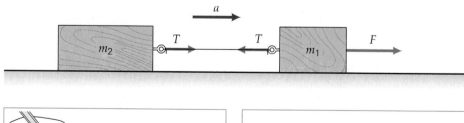

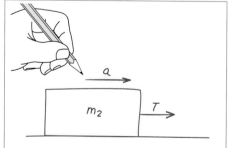
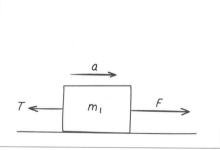

•**FIGURE 4.9 An accelerated system** See Example 4.3.

Isolating the masses

the magnitude of the force acting at any point. For a very light string, the force at the right end of the string has the same magnitude (T) as the force at the left end.]

Thinking It Through.   It is important to remember that the second law may be applied to a total system or any part of it (a subsystem, so to speak). This makes for the analysis of a particular component of a system if desired. Identification of the acting forces is critical, as this Example shows. We then apply $F = ma$ to each subsystem or component.

Solution.   Carefully listing the data and what we want to find, we have

Given: $m_1 = 2.5$ kg   *Find:*   (a) a (acceleration)
          $m_2 = 3.5$ kg   (b) T (tension, a force)
          $F = 12.0$ N

[Since the surface is ideally frictionless, the weight forces of the blocks, which ordinarily affect the frictional forces (Section 4.6), can be ignored.]

    Given an applied force that produces motion, the acceleration of the masses can be found from Newton's second law. In using the second law, it is important to keep in mind that it applies to the total system *or to any part of it*—that is, to the total mass ($m_1 + m_2$), to m_1 individually, or to m_2 individually. However, *we must be sure to identify correctly the appropriate force or forces in each case.* The net force acting on the combined masses, for example, is not the same as the net force acting on m_2 considered separately.

(a)   Drawing an imaginary line around m_1 and m_2, we see that the net force acting on this system is F. Representing the total mass simply as m, we can thus write

$$a = \frac{F_{net}}{m} = \frac{F}{m_1 + m_2} = \frac{12.0 \text{ N}}{2.5 \text{ kg} + 3.5 \text{ kg}} = 2.0 \text{ m/s}^2$$

The acceleration is in the direction of the applied force, as the figure indicates. Note that m is the *total* mass of the system, or all the mass that is accelerated. (The mass of the light string is small enough to be ignored.)

Note: When an object is described as "light," you can ignore its mass in analyzing the problem situation. That is, the mass is negligible relative to the other masses.

(b)   Under tension a force is exerted *on* an object by flexible strings (or ropes or wires) and is directed along the string. Note in the figure that we are assuming the tension to be transmitted *undiminished* through the string. That is, the tension is the same everywhere in the string. Thus, the magnitude of **T** acting on m_2 is the same as that acting on m_1. This is actually true only if the string has zero mass. Only such idealized *light* (negligible mass) strings or ropes will be considered in this book.

    So, there is a force of magnitude T on each of the masses because of tension in the connecting string. The T forces on the masses are equal in magnitude and opposite in direction. Note that the tension forces did not appear in part (a), where the total system of both masses was considered. In this case, the *internal* equal and opposite T forces cancel each other.

    However, each mass may also be considered a separate system to which Newton's second law applies. In these systems, the tension comes into play explicitly. Looking at the sketch of the isolated m_2 in Fig. 4.9, we see that the only force acting to accelerate this mass is T. From the values of m_2 and a, the magnitude of this force is given directly by

$$F_{net} = T = m_2a = (3.5 \text{ kg})(2.0 \text{ m/s}^2) = 7.0 \text{ N}$$

    An isolated sketch of m_1 is also shown in Fig. 4.9, and the second law can equally well be applied to this mass to find T. We must add the forces vectorially to get the net force on m_1 that produces its acceleration. That is,

$$F_{net} = F - T = m_1a \quad \textit{(direction of F taken as positive)}$$

Then, solving for T,

$$T = F - m_1a$$

$$= 12.0 \text{ N} - (2.5 \text{ kg})(2.0 \text{ m/s}^2) = 12.0 \text{ N} - 5.0 \text{ N} = 7.0 \text{ N}$$

Follow-up Exercise.   Suppose there were an additional horizontal force to the left of 3.0 N applied to m_2 in Fig. 4.9. What would be the tension in the connecting string in this case?

The Second Law in Component Form

Not only does Newton's second law hold for any part of a system, it also applies to the components of motion. A force may be expressed in component notation in two dimensions as follows:

$$\mathbf{F} = m\mathbf{a}$$

and

$$F_x\hat{\mathbf{x}} + F_y\hat{\mathbf{y}} = m(a_x\hat{\mathbf{x}} + a_y\hat{\mathbf{y}}) = ma_x\hat{\mathbf{x}} + ma_y\hat{\mathbf{y}} \tag{4.3a}$$

Hence we have the x and y components

$$F_x = ma_x$$
$$F_y = ma_y \tag{4.3b}$$

and Newton's second law applies separately to each component of motion. Note that *both* equations must be true. An example of how the second law is applied to components follows.

Example 4.4 ■ Newton's Second Law: Components of Force

A block of mass 0.50 kg travels with a speed of 2.0 m/s in the x direction on a flat, frictionless surface. On passing through the origin, it experiences a constant force of 3.0 N at an angle of 60° relative to the x-axis for 1.5 s (•Fig. 4.10). What is the velocity of the block at the end of this time?

Thinking It Through. With the force at an angle to the initial motion, it would appear that the solution is complicated. But note in the Fig. 4.10 insert that the force can be resolved into components. The motion can then be analyzed in each component direction.

Solution. First we write down the given data and what is to be found.

Given: $m = 0.50$ kg   *Find:* **v** (velocity at the end of 1.5 s)
$\quad\quad v_0 = v_{x_0} = 2.0$ m/s
$\quad\quad v_{y_0} = 0$
$\quad\quad F = 3.0$ N, $\theta = 60°$
$\quad\quad t = 1.5$ s

Let's find the magnitudes of the forces in the component (x and y) directions:

$$F_x = F\cos 60° = (3.0\text{ N})(0.500) = 1.5\text{ N}$$

$$F_y = F\sin 60° = (3.0\text{ N})(0.866) = 2.6\text{ N}$$

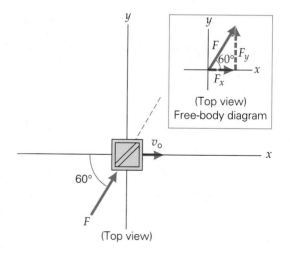

•**FIGURE 4.10   Off the straight and narrow**   A force is applied to a moving block when it reaches the origin, and the block is deviated from its straight-line path. See Example 4.4.

Then, applying Newton's second law to each direction to find the acceleration components, we get

$$a_x = \frac{F_x}{m} = \frac{1.5 \text{ N}}{0.50 \text{ kg}} = 3.0 \text{ m/s}^2$$

$$a_y = \frac{F_y}{m} = \frac{2.6 \text{ N}}{0.50 \text{ kg}} = 5.2 \text{ m/s}^2$$

Next, from the kinematic equation relating velocity and acceleration, the velocity of the block is

$$v_x = v_{x_0} + a_x t = 2.0 \text{ m/s} + (3.0 \text{ m/s}^2)(1.5 \text{ s}) = 6.5 \text{ m/s}$$

$$v_y = v_{y_0} + a_y t = 0 + (5.2 \text{ m/s}^2)(1.5 \text{ s}) = 7.8 \text{ m/s}$$

And, at the end of the 1.5 s, the velocity of the block is

$$\mathbf{v} = v_x \hat{\mathbf{x}} + v_y \hat{\mathbf{y}} = (6.5 \text{ m/s}) \, \hat{\mathbf{x}} + (7.8 \text{ m/s}) \, \hat{\mathbf{y}}$$

Follow-up Exercise. (a) What is the direction of motion at the end of the 1.5 s? (b) If the force were applied at an angle of 30° (rather than 60°) relative to the x axis, how would this change the results of Example 4.4?

4.4 Newton's Third Law of Motion

Objectives: To (a) state and explain Newton's third law of motion and (b) identify action–reaction force pairs.

Newton formulated a third law that is as far-reaching in its physical significance as the first two. For a simple introduction to the third law, consider the forces involved in seat belt safety. When the brakes are suddenly applied in a moving car, you continue to move forward (the frictional force on the seat of your pants is not enough to stop you). In doing so, you exert forces on the seat belt and shoulder strap. The belt and strap exert corresponding reaction forces on you, causing you to slow down with the car. If you haven't buckled up, you may keep on going (Newton's first law) until another applied force, such as that applied by the dashboard or windshield, slows you down.

We commonly think of single forces. However, Newton recognized that it is impossible to have a single force. He observed that in any application of force, there is always a mutual interaction, and forces always occur in pairs. An example given by Newton was this: If you press on a stone with a finger, then the finger is also pressed by, or receives a force from, the stone.

Newton termed the paired forces *action* and *reaction*, and **Newton's third law of motion** is as follows:

For every force (action), there is an equal and opposite force (reaction).

In symbol notation,

$$\mathbf{F}_{12} = -\mathbf{F}_{21}$$

That is, $\mathbf{F}_{12}$ is the force exerted *on* object 1 *by* object 2, and $-\mathbf{F}_{21}$ is the equal and opposite force exerted *on* object 2 *by* object 1. (The minus sign indicates the opposite direction.) *Which force is considered the action or the reaction is arbitrary;* $\mathbf{F}_{21}$ may be the reaction to $\mathbf{F}_{12}$ or vice versa.

Newton's third law may seem to contradict Newton's second law. If there are always equal and opposite forces, how can there be a nonzero net force? An important thing to remember about a force pair of the third law is that *the action–reaction forces do not act on the same object*. The second law is concerned with force(s) acting on a particular object (or system). The opposing forces of the third law act on different objects. Hence, the forces cannot cancel each other out or have a vector sum of zero when we apply the second law to the objects.

Newton's third law—action and reaction

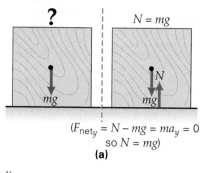

$(F_{net_y} = N - mg = ma_y = 0$
so $N = mg)$
(a)

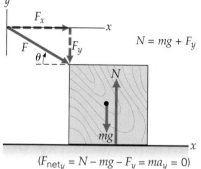

$(F_{net_y} = N - mg - F_y = ma_y = 0)$
(b)

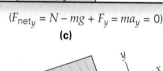

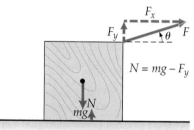

$(F_{net_y} = N - mg + F_y = ma_y = 0)$
(c)

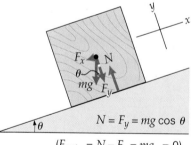

$(F_{net_y} = N - F_y = ma_y = 0)$
(d)

$N - mg = ma_y > 0$
so $N > mg$

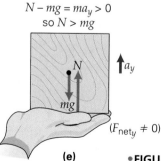

$(F_{net_y} \neq 0)$
(e)

To illustrate this distinction, consider the situations shown in •Fig. 4.11. We often tend to forget the reaction force. For example, in Fig. 4.11a (left), the obvious force that acts on a block sitting on a table is the Earth's gravitational attraction, which is expressed by the weight mg. But, there *has to be another force* acting on it. In fact, the other force must be equal and opposite in weight. Otherwise, by Newton's second law, the block would be accelerating downward.

The *block* exerts a downward force of magnitude mg on the table, and in turn the *table* exerts an equal but opposite force upward on the block (N), as illustrated in Fig. 4.11a (right). The net force *on the block* is then zero, and the block doesn't accelerate; thus, $F_{net_y} = N - mg = 0$. You could easily demonstrate that this upward force on the block is there by placing the block on your hand and holding it stationary—you would exert an upward force on the block. (And you would feel a reaction force of $-N$ on your hand.)

We call the force a surface exerts on an object a *normal* force and use the symbol N to denote this. *Normal* means *perpendicular*. The **normal force** a surface exerts on an object is always perpendicular to the surface. In Fig. 4.11a, the normal force is equal and opposite to the weight of the block.

The normal force is not always equal and opposite to an object's weight, however. The normal force is a "reaction" force; it reacts to the situation. If a force is applied to the block downward at an angle, as shown in Fig. 4.11b, then a downward component (F_y) acts on the block in addition to the block's weight, and the normal force must balance both forces ($F_{net_y} = N - mg - F_y = 0$.) and $N = mg + F_y$.

In Fig. 4.11b, the block could accelerate to the right (why?). However, this component might be balanced by a frictional force, as it often is. We'll bring friction into the picture in a later section; the normal force will be quite important there.

For the situation in Fig. 4.11c, the normal force is reduced. Can you see why? Note that the F_y component of the applied force is opposite in direction to the weight. Hence, in magnitude, $N = mg - F_y$.

Inclined planes are important in physics; for a block on an inclined plane (Fig. 4.11d), the normal force on the block is *perpendicular to the plane surface*. This force is the reaction to the block's force on the plane surface, and since $a_y = 0$ (no motion in the y direction), we have $F_{net_y} = N - F_y = 0$ and $N = F_y = mg \cos \theta$. The weight component (F_x) accelerates the block down the plane in the absence of an equal opposing friction force between the block and the plane surface.

Finally, Fig. 4.11e shows a case in which there is acceleration in the direction of the normal force. Here, a person is accelerating the block vertically, and $N - mg = ma_y$. Hence in this case, $N > mg$.

As another example, consider the situation in •Fig. 4.12a. Two third-law force pairs are acting when the person is holding the briefcase. There is a pair of contact forces: The person's hand exerts an upward force on the handle, and the handle exerts an equal downward force on the hand. That is, $F_1 = -F_1$. This is an action–reaction pair, with the forces acting on different objects. The other third-law force pair consists of action-at-a-distance forces associated with gravitational attraction: The Earth attracts the briefcase (its weight), and the briefcase attracts the Earth, or $F_2 = -F_2$.

Concentrating on the briefcase in isolation, we can see that only two of the four forces in Fig. 4.12a act on it—the upward force on the handle (F_1) and the downward force of the briefcase's weight ($-F_2$). These do *not* form a third-law force pair, however, because they act on the *same* object. Since the briefcase does not accelerate, these forces must be equal and opposite. Thus, the net force on the isolated stationary briefcase is zero, as required by Newton's first and second laws.

•**FIGURE 4.11 Second and third law distinctions** Newton's second law deals with the forces acting on a particular object (or system). The third law deals with the force pair that acts on different objects. This distinction is illustrated for various situations.

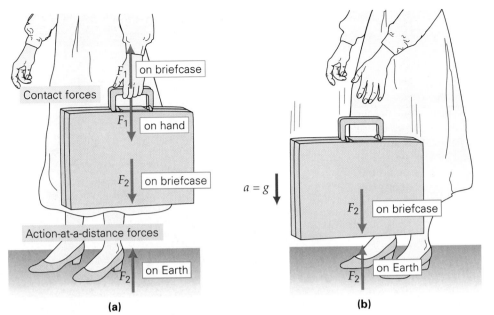

•FIGURE 4.12 **Force pairs of Newton's third law** **(a)** When the person holds the briefcase, there are two force pairs: a contact pair (F_1 and $-F_1$) and an action-at-a-distance (gravity) pair (F_2 and $-F_2$). The net force acting on the briefcase is zero: The upward contact force balances the downward weight force. Note, however, that the upward contact force and downward weight force do *not* form a third-law pair. **(b)** When the briefcase is released, there is an unbalanced force acting on the case (its weight force), and it accelerates downward (at g for free fall).

When the person drops the briefcase (Fig. 4.12b), there is then a nonzero net force (the unbalanced force of gravity) on the case, and it accelerates toward the Earth. But there is still a pair of third law forces *acting on different objects*—the briefcase and the Earth. (The Earth actually accelerates upward toward the briefcase as well, but the amount of the acceleration is infinitesimal. Can you explain why?)

Jet propulsion is yet another example of Newton's third law in action. In the case of a rocket, the rocket and exhaust gases exert equal and opposite forces on each other. As a result, the exhaust gases are accelerated away from the rocket, and the rocket is accelerated in the opposite direction. When big rockets "blast off" as in a Space Shuttle launch, there is a fiery exhaust from the rocket. A common misconception is that the exhaust gases "push" against the launch pad to accelerate the rocket. If this misconception were true, there would be no space travel, since there is nothing to "push" against in space. The correct explanation is one of action (gases exerting a force on the rocket) and reaction (rocket exerting an opposite force on the gases).

Another action–reaction pair is given in the Insight on page 112, and Demonstration 2 shows action–reaction forces in terms of the tension in a string.

4.5 More on Newton's Laws: Free-body Diagrams and Translational Equilibrium

Objectives: To (a) apply Newton's laws in analyzing various situations, using free-body diagrams, and (b) understand the concept of translational equilibrium.

Now that you have been introduced to Newton's laws and some applications in analyzing motion, the importance of these laws should be evident. They are so simply stated, yet so far-reaching. The second law is probably the most often applied, because of its mathematical relationship. However, the first and third laws are often used in qualitative analysis, as our study of physics will reveal.

The simple relationship expressed by Newton's second law, $\mathbf{F} = m\mathbf{a}$, allows the quantitative analysis of force and motion. We can think of it as a cause-and-effect relationship, with force being the cause and acceleration being the motional effect.

In general, you will be concerned with applications involving constant forces. Constant forces result in constant accelerations and allow us to use the kinematic equations from Chapter 2 in analyzing the motion. When there is a variable force, Newton's second law holds for the *instantaneous* force and acceleration, but the

Note: Remember that **F** means the *net* force, or $\Sigma_i \mathbf{F}_i$.

Action and Reaction: The Harrier "Jump" Jet

The action–reaction principle of Newton's third law is used to power jet aircraft (and propeller-driven planes), usually for propulsion rather than directly for lift. However, there is one type of aircraft that uses jet propulsion to take off vertically—no runway needed, like a helicopter: the Harrier "Jump" jet originally developed by the British (Fig. 1). This plane has four engine nozzles that can be swiveled to point at any angle from vertical to horizontal. These engines provide the power for both vertical and horizontal flight.

A Harrier takeoff is illustrated in Fig. 2. (There are other small jets on the plane to help control and stabilize the aircraft that are not shown.) As shown in Fig. 2a, the nozzles are directed downward for takeoff, and the reaction forces raise the plane. Achieving a proper altitude, the nozzles are swiveled and a forward component of the force gives the plane a forward acceleration (Fig. 2b). Finally, with the jets horizontal, the planes flies conventionally (Fig. 2c).

But wait! Did you notice something wrong with Fig. 2c? The plane has weight, or a downward force acting on it, that is not shown. What is the counterbalancing force to this weight that keeps the plane from falling? As you will learn in Chapter 9, this *lift* force results from the curved wing surface and fluid (air) dynamics.

FIGURE 1 "Jump" jet A Harrier "Jump" jet lifting off vertically.

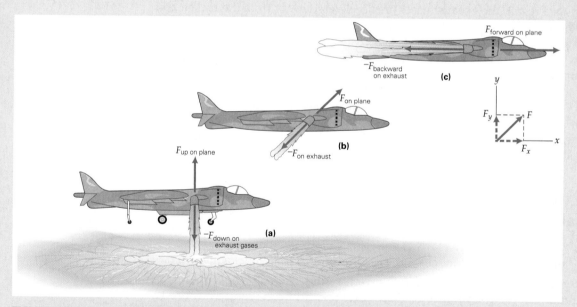

FIGURE 2 Up, up, and away (a) The reaction force to the downward exhaust force from the engines raises the plane. Only one nozzle is shown for simplicity. (b) At the appropriate altitude, the engine nozzles are swiveled and a forward force component gives the plane a forward acceleration (while still gaining altitude—why?). (c) With the nozzles swiveled 90° from their position in (a), the plane flies conventionally.

Demonstration 2 ■ Tension in a String: Action and Reaction Forces

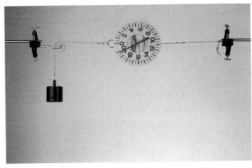

(a) Two suspended 2-kilogram masses are at-tached to opposite sides of a scale (calibrated in newtons). The total suspended weight is $w = mg = (4.00 \text{ kg})(9.80 \text{ m/s}^2) = 39.2 \text{ N}$, yet the scale reads about 20 N. Is something wrong with the scale?

(b) No, think of it in this manner. The effect of the weight of the mass on the right is replaced by fixing the end of the string. The other mass stretches the scale spring, giving a reading of about 20 N [or $w = mg = (2.00 \text{ kg})(9.80 \text{ m/s}^2) = 19.6 \text{ N}$].

(c) Similarly, the other end of the scale can be fixed. (A fixed pulley merely changes the direction of the force, and the scale can be hung vertically with the same effect.) In all cases, the tension in the string is 19.6 N, as the scale shows.

acceleration will vary with time. We will generally limit ourselves to constant ac-celerations and forces.

This section presents several examples of applications of Newton's second law so that you can become familiar with its use. This small but powerful equa-tion will be used again and again throughout this text.

Newton's second law gives the acceleration resulting from an applied force, but it can also be used to find the force from motional effects, as the following Example shows.

Example 4.5 ■ A Braking Car: Finding a Force From Motional Effects

A car traveling at 72.0 km/h along a straight, level road is brought uniformly to a stop in a distance of 40.0 m. If the car weighs 8.80×10^3 N, what is the braking force?

Thinking It Through. It is noted that the car's velocity changes, so there is an acceleration. Given a distance, we might surmise that first the acceleration is found by using a kinematic equation, and then the acceleration is used to compute the force.

Solution. In bringing the car to a stop, the braking force caused an acceleration (actually a deceleration), as illustrated in •Fig. 4.13. Listing what is given and what we must find, we have

Given: $v_0 = 72.0 \text{ km/h} = 20.0 \text{ m/s}$ *Find:* F_b (braking force)
$v = 0$
$x = 40.0 \text{ m}$
$w = 8.80 \times 10^3 \text{ N}$

•**FIGURE 4.13 Finding force from motional effects** See Example 4.5.

We know that $F = ma$, so we can easily calculate F_b if we can find m and a. The car's mass m can be obtained from its given weight. The other given quantities should remind you of a kinematic equation from Chapter 2 from which the acceleration a can be found. Since the car is brought uniformly to a stop, the acceleration is constant and we may use Eq. 2.10, $v^2 = v_0^2 + 2ax$, to find a:

$$a = \frac{v^2 - v_0^2}{2x} = \frac{0 - (20.0 \text{ m/s})^2}{2(40.0 \text{ m})} = -5.00 \text{ m/s}^2$$

The minus sign indicates that the acceleration is opposite to v_0, as expected for a braking force, which slows the car.

The mass of the car is obtained from the weight: $w = mg$, or $m = w/g$. Using this expression along with the acceleration obtained previously gives a braking force of

$$F_b = ma = \left(\frac{w}{g}\right)a = \left(\frac{8.80 \times 10^3 \text{ N}}{9.80 \text{ m/s}^2}\right)(-5.00 \text{ m/s}^2) = -4.49 \times 10^3 \text{ N}$$

Follow-up Exercise. A 1000-kilogram car starting from rest travels in a straight line and uniformly reaches a speed of 10 m/s in 5.0 s. What is the magnitude of the net force that accelerates the car?

Example 4.6 ■ The Atwood Machine: Finding the Acceleration of a System

The *Atwood machine* consists of two masses suspended from a fixed pulley, as shown in •Fig. 4.14. If $m_1 = 0.55$ kg and $m_2 = 0.80$ kg, (a) what is the acceleration of the system? (b) What is the magnitude of the tension in the string? (Consider the pulley to be frictionless and the masses of the string and the pulley to be negligible.)

Thinking It Through. The system consists of the two masses that are acted on by forces, so this appears to be a straightforward application of the second law. To find the tension in the string, the masses will have to be isolated as separate systems so as to have T appear in the second-law equations.

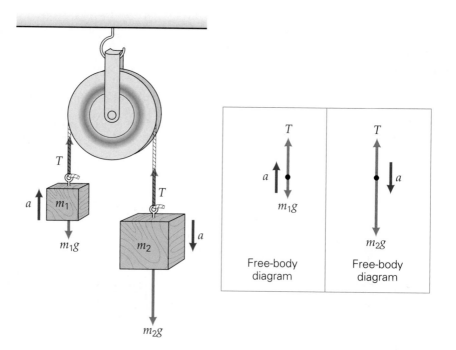

•FIGURE 4.14 **The Atwood machine** A single, fixed pulley is simply a direction changer. Free-body diagrams are shown for the forces in the box. See Example 4.6.

Solution. Listing the two given quantities, we have

Given: $m_1 = 0.55$ kg   *Find:* (a) a (acceleration)
       $m_2 = 0.80$ kg   (b) T (tension)

(a) Since m_2 is greater than m_1, if released from rest, m_2 will fall and m_1 will rise with accelerations of the same magnitude in opposite directions. Looking at the two masses as separate systems, we are free to assign directional plus and minus signs arbitrarily. Here, *the positive direction is taken to be the direction of acceleration for both masses.* That is, for m_2, downward is taken to be the positive direction; for m_1, upward is positive. The pulley is simply a direction changer. By analogy, you might think of the system as two masses that are connected by a horizontal string and are moving to the right (as in Fig. 4.9).

Then, applying Newton's second law to the system as a whole (and thus ignoring the T forces, which are internal) gives

$$F_2 - F_1 = m_2 g - m_1 g$$

$$= (m_1 + m_2)a$$

net force = total mass × acceleration

Solving for a gives

$$a = \frac{(m_2 - m_1)g}{m_1 + m_2}$$

$$= \frac{(0.80 \text{ kg} - 0.55 \text{ kg})(9.8 \text{ m/s}^2)}{0.55 \text{ kg} + 0.80 \text{ kg}} = 1.8 \text{ m/s}^2$$

(b) To find the magnitude of the tension, the second law must be applied to an isolated mass on which the tension acts. This will give an equation with T. Note the boxed sketches in Fig. 4.14, which show all the forces of the system. These sketches are **free-body diagrams**, which represent an object as a "particle" or point-sized mass and shows all the forces acting on that particle. Free-body diagrams, discussed in the following Problem-Solving Strategy, are very helpful in analyzing situations involving several bodies and multiple forces.

Applying the second law to each of the masses, we have

$$T - m_1 g = m_1 a$$

$$m_2 g - T = m_2 a$$

where the direction of the acceleration for each mass (upward for m_1 and downward for m_2) is taken as positive so as to avoid a minus sign for a. The magnitudes of the accelerations of the masses are equal. Eliminating T from the two equations gives the equation derived above:

$$m_2 g - m_1 g = m_2 a + m_1 a = (m_1 + m_2)a$$

Once the acceleration is found, use it in either of the previous two equations, solved for T. Using the first equation gives

$$T = m_1 a + m_1 g = m_1(a + g)$$

$$= (0.55 \text{ kg})(1.8 \text{ m/s}^2 + 9.8 \text{ m/s}^2) = 6.4 \text{ N}$$

Follow-up Exercise. Suppose there were a constant retarding force of friction, f, in the pulley of an Atwood machine when the masses were in motion. What would be the general (symbol) form of the acceleration in this case?

Incidentally, the Atwood machine is named after the British scientist George Atwood (1746–1807), who used the arrangement to study motion and measure the value of g. The masses can be chosen to minimize the acceleration, making it easier to measure the time of fall.

When working with problems in which two or more forces or components of force act on a body, it is convenient and instructive to draw a free-body diagram of the forces, as was done in Example 4.6. In such a diagram, we show all the forces acting on the body or object. If several bodies are involved, we may make a diagram for each body *separately*, showing all of the forces acting on each individual body.

In illustrations of physical situations, sometimes called *space diagrams* (such as the Atwood machine in Fig. 4.14), force vectors may be drawn at different locations to indicate their points of application. However, since we are concerned only with linear motions, vectors in free-body diagrams (such as those boxed in Fig. 4.14) may be shown emanating from a common point, which is chosen as the origin of x–y axes. One of the axes is generally chosen along the direction of the net force acting on a body, since that is the direction in which the body will accelerate. Also, it is often important to resolve force vectors into components, and properly chosen x–y axes simplify this.

In a free-body diagram, the vector arrows do not have to be exactly to scale. However, it should be made apparent if there is a net force, and whether forces balance each other in a particular direction. When the forces aren't balanced, we know from Newton's second law that there must be an acceleration.

In summary, the general steps in constructing and using free-body diagrams are as follows (refer to the accompanying Learn by Drawing as you read):

1. Sketch a space diagram (if one is not already available) and identify the forces acting on each body of the system.

2. Isolate the body for which the free-body diagram is to be constructed. Draw a set of Cartesian axes with the origin at a point through which the forces act and with one of the axes along the direction of the body's acceleration. (This will be in the direction of the net force if there is one.)

3. Draw properly oriented force vectors on the diagram emanating from the origin of the axes. If there is an unbalanced force, assume a direction of acceleration and indicate it with an acceleration vector. Be sure to include only those forces that act on the isolated body.

4. Resolve any forces that are not directed along the x or y axes into x or y components. Use the free-body diagram to analyze the forces in terms of Newton's second law of motion. (Note: If the acceleration is in the direction opposite that selected, indicate this by an acceleration with an opposite sign in the solution. For example, if the motion is taken to be in the positive direction and it is actually in the opposite direction, the acceleration will be negative.)

Free-body diagrams are a particularly useful way of following one of the Suggested Problem-Solving Procedures in Chapter 1: Draw a diagram as an aid in visualizing and analyzing the physical situation of the problem. *Make it a practice to draw free-body diagrams for force problems, as is done in the following Examples.*

LEARN BY DRAWING

Drawing a Free-body Diagram

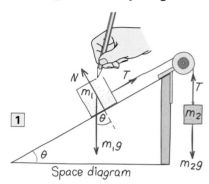

1 Space diagram

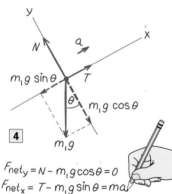

2

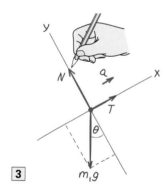

3

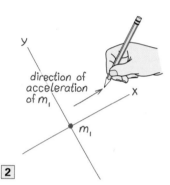

4

$$F_{net_y} = N - m_1 g \cos\theta = 0$$
$$F_{net_x} = T - m_1 g \sin\theta = ma$$

Example 4.7 ■ Components of Force and Free-body Diagrams

A force of 15.0 N is applied at an angle of 30° to the horizontal on a 0.750-kilogram block at rest on a frictionless surface as illustrated in •Fig. 4.15. (a) What is the magnitude of the resulting acceleration of the block? (b) What is the magnitude of the normal force?

Thinking It Through. The applied force may be resolved into components. The horizontal component accelerates the block. The vertical component affects the normal force (review Fig. 4.11).

Solution. First we write down the given data and what is to be found.

Given: $F = 15.0\,\text{N}$ *Find:* (a) a (acceleration)
$\quad\quad m = 0.750\,\text{kg}$ (b) N (normal force)
$\quad\quad \theta = 30°$
$\quad\quad v_o = 0$

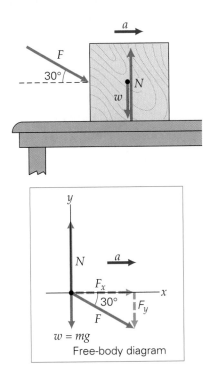

Then we draw a free-body diagram for the block as in Fig. 4.15. Note that the forces act in the same directions as in the space diagram.

(a) The acceleration of the block is given by Newton's second law. We choose our axes so that a is in the $+x$ direction. As the free-body diagram shows, only a component (F_x) of the applied force F acts in this direction. The component of F in the direction of motion is $F_x = F \cos\theta$. We apply Newton's law in the $+x$ direction,

$$F_x = F \cos 30° = ma_x$$

$$a_x = \frac{F \cos 30°}{m} = \frac{(15.0\,\text{N})(0.866)}{0.750\,\text{kg}} = 17.3\,\text{m/s}^2$$

(b) This is the total acceleration of the block, since the block moves only in the x direction (it does not accelerate in the y direction). With $a_y = 0$, the sum of the forces in the y direction must then be zero. That is, the downward component of F acting on the block, F_y, and its downward weight force w must be balanced by the upward normal force N that the surface exerts on the block. If this were not the case, then there would be a net force and an acceleration in the y direction.
We sum the forces in the y direction with upward as positive:

$$N - F_y - w = 0$$

or

$$N - F \sin 30° - mg = 0$$

and

$$N = F \sin 30° + mg = (15.0\,\text{N})(0.500) + (0.750\,\text{kg})(9.80\,\text{m/s}^2) = 14.85\,\text{N}$$

The surface then exerts a force of 14.85 N upward on the block, which balances the downward forces acting on it.

Follow-up Exercise. (a) Suppose the applied force on the block is applied for only a short time. What is the magnitude of the normal force after the applied force is removed? (b) If the block slides off the edge of the table, what would be the net force on the block just after leaving the table (with the applied force removed)?

•**FIGURE 4.15 Newton's second law and components of force**
See Example 4.7.

Problem-Solving Hint

There is no single fixed way to go about solving a problem. However, there are general strategies or procedures that are helpful in solving problems involving Newton's second law. Using our Suggested Problem-Solving Procedures introduced in Chapter 1, you might include the following in force applications:

- Draw a free-body diagram for each individual body, showing all of the forces acting on that body.
- Depending on what is to be found, Newton's second law may be applied to the system as a whole (in which case internal forces cancel) or to a part of the system. Basically, *you want to obtain an equation containing the quantity for which you want to solve.* Review Example 4.3. (If there are two unknown quantities, application of Newton's law to two parts of the system may give you two equations and two unknowns. See Example 4.8.)
- Keep in mind that Newton's second law may be applied to components of motion and that forces may be resolved into components to do this. Review Example 4.7.

Note: Review pp. 21–24.

Example 4.8 ■ The Atwood Machine Revisited: Motion on a Frictionless Inclined Plane

Two masses are connected by a light string running over a light pulley of negligible friction as illustrated in •Fig. 4.16. One mass ($m_1 = 5.0$ kg) is on a frictionless 20° inclined plane, and the other ($m_2 = 1.5$ kg) is freely suspended. What is the acceleration of the masses?

Thinking It Through.   Apply the preceding Problem-Solving Hint.

Solution.   Following our usual procedure, we write

Given: $m_1 = 5.0$ kg   *Find:*   **a** (acceleration)
$\quad\quad\quad m_2 = 1.5$ kg
$\quad\quad\quad \theta = 20°$

To help us visualize the forces involved, we isolate m_1 and m_2 and draw free-body diagrams for each mass. For mass m_1, there are three concurrent forces (forces acting through a common point). These are T, m_1g, and N, where T is the tension force in the string and N is the normal force of the plane on the block. The forces are shown emanating from their common point of action. (Recall that a vector arrow can be moved as long as its direction is not changed.)

We will start by assuming that m_1 accelerates up the plane, which is taken to be in the x direction. (It makes no difference whether it is assumed that m_1 accelerates up or

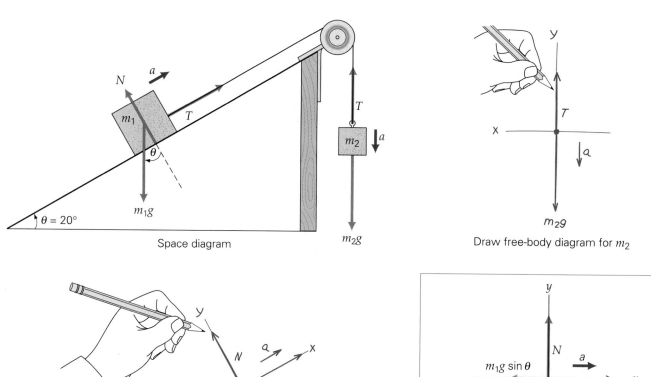

Space diagram

Draw free-body diagram for m_2

Draw free-body diagram for m_1

Free-body diagram for m_1
(reoriented and colored)

•**FIGURE 4.16   Application of Newton's second law**   See Example 4.8. (Not drawn to scale.)

down the plane, as we shall see shortly.) Notice that m_1g (the weight) is broken down into components. The x component is in the assumed direction of acceleration, and the y component acts perpendicularly to the plane and is balanced by the normal force N. (There is no acceleration in the y direction, so there is no net force in this direction.)

Then, applying Newton's second law in component form to m_1, we have

$$F_{x_1} = T - m_1g \sin \theta = m_1a$$

$$F_{y_1} = N - m_1g \cos \theta = 0 \qquad \text{(no net force, forces cancel)}$$

And for m_2,

$$F_{y_2} = m_2g - T = m_2a$$

where the masses of the string and pulley have been neglected.

Adding the first and last equations to eliminate T, we have

$$m_2g - m_1g \sin \theta = (m_1 + m_2)a$$

(net force = *total* mass × acceleration)

(Note that this is the equation that would be obtained by applying the second law to the system as a whole, in which the T forces cancel.)

Then, we solve for a:

$$a = \frac{m_2g - m_1g \sin 20°}{m_1 + m_2}$$

$$= \frac{(1.5 \text{ kg})(9.8 \text{ m/s}^2) - (5.0 \text{ kg})(9.8 \text{ m/s}^2)(0.342)}{5.0 \text{ kg} + 1.5 \text{ kg}}$$

$$= -0.32 \text{ m/s}^2$$

The minus sign indicates that the acceleration is opposite to the assumed direction. That is, m_1 actually accelerates down the plane and m_2 accelerates upward. As this example shows, if you assume the acceleration to be in the wrong direction, the sign on the result will give you the correct direction anyway.

Could you find the tension force T in the string if you were asked to do so? How this could be done should be quite evident from the free-body diagram.

Follow-up Exercise. (a) In Example 4.8, what mass m_2 would cause m_1 to accelerate up the plane? (b) Keeping the masses the same as in the Example, how should the angle of incline be adjusted so that m_1 would accelerate up the plane?

Translational Equilibrium

Forces may act on an object without producing an acceleration. In such a case, with $a = 0$, we know from Newton's second law that

$$\Sigma\mathbf{F} = 0 \qquad (4.4)$$

That is, the vector sum of the forces, or the net force, is zero, so the object either remains at rest (as in •Fig. 4.17) or moves with a constant velocity. In such cases, objects are said to be in **translational equilibrium**. When remaining at rest, an object is said to be in *static translational equilibrium*.

It follows that the sums of the rectangular components of the forces for an object in translational equilibrium are also zero. (Why?)

$$\Sigma\mathbf{F}_x = 0 \qquad \textit{(translational} \qquad (4.5)$$
$$\Sigma\mathbf{F}_y = 0 \qquad \textit{equilibrium only)}$$

For three-dimensional problems, we would add $\Sigma\mathbf{F}_z = 0$. However, we will restrict our discussion of forces to two dimensions.

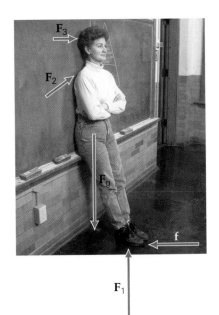

(a)

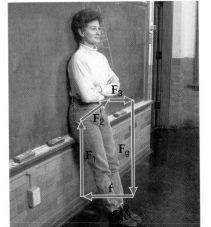

(b)

•**FIGURE 4.17 Many forces, no acceleration** (a) At least five different external forces act on this physics professor. (Here **f** is the force of friction.) Nevertheless, she experiences no acceleration. Why? (b) Adding the force vectors by the polygon method reveals that the vector sum of the forces is zero. The professor is in static translational equilibrium. (She is also in static rotational equilibrium; we'll see why in Chapter 8.)

These equations give what is often referred to as the **condition for translational equilibrium.** (A condition for rotational considerations will be given in Chapter 8.) Let's apply this translational equilibrium condition to a static equilibrium case.

Example 4.9 ■ A Hanging Sign: Static Translational Equilibrium

A 3.0-kilogram sign hangs in a hall in the Physics Department as shown in •Fig. 4.18a. What is the minimum tensile strength necessary for the cord that is used to hang the sign?

Thinking It Through. The sign hangs motionless on the wall, so it is in static translational equilibrium. Hence, the summation of the forces is zero, and here we see that there will be components in two dimensions.

Solution.

Given: $m = 3.0 \text{ kg}$ *Find:* T_1 and T_2
$\theta_1 = \theta_2 = 45°$ (tensions in the cord)

The minimum *tensile strength* of the cord is the amount of tension the cord must be able to support without breaking. With a cord of sufficient strength, the sign will hang in static equilibrium. We want to find the values of T_1 and T_2 that will *just* support the weight (mg) of the sign.

The rectangular components of the tensions are shown in the free-body diagram (Fig. 4.18b). Applying the component conditions for static translational equilibrium (Eq. 4.5), we have

$$\Sigma F_x = T_{1_x} - T_{2_x} = 0$$

or

$$T_{1_x} = T_{2_x}$$

Hence, the magnitudes of the x components of the tensions are equal, as you might have expected. (They are the only forces in the x direction, and they balance each other.) However, this gives no direct information about the magnitude of the tensions.

Since angles θ_1 and θ_2 are equal, we also know that $T_{1_y} = T_{2_y}$. (Why?) We apply the other component condition:

$$\Sigma F_y = T_{1_y} + T_{2_y} - mg = 2T_{1_y} - mg$$
$$= 2(T_1 \sin 45°) - mg = 0$$

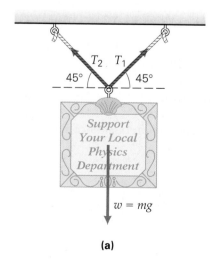

(a)

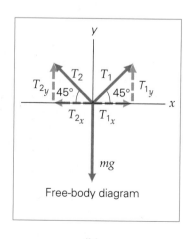

Free-body diagram

(b)

•**FIGURE 4.18 Static translational equilibrium** See Example 4.9.

Solving for T_1,

$$T_1 = \frac{mg}{2 \sin 45°} = \frac{(3.0 \text{ kg})(9.8 \text{ m/s}^2)}{2(0.707)} = 21 \text{ N}$$

With the x and y tension components equal, $T_1 = T_2 = 21$ N, so the sign should be hung from a cord with a tensile strength of at least 21 N.

Follow-up Exercise.   How could a heavier sign be hung in Example 4.9 from a cord with a tensile strength of less than 21 N?

4.6   Friction

Objectives: **To explain (a) the causes of friction and (b) how it is described by using coefficients of friction.**

Friction refers to the ever-present resistance to motion that occurs whenever two materials, or media, are in contact with each other. This resistance occurs for all types of media—solids, liquids, and gases—and is characterized as the **force of friction** (f). We have up to now generally ignored all kinds of friction (including air resistance) in examples and problems for simplicity. Now that you know how to describe motion, we are ready to consider situations that are more realistic, in that the effects of friction are included.

In some real situations, we want to increase friction—for example, by putting sand on an icy road or sidewalk to improve traction. This might seem contradictory, since an increase in friction presumably would increase the resistance to motion. However, consider the forces involved in walking as illustrated in •Fig. 4.19. Without friction, the foot would slip backward. (Think about walking on a slippery surface.) The force of friction prevents this and sometimes needs to be increased on slippery surfaces (•Fig. 4.20a). In other situations, we try to reduce friction (Fig. 4.20b). For instance, we lubricate moving machine parts to allow them to move more freely, lessen wear, and reduce expenditure of energy. Automobiles would not run without friction-reducing oils and greases.

This section is concerned chiefly with friction between solid surfaces. All surfaces are microscopically rough, no matter how smooth they appear or feel. It was originally thought that friction was primarily due to the mechanical interlocking of surface irregularities, or *asperities* (high spots). However, research has shown that the friction between the contacting surfaces of ordinary solids (metals in particular) is mostly due to local adhesion. When surfaces are pressed together, local welding or bonding occurs in a few small patches where the largest asperities make contact. To overcome this local adhesion, a force great enough to pull apart the bonded regions must be applied. Once contacting surfaces are in relative motion, another form of friction may result when the asperities of a harder material dig into a softer material, with a "plowing" effect.

Friction between solids is generally classified into three types: static, sliding (kinetic), and rolling. **Static friction** includes all cases in which the frictional force is sufficient to prevent relative motion between surfaces. Suppose you want to move a large desk. You push it, but the desk doesn't move. The force of static friction between the desk's legs and the floor opposes and equals the horizontal force you are applying, so there is no motion—a static condition.

Sliding friction, or **kinetic friction**, occurs when there is relative (sliding) motion at the interface of the surfaces in contact. In pushing on the desk, you eventually get it sliding, but there is still a great deal of resistance between the desk's legs and the floor—kinetic friction.

Rolling friction occurs when one surface rotates as it moves over another surface but does not slip or slide at the point or area of contact. Rolling friction, such as occurs between a train wheel and a rail, is attributed to small local deformations in the contact region. This type of friction is somewhat difficult to analyze.

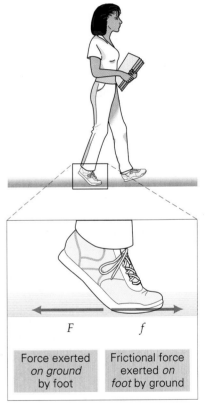

F	f
Force exerted *on ground* by foot	Frictional force exerted *on foot* by ground

•**FIGURE 4.19   Friction and walking**   The force of friction is shown in the direction of the walking motion. This may seem wrong at first glance, but it's not. The force of friction prevents the foot from slipping backward while the other foot is brought forward. If you walk on a deep-pile rug, F is evident in that the pile will be bent backward.

(a)

(b)

•**FIGURE 4.20 Increasing and decreasing friction** **(a)** To get a fast start, drag racers need to make sure that their wheels don't slip when the light goes on and they floor the accelerator. They therefore try to maximize the friction between their tires and the track by "burning in" the tires just before the start of the race. This is done by spinning the wheels with the brakes on until the tires are extremely hot. The rubber becomes so sticky that it almost welds itself to the road surface. **(b)** Water serves as a good lubricant to reduce friction in rides such as this one.

Frictional Forces and Coefficients of Friction

Here, we will consider the forces of friction on stationary and sliding objects. These are called the force of static friction and the force of kinetic (or sliding) friction, respectively. Experimentally, it is found that the force of friction depends on both the nature of the two surfaces and on the *load*, or the force that presses the surfaces together. For an object on a horizontal surface, this force is equal in magnitude to the object's weight. However, as was shown in Fig. 4.16, on an inclined plane only a component of the weight force contributes to the load.

Thus, to avoid confusion, remember that the force of friction is proportional to the normal force N in magnitude; that is, $f \propto N$. As we learned earlier, the *normal force* always acts perpendicular to, and away from, the surface. (It is the force exerted *by* the surface *on* the object.) In the absence of other perpendicular forces, the normal force is equal in magnitude to the component of the weight force acting perpendicular to the surface by the third law.

The force of static friction (f_s) between parallel surfaces in contact acts in the direction that opposes the initiation of relative motion between the surfaces. The magnitude has different values such that

$$f_s \leq \mu_s N \qquad (static\ conditions) \qquad (4.6)$$

where μ_s is the **coefficient of static friction**. (Note that it is a dimensionless constant. Why?)

The less-than-or-equal-to sign ($\leq$) indicates that the force of static friction may have different values or magnitudes up to some maximum. To understand this, look at •Fig. 4.21. In Fig. 4.21a, one person pushes on a file cabinet and it doesn't move. With no acceleration, the net force on the cabinet is zero, and $F - f_s = 0$, or $F = f_s$. Suppose that a second person also pushes and the file cabinet still doesn't budge (Fig. 4.21b). Then f_s must now be larger, since the applied force has been increased. Finally, if the applied force is made large enough to overcome the static friction, motion occurs (Fig. 4.21c). The greatest, or maximum, force of static friction is exerted just before the cabinet starts to slide (Fig. 4.21b), and for this case Eq. 4.6 can be written with an equal sign:

$$f_{s_{max}} = \mu_s N \qquad (4.7)$$

Once an object is sliding, there is a force of kinetic friction (f_k) acting on it. That force acts in the direction opposite to the direction of motion and has a magnitude of

$$f_k = \mu_k N \qquad (sliding\ conditions) \qquad (4.8)$$

where μ_k is the **coefficient of kinetic friction** (sometimes called the coefficient of sliding friction). Generally, the coefficient of kinetic friction is less than the coefficient of static friction ($\mu_k < \mu_s$) for two surfaces, which means that the force of kinetic friction is less than $f_{s_{max}}$ as illustrated in Fig. 4.21. The coefficients of friction between some common materials are listed in Table 4.1.

Note that the force of static friction (f_s) exists in response to an applied force. The magnitude of f_s and its direction depend on the magnitude and direction of the applied force. Up to its maximum value, the force of static friction is equal in magnitude and opposite to the applied force (F), since there is no acceleration ($F - f_s = 0 = ma$). Thus, if the person in Fig. 4.21a pushed on the cabinet in the opposite direction, f_s would also be in the opposite direction. If there were no applied force F, then f_s would be zero. When F exceeds $f_{s_{max}}$, the cabinet slides and kinetic friction comes into effect, with $f_k = \mu_k N$. If F is equal to f_k, the cabinet will slide with a constant velocity; if F is greater than f_k, the cabinet will accelerate.

It has been experimentally determined that the coefficients of friction (and therefore the forces of friction) are nearly independent of the size of the contact area between metal surfaces. This means that the force of friction between a

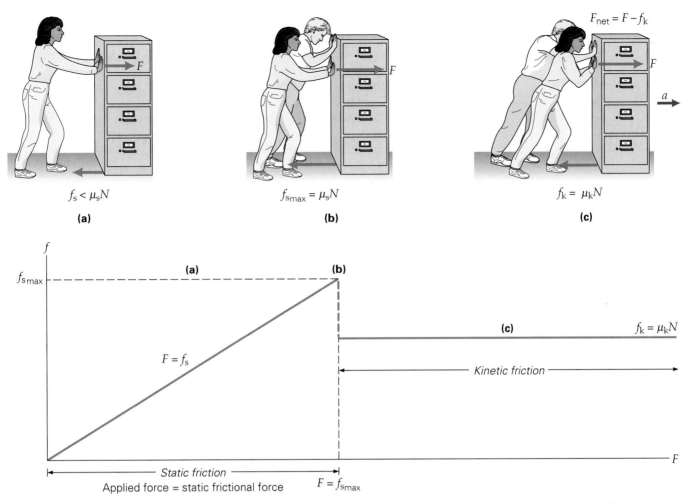

$F_{net} = F - f_k$

$f_s < \mu_s N$

(a)

$f_{s_{max}} = \mu_s N$

(b)

$f_k = \mu_k N$

(c)

$f_{s_{max}}$

(a)   **(b)**

$F = f_s$

(c) $f_k = \mu_k N$

Kinetic friction

Static friction

$F = f_{s_{max}}$

Applied force = static frictional force

•**FIGURE 4.21   Force of friction versus applied force   (a)** In the static region of the graph, as the applied force F increases, so does f_s: That is, $f_s = F$ and $f_s < \mu_s N$. **(b)** When the applied force F exceeds $f_{s_{max}} = \mu_s N$, the heavy file cabinet is set into motion. **(c)** Once the cabinet is moving, the frictional force is decreased, since kinetic friction is less than static friction ($f_k < f_{s_{max}}$).Thus, if the applied force is maintained at $F = f_{s_{max}}$ there is a net force, and the cabinet is accelerated. For the cabinet to move with constant velocity, the applied force must be reduced to equal the kinetic friction force, $f_k = \mu_k N$.

TABLE 4.1   Approximate Values for Coefficients of Static and Kinetic Friction between Certain Surfaces

Friction between Materials	μ_s	μ_k
aluminum on aluminum	1.90	1.40
glass on glass	0.94	0.35
rubber on concrete		
dry	1.20	0.85
wet	0.80	0.60
steel on aluminum	0.61	0.47
steel on steel		
dry	0.75	0.48
lubricated	0.12	0.07
Teflon on steel	0.04	0.04
Teflon on Teflon	0.04	0.04
waxed wood on snow	0.05	0.03
wood on wood	0.58	0.40
lubricated ball bearings	<0.01	<0.01
synovial joints (at the ends of most long bones—for example, elbows and hips)	0.01	0.01

brick-shaped metal block and a metal surface is the same regardless of whether the block is lying on a larger side or a smaller side. The observation is not generally valid for other surfaces, such as wood, and does not apply to plastic or polymer surfaces.

Finally, keep in mind that although the equation $f = \mu N$ holds in general for frictional forces, friction may not be linear over a wide range. That is, μ is not always constant. For example, the coefficient of kinetic friction varies somewhat with the relative speed of the surfaces. However, for speeds up to several meters per second the coefficients are relatively constant. Thus, this discussion will neglect any variations due to speed (or area), and the forces of static and kinetic friction will be assumed to depend only on the load (N) and the nature of the materials as expressed by the given coefficients of friction.

Example 4.10 ■ Pulling a Crate: Static and Kinetic Forces of Friction

(a) If the coefficient of static friction between the 40.0-kilogram crate in •Fig. 4.22 and the floor is 0.650, with what horizontal force must the worker pull to move the crate? (b) If the worker maintains that force once the crate starts to move and the coefficient of kinetic friction between the surfaces is 0.500, what is the magnitude of the acceleration of the crate?

Thinking It Through. This scenario involves applications of the forces of friction. In (a), the maximum force of static friction must be calculated. In (b), if the worker maintains an applied force of this magnitude after the crate is in motion, there will be an acceleration, since $f_k < f_{s_{max}}$.

Solution. Listing the given data and what we want to find, we have

Given: $m = 40.0$ kg   *Find:* (a) F (force necessary to move crate)
$\quad\quad\quad \mu_s = 0.650$ $\quad\quad\quad\quad\quad\quad$ (b) a (acceleration)
$\quad\quad\quad \mu_k = 0.500$

(a) The crate will not move until the applied force F slightly exceeds the maximum static frictional force $f_{s_{max}}$. So we must find $f_{s_{max}}$ to see what force the worker must apply. The weight of the crate and the normal force are equal in magnitude in this case (see the free-body diagram in Fig. 4.22), so the maximum force of static friction is

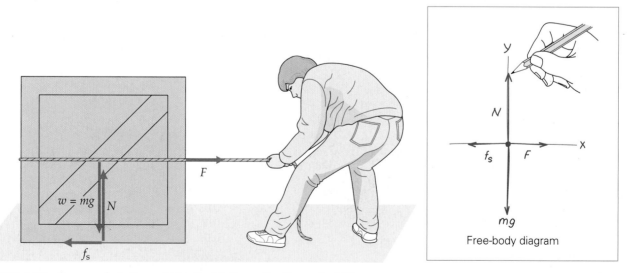

•**FIGURE 4.22   Forces of static and kinetic friction**   See Example 4.10.

$$f_{s_{max}} = \mu_s N = \mu_s(mg)$$
$$= (0.650)(40.0 \text{ kg})(9.80 \text{ m/s}^2) = 255 \text{ N} \quad \text{(about 57 lb)}$$

The crate moves if the applied force exceeds this value: $F > 255$ N.

(b) Now the crate is in motion and the worker maintains a constant applied force $F = f_{s_{max}} = 255$ N. The force of kinetic friction f_k acts on the crate, but this is smaller than F because $\mu_k < \mu_s$. Hence, there is a net force, and the acceleration of the crate may be found by using Newton's second law:

$$F - f_k = F - \mu_k N = ma$$

or

$$a = \frac{F - \mu_k N}{m} = \frac{F - \mu_k(mg)}{m}$$
$$= \frac{255 \text{ N} - (0.500)(40.0 \text{ kg})(9.80 \text{ m/s}^2)}{40.0 \text{ kg}} = 1.5 \text{ m/s}^2$$

Follow-up Exercise. On the average, by what factor does μ_s exceed μ_k for nonlubricated, metal-on-metal surfaces? (See Table 4.1.)

Let's look at the worker and the crate again, but this time assume that he applies the force at an angle (●Fig. 4.23).

Example 4.11 ■ Pulling at an Angle: A Closer Look at the Normal Force

A worker pulling a crate applies a force at an angle of 30° to the horizontal as shown in Fig. 4.23. How large a force must he apply to move the crate? (Before looking at the solution, would you expect that the force needed in this case would be greater or lesser than in Example 4.10?)

Thinking It Through. Here we see from the figure that the applied force is at an angle to the horizontal surface, so the vertical component will affect the normal force (see Fig. 4.11). This will in turn affect the maximum force of static friction.

Solution. The data are the same as in Example 4.10, except that the force is applied at an angle.

Given: $\theta = 30°$ *Find:* F (force necessary to move crate)

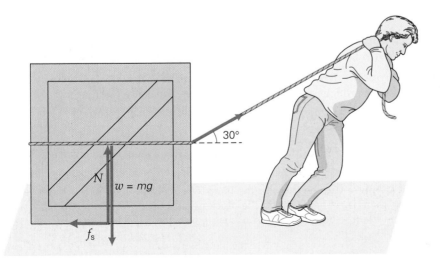
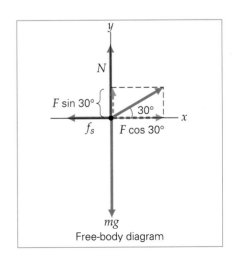

●**FIGURE 4.23 Normal force** See Example 4.11.

In this case, the crate will move when the *horizontal component* of the applied force, $F \cos 30°$, slightly exceeds the maximum static friction force. So, we may write for the maximum friction:

$$F \cos 30° = f_{s_{max}} = \mu_s N$$

However, the magnitude of the normal force is *not* equal to that of the weight of the crate here because of the upward component of the applied force (see the free-body diagram in the figure). By the second law, since $a_y = 0$, we have

$$N + F \sin 30° - mg = 0$$

or

$$N = mg - F \sin 30°$$

In effect, the applied force partially supports the weight of the crate. Substituting this expression for N into the first equation gives

$$F \cos 30° = \mu_s(mg - F \sin 30°)$$

Solving for F gives

$$F = \frac{mg}{(\cos 30°/\mu_s) + \sin 30°}$$

$$= \frac{(40.0 \text{ kg})(9.80 \text{ m/s}^2)}{(0.866/0.650) + 0.500} = 214 \text{ N (about 48 lb)}$$

Thus, less applied force is needed in this case, reflecting the fact that the frictional force is less because of the reduced normal force.

Follow-up Exercise.   Note that in this Example, applying the force at an angle produces two effects. As the angle between the applied force and the horizontal increases, the horizontal component of the applied force is reduced. However, the normal force also gets smaller, resulting in a lower $f_{s_{max}}$. Does one effect always outweigh the other? That is, does the applied force F necessary to move the crate *always decrease* with increasing angle? (*Hint*: Investigate F for different angles. For example, compute F for 20° and 50°. You already have a value for 30°. What do the results tell you?)

Now let's look at ways to determine coefficients of friction experimentally.

Example 4.12 ■ A Sliding Block: Determining the Coefficient of Kinetic Friction

A block slides with a constant velocity down a plane inclined at 37° to the horizontal (●Fig. 4.24). What is the coefficient of kinetic friction between the block and the plane?

Thinking It Through.   If the block slides down the plane with a constant velocity (zero acceleration), then the weight component down the plane is balanced by the force of kinetic friction up the incline. Hence, they may be equated, which may provide a way to find μ_k.

Solution.

Given: $a = 0$ (because v is constant)   *Find:* μ_k (coefficient of kinetic friction)
    $\theta = 37°$

Note that we are not given the mass of the block. We will not need it. Since the acceleration is zero, there is no net force on the block in either the x or y direction. So, noting the forces in the free-body diagram and using Newton's second law,

$$F_x = 0 = mg \sin \theta - f_k$$

$$F_y = 0 = N - mg \cos \theta$$

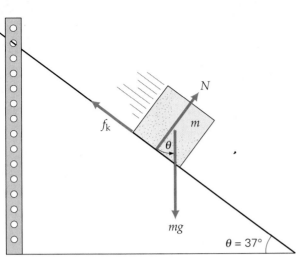

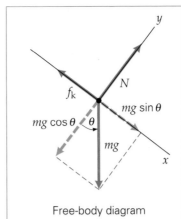

•FIGURE 4.24 **Coefficient of kinetic friction** See Example 4.12.

Free-body diagram

Rearranging the equations, we have

$$f_k = mg \sin \theta$$

$$N = mg \cos \theta$$

Then, since $f_k = \mu_k N$,

$$\mu_k = \frac{f_k}{N} = \frac{mg \sin \theta}{mg \cos \theta} = \tan \theta = \tan 37° = 0.75$$

Thus, adjusting the angle of incline until the velocity of the block sliding down the plane is constant allows μ_k to be determined experimentally from the angle of incline. From the preceding general result, we can write

$$\mu_k = \tan \theta$$

Suppose that μ_s between the plane and the block is 0.90. Can you determine the angle of incline at which the block will start to move down the plane? It will be an angle just slightly greater than that at which the component of the weight force down the plane equals the maximum force of static friction. That is,

$$mg \sin \theta = f_{s_{max}} = \mu_s N = \mu_s (mg \cos \theta) \ .$$

and

$$\frac{\sin \theta}{\cos \theta} = \tan \theta = \mu_s$$

Thus,

$$\theta = \tan^{-1} \mu_s = \tan^{-1}(0.90) = 42°$$

Therefore, the block will move if the angle of incline exceeds 42°. Adjusting the angle of incline until the block just starts to slide down the plane is an experimental way of approximating μ_s. (This critical angle is called the *angle of repose*.)

Notice how the difference in angles in this Example points out the difference between static friction and kinetic friction. You must tilt the incline to 42° to start the block moving but must reduce it to 37° to keep the block's velocity constant.

Follow-up Exercise. What would be the dynamic situation if the plane in Example 4.12 were inclined at an angle greater than 37°? Would this be a convenient way to determine μ_k?

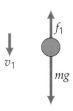

v_1

(a) As v increases, so does f.

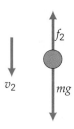

(b) When $f = mg$, the object falls with a constant (terminal) velocity.

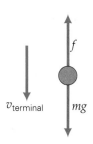

(c)

•**FIGURE 4.26 Air resistance and terminal velocity** **(a)** As the speed of a falling object increases, so does the frictional force of air resistance. **(b)** When this force of friction equals the weight of the object, the net force is zero and the object falls with a constant (terminal) velocity. **(c)** A plot of speed versus time, showing these relationships.

•**FIGURE 4.25 Air foil** The air foil at the top of the truck's cab makes the truck streamlined and therefore reduces air resistance.

Air Resistance

Air resistance refers to the resistance force acting on an object as it moves through air. In other words, air resistance is a frictional force. In analyses of free fall, you can generally ignore the effect of air resistance and still get valid approximations for objects falling relatively short distances. However, for longer distances, air resistance cannot be ignored.

Air resistance arises because a moving object collides with air molecules. Therefore, air resistance depends on the object's shape and size (which determine the area exposed to collisions) as well as its speed. The larger the object and the faster it moves, the more collisions there will be with air molecules. (Air density is also a factor, but this can be assumed to be constant near the Earth's surface.) To reduce air resistance (and fuel consumption), automobiles are made more "streamlined" and air foils are used on trucks and campers (•Fig. 4.25).

Consider now a falling object. Since air resistance depends on speed, as a falling object accelerates under the influence of gravity, the retarding force of air resistance increases (•Fig. 4.26a). Eventually, the magnitude of the retarding force equals that of the object's weight force (Fig. 4.26b), so the net force on the object is zero. The object then falls with a maximum constant velocity, which is called the **terminal velocity**.

This can be easily seen from Newton's second law. For the falling object, we have

$$F_{net} = ma$$

$$mg - f = ma$$

where downward has been taken as positive for convenience. Solving for a,

$$a = g - \frac{f}{m}$$

where a is the magnitude of the *instantaneous* acceleration.

Notice that the acceleration for a falling object when air resistance is included is less than g; that is, $a < g$ or $a < 9.8$ m/s². As the object continues to fall, the force of air resistance f increases until $a = 0$ when $f = mg$ and $f - mg = 0$. The object then falls at its constant terminal velocity.

For a skydiver with an unopened parachute, the terminal velocity is about 200 km/h (about 125 mi/h). To reduce the terminal velocity, so that it can be reached sooner and the time of fall be extended, a skydiver will try to increase exposed body area to a maximum by assuming a spread-eagle position (•Fig. 4.27). Doing this takes advantage of the dependence of air resistance on the size and shape of the falling object. Once the parachute is open (giving a larger exposed area and a shape that catches the air), the additional air resistance slows the diver down to about 40 km/h (or 25 mi/h), which is preferable for landing.

Conceptual Example 4.13

Race You Down: Air Resistance and Terminal Velocity

From a high altitude, a balloonist simultaneously drops two balls of identical size but appreciably different weight. Assuming that both balls reach terminal velocity during the fall, (a) the heavier ball reaches terminal velocity first, (b) the balls reach terminal velocity at the same time, (c) the heavier ball hits the ground first, (d) the balls hit the ground at the same time. *Clearly establish the reasoning and physical principle(s) used in determining your answer before checking it below. That is, **why** did you select your answer?*

Reasoning and Answer. Terminal velocity is reached when the weight of a ball is balanced by the frictional air resistance. Both balls start to fall with the same acceleration, *g*, and their speeds and the retarding forces of air resistance increase at the same rate. The weight of the lighter ball will be balanced first, so (a) and (b) are incorrect. The lighter ball reaches terminal velocity ($a = 0$), but the heavier ball continues to accelerate and pulls ahead of the lighter ball. Hence, the heavier ball hits the ground first and the answer is (c), and (d) does not apply.

Follow-up Exercise. Suppose the heavier ball were much larger than the lighter ball. How might this difference affect the outcome?

•**FIGURE 4.27 Terminal velocity**
Sky divers assume a spread-eagle position to maximize air resistance. This causes them to reach terminal velocity more quickly and prolongs the time of fall.

Chapter Review

Important Terms

force   *100*
net (unbalanced) force   *100*
inertia   *102*
Newton's first law of motion
  (law of inertia)   *102*
newton (unit)   *104*
Newton's second law of
  motion   *104*

weight   *104*
Newton's third law of
  motion   *109*
normal force   *110*
free-body diagram   *115*
translational equilibrium   *119*
condition for translational
  equilibrium   *120*

force of friction   *121*
static friction   *121*
kinetic (sliding) friction   *121*
rolling friction   *121*
coefficient of static friction   *122*
coefficient of kinetic friction   *122*
air resistance   *128*
terminal velocity   *128*

Important Concepts

- A force is something capable of changing an object's state of motion. To produce a change in motion, there must be a net or unbalanced force.

- Newton's first law of motion is also called the law of inertia, where inertia is the natural tendency of an object to maintain its state of motion. It states that in the absence of a net applied force, a body at rest remains at rest and a body in motion remains in motion with constant velocity.

- Newton's second law relates the net force acting on an object or system to the (total) mass and the resulting acceleration. It defines the cause-and-effect relationship between force and acceleration ($\mathbf{F} \propto m\mathbf{a}$).

- Newton's third law states that for every force there is an equal and opposite reaction force. The oppos-

ing forces of a third-law pair always act on different objects.

- Friction is the resistance to motion that occurs between contacting surfaces.

- The frictional force between surfaces is characterized by coefficients of friction (μ), one for the static case and one for the kinetic (moving) case. In many cases, $f = \mu N$, where N is the normal force—the force perpendicular to the surface (the force exerted *by* the surface *on* the object). As a ratio of forces, μ is unitless.

- The force of air resistance on a falling object increases with increasing speed until the object falls at a constant rate called the terminal velocity.

Important Equations

Newton's Second Law:

$$\mathbf{F} = m\mathbf{a} \qquad (4.1)$$

where $\mathbf{F} = \sum_i \mathbf{F}_i = \mathbf{F}_{net}$

Weight:

$$w = mg \qquad (4.2)$$

Component Form of Newton's Second Law:

$$F_x\,\hat{\mathbf{x}} + F_y\,\hat{\mathbf{y}} = ma_x\,\hat{\mathbf{x}} + ma_y\,\hat{\mathbf{y}} \qquad (4.3a)$$

Condition for Translational Equilibrium:

$$\Sigma\mathbf{F} = 0 \qquad (4.4)$$

or

$$\Sigma F_x = 0 \quad \text{and} \quad \Sigma F_y = 0 \qquad (4.5)$$

Force of Static Friction:

$$f_s \le \mu_s N \qquad (4.6)$$

$$f_{s_{max}} = \mu_s N \qquad (4.7)$$

Force of Kinetic (Sliding) Friction:

$$f_k = \mu_k N \qquad (4.8)$$

Exercises

Note: Exercises with a CD-ROM icon have corresponding simulations on the Interactive Journey through physics CD-ROM; a complete cross-reference of text problems to CD-ROM activities appears in the Media Pack.

Note: Unless otherwise stated, all objects are located near the Earth's surface, where $g = 9.80 \text{ m/s}^2$.

4.1 The Concepts of Force and Net Force
and
4.2 Inertia and Newton's First Law of Motion

1. If an object remains at rest, there must be no force acting on it. Is this statement correct? Explain.

2. The tendency of an object to maintain its state of motion is called (a) Newton's first law, (b) Galileo's principle, (c) inertia.

3. If an object is moving with constant velocity, (a) there must be a force in the direction of the velocity, (b) there must be no force in the direction of the velocity, (c) there must be no net force.

4. If the net force on an object is zero, the object must (a) be at rest, (b) be in motion with a constant velocity, (c) have zero acceleration, (d) none of these.

5. An object weighs 300 N on Earth and 50 N on the Moon. Does the object also have less inertia on the Moon?

6. Consider an air-bubble level that is sitting on a horizontal surface (•Fig. 4.28). Initially, the air bubble is in the middle of the horizontal glass tube. (a) If the level is pushed and a force is applied to accelerate it, which way would the bubble move? Which way would the bubble move if the force is then removed and the level slows down due to friction? (b) Such a level is sometimes used as an "accelerometer" to indicate the direction of the acceleration. Explain the principle involved. [Hint: Think about pushing a pan of water.]

7. As a follow-up to Exercise 6, consider the situation of a child holding a helium balloon in a closed car at rest. What would the child observe when the car (a) accelerates from rest and (b) brakes to a stop? (The balloon does not touch the roof of the car.)

8. Objects have no weight in outer space. How can you then distinguish their masses?

9. If a tablecloth is yanked out very quickly, the dishes on top of it will barely move (•Fig. 4.29). Why?

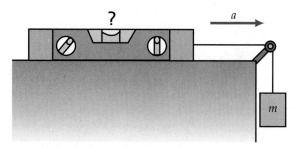

•**FIGURE 4.28   An air-bubble level/accelerometer**   See Exercise 6.

•**FIGURE 4.29   No more dinner?**   See Exercise 9.

10. If your professor's car is traveling west with a constant velocity of 55 mi/h west on a straight highway, what is the net force acting on it?

11. ■ Which has more inertia, 20 cm³ of water or 10 cm³ aluminum, and how many times more? (See Table 9.2.)

12. ■■ 🔘 You are told that an object has zero acceleration. Two forces on the object are $\mathbf{F}_1 = 3.6$ N at 74° below the $+x$ axis and $\mathbf{F}_2 = 3.6$ N at 34° above the $-x$ axis. Is there a third force on the object? If yes, what is it? Can you tell whether the object is at rest or in motion?

13. ■■ 🔘 A 5.0-kilogram block at rest on a frictionless surface is acted on by forces $F_1 = 5.5$ N and $F_2 = 3.5$ N, as illustrated in •Fig. 4.30. What additional horizontal force will keep the block at rest?

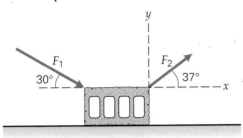

•FIGURE 4.30   Two applied forces   See Exercises 13 and 86.

14. ■■■ 🔘 A 1.5-kilogram object moves up the y axis with a constant speed. When it reaches the origin, the forces $\mathbf{F}_1 = 5.0$ N at 37° above the $+x$ axis, $\mathbf{F}_2 = 2.5$ N in the $+x$ direction, $\mathbf{F}_3 = 3.5$ N at 45° below the $-x$ axis, and $\mathbf{F}_4 = 1.5$ N in the $-y$ direction are applied to it. (a) Will the object continue to move along the y axis? (b) If not, what simultaneously applied force will keep it moving along the y axis with a constant speed?

4.3 Newton's Second Law of Motion

15. The newton unit of force is equivalent to (a) kg·m/s, (b) kg·m/s², (c) kg·m²/s, (d) none of these.

16. In general, this chapter considered forces that were applied to objects of constant mass. What would be the situation if mass were added to or lost from a system while a force was being applied to the system? Give examples of situations in which this might happen.

17. Good football wide receivers usually have "soft" hands for catching balls (•Fig. 4.31). How would you interpret this description on the basis of Newton's second law?

18. ■ A 3.0-newton net force is applied to a 1.5-kilogram mass. What is its acceleration?

19. ■ What is the mass of an object that accelerates at 3.0 m/s² under the influence of a 5.0-newton net force?

20. ■ A worker pushes on a crate, which experiences a net force of 75 N. If the crate also experiences an acceleration of 0.50 m/s², what is its weight?

•FIGURE 4.31   Soft hands   See Exercise 17.

21. ■ An ocean tanker has a gross mass of 7.0×10^7 kg. What constant net force would give the tanker an acceleration of 0.10 m/s²?

22. ■ A 6.0-kilogram object is brought to the Moon, where the acceleration due to gravity is only one-sixth of that on the Earth. What is the mass of the object on the Moon? (a) 1.0 kg (b) 6.0 kg (c) 59 kg (d) 9.8 kg

23. ■ What is the mass of a person weighing 740 N?

24. ■ What is the net force acting on a 1.0-kilogram object in free fall?

25. ■ What is the weight of an 8.0-kilogram mass in newtons? How about in pounds?

26. ■■ What is the weight of a 150-pound person in newtons? What is his mass in kilograms?

27. ■■ (a) Is the product label shown in •Fig. 4.32 correct on the Earth? (b) Is it correct on the Moon, where the acceleration due to gravity is only about one-sixth of that on Earth? If not, what should the label be on the Moon?

•FIGURE 4.32   Correct label?   See Exercise 27.

28. ■■ In a college orientation competition, 18 students lift a sports car. While holding the car off the ground, each student exerts an upward force of 600 N. (a) What is the mass of the car in kilograms? (b) What is its weight in pounds?

29. ■■ A horizontal force of 12 N acts on an object that rests on a level, frictionless surface on the Earth, where the object has a weight of 98 N. (a) What is the magnitude of the acceleration of the object? (b) What would be the acceleration of the same object in a similar situation on the Moon?

30. ■■ The engine of a 1.0-kilogram toy plane exerts a 15-newton forward force on the plane. If the air exerts an 8.0-newton resistive force on the plane, what is the magnitude of the acceleration of the plane?

31. ■■ When a horizontal force of 300 N is applied to a 75.0-kilogram box, the box slides on a level floor, opposed by a force of kinetic friction of 120 N. What is the magnitude of the acceleration of the box?

32. ■■ An ocean-going oil tanker that has a mass of 6.4×10^7 kg and that travels with a speed of 15 knots takes 5.0 km to come to a stop. What is the magnitude of the constant force required to do this? [1 knot = 1.15 (statute) mi/h]

33. ■■ A stalled 1500-kilogram automobile is pushed toward a gas station by a man and a woman on a level road. The applied horizontal forces are 200 N for the woman and 300 N for the man. (a) If there is an effective force of friction of 300 N on the car as it moves, what is the car's acceleration? (b) Once the car is moving appreciably, what would be an appropriate combined applied force, and why?

34. ■■ In an emergency stop to avoid an accident, a shoulder-strap seat belt holds a 60-kilogram passenger firmly in place. If the car was initially traveling at 90 km/h and came to a stop in 5.5 s along a straight, level road, what was the average force applied to the passenger by the seat belt?

35. ■■ A jet catapult on an aircraft carrier accelerates a 2000-kilogram plane uniformly from rest to a launch speed of 320 km/h in 2.0 s. What is the magnitude of the net force on the plane?

36. ■■ In serving, a tennis player accelerates a 56-gram tennis ball horizontally from rest to a speed of 35 m/s. Assuming the acceleration is uniform when the racquet applies a force over a distance of 0.50 m, what is the magnitude of the force exerted on the ball?

4.4 Newton's Third Law of Motion

37. A brick hits a glass window. The brick breaks the glass, so (a) the magnitude of the force of the brick on the glass is greater than the magnitude of the force of the glass on the brick, (b) the magnitude of the force of the brick on the glass is smaller than the magnitude of the force of the

glass on the brick, (c) the magnitude of the force of the brick on the glass is equal to the magnitude of the force of the glass on the brick, (d) none of the above.

38. The force pair of Newton's third law (a) consists of forces that are always opposite but not always equal, (b) always cancel each other when the second law is applied to a body, (c) always act on the same object, (d) consists of forces that are equal and opposite but act on different objects.

39. Is there something wrong with the following statement? A baseball bat hits a baseball, and there are equal and opposite forces on the baseball. The forces then cancel, and there is no motion.

40. Draw a free-body diagram of a book sitting on a horizontal surface. Identify the reaction forces to all the forces in the diagram.

41. By using the right technique, a karate master can exert huge forces on objects. If a fist hits a brick with a force of 800 N, what else do we know according to Newton's third law? Is this something you can try at home?

42. ■ A person pushes on a block of wood that has been placed against a wall. Draw a free-body diagram and identify the reaction forces to all the forces on the block.

43. ■■ In the pairs competition of Olympic figure skating, a male skater with a mass of 60 kg pushes a female skater who has a mass 45 kg, causing her to accelerate at a rate of 2.0 m/s^2. At what rate will the male skater accelerate? What is the direction of his acceleration?

44. ■■ Jane and John, with masses of 50 kg and 60 kg, respectively, stand on a frictionless surface 10 m apart. John pulls on a connecting rope, giving Jane an acceleration of 0.92 m/s^2 toward him. (a) What is John's acceleration? (b) If the pulling force is applied constantly, where will they meet?

4.5 More on Newton's Laws: Free-body Diagrams and Translational Equilibrium

Note: In Exercises with strings and pulleys, "ideal conditions" means that the masses of the string(s) and pulley(s), as well as the friction of the pulley, should be neglected.

45. ■■ Is it possible for a 500-kilogram object on a scale to appear to weigh only 4000 N in a moving elevator? If so, explain and describe the motion.

46. ■■ ⬤ A 75.0-kilogram person is standing on a scale in an elevator. What is the reading of the scale in newtons if the elevator is (a) at rest, (b) moving up with a constant velocity of 2.00 m/s, (c) accelerating up at 2.00 m/s^2?

47. ■■ ⬤ In Exercise 46, what if the elevator is accelerating downward at the same rate?

48. ■■ Two boats pull a 75.0-kilogram water skier as illustrated in •Fig. 4.33. (a) If each boat pulls with a force of 600 N and the skier travels with a constant velocity, what is the magnitude of the retarding force between the water and the skis? (b) Assuming the retarding force remains constant, if each boat pulls with a force of 700 N, what is the magnitude of the acceleration of the skier?

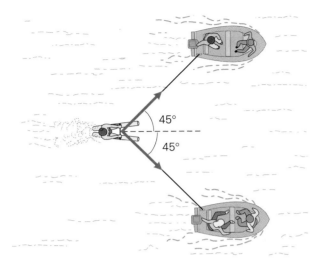

•**FIGURE 4.33   Double tow**   See Exercise 48.

49. ■■ (a) A 65-kilogram water skier is pulled due east by a boat with a horizontal force of 400 N with a water drag of 300 N. A sudden gust of wind supplies another horizontal force of 50 N on the skier at an angle of 60° north of east. What is the skier's acceleration at that instant? (b) What would be the acceleration if the wind force were in the opposite direction of that in part (a)?

50. ■■ A boy pulls a box of mass 30 kg with a force of 25 N in the direction shown in •Fig. 4.34. (a) What is the acceleration of the box? (b) What is the normal force on the box by the ground? Ignore friction.

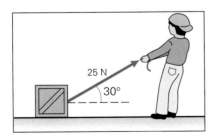

•**FIGURE 4.34   Pulling a box**   See Exercise 50.

51. ■■ A girl pushes a 25-kilogram lawn mower as shown in •Fig. 4.35. If $F = 30$ N and $\theta = 37°$, (a) what is the acceleration of the mower? (b) What is the normal force exerted on the mower by the lawn? Ignore friction.

52. ■■ 🌐 An Olympic skier coasts down a slope with an angle of inclination of 37°. (a) Neglecting friction, what is the acceleration of the skier? (b) If the skier has a speed

•**FIGURE 4.35   Mowing the lawn**   See Exercise 51.

of 5.0 m/s at the top of the slope, what is his speed when he reaches the bottom of the 35-meter-long slope?

53. ■■ A car is coasting (engine off) up a 30° grade. If the speed of the car is 25 m/s at the bottom of the grade, what is the distance traveled by the car before coming to rest?

54. ■■ A horizontal force of 40 N acting on a block on a frictionless level surface produces an acceleration of 2.5 m/s². A second block with a mass of 4.0 kg is dropped onto the first. What is the magnitude of the acceleration of the combination if the same force continues to act? (Assume that the second block does not slide on the first block.)

55. ■■ Two blocks, weighing 80 N and 50 N, respectively, sit side by side in contact with each other on a frictionless horizontal surface. (a) If a constant horizontal force of 40 N is applied to one of the blocks in the direction of the other block, what is the magnitude of the resulting acceleration? (b) If the force is applied instead to the block downward at an angle of 25° to the horizontal, what would be the acceleration?

56. ■■ A 50-kilogram gymnast hangs vertically from a pair of parallel rings. (a) If the ropes supporting the rings are attached to the ceiling directly above, what is the tension in each rope? (b) If the ropes are supported so that they make an angle of 45° with the ceiling, what is the tension in each rope?

57. ■■ A truck of mass 3000 kg tows a 1500-kilogram car by a chain. If the net forward force on the truck by the ground is 3200 N, (a) what is the acceleration of the car? (b) What is the tension in the connecting chain?

58. ■■ 🌐 Three blocks are pulled along a frictionless surface by a horizontal force as shown in •Fig. 4.36. (a) What is the acceleration of the system? (b) What are the tension forces in the light strings? [Hint: Can T_1 equal T_2? Investigate by drawing free-body diagrams of each block separately.]

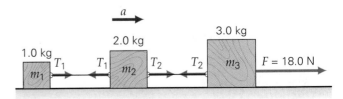

•**FIGURE 4.36   Three-block system**   See Exercises 58 and 87.

59. ■■ Assume ideal conditions for the apparatus illustrated in •Fig. 4.37. What is the acceleration of the system if (a) $m_1 = 0.25$ kg, $m_2 = 0.50$ kg, and $m_3 = 0.25$ kg, and (b) $m_1 = 0.35$ kg, $m_2 = 0.15$ kg, and $m_3 = 0.50$ kg?

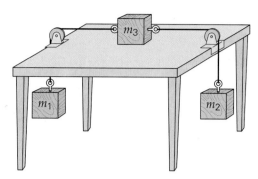

•**FIGURE 4.37 Which way will they accelerate?** See Exercises 59, 89, and 90.

60. ■■ An Atwood machine (Fig. 4.14) has suspended masses of 0.25 kg and 0.20 kg. Under ideal conditions, what will be the acceleration of the smaller mass?

61. ■■ Consider an ideal Atwood machine (Fig. 4.14) in which m_2 (0.15 kg) is released 1.0 m above the floor and m_1 (0.14 kg) rests on the floor. If it takes m_2 2.4 s to reach the floor, what is the experimental value of g?

62. ■■■ One mass, $m_1 = 0.215$ kg, of an ideal Atwood machine rests on the floor 1.10 m below the other mass, $m_2 = 0.255$ kg. (a) If the masses are released from rest, how long does it take m_2 to reach the floor? (b) How high will mass m_1 ascend from the floor? [Hint: When m_2 hits the floor, m_1 continues to move upward.]

63. ■■■ A 0.20-kilogram ball is released from a height of 10 m above the beach; the impression the ball makes in the sand is 5.0 cm deep while the ball stops uniformly. What is the force acting on the ball by the sand?

64. ■■■ In the ideal apparatus shown in •Fig. 4.38, $m_1 = 2.0$ kg. What is m_2 if both masses are at rest? How about if both are moving at a constant velocity?

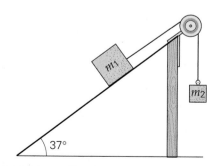

•**FIGURE 4.38 Inclined Atwood machine** See Exercises 64, 65, and 91.

65. ■■■ In the ideal setup shown in Fig. 4.38, $m_1 = 3.0$ kg and $m_2 = 2.5$ kg. (a) What is the acceleration of the masses? (b) What is the tension in the string?

4.6 Friction

Note: Neglect air resistance unless otherwise stated.

66. In general, the frictional force (a) is greater for smooth surfaces, (b) depends on sliding speeds, (c) is proportional to the normal force, (d) depends significantly on the surface area.

67. Identify the direction of the friction force in the following cases: (a) a book sitting on a table, (b) a box sliding on a horizontal surface, (c) a car making a turn on a flat road, (d) the initial motion of a machine part delivered on a conveyor belt in an assembly line.

68. The coefficient of kinetic friction μ_k (a) is usually greater than the coefficient of static friction μ_s, (b) usually equals μ_s, (c) is usually smaller than μ_s, (d) equals the applied force that exceeds the maximum static force.

69. The purpose of anti-lock brakes is to avoid the lock-up of the wheels to keep the car rolling rather than sliding. Why would rolling decrease the stopping distance?

70. Is it easier to push or pull a lawnmower at an angle? Explain.

71. •Figure 4.39 shows the front and rear wings of an Indy car. These wings generate *down force*, the vertical downward force by the air moving over the car. Why is such a down force desired? Will the additions of the wings slow down the car? An Indy car can create a down force equal to twice its weight. Why not simply make the cars heavier?

•**FIGURE 4.39 Down force** See Exercise 71.

72. (a) We commonly say that friction opposes motion. Yet, when we walk, the frictional force is in the direction of our motion (Fig. 4.19). Is there an inconsistency in terms of Newton's second law? Explain. (b) What effects would wind have on air resistance? [Hint: The wind can blow in different directions.]

73. Why are racing tires wide and smooth whereas regular tires are narrower and have tread (•Fig. 4.40)? Are there frictional and/or safety considerations?

•**FIGURE 4.40 Racing tires versus passenger tires: safety or speed?** See Exercise 73.

74. ■ 🌀 In moving a 35.0-kilogram desk from one side of a classroom to the other, a professor finds that a horizontal force of 275 N is necessary to set the desk in motion and a force of 195 N is necessary to keep it in motion with a constant speed. What are the coefficients of (a) static and (b) kinetic friction between the desk and the floor?

75. ■ 🌀 A 40-kilogram crate is at rest on a level surface. If the coefficient of static friction between the crate and the surface is 0.69, what horizontal force is required to get the crate moving?

76. ■ A 20-kilogram box sits on a horizontal surface. The forces required to set it in motion and then to keep it in motion at constant velocity are 150 N and 120 N, respectively. What are the coefficients of static and kinetic friction?

77. ■ The coefficients of static and kinetic friction between a 50-kilogram box and a horizontal surface are 0.60 and 0.40, respectively. (a) What is the acceleration of the object if a 250-newton horizontal force is applied to the box? (b) What if the applied force is 350 N?

78. ■■ 🌀 A 1500-kilogram automobile travels at a speed of 90 km/h along a straight concrete highway. Faced with an emergency situation, the driver jams on the brakes, and the car skids to a stop. What will the stopping distance be for (a) dry pavement and (b) wet pavement?

79. ■■ 🌀 A hockey player hits a puck with her stick, giving the puck an initial speed of 5.0 m/s. If the puck slows uniformly and comes to rest in a distance of 20 m, what is the coefficient of kinetic friction between the ice and the puck?

80. ■■ 🌀 A packing crate is placed on a 20° inclined plane. If the coefficient of static friction between the crate and the plane is 0.65, will the crate slide down the plane? (Justify your answer.)

81. ■■ A wood block with a weight of 3.5 N is placed on a wood board, measuring 0.75 m × 0.75 m, that is elevated at one end to act as an inclined plane. What is the elevation of the higher end of the board when the block, once it is in motion, slides down with a constant velocity?

82. ■■ 🌀 Suppose the slope conditions for the skier shown in •Fig. 4.41 are such that he travels with a constant velocity. From the photo, could you find the coefficient of kinetic friction between the snowy surface and the skis? If so, describe how this would be done.

•**FIGURE 4.41 A downslope run** See Exercise 82.

83. ■■ 🌀 A 5.0-kilogram wooden block is placed on an adjustable wooden inclined plane. (a) What is the angle of incline above which the block will *start* to slide down the plane? (b) At what angle of incline will the block then slide down the plane with a constant speed?

84. ■■ A block that has a mass of 2.0 kg and is 10 cm wide on each side just slides down an inclined plane with a 30° angle of incline (•Fig. 4.42). Another block of the same height and same material has base dimensions of 20 cm × 10 cm and thus a mass of 4.0 kg. (a) At what critical angle will the more massive block start to slide down the plane? (b) Estimate the coefficient of static friction between the block and the plane.

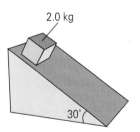
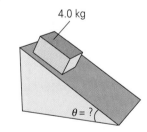

•**FIGURE 4.42 At what angle will it begin to slide?** See Exercise 84.

85. ■■ 🌀 While being unloaded from a moving truck, a 10-kilogram suitcase is placed on a flat ramp inclined at 37°. When released from rest, the suitcase accelerates down the ramp at 0.15 m/s². What is the coefficient of kinetic friction between the suitcase and the ramp?

86. ■■ For the situation shown in Fig. 4.30, what is the minimum coefficient of static friction between the block and the surface that will keep the block from moving? (F_1 = 5.0 N, F_2 = 4.0 N, and m = 5.0 kg.)

87. ■■ 🌀 For the system illustrated in Fig. 4.36, if $\mu_s = 0.45$ and $\mu_k = 0.35$ between the blocks and the surface, what applied forces will (a) set the blocks in motion and (b) move the blocks with a constant velocity?

88. ■■ 🌀 In the apparatus shown in •Fig. 4.43, $m_1 = 10$ kg and the coefficients of static and kinetic friction between m_1 and the table are 0.60 and 0.40, respectively. (a) What mass of m_2 will set the system in motion? (b) With mass m_2 as in part (a), after the system moves, what is the acceleration?

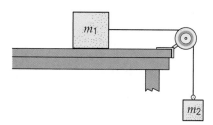

•**FIGURE 4.43   Friction and motion.**   See Exercise 88.

89. ■■■ For the apparatus in Fig. 4.37, what is the minimum value of the coefficient of static friction between the block (m_3) and the table that would keep the system at rest if $m_1 = 0.25$ kg, $m_2 = 0.50$ kg, and $m_3 = 0.75$ kg? (Assume ideal conditions for the string and pulleys.)

90. ■■■ If the coefficient of kinetic friction between the block and the table in Fig. 4.37 is 0.560, and $m_1 = 0.150$ kg and $m_2 = 0.250$ kg, (a) what should m_3 be if the system is to move with a constant speed? (b) If $m_3 = 0.100$ kg, what is the acceleration of the system? (Assume ideal conditions for the string and pulleys.)

91. ■■■ In the apparatus shown in Fig. 4.38, $m_1 = 2.0$ kg and the coefficients of static and kinetic friction between m_1 and the inclined plane are 0.30 and 0.20, respectively. (a) What is m_2 if both masses are to remain at rest? (b) What is m_2 if both masses are moving with constant velocity? Neglect all friction.

Additional Exercises

92. A force acts on a mass. If the force is halved and the mass is doubled, how is the acceleration affected? (Give a factor of change—for example, 2 times as great or 1/2 as great.)

93. A rifle weighs 50.0 N, and its barrel is 0.750 m long. It shoots a 25.0-gram bullet, which leaves the barrel at a speed of 300 m/s (muzzle velocity) after being uniformly accelerated. What is the magnitude of the force exerted on the rifle by the bullet?

94. The maximum load that can safely be supported by a rope in an overhead hoist is 400 N. What is the maximum acceleration that can safely be given to a 25-kilogram object being hoisted vertically upward?

95. The coefficient of static friction between a 9.0-kilogram object and a horizontal surface is 0.45. Would a force of 35 N applied horizontally cause the object to move from rest? If so, what would be its acceleration?

96. An object with a mass of 2.0 kg travels with a constant velocity of 4.8 m/s northward. It is then acted on by a force of 6.5 N in the direction of the motion and a force of 8.5 N to the south, both of which continue even after the mass comes momentarily to rest. (a) How far will the object travel before coming to rest? (b) What will be its position 1.5 s after the object comes momentarily to rest?

97. In the operation of a machine, a 5.0-kilogram steel part is to move on a horizontal steel surface. Force is applied to the part downward at an angle of 30° from the horizontal. (a) What is the magnitude of the applied force required to set the part in motion if the surface is dry? (b) If the surface is lubricated, what is the needed force?

98. 🌀 A crate weighing 9.80×10^3 N is pulled up a 37° incline by a force parallel to the plane. If the coefficient of kinetic friction between the crate and the surface of the plane is 0.750, what is the magnitude of the applied force required to move the crate with a constant velocity?

99. For an Atwood machine with suspended masses of 0.30 kg and 0.40 kg, the acceleration of the masses is measured as 0.95 m/s². What is the effective force of friction for the system?

100. A catapult jet plane with a weight of 2.75×10^6 N is ready for takeoff. If its engines supply 6.35×10^6 N of net thrust, how long a runway will the plane need to reach its minimum take-off speed of 285 km/h?

101. A 0.45-kilogram shuffleboard puck is given an initial speed of 4.5 m/s down the flat playing surface. If the coefficient of sliding friction between the puck and the surface is 0.20, how far will the puck slide before coming to rest?

102. A frictionless ramp 135 m long is to be built for a ski jump. If a skier starting from rest at the top is to have a speed no faster than 24 m/s at the bottom, what should be the maximum angle of inclination? Ignore friction.

103. A school bus pulls into an intersection as a car approaches with a speed of 25 km/h on an icy street. Seeing the bus from 26 m away, the driver of the car locks the brakes, causing the car to slide toward the intersection. If the coefficient of kinetic friction between the car's tires and the icy road is 0.10, does the car hit the bus?

104. While catching a baseball that is traveling horizontally with a speed of 15 m/s, the player's glove and arm move straight backward 25 cm from the time of contact to the time the ball comes to rest. If the ball has a mass of 0.14 kg, what is the magnitude of the average force on the ball during that interval?

Work and Energy

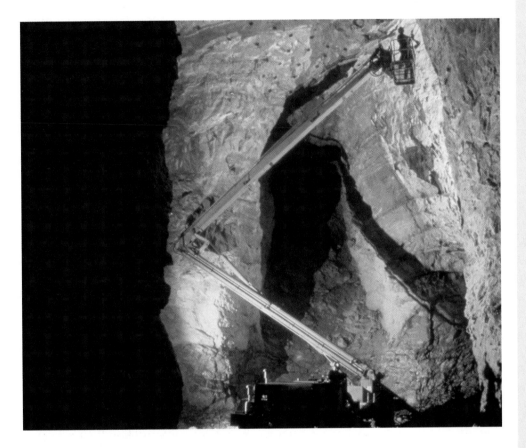

Work and energy can be somewhat slippery concepts. For the most part, our everyday sense of these words accords pretty well with their definitions in physics, but not always. You might think that this crane, for example, must be performing a good deal of work to support the miner and move him around, but that's not necessarily true from a physicist's point of view. Depending on how and where the bucket is moved, the crane might be doing no work at all. It might even be doing *negative* work—a concept that most people can't easily imagine. A physicist would also say that the miner has slightly less energy down here than he would outside the mine—not because he prefers working on his lawn to mining zinc, but because he's a tiny bit closer to the center of the Earth.

This chapter centers on two concepts that are important in both science and everyday life—*work* and *energy*. We commonly think of work as being associated with doing or accomplishing something. Because work makes us physically (and sometimes mentally) tired, we have invented machines and use them to decrease the amount of effort we expend personally. Energy tends to bring to mind the cost of fuel for transportation and heating, or perhaps the food that supplies the energy our bodies need to carry out life processes and to do work.

Although these notions do not really define work and energy, they point us in the right direction. As you might have guessed work and energy are closely related. In physics, as in everyday life, when something possesses energy, it usually has the ability to do work. For example, water rushing through the sluices of a dam has energy of motion, and this energy allows the water to do the work of driving a turbine or dynamo. Conversely, no work can be performed without energy.

Energy exists in various forms: There is mechanical energy, chemical energy, electrical energy, heat energy, nuclear energy, and so on. A transformation from one form to another may take place, but the total amount of energy is *conserved*, or always remains the same. This is the point that makes the concept of energy so useful. When a physically measurable quantity is conserved, it not only gives us an insight that leads to a better understanding of nature, but also usually provides another approach to practical problems. (You will be introduced to other conserved quantities during the course of our study of physics.)

5.1 Work Done by a Constant Force

Objectives: To (a) define mechanical work and (b) compute the work done in various situations.

The word *work* is commonly used in a variety of ways: We go to work; we work on projects; we work at our desks or on computers; we work problems. In physics, however, work has a very specific meaning. Mechanically, **work** involves force and displacement, and we use this word to describe quantitatively what is accomplished when a force moves an object through a distance. In the simplest case of a *constant* force acting on an object:

Work—involves force and displacement

> The work done by a constant force in moving an object is equal to the product of the magnitudes of the displacement and the component of the force parallel to the displacement.

Work then involves moving an object through a distance. A force may be applied, as in •Fig. 5.1a, but *if there is no motion (no displacement), then no work is done.* For a constant force **F** acting *in the same direction* as the displacement **d** (Fig. 5.1b), the work (W) is simply the product of their magnitudes:

$$W = Fd \qquad (5.1)$$

(As you might expect, when work is done in Fig. 5.1b, energy is expended. We shall discuss the relationship between work and energy in Section 5.3.)

In general, work is done on an object only by a force, or force *component*, parallel to the line of motion or displacement of that object (Fig. 5.1c). That is, if the force acts at an angle θ to the object's displacement, then $F_{\parallel} = F \cos \theta$ is the com-

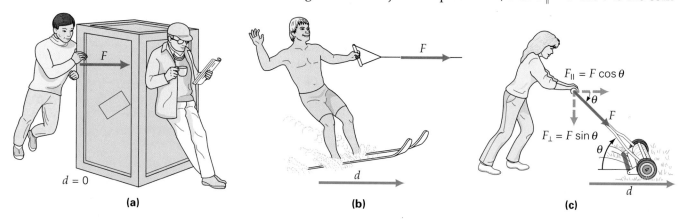

•**FIGURE 5.1   Work done by a constant force—the product of the magnitudes of the parallel component of force and the displacement**   **(a)** If there is no displacement, no work is done: $W = 0$. **(b)** For a constant force in the same direction as the displacement, $W = Fd$. **(c)** For a constant force at an angle to the displacement, $W = (F \cos \theta)d$.

ponent of the force parallel to the displacement. Thus, a more general equation for work done by a constant force is

$$W = F_{\parallel}d = (F \cos \theta)d \quad \textit{work done by a constant force} \quad (5.2)$$

Notice that θ is the angle *between* the force and the displacement vectors; as a reminder, cos θ can be written between the magnitudes of the force and displacement in Eq. 5.2. If $\theta = 0°$ (force and displacement are in the same direction as in Fig. 5.1b), then $W = F(\cos 0°)d = Fd$, so Eq. 5.2 reduces to Eq. 5.1. The perpendicular component of the force, $F_{\perp} = F \sin \theta$, does no work since there is no displacement in this direction.

The units of work can be determined from the equation $W = Fd$. With force in newtons and displacement in meters, work has the SI unit of newton-meters (N·m). This unit is given the special name **joule** (J):

$$Fd = W$$

$$1 \text{ N·m} \equiv 1 \text{ J}$$

For example, the work done by a force of 25 N in moving an object through a parallel displacement of 2.0 m is $W = Fd = (25 \text{ N})(2.0 \text{ m}) = 50 \text{ N·m}$, or 50 J.

From the previous equation, we also see that in the British system work would have the unit pound-foot. However, this name is commonly written in reverse. The British standard unit of work is the *foot-pound* (ft·lb).

Remember that *work is a scalar quantity* and, as such, may have a positive or negative value. In Fig. 5.1b, the work would be positive because the force acts in the same direction as the displacement (cos 0° is positive). The work would also be positive in Fig. 5.1c, because a force component acts in the direction of the displacement (cos θ is positive).

However, if the force, or a force component, acts in the opposite direction of the displacement, the work is negative since the cosine term is negative. For example, for $\theta = 180°$ (force opposite displacement), cos 180° = −1, so the work is negative: $W = Fd = (F \cos 180°)d = -Fd$. An example is a braking force that tends to slow down or decelerate an object. Work is done (and energy expended) to accomplish this action. See Learn by Drawing on p. 140.

The product of two vectors (force and displacement) in this case is a special type of vector multiplication and yields a scalar quantity equal to $(F \cos \theta)d$. Thus, work is a scalar—it does not have direction. It can, however, be positive, zero, or negative, depending on the angle.

The joule (J) (pronounced "jool") was named in honor of James Prescott Joule (1818–1889), a British scientist who investigated work and energy.

Example 5.1 ■ Applied Psychology: Mechanical Work

A student holds her psychology textbook, which has a mass of 1.5 kg, out of a second-story dormitory window until her arm is tired; then she releases it (•Fig. 5.2). (a) How much work is done on the book by the student in simply holding it out the window? (b) How much work will have been done by the force of gravity during the time in which the book falls 3.0 m?

Thinking It Through. Analyze the situations in terms of the definition of work, keeping in mind that force and displacement are the key factors.

Solution. Listing the data, we have

Given: $v_0 = 0$ (initially at rest)   *Find:* (a) W (work in holding)
       $m = 1.5$ kg   (b) W (work in falling)
       $d = 3.0$ m

(a) Even though the student gets tired (because work is performed within the body to maintain muscles in a state of tension), she does *no* mechanical work *on the book* in merely holding it stationary. She exerts an upward force on the book (equal in magnitude to its weight), but the displacement is zero in this case ($d = 0$). Thus, $W = Fd = F \times 0 = 0$ J.

(b) While the book is falling, the net force acting on it is the force of gravity, which is equal in magnitude to the weight of the book: $F = w = mg$ (neglecting air resistance).

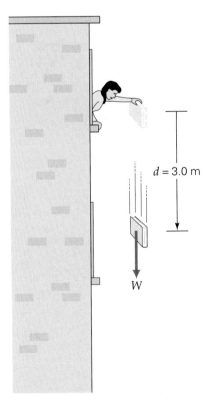

•**FIGURE 5.2 Mechanical work requires motion** See Example 5.1.

The displacement is in the same direction as the force ($\theta = 0°$) and has a magnitude of $d = 3.0$ m, so the work done by gravity is

$$W = Fd \cos 0° = (mg)d = (1.5 \text{ kg})(9.8 \text{ m/s}^2)(3.0 \text{ m}) = +44 \text{ J}$$

Follow-up Exercise. How much work is done on a 0.20-kilogram ball by gravity as the ball rises between heights of 2.0 m and 3.0 m?

LEARN BY DRAWING

Determining the Sign of Work

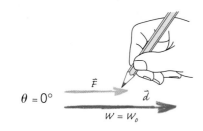

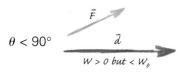

$\theta = 0°$ $\vec{F}$ $\vec{d}$
$W = W_o$

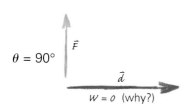

$\theta < 90°$ $\vec{F}$ $\vec{d}$
$W > 0$ but $< W_o$

$\theta = 90°$ $\vec{F}$ $\vec{d}$
$W = 0$ (why?)

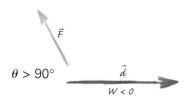

$\theta > 90°$ $\vec{F}$ $\vec{d}$
$W < 0$

$\theta = 180°$ $\vec{F}$ $\vec{d}$
$W = -W_o$

Example 5.2 ■ Yard Work: Parallel Component of a Force

If the person in Fig. 5.1c pushes on the lawn mower with a constant force of 90.0 N at an angle of 40° to the horizontal, how much work does she do in pushing it a horizontal distance of 7.50 m?

Thinking It Through. Only the component of the force parallel to the displacement does work on the mower. Clearly identify the angle between the force and the displacement.

Solution.

Given: $F = 90.0$ N   *Find:* W (work done by person)
$\theta = 40°$
$d = 7.50$ m

Here, the horizontal component of the applied force, $F \cos \theta$, is parallel to the displacement, so Eq. 5.2 applies:

$$W = (F \cos 40°)d = (90.0 \text{ N})(0.766)(7.50 \text{ m}) = +517 \text{ J}$$

Work is done by the horizontal component of the force, but the vertical component does no work because there is no vertical displacement.

Follow-up Exercise. When you wheel a wheelbarrow on a level surface, the force you apply has an upward component. Is work done by this component?

We commonly specify which force is doing work *on* which object. For example, the force of gravity does work on a falling object, such as the book in Example 5.1. Also, when you lift an object, *you* do work *on* the object. We sometimes describe this as doing work *against* gravity, because the force of gravity acts in the direction opposite that of the applied lift force and opposes it.

In both Examples 5.1 and 5.2, work was done by a single constant force. If more than one force acts on an object, the work done by each can be calculated separately. The *total* or *net work* is then the work done by all the forces, or the scalar sum of those quantities of work, as shown in the following Example.

Example 5.3 ■ Total or Net Work

A 0.75-kilogram block slides with a uniform velocity down a 20° inclined plane (•Fig. 5.3). (a) How much work is done by the force of friction on the block as it slides the total length of the plane? (b) What is the net work done on the block? (c) Discuss the net work done if the angle of incline is adjusted so that the block accelerates down the plane.

Thinking It Through. (a) The length of the plane can be found by using trigonometry, so this part boils down to finding the force of friction. (b) The net work is the sum of all the work done by the individual forces. (c) If there is an acceleration, Newton's second law applies, and this involves a net force.

Solution. We list what is given, but equally importantly we want to know specifically what is to be found.

Given: $m = 0.75$ kg   *Find:* (a) W_f (work done by friction)
$\theta = 20°$ (b) W_{net} (net work)
$L = 1.2$ m (from figure)   (c) W (discuss work with block accelerating)

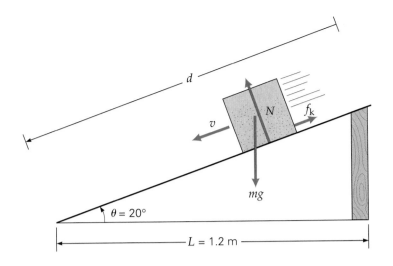

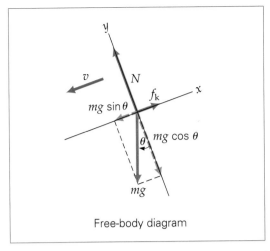

Free-body diagram

•FIGURE 5.3 Total or net work
See Example 5.3.

(a)   Note from Fig. 5.3 that only two forces do work because there are only two forces parallel to the motion: f_k, the force of kinetic friction; and $mg \sin\theta$, the component of the block's weight acting down the plane. The normal force and $mg \cos\theta$, the components of the block's weight acting perpendicular to the plane, do no work on the block. (Why?) We first find the work done by the frictional force:

$$W_f = f_k d \cos 180° = -f_k d = -\mu_k N d$$

where $f_k = \mu_k N$. The angle 180° indicates that the force and displacement are in opposite directions. (It is common in such cases to write $W_f = -f_k d$ directly, since friction typically opposes motion.) The magnitude of the normal force N is equal to that of the component of the block's weight perpendicular to the surface (these are equal and opposite forces. Why?):

$$N = mg \cos\theta$$

The distance d the block slides down the plane can be found by using trigonometry. Note that $\cos\theta = L/d$, so

$$d = \frac{L}{\cos\theta}$$

But what is μ_k? It is not given (nor are the types of surfaces specified). This is an example in which a key piece of data is given to you indirectly in a problem. Note that the block slides at a uniform (constant) velocity down the plane. In Chapter 4, you learned that for uniform motion ($a = 0$) down a plane, $\mu_k = \tan\theta$. Thus,

$$W_f = -\mu_k N d = -(\tan\theta)(mg \cos\theta)\left(\frac{L}{\cos\theta}\right) = -mgL \tan 20°$$

$$= -(0.75 \text{ kg})(9.8 \text{ m/s}^2)(1.2 \text{ m})(0.364) = -3.2 \text{ J}$$

(b)   To find the net work, we need to calculate the work done by gravity and then add it to our result in part (a). Since $F_\parallel$ for gravity is just $mg \sin\theta$, we have

$$W_g = F_\parallel d = (mg \sin\theta)\left(\frac{L}{\cos\theta}\right) = mgL \tan 20° = +3.2 \text{ J}$$

where the calculation is the same as in (a) except for the sign. Then,

$$W_{net} = W_g + W_f = +3.2 \text{ J} + (-3.2 \text{ J}) = 0$$

Remember that work is a scalar quantity, so scalar addition is used to find net work.

(c)   If the block accelerates down the plane, then from Newton's second law, $F = mg \sin\theta - f_k = ma$. The component of the gravitational force is greater than the opposing

Note: Recall the discussion of friction in Section 4.6.

frictional force, so there is net work done on the block because now $|W_g| > |W_f|$. You might be wondering what the effect of nonzero net work is. As you will learn shortly, it causes a change in the amount of energy an object has.

Follow-up Exercise. In part (c) of this Example, is it possible for the frictional work to be greater in magnitude than the gravitational work? What would this mean in terms of the block's speed?

Problem-Solving Hint

Note how in part (a) of Example 5.3, the equation for W_f was simplified by using the algebraic expressions for N and d instead of computing these quantities initially. It is a good rule of thumb not to plug numbers into an equation until you have to. Simplifying an equation through cancellation is easier with symbols and saves computation time.

5.2 Work Done by a Variable Force

Objectives: To (a) differentiate between work done by constant and variable forces and (b) compute the work done by a spring force.

The discussion in the preceding section was limited to work done by constant forces. In general, however, forces are variable; that is, they change with time and/or position. For example, someone pushes harder and harder on an object to overcome the force of static friction, until the applied force exceeds $f_{s_{max}}$. However, the force of static friction does no work because there is no motion or displacement.

An example of a variable force doing work in stretching a spring is illustrated in •Fig. 5.4. As a spring is stretched (or compressed) farther and farther, the restoring force of the spring (the force that opposes the stretching) gets greater and an increasing applied force is required. For most springs, the applied force F is di-

In Fig 5.4, the hand applies a variable force F in stretching the spring. At the same time, the spring exerts an equal and opposite force, F_s, on the hand.

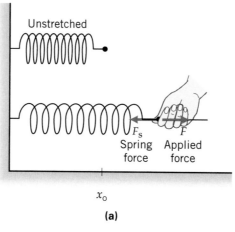

Unstretched

F_s F
Spring   Applied
force   force

x_o

(a)

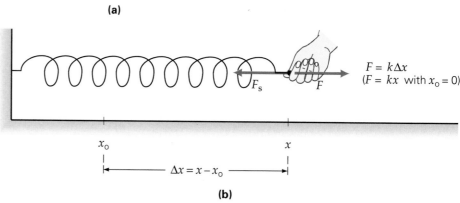

$F = k\Delta x$
($F = kx$ with $x_o = 0$)

F_s F

x_o x

$\Delta x = x - x_o$

(b)

•**FIGURE 5.4 Spring force** (a) An applied force F stretches the spring, and the spring exerts an equal and opposite force F_s on the hand. **(b)** The magnitude of the force depends on the change in the spring's length. This is often referenced to the position of a mass on the end of the spring.

rectly proportional to the change in length of the spring from its unstretched length. In equation form, this relationship is expressed as

$$F = k \Delta x = k(x - x_o)$$

or, if we chose $x_o = 0$,

$$F = kx$$

where x now represents the distance the spring is stretched (or compressed) from its unstretched length. As can be seen, the force varies with x. We describe this by saying that the *force is a function of position.*

The k in this equation is a constant of proportionality and is commonly called the **spring constant,** or **force constant.** The greater the value of k, the stiffer or stronger is the spring. As you should be able to prove to yourself, the SI units of k are newtons per meter (N/m).

k, the spring constant or force constant

The relationship expressed by the equation $F = kx$ holds only for ideal springs. Real springs approximate this linear relationship between force and displacement within certain limits. For example, if a spring is stretched beyond a certain point, called its *elastic limit*, the spring will be permanently deformed and $F = kx$ will no longer apply.

Note that a spring exerts a force that is equal and opposite to the external applied force F. Thus

$$F_s = -k \Delta x = -k(x - x_o)$$

or, with $x_o = 0$,

$$F_s = -kx \qquad \textit{ideal spring force} \qquad (5.3)$$

Hooke's law

The minus sign indicates that this spring force acts in the direction opposite to the displacement whether the spring is stretched or compressed. This equation is a form of what is known as *Hooke's law*, named after Robert Hooke, a contemporary of Newton.

To compute the work done by variable forces generally requires calculus. But we are fortunate that the spring force can be computed by using the average force. An analogous case involving average velocity, $\bar{v} = (v + v_0)/2$, was used in the study of kinematics in Chapter 2, and it is instructive to explore this analogy graphically.

Plots of F versus x and v versus t are shown in •Fig. 5.5. The graphs have straight-line slopes of k and a, respectively, with $F = kx$ and $v = at$. [The applied force F, rather than the spring force F_s, is plotted to avoid a negative slope ($F_s = -kx$) and thereby allow analogous graphs.]

By the same reasoning given for v in Section 2.3, the average force $\bar{F}$ can be expressed as

$$\bar{F} = \frac{F + F_o}{2}$$

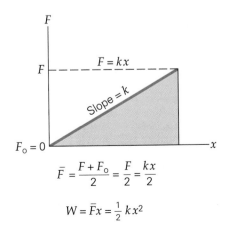

$$\bar{F} = \frac{F + F_o}{2} = \frac{F}{2} = \frac{kx}{2}$$

$$W = \bar{F}x = \frac{1}{2}kx^2$$

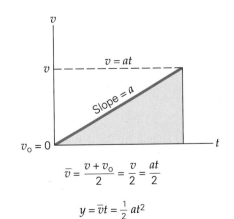

$$\bar{v} = \frac{v + v_0}{2} = \frac{v}{2} = \frac{at}{2}$$

$$y = \bar{v}t = \frac{1}{2}at^2$$

•FIGURE 5.5 **Work done by a uniformly variable force** The work done by a uniformly varying force of the form $F = kx$ is $W = \frac{1}{2}kx^2$. A plot of this special case of a variable force is graphically analogous to a plot of v versus t for a uniformly varying velocity starting at rest ($v_o = 0$): $v = at$. The work (W) and the distance (y) are equal to the areas under the respective straight lines.

5.2 Work Done by a Variable Force **143**

or, since $F_o = 0$ (why?),

$$\overline{F} = \frac{F}{2}$$

Thus, the work done in stretching or compressing the spring an amount x from its unstretched length is

$$W = \overline{F}x = \frac{Fx}{2}$$

Since $F = kx$, the work done is

$$W = \tfrac{1}{2}kx^2 \qquad \begin{array}{l} \textit{work done in stretching} \\ \textit{(or compressing)} \\ \textit{a spring from } x_o = 0 \end{array} \qquad (5.4)$$

Note: We have tried to show the physical basis of Eq. 5.4.

Note that the work done is just the area under the curve in Fig. 5.5.

Example 5.4 ■ Determining the Spring Constant

A 0.15-kilogram mass is suspended from a vertical spring and descends a distance of 4.6 cm, after which it hangs at rest (•Fig. 5.6). An additional 0.50-kilogram mass is then suspended from the first. What is the total extension of the spring? (Neglect the mass of the spring.)

Thinking It Through.   The spring constant k appears in Eq. 5.3. Therefore to find the value of k for a particular instance, the spring force and distance the spring is stretched (or compressed) must be known.

Solution.   The data given are as follows:

Given: $m_1 = 0.15$ kg   *Find:* x (total stretch length)
 $x_1 = 4.6$ cm $= 0.046$ m
 $m_2 = 0.50$ kg

The total stretch distance is given by $F = kx$, where F is the applied force, which in this case is the weight of the mass suspended on the spring. However, the spring constant, k, is not given. This may be found from the data pertaining to the suspension of m_1 and resulting displacement x_1. (This is a common method of determining spring constants.) As seen in Fig. 5.6a, the magnitudes of the weight force and the restoring spring force are equal since $a = 0$, so we may write (with directional minus signs neglected)

$$F_s = m_1 g = kx_1$$

and

$$k = \frac{m_1 g}{x_1} = \frac{(0.15 \text{ kg})(9.8 \text{ m/s}^2)}{0.046 \text{ m}} = 32 \text{ N/m}$$

Then, knowing k, we find the total extension of the spring from the balanced force situation shown in Fig. 5.6b:

$$F = (m_1 + m_2)g = kx$$

Thus,

$$x = \frac{(m_1 + m_2)g}{k} = \frac{(0.15 \text{ kg} + 0.50 \text{ kg})(9.8 \text{ m/s}^2)}{32 \text{ N/m}} = 0.20 \text{ m (or 20 cm)}$$

Follow-up Exercise.   How much work is done by gravity in stretching the spring through both displacements in Example 5.4?

Problem-Solving Hint

The reference position x_o for the change in length of a spring is arbitrary and is usually chosen for convenience. *The important quantity in computing work is the position difference Δx, or the net change in the length of the spring.* As shown in •Fig. 5.7 for a mass suspended on a spring, x_o can be referenced to the unloaded length of the spring

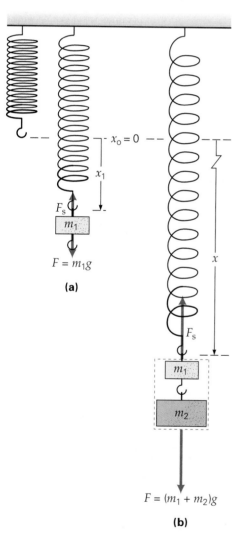

•**FIGURE 5.6 Determining the spring constant and the work done in stretching a spring**
See Example 5.4.

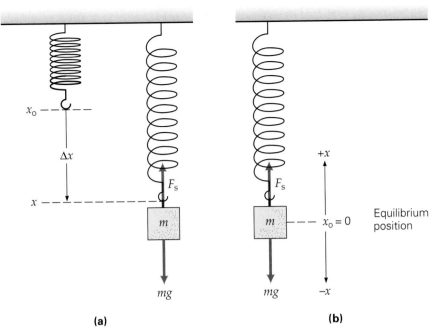

•FIGURE 5.7 Displacement reference
The reference position x_o is arbitrary and is usually chosen for convenience. It may be **(a)** at the end of the spring at its unloaded position or **(b)** at the equilibrium position when a mass is suspended on the spring. The latter is particularly convenient in cases in which the mass oscillates up and down on the spring.

(a) (b)

or to the loaded position, which may be taken as the zero position for convenience. In Example 5.4, x_o was referenced to the end of the unloaded spring. Also, with the displacement only in one direction, downward was designated as the $+x$ direction to avoid minus signs.

When the net force on the suspended mass is zero, the mass is said to be at its equilibrium position (as in Fig. 5.7a with m_1 suspended). This position, rather than the unloaded length, may be taken as a zero reference ($x = x_o = 0$, Fig. 5.7b). The equilibrium position is a convenient reference point for a case in which the mass oscillates up and down on the spring. (We will describe this motion in Chapter 13.) Notice that in general there are both positive and negative directions. The chosen signs are arbitrary but must be adhered to during the whole problem. Also, since the displacement is in the vertical direction, the x's are often replaced by y's.

5.3 The Work–Energy Theorem: Kinetic Energy

Objectives: To (a) explain the work–energy theorem and (b) apply it in solving
problems.

Now that we have an operational definition of work, we are ready to look at how work is related to energy. Energy is one of the most important concepts in science. We describe it as a quantity that objects or systems possess. Basically, work is something that is *done on* objects, whereas energy is something that objects *have*.

One form of energy that is closely associated with work is *kinetic energy*. (Another form of energy, *potential energy*, will be described in Section 5.4.) Consider an object at rest on a frictionless surface. Let a horizontal force act on the object and set it in motion. Work is done *on* the object, but where does the work "go," so to speak? It goes into setting the object into motion or changing its *kinetic* conditions. Because of its motion, we say the object has energy—kinetic energy, which gives it the capability to do work.

For a constant net force doing work on a moving object, as illustrated in •Fig. 5.8, the force does an amount of work, $W = Fx$. But what are the kinematic effects? The force causes the object to accelerate, and from Equation 2.10, $v^2 = v_0^2 + 2ax$,

$$a = \frac{v^2 - v_0^2}{2x}$$

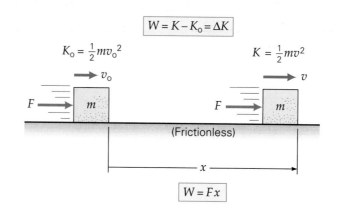

•FIGURE 5.8 The relationship of work and kinetic energy The work done on a block in moving along a horizontal frictionless surface is equal to the change in its kinetic energy, $W = \Delta K$.

where v_o may or may not be zero and Δx can be written as x if $x_o = 0$. Writing the force in its Newton's second law form ($F = ma$) and then substituting the expression for a from the previous equation gives

$$F = ma = m\left(\frac{v^2 - v_o{}^2}{2x}\right)$$

Using this expression in the equation for work, we get

$$W = Fx = m\left(\frac{v^2 - v_o{}^2}{2x}\right)x$$

$$= \tfrac{1}{2}mv^2 - \tfrac{1}{2}mv_o{}^2$$

It is convenient to define $\tfrac{1}{2}mv^2$ as the **kinetic energy** (K).

Kinetic energy—the energy of motion

$$K = \tfrac{1}{2}mv^2 \qquad kinetic\ energy \qquad (5.5)$$

SI unit of energy: joule (J)

Kinetic energy is often called the energy of motion. Note that it is directly proportional to the square of the (instantaneous) speed of a moving object.

Then, in terms of kinetic energy, the previous work expression can be written as follows:

$$W = \tfrac{1}{2}mv^2 - \tfrac{1}{2}mv_o{}^2 = K - K_o = \Delta K$$

or

Work–energy theorem

$$W = \Delta K \qquad (5.6)$$

where it is understood that W is the net work if more than one force acts on the object, as shown in Example 5.3. This equation is called the **work–energy theorem**; it relates the work done on an object to the change in its kinetic energy. That is, *the net work done on a body by an external net force is equal to the change in kinetic energy of the body*. Both work and energy have the units of joules, and both are *scalar* quantities.

Keep in mind that the work–energy theorem is true in general for varying forces and not just for the special case considered in deriving Eq. 5.6. As a concrete example, recall that in Example 5.1 the force of gravity did +44 J of work on a book that fell from rest through a distance of $y = 3.0$ m. At that position and instant, the falling book had 44 J of kinetic energy. Since $v_o = 0$ in this case, $v^2 = 2gy$, and $gy = v^2/2$. Substituting this expression into the equation for the work done on the falling book by gravity, we get

$$W = Fd = mgy = \frac{mv^2}{2} = K$$

where $K_o = 0$. Thus the kinetic energy gained by the book is equal to the net work done on it: 44 J in this instance. (As an exercise, confirm this fact by calculating the speed of the book and evaluating its kinetic energy.)

What the work–energy theorem tells us is that when work is done, there is change in or a transfer of energy. In general then, we might say that *work is a measure of the transfer of kinetic energy.* For example, a force doing work on an object that causes the object to speed up gives rise to an increase in the object's kinetic energy. Conversely, work done by the force of kinetic friction may cause a moving object to slow down and decrease its kinetic energy. So, for a change in or transfer of kinetic energy, there must be a net force and net work, as Eq. 5.6 tells us.

When an object is in motion, it possesses kinetic energy and has the capability to do work. For example, a moving automobile has kinetic energy and can do work in crumpling a fender in a fender-bender—not *useful* work in that case, but still work. Other examples are shown in •Fig. 5.9.

(a)

(b)

•FIGURE 5.9 **Kinetic energy and work** (a) A moving object, such as a wrecking ball, possesses kinetic energy and so can do work. (b) A great deal of kinetic energy was converted to work when a large meteorite made this crater in Arizona, perhaps 10 000 y ago. The crater is 1200 m in diameter and 180 m deep.

Example 5.5 ■ A Game of Shuffleboard: The Work–Energy Theorem

A shuffleboard player (•Fig. 5.10) pushes a 0.25-kilogram puck, initially at rest, such that a constant horizontal force of 6.0 N acts on it through a distance of 0.50 m. (Neglect friction.) (a) What are the kinetic energy and the speed of the puck when the force is removed? (b) How much work would be required to bring the puck to rest?

Thinking It Through. Apply the work–energy theorem. If you can find the work done, you know the change in kinetic energy, and vice versa.

Solution. Listing the given data as usual, we have

Given: $m = 0.25$ kg   *Find:* (a) K (kinetic energy)
$\quad\quad F = 6.0$ N v (speed)
$\quad\quad d = 0.50$ m   (b) W (work done in stopping puck)
$\quad\quad v_o = 0$

(a) Since the speed is not known, we cannot compute the kinetic energy ($K = \frac{1}{2}mv^2$) directly. However, kinetic energy is related to *work* by the work–energy theorem. The work done on the puck is

$$W = Fd = (6.0\ \text{N})(0.50\ \text{m}) = +3.0\ \text{J}$$

Then, by the work–energy theorem,

$$W = \Delta K = K - K_o = 3.0\ \text{J}$$

But $K_o = \frac{1}{2}mv_o^2 = 0$ because $v_o = 0$, so

$$K = 3.0\ \text{J}$$

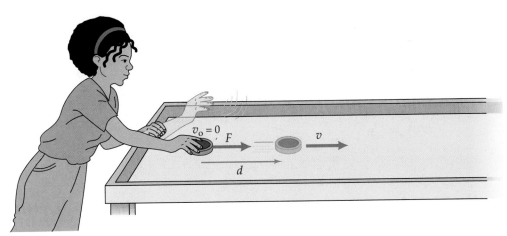

•FIGURE 5.10 **Work and kinetic energy** See Example 5.5.

The speed can be found from the kinetic energy. Since $K = \frac{1}{2}mv^2$, we have

$$v = \sqrt{\frac{2K}{m}} = \sqrt{\frac{2(3.0 \text{ J})}{0.25 \text{ kg}}} = 4.9 \text{ m/s}$$

(b) As you might guess, the work required to bring the puck to rest is equal to its kinetic energy (the amount of energy that must be "removed" from the puck to stop its motion). To confirm this equality, we essentially perform the reverse of the previous calculation, with $v_o = 4.9$ m/s and $v = 0$:

$$W = K - K_o = 0 - K_o = -\frac{1}{2}mv_o^2 = -\frac{1}{2}(0.25 \text{ kg})(4.9 \text{ m/s})^2 = -3.0 \text{ J}$$

The minus sign indicates that the puck loses energy as it slows down. The work is done *against* the motion of the puck; that is, the opposing force is in a direction opposite that of the motion. (In a real situation, the opposing force would be friction.)

Follow-up Exercise. Suppose the puck in this Example had twice the final speed. Would it then take twice as much work to stop the puck?

Problem-Solving Hint

Notice how work–energy considerations were used to find speed in Example 5.5. This could be done in another way. First, the acceleration could be found from $a = F/m$, and then the kinematic equation $v^2 = v_o^2 + 2ax$ could be used to find v (where $x = d$ in the equation).

The point is that many problems can be solved in different ways, and finding the fastest and most efficient way is often the key to success. As our discussion of energy progresses, you will see how useful and powerful the notions of work and energy are, both as theoretical concepts and as practical tools for solving many kinds of problems.

Conceptual Example 5.6

Kinetic Energy: Mass versus Speed

In a football game, a 140-kilogram guard runs at a speed of 4.0 m/s, and a 70-kilogram free safety moves at 8.0 m/s. Then, it is correct to say that (a) both players have the same kinetic energy, (b) the safety has twice as much kinetic energy as the guard, (c) the guard has twice as much kinetic energy as the safety, (d) the safety has four times as much kinetic energy as the guard.

Reasoning and Answer. The kinetic energy of a body depends on both its mass and its speed. You might think that, with half the mass but twice the speed, the safety would have the same kinetic energy as the guard, but this is not the case. Kinetic energy, $K = \frac{1}{2}mv^2$, is directly proportional to the mass but proportional to the *square* of the speed. Thus, halving the mass decreases the kinetic energy by a factor of two; so if the two athletes had equal speeds, the safety would have half as much kinetic energy as the guard.

However, doubling the speed increases the kinetic energy, not by a factor of 2 but by a factor of 2^2, or 4. Thus, the safety, with half the mass but twice the speed, would have $\frac{1}{2} \times 4 = 2$ times as much kinetic energy as the guard, and the answer is (b).

Note that to answer this question it was not necessary to calculate the kinetic energy of each player. We can do so, however, to check our conclusions:

$$K_s = \frac{1}{2}m_s v_s^2 = \frac{1}{2}(70 \text{ kg})(8.0 \text{ m/s})^2 = 2.2 \times 10^3 \text{ J}$$

$$K_g = \frac{1}{2}m_g v_g^2 = \frac{1}{2}(140 \text{ kg})(4.0 \text{ m/s})^2 = 1.1 \times 10^3 \text{ J}$$

Thus we see explicitly that our answer was correct.

Follow-up Exercise. Suppose that the safety's speed were only 50 percent greater than the guard's, or 6.0 m/s. Which athlete would then have the greater kinetic energy, and how many times greater?

Note that the work–energy theorem relates the work done to the *change* in the kinetic energy. Often we have $v_o = 0$ and $K_o = 0$, so $W = \Delta K = K$. But take care! You *cannot* simply use the square of the change in speed $(\Delta v)^2$ to calculate ΔK, as you might at first think. In terms of speed, we have

$$W = \Delta K = K - K_o = \tfrac{1}{2}mv^2 - \tfrac{1}{2}mv_o{}^2 = \tfrac{1}{2}m(v^2 - v_o{}^2)$$

But $v^2 - v_o{}^2$ is *not* the same as $(v - v_o)^2 = (\Delta v)^2$, since $(v - v_o)^2 = v^2 - 2vv_o + v_o{}^2$. Hence the work or change in kinetic energy is **not** equal to $\tfrac{1}{2}m(v - v_o)^2 = \tfrac{1}{2}m(\Delta v)^2$.

What this means is that you must compute the kinetic energy of an object at one point or time (using the instantaneous speed to get the instantaneous kinetic energy) and also at another point or time. Then the quantities are subtracted to find the change in kinetic energy, or the work. Alternatively, you can find the difference of the *squares* of the speeds $(v^2 - v_o{}^2)$ first in computing the change.

Conceptual Example 5.7

An Accelerating Car: Speed and Kinetic Energy

A car traveling at 5.0 m/s speeds up to 10 m/s, with an increase in kinetic energy that requires work W_1. Then, the speed is increased from 10 m/s to 15 m/s, requiring work W_2. How do the amounts of work compare? (a) $W_1 > W_2$, (b) $W_1 = W_2$, (c) $W_2 > W_1$.

Reasoning and Answer. As noted above, the work–energy theorem relates the work done to the *change* in the kinetic energy. Since the speeds have the same increment in each case ($\Delta v = 5.0$ m/s), it might appear that (b) would be the answer. However, keep in mind that the work is equal to the *change* in kinetic energy and involves $(v_2{}^2 - v_1{}^2)$, *not* $(\Delta v)^2 = (v_2 - v_1)^2$.

So, the greater the speed of an object, the greater its kinetic energy, and we would expect the *difference* in kinetic energy in changing speeds (or work required) to be greater for higher speeds for the same Δv. Thus, (c) is the answer.

We can confirm this conclusion by calculating the actual changes in the kinetic energy in each of these cases:

$$W_1 = \tfrac{1}{2}m(v^2 - v_o{}^2) = \tfrac{1}{2}m[(10 \text{ m/s})^2 - (5.0 \text{ m/s})^2] = \tfrac{1}{2}m(75 \text{ m}^2/\text{s}^2)$$

Similarly,

$$W_2 = \tfrac{1}{2}m(v^2 - v_o{}^2) = \tfrac{1}{2}m[(15 \text{ m/s})^2 - (10 \text{ m/s})^2] = \tfrac{1}{2}m(125 \text{ m}^2/\text{s}^2)$$

And since m is the same for each case, we see that $W_2 > W_1$, which confirms our expectations. The main point is that the Δv values are the same, but it requires more work to increase the kinetic energy of an object at higher speeds.

Follow-up Exercise. Suppose the car speeds up a third time, from 15 m/s to 20 m/s, a change requiring work W_3. How does the work done in this increment compare with W_2?

5.4 Potential Energy

Objectives: **To (a) explain how potential energy depends on position and (b) compute values of gravitational potential energy.**

An object in motion has kinetic energy. However, whether an object is in motion or not, it may have another form of energy—potential energy. As the name implies, an object having potential energy has the *potential* to do work. You can probably think of many examples: a compressed spring, a drawn bow, water held back by a dam, a wrecking ball poised to drop. In all such cases, the potential to do work derives from the position or configuration of bodies. The spring has energy because it is compressed, the bow because it is drawn, the water and the ball

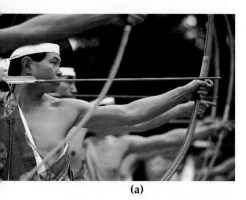

(a)

(b)

•**FIGURE 5.11 Potential energy**
Potential energy has many forms.
(a) Work must be done to bend the
bow, giving it potential energy.
That energy is converted into
kinetic energy when the arrow is
released. **(b)** Gravitational potential
energy is converted into kinetic
energy when an object falls.
(Where did the gravitational
potential energy of the water and
the diver come from?)

because they have been lifted above the surface of the Earth (•Fig. 5.11). Consequently, **potential energy** (U) is often called the energy of position (and/or configuration).

In a sense, potential energy can be thought of as stored work. You have already seen an example of potential energy in Section 5.2 when work was done in compressing a spring from its equilibrium position. Recall that the work done in such a case is $W = \frac{1}{2}kx^2$ (with $x_o = 0$). Note that the amount of work done depends on the amount of compression (x). Because work is done, there is a *change* of position and a *change* in potential energy (ΔU), which is equal to the work done *by the applied force* in compressing (or stretching) the spring:

$$W = \Delta U = U - U_o = \tfrac{1}{2}kx^2 - \tfrac{1}{2}kx_o^2$$

Thus, with $x_o = 0$ and $U_o = 0$, as they are commonly taken to be for convenience, the *potential energy of a spring* is

$$U = \tfrac{1}{2}kx^2 \qquad \begin{array}{l}\textit{potential energy}\\ \textit{of a spring}\end{array} \qquad (5.7)$$

SI unit of energy: joule (J)

[Since the potential energy has x^2 as one term, the previous problem-solving hint applies when $x_o \neq 0$; that is, $x^2 - x_o^2 \neq (x - x_o)^2$.]

Perhaps the most common type of potential energy is **gravitational potential energy**. In this case, position refers to the height of an object above some reference point, such as the floor or the ground. Suppose that an object of mass m is lifted a distance Δh (•Fig. 5.12). Work is done against the force of gravity, and an applied force at least equal to the object's weight is necessary to lift it: $F = w = mg$. The work done in lifting is then equal to the change in potential energy. Expressed in equation form, since there is no overall change in kinetic energy, we have

$$\text{work} = \text{change in potential energy}$$

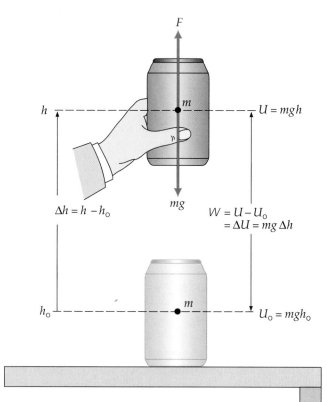

•**FIGURE 5.12 Gravitational potential energy**
The work done in lifting an object is equal to the
change in gravitational potential energy,
$W = F \Delta h = mg(h - h_o)$.

or

$$W = F\,\Delta h = \Delta U = U - U_o = mgh - mgh_o$$

And, with the common choices of $h_o = 0$ and $U_o = 0$, the gravitational potential energy is

$$U = mgh \qquad \textit{gravitational potential energy} \qquad (5.8)$$

SI unit of energy: joule (J)

Because we commonly take upward to be the $+y$ direction, Eq. 5.8 is often written $U = mgy$. (Eq. 5.8 is the gravitational potential energy on or near the Earth's surface where g is considered to be constant. A more general form of gravitational potential energy shall be given in Section 7.5.)

Gravitational potential energy

Example 5.8 ■ A Thrown Ball: Kinetic Energy and Gravitational Potential Energy

A 0.50-kilogram ball is thrown vertically upward with an initial velocity of 10 m/s (•Fig. 5.13). (a) What is the change in the ball's kinetic energy between the starting point and its maximum height? (b) What is the change in the ball's potential energy between its starting point and the maximum height? (Neglect air resistance.)

Thinking It Through. Kinetic energy is lost and gravitational potential energy is gained as the ball travels upward.

Solution. Studying Fig. 5.13 and listing the data, we have

Given: $m = 0.50$ kg $\qquad$ *Find:* (a) ΔK (change in kinetic energy)
$\qquad\quad v_o = 10$ m/s $\qquad\qquad\qquad$ (b) ΔU (the change in potential energy
$\qquad\quad a = g$ $\qquad\qquad\qquad\qquad\qquad$ between y_o and y_{max})

(a) To find the *change* in kinetic energy, we first compute the kinetic energy at each point. We know the initial velocity v_o, and at the maximum height, $v = 0$, so $K = 0$. Thus,

$$\Delta K = K - K_o = 0 - K_o = -\tfrac{1}{2}mv_o^2 = -\tfrac{1}{2}(0.50\ \text{kg})(10\ \text{m/s})^2 = -25\ \text{J}$$

That is, the ball loses 25 J of kinetic energy as negative work is done on it by the force of gravity (force and displacement are in opposite directions).

(b) To find the change in potential energy we need to know the ball's height above its starting point when $v = 0$. Using the equation $v^2 = v_o^2 - 2gy$ to find y_{max},

$$y_{max} = \frac{v_o^2}{2g} = \frac{(10\ \text{m/s})^2}{2(9.8\ \text{m/s}^2)} = 5.1\ \text{m}$$

Then, with $y_o = 0$ and $U_o = 0$, $\Delta U = U - U_o = U - 0$, and

$$\Delta U = U = mgy_{max} = (0.50\ \text{kg})(9.8\ \text{m/s}^2)(5.1\ \text{m}) = +25\ \text{J}$$

The potential energy increases by 25 J, as might be expected. Notice that this is the change in potential energy with respect to the release point, which was taken as the zero reference point ($y_o = 0$).

Follow-up Exercise. In this Example, what are the total changes in the ball's kinetic and potential energies when it returns to the starting point?

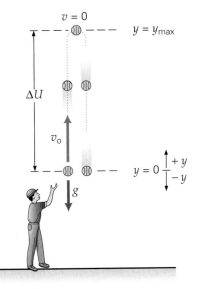

•**FIGURE 5.13 Kinetic and potential energies** See Example 5.8. (The ball is displaced sideways for clarity.)

Zero Reference Point

An important point is illustrated in Example 5.8: namely, the choice of a zero reference point. Potential energy is the energy of *position*, and the potential energy at a particular position (U) is referenced to the potential energy at some other position (U_o). The reference position or point is arbitrary, as is the origin of a set of coordinate axes for analyzing a system. They are usually chosen with convenience in

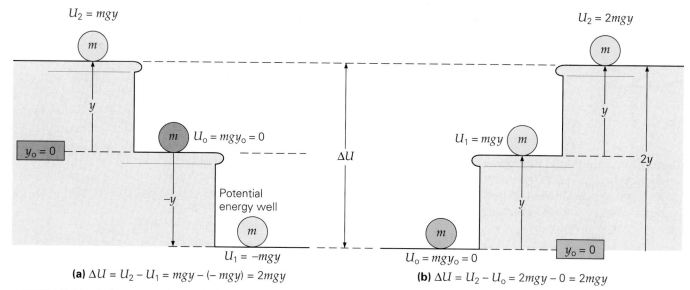

$U_2 = mgy$

$U_0 = mgy_0 = 0$

$y_0 = 0$

$-y$

Potential
energy well

$U_1 = -mgy$

(a) $\Delta U = U_2 - U_1 = mgy - (-mgy) = 2mgy$

$U_2 = 2mgy$

$U_1 = mgy$

$U_0 = mgy_0 = 0$

$y_0 = 0$

(b) $\Delta U = U_2 - U_0 = 2mgy - 0 = 2mgy$

•**FIGURE 5.14   Reference point and change in potential energy** **(a)** The choice of a reference point (zero height) is arbitrary and may give rise to a negative potential energy. An object is said to be in a potential energy well in this case. **(b)** The well may be avoided by selecting a new zero reference. Note that the difference, or *change*, in potential energy (ΔU) associated with the two positions is the same regardless of the reference point. There is no physical difference even though there are two coordinate systems and two different zero reference points.

mind, for example, $y_0 = 0$ or $h_0 = 0$. The value of the potential energy at a particular position depends on the reference point used. However, the *difference or change in potential energy associated with two positions is the same regardless of the reference position.*

In Example 5.8 if ground level had been taken as the zero reference point, then U_0 at the release point would not have been zero. However, U at the maximum height would have been greater and $\Delta U = U - U_0$ would have been the same. This concept is illustrated in •Fig. 5.14. Note that the potential energy can be negative. When an object has a negative potential energy, it is said to be in a potential energy *well*, which is analogous to being in an actual well. Work is needed to raise the object to a higher position in the well or to get it out of the well.

Also, for gravitational potential energy, the path by which an object is raised (or lowered) makes no difference (•Fig. 5.15). That is, *the change in gravitational potential energy is independent of path.* As illustrated in the figure, an object raised to a height h has a change in potential energy of $\Delta U = mgh$ whether it is lifted vertically or moved along an inclined plane. This is because the force of gravity always acts downward and only vertical displacement (or vertical component) is involved in doing work against gravity and changing the potential energy.

•**FIGURE 5.15   Path-independence of the change in gravitational potential energy** When it is resting on the table, the crate has the same potential energy relative to the floor regardless of how it got there. Only the vertical component of the force that moves the crate does work against gravity in lifting it vertically or moving it up an inclined plane or ramp.

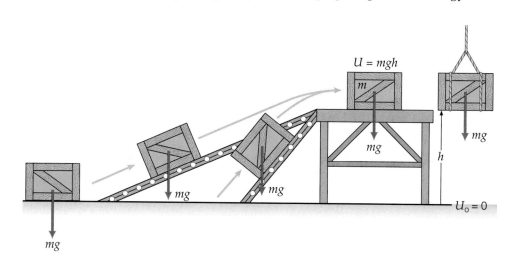

$U = mgh$

mg

h

mg

mg

mg

mg

$U_0 = 0$

5.5 The Conservation of Energy

Objectives: To (a) distinguish between conservative and nonconservative forces and (b) explain their effects on the conservation of energy.

Conservation laws are the cornerstones of physics, both theoretically and practically. Most scientists would probably name the conservation of energy as the most profound and far-reaching of these important laws. When we say something is *conserved*, we mean that it is constant, or has a constant value. Because so many things continually change in physical processes, conserved quantities are extremely helpful in our attempts to understand and describe the universe. Keep in mind, though, that quantities are generally conserved only under special conditions.

One of the most important conservation laws is that concerning conservation of energy. (You may have seen this coming in Example 5.8.) A familiar statement is that the total energy of the universe is conserved. This is true because the whole universe is taken to be a system. A *system* is defined as a definite quantity of matter enclosed by boundaries, either real or imaginary. In effect, the universe is the largest possible closed, or isolated, system we can imagine. Within a *closed system*, particles can interact with each other but have absolutely no interaction with anything outside. In general then, the amount of energy remains constant when no mechanical work is done on or by the system, and no energy is transmitted to or from the system (including thermal energy and radiation).

Thus, the **law of conservation of total energy** may be stated:

> The total energy of an isolated system is always conserved.

Within such a system, energy may be converted from one form to another, but the total amount of all forms of energy is constant or unchanged. Total energy can never be created or destroyed.

Note: A system is a physical situation with real or imaginary boundaries. A classroom might be considered a system, and so might an arbitrary cubic meter of air.

Conservation of total energy

Conservative and Nonconservative Forces

We can make a general distinction among systems by considering two categories of forces that may act within them: conservative and nonconservative forces. You have already been introduced to a couple of conservative forces: the force due to gravity and the spring force. A classic nonconservative force, friction, was considered in Chapter 4. Here's the distinction:

> A force is said to be conservative if the work done by or against it in moving an object is independent of the object's path.

Note: Friction was discussed in Section 4.6.

What this means is that the work done by a **conservative force** depends only on the initial and final positions of an object.

Conservative force—work independent of path

The concept of conservative and nonconservative forces is sometimes difficult to comprehend at first. Because this concept is so important in the conservation of energy, let's consider some illustrative examples to increase our understanding.

First, what does *independent of path* mean? An example was given in Fig. 5.15, where work was done against the conservative force of gravity. The figure illustrates that the work done in moving a crate onto a table does not depend on the *path* but only on the initial and final positions. The magnitude of the work done is equal to the change in potential energy, and in fact, *the concept of potential energy is associated only with conservative forces*. A change in potential energy can be defined in terms of the work done by a conservative force.

Conversely, work done against a **nonconservative force**, such as friction, *does* depend on path. Given the same frictional conditions, it would take more work (against friction) to move the crate up the longer ramp, so the work would depend on path—the longer the path, the more frictional work done. It should therefore be evident that the *total* work done is not equal to a change in potential energy. Some work was done against a nonconservative force, in addition to the conservative gravitational force. In this case, the energy associated with the work done against friction would be converted to heat energy. Hence, in a sense, a conservative force allows you to conserve or store all of the energy as potential energy, whereas a nonconservative force does not.

To further highlight this idea, consider moving the crate around the table top. Here, work is done only against the nonconservative frictional force (why?). Certainly the work done in moving the crate between two points depends on the path taken. It would be greater, for example, if we moved the crate around the table's edges before coming to the destination point than if we took a straight-line path.

Another approach to explain the distinction between conservative and nonconservative forces is through an equivalent statement of the previous definition:

Another way of describing a conservative force

A force is conservative if the work done by or against it in moving an object through a round trip is zero.

Consider a book resting on a table. It has gravitational potential energy $U = mgh$, relative to some reference point $h_o = 0$. You could drop it on the floor, pick it up, and place it back at its original position; or you could pick it up from the table and carry it around with you all day, then place it back at its original position. Both are round trips, and the potential energy of the book is the same when it is returned to its original position. Thus, the change in its potential energy or the work done by the conservative force of gravity is $\Delta U = W = 0$. However, if you pushed the book around on the tabletop and eventually back to its original position, the work done against the nonconservative force of friction would depend on the path (the longer the path, the more work done). The work done would not be stored but lost as heat and sound.

Notice that for the *conservative* gravitational force, the force and displacement are sometimes in the same direction (in which case positive work is done) and sometimes in opposite directions (in which case negative work is done against the force) during a round trip. Think of the simple case of the book falling to the floor and being placed back on the table. With positive and negative work, the total work done by gravity can be zero.

However, for only a *nonconservative* force of kinetic friction, which always opposes the motion or is in the opposite direction to the displacement, the total work done in a round trip can *never* be zero and is always negative (energy lost). But don't get the idea that nonconservative forces only take energy away from a system. On the contrary, we often supply nonconservative pushes and pulls (forces) that add to the energy of a system, such as when you push a stalled car.

Conservation of Total Mechanical Energy

The idea of a conservative force allows us to extend the conservation of energy to the special case of mechanical energy, which greatly helps us better analyze many physical situations. The sum of the kinetic and potential energies is called the **total mechanical energy**:

Mechanical energy—kinetic plus potential

$$E_{\substack{total \\ mechanical \\ energy}} = K_{\substack{kinetic \\ energy}} + U_{\substack{potential \\ energy}} \tag{5.9}$$

For a **conservative system** (one in which only conservative forces do work), the total mechanical energy is constant, or conserved; that is,

$$E = E_o$$

$$K + U = K_o + U_o$$

or

$$\tfrac{1}{2}mv^2 + U = \tfrac{1}{2}mv_o^2 + U_o \tag{5.10}$$

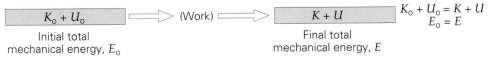

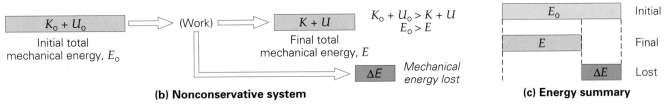

(a) Conservative system

(b) Nonconservative system

(c) Energy summary

•**FIGURE 5.16 Conservative and nonconservative systems** (a) The work done by a conservative force exchanges energy between kinetic and potential forms: that is, $\Delta K + \Delta U = 0$. This means that $K_o + U_o = K + U$, or $E_o = E$, and the total mechanical energy is conserved (none is lost or gained). **(b)** For a nonconservative force, not all the work goes into the mechanical energy exchange, but some energy (ΔE) is generally lost, such as to friction. The mechanical energy is not conserved in this case. **(c)** A graphical energy summary for the nonconservative case. (Keep in mind that a nonconservative force doing work may instead add energy to a system.) How would the energy summary for a conservative system look?

This is a mathematical statement of the **law of the conservation of mechanical energy**:

In a conservative system, the sum of all types of kinetic energy and potential energy is constant and equals the total mechanical energy of the system.

Conservation of mechanical energy

The kinetic and potential energies in a conservative system may change, but their sum is always constant. This is illustrated in •Fig. 5.16a.

Notice for a conservative system, when work is done and energy is transferred, we have

$$K + U = K_o + U_o$$
$$\textit{final energy} \qquad \textit{initial energy}$$

This equation can be rewritten as

$$(K - K_o) + (U - U_o) = 0 \qquad (5.11)$$

or as

$$\Delta K + \Delta U = 0 \quad \textit{(only when nonconservative forces do no work)}$$

This expression tells us that these quantities are related in seesaw fashion: If there is a decrease in potential energy, then the kinetic energy must increase by an equal amount to keep the sum equal to zero. However, in a nonconservative system, some mechanical energy is generally lost (for example, to the heat of friction), and thus $\Delta K + \Delta U < 0$. Such a situation is illustrated in Fig. 5.16b. In terms of the total mechanical energy, $\Delta E = E - E_o$, where ΔE is the amount of energy lost from the system. But, keep in mind, as pointed out above, a nonconservative force may instead add energy to a system.

Examples 5.9 through 5.11 illustrate the conservation of mechanical energy for some conservative systems.

Example 5.9 ■ Look Out Below! Conservation of Mechanical Energy

A painter on a scaffold drops a 1.50-kilogram can of paint from a height of 6.00 m. (a) What is the kinetic energy of the can when it is at a height of 4.00 m? (b) With what speed will the can hit the ground? (Neglect air resistance.)

Thinking It Through. Conservation of energy: The initial total mechanical energy can be found, and potential energy decreases as kinetic energy (and speed) increases.

Solution. Listing what is given and what we are to find, we have

Given: $m = 1.50$ kg   *Find:* (a) K (kinetic energy at h)
$\quad\quad\quad h_o = 6.00$ m $\quad\quad\quad\quad$ (b) v (speed hitting the ground)
$\quad\quad\quad h = 4.00$ m
$\quad\quad\quad v_o = 0$

(a) First, it is convenient to find the can's total mechanical energy, since this quantity is conserved while it is falling (why?). Initially, with $v_o = 0$, the can's total mechanical energy is all potential. Taking the ground as the zero reference point, we have

$$E = K + U = 0 + mgh_o = (1.50 \text{ kg})(9.80 \text{ m/s}^2)(6.00 \text{ m}) = 88.2 \text{ J}$$

The relation $E = K + U$ continues to hold while the can is falling, but now we know what E is. Rearranging the equation, we have $K = E - U$ and can find U at $h = 4.00$ m:

$$K = E - U = E - mgh = 88.2 \text{ J} - (1.50 \text{ kg})(9.80 \text{ m/s}^2)(4.00 \text{ m}) = 29.4 \text{ J}$$

Alternatively, we could have computed the change in (in this case, the loss of) potential energy, ΔU. Whatever potential energy was lost must have been gained as kinetic energy (Eq. 5.11). Then,

$$\Delta K + \Delta U = 0$$

$$(K - K_o) + (U - U_o) = (K - K_o) + (mgh - mgh_o) = 0$$

With $K_o = 0$ (since $v_o = 0$),

$$K = mg(h_o - h) = (1.50 \text{ kg})(9.8 \text{ m/s}^2)(6.00 \text{ m} - 4.00 \text{ m}) = 29.4 \text{ J}$$

(b) Just before the can strikes the ground ($h = 0$, $U = 0$), the total mechanical energy is all kinetic, or

$$E = K = \tfrac{1}{2}mv^2$$

Thus,

$$v = \sqrt{\frac{2E}{m}} = \sqrt{\frac{2(88.2 \text{ J})}{1.50 \text{ kg}}} = 10.8 \text{ m/s}$$

Basically, all of the potential energy of a free-falling object released from some height h is converted into kinetic energy just before the object hits the ground, so

$$|\Delta K| = |\Delta U|$$

$$\tfrac{1}{2}mv^2 = mgh$$

or

$$v = \sqrt{2gh}$$

Note that the mass cancels and is not a consideration. This result is also obtained from the kinematic equation $v^2 = v_o^2 + 2gy$ (Eq. 2.10′), with $v_o = 0$ and $y = h$.

Follow-up Exercise. A fellow painter on the ground wishes to toss a paintbrush vertically upward a distance of 5.0 m to his partner on the scaffold. Use conservation of mechanical energy methods to determine the minimum speed that he must give to the brush.

Conceptual Example 5.10

A Matter of Direction? Speed and Conservation of Energy

Three balls of equal mass m are projected with the same speed in different directions as shown in •Fig. 5.17. If air resistance is neglected, which ball would you expect to strike the ground with the greatest speed? (a) ball 1, (b) ball 2, (c) ball 3, (d) all strike with the same speed.

Reasoning and Answer. All of the balls have the same initial kinetic energy, $K_o = \tfrac{1}{2}mv_o^2$. (Recall that energy is a scalar quantity, and the different directions of projections do not produce any difference in the kinetic energies.) Regardless of their trajectories, all of the balls ultimately descend a distance h relative to their common starting point, so they all

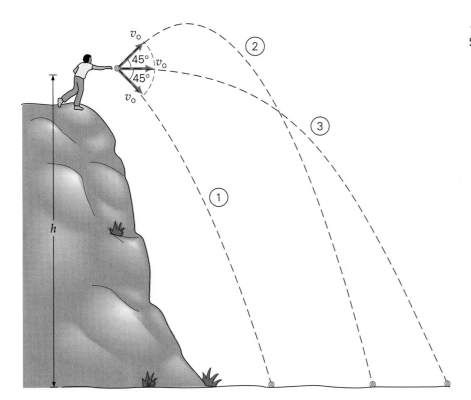

•FIGURE 5.17 **Speed and energy** See Conceptual Example 5.10.

lose the same amount of potential energy. (*U* is energy of *position* and thus is *independent of path*—see Fig. 5.15.)

By the law of conservation of energy, the amount of potential energy each ball loses is equal to the amount of kinetic energy it gains. Since all balls start with the same amount of kinetic energy and gain the same amount of kinetic energy, all three will have equal kinetic energies just before striking the ground. This means that their speeds must be equal, so the answer is (d).

Note that although balls 1 and 2 are projected at 45° angles, this is not a factor. Since the change in potential energy is independent of path, it is independent of the projection angle. The vertical distance between the starting point and the ground is the same (*h*) for projectiles at any angle. (Note: Although the strike speeds are equal, the *time* it takes each ball to reach the ground is different.)

Follow-up Exercise. Would the balls strike the ground with different speeds if their masses were different?

Example 5.11 ■ Conservative Forces: Mechanical Energy of a Spring

A 0.30-kilogram mass sliding on a horizontal frictionless surface with a speed of 2.5 m/s, as depicted in •Fig. 5.18, strikes a light spring that has a spring constant of 3.0×10^3 N/m. (a) What is the total mechanical energy of the system? (b) What is the kinetic energy (K_1) of the mass when the spring is compressed a distance $x_1 = 1.0$ cm? (Assume that no energy is lost in the collision.)

Thinking It Through. (a) All of the total energy is initially kinetic. (b) The total energy is the same as in (a), but it is now divided between kinetic energy and spring potential energy.

Solution.

Given: $m = 0.30$ kg   *Find:* (a) E (total mechanical energy)
$v_o = 2.5$ m/s   (b) K_1 (kinetic energy)
$k = 3.0 \times 10^3$ N/m
$x_1 = 1.0$ cm $= 0.010$ m

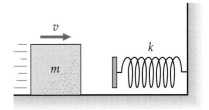

•FIGURE 5.18 **Conservative force and the mechanical energy of a spring** See Example 5.11.

(a) Before the mass makes contact with the spring, the total mechanical energy of the system is all in the form of kinetic energy, and

$$E = K_o = \tfrac{1}{2}mv_o^2 = \tfrac{1}{2}(0.30 \text{ kg})(2.5 \text{ m/s})^2 = 0.94 \text{ J}$$

Since the system is conservative (no energy is lost), this is the total energy at any time.

(b) When the spring is compressed a distance x_1, it has potential energy $U_1 = \tfrac{1}{2}kx_1^2$, and

$$E = K_1 + U_1 = K_1 + \tfrac{1}{2}kx_1^2$$

Solving for K_1, we have

$$K_1 = E - \tfrac{1}{2}kx_1^2$$

$$= 0.94 \text{ J} - \tfrac{1}{2}(3.0 \times 10^3 \text{ N/m})(0.010 \text{ m})^2 = 0.94 \text{ J} - 0.15 \text{ J} = 0.79 \text{ J}$$

Follow-up Exercise. How far will the spring in Example 5.11 be compressed when the mass comes to a stop?

See Learn by Drawing on this page for another example of energy exchange.

LEARN BY DRAWING

Energy Exchanges: A Falling Ball

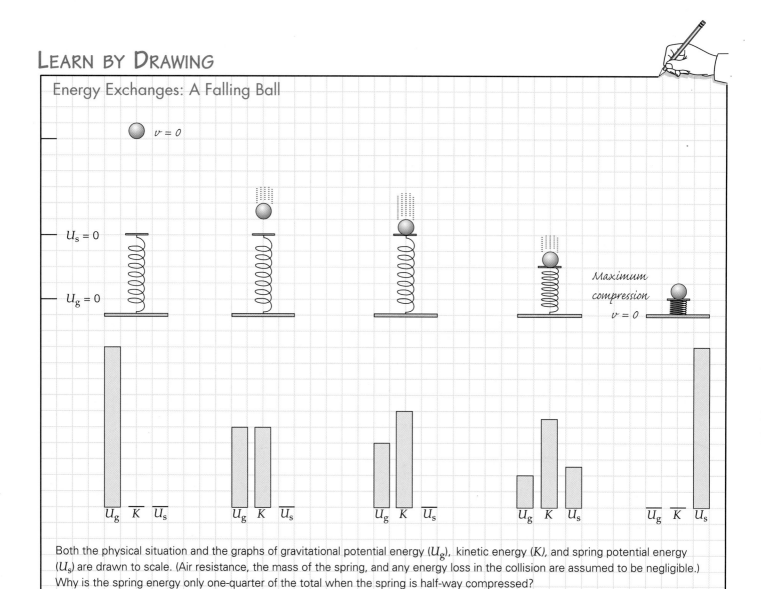

Both the physical situation and the graphs of gravitational potential energy (U_g), kinetic energy (K), and spring potential energy (U_s) are drawn to scale. (Air resistance, the mass of the spring, and any energy loss in the collision are assumed to be negligible.) Why is the spring energy only one-quarter of the total when the spring is half-way compressed?

Total Energy and Nonconservative Forces

In the preceding examples, we ignored the force of friction, which is probably the most common nonconservative force. In general, both conservative and nonconservative forces can do work on objects. However, as you know, when nonconservative forces do work, the mechanical energy is not conserved. Mechanical energy is "lost" through the work done by a nonconservative force such as friction.

You might think that we can no longer use an energy approach to analyze problems involving nonconservative forces, since mechanical energy can be lost or dissipated (•Fig. 5.19). However, in some instances we can use the conservation of total energy to find out how much energy was lost to the work done by a nonconservative force. Suppose an object initially has mechanical energy and that nonconservative forces do an amount of work (W_{nc}) on it, which is equal to the mechanical energy that is lost or not conserved. Starting with the work–energy theorem, we have

$$W = \Delta K = K - K_o$$

In general, the net work (W) may be done by both conservative forces (W_c) and nonconservative forces (W_{nc}), so we may write

$$W_c + W_{nc} = K - K_o \tag{5.12}$$

But recall that the work done by conservative forces is independent of path and depends only on the initial and final positions, as does the change in potential energy. Thus, $W_c = U_o - U$, and Eq. 5.12 then becomes

$$W_{nc} = K - K_o - (U_o - U)$$
$$= (K + U) - (K_o + U_o)$$
$$W_{nc} = E - E_o = \Delta E \tag{5.13}$$

Hence, the work done by the nonconservative forces acting on a system is equal to the change in mechanical energy. Notice that for dissipative forces such as friction, $E_o > E$. Thus the change is negative, indicating a decrease in mechanical energy. The following Example illustrates this concept.

•FIGURE 5.19 **Nonconservative force and energy loss** Friction is a nonconservative force—when it is present, mechanical energy is not necessarily conserved. Can you tell from the photo what is happening to the work being done by the motor on the grinding wheel after it is converted into rotational kinetic energy?

Example 5.12 ■ Downhill Racer: Nonconservative Force

A skier with a mass of 80 kg starts from rest and skis down a slope from an elevation of 110 m (•Fig. 5.20). The speed of the skier at the bottom of the slope is 20 m/s. (a) Show that the system is nonconservative. (b) How much work is done by the nonconservative force of friction?

Thinking It Through.   (a) If the system is nonconservative, then $E_o \neq E$, and these quantities can be computed. (b) We cannot determine the work from force–distance considerations, but W_{nc} is equal to the difference in total energies (Eq. 5.13).

$v = 20$ m/s

110 m

•FIGURE 5.20 **Work done by a nonconservative force** See Example 5.12.

Solution.

Given: $m = 80$ kg
$v_o = 0$
$v = 20$ m/s
$h = 110$ m

Find: (a) Show that E is not constant
(b) W_{nc} (work done by friction)

(a) If the system is conservative, the total mechanical energy is constant. Taking $U_o = 0$ at the bottom of the hill, we find the initial energy at the top of the hill to be

$$E_o = U = mgh = (80 \text{ kg})(9.8 \text{ m/s}^2)(110 \text{ m}) = 8.6 \times 10^4 \text{ J}$$

And at the bottom of the slope,

$$E = K = \tfrac{1}{2}mv^2 = \tfrac{1}{2}(80 \text{ kg})(20 \text{ m/s})^2 = 1.6 \times 10^4 \text{ J}$$

Therefore, $E_o \neq E$, so this system is not conservative.

(b) The amount of work done by the nonconservative force of friction is equal to the change in the mechanical energy, or to the amount of mechanical energy lost (Eq. 5.13):

$$W_{nc} = E - E_o = (1.6 \times 10^4 \text{ J}) - (8.6 \times 10^4 \text{ J}) = -7.0 \times 10^4 \text{ J}$$

This is over 80% of the initial energy. (Where did this energy actually go?)

Note: Free fall is discussed in Section 2.5.

Follow-up Exercise. In free fall, air resistance is neglected; but for skydivers, air resistance has a very practical effect. Typically, a skydiver descends about 450 m before reaching a terminal velocity (Section 4.6) of 60 m/s. (a) What is the percentage energy loss to nonconservative forces during this descent? (b) Show that after terminal velocity is reached, the rate of energy loss is given by $(60mg)$ J/s, where m is the mass of the skydiver.

Note that in a nonconservative system the *total energy* (*not* total mechanical energy) is conserved (including heat and so on), but not all of it is available for mechanical work. For a conservative system, you get back what you put in, so to speak. That is, if you do work on the system, the transferred energy is available to do work. But keep in mind that conservative systems are idealizations, because all real systems are nonconservative to some degree. However, working with ideal conservative systems gives us an understanding of the conservation of energy.

During this course of study, you will be concerned with other forms of energy, such as thermal, electrical, nuclear, and chemical energies. In general, on the microscopic and submicroscopic levels, these forms of energy can be described in terms of kinetic energy and potential energy. Also, you will learn that mass is a form of energy and that the law of the conservation of energy must take this form into account to be applied to the analysis of nuclear reactions.

5.6 Power

Objectives: To (a) define power and (b) describe mechanical efficiency.

A particular task may require a certain amount of work, but that work might be done over different lengths of time or at different rates. For example, suppose that you have to mow a lawn. This takes a certain amount of work, but you might do the job in a half hour, or you might take an hour or two. There's a practical distinction to be made here. There is usually not only an interest in the amount of work done, but also an interest in how fast it is done—that is, the rate at which it is done. *The rate of doing work* is called **power**.

The average power is the work done divided by the time it takes to do the work, or work per unit of time:

Power: the time rate of doing work

$$\overline{P} = \frac{W}{t - t_o}$$

If t_o is taken to be zero,

$$\overline{P} = \frac{W}{t} \tag{5.14}$$

If the work is done by a constant force (magnitude F) while an object moves through a displacement (magnitude d), then

$$\overline{P} = \frac{W}{t} = \frac{Fd}{t} = F\overline{v} \qquad (5.15)$$

SI unit of power: J/s or watt (W)

where it is assumed that the force is in the direction of the displacement. Here $\overline{v}$ is the magnitude of the average velocity. If the velocity is constant, then $\overline{P} = P = Fv$. If the force and displacement are not in the same direction, then we can write

$$\overline{P} = \frac{Fd\cos\theta}{t} = F\overline{v}\cos\theta \qquad (5.16)$$

where θ is the angle between the force and displacement.

As you can see from Eq. 5.15, the SI unit of power is joules per second (J/s), but this unit is given another name:

$$1\ \text{J/s} \equiv 1\ \text{watt (W)}$$

The watt (W), unit of power

The SI power unit is called a **watt (W)** in honor of James Watt (1736–1819), a Scottish engineer who developed one of the first practical steam engines. A common unit of electrical power is the *kilowatt* (kW).

The British unit of power is foot-pounds per second (ft·lb/s). However, a larger unit, the **horsepower (hp)** is more commonly used:

$$1\ \text{hp} \equiv 550\ \text{ft·lb/s} = 746\ \text{W}$$

In Watt's time, steam engines were replacing horses for work in mines and mills. To characterize the performance of his new engine, which was more efficient than other existing ones, Watt used the average rate at which a horse could do work as a unit—a horsepower.

Power tells you how fast work is being done *or* how fast energy is transferred. For example, motors have power ratings (commonly given in horsepower). A 1-horsepower motor can do a given amount of work in half the time that a $\frac{1}{2}$-horsepower motor would take, or twice the work in the same amount of time. That is, a 1-horsepower motor is twice as "powerful" as a $\frac{1}{2}$-horsepower motor (see Example 5.14).

Example 5.13 ■ A Crane Hoist: Work and Power

A crane hoist like the one shown in •Fig. 5.21 lifts a load of 1.0 metric ton a vertical distance of 25 m in 9.0 s at a constant velocity. How much useful work is done by the hoist each second?

Thinking It Through. The useful work done each (per) second is power output, so this is what is to be found.

Solution.

Given: $m = 1.0$ metric ton   *Find:* W per second (= power)
           $= 1.0 \times 10^3$ kg
        $h = 25$ m
        $t = 9.0$ s

•**FIGURE 5.21 Power delivery** See Example 5.13.

Keep in mind that the work per unit time (work per second) is power, so this is what we need to compute. Since the load moves with a constant velocity, $\overline{P} = P$. (Why?) The work is done against gravity, so $F = mg$, and

$$P = \frac{W}{t} = \frac{Fd}{t} = \frac{mgh}{t}$$

$$= \frac{(1.0 \times 10^3\ \text{kg})(9.8\ \text{m/s}^2)(25\ \text{m})}{9.0\ \text{s}} = 2.7 \times 10^4\ \text{W (or 27 kW)}$$

Thus, since a watt (W) is a joule per second (J/s), the hoist did 2.7×10^4 J of work each second. Note that the velocity has a magnitude of $v = d/t = 25$ m/9.0 s $= 2.8$ m/s, and the power is $P = Fv$.

Follow-up Exercise. If the hoist motor of the crane in this Example is rated at 70 hp, what percentage of this power output goes into useful work?

Example 5.14 ■ Cleaning Up: Work and Time

The motors of two vacuum cleaners have net power outputs of 1.00 hp and 0.500 hp, respectively. (a) How much work in joules can each motor do in 3.00 min? (b) How long does it take for each motor to do 97.0 kJ of work?

Thinking It Through. (a) Since power is work/time ($P = W/t$), the work can be computed. Note that power is given in horsepower units. (b) This is another application of Eq. 5.15.

Solution.

Given: $P_1 = 1.00$ hp $= 746$ W   *Find:* (a) W (work)
$P_2 = 0.500$ hp $= 373$ W   (b) t (time)
$t = 3.00$ min $= 180$ s
$W = 97.0$ kJ $= 97.0 \times 10^3$ J

(a) Since $P = W/t$,

$$W_1 = P_1 t = (746 \text{ W})(180 \text{ s}) = 1.34 \times 10^5 \text{ J}$$

$$W_2 = P_2 t = (373 \text{ W})(180 \text{ s}) = 0.67 \times 10^5 \text{ J}$$

Note that in the same time the smaller motor does half the work the larger one does, as you would expect.

(b) The times are given by $t = W/P$, and for the same amount of work,

$$t_1 = \frac{W}{P_1} = \frac{97.0 \times 10^3 \text{ J}}{746 \text{ W}} = 130 \text{ s}$$

$$t_2 = \frac{W}{P_2} = \frac{97.0 \times 10^3 \text{ J}}{373 \text{ W}} = 260 \text{ s}$$

Note that the smaller motor takes twice as long as the larger one to do the same amount of work.

Follow-up Exercise. (a) A 10-horsepower motor breaks down and is temporarily replaced with a 5-horsepower motor. What can you say about the rate of work output? (b) Suppose the situation were reversed, a 5-horsepower motor replaced with a 10-horsepower motor?

A real-life example of power, energy, and physics is given in the Insight on p. 163.

Efficiency

Machines and motors are quite common, and we often talk about their efficiency. Efficiency involves work, energy, and/or power. Simple and complex machines that do work have mechanical parts that move, so some input energy is always lost because of friction or some other cause (perhaps in the form of sound). Thus, not all of the input energy goes into doing useful work.

Mechanical efficiency is essentially a measure of what you get out for what you put in—that is, the *useful* work output for the energy input. **Efficiency (ε)** is given as a fraction (or percentage):

More Broken Records: The Clap Skate

At the 1998 Winter Olympics in Nagano, Japan, speed skating records were broken and new ones set—largely because of an innovation in ice skates (and applied physics). This new skate technology, invented by Dutch researchers in biomechanics, is commonly called the *clap skate*. The skates are designed to increase the amount of time they are in contact with the ice and therefore also the length of the skater's stride.

The new skates have a spring-loaded hinge on the toe that allows the blade to pull away from the heel (Fig. 1). With a longer stride, the amount of work done by the skater's leg muscle is increased, providing more kinetic energy and speed. Toward the end of the stride, the blade returns to the boot with a "clap" sound when the foot is lifted off the ice—hence the name clap skate. (The skates' inventors called them "slap skates," however, because the skates enables a skater to "slap on" an extra amount of work with each stride.)

Traditional skates require skaters to push from side to side. However, with claps, a new skating technique must be learned, with an emphasis on pushing straight back during the stride to keep the blade on the ice longer. Will the clap skate replace the traditional skate? Probably not completely. On a long track against the clock, the clapping is no more than an annoyance. But on a relatively short curved track, where skaters compete against each other, a surprise move to pass another skater would be well announced—clap, clap, clap—no surprise.

FIGURE 1   Clap, clap, clap   The new clap skate is helping skaters set records. A hinge allows the heel to lift off the blade so that the blade stays on the ice longer, thus providing the skater with more kinetic energy and speed.

$$\varepsilon = \frac{\text{work output}}{\text{energy input}} \ (\times 100\%) = \frac{W_{\text{out}}}{E_{\text{in}}} \ (\times 100\%) \qquad (5.17)$$

Efficiency—What you get out for what you put in

Efficiency is a unitless quantity

For example, if a machine has a 100-joule (energy) input and a 50-joule (work) output, then its efficiency is

$$\varepsilon = \frac{W_{\text{out}}}{E_{\text{in}}} = \frac{50 \text{ J}}{100 \text{ J}} = 0.50 \ (\times 100\%) = 50\%$$

An efficiency of 0.50, or 50%, means that half of the energy input is lost because of friction or some other cause and doesn't serve its intended purpose.

You can also write efficiency in terms of power (*P*):

$$\varepsilon = \frac{P_{\text{out}}}{P_{\text{in}}} \ (\times 100\%) \qquad (5.18)$$

Note that if both terms of the ratio in Eq. 5.17 are divided by time (*t*), $W_{\text{out}}/t = P_{\text{out}}$ and $E_{\text{in}}/t = P_{\text{in}}$.

Example 5.15 ■ Home Improvement: Mechanical Efficiency and Work Output

The motor of an electric drill with an efficiency of 80% has a power input of 600 W. How much useful work is done by the drill in a time of 30 s?

Thinking It Through.   This is an application of Eq. 5.18 and the definition of power.

Solution.

Given: $\varepsilon = 80\% = 0.80$ **Find:** W_{out} (work output)
        $P_{in} = 600$ W
         $t = 30$ s

Given the efficiency and power input, we can readily find the power output (P_{out}) from Eq. 5.18, and this is related to the work output ($P_{out} = W_{out}/t$). First, we rearrange Eq. 5.18:

$$P_{out} = \varepsilon P_{in} = (0.80)(600 \text{ W}) = 4.8 \times 10^2 \text{ W}$$

Then,

$$W_{out} = P_{out}t = (4.8 \times 10^2 \text{ W})(30 \text{ s}) = 1.4 \times 10^4 \text{ J}$$

Follow-up Exercise. (a) Is it possible to have a mechanical efficiency of 100%? (b) What would an efficiency of greater than 100% imply?

TABLE 5.1 Typical Efficiencies of Some Machines

Machine	Efficiency (%)
Compressor	85
Electric motor	70–95
Automobile	<15
Human muscle*	20–25
Steam locomotive	5–10

*Technically not a machine, but used to perform work.

The typical efficiencies of some machines are given in Table 5.1. You may be surprised by the relatively low efficiency of the automobile. Much of the energy input (from gasoline combustion) is lost as exhaust heat and through the cooling system (more than 60%), and friction accounts for a great deal more. Less than 15% of the input energy is converted to useful work that goes into propelling the vehicle. Air conditioning, power steering, radio, and tape and CD players are nice, but they use energy and also contribute to the car's decrease in efficiency.

Chapter Review

Important Terms

work *138*
joule (unit) *139*
spring constant (force constant) *143*
kinetic energy *146*
work–energy theorem *146*
potential energy *150*

gravitational potential energy *150*
law of conservation of total energy *153*
conservative force *153*
nonconservative force *153*
total mechanical energy *154*
conservative system *154*

law of conservation of mechanical energy *155*
power *160*
watt (unit) *161*
horsepower (unit) *161*
efficiency *162*

Important Concepts

- Work done by a constant force is the product of the displacement and the component of the force parallel to the displacement.
- Work done by a variable force requires advanced mathematics. A simple example for which we can use average values is the work done by a spring force.
- Kinetic energy is the energy of motion. By the work–energy theorem, the net work done on an object (by a net external force) is equal to the change in the kinetic energy of the object.
- Potential energy is the energy of position and/or configuration. Most common is gravitational poten-

tial energy, associated with the Earth's gravitational attraction.
- Conservation of energy: The total energy of the universe or of an isolated system is always conserved. The total mechanical energy (kinetic plus potential) is constant in a conservative system.
- In systems with nonconservative forces, total mechanical energy is changed. If friction is the nonconservative force, then total mechanical energy is lost.
- Power is the time rate of doing work (or expending energy).
- Efficiency relates work output to energy (work) input.

Important Equations

Work:
$$W = (F \cos \theta)d \qquad (5.2)$$

Hooke's Law (spring force):
$$F_s = -kx \qquad (5.3)$$

Work Done in Stretching or Compressing a Spring (from Equilibrium):
$$W = \tfrac{1}{2}kx^2 \qquad (5.4)$$

Kinetic Energy:
$$K = \tfrac{1}{2}mv^2 \qquad (5.5)$$

Work–Energy Theorem:
$$W = K - K_o = \Delta K \qquad (5.6)$$

Elastic (Spring) Potential Energy:
$$U = \tfrac{1}{2}kx^2 \quad \text{(with } x_o = 0\text{)} \qquad (5.7)$$

Gravitational Potential Energy:
$$U = mgh \quad \text{(with } h_o = 0\text{)} \qquad (5.8)$$

Conservation of Mechanical Energy:
$$\tfrac{1}{2}mv^2 + U = \tfrac{1}{2}mv_o{}^2 + U_o \qquad (5.10)$$

Conservation of Energy (with a Nonconservative Force):
$$W_{nc} = E - E_o = \Delta E \qquad (5.13)$$

Average Power:
$$\overline{P} = \frac{W}{t} = \frac{Fd}{t} = F\overline{v} \qquad (5.15)$$

(constant force in direction of d and v)
$$\overline{P} = \frac{Fd \cos \theta}{t} \qquad (5.16)$$

(force acts at an angle θ between d and v)

Efficiency (percent):
$$\varepsilon = \frac{W_{out}}{E_{in}} (\times 100\%) \qquad (5.17)$$
$$= \frac{P_{out}}{P_{in}} (\times 100\%) \qquad (5.18)$$

Exercises

5.1 Work Done by a Constant Force

1. The units of work are (a) N·m, (b) kg·m²/s², (c) J, (d) all of the preceding.

2. If you push against a wall, are you doing any work? Explain.

3. (a) As a weightlifter strains preparing to lift a barbell (•Fig. 5.22a), is he doing work? Why or why not? (b) In raising the barbell above his head, is he doing work? Explain. (c) In holding the barbell above his head (Fig. 5.22b), is he doing more work, no work, or the same amount of work as in lifting the barbell? Explain. (d) If the weightlifter drops the barbell, is work done on the barbell? Explain what happens in this situation.

| (a) | (b) |

•**FIGURE 5.22** **Man at work?** See Exercise 3.

4. You are carrying a backpack across the campus. What is the work done by your carrying force on the backpack? Explain.

5. A jet plane flies in a vertical circular loop. In what regions of the loop is the work done by the plane's weight positive and/or negative? Is the work constant? If not, are there maximum and minimum instantaneous values? Explain.

6. ■ 🔵 A person does 50 J of work in moving a 30-kilogram box over a 10-meter distance on a horizontal surface. What is the minimum force required?

7. ■ A 5.0-kilogram box slides a 10-meter distance on ice. If the coefficient of kinetic friction if 0.20, what is the work done by the friction force?

8. ■ You carry a 15-kilogram bucket at constant speed from the ground floor to the third floor in the Science building. If the third floor is 10 m above the ground, how much work do you do on the bucket?

9. ■ 🔵 A boy pushing a lawnmower on a level lawn applies a constant force of 250 N at an angle of 30° downward from the horizontal. How far does the boy push the mower in doing 1.44×10^3 J of work?

10. ■■ A 3.00-kilogram block slides down a frictionless plane inclined 20° to the horizontal. If the length of the plane's

surface is 1.50 m, how much work is done, and by what force?

11. ■■ Suppose the coefficient of kinetic friction between the block and the plane in Exercise 10 is 0.275. What would be the net work done in this case?

12. ■■ The formula for work can be written $W = F_\parallel d$, where $F_\parallel$ is the component of the force parallel to the displacement. Show that work is also given by $Fd_\parallel$, where $d_\parallel$ is the component of the displacement parallel to the force.

13. ■■ A hot-air balloon with a mass of 500 kg ascends at a constant rate of 1.50 m/s for 20.0 s. How much work is done by the upward buoyant force? (Neglect air resistance.)

14. ■■ A father pulls his young daughter on a sled with a constant velocity on a level surface through a distance of 10 m, as illustrated in •Fig. 5.23a. If the total mass of the sled and the girl is 35 kg and the coefficient of kinetic friction between the sled runners and the snow is 0.20, how much work does the father do?

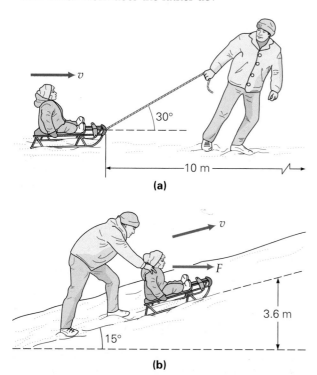

(a)

(b)

•FIGURE 5.23 **Fun and work** See Exercises 14 and 15.

15. ■■ A father pushes horizontally on his daughter's sled to move it up a snowy incline (Fig. 5.23b). If the sled moves with a constant velocity up the hill, how much work does he do in moving it from the bottom to the top of the hill? (Some necessary data are given in Exercise 14.)

16. ■■ A boy pulls a 20-kilogram box with a 50-newton force at 37° above a horizontal surface. If the coefficient of kinetic friction between the box and the horizontal surface is 0.15 and the box is pulled a distance of 25 m, what is (a) the work done by the boy? (b) the work done by the friction force? (c) the net work done on the box?

17. ■■■ A 500-kilogram helicopter ascends from the ground with an acceleration of 2.00 m/s². Over a 5.00-second interval, what is (a) the work done by the lifting force? (b) the work done by the gravitational force? (c) the net work done on the helicopter?

18. ■■■ A 50-kilogram crate slides down a 5.0-meter loading ramp that is inclined at an angle of 25° to the horizontal. A worker pushes on the crate parallel to the ramp surface so that it slides down with a constant velocity. If the coefficient of kinetic friction between the crate and the ramp is 0.33, how much work is done by (a) the worker, (b) the force of friction, and (c) the force of gravity? (d) What is the net work done on the crate?

5.2 Work Done by a Variable Force

19. 💿 The work done by a variable force of the form $F = kx$ is equal to (a) kx^2 (b) kx, (c) the area under the F versus x curve, (d) none of these.

20. If a spring is compressed 2.0 cm from its equilibrium position and then compressed an additional 4.0 cm, how much more work is done in the second compression than in the first? (a) the same amount, (b) twice as much, (c) four times as much, (d) eight times as much.

21. Does it take the same amount of work to stretch the second centimeter as the first centimeter of a spring from its equilibrium position? Explain.

22. ■ To measure the spring constant of a certain spring, a student applies a 4.0-newton force; the spring stretches by 5.0 cm. What is the spring constant?

23. ■ If a 10-newton force is used to compress a spring with a spring constant of 4.0×10^2 N/m, what is the resulting spring compression?

24. ■ A spring has a spring constant of 40 N/m. How much work is required to stretch the spring 2.0 cm from its equilibrium position?

25. ■ If it takes 400 J of work to stretch a spring 8.00 cm, what is the spring constant?

26. ■ A student suspends a 0.25-kilogram mass on a spring, which stretches a distance of 5.0 cm. The student then removes the mass and pulls down the spring, stretching it and doing 10 J of work. How far is the spring stretched as a result of the work done by the student?

27. ■■ When a 75-gram mass is suspended from a vertical spring, the spring is stretched from a length of 4.0 cm to a length of 7.0 cm. If the mass is then pulled downward an additional 10 cm, what is the total work done against the spring force?

28. ■■ A particular spring has a force constant of 2.5×10^3 N/m. (a) How much work is done in stretching the relaxed spring

by 6.0 cm? (b) How much more work is done in stretching the spring an additional 2.0 cm?

29. ■■ For the spring in Exercise 28, how much mass would have to be suspended from the vertical spring to stretch it (a) the first 6.0 cm and (b) the additional 2.0 cm?

30. ■■ A particular force is described by the equation $\mathbf{F} = (60 \text{ N/m})x\,\hat{\mathbf{x}}$. How much work is done when this force pushes a box horizontally between (a) $x_o = 0$ and $x = 0.15$ m, and (b) $x_o = 0.15$ m and $x = 0.25$ m?

31. ■■ Compute the work done by the variable force in the F versus x graph in ●Fig. 5.24. [Hint: The area of a triangle is $A = \frac{1}{2}$altitude $\times$ base.]

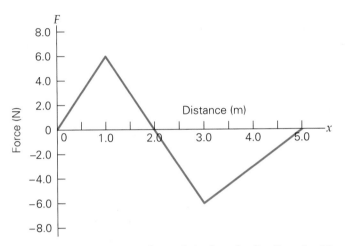

●FIGURE 5.24 **How much work is done?** See Exercise 31.

5.3 The Work–Energy Theorem: Kinetic Energy

32. If the angle between the force and the displacement is greater than 90°, (a) work is positive, (b) work is negative, (c) the change in energy is zero, (d) energy is either gained or lost.

33. ●Figure 5.25 is a close-up of the clap skates discussed in the Insight on p. 163. The skates have a spring-loaded hinged toe at the front of the boot and a plunger system

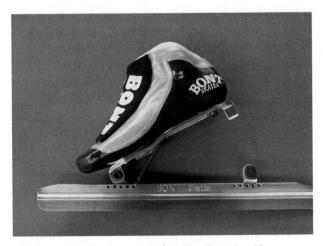

●FIGURE 5.25 **Clap to records** See Exercise 33.

at the heel. As the skater strides, the heel lifts from the blade—unlike traditional skates—lengthening the skater's stride. Explain how these skates can improve performance.

34. ◐ Which of the following objects has the smallest kinetic energy: (a) an object of mass $4m$ and speed v; (b) an object of mass $3m$ and speed $2v$; (c) an object of mass $2m$ and speed $3v$; (d) an object of mass m and speed $4v$?

35. A 0.50-kilogram ball with a speed of 10 m/s hits a wall and bounces back with only half the original speed. How much kinetic energy is lost in the ball's collision with the wall?

36. ■ A large car and a small car, both traveling with the same speed, skid to a stop with the same coefficient of friction. Use the work–energy theorem to determine how their stopping distances compare.

37. ■ A large car of mass $2m$ travels with speed v, and a small car of mass m travels with speed $2v$. Both skid to a stop with the same coefficient of friction. How do their stopping distances compare? (Use the work–energy theorem, not Newton's laws.)

38. ■ ◐ An automobile with a mass of 1200 kg travels at a speed of 90 km/h. (a) What is its kinetic energy? (b) What is the net work that would be required to bring it to rest?

39. ■ A constant net force of 75 N acts on an object initially at rest and acts through a parallel distance of 0.60 m. (a) What is the final kinetic energy of the object? (b) If the object has a mass of 0.20 kg, what is its final speed?

40. ■ ◐ A 3.0-gram bullet traveling at 350 m/s hits a tree and slows down uniformly while penetrating a distance of 12 cm. What was the force exerted on the bullet in bringing it to rest?

41. ■■ ◐ A 0.60-kilogram object with an initial velocity of 3.0 m/s in the positive x direction is acted on by a force in the direction of motion. The force does +2.5 J of work. What is the final velocity of the object?

42. ■■ The stopping distance of a vehicle is an important safety factor. Assuming a constant braking force, use the work–energy theorem to show that a vehicle's stopping distance is proportional to the square of its initial speed. If an automobile traveling at 45 km/h is brought to a stop in 50 m, what would be the stopping distance for an initial speed of 90 km/h?

43. ■■■ If the work required to speed a car up from 10 km/h to 20 km/h is 5.0×10^3 J. What would be the work required to increase its speed from 20 km/h to 30 km/h?

5.4 Potential Energy

44. The potential energy change of a spring is proportional to (a) k^2, (b) $(x - x_o)^2$, (c) x_o^2, (d) $x^2 - x_o^2$.

45. A change in gravitational potential energy (a) is always positive, (b) depends on the reference point, (c) depends on the path, (d) depends only on the initial and final heights.

46. Sketch a plot of U versus x for a mass oscillating on a spring between the limits of $-A$ and $+A$.

47. ■ What is the gravitational potential energy, relative to the ground, of a 1.0-kilogram box at the top of a 50-meter building?

48. ■ To store exactly 1.0 J of potential energy in a spring for which $k = 45$ N/m, how much would the spring have to be stretched beyond its equilibrium length?

49. ■ How much more gravitational potential energy does a 1.0-kilogram hammer have when it is on a shelf 1.5 m high than when it is on a shelf 0.90 m high?

50. ■ You are told that the gravitational potential energy of a 2.0-kilogram object decreased by 10 J. What physically happened to the object? Can you determine its height?

51. ■■ A 0.20-kilogram stone is thrown vertically upward with an initial velocity of 7.5 m/s from a starting point 1.2 m above the ground. (a) What is the potential energy of the stone at its maximum height relative to the ground? (b) What is the change in the potential energy of the stone between its launch point and its maximum height?

52. ■■ 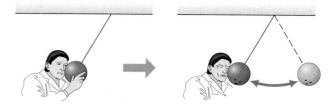 A 60-kilogram diver dives off a board that is 5.0 m above the surface of water; he touches the bottom of a pool 3.0 m below the surface of the water. What are (a) the potential energies of the diver relative to the water surface when he is on the board and at the bottom of the pool? (b) the change in the diver's potential energy relative to the board, to the water surface, and to the bottom of the pool?

53. ■■ The floor of a basement is 3.0 m below ground level, and the floor of the attic is 4.5 m above ground level. (a) What are the potential energies of 1.5-kilogram objects in the basement and attic, relative to ground level? (b) If one such object in the attic were brought to the basement, what would be the change in potential energy relative to the attic, ground, and basement? [Hint: Make a sketch.]

54. ■■■ A student has six textbooks, each with a thickness of 4.0 cm and a weight of 30 N. What is the minimum work the student would have to do to stack the books on top of one another with each initially on the table surface?

5.5 The Conservation of Energy

55. 🔵 If a nonconservative force acts on an object, (a) the object's kinetic energy is always greater than its potential energy, (b) the mechanical energy is conserved, (c) its kinetic energy is conserved, (d) the total energy is conserved.

56. The speed of a pendulum is greatest (a) before the pendulum reaches the bottom of its swing, (b) when its acceleration is the greatest, (c) when its potential energy is the least, (d) when its kinetic energy is a minimum.

57. A bowling ball suspended from a ceiling is displaced from the vertical position to one side and released from rest just in front of a student's nose (•Figure 5.26). If the student doesn't move, why won't the bowling ball hit his nose?

•**FIGURE 5.26   In the face?**   See Exercise 57.

58. Discuss all the different energy conversions involved in a pole vault, as shown in •Fig. 5.27. (Include the vaulter's running start. Where does the energy for this come from?)

59. An energy transfer question: When jumping straight upward from the ground, you can achieve only the same maximum height with each jump. However, on a trampoline, you can jump higher with each bounce. Why?

•**FIGURE 5.27   Energy conversion(s)**   See Exercise 58.

There's a limit on the trampoline too. What determines this limit?

60. ■ 🌐 A person standing on a bridge at a height of 115 m above a river drops a 0.250-kilogram rock. (a) What is the rock's mechanical energy at the time of release relative to the river surface? (b) What are the rock's kinetic, potential, and mechanical energies after it has fallen 75.0 m? (c) Just before the rock hits the water, what are its speed and total mechanical energy? (d) Repeat (a)–(c) but take the reference point ($h = 0$) at the elevation where the rock is released.

61. ■ A 0.300-kilogram ball is thrown vertically upward with an initial speed of 10.0 m/s. If the initial potential energy is called zero, find the ball's kinetic, potential, and mechanical energies (a) at its initial position, (b) at 2.50 m above the initial position, (c) at its maximum height.

62. ■ What is the maximum height reached by the ball in Exercise 61?

63. ■■ A 0.50-kilogram ball thrown vertically upward has an initial kinetic energy of 80 J. What are (a) its kinetic and potential energies when it has traveled three-quarters of the distance to its maximum height? (b) its speed at this point? (c) its potential energy at the maximum height (assume a reference is chosen to be zero at the launch point)?

64. ■■ A girl swings back and forth on a swing whose ropes are 4.00 m long. The maximum height she reaches is 2.00 m above the ground. At the lowest point of the swing, she is 0.500 m above the ground. What is the girl's maximum speed, and where does she attain it?

65. ■■ When a certain rubber ball is dropped from a height of 1.25 m onto a hard surface, it loses 18.0% of its mechanical energy on each bounce. How high will the ball bounce (a) on the first bounce? (b) on the second bounce? (c) With what speed would the ball have to be thrown downward to make it reach its original height on the first bounce?

66. ■■ A skier coasts down a very smooth slope as shown in •Fig. 5.28. If the speed of the skier at point A is 5.0 m/s, what is his speed at point B?

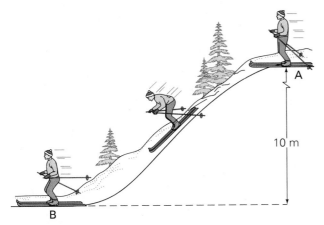

•FIGURE 5.28 **Down the slope** See Exercises 66 and 71.

67. ■■ 🌐 A roller coaster rolls on a track as shown in •Fig. 5.29. If the speed of the roller coaster at point A is 5.0 m/s, (a) what is its speed at point B? (b) Will it reach point C? (c) What speed at A is required for the roller coaster to reach point C?

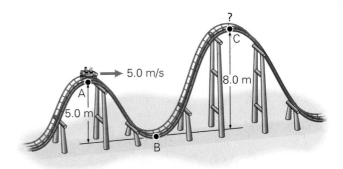

•FIGURE 5.29 **Enough energy?** See Exercise 67.

68. ■■ A simple pendulum has a length of 0.75 m and a bob whose mass is 0.15 kg. The bob is released from an angle of 25° relative to a vertical reference line (•Fig. 5.30). (a) Show that the vertical height of the bob when released is $h = L(1 - \cos 25°)$. (b) What is the kinetic energy of the bob when the string is at an angle of 9.0°? (c) What is the speed of the bob at the bottom of the swing? (Neglect friction and the mass of the string.)

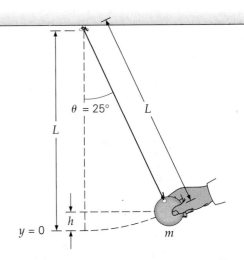

•FIGURE 5.30 **A pendulum swings** See Exercises 68 and 69.

69. ■■ Suppose the simple pendulum in Exercise 68 were released from an angle of 60°. (a) What would be the speed of the bob at the bottom of the swing? (b) To what height would the bob swing on the other side? (c) What angle of release would give half the speed of the 60° release angle at the bottom of the swing?

70. ■■ 🌐 A 1.5-kilogram box sliding with a speed of 12 m/s on a frictionless surface approaches a horizontal spring (see Fig. 5.18). If the spring has a spring constant of 2000 N/m, (a) how far will the spring be compressed in stopping the box? (b) How far will the spring be

compressed when the box's speed is reduced to half of its initial speed?

71. ■■■ 🌐 In Exercise 66, the skier has a mass of 60 kg. If the force of friction retards the motion by doing 2500 J of work, what is his speed at point B?

5.6 Power

72. Which of the following is not a unit of power? (a) J/s, (b) W·s, (c) ft·lb/s, (d) hp

73. If you check your electricity bill, you will note that you are paying the power company for so many kilowatt-hours (kWh). Are you really paying for power? Explain.

74. (a) Does efficiency describe how fast work is done? Explain. (b) Does a machine with a higher efficiency always do work faster than one with a lower efficiency? Explain.

75. Two students who weigh the same start at the same ground floor location at the same time to go to the same classroom on the third floor by different routes. If they arrive at different times, which student will have expended more power? Explain.

76. ■ What is the power in watts of a motor rated at 1/4 hp?

77. ■ Marcia consumes 8.4×10^6 J (2000 food calories) of energy per day while maintaining a constant weight. What is the average power she produces in a day?

78. ■ A 1500-kilogram race car can go from zero to 90 km/h in 5.0 s. What average power is required to do this?

79. ■ The two 0.50-kilogram weights of a cuckoo clock descend 1.5 m in a 3-day period. At what rate is gravitational potential energy decreased?

80. ■ A 60-kilogram woman runs up a staircase 15 m high (vertically) in 20 s. (a) How much power does she expend? (b) What is the horsepower rating?

81. ■■ An electric motor with a 2.0-horsepower output drives a machine with an efficiency of 45%. What is the energy output of the machine per second?

82. ■■ Water is lifted out of a well 30.0 m deep by a motor rated at 1 hp. Assuming 100% efficiency, how many kilograms of water can be lifted in 1 min?

83. ■■ In Exercise 82, if the motor has an efficiency of 83% and 230 kg of water is to be lifted in 1 min, what would be the required horsepower rating?

84. ■■■ A 3250-kilogram aircraft takes 12.5 min to achieve its cruising altitude of 10.0 km and cruising speed of 850 km/h. If the plane's engines deliver, on the average, 1500 hp of power during this time, what is the efficiency of the engines?

85. ■■■ A sled and driver with a total mass of 120 kg are pulled up a hill with a 15° incline by a horse, as illustrated in •Fig. 5.31. (a) If the overall retarding frictional force is 950 N and the sled moves up the hill with a constant velocity of 5.0 km/h, what is the power output of the horse? (Express in horsepower, of course. Note the magnitude of your answer and explain.) (b) Suppose that in a spurt of energy the horse accelerates the sled uniformly from 5.0 km/h to 20 km/h in 5.0 s. What is the horse's maximum instantaneous power output? Assume the same force of friction.

•**FIGURE 5.31 A one-horse open sleigh** See Exercise 85.

Additional Exercises

86. An electron ($m = 9.11 \times 10^{-31}$ kg) has 8.00×10^{-17} J of kinetic energy. What is the speed of the electron?

87. A large electric motor with an efficiency of 75% has a power output of 1.5 hp. If the motor is run steadily and the cost of electricity is $0.12/kWh, how much does it cost, to the nearest penny, to run the motor for 2.0 h?

88. 💿 A 28-kilogram child slides down a playground slide from a height of 3.0 m above the bottom of the slide. If her speed at the bottom is 2.5 m/s, how much energy is lost to friction?

89. In planing a piece of wood 35 cm long, a carpenter applies a force of 40 N to a plane at a downward angle of 25° to the horizontal. How much work is done by the carpenter?

90. A spring with a force constant of 50 N/m is stretched from 0 to 20 cm. Is more work required to stretch the spring from 10 cm to 20 cm than from 0 to 10 cm? If so, how much more? Justify your answer mathematically.

91. A 120-kilogram sleigh is pulled by one horse at a constant velocity for a distance of 0.75 km on a level snowy surface. The coefficient of kinetic friction between the sleigh runners and the snow is 0.25. (a) Calculate the work done by the horse. (b) Calculate the work done by friction.

92. A water slide has a height of 4.0 m. The people coming down the slide shoot out horizontally at the bottom, which is a distance of 1.5 m above the water in the swimming pool. If a person starts down the slide from rest, neglecting frictional losses, how far from a point directly below the bottom of the slide does the person land? Does it make any difference whether the person is a small child or an adult?

93. A hiker plans to swing on a rope across a ravine in the mountains, as illustrated in •Fig. 5.32, and to drop when she is just above the far edge. (a) At what horizontal speed should she be moving when she starts to swing? (b) When would she be in danger of falling into the ravine? Explain.

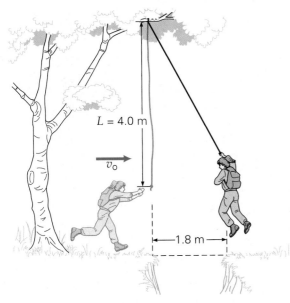

$L = 4.0$ m

v_0

1.8 m

•FIGURE 5.32 Can she make it? See Exercise 93.

94. A tractor pulls a wagon with a constant force of 700 N at a constant speed of 20.0 km/h. (a) How much work is done by the tractor in 3.50 min? (b) What is the tractor's power output?

95. A sports car weighs one-third as much as a large luxury car. (a) If the sports car is traveling at a speed of 90 km/h, at what speed would the larger car have to travel to have the same kinetic energy? (b) Suppose that the large car is traveling at a speed that gives it half the kinetic energy of the sports car. What is the speed of the more massive car?

96. A ball with a mass of 0.360 kg is dropped from a height of 1.20 m above the top of a fixed vertical spring, whose force constant is 350 N/m. (a) What is the maximum distance the spring is compressed by the ball? (Neglect energy loss due to the collision.) (b) What is the speed of the ball when the spring has been compressed 5.00 cm?

97. A student on a sled starts from rest at a vertical height of 25 m above the horizontal base of the hill and slides down. If the sled and the student have a speed of 20 m/s at the bottom of the hill, is mechanical energy conserved? Justify your answer.

98. A constant horizontal force of 30 N is required to move a box with a constant speed along a rough horizontal surface. If the force does work at a rate of 50 W, (a) what is the box's speed? (b) How much work is done by the force in 2.5 s?

99. In 10 s, a 70-kilogram student runs up two flights of stairs whose combined vertical height is 8.0 m. Compute the student's power output in doing work against gravity in (a) watts and (b) horsepower.

Momentum and Collisions

Tomorrow, the sportscasters may say that in this instant the momentum of the entire game changed. As a result of this clutch hit, one team gained momentum while their opponents lost it. But regardless of the effect on the team, it's clear that the momentum of the *ball* must have changed dramatically in the instant before this photograph was taken. The ball was traveling from right to left, probably at a pretty good rate of speed—and thus with lots of momentum. But a collision with several pounds of hardwood—with plenty of momentum of its own—changed its trajectory in a fraction of a second. A fan might say that the batter turned it around; after studying Chapter 4, you might say that the force he applied gave it a large negative acceleration, reversing its velocity vector (and probably increasing its magnitude as well). Yet, if you summed up the momentum of the ball and bat just before the collision and just afterward, you'd discover that although both the ball and the bat had momentum changes, the total momentum never changed!

If you were bowling and the ball bounced off the pins and rolled back toward you, you would probably be very surprised. But why? What leads us to expect that the ball will send the pins flying and continue on its way, rather than rebounding? You might say that the momentum of the ball carries it onward even after the collision (and you would be right)—but what does that really mean? In this chapter, you will study the concept of *momentum* and learn how it is particularly useful in analyzing motion and collisions.

6.1 Linear Momentum

Objective: To compute linear momentum and the components of momentum.

The term *momentum* may bring to mind a football player running down the field, knocking down players who are trying to stop him. Or you might have heard someone say that a team lost its momentum (and so lost the game). Such everyday usages give some insight into the meaning of momentum. They suggest the idea of mass in motion and therefore of inertia. We tend to think of heavy or massive objects in motion as having a great deal of momentum, even if they move very slowly. However, according to the technical definition of momentum, a light object can have just as much momentum as a heavier one, and sometimes more.

Newton referred to what modern physicists term **linear momentum** as "the quantity of motion . . . arising from velocity and the quantity of matter conjointly." In other words, the momentum of a body is proportional to both its mass and its velocity. By definition,

The linear momentum of an object is the product of its mass and velocity:

$$\mathbf{p} = m\mathbf{v} \qquad (6.1)$$

SI unit of momentum: kilogram-meters per second (kg·m/s)

It is common to refer to linear momentum as simply momentum. From Eq. 6.1, you can see that the SI units for momentum are kilogram-meters per second. Momentum is a vector quantity having the same direction as the velocity. Like velocity, it can be resolved into rectangular components: $\mathbf{p}_x = m\mathbf{v}_x$ and $\mathbf{p}_y = m\mathbf{v}_y$.

Equation 6.1 expresses the momentum of a single object or particle. For a system of more than one particle, the **total linear momentum** of the system is the vector sum of the momenta (plural of momentum) of the individual particles:

$$\mathbf{P} = \mathbf{p}_1 + \mathbf{p}_2 + \mathbf{p}_3 + \cdots = \Sigma_i\mathbf{p}_i \qquad (6.2)$$

Note: The momentum vector of a single object is in the direction of the object's velocity.

Total linear momentum—a vector sum

Example 6.1 ■ Momentum: Mass *and* Velocity

A 100-kilogram football player runs with a velocity of 4.0 m/s straight down the field. A 1.0-kilogram artillery shell leaves the barrel of a gun with a muzzle velocity of 500 m/s. Which has the greater momentum (magnitude)—the football player or the shell?

Thinking It Through. Given the mass and velocity of an object, the momentum can be calculated from Eq. 6.1.

Solution. As usual, we first list the given data and what we are to find, using the subscripts "p" and "s" to refer to the player and shell, respectively.

Given: $m_p = 100$ kg   Find: p_p and p_s
        $v_p = 4.0$ m/s   (magnitudes of the momenta)
        $m_s = 1.0$ kg
        $v_s = 500$ m/s

The magnitude of the momentum of the player is

$$p_p = m_pv_p = (100 \text{ kg})(4.0 \text{ m/s}) = 4.0 \times 10^2 \text{ kg·m/s}$$

and that of the shell is

$$p_s = m_sv_s = (1.0 \text{ kg})(500 \text{ m/s}) = 5.0 \times 10^2 \text{ kg·m/s}$$

Thus, the much lighter, or less massive, shell has the greater momentum. Remember: The magnitude of momentum depends on *both* the mass and the magnitude of the velocity.

(a)

(b)

(c)

•**FIGURE 6.1 Three moving objects: a comparison of momenta and kinetic energies** (a) A .22-caliber bullet shattering a ball point pen; (b) a cruise ship, (c) a glacier, Glacier Bay, Alaska. See Example 6.2.

Follow-up Exercise. What would the football player's speed have to be for his momentum to have the same magnitude as the artillery shell's? Would this speed be realistic?

Example 6.2 ■ Momentum versus Kinetic Energy: Some Ballpark Comparisons

(a) Consider the three objects shown in •Fig. 6.1—a .22-caliber bullet, a cruise ship, and a glacier. Assuming each to be moving at its normal speed, which do you think has the greatest linear momentum? The least? Which has the greatest kinetic energy? The least? (List your choices before going on.)

	Greatest	Least
Momentum	_____	_____
Kinetic energy	_____	_____

(b) Using the estimates given below as your data, calculate the magnitudes of the momenta and the kinetic energies of the three objects. How accurate were your original answers—any surprises?

Bullet: A typical .22-caliber bullet would have a weight of about 30 grains and a muzzle velocity of about 1300 ft/s. (A grain, abbreviated gr, is an old British unit used for pharmaceuticals, such as 5-grain aspirin tablets; 1 lb = 7000 gr.)

Ship: A ship like the one shown would have a weight of about 70 000 tons and a speed of about 20 knots. (A knot is another old unit, still commonly used in nautical contexts; 1 knot = 1.15 mi/h).

Glacier: The glacier might be 1 km wide, 10 km long, and 250 m deep, and it might move at a rate of 1 m per day. (There is much variation among glaciers. Therefore these figures must involve more assumptions and rougher estimates than those for the bullet or ship. For example, we are assuming a uniform, rectangular cross-sectional area for the glacier. The depth is particularly difficult to estimate from a photograph; a minimum value is given by the fact that glaciers must be at least 50–60 m thick before they can "flow." Observed speeds range from a few centimeters to as much as 40 m a day for valley glaciers such as the one in Fig. 6.1c. The value we have chosen is considered a typical one.)

Solution. Converting our data to metric units, we have:

Bullet:
$$m_b = 30 \text{ gr} \left(\frac{1 \text{ lb}}{7000 \text{ gr}} \right) \left(\frac{1 \text{ kg}}{2.2 \text{ lb}} \right) = 0.0019 \text{ kg}$$

$$v_b = 1.3 \times 10^3 \text{ ft/s} \left(\frac{0.305 \text{ m/s}}{\text{ft/s}} \right) = 4.0 \times 10^2 \text{ m/s}$$

Ship:
$$m_s = 7.0 \times 10^4 \text{ ton} \left(\frac{2.0 \times 10^3 \text{ lb}}{\text{ton}} \right) \left(\frac{1 \text{ kg}}{2.2 \text{ lb}} \right) = 6.4 \times 10^7 \text{ kg}$$

$$v_s = 20 \text{ knots} \left(\frac{1.15 \text{ mi/h}}{\text{knot}} \right) \left(\frac{0.447 \text{ m/s}}{\text{mi/h}} \right) = 10 \text{ m/s}$$

Glacier: width = 10^3 m, length = 10^4 m, depth = 2.5×10^2 m;

$$v_g = 1 \text{ m/day} \left(\frac{1 \text{ day}}{86\,400 \text{ s}} \right) = 1.2 \times 10^{-5} \text{ m/s}$$

We know the equations for kinetic energy, $K = \frac{1}{2}mv^2$ (Eq. 5.5), and momentum, $p = mv$ (Eq. 6.1). We have all the speeds and masses except for m_g, the mass of the glacier. To calculate this we need to know the density of ice, since $m = \rho V$ (Eq. 1.1). The density of ice is less than that of water, but the two are not very different, so we will use the density of water, 1.0×10^3 kg/m³, to simplify the calculations. (The actual density of ice is 0.92×10^3 kg/m³, but most of our other data in this Example are also approximations or estimates, so this shortcut should produce results that are good enough for our present purposes.)

$$m_g = \rho V = \rho(\ell \times w \times d)$$
$$\approx (1.0 \times 10^3 \, \text{kg/m}^3)(10^4 \, \text{m})(10^3 \, \text{m})(2.5 \times 10^2 \, \text{m}) = 2.5 \times 10^{12} \, \text{kg}$$

Then, calculating the magnitudes of the momenta of the objects, we have

Bullet: $\quad p_b = m_b v_b = (0.0019 \, \text{kg})(4.0 \times 10^2 \, \text{m/s}) = 0.76 \, \text{kg·m/s}$

Ship: $\quad p_s = m_s v_s = (6.4 \times 10^7 \, \text{kg})(10 \, \text{m/s}) = 6.4 \times 10^8 \, \text{kg·m/s}$

Glacier: $\quad p_g = m_g v_g = (2.5 \times 10^{12} \, \text{kg})(1.2 \times 10^{-5} \, \text{m/s}) = 3.0 \times 10^7 \, \text{kg·m/s}$

Thus, the glacier has about 40 million times as much momentum as the bullet, and the ship nearly a billion times as much. The enormous differences in the masses of these objects outweigh the differences in their speeds.

When we turn to the kinetic energies, however, the dependence on the *square* of the speed (Example 5.6) produces results that might surprise you:

Bullet: $\quad K_b = \frac{1}{2}m_b v_b^2 = \frac{1}{2}(0.0019 \, \text{kg})(4.0 \times 10^2 \, \text{m/s})^2 = 1.5 \times 10^2 \, \text{J}$

Ship: $\quad K_s = \frac{1}{2}m_s v_s^2 = \frac{1}{2}(6.4 \times 10^7 \, \text{kg})(10 \, \text{m/s})^2 = 3.2 \times 10^9 \, \text{J}$

Glacier $\quad K_g = \frac{1}{2}m_g v_g^2 = \frac{1}{2}(2.5 \times 10^{12} \, \text{kg})(1.2 \times 10^{-5} \, \text{m/s})^2 = 1.8 \times 10^2 \, \text{J}$

The bullet actually has about the same kinetic energy as the glacier—but the ship has more than 10 million times as much kinetic energy as either of them!

Problem-Solving Strategies: "Ballpark" and Order-Of-Magnitude Calculations

Calculations such as we have performed in Example 6.2 are sometimes known as "ballpark" or "back of the envelope" calculations. In performing such calculations we know that, because the data are only approximate (they are not known precisely, or they are being rounded off to simple whole numbers for the purpose of rapid calculation), the answer can't be taken as very accurate (see Section 1.6). However, it is often precise enough for our needs. In Example 6.2, some of the data were unavoidable approximations (the glacier isn't a simple rectangular object), whereas others, such as the density of ice, were deliberately approximated to simplify calculation. Nevertheless, this method gave us a good idea of the relative momenta and kinetic energies of the three objects.

Physicists often find it convenient to use a formalized version of the "ballpark" approach, commonly known as *order-of-magnitude calculation*. In this method, all values are rounded off to the nearest "order of magnitude," or power of 10: 10^2, 10^3, 10^4, and so on. Since it is very easy to multiply and divide powers of 10, this shortcut makes calculation very fast. However, although sometimes just the powers of 10 can be used in a calculation, it is usually a good idea to retain one significant figure of the prefix. This is particularly true when quantities are squared or raised to higher powers, as in the kinetic energy calculations. For instance, $v_s = 4 \times 10^2 \, \text{m/s}$ or $\approx 10^2 \, \text{m/s}$. Then $v_s^2 = (4 \times 10^2 \, \text{m/s})^2 = 16 \times 10^4 \, \text{m}^2/\text{s}^2 \approx 10^5 \, \text{m}^2/\text{s}^2$. However, if we had used just $10^2 \, \text{m/s}$, we would have obtained $10^4 \, \text{m}^2/\text{s}^2$.

In order-of-magnitude calculations, we can generally expect the answer to be correct to the nearest order of magnitude. In other words, if we get an answer of 1000, the true value might really be 750 or 1200, but we can be pretty confident that it is closer to 1000 than to 100 or to 10 000.

Example: A light-year—a unit used in astronomy—is the distance light travels in 1 y in a vacuum. The speed of light is $c = 3.00 \times 10^8 \, \text{m/s}$. Approximately how many meters are there in a light-year?

Solution: From the Conversion Factors Table inside the front cover (or simple calculation), we know that there are $3.16 \times 10^7 \, \text{s}$ in 1 year. Using the relationship distance = speed × time,

$$d = ct = (3 \times 10^8 \, \text{m/s})(3 \times 10^7 \, \text{s}) \approx 10^{16} \, \text{m}$$

Note that the product of the powers of 10 gives 10^{15}, but $3 \times 3 = 9$, so the prefixes give about another order of magnitude. Thus $(10 \times 10^{15}) = 10^{16}$ is the order of magnitude of the number of meters in a light-year. (The actual value is $9.48 \times 10^{15} \, \text{m}$.)

Example 6.3 ■ Total Momentum: A Vector Sum

What is the total momentum for each of the systems of particles illustrated in •Fig. 6.2?

Thinking It Through. The total momentum is the vector sum of the individual momenta (Eq. 6.2). This can be computed for each component.

Solution.

Given: Magnitudes and directions   *Find:* (a) Total momentum (**P**)
of momenta from Fig. 6.2   (b) Total momentum (**P**)

(a) The total momentum of a system is the vector sum of the momenta of the individual particles, so

$$\mathbf{P} = \mathbf{p}_1 + \mathbf{p}_2 = (2.0 \text{ kg·m/s})\hat{\mathbf{x}} + (3.0 \text{ kg·m/s})\hat{\mathbf{x}} = (5.0 \text{ kg·m/s})\hat{\mathbf{x}} \quad (+x \text{ direction})$$

(b) Computing the total momenta in the x and y directions gives

$$\mathbf{P}_x = \mathbf{p}_1 + \mathbf{p}_2 = (5.0 \text{ kg·m/s})\,\hat{\mathbf{x}} + (-8.0 \text{ kg·m/s})\,\hat{\mathbf{x}}$$

$$= -(3.0 \text{ kg·m/s})\,\hat{\mathbf{x}} \quad (-x \text{ direction})$$

$$\mathbf{P}_y = \mathbf{p}_3 = (4.0 \text{ kg·m/s})\,\hat{\mathbf{y}} \quad (+y \text{ direction})$$

Then

$$\mathbf{P} = \mathbf{P}_x + \mathbf{P}_y = (-3.0 \text{ kg·m/s})\,\hat{\mathbf{x}} + (4.0 \text{ kg·m/s})\,\hat{\mathbf{y}}$$

Follow-up Exercise. In Example 6.3, if $\mathbf{p}_1$ and $\mathbf{p}_2$ in (a) were added to $\mathbf{p}_2$ and $\mathbf{p}_3$ in (b), what would be the total momentum?

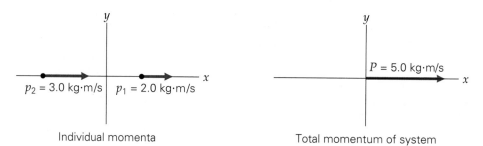

Individual momenta   Total momentum of system

(a) $\mathbf{P} = \mathbf{p}_1 + \mathbf{p}_2$

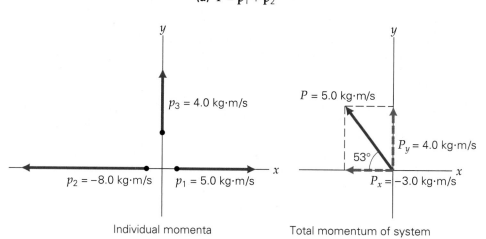

•**FIGURE 6.2   Total momentum**
The total momentum of a system
of particles is the vector sum of
the individual momenta. See
Example 6.3.

Individual momenta   Total momentum of system

(b) $\mathbf{P} = \mathbf{p}_1 + \mathbf{p}_2 + \mathbf{p}_3$

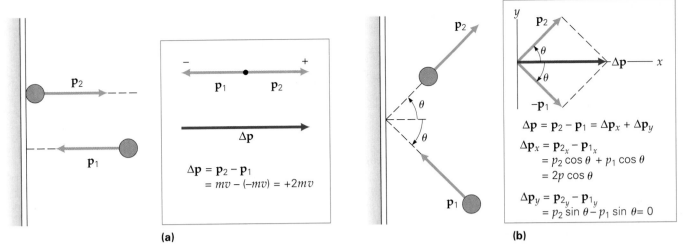

•FIGURE 6.3 Change in momentum The change in momentum is given by the *difference* in the momentum vectors. **(a)** Here the vector sum is zero, but the vector *difference,* or change in momentum, is not. (The particles are displaced for convenience.) **(b)** The change in momentum is found by computing the change in the components.

In Example 6.3a, the momenta were along the coordinate axes and thus added straightforwardly. If the motion of one (or more) of the particles is not along an axis, its momentum may be broken up, or resolved, into rectangular components and added to the component sums as in Example 6.3b—just as you learned to do with force components in Chapter 4. For example, suppose you were given a momentum vector with a magnitude of 5.0 kg·m/s at an angle of 53° relative to the negative x axis (like the total momentum vector in Fig. 6.2b). Then, its rectangular components are

Note: Review Example 4.4 and Fig. 4.10.

$$\mathbf{p}_x = -p\cos 53° \,\hat{\mathbf{x}} = -(5.0 \text{ kg·m/s})(0.60)\,\hat{\mathbf{x}} = (-3.0 \text{ kg·m/s})\,\hat{\mathbf{x}}$$

$$\mathbf{p}_y = p\sin 53° \,\hat{\mathbf{x}} = (5.0 \text{ kg·m/s})(0.80)\,\hat{\mathbf{y}} = (4.0 \text{ kg·m/s})\,\hat{\mathbf{y}}$$

where the signs give the directions. Each momentum vector would be broken down thus and the rectangular components along the axis in each dimension added separately to find the vector sum.

Since momentum is a vector, a change in momentum can result from a change in magnitude and/or direction. Examples of changes in the momenta of particles because of changes of direction on collision are illustrated in •Fig. 6.3. In the figure, the magnitude of a particle's momentum is taken to be the same before and after collision (as indicated by the arrows of equal length). Figure 6.3a illustrates a direct rebound—a 180° change in direction. Note that the change in momentum ($\Delta \mathbf{p}$) is the vector difference and that directional signs for the vectors are important. Figure 6.3b shows a glancing collision, for which the change in momentum is given by the component differences.

Force and Momentum

As you know from Chapter 4, a change in velocity (an acceleration) requires a force. Similarly, since momentum is directly related to velocity (as well as mass), a change in momentum also requires a force. In fact, Newton originally expressed his second law of motion in terms of momentum rather than acceleration. The force-momentum relationship may be seen by starting with $\mathbf{F} = m\mathbf{a}$ and using $\mathbf{a} = (\mathbf{v} - \mathbf{v}_o)/\Delta t$, where the mass is assumed to be constant:

$$\mathbf{F} = m\mathbf{a} = \frac{m(\mathbf{v} - \mathbf{v}_o)}{\Delta t} = \frac{m\mathbf{v} - m\mathbf{v}_o}{\Delta t} = \frac{\mathbf{p} - \mathbf{p}_o}{\Delta t} = \frac{\Delta \mathbf{p}}{\Delta t}$$

•FIGURE 6.4 Change in the momentum of a projectile The total momentum vector of a projectile is tangential to its path (as is its velocity); this vector changes in magnitude and direction because of the action of an external force (gravity). The x component of the momentum is constant. (Why?)

Newton's second law of motion in terms of momentum

$$\mathbf{F} = \frac{\Delta \mathbf{p}}{\Delta t} \qquad (6.3)$$

where $\mathbf{F}$ is the *average* force if the acceleration is not constant or the *instantaneous* force if Δt goes to zero.

Expressed in this form, Newton's second law states that *the net external force acting on an object is equal to the time rate of change of the object's momentum.* It is easily seen from the development of Eq. 6.3 that the equations $\mathbf{F} = m\mathbf{a}$ and $\mathbf{F} = \Delta\mathbf{p}/\Delta t$ are equivalent if the mass is constant. In some situations, however, the mass may vary. This will not be a consideration here in our discussion of particle collisions, but a special case will be given later in the chapter. The more general form of Newton's second law, Eq. 6.3, is true even if the mass varies.

Just as the equation $\mathbf{F} = m\mathbf{a}$ indicates that an acceleration is evidence of a net force, the equation $\mathbf{F} = \Delta\mathbf{p}/\Delta t$ indicates that *a change in momentum is evidence of a net force.* For example, as illustrated in •Fig. 6.4, the momentum of a projectile is tangential to the projectile's parabolic path and changes in both magnitude and direction. The change in momentum indicates that there is a net force acting on the projectile, which you know is the force of gravity. Changes in momentum were illustrated in Fig. 6.3. Can you identify the forces in these cases? Think in terms of Newton's third law.

6.2 Impulse

Objectives: To relate (a) impulse and momentum, and (b) kinetic energy and momentum.

When two objects—such as a hammer and a nail, a golf club and a golf ball, or even two cars—collide, they exert large forces on one another for a short period of time (•Fig. 6.5a). The force is not constant in this case. However, Newton's second law in momentum form is still useful for analyzing such situations by using average values. Written in this form, the law states that the *average* force is equal to the time rate of change of momentum, $\overline{\mathbf{F}} = \Delta\mathbf{p}/\Delta t$ (Eq. 6.3). Rewriting the equation to express the change in momentum, we have

$$\overline{\mathbf{F}}\,\Delta t = \Delta\mathbf{p} = \mathbf{p} - \mathbf{p}_\mathrm{o} \qquad (6.4)$$

The term $\overline{\mathbf{F}}\,\Delta t$ is known as the **impulse** of the force:

$$\text{Impulse} = \overline{\mathbf{F}}\,\Delta t = \Delta\mathbf{p} = m\mathbf{v} - m\mathbf{v}_\mathrm{o} \qquad (6.5)$$

SI unit of impulse and momentum: newton-seconds (N·s)

Thus, *the impulse exerted on a body is equal to the change in the body's momentum.* This statement is referred to as the **impulse–momentum theorem**. Impulse has units of newton-seconds (N·s), which are also units of momentum by Eq. 6.4 and Eq. 6.1 (1 kg·m/s = 1 N·s).

(a)

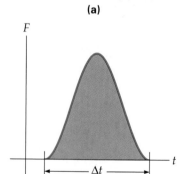

(b)

•FIGURE 6.5 Collision impulse (a) Collision impulse causes the football to be deformed. **(b)** The impulse is the area under the curve of an F versus t graph. Note that the impulse force on the ball is not constant but rises to a maximum.

In Chapter 5 you learned that by the work–energy theorem ($W = F \Delta x = \Delta K$), the area under an F (net force) versus x curve is equal to the net work or change in kinetic energy. Similarly, the area under an F versus t curve is equal to the impulse or the change in momentum (Fig. 6.5b). An impulse force varies with time and is not a constant force, as Eq. 6.4 might suggest. However, in general, it is convenient to talk about the equivalent *constant* average force $\overline{F}$ acting over a time interval Δt to give the same impulse (same area under the force versus time curve), as shown in •Fig. 6.6.

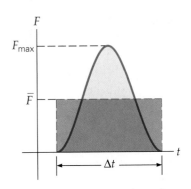

•**FIGURE 6.6   Average impulse force**   The area under the average force curve ($\overline{F} \Delta t$, within the dashed red lines) is the same as the area under the F versus t curve, which is usually difficult to evaluate.

Example 6.4 ■ Teeing Off: The Impulse–Momentum Theorem

A golfer drives a 0.10-kilogram ball from an elevated tee, giving the ball an initial horizontal speed of 40 m/s (about 90 mi/h). The club and the ball are in contact for 1.0 ms (millisecond). What is the average force exerted by the club on the ball during this time?

Thinking It Through.   The average force is equal to the time rate of change of momentum (Eq. 6.5 rearranged).

Solution.

Given: $m = 0.10$ kg   *Find:* $\overline{F}$ (average force)
       $v = 40$ m/s
       $v_0 = 0$
       $\Delta t = 1.0$ ms $= 1.0 \times 10^{-3}$ s

Notice that the mass and the initial and final velocities are given, so the change in momentum can be found. Then, the average force can be computed from the impulse-momentum theorem:

$$\overline{F} \Delta t = p - p_0 = mv - mv_0$$

Thus,

$$\overline{F} = \frac{mv - mv_0}{\Delta t} = \frac{(0.10 \text{ kg})(40 \text{ m/s}) - 0}{1.0 \times 10^{-3} \text{ s}} = 4000 \text{ N} \quad \text{(or 900 lb)}$$

The force is in the direction of the acceleration and is the *average* force. The instantaneous force is even greater near the midpoint of time interval of the collision (Δt in Fig. 6.6).

Follow-up Exercise.   Suppose the golfer in Example 6.4 drives a ball with the same average force but "follows through" on the swing so as to increase the contact time to 1.5 ms. What effect would this have on the initial horizontal speed of the drive?

$\overline{F} \Delta t = mv_0$

(a)

Example 6.4 illustrates the large internal forces that colliding objects can exert on one another during short contact times. However, in some instances the impulse may be manipulated to reduce the force. Suppose there is a fixed change in momentum in a given situation. Then, since $\Delta p = \overline{F} \Delta t$, if Δt could be made longer, the average impulse force $\overline{F}$ would be reduced.

You have probably tried to minimize the impulse force on occasion. For example, when jumping from a height onto a hard surface, you try not to land stiff-legged. The abrupt stop (small Δt) would apply a large impulse force to your leg bones and joints and could cause injury. If you bend your knees as you land, the impulse is constant and vertically upward opposite your velocity ($\overline{F} \Delta t = \Delta p = -mv_0$, with the final velocity zero). Thus increasing the time interval Δt makes the impulse force smaller.

Similarly, in catching a fast-moving, hard ball, you quickly learn not to catch it with your arms rigid but rather to move your hands with the ball. This movement increases the contact time and reduces the impulse force and the "sting" (•Fig. 6.7).

$\overline{F} \Delta t = mv_0$

(b)

•**FIGURE 6.7   Adjust the impulse** (a) The change in momentum in catching the ball is a constant mv_0. If the ball is stopped quickly (small Δt), the impulse force is large (big $\overline{F}$) and stings the catcher's bare hands. (b) Increasing contact time (large Δt) by moving the hands with the ball reduces the impulse force and makes catching more enjoyable.

The Automobile Air Bag

A dark, rainy night—a car goes out of control and hits a big tree head-on! But the driver walks away with only minor injuries because he had his seatbelt buckled and had a car equipped with air bags. Air bags, along with seatbelts, are safety devices designed to prevent injuries to passengers in the front seat in automobile collisions. (Backseat air bags are also available.)

When a car collides with something basically immovable, such as a tree or a bridge abutment, or has a head-on collision with another vehicle, it stops almost instantaneously. If the front-seat passengers have not buckled up (and there are no air bags), they keep moving until acted on by an external force (Newton's first law). For the driver, this force is supplied by the steering wheel and column; for the passenger, by the dashboard and/or windshield.

Even when everyone has buckled up, there can be injuries. Seatbelts absorb energy by stretching and spread the force over a wide area to reduce the pressure. However, if a car is going fast enough and hits something truly immovable, there may be too much energy for the belts to absorb. This is where the air bag comes in. The bag inflates automatically on hard impact (Fig. 1), cushioning the driver (and front-seat passenger if both sides are equipped with air bags). In terms of impulse, the air bag increases the stopping contact time—the fraction of a second it takes your head to sink into the inflated bag is many times longer than the instant in which your head would be stopped by the dashboard. A longer contact time means a reduced impact force and thus much less likelihood of an injury. (Because the bag is large, the total impact force is also spread over a greater area of the body, so the force on any one part of the body is less.)

An interesting point is the inflating mechanism of an air bag. Think of how little time elapses between the front-end impact and the driver hitting the steering column in the collision of a fast-moving automobile. We say such a collision takes place instantaneously, yet during this time the air bag must be inflated! How is this done?

First, the air bag is equipped with sensors that detect the sharp deceleration associated with a head-on collision the instant it begins. If the deceleration exceeds the sensors' threshold settings, an electric current in an igniter in the air bag sets

FIGURE 1 Impulse and safety An automobile air bag increases the contact time, thereby decreasing the impulse force that could cause injury.

off a chemical explosion that generates gas to inflate the bag. The complete process from sensing to full inflation takes only on the order of 25 *thousandths* of a second (0.025 s).

The sensors' signals go first to a control unit, which determines whether a frontal collision rather than a system malfunction is occurring. (Accidental deployment could be dangerous as well as costly.) Typically, the control unit compares signals from two different sensors for collision verification. The unit is equipped with its own power source, since the car's battery and alternator are usually destroyed in a hard front-end collision. Sensing a collision, the control unit completes the circuit to the air bag igniter. It initiates a chemical reaction in a sodium azide (NaN_3) propellant. Gas (mostly nitrogen) is generated at an explosive rate, which inflates the air bag. The bag itself is made of thin nylon that is covered with cornstarch. The cornstarch acts as a lubricant to help the bag unfold smoothly on inflation.

Not only is sodium azide explosive, but it is also poisonous. Thus, you should not open a collapsed air bag, and if a car is junked, undeployed air bags must be disposed of

Another example in which the contact time is increased to decrease the impulse force is given in the Insight on this page.

In other instances, the *applied* impulse force may be relatively constant and the contact time (Δt) deliberately increased to produce a greater impulse, and thus a greater change in momentum ($\overline{F} \Delta t = \Delta p$). This is the principle of "following through" in sports, for example, when hitting a ball with a bat or driving a golf ball. In the latter case (•Fig. 6.8a), assuming that the golfer supplies the same average force with each swing, the longer the contact time, the greater will be the impulse or the momentum the ball receives. That is, with $\overline{F} \Delta t = mv$ (since $v_0 = 0$), the greater the value of Δt, the greater will be the final velocity of the ball. (This was illustrated in Follow-up Exercise 6.4.) As you learned in Section 3.4, a greater projection velocity gives a projectile a greater range. (What

properly. Other, nonpoisonous, compounds are being developed for use as air bag inflators.

In general, air bags offer protection only if the occupants are thrown forward, because the bags are designed to deploy only in front-end collisions. They are of little use in side-impact crashes. Side air bags are now available in some automobiles.

Recent Concerns

In some cases, however, the deployment of air bags has been shown to cause injuries and deaths. An air bag is not a soft, fluffy pillow. When activated, it is ejected out of its compartment at speeds up to 320 km/h (200 mi/h) and could hit a person close by with enough force to cause severe injury and even death. Adults are advised to sit at least 13 cm (10 in.) from the air bag compartment to allow a margin of safety from the 5–8 cm (2–3 in.) injury "risk zone." Seats should be adjusted to allow the proper safety distance.*

The most serious concern involves children. Children may get too close to the dashboard if they are not buckled in or not buckled in securely with an adjustable shoulder harness (Fig. 2a). Using a rear-facing child seat in the front passenger seat is also dangerous if the air bag inflates (Fig. 2b). According to government recommendations, 1) *children 12 years old and younger should ride buckled up in the back seat*, and 2) *a rear-facing child seat should _never_ be put in front of an air bag.*

To avoid the possibility of air bag injury, it is now possible to have a cut-off switch installed in an automobile to turn off one or both of the front air bags, pending approval by the appropriate government safety agency. Air bag deactivation may be authorized for the following reasons:

- A rear-facing child seat must be placed in the front seat because the car has no back seat or one that is too small.
- A child 12 years old or younger must ride in the front seat because of a medical condition that requires frequent monitoring.
- An individual who drives (or rides in the front seat) has a medical condition that would make it safer to have the air bag(s) turned off.

*Guidelines from the National Highway Traffic Safety Administration

- Drivers who must sit within several centimeters of the air bag, typically because of extremely short stature (137 cm—or 4 ft, 6 in.—or less).

There may be specific problems in some instances, but air bags do save many lives. All new passenger cars are now required to have dual air bags. But even if your car is equipped with air bags, _always_ remember to buckle up. (Maybe we should make that Newton's fourth law of motion!)

(a)

(b)

FIGURE 2 The wrong ways (a) Children who must ride in the front seat for medical monitoring should be buckled in securely so as not to get too close to the air bag in the dashboard. (b) Rear-facing child seats should never be used in the front passenger seat because injury could occur if the airbag is activated.

angle of projection should a golfer try to achieve when driving the ball on a level fairway?)

In some instances, a long follow-through that increases the contact time may be used to improve control (Fig. 6.8b).

The word *impulse* implies that the impulse force acts only briefly or quickly (like an "impulsive" person), and this is true in many instances. However, the definition of impulse places no limit on the time interval over which the force may act. Technically, a comet at its closest approach to the Sun is involved in a collision, because in physics collision forces do not have to be contact forces. Basically, a **collision** is an interaction of objects in which there is an exchange of momentum and energy. As you might expect from the work–energy theorem and the impulse–momentum theorem, momentum and kinetic energy are directly related.

Collision—exchange of momentum and energy

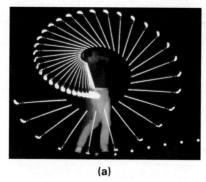

(a)

(b)

•**FIGURE 6.8 Increasing the contact time** **(a)** A golfer follows through on a drive swing. One reason is to increase the contact time so that the ball receives greater impulse and momentum. **(b)** The follow-through on a long putt increases the contact time for greater impulse and momentum, but the main reason is for directional control. Notice that the putter is in contact with the ball for a time equivalent to about 4 flash intervals.

A little algebraic manipulation of the equation for kinetic energy (Eq. 5.5) allows us to express kinetic energy in terms of the *magnitude* of momentum:

Kinetic energy and momentum

$$K = \tfrac{1}{2}mv^2 = \frac{(mv)^2}{2m} = \frac{p^2}{2m} \tag{6.6}$$

Thus, kinetic energy and momentum are intimately related. The conservation of these quantities is important in collisions, as we shall see later in the chapter.

6.3 The Conservation of Linear Momentum

Objectives: To (a) explain the conditions for the conservation of linear momentum and (b) apply it to physical situations.

Like total mechanical energy, the momentum of a body or system is a conserved quantity under certain conditions. This fact allows us to analyze a wide range of situations and solve many problems readily. The conservation of momentum is one of the most important principles in physics. In particular, it is used to analyze the collision of objects ranging from subatomic particles to automobiles in traffic accidents.

For the linear momentum of a particle or an object to be conserved (to remain constant with time), one condition must hold. This condition is apparent from the momentum form of Newton's second law (Eq. 6.3). If the net force acting on a particle is zero, that is,

$$\mathbf{F} = \frac{\Delta \mathbf{p}}{\Delta t} = 0$$

then,

$$\Delta \mathbf{p} = 0 = \mathbf{p} - \mathbf{p}_o$$

where $\mathbf{p}_o$ is the initial momentum and $\mathbf{p}$ is the momentum at some later time. Since they are equal, the momentum is conserved:

$$\mathbf{p} = \mathbf{p}_o \tag{6.7}$$

or

$$m\mathbf{v} = m\mathbf{v}_o$$

Note that this conservation is consistent with Newton's first law: An object remains at rest ($\mathbf{p} = 0$), or in motion with a *uniform* velocity (constant $\mathbf{p}$), unless acted on by a net external force.

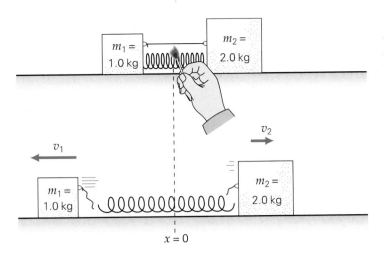

The conservation of momentum can be extended to a system of particles if we write Newton's second law in terms of the sums (resultants) of the forces acting on the system and of the momenta of the particles: $\mathbf{F} = \Sigma\mathbf{F}_i$ and $\mathbf{P} = \Sigma\mathbf{p}_i = \Sigma m\mathbf{v}_i$. A more complicated argument similar to the development of Eq. 6.7 can be made for systems of particles. It turns out that if there is no net external force acting *on the system*, then $\mathbf{P} = \mathbf{P}_o$. This generalized condition is referred to as the law of **conservation of linear momentum**:

Note: A lowercase **p** means individual momentum. A capital **P** means total system momentum. Both are vectors.

Conservation of momentum— no net external force

There are other ways to express this condition. For example, recall from Chapter 5 that a *closed,* or *isolated,* system is one on which no external work is done. Therefore there is no net external force acting on such a system, so the total linear momentum of an isolated system is conserved.

Within a system, internal forces may be acting, for example, when particles collide. These are force pairs of Newton's third law, and there is a good reason such forces are not explicitly referred to in the condition for the conservation of momentum. By Newton's third law, these internal forces are equal and opposite and vectorially cancel each other. Thus, the net internal force of a system is zero.

Note: Third-law force pairs were discussed in Section 4.4.

An important point to understand, however, is that the momenta of *individual* particles or objects within a system may change. But in the absence of a net external force, the *vector sum* of all the momenta remains the same. If the objects are initially at rest (total momentum is zero) and then are set in motion as the result of internal forces, the total momentum must still add to zero. This principle is illustrated in •Fig. 6.9 and analyzed in Example 6.5. Objects in an isolated system may transfer momentum among themselves, but the total momentum after the changes must add up to the initial value, assuming friction is negligible. (See Demonstration 3.)

The conservation of momentum is often a powerful and convenient tool for analyzing situations involving motion. Its application is illustrated in the following Examples.

Example 6.5 ■ Before and After: Conservation of Momentum

Two masses, $m_1 = 1.0$ kg and $m_2 = 2.0$ kg, are held on either side of a light compressed spring by a light string joining them, as shown in Fig. 6.9. The string is burned (negligible external force), and the masses move apart on the frictionless surface, with m_1 having a velocity of 1.8 m/s to the left. What is the velocity of m_2?

Thinking It Through. With no net external force (the weights are canceled by normal forces), the total momentum of the system is conserved. It is initially zero, so afterward the momentum of m_2 is *equal* and opposite that of m_1. (Vector addition gives zero total momentum.) (The term *light* indicates that the masses of the spring and string can be ignored.)

Solution. Listing the masses and speed given, we have

Given: $m_1 = 1.0$ kg   *Find:* v_2 (velocity—speed and direction)
$m_2 = 2.0$ kg
$v_1 = 1.8$ m/s (left)

Here the system consists of the two masses and the spring. Since the spring force is internal to the system, the momentum of the system is conserved. It should be apparent that the initial total momentum of the system ($\mathbf{P}_o$) is zero, and therefore the final momentum must also be zero. Thus, we may write

$$\mathbf{P}_o = \mathbf{P} = 0 \quad \text{and} \quad \mathbf{P} = \mathbf{p}_1 + \mathbf{p}_2 = 0$$

(The spring does not come into the equations because its velocity is zero.) Thus,

$$\mathbf{p}_2 = -\mathbf{p}_1$$

which means that the momenta of m_1 and m_2 are equal and opposite. Using directional signs (with $+x$ to the right in the figure), we get

$$p_2 - p_1 = m_2 v_2 - m_1 v_1 = 0$$

or in terms of magnitudes

$$m_2 v_2 = m_1 v_1$$

and

$$v_2 = \left(\frac{m_1}{m_2}\right) v_1 = \left(\frac{1.0 \text{ kg}}{2.0 \text{ kg}}\right)(1.8 \text{ m/s}) = 0.90 \text{ m/s}$$

Thus, the velocity of m_2 is 0.90 m/s in the positive x direction, or to the right in the figure. This is half v_1, as you might have expected, since m_2 has twice the mass of m_1.

Follow-up Exercise. (a) Suppose that the large block in Fig. 6.9 were attached to the Earth's surface so that the block could not move when the string was burned. Would momentum be conserved in this case? Explain. (b) Two girls, each having a mass of 50 kg, stand at rest on skateboards with negligible friction. The first tosses a 2.5-kilogram ball to the second. If the speed of the ball is 10 m/s, what is the speed of each girl after the ball is caught, and what is the momentum of the ball before it is tossed, while it is in the air, and after it is caught?

Demonstration 3 ■ Internal Forces (and Conservation of Momentum?)

In both cases, the initial total momentum is zero. What would happen after the actions (and reactions) begin? Actually, there is an external force (what is it?), so the momentum is not conserved and the motion is limited.

(a)

(a) Getting ready to push off. Note that the forces are internal to the system.

(b)

(b) Play ball! Here a ball is used to exchange momentum between parts of the system.

Conceptual Example 6.6

How Far? Conservation of Momentum with External Forces

Consider the situation in Fig. 6.9 again. Let's assume that there is friction and that the blocks are made from different materials such that the constant frictional forces acting on the moving blocks have the same magnitude. (Why would they have to be different materials?) Then, under these conditions, on coming to rest, (a) both blocks have traveled the same distance, (b) the lighter block has traveled twice the distance of the heavier one, (c) the lighter block has traveled four times the distance of the heavier one.

Reasoning and Answer.   The spring force is internal and initially the frictional forces cancel within the system, so the linear momentum is conserved and we can get the initial speeds as in Example 6.5. However, let's solve this example algebraically rather than by using numbers. (It's usually simpler to work with symbols.) To define the notation, we will say that $m_1 = m$ and $m_2 = 2m$ and will specify the initial velocities, v_{1_o} and v_{2_o}, to distinguish them from the final velocities, $v_1 = v_2 = 0$. (Properly defined notation is important.) Then, using the conservation of momentum as above, we have

$$v_{1_o} = \left(\frac{m_2}{m_1}\right)v_{2_o} = \left(\frac{2m}{m}\right)v_{2_o} = 2v_{2_o}$$

as shown previously with numbers.

   With the same constant frictional decelerating force, each block has a constant deceleration, so we can use a kinematic equation from Chapter 2: $v^2 = v_o^2 + 2ax$. With $v = 0$, the magnitude of x is given by

$$v^2 = 0 = v_o^2 - 2ax \quad \text{and} \quad x = \frac{v_o^2}{2a}$$

The distance, then, is proportional to the *square* of the initial speed. So, with twice the initial speed, does the lighter block travel four times the distance of the heavier block, making the answer (c)? We had better look more closely. Note that the acceleration also appears in the expression for the distance. The frictional forces (f) on each block have the same magnitude, but since the masses of the blocks are different, so are the accelerations. After the spring releases, the net force on each block is friction, f. Hence the acceleration of each is $a = f/m$, and the general expression for x becomes

$$x = \frac{mv_o^2}{2f}$$

Now we see that the distance is proportional not only to v^2 but also to m. Forming a ratio of x_1 and x_2, and using known ratio values, we get

$$\frac{x_1}{x_2} = \left(\frac{m_1}{m_2}\right)\left(\frac{v_{1_o}}{v_{2_o}}\right)^2 = (\tfrac{1}{2})(2)^2 = 2$$

Thus, $x_1 = 2x_2$, or the lighter block travels twice the distance of the heavier one, so the answer is (b).

Follow-up Exercise.   If the magnitude of the constant frictional force on the lighter block were twice that on the heavier block, would the blocks then travel the same distance? Justify your answer.

Example 6.7 ■ A Glancing Collision: Components of Momentum

A moving shuffleboard puck has a glancing collision with a stationary one of the same mass, as shown in •Fig. 6.10. If friction is negligible, what are the speeds of the pucks after the collision?

Thinking It Through.   Even in a glancing collision, the momentum is conserved (with no net external force; the puck weights are canceled by normal forces). If the total momentum is conserved, then each *component* of the momentum is conserved.

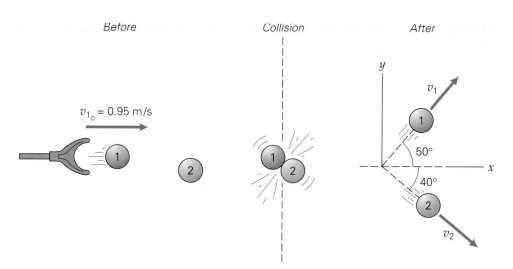

•**FIGURE 6.10 A glancing collision** Momentum is conserved in an isolated system. The motion in two dimensions may be analyzed in terms of the components of momentum, which are also conserved. See Example 6.7. Note that v_1 and v_2 form a 90° angle *only* if masses m_1 and m_2 are equal.

Solution.

Given: $v_{1_0} = 0.95$ m/s   **Find:** v_1 and v_2 (speeds after collision)
$\quad\quad\quad\theta_1 = 50°$
$\quad\quad\quad\theta_2 = 40°$
$\quad\quad\quad m_1 = m_2$

The collision forces of the two-puck system are internal forces, so the total momentum is conserved (the external force of friction is negligible for the very brief collision time). This means that the rectangular components of momentum are also conserved. That is, $\mathbf{P} = \mathbf{P}_x + \mathbf{P}_y$ is a constant, so $\mathbf{P}_x$ and $\mathbf{P}_y$ must be constant. For the x component of the momentum with signs indicating directions (positive axes shown in figure),

$$\overset{before}{P_{x_0}} = \overset{after}{P_{x_1} + P_{x_2}}$$

or

$$m_1 v_{1_0} = m_1 v_1 \cos 50° + m_2 v_2 \cos 40° \tag{1}$$

For the y component,

$$0 = P_{y_1} - P_{y_2}$$

or

$$0 = m_1 v_1 \sin 50° - m_2 v_2 \sin 40° \tag{2}$$

where the minus sign indicates that the component of puck 2 is in the negative y direction. With two equations and two unknowns, the rest is basic algebra. The masses cancel in Eq. 2 since they are equal ($m_1 = m_2$), and

$$v_2 = \left(\frac{\sin 50°}{\sin 40°}\right) v_1 = \left(\frac{0.766}{0.643}\right) v_1 = 1.19 v_1 \tag{3}$$

Substituting for v_2 in Eq. 1 and canceling the masses (as above) gives

$$v_{1_0} = v_1 \cos 50° + (1.19 v_1)(\cos 40°)$$

and

$$v_1 = \frac{v_{1_0}}{\cos 50° + (1.19)(\cos 40°)} = \frac{0.95 \text{ m/s}}{0.643 + (1.19)(0.766)} = 0.61 \text{ m/s}$$

Using this value in Eq. 3 gives

$$v_2 = 1.19 v_1 = 1.19(0.61 \text{ m/s}) = 0.73 \text{ m/s}$$

The directions of v_1 and v_2 are given by θ_1 and θ_2, respectively.

Follow-up Exercise.   (a) Is the kinetic energy conserved in the collision in this Example? Explain. (b) In such a collision, is it possible for both θ_1 and θ_2 to lie above the $+x$ axis? Why or why not?

The conservation of momentum is one of the most important principles in physics. As mentioned previously, it is used to analyze the collisions of objects ranging from subatomic particles to automobiles in traffic accidents. In many instances, external forces may be acting on the objects, which means that the momentum is not conserved.

But, as you will learn in the next section, the conservation of momentum often allows a good approximation *over the short time of a collision* because the internal forces, for which momentum is conserved, are much greater than the external forces. For example, external forces such as gravity and friction also act on colliding objects but are often relatively small compared to the internal forces. Therefore, if the objects interact for only a brief time, the effects of the external forces may be negligible compared to those of the large internal forces.

6.4 Elastic and Inelastic Collisions

Objective: To describe the conditions on kinetic energy and momentum in elastic and inelastic collisions.

Taking a closer look at collisions in terms of the conservation of momentum is simpler if we consider an isolated system, such as a system of particles (or balls) involved in head-on collisions. For simplicity, we will consider only collisions in one dimension. Such collisions can also be analyzed in terms of the conservation of energy. On the basis of what happens to the total kinetic energy, we can define two types of collisions: *elastic* and *inelastic*.

In an **elastic collision** (•Fig. 6.11a), the total kinetic energy is conserved. That is, the *total* kinetic energy of all the objects of the system after the collision is the same as their *total* kinetic energy before the collision. Kinetic energy may be traded between objects of a system, but the total remains constant. That is,

$$\begin{aligned} total\ K\ after = total\ K\ before \\ K_f = K_i \end{aligned} \qquad \begin{aligned} condition\ for\ an \\ elastic\ collision \end{aligned} \qquad (6.8)$$

Elastic collision—total kinetic energy conserved

During such a collision, some or all of the initial kinetic energy is temporarily converted to potential energy as the objects are deformed. But, after the maximum deformations, the objects elastically spring back to their original shapes, and the system regains all of its original kinetic energy. For example, two steel balls or two billiard balls may have a nearly elastic collision with each ball having the same shape afterward as before; that is, there is no permanent deformation.

In an **inelastic collision** (Fig. 6.11b), total kinetic energy is *not* conserved. For example, one or more of the colliding objects may not spring back to its original shape, or heat or sound may be generated. In such interactions, work is done

Note: In reality, only atoms and subatomic particles have truly elastic collisions, but some larger hard objects have nearly elastic collisions in which the kinetic energy is approximately conserved.

Inelastic collision—total kinetic energy not conserved

(a)

(b)

•**FIGURE 6.11 Collisions**
(a) Elastic collisions. (b) An inelastic collision.

by nonconservative forces (Section 5.5), such as friction, and some kinetic energy is lost:

$$\text{total } K \text{ after} < \text{total } K \text{ before}$$
$$K_f < K_i \qquad \qquad \text{condition for an} \qquad (6.9)$$
$$\text{inelastic collision}$$

For example, a hollow aluminum ball that collides with a solid steel ball may be dented. Permanently deforming the ball takes work, and that work is done at the expense of the original kinetic energy of the system.

For isolated systems, momentum is conserved in both elastic and inelastic collisions. For an inelastic collision, only an amount of kinetic energy consistent with the conservation of momentum may be lost. It may seem strange that kinetic energy can be lost and momentum still be conserved, but this fact provides insight into the difference between scalar and vector quantities.

Energy and Momentum in Inelastic Collisions

To see how momentum can remain constant while the kinetic energy changes (decreases) in inelastic collisions, consider the examples illustrated in •Fig. 6.12. In Fig. 6.12a, two balls of equal mass ($m_1 = m_2$) approach each other with equal and opposite velocities ($v_{1_o} = -v_{2_o}$). Hence, the total momentum before the collision is vectorially zero, but the scalar kinetic energy is *not* zero. After the collision the balls are stuck together and stationary, so the total momentum is unchanged—still zero. Momentum is conserved because the forces of collision are internal to the system of the two balls; thus there is no net external force on the system. The total kinetic energy, however, has decreased to zero. In this case, the kinetic energy went into the work done in permanently deforming the balls. Some energy may also have gone into doing work against friction (producing heat) or been lost in some other way (for example, in producing sound).

It should be noted that the balls need not come to rest after collision. In a less inelastic collision, the balls may recoil in opposite directions at reduced but equal speeds. The momentum would be conserved (still equal to zero—why?), but the

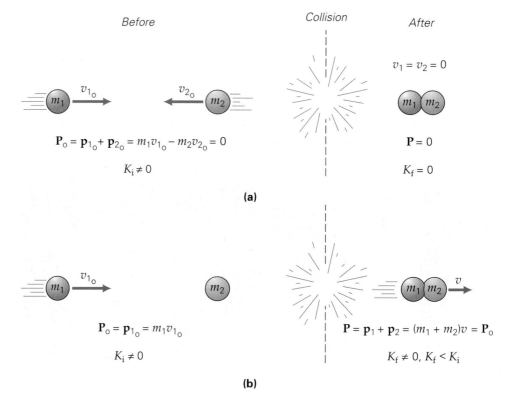

•**FIGURE 6.12 Inelastic collisions** In inelastic collisions, momentum is conserved but kinetic energy is not. Collisions like the ones shown here, in which the objects stick together, are called completely or totally inelastic collisions. The maximum amount of kinetic energy lost is consistent with the conservation of momentum.

kinetic energy would again not be conserved. Under all conditions, the amount of kinetic energy lost must be consistent with momentum conservation.

In Fig. 6.12b, one ball is initially at rest as the other approaches. The balls stick together after collision but are still in motion. Both of these cases are examples of a **completely inelastic collision**, in which the objects stick together and hence have the same velocity after colliding.

Assume that the balls in Fig. 6.12b have different masses. Since the momentum is conserved even in inelastic collisions,

$$\underset{\text{before}}{m_1 v_0} = \underset{\text{after}}{(m_1 + m_2)v}$$

and

$$v = \left(\frac{m_1}{m_1 + m_2}\right)v_0 \qquad \begin{array}{l}\textit{m}_2 \textit{ initially at rest}\\ \textit{(completely inelastic}\\ \textit{collision only)}\end{array} \qquad (6.10)$$

Thus, v is less than v_0, since $m_1/(m_1 + m_2)$ must be less than 1. Now let us consider how much kinetic energy has been lost. Initially, $K_i = \frac{1}{2}m_1 v_0^2$; finally, after the collision,

$$K_f = \frac{1}{2}(m_1 + m_2)v^2$$

Substituting for v from Eq. 6.10 and simplifying the result, we have

$$K_f = \frac{1}{2}(m_1 + m_2)\left(\frac{m_1 v_0}{m_1 + m_2}\right)^2 = \frac{\frac{1}{2}m_1^2 v_0^2}{m_1 + m_2} = \left(\frac{m_1}{m_1 + m_2}\right)\frac{1}{2}m_1 v_0^2 = \left(\frac{m_1}{m_1 + m_2}\right)K_i$$

and

$$\frac{K_f}{K_i} = \frac{m_1}{m_1 + m_2} \qquad \begin{array}{l}\textit{m}_2 \textit{ initially at rest}\\ \textit{(completely inelastic}\\ \textit{collision only)}\end{array} \qquad (6.11)$$

Equation 6.11 gives the fractional amount of the initial kinetic energy that the system has after a completely inelastic collision. For example, if the masses of the balls are equal ($m_1 = m_2$), then $m_1/(m_1 + m_2) = \frac{1}{2}$ and $K_f/K_i = \frac{1}{2}$, or $K_f = K_i/2$. That is, only half of the initial kinetic energy is lost (consistent with the conservation of momentum).

Note that all the kinetic energy cannot be lost in this case no matter what the masses of the balls are. The momentum after collision cannot be zero, since it was not zero initially. Thus, the balls must be moving and must have at least some kinetic energy.

Example 6.8 ■ Stuck Together: Completely Inelastic Collision

A 1.0-kilogram ball with a speed of 4.5 m/s strikes a 2.0-kilogram stationary ball. If the collision is completely inelastic, (a) what are the speeds of the balls after the collision? (b) What percentage of the initial kinetic energy do they have after the collision? (c) What is the total momentum after the collision?

Thinking It Through. Recall the definition of a *completely inelastic collision*. The balls stick together after collision; kinetic energy is *not* conserved, but total momentum is.

Solution. Using the labeling as in the preceding discussion, we have

Given: $m_1 = 1.0$ kg   *Find:* (a) v (speed after collision)
$\qquad\quad m_2 = 2.0$ kg
$\qquad\quad v_0 = 4.5$ m/s
$\qquad\qquad\qquad\qquad\qquad$ (b) $\dfrac{K_f}{K_i}(\times 100\%)$
$\qquad\qquad\qquad\qquad\qquad$ (c) $\mathbf{P}_f$ (total momentum after collision)

(a) The momentum is conserved and

$$\mathbf{P} = \mathbf{P}_0$$

or

$$(m_1 + m_2)v = m_1 v_o$$

The balls stick together and have the same speed after collision. This speed is then

$$v = \left(\frac{m_1}{m_1 + m_2}\right)v_o = \left(\frac{1.0\ \text{kg}}{1.0\ \text{kg} + 2.0\ \text{kg}}\right)(4.5\ \text{m/s}) = 1.5\ \text{m/s}$$

(b) The fractional part of the initial kinetic energy that the balls have after the completely inelastic collision is given by Eq. 6.11. Notice that this fraction, as given by the masses, is the same as that for the speeds (Eq. 6.10) in this special case. By inspection, we can write

$$\frac{K_f}{K_i} = \frac{m_1}{m_1 + m_2} = \frac{1.0\ \text{kg}}{1.0\ \text{kg} + 2.0\ \text{kg}} = \frac{1}{3} = 0.33\ (\times 100\%) = 33\%$$

Let's show this explicitly:

$$\frac{K_f}{K_i} = \frac{\frac{1}{2}(m_1 + m_2)v^2}{\frac{1}{2}m_1 v_o^2} = \frac{\frac{1}{2}(1.0\ \text{kg} + 2.0\ \text{kg})(1.5\ \text{m/s})^2}{\frac{1}{2}(1.0\ \text{kg})(4.5\ \text{m/s})^2} = 0.33\ (= 33\%)$$

Keep in mind that Eq. 6.11 applies *only* to *completely* inelastic collisions. For other types of collisions, the initial and final values of the kinetic energy must be computed explicitly.

(c) The momentum is conserved in all collisions (in the absence of external forces), so the total momentum after collision is the same as before. That value is the momentum of the incident ball, with a magnitude of

$$P_f = p_{1_o} = m_1 v_o = (1.0\ \text{kg})(4.5\ \text{m/s}) = 4.5\ \text{kg·m/s}$$

and in the same direction.

Follow-up Exercise. A small hard-metal ball of mass m collides with a larger, stationary soft-metal ball of mass M. A *minimum* amount of work W is required to make a dent in the larger ball. If the smaller ball initially has kinetic energy $K = W$, will the larger ball be dented in a completely inelastic collision?

___

Energy and Momentum in Elastic Collisions

For a general *elastic* collision of two objects,

$$\text{Conservation of } K: \quad \underset{before}{\tfrac{1}{2}m_1 v_{1_o}^2 + \tfrac{1}{2}m_2 v_{2_o}^2} = \underset{after}{\tfrac{1}{2}m_1 v_1^2 + \tfrac{1}{2}m_2 v_2^2} \qquad (6.12a)$$

Linear momentum is conserved even if the collision is inelastic, so we can write

$$\text{Conservation of } \mathbf{P}: \quad m_1\mathbf{v}_{1_o} + m_2\mathbf{v}_{2_o} = m_1\mathbf{v}_1 + m_2\mathbf{v}_2 \qquad (6.12b)$$

Thus for head-on elastic collisions, knowing the masses of the objects and the initial velocities allows you to find the final velocities.

A common collision situation is that in which one of the objects is initially stationary and then is hit head-on (•Fig. 6.13). Assume that the motion of both of the balls after the collision is to the right, or in the positive direction, as shown in the figure. If this is not the case, the math will tell you. (How?)

This is a one-dimensional situation. The condition for an elastic, head-on collision in this situation is

$$\tfrac{1}{2}m_1 v_{1_o}^2 = \tfrac{1}{2}m_1 v_1^2 + \tfrac{1}{2}m_2 v_2^2 \qquad (6.13a)$$

and conservation of linear momentum gives

$$m_1 v_{1_o} = m_1 v_1 + m_2 v_2 \qquad (6.13b)$$

Rearranging these equations gives

$$m_1(v_{1_o}^2 - v_1^2) = m_2 v_2^2 \qquad (6.13a')$$

and

$$m_1(v_{1_o} - v_1) = m_2 v_2 \qquad (6.13b')$$

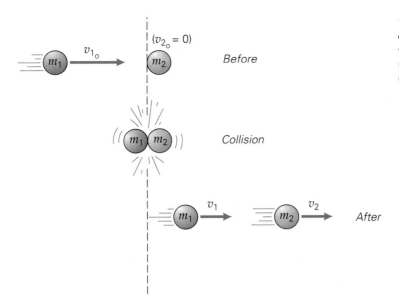

$(v_{2_0} = 0)$

v_{1_0}

m_1 m_2 *Before*

m_1 m_2 *Collision*

m_1 v_1 m_2 v_2 *After*

Then, using the algebraic relationship $x^2 - y^2 = (x + y)(x - y)$ and dividing the first equation by the second, we get

$$\frac{m_1(v_{1_0} + v_1)(v_{1_0} - v_1)}{m_1(v_{1_0} - v_1)} = \frac{m_2 v_2^2}{m_2 v_2} \quad \frac{(\text{Eq. 6.13a}')}{(\text{Eq. 6.13b}')} \; (divide)$$

or

$$v_{1_0} + v_1 = v_2 \tag{6.14}$$

Equation 6.14 can now be used to eliminate v_1 or v_2 from Eq. 6.13b. Thus, each of these final velocities can be expressed in terms of the initial velocity (v_{1_0}):

$$v_1 = \left(\frac{m_1 - m_2}{m_1 + m_2}\right) v_{1_0} \qquad \text{(6.15)}$$

(final velocities for elastic, head-on collision with m_2 initially stationary)

$$v_2 = \left(\frac{2m_1}{m_1 + m_2}\right) v_{1_0} \qquad \text{(6.16)}$$

Note that the final velocities depend on mass *ratios* of the objects. Looking at Eq. 6.15, you can see that if m_1 is greater than m_2, then v_1 is positive, or in the same direction as the initial velocity of the incoming ball, as was assumed in Fig. 6.13. However, if m_2 is greater than m_1, then v_1 is negative, and the incoming ball recoils in the opposite direction after collision. Equation 6.16 shows that v_2 is always in the same direction as the velocity of the incoming ball.

You can also get some general ideas about what happens after such a collision by considering three possible situations for the relative masses of the objects, illustrated in •Fig. 6.14.

Case 1. $m_1 = m_2$ (Fig. 6.14a). From Eqs. 6.15 and 6.16,

$$v_1 = 0 \quad \text{and} \quad v_2 = v_{1_0}$$

Elastic head-on collision, $m_1 = m_2$

That is, if the masses of the colliding objects are equal, the objects simply exchange momentum and energy. The incoming ball is stopped on collision, and the originally stationary ball moves off with the same velocity as that of the incoming ball, thus obviously conserving system kinetic energy and momentum. (A real-world example would be two billiard balls.)

Case 2. $m_1 \gg m_2$ (m_1 very much greater than m_2, Fig. 6.14b). In this case, m_2 can be ignored in the addition and subtraction with m_1 in Eqs. 6.15 and 6.16, and

Elastic head-on collision, $m_1 \gg m_2$

$$v_1 \approx v_{1_0} \quad \text{and} \quad v_2 \approx 2v_{1_0}$$

•**FIGURE 6.14 Special cases of head-on elastic collisions** (a) When a moving object collides elastically with a stationary object of equal mass, there is a complete exchange of momentum and energy. (b) When a very massive moving object collides elastically with a much less massive stationary object, the very massive object continues to move essentially as before, and the less massive object is given a velocity almost twice the initial velocity of the large mass. (c) When a moving object of small mass collides elastically with a very massive stationary object, the incoming object recoils in the opposite direction with approximately the same speed and the very massive object remains essentially stationary.

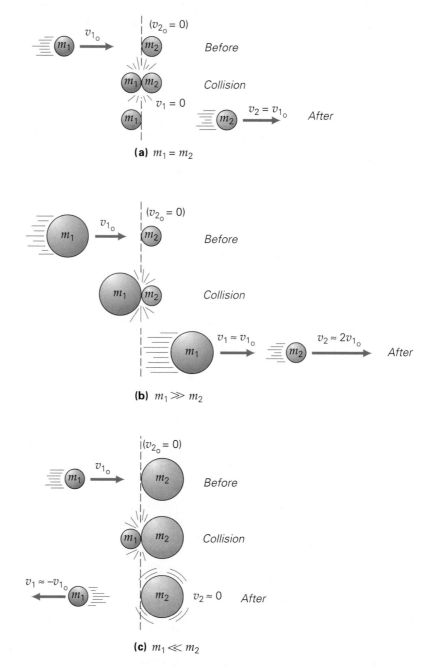

(a) $m_1 = m_2$

(b) $m_1 \gg m_2$

(c) $m_1 \ll m_2$

This tells you that if a very massive object collides with a stationary light object, the massive object is slowed down only slightly by the collision, and the light object is knocked away with a velocity almost twice that of the initial velocity of the massive object. (Think of a bowling ball hitting a pin.)

Elastic head-on collision,
$m_1 \ll m_2$

Case 3. $m_1 \ll m_2$ (m_1 very much less than m_2, Fig. 6.14c). Here, m_1 can be ignored in the addition and subtraction with m_2 in Eqs. 6.15 and 6.16, and

$$v_1 \approx -v_{1_o} \quad \text{and} \quad v_2 \approx 0$$

(In the second equation, the approximation $m_1/m_2 \approx 0$ is made.) Thus, if a light object collides with a massive stationary one, the massive object remains *almost* stationary, and the light object recoils backward with approximately the same speed it had before collision. An extreme case of this type is similar to a particle striking a solid, immovable wall (see Fig. 6.3). If the wall is anchored to the ground, then the Earth and wall recoil together, but this is totally unnoticeable (why?). Total momentum, however is still conserved. For the case in Fig. 6.14c, the massive ball must move a bit after the collision to conserve momentum. (Think of throwing a Superball at a stationary bowling ball.)

Example 6.9 ■ Elastic Collision: Conservation of Momentum and Kinetic Energy

A 0.30-kilogram object with a speed of 2.0 m/s in the positive x direction has a head-on elastic collision with a stationary 0.70-kilogram object located at $x = 0$. What is the distance separating the objects 2.5 s after the collision?

Thinking It Through. The incoming object is less massive than the stationary one. Even though it is not considerably less massive, we might expect the objects to separate in opposite directions after collision. Equations 6.15 and 6.16 will give the velocities; knowing the time, we can find the distances traveled or the separation distance after collision.

Solution. Using the previous notation, we have

Given: $m_1 = 0.30$ kg   *Find:* $\Delta x = x_2 - x_1$ (separation distance
    $v_{1_o} = 2.0$ m/s   2.5 s after collision)
    $m_2 = 0.70$ kg
    $v_{2_o} = 0$
    $t = 2.5$ s

From Eqs. 6.15 and 6.16, the velocities after collision are

$$v_1 = \left(\frac{m_1 - m_2}{m_1 + m_2}\right)v_{1_o} = \left(\frac{0.30 \text{ kg} - 0.70 \text{ kg}}{0.30 \text{ kg} + 0.70 \text{ kg}}\right)(2.0 \text{ m/s}) = -0.80 \text{ m/s}$$

$$v_2 = \left(\frac{2m_1}{m_1 + m_2}\right)v_{1_o} = \left[\frac{2(0.30 \text{ kg})}{0.30 \text{ kg} + 0.70 \text{ kg}}\right](2.0 \text{ m/s}) = 1.2 \text{ m/s}$$

Here m_1 is less than m_2, but not so *much* less that it could be ignored, as was done in part (c) of Fig. 6.14.

The objects are separating after collision, and their positions relative to the collision point ($x_o = 0$) are

$$x_1 = v_1 t = (-0.80 \text{ m/s})(2.5 \text{ s}) = -2.0 \text{ m}$$

$$x_2 = v_2 t = (1.2 \text{ m/s})(2.5 \text{ s}) = 3.0 \text{ m}$$

So

$$\Delta x = x_2 - x_1 = 3.0 \text{ m} - (-2.0 \text{ m}) = 5.0 \text{ m}$$

The objects are 5.0 m apart at that time.

Follow-up Exercise. Suppose the objects in this Example had equal masses. What would be the separation distance 2.5 s after collision in this case?

Example 6.10 ■ Spare! Conservation of Momentum and Kinetic Energy

A 7.1-kilogram bowling ball with a speed of 6.0 m/s has a head-on elastic collision with a stationary 1.6-kilogram pin. (a) What is the velocity of each object after the collision? (b) What is the total momentum after the collision?

Thinking It Through. If the ball and the pin have an elastic collision, then Eqs. 6.15 and 6.16 apply and will give the velocities. (The ball is quite a bit more massive than the pin, so as in Case 2 above, the ball and the pin will go in same direction after collision—as you would expect.) The total momentum is conserved, so we can calculate it from the initial total momentum. (Why?)

Solution. Listing the data with the same notation as before, we have

Given: $m_1 = 7.1$ kg   *Find:* (a) $\mathbf{v}_1$ and $\mathbf{v}_2$ (velocities after collision)
    $v_{1_o} = 6.0$ m/s   (b) $\mathbf{P}$ (total momentum after collision)
    $m_2 = 1.6$ kg
    $v_{2_o} = 0$

(a) The velocities are given directly by Eqs. 6.15 and 6.16, and the direction of the incoming ball is taken as positive:

$$v_1 = \left(\frac{m_1 - m_2}{m_1 + m_2}\right)v_{1_0} = \left(\frac{7.1 \text{ kg} - 1.6 \text{ kg}}{7.1 \text{ kg} + 1.6 \text{ kg}}\right)(6.0 \text{ m/s}) = 3.8 \text{ m/s}$$

$$v_2 = \left(\frac{2m_1}{m_1 + m_2}\right)v_{1_0} = \left[\frac{2(7.1 \text{ kg})}{(7.1 \text{ kg} + 1.6 \text{ kg})}\right](6.0 \text{ m/s}) = 9.8 \text{ m/s}$$

So both objects move in the same direction. Here m_1 is greater than m_2, but not *so much* greater that m_2 could be ignored, as was done in part (b) of Fig. 6.14.

(b) Momentum is conserved in elastic (and inelastic) collisions, so the total momentum afterward is the same as that before the collision. Thus the total momentum is that of the incident ball, in the same direction and with a magnitude of

$$P_f = P_{1_0} = m_1 v_{1_0} = (7.1 \text{ kg})(6.0 \text{ m/s}) = 43 \text{ kg·m/s}$$

Follow-up Exercise. In general, for head-on, elastic collisions with one of the objects initially stationary, is it possible for the velocities to be the same after collision? Explain.

Conceptual Example 6.11

Two In, One Out?

A novelty collision device, as shown in •Fig. 6.15, consists of five identical metal balls. When one ball swings in, after multiple collisions, one ball swings out at the other end of the row of balls. When two balls swing in, two swing out; when three swing in, three swing out, and so on—always the same number out as in.

Suppose that two balls, each of mass m, swing in at velocity v and collide with the next ball. Why doesn't one ball swing out at the other end with a velocity $2v$?

Reasoning and Answer. The collisions along the horizontal row of balls are approximately elastic. Two balls in and one out with twice the velocity wouldn't violate the conservation of momentum: $(2m)v = m(2v)$. However, there's another condition if we assume elastic collisions—the conservation of kinetic energy. Let's check to see if this is so for this case:

$$K_0 = K$$

$$\underset{before}{\tfrac{1}{2}(2m)v^2} \overset{?}{=} \underset{after}{\tfrac{1}{2}m(2v)^2}$$

$$mv^2 \neq 2mv^2$$

Hence, the kinetic energy would *not* be conserved if this happened, and the equation is telling us that this situation violates established physical principles and does not occur. Note that there's a big violation—more energy out than in.

Follow-up Exercise. Suppose the first ball of mass m were replaced with a ball of mass $2m$. When this ball is pulled back and allowed to swing in, how many balls will swing out? [*Hint*: Think about the analogous situation in Fig. 6.14 and remember that the balls in the row are actually colliding. It may help to think of them as being separated.]

•**FIGURE 6.15 In and out** See Conceptual Example 6.11.

6.5 Center of Mass

Objectives: To (a) explain the concept of the center of mass and compute its location for simple systems, and (b) describe how the center of mass and center of gravity are related.

The conservation of total momentum gives us a method of analyzing a "system of particles." Such a system may be virtually anything—a volume of gas, water in a container, or a baseball. Another important concept allows us to analyze the overall motion of a system of particles. It involves representing the whole system as a single particle. This concept will be introduced here and applied in more detail in the upcoming chapters.

If no net external force acts on a particle, we have seen that its linear momentum is constant. Similarly, if no net external force acts on a *system* of particles, the linear momentum of the system is constant. This similarity implies that a system of particles might be represented by an *equivalent* single particle. Moving rigid objects, such as balls, automobiles, and so forth, are essentially systems of particles and can be effectively represented by equivalent single particles when we analyze motion. Such representation is done through the concept of the **center of mass (CM)**:

> The center of mass is the point at which all of the mass of an object or system may be considered to be concentrated, for the purposes of linear or translational motion only.

Even if a rigid object is rotating, an important result (beyond the scope of this text to derive) is that the center of mass moves as though it were a particle (•Fig. 6.16). The center of mass is sometimes described as the *balance point* of a solid object. For example, if you balance a meterstick on your finger, the center of mass of the stick is located directly above your finger and all of the mass (or weight) seems to be concentrated there.

An expression similar to Newton's second law for a single particle applies to a *system* when the center of mass is used:

$$\mathbf{F}_{\text{net external}} = M\mathbf{A}_{\text{CM}} \qquad (6.17)$$

where $\mathbf{F}$ is the net external force on the system, M is the total mass of the system or the sum of the masses of the particles of the system ($M = m_1 + m_2 + m_3 + \cdots + m_n$, where the system has n particles), and $\mathbf{A}_{\text{CM}}$ is the acceleration of the center of mass of the system. In words, Eq. 6.17 says that the *center of mass* of a system of particles moves as though all the mass of the system were concentrated there and acted on by the resultant of the external forces. Note that the movement of the parts of the system is *not* predicted by Eq. 6.17.

It follows that *if the net external force on a system is zero*, the total linear momentum of the center of mass is conserved (stays constant) because

$$\mathbf{F} = M\mathbf{A}_{\text{CM}} = M(\Delta\mathbf{V}_{\text{CM}}/\Delta t) = \Delta(M\mathbf{V}_{\text{CM}})/\Delta t = \Delta\mathbf{P}/\Delta t \qquad (6.18)$$

Thus the total momentum of the system, $\mathbf{P} = M\mathbf{V}_{\text{CM}}$, is constant. Since M is constant (why?), in this case $\mathbf{V}_{\text{CM}}$ is a constant. Thus the center of mass either moves with a constant velocity or remains at rest.

•**FIGURE 6.16 Center of mass** **(a)** The center of mass of this sliding wrench moves as though it were a particle. Note the white dot on the wrench that marks the center of mass. **(b)** Even after exploding, the center of mass of a fireworks projectile would follow a parabolic path in the absence of air resistance.

(a)

(b)

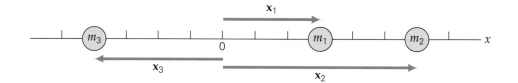

•FIGURE 6.17 **System of particles in one dimension**
Where is the system's center of mass? See Example 6.12.

Although you may more readily visualize the center of mass of a solid object, the concept of the center of mass applies to any system of particles or objects, even a quantity of gas. For a system of n particles arranged in one dimension, along the x axis (•Fig. 6.17), the location of the center of mass is given by

$$X_{CM} = \frac{m_1x_1 + m_2x_2 + m_3x_3 + \cdots + m_nx_n}{m_1 + m_2 + m_3 + \cdots + m_n} \qquad (6.19)$$

That is, X_{CM} is the x coordinate of the center of mass of a system of particles. In shorthand notation (using signs to indicate vector directions),

$$X_{CM} = \frac{\Sigma_i m_i x_i}{M} \qquad (6.20)$$

where Σ_i is the summation of the products $m_i x_i$ for n particles ($i = 1, 2, 3, \ldots, n$). If $\Sigma_i m_i x_i = 0$, then $X_{CM} = 0$, and the center of mass of the one-dimensional system is located at the origin.

Other coordinates of the center of mass for systems of particles are similarly defined. For a two-dimensional distribution of masses, the coordinates of the center of mass are (X_{CM}, Y_{CM}).

Example 6.12 ■ Finding the Center of Mass: A Summation Process

Three masses, 2.0 kg, 3.0 kg, and 6.0 kg, are located at positions (3.0, 0), (6.0, 0), and (−4.0, 0), respectively, in meters from the origin (•Fig. 6.17). Where is the center of mass of this system?

Thinking It Through. The masses and the positions are given, so the definition of X_{CM} (Eq. 6.20) applies. However, keep in mind that the positions are vector displacements from the origin and these are indicated by the appropriate signs (+ or −).

Solution.

Given: $m_1 = 2.0$ kg   *Find:* X_{CM} (CM coordinate)
$m_2 = 3.0$ kg
$m_3 = 6.0$ kg
$x_1 = 3.0$ m
$x_2 = 6.0$ m
$x_3 = -4.0$ m

Then, simply performing the summation as indicated in Eq. 6.20,

$$X_{CM} = \frac{\Sigma_i m_i x_i}{M}$$

$$= \frac{(2.0\text{ kg})(3.0\text{ m}) + (3.0\text{ kg})(6.0\text{ m}) + (6.0\text{ kg})(-4.0\text{ m})}{2.0\text{ kg} + 3.0\text{ kg} + 6.0\text{ kg}} = 0$$

The center of mass is at the origin.

Follow-up Exercise. Describe a single particle that could replace the system in this Example in order to analyze the translational part of the motion of the system as a whole.

Example 6.13 ■ A Dumbbell: Center of Mass Revisited

A dumbbell (•Fig. 6.18) has a connecting bar of negligible mass. Find the location of the center of mass (a) if m_1 and m_2 are each 5.0 kg, and (b) if m_1 is 5.0 kg and m_2 is 10.0 kg.

Thinking It Through. This Example shows how the location of the center of mass depends on mass distribution. In (b), you might expect the center of mass to be located closer to the more massive end of the dumbbell.

Solution.

Given: $x_1 = 0.20$ m *Find:* (a) (X_{CM}, Y_{CM}) (CM coordinates)
$x_2 = 0.90$ m   (b) (X_{CM}, Y_{CM})
$y_1 = y_2 = 0.10$ m
(a) $m_1 = m_2 = 5.0$ kg
(b) $m_1 = 5.0$ kg
$m_2 = 10.0$ kg

Note that each mass is considered to be a particle located at the center of the sphere (its center of mass).

(a) Finding X_{CM} gives

$$X_{CM} = \frac{m_1 x_1 + m_2 x_2}{m_1 + m_2}$$

$$= \frac{(5.0 \text{ kg})(0.20 \text{ m}) + (5.0 \text{ kg})(0.90 \text{ m})}{5.0 \text{ kg} + 5.0 \text{ kg}} = 0.55 \text{ m}$$

Similarly, we find that $Y_{CM} = 0.10$ m. (You might have seen this right away, since each center of mass is at this height.) The center of mass of the dumbbell is then located at $(X_{CM}, Y_{CM}) = (0.55$ m, 0.10 m), or midway between the end masses.

(b) With $m_2 = 10.0$ kg,

$$X_{CM} = \frac{m_1 x_1 + m_2 x_2}{m_1 + m_2}$$

$$= \frac{(5.0 \text{ kg})(0.20 \text{ m}) + (10.0 \text{ kg})(0.90 \text{ m})}{5.0 \text{ kg} + 10.0 \text{ kg}} = 0.67 \text{ m}$$

which is $\frac{2}{3}$ of the way between the masses. (Note that the distance of the CM from the center of m_1 is $\Delta x = 0.67$ m $- 0.20$ m $= 0.47$ m; with the distance $L = 0.70$ m between the centers of the masses, then $\Delta x/L = 0.47$ m$/0.70$ m $= 0.67$, or $\frac{2}{3}$.) You might expect the balance point of the dumbbell in this case to be closer to m_2. The y coordinate of the center of mass is again $Y_{CM} = 0.10$ m, as you can prove for yourself.

Follow-up Exercise. In part (b) of this Example, take the origin of the coordinate axes to be at the point where m_1 touches the x axis. What are the coordinates of the CM in this case, and how does its location compare to that found in the Example?

In Example 6.13, when the value of one of the masses changed, the x coordinate of the center of mass changed. You might have expected the y coordinate to change also. However, the centers of the end masses were still at the same height, and Y_{CM} remained the same. To increase Y_{CM}, one or both of the end masses would have to be in a higher position.

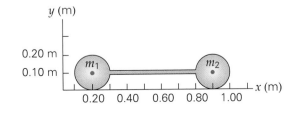

•**FIGURE 6.18 Location of the center of mass** See Example 6.13.

Now let's see how the center of mass can be applied to a realistic situation.

Example 6.14 ■ Internal Motion: Where's the Center of Mass?

A 75.0-kilogram man stands in the far end of a 50.0-kilogram boat 100 m from the shore, as illustrated in •Fig. 6.19. If he walks to the other end of the 6.00-meter-long boat, how far is the man then from the shore? Neglect friction, and assume that the center of mass of the boat is at its midpoint.

Thinking It Through.   The answer is *not* 100 m − 6.00 m = 94.0 m, because the boat moves as the man walks. Why? With no net external force, the acceleration of the center of mass of the man–boat system is zero (Eq. 6.17), and so is the total momentum by Eq. 6.18 ($\mathbf{P} = M\mathbf{V}_{CM} = 0$). Hence, the velocity of the center of mass of the system is zero, or the center of mass is stationary and remains so to conserve system momentum; that is, X_{CM} (initial) = X_{CM} (final).

Solution.   Taking the shore as the origin ($x = 0$), we have

Given: $m_m = 75.0$ kg   *Find:* x_{m_f} (distance of man from shore)
 $x_{m_i} = 100$ m
 $m_b = 50.0$ kg
 $x_{b_i} = 94.0$ m + 3.00 m = 97.0 m
 (where the location of the boat is taken at its center of mass)

Note that if we take the man's final position to be a distance x_{m_f} from the shore, then the final position of the boat's center of mass will be $x_{b_f} = x_{m_f} + 3.00$ m, since he will be at the front of the boat.
   Then initially,

$$X_{CM_i} = \frac{m_m x_{m_i} + m_b x_{b_i}}{m_m + m_b}$$

$$= \frac{(75.0 \text{ kg})(100 \text{ m}) + (50.0 \text{ kg})(97.0 \text{ m})}{75.0 \text{ kg} + 50.0 \text{ kg}} = 98.8 \text{ m}$$

And finally,

$$X_{CM_f} = \frac{m_m x_{m_f} + m_b x_{b_f}}{m_m + m_b}$$

$$= \frac{(75.0 \text{ kg})x_{m_f} + (50.0 \text{ kg})(x_{m_f} + 3.00 \text{ m})}{75.0 \text{ kg} + 50.0 \text{ kg}} = 98.8 \text{ m}$$

Here X_{CM_f} = 98.8 m = X_{CM_i} since the CM does not move. Then, solving for x_{m_f}, we get

$$(125 \text{ kg})(98.8 \text{ m}) = (125 \text{ kg})x_{m_f} + (50.0 \text{ kg})(3.00 \text{ m})$$

and

$$x_{m_f} = 97.6 \text{ m}$$

from the shore.

Follow-up Exercise.   Suppose the man then walks back to his original position at the opposite end of the boat. Would he be 100 m from shore?

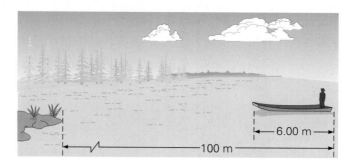

•**FIGURE 6.19 Walking toward shore**
See Example 6.14.

Center of Gravity

As you know, mass and weight are directly related. Closely associated with the center of mass is the concept of the **center of gravity (CG)**, the point where all of the weight of an object may be considered to be concentrated in representing the object as a particle. If the acceleration due to gravity is constant both in magnitude and direction over the extent of the object, Eq. 6.20 can be rewritten as

$$MgX_{CM} = \Sigma_i m_i g x_i \qquad (6.21)$$

Then, the weight, Mg, acts as if it were concentrated at X_{CM}, and the center of mass and the center of gravity coincide. As you may have noticed, the location of the center of gravity was implied in some previous figures in Chapter 4, where the vector arrows for weight ($\mathbf{w} = m\mathbf{g}$) were drawn from a point at or near the center of an object.

In some cases, the center of mass or the center of gravity of an object may be located by symmetry. For example, for a spherical object that is homogeneous (the mass is distributed evenly throughout), the center of mass is at the geometrical center (or center of symmetry). In Example 6.13a, where the end masses of the dumbbell were equal, it was probably apparent that the center of mass was midway between them.

The location of the center of mass of an irregularly shaped object is not so evident and is usually difficult to calculate (even with advanced mathematical methods that are beyond the scope of this book). In some instances, the center of mass may be located experimentally. For example, the center of mass of a flat, irregularly shaped object can be determined experimentally by suspending it freely from different points (•Fig. 6.20). A moment's thought should convince you that the center of mass (or center of gravity) always lies vertically below the point of suspension. Since the center of mass is defined as the point at which all the mass of a body can be considered to be concentrated, this is analogous to a particle of

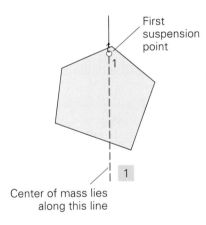

Center of mass lies along this line

Center of mass also lies along this line

(a)

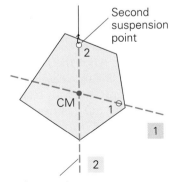

(b)

•**FIGURE 6.20 Location of the center of mass by suspension** (a) The center of mass of a flat, irregularly shaped object can be found by suspending the object from two or more points. The CM (and CG) lies on a vertical line under any point of suspension, so the intersection of two such lines marks its location midway through the thickness of the body. The sheet could be balanced horizontally at this point. Why? **(b)** The process is illustrated with a cutout map of the United States. Note that a plumb line dropped from any other point (third photo) does in fact pass through the CM as located in the first two photos.

(a)

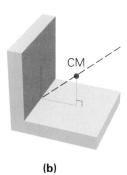

(b)

•**FIGURE 6.21 The center of mass may be located outside of a body** The center of mass may lie either inside or outside of a body, depending on its mass distribution. **(a)** For a uniform ring, the center of mass is at its center. **(b)** For an L-shaped object, if the mass distribution is uniform and the legs are of equal length, the center of mass lies on the diagonal between the legs.

•**FIGURE 6.22 Center of gravity** By arching her body, this high jumper can get over the bar even though her center of gravity passes beneath it.

mass suspended from a string. Suspending the object from two or more points and marking the vertical lines on which the center of mass must lie locates the center of mass as the intersection of the lines.

The center of mass of an object may lie outside the body of the object (•Fig. 6.21). The center of mass of a homogeneous ring is at the ring's center. The mass in any section of the ring is compensated by the mass in an equivalent section directly across the ring, and by symmetry the center of mass is at the center. For an L-shaped object with equal legs, the center of mass lies on a line that makes a 45° angle with both legs. Its location can easily be determined by suspending the L from a point on one of the legs and noting where a vertical line from that point intersects the diagonal line.

Keep in mind that the location of the center of mass or center of gravity of an object depends on the distribution of mass. Therefore, for a flexible object such as the human body, the position of the center of gravity changes as the object changes configuration (mass distribution). For example, when a person raises both arms overhead, his or her center of gravity is raised several centimeters. For a high jumper going over a bar, the center of gravity lies outside the arched body (•Fig. 6.22). In fact, the center of gravity passes *beneath* the bar. This is done purposefully because work must be done to raise the center of gravity, and only the jumper's body has to clear the bar, not the CG.

6.6 Jet Propulsion and Rockets

Objective: To apply the conservation of momentum in the explanation of jet propulsion and the operation of rockets.

The word *jet* is sometimes used to refer to a stream of liquid or gas emitted at a high speed—for example, a jet of water from a fountain or a jet of air from an automobile tire. **Jet propulsion** is the application of such jets to the production of motion. This concept usually brings to mind jet planes and rockets, but squid and octopi propel themselves by squirting jets of water.

You have probably tried the simple application of blowing up a balloon and releasing it. Lacking any guidance or rigid exhaust system, the balloon zig-zags around, driven by the escaping air. In terms of Newton's third law, the air is forced out by the contraction of the stretched balloon—that is, the balloon exerts a force on the air. Thus there must be an equal and opposite reaction force exerted by the air on the balloon. It is this force that propels the balloon on its erratic path.

Jet propulsion is explained by Newton's third law, and in the absence of external forces, the conservation of momentum also applies. You may understand this better by considering the recoil of a rifle, taking the rifle and the bullet as an isolated system (•Fig. 6.23). Initially, the total momentum of this system is zero. When the rifle is fired (by remote control to avoid external forces), the expansion of the gases from the exploding charge accelerates the bullet down the barrel. These gases push backward on the rifle as well, producing a recoil force (the "kick" experienced by a person firing a weapon). Since the initial momentum of the system is zero and the force of the expanding gas is an internal force, the momenta of the bullet and of the rifle must be exactly equal and opposite at any instant. After the bullet leaves the barrel, there is no propelling force, so the bullet and the rifle move with constant velocities (unless acted on by a net external force such as gravity or air resistance).

Similarly, the thrust of a rocket is created by exhausting the gas from burning fuel out the rear of the rocket. The expanding gas exerts a net force on the rocket that propels the rocket in the forward direction (•Fig. 6.24a,b). The rocket exerts a reaction force on the gas, so all of the gas is directed out the exhaust nozzle. If the rocket is at rest when the engines are turned on and there are no external forces (as in deep space, where friction is zero and gravitational forces are negligible), then the instantaneous momentum of the exhaust gas is equal and opposite to that of the rocket. The numerous exhaust gas molecules have small masses and high

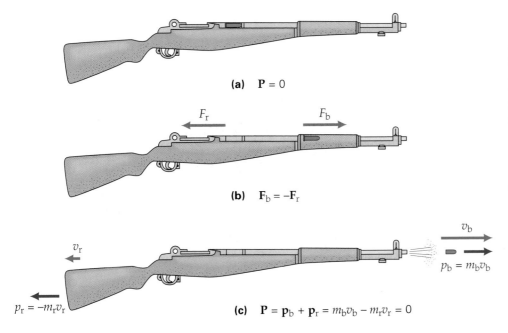

(a) $\mathbf{P} = 0$

(b) $\mathbf{F}_b = -\mathbf{F}_r$

v_b

$p_b = m_b v_b$

v_r

$p_r = -m_r v_r$

(c) $\mathbf{P} = \mathbf{p}_b + \mathbf{p}_r = m_b v_b - m_r v_r = 0$

•**FIGURE 6.23 Conservation of momentum** **(a)** Before the rifle is fired, the total momentum of the rifle and bullet (as an isolated system) is zero. **(b)** During firing, there are equal and opposite internal forces, and the instantaneous total momentum of the rifle-bullet system remains zero (neglecting external forces, such as arise when a rifle is being held). **(c)** When the bullet leaves the barrel, the total momentum of the system is still zero.

velocities, and the rocket has a much larger mass and a smaller velocity. (See Demonstration 4.)

Unlike a rifle firing a single shot, which has negligible mass, a rocket continuously loses mass when burning fuel (it is more like a machine gun). Thus, the rocket is a system for which the mass is not constant. As the mass of the rocket decreases, it accelerates more easily. Multistage rockets take advantage of this fact. The hull of a burnt-out stage is jettisoned to give a further in-flight reduction in mass (Fig. 6.24c). The payload (cargo) is typically a very small part of the initial mass of rockets for space flights.

Suppose that the purpose of a space flight is to land a payload on the Moon. At some point on the journey, the gravitational attraction of the Moon will become greater than that of Earth, and the spacecraft will accelerate toward the Moon. A soft landing is desirable, so the spacecraft must be slowed down enough to go into orbit around the Moon. This is done by using the rocket engines to apply a *reverse thrust*, or braking thrust. The spacecraft is maneuvered through a 180° angle, or turned around, which is quite easy to do in space. The rocket engines are then fired, expelling the exhaust gas toward the Moon and supplying a braking action.

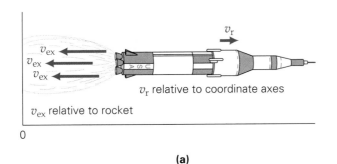

v_r

v_{ex}
v_{ex}
v_{ex}

v_r relative to coordinate axes

v_{ex} relative to rocket

0

(a)

•**FIGURE 6.24 Jet propulsion and mass reduction** **(a)** A rocket burning fuel is continuously losing mass and so becomes easier to accelerate. The resulting force on the rocket (the thrust) depends on the product of the rate of change of its mass with time and the velocity of the exhaust gases: $(\Delta m/\Delta t)\mathbf{v}_{ex}$. Since the mass is decreasing, $(\Delta m/\Delta t)$ is negative and the thrust is opposite $\mathbf{v}_{ex}$. **(b)** The space shuttle uses a multistage rocket. Both the two booster rockets and the huge external fuel tank are jettisoned in flight. **(c)** The first and second stages of a *Saturn V* rocket separating after 148 s of burn time.

(b)

(c)

Demonstration 4 ■ Jet Propulsion

A demonstration of Newton's third law and the conservation of momentum. For a rocket or jet engine, thrust is obtained by burning fuel and exhausting the gas out the rear of the engine. Here a fire extinguisher exhausting CO_2 takes the place of the engine.

(a) Ignition and blast off!

(b) Away we go—equal and opposite forces.

You have experienced a reverse thrust effect if you have flown in a commercial jet. In this instance, however, the craft is not turned around. Instead, after touchdown, the jet engines are revved up and a braking action can be felt. Ordinarily, revving up the engines accelerates the plane forward. The reverse thrust is accomplished by activating thrust reversers in the engines that deflect the exhaust gases forward (•Fig. 6.25). The gas experiences an impulse force and a change in momentum in the forward direction (see Fig. 6.3b), and the engine and the aircraft have an equal and opposite momentum change and braking impulse force.

Question: There are no end-of-chapter exercises for this section, so test your knowledge with this one: Astronauts use hand-held maneuvering devices (small rockets) to move around on space walks. Describe how these rockets would be used. Is there any danger on an untethered space walk?

•FIGURE 6.25 Reverse thrust
Thrust reversers are activated on jet engines on landing to help slow the plane. The gas experiences an impulse force and a change in momentum in the forward direction, and the plane experiences an equal and opposite momentum change and a braking impulse force.

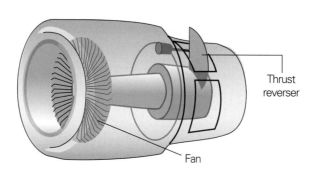

Normal operation

Thrust reverser

Fan

Thrust reverser activated

Chapter Review

Important Terms

linear momentum   *173*
total linear momentum   *173*
impulse   *178*
impulse–momentum theorem   *178*

conservation of linear
   momentum   *183*
elastic collision   *187*
inelastic collision   *187*

completely inelastic collision   *189*
center of mass (CM)   *195*
center of gravity (CG)   *199*
jet propulsion   *200*

Important Concepts

- The linear momentum of a particle is a vector and is defined as the product of mass and velocity.
- The total linear momentum of a system is the vector sum of the momenta of the individual particles.
- In the absence of a net external force, the total linear momentum of a system is conserved.
- The impulse–momentum theorem relates impulse to the change in momentum.

- A collision is elastic if the total kinetic energy is conserved. Momentum is conserved in both elastic and inelastic collisions. In a completely inelastic collision, objects stick together after impact.
- The center of mass is the point at which all of the mass of an object or system may be considered to be concentrated. (The center of gravity is the point where all the weight may be considered to be concentrated.)

Important Equations

Linear Momentum:
$$\mathbf{p} = m\mathbf{v} \tag{6.1}$$

Total Linear Momentum of a System:
$$\mathbf{P} = \mathbf{p}_1 + \mathbf{p}_2 + \mathbf{p}_3 + \cdots = \Sigma_i \mathbf{p}_i \tag{6.2}$$

Newton's Second Law in Terms of Momentum:
$$\mathbf{F} = \frac{\Delta \mathbf{p}}{\Delta t} \tag{6.3}$$

Impulse–Momentum Theorem:
$$\text{Impulse} = \overline{\mathbf{F}}\,\Delta t = \Delta \mathbf{p} = m\mathbf{v} - m\mathbf{v}_\text{o} \tag{6.5}$$

Conditions for an Elastic Collision:
$$\mathbf{P}_\text{f} = \mathbf{P}_\text{i} \tag{6.8}$$
$$K_\text{f} = K_\text{i}$$

Conditions for an Inelastic Collision:
$$\mathbf{P}_\text{f} = \mathbf{P}_\text{i} \tag{6.9}$$
$$K_\text{f} < K_\text{i}$$

Final Velocities in Head-On Two-Body Elastic Collisions ($v_{2_\text{o}} = 0$):
$$v_1 = \left(\frac{m_1 - m_2}{m_1 + m_2} \right) v_{1_\text{o}} \tag{6.15}$$

$$v_2 = \left(\frac{2m_1}{m_1 + m_2} \right) v_{1_\text{o}} \tag{6.16}$$

Coordinate of the Center of Mass (using signs for directions):
$$X_\text{CM} = \frac{\Sigma_i m_i x_i}{M} \tag{6.20}$$

Exercises

6.1 Linear Momentum

1. Linear momentum has units of (a) N/m, (b) kg·m/s, (c) N/s, (d) all of these.

2. Linear momentum is (a) always conserved, (b) a scalar quantity, (c) a vector quantity, (d) unrelated to force.

3. Does a fast-running running back always have more linear momentum than a slow-moving, more massive lineman? Explain.

4. ■ If a woman of mass 60 kg is riding in a car traveling at 90 km/h, what is the magnitude of her linear momentum relative to (a) the ground and (b) the car?

5. ■ The linear momentum of a runner in a 100-meter dash is 7.5×10^2 kg·m/s. If the runner's speed is 10 m/s, what is his mass?

6. ■ Find the magnitude of the linear momentum of (a) a 7.1-kilogram bowling ball traveling at 12 m/s and (b) a 1200-kilogram automobile traveling at 90 km/h.

7. ■ Who has more linear momentum, a 70-kilogram running back running at 8.5 m/s or a 120-kilogram lineman moving at 5.0 m/s?

8. ■ How fast would a 1200-kilogram car travel if it had the same linear momentum as a 1500-kilogram truck traveling at 90 km/h?

9. ■ A ball of mass 3.0 kg has a linear momentum of 12 kg·m/s. What is the kinetic energy of the ball?

10. ■■ A 0.150-kilogram baseball traveling with a horizontal speed of 4.50 m/s is hit by a bat and then moves with a speed of 34.7 m/s in the opposite direction. What is the change in the ball's momentum?

11. ■■ A 15.0-gram rubber bullet hits a wall with a speed of 150 m/s. If the bullet bounces straight back with a speed of 120 m/s, what is the change in momentum of the bullet?

12. ■■ If two protons approach each other with speeds of 340 m/s and 450 m/s, respectively, what is the total

momentum of the two-particle system? [Hint: Find the mass of a proton in the Table inside the back cover.]

13. ■■ If a 0.50-kilogram ball is dropped from a height of 10 m, what is the momentum of the ball (a) 0.75 s after being released and (b) just before it hits the ground?

14. ■■ Taking the density of air to be 1.29 kg/m³, what is the magnitude of the linear momentum of a cubic meter of air moving with a wind speed of (a) 36 km/h and (b) 74 mi/h (the wind speed at which a tropical storm becomes a hurricane).

15. Two runners, of mass 70 kg and 60 kg, respectively, running on a straight track have a total linear momentum of 350 kg·m/s. If the heavier runner is running at 2.0 m/s, what are the possible velocities of the lighter runner?

16. ■■ A 0.20-kilogram billiard ball traveling at a speed of 15 m/s strikes the side rail of the table at an angle of 60° (•Fig. 6.26). If the ball rebounds at the same speed and angle, what is the change in its momentum?

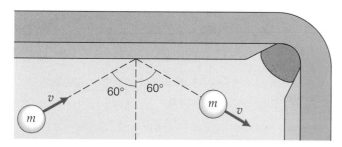

•FIGURE 6.26   Glancing collision   See Exercises 16, 17, and 36.

17. ■■ Suppose that the billiard ball in Fig. 6.26 approaches the rail at a speed of 15 m/s and an angle of 60° but rebounds at a speed of 10 m/s and an angle of 50°. What is the change in momentum now? [Hint: Use components.]

18. ■■ A person pushes a 10-kilogram box as it starts from rest and accelerates to a speed of 4.0 m/s. If the force acts on the box for a time of 2.5 s, what is the force exerted by the person? (Ignore friction.)

19. ■■ A loaded tractor-trailer with a total mass of 5000 kg traveling at 3.0 km/h coasts into a loading dock and collides, coming to a stop in 0.64 s. What is the magnitude of the average force exerted on the truck by the dock?

20. ■■ A 2.0-kilogram mud ball drops from rest at a height of 15 m. If the impact between the ball and the ground lasts 0.50 s, what is the average force exerted by the ball on the ground?

21. ■■■ At a basketball game, a 120-pound cheerleader is tossed vertically upward with a speed of 4.50 m/s by another cheerleader. (a) What is the cheerleader's change in momentum from the time she is released to just before being caught if she is caught at the height at which she

was released? (b) Would there be any difference in this change if she were caught 0.25 m below the point of release? If so, what is the change then?

6.2 Impulse

22. Impulse is equal to (a) $F \Delta x$, (b) the change in kinetic energy, (c) the change in momentum, (d) $\Delta p/\Delta t$.

23. A karate student tries *not* to follow through in order to break a board as shown in •Fig. 6.27. How can the abrupt stoppage of the fist (by not following through) generate so much force?

•FIGURE 6.27   A karate punch   See Exercise 23.

24. Explain the difference for each of the following pairs of actions in terms of impulse: (a) a golfer's long drive and a short chip shot, (b) a boxer's jab and knock-out punches, (c) a baseball player's bunting action and home-run swing.

25. Explain the principle behind (a) the use of Styrofoam as packing material to prevent objects from breaking and (b) the padding of dashboards in cars to prevent injuries.

26. ■ When tossed upward and hit horizontally by a batter, a 0.20-kilogram softball receives an impulse of 3.0 N·s. With what horizontal speed does the ball leave the bat?

27. ■ An automobile with a linear momentum of 3.0×10^4 kg·m/s is brought to a stop in 5.0 s. What is the magnitude of the average braking force?

28. ■ A pool player imparts an impulse of 3.2 N·s to a stationary 0.25-kilogram cue ball with a cue stick. What is the speed of the ball after impact?

29. ■■ For the karate punch in Exercise 23, assume that the fist has a mass of 0.35 kg and that the speeds of the fist just before and after hitting the board are 10 m/s and zero, respectively. What is the average force exerted by the fist on the board if (a) the fist follows through, so the contact time is 3.0 ms, or (b) the fist stops abruptly, so the contact time is only 0.30 ms.

30. ■■ The spout of a water tower used to fill fire-fighting tank trucks is 7.0 m above the ground. If the spout is inadvertently left open and water runs out vertically at a rate of 3.0 kg/s, what is the average force exerted by the water on the ground?

31. ■■ A 0.45-kilogram volleyball is hit with a horizontal velocity of 4.0 m/s over the net. One of the players on the

front line jumps up and hits it back with a horizontal velocity of 7.0 m/s. If the contact time is 0.040 s, what was the average force applied by the player?

32. ■■ A 1.0-kilogram ball is thrown horizontally with a velocity of 15 m/s against a wall. If the ball rebounds with a velocity of 13 m/s and the contact time is 0.020 s, what is the force on the ball by the wall?

33. ■■ In bare hands with his arms rigidly extended, a boy catches a 0.16-kilogram baseball coming directly toward him at a speed of 25 m/s. He emits an audible "ouch!" because the ball stings his hands. He learns quickly to move his hands with the ball as he catches it. If the contact time of the collision is increased from 3.5 ms to 8.5 ms in this way, how do the magnitudes of the average impulse forces compare?

34. ■■ A one-dimensional impulse force acts on a 3.0-kilogram object as diagrammed in •Fig. 6.28. Find (a) the magnitude of the impulse given to the object, (b) the magnitude of the average force, and (c) the final speed if the object had an initial velocity of 6.0 m/s opposite the impulse.

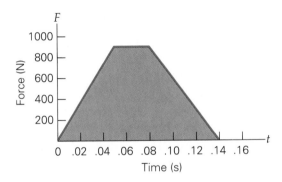

•**FIGURE 6.28   Force versus time graph**   See Exercise 34.

35. ■■ A 0.35-kilogram piece of putty is dropped from a height of 2.5 m above a flat surface. When it hits the surface, the putty comes to rest in 0.30 s. What is the average force exerted on the putty by the surface?

36. ■■ If the billiard ball in Fig. 6.26 is in contact with the rail for 0.010 s, what is the magnitude of the average force exerted on the ball? (See Exercise 16.)

37. ■■■ A mother holds her infant son on her lap rather than putting him in a child seat. In a car crash of only 25 mph (40 km/h), the baby, weighing 12 lb (of mass 5.5 kg) will require his mother's arms to exert a force of 240 lb. to avoid injury. How long does the crash last?

6.3 The Conservation of Linear Momentum

38. The linear momentum of an object is conserved if (a) the force acting on the object is conservative, (b) there is a single, unbalanced internal force acting on the object, (c) the mechanical energy is conserved, (d) none of these.

39. A fan boat of the type used in swampy and marshy areas is shown in •Fig. 6.29. Using linear momentum conser-

vation, explain the principle of its propulsion. What would happen if a sail were installed on the boat behind the fan?

•**FIGURE 6.29   Fan propulsion**   See Exercise 39.

40. Internal forces do not affect the conservation of momentum of a system because (a) they cancel each other, (b) their effects are canceled by external forces, (c) they can never produce a change in velocity, (d) Newton's second law is not applicable to them.

41. ⬤ Imagine yourself standing in the middle of a frozen lake. The ice is so smooth that it is frictionless. How could you get to shore? (You couldn't walk. Why?)

42. ■ ⬤ A 60-kilogram astronaut floating in space outside the capsule throws his 0.50-kilogram hammer such that it moves with a speed of 10 m/s relative to the capsule. What happens to the astronaut?

43. ■ In a pairs figure-skating competition, a 65-kilogram male and his 45-kilogram partner stand together on skates on the ice. If they push apart and the female has a velocity of 1.5 m/s eastward, what is the velocity of the male? (Neglect friction.)

44. ■ To get off a frozen, frictionless lake, a 70.0-kilogram person throws a 0.150-kilogram object horizontally, directly away from the shore, with a speed of 2.00 m/s. If the person is 5.00 m from the shore, how long does it take for him to reach it?

45. ■■ A bullet of mass 100 g is fired horizontally into a 14.9-kilogram block of wood resting on a horizontal surface, and the bullet becomes embedded in the block. If the muzzle speed of the bullet is 250 m/s, what is the velocity of the block containing the embedded bullet immediately after the impact? (Neglect surface friction.)

46. ■■ A 3.0-kilogram object initially at rest explodes and splits into three fragments. One fragment has a mass of 0.50 kg and flies off along the negative x axis at a speed of 2.8 m/s, and another has a mass of 1.3 kg and flies off along the negative y axis at a speed of 1.5 m/s. What are the speed and direction of the third fragment?

47. ■■ Suppose that the 3.0-kilogram object in Exercise 46 is initially traveling at a speed of 2.5 m/s in the positive

x direction. What will the speed and direction of the third fragment be in this case?

48. ■■ Identical cars hit each other and lock bumpers. In each of the following cases, what are the speeds of the cars immediately after coupling? (a) A car moving with a speed of 90 km/h approaches a stationary one. (b) Two cars approach each other with speeds of 90 km/h and 120 km/h, respectively. (c) Two cars travel in the same direction with speeds of 90 km/h and 120 km/h, respectively.

49. ■■ ⬤ A 1200-kilogram car moving to the right with a speed of 25 m/s collides with a 1500-kilogram truck and locks bumpers with the truck. Calculate the velocity of the combination after the collision if the truck is initially (a) at rest, (b) moving to the right with a speed of 20 m/s, (c) moving to the left with a speed of 20 m/s.

50. ■■ A 10-gram bullet moving horizontally at 400 m/s penetrates a 3.0-kilogram wood block resting on a frictionless horizontal surface. If the bullet slows down to 300 m/s after emerging from the block, what is the speed of the block immediately after the bullet emerges (•Fig. 6.30)?

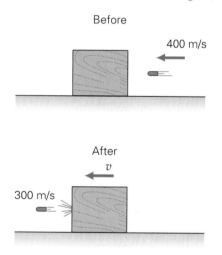

•FIGURE 6.30 Momentum transfer? See Exercise 50.

51. ■■ A 1600-kilogram (empty) truck rolls with a speed of 2.5 m/s under a loading bin, and a mass of 3500 kg is deposited in the truck. What is the truck's speed immediately after loading?

52. ■■■ A projectile that is fired from a gun has an initial velocity of 90.0 km/h at an angle of 60.0° above the horizontal. When the projectile is at the top of its trajectory, an internal explosion causes it to separate into two fragments of equal mass. One of the fragments falls straight downward as though it has been released from rest. How far does the other fragment land from the gun?

53. ■■■ A *ballistic pendulum* is a device used to measure the velocity of a projectile—for example, the muzzle velocity of a rifle bullet. The projectile is shot horizontally into, and becomes embedded in, the bob of a pendulum as illustrated in •Fig. 6.31. The pendulum swings upward to some height (*h*), which is measured. The masses of the

block and the bullet are known. Using the laws of momentum and energy, show that the initial speed of the projectile is given by $v_0 = [(m + M)/m] \sqrt{2gh}$.

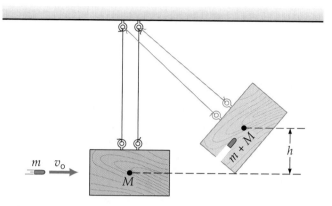

•FIGURE 6.31 A ballistic pendulum See Exercises 53, 70, and 71.

6.4 Elastic and Inelastic Collisions

54. Which of the following is *not* conserved in an inelastic collision: (a) momentum, (b) mass, (c) kinetic energy, (d) total energy?

55. ⬤ In a head-on elastic collision, mass m_1 strikes a stationary mass m_2. The energy transfer is complete if (a) $m_1 = m_2$, (b) $m_1 \gg m_2$, (c) $m_1 \ll m_2$, (d) the masses stick together.

56. A large mass and a small mass have a head-on elastic collision. Which mass receives the greater impulse: (a) the large mass, (b) the small mass, or (c) the impulses are equal?

57. Since $K = p^2/2m$, how can kinetic energy be lost in an inelastic collision yet the total momentum still be conserved? Explain.

58. ■■ ⬤ A 4.0-kilogram ball with a velocity of 4.0 m/s in the +x direction elastically collides head-on with a stationary 2.0-kilogram ball. What are the velocities of the balls after the collision?

59. ■■ A ball with a mass of 0.10 kg is traveling with a velocity of 0.50 m/s in the +x direction and collides head-on with a 5.0-kilogram ball that was at rest. Find the velocities of the balls after the collision; assume that it is elastic.

60. ■■ For the apparatus in Fig. 6.15, show that one ball swinging in with speed v_0 will not cause two balls to swing out with speed $v_0/2$.

61. ■■ A proton of mass m moving with a speed of 3.0×10^6 m/s undergoes a head-on elastic collision with an alpha particle of mass $4m$, which is initially at rest. What are the velocities of the two particles after the collision?

62. ■■ ⬤ Two balls with masses of 2.0 kg and 6.0 kg travel toward each other at speeds of 12 m/s and 4.0 m/s, respectively. If the balls have a head-on inelastic collision

and the 2.0-kilogram ball recoils with a speed of 8.0 m/s, how much kinetic energy is lost in the collision?

63. ■■ A 1500-kilogram truck moving with a speed of 25 m/s runs into the rear of a 1200-kilogram stopped car. If the collision is perfectly inelastic, what is the kinetic energy lost in the collision?

64. ■■ Two balls approach each other as shown in •Fig. 6.32, where $m = 2.0$ kg, $v = 3.0$ m/s, $M = 4.0$ kg, $V = 5.0$ m/s. If the balls collide and stick together at the origin, (a) what are the components of the velocity v' of the balls after collision? (b) What is the angle θ?

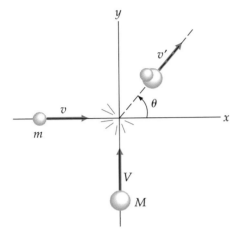

•FIGURE 6.32   A completely inelastic collision   See Exercise 64.

65. ■■ A 1500-kilogram car traveling east at 90.0 km/h and a 3000-kilogram minivan traveling south at 60.0 km/h collide at a perpendicular intersection. Assuming that the collision is completely inelastic, what is the velocity of the vehicles immediately after collision?

66. ■■ A cue ball traveling at 0.75 m/s hits the stationary 8-ball, which moves off with a speed of 0.25 m/s at an angle of 37° relative to the cue ball's initial direction. Assuming that the collision is inelastic, at what angle will the cue ball be deflected, and what will be its speed?

67. ■■ A fellow student states that the total momentum of a three-particle system ($m_1 = 0.25$ kg, $m_2 = 0.20$ kg, and $m_3 = 0.33$ kg) is initially zero and that after an inelastic triple collision, he calculates the particles to have velocities of 4.0 m/s at 0°, 6.0 m/s at 120°, and 2.5 m/s at 230°, respectively, measured from the $+x$ direction. Do you agree with his calculations? If not, assuming the first two answers to be correct, what should be the momentum of the third particle so the total is zero?

68. ■■ A freight car of mass 25 000 kg rolls down an inclined track through a vertical distance of 2.5 m. At the bottom of the incline, on a level track, the car collides and couples with an identical freight car that was at rest. What percentage of the initial kinetic energy is lost in the collision?

69. ■■■ In an elastic head-on collision with a stationary target particle, a moving particle recoils at one-third of its

incident speed. What is (a) the ratio of the particle masses (m_1/m_2)? (b) the speed of the target particle after the collision in terms of the incoming particle's initial speed?

70. ■■■ Show that the fraction of kinetic energy lost in a ballistic pendulum collision (as in Fig. 6.31) is equal to $M/(m + M)$.

71. ■■■ A 10-gram bullet is fired horizontally into, and becomes embedded in, a suspended block of wood with a mass of 0.890 kg (see Fig. 6.31). (a) How does the speed of the block with the embedded bullet immediately after the collision compare with the initial speed (v_0)? (b) If the block with the embedded bullet swings upward and its center of mass is raised 0.40 m, what was the initial speed of the bullet? (c) Was the collision elastic? If not, what percentage of the initial kinetic energy was lost?

72. ■■■ A moving billiard ball collides with an identical stationary one, and the incoming ball is deflected at an angle of 45° from its original direction. Show that if the collision is elastic, both balls will have the same speed afterward and will move at 90° relative to each other.

73. ■■■ (a) For an elastic, two-body head-on collision, show that in general $v_2 - v_1 = -(v_{2_0} - v_{1_0})$. That is, the relative speed of recession after the collision is the same as the relative speed of approach before it. (b) In general, a collision is completely inelastic, completely elastic, or somewhere in between. The degree of elasticity is sometimes expressed as the *coefficient of restitution* (e), defined as the ratio of the relative velocities of recession and approach: $v_2 - v_1 = -e(v_{2_0} - v_{1_0})$. What are the values of e for an elastic collision and a completely inelastic collision?

74. ■■■ The coefficient of restitution (see Exercise 73) for steel colliding with steel is 0.95. If a steel ball is dropped from a height h_0 above a steel plate, to what height will the ball rebound?

6.5 Center of Mass

75. •Figure 6.33 shows a performer walking on a tightrope. The long pole he holds curves down at the ends. What effect does this pole have on the center of mass of the performer–pole system? [In Chapter 8, you will learn why

•FIGURE 6.33   Tightrope walking   See Exercise 75.

this effect and another effect caused by the long pole will make walking on a tightrope easier.]

76. The center of mass of an object (a) always lies at the center of the object, (b) is at the location of the most massive particle in the object, (c) always lies within the object, (d) none of these.

77. The center of mass and center of gravity of an object coincide if (a) the object is flat, (b) there is a uniform mass distribution, (c) they both lie inside the object, (d) the acceleration due to gravity is constant.

78. A spacecraft is initially at rest in free space, and then its rocket engines are fired. Describe the motion of the center of mass of the system.

79. ■ (a) The center of mass of a system consisting of two 0.10-kilogram particles is located at the origin. If one of the particles is at (0, 0.45 m), where is the other? (b) If they are moved so their center of mass is at (0.25 m, 0.15 m), can you tell where the particles are located?

80. ■ The centers of a 4.0-kilogram sphere and a 7.5-kilogram sphere are separated by a distance of 1.5 m. Where is the center of mass of the system?

81. ■ (a) Find the center of mass of the Earth–Moon system. [Hint: Use Appendix I and consider the distance between the two to be measured from their centers.] (b) Where is that center of mass relative to the surface of the Earth?

82. ■■ Find the center of mass of a system composed of three spherical objects with masses of 3.0 kg, 2.0 kg, and 4.0 kg and centers located at $(-6.0$ m, 0), (1.0 m, 0), and (3.0 m, 0), respectively.

83. ■■ For the system described in Exercise 82, where would a fourth sphere with a mass of 0.50 kg have to be located for the center of mass of the system to be at the origin?

84. ■■ A uniform 3.0-kilogram rod of length 5.0 m supports two point masses of 4.0 kg and 6.0 kg, respectively (•Fig. 6.34). Where is the center of mass of the system?

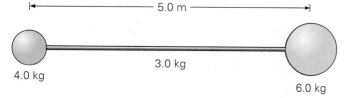

|←—————— 5.0 m ——————→|

3.0 kg

4.0 kg

6.0 kg

•FIGURE 6.34 **Find the center of mass** See Exercise 84.

85. ■■ A uniform sheet of metal measures 25 cm by 25 cm. If a circular piece with a radius of 5.0 cm is cut from the center of the sheet, where is its center of mass now?

86. ■■ Locate the center of mass of the system shown in •Fig. 6.35 (a) if all of the masses are equal; (b) if

$m_2 = m_4 = 2m_1 = 2m_3$; (c) if $m_1 = 1.0$ kg, $m_2 = 2.0$ kg, $m_3 = 3.0$ kg, and $m_4 = 4.0$ kg.

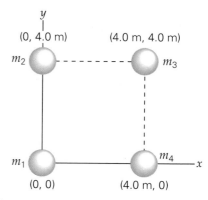

•FIGURE 6.35 **Where's the center of mass?** See Exercise 86.

87. ■■ A system of two masses has a center of mass given by X_{CM_1}. Another system of three masses has a center of mass given by X_{CM_2}. Show that if all five masses are considered to be one system, the center of mass of that combined system is not $X_{CM} = X_{CM_1} + X_{CM_2}$.

88. ■■ A 100-kilogram astronaut (mass includes space gear) on a space walk is 5.0 m from a 3000-kilogram space capsule at the full length of her safety cord. To return to the capsule, she pulls herself along the cord. Where do the astronaut and capsule meet?

89. ■■ Two skaters with masses of 65 kg and 45 kg, respectively, stand 8.0 m apart; each holds one end of a piece of rope. (a) If they pull themselves along the rope until they meet, how far does each skater travel? (Neglect friction.) (b) If only the 45-kilogram skater pulls along the rope until she meets her friend (who just holds onto the rope), how far does each skater travel?

90. ■■■ Three particles, each with a mass of 0.25 kg, are located at $(-4.0$ m, 0), (2.0 m, 0), and (0, 3.0 m) and are acted on by forces $\mathbf{F}_1 = (-3.0$ N$)\hat{\mathbf{y}}$, $\mathbf{F}_2 = (5.0$ N$)\hat{\mathbf{y}}$, and $\mathbf{F}_3 = (4.0$ N$)\hat{\mathbf{x}}$, respectively. Find the acceleration (magnitude and direction) of the center of mass. [Hint: Consider the components of that acceleration.]

Additional Exercises

91. 🌑 Two objects have the same momentum. (a) Will they always have the same kinetic energy? (b) What can you say definitely about their motion?

92. A 1.0-kilogram object moving at 10 m/s collides with a stationary 2.0-kilogram object as shown in •Fig. 6.36. If the collision is perfectly inelastic, how far along the inclined plane will the combined system travel? Neglect friction.

93. Near the end of a chess game, each player has three pieces left: a knight (N), the queen (Q), and the king (K), as shown in •Fig. 6.37 (where B stands for black and W for white). If the queen has 3.0 times the mass of a knight

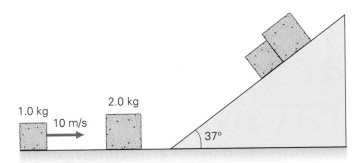

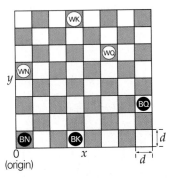

•**FIGURE 6.36   How far up?**   See Exercise 92.

and the king has 4.0 times the mass of a knight, find (a) the location (coordinates X_{CM} and Y_{CM}) of the CM of the black pieces and (b) the location of the CM of the white pieces. (c) Find the location of the CM for the system of all of the pieces, regardless of color. (Taking the lower left corner of the board as the origin of the coordinate system, express the coordinates in terms of d, the length of a side of the board's squares; assume that each piece is located at the center of the square it occupies.)

•**FIGURE 6.37   No checkmate yet**   See Exercises 93 and 94.

94. In Exercise 93, the locations of the centers of mass of the black and white chess pieces were found in (a) and (b). To find the CM for all of the pieces, could these centers of mass be treated as the locations of point particles where the respective total masses (black and white) were considered concentrated? Check and find out.

95 Two balls of equal mass (0.50 kg) approach the origin along the positive x and y axes at the same speed (3.3 m/s). (a) What is the total momentum of the system? (b) Will the balls necessarily collide at the origin? What is the total momentum of the system after both balls have passed through the origin?

96. Two identical billiard balls approach each other at the same speed (2.0 m/s). At what speeds do they rebound after a head-on elastic collision?

97. A truck with a mass of 2400 kg travels at a constant speed of 90 km/h. (a) What is the magnitude of the truck's linear momentum? (b) What average force would be required to stop the truck in 8.0 s?

98. A 15 000-newton automobile travels at a speed of 45 km/h northward along a street, and a 7500-newton sports car travels at a speed of 60 km/h eastward along an intersecting street. (a) If neither driver brakes and the cars collide at the intersection and lock together, what will the velocity of the cars be immediately after the collision? (b) What percentage of the initial kinetic energy will be lost in the collision?

99. ⬤ For a movie scene, a 75-kilogram stunt man drops from a tree onto a 50-kilogram sled moving with a velocity of 10 m/s toward the shore on a frozen lake. (a) What is the speed of the sled after the stunt man is on board? (b) If the sled hits the bank and stops but the stunt man keeps on going, with what speed does he leave the sled? (Neglect friction.)

100. During a snowball fight, a 0.25-kilogram snowball traveling at a speed of 14 m/s hits a student in the back of the head. (a) What is the impulse? Is this an elastic collision? (b) If the contact time is 0.10 s, what is the average force on the student's head?

101. A 2.5-kilogram block sliding with a constant velocity of 6.0 m/s on a frictionless horizontal surface approaches a stationary 6.5-kilogram block. (a) If the blocks suffer a completely inelastic collision, what is their velocity after it? (b) How much mechanical energy is lost in the completely inelastic collision?

102. A 90-kilogram astronaut is stranded in space at a point 6.0 m from his spaceship. To get back, he throws a 0.50-kilogram piece of equipment so that it moves at a speed of 4.0 m/s directly away from the spaceship. How long will it take him to reach the ship?

103. In nuclear reactors, subatomic particles called neutrons are slowed down by allowing them to collide with the atoms of a moderator material, such as a carbon atom, which is 12 times more massive than a neutron. (a) In a head-on elastic collision with a carbon atom, what percentage energy is lost by a neutron? (b) If the neutron has an initial speed of 1.5×10^7 m/s, what will its speed be after collision?

104. A 70-kilogram athlete achieves a height of 2.25 m in a high jump. Considering the jumper and the Earth as an isolated system, with what speed does the Earth initially move as the jumper launches himself upward?

105. A uniform, flat piece of metal is shaped like an equilateral triangle with sides that are 30 cm long. What are the coordinates of the center of mass in the xy plane if one apex is at the origin and one side is along the y axis?

Circular Motion and Gravitation

Insights

- The Centrifuge: Separating Blood Components
- Space Exploration: Gravity Assists
- Space Colonies And Artificial Gravity

People often say that rides like this "defy gravity." Of course, you know that in reality gravity cannot be defied; it commands respect. There is nothing that will shield you from it and no place in the universe where you can go to be entirely free of it. What, then, keeps these people in their seats, despite the tug of Earth's ever-present gravity? You may be surprised to find that if you can answer this question, you'll also be able to understand what keeps satellites from falling to Earth and keeps the Earth in orbit around the Sun.

Circular motion is everywhere, from atoms to galaxies, from flagella of bacteria to Ferris wheels. Two terms are frequently used to describe such motion. In general, we say that an object *rotates* when the axis of rotation lies within the body and that it *revolves* when the axis is outside the body. Thus, the Earth rotates on its axis and revolves about the Sun.

When a body rotates on its axis, all the particles of the body move in circular paths about the body's axis of rotation. For example, all the particles that make up a compact disc travel in circles about the hub of the CD player. In fact, as a "particle" on Earth, you are continuously in circular motion about the Earth's rotational axis.

Circular motion is motion in two dimensions, and so it can be described by rectangular components as used in Chapter 3. However, it is usually more convenient to describe circular motion in terms of angular quantities that will be introduced in this chapter. Being familiar with the description of circular motion will make the study of rotating rigid bodies much easier, as you will find in Chapter 8.

Gravity plays a large role in determining the motions of the planets, since it supplies the force necessary to maintain their nearly circular orbits. This chapter will consider Newton's law of gravitation, which describes this fundamental force, and will analyze planetary motion in terms of this and related basic laws. The same considerations will help you understand the motions of Earth satellites, which include one natural one (the Moon) and many artificial ones.

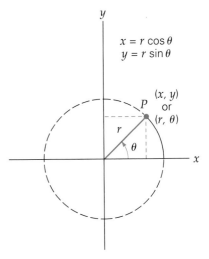

•FIGURE 7.1 Polar coordinates
A point may be described by polar coordinates instead of Cartesian coordinates—(r, θ) instead of (x, y). For a circle, θ is the angular distance and r is the radial distance. The two types of coordinates are related by the transformation equations $x = r \cos \theta$ and $y = r \sin \theta$.

7.1 Angular Measure

Objectives: To (a) define units of angular measure and (b) show how angular measure is related to circular arc length.

Motion is described as a time rate of change of position. As you might guess, *angular* speed and velocity also involve a time rate of change of position, which is expressed by an *angle*. Consider a particle traveling in a circular path, as shown in •Fig. 7.1. At a particular instant, the particle's position (P) may be designated by the Cartesian coordinates x and y. However, the position may also be designated by the polar coordinates r and θ. The distance r extends from the origin, and the angle θ is commonly measured counterclockwise from the x axis. The transformation equations that relate one set of coordinates to the other are

$$x = r \cos \theta \qquad (7.1a)$$

$$y = r \sin \theta \qquad (7.1b)$$

Polar coordinates, r and θ

as can be seen from the x and y components of point P in Fig. 7.1.

Note that r is the same for any point on a given circle. As a particle travels in a circle, the value of r is constant and only θ changes with time. Thus, circular motion can be described by using one polar coordinate (θ) that changes with time, instead of two Cartesian coordinates (x and y), both of which change with time.

Analogous to linear displacement is **angular displacement**, which is given by

$$\Delta\theta = \theta - \theta_\text{o} \qquad (7.2)$$

or simply $\Delta\theta = \theta$ when we choose $\theta_\text{o} = 0°$. A unit commonly used to express angular displacement (or angles) is the degree (°); there are 360° in one complete circle, or revolution. Each degree is divided into 60 minutes and each minute into 60 seconds. (These divisions have no relationship to time units.)

It is important to be able to relate the angular description of circular motion to the orbital or tangential description—that is, to relate the angular displacement to the arc length s. The *arc length* is the distance traveled along the circular path, and the angle θ is said to *subtend* (define) the arc length. A unit that is very convenient for relating angle to arc length is the **radian (rad)**, which is defined as the angle subtending an arc length (s) that is equal to the radius (r) (•Fig. 7.2a).

To get a general relationship between radians and degrees, let's consider the distance around a complete circle (360°). Basically, a radian is the angle subtending a circular arc equal in length to the radius. Thus for one full circle with $s = 2\pi r$ (the circumference), there is a total of $\theta = 2\pi r/r = 2\pi$ rad, or

$$2\pi \text{ rad} = 360°$$

This relationship can be used to obtain convenient conversions of common angles (Table 7.1). Thus,

$$1 \text{ rad} = 360°/2\pi = 57.3°$$

TABLE 7.1 Equivalent Degree and Radian Measures

Degrees	Radians
360°	2π
180°	π
90°	$\pi/2$
60°	$\pi/3$
57.3°	1
45°	$\pi/4$
30°	$\pi/6$

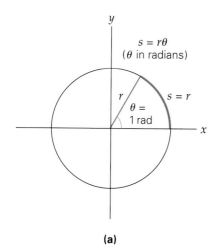

$s = r\theta$
(θ in radians)

$s = r$

$\theta = 1$ rad

(a)

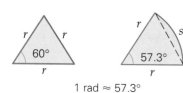

$60°$

$57.3°$

$s = r$

1 rad $\approx 57.3°$

(b)

•**FIGURE 7.2   Radian measure**
(a) Angular displacement may be measured either in degrees or in radians (rad). An angle θ is subtended by an arc length s. When $s = r$, the angle subtending s is defined to be 1 rad. More generally, $\theta = s/r$ with θ in radians. **(b)** A radian is approximately equal to $57.3°$, and the circular sector marked off by this angle is similar to an equilateral triangle with sides r.

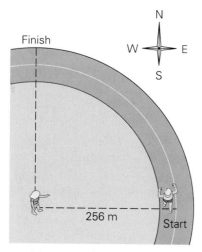

Finish

256 m

Start

N
W   E
S

•**FIGURE 7.3   Arc length—found by means of radians**   See Example 7.1.

(to 3 significant figures). A radian is thus slightly smaller than one of the angles of an equilateral triangle, as Fig. 7.2b shows. Notice in Table 7.1 that the angles in radians are expressed in terms of π explicitly, for convenience.

Similarly, the number of radians subtended by an arbitrary arc length s is equal to the number of radii that will go into s, or the number of radians $\theta = s/r$. Thus we can write

$$s = r\theta \qquad (7.3)$$

which is an important relationship between the circular arc length s and its radius r. Notice that since $\theta = s/r$, the angle in radians is the ratio of two lengths. This means that a radian measure is a pure number—that is, it is dimensionless and so has no units.

Example 7.1 ■ Finding Arc Length: Using Radian Measure

A spectator standing at the center of a circular running track observes a runner start a practice race 256 m due east of her own position (•Fig. 7.3). The runner runs on the same track to the finish line, which is located due north of the observer's position. What is the distance of the run?

Thinking It Through.   Note that the subtending angle of the section of circular track is $\theta = 90°$. The arc length (s) can be found, since the radius r of the circle is known.

Solution.   Listing what is given and what is to be found, we have

Given: $r = 256$ m   *Find:* s (arc length)
$\theta = 90° = \pi/2$ rad

Simply using Eq. 7.3 to find the arc length, we get

$$s = r\theta = (256 \text{ m})\left(\frac{\pi}{2}\right) = 402 \text{ m}$$

Note that the rad unit is omitted, but the equation is dimensionally correct. Why?

Follow-up Exercise.   What would be the distance of one complete lap around the track in this Example?

Example 7.2 ■ How Far Away? A Useful Approximation

A sailor measures the length of a distant tanker as an angular distance of 1° 9′ (1 degree, 9 minutes) with a divided circle, as illustrated in •Fig. 7.4a. He knows from the shipping charts that the tanker is 150 m in length. Approximately how far away is the tanker?

Thinking It Through.   Note in the Learn by Drawing on p. 213 that for small angles, the arc length approximates the vertical leg of the triangle, or $s \approx y$. Hence, if we know the length and the angle, we can find the radial distance, which is approximately equal to the tanker's distance.

Solution.   To approximate the distance, we take the ship's length to be nearly equal to the arc length subtended by the measured angle. This is a good approximation for small angles. The data are then

Given: $\theta = 1° 9′$ *Find:* r (radial distance)
$s = 150$ m

Equation 7.3 can be used to find r, but first we must convert θ to radians. Changing the minutes to degrees, we get $9'\left(\dfrac{1°}{60'}\right) = 0.15°$, and $\theta = 1.15°$. Then,

$$\theta = 1.15°\left(\frac{2\pi \text{ rad}}{360°}\right) = 0.0201 \text{ rad}$$

•FIGURE 7.4 Angular distance
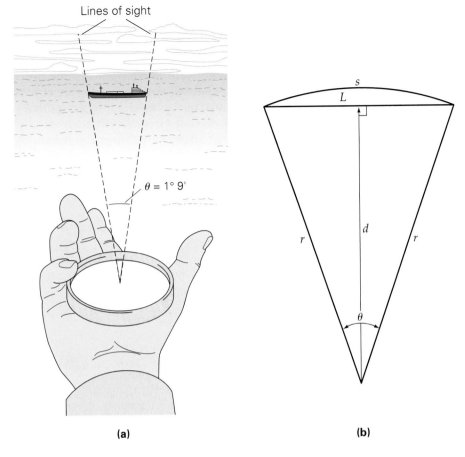

(a)

(b)

For small angles, the arc length is approximately a straight line, or the chord length. Knowing the length of the tanker, we can find how far away it is by measuring its angular size. See Example 7.2.

LEARN BY DRAWING

The Small-Angle Approximation

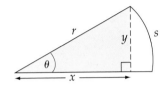

θ *not* small:

$$\theta\ (\text{in rad}) = \frac{s}{r}$$

$$\sin \theta = \frac{y}{r} \qquad \tan \theta = \frac{y}{x}$$

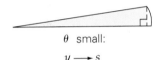

θ small:

$$y \longrightarrow s$$
$$x \longrightarrow r$$

$$\theta\ (\text{in rad}) = \frac{s}{r} \approx \frac{y}{r} \approx \frac{y}{x}$$

$$\theta\ (\text{in rad}) \approx \sin \theta \approx \tan \theta$$

(Note that any of the equivalent relationships in Table 7.1 could be used for the conversion, such as π rad/180° or even 1 rad/57.3°.) Then,

$$r = \frac{s}{\theta} = \frac{150\ \text{m}}{0.0201} = 7.46 \times 10^3\ \text{m} = 7.46\ \text{km}$$

As noted, the distance r is an approximation, obtained by assuming that for small angles the arc length s and the straight-line chord length L are very nearly the same length (Fig. 7.4b). How good an approximation is this? To see, let's compute the perpendicular distance d to the ship. From the geometry, we have $\tan \theta/2 = L/2d$, so

$$d = \frac{L}{2 \tan \theta/2} = \frac{150\ \text{m}}{2 \tan (1.15°/2)} = 7.47 \times 10^3\ \text{m} = 7.47\ \text{km}$$

This is a pretty good approximation—the values are nearly equal.

Follow-up Exercise.   As pointed out, the approximation in this Example was for *small* angles. You might wonder what is small. To investigate this question, what would be the percentage error of the approximated distance to the tanker for an angle of 20°?

Problem-Solving Hint

In computing trigonometric functions such as $\tan \theta$ or $\sin \theta$, the angle may be expressed in degrees or radians; for example, $\sin 30° = \sin (\pi/6\ \text{rad}) = \sin (0.524\ \text{rad}) = 0.500$. When finding trig functions with a calculator, note that there is usually a way to change the angle entry between "deg" and "rad" modes. Calculators commonly are set in the degree mode, so if you want to find the value of, say, $\sin (1.22\ \text{rad})$, first change to the "rad" mode and enter 1.22; $\sin (1.22\ \text{rad}) = 0.939$.

Your calculator may have a third, "grad," mode. The grad is a little-used angular unit. A grad is 1/100 of a right (90°) angle; that is, there are 100 grads in a right angle.

7.2 Angular Speed and Velocity

Objectives: To (a) describe and compute angular speed and velocity, and (b) explain their relationship to tangential speed.

The description of circular motion in angular form is analogous to the description of linear motion. In fact, you'll notice that the equations are almost mathematically identical, with different symbols being used to indicate that the quantities have different meanings. The lower-case Greek letter omega is used to represent **average angular speed** ($\overline{\omega}$), the angular displacement divided by the total time to travel the distance:

$$\overline{\omega} = \frac{\Delta\theta}{\Delta t} = \frac{\theta - \theta_{o}}{t - t_{o}} \qquad (7.4)$$

We say the units of angular speed are radians per second; technically they are s^{-1} since the radian is unitless, but it is useful to keep the rad. The *instantaneous angular speed* is given for an extremely small time interval, Δt approaches zero.

It is common to take θ_{o} and t_{o} to be zero when possible, and we generally write

$$\overline{\omega} = \frac{\theta}{t} \quad \text{or} \quad \theta = \overline{\omega}t \qquad (7.5)$$

SI unit of angular speed: radians per second (rad/s or s^{-1})

As in the linear case, if the angular speed is *constant*, then $\overline{\omega} = \omega$.

Another common descriptive unit for angular speed is revolutions per minute (rpm); for example, a CD (compact disk) rotates at a speed of 200–500 rpm (depending on the location of the track). This nonstandard unit of revolutions per minute is readily converted to radians per second, since 1 revolution = 2π rad.

The **average and instantaneous angular velocities** are analogous to their linear counterparts. Angular velocity is associated with angular displacement. Both are vectors and so have direction; however, this directionality may be specified in a special way. In one-dimensional, or linear, motion, a particle can go only in one direction or the other (+ or −), so the displacement and velocity vectors can have only these two directions. In the angular case, a particle moves one way or the other, but the motion is along its *circular path*. Thus, the angular displacement and angular velocity vectors of a particle in circular motion can have only two directions, which correspond to going around the circular path with either increasing or decreasing angular displacement from θ_{o}—that is, clockwise or counterclockwise. Let's focus on the angular velocity vector, $\boldsymbol{\omega}$. (The direction of the angular displacement will be the same as that of the angular velocity. Why?)

The direction of the angular velocity vector is given by a *right-hand rule*, illustrated in •Fig. 7.5a. When the fingers of your right hand are curled in the direction of circular motion, your extended thumb points in the direction of $\boldsymbol{\omega}$. Note that circular motion can be in only one of two circular *senses*, clockwise or counterclockwise. Plus and minus signs can be used to distinguish circular rotation directions relative to the angular velocity vector. It is customary to take a counterclockwise rotation as positive (+), since positive angular distance (and displacement) is conventionally measured counterclockwise from the positive x axis.

Question: Why not just designate the direction of the angular velocity vector to be either clockwise or counterclockwise?

Answer: Clockwise (cw) and counterclockwise (ccw) are directional senses or indications rather than actual directions. These rotational senses are like right and left. If you faced another person and were asked if some object is on the right or

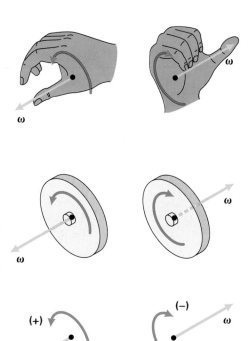

(a)   **(b)**

•FIGURE 7.5 Angular velocity The direction of the angular velocity vector for an object in rotational motion is given by a right-hand rule: When the fingers of the right hand are curled in the direction of the rotation, the extended thumb points in the direction of the angular velocity vector. Circular senses or directions are commonly indicated by **(a)** plus and **(b)** minus signs.

left, your answers would disagree. Similarly, if you held this book up toward a person facing you and rotated it, would it be rotating cw or ccw?

We can use cw and ccw to indicate rotational "directions" when they are specified relative to a reference, for example, the positive *x* axis as in the preceding discussion. Referring to Fig. 7.5, imagine yourself being first on one side of one of the rotating disks and then on the other. Which way is the disk rotating from each vantage point, cw or ccw? Then, apply the right-hand rule on both sides. You should find that the angular velocity vector direction is the same for both locations (because it is referenced to the right hand). Relative to this vector—for example, looking at the tip—there is no ambiguity in using + and − to indicate rotational senses or directions.

Relationship between Tangential and Angular Speeds

A particle moving in a circle has an instantaneous velocity tangential to its circular path. For a constant angular velocity and speed, the particle's orbital or **tangential speed** *v* (the magnitude of the tangential velocity) is also constant. How the angular and tangential speeds are related is revealed by starting with Eq. 7.3 ($s = r\theta$) and Eq. 7.5 ($\theta = \omega t$):

$$s = r\theta = r(\omega t)$$

The arc length, or distance, is also given by

$$s = vt$$

Combining the equations for *s* gives

$$v = r\omega \qquad \begin{array}{l} \textit{tangential speed relative to} \\ \textit{angular speed for circular motion} \end{array} \qquad (7.6)$$

where ω is in radians per second. Equation 7.6 holds in general for instantaneous tangential and angular speeds for solid or rigid body rotation about a fixed axis even when ω might vary with time.

Note that all the particles of an object rotating with constant angular velocity have the same angular speed, but the tangential speeds are different at different distances from the axis of rotation (•Fig. 7.6 and Demonstration 5).

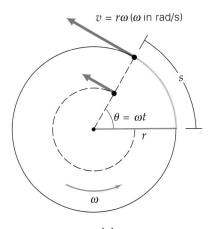

(b)

•FIGURE 7.6 **Tangential and angular speeds** (a) Tangential and angular speeds are related by $v = r\omega$, with ω in radians per second. Note that all of the particles of an object rotating about a fixed axis travel in circles. All the particles have the same angular speed (ω), but particles at different distances from the axis of rotation have different tangential speeds. (b) Sparks from a grinding wheel provide a graphic illustration of instantaneous tangential velocity. (Why do the paths curve slightly?)

Demonstration 5 ■ Constant Angular Velocity, but Not Tangential Velocity

This demonstration of uniform circular motion shows that tangential velocities are different at different radii. Neon bulbs are placed at different radii on a spinning wheel. The bulbs light up every 1/120 s for the same period of time. As seen in the photo, the bulbs at the greater radii have greater tangential speeds since they have longer lighted path lengths for the same time periods.

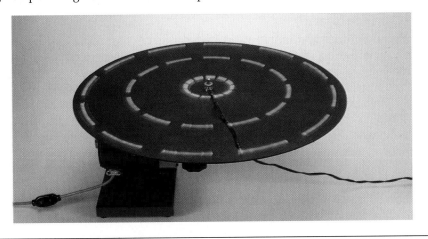

•**FIGURE 7.7 Around they go—
at the same speeds?** See
Example 7.3.

Example 7.3 ■ Merry-Go-Rounds: Do Some Go Faster Than Others?

An amusement park merry-go-round (•Fig. 7.7) at its constant operational speed makes one complete rotation in 45 s. Two children are on horses, one at 3.0 m from the center of the ride and the other farther out, 6.0 m from the center. What are (a) the angular speed and (b) the tangential speed of each child?

Thinking It Through.   The angular speed of each child is the same, since both make a complete rotation in the same time. However, the tangential speeds will be different, because the radii are different. That is, the child at the greater radius travels in a larger circle during the rotation time and so must travel faster.

Solution.

Given: $\theta = 2\pi$ rad (one rotation)   *Find:* (a) ω_1 and ω_2 (angular speeds)
  $t = 45$ s   (b) v_1 and v_2 (tangential speeds)
  $r_1 = 3.0$ m
  $r_2 = 6.0$ m

(a) It should be apparent that $\omega_1 = \omega_2$—that both riders rotate at the same angular speed. All points on the merry-go-round travel through 2π rad in the time it takes to make one rotation. The angular speed can be found from Eq. 7.5 (constant ω), and

$$\omega = \frac{\theta}{t} = \frac{2\pi \text{ rad}}{45 \text{ s}} = 0.14 \text{ rad/s}$$

where the unitless radian is included for clarity. Hence, $\omega = \omega_1 = \omega_2 = 0.14$ rad/s.

(b) The tangential speed is different at different locations on the merry-go-round. All of the "particles" making up the merry-go-round go through one rotation in the same time. Therefore, the farther a point is from the center, the longer its circular path will be, and its tangential speed will be greater, as Eq. 7.6 indicates (see also Fig. 7.6a). Thus,

$$v_1 = r_1\omega = (3.0 \text{ m})(0.14 \text{ rad/s}) = 0.42 \text{ m/s}$$

$$v_2 = r_2\omega = (6.0 \text{ m})(0.14 \text{ rad/s}) = 0.84 \text{ m/s}$$

(Note that the unitless radian has been dropped from the answer.)
  Thus a rider on the outer part of the ride has a greater tangential speed than one closer to the center, as you should expect. The outer rider must travel faster than an inner one to make one rotation in the same time.

Follow-up Exercise.   On an old 45-rpm record album, the beginning track is 8.0 cm from the center, and the end track is 5.0 cm from the center. (a) What are the angular speeds and the tangential speeds at these distances when the record is spinning at 45 rpm? (b) Why on oval race tracks do inside and outside runners have different starting points, such that some runners start "ahead" of others?

Whenever a tangential or linear quantity is calculated from an angular quantity, the angular unit radians is dropped in the final answer. When an angular quantity is asked for, the unit radians is usually included in the final answer for clarity.

Period and Frequency

Some other quantities commonly used to describe circular motion are period and frequency. The **period** (T) is the time it takes for an object in circular motion to make one complete revolution, or *cycle*. For example, the period of revolution of the Earth about the Sun is 1 year, and the period of the Earth's axial rotation is 24 hours. The standard unit of period is the second (s). Descriptively, the period is sometimes given in seconds per revolution (s/rev) or seconds per cycle (s/cycle).

  Closely related to the period is the **frequency** (f), which is the number of revolutions, or cycles, made in a given time, generally a second. For example, if a particle traveling uniformly in a circular orbit makes 5.0 revolutions in 1.0 s, the frequency (of revolution) is $f = 5.0$ rev/1.0 s $= 5.0$ rev/s, or 5.0 cycles/s (cps). *Revolution* and *cycle* are merely descriptive terms and not part of the frequency unit. The unit of frequency is inverse seconds (1/s, or s^{-1}), which is called the **hertz** (Hz) in the SI.

The hertz (Hz), a unit of frequency, is named for Heinrich Hertz (1857–1894), a German physicist and pioneering investigator of electromagnetic waves, which are also characterized by frequency.

Since the descriptive units for frequency and period are inverses of one another (cycles/second and seconds/cycle), it follows that the two quantities are related by

$$f = \frac{1}{T} \qquad \textit{frequency and period} \qquad (7.7)$$

Note: Frequency (f) and period (T) are inversely related.

SI unit of frequency: hertz (Hz, 1/s or s^{-1})

where the period is in seconds and the frequency is in hertz or inverse seconds.

The frequency can also be related to the angular speed. For uniform circular motion, the orbital speed can be written as $v = 2\pi r/T$: that is, the distance traveled in one revolution divided by the time for one revolution (1 period). Similarly, for the angular case, since a distance of 2π rad is traveled in 1 period (by definition of the period), we have

$$\omega = \frac{2\pi}{T} = 2\pi f \qquad \begin{array}{l}\textit{angular speed in terms} \\ \textit{of period and frequency}\end{array} \qquad (7.8)$$

Example 7.4 ■ Frequency and Period: An Inverse Relationship

A compact disc (CD) rotates in a player at a constant speed of 200 rpm. What are the CD's (a) frequency and (b) period of revolution?

Thinking It Through. We can use the relationships among the frequency (f), the period (T), and angular frequency (ω)—Eqs. 7.7 and 7.8.

Solution. The angular speed is not in standard units, and so must be converted. Revolutions per minute (rpm) can be converted to radians per second (rad/s).

Given: $\omega = 200 \text{ rpm} \left(\dfrac{2\pi \text{ rad}}{1 \text{ rev}}\right)\left(\dfrac{1 \text{ min}}{60 \text{ s}}\right)$ *Find:* (a) f (frequency)

$= 20.9 \text{ rad/s}$ (b) T (period)

(A convenient conversion facator is 1 rev/min $= \pi/30$ rad/s. Can you see how this is obtained?)

(a) Rearranging Eq. 7.8 and solving for f, we get

$$f = \frac{\omega}{2\pi} = \frac{20.9 \text{ rad/s}}{2\pi} = 3.33 \text{ Hz}$$

The units of 2π are rad/cycle or revolution, so the result is in cycles/second or inverse seconds, which is the hertz.

(b) Equation 7.8 could be used to find T, but Eq. 7.7 is a bit simpler:

$$T = \frac{1}{f} = \frac{1}{3.33 \text{ Hz}} = 0.300 \text{ s}$$

Thus, it takes 0.300 s for the CD to make one revolution. (Notice that with Hz = 1/s, the equation is dimensionally correct.)

Follow-up Exercise. If the period of a particular CD is 0.500 s, what is its angular speed in revolutions per minute?

7.3 Uniform Circular Motion and Centripetal Acceleration

Objectives: **To (a) explain why there is a centripetal acceleration in constant or uniform circular motion, and (b) compute centripetal acceleration.**

A simple, but important, type of circular motion is **uniform circular motion**, which occurs when an object moves at a constant speed in a circular path. An example of this movement is a car going around a circular track (•Fig. 7.8). The motion of

Uniform circular motion— constant speed but not constant velocity

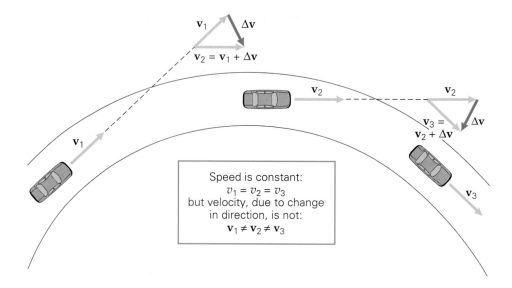

•FIGURE 7.8 Uniform circular motion The speed of an object in uniform circular motion is constant, but the object's velocity changes in the direction of motion. Thus, there is an acceleration.

Speed is constant:
$v_1 = v_2 = v_3$
but velocity, due to change in direction, is not:
$\mathbf{v}_1 \neq \mathbf{v}_2 \neq \mathbf{v}_3$

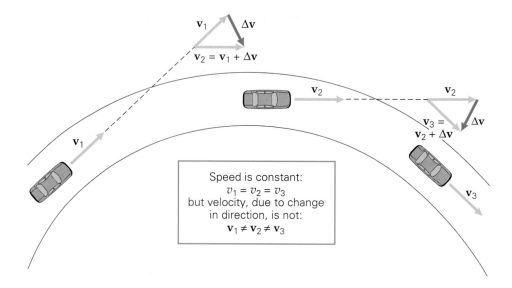

Note: Review the discussion of curvilinear motion in Section 3.1.

the Moon around the Earth is approximated by uniform circular motion. Such motion is curvilinear, so you know from the discussion in Chapter 3 that there must be an acceleration. But what is its magnitude and direction?

Centripetal Acceleration

The acceleration of uniform circular motion is not in the same direction as the instantaneous velocity (which is tangent to the circular path at any point). If it were, the object would speed up and the motion wouldn't be uniform. Recall that acceleration is the time rate of change of velocity and that velocity has both magnitude and direction. In uniform circular motion, the direction of the velocity is continuously changing, which is a clue to the direction of the acceleration (see Fig. 7.8).

This directionality is shown more explicitly in •Fig. 7.9a. The velocity vectors at the beginning and end of a time interval give the change in velocity, or $\Delta\mathbf{v}$, via vector addition (or subtraction). The vector triangles for several instantaneous velocities are shown in the figure. All of the instantaneous velocity vectors have the same magnitude or length (constant speed), but they differ in direction. Note that since $\Delta\mathbf{v}$ is not zero, there must be an acceleration ($\mathbf{a} = \Delta\mathbf{v}/\Delta t$).

As illustrated in Fig. 7.9b, as Δt (or $\Delta\theta$) becomes smaller, $\Delta\mathbf{v}$ points more toward the center of the circular path. As Δt approaches zero, the instantaneous change in the velocity, and therefore the acceleration, is exactly toward the center of the circle. As a result, the acceleration in uniform circular motion is called **centripetal acceleration**, which means center-seeking acceleration (from Latin *centri*, "center," and *petere*, "to fall toward" or "to seek").

Without the centripetal acceleration, the motion would not be in a curved path but in a straight line. The centripetal acceleration must be directed radially inward, that is, with no component in the direction of the (tangential) velocity, or else the magnitude of that velocity would change (•Fig. 7.10). Note that for an object in uniform circular motion, the direction of the centripetal acceleration is continuously changing. In terms of x and y components, a_x and a_y are not constant. Can you describe how this differs from projectile motion?

The magnitude of the centripetal acceleration can be deduced from the small shaded triangles in Fig. 7.9. (For very short time intervals, the arc length of Δs is almost a straight line—the chord.) These two triangles are similar because each has a pair of equal sides surrounding the same angle $\Delta\theta$. (Note that the velocity vectors have the same magnitude.) Thus, Δv is to v as Δs is to r, which we can write as

$$\frac{\Delta v}{v} \approx \frac{\Delta s}{r}$$

Centripetal (center-seeking) acceleration: always directed toward the center of the circle

•**FIGURE 7.9   Analysis of centripetal acceleration**
(a) The velocity vector of an object in uniform circular motion is constantly changing direction. **(b)** As Δt, the time interval for $\Delta\theta$, is taken smaller and smaller and approaches zero, Δv (the change in the velocity, and therefore an acceleration) is directed toward the center of the circle. The centripetal, or center-seeking, acceleration has a magnitude of $a_c = v^2/r$.

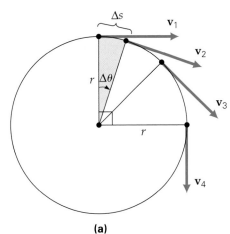

(a)

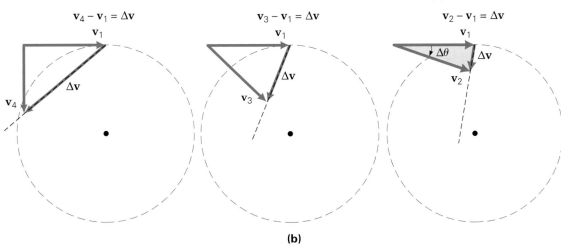

(b)

The arc length Δs is the distance traveled in time Δt, thus $\Delta s = v\,\Delta t$:

$$\frac{\Delta v}{v} \approx \frac{\Delta s}{r} = \frac{v\,\Delta t}{r}$$

and

$$\frac{\Delta v}{\Delta t} \approx \frac{v^2}{r}$$

Then, as Δt approaches zero, this approximation becomes exact. The instantaneous centripetal acceleration, $\Delta v/\Delta t = a_c$, thus has a magnitude

$$a_c = \frac{v^2}{r} \qquad \begin{array}{l}\textit{magnitude of} \\ \textit{centripetal acceleration}\end{array} \qquad (7.9)$$

Using Eq. 7.6 ($v = r\omega$), the centripetal acceleration equation can be written in terms of the angular speed:

$$a_c = \frac{v^2}{r} = \frac{(r\omega)^2}{r} = r\omega^2 \qquad \begin{array}{l}\textit{centripetal acceleration} \\ \textit{in terms of angular speed}\end{array} \qquad (7.10)$$

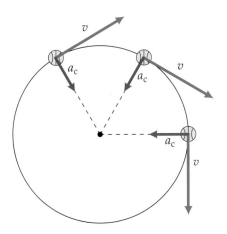

•**FIGURE 7.10   Centripetal acceleration**   For an object in uniform circular motion, the centripetal acceleration is directed radially inward. There is no acceleration component in the tangential direction; if there were, the magnitude of the velocity (tangential *speed*) would change.

Note: Centripetal acceleration depends on tangential speed (v) and radius (r).

Example 7.5 ■ A Satellite in Orbit: Centripetal Acceleration

A space station is in a circular orbit about the Earth at an altitude h of 5.0×10^2 km. If the station makes one revolution every 95 min, what are its (a) orbital speed and (b) centripetal acceleration?

Thinking It Through. From the given data, we can find the length of the circular path (by calculating the radius); and from the period, we can find the orbital speed. The centripetal acceleration can then be computed by using Eq. 7.9.

Solution.

Given: $h = 5.0 \times 10^2$ km   Find: (a) v (tangential, or orbital, speed)
        $= 5.0 \times 10^5$ m   (b) $\mathbf{a}_c$ (centripetal acceleration)
        $t = T = 95$ min $= 5.7 \times 10^3$ s

(a) The radius of the circular orbit is not h but $R_E + h$, where R_E is the radius of the Earth, 6.4×10^6 m (see Appendix III):

$$r = R_E + h = (6.4 \times 10^6 \text{ m}) + (5.0 \times 10^5 \text{ m}) = 6.9 \times 10^6 \text{ m}$$

The station travels the circumference of its circular orbit ($2\pi r$) in the given time, so the tangential speed is

$$v = \frac{2\pi r}{T} = \frac{2\pi(6.9 \times 10^6 \text{ m})}{5.7 \times 10^3 \text{ s}} = 7.6 \times 10^3 \text{ m/s}$$

(b) Then the centripetal acceleration is (from Eq. 7.9)

$$a_c = \frac{v^2}{r} = \frac{(7.6 \times 10^3 \text{ m/s})^2}{6.9 \times 10^6 \text{ m}} = 8.4 \text{ m/s}^2$$

directed toward the center of the orbit. Alternatively, we could have computed the angular speed, $\omega = 2\pi/T$, and used Eq. 7.10, $a_c = r\omega^2$.

As you might suspect, this centripetal acceleration is supplied by the Earth's gravitational force—it is the acceleration due to gravity at an altitude of 500 km.

Follow-up Exercise. (a) How does the acceleration found in this Example compare percentagewise with the acceleration due to gravity on Earth's surface? (b) Would an astronaut in the space station be "weightless"?

A practical application of centripetal acceleration is discussed in the Insight on this page and illustrated in the Example that follows.

Insight   Insight   Insight   Insight   Insight   Insight   Insight

The Centrifuge: Separating Blood Components

The centrifuge is a rotating machine used to separate particles of different sizes and densities suspended in a liquid (or a gas). For example, cream is separated from milk by centrifuging, and blood components are separated in centrifuges in medical laboratories (see Fig. 7.11).

Blood components will eventually settle toward the bottom of a tube in layers—a process called sedimentation—under the influence of normal gravity alone. The viscous drag of the plasma on the particles is analogous to (but much greater than) the air resistance that determines the terminal velocity of falling objects (Section 4.6). Red blood cells settle in the bottom layer of the plasma because they reach a greater terminal velocity than do the white blood cells and platelets and so reach the bottom sooner. However, gravitational sedimentation is generally very slow.

Since clinicians cannot afford to wait a long time to see the fractional volume of red cells in the blood, centrifugation is used to speed up the sedimentation process. Tubes are spun horizontally. The resistance of the fluid medium on the particles supplies the centripetal acceleration that keeps them moving in slowly widening circles as they settle toward the bottom of the tube. The bottom of the tube itself must exert a strong force on the contents as a whole and must be strong enough so as not to break.

Laboratory centrifuges commonly operate at speeds sufficient to produce centripetal accelerations thousands of times larger than g (see Example 7.6). Since the principle of the centrifuge involves centripetal acceleration, perhaps "centripuge" would be a more descriptive name.

Example 7.6 ■ A Centrifuge: Centripetal Acceleration

A laboratory centrifuge like that shown in •Fig. 7.11 operates at a rotational speed of 12 000 rpm. (a) What is the magnitude of the centripetal acceleration of a red blood cell at a radial distance of 8.00 cm from the centrifuge's axis of rotation? (b) How does this acceleration compare to g?

Thinking It Through. Here, the angular speed and the radius are given, so the magnitude of the centripetal acceleration can be computed directly from Eq. 7.10. The result can be compared to g by using $g = 9.80$ m/s².

Solution. The data are as follows:

Given: $\omega = 1.20 \times 10^4$ rpm $\left(\dfrac{\pi/30 \text{ rad/s}}{\text{rpm}}\right)$ *Find:* (a) a_c
(b) a_c compared to g

$= 1.26 \times 10^3$ rad/s
$r = 8.00$ cm $= 0.0800$ m

where we convert the angular speed to radians per second.

(a) The centripetal acceleration is found from Eq. 7.10:

$$a_c = r\omega^2 = (0.0800 \text{ m})(1.26 \times 10^3 \text{ rad/s})^2 = 1.27 \times 10^5 \text{ m/s}^2$$

(b) Using the relationship 1 $g = 9.80$ m/s² to express a_c in terms of g, we get

$$a_c = (1.27 \times 10^5 \text{ m/s}^2)\left(\frac{1 g}{9.80 \text{ m/s}^2}\right) = 1.30 \times 10^4 g \ (= 13\,000 \ g!)$$

Follow-up Exercise. What angular speed in revolutions per minute would give an acceleration of 1 g at the radial distance in this Example?

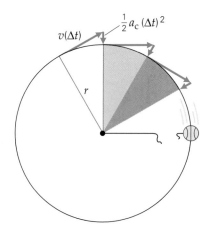

•**FIGURE 7.12 Displacement components, uniform circular motion** In a short time, Δt, an object travels a distance $v \Delta t$ tangentially and a distance $\frac{1}{2}a_c(\Delta t)^2$ radially. Thus, the object is essentially "falling" toward the center of the circle (per the short, inward arrows) as it travels tangentially (per the long, outward arrows). As Δt approaches zero, these components describe a circular path. (What happens if the string breaks? See Conceptual Example 7.7.)

The concept of centripetal acceleration may be clearer if you look at the displacement components of uniform circular motion for a short time interval, Δt (•Fig. 7.12). The rectangular components in this case are directed tangentially and radially. An object travels a distance $v \Delta t$ tangent to the circle, and at the same time it is displaced a distance $\frac{1}{2}a_c(\Delta t)^2$ radially toward the center of the circle. As Δt approaches zero, a circular path is described. Because of its centripetal acceleration, the Moon in a sense is continuously "falling" toward the Earth. Fortunately, it never gets here because it is traveling tangentially at the same time.

Suppose that an automobile moves into a level, circular curve. To negotiate the curve, the car must have a centripetal acceleration determined by the equation $a_c = v^2/r$. This acceleration is supplied by the force of friction between the tires and the road. Tires act radially (outwardly) on the road, and the reaction (friction) force to this outward force acts inward on the tires.

However, this (static, why?) friction has a maximum limiting value. If the speed of the car is high enough, the friction will not be sufficient to supply the necessary centripetal acceleration, and the car will skid outward from the center of the curve. If the car moves onto a wet or icy spot, the friction between the tires and the road may be reduced, allowing the car to skid at an even lower speed. (Banking a curve also helps vehicles negotiate the curve. See Exercises 42 and 55.)

Conceptual Example 7.7

Breaking Away

A ball attached to a string is swung with uniform motion in a horizontal circle above a person's head (•Fig. 7.13a). If the string breaks, which of the trajectories shown in Fig. 7.13b (viewed from above) would the ball follow?

•**FIGURE 7.11 Centrifuge** Centrifuges are used to separate particles of different sizes and densities suspended in liquids. For example, red and white blood cells can be separated from each other and from the plasma that makes up the liquid portion of the blood.

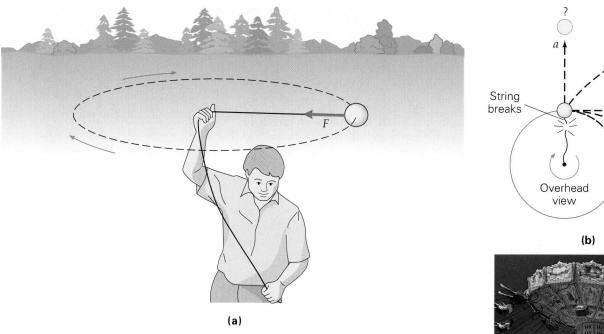

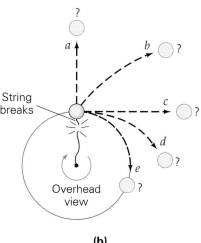

(b)

(c)

•**FIGURE 7.13 Centripetal force** (a) A ball is swung in a horizontal circle. (b) If the string breaks and the centripetal force goes to zero, what happens to the ball? See Conceptual Example 7.7. (c) What happens to the passengers on this amusement park ride as the angular velocity increases?

Reasoning and Answer. When the string breaks, the centripetal force goes to zero. Newton's first law states that if no force acts on an object in motion, the object will continue to move in a straight line. This rules out *b*, *d*, and *e*.

It should be evident from Figs. 7.6, 7.9, 7.10, and 7.11 that at any instant (including the instant when the string breaks), the isolated ball has a horizontal, tangential velocity. The downward force of gravity acts on it, but this force affects only its vertical motion, which is not visible in Fig. 7.13b. The ball thus flies off tangentially and is essentially a horizontal projectile (with $v_{x_o} = v$, $v_{y_o} = 0$, and $a_y = -g$). Viewed from above, the ball would follow the path labeled *c*.

Follow-up Exercise. If you swing a ball in a horizontal circle about your head, can the string be exactly horizontal? (See Fig. 7.13a.) Explain your answer. *Hint:* Analyze the forces acting on the ball.

Centripetal Force

To provide an acceleration, there must be a net force. Thus to produce a centripetal (inward) acceleration, we must have a **centripetal force** (net inward force). Expressing the magnitude of this force in terms of Newton's second law ($F = ma$) and inserting the expression for centripetal acceleration from Eq. 7.9, we can write

$$F_c = ma_c = \frac{mv^2}{r} \qquad \begin{array}{l} \textit{magnitude of} \\ \textit{centripetal force} \end{array} \qquad (7.11)$$

The centripetal force, like the centripetal acceleration, is directed toward the center of the circular path.

Keep in mind that in general a force applied at an angle to the direction of motion of an object produces changes in the magnitude *and* direction of the velocity. However, when a force is continuously applied at an angle of 90° to the direction of motion (as is centripetal force), only the direction of the velocity changes. Also notice that because it is always perpendicular to the direction of motion, a centripetal force does no work (why?). Therefore, by the work-energy theorem, a centripetal force does not change the kinetic energy or speed of the object.

As we have seen in Examples 7.6 and 7.7, a centripetal force can be provided by gravitational attraction or by a tension force in a string. In other situations, it could be a static frictional force, as •Fig. 7.14 and Examples 7.8 and 7.9 show.

Example 7.8 ■ Where the Rubber Meets the Road: Friction and Centripetal Force

A car approaches a level, circular curve with a radius of 45.0 m. If the concrete pavement is dry, what is the maximum speed at which the car can negotiate the curve at a constant speed?

Thinking It Through. The car will be in uniform circular motion on the curve, so there must be a centripetal force. This force is supplied by friction, so the maximum frictional force provides the centripetal (net) force when the car is at its maximum tangential speed.

Solution. Writing down what is given and what is to be found, we have

Given: $r = 45.0$ m   *Find:* v (maximum speed)

To go around the curve at a particular speed, the car must have a centripetal acceleration and therefore a centripetal force must act on it. This force is supplied by static friction between the tires and the road. (The tires are not slipping or skidding relative to the road.)
Recall from Chapter 4 that the maximum frictional force is given by $f_{s_{max}} = \mu_s N$, where here N is the magnitude of the normal force on the car and is equal to the weight of the car, mg, on the level road (why?). We may set this equal in magnitude to the expression for centripetal force ($F_c = mv^2/r$) to find the maximum speed. To find $f_{s_{max}}$ we will need the coefficient of friction between rubber and concrete; from Table 4.1, $\mu_s = 1.20$. Then, we can write

$$f_{s_{max}} = F_c$$

$$\mu_s N = \mu_s mg = \frac{mv^2}{r}$$

So

$$v = \sqrt{\mu_s rg} = \sqrt{(1.20)(45.0 \text{ m})(9.80 \text{ m/s}^2)} = 23.0 \text{ m/s}$$

(about 83 km/h, or 52 mi/h).

Follow-up Exercise. Would the centripetal force be the same for all types of vehicles in this Example?

•**FIGURE 7.14   Frictional centripetal force**   The centripetal force necessary for the skaters to round the curve is supplied by frictional force. If there were no friction between the skates and the ice, what would happen?

The proper safe speed for driving on a highway curve is an important consideration. The coefficient of friction between tires and road may vary, depending on weather, road conditions, the design of the tires, the amount of tread wear, and so on. When a curved road is designed, safety may be promoted by banking, or inclining, the roadway. This reduces the chances of skidding because the normal force exerted on the car by the road then has a component toward the center of the curve that reduces the need for friction. In fact, for a circular curve with a given banking angle and radius, there is one speed for which no friction is required at all. This condition is used in banking design (see Exercise 55).

Let's look at one more centripetal force example with two objects in uniform circular motion. It will help you understand the motions of satellites in circular orbits in a later section.

Example 7.9 ■ Strung Out: Centripetal Force and Newton's Second Law

Suppose that two masses, $m_1 = 2.5$ kg and $m_2 = 3.5$ kg, respectively, are connected by light strings and are in uniform circular motion on a horizontal frictionless surface as illustrated in •Fig. 7.15, where $r_1 = 1.0$ m and $r_2 = 1.3$ m. The forces acting on the masses

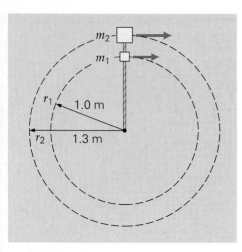

are $T_2 = 2.9$ N and $T_1 = 4.5$ N. Find (a) the centripetal accelerations and (b) the magnitudes of the tangential velocities (tangential speeds) of the masses.

Thinking It Through. The centripetal forces on the masses are supplied by the tensions (T_1 and T_2) in the strings. Isolating the masses, the net force on a mass is equal to its centripetal force ($F_c = ma_c$), and a_c for that mass can be found. The tangential speeds can then be found since the radii are known ($a_c = v^2/r$).

Solution.

Given: $r_1 = 1.0$ m and $r_2 = 1.3$ m   *Find:* (a) $\mathbf{a}_{c_1}$ and $\mathbf{a}_{c_2}$ (centripetal accelerations)
$m_1 = 2.5$ kg and $m_2 = 3.5$ kg   (b) v_1 and v_2 (tangential speeds)
$T_2 = 2.9$ N
$T_1 = 4.5$ N

By isolating m_2 in the figure, you can see that the centripetal force is provided by the tension in the string ($\mathbf{T}_2$ is the only force acting on m_2 toward the center of its circular path). Thus,

$$T_2 = m_2 a_{c_2}$$

and

$$a_{c_2} = \frac{T_2}{m_2} = \frac{2.9 \text{ N}}{3.5 \text{ kg}} = 0.83 \text{ m/s}^2$$

The acceleration is toward the center of the circle. You can find the tangential speed of m_2 from $a_c = v^2/r$:

$$v_2 = \sqrt{a_{c_2} r_2} = \sqrt{(0.83 \text{ m/s}^2)(1.3 \text{ m})} = 1.0 \text{ m/s}$$

The situation is a bit different for m_1. In this case there are two forces acting on m_1: the string tensions $\mathbf{T}_1$ and $-\mathbf{T}_2$. Also, by Newton's second law, in order to have a centripetal acceleration there must be a net force, which is given by the difference in the two tensions. Thus we have

$$F_{\text{net}_1} = +T_1 + (-T_2) = m_1 a_{c_1} = \frac{m_1 v_1^2}{r_1}$$

where the radial direction (toward the center of the circular path) is taken to be positive. Then

$$a_{c_1} = \frac{T_1 - T_2}{m_1} = \frac{4.5 \text{ N} + (-2.9 \text{ N})}{2.5 \text{ kg}} = 0.64 \text{ m/s}^2$$

and

$$v_1 = \sqrt{a_{c_1} r_1} = \sqrt{(0.64 \text{ m/s}^2)(1.0 \text{ m})} = 0.80 \text{ m/s}$$

Follow-up Exercise. Notice in this Example that the centripetal acceleration of m_2 is greater than that of m_1, yet $r_2 > r_1$, and $a_c \propto 1/r$. Is there something wrong here? Explain.

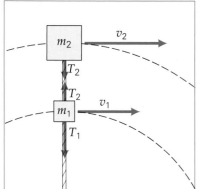

•**FIGURE 7.15 Centripetal force and Newton's second law** See Example 7.9.

7.4 Angular Acceleration

Objectives: To (a) define angular acceleration and (b) analyze rotational kinematics.

As you might have guessed, another type of acceleration in angular motion is angular acceleration. This is the time rate of change of angular velocity. In the case of circular motion, if there were an angular acceleration, the motion would not be uniform because the speed would be changing. Analogous to the linear case, the magnitude of the **average angular acceleration** ($\bar{\alpha}$) is

$$\bar{\alpha} = \frac{\Delta\omega}{\Delta t}$$

where the bar over the alpha indicates that it is an average. With $t_o = 0$ and if the angular acceleration is constant so that $\bar{\alpha} = \alpha$, we have

$$\alpha = \frac{\omega - \omega_o}{t} \qquad \text{(constant angular acceleration)}$$

SI unit of angular acceleration: radians per second squared (rad/s^2)

or

$$\omega = \omega_o + \alpha t \qquad \text{(constant acceleration only)} \qquad (7.12)$$

No boldface symbols are used in Eq. 7.12 because, in general, plus and minus signs will be used to indicate angular directions, as described earlier. As in the case of linear motion, if the angular acceleration increases the angular velocity, both quantities have the same sign, meaning that their vector directions are the same ($\boldsymbol{\alpha}$ is in the same direction as $\boldsymbol{\omega}$ as given by the right-hand rule). If the angular acceleration decreases the angular velocity, the two quantities have opposite signs, meaning that their vectors are opposed ($\boldsymbol{\alpha}$ is in the direction opposite to $\boldsymbol{\omega}$ as given by the right-hand rule, or is an angular deceleration, so to speak).

Example 7.10 ■ A Rotating CD: Angular Acceleration

A compact disc (CD) accelerates uniformly from rest to its operational speed of 500 rpm in 3.50 s. What is the angular acceleration of the CD (a) during this time? (b) after this time? (c) If the CD comes uniformly to a stop in 4.50 s, what is its angular acceleration?

Thinking It Through. (a) We are given the initial and final angular velocities; the constant (uniform) angular acceleration can thus be calculated (Eq. 7.12) since we know how long the CD accelerates. (b) Keep in mind that the operational speed is constant. (c) Everything is given for Eq. 7.12, but a negative result should be expected. (Why?)

Solution.

Given: $\omega_o = 0$

$\omega = 500 \text{ rpm} \left(\dfrac{\pi/30 \text{ rad/s}}{1 \text{ rev/min}} \right) = 52.4 \text{ rad/s}$

$t = 3.50 \text{ s (starting up)}$
$t = 4.50 \text{ s (coming to a stop)}$

Find: (a) α (during startup)
(b) α (in operation)
(c) α (in coming to a stop)

(a) Using Eq. 7.12, we get

$$\alpha = \frac{\omega - \omega_o}{t} = \frac{52.4 \text{ rad/s} - 0}{3.50 \text{ s}} = 15.0 \text{ rad/s}^2$$

in the direction of the angular velocity.

(b) After the CD reaches its operational speed, the angular velocity remains constant, so $\alpha = 0$.

(c) Again using Eq. 7.12, but this time with $\omega_o = 500$ rpm and $\omega = 0$, we get

$$\alpha = \frac{\omega - \omega_o}{t} = \frac{0 - 52.4 \text{ rad/s}}{4.50 \text{ s}} = -11.6 \text{ rad/s}^2$$

where the minus sign indicates that the angular acceleration is in the direction opposite that of the angular velocity.

Follow-up Exercise. What are the directions of the $\boldsymbol{\omega}$ and $\boldsymbol{\alpha}$ vectors in part (a) of this Example if the CD rotates clockwise when viewed from above?

As with arc length and angle ($s = r\theta$) and tangential and angular speeds ($v = r\omega$), there is a relationship between the tangential acceleration and the angular acceleration. The **tangential acceleration** is associated with the tangential speed changes and hence continuously changes direction. The magnitudes of the

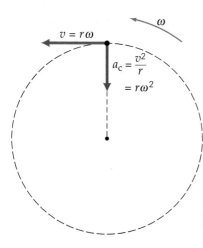

(a) Uniform circular motion
$(a_t = r\alpha = 0)$

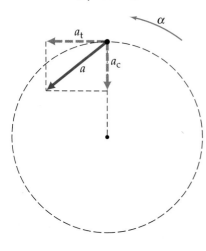

(b) Nonuniform circular motion
$(\mathbf{a} = \mathbf{a}_t + \mathbf{a}_c)$

•**FIGURE 7.16 Acceleration and circular motion** **(a)** In uniform circular motion, there is centripetal acceleration but no angular acceleration ($\alpha = 0$) or tangential acceleration ($a_t = r\alpha = 0$). **(b)** In nonuniform circular motion, there are angular and tangential accelerations, and the total acceleration is the vector sum of the tangential and centripetal components.

tangential and angular accelerations are related by a factor of r. For circular motion with a constant radius r:

$$a_t = \frac{\Delta v}{\Delta t} = \frac{\Delta(r\omega)}{\Delta t} = \frac{r\,\Delta\omega}{\Delta t} = r\alpha$$

so

$$a_t = r\alpha \qquad \textit{magnitude of tangential acceleration} \qquad (7.13)$$

The tangential acceleration (a_t) is written with a subscript t to distinguish it from the radial, or *centripetal, acceleration* (a_c), which is necessary for circular motion.

For uniform circular motion, there is no angular acceleration ($\alpha = 0$) or tangential acceleration, as can be seen from Eq. 7.13. There is only centripetal acceleration (•Fig. 7.16a).

However, when there is an angular acceleration α (and therefore a tangential acceleration, $a_t = r\alpha$), there is a change in *both* the angular and the tangential velocities ($v = r\omega$). As a result, the centripetal acceleration, $a_c = v^2/r = r\omega^2$, must increase or decrease if the object is to maintain the same circular orbit (that is, if r is to stay the same). When there are both tangential and centripetal accelerations, the total instantaneous acceleration is their vector sum (Fig. 7.16b). The tangential acceleration vector and the centripetal acceleration vector are perpendicular to each other at any instant, and the total acceleration is $\mathbf{a} = a_t\hat{\mathbf{t}} + a_c\hat{\mathbf{r}}$, where $\hat{\mathbf{t}}$ and $\hat{\mathbf{r}}$ are unit vectors directed tangentially and radially inward, respectively. You should be able to find the magnitude of $\mathbf{a}$ and the angle it makes relative to $\mathbf{a}_t$ by using trigonometry (Fig. 7.16b).

By now, the direct correspondence between the linear and angular kinematic equations should be apparent to you. The other angular equations can be derived, as was done for the linear ones in Chapter 2. That development will not be shown; the set of angular equations with their linear counterparts for constant accelerations are listed in Table 7.2. A quick review of Chapter 2 (with a change of symbols) will show you how the angular equations are derived.

Example 7.11 ■ Even Cooking: Rotational Kinematics

A microwave oven has a rotating plate 30 cm in diameter for even cooking. From rest, the plate accelerates at a uniform rate of 0.87 rad/s² for 0.50 s before reaching its constant operational speed. (a) How many revolutions does the plate make before reaching its operational speed? (b) What are the constant angular speed of the plate and the tangential speed at its rim?

Thinking It Through. This Example involves the use of the angular kinematic equations (Table 7.2). In (a), the angular displacement (θ) will give the number of revolutions. For (b), we want to find ω and then $v = r\omega$.

Solution. Listing what is given and what we are to find, we have

Given: $d = 30$ cm, $r = 15$ cm $= 0.15$ m (radius) *Find:* (a) θ (in revolutions)
$\omega_o = 0$ (at rest) (b) ω and v (angular and
$\alpha = 0.87$ rad/s² tangential speeds)
$t = 0.50$ s

(a) To find the angular distance (ω) *in radians*, we can use Eq. 4 from Table 7.2:

$$\theta = \omega_o t + \tfrac{1}{2}\alpha t^2 = 0 + \tfrac{1}{2}(0.87\ \text{rad/s}^2)(0.50\ \text{s})^2 = 0.11\ \text{rad}$$

Since 2π rad = 1 rev (or 360°),

$$0.11\ \text{rad}\left(\frac{1\ \text{rev}}{2\pi\ \text{rad}}\right) = 0.018\ \text{rev}$$

so the plate reaches its operational speed in only a small fraction of a revolution.

(b) From Table 7.2, we see that Eq. 3 gives the angular speed, and

$$\omega = \omega_o + \alpha t = 0 + (0.87 \text{ rad/s}^2)(0.50 \text{ s}) = 0.44 \text{ rad/s}$$

Then, Eq. 7.6 gives the tangential speed at the rim radius:

$$v = r\omega = (0.15 \text{ m})(0.44 \text{ rad/s}) = 0.066 \text{ m/s}$$

Follow-up Exercise. When the oven is turned off, the plate makes half of a revolution before stopping, what is its angular acceleration?

7.5 Newton's Law of Gravitation

Objectives: To (a) describe Newton's law of gravitation and how it relates to the acceleration due to gravity, and (b) apply the general formulation of gravitational potential energy.

Another of Isaac Newton's many accomplishments was the formulation of what is called the **universal law of gravitation**. This law is very powerful and fundamental. Without it, for example, we would not understand the cause of tides or know how to put satellites into particular orbits around the Earth. This law allows us to analyze the motions of planets and comets, stars and galaxies. The word *universal* in the name indicates that we believe it to apply everywhere in the universe. (This term highlights the importance of the law, but for brevity it is common to refer simply to Newton's law of gravitation or the law of gravitation.)

Newton's law of gravitation in mathematical form gives a simple relationship for the gravitational interaction between two particles, or point masses, m_1 and m_2, separated by a distance r (●Fig. 7.17). Basically, every particle in the universe has an attractive interaction with every other particle. The forces of mutual interaction are equal and opposite, forming a force pair as described by Newton's third law (Chapter 4).

The gravitational attraction, or force (F), decreases as the the square of the distance (r^2) between two point masses increases; that is, the magnitude of the gravitational force and the distance separating the two particles are related as follows:

$$F \propto \frac{1}{r^2} \tag{7.14}$$

(This type of relationship is called an *inverse-square law*—F is inversely proportional to r^2.)

Newton's law also correctly postulates that the gravitational force, or attraction of a body, depends on the body's mass—the greater the mass, the greater the attraction. However, because gravity is a mutual interaction between masses, it should be directly proportional to both masses—that is, to their product ($F \propto m_1 m_2$).

Hence, Newton's law of gravitation has the form $F \propto m_1 m_2 / r^2$. Expressed as an equation, the magnitude of the mutually attractive gravitational forces between two masses is given by

$$F = \frac{G m_1 m_2}{r^2} \tag{7.15}$$

Newton's universal law of gravitation

where G is a constant called the **universal gravitational constant**:

$$G = 6.67 \times 10^{-11} \text{ N·m}^2/\text{kg}^2$$

This constant is often referred to as "big G" to distinguish it from "little g," the acceleration due to gravity. Note from Eq. 7.15 that F approaches zero only when r is infinitely large. Thus, the gravitational force has, or acts over, an infinite range.

How did Newton come to his conclusions about the force of gravity? Legend has it that his insight came after he observed an apple fall from a tree to

TABLE 7.2 Equations for Linear and Angular Motion with Constant Acceleration*

Linear	Angular	
$x = \bar{v}t$	$\theta = \bar{\omega}t$	(1)
$\bar{v} = \dfrac{v + v_o}{2}$	$\bar{\omega} = \dfrac{\omega + \omega_o}{2}$	(2)
$v = v_o + at$	$\omega = \omega_o + \alpha t$	(3)
$x = v_o t + \frac{1}{2}at^2$	$\theta = \omega_o t + \frac{1}{2}\alpha t^2$	(4)
$v^2 = v_o^2 + 2ax$	$\omega^2 = \omega_o^2 + 2\alpha\theta$	(5)

*For these equations, $x_o = 0$, $\theta_o = 0$, and $t_o = 0$. The first equation in each column is general, that is, is not limited to situations where the acceleration is constant.

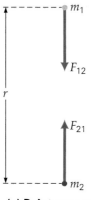

(a) Point masses

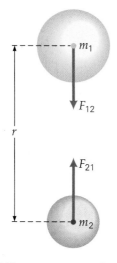

(b) Homogeneous spheres

$$F = \frac{Gm_1 m_2}{r^2}$$

•FIGURE 7.17   Universal law of gravitation   **(a)** Any two particles, or point masses, are gravitationally attracted to each other with a force that has a magnitude given by Newton's universal law of gravitation. **(b)** For homogeneous spheres, the masses may be considered to be concentrated at their centers.

Note: The symbol g is reserved for the acceleration due to gravity at the Earth's surface; a_g is more generally the acceleration due to gravity at some greater radial distance.

the ground. Newton had been wondering what supplied the centripetal force to keep the Moon in orbit and might have had this thought: "If gravity attracts an apple toward the Earth, perhaps it also attracts the Moon, and the Moon is falling, or accelerating toward the Earth, under the influence of gravity" (•Fig. 7.18).

Whether or not the legendary apple did the trick, Newton assumed that the Moon and the Earth were attracted to each other and could be treated as point masses, with their total masses concentrated at their centers. The inverse-square relationship had been speculated on by some of his contemporaries. Newton's achievement was demonstrating that the relationship could be deduced from one of Johannes Kepler's laws of planetary motion (Section 7.6).

Newton expressed Eq. 7.15 as a proportion ($F \propto m_1 m_2 / r^2$) because he did not know the value of G. It was not until 1798 (71 years after Newton's death) that the value of the universal gravitational constant was experimentally determined by an English physicist, Henry Cavendish. Cavendish used a sensitive balance to measure the gravitational force between separated spherical masses. If F, r, and the m's are known, big G can be computed from Eq. 7.15.

As mentioned earlier, Newton considered the nearly spherical Earth and Moon to be point masses located at their respective centers. It took him some years, using mathematical methods he developed, to prove that this is the case for spherical, *homogeneous* objects.* The general concept is illustrated in •Fig. 7.19.

Example 7.12 ■ Gravitational Attraction between the Earth and the Moon: A Centripetal Force

Estimate the magnitude of the mutual gravitational force between the Earth and the Moon. (You can assume that the Earth and the Moon are homogeneous spheres.)

Thinking It Through.   This is an application of Eq. 7.15 for which the masses and the distance must be looked up. (See the Appendices.)

Solution.   No data are given, so they must be available from references.

Given: (from Appendix III)   *Find:* F (gravitational force)
$M_E = 6.0 \times 10^{24}$ kg
$m_M = 7.4 \times 10^{22}$ kg
$r_{EM} = 3.8 \times 10^8$ m

The average distance from the Earth to the Moon (r_{EM}) is taken to be the distance from the center of one to the center of the other. Using Eq. 7.15, we get

$$F = \frac{Gm_1 m_2}{r^2} = \frac{GM_E m_M}{r_{EM}^2}$$

$$= \frac{(6.67 \times 10^{-11} \text{ N·m}^2/\text{kg}^2)(6.0 \times 10^{24} \text{ kg})(7.4 \times 10^{22} \text{ kg})}{(3.8 \times 10^8 \text{ m})^2}$$

$$= 2.1 \times 10^{20} \text{ N}$$

This is the magnitude of the centripetal force that keeps the Moon revolving in its orbit around the Earth. It is a very large force, but our Moon is a very massive object, with a correspondingly large inertia to overcome.

Follow-up Exercise.   With what acceleration is the Moon accelerating toward Earth?

The acceleration due to gravity at a particular distance from a planet can also be investigated by using Newton's second law of motion and his law of gravitation. The magnitude of the acceleration due to gravity, which we will generally write as a_g, at a distance r from the center of a spherical mass M is found by set-

---

*For a homogeneous sphere, the equivalent point mass is located at the center of mass. However, this is a special case. The center of gravitational force and the center of mass of a configuration of particles or an object do not generally coincide.

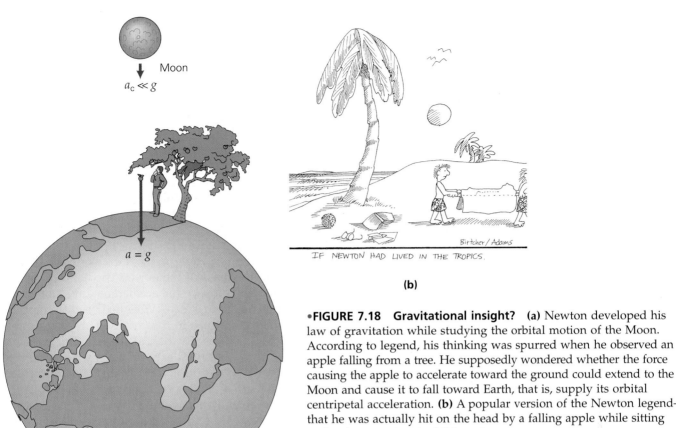

•FIGURE 7.18 Gravitational insight? **(a)** Newton developed his law of gravitation while studying the orbital motion of the Moon. According to legend, his thinking was spurred when he observed an apple falling from a tree. He supposedly wondered whether the force causing the apple to accelerate toward the ground could extend to the Moon and cause it to fall toward Earth, that is, supply its orbital centripetal acceleration. **(b)** A popular version of the Newton legend—that he was actually hit on the head by a falling apple while sitting under the tree—inspired this cartoon.

ting the force of gravitational attraction due to that spherical mass equal to ma_g, the net force on an object at distance r:

$$ma_g = \frac{GmM}{r^2}$$

Then, the acceleration due to gravity at any distance r from the planet's center is

$$a_g = \frac{GM}{r^2} \tag{7.16}$$

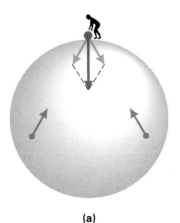
(a)

(b)

•FIGURE 7.19 Uniform spherical masses **(a)** Gravity acts between any two particles. The resultant gravitational force exerted on an object outside a homogeneous sphere by two particles at symmetric locations within the sphere is directed toward the center of the sphere. **(b)** Because of the sphere's symmetry and uniform mass distribution, the net effect is as though all the mass of the sphere were concentrated as a particle at its center. For this special case, the gravitational center of force and center of mass coincide, but this is not generally true for other objects. (Only a few of the red force arrows are shown because of space.)

Notice that a_g is proportional to $1/r^2$, so the farther away an object is from the planet, the smaller its acceleration due to gravity and the smaller the attractive force (ma_g) on the object. The force or weight vector is directed toward the center of the planet.

Equation 7.16 can be applied to the Moon or any planet. For example, taking the Earth to be a point mass M_E located at its center and R_E as its radius, we have the acceleration due to gravity (g) at the Earth's surface by setting the distance $r = R_E$:

$$a_{g_E} = g = \frac{GM_E}{R_E{}^2} \qquad (7.17)$$

This equation has several interesting implications. First, it reveals that taking g to be constant everywhere on the surface of the Earth involves assuming that the Earth has a homogeneous mass distribution and that the distance from the center of the Earth to any location on its surface is the same. These two assumptions are not exactly true. Therefore, taking g to be a constant is only an approximation, but one that works pretty well for most situations.

Also, you can see why the acceleration due to gravity is the same for all free-falling objects, that is, independent of the mass of the object. The mass of the object doesn't appear in Eq. 7.17, so all objects in free fall accelerate at the same rate.

Mass of the Earth

Finally, if you're observant, you'll notice that Eq. 7.17 can be used to compute the mass of the Earth. All of the other quantities in the equation are measurable and their values are known, so M_E can readily be calculated. This is what Cavendish did after he determined the value of G experimentally. Then he also found the average density of the Earth. (What do you think this showed? Think of the difference between the Earth's crust and its metallic core. How does the density of a crustal rock compare to Cavendish's average density of the Earth?)

The acceleration due to gravity does vary with altitude. At a distance h above the Earth's surface, we have $r = R_E + h$. The acceleration is then given by

$$a_g = \frac{GM_E}{(R_E + h)^2} \qquad (7.18)$$

Problem-Solving Hint

When comparing accelerations due to gravity or gravitational forces, you will often find it convenient to work with ratios. For example, comparing a_g to g (Eqs. 7.16 and 7.17) for the Earth gives

$$\frac{a_g}{g} = \frac{GM_E/r^2}{GM_E/R_E{}^2} = \frac{R_E{}^2}{r^2} = \left(\frac{R_E}{r}\right)^2 \quad \text{or} \quad \frac{a_g}{g} = \left(\frac{R_E}{r}\right)^2$$

Note how the constants cancel out. Taking $r = R_E + h$, you can easily compute a_g/g, or the acceleration due to gravity at some altitude h above the Earth compared to g on the Earth's surface (9.80 m/s²).

Because R_E is very large compared to readily attainable everyday altitudes above the Earth's surface, the acceleration due to gravity does not decrease very rapidly as we ascend. At an altitude of 16 km (10 mi), $a_g/g = 0.99$—a_g is still 99% of the value of g at the Earth's surface. At an altitude of 160 km (100 mi), a_g is 95% of g.

Another aspect of the decrease of g with altitude concerns potential energy. In Chapter 5, you learned that $U = mgh$ for an object at a height h above some zero reference point, since g is essentially constant near the Earth's surface. This potential energy is equal to the work done in raising the object a distance h above the Earth's surface in a *uniform* gravitational field. But what if the change in altitude is so large that g cannot be considered constant while work is done in moving an object, such as a satellite? In this case, the equation $U = mgh$ doesn't apply. In general, it can be shown (using mathematical methods that are beyond the

scope of this book) that the **gravitational potential energy** of two point masses
separated by a distance r is given by

$$U = -\frac{Gm_1 m_2}{r} \qquad (7.19)$$

The minus sign in Eq. 7.19 arises from the choice of the zero reference point (the point where $U = 0$), which is $r = \infty$.

In terms of the Earth and a mass m at an altitude h above the Earth's surface,

$$U = -\frac{Gm_1 m_2}{r} = -\frac{GmM_E}{R_E + h} \qquad (7.20)$$

where r is the distance separating the Earth's center and the mass. What this means is that on Earth we are in a negative gravitational potential energy well (•Fig. 7.20) that extends to infinity because the force of gravity has an infinite range. As h increases, so does U. That is, U becomes *less negative*, or gets closer to zero, corresponding to a higher position in the potential well. The same is true for the finite well approximation given by $U = mgh$, but over long distances, for which Eq. 7.20 applies, the change in U is not linear; it varies as $1/r$.

Note: Review Fig. 5.14, which illustrates negative potential energy.

Thus, when gravity does negative work (an object moves higher in the well) or gravity does positive work (an object falls lower in the well), there is a *change* in potential energy. As with finite potential energy wells, this change in energy is usually what's important in analyzing situations.

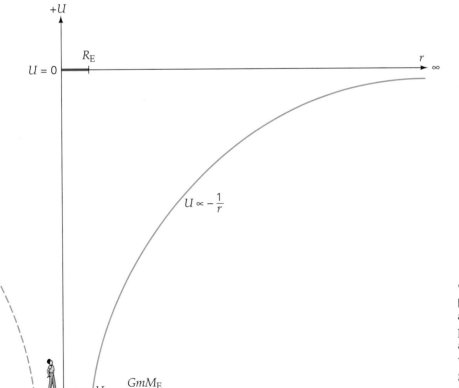

•**FIGURE 7.20   Gravitational potential energy well**   On Earth, we are in a negative gravitational potential energy well. As with an actual well or hole in the ground, work must be done against gravity to get higher in the well. The potential energy of an object increases as it moves higher in the well. This means that the value of U becomes less negative. The top of the Earth's gravitational well is at infinity, where the gravitational potential energy is, by choice, zero.

Example 7.13 ■ Different Orbits: Change in Gravitational Potential Energy

Two 50-kilogram satellites are in circular orbits about the Earth at altitudes of 1000 km (about 620 mi) and 36000 km (about 22300 mi), respectively. The lower one monitors particles about to enter the atmosphere, and the higher geosynchronous one (synchronous with the Earth's rotation) takes weather pictures from its stationary position with respect to the Earth's surface. What is the difference in the gravitational potential energies of the two satellites in their respective orbits?

Note: For communications, many satellites are launched into a circular orbit above the equator at an altitude of about 36000 km. Satellites there are *synchronous* to Earth's rotation. That is, they remain "fixed" over one point on the equator; to an observer on Earth, they are always seen in the same position in the sky.

Thinking It Through. The potential energies of the satellites are given by Eq. 7.19, where the greater the radial distance (r), the smaller or less negative is U. Thus the satellite with the greater r is higher in the gravitational potential energy well and has more potential energy.

Solution. Listing the data so that we can better see what's given, we have

Given: $m = 50$ kg
$h_1 = 1000$ km $= 1.0 \times 10^6$ m
$h_2 = 36000$ km $= 36 \times 10^6$ m
$M_E = 6.0 \times 10^{24}$ kg (from table inside back cover)
$R_E = 6.4 \times 10^6$ m

Find: ΔU (potential energy difference)

The difference in the gravitational potential energy can be computed directly from Eq. 7.20. Keep in mind that the potential energy is the energy of position, so we compute the potential energies for each position or altitude and subtract. Thus we have

$$\Delta U = U_2 - U_1 = -\frac{GmM_E}{R_E + h_2} - \left(-\frac{GmM_E}{R_E + h_1}\right) = GmM_E\left(\frac{1}{R_E + h_1} - \frac{1}{R_E + h_2}\right)$$

$$= (6.67 \times 10^{-11}\,\text{N·m}^2/\text{kg}^2)(50\,\text{kg})(6.0 \times 10^{24}\,\text{kg})$$

$$\times \left[\frac{1}{(6.4 \times 10^6\,\text{m}) + (1.0 \times 10^6\,\text{m})} - \frac{1}{(6.4 \times 10^6\,\text{m}) + (36 \times 10^6\,\text{m})}\right]$$

$$= +2.2 \times 10^9\,\text{J}$$

Because ΔU is positive, m_2 is higher in the gravitational potential energy well than m_1. Note that even though both U_1 and U_2 are negative, U_2 is "more positive" or "less negative" and closer to zero.

Follow-up Exercise. (a) Suppose that the altitude of the higher satellite in this Example were doubled, to 72000 km. Would the difference in the gravitational potential energies of the two satellites then be twice as great? Justify your answer. (b) Note that U_2 has a *smaller* negative value than U_1 (since $h_2 > h_1$). What does this mean?

Using the gravitational potential energy (Eq. 7.19) gives the equation for the total mechanical energy a different form than it had in Chapter 5. For example, the total mechanical energy of a mass m_1 moving near a stationary mass m_2 is

$$E = K + U = \tfrac{1}{2}m_1v^2 - \frac{Gm_1m_2}{r} \qquad (7.21)$$

(Note that the potential energy $U = -Gm_1m_2/r$ is negative in this case and *not* written as mgh.) This equation and the principle of the conservation of energy can be applied to the Earth moving about the Sun, by neglecting other gravitational forces. The Earth's orbit is not quite circular but slightly elliptical. At *perihelion* (the point of the Earth's closest approach to the Sun), the mutual gravitational potential energy is less (a larger negative number) than it is at *aphelion* (the point farthest from the Sun). Therefore, as can be seen from Eq. 7.21 in the form $\tfrac{1}{2}m_1v^2 = E + Gm_1m_2/r$, where E is constant, the Earth's kinetic energy and orbital speed are greatest at perihelion (the smallest value of r) and least at aphelion (the greatest value of r). Or, in general, the Earth's orbital speed is greater when it is nearer the Sun than when it is farther away.

Mutual gravitational potential energy also applies to a group, or *configuration*, of more than two masses. That is, there is gravitational potential energy due to the masses being in a configuration. This is because work was done in bringing them together. Suppose that there is a single fixed mass m_1, and another mass m_2 is brought close to m_1 from an infinite distance (where $U = 0$). The work done against the attractive force of gravity is negative (why?) and equal to the change in the mutual potential energy of the masses, which are now separated by a distance r_{12}; that is, $U_{12} = -Gm_1m_2/r_{12}$.

If a third mass m_3 is brought close to the other two fixed masses, there are then two forces of gravity acting on m_3, so $U_{13} = -Gm_1m_3/r_{13}$ and $U_{23} = -Gm_2m_3/r_{23}$. The total gravitational potential energy of the configuration is therefore

$$U = U_{12} + U_{13} + U_{23}$$

$$= -\frac{Gm_1m_2}{r_{12}} - \frac{Gm_1m_3}{r_{13}} - \frac{Gm_2m_3}{r_{23}} \qquad (7.22)$$

A fourth mass could be brought in, but this development should be sufficient to suggest that the total gravitational potential energy of a configuration of particles is equal to the sum of the individual potential energies for all pairs of particles.

Example 7.14 ■ Total Gravitational Potential Energy: Energy of Configuration

Three masses are in a configuration as shown in •Fig. 7.21. What is their total gravitational potential energy?

Thinking It Through. Equation 7.22 applies, but be sure to keep your masses and their distances distinct.

Solution. From the figure we have,

Given: $m_1 = 1.0$ kg $\qquad$ *Find:* U (total gravitational potential)
$m_2 = 2.0$ kg
$m_3 = 2.0$ kg
$r_{12} = 3.0$ m
$r_{13} = 4.0$ m
$r_{23} = 5.0$ m (3-4-5-right triangle)

We can use Eq. 7.22 directly since only three masses are used in this example. (Note that Eq. 7.22 can be extended to four, five, . . . or any number of masses.)

$$U = U_{12} + U_{13} + U_{23}$$

$$= -\frac{Gm_1m_2}{r_{12}} - \frac{Gm_1m_3}{r_{13}} - \frac{Gm_2m_3}{r_{23}}$$

$$= (6.67 \times 10^{-11} \text{ N·m}^2/\text{kg}^2)$$

$$\times \left[-\frac{(1.0 \text{ kg})(2.0 \text{ kg})}{3.0 \text{ m}} - \frac{(1.0 \text{ kg})(2.0 \text{ kg})}{4.0 \text{ m}} - \frac{(2.0 \text{ kg})(2.0 \text{ kg})}{5.0 \text{ m}} \right]$$

$$= -1.3 \times 10^{-10} \text{ J}$$

Follow-up Exercise. Explain what the *negative* potential energy in this Example means physically.

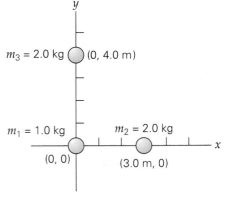

•**FIGURE 7.21 Total gravitational potential energy** See Example 7.14.

The effects of gravity are familiar to us, perhaps indirectly. When we lift an object, we may think of it as being heavy, but we are working against gravity. Gravity causes rocks to tumble down and causes mudslides. But we often put gravity to use. For example, fluids from bottles for intravenous injections flow

Space Exploration: Gravity Assists

As you read this, the *Cassini* spacecraft is on its way to Saturn via Jupiter, after having made two Venus flybys and one Earth flyby (Fig. 1).* What is going on here? Why was the spacecraft launched toward Venus, an inner planet in order to go to Saturn, an outer planet?

Although space probes can be launched from Earth with the current rocket technology, there are limitations—namely, fuel versus payload: The more fuel, the smaller the payload. Using rockets alone, planetary spacecraft are realistically limited to visiting Venus, Mars, and Jupiter. The other planets could not be reached by a spacecraft of reasonable size without taking decades to get there.

So, how will *Cassini* get to Saturn in 2004, almost 7 years after its 1997 launch? By using gravity in a clever scheme called *gravity assist*. Using gravity assists, missions to all of

*Cassini was the French-Italian astronomer who studied Saturn, discovering four of its moons and that Saturn's rings are split into two parts by a narrow gap, now called the Cassini Division. The *Cassini* spacecraft will release a "Huygens" probe to Saturn's moon Titan. This moon was discovered by the Dutch scientist Christian Huygens.

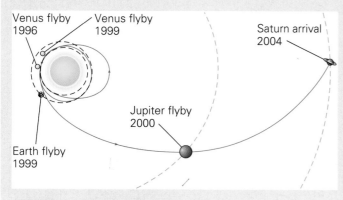

FIGURE 1   Cassini trajectory

the planets in our solar system are possible. Rocket energy is needed to get a spacecraft to the first planet, and after that the energy is more or less "free." Basically, during a planetary swingby there is an exchange of energy between the planet and the spacecraft, which enables the spacecraft to increase its velocity relative to the Sun. (This is sometimes called a slingshot effect.)

Gravity assists, or swingbys, are a key factor that has made exploration of the solar system possible. The first swingby was in 1973 for the *Mariner 10* mission, which flew by Venus on its way to Mercury. Since then, there have been many planetary missions with gravity assists—for example, *Voyagers* to the outer planets and *Galileo* to Jupiter. Perhaps most spectacular of all was the *Voyager 2* spacecraft, which made three planetary swingbys (of Jupiter, Saturn, and Uranus) on the way to its encounter with Neptune in 1989. After reaching Neptune, there was another swingby in order to gain momentum and trajectory that sent the spacecraft out of the solar system—and it is still going (Newton's first law).

Let's take a brief look at the physics of this ingenious use of gravity. Imagine the *Cassini* spacecraft making a swingby of Jupiter as illustrated in Fig. 2, where the momenta vectors of the motion are shown. Recall from Chapter 6 that a collision is an interaction of objects in which there is an exchange of momentum and energy. Technically, in a swingby, a spacecraft is having a "collision" with the planet.

As shown in Fig. 2, when the spacecraft approaches from "behind" the planet and leaves in "front" (relative to the planet's motional direction), the gravitational interaction gives rise to a change in momentum—that is, $|\mathbf{p}_2| > |\mathbf{p}_1|$ and the direction is different. Looking at the vector triangle, we see there is a $\Delta\mathbf{p}$ in the general "forward" direction of the spacecraft. Since $\Delta\mathbf{p} \propto \mathbf{F}$, there is a force acting on the craft that gives it a "kick" of energy in that direction. So, there is positive work done ($\Delta W > 0$) and a change in ki-

because of gravity. An extraterrestrial application of putting gravity to work is given in the Insight on this page.

7.6   Kepler's Laws and Earth Satellites

Objectives: To (a) state and explain Kepler's laws of planetary motion, and
(b) describe the orbits and motions of satellites.

The force of gravity determines the motions of the planets and satellites and holds the solar system (and galaxy) together. A general description of planetary motion had been set forth shortly before Newton's time by the German astronomer and mathematician Johannes Kepler (1571–1630). Kepler was able to formulate three empirical laws from observational data gathered during a 20-year period by the Danish astronomer Tycho Brahe (1546–1601). Brahe's extensive observations of stars and planets were done without the benefit of a telescope, which had not yet been invented. But the observations Brahe made were more accurate than those of most of his contemporaries because he had developed better instrumentation. As a result, he is considered to have been one of the greatest practical astronomers.

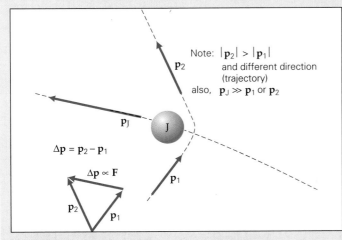

FIGURE 2   **Spacecraft on a planetary swingby**   The gravitational interaction of Jupiter (J) gives the spacecraft a change in momentum (greater magnitude and different direction), as the vector diagram shows. With $\Delta\mathbf{p} \propto \mathbf{F}$, positive work is done on the craft: It has more energy and a greater speed on leaving the region than on entering. By the conservation of momentum, the planet also has a change in momentum, but the motional effects on it are negligible because of its large mass.

netic energy ($\Delta K > 0$, from the work–energy theorem), and the spacecraft leaves with more energy, a greater speed, and a new direction. (If the flyby occurred in the opposite direction, the spacecraft would slow down.)

Momentum and energy are conserved in this elastic collision, and the planet gets an equal and opposite change in momentum, giving a retarding effect. But because the planet's mass is so much larger than the spacecraft, the effect is negligible.

To help you grasp the idea of a gravity assist, consider the analogous roller derby "slingshot maneuver" illustrated in Fig. 3. The skaters interact, and S comes out of the "flyby" with increased speed. Here, the change in momentum on the "slinger," skater J, would probably be noticeable, but that would not be so on Jupiter or any other planet.

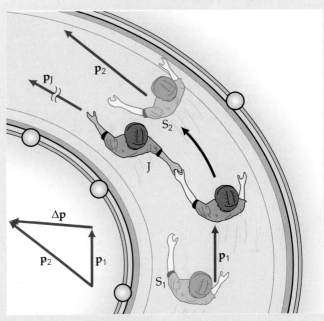

FIGURE 3   **Skating swingby**   As an analogy to a planetary swingby, consider a roller derby "slingshot maneuver." Skater J slings skater S, who comes out of the "flyby" with greater speed than she had before. In this case, the change in momentum on skater J, the slinger, would probably be noticeable. (Why?)

Kepler went to Prague to assist Brahe, who was the official mathematician at the court of the Holy Roman Emperor. Brahe died the next year and Kepler succeeded him, inheriting his records of the positions of the planets. Analyzing these data, Kepler announced the first two of his three laws in 1609 (the year Galileo built his first telescope). These laws were applied initially only to Mars. Kepler's third law came 10 y later.

Interestingly enough, Kepler's laws of planetary motion, which took him about 15 y to deduce from observed data, can now be derived theoretically with a page or two of calculations. These three laws apply not only to planets, but to any system composed of a body revolving about a more massive body, to which the inverse-square law of gravitation applies (such as the Moon, artificial Earth satellites, and solar-bound comets).

Kepler's first law (the law of orbits) says that

Planets move in elliptical orbits with the Sun at one of the focal points.

Kepler's first law: the law of orbits

An ellipse, shown in •Fig. 7.22a, has, in general, an oval shape, resembling a flattened circle. In fact, a circle is a special case of an ellipse in which the focal points,

•FIGURE 7.22 Kepler's first and second laws of planetary motion
(a) In general, an ellipse has an oval shape. The sum of the distances from the focal points F to any point on the ellipse is constant: $r_1 + r_2 = 2a$, where $2a$ is the length of the line joining the two points on the ellipse at the greatest distance from its center, called the major axis. (The line joining the two points closest to the center is $2b$, the minor axis.) Planets revolve about the Sun in elliptical orbits for which the Sun is at one of the focal points and nothing is at the other. **(b)** A line joining the Sun and a planet sweeps out equal areas in equal times. Since $A_1 = A_2$, a planet travels faster along s_1 than along s_2.

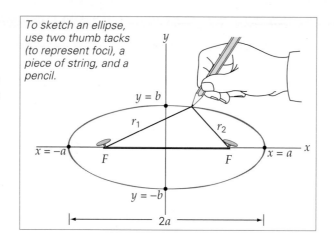

To sketch an ellipse, use two thumb tacks (to represent foci), a piece of string, and a pencil.

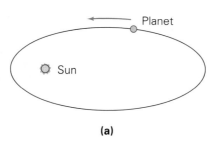

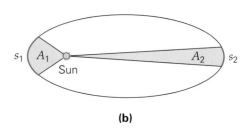

(a)

(b)

or *foci* (plural of focus), are at the same point (the center of the circle). Although the orbits of the planets are elliptical, most do not deviate very much from circles (Mercury and Pluto are notable exceptions; see Appendix III). For example, the difference between the perihelion and aphelion of the Earth (its closest and farthest distances from the Sun, respectively) is about 5 million km. This may sound like a lot, but it is only a little over 3 percent of 150 million km, which is the average distance between the Earth and the Sun.

Kepler's second law (the law of areas) says that

Kepler's second law: the law of areas

A line from the Sun to a planet sweeps out equal areas in equal lengths of time.

This law is illustrated in Fig. 7.22b. Since the time to travel the different orbital distances (s_1 and s_2) is the same, this law tells you that the orbital speed of a planet varies in different parts of its orbit. Because a planet's orbit is elliptical, its orbital speed is greater when it is closer to the Sun than when it is farther away. We used the conservation of energy in Section 7.5 (Eq. 7.21) to deduce this relationship for the Earth.

Kepler's third law (the law of periods) says that

Kepler's third law: the law of periods

The square of the orbital period of a planet is directly proportional to the cube of the average distance of the planet from the Sun; that is, $T^2 \propto r^3$.

Kepler's third law is easily derived for the special case of a planet with a circular orbit, using Newton's law of gravitation. Since the centripetal force is supplied by the force of gravity, the expressions for these forces can be set equal:

$$\frac{m_p v^2}{r} = \frac{Gm_p M_S}{r^2}$$

centripetal   *gravitational*
force   *force*

and

$$v = \sqrt{\frac{GM_S}{r}}$$

where m_p and M_S are the masses of the planet and the Sun, respectively, and v is the planet's orbital speed. But, $v = 2\pi r/T$ (circumference/period), so

$$\frac{2\pi r}{T} = \sqrt{\frac{GM_S}{r}}$$

Squaring both sides and solving for T^2 gives

$$T^2 = \left(\frac{4\pi^2}{GM_S}\right) r^3$$

or

$$T^2 = Kr^3 \qquad\qquad (7.23) \qquad \textbf{The law of periods}$$

The constant K for solar system planetary orbits is easily evaluated from orbital data (T and r) for the Earth: $K = 2.97 \times 10^{-19}$ s^2/m^3. (As an exercise, you might wish to convert K to the more useful units of y^2/km^3.) Notice that Eq. 7.23 can be used to find the value of K, and if G is known, then the mass of the Sun can be determined. (Note: This value of K does *not* apply to Earth satellites. Why?)

Earth Satellites

We are only about a half a century into the space age. Since the 1950s, numerous unmanned satellites have been put into orbit about the Earth, and now astronauts regularly spend weeks or months in orbiting space laboratories.

Putting a spacecraft into orbit about the Earth (or any planet) is an extremely complex task. However, you can get a basic understanding of the problem from fundamental principles. First, suppose that a projectile could be given the initial speed required to take it just to the top of the Earth's potential energy well. At the exact top of the well, which is an infinite distance away ($r = \infty$), the potential energy is zero. By the conservation of energy and Eq. 7.19,

$$\underset{inital}{K_o + U_o} = \underset{final}{K + U}$$

$$\tfrac{1}{2}mv_{esc}^2 - \frac{GmM_E}{R_E} = 0 + 0$$

where v_{esc} is the **escape speed**: the initial speed needed to escape from the surface of a planet or moon. The final energy is zero since the projectile stops at the top of the well (at very large distances, it is barely moving), and $U = 0$ there. Solving for v_{esc} gives

Definition of:
Escape speed

$$\tfrac{1}{2}mv_{esc}^2 = \frac{GmM_E}{R_E}$$

and

$$v_{esc} = \sqrt{\frac{2GM_E}{R_E}} \qquad\qquad (7.24)$$

Since $g = GM_E/R_E^2$ (Eq. 7.17), it is convenient to write

$$v_{esc} = \sqrt{2gR_E} \qquad\qquad (7.25)$$

Although derived for Earth, this equation may be used generally to find the escape speeds for other planets and our Moon.

Example 7.15 ■ No Return: Escape Speed

What is the escape speed at the Earth's surface?

Thinking It Through.   This is a direct application of Eq. 7.25, with the Earth's radius found inside the back cover (or Appendix III).

Solution. Equation 7.25 can be used with known data:

$$v_{esc} = \sqrt{2gR_E} = \sqrt{2(9.80 \text{ m/s}^2)(6.4 \times 10^6 \text{ m})} = 11 \times 10^3 \text{ m/s} = 11 \text{ km/s} \quad (\text{about 7 mi/s})$$

Thus, if a projectile or spacecraft could be given an initial upward speed of 11 km/s (about 40 000 km/h, or 25 000 mi/h), it would leave the Earth and not return. Note from the equation that the escape speed is independent of the mass of the spacecraft; this is the initial speed required to send *any* object to the top of the Earth's potential energy well.

Such an initial speed is not attainable instantly at launch but can be acheived in multi-stage steps. For example, in the 1980s probes (*Voyagers*) to the outer planets left Earth never to return and are currently beyond the limits of the solar system.

Follow-up Exercise. How many times greater is the escape speed of the Earth than that of the Moon?

A tangential speed less than the escape speed is required for a satellite to orbit. Consider the centripetal force of a satellite in circular orbit about the Earth. Since the centripetal force on the satellite is supplied by the gravitational attraction between the satelite and the Earth, we may again write

$$F = \frac{mv^2}{r} = \frac{GmM_E}{r^2}$$

Then

$$v = \sqrt{\frac{GM_E}{r}} \tag{7.26}$$

where $r = R_E + h$. For example, suppose that a satellite is in a circular orbit at an altitude of 500 km (about 300 mi); its tangential speed is

$$v = \sqrt{\frac{GM_E}{r}} = \sqrt{\frac{GM_E}{R_E + h}} = \sqrt{\frac{(6.67 \times 10^{-11} \text{ N·m}^2/\text{kg}^2)(6.0 \times 10^{24} \text{ kg})}{(6.4 \times 10^6 \text{ m}) + (5.0 \times 10^5 \text{ m})}}$$

$$= 7.6 \times 10^3 \text{ m/s} = 7.6 \text{ km/s} \quad (\text{about 4.7 mi/s})$$

This is about 27 000 km/h, or 17 000 mi/h. As can be seen from Eq. 7.26, the required circular orbital speed *decreases* with altitude.

In practice, a satellite is given a tangential speed by a component of the thrust from a rocket stage (•Fig. 7.23a). The inverse-square relationship of Newton's law of gravitation means that the satellite orbits that are possible about a large mass are ellipses, of which a circular orbit is a special case. This is illustrated in Fig. 7.23b for Earth, using the previously calculated values. If a satellite is not given a sufficient tangential speed, it will fall back to Earth (and possibly be burned up while falling through the atmosphere). If the tangential speed reaches the escape speed, the satellite will leave its orbit and go off into space.

Finally, the total energy of an orbiting satellite in circular orbit is

$$E = K + U = \tfrac{1}{2}mv^2 - \frac{GmM_E}{r} \tag{7.27}$$

Substituting the expression for v from Eq. 7.26 into the kinetic energy term gives

$$E = \frac{GmM_E}{2r} - \frac{GmM_E}{r}$$

Thus,

$$E = -\frac{GmM_E}{2r} \qquad \begin{array}{l} \textit{total energy of} \\ \textit{an orbiting satellite} \end{array} \tag{7.28}$$

Note that the total energy of the satellite is negative: More work is required to put a satellite into a higher orbit, where it has more potential and total energy. The total energy E increases as its *numerical value* becomes smaller—less negative—in go-

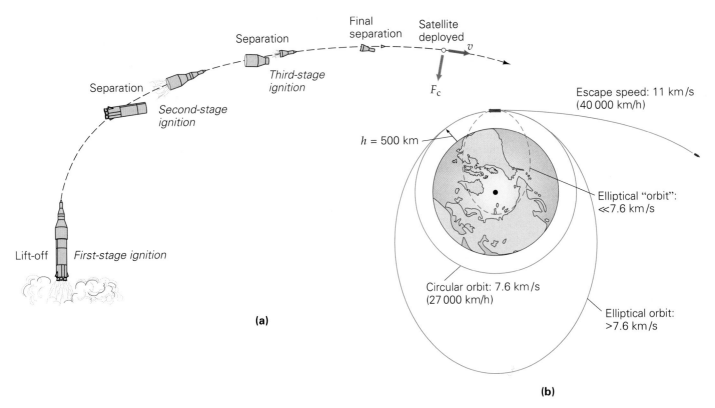

(a)

(b)

•**FIGURE 7.23 Satellite orbits** (a) A satellite is put into orbit by giving it a tangential speed sufficient for maintaining an orbit at a particular altitude. The higher the orbit, the smaller the tangential speed. (b) At an altitude of 500 km, a tangential speed of 7.6 km/s is required for a circular orbit. With a tangential speed between 7.6 km/s and 11 km/s (the escape speed), the satellite would move out of the circular orbit. Since it would not have the escape speed, it would "fall" around the Earth in an elliptical orbit with the Earth's center at one focal point. A tangential speed less than 7.6 km/s would also give an elliptical path about the center of the Earth; but because the Earth is not a point mass, a certain minimum speed is needed to keep the satellite from striking the Earth's surface.

ing to a higher orbit toward the zero potential at the top of the well. That is, the farther a satellite is from Earth, the greater its total energy.

To help understand why the total energy increases when its value becomes less negative, think of a change in energy from, say, 5.0 J to 10 J. This would be considered an increase in energy. Similarly, a change from -10 J to -5.0 J would be an increase in energy, even though the *absolute* value has decreased:

$$\Delta U = U - U_{\text{o}} = -5.0\,\text{J} - (-10\,\text{J}) = +5.0\,\text{J}$$

The relationship of speed and energy to orbital radius is summarized in Table 7.3.

Also note from the development of Eq. 7.28 that the kinetic energy of an orbiting satellite is equal to the absolute value of its total energy:

$$K = \frac{GmM_{\text{E}}}{2r} = |E| \tag{7.29}$$

TABLE 7.3 Relationship of Radius, Speed, and Energy for Circular Orbital Motion

	Increasing r (larger orbit)	*Decreasing r (smaller orbit)*
ω	decreases	increases
v	decreases	increases
K	decreases	increases
U	increases (smaller negative value)	decreases (larger negative value)
$E\ (= K + U)$	increases (smaller negative value)	decreases (larger negative value)

Conceptual Example 7.16

Applying the Brakes to Speed Up: Conservation of Energy

A spacecraft is in a circular orbit about the Earth, well outside the atmosphere. Describe what happens to the spacecraft if its retrorockets are fired. (Retrorockets point in the direction of the spacecraft's motion and produce a reverse thrust.)

Note: The work–energy theorem is discussed in Section 5.3.

Reasoning and Answer. The retrorockets apply a braking force, or reverse thrust, to the spacecraft. *Negative* work is done (because the force and displacement are in opposite directions), and the spacecraft loses some energy (the work–energy theorem). That is, the total energy of the spacecraft,

$$E = -\frac{GmM_E}{2r}$$

is reduced. If E decreases (that is, takes on a greater negative value), we can see from the above equation that r must also decrease. Thus, the spacecraft will spiral inward toward the Earth. From Eq. 7.29,

$$K = \tfrac{1}{2}mv^2 = \frac{GmM_E}{2r} \quad \text{or} \quad v^2 = \frac{GM_E}{r}$$

So, if r decreases, v increases, and the spacecraft actually speeds up—even though both U and E decrease, K increases!

Follow-up Exercise. Suppose the retrorockets of an orbiting spacecraft were fired until the craft was within the top layers of the Earth's atmosphere, where air resistance is appreciable. If the rockets were then shut off, what would happen to the spacecraft?

Reverse thrust, provided by the engines of docked cargo ships, is being used to put the Russian space station *Mir* into lower orbits and will ultimately lead to its final destruction. A final thrust will send the station into a decaying orbit and into our atmosphere. Most of the 120-ton *Mir* will burn up in the atmosphere; however, some large pieces are expected to fall into the Pacific Ocean.

Conceptual Example 7.17

Forward Thrust: A Change in Energy

The engines of a spacecraft in circular orbit about the Earth are fired so as to give the craft a *forward* thrust. How will the spacecraft's energy be changed after the firing, assuming another circular orbit is achieved? (a) K increased and U decreased, (b) K decreased and U increased, (c) both K and U increased, (d) the total energy decreased.

Reasoning and Answer. You might think that a forward thrust would cause the spacecraft to speed up and increase its kinetic energy. However, an increase in v would require an increased centripetal force to maintain the spacecraft in orbit. This force is supplied by gravitational attraction, and so could increase only if r were reduced (that is, if the spacecraft moved to a smaller orbit)—something that would not result from the application of a forward thrust. (Draw a diagram to confirm this; in what direction would the thrust act with respect to the circular orbit?)

A forward thrust (in contrast to a reverse thrust in Example 7.16) will send the craft into a larger rather than a smaller orbit, because the forward thrust does positive work on the spacecraft, causing it to rise higher in its potential energy well and increasing its potential energy. With an increase in r, the kinetic energy *decreases* (Eq. 7.29) while the total energy increases by becoming less negative (Eq. 7.28). Hence, the answer is (b). (In actuality, if the engine "burn" is brief, the spacecraft would go into an elliptical orbit with its lowest altitude equal to the radius of the initial circular orbit.)

Follow-up Exercise. Imagine yourself in a spacecraft in a circular orbit, well behind a space station in the same radial orbit. Wishing to dock with the station, what maneuvers would you execute to catch up with it?

The advent of the space age and orbiting satellites has brought us the terms *weightlessness* and *zero gravity,* because astronauts appear to "float" about in orbiting spacecraft (•Fig. 7.24a). However, the terms are misnomers. As mentioned earlier in the chapter, gravity is an infinite-range force, and Earth's gravity acts on a spacecraft and astronauts, supplying the centripetal force necessary to keep them in orbit. Gravity there is not zero, so there must be weight.

A better term to describe the floating effect of astronauts in orbiting spacecraft would be *apparent weightlessness.* They "float" because both the astronauts and the spacecraft are centripetally accelerating (or falling) toward the Earth at the same rate. To help you understand this effect, consider an analogous situation of a person standing on a scale in an elevator (Fig. 7.24b). The "weight"

(a)

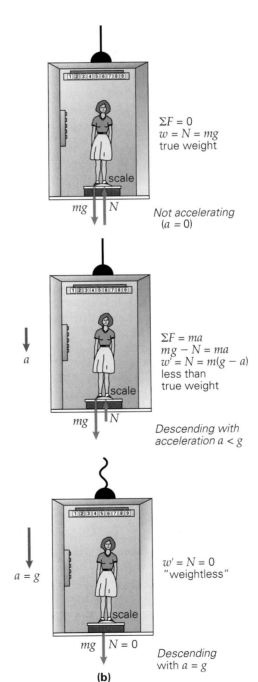

$\Sigma F = 0$
$w = N = mg$
true weight

Not accelerating
(a = 0)

$\Sigma F = ma$
$mg - N = ma$
$w' = N = m(g - a)$
less than
true weight

Descending with
acceleration a < g

$w' = N = 0$
"weightless"

Descending
with a = g

(b)

•**FIGURE 7.24 Apparent weightlessness (a)** An astronaut "floats" in a spacecraft, seemingly in a weightless condition. **(b)** In a stationary elevator (top), a scale reads true weight. The weight reading is the reaction force N of the scale on the person. If the elevator is descending with an acceleration $a < g$ (middle), the reaction force and apparent weight are less than the true weight. If the elevator were in free fall, $a = g$, the reaction force and indicated weight would be zero since the scale would be falling as fast as the person. **(c), (d)** "Weightlessness" on Earth.

(c)

(d)

Space Colonies and Artificial Gravity

Space colonies, long a staple of science fiction stories and films, are now in the planning stage. However, humans cannot live for long periods of time in a "zero-gravity" environment (see Exercise 71) without detrimental physical effects. It has been suggested that this problem could be avoided if the colony were housed in a huge rotating wheel, which would supply "artificial gravity" (Fig. 1).

As you know, centripetal force is necessary to keep an object in rotational circular motion. On the rotating Earth, that force is supplied by gravity, and we refer to it as weight. Because of our weight, we exert a force on the ground, and the normal force (by Newton's third law) upward on our feet gives us the feeling of "having our feet on solid ground." In a rotating space colony, the situation is somewhat reversed. The rotating colony would supply the centripetal force on the inhabitants, which would be perceived as a weight sensation on the soles of our feet, or artificial gravity (Fig. 2a). Rotation at the proper speed would produce a simulation of normal gravity ($g = 9.80$ m/s^2) within the colony wheel. Note that in the colonists' world, "down" would be outward, toward the periphery of the space station, and "up" would always be inward, toward the axis of rotation.

The inhabitants would experience normal gravitational effects. For example, in Fig. 2a, suppose that the person in

FIGURE 1   Space colony and artificial gravity   It has been suggested that a space colony could be housed in a huge, rotating wheel as in this artist's conception. The rotation would supply the "artificial gravity" for the colonists.

measurement that the scale registers is actually the normal force N of the scale on the person. In a non-accelerating elevator ($a = 0$), we have $N = mg = w$, and N is equal to the true weight of the individual. However, suppose the elevator is descending with an acceleration a, and $a < g$. As the vector diagram in the figure shows,

$$mg - N = ma$$

and the *apparent* weight w' is

$$w' = N = m(g - a) < mg$$

where the downward direction is taken as positive in this instance. With a downward acceleration a, we see that N is less than mg, hence the scale indicates that the person weighs less than the true weight. Note that the *apparent acceleration* due to gravity is $g' = g - a$.

Now suppose the elevator were in free fall with $a = g$. As you can see, N and the apparent weight w' would be zero. Essentially, the scale is accelerating or falling at the same rate as the person. The scale may indicate a "weightless" condition, but gravity still acts, as would be noted by the sudden stop at the bottom of the shaft.

(Note that you do not have to become an astronaut or be caught in a falling elevator to experience this condition. Some more familiar examples are shown in Fig. 7.24c,d.)

Space has been called the final frontier. Some day, instead of brief stays in Earth-orbiting spacecraft, we may have permanent space colonies (see the Insight on this page).

green let go of the ball. Viewed from outside the rotating wheel, you would say the ball has a tangential velocity v as a result of the rotation, and it would go off in a straight line (Newton's first law) until it hits the side of the wheel (as in-dicated by the dashed line in the figure). However, the in-habitant rotates with the wheel; from that perspective, the ball merely falls to the floor as it would on Earth. Think about it.

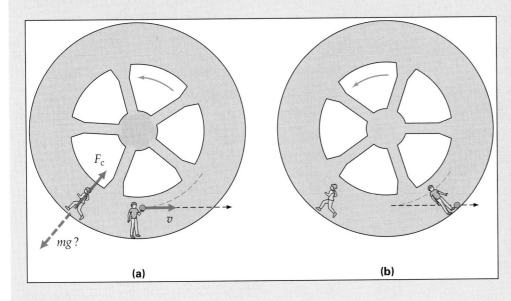

(a)　　　　　　　　　　　　**(b)**

FIGURE 2 Rotating space colony **(a)** In the frame of reference of someone in a rotating space colony, centripetal force, coming from the normal force N of the floor, would be perceived as weight sensation, or artificial gravity, since we are used to feeling N upward on our feet to balance gravity. Rotation at the proper speed would simulate normal gravity. To an outside observer, a dropped ball would follow a tangential straight-line path (shown as dashed). **(b)** A colonist would observe the ball to fall downward as in a normal gravitational situation.

Chapter Review

Important Terms

angular displacement *211*
radian (rad) *211*
average angular speed *214*
average angular velocity *214*
instantaneous angular velocity *214*
tangential speed *215*
period *216*
frequency *216*

hertz (Hz) *216*
uniform circular motion *217*
centripetal acceleration *218*
centripetal force *222*
average angular acceleration *224*
tangential acceleration *225*
universal law of gravitation *227*
universal gravitational constant *227*

gravitational potential energy *231*
Kepler's first law (the law of orbits) *235*
Kepler' second law (the law of areas) *236*
Kepler's third law (the law of periods) *236*
escape speed *237*

Important Concepts

- The radian (rad) is a measure of angle; 1 rad is the angle of a circle subtended by an arc length (s) equal to the radius (r).
- Tangential speed (v) and angular speed (ω) for circular motion are directly proportional, with the radius r being the constant of proportionality ($v = r\omega$).
- In uniform circular motion, a centripetal acceleration is required and is always directed toward the center of the circular path.
- A centripetal force (net force directed toward the center of a circle) is a requirement for circular motion.

- Angular acceleration (α) is the time rate of change of angular velocity.
- According to Newton's law of gravitation, every particle attracts every other particle in the universe with a force that is proportional to the masses of both particles and inversely proportional to the square of the distance between them.
- An Earth satellite is in a negative potential energy well. The higher in the well, the greater the potential energy and the less the kinetic energy.

Important Equations

Arc Length (angle in radians):
$$s = r\theta \tag{7.3}$$

Angular Kinematic Equations (for $\theta = 0$ and $t_o = 0$; see Table 7.2 for linear analogues):

$$\theta = \bar{\omega}t \quad \begin{array}{l} \text{(general, not limited} \\ \text{to constant acceleration)} \end{array} \tag{7.5}$$

$$\left.\begin{array}{l} \bar{\omega} = \dfrac{\omega + \omega_o}{2} \\[2mm] \omega = \omega_o + \alpha t \\[2mm] \theta = \omega_o t + \frac{1}{2}\alpha t^2 \\[2mm] \omega^2 = \omega_o^2 + 2\alpha\theta \end{array}\right\} \begin{array}{l} \\ \text{constant} \\ \text{acceleration} \\ \text{only} \end{array} \quad \begin{array}{r} \text{(2, Table 7.2)} \\[2mm] \text{(7.12)} \\[2mm] \text{(4, Table 7.2)} \\[2mm] \text{(5, Table 7.2)} \end{array}$$

Tangential and Angular Speeds:
$$v = r\omega \tag{7.6}$$

Angular Speed (with uniform circular motion):
$$\omega = \frac{2\pi}{T} = 2\pi f \tag{7.8}$$

Frequency and Period:
$$f = \frac{1}{T} \tag{7.7}$$

Centripetal Acceleration:
$$a_c = \frac{v^2}{r} = r\omega^2 \tag{7.10}$$

Tangential and Angular Accelerations:
$$a_t = r\alpha \tag{7.13}$$

Centripetal Force and Acceleration:
$$F_c = ma_c = \frac{mv^2}{r} \tag{7.11}$$

Newton's Law of Gravitation:
$$F = \frac{Gm_1m_2}{r^2} \tag{7.15}$$
$$G = 6.67 \times 10^{-11}\,\text{N}\cdot\text{m}^2/\text{kg}^2$$

Acceleration Due to Gravity at an Altitude h:
$$a_g = \frac{GM_E}{(R_E + h)^2} \tag{7.18}$$

Gravitational Potential Energy of Two Particles:
$$U = -\frac{Gm_1m_2}{r} \tag{7.19}$$

Kepler's Law of Periods:
$$T^2 = Kr^3 \tag{7.23}$$
(K depends on the mass of the object being orbited; for objects orbiting the Sun, $K = 2.97 \times 10^{-19}\,\text{s}^2/\text{m}^3$.)

Escape Speed (from Earth):
$$v_{esc} = \sqrt{\frac{2GM_E}{r_E}} = \sqrt{2gR_E} \tag{7.24}$$

Energy of Orbiting Satellite Orbiting Earth:
$$E = -\frac{GmM_E}{2r} \tag{7.28}$$
$$K = |E| \tag{7.29}$$

Exercises

7.1 Angular Measure

1. The radian unit is equivalent to a ratio of (a) degree/time, (b) length, (c) length/length, (d) length/time.

2. A proper conversion factor for degrees to radians would be (a) $(\pi/7\text{ rad})/270°$, (b) $(\pi/2\text{ rad})/60°$, (c) $(3\pi\text{ rad})/540°$, (d) $(0.50\text{ rad})/180°$.

3. The Cartesian coordinates of a point on a circle are (1.5 m, 2.0 m). What are the point's polar coordinates (r, θ)?

4. ■ In Cartesian coordinates, the equation of a circle is $x^2 + y^2 = a^2$, where a is the radius of the circle. What is that equation in polar coordinates?

5. ■ Convert the following angles from degrees to radians. Report to 3 significant figures. (a) 15°, (b) 45°, (c) 90°, and (d) 120°.

6. ■ Convert the following angles from radians to degrees. (a) $\pi/6$ rad, (b) $5\pi/12$ rad, (c) $3\pi/4$ rad, and (d) π rad.

7. ■ What is the arc length subtended by an angle of $\pi/4$ rad on a circle with a radius of 6.0 cm?

8. ■ You measure the length of a distant car as an angular distance of 2.0°. If the car is actually 5.0 m long, approximately how far away is the car?

9. ■ A jogger on a circular track that has a radius of 0.250 km runs a distance of 1.00 km. What angular distance does the jogger cover in (a) radians and (b) degrees?

10. ■ Assume that the Earth's orbit around the Sun is circular. In kilometers, what is the approximate orbital distance the Earth travels in 3 months?

11. ■ Using the ratio equation $s/\theta = c/360°$, where c is the circumference of a circle, (a) show that by definition 1 rad is equivalent to 57.3°. (b) Show that 2π rad is equivalent to 360°.

12. ■■ The hour, minute, and second hands on a clock are 0.25 m, 0.30 m, and 0.35 m long, respectively. What are the distances traveled by the tips of the hands in a 30-minute interval?

13. ■■ In Europe, a large circular walking track with a diameter of 0.900 km is marked in angular distances in radians. An American tourist who walks 3.00 mi daily goes to the track. How many radians should he walk?

14. ■■ You ordered a 12-inch-diameter pizza for a party of five. To distribute the pizza evenly, how should you cut it (●Fig. 7.25)?

●FIGURE 7.25 **Tough pizza to cut** See Exercise 14.

15. ■■ (a) In degrees and radians, what is the angular width of a full Moon as viewed from Earth? (b) of a full Earth as viewed from the Moon? [Hint: Use Appendix III.]

16. ■■ To attend the 1996 Summer Olympics, a fan flew from Los Angeles (32° N, 119° W) to Atlanta (33° N, 84° W). Using your knowledge of angular measure, determine the approximate shortest flight distance in kilometers.

17. ■■■ Could a circular pie be cut such that all of the wedge-shaped pieces have an arc length along the outer crust equal to the pie's radius? If not, how many such pieces could you cut, and what would be the angular dimension of the final piece?

18. ■■■ Electrical wire of diameter 0.75 cm is wound on a spool with a radius of 30 cm and a length of 24 cm. (a) Through how many radians must the spool be turned to wrap one even layer of wire? (b) What is the length of this wound wire?

7.2 Angular Speed and Velocity

19. Viewed from above, a turntable is observed to rotate counterclockwise. The angular velocity vector then is (a) tangential to the turntable rim, (b) out of the plane of the turntable, (c) counterclockwise, (d) none of the preceding.

20. The frequency unit of hertz is equivalent to (a) that of the period, (b) cycle, (c) radian/s, (d) s^{-1}.

21. ● Do all points on a wheel rotating about a fixed axis have the same angular velocity? the same tangential speed? Explain.

22. When clockwise or counterclockwise is used to describe rotational motion, why is a phrase such as "viewed from above" added?

23. ● A wheel is spinning at a constant angular velocity about an axis through its center. Which point(s) on the wheel have the greatest and smallest tangential speeds?

24. ■ A wheel spins at 30 rpm. What is this angular speed expressed in radians per second?

25. ■ A satellite in a circular orbit has a period of 10 h. What is the satellite's frequency in revolutions per day?

26. ■ A race car makes two and a half laps around a circular track in 3.0 min. What is the car's average angular speed?

27. ■ If a ball is rotating with an angular speed of 3.5 rad/s about a central axis, how long does it take for the ball to go through one revolution?

28. ■ What is the period of revolution for (a) a centrifuge rotating at 12 000 rpm and (b) a computer hard disk drive rotating at 10 000?

29. ■■ Determine which has the greater angular speed: particle A, which travels 160° in 2.00 s, or particle B, which travels 4π rad in 8.00 s.

30. ■■ ● The tangential speed of a particle on a rotating wheel is 3.0 m/s. If the particle is 0.20 m from the axis of rotation, how long will it take for the particle to go through one revolution?

31. ■■ ● A merry-go-round makes 24 revolutions in a 3.0-minute ride. (a) What is its average angular speed? (b) What are the tangential speeds of persons 4.0 m and 5.0 m from the center or axis of rotation?

32. ■■ In 1.5 min, a car moving at constant speed gets halfway around a circular track that has a diameter of 1.0 km. What are the car's (a) angular speed and (b) tangential speed?

33. ■■ Calculate the angular speeds in radians per second of the Earth (a) rotating on its axis and (b) revolving around the Sun.

34. ■■ A CD-ROM in a computer rotates at 500 rpm. How long will it take the disc to make one revolution?

35. ■■■ The driver of a car sets the cruise control and ties the steering wheel so that the car travels at a uniform speed of 15 m/s in a circle with a diameter of 120 m. (a) Through what angular distance does the car move in 4.00 min? (b) What arc distance does it travel in this time?

7.3 Uniform Circular Motion and Centripetal Acceleration

36. In uniform circular motion, there is a (a) constant velocity, (b) constant angular velocity, (c) zero acceleration, (d) net tangential acceleration.

37. If the centripetal force on a particle in uniform circular motion is increased, (a) the tangential speed will remain constant, (b) the tangential speed will decrease, (c) the radius of the circular path will increase, (d) the tangential speed will increase and/or the radius will decrease.

38. Explain why mud flies off a fast-spinning wheel.

39. The spin cycle of a washing machine is used to extract water from recently washed clothes. Explain the physical principle(s) involved.

40. The apparatus illustrated in •Fig. 7.26 is used to demonstrate forces in a rotating system. The floats are in jars of water. When the arm is rotated, which way will the floats move? (Compare with the accelerometer in Fig. 4.28.) Does it make a difference which way the arm is rotated?

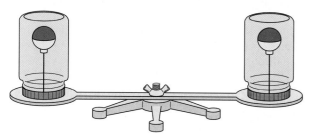

•**FIGURE 7.26   When set in motion, a rotating system** See Exercise 40.

41. When rounding a curve in a fast-moving car, we experience a feeling of being thrown outward (•Fig. 7.27). It is sometimes said that this is because of an outward centrifugal (center-fleeing) force. However, in terms of Newton's laws, this *pseudo*, or *false*, force doesn't really exist. Analyze the situation in the figure to show that this is the case. [Hint: Start with Newton's first law.]

•**FIGURE 7.27   A center-fleeing force?**   See Exercise 41.

42. Many race tracks have banked curves, which allow the cars to travel faster than if the curves were flat. Actually, cars could make turns on these banked curves if there were no friction at all. Explain this with the free-body diagram shown in •Fig. 7.28.

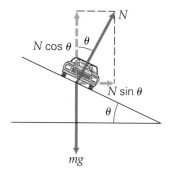

•**FIGURE 7.28   Banking safety**   See Exercises 42 and 55.

43. ■ A rotating cylinder about 10 mi in length and 5.0 mi in diameter is designed to be used as a space colony. With what angular speed must it rotate so the residents on it will experience the same acceleration as that due to gravity on Earth?

44. ■ An Indy car with a speed of 120 km/h goes around a level, circular track with a radius of 1.00 km. What is the centripetal acceleration of the car?

45. ■ 🌐 A wheel of radius 1.5 m rotates at a uniform speed. If a point on the rim of the wheel has a centripetal acceleration of 1.2 m/s², what is the point's tangential speed?

46. ■■ The Moon revolves around the Earth in 29.5 days in a nearly circular orbit with a radius of 3.8×10^5 km. Assume that the Moon's motion is uniform. With what acceleration is the Moon falling toward the Earth?

47. ■■ A car with a constant speed of 83.0 km/h enters a circular flat curve with a radius of curvature of 0.400 km. If the friction between the road and the car's tires can supply a centripetal acceleration of 1.25 m/s², does the car negotiate the curve safely? Justify your answer.

48. ■■ 🌐 Imagine that you swing about your head a 0.250-kilogram ball attached to the end of a string. The ball moves at a constant speed in a horizontal circle with a radius of 1.50 m. If it takes 1.20 s for the ball to make one revolution, (a) what is the ball's tangential speed? (b) What centripetal force are you imparting to the ball via the string? (c) Can the string be exactly horizontal?

49. ■■ In Exercise 48, what angle would the string make relative to the horizontal?

50. ■■ A student is to swing a bucket of water in a vertical circle without spilling any (•Fig. 7.29). If the distance from his shoulder to the center of mass of the bucket of water is 1.0 m, what is the minimum speed required to keep the water from coming out of the bucket at the top of the swing?

•**FIGURE 7.29   Weightless water?**   See Exercise 50.

51. ■■ A car of mass 1.5×10^3 kg rounds a circular turn of radius 20 m. If the road is flat and the coefficient of friction is 0.50 between the tires and the road, how fast can the car travel without skidding?

52. ■■ A jet pilot puts an aircraft with a constant speed of 700 km/h into a vertical circular loop with a radius of 2.0 km. (a) At the bottom of the loop, what is the normal force on the seat by the pilot? (b) What is the normal force at the top of the loop?

53. ■■■ A block of mass m slides down an inclined plane into a loop-the-loop of radius r (•Fig. 7.30). (a) Neglecting friction, what is the minimum speed the block must have at the highest point of the loop to stay in the loop? [Hint: What force must act on the block at the top of the loop to keep the block on a circular path?] (b) At what vertical height on the inclined plane (in terms of the radius of the loop) must the block be released if it is to have the required minimum speed at the top of the loop?

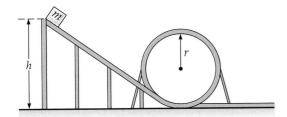

•FIGURE 7.30 **Loop-the-loop** See Exercise 53.

54. ■■■ For a scene in a movie, a stunt driver drives a pickup truck with a length of 4.25 m and a mass of 1.5×10^3 kg around a circular curve with a radius of curvature of 0.333 km (•Fig. 7.31). The truck is to curve off the road, jump across a gully 10 m wide, and land on the other side 2.96 m below the initial side. What is the minimum centripetal acceleration the truck must have in going around the circular curve to clear the gully and land on the other side?

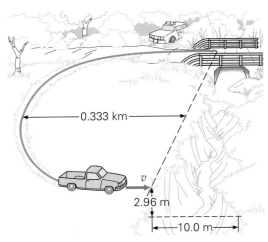

•FIGURE 7.31 **Over the gully** See Exercise 54.

55. ■■■ For the banked curve in Fig. 7.28, the horizontal component of the normal force toward the center of the curve reduces the need for friction to prevent skidding. In fact, for a circular curve banked at a given angle, there is one speed for which no frictional force is required—the centripetal force is supplied completely by the inward component of the normal force. This relationship is used in designing banked curves. (a) Show that the banking angle θ for which no friction is required for a car to negotiate a circular curve safely is given by $\tan \theta = v^2/gr$, where v is the car's speed and r is the radius of curvature. (b) How would the banking angle differ for a more massive truck? (c) Show how friction affects the situation.

56. ■■■ In a theme park, you walk into a giant can-shaped room that can rotate. When the angular speed of the "room" reaches a certain value, the bottom "floor" drops out (•Fig. 7.32). However, you don't fall out! Explain why. If the coefficient of static friction between you and the side wall is 0.30 and the radius of the room is 2.5 m, what is the minimum rotational speed of the room so that you can stay in the bottomless room?

•FIGURE 7.32 **Bottomless room** See Exercise 56.

7.4 Angular Acceleration

57. The angular acceleration in circular motion (a) is equal in magnitude to the tangential acceleration divided by the radius, (b) increases the angular velocity if in the same direction, (c) has units of s^{-2}, (d) all of the preceding.

58. Can you think of an example of a car having both centripetal acceleration and angular acceleration?

59. Is it possible for a car in circular motion to have angular acceleration but not centripetal acceleration?

60. ■ ⊙ During an acceleration, the angular speed of an engine increases from 700 rpm to 3000 rpm in 3.0 s. What is the average angular acceleration of the engine?

61. ■ ⊙ A merry-go-round accelerating uniformly from rest achieves its operating speed of 2.5 rpm in five revolutions. What is the magnitude of its angular acceleration?

62. ■■ A $33\frac{1}{3}$-rpm record on a turntable uniformly reaches its operating speed in 2.45 s once the record player is turned on. (a) What is the angular distance traveled during this time? (b) What is the corresponding arc length in feet on the circumference of a 12-inch-diameter record?

63. ■■ 🌀 A bicycle being repaired is turned upside down, and one wheel is rotated at a rate of 60 rpm. If the wheel slows uniformly to a stop in 15 s, how many revolutions does it make during this time?

64. ■■ A Ferris wheel with a diameter of 35.0 m starts from rest and achieves its maximum operational tangential speed of 2.20 m/s in a time of 15.0 s. (a) What is the magnitude of the wheel's angular acceleration? (b) What is the magnitude of the tangential acceleration after the maximum operational speed is reached?

65. ■■ 🌀 The blades of a fan running at low speed turn at 250 rpm. When the fan is switched to high speed, the rotation rate increases uniformly to 350 rpm in 5.75 s. (a) What is the magnitude of the angular acceleration of the blades? (b) How many revolutions do the blades go through while the fan is accelerating?

66. ■■ 🌀 In landing, the rotor of a helicopter slows from an angular speed of 4500 rpm to rest in 5.0 s. How many revolutions does the rotor travel during this time?

67. ■■ A car on a circular track with a radius of 0.30 km accelerates from rest with a constant angular acceleration of magnitude 4.5×10^{-3} rad/s². (a) How long does it take the car to make one lap around the track? (b) What is the total (vector) acceleration of the car when it has completed half of a lap?

68. ■■■ A student investigating circular motion places a dime 10 cm from the center of a $33\frac{1}{3}$-rpm record on a turntable. The record player can accelerate at 1.42 rad/s². The student notes that the dime slides outward 2.25 s after she has switched on the turntable. (a) Why does the dime slide outward? (b) What is the coefficient of static friction between the dime and the record?

7.5 Newton's Law of Gravitation

69. The gravitational force is (a) a linear function of distance, (b) an infinite-range force, (c) applicable only to our solar system, (d) sometimes repulsive.

70. The acceleration due to gravity of an object on the Earth's surface (a) is a universal constant like G, (b) does not depend on the Earth's mass, (c) is directly proportional to the Earth's radius, (d) does not depend on the object's mass.

71. Astronauts in a spacecraft orbiting the Earth or out for a "spacewalk" (•Fig. 7.33) are seen to "float" in midair. This is sometimes referred to as weightlessness or zero gravity (zero g). Are these terms correct? Explain why an astronaut appears to float in or near an orbiting spacecraft.

•**FIGURE 7.33   Out for a walk**   Why does this astronaut seem to "float"? See Exercise 71.

72. If the cup in •Fig. 7.34 were dropped, no water would run out. Explain.

•**FIGURE 7.34   Let it go** See Exercise 72.

73. Can you determine the mass of the Earth simply by measuring the gravitational acceleration near the Earth's surface? If yes, give the details.

74. (a) A spring scale calibrated in kilograms on the Earth is taken to the Moon to make measurements. Will the scale read correctly? Explain. (b) For a given mass, will the scale reading be more or less at the Earth's equator than at the North Pole? Explain.

75. ■ From the known mass and radius of the Earth, compute the value of the acceleration due to gravity g at the surface of the Earth.

76. ■ Calculate the gravitational force between the Earth and the Moon.

77. ■■ Approximate the gravitational force exerted on the Moon by the Earth and the Sun during (a) a solar eclipse (the Moon is between the Sun and the Earth) and (b) a lunar eclipse (the Earth is between the Sun and the Moon).

78. ■■ Two objects are attracting each other with a 0.90-newton gravitational force. If the distance between them is tripled, what is the new gravitational force between them?

79. ■■ Four identical masses of 2.5 kg each are located at the corners of a square with 1.0 m sides. What is the net force on any one of the masses?

80. ■■ What is the acceleration due to gravity on the top of Mt. Everest? The summit is about 8.80 km above sea level. (Report to 3 significant figures.)

81. ■■ A 75-kilogram man weighs 735 N on Earth's surface. How far above the surface would he have to go to "lose" 10% of his body weight?

82. ■■ For a spacecraft going directly from the Earth to the Moon, beyond what point will lunar gravity begin to dominate; that is, where will the lunar gravitational force be equal in magnitude to the Earth's gravitational force? Are the astronauts on board truly weightless at this point?

83. ■■ Which is greater, the gravitational force exerted on the Earth by the Sun or by the Moon? (Compare these attractions by forming a ratio and giving a factor of how many times greater or smaller one is than the other.)

84. ■■ Assuming that we could get more soil from somewhere, approximately how thick would an added uniform outer layer on the Earth have to be to have $g = 10.0$ m/s^2 exactly? (Take the average density of the Earth to be 5.52 g/cm^3 and $R_E = 6.38 \times 10^3$ km.)

85. ■■■ (a) What is the mutual gravitational potential energy of the configuration shown in •Fig. 7.35 if all the masses are 1.0 kg? (b) What is the gravitational force per unit mass at the center of the configuration?

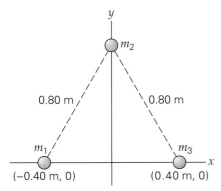

•FIGURE 7.35 **Gravitational potential, gravitational force, and center of mass** See Exercise 85.

7.6 Kepler's Laws and Earth Satellites

86. A new planet is discovered and its period determined. The new planet's distance from the Sun could then be found by using Kepler's (a) first law, (b) second law, (c) third law.

87. 🌐 A planet moves in its orbit about the Sun. (a) Its speed is constant. (b) Its distance from the Sun is constant. (c) It moves faster when it is closer to the Sun. (d) It moves slower when it is closer to the Sun.

88. (a) In one revolution, how much work does the centripetal force do on a satellite in circular orbit about the Earth? (b) A person in a freely falling elevator thinks he can avoid

injury by jumping upward just before the elevator strikes the floor. Would this work?

89. (a) In putting satellites into orbits, rockets are launched eastward. Why? (b) In the United States, satellites are launched from Florida. Why not from California, which generally has better weather conditions?

90. ■ 🌐 An instrument package is projected vertically upward to collect data at the top of the Earth's atmosphere (at an altitude of about 800 km). (a) What initial speed is required at the Earth's surface for the package to reach this height? (b) What percentage of the escape speed is this initial speed?

91. ■■ Compute the constant K of Kepler's third law for the orbits of (a) Earth and (b) Venus. (See Appendix III for data.)

92. ■■ Using a development similar to Kepler's law of periods for planets orbiting the Sun, find the required altitude of geosynchronous satellites above the Earth. [Hint: The period of such satellites is the same as that of the Earth.]

93. ■■ Venus has a rotational period of 243 d. What would be the altitude of a geosynchronous satellite for this planet?

94. ■■ The asteroid belt that lies between Mars and Jupiter may be the debris of a planet that broke apart or that was not able to form as a result of Jupiter's strong gravitation. The asteroid belt has a period of 5.0 y. How far from the Sun would this "fifth" planet have been?

Additional Exercises

95. At sunset, the Sun has an angular width of about 0.50°. From the time the lower edge of the Sun just touches the horizon, about how long does it take in minutes to disappear?

96. Ocean tides are primarily produced by the gravitational attraction of the Moon. Explain how this attraction gives rise to two tidal "bulges" on opposite sides of the Earth, resulting in two daily high tides and two daily low tides (•Figure 7.36). [Hint: Consider the inverse-square relation between the distance and the gravitational force acting on

•FIGURE 7.36 **Tides** (a) Low tide and (b) high tide at the same location off the California coast. See Exercise 96.

the water on opposite sides of the Earth and on the Earth itself.] (You may want to go to the library for a little help on this one.)

97. A 60-kilogram woman prepares to swing out over a pond on a rope (•Fig. 7.37). What tension in the rope is required if she is to maintain a circular arc (a) when she starts her swing by stepping off the platform, (b) when she is 5.0 m above the pond surface, and (c) at the lowest point of her swing? [Hint: Treat the woman and the rope as a simple pendulum.]

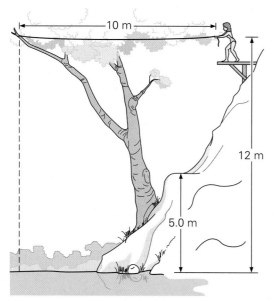

•FIGURE 7.37 **A swinging time** See Exercises 97 and 98.

98. Suppose that the woman swinging on the rope in Fig. 7.37 (see Exercise 97) starts her swing at a lower point and has a tangential speed of 8.5 m/s when the rope is at an angle of 30° to the vertical. (a) What is her tangential acceleration at that time? (b) What is the magnitude of the vector sum of the tangential and centripetal accelerations at that time? (c) From what height above the pond surface did she start her swing?

99. An ultracentrifuge (a very fast centrifuge) operates at a speed of 5.00×10^6 rpm. It contains test tubes filled with viruses. (a) What is the centripetal acceleration on a virus at a radial distance of 4.00 cm from the centrifuge's axis

of rotation? (b) How does this acceleration compare with g, the acceleration due to gravity?

100. A garden hose wound on a rotating storage cylinder has an outer layer at a distance of 0.45 m from the center of the cylinder. As this layer of wound hose is pulled out, the cylinder makes three revolutions. What is the length of the unwound hose?

101. What is the escape speed of a satellite at an altitude of 750 km in a circular orbit about the Earth ?

102. An astronaut has a weight of 735 N on Earth. What would be the force of gravity on the astronaut in a spacecraft in a circular orbit at an altitude of 450 km?

103. The value of the Kepler constant is $K = 2.97 \times 10^{-19}\,\text{s}^2/\text{m}^3$, and $T^2 = (2.97 \times 10^{-19}\,\text{s}^2/\text{m}^3)r^3$. With nonstandard units, K can be made equal to 1, and the equation can be written as $T^2 = r^3$. Using Earth data, find the units of K that make it equal to 1. [Hint: The average, or mean, distance of the Earth from the Sun is used in astronomy as a unit called an *astronomical unit* (AU).]

104. A pendulum swinging in a circular arc under the influence of gravity, as shown in •Fig. 7.38, has both centripetal and tangential components of acceleration. (a) If the pendulum bob has a speed of 2.7 m/s when the cord makes an angle of $\theta = 15°$ with the vertical, what are the magnitudes of the components at this time? (b) When is the centripetal acceleration a maximum? What is the value of the tangential acceleration at that time?

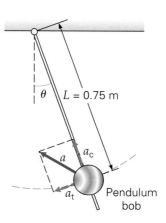

•FIGURE 7.38 **A swinging pendulum** See Exercise 104.

Rotational Motion and Equilibrium

INSIGHTS

- The Leaning Tower of Pisa
- Stability in Action
- Slide or Roll to a Stop? — Antilock Brakes

I t's always a good idea to keep your equilibrium—but it's more important in some situations than in others! Looking at a photo like this, your first reaction is probably to wonder how these acrobats keep from falling. Presumably, the poles must help—but in what way? If you think about the picture and about what you have already learned, a less obvious but in some ways more interesting question may come to mind: Why do we think they're in danger in the first place? After all, the wire is strong enough to support their weight.

We might say they are in equilibrium. Translational equilibrium ($\Sigma \mathbf{F} = 0$) was discussed in Chapter 4, but here there is another consideration—rotation. Should the performers start to fall (which hopefully they won't), there would initially be a sideways rotation about the wire. To avoid this calamity, another condition must be met—rotational equilibrium, which we will discuss in this chapter.

These tightrope riders are striving to elude rotational motion. But rotational motion is very important in physics because rotating objects are all around us: wheels on vehicles (as on the bicycles in the photo); gears and pulleys in machinery; planets in our solar system; and even many bones in the human body. (Can you think of a few bones that rotate in sockets?)

Fortunately, the equations describing rotational motion can be written as almost direct analogues of those for translational (linear) motion. In Chapter 7, this similarity was pointed out for the kinematic equations. With the addition of equations describing rotational dynamics, you will be able to analyze the general motions of real objects.

8.1 Rigid Bodies, Translations, and Rotations

Objectives: To (a) distinguish between pure translational and pure rotational motions of a rigid body and (b) state the condition(s) for rolling without slipping.

It was convenient initially to consider motions of objects with the understanding that the motion of an object can be represented by a particle located at its center of mass. Rotation, or spinning, was not a consideration then because a particle, a point mass, has no physical dimensions. Rotational motion becomes relevant when we analyze the motion of solid extended objects or rigid bodies, the focus of this chapter.

Definition of:
A rigid body

> A **rigid body** is an object or system of particles in which the distances between particles are fixed and remain constant.

Note: The words "rotation" and "revolution" are commonly used synonymously. In general, this book uses rotation when the axis of rotation goes through the body (the Earth's rotation on its axis, in a period of 24 h) and revolution when the axis is outside the body (the revolution of the Earth about the Sun, in a period of 365 d).

A quantity of liquid water is not a rigid body, but the ice that would form if the water were frozen would be. The discussion of rigid body rotation is thus conveniently restricted to solids. Actually, the concept of a rigid body is an idealization. In reality, the particles (atoms and molecules) of a solid vibrate constantly. Also, solids can undergo elastic (and inelastic) deformations (Chapter 6). Even so, most solids can be considered to be rigid bodies for purposes of analyzing rotational motion.

A rigid body may be subject to either or both of two types of motions: translational and rotational. Translational motion is basically the linear motion we studied in previous chapters. If an object has only **translational motion** (•Fig. 8.1a),

•**FIGURE 8.1 Rolling—a combination of translational and rotational motions** (a) In pure translational motion, all the particles of an object have the same instantaneous velocity. (b) For pure rotational motion, all the particles of an object have the same instantaneous angular velocity. (c) Rolling is a combination of translational and rotational motions. Summing the velocity vectors for these two motions shows that the point of contact (for a sphere) or the line of contact (for a cylinder) is instantaneously at rest. (d) The line of contact (or, for a sphere, a line through the point of contact) is called the instantaneous axis of rotation. Note that the center of mass of a rolling object on a level surface moves linearly and remains over the point or line of contact.

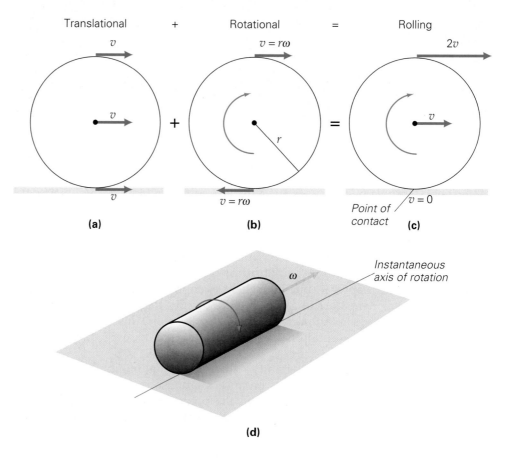

every particle has the same instantaneous velocity, which means there is no rotation. (Why?)

An object may have only **rotational motion**, motion about a fixed axis (Fig. 8.1b). In this case, all the particles of an object have the same instantaneous angular velocity and travel in circles about the axis of rotation. Although the axis of rotation of an object is commonly taken through its center of mass, this is not always the case. For example, you might pivot a meterstick through one end and have rotational motion about an axis through that end.

General rigid body motion is a combination of translational and rotational motions. When you throw a ball, the translational motion is described by the motion of its center of mass (as in projectile motion). But the ball may also spin, or rotate, and usually does. A common example of rigid body motion involving both translation and rotation is rolling, as illustrated in Fig. 8.1c. The combined motion of any point or particle is given by the vector sum of its instantaneous velocity vectors. (Three points or particles are shown in the figure—top, middle, and bottom). At each instant, a rolling object rotates about an **instantaneous axis of rotation** through its point of contact with the surface (for a sphere) or along its line of contact with the surface (for a cylinder; Fig. 8.1d). The location of this axis changes with time. However, note in Fig. 8.1c that the point or line of contact of the body with the surface is instantaneously at rest (zero velocity), as can be seen from the vector addition of the combined motions at that point.

When an object rolls without slipping, the translational and rotational motions are related simply as discussed in Chapter 7. For example, when a uniform ball (or cylinder) rolls in a straight line on a flat surface, it turns through an angle θ, and a point on the object that was initially in contact with the surface moves through an arc distance s (•Fig. 8.2). From Chapter 7, you know that $s = r\theta$. The center of mass of the ball is directly over the point of contact and moves a linear distance s. Then

$$v_{CM} = \frac{s}{t} = \frac{r\theta}{t} = r\omega$$

In terms of the speed of the center of mass and the angular speed ω, the *condition for rolling without slipping* is

$$v_{CM} = r\omega \qquad \begin{array}{c}\text{condition for rolling}\\\text{without slipping}\end{array} \qquad (8.1)$$

(This expression is generally written as $v = r\omega$, with the understanding that v is the magnitude of the velocity of the center of mass.) The condition is also expressed by

$$s = r\theta \qquad (8.1a)$$

itself, where s is the distance an object rolls, or the distance the center of mass moves.

Carrying Eq. 8.1 one step further, we can write an expression for the time rate of change of the velocity: $v_{CM}/t = r\omega/t$ (starting from rest). This yields an equation for *accelerated rolling without slipping*:

$$a_{CM} = \frac{v_{CM}}{t} = \frac{r\omega}{t} = r\alpha \qquad (8.1b)$$

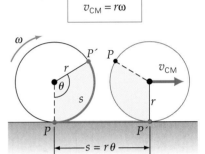

$$v_{CM} = r\omega$$

•**FIGURE 8.2 Rolling without slipping** As an object rolls without slipping, the length of the arc between two points of contact on the circumference is equal to the linear distance traveled—think of paint coming off a roller. This distance is $s = r\theta$. The speed of the center of mass is $v_{CM} = r\omega$.

Example 8.1 ■ Slipping or Not Slipping?: That Is the Question

A cylinder with a radius of 12 cm rolls with an instantaneous angular speed of 0.75 rad/s down an inclined plane. (a) If the center of mass of the cylinder travels at a speed of 0.10 m/s at this time, does the cylinder roll without slipping? (b) The cylinder goes onto a level surface and rolls without slipping with the same speed of the center of mass as in part (a). Assuming this speed remains constant for 2.0 s, through what angle does the cylinder turn during this time?

Thinking It Through. Part (a) is a simple test of Eq. 8.1, which must be satisfied for rolling without slipping. In part (b), Eq. 8.1 is valid and can be used to find the angular speed. With the speed and time, we can find the angular distance or the angle through which the cylinder rotated.

Solution. We list the data as usual:

Given: $r = 12$ cm $= 0.12$ m   *Find:* (a) if cylinder is slipping
        $\omega = 0.75$ rad/s   (b) θ (angle of rotation)
        $v_{CM} = 0.10$ m/s
        $t = 2.0$ s

(a) If the data satisfy Eq. 8.1, the cylinder will be rolling without slipping:

$$v_{CM} = r\omega = (0.12 \text{ m})(0.75 \text{ rad/s}) = 0.090 \text{ m/s} \neq 0.10 \text{ m/s}$$

Hence, the cylinder is slipping as it goes down the plane. The center of mass is traveling too fast for the nonslipping, rolling condition.

(b) For rolling without slipping on the level surface, the angular speed is given by Eq. 8.1:

$$\omega = \frac{v_{CM}}{r} = \frac{0.10 \text{ m/s}}{0.12 \text{ m}} = 0.83 \text{ rad/s}$$

(Note that the unit of angular speed is 1/s, or s^{-1}, but we add the unitless radian, or rad, for clarity.) Since this speed is constant for the time period, we have

$$\theta = \omega t = (0.83 \text{ rad/s})(2.0 \text{ s}) = 1.7 \text{ rad}$$

The cylinder makes a little over one-quarter of a rotation (right?). Check it yourself.

Follow-up Exercise. How far does the cylinder travel linearly in part (b) of this Example? Find this distance by using two different methods.

8.2 Torque, Equilibrium, and Stability

Objectives: To (a) define torque, (b) apply the conditions for mechanical equilibrium, and (c) describe the relationship between the location of the center of gravity and stability.

Torque

As with translational motion, a force is necessary to produce a change in rotational motion. The rate of change of motion depends not only on the magnitude of the force, but also on the perpendicular distance of its line of action from the axis of rotation, $r_\perp$ (•Fig. 8.3a,b). The line of action of a force is a line extending through the force vector arrow—that is, the line along which the force acts.

Figure 8.3 shows that $r_\perp = r \sin\theta$, where r is the straight-line distance between the axis of rotation and the point at which the force acts, and θ is the angle between the line of r and the force vector **F**. This perpendicular distance $r_\perp$ is called the **moment arm** or **lever arm**.

Torque: the product of a lever arm and a force

The product of the force and the lever arm is called **torque**, τ (from the Latin *torquere*, meaning "to twist"). The magnitude of the torque provided by the force is

$$\tau = r_\perp F = rF \sin\theta \tag{8.2}$$

SI unit of torque: meter-newton (m·N)

The SI units of torque are meters times newtons (m·N). These are the same as the units of work, $W = Fd$ (N·m, or J). However, we will write the units of torque in reverse order as m·N to avoid confusion.

Note that rotational motion is not *always* produced when a force acts on a stationary rigid body. Notice from Eq. 8.2 that $\tau = 0$ when $\theta = 0°$—that is, when the force acts through the axis of rotation (Fig. 8.3c). Also, when $\theta = 90°$, the torque is maximum and the force acts perpendicularly to r, as when you push a door open. Keep in mind that the motion or angular acceleration depends on *where* a perpendicular force is applied (and therefore on the length of the lever arm). As a practical example, think of applying a force to a heavy glass door that swings in and out. Where you apply the force makes a great difference in how easily the door opens, or rotates (through the hinges) on its axis. Have you ever tried to open such a door and inadvertently pushed on the side near the hinges? This force produces a small torque and little or no rotational motion.

We can think of torque in rotational motion as the analogue of force in translational motion. An unbalanced or net force changes translational motion, and an unbalanced or net torque changes rotational motion. Torque, the product of a force and moment arm—both vectors—is itself a vector. Its direction is always perpendicular to the plane of the force and moment arm vectors and is given by a *right-hand rule* similar to that for angular velocity given in Section 7.2. If the fingers of the right hand are curled around the axis of rotation in the direction that the torque would produce a rotational (angular) acceleration, the extended thumb points in the direction of the torque. A sign convention, as in the case of linear motion, can be used to represent torque directions, as we will soon see.

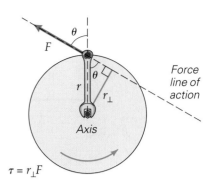

$$\tau = r_\perp F$$

(a) Counterclockwise torque

Example 8.2 ■ Lifting and Holding: Muscle Torque at Work

In our bodies, torques produced by the contraction of our muscles cause some bones to rotate at joints. For example, when you lift something with your forearm, a torque is applied by the biceps muscle on the lower arm (•Fig. 8.4). With the axis of rotation through the elbow joint and the muscle attached 4.0 cm from the joint, what are the magnitudes of the muscle torques for cases (a) and (b) in Fig. 8.4 if the muscle exerts a force of 600 N?

Thinking It Through. As in many rotational situations, it is important to know the orientations of the r and F vectors so that we can find the angle *between* them to determine the lever arm. Note in the insert in Fig. 8.4a that if the tails of the **r** and **F** vectors were put together, the angle between them would be greater than 90°.

Solution. First we list the data given here and in the figure. This Example demonstrates an important point. Recall that θ is the angle *between* the radial vector **r** and the force **F**.

Given: $r = 4.0$ cm $= 0.040$ m   **Find:** (a) τ (muscle torque magnitude)
         $F = 600$ N   (b) τ (muscle torque magnitude)
         (a) $\theta = 30° + 90°$
         (b) $\theta = 90°$

(a) In this case, **r** is directed along the forearm, so the angle between the **r** and **F** vectors is $\theta = 30° + 90°$. Thus, $r \sin \theta = r \sin (30° + 90°) = r \cos 30°$. (See Appendix I for

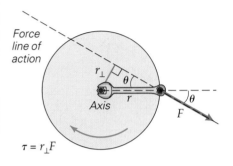

$$\tau = r_\perp F$$

(b) Smaller clockwise torque

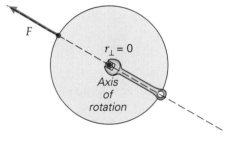

(c) Zero torque

•**FIGURE 8.3 Torque and moment arm** (a) The perpendicular distance $r_\perp$ from the axis of rotation to the line of action of a force is called the moment arm (or lever arm) and is equal to $r \sin \theta$. The torque, or twisting force, that produces rotational motion is given by $\tau = r_\perp F$. **(b)** The same force in the opposite direction with a smaller moment arm produces a smaller torque in the opposite direction. Note that $r_\perp F = rF_\perp$, or $(r \sin \theta)F = r(F \sin \theta)$. **(c)** When a force acts through the axis of rotation, $r_\perp = 0$ and $\tau = 0$.

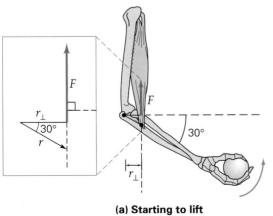

(a) Starting to lift   **(b) Holding**

•**FIGURE 8.4 Human torque** See Example 8.1.

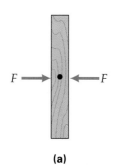

(a)

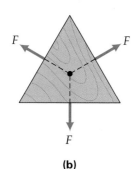

(b)

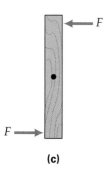

(c)

•**FIGURE 8.5 Equilibrium and forces** Forces with lines of action through the same point are said to be concurrent. The resultants of the concurrent forces acting on the objects in **(a)** and **(b)** are zero, and the objects are in equilibrium, because the net torque *and* net force are zero. In **(c)**, the object is in *translational* equilibrium, but it will undergo angular acceleration; it is *not* in rotational equilibrium.

Conditions for mechanical equilibrium

this and other useful trigonometric relationships.) Or, just notice from the figure inset that $r_\perp = r \cos 30°$ in this case. Using Eq. 8.2, we have

$$\tau = r_\perp F = rF \sin (30° + 90°) = rF \cos 30°$$

$$= (0.040 \text{ m})(600 \text{ N}) \cos 30° = 21 \text{ m·N}$$

at that instant.

(b) Here, the lever arm and the line of action for the force are perpendicular ($\theta = 90°$), and $r_\perp = r \sin 90° = r$. Then,

$$\tau = r_\perp F = rF = (0.040 \text{ m})(600 \text{ N}) = 24 \text{ m·N}$$

Follow-up Exercise. In (a) of this Example, there must have been a net torque, since the ball was lifted by a rotation of the forearm. In (b), the ball is just being held and there is no rotational acceleration, so there is no net torque on the system. Identify the other torque(s) in each case.

Before considering rotational dynamics with net torques and rotational motions, let's first look at the situation in which the forces and torques acting on a body are balanced or in equilibrium.

Equilibrium

In general, equilibrium means that things are in balance or are stable. This definition applies in the mechanical sense to forces and torques. Unbalanced forces produce translational accelerations, but *balanced* forces produce the condition we call *translational equilibrium*. Similarly, unbalanced torques produce rotational accelerations, and *balanced* torques produce *rotational equilibrium*.

According to Newton's first law of motion, when the sum of the forces acting on a body is zero, the body remains either at rest (static) or in motion with a constant velocity. In either case, the body is said to be in **translational equilibrium**. Stated another way, the *condition for translational equilibrium* is that the net force on a body is zero, $\Sigma \mathbf{F}_i = 0$. In other words, the vector sum of all forces equals zero: $\Sigma \mathbf{F}_i = \mathbf{F}_1 + \mathbf{F}_2 + \mathbf{F}_3 + \cdots = 0$. It should be apparent that this condition is satisfied for the situations illustrated in •Fig. 8.5a,b. Forces with lines of action through the same point are called **concurrent forces**. When these forces vectorially add to zero as in (a) and (b), the body is in translational equilibrium.

But what about the situation pictured in Fig. 8.5c? Here $\Sigma \mathbf{F}_i = 0$, but the opposing forces will cause the object to rotate, and it will clearly not be in a state of static equilibrium. (Such a pair of equal and opposite forces not having the same line of action is called a *couple*.) Thus, the condition $\Sigma \mathbf{F}_i = 0$ is a necessary, but *not sufficient*, condition for static equilibrium.

Since $\Sigma \mathbf{F}_i = 0$ is the condition for translational equilibrium, you might predict (and correctly so) that $\Sigma \boldsymbol{\tau}_i = 0$ is the *condition for rotational equilibrium*. That is, if the sum of the *torques* acting on an object is zero, then the object is in **rotational equilibrium**—it is rotationally at rest or rotates with a constant angular velocity.

Thus we see that there are two equilibrium conditions. Taken together, they define what is called **mechanical equilibrium**:

A body is said to be in mechanical equilibrium when the conditions for both translational and rotational equilibrium are satisfied:

$$\Sigma \mathbf{F}_i = 0 \quad \text{(for translational equilibrium)}$$

$$\Sigma \boldsymbol{\tau}_i = 0 \quad \text{(for rotational equilibrium)}$$

(8.3)

A rigid body in mechanical equilibrium may be either at rest or moving with a constant linear and/or angular velocity. An example of the latter is an object rolling without slipping on a level surface with the center of mass having a constant velocity. This is an ideal condition, however, because there is always some friction in reality.

Of greater practical interest is **static equilibrium**, the condition that exists when a rigid body remains at rest. There are many instances in which we do not want things to move, and this absence of motion can occur only if the equilibrium conditions are satisfied. It is particularly comforting to know, for example, that a bridge you are crossing is in static equilibrium and not subject to translational or rotational motions.

Let's consider examples of static translational equilibrium and static rotational equilibrium separately, and then one in which both apply.

Example 8.3 ■ Translational Static Equilibrium: No Translational Motion

A picture hangs motionless on a wall as shown in •Fig. 8.6a. If it has a mass of 5.0 kg, what are the tension forces in the wires?

Thinking It Through. Since the picture remains motionless, it must be in static equilibrium, so applying the conditions for mechanical equilibrium should give equations that yield the tensions. Note that all the forces (tension and weight forces) are concurrent—that is, their lines of action pass through a common point, the nail. Because of this, the condition for rotational equilibrium ($\Sigma \tau_i = 0$) is automatically satisfied—with respect to an axis of rotation along the nail, the moment arms ($r_\perp$) of the forces are zero. Thus, we need consider only translational equilibrium.

Solution.

Given: $\theta = 45°$ *Find:* T_1 and T_2
$m = 5.0$ kg

You will find it helpful to isolate the forces acting on the picture in a free-body diagram as was done in Chapter 4 for force problems (Fig. 8.6b). The diagram shows the concurrent forces acting through their common point. Note that we have moved all the force vectors to the common point, which is taken as the origin of the coordinate axes. The weight force mg acts downward.

With the system in static equilibrium, the net force is zero; that is, $\Sigma \mathbf{F}_i = 0$. Thus, the sums of the rectangular components are also zero: $\Sigma \mathbf{F}_{x_i} = 0$ and $\Sigma \mathbf{F}_{y_i} = 0$. Then (using ± for direction), we get

$$\Sigma \mathbf{F}_{x_i}: \quad T_1 \cos 45° - T_2 \cos 45° = 0$$

This just tells you that the *x* components of the forces are equal and opposite and that $T_1 = T_2 = T$. For the *y* components, with two upward components of $T \sin 45°$, we have

$$\Sigma \mathbf{F}_{y_i}: \quad T \sin 45° + T \sin 45° - mg = 0$$

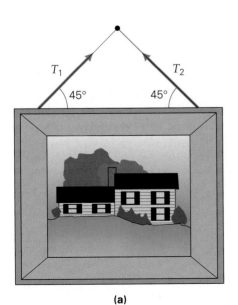

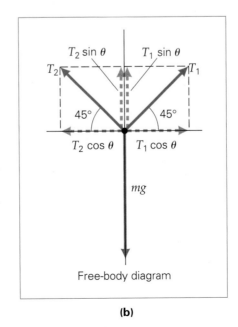

Free-body diagram

(a)   (b)

•**FIGURE 8.6 Translational static equilibrium** (a) Since the picture hangs motionless on the wall, the sum of the forces acting on it must be zero. The forces are concurrent, with their lines of action passing through a common point at the nail. (b) This is shown in the free-body diagram, where all the forces have been represented as acting at this point. T_1 and T_2 have been moved to the nail point for convenience. Note, however, that these are the forces acting on the *picture*, not on the nail. See Example 8.3.

or

$$2T \sin 45° - mg = 0$$

Thus,

$$T = \frac{mg}{2 \sin 45°} = \frac{(5.0 \text{ kg})(9.8 \text{ m/s}^2)}{2(0.707)} = 35 \text{ N}$$

Follow-up Exercise. Analyze the situation in Fig. 8.6 that would result if the wires were shortened such that the angles were decreased. Carry this to the limit where the angles approach zero.

As pointed out earlier, torque is a vector and direction is important. Similar to linear motion (Chapter 2), in which we used plus and minus signs to express opposite directions (for example, $+x$ and $-x$), we designate torque directions as being plus and minus depending on the rotational acceleration they tend to produce. The rotational "directions" are taken as clockwise or counterclockwise around the axis of rotation. A torque that tends to produce a counterclockwise rotation will be taken as positive ($+$), and a torque that tends to produce a clockwise acceleration will be taken as negative ($-$).

This convention is arbitrary. Remember only that the magnitude of the sum of the counterclockwise torques must equal that of the clockwise torques to have rotational equilibrium. To illustrate, let's apply our convention to the situation shown in •Fig. 8.7.

Example 8.4 ■ Rotational Static Equilibrium: No Rotational Motion

Three masses are suspended from a meterstick as shown in Fig. 8.7a. How much mass must be suspended on the right side for the system to be in static equilibrium? (Neglect the mass of the meterstick.)

Thinking It Through. As the free-body diagram (Fig. 8.7b) shows, the translational equilibrium condition will be satisfied with the upward reaction force N balancing the downward weight forces, so long as the stick is horizontal. But N is not known if m_3 is unknown, so applying the condition for rotational equilibrium should give the required value of m_3. (Note the lever arms are measured from the pivot point, or the center of the meterstick.)

Solution. From the figure, we have

Given: $m_1 = 25$ g   *Find:* m_3 (unknown mass)
$\quad\quad\quad r_1 = 50$ cm
$\quad\quad\quad m_2 = 75$ g
$\quad\quad\quad r_2 = 30$ cm
$\quad\quad\quad r_3 = 35$ cm

Because the condition for translational equilibrium ($\Sigma \mathbf{F}_i = 0$) is satisfied, $N - Mg = 0$, or $N = Mg$, where M is the total mass. This is true no matter what the total mass may be—that is, regardless of how much mass we add for m_3. However, unless the proper mass for m_3 is placed on the right side, the stick will experience a net torque and rotate.

Notice that the masses on the left side produce torques that tend to rotate the stick counterclockwise, and the mass on the right side produces a torque that tends to rotate it clockwise. We apply the condition for rotational equilibrium by summing the torques about an axis. Let's take this to be through the center of the stick at the 50-centimeter position, or point A in Fig. 8.7b. Then, noting that N passes through the axis of rotation ($r_\perp = 0$) and produces no torque, we have

$$\Sigma \tau_i: \quad \tau_1 + \tau_2 + \tau_3 = + r_1 F_1 + r_2 F_2 + r_3 F_3$$

$$= r_1(m_1 g) + r_2(m_2 g) - r_3(m_3 g) = 0$$

(using our sign convention for torque vectors)

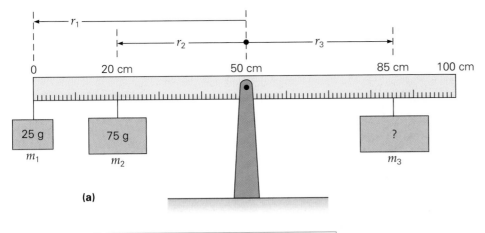

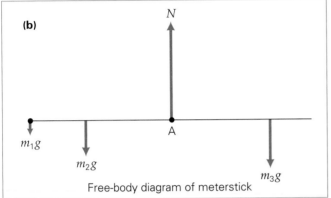

(a)

(b)

N

A

m_1g

m_2g

m_3g

Free-body diagram of meterstick

y

$-\tau$ $+\tau$

x

Sign convention

•**FIGURE 8.7 Rotational static equilibrium** For the meterstick to be in rotational equilibrium, the sum of the torques acting about any selected axis must be zero. (The mass of the meterstick is considered negligible.) See Example 8.4.

Noting that the g's cancel out and solving for m_3, we have

$$m_3 = \frac{m_1r_1 + m_2r_2}{r_3} = \frac{(25 \text{ g})(50 \text{ cm}) + (75 \text{ g})(30 \text{ cm})}{35 \text{ cm}} = 100 \text{ g}$$

where it was convenient not to convert to standard units. (The mass of the stick was neglected. If the stick is uniform, however, its mass will not affect the equilibrium as long as the pivot point is at the 50-centimeter mark. Why?)

Follow-up Exercise.   The axis of rotation could have been taken through any point along the stick. That is, if a system is in static rotational equilibrium, the condition $\Sigma\tau_i = 0$ holds for *any* axis of rotation. Demonstrate this for the system in this Example by taking the axis of rotation through the left end of the stick ($x = 0$).

In general, the conditions for both translational and rotational equilibrium need to be written explicitly to solve a statics problem. Example 8.5 is one such case.

Example 8.5 ■ Static Equilibrium: No Translation, No Rotation

A ladder with a mass of 15 kg rests against a smooth wall (•Fig. 8.8a). A painter who has a mass of 78 kg stands on the ladder as shown in the figure. What frictional force must act on the bottom of the ladder to keep it from slipping?

Thinking It Through.   Here we have a variety of forces and torques. However, the ladder will not slip as long as the conditions for static equilibrium are satisfied. Summing both the forces and torques to zero should enable us to solve for the necessary frictional force. Also, as we will see, choosing a convenient axis of rotation for summing the torques can simplify the torque equation.

•FIGURE 8.8 Static equilibrium
For the painter's sake, the ladder
has to be in static equilibrium; that
is, both the sum of the forces and
the sum of the torques must be
zero. See Example 8.5.

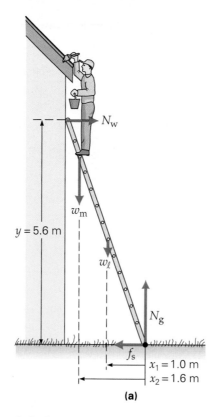

(a)

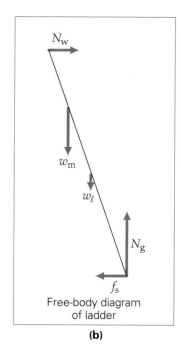

Free-body diagram
of ladder

(b)

Solution.

Given: $m_\ell = 15$ kg   *Find:* f_s (force of static friction)
     $m_m = 78$ kg
     Distances given in figure

Because the wall is smooth, there is negligible friction between it and the ladder, and only the normal reaction force of the wall (N_w) acts on the ladder at this point (Fig. 8.8b).

In applying the conditions for static equilibrium, you are free to choose any axis of rotation for the rotational condition. (The conditions must hold for any part of a system for static equilibrium; that is, there can't be motion in any part of the system.) Note that choosing an axis at the end of the ladder where it touches the ground eliminates the torques due to f_s and N_g since the moment arms are zero. Then you can write three equations (using mg for w):

$$\Sigma F_{x_i}: \quad N_w - f_s = 0$$
$$\Sigma F_{y_i}: \quad N_g - m_m g - m_\ell g = 0$$

and

$$\Sigma \tau_i: \quad (m_\ell g)x_1 + (m_m g)x_2 + (-N_w y) = 0$$

The weight of the ladder is considered to be concentrated at its center of gravity. Solving the third equation for N_w and substituting the given values for the masses and distances gives

$$N_w = \frac{(m_\ell g)x_1 + (m_m g)x_2}{y}$$

$$= \frac{(15 \text{ kg})(9.8 \text{ m/s}^2)(1.0 \text{ m}) + (78 \text{ kg})(9.8 \text{ m/s}^2)(1.6 \text{ m})}{5.6 \text{ m}} = 2.4 \times 10^2 \text{ N}$$

From the first equation then,

$$f_s = N_w = 2.4 \times 10^2 \text{ N}$$

Follow-up Exercise. In this Example, would the frictional force between the ladder and the ground (call it f_{s_1}) remain the same if there were friction between the wall and the ladder (call it f_{s_2})? Justify your answer.

As the preceding Examples have shown, a good procedure to follow in working problems involving static equilibrium is as follows:

1. Sketch a space diagram of the problem.
2. Draw a free-body diagram, showing and labeling all external forces and normally resolving the forces into x and y components.
3. Apply the equilibrium conditions. Sum the forces: $\Sigma \mathbf{F}_i = 0$, or usually in component form $\Sigma \mathbf{F}_{x_i} = 0$ and $\Sigma \mathbf{F}_{y_i} = 0$. Sum the torques: $\Sigma \boldsymbol{\tau}_i = 0$, remembering to select an appropriate axis of rotation to reduce the number of terms as much as possible. (But be sure to include the unknown force.) Use $\pm$ sign conventions for both $\mathbf{F}$ and $\boldsymbol{\tau}$.
4. Solve for the unknown quantities.

Stability and Center of Gravity

The equilibrium of a particle or a rigid body can also be described as being stable or unstable in a gravitational field. For rigid bodies, these categories of equilibria are conveniently analyzed in terms of the center of gravity. Recall from Chapter 6 that the **center of gravity** is the point at which all the weight of an object may be considered to be concentrated in representing it as a particle. When the acceleration due to gravity is constant, the center of gravity and the center of mass coincide.

Note: Review Section 6.5.

If an object is in **stable equilibrium**, any small displacement results in a restoring force or torque, which tends to return the object to its original equilibrium position. As illustrated in •Fig. 8.9a, a ball in a bowl is in stable equilibrium. Analogously, the center of gravity of an extended body in stable equilibrium is

Stable equilibrium—a restoring torque

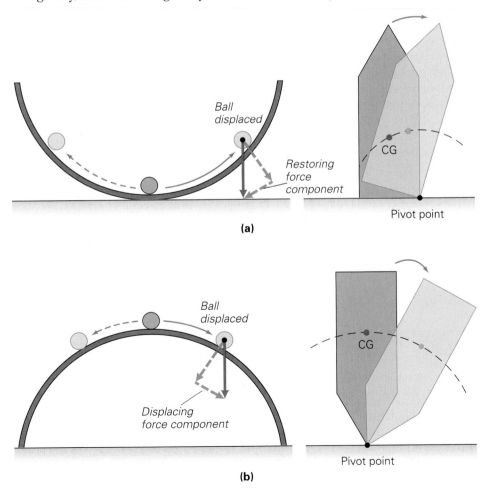

•**FIGURE 8.9   Stable and unstable equilibria   (a)** When an object is in stable equilibrium, any small displacement from an equilibrium position results in a force or torque that tends to return the object to that position. A ball in a bowl (left) returns to the bottom after being displaced. Analogously, the center of gravity (CG) of an extended object (right) can be thought of as being in a potential energy bowl—a small displacement raises the CG, increasing the object's potential energy. **(b)** For an object in unstable equilibrium, any small displacement from its equilibrium position results in a force or torque that tends to take it farther away from that position. The ball on top of an overturned bowl (left) is in unstable equilibrium. For an extended object (right), the CG can be thought of as being on an inverted potential energy bowl—a small displacement lowers the CG, decreasing the object's potential energy.

essentially in a potential energy bowl. Any slight displacement raises its center of gravity, and a restoring force tends to return it to the position of minimum potential energy. This force actually produces a restoring torque that is due to a component of the weight force and that tends to rotate the object about a pivot point back to its original position.

For an object in **unstable equilibrium**, any small displacement from equilibrium results in a force (and a torque) that tends to rotate the object farther away from its equilibrium position. This situation is illustrated in Fig. 8.9b. Note that the center of gravity is at the top of an overturned, or inverted, potential energy bowl; that is, the potential energy is at a maximum in this case.

Small displacements or slight disturbances have profound effects on objects in unstable equilibrium—it doesn't take much to cause such an object to change its position. Yet, even if the angular displacement of an object in stable equilibrium is quite substantial, the object will still be restored to its equilibrium position. An object lying on its long side can be rotated through quite a distance and will still fall back to that original position. This is another way of expressing the *condition of stable equilibrium:*

> An object is in stable equilibrium as long as its center of gravity after a small displacement still lies above and inside its original base of support. That is, the line of action of the weight force of the center of gravity intersects the original base of support.

When this is the case, there will always be a restoring torque (•Fig. 8.10a). However, when the center of gravity or center of mass falls outside the base of support, over goes the object—because of a gravitational torque that rotates it away from its equilibrium position (Fig. 8.10b).

Rigid bodies with wide bases and low centers of gravity are therefore most stable and least likely to tip over. This relationship is evident in the design of high-speed race cars, which have wide wheel bases and centers of gravity close to the ground (•Fig. 8.11).

The location of the center of gravity of the human body has an effect on certain physical abilities. For example, women can generally bend over and touch their toes or touch their palms to the floor more easily than can men, who often fall over trying. On the average, men have higher centers of gravity (larger shoulders) than do women (larger pelvises), so it is more likely that a man's center of gravity will be outside his base of support when he bends over. (See Demonstration 6.) See the Insight on p. 264 for another real-life example of equilibrium and stability.

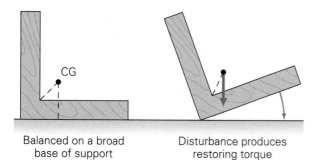

Balanced on a broad
base of support

Disturbance produces
restoring torque

(a) Stable Equilibrium

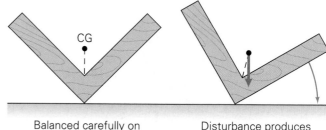

Balanced carefully on
narrow base of support (point)

Disturbance produces
displacing torque

(b) Unstable Equilibrium

•**FIGURE 8.10 Examples of stable and unstable equilibria** (a) When the center of gravity is above and inside an object's base of support, the object is in stable equilibrium (there is a restoring torque). Note how the line of action of the weight of the center of gravity (CG) intersects the original base of support after the displacement. **(b)** When the center of gravity lies outside the base of support or the line of action of the weight does not intersect the original base of support, the object is unstable (there is a displacing torque).

(a)　　　　　　　　　　**(b)**　　　　　　　　　　**(c)**

•**FIGURE 8.11 Stable, stabler, stablest** **(a)** The acrobat's base of support is very narrow—the small area of head-to-head contact. As long as his center of gravity remains above this area, he is in equilibrium, but a displacement of only a few inches would probably be enough to topple him. (Why he is in a spread-eagle position will become clearer in Section 8.3.) **(b)** This balancing rock in Idaho has a slightly larger base of support than the acrobat in (a); still, a displacement of its center of gravity by a few feet would send it crashing down. **(c)** Race cars are very stable because of their wide wheel bases and low center of gravity.

Demonstration 6 ■ Stability and Center of Gravity

A demonstration to show that women generally have more stability than men—because of their mass distribution and the location of their center of gravity.

(a) An object is placed just beyond the fingertips of a man kneeling with elbows against his knees. Maintaining his balance, he can topple the object with his nose.

(b) With hands now clasped behind his back, the man's center of gravity shifts beyond his base of support (knees) as he bends over to topple the object, and he will topple instead.

(c) A woman can generally do it both ways.

(d) Because her center of gravity is positioned lower in the body, it does not move outside the base of support (past her knees) as she bends with hands clasped behind her back.

The Leaning Tower of Pisa: Stability in Question

The Leaning Tower of Pisa is a famous Italian landmark from which Galileo allegedly performed experiments (see the Chapter 2 Insight on p. 52). Recently, the Tower of Pisa has been in the spotlight—in attempts to keep it from falling. The Tower started leaning before its completion in A.D. 1350 because of soft subsoil. It was closed to the public in 1990, with a lean of about 5.5° off the perpendicular (about 5 m at the top) and an average lean increase of about 1.2 mm a year.

Cement was injected into the base in the 1930s, but the lean accelerated. There has been an international effort to save the Tower. From 1993 to 1995, a concrete ring was built at the base and 620 tons of lead counterweight was placed on it (Fig. 1). Within months, the incline was corrected by 2 to 3 cm. But the construction of a second ring jolted the tower, and overnight it tilted 2.5 mm. With that big scare, an additional 230 tons of lead weight was added.

Current plans call for soil to be removed from beneath the high side of the Tower to balance the natural settling (and to keep the center of gravity above the base of support). However, plastic-coated steel cables will be attached to the Tower to support the structure should there be any sudden shifts (Fig. 2). Should the settling continue, long steel cables connected to ground anchors would be installed to prevent further tilting. It is hoped that these measures will correct the tilt by about 0.5°—enough to save the Tower but still have the leaning attraction.

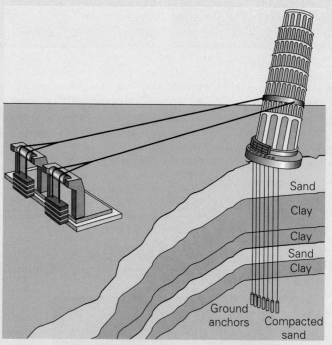

FIGURE 2 Keep the center of gravity over the base of support. Efforts will be made to stabilize the Tower by removing soil from beneath the high side. For safety in case of sudden shifts, steel support cables will be attached.

FIGURE 1 Hold it stable! Tons of lead counterweight being used to help correct the Tower's lean.

Example 8.6 ■ Stack Them Up: Center of Gravity

Uniform, identical bricks 20 cm long are stacked so that 4.0 cm of each brick extends beyond the brick beneath, as shown in •Fig. 8.12a. How many bricks can be stacked in this way before the stack falls over?

Thinking It Through.   As each brick is added, the center of mass (or center of gravity) of the stack moves to the right. The stack will be stable as long as the combined center of mass (CM) is over the base of support—the bottom brick. All of the bricks have the same mass, and the center of mass of each is located at its midpoint. So, the horizontal location of the stack's CM must be computed as bricks are added. The location of the CM was discussed in Chapter 6 (see Eq. 6.19).

Solution.

Given: length = 20 cm   *Find:*   maximum number of bricks
            displacement, each brick = 4.0 cm   with stability

Taking the origin to be at the center of the bottom brick, the horizontal coordinate of the center of mass (or center of gravity) for the first two bricks in the stack is given by Eq. 6.19, where $m_1 = m_2 = m$ and x_2 is the displacement of the second brick:

$$X_{CM_2} = \frac{mx_1 + mx_2}{m + m} = \frac{m(x_1 + x_2)}{2m} = \frac{x_1 + x_2}{2} = \frac{0 + 4.0 \text{ cm}}{2} = 2.0 \text{ cm}$$

The masses of the bricks cancel out (since they are all the same). For three bricks,

$$X_{CM_3} = \frac{m(x_1 + x_2 + x_3)}{3m} = \frac{0 + 4.0 \text{ cm} + 8.0 \text{ cm}}{3} = 4.0 \text{ cm}$$

For four bricks,

$$X_{CM_4} = \frac{m(x_1 + x_2 + x_3 + x_4)}{4m} = \frac{0 + 4.0 \text{ cm} + 8.0 \text{ cm} + 12.0 \text{ cm}}{4} = 6.0 \text{ cm}$$

and so on.

This series of results shows that the center of mass of the stack moves horizontally 2.0 cm for each brick added to the bottom one. For a stack of six bricks, the center of mass is 10 cm from the origin and directly over the edge of the bottom brick (2.0 cm × 5 *added* bricks = 10 cm, which is half the length of the bottom brick), so the stack is in unstable equilibrium. The stack may not topple if the sixth brick is positioned very carefully, but it is doubtful that this could be done in practice. A seventh brick would definitely cause the stack to fall.

Follow-up Exercise. If the bricks in this Example were stacked so that alternately 4.0 cm and 6.0 cm extended beyond the brick beneath, how many bricks could be stacked in this manner before the stack toppled?

For another case of stability, see the Insight on p. 266.

8.3 Rotational Dynamics

Objectives: To (a) describe the moment of inertia of a rigid body and (b) apply the rotational form of Newton's second law to physical situations.

Moment of Inertia

Torque is the rotational analogue of force in linear motion, and a net torque produces rotational motion. To analyze this relationship, consider a constant net force acting on a particle of mass m (•Fig. 8.13). The magnitude of the torque on the particle is

$$\tau = r_\perp F = rF_\perp = rma_\perp = mr^2\alpha \qquad torque\ on\ a\ particle \qquad (8.4)$$

where $a_\perp = r\alpha$ is the tangential acceleration (a_t, Eq. 7.13). For the rotation of a rigid body about a fixed axis, this equation can be applied to each particle and the results summed over the entire body (n particles) to find the total torque. Since all the particles of a rotating body have the same angular acceleration, we can simply add the individual torque magnitudes:

$$\Sigma\tau = \tau_1 + \tau_2 + \tau_3 + \cdots + \tau_n$$
$$= m_1r_1^2\alpha + m_2r_2^2\alpha + m_3r_3^2\alpha + \cdots + m_nr_n^2\alpha$$
$$= (m_1r_1^2 + m_2r_2^2 + m_3r_3^2 + \cdots + m_nr_n^2)\alpha$$
$$\Sigma\tau = \left(\sum_{i=1}^{n} m_ir_i^2\right)\alpha \qquad (8.5)$$

But for a rigid body, the masses (m_i's) and the distances from the axis of rotation (r_i's) are constant. Therefore, the quantity in the parentheses in Eq. 8.5 is constant and it is called the **moment of inertia**, I:

$$I = \Sigma m_ir_i^2 \qquad moment\ of\ inertia \qquad (8.6)$$

SI unit of moment of inertia: kilogram-meters squared (kg·m²)

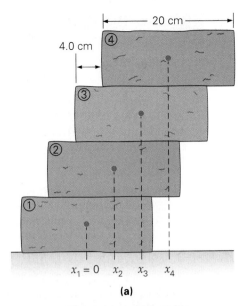

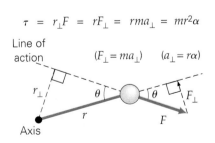

(b)

•FIGURE 8.12 **Stack them up!** (a) How many bricks can be stacked like this before the stack falls? See Example 8.6. (b) Try a similar experiment with books.

Torque on a particle

•FIGURE 8.13 **Torque on a particle** The magnitude of the torque on a particle of mass m is $\tau = mr^2\alpha$.

Stability in Action

When riding a bicycle and going around a curve or making a circular turn on a level surface, a rider instinctively leans into the curve. Why? We might think that leaning over, rather than remaining upright, is more likely to cause a spill. However, leaning really does increase stability—it's all a matter of torques.

When a bicycle or vehicle goes around a level circular curve, a centripetal force is needed, as we learned in Chapter 7. This force is generally supplied by the force of static friction between the tires and the road. As illustrated in Fig. 1a, the reaction force **R** of the ground on the bicycle provides the required centripetal force $\mathbf{F_c}$ to round the curve and the normal force **N**.

Suppose the rider tried to go around the curve with these forces while remaining upright as shown in Fig. 1a. Note that the line of action of **R** does not go through the system's center of gravity (indicated by a dot). Considering an axis of rotation through this point, there would be a coun-

terclockwise torque that would tend to rotate the bicycle in such a way that the wheels would slide inward underneath the rider. However, if the rider leans inward at the proper angle (Fig. 1b), both the line of action of **R** and the weight force go through the center of gravity and there is no rotational instability (as the gentleman on the bicycle well knew).

(Being alert, you might observe there is still a torque on the rider. Indeed, when the rider leans into the curve, the weight force gives rise to a torque about an axis through the point of contact with the ground. This, along with the turning of the handle bars, causes the bicycle to turn. If the bicycle were not moving, there would be a rotation about this axis and the bicycle and rider would fall over.)

The need to lean into a curve is readily apparent in bicycle and motorcycle races on level tracks. Things can be made easier for the riders if tracks or roadways are banked to provide a natural lean (recall Section 7.3 and Exercise 7.55).

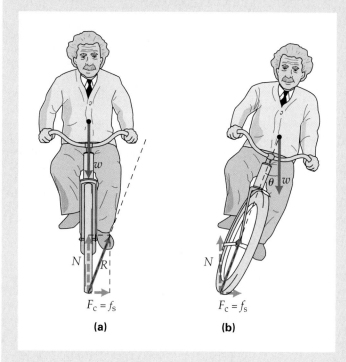

(a)   (b)

FIGURE 1 Leaning into a curve When rounding a curve or making a turn, a bicycle rider must lean into the curve. (This rider could have told you why.)

Thus, the magnitude of the net torque can be conveniently written as

Rotational form of Newton's second law

Note: τ means τ_{net}

$$\tau = I\alpha \qquad \text{torque on a rigid body} \qquad (8.7)$$

This is the *rotational form of Newton's second law* (**F** = $m\mathbf{a}$ and $\boldsymbol{\tau} = I\boldsymbol{\alpha}$, in vector form). Keep in mind that *net* forces and torques ($\mathbf{F}_{net}$ and $\boldsymbol{\tau}_{net}$) are necessary to produce accelerations, although this is not indicated explicitly in Eq. 8.7; τ means *net* torque.

As you might infer from a comparison of these two forms of Newton's law, the moment of inertia I is a measure of *rotational inertia*, or a body's tendency to resist change in its rotational motion. Although I is constant for a rigid body and is the rotational analogue of mass, you must keep in mind that, unlike the mass of a particle, the moment of inertia is referenced to a particular axis and can have different values for different axes.

The moment of inertia also depends on the mass distribution *relative* to the axis of rotation. It is easier (that is, it takes less torque) to give an object an angular acceleration about some axes than about others. The following Example illustrates this point.

Example 8.7 ■ Rotational Inertia: Mass Distribution and Axis of Rotation

Find the moment of inertia about the axis indicated for each of the one-dimensional dumbbell configurations in •Fig. 8.14. (Consider the mass of the connecting bar to be negligible, and report to three significant figures.)

Thinking It Through. This is a direct application of Eq. 8.6 for cases with different masses and distances. It will show that the moment of inertia of an object depends on the axis of rotation and on the mass distribution relative to the axis of rotation. The sum for I will include only two terms (two masses).

Solution.

Given: Values of m and r from the figure.   *Find:* $I = \Sigma m_i r_i^2$

With $I = m_1 r_1^2 + m_2 r_2^2$,

(a) $I = (30 \text{ kg})(0.50 \text{ m})^2 + (30 \text{ kg})(0.50 \text{ m})^2 = 15 \text{ kg·m}^2$

(b) $I = (40 \text{ kg})(0.50 \text{ m})^2 + (10 \text{ kg})(0.50 \text{ m})^2 = 12.5 \text{ kg·m}^2$

(c) $I = (30 \text{ kg})(1.5 \text{ m})^2 + (30 \text{ kg})(1.5 \text{ m})^2 = 135 \text{ kg·m}^2$

(d) $I = (30 \text{ kg})(0 \text{ m})^2 + (30 \text{ kg})(3.0 \text{ m})^2 = 270 \text{ kg·m}^2$

(e) $I = (40 \text{ kg})(0 \text{ m})^2 + (10 \text{ kg})(3.0 \text{ m})^2 = 90 \text{ kg·m}^2$

This Example clearly shows how the moment of inertia depends on the mass *and* its distribution relative to a particular axis of rotation. In general, the moment of inertia is larger the farther the mass is from the axis of rotation. This principle is important in the design of flywheels, which are used in automobiles to keep the engine running smoothly between cylinder firings. The mass of a flywheel is concentrated near the rim, giving a large moment of inertia, which resists changes in motion.

Follow-up Exercise. In parts (d) and (e) of this Example, would the moments of inertia be different if the axis of rotation were through m_2? Explain.

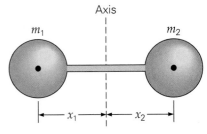

(a) $m_1 = m_2 = 30$ kg
    $x_1 = x_2 = 0.50$ m

(b) $m_1 = 40$ kg, $m_2 = 10$ kg
    $x_1 = x_2 = 0.50$ m

(c) $m_1 = m_2 = 30$ kg
    $x_1 = x_2 = 1.5$ m

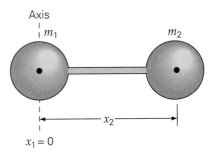

(d) $m_1 = m_2 = 30$ kg
    $x_1 = 0$, $x_2 = 3.0$ m

(e) $m_1 = 40$ kg, $m_2 = 10$ kg
    $x_1 = 0$, $x_2 = 3.0$ m

•FIGURE 8.14 Moment of inertia The moment of inertia depends on the distribution of mass relative to a particular axis of rotation and, in general, has a different value for each axis. This difference reflects the fact that objects are easier or more difficult to rotate about certain axes. See Example 8.7.

Balancing Act: Locating the Center of Gravity

A rod with a movable ball, like that shown in •Fig. 8.15, is more easily balanced if the ball is in a higher position. This is because with the ball in a higher position, (a) the system has a higher center of gravity and more stability; (b) when the center of gravity is off the vertical, there is less torque and smaller angular acceleration; (c) the moment of inertia about the axis of rotation is larger; (d) the center of gravity is closer to the axis of rotation.

Reasoning and Answer. With the ball at any position and the rod vertical, the system is in unstable equilibrium. We saw in Section 8.2 that rigid bodies with wide bases and *low* centers of gravity are more stable, so answer (a) isn't correct. Any slight movement will cause the rod to rotate about an axis through its point of contact. With the center of gravity (CG) at a higher position and off the vertical, there would be a greater lever arm (and thus a *greater* torque), so (b) too is incorrect. Also, with the ball in a higher position, the center of gravity is *farther* from the axis of rotation, which eliminates (d). However, moving the CG farther from the axis of rotation has an interesting consequence—a greater moment of inertia or resistance to change in rotational motion.

    With the ball in a higher position, as the rod starts to rotate there is a greater torque, but the increased moment of inertia produces an even greater resistance to rotational motion. With the rod rotationally accelerating more slowly, you have more time to adjust your

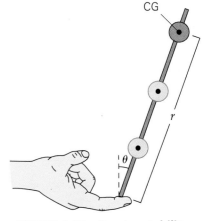

•FIGURE 8.15 Greater stability with a higher center of gravity? See Conceptual Example 8.8.

hand and bring your finger and the axis of rotation under the center of gravity. The torque is then zero and the rod is again in equilibrium, albeit unstable. Thus, the answer is (c).

We can understand this result more clearly if we consider the angular acceleration of the rod as it starts to tilt. The smaller the angular acceleration, the more time is available to adjust your hand under the rod to balance it. The angular acceleration of the rod is given by Eq. 8.7: $\alpha = \tau/I$.

Now let us look at how the torque, τ, and moment of inertia, I, vary with r, the distance of the system's CG from the axis of rotation (through the balancing finger). Neglecting the mass of the rod and considering the CG to be simply that of the balls, $\tau = rF = rmg \sin \theta$ (Eq. 8.2) and $I = mr^2$ (Eq. 8.5). In other words, τ increases with r while I increases with r^2. Combining these relationships, we can see that α is proportional to $1/r$. Thus we have our paradoxical result: The farther the rod's CG from the axis of rotation, the more slowly it tilts and the easier it is to balance.

Follow-up Exercise. When walking on a thin bar or rail, such as a railroad rail, you have probably found that it helps to hold your arms outstretched. Similarly, tightrope walkers often carry long poles, as in the chapter-opening photo. How does this help a performer to maintain balance?

As Example 8.8 shows, the moment of inertia is an important consideration in rotational motion. By changing the axis of rotation and the relative mass distribution, the value of I can be changed and the motion affected. You were probably told to do this if you played softball or baseball as a child. When at bat, children are often instructed to "choke up" on the bat—to move their hands farther up on the handle.

Now you know why. In doing so, the axis of rotaion is moved closer to the more massive end of the bat (or its center of mass). Hence the moment of inertia is decreased (smaller r in the mr^2 term). Then when a swing is taken, the angular acceleration is greater. The bat gets around quicker, and the chance of hitting the ball before it goes past is greater. A batter has only a fraction of second to swing, and with $\theta = \frac{1}{2}\alpha t^2$, a larger α allows the quicker rotation of the bat through an angle of approximately 90°.

Parallel Axis Theorem

The calculations for the moments of inertia of most extended rigid bodies require math that is beyond the scope of this book. The results for some common shapes are given in •Fig. 8.16. The rotational axes are generally taken along axes of symmetry, that is, running through the center of mass so as to give a symmetrical mass distribution. An exception is the rod with an axis of rotation through one end (Fig. 8.16c). This axis is parallel to an axis of rotation through the center of mass of the rod (Fig. 8.16b). The moment of inertia about such a parallel axis is given by a useful theorem called the **parallel axis theorem**:

Note: $I = I_{CM}$ when $d = 0$, which is the minimum I value.

$$I = I_{CM} + Md^2 \tag{8.8}$$

where I is the moment of inertia about an axis that is parallel to one through the center of mass, I_{CM}, and at a distance d from it, and M is the total mass of the body (•Fig. 8.17). For the axis through the end of the rod (Fig. 8.16c), the moment of inertia is obtained by applying the parallel axis theorem to the thin rod in Fig. 8.16b:

$$I = I_{CM} + Md^2 = \tfrac{1}{12}ML^2 + M\left(\frac{L}{2}\right)^2 = \tfrac{1}{12}ML^2 + \tfrac{1}{4}ML^2 = \tfrac{1}{3}ML^2$$

Applications of Rotational Dynamics

The rotational form of Newton's second law allows us to analyze dynamic rotational situations. Examples 8.9 and 8.10 illustrate how this is done. In such situations, it is very important to make certain that all the data are properly listed to help with the increasing number of variables.

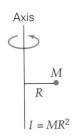

(a) Particle

$$I = MR^2$$

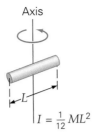

(b) Thin rod

$$I = \frac{1}{12} ML^2$$

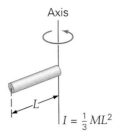

(c) Thin rod

$$I = \frac{1}{3} ML^2$$

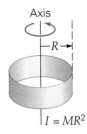

(d) Thin cylindrical shell, hoop, or ring

$$I = MR^2$$

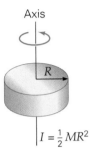

(e) Solid cylinder or disk

$$I = \frac{1}{2} MR^2$$

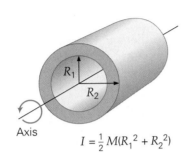

(f) Annular cylinder

$$I = \frac{1}{2} M(R_1^2 + R_2^2)$$

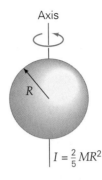

(g) Solid sphere about any diameter

$$I = \frac{2}{5} MR^2$$

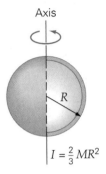

(h) Thin spherical shell

$$I = \frac{2}{3} MR^2$$

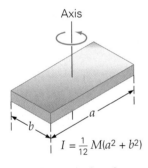

(i) Rectangular plate

$$I = \frac{1}{12} M(a^2 + b^2)$$

(j) Thin rectangular sheet

$$I = \frac{1}{12} ML^2$$

(k) Thin rectangular sheet

$$I = \frac{1}{3} ML^2$$

•**FIGURE 8.16   Moments of inertia for some uniform objects with common shapes**

Example 8.9 ■ Opening the Door: Torque In Action

A student opens a 12-kilogram door by applying a constant force of 40 N at a perpendicular distance of 0.90 m from the hinges (•Fig. 8.18). If the door is 2.0 m in height and 1.0 m wide, what is the magnitude of the angular acceleration of the door? (Assume that the door rotates freely on its hinges.)

Thinking It Through.   From the given information we can calculate the applied torque. To find the angular acceleration of the door, we need to know its moment of inertia. This can be calculated since we know mass and dimensions.

Solution.   From the problem, we can list the following:

Given: $M = 12$ kg
   $F = 40$ N
   $r_\perp = r = 0.90$ m
   $h = 2.0$ m (door height)
   $w = 1.0$ m (door width)

Find: α (magnitude of angular acceleration)

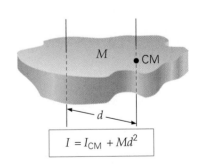

$$I = I_{CM} + Md^2$$

•**FIGURE 8.17   Parallel axis theorem**   The moment of inertia about an axis parallel to another through the center of mass of a body is $I = I_{CM} + Md^2$, where M is the total mass of the body and d is the distance between the two axes.

We need to apply the rotational form of Newton's second law (Eq. 8.7): $\tau = r_\perp F = I\alpha$. The problem boils down to finding the moment of inertia of the door.

Looking at Fig. 8.16, we see that part (k) applies to a door (treated as a uniform rectangle) rotating on hinges, so $I = \frac{1}{3}ML^2$, where $L = w$, the width of the door. We must simply do the calculations:

$$\tau = I\alpha$$

or

$$\alpha = \frac{\tau}{I} = \frac{r_\perp F}{\frac{1}{3}ML^2} = \frac{3rF}{Mw^2} = \frac{3(0.90\text{ m})(40\text{ N})}{(12\text{ kg})(1.0\text{ m})^2} = 9.0\text{ rad/s}^2$$

Follow-up Exercise. In this Example, if the constant torque were applied through an angular distance of 45° and then removed, how long would it take the door to swing completely open (90°)?

In problems involving pulleys in Chapter 4, the mass (and inertia) of the pulley was always neglected to simplify things. Now we know how to include this and can treat pulleys more realistically.

•FIGURE 8.18   Torque in action
See Example 8.9.

Example 8.10 ■ Pulleys Have Mass Too: Taking Account of Pulley Inertia

A block of mass m hangs from a string wrapped around a frictionless, disk-shaped pulley of mass M and radius R, as shown in •Fig. 8.19. If the block descends from rest under the influence of gravity, what is the magnitude of the linear acceleration of the block? (Neglect the mass of the string.)

Thinking It Through. Real pulleys have mass and rotational inertia, which affect the motion. The suspended mass via the string applies a torque to the pulley. Here we use the rotational form of Newton's second law to find the angular acceleration of the pulley and then its tangential acceleration, which is the same in magnitude as the linear acceleration of the block. (Why?)

Solution. The linear acceleration of the block depends on the angular acceleration of the pulley, so we look at the pulley system first. The pulley is treated as a disk and thus has a moment of inertia of $I = \frac{1}{2}MR^2$ (Fig. 8.16e). It experiences a torque due to the tension force in the string (T). With $\tau = I\alpha$ (considering only the upper dashed box in Fig. 8.19),

$$\tau = r_\perp F = RT = I\alpha = \frac{1}{2}MR^2\alpha$$

so

$$\alpha = \frac{2T}{MR}$$

The linear acceleration of the block and the angular acceleration of the pulley are related by $a = R\alpha$, so

$$a = R\alpha = \frac{2T}{M} \tag{1}$$

But T is unknown. Looking at the descending mass (the lower dashed box) and summing the forces in the vertical direction (positive in the direction of motion) gives

$$mg - T = ma$$

or

$$T = mg - ma \tag{2}$$

Using Eq. 2 to eliminate T from Eq. 1 gives

$$a = \frac{2T}{M} = \frac{2(mg - ma)}{M}$$

Solving for a, we get

$$a = \frac{2mg}{(2m + M)} \tag{3}$$

Note that if $M \to 0$ (as for ideal, massless pulleys in previous chapters), then $I \to 0$, and $a = g$ (from Eq. 3). Here, however, $M \neq 0$, so we have $a < g$. (Why?)

Follow-up Exercise. Pulleys can be analyzed even more realistically. In Example 8.10, friction was neglected, but practically a frictional torque (τ_f) exists and could be included. What would be the form of the angular acceleration in this case? Show your result to be dimensionally correct.

In pulley problems, we also neglect the mass of the string, which still gives a good approximation if the string is relatively light. Taking the mass of the string into account would give a continuously varying mass hanging on the pulley, thus producing a variable torque. Such a problem is beyond the scope of this book.

Suppose you had masses suspended from each side of a pulley. Here, you'd have to compute the net torque. If you didn't know the values of the masses, or which way the pulley would rotate, then you could simply assume a direction. As in the linear case, if the result came out with the opposite sign, this would indicate that you had assumed the wrong direction.

Problem-Solving Hint

For problems such as those of Examples 8.9 and 8.10, dealing with coupled rotational and translational motions, keep in mind that, with no rope slippage, the magnitudes of the accelerations are usually related by $a = r\alpha$, while $v = r\omega$ relates the magnitudes of the velocities at any instant of time. Applying Newton's second law (in rotational or linear form) to different parts of the system gives equations that can be combined by using such relationships. Also, for rolling without slipping, $a = r\alpha$ and $v = r\omega$ relate the angular quantities to the linear motion of the center of mass.

Another application of rotational dynamics is the analysis of motion on an inclined plane. Previously, we worked with block-shaped objects sliding down planes. Now we can generalize to include objects that can roll.

Example 8.11 ■ Rolling Down: Torque and Acceleration

A solid, rigid spherical ball of mass M and radius R is released at the top of a hard-surfaced inclined plane. The ball rolls without slipping, with only static friction between it and the plane (•Fig. 8.20a). What is the acceleration of the ball's center of mass?

Thinking It Through. Since the ball rolls without slipping, there is no relative motion between it and the plane at the point of contact (through which the instantaneous axis of rotation passes). At this point the force of static friction (f_s) acts on the ball. Without this force, the ball would slide down the plane. This force is perpendicular to the instantaneous axis of rotation, or directed up the plane. Detailed analyses are given in the solution for *two axes of rotaion*, just to show that the results are the same, as they should be.

Solution. You can analyze the motion by using two different axes of rotation. First, let's consider an axis through the center of mass of the ball (the central dot in Fig. 8.20a) and perpendicular to the page so that we can see the effect of the frictional torque. The acceleration of the ball's center of mass is down the plane, and the ball's angular acceleration is about an axis through that center.

Since the reaction force (the normal force N) and the weight force act through the axis of rotation (see the vector diagram in Fig. 8.20b), they produce no torques, and the angular acceleration is produced entirely by the frictional torque due to f_s.

Applying Newton's second law, $\tau = I\alpha$, and using the moment of inertia about an axis through the ball's center of mass as found in Fig. 8.16g ($I = \frac{2}{5}MR^2$), we have

$$\tau = Rf_s = I\alpha = (\tfrac{2}{5}MR^2)\left(\frac{a}{R}\right)$$

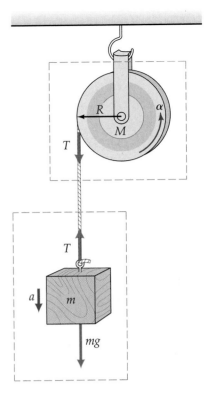

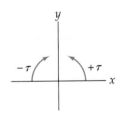

•**FIGURE 8.19 Pulley with inertia** Taking the mass, or rotational inertia, of a pulley into account allows a more realistic description of the motion. See Example 8.10.

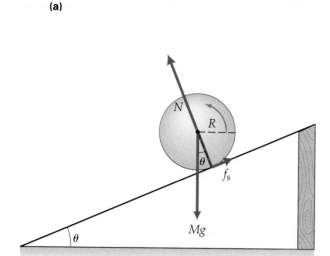

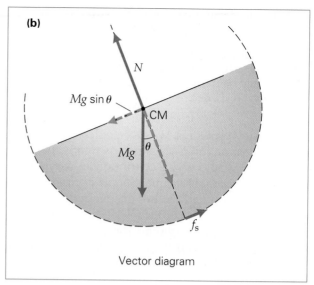

Vector diagram

•**FIGURE 8.20 Rolling down an inclined plane** For a rigid ball rolling down a hard-surfaced inclined plane, the force of static friction acts on the ball in response to the parallel component of the weight force. See Example 8.11.

where we substituted a/R for α. Then, solving for f_s,

$$f_s = \tfrac{2}{5}Ma$$

Looking at the translational motion of the center of mass and summing the forces acting parallel to the plane (see the vector diagram; positive in the direction of motion) gives

$$Mg\sin\theta - f_s = Ma$$

or

$$Mg\sin\theta - \tfrac{2}{5}Ma = Ma$$

Thus,

$$a = \tfrac{5}{7}g\sin\theta$$

Here is another situation in which the acceleration does not depend on M or R. Therefore, uniform spheres of any size or density will roll down the same plane with the same acceleration (assuming no slippage).

Before leaving this Example, let's look at the situation from another perspective. Consider analyzing the rotational motion about the instantaneous axis of rotation, which runs through the point of contact of the sphere and the plane's surface. This point is instantaneously at rest, as noted earlier. As can be seen from Fig. 8.20, both the normal force and the frictional force act through this point and so produce no torque. The torque on the ball from this perspective comes from the component of the weight force parallel to the plane, $Mg\sin\theta$ (see the vector diagram). The torque is then

$$\tau = rF_\perp = R(Mg\sin\theta) = I_i\alpha \tag{1}$$

where I_i is the moment of inertia *about the instantaneous axis of rotation*. Note that this is *not* equal to the moment of inertia used above, $I_{CM} = \tfrac{2}{5}MR^2$, which is the moment of inertia about an axis *through the center of mass* of the ball. You can find the moment of inertia about the instantaneous axis from the parallel axis theorem:

$$I_i = I_{CM} + Md^2 = \tfrac{2}{5}MR^2 + MR^2 = \tfrac{7}{5}MR^2 \tag{2}$$

Substituting for I_i from Eq. 2 into Eq. 1 gives

$$R(Mg\sin\theta) = I_i\alpha = (\tfrac{7}{5}MR^2)\left(\frac{a}{R}\right)$$

and

$$a = \tfrac{5}{7}g\sin\theta$$

This is the same as the acceleration found for the other axis of rotation (as it should be).

Follow-up Exercise. That uniform spheres of any size or density will roll down the same plane with the same acceleration is shown in this Example. Will different uniform objects with different shapes (for example, cylinders or hoops) roll down this incline with the same acceleration as a sphere?

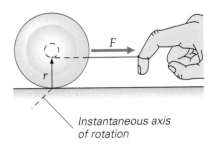

Applying a Torque: Which Way Does the Yo-yo Roll?

The string of a yo-yo sitting on a level surface is pulled as shown in •Fig. 8.21. Will the yo-yo roll (a) toward the person or (b) away from the person?

Reasoning and Answer. Apply the physics we have just studied to the situation. Note that the instantaneous axis of rotation is along the line of contact of the yo-yo with the surface. If you had a stick standing vertically in place of the **r** vector and pulled on a string attached to its top in the direction of **F**, which way would the stick rotate? Clockwise (about its instantaneous axis of rotation). The yo-yo reacts similarly—that is, it rolls in the direction of the pull, so the answer is (a). (Get a yo-yo and try it if you're a nonbeliever.)

There is more interesting physics in our yo-yo situation. The pull force is not the only force acting on the yo-yo; there are three others. Do they contribute torques? Let's identify these forces. There's the vertical weight of the yo-yo (gravitational attraction) and the normal force from the surface. Also, there is a horizontal force of static friction between the yo-yo and the surface (otherwise the yo-yo would slide rather than roll). But, all these forces act through the line of contact or through the instantaneous axis of rotation, so there are no torques here (why?).

What would happen if we increased the angle of the string or pull force (relative to the horizontal) as illustrated in •Fig. 8.22a? The yo-yo would still roll to the right. But note in Fig. 8.22b that, at some critical angle θ_c, the line of force goes through the axis of rotation and the net force and net torque on the yo-yo are zero, so it is in equilibrium.

If we exceed this critical angle (Fig. 8.22c), the yo-yo will begin to roll counterclockwise, or to the left. Note that the line of action of the force is on the other side of the axis of rotation than that in Fig. 8.22a and that the lever arm ($r_\perp$) has changed directions, resulting in a reversed net torque direction. (There is a lot of physics in a yo-yo.)

•**FIGURE 8.21 Pulling the yo-yo's string** See Conceptual Example 8.12.

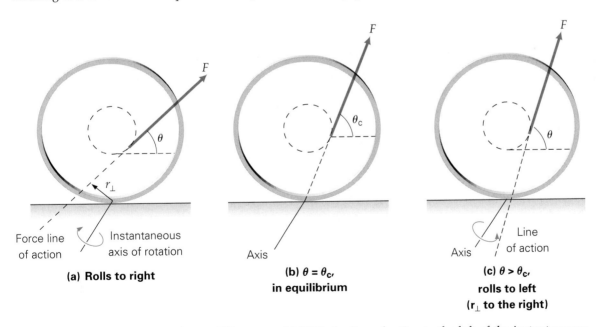

(a) Rolls to right

(b) $\theta = \theta_c$, in equilibrium

(c) $\theta > \theta_c$, rolls to left ($r_\perp$ to the right)

•**FIGURE 8.22 The angle makes a difference** (a) With the line of action to the left of the instantaneous axis, the yo-yo rolls to the right. (b) At a critical angle θ_c, the line of action passes through the axis and the yo-yo is in equilibrium. (c) When the line of action is to the right of the axis, the yo-yo rolls to the left. See Conceptual Example 8.12.

Follow-up Exercise. Suppose you set the yo-yo string at the critical angle, with the string over a round, horizontal bar at the appropriate height, and you suspended a weight on the end of the string to supply the force for the equilibrium condition. What would happen if you then pulled the yo-yo toward you, away from its equilibrium position, and released it?

8.4 Rotational Work and Kinetic Energy

Objectives: To discuss, explain, and use the rotational forms of (a) work, (b) kinetic energy, and (c) power.

This section gives the rotational analogues associated with work and kinetic energy for constant torques. Because their development is very similar to that given for their linear counterparts, detailed discussion is not needed.

Rotational Work. We can go directly from work done by a force to work done by a torque since they are related ($\tau = r_\perp F$). For rotational motion, **rotational work** $W = Fs$ for a single force F acting tangentially along an arc length s:

$$W = Fs = F(r_\perp \theta) = \tau\theta$$

where θ is in radians. Thus, for a single torque acting through an angle of rotation θ,

Rotational work

$$W = \tau\theta \tag{8.9}$$

In this book, both torque (τ) and angular displacement (θ) vectors are almost always along the fixed axis of rotation, so you will not need to be concerned about parallel components as you were for translational work. The torque and angular displacement may be in opposite directions, in which case the torque does negative work and slows down the rotation of the body. Negative rotational work is analogous to F and d being in opposite directions for translational motion.

Rotational Power. An expression for **rotational power**, the rotational analogue of power (the time rate of doing work), is easily obtained from Eq. 8.9:

Rotational power

$$P = \frac{W}{t} = \tau\left(\frac{\theta}{t}\right) = \tau\omega \tag{8.10}$$

The Work–Energy Theorem and Kinetic Energy

The relationship between the net rotational work and change in rotational kinetic energy can be derived as follows, starting with the equation for rotational work:

$$W = \tau\theta = I\alpha\theta$$

Since our torques are due only to constant forces, we get a constant α. For a constant angular acceleration, $\omega^2 = \omega_0^2 + 2\alpha\theta$, and therefore

$$W = I\left(\frac{\omega^2 - \omega_0^2}{2}\right) = \tfrac{1}{2}I\omega^2 - \tfrac{1}{2}I\omega_0^2$$

Thus, using Eq. 5.6, we get

Rotational analogue of work–energy theorem

$$W = \tfrac{1}{2}I\omega^2 - \tfrac{1}{2}I\omega_0^2 = K - K_0 = \Delta K \tag{8.11}$$

Here, the expression for **rotational kinetic energy**, K, is

Rotational kinetic energy

$$K = \tfrac{1}{2}I\omega^2 \tag{8.12}$$

Thus, the net rotational work is equal to the change in the rotational kinetic energy. Consequently, to change the *rotational* kinetic energy of an object, a net torque must be applied.

TABLE 8.1 Translational and Rotational Quantities and Equations

Translational		Rotational	
Force:	$\mathbf{F}$	Torque (magnitude):	$\tau = rF \sin \theta$
Mass (inertia):	m	Moment of inertia:	$I = \Sigma m_i r_i^2$
Newton's second law:	$\mathbf{F} = m\mathbf{a}$	Newton's second law:	$\boldsymbol{\tau} = I\boldsymbol{\alpha}$
Work:	$W = Fd$	Work:	$W = \tau\theta$
Power:	$P = Fv$	Power:	$P = \tau\omega$
Kinetic energy:	$K = \frac{1}{2}mv^2$	Kinetic energy:	$K = \frac{1}{2}I\omega^2$
Work–energy theorem:	$W = \frac{1}{2}mv^2 - \frac{1}{2}mv_0^2 = \Delta K$	Work–energy theorem:	$W = \frac{1}{2}I\omega^2 - \frac{1}{2}I\omega_0^2 = \Delta K$
Momentum:	$\mathbf{p} = m\mathbf{v}$	Momentum:	$\mathbf{L} = I\boldsymbol{\omega}$

It is possible to derive the kinetic energy of a rotating rigid body (on a fixed axis) directly. Summing the instantaneous tangential velocities and kinetic energies of the body's individual particles relative to the fixed axis gives

$$K = \tfrac{1}{2}\Sigma m_i v_i^2 = \tfrac{1}{2}(\Sigma m_i r_i^2)\omega^2 = \tfrac{1}{2}I\omega^2$$

where $v_i = r_i\omega$. Thus, Eq. 8.12 doesn't represent a new form of energy. It is simply another expression for kinetic energy, in a form that is more convenient for rigid-body rotation.

A summary of the analogous equations for translational and rotational motions is given in Table 8.1. (The table also contains momentum, which we shall discuss in Section 8.5.)

When an object has both translational and rotational motions, its total kinetic energy may be divided into parts to reflect this. For example, for a cylinder rolling without slipping on a level surface, the motion is purely rotational relative to the instantaneous axis of rotation, which is instantaneously at rest (see Example 8.12). The kinetic energy is

$$K = \tfrac{1}{2}I_i\omega^2$$

where I_i is the moment of inertia about the instantaneous axis. This moment of inertia is given by the parallel axis theorem (Eq. 8.8): $I_i = I_{CM} + MR^2$, where R is the radius of the cylinder. Then

$$K = \tfrac{1}{2}I_i\omega^2 = \tfrac{1}{2}(I_{CM} + MR^2)\omega^2 = \tfrac{1}{2}I_{CM}\omega^2 + \tfrac{1}{2}MR^2\omega^2$$

But since there is no slipping, $v_{CM} = R\omega$, and

$$\underset{\substack{total \\ KE}}{K} = \underset{\substack{rotational \\ KE}}{\tfrac{1}{2}I_{CM}\omega^2} + \underset{\substack{translational \\ KE}}{\tfrac{1}{2}Mv_{CM}^2} \qquad \text{(rolling, no slipping)} \quad (8.13)$$

Rolling kinetic energy

Note that although a cylinder was used as an example here, this is a general result and applies to any object that is rolling without slipping. *Thus, the total kinetic energy of such an object is the sum of two contributions: the translational kinetic energy of the center of mass and the rotational kinetic energy relative to a horizontal axis through the center of mass.*

Note: A rolling body has both translational kinetic energy and rotational kinetic energy.

Example 8.13 ■ Division of Energy: Rotational and Translational

A uniform, solid 1.0-kilogram cylinder rolls without slipping at a speed of 1.8 m/s on a flat surface. (a) What is the total kinetic energy of the cylinder? (b) What percentage of this total is rotational kinetic energy?

Thinking It Through. The cylinder has both rotational and translational kinetic energies, so Eq. 8.13 applies, and its terms are related by the condition of rolling without slipping.

Solution.

Given: $M = 1.0$ kg   *Find:* (a) K (total kinetic energy)
$\quad\quad v_{CM} = 1.8$ m/s
$\quad\quad I_{CM} = \frac{1}{2}MR^2$ (from Fig. 8.16e)   (b) $\dfrac{K_r}{K}$ ($\times$ 100%) (percentage
$\quad\quad\quad\quad\quad\quad\quad\quad\quad\quad\quad\quad\quad\quad\quad\quad\quad\quad$ of rotational energy)

(a) The cylinder rolls without slipping, so the condition $v_{CM} = R\omega$ applies. Then, the total kinetic energy is the sum of the rotational kinetic energy K_r and the translational kinetic energy of the center of mass K_{CM} (Eq. 8.13):

$$K = \frac{1}{2}\left(\frac{1}{2}MR^2\right)\left(\frac{v_{CM}}{R}\right)^2 + \frac{1}{2}Mv_{CM}^2 = \frac{1}{4}Mv_{CM}^2 + \frac{1}{2}Mv_{CM}^2$$

$$= \frac{3}{4}Mv_{CM}^2 = \frac{3}{4}(1.0\text{ kg})(1.8\text{ m/s})^2 = 2.4\text{ J}$$

(b) The rotational kinetic energy K_r of the cylinder is the first term of the preceding equation, so, forming a ratio in symbol form, we get

$$\frac{K_r}{K} = \frac{\frac{1}{4}Mv_{CM}^2}{\frac{3}{4}Mv_{CM}^2} = \frac{1}{3}(\times\ 100\%) = 33\%$$

Thus, the total kinetic energy of the cylinder is made up of rotational and translational parts, with one-third being rotational.

     Note that the radius of the cylinder was not needed and that neither the mass nor any numerical values were needed in part (b). Because we used a ratio, these quantities canceled out. *Don't* think that this exact division of energy is a general result, however. It is easy to show that the percentage is different for objects with different moments of inertia. For example, you should expect a rolling sphere to have a smaller percentage of rotational kinetic energy than a cylinder has because the sphere has a smaller moment of inertia ($I = \frac{2}{5}MR^2$).

Follow-up Exercise. Potential energy can be brought into the act by applying the conservation of energy to an object rolling up or down an inclined plane. In this Example, suppose that the cylinder rolled up a 20° inclined plane without slipping. (a) At what vertical height on the plane does the cylinder stop? (b) To find the height in part (a), you probably equated the initial total kinetic energy to the final gravitational potential energy. That is, the total kinetic energy was reduced by the work done by gravity. However, a frictional force also acts (to prevent slipping—see Fig. 8.20). Is there not work done here too?

Example 8.14 ■ Rolling Down versus Sliding Down: Which Is Faster?

A uniform cylindrical hoop is released from rest at a height of 0.25 m near the top of an inclined plane (•Fig. 8.23). If the cylinder rolls down the plane without slipping and there is no energy loss due to friction, what is the linear speed of the cylinder's center of mass at the bottom of the incline?

Thinking It Through. Here, potential energy is converted into kinetic energy—both rotational and translational. The conservation of (mechanical) energy applies.

Solution.

Given: $h = 0.25$ m   *Find:* v_{CM} (speed of CM)
$\quad\quad I_{CM} = MR^2$ (from Fig. 8.16d)

Because the total mechanical energy of the cylinder is conserved, you can write

$$E_o = E$$

or, since $v_o = 0$ at the top of the incline, we have

$$\underset{\text{at rest}}{Mgh} = \frac{1}{2}I_{CM}\omega^2 + \underset{\text{at bottom of incline}}{\frac{1}{2}Mv_{CM}^2}$$

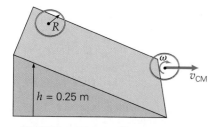

•**FIGURE 8.23 Rolling motion and energy** When an object rolls down an inclined plane, potential energy is converted to translational *and* rotational kinetic energies. This makes the rolling slower than frictionless sliding. See Example 8.14.

Using the rolling condition $v_{CM} = R\omega$ gives

$$Mgh = \tfrac{1}{2}(MR^2)\left(\frac{v_{CM}}{R}\right)^2 + \tfrac{1}{2}Mv_{CM}^2 = Mv_{CM}^2$$

Solving for v_{CM}, we get

$$v_{CM} = \sqrt{gh} = \sqrt{(9.80 \text{ m/s}^2)(0.25 \text{ m})} = 1.6 \text{ m/s}$$

Again, not much numerical information was needed here. Note that the hoop rolls down from the same height as the solid cylinder rolled up in the Example 8.13 Follow-up Exercise, yet the speed of the hoop is less than that of the cylinder at the bottom of the incline. You should know why—differences in moment of inertia.

Follow-up Exercise. Suppose the inclined plane in this Example were frictionless and the hoop slid down the plane instead of rolling. How would the speed at the bottom compare in this case? Why are the speeds different?

As Example 8.14 shows, v_{CM} is independent of M and R. The masses and radii cancel out, so all objects of a particular shape (with the same formula for the moment of inertia) roll with the same speed, regardless of size or density. But the rolling speed does vary with the moment of inertia, which varies with the shape. Therefore, rigid bodies with different shapes roll with different speeds. For example, if you released a cylindrical hoop, a solid cylinder, and a uniform sphere at the same time from the top of an inclined plane, the sphere would win the race to the bottom, followed by the cylinder, with the hoop coming in last—every time!

You can try this experiment with a couple of food cans or other cylindrical containers—one full of some solid material (a rigid body) and one empty with the ends cut out—and a smooth, solid ball. Remember that the masses and the radii make no difference. You might think that an annular cylinder (a hollow cylinder with inner and outer radii that vary appreciably—Fig. 8.16f) would be a possible front-runner, or front-roller, in such a race, but it wouldn't win. The rolling race down an incline is fixed even when you vary the masses and the radii. (See Exercise 76.)

Another aspect of rolling is discussed in the Insight on p. 278.

8.5 Angular Momentum

Objectives: To (a) define angular momentum and (b) apply the conservation of angular momentum to physical situations.

Another important quantity in rotational motion is angular momentum. Recall from Section 6.1 how the linear momentum of an object is changed by a force. Similarly, angular momentum is associated with torque. As we have seen, torque is the product of a moment arm and a force. In a similar manner, **angular momentum (L)** is the product of a moment arm and a linear momentum. For a particle of mass m, the magnitude of the linear momentum is $p = mv$ and $v = r\omega$. The magnitude of the angular momentum is

$$L = r_\perp p = mr_\perp v = mr_\perp^2\omega \qquad \begin{array}{c}\textit{single-particle}\\ \textit{angular momentum}\end{array} \qquad (8.14)$$

SI unit of angular momentum: kilogram-meters squared per second $(\text{kg}\cdot\text{m}^2/\text{s})$

where v is the tangential speed of the particle, $r_\perp$ the moment arm, and ω the angular speed. For circular motion, $r_\perp = r$, since $\mathbf{v}$ is perpendicular to $\mathbf{r}$.

For a system of particles comprising a rigid body, the total magnitude of the angular momentum is

$$L = (\Sigma m_i r_i^2)\omega = I\omega \qquad \begin{array}{c}\textit{rigid-body}\\ \textit{angular momentum}\end{array} \qquad (8.15)$$

Slide or Roll to a Stop?—Antilock Brakes

In an emergency situation while driving, you may instinctively jam on the brakes, trying to come to a quick stop—that is, to stop in the shortest distance. But with the wheels locked, the car skids, or slides, to a stop, often out of control. In this case, the force of sliding friction is acting on the wheels.

To prevent this, you may have learned to pump the brakes in order to roll rather than slide to a stop, particularly on wet or icy roads. Many newer automobiles have computerized antilock braking systems (ABS) that do this automatically. When the brakes are applied firmly and the car begins to slide, sensors in the wheels note this and a computer takes over control of the braking system. It momentarily releases the brakes and then varies the brake-fluid pressure with a pumping action (up to thirteen times per second!) so that the wheels will continue to roll without slipping.

In the absence of sliding, both rolling friction and static friction act. In many cases, however, the force of rolling friction is small and only static friction need be taken into account, as will be done here. The ABS works to keep the static friction near the maximum, $f_s \approx f_{s_{max}}$, which you can't do easily by foot.

Does sliding instead of rolling make a big difference in the stopping distance for an automobile? We can calculate the difference by assuming that rolling friction is negligible. Although the external force of static friction does no work to dissipate energy in slowing a car (this is done internally by friction on the brake pads), it does determine whether the wheels roll or slide.

In Example 2.9, a vehicle stopping distance was given by

$$x = \frac{v_o^2}{2a}$$

By Newton's second law, the net force in the horizontal direction is

$$F = f = \mu N = \mu mg = ma$$

and the stopping acceleration is then

$$a = \mu g$$

Thus,

$$x = \frac{v_o^2}{2\mu g} \qquad (1)$$

But, as was noted in Chapter 4, the coefficient of sliding (kinetic) friction is generally less than that of static friction; that is, $\mu_k < \mu_s$. The general difference between rolling and sliding stops can be seen by using the same initial velocity (v_o for both cases). Then, using Eq. 1 to form a ratio,

$$\frac{x_{roll}}{x_{slide}} = \frac{\mu_k}{\mu_s} \quad or \quad x_{roll} = \left(\frac{\mu_k}{\mu_s}\right) x_{slide}$$

From Table 4.1, the value of μ_k for rubber on wet concrete is 0.60, and the value of μ_s for these surfaces is 0.80. Using these values for a comparison of the stopping distances gives

$$x_{roll} = \left(\frac{0.60}{0.80}\right) x_{slide} = (0.75) x_{slide}$$

The car comes to a rolling stop in 75 percent of the distance required for a sliding stop, for example, 15 m instead of 20 m. Although this may vary for different conditions, it could be an important, perhaps life-saving, difference.

---

which, for rotation about a fixed axis, in vector notation, is

Rigid body angular momentum

$$\mathbf{L} = I\boldsymbol{\omega} \qquad (8.16)$$

Thus, **L** is in the direction of the angular velocity vector ($\boldsymbol{\omega}$) as given by the right-hand rule.

Note: Review Eq. 6.3, Section 6.1.

For linear motion, the momentum of a system is related to force by $\mathbf{F} = \Delta \mathbf{P}/\Delta t$. Angular momentum is analogously related to net torque:

$$\boldsymbol{\tau} = I\boldsymbol{\alpha} = \frac{I\Delta\boldsymbol{\omega}}{\Delta t} = \frac{\Delta(I\boldsymbol{\omega})}{\Delta t} = \frac{\Delta \mathbf{L}}{\Delta t}$$

That is,

Net torque: the time rate of change of angular momentum

$$\boldsymbol{\tau} = \frac{\Delta \mathbf{L}}{\Delta t} \qquad (8.17)$$

Thus, the net torque is equal to *the time rate of change of angular momentum*. In other words, a net torque causes a *change* in angular momentum.

Conservation of Angular Momentum

Equation 8.17 was derived by using $\boldsymbol{\tau} = I\boldsymbol{\alpha}$, which applies to a rigid system of particles or a rigid body having a constant moment of inertia. However, Eq. 8.17 is a general one that also applies to a nonrigid system of particles. In such a system, there may be a change in the mass distribution and a change in the moment of inertia. As a result, there may be an angular acceleration even in the absence of a net torque. How can this be?

If the net torque on a system is zero, then, by Eq. 8.17, $\tau = \Delta L / \Delta t = 0$, and

$$\Delta L = L - L_o = I\omega - I_o\omega_o = 0$$

or

$$I\omega = I_o\omega_o \qquad (8.18)$$

Thus, the condition for the **conservation of angular momentum** is as follows:

> In the absence of an external, unbalanced torque, the total (vector) angular momentum of a system is conserved (remains constant).

As with total linear momentum, the internal torques arising from internal forces cancel out.

For a rigid body with a constant moment of inertia (that is, $I = I_o$), the angular speed remains constant ($\omega = \omega_o$) in the absence of a net torque. But it is possible for the moment of inertia to change in some systems, giving rise to a change in the angular speed, as the following Example illustrates.

Conservation of angular momentum

Note: Angular momentum is conserved when the net torque is zero. (**L** stays fixed.) This is the third conservation law in mechanics.

Example 8.15 ■ Pull It Down: Conservation of Angular Momentum

A small ball at the end of a string that passes through a tube is swung in a circle, as illustrated in •Fig. 8.24. When the string is pulled downward through the tube, the angular speed of the ball increases. (a) Is this caused by a torque due to the pulling force? (b) If the ball is initially swung at a speed of 2.8 m/s in a circle with a radius of 0.30 m, what will be its tangential speed if the string is pulled down far enough to reduce the radius of the circle to 0.15 m?

Thinking It Through. (a) A force is applied to the ball via the string, but consider the axis of rotation. (b) In the absence of a net torque, the angular momentum is conserved (Eq. 8.18), and the tangential speed is related to the angular speed by $v = r\omega$.

Solution.

Given: $r_1 = 0.30$ m   *Find:* (a) Cause of the increase in angular speed
        $r_2 = 0.15$ m   (b) v_2 (final tangential speed)
        $v_1 = 2.8$ m/s

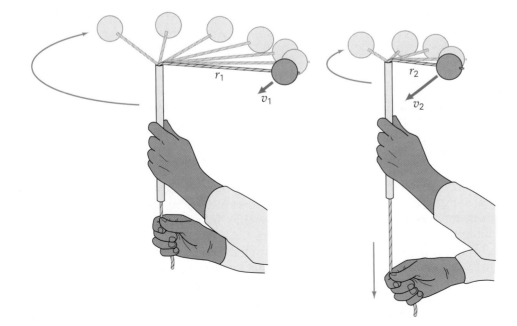

•**FIGURE 8.24 Conservation of angular momentum** When the string is pulled downward through the tube, the revolving ball speeds up. See Example 8.15.

(a) The change in the angular velocity, or the angular acceleration, is not caused by a torque due to the pulling force. The force on the ball as transmitted by the string acts through the axis of rotation, and therefore the torque is zero. As the rotating portion of the string is shortened, the moment of inertia of the ball ($I = mr^2$, from Fig. 8.16a) decreases. This causes the ball to speed up, since in the absence of an external torque, the angular momentum of the ball is conserved.

(b) Since the angular momentum is conserved, we have

$$I_o\omega_o = I\omega$$

Then, using $I = mr^2$ and $\omega = v/r$ gives

$$mr_1v_1 = mr_2v_2$$

and

$$v_2 = \left(\frac{r_1}{r_2}\right)v_1 = \left(\frac{0.30 \text{ m}}{0.15 \text{ m}}\right)2.8 \text{ m/s} = 5.6 \text{ m/s}$$

When the radial distance is shortened, the ball speeds up.

Follow-up Exercise. Let's look at the situation in this Example in terms of work and energy. If the inital speed is the same and the vertical pulling force is 7.8 N, what is the final speed of the 0.10-kilogram ball?

Example 8.15 should help you understand Kepler's law of areas (Chapter 7). A planet's angular momentum is essentially conserved, since the Sun exerts no torque on it (why?) and we can neglect the torques from other planets (why?). Thus, when a planet is closer to the Sun in its elliptical orbit and so has a shorter moment arm, its speed is greater by the conservation of angular momentum. Similarly, when an orbiting satellite's altitude varies during the course of an elliptical orbit about a planet, the satellite speeds up or slows down in accordance with the same principle.

Real-Life Angular Momentum

A popular lecture demonstration of the conservation of angular momentum is shown in •Fig. 8.25a,b. A person sitting on a stool that rotates holds weights with his arms outstretched and is started slowly rotating. An external torque to start this rotation must be supplied by someone else, because the person on the stool cannot initiate the motion by himself. (Why not?) Once rotating, if the person brings his arms inward, the angular speed increases and he spins much faster. Extending his arms again slows him down. Can you explain this?

If L is constant, what happens to ω when I is made smaller by reducing r? The angular speed must increase to compensate and keep L constant. Ice skaters do dizzying spins by pulling in their arms to reduce their moment of inertia. Similarly, a diver spins during a high dive by tucking in, greatly decreasing his or her moment of inertia. The enormous wind speeds of tornadoes and hurricanes represent another example of the same effect (Fig. 8.25c).

Angular momentum, **L**, is a vector, and when it is conserved or constant, its magnitude *and* direction must remain unchanged. Thus, when no external torques act, the direction of **L** is fixed in space. This is the principle of passing a football accurately as well as of the gyrocompass (•Fig. 8.26). A football is normally passed with a spiraling rotation. This spin, or gyroscopic action, stabilizes the ball's spin axis in the direction of motion. Similarly, rifle bullets are set spinning by the rifling in the barrel for directional stability.

The **L** vector of a spinning gyroscope in the compass is set in a particular direction (usually north). In the absence of external torques, the compass direction remains fixed, even though its carrier (an airplane or ship, for example) changes directions. You may have played with a toy gyroscope that is set spinning and placed on a pedestal. In a "sleeping" condition, the gyro stands straight up with its angular

(a)

(b)

(c)

•**FIGURE 8.25 Moment of inertia change** (a) Spinning slowly with masses in outstretched arms, the man's moment of inertia is relatively large (masses are farther from the axis of rotation). Note that he is isolated, with no external torques (neglecting friction), so angular momentum, $L = I\omega$, is conserved. (b) Pulling his arms inward decreases the moment of inertia (why?). Consequently ω must increase, and he goes into a dizzy spin. (c) The same principle helps explain the violence of the winds that spiral around the center of a hurricane. As air rushes in toward the center of the storm, where the pressure is low, its rotational velocity must increase for angular momentum to be conserved.

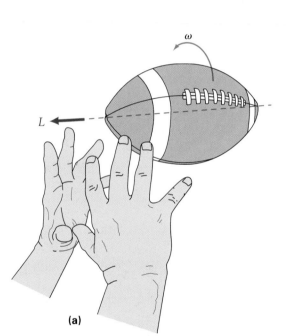
(a)

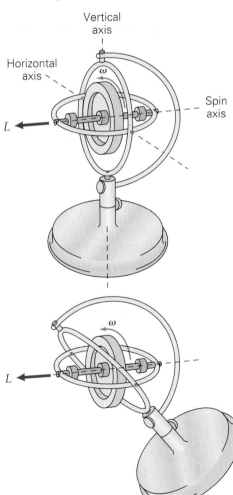
(b)

•**FIGURE 8.26 Constant direction of angular momentum** When angular momentum is conserved, its direction is constant in space. (a) This can be demonstrated by a passed football. (b) Gyroscopic action also occurs in a gyroscope, a rotating wheel that is universally mounted on gimbals (rings) so that it is free to turn about any axis. When the frame moves, the wheel maintains its direction. This is the principle of the gyrocompass.

momentum vector fixed in space for some time. The gyro's center of gravity is on the axis of rotation, so there is no net torque due to its weight.

However, the gyroscope eventually slows down because of friction, causing **L** to tilt. In doing so, you may have noticed how the spin axis revolves, or *precesses*, about the vertical axis. It revolves tilted over, so to speak (•Fig. 8.27a). Since the gyroscope precesses, the angular momentum vector **L** is no longer constant in direction, indicating that a torque must be acting to give a change (Δ**L**) with time. As can be seen from the figure, the torque arises from the vertical component of the weight force, since the center of gravity no longer lies directly above the point of support or on the vertical axis of rotation. The instantaneous torque is such that the gyroscope's axis moves or precesses about the vertical axis.

In a similar manner, the Earth's rotational axis precesses. The Earth's spin axis is tilted $23\frac{1}{2}°$ with respect to a line perpendicular to the plane of its revolution about the Sun; the axis precesses about this line (Fig. 8.27b). The precession is due to slight gravitational torques exerted on the Earth by the Sun and the Moon. The torques arise because the Earth is not perfectly spherical—it has an equatorial bulge.

The period of the precession of the Earth's axis is about 26 000 years, so it has little day-to-day effect. However, it does have an interesting long-term effect. Polaris will not always be (nor has it always been) the North Star, that is, the star toward which the Earth's axis of rotation points. About 5000 years ago, Alpha Draconis was the North Star, and 5000 years from now it will be Alpha Cephei, which is at an angular distance of about 68° away from Polaris on the circle described by the precession of the Earth's axis.

There are some other long-term torque effects on the Earth and the Moon. Did you know that the Earth's daily spin rate is slowing down and hence the days are getting longer? Also, that the Moon is receding, or getting farther away, from the Earth? This is primarily due to ocean tidal friction, which gives rise to a torque. As a result, the Earth's spin angular momentum, and therefore its rate of rotation, is changing. The slowing rate of rotation causes the average day to be longer; this century will be about 25 seconds longer than the previous.

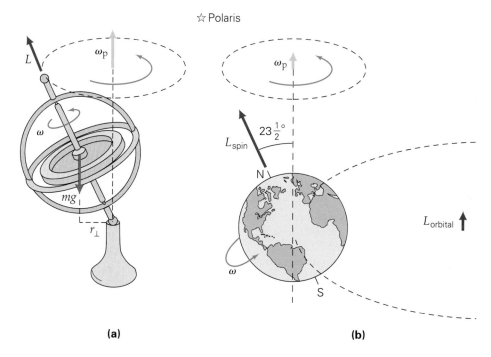

•**FIGURE 8.27 Precession** An external torque causes a change in angular momentum. **(a)** For a spinning gyroscope, this change is directional, and the axis of rotation precesses at angular acceleration ω_p about a vertical line. (The torque due to the weight force would point out of the page as drawn here, as would Δ**L**.) Note that although there is a torque that would topple a nonspinning gyroscope, a spinning gyroscope doesn't fall. **(b)** Similarly, the Earth's axis precesses because of gravitational torques caused by the Sun and the Moon. We don't notice this motion because the period of precession is about 26 000 y.

(a)

(b)

But this is an average rate. At times, the Earth's rotation speeds up for relatively short periods. This increase is thought to be associated with the rotational inertia of the liquid layer of the Earth's core (see the Insight in Chapter 13).

The tidal torque on the Earth results chiefly from the Moon's gravitational attraction, which is the main cause of ocean tides. This torque is *internal* to the Earth–Moon system, so the total angular momentum of that system is conserved. Since the Earth is losing angular momentum, the Moon must be gaining angular momentum to keep the total angular momentum of the system constant. The Earth loses rotational (spin) angular momentum, and the Moon gains orbital angular momentum. As a result, the Moon drifts slightly farther from Earth and its orbital speed decreases. The Moon moves away from the Earth at about 1 cm per lunar revolution, or lunar month (calculated from the decrease rate of the Earth's rotation). Thus, the Moon moves in a slowly widening spiral.

Finally, a common example in which angular momentum is an important consideration—the helicopter. What would happen if a helicopter had a single rotor? Since the motor supplying the torque is internal, the angular momentum would be conserved. Initially, $\mathbf{L} = 0$, hence to conserve the total system (rotor plus body) angular momentum, the separate angular momenta of the rotor and body would have to be in opposite directions to cancel. Thus on takeoff, the rotor would rotate one way and the helicopter body the other, which is not a desirable situation.

To prevent this situation, helicopters have two rotors. Large helicopters have two overlapping rotors (•Fig. 8.28a). The oppositely rotating rotors cancel each other's angular momenta, so the helicopter body does not have to rotate to provide canceling angular momentum. The rotors are offset at different heights so that they do not collide.

Small helicopters with a single overhead rotor have small "anti-torque" tail rotors (Fig. 8.28b). The tail rotor produces a thrust like a propeller and supplies the torque to counterbalance the torque produced by the overhead rotor. The tail rotor also helps in steering the craft. By increasing or decreasing the tail rotor's thrust, the helicopter turns (rotates) one way or the other.

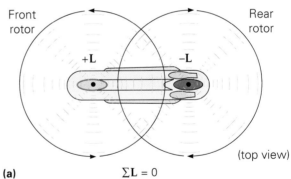

(a) $\Sigma \mathbf{L} = 0$

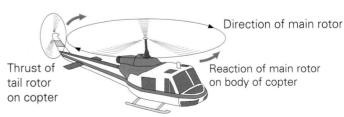

(b)

•**FIGURE 8.28   Different rotors**

Chapter Review

Important Terms

rigid body 252
translational motion 252
rotational motion 253
instantaneous axis of rotation 253
moment (lever arm) 254
torque 254
translational equilibrium 256
concurrent forces 256

rotational equilibrium 256
mechanical equilibrium 256
static equilibrium 257
center of gravity 261
stable equilibrium 261
unstable equilibrium 262
moment of inertia 265

parallel axis theorem 268
rotational work 274
rotational power 274
rotational kinetic energy 274
angular momentum 277
conservation of angular
   momentum 279

Important Concepts

- In pure translational motion, all the particles of an object have the same instantaneous velocity.
- In pure rotational motion (about a fixed axis), all the particles of an object have the same instantaneous angular velocity.
- Torque, the rotational analogue of force, is the product of a force and a moment arm, or lever arm.
- Mechanical equilibrium requires that the summation of the forces be zero (translational equilibrium) and that the summation of the torques be zero (rotational equilibrium).

- An object is in stable equilibrium as long as its center of gravity, upon small displacement, lies above and inside its original base of support.
- Moment of inertia (I) is the rotational analogue of mass in the rotational form of Newton's second law, $\boldsymbol{\tau} = I\boldsymbol{\alpha}$.
- Angular momentum, the product of a moment arm and linear momentum as well as the product of a moment of inertia and angular velocity, is conserved in the absence of an external, unbalanced torque.

Important Equations

Condition for Rolling without Slipping:
$$v_{CM} = r\omega \tag{8.1}$$
$$(\text{or } s = r\theta \quad \text{or} \quad a_{CM} = r\alpha)$$

Torque (magnitude):
$$\tau = r_\perp F = rF \sin \theta \tag{8.2}$$

Conditions for Translational and Rotational Mechanical Equilibria, Respectively:
$$\Sigma \mathbf{F}_i = 0 \quad \text{and} \quad \Sigma \boldsymbol{\tau}_i = 0 \tag{8.3}$$

Torque on a Particle (magnitude):
$$\tau = r_\perp F = rF_\perp = rma_\perp = mr^2\alpha \tag{8.4}$$

Moment of Inertia:
$$I = \Sigma m_i r_i^2 \tag{8.6}$$

Rotational Form of Newton's Second Law (magnitude):
$$\tau = I\alpha \tag{8.7}$$

Parallel Axis Theorem:
$$I = I_{CM} + Md^2 \tag{8.8}$$

Rotational Work:
$$W = \tau\theta \tag{8.9}$$

Rotational Power:
$$P = \tau\omega \tag{8.10}$$

Work–Energy Theorem:
$$W = \Delta K = \tfrac{1}{2}I\omega^2 - \tfrac{1}{2}I\omega_0^2 \tag{8.11}$$

Rotational Kinetic Energy:
$$K = \tfrac{1}{2}I\omega^2 \tag{8.12}$$

Kinetic Energy of a Rolling Object:
$$K = \tfrac{1}{2}I_{CM}\omega^2 + \tfrac{1}{2}Mv_{CM}^2 \tag{8.13}$$

Angular Momentum of a Particle in Circular Motion (magnitude):
$$L = r_\perp p = mr_\perp v = mr_\perp^2\omega \tag{8.14}$$

Angular Momentum of a Rigid Body:
$$\mathbf{L} = I\boldsymbol{\omega} \tag{8.16}$$

Torque in Terms of Angular Momentum:
$$\boldsymbol{\tau} = \frac{\Delta \mathbf{L}}{\Delta t} \tag{8.17}$$

Conservation of Angular Momentum (with $\tau = 0$):
$$I\boldsymbol{\omega} = I_0\boldsymbol{\omega}_0 \tag{8.18}$$

Exercises

8.1 Rigid Bodies, Translations, and Rotations

1. In pure rotational motion of a rigid body, (a) all the particles of the body have the same angular velocity, (b) all the particles of the body have the same tangential velocity, (c) the acceleration is always zero, (d) there are always two simultaneous axes of rotation.

2. The condition for rolling without slipping is (a) $a = r\omega^2$, (b) $v = r\omega$, (c) $F = ma$, (d) $a_c = v^2/r$.

3. Suppose someone in your physics class says that it is possible for a rigid body to have translational motion and rotational motion at the same time. Would you agree? Why?

4. For a rolling cylinder, what would happen if v were less than $r\omega$? Is it possible for v to be greater than $r\omega$? Explain.

5. For the tires on your skidding car, (a) $v = r\omega$, (b) $v > r\omega$, (c) $v < r\omega$.

6. ■ A rope goes over a circular pulley with a radius 6.5 cm. If the pulley makes four revolutions without the rope slipping, what length of rope passes over the pulley?

7. ■ A wheel rolls five revolutions on a horizontal surface without slipping. If the center of the wheel moves 3.2 m, what is the radius of the wheel?

8. ■ A wheel rolls uniformly on level ground without slipping. A piece of mud on the wheel flies off when it is at the 9 o'clock position (rear of wheel). Describe the subsequent motion of the mud.

9. ■ A circular disk with a radius of 0.25 m rolls without slipping with an angular speed of 2.0 rad/s on a level surface. (a) What is the linear speed of the center of mass of the disk? (b) What is the instantaneous tangential speed of the top of the disk?

10. ■■ A ball with a radius of 15 cm rolls on a level surface; the translational speed of the center of mass is 0.25 m/s. What is the angular speed about the center of mass if the ball rolls without slipping?

11. ■■ A disk with a radius of 0.15 m rotates through 270° as it travels 0.71 m. Does the disk roll without slipping?

12. ■■ Show that $a = r\alpha$ is also a condition for rolling without slipping.

13. ■■■ A cylinder with a diameter of 20 cm rolls with an angular speed of 0.50 rad/s on a level surface. If the cylinder experiences a uniform tangential acceleration of 0.018 m/s² without slipping until its angular speed is 1.25 rad/s, through how many complete revolutions does the cylinder rotate during the acceleration time?

8.2 Torque, Equilibrium, and Stability

14. A net torque is possible when (a) all forces act through the axis of rotation, (b) $\Sigma F_i = 0$, (c) a body is in rotational equilibrium, (d) a body remains in unstable equilibrium.

15. If an object in unstable equilibrium is slightly displaced, (a) its potential energy will decrease, (b) the center of gravity is directly above the axis of rotation, (c) no gravitational work is done, (d) stable equilibrium follows.

16. Explain the balancing acts in •Fig. 8.29. Where are the centers of gravity?

•**FIGURE 8.29 Balancing acts** See Exercise 16. *Left:* A toothpick on the rim of the glass supports a fork and spoon. *Right:* (*top*) A hinged board is mounted on a pole. (*bottom*) The handle of a hammer is placed in a loop that runs around the board; the hinged part of the board remains horizontal.

17. •Figure 8.30 shows a toy clown walking on a tightrope. Explain how this brainless toy can perform such a difficult task. Is the clown in stable equilibrium? What happens if

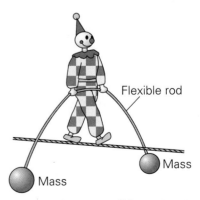

•**FIGURE 8.30 Tightrope walking** See Exercise 17.

the weights are removed? [Hint: Think about the center of gravity.]

18. ■ The drain plug on a car's engine is required to be tightened to a torque of 25 m·N. If you are using a 0.15-meter-long wrench to change the oil of that engine, what is the minimum force you need to apply to loosen the plug?

19. ■ In Exercise 18, due to space limitations (you are crawling under the car), you cannot apply the force perpendicularly to the length of the wrench. If your force makes a 30° angle to that length, what is the force required?

20. ■ In Fig. 8.4a, if the arm makes a 37° angle to the horizontal and a torque of 18 m·N is to be produced, what is the force required by the biceps muscle?

21. ■ How many different positions of stable equilibrium and unstable equilibrium are there for a cube? Consider each surface, edge, and corner to be a different position.

22. ■ The pedals of a bicycle rotate in a circle with a diameter of 40 cm. What is the maximum torque a 55-kilogram rider can apply by putting all of her weight on one pedal?

23. ■ A 35-kilogram child sits on a uniform seesaw of negligible mass; she is 2.0 m from the pivot point (or fulcrum). How far from the pivot point on the other side will her 30-kilogram playmate have to sit for the seesaw to be in equilibrium?

24. ■ A uniform meterstick pivoted at its center, as in Example 8.4, has a 100-gram mass suspended at the 25.0-centimeter position. (a) At what position should a 75.0-gram mass be suspended to put the system in equilibrium? (b) What mass would have to be suspended at the 90.0-centimeter position for the system to be in equilibrium?

25. ■■ Show that the balanced meterstick in Example 8.4 is in static rotational equilibrium about a horizontal axis through the zero end of the stick.

26. ■■ Telephone and electrical lines are allowed to sag between poles so that the tension will not be too great when something hits or sits on the line. Suppose that a line were stretched almost perfectly horizontally between two poles that are 30 m apart. If a 0.25-kilogram bird perches on the wire midway between the poles and the wire sags 1.0 cm, what would the tension in the wire be?

27. ■■ In •Fig. 8.31, what is the force F_m supplied by the deltoid muscle so as to hold up the outstretched arm if the mass of the arm is 3.0 kg? (F_j is the joint force on the bone of the upper arm—the humerus.)

28. ■■ A variation of *Russell traction* (•Fig. 8.32) supports the lower leg when in a cast. Suppose that the patient's leg and cast have a combined mass of 15.0 kg and m_1 is 4.50 kg. (a) What is the reaction force of the leg muscles to the traction? (b) What must m_2 be to keep the leg horizontal?

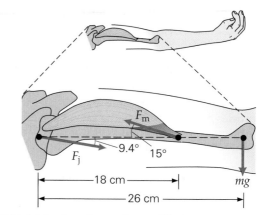

•FIGURE 8.31 **Arm in static equilibrium** See Exercise 27.

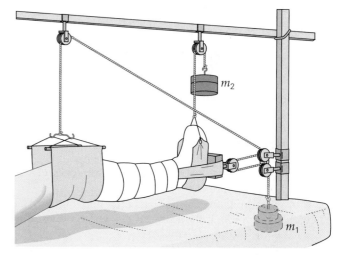

•FIGURE 8.32 **Static traction** See Exercise 28.

29. ■■ A uniform meterstick weighing 5.0 N is pivoted so it can rotate about a horizontal axis through one end (•Fig. 8.33). If a 0.15-kilogram mass is suspended 75 cm from the pivoted end, what is the tension in the string?

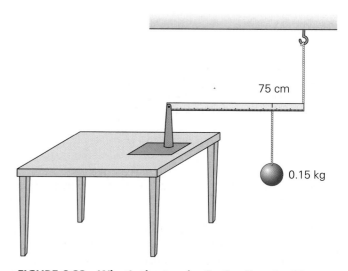

•FIGURE 8.33 **What's the tension?** See Exercise 29.

30. ■■ In Example 8.5, if the mass of the painter is 65 kg, what frictional force must act on the bottom of the ladder to keep it from slipping?

31. ■■ An artist wishes to construct a birds-and-bees mobile, as shown in •Fig. 8.34. If the mass of the bee on the lower left is 0.10 kg and each vertical support string has a length of 30 cm, what are the masses of the other birds and bees? (Neglect the masses of the bars and strings.)

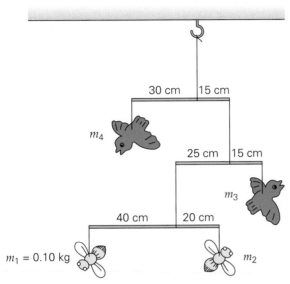

•FIGURE 8.34 **Birds and bees** See Exercise 31.

32. ■■ (a) How many uniform, identical textbooks of width 25.0 cm can be stacked on top of each other on a level surface without the stack falling over if each successive book is displaced 3.00 cm in width relative to the book below it? (b) If the books are 5.00 cm thick, what will be the height of the center of mass of the stack above the level surface?

33. ■■ If four metersticks were stacked on a table with 10 cm, 15 cm, 30 cm, and 50 cm, respectively, hanging over the edge, as shown in •Fig. 8.35, would the metersticks remain on the table?

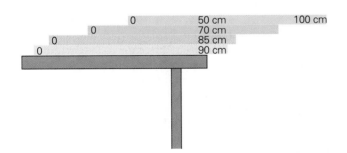

•FIGURE 8.35 **Will they fall off?** See Exercise 33.

34. ■■ A 10.0-kilogram solid uniform cube with 0.500-meter sides rests on a level surface. What is the minimum amount of work necessary to put the cube into an unstable equilibrium position?

35. ■■■ While standing on a long board resting on a scaffold, a 70-kilogram painter paints the side of a house (•Fig. 8.36). If the mass of the board is 15 kg, how close to the end can the painter stand without tipping the board over?

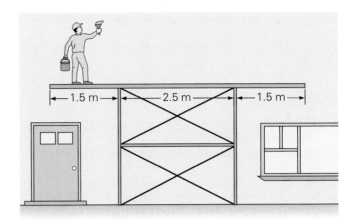

•FIGURE 8.36 **Not too far!** See Exercises 35 and 36.

36. ■■■ Suppose that the board in Fig. 8.36 were suspended from vertical ropes attached to each end instead of resting on scaffolding. If the painter stood 1.5 m from one end of the board, what would the tensions in the ropes be? (See Exercise 35 for additional data.)

37. ■■■ The forces acting on Einstein and the bicycle (Fig. 1 of the Insight on p. 266) are the total weight of Einstein and the bicycle (mg) at the center of gravity of the system, the normal force (N) by the road, and the force of static friction (f_s) on the tires exerted by the road. (a) Explain why the angle of lean θ the bicycle makes with the vertical must be given by $\tan \theta = f_s/N$ if Einstein is to maintain balance. (b) The angle θ in the picture is about 11°. What is the minimum coefficient of static friction μ_s between the road and the tires? (c) If the radius of the circular path is 6.5 m, what is the maximum speed of Einstein's bicycle? [Hint: The net torque about the center of gravity must be zero for rotational equilibrium.]

8.3 Rotational Dynamics

38. The moment of inertia of a rigid body (a) depends on the axis of rotation, (b) cannot be zero, (c) depends on mass distribution, (d) all of the preceding.

39. Which of the following best describes the physical quantity called torque: (a) rotational analogue of force, (b) energy due to rotation, (c) rate of change of linear momentum, (d) force that is tangent to a circle?

40. (a) Does the moment of inertia of a rigid body depend in any way on the center of mass? Explain. (b) Can a moment of inertia have a negative value? If so, explain what this would mean.

41. Why does the moment of inertia of a rigid body have different values for different axes of rotation? What does this mean physically?

42. When a hard-boiled egg laying on a table is given a quick torque (spin), it will rise up and spin on one end like a top. A raw egg will not. Why the difference?

43. Why does jerking a paper towel from a roll cause the paper to tear better than pulling it smoothly? Will the amount of paper on the roll affect the results?

44. (a) Why is it easier to balance a meterstick vertically on your finger than on a pencil? (b) A softball, a volleyball, and a basketball are released at the same time from the top of an inclined plane. Give the results of the race.

45. ■ For the system of masses shown in •Fig. 8.37, find the moment of inertia about (a) the x axis, (b) the y axis, (c) an axis through the origin perpendicular to the page (z axis). Neglect the masses of the connecting rods.

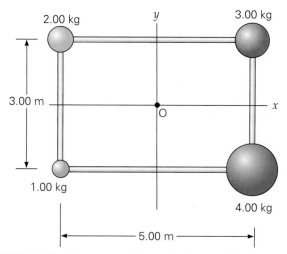

•**FIGURE 8.37   Moments of inertia about different axes**
See Exercise 45.

46. ■ 🌐 A fixed 0.15-kilogram solid-disk pulley with a radius of 0.075 m is acted on by a net torque of 6.4 m·N. What is the angular acceleration of the pulley?

47. ■ 🌐 What net torque is needed to give a uniform 20-kilogram solid ball with a radius of 0.20 m and angular acceleration of 2.0 rad/s²?

48. ■ A light meterstick is loaded with bodies of masses 2.0 kg and 4.0 kg at the 30 cm and 75 cm positions, respectively. What is (a) the moment of inertia about an axis through the 0-centimeter end of the meterstick? (b) the moment of inertia about an axis through the center of mass of the system? (c) Use the parallel axis theorem to find the moment of inertia about the axis through the 0-centimeter end of the stick, and compare the result with that of part (a).

49. ■■ Two masses are joined by a light rod as shown in •Fig. 8.38. Find the moment of inertia of the system about an axis perpendicular to the rod through (a) the center of the rod and (b) the center of mass. (c) Is the moment of inertia about the center of mass the minimum?

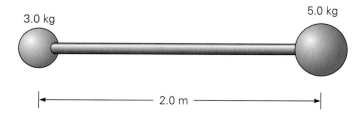

•**FIGURE 8.38   Moments of inertia about axes through different centers**   See Exercise 49.

50. ■■ A 2000-kilogram Ferris wheel accelerates from rest to an angular speed of 2.0 rad/s in 12 s. Approximating the Ferris wheel as a circular disk with a radius of 30 m, what is the net torque on the wheel?

51. ■■ A 15-kilogram uniform sphere with a radius of 15 cm rotates about an axis tangent to its surface at 3.0 rad/s. A constant torque of 10 m·N then increases the rotational speed to 7.5 rad/s. Through what angle does the sphere rotate while accelerating?

52. ■■ A 10-kilogram solid disk of radius 0.50 m is rotated about an axis through its center. If the disk accelerates from rest to an angular speed of 3.0 rad/s while rotating 2.0 revolutions, what net torque is required?

53. ■■ If the Earth rotated about a north–south axis tangent to the Earth at the Equator, how much more moment of inertia would it have compared to that about an axis through its center?

54. ■■ A 0.25-kilogram disk-shaped machine part with a radius of 6.0 cm rotates eccentrically (off-centered) about an axis that is normal to its flat surface and located two-thirds of the distance from the center to the circumference. (a) What torque is required to rotate the part about this off-center axis with an angular acceleration of 2.0 rad/s²? (b) How much greater is this torque than the torque required to rotate the part with the same angular acceleration about a parallel axis through the center?

55. ■■ Two masses are suspended from a pulley as shown in •Fig. 8.39 (the Atwood machine revisited, Chapter 4). The pulley has a mass of 0.20 kg, a radius of 0.15 m, and a constant torque of 0.35 m·N due to the friction between the rotating pulley and its axle. What is the magnitude of the acceleration of the suspended masses if $m_1 = 0.40$ kg and $m_2 = 0.80$ kg? (Neglect the mass of the string.)

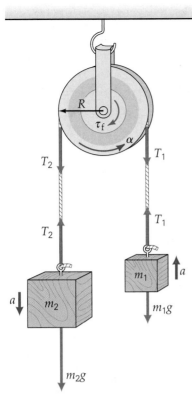

•FIGURE 8.39 **The Atwood machine revisited** See Exercise 55.

56. ■■ For the system in •Fig. 8.40, $m_1 = 8.0$ kg, $m_2 = 3.0$ kg, $\theta = 30°$, and the radius and mass of the pulley are 0.10 m and 0.10 kg, respectively. (a) What is the acceleration of the masses? (Neglect friction and the string's mass.) (b) The pulley has a constant frictional torque of 0.050 m·N when the system is in motion, what is the acceleration? [Hint: Isolate the forces. The tensions in the strings are different. Why?]

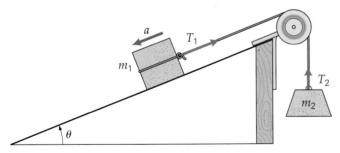

•FIGURE 8.40 **Inclined plane and pulley** See Exercise 56.

57. ■■ What is the magnitude of the tangential force that must be applied to the rim of a 2.0-kilogram wheel that is disk-shaped and has a radius of 0.50 m in order to give it an angular acceleration of 4.8 rad/s²? (Neglect friction.)

58. ■■ A man starts his lawn mower by applying to the starter rope a constant tangential force of 150 N to the 0.30-kilogram disk-shaped flywheel. The diameter of the flywheel is 18 cm. What is the angular speed of the wheel after it has turned through one revolution? (Neglect friction and motor compression.)

59. ■■ A meterstick, pivoted about a horizontal axis through the 0-centimeter end, is held in a horizontal position and let go. (a) What is the tangential acceleration of the 100-centimeter position? Are you surprised by this result? (b) Which position has a tangential acceleration equal to the acceleration due to gravity?

60. ■■ Pennies are placed every 10 cm on a meterstick. One end is put on a table and the other end is held horizontally with a finger as shown in •Fig. 8.41. If the finger is pulled away, what happens to the pennies? [Hint: See Exercise 59.]

•FIGURE 8.41 **Money left behind?** See Exercise 60.

61. ■■■ A uniform 2.0-kilogram cylinder of radius 0.15 m is suspended by two strings wrapped around it (•Fig. 8.42). As the cylinder descends, the strings unwind from it. What is the acceleration of the center of mass of the cylinder? (Neglect the mass of the string.)

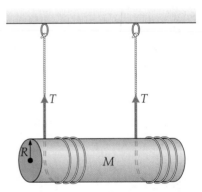

•FIGURE 8.42 **Unwinding with gravity** See Exercise 61.

62. ■■■ A uniform hoop rolls without slipping down a 15° inclined plane. What is the acceleration of the hoop's center of mass?

63. ■■■ In Fig. 8.20, what would be the maximum angle of incline (in terms of the coefficient of static friction) for the ball to roll without slipping down the incline?

8.4 Rotational Work and Kinetic Energy

64. From $W = \tau\theta$, the unit(s) of rotational work is (a) watt, (b) N·m, (c) kg·rad/s², (d) N·rad.

65. A bowling ball rolls without slipping on a flat surface. The ball has (a) rotational kinetic energy; (b) translational kinetic energy; (c) both translational and rotational kinetic energies.

66. ■ A constant retarding torque of 12 m·N stops a rolling wheel of diameter 0.80 m in a distance of 15 m. How much work is done by the torque?

67. ■ A person opens a door by applying a 15-newton force perpendicular to it at a distance 0.90 m from the hinges. The door is pushed wide open (to 120°) in 2.0 s. (a) How much work was done? (b) What was the average power delivered?

68. ■ A constant torque of 10 m·N is applied to a 10-kilogram uniform disk of radius 0.20 m. What is the angular speed of the disk about an axis through its center after it rotates 2.0 revolutions from rest?

69. ■ A 2.5-kilogram pulley of radius 0.15 m is pivoted about an axis through its center. What constant torque is required so the pulley will reach an angular speed of 25 rad/s after rotating 3.0 revolutions, starting from rest?

70. ■ Use the conservation of mechanical energy to find the linear speed of the descending mass ($m = 1.0$ kg) of Fig. 8.19 after it has descended a vertical distance of 2.0 m from rest. (For the pulley, $M = 0.30$ kg and $R = 0.15$ m. Neglect friction and the mass of the string.)

71. ■■ A sphere with a radius of 15 cm rolls on a level surface with a constant angular speed of 10 rad/s. To what height on a 30° inclined plane will the sphere roll before coming to rest? (Neglect frictional losses.)

72. ■■ A thin 1.0-meter-long rod pivoted at one end falls (rotates) from a horizontal position, starting from rest and with no friction. What is the angular speed of the rod when it is vertical? [Hint: Consider the center of mass and use the conservation of mechanical energy.]

73. ■■ A uniform sphere and a uniform cylinder with the same mass and radius roll at the same velocity side by side on a level surface without slipping. If the sphere and the cylinder approach an inclined plane and roll up it without slipping, will they be at the same height on the plane when they come to a stop? If not, what will be the percentage height difference?

74. ■■ 💿 A hoop starts from rest at a height 1.2 m above the base of an inclined plane and rolls down under the influence of gravity. What is the linear speed of the hoop's center of mass just as the hoop leaves the incline and rolls onto a horizontal surface? (Neglect friction.)

75. ■■ An industrial flywheel with a moment of inertia of 4.25×10^2 kg·m² rotates with a speed of 7500 rpm. (a) How much work is required to bring it to rest? (b) If this work is done uniformly in 1.5 min, how much power is expended?

76. ■■ 💿 A cylindrical hoop, a cylinder, and a sphere of equal radius and mass are released at the same time from the top of an inclined plane. Using the conservation of mechanical energy, show that the sphere always gets to the bottom of the incline first with a faster speed and that the hoop always arrives last with the slowest speed.

77. ■■ 💿 For the following objects, which roll without slipping, determine the rotational kinetic energy about the center of mass as a percentage of the total instantaneous kinetic energy: (a) solid sphere, (b) a thin spherical shell, (c) a thin cylindrical shell.

78. ■■ A 0.050-kilogram phonograph record with a radius 0.15 m drops onto a turntable and is soon rotating at $33\frac{1}{3}$ rpm. How much work must be supplied to get the record to rotate at this speed, and what supplies it?

79. ■■■ A steel ball rolls down an incline into a loop-the-loop of radius R (•Fig. 8.43a). (a) What minimum speed must the ball have at the top of the loop in order to stay on the track? (b) At what vertical height (h) on the incline, in terms of the radius of the loop, must the ball be released in order for it to have the required minimum speed at the top of the loop? (Neglect frictional losses.) (c) Figure 8.43b shows the loop-the-loop of a roller coaster. What are the sensations of the riders if the roller coaster has the minimum speed or greater at the top of the loop? [Hint: In case the speed is below the minimum, seat and shoulder straps hold the riders in.]

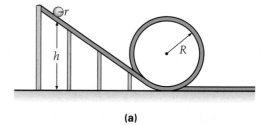

(a)

(b)

•**FIGURE 8.43 Loop-the-loop and rotational speed** See Exercise 79.

8.5 Angular Momentum

80. The units of angular momentum are (a) N·m, (b) kg·m/s², (c) kg·m²/s, (d) J·m.

81. A child stands on the edge of a rotating (hand-driven) merry-go-round. He then starts to walk toward the center of the merry-go-round. A dangerous situation results. Why?

82. The release of vast amounts of carbon dioxide may increase the Earth's average temperature through the greenhouse effect and cause the polar ice caps to melt. If this occurred and the ocean level rose substantially, what effect would it have on the Earth's rotation and on the length of the day?

83. An ice skater goes into a fast spin by tucking her arms in (•Fig. 8.44a), and a high-platform diver often draws her legs up against her chest (Fig. 8.44b). Why are the arms and legs put into these positions in each case?

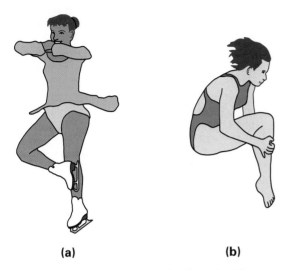

(a)   **(b)**

•**FIGURE 8.44 Faster rotation** See Exercise 83.

84. In the classroom demonstration illustrated in •Fig. 8.45, a person on a rotating stool holds a rotating bicycle wheel by handles attached to the wheel. When the wheel is held

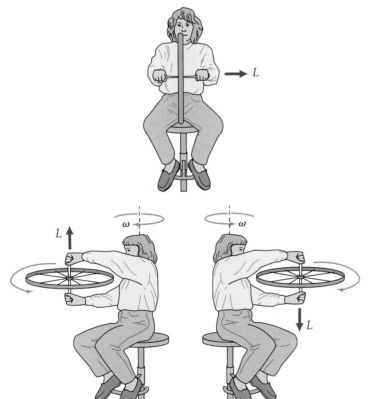

•**FIGURE 8.45 A double rotation** See Exercise 84.

horizontally, she rotates one way (clockwise as viewed from above). When the wheel is turned over, she rotates in the opposite direction. Explain why this occurs. [Hint: Consider angular momentum vectors.]

85. Cats usually land on their feet when they fall, even if held upside down when dropped (•Fig. 8.46). While a cat is falling, there is no external torque and its center of mass falls as a particle. How can cats turn themselves over while falling?

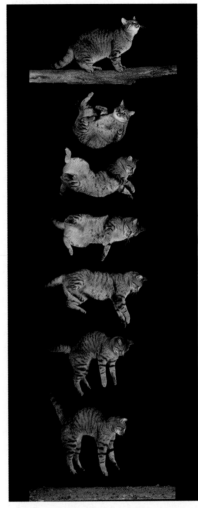

•**FIGURE 8.46 One down, eight to go** See Exercise 85.

86. ■ What is the angular momentum of a 2.0-gram particle moving counterclockwise (as viewed from above) with an angular speed of 5π rad/s in a horizontal circle of radius 15 cm? (Give magnitude and direction.)

87. ■ A 10-kilogram rotating disk of radius 0.25 m has an angular momentum of 0.45 kg·m²/s. What is its angular speed?

88. ■■ Compute the ratio of the Earth's orbital angular momentum and its rotational angular momentum. Are these momenta in the same direction?

Exercises **291**

89. ■■ The period of the Moon's rotation is the same as the period of its revolution: 27.3 days (sidereal). What is the angular momentum for each, rotation and revolution? (Because the periods are equal, we see only one side of the Moon from Earth.)

90. ■■ Circular disks are used in automobile clutches and transmissions. When a rotating disk couples with a stationary one through friction, energy from the rotating disk is transferred to the stationary one. If a disk rotating at 800 rpm couples with a stationary disk with three times the moment of inertia, what is the angular speed of the combination?

91. ■■ 🌀 A rotating spherical star in part of its life cycle expands to six times its normal volume. Assuming the mass remains constant and uniformly distributed inside the star, how is the period of rotation affected?

92. ■■ 🌀 A skater has a moment of inertia of 100 kg·m² when his arms are outstretched and a moment of inertia of 75 kg·m² when his arms are tucked in close to his chest. If he starts to spin at an angular speed of 2.0 rps (revolutions per second) with his arms outstretched, what will his angular speed be when they are tucked in?

93. ■■ 🌀 An ice skater spinning with outstretched arms has an angular speed of 4.0 rad/s. She then tucks in her arms, decreasing her moment of inertia by 7.5%. (a) What is the resulting angular speed? (b) By what factor does the skater's kinetic energy change? (Neglect any frictional effects.) (c) Where does the extra kinetic energy come from?

94. ■■■ A comet approaches the Sun as illustrated in •Fig. 8.47 and is deflected by the Sun's gravitational attraction. This event is considered a collision, and b is called the impact parameter. Find the distance of closest approach (d) in terms of the impact parameter and the velocities (v_o at large distances and v at closest approach). Consider the radius of the Sun to be negligible compared to d. (As the figure shows, the tail of a comet always "points" away from the Sun.)

95. ■■■ A 0.50-kilogram kitten stands on the edge of a lazy Susan (a turntable) of mass 1.5 kg and radius 0.30 m. Assume that the lazy Susan has frictionless bearings and is initially at rest. (a) What will happen if the kitten starts walking around the edge of the lazy Susan? (b) If the kitten walks at a speed of 0.25 m/s relative to the ground, what will be the angular speed of the lazy Susan? (c) When the kitten has walked completely around the edge and is back at its starting point, will that point be above the same point on the ground as it was at the start? If not, where is the kitten relative to the starting ground point? (Speculate on what might happen if everyone on Earth suddenly started to run eastward. What effect might this have on the length of a day?)

Additional Exercises

96. A 10 000-kilogram bridge of length 10 m is supported at both ends. If a 2000-kilogram car is parked on the bridge 3.0 m from the left support, what are the supporting forces at the left and right ends?

97. A spring is twisted 60° about its linear axis (axis of symmetry) by a torque of 100 m·N. How much torsional, or rotational, energy is stored in the spring? [Hint: Let the restoring torque of the spring be $\tau = k\theta$, the rotational analogue of $F = kx$ (Hooke's law—see Chapter 5) with the minus sign omitted.]

98. 🌀 Prove that a thin hoop or ring starting from rest will roll more slowly down an inclined plane than an annular cylinder. (Remember that masses and radii do not have to be taken into account.)

99. A 10-kilogram pulley of radius 0.50 m is connected to a 5.0-kilogram mass through a string as shown in •Fig. 8.48. If the mass is let go from rest, (a) what is its linear acceleration? (b) What is the angular acceleration of the pulley?

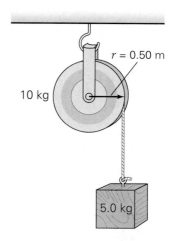

•**FIGURE 8.48 Linear and angular accelerations** See Exercise 99.

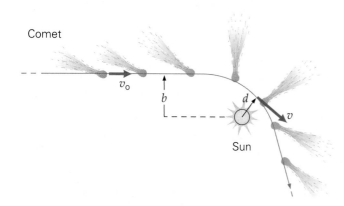

•**FIGURE 8.47 A comet "collision"** See Exercise 94.

100. A bicycle wheel has a diameter of 0.80 m and a mass of 2.0 kg. (a) If the bicycle travels with a linear speed of 1.5 m/s, what is the angular speed of the wheel? (b) What is the angular momentum of the wheel? (Consider the wheel to be a thin ring.)

101. A piece of machinery with a moment of inertia of 2.6 kg·m² rotates with a constant angular speed of 4.0 rad/s when experiencing a frictional torque of 0.56 m·N. What is the net torque acting on the piece?

102. The location of a person's center of gravity relative to his or her height can be found by using the arrangement shown in •Fig. 8.49. The scales are initially adjusted to zero with the board alone. Locate the center of gravity of the person relative to the horizontal dimension. Would you expect the location of the center of gravity in other dimensions to be exactly at the midway points? Explain.

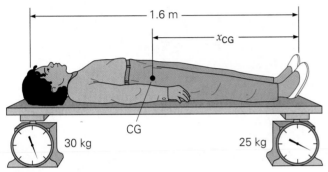

•**FIGURE 8.49   Locating the center of gravity**   See Exercise 102.

103. Using bricks identical to those in Example 8.6, you are asked to create a stack of nine bricks (maximum) without its toppling, with each added brick displaced an equal distance horizontally in the same direction. (a) What is the maximum displacement that will allow this? (b) What is the height of the center of mass of the stack if the uniform bricks are 8.0 cm thick?

104. The *radius of gyration* (k) is defined as the distance from an axis of rotation at which all the mass of an object would have to be concentrated to give the same moment of inertia as the actual mass distribution would have—that is,

$I = Mk^2$. Take the object to be a rotating particle of mass M. From the axes shown in Fig. 8.16, determine the radius of gyration for (a) a uniform sphere, (b) a uniform cylinder, and (c) a particle.

105. A 25-kilogram child jumps onto the rim of a 100-kilogram rotating disk of radius 2.0 m. If the angular speed of the disk before the child's jump was 2.0 rad/s, what is the angular speed of the disk–child system?

106. A mass is suspended by two cords as shown in •Fig. 8.50. What are the tensions in the cords?

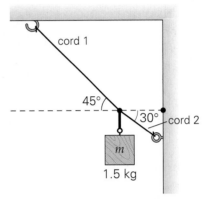

•**FIGURE 8.50   Strung-out equilibrium**   See Exercises 106 and 107.

107. If the cord attached to the vertical wall in Fig. 8.50 were horizontal (instead of at a 30° angle), what would the tensions in the cords be?

108. The outer back wheels of an oversize tractor-trailer rig are 3.66 m apart. When the trailer is loaded, its center of gravity is equidistant from its sides and 3.58 m above the road surface. At what road angle will the parked trailer be in unstable equilibrium?

109. A 0.50-kilogram thin rectangular plate has dimensions of 30 cm by 40 cm. For which axis of rotation does the plate have the greater rotational inertia, an axis tangent to a smaller side at its midpoint or an axis tangent to a larger side at its midpoint? How many times more? (Both axes are perpendicular to the plate surface.)

Solids and Fluids

L ike the person hang-gliding, we exist at the intersection of three realms. We walk on the solid surface of the Earth, and in our daily lives use solid objects of all sorts, from scissors to computers. But we are surrounded by fluids—liquids and gases—on which we are completely dependent. Without the water that we drink, we could survive for a few days at most; without the oxygen in the air that we breathe, we could not live for more than a few minutes. Indeed, we ourselves are not nearly as solid as we like to think. By far the most abundant substance in our bodies is water, and it is in the watery environment of our cells that all the chemical processes on which life depends take place.

On the basis of general physical distinctions, matter is commonly divided into these three phases: solid, liquid, and gas. A solid has a definite shape and volume. A liquid has a fairly definite volume but assumes the shape of its container. A gas takes on the shape and volume of its container. Solids and liquids are sometimes called *condensed matter*. We will utilize a different classification scheme and consider matter in terms of solids and fluids. Liquids and gases are referred to collectively as fluids. A **fluid** is a substance that can flow; liquids and gases qualify, but solids do not.

A simplistic description of solids is that they are made up of particles called atoms that are held rigidly together by interatomic forces. This concept of an ideal rigid body was used in Chapter 8 to describe rotational motion. Real solid bodies are not absolutely rigid and can be elastically deformed by external forces. Elasticity usually brings to mind a rubber band or spring that will resume its original dimensions even after being greatly deformed. In fact, all materials are elastic to some degree, even very hard steel. But, as you will learn, there's a limit to such deformation—an *elastic limit*.

Fluids, however, have little or no elastic response to force—a force merely causes an unconfined fluid to flow. This chapter pays particular attention to the behavior of fluids, shedding light on such questions as why icebergs and ocean liners float, why paper towels absorb spills, and what the 10W–40 on a can of motor oil means. You'll also discover why the person in the photo can neither float like a balloon nor fly like a hummingbird yet can, with the aid of a suitably shaped piece of plastic, soar like an eagle.

Because of their fluidity, liquids and gases have many properties in common, and it is convenient to study them together. There are important differences as well: For example, liquids are not very compressible, whereas gases are easily compressed, as we will shortly see.

9.1 Solids and Elastic Moduli

Objectives: To (a) distinguish between stress and strain and (b) use elastic moduli to compute dimensional changes.

As stated previously, all solid materials are elastic to some degree. That is, a body that is slightly deformed by an applied force will return to its original dimensions or shape when the force is removed. The deformation may not be noticeable for many materials, but it's there.

You may be able to visualize why materials are elastic if you think in terms of the simplistic model of a solid in •Fig. 9.1. The atoms of the solid substance are imagined to be held together by springs. The elasticity of the springs represents the resilient nature of the interatomic forces. The springs resist permanent deformation, as do the forces between atoms.

The elastic properties of solids are commonly discussed in terms of stress and strain. **Stress** is a measure of the force causing a deformation. **Strain** is a relative measure of the deformation a stress causes. Quantitatively, *stress is the applied force per unit cross-sectional area:*

$$\text{stress} = \frac{F}{A} \tag{9.1}$$

SI unit of stress: newtons per square meter (N/m²)

Here F is the magnitude of the applied force normal (perpendicular) to the cross-sectional area. Equation 9.1 shows that the SI units for stress are newtons per square meter (N/m²).

As illustrated in •Fig. 9.2, a force applied to the ends of a rod gives rise to either a *tensile stress* (an elongating tension) or a *compressional stress* (a shortening tension), depending on the direction of the force. In both these cases, the *tensile strain* is the ratio of the change in length to the original length:

$$\text{strain} = \frac{\text{change in length}}{\text{original length}} = \frac{\Delta L}{L_\text{o}} \tag{9.2}$$

Strain is a unitless quantity

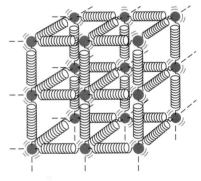

•**FIGURE 9.1 A springy solid**
The elastic nature of interatomic forces is indicated by simplistically representing them as springs, which, like the forces, resist deformation.

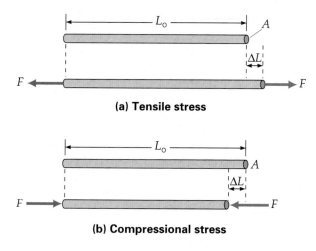

•**FIGURE 9.2 Tensile and compressional stress and strain** Tensile and compressional stresses are due to forces applied normally to the surface area of the ends of bodies. **(a)** A tension, or tensile stress, tends to increase the length of an object. **(b)** A compressional stress tends to shorten the length.

(a) Tensile stress

(b) Compressional stress

where $\Delta L = L - L_o$. Note that strain is a unitless quantity (length/length). It is the *fractional change* in length. For example, if the strain is 0.05, the material has changed in length by 5% of its original length.

As might be expected, the resulting strain is proportional to the applied stress; that is, strain $\propto$ stress. For relatively small stresses, this is a direct proportion. The constant of proportionality, which depends on the nature of the material, is called the **elastic modulus**. Thus,

$$\text{stress} = \text{elastic modulus} \times \text{strain}$$

or

Definition of:
Elastic modulus

$$\text{elastic modulus} = \frac{\text{stress}}{\text{strain}} \qquad (9.3)$$

SI unit of elastic modulus: newtons per square meter (N/m²)

That is, the elastic modulus is the stress divided by the strain, or the ratio of stress to strain. The elastic modulus has the same units as stress. Why?

Three general types of elastic moduli (plural of modulus) are associated with stresses that produce changes in length, shape, or volume. These are called *Young's modulus, shear modulus,* and *bulk modulus,* respectively.

Change in Length: Young's Modulus

•Figure 9.3 is a graph of the tensile stress versus the strain for a typical metal rod. The curve is a straight line up to a point called the *proportional limit*. Beyond this point, the strain begins to increase more rapidly to another critical point called the **elastic limit**. If the tension is removed at this point, the material will return

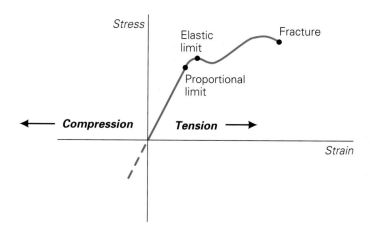

•**FIGURE 9.3 Stress versus strain** A plot of stress versus strain for a typical metal rod is a straight line up to the proportional limit. Then elastic deformation continues until the elastic limit is reached. Beyond that, the rod will be permanently deformed and will eventually fracture or break.

TABLE 9.1 Elastic Moduli for Various Materials (in N/m²)

Substance	Young's Modulus (Y)	Shear Modulus (S)	Bulk Modulus (B)
Solids			
Aluminum	7.0×10^{10}	2.5×10^{10}	7.0×10^{10}
Bone (limb)	1.5×10^{10}	8.0×10^{10}	
Brass	9.0×10^{10}	3.5×10^{10}	7.5×10^{10}
Copper	11×10^{10}	3.8×10^{10}	12×10^{10}
Glass	5.7×10^{10}	2.4×10^{10}	4.0×10^{10}
Iron	15×10^{10}	6.0×10^{10}	12×10^{10}
Steel	20×10^{10}	8.2×10^{10}	15×10^{10}
Liquids			
Alcohol, ethyl			1.0×10^{9}
Glycerin			4.5×10^{9}
Mercury			26×10^{9}
Water			2.2×10^{9}

to its original length. If the tension is applied beyond the elastic limit and then removed, the material will recover somewhat but will retain some permanent deformation.

The straight-line part of the graph shows a direct proportionality between stress and strain. This relationship, first formalized by Robert Hooke in 1678, is now known as *Hooke's law*. (It is the same general relationship as that given for a spring in Section 5.2—see Fig. 5.5.) The elastic modulus for a tension or a compression is called **Young's modulus** (Y):

$$\underbrace{\frac{F}{A}}_{stress} = Y \underbrace{\left(\frac{\Delta L}{L_{o}}\right)}_{strain} \quad \text{or} \quad Y = \frac{F/A}{\Delta L/L_{o}} \qquad (9.4)$$

SI unit of Young's modulus: newtons per square meter (N/m²)

Thomas Young (1773–1829) was a British physicist who investigated the mechanical properties of materials and optical phenomena.

The units of Young's modulus are the same as of stress, newtons per square meter (N/m²), since the ratio of lengths is unitless. Some typical values of Young's modulus are given in Table 9.1.

To obtain a conceptual or physical understanding of Young's modulus, let's solve Eq. 9.4 for ΔL:

$$\Delta L = \left(\frac{FL_{o}}{A}\right)\frac{1}{Y} \quad \text{or} \quad \Delta L \propto \frac{1}{Y}$$

Hence we see that the larger the modulus of a material, the smaller its relative change in length (with other parameters being equal).

Example 9.1 ■ Pulling My Leg: Under a Lot of Stress

The femur (upper leg bone) is the longest and strongest bone in the body. Taking a typical femur to be approximately circular with a radius of 2.0 cm, how much force would be required to extend the bone by 0.010%?

Thinking It Through. We can see that Eq. 9.4 should apply, but where does the percentage increase fit in? Recognize that the $\Delta L/L_{o}$ term is the *fractional* increase in length. For example, if you had a spring with a length of 10 cm (L_{o}) and you stretched it 1.0 cm (ΔL), then $\Delta L/L_{o} = 1.0$ cm/10 cm $= 0.10$. This ratio can readily be changed to a percentage, and we would say the spring's length was increased be 10%. So, the percentage increase is really just the value of the $\Delta L/L_{o}$ term (multiplied by 100%).

Solution. Listing the data, we have

Given: $r = 2.0$ cm $= 0.020$ m **Find:** tensile force F
$\Delta L/L_o = 0.010\% = 1.0 \times 10^{-4}$
$Y = 1.5 \times 10^{10}$ N/m² (from Table 9.1)

Using Eq. 9.4, we have

$$F = Y(\Delta L/L_o)A = Y(\Delta L/L_o)\pi r^2$$
$$= (1.5 \times 10^{10} \text{ N/m}^2)(1.0 \times 10^{-4})\pi(0.020 \text{ m})^2 = 1.9 \times 10^3 \text{ N}$$

How much force is this? Quite a bit—more than 400 lb.; the femur is a pretty strong bone.

Follow-up Exercise. A total mass of 16 kg is suspended from a steel wire with a diameter of 0.10 cm. (a) By what percentage does the length of the wire increase? (b) The tensile or ultimate strength is the maximum stress a material can support before breaking or fracturing. If the tensile strength of the steel wire in (a) is 4.9×10^8 N/m², how much mass could be suspended before it would break?

Change in Shape: Shear Modulus

Another way an elastic body can be deformed is by a *shear stress*. In this case, the deformation is due to an applied force that is tangential to the surface area (•Fig. 9.4). A change in shape results without a change in volume. The *shear strain* is given by x/h, where x is the relative displacement of the faces and h is the distance between them.

The shear strain is sometimes defined in terms of the *shear angle* (ϕ). As Fig. 9.4 shows, $\tan \phi = x/h$. But the shear angle is usually quite small, so a good approximation is $\tan \phi \approx \phi \approx x/h$, where the angle ϕ is in radians. (If $\phi = 10°$, for example, there is only 1.0% difference between ϕ and $\tan \phi$.) The **shear modulus** (sometimes called the modulus of rigidity) is then

$$S = \frac{F/A}{x/h} \approx \frac{F/A}{\phi} \qquad (9.5)$$

SI unit of shear modulus: newtons per square meter (N/m²)

Note in Table 9.1 that the shear modulus is generally less than Young's modulus; S is approximately $Y/3$ for many materials, which indicates that there is a greater response to a shear stress than to a tensile stress. Also, note the inverse relationship of $\phi \propto 1/S$, similar to that pointed out previously for Young's modulus.

A shear stress may also be of the torsional type, resulting from the twisting action of a torque. For example, a torsional shear stress may shear off the head of a bolt that is being tightened.

Liquids do not have shear moduli (or Young's moduli), hence the gaps in Table 9.1. A shear stress cannot be effectively applied to a liquid or a gas. It is often said that *fluids cannot support a shear*. Why?

Change in Volume: Bulk Modulus

Suppose that a force directed inward acts over the entire surface of a body (•Fig. 9.5). Such a *volume stress* is often applied by pressure transmitted by a fluid (Section 9.2). An elastic material will be compressed by a volume stress; that is, it will show a change in volume but not in general shape in response to a pressure change Δp. (Pressure is the force per unit area, as we shall see in Section 9.2.) The change in pressure is equal to the volume stress, or $\Delta p = F/A$. The *volume strain* is the ratio of the volume change (ΔV) to the original volume (V_o). The **bulk modulus** (B) is then

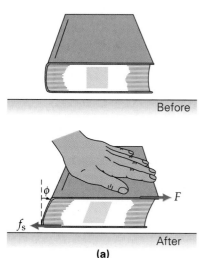

Before

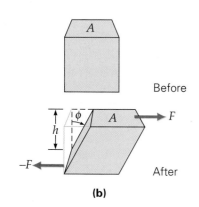

After

(a)

A

Before

A

After

(b)

•**FIGURE 9.4 Shear stress and strain** A shear stress is produced when a force is applied tangentially to a surface area. The strain is measured in terms of the relative displacement of the object's faces or the shear angle ϕ.

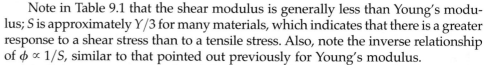

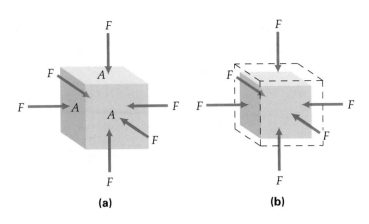

$$B = \frac{F/A}{-\Delta V/V_o} = -\frac{\Delta p}{\Delta V/V_o} \qquad (9.6)$$

SI unit of bulk modulus: newtons per square meter (N/m^2)

The minus sign is introduced to make B a positive quantity, since $\Delta V = V - V_o$ is negative for an increase in external pressure. And, similar to the previous moduli relationships, $\Delta V \propto 1/B$.

Bulk moduli are listed for solids *and* liquids in Table 9.1. Gases have bulk moduli too, since they can be compressed. For a gas, it is common to talk about the reciprocal of the bulk modulus, which is called the **compressibility** (k):

$$k = \frac{1}{B} \qquad (\textit{for gases}) \qquad (9.7)$$

The change in volume ΔV is thus directly proportional to the compressibility k.

Solids and liquids are relatively incompressible and have small values of compressibility. Conversely, gases are easily compressed and have large compressibilities, which vary with pressure and temperature.

Example 9.2 ■ Compressing a Liquid: Volume Stress and Bulk Modulus

By how much should the pressure on a liter of water be changed to compress it by 0.10%?

Thinking It Through. Similar to the fractional change in length $\Delta L/L_o$, the fractional change in volume is given by $-\Delta V/V_o$, which may be expressed as a percentage. The pressure change can then be found from Eq. 9.6. Compression implies a negative ΔV.

Solution. Listing the data, we have

Given: $-\Delta V/V_o = 0.0010$ (or 0.10%)   *Find:* $\Delta p = F/A$
   $V_o = 1.0\,L = 1000\,cm^3$
   $B_{H_2O} = 2.2 \times 10^9\,N/m^2$ (from Table 9.1)

Note that $-\Delta V/V_o$ is the *fractional* change in the volume; for a compression of 0.10%, it is 0.0010 (no units, a ratio of volumes). With $V_o = 1000\,cm^3$, the volume reduction is

$$-\Delta V = 0.0010V_o = 0.0010(1000\,cm^3) = 1.0\,cm^3$$

However, the change in volume is not needed. The fractional change, as listed in the Given data, can be used directly in Eq. 9.6 to find the pressure increase:

$$\Delta p = B\left(\frac{-\Delta V}{V_o}\right) = (2.2 \times 10^9\,N/m^2)(0.0010) = +2.2 \times 10^6\,N/m^2$$

(This increase is about 22 times normal atmospheric pressure!)

Follow-up Exercise. If an extra $1.0 \times 10^6 N/m^2$ of pressure above normal atmospheric pressure is applied to a half liter of water, what is the change in the water's volume?

Feat of Strength or Knowledge of Materials?

The breaking of wooden boards, concrete blocks, or similar materials with a bare hand or foot is an impressive demonstration often done by karate experts (Fig. 1). The physics of this feat can be analyzed in terms of the properties of the materials involved.

In delivering the blow, the expert imparts a large impulse force (Chapter 6) to the top board or block, which bends under the pressure. (Note that the objects are struck midway between the end supports.) At the same time, the bones of the hand are being compressed as well. Fortunately, human bone can withstand more compressive force than can wood, tile, or concrete, which is why the expert's bones aren't

FIGURE 1   Feat of strength or knowledge of materials? Breaking wooden boards, concrete blocks, or roofing tiles (as shown here) with a karate blow depends on both the physical strength of the expert and the strength of the material. Wood, concrete, and tile have different maximum tensile and compressional stresses, but the maximum compressional strength of bone is greater than any of these. (This photo, from a high-speed photography sequence, was made toward the end of the blow.)

damaged. (The ultimate compressive strength of bone is at least four times greater than that of concrete.*)

When each board or block is hit, the upper surface is compressed and the lower surface is elongated, or subjected to a tension force. Wood, tile, and concrete are weaker under tension than under compression. (The ultimate tensile strength of concrete is only about a twentieth of its ultimate compressive strength.) Thus, the board or block begins to crack at the bottom surface first. The crack propagates from the underside *toward* the hand, widens, and becomes a complete break. Thus the hand never actually "cuts through" the board or block.

The amount of force required to break a board or block in this way depends on several factors. For a wooden board, these are the type of wood, the width and thickness, and the distance between the end supports. Also, the edge of the hand must strike the board parallel to the grain of the wood. Similar considerations apply for a tile or a concrete block. Because tile and concrete are more rigid than wood, more force is generally required to break these substances.

Some karate experts are able to break through a stack of boards or tiles and more than one concrete block. The force required does not increase by a factor equal to the number of objects. High-speed photography shows that the hand makes contact with only one or two boards or tiles at the top of the stack. Then, as each object breaks, it collides with and breaks through the one below it.

Even though the properties of the materials seem to guarantee the success of this demonstration, you should not attempt the feat unless you are an expert and know what you're doing. The board or block might win.

*Ultimate strength is the maximum strength a material can stand before it breaks or fractures.

9.2   Fluids: Pressure and Pascal's Principle

Objectives: To (a) explain the pressure–depth relationship and (b) state Pascal's principle and describe how it is used in practical applications.

A force can be applied to a solid at a point of contact, but this won't work with a fluid, as we saw in the preceding discussion on bulk moduli. With fluids, a force must be applied over an area. Such an application of force is expressed in terms of **pressure**, or the *force per unit area*:

Pressure—force per unit area

$$p = \frac{F}{A} \tag{9.8a}$$

SI unit of pressure: newtons per square meter (N/m^2) or pascal (Pa)

The force in this equation is understood to be acting normally (perpendicularly) to the surface area. F may be the perpendicular component of a force that acts at an angle to the surface (•Fig. 9.6).

As Fig. 9.6 shows, in the more general case we should write

$$p = \frac{F_\perp}{A} = \frac{F \cos \theta}{A} \tag{9.8b}$$

Pressure is a scalar quantity (with magnitude only), even though the force producing it or the force produced within a fluid by a pressure does have direction and so is a vector.

Pressure has SI units of newtons per square meter (N/m²). This combined unit is given the special name **pascal (Pa)** in honor of the French scientist and philosopher Blaise Pascal (1623–1662), who studied fluids and pressure:

$$1 \, \text{Pa} = 1 \, \text{N/m}^2$$

In the British system, a common unit of pressure is pounds per square inch (lb/in², or psi). Other units, some of which will be introduced later, are used in special applications.

Before going on, it may be helpful to review density, which is an important consideration in the study of fluids. Recall from Chapter 1 that the density (ρ) of a substance is defined (in Eq. 1.1) as

$$\text{density} = \frac{\text{mass}}{\text{volume}}$$

$$\rho = \frac{m}{V}$$

SI unit of density: kilograms per cubic meter (kg/m³)

(common cgs unit: grams per cubic centimeter, or g/cm³)

That is, density is the mass per *unit* volume. The densities of some common substances are given in Table 9.2.

Water has a density of $1.00 \times 10^3 \, \text{kg/m}^3$ (or 1.00 g/cm³), from the original definition of the kilogram (Chapter 1). Mercury has a density of $13.6 \times 10^3 \, \text{kg/m}^3$ (or 13.6 g/cm³). Hence, mercury is denser than water—13.6 times more dense. Gasoline is less dense than water. (See Table 9.2.)

We say that density is a measure of the compactness of the matter of a substance—the greater the density, the more matter or mass in a given volume. Notice how density quantifies the amount or mass per volume. It would be incorrect to say that mercury is "heavier" than water. Why? Because you could have a large volume of water that would be heavier than some much smaller volume of mercury.

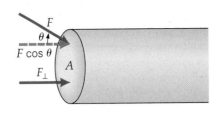

$$p = \frac{F_\perp}{A} = \frac{F \cos \theta}{A}$$

•**FIGURE 9.6 Pressure** Pressure is usually written $p = F/A$, where it is understood that F is the force or component of force normal to the surface. In general, then, $p = (F \cos \theta)/A$.

Pressure and Depth

If you have gone diving, you well know that pressure increases with depth, having felt the increased pressure on your eardrums. An opposite effect is commonly felt when you fly in a plane or ride in a car going up a mountain. With increasing altitude, your ears may "pop" because of reduced external air pressure.

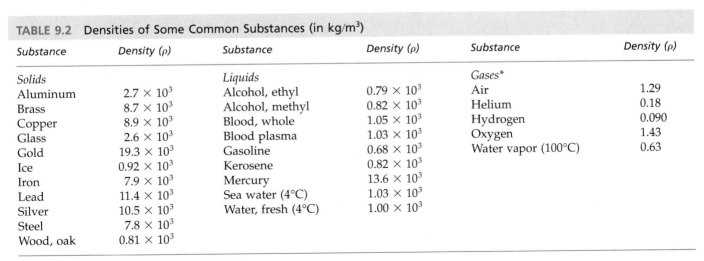

TABLE 9.2 Densities of Some Common Substances (in kg/m³)

Substance	Density (ρ)	Substance	Density (ρ)	Substance	Density (ρ)
Solids		*Liquids*		*Gases**	
Aluminum	2.7×10^3	Alcohol, ethyl	0.79×10^3	Air	1.29
Brass	8.7×10^3	Alcohol, methyl	0.82×10^3	Helium	0.18
Copper	8.9×10^3	Blood, whole	1.05×10^3	Hydrogen	0.090
Glass	2.6×10^3	Blood plasma	1.03×10^3	Oxygen	1.43
Gold	19.3×10^3	Gasoline	0.68×10^3	Water vapor (100°C)	0.63
Ice	0.92×10^3	Kerosene	0.82×10^3		
Iron	7.9×10^3	Mercury	13.6×10^3		
Lead	11.4×10^3	Sea water (4°C)	1.03×10^3		
Silver	10.5×10^3	Water, fresh (4°C)	1.00×10^3		
Steel	7.8×10^3				
Wood, oak	0.81×10^3				

*At 0°C and 1 atm unless otherwise specified.

•FIGURE 9.7 **Pressure and depth** The extra
pressure at a depth h in a liquid is due to the
weight of the liquid above: $p = \rho g h$, where ρ is the
density of the liquid (assumed to be constant). This
is shown for an imaginary rectangular column of
liquid.

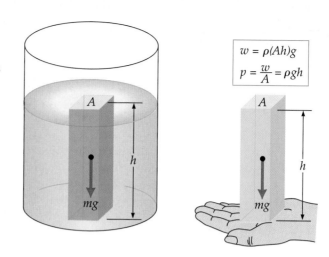

$$w = \rho(Ah)g$$
$$p = \frac{w}{A} = \rho g h$$

How the pressure in a fluid varies with depth can be demonstrated by considering a container of liquid at rest. Imagine you can isolate a rectangular column of water as shown in •Fig. 9.7. Then, the force on the bottom of the container
below the column (or the hand) is equal to the weight of the liquid making up
the column: $F = w = mg$. Since density is $\rho = m/V$, the mass in the column is
equal to the density times the volume; that is, $m = \rho V$. (The liquid is assumed incompressible, so ρ is constant.)

The volume of the isolated liquid column is equal to the height of the column
times the area of its base, $V = hA$. Thus we can write

$$F = w = mg = \rho V g = \rho g h A$$

With $p = F/A$, the pressure at a depth h due to the weight of the column is

$$p = \rho g h \qquad (9.9)$$

This is a general result for incompressible liquids. The pressure is the same everywhere on a horizontal plane at a depth h (with ρ and g constant). Also note that
Eq. 9.9 is independent of the base area of the rectangular column. We could have
taken the whole cylindrical column of the liquid in the container in Fig. 9.7 and
gotten the same result.

The derivation of Eq. 9.9 did not take into account pressure being applied to
the open surface of the liquid. This factor adds to the pressure at a depth h to give
a *total* pressure of

Pressure–depth relationship

$$p = p_o + \rho g h \qquad \begin{array}{l}\textit{(incompressible fluid} \\ \textit{at constant density)}\end{array} \qquad (9.10)$$

where p_o is the pressure applied to the liquid surface (that is, at $h = 0$).

For an open container, p_o is atmospheric pressure, or the weight (force) per
area due to the gases in the atmosphere above the liquid's surface. The average
atmospheric pressure at sea level is sometimes used as a unit, called an **atmosphere (atm)**:

$$1 \text{ atm} \equiv 101.325 \text{ kPa} = 1.01325 \times 10^5 \text{ N/m}^2 \approx 14.7 \text{ lb/in}^2$$

How atmospheric pressure is measured will be described shortly.

Example 9.3 ■ A Scuba Diver: Pressure and Force

(a) What is the total pressure on the back of a scuba diver in a lake at a depth of 8.00 m?
(b) What is the force on the diver's back due to the water alone, taking the surface of
the back to be a rectangle 60.0 cm by 50.0 cm?

Thinking It Through. (a) This is a direct application of Eq. 9.10 in which p_o is taken as the atmospheric pressure p_a. (b) Knowing the area and the pressure due to the water, the force can be found from the definition of pressure, $p = F/A$.

Solution.

Given: $h = 8.00$ m
$\quad A = 60.0$ cm $\times$ 50.0 cm
$\qquad = 0.600$ m $\times$ 0.500 m $= 0.300$ m^2
$\quad \rho_{H_2O} = 1.00 \times 10^3$ kg/m^3 (from Table 9.2)
$\quad p_a = 1.01 \times 10^5$ N/m^2

Find: (a) p (total pressure)
$\qquad$ (b) F (force due to water)

(a) The total pressure is the sum of the pressure due to the water and the atmospheric pressure (p_a). By Eq. 9.10, this is

$$p = p_a + \rho g h$$
$$= (1.01 \times 10^5 \text{ N/m}^2) + (1.00 \times 10^3 \text{ kg/m}^3)(9.80 \text{ m/s}^2)(8.00 \text{ m})$$
$$= (1.01 \times 10^5 \text{ N/m}^2) + (0.784 \times 10^5 \text{ N/m}^2) = 1.79 \times 10^5 \text{ N/m}^2 \text{ (or Pa)}$$
$$\text{(expressed in atmospheres)} \approx 1.8 \text{ atm}$$

(b) The pressure p_w due to the water alone is the $\rho g h$ portion of the preceding equation, so $p_w = 0.784 \times 10^5$ N/m^2. Then, $p_w = F/A$, and

$$F = p_w A = (0.784 \times 10^5 \text{ N/m}^2)(0.300 \text{ m}^2)$$
$$= 2.35 \times 10^4 \text{ N} \quad \text{(or } 5.29 \times 10^3 \text{ lb—about 2.6 tons!)}$$

Follow-up Exercise. You might question the answer to part (b) of this Example—how could the diver support such a force? To get a better idea of the forces our bodies can support, what would be the force on the diver's back at the water surface from atmospheric pressure alone? How do you suppose our bodies can support such forces or pressures?

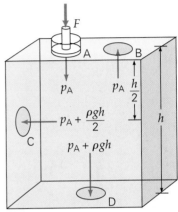

•**FIGURE 9.8 Pascal's principle**
The pressure applied at point A is fully transmitted to all parts of the fluid and to the walls of the container. There is also pressure due to the weight of the fluid above at different depths (for instance, $\rho g h/2$ at C and $\rho g h$ at D).

Pascal's Principle

When the pressure (for example, air pressure) is increased on the entire open surface of an incompressible liquid at rest, the pressure at any point in the liquid or on the boundary surfaces increases by the same amount. The effect is the same if pressure is applied by means of a piston to any surface of an enclosed fluid (•Fig. 9.8). The transmission of pressure in fluids was studied by Blaise Pascal (after whom the SI pressure unit is named), and the observed effect is called **Pascal's principle**:

Pascal's principle

> Pressure applied to an enclosed fluid is transmitted undiminished to every point in the fluid and to the walls of the container.

For an incompressible liquid, the pressure change is transmitted essentially instantaneously. For a gas, a pressure change will generally be accompanied by a change in volume and/or temperature, but after equilibrium has been reestablished, Pascal's principle remains valid.

Common practical applications of Pascal's principle include the hydraulic braking systems used on automobiles. A force on the brake pedal transmits a force to the wheel brake cylinder. Similarly, hydraulic lifts and jacks are used to raise automobiles and other heavy objects (•Fig. 9.9).

Using Pascal's principle, we can show how such systems not only allow us to transmit force from one place to another, but also to multiply that force. The input pressure p_i, supplied by compressed air for a garage lift, for example, gives an input force F_i on a small piston area A_i (Fig. 9.9). The full magnitude of the pressure is transmitted to the output piston, which has an area A_o. Since $p_i = p_o$,

$$\frac{F_i}{A_i} = \frac{F_o}{A_o}$$

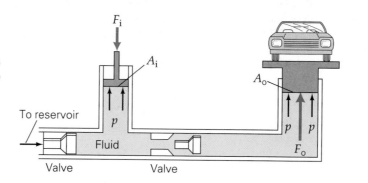

•FIGURE 9.9 The hydraulic lift Because the input and output pressures are equal (Pascal's principle), a small input force gives a large output force proportional to the ratio of the piston areas.

$$F_o = \left(\frac{A_o}{A_i}\right)F_i$$

and

$$F_o = \left(\frac{A_o}{A_i}\right)F_i \qquad \begin{array}{l}\textit{hydraulic}\\\textit{force multiplication}\end{array} \qquad (9.11)$$

With A_o larger than A_i, then F_o will be larger than F_i. The input force is greatly multiplied.

Example 9.4 ■ The Hydraulic Lift: Pascal's Principle

A garage lift has input and lift (output) pistons with diameters of 10 cm and 30 cm, respectively. The lift is used to hold up a car with a weight of 1.4×10^4 N. (a) What is the force on the input piston? (b) What pressure is applied to the input piston?

Thinking It Through. (a) Pascal's principle, as expressed in the hydraulic Eq. 9.11, has four variables, and three are given (areas via diameters). (b) The pressure is simply $p = F/A$.

Solution.

Given: $d_i = 10$ cm $= 0.10$ m   *Find:* (a) F_i (input force)
$\quad\quad\quad d_o = 30$ cm $= 0.30$ m $\quad\quad\quad$ (b) p_i (input pressure)
$\quad\quad\quad F_o = 1.4 \times 10^4$ N

(a) Rearranging Eq. 9.11 and using $A = \pi r^2 = \pi d^2/4$ for the circular piston ($r = d/2$) gives

$$F_i = \left(\frac{A_i}{A_o}\right)F_o = \left(\frac{\pi d_i^2/4}{\pi d_o^2/4}\right)F_o = \left(\frac{d_i}{d_o}\right)^2 F_o$$

$$F_i = \left(\frac{0.10\ \text{m}}{0.30\ \text{m}}\right)^2 F_o = \frac{F_o}{9} = \frac{1.4 \times 10^4\ \text{N}}{9} = 1.6 \times 10^3\ \text{N}$$

The input force is one-ninth of the output force, or there was a force multiplication of 9 (that is, $F_o = 9F_i$).

(Note that we didn't really need to write the complete expressions for the areas. We know that the area of a circle is proportional to the square of its diameter. If the ratio of the piston diameters is 3 to 1, the ratio of their areas must therefore be 9 to 1, and we could have used this ratio directly in Eq. 9.11.)

(b) Then we apply Eq. 9.8a:

$$p_i = \frac{F_i}{A_i} = \frac{F_i}{\pi r_i^2} = \frac{F_i}{\pi(d_i/2)^2} = \frac{1.6 \times 10^3\ \text{N}}{\pi(0.10\ \text{m})^2/4}$$

$$= 2.0 \times 10^5\ \text{N/m}^2\ (= 200\ \text{kPa})$$

This pressure is about 30 lb/in², a common pressure used in automobile tires and about twice atmospheric pressure (about 100 kPa, or 15 lb/in²).

Follow-up Exercise. Pascal's principle is used in shock absorbers on automobiles and on the landing gear of airplanes. (The polished steel piston rods can be seen above the wheels on aircraft.) In these devices, a large force (the shock produced on hitting a bump in the road or an airport runway at high speed) must be reduced to a safe level. Basically, fluid is forced from a larger diameter cylinder into a smaller diameter cylinder (the reverse of the situation in Fig. 9.9). Suppose that the input piston of a shock absorber on a jet plane has a diameter of 8.0 cm. What would be the diameter of the output cylinder that would reduce the force by a factor of 10?

As Example 9.4 shows, we can relate the force directly to the diameters of the pistons: $F_i = (d_i/d_o)^2 F_o$, or $F_o = (d_o/d_i)^2 F_i$. By making $d_o \gg d_i$ we can get huge factors of force multiplication, as is typical for hydraulic presses, jacks, and earth-moving equipment. (The shiny input piston rods are often visible on front loaders and backhoes.) Inversely, we can get a force reduction by making $d_i > d_o$, as in Follow-up Exercise 9.4.

However, don't think that you are getting something for nothing with large force multiplications. Energy is still a factor, and it can never be multiplied by a machine. (Why not?) Looking at the work involved and assuming the work output is equal to the work input, $W_o = W_i$ (an ideal condition—why?), we have from Eq. 5.1

$$F_o x_o = F_i x_i$$

or

$$F_o = \left(\frac{x_i}{x_o}\right) F_i$$

where x_o and x_i are the output and input distances moved by the respective pistons.

Thus, the output force can be much greater than the input force only if the input distance is much greater than the output distance. For example, if $F_o = 10F_i$, then $x_i = 10x_o$, and the input piston must travel ten times the distance of the output piston. We say that force is multiplied at the expense of distance. Specifically, with the preceding numbers, suppose the brake pads of a simple hydraulic braking system of a car move 5.0 mm when the brakes are applied. Then, the input distance of the pedal is $x_i = 10x_o = 10(5.0 \text{ mm}) = 50$ mm—a greater input distance, but the brake pads are applied with 10 times the force applied to the pedal, $F_o = 10F_i$.

Pressure Measurement

Pressure can be measured by a variety of mechanical devices that are often spring-loaded. Another type of instrument, called a manometer, uses a liquid—usually mercury. An *open-tube manometer* is illustrated in •Fig. 9.10a. One end of the U-shaped tube is open to the atmosphere, and the other is connected to the container of gas whose pressure is to be measured. The liquid in the U-tube acts as a reservoir through which pressure is transmitted according to Pascal's principle.

The pressure of the gas (p) is balanced by the weight of the column of liquid (of height h, the difference in the heights of the columns) and the atmospheric pressure (p_a) on the open liquid surface:

$$p = p_a + \rho g h \tag{9.12}$$

The pressure p is called the **absolute pressure**.

You may have experience measuring pressure by means of pressure gauges; a tire gauge used to measure air pressure in automobile tires is a common example (Fig. 9.10b). Such gauges, quite appropriately, measure **gauge pressure**: A pressure gauge registers only the pressure *above* atmospheric pressure. Hence, to get the absolute pressure (p), you have to add the atmospheric pressure (p_a) to the gauge pressure (p_g):

$$p = p_a + p_g$$

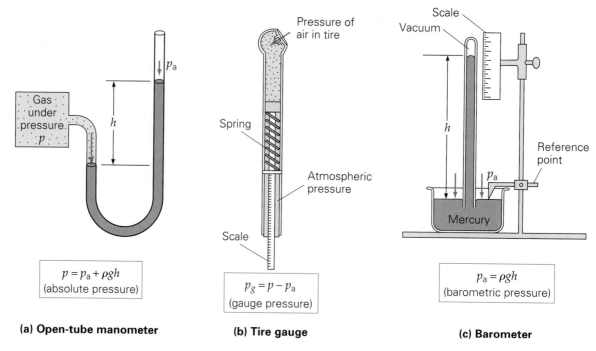

(a) Open-tube manometer

$$p = p_a + \rho g h$$
(absolute pressure)

(b) Tire gauge

$$p_g = p - p_a$$
(gauge pressure)

(c) Barometer

$$p_a = \rho g h$$
(barometric pressure)

•**FIGURE 9.10 Pressure measurement** (a) For an open-tube manometer, the pressure of the gas in the container is balanced by the pressure of the liquid column and atmospheric pressure acting on the open surface of the liquid. The absolute pressure of the gas equals the sum of the atmospheric pressure (p_a) and $\rho g h$, the gauge pressure. (b) A tire gauge measures gauge pressure, the difference between the pressure in the tire and atmospheric pressure: $p_{gauge} = p - p_a$. Thus, if a tire gauge reads 200 kPa (30 lb/in²), the actual pressure within the tire is 1 atm higher, or 300 kPa. (c) A barometer is a closed-tube manometer that is exposed to the atmosphere and thus reads only atmospheric pressure.

For example, suppose your tire gauge reads a pressure of 200 kPa ($\approx$ 30 lb/in²). The absolute pressure within the tire is then $p = p_a + p_g = 101\ \text{kPa} + 200\ \text{kPa} = 301\ \text{kPa}$, where normal atmospheric pressure is about 101 kPa (14.7 lb/in²), as will be shown shortly.

It is the gauge pressure of a tire that keeps it rigid or operational. In terms of the more familiar pounds per square inch (psi, or lb/in²), a tire with a gauge pressure of 30 psi has an absolute pressure of about 45 psi (30 + 15, with atmospheric pressure $\approx$ 15 psi). Hence, the pressure on the inside of the tire is 45 psi, and that on the outside is 15 psi. The Δp of 30 psi keeps the tire inflated. If you open the valve or get a puncture, the internal and external pressures equalize and you have a flat!

Atmospheric pressure itself can be measured with a *barometer*. The principle of a mercury barometer is illustrated in Fig. 9.10c. The device was invented by Evangelista Torricelli (1608–1647), Galileo's successor as professor of mathematics at the university in Florence. A simple barometer consists of a tube filled with mercury that is inverted into a reservoir. Some mercury runs out, but as much as can be supported by the air pressure on the surface of the reservoir pool remains in the tube. This device can be considered to be a *closed-tube manometer*, and the absolute pressure is just the atmospheric pressure, since the gauge pressure (the pressure *above* atmospheric pressure) is zero.

The atmospheric pressure is then equal to the pressure due to the weight of the column of mercury, or

$$p_a = \rho g h \qquad (9.13)$$

A *standard atmosphere* is defined as the pressure supporting a column of mercury exactly 76 cm in height at sea level and at 0°C.

Example 9.5 ■ Standard Atmospheric Pressure: Converting to Pascals

If a standard atmosphere supports a column height of exactly 76 cm of mercury (chemical symbol Hg), what is the standard atmospheric pressure in pascals? (The density of mercury is $13.5951 \times 10^3 \text{ kg/m}^3$ at 0°C, and $g = 9.80665 \text{ m/s}^2$.)

Thinking It Through. This is a direct application of Eq. 9.13 to find standard atmospheric pressure in metric units (pascals or newtons per square meter). Note that the height of the column is given in centimeters.

Solution.

Given: $h = 76 \text{ cm} = 0.76 \text{ m}$ (exact)   *Find:* p_a (atmospheric pressure)
$\rho_{Hg} = 13.5951 \times 10^3 \text{ kg/m}^3$
$g = 9.80665 \text{ m/s}^2$

Using Eq. 9.13, we have

$$p_a = \rho_{Hg}gh = (13.5951 \times 10^3 \text{ kg/m}^3)(9.80665 \text{ m/s}^2)(0.760000 \text{ m})$$
$$= 101325 \text{ N/m}^2 = 1.01325 \times 10^5 \text{ Pa} \quad \text{(or 101.325 kPa)}$$

Follow-up Exercise. What would be the height of a barometer column for one standard atmosphere if water were used instead of mercury?

Changes in atmospheric pressure can be observed as changes in the height of the mercury column. Atmospheric pressure is commonly reported in terms of the height of the barometer column, and weather forecasters say the barometer is rising or falling. That is, 1 standard atmosphere (atm) of pressure has equivalent values of

$$1 \text{ atm} = 76 \text{ cm Hg} = 760 \text{ mm Hg}$$
$$= 29.92 \text{ in. Hg} \text{ (about 30 in. Hg)}$$

In honor of Torricelli, a pressure supporting 1 mm of mercury is given the name *torr*:

$$1 \text{ mm Hg} \equiv 1 \text{ torr}$$

and

$$1 \text{ atm} = 760 \text{ torr}$$

Because mercury is very toxic, it is sealed inside a barometer. A safer and less expensive device widely used to measure atmospheric pressure is the *aneroid* (without fluid) *barometer*. In an aneroid barometer, a sensitive metal diaphragm on an evacuated container (something like a drum head) responds to pressure changes, which are indicated on a dial. This is the kind of barometer you frequently find in homes in decorative wall mountings.

Since air is compressible, the atmospheric density and pressure are greatest at the Earth's surface and decrease with altitude. We live at the bottom of the atmosphere but don't notice its pressure very much in our daily activities. Remember that our bodies are composed largely of fluids, which exert a matching outward pressure. Indeed, the external pressure of the atmosphere is so important to our normal functioning that we take it with us wherever we can. The pressurized suits worn by astronauts in space or on the Moon are needed not only to supply oxygen, but also to provide an external pressure similar to that on the Earth's surface.

A very important gauge pressure reading is discussed in the Insight on p. 308. Read it before going on to Example 9.6.

Note: Another unit sometimes used in weather reports is the millibar (mb). By definition, $1 \text{ atm} = 1.01325 \times 10^5 \text{ N/m}^2$ $= 1.01325 \text{ bar} = 1013.25 \text{ mb}$. Normal atmospheric pressures are around 1000 mb.

Blood Pressure

Basically, a pump is a machine that transfers mechanical energy to a fluid, thereby increasing the pressure and causing the fluid to flow. There are a wide variety of pumps, but one of interest to everyone is the heart. The heart is a muscular pump that drives blood through the body's circulatory network of arteries, capillaries, and veins. With each pumping cycle, the human heart's interior chambers enlarge and fill with freshly oxygenated blood from the lungs (Fig. 1).

On contraction of chambers called ventricles, blood is forced out through the arteries. Smaller and smaller arteries branch off from the main ones, until the very small capillaries are reached. There food and oxygen being carried by the blood are exchanged with the surrounding tissues, and wastes are picked up. The blood then flows into the veins to complete the circuit back to the heart.

The arterial blood pressure rises and falls in response to the cardiac cycle. That is, when the ventricles contract, forcing blood into the arterial system, the pressure in the arteries increases sharply. The maximum pressure achieved during the ventricular contraction is called the *systolic pressure*. When the ventricles relax, the arterial pressure drops, and the lowest pressure before the next contraction is called *diastolic pressure*. (These pressures are named after two parts of the pumping cycle, the *systole* and the *diastole*.)

The walls of the arteries have considerable elasticity and expand and contract with each pumping cycle. This alternate expansion and contraction can be felt as a *pulse* in an artery near the body surface. For example, the radial artery near the surface of the wrist is commonly used to measure a person's pulse. The pulse rate is equal to the ventricle contraction rate, and hence the pulse rate indicates the heart rate.

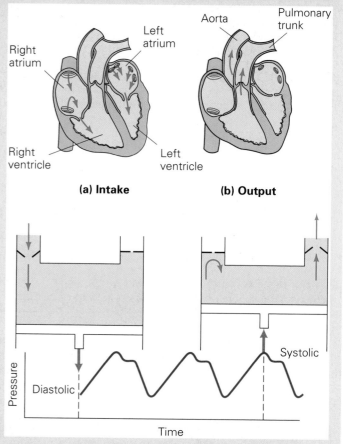

FIGURE 1 The heart as a pump The human heart is analogous to a mechanical force pump. Its pumping action, consisting of **(a)** intake and **(b)** output, gives rise to variations in blood pressure.

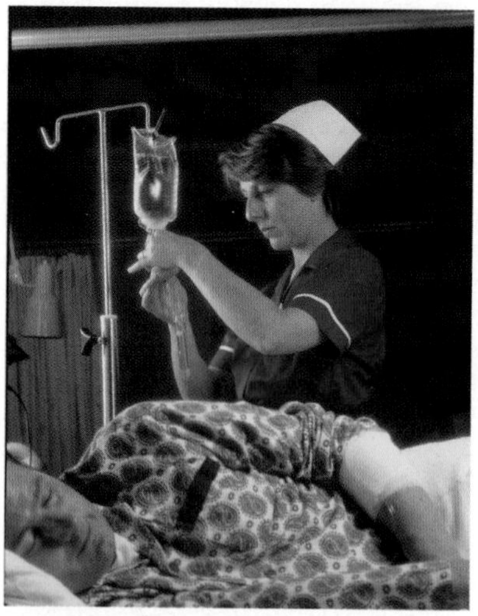

•**FIGURE 9.11 What height is needed?** See Example 9.6.

Example 9.6 ■ An IV: Gravity Assist

An IV ("*intravenous*" injection) is a quite different type of gravity assist from that discussed for space probes in Chapter 7. Consider a hospital patient who receives an IV under gravity flow as seen in •Fig. 9.11. If the blood gauge pressure in the vein is 20.0 mm Hg, above what height should the bottle be placed for the IV to function properly?

Thinking It Through. The fluid gauge pressure at the bottom of the IV tube must be greater than the pressure in the vein and can be computed from Eq. 9.9. (The liquid is assumed to be incompressible.)

Solution.

Given: $p_v = 20.0$ mm Hg (vein gauge pressure)   *Find:* h (height for
$\rho = 1.05 \times 10^3$ kg/m^3 (whole blood $p_v > 20$ mm Hg)
        from Table 9.2)

First, we need to change the common medical unit of mm Hg (or torr) to the SI unit of pascals (Pa, or N/m^2):

$$p_v = (20.0 \text{ mm Hg})[133 \text{ Pa/(mm Hg)}] = 2.66 \times 10^3 \text{ Pa}$$

Then, for the condition of $p > p_v$,

Taking a person's blood pressure involves measuring the pressure of the blood on the arterial walls. This is done with a *sphygmomanometer* (the Greek word *sphygmo* means "pulse"). An inflatable cuff is used to shut off the blood flow temporarily. The cuff pressure is slowly released, and the artery is monitored with a stethoscope (Fig. 2). A point is reached

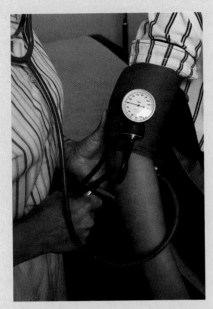

FIGURE 2   Measuring blood pressure   The pressure is indicated on the gauge in millimeters Hg or torr.

where blood is just forced through the constricted artery. This flow is turbulent and gives rise to a specific sound with each heartbeat.

When the sound is first heard, the systolic pressure is noted, which is normally about 120 mm Hg. When the turbulent beats disappear because of smooth blood flow, the diastolic reading is taken. The pressure at this point is normally 80 mm Hg. Blood pressure is commonly reported by giving the systolic and diastolic pressures separated by a slash, for example, 120/80 (read as "120 over 80"). Normal blood pressure ranges between 100 and 140 for the systolic pressure and between 70 and 90 for the diastolic pressure. (Blood pressure is a gauge pressure. Why?)

Away from the heart, blood vessels branch to smaller and smaller diameters. As you might expect, pressure in the blood vessels decreases as their diameter decreases. In small veins, such as those in the arm, the blood pressure is on the order of 10 to 20 mm Hg, and there is no systolic–diastolic variation.

High blood pressure is a common health problem. The elastic walls of the arteries expand under the hydraulic force of the blood pumped from the heart. Their elasticity may diminish with age, however. Fatty deposits (of cholesterols) can narrow and roughen the arterial passageways, impeding the blood flow and giving rise to a form of arteriosclerosis, or hardening of the arteries. Because of these defects, the driving pressure must increase to maintain a normal blood flow. The heart must work harder, which places a greater demand on its muscles. A relatively slight decrease in the effective cross-sectional area of a blood vessel has a rather large effect on the flow rate.

$$p = \rho g h > p_v$$

or

$$h > \frac{p_v}{\rho g} = \frac{(2.66 \times 10^3 \text{ Pa})}{(1.05 \times 10^3 \text{ kg/m}^3)(9.80 \text{ m/s}^2)} = 0.259 \text{ m } (\approx 26 \text{ cm})$$

The IV bottle needs to be at least 26 cm above the injection site.

Follow-up Exercise.   The normal (gauge) blood pressure range is commonly reported as 120/80 (in millimeters Hg). Why is the blood pressure of 20 mm Hg in this Example so low?

9.3   Buoyancy and Archimedes' Principle

Objectives: To (a) relate the buoyant force and Archimedes' principle and (b) tell whether an object will float in a fluid based on relative densities.

When placed in a fluid, an object will either sink or float. This is most commonly observed with liquids—for example, objects floating or sinking in water. But the same effect occurs in gases. A falling object sinks in the atmosphere, and other bodies float (•Fig. 9.12).

Things float because they are buoyant, or are buoyed up. For example, if you immerse a cork in water and release it, the cork will be buoyed up to the surface and float there. From your knowledge of forces, you know that such motion requires an upward net force on the object. That is, there must be an upward force acting on the object that is greater than its downward weight force. The forces are

•**FIGURE 9.12   Fluid buoyancy** The air is a fluid in which objects such as these dirigibles float. The helium inside the blimps is lighter or less dense than the surrounding air. The blimps are supported by the resulting buoyant forces.

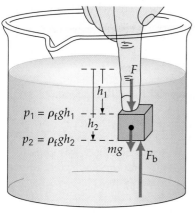

$$\Delta p = \rho_f g(h_2 - h_1)$$

(a)

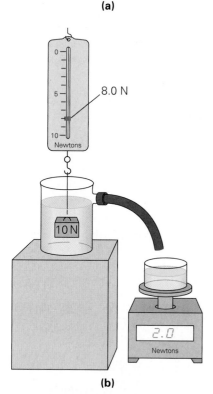

(b)

•**FIGURE 9.13 Buoyancy and Archimedes' principle** (a) A buoyant force arises from the pressure difference between different depths. The pressure on the bottom of the submerged block (p_2) is greater than that on the top (p_1), so there is a (buoyant) force directed upward. (It is shifted for clarity.) (b) Archimedes' principle: The buoyant force on the object is equal to the weight of the volume of fluid it displaces. (The scale is set to read zero when the container is empty.)

equal when the object floats, moves at constant velocity, or remains stationary. The upward force resulting from an object being wholly or partially immersed in a fluid is called the **buoyant force**.

How the buoyant force comes about can be seen by considering a buoyant object being held under the surface of a fluid (•Fig. 9.13a). The pressures on the upper and lower surfaces of the block are $p_1 = \rho_f g h_1$ and $p_2 = \rho_f g h_2$, respectively, where ρ_f is the fluid density. Thus, there is a pressure difference $\Delta p = p_2 - p_1 = \rho_f g (h_2 - h_1)$ between the top and bottom of the block, which gives an upward force (the buoyant force), F_b. This force is balanced by the applied force and the weight of the block.

It is not difficult to derive an expression for the magnitude of the buoyant force. We know that pressure is force per unit area. Thus, if both the top and bottom areas of the block are A, the magnitude of the net buoyant force in terms of the pressure difference is

$$F_b = p_2 A - p_1 A = (\Delta p)A = \rho_f g(h_2 - h_1)A$$

Since $(h_2 - h_1)A$ is the volume of the block, and hence the volume of fluid displaced by the block, V_f, we can write this expression as

$$F_b = \rho_f g V_f$$

But $\rho_f V_f$ is simply the mass of the fluid displaced by the block, m_f. Thus we can write the expression for the buoyant force $F_b = m_f g$: The magnitude of the buoyant force is equal to the weight of the fluid displaced by the block (Fig. 9.13b). This general result is known as **Archimedes' principle**:

> A body immersed wholly or partially in a fluid is buoyed up by a force equal in magnitude to the weight of the *volume of fluid* it displaces:

$$F_b = m_f g = \rho_f g V_f \qquad (9.14)$$

Archimedes (287–212 B.C.) was given the task of determining whether a crown made for a certain king was pure gold or contained some other, cheaper metal. Legend has it that the solution to the problem came to him when he was bathing, perhaps from seeing the water level rise when he got into the tub and experiencing the buoyant force on his limbs. In any case, it is said that he was so excited that he ran through the streets of the city shouting "Eureka!" (Greek for "I have found it"). Although Archimedes' solution to the problem involved density and volume (see Exercises 53 and 100), it presumably got him thinking about buoyancy.

Example 9.7 ■ Lighter Than Air: Buoyant Force

What is the buoyant force in air on a spherical helium balloon with a radius of 30 cm if $\rho_{air} = 1.29$ kg/m³? (Neglect the weight of the balloon material.)

Thinking It Through. Helium is termed "lighter" (less dense) than air, so there is a buoyant force for the balloon in air. This force is equal to the weight of the volume of air that balloon displaces. To find this weight, we first find the balloon's volume and then use the density of air to find the air's mass and weight.

Solution.

Given: $r = 30$ cm $= 0.30$ m   *Find:* F_b (buoyant force)
$\rho_{air} = 1.29$ kg/m³

The volume of the air in the balloon is

$$V = \tfrac{4}{3}\pi r^3 = \tfrac{4}{3}\pi(0.30 \text{ m})^3 = 0.11 \text{ m}^3$$

Then, by Eq. 9.14, the weight of the air displaced by the balloon's volume, or the magnitude of the upward buoyant force, is

$$F_b = m_{air}g = (\rho_{air}V)g = (1.29 \text{ kg/m}^3)(0.11 \text{ m}^3)(9.8 \text{ m/s}^2) = 1.4 \text{ N}$$

Note that the buoyant force depends on the *density of the fluid* and *the volume of the body*. Shape makes no difference.

Follow-up Exercise. Would the buoyant force on the balloon be greater or lesser if the balloon were submerged in water, and by what factor? (Assume the same volume.)

Conceptual Example 9.8

Weight and Buoyant Force: Archimedes' Principle

An overflow container filled with water, like that shown on the left in Fig. 9.13b, sits on a scale that reads 40 N. (The water level is just below the exit tube in the side of the container.) A 5.0-newton object that is less dense than water (such as a block of wood) is placed in the container. The water it displaces runs out the exit tube into another container that is not on the scale. Will the scale reading then be (a) exactly 45 N, (b) between 40 N and 45 N, (c) exactly 40 N, (d) less than 40 N?

Reasoning and Answer. By Archimedes' principle, the block is buoyed upward with a force equal in magnitude to the weight of the water displaced. Since the block floats, the upward buoyant force must balance the weight of the block and has a magnitude of 5.0 N. Thus a volume of water weighing 5.0 N is displaced from the container as a 5.0-newton weight is added to the container. The scale still reads 40 N, so the answer is (c).

Note that the upward buoyant force and the block's weight force act *on the block*. The reaction force (pressure) of the block *on the water* is transmitted to the bottom of the container (Pascal's principle) and is registered on the scale.

Follow-up Exercise. Would the scale reading still be 40 N if the object had a density greater than that of water?

Buoyancy and Density

We commonly say that helium and hot-air balloons float because they are lighter than air. To be correct, technically we should say they are *less dense than air*. An object's density will tell you whether it will sink or float in a fluid as long as you also know the density of the fluid. Consider a solid uniform object totally immersed in a fluid. The weight of the object is

$$w_o = m_o g = \rho_o V_o g$$

The weight of the volume of fluid displaced, or the magnitude of the buoyant force, is

$$F_b = w_f = m_f g = \rho_f V_f g$$

If the object is *completely submerged*, $V_o = V_f$. Dividing the second equation by the first gives

$$\frac{F_b}{w_o} = \frac{\rho_f}{\rho_o} \quad \text{or} \quad F_b = \left(\frac{\rho_f}{\rho_o}\right) w_o \qquad (9.15)$$

Thus, if ρ_o is less than ρ_f, then F_b will be greater than w_o and the object will be buoyed to the surface and float. If ρ_o is greater than ρ_f, then F_b will be less than w_o and the object will sink. If ρ_o equals ρ_f, then F_b will be equal to w_o and the object will remain in equilibrium at any submerged depth (as long as the density of the fluid is constant). If the object is not uniform, such that its density varies over its volume, then the density of the object is its average density.

These three conditions expressed in words are as follows:

An object will float in a fluid if the average density of the object is less than the density of the fluid.

An object will sink in a fluid if the average density of the object is greater than the density of the fluid.

An object will be in equilibrium at any submerged depth in a fluid if the average density of the object and the density of the fluid are equal.

Float or sink? Depends on densities of object and fluid

This demonstration of buoyancy shows that the overall density of a can of Diet Coke is less than that of water, while the density of a can of Classic Coke is greater. Consider the following questions: Does one can have a greater volume of metal? higher gas pressure inside? more fluid volume? Do calories make a difference? Investigate the possibilities to determine the reason(s) for the different densities.

(a) Unopened cans of Coke are dropped into a container of water.

(b) The can of Classic Coke sinks, and the can of Diet Coke floats.

•**FIGURE 9.14   Equal densities and buoyancy**   This soft drink contains colored gelatin beads that remain suspended for months with no change.

See Demonstration 7 for examples of the first two conditions and •Fig. 9.14 for the last one.

The densities of some solids and fluids are given in Table 9.2 (p. 301). A quick look will tell you if an object will float in a fluid, regardless of the shape or volume of the object. The conditions stated above also apply to a fluid in a fluid, provided the two are immiscible (do not mix).

In general, the densities of objects or fluids will be assumed to be uniform and constant in this book. (The density of the atmosphere varies with altitude but is relatively constant near the surface of the Earth.) However, in practical applications, as we have seen above, it is the *average* density of an object that often matters with regard to floating and sinking. An ocean liner is overall less dense than water, even though it is made of steel. Most of its volume is occupied by air, so the liner's average density is less than that of water. Similarly, the human body has air-filled spaces, so we float in water.

In some instances, the overall density of an object is purposefully varied. For example, a submarine submerges by flooding its tanks with sea water (called "taking on ballast"), which increases its average density while reducing the buoyant force (less water is displaced). When the sub is to surface, the water is pumped out of the tanks, so the average density of the sub becomes less than that of the surrounding sea water.

Example 9.9 ■ Float or Sink? Comparison of Densities

A uniform solid cube of material 10 cm on each side has a mass of 700 g. (a) Will the cube float in water? (b) If so, how much of its volume will be submerged?

Thinking It Through.   (a) The question is whether the density of the cube material is greater or less than that of water, so we compute the cube's density. (b) If the cube floats, then the buoyant force and the cube's weight are equal. Both of these forces are related to volume, so we can write them in terms of volume and equate them.

Solution.

Given: $m = 700\,\text{g} = 0.700\,\text{kg}$
$L = 10\,\text{cm}$
$\rho_w = 1.00 \times 10^3\,\text{kg/m}^3 = 1.00\,\text{g/cm}^3$
(Table 9.2)

Find: (a) Whether the cube will float in water
(b) The percentage of the volume submerged if it does float

It is sometimes convenient to work in cgs units in comparing quantities, particularly when working with ratios. For densities in g/cm^3, drop the "$\times 10^3$" from the values given in Table 9.2 for solids and liquids and add "$\times 10^{-3}$" for gases.

(a) The density of the cube material is

$$\rho_c = \frac{m}{V_c} = \frac{m}{L^3} = \frac{700\,\text{g}}{(10\,\text{cm})^3} = 0.70\,\text{g/cm}^3 < \rho_w = 1.00\,\text{g/cm}^3$$

Since ρ_c is less than ρ_w, the cube will float.

(b) The weight of the cube is $w_c = \rho_c g V_c$. When the cube is floating (in equilibrium), its weight is balanced by the buoyant force. That is, $F_b = \rho_w g V_w$, where V_w is the volume of water the submerged part of the cube displaces. Equating the expressions for weight and buoyant force gives

$$\rho_w g V_w = \rho_c g V_c$$

or

$$\frac{V_w}{V_c} = \frac{\rho_c}{\rho_w} = \frac{0.70\,\text{g/cm}^3}{1.00\,\text{g/cm}^3} = 0.70$$

Thus, $V_w = 0.70 V_c$, and 70% of the cube is submerged.

Follow-up Exercise. Most of an iceberg floating in the ocean (•Fig. 9.15) is submerged. What is seen is the proverbial "tip of the iceberg." What percentage of an iceberg's volume is seen above the surface? [Note: Icebergs are frozen *fresh* water floating in cold (salty) water.]

•**FIGURE 9.15 The tip of the iceberg** The vast majority of an iceberg's bulk is underneath the water. This large "berg" was photographed off the Antarctic Peninsula. (Does the submerged ice have to be directly below the exposed tip of the iceberg?)

Conceptual Example 9.10

A Floating Candle: Archimedes' Principle and Buoyant Force

A dripless candle is weighted slightly on the bottom so that it floats upright in a container of water as illustrated in •Fig. 9.16. As the candle burns, (a) the portion of the candle above the water burns down and is extinguished, (b) the candle rises relative to the base of the container, (c) the candle sinks lower relative to the base of the container.

Reasoning and Answer. To answer this, we need to look at the situation in terms of Archimedes' principle and buoyant force. Initially, the buoyant force balances the candle's original weight, so it floats at a certain level. As the candle burns, it gradually loses mass (weight). This loss gives rise to a net upward force, and the candle rises until the buoyant force and weight are again equal (with less candle submerged and a smaller buoyant force). Hence, the answer is (b).

The candle continuously rises and stays lit. To understand the process better, you may find it helpful to think of the candle as being burned in increments or segments and then extrapolate this to continuous burning.

Follow-up Exercise. A person sitting in a small boat floating on a large swimming pool notes the water level on the side of the pool. He then drops several concrete blocks over the side of the boat into the water. Does the level of the pool change? If so, how?

•**FIGURE 9.16 Rise or sink?** See Conceptual Example 9.10.

A quantity called specific gravity is related to density. It is commonly used for liquids but also applies to solids. Basically, it is a comparison of the weight of a volume of a substance with the weight of an equal volume of water. The **specific gravity** of a substance is equal to the ratio of the density of the substance (ρ_s) to the density of water (ρ_w at 4°C, the temperature for maximum density):

Specific gravity: a comparison to water

$$\text{sp. gr.} = \frac{\rho_s}{\rho_w}$$

Because it is a ratio of densities, specific gravity has no units. In cgs units $\rho_w = 1.00 \text{ g/cm}^3$, so

$$\text{sp. gr.} = \frac{\rho_s}{1.00} = \rho_s$$

That is, the specific gravity of a substance is equal to the numerical value of its density *in cgs units*. For example, if a liquid has a density of 1.5 g/cm³, its specific gravity is 1.5, which tells you that it is 1.5 times denser than water. To get density values in grams per cubic centimeter, divide the value in Table 9.2 by 10^3. The specific gravities of automobile coolant and battery electrolyte (both water solutions) are commonly measured so as to determine their relative concentrations of antifreeze and sulfuric acid, respectively.

9.4 Fluid Dynamics and Bernoulli's Equation

Objectives: To (a) identify the simplifications used in describing ideal fluid flow and (b) use the continuity equation and Bernoulli's equation to explain common effects.

In general, fluid motion is difficult to analyze. For example, think of trying to describe the motion of a particle (a molecule) of water in a rushing stream. The overall motion of the stream may be apparent, but a mathematical description of the motion of any one particle of it may be virtually impossible because of eddy currents (small whirlpool motions), the gushing of water over rocks, frictional drag on the stream bottom, and so on. A basic description of fluid flow is conveniently obtained by ignoring such complications and considering an ideal fluid. Actual fluid flow can then be approximated in reference to this simpler theoretical model.

In this simplified approach to fluid dynamics, it is customary to consider four characteristics of an **ideal fluid**. In such a fluid, flow is (1) *steady*, (2) *irrotational*, (3) *nonviscous*, and (4) *incompressible*.

Condition 1: *Steady flow* means that all the particles of a fluid have the same velocity as they pass a given point.

Steady flow might be called smooth or regular flow. The path of steady flow can be depicted in the form of **streamlines** (•Fig. 9.17). Every particle that passes a particular point moves along a streamline. That is, every particle moves along the same path (streamline) as particles that passed by earlier. Streamlines never cross. If they did, a particle would have alternate paths and abrupt changes in its velocity, in which case the flow would not be steady.

Steady flow requires low velocities. For example, steady flow is approximated by the flow relative to a canoe that is gliding slowly through still water. When the flow velocity is high, eddies tend to appear, especially near boundaries, and the flow becomes turbulent.

Streamlines also indicate the relative magnitude of the fluid velocity. The velocity is greater where the streamlines are closer together. Notice this effect in Fig. 9.17a. The reason for this will be explained shortly.

Condition 2: *Irrotational flow* means that a fluid element (a small volume of the fluid) has no net angular velocity, which eliminates the possibility of whirlpools and eddy currents (the flow is nonturbulent).

Consider the small paddle wheel in Fig. 9.17a. With a zero net torque, it does not rotate. Thus, the flow is irrotational.

Condition 3: *Nonviscous flow* means that viscosity is neglected.

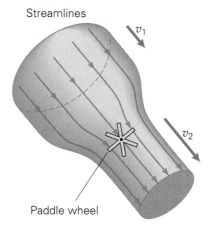

Streamlines

v_1

v_2

Paddle wheel

(a)

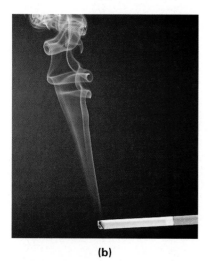

(b)

•**FIGURE 9.17 Streamline flow**
(a) Streamlines never cross and are closer together in regions of greater fluid velocity. The stationary paddle wheel indicates that the flow is irrotational, or without whirlpools and eddy currents. **(b)** The smoke begins to rise in nearly streamline flow but quickly becomes rotational and turbulent.

Viscosity refers to a fluid's internal friction, or resistance to flow. A truly nonviscous fluid would flow freely with no energy lost within it. Also, there would be no frictional drag between the fluid and the walls containing it. In reality, when a liquid flows through a pipe, the speed is lesser near the walls because of frictional drag and is greater toward the center of the pipe. (Viscosity is discussed in more detail in Section 9.5.)

Condition 4: *Incompressible flow* means that the fluid's density is constant.

Liquids can usually be considered incompressible. Gases, conversely, are quite compressible. Sometimes, however, gases approximate incompressible flow—for example, air flowing relative to the wings of an airplane traveling at low speeds.

Theoretical or ideal fluid flow is not characteristic of most real situations. But the analysis of ideal flow provides results that approximate, or generally describe, a variety of applications. This analysis is not usually derived from Newton's laws but instead from two basic principles: conservation of mass and conservation of energy.

Equation of Continuity

If there are no losses of fluid within a uniform tube, the mass of fluid flowing into the tube in a given time must be equal to the mass flowing out of the tube in the same time (by the conservation of mass). For example, in •Fig. 9.18, the mass (Δm_1) entering the tube during a short time (Δt) is

$$\Delta m_1 = \rho_1 \, \Delta V_1 = \rho_1 (A_1 v_1 \, \Delta t)$$

Equation of continuity

where A_1 is the cross-sectional area of the tube at the entrance and, in a time Δt, a fluid particle moves a distance equal to $v_1 \Delta t$. Similarly, the mass leaving the tube in the same time interval is

Flow rate equation

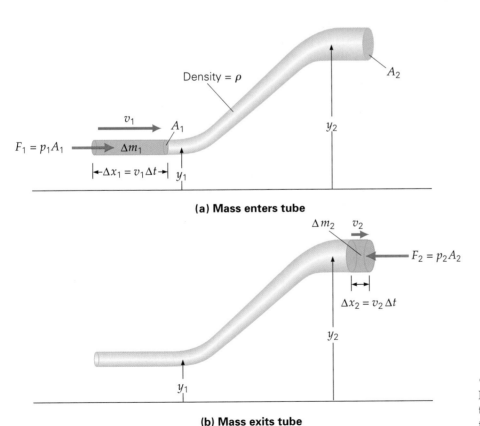

(a) Mass enters tube

(b) Mass exits tube

•**FIGURE 9.18 Flow continuity** Ideal fluid flow can be described in terms of the conservation of mass by the equation of continuity.

$$\Delta m_2 = \rho_2\,\Delta V_2 = \rho_2(A_2 v_2\,\Delta t)$$

Since the mass is conserved, $\Delta m_1 = \Delta m_2$, and

Equation of continuity

$$\rho_1 A_1 v_1 = \rho_2 A_2 v_2 \quad \text{or} \quad \rho A v = \text{constant} \qquad (9.16)$$

This general result is called the **equation of continuity.**

For an incompressible fluid, the density ρ is constant, so

Flow rate equation

$$A_1 v_1 = A_2 v_2 \quad \text{or} \quad Av = \text{constant} \quad \textit{(for an incompressible fluid)} \quad (9.17)$$

This is sometimes called the **flow rate equation**, since Av has the standard units of cubic meters per second (m³/s, or volume/time). (In the British system, gallons per minute is often used.) Note that the flow rate equation shows that the fluid velocity is greater where the cross-sectional area of the tube is smaller. That is,

$$v_2 = \left(\frac{A_1}{A_2}\right) v_1$$

and v_2 is greater than v_1 if A_2 is less than A_1. This effect is evident in the common experience that the speed of water is greater from a hose fitted with a nozzle than from the same hose without a nozzle (•Fig. 9.19).

The flow rate equation can be applied to the flow of blood in your body. Blood flows from the heart into the aorta. It then makes a circuit through the circulatory system, passing through arteries, arterioles (small arteries), capillaries, venules (small veins) and back to the heart through veins. The speed is lowest in the capillaries—a contradiction? No. The *total* area of the capillaries is much larger than that of the arteries or veins, so the flow rate equation holds.

•**FIGURE 9.19 Flow rate** By the flow rate equation, the fluid speed is greater when the cross-sectional area of a tube is smaller. Think of a hose that is equipped with a nozzle such that the cross-sectional area is made smaller.

Example 9.11 ■ Blood Flow: Cholesterol and Plaque

High cholesterol in the blood can cause fatty deposits called plaques to form on the walls of blood vessels. Suppose a plaque reduces the effective radius of an artery by 25%. How does this partial blockage affect the speed of blood through the artery?

Thinking It Through. We will use the flow rate equation (Eq. 9.17), but note that no values of area or speed are given. This indicates that we should use ratios.

Solution. If we take the unclogged artery to have a radius r_1, the plaque then reduces the effective radius to r_2.

Given: $r_2 = 0.75 r_1$ (for a 25% reduction)   *Find:* v_2

Writing the flow rate equation in terms of the radii, we have

$$A_1 v_1 = A_2 v_2$$
$$(\pi r_1^2) v_1 = (\pi r_2^2) v_2$$

Rearranging and canceling, we get

$$v_2 = (r_1/r_2)^2\, v_1$$

From the given, $r_1/r_2 = 1/0.75$, so

$$v_2 = (1/0.75)^2\, v_1 = 1.8 v_1$$

Hence, the speed through the clogged artery increases by 80%.

Follow-up Exercise. By how much would the effective radius of an artery have to be reduced to have a 50% increase in the blood speed through it?

Bernoulli's Equation

The conservation of energy or the general work-energy theorem leads to another relationship that has great generality for fluid flow. This relationship was first derived in 1738 by the Swiss mathematician Daniel Bernoulli (1700–1782) and is named for him.

Let's look again at the ideal fluid flowing in the tube in Fig. 9.18. Work is done by the external forces at the ends of the tube; F_1 does positive work (in the same direction as the fluid's motion) and F_2 does negative work (opposite to the fluid motion). The net work done on the system by these forces is then

$$W = F_1 \Delta x_1 - F_2 \Delta x_2 = (p_1 A_1)(v_1 \Delta t) - (p_2 A_2)(v_2 \Delta t)$$

The flow rate equation (Eq. 9.17) requires that $A_1 v_1 = A_2 v_2$, so we may write the work equation as

$$W = A_1 v_1 \Delta t (p_1 - p_2)$$

Recall from the preceding derivation of the equation of continuity that $\Delta m_1 = \rho_1 \Delta V_1 = \rho_1 (A_1 v_1 \Delta t)$, and $\Delta m_1 = \Delta m_2$, so we may write in general

$$W = \frac{\Delta m}{\rho}(p_1 - p_2)$$

The net work done on the system by the external forces (nonconservative work) must be equal to the change in total mechanical energy. That is, $W = \Delta E = \Delta K + \Delta U$. Looking at the change in kinetic energy of an element of mass Δm, we have

$$\Delta K = \tfrac{1}{2}\Delta m(v_2{}^2 - v_1{}^2)$$

The corresponding change in gravitational potential energy is

$$\Delta U = \Delta m g(y_2 - y_1)$$

Thus,

$$W = \Delta K + \Delta U$$

$$\frac{\Delta m}{\rho}(p_1 - p_2) = \tfrac{1}{2}\Delta m(v_2{}^2 - v_1{}^2) + \Delta m g(y_2 - y_1)$$

Canceling each Δm and rearranging gives the common form of **Bernoulli's equation**:

or

$$p_1 + \tfrac{1}{2}\rho v_1{}^2 + \rho g y_1 = p_2 + \tfrac{1}{2}\rho v_2{}^2 + \rho g y_2$$

$$p + \tfrac{1}{2}\rho v^2 + \rho g y = \text{constant}$$

(9.18)

Note that in working with a fluid, the terms in Bernoulli's equation are work or energy *per unit volume* (J/m^3). That is, $W = F \Delta x = p(A \Delta x) = p \Delta V$, and $p = W/\Delta V$ (work/volume). Similarly, with $\rho = m/V$, we have $\tfrac{1}{2}\rho v^2 = \tfrac{1}{2}mv^2/V$ (energy/volume) and $\rho g y = mgy/V$ (energy/volume).

Note: Compare the derivation of Eq. 5.10 in Section 5.5.

Bernoulli's equation, or principle, can be applied to many situations. If there is horizontal flow ($y_1 = y_2$), then $p + \tfrac{1}{2}\rho v^2 = \text{constant}$, which indicates that the pressure decreases if the fluid speed increases (and vice versa). This effect is illustrated in •Fig. 9.20, where the difference in flow heights through the pipe is considered negligible (so the $\rho g y$ term drops out). (See Demonstration 8.)

Bernoulli effects

Chimneys and smokestacks are tall in order to take advantage of the more consistent and higher wind speeds at greater heights. The faster the wind blows over the top of a chimney, the lower the pressure, and the greater the pressure difference between the bottom and top of the chimney. Thus, the chimney draws

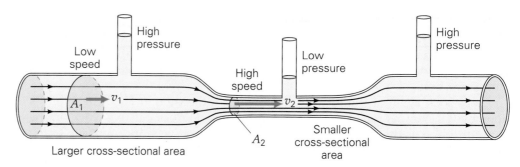

•**FIGURE 9.20 Flow rate and pressure** Taking the horizontal difference in flow heights to be negligible in a constricted pipe, Bernoulli's equation becomes $p + \frac{1}{2}\rho v^2 = $ constant. In a region of smaller cross-sectional area, the flow speed is greater (flow rate equation); from Bernoulli's equation, the pressure in that region is lower than in other regions.

Demonstration 8 ■ Bernoulli-trapped Ball

How does blowing downward through a funnel keep a Ping Pong ball from falling? In this demonstration, seeing is believing in Bernoulli. The air stream above and around the ball (along the wall of the funnel) is moving faster than the air below the ball, and the faster the air moves, the less pressure it exerts. Thus the greater pressure of the slower air below the ball holds it up. Without an air stream, the pressures are equalized (both atmospheric), and the ball falls.

(a) The demonstrator prepares to place the ball in an inverted funnel and to blow downward through the funnel.

(b) With air blowing through, the ball remains "trapped" in the funnel.

(c) When the demonstrator is out of breath, the ball falls.

---

exhaust out better. Bernoulli's equation and the continuity equation ($Av = $ constant) also tell you that if the cross-sectional area of a pipe is reduced, so the velocity of the fluid passing through it is increased, the pressure is reduced.

The Bernoulli effect (as it is sometimes called) is also partially responsible for the lift of an airplane. Ideal air flow over an airfoil or wing is shown in •Fig. 9.21 (turbulence is neglected). The wing is curved on the top side and angled relative to the incident streamlines. As a result, the streamlines above the wing are closer together than those below, which means a higher air speed and lower pressure above the

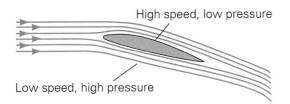

High speed, low pressure

Low speed, high pressure

•FIGURE 9.21 **Airplane lift—Bernoulli in action** Because of the shape and orientation of an airfoil or airplane wing, the air streamlines are closer together, and the air speed is greater, above the wing than below it. By Bernoulli's principle, a pressure difference results that supplies an upward force, or lift.

wing. With a higher pressure on the bottom of the wing, there is a net upward force, or *lift*. Note in the figure that the streamlines leaving the wing curve downward. This curving reflects the fact that since the wing is acquiring upward momentum, the air molecules must acquire an equal downward component of momentum.

Suppose a fluid is at rest ($v_2 = v_1 = 0$). Bernoulli's equation then becomes

$$p_2 - p_1 = \rho g (y_1 - y_2)$$

This is the pressure–depth relationship derived earlier (Eq. 9.10).

Example 9.12 ■ Flow Rate from a Tank: Bernoulli's Equation

Water escapes through a small hole in a tank as depicted in •Fig. 9.22. What is the initial flow rate of the water out of the tank?

Thinking It Through. There is an appreciable difference in flow heights (y) across the tank. Approximations can be made to delete other variables in Bernoulli's equation; for example, the atmospheric pressures on the tank water surface and the hole are nearly equal.

Solution. By Bernoulli's equation,

$$p_1 + \tfrac{1}{2}\rho v_1{}^2 + \rho g y_1 = p_2 + \tfrac{1}{2}\rho v_2{}^2 + \rho g y_2$$

where $y_2 - y_1$ is the height of the liquid surface above the hole. The atmospheric pressures acting on the open surface and at the hole, p_1 and p_2, respectively, are essentially equal and cancel from the equation, as does the density, giving

$$v_1{}^2 - v_2{}^2 = 2g(y_2 - y_1)$$

The equation of continuity says that $A_1 v_1 = A_2 v_2$, where A_2 is the cross-sectional area of the tank and A_1 is that of the hole. Since A_2 is much greater than A_1, then v_1 is much greater than v_2. So a good approximation is

$$v_1{}^2 = 2g(y_2 - y_1) \qquad \text{or} \qquad v_1 = \sqrt{2g(y_2 - y_1)}$$

The flow rate (volume/time) is then

$$\text{flow rate} = A_1 v_1 = A_1 \sqrt{2g(y_2 - y_1)}$$

Given the area of the hole and the height of the liquid above it, you could find the instantaneous speed of the water coming from the hole and the instantaneous flow rate.

Follow-up Exercise. What would be the percentage change in the initial flow rate from the tank in this Example if the diameter of the small circular hole were increased by 30.0%?

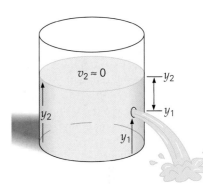

•FIGURE 9.22 **Fluid flow from a tank** The flow rate is given by Bernoulli's equation. See Example 9.12.

Another example of the Bernoulli effect is given in the Insight on p. 320.

Conceptual Example 9.13

A Stream of Water: Smaller and Smaller

You have probably observed that a steady stream of water flowing out of a faucet gets smaller the farther it gets from the faucet. Why does this happen?

Reasoning and Answer. This effect can be explained by Bernoulli's principle. As the water falls, it accelerates and its speed increases. Then, by Bernoulli's principle, the internal liquid pressure inside the stream decreases (see Fig. 9.20). A pressure difference

Throwing a Curveball

In baseball, a pitcher can cause a baseball to curve as it moves toward home plate by giving it an appropriate spin. Why a curveball curves can be understood in terms of Bernoulli's equation and the viscous properties of air.

If air were an ideal fluid, the spin of a pitched ball would have no effect in changing the direction of its motion. (In Fig. 1a, the streamlines are those that would be seen by an observer moving with the ball.) However, because air is viscous, friction between it and the ball causes a thin (boundary) layer of air to be dragged around by the spinning ball. This is illustrated in Fig. 1b for a relatively slow airstream in which the flow is generally smooth. The speed of the ball relative to the air is thus greater on one side than on the other because the velocities of the ball and of the air are in the same direction on one side (top in Fig. 1b) and in opposite directions on the other. By Bernoulli's equation, the low-velocity side (v') has a greater pressure than the high-velocity side (v''). Thus, the ball experiences a net force toward the low-pressure side and is deflected (curves).

However, baseballs are usually pitched very fast, and another effect applies. With a high air flow speed, the rotation of the ball causes the boundary layer to separate at different points on either side of the ball. This separation gives rise to a turbulent wake that is deflected in the spin direction (Fig. 2). A pressure difference results, and the ball is deflected or curves the opposite way. This is called the Magnus effect (after Gustav Magnus, who observed the effect in 1852). In terms of Newton's third law, the turbulent wake is deflected in one direction and the reaction force deflects the ball in the opposite direction.

Note that the ball in Fig. 2 is rotating counterclockwise and is deflected to the left (toward the top, in this overhead view). To have the ball curve to the right requires a clockwise rotation. In general, the curve is in the direction that the front part of the ball is rotating. With the spin axis in the horizontal plane, the deflecting force may be up or down, which can cause the ball

not to drop as much under the influence of gravity, or to "sink" even faster. The spin axis can be oriented in any direction to produce a variety of effects with the deflecting force and gravity.

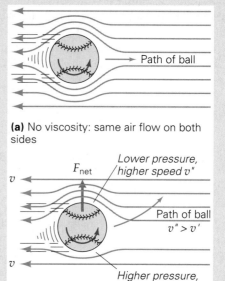

(a) No viscosity: same air flow on both sides

(b) With viscosity: faster air flow on one side of the ball creates a pressure difference

FIGURE 1   Bernoulli effect   (a) If air were an ideal fluid (without viscosity) and spin were applied, air flow would be the same on both sides of the ball. **(b)** Because air has viscosity, some air is dragged around with the ball, so the speed of the ball with respect to the air is greater on one side (top view shown). Pressure on that side is lower, resulting in a net force in that direction.

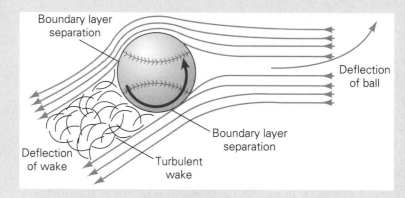

FIGURE 2   Magnus effect
The turbulent wake of a fast-moving ball gives rise to a curving of the ball. In terms of Newton's third law, the wake is deflected in one direction and the ball is deflected in the opposite direction by the reaction force.

between that inside stream and the atmospheric pressure on the outside has been created. As a result, there is an increasing inward force as the stream falls, so it becomes smaller. Eventually, the stream may get so thin that it breaks up into individual droplets.

Follow-up Exercise.   The equation of continuity can also be used to explain this stream effect. Give this explanation.

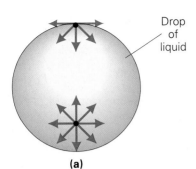

(a)

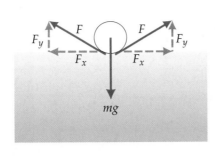

(b)

•**FIGURE 9.23 Surface tension** **(a)** The net force on a molecule in the interior of a liquid is zero, because it is surrounded by other molecules. However, a molecule at the surface experiences a nonzero net force due to the attractive forces of the neighboring molecules just below the surface. **(b)** For an object such as a needle to form a surface depression, work must be done, since more interior molecules must be brought to the surface to increase the area. As a result, the surface area acts like a stretched elastic membrane, and the weight force of the object is supported by the upward components of the surface tension. Insects such as this water strider can walk on water because of the upward components of the surface tension, much as you might walk on a large trampoline. Note the depressions in the surface of the liquid where the legs touch it.

*9.5 Surface Tension; Viscosity and Poiseuille's Law

Objectives: **To (a) describe the source of surface tension and its effects and (b) discuss fluid viscosity.**

Surface Tension

The molecules of a liquid exert small attractive forces on each other. Even though molecules are electrically neutral overall, there is often some slight asymmetry of charge that gives rise to attractive forces between molecules (called *van der Waals forces*). Within a liquid, where any molecule is completely surrounded by other molecules, the net force is zero (•Fig. 9.23a). However, for molecules at the surface of the liquid, there is no attractive force acting from above the surface. (The effect of air molecules is small and considered negligible.) As a result, the molecules of the surface layer experience net forces due to the attraction of neighboring molecules just below the surface. This inward pull on the surface molecules causes the surface of the liquid to contract and to resist being stretched or broken, a property called **surface tension**.

Surface tension: resistance to stretching of surface

If a sewing needle is carefully placed on the surface of a bowl of water, the surface acts like an elastic membrane under tension. There is a slight depression in the surface, and molecular forces along the depression act at an angle to the surface (Fig. 9.23b). The vertical components of these forces balance the weight (mg) of the needle, and it "floats" on the surface. Similarly, surface tension supports the weight of a water strider.

The net effect of surface tension is to make the surface area of a liquid as small as possible. That is, a given volume of liquid tends to assume the shape that has the least surface area. As a result, drops of water and soap bubbles have spherical shapes because a sphere has the smallest surface area for a given volume (•Fig. 9.24). In forming a drop or bubble, surface tension pulls the molecules together to minimize the surface area.

Viscosity

All real fluids have an internal resistance to flow, which is described as **viscosity**. Viscosity can be considered to be friction between the molecules of a fluid. In liquids, it is caused by short-range cohesive forces and, in gases, by collisions

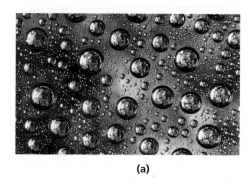

(a)   **(b)**

between molecules (see the discussion of air resistance in Section 4.6). The viscous drag for both liquids and gases depends on their velocity and may be directly proportional to it in some cases. However, the relationship varies with the conditions; for example, the drag is approximately proportional to v^2 or v^3 for turbulent flow.

Internal friction causes fluid layers to move relative to each other in response to a shear stress. This layered motion, called *laminar flow*, is characteristic of steady flow for viscous liquids at low velocities (•Fig. 9.25a). At higher velocities, the flow becomes rotational, or *turbulent*, and difficult to analyze.

Since there are shear stresses and shear strains (deformation) in laminar flow, the viscous property of a fluid can be described by a coefficient, as elastic moduli were used in Section 9.1. Viscosity is characterized by a *coefficient of viscosity*, η (the Greek letter eta), commonly referred to as simply the **viscosity**.

Viscosity: resistance to flow

The coefficient of viscosity is, in effect, the ratio of the shear stress to the rate of change of the shear strain (since motion is involved). When the equation for the ratio is written, unit analysis shows that the SI unit of viscosity is pascal-seconds (Pa·s). This combined unit is called the *poiseuille* (Pl), in honor of the French scientist Jean Poiseuille (1799–1869), who studied the flow of liquids, particularly blood. The cgs unit of viscosity is the *poise* (P). A smaller multiple, the *centipoise* (cP), is widely used because of its convenient size; 1 Pl = 10^3 cP.

The viscosities of some fluids are listed in Table 9.3. The larger the viscosity of a liquid, which is easier to visualize than that of a gas, the greater the shear stress required to get the liquid layers to slide along each other. Note, for example, the large viscosity of glycerin compared to that of water.

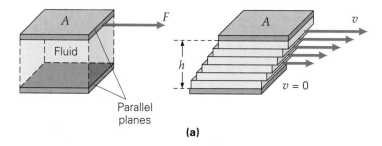

(a)

•FIGURE 9.25 **Laminar flow** (a) A shear stress causes fluid layers to move over each other in laminar flow. The shear force and the flow rate depend on the viscosity of the fluid. (b) For laminar flow through a pipe, the fluid speed is less near the walls of the pipe than near the center because of frictional drag between the walls and the fluid.

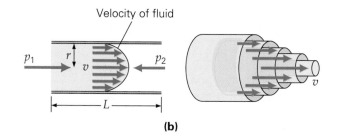

(b)

TABLE 9.3   Viscosities of Various Fluids*

Fluid	Viscosity (η)	
	Poisueille (Pl)	Centipoise (cP)
Liquids		
Alcohol, ethyl	1.2×10^{-3}	1.2
Blood, whole (37°C)	1.7×10^{-3}	1.7
Blood plasma (37°C)	2.5×10^{-3}	2.5
Glycerin	1.5	1.5×10^3
Mercury	1.55×10^{-3}	1.55
Oil, light machine	1.1	1.1×10^3
Water	1.00×10^{-3}	1.00
Gases		
Air	1.9×10^{-5}	1.9×10^{-2}
Oxygen	2.2×10^{-5}	2.2×10^{-2}

*At 20°C unless otherwise indicated. 1 poiseuille (Pl) = 10^3 centipoise (cP).

As you might expect, viscosity, and thus fluid flow, varies with temperature, which is evident from the old saying "slow as molasses in January." A familiar application is the viscosity grading of motor oil used in automobiles. In winter, a low-viscosity, or relatively thin, oil should be used (such as SAE grade 10W or 20W) because it will flow more readily, particularly when the engine is cold at start-up. In summer, a higher-viscosity, or thicker, oil is used (SAE 30, 40, or even 50).

Seasonal changes in the grade of motor oil are not necessary if you use the multigrade year-round oils. These oils contain additives called viscosity improvers, which are polymers whose molecules are long, coiled chains. A temperature increase causes the molecules to uncoil and intertwine with each other. Thus, the normal decrease in viscosity is counteracted. The action is reversed on cooling, and the oil maintains a relatively small viscosity range over a large temperature range. Such motor oils are graded as, for example, SAE 10W-40 (or 10W-40 for short).

Poiseuille's Law   Viscosity makes analyzing fluid flow difficult. For example, when a fluid flows through a pipe, there is frictional drag between the liquid and the walls, and the fluid velocity is greater toward the center of the pipe (Fig. 9.25b).

In practice, this effect makes a difference in the *average flow rate Q* of a fluid, which is given by $Q = A\bar{v} = \Delta V/\Delta t$ (see Eq. 9.17) and describes the volume (ΔV) of fluid flowing past a given point in a time Δt. The SI unit of flow rate is cubic meters per second (m³/s). The flow rate depends on the properties of the fluid and the dimensions of the pipe as well as on the pressure difference (Δp) between the ends of the pipe.

Jean Poiseuille studied flow in pipes and tubes, assuming constant viscosity and steady or laminar flow. He derived the following relationship, known as **Poiseuille's law**, for the flow rate:

$$Q = \frac{\pi r^4 \Delta p}{8\eta L}$$

(9.19)   Poiseuille's law—viscous flow

where r is the radius of the pipe and L is its length.

As should be expected, the flow rate is inversely proportional to the viscosity (η) and the length of the pipe. Also as expected, the flow rate is directly proportional to the pressure difference, Δp, between the ends of the pipe. Somewhat

Note: SAE stands for Society of Automotive Engineers, an organization that designates the grades of motor oils based on flow rate or viscosity.

surprisingly, however, the flow rate is proportional to r^4, which makes it more highly dependent on the radius of the tube than we might have thought.

An application of fluid flow was considered for a medical "IV" in Example 9.6. However, Poiseuille's law, which incorporates flow rate, provides more reality in this application, as the following Example shows.

Example 9.14 ■ Poiseuille's Law: A Blood Transfusion

A hospital patient needs a blood transfusion, which will be administered through a vein in the arm via a gravity IV. The physician wishes to have 500 cc of whole blood delivered over a period of 10 min by using an 18-gauge needle with a length of 50 mm and an inner diameter of 1.0 mm. At what height above the arm should the bag of blood be hung? (Assume a venous blood pressure of 15 mm Hg.)

Thinking It Through. This is an application of Poiseuille's law (Eq. 9.19) to find the pressure needed at the inlet of the needle that will provide the required flow rate (Q). Note that $\Delta p = p_i - p_o$ (inlet pressure minus outlet pressure). Knowing the inlet pressure, we can find the required height of the bag as in Example 9.6. *Caution:* There are a lot of nonstandard units, and some quantities are assumed to be known from tables.

Solution. First we write the given (and known) quantities, converting to standard SI units as we go:

Given: $\Delta V = 500$ cc $= 500$ cm^3 $(1$ m$^3/10^6$ cm$^3) = 5.00 \times 10^{-4}$ m^3 *Find:* h (height of bag)
$\Delta t = 10$ min $= 600$ s $= 6.00 \times 10^2$ s
$L = 50$ mm $= 5.0 \times 10^{-2}$ m
$d = 1.0$ mm, or $r = 0.50$ mm $= 5.0 \times 10^{-4}$ m
$p_o = 15$ mm Hg $= 15$ torr $(133$ Pa/torr$) = 2.0 \times 10^3$ Pa
$\eta = 1.7 \times 10^{-3}$ Pl (whole blood, from Table 9.3)

The flow rate Q is

$$Q = \Delta V/\Delta t = 5.00 \times 10^{-4}\ \text{m}^3/(6.00 \times 10^2\ \text{s}) = 8.33 \times 10^{-7}\ \text{m}^3/\text{s}$$

We apply this value to Eq. 9.19 and solve for Δp:

$$\Delta p = \frac{8\eta L Q}{\pi r^4} = \frac{8(1.7 \times 10^{-3}\ \text{Pl})(5.0 \times 10^{-2}\ \text{m})(8.33 \times 10^{-7}\ \text{m}^3/\text{s})}{\pi(5.0 \times 10^{-4}\ \text{m})^4} = 2.9 \times 10^3\ \text{Pa}$$

With $\Delta p = p_i - p_o$, we have

$$p_i = \Delta p + p_o = (2.9 \times 10^3\ \text{Pa}) + (2.0 \times 10^3\ \text{Pa}) = 4.9 \times 10^3\ \text{Pa}$$

Then, to find the height of the bag that will deliver this pressure, we use $p_i = \rho g h$ (where $\rho_{\text{blood}} = 1.05 \times 10^3$ kg/m^3, from Table 9.2), so

$$h = \frac{p_i}{\rho g} = \frac{4.9 \times 10^3\ \text{Pa}}{(1.05 \times 10^3\ \text{kg/m}^3)(9.80\ \text{m/s}^2)} = 0.48\ \text{m}$$

Hence, for the prescribed flow rate, the "pint" of blood should be hung about 48 cm above the needle in the arm.

Follow-up Exercise. Suppose the physician wants to follow up the blood transfusion with 500 cc of saline solution at the same rate of flow. At what height should the saline bag be placed? (The "isotonic" saline solution administered by IV is a 0.85% aqueous salt solution, which has the same salt concentration as do body cells. To a good approximation, saline has the same density as water.)

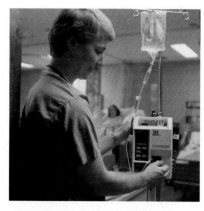

•**FIGURE 9.26 IV technology** The mechanism of intravenous injection is still a gravity assist, but IV flow rates are now commonly controlled and monitored by machines.

Gravity-flow IVs are still used, but with modern technology, the flow rates of IVs are now often controlled and monitored by machines (•Fig. 9.26).

Chapter Review

Important Terms

fluid *294*
stress *295*
strain *295*
elastic modulus *296*
elastic limit *296*
Young's modulus *297*
shear modulus *298*
bulk modulus *298*
compressibility *299*

pressure *300*
pascal (Pa) *301*
atmosphere (atm) *302*
Pascal's principle *303*
absolute pressure *305*
gauge pressure *305*
buoyant force *310*
Archimedes' principle *310*
specific gravity *313*

ideal fluid *314*
streamlines *314*
equation of continuity *316*
flow rate equation *316*
Bernoulli's equation *317*
*surface tension *321*
*viscosity *322*
*Poiseuille's law *323*

Important Concepts

- Elastic modulus is the ratio of stress to strain.
- Young's modulus is the ratio of tensile stress to strain. (The stress may also be compressional.)
- Pressure applied to an enclosed fluid is transmitted undiminished to every point in the fluid and to the walls of the container (Pascal's principle).
- A body immersed wholly or partially in a fluid is buoyed up by a force equal in magnitude to the weight of the volume of fluid it displaces (Archimedes' principle).
- An object will float (sink) in a fluid if the average density of the object is less (greater) than the density of the fluid.
- Bernoulli's equation is basically a mathematical statement of the conservation of energy for a fluid.

Important Equations

Stress:

$$\text{stress} = \frac{F}{A} \tag{9.1}$$

Strain (tensile):

$$\text{strain} = \frac{\Delta L}{L_\text{o}} = \frac{L - L_\text{o}}{L_\text{o}} \tag{9.2}$$

Young's Modulus:

$$Y = \frac{F/A}{\Delta L/L_\text{o}} \tag{9.4}$$

Shear Modulus:

$$S = \frac{F/A}{x/h} \approx \frac{F/A}{\phi} \tag{9.5}$$

Bulk Modulus:

$$B = \frac{F/A}{-\Delta V/V_\text{o}} = -\frac{\Delta p}{\Delta V/V_\text{o}} \tag{9.6}$$

Compressibility:

$$k = \frac{1}{B} \tag{9.7}$$

Pressure:

$$p = \frac{F}{A} \tag{9.8a}$$

Pressure–Depth Equation (for an incompressible fluid at constant density):

$$p = p_\text{o} + \rho g h \tag{9.10}$$

Archimedes' Principle:

$$F_\text{b} = m_\text{f} g = \rho_\text{f} g V_\text{f} \tag{9.14}$$

Equation of Continuity:

$$\rho_1 A_1 v_1 = \rho_2 A_2 v_2 \quad \text{or} \quad \rho A v = \text{constant} \tag{9.16}$$

Flow Rate Equation (for an incompressible fluid):

$$A_1 v_1 = A_2 v_2 \quad \text{or} \quad A v = \text{constant} \tag{9.17}$$

Bernoulli's Equation:

$$p_1 + \tfrac{1}{2}\rho v_1^2 + \rho g y_1 = p_2 + \tfrac{1}{2}\rho v_2^2 + \rho g y_2$$

or

$$p + \tfrac{1}{2}\rho v^2 + \rho g y = \text{constant} \tag{9.18}$$

*Poiseuille's Law:

$$Q = \frac{\pi r^4 \Delta p}{8 \eta L} \tag{9.19}$$

Exercises

9.1 Solids and Elastic Moduli

(Use as many significant figures as you need to show small changes.)

1. The pressure on an elastic body is described by (a) a modulus, (b) work, (c) stress, (d) strain.

2. Shear moduli are not zero for (a) solids, (b) liquids, (c) gases, (d) all of these.

3. One material has a greater Young's modulus than another. What does this tell you?

4. Why are scissors sometimes called shears? Is this a descriptive name in the physical sense?

5. Write the general form of Hooke's law, and find the units of the "spring constant" for elastic deformation.

6. ■ A 50-kilogram mass is suspended from a cable with a diameter of 2.0 cm. What is the stress on the cable?

7. ■ A 5.0-meter-long rod is stretched 0.10 m by a force. What is the strain in the rod?

8. ■ A 250-newton force is applied at a 37° angle to the surface of the end of a square bar. That surface is 4.0 cm on a side. What are (a) the compressional stress and (b) the shear stress on the bar?

9. ■■ A metal wire 1.0 mm in diameter and 2.0 m long hangs vertically with a 6.0-kilogram mass suspended from it. If the wire stretches 1.4 mm under the tension, what is the value of Young's modulus for the metal?

10. ■■ A 5.0-kilogram mass is supported by an aluminum wire of length 2.0 m and diameter 2.0 mm. How much will the wire stretch?

11. ■■ A copper wire has a length of 5.0 m and a diameter of 3.0 mm. Under what compressional load will its length decrease by 0.30 mm?

12. ■■ Each steel rail in a railroad is 8.0 m long and has a cross-sectional area of 0.0025 m². No gap has been left between the rails, and on a hot day, each rail tries to thermally expand 3.0×10^{-3} m. What is the force on the ends of a rail? (Is this a good reason to leave expansion gaps between the rails?)

13. ■■ A rectangular steel column (20.0 cm × 15.0 cm) supports a load of mass 12.0 metric tons. If the column was 2.00 m in length before being stressed, what is the decrease in length?

14. ■■ A bimetallic rod as illustrated in •Fig. 9.27 is composed of brass and copper. If the rod is subjected to a compressive force of 5.00×10^4 N, which way will the rod bend? (Justify your answer mathematically.)

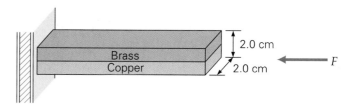

•**FIGURE 9.27 Bimetallic rod and mechanical stress** See Exercise 14.

15. ■■ A 500-newton shear force is applied to one face of a cube of aluminum measuring 10 cm on each side. What is the displacement of that face relative to the opposite face?

16. ■■ Two metal posts of aluminum and copper, respectively, of the same size are subjected to equal shear stresses. Which post will show the larger deformation angle and by what factor?

17. ■■ A rectangular block of gelatin of length, width, and height 10 cm, 8.0 cm, and 4.0 cm, respectively, is subjected to a 0.40-newton shear force on its upper surface. If the top surface is displaced 0.30 mm relative to the bottom surface, what is the shear modulus of the gelatin?

18. ■■ Two metal plates are held together by two steel rivets, each of diameter 0.20 cm and length of 1.0 cm. How much force must be applied parallel to the plates to shear off both rivets?

19. ■■ (a) Which of the liquids in Table 9.1 has the greatest compressibility? (b) For equal volumes of ethyl alcohol and water, which would require more pressure to be compressed by 0.10%, and how many times more?

20. ■■■ A brass cube 6.0 cm on each side is placed in a pressure chamber and subjected to a pressure of 1.2×10^7 N/m² on all its surfaces. By how much will each side be compressed under this pressure?

21. ■■■ A 45-kilogram traffic light is suspended from two steel cables of equal length and radii 0.50 cm. If each cable makes a 15° angle with the horizontal, what is the fractional increase in their length due to the weight of the light?

9.2 Fluids: Pressure and Pascal's Principle

22. •Figure 9.28 shows a famous magician's "trick." She lies on a bed of thousands of nails and a sledge hammer is used to pound on an anvil on her chest. The nails will not pierce her skin. Explain why.

•FIGURE 9.28 **The nail mattress** See Exercise 22.

23. For the pressure–depth relationship for a fluid ($p = p_o + \rho g h$), it is assumed that (a) the pressure decreases with depth, (b) a pressure difference depends on the reference point, (c) the fluid density is constant, (d) this applies only to liquids.

24. Two dams form artificial lakes of equal depth. However, one lake backs up 15 km behind the dam, and the other backs up 50 km behind. What effect does the difference in length have on the pressures on the dams?

25. A water dispenser for pets has an inverted plastic bottle, as shown in •Fig. 9.29. (The water is dyed blue for contrast.) When a certain amount of water is drunk from the bowl, more water flows automatically from the bottle into the bowl. The bowl never overflows. Explain the operation of the dispenser. Does the height of the water in the bottle depend on the surface area of the water in the bowl?

•FIGURE 9.29 **Pet barometer** See Exercise 25.

26. (a) Liquid storage cans, such as gasoline cans, generally have capped vent holes. What is the purpose of the vent, and what happens if you forget to remove the cap before you pour the liquid? (b) Explain how a medicine dropper works. (c) Explain how we breathe (inhalation and exhalation).

27. Automobile tires are inflated to about 30 lb/in^2, whereas thin bicycle tires are inflated to 90 to 115 lb/in^2—at least three times higher pressure! Why?

28. Blood pressure is usually measured at the arm. However, suppose the pressure reading were taken on the calf of the leg of a standing person. Would there be a difference in principle? Explain.

29. ■ In his original barometer, Pascal used water instead of mercury. How high would the water column have been? Is this practical?

30. ■ If you dive to 15 m below the surface of a lake, (a) what is the pressure due to the water alone? (b) What is the total or absolute pressure at that depth?

31. ■ In an open U-tube, the pressure of a 15-centimeter water column on one side is balanced by a column of gasoline on the other side. What is the height of the gasoline column?

32. ■ A 75-kilogram athlete does a single-hand handstand. If the area of the hand in contact with the floor is 125 cm^2, what pressure is exerted on the floor?

33. ■ The gauge pressure in both tires of a bicycle is 690 kPa. If the bicycle and the rider have a combined mass of 90.0 kg, what is the area of contact of *each* tire with the ground if each tire supports half the total weight?

34. ■■ What is the pressure exerted on a 15° ski slope by a 90-kilogram skier going down it? The area of contact of the skis with the snow is 0.40 m^2.

35. ■■ In a lecture demonstration, an empty can of dimensions 0.24 m × 0.16 m × 0.10 m is used to demonstrate the force exerted by air pressure. A small quantity of water is poured into the can, and the water is brought to a boil. Then, the can is capped and dipped in cold water. To your surprise, the can is crushed (•Fig. 9.30). Explain. Assume the inside of the can is in a perfect vacuum. What is the force exerted on the can by air pressure?

•FIGURE 9.30 **Air pressure** See Exercise 35.

36. ■■ What is the fractional decrease in pressure when a barometer is raised 35 m to the top of a building? (Assume that the density of air is constant over that distance.)

37. ■■ A student decides to compute the standard barometric reading on top of Mt. Everest (29 028 ft) by assuming the density of air has the same constant density as at sea level. Try this yourself. What does the result tell you?

38. •Figure 9.31 shows a method Pascal used to demonstrate the importance of fluid pressure on fluid depth. An oak barrel with lid of area 0.20 m² is filled with water. A long, thin tube of cross-sectional area 5.0×10^{-5} m² is inserted into a hole at the center of the lid, and water is poured into the tube. When the water reaches 12 m high, the barrel burst. What was (a) the weight of the water in the tube? (b) the pressure of the water on the lid of the barrel? (c) the upward net force on the lid due to the water pressure?

•**FIGURE 9.31   Pascal and the bursting barrel**   See Exercise 38.

39. ■■ In a sample of seawater taken from an oil spill region, it is found that an oil layer 4.0 cm thick floats on 55 cm of water. If the density of the oil is 0.75×10^3 kg/m³, what is the absolute pressure on the bottom of the container?

40. ■■ The pressure exerted by a person's lungs can be measured by having the person blow as hard as possible into one side of a manometer. If a person blowing into one side of an open-tube manometer produces an 80-centimeter difference in the heights of the columns of water in the manometer arms, what is the lung gauge pressure?

41. ■■ In 1960, the U.S. Navy's bathyscaphe *Trieste* (a submersible) descended to a depth of 10 912 m (about 35 000 ft) into the Marianas Trench in the Pacific Ocean. (a) What was the pressure at that depth? (Assume that sea water is incompressible.) (b) What was the force on a circular observation window with a diameter of 15 cm?

42. ■■ The output piston of a hydraulic press has a cross-sectional area of 0.20 m². (a) How much pressure on the input piston is required for the press to generate a force of 1.5×10^6 N? (b) What force is applied to the input piston if that piston has a diameter of 5.0 cm?

43. ■■ A garage hydraulic lift has two pistons, one of cross-sectional area 4.00 cm² and one of cross-sectional area 250 cm². If this lift is designed to lift a 3000-kilogram car, what minimum force must be applied to the small piston?

44. ■■ The force required in Exercise 43 is usually provided by compressed air. If the small piston is circular, what must be the minimum air pressure applied to the small piston?

45. ■■■ A hydraulic balance used to detect small change in mass is shown in •Fig. 9.32. If a mass *m* of 0.25 g is placed on the balance platform, by how much will the height of the water in the smaller cylinder of diameter 1.0 cm have changed when the balance comes to equilibrium?

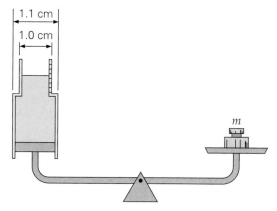

•**FIGURE 9.32   A hydraulic balance**   See Exercise 45.

9.3 Buoyancy and Archimedes' Principle

46. A wood block floats in a swimming pool. The buoyant force on the block by water depends on (a) the volume of water in the pool, (b) the volume of the wood block, (c) the volume of the wood block under water, (d) all of the above.

47. If a submerged object displaces an amount of liquid of greater weight than its own and is then released, the object will (a) rise to the surface and float, (b) sink, (c) remain in equilibrium at its submerged position.

48. ■ (a) What is the most important factor in constructing a life jacket that will keep a person afloat? (b) Why is it so easy to float in the Great Salt Lake in Utah?

49. An ice cube floats in a glass of water. As the ice melts, how does the level of the water in the glass change? Would it make any difference if the ice cube were hollow? Explain.

50. Ocean-going ships in port are loaded to the *Plimsoll mark*, a line indicating the maximum safe loading depth. However, in New Orleans, which is located at the mouth of the Mississippi River and is noted for its brackish water (partly salt, partly fresh), ships are loaded until the Plimsoll mark is somewhat below the water line. Why?

51. ■ A submarine has a mass of 10 000 metric tons. What mass of water must be displaced for the sub to be in equilibrium just below the ocean surface?

52. ■ A cube 8.5 cm on each side has a mass of 0.65 kg. Will the cube float in water?

53. ■ Suppose that Archimedes found that the king's crown had a mass of 0.750 kg and a volume of 3.980×10^{-5} m³. (a) How did Archimedes determine the crown's volume? (b) Was the crown pure gold?

54. ■ A rectangular boat as illustrated in •Fig. 9.33 is overloaded such that the water level is just 1.0 cm below the top of the boat. What is the combined mass of the people and the boat?

•FIGURE 9.33 **An overloaded boat** (Not drawn to scale.) See Exercise 54.

55. ■ An aluminum cube 0.15 m on each side is completely submerged in water. What is the buoyant force on it? How does the answer change if the cube is made of steel?

56. ■■ An object has a weight of 8.0 N when in air. However, it apparently weighs only 4.0 N when it is completely submerged in water. What is the density of the object?

57. ■■ When a crown of mass 0.80 kg is submerged in water, the crown's apparent weight is measured to be 7.3 N. Is the crown pure gold?

58. ■■ An irregularly shaped piece of metal has a weight of 0.882 N. It is suspended from a scale, which reads 0.735 N when the piece is submerged in water. What are the volume and density of the piece of metal?

59. ■■ A flat-bottomed rectangular boat is 4.0 m long and 1.5 m wide. If the load is 2000 kg (including the mass of the boat), how much of the boat will be submerged when it floats in a lake?

60. ■■ A block of iron quickly sinks in water, but ships constructed of iron float. A solid cube of iron 1.0 m on each side is made into sheets. From these sheets, to make a hollow cube that will not sink, what should the minimum length of the sides be?

61. ■■ Plans are being made to bring back the zeppelin—a lighter-than-air airship, like the Goodyear blimp, that carries passengers and cargo but is filled with helium, not flammable hydrogen (as was used in the ill-fated *Hindenburg*). One design calls for the ship to be 110 m long and to have a mass comprised of 30.0 metric tons of structure plus the mass of the helium needed to fill it. Assuming the ship's "envelope" to be cylindrical, what would its diameter have to be so as to lift the total weight?

62. ■■■ A girl floats in a lake with 97% of her body beneath the water. What are (a) her mass density and (b) her weight density?

63. ■■■ A bar of 24-karat (pure) gold is supposed to be completely solid. To check this, a scientist takes the bar into the lab. Using a scale, she finds that it has a weight of 9.8 N and an apparent weight of only 9.1 N when completely immersed in water. Is the bar solid?

9.4 Fluid Dynamics and Bernoulli's Equation

64. The speed of blood flow is greater in arteries than in capillaries. However, the flow rate equation (Av = constant) seems to predict that the speed should be greater in the smaller capillaries. Can you resolve this apparent inconsistency?

65. If the speed at some point in a fluid changes with time, the fluid flow is *not* (a) steady, (b) irrotational, (c) incompressible, (d) nonviscous.

66. According to Bernoulli's equation, if the pressure on a liquid is increased, (a) the flow speed always increases, (b) the height of the liquid always increases, (c) both the flow speed and the height of the liquid may increase, (d) none of these.

67. When you suddenly turn on a water faucet in a shower, the shower curtain moves inward. Why?

68. If an Indy car had a flat bottom, it would be very unstable due to the lift it gets when it moves fast (like an airplane wing). To increase friction and stability, the bottom has a concave section called the *Venturi tunnel* (•Fig. 9.34). In

•FIGURE 9.34 **Venturi tunnel** See Exercise 68.

terms of Bernoulli's equation, explain how this concavity supplies extra down force to the car in addition to the front and rear wings (Exercise 4.71).

69. Two common demonstrations of Bernoulli effects: (a) If you hold a narrow strip of paper in front of your mouth and blow over the top surface, the strip will rise (•Fig. 9.35a). (Try it.) Why? (b) A plastic egg is supported vertically by a stream of air from a tube (Fig. 9.35b). The egg will not move away from the midsteam position. Why not?

(a)

(b)

•**FIGURE 9.35   Bernoulli effects**   See Exercise 69.

70. ■ An ideal fluid is moving at 3.0 m/s in a section of a pipe of radius 0.20 m. If the radius in another section is 0.35 m, what is the flow speed there?

71. ■ If the radius of a pipe narrows to half of its original size, what is the ratio of the flow speed in the narrow section to that in the wide section, assuming the same flow rate?

72. ■■ Show that the static pressure–depth equation can be derived from Bernoulli's equation.

73. ■■ The speed of blood in a major artery of diameter 1.0 cm is 4.5 cm/s. (a) What is the flow rate in the artery? (b) If the capillary system has a total cross-sectional area of 2500 cm², the average speed of blood through the capillaries is what percentage of that through the major artery? (c) Why must blood flow at low speed through the capillaries?

74. ■■ A room measures 3.0 m by 4.5 m by 6.0 m. If the heating and air conditioning ducts to and from the room are circular, with diameter 0.30 m, and all the air in the room is to be exchanged every 12 min, (a) what is the average flow rate? (b) What is the necessary flow speed in the duct? (Assume that the density of the air is constant.)

75. ■■ The spout heights for the container in •Fig. 9.36 are 10 cm, 20 cm, 30 cm, and 40 cm. The water level is maintained at a 45 cm height by an outside supply. (a) What is the speed of the water out of each hole? (b) Which water stream has the greatest range relative to the base of the container? Justify your answer.

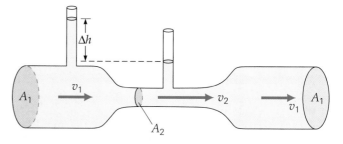

•**FIGURE 9.36   Streams as projectiles**   See Exercise 75.

76. ■■■ In an industrial cooling process, water is circulated through a system. If water is pumped with a speed of 0.45 m/s under a pressure of 400 torr from the first floor through a 6.0-centimeter-diameter pipe , what will be the pressure on the next floor 4.0 m above in a pipe with a diameter of 2.0 cm?

77. ■■■ Water flows at a rate of 25 L/min through a horizontal 7.0-centimeter-diameter pipe under a pressure of 6.0 Pa. At one point, calcium deposits reduce the cross-sectional area of the pipe to 30 cm². What is the pressure at this point? (Consider the water to be an ideal fluid.)

78. ■■■ A Venturi meter can be used to measure the flow speed of a liquid. A simple one is shown in •Fig. 9.37.

•**FIGURE 9.37   A flow speed meter**   See Exercise 78.

Show that the flow speed of an ideal fluid is given by

$$v_1 = \sqrt{\frac{2g\,\Delta h}{(A_1{}^2/A_2{}^2) - 1}}$$

*9.5 Surface Tension; Viscosity and Poiseuille's Law

79. Water droplets and soap bubbles tend to assume the shape of a sphere. This effect is due to (a) viscosity, (b) surface tension, (c) laminar flow, (d) none of the above.

80. Some insects can walk on water because (a) the density of water is greater than that of the insect, (b) water is viscous, (c) water has surface tension, (d) none of the above.

81. The viscosity of a fluid is due to (a) forces causing friction between the molecules, (b) surface tension, (c) density, (d) none of the above.

82. A motor oil is labeled 10W-40. The numbers 10 and 40 specify the (a) density, (b) surface tension, (c) viscosity, (d) weight of the oil.

83. ■■ A horizontal pipe with an inside diameter of 3.0 cm and a length of 6.0 m carries water, which has a flow rate of 40 L/min. What is the required pressure difference between the ends of the pipe?

84. ■■ A hospital patient receives a transfusion of 500 cc of blood through a needle with a length of 5.0 cm and an inner diameter of 1.0 mm. If the blood bag is suspended 0.85 m above the needle, how long does the transfusion take? (Neglect the viscosity of the blood flowing in the plastic tube between the bag and the needle.)

Additional Exercises

85. A copper wire 100.00 cm long is stretched to 100.02 cm when it supports a certain load. If an aluminum wire of the same diameter is used to support the same load, what should its initial length be if its stretched length is to be 100.02 cm also?

86. A steel cube 0.25 m on each side is suspended from a scale and immersed in water. What will the scale read?

87. A wood cube 0.30 m on each side has a density of 700 kg/m³ and floats level in water. (a) What is the distance from the top of the wood to the water surface? (b) What mass has to be placed on top of the wood so that its top is just at the water level?

88. A 60-kilogram woman balances herself on the heel of one of her high-heeled shoes. If the heel is a square of sides 1.5 cm, what is the pressure exerted on the floor? (Express your answer in pascals, pounds per square inch, and atmospheres.)

89. What total inward force does the atmosphere exert on the body of a person with a surface area of 1.30 m²? (Express your answer in both newtons and pounds.)

90. A scuba diver dives to a depth of 12 m in a lake. If the circular glass plate on the diver's face mask has a diameter of 18 cm, what is the force on it due to the water only?

91. Oil is poured into the open side of an open-tube manometer containing mercury. What is the density of the oil if a column of mercury 5.0 cm high supports a column of oil 80 cm high?

92. If the atmosphere had a constant density equal to its density at the Earth's surface, how high would it extend? What does this tell you?

93. A cylinder has a diameter of 15 cm (•Fig. 9.38). The water level in the cylinder is maintained at a constant height of 0.45 m. If the diameter of the spout pipe is 0.50 cm, how high is h, the vertical stream of water? (Assume the water to be an ideal fluid.)

•FIGURE 9.38 **How high a fountain?** See Exercise 93.

94. A vertical steel beam with a rectangular cross-sectional area of 24 cm² is used to support a sagging floor in a building. If the beam supports a load of 10 000 N, by what percentage is the beam compressed?

95. A container 100 cm deep is filled with water. If a small hole is punched in its side 25 cm from the top, at what initial speed will the water flow from the hole?

96. Show that specific gravity is equivalent to the ratio of densities, given that its strict definition is the ratio of the weight of a given volume of a substance to the weight of an equal volume of water.

97. The flow rate of water through a garden hose is 66 cm³/s, and the hose and nozzle have cross-sectional areas of 6.0 cm² and 1.0 cm², respectively. (a) If the nozzle is held 10 cm above the spigot, what are the flow speeds through the spigot and the nozzle? (b) What is the pressure

difference between these points? (Consider the water to be an ideal fluid.)

98. Ancient stonemasons sometimes split huge blocks of rock by inserting wooden pegs into holes drilled in the rock and then pouring water on the pegs. Can you explain the physics that underlies this technique? [Hint: Think about sponges and paper towels.]

99. A crane lifts from the bottom of a lake a rectangular iron bar whose dimensions are 0.25 m × 0.20 m × 10 m. What is the minimum upward force the crane must supply when the bar is (a) in the water and (b) out of the water?

100. •Figure 9.39 shows a simple laboratory experiment. Calculate the (a) volume and (b) density of the suspended sphere. (Assume that the density of the sphere is uniform and that the liquid in the beaker is water.) (c) Would you be able to make the same determinations if the liquid in the beaker were mercury (see Table 9.2)? Explain.

•**FIGURE 9.39   Dunking a sphere**   See Exercise 100.

Temperature

Like sailboats, hot-air balloons are low-tech devices in a high-tech world. You can equip a balloon with the latest satellite-linked, computerized navigational system and attempt to fly across the Pacific, but the basic principles that keep you aloft were known and understood centuries ago.

But why must the air in a hot-air balloon be hot? More fundamentally, what do we mean by "hot"? How does hot air differ from cool air? Temperature and heat are frequent subjects of conversation, but if you had to state what the words really mean, you might find yourself at a loss. We use thermometers of all sorts to record temperatures, which provide an objective equivalent for our sensory experience of hot and cold. And we know that a temperature change generally results from the application or removal of heat. Temperature, therefore, is related to heat. But what is heat? In this chapter you'll find that the answers to such simple questions lead to an understanding of some far-reaching physical principles.

An early theory of heat considered it to be a fluidlike substance called caloric (from the Latin word *calor,* meaning "heat") that could be made to flow into and out of a body. Even though this theory has been abandoned, we still speak of heat as flowing from one object to another. Heat is now characterized as energy, and temperature and thermal properties are explained by considering the atomic and molecular behavior of substances. This and the next two chapters examine the nature of temperature and heat in terms of microscopic (molecular) theory and macroscopic observations. Here, you'll explore the nature of heat and the ways in which we measure temperature. You'll also encounter the gas laws, which explain not only the behavior of balloons, but also more important phenomena, such as how our lungs supply us with the oxygen we need to live.

10.1 Temperature and Heat

Objective: To distinguish between temperature and heat.

A good way to begin studying thermal physics is with definitions of temperature and heat. **Temperature** is a relative measure, or indication, of hotness or coldness. A hot stove is said to have a high temperature, and an ice cube to have a low temperature. An object that has a higher temperature than another object does is said to be hotter. Note that hot and cold are relative terms, like tall and short. We can perceive temperature by touch. However, this temperature sense is somewhat unreliable, and its range is too limited to be useful for scientific purposes.

Heat is related to temperature and describes the process of energy transfer from one object to another. That is, **heat** is *the net energy transferred from one object to another because of a temperature difference.* Generally, heat transfer is the result of a temperature difference (though not always, as we shall find in Section 11.4). Thus, heat is energy in transit, so to speak. Once transferred, the energy becomes part of the total energy of the molecules of the object or system, its **internal energy**.

On a microscopic level, temperature is associated with molecular motion. We will show that in kinetic theory (Section 10.5), which treats gas molecules as point particles, temperature is a measure of the average random *translational* kinetic energy of the molecules. However, diatomic molecules and other real substances, besides having such translational "temperature" energy, also may have kinetic energy due to linear vibration and rotation as well as potential energy due to the attractive forces between molecules. These energies would not contribute to the gas's temperature measurement but are definitely part of the internal energy, which is the sum total of all such energies (•Fig. 10.1).

Note that a higher temperature does not necessarily mean that one system has a greater internal energy than another. For example, in a classroom on a cold day, the air temperature is relatively high compared to that of the outdoor air. But all

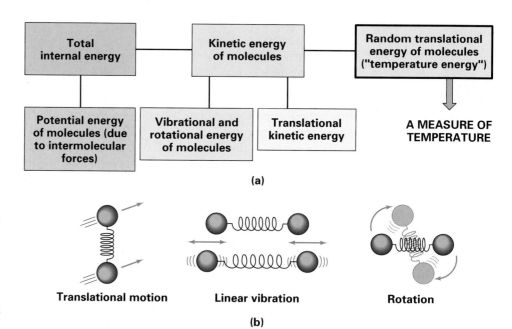

•**FIGURE 10.1 Molecular motions** **(a)** Temperature is associated with random translational motion; the internal energy of a system is the total energy. **(b)** A molecule may move as a whole in translational motion and/or have vibrational and rotational motions.

that cold air outside the classroom has far more internal energy as a whole than does the warm air inside, simply because there is so much *more* of it. If this were not the case, heat pumps would not be practical (Chapter 12). The internal energy of a system also depends on its mass, or the number of molecules in the system.

When heat is transferred between two objects, whether or not they are touching, they are said to be in *thermal contact*. When there is no longer a net heat transfer between objects in thermal contact, they have come to the same temperature and are said to be in *thermal equilibrium*.

(a)

10.2 The Celsius and Fahrenheit Temperature Scales

Objectives: To (a) explain how a temperature scale is constructed and (b) convert temperatures from one scale to another.

A measure of temperature is obtained by using a **thermometer**, a device constructed to make evident some property of a substance that changes with temperature. Fortunately, many physical properties of materials change sufficiently with temperature to be used as the bases for thermometers. By far the most obvious and commonly used property is **thermal expansion**, a change in the dimensions or volume of a substance that occurs when the temperature changes.

Almost all substances expand with increasing temperature, but they do so to different extents. Most substances also contract with decreasing temperature. (Thermal expansion refers to both expansion and contraction; contraction is considered to be a negative expansion.) Because some metals expand more than others, a bimetallic strip (strips of two different metals bonded together) can be used to measure temperature changes. As heat is applied, the composite strip will bend away from the side made of the metal that expands more (•Fig. 10.2). Coils formed from such strips are used in dial thermometers and in common household thermostats (•Fig. 10.3).

A common thermometer is the liquid-in-glass type, which is based on the thermal expansion of a liquid. A liquid in a glass bulb expands into a glass stem, rising in a capillary bore (a thin tube). Mercury and alcohol (usually dyed red to make it more visible) are the liquids used in most liquid-in-glass thermometers. These substances were chosen because of their relatively large thermal expansion and because they remain liquids over normal temperature ranges.

Thermometers are calibrated so that a numerical value can be assigned to a given temperature. For the definition of any standard scale or unit, two fixed reference points are needed. Since all substances change dimensions with temperature, an absolute reference for expansion is unavailable. However, the necessary fixed points can be correlated to physical phenomena that always occur at the same temperatures.

(b)

•**FIGURE 10.3 Bimetallic coil**
Bimetallic coils are used in **(a)** dial thermometers and **(b)** household thermostats. In a thermostat, the expansion and contraction of the coil are used to regulate the heating or cooling system, turning it off and on as the room temperature changes.

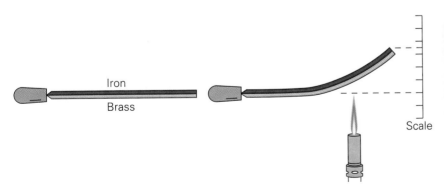

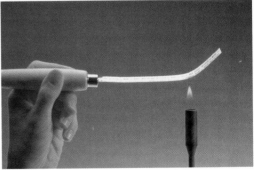

(a) Initial condition   (b) Heated condition

•**FIGURE 10.2 Thermal expansion**   (a) A bimetallic strip is made of two strips of different metals bonded together. (b) When such a strip is heated, it bends because of unequal expansions of the two metals. Here brass expands more than iron, so the deflection is toward the iron. The deflection of the end of a strip could be used to measure temperature.

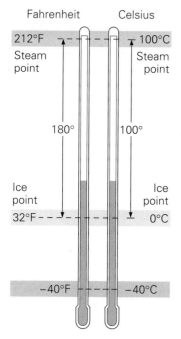

•FIGURE 10.4 **Celsius and Fahrenheit temperature scales** Between the ice and steam fixed points, there are 100 degrees on the Celsius scale and 180 degrees on the Fahrenheit scale. Thus, a Celsius degree is 1.8 times larger than a Fahrenheit degree.

Note: For distinction, a particular temperature measurement, such as $T = 20°C$, is written with °C (pronounced 20 degrees Celsius), whereas a temperature interval, such as $\Delta T = 80°C - 60°C = 20$ C°, is written with C° (pronounced 20 Celsius degrees).

The ice point and the steam point of water are two convenient fixed points. Also commonly known as the freezing and boiling points, these two points are the temperatures at which pure water freezes and boils under a pressure of 1 atm (standard pressure).

The two familiar temperature scales are the **Fahrenheit temperature scale** and the **Celsius temperature scale**. As shown in •Fig. 10.4, the ice and steam points have values of 32°F and 212°F, respectively, on the Fahrenheit scale and 0°C and 100°C on the Celsius scale. On the Fahrenheit scale, there are 180 equal intervals, or degrees, (F°), between the two reference points; on the Celsius scale, there are 100 (C°). Therefore, a Celsius degree is larger than a Fahrenheit degree. Since $180/100 = 1.8$, it is almost twice as large.

A relationship for converting between the two scales can be obtained from a graph of Fahrenheit temperature (T_F) versus Celsius temperature (T_C), such as the one in •Fig. 10.5. The equation of the straight line (in slope–intercept form, $y = mx + b$) is $T_F = (180/100)T_C + 32$, and

$$T_F = \tfrac{9}{5}T_C + 32 \quad \text{or} \quad T_F = 1.8T_C + 32 \qquad \begin{array}{l}\textit{Celsius-to-Fahrenheit} \\ \textit{conversion}\end{array} \quad (10.1)$$

where $\tfrac{9}{5}$ or 1.8 is the slope of the line and 32 is the intercept on the vertical axis. Thus, to change from a Celsius temperature (T_C) to its equivalent Fahrenheit temperature (T_F), you simply multiply the Celsius reading by $\tfrac{9}{5}$ and add 32.

The equation can be solved for T_C to convert from Fahrenheit to Celsius:

$$T_C = \tfrac{5}{9}(T_F - 32) \qquad \begin{array}{l}\textit{Fahrenheit-to-Celsius} \\ \textit{conversion}\end{array} \quad (10.2)$$

Example 10.1 ■ Converting Temperature Scale Readings: Fahrenheit and Celsius

What are (a) 20°C on the Fahrenheit scale and (b) normal body temperature, 98.6°F, on the Celsius scale?

Thinking It Through. This is a direct application of Eqs. 10.1 and 10.2.

Solution. We wish to make the following conversions:

Given: (a) $T_C = 20°C$ *Find:* (a) T_F
  (b) $T_F = 98.6°F$ (b) T_C

•FIGURE 10.5 **Fahrenheit versus Celsius** (a) A plot of Fahrenheit temperature versus Celsius temperature gives a straight line of the general form $y = mx + b$, where $T_F = \tfrac{9}{5}T_C + 32$. (b) The temperature is given here in degrees Celsius. What is the equivalent temperature in degrees Fahrenheit?

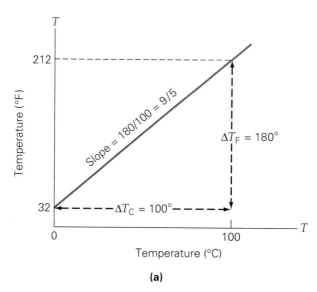

(a)

(b)

(a) Equation 10.1 is for changing Celsius readings to Fahrenheit:

$$T_F = \tfrac{9}{5} T_C + 32 = \tfrac{9}{5}(20) + 32 = 68°F$$

This temperature, often taken to be room temperature, is a good one to remember.

(b) Equation 10.2 changes Fahrenheit to Celsius:

$$T_C = \tfrac{5}{9}(T_F - 32) = \tfrac{5}{9}(98.6 - 32) = \tfrac{5}{9}(66.6)$$

$$= 37.0°C$$

On the Celsius scale, normal body temperature has a whole-number value. Keep in mind that a Celsius degree is 1.8 times (almost twice) as large as a Fahrenheit degree, so a temperature elevation of several degrees on the Celsius scale makes a big difference. For example, a temperature of 40.0°C represents an elevation of 3.0 C° over normal body temperature. This is an increase of $3.0 \times 1.8 = 5.4°$ on the Fahrenheit scale, or a temperature of $98.6 + 5.4 = 104.0°F$. For more on "normal" body temperature, see the Insight on this page.

Follow-up Exercise.   Convert the following temperatures: (a) −40°F to Celsius and (b) −40°C to Fahrenheit.

Insight   Insight   Insight   Insight   Insight   Insight   Insight

Human Body Temperature

We commonly take "normal" human body temperature to be 98.6°F (or 37.0°C). The source of this value is a study of human temperature readings done in 1868—more than 130 years ago! A more recent study, done in 1992, notes that the 1868 study used thermometers that were not as accurate as modern electronic thermometers. The new study has some interesting results.

The "normal" human body temperature from oral measurements varies with individuals over a range of about 96°F to 101°F, with an average temperature of 98.2°F. After strenuous exercise, the oral temperature can rise as high as 103°F. When the body is exposed to cold, oral temperatures can fall below 96°F. A rapid drop in temperature of 2 to 3 F° produces uncontrollable shivering. There is not only a contraction of the skeletal muscles, but also of the tiny muscles attached to the hair follicles. The result is "goose bumps."

Your body temperature is typically lower in the morning, after you have slept and your digestive processes are at a low point. "Normal" body temperature generally rises during the day to a peak and then recedes. The study also indicated that women have a slightly higher average temperature than men do (98.4°F versus 98.1°F).

What about the extremes? A fever temperature is typically between 102°F and 104°F. A body temperature above 106°F is extremely dangerous. The enzymes for certain chemical reactions in the body begin to be inactivated, and a total breakdown of body chemistry can result.

On the cold side, decreasing body temperature results in memory lapse and slurred speech, muscular rigidity, erratic heartbeats, loss of consciousness, and, below 78°F, death due to heart failure. However, mild hypothermia (lower-than-normal body temperature) can be beneficial. A decrease in body temperature slows down the body's chemical reactions, and cells use less oxygen than they normally do. This fact is applied in some surgeries (Fig. 1). A patient's body temperature may be lowered significantly to avoid damage to the heart, which must be stopped during such procedures, and to the brain.

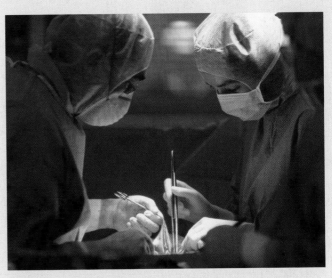

FIGURE 1   Lower than normal   During some surgeries, the patient's body temperature is lowered to slow down the body's chemical reactions and to reduce the need for blood to supply oxygen to the tissues.

Because Eqs. 10.1 and 10.2 are so similar, it is easy to miswrite them. Since they are equivalent, you need to know only one of the equations—say, Celsius to Fahrenheit (Eq. 10.1, $T_F = \frac{9}{5}T_C + 32$). Solving it for T_C algebraically gives Eq. 10.2. A good way to make sure you have written the conversion equation correctly is to test it with a known temperature, such as the boiling point of water. For example, $T_C = 100°$, so

$$T_F = \frac{9}{5}T_C + 32 = \frac{9}{5}(100) + 32 = 212°F$$

Thus we know the equation is correct.

Liquid-in-glass thermometers are adequate for many temperature measurements, but problems arise when very accurate determinations are needed. A material may not expand uniformly over a wide temperature range. When calibrated to the ice and steam points, an alcohol thermometer and a mercury thermometer have the same readings at these points. But because alcohol and mercury have different expansion properties, the thermometers will not have exactly the same reading at an intermediate temperature, such as room temperature. For very sensitive temperature measurements and to define intermediate temperatures precisely, some other type of thermometer must be used. One such thermometer is discussed next.

10.3 Gas Laws and Absolute Temperature

Objectives: **To (a) describe the ideal gas law and (b) explain how it is used to determine absolute zero.**

Whereas different liquid-in-glass thermometers show slightly different readings for temperatures other than fixed points because of the liquids' different expansion properties, a thermometer that uses a gas gives the same readings regardless of the gas used. All gases at very low densities exhibit the same expansion behavior.

The variables that describe the behavior of a given quantity (mass) of gas are pressure, volume, and temperature (p, V, and T). When temperature is held constant, the pressure and volume of a quantity of gas are related as follows:

$$pV = \text{constant} \quad \text{or} \quad p_1V_1 = p_2V_2 \qquad \textit{(at constant temperature)} \quad (10.3)$$

That is, the product of the pressure and volume is a constant. This relationship is known as *Boyle's law*, after Robert Boyle (1627–1691), the English chemist who discovered it. (See Demonstration 9.)

When the pressure is held constant, the volume of a quantity of gas is related to the *absolute* temperature (to be defined shortly):

$$\frac{V}{T} = \text{constant} \quad \text{or} \quad \frac{V_1}{T_1} = \frac{V_2}{T_2} \qquad \textit{(at constant pressure)} \quad (10.4)$$

That is, the ratio of the volume and the temperature is a constant. This relationship is known as *Charles' law*, named for the French scientist Jacques Charles (1747–1823), who made early hot-air balloon flights and was therefore quite interested in the relationship between the volume and temperature of a gas. A popular demonstration of Charles' law is shown in •Fig. 10.6.

Low-density gases obey these laws, which may be combined into a single relationship. Since $pV = \text{constant}$ and $V/T = \text{constant}$ for a given quantity of gas, then pV/T must also equal a constant. This relationship is the **ideal gas law**:

$$\frac{pV}{T} = \text{constant} \quad \text{or} \quad \frac{p_1V_1}{T_1} = \frac{p_2V_2}{T_2} \qquad \begin{array}{l}\textit{ideal gas law}\\\textit{(ratio form)}\end{array} \quad (10.5)$$

•**FIGURE 10.6   Charles' law in action**   Demonstrations of the relationship between volume and temperature of a quantity of gas. A weighted balloon, initially at room temperature, is placed in a beaker of water. **(a)** When ice is placed in the beaker and the temperature falls, the balloon's volume is reduced. **(b)** When the water is heated and the temperature rises, the balloon's volume increases.

Demonstration 9 ■ Boyle's Shaving Cream

A demonstration that shows Boyle's law, $p \propto 1/V$, the inverse relationship of pressure (p) and volume (V). Note: A similar effect occurs for a Boyle's marshmallow.

(a) A swirl of shaving cream is placed in a transparent vacuum chamber.

(b) A vacuum pump begins to evacuate the chamber.

(c) As the pressure is reduced, the swirl of cream grows in volume from the expansion of the air bubbles trapped in the cream.

(d) In a dramatic volume reduction, the shaving cream cannot stand up to the sudden increase in pressure when the chamber is vented to the atmosphere.

That is, the ratio pV/T at one time (t_1) is the same as at another time (t_2), or at any other time, as long as the quantity of gas does not change.

This relationship can be written in a more general form that applies not just to a given quantity of a single gas but to any quantity of any dilute gas:

$$\frac{pV}{T} = Nk_B \quad \text{or} \quad pV = Nk_BT \qquad \textit{ideal gas law} \qquad (10.6)$$

Here N is the number of molecules in the sample of gas and k_B is a number known as *Boltzmann's constant*: $k_B = 1.38 \times 10^{-23}$ J/K. The K stands for temperature on the Kelvin scale, discussed below. (Can you show that the units are correct?) Note that the mass of the sample does not appear explicitly in Eq. 10.6. However, a given number of molecules of a gas is proportional to the total mass of the gas. The ideal gas law, sometimes called the *perfect gas law*, applies to all gases at low densities and describes fairly accurately the behavior of most gases at normal densities.

Macroscopic Form of the Ideal Gas Law

Equation 10.6 is a "microscopic" (*micro* means small) form of the ideal gas law in that it refers to the tiny molecules of a gas. (The equation contains N, the number of molecules.) However, the law can be written in a "macroscopic" (*macro* means large) form, which involves quantities that can be measured with everyday laboratory equipment. In this form, we have

$$pV = nRT \qquad \textit{ideal gas law} \qquad (10.7)$$

Ideal gas law (with T on the Kelvin scale)

with the constants nR rather than Nk_B used for convenience. Here, n is the number of moles (mol) of the gas, a quantity defined below, and R is called the *universal gas constant*:

$$R = 8.31 \text{ J/(mol·K)}$$

In chemistry a **mole** (abbreviated mol) of a substance is defined as the quantity that contains **Avogadro's number** (N_A) of molecules:

$$N_A = 6.02 \times 10^{23} \text{ molecules/mol}$$

Thus, the n and N in the two forms of the ideal gas law are related by $N = nN_A$. From Eq. 10.7 it is easy to show that 1 mol of *any* gas occupies 22.4 L at 0°C and 1 atm. These conditions are known as *standard temperature and pressure (STP)*.

Equation 10.7 is a practical form of the gas law, because we generally work with measured quantities or moles (n) of gases rather than the number of molecules (N). To use Eq. 10.7, we need to know the number of moles of a gas. The mass of 1 mol of any substance is simply its formula mass expressed in grams. The formula mass is determined from the chemical formula and the atomic masses of the atoms (these are listed in Appendix IV and are commonly rounded to the nearest one-half). For example, water, H_2O, with two hydrogen atoms and one oxygen atom, has a formula mass of $2.0 + 16.0 = 18.0$, because the atomic mass of each hydrogen atom is 1.0 g and that of an oxygen atom is 16.0 g. Thus 1 mol of water has a formula mass of 18.0 g. Similarly, the oxygen we breathe, O_2, has a formula mass of $2 \times 16.0 = 32.0$. Thus a mole of oxygen, 32 g, would occupy 22.4 L at STP.

It is interesting to note that Avogadro's number allows you to compute the mass of a particular type of molecule. For example, suppose you want to know the mass of a water molecule (H_2O). As we have just seen, the formula mass of 1 mol of water is 18.0 g, or 18.0 g/mol. The molecular mass (m) is then given by

$$m = \frac{\text{formula mass (in kilograms)}}{N_A}$$

and, converting grams to kilograms, we have

$$m_{H_2O} = \frac{(18.0 \text{ g/mol})(10^{-3} \text{ kg/g})}{6.02 \times 10^{23} \text{ molecules/mol}} = 2.99 \times 10^{-26} \text{ kg/molecule}$$

The *formula mass* (in kilograms) is equivalent to the mass of 1 mol (in kilograms).

Absolute Zero and the Kelvin Temperature Scale

The important point for the purpose of this section is that the pressure and volume are directly proportional to temperature: $pV \propto T$. This relationship allows a gas to be used to measure temperature in a *constant-volume gas thermometer*. Holding the volume of a gas constant, which can be done easily in a rigid container (•Fig. 10.7), gives $p \propto T$. Thus, with a constant-volume gas thermometer, tem-

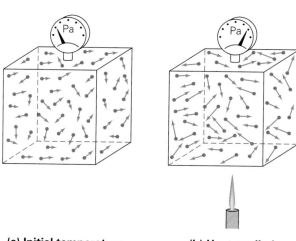

•FIGURE 10.7 Constant-volume gas thermometer Such a thermometer indicates temperature as a function of pressure, since for a low-density gas, $p \propto T$. **(a)** At some initial temperature, the pressure reading has a certain value. **(b)** When the gas thermometer is heated, the pressure (and temperature) reading is higher, because on the average, the gas molecules are moving faster.

(a) Initial temperature   **(b) Heat applied**

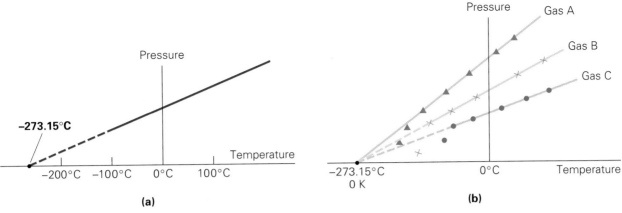

•FIGURE 10.8 Pressure versus temperature **(a)** A low-density gas kept at a constant volume gives a straight line on a p versus T graph: $p = (Nk_B/V)T$. When the line is extended to the zero pressure value, a temperature of $-273.15°C$ is obtained, which is taken to be absolute zero. **(b)** Extrapolation of lines for all low-density gases indicates the same absolute zero temperature. The actual behavior of gases deviates from this straight-line relationship at low temperatures because they start to liquefy.

perature is read in terms of pressure. A plot of pressure versus temperature gives a straight line in this case (•Fig. 10.8a).

As can be seen in Fig. 10.8b, measurements of real gases (plotted data points) deviate from the values predicted by the ideal gas law at low temperatures. This is because the gases liquefy at low temperatures. However, the relationship is linear over a large temperature range, and it looks as though the pressure might reach zero with decreasing temperature if the gas continued to be gaseous (ideal or perfect).

The absolute minimum temperature for an ideal gas is therefore inferred by extrapolating, or extending the straight line to the axis, as in Fig. 10.8b. This temperature is found to be $-273.15°C$ and is designated as **absolute zero**. Absolute zero is believed to be the lower limit of temperature, but it has never been attained. In fact, there is a physical law that says it never can be (Chapter 12).

There is no known upper limit to temperature. For example, the temperatures at the centers of some stars are estimated to be greater than 100 million degrees.

Absolute zero is the foundation of the **Kelvin temperature scale**, named after the British scientist Lord Kelvin.* On this scale, $-273.15°C$ is taken as the zero point—that is, as 0 K (•Fig. 10.9). The size of a single unit of Kelvin temperature is the same as the Celsius degree, so temperatures on these scales are related by

$$T_K = T_C + 273.15 \qquad \begin{array}{l} \textit{Celsius-to-Kelvin} \\ \textit{conversion} \end{array} \qquad (10.8)$$

Here T_K is the temperature in **kelvins** (*not* degrees Kelvin; for example, 300 kelvins). The kelvin is abbreviated as K (*not* °K). In most instances, a rounded value can be used in Eq. 10.8: $T_K = T_C + 273$.

The absolute Kelvin scale is the official SI temperature scale; however, the Celsius scale is used in most parts of the world for everyday temperature readings. The absolute temperature in kelvins is used primarily in scientific applications.

Caution: Keep in mind that Kelvin temperatures *must* be used with the ideal gas law. It is a common mistake to use Celsius or Fahrenheit temperatures in that equation. Suppose you used a Celsius temperature of $T = 0°C$ in the gas law. You would have $pV = 0$, which makes no sense.

*Lord Kelvin, born William Thomson (1824–1907), developed devices to improve telegraphy and the compass and was involved in the laying of the first transatlantic cable. When he received his title, it is said that he considered choosing Lord Cable or Lord Compass as his title but decided on Lord Kelvin, after a river that runs near the University of Glasgow in Scotland, where he was a professor of physics for 50 years.

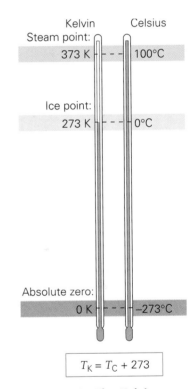

•FIGURE 10.9 The Kelvin temperature scale The lowest temperature on the Kelvin scale (corresponding to $-273.15°C$) is absolute zero. A unit interval on the Kelvin scale, called a kelvin and abbreviated K, is equivalent to a temperature change of 1 C°; thus $T_K = T_C + 273.15$. (The constant is usually rounded to 273 for convenience. For example, a temperature of 0°C is equal to 273 kelvins.)

Also note that there can be no negative temperatures on the Kelvin scale if absolute zero is the lowest possible temperature. That is, the Kelvin scale doesn't have an arbitrary zero temperature somewhere within the scale—0 K is absolute zero, period.

Example 10.2 ■ Deepest Freeze: Absolute Zero on the Fahrenheit Scale

What is absolute zero on the Fahrenheit scale?

Thinking It Through. We must convert 0 K to the Fahrenheit scale. However, the first conversion is to the Celsius scale. (Why?)

Solution.

Given: $T_K = 0$ K   *Find:* T_F

Temperatures on the Kelvin scale are related directly to Celsius temperatures by $T_K = T_C + 273.15$, so first we convert 0 K to a Celsius value:

$$T_C = T_K - 273.15 = 0 - 273.15 = -273.15°C$$

Then converting to Fahrenheit (Eq. 10.1) gives

$$T_F = \tfrac{9}{5}T_C + 32 = \tfrac{9}{5}(-273.15) + 32 = -459.67°F$$

Thus, absolute zero is about $-460°F$.

Follow-up Exercise. There is an absolute temperature scale associated with the Fahrenheit temperature called the Rankine scale. A Rankine degree is the same size as a Fahrenheit degree, and absolute zero is taken as 0°R (zero degrees Rankine). Write the conversion equations between (a) the Rankine and the Fahrenheit scales, (b) the Rankine and the Celsius scales, and (c) the Rankine and the Kelvin scales.

Initially, gas thermometers were calibrated by using the ice and steam points. The Kelvin scale uses absolute zero and a second fixed point adopted in 1954 by the International Committee on Weights and Measures. This second fixed point is the **triple point of water**, which represents a unique set of conditions where water co-exists simultaneously in equilibrium as a solid, a liquid, and a gas. The conditions for the triple point are a pressure of 4.58 mm Hg, or about 610 Pa (760 mm Hg is normal atmospheric pressure), and a temperature taken to be 273.16 K. The kelvin is then defined as 1/273.16 of the temperature at the triple point of water.*

Now let's use the ideal gas law, which requires absolute temperatures.

Example 10.3 ■ The Ideal Gas Law: Using Absolute Temperatures

A quantity of low-density gas in a rigid container is initially at room temperature (20°C) and a particular pressure (p_1). If the gas is heated to a temperature of 60°C, by what factor does the pressure change?

Thinking It Through. A "factor" of change implies a ratio (p_2/p_1), so Eq. 10.5 should apply. Note that the container is rigid, which implies that $V_1 = V_2$.

Solution.

Given: $T_1 = 20°C$ *Find:* p_2/p_1 (pressure ratio or factor)
$T_2 = 60°C$
$V_1 = V_2$

Since we want the factor of pressure change, we write p_2/p_1 as a ratio. For example, if $p_2/p_1 = 2$, then $p_2 = 2p_1$, or the pressure would change (increase) by a factor of 2. The

---

*The 273.16 value given here for the triple point temperature and the -273.15 value in Eq. 10.8 indicate different things. The number in the equation comes from the fact that $-273.15°C$ is taken as 0 K. The 273.16 K (or 0.01°C) is a different reading on a different temperature scale.

ratio also indicates that we should use the ideal gas law in ratio form. The gas law re-quires *absolute* temperatures, so we first change the Celsius temperatures to kelvins:

$$T_1 = 20°C + 273 = 293 \text{ K}$$

$$T_2 = 60°C + 273 = 333 \text{ K}$$

where a rounded value of 273 was used in Eq. 10.8 for convenience. Then, using the ideal gas law (Eq. 10.5) in the form $p_2V_2/T_2 = p_1V_1/T_1$, and since $V_1 = V_2$, we have

$$p_2 = \left(\frac{T_2}{T_1}\right)p_1 = \left(\frac{333 \text{ K}}{293 \text{ K}}\right)p_1 = 1.14p_1$$

Thus, p_2 is 1.14 times p_1; the pressure increases by a factor of 1.14, or by 14 percent. (What would the factor be if the Celsius temperatures were *incorrectly* used? It would be much larger: $60°C/20°C = 3$.)

Note: *Always* use Kelvin (absolute) temperatures with the ideal gas law.

Follow-up Exercise.   If the gas in this Example is heated when at room temperature so the pressure increases by a factor of 1.26, what is the final Celsius temperature?

Because of its absolute nature, the Kelvin temperature scale has special signifi-cance. As we shall see in Section 10.5, the absolute temperature is directly propor-tional to the internal energy of an ideal gas and so can be used as an indication of that energy. There are no negative values on the absolute scale. Negative absolute tem-peratures would imply negative internal energy for the gas—a meaningless concept.

Suppose you were asked to double the temperatures of, say, $-10°C$ and $0°C$. What would you do? The following Example should help.

Conceptual Example 10.4

Some Like It Hot: Doubling the Temperature

The evening weather report gives the day's high temperature as 10°C and predicts the next day's high to be 20°C. John says that this means it will be twice as warm tomor-row; Mary says it does not. With whom do you agree?

Reasoning and Answer.   Keep in mind that temperature gives a relative *indication* of hotness or coldness. Certainly, 20°C would be warmer than 10°C, but just because the numerical value of the temperature is twice as great does not necessarily mean it is twice as warm—only that the air temperature is 10 degrees higher and therefore relatively warmer.

The *absolute* temperature, with an absolute zero, is not doubled: $T_{K_1} = T_{C_1} + 273 = 10°C + 273 = 283 \text{ K}$ and $T_{K_2} = 20°C + 273 = 293 \text{ K}$. So the absolute temperature is increased only by a factor of $\Delta T_K/T_{K_1} = 10 \text{ K}/283 \text{ K} = 0.035$ (or 3.5%—not 100%), and we should agree with Mary.

Follow-up Exercise.   The weather report gives the day's high temperature as 0°C. If the next day's temperature were double that, what would the temperature be in degrees Celsius? Would this be environmentally possible?

10.4   Thermal Expansion

Objective: To calculate the thermal expansions of solids and liquids.

Changes in the dimensions and volumes of materials are common thermal effects. As you learned earlier, thermal expansion provides a means of temperature mea-surement. The thermal expansion of gases is generally described by the ideal gas law and is very obvious. Less dramatic, but by no means less important, is the thermal expansion of solids and liquids.

Note: Solids are discussed in Section 9.1.

Thermal expansion results from a change in the average distance separating the atoms of a substance. The atoms are held together by bonding forces, which can be simplistically represented as springs in a simple model of a solid (see Fig. 9.1). The atoms vibrate back and forth; with increased temperature (that is,

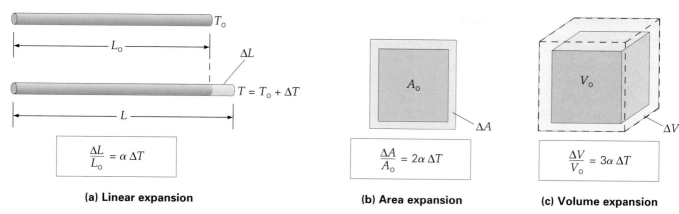

(a) Linear expansion

$$\frac{\Delta L}{L_o} = \alpha \Delta T$$

(b) Area expansion

$$\frac{\Delta A}{A_o} = 2\alpha \Delta T$$

(c) Volume expansion

$$\frac{\Delta V}{V_o} = 3\alpha \Delta T$$

•**FIGURE 10.10 Thermal expansion** (a) Linear expansion is proportional to the temperature change; that is, the change in length ΔL is proportional to ΔT, and $\Delta L/L_o = \alpha \Delta T$, where α is the thermal coefficient of linear expansion. **(b)** For isotropic expansion, the thermal coefficient of area expansion is approximately 2α. **(c)** The thermal coefficient of volume expansion for solids is about 3α.

more internal energy), they become increasingly active and vibrate over greater distances. With wider vibrations in all dimensions, the solid expands as a whole.

The change in one dimension of a solid (length, width, or thickness) is called *linear* expansion. For small temperature changes, linear expansion is approximately proportional to ΔT, or $T - T_o$ (•Fig. 10.10a). The fractional change in length is $(L - L_o)/L_o$, or $\Delta L/L_o$, where L_o is the original length (at the initial temperature). This ratio is related to the change in temperature by

$$\frac{\Delta L}{L_o} = \alpha \Delta T \quad \text{or} \quad \Delta L = \alpha L_o \Delta T \qquad (10.9)$$

where α is the **thermal coefficient of linear expansion**. Note that the unit of α is inverse temperature: inverse Celsius degrees ($1/C^\circ$, or $C^{\circ -1}$). Values of α for some materials are given in Table 10.1.

A solid may have different coefficients of linear expansion for different directions, but for simplicity this book will assume that the same coefficient applies to all directions (in other words, that solids show *isotropic* expansion). Also, the coefficient of expansion may vary slightly for different temperature ranges. Since this variation is negligible for most common applications, α will be considered to be constant and independent of temperature.

Equation 10.9 can be rewritten to give the final length (L) after a temperature change:

TABLE 10.1 Values of Thermal Expansion Coefficients (in $C^{\circ -1}$) for Some Materials at 20°C

Material	Coefficient of linear expansion (α)	Material	Coefficient of volume expansion (β)
Aluminum	24×10^{-6}	Alcohol, ethyl	1.1×10^{-4}
Brass	19×10^{-6}	Gasoline	9.5×10^{-4}
Brick or concrete	12×10^{-6}	Glycerin	4.9×10^{-4}
Copper	17×10^{-6}	Mercury	1.8×10^{-4}
Glass, window	9.0×10^{-6}	Water	2.1×10^{-4}
Glass, Pyrex	3.3×10^{-6}		
Gold	14×10^{-6}		
Ice	52×10^{-6}	Air (and most other gases at 1 atm)	3.5×10^{-3}
Iron and steel	12×10^{-6}		

$$\Delta L = \alpha L_o \Delta T$$
$$L - L_o = \alpha L_o \Delta T$$
$$L = L_o + \alpha L_o \Delta T$$

or

$$L = L_o(1 + \alpha \Delta T) \qquad (10.10)$$

We can then use Eq. 10.10 to compute the thermal *area* expansion of flat objects. Since area (A) is length squared (L^2) for a square, we get

$$A = L^2 = L_o^2(1 + \alpha \Delta T)^2 = A_o(1 + 2\alpha \Delta T + \alpha^2 \Delta T^2)$$

where A_o is the original area. Because the values of α for solids are much less than 1 ($\sim 10^{-5}$, as shown in Table 10.1), the second-order term (containing α^2) can be dropped with negligible error. As a first-order approximation, then,

$$A = A_o(1 + 2\alpha \Delta T) \quad \text{or} \quad \frac{\Delta A}{A_o} = 2\alpha \Delta T \qquad (10.11)$$

Thus, the **thermal coefficient of area expansion** (Fig. 10.10b) is twice as large as the coefficient of linear expansion (that is, it is equal to 2α). See Learn by Drawing.
    Similarly, a first-order expression for thermal *volume* expansion is

$$V = V_o(1 + 3\alpha \Delta T) \quad \text{or} \quad \frac{\Delta V}{V_o} = 3\alpha \Delta T \qquad (10.12)$$

The **thermal coefficient of volume expansion** (Fig. 10.10c) is equal to 3α.
    The equations for thermal expansions are approximations. Even though an equation is a description of a physical relationship, always keep in mind that it may be only an approximation of physical reality and/or may apply only in certain situations. Because calculations of thermal expansion involve such small numbers, you may assume that all quantities are exact.
    The thermal expansion of materials is an important consideration in construction. Seams are put in concrete highways and sidewalks to allow room for expansion and prevent cracking. Expansion gaps in large bridge structures and between railroad rails are necessary to prevent damage. Similarly, expansion loops are found in oil pipelines (•Fig. 10.11). The thermal expansion of steel beams and girders can produce tremendous pressures, as the following Example shows.

LEARN BY DRAWING
Thermal Area Expansion

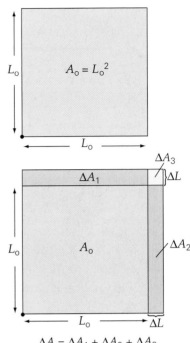

$$\Delta A = \Delta A_1 + \Delta A_2 + \Delta A_3$$
$$\Delta A_1 = \Delta A_2 = L_o \, \Delta L$$
$$= L_o \,(\alpha L_o \, \Delta T\,) = \alpha A_o \, \Delta T$$
Since ΔA_3 is very small
compared to ΔA_1 and ΔA_2,
$$\Delta A \approx 2\alpha A_o \, \Delta T$$

Example 10.5 ■ Temperature Rising: Thermal Expansion and Stress

A steel beam is 5.0 m long at a temperature of 20°C (68°F). On a hot day, the temperature rises to 40°C (104°F). (a) What is the change in the beam's length due to thermal expansion? (b) Suppose that the ends of the beam are initially in contact with rigid vertical supports. How much force will the expanded beam exert on the supports if the beam has a cross-sectional area of 60 cm²?

Thinking It Through.   (a) This is a direct application of Eq. 10.9. (b) As the constricted beam expands, it applies a stress, and hence a force, to the supports. For a linear expansion, the Young's modulus relationship (Section 9.1) should come into play.

Solution.

Given: $L_o = 5.0$ m   *Find:*   (a) ΔL (change in length)
$\qquad\quad T_o = 20°C$ $\qquad\qquad\qquad\quad$ (b) F (force)
$\qquad\quad T = 40°C$
$\qquad\quad \alpha = 12 \times 10^{-6} \; C°^{-1}$ (from Table 10.1)
$\qquad\quad A = 60 \; cm^2 \left(\dfrac{1 \; m}{100 \; cm}\right)^2 = 6.0 \times 10^{-3} \; m^2$

(a)

(b)

•**FIGURE 10.11 Expansion gaps**
(a) Expansion gaps are built into bridge roadways to prevent contact stresses produced by thermal expansion. **(b)** These loops in oil pipelines serve a similar purpose. As hot oil passes through them, the pipes expand; loops take up the extra length.

(a) Using Eq. 10.9 to find the change in length with $\Delta T = T - T_{\mathrm{o}} = 40°C - 20°C = 20\ C°$, we get

$$\Delta L = \alpha L_{\mathrm{o}} \Delta T = (12 \times 10^{-6}\ C^{°-1})(5.0\ \mathrm{m})(20\ C°) = 1.2 \times 10^{-3}\ \mathrm{m} = 1.2\ \mathrm{mm}$$

This may not seem like much of an expansion, but it can give rise to a great deal of force if the beam is constrained and kept from expanding, as part (b) will show.

(b) By Newton's third law, if the beam is kept from expanding, the force the beam exerts on its constraint supports is equal to the force exerted by the supports to prevent the beam from expanding by a length ΔL. This is the same as the force that would be required to compress the beam by that length. Using the Young's modulus form of Hooke's law (Chapter 9) with $Y = 20 \times 10^{10}\ \mathrm{N/m^2}$ (Table 9.1), we calculate the stress on the beam as

$$\frac{F}{A} = \frac{Y\,\Delta L}{L_{\mathrm{o}}} = \frac{(20 \times 10^{10}\ \mathrm{N/m^2})(1.2 \times 10^{-3}\ \mathrm{m})}{5.0\ \mathrm{m}} = 4.8 \times 10^{7}\ \mathrm{N/m^2}$$

The force is then

$$F = (4.8 \times 10^{7}\ \mathrm{N/m^2})A = (4.8 \times 10^{7}\ \mathrm{N/m^2})(6.0 \times 10^{-3}\ \mathrm{m^2})$$

$$= 2.9 \times 10^{5}\ \mathrm{N} \quad (\text{about } 65\,000\ \mathrm{lb},\ \text{or } 32.5\ \mathrm{tons!})$$

Follow-up Exercise. Expansion gaps between identical steel beams laid end-to-end are specified to be 0.060% of the length of a beam at the installation temperature. With this specification, what is the temperature range for noncontact expansion?

Conceptual Example 10.6

Larger or Smaller? Area Expansion

A circular piece is cut from a flat metal sheet (•Fig. 10.12a). If the sheet is then heated in an oven, the size of the hole will (a) become larger, (b) become smaller, (c) be unchanged.

Reasoning and Answer. It is a common misconception to think that the area of the hole will shrink because of expansion of the metal around it. Think of the piece of metal removed from the hole rather than of the hole itself. This piece would expand with increasing temperature. The metal in the heated sheet reacts as if the piece removed were still part of it. (Think of putting the piece of metal back into the hole and heating, as in Fig. 10.12b. Or consider drawing a circle on an uncut metal sheet and heating it.) So, the answer is (a).

Follow-up Exercise. A circular ring of iron has a tight-fitting metal bar inside it, across its diameter. If heated in an oven to a high temperature, would the circular ring be distorted or bent out of shape, or would it remain circular?

Fluids (liquids and gases), like solids, normally expand with increasing temperature. Because fluids have no definite shape, only volume expansion (and not linear or area expansion) is meaningful. The expression is

$$\frac{\Delta V}{V_{\mathrm{o}}} = \beta\,\Delta T \qquad \textit{fluid volume expansion} \qquad (10.13)$$

where β is the coefficient of volume expansion for fluids. Note in Table 10.1 that the values of β for fluids are typically larger than the values of 3α for solids.

Water exhibits an anomalous volume expansion near its freezing point. The volume of a given amount of water decreases as it is cooled from room temperature, until its temperature reaches 4°C (•Fig. 10.13a). Below 4°C, the volume increases, and therefore the density decreases (Fig. 10.13b). This means that water has a maximum density ($\rho = m/V$) at 4°C (actually, 3.98°C).

When water freezes, the molecules form a hexagonal (six-sided) lattice pattern. (This is why snowflakes have hexagonal shapes.) It is the open structure of this lattice that gives water its almost unique property of expanding on freezing and being less dense as a solid than as a liquid. (This is why ice floats in water and frozen water pipes burst.) The variation of the density of water over the temperature range of 4°C to 0°C indicates that the open lattice structure is beginning to form at about 4°C rather than arising instantaneously at the freezing point.

This property has an important environmental effect: Bodies of water, such as lakes and ponds, freeze at the top first. As a lake cools toward 4°C, water near the surface loses energy to the atmosphere, becomes denser, and sinks. The warmer, less dense water near the bottom rises. However, once the colder water on top reaches temperatures below 4°C, it becomes less dense and remains at the surface, where it freezes. If water did not have this property, lakes and ponds would freeze from the bottom up, which would destroy much of their animal and plant life (and would make ice skating a lot less popular). There would also be no oceanic ice caps at the polar regions. Rather, there would be a thick layer of ice at the bottom of the ocean, covered by a layer of water.

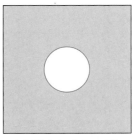

(a) Metal plate with hole

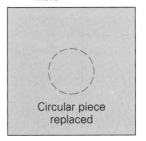

(b) Metal plate without hole

•FIGURE 10.12 **A larger or smaller hole?** See Conceptual Example 10.6.

Conceptual Example 10.7

Quick Chill: Temperature and Density

Ice is put into a container of water at room temperature. For faster cooling, the ice (a) should be allowed to float naturally in the water or (b) be pushed to the bottom of the container with a stick and held there.

Reasoning and Answer. When the ice melts, the water in the vicinity is cooled and hence becomes denser (Fig. 10.13b). If the ice is permitted to float at the top, the denser water sinks and the warmer, less dense water rises. This mixing causes the water to cool quickly. If the ice were at the bottom of the container, however, the colder, denser water would remain there, and cooling would be slowed, so the answer is (a).

Follow-up Exercise. Suppose that the density-versus-temperature curve for water (Fig. 10.13b) were inverted, so that it dipped downward. What would this imply for the situation in this Example and the freezing of lakes?

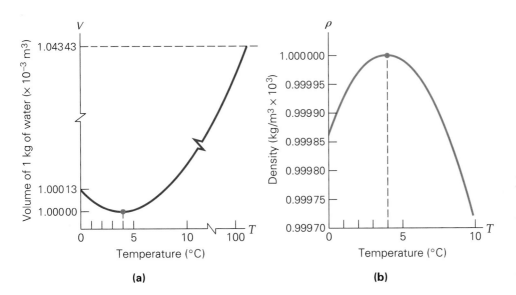

(a)

(b)

•FIGURE 10.13 **Thermal expansion of water** Water exhibits nonlinear expansion behavior near its freezing point. **(a)** Above 4°C (actually, 3.98°C), water expands with increasing temperature, but from 4°C down to 0°C, it expands with decreasing temperature. **(b)** As a result, water has its maximum density near 4°C.

10.5　The Kinetic Theory of Gases

Objectives: To (a) relate kinetic theory and temperature and (b) explain the process of diffusion.

If the molecules of a gas sample are viewed as colliding particles, the laws of mechanics can be applied to each molecule of that gas. We should then be able to describe its characteristics, such as pressure, internal energy, and so on, in terms of molecular motion. Because of the large number of particles involved, however, a statistical approach is employed for such a description.

One of the major accomplishments of theoretical physics was to do exactly this: derive the ideal gas law from mechanical principles. This derivation led to a new interpretation of temperature in terms of the translational kinetic energy of the gas molecules. As a theoretical starting point, the molecules of an ideal gas are viewed as point masses in random motion with relatively large distances separating them.

Our discussion in this book involves only two types of gases: (1) *monatomic* (single-atom gases), such as He, and (2) *di*atomic (two-atom gas molecules), such as O_2. In either case, the vibrational and rotational motions can be ignored in terms of temperature and pressure, since these quantities depend only on *linear* motion.

Note: Elastic collisions are discussed in Section 6.4.

According to the **kinetic theory of gases**, the molecules undergo perfectly elastic collisions with each other and with the walls of the container. The forces between molecules are considered to act only over a short range, so the molecules interact with each other only *during* collisions. And since the molecules of an ideal gas interact only during collisions, such a gas would remain gaseous at low temperatures; it would not liquefy as real gases do. This gives the theoretical ideal gas its idealness. Colliding molecules lose and gain kinetic energy, with corresponding changes in their potential energy. We ignore this potential energy in summing the total energy of the system because molecules spend a negligible amount of time in collisions.

From Newton's laws of motion, the force on the walls of the container can be calculated from the change in momentum of the gas molecules when they collide with the walls (•Fig. 10.14). If this force is expressed in terms of pressure (force/area), the following equation is obtained (see Appendix II for derivation):

$$pV = \tfrac{1}{3}Nmv_{rms}^2 \tag{10.14}$$

where V is the volume of the container or gas, N the number of gas molecules in the closed container, and m the mass of a gas molecule. In this expression, v_{rms} is the average speed, but a special kind of average. It is obtained by averaging the squares of the speeds and then taking the square root of the average—that is, $\sqrt{\overline{v^2}} = v_{rms}$. As a result, v_{rms} is called the *root-mean-square* (*rms*) speed.

Solving Eq. 10.6 for pV and equating that expression with Eq. 10.14 shows how temperature came to be interpreted as a measure of translational kinetic energy:

Average molecular kinetic energy and temperature

$$pV = Nk_BT = \tfrac{1}{3}Nmv_{rms}^2 \quad \text{or} \quad \tfrac{1}{2}mv_{rms}^2 = \tfrac{3}{2}k_BT \quad \begin{array}{l}\textit{(for all}\\ \textit{ideal gases)}\end{array} \tag{10.15}$$

Thus, the temperature of a gas (and that of the walls of the container or a thermometer bulb in thermal equilibrium with the gas) is directly proportional to its average random kinetic energy (per molecule), since $\overline{K} = \tfrac{1}{2}mv_{rms}^2 = \tfrac{3}{2}k_BT$. (Don't forget that T is the absolute temperature in kelvins.)

Example 10.8　■　Molecular Speed: Relation to Absolute Temperature

What is the average (rms) speed of a helium atom (He) in a helium balloon at room temperature? (Take the mass of the helium atom to be 6.65×10^{-27} kg.)

Thinking It Through. All the data we need to solve for the average speed in Eq. 10.15 are known.

Solution.

Given: $m = 6.65 \times 10^{-27}$ kg   *Find:* v_{rms} (rms speed)
$T = 20°C$ (room temperature)
$k_B = 1.38 \times 10^{-23}$ J/K (known)

We will use Eq. 10.15, so we list k_B among the given quantities.
    We must change the Celsius temperature to kelvins, but note that k_B has units of J/K. So,

$$T_K = T_C + 273 = 20°C + 273 = 293 \text{ K}$$

where the 273.15 in Eq. 10.8 was rounded to 273 for convenience.
    Rearranging Eq. 10.15, we have

$$v_{rms} = \sqrt{\frac{3k_B T}{m}} = \left[\frac{3(1.38 \times 10^{-23} \text{ J/K})(293 \text{ K})}{6.65 \times 10^{-27} \text{ kg}}\right]^{1/2} = 1.35 \times 10^3 \text{ m/s} = 1.35 \text{ km/s}$$

This is over 3000 mi/h—pretty fast!

Follow-up Exercise. In this Example, if the temperature of the gas were increased by 10 C°, what would be the corresponding percentage increases in the average (rms) speed and in the average kinetic energy?

Interestingly, Eq. 10.15 predicts that at absolute zero ($T = 0$ K), all translational molecular motion of a gas would cease. According to classical theory, this would correspond to absolute zero energy. However, modern quantum theory says that there would still be some zero-point motion, and a corresponding minimum *zero-point energy*. Basically, absolute zero is the temperature at which all the energy that *can* be removed from an object has been removed.

Internal Energy of Monatomic Gases

Because the "particles" in an ideal monatomic gas do not vibrate or rotate, as explained previously, the total kinetic energy of all the molecules is equal to the total internal energy of the gas. That is, the gas's internal energy is all "temperature" energy (Section 10.1). With N molecules in a system, we can use Eq. 10.15, the energy per molecule, to write an expression for the total internal energy U:

$$U = N(\tfrac{1}{2}mv_{rms}^2) = \tfrac{3}{2}Nk_B T = \tfrac{3}{2}nRT \qquad \begin{array}{l}\textit{(for ideal monatomic}\\ \textit{gases only)}\end{array} \qquad (10.16)$$

Thus, we see that the internal energy of an ideal monatomic gas is directly proportional to its absolute temperature. (In Section 10.6 we will see this is true regardless of the molecular structure of the gas. However, the expression for U will be a bit different for gases that are not monatomic.) This means that if the absolute temperature of a gas is doubled (by heat transfer), for example, from 200 K to 400 K, then its internal energy is also doubled. This does not apply to the Celsius and Fahrenheit temperatures, since their zero points are not referenced to the zero-point energy. If a gas is at a temperature of 0°C (or 0°F) and its internal energy is doubled, the Celsius (or Fahrenheit) temperature is not doubled—doubling zero gives zero, which is not a new temperature.

Diffusion

We depend on our sense of smell to detect odors, such as the smell of smoke from something burning. That you can smell something from a distance implies that molecules get from one place to another in the air, from the source to your nose. This process of random molecular mixing in which particular molecules move from a region where they are present in higher concentration to one where they are in lower concentration is called **diffusion**. Diffusion also occurs readily in liquids; think about what happens to a drop of ink in a glass of water (•Fig. 10.15). It even occurs to some degree in solids.

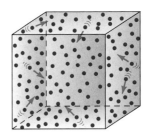

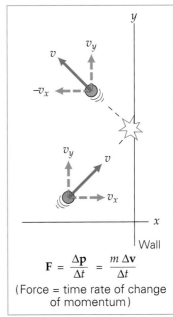

$$\mathbf{F} = \frac{\Delta \mathbf{p}}{\Delta t} = \frac{m \, \Delta \mathbf{v}}{\Delta t}$$

(Force = time rate of change of momentum)

•**FIGURE 10.14   Kinetic theory of gases**   The pressure a gas exerts on the walls of a container is due to the force resulting from the change in momentum of the gas molecules that collide with the wall. The force exerted by an individual molecule is equal to the time rate of change of momentum: $\mathbf{F} = \Delta \mathbf{p}/\Delta t = m \, \Delta \mathbf{v}/\Delta t$, where $\mathbf{p} = m\mathbf{v}$. The sum of the instantaneous normal components of the collision forces gives rise to the average pressure on the wall.

Note: Eq. 10.16 applies only to ideal gases or very dilute real gases.

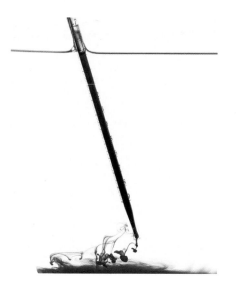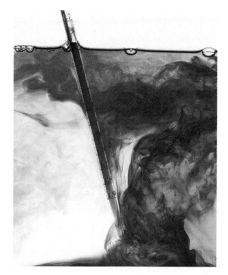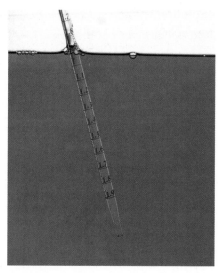

•**FIGURE 10.15 Diffusion in liquids** Random molecular motion would eventually distribute the dye throughout the water. Here there is some distribution due to mixing, and the ink colors the water after a few minutes. The distribution would take more time by diffusion only.

The rate of diffusion for a particular gas depends on the rms speed of its molecules. Even though gas molecules have large average speeds (Example 10.8), the molecules do not fly from one side of a room to the other. There are frequent collisions, and as a result the molecules "drift" rather slowly. For example, suppose someone opened a bottle of ammonia on the other side of a closed room. It would take some time for the ammonia to diffuse across the room until you could smell it. (Much of the movement that people commonly attribute to diffusion is actually due to air currents.)

Gases can also diffuse through porous materials or permeable membranes. (This process is sometimes referred to as *effusion*.) Energetic molecules enter the material through the pores (openings), and, colliding with the pore walls, they slowly meander through the material. Such gaseous diffusion can be used to separate different gases from a mixture physically.

The kinetic theory of gases says that the average translational kinetic energy (per molecule) is proportional to the absolute temperature of a gas, $\frac{1}{2} m v_{\text{rms}}^2 = \frac{3}{2} k_B T$. So, on the average, the molecules of different gases (having different masses) move at different speeds at a given temperature. As you might expect, because they move faster, lighter gas molecules diffuse through the tiny openings of a porous material faster than heavier gas molecules do.

For example, at a particular temperature, molecules of oxygen (O_2) move faster on the average than the more massive molecules of carbon dioxide (CO_2) do. Because of this difference in molecular speed, oxygen can diffuse through a barrier faster than carbon dioxide can. Suppose that a mixture of equal volumes of oxygen and carbon dioxide is contained on one side of a porous barrier (•Fig. 10.16). After a while, some O_2 molecules and CO_2 molecules will have diffused through

•**FIGURE 10.16 Separation by gaseous diffusion** The molecules of both gases diffuse (or effuse) through the porous barrier. But, because oxygen molecules have the greater average speed, more of them pass through. Thus, over time there is a greater concentration of oxygen molecules on the other side of the barrier.

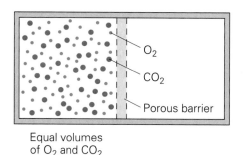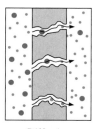

Equal volumes
of O_2 and CO_2

Diffusion through barrier

Physiological Diffusion

Diffusion plays a central role in many life processes. For example, consider a cell membrane in the lung. Such a membrane is permeable to a number of substances, any of which will diffuse through the membrane from a region where its concentration is high to a region where its concentration is low. Most importantly, the lung membrane is permeable to oxygen (O_2), and the transfer of O_2 across the membrane occurs because of a concentration gradient.

The blood carried to the lungs is low in O_2, having given up the oxygen to tissues for metabolism during its circulation through the body. Conversely, the air in the lungs is high in O_2 because there is a continuous exchange of fresh air in the breathing process. As a result of this concentration difference, or gradient, O_2 diffuses from the lung volume into the blood that flows through the lung tissue, and the blood leaving the lungs is high in O_2.

Exchanges between the blood and the tissues occur across capillary walls, and diffusion again is a major factor. The chemical composition of arterial blood is regulated to maintain the proper concentrations of particular solutes (substances dissolved in the blood solution), so diffusion takes place in the appropriate directions across capillary walls. For example, as cells take up O_2 and glucose (blood sugar), the blood continuously brings in fresh supplies of the substances to maintain the concentration gradient needed for diffusion to the cells. The continuous production of carbon dioxide

(CO_2) and metabolic wastes in the cells produces concentration gradients in the opposite direction for these substances. They therefore diffuse out of the cells into the blood, to be carried away from the tissue by the circulatory system.

During periods of physical exertion, there is an increase in cellular activity. More O_2 is used up and more CO_2 is produced, thereby increasing the concentration gradients and the diffusion rates. With an increased demand for O_2, how do the lungs respond to provide this to the blood? As you might expect, the rate of diffusion depends on the surface area and thickness of the lung membrane. Deeper breathing during exercise causes the alveoli (small air sacs in the lungs) to increase in volume. Such stretching increases the alveolar surface area and decreases the thickness of the membrane wall.

Also, the heart works harder during exercise, and the blood pressure is raised. The increased pressure forces open capillaries that are normally closed during rest or normal activity. As a result, the total exchange area between the blood and cells is increased. Each of these changes helps expedite the exchange of gases during exercise.

Now you can understand why people with emphysema (a disease involving the breakdown of alveoli walls) or pneumonia (characterized by fluid accumulations within or around the lungs) have difficulty providing enough oxygen to their tissues.

the barrier. The diffused gases would still be mixed. After another diffusion stage, however, the oxygen concentration would become greater on the far side of the barrier. Almost pure oxygen can be obtained by repeating the separation process many times. Separation by gaseous diffusion is a key process in obtaining enriched uranium, which was used in the atomic bomb and in nuclear reactors that generate electricity (Chapter 30).

Fluid diffusion is very important to organisms. In plant photosynthesis, carbon dioxide from the air diffuses into leaves, and oxygen and water vapor diffuse out. The diffusion of liquid water across a permeable membrane down a concentration gradient is called **osmosis**; this process is vital in living cells. Osmotic diffusion is also important to kidney functioning: Tubules in the kidneys concentrate waste matter from the blood in much the same way oxygen is removed from mixtures. (See the Insight on this page for other examples of diffusion.)

*10.6 Kinetic Theory, Diatomic Gases, and the Equipartition Theorem

Objectives: To understand (a) the difference between monatomic and diatomic gases, (b) the meaning of the equipartition theorem, and (c) the expression for the internal energy of a diatomic gas.

In the real world, most of the gases we deal with are *not* monatomic gases. Recall from chemistry that monatomic gases are elements known as *noble* or *inert* gases because they do not readily combine with other atoms. These elements are found on the far right side of the periodic chart: helium, neon, argon, krypton, xenon, and radon.

However, the mixture of gases we breathe (collectively known as "air") consists mainly of diatomic molecules of nitrogen (N_2, 78% by volume) and oxygen (O_2, 21% by volume). Both of these gases have two identical atoms chemically bonded together to form a single molecule. How do we deal with these realistic, more complicated molecules in terms of the kinetic theory of gases? There are even more complicated gas molecules consisting of more than two atoms, such as carbon dioxide (CO_2). However, because of their complexity, we will limit our discussion to diatomic molecules.

The Equipartition Theorem

As we saw in Section 10.5, it is only the translational kinetic energy of a gas that determines the gas's temperature. Thus, for any type of gas, regardless of how many atoms make up its molecules, it is *always true* that the average *translational* kinetic energy per molecule is still proportional to the temperature of the gas (Eq. 10.15): $\frac{1}{2}mv_{rms}^2 = \frac{3}{2}k_BT$ (for all gases).

Recall that for monatomic gases, the total internal energy, U, consists solely of translational kinetic energy. For diatomic molecules, this is not true, because a diatomic molecule is free to rotate and vibrate in addition to moving linearly. Therefore, these extra forms of energy must be taken into account. The expression given in Eq. 10.16 ($U = \frac{3}{2}Nk_BT$) for monatomic gases, which assumes that the total energy is due only to translational kinetic energy, therefore does *not* hold for diatomic gases.

Scientists wondered exactly how the expression for the internal energy of a diatomic gas might differ from that for a monatomic gas. In looking at the kinetic theory derivation that led up to Eq. 10.16, they realized that the factor of 3 in this equation was due to the fact that the gas molecules had three independent linear ways (dimensions) of moving. Thus for each molecule, there were three independent ways of possessing kinetic energy: with x, y, or z linear motion. Each independent way a molecule has for possessing energy is called a **degree of freedom**.

According to this scheme, a monatomic gas has only three degrees of freedom, since its molecules can move only linearly and possess kinetic energy in three dimensions. It was reasoned that quite possibly a diatomic gas could vibrate (see Fig. 10.1), thus having vibrational kinetic and potential energies (two additional degrees of freedom). A diatomic molecule might also rotate (•Fig. 10.17).

However, the rotational about the molecular axis (x in the figure) possesses very little energy compared to the other axes. As illustrated in Fig. 10.17, if we assume the atoms to be small spheres on the end of a stick, the moment of inertia (I) about the stick (the molecular axis) is very small and taken to be negligible. Since rotational kinetic energy depends directly on the moment of inertia, rotational motion about this axis provides negligible rotational kinetic energy. Thus, there are only two degrees of freedom associated with diatomic molecule rotation.

On the basis of the new understanding of monatomic gases and their three degrees of freedom, the **equipartition theorem** was proposed. (As the name implies, the total energy of a gas or molecule is "partitioned," or divided, equally for each degree of freedom.) That is,

On average, the total internal energy (U) of an ideal gas is divided equally among each degree of freedom its molecules possess. Furthermore, each degree of freedom contributes $\frac{1}{2}Nk_BT$ (or $\frac{1}{2}nRT$) to the total internal energy.

The equipartition theorem fits the special case of a monatomic gas, since this theorem predicts that $U = \frac{3}{2}Nk_BT$, which we know to be true. With three degrees of freedom, we have $U = 3(\frac{1}{2}Nk_BT)$, in agreement with the monatomic result given earlier (Eq. 10.16).

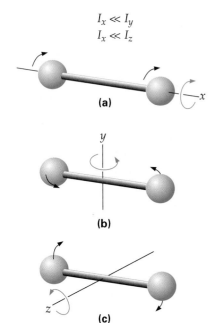

$I_x \ll I_y$
$I_x \ll I_z$

(a)

(b)

(c)

•**FIGURE 10.17   Model of a diatomic gas molecule**   If we consider a diatomic molecule as two spheres on the end of a stick, the rotation about (a) the molecular axis is very small or negligible compared to (b) and (c), the other axes. That is, the moment of inertia about the x axis (I_x) is relatively small and so is the rotational kinetic energy ($K = \frac{1}{2}I\omega^2$).

Definition of:
Equipartition theorem

The Internal Energy of a Diatomic Gas

Exactly how does the equipartition theorem enable us to calculate the internal energy of a diatomic gas such as oxygen? To do this, we must realize that U must now include all the available degrees of freedom. In addition to the translational degrees of freedom, what other motions are the molecules undergoing? The analysis is complicated and beyond the scope of this text, so we give only the general results. *Typically for normal (room) temperatures, quantum theory predicts (and experiment verifies) that only the rotational motions are important in the degrees of freedom. Diatomic molecules do NOT vibrate at normal temperatures.*

So, the total internal energy of a diatomic gas is comprised of the internal energies due to the three linear degrees of freedom and two rotational degrees of freedom, for a total of five. Hence, we can write

$$U = K_{\text{trans}} + K_{\text{rot}} = 3(\tfrac{1}{2}nRT) + 2(\tfrac{1}{2}nRT)$$

$$= \tfrac{5}{2}nRT = \tfrac{5}{2}Nk_BT \qquad \text{\textit{(for diatomic gases at room temperature)}} \qquad (10.17)$$

Thus a monatomic gas sample at normal room temperature has 40% less internal energy than a diatomic gas sample at the same temperature, or equivalently possesses only 60% of the internal energy of the diatomic gas sample.

Example 10.9 ■ Monatomic versus Diatomic: Are Two Atoms Better Than One?

More than 99% of the air we breathe consists of diatomic gases, mainly nitrogen (N_2, 78%) and oxygen (O_2, 21%). There are traces of other gases, one of which is radon (Ra), a monatomic gas arising from radioactive decay of uranium in the ground. [Radon itself is radioactive, which is irrelevant here, but this fact does make radon a possible health danger when concentrated inside a house (as we shall see in Chapter 29).] (a) Calculate the total internal energy of 1.00-mole samples each of oxygen and radon at room temperature (20°C). (b) For each sample, calculate the amount of internal energy associated with molecular *translational* kinetic energy.

Thinking It Through. (a) We have to consider the number of degrees of freedom in a monatomic gas and a diatomic gas in computing the internal energy (U). (b) Only three linear degrees of freedom contribute to the translational kinetic energy portion (U_{trans}) of the internal energy.

Solution. We list the data and convert to kelvins right away because we know that the internal energy is expressed in terms of absolute temperature:

Given: $n = 1.00$ mol *Find:* (a) U for O_2 and Ra samples
  $T = 20°C + 273 = 293$ K   at room temperature
       (b) U_{trans} for O_2 and Ra samples
       at room temperature

(a) Let's compute the total internal energy of the (monatomic) radon sample first, using Eq. 10.16:

$$U_{\text{Ra}} = \tfrac{3}{2}nRT = \tfrac{3}{2}(1.00 \text{ mol})[8.31 \text{ J}/(\text{mol·K})](293 \text{ K}) = 3.65 \times 10^3 \text{ J}$$

Since this sample is at room temperature, the (diatomic) oxygen will also include internal energy stored as two extra degrees of freedom, due to rotation. Thus we have

$$U_{\text{O}_2} = \tfrac{5}{2}nRT = \tfrac{5}{2}(1.00 \text{ mol})[8.31 \text{ J}/(\text{mol·K})](293 \text{ K}) = 6.09 \times 10^3 \text{ J}$$

As we have seen, even though there is the same number of molecules in each sample and the temperature is the same, the oxygen sample has almost 70% more energy.

(b) For (monatomic) radon, all the internal energy is in the form of translational kinetic energy; thus the answer is the same as in (a):

$$U_{\text{trans}} = U_{\text{Ra}} = 3.65 \times 10^3 \text{ J}$$

For (diatomic) oxygen, only $\frac{3}{2}nRT$ of the total internal energy ($\frac{5}{2}nRT$) is from translational kinetic energy, so the answer is the same as for radon; that is, $U_{\text{trans}} = 3.65 \times 10^3$ J for both gas samples.

Follow-up Exercise. In this Example, how much energy is associated with the rotational motion of the oxygen molecules? Which sample has the highest rms speed? (Note: The mass of one radon atom is about 7 times the mass of an oxygen molecule.) Explain your reasoning.

Chapter Review

Important Terms

temperature *334*
heat *334*
internal energy *334*
thermometer *335*
thermal expansion *335*
Fahrenheit temperature scale *336*
Celsius temperature scale *336*
ideal (perfect) gas law *338*
mole *340*

Avogadro's number *340*
absolute zero *341*
Kelvin temperature scale *341*
kelvins *341*
triple point of water *342*
thermal coefficient of linear
  expansion *344*
thermal coefficient of area
  expansion *345*

thermal coefficient of volume
  expansion *345*
kinetic theory of gases *348*
diffusion *349*
osmosis *351*
*degree of freedom *352*
*equipartition theorem *352*

Important Concepts

- Heat is the net energy transferred from one object to another because of temperature differences. Once transferred, the energy becomes part of the internal energy of the object (or system).
- The ideal gas law relates pressure, volume, and absolute temperature.
- Absolute zero (0 K) corresponds to $-273.15°C$.

- Thermal coefficients of expansion relate the fractional change in dimension(s) to a change in temperature.
- According to the kinetic theory of gases, the absolute temperature of a gas is directly proportional to the average random kinetic energy per molecule.

Important Equations

Celsius–Fahrenheit Conversion:

$$T_F = \tfrac{9}{5}T_C + 32 \tag{10.1}$$

$$T_C = \tfrac{5}{9}(T_F - 32) \tag{10.2}$$

Ideal (or perfect) Gas Law (always absolute temperature):

$$pV = Nk_BT \quad \text{or} \quad \frac{p_1V_1}{T_1} = \frac{p_2V_2}{T_2} \tag{10.5, 10.6}$$

or

$$pV = nRT \tag{10.7}$$

$$k_B = 1.38 \times 10^{-23} \text{ J/K}$$

$$R = 8.31 \text{ J/(mol·K)}$$

Avogadro's number: $N_A = 6.02 \times 10^{23}$ molecules/mol

Kelvin–Celsius Conversion:

$$T_K = T_C + 273.15 \tag{10.8}$$

Thermal Expansion of Solids:

$$\text{linear:} \quad \frac{\Delta L}{L_o} = \alpha \,\Delta T \quad \text{or} \quad L = L_o(1 + \alpha \,\Delta T) \tag{10.9, 10.10}$$

$$\text{area:} \quad \frac{\Delta A}{A_o} = 2\alpha \,\Delta T \quad \text{or} \quad A = A_o(1 + 2\alpha \,\Delta T) \tag{10.11}$$

$$\text{volume:} \quad \frac{\Delta V}{V_o} = 3\alpha \,\Delta T \quad \text{or} \quad V = V_o(1 + 3\alpha \,\Delta T) \tag{10.12}$$

Thermal Volume Expansion of Fluids:

$$\frac{\Delta V}{V_o} = \beta \,\Delta T \tag{10.13}$$

Results of Kinetic Theory of Gases:

$$pV = \tfrac{1}{3}Nmv_{\text{rms}}^2 \tag{10.14}$$

$$\tfrac{1}{2}mv_{\text{rms}}^2 = \tfrac{3}{2}k_BT \quad \text{(all gases)} \tag{10.15}$$

$$U = \tfrac{3}{2}Nk_BT = \tfrac{3}{2}nRT \quad \begin{matrix}\text{(ideal monatomic}\\\text{gases only)}\end{matrix} \tag{10.16}$$

$$U = \tfrac{5}{2}Nk_BT = \tfrac{5}{2}nRT \quad \text{(diatomic gases)} \tag{10.17}$$

Exercises*

10.2 The Celsius and Fahrenheit Temperature Scales

1. 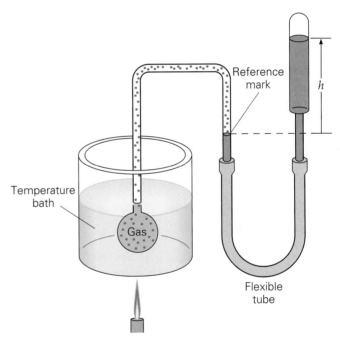 Which one of the following is the closest to 15°C? (a) 8.3°F, (b) 27°F, (c) 40°F, (d) 50°F?

2. Which of the following temperature scales has the smaller degree interval, Celsius or Fahrenheit?

3. Heat flows spontaneously from a body at a higher temperature to one at a lower temperature that is in thermal contact with it. Does it always flow from a body with more internal energy to one with less internal energy? Explain.

4. What is the hottest (highest temperature) item in a home? [Hint: Think about this, and maybe a light will come on.]

5. ■ Convert the following to Celsius readings: (a) 1000°F, (b) 0°F, (c) −20°F, (d) −40°F.

6. ■ Convert the following to Fahrenheit readings: (a) 150°C, (b) 32°C, (c) −25°C, (d) −273°C.

7. ■ The boiling point of alcohol is 172°F. What is this temperature on the Celsius scale?

8. ■ A person running a fever has a body temperature of 39.4°C. What is this temperature on the Fahrenheit scale?

9. ■ The highest and lowest recorded air temperatures in the United States are 134°F (Death Valley, California, 1913) and −80°F (Prospect Creek, Alaska, 1971), respectively. What are these temperatures on the Celsius scale?

10. ■ Which is the lower temperature, (a) 245°C or 245°F? (b) 200°C or 375°F?

11. ■ The highest and lowest recorded air temperatures in the world are 58°C (Libya, 1922) and −89°C (Antarctica, 1983). What are these temperatures on the Fahrenheit scale?

12. ■■ Suppose you didn't know that −40°C = −40°F (see Follow-up Exercise 10.1) and are asked if the Celsius and Fahrenheit temperatures are ever equal. How could you find out algebraically? Do it.

13. ■■■ (a) The largest temperature drop recorded in the United States in one day occurred in Browning, Montana, in 1916, when the temperature went from 7°C to −49°C. What is the corresponding change on the Fahrenheit scale? (b) The largest recorded temperature rise occurred in 2 min from −4°F to 45°F (Spearfish, South Dakota, 1943). What is the corresponding change on the Celsius scale?

*Assume all temperatures to be exact, and neglect significant figures for small dimension changes.

14. ■■■ (a) On the plot of Fahrenheit temperature versus Celsius temperature in Fig. 10.5a, the y-intercept is 32°F. What would be the x-intercept value? (b) What would the slope and y-intercept be if the graph were plotted in the opposite way (Celsius versus Fahrenheit)?

10.3 Gas Laws and Absolute Temperature

15. The temperature used in the ideal gas law is (a) Celsius, (b) Fahrenheit, (c) Kelvin, (d) any of the preceding.

16. When the temperature of a quantity of gas is increased, (a) the pressure must increase, (b) the volume must increase, (c) both the pressure and volume must increase, (d) none of the preceding.

17. The temperature of a quantity of gas is decreased. How is the density affected (a) if the pressure is held constant? (b) if the volume is held constant?

18. A type of constant-volume gas thermometer is shown in •Fig. 10.18. Describe how it operates.

•FIGURE 10.18   A type of constant-volume gas thermometer   See Exercise 18.

19. Describe how a constant-pressure gas thermometer might be constructed.

20. In terms of the ideal gas law, what would a temperature of absolute zero imply? a negative absolute temperature?

21. Excited about a New Year's Eve party in Times Square, you pumped up 10 balloons in your warm apartment and took them to the cold square. However, you were very disappointed. What happened? Why?

22. ■ 🌐 Convert the following temperatures to absolute temperatures in kelvins: (a) 0°C, (b) 100°C, (c) 20°C, and (d) −35°C.

23. ■ Convert the following temperatures to degrees Celsius: (a) 0 K, (b) 250 K, (c) 273 K, and (d) 325 K.

24. ■ What are the absolute temperatures in kelvins for (a) −40°F and (b) −40°C?

25. ■ Which is the lower temperature, 300°F or 300 K?

26. ■ The surface temperature of the Sun is about 6000 K. (a) What is this temperature on the Fahrenheit and Celsius scales? (b) The surface temperature is sometimes reported to be 6000°C. Assuming that 6000 K is correct, what is the percentage error of this Celsius value?

27. ■ How many moles are there in (a) 40 g of water, (b) 245 g of H_2SO_4 (sulfuric acid), (c) 138 g of NO_2 (nitrogen dioxide), (d) 56 L of SO_2 (sulfur dioxide) at STP?

28. ■ 🌐 A constant-volume gas thermometer exerts an absolute pressure of 1000 Pa at room temperature (20°C). If the pressure increases to 1500 Pa, what is the new Celsius temperature?

29. ■ The air in a balloon of volume 0.10 m³ exerts a pressure of 1.4×10^5 Pa. If the volume of the balloon increases to 0.12 m³ at a constant temperature, what is the pressure?

30. ■ 🌐 Show that 1.00 mol of ideal gas under STP (standard temperature of exactly 0°C and pressure of exactly 1 atm) occupies a volume of 0.0224 m³ = 22.4 L.

31. ■ 🌐 What is the volume occupied by 4.00 g of hydrogen under a pressure of 2.00 atm and a temperature of 300 K?

32. ■ How many molecules are there in (a) 1.00 L and (b) 1.00 cm³ of a gas at STP?

33. ■■ The Kelvin temperature of an ideal gas is doubled and its volume is halved. How is the pressure affected?

34. ■■ On a warm day (92°F), the air in a balloon occupies a volume of 0.20 m³ and exerts a pressure of 20.0 lb/in². If the balloon is cooled to 32°F in a refrigerator, the pressure decreases to 14.7 lb/in². What is the volume of the balloon? (Assume that the air behaves as an ideal gas.)

35. ■■ A steel-belted radial automobile tire is inflated to a gauge pressure of 30.0 lb/in² when the temperature is 61°F. Later in the day, the temperature rises to 100°F. Assuming the volume of the tire remains constant, what is the pressure at the elevated temperature? [Hint: Remember that the ideal gas law uses absolute pressure.]

36. ■■ If 2.4 m³ of a gas initially at STP is compressed to 1.6 m³ and its temperature raised to 30°C, what is the final pressure?

37. ■■ 🌐 The pressure on a low-density gas in a cylinder is kept constant as the temperature of the gas is increased from 10°C to 40°C. (a) Does the gas expand or compress? (b) What is the percentage change in the volume of the gas?

38. ■■■ A diver releases an air bubble of volume 2.0 cm³ from a depth of 15 m below the surface of a lake; there the temperature is 7.0°C. What is the volume of the bubble when it reaches the surface of the lake, where the temperature is 20°C?

10.4 Thermal Expansion

39. The units of the thermal coefficient of linear expansion are (a) m/C°, (b) m²/C°, (c) m·C°, (d) 1/C°.

40. The thermal coefficient of volume expansion for a solid is (a) α, (b) 2α, (c) 3α, (d) α^3.

41. A cube of ice sits on a bimetallic strip at room temperature (•Fig. 10.19). What will happen if (a) the upper strip is aluminum and the lower strip brass or (b) the upper strip is iron and the lower strip copper? (c) If the cube is made of a hot metal rather than ice and the two strips are brass and copper, should the brass or copper be on top to keep the cube from falling off?

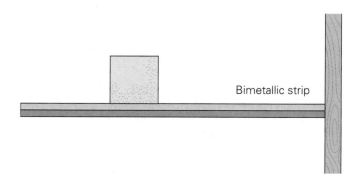

•FIGURE 10.19 **Which way will the cube go?** See Exercise 41.

42. A demonstration of thermal expansion is shown in •Fig. 10.20. Initially, the ball goes through the ring made of the same metal. (a) When the ball is heated, it does not go through the ring. (b) If both the ball and the ring are heated, the ball again goes through the ring. Explain what is being demonstrated.

43. What happens to a volume of water if it is cooled from 4°C to 2°C?

44. A solid metal disk rotates freely, so the conservation of angular momentum applies (Chapter 8). If the disk is heated while it is rotating, will there be any effect on the rate of rotation (the angular speed)?

45. A circular ring of iron has a tight-fitting iron bar across its diameter as illustrated in •Fig. 10.21. If the ring-and-bar arrangement is heated in an oven to a high temperature,

(a)   (b)   (c)

•**FIGURE 10.20   Ball-and-ring expansion**   See Exercises 42 and 53.

will the circular ring be distorted? What if the bar is made of aluminum?

•**FIGURE 10.21   Stress out of shape?**   See Exercise 45.

46. ■ 🔘 A steel beam 10 m long is installed in a structure at 20°C. What are its changes in length in millimeters at the temperature extremes of −30°C to 45°C?

47. ■ An aluminum tape measure is accurate at 20°C. If the tape measure is placed in a deep freeze at −5.0°C, what would be the tape measure's percentage error because of thermal contraction? Would it read high or low?

48. ■ Concrete highway slabs are poured in lengths of 10.0 m. How wide in millimeters should the expansion gaps between the slabs be at a temperature of 20°C to ensure that there will be no contact between adjacent slabs over a temperature range of −25°C to 45°C?

49. ■ A man's gold wedding ring has an inner diameter of 2.4 cm at 20°C. If the ring is dropped into boiling water, what will be the change in the inner diameter of the ring?

50. ■■ What temperature change would cause a 0.10% increase in the volume of a quantity of water that was initially at 20°C?

51. ■■ A piece of copper tubing used in plumbing has a length of 60.0 cm and an inner diameter of 1.50 cm at 20°C. When hot water at 85°C flows through the tube, what are (a) the tube's new length and (b) the change in its cross-sectional area? Can the latter affect the flow speed?

52. ■■ A circular piece of diameter 8.00 cm is cut from an aluminum sheet at 20°C. If the sheet is then placed in an oven and heated to 150°C, (a) will the hole be larger or smaller? (b) What will the new area of the hole be?

53. ■■ In Fig. 10.20, the steel ring of diameter 2.5 cm is 0.10 mm smaller in diameter than the steel ball at 20°C. If you have a torch, which should you heat so that the ball will go through the ring? What is the required temperature?

54. ■■ A circular steel plate of radius 0.10 m is cooled from 350°C to 20°C. By what percentage does the plate's area decrease?

55. ■■ One morning when the temperature is 10°C, an employee at a rent-a-car company fills the 25-gallon steel gas tank of an automobile to the top and then parks the car a short distance away. That afternoon, when the temperature is 30°C, gasoline drips from the tank onto the pavement. How much gas will be lost? (Neglect the expansion of the tank.) What is the lesson as to the best time to buy gas (if you drive right away and don't lose it!)?

56. ■■ Show that the density of a solid substance varies with temperature as $\rho = \rho_{\mathrm{o}}(1 - 3\alpha\,\Delta T)$ or $\rho = \rho_{\mathrm{o}}(1 - \beta\,\Delta T)$.

57. ■■ The density of mercury is 13.6×10^3 kg/m³ at 0°C. Calculate its density at 100°C.

58. ■■ A copper block has an internal spherical cavity of diameter 10 cm (•Fig. 10.22). The block is heated in an oven from 20°C to 500 K. (a) Does the cavity get larger or smaller? (b) What is the change in the cavity's volume?

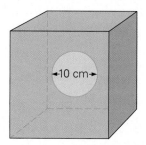

•**FIGURE 10.22   A hole in a block**   See Exercise 58.

59. ■■■ A brass rod has a circular cross section of radius 0.500 cm. It fits into a circular hole in a copper sheet with a clearance of 0.010 mm completely around it when both it and the sheet are at 20°C. (a) At what temperature will the clearance be zero? (b) Would such a tight fit be possible if the sheet were brass and the rod were copper?

60. ■■■ A Pyrex beaker that has a capacity of 1000 cm³ at 20°C contains 990 cm³ of mercury at that temperature. Is there some temperature at which the mercury will completely fill the beaker? Justify your answer. (Assume no mass loss by vaporization.)

10.5 The Kinetic Theory of Gases

61. If the kinetic energy of an average ideal gas molecule in a sample at 20°C doubles, its final temperature must be (a) 10°C, (b) 40°C, (c) 313°C, (d) none of the preceding.

62. 💿 If the temperature of a quantity of ideal gas is raised from 20°C to 40°C, its internal energy is (a) doubled, (b) tripled, (c) unchanged, (d) none of the preceding.

63. Equal volumes of helium gas (He) and neon gas (Ne) at the same temperature (and pressure) are on opposite sides of a porous membrane (•Fig. 10.23). Describe what happens after a period of time, and why.

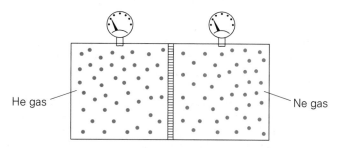

•**FIGURE 10.23   What happens as time passes?**   See Exercise 63.

64. Natural gas is odorless; to alert people to gas leaks, the gas company adds an additive with smell. When there is a gas leak, the additive reaches your nose before the gas

does. What can you conclude about the masses of the additive molecules and gas molecules?

65. ■ 💿 What is the average kinetic energy per molecule of an ideal gas at (a) 20°C and (b) 100°C?

66. ■ 💿 If the temperature of an ideal gas is raised from 20°C to 50°C, by how much will the internal energy of 2.0 mol of gas change?

67. ■ 💿 (a) What is the average kinetic energy per molecule of an ideal gas at a temperature of 27°C? (b) What is the average (rms) speed of the molecules if the gas is helium? (A helium molecule consists of a single atom of mass 6.65×10^{-27} kg.)

68. ■ 💿 What is the average speed of the molecules in low-density oxygen gas at 0°C? (The mass of an oxygen molecule, O_2, is 5.31×10^{-26} kg.)

69. ■■ 💿 At a given temperature, which would be greater, the rms speed of oxygen (O_2) or of ozone (O_3), and how many times greater?

70. ■■ If the temperature of an ideal gas is raised from 25°C to 100°C, how many times faster is the new average (rms) speed of the gas molecules?

71. ■■ A quantity of an ideal gas is at 0°C. An equal quantity of another ideal gas is twice as hot. What is its temperature?

*10.6 Kinetic Theory, Diatomic Gases, and the Equipartition Theorem

72. The temperature of a diatomic molecule such as O_2 is a measure of its (a) translational kinetic energy, (b) rotational kinetic energy, (c) vibrational kinetic energy, (d) all of the above.

73. On the average, the total internal energy of a gas is divided equally among (a) each atom, (b) each degree of freedom, (c) linear motion, rotational motion, and vibrational motion, (d) none of the above.

74. ■ If 1.0 mol of an ideal monatomic gas has a total internal energy of 5.0×10^3 J at a certain temperature, what is the total internal energy of 1.0 mol of a diatomic gas at the same temperature?

75. ■ What is the total internal energy of 1.0 mol of O_2 gas at 20°C?

76. ■■ For the average molecule in a sample of N_2 gas at 0°C, what is its (a) translational kinetic energy? (b) rotational kinetic energy? (c) total internal energy?

Additional Exercises

77. In Exercise 55, if the expansion of the steel gas tank is *not* ignored, what is the volume of gas spilled?

78. To conserve energy, thermostats in an office building are set at 78°F in the summer and 65°F in the winter. What would the settings be if the thermostats had a Celsius scale?

79. A copper wire 0.500 m long is at 20°C. If the temperature is increased to 100°C, what is the change in the wire's length?

80. If 2.0 mol of oxygen gas is confined in a 10-liter bottle under a pressure of 6.0 atm, what is the average translational kinetic energy of an oxygen molecule?

81. A mercury thermometer has a uniform capillary bore whose cross-sectional area is 0.012 mm^2. The volume of the mercury in the thermometer bulb at 10°C is 0.130 cm^3. If the temperature is increased to 50°C, how much will the height of the mercury column in the capillary bore change? (Neglect the expansion of the mercury in the bulb and of the glass of the thermometer.)

82. An aluminum cube is 10.0 cm on each side at 20°C. It is heated in an oven to 300°C. (a) What is the cube's new volume? (b) Using this volume, compute the length of each edge of the cube.

83. What temperature increase would produce a stress of 8.0×10^7 N/m^2 on a rigidly held steel beam?

84. Using the ideal gas law, express the coefficient of volume expansion in terms of temperature and pressure. [Hint: Use different temperatures and pressures.]

85. An ideal gas occupies a container of volume 0.75 L at STP. Find (a) the number of moles and (b) the number of molecules of the gas. (c) If the gas is carbon monoxide (CO), what is its mass?

86. An aluminum tape measure is used to measure a 1.0-meter-long steel beam. At 0°C, the measurement is accurate. What will the measurement be if the temperature increases to 40°C? (Ignore significant figures.)

87. (a) If the temperature drops by 10 C°, what is the corresponding temperature change on the Fahrenheit scale? (b) If the temperature rises by 10 F°, what is the corresponding change on the Celsius scale?

88. A new metal alloy is made into a 50-centimeter rod. (a) Determine the alloy's coefficient of linear expansion if, after being heated from 20°C to 250°C, the rod has elongated by 0.44 mm. (b) Is this a potentially valuable alloy? Explain.

89. Calculate the number of gas molecules in a container of volume 0.20 m^3 filled with gas under a partial vacuum of pressure 20 Pa at 20°C.

90. Is there a temperature that has the same numerical value on the Kelvin and Fahrenheit scales? Justify your answer.

91. Show that the coefficient of area expansion for solids is approximately equal to 2α.

92. In the troposphere (the lowest part of the atmosphere), the temperature decreases rather uniformly with altitude at a rate of about 6.5 C°/km; this decrease is called the *lapse rate*. What are the temperatures (a) near the top of the troposphere (which has an average thickness of 11 km) and (b) outside a commercial aircraft flying at a cruising altitude of 34 000 ft? (Assume that the ground temperature is normal room temperature.)

93. A mercury thermometer has a bulb volume of 0.200 cm^3 and a capillary bore diameter of 0.065 mm. How far up the bore will the column of mercury move if the overall temperature of the thermometer is increased by 30 C°? (Neglect the expansion of the glass of the thermometer.)

94. On a hot (40°C) day, copper electric wires are used to connect a power box to a home 150 m away. On a cold (−10°C) winter night, what should the minimum extra length of the wire be in centimeters to avoid snapping?

95. ⚫ A quantity of an ideal gas initially at atmospheric pressure is maintained at a constant temperature while it is compressed to half of its volume. What is the final pressure of the gas?

96. Use algebraic equations (not numbers) to demonstrate the discrepancy shown in Example 10.5.

Heat

INSIGHTS

- Phase Changes and Ice Skating
- The Greenhouse Effect
- Heat Transfer by Radiation:
  The Microwave Oven

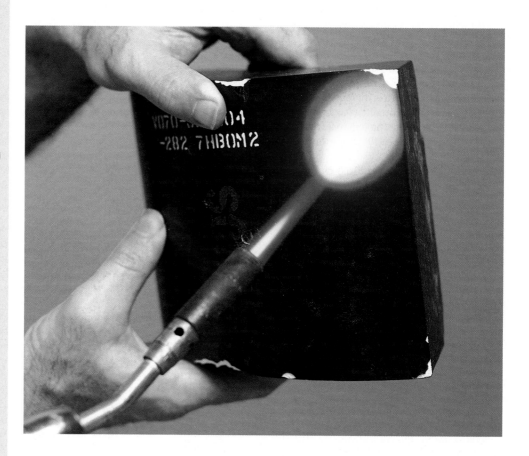

Most of us know to be very careful about handling anything that has recently been in contact with a flame or other source of heat. Yet, while the metal handle of a pot on the stove may be almost as hot as the bottom, a wooden pot handle is probably safe to touch. In other words, not only does heat flow through objects, but the rate of flow can vary widely (as the photograph shows in dramatic fashion). Direct physical contact isn't always necessary for heat to be transmitted: Think of touching the dashboard or the metal seat-belt buckle of a car left in direct sunlight for a few hours.

Think for a moment how crucial heat is in our existence. Our bodies must precisely balance heat loss and heat gain to stay within the narrow temperature range necessary for life. On a larger scale, the average temperature of the Earth, so critical to our environment and the survival of the organisms that inhabit it, is maintained through a similar balance. We are warmed by energy that comes to us across a 150-million-kilometer void. Each day, vast quantities of energy reach our planet's atmosphere and surface from the Sun, only to be radiated away into the cold of space.

These thermal balances are delicate, and any disturbance can have serious consequences. In an individual, sickness can disrupt the balance, producing a chill or fever. Similarly, we worry that an atmospheric buildup of "greenhouse" gases, a product of our industrial society, could give our whole planet a "fever" that would affect all life.

In this chapter you'll learn what heat is and how it is measured, and will study the various processes by which heat passes from one body to another. This knowledge will allow you to explain many everyday things as well as provide a basis for understanding the conversion of thermal energy into useful mechanical work.

11.1 Units of Heat

Objectives: **To (a) distinguish the various units of heat and (b) define the mechanical equivalent of heat.**

Like work, heat involves a transfer of energy. It is commonly thought that heat is a form of energy, but this is not true in the strictest sense. Rather, **heat** is the name we use to describe a type of net energy *transfer*. When we refer to "heat" or "heat energy" in this book, we mean the addition (or removal) of internal energy to a body or system.

Because heat is energy *in transit*, it is measurable as energy losses or gains and is described by standard energy units like any other quantity of energy. Recall that the SI standard unit of energy is the joule (J), or newton-meter (N·m). Thus, it is correct to say, for example, that 20 J of heat energy is transferred from one body to another.

As usual, we will use SI units, but we define for completeness other commonly used units of heat. A chief one is the **kilocalorie (kcal)** (•Fig. 11.1a):

> One kilocalorie (kcal) is defined as the amount of heat needed to raise the temperature of 1 kg of water 1 C° (from 14.5°C to 15.5°C).

Definition of:
The kilocalorie

The kilocalorie is technically known as the 15° kilocalorie. The temperature range is specified because the energy needed to raise 1 kg of water by 1 C° varies slightly with temperature. It is a minimum between 14.5°C and 15.5°C at a pressure of 1 atm. Over the temperature range in which water is a liquid, the variation is small and can be ignored for most purposes, as we will do.

For smaller quantities, the **calorie (cal)** is sometimes used (1 kcal = 1000 cal). *One calorie is the amount of heat needed to raise the temperature of 1 g of water by 1 C° (from 14.5°C to 15.5°C)* (Fig. 11.1b).

A familiar use of the larger kilocalorie is for specifying the energy values of foods. In this context the word is usually shortened to *Calorie* (Cal). That is, people on diets really count kilocalories; a piece of cake may contain 400 Cal (kcal), or 400 000 cal. The capital C distinguishes the larger kilogram-Calorie, or kilocalorie, from the smaller gram-calorie, or calorie. They are sometimes referred to as "big Calorie" and "little calorie." (In some countries, the joule is used for food values—see •Fig. 11.2.)

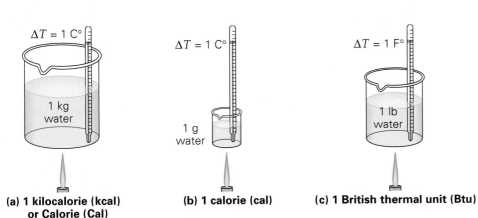

•**FIGURE 11.1 Units of heat**
(a) A kilocalorie raises the temperature of 1 kg of water by 1 C°. **(b)** A calorie raises the temperature of 1 g of water by 1 C°. **(c)** A Btu raises the temperature of 1 lb of water by 1 F°. (Not drawn to scale.)

•FIGURE 11.2   It's a joule   In Australia, diet drinks are labeled as being "low joule." In Germany, the labeling is a bit more specific: "Less than 4 kilojoules (1 kcal) in 0.3 Liter." How does this compare to our diet drinks?

Benjamin Thompson (1753–1814) was born in New England but went to England at the time of the American Revolution. He later worked in Bavaria, where he was made a count. He took the name Count Rumford after his birthplace (now known as Concord, New Hampshire).

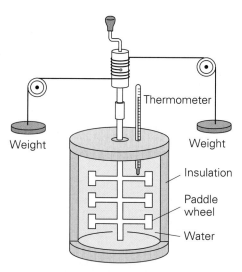

•FIGURE 11.3   Joule's apparatus for measuring the mechanical equivalent of heat   As the weights descend, the paddle wheels churn the water and mechanical energy or work is converted into heat energy. For every 4186 J of work done, the temperature of the water rises 1 C° per kilogram, and 4186 J is equivalent to 1 kilocalorie.

A unit of heat that is commonly used in industry is the **British thermal unit (Btu)**. *One Btu is the amount of heat needed to raise the temperature of 1 lb of water by 1 F° (from 63°F to 64°F)* (Fig. 11.1c), and 1 Btu = 252 cal = 0.252 kcal. If you buy an air conditioner or an electric heater, you will find it rated in Btus; for example, window air conditioners range from 4000 to 25 000 Btus. This number is actually Btus *per hour* and specifies how much heat energy the unit will remove (or deliver, in the case of a heater) during that time.

The Mechanical Equivalent of Heat

The idea that heat is a transfer of energy is the result of work by many scientists on the relationship of heat and energy. Some early observations were made by Benjamin Thompson, Count Rumford, while he was supervising the boring of cannon barrels in Germany. Rumford noticed that water put into the bore of the cannon to prevent overheating during drilling boiled away and had to be frequently replenished. He did several experiments, including ones in which he tried to detect "caloric fluid" by changes in the weights of heated substances. He eventually concluded that mechanical work was responsible for the heating of the water.

This conclusion was later proven quantitatively by James Joule, the English scientist after whom the SI unit for work and energy is named. Using the apparatus shown in •Fig. 11.3, Joule demonstrated that when a given amount of mechanical work was done, the water was heated, as indicated by an increase in temperature. He found that for every 4186 J of work done, the temperature of the water rose 1 C° per kg, or that 4186 J was equivalent to 1 kcal:

$$1 \text{ kcal} = 4186 \text{ J} = 4.186 \text{ kJ}$$

This relationship is called the **mechanical equivalent of heat**. It provides a conversion factor between heat and mechanical (standard) units, such as

$$1 \text{ cal} = 4.186 \text{ J}$$

Example 11.1 ■ Stirred, Not Shaken: Mechanical Equivalent of Heat

How much mechanical work in joules would have to be done in Joule's apparatus to raise the temperature of a liter of water from room temperature (20°C) to 25°C?

Thinking It Through.   We learned that temperature rises 1 C° per kilogram for every 4186 J of work. The relationship of a liter of water to mass is the key. Temperatures are assumed to be exact.

Solution.   Listing the given data and what we want to find, we have

Given: $T_i = 20°C$ *Find:* W (work in joules)
            $T_f = 25°C$
            $V = 1.0$ L

where T_i and T_f are the initial and final temperatures, respectively.

From Joule's mechanical equivalent of heat, we know how much heat energy (symbolized Q) it takes to raise the temperature of a certain amount of water 1 C°. Here, the amount of water is given in terms of its volume. However, we know from Chapter 1 that 1.0 L of water has a mass of 1.0 kg.

Since 4.186 kJ raises the temperature of 1 kg of water 1 C°, to raise the temperature of 1 kg of water by $\Delta T = T_f - T_i = 25°C - 20°C = 5$ C° would require five times this amount of heat energy, if we assume no heat loss. This energy is supplied by mechanical work, and

$$W = Q = 4.186 \text{ kJ/C°} \times 5 \text{ C°} = 20.93 \text{ kJ} = 2.093 \times 10^4 \text{ J}$$

Follow-up Exercise.   To get an idea of the mechanical equivalent of 5.00 kcal of heat, to what height could a 100-kilogram object (of equivalent weight 220 lb) be lifted by using this amount of energy?

11.2 Specific Heat

Objectives: To (a) describe specific heat and (b) explain how the specific heats of materials are obtained from calorimetry.

Note: Review Section 10.1 and Fig. 10.1.

Recall from Chapter 10 that when heat is added to a substance, the energy may go to increase the random molecular motion, which results in a temperature change, and also to increase the potential energy associated with the molecular bonds. Different substances have different molecular configurations and bonding patterns. Thus, if equal amounts of heat are added to equal masses of different substances, the resulting temperature changes will *not* generally be the same. For example, suppose that you have 1 kg of water and 1 kg of aluminum and add 1 kcal of heat to each. From the definition of the kilocalorie, you know that the temperature of the water will rise 1 C°. But the temperature of the aluminum will increase by 4.5 C°. The reason for this difference is that more of the added energy goes into the nontranslational ("nontemperature") part of the internal energy of the water than of the aluminum.

The amount of heat (Q) required to change the temperature of a substance is proportional to the mass (m) of the substance and to the change in the temperature (ΔT). That is, $Q \propto m \Delta T$, or, in equation form,

$$Q = mc\,\Delta T \quad \text{or} \quad c = \frac{Q}{m\,\Delta T} \qquad (11.1)$$

**Definition of:
Specific heat**

Here $\Delta T = T_f - T_i$ is the temperature change, or the difference between final temperature (T_f) and the initial temperature (T_i), and c is the *specific heat capacity*, or simply the **specific heat**. It is characteristic of, or *specific* for, a given substance and is independent of m. Thus, the specific heat gives us an indication of a material's internal molecular configuration and bonding.*

Writing Eq. 11.1 as $c = Q/(m\,\Delta T)$ shows us that the units of specific heat are J/(kg·K) [or kcal/(kg·C°)]. The standard SI unit of specific heat is the one with the temperature in kelvins. However, the use of Celsius temperature in this unit [J/(kg·C°)] is commonly accepted because the size of a Celsius degree is the same as the size of a kelvin.

Note that the *specific heat is the amount of heat energy required to raise the temperature of 1 kg of a substance by 1 C°*. The specific heats of some common substances are given in Table 11.1. Specific heat depends somewhat on temperature (and pressure); but over normal temperature intervals, such variations can be neglected and c can be considered constant.

The greater the specific heat of a substance, the more energy must be transferred to it or taken from it to change the temperature of a given mass of it. That is, a substance with a higher specific heat has a greater capacity for heat, or accepts or yields more heat for a given temperature change (and mass).

Water has a relatively large specific heat of 4186 J/(kg·C°), or 1.00 kcal/(kg·C°). (The kilocalorie value is exactly 1.00 because the definition of the kilocalorie states that 1 kcal raises the temperature of 1 kg of water by 1 C°.)

You have been the victim of the high specific heat of water if you have ever burned your mouth on a baked potato or the cheese of a pizza. These foods have a high water content and a large heat capacity, so they don't cool off as quickly as some other foods do.

Note from Eq. 11.1 that when there is a temperature increase, ΔT is positive ($T_f > T_i$) and Q is positive. This corresponds to energy being *added to* a system. Conversely, ΔT and Q are negative when energy is *removed from* a system.

$+Q$ = heat added
$-Q$ = heat removed

---

*This may also be expressed in terms of *molar heat capacity*, in which the mass is expressed in term of moles (Section 10.3) rather than kilograms. This is common in chemical analyses, particularly when working with gases.

TABLE 11.1 Specific Heats of Various Substances (at 20°C and 1 atm)

Substance	Specific heat (c)	
	J/(kg·C°)	kcal/(kg·C°) [or cal/(g·C°)]
Air (50°C)	1050	0.25
Alcohol, ethyl	2430	0.58
Aluminum	920	0.22
Copper	390	0.093
Glass	840	0.20
Ice (−5°C)	2100	0.50
Iron or steel	460	0.11
Lead	130	0.031
Mercury	140	0.033
Soil (average)	1050	0.25
Steam (110°C)	2010	0.48
Water (15°C)	4186	1.00
Wood (average)	1680	0.40

Example 11.2 ■ Warming Up: Specific Heat

How much heat is required to raise the temperature of 0.20 kg of water from 15°C to 45°C?

Thinking It Through. This is a direct application of Eq. 11.1. All quantities are known, including the specific heat of water (Table 11.1).

Solution.

Given: $m = 0.20$ kg $\quad\quad\quad\quad\quad$ *Find:* Q (heat)
$\quad\quad\Delta T = T_f - T_i = 45°C - 15°C = 30\ C°$
$\quad\quad\quad c = 4186$ J/(kg·C°)
$\quad\quad\quad\quad$ (from Table 11.1)

Using Eq. 11.1, we get

$$Q = mc\,\Delta T = (0.20\ \text{kg})[4186\ \text{J/(kg·C°)}](30\ \text{C°}) = +2.5 \times 10^4\ \text{J}$$

where the + sign indicates that heat was added to the system.

Follow-up Exercise. How much heat (in joules) would have to be removed from 0.20 kg of water at 50°C in order to lower its temperature to 20°C?

Example 11.3 ■ Cooling Down: Specific Heat Again

A half-liter of water at 30°C is cooled, with the removal of 63 kJ of heat. What is the final temperature of the water?

Thinking It Through. Recall that 1.0 L of water has a mass of 1.0 kg; hence the mass of the water is half that, or 0.50 kg. We want T_f; note that it is part of ΔT in Eq. 11.1.

Solution.

Given: $m = 0.50$ kg $\quad\quad\quad\quad\quad$ *Find:* T_f (final temperature)
$\quad\quad\quad T_i = 30\ C°$
$\quad\quad\quad Q = -63$ kJ
$\quad\quad\quad\quad$ (negative because heat
$\quad\quad\quad\quad$ energy is removed)
$\quad\quad\quad c = 4.186$ kJ/(kg·C°)
$\quad\quad\quad\quad$ (from Table 11.1)

With relatively large numbers, it can be convenient to work in kilojoules rather than the standard joules. Writing out the ΔT term of Eq. 11.1 gives

$$Q = mc\,\Delta T = mc\,(T_f - T_i)$$

Solving for T_f gives

$$T_f = \frac{Q}{mc} + T_i = \frac{-63\text{ kJ}}{(0.50\text{ kg})[4.186\text{ kJ}/(\text{kg}\cdot\text{C}°)]} + 30°\text{C} = -30°\text{C} + 30°\text{C} = 0°\text{C}$$

The (liquid) water is at its freezing point; the removal of more heat would cause the water to freeze. However, Eq. 11.1 *does not apply* when the temperature change (ΔT) is an interval that includes a change of phase, as you will learn in Section 11.3.

Follow-up Exercise. Suppose in this Example that 30 kJ of heat were initially removed, then 60 kJ added. What would be the final temperature of the water?

Example 11.4 ■ Heat Capacity: Different Materials

Equal masses of aluminum (Al) and copper (Cu) are at the same temperature. Which will require the greater heat to raise its temperature by a given amount, and how many times greater is this than the heat that would have to be added to the other metal?

Thinking It Through. The first part of the question can be answered from the definition and values of specific heat. The second part is more detailed, but "how many times greater" implies a factor or ratio.

Solution.

Given: $m_{Al} = m_{Cu}$
$\Delta T_{Al} = \Delta T_{Cu}$
$c_{Al} = 920\text{ J}/(\text{kg}\cdot\text{C}°)$
$c_{Cu} = 390\text{ J}/(\text{kg}\cdot\text{C}°)$
(from Table 11.1)

Find: Which material requires the greater value of Q and how many times greater it is

Since aluminum has a larger specific heat, it has a higher heat capacity, and more heat is required to raise its temperature by a given amount.

If we find $x = Q_{Al}/Q_{Cu}$, we can write $Q_{Al} = xQ_{Cu}$, or Q_{Al} is x times greater than Q_{Cu}. To find the ratio we use Eq. 11.1 for each metal and divide one by the other:

$$\frac{Q_{Al}}{Q_{Cu}} = \left(\frac{m_{Al}}{m_{Cu}}\right)\left(\frac{c_{Al}}{c_{Cu}}\right)\left(\frac{\Delta T_{Al}}{\Delta T_{Cu}}\right) = \frac{c_{Al}}{c_{Cu}}$$

where quantities that are given as being equal (mass and ΔT) have been grouped in parentheses so that it can be easily seen that they cancel out. Then,

$$Q_{Al} = \left(\frac{c_{Al}}{c_{Cu}}\right)Q_{Cu} = \left[\frac{920\text{ J}/(\text{kg}\cdot\text{C}°)}{390\text{ J}/(\text{kg}\cdot\text{C}°)}\right]Q_{Cu}$$

$$Q_{Al} = 2.36Q_{Cu}$$

That is, 2.36 times more heat is required for the aluminum than for the copper.

Follow-up Exercise. To illustrate that the specific heat of water is relatively large, let equal amounts of heat be added to equal masses of water and mercury initially at the same temperature. Which liquid has the greater temperature change, and how many times greater?

Problem-Solving Hint

When the quantities given in a problem are equal, such as the masses and temperature intervals in Example 11.4, it is a good indication that a ratio can be used to cancel out the equal quantities. Also keep in mind that phrases such as "how many times more" imply a factor derived from a ratio.

•FIGURE 11.4   Calorimetry apparatus The calorimeter cup (center, with black insulating ring) goes into the larger container. The cover with the thermometer and stirrer is seen at the right. Metal shot or pieces of metal are heated in the small cup with handle that is inserted into the hole at the top of the steam generator on the tripod.

Calorimetry

Calorimetry is the quantitative measure of heat exchange, which allows us to determine the specific heats of substances. Such measurements are made by using a *calorimeter* (cal-oh-RIM-i-ter), an insulated container that allows little heat loss (ideally none). A simple laboratory calorimeter is shown in •Fig. 11.4.

The specific heat of a substance is determined by measuring the quantities in Eq. 11.2 ($Q = mc\,\Delta T$) other than c. A substance of known mass and temperature is put into a quantity of water in a calorimeter. The water is at a different temperature, usually a lower one. The principle of the conservation of energy is then applied to determine c. This procedure is sometimes called the *method of mixtures*. An example follows. Keep in mind that such heat exchange problems are just a matter of "thermal accounting." If something loses heat, something else must gain heat—that is, $\Sigma Q_i = 0$—or the conservation of energy would be violated.

Example 11.5 ■ Calorimetry: The Method of Mixtures

Students in a physics lab are to determine the specific heat of copper experimentally. They heat 0.150 kg of copper shot to 100°C and then carefully pour the hot shot into a calorimeter cup (Fig. 11.4) containing 0.200 kg of water at 20°C. The final temperature of the mixture in the cup is measured to be 25°C. If the aluminum cup has a mass of 0.045 kg, what is the specific heat of copper? (Assume that there is no heat loss to the surroundings.)

Thinking It Through.   The conservation of energy is involved: The sum of heats lost = the sum of heats gained. In calorimetry problems, it is important to identify and label all of the quantities so as to keep them straight. Identification of the heat gains and losses is thus very important. You will probably use this method in the laboratory.

Solution.   We will use subscripts Cu, H_2O, and Al to refer to the copper, water, and aluminum calorimeter cup, respectively, and the subscripts h, i, and f to refer to the temperatures of the hot metal shot, the water (and cup) initially at room temperature, and the final temperature of the system, respectively. With this notation, we have,

Given: $m_{Cu} = 0.150$ kg   *Find:* c_{Cu} (specific heat)
$\quad\quad m_{H_2O} = 0.200$ kg
$\quad\quad c_{H_2O} = 4186$ J/(kg·C°)
$\quad\quad\quad$ (from Table 11.1)
$\quad\quad m_{Al} = 0.045$ kg
$\quad\quad c_{Al} = 920$ J/(kg·C°)
$\quad\quad T_h = 100°C, T_i = 20°C,$
$\quad\quad\quad$ and $T_f = 25°C$

If no heat is lost from the system, its energy is conserved: $\Sigma Q_i = 0$, and

$$\Sigma Q_i = Q_{H_2O} + Q_{Al} + Q_{Cu} = 0$$

or

$$Q_{H_2O} + Q_{Al} = -Q_{Cu}$$

(heats gained by water and cup) = (heat lost by metal)

Substituting Eq. 11.1 ($Q = mc\,\Delta T$) for these heats, we have

$$m_{H_2O}c_{H_2O}\,\Delta T_{H_2O} + m_{Al}c_{Al}\,\Delta T_{Al} = -m_{Cu}c_{Cu}\,\Delta T_{Cu}$$

or

$$m_{H_2O}c_{H_2O}(T_f - T_i) + m_{Al}c_{Al}(T_f - T_i) = -m_{Cu}c_{Cu}(T_f - T_h)$$

Here the water and aluminum cup initially at T_i are heated to T_f, so $\Delta T_{H_2O} = \Delta T_{Al} = (T_f - T_i)$; and the copper initailly at T_h is cooled to T_f, so $\Delta T_{Cu} = (T_f - T_h)$. Solving for c_{Cu} gives

$$c_{Cu} = \frac{(m_{H_2O}c_{H_2O} + m_{Al}c_{Al})(T_f - T_i)}{-m_{Cu}(T_f - T_h)}$$

$$= \frac{\{(0.200 \text{ kg})[4186 \text{ J}/(\text{kg} \cdot \text{C}°)] + (0.045 \text{ kg})[920 \text{ J}/(\text{kg} \cdot \text{C}°)]\}(25°C - 20°C)}{-0.150 \text{ kg}(25°C - 100°C)}$$

$$= 390 \text{ J}/(\text{kg} \cdot \text{C}°)$$

Follow-up Exercise. In actual calorimetry experiments, some heat is lost from the system. What effect would this have on the experimentally determined specific heat?

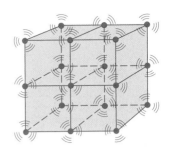

(a) Solid

11.3 Phase Changes and Latent Heat

Objectives: To (a) compare and contrast the three common phases of matter and (b) relate latent heat to phase changes.

Matter normally exists in one of three *phases*: solid, liquid, or gas (•Fig. 11.5). The phase that a substance is in depends on the substance's internal energy (as indicated by its temperature) and the pressure on it. You probably think more readily of adding or removing heat to change the phase of a substance because most of your experience with phase changes has been at normal atmospheric pressure, which is relatively constant.

In the **solid phase**, molecules are held together by attractive forces, or bonds. Simplistically, these bonds can be represented as springs (Fig. 11.5a), as was done in Section 9.1. Adding heat causes increased motion about the molecular equilibrium positions. If enough heat is added to provide sufficient energy to break the intermolecular bonds, the solid undergoes a phase change and becomes a liquid. The temperature at which this occurs is called the **melting point**. The temperature at which a liquid becomes a solid is called the **freezing point**. In general, these temperatures are the same for a given substance, but they can differ slightly.

Ice, table salt, and most metals are *crystalline* solids. That is, they have orderly molecular or atomic arrangements. Other substances, however, such as glass, are noncrystalline, or *amorphous*. Instead of melting at a particular temperature, these substances melt over a temperature range. We will be concerned primarily with substances that have definite melting points.

In the **liquid phase**, molecules of a substance are relatively free to move, and a liquid assumes the shape of its container (Fig. 11.5b). In certain liquids, there may be some locally ordered structure over several molecular lengths, giving rise to so-called liquid crystals, such as are used in LCDs (liquid crystal displays) of calculators and clocks (Chapter 24). Adding heat increases the motion of the molecules of a liquid. When the molecules have enough energy to become separated by large distances (compared to their diameters), the liquid changes to the **gaseous phase**, or **vapor phase** (Fig. 11.5c). (The distinction between a gas and a vapor will be made shortly.) This change may occur slowly by the process of evaporation or rapidly at a particular temperature called the **boiling point**. The temperature at which a gas condenses and becomes a liquid is the **condensation point**.

Some solids, such as dry ice (solid carbon dioxide), moth balls, and certain air fresheners, change directly from the solid to the gaseous phase. This process is called **sublimation**. Like the rate of evaporation, the rate of sublimation increases with temperature. A phase change from a gas to a solid is called *deposition*. Frost, for example, is solidified water vapor deposited directly on grass, car windows, and other objects. Frost is not frozen dew, as is sometimes mistakenly assumed.

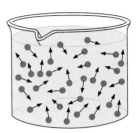

(b) Liquid

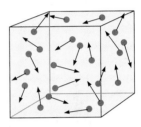

(c) Gas

•**FIGURE 11.5 Three phases of matter** (a) The molecules of a solid are held together by bonds; consequently, a solid has definite shape and volume. (b) The molecules of a liquid can move more freely, so a liquid has a definite volume and assumes the shape of its container. (c) The molecules of a gas interact weakly and are separated by relatively large distances; thus, a gas has no definite shape or volume.

Note: Solid, liquid, and gas are sometimes referred to as states of matter rather than phases of matter, but the *state* of a system has a different meaning in physics, as you will learn in Chapter 12.

Latent Heat

In general, when heat energy is transferred to a substance, the substance's temperature increases. However, when heat that is added (or removed) causes only a phase change, the temperature of the substance does *not* change. For example, if heat is added to a quantity of ice [$c = 2100$ J/(kg·C°) or 0.50 kcal/(kg·C°)] at $-10°C$, the

Note: Keep in mind that ice can be colder than 0°C.

Substance	Melting point	L_f		Boiling point	L_v	
		J/kg	kcal/kg		J/kg	kcal/kg
Alcohol, ethyl	−114°C	1.0×10^5	25	78°C	8.5×10^5	204
Gold	1063°C	0.645×10^5	15.4	2660°C	15.8×10^5	377
Helium*	—	—	—	−269°C	0.21×10^5	5
Lead	328°C	0.25×10^5	5.9	1744°C	8.67×10^5	207
Mercury	−39°C	0.12×10^5	2.8	357°C	2.7×10^5	65
Nitrogen	−210°C	0.26×10^5	6.1	−196°C	2.0×10^5	48
Oxygen	−219°C	0.14×10^5	3.3	−183°C	2.1×10^5	51
Tungsten	3410°C	1.8×10^5	44	5900°C	48.2×10^5	1150
Water	0°C	3.33×10^5	80	100°C	22.6×10^5	540

*Not a solid at 1 atm pressure; melting point −272°C at 26 atm.

temperature of the ice increases until it reaches its melting point of 0°C. At this point, the addition of more heat does not increase the temperature but causes the ice to melt, or change phase. (The heat must be added slowly so that the ice and melted water remain in thermal equilibrium.) Once the ice is melted, adding more heat will again cause the temperature of the water to rise. A similar situation occurs during the liquid–gas phase change at the boiling point. Adding more heat to boiling water only causes more vaporization, not a temperature increase.

From the earlier description of the molecular nature of the three phases of matter, you can see that during a phase change, the heat energy goes into the work of breaking bonds and separating molecules rather than into increasing the temperature. The heat energy involved in a phase change is called the **latent heat** (L)*, and

Latent heat and phase changes

$$Q = mL \quad \text{or} \quad L = \frac{Q}{m} \tag{11.2}$$

where m is the mass of the substance. As this equation shows, the latent heat has units of joules per kilogram (J/kg) in the SI, or kilocalories per kilogram (kcal/kg).

The latent heat for a solid–liquid phase change is called the **latent heat of fusion** (L_f), and that for a liquid–gas phase change is called the **latent heat of vaporization** (L_v). These are often referred to as simply the heat of fusion and the heat of vaporization. The latent heats of some substances, along with their melting and boiling points, are given in Table 11.2. (The latent heat for the less common solid–gas phase change is called the *latent heat of sublimation* and is symbolized by L_s.) As you might expect from the conservation of energy, the latent heat (in joules per kilogram) is the amount of energy per kilogram given up when the phase change is in the opposite direction, from liquid to solid or gas to liquid.

It is helpful to focus on the fusion and vaporization of water. A plot of temperature versus heat energy for a quantity of water is shown in •Fig. 11.6. Note that when heat is added (or removed) at the temperature of a phase change, 0°C or 100°C, the temperature remains constant. Once the phase change is complete, adding more heat causes the temperature to increase. The slopes of the phase lines in Fig. 11.6 are not all the same, which indicates that the specific heats of the various phases are not all the same. (Why?)

For water, the latent heats of fusion and vaporization are

$$L_f = 3.33 \times 10^5 \text{ J/kg} \quad \text{(or about 80 kcal/kg)}$$

$$L_v = 22.6 \times 10^5 \text{ J/kg} \quad \text{(or about 540 kcal/kg)}$$

latent heats for water

*Latent is Latin for "hidden"; Q seems to be disappearing with no change in temperature.

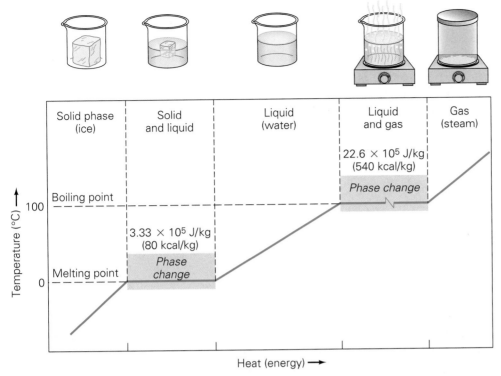

•FIGURE 11.6 **Temperature versus heat for water** As heat is added to the various phases of water, the temperature increases. During a phase change, however, the heat energy does the work of separating the molecules and the temperature remains constant. Note the different slopes of the phase lines, which indicate different values of specific heat. (Not drawn to scale.)

That is, 3.33×10^5 J (or 80 kcal) of heat is needed to melt 1 kg of ice at 0°C, and 22.6×10^5 J (or 540 kcal) of heat is needed to convert 1 kg of water to steam at 100°C (•Fig. 11.7). Note that the latent heat of vaporization is almost 7 times the latent heat of fusion. This indicates that more energy is needed to separate the molecules in going from water to steam than to break up the lattice structure in going from ice to water.

The term "latent" may be understood by considering a situation involving human skin. Since 22.6×10^5 J of heat energy is required to convert 1 kg of boiling water into steam, the conservation of energy tells you that when 1 kg of steam

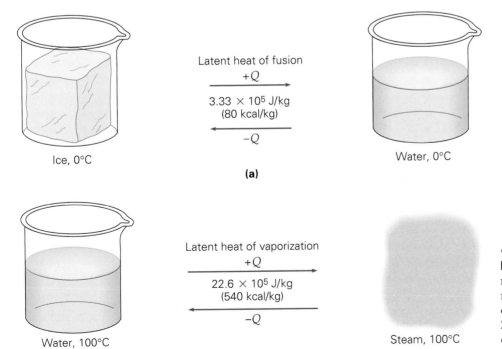

•FIGURE 11.7 **Phase changes and latent heats** (a) At 0°C, 3.33×10^5 J must be added to 1 kg of ice or removed from 1 kg of water to change its phase. (b) At 100°C, 22.6×10^5 J must be added to 1 kg of water or removed from 1 kg of steam to change its phase.

condenses into water, 22.6×10^5 J of energy must be given up. As a result, burns from steam are usually more serious than those from boiling water (for equal masses). The condensing of the steam on the skin provides an additional 22.6×10^5 J/kg of heat that is seemingly hidden until contact.

Example 11.6 ■ Temperature Changes and Phase Changes: Specific Heat and Latent Heat

Heat is added to 0.500 kg of water at room temperature (20°C). How much heat in joules is required to change the water to steam at 110°C?

Thinking It Through. Three steps are involved: heating water (specific heat), a phase change (latent heat), and heating steam (specific heat).

Solution.

Given: $m = 0.500$ kg   *Find:* Q (total heat in joules)
$\quad\quad\quad T_i = 20°C, T_f = 110°C$
$\quad\quad\quad L_v = 22.6 \times 10^5$ J/kg (from Table 11.2)
$\quad\quad\quad$ (specific heats from Table 11.1)

The temperature intervals and the heats required in the process are shown in •Fig. 11.8. First, the water is heated to the boiling point (100°C), where a phase change occurs, and then the steam is heated to 110°C. The Q's for the various steps are

(heating water) $Q_1 = mc_w \Delta T_1 = (0.500$ kg$)[4186$ J/(kg·C°)$](100°C - 20°C) = 1.67 \times 10^5$ J

(phase change) $Q_{L_v} = mL_v = (0.500$ kg$)(22.6 \times 10^5$ J/kg$) = 11.3 \times 10^5$ J

(heating steam) $Q_2 = mc_s \Delta T_2 = (0.500$ kg$)[2010$ J/(kg·C°)$](110°C - 100°C) = 0.101 \times 10^5$ J

Then, rounding, we get

$$Q_{total} = \Sigma_i Q_i = (1.7 + 11.3 + 0.1) \times 10^5 \text{ J} = +13.1 \times 10^5 \text{ J}$$

Follow-up Exercise. How many joules of heat would have to be removed to condense 1.00 kg of steam at 110°C?

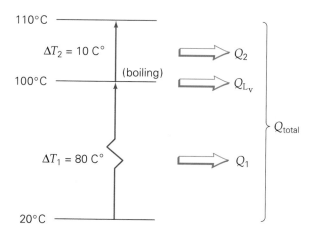

•**FIGURE 11.8   Steps in changing phases**   In going from water at 20°C to steam at 110°C, three heat-adding steps are involved. See Example 11.6.

Problem-Solving Hint

Note that you must compute the latent heat at each phase change. It is a common error to use the specific heat equation with a temperature interval that includes a phase change. Also, you cannot assume a complete phase change until you have checked it numerically.

Example 11.7 ■ Water and Ice: Thermal Equilibrium

A 0.30-kilogram piece of ice at 0°C is placed in a liter of water at room temperature (20°C) in an insulated container. Assuming that no heat is lost to the container, what is the final temperature of the water?

Thinking It Through. The final water temperature will be less than room temperature because heat is removed from the water to melt the ice (latent heat). Some ice will melt, but will all of it? To find out, we compare how much latent heat would be required to melt the total mass to how much energy the water has to give up in going to 0°C. Once this is determined, then we will know how to proceed—a mixture of ice and water at 0°C or melting and warming of the ice water to a final temperature that we must determine.

Solution. We list the data as usual:

Given: m_{ice} = 0.30 kg   *Find:* T_f (final temperature)
T_{ice} = 0°C
V_{H_2O} = 1.0 L, so m_{H_2O} = 1.0 kg
T_{H_2O} = 20°C
(latent heats from Table 11.2)

We know that 1.0 L of water has a mass (m_{H_2O}) of 1.0 kg and that the water supplies the heat to melt the ice. If all of the ice melts, we could then view the system as being two masses of water at different temperatures, which come to equilibrium at some intermediate temperature.

Since there is no heat loss, it is tempting to write an expression equating the amount of heat lost by the water to the amounts of heat used to melt the ice and to warm up the ice water. But does *all* the ice melt? To melt 0.30 kg of ice requires

$$Q_{ice} = m_{ice}L_f = (0.30 \text{ kg})(3.33 \times 10^5 \text{ J/kg}) = 1.0 \times 10^5 \text{ J}$$

Then, looking at the water, you must ask how much heat it can supply to the ice. The maximum amount would be that given up in lowering the temperature of the water to 0°C, which would be a temperature decrease of ΔT_{H_2O} = −20 C°. This *maximum* amount of heat would be

$$Q_{H_2O} = m_{H_2O}c_{H_2O}\Delta T_{H_2O} = (1.0 \text{ kg})[4186 \text{ J/(kg·C°)}](-20 \text{ C°}) = -8.4 \times 10^4 \text{ J}$$

Thus, all of the ice does *not* melt, since the water does not have enough energy to do so. The final temperature of the water is therefore 0°C.

The heat given up by the water (Q_{H_2O} = 8.4 × 10⁴ J) will melt a mass of ice equal to

$$m_{ice} = \frac{Q_{H_2O}}{L_f} = \frac{8.4 \times 10^4 \text{ J}}{3.33 \times 10^5 \text{ J/kg}} = 0.25 \text{ kg}$$

Thus, the final mixture would be 0.30 kg − 0.25 kg = 0.05 kg of ice in thermal equilibrium with 1.25 kg (or 1.25 L) of water at T_f = 0°C.

Follow-up Exercise. In this Example, what would the initial temperature of the water have to be to melt all of the ice but heat up no further?

Note: This is a matter of "thermal accounting"—if something loses, something else gains.

Note that the freezing and boiling points of 0°C and 100°C, respectively, for water apply at 1 atm of pressure. The phase-change temperatures vary with pressure. For example, the boiling point of water decreases with decreasing pressure. In fact, a container of water in a vacuum chamber will boil at room temperature. The cooling effect of the boiling (the removal of latent heat) will eventually cause the remaining water to freeze if the vacuum is maintained.

At high altitudes, where there is lower atmospheric pressure, the boiling point of water is lowered. For example, at Pike's Peak, Colorado, at an elevation of about 4300 m, the atmospheric pressure is about 600 torr, and water boils at about 94°C rather than at 100°C. The lower temperature lengthens the cooking time of food. A pressure cooker can be used to reduce the cooking time (at Pike's Peak or at sea level)—by increasing the pressure, a pressure cooker raises the boiling point.

Phase Changes and Ice Skating

It was once thought that the lowering of the freezing point of water by pressure provided the mechanism that makes ice skating possible. Supposedly, the pressure of the narrow skate blade on the ice would lower the melting point below the ambient temperature. The skater would thus glide on a thin film of water, which quickly refroze when the pressure was removed. However, for water, a pressure of about 120 atm is needed to lower the freezing point from 0°C to −1°C.

A typical skate blade has an area of about 27 cm × 0.40 cm = 11 cm² = 11 × 10⁻⁴ m². For a 70-kilogram skater with all the weight on one blade, the pressure would be $p = F/A = mg/A = (70 \text{ kg})(9.8 \text{ m/s})(11 \times 10^{-4} \text{ m}^2) = 6.2 \times 10^5 \text{ N/m}^2$, or approximately 6 atm. This is hardly enough, and the outdoor ice skating temperature is usually well below −1°C.

Frictional heating between the skate blade and the ice does contribute to melting. For high-friction surfaces, such as skis on snow, this is the main mechanism. However, for ice skating another factor contributes to the low coefficient of friction of ice. This factor is called *surface melting*, proposed by the English scientist Michael Faraday (1791–1867; Chapter 20). He suggested that a thin layer of liquid normally exists on the surface of a solid even at temperatures well below the solid's melting point. Modern techniques have shown this to be the case for most solids. For ice, the thickness of the water film is about 40 nm near 0°C and about 0.50 nm near −35°C. (Recall that 1 nm = 10⁻⁹ m.)

Conversely, the freezing point of water decreases with increasing pressure. This inverse relationship is characteristic of only a very few substances, including water (Section 10.4), that expand when they freeze. Pressure and ice skating are discussed in the Insight on this page.

Evaporation

The **evaporation** of water from an open container becomes evident only after a relatively long period of time. This phenomenon can be explained in terms of the kinetic theory of gases (Section 10.5). The molecules in a liquid are in motion at different speeds. A faster-moving molecule near the surface may momentarily leave the liquid. If its speed is not too large, the molecule will return to the liquid because of the attractive forces exerted by the other molecules. Occasionally, however, a molecule has a large enough speed to leave the liquid entirely and become part of the air. The higher the temperature of the liquid, the more likely this is to occur.

The escaping molecules take their energy with them. Since those molecules with greater-than-average energy are the ones most likely to escape, the energy and temperature of the remaining liquid will be reduced. Thus, *evaporation is a cooling process* for the object from which the molecules escape. You have probably noticed this when drying off after a bath or shower. About 2500 kJ is needed to evaporate 1 kg of water from the skin. This energy requirement can be estimated by considering a different process. The amount of heat per kilogram (Q/m) needed to raise the temperature of water from 35°C (approximately skin temperature) to 100°C ($\Delta T = 100°C − 35°C = 65 \text{ C}°$) is $c \Delta T$ (since $Q = mc \Delta T$). Adding the latent heat of vaporization, which is also the heat per unit mass ($Q = mL_v$ and $L_v = Q/m$), gives

$$\frac{Q}{m} = c \Delta T + L_v = [4186 \text{ J/(kg·C}°)](65°C) + (22.6 \times 10^5 \text{ J/kg}) \approx 2500 \text{ kJ/kg}$$

Evaporation will never lead to heating to the boiling point (100°C), but this estimate gives the maximum possible cooling effect.

Although evaporation is a relatively slow process, it is often important in preventing our bodies from overheating. Usually, radiation and conduction (discussed in Section 11.4) are sufficient to maintain a rate of heat loss that keeps us comfortable, given the temperature difference between our bodies and our surroundings. However, when the air gets hot and the temperature difference narrows (or disappears), we start to perspire. The evaporation of perspiration helps cool our bodies. On a hot summer day, a person may stand in front of a fan and remark how cool the blowing air feels. But the fan is merely blowing hot air from one place to another. The air feels cool because its flow promotes evaporation, which removes heat energy. Of course, evaporation depends on the humidity (the amount of moisture already in the air). We feel less comfortable on hot humid days because evaporation is reduced. (Why?)

11.4 Heat Transfer

Objectives: **To (a) describe the three methods of heat transfer and (b) give practical and/or environmental examples of each.**

Since heat is defined as energy in transit, how the transfer takes place is important. Heat moves from place to place by one of three mechanisms: conduction, convection, or radiation.

Conduction

You can keep a pot of coffee hot on a hot stove because heat is conducted through the coffee pot from the burner. The process of **conduction** is visualized as resulting from molecular interactions. Molecules in one part of a body at a higher temperature vibrate faster. They collide with, and transfer some of their energy to, the less energetic molecules located toward the cooler part of the body. In this way, energy is conductively transferred from a higher-temperature region to a lower-temperature region—transfer as a result of a temperature difference.

Solids can be divided into two general categories: metals and nonmetals. *Metals* are good conductors of heat, or **thermal conductors.** Modern theory views metals as having a large number of electrons that are free to move around (not permanently bound to a particular molecule or atom). These free electrons are believed to be primarily responsible for the heat conduction of metals. *Nonmetals,* such as wood or cloth, have relatively few free electrons and are poor heat conductors. A poor heat conductor is called a **thermal insulator.**

In general, the ability of a substance to conduct heat depends on the substance's phase. Gases are poor thermal conductors: Their molecules are relatively far apart, and collisions are therefore infrequent. Liquids are better thermal conductors than gases are because their molecules are closer together and can interact more readily.

Heat conduction can be described quantitatively as the time rate of heat flow $(\Delta Q/\Delta t)$ in a material for a given temperature difference (ΔT), as illustrated in •Fig. 11.9. It has been established through experiments that the rate of heat flow through a substance depends on the temperature difference between its boundaries. Heat conduction also depends on the size and shape of the object. In our analysis of heat flow, we will use a uniform slab of the substance.

A moment's thought should convince you that the heat flow through a slab of material is directly proportional to its surface area (A) and inversely proportional to its thickness (d). That is,

$$\frac{\Delta Q}{\Delta t} \propto \frac{A \, \Delta T}{d}$$

The term $\Delta T/d$ is called the *thermal gradient* (the change in temperature per unit of length). Using a constant of proportionality allows us to write the relation as an equation:

$$\frac{\Delta Q}{\Delta t} = \frac{kA \, \Delta T}{d} \qquad \text{(conduction only)} \qquad (11.3)$$

The constant k, called the **thermal conductivity**, characterizes the heat-conducting ability of a material. The greater the value of k for a material, the more rapidly it will conduct heat. The units of k are joules per meter-second-Celsius degree [J/(m·s·C°),

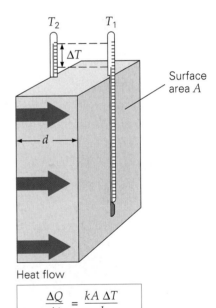

$$\frac{\Delta Q}{\Delta t} = \frac{kA \, \Delta T}{d}$$

•**FIGURE 11.9   Thermal conduction**   Heat conduction is characterized by the time rate of heat flow $(\Delta Q/\Delta t)$ in a material with a temperature difference ΔT. For a slab of material, $\Delta Q/\Delta t$ is directly proportional to the cross-sectional area (A) and the thermal conductivity of the material (k); it is inversely proportional to the slab thickness (d).

TABLE 11.3 Thermal Conductivities of Some Substances

Substance	Thermal conductivity (k)	
	$J/(m \cdot s \cdot C°)$	$kcal/(m \cdot s \cdot C°)$
Metals		
Aluminum	240	5.73×10^{-2}
Copper	390	9.32×10^{-2}
Iron and steel	46	1.1×10^{-2}
Silver	420	10×10^{-2}
Liquids		
Transformer oil	0.18	4.3×10^{-5}
Water	0.57	14×10^{-5}
Gases		
Air	0.024	0.57×10^{-5}
Hydrogen	0.17	4.1×10^{-5}
Oxygen	0.024	0.57×10^{-5}
Other materials		
Brick	0.71	17×10^{-5}
Concrete	1.3	31×10^{-5}
Cotton	0.075	1.8×10^{-5}
Fiberboard	0.059	1.4×10^{-5}
Floor tile	0.67	16×10^{-5}
Glass (typical)	0.84	20×10^{-5}
Glass wool	0.042	1.0×10^{-5}
Human tissue (average)	0.20	4.8×10^{-5}
Ice	2.2	53×10^{-5}
Styrofoam	0.042	1.0×10^{-5}
Wood, oak	0.15	3.6×10^{-5}
Wood, pine	0.12	2.9×10^{-5}
Vacuum	0	0

or $kcal/(m \cdot s \cdot C°)$]. The thermal conductivities of various substances are listed in Table 11.3. These values vary slightly over different temperature ranges but can be considered to be constant over normal temperature ranges and differences.

Compare the relatively large thermal conductivities of the good thermal conductors, the metals, with the relatively small thermal conductivities of some good thermal insulators, such as Styrofoam and wood. Some cooking pots have copper coatings on the bottoms (•Fig. 11.10). Being a good conductor of heat, the copper promotes the distribution of heat over the bottom of a pot for even cooking. Conversely, plastic foams are good insulators mainly because they contain pockets of air. Recall that gases are poor conductors, and note the low thermal conductivity of air in Table 11.3.

•FIGURE 11.10 Copper-bottomed pots A layer of copper is found on the bottoms of some pots and saucepans. The high heat conductivity of this metal ensures the rapid and even spread of heat from the burner, reducing the likelihood of local "hot spots." (Contrast the chapter-opening photograph.)

Example 11.8 ■ Insulation: Controlling Thermal Conductivity

A room with a pine ceiling that measures 3.0 m × 5.0 m × 2.0 cm thick has a layer of glass wool insulation above it that is 6.0 cm thick (•Fig. 11.11a). On a cold day, the temperature inside the room is 20°C and the temperature in the attic above the room is 8.0°C. Assuming that the temperatures remain constant with a steady heat flow, how much energy does the layer of insulation save in 1.0 h? Assume losses are due to conduction only.

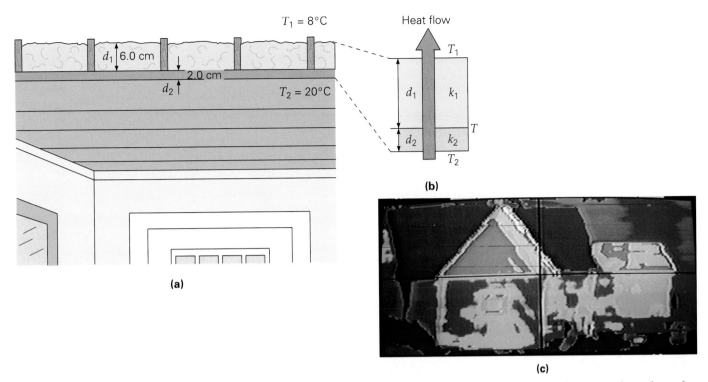

$T_1 = 8°C$ Heat flow

d_1 6.0 cm

2.0 cm

d_2 $T_2 = 20°C$

T_1

d_1 k_1

T

d_2 k_2

T_2

(a)

(b)

(c)

•**FIGURE 11.11**   **Insulation and thermal conductivity**   **(a, b)** Attics should be insulated to prevent the loss of heat through conduction. See Example 11.8. **(c)** This thermogram of a house allows us to visualize heat loss. (Blue represents the areas that lose the least heat; white, pink, and red, the most.) What conclusions can you draw from it? (Compare Fig. 11.15.)

Thinking It Through.   Here we have two materials, so we must consider Eq. 11.3 for two different thermal conductivities (k). We want to find $\Delta Q / \Delta t$ for the combination so that we can get ΔQ for $\Delta t = 1.0$ h. The situation is a bit complicated because the heat flows through two materials. But we know that at a steady rate, the heat flows must be the same through both. (Why?) To find the energy saved in 1.0 h, we need to compute how much heat is conducted in this time with and without the layer of insulation. Then we will derive a general equation for multimaterial insulations.

Solution.   Computing some of the quantities and making conversions as we list the data, we have

Given: $A = 3.0$ m $\times$ 5.0 m $= 15$ m^2 *Find:*   Energy saved
$d_1 = 6.0$ cm $= 0.060$ m   in 1.0 h
$d_2 = 2.0$ cm $= 0.020$ m
$\Delta T = T_2 - T_1 = 20°C - 8.0°C = 12$ C°
$\Delta t = 1.0$ h $= 3.6 \times 10^3$ s
$k_1 = 0.042$ J/(m·s·C°) (glass wool)
$k_2 = 0.12$ J/(m·s·C°) (wood)   (from Table 11.3)

(In working such problems with lots of given quantities, it is especially important to label all the data correctly.)

First, let's consider how much heat would be conducted in 1.0 h through the wooden ceiling with no insulation present. Since we know Δt, we can rearrange Eq. 11.3 to find ΔQ_c (heat conducted through the wooden ceiling):

$$\Delta Q_c = \left(\frac{k_2 A \, \Delta T}{d_2}\right)\Delta t$$

$$= \left\{\frac{[0.12 \text{ J/(m·s·C°)}](15 \text{ m}^2)(12 \text{ C°})}{0.020 \text{ m}}\right\}(3.6 \times 10^3 \text{ s}) = 3.9 \times 10^6 \text{ J}$$

Now we need to find the heat conducted through the ceiling *and* the insulation layer together. Let T be the temperature at the interface of the materials and T_2 and T_1 be the warmer and cooler temperatures, respectively (Fig. 11.11b). Then

$$\frac{\Delta Q_1}{\Delta t} = \frac{k_1 A(T - T_1)}{d_1} \quad \text{and} \quad \frac{\Delta Q_2}{\Delta t} = \frac{k_2 A(T_2 - T)}{d_2}$$

We don't know T, but when the conduction is steady, the flow rates are the same for both; that is, $\Delta Q_1/\Delta t = \Delta Q_2/\Delta t$, or

$$\frac{k_1 A(T - T_1)}{d_1} = \frac{k_2 A(T_2 - T)}{d_2}$$

The A's cancel, and solving for T gives

$$T = \frac{k_1 d_2 T_1 + k_2 d_1 T_2}{k_1 d_2 + k_2 d_1}$$

Substituting into either of the flow rate equations and rearranging, we get a general expression for $\Delta Q/\Delta t$ for the combined layers:

$$\frac{\Delta Q}{\Delta t} = \frac{A(T_2 - T_1)}{(d_1/k_1) + (d_2/k_2)} \tag{11.4*}$$

$$= \frac{(15 \text{ m}^2)(12 \text{ C}°)}{[0.060 \text{ m}/0.042 \text{ J}/(\text{m·s·C}°)] + [0.020 \text{ m}/0.12 \text{ J}/(\text{m·s·C}°)]} = 1.1 \times 10^2 \text{ J/s}$$

In $1.0 \text{ h} = 3.6 \times 10^3 \text{ s}$,

$$\Delta Q = (1.1 \times 10^2 \text{ J/s}) \Delta t = (1.1 \times 10^2 \text{ J/s})(3.6 \times 10^3 \text{ s}) = 4.0 \times 10^5 \text{ J}$$

Thus, the heat loss is decreased by

$$\Delta Q_c - \Delta Q = (39 \times 10^5 \text{ J}) - (4.0 \times 10^5 \text{ J}) = 35 \times 10^5 \text{ J}$$

Ideally, this represents a savings of

$$\frac{35 \times 10^5 \text{ J}}{39 \times 10^5 \text{ J}} (\times 100\%) = 90\%$$

Follow-up Exercise.   Suppose that in this Example there is an upstairs room (in the attic) with a pine floor, similar to the lower ceiling, covered with floor tile 0.50 cm thick. What would be the flow rate of heat between floors in this case if the upstairs room temperature is 15°C?

Convection

In general, compared to solids, liquids and gases are not good thermal conductors. However, the mobility of molecules in fluids permits heat transfer by another process—convection. **Convection** is heat transfer as a result of mass transfer, which can be natural or forced.

Natural convection cycles occur in liquids and gases. For example, when cold water is in contact with a hot object, such as the bottom of a pot on a stove, the object transfers heat to the water adjacent to the pot by conduction. But the water carries the heat away with it by natural convection, and a cycle is set up in which upper, cold water replaces the rising warm water. Such cycles are important in atmospheric processes, as illustrated in •Fig. 11.12. During the day, the ground heats up more quickly than do large bodies of water, as you may have noticed if you have been to the beach. This occurs both because the water has a greater specific heat than the land and because convection currents disperse the absorbed heat throughout the great volume of water. The air in contact with the warm ground is heated by conduction. That air expands, becoming less dense than in surrounding cooler air. As a result, the warm air rises (air currents) and, to fill the space, other air moves horizontally (winds)—creating a sea breeze near a large body of water. Cooler air descends, and a thermal convection cycle is set up, which transfers heat away from the land. At night, the ground loses its heat more quickly than the water, and the water surface is warmer than the land. As a result, the cycle is reversed.

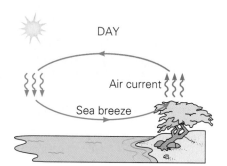

DAY

Air current

Sea breeze

Land warmer than water

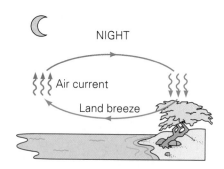

NIGHT

Air current

Land breeze

Water warmer than land

•FIGURE 11.12   Convection cycles   During the day, natural convection cycles give rise to sea breezes near large bodies of water. At night, the pattern of circulation is reversed and land breezes blow. The temperature differences between land and water are the result of their different specific heats. Water has a much greater specific heat, so the land warms up more quickly during the day. At night, the land cools more quickly, while the water remains warmer because of its greater heat capacity.

*This equation can be extended to any number of layers or slabs of materials by adding additional d/k terms to the denominator: $\Delta Q/\Delta t = A(T_2 - T_1)/\Sigma(d_i/k_i)$.

You can see the evidence of convection currents in the air above a hot road surface in the summer and in transparent liquids, such as heated water in a glass container. This is because regions of different temperatures have different densities, which cause a bending, or refraction, of the light that passes through the regions (Chapter 22).

In *forced convection*, the medium of heat transfer is moved mechanically. Note that we can have heat transfer without a temperature difference. In fact, we can have energy transferred from a low-temperature region to a high-temperature region, as in the case of the forced convection of a refrigerator coolant removing energy from the inside of the refrigerator. (The circulating coolant carries heat energy from the inside of the refrigerator, and this heat is given up to the environment, as we will see in Section 12.4.)

Other common examples of forced convection systems are forced-air heating systems in homes (•Fig. 11.13), the human circulatory system, and the cooling system of an automobile engine. The human body does not use all of the energy obtained from food; a great deal is lost. (There's usually a temperature difference between your body and your surroundings.) So that body temperature will stay normal, the internally generated heat energy is transferred close to the skin surface by blood circulation. From the skin, that energy is conducted to the air or lost by radiation (the other heat-transfer mechanism, to be discussed shortly).

Water or some other coolant is circulated (pumped) through most automobile cooling systems (some engines are air-cooled). The fluid medium carries heat to the radiator (a heat exchanger), where forced air flow produced by the fan carries it away. The *radiator* of an automobile is actually misnamed—most of the heat is transferred from it by forced convection rather than by radiation.

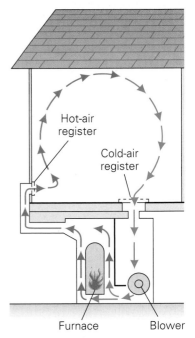

•FIGURE 11.13 Forced convection Houses are commonly heated by forced convection. Registers or gratings in the floors or walls allow heated air to enter and cooler air to return to the heat source. In older homes, natural convection was (and still is) used.

Conceptual Example 11.9

Foam Insulation: Better Than Air?

Polymer foam insulation is sometimes blown into the space between the inner and outer walls of a house. Since air is a good thermal insulator, why is the foam insulation needed? (a) To prevent loss of heat by conduction, (b) to prevent loss of heat by convection, (c) for fireproofing.

Reasoning and Answer. Polymer foams will generally burn, so (c) isn't likely to be the answer. Air is a poor thermal conductor, even poorer than plastic foam (Styrofoam, Table 11.3), so the answer can't be (a). However, as a gas, it is subject to convection within the wall space. In the winter, the air near the warm inner wall is heated and rises, thus setting up a convection cycle in the space and transferring heat to the cold outer wall. In the summer, with air conditioning, the heat-loss cycle is reversed. Foam blocks such convection cycles. Hence, the answer is (b).

Follow-up Exercise. Thermal underwear and thermal blankets are loosely knitted with lots of small holes. Wouldn't they be more effective if the material were denser?

Radiation

Conduction and convection require some material as a transport medium. The third mechanism of heat transfer needs no medium; it is called **radiation**, which refers to energy transfer by electromagnetic waves (Chapter 20). Heat is transferred to the Earth from the Sun through empty space by radiation. Visible light and other forms of electromagnetic radiation are commonly referred to as radiant energy.

You have experienced heat transfer by radiation if you've ever stood near an open fire (•Fig. 11.14). You can feel the heat on your exposed hands and face. This heat transfer is not due to convection or conduction, since heated air rises and air is a poor conductor. Visible radiation is emitted from the burning material, but most of the heating effect comes from the invisible **infrared radiation** emitted by

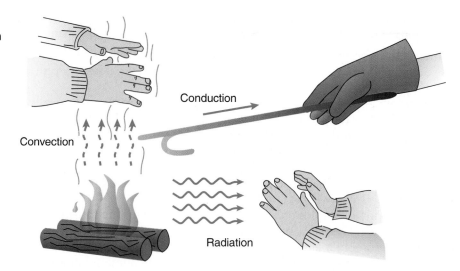

•FIGURE 11.14 Heating by conduction, convection, and radiation

The hands at left are warmed by the convection of rising hot air (and some radiation). The gloved hand at upper right is warmed by conduction. The hands at lower right are warmed by radiation.

Conduction

Convection

Radiation

Note: Resonance is discussed in Section 13.5.

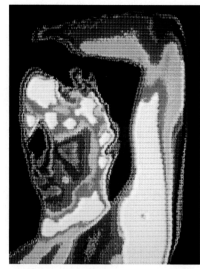

•FIGURE 11.15 Thermography

A thermogram of a man holding his arm above his head. The skin temperature varies about 1 C° for each color, from white (hottest) to black (coldest). Note the high temperature of the armpit, due to the proximity of blood vessels to the skin, and the cooler nose and hair.

the glowing embers or coals. You feel this radiation because it is absorbed by water molecules in your skin. (Body tissue is about 85% water.) The water molecule has an internal vibration whose frequency coincides with that of infrared radiation, which is therefore absorbed readily. (This effect is called *resonance absorption*. The electromagnetic wave drives the molecular vibration, and energy is transferred to the molecule, somewhat like pushing a swing.)

Infrared radiation is sometimes referred to as "heat rays." You may have noticed the red infrared lamps used to keep food warm in cafeterias. Heat transfer by infrared radiation is also important in maintaining our planet's warmth by a mechanism known as the *greenhouse effect*. This important environmental topic and another common example of heating by radiation absorption are discussed in the Insights on pp. 379 and 380.

Although infrared radiation is invisible to the human eye, it can be detected by other means. The frequency of the infrared radiation is proportional to the temperature of its source. This relationship is the basis for infrared thermometers, which, using infrared detectors, can measure temperature remotely. Also, some cameras can use special infrared film. A picture taken with this film is an image consisting of contrasting light and dark areas corresponding to regions of higher and lower temperatures. Special instruments that apply such *thermography* are used in industry and medicine; the images they produce are called *thermograms* (•Fig. 11.15).

A new application of thermograms is for security. The system consists of an infrared camera and a computer that identifies an individual by means of the unique heat pattern emitted by the facial blood vessels. The camera takes a picture of the radiation from a person's face, which is compared with an earlier image stored in the computer memory. It is reported that the system can even distinguish between identical twins, whose facial features are slightly different. Also, changes in body temperature from weather conditions or a fever do not affect the identification, as the relative patterns of radiation remain the same.

The rate at which an object radiates energy has been found to be proportional to the fourth power of the absolute temperature (T^4). This relationship is expressed in an equation known as **Stefan's law:**

$$P = \frac{\Delta Q}{\Delta t} = \sigma A e T^4 \tag{11.5}$$

where P is the power radiated in watts (W) or joules per second (J/s). The symbol σ (the Greek letter sigma) is the *Stefan-Boltzmann constant*: $\sigma = 5.67 \times 10^{-8}$ W/(m²·K⁴). The radiated power is also proportional to the surface area (A) of the object. The **emissivity** (e) is a unitless number between 0 and 1 that is characteristic of the

The Greenhouse Effect

You may have heard or read about the greenhouse effect in connection with global warming. The *greenhouse effect* helps regulate the Earth's long-term average temperature, which has been fairly constant. A portion of the solar radiation we receive reaches and warms the Earth's surface. The Earth in turn reradiates energy in the form of infrared radiation. The balance between absorption and radiation is a major factor in stabilizing the Earth's temperature.

This balance is affected by the concentration of *greenhouse gases*—primarily water vapor and carbon dioxide (CO_2)—in the atmosphere. As the infrared radiation passes through the atmosphere, some of the radiation is absorbed by the greenhouse gases. These gases are selective absorbers: They absorb radiation at certain wavelengths but not at others (Fig. 1a).

If terrestrial infrared radiation is absorbed, the atmosphere warms, warming the Earth (heat transfer by radiation). This rise in surface temperature causes a shift in the wavelength of the emitted radiation. The wavelength is eventually shifted to a "window" in the absorption spectrum where little or no absorption takes place, and the terrestrial radiation passes through the atmosphere into space. Thus, the Earth loses energy, and its surface temperature decreases.

But with a temperature decrease, the terrestrial radiation shifts to a longer wavelength and is again absorbed by the greenhouse gases. We have a turning on and off, so to speak, similar to the action of a thermostat. Hence, the selective absorption of atmospheric gases plays an important role in maintaining the Earth's average temperature

Why is this phenomenon called the greenhouse effect? The reason is that the atmosphere functions like the glass in a greenhouse. That is, the absorption and transmission properties of glass are similar to those of the atmospheric greenhouse gases—in general, visible radiation is transmitted, but infrared radiation is selectively absorbed (Fig. 1b). We have all observed the warming effect of sunlight passing through glass, for ex-

ample, in a closed car on a sunny, cold day. Similarly, a greenhouse heats up by absorbing sunlight and trapping the reradiated infrared radiation. Thus, it is quite warm inside on a sunny day, even in winter. (The glass enclosure also keeps warm air from escaping upward, as it does in the atmosphere. In practice, this elimination of heat loss by convection is the chief factor in maintaining an elevated temperature. The temperature of a greenhouse in the summer is controlled by painting the glass panels white so that some of the sunlight is reflected and by opening panels to let some hot air escape.)

The problem is that on Earth, human activities may accentuate greenhouse warming. With the combustion of fuels for heating and industrial processes, vast amounts of CO_2 and other greenhouse gases are vented into the atmosphere. There is concern that the result of this trend will be—or already is—*global warming*: an increase in the Earth's average temperature that could dramatically affect the environment. For example, the climate in many parts of the globe would be altered, with effects on agricultural production and world food supplies that are very difficult to predict. A general rise in temperature could cause partial melting of the polar ice caps. Sea levels would rise, flooding low-lying regions and endangering coastal ports and population centers.

In an actual greenhouse, the visible portion of the sunlight is transmitted through the glass, but the short-wavelength ultraviolet radiation is absorbed. The ultraviolet spectrum from 280 nm to 400 nm is divided into two regions, called UV-A (320–400 nm) and UV-B (280–320 nm). Ordinary window glass does not appreciably transmit UV-B radiation, which is the primary cause of suntans and sunburns (Section 20.4). Hence, although a considerable warming effect occurs through glass from visible radiation, you do not receive a tan or severe sunburn through ordinary window glass because of UV-B absorption. Slight reddening may occur for sensitive skin from the UV-A radiation that is transmitted.

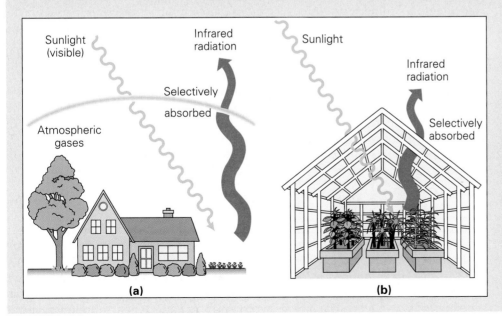

FIGURE 1   The greenhouse effect   **(a)** The greenhouse gases of the atmosphere, particularly water vapor and carbon dioxide, are selective absorbers with absorption properties similar to those of glass used in greenhouses. Visible light is transmitted and heats the surface, while some of the infrared radiation that is re-emitted is absorbed. **(b)** A greenhouse operates in a similar way.

Heat Transfer by Radiation: The Microwave Oven

The microwave oven quickly became a common kitchen appliance. It is both time-saving and energy-saving, since the oven doesn't have to be warmed up as does a conventional oven. The principle of operation of the microwave oven is heat transfer by radiation.

Microwaves are a form of electromagnetic radiation; they have a frequency range just below that of infrared radiation (Chapter 21). Like infrared radiation, microwaves are absorbed chiefly by water molecules (in a molecular resonance). Fats and sugars also absorb mcrowaves.

In a microwave oven, the microwaves are generated electronically and distributed by reflection from a metal stirrer or fan and off the metal walls (Fig. 1). Because the walls reflect the radiant energy, they do not get hot.

Microwaves pass through plastic wrap, glass, or dishes made of other "microwave-safe" materials and are absorbed by water molecules in the food, producing rapid heating. (Metal utensils or objects cannot be used in microwave ovens because the microwaves can dislodge electrons from metals, causing sparking and possibly damage.) The microwaves do not penetrate the food completely but are absorbed within 2 to 3 cm of the surface. Heat is then conducted to the interior of the food, just as it is in conventional oven heating. This is why it is advisable to let large items or portions sit for a time after the microwave oven has shut off, so that they will be warmed or cooked throughout.

Since microwaves could be absorbed by water molecules in the skin, causing burns, microwave ovens have several

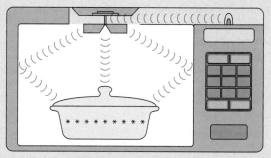

FIGURE 1   Heat transfer by radiation   In a microwave oven, microwaves (a type of electromagnetic radiation) are absorbed by molecules, chiefly those of water (and to a lesser extent, fats and sugars), raising their temperature.

important safety features. The door is tight-fitting so that microwaves cannot leak out. The glass in the door is fitted with a metal shield that has small holes through which you can view the food without opening the door. Microwaves are reflected by this shield and prevented from coming through the glass (essentially, the waves are larger than the holes). Also, there is a mechanism that automatically shuts off the oven when the door is opened. In fact, the oven cannot be turned on when the door is open.

You may be hearing more about microwaves soon. Developmental work is being done on microwave clothes dryers, and small prototype versions are now available.

---

material. Dark surfaces have emissivities close to 1, and shiny surfaces have emissivities close to 0. The emissivity of human skin is about 0.70.

Dark surfaces are not only better emitters of radiation, they are also good absorbers. In general, *a good absorber is also a good emitter*. That is, a body that is a good absorber of certain wavelengths of radiation is a good emitter of those same wavelengths. An ideal, or perfect, absorber (and emitter) is referred to as a **blackbody** ($e = 1.0$). Shiny surfaces are poor absorbers, since most of the incident radiation is reflected. This fact can be demonstrated easily as shown in •Fig. 11.16. (Can you see why it is better to wear light-colored clothes in the summer and dark-colored clothes in the winter?)

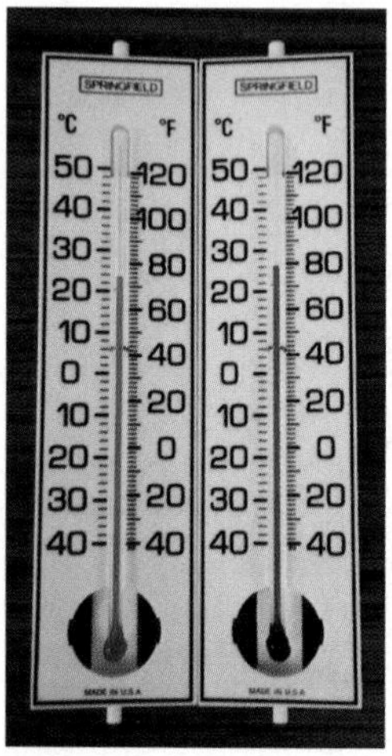

When an object is in thermal equilibrium with its surroundings, its temperature is constant; thus, it must be emitting and absorbing radiation at the same rate. If the temperatures of the object and its surroundings are different, there must be a net flow of radiant energy. If an object is at a temperature T and its surroundings are at a temperature T_s, the net rate of energy loss or gain per unit time (power) is given by

$$P_{net} = \sigma A e(T_s^4 - T^4) \qquad (11.6)$$

Note that if T_s is less than T, then P (or $\Delta Q/\Delta t$) will be negative, indicating a net energy loss. *Keep in mind that the temperatures used in calculating radiated power are the absolute temperatures in kelvins.*

You may have noticed in Chapter 10 that we defined *heat* as the net energy transfer because of a temperature difference. The word *net* here is important. It is possible to have energy transfer between an object and its surroundings, or be-

•**FIGURE 11.16   Good absorber** Black objects are generally good absorbers. The bulb of the thermometer on the right has been painted black. Note the difference in the temperature readings.

tween objects, at the same temperature. Note that if $T_s = T$ (no temperature difference) in Eq. 11.6, there is a continuous exchange of radiant energy (Eq. 11.5 still holds), but there is no *net* change of internal energy. Suppose you had two identical objects relatively near each other with the same temperatures and emissivities. They would constantly be exchanging energy via radiation, even though there is no temperature difference. But on the average, there would be no *net* energy transfer between them.

Example 11.10 ■ Body Heat: Radiant Heat Transfer

Suppose that your skin has an emissivity of 0.70 and that its exposed area is 0.27 m². How much net energy will be radiated per second from this area if the air temperature is 20°C? Assume your skin temperature to be the same as normal body temperature, 37°C. (The average skin temperature is actually less. Why?)

Thinking It Through. Everything is given for us to find P_{net} from Eq. 11.6. We must remember to work in kelvins.

Solution.

Given: $T_s = 20°C + 273 = 293$ K $\qquad$ *Find:* P_{net} (net power)
$\qquad\quad T = 37°C + 273 = 310$ K
$\qquad\quad e = 0.70$
$\qquad\quad A = 0.27$ m²
$\qquad\quad \sigma = 5.67 \times 10^{-8}$ W/(m²·K⁴) (known)

Using Eq. 11.6 directly, we get

$$P_{net} = \sigma A e (T_s^4 - T^4)$$

$$= [5.67 \times 10^{-8} \text{ W/(m}^2\text{·K}^4)](0.27 \text{ m}^2)(0.70)[(293 \text{ K})^4 - (310 \text{ K})^4]$$

$$= -20 \text{ W (or } -20 \text{ J/s)}$$

Since $P_{net} = \Delta Q / \Delta t$, 20 J of energy is radiated, or *lost* (as indicated by the minus sign), each second.

Follow-up Exercise. A piece of cake contains 100 Calories. If all of this energy were converted to body heat, how long would it supply the radiative heat loss computed in this Example?

Problem-Solving Hint

Note that in Example 11.10 the fourth powers of the temperatures were found first and then their difference. It is *not* correct to find the temperature difference and then raise it to the fourth power: $T_s^4 - T^4 \neq (T_s - T)^4$.

Let's consider a thoughtful example of heat transfer.

Conceptual Example 11.11

Solar Panels: Reducing the Heat Transfer

Solar panels are used to collect solar energy to heat water, which may then be used directly to heat a home at night. The panel boxes have black interiors (why?) through which the piping runs to carry the water, and the top side is covered with glass. However, ordinary glass absorbs most of the Sun's ultraviolet radiation. This absorption reduces the heating effect. Wouldn't it be better to leave the glass off the panel boxes?

Reasoning and Answer. Some energy is absorbed by the glass, but the glass serves a useful purpose and saves a lot more energy than it absorbs. As the black interior and piping of the panel box heat up, there could be heat loss by radiation (infrared) and convection. The glass prevents this via the greenhouse effect. It absorbs much of the infrared radiation and keeps the convection inside the solar panel box. (See the Insight on the greenhouse effect, p. 379.)

Follow-up Exercise. Is there any practical reason for window drapes (other than privacy)?

Let's look at a few more real-life examples of heat transfer. In the spring, a late frost could kill the buds on fruit trees in an orchard. To reduce heat transfer, some growers spray water on the trees to form ice before a hard frost occurs. Using ice to save buds? Ice is a relatively poor (and inexpensive) conductor of heat, so it has an insulating effect. It will maintain the bud temperature at 0°C, not going below, and therefore protects the buds.

Another method to protect orchards from freezing is the use of smudge pots, containers in which material is burned to create a dense cloud of combustion products. At night, when the Sun-warmed ground cools off by radiation, the cloud absorbs this heat and reradiates it back to the ground. Thus the ground takes longer to cool, hopefully without reaching freezing temperatures before the Sun comes up. (Recall that frost is the direct condensation of water vapor in the air to ice—not frozen dew.)

Lastly, check out •Fig. 11.17. Why would anyone wear a dark robe in the desert? We learned that dark objects absorb radiation (Fig. 11.16). Wouldn't a white robe be better? A dark robe definitely absorbs more radiant energy and warms the air inside near the body. But note that the robe is open at the bottom. The warm air rises (by buoyancy) and exits at the neck area, and outside cooler air enters the robe at the bottom—a natural-convection air circulation!

•**FIGURE 11.17 A dark robe in the desert?** Dark objects absorb more radiation than do lighter ones, and they are hotter. What's going on here?

Chapter Review

Important Terms

heat *361*
kilocalorie (kcal) *361*
calorie (cal) *361*
British thermal unit (Btu) *362*
mechanical equivalent of heat *362*
specific heat *363*
solid phase *367*
melting point *367*
freezing point *367*
liquid phase *367*

gaseous phase (vapor phase) *367*
boiling point *367*
condensation point *367*
sublimation *367*
latent heat *368*
latent heat of fusion *368*
latent heat of vaporization *368*
evaporation *372*
conduction *373*

thermal conductor *373*
thermal insulator *373*
thermal conductivity *373*
convection *376*
radiation *377*
infrared radiation *377*
Stefan's law *378*
emissivity *378*
blackbody *380*

Important Concepts

• As a form of energy in transit, heat has the SI unit of joule (J). Common nonstandard units are the kilocalorie (kcal) and the British thermal unit (Btu).

• The mechanical equivalent of heat relates joules to kilocalories.

• The amount of heat required to raise the temperature of 1 kg of a substance by 1 C° is called the spe-

cific heat. The greater the specific heat of a substance, the more heat energy necessary to change the temperature of a given mass.

• Latent heat is the heat energy, in joules per kilogram, associated with a phase change.

• Heat transfer is effected by three methods: conduction, convection, and radiation.

Important Equations

Mechanical Equivalent of Heat:

$$1 \text{ kcal} = 4186 \text{ J} = 4.186 \text{ kJ}$$
$$1 \text{ cal} = 4.186 \text{ J}$$

Specific Heat (specific heat capacity):

$$Q = mc\,\Delta T \quad \text{or} \quad c = \frac{Q}{m\,\Delta T} \qquad (11.1)$$

Latent Heat:

$$Q = mL \quad \text{or} \quad L = \frac{Q}{m} \qquad (11.2)$$

(for water)

$$L_f = 3.33 \times 10^5 \text{ J/kg (or 80 kcal/kg)}$$
$$L_v = 22.6 \times 10^5 \text{ J/kg (or 540 kcal/kg)}$$

Thermal Conduction:

$$\frac{\Delta Q}{\Delta t} = \frac{kA\,\Delta T}{d} \qquad (11.3)$$

Stefan's Law:

$$P = \frac{\Delta Q}{\Delta t} = \sigma A e T^4 \qquad (11.5)$$

Net Radiant Power (loss or gain):

$$P_{\text{net}} = \sigma A e (T_s{}^4 - T^4) \qquad (11.6)$$

Exercises*

11.1 Units of Heat

1. The SI unit of heat energy is the (a) calorie, (b) kilocalorie, (c) Btu, (d) joule.

2. Which of the following is the largest unit of heat energy: (a) calorie, (b) Btu, (c) joule, (d) kilojoule?

3. ■ A heater supplies 240 Btu of energy. What is this in joules?

4. ■ A person goes on a 1500-Cal-per-day diet to lose weight. What is the equivalent daily allowance expressed in joules?

5. ■ A window air conditioner has a rating of 20 000 Btu (per hour). What would this rating be in watts?

6. ■ What is the mechanical equivalent of heat expressed in British thermal units?

7. ■■ A student eats a Thanksgiving dinner that totaled 2800 Cal. He wants to use up all that energy by lifting a 20-kilogram mass a distance of 1.0 m. (a) How many times must he lift the mass? (b) If he can lift the mass once every 5.0 s, how long does this exercise take (neglecting lowering)?

11.2 Specific Heat

8. The amount of heat necessary to change the temperature of 1 kg of a substance by 1 C° is called the substance's (a) specific heat, (b) latent heat, (c) heat of combustion, (d) mechanical equivalent of heat.

*Neglect heat losses in the exercises unless instructed otherwise, and consider all temperatures to be exact.

9. When you swim in the ocean or a lake at night, the water may feel pleasantly warm even when the air is quite cool. Why?

10. Is it possible to have a negative specific heat? Explain.

11. ● Is it possible to have negative heat? Explain.

12. Near a large body of water, what would you expect the direction of a breeze to be during the night?

13. ● Does heat flow depend on the temperatures of two locations or only on the temperature difference?

14. An astronaut in deep space says, "I'm going to perform an experiment to see how many Btus it takes to increase the temperature of 3.0 lb of water from 60°F to 212°F." What is wrong with this statement, and what does it tell you?

15. ■ ● The temperature of a lead block and a copper block, both 1.0 kg and both at 110 K, is to be raised to 190 K. (a) Which one requires more heat? (b) How much more?

16. ■ ● A 5.0-gram pellet of aluminum at 20°C gains 200 J of heat. What is its final temperature?

17. ■ How many joules heat must be added to 5.0 kg of water at 20°C to bring it to the boiling point?

18. ■■ A 0.250-kilogram cup at 20°C is filled with 0.250 kg of boiling coffee. The cup and the coffee come to thermal equilibrium at 80°C. If no heat is lost, what is the specific heat of the cup material? [Hint: Consider the coffee to be essentially boiling water.]

19. ■■ An aluminum serving spoon at 100°C is placed in a Styrofoam cup containing 0.200 kg of water at 20°C. If the final equilibrium temperature is 30°C and no heat is lost to the cup itself, what is the mass of the spoon?

20. ■■ ⬤ How many kilograms of aluminum will experience the same temperature rise as 3.00 kg of copper when the same amount of heat is added to each?

21. ■■ A waterfall is 75 m high. If all of the gravitational potential energy of the water were converted into heat energy, by how much would the temperature of the water increase in going from the top to the bottom of the falls? [Hint: Consider a kilogram of water going over the falls.]

22. ■■ A cube of aluminum 10 cm on each side is cooled from 100°C to 20°C. If the energy removed from the aluminum cube were added to a copper cube of the same size at 20°C, what would be the final temperature of the copper cube?

23. ■■ A camper heats 30 L of water to boiling to get hot water for bathing. What volume of water from a 15°C stream must he then add to get the bath water to 45°C? (Neglect any losses.)

24. ■■ To determine the specific heat of a new metal alloy, 0.150 kg of the substance is heated to 400°C and then placed in a 0.200-kilogram aluminum calorimeter cup containing 0.400 kg of water at 10.0°C. If the final temperature of the mixture is 30.5°C, what is the specific heat of the alloy? (Ignore the calorimeter stirrer and thermometer.)

25. ■■ In a calorimetry experiment, 0.50 kg of a metal at 100°C is added to 0.50 kg of water at 20°C in an aluminum calorimeter cup. The cup has a mass of 0.250 kg. (a) If the final temperature of the mixture is 25°C, what is the specific heat of the metal? (b) What would be the effect on the calculated specific heat value if some water splashed out of the cup when the metal was added?

26. ■■■ An electric immersion heater has a power rating of 1500 W. If the heater is placed in a liter of water at 20°C, how many minutes will it take to bring the water to a boil? (Assume that there is no heat loss except to the water itself.)

27. ■■■ A 0.100-kilogram piece of aluminum at 90°C is immersed in 1.00 kg of water at 20°C. Assuming no heat is lost to the surroundings or container, what is the temperature of the metal and water when they reach thermal equilibrium?

28. ■■■ At what average rate would heat have to be removed from 1.5 L of (a) water and (b) mercury to reduce the liquid's temperature from 20°C to its freezing point in 3.0 min?

11.3 Phase Changes and Latent Heat

29. The SI units of latent heat are (a) $1/C°$, (b) $J/(kg·C°)$, (c) $J/C°$, (d) J/kg.

30. Latent heat is always (a) part of the specific heat, (b) related to the specific heat, (c) the same as the mechanical equivalent of heat, (d) involved in a phase change.

31. Why do different substances have different freezing and boiling points? Would you expect the latent heats to be different for different substances? Explain.

32. You are monitoring the temperature of some cold ice cubes (−5.0°C) in a closed, insulated cup as the ice and cup are heated. Initially, the temperature rises, but it stops at 0°C. After a while it begins rising again. Is there anything wrong with the thermometer? Explain.

33. In general, you can get a more severe burn from steam at 100°C than from an equal mass of hot water at 100°C. Why?

34. ■ How much heat is required to boil away 0.500 kg of water that is initially at 100°C?

35. ■ How much more heat is needed to convert 1.0 kg of water at 100°C to steam at 100°C than to raise the temperature of 1.0 kg of water from 0°C to 100°C?

36. ■ How much heat must be added to 0.75 kg of lead at 20°C to cause it to melt completely?

37. ■ How much energy in joules is required to boil away 0.50 L of liquid nitrogen at −210°C? (Take the density of liquid nitrogen to be 0.80×10^3 kg/m³.)

38. ■ How much heat in joules is required to boil away 0.50 kg of water initially at 50°C?

39. ■ A mercury diffusion vacuum pump contains 0.015 kg of mercury vapor at a temperature of 630 K. Suppose that you wanted to condense the vapor. How much heat would have to be removed?

40. ■■ (a) A 0.500-kilogram piece of ice at −20°C is converted to steam at 115°C. How much heat must be supplied to do this? (b) To convert the steam back to ice at −5°C, how much heat would have to be removed?

41. ■■ How much ice (at 0°C) must be added to 1.0 kg of water at 100°C so as to end up with all liquid at 20°C?

42. ■■ A 0.60-kilogram piece of ice at −10°C is placed in 0.30 kg of water at 50°C. How much water is there when the system reaches thermal equilibrium?

43. ■■ If 0.050 kg of ice at 0°C is added to 0.300 kg of water at 25°C in a 0.100-kilogram aluminum calorimeter cup, what is the final temperature of the water?

44. ■■ Steam at 100°C is bubbled into 0.250 kg of water at 20°C in a calorimeter cup. How much steam will have been added when the water in the cup reaches 60°C? (Ignore the effect of the cup.)

45. ■■ Ice is added to 0.75 L of tea at 20°C to make very cold iced tea. If enough 0°C ice is added so that the mixture is all liquid, how much liquid is in the pitcher when this occurs?

46. ■■ A volume of 0.50 L of water at 16°C is put into an aluminum ice cube tray of mass 0.250 kg at the same temperature. How much energy must be removed from this system by the refrigerator to turn the water into ice at −8.0°C?

47. ■■■ Suppose 3.0 cm of rain at 10°C falls on a rectangular region of dimensions 2.0 km × 3.0 km. How much energy was released when water vapor condensed to form this amount of rain? Is it a large amount of energy?

48. ■■■ A kilogram of a substance gives a T-versus-Q graph as shown in •Fig. 11.18. What are the (a) melting and boiling points? In SI units, what are (b) the specific heats of the various phases and (c) the latent heats at the various phase changes?

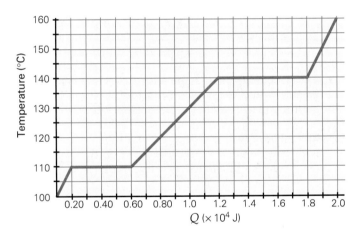

•**FIGURE 11.18   Temperature versus heat**   See Exercise 48.

49. ■■■ In an experiment, a 0.150-kilogram piece of ceramic superconducting material at 20°C is placed in liquid nitrogen (N_2) at its boiling point to cool. The N_2 is in a perfectly insulated flask that allows the gaseous N_2 to escape immediately. How many liters of liquid nitrogen will be boiled away? (Take the specific heat of the ceramic material to be the same as that of glass, and take the density of liquid nitrogen to be 0.80×10^3 kg/m³.)

11.4 Heat Transfer

50. The warming of the atmosphere involves (a) conduction, (b) convection, (c) radiation, (d) all of these.

51. Water is a very poor heat conductor. Why can it be thoroughly heated relatively quickly?

52. A plastic ice tray and a metal ice tray are removed from the same freezer (same initial temperature). However, once your hands touch both, the metal one feels cooler. Why?

53. (a) You blow on a spoonful of hot soup to cool it. Yet, your breath is warm. How does blowing cause the soup to cool? (b) On hot, humid days, ice can build up on the cooling coils of a window air conditioner. Does this help with the room cooling? Explain.

54. Why is the warning shown on the highway road sign in •Fig. 11.19 necessary?

•**FIGURE 11.19   A cold warning**   See Exercise 54.

55. What is the purpose of the fins on a motorcycle radiator or an automobile radiator (•Fig. 11.20)?

•**FIGURE 11.20   Radiator with fins**   See Example 55.

56. Newton's law of cooling (Sir Isaac was a busy man) states that, in general, the rate of heat loss from an object is proportional to the temperature difference between it and the surroundings: $\Delta Q / \Delta t = K \Delta T$, where K is a constant that includes losses by conduction, convection, and radiation. Apply this law to the following question: Would a cup of hot coffee stay hotter longer if you put cream in it right away or waited until you were ready to drink it, at a later time?

57. A Thermos bottle (•Fig. 11.21) keeps cold beverages cold and hot ones hot. It consists of a double-walled, partially evacuated container with silvered walls (mirrored

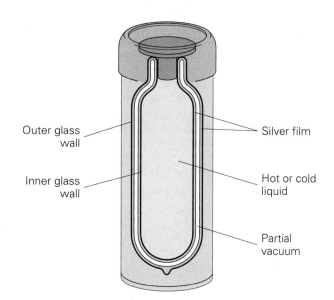

•FIGURE 11.21   **Thermal insulation**   The Thermos bottle uses all three methods of heat transfer. See Exercise 57.

interior). The bottle is constructed to minimize all three mechanisms of heat transfer. Explain how.

58. If you put your hand in a hot oven, the heat is uncomfortable but tolerable. Yet, if you touch a pan in the oven, you burn your hand. Why the difference?

59. ■ How many times faster would heat be conducted from your bare feet by a tile floor than by an oak floor? (Assume that both floors have the same temperature and thickness.)

60. ■ If an object has an emissivity of 0.75 and an area of 0.20 m², how much energy does it radiate outward per second at 20°C?

61. ■■ The glass pane in a window has dimensions of 2.00 m × 1.50 m and is 4.00 mm thick. How much heat in joules will flow through the glass in 1.00 h if there is a temperature difference of 2 C° between the inner and outer surfaces? (Consider conduction only.)

62. ■■ Assuming the human body has a 4.0-centimeter-thick layer of skin tissue and a surface area of 1.5 m², estimate the rate at which heat is conducted from inside the body to the surface if the skin temperature is 33°C. (Assume normal body temperature of 37°C for the interior temperature, and neglect insulating clothes.)

63. ■■ A copper teakettle with a circular bottom 30 cm in diameter has a uniform thickness of 2.5 mm. It sits on a burner whose temperature is 150°C. (a) If the teakettle is full of boiling water, what is the rate of heat conduction through its bottom? (b) Assuming that the heat from the burner is the only heat input, how much water is boiled away in 5.0 min? Is your answer reasonable? If not, explain why.

64. ■■ An aluminum bar and a copper bar of identical cross-sectional area have the same temperature difference

between their ends and conduct heat at the same rate. Which bar is longer, and how many times longer?

65. ■■ A Styrofoam box has a surface area of 1.0 m² and a thickness of 2.5 cm. If 5.0 kg of ice at 0°C is stored inside and the outside temperature is 35°C, how long does it take for all the ice to melt? (Consider conduction only.)

66. ■■ The thermal insulation used in building is commonly rated in terms of its *R-value*, defined as d/k, where d is the thickness of the insulation in inches and k is the thermal conductivity. For example, 3.0 in. of foam plastic would have an R-value of 3.0/0.30 = 10, where, in British units, $k = 0.30$ Btu·in./(ft²·h·F°). This value is expressed as R-10. (a) What does the R-value tell you; that is, how does thermal insulation vary with R-value? (b) What thicknesses of (1) fiberboard and (2) brick would give an R-value of R-10? [Hint: Use known ratios for unit conversion.]

67. ■■ Pine wood 14 in. thick has an R-value of 19. What thickness of (a) glass wool and (b) fiberboard would have the same R-value? (See Exercise 66.)

68. ■■ A large picture window measures 2.0 m by 3.0 m. At what rate will heat be conducted through the window when the room temperature is 20°C and the outside temperature is 0°C (a) if the window has a single pane of glass 4.0 mm thick and (b) if the window has a double pane of glass (a thermopane), and each pane is 2.0 mm thick with an intervening air space of 1.0 mm? (Assume that there is a constant temperature difference.)

69. ■■ The emissivity of an object is 0.60. How many times greater would the outward radiation be from a similar perfectly blackbody at the same temperature?

70. ■■ A house wall is composed of a solid concrete block with outside brick veneer and faced on the inside with fiberboard as illustrated in •Fig. 11.22. If the outside

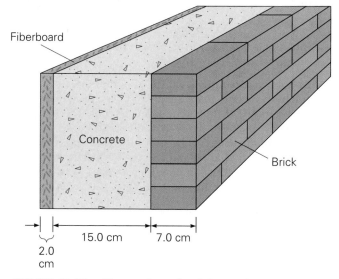

•FIGURE 11.22   **Thermal conductivity and heat loss**   See Exercise 70.

temperature on a cold day is −10°C and it is 20°C on the inside, how much energy in joules is conducted through a wall with dimensions of 3.5 m × 5.0 m in 1.0 h?

71. ■■ Suppose you wished to cut the heat loss through the wall in Exercise 70 by 50% by installing insulation. What thickness of Styrofoam would have to be placed between the fiberboard and concrete block to accomplish this?

72. ■■ If the temperature of an object is increased from 20°C to 40°C, how are the object's (a) emissivity and (b) output radiation rate affected? (c) What are the effects if instead the object's temperature in kelvins is doubled?

73. ■■ A steel cylinder of radius 5.0 cm and length 4.0 cm is placed in end-to-end thermal contact with a copper cylinder of the same dimensions. If the free ends of the two cylinders are maintained at constant temperatures of 95°C (steel) and 15°C (copper), how much heat will flow through the cylinders in 20 min?

74. ■■■ For the metal cylinders in Exercise 73, what is the temperature at the interface of the cylinders?

75. ■■■ Solar heating takes advantage of solar collectors such as the type shown in •Fig. 11.23. About 1000 W/m² of incoming solar radiation reaches the Earth's surface during daylight hours on a clear day. (The rest is reflected, scattered, absorbed, and so on.) How much heat energy would be collected, on the average, by the cylindrical collector shown in the figure during 10 h of collection?

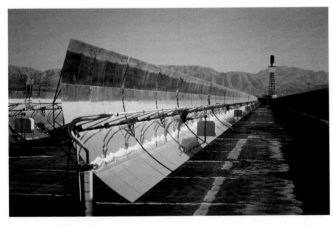

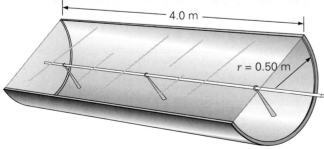

•FIGURE 11.23   Solar collector and solar heating   See Exercise 75.

Additional Exercises

76. A modern engine of alloy construction consists of 25 kg of aluminum and 80 kg of iron. How much heat does it absorb as its temperature increases from 20°C to 120°C as it warms up to operating temperature?

77. In a calorimetry experiment, 0.35 kg of aluminum shot at 100°C is carefully poured into 50 mL of water that has been chilled to 10°C. Neglecting any losses to the container (or otherwise), what is the final equilibrium temperature of the mixture?

78. A lamp filament radiates 100 W of power when the temperature of the surroundings is 20°C and 99.5 W of power when the temperature of the surroundings is 30°C. If the temperature of the filament is the same in each case, what is that temperature on the Celsius scale?

79. A student doing an experiment pours 0.150 kg of heated copper shot into a 0.375-kilogram aluminum calorimeter cup containing 0.200 kg of water at 25°C. The mixture (and the cup) comes to thermal equilibrium at 28°C. What was the initial temperature of the shot?

80. A student mixes 1.0 L of water at 40°C with 1.0 L of ethyl alcohol at 20°C. Assuming that there is no heat loss to the container or the surroundings, what is the final temperature of the mixture? [Hint: See Table 11.1.]

81. After a barrel jump, a 65-kilogram ice skater traveling at 25 km/h glides to a stop. If 40% of the frictional heat generated by the skate blades goes into melting ice at 0°C, how much ice is momentarily melted? Where does the other 60% of the energy go?

82. A 0.030-kilogram lead bullet hits a steel plate, both initially at 20°C. The bullet melts and splatters on impact. (This action has been photographed.) Assuming that the bullet receives 80% of its kinetic energy as heat energy, at what minimum speed must it be traveling to melt on impact?

83. ⦿ Two metals, 0.50 kg of aluminum and 0.50 kg of iron, initially at 20°C are heated to 100°C. Which metal gains more heat, and how much more?

84. Equal amounts of heat are added to different quantities of copper and lead. The temperature of the copper increases by 5.0 C°, and the temperature of the lead by 10 C°. Which piece of metal has the greater mass, and how much greater?

85. How much heat is required to convert 0.75 kg of ice at −10°C to steam at 120°C?

Thermodynamics

At certain locations on Earth, hot springs from deep in the interior rise to the surface. In Yellowstone National Park, they produce boiling pools and geysers such as Old Faithful, shown here. On the coast of Iceland, they warm the ocean, creating a tropical lagoon surrounded by glaciers. Such intriguing settings are popular vacation spots.

But the uses of such hot springs extend beyond recreation. Wherever there is a temperature difference, the potential exists for obtaining useful work. For instance, geothermal power plants draw on geysers as a renewable resource and produce no pollution. In this chapter you'll learn under what conditions, and with what efficiency, heat can be exploited to perform work, in machines as different as automobile engines and home freezers. You'll also find that the laws governing such energy conversions include some of the most general and far-reaching in all of physics.

As the word implies, **thermodynamics** deals with the transfer or the actions (dynamics) of heat (the Greek word for "heat" is *therme*). The development of thermodynamics started about 200 years ago and grew out of efforts to develop heat engines. The steam engine was one of the first of such devices, which convert heat energy to mechanical work. Steam engines in factories and locomotives powered the industrial revolution that changed the world. Although this course of study is primarily concerned with heat and work, thermodynamics is a broad and comprehensive science that includes a great deal more than heat engine theory. In this chapter, you will learn about the two general laws on which thermodynamics is based as well as the concept of entropy.

12.1 Thermodynamic Systems, States, and Processes

Objectives: To (a) define thermodynamic systems and states of systems, and (b) explain how thermal processes affect such systems.

Definition of:
Thermodynamics system

Thermodynamics is a field that makes use of many special terms, or everyday terms with special meanings, and you will find it useful to become familiar with some of them at the onset. The term **system**, as used in thermodynamics, refers to a definite quantity of matter enclosed by boundaries or surfaces, either real or imaginary. For example, a quantity of gas in the piston-cylinder of an engine has real boundaries, and imaginary boundaries enclose a cubic meter of air in a room. The enclosing boundaries of a system need have no definite shape, nor do they have to enclose a fixed volume. For example, a cylinder of gas could be compressed by advancing the piston.

Occasionally it is necessary to consider systems for which there is a transfer of matter across boundaries. However, for the most part, we will consider systems of constant mass. An important thermodynamic consideration is the interchange of energy between a system and its surroundings. This exchange may occur through a transfer of heat and/or the performance of mechanical work. For example, if a balloon is heated, it expands and work is done against the surrounding atmospheric pressure (and in stretching the balloon).

If heat transfer into or out of the system is impossible, the system is said to be a **thermally isolated system**. Work may be done on a thermally isolated system. For example, a thermally isolated balloon (perhaps surrounded by insulation) can be compressed by an external force or pressure, so work is done on the system; as we know, work results in a transfer of energy.

When heat does enter or leave a system, it is absorbed from or given up to the surroundings or to heat reservoirs. A **heat reservoir** is a system assumed to have an unlimited heat capacity. Thus any amount of heat can be withdrawn from or added to a heat reservoir without appreciably changing its temperature. This is like taking a cup of water from or adding a cup of water to the ocean—you don't change the water level appreciably. Similarly, pouring a warm bottle of water into a cold lake does not noticeably raise the lake's temperature.

State of a System

Just as there are kinematic equations to describe aspects of the motion of an object, there are **equations of state** to describe the conditions of thermodynamic systems. Such an equation expresses a mathematical relationship of the thermodynamic variables of a system. The ideal gas law, $pV = Nk_BT$, is a simple equation of state. This expression establishes a relationship among the pressure (p), volume (V), absolute temperature (T), and mass (N, the number of molecules) of a gas. These quantities are called *state variables*. Different gaseous states have different sets of values for these variables. A set of these variables that satisfies the ideal gas law specifies a state of a system of ideal gas completely as long as the system is in thermal equilibrium and has a uniform temperature.

The state of a given mass of gas in a closed system can be specified by the variables p, V, and T. It is convenient to plot the states according to the thermodynamic coordinates (p, V, T), much as we plot graphs with Cartesian coordinates (x, y, z). A general two-dimensional illustration of such a plot is shown in •Fig. 12.1. (This is not a p-V graph for an ideal gas. Several of these will be shown shortly.)

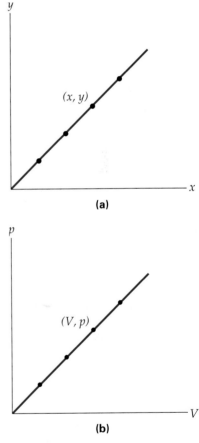

•FIGURE 12.1 Graphing states (a) On a common Cartesian graph, the coordinates (x, y) represent an individual point. (b) Similarly, on a p-V graph or diagram, the coordinates (V, p) represent a particular state of a system. It is common to say p-V diagram, rather than V-p, to match the order of the coordinates in the ideal gas law (p, V, T). The graph is not that described by the ideal gas law equation of state.

Just as the coordinates (x, y) specify individual points on a Cartesian graph, the coordinates (V, p) specify individual *states* on the p-V graph or diagram.

Using the ideal gas law, $pV = Nk_BT$ (Eq. 10.6), we can determine the temperature of a gas if we know its pressure and volume (and the number of molecules or moles in the sample). In other words, on a p-V diagram for a gas, each "coordinate" gives the pressure and volume directly, as well as indirectly the unique temperature for that condition or state of the gas. Thus for most situations involving a gas, only a p-V plot is necessary. In some cases, however, it can be instructive to refer to other coordinate plots such as p-T or T-V.

Processes

A **process** is a change in the state, or the thermodynamic coordinates, of a system. That is, when a system undergoes a process, the set of variables p, V, and T that describe its state completely will, in general, change. For example, we go from state (p_1, V_1, T_1) to state (p_2, V_2, T_2) by a thermodynamic process. Processes are said to be either reversible or irreversible.

Suppose that a system of gas in equilibrium (known p, V, T values) is allowed to expand quickly. The state of the system will change rapidly and unpredictably but will eventually return to equilibrium with another set of thermodynamic coordinates, in another state. On a graph such as a p-V diagram (•Fig. 12.2), we could show the initial state and the final state, but not what happened in between. Since the intermediate states changed so fast, there would be no data describing them. This is called an **irreversible process**—a process for which the intermediate steps are nonequilibrium states. "Irreversible" does not mean that the system can't be taken back to the initial state; it means only that the process path can't be retraced because of the nonequilibrium conditions.

If, however, the gas expands very, very slowly, passing from one known equilibrium state to a neighboring one and eventually arriving at the final state (see Fig. 12.2), then the process path between the initial and final states would be known. Such a process is called a **reversible process**—one whose path is known. In fact, a perfectly reversible process cannot be achieved. All real thermodynamic processes are irreversible to some degree because they follow complicated paths with many intermediate states. However, the concept of an ideal reversible process is useful.

These thermodynamic terms and ideas will now be applied in the laws of thermodynamics.

12.2 The First Law of Thermodynamics

Objectives: To (a) explain the relationship among internal energy, heat, and work as expressed by the first law, and (b) analyze various thermal processes.

The conservation of energy is valid for any system. Thus, the **first law of thermodynamics** is simply a statement of the conservation of energy as applied to thermodynamic systems. Heat, internal energy, and work are the quantities involved in a thermodynamic system. Suppose that some heat (Q) is added to a system. Where does it go? That it serves to increase the system's internal energy (ΔU; see Chapter 10) is certainly a possibility, as when the total molecular kinetic energy of an ideal gas increases with a temperature rise. Another possibility is that it could result in work (W) being done by the system. For example, when a heated gas expands, it does work on its surroundings (think of expanding air in a balloon).

Thus, it is possible for the added heat to go into internal energy, work, or both. Writing this as an equation gives a mathematical expression for the first law:

$$Q = \Delta U + W \qquad (12.1)$$

Our sign convention will be as follows: A positive value of heat ($+Q$) means that heat is *added to* the system, and a positive value of work ($+W$) means that work

Note: A change of state (of a system) does not necessarily mean a change of phase, such as from liquid to solid. To avoid confusion, always refer to solid, liquid, and gas as *phases* rather than states of matter.

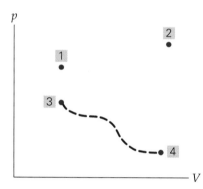

•**FIGURE 12.2 Paths of reversible and irreversible processes** If a gas quickly goes from state 1 to state 2, the process is irreversible since we do not know the "path." If, however, the gas is taken through many closely spaced states (as in going from state 3 to state 4), the process is reversible, at least in principle. Reversible means "exactly retraceable."

First law of thermodynamics: conservation of energy

is *done by* the system (for example, the work done by an expanding gas). Negative quantities mean that heat *is removed from the system* ($-Q$) and work *is done on the system* ($-W$), for example, when a gas is compressed.

Before going on with some examples of the first law, let's make certain we understand a fundamental difference among U, Q, and W. Any given system in a particular state will have a certain amount of internal energy U. Yet, a system does not possess certain amounts of heat or work. These are what *change* the state of the system and generally change the internal energy. When heat is added to or removed from a system or work is done on or by a system, *thermodynamic processes* occur that can change the system from one state to another, each having a particular internal energy U. That is, the internal energy depends only on the state of the system and not on how it got there.

Hence, to find the change in internal energy, ΔU, in going from one state to another, we need to know only the internal energy of the two states, and $\Delta U = U_2 - U_1$. In other words, the change in the internal energy, ΔU, is independent of the process path; it depends only on the initial and final states. This is analogous to the change in gravitational potential energy (mgh) being independent of path (Section 5.4).

The first law can be applied to several processes for a closed system of an ideal gas. In some of these processes, one thermodynamic variable is kept constant. Such processes have names beginning with iso- (from the Greek *isos*, meaning "equal").

Isobaric Process. A constant-pressure process is called an **isobaric process** (*iso* for equal, *bar* for pressure). An isobaric process for an ideal gas is illustrated in •Fig. 12.3. On a *p-V* diagram, the path of an isobaric process is along a horizontal line called an *isobar*. When heat is added to the gas in the cylinder, the ratio V/T must remain constant ($V/T = Nk_B/p$ = a constant). The heated gas expands; there is an increase in its volume. The temperature must therefore also increase proportionally, which means the internal energy of the gas increases. (Recall from Chapter 10 that kinetic theory says that the internal energy of an ideal gas is directly proportional to its absolute temperature.)

Work is done by the gas as it expands, in moving the cylindrical piston in Fig. 12.3. From the definition of work in Section 5.1,

Note: Keep in mind that, unless otherwise stated, we are working with constant-mass systems and with ideal gases.

Note: *Isobaric* means constant *pressure*.

Note: Review Sections 10.1 and 10.5.

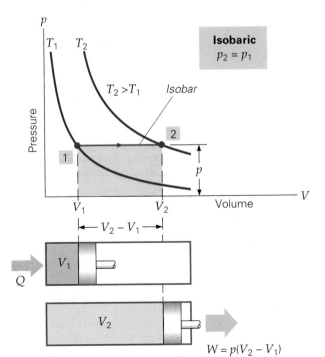

•**FIGURE 12.3 Isobaric (constant-pressure) process** The heat added to the gas in the frictionless piston goes into increasing the internal energy of the gas and doing work (the expanding gas moves the piston): $Q = \Delta U + W$. The work is equal to the area under the process path (from state 1 to state 2 here) on the *p-V* diagram.

$$W = F \Delta x$$

In terms of pressure ($p = F/A$), the force can be written $F = pA$, where A is the area of the piston. Then, the previous equation becomes

$$W = pA \Delta x$$

The change in volume is a right cylinder with end area A and height Δx. Hence $A \Delta x$ is simply the change in the volume of the gas, $A \Delta x = \Delta V = V_2 - V_1$. Thus,

$$W = p \Delta V = p(V_2 - V_1) \quad \text{(for an isobaric process)} \quad (12.2)$$

In Fig. 12.3, you can see that $p \Delta V$ is equal to area of the rectangle under the isobar on the p-V diagram. For a nonisobaric process (one in which the pressure changes), the work is also equal to the area under the line showing the process path, but the shape is not rectangular. Thus, the work depends on the process path as well as on the initial and final states: There will be different areas under different paths.

The internal energy of a quantity of an ideal gas depends only on its (absolute) temperature, however. Therefore a change in this internal energy is *independent* of the process path and depends only on the initial and final states, or on the difference between the temperatures of these states ($\Delta U = U_2 - U_1 \propto T_2 - T_1$).

Since V_2 is greater than V_1 for an expanding gas, work is done by the system ($+W$). In terms of the first law, then,

$$Q = \Delta U + W = \Delta U + p \Delta V \quad \text{(for an isobaric process)} \quad (12.3)$$

That is, the heat added to the system goes into increasing the internal energy *and* into work done by the system. If the process were reversed and the gas were being compressed by an external force doing work *on* the system, all the quantities would be negative. Heat would flow out of the system ($-Q$), and the internal energy, and therefore the temperature, of the gas would decrease ($-\Delta U$).

Note: *Isometric* means constant *volume*.

Isometric Process. An **isometric process** (short for iso*volu*metric, or equal-volume, process) is a constant-volume process, sometimes called an *isochoric process*. As illustrated in •Fig. 12.4, the process path on a p-V diagram is along a vertical line, commonly called an *isomet*. No work is done ($W = p \Delta V = 0$, since $\Delta V = 0$), so if heat is added, it goes into increasing the internal energy, and therefore the temperature, of the gas. By the first law,

$$Q = \Delta U + W = \Delta U + 0$$

and thus

$$Q = \Delta U \quad \text{(for an isometric process)} \quad (12.4)$$

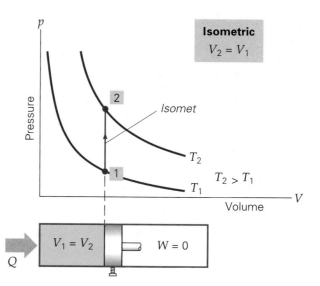

•**FIGURE 12.4 Isometric (constant-volume) process** All of the heat added to the gas goes into increasing the internal energy when the volume is held constant: $Q = \Delta U$. This causes an increase in temperature.

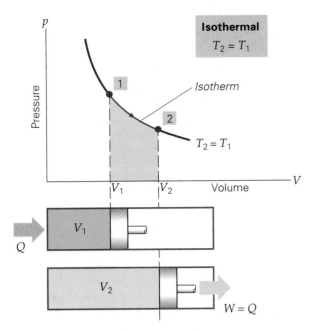

•**FIGURE 12.5 Isothermal (constant-temperature)** **process**   All of the heat added to the gas goes into doing work (the expanding gas moves the piston): since $\Delta T = 0$, then $\Delta U = 0$ and $Q = W$. As always, the work is equal to the area under the process path on the p-V diagram.

Isothermal Process.   An **isothermal process** is a constant-temperature process (•Fig. 12.5). In this case, the process path is along an *isotherm*, or a line of constant temperature. Since $p = (Nk_BT)/V = (\text{constant})/V$ for an ideal-gas isothermal process, an isotherm is a hyperbola on a p-V diagram. (The general form of the equation for a hyperbola is $y = a/x$.)

Note: *Isothermal* means constant temperature.

In going from state 1 to state 2 in Fig. 12.5, heat is added to the system, and both the pressure and volume change in order to keep the temperature constant (pressure decreases and volume increases). The work done by the expanding system $(+W)$ is again equal to the area under the process path.

For an isothermal process, the internal energy of the ideal gas remains constant ($\Delta U = 0$), because the temperature is constant. By the first law,

$$Q = \Delta U + W = 0 + W$$

and

$$Q = W \qquad \textit{(for an isothermal process)} \qquad (12.5)$$

Thus, for an ideal gas, in an isothermal process heat energy is converted to mechanical work (or vice versa for the reverse path). See Learn by Drawing on p. 394.

Adiabatic Process.   There's one more type of process in which a thermodynamic condition remains constant (•Fig. 12.6). In an **adiabatic process**, no heat is transferred into or out of the system; that is, $Q = 0$ (the Greek word *adiabatos* means "impassable.") The condition $Q = 0$ is satisfied for a thermally isolated system. This is an ideal situation, since there is always some heat transfer in actual system processes. Under actual conditions, we can only approximate adiabatic processes. For example, processes that are nearly adiabatic can take place in a system that is not thermally isolated if they occur rapidly enough that there isn't time for much energy to be transferred into or out of the system.

Adiabatic process: no heat transfer

Note: In practice, adiabatic means *quickly*, before significant heat can flow in or out ($Q = 0$).

As such a system follows the process path, a curve called an *adiabat*, all three thermodynamic coordinates change. For example, suppose a quantity of ideal gas were compressed in a thermally isolated cylindrical piston. If the piston were suddenly released, the gas would expand (that is, p and V would change). Work would be done at the expense of the internal energy of the gas, so the temperature of the gas would drop. By the first law,

$$Q = 0 = \Delta U + W$$

and

$$W = -\Delta U \qquad \textit{(for an adiabatic expansion)} \qquad (12.6)$$

Leaning on Isotherms

When you are analyzing the various thermodynamic processes dicussed in this chapter, it is sometimes hard to keep track of the heat flow (Q), work (W), and internal energy change (ΔU), together with their correct signs. One trick that can help with this bookkeeping is to superimpose a series of isotherms on the p-V plot you are working with (as in Figs. 12.2 and 12.4–12.6). This is generally useful even if the situation you are studying does not involve an isothermal process.

Before starting, recall the two important properties of an isothermal process. By definition, an isothermal process is one in which the temperature remains constant:

1. In an isothermal process, ΔU is zero. (Why?)
2. Since T is constant, the ideal gas law (Eq. 10.3) tells us that in an isothermal process, pV must also be constant: $pV = k$. You may recall from algebra that this is the equation of a hyperbola; on a p-V diagram, an isotherm is a hyperbola. The farther from the axes the hyperbola is, the higher the temperature it represents (Fig. 1).

To take advantage of these properties, follow these steps:

- Sketch a set of isotherms for a series of different temperatures on the p-V plot (Fig. 1).
- Then sketch the actual process or processes you are analyzing—for example, those shown in Fig. 2 and Fig. 3. (Take a moment to recall what the names mean: isomet = constant volume, isobar = constant pressure, and adiabat = no heat flow.)
- Finally, use the first law of thermodynamics, $Q = \Delta U + W$, to calculate the final results, including signs. W is the area under the p-V curve for the process represented, and its sign is determined by

expansion ($+$) or compression ($-$). The sign of ΔT will be clear from the isotherms; they serve as a temperature scale, with progressively higher temperatures represented by isotherms at greater distances from the axes. A rise in T implies a positive value of ΔU, since the internal energy of the gas increases with its temperature. You will thus have all you need to determine the sign of Q and whether heat flows into ($Q > 0$) or out of the system ($Q < 0$).

The following two examples indicate the power of this approach. In Fig. 2, we can decide whether heat flows into or out of the gas during an isobaric expansion. Expansion implies positive work. But what is the direction of the heat flow (or is it zero)? Sketching the isobar, we see that it crosses from lower-temperature isotherms to higher-temperature ones. Hence there is a temperature increase, and ΔU must be positive. Since the gas has both gained internal energy and done positive work, both terms on the right-hand side of the first-law equation are positive, so it follows that Q must be positive. Positive heat flow means "into the gas," so we have our answer.

An isometric process involving a pressure drop is analyzed in Fig. 3. Again, starting with lightly drawn isotherms, we plot a vertical line (constant volume) with an arrow pointing downward to show pressure reduction. Since there is no volume change, the gas does no work. However, we can see that its temperature drops (why?), so its internal energy must decrease. This means that ΔU is negative, hence Q must be negative. Thus heat flows "out of the gas."

As an exercise, try analyzing an adiabatic process via this graphical approach to determine where the work comes from if there is no heat flow into or out of the gas.

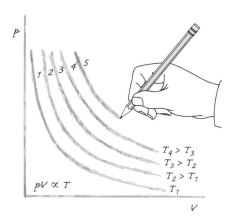

FIGURE 1 Isotherms on a _p-V_ plot

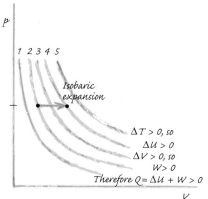

FIGURE 2 An isobaric expansion

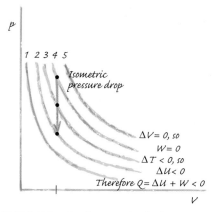

FIGURE 3 An isometric decline in pressure

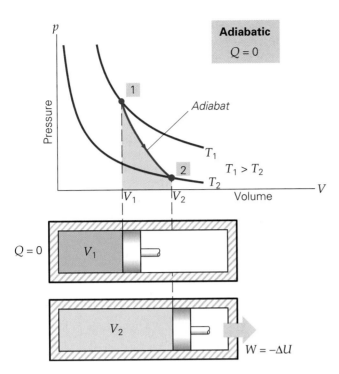

•FIGURE 12.6 Adiabatic (no heat transfer) process In an adiabatic process (shown here as a cylinder with heavy insulation), no heat is added to or removed from the system: $Q = 0$. Work is done at the expense of the internal energy: $W = -\Delta U$. The pressure, volume, and temperature all change in the process.

Thus, in an adiabatic expansion, work (the area under the process path) is done by the system and is therefore positive. In turn, ΔU must be negative, corresponding to a decrease in the system's internal energy. Since the internal energy, U, of an ideal gas depends only on its temperature, this decrease in internal energy is evidenced by a decrease in temperature.

Conceptual Example 12.1

Exhaling: Blowing Hot and Cold?

The air in your lungs is warm. This can be shown by putting your bare upper arm near your mouth and blowing air on your arm with your mouth opened widely. If you were to repeat this with your lips puckered, the air would feel (a) warmer, (b) cooler, (c) the same.

Reasoning and Answer.   We'll go from answer to reasoning on this one, since it is easy to determine the correct answer experimentally. Try it, and you might be surprised to find that the answer is (b).

   The interesting thing is why this occurs. When you blow on your arm with mouth open, you get a gush of warm air (roughly at body temperature). However, when you blow with your lips puckered, the stream of air is compressed. Then, as it emerges, the air expands, doing work against the atmosphere. This work is done at the expense of the internal energy of the air, so the temperature of the air is decreased and it feels cooler. This takes place quickly and is an essentially adiabatic process, at least for a few seconds. Eventually, the cooled air absorbs heat from the surrounding air and its temperature rises.

Follow-up Exercise.   After repairing a flat tire in his cool garage, a bicyclist pumps the tire up to full pressure. He then feels the pump and tire, and both feel warmer than before. (Neglect friction. The pump sides and plunger are assumed to be slippery.) Explain why this happens in terms of thermodynamic processes.

Example 12.2 ■ Isobaric Expansion: Change in Internal Energy

An ideal gas occupies a volume of 22.4 L at STP (standard temperature and pressure; see Chapter 10). While absorbing 2.53 kJ of heat from the surroundings, the gas expands isobarically to 32.4 L. What is the change in the internal energy of the gas?

12.2 The First Law of Thermodynamics **395**

Thinking It Through. Heat is absorbed, and hence Q is positive. Also, for an isobaric process, we can compute the work, W, as the area of a rectangle. Here W will be positive (why?). Lastly, the first law will allow us to compute ΔU.

Solution. Listing the data, along with appropriate conversions, we have

Given: $p_1 = p_2 = 1$ atm $\qquad\qquad$ *Find:* ΔU (change in internal energy)
$\qquad\quad = 1.01 \times 10^5 \, \text{N/m}^2$ (Pa)
$\qquad V_1 = 22.4 \, \text{L} = 22.4 \times 10^{-3} \, \text{m}^3$
$\qquad V_2 = 32.4 \, \text{L} = 32.4 \times 10^{-3} \, \text{m}^3$
$\qquad T_1 = 0°\text{C} = 273 \, \text{K}$
$\qquad Q = 2.53 \, \text{kJ} = 2.53 \times 10^3 \, \text{J}$

The change in internal energy is given by the first law: $\Delta U = Q - W$. We know Q, but we must first compute W, the work done by the expanding gas, to find ΔU. Using Eq. 12.2,

$$W = p\,\Delta V = p(V_2 - V_1)$$
$$= (1.01 \times 10^5 \, \text{N/m}^2)[(32.4 \times 10^{-3} \, \text{m}^3) - (22.4 \times 10^{-3} \, \text{m}^3)] = 1.01 \times 10^3 \, \text{J}$$

Then, by the first law,

$$\Delta U = Q - W = (2.53 \times 10^3 \, \text{J}) - (1.01 \times 10^3 \, \text{J}) = 1.52 \times 10^3 \, \text{J}$$

Follow-up Exercise. In this Example, what is the equilibrium temperature, in degrees Celsius, of the gas after the expansion?

12.3 The Second Law of Thermodynamics and Entropy

Objectives: To (a) state and explain the second law of thermodynamics in several forms and (b) explain the concept of entropy.

Suppose that a piece of hot metal is placed in an insulated container of cool water. Heat will be transferred from the metal to the water, and the two will come to thermal equilibrium at some intermediate temperature. For a thermally isolated system, the system's total energy remains constant. Could there be a reverse process in which heat would be transferred from the cooler water to the hot metal? This process would not happen naturally. But if it did, the total energy of the system would remain constant, and the energy transfer would *not* violate energy conservation.

Clearly, there must be another principle that specifies the *direction* in which a process can take place. This principle is embodied in the **second law of thermodynamics**, which says that certain processes do not take place, or have never been observed to take place, even though they are consistent with the first law.

There are many equivalent statements of the second law, which are worded differently according to their application. One applicable to the above situation is

Heat will not flow spontaneously from a colder body to a warmer body.

An alternative statement of the second law involves *thermal cycles*. A thermal cycle typically consists of several separate thermal processes, connected such that they end up back at the starting conditions. If a gas is used as the material, this means that the gas comes back to the same *p-V-T* state in which it started. The second law, stated in terms of a thermal cycle (operating as a heat engine, which we shall discuss in Section 12.4), is as follows:

In a thermal cycle, heat energy cannot be completely transformed into mechanical work.

In general, the second law applies to all forms of energy. It is considered to be true because no one has ever found an exception to it. If it were not valid, a perpetual motion machine could be built. Such a machine could transform heat completely into work and motion (mechanical energy), with no energy loss. The mechanical energy could then be transformed back into heat and be used to reheat the reservoir (again with no loss). Since the processes could be repeated indefinitely, the machine would run perpetually. All of the energy is accounted for, so this

Second law of thermodynamics

Note: Real machines can be made virtually frictionless, but their efficiency is still less than 100%. The second law limit does *not* refer to frictional losses.

396 **CHAPTER 12** Thermodynamics

situation does not violate the first law. However, it is obvious that real machines are always less than 100% efficient, that the work output is always less than the energy input. Another statement of the second law is therefore:

It is impossible to construct an operational perpetual motion machine.

Many attempts have been made to construct perpetual motion machines, but there have been no known successes.*

It would be convenient to have some way of expressing the direction of a process in terms of the thermodynamic properties of a system. One such property is temperature. In analyzing a conductive heat transfer process, you need to know the temperatures of the system and its surroundings. Knowing the temperature difference allows you to state the direction in which the heat transfer will spontaneously take place, into or out of the system.

Entropy

A property that indicates the *natural direction* of a process was first described by Rudolf Clausius (1822–1888), a German physicist. This property is called **entropy**, a name coined by Clausius. Physically, entropy is a particularly rich, multifaceted concept. Its interpretations are diverse and intriguing. We might draw an analogy to the fable of the Blind Men and the Elephant, in which three blind men describe what each thinks an elephant to be after touching different parts. Thus, a discussion of entropy is likely to resemble a debate:

"Entropy is a measure of a system's ability to do useful work. As a system loses the ability to do work, its entropy increases."

"No, entropy determines the direction of time. It's 'time's arrow' that points out the forward flow of events, thereby distinguishing past events from future ones."

"You're both wrong. Entropy is a measure of disorder. A system naturally moves toward a state of greater disorder or disarray. The more order, the less the system's entropy is."

As you see, entropy has several equally valid physical interpretations. Here we are concerned with the mathematical definition of entropy in terms of thermodynamic properties. The change in a system's entropy (ΔS) when an amount of heat (Q) is added or removed by a reversible process at a constant temperature is

Note: ΔS is positive if a system absorbs heat ($+Q$) and negative if heat leaves ($-Q$). The sign of ΔS is determined by the sign of Q.

$$\Delta S = \frac{Q}{T} \qquad \begin{array}{l}\textit{change in entropy}\\ \textit{(constant temperature only)}\end{array} \qquad (12.7)$$

SI unit of entropy: joule per kelvin (J/K)

where T is the Kelvin temperature and the units of entropy are joules per kelvin (J/K).

If the temperature changes during the process, the change in entropy can still be calculated by using advanced mathematics. Our discussions will be limited to isothermal processes or ones with relatively small temperature changes. For the latter, reasonable approximations of entropy changes may be obtained by using average temperatures, as we will see in Example 12.4. First, however, let's look at a relatively simple calculation of entropy change.

Always use kelvins! What would happen if you used the Celsius temperature for a phase change from ice to water at 0°C?

Example 12.3 ■ Change in Entropy: An Isothermal Process

What is the change in entropy of ethyl alcohol when 0.25 kg of it vaporizes at its boiling point of 78°C (latent heat of vaporization $L_v = 1.0 \times 10^5$ J/kg)?

Thinking It Through. A phase change occurs at constant temperature; hence we can apply Eq. 12.7 after converting to kelvins. From Eq. 11.2, we can compute the amount of heat added.

---

*Although perpetual motion *machines* do not exist, perpetual motion is known to exist—for example, planets in orbit around the Sun.

Solution. From the problem, we have

Given: $m = 0.25$ kg   *Find:* ΔS (change in entropy)
$\quad\quad T = 78°C + 273 = 351$ K
$\quad\quad L_v = 1.0 \times 10^5$ J/kg

Since this is a phase change and we are given the latent heat,

$$Q = mL_v = (0.25 \text{ kg})(1.0 \times 10^5 \text{ J/kg}) = 2.5 \times 10^4 \text{ J}$$

Then

$$\Delta S = \frac{Q}{T} = \frac{+2.5 \times 10^4 \text{ J}}{351 \text{ K}} = +71 \text{ J/K}$$

Note that Q is positive because heat is added to the system. The change in entropy, then, is also positive, and the entropy of the alcohol increases.

Follow-up Exercise. What is the change in entropy of water when 1.00 kg of it freezes to form ice at 0°C?

Example 12.4 ■ Warm Spoon into Cool Water: Entropy Change?

A metal spoon at 24°C is placed in 1.00 kg of water at 18°C. The thermally isolated system comes to equilibrium at a temperature of 20°C. Find the approximate change in the entropy of the system.

Thinking It Through. The system is isolated, so we know that $Q_w + Q_m = 0$, where the subscripts w and m stand for water and metal, respectively. We can determine Q_w from the given data, so we have both Q values. Strictly speaking, we cannot use Eq. 12.7 (why?), but since the temperature changes are small, we can get a good approximation for ΔS by using each object's *average* temperature, $\overline{T}$.

Solution. Using subscripts i and f to stand for initial and final, respectively, we have

Given: $T_{m_i} = 24°C$ *Find:* ΔS (change in entropy of the system)
$\quad\quad T_{w_i} = 18°C$
$\quad\quad m_w = 1.00$ kg
$\quad\quad T_f = 20°C$
$\quad\quad c_w = 4186$ J/(kg·C°)
$\quad\quad\quad$ (from Table 11.1)

As in Example 12.3, we need to find the amount of heat (Q) to solve for ΔS. However, in this system there are two Q values: Q_w, the heat gained by the water, and Q_m, the heat lost by the metal.

With $\Delta T_w = T_f - T_{w_i} = 20°C - 18°C = 2.0 \text{ C}°$, the heat gained by the water is

$$Q_w = m_w c_w \Delta T = (1.00 \text{ kg})[4186 \text{ J/(kg·C°)}](2.0 \text{ C°}) = 8.37 \times 10^3 \text{ J}$$

from Eq. 11.1. This is also the amount of heat *lost* by the metal, since by the conservation of energy, heat gained must equal heat lost. Therefore, $Q_m = -8.37 \times 10^3$ J, where the minus sign indicates a loss of heat.

In this case, we have small temperature changes, so we can approximate the entropy changes by using average temperatures in kelvins:

$$\overline{T}_w = \frac{T_f + T_{w_i}}{2} = \frac{20°C + 18°C}{2} = 19°C = 292 \text{ K}$$

$$\overline{T}_m = \frac{T_{m_i} + T_f}{2} = \frac{24°C + 20°C}{2} = 22°C = 295 \text{ K}$$

We can then use these average temperatures to compute the approximate entropy changes for both the water and the metal:

$$\Delta S_w = \frac{Q_w}{\overline{T}_w} = \frac{+8.37 \times 10^3 \text{ J}}{292 \text{ K}} = +28.7 \text{ J/K}$$

$$\Delta S_m = \frac{Q_m}{\overline{T}_m} = \frac{-8.37 \times 10^3 \text{ J}}{295 \text{ K}} = -28.4 \text{ J/K}$$

The change in the entropy of the *system* is the sum of these results:

$$\Delta S = \Delta S_w + \Delta S_m = +28.7 \text{ J/K} - 28.4 \text{ J/K} = +0.3 \text{ J/K}$$

The entropy of the metal decreased because heat flowed out of it. The entropy of the water increased by a greater amount, so the total entropy of the system overall increased (a positive change).

Follow-up Exercise. What would be the initial temperatures in this Example if the overall entropy change were zero ($\Delta S = 0$)?

Note that the total entropy change of the system in Example 12.4 is positive. In general, the direction of any process is toward an increase in total system entropy. That is, *the entropy of an isolated system never decreases.* Another way to state this observation about entropy is to say that *the entropy of an isolated system increases for every natural process* ($\Delta S > 0$). In coming to an intermediate temperature, the water and spoon in Example 12.4 are undergoing a natural process, so-called because this is what is always observed to occur.

If, in coming to thermal equilibrium with the water, the spoon somehow spontaneously became hotter while the water became cooler, this would certainly be an *unnatural* process ($\Delta S < 0$). Similarly, water at room temperature in an isolated ice cube tray will not naturally turn into ice.

If a system is not isolated, it may undergo a decrease in entropy. For example, if an ice cube tray filled with water, is put into a freezer compartment, the water will freeze, undergoing a decrease in entropy. But there will be a *larger increase in entropy somewhere else* in the environment or the universe. Work has to be done or energy expended to bring about the transfer of heat involved in the change from water to ice. Thus, a more general statement of the second law of thermodynamics in terms of entropy is:

> The total entropy of the universe increases in every natural process.

The entropy of a system is a function of its state. Each state of a system has a particular value of entropy, and a change in entropy depends only on the initial and final states for a process, or $\Delta S = S_f - S_i$. (This is analogous to the change in internal energy for an ideal gas, $\Delta U = U_f - U_i$, which depends only on the initial and final states and *not* on the details of the process.) An entropy change can be represented on a T-S diagram, as shown in •Fig. 12.7. Just as the area under a curve on a p-V diagram for a gas is equal to the work ($W = p \Delta V$), the area under a T-S curve is equal to the heat flow, which is easily seen for the isothermal process in the figure ($Q = T \Delta S$). Like work for a process with variable pressure, heat transferred in a process with variable temperature is also equal to the area under the T-S curve (which, once again, requires advanced mathematics to calculate).

There is a theoretical process for which the entropy is constant. Recall that for an adiabatic process $Q = 0$. Since in this case $\Delta S = Q/T = 0$, there is no change in entropy. Hence, an adiabatic process is a constant-entropy process, or an **isentropic process**. With the inclusion of an ideal adiabatic process, we may say that during any process the entropy of the universe can only increase or remain constant ($\Delta S \geq 0$). How would an isentropic process appear on a T-S diagram?

Note: The entropy of a closed system increases for every natural process.

Note: Total energy can neither be created nor destroyed; total entropy can be created but not destroyed.

•FIGURE 12.7 *T-S*
(temperature–entropy) diagrams
When the temperature (*T*) is plotted versus entropy (*S*), the area under the process path is equal to the heat (*Q*) transferred in the process. This is similar to work being equal to the area under a process path on a *p-V* diagram. **(a)** This *T-S* diagram is for an isothermal process. (How can you tell?) Note that the area under the path is that of a rectangle. **(b)** This *T-S* diagram is for a general process in which the temperature varies.

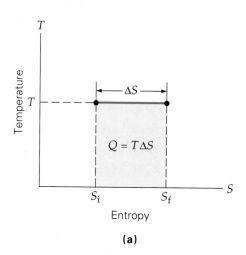

(a)

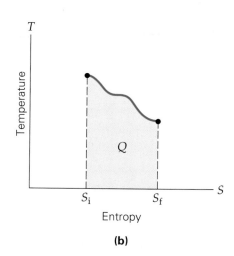

(b)

Conceptual Example 12.5

More Order in the System: Entropy Decreasing?

Which of the following involves a *decrease* in the system's entropy: (a) A hot fudge sundae is left uneaten too long, so the ice cream melts and the fudge solidifies; (b) a green plant combines water and carbon dioxide molecules in photosynthesis to make a larger, more complex molecule of sugar; (c) you drop your term paper on the way up the library stairs and find that, when you have gathered up all the pages, they are no longer in sequence; (d) perfume in an opened bottle evaporates and fills the room with scent; (e) a shortage of library personnel makes it increasingly difficult to find the books you want in their proper locations on the shelves?

Reasoning and Answer. It is not difficult to see that both cases (c) and (e) involve a decrease in order and thus an increase in entropy. The same is true, in a slightly less obvious way, of case (d). When the perfume molecules are no longer confined within the restricted space of the bottle but instead are distributed at random throughout the larger volume of air in the room, there is a loss of order—the positions of individual perfume molecules cannot be specified as accurately. Similarly in case (a), the system (fudge plus ice cream) is less highly ordered when the speeds of all the molecules are randomized than when they are divided into two distinct populations, one hot and one cold. (We showed in another way in Example 12.4 how the flow of heat from a warmer body to a colder one entails an increase in entropy.) Hence the answer is (b). In this instance a more complex, highly ordered structure was created from simpler components—a process that will not occur naturally without an input of energy and a larger increase in entropy somewhere else in the universe (see the Insight on p. 401).

Follow-up Exercise. In a garden compost pile, newly clipped grass blades are gradually broken down into a natural fertilizer. Does this process represent an increase or decrease in the total entropy of the grass blades? of the universe?

12.4 Heat Engines and Thermal Pumps

Objectives: To (a) explain the concept of a heat engine and compute thermal efficiency, and (b) explain the concept of a thermal pump and compute coefficient of performance.

Heat engine: converts heat → work

A **heat engine** is any device that converts heat energy to work. Since the second law says that perpetual motion machines are impossible, some of the heat supplied to a heat engine will necessarily be lost. In studying thermodynamics, there is no need to be concerned with the usual mechanical components of an engine, such as pistons, cylinders, and gears. For our purposes, a heat engine is simply a device that takes heat from a high-temperature source (a hot reservoir), converts some of it to useful work, and transfers the rest to its surroundings

Life, Order, and the Second Law

The second law of thermodynamics is one of the foundations of physics and operates universally—the total entropy of the universe increases in every natural process. Certainly life is a natural process. Yet clearly, there is an increase in order during the development of the human embryo into a mature adult. Moreover, in virtually all life processes larger, more complex molecules and cellular structures are continuously being synthesized from simpler components. Does this mean that living things are exempt from the second law?

To answer this question, we must realize that cells are not closed systems. They are involved in continuous exchanges of matter and energy with their surroundings. Living cells are constantly occupied in the exchange and conversion of energy from one form to another, which is governed by the first law of thermodynamics. For example, cells convert chemical potential energy into kinetic energy (movement) and electrical energy (the basis of nerve impulses). Many plant cells convert sunlight energy into chemical energy through photosynthesis (Fig. 1a). We have taken a hint from nature in attempting to harness the energy in sunlight (Fig. 1b).

In all biological processes, living systems maintain or even increase their organization by causing a larger decrease in the order of their surroundings. The second law is not violated by a *local* decrease in entropy as long as the *total* entropy of the universe increases in the process.

As an example of how cells create order while remaining within the confines of the second law, consider how they use chemical raw materials such as the sugar glucose ($C_6H_{12}O_6$). Cells use carbon atoms from glucose (and other organic compounds) to build up more complex molecules—a decrease in entropy. But they get the energy for these reactions by "burning" a great many glucose molecules as well—oxidizing them to produce many smaller, simpler molecules of water (H_2O) and carbon dioxide (CO_2). This process involves an increase in entropy. The product molecules, being less highly structured than glucose, have greater entropy. Moreover, heat is released by the oxidation process, raising the entropy of the cell's surroundings. Thus a cell "pays" for its increase in order by increasing the disorder of the rest of the universe.

But, you may ask, where did the complex glucose molecule, with its chemical potential energy, come from in the first place? As we mentioned above, photosynthetic cells use the energy of sunlight to make glucose and related compounds. Thus we can look at the flow of energy through organisms in a more general way. As we will see in the Chapter 20 Insight on the greenhouse effect, life on Earth is dependent on receiving radiant energy in the form of sunlight and reradiating infrared energy back into space. In effect, the Sun is a high-temperature source and space is a low-temperature

(a)

(b)

FIGURE 1 Solar collectors (a) Green plants capture trillions of joules of solar energy every day and have been doing so for billions of years. The chemical energy they store in the form of sugars and other complex molecules is used by nearly all organisms on Earth to maintain and increase the highly organized state that we call life. (b) Humans have developed devices (such as these reflective solar energy concentrators) to capture some of the energy of sunlight, although the technology is not yet in widespread use.

sink. The flow of energy from one to another allows living things to decrease entropy locally, creating this island of order that we call the biosphere. Thus biological organisms do not violate the second law, as is sometimes thought; they merely take advantage of a loophole in it.

•FIGURE 12.8 Work from heat
At this geothermal energy facility in the San Francisco area, natural heat is exploited to do useful work (generating electricity) without the need to "burn" fuel. Water heated by thermal processes beneath the Earth's surface serves as the hot reservoir. (What do you think is the cold reservoir?)

(a cold, or low-temperature, reservoir). For example, most of the turbines that generate the electricity we use (Chapter 20) are heat engines, using input heat from various sources: the burning of chemical fuels (oil, gas, or coal); nuclear reactions (Chapter 30); or the thermal energy already present beneath the Earth's surface (•Figure 12.8). They might be cooled by cold river water, thus losing heat to this low-temperature reservoir. A generalized heat engine is usually represented as shown in •Fig. 12.9a.

Simply adding heat to a gas in a cylindrical piston can produce work in a single process. But since a continuous output is usually wanted, practical heat engines usually operate in a **thermal cycle,** or a series of processes, which brings the engine or system back to its original condition. Cyclic heat engines include steam engines and internal combustion engines, such as automobile engines. An idealized, rectangular thermodynamic cycle is shown in Fig. 12.9b. It consists of two isobars and two isomets. When these processes occur in the sequence indicated, the system goes through a cycle (1-2-3-4-1), returning to its original condition. Recall that for an isobaric process, the work done is equal to the area under the isobar, $W = p \Delta V$, on a p-V diagram (see Fig. 12.3). Here, the work in the 1-2 process is positive and the work in the 3-4 process is negative (why?). Hence, the *net work* (W_{net}) is the area of the rectangle formed by the isobars and isomets.

For a general heat engine cycle, the net work done per cycle is the area enclosed by the process paths that compose the cycle as shown in Fig. 12.9c. When the paths are not straight lines, advanced mathematics is needed to calculate the area. The idea is the same regardless of the details: For a heat engine, the object is to get more positive work out ($+W$, expansion) than must be put back ($-W$, compression). This means a *net positive work* output per cycle.

Conceptual Example 12.6

The Drinking Bird: A Heat Engine

An unusual example of a cyclic heat engine is a novelty item called a drinking bird, shown in •Fig. 12.10. The glass portion of the bird contains a volatile liquid (usually ether). When the absorbent material on the head and beak is initially wet and the bird is released, it begins a slow, periodic motion: dunking its head in the water, bobbing back out for a while, and then repeating the dunking/bobbing cycle. Using the principles of thermodynamics, expain how this "heat engine" works.

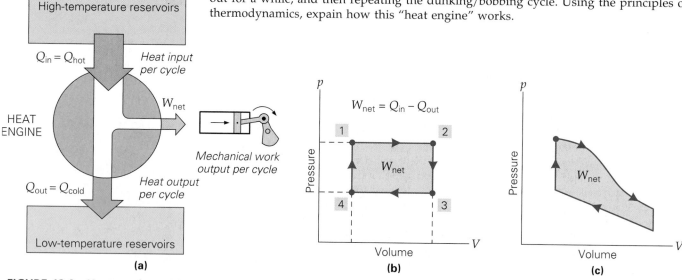

•FIGURE 12.9 Heat engine (a) Energy flow for a generalized cyclic heat engine. Note that the width of the arrow representing Q_{in} is equal to the combined widths of the arrows representing W_{net} and Q_{out}, reflecting the conservation of energy. (b) This cyclic process consists of two isobars and two isomets. The net work output per cycle is the area of the rectangle formed by the process paths. (c) For a general heat engine cycle, the net work (W_{net}) per cycle is still represented by the area inside the cycle's separate paths.

Reasoning and Answer. When the head and beak are initially wet, evaporation occurs. Thus heat is removed from the head, which lowers its temperature and internal vapor pressure. The higher vapor pressure is now in the body, forcing the liquid into the head. This raises the bird's center of gravity above the pivot point, and a torque causes a forward rotation. The bird then rewets its beak in the glass of water.

The liquid from the body runs into the head, warming it. With the tube not completely filled with liquid and partially open, the pressures in the head and body equalize, so there is no longer a pressure difference. The liquid then drains back into the body. With the lowering of its center of gravity below the pivot point, the bird swings back to the vertical position, and the cycle begins again. The net effect is that heat is transferred from a high-temperature reservoir (the body) to a low-temperature reservoir (the head), with work being done (evidenced by the rotational motion). Gravity and the atmosphere also play roles in the operation of this specialized heat engine.

Follow-up Exercise. Describe how changes in atmospheric conditions such as humidity and pressure would affect the operation of the drinking bird.

Thermal Efficiency

Thermal efficiency is used to rate heat engines. The **thermal efficiency** (ϵ_{th}) of a heat engine is defined as

$$\epsilon_{th} = \frac{\text{net work out}}{\text{heat in}} = \frac{W_{net}}{Q_{in}} \qquad (12.8)$$

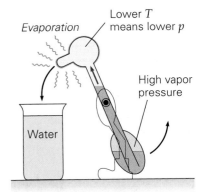

The efficiency tells you what you get out (net) for what you put in, so to speak. The higher compression and pressure present in an automotive diesel engine, for example, give it a higher thermal efficiency than a gasoline engine—about 25% as opposed to 15%.* For more about engines, see the Insight on p. 404.

For one cycle of a cyclic heat engine, the work output per cycle W_{net} is determined by the first law as applied to the whole cycle. We now take the symbol Q to mean the *magnitude* of each heat flow and show Q_{out} explicitly as $-Q_{cold}$ (flow out) and Q_{in} as $+Q_{hot}$ (flow in):

$$Q = \Delta U + W \quad \text{or} \quad Q_{hot} - Q_{cold} = 0 + W_{net} \quad \text{(full cycle)}$$

Here we have set ΔU to zero, since the system returns to its original state in completing a cycle ($\Delta T = 0$). The thermal efficiency (per cycle) of a cyclic heat engine is thus given by

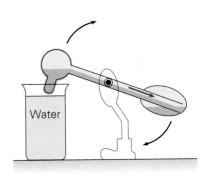

•**FIGURE 12.10   A heat engine?** See Conceptual Example 12.6.

$$\epsilon_{th} = \frac{W_{net}}{Q_{hot}} = \frac{Q_{hot} - Q_{cold}}{Q_{hot}} = 1 - \frac{Q_{cold}}{Q_{hot}} \qquad (12.9)$$

Thermal efficiency of a heat engine

In general, each cycle of an engine has the same efficiency.

Like mechanical efficiency, thermal efficiency is a dimensionless fraction and is commonly expressed as a percentage. Equation 12.9 indicates that a heat engine could have 100% efficiency if Q_{cold} were zero. This would mean that no heat energy would be lost and all the heat input would be converted to useful work, which is impossible according to the second law of thermodynamics. In 1851, Lord Kelvin stated this observation, yet another form of the second law:

Note: Here all Q's are positive; heat flow in or out will be shown by a + or − sign.

> No heat engine operating in a cycle can convert its heat input completely to work.

The second law, in terms of heat engines

Practically, we see from Eq. 12.9 that to maximize the work output per cycle of a heat engine, we must work to minimize the ratio of Q_{cold}/Q_{hot}, which increases the efficiency.

---

*The thermal efficiency of a heat engine in an automobile depends on many parameters, including the operation of features such as air conditioning. These values are only approximate.

The Otto Cycle

The internal combustion engine is a cyclic heat engine employed in automobiles and many other devices. Two different cycles of operation are in general use: a two-stroke cycle and a four-stroke cycle. (The number of strokes tells how many up and down motions a piston makes in a cycle; for example, a two-stroke cycle has one upstroke and one downstroke per cycle.) Engines with a two-stroke cycle are commonly used in motorcycles, outboard motors, chain saws, and some lawn mowers.

Most automobile engines utilize a four-stroke cycle. The steps in this cycle are shown in Fig. 1, along with a *p-V* diagram of the theoretical thermodynamic pro-cesses each cycle approximates. The theoretical cycle is called the Otto cycle, for the German engineer Nickolas Otto (1832–1891), who built one of the first successful gasoline engines.

During the intake stroke (1–2), an isobaric expansion, the air–fuel mixture is admitted at atmospheric pressure through the open valve as the piston drops. This mixture is compressed adiabatically (very quickly) on the compression stroke (2–3), which is followed by ignition (isometric pressure rise) (3–4) and an adiabatic expansion during the power stroke (4–5). Next is an isometric cooling of the system when the piston is at its lowest position (5–2). The final, exhaust stroke is along the isobaric leg of the Otto cycle (2–1). Notice that it takes two cycles of the piston to produce one power stroke.

A two-stroke engine does not have an isobaric leg on its theoretical cycle. The air–fuel mixture is added to the cylinder at the same time as the previously combusted gases are expelled, during the isometric cooling process (5–2). This results in some loss of fuel (in the exhaust gases), which is the major disadvantage of the two-stroke engine. An advantage is that there is one power stroke for every crankshaft revolution in a two-stroke engine, compared to only *one* power stroke for *two* crankshaft revolutions in a four-stroke engine, so a two-stroke engine requires less work and can have higher efficiency.

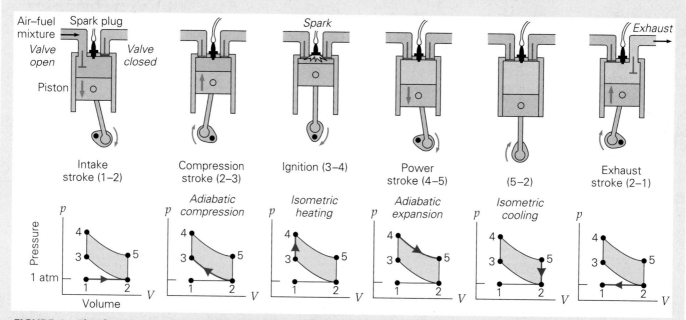

FIGURE 1   The four stroke cycle of a heat engine   The process steps for the four-stroke Otto cycle. The piston moves up and down twice each cycle, making a total of four strokes per cycle.

Example 12.7 ■ Thermal Efficiency: What You Get Out of What You Put In

The small, gasoline-powered engine of a leaf blower removes 800 J of heat energy from a high-temperature reservoir and exhausts 700 J to a low-temperature reservoir. What is the engine's thermal efficiency?

Thinking It Through.   We can apply the definition of thermal efficiency of a heat engine (Eq. 12.8). We first need W_{net} to determine ϵ_{th}.

Solution.

Given: $Q_{hot} = 800$ J   *Find:* ϵ_{th} (thermal efficiency)
    $Q_{cold} = 700$ J

Let's calculate W_{net} first from the conservation of energy (first law) applied over one full cycle (so $\Delta U = 0$). Recall that our notation for heat now indicates magnitude only, so we show the minus sign explicitly for heat flow out:

$$W_{net} = Q_{hot} - Q_{cold} = 800\ J - 700\ J = 100\ J$$

Therefore, from Eq. 12.8 the thermal efficiency is

$$\epsilon_{th} = \frac{W_{net}}{Q_{hot}} = \frac{100\ J}{800\ J} = 0.125\ (\times 100\%) = 12.5\%$$

Follow-up Exercise. What would be the net output work of the engine in this Example if the efficiency were raised to 15% and the input heat were raised to 1000 J/cycle?

Thermal Pumps: Refrigerators, Air Conditioners, and Heat Pumps

The function performed by a thermal pump is basically the reverse of that of a heat engine. That is, the name **thermal pump** is a generic term for *any* device, including refrigerators, air conditioners, and heat pumps, that transfers heat energy from a low-temperature reservoir to a high-temperature reservoir (•Fig. 12.11a). To do this, there must be work input, since the second law says that heat will not spontaneously flow from a cold body to a hot body.

Note: As used here, the term *thermal pump* is general and does not refer specifically to the devices that are commonly used for heating and cooling of buildings. These will be discussed shortly and are called "heat pumps."

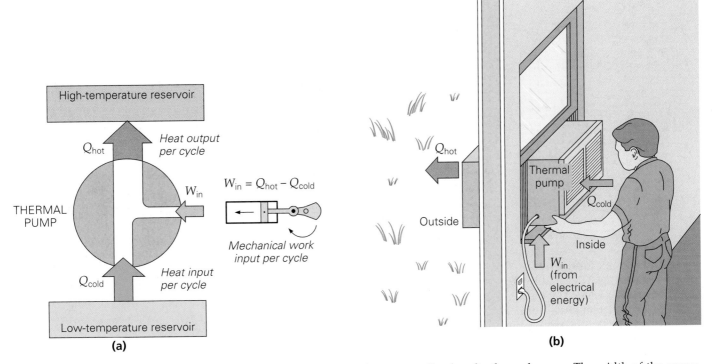

•**FIGURE 12.11 Thermal pumps** (a) An energy-flow diagram for a generalized cyclic thermal pump. The width of the arrow representing Q_{hot}, the heat transferred into the high-temperature reservoir, is equal to the combined widths of the arrows representing W_{in} and Q_{cold}, reflecting the conservation of energy. (b) An air conditioner is an example of a thermal pump. With work input, it transfers heat (Q_{cold}) from a low-temperature reservoir (inside the house) to a high-temperature reservoir (outside). Thermodynamically, it operates like a refrigerator.

A familiar example of a thermal pump is a refrigerator. With work input (from electrical energy), heat is transferred from inside the refrigerator (low-temperature reservoir) to the surroundings (high-temperature reservoir). Have you ever wondered how a refrigerator (or an air conditioner, another thermal pump; Fig. 12.11b) operates? The knowledge you have gained about thermodynamic processes will allow you to understand the basic operation (•Fig. 12.12).

The refrigerant, or heat-transferring medium, is a substance with a relatively low boiling point. Ammonia (boiling point $-33.3°C$ at 1 atm), sulfur dioxide ($-10.1°C$), and Freon ($-29.8°C$) can be used as refrigerants. Until recently Freon was used in most domestic refrigerators. The other compounds were once used, but they were quite dangerous if the system happened to leak. Freon has been replaced by still other refrigerants due to its destruction of the Earth's ozone layer (see the Insight in Chapter 19).

It is convenient to think of the system diagrammed in Fig. 12.12 as having high and low (temperature and pressure) sides. On the low side, heat is transferred to the evaporator coils from inside the refrigerator. The heat causes the refrigerant to boil and is carried away as latent heat of vaporization. The gaseous vapor is drawn into the compressor chamber on the downstroke of the piston and is compressed on the upstroke (work input by compressor motor). The compression increases the temperature of the gas, which is discharged from the compressor as a superheated vapor (on the high side). This vapor condenses in the cooler condenser unit, where circulating air (or water in larger units) carries away the latent heat of condensation and the heat of compression. The condensed liquid collects in a receiver.

An expansion valve maintains a balance between the high and low sides, thereby controlling the rate at which the refrigerant passes back to the low side. On being admitted to the low side, the liquid immediately boils and vaporizes because of the low pressure. This is a cooling process, because the heat of vaporization is supplied by the internal energy of the liquid. The cooled refrigerant is drawn into the evaporator, and the cycle begins again.

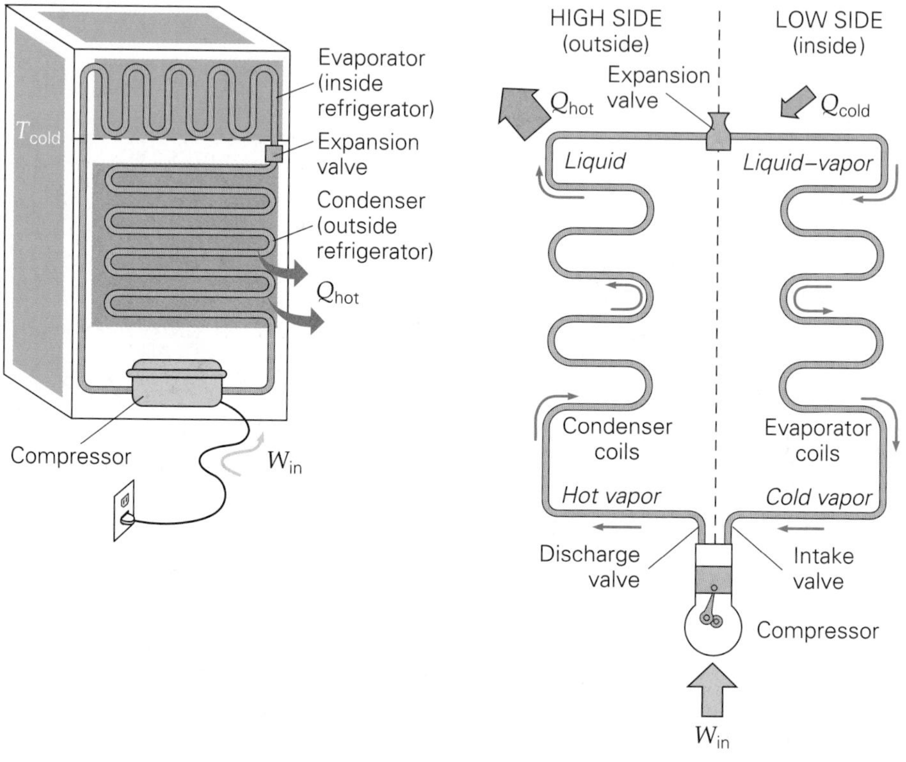

Refrigerator operation

•**FIGURE 12.12 Refrigerator operation** Heat (Q_{cold}) is carried away from the interior by the refrigerant as latent heat. This heat energy and that of the work input (W_{in}) are discharged from the condenser to the surroundings (Q_{hot}). A refrigerator can be thought to remove heat (Q_{cold}) from an already cold region *or* as a heat pump adding heat (Q_{hot}) to an already warm area (the kitchen).

In essence, a refrigerator (or air conditioner) pumps heat up a temperature gradient, or "hill." (Think of pumping water up an actual hill against the force of gravity.) The cooling efficiency of this operation depends on the amount of heat extracted from the cold temperature reservoir (the freezer compartment or the inside of a house), Q_{cold}, compared to the work W_{in} needed to do so. Since a practical refrigerator operates in a cycle to provide a continuous removal of heat, $\Delta U = 0$ for the cycle. Then, by the conservation of energy (or first law), $Q_{cold} + W_{in} = Q_{hot}$, where Q_{hot} is the heat ejected to the high temperature reservoir or the outside.

The measure of a refrigerator or air conditioner's performance is defined differently from that of a heat engine, because of the difference in their functions. For these appliances, the efficiency is expressed as a **coefficient of performance (COP)**. Since the purpose is to extract the most heat (Q_{cold}) per unit work input (W_{in}), the coefficient of performance for a refrigerator (or air conditioner), COP_{ref}, is expressed as their ratio:

$$COP_{ref} = \frac{Q_{cold}}{W_{in}} = \frac{Q_{cold}}{Q_{hot} - Q_{cold}} \qquad \text{(for refrigerator or air conditioner)} \qquad (12.10)$$

Cooling with a refrigerator or air conditioner

where the first-law conservation relationship given above was used to express the input work in terms of the two heats.

Thus, the greater the COP, the better the performance—more heat is extracted for each unit of work done. For normal refrigerator operation, the work input is less than the heat removed, so the COP is greater than 1. The COP's of typical refrigerator cycles range from 3 to 5, depending on operating conditions. This means that the heat removed from the cold reservoir (the refrigerator freezer or the inside of a house) is 3 to 5 times the work needed to remove it.

Refrigerators and air conditioners are generically referred to as thermal pumps because they basically "pump" heat uphill. However, the term **heat pump** is now more specifically applied to the common commercial devices used to cool homes and offices in the summer and to heat them in the winter. The summer operation is that of an air conditioner, as described above. In this mode it cools the inside (of a house) and heats the outdoors. Operating in the heating mode, a heat pump heats the inside and cools the outdoors, usually by taking heat energy from the air. This may seem hard to believe, as it may be quite cold outside. However, keep in mind that the air has internal energy regardless of its temperature.

For a heat pump in its heating mode, the heat input is the major interest, so the COP is defined differently from that of a refrigerator or air conditioner. As you might guess, it is the ratio of Q_{hot} to W_{in} (heat you get out for work you put in):

$$COP_{hp} = \frac{Q_{hot}}{W_{in}} = \frac{Q_{hot}}{Q_{hot} - Q_{cold}} \qquad \text{(for heat pump in heating mode)} \qquad (12.11)$$

Heating with a heat pump

where again $Q_{cold} + W_{in} = Q_{hot}$. Typical COPs for heat pumps range between 2 and 4, depending on operating conditions.

Compared to electrical heating, heat pumps are efficient. For each unit of electric energy consumed, a heat pump typically pumps in from one and one-half to three times as much heat as direct electric heating systems provide. However, if the outside temperature is very low, a heat pump may not be efficient enough to heat a building adequately. Then it must be supplemented by a conventional heating system. Some heat pumps use water from underground reservoirs, wells, or buried loops of pipe as a heat source and sink. These are more efficient than the ones that use the outside air because water has a larger specific heat than air, and the average temperature difference between the water and the inside air is usually smaller.

Example 12.8 ■ **Air Conditioner/Heat Pump: Thermal Switch Hitting**

An air conditioner operating in summer extracts 100 J of heat from the interior of the house for every 40 J of electric energy required to operate it. Determine (a) its COP and (b) its COP if it is reversed in the winter, running as a heat pump to move the same amount of heat for the same amount of electric energy.

Thinking It Through. We know the input work and input heat in part (a), so we apply the definition of COP for a refrigerator (Eq. 12.10). Operating in reverse, it is the output heat that is important, so we must find this from the conservation of energy before we can compute the COP for operation as a heat pump.

Solution.

Given: $Q_{cold} = 100$ J *Find:* (a) COP_{ref}
$\quad\quad\quad W_{in} = 40$ J $\quad\quad\quad$ (b) COP_{hp}

(a) Using Eq. 12.10, we determine the COP for this engine operating as an air conditioner:

$$COP_{ref} = \frac{Q_{cold}}{W_{in}} = \frac{100 \text{ J}}{40 \text{ J}} = 2.5$$

(b) When the engine operates as a heat pump, the relevant heat is the output heat, which we can calculate from the conservation of energy:

$$Q_{hot} = Q_{cold} + W_{in} = 100 \text{ J} + 40 \text{ J} = 140 \text{ J}$$

Thus the COP for this engine operating as a heat pump in winter is, from Eq. 12.11,

$$COP_{hp} = \frac{Q_{hot}}{W_{in}} = \frac{140 \text{ J}}{40 \text{ J}} = 3.5$$

Follow-up Exercise. (a) Suppose you redesigned the engine in this Example to perform the same operation but with 25% less work input. What would be the new values of the two COPs? (b) Which COP would have the larger percentage increase?

12.5 The Carnot Cycle and Ideal Heat Engines

Objectives: To (a) explain how the Carnot cycle applies to heat engines, (b) compute the ideal Carnot efficiency, and (c) state the third law of thermodynamics.

Lord Kelvin's statement of the second law of thermodynamics says that any *cyclic* heat engine, regardless of its design, must always exhaust some heat energy (Section 12.4). But how much heat must be lost in the process? In other words, what is the *maximum* possible efficiency of a heat engine? In designing heat engines, engineers strive to make them as efficient as possible, but there must be some theoretical limit, and, according to the second law, it must be less than 100%.

Sadi Carnot (1796–1832), a French engineer, studied this limit. The first thing he sought was the thermodynamic cycle an *ideal* heat engine would use, that is, the most efficient cycle. Carnot found that the ideal heat engine absorbs heat from a *constant* high-temperature reservoir (T_{hot}) and exhausts it to a *constant* low-temperature reservoir (T_{cold}). These are ideally reversible isothermal processes and may be represented as two isotherms on a *p*-*V* diagram. But what are the processes that complete the cycle and make it the most efficient cycle operating between these temperature extremes? Carnot showed that these are reversible adiabatic processes. As we saw in Section 12.2, they are called adiabats when represented on a graph (•Fig. 12.13a).

Thus, the ideal **Carnot cycle** consists of two isotherms and two adiabats and is conveniently represented on a *T*-*S* diagram, where it forms a rectangle (Fig. 12.13b). The area under the upper isotherm (1–2) is the heat added to the system from the high-temperature reservoir: $Q_{hot} = T_{hot} \Delta S$. Similarly, the area

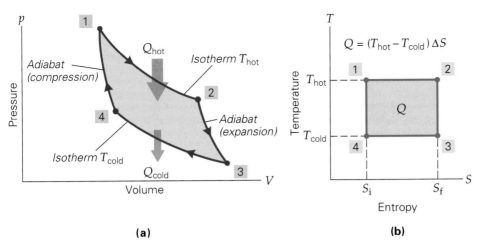

•FIGURE 12.13 The Carnot cycle
(a) The Carnot cycle consists of two isotherms and two adiabats. Heat is absorbed during the isothermal expansion and exhausted during the isothermal compression. (b) On a T-S diagram, the Carnot cycle forms a rectangle, the area of which is equal to Q.

(a)   **(b)**

under the lower isotherm (3–4) is the heat exhausted: $Q_{cold} = T_{cold} \Delta S$. Here, Q_{hot} and Q_{cold} are magnitudes of heat transfers at *constant* temperatures (T_{hot} and T_{cold}). There is no heat transfer ($Q = 0$) in the adiabatic legs of the cycle. (Why?) This is not typical of practical heat-engine cycles. See Exercises 68 and 69.

The difference between these heat transfers is the work output, which is equal to the area of the rectangle enclosed by the process paths (the shaded area on the diagram):

$$W_{net} = Q_{hot} - Q_{cold} = (T_{hot} - T_{cold}) \Delta S$$

Since ΔS is the same for the areas under both isotherms, these expressions can be used to relate the temperatures and heats. That is, since

$$Q_{hot} = T_{hot} \Delta S \quad \text{and} \quad Q_{cold} = T_{cold} \Delta S$$

then

$$\frac{Q_{hot}}{T_{hot}} = \frac{Q_{cold}}{T_{cold}} \quad \text{or} \quad \frac{Q_{hot}}{Q_{cold}} = \frac{T_{hot}}{T_{cold}} \tag{12.12}$$

This equation can be used to express the efficiency of an ideal heat engine in terms of temperature. From Eq. 12.9, this ideal **Carnot efficiency** (ϵ_C) is

$$\epsilon_C = 1 - \frac{Q_{cold}}{Q_{hot}} = 1 - \frac{T_{cold}}{T_{hot}}$$

or

$$\epsilon_C = 1 - \frac{T_{cold}}{T_{hot}} = \frac{T_{hot} - T_{cold}}{T_{hot}} \qquad \begin{array}{l} \textit{Carnot efficiency} \\ \textit{(for an ideal} \\ \textit{heat engine)} \end{array} \tag{12.13}$$

where the fractional efficiency is often expressed as a percentage. Note that T_{cold} and T_{hot} are Kelvin temperatures.

The Carnot efficiency expresses the theoretical upper limit on the thermodynamic efficiency of a cyclic heat engine; this limit can never be achieved. A true Carnot engine cannot be built because the necessary reversible processes can only be approximated. Although reversible adiabatic processes can be approached, reversible isothermal processes are virtually impossible to approximate during the heat transfer processes in a real engine.

However, the Carnot efficiency does show that the greater the difference in the temperatures of the heat reservoirs, the greater the efficiency. For example, if T_{hot} is twice T_{cold}, or $T_{cold}/T_{hot} = \frac{1}{2} = 0.50$, the Carnot efficiency will be

$$\epsilon_C = 1 - \frac{T_{cold}}{T_{hot}} = 1 - 0.50 = 0.50 \, (\times 100\%) = 50\%$$

Note: Keep in mind that the temperature in the Carnot efficiency is the absolute, or Kelvin, temperature.

Note: The Carnot efficiency, which can never be attained, gives an ideal upper limit.

But if T_{hot} is four times T_{cold}, or $T_{cold}/T_{hot} = \frac{1}{4} = 0.25$, then

$$\epsilon_C = 1 - \frac{T_{cold}}{T_{hot}} = 1 - 0.25 = 0.75 \, (\times 100\%) = 75\%$$

Since a heat engine can never attain 100% thermal efficiency, it is useful to compare its actual efficiency, ϵ_{th}, to its theoretical maximum efficiency, that of a Carnot cycle, ϵ_C. The ratio of these two efficiencies is sometimes called the engine's **relative efficiency**, ϵ_{rel}, defined as

$$\epsilon_{rel} = \frac{\epsilon_{th}}{\epsilon_C} \quad \text{(relative efficiency of a heat engine)} \quad (12.14)$$

We can express this quantity as a ratio or a percentage by multiplying by 100%. To see the importance of this concept in more detail, study the next Example carefully.

Example 12.9 ■ Carnot Efficiency: The Yardstick of Any Real Engine

An engineer is designing a cyclic heat engine to operate between temperatures of 150°C and 27°C. (a) What is the maximum theoretical efficiency that can be achieved? (b) Suppose the engine, when built, does 100 J of work per cycle for every 500 J of input heat per cycle. What is its relative efficiency?

Thinking It Through. The maximum efficiency for certain high and low temperatures is given by Eq. 12.13. We must convert temperatures to the Kelvin scale. In part (b), we need to calculate the actual efficiency and compare it to our answer in (a).

Solution.

Given: $T_{hot} = 150°C + 273 = 423$ K $\qquad$ *Find:* (a) ϵ_C (Carnot efficiency)
$\qquad\quad T_{cold} = 27°C + 273 = 300$ K $\qquad\qquad\quad$ (b) ϵ_{rel} (relative efficiency)
$\qquad\quad W_{net} = 100$ J
$\qquad\quad Q_{hot} = 500$ J

(a) Using Eq. 12.13, we get

$$\epsilon_C = 1 - \frac{T_{cold}}{T_{hot}} = 1 - \frac{300 \text{ K}}{423 \text{ K}} = 0.291 \, (\times 100\%) = 29.1\%$$

(b) The actual efficiency is, from Eq. 12.8,

$$\epsilon_{th} = \frac{W_{net}}{Q_{hot}} = \frac{100 \text{ J}}{500 \text{ J}} = 0.200, \quad \text{or} \quad 20.0\%$$

Thus

$$\epsilon_{rel} = \frac{\epsilon_{th}}{\epsilon_C} = \frac{0.200}{0.291} = 0.687, \quad \text{or} \quad 68.7\%$$

In other words, the engine is operating at almost 70% of its theoretical maximum. Not bad!

Follow-up Exercise. If the operating high temperature of the engine in this Example were increased to 200°C, what would be the change in the theoretical efficiency?

There are also Carnot COPs for refrigerators and heat pumps. (See Exercise 88.)

The Third Law of Thermodynamics

Another interesting conclusion can be drawn from the expression for the Carnot efficiency (Eq. 12.13). To have ϵ_C equal to 100%, T_{cold} would have to be absolute zero (see Section 10.3). Since a heat engine with 100% efficiency is impossible by the second law, the **third law of thermodynamics** is as follows:

It is impossible to reach a temperature of absolute zero.

Absolute zero has never been observed experimentally. If it were, this would in effect violate the second law. In cryogenic (low-temperature) experiments, scientists have come close to absolute zero—to about 20 nK (2.0×10^{-8} K)—but have never reached it.

Chapter Review

Important Terms

Important Concepts

- The state of a system is described by an equation of state. A process is a change in the state of a system. There are ideal reversible processes (changes through equilibrium states) and irreversible processes (changes through nonequilibrium states).

- The first law of thermodynamics is a statement of the conservation of energy for a thermodynamic system.

- The second law of thermodynamics specifies whether a process can take place naturally or the direction of a process.

- The total entropy of the universe increases in every natural process.

- A heat engine is a device that converts heat energy into work. Thermal efficiency is the ratio of work output to heat input.

- A thermal pump is a device that transfers heat energy from a low-temperature reservoir to a high-temperature reservoir. The coefficient of performance (COP) is the ratio of useful heat transferred to the work input required to do this.

Important Equations

First Law of Thermodynamics:

$$Q = \Delta U + W \tag{12.1}$$

$$\text{sign convention} \begin{cases} +Q, \text{ heat } added \text{ to system} \\ -Q, \text{ heat } removed \text{ from system} \\ +W, \text{ work done } by \text{ system} \\ -W, \text{ work done } on \text{ system} \end{cases}$$

Work Done by an Expanding Gas (constant pressure):

$$W = p\,\Delta V = p(V_2 - V_1) \tag{12.2}$$

Change in Entropy (constant temperature):

$$\Delta S = \frac{Q}{T} \tag{12.7}$$

Thermal Efficiency of a Heat Engine:

$$\epsilon_{th} = \frac{W_{net}}{Q_{hot}} = \frac{Q_{hot} - Q_{cold}}{Q_{hot}} = 1 - \frac{Q_{cold}}{Q_{hot}} \tag{12.9}$$

Coefficients of Performance:

$$COP_{ref} = \frac{Q_{cold}}{W_{in}} = \frac{Q_{cold}}{Q_{hot} - Q_{cold}} \tag{12.10}$$

(for refrigerator or air conditioner)

$$COP_{hp} = \frac{Q_{hot}}{W_{in}} = \frac{Q_{hot}}{Q_{hot} - Q_{cold}} \tag{12.11}$$

(for heat pump in heating mode)

Carnot Efficiency of an Ideal Heat Engine:

$$\epsilon_C = 1 - \frac{T_{cold}}{T_{hot}} = \frac{T_{hot} - T_{cold}}{T_{hot}} \tag{12.13}$$

Exercises*

12.1 Thermodynamic Systems, States, and Processes

1. There may be an exchange of heat with the surroundings for (a) a thermally isolated system, (b) a completely isolated system, (c) a heat reservoir, (d) none of these.

2. Only initial and final states are known for irreversible processes on (a) p-V diagrams, (b) p-T diagrams, (c) V-T diagrams, (d) all of these.

3. On a p-V diagram, a reversible process is one (a) whose path is known, (b) whose path is unknown, (c) for which the intermediate steps are nonequilibrium states, (d) none of these.

4. 🔘 Explain why the p-V diagram in Fig. 12.1b is not that of an ideal gas.

12.2 The First Law of Thermodynamics

5. 🔘 On a p-V diagram, sketch a general cyclic process that consists of an isothermal expansion, an isobaric compression, and an isometric process—in that order.

6. 🔘 On a p-T diagram, sketch the general paths for the following reversible processes for an ideal gas: (a) isothermal, (b) isobaric, and (c) isometric.

7. There is no exchange of heat in an (a) isothermal process, (b) adiabatic process, (c) isobaric process, (d) isometric process.

8. 🔘 According to the first law of thermodynamics, if work is done on a system, then (a) the internal energy of the system must change, (b) heat must be transferred from the system, (c) the internal energy of the system changes and/or heat is transferred from the system, (d) heat is transferred to the system.

9. 🔘 If heat is added to a system of ideal gas during an isothermal expansion process, (a) work is done on the system, (b) the internal energy decreases, (c) the effect is the same as for an isometric process, (d) none of these.

10. (a) Why does the body of a hand pump become hot when a tire is pumped up? (Neglect friction. The plunger is usually lubricated to prevent air leakage.) (b) Why does the valve of a tire become cold when air is released from the tire?

11. In •Fig. 12.14, the plunger is pushed in quickly in a syringe and the small pieces of paper in the syringe catch fire. Explain by using the first law of thermodynamics.

*Take temperatures to be exact—that is, with no uncertainty.

(Similarly, in a diesel engine, there are no spark plugs. How can the air–fuel mixture be ignited?)

•**FIGURE 12.14 Fire syringe** See Exercise 11.

12. ■ 🔘 A rigid container contains 1 mol of ideal gas that slowly receives 2.0×10^4 J of heat. What is the change in the internal energy of the gas in joules?

13. ■ A quantity of ideal gas goes through a cyclic process and does 400 J of net work. (a) Is a net amount of heat added to or removed from the system, and how much heat is involved? (b) Does the temperature of the gas increase or decrease over the cycle?

14. ■ 🔘 The internal energy of an ideal gas system increases by 20 J, while 50 J of heat is removed from the system. Does the system do work, or is work done on the system? How much work is involved?

15. ■ 🔘 While doing 500 J of work, a system of ideal gas is expanded adiabatically to 1.5 times its volume. (a) Does the temperature of the gas increase or decrease? (b) How much heat is transferred? (c) What is the change in the internal energy of the gas?

16. ■ An ideal gas system expands from 1.0 m³ to 3.0 m³ at STP (standard temperature and pressure). If the system absorbs 5.0×10^4 J of heat in the process, what is the change in internal energy of the system?

17. ■■ 🔘 A system of gas at low density has an initial pressure of 1.65×10^4 Pa and occupies a volume of 0.20 m³. The slow addition of 1000 J of heat to the system causes it to expand isobarically to a volume of 0.40 m³. (a) How much work is done by the system in the process? (b) Does the internal energy of the system change? If so, by how much?

18. ■■ An Olympic weight lifter lifts 145 kg a vertical distance of 2.1 m. In doing so, his internal energy decreases by 6.0×10^4 J. Consider the weight lifter to be a thermodynamic system. How much heat flows and in what direction?

19. ■■ A system of ideal gas is taken through the reversible processes shown in •Fig. 12.15. In terms of state variables, (a) what is the overall change in the internal energy of the gas? (b) How much work is done by or on the gas? (c) What is the amount of heat transfer in the overall process?

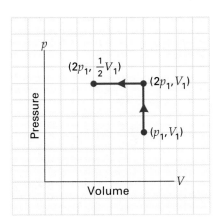

•**FIGURE 12.15   A *p-V* diagram for an ideal gas**   See Exercise 19.

20. ■■ The temperature of 2.0 mol of ideal gas is increased from 150°C to 250°C by two different process paths. In one, 2500 J of heat is added to the system; in the other, 3000 J of heat is added. (a) What is the change in the internal energy of the gas in each case? [Hint: See Eq. 10.16.] (b) In which case is more work done, and how much more?

21. ■■ A gram of water (1.00 cm³) at 100°C is converted to a volume of 1671 cm³ of steam at atmospheric pressure. What is the change in the internal energy of the system?

22. ■■ ⊙ A fixed quantity of gas undergoes the reversible changes illustrated in the *p-V* diagram in •Fig. 12.16. How much work is done in each process?

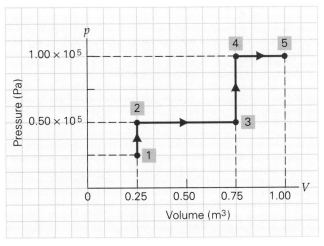

•**FIGURE 12.16   A *p-V* diagram and work**   See Exercises 22 and 23.

23. ■■ ⊙ Suppose that after the final process shown in Fig. 12.16 (see Exercise 22), the pressure of the gas is decreased isometrically from 1.0×10^5 Pa to 0.70×10^5 Pa, and then the gas is compressed isobarically from 1.0 m³ to 0.80 m³. What is the total work done in all of these processes, including 1 through 5?

24. ■■■ A quantity of gas is compressed as shown on the *p-V* diagram in •Fig. 12.17. How much work is done on the system?

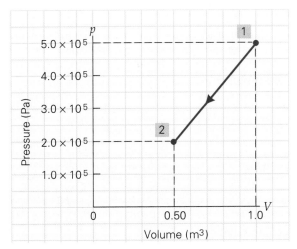

•**FIGURE 12.17   A variable *p-V* process and work**   See Exercise 24.

25. ■■■ ⊙ One mole of ideal gas is taken through the cyclic process shown in •Fig. 12.18. (a) Compute the work involved (*W*) for each of the four processes. (b) Find ΔU, *W*, and *Q* for the complete cycle. (c) What is T_3?

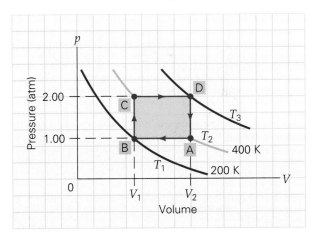

•**FIGURE 12.18   A cyclic process**   See Exercise 25.

12.3 The Second Law of Thermodynamics and Entropy

26. The second law of thermodynamics (a) describes the state of a system, (b) applies only when the first law is satisfied, (c) precludes perpetual motion machines, (d) does not apply to an isolated system.

27. In any natural process, the change in the entropy of the universe is (a) negative, (b) positive, (c) zero, (d) the same as the change in the internal energy.

28. Do the entropies of the systems in the following processes increase or decrease? (a) *ice* melts, (b) *water vapor* condenses, (c) *water* is heated on a stove, (d) *food* is cooled in a refrigerator.

29. In Example 12.4, what would be the implication if the total change in entropy had been negative?

30. When a quantity of hot water is mixed with a quantity of cold water, the combined system comes to thermal equilibrium at some intermediate temperature. How does the entropy of the system change?

31. ■ What change in entropy is associated with the reversible phase change of 1.0 kg of ice to water at 0°C?

32. ■ What is the change in entropy when 0.50 kg of steam condenses to water at 100°C?

33. ■ What is the change in entropy when 0.50 kg of mercury vapor ($L_v = 2.7 \times 10^5$ J/kg) condenses to a liquid at its boiling point of 357°C?

34. ■ Which process has the greater change in entropy: 0.75 kg of ice changing to water at 0°C, or 0.25 kg of water changing to steam at 100°C?

35. ■ One mole of ideal gas goes through an isothermal compression at room temperature. If 7.5×10^3 J of work is done in compressing the gas, what is the change in entropy of the system?

36. ■■ A quantity of an ideal gas initially at STP undergoes a reversible isothermal expansion and does 3.0×10^3 J of work on its surroundings in the process. What is the change in the entropy of the gas?

37. ■■ During a liquid-to-solid phase change of a quantity of a substance, the change in entropy is -4.19×10^3 J/K. If 1.67×10^6 J of heat is removed in the process, what is the freezing point of the substance in degrees Celsius?

38. ■■ An isolated system consists of two reservoirs with temperatures 373 K and 273 K, respectively. If 1000 J of heat were to flow from the cold reservoir to the hot reservoir spontaneously, (a) what would be the change in entropy of the hot reservoir? (b) of the cold reservoir? (c) What would be the total change in entropy of the isolated system? (d) Could the process take place naturally?

39. ■■ Two heat reservoirs at temperatures 200°C and 60°C, respectively, are brought into thermal contact, and 1.50×10^3 J of heat spontaneously flows from one to the other. What is the change in the entropy of the two-reservoir system?

40. ■■ A system of ideal gas at 20°C undergoes an isometric process with an internal energy decrease of 4.50×10^3 J to a final temperature of 16°C. What is the approximate change in entropy in joules per kelvin?

41. ■■ (a) How much heat is transferred in the processes shown on the T-S diagram in •Fig. 12.19? (b) What type of process takes place as the system goes from 2 to 3?

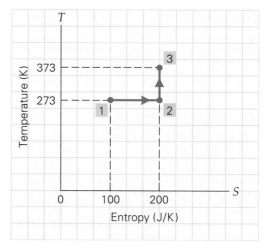

•**FIGURE 12.19 Entropy and heat** See Exercises 41 and 42.

42. ■■ Suppose that the system described by the T-S diagram in Fig. 12.19 is returned to its original state, 1, by a reversible process depicted by a straight line from 3 to 1. What is the change in entropy of the total cycle? How much heat is transferred in the cyclic process?

43. ■■■ A 50.0-gram ice cube at 0°C is placed in 500 mL of water at 20°C. Estimate the change in entropy when all the ice has melted (a) for the ice, (b) for the water, and (c) for the ice–water system.

12.4 Heat Engines and Thermal Pumps*

44. For a cyclic heat engine, (a) $\epsilon_{th} > 1$, (b) $Q_{hot} = W_{out}$, (c) $\Delta U = W_{out}$, (d) $Q_{hot} > Q_{cold}$.

45. A thermal pump (a) is rated by thermal efficiency, (b) requires work input, (c) is not consistent with the second law, (d) violates the first law.

46. Is leaving a refrigerator door open a practical way to air condition a room? Explain.

47. Lord Kelvin's statement of the second law ("No heat engine operating in a cycle can convert its heat input completely to work") as applied to heat engines refers to their operating *in a cycle*. Why is the italicized phrase included?

48. The energy output of a thermal pump is greater than the energy used to operate the pump. Is this device a violation of the first law of thermodynamics?

*Consider efficiencies to be exact.

49. In normal atmospheric convection cycles, colder air from a higher altitude is transferred to a lower, warmer level. Does this process violate the second law? Explain.

50. ■ A gasoline engine is 28% efficient. If the engine absorbs 2000 J of heat in each cycle, (a) what is the work output in each cycle? (b) How much heat is exhausted in each cycle?

51. ■ If an engine does 200 J of work and exhausts 600 J of heat per cycle, what is its efficiency?

52. ■ A heat engine with an efficiency of 40% does 800 J of work each cycle. How much heat per cycle is lost to the surroundings (the low-temperature reservoir)?

53. ■ An internal combustion engine with a thermal efficiency of 15.0% does 2.60×10^4 J of work each cycle. How much heat is lost by the engine in each cycle?

54. ■ The heat output of a particular engine is 7.5×10^3 J per cycle, and the work output is 4.0×10^3 J per cycle. What is the thermal efficiency of the engine?

55. ■■ A steam engine does 4500 J of useful work each cycle but loses 500 J to friction and exhausts 6300 J of heat energy. What is the engine's efficiency?

56. ■■ An engineer redesigns a heat engine and improves its thermal efficiency from 20% to 25%. (a) Does the ratio of the heat output to heat input increase or decrease, and why? (b) What is the change in Q_{cold}/Q_{hot}?

57. ■■ A gasoline engine burns gas as it releases 3.3×10^8 J of heat per hour. (a) What is the energy input during a period of 2.0 h? (b) If the engine delivers 25 kW of power during this time, what is its efficiency?

58. ■■ A refrigerator takes heat from its cold interior at a rate of 7.5 kW when the required work is done at a rate of 2.5 kW. At what rate is heat exhausted to the kitchen?

59. ■■ A refrigerator with a COP of 2.2 removes 4.2×10^5 J of heat from its storage area each cycle. (a) How much heat is exhausted each cycle? (b) What is the total work input in joules for 10 cycles?

60. ■■ A heat pump removes 2.0×10^3 J of heat from the outdoors and delivers 3.5×10^3 J of heat to the inside of a house each cycle. (a) How much work is required per cycle? (b) What is the COP of the pump?

61. ■■ An air conditioner has a COP of 2.75. What is its power rating if it is to remove 1.00×10^7 J of heat in 20 min?

62. ■■ A heat engine has a thermal efficiency of 30.0%. Its heat input each cycle is supplied by the condensation of 8.00 kg of steam at 100°C. (a) What is the work output per cycle? (b) How much heat is lost to the surroundings each cycle?

63. ■■ A heat pump that uses an underground water reservoir as a heat source extracts 2.1×10^5 J each cycle while requiring 3.0×10^4 J of work. (a) How much heat is delivered to the inside of the house? (b) What is the temperature drop of the water when returned to the reservoir if 5.0 kg is used each cycle?

64. ■■■ A coal-fired power plant produces 900 MW of electricity and operates at an overall thermal efficiency of 35%. (a) What is the thermal input (power) to the plant? (b) What is the rate of heat discharge from the plant? (c) Why is the water heated by the discharge cooled in a cooling tower before being ejected into a nearby river?

65. ■■■ A four-stroke engine runs on the Otto cycle and delivers 150 hp. If the efficiency of the engine is 20%, what is (a) the rate of energy input? (b) the rate of heat energy exhausted to the environment?

12.5 The Carnot Cycle and Ideal Heat Engines

66. ◉ The Carnot cycle consists of (a) two isobaric and two isothermal processes, (b) two isometric and two adiabatic processes, (c) two adiabatic and two isothermal processes, (d) four arbitrary processes that return the system to its initial state.

67. Which of the following temperature reservoir relationships would have the highest efficiency for a Carnot engine? (a) $T_{cold} = 0.15 T_{hot}$, (b) $T_{cold} = 0.25 T_{hot}$, (c) $T_{cold} = 0.50 T_{hot}$, (d) $T_{cold} = 0.90 T_{hot}$

68. An idealized cycle for a two-stroke internal combustion engine is shown in •Fig. 12.20. It consists of an adiabatic expansion and compression (legs 1 and 3, respectively) and an isometric decompression and compression (legs 2 and 4, respectively). The text stated that in a Carnot cycle, heat transfers occur at constant temperatures. For the cycle in Fig. 12.20, (a) is heat taken in at a constant temperature? (b) Is heat rejected at a constant temperature? (c) Draw a sketch that indicates the areas representing work (W, with

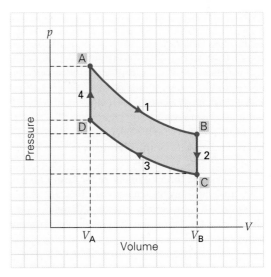

•**FIGURE 12.20 Non-Carnot cycle** Two adiabats and two isomets. See Exercises 68 and 69. (Not drawn to scale.)

signs) during expansion and compression, and draw another sketch for the net work.

69. In Fig. 12.20, legs 1 and 3 are adiabatic processes. Suppose these were isothermal processes instead. (a) Does this change make the cycle a Carnot cycle? Explain. (b) In which legs are there heat transfers? [Indicate directions by signs.] Do these transfers occur at constant temperatures?

70. Automobile engines can be either air-cooled or water-cooled. Which type of engine would you expect to be more efficient, and why?

71. ■ Diesel engines are more efficient than gasoline engines. Which type of engine runs hotter? Why?

72. ■ A steam engine operates between 100°C and 20°C. What is its Carnot efficiency?

73. ■ A Carnot engine with an efficiency of 35% takes in heat from a high-temperature reservoir at 147°C. What is the Celsius temperature of the low-temperature reservoir?

74. ■ What is the temperature of the hot reservoir of a 30% efficient Carnot engine with a 20°C cold reservoir?

75. ■ It has been proposed that the temperature difference in the ocean could be utilized to run a heat engine to generate electricity. In tropical regions, the water temperature is about 25°C at the surface and about 5°C at depth. (a) What would be the maximum theoretical efficiency of such an engine? (b) Would a heat engine with such a low efficiency be practical? Explain.

76. ■ An engineer wants to run a heat engine with an efficiency of 40% between a high-temperature reservoir at 350°C and a low-temperature reservoir. Below what temperature must the low-temperature reservoir be for practical operation of the engine?

77. ■■ An ideal heat engine takes 2.7×10^4 J of heat per cycle from a high-temperature reservoir at 320°C and exhausts some of it to a low-temperature reservoir at 120°C. How much net work is done by the engine per cycle?

78. ■■ A Carnot engine of efficiency 40% operates with a low-temperature reservoir at 50°C and exhausts 1200 J of heat each cycle. What are (a) the heat input per cycle and (b) the Celsius temperature of the high-temperature reservoir?

79. ■■ An ideal heat engine takes in heat from a reservoir at 327°C and has an efficiency of 30%. If the exhaust temperature does not vary and the efficiency is increased to 40%, what would be the increase in the temperature of the hot reservoir?

80. ■■ An inventor claims to have developed a heat engine that, on each cycle, takes in 5.0×10^5 J of heat from a high-temperature reservoir at 400°C and exhausts 2.0×10^5 J

to the surroundings at 125°C. Would you invest your money in the production of this engine? Explain.

81. ■■ Equation 12.13 shows that the greater the temperature difference between the reservoirs of a heat engine, the greater the Carnot efficiency. Suppose you had the choice of raising the high-temperature reservoir a certain number of kelvins or lowering the low-temperature reservoir the same number of kelvins. Which would you choose to increase the efficiency? Prove your answer numerically.

82. ■■ A heat engine operates at a relative efficiency of 45% (ϵ_{rel}). If the temperature of the high-temperature and low-temperature reservoirs are 400°C and 100°C, respectively, what are the Carnot efficiency ϵ_C and the thermal efficiency ϵ_{th} of the engine?

83. ■■ A heat engine has half the thermal efficiency (ϵ_{th}) of a Carnot engine operating between temperatures of 100°C and 375°C. (a) What is the relative efficiency of the heat engine? (b) If the heat engine absorbs heat at a rate of 50 kW, at what rate is heat exhausted?

84. ■■ The working substance of a cyclic heat engine is 0.75 kg of an ideal gas. The cycle consists of two isobaric processes and two isometric processes as shown in •Fig. 12.21. What would be the efficiency of a Carnot engine operating with the same high-temperature and low-temperature reservoirs?

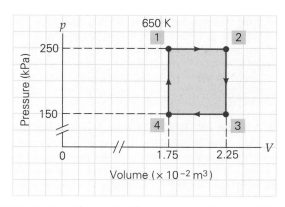

•FIGURE 12.21 **Thermal efficiency** See Exercise 84.

85. ■■ In each cycle, a Carnot engine takes 800 J of heat from a high-temperature reservoir and discharges 600 J to a low-temperature reservoir. (a) What is the Carnot efficiency of the engine? (b) What is the temperature ratio of the high-temperature reservoir to the low-temperature reservoir?

86. ■■ A Carnot engine operating between reservoirs at 27°C and 227°C does 1500 J of work each cycle. (a) What is the efficiency of the engine? (b) What is the change in entropy of the engine each cycle?

87. ■■ Because of material limitations, the maximum temperature of the superheated steam used in a turbine for electrical generation is about 540°C. (a) If the steam condenser operates at 20°C, what is the ideal efficiency? (b) The actual efficiency is about 35–40%. What does this tell you?

88. ■■■ There is a Carnot coefficient of performance (COP_C) for an ideal or Carnot heat pump. (a) Show that this quantity is given by

$$COP_C = \frac{T_{hot}}{T_{hot} - T_{cold}}$$

(b) What does this quantity tell you about adjusting the temperatures for maximum performance of a heat pump? (Could you guess the equation for the COP_C for a refrigerator?)

89. ■■■ A salesperson tells you that a new refrigerator with a high COP removes 2.6×10^3 J each cycle from the inside of the refrigerator at a temperature of 5.0°C and rejects 2.8×10^3 J into the 30°C kitchen. (a) What is the refrigerator's COP? (b) Is this scenario possible? Justify your answer. (See Exercise 88.)

90. ■■■ A Carnot engine has an efficiency of 40%. Suppose it could be run in reverse as a heat pump. What would be the COP_C of the pump? (See Exercise 88.)

91. ■■■ An ideal heat pump is equivalent to a Carnot engine running in reverse. Show that the input work required for an ideal heat pump is given by

$$W_{in} = Q_{cold}\left[\left(\frac{T_{hot}}{T_{cold}}\right) - 1\right]$$

Additional Exercises

92. 🔘 A thermally isolated quantity of gas with an initial volume of 10 L is compressed isobarically at a pressure of 300 kPa. If 900 J of work is done on the system, what is the final volume of the gas?

93. For the two-step reversible process illustrated in •Fig. 12.22, the same amount of heat is added to and removed from the system. Prove that the total entropy (of the universe) increases in the process.

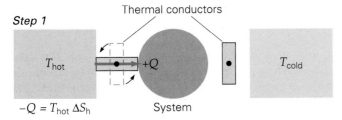

Step 1

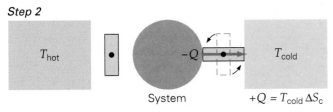

Step 2

•**FIGURE 12.22   Entropy increase?**   See Exercise 93.

94. Use the general definition of efficiency of a heat engine, $\epsilon_{th} = W_{net}/Q_{in}$, to derive the Carnot efficiency. [Hint: Consider the cycle in terms of entropy, and see Fig. 12.13b.]

95. 🔘 A gas is enclosed in a cylindrical piston with a radius of 12.0 cm. Heat is slowly added to the system while the pressure is maintained at 1.00 atm, and the piston moves 6.00 cm. (a) What type of process is this? (b) If the quantity of heat transferred to the system during the expansion is 420 J, what is the change in the internal energy of the gas?

96. A heat engine operates between a reservoir at 20°C and one at 250°C. What is the maximum theoretical efficiency of the engine? Could this efficiency ever be achieved?

97. Two automobiles with equal masses of 1500 kg are traveling at 90 km/h when they have a head-on, totally inelastic collision at 20°C. Estimate the change in entropy of this process. Assume all the kinetic energy is converted to heat.

98. The energy or useful work lost as a result of a change in the entropy of the universe is given by $W = T_{cold}\Delta S_u$, where T_{cold} is the temperature of the cold reservoir for the process and S_u is the entropy of the universe. (a) Show that for heat (Q) transferred from a high-temperature reservoir to a low-temperature reservoir, $W = Q[1 - (T_{cold}/T_{hot})]$. (b) What is the quantity within the brackets?

99. In an isothermal expansion at 27°C, an ideal gas does 30 J of work. What is the change in entropy of the gas?

100. An ideal gas in a cylinder is compressed adiabatically while 1500 J of work is done in moving a piston in the cylinder. What is the change in the internal energy of the gas? Does the temperature of the gas increase or decrease?

101. In a thermodynamic process, 2500 J of heat is transferred to a system and 1000 J of work is done on the system. What is the change in the internal energy of the system?

102. A heat reservoir supplies heat to melt 0.75 kg of ice at 0°C. Once all of the ice has melted, what is the entropy change?

103. In 1834 Joule performed an important experiment on the internal energy of a gas. Two containers connected by a valve are submerged in a water bath (•Fig. 12.23). Container A is filled with a gas at high pressure, and container B is empty (evacuated). If the valve is opened quickly, the gas in container A undergoes "free expansion" into the evacuated container (in other words, the gas does no work in pushing against "a vacuum"). The temperature of the water bath is the same after the gas has again come to equilibrium. Does the internal energy of the gas change?

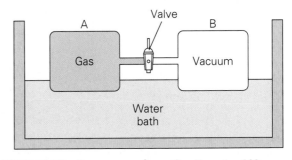

•**FIGURE 12.23   Free expansion**   See Exercise 103.

Vibrations and Waves

Insights

- Earthquakes, Seismic Waves, and Seismology

- Desirable and Undesirable Resonances

This is what a lot of people probably think of first when they hear the word *wave*. We're all familiar with ocean waves or their smaller relatives, the ripples that form on the surface of a lake or pond when something disturbs the surface. Yet in many ways, the waves that are most important to us, as well as most interesting to physicists, are either invisible or don't look like waves. Sound, for example, is a wave. As a wave traveling through a medium, it shares many properties with water waves. Perhaps most surprisingly, light is a wave. In fact, all electromagnetic radiations are waves—radio waves, microwaves, X-rays, and so on. Whenever you peer through a microscope or put on a pair of glasses or look up at a rainbow, you are experiencing wave energy in the form of light. In Chapter 28 you'll learn how even moving particles have wavelike properties. But first, we need to look at the basic description of waves.

What is a wave, and what properties do waves of *all* kinds have in common? You will find that to understand wave motion, it is necessary to study some other kinds of motion that at first might not seem closely related to waves. It turns out, however, that the motion of a weight bouncing up and down on a spring or the swinging of a pendulum in a grandfather clock provides important keys to an understanding of wave motion, whether it be that of an electron, a Slinky®, or the perfect wave that surfers dream about.

A vibration or oscillation involves back-and-forth motion, such as that of a swinging pendulum. Another example is a ball in a round-bottomed bowl. If the ball is displaced from its equilibrium position, it will roll back and forth, or oscillate, and finally come to rest at the equilibrium position, where its potential energy is lowest.

Recall from Chapter 8 that the ball must be oscillating about a point of stable equilibrium, for which there is a restoring force or torque. This is true in general for objects that undergo vibrating or oscillating motions. In a material medium, the restoring force is provided by intermolecular forces. If a molecule is disturbed, restoring forces exerted by its neighbors tend to return it to its original position, and it begins to oscillate. In so doing, it affects the adjacent molecules, which are in turn set into oscillation.

In this situation, we can see that something happens at point A at time t_1, and this causes a similar happening at a point B at a later time t_2, and so on. This is referred to as *propagation*. But what is propagated by the molecules in a material? It is energy. A single disturbance, which happens when you give the end of a stretched rope a quick shake, gives rise to what is referred to as a *wave pulse*. A continuous, repetitive disturbance gives rise to a continuous propagation of energy that we call wave motion.

13.1 Simple Harmonic Motion

Objectives: **To (a) describe simple harmonic motion and (b) relate energy and speed in such motion.**

The motion of an oscillating object depends on the restoring force that makes it go back and forth. It is convenient to begin to study such motion by considering the simplest type of force acting along the x axis: a force that is directly proportional to the displacement from equilibrium. One such force is the (ideal) spring force, described by **Hooke's law**,

$$F_s = -kx \qquad (13.1)$$

where k is the spring constant. Recall from Chapter 5 that the minus sign indicates that the force always points in the opposite direction to the displacement, $F_s = k(-x)$. That is, it always tends to restore the spring to its equilibrium position.

Suppose that an object on a horizontal frictionless surface is connected to a spring as shown in •Fig. 13.1. When the object is displaced to one side of its equilibrium position and released, it will move back and forth—that is, it will vibrate or oscillate. Here, an oscillation or a vibration is clearly a *periodic motion*, a motion that repeats itself again and again along the same path. For linear oscillations, like those of an object attached to a spring, the path may be back and forth or up and down. For the angular oscillation of a pendulum, the path is back and forth along a circular arc.

Motion under the influence of the type of force described by Hooke's law is called **simple harmonic motion (SHM)**, because the force is the simplest restoring force and because the motion can be described by harmonic functions (sines and cosines), as you will see later in the chapter. The directed distance of the object in SHM from its equilibrium position is its **displacement**. Note in Fig. 13.1 that the displacement can be either positive or negative ($+x$ or $-x$), which indicates direction. The maximum displacements are $+A$ and $-A$ (Fig. 13.1b,d).

The magnitude of the maximum displacement, or the maximum distance of an object from its equilibrium position, is called the **amplitude** (A). It is a scalar quantity that expresses the distance for both extreme displacements.

Besides the amplitude, two other important quantities for describing an oscillation are its period and frequency. The **period** (T) is the time needed to complete one cycle of motion. A cycle is a *complete* round trip, or motion through a complete

Note: Hooke's law (for an ideal spring) is discussed in Section 5.2.

Note: Here $+x$ signifies spring extension and $-x$ means compression.

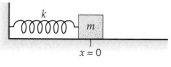

(a) Equilibrium

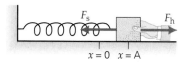

(b) $t = 0$ Just before release

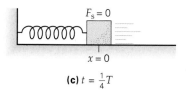

(c) $t = \frac{1}{4}T$

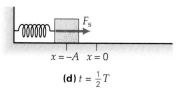

(d) $t = \frac{1}{2}T$

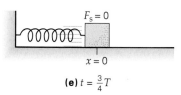

(e) $t = \frac{3}{4}T$

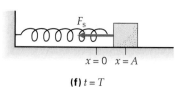

(f) $t = T$

•**FIGURE 13.1 Simple harmonic motion (SHM)** When displaced from **(a)** its equilibrium position $x = 0$ and **(b)** released, the object on a spring undergoes SHM (assuming no frictional losses). The time it takes to complete one cycle is the period of oscillation (T). (Here F_s is the spring force and F_h is the force the hand applies.) **(c)** At $t = T/4$, the object is back at its equilibrium position; **(d)** at $t = T/2$, it is at $x = -A$. **(e)** During the next half cycle, the motion is to the right; **(f)** at $t = T$, the object is back at its initial ($t = 0$) starting position.

TABLE 13.1 Terms Used to Describe SHM

displacement—directed distance of an object ($\pm x$) from its equilibrium position.
amplitude (A)—magnitude of the maximum displacement or the maximum distance of an object from its equilibrium position.
period (T)—the time for one *complete* cycle of motion.
frequency (f)—the number of cycles per second (in hertz or inverse seconds, where $f = 1/T$).

---

oscillation. For example, if an object starts at $x = A$ (Fig. 13.1b), when it returns to $x = A$ (as in Fig. 13.1f), it will have completed one cycle in a time we call one period. If an object were initially at $x = 0$ when disturbed, then its second return to this point would mark a cycle. In either case, the object would travel a distance of $4A$ during one cycle. Can you show this? (Why a *second* return?)

The **frequency** (f) is the number of cycles per second. The frequency and the period are related by

$$f = \frac{1}{T} \qquad \begin{array}{l}\textit{frequency}\\\textit{and period}\end{array} \qquad (13.2)$$

SI unit of frequency: hertz (Hz), or cycle per second (cycle/s)

The inverse relationship is reflected in the units: The period is the number of seconds per cycle, and the frequency is the number of cycles per second. For example, if $T = \frac{1}{2}$ s/cycle, then $f = 2$ cycles/s.

The standard unit of frequency is the **hertz (Hz)**, which is one cycle per second.* From Eq. 13.2, frequency has the unit inverse seconds (1/s, or s^{-1}), since the period is a measure of time. Although cycle is not really a unit, you might find it convenient at times to express frequency in cycles per second to help with unit analysis. This is similar to the way the radian (rad) is used in the description of circular motion in Sections 7.1 and 7.2.

The preceding terms used to describe SHM are summarized in Table 13.1.

Energy and Speed of a Spring–Mass System in SHM

Recall from Chapter 5 that the potential energy stored in a spring stretched or compressed a distance $\pm x$ from equilibrium is given by

$$U = \tfrac{1}{2}kx^2 \qquad (13.3)$$

The *change* in potential energy of an object oscillating on a spring is related to the work done by the spring force, as illustrated and explained in •Fig. 13.2. (Compare Fig. 5.5, Section 5.2.) An object with mass m oscillating on a spring also has kinetic energy. The kinetic and potential energies together give the total mechanical energy of the system:

$$E = K + U = \tfrac{1}{2}mv^2 + \tfrac{1}{2}kx^2 \qquad (13.4)$$

When the object is at one of its maximum displacements, $+A$ or $-A$, it is instantaneously at rest ($v = 0$). Thus, all the energy is in the form of potential energy at this location; that is,

$$E = \tfrac{1}{2}m(0)^2 + \tfrac{1}{2}k(\pm A)^2 = \tfrac{1}{2}kA^2$$

or

$$E = \tfrac{1}{2}kA^2 \qquad \begin{array}{l}\textit{total energy in}\\\textit{SHM of a spring}\end{array} \qquad (13.5)$$

---

*The unit is named for Heinrich Hertz (1857–1894), a German physicist and early investigator of electromagnetic waves.

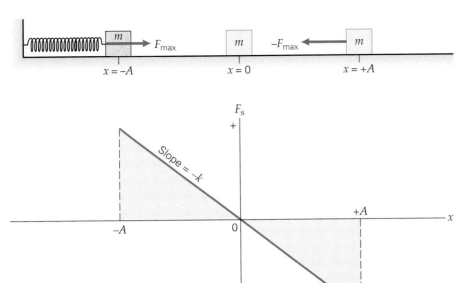

•FIGURE 13.2 Work and potential energy   The work done *by the spring force* on the oscillating object is equal to the negative change in potential energy $(-\Delta U)$. The magnitude of this change is equal to the area under the spring force (F_s) versus displacement (x) curve (where $F_s = -kx$). The work can be positive or negative. For example, in going from $x = 0$ to $x = A$, the work is negative (F_s and x in opposite directions) and U increases. (Hence the minus sign in $W_F = -\Delta U$.) But, going back from $x = A$ to $x = 0$, the work done by the spring is positive (why?), and U decreases by the same amount. Thus, the overall ΔU is zero. (How about going from $x = 0$ to $x = -A$ and back—or halfway back?)

This result is a general result for SHM:

> The total energy of an object in simple harmonic motion is directly proportional to the square of the amplitude of the motion.

Equation 13.5 allows us to express the velocity of an object oscillating on a spring as a function of position:

$$E = K + U \quad \text{or} \quad \tfrac{1}{2}kA^2 = \tfrac{1}{2}mv^2 + \tfrac{1}{2}kx^2$$

Rearranging, we get

$$v^2 = \frac{k}{m}(A^2 - x^2)$$

so

$$v = \pm\sqrt{\frac{k}{m}(A^2 - x^2)} \qquad (13.6)$$

Note: This discussion will be limited to light springs, the mass of which can be considered negligible.

Velocity of an object in SHM

where the $\pm$ indicates velocity direction. Note that at $x = \pm A$, the velocity is zero since the object is instantaneously at rest when at its maximum displacement from equilibrium.

Note that when the oscillating object passes through the equilibrium position $(x = 0)$, the potential energy is zero. At that instant, all the energy is kinetic and the object is traveling at its maximum speed, v_{max} (Fig. 13.1c,e). The energy expression for this case is

$$E = \tfrac{1}{2}kA^2 = \tfrac{1}{2}mv_{max}^2$$

and thus

$$v_{max} = \sqrt{\frac{k}{m}}\,A \qquad \begin{array}{l}\textit{maximum}\\ \textit{speed of mass}\\ \textit{on a spring}\end{array} \qquad (13.7)$$

Consider the use of energy methods in the following Example. You can visualize the continuous tradeoff between kinetic and potential energy in Learn by Drawing on p. 422.

Oscillating in a Parabolic Potential Well

A way to visualize conservation of energy for simple harmonic motion is shown in Fig. 1. The potential energy of a spring–mass system can be sketched on a plot of energy (E) versus position (x). Since $U = \frac{1}{2}kx^2 \propto x^2$, the graph is a *parabola*.

In the absence of nonconservative forces, the total energy of the system, E, is constant. But E is the sum of the kinetic and potential energies. During the oscillations, there is a continuous tradeoff between the two types of energies, but their sum remains constant. Mathematically this is written as $E = K + U$. In Fig. 2, U (shown as a blue arrow) is represented by the vertical distance from the x axis. Since E is constant, independent of x, it will plot as a horizontal line (shown in green). The kinetic energy is the piece of the total energy that is *not* potential energy, that is, $K = E - U$; it can be graphically interpreted (purple arrow) as the vertical distance between the potential energy parabola and the horizontal green total energy line. As the object oscillates on the x axis, the energy tradeoffs can be visualized as the lengths of the two arrows change.

A general location, x_1, is shown in Fig. 2. Neither the kinetic energy nor the potential energy is at its maximum value of E there. These maximum values occur instead at $x = 0$ and $x = \pm A$, respectively. The motion cannot exceed $x = \pm A$ because that would imply a negative kinetic energy, which is physically impossible (why?). The amplitude positions are sometimes called the *endpoints* of the motion because they are the locations where the speed is instantaneously zero and the object reverses direction.

Try using the graphical approach to answer the following questions (and make up some of your own): What do you have to do to E to increase the amplitude of oscillation, and how might you do this? What happens to the amplitude of a real-life system in SHM in the presence of a force such as friction when E decays with time?

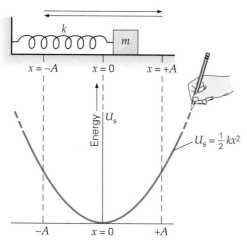

FIGURE 1 The potential energy "well" of a spring–mass system The potential energy of a spring stretched or compressed ($+x$) from its equilibrium position ($x = 0$) is a parabola, since $U_s \propto x^2$. At $x = \pm A$, all of the system's energy is potential.

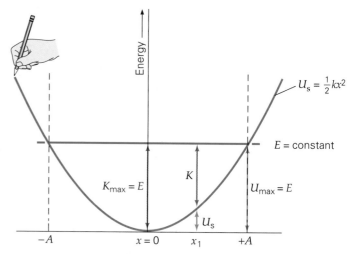

FIGURE 2 Energy transfers as the spring–mass system oscillates The vertical distance from the x axis to the parabola is the system's potential energy. The remainder—the vertical distance between the parabola and the horizontal line representing the system's constant total energy E—is the system's kinetic energy.

Example 13.1 ■ A Block and a Spring: Simple Harmonic Motion

A block with a mass of 0.25 kg sitting on a frictionless surface is connected to a light spring that has a spring constant of 180 N/m (see Fig. 13.1). If the block is displaced 15 cm from its equilibrium position and released, what are (a) the total energy of the system and (b) the speed of the block when it is 10 cm from its equilibrium position?

Thinking It Through. The total energy depends on the spring constant (k) and amplitude (A), which are given. At $x = 10$ cm, we expect the speed to be less than the maximum speed (why?).

Solution. First we list the given data, as usual, and what it is we are to find. The initial displacement *is* the amplitude (why?).

Given: $m = 0.25$ kg
$k = 180$ N/m
$A = 15$ cm $= 0.15$ m
$x = 10$ cm $= 0.10$ m

Find: (a) E (total energy)
(b) v (speed)

(a) The total energy is given directly by Eq. 13.5:

$$E = \tfrac{1}{2}kA^2 = \tfrac{1}{2}(180 \text{ N/m})(0.15 \text{ m})^2 = 2.0 \text{ J}$$

(b) The instantaneous speed of the block at a distance of 10 cm from the equilibrium position is given by Eq. 13.6 without directional signs:

$$v = \sqrt{\frac{k}{m}(A^2 - x^2)} = \sqrt{\frac{180 \text{ N/m}}{0.25 \text{ kg}}[(0.15 \text{ m})^2 - (0.10 \text{ m})^2]} = \sqrt{9.0 \text{ m}^2/\text{s}^2} = 3.0 \text{ m/s}$$

What would the speed be at $x = -10$ cm?

Follow-up Exercise. In part (b) of this Example, the block at $x = 10$ cm is at two-thirds or 67% of its maximum displacement. Is its speed at that position therefore 67% of its maximum speed? Prove your answer mathematically.

The spring constant is commonly determined by placing an object of known mass on the end of the spring and letting it settle vertically into a new equilibrium position. The next Example shows some typical results.

Example 13.2 ■ The Spring Constant: Experimental Determination

When a 0.50-kilogram mass is suspended from a spring, the spring stretches a distance of 10 cm to a new equilibrium position (•Fig. 13.3a). (a) What is the spring constant of the spring? (b) The mass is then pulled down another 5.0 cm and released. What is the highest position of the oscillating mass?

Thinking It Through. At the equilibrium position, the net force on the mass is zero because $\mathbf{a} = 0$. In (b) we will use $-y$ to designate "downward," as is customary in vertical problems.

Solution.

Given: $m = 0.50$ kg
$y_o = 10$ cm $= 0.10$ m
$y = -5.0$ cm $= -0.050$ m

Find: (a) k (spring constant)
(b) A (amplitude)

(a) When the suspended mass and the stretched spring are in equilibrium (Fig. 13.3a), the net force on the mass is zero. Thus the weight of the mass and the spring force are equal and opposite. Then, equating their magnitudes,

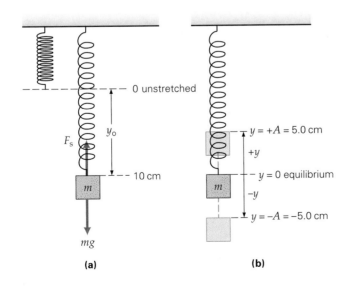

(a)   (b)

•**FIGURE 13.3 Determination of the spring constant** (a) When an object suspended on a spring is in equilibrium, the two forces on it cancel: $F_s = w$, or $ky_o = mg$. Thus the spring constant k can be computed: $k = mg/y_o$. (b) The zero reference point of SHM of an object suspended on a spring is conveniently taken as the new equilibrium position, as the motion is symmetric about this point. See Example 13.2.

$$F_s = w$$

or

$$ky_o = mg$$

Thus,

$$k = \frac{mg}{y_o} = \frac{(0.50\ \text{kg})(9.8\ \text{m/s}^2)}{0.10\ \text{m}} = 49\ \text{N/m}$$

(b) Once set into motion, the mass oscillates up and down through the equilibrium position. Since the motion is symmetric about this point, it is designated as the zero reference point of the oscillation (Fig. 13.3b). The initial displacement is $-A$, so the highest position of the mass will be 5.0 cm above the equilibrium position $(+A)$.

Follow-up Exercise. How much more potential energy does the spring in this Example have at the bottom of its oscillation that at the top?

13.2   Equations of Motion

Objectives: To (a) understand the equation of motion for SHM and (b) explain what is meant by phase and phase differences.

We refer to the **equation of motion** of an object as the equation that gives the object's position as a function of time. For example, the equation of motion with a constant linear acceleration is $x = v_o t + \frac{1}{2}at^2$, where v_o is the initial velocity (Chapter 2). However, the acceleration is not constant for simple harmonic motion, so the kinematic equations of Chapter 2 do not apply to this case.

The equation of motion for an object in simple harmonic motion can be derived from a relationship between simple harmonic and uniform circular motions. SHM can be simulated by a component of uniform circular motion, as illustrated in •Fig. 13.4. As the illuminated object moves in uniform circular motion (with constant angular speed ω) in a horizontal plane, its shadow moves back and forth horizontally, following the same path as the object on the spring, which is in simple harmonic motion. Since the shadow and the object have the same position at any time, it follows that the equation of *horizontal* motion for the object in circular motion is the same as the equation of motion for a horizontally oscillating object on the spring.

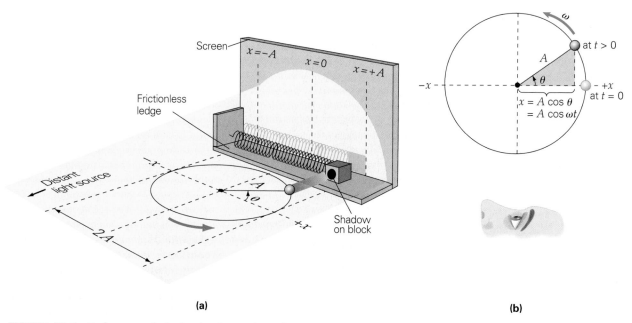

(a)

(b)

•**FIGURE 13.4   Reference circle for horizontal motion** (a) The shadow of the object in uniform circular motion has the same horizontal motion as the object on the spring in simple harmonic motion. (b) The motion can thus be described by $x = A \cos \theta = A \cos \omega t$ (assuming $x = +A$ at $t = 0$).

From the reference circle in Fig. 13.4b, the x coordinate (position) of the object is given by

$$x = A \cos \theta$$

But the object moves with a constant angular velocity with a magnitude ω. In terms of angular distance θ, assuming $\theta = 0$ at $t = 0$, we have $\theta = \omega t$, so

$$x = A \cos \omega t \qquad \text{(SHM for } x_o = +A) \qquad (13.8)$$

The angular speed ω (in rad/s) of the *reference circle object* is called the *angular frequency* of the oscillating object, since $\omega = 2\pi f$, where f is the frequency of revolution or rotation (Section 7.2). Figure 13.4 shows that the frequency of the "orbiting" object is the same as the frequency of oscillation of the object on the spring. Thus, we can write Eq. 13.8 alternatively (using $f = 1/T$) as

Note: See Eq. 7.8, Section 7.2.

$$x = A \cos(2\pi ft) = A \cos\left(\frac{2\pi t}{T}\right) \qquad \text{(SHM for } x_o = +A) \quad (13.9)$$

Equations 13.8 and 13.9 give three equivalent forms of the equation of motion for an object in simple harmonic motion. Any one of them can be used for convenience, depending on the known parameters. For example, suppose you are given the time t in terms of the period T—say, $t_o = 0$, $t_1 = T/2$, and $t_2 = T$—and are asked to find the position of an object in SHM at these times. In this case, it is convenient to use Eq. 13.9, and

Different forms of the equation of motion

$$t_o = 0 \qquad x_o = A \cos\left[2\pi(0)/T\right] = A \cos 0 = A$$

$$t_1 = \frac{T}{2} \qquad x_1 = A \cos\left[2\pi(T/2)/T\right] = A \cos \pi = -A \qquad (13.10)$$

$$t_2 = T \qquad x_2 = A \cos\left[2\pi T/T\right] = A \cos 2\pi = A$$

Hence, the results tell us that the object was initially at $x = A$, as we knew. One-half period later it was at $x = -A$, or the opposite extreme of its oscillation; and at a time of one period (T), it was back where it started, which is to be expected since the motion is periodic.

The period of an object oscillating on a spring can be expressed in terms of the mass m (or inertia) of the object and the spring constant (or stiffness) k. Our algebraic equation of motion describes the motion once you know the numerical values.

To show the power of the reference circle, let us use it to compute the period of the spring–object system. Note that the time for the object in the reference circle to make one complete "orbit" is exactly the time it takes for the oscillating object to make one complete cycle (see Fig. 13.4). Thus all we need is the time for one reference circle orbit and we have the period of oscillation. Because the reference circle object is in uniform circular motion at a constant speed equal to the maximum speed of oscillation, the object travels a distance of one circumference in one period. Because $t = d/v$, where $t = T$, d is the circumference, and v is v_{max} given by Equation 13.7,

$$T = \frac{d}{v_{max}} = \frac{2\pi A}{\sqrt{k/m}A}$$

or

$$T = 2\pi\sqrt{\frac{m}{k}} \qquad \begin{array}{l} \textit{period of object} \\ \textit{oscillating on a spring} \end{array} \qquad (13.11)$$

Because the amplitudes canceled out in Eq. 13.11, *the period (and frequency) are independent of the amplitude of the motion.* This statement is a general feature of simple harmonic oscillators—that is, for oscillators driven by a linear restoring force, such as a spring obeying Hooke's law. For example, a pendulum-driven clock that is not properly rewound and is running down would still keep correct time because the period would remain unchanged as its amplitude decreased.

Note: Period and frequency are independent of amplitude for SHM.

We see from Eq. 13.11 that the greater the mass, the longer the period; and the greater the spring constant (the stiffer the spring), the shorter the period. It is the *ratio* of mass to stiffness that determines the period. Thus you can offset a mass increase by using a stiffer spring.

Since $f = 1/T$,

$$f = \frac{1}{2\pi}\sqrt{\frac{k}{m}} \qquad \begin{array}{l}\textit{frequency of object}\\ \textit{oscillating on a spring}\end{array} \qquad (13.12)$$

Thus, the greater the spring constant (the stiffer the spring), the more frequently the system vibrates, as you might expect.

Also, note that since $\omega = 2\pi f$, we may write

$$\omega = \sqrt{\frac{k}{m}} \qquad \begin{array}{l}\textit{anguluar frequency of object}\\ \textit{oscillating on a spring}\end{array} \qquad (13.13)$$

As another example, a simple pendulum (a small, heavy object on a string) will undergo simple harmonic motion for small angles of oscillation. Advanced mathematics is needed to show that the period of a simple pendulum oscillating through a small angle ($\theta \leq 10°$) is given, to a good approximation, by

$$T = 2\pi\sqrt{\frac{L}{g}} \qquad \begin{array}{l}\textit{period of a}\\ \textit{simple pendulum}\end{array} \qquad (13.14)$$

where L is the length of the pendulum and g is the acceleration due to gravity.

An important difference between the spring–mass period and the pendulum period is that the latter is independent of the mass of the bob. Can you explain why? Think about what supplies the restoring force for the pendulum's oscillations. It is gravity. Hence, we would expect the acceleration (along with the velocity and period) to be independent of mass. That is, the gravitational force automatically provides the same acceleration to different bob masses of pendulums with the same length. Thus the acceleration doesn't depend on the mass. We have seen that similar effects occur in free fall (Chapter 2) and for blocks sliding and cylinders rolling down inclines (Chapters 4 and 8, respectively). The following Example demonstrates the usage of the equation of motion for SHM.

Example 13.3 ■ Equation of Motion for a Practical Oscillation: Shaken, Not Stirred

To stir a mixture of liquids consistently and gently, a biologist designs a simple mechanical system consisting of a horizontal spring attached to a beaker on the slippery Teflon® surface. She pours two liquids with a total mass of 100 g into the beaker, which has a mass of 50.0 g. She then pulls on the beaker, stretching the spring. On releasing the beaker from rest ($t = 0$), she sets the system in motion with an amplitude of 5.00 cm. To avoid spilling and to promote gentle mixing, the frequency is designed to be exactly 0.300 Hz. (a) What is the required spring constant? (b) What is the displacement of the beaker from equilibrium at time $t = 2.00$ s?

Thinking It Through. (a) Since we know the frequency and total mass (beaker plus liquids), we can use Eq. 13.12 to solve for the spring constant. (b) The system has an equation of motion given by Eq. 13.8. We must be careful to express the angle in radians.

Solution.

Given: $m = 50.0$ g $+ 100$ g $= 150$ g $= 0.150$ kg (total mass)   *Find:* (a) k (spring constant)
    $A = 5.00$ cm $= 5.00 \times 10^{-2}$ m   (b) x (displacement)
    $f = 0.300$ Hz
    $t = 2.00$ s

(a) We can rearrange Eq. 13.12 to solve for the spring constant:

$$k = 4\pi^2 m f^2 = 4\pi^2 (0.150 \text{ kg})(0.300 \text{ Hz})^2 = 0.533 \text{ N/m}$$

(b) First we calculate the angular frequency, ω, from the relationship between angular frequency and frequency (Eq. 7.8):

$$\omega = 2\pi f = 2\pi(0.300 \text{ Hz}) = 1.88 \text{ rad/s}$$

To find the location of the beaker (its displacement from equilibrium), we use the equation of motion, Eq. 13.8:

$$x = A \cos \omega t = (0.0500 \text{ m}) \cos[(1.88 \text{ rad/s})(2.00 \text{ s})] = (0.0500 \text{ m}) \cos(3.76 \text{ rad})$$

$$= -0.0407 \text{ m} = -4.07 \text{ cm}$$

The minus sign indicates that at this time the spring is compressed 4.07 cm from equilibrium.

Follow-up Exercise. The cone of a loudspeaker moves in SHM in and out a distance of 0.20 mm while vibrating at a frequency of 500 Hz. **(a)** What is its maximum speed? **(b)** Write the equation of motion in the form of Eq. 13.8 for the cone, and use it to find the cone's displacement at $t = 1.1 \times 10^{-3}$ s.

Problem-Solving Hint

Note that in the part (b) calculation of Example 13.3, where we have cos (3.76), the angle is in radians, *not* degrees. Don't forget to set your calculator to radians (rather than degrees) when finding the value of a trigonometric function in equations for simple harmonic or circular motion.

Example 13.4 ■ Fun with a Pendulum: Frequency and Period

A helpful older brother takes his sister to play on the swings in the park. He pushes her from behind on each return. Assuming that the swing behaves as a simple pendulum with a length of 2.50 m, (a) what would be the frequency of the oscillations, and (b) what would be the interval between the brother's pushes?

Thinking It Through. (a) The period is given by Eq. 13.14, and the frequency and period are inversely related: $f = 1/T$. (b) Since the brother pushes from one side on each return, he must push once every completed cycle, so the time between his pushes is equal to the swing's period.

Solution.

Given: $L = 2.50$ m   *Find:* (a) f (frequency)
                                     (b) T (period)

(a) We can invert Eq. 13.14 to solve directly for the frequency:

$$f = \frac{1}{T} = \frac{1}{2\pi}\sqrt{\frac{g}{L}} = \frac{1}{2\pi}\sqrt{\frac{9.80 \text{ m/s}^2}{2.50 \text{ m}}} = 0.315 \text{ Hz}$$

(b) The period is then found from the frequency:

$$T = 1/f = 1/(0.315 \text{ Hz}) = 3.17 \text{ s}$$

The brother must push every 3.17 s to maintain a steady swing (and to keep his sister from complaining).

Follow-up Exercise. In this Example, the older brother, a physics buff, carefully measures the period of the swing to be 3.18 s, not 3.17 s. If the length is accurate at 2.50 m, what is the acceleration due to gravity at their location? Using this accurate value of g, do you think the park is at sea level?

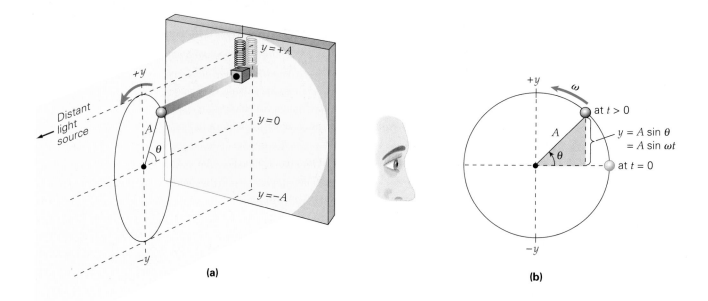

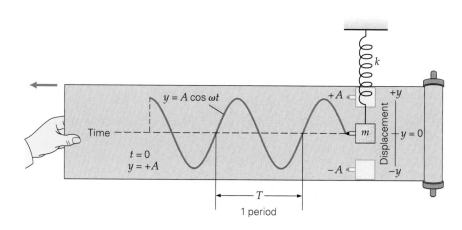

•FIGURE 13.5 Reference circle for vertical motion (a) The shadow of the object in uniform circular motion has the same vertical motion as the object oscillating on the spring in simple harmonic motion. (b) The motion can thus be described by $y = A \sin \theta = A \sin \omega t$ (assuming $y = 0$ at $t = 0$).

A vertical version of the reference circle can be used to describe vertical simple harmonic motion, as illustrated in •Fig. 13.5 for a suspended object oscillating on a spring. Suppose that, having been given an upward push, the object initially started its upward motion from its equilibrium position $y = 0$ (recall that $+y$ means upward). In this case, the vertical displacement is given by $y = A \sin \theta$. A development similar to that for the horizontal case leads to this equation of motion:

$$y = A \sin \omega t \qquad \begin{array}{l}\textit{initially upward}\\\textit{motion with } y_o = 0\end{array}$$

Notice that at $t = 0$, the object is indeed at its equilibrium point ($y = 0$).

However, suppose the object starts initially from its maximum positive displacement position. As shown in •Fig. 13.6, the resulting curve must instead be a cosine function

$$y = A \cos \omega t \qquad \begin{array}{l}\textit{vertical motion}\\\textit{with } y_o = A, v_o = 0\end{array}$$

This is because only a cosine equation correctly describes the initial conditions— $y = +A$ at $t = 0$.

Thus, the equation of motion for an oscillating object may be either a sine or a cosine function. Both of these functions are referred to as being *sinusoidal*. That is, simple harmonic motion is described by a sinusoidal function of time.

•FIGURE 13.6 Sinusoidal equation of motion As time passes, the oscillating object traces out a sinusoidal curve on the moving paper. In this case, $y = A \cos \omega t$, because the object's initial displacement is $y_o = +A$.

Initial Conditions and Phase

You may be wondering how to decide whether to use a sine or cosine function to describe a particular case of simple harmonic motion. In general, the form of the function is determined by the initial displacement and velocity of the object: the *initial conditions* of the system. These initial conditions are the values of the displacement and velocity at $t = 0$; when, taken together, they tell how the system is initially set into motion.

Let's look at four special cases. If an object in vertical SHM has an initial displacement of $y = 0$ at $t = 0$ and moves initially upward, the equation of motion is $y = A \sin \omega t$ (•Fig. 13.7a). Note that $y = A \cos \omega t$ does not satisfy the initial condition, that is, $y_o = A \cos \omega t = A \cos \omega(0) = A$, since $\cos 0 = 1$.

Suppose that the object is initially released ($t = 0$) from its positive amplitude position ($+A$), as in the case of the object on a spring shown in Fig. 13.6. Here, the equation of motion is $y = A \cos \omega t$ (Fig. 13.7b). This expression satisfies the initial condition: $y_o = A \cos \omega(0) = A$.

The other two cases are $y = 0$ at $t = 0$, with motion initially downward (for an object on a spring) or in the negative direction (for horizontal SHM); and $y = -A$ at $t = 0$, meaning that the object is initially at its negative amplitude

Note: Initial conditions include both x_o and v_o, the displacement and velocity at $t = 0$.

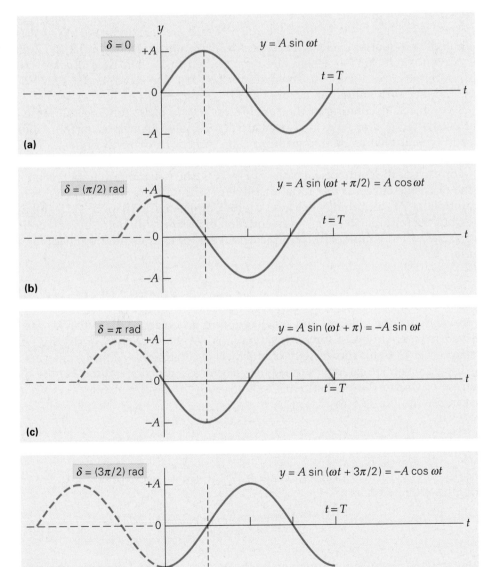

(a)

(b)

(c)

(d)

•**FIGURE 13.7   Phase differences** For the general SHM equation $y = A \sin (\omega t + \delta)$, oscillations are described by either sine or cosine terms for the phase constants shown. The initial displacement and velocity direction determine δ. Each curve is 90° (or $\pi/2$ rad) out of phase with the preceding one. This is equivalent to shifting each successive curve a quarter of a cycle relative to the previous one. Everthing else, including the period, is identical for each curve (why?).

position. These motions are described by $y = -A \sin \omega t$ and $y = -A \cos \omega t$, respectively, as illustrated in Fig. 13.7c and 13.7d.

Note in Fig. 13.7 that if the curves are extended in the negative direction to the horizontal axis (dashed purple lines in figure), they all have the same shape but have been "shifted," so to speak. We describe this change as a shift in *phase*. For the general case, we may write the equation of simple harmonic motion as

General equation of motion for simple harmonic motion

$$y = A \sin (\omega t + \delta) \qquad (13.15)$$

where $(\omega t + \delta)$ is the *phase angle* and δ is the **phase constant**. The phase constant essentially shifts the curve to match the appropriate sinusoidal function to the motion and is determined by the initial conditions.

The phase constants for the cases in Fig. 13.7 are given in the figure. For example, for $\delta = 90°$ (or $\pi/2$ rad), the equation of motion is $y = A \cos \omega t$, and an object oscillating up and down on a spring would be initially released at the $+A$ position. You probably recognize that the curve for the $\delta = 90°$ is a cosine. This can be shown algebraically by using the following trigonometric formula: $\sin (a + b) = \sin a \cos b + \cos a \sin b$. Then, with $\delta = 90°$,

$$y = A \sin (\omega t + 90°) = A[\sin \omega t \cos 90° + \cos \omega t \sin 90°] = A \cos \omega t$$

since $\cos 90° = 0$ and $\sin 90° = 1$.

The curves for $\delta = 0$ and $\delta = 90°$ (or $\pi/2$ rad) are said to be $90°$ *out of phase*, or shifted by a quarter cycle, with respect to one another. Notice in Fig. 13.7 that the curve for $\delta = 90°$ has essentially been shifted to the left by a quarter cycle ($90°$) from the curve for $\delta = 0$. The cases for $\delta = 180°$ (or π rad) and $\delta = 270°$ (or $3\pi/2$ rad) are each shifted an additional $90°$ out of phase.

Two objects oscillating at the same frequency *in phase* (having the same δ) will oscillate together. Two that oscillate completely out of phase ($180°$ difference in δ) will always be going in opposite directions or be at opposite maximum displacements.

For each of the four special cases in Fig. 13.7, the equation of motion is either a sine or cosine curve. If the initial displacement is not zero or $\pm A$, then δ is not a multiple of $90°$ and things get a bit more complicated. Equation 13.15 still applies, but the formula above gives an equation of motion with both sine and cosine terms. See Demonstration 10 for a pictorial view of a sinusoidal oscillation and phase.

Velocity and Acceleration in SHM

Expressions for the velocity and acceleration of an object in SHM can be easily derived from energy and force considerations. We have already derived an expression (Eq. 13.6) for the velocity of an object oscillating on a spring in terms of the displacement. We can now express the velocity as a function of time. For the special case of vertical simple harmonic motion, with the displacement $y = A \sin \omega t$, we can rewrite Eq. 13.6 as

$$v = \sqrt{\frac{k}{m}(A^2 - y^2)} = \sqrt{\frac{k}{m}(A^2 - A^2 \sin^2 \omega t)} = \sqrt{\frac{k}{m}} A \sqrt{1 - \sin^2 \omega t}$$

Since $\omega = \sqrt{k/m}$ (Eq. 13.13) and $\cos \theta = \sqrt{1 - \sin^2 \theta}$ (see Appendix I), we can write this expression as

Note: Maximum speed $v = \omega A$.

$$v = \omega A \cos \omega t \quad \begin{matrix} \textit{vertical velocity} \\ \textit{if } v_o \textit{ upward, } y_o = 0 \end{matrix} \qquad (13.16)$$

where the directional signs are given by the cosine function. Using Newton's second law to find the acceleration with the spring force $F_s = -ky$, we have

A demonstration to show that SHM can be represented by a sinusoidal function.
A "graph" of the function is generated with an analogue of a strip chart recorder.

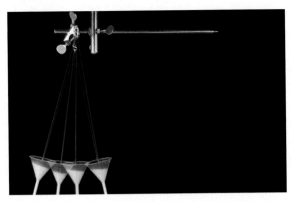

(a) A salt-filled funnel oscillates, suspended from two strings.

(b) The salt falls on a black-painted poster board that will be pulled in a direction perpendicular to the plane of the funnel's oscillation.

(c) Away we go.

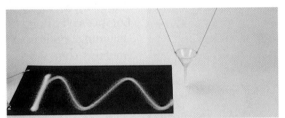

(d) The salt trail traces out a plot of displacement versus time, or $y = A \sin (\omega t + \delta)$. Note that in this case the phase constant is about $\delta = 90°$ and $y = A \cos \omega t$. (Why?)

$$a = \frac{F_s}{m} = \frac{-ky}{m} = -\frac{k}{m} A \sin \omega t$$

and since $\omega = \sqrt{k/m}$,

$$a = -\omega^2 A \sin \omega t = -\omega^2 y \qquad \begin{array}{l} \textit{vertical acceleration} \\ \textit{if } v_0 \textit{ upward, } y_0 = 0 \end{array} \qquad (13.17)$$

Note that the functions for the velocity and acceleration are out of phase with that for the displacement. Since the velocity is 90° out of phase with the displacement, the speed is greatest when $\cos \omega t = \pm 1$ at $y = 0$—when the oscillating object is passing through its equilibrium position. The acceleration is 180° out of phase with the displacement (as indicated by the minus sign on the right-hand side of Eq. 13.17). Therefore, the magnitude of the acceleration is a maximum when $\sin \omega t = \pm 1$ at $y = \pm A$—when the displacement is a maximum, or when the object is at an amplitude position. At any position except the equilibrium position, the directional sign of the acceleration is opposite that of the displacement, as it should be for an acceleration resulting from a restoring force. At the equilibrium position, both the displacement and acceleration are zero. (Can you see why?)

Note: Maximum acceleration magnitude $a = \omega^2 A$.

Also note that the acceleration for SHM is not constant with time. Hence, the kinematic equations for acceleration (Chapter 2) *cannot* be used, since they are for constant accelerations.

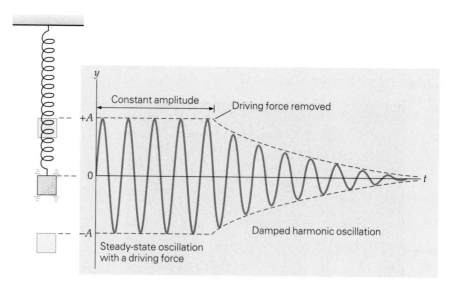

•FIGURE 13.8 Damped harmonic motion When a driving force adds energy to a system equal to its energy losses, the oscillation is steady with a constant amplitude. When the driving force is removed, the oscillations decay (that is, are damped), and the amplitude decreases nonlinearly with time.

Damped Harmonic Motion

Simple harmonic motion with a constant amplitude implies that there are no losses of energy, but in practical applications there are always some frictional losses. Therefore, to maintain a constant-amplitude motion, energy must be added to the system by some external driving force, such as someone pushing a swing. Without a driving force, the amplitude and the energy of an oscillator decrease with time, giving rise to **damped harmonic motion** (•Fig. 13.8). The time required for the oscillations to cease, or to damp out, depends on the magnitude and type of the damping force (for example, air resistance).

In many applications involving continuous periodic motion, damping is unwanted and necessitates energy input. However, in some instances, damping is desirable. For example, the dial in a spring-operated bathroom scale oscillates briefly before stopping at a weight reading. If not properly damped, these oscillations would continue for some time and you would have to wait before you could read your weight. Damping is also required for shock absorbers on automobiles and needle indicators on instruments measuring electrical quantities. In California many new buildings incorporate damping mechanisms (giant shock absorbers) to dampen their oscillatory motion after they are set in motion by earthquake waves.

13.3   Wave Motion

Objectives: To (a) describe wave motion in terms of various parameters and
(b) identify different types of waves.

The world is full of waves of various types—some examples are water waves, sound waves, waves generated by earthquakes, and light waves. All waves result from a disturbance, the source of the wave. In this chapter we will be concerned with mechanical waves, or those that are propagated in some medium. (Light waves, which do not require a propagating medium, will be considered in more detail in later chapters.)

When a medium is disturbed, energy is imparted to it. Suppose that energy is added to a material mechanically, such as by impact or (in the case of a gas) by compression. This addition sets some of the particles vibrating. Because the particles are linked by intermolecular forces, the oscillation of each particle affects that of its neighbors. The added energy propagates, or spreads, by means of interactions among the particles of the medium. An analogy for this process is shown in •Fig. 13.9, where the "particles" are dominoes. As each domino falls, it topples

•FIGURE 13.9   Energy transfer The propagation of a disturbance, or a transfer of energy through space, is seen in a row of falling dominoes.

the one next to it. Thus energy is transferred from domino to domino, and the disturbance propagates through the medium.

In this case, there is no restoring force between the dominoes, so they do not oscillate as do particles in a continuous material medium. Therefore the disturbance moves in space, but it does not repeat itself in time at any one location.

Similarly, if the end of a stretched rope is given a quick shake, the disturbance transfers energy from the hand to the rope, as illustrated in •Fig. 13.10. The forces acting between the rope "particles" cause them to move in response to the motion of the hand, and a *wave pulse* travels down the rope. Each "particle" goes up and then back down as the pulse passes by. This motion of individual particles and that of the wave pulse propagation as a whole can be observed by tying pieces of ribbon onto the rope (at x_1 and x_2 in Fig. 13.10). As the disturbance passes point x_1, the ribbon rises and falls, as do the rope "particles." Later, the same occurs for the ribbon at x_2, which indicates that the disturbance energy is propagating or traveling along the rope.

In a continuous material medium, particles interact with their neighbors and restoring forces cause them to oscillate when they are disturbed. Thus any disturbance not only propagates through space but may be repeated over and over in time at each position. Such a regular, rhythmic disturbance in both time and space is called a **wave**, and the transfer of energy is said to take place by means of **wave motion**.

A continuous wave motion, or *periodic wave*, requires a disturbance from an oscillating source (•Fig. 13.11). In this case, the particles move up and down continuously. If the driving source is such that a constant amplitude is maintained (the source oscillates in simple harmonic motion), the resulting particle motion is also simple harmonic.

Such periodic wave motion will have sinusoidal forms (sine or cosine) in both time and space. Being sinusoidal in space means that if you took a photograph of the wave at any instant (freezing it in time), you would see a sinusoidal waveform (such as one of the curves in Fig. 13.11). However, if you looked at a single point in space as a wave passed by, you would see a particle of the medium oscillating up and down sinusoidally with time, like the mass on a spring discussed in Section 13.2. (For example, imagine looking through a thin slit at a fixed location on the moving paper in Fig. 13.6. The wave trace would be seen rising and falling like a particle.)

Wave Characteristics

Specific characteristics of sinusoidal waves are used to describe them. As for a particle in simple harmonic motion, the *amplitude* (A) of a wave is the magnitude of the maximum displacement or the maximum distance from the particle equilibrium position (Fig. 13.11). This quantity corresponds to the height of a wave crest or the depth of a trough. Recall from Section 13.2 that, in SHM, the total energy of the oscillator is proportional to the square of the amplitude. Similarly, the energy *transported* by a wave is proportional to the square of its amplitude ($E \propto A^2$). Note the difference—a wave is one way of *transmitting* energy through space, whereas an oscillator's energy is localized in space.

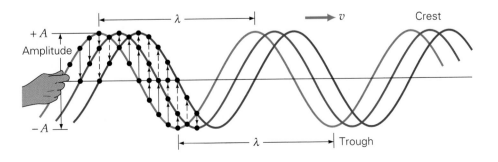

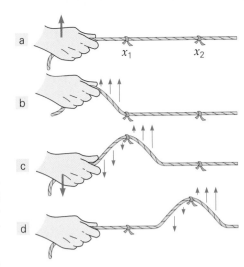

•**FIGURE 13.10 Wave pulse** The hand disturbs the stretched rope in a quick up and down motion, and a wave pulse propagates along the rope. (The red arrows represent the velocities of the hand and of pieces of the rope at different times and locations.) The rope "particles" move up and down as the pulse passes. The energy in the pulse is thus *both* kinetic and potential (elastic and gravitational).

Note: A wave is a combination of oscillations in space and time.

•**FIGURE 13.11 Periodic wave** A continuous harmonic disturbance can set up a sinusoidal wave in stretched rope, and the wave travels down the rope with wave speed v. Note that the rope "particles" oscillate vertically in simple harmonic motion. The distance between two successive points that are in phase (for example, at two crests) on the waveform is the wavelength λ of the wave. Can you tell how much time has elapsed, as a fraction of the period T, between the first (red) and last (blue) waves?

For a periodic wave, the distance between two successive crests (or troughs) is called the **wavelength** (λ) (Fig. 13.11). Actually, it is the distance between any two successive parts of the wave that are in phase (that is, at identical points on the waveform). The crest and trough positions are usually used for convenience. Note that a wavelength corresponds spatially to one cycle. Keep in mind that the wave, not the medium or material, is traveling.

The *frequency* (f) of a wave is the number of cycles per second—that is, the number of complete waveforms, or wavelengths, that pass by a given point during each second. The wave frequency is the same as the frequency of the SHM source that created it.

A periodic wave is said to possess a *period* (T). The period $T = 1/f$ is the time for one complete waveform (a wavelength) to pass by a given point. Since a wave moves, it also has a **wave speed** (or velocity, if direction is specified). Any particular point on the wave (for example, a crest) travels a distance of one wavelength λ in a time of one period T. Then, since $v = d/t$ and $f = 1/T$, we have

$$v = \frac{\lambda}{T} = \lambda f \qquad \text{wave speed} \qquad (13.18)$$

Note that the dimensions are correct (length/time). In general, the wave speed depends on the nature of the medium in addition to the source frequency f.

Example 13.5 ■ Dock of the Bay: Finding Wave Speed

A person on a pier observes a set of incoming waves that have a sinusoidal form with a distance of 1.6 m between the crests. If a wave laps against the pier every 4.0 s, what are (a) the frequency and (b) the speed of the waves?

Thinking It Through. We know the period and wavelength, so we can use the definition of frequency and Eq. 13.18 for wave speed.

Solution. The distance between crests is the wavelength, so we have

Given: $\lambda = 1.6$ m $\qquad$ *Find:* (a) f (frequency)
$\qquad\quad\; T = 4.0$ s $\qquad\qquad\qquad$ (b) v (wave speed)

(a) The lapping indicates the arrival of a wave crest, so 4.0 s is the wave period—the time it takes to travel one wavelength (the crest-to-crest distance). Then

$$f = \frac{1}{T} = \frac{1}{4.0 \text{ s}} = 0.25 \text{ s}^{-1} = 0.25 \text{ Hz}$$

(b) The frequency or the period can be used in Eq. 13.18 to find the wave speed:

$$v = \lambda f = (1.6 \text{ m})(0.25 \text{ s}^{-1}) = 0.40 \text{ m/s}$$

Alternatively,

$$v = \frac{\lambda}{T} = \frac{1.6 \text{ m}}{4.0 \text{ s}} = 0.40 \text{ m/s}$$

Follow-up Exercise. On another day, the person measures the speed of sinusoidal water waves at 0.25 m/s. (a) How far does a wave crest travel in 2.0 s? (b) If the distance between successive crests is 2.5 m, what is the frequency of these waves?

Types of Waves

Waves may be divided into two types based on the direction of the particles' oscillations relative to the wave velocity. In a **transverse wave**, the particle motion is perpendicular to the direction of the wave velocity. The wave produced in a stretched string (Fig. 13.11) is an example of a transverse wave, as is the wave shown in •Fig. 13.12a. A transverse wave is sometimes called a *shear wave* because the disturbance supplies a force that tends to shear the medium—to separate layers of that medium at a right angle to the direction of the wave velocity. Shear waves can propagate only in solids, since a liquid or a gas cannot support a shear. That is, a liquid or a gas does not have sufficient restoring forces between its particles to propagate a transverse wave.

In a **longitudinal wave**, the particle oscillation is parallel to the direction of the wave velocity. A longitudinal wave can be produced in a stretched spring by moving the coils back and forth along the spring axis (Fig. 13.12b). Alternating pulses of compressions and relaxations move along the spring. A longitudinal wave is sometimes called a *compressional wave*.

Sound waves in air are another example of longitudinal waves. A periodic disturbance produces compressions in the air. Between the compressions are *rarefactions*, regions where the density of the air is reduced, or rarefied. A loudspeaker oscillating back and forth, for example, can create these compressions and rarefactions, which travel out into the air as sound waves. Sound will be discussed in detail in Chapter 14.

Longitudinal waves can propagate in solids, liquids, and gases, since all phases of matter can be compressed to some extent. The propagations of transverse and longitudinal waves in different media give information about the Earth's interior structure, as discussed in the Insight on p. 436.

The sinusoidal profile of water waves might make you think that these are transverse waves. Actually, they reflect a combination of longitudinal and transverse motions (•Fig. 13.13). The particle motion may be nearly circular at the surface but becomes more elliptical with depth, eventually becoming longitudinal. A hundred meters or so below the surface of a large body of water, the wave disturbances have little effect. For example, a submarine at these depths is undisturbed by large waves on the ocean's surface. As a wave approaches shallower water near shore, the water particles have difficulty completing their elliptical paths. When the water becomes too shallow, the particles can no longer move through the bottom parts of their paths and the wave breaks. Its crest falls forward to form breaking surf as the waves' kinetic energy is transformed into potential energy—a water "hill" that eventually topples over.

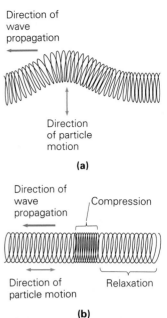

•**FIGURE 13.12 Transverse and longitudinal waves** (a) In a transverse wave, the particle motion is perpendicular to the direction of the wave velocity, as shown here in a spring for a wave moving to the left. (b) In a longitudinal wave, the particle motion is parallel to (or *along*) the direction of the wave velocity. Here a wave pulse also moves to the left. Can you explain the motion of the wave *source* for both types of waves?

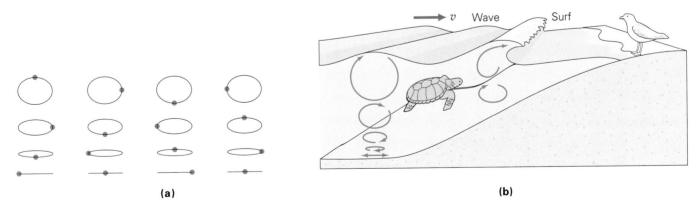

•**FIGURE 13.13 Water waves** Water waves are a combination of longitudinal and transverse motions. **(a)** At the surface, the water particles move in circles, but their motions become more longitudinal with depth. **(b)** When a wave approaches the shore, the lower particles are forced into steeper paths until finally the wave breaks or falls over to form surf.

Earthquakes, Seismic Waves, and Seismology

The Earth's interior structure is still something of a mystery. The deepest mine shafts and drillings extend only a few kilometers into the Earth. Using waves to probe the Earth's structure is one way to investigate it further. Waves generated by earthquakes have proved to be especially useful for this purpose. Seismology is the study of these waves, called *seismic waves*.

Earthquakes are caused by the sudden release of built-up stress along cracks and faults, such as the famous San Andreas Fault in California (Fig. 1). According to the geological theory of plate tectonics, the outer layer of the Earth consists of rigid plates, huge slabs of rock that move very slowly relative to one another. Stresses are continuously built up, particularly along boundaries between plates.

When slippage finally occurs, the energy from this stress-relieving event propagates outward as (seismic) waves from a site below the surface called the *focus*. These are of two general types: surface waves and body waves. *Surface waves*, which move along the Earth's surface, account for most earthquake damage (Fig. 2). *Body waves* travel through the Earth. There are both longitudinal and transverse body waves. The compressional (longitudinal) waves are called *P waves,* and shear (transverse) waves are called *S waves* (Fig. 3). The P and S stand for primary and secondary and indicate the waves' relative speeds (actually, their arrival times at monitoring stations). In general, primary waves travel through materials faster than do secondary waves and are detected first. An earthquake's rating on the Richter scale is related to the energy released in the form of seismic waves.

Seismic stations around the world monitor these waves with sensitive detecting instruments called seismographs (Fig. 4). From the data gathered, we can map the paths of the waves through the Earth and thereby learn about the interior structure. The Earth's interior seems to be divided into three

FIGURE 2   Bad vibrations   Earthquake damage caused by the major shock that struck Kobe, Japan, in January 1995.

general regions: the crust, the mantle, and the core, which itself has a solid inner region and a liquid outer region.*

The locations of their boundaries are determined in part by the shadow zones. Shadow zones are regions where no

*The crust is in most places about 24–30 km (15–20 mi) thick; the mantle is 2900 km (1800 mi) thick; and the core has a radius of 3450 km (2150 mi). The solid inner core has a radius of about 1200 km (750 mi).

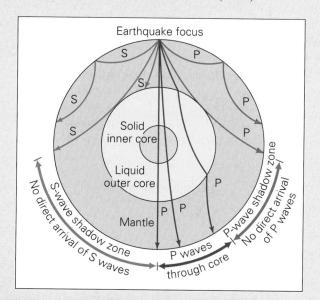

FIGURE 3   Compressional and shear waves
Earthquakes produce waves that travel through the Earth. Because transverse S waves are not detected on the opposite side of the Earth, scientists believe that at least part of the Earth's core is a viscous liquid under high pressures and temperatures. The waves bend continuously, or refract, because their speed varies with depth.

FIGURE 1   The San Andreas Fault   A small section of the fault, which runs through the San Francisco Bay area as well as the more rural regions of California shown here.

waves of a particular type are detected. These zones appear because although longitudinal waves can travel through solids *or* liquids, transverse waves can travel only through solids. When an earthquake occurs at a particular location, P waves are detected on the other side of the Earth but S waves are not (see Fig. 3). The absence of S waves in a shadow zone leads to the conclusion that the Earth must have a region near its center that is in the liquid phase. This region is a highly viscous metallic liquid—but definitely a liquid, since it does not support a shear (transverse waves are not propagated).

When the transmitted P waves enter and leave the liquid region, they are refracted (bent). This refraction gives rise to a P wave shadow zone, which indicates that only the outer part of the core is liquid. As you will learn in Chapter 19, the combination of a liquid outer core and rotation may be responsible for the Earth's magnetic field.

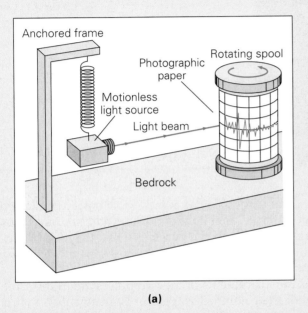

(a)

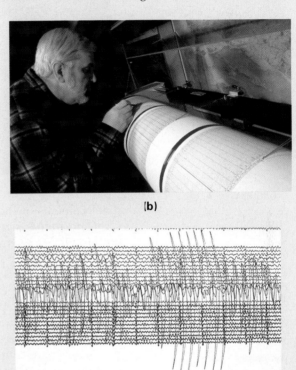

(b)

(c)

FIGURE 4   The seismograph   **(a)** A simple seismograph. The device records the amplitude of the ground vibrations. The energy of the seismic waves is proportional to the square of their amplitude. **(b)** A U.S. Geological Survey scientist monitors seismograph readings. **(c)** Seismograph trace from the destructive Kobe, Japan, earthquake of 1995.

13.4   Wave Properties

Objective: **To explain various wave properties and resulting phenomena.**

Among the properties exhibited by all waves are interference, superposition, reflection, refraction, dispersion, and diffraction. Particles do not share these properties.

Interference and Superposition

When two or more waves meet or pass through the same region of a medium, they pass through each other and proceed without being altered. While they are in the same region, the waves are said to be interfering.

What happens during interference; that is, what does the combined waveform look like? The relatively simple answer is given by the **principle of superposition**:

> At any time, the combined waveform of two or more interfering waves is given by the sum of the displacements of the individual waves at each point in the medium.

The principle of **interference** is illustrated in •Fig. 13.14. The displacement of the combined waveform at any point is given by $y = y_1 + y_2$, where y_1 and y_2 are the

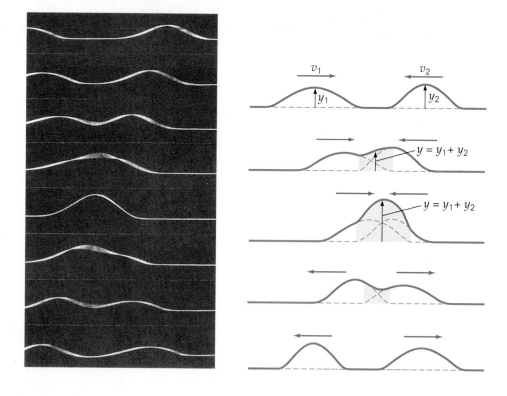

•FIGURE 13.14 Principle of superposition When two waves meet, they interfere. The beige tint marks the area where the two waves, moving in opposite directions, overlap and combine. The displacement at any point on the combined wave is equal to the sum of the displacements on the individual waves: $y = y_1 + y_2$.

displacements of the individual pulses at that point (directions are indicated by plus and minus signs). Interference, then, is the physical addition of waves. In adding waves, we must take into account the possibility that they are producing disturbances in opposite directions. In other words, we must treat the disturbances in terms of vector addition.

In Fig. 13.14, the vertical displacements of the two pulses are in the same direction, and the amplitude of the combined waveform is greater than that of either pulse. This situation is called **constructive interference**. Conversely, if one pulse has a negative displacement, the two pulses tend to cancel each other when they overlap, and the amplitude of the combined waveform is smaller than that of either pulse. This situation is called **destructive interference**.

Special cases of total constructive and destructive interference are shown in •Fig. 13.15. These are shown for traveling wave pulses of the same width and amplitude. At the instant these interfering waves are exactly in phase (crest coincid-

•FIGURE 13.15 Interference
(a) When two wave pulses of the same amplitude meet and are in phase, they interfere constructively. When the pulses are exactly superimposed (3), total constructive interference occurs.
(b) When the interfering pulses are completely out of phase (180°) and are exactly superimposed (3), total destructive interference occurs.

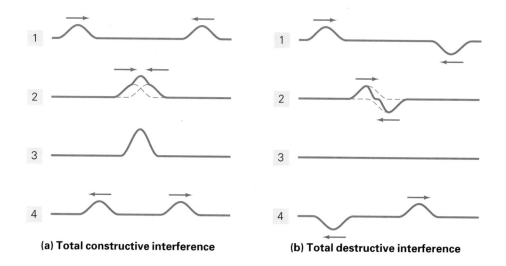

(a) Total constructive interference   **(b) Total destructive interference**

ing with crest), the amplitude of the combined waveform is twice that of either individual wave. This case is referred to as **total constructive interference**. When these interfering pulses are completely out of phase (180° difference, or crest coinciding with trough), the waveforms momentarily disappear; that is, the amplitude of the combined wave is zero. This case is called **total destructive interference**.

The word *destructive* unfortunately tends to imply that the energy as well as the form of the waves is destroyed. This is not the case. At the point of total destructive interference, when the net wave shape and hence potential energy are zero, the wave energy is stored in the medium completely in the form of kinetic energy.

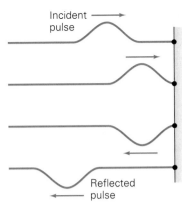

(a) Phase shift of 180°

Reflection, Refraction, Dispersion, and Diffraction

Besides meeting other waves, waves can (and do) meet objects or a boundary with another medium. In such cases, several things may occur. One of these is reflection. **Reflection** occurs when a wave strikes an object or comes to a boundary of another medium and is at least partly diverted back into the original medium. An echo is the reflection of sound waves, and mirrors reflect light waves.

Two cases of reflection are illustrated in •Fig. 13.16. If the end of the string is fixed, the reflected pulse is inverted, or undergoes a 180° phase shift (Fig. 13.16a). This is because the pulse causes the string to exert an upward force on the wall, and the wall exerts an equal and opposite downward force on the string (by Newton's third law). This downward force creates the downward or inverted reflected pulse. If the end of the string is free to move, then the reflected pulse is not inverted (zero or no phase shift). This is illustrated in Fig. 13.16b, where the string is attached to a light ring that can move freely on a smooth pole. The ring is accelerated upward by the front portion of the incoming pulse and then comes back down, thus creating a noninverted reflected pulse.

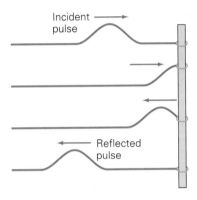

(b) No phase shift

•FIGURE 13.16   Reflection (a) When a wave (pulse) on a string is reflected from a fixed boundary, the reflected wave is inverted, or undergoes a 180° phase shift. (b) If the string is free to move at the boundary, there is no phase shift of the reflected wave.

More generally, when a wave strikes a boundary, the wave is not completely reflected. Instead, some of the wave's energy is reflected and some is transmitted or absorbed. When a wave crosses a boundary into another medium, its speed generally changes because the new material has different characteristics. Entering the medium obliquely (at an angle), the transmitted wave moves in a direction different from that of the incident wave. This phenomenon is called **refraction** (•Fig. 13.17a).

Since refraction depends on changes in wave speed, you might be wondering which physical parameters determine the wave speed. Generally, there are two types of situations. The simplest kind of wave is one whose speed does *not* depend on the wavelength (or frequency) of the wave. All such waves travel at the same speed determined solely by the properties of the medium. These waves are called *nondispersive waves*, because they do not disperse, or spread apart from one another. An example of a nondispersive transverse wave is a wave on a string, where, as we

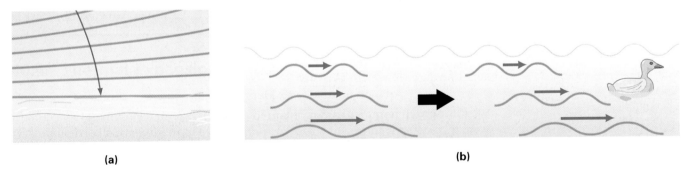

(a)   (b)

•FIGURE 13.17   Refraction and dispersion   (a) Refraction of water waves is shown from overhead. As the crests approach the beach, their left edge slows as it enters shallow water first. Thus the whole crest rotates, approaching the beach more or less head on. (b) For dispersive waves, wave speed depends on frequency or wavelength, as shown for surface waves traveling in deep water.

•**FIGURE 13.18 Diffraction**

Diffraction effects are greatest when the opening (or object) is about the same size as or smaller than the wavelength of the waves. **(a)** With an opening much larger than the wavelength of these plane water waves, diffraction is noticeable only near the edges. **(b)** With an opening about the same size as the wavelength of the waves, diffraction produces nearly semi-circular waves.

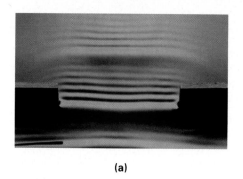

(a)

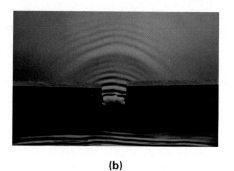

(b)

shall see, the speed is determined only by the tension and mass density of the string (Section 13.5). Sound is a nondispersive longitudinal wave; the speed of sound (in air) is determined only by the compressibility and density of the air. Indeed, if sound speed did depend on the frequency, at the back of the symphony hall you might hear the violins well before the clarinets, even though the two sound waves were in perfect synchronization when they left the orchestra pit.

When the wave speed *does* depend on wavelength (or frequency), the waves are said to exhibit **dispersion**—waves of different frequencies spread apart from one another. Although waves of light are nondispersive in a vacuum, when they enter water they become dispersive. Thus in water the different wavelengths of light (seen as colors) separate. This is the basis for rainbow formation, as we shall see in Chapter 22. Water waves are dispersive but only in very deep water (Fig. 13.17b). Under such conditions, longer wavelengths travel faster than shorter ones. Thus if waves are created in a storm region in the deep ocean, the long-wavelength, low-frequency waves reach the shore first. Dispersion will be most important for us in our study of light, but you should remember that waves other than light can also be dispersive under the right conditions.

Diffraction refers to the bending of waves around an edge of an object but is unrelated to refraction. For example, if you stand along an outside wall of a building near the corner, you can hear people talking around the corner. Assuming there are no reflections or air motion (wind), this would not be possible if the sound waves traveled in a straight line. As the sound waves pass the corner, instead of being sharply cut off, they "wrap around" the edge; thus you can hear the sound.

In general, the effects of diffraction are evident only when the size of the diffracting object or opening is about the same as or smaller than the wavelength of the waves. The dependence of diffraction on wavelength and size of the object or opening is illustrated in •Fig. 13.18. For many waves, diffraction is negligible under normal circumstances. For instance, visible light has wavelengths on the order of 10^{-6} m. Such wavelengths are much too small to exhibit diffraction when passing through common-sized openings, such as an eyeglass lens.

Reflection, refraction, dispersion, and diffraction will be considered in more detail for light waves in Chapter 22.

13.5 Standing Waves and Resonance

Objectives: To (a) describe the formation and characteristics of standing waves and (b) explain the phenomenon of resonance.

If you shake one end of a stretched rope, waves travel down it to the fixed end and are reflected back. The waves going down and back interfere. In most cases, the combined waveforms have a changing, jumbled appearance. But if the rope is shaken at just the right frequency, a steady waveform, or series of uniform loops, appears to stand in place along the rope. Appropriately, this phenomenon is called

•FIGURE 13.19 Standing waves (a) Standing waves are formed by interfering waves traveling in opposite directions. (b) Conditions of destructive and constructive interference recur as each wave travels a distance of $\lambda/4$ in a time of $t = T/4$. The velocities of the rope particles are indicated by the arrows. This gives rise to standing waves with stationary nodes and maximum amplitude antinodes.

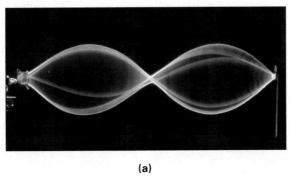

(a)

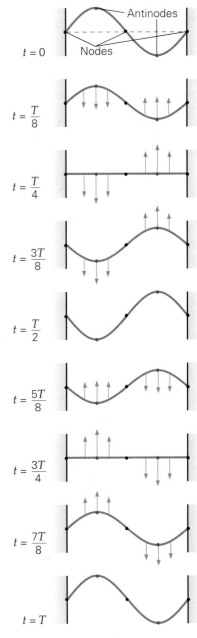

(b)

a **standing wave** (•Fig. 13.19a; see Demonstration 11 on p. 442). It arises because of interference with the reflected waves, which have the same wavelength, amplitude, and speed. Since the two identical waves travel in opposite directions, the net energy flow down the rope is zero. The energy is "standing" in the loops.

Some points on the rope remain stationary at all times and are called **nodes**. At these points, the displacements of the interfering waves are *always* equal and opposite. Thus, by the principle of superposition, the interfering waves must cancel each other completely at these points, and the rope does not undergo displacement there. At all other points, the rope oscillates back and forth at the same frequency. The points of maximum amplitude, where constructive interference is greatest, are called **antinodes**. As you can see in Fig. 13.19b, adjacent antinodes are separated by a half wavelength ($\lambda/2$), or one loop; adjacent nodes are also separated by a half wavelength.

Standing waves can be generated in a rope by more than one driving frequency; the higher the frequency, the more oscillating half-wavelength loops in the rope. The only requirement is that the half wavelengths "fit" the rope length. The frequencies at which large-amplitude standing waves are produced are called **natural frequencies**, or **resonant frequencies**. The resulting standing wave patterns are called *normal*, or *resonant*, *modes of vibration*. In general, all systems that oscillate have one or more natural frequencies, which depend on such factors as mass, elasticity or restoring force, and geometry (boundary conditions). The natural frequencies of a system are sometimes called its *characteristic* frequencies.

A stretched string or rope can be analyzed to determine its natural frequencies. The boundary conditions are that the ends are fixed; thus, there must be a node at each end. The number of closed segments or loops of a standing wave that will fit between the nodes at the ends (along the length of the string) is equal to an integral number of *half*-wavelengths (•Fig. 13.20). Note that $L = \lambda_1/2$, $L = 2(\lambda_2/2)$, $L = 3(\lambda_3/2)$, $L = 4(\lambda_4/2)$, and so on. In general,

$$L = n\left(\frac{\lambda_n}{2}\right) \quad \text{or} \quad \lambda_n = \frac{2L}{n} \quad (\text{for } n = 1, 2, 3, \ldots)$$

The natural frequencies of oscillation, where v is the speed of waves on a string, are

$$f_n = \frac{v}{\lambda_n} = n\left(\frac{v}{2L}\right) = nf_1 \quad \text{for } n = 1, 2, 3, \ldots \qquad \begin{array}{l}\textit{natural frequencies for} \\ \textit{a stretched string}\end{array} \quad (13.19)$$

The lowest natural frequency ($f_1 = v/2L$ for $n = 1$,) is called the **fundamental frequency**. All of the other natural frequencies are integral multiples of the fundamental frequency: $f_n = nf_1$ (for $n = 1, 2, 3, \ldots$). The set of frequencies $f_1, f_2 = 2f_1$, $f_3 = 3f_1, \ldots$, is called a **harmonic series**: f_1 (the fundamental frequency) is the *first harmonic*, f_2 the *second harmonic*, and so on.

Strings fixed at each end are found in stringed musical instruments such as violins, pianos, and guitars. When such a string is excited, the resulting vibration

Demonstration 11 ■ Flame Standing Wave Pattern

Demonstration of a compressional standing wave: a sound wave in a closed tube. The distance between successive high (or low) flames is equal to a half wavelength of the standing sound (compressional) wave in the pipe. What would happen to the waveform if the next lower resonant frequency were tuned in?

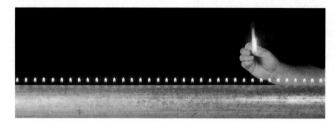 

(a) Regularly spaced holes in a piece of pipe allow a line of uniformly sized flames when flammable gas is supplied to the pipe. One end of the pipe is closed by a metal plate, and the other end is fitted with a rubber diaphragm and loudspeaker capable of sending sound waves down the tube to reflect off the closed end.

(b) When an audio oscillator driving the speaker is tuned to a resonant frequency of the pipe, standing sound (compressional) waves are created. The highest flames are spaced a half wavelength apart. [See *The Physics Teacher, 17*, 307 (1979).] Note: Some fast, contemporary music with good bass instead of an oscillator output produces a rapidly varying, dancing flame pattern.

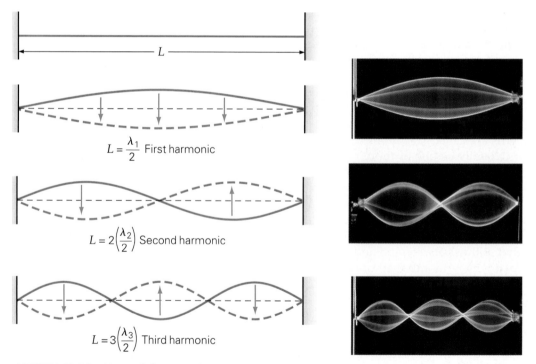

•**FIGURE 13.20 Natural frequencies** A stretched string can have standing waves only at certain frequencies. These correspond to the numbers of half-wavelength loops that will fit along the length of string between the nodes at the fixed ends.

generally includes several harmonics in addition to the fundamental frequency. The number of harmonics depends on how and where the string is excited—that is, plucked, struck, or bowed. It is the combination of harmonic frequencies that gives a particular instrument its characteristic sound quality. (More on this in Chapter 14.) As Eq. 13.19 shows, the fundamental frequency of a stretched string, as well as the other harmonics, depends on the length of the string. Think of how different notes are obtained on a particular string of a violin or a guitar (•Fig. 13.21).

Natural frequencies also depend on other parameters, such as mass and force, which affect the wave speed in the string. For a stretched string, the wave speed (v) can be shown to be

$$v = \sqrt{\frac{F_T}{\mu}} \tag{13.20}$$

where F_T is the tension in the string and μ is the linear mass density (mass per unit length, $\mu = m/L$). (We use F_T, rather than T of previous chapters, so as not to confuse the tension with period T.) Thus Eq. 13.19 can be written as

$$f_n = n\frac{v}{2L} = \frac{n}{2L}\sqrt{\frac{F_T}{\mu}} = nf_1 \quad \text{(for } n = 1, 2, 3, \ldots) \tag{13.21}$$

Note that the greater the linear mass density of a string, the lower its natural frequencies. As you may know, the low-note strings on a violin or guitar are thicker, or more massive, than the high-note strings. By tightening a string, we increase all frequencies of that string. Changing string tension is how violinists, for example, tune their instruments before a performance.

•**FIGURE 13.21 Fundamental frequencies** Performers on string instruments such as the violin or guitar use their fingers to stop or fret the strings. By pressing a string against the fingerboard, the player reduces the amount of its length that is free to vibrate. This reduction changes the resonant frequency of the string and thus the pitch of the tone it produces.

Example 13.6 ■ A Piano String: Fundamental Frequency and Harmonics

A piano string with a length of 1.15 m and a mass of 20.0 g is under a tension of 6.30×10^3 N. (a) What is the fundamental frequency of the string when it is struck? (b) What are the frequencies of the next two harmonics?

Thinking It Through. We have the tension and can calculate the linear mass density from the data. This will give us the fundamental frequency, and from that we can get the harmonics.

Solution.

Given: $L = 1.15$ m   *Find:* (a) f_1 (fundamental frequency)
 $m = 20.0$ g $= 0.0200$ kg   (b) f_2 and f_3 (frequencies of next
 $F_T = 6.30 \times 10^3$ N   two harmonics)

(a) The linear mass density of the string is

$$\mu = \frac{m}{L} = \frac{0.0200 \text{ kg}}{1.15 \text{ m}} = 0.0174 \text{ kg/m}$$

Then, using Eq. 13.21, we get

$$f_1 = \frac{1}{2L}\sqrt{\frac{F_T}{\mu}} = \frac{1}{2(1.15 \text{ m})}\sqrt{\frac{6.30 \times 10^3 \text{ N}}{0.0174 \text{ kg/m}}} = 262 \text{ Hz}$$

This is approximately the frequency of middle C (C_4) on a piano.

(b) Since $f_2 = 2f_1$ and $f_3 = 3f_1$,

$$f_2 = 2f_1 = 2(262 \text{ Hz}) = 524 \text{ Hz}$$

and

$$f_3 = 3f_1 = 3(262 \text{ Hz}) = 786 \text{ Hz}$$

The second harmonic corresponds approximately to C_5 on a piano, since by definition the frequency doubles with each octave (every eighth white key).

Follow-up Exercise. A musical note is referenced to the fundamental vibrational frequency, or first harmonic. In musical terms, the second harmonic is the first overtone, the third harmonic is the second overtone, and so on. If an instrument has a third overtone with a frequency of 880 Hz, what is the frequency of the first overtone?

Note: Don't be confused by the language: *First* overtone means the first frequency above the fundamental frequency—that is, the *second* harmonic.

Tuning Up: Raising the Frequency of a Guitar String

You wish to raise the fundamental frequency of a guitar string from the A note (220 Hz) below middle C to the A note (440 Hz) above middle C. Would you (a) loosen the string to halve its tension, (b) tighten the string to double the tension, (c) use another string of the same material with half the diameter at the same tension, (d) use another string of the same material with twice the diameter at the same tension?

Reasoning and Answer. The fundamental frequency of a stretched string is given by Eq. 13.21:

$$f = \frac{1}{2L}\sqrt{\frac{F_T}{\mu}} \quad \text{(for } n = 1\text{)}$$

The frequency of the string is thus proportional to the *square root* of the tension force F_T, so loosening the string—that is, decreasing F_T—would not increase the frequency. Nor would doubling the tension double the frequency (because $\sqrt{2F_T} \neq 2\sqrt{F_T}$). Thus, neither (a) nor (b) is the correct answer.

The question then is, how does the frequency vary with the diameter of the string? Clearly, if we stay with the same string material (same density ρ), the greater the diameter of the string, the greater its mass per unit length (μ). Hence, a thinner string will vibrate at a higher frequency, and the answer can only be (c).

We must also confirm that halving the diameter doubles the frequency. Remember that doubling the tension did *not* double the frequency! So let's be careful.

Here μ is the mass per unit length ($\mu = m/L$), and the mass can be written in terms of the volume density (ρ) and the geometry of the string, $m = \rho V$, where V is the volume of the string. The volume can be written in terms of its circular cross-sectional area (assumed constant) and the length of the string, which we treat as a long cylinder:

$$V = AL = \frac{\pi d^2 L}{4}$$

where d is the diameter of the string ($A = \pi r^2 = \pi d^2/4$).

Writing the frequency in terms of these parameters, we have

$$f = \frac{1}{2L}\sqrt{\frac{F_T}{\mu}} = \frac{1}{2L}\sqrt{\frac{F_T L}{m}} = \frac{1}{2L}\sqrt{\frac{4F_T}{\rho \pi d^2}} = \left(\frac{1}{L}\sqrt{\frac{F_T}{\rho \pi}}\right)\frac{1}{d}$$

So, with constant F_T, ρ, and L (the active length of the string between the bridge and neck of a stringed instrument), the frequency is inversely proportional to the diameter of the string:

$$f \propto \frac{1}{d}$$

We need to show explicitly that using a string with half the diameter ($d_2 = d_1/2$) would double the frequency. Forming a ratio shows this quickly, since

$$\frac{f_2}{f_1} = \frac{d_1}{d_2} \quad \text{or} \quad f_2 = \left(\frac{d_1}{d_2}\right)f_1$$

For $f_2 = 2f_1$ (that is, 440 Hz $= 2 \times 220$ Hz), we must have $d_1/d_2 = 2$ or $d_2 = d_1/2$, as expected.

Follow-up Exercise. The fundamental frequency of a violin string is A below middle C (220 Hz). How could you tune this string to middle C (264 Hz) *without* changing strings as was done in this Example?

When an oscillating system is driven at one of its natural, or resonant, frequencies, maximum energy transfer to the system occurs. The system is physically suited to any of its natural frequencies. These are the frequencies at which it "wants" to vibrate, so to speak. The condition of driving a system at a natural frequency is referred to as **resonance**.

A common example of a system in mechanical resonance is someone being pushed on a swing. (For some typical numerical values, see Example 13.4.) Basically, a swing is a simple pendulum and has only one resonant frequency for a given length [$f = (1/2\pi)\sqrt{g/L}$]. If you push the swing with this frequency and in phase with its motion, the amplitude and energy increase (•Fig. 13.22). If you push at a slightly different frequency, the energy transfer is no longer a maximum. (What do you think happens if you push with the resonant frequency but 180° out of phase with the swing's motion?)

Unlike a simple pendulum, a stretched string has many natural frequencies. Almost any driving frequency will cause a disturbance in the string. However, if the frequency of the driving force is not equal to one of the natural frequencies, the resulting wave will be relatively small and jumbled. However, when the frequency of the driving force matches one of the natural frequencies, the maximum amount of energy is transferred to the string. A steady standing wave pattern results with the amplitude at the antinodes becoming relatively large.

Mechanical resonance is not the only type of resonance. When you tune a radio, you are changing the resonant frequency of an electrical circuit (Chapter 21) so that it will be driven by, or will pick up, the frequency signal of the station you want. Other examples of resonance are described in the Insight on this page.

•**FIGURE 13.22 Resonance in the playground** The swing behaves like a pendulum in SHM. To transfer energy efficiently, the woman must time her pushes to its natural frequency.

Insight

Desirable and Undesirable Resonances

When we hold a seashell to our ear, we say we hear the ocean, or a sound like that of the ocean. What causes this is a resonance effect—a desirable one because of the pleasant sound. Soft sounds enter the shell from the environment, and some of these are at the resonant frequencies of the shell. Standing sound waves are set up that can be heard when the shell is held close to the ear. (We shall learn more about sound in Ch. 14.)

The mixture of sound frequencies is not harmonic because of the complicated shape of the shell's interior. As a result, the sound has a variety of frequencies—"white noise," so to speak (analogous to "white light," which is a mixture of all visible frequencies). The resulting sound is similar to that of the ocean.

When a large number of soldiers march over a small bridge, they are generally ordered to break step. The reason is that the marching frequency may correspond to one of the natural frequencies of the bridge and set it into resonant vibration, which could cause it to collapse. This actually occurred on a suspension bridge in England in 1831. The bridge was weak and in need of repair, but the resonance vibrations induced by the marching soldiers crossing the bridge helped it fail sooner—and caused some injuries.

Another incidence of bridge vibration wasn't due to marching soldiers but to the driving force of the wind. On the morning of November 7, 1940, winds with speeds of 40 to 45 mi/h started the main span of the Tacoma Narrows Bridge (in Washington state) vibrating. The bridge, 2800 ft (855 m) long and 39 ft (12m) wide, had been first opened to traffic only 4 months earlier.

During the first month of use, small transverse modes of vibration had been observed. But on November 7, special wind effects drove the bridge in near resonance, and the main span vibrated at a frequency of 36 vib/min and an amplitude of 1.5 ft. At 10 A.M., the main span began to vibrate in a torsional (twisting) mode in two segments at a frequency of 14 vib/min. The wind continued to drive the bridge in res-

onance, and the vibrational amplitude increased. Shortly after 11 A.M., the main span collapsed (Fig. 1).*

"Galloping Gertie" (the nickname given to the bridge) was rebuilt on the same tower foundations. However, the new design made the structure stiffer to increase its resonant frequency so that high winds could not produce unwanted resonance.

*It is doubtful that the gusting of the wind set the bridge into vibration. The wind velocity was moderately steady, and gust fluctuations are normally random. One explanation for the driving source of the oscillations involves the formation of vortices as the wind blew past the bridge. Vortices are like the eddies that form in water at the end of oars when a boat is rowed. The wind blowing over and under the bridge formed vortices that rotated in opposite directions. The formation and "shedding" of the vortices (like eddies coming off oars) would have imparted energy to the bridge, and if the frequency of this action were near a natural frequency, a standing wave would have been set up.

FIGURE 1 Galloping Gertie The collapse of the Tacoma Narrows Bridge on November 7, 1940, is captured in this frame from a movie camera.

Chapter Review

Important Terms

Important Concepts

- Simple harmonic motion (SHM) requires a restoring force directly proportional to the displacement, such as the spring force.

- In general, the total energy of an object in SHM is directly proportional to the square of the amplitude.

- The form of an equation of motion for an object in SHM depends on the object's initial displacement and velocity, or initial conditions expressed in terms of a phase constant.

- A wave is a disturbance in time and space; energy is transferred or propagated by wave motion.

- At any time, the combined waveform of two or more interfering waves is given by the sum of the displacements of the individual waves at each point in the medium.

- At natural frequencies, standing waves can result from the interference of two waves of identical wavelength, amplitude, and speed traveling in opposite directions on a string.

Important Equations

Hooke's Law:

$$F_s = -kx \qquad (13.1)$$

Frequency and Period for SHM:

$$f = \frac{1}{T} \qquad (13.2)$$

Total Energy of a Spring and Mass in SHM:

$$E = \tfrac{1}{2}kA^2 = \tfrac{1}{2}mv^2 + \tfrac{1}{2}kx^2 \qquad (13.4\text{–}5)$$

Velocity of Oscillating Mass on a Spring:

$$v = \pm\sqrt{\frac{k}{m}(A^2 - x^2)} \qquad (13.6)$$

Period of Mass Oscillating on a Spring:

$$T = 2\pi\sqrt{\frac{m}{k}} \qquad (13.11)$$

Angular Frequency of a Mass Oscillating on a Spring:

$$\omega = 2\pi f = \sqrt{\frac{k}{m}} \qquad (13.13)$$

Period of a Simple Pendulum (small-angle approximation):

$$T = 2\pi\sqrt{\frac{L}{g}} \qquad (13.14)$$

Displacement of a Mass in SHM:

$$y = A\sin(\omega t + \delta) \qquad (13.15)$$

$$(\text{with } \delta = 0) \quad y = A\sin\omega t = A\sin 2\pi f t = A\sin\frac{2\pi t}{T}$$

Velocity of a Mass in SHM (δ = 0):

$$v = \omega A\cos\omega t \qquad (13.16)$$

Acceleration of a Mass in SHM (δ = 0):

$$a = -\omega^2 A\sin\omega t = -\omega^2 y \qquad (13.17)$$

Wave Speed:

$$v = \frac{\lambda}{T} = \lambda f \qquad (13.18)$$

Natural Frequencies in a Stretched String:

$$f_n = n\left(\frac{v}{2L}\right) = \frac{n}{2L}\sqrt{\frac{F_T}{\mu}} \quad (\text{for } n = 1, 2, 3, \dots) \quad (13.21)$$

Exercises

13.1 Simple Harmonic Motion

1. A particle in SHM (a) has variable amplitude, (b) has a restoring force in the form of Hooke's law, (c) has a frequency directly proportional to its period, (d) has its position represented graphically by $x(t) = at + b$.

2. ● The maximum kinetic energy of a spring–mass system in SHM is equal to (a) A, (b) A^2, (c) kA, (d) $kA^2/2$.

3. ● If the amplitude of an object in SHM is doubled, how are (a) the energy and (b) the maximum speed affected?

4. How does the speed of an object in SHM change as it approaches its equilibrium position? Explain.

5. If the period of a system in SHM is doubled, its frequency is (a) doubled, (b) halved, (c) four times as large, (d) one-quarter as large.

6. ● When a particle in SHM is at the equilibrium position, the potential energy of the system is (a) zero, (b) maximum, (c) negative, (d) none of the above.

7. ■ A particle oscillates in SHM with an amplitude A. What is the total *distance* it travels in one period?

8. ■ A 1.0-kilogram object oscillating on a spring completes a cycle every 0.50 s. What is the frequency of this oscillation?

9. ■ A particle in simple harmonic motion has a frequency of 40 Hz. What is the period of this oscillation?

10. ■ The frequency of a simple harmonic oscillator is doubled from 0.25 Hz to 0.50 Hz. What is the change in its period?

11. ■ What is the spring constant of a spring that stretches 6.0 cm when a 0.25-kilogram mass is suspended from it?

12. ■ ● An object of mass 0.50 kg is attached to a spring, with spring constant 10 N/m. If the object is stretched 0.050 m from the equilibrium position and released, what is its maximum speed?

13. ■■ Atoms in a solid are in continuous vibrational motion due to thermal energy. At room temperature, the amplitude of these atomic vibrations is typically about 10^{-9} cm, and their frequency is on the order of 10^{12} Hz. (a) What is the approximate period of oscillation of a typical atom? (b) What is its maximum speed?

14. ■■ A 0.500-kilogram object resting on a horizontal frictionless surface is attached to a spring with a spring constant of 150 N/m. If the object is pulled 0.150 m from its equilibrium position and released, what are the magnitude of force on the object and its acceleration at (a) $t = 0$, (b) $x = 0.050$ m, and (c) $x = 0$?

15. ■■ (a) At what position is the magnitude of the force on an object in a spring–object system maximum: $x = 0$, $x = -A$, or $x = +A$? (b) How about its speed? (c) the magnitude of its acceleration?

16. ■■ In a lab experiment, you are given a spring with a spring constant of 12 N/m. What mass would you suspend on the spring to have an oscillation period of 0.91 s when in SHM?

17. ■■ If the oscillating spring in Exercise 16 had an amplitude of 15 cm, (a) what would be its maximum speed, and (b) where would this occur? (c) What would be the speed at a half-amplitude position?

18. ■■ ● If the mass of an object attached to a spring is doubled, what is the ratio of the new period to the old one?

19. ■■ ● If the spring constant in a spring–mass system is doubled, what is the ratio of the new period to the old one?

20. ■■ A 0.25-kilogram mass on a vertical spring oscillates at a frequency of 1.0 Hz. If a 0.50-kilogram mass were instead suspended from the same spring, what would be the frequency of oscillation?

21. ■■ The oscillations of two oscillating spring–object systems are graphed in •Fig. 13.23. If the mass of the object in (a) is 4 times that in (b), which system has more energy, and how many times more?

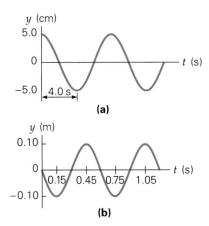

•FIGURE 13.23 **Wave energy and equation of motion**
See Exercises 21, 47, and 48.

22. ■■ ● A 0.25-kilogram object suspended on a light spring is released from a position 15 cm above the stretched equilibrium position. The spring has a spring constant of 80 N/m. (a) What is the total (spring) energy of the system? (Neglect gravitational potential energy.) (b) Does this energy depend on the mass of the object? Explain.

23. ■■ What is the speed of the object in Exercise 22 when the object is (a) 5.0 cm above its equilibrium position and (b) 5.0 cm below its equilibrium position? (c) What is the object's maximum speed, and where does this occur?

24. ■■■ A 75-kilogram circus performer jumps from a height 5.0 m onto a trampoline and stretches it 0.30 m. Assume that the trampoline obeys Hooke's law. (a) How far will it stretch if the performer jumps from a height 8.0 m? (b) How far will it stretch if the performer stands still on it while taking a bow?

25. ■■■ A 0.250-kilogram ball is dropped from a height of 10.0 cm onto a spring as illustrated in •Fig. 13.24. If the spring has a spring constant of 60.0 N/m, (a) what distance will the spring be compressed? (Neglect energy loss during collision.) (b) On recoiling upward, how high will the ball go?

•FIGURE 13.24 How far down? See Exercise 25.

13.2 Equations of Motion

26. The equation of motion for a particle in SHM (a) is always a cosine function, (b) reflects damping action, (c) is independent of the initial conditions, (d) gives the position of the particle as a function of time.

27. ● If the length of a pendulum is doubled, what is the ratio of the new period to the old one?

28. The apparatus in Fig. 13.6 demonstrates that the motion of an object on a spring can be described by a sinusoidal function of time. How could this be demonstrated for a pendulum?

29. Could simple harmonic motion be described by a tangent function? Explain.

30. ● Would the period of a pendulum in an upward accelerating elevator be increased or decreased from that of the period in a non-accelerating elevator? Explain.

31. ■ What mass on a spring with a spring constant of 10 N/m will oscillate with a period of 2.0 s?

32. ■ A 0.50-kilogram mass oscillates in simple harmonic motion on a spring with a spring constant of 200 N/m. What are (a) the period and (b) the frequency of the oscillation?

33. ■ ● The simple pendulum in a grandfather clock is 1.0 m long. What are (a) the period and (b) the frequency of this pendulum?

34. ■ A breeze sets into oscillation a lamp suspended from the ceiling. If the period is 1.0 s, what is the distance from the ceiling to the lamp at the lowest point? Assume the lamp acts as a simple pendulum.

35. ■ ● Write the general equation of motion for an object that is on a horizontal frictionless surface and is connected to a spring at equilibrium (a) if the object is initially given a quick push outward and (b) if the object is pulled away from the spring and released.

36. ■ ● The equation of motion for a vertical SHM oscillator is given by $y = (0.10\ m)\sin 100t$. What are the (a) amplitude, (b) frequency, and (c) period of this motion?

37. ■ ● The displacement of an object in SHM is given by $y = (5.0\ cm)\sin 20\pi t$. What are the (a) amplitude, (b) frequency, and (c) period of oscillation?

38. ■ ● If a SHM oscillator's displacement obeys the equation $y = (0.25\ m)\cos 314t$, where y is in meters and t in seconds, what is the position of the oscillator at (a) $t = 0$, (b) $t = 5.0\ s$, and (c) $t = 15\ s$?

39. ■■ Show that the total energy of a spring–object system in simple harmonic motion is given by $\frac{1}{2}m\omega^2 A^2$.

40. ■■ Show that for a pendulum to oscillate at the same frequency as an object on a spring, the pendulum's length must be $L = mg/k$.

41. ■■ ● Students use a simple pendulum with a length of 36.90 cm to measure the acceleration of gravity at the location of their school. If the period of the pendulum is 1.220 s, what is the experimental value of g at the school?

42. ■■ ● The equation of motion of a particle in vertical simple harmonic motion is given by $y = (10\ cm)\sin 0.50t$. What are the particle's (a) displacement, (b) velocity, and (c) acceleration at $t = 1.0\ s$?

43. ■■ A 0.15-kilogram block oscillates vertically in simple harmonic motion on a spring with a spring constant 6.0 N/m. The block was initially raised 8.0 cm from equilibrium. (a) Write the equation of motion. (b) Determine the phase constant. (c) What is the displacement of the block at $t = 0.50\ s$?

44. ■■ Two objects of equal mass oscillate on light springs, the second with a spring constant twice that of the first. Which system will have the greater period, and how many times greater?

45. ■■ 🌑 If a grandfather clock were taken to the Moon, where the acceleration due to gravity is only one-sixth (assume exact) of that on the Earth, by what factor will the period of oscillation change? Does it increase or decrease? (See Exercise 33.)

46. ■■ A 0.075-kilogram mass oscillates on a light spring, and another 0.075-kilogram mass oscillates as the bob of a simple pendulum in SHM. If the periods of the oscillations are equal and the pendulum length is 0.30 m, what would be the spring constant of the spring?

47. ■■ The motion of a particle is described by the curve in Fig. 13.23a. Write the equation of motion in terms of a cosine function.

48. ■■ The motion of a 0.35-kilogram mass oscillating on a light spring is described by the curve in Fig. 13.23b. (a) Write the equation for its displacement as a function of time. (b) What is the spring constant of the spring?

49. ■■■ The forces acting on a simple pendulum are shown in •Fig. 13.25. (a) Show that for the small-angle approximation ($\sin \theta \approx \theta$), the force producing the motion has the same form as Hooke's law. (b) Show by analogy with an object on a spring that the period of a simple pendulum is given by $T = 2\pi\sqrt{L/g}$. [Hint: Think of the effective spring constant.]

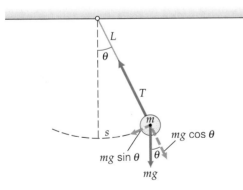

•FIGURE 13.25 **SHM of a pendulum**   See Exercise 49.

50. ■■■ A grandfather clock uses a pendulum that is 75 cm long. It is accidentally broken, and when repaired the length is shorter by 2.0 mm. Considering the clock's pendulum to be a simple pendulum, (a) will the repaired clock gain or lose time? (b) By how much will the repaired clock differ from the correct time (taken to be the time determined by the original pendulum in 24 h)? (c) If the pendulum string were metal, would temperature make a difference in the timekeeping of the clock? Explain.

13.3 Wave Motion

51. Wave motion in a material involves (a) the propagation of a disturbance, (b) interparticle interactions, (c) the transfer of energy, (d) all of the preceding.

52. For a periodic wave propagating at a speed v, the following relationship holds: (a) $\lambda = v/f$, (b) $v = \lambda/f$, (c) $v = \lambda f^2$, (d) $f = \lambda/v$.

53. What type(s) of wave(s) will propagate through (a) solids, (b) liquids, and (c) gases?

54. •Figure 13.26 shows snapshots of two waves. Identify whether each is transverse or longitudinal.

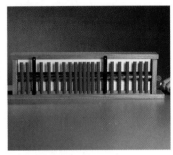

•FIGURE 13.26 **Transverse or longitudinal?** See Exercise 54.

55. ■ A longitudinal sound wave has a speed of 340 m/s in air. If this wave produces a tone with a frequency of 1000 Hz, what is the wavelength?

56. ■ A transverse wave has a wavelength of 0.50 m and a frequency of 20 Hz. What is the wave speed?

57. ■ A student reading his physics book on a lake dock notices that the distance between two incoming wave crests is about 2.4 m, and he then measures the time of arrival between wave crests to be 1.6 s. What is the approximate speed of the waves?

58. ■ Light waves travel in a vacuum at a speed of 300 000 km/s. The frequency of visible light is about 5×10^{14} Hz. What is the approximate wavelength of the light?

59. ■ A certain laser emits light of wavelength 6.33×10^{-7} m. What is the frequency of this light in a vacuum?

60. ■■ The range of sound frequencies audible to the human ear extends from about 20 Hz to 20 kHz. If the speed of sound in air is 345 m/s, what are the limits of this audible range expressed in wavelengths?

61. ■■ The AM frequencies on a radio dial range from 550 kHz to 1600 kHz, and the FM frequencies range from 88.0 MHz to 108 MHz. All of these radio waves travel at a speed of 3.00×10^8 m/s (speed of light). What are the wavelength ranges of (a) the AM band and (b) the FM band in terms of wavelengths?

62. ■■ A sonar generator on a submarine on the ocean surface produces periodic ultrasonic waves at a frequency of 2.50 MHz. The wavelength of the waves in sea water is 4.80×10^{-4} m. When the generator is directed downward, an echo reflected from the ocean floor is received 10.0 s later. How deep is the ocean at that point? (Assume wavelength is constant at all depths.)

63. ■■ In watching a transverse wave go by, a woman notes that 13 crests go by in a time of 3.0 s. If she measures the distance between two successive crests to be 0.75 m and if the first and the last points that pass her are crests, what is the speed of the wave?

64. ■■ A wave traveling in the $+x$ direction at a certain time is shown in •Fig. 13.27a. The particle displacement at a particular location in the medium through which the wave travels is shown in Fig. 13.27b. (a) What is the amplitude of the traveling wave? (b) What is the wave speed?

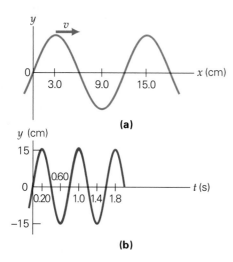

•**FIGURE 13.27   How high and how fast?**   See Exercise 64.

65. ■■ Assume that P and S (primary and secondary) waves from an earthquake with a focus near the Earth's surface travel through the Earth at nearly constant average speeds of 8.0 km/s and 6.0 km/s, respectively. Assume that there is no deflection or refraction of the waves. (a) How long is the delay between the arrivals of successive waves at a seismic monitoring station located 90° in latitude from the epicenter (the spot on the surface directly above the focus) of the quake? (b) Do the waves cross the boundary of the mantle? (c) How long does it take for the waves to arrive at a monitoring station on the opposite side of the Earth?

66. ■■■ The speed of longitudinal waves traveling in a long solid rod is given by $v = \sqrt{Y/\rho}$, where Y is Young's modulus and ρ is the density of the solid. If a disturbance has a frequency of 40 Hz, what is the wavelength of the waves it produces in (a) an aluminum rod and (b) a copper rod? [Hint: See Tables 9.1 and 9.2.]

67. ■■■ As noted in Exercise 66, the speed of longitudinal waves in a solid rod is given by $v = \sqrt{Y/\rho}$. Fred strikes a steel train rail with a hammer at a frequency of 0.50 Hz, and Wilma puts her ear to the rail 1.0 km away. (a) How long after the first strike does Wilma hear the sound? (b) What is the time interval between successive sound pulses she hears? [Hint: See Tables 9.1 and 9.2]

13.4 Wave Properties

68. When waves meet each other and interfere, the resultant waveform is determined by (a) reflection, (b) refraction, (c) diffraction, (d) superposition.

69. Refraction (a) involves constructive interference, (b) refers to a change in direction at media interfaces, (c) is synonymous with diffraction, (d) occurs only for mechanical waves, not for light.

70. What is destroyed when destructive interference occurs?

71. Dolphins determine the location of prey by emitting ultrasonic sound (this is called *echolocation*). Which wave phenomenon is involved?

72. If sound waves were dispersive (that is, the speed of sound depended on its frequency), what would be the consequences if you were in a concert hall?

13.5 Standing Waves and Resonance

73. For two traveling waves to form standing waves, the waves must have the same (a) frequency, (b) amplitude, (c) speed, (d) all of the preceding.

74. When a stretched violin string oscillates in its third harmonic mode, then the standing wave in the string will exhibit (a) 3 wavelengths, (b) 1/3 wavelength, (c) 3/2 wavelengths, (d) 2 wavelengths.

75. A child's swing (a pendulum) has only one natural frequency f_o, yet it can be kept going by pushes at a frequency of $f_o/2$. How is this possible?

76. A guitar string fixed at both ends is vibrating in its fourth harmonic. How many nodes are there on the string?

77. ■ The fundamental frequency of a stretched string is 100 Hz. What are the frequencies of (a) the second harmonic and (b) the third harmonic?

78. ■ If the frequency of the third harmonic of a vibrating string is 450 Hz, what is the frequency of the first harmonic?

79. ■ 💿 A standing wave is formed in a stretched string that is 3.0 m long. What are the wavelengths of (a) the first harmonic and (b) the third harmonic?

80. ■■ 💿 Will a standing wave be formed in a 4.0-meter length of stretched string that transmits waves at a speed of 12 m/s if it is driven at a frequency of (a) 15 Hz or (b) 20 Hz?

81. ■■ Two waves of equal amplitude and wavelength of 0.80 m travel with a speed of 250 m/s in opposite directions in a string. If the string is 2.0 m long, which harmonic mode is the standing wave set up in the string?

82. ■■ A piece of rubber tubing with a linear mass density of 0.125 kg/m is stretched by a force of 9.00 N. (a) What will be the transverse wave speed in the tubing? (b) If the stretched tubing has a length of 10.0 m, what are its natural frequencies?

83. ■■ Find the first four harmonics for a string that is 2.0 m long, has a linear mass density of 2.5×10^{-2} kg/m, and is under a tension of 40 N.

84. ■■ Two stretched strings have the same tension and linear mass density. Are any of the first six harmonics of the strings equal if the string lengths are (a) 1.0 m and 3.0 m or (b) 1.5 m and 2.0 m, respectively?

85. ■■ A violin string is tuned to 440 Hz (fundamental frequency, or first harmonic). When playing the instrument, the violinist puts a finger down on the string one-eighth of the string length from the neck end. What is the frequency of the string when played like this?

86. ■■■ A thin, flexible metal rod is 1.0 m long. It is clamped at one end to a table, and the other end can vibrate freely. What are the natural frequencies of the rod if the wave speed in the material is 3.5×10^3 m/s?

87. ■■■ In a common laboratory experiment on standing waves, the waves are produced in a stretched string by an electrical vibrator that oscillates at 60 Hz (•Fig. 13.28). The string runs over a pulley, and a hanger is suspended from the end. The tension in the string is varied by adding weights to the hanger. If the active length of string (the part that vibrates) is 1.5 m and this length of string has a mass of 0.10 g, what masses must be suspended to produce the first four harmonics in that length?

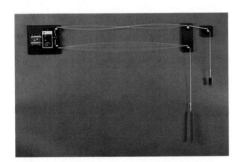

•FIGURE 13.28 **Standing waves in strings** Twin vibrating strings with standing waves. This demonstration model allows you to vary string tension, string length, and string type (linear mass density). Also, the vibration frequency can be adjusted. See Exercise 87.

Additional Exercises

88. A 0.10-kilogram block suspended on a spring is pulled to 8.0 cm below its equilibrium position and released. When the block passes through the equilibrium position, it has a speed of 0.40 m/s. What is the speed of the block when it is 4.0 cm from the equilibrium position?

89. 💿 A standing wave has nodes at $x = 0$ cm, $x = 6.0$ cm, $x = 12$ cm, and $x = 18$ cm. (a) What is the wavelength of the waves that are interfering to produce this standing wave? (b) At what positions are the antinodes?

90. A simple pendulum makes 5.0 complete cycles in 10 s. (a) What are the frequency and period of this pendulum? (b) What is its length?

91. 💿 The motion of an object is $x = (0.10 \text{ m}) \cos 50t$. What are the magnitudes of the object's maximum velocity and maximum acceleration?

92. A steel piano wire is 60 cm long and has a mass of 3.0 g. If the tension in the wire is 550 N, what are the fundamental frequency and wavelength?

93. 💿 A stretched string exhibits four equal loops in a standing wave when driven at a frequency of 420 Hz. What driving frequency will set up a wave with two equal loops?

94. On a violin, a correctly tuned A string has a frequency of 440 Hz. If an A string produces sound at 450 Hz when under a tension of 500 N, what should the tension be to produce the correct frequency?

95. If the tension in a string fixed at both ends is doubled, what will happen to the (a) wavelength and (b) frequency of the first harmonic?

96. What is the length of a string whose third harmonic standing wave has a frequency of 27 Hz and along which waves travel at 45 m/s?

97. 💿 An object in vertical simple harmonic motion is described by $y = (0.20 \text{ cm}) \sin 1.8\pi t$. What is the speed of the object at $t = 10$ s?

98. During an earthquake, the corner of a tall building oscillates with an amplitude of 20 cm at 0.50 Hz. What are the magnitudes of (a) the maximum displacement, (b) the maximum velocity, and (c) the maximum acceleration of the corner of that building?

99. An object resting on a horizontal frictionless surface is connected to a fixed spring. The object is displaced 16 cm from its equilibrium position and released. At $t = 0.50$ s, it is 8.0 cm from its equilibrium position (and has not passed through it yet). What is the object's period of oscillation?

Sound

Insights

- Speech and Hearing
- Doppler Radar

Good vibrations, clearly! We owe a lot to sound waves. Not only do they provide us with one of our main sources of enjoyment in the form of music, they also bring us a wealth of vital information about our environment, from the chime of a doorbell to the warning shrill of a police siren to the song of a mockingbird. Indeed, they are the basis for our major form of communication: speech. They can also constitute a highly irritating distraction (noise). But sound waves become music, speech, or noise only when our ears perceive them. Physically, sound is simply waves that propagate in solids, liquids, and gases. Without a medium, there can be no sound; in a vacuum such as outer space, there is utter silence.

This distinction between the sensory and physical meanings of sound gives you a way to answer the old philosophical question, If a tree falls in the forest where there is no one to hear it, does it make a sound? The answer depends on how sound is defined—it is no if we are thinking in terms of sensory hearing, but yes if we are considering only the physical waves. Since sound waves are all around us most of the time, we are exposed to many interesting sound phenomena. You'll explore some of the most important of these in this chapter.

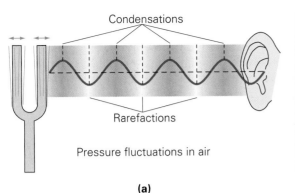

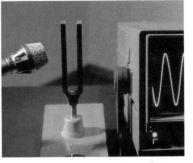

(a)

(b)

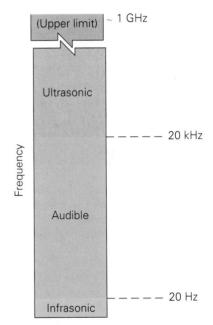

•FIGURE 14.1 **Vibrations make waves** **(a)** A vibrating tuning fork disturbs the air, producing alternating high-pressure regions (condensations) and low-pressure regions (rarefactions), which form sound waves. **(b)** After being picked up by a microphone, the pressure variations are converted to electrical signals. When these signals are displayed on an oscilloscope, the sinusoidal waveform is evident.

14.1 Sound Waves

Objectives: **To (a) define sound and (b) explain the sound frequency spectrum.**

For sound waves to exist, there must be a disturbance or vibrations in some medium. This disturbance may be the clapping of hands or the skidding of tires as a car comes to a sudden stop. Underwater you can hear the click of rocks against one another. If you put your ear to a thin wall, you can hear sounds from the other side through it. **Sound waves** in gases and liquids (fluids) are primarily longitudinal waves. However, sound disturbances moving through solids can have both longitudinal and transverse components. The intermolecular interactions in solids are much stronger than in fluids and allow transverse components to propagate.

The characteristics of sound waves can be visualized by considering those produced by a tuning fork (•Fig. 14.1). A tuning fork is essentially a metal bar bent into a U shape. The prongs, or tines, vibrate when struck. The fork vibrates at its fundamental frequency (with an antinode at the end of each tine), so a single tone is heard. (A *tone* is sound with a definite frequency.) The vibrations disturb the air, producing alternating high-pressure regions called *condensations* and low-pressure regions called *rarefactions*. As the fork vibrates, these disturbances propagate outward, and a series of them can be described by a sinusoidal wave.

As the disturbances traveling through the air reach the ear, the eardrum (a thin membrane) is set into vibration by the pressure variations. On the other side of the eardrum, tiny bones (the hammer, anvil, and stirrup) carry the vibrations to the inner ear, where they are picked up by the auditory nerve (see the Insight on p. 458).

Characteristics of the ear limit the perception of sound. Only sound waves with frequencies between about 20 Hz and 20 kHz (kilohertz) initiate nerve impulses that are interpreted by the human brain as sound. This frequency range is called the **audible region** of the **sound frequency spectrum** (•Fig 14.2).

Frequencies lower than 20 Hz are in the **infrasonic region**. Waves in this region, which we can't hear, are found in nature. Longitudinal waves generated by earthquakes have infrasonic frequencies, and we use these waves to study the Earth's interior (Chapter 13). Infrasonic waves, or *infrasound*, are also generated by wind and weather patterns. Elephants and cattle have hearing response in the infrasonic region and may even give early warnings of earthquakes and weather disturbances. Aircraft, automobiles, and other rapidly moving objects can also produce infrasound.

Above 20 kHz is the **ultrasonic region**. Ultrasonic waves can be generated by high-frequency vibrations in crystals. Ultrasonic waves, or *ultrasound*, cannot be detected by humans but can be by other animals. The audible region for dogs extends to about 45 kHz, so ultrasonic whistles can be used to call dogs without

•FIGURE 14.2 **Sound frequency spectrum** The audible region of sound for humans lies between about 20Hz and 20kHz. Below this is the infrasonic region, and above it is the ultrasonic region. The upper limit is about 1 GHz because of the elastic limitations of materials.

disturbing people. Cats and bats have even higher ranges, up to about 70 kHz and 100 kHz, respectively.

There are many other practical applications of ultrasound. Since ultrasound can travel for kilometers in water, it is used in sonar. Sonar is the ultrasound counterpart of radar, which uses radio waves for ranging and detection. Sound pulses generated by the sonar apparatus are reflected by underwater objects, and the resulting echoes are picked up by a detector. The time required for a sound pulse to make one round trip, together with the speed of sound in water, gives the distance of the reflecting object. In a similar manner, ultrasound is used in autofocus cameras. Distance measurement allows focus adjustments to be made. Sonar can be used not only for ranging but also to create images of the seafloor and objects resting on it, such as shipwrecks (•Fig. 14.3a).

Sonar appeared in the animal kingdom long before it was developed by human engineers. On their nocturnal hunting flights, bats use a kind of natural sonar to navigate in and out of their caves and to locate and catch flying insects. The bats emit pulses of ultrasound and track their prey by means of the reflected echoes (Fig. 14.3b). Certain species of cave-dwelling birds have evolved the same ability.

In medicine, ultrasound is used to examine internal tissues and organs that are nearly invisible to X-rays. Perhaps the best known medical application of ultrasound is its use to view a fetus without exposing it to the dangerous effects of X-rays. Ultrasonic generators (transducers) made of quartz crystals produce high-frequency waves that are used to scan a designated region of the body from several angles. Reflections from the scanned areas are monitored, and a computer constructs an image from the reflected signals. Images are recorded sev-

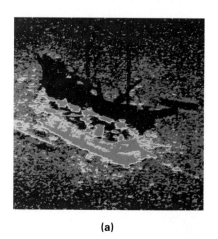

(a)

(b)

•**FIGURE 14.3 Some uses of ultrasound** **(a)** This computer-enhanced false-color image, constructed from sonar signals, shows the wreck of the American sailing ship *Hamilton*. The schooner, sunk in Lake Ontario during the War of 1812, lies under 91 m (300 ft) of water. **(b)** With the aid of their own natural sonar systems, bats hunt flying insects. They emit pulses of ultrasonic waves, which lie within their audible region, and use the echoes reflected from their prey to guide their attack. (Notice the size of the animal's ears.) **(c)** Ultrasound generated by crystal transducers, which convert electrical oscillations into mechanical vibrations, is transmitted through tissue and is reflected from internal structures. The reflected waves are detected by the transducers, and the signals are used to construct an image, or echogram, as shown here for a well-developed fetus.

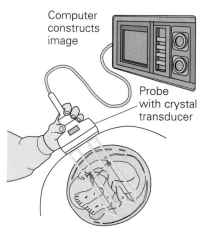

Computer constructs image

Probe with crystal transducer

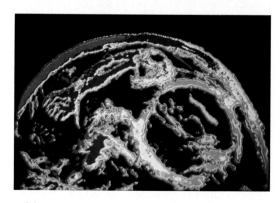

(c)

eral times each second. The series of images provides a "moving picture" of an internal structure, such as the heart of a fetus. A still shot, or echogram, is shown in Fig. 14.3c.

In industrial and home applications, ultrasonic baths are used to clean metal machine parts, dentures, and jewelry. The high-frequency (short-wavelength) ultrasound vibrations loosen particles in otherwise inaccessible places.

Ultrasonic frequencies extend into the megahertz (MHz) range, but the sound frequency spectrum does not continue indefinitely. There is an upper limit of about 10^9 Hz, or 1 GHz (gigahertz), which is determined by the upper limit of the elasticity of the materials through which the sound propagates.

14.2 The Speed of Sound

Objectives: To (a) tell how the speed of sound differs in different media and
(b) describe the temperature dependence of the speed of sound in air.

In general, the speed at which a disturbance moves through a medium depends on the elasticity and density of the medium. For example, as you learned in Chapter 13, the wave speed in a stretched string is given by $v = \sqrt{F_T/\mu}$, where F_T is the tension in the string and μ is the linear mass density.

Similar expressions describe wave speeds in solids and liquids, for which the elasticity is expressed in terms of moduli (Chapter 9). In general, the speeds of sound in solids and liquids are given by $v = \sqrt{Y/\rho}$ and $v = \sqrt{B/\rho}$, respectively, where Y is Young's modulus, B is the bulk modulus, and ρ is the density. The speed of sound in a gas is inversely proportional to the square root of the molecular mass, but the equation is more complicated and will not be presented here.

Note: Review Section 9.1.

Solids are generally more elastic than liquids, which in turn are more elastic than gases. In a highly elastic material, the restoring forces between the atoms or molecules cause a disturbance to propagate faster. Thus, the speed of sound is generally about 2 to 4 times faster in solids than in liquids and about 10 to 15 times faster in solids than in gases such as air (Table 14.1).

Although not expressed explicitly in the preceding equations, the speed of sound also depends on the temperature of the medium. In dry air, for example, the speed of sound is 331 m/s (about 740 mi/h) at 0°C. As the temperature increases, so do the speeds of the gas molecules. As a result, the molecules collide with each other more frequently, and a disturbance is transmitted more quickly. Thus the speed of sound in air increases with increasing temperature. For *normal environmental temperatures*, the speed of sound in air increases by about 0.6 m/s for each degree Celsius above 0°C. Thus, a good approximation of the speed of sound in air for a particular (environmental) temperature is given by

$$v = (331 + 0.6T_C)\ \text{m/s} \qquad \begin{array}{l}\textit{speed of sound}\\ \textit{in air}\end{array} \qquad (14.1)$$

where T_C is the air temperature in degrees Celsius.* Although not written explicitly, the units associated with the factor 0.6 are meters per second per Celsius degree $[\text{m}/(\text{s}\cdot\text{C}°)]$.

Let's take a comparative look at the speeds of sound in different media.

TABLE 14.1 The Speed of Sound in Various Media (typical values)

Medium	Speed (m/s)
Solids	
Aluminum	5100
Copper	3500
Iron	4500
Glass	5200
Polystyrene	1850
Zinc	3200
Liquids	
Alcohol, ethyl	1125
Mercury	1400
Water	1500
Gases	
Air (0°C)	331
Air (100°C)	387
Helium (0°C)	965
Hydrogen (0°C)	1284
Oxygen (0°C)	316

*A better approximation of these and higher temperatures is given by the expression

$$v = \left(331\ \sqrt{1 + \frac{T_C}{273}}\right)\ \text{m/s}$$

In Table 14.1, see v for air at 100°C, which is outside the normal environmental temperature range.

Example 14.1 ■ Solid, Liquid, Gas: Speed of Sound in Different Media

From their material properties, find the values of the speeds of sound in (a) a solid copper rod, (b) liquid water, and (c) air at room temperature (20°C).

Thinking It Through. We know that the speeds of sound in solids and liquids depend on elastic moduli and densities. These values are available in Tables 9.1 and 9.2. The speed of sound in air is given by Eq. 14.1.

Solution.

Given: $Y_{Cu} = 11 \times 10^{10}$ N/m^2 *Find:* (a) v_{Cu} (speed in copper)
        $B_{H_2O} = 2.2 \times 10^9$ N/m^2 (b) v_{H_2O} (speed in water)
        $\rho_{Cu} = 8.9 \times 10^3$ kg/m^3 (c) v_{air} (speed in air)
        $\rho_{H_2O} = 1.0 \times 10^3$ kg/m^3
              (all values from Tables 9.1 and 9.2)
        $T_C = 20°C$ (for air)

(a) To find the speed of sound in a copper rod, we use the expression $v = \sqrt{Y/\rho}$:

$$v_{Cu} = \sqrt{Y/\rho} = \sqrt{(11 \times 10^{10} \text{ N/m}^2)/(8.9 \times 10^3 \text{ kg/m}^3)} = 3.5 \times 10^3 \text{ m/s}$$

(b) For water, $v = \sqrt{B/\rho}$:

$$v_{H_2O} = \sqrt{B/\rho} = \sqrt{(2.2 \times 10^9 \text{ N/m}^2)/(1.0 \times 10^3 \text{ kg/m}^3)} = 1.5 \times 10^3 \text{ m/s}$$

(c) For air at 20°C, by Eq. 14.1, we have

$$v_{air} = (331 + 0.6T_C) \text{ m/s} = [331 + 0.6(20)] \text{ m/s} = 343 \text{ m/s}$$

Follow-up Exercise. In this Example, how many times faster is the speed of sound in copper (a) than in water and (b) than in air (at room temperature)?

A generally useful value for the speed of sound in air is $\frac{1}{3}$ km/s (or $\frac{1}{5}$ mi/s). Using this value, you can, for example, estimate how far away lightning has struck by counting the number of seconds between the time you observe the flash and the time you hear the associated thunder. Because the speed of light is so fast, you see the lightning flash almost instantaneously. The sound waves of the thunder travel relatively slowly, at about $\frac{1}{3}$ km/s. For example, if the interval between the two events is measured to be 6 s (often by counting "one thousand one, one thousand two, . . ."), the lightning stroke was about 2 km away ($\frac{1}{3}$ km/s × 6 s).

You may also have noticed the delay of sound relative to light at a baseball game. If you're sitting in the outfield stands, you see the batter hit the ball before you hear the crack of the bat.

Example 14.2 ■ Safe or Out? You Make the Call

On a cool October afternoon (air temperature = 15°C), from your seat in the centerfield stands 113 m from first base, you witness the play that will decide the World Series. You see the runner's foot touch the bag; half a second later, straining your ears, you hear the faint thud of the ball in the first baseman's glove. The umpire signals safe; half the fans boo loudly. As a student of physics, you make the call—did the ump blow it?

Thinking It Through. This Example has to do with the delay of hearing the sound, and hence the speed of sound. You can see the action almost instantaneously yet the sound takes an appreciable time to reach you, because the speed of light is much faster than that of sound. If the sound travel time is less than $\frac{1}{2}$ s, then the runner had already touched the base when the first baseman caught the ball, and the runner is safe. If that time is greater that $\frac{1}{2}$ s, he's out. (Why?)

Solution. Listing the data, we have

Given: $d = 113$ m   *Find:* t (sound travel time)
$T = 15°C$
$t = \frac{1}{2}$ s

With a temperature of 15°C, the speed of sound is, from Eq. 14.1,

$$v = (331 + 0.6 T_C) \text{ m/s} = [331 + 0.6(15°C)] \text{ m/s} = 340 \text{ m/s}$$

For a constant speed, the general distance–time relation is $d = vt$, so the time for the sound to travel the distance to your seat is

$$t = \frac{d}{v} = \frac{113 \text{ m}}{340 \text{ m/s}} = 0.332 \text{ s}$$

or just under one-third of a second. The runner reached the base about $\frac{1}{2}$ s $- \frac{1}{3}$ s $= \frac{1}{6}$ s before the ball arrived. The ump was right!

To show why it is justified to say that the visual observation takes place almost instantaneously, let's see how long it takes for light to travel from first base to your seat. For practical purposes, the speed of light in air is the same as that in a vacuum, $c = 3.00 \times 10^8$ m/s, which is on the order of 10^6 (a million) times greater than the speed of sound. Then, the time for light to travel 113 m is

$$t = \frac{d}{c} = \frac{113 \text{ m}}{3.00 \times 10^8 \text{ m/s}} = 3.77 \times 10^{-7} \text{ s}$$

or 0.377 μs—not a very long time.

Follow-up Exercise. On a day when the air temperature is 20°C, a hiker shouts and hears the echo from the vertical stone cliff face 5.00 s later. (a) How far away is the cliff? (b) If the air temperature were 15°C, what would be the difference in the times of hearing the echoes?

Conceptual Example 14.3

Speed of Sound: Sound Traveling Far and Wide

Note that the speed of *dry* air for a given temperature is given to a good approximation by Eq. 14.1. However, the moisture content of the air (humidity) varies, and this variation affects the speed of sound. At the same temperature, would sound travel faster in (a) dry air or (b) moist air?

Reasoning and Answer. According to an old folklore saying, "Sound traveling far and wide, a stormy day will betide" (happen or predict). This saying implies that sound travels faster on a highly humid day when a storm or precipitation is likely. But is this saying valid?

Near the beginning of this section, we learned that the speed of sound in a gas is inversely proportional to the square root of the gas's molecular mass. So, at constant pressure, is moist air more or less dense than dry air?

In a volume of moist air, a large number of water (H_2O) molecules occupy the space normally occupied by either nitrogen (N_2) or oxygen (O_2) molecules, which make up 98% of the air. Water molecules are less massive than both nitrogen and oxygen molecules. [From Section 10.3, the molecular (formula) masses are H_2O, 18 g; N_2, 28 g; and O_2, 32 g.] Thus, the average molecular mass of moist air is less than that of dry air, and the speed of sound is greater in moist air.

We can look at this situation another way: Since water molecules are less massive, they have less inertia and respond to the sound wave faster than nitrogen or oxygen molecules do. The water molecules therefore propagate the disturbance faster.

Humidity was included here as an interesting consideration for the speed of sound in air. However, henceforth in computing the speed of sound in air at a certain temperature, we will consider only *dry* air (Eq. 14.1) unless otherwise stated.

Follow-up Exercise. Considering only molecular masses, would you expect the speed of sound to be greatest in nitrogen, oxygen, or helium (at the same temperature and pressure)? Explain.

Sound is one of our major means of communicating. For more on our speech and hearing, see the Insight on p. 458.

Speech and Hearing

Sound is one of our most important means of interpersonal communication, and our most important sound source is the human voice. Let's look at the anatomy and physics of the voice and of the ear, which serves as our receiver for sound.

The Human Voice

The energy for sounds associated with the human voice originates in the muscle action of the diaphragm, which forces air up from the lungs. To produce variations in sound, this steady stream of air must be periodically disturbed, or "modulated." The fundamental modulating organ is the larynx (the "voice box"), across which are stretched membrane-like folds called the vocal cords. The opening and closing of the vocal cords modulates the air stream to produce sounds (Fig. 1).

As illustrated in the figure, there are two sets of vocal cords: (1) *false vocal cords*, which do not produce sound but help close the larynx during swallowing, and (2) *true vocal cords*, which are elastic and are responsible for vocal sounds. When air is forced between the true vocal cords, they vibrate from side to side, producing sound. During normal breathing, these vocal cords remain relaxed and open. The opening between them is called the *glottis*.

Changing the tension of the vocal cords controls the pitch (perceived frequency) of the vocal sounds (as we shall see in Section 14.6). The loudness of the sound is related to the pressure of the air passing over the vocal cords. A strong blast of air results in greater amplitude of the vibrations and louder sound.

The sound waves from the larynx are further modulated in the numerous resonance cavities in the throat, mouth, and nose, where standing waves are set up. Some of the vocal cavities can be altered by means of controllable structures, such as the tongue and lips, so as to produce a wide variety

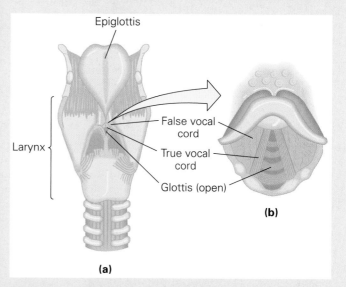

FIGURE 1   Anatomy of the "voice box"   (a) A cross section of the larynx. **(b)** A down-the-throat view of the vocal cords.

of sounds and words. (Sound out the vowel letters *a, e, i, o,* and *u,* and notice the positions of your tongue and lips.)

The waveforms of voice sounds are quite specific for individuals and provide a "voice print" that can be used for identification, just as fingerprints are. The validity of voice prints for legal identification, however, is highly controversial.

Hearing

The anatomy of the human ear is illustrated in Fig. 2. Sound enters the outer ear and travels through the ear (or auditory) canal to the *eardrum* (the tympanum), which separates the

14.3   Sound Intensity and Sound Intensity Level

Objectives: To **(a)** define sound intensity and explain how it varies with distance from a point source and **(b)** calculate sound intensity levels on the decibel scale.

Wave motion involves the propagation of energy. The rate of the energy transfer is expressed in terms of **intensity**, which is the energy transported per unit time across a unit area. Since energy/time is power, intensity is power/area:

$$\text{intensity} = \frac{\text{energy/time}}{\text{area}} = \frac{\text{power}}{\text{area}}$$

The standard units of intensity (power/area) are watts per square meter (W/m^2).

Consider a point source that sends out spherical sound waves, as shown in •Fig. 14.4. If there are no losses, the sound intensity at a distance R from the source is

$$I = \frac{P}{A} = \frac{P}{4\pi R^2} \tag{14.2}$$

where P is the power of the source and $4\pi R^2$ is the area of a sphere of radius R, through which the sound energy passes perpendicularly.

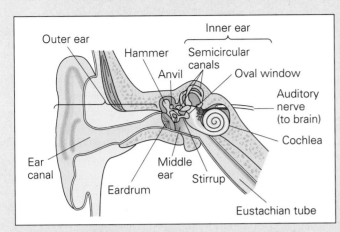

FIGURE 2   Anatomy of the human ear   The ear converts pressure in the air into electrical nerve impulses that are interpreted by the brain as sounds.

outer ear from the middle ear. The eardrum is a membrane that vibrates in response to the impinging sound waves. The vibrations are transmitted through the middle ear, which contains an intricate set of connected bones, commonly called the *hammer* (malleus), *anvil* (incus), and *stirrup* (stapes), because of their shapes.

The bones of the middle ear form a delicate set of levers with a force multiplication factor (mechanical advantage) of about 2. However, the amplification of the pressure in a wave is much greater because the area of the eardrum is about 20 times that of the *oval window*, the membrane-covered opening to the inner ear. (The stirrup is in contact with the membrane.) Thus, pressure is amplified by a factor of about 40.

The inner ear includes the semicircular canals, which are important in controlling balance, and the *cochlea*. It is in the cochlea that sound waves are translated into nerve impulses and that pitch or frequency discrimination is made. The cochlea consists of a series of liquid-filled tubes, coiled into a spiral shape that resembles a snail shell. Supported by a membrane, called the *basilar membrane*, within the tubes are thousands of special receptor cells called *hair cells*. The hair cells are specialized nerve receptors. When a region of the basilar membrane is set into vibration by sound of a particular frequency, hair cells in that region are stimulated and nerve impulses are sent to the brain, where they are interpreted as sound. The brain translates this positional information (impulses from particular fibers originating in specific regions of the basilar membrane) back into frequency information (the subjective physiological sensation of pitch).

The sensation of loudness is determined by the amplitude of vibration of the basilar membrane. The greater the amplitude, the more strongly the hair cells are stimulated and the more impulses they send to the brain in a given period of time. The brain translates this information into degrees of loudness.

Incidentally, the middle ear is connected to the throat by the Eustachian tube, the end of which is normally closed. It opens during swallowing and yawning to permit air to enter and leave, so internal and external pressures are equalized. You have probably experienced a "stopping up" of your ears with a sudden change in atmospheric pressure (for example, during rapid ascents or descents in elevators or airplanes). Swallowing opens the Eustachian tubes and relieves the excess pressure difference on the middle ear.

The intensity for a point source is therefore *inversely proportional to the square of the distance from the source* (an inverse-square relationship). Two intensities at different distances from a source of constant power can be compared as a ratio:

$$\frac{I_2}{I_1} = \frac{P/4\pi R_2^2}{P/4\pi R_1^2} = \frac{R_1^2}{R_2^2}$$

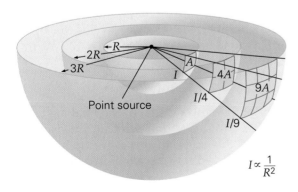

$$I \propto \frac{1}{R^2}$$

•FIGURE 14.4   Intensity of a point source   The energy emitted from a point source spreads out equally in all directions. Since intensity is power/area, $I = P/A = P/4\pi R^2$, where the area is that of a spherical surface. The intensity then decreases with the distance from the source as $1/R^2$.

or

$$\frac{I_2}{I_1} = \left(\frac{R_1}{R_2}\right)^2 \qquad (14.3)$$

Suppose that the distance from a point source is doubled; that is, $R_2 = 2R_1$, or $R_1/R_2 = \frac{1}{2}$. Then

$$\frac{I_2}{I_1} = \left(\frac{R_1}{R_2}\right)^2 = \left(\frac{1}{2}\right)^2 = \frac{1}{4}$$

and

$$I_2 = \frac{I_1}{4}$$

Since the intensity decreases by a factor of $1/R^2$, doubling the distance decreases the intensity to a quarter of its original value.

A good way to understand this inverse-square relationship intuitively is to look at the geometry of the situation. As Fig. 14.4 shows, the greater the distance from the source, the larger the area over which a given amount of sound energy is spread, and thus the lower its intensity. (Imagine having to paint two walls of different areas. If you had the same amount of paint to use on each, you'd have to spread it more thinly over the larger wall.) Since this area increases as the square of the radius R, the intensity decreases accordingly—that is, as $1/R^2$.

Sound intensity is perceived by the ear as **loudness**. On the average, the human ear can detect sound waves (at 1 kHz) with an intensity as low as 10^{-12} W/m^2. This intensity (I_o) is referred to as the *threshold of hearing*. Thus, for us to hear a sound, it must not only have a frequency in the audible range, but also be of sufficient intensity. As the intensity is increased, the perceived sound becomes louder. At an intensity of 1.0 W/m^2, the sound is uncomfortably loud and may be painful to the ear. This intensity (I_p) is called the *threshold of pain*.

Note that the thresholds of pain and hearing differ by a factor of 10^{12}:

$$\frac{I_p}{I_o} = \frac{1.0 \text{ W/m}^2}{10^{-12} \text{ W/m}^2} = 10^{12}$$

That is, the intensity at the threshold of pain is a *trillion* times greater than that at the threshold of hearing. Within this enormous range, the perceived loudness is not directly proportional to the intensity. That is, if the intensity is doubled, the perceived loudness does not double. In fact, doubling of perceived loudness corresponds approximately to an increase in intensity by a factor of 10. For example, a sound with an intensity of 10^{-5} W/m^2 would be perceived to be twice as loud as one with an intensity of 10^{-6} W/m^2 (the smaller the negative exponent, the larger the number).

Sound Intensity Level—The Bel and the Decibel

It is convenient to compress the large range of sound intensities by using a logarithmic scale (base-10) to express intensity levels. The intensity level of a sound must be referenced to a standard intensity, which is taken to be that of the threshold of hearing, $I_o = 10^{-12}$ W/m^2. Then, for any intensity I, the intensity level is the log of the ratio of I to I_o: $\log I/I_o$. For example, if a sound has an intensity of $I = 10^{-6}$ W/m^2,

$$\log \frac{I}{I_o} = \log \frac{10^{-6} \text{ W/m}^2}{10^{-12} \text{ W/m}^2} = \log 10^6 = 6 \text{ B}$$

(Recall that $\log_{10} 10^x = x$.) The exponent of the power of ten in the final log term is taken to have a unit called the **bel** (B). Thus, a sound with an intensity of 10^{-6} W/m^2 has an intensity level of 6 B on this scale. In this way, the intensity range from 10^{-12} W/m^2 to 1.0 W/m^2 is compressed into a scale of intensity levels ranging from 0 B to 12 B.

Threshold of hearing:
$I_o = 10^{-12}$ W/m^2

Threshold of pain:
$I_p = 1.0$ W/m^2

Note: The bel was named in honor of Alexander Graham Bell, the inventor of the telephone.

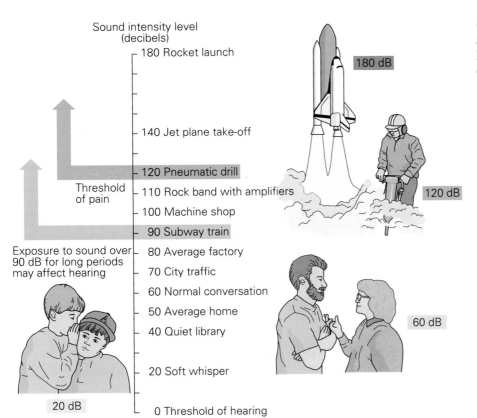

Sound intensity level
(decibels)

- 180 Rocket launch
- 140 Jet plane take-off
- 120 Pneumatic drill
- 110 Rock band with amplifiers
- 100 Machine shop
- 90 Subway train
- 80 Average factory
- 70 City traffic
- 60 Normal conversation
- 50 Average home
- 40 Quiet library
- 20 Soft whisper
- 0 Threshold of hearing

Threshold
of pain

Exposure to sound over
90 dB for long periods
may affect hearing

180 dB

120 dB

60 dB

20 dB

•FIGURE 14.5 Sound intensity levels
and the decibel scale The intensity
levels of some common sounds on the
decibel (dB) scale.

A finer intensity scale is obtained by using a smaller unit, the **decibel** (dB), which is a tenth of a bel. The 0 to 12 B range corresponds to 0 to 120 dB. In this case, the equation for the relative **sound intensity level**, or **decibel level** (β), is

$$\beta = 10 \log \frac{I}{I_\mathrm{o}} \qquad \text{where } I_\mathrm{o} = 10^{-12}\,\mathrm{W/m^2} \quad (14.4)$$

Sound intensity level in decibels

Note that the sound intensity level (in decibels) is *not* the same as the sound intensity (in watts per square meter).

The decibel intensity scale and familiar sounds at some intensity levels are shown in •Fig. 14.5. Sound intensities can have detrimental effects on hearing, and because of this the U.S. government has set occupational noise-exposure limits.

Example 14.4 ■ Sound Intensity Levels: Using Logarithms

What are the intensity levels of sounds with intensities of (a) $10^{-12}\,\mathrm{W/m^2}$ and (b) $5.0 \times 10^{-6}\,\mathrm{W/m^2}$?

Thinking It Through. We can find the sound intensity levels directly by using Eq. 14.4.

Solution.

Given: (a) $I = 10^{-12}\,\mathrm{W/m^2}$ *Find:* (a) β (sound intensity level)
(b) $I = 5.0 \times 10^{-6}\,\mathrm{W/m^2}$ (b) β

(a) Using Eq. 14.4 we have,

$$\beta = 10 \log \frac{I}{I_\mathrm{o}} = 10 \log \left(\frac{10^{-12}\,\mathrm{W/m^2}}{10^{-12}\,\mathrm{W/m^2}} \right) = 10 \log 1 = 0\,\mathrm{dB}$$

The intensity is the same as that at the threshold of hearing. (Recall that $\log 1 = 0$, since $1 = 10^0$ and $\log 10^0 = 0$.)

(b) $\beta = 10 \log \dfrac{I}{I_\mathrm{o}} = 10 \log \left(\dfrac{5.0 \times 10^{-6}\,\mathrm{W/m^2}}{10^{-12}\,\mathrm{W/m^2}} \right)$

$= 10 \log (5.0 \times 10^6) = 10(\log 5.0 + \log 10^6) = 10(0.70 + 6.0) = 67\,\mathrm{dB}$

Follow-up Exercise.   Note in this Example that the intensity of $5.0 \times 10^{-6}\,\mathrm{W/m^2}$ is halfway between 10^{-6} and 10^{-5} (or 60 and 70 dB), yet this intensity does not correspond to a midway value of 65 dB. (a) Why? (b) What intensity *does* correspond to 65 dB? (Compute to 3 significant figures.)

Example 14.5 ■ Intensity Level Factors: Using Ratios

(a) What is the difference in the intensity levels if the intensity of a sound is doubled? (b) By what factors does the intensity increase for intensity level *differences* of 10 dB and 20 dB?

Thinking It Through.   (a) If the intensity is doubled, we know the increase ratio is 2. That is, $I_2 = 2I_1$, and $I_2/I_1 = 2$. We can then use Eq. 14.4 to find the intensity difference. Recall that $\log a - \log b = \log a/b$. (b) Here it is important to note that these values are intensity level *differences*, $\Delta \beta = \beta_2 - \beta_1$, *not* intensity *levels*. The equation developed in (a) will work. (Why?)

Solution.   Listing the data, we have

Given: (a) $I_2 = 2I_1$ *Find:* (a) $\Delta \beta$ (intensity level difference)
          (b) $\Delta \beta = 10$ dB   (b) I_2/I_1 (factors of increase)
              $\Delta \beta = 20$ dB

(a)   With $\log a - \log b = \log a/b$, using Eq. 14.4 and this relationship, we have for the difference $\Delta \beta = \beta_2 - \beta_1 = 10[\log (I_2/I_\mathrm{o}) - \log (I_1/I_\mathrm{o})] = 10 \log (I_2/I_\mathrm{o})/(I_1/I_\mathrm{o}) = 10 \log I_2/I_1$. Then,

$$\Delta \beta = 10 \log \dfrac{I_2}{I_1} = 10 \log 2 = 3\,\mathrm{dB}$$

Thus, doubling the intensity increases the intensity level by 3 dB (for example, an increase from 55 dB to 58 dB).

(b)   For a 10-decibel difference,

$$\Delta \beta = 10\,\mathrm{dB} = 10 \log \dfrac{I_2}{I_1} \quad \text{and} \quad \log \dfrac{I_2}{I_1} = 1.0$$

Since $\log 10^1 = 1$,

$$\dfrac{I_2}{I_1} = 10^1 \quad \text{and} \quad I_2 = 10\,I_1$$

Similarly, for a 20-decibel difference,

$$\Delta \beta = 20\,\mathrm{dB} = 10 \log \dfrac{I_2}{I_1} \quad \text{and} \quad \log \dfrac{I_2}{I_1} = 2.0$$

Since $\log 10^2 = 2$,

$$\dfrac{I_2}{I_1} = 10^2 \quad \text{and} \quad I_2 = 100\,I_1$$

Thus, an intensity level difference of 10 dB corresponds to increasing (or decreasing) the intensity by a factor of 10. An intensity level difference of 20 dB corresponds to increasing (or decreasing) the intensity by a factor of 100.

You should be able to guess the factor that corresponds to an intensity level difference of 30 dB. In general, the factor of the intensity change is $10^{\Delta B}$, where ΔB is the level difference in bels. Since 30 dB = 3 B and $10^3 = 1000$, the intensity changes by a factor of 1000 for an intensity level difference of 30 dB.

Follow-up Exercise.   A $\Delta \beta$ of 20 dB and a $\Delta \beta$ of 30 dB correspond to factors of 100 and 1000, respectively, in intensity changes. Does a $\Delta \beta$ of 25 dB correspond to an intensity change factor of 500? Explain.

Example 14.6 ■ Combined Sound Levels: Adding Intensities

Sitting at a sidewalk restaurant table, a friend talks to you in normal conversation (60 dB). At the same time, the intensity level of the street traffic reaching you is also 60 dB. What is the total intensity level of the combined sounds?

Thinking It Through.   It is tempting simply to add the two sound intensity levels together and say the total is 120 dB. But intensity levels in decibels are logarithmic, so you can't add them in the normal way. However, intensities (I) can be added arithmetically, since energy and power are scalar quantities. Then the combined intensity level can be found from the sum of the intensities.

Solution.   We have

Given: $\beta_1 = 60 \text{ dB}$ *Find:*   Total β
$\quad\quad\quad\beta_2 = 60 \text{ dB}$

Let's find the intensities associated with the intensity levels:

$$\beta_1 = 60 \text{ dB} = 10 \log \frac{I_1}{I_o} = 10 \log \left(\frac{I_1}{10^{-12} \text{ W/m}^2} \right)$$

By inspection, we have

$$I_1 = 10^{-6} \text{ W/m}^2$$

Similarly, $I_2 = 10^{-6} \text{ W/m}^2$, since both intensity levels are 60 dB. So, the total intensity is

$$I_{total} = I_1 + I_2 = 1.0 \times 10^{-6} \text{ W/m}^2 + 1.0 \times 10^{-6} \text{ W/m}^2 = 2.0 \times 10^{-6} \text{ W/m}^2$$

Then, converting back to intensity level, we get

$$\beta = 10 \log \frac{I_{total}}{I_o} = 10 \log \left(\frac{2.0 \times 10^{-6} \text{ W/m}^2}{10^{-12} \text{ W/m}^2} \right) = 10 \log (2.0 \times 10^6)$$

$$= 10(\log 2.0 + \log 10^6) = 10(0.30 + 6.0) = 63 \text{ dB}$$

This value is a long way from 120 dB! Notice that the combined intensities doubled the value and the intensity level increased by 3 dB, in agreement with our finding in part (a) of Example 14.5.

Follow-up Exercise.   In this Example, suppose the added noise gave a total that *tripled* the value of the conversation intensity. What would be the total combined intensity level in this case?

14.4   Sound Phenomena

Objectives: **To (a) explain sound reflection, refraction, and diffraction and (b) distinguish between constructive and destructive interference.**

Reflection, Refraction, and Diffraction

A sound wave can be reflected; an echo is a familiar example of sound *reflection*. Sound *refraction* is less common than reflection, but you may have experienced this effect on a calm summer evening. Then it is sometimes possible to hear distant voices or other sounds that ordinarily would not be audible. This effect is due to the refraction, or bending, of the sound waves as they pass from one region into a region where the air density is different. The effect is similar to what would happen if the sound passed into another medium.

The required conditions for sound refraction are a layer of cooler air near the ground or water and a layer of warmer air above it. These conditions occur frequently over bodies of water because of cooling after sunset (•Fig. 14.6). As a result, the waves are refracted in an arc that may allow a distant person to receive an increased intensity of sound.

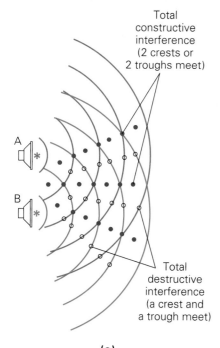

Total constructive interference (2 crests or 2 troughs meet)

A

B

Total destructive interference (a crest and a trough meet)

(a)

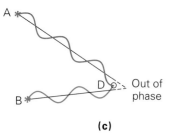

A

λ_{AC}

L_{AC}

B

λ_{BC}

L_{BC}

C

In phase

(b)

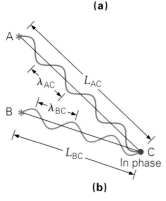

A

D

Out of phase

B

(c)

•**FIGURE 14.7   Interference**
(a) Sound waves from two point sources spread out and interfere. **(b)** At points where the waves arrive in phase (zero phase difference), such as point C, constructive interference occurs. **(c)** At points where the waves arrive completely out of phase (phase difference of 180°), such as point D, destructive interference occurs. The phase difference at a particular point depends on the path lengths the waves travel to reach that point.

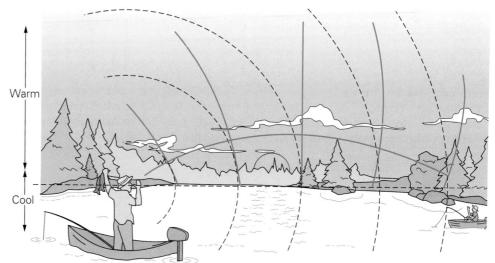

•**FIGURE 14.6   Sound refraction**   Sound travels slower in the cool air near the water surface than in the upper, warmer air. As a result, the waves are refracted, or bent. This bending increases the intensity of the sound at a distance where it otherwise might not be heard.

Another bending phenomenon is *diffraction,* a bending of sound around corners or around an object. We usually think of waves as traveling in straight lines. However, you can hear someone who is standing around a corner but whom you cannot see. This bending is different from that of refraction, in which no obstacle causes the bending.

Reflection, refraction, and diffraction are described in a general sense here for sound. These phenomena are important considerations for light waves and will be discussed more fully in Chapters 22 and 24.

Interference

Like waves of any kind, sound waves *interfere* when they meet. Suppose that two loudspeakers separated by some distance emit sound waves in phase at the same frequency. If we consider the speakers to be point sources, then the waves will spread out spherically and interfere (•Fig. 14.7a). The lines from a particular speaker represent wave crests (or condensations), and the troughs (or rarefactions) lie in the intervening white areas.

In particular regions of space, there will be constructive or destructive interferences. For example, if the waves meet in a region where they are exactly in phase (where two crests or two troughs coincide), there will be total **constructive interference** (Fig. 14.7b). Notice that the waves have the same motion at point C in the figure. Conversely, if the waves meet such that the crest of one coincides with the trough of the other (at point D), the two waves will cancel each other out (Fig. 14.7c). The result will be total **destructive interference**.

It is convenient to describe the path lengths traveled by the waves in terms of wavelength (λ) to determine whether they arrive in phase. In such analysis we work with points of zero displacement rather than crests or troughs. Consider the waves arriving at point C in Fig. 14.7b. The path lengths in this case are AC = 4λ and BC = 3λ. The **phase difference** ($\Delta\theta$) is related to the **path-length difference** (ΔL) by the simple relationship

$$\Delta\theta = \frac{2\pi}{\lambda}(\Delta L) \qquad \begin{array}{l}\textit{phase difference and}\\ \textit{path-length difference}\end{array} \qquad (14.5)$$

Since 2π rad is equivalent, in angular terms, to a full wave cycle or wavelength, multiplying the path-length difference by $2\pi/\lambda$ gives the phase difference in radians. For the example illustrated in Fig. 14.7b, we have

$$\Delta\theta = \frac{2\pi}{\lambda}(AC - BC) = \frac{2\pi}{\lambda}(4\lambda - 3\lambda) = 2\pi \text{ rad}$$

When $\Delta\theta = 2\pi$ rad, the waves are shifted by one wavelength. This is the same as $\Delta\theta = 0°$, so the waves are actually in phase. Thus, the waves interfere constructively in the point C region, increasing the intensity, or loudness, of the sound detected there.

From Eq. 14.5, we see that the sound waves are in phase at any point where the path-length difference is zero or an integral multiple of the wavelength. That is,

$$\Delta L = n\lambda \qquad (n = 0, 1, 2, 3, \dots) \qquad \begin{array}{l} \textit{condition for} \\ \textit{constructive interference} \end{array} \qquad (14.6)$$

A similar analysis for the situation in •Fig. 14.7c, where $AD = 2\frac{3}{4}\lambda$ and $BD = 2\frac{1}{4}\lambda$, gives

$$\Delta\theta = \frac{2\pi}{\lambda}(2\tfrac{3}{4}\lambda - 2\tfrac{1}{4}\lambda) = \pi \text{ rad}$$

or $\Delta\theta = 180°$. At point D, the waves are completely out of phase, and destructive interference occurs in this region.

Sound waves will be out of phase at any point where the path-length difference is an odd number of half-wavelengths ($\lambda/2$), or

$$\Delta L = m\left(\frac{\lambda}{2}\right) \qquad (m = 1, 3, 5, \dots) \qquad \begin{array}{l} \textit{condition for} \\ \textit{destructive interference} \end{array} \qquad (14.7)$$

At these points, a softer or less intense sound will be heard or detected. If the amplitudes of the waves are exactly equal, the destructive interference is total and no sound is heard!

An application of destructive interference for sound waves is illustrated in •Fig. 14.8. Engine noise in commercial jet aircraft cabins can be distracting. In a new technique, microphones throughout the cabin pick up the various sounds; a computer analyzes them and generates sound waves that are 180° out of phase with the undesirable noise. Destructive interference then reduces the cabin noise.

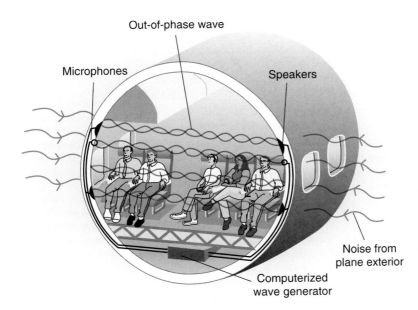

Out-of-phase wave

Microphones

Speakers

Noise from plane exterior

Computerized wave generator

•FIGURE 14.8 **Destructive interference** Destructive interference is used to reduce the cabin noise in commercial jet aircraft. Microphones route noise to a computer, which then drives speakers to produce sound waves 180° out of phase with the noise. The resulting destructive interference makes the cabin quieter.

Example 14.7 ■ Pump up the Volume: Sound Interference

At an open-air concert on a hot day (air temperature of 25°C), you sit 7.00 m and 9.10 m, respectively, from a pair of speakers, one at each side of the stage. A musician, warming up, plays a single 494-hertz tone. What do you hear? (Consider the speakers to be point sources.)

Thinking It Through. The sound waves from the speakers interfere. Is it constructive or destructive interference or something in between? This depends on the path-length difference, which we can compute from the given distances.

Solution.

Given: $d_1 = 7.00$ m and $d_2 = 9.10$ m   *Find:* ΔL (path-length difference
$\qquad f = 494$ Hz $\qquad\qquad\qquad\qquad\qquad\qquad$ in wavelength units)
$\qquad T = 25°C$

The path-length difference (2.10 m) between the waves arriving at your location must be expressed in terms of the wavelength of the sound. To do this, we first need to know the wavelength. Given the frequency, we can find the wavelength from the relationship $\lambda = v/f$, provided we know the value of the speed of sound v at the given temperature. We calculate v from Eq. 14.1:

$$v = 331 + 0.6T_C = 331 + 0.6(25) = 346 \text{ m/s}$$

The wavelength of the sound waves is then

$$\lambda = \frac{v}{f} = \frac{346 \text{ m/s}}{494 \text{ Hz}} = 0.700 \text{ m}$$

Thus, the distances in terms of wavelength are

$$d_1 = (7.00 \text{ m})\left(\frac{\lambda}{0.700 \text{ m}}\right) = 10.0\lambda \quad \text{and} \quad d_2 = (9.10 \text{ m})\left(\frac{\lambda}{0.700 \text{ m}}\right) = 13.0\lambda$$

The path-length difference in terms of wavelengths is

$$\Delta L = d_2 - d_1 = 13.0\lambda - 10.0\lambda = 3.0\lambda$$

This is an integral number of wavelengths ($n = 3$), so constructive interference occurs. The sounds of the two speakers reinforce each other, and you hear an intense tone at 494 Hz.

Follow-up Exercise. Suppose that in this Example the tone traveled to a person sitting 7.00 m and 8.75 m, respectively, from the two speakers. What would be the situation in this case?

Another interesting interference effect occurs when two tones of nearly the same frequency ($f_1 \approx f_2$) are sounded simultaneously. The ear senses pulsations in loudness known as **beats**. The human ear can detect as many as 7 beats per second before they sound "smooth" or unpulsating.

Suppose that two sinusoidal waves with the same amplitude but slightly different frequencies interfere (•Fig. 14.9a). Figure 14.9b represents the resulting sound wave. The amplitude of the combined wave varies sinusoidally, as shown by the black curves (known as *envelopes*) that outline the wave.

What does this mean in terms of listener perception? A listener will hear a pulsating sound (beats) as determined by the envelopes. The maximum amplitude is $2A$ (at the point where the maxima of the two original waves interfere constructively). Detailed mathematics shows that a listener will hear these beats at a frequency called the **beat frequency** (f_b), given by

$$f_b = |f_1 - f_2| \qquad\qquad\qquad (14.8)$$

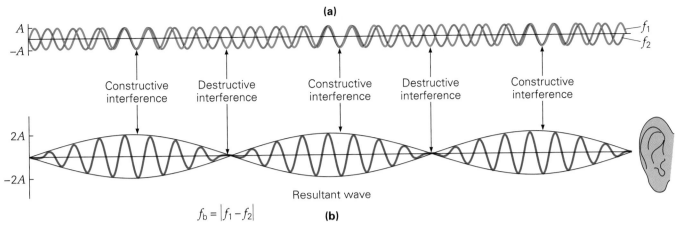

(a)

Constructive interference   Destructive interference   Constructive interference   Destructive interference   Constructive interference

Resultant wave

$$f_b = |f_1 - f_2|$$

(b)

•**FIGURE 14.9 Beats** Two traveling waves of equal amplitude and slightly different frequencies interfere and give rise to pulsating tones called beats. The beat frequency is given by $f_b = |f_1 - f_2|$.

The absolute value is taken since the frequency f_b cannot be negative, even if $f_2 > f_1$. A negative beat frequency would be meaningless.

Beats can be produced when tuning forks of nearly the same frequency are vibrating at the same time. For example, using forks with frequencies of 516 Hz and 513 Hz, the beat frequency is $f_b = 516$ Hz $- 513$ Hz $= 3$ Hz, and three beats are heard each second. Musicians tune two stringed instruments to the same note by adjusting the tensions in the strings until the beats disappear ($f_1 = f_2$).

14.5 The Doppler Effect

Objectives: **To (a) describe and explain the Doppler effect and (b) give some examples of its occurrences and applications.**

If you stand along a highway and a car or truck approaches you with its horn blowing, the **pitch** (the perceived frequency) of the sound is higher as the vehicle approaches and lower as it recedes. You can also hear variations in the frequency of the motor noise when you watch a race car going around a track. A variation in the perceived sound frequency due to the motion of the sound source is an example of the **Doppler effect**.

As •Fig. 14.10 shows, the sound waves emitted by a moving source tend to bunch up in front of the source and spread out in back. The Doppler shift in frequency can be found by assuming that the air is at rest in a reference frame such as that depicted in •Fig. 14.11. The speed of sound in air is v, and the speed of the moving source is v_s. The frequency of the sound produced by the source is f_s. In one period, $T = 1/f_s$, a wave crest moves a distance $d = vT = \lambda$. (The sound wave would travel this distance in still air in any case, whether or not the source is moving.) But, in one period, the source travels a distance $d_s = v_s T$ before emitting another wave crest. The distance between the successive wave crests is thus shortened to a wavelength λ':

$$\lambda' = d - d_s = vT - v_s T = (v - v_s)T = \frac{v - v_s}{f_s}$$

The frequency heard by the observer (f_o) is related to the shortened wavelength by $f_o = v/\lambda'$, and substituting v/λ' gives

$$f_o = \frac{v}{\lambda'} = \left(\frac{v}{v - v_s}\right)f_s$$

The Austrian physicist Christian Doppler (1803–1853) first described what we now call the Doppler effect.

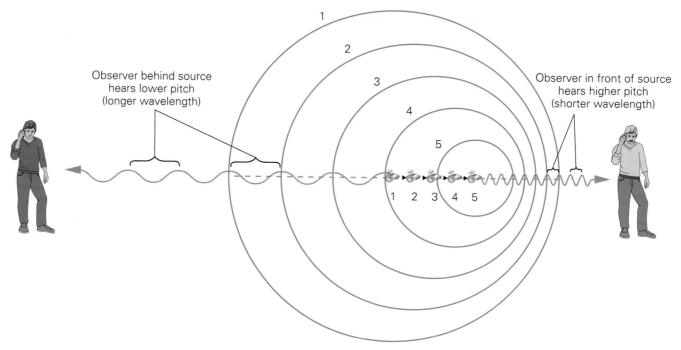

Observer behind source hears lower pitch (longer wavelength)

Observer in front of source hears higher pitch (shorter wavelength)

•FIGURE 14.10 **The Doppler effect** The sound waves bunch up in front of a moving source—the whistle—giving a higher frequency there. They trail out behind the source, giving a lower frequency there. (Not drawn to scale. Why?)

or

$$f_o = \left(\frac{1}{1 - \dfrac{v_s}{v}} \right) f_s \qquad \begin{array}{l} \textit{source moving} \\ \textit{toward a stationary observer} \end{array} \qquad (14.9)$$

$$\textit{where} \quad v_s = \textit{speed of source}$$
$$\textit{and} \quad v = \textit{speed of sound}$$

Since $1 - (v_s/v)$ is less than 1, f_o is greater than f_s in this situation. For example, suppose that the speed of the source is a tenth of the speed of sound: $v_s = v/10$, or $v_s/v = \frac{1}{10}$. Then, by Eq. 14.9, $f_o = \frac{10}{9}f_s$.

Similarly, when the source is moving away from the observer ($\lambda' = d + d_s$), the observed frequency is given by

$$f_o = \left(\frac{v}{v + v_s} \right) f_s = \left(\frac{1}{1 + \dfrac{v_s}{v}} \right) f_s \qquad \begin{array}{l} \textit{source moving} \\ \textit{away from a stationary observer} \end{array} \qquad (14.10)$$

Here f_o is less than f_s. (Why?)

•FIGURE 14.11 **The Doppler effect and wavelength** Sound from a moving car's horn travels a distance d in a time T. During this time, the car (the source) travels a distance d_s before putting out a second pulse, thereby shortening the observed wavelength of the sound in the approaching direction.

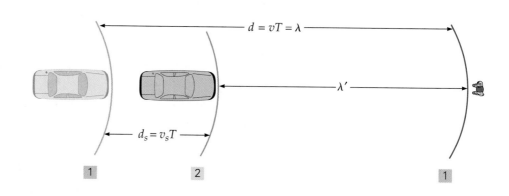

Combining Eqs. 14.9 and 14.10 yields a general equation for the observed frequency with a moving source and a stationary observer:

$$f_o = \left(\frac{v}{v \pm v_s}\right)f_s = \left(\frac{1}{1 \pm \dfrac{v_s}{v}}\right)f_s \quad \begin{cases} - \textit{ for source moving} \\ \textit{toward stationary observer} \\ + \textit{ for source moving} \\ \textit{away from stationary observer} \end{cases} \quad (14.11)$$

As you might expect, the Doppler effect also occurs with a moving observer and a stationary source. This situation is a bit different. As the observer moves toward the source, the distance between successive wave crests is the normal wavelength (or $\lambda = v/f_s$), but the measured wave speed is different. Relative to the approaching observer, the sound from the stationary source has a wave speed of $v' = v + v_o$, where v_o is the speed of the observer and v is the speed of sound in still air. (The observer moving toward the source is moving in a direction opposite that of the propagating waves and thus meets more wave crests in a given time.) The observed frequency (f_o) is then (with $\lambda = v/f_s$):

$$f_o = \frac{v'}{\lambda} = \left(\frac{v + v_o}{v}\right)f_s$$

or

$$f_o = \left(1 + \frac{v_o}{v}\right)f_s \quad \begin{matrix} \textit{observer moving} \\ \textit{toward a stationary source} \end{matrix} \quad (14.12)$$

$$\textit{where} \quad v_o = \textit{speed of observer}$$
$$\textit{and} \quad v = \textit{speed of sound}$$

Similarly, for an observer moving away from a stationary source, the perceived wave speed is $v' = v - v_o$, and

$$f_o = \frac{v'}{\lambda} = \left(\frac{v - v_o}{v}\right)f_s$$

or

$$f_o = \left(1 - \frac{v_o}{v}\right)f_s \quad \begin{matrix} \textit{observer moving} \\ \textit{away from a stationary source} \end{matrix} \quad (14.13)$$

Equations 14.12 and 14.13 can be combined into a general equation for a moving observer and a stationary source:

$$f_o = \left(\frac{v \pm v_o}{v}\right)f_s = \left(1 \pm \frac{v_o}{v}\right)f_s \quad \begin{cases} + \textit{ for observer moving} \\ \textit{toward stationary source} \\ - \textit{ for observer moving} \\ \textit{away from stationary source} \end{cases} \quad (14.14)$$

There are also the cases when both the source and the observer are moving, either toward one another or away from one another. We will not consider them mathematically (but will do so conceptually in Example 14.9).

Example 14.8 ■ On the Road Again: The Doppler Effect

As a truck traveling at 96 km/h approaches and passes a person standing along the highway, the driver sounds the horn. If the horn has a frequency of 400 Hz, what are the frequencies of the sound waves heard by the person (a) as the truck approaches and (b) after it has passed? (Assume that the speed of sound is 346 m/s.)

Thinking It Through. This is an application of the Doppler effect, Eq. 14.11, with a moving source and a stationary observer. In such problems it is important to identify the data correctly.

Solution.

Given: $v_s = 96 \text{ km/h} = 27 \text{ m/s}$ *Find:* (a) f_o (observed frequency approaching)
$f_s = 400 \text{ Hz}$ (b) f_o (observed frequency moving away)
$v = 346 \text{ m/s}$

(a) From Eq. 14.11 with a minus sign (source approaching stationary observer),

$$f_o = \left(\frac{v}{v - v_s}\right) f_s = \left(\frac{346 \text{ m/s}}{346 \text{ m/s} - 27 \text{ m/s}}\right)(400 \text{ Hz}) = 434 \text{ Hz}$$

(b) A plus sign is used in Eq. 14.11 when the source is moving away:

$$f_o = \left(\frac{v}{v + v_s}\right) f_s = \left(\frac{346 \text{ m/s}}{346 \text{ m/s} + 27 \text{ m/s}}\right)(400 \text{ Hz}) = 371 \text{ Hz}$$

Follow-up Exercise. Suppose that the observer in this Example was initially moving toward and then past a stationary 400-hertz source at a speed of 96 km/h. What would be the observed frequencies? (Will they differ from those for the moving source?)

Conceptual Example 14.9

It's All Relative: Moving Source and Moving Observer

Suppose a sound source and an observer are moving away from one another in opposite directions, each at half the speed of sound in air. Then, the observer would (a) receive sound with a frequency higher than the source frequency, (b) receive sound with a frequency lower than the source frequency, (c) receive sound with the same frequency as the source frequency, (d) receive no sound from the source.

Reasoning and Answer. As we know, when a source moves away from a stationary observer, the observed frequency is lower (Eq. 14.10). Similarly, when an observer moves away from a stationary source, the observed frequency is also lower (Eq. 14.13). With both source and observer moving away from each other in opposite directions, the combined effect would make the observed frequency even less, so the answer is not (a) or (c).

It would appear that (b) is the correct answer, but we must logically eliminate (d) for completeness. Remember that the speed of sound relative to the air is constant. Therefore, (d) would be correct *only if the observer is moving faster than the speed of sound* relative to the air. Since the observer is moving at only half the speed of sound, (b) is the correct answer.

Think about it this way. Regardless of how fast the source is moving, the sound from the source is moving at the speed of sound through the air toward the observer. The observer is moving at only half the speed of sound through the air, so the sound from the source can easily reach and pass the observer.

Follow-up Exercise. What would be the situation if both the source and the observer were traveling in the same direction with the same subsonic speed? (Subsonic, as opposed to supersonic, refers to a speed that is less than the speed of sound in air.)

Problem-Solving Hint

You may find it difficult to remember whether a plus or minus sign is used in the general equations for the Doppler effect. Let your experience help you. For the commonly experienced case of being a stationary observer, the sound frequency increases when the source approaches, so the denominator in Eq. 14.11 must be smaller than the numerator. For this case, you use the minus sign. When the source is receding, the frequency is lower. The denominator in Eq. 14.11 must then be larger than the numerator, and you use the plus sign for this case. Similar reasoning will help you choose a plus or minus sign for the numerator in Eq. 14.14.

The Doppler effect also occurs for light waves, although the formulas describing it are different from those given above. When a distant light source such as a star moves away from us, the frequency of the light we receive from it is lowered. That is, the light is shifted toward the red (long-wavelength) end of the spectrum, an effect known as a *Doppler red shift*. Similarly, the frequency of light from an object approaching us is increased—the light is shifted toward the blue (short-wavelength) end of the spectrum, producing a *Doppler blue shift*. The magnitude of the frequency shift is related to the speed of the source.

The Doppler shift of light from astronomical objects is very useful to astronomers. The rotation of a planet, star, or other body can be established by looking at the Doppler shifts of light from opposite sides of the object—because of the rotation, one side is receding (and hence red-shifted), and the other approaching (blue-shifted). Similarly, the Doppler shifts of light from stars in different regions of our galaxy, the Milky Way, indicates that the galaxy is rotating.

You have been subjected to a practical application of the Doppler effect if you have ever been caught speeding in your car by police radar, which uses reflected radio waves. (Radar stands for *ra*dio *d*etecting *a*nd *r*anging and is similar to underwater sonar, which uses ultrasound.) If the radio waves are reflected from a parked car, the reflected waves return to the source with the same frequency. But for a car that is moving toward a patrol car, the reflected waves have a higher frequency, or are Doppler-shifted. Actually, there is a double Doppler shift. The moving car acts like a moving observer in receiving the wave (first Doppler shift), and in reflecting it the car acts like a moving source emitting a wave (second Doppler shift). The magnitudes of the shifts depend on the speed of the car. A computer quickly calculates this speed and displays it for the police officer.

For another important application of the Doppler effect in weather and airplane safety, see the Insight on p. 472.

Sonic Booms

Consider a jet plane that can travel at supersonic speeds. As the speed of a moving source of sound approaches the speed of sound, the waves ahead of the source come close together (•Fig. 14.12). When a plane is traveling at the speed of sound, the waves can't outrun it, and they pile up in front. At supersonic speeds, the waves overlap. This overlapping of a large number of waves produces many points of constructive interference, forming a large pressure ridge, or *shock wave*. This is sometimes called a *bow wave* because it is analogous to the wave produced by the bow of a boat moving through water at a speed greater than the speed of the water waves (Fig. 14.12c).

From aircraft traveling at supersonic speed, the shock wave trails out to the sides and downward. When this pressure ridge passes over an observer on the ground, the large concentration of energy produces what is known as a **sonic boom**. There is really a double boom because shock waves are formed at both ends of the aircraft. Under certain conditions, the shock waves can break windows and cause other damage. (Sonic booms are no longer heard as frequently as in the past. Pilots are now instructed to fly supersonically only at high altitudes and away from populated areas.)

On a smaller scale, you have probably heard a "mini" sonic boom. The "crack" of a whip is actually a sonic boom created by the transonic speed of the whip's tip—that is, the speed of the tip changes from subsonic to supersonic (and back).

A common misconception is that a sonic boom occurs only when a plane breaks the sound barrier. As an aircraft approaches the speed of sound, the pressure ridge in front of it is essentially a barrier that must be overcome with extra power. However, once supersonic speed is reached, this barrier is no longer there, and the shock wave, continuously created, trails behind the plane, producing booms along its ground path.

Doppler Radar

Radar has been used since the early 1940s to provide information about rainstorms and other forms of precipitation. This information is obtained from the intensity of the reflected signal. Such conventional radars can also detect the hooked (rotational) "signature" of a tornado, but only after the storm is well developed.

A major improvement in weather forecasting came about with the development of a radar system that could measure the Doppler frequency shift in addition to the magnitude of the echo signal reflected from precipitation. The Doppler shift is related to the *radial* velocity of the precipitation blown by the wind (the component of the wind velocity moving either toward or away from the radar installation).

A Doppler-based radar system (Fig. 1a) can penetrate a storm and monitor its wind speeds. The direction of a storm's wind-driven rain gives a wind "field" map of the storm region. Such maps provide strong clues of developing tornadoes, so meteorologists can detect them much earlier than was ever before possible (Fig. 1b). With Doppler radar, forecasters have been able to predict tornadoes as much as 20 min before they touch down, as compared to just over 2 min for

conventional radar. Doppler radar has saved many lives with this increased warning time. The National Weather Service has a network of Doppler radars around the United States, and Doppler radar scans are now common on TV weather forecasts and on the Internet.

Doppler radars have also been installed at major airports for another use. Several airplane crashes and near crashes have been attributed to downward wind bursts (also known as microbursts and down bursts). Such strong down drafts cause wind shears capable of forcing an aircraft to the ground. The Doppler radar used at airports is designed to detect dangerous wind bursts in the vicinity of the landing approaches.

Wind bursts generally result from high-speed down drafts in the turbulence of thunderstorms, but they can also occur in clear air when rain evaporates high above ground. A down draft spreads out when it hits the ground and then circulates upward. An airplane entering this pattern experiences an unexpected upward headwind that lifts the plane. The pilot often cuts speed and lowers the plane's nose to compensate. Further into the circular pattern, the wind

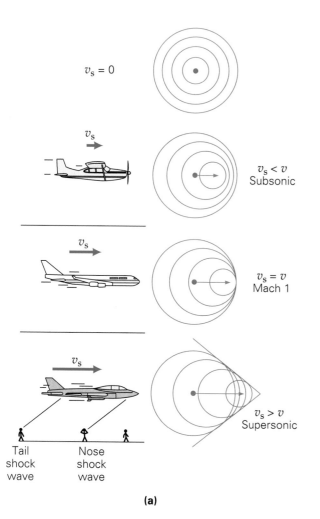

(a)

(b)

(c)

●**FIGURE 14.12 Bow waves and sonic booms** **(a)** When an aircraft exceeds the speed of sound in air, the sound waves form a pressure ridge, or shock wave. As the trailing shock wave passes over the ground, observers hear a sonic boom—actually two booms, because shock waves are formed at the front and tail of the plane. **(b)** A bullet traveling at a speed of 500 m/s. Note the shock waves produced (and the turbulence behind the bullet). The image was made by using interferometry with polarized light and a pulsed laser, with an exposure time of 20 ns. **(c)** A shock wave in water. Bow waves like these are produced whenever a boat travels faster than the speed of the water waves that it creates.

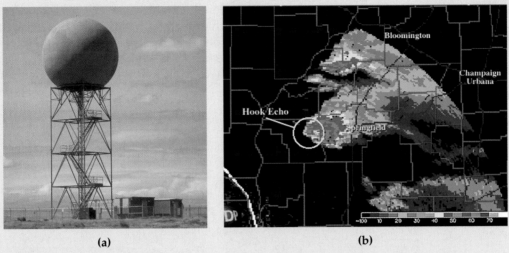

FIGURE 1   Doppler radar   **(a)** A Doppler radar installation. **(b)** Doppler radar depicts the precipitation inside a thunderstorm. A hook echo is a signature of a possible tornado.

quickly turns downward. The airplane can suddenly lose altitude and, when near the ground, as on landing, may crash. Since Doppler radar can detect the wind speed and the direction of raindrops in clouds, as well as dust and other objects floating in the air, it can provide an early warning against dangerous wind shear conditions.

Ideally, the sound waves produced by a supersonic aircraft form a cone-shaped shock wave (•Fig. 14.13). The waves travel outward with a speed v, and the speed of the plane (the source) is v_s. Note from the figure that the angle between a line tangent to the spherical waves and the line along which the plane is moving is given by

$$\sin \theta = \frac{vt}{v_s t} = \frac{v}{v_s} \tag{14.15}$$

The inverse ratio of the speeds is called the **Mach number** (*M*), named after Ernst Mach (1838–1916), an Austrian physicist, who used it in studying supersonics:

$$M = \frac{v_s}{v} \tag{14.16}$$

If v equals v_s, the plane is flying at the speed of sound, and the Mach number is 1 (that is, $v_s/v = 1$). Therefore, a Mach number less than 1 indicates a subsonic speed, and a Mach number greater than 1 indicates a supersonic speed. In the latter case, the Mach number tells the speed of the aircraft in terms of a multiple of the speed of sound. A Mach number of 2, for instance, indicates a speed twice the speed of sound.

14.6   Musical Instruments and Sound Characteristics

Objective: To explain some of the sound characteristics of musical instruments in physical terms.

Musical instruments provide good examples of standing waves and boundary conditions. On some stringed instruments, different notes are produced by using finger pressure to vary the lengths of the strings (•Fig. 14.14). As you learned in Chapter 13, the natural frequencies of a stretched string (fixed at each end, as is the case for the strings on an instrument), are $f_n = n(v/2L)$, from Eq. 13.21, where

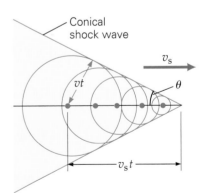

•**FIGURE 14.13   Shock wave cone and Mach number**   When the speed of the source (v_s) is greater than the speed of sound in air (v), the interfering spherical sound waves form a conical shock wave that appears as a V-shaped pressure ridge when viewed in 2 dimensions. The angle θ is given by $\sin \theta = v/v_s$, and the inverse ratio v_s/v is called the Mach number.

•**FIGURE 14.14 A shorter vibrating string, a higher frequency** Different notes are produced on stringed instruments such as guitars, violins, or cellos by placing a finger on a string to change its effective, or vibrating, length.

$v = \sqrt{F_T/\mu}$. Initially adjusting the tension in a string tunes it to a particular (fundamental) frequency. Then the effective length of the string is varied by finger location and pressure.

Standing waves are also set up in wind instruments. For example, consider a pipe organ with fixed pipe lengths, which may be open or closed (•Fig. 14.15). An open pipe is open at both ends, and a closed pipe is closed at one end and open at the other (the antinode end). Analysis similar to that done in Chapter 13 for a stretched string with the proper boundary conditions shows that the natural frequencies for the pipes (where v is the speed of sound in air) are

$$f_n = \frac{v}{\lambda_n} = n\left(\frac{v}{2L}\right) = nf_1 \quad n = 1, 2, 3, \ldots \quad \begin{array}{l}\textit{natural frequencies for} \\ \textit{a pipe open on both ends}\end{array} \quad (14.17)$$

and

$$f_m = \frac{v}{\lambda_m} = m\left(\frac{v}{4L}\right) = mf_1 \quad m = 1, 3, 5, \ldots \quad \begin{array}{l}\textit{natural frequencies for} \\ \textit{a pipe closed on one end}\end{array} \quad (14.18)$$

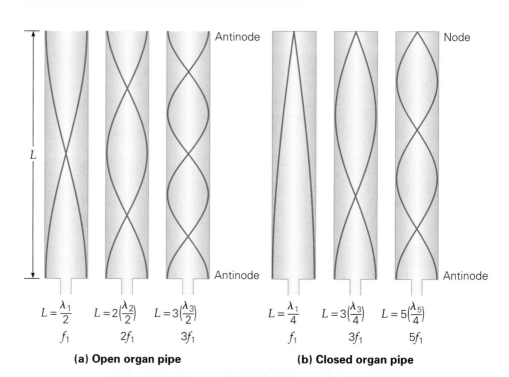

(a) **Open organ pipe**   (b) **Closed organ pipe**

(c)

•**FIGURE 14.15 Organ pipes** Longitudinal standing waves are formed in vibrating air columns in pipes (illustrated here as sinusoidal curves). **(a)** An open pipe has antinodes at both ends. **(b)** A closed pipe has a closed (node) end and an open (antinode) end. **(c)** A modern pipe organ. The pipes can be open or closed.

Note that the natural frequencies depend on the length of a pipe. This is an important consideration in a pipe organ (Fig. 14.15c), particularly in selecting the dominant or fundamental frequency. (Pipe diameter is also a factor but is not considered in this simple analysis.)

The same physical principles apply to wind and brass instruments. In all of these, the human breath is used to create standing waves in an open tube. Most such instruments allow the player to vary the effective length of the tube, and thus the pitch produced—either by opening and closing holes in the tube, as in woodwinds, or with the help of slides or valves that vary the actual length of tubing in which the air can resonate, as in most brasses (•Fig. 14.16).

Recall from Chapter 13 that a musical note or tone is referenced to the fundamental vibrational frequency. In musical terms, the first overtone is the second harmonic, the second overtone is the third harmonic, and so on. Note that for a closed organ pipe (Eq. 14.18), the even harmonics are missing.

(a)

Example 14.10 ■ Pipe Dreams: Fundamental Frequency

A particular open organ pipe has length of 0.653 m. Taking the speed of sound in air to be 345 m/s, what is the fundamental frequency of this pipe?

Thinking It Through. The fundamental frequency ($n = 1$) for an open pipe is given directly by Eq. 14.17. Physically, there is a half wavelength ($\lambda/2$) in the length of the pipe, so $\lambda = 2L$.

Solution.

Given: $L = 0.653$ m
$v = 345$ m/s (speed of sound)

Find: f_1 (fundamental frequency)

With $n = 1$,

$$f_1 = \frac{v}{2L} = \frac{345 \text{ m/s}}{2(0.653 \text{ m})} = 264 \text{ Hz}$$

This frequency is middle C (C_4).

Follow-up Exercise. A closed organ pipe has a fundamental frequency of 256 Hz. What would be the frequency of its first overtone? Is this audible?

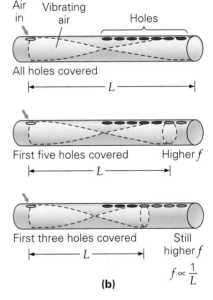

(b)

•**FIGURE 14.16   Wind instruments**
(a) The saxophone, like all wind instruments, is essentially an open tube. (b) The effective length of the air column, and hence the pitch of the sound, is varied by opening and closing holes along the tube. The frequency f is inversely proportional to the effective length L of the air column.

Perceived sounds are described by terms whose meanings are similar to those used to describe the physical properties of sound waves. Physically, a wave is generally characterized by intensity, frequency, and waveform (harmonics). The corresponding terms used to describe the sensations of the ear are loudness, pitch, and quality (or timbre). These general correlations are shown in Table 14.2. However, the correspondence is not perfect. The physical properties are objective and can be measured directly. The sensory effects are subjective and vary from person to person. (Think of temperature as measured by a thermometer and by the sense of touch.)

Sound intensity and its measurement on the decibel scale were covered in Section 14.3. Loudness is related to intensity, but the human ear responds differently to sounds of different frequencies. For example, two tones of the same intensity (in watts per square meter) but different frequencies might be judged by the ear to have different loudnesses.

Frequency and pitch are often used synonymously, but again there is an objective–subjective difference. If the same low-frequency tone is sounded at two intensity levels, most people will say that the more intense sound has a lower pitch, or perceived frequency.

TABLE 14.2 General Correlation between Perception and Physical Characteristics of Sound

Sensory Effect	Physical Wave Property
Loudness	Intensity
Pitch	Frequency
Quality (timbre)	Waveform (harmonics)

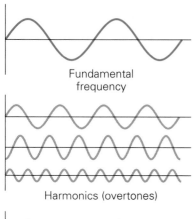

Fundamental frequency

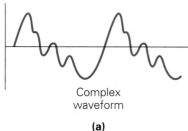

Harmonics (overtones)

Complex waveform

(a)

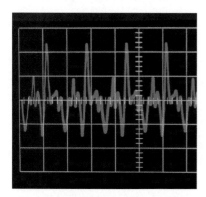

(b)

•**FIGURE 14.18 Waveform and quality (a)** The superposition of sounds of different frequencies and amplitudes gives a complex waveform. The harmonics, or overtones, determine the quality of the sound. **(b)** The waveform of a violin tone is displayed on an oscilloscope.

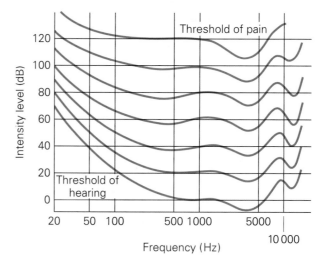

•**FIGURE 14.17 Equal loudness contours** The curves indicate tones that are judged to be equally loud, though they have different frequencies and intensity levels. For example, on the lowest contour, a 1000-hertz tone at 0 dB sounds as loud as a 50-hertz tone at 40 dB. Note that the frequency scale is logarithmic to compress the large frequency range.

The curves in the graph of intensity level versus frequency shown in •Fig. 14.17 are called equal *loudness contours* (or Fletcher-Munson curves, after the researchers who generated them). They join points representing intensity–frequency combinations that a person with average hearing judges to be equally loud. The top curve shows that the decibel level of the threshold of pain does not vary much from 120 dB, regardless of the frequency of the sound. However, the threshold of hearing, represented by the lowest contour, varies widely with frequency. For a tone with a frequency of 2000 Hz, the threshold of hearing is 0 dB. But a 20-hertz tone would have to have an intensity level of over 70 dB just to be heard (the extrapolated *y*-intercept of the lowest curve).

It is interesting to note the dips (or minima) in the curves. These indicate that the ear is most sensitive to sounds with frequencies around 4000 Hz and 12 000 Hz. Note that a tone with a frequency of 4000 Hz can be heard at intensity levels - below 0 dB. These minima occur as a result of resonance in a closed cavity in the auditory canal (similar to a closed pipe). The length of the cavity is such that it has a fundamental resonance frequency of about 4000 Hz, resulting in extra sensitivity. As in a closed cavity, the next natural frequency is the third harmonic (see Eq. 14.18), which is three times the fundamental frequency, or about 12 000 Hz.

The **quality** of a tone is the characteristic that enables it to be distinguished from another of basically the same intensity and frequency. Tone quality depends on the waveform—specifically, on the number of harmonics (overtones) present and their relative intensities (•Fig. 14.18). The tone of a voice depends in large part on the vocal resonance cavities. One person can sing a tone with the same basic frequency and intensity as another, but different combinations of overtones give the two voices different qualities.

The notes of a musical scale correspond to certain frequencies; as we saw in Example 14.10, middle C (C_4) has a frequency of 264 Hz. When a note is played on an instrument, its assigned frequency is that of the first harmonic, which is the fundamental frequency. (The second harmonic is the first overtone; the third harmonic is the second overtone, and so on.) This frequency is dominant over the accompanying overtones that determine the sound quality of the instrument. Recall from Chapter 13 that the overtones produced depend on how an instrument is played. Whether a violin string is plucked or bowed, for example, can be discerned from the quality of identical notes.

Chapter Review

Important Terms

sound waves   *453*
audible region   *453*
sound frequency spectrum   *453*
infrasonic region   *453*
ultrasonic region   *453*
intensity   *458*
loudness   *460*
bel   *460*

decibel   *461*
sound intensity level (decibel
   level)   *461*
constructive interference   *464*
destructive interference   *464*
phase difference   *464*
path-length difference   *464*

beats   *466*
beat frequency   *466*
pitch   *467*
Doppler effect   *467*
sonic boom   *471*
Mach number   *473*
quality   *476*

Important Concepts

- The sound frequency spectrum is divided into infrasonic ($f < 20$ Hz), audible (20 Hz $< f <$ 20 kHz), and ultrasonic ($f > 20$ kHz) frequency regions.
- The speed of sound in a medium depends on the elasticity of the medium and its density. In general, $v_{\text{solids}} > v_{\text{liquids}} > v_{\text{gases}}$.
- The intensity of a point source is inversely proportional to the square of the distance from the source.
- The sound intensity level is a logarithmic function of the sound intensity and is expressed in decibels (dB).
- Sound wave interference of two point sources depends on phase difference as related to path-length

difference. Sound waves that arrive at a point in phase reinforce each other (constructive interference); sound waves that arrive at a point out of phase cancel each other (destructive interference).

- The Doppler effect depends on the velocities of the sound source and observer relative to still air. When the relative motion of the source and observer is toward each other, the observed pitch increases; when the relative motion of source and observer is away from each other, the observed pitch decreases.

Important Equations

Speed of Sound (in meters per second):

$$v = (331 + 0.6T_C) \text{ m/s} \tag{14.1}$$

Intensity of a Point Source:

$$I = \frac{P}{4\pi R^2} \quad \text{and} \quad \frac{I_2}{I_1} = \left(\frac{R_1}{R_2}\right)^2 \tag{14.2, 14.3}$$

Intensity Level (in decibels):

$$\beta = 10 \log \frac{I}{I_o} \quad \text{where } I_o = 10^{-12} \text{ W/m}^2 \tag{14.4}$$

Phase Difference (where ΔL is path-length difference):

$$\Delta\theta = \frac{2\pi}{\lambda}(\Delta L) \tag{14.5}$$

Condition for Constructive Interference:

$$\Delta L = n\lambda \quad (n = 0, 1, 2, 3, \dots) \tag{14.6}$$

Condition for Destructive Interference:

$$\Delta L = m\left(\frac{\lambda}{2}\right) \quad (m = 1, 3, 5, \dots) \tag{14.7}$$

Beat Frequency:

$$f_b = |f_1 - f_2| \tag{14.8}$$

Doppler Effect:

$$f_o = \left(\frac{v}{v \pm v_s}\right)f_s = \left(\frac{1}{1 \pm \dfrac{v_s}{v}}\right)f_s \tag{14.11}$$

where v_s = *speed of source*
and v = *speed of sound*
$\begin{cases} - \textit{for source moving toward stationary observer} \\ + \textit{for source moving away from stationary observer} \end{cases}$

$$f_o = \left(\frac{v \pm v_s}{v}\right)f_s = \left(1 \pm \frac{v_o}{v}\right)f_s \tag{14.14}$$

where v_o = *speed of observer*
and v = *speed of sound*
$\begin{cases} + \textit{for observer moving toward stationary source} \\ - \textit{for observer moving away from stationary source} \end{cases}$

Mach Number:

$$M = \frac{v_s}{v} \tag{14.16}$$

Natural Frequencies of Organ Pipe Open on Both Ends:

$$f_n = n\left(\frac{v}{2L}\right) = nf_1 \quad (n = 1, 2, 3, \dots) \tag{14.17}$$

Natural Frequencies of Organ Pipe Closed on One End:

$$f_m = m\left(\frac{v}{4L}\right) = mf_1 \quad (m = 1, 3, 5, \dots) \tag{14.18}$$

Exercises

14.1 Sound Waves
and
14.2 The Speed of Sound

1. A sound wave with a frequency of 15 Hz is in what region of the sound spectrum: (a) audible, (b) infrasonic, (c) ultrasonic, (d) supersonic?

2. A sound wave in air (a) is longitudinal, (b) is transverse, (c) has longitudinal and transverse components, (d) travels faster than a sound wave through a liquid.

3. The speed of sound is generally greatest in (a) solids, (b) liquids, (c) gases, (d) a vacuum.

4. The speed of sound in air (a) is about 1/3 km/s, (b) is about 1/5 mi/s, (c) depends on temperature, (d) all of these.

5. Suggest a possible explanation of why some flying insects produce buzzing sounds and some do not.

6. Explain why sound travels faster in warmer air than in colder air.

7. Two different-frequency sounds are emitted from a single loudspeaker. Which one will reach your ear first?

8. The speed of sound in air is temperature dependent. What effect should humidity have, if any?

9. The wave speed in a liquid is given by $v = \sqrt{B/\rho}$, where B is the bulk modulus and ρ is the density. Show that this equation is dimensionally correct. What about $v = \sqrt{Y/\rho}$ for a solid (Y is Young's modulus)?

10. ■ What is the speed of sound in air at (a) 10°C and (b) 20°C?

11. ■ The speed of sound in air on a summer day is 350 m/s. What is the air temperature?

12. ■ The thunder from a lightning flash is heard by an observer 3.0 s after she sees the flash. What is the approximate distance to the lightning strike in (a) kilometers and (b) miles?

13. ■ Sonar is used to map the ocean floor. If an ultrasonic signal is received 2.0 s after it is emitted, how deep is the ocean floor?

14. ■■ Particles approximately 3.0×10^{-2} cm in diameter are to be scrubbed loose in an aqueous ultrasonic cleaning bath. Above what frequency should the bath be operated to produce wavelengths of this size and smaller?

15. ■■ A tuning fork vibrates at a frequency of 256 Hz. What is the wavelength of the sound coming from the fork when the air temperature is (a) 0°C and (b) 20°C?

16. ■■ Brass is an alloy of copper and zinc. Does the addition of zinc to copper cause an increase or a decrease in the speed of sound in brass rods as compared to copper rods?

17. ■■ The speed of sound in steel is about 4.5 km/s. A steel rail is struck with a hammer, and there is an observer 0.30 km away with one ear to the rail. (a) How much time will elapse from the time the sound is heard through the rail until the time it is heard through the air? Assume that the air temperature is 20°C and that no wind is blowing. (b) How much time would elapse if the wind were blowing toward the observer at 36 km/h from where the rail was struck?

18. ■■ At a baseball game on a cool day (air temperature of 16°C), a fan hears the crack of the bat 0.25 s after observing the batter hit the ball. How far is the fan from home plate?

19. ■■ A 2000-hertz tone is sounded when the air temperature is 20°C and then again at 10°C. What is the percentage change in the wavelength of the sound between the higher and the lower temperature?

20. ■■ A person holds a rifle horizontally and fires at a target. The bullet has a muzzle speed of 200 m/s, and the person hears the bullet strike the target 1.00 s after firing it. The air temperature is 72°F. What is the distance to the target?

21. ■■ A hiker shouts and hears the echo reflected from a cliff wall 3.40 s later. If the air temperature is 20°C, how far is the hiker from the cliff? (Assume no wind.)

22. ■■ The driver of a truck parked on a mountain road gives the air horn a quick blast toward a rock cliff 0.760 km away. If he hears the echo 4.50 s later, what is the approximate air temperature at this mountain elevation?

23. ■■■ Sound propagating through air at 30°C passes through a vertical cold front into air that is 4.0°C. If the sound has a frequency of 2400 Hz, by what percentage does its wavelength change in crossing the boundary?

14.3 Sound Intensity and Sound Intensity Level

24. If the air temperature increases, would the sound intensity from a constant output point source (a) increase, (b) decrease, or (c) remain unchanged?

25. The decibel scale is referenced to a standard intensity of (a) 1.0 W/m², (b) 10^{-12} W/m², (c) normal conversation, (d) the threshold of pain.

26. The Richter scale, used to measure the intensity level of earthquakes, is a logarithmic scale, as is the decibel scale. Why are such scales used?

27. Can there be negative decibel levels, such as −10 dB? If so, what would these mean?

28. ■ Calculate the intensity generated by a 1.0-watt point source of sound at a location (a) 3.0 m and (b) 6.0 m from it.

29. ■ What happens to the intensity of sound at a location if the distance from the point source of sound is halved?

30. ■ By how many times must the distance from a point source be increased to reduce the sound intensity by half?

31. ■ Calculate the intensity level for (a) the threshold of hearing and (b) the threshold of pain.

32. ■ Find the intensity levels in decibels for sounds with intensities of (a) 10^{-2} W/m^2, (b) 10^{-6} W/m^2, and (c) 10^{-15} W/m^2.

33. ■ If the intensity of one sound is 10^{-4} W/m^2 and the intensity of another is 10^{-2} W/m^2, what is the difference in their intensity levels?

34. ■ Is a 100-watt speaker twice as loud as a 50-watt one at the same distance?

35. ■■ What is the intensity of a sound that has an intensity level of (a) 50 dB and (b) 90 dB?

36. ■■ Noise levels for some commercial jet aircraft are given in Table 14.3. Which are the lowest and highest intensities for (a) takeoff and (b) landing?

TABLE 14.3 Takeoff and Landing Noise Levels for Some Common Commercial Jet Aircraft *

Aircraft	Takeoff noise (dB)	Landing noise (dB)
737	85.7–97.7	99.8–105.3
747	89.5–110.0	103.8–107.8
DC-10	98.4–103.0	103.8–106.6
L-1011	95.9–99.3	101.4–102.8

*Noise level readings taken from 198 m (650 ft). Range depends on the aircraft model and the type of engine used.

37. ■■ What would be the intensity level of a 23-decibel sound after being amplified (a) 10 thousand times, (b) a million times, (c) a billion times?

38. ■■ A tape player has a signal-to-noise ratio of 53 dB. How many times larger is the intensity of the signal than that of the background noise?

39. ■■ The sound intensity levels for a machine shop and a quiet library are 90 dB and 40 dB, respectively. (a) What is the intensity of each? (b) How many times greater is the intensity of the sound in the machine shop than that in the library?

40. ■■ A dog's barking has a sound intensity level of 40 dB. What would the intensity level be if two identical dogs were barking?

41. ■■ At a rock concert, the average sound intensity level for a person in a front-row seat is 110 dB for a single band. If all the bands scheduled to play produce sound of that same intensity, how many of them would have to play simultaneously for the sound level to be at or above the threshold of pain?

42. ■■ At a distance of 10.0 m from a point source, the intensity level is measured to be 70 dB. At what distance from the source will the intensity level be 40 dB?

43. ■■ At a 4th of July celebration, a firecracker explodes (•Fig. 14.19). Considering the firecracker to be a point source, what are the intensities heard by observers at points B, C, and D relative to that for the observer at A?

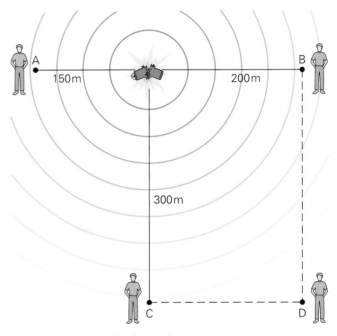

•FIGURE 14.19 **A big bang** See Exercise 43.

44. ■■ A person standing 4.0 m from a wall shouts such that the sound strikes the wall with an intensity of 2.5×10^{-4} W/m^2. Assuming that the wall absorbs 20% of the incident energy and reflects the rest, what is the sound intensity level just before and after the sound is reflected?

45. ■■ A gas-powered lawnmower is rated at 95 dB. (a) What is the sound intensity for this mower? (b) How many times more intense is the sound of this mower than that of an electric-powered mower rated at 83 dB?

46. ■■■ A 1000-hertz tone issuing from a loudspeaker has an intensity level of 100 dB at a distance of 2.5 m. If the speaker is assumed to be a point source, how far from the speaker will the sound have undiminished intensity levels (a) of 60 dB and (b) barely enough to be heard?

47. ■■■ A bee produces a buzzing sound that is barely audible to a person 3.0 m away. How many bees would

have to be buzzing at that distance to produce a sound with an intensity level of 50 dB?

14.4 Sound Phenomena
and
14.5 The Doppler Effect

48. Beats are the direct result of (a) interference, (b) refraction, (c) diffraction, (d) the Doppler effect.

49. Traffic radar is based on (a) beats, (b) the Doppler effect, (c) interference, (d) the sonic boom.

50. Do interference beats have anything to do with the "beat" of music? Explain.

51. (a) Is there a Doppler effect if a sound source and an observer are moving with the same velocity? (b) What would be the effect if a moving source accelerated toward a stationary observer?

52. How fast would a "jet fish" have to swim to create an aquatic sonic boom?

53. As a person walks *between* a pair of loudspeakers that produce tones of the same amplitude and frequency, he hears a varying sound intensity. Explain.

54. Two sound waves with the same wavelength, 0.50 m, arrive at a point after having traveled (a) 2.50 m and 3.75 m and (b) 3.25 m and 8.25 m, respectively. What type of interference occurs in each case?

55. ■ Two adjacent point sources, A and B, are directly in front of an observer and emit identical 1000-hertz tones. To what closest distance behind source B would source A have to be moved for the observer to hear no sound? (Assume that the air temperature is 20°C and ignore intensity fall-off with distance.)

56. ■ A violinist and a pianist simultaneously sound notes with frequencies of 436 Hz and 440 Hz, respectively. What beat frequency will the musicians hear?

57. ■ A violinist tuning an instrument to a piano note of 264 Hz detects three beats per second. What are the possible frequencies of the violin tone?

58. ■ 🌐 On a 20°C day with no wind blowing, the frequency heard by a moving person from a 500-hertz stationary siren is 520 Hz. (a) Is the person moving toward the siren or away from the siren? (b) What is the person's speed?

59. ■ 🌐 What is the frequency heard by a person driving 50 km/h directly towards a factory whistle (f = 800 Hz) if the air temperature is 0°C?

60. ■■ 🌐 While standing near a railroad crossing, you hear a train horn. The frequency emitted by the horn is 400 Hz. If the train is traveling at 90.0 km/h and the air temperature is 25°C, what is the frequency you hear (a) when the train is approaching and (b) after it has passed?

61. ■■ Two identical strings on different cellos are tuned to the 440-hertz A note. The peg holding one of the strings slips, so its tension is decreased by 1.5%. What is the beat frequency heard when the strings are then played together?

62. ■■ 🌐 How fast, in kilometers per hour, must a sound source be moving toward you to make the observed frequency 5.0% greater than the true frequency? (Assume that the speed of sound is 340 m/s.)

63. ■■ What is the half-angle of the shock wave of a jet aircraft just as it breaks the sound barrier?

64. ■■ On transatlantic flights, the supersonic transport (SST) *Concorde* flies at a speed of Mach 1.5. What is the half-angle of the conical shock wave formed by the *Concorde* at this speed?

65. ■■ The half-angle of the conical shock wave formed by a supersonic jet is 35°. What are (a) the Mach number of the aircraft and (b) the actual speed of the aircraft if the air temperature is −20°C?

66. ■■■ Two point-source loudspeakers are a certain distance apart and a person stands 12.0 m in front of one of them on a line perpendicular to the baseline of the speakers. If the speakers emit identical 1000-hertz tones, what is their minimum nonzero separation so the observer hears little or no sound? (Take the speed of sound as exactly 340 m/s.)

67. ■■■ 🌐 A bystander hears a siren vary in frequency from 476 Hz to 404 Hz as a fire truck approaches, passes by, and moves away on a straight street (•Fig. 14.20). What is the speed of the truck? (Take the speed of sound in air to be 343 m/s.)

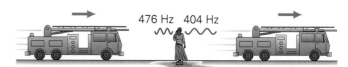

•**FIGURE 14.20   The siren's wail**   See Exercise 67.

14.6 Musical Instruments and Sound Characteristics

68. The human ear can hear tones best at (a) 1000 Hz, (b) 4000 Hz, (c) 6000 Hz, (d) all frequencies.

69. The quality of sound depends on its (a) waveform, (b) frequency, (c) speed, (d) intensity.

70. (a) After a snowfall, why does it seem particularly quiet? (b) Why do empty rooms sound hollow? (c) Why do people's voices sound fuller or richer when they sing in the shower?

71. Why aren't the frets on a guitar evenly spaced?

72. Is it possible for an open and a closed organ pipe of the same length to produce notes of the same frequency? Justify your answer.

73. When you blow across the top of a bottle with water in it, why does the frequency of the sound increase with increasing levels of water?

74. ■ The first three natural frequencies of an organ pipe are 126 Hz, 378 Hz, and 630 Hz. (a) Is the pipe an open or closed one? (b) Taking the speed of sound in air to be 340 m/s, find the length of the pipe.

75. ■ A closed organ pipe has a fundamental frequency of 528 Hz (a C note) at 20°C. What is the fundamental frequency of the pipe when the temperature is 0°C?

76. ■ The human ear canal is about 2.5 cm long. It is open at one end and closed at the other (see Fig. 1 in the Insight on p. 458). What is the fundamental frequency of the ear canal (at 20°C)? To what frequency is the ear most sensitive?

77. ■■ A closed organ pipe has a length of 0.800 m. At 20°C, what are the frequencies of (a) the second harmonic and (b) the third harmonic?

78. ■■ Both an open organ pipe and a closed organ pipe have lengths of 0.52 m at 20°C. What is the fundamental frequency of each pipe?

79. ■■ Using the sign convention of Eqs. 14.11 and 14.14, show that the general equation of the Doppler effect for a moving source and a moving observer is given by

$$f_o = f_s \left(\frac{v \pm v_o}{v \mp v_s} \right)$$

80. ■■ A fire truck travels at a speed of 90 km/h with its siren emitting sound at a frequency of 500 Hz. What is the frequency heard by a passenger in a car traveling at 65 km/h in the opposite lane to the fire truck, (a) approaching it and (b) moving away from it? [Hint: See Exercise 79, and take the speed of sound to be 354 m/s (a hot day).]

81. ■■ A tuning fork with a frequency of 440 Hz is held above a resonance tube partially filled with water. Assuming that the speed of sound in air is 342 m/s, for what three smallest heights of the air column will resonance occur? [Hint: For resonance to occur, the frequency of the tuning fork must match that of the tube.]

82. ■■■ A closed organ pipe has a fundamental frequency of 660 Hz in air at 0°C. What is the fundamental frequency when the pipe is filled with the helium?

Additional Exercises

83. Two identical sources producing 440-hertz tones are located 6.97 m and 8.90 m from an observation point. If the

air temperature that day is 15°C, how do the waves interfere at the point?

84. How long does it take sound to travel 3.5 km in air if the temperature is 30°C?

85. If a person standing 30.0 m from a 550-hertz point source moves 5.0 m closer to the source, by what factor does the sound intensity change?

86. An office of a mail-order company has 50 computers, which generate a sound intensity level of 40 dB (from the keyboards). The office manager tries to cut the noise in half by removing 25 computers. Does he achieve his goal? What is the intensity level generated by 25 computers?

87. How fast must an observer be moving toward a 500-hertz siren so she hears double the frequency?

88. The intensity levels of two people's speech are 60.0 dB and 65.0 dB, respectively. What is the intensity level of the combined sounds?

89. The intensity level of a sound is 90 dB at a certain distance from its source. How much energy falls on an area of 1.5 m^2 in 5.0 s?

90. A jet flies at a speed of Mach 2.0. What is the half-angle of the conical shock wave formed by the aircraft? Can you tell its speed from this information?

91. The note A (440 Hz) is produced by a stringed musical instrument. The air temperature is 20°C. Approximately how many vibrations does the string make before the sound reaches a person 30 m away?

92. A person drops a stone into a deep well and hears the splash 3.16 s later. How deep is the well? (Assume the air temperature in the well to be 10°C.)

93. The frequency of an ambulance siren is 700 Hz. What are the frequencies heard by a stationary pedestrian as the ambulance approaches and then passes her at a speed of 90.0 km/h? (Assume the air temperature is 20°C.)

94. One hunter sees another who is 300 m away fire his rifle (smoke comes out of the barrel). If the air temperature is 5°C, how long will it be until the first hunter hears the shot? (Assume no wind.)

95. An open organ pipe has a length of 0.75 m. What would be the length of a closed organ pipe whose third harmonic ($m = 3$) is the same as the fundamental frequency of the open pipe?

CHAPTER 5

Electric Charge, Forces, and Fields

Very few processes in nature deliver such an enormous amount of energy to a small area in such a tiny fraction of a second as a bolt of lightning from a dark storm cloud. Yet not many people have experienced its tremendous power at close range; only a few hundred people are actually struck by lightning each year in the United States.

It might surprise you, therefore, to realize that you have almost certainly had an experience that is practically identical, from a physicist's point of view. Have you ever walked across a carpeted room and then gotten a tiny shock when you reached for a brass doorknob? Though the scale is dramatically different, the physical processes involved are essentially the same.

Electricity seems to have a flair for the dramatic. To most of us, the word conjures up striking images: a blaze of floodlights, crackling live wires, the skyline of a great city at night, or a flash of lightning. There is often a hint of anxiety in our associations, for we know that electricity can be dangerous. We also know, however, that electricity can be tamed, even domesticated. In the home or the office, we have come to take it for granted. Indeed, the extent to which we depend on it becomes evident only when the power goes off unexpectedly, giving us a dramatic reminder of the role that it plays in our daily lives. Yet less than a century ago there were no power lines crossing the land, no electric lights or appliances—none of the seemingly endless electrical applications surrounding us today.

We now know that electricity and magnetism are in fact related. Today, this "unified" electromagnetic force is recognized as one of the four fundamental forces (along with gravity, discussed in Chapter 7, and the strong and weak nuclear forces, discussed in Chapters 29 and 30). It is customary in introductory physics courses to consider the electrical part of the electromagnetic force before the magnetic part and only then to combine electricity and magnetism into electromagnetism. This will be our approach. Among other things, in this chapter you'll learn what the lightning bolt and the spark from the doorknob have in common.

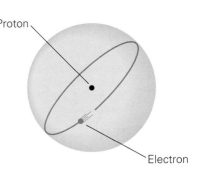

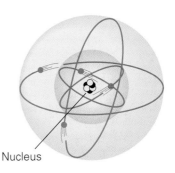

(a) Hydrogen atom

15.1 Electric Charge

Objectives To (a) distinguish between the two types of electric charge, (b) state the force law that operates between charged objects, and (c) understand and use the law of charge conservation.

What is electricity? Perhaps the broadest answer is that electricity is a collective term describing phenomena associated with the interaction between *electrically charged* objects. We start our study of electricity by investigating electric forces between charged objects *at rest*. This area of physics is called electro*statics*, because all the charges remain at rest.

Like mass, **electric charge** is a fundamental property of matter. Electric charge is fundamentally associated with atomic particles, the electron and the proton. The simplistic solar system model of the atom, shown in •Fig. 15.1, likens its structure to planets orbiting the Sun. The *electrons* are viewed as orbiting a nucleus, a core containing most of the atom's mass in the form of *protons* and electrically neutral particles called *neutrons*. As we saw in Section 7.5, the centripetal force that keeps the planets in their orbits about the Sun is supplied by gravity. Similarly, the force that keeps the electrons orbiting the nucleus is supplied by electrical attraction. However, there is an important distinction between the gravitational and electric forces.

There is apparently only one type of mass, and this gives rise to attractive gravitational forces only. Electric charge, however, comes in two types, distinguished by the labels positive (+) and negative (−). Protons carry a positive charge, and electrons carry a negative charge. Different combinations of the two kinds of charges can produce *either* attractive *or* repulsive forces.

The directions of the electric forces when charges interact with one another are given by the following simple principle, sometimes called the **law of charges** or the **charge–force law:**

Like charges repel each other, and unlike charges attract each other.

That is, two negatively charged particles or two positively charged particles experience mutually repulsive forces, whereas particles with opposite charges are attracted to each other (•Fig. 15.2). These forces are equal and opposite, and they act on different objects, in keeping with Newton's action–reaction law for forces.

(b) Beryllium atom

•**FIGURE 15.1 Simplistic model for atoms** The so-called solar system model of **(a)** a hydrogen atom and **(b)** a beryllium atom views the electrons (negatively charged) as orbiting the nucleus (positively charged), analogous to the planets orbiting the Sun. The electronic structure of atoms is actually much more complicated than this.

Note: Recall the discussion of Newton's third law in Section 4.4.

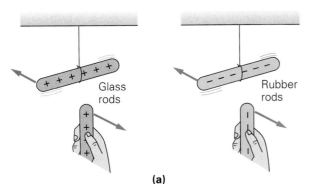

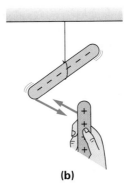

(a)

(b)

•**FIGURE 15.2 The charge–force law, law of charges** **(a)** Like charges repel. **(b)** Unlike charges attract.

TABLE 15.1 Particles and Electric Charge

Particle	Electric Charge*	Mass*
Electron	-1.602×10^{-19} C	$m_e = 9.109 \times 10^{-31}$ kg
Proton	$+1.602 \times 10^{-19}$ C	$m_p = 1.672 \times 10^{-27}$ kg
Neutron	0	$m_n = 1.674 \times 10^{-27}$ kg

*Even though the values are displayed to four significant figures, we will usually use only two or three in our calculations.

The charge of an electron and the charge of a proton are equal in magnitude, though opposite in sign. Abbreviated as e, the charge of the electron is taken as the fundamental unit of charge, since it is the smallest charge that has been observed in nature.* Thus the electric charge of an object is an integral multiple of fundamental charges (either positive or negative). Our general symbol for charge will be the letter q. For the charge of an object we can write

$$q = ne \qquad (15.1)$$

SI unit of charge: coulomb (C)

where n is an integer. We sometimes say that charge is "quantized," which means that it occurs only in integral multiples of the fundamental electronic charge. Conversely, mass does not appear to be quantized.

The SI unit of charge is the **coulomb** (**C**), named for the French physicist Charles A. Coulomb (1736–1806), who discovered a relationship between electric force and charge (to be considered later in this chapter). The charges and masses of the electron, proton, and neutron are given in Table 15.1.

We frequently use several terms when we discuss charged objects. Saying that an object has a **net charge** means that it has an excess of either positive or negative charges. As you will learn in Section 15.2, excess charges are commonly produced by a transfer of electrons. For example, if an object has a net charge of $+1.60 \times 10^{-19}$ C, it could be that one of its atoms has lost an electron. This means that the atom has one proton whose charge is no longer canceled by an electron, so the atom is no longer electrically neutral. (Such charged atoms are called *ions.*) The charge on the electron is a tiny fraction of a coulomb, and a coulomb's worth of net charge on an object is rarely seen in everyday situations. More typically we deal with net charges in the *microcoulomb* (μC, or 10^{-6} C) to *picocoulomb* (pC, or 10^{-12} C) range.

In dealing with any electrical phenomena, another important principle is **conservation of charge**:

The net charge of an isolated system remains constant.

The net charge may be other than zero, but it remains constant. Suppose, for example, that a system consists initially of two electrically neutral objects, and electrons are transferred from one to the other. The object with added electrons will then have a net negative charge, and the object with fewer electrons will have a net positive charge of equal magnitude. Thus, the net charge of the *system* remains zero. If we think about the universe as a whole, conservation of charge means that the net charge *of the universe* is constant, though no one knows what the numerical value of that net charge might be.

This principle does *not* mean that charged particles cannot be created or destroyed. Physicists have been studying this since the early part of the twentieth

*According to recent experiments, protons (as well as neutrons and other particles) are made up of particles called *quarks*, which carry charges of $\pm\frac{1}{3}$ and $\pm\frac{2}{3}$ of the electronic charge. There is experimental evidence of the existence of quarks within the nucleus, but free quarks have not been detected. According to current theory, that may be impossible (Chapter 30).

century, as you will learn in the chapters on modern physics. However, by conservation of charge, charged particles are created or destroyed only in pairs with equal and opposite charges. Like the other conservation laws of physics, the concept of charge conservation is called a law because no violation of it has ever been observed experimentally.

Example 15.1 ■ On the Carpet: Quantized Charge

(a) If you shuffle across a carpeted floor on a dry day and acquire a net charge of $-2.0\ \mu C$ (1 microcoulomb = $1\ \mu C = 10^{-6}$ C), will you have a deficiency or excess of electrons? (b) How many missing or extra electrons will you have?

Thinking It Through. (a) From the sign of the net charge, we can tell whether there is an excess of electrons. (b) Once established, since the charge of one electron is known, we can quantify their deficiency or excess.

Solution. First, we express the charge in coulombs and list what we are to find:

Given: $q = -2.0\ \mu C$ ———

$= -2.0 \times 10^{-6}$ C

$e = -1.60 \times 10^{-19}$ C (Table 15.1)

Find: (a) whether you have lost or gained electrons

(b) number of missing or excess electrons

(a) Since the sign of your net charge is negative and electrons carry a negative charge, you have acquired an excess of electrons.

(b) The net charge is made up of an integral number of electronic charges. Using Eq. 15.1, we have

$$n = \frac{q}{e} = \frac{-2.0 \times 10^{-6}\ C}{-1.60 \times 10^{-19}\ C/electron} = 1.3 \times 10^{13}\ electrons$$

As you can see, net charges usually involve huge numbers of electrons. This fact makes the addition of one electron, or even a million electrons, virtually impossible to detect. Since charge is quantized in extremely small packages, we can think of the electrons as flowing like a charged liquid and, in most cases, can ignore that the transferred charge is composed of individual particles. This is similar to our picture of a river as a moving continuous fluid, even though we know it is composed of many individual water molecules.

Follow-up Exercise. In this Example, (a) what type of charge does the carpet acquire? (b) How much charge does the carpet acquire? (c) Does it have an excess or deficiency of electrons? (d) How many electrons?

15.2 Electrostatic Charging

Objectives To (a) distinguish between conductors and insulators, (b) explain the operation of the electroscope, and (c) distinguish among charging by friction, conduction, induction, and polarization.

That there are two types of electric charges and attractive and repulsive forces can be demonstrated easily. Before learning how this is done, you need to be able to distinguish between electrical conductors and insulators. What distinguishes these broad groups of substances is their ability to conduct, or transmit, electric charge. Some materials, particularly metals, are good **conductors** of electric charge. Others, such as glass, rubber, and most plastics, are **insulators**, or poor electrical conductors. A comparison of the relative magnitudes of the conductivities of some materials is given in ●Fig. 15.3.

A general picture is that in conductors, *valence* electrons of atoms—the ones in the outermost orbits—are loosely bound. As a result, valence electrons can be easily removed from the atom and moved about in the conductor. That is, they are not permanently bound to a particular atom. (This electron mobility also plays

•**FIGURE 15.3 Conductors, semiconductors, and insulators** A comparison of the relative magnitudes of the electrical conductivities of various materials. (Not drawn to scale.)

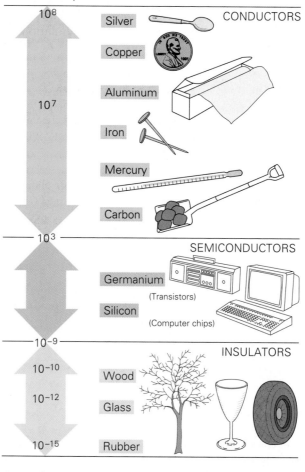

Relative magnitude   Material
of conductivity

CONDUCTORS

10^8 Silver

Copper

10^7 Aluminum

Iron

Mercury

Carbon

10^3 ———————— SEMICONDUCTORS

Germanium

(Transistors)

Silicon

(Computer chips)

10^{-9} ———————— INSULATORS

10^{-10} Wood

10^{-12} Glass

10^{-15} Rubber

Note: An electroscope can detect whether an object is electrically charged; if so, the electroscope can detect the sign of its charge.

a major role in thermal conduction, Section 11.4.) In insulators, however, the valence electrons are more tightly bound and aren't readily moved. (*Continuous* conduction of electric charge is called an *electric current*, which we will consider in more detail in Chapter 17.)

As Fig. 15.3 shows, there is also an intermediate class of materials called **semiconductors**. Their ability to conduct charge is much less than that of metals but much greater than that of insulators. The conductivity of semiconductors can be adjusted by adding certain types of atomic impurities in varying concentrations. Semiconductors form the basis of the transistors, solid-state circuits, and chips that have become the backbone of the modern computer industry.

The *electroscope* is a device that can be used to demonstrate the characteristics of electric charge (•Fig. 15.4). In its simplest form, it consists of a metal rod with a metallic bulb at one end and a pair of hanging foil "leaves," usually made of gold or aluminum, at the other end. This arrangement is insulated from its protective glass container by a rubber cork. When charged objects are brought close to the bulb, electrons in the bulb are either attracted or repelled, according to the charge–force law. For example, if a negatively charged rod is brought near the bulb, electrons are repelled, and the bulb is left with a positive charge. The electrons are conducted to the metal foil leaves, which separate because of a mutually repulsive electric force between them (Fig. 15.4b). Similarly, if a positively charged rod is brought near the bulb, the leaves also diverge. (Can you explain why?)

Notice that the net charge of the electroscope remains zero in these instances because it is isolated (insulated)—only the *distribution* of charge is altered. However, it is possible to give an electroscope (and other objects) a nonzero net charge by electrostatic charging.

Bulb

(a) Neutral electroscope has charges evenly distributed; leaves are close together.

Negatively charged rod

Positively charged rod

(b) Electrostatic forces cause leaves to separate.

•**FIGURE 15.4 The electroscope**
An electroscope can be used to determine whether an object is electrically charged. When a charged object is brought near the bulb, the leaves separate.

Charging by Friction

In general, **electrostatic charging** is a process by which an insulator or an insulated conductor receives a net charge. In one such charging process, when certain insulator materials are rubbed with cloth or fur, they become electrically charged. For example, if a hard rubber rod is rubbed with fur, the rod will acquire a net negative charge; rubbing a glass rod with silk will give the rod a net positive charge. This is called **charging by friction**. The transfer of charge is due to the contact between the materials and depends on the nature of the materials; the charges are not merely rubbed off by friction.

You have almost certainly experienced a result of frictional charging when, after walking across a carpet on a dry day, you get "zapped" by a spark when you reach for a metal object, such as a doorknob. This happens because you have become electrostatically charged; that is, you have picked up a net charge from the carpet. The charge produces an electric force great enough to *ionize* (free electrons from) the air molecules between your hand and the knob when your hand comes close to the metal knob. The resulting flow of charges gives rise to a spark discharge between hand and metal. This doesn't occur on humid days. With adequate humidity, a thin film of moisture on objects prevents the build-up of charge by conducting it away.

As with heat, when an electric charge moves in a conductor, it is common to talk about a flow. Like heat flow, the flow of electricity was once thought to be due to some type of fluid transfer. Ben Franklin proposed a single-fluid theory of electricity. He assumed that all bodies had some normal amount of "electrical fluid." When some of this was transferred, for example, by rubbing two bodies together, one body would then have an excess of it and the other a deficiency. Franklin indicated these conditions by plus and minus signs, respectively, which is the origin of the sign convention for charge.

Note: From an external viewpoint you cannot tell whether the rubber rod gained negative charges or the fur gained positive charges. In other words, moving electrons to the rubber rod results in the same physical situation as moving positive charges to the fur. However, because the rubber is an insulator and its electrons are therefore tightly bound, we might suspect that the fur lost electrons and the rubber gained them. In solids, the protons, being in the nuclei of the atoms, do not move; only electrons move. It is just a question of which material most easily loses electrons.

Charging by Contact or by Conduction

Bringing a charged rod close to an electroscope will reveal that the rod is charged, but it won't tell you *how* the rod is charged (positively or negatively). This distinction can be made, however, if the electroscope is first given a known type of charge. For example, electrons can be transferred from one object to another if the two objects touch, as illustrated in •Fig. 15.5a for an electroscope bulb and a negatively charged rod. The electrons in the rod are mutually repelled by one another. Some will be able to "escape" and move onto the electroscope. In this case we say the electroscope has been **charged by contact** or by **conduction** (Fig. 15.5b). "Conduction" refers to the flow of charge during the short period of time the electrons are in motion.

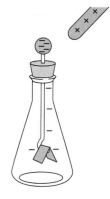

(a) Neutral electroscope is touched with negatively charged rod.

(b) Charges are transferred to bulb; electroscope has net negative charge.

(c) Negatively charged rod repels electrons; leaves separate further.

(d) Positively charged rod attracts electrons; leaves collapse.

•**FIGURE 15.5   Charging by contact or conduction**   **(a)** The electroscope is initially neutral (but the charges are separated), and a nearby charged rod is then touched to the bulb. **(b)** Charge is transferred to the electroscope. **(c)** When a rod of the same charge is brought near the bulb, the leaves separate further. **(d)** When an oppositely charged rod is brought nearby, the leaves collapse or move closer together.

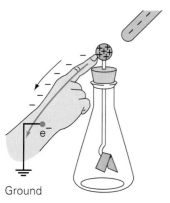

(a) Repelled by the negatively charged rod, electrons are transferred to ground through hand.

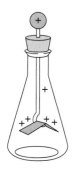

(b) Electroscope is left positively charged.

•**FIGURE 15.6   Charging by induction**   **(a)** Touching the bulb with a finger provides a path to ground for charge transfer. The symbol e⁻ stands for electron. **(b)** When the finger is removed, the electroscope has a net positive charge, opposite that of the rod.

If a negatively charged rod is brought close to the now negatively charged electroscope, the leaves will separate even further as more electrons are repelled onto them from the bulb (Fig. 15.5c). An oppositely (positively) charged rod will cause the leaves to collapse by attracting some electrons up to the bulb and away from the leaves (Fig. 15.5d).

Charging by Induction

Since it is electrons that are transferred in conductors, how can an electroscope ever become positively charged? This can be done by a process called **charging by induction**. Touching the bulb with a finger *grounds* the electroscope—that is, provides a path by which electrons can escape from the bulb (•Fig. 15.6). Then when a negatively charged rod is brought close to (but not touching) the bulb, the rod repels electrons from the bulb into the finger and down into the Earth. Removing the finger while the charged rod is kept nearby leaves the electroscope with a net positive charge, because the removed electrons have no way of flowing back once the ground path is removed.

Charge Separation by Polarization

Both charging by contact and induction involve a removal of charge from an object. However, an object can have some charge moved *within* it to give different regions of charge yet keep a net charge of zero. In this case, induction brings about **polarization**, or separation of charge (•Fig. 15.7). Now you can understand why, when you rub a balloon on your hair or sweater, the balloon will stick to the wall or ceiling. The balloon is charged by friction, and the charged balloon induces an opposite charge on the surface, creating an attractive electric force.

Electrostatic charging can be annoying (even dangerous), but it can also be beneficial in a variety of practical applications. Clothes and papers often stick together because of static cling, and an electrostatic spark discharge can start a fire or cause an explosion in the presence of a flammable gas. But the air we breathe is cleaner because of electrostatic precipitators used in smokestacks. In these devices, electrical discharges cause the particles that are a byproduct of fuel combustion to acquire a charge. This particulate matter is then removed from the flue gases by means of electric force. On a smaller scale, electrostatic air cleaners are available for the home. Another, now almost indispensable application is the electrostatic copier (see the Insight on p. 490).

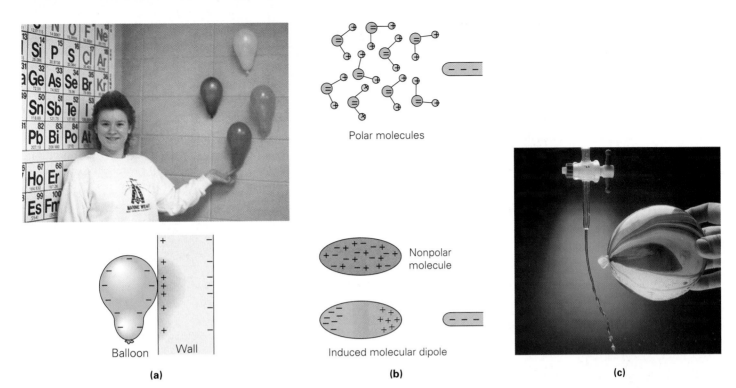

•**FIGURE 15.7 Polarization** (a) When the balloons are charged by friction and placed in contact with the wall, an opposite charge is induced on the wall's surface, to which the balloons then stick by the force of electrostatic attraction. (b) Some molecules, such as those of water, are polar in nature—that is, they have separated regions of positive and negative charge. But even some molecules that are not normally polar can be polarized by the presence of a nearby charged object. The electric force induces a separation of charge, making the molecules into induced molecular dipoles. (c) A stream of water bends toward a charged balloon. No charge needs to be induced in the water molecules, because they are already natural dipoles. The charged balloon simply attracts the ends of the molecules that carry the opposite charge.

15.3 Electric Force

Objectives: To (a) understand Coulomb's law and (b) use it to calculate the electric force between charged particles.

The relative directions of the electric forces on mutually interacting charges are given by the charge–force law. However, what about the *magnitude* or strength of the electric force? This was investigated by Charles de Coulomb, using a delicate balance to measure the force. He found that the magnitude of the electric force between two point charges (q_1 and q_2) depended on the product of the charges and varied inversely as the square of the distance between them. That is, $F_e \propto q_1 q_2 / r^2$. (Note that this is an inverse-square relationship, mathematically similar to the one for the force of gravity between two point masses, $F_g \propto m_1 m_2 / r^2$.) Here the subscripts "e" and "g" help us distinguish electric force from gravitational force.

Like Cavendish's measurements for the determination of the universal gravitational constant G, where $F_g = G m_1 m_2 / r^2$ (Section 7.4), Coulomb's measurements provided a constant of proportionality, so the electric force can be written in equation form. Thus, the magnitude of the electric force between two electric point charges is described by an equation called **Coulomb's law:**

$$F_e = \frac{k q_1 q_2}{r^2} \qquad (\textit{point charges only}) \qquad (15.2)$$

where r is the distance between the charges (•Fig. 15.8a) and k is a constant:

$$k = 8.988 \times 10^9 \, \text{N·m}^2/\text{C}^2 \approx 9.00 \times 10^9 \, \text{N·m}^2/\text{C}^2$$

Note: Compare Eq. 7.15.

Note: Coulomb's law gives the electric force, but only between *point charges*.

Note: In calculations, we will take k to be *exact* at 9.00×10^9 N·m^2/C^2 for significant figure purposes.

Xerography, Electrostatic Copiers, and Laser Printers

Xerography, coined from the Greek words *xeros* (meaning "dry") and *graphein* (meaning "to write"), refers to a dry process by which almost any printed material can be copied. This process makes use of a photoconductor, which is a light-sensitive semiconductor, such as selenium. When kept in darkness, a photoconductor is a good insulator that can be electrostatically charged. However, when light strikes the material, it becomes conductive and the electric charge can be removed.

In transfer xerography, a photoconductor-coated plate, drum, or belt is electrostatically charged and then receives a projected image of the page to be copied (Fig. 1). The illuminated portions of the photoconductor coating become conducting and are discharged, but the areas corresponding to

FIGURE 1 Printing by xerography (a) A typical xerographic copier in cutaway front view. ① The drum is charged positively at charging electrode C_1. ② The light reflected from the original page is focused onto the drum, creating a "positive charge" image. ③ Then negatively charged toner is sprayed onto the drum, making a true ink positive of the original. The paper is carried by conveyor belt from the blank paper feed to a positive charging electrode C_2 and then on to the selenium-coated drum. ④ The positive toner image on the drum is transferred to the paper. ⑤ The paper is then carried through hot rollers, which press and "set" the toner, and the copy is delivered to the output tray. **(b)** Detail of the processes occurring at the numbered locations in (a).

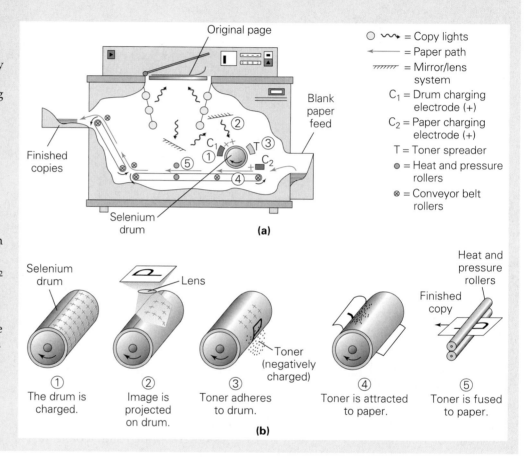

(a)

= Copy lights
= Paper path
= Mirror/lens system
C_1 = Drum charging electrode (+)
C_2 = Paper charging electrode (+)
T = Toner spreader
= Heat and pressure rollers
= Conveyor belt rollers

(b)

① The drum is charged.
② Image is projected on drum.
③ Toner adheres to drum.
④ Toner is attracted to paper.
⑤ Toner is fused to paper.

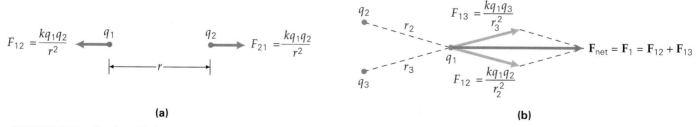

(a)   **(b)**

•**FIGURE 15.8 Coulomb's law (a)** The mutual electrostatic forces on two point charges are equal and opposite. **(b)** For a configuration of two or more point charges, the force on a particular charge is the vector sum of the forces on it due to all the other charges.

the dark print remain charged. Essentially, this creates an electric charge copy of the original sheet. Then the photoconductor copy comes into contact with a negatively charged powder called toner, or dry ink. The toner is attracted to and adheres to the charged regions. Paper is placed over the inked photoconductor and given a positive charge. The toner is then attracted to the paper, and heating causes it to be permanently fused to the paper. All of this takes place very quickly—out comes your copy!

The laser printer used with computers is basically a xerographic machine. In this case, there is no "original copy" per se; the information to be printed is stored in the computer. A laser (Chapter 27) scans back and forth across a rotating, charged drum in the printer (Fig. 2). The laser beam passes through a device called a modulator, which turns the beam on or off, according to the signals received from the computer. Wherever the fine light beam strikes the drum, that point is discharged, while unilluminated areas, corresponding to the letters, remain charged. In this way a charged image of what is to be printed is produced. The rest of the printing process then takes place as in xerography.

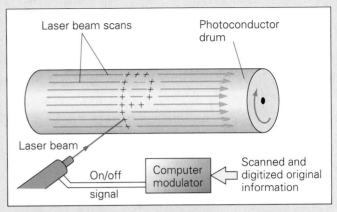

FIGURE 2 Laser printer   A computer-controlled laser scans the charged photoconductor drum, causing the charge to bleed off where the beam strikes. When the laser beam is turned off, charged regions remain, which can be reproduced as in the xerography process.

There are equal and opposite forces on the charges (Fig. 15.8a). These forces are sometimes referred to as electro*static* forces, emphasizing the fact that Coulomb's law applies to fixed, or *static*, charges.

In some instances, we are concerned with the force on a particular charge in a configuration of two or more charges. The net electric force on a particular charge is simply the vector sum of the forces on it due to all the other charges (Fig. 15.8b), as we shall see in the next two Examples.

Conceptual Example 15.2

Free of Charge: Electric Forces

A rubber comb pulled through dry hair can acquire a net negative charge. The comb can then be used to pick up small pieces of uncharged paper. This seems to violate Coulomb's force law: Since the paper has zero net charge, you might expect there to be no electric force. Explain how the attraction comes about.

Reasoning and Answer.   When the charged comb is near the paper, the paper becomes polarized (•Fig. 15.9). The positive end of the paper is closer to the comb than the negative end. Since the electric force varies *inversely* with the square of the distance, the attraction (F_1) between the comb and the positive end is greater than the repulsion (F_2) between the comb and the negative end, and the vector sum of the forces on the paper is toward the comb. Instead of contradicting Coulomb's force law, this phenomenon depends crucially on it and illustrates clearly that the electrostatic forces get weaker with greater distance!

Follow-up Exercise.   Does this phenomenon tell you the sign of the charge on the comb? Why or why not?

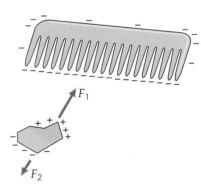

•**FIGURE 15.9 Comb and paper**
See Conceptual Example 15.2.

Example 15.3 ■ Coulomb's Law: Reviewing Vector Addition

(a) Two point charges of $-1.0\ \mu C$ and $+2.0\ \mu C$ are separated by a distance of 0.30 m (•Fig. 15.10a). What is the electrostatic force on each particle? (b) A configuration of three charges is shown in Fig. 15.10b. What is the electrostatic force on q_3?

Thinking It Through. (a) For the two point charges, we use Coulomb's law (Eq. 15.2), noting that the forces are attractive (why?). (b) Here we must add the two (vector) forces acting on q_3 due to q_1 and q_2. The angle θ is known (how?), which will enable us to calculate the x and y force components.

Solution. Listing the data and converting microcoulombs to coulombs, we have

Given: (a) $q_1 = -1.0\ \mu C$ ——— $= -1.0 \times 10^{-6}$ C   Find: (a) $\mathbf{F}_{12}$ and $\mathbf{F}_{21}$
(b) $\mathbf{F}_3$

$q_2 = +2.0\ \mu C$ ——— $= +2.0 \times 10^{-6}$ C

$r = 0.30$ m

(b) Data given in Figure 15.10b. Convert charges to coulombs as in (a).

(a) Equation 15.2 gives the magnitude of the force acting on each point charge:

$$F_{12} = F_{21} = F = \frac{kq_1q_2}{r^2} = \frac{(9.00 \times 10^9\ \text{N·m}^2/\text{C}^2)(1.0 \times 10^{-6}\ \text{C})(2.0 \times 10^{-6}\text{C})}{(0.30\ \text{m})^2}$$

$$= 0.20\ \text{N}$$

Only the magnitude of the charges is used since Coulomb's law gives only the force magnitude. However, because the charges are of opposite sign, we know that the force is attractive. By Newton's third law, F_{21} is equal to F_{12} but in the opposite direction (see the Problem-Solving Hint on p. 493).

(b) We must vectorially add the forces F_{31} and F_{32}. Since all the charges are positive, the forces are repulsive as shown in the vector diagram in Fig. 15.10b. Since $q_1 = q_2$ and the charges are equidistant from q_3, then F_{31} and F_{32} are of equal magnitude.

Note from the figure that $r_{31} = r_{32} = 0.50$ m. (Why?) With data from the figure we again use Eq. 15.2:

$$F_{31} = F_{32} = \frac{kq_2q_3}{r_{32}^{\ 2}}$$

$$= \frac{(9.00 \times 10^9\ \text{N·m}^2/\text{C}^2)(2.5 \times 10^{-6}\ \text{C})(3.0 \times 10^{-6}\ \text{C})}{(0.50\ \text{m})^2} = 0.27\ \text{N}$$

Taking into account the directions of F_{31} and F_{32}, we see by symmetry that the y components of the vectors cancel. Thus F_3 acts along the x axis:

$$F_3 = F_{31}\cos\theta + F_{32}\cos\theta = 2F_{32}\cos\theta$$

$q_1 = -1.0\ \mu C$ $q_2 = +2.0\ \mu C$

$\xrightarrow{}\ F_{12}$ $F_{21}\ \longleftarrow$

|← 0.30 m →|

(a)

•**FIGURE 15.10 Coulomb's law and electrostatic forces** See Example 15.3.

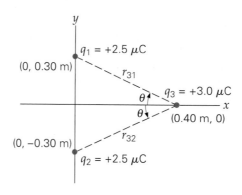

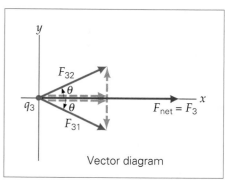

Vector diagram

(b)

since $F_{31} = F_{32}$. The angle θ can be determined from the distance triangles; that is,

$$\tan \theta = \frac{y}{x} = \frac{0.30 \text{ m}}{0.40 \text{ m}} = 0.75, \quad \text{and} \quad \theta = \tan^{-1}(0.75) = 37°$$

So

$$F_3 = 2F_{32}\cos\theta = 2(0.27 \text{ N})\cos 37° = 0.43 \text{ N}$$

in the $+x$ direction.

Follow-up Exercise. In part (b) of this Example, what is the force on q_3 if that charge is instead located at (0.80 m, 0)? Why is this answer closer to the sum of the two force magnitudes ($F_{31} + F_{32}$) than that in the Example?

Problem-Solving Hint

The signs of the charges can be used explicitly in Eq. 15.2. If they are, a positive value for F indicates a repulsive force and a negative value an attractive force. However, it is usually less confusing to calculate the magnitude of the force by using only the magnitude of the charges (as was done in Example 15.3) and to apply the charge–force law to determine whether the force is attractive or repulsive. This latter approach will be the one we use.

The forces in Example 15.3 are not large by everyday standards. But on the atomic scale, even these tiny forces can produce huge accelerations on very light particles such as electrons. Also, the inverse-square relationship can make a huge difference, since atomic and nuclear distances are so small. Consider the following Example involving protons.

Example 15.4 ■ Inside the Nucleus: Repulsive Electrostatic Forces

(a) What is the magnitude of the repulsive electrostatic force between two protons in a nucleus? Take the distance between the centers of nuclear protons to be 3.0×10^{-15} m. (b) If these protons were released from rest, how would the magnitude of their initial acceleration compare to that of the acceleration due to gravity on the Earth's surface, g?

Thinking It Through. (a) We must apply Coulomb's law to find the repulsive force. (b) To find the initial acceleration, we use Newton's second law ($\mathbf{F} = m\mathbf{a}$).

Solution. Listing the known quantities, we have

Given: $r = 3.0 \times 10^{-15}$ m
$\quad\quad q_1 = q_2 = +1.60 \times 10^{-19}$ C (Table 15.1)
$\quad\quad m_p = 1.67 \times 10^{-27}$ kg (Table 15.1)

Find: (a) F_e (magnitude of force)
$\quad\quad$ (b) a/g (magnitude of acceleration compared to g)

(a) Using Coulomb's law (Eq. 15.2), we have

$$F_e = \frac{kq_1q_2}{r^2} = \frac{(9.00 \times 10^9 \text{ N·m}^2/\text{C}^2)(1.60 \times 10^{-19} \text{ C})(1.60 \times 10^{-19} \text{ C})}{(3.0 \times 10^{-15} \text{ m})^2} = 26 \text{ N}$$

(This is about 6 lb of force, so we might expect a huge acceleration in part [b].)

(b) If it acted alone on a proton, this force would produce an acceleration

$$a = \frac{F_e}{m_p} = \frac{26 \text{ N}}{1.67 \times 10^{-27} \text{ kg}} = 1.6 \times 10^{28} \text{ m/s}^2$$

Then

$$\frac{a}{g} = \frac{1.6 \times 10^{28} \text{ m/s}^2}{9.8 \text{ m/s}^2} = 1.6 \times 10^{27}$$

That is, $a = (1.6 \times 10^{27})g \approx 10^{27} g$! (*A whole lot bigger!*)

Large atoms contain many protons in their nuclei, so the repulsive force on any of these protons would be even larger, which should tend to cause the nucleus to fly apart.

Since this doesn't generally happen, there must be a stronger attractive force holding the nucleus together. This is the nuclear (or strong) force, which will be discussed in Chapters 29 and 30.

Follow-up Exercise. Suppose you could place a free proton on the ground, and wished to place a second one directly above the first so that the weight of the second proton would be exactly balanced by the electric repulsion between them. How far apart must the protons be?

Although there is a striking similarity between the mathematical form of the expressions for the electric and gravitational forces, there is a huge difference in their relative strengths, as is shown in the following Example.

Example 15.5 ■ Inside the Atom: Electric versus Gravitational Force

Compare the magnitudes of the electric and gravitational forces between a proton and an electron. Express your answer as a ratio of electric force to gravitational force. In other words, how many times larger is the electric force than the gravitational force?

Thinking It Through. We use the charges to compute the electric force and the masses for the gravitational force. The separation distances are not given, but since both the electric and gravitational forces are inverse-square laws, when we compute the ratio, r will cancel out.

Solution. The charges and masses of the particles are known (Table 15.1), and we also know the values of the constants in the equations for the electric and gravitational forces. Thus, we have

Given: $q_e = -1.60 \times 10^{-19}$ C   *Find:* $\dfrac{F_e}{F_g}$ (ratio of forces)
$q_p = +1.60 \times 10^{-19}$ C
$m_e = 9.11 \times 10^{-31}$ kg
$m_p = 1.67 \times 10^{-27}$ kg

The expressions for the forces are

$$F_e = \frac{k q_e q_p}{r^2} \quad \text{and} \quad F_g = \frac{G m_e m_p}{r^2}$$

Forming a ratio of magnitudes for comparison purposes (and to cancel r) gives

$$\frac{F_e}{F_g} = \frac{k q_e q_p}{G m_e m_p}$$

$$= \frac{(9.00 \times 10^9 \text{ N·m}^2/\text{C}^2)(1.60 \times 10^{-19} \text{ C})^2}{(6.67 \times 10^{-11} \text{ N·m}^2/\text{kg}^2)(9.11 \times 10^{-31} \text{ kg})(1.67 \times 10^{-27} \text{ kg})} = 2.27 \times 10^{39}$$

or

$$F_e = (2.27 \times 10^{39})F_g$$

The magnitude of the electrostatic force between a proton and an electron is more than 10^{39} times greater than the gravitational force! For this reason, the gravitational force between charged particles is usually neglected. (Note that 10^{39} is 1000 trillion trillion trillion. In comparison, our national debt is about 6 trillion dollars.)

Follow-up Exercise. Show that gravity is even more negligible for the force between two electrons by repeating the calculation. Explain why this is so.

15.4 Electric Field

Objectives: To (a) understand the definition of the electric field and (b) plot electric field lines and calculate electric fields for simple charge distributions.

The electric force, like the gravitational force, is an action-at-a-distance force. In fact, we can see that the range of the electric force between two point charges is infinite, since $F_e \propto 1/r^2$ and so approaches zero only if r approaches infinity. Thus, a particular arrangement, or configuration, of charges can have an effect on an additional charge placed anywhere nearby (or possibly anywhere in space).

The idea of a force acting across space was a difficult one for early investigators, and the concept of a *force field*, or simply a field, was introduced. Conceptually, an *electric field* surrounds every arrangement of charges. Thus, the field represents the *physical effect* of a particular configuration of charges on the nearby space. We think of the field as representing what is different about the nearby space because those charges are there. This conception allows us to see other charges as interacting with the electric field rather than with the charge configuration that created the field.

Note: Charges set up an electric field, which then acts on other charges placed in the same region.

The electric field is a *vector field* and can tell you what force a charge experiences at a particular position in space. The strength or magnitude of the electric field is expressed as *force per unit charge*. We think of mapping an electric field by placing a *test charge* at various locations and calculating the force per unit charge. With these values, the electric field can be plotted. Once you know the electric field pattern due to a charge or charge configuration throughout space, you can then ignore the charge(s) and talk solely in terms of the field. This procedure often greatly facilitates calculations.

In investigating the electric field around a charge or configuration of charges, it is important for the test charge to be small enough that the force it exerts on the field-producing charge or charges is negligible. This will ensure that the introduction of the test charge does not change the field-producing charge arrangement. Since the direction of the force on the charge depends on whether the charge is positive or negative, this too must be specified in a field representation. By convention, a *positive* test charge (q_0) is used for measuring the field. The **electric field** is then defined as

Note: A *test charge* (q_0) is small and positive.

$$E = \frac{\mathbf{F}_{on\,q_0}}{q_0} \qquad \textit{electric field definition} \qquad (15.3)$$

Note: Think of the electric field definition as being useful in the same way that the price *per pound* is for food items. Knowing how much you want of an item, you can compute how much it will cost if you know the price *per pound*. Similarly, given the magnitude of a charge placed in an electric field, you can compute the force on it if you know the field strength in newtons *per coulomb*.

SI unit of electric field: newton/coulomb (N/C)

E is a vector field and at any point has the direction of the force experienced by a *positive* (test) charge placed there. The SI unit of the electric field is the newton per coulomb (N/C).

The magnitude of the electric field (E), or the force per unit charge, at a distance of r meters from a point charge of q coulombs is then easily computed from Coulomb's force law:

$$E = \frac{F_e}{q_0} = \frac{1}{q_0}\left(\frac{kq_0q}{r^2}\right) = \frac{kq}{r^2}$$

That is,

$$E = \frac{kq}{r^2} \qquad \begin{array}{l}\textit{magnitude of electric field}\\ \textit{due to point charge } q\end{array} \qquad (15.4)$$

The field direction is given by the charge–force law, with q_0 taken as positive.

It is important to note that in the derivation of Eq. 15.4, q_0 canceled out. This must always happen since the field is produced by the nearby arrangement of charges, *not* by the test charge. In the case of a point charge, the electric field is produced by q, and the expression for E should not have the test charge in it.

Some electric field vectors in the vicinity of a positive charge are illustrated in •Fig. 15.11a. Note that the vectors point away from the positive charge, which is the direction of the force a positive test charge experiences, and that the magnitude of the vectors decreases with distance from the charge (an inverse-square relationship). For a *configuration* of charges, the total, or net, electric field at any point is the vector sum of the electric fields due to the individual charges. This is called the *superposition principle*. The use of this principle is shown in the next two Examples.

Note: Total electric field: $\mathbf{E} = \sum \mathbf{E}_i$

•**FIGURE 15.11 Electric field**
(a) Electric field direction is determined by using a positive test charge. Note that the magnitude of the field (lengths of vectors) becomes smaller as the distance from the source charge increases, reflecting the inverse-square relationship characteristic of the field from a point charge. **(b)** The vectors are connected to give electric field lines, or lines of force.

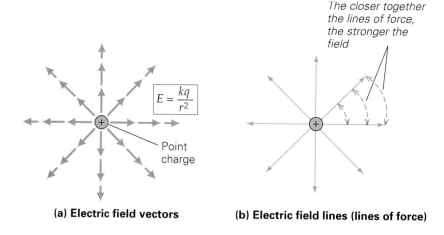

$$E = \frac{kq}{r^2}$$

Point charge

The closer together the lines of force, the stronger the field

(a) Electric field vectors   **(b) Electric field lines (lines of force)**

Example 15.6 ■ Electric Fields in One Dimension: Zero Net Field

Two point charges are placed on the x axis as shown in •Fig. 15.12. Find the locations on the axis where the electric field is zero.

Thinking It Through. Each point charge produces its own electric field. The superposition principle tells us how to find the total, or net, electric field (vector sum). We are looking for places on the x axis where these fields are equal and opposite so as to cancel and give no *net* electric field. Since both charges are positive, their fields point to the right at all points right of q_2 (the test charge gives the field direction as positive). Therefore they cannot cancel in that region. Similarly, to the left of q_1 both fields point to the left and cannot cancel. The only possibility is *between* the charges.

Solution. Let us specify the location by x relative to q_1. (Convert charges from microcoulombs to coulombs.)

Given: $d = 0.60$ m (distance between charges)   *Find:* x [the location(s) of zero E]
  $q_1 = +1.5\ \mu C = +1.5 \times 10^{-6}$ C
  $q_2 = +6.0\ \mu C = +6.0 \times 10^{-6}$ C

If a small positive test charge is placed on the axis between the charges, the two forces on it will oppose one another. The two fields will cancel if their two magnitudes are equal. We use Eq. 15.4, the expression for the electric field from a point charge:

$$E_1 = E_2 \quad \text{or} \quad \frac{kq_1}{x^2} = \frac{kq_2}{(d-x)^2}$$

Rearranging this expression and canceling the constant k, we can write

$$\frac{1}{x^2} = \frac{(q_2/q_1)}{(d-x)^2}$$

Since $q_2/q_1 = 4$, we can take the square root of both sides:

$$\sqrt{\frac{1}{x^2}} = \sqrt{\frac{q_2/q_1}{(d-x)^2}} = \sqrt{\frac{4}{(d-x)^2}} \quad \text{or} \quad \frac{1}{x} = \frac{2}{d-x}$$

Solving for x, we get $x = d/3 = 0.60$ m$/3 = 0.20$ m. This makes sense physically: Because q_2 is the larger of the two positive charges, the only way to make the two fields equal in magnitude is to be closer to q_1.

Follow-up Example. Repeat this Example, changing the sign of the right-hand charge. How is this situation physically different?

•**FIGURE 15.12 Electric field in one dimension** See Example 15.6.

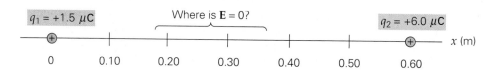

$q_1 = +1.5\ \mu C$ Where is **E** = 0? $q_2 = +6.0\ \mu C$

0   0.10   0.20   0.30   0.40   0.50   0.60 x (m)

Example 15.7 ■ Electric Fields in Two Dimensions: Using Components

What is the electric field at the origin for the three-charge configuration in •Fig. 15.13?

Thinking It Through. According to the superposition principle, the electric field at the origin will be the vector sum $\mathbf{E}_1 + \mathbf{E}_2 + \mathbf{E}_3$. We know the directions of each vector ($\mathbf{E}$ points away from positive charges and toward negative charges). We also know the magnitudes, since each is produced by a point charge (hence $E = kq/r^2$).

Solution. From the figure, we have

Given: $q_1 = -1.00\ \mu C = -1.00 \times 10^{-6}\ C$ *Find:* $\mathbf{E}$ (total electric field at origin)
$q_2 = +2.00\ \mu C = +2.00 \times 10^{-6}\ C$
$q_3 = -1.50\ \mu C = -1.50 \times 10^{-6}\ C$
$r_1 = 3.50\ m$
$r_2 = 5.00\ m$
$r_3 = 4.00\ m$

By the law of charges, an imaginary *positive* test charge q_0 at the origin tells us the direction of the electric field due to each charge. The directions are indicated in the vector diagram in Fig. 15.13. Then we use Eq. 15.4 to find the magnitude of each electric field vector:

$$E_1 = \frac{kq_1}{r_1^2} = \frac{(9.00 \times 10^9\ N{\cdot}m^2/C^2)(1.00 \times 10^{-6}\ C)}{(3.50\ m)^2} = 7.35 \times 10^2\ N/C$$

$$E_2 = \frac{kq_2}{r_2^2} = \frac{(9.00 \times 10^9\ N{\cdot}m^2/C^2)(2.00 \times 10^{-6}\ C)}{(5.00\ m)^2} = 7.20 \times 10^2\ N/C$$

$$E_3 = \frac{kq_3}{r_3^2} = \frac{(9.00 \times 10^9\ N{\cdot}m^2/C^2)(1.50 \times 10^{-6}\ C)}{(4.00\ m)^2} = 8.44 \times 10^2\ N/C$$

The magnitudes of the x and y components of the *total field* are then

$$E_x = E_1 + E_2 = 7.35 \times 10^2\ N/C + 7.20 \times 10^2\ N/C = 1.46 \times 10^3\ N/C$$

$$E_y = E_3 = 8.44 \times 10^2\ N/C$$

So, in component form,

$$\mathbf{E} = E_x\hat{\mathbf{x}} + E_y\hat{\mathbf{y}} = (1.46 \times 10^3\ N/C)\hat{\mathbf{x}} + (8.44 \times 10^2\ N/C)\hat{\mathbf{y}}$$

You should be able to show that in magnitude-angle form this is

$$E = 1.69 \times 10^3\ N/C \text{ at } \theta = 30.0° \qquad \text{relative to the } +x \text{ axis}$$
$$\text{(in the first quadrant)}$$

Follow-up Exercise. In this Example, remove q_1 and find the electric field at its location due to the other two charges.

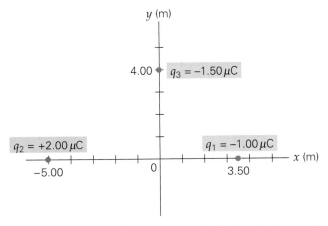
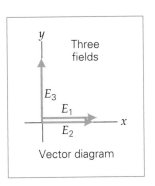
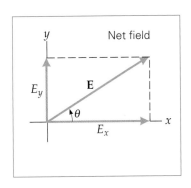

•**FIGURE 15.13** **Finding the electric field** See Example 15.7.

There is a convenient way of *graphically* representing the electric field pattern in space through the use of **electric lines of force**, or *electric field lines*. Consider the electric field vectors near a positive point charge, as in Fig. 15.11a. If we connect the vectors as in Fig. 15.11b, we are constructing the lines of force. The electric field is stronger nearer a charge, where the lines of force are closer together. And at any point on a line of force, the electric field direction is along the tangent to the line. The lines of force also have arrows attached to them indicating the field direction. These are the general "rules" for sketching and interpreting electric field lines:

1. The closer together the lines of force, the stronger the electric field;
2. The direction of the electric field is tangent to the lines of force;
3. The electric field lines start at positive charges and end at negative charges;
4. The number of lines leaving or entering a charge is proportional to the magnitude of that charge.

These rules enable us to "map" the pattern of electric lines of force due to various charge configurations. See the accompanying Learn by Drawing feature.

An **electric dipole** consists of two separate electric charges (or "poles," as they were known historically). One reason the study of electric dipoles is important is that they give us a model for permanently polarized molecules, like the water molecule. Even though the net charge on the dipole is zero, it creates an electric field because the charges are separated. If they were not, the charges would cancel, the net field would drop to zero everywhere. In •Fig. 15.14a, we show the electric field pattern of an electric dipole. The construction of electric field vectors and then electric lines of force is an exercise in artistic vector addition.

In the following Example, you can see that the field from a dipole, although not zero, certainly decreases much faster than that due to a point charge as we move away from it.

LEARN BY DRAWING

Electric Lines of Force

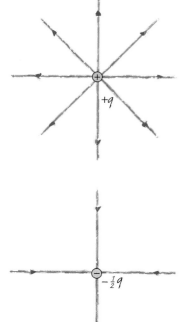

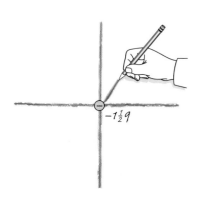

How many lines should be drawn for $-1\frac{1}{2}\,q$, and what should their direction be?

Example 15.8 ■ Fast Fade: Electric Field Due to a Dipole

Using Fig. 15.14b, show that the electric field far from a dipole, on its perpendicular bisector (the x axis), is given by kqd/x^3.

Thinking It Though. We need to add the two fields (one from each end of the dipole) vectorially to find the net field.

Solution. We are given the two charges and the requirement of being "very distant," which means that d is much smaller than x:

Given: $q_1 = +q$ *Find:* E (electric field on the x axis)
$q_2 = -q$
$x \gg d$

Since we are far from each charge, we know that $r = \sqrt{x^2 + (d/2)^2} \approx x$ since $x \gg d$. The net field is the vector sum of the two fields. The two fields have the same magnitude (why?), each given by $E \approx kq/x^2$ since $r \approx x$. By symmetry, the x components of these two fields cancel and the y components add to give a field pointing down. Since $\sin\theta = (d/2)/r \approx (d/2)/x$ from the geometry, the magnitude of the net field is

$$E_{net} = 2E_y = 2E\sin\theta$$

$$\approx 2(kq/x^2)\left(\frac{d/2}{x}\right) = \frac{kqd}{x^3} \propto \frac{1}{x^3}$$

At large distances the electric field decreases *more rapidly* than that of a point charge ($1/x^3$ versus $1/x^2$). If you looked back at the charges from the very distant field point, you would see the charges almost on top of one another, thereby almost canceling out. No wonder the net field decreases more rapidly!

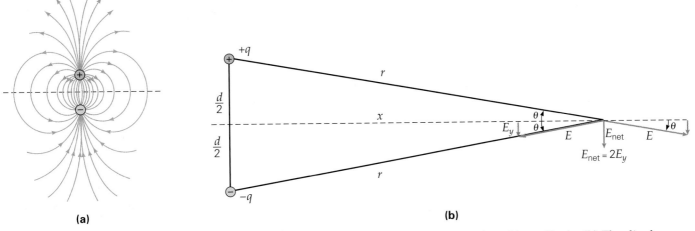

(a)   (b)

•**FIGURE 15.14 The electric field from a dipole** **(a)** The electric field line pattern produced by a dipole. **(b)** The dipole electric field is constucted on the x axis. It points down, in agreement with the overall pattern in part (a).

Follow-up Exercise. After changing the negative charge to a positive one, repeat the calculation in this Example. Show that E points in the $+x$ direction and that $E = k(2q)/x^2$. Why is d *not* in your final answer, and why is it an inverse-*square* decrease, unlike the dipole result?

•Figure 15.15 shows the vector construction of the electric field for a single, large charged plate and then between two oppositely charged parallel plates. Once we understand what the field looks like near one large plate, putting two together is easy.

If we stay away from the plate edges, the field is constant in direction and magnitude. The mathematical expression for the electric field magnitude between two closely spaced plates is too complicated to derive here, but the result is

$$E = \frac{4\pi kQ}{A} \qquad \textit{(between parallel plates)} \qquad (15.5)$$

where Q is the magnitude of the total charge on one of the plates and A is the area of one plate. Parallel plates are very common in electronic applications. For example, your TV screen displays a picture when it is impacted by a stream of electrons. These electrons are accelerated by an oppositely charged arrangement of (approximately) parallel plates. In Chapter 16 we shall see that an important element in many electronic circuits is a device called a capacitor, which, in its simplest form, is just a set of parallel plates.

The formation of cloud-to-ground lightning strokes can be modeled approximately as closely spaced "parallel plates." (See the Insight on p. 500.) With that model, we can estimate the charge on the clouds, as the next Example shows.

Example 15.9 ■ Parallel Plates: Determining the Charge on Storm Clouds

The electric field E required to ionize air is about 1.0×10^4 N/C. When the field reaches this value, the least bound electrons begin leaving their molecules, eventually creating lightning. Assume that this value of field E exists between a negatively charged lower cloud surface and the positively charged ground (Insight Fig. 1a). If we take the storm clouds to be squares 10 mi on each side, estimate the total negative charge on the lower cloud surface.

Thinking It Through. The electric field is given, so we can use Eq. 15.5 to estimate Q. First we must convert the cloud area A (one of the "plates") into square meters.

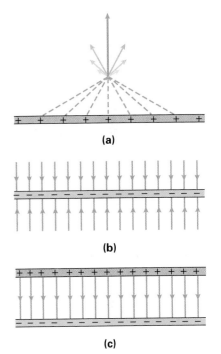

•**FIGURE 15.15 Electric field due to very large parallel plates** **(a)** Above a positively charged plate, the net electric field points up. Here the horizontal components of the electric fields from various locations on the plate cancel out. Below the plate, E points down (not shown). **(b)** For a negatively charged plate, the electric field direction (shown on both sides of the plate) is reversed. **(c)** Superimposing the fields from both plates results in cancellation outside the plates and an approximately uniform field between them.

Lightning and Lightning Rods

We are all familiar with the violent release of electrical energy in the form of lightning. Although it is a common occurrence, we still have a lot to learn about the formation of lightning. We do know that during the development of a cumulonimbus (storm) cloud, a separation of charge occurs. The cloud acquires regions of different charge, with the bottom of the cloud generally negatively charged. As a result, an opposite charge is induced on the surface of the Earth

(Fig. 1a). Eventually, lightning may reduce this charge difference by ionizing the air, allowing a flow of charge between cloud and ground. However, air is a good insulator; the electric field must be quite strong for this ionization to occur.

How the separation of charge takes place in a cloud is not fully understood, but it must be associated somehow with the rapid vertical movement of air and moisture within storm clouds. Water is a polar molecule—it has regions of

(a)   (b)   (c)

FIGURE 1   Lightning and lightning rods   (a) Cloud polarization induces a charge on the Earth's surface. (b) When the field becomes large enough, an electrical discharge results, which we call lightning. (c) A lightning rod provides a path to ground so as to prevent damage.

Solution.

Given: $E = 1.0 \times 10^4$ N/C
$d = 10$ mi $\approx 1.6 \times 10^4$ m
(to two significant figures)

Find: Q (the magnitude of the charge on the lower cloud surface)

Using $A = d^2$ for the area of a square, we solve Eq. 15.5 for the charge:

$$Q = \frac{EA}{4\pi k} = \frac{(1.0 \times 10^4 \text{ N/C})(1.6 \times 10^4 \text{ m})^2}{4\pi(9.0 \times 10^9 \text{ N·m}^2/\text{C}^2)} = 23 \text{ C}$$

We are justified in using this expression only if the distance between the clouds and the ground is much less than their size. (Why?) This is equivalent to assuming that they are less than several miles from the Earth's surface.

Note that 23 C is a huge amount of charge compared to the static charge a person might get from shuffling her feet on a rug. However, this charge is spread out over a large area, and any one small area does not contain a lot of charge.

Follow-up Exercise.   In this Example, (a) what is the direction of the electric field? (b) How many excess electrons does the lower cloud surface have per square meter?

The complete electric field patterns are shown for some charge configurations and charged conductors in •Fig. 15.16. Note that the electric field lines begin on charge, and under certain circumstances water molecules can break apart to produce positively charged and negatively charged ions. It is thought that some ionization may occur as a result of frictional forces between water droplets. However, a more plausible theory describes the separation as taking place during ice pellet formation. It has been shown experimentally that as water droplets freeze, positively charged ions are concentrated in the colder, outer regions of the droplets, whereas negatively charged ions are concentrated in the warmer, interior regions. Thus, a freezing droplet has a positively charged outer ice shell and a negatively charged liquid interior.

As the interior of a droplet begins to freeze, it expands and shatters the outer shell. This gives rise to positively charged ice fragments, which are carried upward by the internal cloud turbulence. This occurs on a large scale, and the remaining, relatively heavy droplets with their negative charge eventually settle to the base of the cloud.

Most lightning occurs entirely within a cloud (intracloud discharges), where it cannot be seen directly. However, familiar visible discharges do take place between clouds (cloud-to-cloud discharges) and between a cloud and the Earth (cloud-to-ground discharges). Pictures of cloud-to-ground discharges taken with special high-speed cameras reveal a nearly invisible downward ionization path. This occurs in a series of jumps or steps and so is called a *stepped leader*. As the leader nears the ground, positively charged ions in the form of a *streamer* rise from trees, tall buildings, or the ground to meet it.

When a streamer and leader make contact, the electrons along the leader channel begin to flow downward. The initial flow is near the ground, and as it continues, electrons positioned successively higher begin to migrate downward. Hence, the path of electron flow is continuously extended upward in what is called a *return stroke*. The surge of charge flow in the return stroke causes the conductive path to be illuminated, producing the bright flash seen by the eye and recorded in time-exposure photographs of lightning (Fig. 1b). Most lightning flashes have a duration of less than 0.50 s. Usually after the initial discharge, ionization again takes place along the original channel and another return stroke occurs. Typical lightning events have 3 or 4 return strokes.

Ben Franklin is often said to have been the first to demonstrate the electrical nature of lightning. In 1750 he suggested an experiment using a metal rod on a tall building. However, a Frenchman named d'Alibard set up the experiment and drew sparks from a rod during a thunderstorm. Franklin later performed a similar experiment with a kite he flew during a thunderstorm. He drew sparks to his knuckles from a key hanging at the end of a conductive kite cord. Ben was extremely lucky that he wasn't electrocuted. *Under no circumstances should you try to duplicate this experiment.* On the average, lightning kills 200 people a year in the United States and injures another 550.

A practical outcome of Franklin's work with lightning was the lightning rod, which was described in *Poor Richard's Almanac* in 1753. It consists simply of a pointed metal rod connected by a wire to a metal rod that has been driven into the Earth, or grounded (Fig. 1c). Franklin wrote that the rod "either prevents the stroke from the cloud or, if the stroke is made, conducts the stroke to Earth with safety of the building."

The latter is the general principle of operation of a lightning rod. The elevated rod intercepts the downward ionized stepped leader from a cloud, harmlessly discharging it to the ground before it reaches the structure or makes contact with an upward streamer. This prevents the formation of the damaging electrical surge associated with the return stroke.

positive charges and end on negative charges (or at infinity when there is no nearby negative charge). Also note that the lines do not cross (why not?). When necessary, we should choose the number of lines emanating from, or ending at, a charge, in proportion to the magnitude of that charge. Thus if we draw six lines *leaving* a point charge of $+2.0$ μC, we must show three lines *ending on* a nearby point charge of -1.0 μC in order to be consistent.

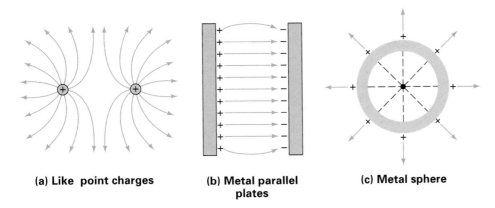

(a) Like point charges   **(b) Metal parallel plates**   **(c) Metal sphere**

•**FIGURE 15.16 Electric fields** Electric fields for various charge configurations: **(a)** Like point charges. **(b)** Oppositely charged parallel plates. The field is relatively uniform between the plates. **(c)** A charged metal sphere. The electric field outside the sphere is as though all the charge on the sphere were concentrated at its center. The electric field inside the sphere is zero.

15.5 Conductors and Electric Fields

Objectives: To (a) describe the electric field near the surface and in the interior of a conductor, (b) determine where the highest concentration of excess charge accumulates on a charged conductor, and (c) sketch the electric field line pattern outside a charged conductor.

The electric fields associated with charged conductors (that are isolated or insulated) have several interesting properties. By definition, in a static situation, charges that remain at rest must experience no electric force, so it follows that

> The electric field is zero everywhere *inside* a charged conductor.

Thus, to shield a region from stray electric fields, we can surround it with metal screening in the laboratory or, in everyday situations, with solid metal (see Exercises 60 and 62).

Also, as you might expect, excess charges on a conductor tend to get as far away from each other as possible since they are very mobile. Thus,

> Any *excess* charge on an isolated conductor resides entirely on the surface of the conductor.

Another property of static electric fields and conductors is that there cannot be any tangential component of the electric field at the conductor surface. If this were not true, charges would move along the surface, contrary to our assumption of a static situation. Thus it must be true that

> The electric field at the surface of a charged conductor is perpendicular to the surface.

Lastly, the excess charge on a conductor of irregular shape is most closely packed where the surface is highly curved, and thus E will be largest there. That is,

> Excess charge tends to accumulate at sharp points, or locations of greatest curvature, on charged conductors, so the highest charge accumulations occur where the electric field from the conductor is the largest.

These results are illustrated in •Fig. 15.17. Note that these results are true only for good conductors, and even then only under static conditions. Electric fields can exist inside a nonconductor, for example, and even inside conductors when conditions vary over time.

To understand *why* most of the charge accumulates in the highly curved surface regions, consider the forces acting *between* charges on the surface of the conductor. Where the surface is slightly curved, these forces will be directed nearly parallel to the surface (they would be parallel for a completely flat surface). The charges will spread out until the parallel forces in opposite directions cancel out. At a sharp end, the forces between charges will be directed more nearly perpendicular to the surface. Here there will be less tendency for the charges to move parallel to the surface since the force component parallel to the surface will be re-

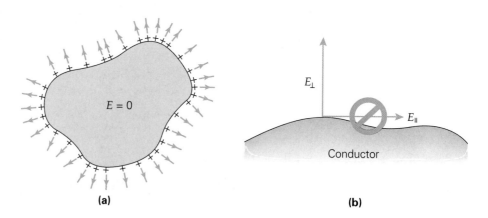

•**FIGURE 15.17 Electric fields and conductors** **(a)** The electric field is zero inside a conductor. Any excess charge resides entirely on the conductor's surface. For an irregularly shaped conductor, the excess charge accumulates in the regions of highest curvature. The electric field near the surface is perpendicular to that surface and strongest where the charge is densest. **(b)** Under static conditions, the electric field must not have a component tangential to the conductor's surface.

duced considerably. Thus the highly curved regions of the surface have a higher concentration of charge. (Draw some surfaces and force vectors to prove this to yourself.) Examples of both situations are shown in •Fig. 15.18.

Let's look more closely at some of these results for a spherical conductor. Suppose that a metal sphere is given a charge $+Q$, as shown in Fig. 15.16c. Because of the mutually repelling forces between them, the like charges will be distributed evenly over the sphere and will eventually come to rest on its outside surface. At this point, the conductor is in electrostatic equilibrium. Because the charge is distributed symmetrically, the electric field outside a charged sphere is as though all the excess charge on the sphere were concentrated at its center; that is, $E = kQ/r^2$, where r is greater than the radius R of the sphere. Inside the charged sphere, the electric field is zero. (Notice the analogy to the gravitational case, in which all the mass of a uniform sphere can be thought of as concentrated at its center.)

The sphere's electric field is mathematically identical to that of a point charge, if we are concerned only about the region outside the sphere. Hence we can represent the electric field *just outside* the sphere of radius R (with $r \approx R$) by rewriting Eq. 15.4. After multiplying both the numerator and denominator by 4π, we have

$$E_{\text{surface}} = 4\pi k(Q/4\pi R^2)$$

But the expression $Q/4\pi R^2$ is simply the *surface charge density* (total surface charge divided by surface area of a sphere in coulombs per square meter) on the surface of the sphere; $4\pi k$ is a constant. Thus, we see that if you place the same excess charge on spheres of different size, the larger spheres will have smaller surface electric field strengths. This is because the surface charge density is less on the larger spheres, since they have the same charge $(+Q)$ spread over more surface area. This special case illustrates the general result that electric fields at a conductor surface are highest where the radius of curvature (R) is smallest (that is, the areas with the greatest curvature).

An interesting situation occurs if there is a large concentration of charge on a conductor with a sharp point. The electric field strength in that region may be large enough to ionize molecules of air (that is, to pull electrons off the molecules) and a spark discharge may occur. More charge can be placed on a gently curved conductor, such as a sphere, before a spark discharge will occur. The concentration of charge at the sharp point of a conductor is one reason for the effectiveness of lightning rods. (See the Insight on p. 500.) As an example of some of these charged conductor properties, consider the following classic electrostatic experiment.

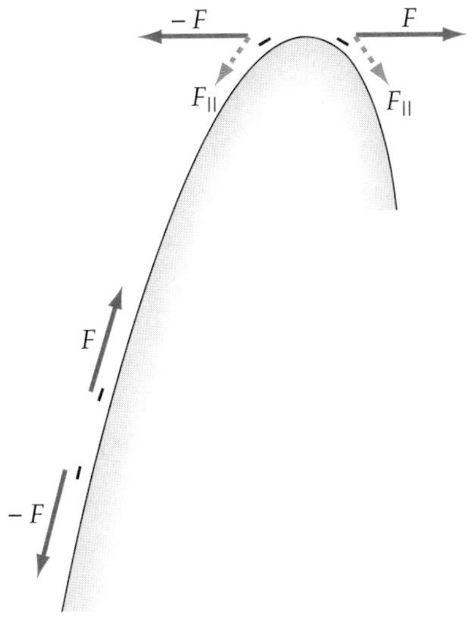

•**FIGURE 15.18 Concentration of charge on a curved surface** On a flat surface, the repulsive forces between excess charges are parallel to the surface and tend to push the charges apart. On a sharply curved surface, in contrast, these forces are directed at an angle to the surface. Their components parallel to the surface are smaller, allowing more charge to concentrate in such areas.

Conceptual Example 15.10

An Ice Pail Experiment

A positively charged rod is held inside an isolated metal container that has uncharged electroscopes conductively attached to its inside and outside surfaces (•Fig. 15.19a). With the charged rod held inside the container, (a) neither electroscope would show a deflection, (b) only the outside-connected electroscope would be deflected, (c) only the inside-connected electroscope would be deflected, (d) both electroscopes would be deflected.

Note: Compare Section 7.5.

•**FIGURE 15.19 An ice pail experiment** (a) See Conceptual Example 15.10. (b) See Follow-up Exercise 15.10.

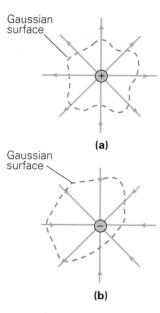

(a)

(b)

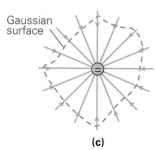

(c)

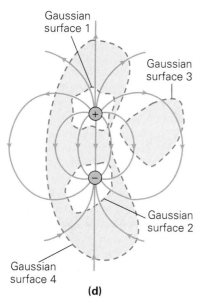

(d)

•**FIGURE 15.20 Various Gaussian surfaces and lines of force**
(a) Surrounding a single positive point charge. **(b)** Surrounding a single negative point charge. **(c)** Surrounding a larger negative point charge. **(d)** Four different surfaces surrounding various parts of an electric dipole.

Reasoning and Answer. The positively charged rod would attract negative charges, causing the inside of the metal container to become negatively charged. The outside would thus acquire a positive charge. Hence, both electroscopes would be charged (though with opposite signs) and would show deflections, so the answer is (d). A similar experiment was performed by the nineteenth-century English scientist Michael Faraday using ice pails, so this setup is often called Faraday's ice pail experiment.

Follow-up Exercise. Suppose that the positively charged rod *touched* the metal container as shown in Fig. 15.19b. What would be the effect on the electroscopes?

*15.6 Gauss's Law for Electric Fields: A Qualitative Approach

Objectives: To (a) state the physical basis of Gauss's Law and (b) use the law to make qualitative predictions.

A fundamental law of electric fields was discovered by Karl Friedrich Gauss (1777–1855), a German mathematician. In its mathematical form, this relationship can be used to calculate electric field strengths in cases of high symmetry. However, even a purely qualitative account of Gauss's law can teach us much interesting physics.

Consider the single positive electric charge in •Fig. 15.20a. Let us picture an imaginary closed surface surrounding it. Such surfaces are called **Gaussian surfaces.** Now, as before, let us designate outward-pointing electric field lines as positive and inward-pointing field lines as negative. If we count the lines of both types and total them up (that is, subtract the number of negative lines from the number of positive ones), we find that the total is positive (since in this simple case there are *only* positive lines). That is, there is a certain net number of outward-pointing electric field lines passing through the surface. Similarly, for a negative charge, as shown in Fig. 15.20b, the count would yield a negative total—a certain net number of inward-pointing lines passing through the surface. This would be true for *any* surface surrounding the charge. If we double the magnitude of the negative charge (Fig. 15.20c), our negative field line count also doubles.

Figure 15.20d shows a dipole with four different imaginary surfaces. Surface 1, which encloses a net positive charge, has a positive field line count, and Surface 2, which encloses a net negative charge, has a negative field line count. The more interesting cases are Surfaces 3 and 4. Note that both include zero net charge—Surface 3 because it includes no charges and Surface 4 because it includes equal and opposite charges. And we find that both Surfaces 3 and 4 have a net field line count of zero.

These situations illustrate the fundamental physical idea of **Gauss's law**[†]:

> The net number of electric field lines passing through an imaginary closed surface is proportional to the amount of net charge enclosed within that surface.

An everyday analogy illustrated in •Fig. 15.21 may help you understand this principle. If you surround a lawn sprinkler with an imaginary surface, you will find that there is a net flow of water out through that surface—clear evidence that what you have inside is a "source" of water (disregarding the pipe bringing water into the sprinkler). In an analogous way, a net outward-pointing electric field indicates the presence of a net positive charge inside the surface, since positive charges, as we know, are "sources" of the electric field. Similarly, a drain inside our imaginary surface would give itself away because there would be a net inward flow of water through the surface (again neglecting the pipe carrying the water away). The following Example illustrates the power of Gauss's law, even qualitatively.

[†]Strictly speaking, this is Gauss's law for electric fields. There is also a version of Gauss's law for magnetic fields, which we will not discuss in this text.

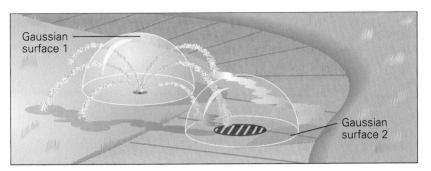

Conceptual Example 15.11

Charged Conductors Revisited: Gauss's Law

A net charge Q is placed on a conductor of arbitrary shape (•Fig. 15.22). Use the qualitative version of Gauss's law to prove that all the charge must lie on the conductor's surface under electrostatic conditions.

Reasoning and Answer.   Since the situation is static equilibrium, there can be no electric field inside the volume of the conductor; otherwise, the almost-free electrons would be moving around. Let us take a Gaussian surface that follows the shape of the actual conductor but is just barely inside the actual surface. Since there are no electric field lines inside the conductor, there are no electric field lines passing through our imaginary surface in either direction. So the net result is zero electric field lines penetrating the Gaussian surface. But, by Gauss's law, the net number of field lines is proportional to the amount of charge enclosed inside the surface. Thus, there must be no net charge within the surface!

Since our imaginary surface can be made as close as we wish to the true surface, it follows that the excess charge, if it cannot be inside the volume of the conductor, must be on the conductor's surface. This is a very elegant proof of what we determined before by considering repulsion and forces between charges.

Follow-up Exercise.   In this Example, if the net charge on the conductor is negative, what is the sign of the net number of lines through a Gaussian surface that completely encloses the conductor, excess charge and all? Explain your reasoning.

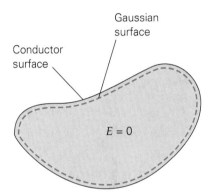

•**FIGURE 15.22   Gauss's law: Excess charge on a conductor**
See Conceptual Example 15.11.

Chapter Review

Important Terms

electric charge   *483*
law of charges *or the*
  charge–force law   *483*
coulomb   *484*
net charge   *484*
conservation of charge   *484*
conductors   *485*

insulators   *485*
semiconductors   *486*
electrostatic charging   *487*
charging by friction   *487*
charging by contact *or by*
  conduction   *487*
charging by induction   *488*

polarization   *488*
Coulomb's law   *489*
electric field   *495*
electric lines of force   *498*
electric dipole   *498*
Gaussian surface   *504*
Gauss's law   *504*

Important Concepts

• Electric charge is the property of an object that determines its electrical behavior: the electric force it can exert and the electric force it can experience.

• Opposite charges attract and like charges repel one another.

• The law of conservation of charge states that the net charge on an isolated system remains constant.

• Coulomb's law gives the force between two point charges.

- The electric field is a vector field that describes how nearby charges modify the space around them. It is defined as the electric force per unit positive charge.
- Electric lines of force are the imaginary lines formed by connecting electric field vectors. Their closeness and direction indicate the magnitude and direction of the electric field at any point.

*• A Gaussian surface is an imaginary closed surface. The net number of lines of force passing through it indicates the amount and sign of net charge enclosed by it.

Important Equations

Quantization of Electric Charge ($|e| = 1.60 \times 10^{-19}$ C):

$$q = ne \qquad (15.1)$$

Coulomb's Law ($k \approx 9.00 \times 10^{9}$ N·m²/C²):

$$F_e = \frac{kq_1q_2}{r^2} \quad \begin{array}{l}\textit{(two point charges,}\\ \textit{magnitude only)}\end{array} \qquad (15.2)$$

Electric Field (definition):

$$\mathbf{E} = \frac{\mathbf{F}_{\text{on } q_0}}{q_0} \qquad (15.3)$$

Electric Field Due to a Point Charge q:

$$E = \frac{kq}{r^2} \quad \textit{(magnitude only)} \qquad (15.4)$$

Electric Field Between Two Closely Spaced Parallel Plates:

$$E = \frac{4\pi kQ}{A} \qquad (15.5)$$

Exercises†

15.1 Electric Charge

1. A combination of two electrons and three protons would have a net charge of (a) +1, (b) −1, (c) +1.6 × 10⁻¹⁹ C, (d) −1.6 × 10⁻¹⁹ C.

2. The directions of the electric forces on two interacting charges are given by (a) the conservation of charge, (b) the charge–force law, (c) the magnitude of the charges, (d) none of these.

3. (a) How do we know that there are two types of electric charge? (b) What would be the effect of designating the charge on the electron as positive and the charge on the proton as negative?

4. An electrically neutral object can be given a net charge by several means. Does this violate the conservation of charge? Explain.

5. If a neutral object becomes positively charged, does its mass increase or decrease? What if it becomes negatively charged?

6. ■ What net electric charge would an object with 1.0 million excess electrons have?

7. ■ In walking across a carpet, you acquire a net negative charge of 50 μC. How many excess electrons do you have?

†Take k to be exact at 9.00 × 10⁹ N·m²/C² and e to be exact at 1.60 × 10⁻¹⁹ C in terms of working with significant figures.

8. ■ A glass rod rubbed with silk acquires a charge of +8.0 × 10⁻¹⁰ C. (a) What is the charge on the silk? (b) How many electrons have been transferred to the silk?

9. ■ A rubber rod rubbed with fur acquires a charge of −4.8 × 10⁻⁹ C. (a) What is the charge on the fur? (b) How much mass is transferred to the rod?

10. ■■ An alpha particle is the nucleus of a helium atom with no electrons. What would be the charge on two alpha particles?

15.2 Electrostatic Charging

11. A rubber rod is rubbed with fur. The fur is then quickly brought near the bulb of an uncharged electroscope. What is the sign of the charge on the leaves of the electroscope?

12. When a positively charged rod is brought near the bulb of a negatively charged electroscope, the leaves (a) separate further, (b) collapse, (c) remain unchanged.

13. A balloon is charged by rubbing and then clings to a wall. Does this mean the wall is also charged?

14. ■ How could an electroscope be positively charged by induction? How could you prove it was positively charged?

15. ■■ Two metal spheres mounted on insulated supports are in contact. How could both spheres be given a net electric charge without touching them? Would the spheres be charged positively or negatively?

15.3 Electric Force

16. The magnitude of the electric force between two point charges is given by (a) the charge–force law, (b) conservation of charge, (c) Coulomb's law, (d) both (a) and (b).

17. Compared to that of the electric force, the strength of the gravitational force between two protons is (a) about the same, (b) somewhat larger, (c) very much larger, (d) very much smaller.

18. The Earth is constantly attracting us by its gravitational force, but the electric force is much greater than the gravitational force. Why don't we experience the electric force?

19. When calculating planetary orbits around the Sun, why can astronomers safely ignore the electric force?

20. ■ 🌐 An electron is a certain distance from a proton. How would the electric force be affected if the electron were moved (a) twice that distance away from the proton and (b) one-third that distance toward the proton?

21. ■ 🌐 Two identical point charges are at a fixed distance from one another. How would the electric force be affected if (a) one of their charges was doubled and the other was halved, (b) both their charges were halved and (c) one charge was halved and the other was unchanged?

22. ■ In a certain organic molecule, the nuclei of two carbon atoms are separated by a distance of 0.25 nm. What is the magnitude of the electric repulsion between them?

23. ■ An electron and a proton are separated by 2.0 nm. (a) What is the magnitude of the force on the electron? (b) What is the net force on the system?

24. ■ 🌐 Two charges originally separated by 30 cm are moved farther apart, until the force between them has decreased by a factor of 10. How far apart are the charges then?

25. ■ 🌐 Two charges are separated to a distance of 100 cm, causing the electric force between them to increase by a factor of exactly 5. What was their initial separation distance?

26. ■■ 🌐 Two charges are attracted by a force of 25 N when separated by 10 cm. What is the force between the charges when the distance between them is 50 cm?

27. ■■ Two charges of $-2.0 \ \mu C$ are placed at opposite ends of a meterstick. Where on the meterstick could (a) a free electron and (b) a free proton be in electrostatic equilibrium?

28. ■■ Charges of $-1.0 \ \mu C$ and $+1.0 \ \mu C$ are placed at opposite ends of a meterstick. Where could (a) a free electron and (b) a free proton be in electrostatic equilibrium?

29. ■■ Two charges, q_1, and q_2, are located at the origin and at (0.50 m, 0), respectively. Where on the x axis must a third charge, q_3, of arbitrary sign be placed to be in electrostatic equilibrium if (a) q_1 and q_2 are like charges of equal magnitude, (b) q_1 and q_2 are unlike charges of equal magnitude, and (c) $q_1 = +3.0 \ \mu C$ and $q_2 = -7.0 \ \mu C$?

30. ■■ On average, the electron and proton in a hydrogen atom are separated by a distance of 5.3×10^{-11} m (●Fig. 15.23). Assuming the orbit of the electron to be circular, (a) what is the electric force on the electron? (b) What is the electron's orbital speed? (c) What is the magnitude of the electron's centripetal acceleration in units of g?

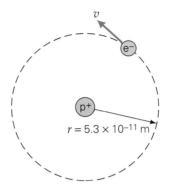

●FIGURE 15.23 Hydrogen atom See Exercises 30 and 31.

31. ■■ Compute the ratio of the magnitudes of the electric force to the gravitational force between the electron and proton in the hydrogen atom (see Fig. 15.23).

32. ■■■ Three charges are located at the corners of an equilateral triangle, as depicted in ●Fig. 15.24. What are the magnitude and the direction of the force on q_1?

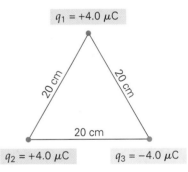

●FIGURE 15.24 Charge triangle See Exercises 32, 51, and 52.

33. ■■■ Four charges are located at the corners of a square as illustrated in ●Fig. 15.25. What are the magnitude

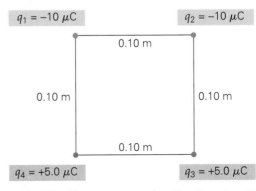

●FIGURE 15.25 Charge rectangle See Exercises 33, 53, 57, 82, and 84.

and the direction of the force (a) on charge q_2 and (b) on charge q_4?

34. ■■■ Two 0.10-gram pith balls are suspended from the same point by threads 30 cm long. (Pith is a light insulating material once used to make helmets worn in tropical climates.) When the balls are given equal charges, they come to rest 18 cm apart as shown in •Fig. 15.26. What is the magnitude of the charge on each ball? (Neglect the mass of the thread.)

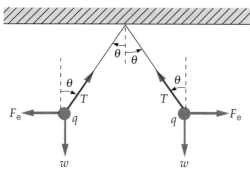

•**FIGURE 15.26 Repelling pith balls** See Exercise 34.

15.4 Electric Field

35. The electric field due to a negative point charge (a) varies as $1/r$, (b) points toward the charge, (c) has a finite range, (d) is the same as that of a positive charge.

36. The units of electric field are (a) C, (b) N/C, (c) N, (d) J.

37. 🌀 At a point in space, an electric force acts vertically downward on an electron. The direction of the electric field at that point is (a) down, (b) up, (c) zero, (d) undetermined by the data.

38. How is the relative magnitude of the electric field in different regions determined from a field vector diagram?

39. How can the relative magnitudes of the field in different regions be determined from an electric field line diagram?

40. Can electric field lines ever cross?

41. A positive charge is inside an isolated metal sphere as shown in •Fig. 15.27. Describe the situation in terms of the electric field and the charge on the sphere. How would the situation change if the charge were negative?

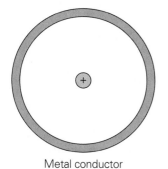

Metal conductor

•**FIGURE 15.27 A point charge inside a thick metal spherical shell** See Exercise 41.

42. ■ If the distance from a location to a charge is doubled, what happens to the magnitude of the electric field produced by the charge at that location?

43. ■ An isolated electron experiences an electric force of 3.2×10^{-14} N. What is the magnitude of the electric field at the electron's location?

44. ■ What is the magnitude of the electric field at a point 0.25 cm away from a point charge of $+2.0\ \mu C$?

45. ■ At what distance from a proton will the magnitude of electric field be 1.0×10^5 N/C?

46. ■■ 🌀 Two charges, $-4.0\ \mu C$ and $-5.0\ \mu C$, are separated by a distance of 20 cm. What is the electric field halfway between the charges?

47. ■■ 🌀 What are the magnitude and the direction of a vertically oriented electric field that would just support the weight of a proton? of an electron?

48. ■■ 🌀 Two charges, $-3.0\ \mu C$ and $-4.0\ \mu C$, are located at $(-0.50\ m, 0)$ and $(0.75\ m, 0)$, respectively. (a) Where on the x axis is the electric field zero? (b) Is there a position (or positions) where the electric field has only a y component? Explain.

49. ■■ Three charges, $+2.5\ \mu C$, $-4.8\ \mu C$, and $-6.3\ \mu C$, are located at $(-0.20\ m, 0.15\ m)$, $(0.50\ m, -0.35\ m)$, and $(-0.42\ m, -0.32\ m)$ respectively. What is the electric field at the origin?

50. ■■ 🌀 Two charges of $+4.0\ \mu C$ and $+9.0\ \mu C$ are 30 cm apart. Where on the line joining the charges is the electric field zero?

51. ■■ What is the electric field at the center of the triangle in Fig. 15.24?

52. ■■ 🌀 Compute the electric field at a point midway between charges q_1 and q_2 in Fig. 15.24.

53. ■■ 🌀 What is the electric field at the center of the square in Fig. 15.25?

54. ■■ 🌀 A particle with a mass of 2.0×10^{-5} kg and a charge of $+2.0\ \mu C$ is released in a (parallel-plate) uniform electric field of 12 N/C. (a) How far does the particle travel in 0.50 s? (b) What is its speed at that point? (c) If the plates are 5.0 cm on each side, how much charge is on each?

55. ■■ Two large parallel plates are oppositely and uniformly charged. If the field between them is 1.7×10^6 N/C, how dense is the charge on each plate (in microcoulombs per square meter)?

56. ■■ Two square, oppositely charged conducting plates measure 20 cm on each side. They are placed close together and parallel to each other. They have charges of $+4.0$ nC

and −4.0 nC, respectively. (a) What is the electric field between the plates? (b) What force does an electron in this region experience?

57. ■■■ 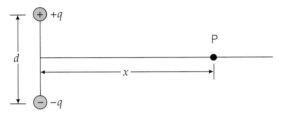 Compute the electric field at a point 4.0 cm from q_2 along a line running toward q_3 in Fig. 15.25.

58. ■■■ Two equal but opposite charges form a dipole as shown in •Fig. 15.28. (a) What are the electric field direction and magnitude at point P? (b) Compute your answer (in terms of k, q, d, and x) by using exact vector addition, and (c) show that it gives the distant field approximation of kqd/x^3 (see Example 15.8).

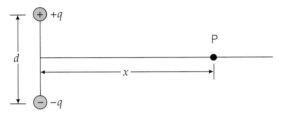

•**FIGURE 15.28   Electric dipole field**   See Exercise 58.

15.5 Conductors and Electric Fields

59. In electrostatic equilibrium, the electric field just below the surface of a charged conductor is (a) the same value as the field just above the surface (b) zero (c) kq/R^2.

60. Is it safe to stay in a car in a lightning storm (•Fig. 15.29)? Explain.

•**FIGURE 15.29   Safe inside a car?**   See Exercise 60.

61. Under electrostatic conditions, the excess charge on a conductor is on its surface. Does this mean that all of the conduction electrons in a conductor are on the surface?

62. An uncharged, thin metal slab is placed in an external electric field that points horizontally to the left. If the slab is oriented such that its large-area side is perpendicular to the field, what is the electric field inside the slab? (a) zero (b) same value as the original external field (c) somewhere between zero and the original external field value.

63. ■ A solid conducting sphere is surrounded by a thick, spherical conducting shell. A total charge $+Q$ is initially placed and released at the center of the inner sphere. After

equilibrium is reached, how much charge is on (a) the interior of the solid sphere, (b) the surface of the solid sphere, (c) the inner surface of the shell, and (d) the outer surface of the shell?

64. ■ In Exercise 63, what is the electric field direction (a) in the interior of the solid sphere, (b) between the sphere and the shell, (c) inside the shell, and (d) outside the shell?

65. ■■ In Exercise 63, write expressions for the electric field magnitude (a) in the interior of the solid sphere, (b) between the sphere and the shell, (c) inside the shell, and (d) outside the shell. Your answer should be in terms of Q, r (the distance from the center of the sphere), and k.

66. ■■ A flat, triangular piece of metal with rounded corners has a net positive charge on it. Sketch the charge distribution on the surface and the electric field lines near the surface of the metal (including direction).

67. ■■■ Approximate a metal needle as a long cylinder with a very pointed but slightly rounded end. Sketch the charge distribution and outside electric field lines if the needle has an excess of electrons on it.

*15.6 Gauss's Law for Electric Fields: A Qualitative Approach

68. How many net electric field lines would pass through a Gaussian surface located totally within a set of oppositely charged parallel plates?

69. A Gaussian surface surrounds an object with a net charge of $+5.0$ μC. Which of the following is true? (a) More electric field lines will point outward than inward. (b) More electric field lines will point inward than outward. (c) The net number of field lines through the surface is zero.

70. Two concentric spherical surfaces enclose a point charge. The radius of the outer sphere is twice that of the inner one. Which sphere will have more electric field lines penetrating?

71. ■ The same Gaussian surface is used to surround two charged objects separately. The net number of field lines penetrating the surface is the same in both cases, but they are oppositely directed. What can you say about the net charges on the two objects?

72. ■■ If a Gaussian surface has 16 field lines leaving it when it surrounds a point charge of $+10.0$ μC and 75 field lines entering it when it surrounds an unknown point charge, what is the amount of charge on the unknown?

73. ■■ If 10 field lines leave a Gaussian surface when it completely surrounds the positive end of an electric dipole, what is the count if it surrounds just the other end?

74. ■■ Suppose a Gaussian surface encloses both a positive point charge that has 6 field lines leaving it and a negative point charge with twice the magnitude of charge as the positive one. What is the net number of field lines passing through the Gaussian surface?

75. ■■ If a net number of electric field lines point out of a Gaussian surface, does that necessarily mean there are no negative charges in the interior?

Additional Exercises

76. An electron is placed in a constant electric field of 3.5×10^3 N/C directed along the $-y$ axis. (a) What is the force on the electron? (b) If the electron is released, how much kinetic energy does it have after it has moved 10 cm?

77. How many electrons are required to make up a net charge of -0.50 μC?

78. Two charges, -3.0 μC and -5.0 μC, are 0.40 m apart. (a) Where should a third charge of -1.0 μC be placed to put it in electrostatic equilibrium? (b) a third charge of $+1.0$ μC?

79. An electron is released from rest at the origin in a uniform electric field that has a magnitude of 450 N/C in the $+x$ direction. (a) How long will it take the electron to travel 1.0 m? (b) What will its position be (in x-y coordinates) at half that time?

80. Find the electric field at point O for the charge configuration shown in •Fig. 15.30.

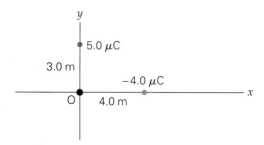

•**FIGURE 15.30 Electric field** See Exercise 80.

81. A charge of $+6.0$ μC at the origin of a set of coordinate axes experiences a force of 1.8 N in the $+y$ direction due to another charge located 0.30 m away. What are the magnitude and the position of the other charge?

82. Compute the electric field at a point midway between charges q_1 and q_4 in Fig. 15.25.

83. A solid spherical conductor (uncharged) is placed in a uniform electric field pointing to the right. Sketch the charges on the conductor and the electric field after electrostatic conditions are attained.

84. If a Gaussian surface completely surrounds the charge arrangement in Fig. 15.25, what is the sign of the net number of lines passing through the surface?

85. In the dipole electric field calculation (Example 15.8), the product of the magnitude of the charge on one end of the dipole and the distance of separation appears as qd in the numerator. Explain why the electric field on the axis perpendicular to the dipole goes to zero if d is zero.

86. For an electric dipole, the product qd is called the *dipole moment*. It is taken to be the vector **p** by assuming it points from the negative end of the dipole toward the positive end (•Fig. 15.31a). With a sketch (Fig. 15.31b), show that a dipole placed in a uniform field will tend to rotate until its dipole moment lines up with the field.

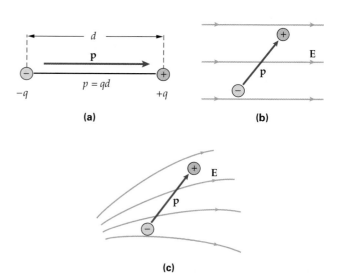

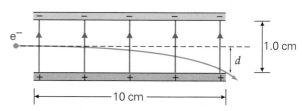

•**FIGURE 15.31 Dipole moment, dipoles in electric fields**
(a) The dipole moment vector has a magnitude of qd and points from the negative to the positive end. (b) An electric dipole placed in a uniform electric field. (c) An electric dipole placed in a nonuniform electric field. See Exercises 86 and 87.

87. What happens to the electric dipole in Exercise 86 if the field is not uniform (Fig. 15.31c), assuming the dipole is free to move?

88. A uniform metal slab is inserted between a pair of oppositely charged parallel plates, itself parallel to the plates. Sketch the resulting electric field.

89. At a speed of 6.0×10^7 m/s, an electron in a computer monitor enters midway between two parallel oppositely charged plates as shown in •Fig. 15.32. If the electric field between the plates is 2.0×10^4 N/C, what is the vertical deflection d of the electron when it leaves the plates?

•**FIGURE 15.32 Electron in a computer monitor** See Exercise 89.

90. In Exercise 89, how much charge is on each plate if each is a square?

Electric Potential, Energy, and Capacitance

Insight

■ Electric Potential and Nerve-Signal Transmission

The girl in the photo is experiencing the effects of electricity. In fact, she is charged to a potential of several thousand volts! Conversely, household circuits, which carry only 120 volts, can give you a nasty and potentially dangerous shock. Yet the girl doesn't seem to be in any trouble or danger. What's going on? You'll find the explanation of this and many other electrical phenomena in this chapter.

16.1 Electric Potential Energy and Potential Difference

Objectives: To (a) understand the concept of electric potential difference ("voltage") and its relationship to electric potential energy and (b) calculate electric potential differences and electric potential energies.

In Chapter 15 we analyzed electrical effects in terms of electric field vectors and electric lines of force. Recall that in mechanics, we first used Newton's laws with free-body diagrams and vectors. Then, in search of a simpler approach to many situations, we introduced *scalar* quantities such as work, kinetic energy, and potential energy. With these concepts, we could solve many problems by using the work–energy theorem and the principle of energy conservation. It is extremely useful, both conceptually and from a problem-solving standpoint, to extend these energy methods to the study of electric fields.

Electric Potential Energy Difference

Let's start with one of the simplest electric field patterns, the field between two large, oppositely charged parallel plates. As we saw in Chapter 15, near the center of the plates the field is uniform in magnitude and direction (•Fig. 16.1a). Suppose we take a small positive charge, q_o, and move it at constant velocity against the electric field E, from the negative plate (A) to the positive plate (B). An external force with the same magnitude as the electric force, $F_e = q_o E$, is required. The work done by the external force is positive (force and displacement in the same direction) and equal to $F_e d \cos 0° = +F_e d$, or $+q_o E d$. At the same time, negative work is done *by the field* (force and displacement in opposite directions), equal to $F_e d \cos 180° = -F_e d$, or $-q_o E d$.

Moreover, if the test charge is released, it will accelerate back toward the negative plate, gaining kinetic energy. Therefore, by moving the charge from plate A to plate B, the external force increases the charge's **electric potential energy, U_e,** by an amount equal to the work done on the charge:

$$\Delta U_e = U_B - U_A = q_o E d$$

The gravitational analogy to the parallel-plate electric field is the Earth's gravitational field near the surface. When an object is raised a vertical distance h, the change in its potential energy is positive and equal to the work done by the external force $F = w = mg$ (Fig. 16.1b):

$$\Delta U_g = U_B - U_A = Fh = mgh$$

For clarity, we use different symbols, h and d, to represent distances in gravitational and electric fields, respectively.

Recall from Chapter 5 that the value of the gravitational potential energy at a point A is arbitrary; we can assign it any value we desire. Similarly, in the parallel-plate case, the value of the electric potential energy at A or B (or any point between) is also arbitrary. It is common (though *not* required) to designate the zero point for electrical potential energy as the negative plate. However, in both cases the only measurable physical quantity is the potential energy *difference* relative to some (arbitrary) reference point.

Electric Potential Difference

In dealing with electric forces, we eliminated the dependence on the test charge by defining the electric field as the *electric force per unit charge*. Then, knowing the electric field, we could determine the force on *any* point charge placed at that location (from $F_e = q_o E$). Similarly, we define the **electric potential difference** between any two points in space as the change in potential energy *per unit positive test charge*. In terms of the work done on the charge by an external force,

•**FIGURE 16.1 Potential energy changes in uniform electric and gravitational fields** **(a)** Moving a positive charge q_o against the electric field requires positive work and increases the electric potential energy. **(b)** Moving a mass m against the gravitational field requires positive work and increases the gravitational potential energy.

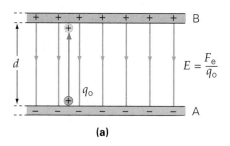

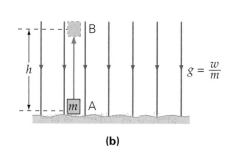

(a)

(b)

$$\Delta V = \frac{\Delta U_e}{q_0} = \frac{W}{q_0} \qquad (16.1)$$

SI unit of electric potential difference: joule/coulomb (J/C) or volt (V)

The SI unit of electric potential difference, or simply potential difference, is the joule per coulomb. This unit has been renamed the **volt** (V) in honor of Alessandro Volta, an Italian scientist (1745–1827). Thus, 1 V = 1 J/C. Potential difference is sometimes called **voltage**, and the symbol for potential difference is commonly changed from ΔV to just V, as we will do later in this chapter.

Potential difference, although based on the potential energy difference, is *not* the same thing. Since potential difference is defined per unit charge, it does not depend on the amount of charge moved, a very useful property, since from ΔV we can calculate ΔU_e for any amount of charged moved. To illustrate this point, let us calculate the potential difference for a uniform field between two parallel plates:

$$\Delta V = \frac{\Delta U_e}{q_0} = \frac{q_0 Ed}{q_0} = Ed \qquad \begin{array}{l}\textit{potential difference}\\ \textit{(parallel plates)}\end{array} \quad (16.2)$$

Note: Use Eq. 16.2 for parallel plates only.

The amount of charge moved, q_0, cancels. Thus the potential difference ΔV depends only on the charged plate characteristics: that is, the field produced (E) and the separation (d). We say that the positively charged plate is at a *higher* **electric potential** than the negatively charged one by an amount ΔV.

Notice that we defined electric potential *difference* without first defining electric potential. Although this may seem backward, there is a good reason for it. Electric potential difference is a physically meaningful quantity that we can actually measure. The electric potential value at a location, in contrast, isn't definable in an absolute way—it depends entirely on our choice of a reference point. We can add or subtract a constant from all potentials and not alter the potential *differences*.

Note: We commonly assign a value of zero for the electric potential of the negative plate, but this is arbitrary. Only potential *differences* are meaningful.

Recall that when we studied mechanical forms of potential energy associated with springs and gravitation in Chapter 5, we dealt only with *changes* in potential energy. Definite values of potential energy could be specified, but only with respect to an arbitrary reference point. Thus, in the case of gravity, we sometimes defined the zero point as the Earth's surface for convenience. However, it is also correct to define the zero point as an infinite distance from Earth, as was done in Figure 7.20.

Note: Review Section 5.4.

The same is true of electric potential energy—and therefore of electric potential, since this is just the potential energy per unit charge. We may want to define the potential as zero at the negative plate of a pair of charged parallel plates. But it is sometimes more convenient to locate the zero point at infinity, as we shall see shortly for the case of a point charge. Either way, the *difference* between two given points will be unaffected. (See the accompanying Learn by Drawing.) Suppose that with one choice of the zero value, point A is at a potential of 100 V and point B at 200 V. With a different zero point, the potential at A might be 1100 V, but the potential at B would then be 1200 V. Or we might designate a zero point that would give A a potential of -500 V; the potential at B would then be -400 V. The *difference* $V_B - V_A$ would always be $+100$ V, regardless of where we assign the zero reference point.

Thus, the concept of a unique value for the potential V is meaningless. Only *changes* in V, which is called *voltage,* are meaningful, and the change is independent of the choice of reference (zero) value. In the following Example, we implicitly assign zero potential to the negative plate, thereby ensuring that the positive plate has a positive value of V. After studying the Example, ask yourself

LEARN BY DRAWING

ΔV Is Independent of
Reference Point

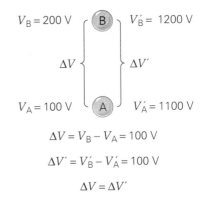

$$\Delta V = V_B - V_A = 100\ \text{V}$$

$$\Delta V' = V'_B - V'_A = 100\ \text{V}$$

$$\Delta V = \Delta V'$$

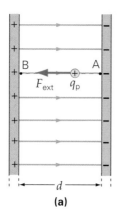

(a)

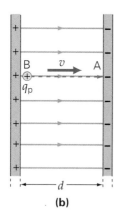

(b)

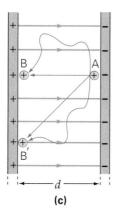

(c)

•FIGURE 16.2 Accelerating a charge (a) Moving a proton from the negative to the positive plate increases the proton's potential energy. See Example 16.1. (b) When it is released from the positive plate, the proton accelerates back toward the negative plate, gaining kinetic energy at the expense of the electric potential energy. (c) The work done to move a proton between any two points in an electrostatic field, such as points A and B or A and B', is independent of the path.

what the potential of the positive plate would be if the negative plate were assigned a potential of +5.0 V, and how this would affect the results.

Example 16.1 ■ Moving a Proton: Parallel Plates and Potential Difference

A proton is moved from the negative to the positive plate of a parallel-plate arrangement in •Fig. 16.2a. The plates are 1.50 cm apart, and the electric field is uniform with a magnitude of 1500 N/C. (a) What is the proton's potential energy change? (b) What is the potential difference between the plates? (c) between the negative plate and a point midway between the plates? (d) If the proton is released from rest at the positive plate (Fig. 16.2b), what speed will it have just before it hits the negative plate?

Thinking It Through. (a) Since the field is constant, the change in potential energy, or work done, is $Fd = (qE)d$. (b) The potential difference can then be computed by dividing by the charge moved. (c) These observations hold true whether the proton is moved completely from one plate to the other or only part way, allowing us to calculate the midpoint. (d) When the proton is released, stored potential energy is converted into kinetic energy, enabling us to find the proton's speed.

Solution. We are given E, and since we are moving a proton, we know its mass and charge.

Given: $E = 1500$ N/C
$q_p = +1.60 \times 10^{-19}$ C
$m_p = 1.67 \times 10^{-27}$ kg
$d = 1.50$ cm $= 1.50 \times 10^{-2}$ m

Find: (a) ΔU_e (potential energy change)
(b) ΔV (plate potential difference)
(c) ΔV_{mid} (potential difference: negative plate to midpoint)
(d) v (speed of released proton just before it reaches negative plate)

(a) Electric potential energy is gained, because positive work is done on the proton:

$$\Delta U_e = (q_p E)d = (+1.60 \times 10^{-19}\,C)(1500\,N/C)(1.50 \times 10^{-2}\,m)$$

$$= +3.60 \times 10^{-18}\,J$$

(b) The potential difference, or plate voltage, is the potential energy change per unit charge, as expressed in Eq. 16.1:

$$\Delta V = \frac{\Delta U_e}{q_p} = \frac{+3.60 \times 10^{-18}\,J}{1.60 \times 10^{-19}\,C}$$

$$= +22.5\,V$$

The positive plate is 22.5 V higher in electric potential than the negative plate.

(c) Since the potential energy gain at the midpoint, ΔU_{mid}, is proportional to the distance moved, the potential difference, ΔV_{mid}, between the negative plate and the midpoint is half the full difference, or about 11.3 V.

(d) The gain in kinetic energy equals the loss in potential energy. Since $K_o = 0$, $\Delta K = K - K_o = K$. Using the absolute value since ΔU_e is negative, we can write

$$\Delta K = |\Delta U_e|$$

$$K = \tfrac{1}{2}m_p v^2 = |\Delta U_e|$$

$$v = \sqrt{\frac{2|\Delta U_e|}{m_p}} = \sqrt{\frac{2(3.60 \times 10^{-18}\,J)}{1.67 \times 10^{-27}\,kg}} = 6.57 \times 10^4\,m/s$$

Even though the amount of kinetic energy gained is very small, the proton acquires considerable speed (about 0.02% the speed of light) because its mass is extremely small.

Follow-up Exercise. How would your answers change in this Example if an alpha particle were moved instead of a proton? (An alpha particle, the nucleus of a helium atom, has a charge of $+2e$ and a mass approximately four times that of a single proton.)

Example 16.1 can be used to show that the potential energy gain and the potential difference are independent of the path taken. This is true of all conservative forces, of which the electrostatic force is one. In Fig. 16.2c, the work done in moving the proton from A to B is the same *regardless of the route* taken. The alternative wiggly paths from A to B and A to B′ require the same work as do their respective straight-line paths. This is because movement at right angles to the field requires no work (why?).

Consider how our energy discussion would differ if we had used an electron instead of a proton. The electron, being negatively charged would be attracted to B, so the external force would have to act in a direction *opposite* the displacement to keep the electron moving at a constant velocity (that is, to prevent the electron from gaining speed). The force would thus do *negative* work on the electron, so the potential energy of the electron would be *decreased*:

$$\Delta U_{electron} = q_e \Delta V = U_B - U_A < 0$$

Unlike the proton, the electron is attracted to the positive plate, which is at a higher potential than the negative plate. If allowed to move freely, it would "fall" toward the region of higher potential (just as the proton "fell" toward the region of lower potential), losing potential energy and gaining kinetic energy. This is a general result for charges in electric fields:

Positive charges, when released, tend to move toward regions of low potential, and negative charges tend to move toward regions of high potential.

Note: Potential differences, like electric fields, are defined in terms of positive charges. Electrons experience the same potential difference but the opposite potential *energy* change. To determine whether potential increases or decreases, decide whether an external force does positive or negative work on a positive test charge.

Example 16.2 ▪ Zero to 1860 mi/s in $2\frac{1}{2}$ mm: Accelerating an Electron

We want to accelerate an electron to 1.00% of the speed of light (1860 mi/s) in a space of 2.50 mm between a pair of large horizontal parallel plates. If the top plate is positively charged, (a) what voltage between the plates is required? (b) What are the direction and magnitude of the electric field? (Disregard the effects of gravity—why?)

Thinking It Through. (a) Using the energy approach, we first calculate the kinetic energy gained by the electron and equate that to the magnitude of its loss of electric potential energy (why?). From the change in potential energy, we can calculate the voltage, or potential difference (Eq. 16.1). (b) Since it is an electron, it will accelerate "uphill," or opposite the electric field direction—equivalently, in the direction of increasing electric potential. We can then calculate the electric field E from the parallel-plate expression (Eq. 16.2).

Solution. One percent of the speed of light is $(0.0100)(3.00 \times 10^8 \text{ m/s}) = 3.00 \times 10^6 \text{ m/s}$ in SI units, and we know the mass and charge of an electron.

Given: $q = -1.60 \times 10^{-19} \text{ C}$
$m = 9.11 \times 10^{-31} \text{ kg}$
$d = 2.50 \text{ mm} = 2.50 \times 10^{-3} \text{ m}$
$v = 3.00 \times 10^6 \text{ m/s}$

Find: (a) ΔV (plate voltage)
(b) E (magnitude and direction)

(a) The electron must be released from the negatively charged plate (why?). The gain in kinetic energy (since $K_o = 0$) is

$$\tfrac{1}{2} m v^2 = \tfrac{1}{2}(9.11 \times 10^{-31} \text{ kg})(3.00 \times 10^6 \text{ m/s})^2 = 4.10 \times 10^{-18} \text{ J}$$

This gain in kinetic energy comes at the expense of electric potential energy, thus $\Delta U_e = -\Delta K = -4.10 \times 10^{-18}$ J. Thus the voltage (potential difference) across the plates is

$$\Delta V = \frac{\Delta U_e}{q} = \frac{-4.10 \times 10^{-18} \text{ J}}{-1.60 \times 10^{-19} \text{ C}} = 25.6 \text{ V}$$

(b) Electric field lines start on the positively charged plate and end on the negatively charged plate; thus E points down. Its magnitude is calculated from the rearranged parallel-plate expression, Eq. 16.2:

$$E = \frac{\Delta V}{d} = \frac{25.6 \text{ V}}{2.50 \times 10^{-3} \text{ m}} = 1.02 \times 10^4 \text{ V/m}$$

Follow-up Exercise. Use energy methods to determine the direction and magnitude of the uniform electric field required to stop a proton initially traveling in the $+x$ direction at 1.00% of the speed of light in the space of 50 cm.

Potential Difference Due to Point Charges

For nonuniform electric fields, the potential difference between two points is determined by applying the definition given in Eq. 16.1. That is, we calculate the work per unit charge in moving a small positive test charge between the two points at constant speed. However, when the field strength varies with location, the mathematics is no longer simple because the work is done by a varying force. The only nonuniform field we will consider in any detail is that due to a point charge (•Fig. 16.3). The expression for the potential difference (voltage) between two points at distances r_A and r_B from a point charge (q) is

$$\Delta V = \frac{kq}{r_B} - \frac{kq}{r_A} \qquad \begin{array}{l} \textit{electric potential difference} \\ \textit{(point charge only)} \end{array} \qquad (16.3)$$

In Fig. 16.3, the point charge is positive. If B is closer to the point charge than A, the potential difference is positive: B is at a higher potential than A. This is because it takes *positive* external work to move a positive test charge from A to B. This situation is analogous to compressing a spring, since the two positive charges naturally repel one another—we must push them together, which requires positive work W.

Thus the potential increases as we consider locations nearer to a positive charge. Again, the potential difference is the same regardless of the path. The work done along path II from A to B is the same as along the direct route, path I.

If the point charge were negative, B would be at a lower potential than A. (Check: Is W negative?) Thus, in going from A to B, a positive charge would lose potential energy. In summary, the potential changes according to the following general rules:

Electric potential increases ($+\Delta V$) as we consider locations nearer to positive charges or farther from negative charges.

Electric potential decreases ($-\Delta V$) as we consider locations farther from positive charges or nearer to negative charges.

The potential at a very large distance from a point charge is usually defined to be zero; this reference point is convenient but not required. For this choice of reference, the *electric potential* at a distance r from the charge is

$$V = \frac{kq}{r} \qquad \begin{array}{l} \textit{electric potential due to} \\ \textit{point charge only} \end{array} \qquad (16.4)$$

Keep in mind, however, that only electric potential *differences* are important, as the next Example illustrates.

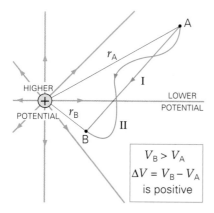

FIGURE 16.3 Electric field and potential due to a point charge Electric potential increases as you move closer to a positive charge. Thus, B is at a higher potential than A, and $\Delta V = V_B - V_A$ is positive.

In the figure:

HIGHER POTENTIAL

LOWER POTENTIAL

$V_B > V_A$
$\Delta V = V_B - V_A$
is positive

Example 16.3 ■ The Hydrogen Atom Revisited: Potential Differences near a Proton

In the Bohr model of the hydrogen atom (Chapter 27), the electron in orbit around the proton can exist only in certain circular orbits. The smallest has a radius of 0.0529 nm, and the next largest has a radius of 0.212 nm. (a) What is the potential difference between the two orbits? (b) Which orbit is at the higher potential? Back up your answer with numerical calculations of the two potentials.

Thinking It Through. (a) Because the proton behaves like a point charge, the potential difference is calculated from the point-charge expression (Eq. 16.3). (b) For the absolute potential values, we use Eq. 16.4 (zero potential at infinity) and compare our answers for the two different distances from the proton.

Solution. The electron orbits in the field of a proton, whose charge we know. First converting the radii into meters, we have:

Given: $q_p = +1.60 \times 10^{-19}$ C   **Find:** (a) ΔV (potential difference)
        $r_1 = 0.0529$ nm $= 5.29 \times 10^{-11}$ m   (b) Orbit at higher potential
        $r_2 = 0.212$ nm $= 2.12 \times 10^{-10}$ m

(a) The difference in potential is given by Eq. 16.3:

$$\Delta V = \frac{kq_p}{r_1} - \frac{kq_p}{r_2} = kq_p \left(\frac{1}{r_1} - \frac{1}{r_2} \right)$$

$$= (9.00 \times 10^9 \text{ N·m}^2/\text{C}^2)(+1.60 \times 10^{-19} \text{ C})\left(\frac{1}{5.29 \times 10^{-11} \text{ m}} - \frac{1}{2.12 \times 10^{-10} \text{ m}} \right)$$

$$= +20.4 \text{ V}$$

(b) The smallest orbit has the higher electric potential because the electron is then closest to the positively charged proton. To check this out numerically, let's use Eq. 16.4 to calculate the value of the potential at both distances. For the smallest orbit,

$$V_1 = \frac{kq_p}{r_1} = \frac{(9.00 \times 10^9 \text{ N·m}^2/\text{C}^2)(+1.60 \times 10^{-19} \text{ C})}{5.29 \times 10^{-11} \text{ m}} = +27.2 \text{ V}$$

For the next largest orbit, we expect a lower value:

$$V_2 = \frac{kq_p}{r_2} = \frac{(9.00 \times 10^9 \text{ N·m}^2/\text{C}^2)(+1.60 \times 10^{-19} \text{ C})}{2.12 \times 10^{-10} \text{ m}} = +6.79 \text{ V}$$

Notice that $\Delta V = 27.2$ V $- 6.8$ V $= +20.4$ V, in agreement with the result of (a).

Follow-up Exercise. In this Example, (a) what is the change in the potential energy if the electron is moved from the smallest orbit to a position very far away from the proton (to "ionize" the atom)? (b) Does the electron move to a region of higher or lower electrical potential?

Electric Potential Energy of Various Charge Configurations

In discussing gravitational energy, we considered the gravitational potential energy of systems of masses. The expressions for electric force and gravitational force are mathematically similar, and so are those for the corresponding potential energies, except for the use of charge in place of mass and the two signs associated with charge. In the gravitational case of two masses, the mutual gravitational potential energy was negative because the force is always attractive. For electric potential energy, the result can be either positive or negative because the electric force can be either attractive or repulsive.

If a positive point charge q_1 is fixed in space and a second positive charge q_2 is pushed toward it from a very large distance ($r = \infty$) to a distance r_{12} (•Fig. 16.4a), the work done is positive. The system of two charges gains electric potential energy:

$$\Delta U_e = q_2 \Delta V = q_2 \left(\frac{kq_1}{r_{12}} - \frac{kq_1}{\infty} \right) = \frac{kq_1 q_2}{r_{12}} - \frac{kq_1 q_2}{\infty} = \frac{kq_1 q_2}{r_{12}}$$

(Recall that dividing any number by infinity yields zero.)

We can choose our zero electric potential energy at any location. For point charges, it is convenient to choose that location to be at very large (infinite) distances. With this choice, we have

$$U_{12} = \frac{kq_1 q_2}{r_{12}} \qquad \begin{array}{l} \textit{electric potential energy} \\ \textit{(two charges)} \end{array} \qquad (16.5)$$

For *unlike* charges, the electrical force is attractive. In that case, by analogy with an attractive gravitational force, the electrostatic potential energy is negative. For *like* charges, the electrical force is repulsive, and the potential energy is positive. Thus if the two charges are of the same sign, it takes positive work to move them

•**FIGURE 16.4   Mutual electric potential energy of point charges** (a) If a positive charge is moved from a very great distance to a distance r_{12} from another positive charge, there is an increase in potential energy because positive work must be done to bring the mutually repelling charges closer together. (b) For more than two charges, the system's electric potential energy is the sum of the energies of each pair.

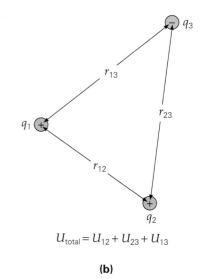

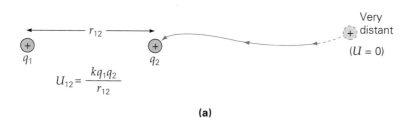

$$U_{12} = \frac{kq_1q_2}{r_{12}}$$

(a)

$$U_{\text{total}} = U_{12} + U_{23} + U_{13}$$

(b)

together; when released, they will move apart, gaining kinetic energy as they lose potential energy. Conversely, it takes positive work to increase the separation of two opposite charges, such as the proton and the electron in a hydrogen atom.

The signs of the charges can be used to keep things straight mathematically, as Example 16.4 shows. Since energy is a scalar quantity, for a static configuration of charges, the *total* potential energy is the algebraic sum of the mutual potential energies of all pairs of charges. For a configuration of many point charges, the total potential energy (U_{total}) is

$$U_{\text{total}} = U_{12} + U_{23} + U_{13} + \cdots \tag{16.6}$$

Only the first three terms of Eq. 16.6 would be needed for the configuration shown in Fig. 16.4b.

Example 16.4 ■ The Water Molecule: Electrostatic Potential Energy

The water molecule is the foundation of life as we know it. Many of its important properties (for example, the reason it is a liquid on the Earth's surface) are related to the fact that it is a polar molecule. Although the molecule's net charge is zero, it is separated into positive and negative regions. A very simple picture of the water molecule, along with the charges, is given in •Fig. 16.5. The distance from each hydrogen atom to the oxygen atom is 9.60×10^{-11} m, and the angle (θ) between the two hydrogen–oxygen bond directions is 104°. What is the total electrostatic energy of the water molecule?

Thinking It Through. Our model of the water molecule involves three charges. Thus the sum for Eq. 16.6 will have three terms. The charges are given, but we must calculate the distance between the hydrogen atoms by trigonometry. The total electrostatic potential energy is the algebraic sum of the mutual potential energies of the three pairs.

Solution. We list the given data.

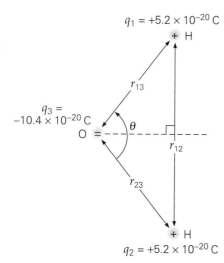

•**FIGURE 16.5 Electrostatic potential energy of a water molecule** A charge configuration has electric potential energy due to the work done in bringing the charges together from very great distances. See Example 16.4.

Given: $q_1 = q_2 = +5.20 \times 10^{-20}$ C (from figure)   *Find:* U_{total} (total electrostatic
$q_3 = -10.4 \times 10^{-20}$ C (from figure)   potential of water molecule)
$r_{13} = r_{23} = 9.60 \times 10^{-11}$ m
$\theta = 104°$

Notice that $(r_{12}/2)/r_{13} = \sin(\theta/2)$. Thus we can solve for r_{12} directly, and we have

$$r_{12} = 2r_{13}\sin\frac{\theta}{2} = 2(9.60 \times 10^{-11}\text{ m})(\sin 52°) = 2(9.60 \times 10^{-11}\text{ m})(0.788)$$

$$= 1.51 \times 10^{-10}\text{ m}$$

To get the algebraic sum of the mutual potential energies of the three pairs, let's calculate separately each pair's contribution to the total. Note that $U_{13} = U_{23}$ (why?). Applying Eq. 16.5, we have

$$U_{12} = \frac{kq_1q_2}{r_{12}} = \frac{(9.00 \times 10^9 \text{ N·m}^2/\text{C}^2)(+5.20 \times 10^{-20} \text{ C})(+5.20 \times 10^{-20} \text{ C})}{1.51 \times 10^{-10} \text{ m}}$$

$$= +1.61 \times 10^{-19} \text{ J}$$

$$U_{13} = U_{23} = \frac{kq_2q_3}{r_{23}} = \frac{(9.00 \times 10^9 \text{ N·m}^2/\text{C}^2)(+5.20 \times 10^{-20} \text{ C})(-10.4 \times 10^{-20} \text{ C})}{9.60 \times 10^{-11} \text{ m}}$$

$$= -5.07 \times 10^{-19} \text{ J}$$

The total electrostatic potential energy is then

$$U_{\text{total}} = U_{12} + U_{13} + U_{23} = (+1.61 \times 10^{-19} \text{ J}) + (-5.07 \times 10^{-19} \text{ J}) + (-5.07 \times 10^{-19} \text{ J})$$

$$= -8.53 \times 10^{-19} \text{ J}$$

The net result is a system with a negative total electrostatic potential energy. Thus positive work is needed to break apart the molecule.

Follow-up Exercise. Another common polar molecule is carbon monoxide, CO, a toxic gas produced commonly in automobiles when the combustion of gasoline molecules is incomplete. The carbon atom is positively charged, and the oxygen atom negatively charged. The distance between the carbon atom and the oxygen atom is 1.20×10^{-10} m, and the charge on each is 6.60×10^{-20} C. Determine the total electrostatic energy of this molecule and the work needed to break it apart completely.

Electric potential energy is a particularly useful quantity when no nonconservative forces are involved. Suppose that two positively charged particles are initially at rest near one another. If we release them, they will fly apart. As a result of the conservation of energy, they will gain kinetic energy at the expense of potential energy. The situation is analogous to that of two masses connected by a compressed spring.

16.2 Equipotential Surfaces and the Electric Field

Objectives: To (a) explain what is meant by an equipotential surface, (b) sketch equipotential surfaces for simple charge configurations, and (c) explain the relationship between equipotential surfaces and electric fields.

Suppose that a positive charge is moved at constant speed perpendicularly to an electric field (such as path I of •Fig. 16.6a). As the charge moves from A to A′ between parallel charged plates, *no work* is done on it by the electric field. The angle between the force due to the field and the displacement is 90°. Thus the work done is zero (cos 90° = 0). If no work is done, then the potential energy of the charge does not change between A and A′, so $\Delta U_{AA'} = 0$. We can conclude that these two points—and *all* other points on path I—are at the same potential:

$$\Delta V_{AA'} = V_{A'} - V_A = \frac{\Delta U_{AA'}}{q} = 0 \quad \text{or} \quad V_A = V_{A'}$$

This is also true for all points on the *plane* parallel to the plates and containing path I. Such a plane, on which the value of the electric potential is constant, is an **equipotential surface** (sometimes referred to simply as an *equipotential*). The word *equipotential* means "same potential." An equipotential surface need not be a plane, nor is it necessarily parallel to the plates or charges that created the field, as is true in the special parallel-plate case we considered here.

Since no work is required to move a charge along an equipotential surface, then it must be generally true that

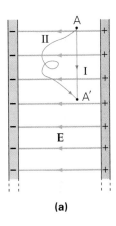

(a)

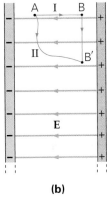

(b)

•**FIGURE 16.6 Construction of equipotential surfaces between parallel plates** (a) The equipotential surfaces are planes parallel to the plates. The work done in moving a charge is zero as long as you start and stop on the same equipotential (compare paths I and II). (b) Once you move to a higher potential (A to B), you can stay on that new equipotential by moving perpendicularly to the electric field (B to B′). The change in potential is independent of path, since the same change occurs via path I as via path II.

Moreover, since the electric field is a conservative field, the work done on the charge will be the same whether we take path I, path II, or any other path from A to A' (Fig. 16.6a). The exact route is irrelevant. As long as the charge returns to the same equipotential surface, the net work done to move it is zero. Thus its electric potential energy remains constant.

If we move the charge opposite the direction of E (path I in Fig. 16.6b)—at right angles to the equipotentials—the electric potential energy, and hence the electric potential, will increase. (Why? Does this require work?) When point B is reached, the charge is on another equipotential surface, but one of a higher potential. If we had instead moved the charge from A to B', the work required would be the same as in moving from A to B (why?). Thus B and B' are on an equipotential.

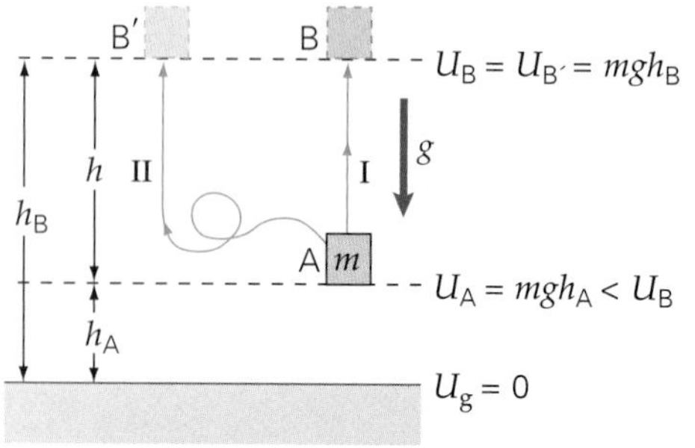

To help us understand the concept of an equipotential surface, consider a gravitational analogy. If we designate the gravitational potential energy as zero at ground level and raise an object a height $h = h_B - h_A$ (from A to B in •Fig. 16.7), the work done by an external force is mgh and is positive. For horizontal movement, the potential energy does not change. For example, the dashed plane at height h_B is a gravitational equipotential surface. So is the plane at h_A—but at a lower potential. Thus, surfaces of constant gravitational potential energy are planes parallel to the Earth's surface. Topographic maps, which display land contours by plotting lines of constant elevation (usually relative to sea level), are actually maps of constant gravitational potential (•Fig. 16.8a,b). Thus, topo-

•FIGURE 16.7 Gravitational potential energy analogy Raising an object in a uniform gravitational field results in an increase in potential energy. At a given height, its potential energy is constant as long as it remains on that equipotential surface. Here **g** points downward, like **E** in Fig. 16.6. Thus $U_B > U_A$.

$U_B = U_{B'} = mgh_B$

$U_A = mgh_A < U_B$

$U_g = 0$

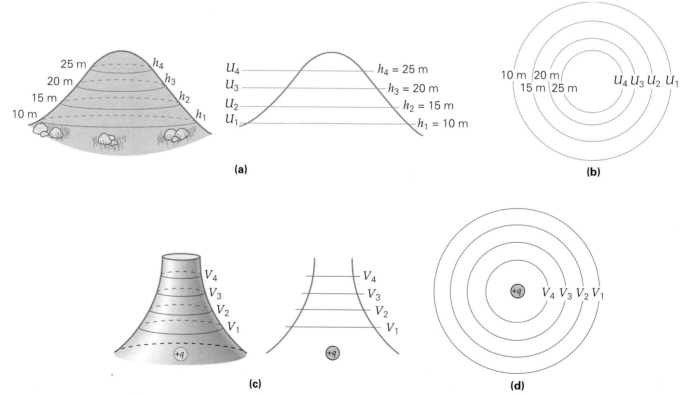

(a)

(b)

(c)

(d)

•FIGURE 16.8 Topographic maps—a gravitational analogy for equipotential surfaces (a) A symmetrical hill with slices at different elevations. Each slice is a plane of constant gravitational potential. (b) A topographic map of the slices in (a). The contours, where the planes intersect the surface, represent gravitational equipotentials. (c) The potential V around a point charge forms a similar symmetrical hill. V is constant at fixed distances from q. (d) Electrical equipotentials around a point charge are spherical (in two dimensions, circular slices).

graphic maps could be called maps of gravitational equipotential surfaces. Note how the equipotentials near a point charge are qualitatively similar to the contours of a hill (Fig. 16.8c,d).

It is extremely useful to know how to sketch equipotential surfaces, because they are intimately related to the electric field and to practical aspects such as voltage. In Learn by Drawing on p. 522, we summarize the method used for sketching equipotential surfaces, given an electric field-line pattern, and conversely for sketching the electric field lines when equipotential surfaces are given. Can you see how the equipotentials of an electric dipole were constructed in •Fig. 16.9?

For the special case of the uniform electric field (•Fig. 16.10), the magnitude of the potential difference (ΔV) between any two equipotential planes separated by a distance Δx is

$$\Delta V = V_3 - V_1$$
$$= E \Delta x$$

Thus if you started on equipotential surface 1 (denoted by potential V_1 in Fig. 16.10) and moved *perpendicularly* away to equipotential surface 3, you would measure a potential increase that depended solely on the electric field strength and the distance traveled.

For a given travel distance Δx, this perpendicular movement yields the *maximum possible* gain in potential. Think of taking one step of length Δx in *any* direction starting from surface 1. The way to maximize the increase in potential would be to step onto surface 3. A step in any other direction, not perpendicular to surface 1 (for example, ending on surface 2 in Fig. 16.10), would yield a smaller change in potential because you would end up on an equipotential surface closer to the one on which you started on (see the three steps, each Δx long, in Fig. 16.10).

By finding the *direction of the maximum potential change*, we are finding the direction *opposite* that of **E**. In one dimension, using plus and minus to indicate directions (here + means toward the positive plate), the expression for the electric field (including direction) in terms of the potential is

$$E = -\left(\frac{\Delta V}{\Delta x}\right)_{max} \qquad \textit{electric field from potential} \qquad (16.7)$$

Thus

E is in the direction opposite that in which V increases most rapidly, or in the direction in which V decreases the most rapidly.

The units of electric field are volts per meter (V/m). In Chapter 15 we saw that E can be expressed in units of newtons per coulomb (N/C; see Section 15.4). You should be able to show, through unit analysis, that these two units are equivalent.

As mentioned above, the potential difference, rather than the electric field, is usually specified for practical applications. For example, a D-cell flashlight battery has a terminal voltage of 1.5 V, meaning it can maintain a potential difference of 1.5 V between its terminals.

Whether you know it or not, you live in an electric field created by the Earth. This field varies with weather conditions and, as such, can be an indicator of approaching lightning and thunderstorms, for instance. Consider Example 16.5 as a practical application of the equipotential surface concept for understanding the Earth's electric field.

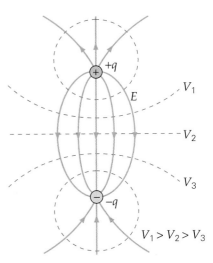

•**FIGURE 16.9 Equipotentials of an electric dipole** Equipotentials are perpendicular to electric field lines. Notice that $V_1 > V_2$, since equipotential surface 1 is closer to the positive charge than is surface 2. To understand how equipotentials are constructed from electric field lines, see Learn by Drawing on p. 522.

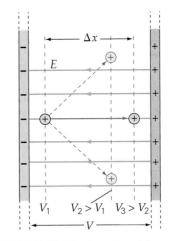

•**FIGURE 16.10 Relationship between the potential change and E** The electric field direction is that of maximum decrease in potential, or opposite the direction of maximum increase in potential (here, in the direction of the solid blue arrow, not the dashed ones; why?) Its magnitude is given by the rate at which the potential changes in volts per meter.

Graphical Relationship between Electric Field Lines and Equipotentials

Since it takes no work to move a charge along an equipotential, equipotential surfaces and electric field lines must be perpendicular to one another. Also, the electric field at any location points in the direction in which the potential decreases most rapidly and has a magnitude equal to the change in potential per meter. We can use these facts to construct equipotentials for a given pattern of electric field lines. The reverse is also true: Given the equipotentials, we can construct the electric field lines. If we know the potential value (in volts) associated with each equipotential, we can estimate the strength and direction of the field.

A couple of examples should provide graphical insight into the close connection between equipotentials and fields. Consider Fig. 1, in which you are given the electric field lines and want to determine the shape of the equipotentials. Pick any point, such as A, and begin moving at right angles to the electric field (mapped by the field lines) at that point. Keep moving, always maintaining this same perpendicular orientation to the field lines. Between lines you may have to approximate, but plan ahead to the next field line so that you can cross it at a right angle. To find another equipotential, start at another point, such as B, and proceed the same way. Sketch as many equipotentials as you need to map the area of interest. Figure 1 shows the result of sketching four such equipotentials, from A (at the highest potential value—can you tell why?) to D (at the lowest potential).

Suppose that you are given the equipotentials instead of the field lines (Fig. 2). The electric field lines point in the direction of decreasing V and are perpendicular to the equipotential surfaces. Thus, to map the field, start at any point and

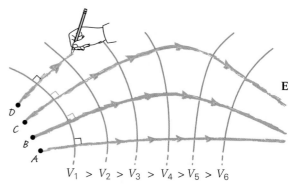

$$V_1 > V_2 > V_3 > V_4 > V_5 > V_6$$

FIGURE 2   Mapping the electric field from equipotentials   Start at a convenient point and trace a line that crosses each equipotential at a right angle. Repeat the process as often as needed to reveal the field pattern, adding arrows to indicate the direction of the field lines from high to low potential.

move in such a way that your path intersects each equipotential surface at a right angle. The resulting field line starting at point A in Fig. 2 is shown. Starting at points B, C, and D gives rise to additional field lines that suggest the complete electric field pattern; you need only add the arrows in the direction of decreasing potential.

Suppose that you want to estimate the magnitude of **E** at some point P, knowing the values of the equipotentials 1.0 cm on either side of it (Fig. 3). From this information you can easily tell that the field points in the direction from A to B (why?). Its approximate magnitude is

$$|\mathbf{E}| = \frac{\Delta V}{\Delta x} = \frac{(1000\ \text{V} - 950\ \text{V})}{2.0 \times 10^{-2}\ \text{m}}$$
$$= 2.5 \times 10^3\ \text{V/m}$$

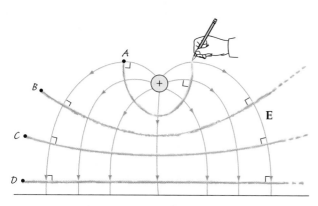

FIGURE 1   Sketching equipotentials from electric field lines   Given the electric field pattern, pick a point in the region of interest and move so that your path is always perpendicular to the next field line. Keep your path as smooth as possible, since kinks and loops in the equipotentials are not allowed. (Why?) To map a higher potential value, move opposite the direction of the electric field and repeat the process. Here $V_A > V_B$ and so on.

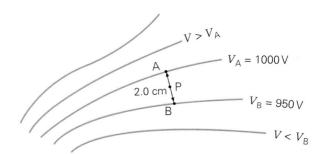

FIGURE 3   Estimating the magnitude of the electric field   The magnitude of the potential change per meter at any point gives the strength of the electric field at that point.

Example 16.5 ■ Electric Barometers: Equipotentials and the Earth's Electric Field

Under normal atmospheric conditions, the Earth is electrically charged and creates an approximately constant electric field of about 150 V/m pointing down near its surface. (a) What is the shape of the equipotentials, and in what direction does the electric potential increase most rapidly? (b) How far apart are two equipotential surfaces that have a 1000-volt difference between them? Which has a higher potential—the one farther from the Earth or the one closer? (c) During an approaching lightning storm, the polarity changes and the field can rise to many times the normal value (see the Insight "Lightning and Lightning Rods" in Chapter 15). Under these conditions, if the field is 900 V/m and points up, now how far apart are the two surfaces in (b)?

Thinking It Through. (a) The field is uniform, so the equipotentials are analogous to the parallel-plate situation. The discussion surrounding Eq. 16.7 will enable us to determine which way the potential increases. (b) We can use Eq. 16.7 to determine how far apart the potentials are. (c) When the electric field reverses, so will the direction in which the potential increases. We can use the same approach as in (b).

Solution. Listing the data, we have

Given: $E = 150$ V/m, down (normal conditions)
   $\Delta V = 1000$ V
   $E' = 900$ V/m, up (storm conditions)

Find: (a) Shape of equipotential surfaces and direction of potential increase
(b) Δx (distance between equipotentials under normal conditions)
(c) $\Delta x'$ (distance between equipotentials under storm conditions)

(a) As we have seen before, uniform electric fields are associated with equipotential *planes*. In this case the planes are parallel to the Earth's surface. The electric field points down. This is the direction *opposite* to the way the potential increases. Thus the potential increases upward.

(b) To determine the separation distance between these two equipotentials, we think of moving vertically so that $\Delta V/\Delta x$ is the maximum value (why?). We solve Eq. 16.7 for Δx (magnitude only, so we ignore the minus sign):

$$\Delta x = \frac{\Delta V}{E} = \frac{1000 \text{ V}}{150 \text{ V/m}} = 6.67 \text{ m}$$

Since the potential increases upward, the higher potential is associated with the surface that is 6.67 m farther from the ground. In other words, the potential decreases as we get closer to the Earth's surface, and that is the direction of **E**.

(c) Again, we solve Eq. 16.7 for Δx (magnitude) but with a different potential difference:

$$\Delta x' = \frac{\Delta V}{E'} = \frac{1000 \text{ V}}{900 \text{ V/m}} = 1.11 \text{ m}$$

The equipotentials are closer together than in (b), indicating a larger electric field strength.

Follow-up Exercise. (a) In part (a) of this Example, can you tell *where* the two surfaces are located relative to the ground? Explain. (b) In part (c) of this Example, which surface is at a higher potential, the one closer to the Earth or the one farther away?

The concept of electric potential provides a convenient unit of energy that is particularly useful in molecular, atomic, nuclear, and elementary particle physics. The **electron volt** (eV) is defined as the kinetic energy acquired by an electron accelerated through a potential difference of exactly 1 V. Because the electron's gain in kinetic energy is equal to the magnitude of its loss in electric potential energy, we can write

$$\Delta K = -\Delta U_e = -(e\,\Delta V) = -(-1.60 \times 10^{-19}\,\text{C})(1.00\,\text{V})$$
$$= 1.60 \times 10^{-19}\,\text{J}$$

This relationship provides us with a conversion factor between the electron volt and the joule to three significant figures:

$$1\,\text{eV} = 1.60 \times 10^{-19}\,\text{J}$$

The electron volt is on the order of typical energies on the atomic scale. Thus it is more convenient to express atomic energies in terms of electron volts than in joules. The energy of *any* charged particle accelerated through *any* potential difference can be expressed in electron volts. For example, if an electron is accelerated from rest through a potential difference of 1000 V, its gain in kinetic energy (K) is

$$K = |e\,\Delta V| = (1\,e)(1000\,\text{V}) = 1000\,\text{eV} = 1\,\text{keV}$$

The "keV" is the abbreviation for the *kilo*electron volt.

The electron volt is conceptually defined in terms of an electron. However, the energy of a particle with any charge, positive or negative, can be expressed in electron volts. Thus, if a particle with a charge of $+2e$, such as an alpha particle (a helium nucleus, consisting of two protons and two neutrons), were accelerated through this voltage, it would gain a kinetic energy of $K = (2e)(1000\,\text{V}) = 2000\,\text{eV} = 2\,\text{keV}$. Note how easy it is to compute the kinetic energy of an accelerated charged particle if you work in electron volts.

Larger units of energy are sometimes needed. For example, in nuclear and elementary particle physics it is not uncommon to find particles with energies of *mega*electron volts (MeV), and *giga*electron volts (GeV); $1\,\text{MeV} = 10^6\,\text{eV}$, and $1\,\text{GeV} = 10^9\,\text{eV}$, respectively.*

Be aware that the electron volt is *not* an SI unit. Thus in using energies to calculate speeds, you must first convert from electron volts back to joules. For example, to calculate the speed of an electron accelerated from rest through 10.0 V, first convert its kinetic energy (10.0 eV) to joules: $K = (10.0\,\text{eV})(1.60 \times 10^{-19}\,\text{J/eV}) = 1.60 \times 10^{-18}\,\text{J}$. Then, using its mass in kg (why?), its speed is

$$v = \sqrt{2K/m} = \sqrt{2(1.60 \times 10^{-18}\,\text{J})/(9.11 \times 10^{-31}\,\text{kg})} = 1.87 \times 10^6\,\text{m/s}$$

16.3 Capacitance

Objectives: To (a) define capacitance and explain what it means physically, and (b) calculate the charge, voltage, electric field, and energy storage for parallel-plate capacitors.

The electric field caused by charged conductors stores electrical energy. Consider the parallel metal plates shown in •Fig. 16.11. Such an arrangement of conductors is a typical example of a **capacitor**. (In theory, any two nearby conductors can be thought of as a capacitor.) Work must be done to transfer charge from one plate to the other. Suppose that one electron is moved between those plates. This requires some external work, since the net positive charge left behind would attract this electron back. Once this work is done, transferring the second electron would be more difficult because it would be repelled by the original electron *and* attracted by a double positive charge left behind. Thus, to separate the charges requires more and more work as more and more charge accumulates on the plates.

*At one time, a billion electron volts was referred to as BeV, but this was abandoned because confusion arose. In some countries, such as Great Britain and Germany, a billion means 10^{12} (which is called a trillion in the United States).

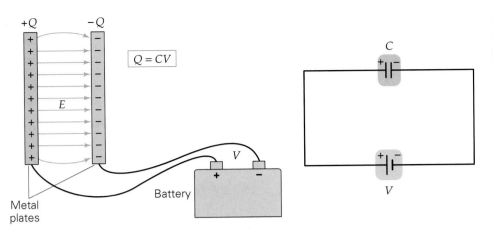

$Q = CV$

+Q   −Q

E

Metal
plates

Battery

+   −

V

(a) Parallel-plate capacitor

C

+ | | −

V

+ | | −

V

(b) Schematic circuit diagram

•**FIGURE 16.11 Capacitor and circuit diagram** (a) Two parallel metal plates are charged by a battery to a charge $Q = CV$, where C is the capacitance. Work is done in charging the capacitor, and energy is stored in the electric field. **(b)** This circuit diagram shows the symbols used for a battery (V) and a capacitor (C). The longer line of the battery symbol is taken as the positive terminal, and the shorter line as the negative terminal. The symbol for a capacitor is similar but has lines of equal length.

(This is analogous to stretching a spring: The more you stretch it, the harder it is to stretch it further.)

The work needed to charge parallel plates can be quickly done (usually in a few microseconds) by a battery. The battery removes electrons from the positive plate and transfers or "pumps" electrons through the wire to the negative plate, doing work on them in the process. The result is a separation of charge and the creation of an electric field in the capacitor. The battery charges the capacitor until the potential difference between the plates is equal to the terminal voltage of the battery. When the capacitor is disconnected from the battery, it remains a kind of reservoir of electrical energy. In accordance with the conservation of energy, the work done by the battery is stored as potential energy in the electric field. This stored energy can then be used to do work.

In most everyday capacitors, the potential difference across the plates is proportional to the charge Q:

$$Q \propto V$$

(Here Q means the magnitude of the charge on *either* plate, since energy is stored even though the net charge on the capacitor is zero.) We can make this into an equation by introducing a constant of proportionality, C:

$$Q = CV \quad \text{or} \quad C = \frac{Q}{V} \qquad (16.8)$$

SI unit of capacitance: coulomb per volt (C/V), or farad (F)

The constant C is called the **capacitance**; it expresses the charge stored per volt, as can be seen from its units, coulombs *per volt*. When we say a capacitor has a "large capacitance," we mean that it has been constructed to hold more charge *per volt* than one of "smaller capacitance." In other words, if you connected the same battery to two different capacitors, thus providing identical voltages across their plates, the one with the larger capacitance would store more charge and therefore more energy.

The combined unit of capacitance is the coulomb per volt (C/V), called a **farad** (F). The farad is a rather large unit (see Example 16.6), so the *microfarad* ($1\,\mu\text{F} = 10^{-6}$ F) or *picofarad* ($1\,\text{pF} = 10^{-12}$ F) is frequently used.

Capacitance depends *only* on the geometry (size, shape, spacing) of the plate arrangement (and the material between the plates, as we shall see). Consider the parallel-plate capacitor, which has a uniform electric field given by Eq. 15.5:

$$E = \frac{4\pi kQ}{A}$$

Note: The net charges on the plates are $+Q$ and $−Q$, but it is customary to refer in general to the charge Q on a capacitor.

Note: From here on, we use V instead of ΔV as our notation for potential difference, or *voltage*. It is still, however, a *difference*.

Definition of:
Capacitance

The farad was named for the English scientist Michael Faraday (1791–1867), an early investigator of electrical phenomena who first introduced the concept of the electric field.

Thus the voltage across the plates can be computed from Eq. 16.2, $V = Ed$, giving

$$V = Ed = \frac{4\pi k Q d}{A}$$

The capacitance of a parallel-plate arrangement is then

$$C = \frac{Q}{V} = \left(\frac{1}{4\pi k}\right)\frac{A}{d} \quad \text{(parallel plates only)} \quad (16.9)$$

Notice that the voltage V depends directly on the charge Q. Thus Q cancels out and the capacitance depends only on the geometrical arrangement of the plates, as we had anticipated. It is common to replace the expression in the parentheses in Eq. 16.9 with a single symbol, ϵ_o, called the *permittivity of free space* (in vacuum). The value of this constant (to three significant figures) is

$$\epsilon_o = \frac{1}{4\pi k} = 8.85 \times 10^{-12}\, C^2/(N \cdot m^2) \quad \begin{array}{l}\textit{permittivity}\\ \textit{of free space}\end{array} \quad (16.10)$$

The constant ϵ_o describes the electrical properties of free space—a vacuum. Its value in air is essentially the same. Along with its magnetic counterpart (Chapter 19), it determines the speed of light.

It is also common to rewrite Eq. 16.9 in terms of ϵ_o:

$$C = \frac{\epsilon_o A}{d} \quad \text{(parallel plates only)} \quad (16.11)$$

Let us use Eq. 16.11 to show just how unrealistically large an air-filled capacitor with a capacitance of 1.0 F would be.

Example 16.6 ■ Parallel-Plate Capacitors: How Large Is a Farad?

What would be the plate area of an air-filled 1.0-farad parallel-plate capacitor if the plate separation were 1.0 mm? Would it be realistic to consider building one?

Thinking It Through. The area can be calculated from Eq. 16.11. Remember to keep every quantity in SI units, so that the answer will be in square meters. We can use the vacuum value for ϵ_o for air.

Solution.

Given: $C = 1.0$ F   *Find:* A (area of one of the plates)
$\quad\quad\quad d = 1.0$ mm $= 1.0 \times 10^{-3}$ m

Solving Eq. 16.11 for the area gives

$$A = \frac{Cd}{\epsilon_o} = \frac{(1.0\,F)(1.0 \times 10^{-3}\,m)}{8.85 \times 10^{-12}\,C^2/(N \cdot m^2)} = 1.1 \times 10^8\,m^2$$

This is about 100 km² (about 40 mi²), or a square 10 km (over 6 mi) on a side. It is unrealistic to build a capacitor this size; 1.0 F is therefore a very large value of capacitance! There are ways, however, to make high-capacity capacitors, as we shall see in Section 16.4.

Follow-up Exercise. In this Example, what would be the plate spacing if you wanted the capacitor to have a plate area roughly the area of the roof of a car? (Assume a rectangular area 1.5 m × 2.0 m.) Compare your answer to a typical atomic diameter of 10^{-9} to 10^{-10} m. Do you think it is feasible to build this?

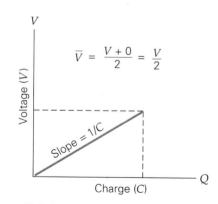

•FIGURE 16.12 **Voltage versus charge** A plot of voltage versus charge for a charging capacitor is a straight line with slope $1/C$ (since $Q = CV$). The average voltage is $\overline{V} = V/2$, and the total work done is equivalent to transferring the charge through $\overline{V}$. Thus $U_C = W = \frac{1}{2}QV$, which is the area under the curve.

The expression for the energy stored in a charged capacitor can be obtained through graphical analysis, since both Q and V vary during *charging*—the process of applying a charge, such as by a battery. A plot of voltage versus charge for charging a capacitor is a straight line with a slope of $1/C$ since $V = (1/C)Q$ (•Fig. 16.12). The graph represents the charging of an initially uncharged capacitor ($V_o = 0$) to some final voltage (V). The work done is equivalent to transferring the total charge through the average voltage $\overline{V}$. Since the voltage varies linearly with the charge, the average is

$$\overline{V} = \frac{V_{final} + V_{initial}}{2} = \frac{V + 0}{2} = \frac{V}{2}$$

Thus, the energy stored in the capacitor (equivalent to the work done by the battery) is

$$U_C = W = Q\overline{V} = \tfrac{1}{2}QV$$

Note: Practice using the various forms of capacitor energy. In a pinch, you need recall only one and the definition of capacitance, $C = Q/V$.

Since $Q = CV$, this equation can be written in several equivalent forms:

$$U_C = \tfrac{1}{2}QV = \frac{Q^2}{2C} = \tfrac{1}{2}CV^2 \qquad \begin{array}{l}\textit{energy storage}\\ \textit{in a capacitor}\end{array} \qquad (16.12)$$

Note: Do not confuse U_C, the energy stored in the capacitor, with the energy gained or lost by a *single charge q* moved from one plate to the other.

The form $U_C = \tfrac{1}{2}CV^2$ is usually the most practical, since the capacitance and the applied voltage are often known. As a very practical application, consider the following life-saving potential of the energy stored in a capacitor in a *cardiac defibrillator*.

Example 16.7 ■ Capacitors to the Rescue: Energy Storage in a Cardiac Defibrillator

Following a heart attack, the heart sometimes beats in erratic (and useless) fashion, called fibrillation. One way to get the heart back to its normal rhythm is to shock it with electrical energy supplied by an instrument called a cardiac defibrillator (•Fig. 16.13). Experiments show that about 300 J of energy is required to produce the desired effect. Typically, a defibrillator stores this energy in a capacitor charged by a 5000-volt power supply. (a) What capacitance is required? (b) What is the charge on the capacitor's plates?

Thinking It Through. (a) To find the capacitance, we solve for C in Eq. 16.12. (b) From the capacitance, the charge follows from the definition of capacitance (Eq. 16.8).

Solution. We list the given data.

Given: $U_C = 300$ J   *Find:*   (a) C (the capacitance)
         $V = 5000$ V   (b) Q (charge on capacitor)

(a) The most appropriate form of Eq. 16.12 is $U_C = \tfrac{1}{2}CV^2$ (why?). Solving for C, we get

$$C = \frac{2U_C}{V^2} = \frac{2(300\text{ J})}{(5000\text{ V})^2} = 2.40 \times 10^{-5}\text{ F} = 24.0\ \mu\text{F}$$

(b) The charge is then

$$Q = CV = (2.40 \times 10^{-5}\text{ F})(5000\text{ V}) = 0.120\text{ C}$$

Follow-up Exercise. In this Example, if the maximum allowable energy for any one defibrillation attempt is 750 J, what is the maximum allowable voltage?

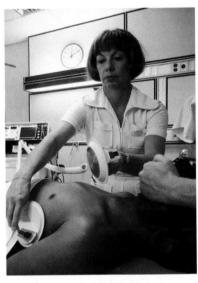

•**FIGURE 16.13 Defibrillator** A burst of electric current (charge flow) from a defibrillator may restore a normal heartbeat in persons who have suffered cardiac arrest. Capacitors are used to store the electrical energy that the device depends on.

Sometimes capacitors provide models for real-life applications. For example, a lightning storm can be considered to be the discharge of a negatively charged cloud to the ground (see the Insight in Chapter 15). Another interesting application of electric potential treats nerve membranes as cylindrical capacitors to explain nerve impulse transmission (see the Insight on p. 528).

16.4 Dielectrics

Objectives: To (a) understand what a dielectric is and (b) understand how it affects the physical properties of a capacitor.

In most capacitors, a sheet of insulating material, such as paper or plastic, is placed between the plates. Such an insulating material is called a **dielectric** and serves several purposes. For one, it keeps the plates from coming into contact, which

Electric Potential and Nerve Signal Transmission

One of the most important systems in the human body is the nervous system. This system is responsible for our reception of external stimuli through our senses (for example, touch). Nerves also provide communication between the brain and our organs and muscles. If you touch something hot, your brain sends the signal "move!" through the nervous system to your hand. But what is this signal, and how does it work?

A typical nerve consists of a bundle of nerve cells called *neurons*, much like individual telephone wires bundled into a single cable. The structure of a typical neuron is shown in Fig. 1a. The cell body, or *soma*, has long attaching structures called *dendrites*, which receive the input signal. The soma is responsible for processing this signal and transmitting it down the *axon*. At the other end of the axon are projections with knobs called *synaptic terminals*. At these knobs, the electrical signal is transmitted to another neuron across a gap called the *synapse*. The human body contains on the order of 100 billion neurons, and each neuron can have several hundred connections! Running the nervous system costs the body about 25% of its energy intake each day.

To understand the electrical nature of nerve signal transmission, let us focus on the axon. A vital component of the axon is its cell membrane, which is typically about 10 nm thick and consists of lipids (polarized hydrocarbon molecules) and embedded protein molecules (Fig. 1b). The membrane has pores called *ion channels* at which large protein molecules regulate the flow of ions across the membrane.

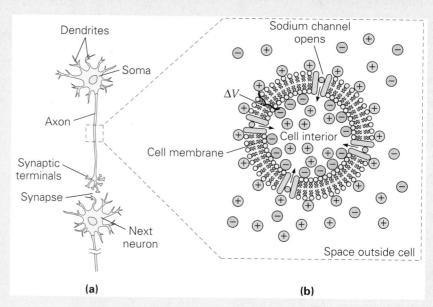

FIGURE 1 **(a)** The structure of a typical neuron. **(b)** An enlargement of the axon membrane, showing the membrane (about 10 nm thick) and the concentration of ions inside and outside the cell. The charge polarization across the membrane leads to a voltage, or membrane potential. When triggered by an external stimulus, the sodium channels open, allowing sodium ions into the cell. This influx changes the membrane potential.

TABLE 16.1 Dielectric Constants for Some Materials

Material	Dielectric Constant (K)
Vacuum	1.000 00
Air	1.000 54
Paper	3.5
Polyethylene	2.3
Polystyrene	2.6
Teflon	2.1
Glass	4.6
Silicone oil	2.6

would allow the electrons to flow back onto the positive plate, thereby neutralizing the charge on the capacitor and energy stored. It also allows flexible plates of aluminum foil to be rolled into a cylinder, giving the capacitor a practical size. Finally, it increases the charge storage capacity of the capacitor and therefore, under the right conditions, the energy stored in the capacitor. This capability depends on the chosen material and is characterized by the **dielectric constant** (K). Values of the dielectric constant for some materials are given in Table 16.1.

How a dielectric affects the electrical properties of a capacitor is illustrated in •Fig. 16.14 (p. 530). The capacitor is fully charged, then carefully disconnected from the battery, and a dielectric is inserted (Fig. 16.14a). In the dielectric material, work is done on molecular dipoles by the existing electric field as they experience electrical forces that tend to align them with that field (Fig. 16.14b). (The molecular polarization may be permanent or temporarily induced by the electric field. In either case the effect is the same.) Work is also done on the dielectric slab as a whole, since the charged plates attract it into the space between them. Both effects *decrease* the potential energy stored in the capacitor *as long as the charge on the plates stays the same*. The dielectric creates its own "reverse" electric field (Fig. 16.14c) and partially cancels the field between the plates.

The key to nerve signal transmission is the fact that these ion channels are selective—they allow only certain types of ions to cross the membrane; others cannot.

The fluid outside the axon, although electrically neutral, contains sodium ions (Na^+) and chlorine ions (Cl^-) in solution. Conversely, the axon's internal fluid is rich in potassium ions (K^+) and negatively charged protein molecules. If it were not for the selective nature of the cell membrane, the Na^+ concentration would be equal on both sides. Under normal (or *resting*) conditions, it is difficult for Na^+ to penetrate the interior of the nerve cell. This process gives rise to a polarization of charge across the membrane. The exterior is positive (with Na^+ trying to enter the region of lower concentration), which in turn attracts the negative proteins to the inner surface of the membrane (Fig. 1b). Thus a cylindrical capacitor-like charge-storage system exists across an axon membrane when it is resting. The *resting membrane potential* (the voltage across the membrane) is defined as $\Delta V = V_{in} - V_{out}$. Since the outside is positively charged, this potential is negative; it ranges from about −40 to −90 mV (millivolts), with a typical value of −70 mV.

Signal conduction occurs when the cell membrane receives a stimulus from the dendrites. Only then does the membrane potential change, and this change is propagated down the axon. The stimulus triggers Na^+ channels in the membrane (closed while resting, like a gate) to open and temporarily allows sodium ions to enter the cell (Fig. 1b). These positive ions are attracted to the negative charge layer on the interior and are driven by the difference in concentration. In about 0.001 s, enough sodium ions have passed through the gated channel to cause a reversal of polarity, and the membrane potential rises, typically to +30 mV. This time sequence of the change in membrane potential is shown in Fig. 2. When the Na^+ concentration difference causes the membrane voltage to become positive, the sodium gates close. A chem-

ical process known as the *Na/K ATPase molecular pump* then reestablishes the resting potential at −70 mV by selectively transporting the excess Na^+ back out to the cell exterior.

This variation in potential (100 mV—from −70 mV to +30 mV) is called the cell's *action potential*. The action potential *is* the signal transmitted down the axon. This voltage "wave" travels at speeds of from 1 to 100 m/s on its way to triggering another such pulse in the adjacent neuron. This speed, along with other factors such as time delays in the synapse region, is responsible for the typical human reaction times totaling a few tenths of a second.

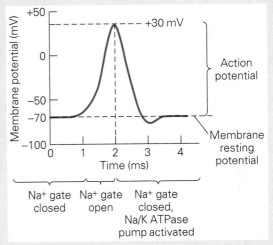

FIGURE 2 As the sodium channels open like gates and sodium ions rush to the cell's interior, the membrane potential changes quickly from its resting value of −70 mV to about +30 mV. The resting potential is restored (between 2 ms and 4 ms) by a pump process that chemically removes the excess Na^+ after the sodium gates have closed (at 2 ms).

This means that the *net* field between the plates, and thus the voltage across the plates (because $V = Ed$), are reduced from their vacuum values. If we designate the vacuum values by E_o and V_o, then the dielectric constant of the material is defined as the ratio of the voltage with the material in place divided by the vacuum voltage. Since V is proportional to E for parallel plates, we have

$$K = \frac{V_o}{V} = \frac{E_o}{E} \qquad (16.13)$$

Definition of:
Dielectric constant

Note that K is dimensionless and is always greater than 1. The dielectric constant can most easily be determined by measuring the two voltages. (We will see how a voltmeter works in Chapter 18.) Since the battery was disconnected and the capacitor isolated, the charge on the plates, Q_o, is unaffected. Since $V = V_o/K$, the new value of the capacitance is larger than the vacuum value by a factor of K:

Note: Equation 16.13 holds only if the battery is disconnected.

$$C = \frac{Q}{V} = \frac{Q_o}{V_o/K} = K\frac{Q_o}{V_o} \qquad or \qquad C = KC_o \qquad (16.14)$$

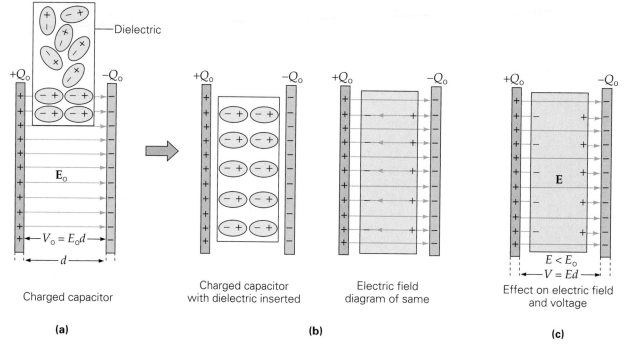

•FIGURE 16.14   The effects of a dielectric on an isolated capacitor   (a) A dielectric with randomly oriented permanent molecular dipoles is inserted between the plates of an isolated charged capacitor. As the dielectric slab is inserted, the capacitor does work on it due to the attractive forces between charges on the capacitor and those induced on the dielectric surfaces. We thus expect the energy storage to decrease if the battery has been disconnected. **(b)** If the dielectric is put into the capacitor, the dipoles orient (or partially orient) themselves with the field, giving rise to an opposing electric field. The dipole field effectively cancels some of the field due to the plate charges. **(c)** The net effects are decreases in the electric field and voltage.

We have seen that inserting a dielectric into an isolated capacitor results in a larger capacitance (the same charge stored at a lower voltage). But what about energy storage? Since there is no energy input (the battery is disconnected) and the capacitor does work on the dielectric, the stored energy *drops* by a factor of K (•Fig. 16.15a):

$$U_C = \frac{Q^2}{2C} = \frac{Q_o^2}{2KC_o} = \frac{Q_o^2/2C_o}{K} = \frac{U_o}{K} < U_o \quad \textit{(battery disconnected)}$$

But if the dielectric is inserted *and the battery remains connected*, the original voltage is maintained and the battery must supply more charge—and therefore do work (Fig. 16.15b). We expect the energy storage to go up in this situation. Here it is the charge on the plates that increases by the factor of K, $Q = KQ_o$. Once again the capacitance increases, since more charge is now stored at the same voltage $C = Q/V = KQ_o/V_o = K(Q_o/V_o) = KC_o$. *Thus, a dielectric increases the capacitance by a factor of K regardless of the conditions under which the dielectric is inserted.* However, in this case (constant V), as we have seen, the energy storage of the capacitor goes up at the expense of the battery:

$$U_C = \tfrac{1}{2}CV^2 = \tfrac{1}{2}KC_oV_o^2 = K(\tfrac{1}{2}C_oV_o^2) = KU_o > U_o \quad \textit{(battery connected)}$$

For a parallel-plate capacitor with a dielectric, the capacitance is given by

$$C = KC_o = \frac{K\epsilon_o A}{d} \quad \textit{(parallel plates only)} \quad (16.15)$$

(using Eq. 16.11 for the parallel-plate capacitance in a vacuum). This is sometimes written as $C = \epsilon A/d$, where $\epsilon = K\epsilon_o$ is the **dielectric permittivity** and is always greater than ϵ_o, the permittivity of the vacuum (free space), by a factor of K.

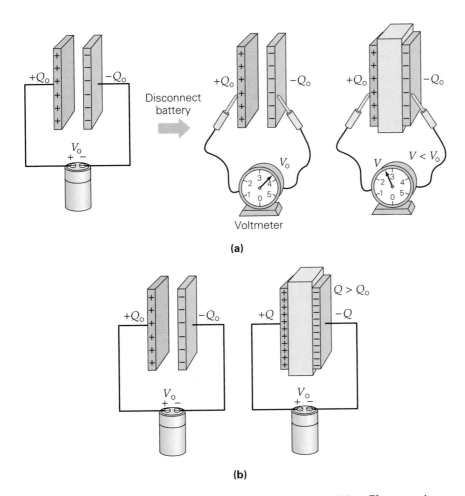

•FIGURE 16.15 **Dielectrics and capacitance** **(a)** A parallel-plate capacitor in air (that is, with no dielectric) is charged by a battery to a charge Q_o and a voltage V_o (left). If the battery is disconnected and the potential across the capacitor measured by a voltmeter, a reading of V_o is obtained (center). But if a dielectric is inserted between the capacitor plates, the voltage drops to $V = V_o/K$ (right), so the amount of energy stored is less. (Can you estimate the dielectric constant from the voltages?) **(b)** A capacitor is charged as in part (a), but the battery is left connected. When a dielectric is inserted into the capacitor, the voltage is maintained at V_o by the battery, but the amount of charge that the capacitor accepts increases to $Q = KQ_o$, so more energy is stored. Both of these observations reflect the increase in capacitance produced by the dielectric.

An assortment of capacitors is shown in •Fig. 16.16a. Changes in capacitance can be put to use in everyday life. *Transducers*, instruments that convert movement into detectable electrical signals (and vice versa), are all around us (Fig. 16.16b), as the next Example shows.

Example 16.8 ■ Electrical Motion Detectors: Capacitors as Transducers

Consider a capacitor (with dielectric) underneath a computer key (Fig. 16.16b). It is connected to a 12.0-volt power supply, has a normal (uncompressed) plate separation of 3.00 mm, and a plate area of 0.750 cm². (a) What is the required dielectric constant if

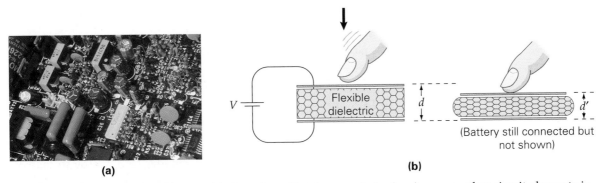

•FIGURE 16.16 **Capacitors in use** **(a)** Capacitors (flat, brown circles here) among other circuit elements in a microcomputer. **(b)** Transducers convert movement into electrical signals that can be measured and analyzed by computer. Some computer keyboards themselves operate in this way, as do other instruments such as seismographs (see Chapter 13). In this schematic representation of a charged capacitor used as a transducer, the plate distance changes. (See Example 16.8.)

the capacitance is 1.10 pF? (b) How much charge is stored on the plates under normal conditions? (c) How much charge flows onto the plates (that is, what is the change in charge) if they are compressed to a separation of only 2.00 mm?

Thinking It Through. (a) We can find the capacitance of air-filled plates from Eq. 16.11 and can then determine the dielectric constant from Eq. 16.14. (b) The charge follows from the definition of capacitance (Eq. 16.8). (c) We must use the compressed-plate separation distance to recompute the capacitance. Then the new charge can be found as in (b). By subtracting the new charge from the old, we get the change in the capacitor's charge (the charge that flowed). If the plates are compressed, we expect charge to flow onto (not away from) the plates (why?).

Solution. We list the given data.

Given: $V = 12.0$ V *Find:* (a) K (dielectric constant)
$d = 3.00$ mm $= 3.00 \times 10^{-3}$ m (b) Q (initial capacitor charge)
$A = 0.750$ cm$^2 = 7.50 \times 10^{-5}$ m^2 (c) ΔQ (change in capacitor charge)
$C = 1.10$ pF $= 1.10 \times 10^{-12}$ F
$d' = 2.00$ mm $= 2.00 \times 10^{-3}$ m

(a) Using Eq. 16.11, the capacitance if the plates were separated by air would be

$$C_o = \frac{\epsilon_o A}{d} = \frac{(8.85 \times 10^{-12}\ \text{C}^2/\text{N·m}^2)(7.50 \times 10^{-5}\ \text{m}^2)}{3.00 \times 10^{-3}\ \text{m}} = 2.21 \times 10^{-13}\ \text{F}$$

Since the dielectric increases the capacitance, it must have a value of

$$K = \frac{C}{C_o} = \frac{1.10 \times 10^{-12}\ \text{F}}{2.21 \times 10^{-13}\ \text{F}} = 4.98$$

(b) The initial charge is then

$$Q = CV = (1.10 \times 10^{-12}\ \text{F})(12.0\ \text{V}) = 1.32 \times 10^{-11}\ \text{C}$$

(c) Under compressed conditions, the capacitance is

$$C' = \frac{K\epsilon_o A}{d'} = \frac{(4.98)(8.85 \times 10^{-12}\ \text{C}^2/\text{N·m}^2)(7.50 \times 10^{-5}\ \text{m}^2)}{2.00 \times 10^{-3}\ \text{m}} = 1.65 \times 10^{-12}\ \text{F}$$

Since the voltage is the same, the new charge will be

$$Q' = C'V = (1.65 \times 10^{-12}\ \text{F})(12.0\ \text{V}) = 1.98 \times 10^{-11}\ \text{C}$$

Since the capacitance increased, the charge increased (as we expected, because the battery is always connected). The increase is

$$\Delta Q = Q' - Q = (1.98 \times 10^{-11}\ \text{C}) - (1.32 \times 10^{-11}\ \text{C}) = 6.60 \times 10^{-12}\ \text{C}$$

As the key is depressed, a charge whose magnitude is related to the displacement flows onto the capacitor, providing a way of measuring the movement electrically. (Where does the charge come from?)

Follow-up Exercise. In this Example, suppose you increased the plate spacing by 1.00 mm from its normal value of 3.00 mm. Would charge flow onto or away from the capacitor? How much charge would flow?

16.5 Capacitors in Series and in Parallel

Objectives: **To (a) find the equivalent capacitance of capacitors connected in series and in parallel, (b) calculate the charges, voltages, and energy storage of individual capacitors in series and parallel configurations, and (c) analyze capacitor networks that include both series and parallel arrangements.**

Capacitors can be connected in two basic ways: *in series* or *in parallel*. In series, the capacitors are connected head to tail, so to speak (•Fig. 16.17a). When they are connected in parallel, all the leads on one side of the capacitors have a common

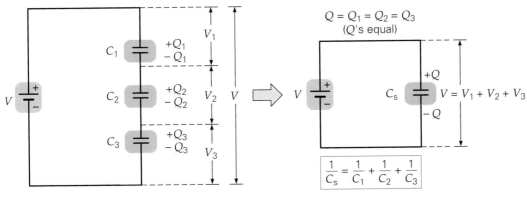

•FIGURE 16.17
Capacitors in series and parallel (a) All capacitors connected in series have the same charge, and the sum of the voltage drops is equal to the voltage of the battery. The total capacitance is equivalent to C_s. (b) When capacitors are connected in parallel, the voltage drops across the capacitors are the same, and the total charge is equal to the sum of the charges on the individual capacitors. The total capacitance is equivalent to C_p.

(a) Capacitors in series

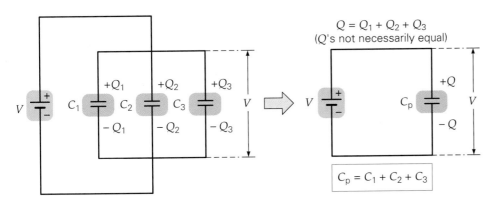

(b) Capacitors in parallel

connection (think of all the "tails" hooked together, and all the "heads" hooked together, as shown in Fig. 16.17b).

Capacitors in Series

When capacitors are wired in series, the charge (magnitude Q) is the same on all the plates:

$$Q = Q_1 = Q_2 = Q_3 = \cdots$$

To see why this must be true, refer to •Fig. 16.18. Note that only plates A and F are actually connected to the battery. Since the plates labeled B and C are isolated, the total charge on them must always be zero. Thus, if the battery puts a charge of $+Q$ on plate A, then $-Q$ is induced on B at the expense of plate C, which acquires a charge of $+Q$. This charge in turn induces $-Q$ on D, and so on down the line.

As we have seen, "voltage drop" is just another name for "change in electrical potential energy per unit charge." Thus, when we add up all the capacitor voltage drops (see Fig 16.17a), going from high to low potential, we get the same value as the drop across the terminals of the battery. (Why?) Thus, the sum of the individual voltage drops across all the capacitors is equal to the voltage of the source:

$$V = V_1 + V_2 + V_3 + \cdots$$

The individual voltages are related to the individual charges by

$$V_1 = \frac{Q}{C_1}, \qquad V_2 = \frac{Q}{C_2}, \qquad V_3 = \frac{Q}{C_3} \cdots$$

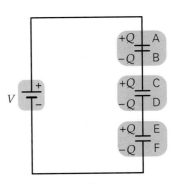

•FIGURE 16.18 **Charges on capacitors in series** Plates B and C together had zero net charge to start. When the battery placed $+Q$ on plate A, charge $-Q$ was induced on B; thus C must have acquired $+Q$ for the BC combination to remain neutral. Continuing this way through the string, we see that all the charges must be the same in magnitude.

16.5 Capacitors in Series and in Parallel **533**

We define the **equivalent series capacitance**, C_s, as the value of a single capacitor that could replace the series combination. Since the combination of capacitors stores a charge of Q at a voltage of V, we have $C_s = Q/V$, or $V = Q/C_s$. Substituting these expressions into the previous equation, we have

$$\frac{Q}{C_s} = \frac{Q}{C_1} + \frac{Q}{C_2} + \frac{Q}{C_3} + \cdots$$

Canceling the common Q's, we get

$$\frac{1}{C_s} = \frac{1}{C_1} + \frac{1}{C_2} + \frac{1}{C_3} + \cdots \qquad \begin{array}{l}\textit{equivalent series} \\ \textit{capacitance}\end{array} \qquad (16.16)$$

It is interesting to note that the value of C_s is always smaller than the smallest capacitance in the combination. The charge stored by the series is $Q = C_i V_i$ (where the subscript i refers to any one capacitor in the string). Since $V_i < V$, this arrangement stores *less* charge than any individual capacitor would if it were connected by itself to the same battery, $Q = C_i V$ (Fig. 16.17a).

It makes *physical* sense that in series the smallest capacitance will get the largest voltage. A small value of C means less charge stored per volt. Since the charges are the same in series, the smaller capacitors require a larger fraction of the total voltage.

Capacitors in Parallel

With a parallel arrangement (Fig. 16.17b), the voltages across the capacitors are the same (why?), and each individual voltage is equal to that of the battery:

$$V = V_1 = V_2 = V_3 = \cdots$$

The total stored charge Q_{total} is equal to the sum of the charges on the individual capacitors (Fig 16.17b):

$$Q_{total} = Q_1 + Q_2 + Q_3 + \cdots$$

We expect the equivalent capacitance in parallel to be larger than the largest one, because more charge per volt can be stored in this way than if any one capacitor were connected to the battery by itself. The individual charges are given by $Q_1 = C_1 V$, $Q_2 = C_2 V$, ... A capacitor with the **equivalent parallel capacitance**, C_p, would hold this same total charge when connected to the battery, so $C_p = Q_{total}/V$ or $Q_{total} = C_p V$. Substituting these expressions into the previous equation, we have

$$C_p V = C_1 V + C_2 V + C_3 V + \cdots$$

and, canceling the common V,

$$C_p = C_1 + C_2 + C_3 + \cdots \qquad \begin{array}{l}\textit{equivalent parallel} \\ \textit{capacitance}\end{array} \qquad (16.17)$$

In the parallel case, the equivalent capacitance C_p is the sum of the individual capacitances. The equivalent capacitance in parallel is larger than the largest individual capacitance, as expected. Since capacitors in parallel have the same voltage, the largest capacitance will store the most charge. As a comparsion of capacitors connected in series and in parallel, consider the next Example.

Example 16.9 ■ Which Stores More Charge: Capacitors in Series or in Parallel?

You have two capacitors, one 2.50 μF and the other 5.00 μF. What are the charge on each and the total charge stored if they are connected across a 12.0-volt battery (a) in series and (b) in parallel?

Thinking It Through. (a) Capacitors in series have the same charge; therefore the smaller capacitor will have most of the voltage ($Q = CV$ = constant; as C goes up, V goes down). With Eq. 16.16 we can find the equivalent capacitance and thus Q on each. (b) Capacitors in parallel have the same voltage (12.0 V; why?); hence the larger capacitor will have the most charge (as Q goes up, C goes up).

Solution. Listing the data, we have

Given: $C_1 = 2.50 \ \mu\text{F} = 2.50 \times 10^{-6} \ \text{F}$
$C_2 = 5.00 \ \mu\text{F} = 5.00 \times 10^{-6} \ \text{F}$
$V = 12.0 \ \text{V}$

Find: (a) Q on each capacitor in series and Q_{total} (total charge)
(b) Q on each capacitor in parallel and Q_{total} (total charge)

(a) In series the total (equivalent) capacitance is given by $1/C_s = 1/C_1 + 1/C_2$. Thus,

$$\frac{1}{C_s} = \frac{1}{2.50 \times 10^{-6} \ \text{F}} + \frac{1}{5.00 \times 10^{-6} \ \text{F}} = \frac{3}{5.00 \times 10^{-6} \ \text{F}}$$

so

$$C_s = 1.67 \times 10^{-6} \ \text{F}$$

(Notice that this result is less than the smallest capacitance in the chain.) Since the charge on each capacitor is the same in series and the same as the total, we have

$$Q_{\text{total}} = Q_1 = Q_2 = C_s V = (1.67 \times 10^{-6} \ \text{F})(12.0 \ \text{V}) = 2.00 \times 10^{-5} \ \text{C}$$

(b) Here we use the equivalent capacitance in parallel relationship, $C_p = C_1 + C_2 = 2.50 \times 10^{-6} \ \text{F} + 5.00 \times 10^{-6} \ \text{F} = 7.50 \times 10^{-6} \ \text{F}$. (Note that this result is greater than the larger individual value of C in the arrangement.) Therefore,

$$Q_{\text{total}} = C_p V = (7.50 \times 10^{-6} \ \text{F})(12.0 \ \text{V}) = 9.00 \times 10^{-5} \ \text{C}$$

In parallel, each capacitor has the full 12.0 V across it, hence

$$Q_1 = C_1 V = (2.50 \times 10^{-6} \ \text{F})(12.0 \ \text{V}) = 3.00 \times 10^{-5} \ \text{C}$$

$$Q_2 = C_2 V = (5.00 \times 10^{-6} \ \text{F})(12.0 \ \text{V}) = 6.00 \times 10^{-5} \ \text{C}$$

The total stored charge is the sum of the two charges, as we would expect in parallel.

Follow-up Exercise. In part (a) of this Example, what is the voltage across each capacitor, and how much energy is stored in each capacitor?

Capacitor arrangements typically involve both series and parallel types, as shown in the following Example.

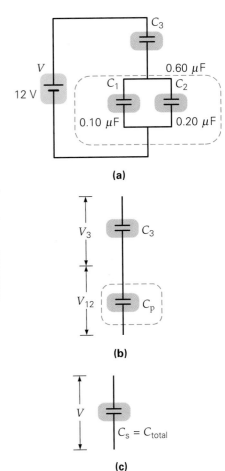

(a)

(b)

(c)

•**FIGURE 16.19 Circuit reduction** By combining capacitances, the capacitor combination is reduced to a single equivalent capacitance. See Example 16.10.

Example 16.10 ■ One Step at a Time: Capacitors in Series–Parallel Combination

Three capacitors are connected in a circuit as shown in •Fig. 16.19a. What is the voltage across each capacitor?

Thinking It Through. The voltage across each capacitor could be found from $V = Q/C$ if the charge on each capacitor were known. The total charge on the capacitors is found by reducing the series–parallel combination to a single equivalent capacitance. Two of the capacitors are in parallel. We first find their single equivalent capacitance (C_p). Then we can treat that equivalent capacitance as being in series with the last capacitor to find the total capacitance of the circuit: that of a single capacitor that would replace the three capacitors with no change in total Q or in total energy stored.

Solution.

Given: Values of capacitance and voltage from figure

Find: V_1, V_2, and V_3 (voltages across capacitors)

Starting with the parallel combination, we have

$$C_p = C_1 + C_2 = 0.10 \ \mu\text{F} + 0.20 \ \mu\text{F} = 0.30 \ \mu\text{F}$$

and the grouping is partially reduced as shown in •Fig. 16.19b. For the series combination of C_p and C_3,

$$\frac{1}{C_s} = \frac{1}{C_3} + \frac{1}{C_p} = \frac{1}{0.60\,\mu F} + \frac{1}{0.30\,\mu F} = \frac{1}{0.60\,\mu F} + \frac{2}{0.60\,\mu F} = \frac{1}{0.20\,\mu F}$$

and

$$C_s = 0.20\,\mu F = 2.0 \times 10^{-7}\,F$$

Using this value as the total equivalent capacitance (Fig. 16.19c), we can find the charge on that equivalent capacitance:

$$Q = C_s V = (2.0 \times 10^{-7}\,F)(12\,V) = 2.4 \times 10^{-6}\,C$$

This is the charge on C_3, since it is in series with the battery. The voltage across C_3 is then

$$V_3 = \frac{Q}{C_3} = \frac{2.4 \times 10^{-6}\,C}{6.0 \times 10^{-7}\,F} = 4.0\,V$$

The sum of the voltages across the capacitors equals the voltage across the battery terminals. Notice that it does not matter if you add the voltage drops by going across C_1 and then across C_3 or by going across C_2 and then across C_3, since the drops across C_1 and C_2 are the same. Thus we have $V = V_{12} + V_3 = 12\,V$ (see Fig. 16.19c):

$$V_{12} = V - V_3 = 12\,V - 4.0\,V = 8.0\,V$$

There is 8.0 V across both C_1 and C_2 since they are connected in parallel.

Follow-up Exercise. In this Example, find (a) the charge stored on each capacitor and (b) the energy stored in each.

Chapter Review

Important Terms

Important Concepts

- The electric potential difference between two points is the work per unit positive charge done by an external force in moving charge between those two points, or the change in electric potential energy per unit positive charge.

- Voltage is synonymous with electric potential *difference*.

- Equipotential surfaces (equipotentials) are surfaces on which a charge has a constant electric potential energy. They are perpendicular to the electric field at all points. It takes no work to move a charge along an equipotential.

- An electron volt (eV) is the kinetic energy gained by an electron accelerated from rest through a potential difference of 1 V.

- A capacitor stores charge, and therefore electric energy, in the form of an electric field. Capacitance is a quantitative measure of how effective a capacitor is in storing charge.

- A dielectric is any nonconducting material capable of being partially polarized when placed in an electric field. When inserted between the plates of a capacitor, a dielectric raises the capacitance by a factor K, the dielectric constant.

- The equivalent series capacitance is always less than that of the smallest capacitor of the series combination.

- The equivalent parallel capacitance is always larger than that of the largest capacitor in the parallel combination.

Important Equations

Electric Potential Difference (Voltage) (definition):

$$\Delta V = \frac{\Delta U_e}{q_o} = \frac{W}{q_o} \qquad (16.1)$$

Electric Potential Difference between Parallel Plates:

$$\Delta V = Ed \qquad (16.2)$$

Electric Potential Due to a Point Charge (V = 0 at r = ∞):

$$V = \frac{kq}{r} \qquad (16.4)$$

Electric Potential Energy of a Configuration of Point Charges:

$$U = U_{12} + U_{23} + U_{13} + \cdots \qquad (16.6)$$

Relationship between Potential and Electric Field:

$$E = -\left(\frac{\Delta V}{\Delta x}\right)_{max} \qquad (16.7)$$

Capacitance (definition):

$$Q = CV \quad \text{or} \quad C = \frac{Q}{V} \qquad (16.8)$$

Capacitance of a Parallel-Plate Capacitor (in air):

$$C = \frac{\epsilon_o A}{d} \qquad (16.11)$$

[where $\epsilon_o = 8.85 \times 10^{-12}$ C^2/(N·m^2)]

Energy in a Charged Capacitor:

$$U_C = \tfrac{1}{2}QV = \frac{Q^2}{2C} = \tfrac{1}{2}CV^2 \qquad (16.12)$$

Dielectric Effect on Capacitance:

$$C = KC_o \qquad (16.14)$$

Equivalent Capacitance for Capacitors in Series:

$$\frac{1}{C_s} = \frac{1}{C_1} + \frac{1}{C_2} + \frac{1}{C_3} + \cdots \qquad (16.16)$$

Equivalent Capacitance for Capacitors in Parallel:

$$C_p = C_1 + C_2 + C_3 + \cdots \qquad (16.17)$$

Exercises

16.1 Electric Potential Energy and Potential Difference

1. The SI unit of electric potential difference is the (a) joule, (b) newton, (c) newton-meter, (d) joule per coulomb.

2. What is the difference (a) between electrostatic potential energy and electric potential, and (b) between electric potential difference and voltage?

3. The electrostatic potential energy of two point charges (a) is inversely proportional to their separation distance, (b) is a vector quantity, (c) is always positive, (d) has units of newton per coulomb.

4. When a proton approaches another proton, the electric potential energy (a) increases, (b) decreases, (c) remains constant.

5. In terms of the way they react to differences in electrical potential, describe why positive charges speed up as they approach negative charges.

6. An electron is released in a region where there is a varying electric potential. The electron will (a) move toward the lower potential region, (b) move toward the higher potential region, (c) remain at rest.

7. In Exercise 6, will the electron gain or lose potential energy?

8. For charged parallel plates, where does a proton have the lowest electric potential energy? (a) near the positive plate, (b) near the negative plate, (c) midway between the plates.

9. For charged parallel plates, where does an electron have the highest electric potential energy? (a) near the positive plate, (b) near the negative plate, (c) midway between the plates.

10. ■ A pair of parallel plates is charged by a 24-volt battery. How much work is required to move a charge of $-4.0\,\mu$C from the positive to the negative plate?

11. ■ If it takes $+1.6 \times 10^{-5}$ J to move a positive charge between two charged parallel plates, (a) what is the magnitude of the charge if the plates are connected to a 6.0-volt battery? (b) Taking the charge to be negative, was it moved from the negative to the positive plate or from the positive to the negative plate?

12. ■ What are the magnitude and direction of the electric field between the two charged parallel plates in Exercise 11 if the plate separation is 4.0 mm?

13. ■ 🌐 A proton is accelerated by a potential difference of 10 kV. How fast is the proton moving if it started from rest?

14. ■ An electron at rest is accelerated by a uniform electric field (1000 V/m) pointing vertically upward. Using Newton's laws, find the electron's velocity after it moves 0.50 cm.

15. ■ (a) Repeat Exercise 14, but find the speed by using energy methods. Get the direction to your answer by considering electric potential. (b) Does the electron gain or lose potential energy?

16. ■ 🌐 (a) What is the difference in potential between two points, one at 20 cm and the other at 40 cm from a charge of 5.5 μC? (b) Which point is at a higher potential?

17. ■■ (a) How far from a charge of $+1.0 \mu C$ is a point with an electric potential value of 10 kV? (b) If the point were moved to three times that distance, how much of a potential change would occur? Does it increase or decrease?

18. ■■ In the Bohr model of the hydrogen atom, the electron can exist only in circular orbits of certain radii, including 0.21 nm and 0.48 nm. (a) Determine the potential difference between these two orbits. (b) Which one is at a higher potential?

19. ■■ In Exercise 18, by how much does the potential energy of the atom change if the electron goes (a) from the lower to the higher orbit, (b) from the higher orbit to the lower orbit, and (c) from the higher orbit to a very large distance.

20. ■■ How much work is required to separate two charges (each $-1.4 \mu C$) completely if they were initially 8.0 mm apart?

21. ■■ In Exercise 20, if the two charges are released at their initial separation distance, how much kinetic energy would each have when they are very distant from one another?

22. ■■ It takes $+5.5$ J of work to move two charges from a large distance to 1.0 cm from one another. If the charges have the same magnitude, (a) how large is each charge, and (b) what can you tell about their signs?

23. ■■ A charge of $+2.0 \mu C$ is initially 0.20 m from a fixed charge of $-5.0 \mu C$ and is then moved to a position 0.50 m from the fixed charge. (a) What is the work required? (b) Does the work depend on the path through which the charge is moved?

24. ■■ An electron is moved from point A to B and then to C along two legs of an equilateral triangle with sides of length 0.25 m (•Fig. 16.20). If the horizontal electric field

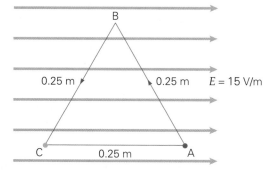

•**FIGURE 16.20 Work and energy** See Exercise 24.

is 15 V/m, (a) what is the magnitude of the work required? (b) What is the potential difference between points A and C? (c) Which point is at a higher potential?

25. ■■ Compute the energy necessary to bring together the charges in the configuration shown in •Fig. 16.21.

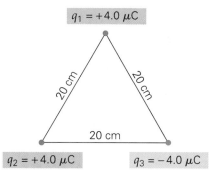

•**FIGURE 16.21 A charge triangle** See Exercises 25, 27, and 28.

26. ■■ Compute the energy necessary to bring together the charges in the configuration shown in •Fig. 16.22.

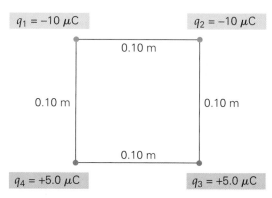

•**FIGURE 16.22 A charge rectangle** See Exercises 26, 29, 30, and 32.

27. ■■ What is the electric potential at the center of the triangle in Fig. 16.21?

28. ■■ Compute the electric potential at a point midway between q_2 and q_3 in Fig. 16.21.

29. ■■■ What is the electric potential at the center of the square in Fig. 16.22?

30. ■■■ Compute the electric potential at a point midway between q_1 and q_4 in Fig. 16.22.

31. ■■■ In a computer monitor, electrons are accelerated from rest through a potential difference of 10 kV in an electron gun (•Fig. 16.23). (a) What is the "muzzle speed" of the electrons emerging from the gun? (b) If the gun is directed at a screen 35 cm away, how long does it take the electrons to reach the screen?

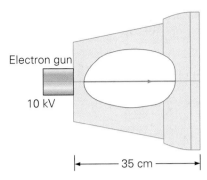

Electron gun

10 kV

|← —— 35 cm —— →|

•**FIGURE 16.23** **Electron speed** See Exercise 31.

32. ■■■ Compute the electric potential at a point 4.0 cm above q_4 along the line joining it to q_1 in Fig. 16.22.

16.2 Equipotential Surfaces and the Electric Field

33. Equipotential surfaces are those surfaces on which (a) the potential is constant, (b) the electric field is zero, (c) the potential is zero.

34. Equipotential surfaces are (a) parallel to the electric field, (b) perpendicular to the electric field, (c) at any angle with respect to the electric field.

35. Sketch the topographic map you would expect from the ocean up to a gently sloping uniform beach area. Label the gravitational equipotentials as to relative height and potential value. Show how to predict, from the map, which way a ball would accelerate if it were initially rolled up the beach away from the water.

36. Explain why two equipotential surfaces cannot intersect.

37. Suppose that you start with a charge at rest on an equipotential surface, move it off that surface, then return it to the same surface and bring it to rest. Is it still true that the net work done on that charge is zero? Explain.

38. What geometrical shape are the equipotential surfaces between two charged parallel plates?

39. Between charged parallel plates, which equipotentials have a higher potential: (a) the ones near the positive plate, (b) the ones near the negative plate, (c) the ones near the middle?

40. The equipotential surfaces outside a point charge are approximately what geometrical shape?

41. For a positive point charge, if you go from one equipotential surface to a smaller one, what happens to the potential?

42. Show that the volt per meter is the same as the newton per coulumb.

43. Explain why an electron volt is a unit of energy, not of voltage. Which is larger, a gigaelectron volt or a megaelectron volt?

44. At a given point on an equipotential surface, the electric field points directly to the (a) next highest equipotential, (b) the next lowest equipotential, (c) parallel to the equipotential surface?

45. Can electric field lines ever cross? Explain, using the result of Exercise 36.

46. Can the electric field at a point be zero yet the electric potential be nonzero at that point? Explain.

47. ■ A uniform electric field of 10 kV/m points vertically upward. How far apart are the equipotential planes that differ by 100 V?

48. ■ In Exercise 47, if the ground is called zero potential, how far above the ground is the equipotential surface corresponding to 7.0 kV?

49. ■ Determine the value of the potential 2.5 mm from the negative plate of a pair of parallel plates separated by 10 mm and connected to a 24-volt battery.

50. ■ Relative to the positive plate in Exercise 49, where is the point with a potential value of 20 V?

51. ■ For a point charge of $+3.50 \ \mu C$, what is the radius of the equipotential surface that is at a potential of 2.50 kV?

52. ■ If the radius of the equipotential surface that is at a potential of 2.20 kV is 14.3 m, what is the magnitude of the point charge creating the potential?

53. ■ Using your results from Exercises 51 and 52, calculate the amount of work (in electron volts and in joules) it would take to move an electron from the closer to the more distant equipotential.

54. ■ The potential difference in a typical lightning discharge can be up to 100 million V. What would be the gain in kinetic energy of an electron after moving through this potential difference, if it loses 80% of its energy to atomic collisions, in (a) electron volts and (b) joules?

55. ■ In a linear accelerator, protons are accelerated through a potential difference of 20 MV. What is their kinetic energy if they started from rest in (a) eV, (b) keV, (c) MeV, (d) GeV, (e) joules?

56. ■■ In Exercise 55, how do your answers change if a doubly charged $(+2e)$ alpha particle is accelerated instead? (Recall that an alpha particle consists of two neutrons and two protons.)

57. ■■ In Exercises 55 and 56, compute the speed of the proton and alpha particle after each is accelerated through the potential.

58. ■■ Calculate the voltage required to give a proton the following kinetic energies. In each case, find the proton's speed. (a) 3.5 eV, (b) 4.1 keV, and (c) 8.0×10^{-16} J

59. ■■ Repeat Exercise 58 for electrons instead of protons.

60. ■■■ Two large parallel plates are separated by 3.0 cm and connected to a 12-volt battery. Starting at the negative plate and moving 1.0 cm toward the positive plate at a 45° angle (•Fig. 16.24), what value of potential would be reached, assuming the negative plate was defined as zero potential?

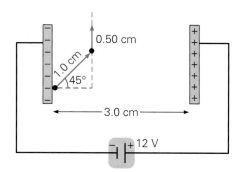

•FIGURE 16.24 **Reaching our potential** See Exercises 60 and 61.

61. ■■■ In Exercise 60, what would be the value of the potential if you then moved 0.50 cm parallel to the plates?

16.3 Capacitance

62. Capacitance has units of (a) farads, (b) joules, (c) coulombs per volt, (d) both (a) and (c).

63. To increase the capacitance and the energy-storage capability of a parallel-plate capacitor, we can (a) increase the plate separation distance, (b) increase the plate area, (c) evacuate the space between the plates, (d) none of the above.

64. If the plates of an isolated parallel-plate capacitor are moved closer to each other, does the energy storage increase, decrease, or remain the same?

65. ■ How much charge flows through a 12-volt battery when a capacitor of 2.0 μF is connected across the battery terminals?

66. ■ A parallel-plate capacitor has a plate area of 0.50 m^2 and a plate separation of 2.0 mm. What is the capacitance of the capacitor?

67. ■ What plate separation is required for a parallel-plate capacitor to have a capacitance of 5.0 × 10^{-9} F if the plate area is 0.40 m^2?

68. ■ A parallel-plate capacitor with a capacitance of 2.5 × 10^{-9} F has a plate separation of 3.0 mm. What is its plate area?

69. ■■ A 12-volt battery is connected to a parallel-plate capacitor with plate areas of 0.20 m^2 each and a plate

separation of 5.0 mm. (a) What is the resulting charge on the capacitor? (b) How much energy is stored in the capacitor?

70. ■■ If the plate separation of the capacitor in Exercise 69 changed to 10 mm after being disconnected from the battery, how do your answers change?

71. ■■■ Assume that a high-tech, fully charged capacitor of capacitance 1.0 F is able to light a 0.50-watt bulb at constant full power for 5.0 s before it quits. What was the terminal voltage of the battery that charged the capacitor?

16.4 Dielectrics

72. Putting a dielectric in an isolated, charged, parallel-plate capacitor (a) decreases the capacitance, (b) decreases the voltage, (c) increases the charge, (d) causes a discharge because the dielectric is a conductor.

73. A parallel-plate capacitor is connected to a battery. If a dielectric is inserted between the plates, (a) the capacitance decreases, (b) the voltage increases, (c) the voltage decreases, (d) the charge increases.

74. Give several reasons a conductor would not be a good choice as a capacitor dielectric.

75. Explain why a stream of water is deflected, or bent, when an electrically charged object is brought close to it. (Try this yourself with a plastic pen, hard rubber comb, or balloon.)

76. ■ A capacitor has a capacitance of 50 pF, which increases to 150 pF when a dielectric material is between its plates. What is the dielectric constant of the material?

77. ■ A capacitor of capacitance 50 pF is immersed in silicone oil ($K = 2.6$). While the capacitor is connected to a 24-volt battery, what will be the charge on the capacitor and the amount of stored energy?

78. ■ The dielectric of a parallel-plate capacitor is to be constructed from glass. The area of each plate is 0.50 m^2. (a) What thickness should the glass have if the capacitance is to be 0.10 μF? (b) What is the charge on the capacitor if it is connected to a 12-volt battery?

79. ■■ A parallel-plate capacitor has a capacitance of 1.5 μF with air between the plates. The capacitor is connected to a 12-volt battery and charged. The battery is then removed. When a dielectric is placed between the plates, a potential difference of 5.0 V is measured across the plates. What is the dielectric constant of the material?

80. ■■■ To store energy to trigger an electronic flash unit for photography, a capacitor is connected to a 400-volt voltage source. The flash requires 20.0 J. The capacitor initially is air-filled and has a capacitance of 100 μF. You then insert a dielectric material, leaving everything else the same. What must be the dielectric constant to store the 20.0 J?

16.5 Capacitors in Series and in Parallel

81. Capacitors in series have the same (a) voltage, (b) charge, (c) energy storage.

82. Capacitors in parallel have the same (a) voltage, (b) charge, (c) energy storage.

83. Under what conditions would two capacitors in series have the same voltage?

84. ■ 🖸 What is the equivalent capacitance of two capacitors, 0.40 μF and 0.60 μF, when they are connected (a) in series and (b) in parallel?

85. ■ 🖸 When a series combination of two uncharged capacitors is connected to a 12-volt battery, 173 μJ of energy is drawn from the battery. If one of the capacitors has a capacitance of 4.0 μF, what is the capacitance of the other?

86. ■■ For the arrangement of three capacitors in •Fig. 16.25, what value of C_1 will give a total equivalent capacitance of 1.7 μF?

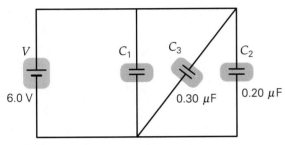

•**FIGURE 16.25   A capacitor triad**   See Exercises 86 and 90.

87. ■■ Three capacitors, 0.25 μF each, are connected in parallel to a 12-volt battery. (a) What is the charge on each capacitor? (b) How much charge is drawn from the battery?

88. ■■ If you are given three identical capacitors of capacitance 1.0 μF each, how many different capacitance values can you obtain?

89. ■■ What are the maximum and minimum equivalent capacitances that can be obtained by combinations of three capacitors of 1.5 μF, 2.0 μF, and 3.0 μF?

90. ■■■ If the capacitance $C_1 = 0.10$ μF in Fig. 16.25, what is the charge on each of the capacitors in the circuit?

91. ■■■ Four capacitors are connected in a circuit as illustrated in •Fig. 16.26. Find the charge on and the voltage difference across each of the capacitors.

Additional Exercises

92. A proton moves 10 mm straight up in the direction in which the electric potential is decreasing most rapidly. (a) If the change in potential is 200 V, estimate the direction and mag-

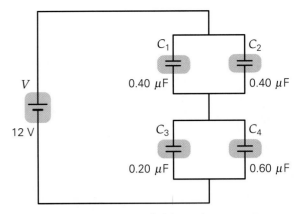

•**FIGURE 16.26   Double parallel in series**   See Exercise 91.

nitude of the electric field. (b) If the initial location has a potential of 116 kV, what is the value of the potential at the ground, 10 cm below?

93. An electric dipole consists of two point charges, $\pm q$, separated by a distance d. How much work (in terms of k, q and d) did it take to bring these charges together from a large distance?

94. Sketch the equipotential surfaces and the electric field line pattern outside a uniformly (negatively) charged long wire. Label the surfaces as to relative potential value, and indicate the electric field direction.

95. A helium atom with one electron already removed (a positive helium ion) consists of a single orbiting electron and a nucleus of two protons and two neutrons. If the electron is in its minimum orbital radius of 0.027 nm, what is the potential energy of the system, in electron volts?

96. Suppose that the three capacitors in Figure 16.19 have the following values: $C_1 = 0.15$ μF, $C_2 = 0.25$ μF, and $C_3 = 0.30$ μF. (a) What is the equivalent capacitance of this arrangement? (b) How much charge will be drawn from the battery? (c) What is the voltage across each capacitor?

97. Two horizontal, conductive parallel plates are separated by a distance of 1.5 cm. What voltage across the plates would be required to suspend (a) an electron or (b) a proton in midair between them? (c) What would be the signs of the charges on the plates in each case?

98. A 1.0-farad parallel-plate capacitor with a distance of 0.50 mm between the plates is to be built. (a) What is the required area of the plates without a dielectric between them? (b) What is the required area of the plates if a polystyrene film ($K = 2.6$) 0.50 mm thick is used between them?

99. The electric field between two parallel plates separated by 3.0 cm has a magnitude of 5.5×10^4 V/m. If the electron is released from the negative plate, how much kinetic energy has it gained when it reaches the positive plate?

Electric Current and Resistance

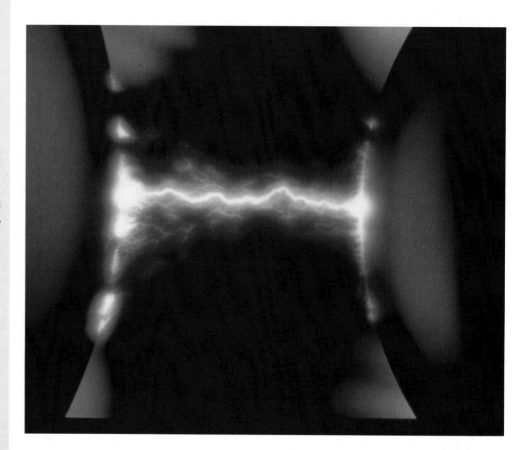

If you were asked to think of electricity and its uses, many images would probably come to mind, including such diverse applications as the lamps you study by, television remotes, and electric leaf blowers. You might also think of some negative images, such as dangerous lightning or sparks you experienced from an overloaded electric outlet.

Common to all of these images is the concept of electric energy. For a light bulb or appliance, electric energy is delivered by electric current in wires; for lightning or a spark, it is delivered through the air. The light, heat, or mechanical energy given off is simply electric energy converted to a different form. In this photograph, for example, the light given off by the spark is energy emitted by molecules in the air. The atoms of those molecules initially received this energy by colliding with the charges that carry the electric current between the two metallic surfaces shown.

But just what is electricity? What is electric current, and how does it travel? What causes it to move through an appliance when we flick a switch? Why does the electric current cause the filament of a light bulb to heat and glow brightly, but not the connecting wires as well? Such questions are basic to our understanding of not only everyday appliances, but also electricity in general. In this chapter you'll learn the basic principles that explain the operation of many common electric appliances and also are responsible for untamed electric currents, such as sparks and those currents we see in Nature's spectacular own fireworks—lightning.

17.1 Batteries and Direct Current

Objectives: To (a) summarize the basic features of a battery and (b) explain how a battery produces a direct current in a circuit.

After studying electric force and energy in Chapters 15 and 16, you can probably guess what is required to produce an *electric current*, or a flow of charge. A flow of mass may occur when there is a gravitational potential energy *difference*. For example, the water in a stream flows downhill, from areas of higher to lower gravitational potential. So, too, spontaneous heat flow occurs only when there is a temperature *difference*, as when a cold object comes in contact with a warm one. Similarly, a flow of electric charge is dependent on a *difference* in electric potential—what we call "voltage."

In solid conductors, particularly metals, the outer electrons of atoms are relatively free to move, while the positively charged nuclei and most of the tightly bound (inner) electrons are fixed in the lattice structure of the material. (In liquid conductors and charged gases called plasmas, both positive and negative ions as well as electrons can move.) Energy is required to move electric charge. This electric energy is produced, or generated, through the conversion of other forms of energy, giving rise to a potential difference, or voltage.

A **battery** is a device that converts chemical energy into electrical energy. The Italian scientist Allesandro Volta is credited with constructing one of the first practical batteries. A simple battery consists of two *electrodes* (either of two inserted metals) in an *electrolyte*, a solution that conducts electricity. With the appropriate electrodes and electrolyte (Volta used zinc and copper electrodes in dilute sulfuric acid), a potential difference develops across the electrodes as a result of chemical action. As you may recall from chemistry, different metals dissolve at different rates in a given acid. When the two metal electrodes dissolve, their atoms move into solution as positively charged ions, leaving behind electrons. The result is a solution of acid and positive ions of both metal types. Both electrodes will accumulate an excess of electrons, but because of differences in the chemical properties of the metals, one will have a larger excess (•Fig. 17.1).

A *circuit* is any complete loop consisting of wires and electrical devices (such as batteries and light bulbs). When a complete circuit is formed—for example, by connecting a light bulb and wires as shown in Fig. 17.1—electrons from the more negative electrode (B) move through the wire and bulb to the less negative electrode (A). The result is a flow of electrons, or an electric current. As it moves through the filament of the bulb, the current can heat the filament to a sufficient temperature to give off visible light (glow).

At this point, you might think the flow would stop, but that is not the case. As electrode A receives more electrons than normal, it attracts the A ions from solution. (A chemical membrane can be placed in the solution to prevent the A and B ions from mixing.) The A ions reattach to electrons and "plate out," or deposit, onto the A electrode as neutral atoms once again. Meanwhile, the electrons are leaving the B electrode through the wire. To generate more electrons at the B electrode, more B ions are released into solution. Thus, negative electrons flow from B to A through the wire and bulb while positive ions flow from B to A in the solution. Eventually this chemical action will cease when the B electrode is entirely dissolved into solution. This process results in an overall loss of chemical potential energy from the battery; that energy is converted into light (energy) and heat (energy) at the bulb.

Note: Recall that the term *voltage* in this text is used to mean "difference in electric potential."

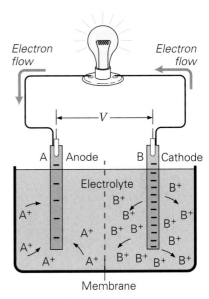

•**FIGURE 17.1   A simple chemical battery or cell**   Chemical processes involving the electrolyte and two unlike metal electrodes cause ions of both metals to dissolve into the solution. Since they dissolve at different rates and thus leave different quantities of excess electrons behind, one electrode (the cathode) becomes more negatively charged than the other (the anode). Thus the anode is at a higher potential than the cathode. The electric potential difference, or voltage, across the electrodes can cause a current, or a flow of charge (electrons), in the wire. There is also a current in the solution as positive ions migrate, completing the circuit. (A membrane is necessary to prevent mixing of ion types.) By convention, the anode is designated the positive terminal and the cathode the negative terminal.

Note: A battery's emf is sometimes called its *open circuit terminal voltage*—when it is connected to nothing.

The electrode that has the larger number of excess electrons is named the **cathode** and designated the negative (−) terminal of the battery. The other electrode, although still having an excess negative charge, has a smaller excess than the cathode and thus is designated the positive (+) terminal. It is called the **anode**. A color code that assigns red to the anode and black to the cathode is sometimes used—for example, on battery jumper cables for automobiles.

Because of the difference in charge on the battery terminals, a potential difference exists between them. The anode, having less negative charge, is at a higher potential than the cathode. The potential difference across the terminals of a battery *when it is not connected* to a circuit is called the battery's **electromotive force (emf)**. This name is somewhat misleading because the electromotive force is not a force but a voltage; as such, it is measured in volts. The emf ($\mathcal{E}$) is the maximum possible potential difference between the battery terminals (•Fig. 17.2a). When a battery is connected to a circuit and charge flows, the voltage across the terminals is slightly less than the emf because the battery has internal resistance, or opposition to charge flow. (*Resistance, R*, as we shall see in Section 17.3, is a quantitative measure of how easily charge flows through a material. A battery's internal resistance depends on such factors as the type of electrolyte and electrode material and their age.) This operating voltage (V) of a battery is called its **terminal voltage** (Fig. 17.2b). We will be chiefly concerned with terminal voltage.

The internal resistance, r, of the battery in Fig. 17.2b is represented explicitly in the *circuit diagram*. Internal resistances are typically small, and the terminal voltage of a battery is usually only slightly less than its emf. However, when a battery supplies a large current to a circuit, the terminal voltage V may drop appreciably below the maximum emf value. The difference between the two voltages $\mathcal{E}$ and V is given by $\mathcal{E} - V = Ir$, where I is the electric current in the battery (covered in Section 17.2). Thus the emf is not a true indication of the state of the battery. Even if the battery's emf, $\mathcal{E}$, is the expected terminal voltage, when the battery delivers current (and thus energy) to a light bulb, for example, the terminal voltage can be considerably less than $\mathcal{E}$, especially if the internal resistance is high. This is a common fault of used or old batteries.

As long as the internal chemical action maintains a potential difference across its terminals, current is supplied to a circuit as in Fig. 17.2b. The battery is said

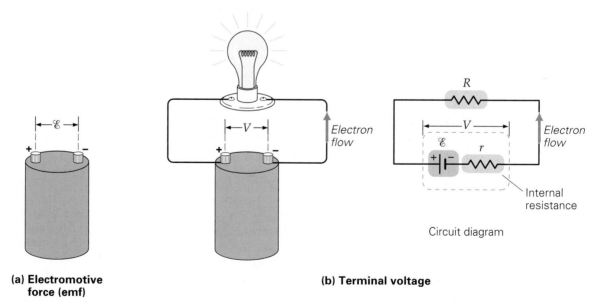

(a) Electromotive force (emf)

(b) Terminal voltage

•**FIGURE 17.2   Electromotive force (emf) and terminal voltage   (a)** The emf ($\mathcal{E}$) of a battery is the maximum potential difference across its terminals. This maximum occurs when the battery is not connected to an external circuit. **(b)** Because of internal resistance (r), the terminal voltage V when the battery is in operation is less than the emf, $\mathcal{E}$. Here R is the resistance of the light bulb.

to *deliver* current to a circuit. Alternatively, we say that the circuit (or its components) *draws* current from the battery. Because the battery's terminal voltage pushes the electrons in one direction only, they can flow only in one direction in such a circuit, from the negative (−) terminal (cathode) to the positive (+) terminal (anode). This type of flow is called **direct current (dc)**. Notice the circuit diagram in the figure. The symbol for a battery is two parallel lines, the longer of which represents the positive (+) terminal and the shorter the negative (−) terminal. Any circuit element, such as the lamp, that opposes charge flow is generally represented by the symbol (−$\wedge\!\wedge\!\wedge$−), which stands for any resistance (R). (Electrical resistance—resistance to charge flow—will be considered in detail in later sections of this chapter. Here, we merely introduce the circuit symbol.)

Note: $\dashv\!\vdash$ = battery
−$\wedge\!\wedge\!\wedge$− = resistance

There are a wide variety of batteries now in use. One of the most common is the 12-volt automobile battery. It consists of six 2-volt cells connected in series. That is, the positive terminal of each cell is connected to the negative terminal of the next cell (shown for three cells in •Fig. 17.3a). When batteries or cells are connected in this fashion, their voltages add.*

If cells are connected in parallel, all their positive terminals have a common connection, as do their negative terminals (Fig. 17.3b). When identical batteries are connected in this way, the potential difference is the same across all of them, and each one supplies a fraction of the current to the circuit (for three batteries with equal voltages, each one supplies one-third of the current). Parallel connections of batteries are familiar to people who have had their car jump started. In that situation, the weak (high r) battery is connected in parallel to a healthy (low r) battery, which delivers most of the current to get the car started. If sources of different voltages are connected in parallel, the situation is more complex.

Other sources of voltage, such as generators and photocells, will be considered in later chapters. One of the more common sources is the "power supply" that plugs into a wall socket and converts the alternating voltage delivered by the power company into direct (and usually adjustable) voltage. You will likely use one in your laboratory. It can be thought of as a battery capable of delivering an adjustable direct current.

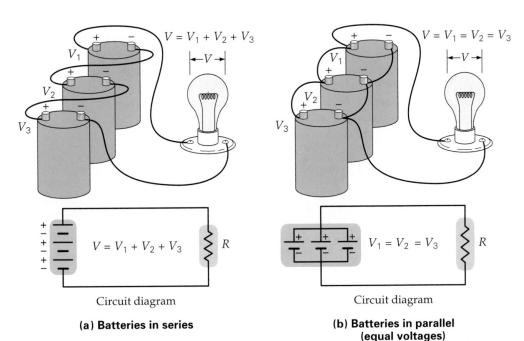

(a) Batteries in series

(b) Batteries in parallel (equal voltages)

•**FIGURE 17.3 Batteries in series and in parallel** **(a)** When batteries are connected in series, their voltages add and the voltage across the resistance R is the sum of the voltages. **(b)** When batteries of the same voltage are connected in parallel, the voltage across the resistance is the same, as if only a single battery were present. In this case, each battery supplies a fraction of the total current.

*Chemical energy is converted to electrical energy in a chemical *cell*. The term *battery* generally refers to a collection, or "battery," of cells.

17.2 Current and Drift Velocity

Objectives: **To (a)** define electric current, **(b)** distinguish between electron flow and conventional current, and **(c)** explain the concept of drift velocity and electric energy transmission.

A battery or some other voltage source connected to a continuous conducting path forms a **complete circuit**. In ordinary applications, *a sustained electric current requires a voltage source and a complete circuit.* A circuit may have a switch (symbolized by S), which is used to open or close the circuit. When the switch is open, the current or circuit component is turned off; when the switch is closed, there is current in the circuit.

Electric Current

Conventional current: in direction opposite that of electron flow

Since it is electrons that move in the wires of a circuit, the net flow of charge is away from the negative terminal of the battery and toward the positive terminal. (Recall from Chapter 15 the charge–force law or from Chapter 16 the fact that electrons flow from areas of low to high potential.) Historically, however, circuit analysis has generally been done in terms of **conventional current**, which is in the direction in which positive charges would flow, or the direction opposite to the electron flow (•Fig. 17.4). (Some situations exist in which positive charge flow *is* responsible for the current—for example, in semiconductors.)

In general we say electric current is characterized by a flow of charge. Quantitatively, the **electric current** (*I*) in a wire is defined as the time rate of flow of net charge. If a net charge *q* passes through the wire's cross-sectional area at a given point in time *t* at a constant rate (•Fig. 17.5), the electric current *I* is defined as

$$I = \frac{q}{t} \qquad \textit{electric current} \qquad (17.1)$$

SI unit of current: coulomb per second (C/s) or ampere (A)

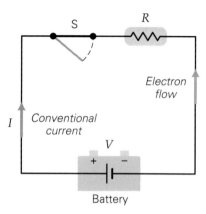

•**FIGURE 17.4 Conventional current** For historical reasons, circuit analysis is usually done with conventional current. Conventional current is in the direction in which positive charges would flow, or opposite to the electron flow in metals and many other common materials.

As this definition indicates, the unit of current is coulombs per second (C/s). This combination of units is called an **ampere (A)** in honor of the French physicist Andre Ampère (1775–1836), an early investigator of electrical and magnetic phenomena. The ampere is commonly shortened to "amp"; for example, a current of 10 A is read as "ten amps." Small currents are expressed in *milliamperes* (mA, or 10^{-3} A) or *microamperes* (μA, or 10^{-6} A), which are also shortened to milliamps and microamps. In a typical household circuit, it is not unusual for the wires to carry several amps of current.

Example 17.1 ■ Counting Electrons: Current and Charge

There is a steady current of 0.50 A in a flashlight bulb for 2.0 min. How much charge passes through the bulb during this time? How many electrons does this represent?

Thinking It Through. We are told the current (rate) and time elapsed. We will use the definition of current (Eq. 17.1) to find *q*. Since each electron has a charge magnitude of 1.6×10^{-19} C, we can convert to get the number of electrons.

Solution. We list the data given, converting time into seconds:

Given: $I = 0.50$ A   *Find:*   *q* (amount of charge)
       $t = 2.0$ min $= 1.2 \times 10^2$ s   *n* (number of electrons)

By Eq. 17.1, $I = q/t$, so the magnitude of the charge is given by

$$q = It = (0.50 \text{ A})(1.2 \times 10^2 \text{ s}) = (0.50 \text{ C/s})(1.2 \times 10^2 \text{ s}) = 60 \text{ C}$$

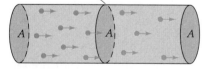

Cross-sectional area *A*

•**FIGURE 17.5 Electric current** Electric current (*I*) in a wire is defined as the net charge (*q*) that passes through a cross-sectional area of the wire at a given location per time: $I = q/t$. *I* has the units of amperes (A), or "amps" for short.

The charge q is made up of n electron charges; that is, $q = ne$ (Eq. 15.1). Solving for n and using the magnitude of the charge on the electron, we have

$$n = \frac{q}{e} = \frac{60\ C}{1.6 \times 10^{-19}\ C/electron} = 3.8 \times 10^{20}\ electrons$$

This may seem like a large number of electrons, but keep in mind that in an aluminum wire, for example, each atom has 13 electrons. From chemical principles, we know that there are about 6×10^{23} aluminum atoms in every 27 g (about an ounce in weight) of aluminum. Thus the percentage of electrons involved in the current is only a small fraction of the total number of electrons in the wire.

Follow-up Exercise.   Many laboratory instruments can measure currents in the nanoampere range. How long would it take for 1 C of charge to flow past a given point in a wire that carries a current of 1.0 nA?

Drift Velocity, Electron Flow, and Electric Energy Transmission

Although we frequently mention charge flow, electric charge does not flow in a conductor in the same way that water flows through a pipe. In the absence of a potential difference, the free electrons in a metal wire move randomly at high speeds, colliding many times per second with the atoms of the lattice. As a result, there is no net flow of charge since, on average, the same amounts of charge pass through a cross-sectional area in opposite directions in a given time.

When a potential difference is applied across the ends of the wire (for example, by a battery), an electric field in one direction, and thus a *net flow* of electrons opposite that direction, is generated. This does *not* mean electrons are moving directly from one end of the wire to the other. They still move about in all directions as they collide with the atoms of the conductor (•Fig. 17.6), but there is an added component (in one direction) to their velocities. The overall result is that their motions are less random, and more of them move toward the positive terminal of the battery than away from it.

This net electron flow is characterized by an average velocity called the **drift velocity**, which is much smaller than the random velocities of the electrons themselves. The magnitude of the drift velocity is on the order of 1.0 mm/s. The net movement of electrons at this speed is relatively slow. A quick calculation using this drift velocity would show that it would take an electron about 17 min to travel 1 m along a wire. Yet a lamp comes on almost instantaneously when you flip the switch (close the circuit), and the electronic signals carrying telephone conversations travel almost instantaneously over miles of wire.

Evidently, *something* must be moving faster than the "drifting" electrons. This is the electric field. When a potential difference is applied, the associated electric field in the conductor travels at a speed close to that of light (roughly 10^8 m/s). The electric field therefore influences the motion of electrons *throughout the conductor* almost instantaneously. Thus there is current everywhere in the circuit almost simultaneously. You don't have to wait for electrons to "get there" from elsewhere. The electrons already in the bulb filament begin to move almost immediately as they feel the electric force, thus delivering energy and creating light essentially with no delay.

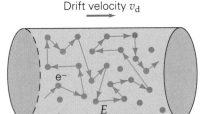

Drift velocity v_d

•**FIGURE 17.6   Drift velocity** Because of collisions with the atoms of the conductor, electron motion is random. However, when the conductor is connected—say, to a battery to form a complete circuit—there is a relatively small net motion in the direction opposite the electric field (or toward the positive high potential terminal of the battery). This net motion is characterized by a drift velocity.

17.3   Resistance and Ohm's Law

Objectives: To (a) define electrical resistance and explain what is meant by an ohmic resistor, (b) summarize the factors that determine resistance, and (c) calculate the effect of these factors in simple situations.

Given that an applied voltage (potential difference) across the two ends of any conducting material causes an electric current, how much charge flows for a given voltage? As you might expect, it is usually true that the greater the voltage, the greater the current. This situation is analogous to more water flowing out of the spigot in a tank per second when the water level in the tank is raised.

However, another factor besides voltage influences the current. Just as internal friction (viscosity) affects fluid flow, the internal resistance of materials affects the flow of electric charge. Any object that offers significant resistance to the flow of charge is called a *resistor* and is represented by the zig-zagged resistor symbol in circuit diagrams. In this book, that symbol is used for a wide range of "resistors," from the cylindrical color-coded ones on printed circuit boards (•Fig. 17.7) to electrical devices such as hairdryers and light bulbs.

Resistance (R) is defined as the ratio of the voltage to the resulting current. This makes physical sense, because if we need to apply a huge voltage to a resistor and still get only a small current, we would say the resistor has a high resistance. Conversely, if a small voltage resulted in a huge current, we would call this a low resistance. Resistance is logically defined as

Definition of:
Electrical resistance

$$R = \frac{V}{I} \quad \text{or} \quad V = IR \qquad (17.2)$$

SI unit of resistance: volt per ampere (V/A), or ohm (Ω)

For some materials, the resistance is constant, or at least approximately so over a range of voltages. A resistor that has constant resistance is said to obey **Ohm's law,** or to be *ohmic*. This law was named after Georg Ohm (1789–1851), a German physicist who investigated the relationship between current and voltage and found cases in which the resistance did not change with voltage. The units of resistance are volts per ampere (V/A), from $R = V/I$ (•Fig. 17.8a). This combined unit is called an **ohm** (Ω). A plot of voltage versus current for an ohmic resistance gives a straight line with a slope equal to its resistance, R (Fig. 17.8b). A common and practical form of Ohm's law is $V = IR$ where R is constant.

Note: "Ohmic" means having constant resistance.

Note: Remember that V stands for ΔV.

Ohm's law is not a fundamental law in the sense that Newton's laws of motion and the first law of thermodynamics are. There is no law that states that materials *must* have constant resistance. Indeed, many of our technological advances in computers and communication are based on semiconductors, which can have highly *nonlinear* voltage–current behaviors. Rather, Ohm's law is an extremely useful generalization that applies to a wide range of materials, particularly metals.

Unless we specify otherwise, we will assume resistors are ohmic. But keep in mind that many everyday appliances, such as light bulbs, have variable resistances. For instance, the tungsten filaments of these bulbs have more resistance at their high operating temperatures than at room temperature. The following Example shows how resistance can be calculated.

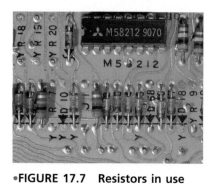

•FIGURE 17.7   Resistors in use A printed circuit board, typically used in computers, includes resistors of different values. The large, striped cylinders are resistors; their four-band color code indicates their resistance in ohms. The smaller cylinders are one-way current "valves" called diodes; the dark rectangle at the top is an integrated circuit chip. The silver paths are printed circuits, which take the place of wires in connecting the components.

Example 17.2   ■   A Car Battery: Electrical Resistance

When a 12-volt automobile battery is connected across a resistor of unknown resistance, a current of 2.5 mA results. What is the value of the resistor?

Thinking It Through.   Resistors are used in circuits to limit current and are manufactured in a wide variety of resistance values. Here we can use the definition of resistance (Eq. 17.2).

Solution.   Listing the data and converting the current to amps, we have

Given: $V = 12$ V   *Find:* R (resistance)
        $I = 2.5$ mA $= 2.5 \times 10^{-3}$ A

Using Eq. 17.2, we get

$$R = \frac{V}{I} = \frac{12 \text{ V}}{2.5 \times 10^{-3} \text{ A}} = 4.8 \times 10^3 \ \Omega$$

That is, the resistor is a 4.8-kilohm (kΩ) resistor. Resistors in the kilohm and megohm (MΩ) ranges are quite common.

Follow-up Exercise. Estimate the resistance of the following common household appliances, assuming a voltage of 120 V is applied across them: (a) a hairblower that has a current of 10 A, (b) a 30-watt VCR that carries a current of 0.25 A, and (c) a corded electric razor that carries a current of 0.10 A.

Factors That Influence Resistance

On the atomic level, resistance arises when electrons collide with lattice atoms or ions making up a material. Thus, resistance is a material property; that is, it depends on the type of material. Geometrical factors also influence an object's resistance. The resistance of a particular conductor of uniform cross section, such as a length of wire, depends on the following factors (•Fig. 17.9): (1) type of material (including any impurities); (2) length; (3) cross-sectional area; and (4) temperature.

As you might expect, the resistance of an object (such as an everyday strand of wire) is *directly* proportional to its length (*L*) and *inversely* proportional to its cross-sectional area (*A*):

$$R \propto \frac{L}{A}$$

For example, a uniform metal wire 4 m long offers twice as much resistance as a similar wire 2 m long. But a wire with a cross-sectional area of 0.50 mm^2 has only half the resistance of one with an area of 0.25 mm^2. These geometrical conditions are analogous to those for liquid flow in a pipe. The longer the pipe, the more resistance (drag); but the larger the diameter (or cross-sectional area) of the pipe, the more liquid it can carry.

Resistivity and Conductivity

As we have noted, the resistance of a material is partly determined by intrinsic atomic properties, which are characterized by the **resistivity** (ρ). This quantity acts as a constant of proportionality for the relationship of resistance to length and area:

$$R = \rho \frac{L}{A} \quad \text{or} \quad \rho = \frac{RA}{L} \tag{17.3}$$

SI unit of resistivity: ohm-meter ($\Omega \cdot$m)

From $\rho = RA/L$, we can see that the units of resistivity are ohm-square meters per meter, or ohm-meters ($\Omega \cdot$m). Resistivity is an atomic property of the material, not the geometry of the actual resistor. Thus it is of practical importance, since it allows us to calculate the resistance of any cylindrical resistor or wire from Eq. 17.3.

The resistivities of some common conductors (and semiconductors and insulators) are given in Table 17.1. The values apply for a particular temperature (20°C, or room temperature), because resistivity can be temperature-dependent. For most metallic conductors, resistivity increases as temperature rises.

A related quantity is the electrical conductivity of a material. As you might guess, the **conductivity** (σ) and the resistivity (ρ) are inverses of one another:

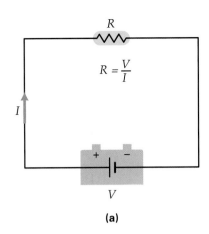

$$R = \frac{V}{I}$$

(a)

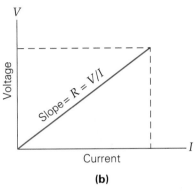

(b)

•**FIGURE 17.8 Resistance and Ohm's law** (a) In principle, any circuit element's resistance is determined by dividing the voltage across it by the resulting current through it. (b) If the element obeys Ohm's law (or has constant resistance and is "ohmic"), then a plot of voltage versus current gives a straight line with a slope equal to R, its resistance.

Resistivity—a material property

Note: Do not confuse resistivity with mass density, which has the same symbol (ρ).

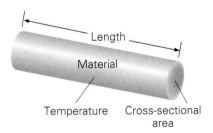

•**FIGURE 17.9 Resistance factors** Factors directly affecting the electrical resistance of a cylindrical conductor are the type of material, the length (L), the cross-sectional area (A), and the temperature.

TABLE 17.1 Resistivities (at 20°C) and Temperature Coefficients of Resistivity for Various Materials*

	ρ $(\Omega \cdot m)$	α $(C^{\circ -1})$		ρ $(\Omega \cdot m)$	α $(C^{\circ -1})$
Conductors			*Semiconductors*		
Aluminum	2.82×10^{-8}	4.29×10^{-3}	Carbon	3.6×10^{-5}	-5.0×10^{-4}
Copper	1.70×10^{-8}	6.80×10^{-3}	Germanium	4.6×10^{-1}	-5.0×10^{-2}
Iron	10×10^{-8}	6.51×10^{-3}	Silicon	2.5×10^{2}	-7.0×10^{-2}
Mercury	98.4×10^{-8}	0.89×10^{-3}			
Nichrome alloy of nickel	100×10^{-8}	0.40×10^{-3}	*Insulators*		
and chromium			Glass	10^{12}	
Nickel	7.8×10^{-8}	6.0×10^{-3}	Rubber	10^{15}	
Platinum	10×10^{-8}	3.93×10^{-3}	Wood	10^{10}	
Silver	1.59×10^{-8}	6.1×10^{-3}			
Tungsten	5.6×10^{-8}	4.5×10^{-3}			

*Values for semiconductors are general values, and resistivities for insulators are typical orders of magnitude.

Conductivity—the reciprocal of resistivity

$$\sigma = \frac{1}{\rho} \qquad conductivity \qquad (17.4)$$

SI unit of conductivity: inverse ohm-meter $[(\Omega \cdot m)^{-1}]$

The unit of conductivity is the inverse of the unit of resistivity: the inverse ohm-meter, or $(\Omega \cdot m)^{-1}$. You can describe resistance by using either resistivity or conductivity, since they are related by Eq. 17.4.

Example 17.3 ■ A Longer Cord: Resistance of a Wire

A 1.5-meter insulated extension cord, which consists of two wires to make a complete circuit, is used to connect a lamp to an outlet. If the copper wire of the cord has a diameter of 2.3 mm (equivalent to American Wire Gauge, or AWG, No. 12), what is the resistance of *one* of the wires at room temperature?

Thinking It Through. A wire is just a long cylinder, and we know the necessary dimensions and material type. Thus we can apply Eq. 17.3 after we look up the resistivity of copper at room temperature in Table 17.1.

Solution. We list the geometrical data and the resistivity we found:

Given: $L = 1.5$ m
$\qquad d = 2.3$ mm $= 2.3 \times 10^{-3}$ m
$\qquad \rho_{Cu} = 1.70 \times 10^{-8}$ $\Omega \cdot$m (Table 17.1)

Find: R (resistance)

The cross-sectional area of the wire is

$$A = \pi r^2 = \pi \left(\frac{d}{2}\right)^2 = \frac{\pi (2.3 \times 10^{-3}\,m)^2}{4} = 4.2 \times 10^{-6}\,m^2$$

Then, using Eq. 17.3, we get

$$R = \frac{\rho L}{A} = \frac{(1.70 \times 10^{-8}\,\Omega \cdot m)(1.5\,m)}{4.2 \times 10^{-6}\,m^2} = 6.1 \times 10^{-3}\,\Omega$$

This result demonstrates why the resistances of the connecting wires in an electrical circuit can usually be neglected: The resistances of the circuit *components* are generally much larger than this. We can thus ignore the resistances of the connecting wires unless the circuit components have comparable resistances.

Follow-up Exercise. Old-fashioned light bulbs used to be constructed from carbon filaments. Estimate the length of the cylindrical carbon filament with a diameter of 0.50 mm that would have a total resistance of 150 Ω at room temperature.

The temperature dependence of resistivity is nearly linear if the temperature change is not too great. An expression similar to the one for thermal linear expansion (Chapter 10) can be written for this relationship. That is, the resistivity (ρ) at a temperature T after a temperature change $\Delta T = T - T_o$ is given by

Note: Compare Eq. 10.10, Secton 10.4, for the linear expansion of a solid.

$$\rho = \rho_o(1 + \alpha \Delta T) \qquad \begin{array}{l}\text{\textit{temperature variation}}\\ \text{\textit{of resistivity}}\end{array} \qquad (17.5)$$

Here α is a constant (over a small temperature range) called the **temperature coefficient of resistivity**, and ρ_o is a reference resistivity at T_o (usually 20°C or 0°C). We can rewrite Eq. 17.5 as

$$\Delta \rho = \rho_o \alpha \Delta T \qquad (17.6)$$

where $\Delta \rho = \rho - \rho_o$ is the change in resistivity that occurs over a given change in temperature (ΔT). The ratio $\Delta \rho / \rho_o$ is dimensionless, so α must have units of inverse Celsius degrees ($C^{\circ -1}$, or $1/C^\circ$). Physically, α is the fractional change in resistivity ($\Delta \rho / \rho_o$) per Celsius degree. The temperature coefficients of resistivity for some materials are listed in Table 17.1. These coefficients are assumed to be constant over the temperature ranges used in the Examples and Exercises.

Temperature dependence of resistivity and resistance

Since resistance is directly proportional to the resistivity (Eq. 17.3), we can use this fact together with Eqs. 17.5 and 17.6 to derive an expression for the resistance of a conductor of uniform cross section as a function of temperature:

$$R = R_o(1 + \alpha \Delta T) \quad \text{or} \quad \Delta R = R_o \alpha \Delta T \qquad \begin{array}{l}\text{\textit{temperature variation}}\\ \text{\textit{of resistance}}\end{array} \qquad (17.7)$$

Here $\Delta R = R - R_o$, the change in resistance relative to its value at 20°C or 0°C. The variation of resistance with temperature provides a means of measuring temperature in the form of an electrical resistance thermometer, as shown in the following Example.

Example 17.4 ■ An Electrical Thermometer: Variation of Resistance with Temperature

A platinum wire has a resistance of 0.50 Ω at 0°C. It is placed in a water bath, where its resistance rises to a final value of 0.60 Ω. What is the temperature of the bath? (Assume α is constant over the temperature range.)

Thinking It Through. Once we get the temperature coefficient of resistivity for platinum from Table 17.1, we can solve for ΔT from Eq. 17.7 and add it to 0°C, the initial temperature to find the bath temperature.

Solution.

Given: $T_o = 0°C$ *Find:* T (temperature of the bath)
         $R_o = 0.50 \Omega$
         $R = 0.60 \Omega$
         $\alpha = 3.93 \times 10^{-3} \, C^{\circ -1}$ (Table 17.1)

The ratio $\Delta R / R_o$ is the fractional change in the initial resistance R_o (at 0°C). We solve Eq. 17.7 for ΔT, using the given values:

$$\frac{\Delta R}{R_o} = \alpha \Delta T$$

or

$$\Delta T = \frac{\Delta R}{\alpha R_o} = \frac{R - R_o}{\alpha R_o} = \frac{0.60 \, \Omega - 0.50 \, \Omega}{(3.93 \times 10^{-3} \, C^{\circ -1})(0.50 \, \Omega)} = 51 \, C^\circ$$

Thus the bath is at $T = T_o + \Delta T = 0°C + 51 \, C^\circ = 51°C$.

Follow-up Exercise. In this Example, if the material had been copper, for which $R_0 = 0.50$ Ω, rather than platinum, what would its resistance be at the same bath temperature? What conclusions can you draw about whether you would choose a material with a high or low temperature coefficient of resistivity to make a sensitive "resistance" thermometer?

Carbon and other semiconductors have negative temperature coefficients of resistivity (see Table 17.1). This means that the resistance of a semiconductor decreases with increasing temperature or increases with decreasing temperature. However, many materials have positive temperature coefficients of resistivity, and their resistances decrease with decreasing temperature. How far can electrical resistance be reduced by lowering the temperature? In certain cases, the resistance can be reduced to zero—not just close to zero, but, as accurately as we can measure, exactly to zero! This condition of **superconductivity**, which has led to technological applications of great importance, is discussed in the Insight on p. 553.

17.4 Electric Power

Objectives: To (a) define electric power, (b) calculate the power delivery of simple electric circuits, and (c) explain joule heating and its significance.

When a sustained current exists in a circuit, the electrons are given energy by the voltage source, such as a battery. As these charge carriers pass through circuit components, they collide with the atoms of the material (experience resistance) and lose energy because of these collisions (Chapter 6). The energy transferred in the collisions can result in a temperature increase of the component. In this way electrical energy can be transformed, at least partially, into thermal energy.

Note: Recall from Chapter 10 that temperature is a measure of the average translational kinetic energy per molecule of a sample.

As you know, electrical energy is commonly converted into various forms for everyday use: not only heat (as in electric stoves), but also light (as in light bulbs) and mechanical motion or work (as in electric drills). According to the law of conservation of energy, however, all of the energy delivered to the charge carriers by the battery must be lost in the circuit. That is, a charge carrier traversing the circuit must lose all the electric potential energy it gained from the battery when that carrier returns to the negative terminal of the battery.

The energy gained by a charge q from a voltage source having a terminal voltage V is qV. This is the work done by the source on the charge, $W = qV$. Over time, the *rate* at which energy is delivered to the external circuit by the battery is called the (average) **electric power**, $\overline{P}$, and is given by

$$\overline{P} = \frac{W}{t} = \frac{qV}{t}$$

If the current and voltage are steady with time (direct current, or dc, conditions, as with a battery), then the average power is the same as the power at any instant. However, if the voltage varies with time (alternating current, or ac, conditions, as we shall see in Chapter 21), so will the current, and we can compute only the average power. We assume steady (dc) currents for the rest of this chapter, so $I = q/t$ (Eq. 17.1). We can rewrite the power equation in terms of the two commonly measured quantities, voltage and current:

$$P = IV$$

As you learned in Chapter 5, the SI unit of power is the watt (W): The ampere (the unit of current I) times the volt (the unit of voltage V) gives the joule per second (J/s), or watt.

Superconductivity

The electrical resistance of metals and alloys generally decreases with decreasing temperature. At relatively low temperatures, some materials exhibit what is called *superconductivity*; that is, the electrical resistance drops to zero.

Superconductivity was discovered in 1911 by Heike Kamerlingh Onnes, a Dutch physicist. The phenomenon was first observed in solid mercury at a temperature of about 4 K (−269°C). The mercury was cooled to this temperature by using liquid helium. The boiling point of helium (the temperature at which it condenses to a liquid) is about −267°C at 1 atm. Lead also exhibits superconductivity when cooled to such a temperature.

An electric current established in a superconducting loop should theoretically persist *indefinitely* with no resistive losses. Currents introduced in superconducting loops have been observed to remain constant for several years. In 1957, a theory was presented by the American physicists John Bardeen (a co-inventor of the transistor), Leon Cooper, and Robert Schrieffer in an attempt to explain various aspects of metallic superconductors. Known as the BCS theory, it presents a quantum-mechanical model in which electrons are viewed as waves traveling through a material. According to this theory, under normal conditions lattice vibrations and imperfections, scatter electrons and give rise to resistance and energy loss. The theory predicts that these mechanisms have no effect on the electrons in superconductors. In the absence of this scattering, the resistance is zero and a current can persist as long as the metal is in a superconducting state.

Keeping a superconductor at the low temperatures of liquid helium is somewhat difficult and costly. Such low temperatures have continued to preclude the use of superconductors in everyday life. However, an ongoing search has revealed materials that become superconducting at higher temperatures. Other superconducting metals and alloys were found for which the critical temperature (the temperature at which the material becomes superconductive) was about 18 K (−255°C). In 1973, a material with a critical temperature of 23 K (−250°C) was discovered.

In 1986, there was a major breakthrough—a new class of superconductors was discovered. These were the so-called ceramic alloys of rare earth elements, such as lanthanum and yttrium. These superconductors were prepared by grinding the mixture of metallic elements together and heating it to a high temperature to produce a ceramic material. One such mixture consisted of barium, yttrium, and copper oxide. The critical temperature for these ceramic mixtures was about 57 K (−216°C).

In 1987, a ceramic material with a critical temperature of 98 K (−175°C) was discovered. This was a major advance because it meant that superconductivity could be obtained by using liquid nitrogen, which has a boiling point of 77 K (−196°C). Liquid nitrogen is relatively plentiful (nitrogen is the chief constituent of air) and inexpensive—it costs cents per liter compared to dollars per liter for liquid helium.

There have been more recent reports of higher critical temperatures and even suggestions that certain copper oxide compounds may lose their resistance to electrical current at room temperature and above. As scientists study the new superconductors, better understanding will be gained, and no doubt the critical temperature will rise. One problem is the difficulty of confirming a superconducting state in small samples or regions of samples. Another is related to the BCS theory. Although many fundamental elements of this theory may help explain high-temperature superconductivity, some reinterpretation and modification will likely be necessary.

There are many possible applications for superconductors. One is superconducting magnets. The strength of an electromagnet depends on the magnitude of the current in the wire coil (Chapter 19). If there were no resistance, there would be greater current and no losses. Used in motors or engines, superconducting electromagnets would provide more power for the same energy input. (Superconducting magnets cooled by liquid helium have been used in ships' engines and particle accelerators.) Such magnets have been used to levitate (Fig. 1) and propel trains. Another application of superconductors might be underground transmission cables with no resistive losses. However, among the many technological problems still to be overcome is how to form wires out of ceramic materials which are generally brittle. The application most likely to be realized soon will be in making faster computer circuits.

Imagine what it would mean if someone discovered a material that was superconducting at refrigerator or room temperature. The absence of electrical resistance opens many possibilities. You're likely to hear more about superconductor applications in the near future.

FIGURE 1 Magnetic levitation A magnetic cube levitates above a piece of "high-temperature" superconducting material that is cooled with liquid nitrogen. When the material becomes cold enough to superconduct, it acquires certain magnetic properties that cause a magnet to be repelled and to levitate.

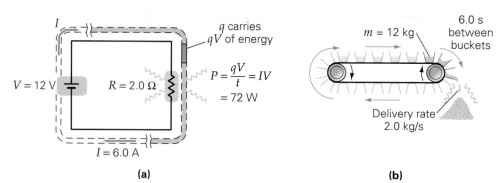

(a)

(b)

•FIGURE 17.10 **Electric power analogy** A simple electric circuit can be considered to be an energy delivery system much like a conveyor belt. **(a)** Think of current as equal continuous segments, each of charge $q = 1.0$ C and each carrying $qV = 12$ J of energy supplied by the battery. Here the current is $I = V/R = 6.0$ A, or 6.0 C/s. Thus the power (energy delivery rate) to the resistor is (6.0 C/s)(12 J/C) = 72 J/s = 72 W. **(b)** On the conveyor belt, assume each bucket carries 12 kg of sand, analogous to $qV = 12$ J in (a). If the buckets arrive at 6.0-second intervals, the "bucket current" is 1/6 bucket/s. Thus the rate of sand delivery (the "sand power") is ($\frac{1}{6}$ bucket/s)(12 kg/bucket) = 2.0 kg/s, analogous to $P = 72$ J/s. Increasing either the amount of sand in each bucket or the rate at which the buckets move results in more sand delivered per second. This analogy can help us understand why increasing *either* current or the voltage results in a power increase ($P = IV$) in the electric circuit.

(a)

(b)

•FIGURE 17.11 **Power ratings**
(a) Light bulbs are rated in watts. If operated at 120 V, this 60-watt bulb uses 60 J of energy each second. **(b)** Appliance ratings list either voltage and power or voltage and current. From either, the current, power, and effective resistance can be found. Here, one appliance is rated at 120 V and 18 W and the other at 120 V and 300 mA. Can you compute the current and resistance for the former and the power required and resistance for the latter?

A visual mechanical analogy to help explain $P = IV$ is shown in •Fig. 17.10. We depict a simple electric circuit as a system for transferring (converted) energy from one place to another and compare it to a conveyor belt system.

Since $R = V/I$, power can be rewritten in two other forms:

$$P = IV = \left(\frac{V}{R}\right)V = \frac{V^2}{R}$$

and

$$P = IV = I(IR) = I^2R$$

Thus there are three equivalent expressions for electric power calculations:

$$P = IV = \frac{V^2}{R} = I^2R \qquad \text{electric power} \qquad (17.8)$$

Usually you choose the one that is most convenient to the problem at hand.

Joule Heat

The thermal energy expended in a current-carrying resistor is referred to as **joule heat**, or **I^2R losses** (pronounced I squared R). In many instances, joule heating is an undesirable side effect: for example, in electrical transmission lines. However, in other situations, the conversion of electrical energy to thermal energy is the main purpose. Heating applications include the heating elements (burners) of electric stoves, hair dryers, immersion heaters, and toasters. A novel application of joule heat in battery testers is discussed in the Insight on page 555.

Electric light bulbs are rated in watts (power)—for example, 100 W or 60 W (•Fig. 17.11a). (These power ratings give total output power in *all* forms, heat and light, for bulbs connected to a voltage of 120 V.) In such bulbs, electrical energy excites the atoms of the filament, and the excited atoms then radiate away the extra energy as a combination of light and heat. The greater the wattage of a bulb, the greater the energy consumption per unit time (joules per second). Incandescent lamps are relatively inefficient as light sources—typically less than 5% of the

Joule Heat and Battery Testing

You may have noticed the individual battery testers on some battery packages (Fig. 1). A battery is placed between the terminals of the tester, and a rising yellow stripe indicates the condition of the battery. Joule heat is one of the ingredients of the operation.

The tester consists of a polyester strip with a printed circuit on one side and a liquid crystal on the other. (Many organic substances exhibit an "intermediate" phase of matter reminiscent of both a solid and a liquid; hence the name *liquid crystal*. These substances have unique optical properties that are affected by heat and voltage, as we shall see in the Insight on LCDs—liquid crystal displays—in Chapter 24.)

The printed circuit is a combination of silver and graphite that appears as a dark film. It is designed to produce graduated resistances. For example, for the AA battery tester, the circuit film is tapered, being narrow at the bottom and wide at the top. The resistance per unit length $(R/L = \rho/A)$ varies accordingly, with the greatest resistance at the narrow (small A) bottom.

You test a battery by inserting it into the back of the package so as to make contact with the ends of the tapered resistor (Fig. 2). The printed circuit strip then carries a current I, the magnitude of which depends on the condition of the battery, and joule heat raises the temperature of the resistor. Since $P = I^2R$, the heating for a given current is greatest at the bottom, where the resistance is highest, and least at the top, where the resistance is lowest.

The heating causes a reaction in the liquid crystal, which results in the thermometer effect on the display side of the tester. The black liquid crystal becomes transparent at a critical temperature. The "color change" occurs when the underlying yellow strip becomes visible through the transparent overcoat. The scale rates the battery as "good" if it supplies enough current to heat the complete liquid crystal strip above the critical temperature; it indicates "replace" if the critical heating occurs only at the bottom of the strip. The same principle applies to batteries to which the tester is attached.

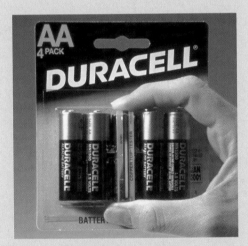

FIGURE 1   Battery tester   The test battery is inserted into the back of the package, and contact is made by pressing with the fingers. The rising yellow stripe indicates the condition of the battery—the longer the stripe, the more life the battery has.

FIGURE 2   Resistance and joule heat   As seen in the back of the test package, a printed circuit provides a calibrated resistance. In this AA battery tester, the circuit film is tapered, being narrow at the bottom and wide at the top. It thus has the greatest resistance at the bottom.

electrical energy is converted to visible light. Most of the energy produced is invisible infrared radiation and heat (Sections 11.4 and 20.4).

Electrical appliances are tagged or stamped with their power ratings. Either the voltage and power requirements or the voltage and current requirements are given (Fig. 17.11b). In either case, the current, power, and effective resistance can be calculated from Eq. 17.8. The power requirements of some household appliances are given in Table 17.2. Keep in mind that lamps and appliances generally operate on *alternating current*, which will be discussed in Chapters 20 and 21. However, the current–voltage equations for alternating current have the same form as those for direct current as long as we are calculating average quantities, as the following Example shows.

TABLE 17.2 Typical Power and Current Requirements for Household Appliances (120 V)

Appliance	Power	Current	Appliance	Power	Current
Air conditioner, room	1500 W	12.5 A	Microwave oven	625 W	5.2 A
Air conditioning, central	5000 W	41.7 A*	Radio–cassette player	14 W	0.12 A
Blender	800 W	6.7 A	Refrigerator, frost-free	500 W	4.2 A
Coffee maker	1625 W	13.5 A	Stove, top burners	6000 W	50.0 A*
Dishwasher	1200 W	10.0 A	Stove, oven	4500 W	37.5 A*
Electric blanket	180 W	1.5 A	Television, color	100 W	0.83 A
Hair dryer	1200 W	10.0 A	Toaster	950 W	7.9 A
Heater, portable	1500 W	12.5 A	Water heater	4500 W	37.5 A*

*High-power appliances such as these are typically wired to 240-volt house supply to reduce the current to half these values (Section 18.5).

Example 17.5 ■ A Modern "Appliance": The Power to Compute

A desktop computer system includes a color monitor and the central processing unit (CPU) with keyboard, each operating at an average voltage of 120 V. A typical 17″ color monitor has a power requirement of 130 W, and a CPU's requirement is about 960 W. (a) Calculate the average current that each component carries. (b) What is the resistance of each component under these operating conditions?

Thinking It Through. (a) We can use $P = IV$ from Eq. 17.8 to find the average current through each component. (b) Once we know the current, $R = V/I$ gives the resistances.

Solution. We are given the following data:

Given: $P_{monitor} = 130$ W   *Find:* (a) I (average current through each)
    $P_{CPU} = 960$ W   (b) R (resistance of each)

(a) Solving for the current from $P = IV$, we get

$$I_{monitor} = \frac{P_{monitor}}{V} = \frac{130\ \text{W}}{120\ \text{V}} = 1.08\ \text{A}$$

and

$$I_{CPU} = \frac{P_{CPU}}{V} = \frac{960\ \text{W}}{120\ \text{V}} = 8.00\ \text{A}$$

(b) The resistances then follow:

$$R_{monitor} = \frac{V}{I_{monitor}} = \frac{120\ \text{V}}{1.08\ \text{A}} = 111\ \Omega$$

and

$$R_{CPU} = \frac{V}{I_{CPU}} = \frac{120\ \text{V}}{8.00\ \text{A}} = 15.0\ \Omega$$

Follow-up Exercise. In addition to the components given in this Example, many desktop computer systems employ a black-and-white laser printer. While printing, such printers typically have a power requirement of 175 W. When they are not printing, they are in "standby mode," which takes only 8.0 W to maintain. Determine the resistance of the printer in each of its two modes.

Conceptual Example 17.6

An Electric Heater: Resistance and Joule Heat

An electric bathroom heater is constructed from a coil of wire. For efficient operation, the resistance of the heater coil (a) should be low, (b) be high, or (c) doesn't matter.

Reasoning and Answer. This problem is simplified by choosing the most convenient form of Eq. 17.8. If we use the form $P = I^2R$, we might be fooled into thinking that *increasing* the resistance would increase the power consumption and thus the joule heating. However, note that power also increases with the square of the current—and current, being inversely proportional to resistance ($I = V/R$), increases as resistance *drops*.

We might conclude (correctly) that because of the I^2 dependence, this second effect will outweigh the first. However, a more direct approach is to use the form of Eq. 17.8 that eliminates the current (which is variable here) and gives the power in terms of the resistance and the voltage (which we can assume to be household voltage and therefore fixed): $P = V^2/R$. From this form we can see clearly that, for a circuit with fixed voltage, power is inversely proportional to resistance. Thus, to promote joule heating, the heating element should have a relatively low resistance, and the answer is (a).

Follow-up Exercise. An ohmic resistor is connected to a battery with adjustable terminal voltage. Use all three forms of electric power (Eq. 17.8) to show that the power output quadruples when the voltage is doubled.

Example 17.7 ■ A Potentially Dangerous Repair: Joule Heat Revisited

A hair dryer is rated at 1200 W for 115 V (average, assumed constant here). The uniform wire filament breaks near one end, and the owner repairs it by removing one section near the break and simply reconnecting it. The filament is then 10.0% shorter than its original length. What will be the change in the heater's power output?

Thinking It Through. Both before and after the breakage, the wire will have the same voltage. However, shortening it will decrease its resistance, resulting in a larger current. Since $P = I^2R$, the increase in current will more than offset the decrease in resistance. Thus we expect the power output to go up.

Solution. Listing the numerical values, we have

Given: $P_{\text{o}} = 1200$ W $\qquad\qquad$ *Find:* P (new power output)
$\quad\quad\quad V = 115$ V
$\quad\quad\quad L = $ new length $= 0.900L_{\text{o}}$

Because 10.0% of the filament's length is gone, it will have 90.0% of its original resistance (R_{o}). From Eq. 17.3, the original resistance is

$$R_{\text{o}} = \rho \frac{L_{\text{o}}}{A}$$

Thus

$$R = \rho \frac{L}{A} = \rho \frac{0.900\, L_{\text{o}}}{A} = 0.900\left(\rho \frac{L_{\text{o}}}{A}\right) = 0.900 R_{\text{o}}$$

This means that the current will increase, since the applied voltage is the same ($V = V_{\text{o}}$). We can write this as $V = IR = I_{\text{o}}R_{\text{o}} = V_{\text{o}}$ and therefore determine that the current increases by about 11% since

$$I = \left(\frac{R_{\text{o}}}{R}\right)I_{\text{o}} = \left(\frac{1}{0.900}\right)I_{\text{o}} = 1.11 I_{\text{o}}$$

Under the initial conditions, the current was $I_{\text{o}} = P_{\text{o}}/V = 1200$ W/115 V $= 10.4$ A. Then the new current will be

$$I = 1.11 I_{\text{o}} = (1.11)(10.4 \text{ A}) = 11.5 \text{ A}$$

The power output could be computed directly from $P = IV$ or by computing the resistance and using $P = I^2R$ in Eq. 17.8. By the latter method, the original resistance was

$$R_{\text{o}} = \frac{V}{I_{\text{o}}} = \frac{115 \text{ V}}{10.4 \text{ A}} = 11.1 \ \Omega$$

The new resistance is therefore

$$R = 0.900 R_{\text{o}} = (0.900)(11.1 \ \Omega) = 9.99 \ \Omega$$

For the repaired dryer, then,

$$P = I^2R = (11.5 \text{ A})^2(9.99 \ \Omega) = 1.32 \times 10^3 \text{ W}$$

The power output of the dryer has been increased by 140 W. *Do not attempt such a repair job.* With reduced resistance and increased current, the filament could overheat and start a fire.

Follow-up Exercise. Suppose that because of a parts shortage, the owner of the hair dryer in this Example replaced the coil with one of the same length and material but made with wire that was 10.0% thicker. How would this affect the power of the dryer? Would such a repair create a potentially dangerous situation? Explain.

We often complain about our electric bills. But what do we actually buy from the electric or power company? What we pay for is electricity measured in **kilowatt-hours (kWh)**. Let's look at our defining equation of work and power to see what quantity this is a unit of. Power is the time rate of doing work, $P = W/t$ or $W = Pt$, so work has the units of watt-seconds (power × time). Converting this to the larger derived unit of kilowatt-hours (kWh), we see that the kilowatt-hour is a unit of work (or energy), equivalent to 3.6 MJ:

$$1 \text{ kWh} = (1000 \text{ W})(3600 \text{ s}) = (1000 \text{ J/s})(3600 \text{ s}) = 3.6 \times 10^6 \text{ J}$$

Thus, we pay the "power" company for electrical energy that we use to do work. We don't really pay for power because that is only the rate of energy delivery. (Similarly, we do not pay the water company for the rate at which they deliver water, only for the total number of gallons delivered.) Let's take a look at the cost of electricity.

Example 17.8 ■ The Price of Coolness: Electric Power Cost

If the motor of a frost-free refrigerator runs 15% of the time, how much does it cost to operate per month (to the nearest penny) if the power company charges 11¢ per kilowatt-hour? (Assume 30 days in a month.)

Thinking It Through. From the power (*rate* of energy usage) and the amount of time the motor is on per day, we can figure out the electrical energy it consumes *daily*. We can project that value to a 30-day month and then determine the cost in dollars on the basis of the typical cost of electrical energy.

Solution. When talking about practical electrical energy delivery, we will work in kilowatt-hours because the joule is a tiny unit.

Given: $P = 500$ W (Table 17.2)   *Find:* Operating cost per month
     Cost = $0.11/kWh

Since the refrigerator motor operates 15% of the time, in one day it runs

$$t = (0.15)(24 \text{ h}) = 3.60 \text{ h}$$

Since $P = W/t$, the electrical work done, or the average energy expended *per day*, is

$$W = Pt = (500 \text{ W})(3.60 \text{ h}) = 1.80 \times 10^3 \text{ Wh} = 1.80 \text{ kWh}$$

Then as a per-day cost we have

$$\left(\frac{1.80 \text{ kWh}}{\text{day}}\right)\left(\frac{\$0.11}{\text{kWh}}\right) = \frac{\$0.20}{\text{day}} \text{ or } 20¢ \text{ per day}$$

For a 30-day month, we have a total cost of

$$\left(\frac{\$0.20}{\text{day}}\right)\left(\frac{30 \text{ day}}{\text{month}}\right) = \$6.00 \text{ per month}$$

Follow-up Exercise. How long would you have to leave a 60-watt light bulb on to use the same amount of electrical energy that the refrigerator motor in this Example does each hour it is on?

The cost of electricity varies with location. On the average, it ranges from about 8¢ to 18¢ per kilowatt-hour in the United States. Do you know the price of electricity in your locality? Check an electric bill to find out. The Insight on electrical costs (p. 560) gives a comparison you might find interesting.

Electrical Efficiency and Natural Resources

In our electrical society, the use of and demand for electricity are continuously increasing. About 25% of the electricity generated in the United States goes into lighting (•Fig. 17.12). This is roughly equivalent to the output of 100 electrical generating (power) plants. Refrigerators consume about 7% of the electricity produced in the United States (the output of about 28 power plants).

This huge consumption of electricity has prompted the federal and many state governments to set minimum efficiency limits for refrigerators, freezers, air conditioners, water heaters, dishwashers, heat pumps, and so forth (•Fig. 17.13). Also, more efficient fluorescent lighting has been developed. The most efficient fluorescent lamp now in use consumes about 25–30% less energy than the average fluorescent lamp and roughly 75% less energy than incandescent lamps with an equivalent illumination output. Researchers are using new techniques in an effort to develop fluorescent lamps with even better efficiencies. The result has been significant energy savings as new, more efficient appliances are gradually replacing inefficient older models. Energy saved translates directly into savings of fuels and other natural resources and a reduction in environmental hazards such as global warming (see the Insight box on p. 379). To see the magnitude of the results of some of these energy efficiency standards, consider the next Example.

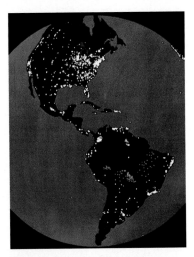

•**FIGURE 17.12 All lit up** A satellite image of the Americas at night. Can you identify the major population centers in the United States and elsewhere? The red spots across part of South America indicate large-scale burning of vegetation. The small yellow spot in Central America shows burning gas flares at oil production sites. At the top right edge you can just glimpse a few of the white city lights of Europe. This image was recorded by a visible/infrared system.

Example 17.9 ■ What We Can Save: Increasing Electrical Efficiency

Most modern power plants produce electricity at a rate of about 1.0 GW (electric power output). Estimate to two significant figures how many fewer power plants (running continuously) the state of California would need if all its households switched from the 500-watt refrigerators of Example 17.8 to more efficient 400-watt refrigerators. (Assume exactly 6 million homes with an average of 1.2 refrigerators per home.)

Thinking It Through. We would like a "ballpark" estimate of whether the refrigerator change would save just a negligible fraction of a plant's energy or a significant amount. We can use the results of Example 17.8 to calculate the effects of many refrigerators operating together.

Solution.

Given: Plant rate = 1.0 GW = 1.0×10^9 W
   = 1.0×10^6 kW
Energy requirement, 500-watt model
   = 1.80 kWh/day (Example 17.8)
Number of homes = 6.0×10^6 homes
Number of refrigerators per home
   = 1.2 refrig/home

Find: How many fewer power plants are required

For the entire state,

$$\left(\frac{1.80 \text{ kWh/day}}{\text{refrig}}\right)(6.0 \times 10^6 \text{ homes})\left(\frac{1.2 \text{ refrig}}{\text{home}}\right) = 1.3 \times 10^7 \frac{\text{kWh}}{\text{day}}$$

The more energy-efficient refrigerators would use only 80% (400 W/500 W = 0.80) of this, or approximately 1.0×10^7 kWh/day. The difference, 3.0×10^6 kWh/day, is the rate at which electric energy is saved. One 1.0-gigawatt power plant produces

$$(1.0 \times 10^6 \text{ kW/plant})\left(24 \frac{\text{h}}{\text{day}}\right) = 2.4 \times 10^7 \frac{(\text{kWh/plant})}{\text{day}}$$

So the replacement refrigerators would save about

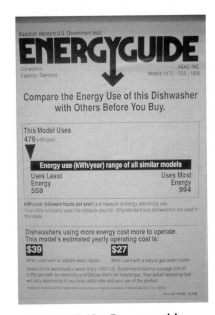

•**FIGURE 17.13 Energy guide** Consumers are made aware of appliance efficiencies in terms of the average yearly cost of operation. Sometimes the yearly cost is given for different kilowatt-hour (kWh) rates, which vary around the United States.

$$\frac{3.0 \times 10^6 \text{ kWh/day}}{2.4 \times 10^7 \text{ kWh/plant-day}} = 0.13 \text{ plants}$$

or about one-eighth of the output of a typical modern plant. Note that this saving would result from a change in just a single domestic appliance. Developing (and using!) more efficient electrical appliances is one way to avoid having to build new electric generating plants as the population and electrical usage increase.

Follow-up Exercise. Electric and gas water heaters are often said to be equally efficient—typically, about 95%. In reality, while gas water heaters are capable of 95% efficiency, it might be more accurate to describe electric water heaters as only about 33% efficient, even though about 95% of the electrical energy they consume is transferred to the water in the form of heat. Explain. [*Hint: Where does the energy for an electric water heater come from? Compare this to the energy delivery of natural gas. Recall the discussion of electrical generation in Section 12.4 and Carnot efficiency in Section 12.5.*]

Insight

The Cost of Portable Electricity

In the United States, the cost of household electricity ranges from about 8¢ to 18¢ per kilowatt-hour, depending on region. Electrical energy prices are still a bargain, considering the work that electricity does and the convenience it provides.

Batteries are widely and increasingly used to supply electrical energy. They power toys, radios, TV remote controls, and a variety of other portable devices. Have you ever thought about the cost of this "portable" electricity, as compared to that of household electricity? Let's take a look and make a general cost comparison. To do this, we need some data on batteries.

For one thing, we need to know the battery cost. This varies a great deal with location and quantity purchased. Focusing on just the AA-cell and the D-cell, a brief survey in the Greenwood, South Carolina, area showed average costs to be about those listed in Table 1.

TABLE 1   Battery Data

	Cost	Rating	Voltage
AA-cell	$0.95	2450 mA·h = 2.45 A·h	1.5 V
D-cell	$1.15	14 250 mA·h = 14.24 A·h	1.5 V

Then, we need to know how much "electricity" we get out of a battery. Batteries are rated in ampere-hours (A·h). This allows you to figure the lifetime of a battery for a given current (rate) output—for example, that needed to operate a flashlight: battery life (h) = amp-hour rating/current output.

Thus, the greater the current output, the shorter the battery life, as we would expect. The amp-hour ratings are not usually listed on batteries or their packages but are available from manufacturers. Some typical values are given in Table 1.

Now, notice that the unit amp-hour is current × time (*It*), and recall that $q = It$. Hence, the battery rating is a measure of the charge supplied over the chemical lifetime of a battery. Converting the ratings to amp-seconds, which is equivalent to 1 C, we have

AA-cell: $q = It = (2.45 \text{ A·h})(3.60 \times 10^3 \text{ s/h}) = 8.82 \times 10^3 \text{ C}$

D-cell: $q = It = (14.25 \text{ A·h})(3.60 \times 10^3 \text{ s/h}) = 5.13 \times 10^4 \text{ C}$

The energy supplied by a battery is given by $W = qV$, so the energy outputs over the lifetimes of the batteries are

AA-cell: $qV = (8.82 \times 10^3 \text{ C})(1.5 \text{ V})$

$= 1.3 \times 10^4 \text{ J} = 0.0036 \text{ kWh}$

D-cell: $qV = (5.13 \times 10^4 \text{ C})(1.5 \text{ V})$

$= 7.7 \times 10^4 \text{ J} = 0.021 \text{ kWh}$

where the conversions to kilowatt-hours are made for comparison. The cost of portable (battery) electricity is then about

AA-cell: price/kWh = $0.95/0.0036 kWh = $260/kWh

D-cell: price kWh = $1.15/0.021 kWh = $55/kWh

Compared to $0.08–$0.18/kWh for household electrical energy costs, we pay a great deal for battery convenience.

Chapter Review

Important Terms

battery   *543*
cathode   *543*
anode   *544*

electromotive force (emf)   *544*
terminal voltage   *544*
direct current (dc)   *545*

complete circuit   *546*
conventional current   *546*
electric current   *546*

Important Concepts

- Electric current is the net rate at which charge flows past a given point.
- The electromotive force of a battery (or any power supply) is the voltage between the two terminals when there is no current.
- Terminal voltage is the voltage across the terminals of a battery or power supply when there is a current. It is less than the emf because of internal resistance.
- The direction of current is that in which positive charge would move. In metals, current is actually electrons moving in the opposite direction.
- The electrical resistance of a circuit element is defined as the voltage across the element divided by the resulting current ($R = V/I$).

- Ohm's law applies to a circuit element if that element exhibits a constant electrical resistance. Ohm's law is commonly written $V = IR$, where R is constant.
- The electrical resistance of a circuit element depends on its geometry (cross-sectional area and length) as well as the intrinsic electrical resistive properties of its material (resistivity). For most metallic conductors, resistivity increases with increasing temperature.
- Electric power is the rate at which work is done or energy is transferred. The power delivered to a circuit element depends on the element's resistance and the voltage applied to it.
- Joule heat is the heat energy expended in current-carrying circuit elements.

Important Equations

Electric Current (definition):
$$I = \frac{q}{t} \tag{17.1}$$

Electrical Resistance (definition):
$$R = \frac{V}{I} \tag{17.2}$$

Ohm's Law:
$$V = IR \quad (R = a\ constant) \tag{17.2}$$

Resistivity:
$$\rho = \frac{RA}{L} \tag{17.3}$$

Conductivity:
$$\sigma = \frac{1}{\rho} \tag{17.4}$$

Temperature Variation of Resistivity (α constant):
$$\rho = \rho_0(1 + \alpha\,\Delta T) \quad \text{or} \quad \Delta\rho = \rho_0\alpha\,\Delta T \tag{17.5,}$$
$$(\text{where } \Delta\rho = \rho - \rho_0) \tag{17.6}$$

Temperature Dependence of Resistance (α constant):
$$R = R_0(1 + \alpha\,\Delta T) \quad \text{or} \quad \Delta R = R_0\alpha\,\Delta T \tag{17.7}$$
$$(\text{where } \Delta R = R - R_0)$$

Electric Power:
$$P = IV = \frac{V^2}{R} = I^2R \tag{17.8}$$

Exercises

Unless otherwise indicated, in this chapter assume that all batteries have negligible internal resistance.

17.1 Batteries and Direct Current

1. When a battery is placed into a complete circuit, the voltage across its terminal is its (a) emf, (b) terminal voltage, (c) power, (d) all of these.

2. As a battery gets old, its (a) emf increases, (b) emf decreases, (c) terminal voltage increases, (d) terminal voltage decreases.

3. When four 1.5-volt batteries are connected in parallel, the output voltage of the combination is (a) 1.5 V, (b) 3.0 V, (c) 6.0 V, (d) none of these.

4. Why does the battery design described on p. 543 require a chemical membrane?

5. A battery has a small internal resistance r (•Fig. 17.14). Explain why the terminal voltage drops as the internal resistance increases even though the emf remains constant.

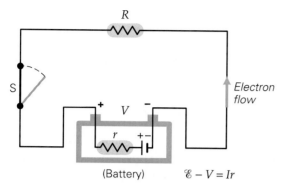

•**FIGURE 17.14   Emf and terminal voltage**   See Exercises 5 and 23.

6. Is it possible for a 12-volt car battery to measure exactly 12 V when it is in use? Explain.

7. ■ (a) Two 1.5-volt dry cells are connected in series. What is the total voltage of the combination? (b) What would be the total voltage if the cells were connected in parallel?

8. ■ What is the voltage across six 1.5-volt batteries when they are connected (a) in series and (b) in parallel?

9. ■■ Two 6.0-volt batteries and one 12-volt battery are connected in series. (a) What is the voltage across the whole arrangement? (b) What arrangement of these three batteries would give a total voltage of 12 V?

10. ■■ Given three batteries with voltages of 1.0 V, 3.0 V, and 12 V, respectively, how many different voltages could be obtained by connecting one or more of the batteries in series, and what are these voltages?

17.2 Current and Drift Velocity

11. The unit of electrical current is (a) C, (b) C/s, (c) A, (d) both (b) and (c).

12. ■ A net charge of 30 C passes a point on a wire in 2.0 min. What is the current in the wire?

13. ■ How long would it take for a net charge of 2.5 C to pass a location in a wire so as to produce a steady current of 5.0 mA?

14. ■ A small toy car draws 0.50 mA of current from a 3.0-volt nicad (nickel–cadmium) battery. In 10 min of operation, (a) how much net charge flows in the circuit and (b) how much energy is lost by the battery?

15. ■■ A car motor draws 50 A when starting up. If the start-up time is 1.5 s, how many electrons pass a given spot in the circuit during this time?

16. ■■ A net charge of 20 C passes a spot in a wire of cross-sectional area 0.30 cm² in 1.25 min; a net charge of 30 C passes a location in another wire of cross-sectional area 0.45 cm² in 1.52 min. Which wire carries more current, and how much more?

17. ■■ A fully charged, heavy-duty battery is rated at 100 A·h at 5.0 A. Thus the life of the battery when delivering 5.0 A is 20 h (100 A·h/5.0 A). How much charge will the battery deliver in this time?

18. ■■ In 45 s, if a charge of 6.7 C of electrons flows to the right past a point at the same time a charge of 8.3 C of protons flows to the left, what are the direction and magnitude of the current at that point?

19. ■■ In a proton linear accelerator, a proton current of 9.5 mA hits the target. (a) How many protons hit the target each second? (b) What is the energy delivered to the target each second if the protons have a kinetic energy of 20 MeV and completely stop in the target?

17.3 Resistance and Ohm's Law*

20. The unit of resistance is the (a) V/A, (b) A/V, (c) W, (d) V.

21. For an ohmic resistor, current and resistance (a) vary with temperature, (b) are directly proportional, (c) are independent of voltage, (d) none of these.

22. ◯ If the voltage (V) were plotted on the same graph versus current (I) for two ohmic conductors with different resistances, how could you tell the less resistive one?

23. ■ A battery labeled 12.0 V supplies 1.90 A to a 6.00-ohm resistor (Fig. 17.14). (a) What is the terminal voltage of the battery? (b) What is its internal resistance?

24. ■ What is the emf of a battery with an internal resistance of 0.15 Ω if the battery delivers 1.5 A to an externally connected 5.0-ohm resistor?

25. ■ A wire is connected across a steady voltage source. (a) If that wire is replaced with one of the same material that is twice as long and has twice the cross-sectional area, how will the current in the wire be affected? (b) How will the current be affected if instead the new wire has the same length but half the diameter of the old one?

26. ■ In a semiconductor, the number of conducting charges can vary drastically with temperature. What does a negative temperature coefficient of resistivity for a semiconductor imply about how the number of charge carriers varies with temperature in such a material?

27. ■ Some states allow the use of aluminum wire in houses in place of copper. If you wanted the resistance of your

*Assume that the temperature coefficients of resistivity given in Table 17.1 apply over large temperature ranges.

wires to be the same as that of copper, how much more thickness of aluminum wire (expressed as a percentage) would you need compared to the copper wire?

28. ■ How much current is drawn from a 12-volt battery when a 15-ohm resistor is connected across its terminals?

29. ■ What voltage must a battery have to produce 0.50 A of current through a 2.0-ohm resistor?

30. ■ 🌐 When a voltage of 6.0 V is applied to an ohmic resistor, the current through the resistor is 0.25 A. What is the current if the applied voltage is 10 V?

31. ■■ A copper wire is 0.60 m long and has a diameter of 0.10 cm. What is the resistance of the wire?

32. ■■ A 100-ohm resistor is placed in a circuit with two identical batteries connected in series. If the resistor draws 1.5 A, what is the terminal voltage of each battery?

33. ■■ An automobile starter is connected to a 12-volt battery. If the starter has an effective resistance of 2.5 Ω, what is the rate at which electrons drift past a point in the circuit?

34. ■■ A certain material is formed into a long rod with a square cross section 0.50 cm on each side. When a voltage of 100 V is applied across a length of the rod 2.0 m long, 5.0 A of current is carried. What are (a) the resistivity and (b) the conductivity of the material? (c) Is it a conductor, insulator, or semiconductor?

35. ■■ Two copper wires have equal cross-sectional areas and lengths of 2.0 m and 0.50 m, respectively. What is the ratio of the current through the shorter wire to that through the longer if they are connected to the same power supply?

36. ■ Two copper wires have equal lengths, but the diameter of one is triple that of the other. What is the ratio of the resistance of the thicker wire to that of the thinner wire?

37. ■■ The wire in a heating element of an electric stove burner has an effective length of 0.75 m and a cross-sectional area of 2.0×10^{-6} m^2. If the wire is made of iron and operates at a temperature of 380°C, what is its resistance?

38. ■■ What is the percentage variation of the resistivity of copper over the range from room temperature (20°C) to 100°C?

39. ■■ A copper wire has a resistance of 25 mΩ at 20°C. When the wire is carrying a current, heat produced by the current causes its temperature to increase by 27 C°. What is the change in the wire's resistance?

40. ■■ What is the change in resistance of a 2000-meter-long aluminum wire of resistance 10.00 Ω (at 0°C) as its temperature changes from the freezing point to the boiling point of water?

41. ■■ When a resistor is connected to a 12-volt source, it draws 185 mA of current. The same resistor, when connected to a 90-volt source, draws 1.25 A of current. Is the resistor ohmic? Justify your answer mathematically.

42. ■■ What length of wire is required to make a 20.0-ohm resistor by winding AWG (American Wire Gauge) No. 10 nichrome wire in a coil? (AWG No. 10 wire has a diameter of 2.588 mm.)

43. ■■ A 1.0-meter length of copper wire with a diameter of 2.0 mm is stretched out; its length increases by 25% while its cross-sectional area decreases but remains uniform. Is the resistance of the wire the same before and after? Justify your answer mathematically.

44. ■■ A particular application requires a 20-meter length of aluminum wire to have a resistance of 0.25 mΩ at 20°C. What diameter should the wire have?

45. ■■ If the resistance of the wire in Exercise 44 cannot vary by more than ±5.0%, what is the wire's operating temperature range?

46. ■■ 🌐 •Figure 17.15 shows data on the dependence of current through a resistor on voltage across that resistor. (a) Is the resistor ohmic? (b) What is its resistance?

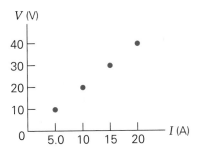

•FIGURE 17.15   An ohmic resistor?   See Exercise 46.

47. ■■■ At 20°C a silicon rod is connected to a battery with a terminal voltage of 6.0 V. A current of 0.50 A results. If the temperature of the rod is increased to 25°C, how much current does the rod then carry?

48. ■■■ An electrical resistance thermometer made of platinum has a resistance of 5.0 Ω at 20°C and is alone in a circuit with a 1.5-volt battery. What is the change in the current in the circuit when the thermometer is heated to 2020°C?

17.4 Electric Power

49. Electric power has units of (a) A^2·Ω, (b) J/s, (c) V^2/Ω, (d) all of these.

50. If the voltage across an ohmic resistor is doubled, the power expended in the resistor (a) increases by a factor of 2, (b) increases by a factor of 4, (c) decreases by half, (d) none of these.

51. If the current through an ohmic resistor is halved, the power expended in the resistor (a) increases by a factor

of 2, (b) increases by a factor of 4, (c) decreases by half, (d) decreases by a factor of 4.

52. Assuming your hair dryer obeys Ohm's law, what would happen if you plugged it directly into a 240-volt outlet in Europe if it is designed to be used in the 120-volt outlets of the United States?

53. Most light-bulb filaments are made of tungsten and are about the same length. What would be different about the filament in a 60-watt bulb compared to a 40-watt bulb?

54. ■ A freezer of resistance 10 Ω is connected to a 110-volt source. What is the power consumed by this freezer?

55. ■ The current through a refrigerator of resistance 12 Ω is 13 A. What is the power consumed by the refrigerator?

56. ■ A digital video disc (DVD) player is rated at 100 W at 120 V. What is its resistance?

57. ■ Show that the quantity volts squared per ohm has SI units of power.

58. ■ An electric heater is designed to produce 50 kW of heat when connected to a 240-volt source. What must the resistance of the heater be?

59. ■■ Assuming it is 90% efficient, how long would the heater in Exercise 58 take to heat 50 gal of water from 20°C to 80°C ?

60. ■■ An electric toy with a resistance of 15 Ω is operated by four 1.5-volt batteries connected in series. (a) What current does the toy draw? (b) What is the power consumed by the toy?

61. ■■ A welding machine draws 18 A of current at 240 V. (a) How much energy does the machine use each second? (b) What is its resistance?

62. ■■ On average, an electric water heater operates for 2.0 h each day. (a) If the cost of electricity is $0.15/kWh, what is the cost of operating the heater during a 30-day month? (b) What is the resistance of a typical water heater? [Hint: See Table 17.2.]

63. ■■ What is the resistance of a heating coil if it is to generate 15 kJ of heat per minute when connected to a 120-volt source?

64. ■■ A 200-watt computer power supply is on 10 h per day. If the cost of electricity is $0.15/kWh, what is the cost (to the nearest dollar) of using the computer for a year (365 days)?

65. ■■ A 120-volt air conditioner unit draws 15 A of current. If it operates for a period of 20 min, (a) how much energy in kilowatt-hours does it use? (b) If the cost of electricity is $0.15/kWh, what is the cost (to the nearest penny) of operating the unit for this time?

66. ■■ Two resistors, 100 Ω and 25 kΩ, are rated for maximum wattages of 1.5 W and 0.25 W, respectively. What is the maximum voltage that can be safely applied to each resistor?

67. ■■ A wire 5.0 m long and 3.0 mm in diameter has a resistance of 100 Ω. A potential difference of 15 V is applied across the wire. Find (a) the current in the wire, (b) the resistivity of its material, and (c) the rate at which heat is being produced in it.

68. ■■ When connected to a voltage source, a coil of tungsten wire initially dissipates 500 W of power. In a short time, the temperature of the coil increases by 150 C° because of joule heat. What is the corresponding change in the power?

69. ■■ A 20-ohm resistor is connected to four 1.5-volt batteries. What is the joule heat loss per minute in the resistor if the batteries are connected (a) in series and (b) in parallel?

70. ■■ A 5.5-kilowatt water heater operates at 240 V. (a) Should the heater circuit have a 20-amp or a 30-amp circuit breaker? (A circuit breaker is a safety device that opens the circuit at its rated current.) (b) Assuming 85% efficiency, how long will the heater take to heat the water in a 55-gallon tank from 20° to 80°C?

71. ■■ A student uses an immersion heater to heat 0.30 kg of water from 20°C to 80°C for tea. If the heater is 75% efficient and takes 2.5 min, what is its resistance? (Assume 120-volt household voltage.)

72. ■■ An ohmic appliance is rated at 100 W when it is connected to a 120-volt source. If the power company cuts the voltage by 5.0% to conserve energy, what is the power consumed by the appliance after the voltage drop?

73. ■■ In an electrical brownout, a light bulb's output drops from its usual 60 W at 120-volt operating voltage. If the voltage was cut in half and the power dropped to 20 W during the brownout, what is the ratio of the bulb's resistance at full power to its resistance during the brownout?

74. ■■ To empty a flooded basement, a water pump must do work (lift the water) at a rate of 2.00 kW. If the pump is wired to a 240-volt source and is 84% efficient, (a) how much current does it draw, and (b) what is its resistance?

75. ■■■ Find the total monthly (30-day) electric bill (to the nearest dollar) for the following household appliance usage if the utility rate is $0.12/kWh: Central air conditioning runs 30% of the time; a blender is used 0.50 h/month; a dishwasher is used 8.0 h/month; a microwave oven is used 15 min/day; the motor of a frost-free refrigerator runs 15% of the time; a stove (range and oven) is used a total of 10 h/month; and a color television is operated 120 h/month. (Use the information given in Table 17.2.)

Additional Exercises

76. Pieces of carbon and copper have uniform cross sections and the same resistance at room temperature. (a) If the temperature of each piece is increased by 10.0 C°, which will have the greater resistance? (b) What is the ratio of the resistances at the higher temperature?

77. A resistance thermometer made of platinum shows a 25% change in resistance over a certain temperature range. What is the change in temperature for this range?

78. An iron conductor with a resistance of 1.75 Ω at 20°C is connected to a 12-volt source. If the conductor is heated in an oven to 300°C, what is the change in the current through it?

79. A computer CD-Rom drive that operates on 120 V is rated at 40 W. (a) How much current does the drive draw? (b) What is the drive's resistance?

80. If the price of electricity is $0.15/kWh, how much would it cost (to the nearest penny) to operate ten 75-watt light bulbs for 8.0 h?

81. A toaster oven is rated for 1.60 kW at 120 V. When the filament ribbon of the oven burns out, the owner makes a repair that reduces the length of the filament by 10%. How does this affect the power output of the oven, and what is the new value?

82. The tungsten filament of an incandescent lamp has a resistance of 200 Ω at room temperature. What would the filament's resistance be at an operating temperature of 1600°C?

83. Pieces of aluminum and copper wire are identical in length and diameter. At some temperature, one of the wires will have the same resistance as the other has at 20°C. What is the temperature? (Is there more than one temperature?)

84. What is the current through an aluminum wire of length 100 m and radius 1.0 mm if the wire is connected to a 1.5-volt battery?

85. An electric iron with a 15-ohm heating element operates at 120 V. How many joules of energy does the iron convert to heat in 1.0 h?

86. When a wire 1.0 mm in diameter and 2.0 m in length is connected to a 3.00-volt source, a current of 11.8 A flows. (a) What is the resistivity of the wire? (b) What material is the wire likely to be made of?

87. A 100-ohm resistor is rated at 0.25 W. (a) What is the maximum current the resistor can carry? (b) What is the maximum operating voltage that should be applied across the resistor?

88. A wire carries a current of 75 mA. How many electrons have to pass a given point in the wire in 2.0 s to produce this current?

89. A current of 0.10 A exists in a single resistor in a circuit with a 40-volt voltage source. (a) What is the resistance of the resistor? (b) How much power is dissipated by the resistor? (c) How much energy is dissipated in 2.0 min?

90. Find the ratio of the resistance of a copper wire to the resistance of an aluminum wire (a) if the wires have the same length and diameter and (b) if the copper wire is twice as long as the aluminum wire and has a diameter half that of the aluminum wire.

91. At 120 V (dc), a 60-watt light bulb is on for 1.0 h. How many electrons move through it in that time?

92. A copper bus bar, used to conduct large amounts of current at a power station, is 1.5 m long and 8.0 cm by 4.0 cm in cross section. What voltage applied to the ends of the bar will produce a current of 3000 A through it?

93. Why is electric power delivered over long distances at high voltages when we know that high voltages can be very dangerous?

94. • Figure 17.16 shows free charge carriers that each have a charge q and move with a speed v_d (drift speed) in a conductor of cross-sectional area A. Let n be the number of free charge carriers per unit volume. (a) Show that the total charge (ΔQ) free to move in the volume element shown is given by $\Delta Q = (nAx)q$. (b) Show that the current in the conductor is given by $I = nqv_d A$.

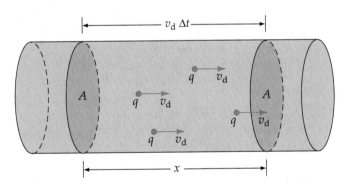

•FIGURE 17.16 **Total charge and current** See Exercises 94 and 95.

95. A copper wire with a cross-sectional area of 13.3 mm^2 (AWG No. 6) carries a current of 1.2 A. If the wire contains 8.5×10^{22} free electrons per cubic centimeter, what is the drift velocity of the electrons? [Hint: See Exercise 94 and Fig. 17.16.]

Basic Electric Circuits

Insights

- An RC Circuit in Action
- Electricity and Personal Safety

The term "electric circuit" probably brings to mind wires. But metal wires are not the only thing that can conduct electricity, as the photo above shows. Since the bulb is lit, the circuit is complete. We can conclude, therefore, that the "lead" in a pencil (actually a form of carbon called graphite) is a good conductor of electricity. The same must be true for the liquid in the beaker—in this case, a solution of ordinary table salt. (The results would be very different if the electrodes consisted of the wooden part of the pencil or the liquid were salad oil.)

Electric circuits are of many kinds and can be designed for many purposes, from boiling water for coffee to lighting a Christmas tree. Circuits of the type in the photo above have practical applications in the laboratory and in industry: They are used in synthesizing or purifying chemical substances and in electroplating metals (for example, making silver plate). Armed with the principles learned in Chapters 15, 16, and 17, you are now ready to analyze some relatively simple circuits. This will give you an appreciation of how electricity actually works.

Circuit analysis most often deals with voltage, current, and power requirements. A circuit may be analyzed theoretically before being assembled. The analysis might show that the circuit would not function properly as designed or that there could be a safety problem (such as overheating due to joule heat).

Circuit diagrams are used to visualize circuits in order to understand their functioning. A few of these diagrams were included in Chapter 17. In actuality, the wires of the circuit are not ordinarily placed in the rectangular pattern of a circuit diagram. The rectangular form is simply a convention that provides a neater presentation and easier visualization of the circuit arrangement.

We'll begin our analysis of circuits by looking at arrangements of resistive elements.

18.1 Resistances in Series, Parallel, and Series–Parallel Combinations

Objectives: To (a) determine the equivalent resistance of resistors in series, parallel, and series–parallel combinations and (b) use equivalent resistances to analyze simple circuits.

The resistance symbol —$\wedge\!\wedge\!\wedge$— in a circuit diagram can represent any resistive element—a commercial resistor, a light bulb, an appliance, and so on. Here we will consider all of these to be ohmic (that is, to have a constant operating resistance), unless otherwise stated. Also, the resistance of each connecting wire in a circuit will be considered to be negligible in comparison to those of the circuit elements.

Resistors in Series

In analyzing a circuit, keep in mind that the *sum of the voltages around a circuit loop* (that is, the gains and losses with + and − signs, respectively) *is zero*. For example, for the circuit in •Fig. 18.1a all of the voltage drops (V_i) across the three resistors combined equal the voltage rise, or increase, V across the battery terminals. In general, for a string of any number of resistors in series, we would write

Recall from Ohm's law (Eq. 17.2) that for resistors, $V = IR$.

$$V - \Sigma_i (IR_i) = 0 \qquad (18.1)$$

or

$$V = \Sigma_i V_i = \Sigma_i (IR_i)$$

where the i summation is taken over all the individual resistances. (The subscript i is often omitted as being understood.)

The fact that the sum of the voltages around the circuit loop is zero is essentially a statement of the conservation of energy. We saw in Chapter 17 that for the charge carriers traversing the circuit, the energy given to them by a voltage source must be lost in the circuit. That is, they must have the same electric potential (conventionally designated zero) on returning to the negative terminal of the source

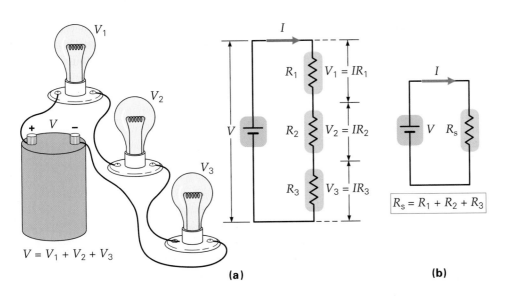

•**FIGURE 18.1   Resistors in series**
(a) When resistors (here, representing the resistances of light bulbs) are connected in series, the current through each of them is the same. Also, $\Sigma_i V_i$, the sum of the voltage drops across the resistors, is equal to V, the voltage of the battery. (b) The equivalent resistance R_s of the resistors in series.

as they had when starting. If not, the charge carriers would gain energy with each round trip of the circuit, thus violating the law of the conservation of energy.

The three resistances (or resistors) in Fig. 18.1a are said to be connected in series, or connected end-to-end. *When resistors are connected in series, the current is the same through all of them,* as required by the conservation of charge. Otherwise, there would be a build-up of charge somewhere along the way. The electrons entering one region of the wire repel the electrons ahead of them, and the whole "stream" moves along without charge pile-up. •Figure 18.2 shows the analogous flow of water along a smooth streambed punctuated by a series of rapids. The rapids can be thought of as gravitational "resistors," with the rocks and pebbles impeding the flow of water. The water current through each course of rapids is the same, and the sum of the potential energies lost in each vertical drop is equal to the work that would have to be done to bring the water back to the top of the hill.

The total voltage drop around a circuit equals the sum of the individual voltage drops across the resistors, which is the same as the battery's voltage. If we label their common current I, Eq. 18.1 can be written explicitly for three resistors (see Fig. 18.1a):

$$V = V_1 + V_2 + V_3$$
$$= IR_1 + IR_2 + IR_3 = I(R_1 + R_2 + R_3)$$

To replace the three resistors by a single resistor R_s (called the **equivalent series resistance**) and maintain the same current, we would need $V = IR_s$, or $R_s = V/I$. Hence the three resistors in series have an equivalent resistance R_s:

$$R_s = \frac{V}{I} = R_1 + R_2 + R_3$$

That is, *the equivalent resistance of resistors in series is the sum of the individual resistances.* The three resistors of Fig. 18.1a could be replaced with a single resistor of resistance R_s (Fig. 18.1b) without affecting the current. For example, if each resistor in Fig. 18.1a had a value of 10 Ω, then R_s would be 30 Ω. This makes physical sense—the more resistors in a row, the more the overall resistance. *The equivalent series resistance is larger than the resistance of the largest resistor in the series.*

This result can be extended to any number of resistors in series:

$$R_s = R_1 + R_2 + R_3 + \cdots = \Sigma_i R_i \qquad \begin{array}{l} equivalent \\ series\ resistance \end{array} \qquad (18.2)$$

Note: R_s is larger than the largest resistance in the series.

Series connections are not normally used in everyday circuits such as house wiring, because they share one major disadvantage. Suppose that one of the light bulbs in the circuit in Fig. 18.1a burned out. What would happen? *All* of the bulbs would go out because the circuit would no longer be complete. If the filament of one of the bulbs is broken or blows out, the circuit is said to be *open*. An open circuit is like a drawbridge stuck in the up position: No traffic can flow across it. Think of an open circuit as equivalent to an infinite resistance; thus the current through it is zero.

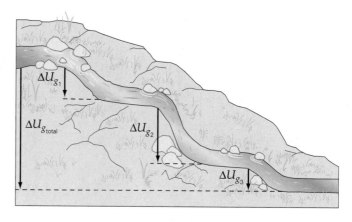

•**FIGURE 18.2 Water-flow analogy to resistors in series** Even though a different amount of gravitational potential energy is lost as the water flows through each set of rapids, the current is the same in each. The total loss of gravitational potential energy is the sum of the three losses. (To make this a complete circuit, some external agent, such as a pump, would have to do work to return the water to the top of the hill, restoring its original gravitational potential energy.)

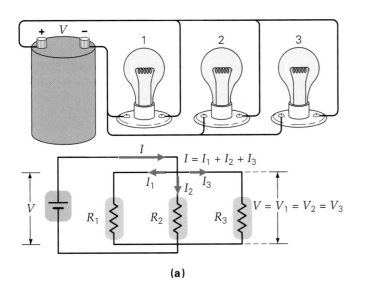

$$V = V_1 = V_2 = V_3$$

$$I = I_1 + I_2 + I_3$$

$$V = V_1 = V_2 = V_3$$

(a)

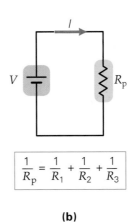

$$\frac{1}{R_p} = \frac{1}{R_1} + \frac{1}{R_2} + \frac{1}{R_3}$$

(b)

•**FIGURE 18.3   Resistors in parallel   (a)** When resistors are connected in parallel, the voltage drop across each of the resistors is the same. The current from the battery divides among the resistors. **(b)** The equivalent resistance R_p of the resistors in parallel.

Resistors in Parallel

Another basic way of connecting resistors in a circuit is in parallel (•Fig. 18.3a). In this case, all the resistors have common connections—that is, all the leads on one side of the resistors are attached together, as are all the leads on the other side. *When resistors are connected in parallel across a battery, the voltage drop across each resistor is the same* (and in this case equal to the voltage of the battery).

However, the current from the battery divides among the different paths, as shown in Fig. 18.3a. Think of how the flow of charge must divide when it comes to a junction in the circuit, like traffic on a highway that divides into two different forks (•Fig. 18.4a). Since charge cannot pile up at the junctions, the total current out of the battery breaks up into smaller currents whose sum equals the total current. Specifically, for three resistors in parallel, we have

$$I = I_1 + I_2 + I_3$$

If the individual resistances are equal, the current divides equally. But in general the resistances are not equal, so the current divides among the resistors in inverse

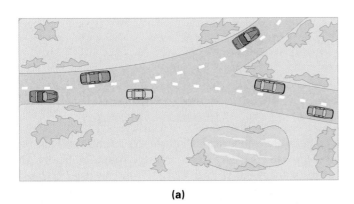

(a)

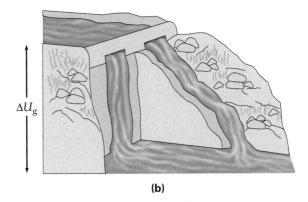

ΔU_g

(b)

•**FIGURE 18.4   Analogies to resistors in parallel   (a)** When a road forks, the total number of cars entering the two branches each second is equal to the number arriving at the fork each second, as does charge at a junction. **(b)** When a given mass of water flows from behind a dam at a given height, the amount of gravitational potential energy that it loses in falling to the streambed below is the same regardless of the path of descent. This situation is analogous to voltages across parallel resistors.

proportion to their resistances—that is, the greatest current takes the path of least resistance.

We define the **equivalent parallel resistance,** R_p, as the value of a single resistor that could replace these three resistors and maintain the same current through the battery. Thus, $R_p = V/I$, or $I = V/R_p$. The voltage drop (V) must be the same across each resistor. Imagine two separate paths for water, each leading from the top of a dam to the bottom. The water loses the same amount of gravitational potential energy regardless of the path it takes (Fig. 18.4b).

Thus for each individual resistor we can write $I_i = V/R_i$. Then,

$$I = I_1 + I_2 + I_3$$

$$= \frac{V}{R_1} + \frac{V}{R_2} + \frac{V}{R_3} = V\left(\frac{1}{R_1} + \frac{1}{R_2} + \frac{1}{R_3}\right) = \frac{V}{R_p}$$

Thus, the equivalent resistance R_p of three resistors in parallel is given (in reciprocal form) by

$$\frac{1}{R_p} = \frac{1}{R_1} + \frac{1}{R_2} + \frac{1}{R_3}$$

That is, the reciprocal of the equivalent resistance is equal to the sum of the reciprocals of all of the individual resistors connected in parallel. The three resistors could therefore be replaced by a single resistor with a resistance value of R_p (see Fig. 18.3b).

This result can be generalized to include any number of resistors in parallel:

$$\frac{1}{R_p} = \frac{1}{R_1} + \frac{1}{R_2} + \frac{1}{R_3} + \cdots = \Sigma_i \frac{1}{R_i} \qquad \begin{array}{l}\textit{equivalent} \\ \textit{parallel resistance}\end{array} \qquad (18.3)$$

Problem-Solving Hint

Equation 18.3 gives $1/R_p$. Don't forget to take the reciprocal of this value to get R_p. The units are not ohms until you invert for the final answer. (If you always carry units along in your calculations, you will be less likely to make this mistake.)

If there are only two resistors in parallel, the formula for the equivalent resistance can be written in a nonreciprocal form, which may be more convenient:

$$\frac{1}{R_p} = \frac{1}{R_1} + \frac{1}{R_2} = \frac{R_1 + R_2}{R_1 R_2}$$

Note: Do *not* generalize Eq. 18.3a. For more than two resistors, you must use Eq. 18.3 or apply 18.3a more than once.

or

$$R_p = \frac{R_1 R_2}{R_1 + R_2} \qquad \begin{array}{l}\textit{equivalent resistance} \\ \textit{(two resistors in parallel)}\end{array} \qquad (18.3a)$$

Note: R_p is smaller than the least resistor in parallel.

The equivalent resistance of a parallel arrangement of resistors is always less than the smallest resistance in the arrangement. This is easily demonstrated mathematically for two parallel resistors—say, of resistances 6.0 Ω and 12.0 Ω. You can use Eq. 18.3a to show that these are equivalent to a single resistor with a resistance of 4.0 Ω.

Physically, the reason the equivalent resistance is smaller when you combine resistors in parallel is as follows. Consider first a 12-volt battery and the 6.0-ohm resistor. The current through the battery is 2.0 A ($I = V/R$). Now connect the 12.0-ohm resistor in parallel to the first. The current through the first will be unaffected (why?), but the new resistor will carry a current of 1.0 A. Thus the *total* current through the battery is 3.0 A. From the battery's point of view, the circuit resistance has dropped. Since the current has increased but the voltage remains the same, the equivalent resistance across the battery terminals must be less than the initial value of 6.0 Ω.

Notice that this argument does not depend on the value of the added resistor. All that matters is that some resistance is added. To see this, consider what would have happened if we had connected a 24-ohm resistor instead of the 12-ohm resistor. It would carry a current of 0.5 A, and the total current would increase to only 2.5 A. The equivalent resistance would now be $R = V/I = (12 \text{ V})/(2.5 \text{ A}) = 4.8 \text{ }\Omega$. Although this result is different from the equivalent resistance in the first case, it is still less than the initial single 6.0-ohm resistor—the overall or equivalent resistance decreases no matter what the value of the added resistor (in parallel)!

Think of the analogy of letting water out of a reservoir dam. If you punch only one hole, there is a certain water current driven by the water pressure behind the dam. If you then punch even the tiniest second hole, the total water flow out of the reservoir will increase. In effect, the overall resistance of the dam to the flow of water will have been reduced.

Thus the general rules of thumb are:

Series connections provide a way to increase total resistance.

Parallel connections provide a way to decrease total resistance.

Example 18.1 ■ Connections Count: Resistors in Series and in Parallel

What is the equivalent resistance of three resistors (1.0 Ω, 2.0 Ω, and 3.0 Ω) when they are connected (a) in series (Fig. 18.1a) and (b) in parallel (Fig. 18.3a)? (c) What current will be delivered from a 12-volt battery for each of these arrangements?

Thinking It Through.   To find the equivalent resistances for (a) and (b), we can apply Eqs. 18.2 and 18.3, respectively. To find the series current for (c), we calculate the total current through the battery as though it were connected to a single resistor with the equivalent resistance. For the parallel currents, we use the fact that each resistor has the same voltage across it—that of the battery.

Solution.   Listing the data, we have

Given: $R_1 = 1.0 \text{ }\Omega$ *Find:*   (a) R_s (series resistance)
              $R_2 = 2.0 \text{ }\Omega$ (b) R_p (parallel resistance)
              $R_3 = 3.0 \text{ }\Omega$ (c) I (current for each case)
              $V = 12 \text{ V}$

(a)   The equivalent series resistance (Eq. 18.2) is:

$$R_s = R_1 + R_2 + R_3 = 1.0 \text{ }\Omega + 2.0 \text{ }\Omega + 3.0 \text{ }\Omega = 6.0 \text{ }\Omega$$

The result is larger than the largest resistance, as expected.

(b)   The equivalent parallel resistance (Eq. 18.3) is

$$\frac{1}{R_p} = \frac{1}{R_1} + \frac{1}{R_2} + \frac{1}{R_3} = \frac{1}{1.0 \text{ }\Omega} + \frac{1}{2.0 \text{ }\Omega} + \frac{1}{3.0 \text{ }\Omega}$$

With a common denominator,

$$\frac{1}{R_p} = \frac{6.0}{6.0 \text{ }\Omega} + \frac{3.0}{6.0 \text{ }\Omega} + \frac{2.0}{6.0 \text{ }\Omega} = \frac{11}{6.0 \text{ }\Omega}$$

or

$$R_p = \frac{6.0 \text{ }\Omega}{11} = 0.55 \text{ }\Omega$$

Note that R_p is found by inverting $1/R_p$ and that it is less than the smallest resistance. If you do not get such a result in parallel (for example, if the result were 1.5 Ω), then you must have made a mistake!

(c) Knowing the equivalent resistance for the series arrangement gives the current:

$$I = \frac{V}{R_s} = \frac{12\text{ V}}{6.0\text{ Ω}} = 2.0\text{ A}$$

Consider the individual voltages. In series, the fractional share of the total voltage is in proportion to the resistance of each resistor. That is, the resistors with the largest resistance get the largest voltage drops. This makes sense since voltage drops cause current. Here the currents are all the same, thus *the larger resistors require the most voltage*. To see this, calculate the voltage drop across each resistor:

$$V_1 = IR_1 = (2.0\text{ A})(1.0\text{ Ω}) = 2.0\text{ V}$$

$$V_2 = IR_2 = (2.0\text{ A})(2.0\text{ Ω}) = 4.0\text{ V}$$

$$V_3 = IR_3 = (2.0\text{ A})(3.0\text{ Ω}) = 6.0\text{ V}$$

The sum of the voltage drops thus equals the battery voltage, as must be the case.
Similarly, for the parallel arrangement, the total current is

$$I = \frac{V}{R_p} = \frac{12\text{ V}}{0.55\text{ Ω}} = 22\text{ A}$$

Note that the current for the parallel combination is much larger than that for the series combination. (Why?) Here the currents through the three resistors are

$$I_1 = \frac{V}{R_1} = \frac{12\text{ V}}{1.0\text{ Ω}} = 12\text{ A}$$

$$I_2 = \frac{V}{R_2} = \frac{12\text{ V}}{2.0\text{ Ω}} = 6.0\text{ A}$$

$$I_3 = \frac{V}{R_3} = \frac{12\text{ V}}{3.0\text{ Ω}} = 4.0\text{ A}$$

The total of these three currents is equal to the current through the battery.
The resistor with the smallest resistance gets most of the current, giving rise to the adage "current takes the path of least resistance." This makes sense because all resistors in parallel have the same voltage. Thus the smallest resistor will have the largest current.

Follow-up Exercise. (a) Calculate the power dissipated in each resistor for each arrangement in this Example. (b) What generalizations can you make? For instance, which resistor gets the most power when the resistors are wired in series? in parallel? (c) Does the series arrangement require more or less power than the parallel case?

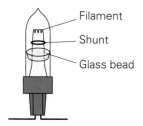

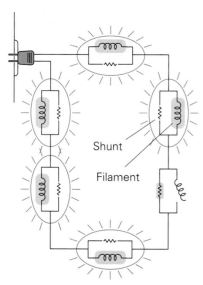

•**FIGURE 18.5 Shunt-wired Christmas tree lights** A shunt, or jumper, in parallel with the bulb filament reestablishes a complete circuit when one of the bulb filaments burns out. Without the shunt, if one bulb burned out, all of the bulbs would go out.

As a wiring application, consider strings of Christmas tree lights. In old strings of lights with large bulbs connected in series, when one bulb blew out, all the others on that string also went out, leaving you to hunt for the faulty bulb. You had a real problem if more than one bulb blew out at the same time! With newer strings having smaller bulbs, one or more bulbs may burn out, but the others remain lit. Does this mean the bulbs are now wired in parallel? No, parallel wiring would make the total equivalent resistance very small, giving a large current in the circuit, which is dangerous.

Instead, an insulated jumper, or "shunt," is wired in parallel with each bulb filament (•Fig. 18.5). When a bulb is in operation, the shunt is insulated from the filament wires and thus does not conduct current. When the filament breaks and the bulb "burns out," there is momentarily no current anywhere in the string. Thus the voltage across the open circuit at the broken filament will be the full 120-volt household voltage. This voltage then causes a sparking that burns off the shunt insulation material. Making contact with the filament wires, the shunt now completes the circuit and the rest of the lights in the string continue to glow. (The shunt, a wire with very little resistance compared to a bulb filament, is indicated by the small resistance symbol in the circuit diagram of Fig. 18.5.) To see the effect (if any) on the string, consider the following Example.

Conceptual Example 18.2

Oh, Tannenbaum! Christmas Tree Lights Revisited

In a string of Christmas tree lights composed of bulbs with jumper shunts, if the filament of one bulb burns out, will the other bulbs (a) glow a little more brightly, (b) glow a little more dimly, or (c) be unaffected?

Reasoning and Answer. If one bulb filament burns out and its shunt completes the circuit, there will be less total resistance in the circuit because the shunt resistance is much less than the filament resistance. (Note that the filaments of the good bulbs and the shunt of the burned-out bulb are in series, so the resistances just add.)

With less total resistance, there will be more current in the circuit and the remaining good bulbs will glow a little brighter since $P = I^2R$, so the answer is (a). For example, suppose a string of lights had 18 identical bulbs. Then the voltage drop across each would be $(120 \text{ V})/18 = 6.7$ V. (The voltage drops across the bulbs must add up to the wall voltage, 120 V.) If one bulb is burned out, the voltage across each lighted bulb would be $(120 \text{ V})/17 = 7.1$ V. This increased voltage causes the current through each bulb to increase. Both increases contribute to a power increase (brighter lights).

Follow-up Exercise. In this Example, by what percentage does the power output of each remaining lighted bulb increase if one bulb burns out? Assume that the bulbs are ohmic.

Series–Parallel Resistor Combinations

Resistors may instead be connected in a circuit in any of a variety of series–parallel combinations. As shown in •Fig. 18.6, circuits with only one voltage source can sometimes be reduced to a single equivalent loop, containing just the voltage source and one equivalent resistance, by applying the series and parallel results.

A procedure for analyzing circuits with different series–parallel combinations of resistors allows us to find the voltages across and the currents in the various resistors. The steps in this procedure are as follows:

1. Starting with the resistor combination farthest from the voltage source, find the equivalent series and parallel resistances.
2. Reduce the circuit until there is a single loop with one total (equivalent) resistance.
3. Find the current delivered to the reduced circuit by using $I = V/R$.
4. Expand the circuit by reversing the reduction steps, and use the current for the reduced circuit to find the currents and voltages for the resistors in each step.

The following Example illustrates this procedure. Analyze the various steps carefully in terms of the series and parallel relationships you have learned previously.

Example 18.3 ■ Series–Parallel Combination of Resistors: Same Voltage or Same Current?

(a) What are the voltage across and the current in each of the resistors R_1 through R_5 in Fig. 18.6a? (b) How much power is dissipated in R_4?

Thinking It Through. We apply the steps described above, calculating the equivalent resistance of each loop of the circuit. Eventually we will have one total (equivalent)

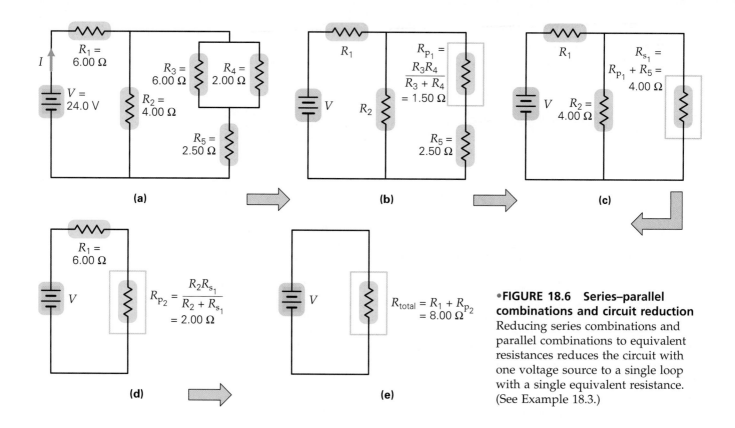

•FIGURE 18.6 Series–parallel combinations and circuit reduction Reducing series combinations and parallel combinations to equivalent resistances reduces the circuit with one voltage source to a single loop with a single equivalent resistance. (See Example 18.3.)

resistance for the whole circuit. From this value we can get the total current and then work backward, expanding to the actual circuit. Each time we encounter a resistor, we calculate the current and voltage for it.

Solution. To verify that the sum of the voltages equals the voltage of the battery, we will calculate to two places after the decimal.

Given: Values in Fig. 18.6a *Find:* (a) V and I for all the resistors
(b) P_4

(a) The parallel combination at the right-hand side of the circuit diagram (farthest from the battery) is first reduced to the equivalent resistance R_{p_1} (see Fig. 18.6b). For convenience we use the two-resistor form for parallel resistors (Eq. 18.3a):

$$R_{p_1} = \frac{R_3 R_4}{R_3 + R_4} = \frac{(6.00\ \Omega)(2.00\ \Omega)}{6.00\ \Omega + 2.00\ \Omega} = 1.50\ \Omega$$

This leaves a series combination along that side, which is reduced to R_{s_1} (Fig. 18.6c):

$$R_{s_1} = R_{p_1} + R_5 = 1.50\ \Omega + 2.50\ \Omega = 4.00\ \Omega$$

Then, R_2 and R_{s_1} are in parallel and can be reduced (again by using Eq. 18.3a for two resistors) to R_{p_2} (Fig. 18.6d):

$$R_{p_2} = \frac{R_2 R_{s_1}}{R_2 + R_{s_1}} = \frac{(4.00\ \Omega)(4.00\ \Omega)}{4.00\ \Omega + 4.00\ \Omega} = 2.00\ \Omega$$

This leaves two resistances in series, which can be combined to give the *total* equivalent resistance (R_{total}) of the circuit (Fig. 18.6e):

$$R_{total} = R_1 + R_{p_2} = 6.00\ \Omega + 2.00\ \Omega = 8.00\ \Omega$$

Then, the battery delivers a current of

$$I = \frac{V}{R_{total}} = \frac{24.0\ \text{V}}{8.00\ \Omega} = 3.00\ \text{A}$$

This is the same as current through R_1 and R_{p_2} since they are in series (Fig. 18.6d). Their voltages are

$$V_1 = IR_1 = (3.00 \text{ A})(6.00 \ \Omega) = 18.0 \text{ V}$$

and

$$V_{p_2} = IR_{p_2} = (3.00 \text{ A})(2.00 \ \Omega) = 6.00 \text{ V}$$

Since R_{p_2} is made up of R_2 and R_{s_1} (Fig. 18.6c), there is a 6.00-volt drop across each of these resistors. Since $R_2 = R_{s_1}$, the current must divide equally:

$$I_2 = 1.50 \text{ A}$$

and

$$I_{s_1} = 1.50 \text{ A}$$

Then I_{s_1} is also the current through R_{p_1} and R_5 in series (Fig. 18.6b). Their voltages are therefore

$$V_{p_1} = I_{s_1} R_{p_1} = (1.50 \text{ A})(1.50 \ \Omega) = 2.25 \text{ V}$$

$$V_5 = I_{s_1} R_5 = (1.50 \text{ A})(2.50 \ \Omega) = 3.75 \text{ V}$$

Note that these do add up to 6.00 V.
Finally, the voltage across R_3 and R_4 is the same as V_{p_1} (why?):

$$V_3 = V_4 = 2.25 \text{ V}$$

The current of 1.50 A (I_{s_1}) divides at the R_3–R_4 junction:

$$I_3 = \frac{V_3}{R_3} = \frac{2.25 \text{ V}}{6.00 \ \Omega} = 0.38 \text{ A}$$

$$I_4 = \frac{V_4}{R_4} = \frac{2.25 \text{ V}}{2.00 \ \Omega} = 1.13 \text{ A}$$

Note that $I_3 + I_4 = I_{s_1}$ within rounding errors.

(b) The power expended in R_4 is

$$P_4 = I_4 V_4 = (1.13 \text{ A})(2.25 \text{ V}) = 2.54 \text{ W}$$

Circuit analysis like that done here may seem involved, but it is simply repeated applications of equations for equivalents of series and parallel resistor combinations, along with the basic relationship $V = IR$.

Follow-up Exercise. In this Example, verify that the total power expended in all of the resistors is the same as the power output of the battery.

18.2 Multiloop Circuits and Kirchhoff's Rules

Objectives: **To (a) understand the physical principles that underlie Kirchhoff's circuit rules and (b) apply these rules in the analysis of actual circuits.**

Simple series–parallel circuits with a single voltage source that can be reduced to a single loop can be analyzed simply, as you learned in Section 18.1. However, in general, circuits may contain several loops, each one having several voltage sources, resistances, or both (collectively referred to as *circuit elements* or *components*). A simple multiloop circuit is shown in •Fig. 18.7a. Some combinations of resistors can be replaced by equivalent resistances (Fig. 18.7b), but this circuit can be reduced only so far by using the procedures illustrated in Section 18.1.

A general method for analyzing multiloop circuits is the application of **Kirchhoff's rules.** These rules embody the conservation of charge and the conservation of energy. (Although they were not stated specifically, Kirchhoff's rules were applied to the simple circuits analyzed in Section 18.1.) Before we present these rules, it is useful to introduce some terminology that will help us describe complex circuits:

Kirchhoff's rules were developed by the German physicist Gustav Kirchhoff (1824–1887).

•FIGURE 18.7 Multiloop circuit

In general, a circuit that contains voltage sources in more than one loop cannot be further reduced by series and parallel reductions. However, some reductions within each loop may be possible, such as from **(a)** to **(b)**. At a circuit junction, where three or more wires come together, the current divides or currents come together, as at junctions A and B in part (b), respectively. The path between two junctions is called a branch. In (b) there are three branches—that is, three different paths between junctions A and B.

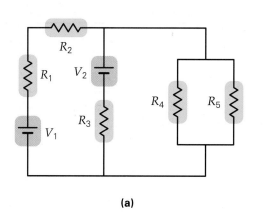

(a)

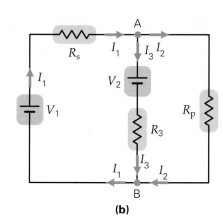

(b)

• A point at which three or more wires are joined is called a **junction** or **node**—for example, point A in Fig. 18.7b. Current either divides or merges at a junction.

• A path connecting two junctions is called a **branch**. A branch may contain one or more circuit elements.

Kirchhoff's Junction Theorem

Kirchhoff's first rule, or **junction theorem**, states that the algebraic sum of the currents at any junction is zero.

$$\sum_i I_i = 0 \qquad \text{\textit{sum of currents at junction}} \qquad (18.4)$$

This rule means that the sum of the currents entering a junction (taken as positive) is equal to the sum of the currents leaving the junction (taken as negative), or that charge is conserved. For the junction at A in Fig. 18.7b, the algebraic sum of the currents is $I_1 - I_2 - I_3 = 0$; equivalently,

$$I_1 = I_2 + I_3$$
$$\textit{current in} = \textit{current out}$$

(This rule was applied in analyzing parallel resistances in Section 18.1.)

You sometimes cannot tell whether a particular current is directed into or out of a junction simply by looking at a multiloop circuit diagram. Generally, you must arbitrarily *assume* current directions at a particular junction. If these assumptions are opposite the actual current direction, a negative current will result. This means that the current is opposite the arbitrarily chosen direction. (Example 18.5 will demonstrate this situation.)

Kirchhoff's Loop Theorem

Kirchhoff's second rule, or **loop theorem**, states that the algebraic sum of the potential differences (voltages) across all of the elements of any *closed* loop is zero:

$$\sum_i V_i = 0 \qquad \text{\textit{sum of voltages around closed loop}} \qquad (18.5)$$

This means that the sum of the voltage rises equals the sum of the voltage drops around a closed loop, which must be true if energy is conserved. (This rule was used in analyzing series resistances in Section 18.1.)

Notice that traversing a circuit loop in different directions will yield either a voltage rise or a voltage drop across each circuit element. Thus it is important to establish a sign convention for voltages. We will use the convention illustrated in •Fig. 18.8. The voltage across a battery is taken to be positive (a voltage rise) if the loop is traversed from the negative terminal toward the positive terminal of the battery, as shown in Fig. 18.8a. The voltage across a battery is taken to be negative if the loop is traversed in the opposite direction, toward the negative terminal. (The assigned branch currents have nothing to do with determining the sign of the voltage across a *battery*. The sign of the voltage depends only on the direction in which we choose to cross the battery.)

The voltage across a resistor is taken to be negative (a drop) if the loop is traversed in the direction of the assigned current in that branch (Fig. 18.8b) and positive if the loop is traversed in the opposite direction. Together, these sign conventions allow you to sum the voltages around a loop either clockwise or counterclockwise. Either way, Eq. 18.5 remains mathematically the same since reversing the direction amounts to multiplying that equation by -1.

A graphical interpretation of Kirchhoff's loop theorem that you may find helpful is presented in Learn by Drawing on pp. 578–579. To see that Kirchhoff's rules are equivalent to our previous series–parallel considerations, examine the following simple circuit analysis.

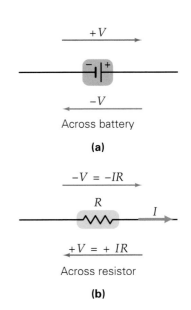

•**FIGURE 18.8 Sign convention for Kirchhoff's rules (a)** When Kirchhoff's rules are applied in going around a circuit loop, the voltage across a battery is taken to be positive (+) if a battery is traversed from the negative to the positive terminal and negative (−) if the battery is traversed from the positive to the negative terminal. **(b)** The voltage across a resistor is taken to be negative (−) if the resistance is traversed in the direction of the assigned branch current and positive (+) if the resistance is traversed in the direction opposite that of the assigned branch current.

Example 18.4 ■ A Simple Circuit: Using Kirchhoff's Rules

Two resistors ($R_1 = 6.00 \ \Omega$ and $R_2 = 3.00 \ \Omega$) are connected in parallel across a 12.0-volt battery. Use Kirchhoff's rules to calculate (a) the current through each resistor and the battery and (b) the equivalent resistance of the pair.

Thinking It Through.   We want to show that Kirchhoff's rules give the same results as do the series and parallel equations. (a) The current divides between the two resistors, so we will use the junction theorem at that point. The loop theorem will enable us to show that both resistors have the same voltage as the battery. (b) The equivalent resistance is then determined from the battery voltage and the total current.

Solution.

Given: $R_1 = 6.00 \ \Omega$ *Find:* (a) current through R_1, R_2, and battery
        $R_2 = 3.00 \ \Omega$ (b) R_p
        $V = 12.0$ V

(a)   For the total current (I) through the battery, which divides between the two resistors, the junction the theorem gives

$$I = I_1 + I_2$$

Using the loop theorem to travel across the battery (from negative to positive terminals) and each resistor in turn (in the direction of the current), we have two equations

$$+V - I_1 R_1 = 0 \quad \text{and} \quad +V - I_2 R_2 = 0$$

From these last two equations we can calculate the currents:

$$I_1 = \frac{V}{R_1} = \frac{12.0 \text{ V}}{6.00 \ \Omega} = 2.00 \text{ A}$$

and

$$I_2 = \frac{V}{R_2} = \frac{12.0 \text{ V}}{3.00 \ \Omega} = 4.00 \text{ A}$$

Thus the current through the battery is $I = I_1 + I_2 = 2.00 \text{ A} + 4.00 \text{ A} = 6.00 \text{ A}$.

(b)   In effect, there is one equivalent resistance hooked across the battery, and it draws 6.00 A when the 12.0 V is applied. Hence the equivalent resistance is

$$R_p = \frac{V}{I} = \frac{12.0 \text{ V}}{6.00 \text{ A}} = 2.00 \ \Omega$$

You should be able to verify this result by using Eq. 18.3.

Follow-up Exercise.   Suppose that the resistors in this Example are connected in series across the battery rather than in parallel. Use Kirchhoff's rules to determine (a) the current through each and (b) the voltage across each.

Kirchhoff Plots: A Graphical Interpretation of Kirchhoff's Loop Theorem

Kirchhoff's loop theorem is usually stated in mathematical terms. However, there is a geometrical interpretation of it that may help you develop better insight into its meaning. This graphical approach allows you to visualize the potential changes in a circuit, either to anticipate the results of mathematical analysis or to confirm the order of magnitude of your results. In relatively simple circuits, this approach can give you much insight into the numerical answers.

The idea is to make a three-dimensional plot out of the circuit wiring diagram. The wires and elements of the circuit form the basis for the x–y plane, or "floor" of the diagram. Plotted perpendicularly to this plane, along the z axis, is the value of the electric potential, with an appropriate choice for zero. Such a diagram is called a *Kirchhoff plot* (Fig. 1).

The rules for constructing a Kirchhoff plot are simple: Start at a known potential value and form a complete loop, finishing at the starting location. The fact that you must come back to the same starting location means that all the potential rises (positive voltages) must be balanced out by the drops (negative voltages). This requirement is the geometrical expression of the conservation of energy, as embodied mathematically by Kirchhoff's loop theorem.

Thus, if the potential increases (say, in traversing a battery from cathode to anode—that is, from the negative terminal to the positive), draw a rise in the z direction. The rise, of course, is just the terminal voltage of the battery. Similarly, if the potential decreases (for example, in traversing a resistor in the direction of the current), draw a drop. If possible, try to make the sizes of the rises and drops (the voltages) to scale. That is, if there is a large rise in potential (such as you would have across a high-voltage battery), then draw that rise large in proportion to the others on the diagram.

You may find it helpful if, before attempting a mathematical analysis of a circuit, you use this graphical method to plan your attack and to estimate the results qualitatively (for example, to reason which resistor might get most of the volt-age). For elaborate circuits, this plotting may be too complicated for practical use. Nevertheless, it is always good to keep this concept in mind, as it illustrates the fundamental idea behind the loop theorem.

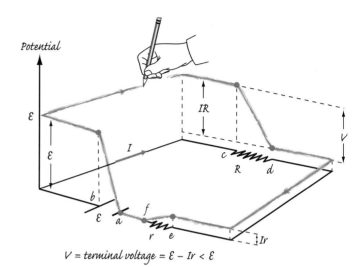

$$V = \text{terminal voltage} = \mathcal{E} - Ir < \mathcal{E}$$

FIGURE 1   Kirchhoff plots: A graphical problem-solving strategy   The schematic of the circuit is laid out in the x–y plane, and the electric potential is plotted perpendicularly along the z axis. Usually, the zero of the potential is taken to be the negative terminal of the battery. A direction for current is assigned, and the value of the potential is plotted around the circuit, following the rules for potential gains and losses. The wires are assumed resistanceless. (How can you tell?) The Kirchhoff plot shows a rise in potential when the battery is traversed from cathode to anode, a drop in potential across the external resistor, and a final (smaller) drop across the internal resistance of the battery.

Application of Kirchhoff's Rules

Example 18.4 is a relatively simple application of Kirchhoff's rules. More complicated, multiloop circuits require a more structured approach. In this book we will use the following general steps in applying Kirchhoff's rules:

1. Assign a current and current direction for each branch in the circuit. This is done most conveniently at junctions.

2. Indicate the loops and the arbitrarily chosen directions in which they are to be traversed (•Fig. 18.9, p. 580). Every branch *must* be in at least one loop.

3. Apply Kirchhoff's first rule to write the equations for the currents, one for each junction that gives a unique equation. (In general, this step gives a set of equations that includes all branch currents, and you may have redundant junctions.)

4. Traverse the number of loops necessary to include all branches. In traversing a loop, apply Kirchhoff's second rule (remember that $V = IR$ for each resistor) and write the equations with the adopted sign convention.

If this procedure is applied properly, steps 3 and 4 give a set of N equations with N unknown currents. These may then be solved for the currents. If more loops

As an example of this method, consider the circuit in Fig. 1: a battery with internal resistance r wired to a single external resistor R. The current is from the anode to the cathode through the external resistor as shown. We start, as usual, by calling the potential of the battery cathode zero. Then, traversing the circuit in the direction of the arrows, we must show a rise in potential going from the cathode to the anode. Next, we show that the potential is constant (flat) as we follow the current through the wires until it meets the external resistor. That is, we indicate no voltage *drop* along connecting wires (why?).

When we reach the resistor, there must be a significant drop in potential. Note, however, that we cannot make the drop across the external resistor equal in magnitude to the rise across the battery because there must be some voltage left to produce a current through the internal resistance. Thus, the terminal voltage of the battery, V, must be less than its emf (the rise between points a and b).

Figure 2 shows a more complicated situation: two resistors in series, and that combination in parallel with a third resistor. All three resistors have the same resistance, R. For simplicity, we neglect the internal resistance of the battery. Starting at point a, there is a rise in potential corresponding to the battery voltage. Then, following the loop through the single resistor, there is a potential drop of equal magnitude across the resistor.

If we follow the loop with the two resistors, we see that each gets only half the total voltage drop. Thus we would expect each to carry only half the current of the single resistor. We know from the formulas developed in Section 18.1 that in parallel, the largest resistance is expected to carry the least current. But notice how this geometrical approach can help develop your intuition and allow you to anticipate numerical results. Here, since all the resistances are equal, we would expect one-third of the total current through the two resistors

and two-thirds through the single resistor. (Can you show this mathematically?)

As an exercise, try redrawing Fig. 2 as it would appear if the two resistors in series had resistances of $0.5R$ and $1.5R$. Can you predict what changes would occur in the circuit? Which resistor now has the largest voltage across it, and how do the currents in the branches compare? Analyze the circuit mathematically to see if your expectations are confirmed.

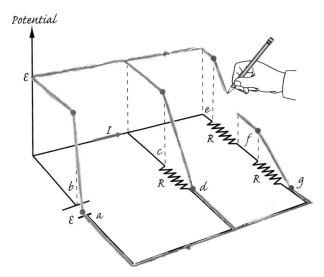

FIGURE 2 Kirchhoff plot of a more complex circuit
Imagine how the plot would change if you varied the values of the three resistors. Then analyze the circuit mathematically to see whether your plots allowed you to anticipate the voltages and currents in the various circuit elements.

are traversed than necessary, you will have redundant equations. Only the number of loops that includes each branch once is needed.

This procedure may seem complicated, but it's generally straightforward, as the following Example shows.

Example 18.5 ■ Branch Currents: Using Kirchhoff's Rules

For the circuit diagrammed in Fig. 18.9, find the branch currents.

Thinking It Through. We cannot use simple series or parallel calculations here (why?). Instead we begin by arbitrarily assigning current directions in each loop, then use the junction theorem and the loop theorem (twice, once each around the two inner loops) to generate three equations, since there are three currents.

Solution.

Given: Values in Fig. 18.9   *Find:* The three branch currents

The branch currents and their directions as well as the loops have been arbitrarily assigned in the diagram. Note that there is a current in every branch and that every branch is in at least one loop. (Some branches are in more than one loop, which is acceptable.)

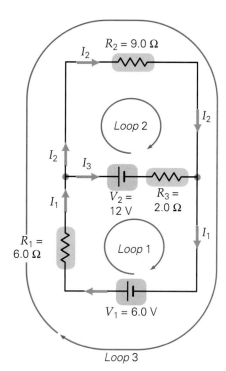

•**FIGURE 18.9 Application of Kirchhoff's rules** To analyze a circuit such as the one shown, assign a current and current direction for each branch in the circuit (most conveniently, at junctions). Identify each loop and the arbitrary direction of traversal. Then write equations for each independent junction (using Kirchhoff's first rule) and for as many loops as needed to include every branch (using Kirchhoff's second rule). Be careful to observe sign conventions. See Example 18.5.

Applying Kirchhoff's first rule at the left-hand junction gives

$$I_1 = I_2 + I_3 \tag{1}$$

For the other junction, we could write $I_2 + I_3 = I_1$ (currents in = currents out), but this equation is equivalent to Eq. 1. (If there were another branch, it would have a different current equation for its junctions.)

Going around loop 1 as in Fig. 18.9 and applying Kirchhoff's second rule with the sign convention gives

$$+V_1 - I_1R_1 - V_2 - I_3R_3 = 0$$

Then, putting in the numerical values from the figures gives

$$+6 - I_1(6) - 12 - I_3(2) = 0$$

or, on rearranging,

$$6I_1 + 2I_3 = -6 \quad \text{or} \quad 3I_1 + I_3 = -3 \tag{2}$$

For convenience, units are omitted and resistances are written to one significant figure.

Similarly, for loop 2,

$$+V_2 - I_2R_2 + I_3R_3 = 0$$

and

$$12 - I_2(9) + I_3(2) = 0$$

Thus,

$$9I_2 - 2I_3 = 12 \tag{3}$$

Equations 1, 2, and 3 form a set of three equations with three unknowns. Physically, the problem is solved; only the math is left to do. You can solve for the I's in several ways. For example, first substitute Eq. 1 into Eq. 2 to eliminate I_1:

$$3(I_2 + I_3) + I_3 = -3$$

This simplifies to

$$3I_2 + 4I_3 = -3 \quad \text{or} \quad I_2 = -1 - \tfrac{4}{3}I_3 \tag{4}$$

Then, substituting Eq. 4 into Eq. 3 eliminates I_2:

$$9(-1 - \tfrac{4}{3}I_3) - 2I_3 = 12$$

That is,

$$-14I_3 = 21 \quad \text{or} \quad I_3 = -1.5 \text{ A}$$

The minus sign on the result tells you that the wrong direction was assumed for I_3.

Putting the value of I_3 into Eq. 4 gives I_2:

$$I_2 = -1 - \tfrac{4}{3}(-1.5 \text{ A}) = 1.0 \text{ A}$$

Then, by Eq. 1,

$$I_1 = I_2 + I_3 = 1.0 \text{ A} - 1.5 \text{ A} = -0.5 \text{ A}$$

The minus sign indicates that I_1 was also assigned the opposite of its true direction. So, at the left-hand junction in Fig. 18.9, I_3 should go into the junction and I_1 and I_2 go out. (You might have suspected this because of the larger 12-volt battery in the I_3 branch.)

Note that loop 3 was not used in this analysis. The equation for this loop would be redundant, giving four equations and three unknowns. This application of Kirchhoff's rules is called the *branch current method*.

Follow-up Exercise. Rework this Example, using the junction theorem and loops 3 and 1 instead of loops 1 and 2.

18.3 RC Circuits

Objectives: To (a) understand the charging and discharging of a capacitor through a resistor, and (b) calculate the current and voltage at specific times during these processes.

The previous sections dealt with circuits having constant currents. In some direct current (dc) circuits, the current can vary with time while maintaining a constant direction (thus still "dc"). This is the case with **RC circuits**, which have a resistor (R), or other resistive components, and a capacitor (C) in series.

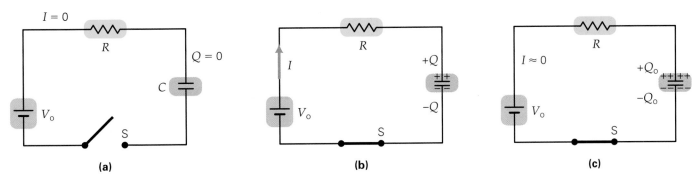

(a)　　　　　　　**(b)**　　　　　　　**(c)**

•**FIGURE 18.10 A series RC circuit (a)** Even though the circuit is open because of the space between the plates of the capacitor, **(b)** there is a current in the circuit when the switch is closed until the capacitor is charged to its maximum value. The rate of charging (and discharging) depends on the product of the values of the resistance and capacitance, which is called the time constant (τ) of the circuit: $\tau = RC$. **(c)** For times much larger than RC, the current is essentially zero and the capacitor is essentially fully charged.

Charging a Capacitor through a Resistor

The charging of an initially uncharged capacitor is depicted in •Fig. 18.10. You may notice that even after the switch is closed, the circuit is still open, because of the separation (air gap) between the plates of the capacitor. Nevertheless, when the switch is closed, charge does flow for a short time while the capacitor is charging.

The maximum charge (Q_o) built up on the capacitor depends on the capacitance (C) and the voltage of the battery (V_o). Recall from Chapter 16 that this charge is $Q_o = CV_o$. Immediately after the switch is closed, at $t = 0$, there is no charge on the capacitor and thus no voltage across it. Thus the full battery voltage appears across the resistor, resulting in an initial current $I_o = V_o/R$ through the resistor.

Recall that as charge increases on the capacitor's plates, so does the voltage across the plates. It takes more work to place more charge on the plates because of the repulsion between like charges. Eventually, the capacitor is charged to the maximum, and the current diminishes to zero as the resistor voltage drops to zero.

The resistance helps determine how fast the capacitor is charged, since the larger its value, the greater the resistance to charge flow. The capacitance also influences the speed of charging—it takes longer to charge a larger capacitor. The voltage across the capacitor increases exponentially with time according to the equation

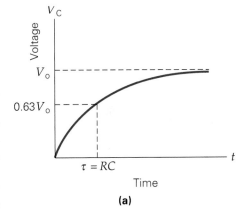

$$V_C = V_o(1 - e^{-t/RC}) \quad \begin{array}{l} \textit{charging capacitor voltage} \\ \textit{in an RC circuit} \end{array} \quad (18.6)$$

where e has an approximate value of 2.718. (The irrational number e is the base of the system of *natural logarithms*.) A graph of V_C versus t is presented in •Fig. 18.11. As expected, V_C approaches V_o at very large times.

A graph of I versus t is given in Fig. 18.11b. The current varies with time according to the equation

$$I = I_o e^{-t/RC} \quad (18.7)$$

The current decreases exponentially with time; it is greatest initially. (Why?)

According to Eq. 18.6, it would theoretically take an infinite amount of time for the capacitor in an RC circuit to become fully charged (to reach the battery voltage, V_o). However, in practice, capacitors become almost completely charged in relatively short times. It is customary to use a special value to express the charging time. This value, called the **time constant** (τ) for an RC circuit, is

$$\tau = RC \quad (18.8)$$

After an elapsed time equal to one time constant, $t = \tau = RC$, the voltage across the charging capacitor has risen to

$$V_C = V_o(1 - e^{-t/\tau}) = V_o(1 - e^{-\tau/\tau}) = V_o(1 - e^{-1})$$

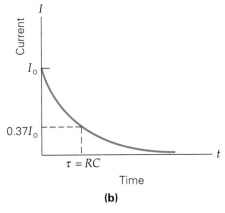

•**FIGURE 18.11 Capacitor charging (a)** In a series RC circuit, the voltage across the capacitor increases nonlinearly with time, reaching 63% of its maximum voltage (V_o) in one time constant, $t = \tau = RC$. **(b)** The current in the circuit is initially a maximum ($I_o = V_o/R$) and decays exponentially with time, falling to 37% of its initial value in one time constant.

Note: Most calculators can raise any number (x) to any power (y) by using the x^y button. For exponential calculations, practice raising e (≈ 2.718) to various negative powers. Use your calculator to check that $e^{-1} \approx 0.37$.

Since $e^{-1} \approx 0.37$, we have

$$V_C \approx 0.63 V_o$$

That is, after one time constant, the voltage across the capacitor has risen to 63% of its maximum value of V_o (or the capacitor is 63% charged, since $Q = CV_C$). Note that at that time the current has decayed to 37% of its initial maximum value (I_o).

At the end of two time constants, $t = 2\tau = 2RC$, the capacitor is charged to more than 86% of its maximum value; at $t = 3\tau = 3RC$, to 95%, and so on. Thus, the capacitor can be considered fully charged after only a few time constants.

Discharging a Capacitor through a Resistor

When a fully charged capacitor is *discharged* through a resistor, the voltage across the capacitor "decays" (decreases) exponentially with time, as does the current:

$$V_C = V_o e^{-t/RC} = V_o e^{-t/\tau} \qquad \begin{array}{l}\textit{discharging capacitor voltage} \\ \textit{in an RC circuit}\end{array} \qquad (18.9)$$

For example, in one time constant, the voltage across the capacitor falls to 37% of its original value (•Fig. 18.12). The current in the circuit decays exponentially, following Eq. 18.7.

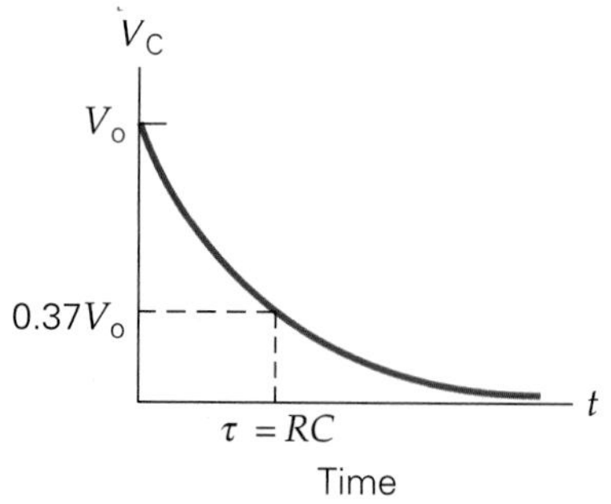

•**FIGURE 18.12 Capacitor discharging** When a charged capacitor is discharged through a resistance, the voltage across the capacitor (and the current in the circuit) decays exponentially with time, falling to 37% of its initial value in one time constant, $t = \tau = RC$.

Example 18.6 ■ Charging a Capacitor: Time Constant

The capacitance and resistance in the RC circuit in Fig. 18.10 are 6.0 μF and 0.25 MΩ, respectively, and the battery has a 12-volt terminal voltage. (a) The capacitor was initially uncharged. What is the voltage across it one time constant after the switch is closed? (b) What are the voltage across the capacitor and the capacitor's charge at $t = 5.0$ s?

Thinking It Through. After one time constant, the capacitor is charged to 63% of its maximum charge, and thus the voltage will have risen to about 63% of its maximum value of 12 V. To find the voltage after a certain time has elapsed, we can use the expression for exponential voltage rise, Eq. 18.6, after calculating the time constant from Eq. 18.8.

Solution. We list the data and apply the charging capacitor expressions:

Given: $C = 6.0\ \mu\text{F} = 6.0 \times 10^{-6}$ F   *Find:* (a) V_C at $t = \tau$ (capacitor voltage)
$R = 0.25\ \text{M}\Omega = 2.5 \times 10^5\ \Omega$ (b) V_C and Q at $t = 5.0$ s
$V_o = 12$ V   (capacitor voltage and charge)

(a)   In one time constant,

$$V_C = 0.63 V_o = 0.63(12\ \text{V}) = 7.6\ \text{V}$$

(b)   When a specific time is given, we need to know the time constant for the circuit:

$$\tau = RC = (2.5 \times 10^5\ \Omega)(6.0 \times 10^{-6}\ \text{F}) = 1.5\ \text{s}$$

Then, for $t = 5.0$ s, using Eq. 18.6 we get

$$V_C = V_o(1 - e^{-t/\tau}) = (12\ \text{V})[1 - e^{(-5.0\,\text{s})/(1.5\,\text{s})}] = (12\ \text{V})(1 - 0.036) = (12\ \text{V})0.964 = 11.6\ \text{V}$$

(Significant figure rules would round this to 12 V, but we carry three figures here to make the point that the voltage is not exactly 12 V.)

Since the charge on a capacitor is proportional to the capacitor's voltage, we have

$$Q = CV_C = (6.0 \times 10^{-6}\ \text{F})(11.6\ \text{V}) = 7.0 \times 10^{-5}\ \text{C}$$

The maximum charge on the capacitor is 7.2×10^{-5} C. After 5.0 s $\approx 3.3\tau$, the capacitor's charge is above 95% of the maximum value. Essentially, it is fully charged.

Follow-up Exercise.   In this Example, once the capacitor is fully charged, it is then discharged through a resistor with a resistance of 1.0 MΩ. How long does it take for half of its charge to leave?

An application of an RC circuit is diagrammed in •Fig. 18.13a. This circuit is called a blinker circuit (or, more impressively, a neon-tube relaxation oscillator). The resistor and capacitor are initially wired in series, and then a small neon tube is connected in parallel with the capacitor. (These neon tubes are about the size of miniature Christmas tree lights.)

When the circuit is closed, the voltage across the capacitor (and the neon tube) rises from 0 to V_b, which is the breakdown voltage of the neon gas in the tube (about 80 V). At that voltage, the gas is ionized and begins to conduct electricity, and the tube lights up. When the tube is in a conducting state, the capacitor discharges through it, and the voltage falls rapidly (Figure 18.13b). When the voltage drops below V_m, called the maintaining voltage, the tube discharge can no longer be sustained, and the tube stops conducting. The capacitor then begins charging again, the voltage rises from V_m to V_b, and the cycle repeats. The continual repetition of this cycle causes the tube to blink on and off. The period of the oscillator, or the time between flashes, depends on the RC time constant.

Other applications of RC circuits are discussed in the Insight on p. 584.

18.4 Ammeters and Voltmeters

Objectives: **To understand (a)** how galvanometers are used as ammeters and voltmeters, **(b)** how multirange versions of these devices are constructed, and **(c)** how they are connected to measure current and voltage in real circuits.

As the names imply, an **ammeter** measures current *through* circuit elements (in amps) and a **voltmeter** measures voltages *across* circuit elements (in volts). A basic component of both of these types of meters is a **galvanometer** (•Fig. 18.14). The galvanometer operates on magnetic principles that will be covered in Chapter 19. In this chapter, it will simply be treated as a circuit element that has an internal resistance (r) and whose needle deflection is directly proportional to the current through it.

Ammeters

The resistance of a galvanometer coil (r) is relatively small, since the coil consists of metal wire. In effect, a galvanometer measures current, but because of the small coil resistance, only currents in the microamp range can be measured without

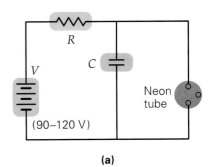

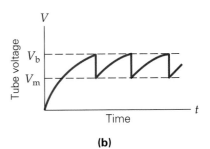

(a)

(b)

•**FIGURE 18.13   Blinker circuit (a)** When a neon tube is connected across the capacitor in a series RC circuit having the proper voltage source, the voltage across the tube (and capacitor) will relax, or oscillate, with time. As result, the tube periodically flashes or blinks. **(b)** A graph of voltage versus time shows the oscillating effect between V_b, the breakdown voltage, and V_m, the maintaining voltage.

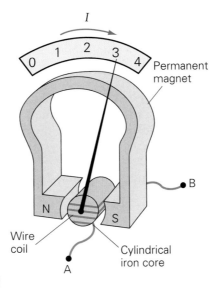

(a)

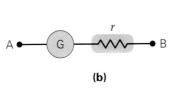

(b)

•**FIGURE 18.14   The galvanometer   (a)** A galvanometer is a current-sensitive device whose needle deflection is proportional to the current through its coil. **(b)** The circuit symbol for a galvanometer is a circle containing a G. The internal resistance (r) of the meter is indicated explicitly.

An RC Circuit in Action

To install wall shelving or to hang a picture securely, it is usually necessary to locate the studs in the walls. Studs are the vertical wooden supports to which plasterboard or paneling is nailed, and they can be very elusive. The traditional way of finding them, by tapping on the wall, generally leads to more frustration than success. Commercially available stud finders employ pivoted magnets, which move when they pass over a nail head in the stud behind the wall, but these too are far from infallible.

An electronic device called Studsensor™ is much easier to use and relies on capacitive sensing. The capacitor forms part of a simple RC charging circuit. However, for this application the metallic plates of the capacitor are arranged end to end rather than facing one another as in a parallel-plate capacitor (Fig. 1). In this configuration the fringe electric field extends a considerable distance beyond the plane of the plates, and any nearby material acts as a dielectric in this field.

The unit is initially calibrated by placing it against the wall and balancing the charging time of the RC circuit against a reference standard. When the sensor containing the capacitor plates is moved over a stud, the character of the dielectric changes, and the capacitance of the plates increases. The greater the capacitance, the longer the charging time of the circuit relative to the calibration standard. The changes in charging time are sensed electronically and conveyed to the user by lighted indicators (Fig. 2).

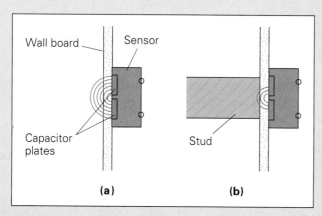

FIGURE 1 RC sensing (a) The fringe electric field of the capacitor plates in a Studsensor permeates the wall board, which acts as a dielectric. (b) When the sensor is over a stud, the capacitance changes. The sensor detects this change as a change in an RC circuit.

FIGURE 2 Studsensor™ The lower green light indicates that the unit is calibrated. Vertical red lights signal the approach to a stud, and the top red light indicates that the sensor is directly over a stud. Some models have both light and sound indicators.

burning out the coil. An ammeter that can be used to measure larger currents therefore must have a small *shunt resistor* (with resistance R_s) in parallel with a galvanometer to take most of the current (•Fig. 18.15). The shunt provides an alternate path by which part of a large current (I) can bypass the galvanometer. The shunt and the galvanometer resistance are in parallel; thus, to carry most of the current the shunt must have the smaller resistance of the two. Because the voltages across the galvanometer and the shunt resistor are equal, we can write

$$V_g = V_s \quad \text{or} \quad I_g r = I_s R_s$$

Using Kirchhoff's junction rule, we see that $I = I_g + I_s$, so the previous equation can be rewritten as

$$I_g r = (I - I_g)R_s$$

Solving for I_g, we get

$$I_g = \frac{I R_s}{r + R_s} \qquad \begin{array}{l} \textit{galvanometer current} \\ \textit{for dc ammeter} \end{array} \qquad (18.10)$$

This equation allows you to select the proper shunt resistance for a given current range and galvanometer. The following Example illustrates how to do this.

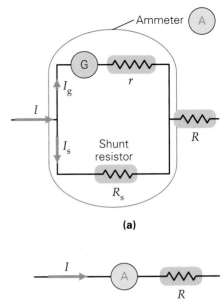

Example 18.7 ■ Designing an Ammeter: Choosing a Shunt Resistor

A galvanometer that can safely carry a maximum coil current of 200 μA (called the full-scale sensitivity) has a coil resistance of 50 Ω. It is to be used in an ammeter designed to read currents up to 3.0 A (at full scale). What is the required shunt resistance?

Thinking It Through. Since the galvanometer can carry only a very small current, most of the current will have to be diverted, or "shunted," through the shunt resistor. We can apply Eq. 18.10, solve for the shunt resistance, R_s. We expect a value much smaller than the coil resistance of the galvanometer (why?).

Solution. Listing the data, we have

Given: $I_g = 200 \ \mu A = 2.00 \times 10^{-4}$ A   *Find:* R_s (shunt resistance)
$r = 50 \ \Omega$
$I_{max} = 3.0$ A

A full-scale reading of 3.0 A means that when a current of 3.0 A enters the ammeter, I_g should be 200 μA. From Eq. 18.10 we get

$$R_s = \frac{I_g r}{I_{max} - I_g}$$

$$= \frac{I_g r}{I_{max} - I_g} = \frac{(2.0 \times 10^{-4} \text{ A})(50 \ \Omega)}{3.0 \text{ A} - (2.00 \times 10^{-4} \text{ A})} = 3.3 \times 10^{-3} \ \Omega$$

Note the small size of the shunt resistance compared to the coil resistance $r = 50 \ \Omega$, as expected. This disparity allows most of the current (2.9998 A at full scale) to pass through the shunt resistor branch. The shunt resistor is made of a material that does not burn out as readily as the thin wire of the galvanometer coil. The ammeter will read currents linearly up to 3.0 A. For example, if a current of 1.5 A flowed into the ammeter, there would be a current of 100 μA in the coil of the galvanometer, which would give a half-scale reading.

Follow-up Exercise. In this Example, if we had used a stunt resistance of 1.0 mΩ, what would be the full-scale reading of the ammeter?

•FIGURE 18.15   A dc ammeter
Here R is the resistance of the resistor whose current is being measured. **(a)** A galvanometer in parallel with a shunt resistor (R_s) is an ammeter capable of measuring various current ranges, depending on the value of R_s. **(b)** The circuit symbol for an ammeter is a circle with an A inside it.

Voltmeters

A voltmeter that is capable of reading voltages higher than the microvolt range is constructed by connecting a large *multiplier resistor* in series with a galvanometer (•Fig. 18.16). Because the voltmeter has a large internal resistance due to the multiplier resistor, it draws little current from the main circuit. When connected *across* a circuit element, the voltmeter (or galvanometer branch) experiences a voltage drop of $V = V_g + V_m$, and most of this drop is across the multiplier resistor rather than the galvanometer coil. By Kirchhoff's loop theorem,

$$V = V_g + V_m$$

$$= I_g r + I_g R_m = I_g(r + R_m)$$

and

$$I_g = \frac{V}{r + R_m} \quad \begin{array}{l} \text{galvanometer current} \\ \text{for dc voltmeter} \end{array} \quad (18.11)$$

Note: Ammeters are galvanometers with *small resistors in parallel.* Voltmeters are galvanometers with *large resistors in series.*

The voltage V is also the potential difference across the circuit element having a resistance R because of the parallel connection (see Fig. 18.16b). If the galvanometer scale is calibrated in volts (instead of amps), the voltage drop across that circuit element can be read from the meter.

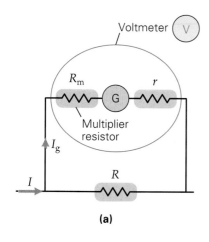

Voltmeter V

R_m

G

r

Multiplier
resistor

I_g

I

R

(a)

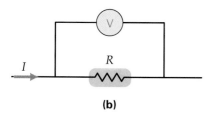

V

I

R

(b)

•**FIGURE 18.16** **A dc voltmeter**
Here R is the resistance of the
resistor whose voltage is being
measured. **(a)** A galvanometer in
series with a multiplier resistor (R_m)
is a voltmeter capable of measuring
various voltage ranges, depending
on the value of R_m. **(b)** The circuit
symbol for a voltmeter is a circle
with a V inside it.

Example 18.8 ■ Designing a Voltmeter: Choosing a Multiplier Resistor

Suppose that the galvanometer in Example 18.7 is to be used instead in a voltmeter with
a full-scale reading of 3.0 V. What is the required multiplier resistance?

Thinking It Through. To turn a galvanometer into a voltmeter, most of the current
should not pass through it; a reduction of current is accomplished by adding a multiplier resistor in series. All the data necessary to calculate the multiplier resistance from
Eq. 18.11 are given.

Solution. First, we list the data given:

Given: $I_g = 200\ \mu A = 2.00 \times 10^{-4}\ A$ *Find:* R_m (multiplier resistance)
        (from Example 18.7)
        $r = 50\ \Omega$ (from Example 18.7)
        $V = 3.0\ V$

From Eq. 18.11, we can find R_m:

$$R_m = \frac{V - I_g r}{I_g}$$

$$= \frac{3.0\ V - (2.00 \times 10^{-4}\ A)(50\ \Omega)}{2.00 \times 10^{-4}\ A} = 1.5 \times 10^4\ \Omega = 15\ k\Omega$$

Follow-up Exercise. The voltmeter in this Example is used to measure the voltage
of a resistor in a circuit. A current of 3.00 A flows through the resistor (1.00 Ω) *before*
the voltmeter is connected. Assuming the total *incoming* current (*I* in Fig. 18.16b)
remains the same after the voltmeter is connected, calculate the current in the
galvanometer.

For versatility, ammeters and voltmeters may be multiranged. This is accomplished by providing the user with a choice of several shunt or multiplier resistors (•Fig. 18.17). Such meters can also be combined into *multimeters*, which
measure voltage, current, and often resistance as well. Electronic digital multimeters are now common (Fig. 18.17c). In place of mechanical galvanometers, these
devices use electronic circuits that analyze digital signals to calculate voltages,
currents, and resistances.

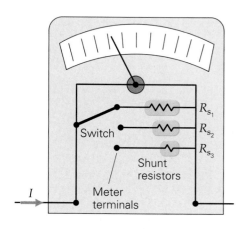
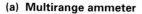

(a) Multirange ammeter

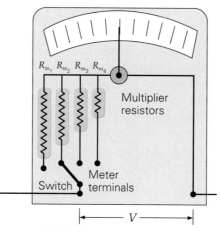

(b) Multirange voltmeter

(c)

•**FIGURE 18.17 Multirange meters** **(a)** An ammeter or **(b)** a voltmeter can be used to measure several ranges of current
and voltage by switching among different shunt and multiplier resistors, respectively. (Instead of a switch, there may be an
exterior terminal for each range.) **(c)** Both functions can be combined in a single multimeter. This multimeter has an electronic
digital readout.

18.5 Household Circuits and Electrical Safety

Objectives: To understand (a) how household circuits are wired, and (b) the underlying principles that govern electrical safety devices.

Although household circuits generally use alternating current, which has not yet been discussed, they include practical applications of some of the principles already studied.

For example, would you expect the elements in a household circuit (lamps, appliances, and so on) to be connected in series or in parallel? From the discussion of Christmas tree lights in Section 18.1, it should be apparent that household elements must be connected in parallel. For example, when the bulb in a lamp blows out, other elements in the circuit continue to work. This would not be the case for a series circuit. Moreover, household appliances and lamps are generally rated for 120 V. If these elements were connected in series, the voltage across them would add up to a *total* of 120 V. None of the individual circuit elements would have their required 120 V.

Power is supplied to a house by a three-wire system (•Fig. 18.18). There is a potential difference of 240 V between the two hot, or high-potential, wires, and each of these has a 120-volt potential difference with respect to the ground. The third wire is grounded at the point where the wires enter the house, usually by a metal rod driven into the ground. This wire is defined as the zero potential and is called the *ground*, or *neutral, wire.*

The potential difference of 120 V needed for most household appliances is obtained by connecting them between the ground wire and either high-potential wire: $\Delta V = 120\,V - 0\,V = 120\,V$ or $\Delta V = 0\,V - (-120\,V) = 120\,V$ (see Fig. 18.18). Even

Note: Household voltage can fluctuate, under normal conditions, between 110 and 130 V. Similarly, 240-volt connections can be as low as 220 V and still be considered normal.

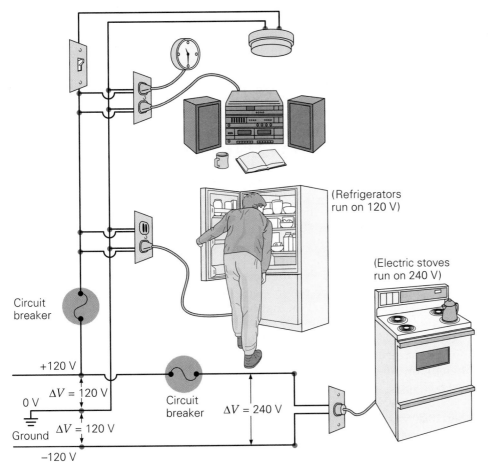

(Refrigerators run on 120 V)

(Electric stoves run on 240 V)

•**FIGURE 18.18   Household wiring** A 120-volt circuit is obtained by connecting between the +120-volt (or −120-volt) line and the ground line. A potential difference of 240 V for large appliances such as electric stoves, central air conditioners, and hot water heaters is obtained by connecting the +120-volt line and the −120-volt line.

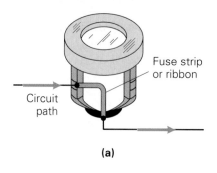

(a)

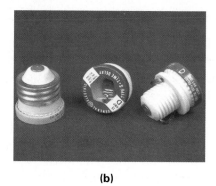

(b)

(c)

•**FIGURE 18.19 Fuses** **(a)** A fuse contains a metallic strip, or ribbon, that melts when the current exceeds a rated value. This opens the circuit and prevents overheating. **(b)** Edison-base fuses (left) have threads similar to those on light bulbs; fuses with different ampere ratings can be interchanged. Type-S fuses (right) have different threads for different ratings that fit specific adapters in the fuse sockets and so cannot be interchanged. **(c)** A type of small fuse commonly found in electrical equipment. The fuse on the right is "blown."

though the ground wire has zero potential, it is a *current-carrying* wire because it is part of the complete circuit. Large appliances such as central air conditioners, ovens, and hot water heaters need 240 V for operation, and this voltage is obtained by connecting them between the two hot wires: $\Delta V = 120 \text{ V} - (-120 \text{ V}) = 240 \text{ V}$.

Although the current through an appliance may be given on a rating tag, it can also be determined from the power rating (using $P = IV$). For example, a stereo rated at 180 W at 120 V would draw an average current of 1.5 A ($I = P/V$).

There are limitations to the number of elements that can be put in a circuit and the total current in that circuit. In particular, the joule heat (or I^2R loss) in the wires must be considered. Generally, the more elements (appliances, and therefore resistances) connected in parallel, the smaller the equivalent resistance of the whole circuit. Thus adding more elements increases the total current flow. By adding too many current-carrying elements, it is possible to overload a household circuit such that it draws too much current and produces too much heat *in the wires*. This heat in the circuit wires could melt the insulation and start a fire.

Overloading is prevented by limiting the current in a circuit by means of two types of devices: fuses and circuit breakers. **Fuses** (•Fig. 18.19) are common in older homes. An Edison-base fuse has threads like those on the base of a light bulb (see Fig. 18.19b). Inside the fuse is a metal strip that melts because of joule heat when the current is larger than the rated value (which is typically 15 A for a 120-volt circuit). The melting of the strip breaks the complete circuit, and the current drops to zero. This "broken" circuit is commonly called an *open circuit*—one with an effectively infinite resistance and zero current.

There are several problems associated with fuses. One problem with Edison-base fuses in particular is that they are interchangeable. For example, a 30-amp fuse can unintentionally be put in a circuit that is rated for 15 A. This would allow a current twice as great as that for which the circuit was designed. Such a problem is avoided with another type of fuse, called a Type-S fuse (Fig. 18.19b). A nonremovable adapter is installed in the socket for a fuse; the adapter is specific for a particular rated value (the fuses and adapters for different ratings have different threads). That is, a 30-amp Type-S fuse cannot be screwed into a 15-amp Type-S adapter, for example, so the wrong fuse can't be used.

Circuit breakers are now used exclusively in wiring new homes. One type (•Fig. 18.20) uses a bimetallic strip (see Chapter 10). As the current through the strip increases, it becomes warmer and bends. At the rated current value, the strip will be bent sufficiently to open the circuit mechanically. The strip quickly cools, so the breaker can be reset. However, a blown fuse or a tripped circuit breaker indicates

Open circuit: Current is zero.

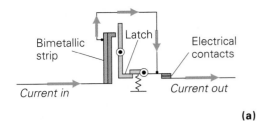

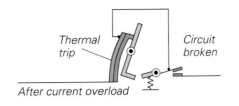

Bimetallic strip — Latch — Electrical contacts

Current in — Current out

(a)

Thermal trip — Circuit broken

After current overload

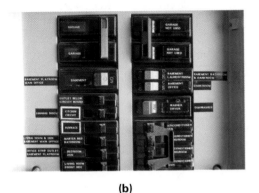

(b)

(c)

•**FIGURE 18.20 Circuit breaker**
(a) A diagram of a thermal trip element. With increased current and joule heating, the element bends until it opens the circuit at some preset current value. Magnetic trip elements are also used. **(b)** A bank of typical household circuit breakers. **(c)** These heavy-duty circuit breakers at an electrical substation must be capable of cutting off current almost instantaneously in power lines operating at very high voltages.

that the circuit is drawing or attempting to draw too much current. *Always find and correct the problem before replacing the fuse or resetting the circuit breaker.*

Switches, fuses, and circuit breakers are placed in the hot (high-potential) side of the line. They would work in the grounded side as a heating overload sensor—but even when the switch was open, the fuse blown, or the breaker tripped, the circuit and any elements in it would still be connected to a high potential, which could be dangerous if a person made electrical contact (•Fig. 18.21a). Even with fuses or circuit breakers in the hot side of the line, there is still a possibility of getting an electrical shock from a defective appliance that has a metal casing, such as a hand drill. If a wire comes loose inside, it could make contact with the casing, which would then be "hot," or at a high potential (Fig. 18.21b). A person could provide a path to the ground and become part of the circuit, thus receiving a shock. For a discussion of the effects of electric shock, see the Insight on p. 590.

To prevent a shock, a third, dedicated grounding wire is added to the circuit that grounds the metal casing (•Fig. 18.22). This wire provides a very low resistance path, bypassing the tool. Such a bypass route is a type of *short circuit*, analogous to taking a shortcut on a trip. This wire does not normally carry current. If a hot wire comes in contact with the casing, the circuit is completed to this grounded or shorting wire. The fuse is blown or the circuit breaker tripped as most of the current is in the third wire to the ground (a low-resistance path) rather than in you (a high-resistance path).

On three-prong **grounded plugs**, the large round prong connects with the dedicated grounding wire. Adapters can be used between a three-prong plug and a two-prong socket. Such an adapter has a grounding lug or grounding

Short circuit: Resistance is zero, current is large.

•**FIGURE 18.21 Electrical safety**
(a) Switches and fuses or circuit breakers should always be wired in the hot side of the line, *not* in the grounded side as shown here. If these elements are wired in the grounded side, the line is at a high potential even when the fuse is blown or a switch is open. **(b)** Even if the fuse or circuit breaker is in the hot side, a potentially dangerous situation exists. If an internal wire comes in contact with the metal casing of an appliance or power tool, a person touching the casing, which is at high voltage, can get a shock.

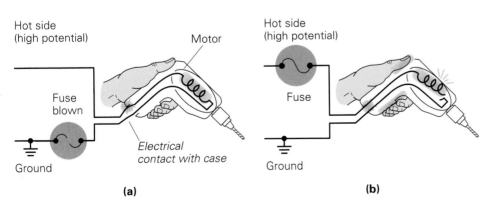

Hot side (high potential) — Motor

Fuse blown

Ground

Electrical contact with case

(a)

Hot side (high potential)

Fuse

Ground

(b)

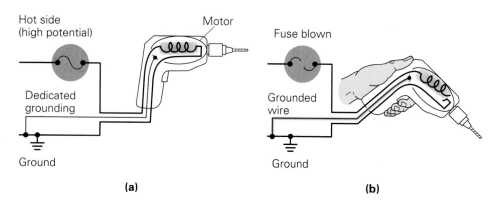

Hot side
(high potential)

Motor

Dedicated
grounding

Ground

(a)

Fuse blown

Grounded
wire

Ground

(b)

•**FIGURE 18.22   Dedicated grounding** **(a)** For safety, a third wire is connected from an appliance or power tool to ground. This dedicated grounding wire normally carries no current (as opposed to the grounded wire of the circuit). **(b)** If a loose internal wire comes in contact with the grounded casing, the shorting to ground through the dedicated grounding wire blows the protective fuse or circuit breaker, and the appliance is not at a high potential as in Fig. 18.21b.

Insight

Electricity and Personal Safety

Safety precautions are necessary to prevent injuries when people work with and use electricity. Electrical conductors (such as wires) are coated with insulating materials so they can be handled safely. However, when a person comes in contact with a charged conductor, a potential difference may exist across part of the body. A bird can sit on a high-voltage line without any problem because the bird's two feet are at essentially the same potential, and thus there is no potential difference to generate a current through the bird. But if a person carrying an aluminum (conducting) ladder touches it to an electrical line, a potential difference exists from the line to the ground, and the ladder and the person are then part of a current-carrying circuit.

The extent of personal injury in such a case depends on the amount of electric current through the body and on the exact circuit path through the body. The current is given by

$$I = \frac{V}{R_{body}}$$

where R_{body} is the resistance of the body. Thus, for a given voltage, the current depends on the body resistance.

Body resistance varies. If the skin is dry, the resistance can be 0.50 MΩ (0.50×10^6 Ω) or more. For a potential difference of 120 V, there would be a current of

$$I = \frac{V}{R_{body}} = \frac{120\ \text{V}}{0.50 \times 10^6\ \Omega} = 0.24 \times 10^{-3}\ \text{A} = 0.24\ \text{mA}$$

This current is almost too weak to be felt.

Suppose, however, that the skin is wet with perspiration. Then R_{body} is only about 5.0 kΩ (5.0×10^3 Ω), and the current is

$$I = \frac{V}{R_{body}} = \frac{120\ \text{V}}{5.0 \times 10^3\ \Omega} = 24 \times 10^{-3}\ \text{A} = 24\ \text{mA}$$

which could be very dangerous.

A basic precaution is to avoid contact with an electrical conductor that might cause a potential difference to exist across the body or part of it. The physical damage resulting from such contact depends on the path of the current. If that path is from the little finger to the thumb on one hand, probably only a burn would result from a large current. However, if the path is from hand to hand through the chest, the effect can be much worse. Some of the possible effects of this latter circuit path are given in Table 1.

Injury results because the current interferes with muscle functions and/or causes burns. Muscle functions are regulated by electrical impulses through the nerves, and these impulses can be influenced by external currents. Muscle reaction and pain can occur from a current of a few milliamps. At about 10 mA, muscle paralysis can prevent a person from releasing the conductor. At about 20 mA, contraction of the chest muscles occurs, which can cause breathing to be impaired or to stop. Death can occur in a few minutes. At 100 mA, there are rapid uncoordinated movements of the heart muscles (called ventricular fibrillation), which prevent the proper pumping action.

Working safely with electricity requires a knowledge of fundamental electrical principles *and* common sense. Electricity must be treated with respect.

TABLE 1   Effects of Electric Current on the Human Body*

Current (approximate)	Effect
1.0 mA (0.001 A)	Mild shock or heating
10 mA (0.01 A)	Paralysis of motor muscles
20 mA (0.02 A)	Paralysis of chest muscles, causing respiratory arrest; fatal in a few minutes
100 mA (0.1 A)	Ventricular fibrillation, preventing coordination of the heart's beating; fatal in a few seconds
1000 mA (1 A)	Serious burns; fatal almost instantly

*The effect on the human body of a given amount of current depends on a variety of conditions. This table gives only general and relative descriptions. The descriptions assume a circuit path that includes the upper chest.

(a)   (b)

•**FIGURE 18.23 Plugging into ground** (a) A three-prong plug and adapter for a two-prong socket. The grounding lug (loop) on the adapter should be connected to the plate-fastening screw on the grounded receptacle box—otherwise, the safety feature is lost. **(b)** A polarized plug. The differently sized prongs permit prewired identification of the high and ground sides of the line.

wire (•Fig. 18.23a) that should be fastened to the receptacle box by the plate-fastening screw or some other means. The receptacle box is grounded by means of the grounding wire. If the adapter lug or wire is not connected, the system is left unprotected, which defeats the purpose of the dedicated grounding safety feature.

You may have noticed another type of plug, a two-prong plug that fits in the socket only one way because one prong is wider than the other and one of the slits of the receptacle is also larger (Fig. 18.23b). This is called a **polarized plug**. *Polarizing* in the electrical sense is a method of identifying the hot and grounded sides of the line so that particular connections can be made.

Such polarized plugs and sockets are an older safety feature. Wall receptacles are wired so that the small slit connects to the hot side and the large slit connects to the neutral, or ground, side. Having the hot side identified in this way makes two safeguards possible. First, the manufacturer of an electrical appliance can design it so that the switch is always in the hot side of the line. Thus, all of the wiring of the appliance beyond the switch is safely neutral when the switch is open and the appliance off. Moreover, the casing of an appliance is connected by the manufacturer to the ground side by means of a polarized plug. Should a hot wire inside the appliance come loose and contact the metal casing, the effect would be similar to that with a dedicated grounding system. The hot side of the line would be shorted to the ground, which would blow a fuse or trip a circuit breaker.

However, the three-wire dedicated grounding system is preferred over polarized plugs and sockets. A wiring error in an appliance or receptacle might cause a problem even with a polarized plug. (What would the problem be?) Another type of electrical safety device, the ground fault interrupter, is discussed in Chapter 20.

Chapter Review

Important Terms

equivalent series resistances   *568*
equivalent parallel resistance   *570*
Kirchoff's rules   *575*
junction (node)   *576*
branch   *576*
Kirchhoff's first rule (junction
   theorem)   *576*

Kirchhoff's second rule (loop
   theorem)   *576*
RC circuit   *580*
time constant   *581*
ammeter (dc)   *583*
voltmeter (dc)   *583*
galvanometer   *583*

fuse   *588*
circuit breaker   *588*
grounded plug   *589*
polarized plug   *591*

Important Concepts

- When resistors are wired in series, the current through all of them must be the same. The equivalent resistance of resistors in series is always greater than that of the largest individual resistance.

- When resistors are wired in parallel, the voltage across all of them must be the same. The equivalent resistance of resistors in parallel is always less than that of the smallest individual resistance.

- Kirchhoff's junction theorem states that the total current into a given junction equals the total current out of that junction (conservation of electric charge).
- Kirchhoff's loop theorem says that in traversing a complete circuit loop, the algebraic sum of the voltage gains and losses is zero, or the sum of the voltage gains equals the sum of the voltage losses (conservation of energy in an electric circuit).
- The time constant for an RC circuit is a "characteristic time" by which we measure the capacitor's charge and discharge.

- An ammeter is a device for measuring current; it consists of a galvanometer and a shunt resistor in parallel. Ammeters are connected in series with the circuit element carrying the current to be measured.
- A voltmeter is a device for measuring voltage; it consists of a galvanometer and a multiplier resistor wired in series. Voltmeters are connected in parallel with the circuit element experiencing the voltage to be measured.

Important Equations

Equivalent Series Resistance:

$$R_s = R_1 + R_2 + R_3 + \cdots = \Sigma_i R_i \qquad (18.2)$$

Equivalent Parallel Resistance:

$$\frac{1}{R_p} = \frac{1}{R_1} + \frac{1}{R_2} + \frac{1}{R_3} + \cdots = \Sigma_i \frac{1}{R_i} \qquad (18.3)$$

Kirchhoff's Rules:

(1) $\qquad \Sigma_i I_i = 0 \quad$ *(junction theorem)* $\quad$ (18.4)
(2) $\qquad \Sigma_i V_i = 0 \quad$ *(loop theorem)* $\quad$ (18.5)

Charging Voltage across a Capacitor in an RC Circuit:

$$V_C = V_0(1 - e^{-t/RC}) \qquad (18.6)$$

Time Constant for an RC Circuit:

$$\tau = RC \qquad (18.8)$$

Discharging Voltage across a Capacitor in an RC Circuit:

$$V_C = V_0 e^{-t/RC} = V_0 e^{-t/\tau} \qquad (18.9)$$

Galvanometer Current (used as ammeter):

$$I_g = \frac{IR_s}{r + R_s} \qquad (18.10)$$

Galvanometer Current (used as voltmeter):

$$I_g = \frac{V}{r + R_m} \qquad (18.11)$$

Exercises

Assume throughout that all resistors are ohmic unless otherwise stated.

18.1 Resistances in Series, Parallel, and Series–Parallel Combinations

1. ■ 💿 Which of the following is always the same for resistors in series? (a) voltage, (b) current, (c) power, (d) energy.

2. ■ 💿 Which of the following is always the same for resistors in parallel? (a) voltage, (b) current, (c) power, (d) energy.

3. ■ 💿 Are the voltages across resistors in series generally the same? If not, could they be so in certain cases?

4. ■ 💿 Are the currents in resistors in parallel generally the same? If not, could they be so in certain cases?

5. ■ 💿 Three resistors that have values of 5 Ω, 2 Ω, and 1 Ω are connected in series to a battery. Which gets the most power, and why?

6. ■ 💿 Three resistors that have values of 5 Ω, 2 Ω, and 1 Ω are connected in parallel to a battery. Which gets the most power, and why?

7. ■ 💿 Three resistors with values of 10 Ω, 20 Ω, and 30 Ω are to be connected. (a) What is the maximum equivalent resistance? (b) What is the minimum equivalent resistance?

8. ■ Two identical resistors (R) are connected in series and then wired in parallel to a 20-ohm resistor. If the total equivalent resistance is 10 Ω, what is the value of R?

9. ■ Two identical resistors (R) are connected in parallel and then wired in series to a 40-ohm resistor. If the total equivalent resistance is 55 Ω, what is the value of R?

10. ■ Three 4.0-ohm resistors are connected in all possible ways to produce different equivalent resistances. Draw a circuit diagram of each possible way. Find the total resistance for each.

11. ■ Three resistors with values of 5.0 Ω, 10 Ω, and 15 Ω are connected in series in a circuit with a 9.0-volt battery. (a) What is the total equivalent resistance? (b) the current in each resistor? (c) At what rate is energy dissipated in the 15-ohm resistor?

12. ■ Find the equivalent resistances for all possible combinations of two or more of the three resistors in Exercise 11.

13. ■ Three resistors with values 1.0 Ω, 2.0 Ω, and 4.0 Ω are connected in parallel in a circuit with a 6.0-volt battery. What are

(a) the total equivalent resistance, (b) the voltage across each resistor, and (c) the power dissipated in the 4.0-ohm resistor?

14. ■■ With an infinite supply of only 1-ohm resistors, what would be the simplest way to make an equivalent resistance of 1.5 Ω?

15. ■■ A length of wire with a resistance of 27 $\mu\Omega$ is cut into three equal segments. The segments are then bundled together to form a conductor one-third as long as the original wire. What is the resistance of the bundled combination?

16. ■■ You are given four 5.00-ohm resistors. Is it possible to connect all of the resistors to produce an effective total resistance of 3.75 Ω?

17. ■■ Three resistors with values of 2.0 Ω, 4.0 Ω, and 6.0 Ω are connected in series in a circuit with a 12-volt battery. (a) How much current is delivered to the circuit by the battery? (b) What is the current through each resistor? (c) How much power is dissipated by each? (d) Compare their total power with the power dissipated by the total equivalent resistance.

18. ■■ Suppose that the resistors in Exercise 17 are connected in parallel. How much current (a) is drawn from the battery? (b) is in each resistor? (c) How much power is dissipated by each resistor? (d) How does their total power compare with the power dissipated by the total equivalent resistance?

19. ■■ Two 8.0-ohm resistors are connected in parallel, as are two 4.0-ohm resistors. These combinations are then connected in series in a circuit with a 12-volt battery. What are the currents in and the voltages across each resistor?

20. ■■ What is the equivalent resistance for the resistors in •Fig. 18.24?

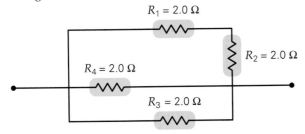

•**FIGURE 18.24 Series–parallel combination** See Exercises 20 and 30.

21. ■■ What is the equivalent resistance between points A and B in •Fig. 18.25?

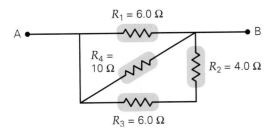

•**FIGURE 18.25 Series–parallel combination** See Exercises 21 and 32.

22. ■■ Find the equivalent resistance for the arrangement of resistors shown in •Fig. 18.26.

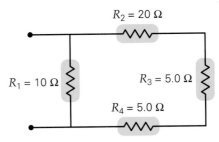

•**FIGURE 18.26 Series–parallel combination** See Exercise 22.

23. ■■ Several 60-watt light bulbs are connected in parallel with a 120-volt source. The very last one blows a 15-amp fuse in the circuit. (a) Sketch a schematic circuit diagram to show the fuse in relation to the bulbs. (b) How many light bulbs are in the circuit (including the very last one)?

24. ■■ Use three 50-ohm resistors and a 120-volt source in a circuit in any arrangement. (a) What is the maximum power for the circuit? Sketch the arrangement. (b) What is the minimum power? Sketch the arrangement.

25. ■■ Find the current and the voltage across the 10-ohm resistor in •Fig. 18.27.

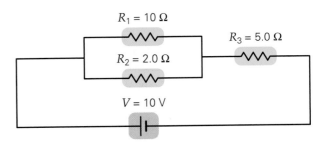

•**FIGURE 18.27 Current and voltage drop of a resistor** See Exercises 25 and 44.

26. ■■ A three-way light bulb can produce 50 W, 100 W, or 150 W of power at 120 V. For each power setting, find (a) the current and (b) the resistance.

27. ■■ For the circuit shown in •Fig. 18.28, find (a) the current in each resistor, (b) the voltage across each resistor, and (c) the total power dissipated.

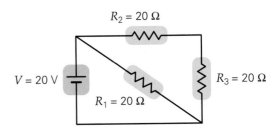

•**FIGURE 18.28 Circuit reduction** See Exercises 27 and 45.

28. ■■ A 120-volt circuit has a circuit breaker rated to trip at 15 A. How many 300-ohm resistors could be connected in parallel in the circuit and not trip the breaker?

29. ■■ In your dorm room, you have two 100-watt lights, a 150-watt color TV, and a 300-watt refrigerator, a 900-watt hairdryer, and a 200-watt computer (with monitor). If there is a 15-amp circuit breaker in the 120-volt power line, will the breaker trip?

30. ■■ Suppose that the resistor arrangement in Fig. 18.24 is connected to a 12-volt battery. What will be (a) the current in each resistor? (b) the voltage across each resistor? (c) the total power dissipated?

31. ■■ To make some hot tea, you use a 500-watt heater in a 120-volt line to heat 0.20 kg of water from 20°C to 80°C. Assume that there is no heat loss, how long does it take?

32. ■■ The terminals of a 6.0-volt battery are connected to points A and B in Fig. 18.25. (a) How much current is in each resistor? (b) How much power is dissipated by each resistor? (c) Compare the sum of the individual power dissipations to the power dissipation of the equivalent resistance for the circuit.

33. ■■ Light bulbs with the power ratings (at 120 V) shown in •Fig. 18.29 are connected in a circuit. (a) What current does the voltage source deliver to the circuit? (b) Find the power dissipated by each bulb. (Take the bulb resistances to be the values when each bulb has a voltage of 120 V.)

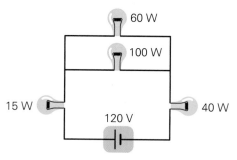

•**FIGURE 18.29 Watt's up?** See Exercise 33.

34. ■■■ For the circuit in •Fig. 18.30, find (a) the current in each resistor and (b) the voltage across each resistor.

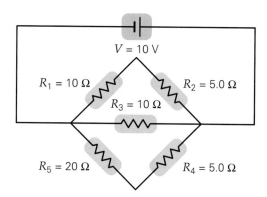

•**FIGURE 18.30 Resistors and currents** See Exercise 34.

35. ■■■ What is the total power dissipated in the circuit shown in •Fig. 18.31?

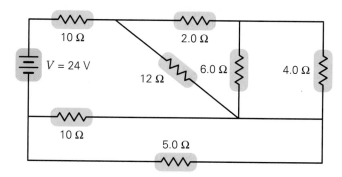

•**FIGURE 18.31 Power dissipation** See Exercise 35.

36. ■■■ What is the equivalent resistance of the arrangement in •Fig. 18.32?

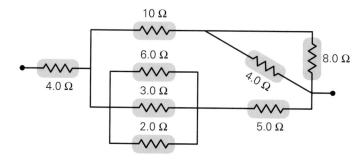

•**FIGURE 18.32 Equivalent resistance** See Exercise 36.

37. ■■■ Find the current in each of the resistors in the circuit in •Fig. 18.33.

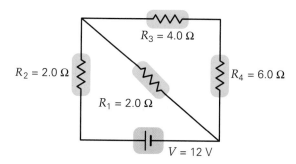

•**FIGURE 18.33 How much current?** See Exercise 37.

38. ■■■ The circuit shown in •Fig. 18.34, named *Wheatstone bridge* after Sir Charles Wheatstone (1802–1875) is used to measure resistance. The resistances R_1, R_2, and R_s are assumed to be known, and R_x is the unknown resistance. R_s is variable and is adjusted until the bridge circuit is balanced, that is, until the galvanometer (G) shows a zero reading (no current in the galvanometer branch). Show that when the bridge is balanced, R_x is given by

$$R_x = \left(\frac{R_2}{R_1}\right) R_s$$

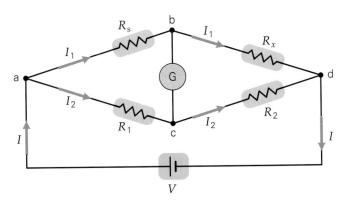

•**FIGURE 18.34** **Wheatstone bridge** See Exercise 38.

18.2 Multiloop Circuits and Kirchhoff's Rules

39. A multiloop circuit has more than one (a) junction, (b) branch, (c) current, (d) all of these.

40. By our sign convention, if a resistor is traversed in the direction of the current in it, (a) the current is negative, (b) the current is positive, (c) the voltage is negative, (d) the voltage is positive.

41. If you traversed a battery from the negative to the positive terminal, is the voltage positive or negative?

42. ■ Traverse loop 3 *opposite* the direction given in Fig. 18.9, and show that the resulting equation is the same as if you followed the arrow direction.

43. ■ For the circuit in Fig. 18.9, reverse the directions of loops 1 and 2 and show that the equations equivalent to those in Example 18.5 are obtained.

44. ■ Use Kirchhoff's loop theorem to find the current through each resistor in Fig. 18.27.

45. ■ Apply Kirchhoff's rules to the circuit in Fig. 18.28 to find the current through each resistor.

46. ■■ Two batteries (terminal voltages 10 V and 4 V) are connected with positive terminals together. A 12-ohm resistor is wired between their negative terminals. (a) Use Kirchhoff's loop theorem to find the current in the circuit and the power dissipated in the resistor. (b) Compare this result to the power output of each battery.

47. ■■ Using Kirchhoff's rules, find the current in each resistor in •Fig. 18.35.

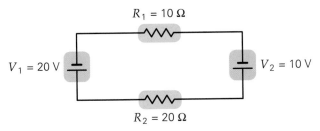

•**FIGURE 18.35** **Single-loop circuit** See Exercise 47.

48. ■■ Apply Kirchhoff's rules to the circuit in •Fig. 18.36, and find (a) the current in each resistor and (b) the rate at which energy is being dissipated in the 8.0-ohm resistor.

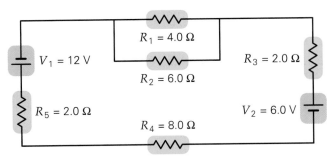

•**FIGURE 18.36** **A loop in a loop** See Exercise 48.

49. ■■ Find the current in each resistor in the circuit in •Fig. 18.37.

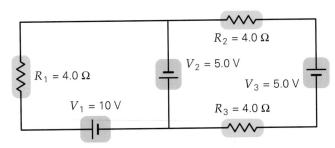

•**FIGURE 18.37** **Double-loop circuit** See Exercise 49.

50. ■■ Find the currents in each resistor in •Fig. 18.38.

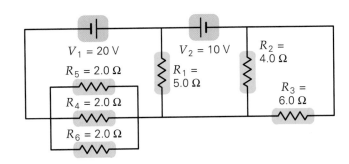

•**FIGURE 18.38** **How many loops?** See Exercise 50.

51. ■■■ For the multiloop circuit shown in •Fig. 18.39, what is the current in each branch?

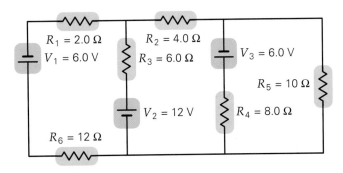

•**FIGURE 18.39** **Triple-loop circuit** See Exercise 51.

18.3 RC Circuits

52. ○ When a capacitor discharges through a resistor, the voltage across the capacitor is a maximum (a) at the beginning of the discharging process, (b) near the middle of the discharging process, (c) at the end of the discharging process, (d) after one time constant.

53. ○ When a capacitor charges through a resistor, the minimum current occurs (a) at the beginning of the charging process, (b) near the middle of the charging process, (c) at the end of the charging process, (d) after one time constant.

54. To increase the time constant of an RC series circuit, you can (a) increase the capacitance, (b) increase the resistance, (c) decrease the capacitance, (d) both (a) and (b).

55. ■ ○ In Fig. 18.10b, the switch is closed at $t = 0$ and the capacitor begins to charge. What is the voltage across the resistor and across the capacitor, expressed in terms of V_o (to two significant figures), (a) just after the switch is closed, (b) after one time constant has elapsed, and (c) after many times constants?

56. ■ ○ A capacitor in a single-loop RC circuit is charged to 63% of its final voltage in 1.5 s. Find (a) the time constant for the circuit and (b) the voltage after 15 s expressed as a percentage of its final voltage. (Use five significant figures to show how close to 100% this percentage is.)

57. ■ In a flashing neon light circuit, a time constant of 2.0 s is desired. If you have a 1.0-microfarad capacitor, what resistance should you use in the circuit?

58. ■■ ○ How many time constants will it take for an initially charged capacitor to be discharged to half of its initial voltage?

59. ■■ ○ A 1.00-microfarad capacitor in series with a resistor is connected to a 12.0-volt source and charged. (a) What resistance is necessary to cause the capacitor to have only 37% of its initial charge 1.50 s after starting to discharge? (b) What is the voltage across the capacitor at $t = 3\tau$ when the capacitor is charging?

60. ■■ ○ An RC circuit with $C = 40$ μF and $R = 6.0$ Ω includes a 24-volt source. With the capacitor initially uncharged, an open switch in the circuit is closed. (a) What is the voltage across the resistor immediately afterward? (b) What is the voltage across the capacitor at that time? (c) What is the current in the resistor at that time?

61. ■■ ○ (a) For the circuit in Exercise 60, after the switch has been closed for $t = 4\tau$, what is the charge on the capacitor? (b) After a long time, what are the voltages across the capacitor and the resistor?

62. ■■■ ○ An RC circuit with a resistance of 5.0 MΩ and a capacitance of 0.40 μF is connected to a 12-volt source. If the capacitor is initially uncharged, what is the change in voltage across it between $t = 2\tau$ and $t = 4\tau$?

63. ■■■ A 3.0-megohm resistor is connected in series with a 0.28-microfarad capacitor. If this arrangement is wired across four 1.5-volt batteries connected in series, (a) what are the initial voltage across the capacitor and the initial current in the circuit? (b) How much charge is on the capacitor after 4.0 s?

18.4 Ammeters and Voltmeters

64. A voltmeter has a (a) large shunt resistance, (b) large multiplier resistance, (c) small shunt resistance, (d) small multiplier resistance.

65. An ammeter has a (a) large shunt resistance, (b) large multiplier resistance, (c) small shunt resistance, (d) small multiplier resistance.

66. To measure the voltage across a circuit element, a voltmeter is connected (a) in series with the element, (b) in parallel with the element, (c) between the high potential side of the element and ground, (d) none of these.

67. What would happen in a circuit if (a) an ammeter were connected in parallel to a resistor and (b) a voltmeter were connected in series to a resistor?

68. ■ A galvanometer with a full-scale sensitivity of 2000 μA has a coil resistance of 100 Ω. If it is to be used in an ammeter with a full-scale reading of 30 A, what is the necessary shunt resistance?

69. ■ If the galvanometer in Exercise 68 were to be used to build a voltmeter with a full-scale reading of 15 V, what would be the required multiplier resistance?

70. ■ A galvanometer with a full-scale sensitivity of 600 μA and a coil resistance of 50 Ω is to be used in an ammeter designed to read 5.0 A at full scale. What is the required shunt resistance?

71. ■■ A galvanometer has a coil resistance of 20 Ω. A current of 200 μA deflects the needle through 10 divisions at full-scale. What resistance is needed to convert the galvanometer to a full-scale 10-volt voltmeter?

72. ■■ An ammeter has a resistance of 1.0 mΩ. Find the current in the ammeter when it is properly connected to a 10-ohm resistor and a 6.0-volt source. (Express your answer to five significant figures to show how it differs from 0.60 A, the value of the current in the absence of the ammeter.)

73. ■■ A voltmeter has a resistance of 30 kΩ. What is the current through the meter when it is connected across a 10-ohm resistor that is wired to a 6.0-volt source?

74. ■■■ An ammeter and a voltmeter together can be used to measure the resistance in a circuit. The ammeter is connected in series with the resistor and that the voltmeter is placed across the resistor only. (a) Show that the resistance is

$$R = \frac{V}{I - (V/R_V)}$$

where V is the voltage measured by the voltmeter, I is the current measured by the ammeter, and R_V is the resistance of the voltmeter. [Hint: Draw a circuit diagram to help analyze the problem.] (b) Use this result to explain why the "perfect" voltmeter would have infinite resistance.

75. ■■■ An ammeter and a voltmeter together can be used to measure the value of a resistor in a circuit. The ammeter is connected in series with the resistor, and the voltmeter is placed across *both* the ammeter and the resistor. (a) Show that the resistance is then given by $R = (V/I) - R_A$, where V is the voltage measured by the voltmeter, I is the current measured by the ammeter, and R_A is the resistance of the ammeter. [Hint: Draw a circuit diagram to help you analyze the problem.] (b) Use this result to explain why the "perfect" ammeter would have zero resistance.

18.5 Household Circuits and Electrical Safety

76. The ground wire in household wiring (a) is a current-carrying wire, (b) is at a voltage of 240 V relative to a "hot" wire, (c) carries no current, (d) none of these.

77. A dedicated grounding wire (a) is the basis for the polarized plug, (b) is necessary for a circuit breaker, (c) normally carries no current, (d) none of these.

78. ■ In terms of electrical safety, what is wrong with the circuit in •Fig. 18.40, and why?

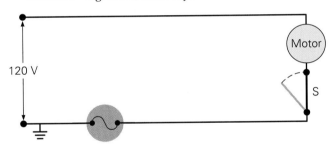

•FIGURE 18.40 **A safety problem?** See Exercise 78.

79. ■ Bodily injury from electrocution depends on the magnitude of the current and its path, yet you commonly see signs that warn "Danger: High Voltage" (•Fig. 18.41). Shouldn't such signs refer to high current? Explain.

•FIGURE 18.41 **Danger—high voltage** Shouldn't it be high current? See Exercise 79.

80. ■ Why is it safe for birds to perch with both feet on the same high-voltage wire, even if the insulation is worn through?

81. ■ After a collision with a power pole, you are fine but are trapped in a car with a bare high-voltage line (with frayed insulation) in contact with the hood of your car. Is it safer to step out of the car one foot at a time or to jump with both feet leaving the car at the same time? Explain your reasoning.

82. ■ Most electrica l codes require the metal case of an electric clothes dryer to have a wire running from the case to a nearby faucet (or metal piece of plumbing). Explain why.

Additional Exercises

83. Two 50-ohm resistors are connected in parallel. That arrangement is then connected to a third 50-ohm resistor in series. What is the effective resistance?

84. Two 100-ohm resistors are connected in series. That arrangement is then connected to a third 100-ohm resistor in parallel. What is the effective resistance?

85. A 4.0-ohm resistor and a 6.0-ohm resistor are connected in series. A third resistor is connected in parallel with the 6.0-ohm resistor. The whole arrangement has an equivalent resistance of 7.0 Ω. What is the value of the third resistor?

86. Find the curre.nt through each resistor in the circuit in •Fig. 18.42.

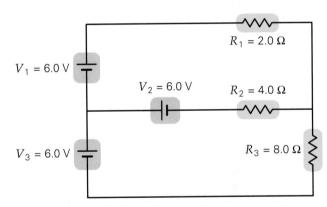

•FIGURE 18.42 **Kirchhoff"s rules** See Exercise 86.

87. Resistors of 30 Ω and 15 Ω are connected in parallel with a 9.0-volt battery. How much energy is dissipated each second by both resistors?

88. Four resistors are connected to a 90-volt source (•Fig. 18.43). (a) Which resistor(s) dissipate the most power, and how much? (b) What is the total power dissipated in the circuit?

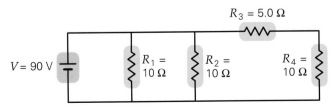

•FIGURE 18.43 **How much power is dissipated?** See Exercise 88.

89. Four resistors are connected in a circuit with a 110-volt source (•Fig. 18.44). (a) What is the current through each resistor? (b) How much power is dissipated by each resistor?

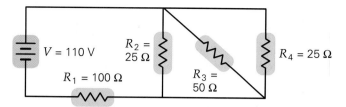

•**FIGURE 18.44   Joule heat losses**   See Exercise 89.

90. Nine resistors, each of the same value R, are connected in a "ladder" fashion (•Fig. 18.45). What is the effective resistance of this arrangement between points A and B?

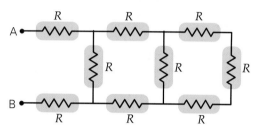

•**FIGURE 18.45   A resistance ladder**   See Exercise 90.

91. In Exercise 90, if a 12.0-volt battery is connected from point A to point B, how much current is in each resistor assuming each has a resistance of 10.0 Ω?

92. The time between flashes of the neon tube, or the period of the oscillator circuit, in Fig. 18.13 is the time it takes for the voltage to rise from V_m to V_b. Show that the period T is

$$T = t_b - t_m = RC \ln\left(\frac{V_o - V_m}{V_o - V_b}\right)$$

where V_o is the maximum voltage, or the voltage of the battery in the circuit.

93. •Figure 18.46 shows the workings of a *potentiometer*, a very accurate device for determining emf. It consists of three batteries, an ammeter, and several resistors, including a uniform long wire that can be "tapped" for a specific

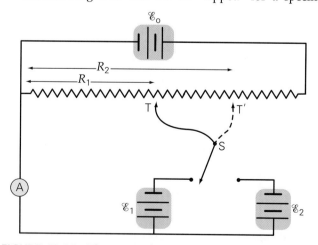

•**FIGURE 18.46   The potentiometer**   See Exercise 93.

fraction of its total resistance. $\mathcal{E}_o$ is the emf of working battery, $\mathcal{E}_1$ designates a battery with a well-known emf, and $\mathcal{E}_2$ denotes the unknown battery. Switch S is thrown toward battery 1, and the point T (for "tapped") is moved along the resistor until the ammeter reads zero. The resistance of this arrangement is R_1. This procedure is repeated with the switch thrown toward battery 2, and the point T is moved to T' until the ammeter again reads zero. The resistance of this arrangement is R_2. Show that

$$\mathcal{E}_2 = \frac{R_2}{R_1}\mathcal{E}_1$$

94. Two 100-watt light bulbs (assumed ohmic) are connected to a 110-volt power source. What are the current in each bulb and the power dissipated in each if the bulbs are connected (a) in series and (b) in parallel?

95. If a 120-volt source were connected across the leads (open ends) of the circuit shown in •Fig. 18.47, how much current would be drawn from it?

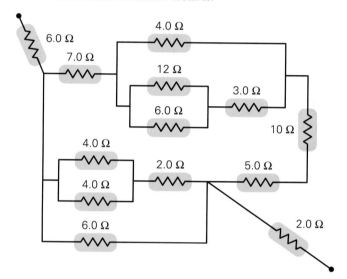

•**FIGURE 18.47   Connect to voltage source**   See Exercise 95.

96. If a combination of three 30-ohm resistors dissipates 3.2 W when connected to a 12-volt battery, how are the resistors connected in the circuit?

97. A battery has three cells, each with an internal resistance of 0.020 Ω and an emf of 1.50 V. The battery is connected in parallel with a 10.0-ohm resistor. (a) Determine the voltage across the resistor. (b) How much current passes through each cell? (The cells in a battery are connected in series.)

98. A galvanometer with an internal resistance of 50 Ω and a full-scale sensitivity of 200 μA is used to construct a multirange voltmeter. What values of multiplier resistors allow for three full-scale voltage readings of 20 V, 100 V, and 200 V? (See Fig. 18.17b and neglect significant figure rules.)

99. A galvanometer with an internal resistance of 100 Ω and a full-scale sensitivity of 100 μA is used in a multirange ammeter. What values of shunt resistors allow for three full-scale current readings of 1.0 A, 5.0 A and 10 A? (See Fig. 18.17a.)

Magnetism

When we think of magnetism, we tend to think of attraction. We know that you can pick things up with a magnet—at least, certain kinds of things. Most of us have probably encountered magnetic latches that hold cabinet doors shut or used magnets to stick notes to the refrigerator. We're less likely to think of repulsion. Yet, wherever there are magnetic attractive forces, there must also be magnetic repulsive forces—and they can be just as useful.

The photo above shows an interesting example. At first glance, this looks like a perfectly ordinary locomotive. But—where are the wheels? In fact, this isn't a conventional train but a high-speed magnetically-levitated train. This means that the train doesn't touch the "rails" at all: It "floats" above them, supported by repulsive forces that are the product of powerful magnets. The advantages are obvious—with no wheels, there is no rolling friction, no bearings to lubricate—in fact, very few moving parts of any kind. Magnetic forces are also used to accelerate and decelerate the train.

But where do the magnetic forces come from? For centuries, the attractive properties of magnets were attributed to supernatural forces. We now associate magnetism with electricity (electromagnetism) as both turn out to be aspects of a single fundamental force or interaction (the electromagnetic force). We have

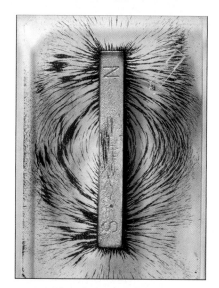

•FIGURE 19.1 **Bar magnet** The iron filings indicate the poles, or centers of force, of a common bar magnet. The compass direction designates these poles as north (N) and south (S). (See Fig. 19.3.)

put electromagnetism to use in motors, generators, radios, telephones, high-speed trains levitated and propelled by magnetic fields, and many other familiar applications. The development of high-temperature superconducting materials (Chapter 17) may soon open the way for practical exploitation of many more devices that are now found only in the laboratory or as experimental prototypes.

Although electricity and magnetism are manifestations of the same fundamental force, it is instructive first to consider them individually and then put them together, so to speak. This and the next chapter will investigate magnetism and its intimate relationship to electricity.

19.1 Magnets, Magnetic Poles, and Magnetic Field Direction

Objectives: To (a) state the force rule between magnetic poles and (b) explain how the magnetic field direction is determined with a compass.

One of the first things anyone notices in examining a common bar magnet is that it has two "centers" of force called *poles*, one at or near each end of the magnet (•Fig. 19.1). Rather than being distinguished as positive and negative, these poles are called north (N) and south (S). This terminology comes from the early use of the magnetic compass. The north pole of a compass magnet is the *north-seeking* end—that is, the end that tends to point *north* on Earth. With two bar magnets, we notice a pattern of attraction and a repulsion between their various ends: Each end is attracted to one end of a second magnet and repelled by the other end.

The attraction and repulsion between poles of magnets is similar to the behavior exhibited by like and unlike electric charges. Analogous to the charge–force law is the **pole–force law**, or **law of poles**:

> Like magnetic poles repel each other, and unlike magnetic poles attract each other.

That is, north and south poles (N–S) attract each other, and north and north poles (N–N) or south and south (S–S) poles repel each other (•Fig. 19.2).

Magnetic poles have always been observed to occur in pairs, never singly. Two opposite poles form what is called a magnetic *dipole*. You might think you could break a bar magnet in half and get two single isolated poles. However, you would find that the resulting pieces are just two shorter magnets, *each with a north and a south pole*. Although an isolated, single magnetic pole (a magnetic *monopole*) has been postulated to exist, it has yet to be found experimentally.

The fact that a magnet always has two poles may make you wonder about the nature of magnetism. Although magnetism appears to have many similarities to electricity, there are many differences between the two. For example, the source of magnetism, although ultimately derived from electric charge, is different. As will be discussed shortly, magnetism is produced by *electric charges in motion*, such as electric currents and orbiting atomic electrons. Stationary electric charges produce only electric fields, not magnetic fields. This understanding of the source of magnetism and some knowledge of magnetic materials (to be covered in Section 19.4) will allow you to see why pieces of a dipole magnet are themselves dipole magnets.

•FIGURE 19.2 **The pole–force law, or law of poles** Like poles (N–N or S–S) repel, and unlike poles (N–S) attract.

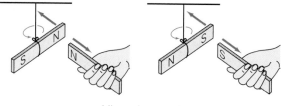

Like poles repel

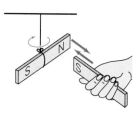

Unlike poles attract

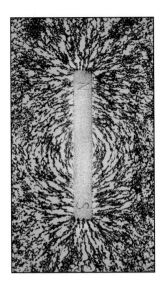

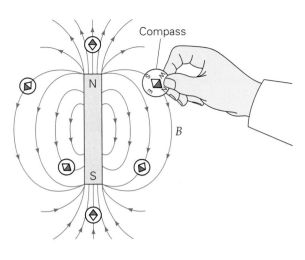

(a)

•FIGURE 19.3 Magnetic fields
(a) Magnetic field lines can be traced and outlined by using iron filings or a compass, as shown for the magnetic field of a single bar magnet. The filings behave like tiny compasses and line up with the field. The closer together the field lines, the stronger the magnetic field. **(b)** Iron filing pattern for the magnetic field between unlike poles; the field lines converge. **(c)** Iron filing pattern for the magnetic field between like poles; the field lines diverge.

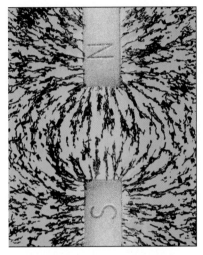

(b)

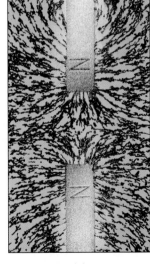

(c)

Defining the Magnetic Field Direction

From your knowledge of electrostatics, you may be tempted to think of a pole as a magnetic "charge" and to wonder if the magnetic force can be expressed in a form similar to Coulomb's law for electric charges, given in Chapter 15. In fact, such a law was developed by Coulomb using the magnetic pole strengths in place of the electric charges. However, this law is now rarely used because it doesn't match our present understanding of magnetism and because it is more convenient to work with magnetic *fields*.

In Section 15.4, you learned how convenient it is to describe the interaction of electrically charged objects in terms of electric fields. An arrangement of electric charges produces an electric field, which is represented by electric field lines, or lines of force. Mathematically, the electric field is defined as the force per unit charge, $\mathbf{E} = \mathbf{F}_e/q_o$. Similarly, it is convenient to describe magnetic interactions in terms of magnetic fields. Like an electric field, a **magnetic field** is a vector quantity; it is represented by the symbol $\mathbf{B}$. A magnetic field surrounds every magnet. The pattern of the magnetic field lines surrounding any magnet can be demonstrated by sprinkling iron filings over the magnet covered with a piece of paper or glass (Fig. 19.1). As you will learn, the iron filings become induced magnets and line up with the field much as compass needles line up with the Earth's magnetic field.

To describe any vector field, we must specify both the magnitude, or strength, and direction of the field at various points. The direction of a magnetic field (which we refer to simply as a "*B* field") is defined as follows, once we have a compass calibrated for direction by the Earth's field:

The direction of a magnetic field (**B**) at any location is the direction that the north pole of a compass at that location would point.

This definition provides another method of mapping a magnetic field: placing a small compass near a magnet. The magnetic forces on the two opposite poles of the needle provide a torque that causes the needle to rotate and line up in the direction of the existing *B* field. If the compass is moved in the direction in which the needle points, the path of the needle traces out a magnetic field line, as illustrated in •Fig. 19.3a.

The direction of the field outside a bar magnet, by the direction definition (above) and the pole–force law, thus points away from the north pole of a magnet and toward the south pole. As with an electric field, the closer together the field lines, the stronger the field. Note how this is indicated by the concentration of iron filings in the pole regions in Fig. 19.3b,c.

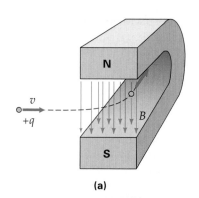

(a)

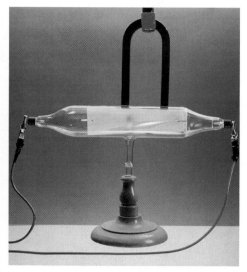

(b)

•**FIGURE 19.4 Force on a moving charged particle (a)** When a charged particle enters a magnetic field, the particle experiences a force that is evident because the charge is deflected from its original path. **(b)** The electron beam in a cathode ray tube (made visible by fluorescent paper) is normally horizontal between the end electrodes but is deflected here because of the magnet.

Note: Like the tesla, the weber and the gauss are named after early investigators of magnetic phenomena.

Since the magnetic north pole of a compass can be used to map out a magnetic field, you might think that the magnitude of **B** would be defined as the magnetic force per unit pole (as **E** is the electric force per unit charge). The magnetic field was once thought of in this manner. However, the magnetic field magnitude, or strength, is now defined in terms of the force on a moving electric charge, as will be discussed in Section 19.2.

19.2 Magnetic Field Strength and Magnetic Force

Objectives: **To (a) define the magnetic field strength in terms of the force exerted on a moving charged particle and (b) determine the magnetic force exerted by a magnetic field on such a particle.**

Experiments indicate that one of the important quantities in determining the *magnetic* force on a particle is the *electric* charge of that particle. That is, there seems to be some connection between *electrical* properties of objects and how they respond to *magnetic* fields. The study of interactions between electrically charged particles and magnetic fields is called **electromagnetism**.

A specific example of such an electromagnetic interaction occurs when a particle with a positive charge (q) moving with a constant velocity enters a region with a uniform magnetic field (for example, between two closely placed pole faces of a large permanent magnet) in a path such that the velocity and the magnetic field are at right angles (•Fig. 19.4). When the charged particle enters the field, the particle is deflected into a curved path, which, if carefully measured, turns out to be a circular arc.

From the study of dynamics, you know that circular motion is caused by a force that is always perpendicular to the particle's velocity. But what gives rise to this force? No electric field is present. The force of gravity is too weak to cause this deflection; furthermore, it would deflect the particle *downward* rather than sideways and into a *parabolic* arc, not a circular one. Evidently, the force is due to the interaction of the moving charge and the magnetic field. That is, *an electrically charged particle moving in a magnetic field may experience a magnetic force.*

It is found experimentally that the deflecting magnetic force is directly proportional to the particle's charge and speed. For the special case in which the velocity vector (**v**) is perpendicular to the magnetic field vector (**B**), we can conveniently define the magnetic field strength (magnitude) as

$$B = \frac{F}{qv} \qquad \text{(definition of B, only for \textbf{v} perpendicular to \textbf{B})} \qquad (19.1)$$

SI unit of magnetic field: newton per ampere-meter [N/(A·m), or tesla (T)]

That is, B is interpreted as the magnetic force per unit charge per unit speed. From Eq. 19.1, the magnetic field strength has units of N/(C·m/s) or N/(A·m), since the ampere is the coulomb per second. This combination of units is given the name **tesla (T)** after a famous early researcher in magnetic fields, Nikola Tesla (1856–1943). Thus 1 T = 1 N/(A·m). The magnetic field is sometimes given in webers per square meter (Wb/m²) when magnetic *flux* (Chapter 20) is relevant. Most everyday magnetic field strengths are much smaller than 1 T. In such situations, it is common to express the field strength in terms of the *gauss* (G), which is 0.0001 T: 1 G = 10^{-4} T. For example, the Earth's magnetic field is on the order of several tenths of a gauss.

Once the magnetic field strength in a location has been determined from Eq. 19.1, then the force on *any* charged particle moving at *any* speed can be found (as long as its direction remains perpendicular to the B field) by solving for F:

$$F = qvB \qquad \text{(force on a charged particle, only for \textbf{v} perpendicular to \textbf{B})} \qquad (19.2)$$

However, if the direction of the charged particle's velocity is *not* perpendicular to the magnetic field, the magnitude of the magnetic force on the particle is not given by Eq. 19.2. In this case, the magnitude of this force depends also on the angle (θ) between the velocity vector and the magnetic field vector—specifically, on the sine of that angle ($\sin \theta$). That is, the force is zero when **v** and **B** are parallel and a maximum when those two vectors are perpendicular. In general, the magnitude of the force is

$$F = qvB \sin \theta \qquad \begin{array}{l}\textit{magnetic force on} \\ \textit{a charged particle}\end{array} \qquad (19.3)$$

When **v** and **B** are perpendicular ($\theta = 90°$), the equation reduces to Eq. 19.2, $F = qvB \sin 90° = qvB$ (maximum force). However, when the **v** and **B** vectors are parallel ($\theta = 0°$ or $180°$), the force on the moving charge is zero, $F = qvB \sin 0° = 0$.

The Right-Hand Force Rule for Moving Charges

The *direction* of the magnetic force on a *positively* charged particle is determined by considering the velocity and magnetic field vectors. This method is stated in the form of a **right-hand force rule** (•Fig. 19.5a):

> When the fingers of the right hand are pointed in the direction of **v** and then curled toward the vector **B**, the extended thumb points in the direction of **F** for a *positive* charge. (For a negative charge, the force is in the opposite direction.)

You might imagine the fingers of the right hand to be physically turning or rotating the **v** vector into **B** so that **v** and **B** are aligned. The magnetic force is always *perpendicular to the imaginary plane formed by the vectors* **v** *and* **B** (Fig. 19.5b). If the moving charge is negative, the right thumb points in the direction opposite that of the force. You can start by assuming that the negative charge is positive, apply the right-hand rule, and then reverse the direction given by the rule.

Some people prefer an equivalent alternative called the right-hand "three finger" rule (Fig. 19.5c). With the forefinger of the right hand pointing in the direction of **v** and the middle finger flexed in the direction of **B**, the extended right thumb points in the direction of **F** on a positive charge. Use whichever method you find easier.

Note: *B* fields that point into the plane of the page are designated by ×. *B* fields that point out of the plane are indicated by •. You can visualize these symbols as though observing the feathered end and tip of an arrow, respectively. This is the general way of showing vectors that point into or out of the page.

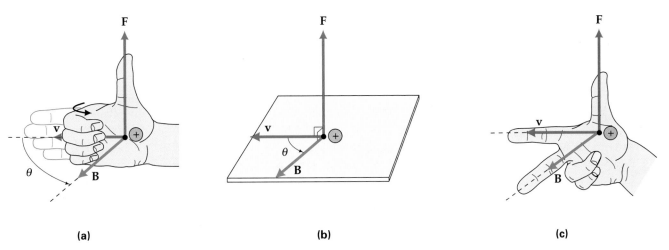

(a)   (b)   (c)

•**FIGURE 19.5 Right-hand rule for magnetic force** **(a)** When the fingers of the right hand are pointed in the direction of **v** and then curled toward the direction of **B**, the extended thumb points in the direction of the force **F** on a *positive* charge. **(b)** The magnetic force is always perpendicular to the plane of **B** and **v** and thus is always perpendicular to the direction of the particle's motion. **(c)** A common equivalent method: When the extended forefinger of the right hand points in the direction of **v** and the middle finger points in the direction of **B**, the extended right thumb points in the direction of **F** on a positive charge. (Reverse the direction for a negative charge.)

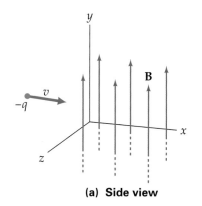

(a) Side view

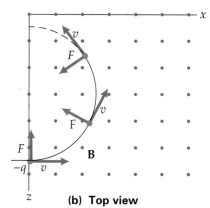

(b) Top view

•**FIGURE 19.6   Path of a charged particle in a magnetic field**   **(a)** A charged particle entering a uniform magnetic field will be deflected— here toward the xy plane by the right-hand rule, because the charge is negative. **(b)** In the field, the force is always perpendicular to the particle's velocity. The particle moves in a circular path if the field is constant and the particle enters perpendicularly to the field direction. See Example 19.2.

Even for Lefties: Using the Right-Hand Rule

In a linear accelerator, a beam of protons travels horizontally northward. To deflect the protons eastward in a uniform magnetic field, the field should point in which direction? (a) vertically downward, (b) west, (c) vertically upward, (d) south.

Reasoning and Answer.   Since the force on a moving charged particle is perpendicular to the plane of **v** and **B**, the magnetic field could not be in the horizontal plane; this would deflect the protons down or up (verify this with the right-hand force rule). Thus we can eliminate choices (b) and (d). Using the right-hand force rule for a positively charged particle to see if **B** is downward (answer a), we find that the force would be to the west. Hence, the answer is (c) for an easterly deflection. Prove this to yourself, using the right-hand force rule.

Follow-up Exercise.   To experience the same force *direction* as the protons in the magnetic field (vertically upward) of this Example, would a beam of *electrons* have to travel north or south?

Example 19.2 ■ Moving in Circles: Force on a Moving Charge

A particle with a charge of -5.0×10^{-4} C moves at a speed of 1.0×10^3 m/s in the $+x$ direction toward a uniform magnetic field of 0.20 T in the $+y$ direction (•Fig. 19.6a). (a) What is the force on the particle just as it enters the magnetic field? (b) Describe the path of the particle while it is in the field.

Thinking It Through.   The magnitude of the magnetic force on a single charge is given by Eq. 19.3. We expect a circular trajectory since the magnetic force is always perpendicular to the particle's velocity. Note that the particle is negatively charged.

Solution.   We are given the following data:

Given: $q = -5.0 \times 10^{-4}$ C   *Find:* (a) **F** (magnetic force on the particle)
  $v = 1.0 \times 10^3$ m/s $(+x)$ (b) Path of the particle in the field
  $B = 0.20$ T $(+y)$

(a)   From Eq. 19.3, we get the magnitude of the force (so we can neglect the minus sign on q):

$$F = qvB \sin \theta$$

$$= (5.0 \times 10^{-4}\ \text{C})(1.0 \times 10^3\ \text{m/s})(0.20\ \text{T})(\sin 90°) = 0.10\ \text{N}$$

The direction of the force just as the particle enters the magnetic field is in the $-z$ direction in Fig. 19.6a. By the right-hand rule, the force on a positive charge would be in the $+z$ direction. But since the charge is negative, the force is in the opposite direction.

(b)   Since the magnetic force is always perpendicular to the velocity of the particle, it supplies the required centripetal force that causes the particle to move in a circular path (Fig. 19.6b). The radius of the path (r) can be found by using Newton's second law:

$$F = ma_c$$

That is, in this case, the centripetal force is supplied by the magnetic force, so

$$qvB = \frac{mv^2}{r} \qquad \text{or} \qquad r = \frac{mv}{qB}$$

If we knew the mass of the particle, we could compute the radius of its circular path.

Follow-up Exercise.   In this Example, if the particle had been a proton, (a) in what direction would it be deflected? (b) If the radius of its circular path were 10 cm and its speed were 1.0×10^6 m/s, what would be the magnetic field strength?

19.3 Electromagnetism—The Source of Magnetic Fields

Objectives: To (a) understand the origin of the magnetic field and calculate its strength for simple cases, and (b) use the right-hand force rule to determine the direction of the magnetic field from the direction of the current that produces it.

Electric and magnetic phenomena, although clearly different in detail, are closely and fundamentally related. In fact, *magnetic* fields are produced by *electric currents* (moving charges) and in turn can exert forces on *moving electric charges*. The fact that magnetic fields are produced by electric currents was discovered by the Danish physicist Hans Christian Oersted in 1820. He noted that an electric current could produce a deflection of a compass needle. This property can be demonstrated with an arrangement like that in •Fig. 19.7. When the circuit is open and there is no current in it, the compass needle points in the northerly direction due to the Earth's magnetic field (to be discussed in Section 19.7). However, when the switch is closed and there is current in the circuit, the compass needle is deflected, indicating that an additional magnetic field (due to the current) is affecting the needle. When the switch is opened and the current drops to zero, the compass needle goes back to pointing north.

Developing expressions for the magnitude of the magnetic field near a current-carrying wire requires mathematics beyond the scope of this book. This section will simply present the results for the magnitude of the **B** field for several arrangements that are used in practical applications.

Magnetic Field near a Long, Straight, Current-Carrying Wire. At a perpendicular distance d from a long, straight wires carrying a current I (•Fig. 19.8), the magnitude of **B** is

$$B = \frac{\mu_o I}{2\pi d} \qquad \textit{(long, straight wire)} \qquad (19.4)$$

where $\mu_o = 4\pi \times 10^{-7}$ T·m/A is a proportionality constant called the *magnetic permeability of free space*. For long, straight wires only, the field lines are closed circles centered on the wire.

Note in Fig. 19.8 that the direction of **B** is given by another right-hand rule, which we call the **right-hand source rule:**

If a current-carrying wire is grasped with the right hand with the extended thumb pointing in the direction of the current (I), the curled fingers indicate the circular sense of the magnetic field direction.

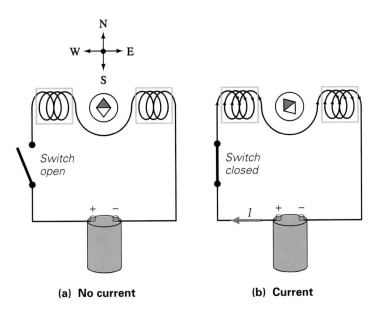

(a) **No current**   (b) **Current**

•FIGURE 19.7 **Electric current and magnetic field** **(a)** With no current in the wire, the compass needle points north. **(b)** With a current in the wire, the compass needle is deflected, indicating the presence of an additional magnetic field superimposed on that of the Earth. In this case, the strength of the additional field is roughly equal in magnitude to that of the Earth. How can you tell?

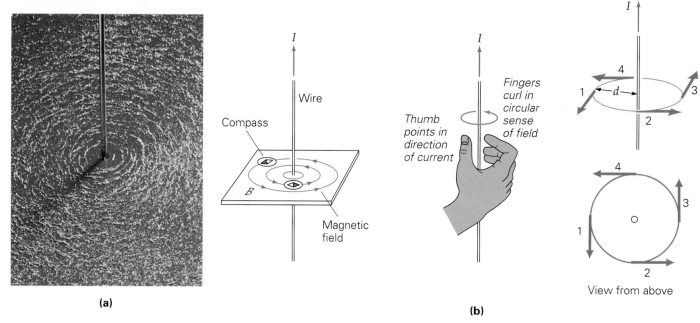

(a)

(b)

•**FIGURE 19.8   Magnetic field around a long, straight current-carrying wire**   (a) The field lines form concentric circles around the wire, as revealed by the iron filing pattern. (b) The circular sense of the field lines is given by the right-hand source rule, and the magnetic field vector is tangent to the circular field line at any point.

Note: The magnetic field vector is tangent to a field line at any point on the circle. The field lines, although typically curved, are true circles only near very long straight wires.

Magnetic Field at the Center of a Circular Current-Carrying Wire Loop.   At the *center* of a circular loop of wire of radius r carrying a current I (•Fig. 19.9), the magnitude of **B** is

$$B = \frac{\mu_{o}I}{2r} \qquad \text{(center of circular loop)} \qquad (19.5)$$

The direction of **B** is given by the right-hand source rule and is perpendicular to the plane of the loop at its center (Fig. 19.9b). Notice that the overall field of the loop (Fig. 19.9c) is geometrically similar to that of a bar magnet (magnetic dipole).

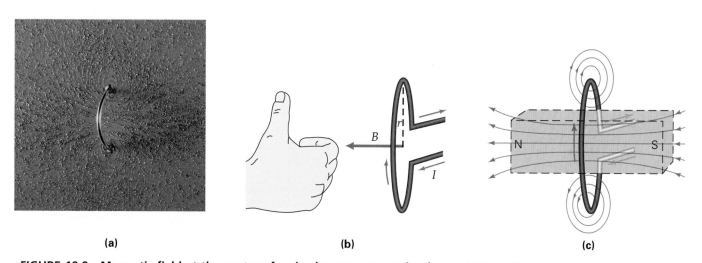

(a)

(b)

(c)

•**FIGURE 19.9   Magnetic field at the center of a circular current-carrying loop**   (a) Iron filing pattern for a current-carrying loop. The magnetic field is perpendicular to the plane of the loop. (b) The direction of the field is given by the right-hand source rule. With the thumb of the right hand in the direction of the conventional current, the curled fingers indicate the direction of **B**. (c) The overall magnetic field of a current-carrying circular loop is similar to that of a bar magnet.

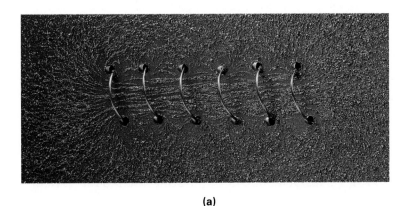

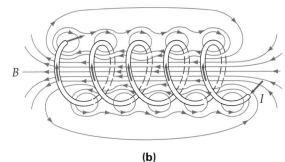

(a) (b)

•FIGURE 19.10 **Magnetic field of a solenoid** (a) The magnetic field of a current-carrying solenoid is fairly uniform near the central axis of the solenoid, as seen in this iron filing pattern. (b) The direction of the field can be determined by applying the right-hand source rule to any of the loops.

Magnetic Field in a Current-Carrying Solenoid. A *solenoid* is a long wire wound in a tight coil, or helix (many circular loops, as shown in •Fig. 19.10). If the radius of the loops is small compared to the length (L) of the coil, **B** is parallel to the solenoid's longitudinal axis. If the solenoid has N turns (loops) and carries a current I, the magnitude of the magnetic field at the center is given by

$$B = \frac{\mu_o NI}{L} \qquad \text{(at center of a solenoid)} \qquad (19.6)$$

The direction of **B** is given by the right-hand source rule as applied to any coil loop.

The expression $n = N/L$, or the number of turns *per meter*, is called the *linear turn density*. Using it allows us to write Eq. 19.6 in simpler form:

$$B = \mu_o nI \qquad \text{(at center of a solenoid)} \qquad (19.7)$$

The longer the solenoid, the more uniform the magnetic field across the cross-sectional area within its coil. An ideal, or infinitely long, solenoid would have a uniform internal field. Although this ideal condition cannot be realized, a long solenoid does produce a relatively uniform magnetic field. Note how the pattern of the field lines for a loop (Fig. 19.9) and especially that for a solenoid (Fig. 19.10) resembles that of the magnetic field of a permanent bar magnet.

Example 19.3 ■ Common Fields: Magnetic Field from a Current-Carrying Wire

The maximum household current in a wire is about 15 A. For simplicity, assume that it is a steady (dc) current (even though it is not). Assume also that this current is in a west-to-east direction (•Fig. 19.11). What are the magnitude and direction of the magnetic field it produces 1.0 cm directly below the wire?

Thinking It Through. This situation is an application of Eq. 19.4, for a long, straight wire with the maximum current.

Solution. Listing the data, we have

Given: I = 15 A *Find:* **B** (magnitude and direction)
  d = 1.0 cm = 0.010 m

From Eq. 19.4, the magnitude of the field at the point 1.0 cm directly below the line is

$$B = \frac{\mu_o I}{2\pi d} = \frac{(4\pi \times 10^{-7}\ \text{T·m/A})(15\ \text{A})}{2\pi(0.010\ \text{m})} = 3.0 \times 10^{-4}\ \text{T}$$

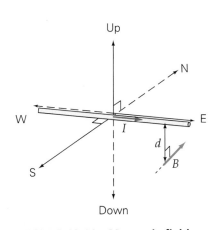

•FIGURE 19.11 **Magnetic field** Finding the magnetic field produced by a straight current-carrying wire. See Example 19.3.

By the right-hand source rule, you should be able to show that the magnetic field vector points northward as shown in Fig. 19.11.

Follow-up Exercise. Estimate the electric current needed to produce a magnetic field of the same strength as that for the long, straight wire in the Example but at a distance of 1.0 m.

To see why the solenoid might be better suited for magnetic applications, consider the next Example.

Example 19.4 ■ Wire versus Solenoid: Concentrating the Magnetic Field

A solenoid is 0.30 m long with 10^3 turns per meter and carries a current of 5.0 A. What is the magnitude of the magnetic field at the center of this solenoid? Compare your result with the field near the single wire in Example 19.3 (notice that this wire carries *three times* the current of the solenoid in this Example!), and comment.

Thinking It Through. The solenoid length is given, but it is not needed. The B field depends only on the linear turn density (n) and on the current (I), as shown in Eq. 19.7.

Solution.

Given: $I = 5.0$ A   *Find:* B (magnitude of magnetic field)
          $n = 10^3$ turns/m

From Eq. 19.7, we get

$$B = \mu_o nI = (4\pi \times 10^{-7}\text{ T·m/A})(10^3\text{ m}^{-1})(5.0\text{ A}) = 2\pi \times 10^{-3}\text{ T} \approx 6.3\text{ mT}$$

Notice that this is about *twenty* times larger than the field near the wire in Example 19.3 and *several hundred* times larger than the Earth's field. This is because the field in the solenoid is the vector sum of all the fields from each coil that makes up the solenoid, and three hundred separate coils make up this solenoid.

Follow-up Exercise. In Example 19.4, if the current was reduced to 1.0 A, how many turns in the same length would be needed to create the same magnetic field near the center of the solenoid?

19.4 Magnetic Materials

Objectives: To explain (a) how ferromagnetic materials enhance external magnetic fields, (b) how "permanent" magnets are produced, and (c) how "permanent" magnetism can be destroyed.

Why are some materials magnetic or easily magnetized, whereas others are not? How can a bar magnet create a magnetic field when there is no obvious current in it? Knowing that a current produces a magnetic field and comparing the magnetic fields of a bar magnet and a current-carrying loop (see Figs. 19.1 and 19.9), you might begin to think that the magnetic field of the bar magnet is in some way due to *internal* currents or moving charges.

Note: Review Fig. 15.1.

Recall that, according to the simplistic solar system model of the atom presented in Chapter 15, the electrons orbit the nucleus. They are thus charges in motion, and each orbiting electron is in effect a current loop that produces its own magnetic field. This model depicts a material as having many internal atomic magnets. However, detailed analysis of atomic structure shows that the net magnetic field produced by the orbiting atomic electrons is much smaller than what would be expected from this simplistic model. Also, the atoms are sometimes randomly arranged. As a result, the net magnetic effect due to orbiting electrons for most materials is either zero or very small.

What then is the source of the magnetism of magnetic materials? Modern theory says that this permanent type of magnetism is an atomic effect due to elec-

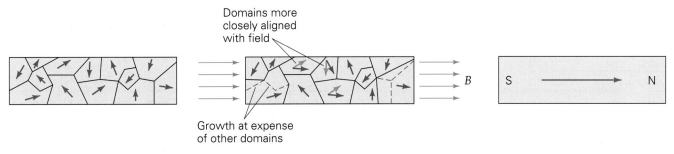

Domains more
closely aligned
with field

B

S ———————→ N

Growth at expense
of other domains

(a) No external magnetic field   **(b) With external magnetic field**   **(c) Resulting bar magnet**

•**FIGURE 19.12   Magnetic domains   (a)** With no external magnetic field, the magnetic domains of a ferromagnetic material are randomly oriented and the material is unmagnetized. **(b)** In an external magnetic field, domains with orientations parallel to the field may grow at the expense of other domains, and the orientations of some domains may become more aligned with the field. **(c)** As a result, the material becomes magnetized, or exhibits magnetic properties.

tron spin. The word "spin" is used because of an early idea that an additional contribution to the atomic magnetic field might come from an electron spinning on its axis. The classical scenario likens a spinning electron to the Earth rotating on its axis. However, this is *not* the case. Electron spin is strictly a quantum-mechanical effect with no direct classical analogue. Nonetheless, this scenario of spinning electrons creating magnetic fields is useful for qualitative thinking and reasoning. (We will revisit the topic of electron spin in relation to the magnetic field it creates later in this chapter.)

In atoms with two or more electrons, the electrons usually are arranged in pairs with their spins oppositely aligned. The magnetic fields then cancel each other, and the material is not magnetic. Aluminum is such a material. However, in certain strongly magnetic materials, known as **ferromagnetic materials,** electron spins do not pair, or cancel, completely. As a result, there is a strong coupling or interaction between neighboring atoms, leading to the formation of large groups of atoms called **magnetic domains.** In a given domain, many of the electron spins are aligned, thus producing a net magnetic field. There aren't very many naturally occurring ferromagnetic materials. The most common are iron, nickel, and cobalt. Gadolinium and certain manufactured alloys are also ferromagnetic.

In an unmagnetized ferromagnetic material, the domains are randomly oriented and there is no net magnetization (•Fig. 19.12a). But, when a ferromagnetic material is placed in an external magnetic field, two things can happen (Fig. 19.12b):

1. Domain boundaries change, and the domains with magnetic orientations in the direction of the external field grow at the expense of the other domains.

2. The magnetic orientation of some domains may change slightly so as to be more aligned with the field.

You now can understand why an unmagnetized piece of iron is attracted to a magnet and why iron filings line up with a magnetic field. Essentially, the pieces of iron become induced magnets (Fig. 19.12c).

Electromagnets and Magnetic Permeability

Ferromagnetic materials are used to make electromagnets, usually by wrapping a wire around an iron core (•Fig. 19.13). The current in the coils creates a magnetic field in the iron, which in turn creates its own field many times larger than the coil field. Turning the current on and off, we can turn the magnet (strictly speaking, its magnetic field) on and off at will. The iron used in an electromagnet is called *soft iron.* It is treated so that when the external field is removed, the magnetic

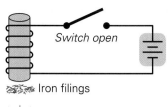

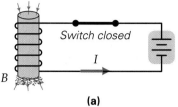

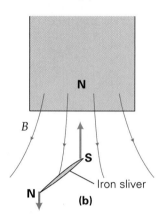

(c)

•**FIGURE 19.13 Electromagnet**
(a) (*top*) With no current in the circuit, there is no magnetic force. (*bottom*) However, with a current in the coil, there is a magnetic field and the iron core becomes magnetized. **(b)** Detail of the lower end of the electromagnet in (a). The sliver of iron is attracted to the end of the electromagnet. **(c)** An electromagnet picking up scrap metal.

domains quickly become unaligned and the iron is said to be *demagnetized*. ("Soft" refers not to the metal's mechanical hardness but to its magnetic properties.)

When an electromagnet is on (Fig. 19.13a), the iron core becomes magnetized and adds to the field of the solenoid. The total field is expressed as

$$B = \mu n I \quad \text{(at the center of the iron-core solenoid)} \quad (19.8)$$

Notice that this equation is similar to that for the magnetic field of an air-core solenoid (Eq. 19.7) but contains μ instead of μ_o, the permeability of free space. Here, μ expresses the **magnetic permeability,** not of free space but of the *core material*. The role permeability plays in magnetism is similar to that of the permittivity, ϵ, in electricity, discussed in Chapter 16. For magnetic materials,

$$\mu = K_m \mu_o \quad (19.9)$$

Here K_m, called the *relative permeability*, is the magnetic analogue of the dielectric constant K.

Both K and K_m are equal to one in a vacuum. Magnetic permeability is a useful material property. A core of a ferromagnetic material with a large permeability in an electromagnet can enhance its field thousands of times (compared to an air core).

Note that the strength of the field of an electromagnet depends on the current ($B = \mu n I$). Large currents produce large fields, but this field generation is accompanied by much greater joule heating (I^2R losses) in the wires. Large electromagnets need water-cooling coils to remove this heat. The problem can be alleviated by using a superconductor, which has zero resistance and thus no joule heating. However, superconducting magnets now in use require tremendous cooling just to keep the conductors in the superconducting state, and this cooling does take energy and itself produces heat. Perhaps someday electromagnets will be made with new high-temperature superconductors (Section 17.3), but today ceramic materials are used that are not easily made into wires that can be wound into an electromagnet.

Iron that retains some magnetism after being in an external magnetic field is called *hard iron* and is used to make so-called permanent magnets. You may have noticed that a paper clip or a screwdriver blade becomes slightly magnetized after being near a magnet. Permanent magnets are produced by heating pieces of some ferromagnetic material in an oven and then cooling them in a strong magnetic field to get the maximum effect.

A *permanent* magnet need not be truly permanent, however—its magnetism can be lost. Hitting such a magnet with a hard object or dropping it on the floor can cause it to lose its domain alignment, thus reducing or eliminating its overall magnetic field. Heating can also cause a loss of magnetism, because the resulting increase in random motions of atoms tends to unalign the domains. One of the worst things you can do to an audio or video magnetic tape cassette is to leave it on the dashboard of a car on a hot day. The thermal motion caused by the heat partially destroys the magnetic voice and/or video signal imprinted on the tape. Above a certain critical temperature, called the **Curie temperature** (or Curie point), domain coupling is destroyed by the increased thermal oscillations and a ferromagnetic material loses its ferromagnetism. This effect was discovered by the French physicist Pierre Curie (1859–1906), husband of (Madame) Marie Curie. The Curie temperature for iron is 770°C.

19.5 Magnetic Forces on Current-Carrying Wires

Objectives: To (a) calculate the magnetic force on a current-carrying wire and the torque on a current-carrying loop and (b) explain the concept of a magnetic moment for such a loop.

We have seen that an electric charge moving in a magnetic field experiences a force unless it is moving parallel to the field lines. Thus, a current-carrying wire in a magnetic field should also experience such a force, because a current is made up of moving charges. That is, the magnetic forces on the moving charges should give a resultant force on the wire conducting them, and in fact this is found to be the case.

A current can be thought of as positive charges moving in a straight wire, as depicted in •Fig. 19.14. Here the magnetic force is at its maximum since $\theta = 90°$. In a time t, a charge q would move on the average through a length $L = vt$, where v is the average drift velocity and is perpendicular to the B field. Since all the moving charges ($\Sigma_i q_i$) in this length of wire experience a force due to the magnetic field, the magnitude of the total maximum force on the length of wire, from Eq. 19.2, is

Note: Remember that thinking of a current in terms of a positive charge is simply a convention. In reality, negative electrons are the charge carriers for ordinary electric current.

$$F_{\max} = (\Sigma_i q_i) vB$$

Substituting L/t for v and rearranging gives

$$F_{\max} = (\Sigma_i q_i)\left(\frac{L}{t}\right) B = \left(\frac{\Sigma_i q_i}{t}\right) LB$$

But $\Sigma_i q_i/t$ is the current (I), since it represents the charge passing through a cross-sectional area of the wire per unit time. Therefore,

$$F_{\max} = ILB \qquad (19.10)$$

Equation 19.10 gives the maximum force on the wire because the current is at an angle of 90° to the magnetic field. If the wire is *not* perpendicular to the field but at some angle θ, then the force on the wire will be less, as in the case of the force on a moving charge. In general, the force on a length of current-carrying wire in a uniform magnetic field is

$$F = ILB \sin \theta \qquad \begin{array}{l} \textit{magnetic force on} \\ \textit{a current-carrying wire} \end{array} \qquad (19.11)$$

Note that if the wire is parallel to the field, the force on it is zero (as is the force on a charge moving parallel to a magnetic field).

The direction of the magnetic force on a current-carrying wire is also given by a right-hand rule. The length of wire is treated as a vector **L** in the direction of the current. If it is imagined to be turned by the fingers of the right hand toward **B**, the extended thumb points in the direction of **F**. This **right-hand force rule for a current-carrying wire** is often stated as follows:

When the fingers of the right hand are pointed in the direction of the conventional current I and then curled toward the vector **B**, the extended thumb points in the direction of the magnetic force on the wire.

Note that this rule is simply an extension of the right-hand force rule on individual charges.

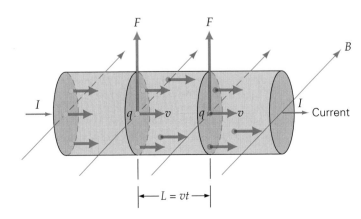

•**FIGURE 19.14 Force on a wire segment** The diagram helps visualize and determine the force on a length of current-carrying wire in a magnetic field. Charge carriers are considered to be positive for convenience only.

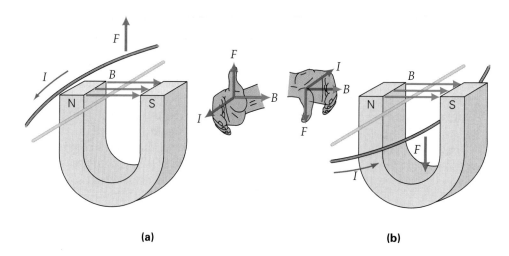

•FIGURE 19.15 Force on a current-carrying wire The direction of the force is given by pointing the fingers of the right hand in the direction of the conventional current I and then curling them toward **B**. The extended thumb points in the direction of **F**. The force is (a) upward and (b) downward.

(a)

(b)

Two examples of magnetic forces on current-carrying wires in a magnetic field are given in •Fig. 19.15. Apply the right-hand force rule in each case to see how the direction of the force is found. To see how two current-carrying wires interact magnetically, consider the following Example.

Example 19.5 ■ Attraction or Repulsion: Magnetic Force between Two Parallel Wires

Two long, parallel wires are separated by a distance d and carry currents I_1 and I_2 in the same direction, as illustrated in •Fig. 19.16. (a) Derive an expression for the magnetic force per unit length on one of the wires due to the other and show that, in keeping with Newton's third law, the forces on the two wires are equal in magnitude. (b) Show that for these current directions, the forces are mutually attractive, in keeping with Newton's third law, that the forces of an action–reaction pair must be opposite in direction.

Thinking It Through. (a) Each wire experiences a force due to the magnetic field of the other. Thus we expect the answer to depend on the current in each wire and the distance between them. We can get the field magnitude from Eq. 19.4, then we can calculate the force from the expression for the force on a current-carrying wire (Eq. 19.10). To find the force *per unit length*, we divide the force by the length of the wire. (b) To find the force direction, we determine the direction of the field created by the other wire, and then use the right-hand force rule for current-carrying wires.

Solution. The symbols are given in Fig. 19.16.

(a) By the right-hand source rule, the magnetic field due to I_1 at wire 2 is in the direction shown; its magnitude (Eq. 19.4) is

$$B_1 = \frac{\mu_o I_1}{2\pi d}$$

The direction of B_1 is perpendicular to the wire carrying I_2, so from Eq. 19.10 the magnetic force on wire 2 is $F_2 = I_2 L B_1$. Thus the magnitude of the force *per unit length* (F_2/L) is

$$\frac{F_2}{L} = I_2 B_1 = \frac{\mu_o I_1 I_2}{2\pi d}$$

The magnetic field due to I_2 exerts a force per unit length on the wire carrying I_1, which similarly is given by

$$\frac{F_1}{L} = \frac{\mu_o I_2 I_1}{2\pi d}$$

This is clearly the same magnitude as the force on wire 2.

(b) By the right-hand force rule, we can verify that each force is *toward* the other wire. Thus, the forces are equal and opposite, in keeping with Newton's third law. Therefore, two parallel wires carrying current in the same direction are attracted to one another.

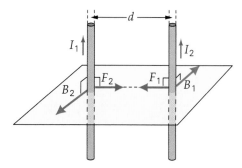

•FIGURE 19.16 Mutual interaction For two parallel current-carrying wires, the magnetic field due to each current interacts with the other current, and there is a force on each wire. If the currents are in the same direction, the forces are attractive. This effect is used to define the ampere unit of current.

Follow-up Exercise. In a household wire cord, the centers of two wires are separated by 1.5 mm. Assume that at some instant, each carries a current of 5.0 A, but in *opposite* directions. (a) Calculate the magnetic force on 1.0 m of one of the wires. (b) Is the force between the wires attractive or repulsive? Explain how you determined the directions of the forces.

The magnetic force between parallel wires in an arrangement like that analyzed in Example 19.5 provides the basis for the definition of the ampere. According to the National Institute of Standards and Technology (NIST),

> The ampere is defined as that current which, if maintained in each of two long parallel wires separated by a distance of 1 m in free space, would produce a magnetic force between the wires of exactly 2×10^{-7} N for each meter of wire.

Torque on a Current-Carrying Loop

Another important instance of magnetic force on a current is that exerted on a current-carrying loop, such as the rectangular loop in •Fig. 19.17a. (The wires connecting the loop to a voltage source are not shown.) Suppose that the loop is free to rotate about an axis passing through opposite sides, as shown in the figure. There are no net forces or torques on the pivot sides of the loop (the sides through which the axis of rotation passes). When these sides are parallel to the B field, the force on them is zero. (Why?) At any other angle to the field, the forces on them are equal and opposite in the plane of the loop (can you see why?) and so produce no net torque or net force. However, the equal and opposite forces on the other two sides of the loop (the sides parallel to the axis of rotation) *do* produce a net torque, which tends to rotate the loop.

To see how this works, consider Fig. 19.17b, which shows a side view of Fig. 19.17a. The magnitude of the magnetic force on each of the nonpivoted sides, whose length is L, is given by $F = ILB$ since L and B are at right angles to each other. Recall that the torque produced by a force is given by $\tau = r_{\perp}F$, where $r_{\perp}$ is the perpendicular lever arm from the axis of rotation to the line of the force. From Fig. 19.17b, we see that $r_{\perp} = \frac{1}{2}w \sin \theta$, where w is the width of the loop and θ is the angle between the normal to the plane of the loop and the direction of the magnetic field. The net torque on the loop from both forces is the sum of the two torques, or

$$\tau = r_{\perp}F + r_{\perp}F = (\tfrac{1}{2}w \sin \theta)F + (\tfrac{1}{2}w \sin \theta)F = wF \sin \theta = w(ILB) \sin \theta$$

Then, since wL is the area (A) of the loop, we can express the magnitude of the torque on a single pivoted, current-carrying loop as

$$\tau = IAB \sin \theta \qquad \text{torque on current-carrying loop} \qquad (19.12)$$

Although derived for a rectangular loop, Eq. 19.12 is valid for a flat loop of *any* shape.

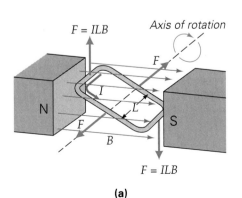

(a)

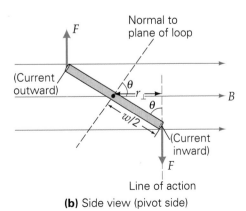

(b) Side view (pivot side)

Note: Torque is discussed in Section 8.2.

•**FIGURE 19.17 Force and torque on a current-carrying pivoted loop** **(a)** A current-carrying rectangular loop oriented in a magnetic field as shown experiences a force on each of its sides. The forces on the sides parallel to the axis of rotation produce a torque that causes the loop to rotate. **(b)** A side view shows the geometry for determining the torque.

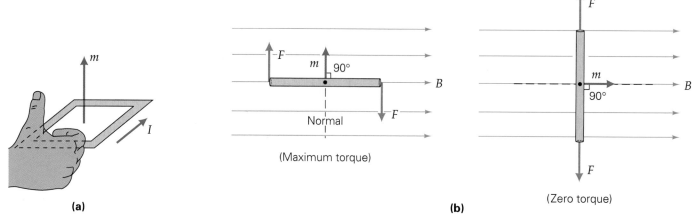

•FIGURE 19.18 Magnetic moment of a current-carrying loop (a) A right-hand rule determines the direction of the loop's magnetic moment vector, **m**. The fingers follow the current direction, and the thumb gives the direction of **m**. (b) Conditions of the maximum and zero torque. The magnetic moment vector of the loop will tend to align with the external magnetic field direction.

The quantity IA is referred to as the magnitude of the **magnetic moment** vector. The symbol for its magnitude is m. Thus $m = IA$. The magnetic moment vector *direction* is determined by circling the fingers of your right hand in the direction of the (conventional) current flow. Your thumb then gives the direction of the magnetic moment vector, **m** (•Fig. 19.18a). Thus we can rewrite Eq. 19.12 in terms of the loop's magnetic moment:

$$\tau = mB \sin \theta \qquad (19.13)$$

Note: A *coil* consists of N loops of the same size, all carrying the same current, I, in series.

Note that if more than one loop makes up the coil—in general, N loops—the torque is N times that due to one loop (since they all have the same current), or

$$\tau = NIAB \sin \theta \qquad \begin{array}{l}\textit{torque on}\\\textit{current-carrying coil}\end{array} \qquad (19.14)$$

The torque tends to align the magnetic moment vector (**m**) with the magnetic field direction. Thus, a loop in a magnetic field will experience a torque until $\sin \theta = 0$ (at which point the forces producing the torque are parallel to the plane of the loop; see Fig. 19.18b). This occurs when the plane of the loop is perpendicular to the field. If the loop is started at rest such that its magnetic moment makes some nonzero angle with the magnetic field direction, the loop will undergo an angular acceleration that will rotate it to the zero angle position. Rotational inertia will cause it to coast through the equilibrium position (zero angle, zero torque) on to the other side. On that side, the torque will slow it down, stop it, then reaccelerate it back toward equilibrium. In the absence of frictional forces, this rotational oscillation would go on forever. In other words, the torque on the current loop is restoring and tends to cause the loop's magnetic moment to oscillate about the field direction.

There are many interesting applications of the concept of magnetic moment at the atomic level. One arises when we consider a spinning electron as a series of concentric current loops (•Fig. 19.19a). Imagined in this way, the electron evi-

•FIGURE 19.19 Magnetic moment of the electron (a) A spinning electron can be depicted as a spinning sphere of negative charge, having its own magnetic moment. (b) When the electron is placed in a magnetic field, its magnetic moment will tend to line up with that field. (c) When lined up with the external field, the electron's own field enhances the external field. In ferromagnets, many electrons doing this in unison produce a net field that can be thousands of times stronger than the external field.

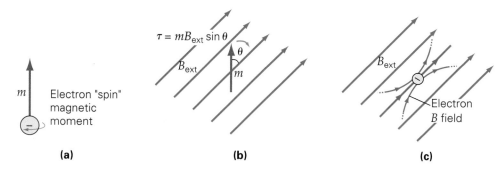

dently must possess a magnetic moment. When placed in an external magnetic field, this moment will tend to line up with the field (Fig. 19.19b), thus enhancing the total field by adding its own field to the external field. In ferromagnetic materials, this enhancement can produce a net magnetic field thousands of times larger than the external field, as we saw previously (Fig. 19.19c).

In addition to the magnetic moment (and field) produced by its spin, an atomic electron has a magnetic moment and associated field as a result of its orbital motion around the nucleus, as the following Example shows.

Example 19.6 ■ The Hydrogen Atom Revisited: Orbital Magnetic Moment of the Electron

An orbiting electron possesses an *orbital* magnetic moment, and the motion can produce a magnetic field. In the Bohr model of the hydrogen atom (discussed in Chapter 27), the single electron travels in a circle with a radius of 0.0529 nm and has a period of 1.5×10^{-16} s. (a) Derive a *general* expression for the orbital magnetic moment in terms of the orbital radius (r), the charge on the electron (e), and the orbital period (T). (b) What are the magnitude and direction (in comparison to its orbital rotation sense) of the orbital magnetic moment of the electron in the hydrogen atom?

Thinking It Through. To compute the magnetic moment of this "atomic loop," we need the area and the current ($m = IA$). For a circular loop, the area is πr^2. To calculate the current, recall the definition, $I = q/t$. In this case, a charge of $q = |e|$ passes any given point in the orbit in each period (T).

Solution. We have the following numerical data:

Given: $|e| = 1.60 \times 10^{-19}$ C (from Table 15.1)
$T = 1.5 \times 10^{-16}$ s
$r = 0.0529$ nm $= 5.29 \times 10^{-11}$ m

Find: (a) General expression for orbital magnetic moment (m)
(b) Orbital magnetic moment of the electron

(a) Here we let e represent the magnitude of the charge on the electron, so the equivalent current is $I = q/t = e/T$. Thus the expression for the orbital magnetic moment is

$$m = IA = \left(\frac{e}{T}\right)\pi r^2 = \frac{e\pi r^2}{T}$$

(b) Inserting the given data, we have

$$m = \frac{e\pi r^2}{T} = \frac{(1.60 \times 10^{-19}\text{ C})\pi(5.29 \times 10^{-11}\text{ m})^2}{1.5 \times 10^{-16}\text{ s}} = 9.4 \times 10^{-24}\text{ A·m}^2$$

An electron orbiting clockwise as we look down on its orbital plane is equivalent to a counterclockwise conventional current (since the electron is negatively charged). If you curl the fingers of your right hand counterclockwise, your thumb will give an orbital magnetic moment direction out of the plane of the orbit (●Fig. 19.20).

Follow-up Exercise. Compare the magnetic moment of the electron in this Example to that of an ordinary coil of 20 closely packed wire loops, each having a diameter of 4.0 cm and carrying a current of 5.0 A.

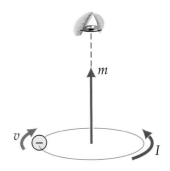

●**FIGURE 19.20 An electron in orbit** As viewed from above, an electron orbiting a proton clockwise in a hydrogen atom is equivalent to a conventional current (I) in the counterclockwise sense. By the right-hand rule for the magnetic moment of a current loop (Fig. 19.18a), the moment (**m**) is directed toward you. See Example 19.6.

On a typical scale, current-carrying loops, even ones with such limited motion, have important applications. Since the current is directly related to the torque, such a loop provides a means to measure current. Also, there are ways to keep a loop rotating, allowing continuous conversion of electrical energy to mechanical energy. This is the operating principle of motors. These and other applications of electromagnetism are considered in Section 19.6.

19.6 Applications of Electromagnetism

Objective: To explain the operation of various instruments whose functions depend on electromagnetic interactions.

With the principles you have learned so far about electromagnetic interactions, you can understand the operation of several widely used scientific instruments. Another very important application, magnetic resonance imaging (MRI), is described in Chapter 28.

The Galvanometer

Note: See Section 18.4.

In the discussion of ammeters and voltmeters in Chapter 18, the galvanometer was considered to be simply a circuit element that measured small currents. Now you are ready to see how it does this. As •Fig. 19.21a shows, a galvanometer consists of a coil of wire loops on an iron core that pivots between the pole faces of a permanent magnet. When a current exists in the coil, the coil experiences a torque. A small spring supplies a counter (restoring) torque, and in equilibrium a pointer indicates a deflection ϕ that is proportional to the current in the coil.

As you learned in Section 19.5, the torque on a single current-carrying loop in a magnetic field is given by $\tau = IAB \sin \theta$. The dependence of the torque on the angle θ between the field lines and a line normal to the plane of the loop would cause a problem in measuring current with a galvanometer coil. If the coil rotated from its position of maximum torque ($\theta = 90°$, or the B field perpendicular to the nonpivoted sides of the coil), the torque would become less, and the pointer deflection ϕ would not be proportional to the current alone. This problem is avoided by making the pole faces curved and wrapping the coil on a cylindrical iron core. The core tends to concentrate the field lines such that **B** is always perpendicular to the nonpivoted side of the coil (Fig. 19.21b).

The magnetic force is then always perpendicular to the plane of the coil, and the torque therefore does *not* vary with the angle θ. We assume that the galvanometer coil is made of N loops, each carrying a current I. Thus, the total torque on the coil is

$$\tau = NIAB \tag{19.15}$$

The restoring torque of the spring (τ_s) in a galvanometer is given by a rotational form of Hooke's law ($F_s = kx$ for linear restoring force):

$$\tau_s = k\phi \tag{19.16}$$

Here ϕ is the deflection angle as measured by the pointer. With a current in the coil, the coil and the pointer will rotate until the torque on the coil is balanced by that on the spring. In equilibrium, the magnitudes of the torques are equal, and by combining Eqs. 19.15 and 19.16 we obtain

$$\phi = \frac{NIAB}{k} \tag{19.17}$$

Thus, the pointer deflection (ϕ) is directly proportional to the current (I).

A properly calibrated galvanometer can measure very small currents (in the microamp range). As we saw in Section 18.4, with an appropriate shunt resistor, a galvanometer forms the basis of an ammeter for measuring larger currents. Also, with an appropriate series multiplier resistor, the galvanometer can be made into a voltmeter.

The dc Motor

In general, *an electrical motor is a device that converts electrical energy into mechanical energy.* Such a conversion occurs with a galvanometer: A current causes the coil to rotate mechanically. However, a galvanometer is not considered to be a motor. A practical **dc motor** must have continuous rotation for continuous energy output.

A pivoted, current-carrying coil (with N loops) in a magnetic field will rotate freely, but for only a half-cycle, or through a maximum angle of 180°. Recall that

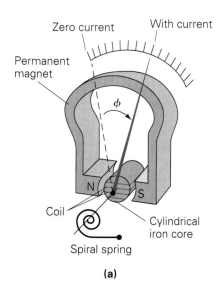

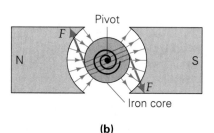

•FIGURE 19.21 The galvanometer (a) The deflection (ϕ) of the needle from its zero-current position is proportional to the current in the coil. A galvanometer can therefore detect and measure currents. (b) A magnet with curved pole faces is used so that the field lines are always perpendicular to the core surface and the torque does not vary with ϕ.

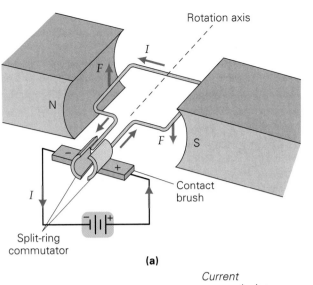

Rotation axis

•FIGURE 19.22   A dc motor
(a) A split-ring commutator
reverses the polarity and current
each half-cycle, so the coil rotates
continuously. (b) An end view
shows the forces on the coil and its
orientation during a half-cycle. (For
simplicity we show a single loop,
but it is actually a coil of N loops.)

(a)

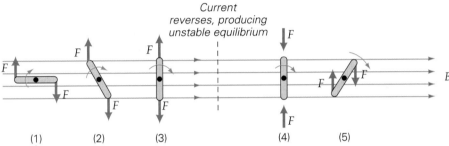

Current
reverses, producing
unstable equilibrium

(b) End view of loop, showing clockwise loop rotation sequence

$\tau = NIAB \sin \theta$, and when the magnetic field is perpendicular to the plane of the coil ($\sin \theta = 0$), the torque is zero and the coil is in equilibrium.

To provide for continuous rotation, the current is reversed every half turn so that the torque-producing forces are reversed. This is done by means of a *split-ring commutator,* an arrangement of two metal half-rings insulated from each other (•Fig. 19.22a). The ends of the wire that forms the coil are fixed to the half-rings, and the current is supplied to the coil through the commutator by means of contact brushes. Then, with one half-ring electrically positive (+) and the other negative (−), the coil and ring rotate. When they have gone through half a rotation, the half-rings come in contact with the opposite brushes. Their polarity is thus reversed, and the current in the coil is in the opposite direction. This changes the directions of the torque (Fig. 19.22b). The torque is zero at the equilibrium position, but the coil is in unstable equilibrium. The coil has enough rotational momentum to continue through this point, and it rotates another half-cycle. This process repeats in continuous operation. For a real motor the rotating shaft is called the *armature.*

The Cathode Ray Tube (CRT)

The **cathode ray tube (CRT)** is a vacuum tube that is used in an oscilloscope, such as those in some laboratories (•Fig. 19.23). Electrons, negatively charged, are "boiled off" a hot filament in an electron gun and accelerated by a voltage applied between the cathode (−) and anode (+). The picture tube in most television sets and computer monitors is also a cathode ray tube (•Fig. 19.24). Magnetic coils are usually used there to deflect the electron beam. The beam scans the fluorescent screen in a 525-line pattern in a fraction of a second. For a black-and-white TV, the signals transmitted from a video camera reproduce an image on the screen as a mosaic of light and dark dots, depending on whether the intermittent beam is on or off at a particular instant.

Producing the images on a color TV or color monitor is somewhat more involved. A common color picture tube has three beams, one for each of the primary

•FIGURE 19.23   Cathode ray
tube (CRT)   The motion of the
deflected beam traces a pattern on
a fluorescent screen.

•FIGURE 19.24 Television tube

(a) A television picture tube is a cathode ray (electron) tube, or CRT. The electrons are accelerated by a high voltage between the cathode and anode and then deflected to the proper location on a fluorescent screen by magnetic fields produced by current-carrying coils. **(b)** The beam scans every other line on the screen in one downward pass that takes $\frac{1}{60}$ s and then scans the lines in between in a second pass of $\frac{1}{60}$ s. This yields a complete picture of 525 lines in $\frac{1}{30}$ s.

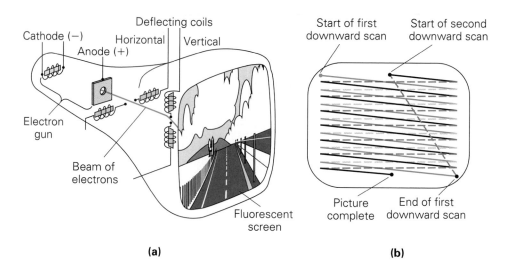

(a)   (b)

colors (red, green, and blue; Chapter 25). Phosphor dots on the screen are arranged in groups of three (triads) with one dot for each primary color. The excitation of the appropriate dots and the resulting combination of colors produce a color picture.

The fact that a TV picture tube is a CRT can be demonstrated by bringing a magnet near a black-and-white TV screen. The magnetic field will deflect the moving electrons and distort the picture. *Do not try this with a color TV!* Internal metallic parts may become magnetized, distorting the picture permanently.

The Mass Spectrometer

Have you ever thought about how the mass of an atom or molecule is measured? Electric and magnetic fields provide a way in one form of **mass spectrometer** (often called a "mass spec" for short). Actually, the masses of *ions* are measured since electric and magnetic fields have motional effects only on charged particles. (Recall that an ion is an atom or molecule with a net electric charge.)

Ions with a known charge ($+q$) are produced by heating the substance to be analyzed. The resulting beam of ions introduced into the mass spec has a distribution of speeds. Ions with a particular velocity are selected by means of a *velocity selector*, made up of charged plates and a magnetic field that allow particles traveling at only that velocity to go undeflected. The values of the electric and magnetic fields between the plates of the velocity selector determine the velocity (•Fig. 19.25). For a positively charged ion, the electric field produces a downward force (magnitude $F = qE$), and the magnetic field produces an upward force (magnitude $F = qvB_1$).

Note: The difference in mass between a neutral atom and its charged (ionic) version amounts to only an electron mass or two. This mass is negligible compared to the mass of the atom, which is dominated by the protons and neutrons in the nucleus.

•FIGURE 19.25 Principles of the mass spectrometer

Ions pass through the velocity selector; those with a particular velocity ($v = E/B_1$) enter a magnetic field (B_2). The ions are deflected, with the radius of the circular path depending on the mass and net charge of the ion. Paths of two different radii indicate that the beam contains ions of two different masses (assuming they have the same charge).

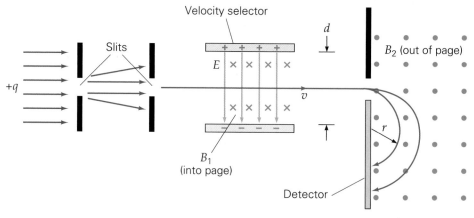

Top view

If the beam is not deflected, the resultant force must be zero, so

$$qE = qvB_1$$

or

$$v = \frac{E}{B_1}$$

If the plates are parallel, then $E = V/d$. Since the voltage and the plate separation distance d are the controllable or measured quantities, a more practical version of the previous equation is

$$v = \frac{V}{B_1 d} \qquad \text{ion speed in velocity selector} \qquad (19.18)$$

Thus, a beam with a known speed can be selected by choosing the appropriate values of V, B_1, and d.

Beyond the velocity selector, the beam passes through a slit into another magnetic field ($\mathbf{B}_2$), which is perpendicular to the direction of the beam. The force due to this magnetic field (magnitude $F = qvB_2$) is always perpendicular to the velocity of the ions, which are thus deflected along a circular arc. The magnetic force supplies the centripetal force (Eq. 7.11) for this motion, and

$$\frac{mv^2}{r} = qvB_2$$

Using the expression for the selected ion velocity (Eq. 19.18), we get

$$m = \left(\frac{qdB_1B_2}{V} \right) r \qquad \begin{array}{l} \textit{particle mass via} \\ \textit{mass spectrometer} \end{array} \qquad (19.19)$$

The quantity inside the parentheses is a constant (assuming all ions have the same net charge q). Hence the greater the mass of an ion, the greater the radius of the circular path. Two circular paths of different radii are shown in Fig. 19.25, which indicates that the beam contains ions of two different masses. If the radius of curvature r for an ion beam is measured (for instance, by recording the position of the beam with a detector), the mass of the ion can be calculated from the other known quantities in Eq. 19.19. Consider the following Example of a mass spec arrangement.

Example 19.7 ■ Molecules Moving in Big Circles: A Mass Spectrometer Setup

One electron is removed from a methane molecule prior to entering the mass spec arrangement shown in Fig. 19.25. After passing through the velocity selector, the charged molecule has a speed of 1.0×10^3 m/s. It then enters the final magnetic field region, in which the field is 6.7×10^{-3} T, and follows a circular path. The molecule lands on the detector 5.0 cm from the entrance to the field. Determine the mass of this molecule. (Neglect the mass of the removed electron.)

Thinking It Through. The centripetal force around the circular arc is provided by the magnetic force. The centripetal force on the charged molecule is given by Eq. 7.11, and the magnetic force by Eq. 19.3. The only unknown is the mass of the molecule, which can then be found.

Solution. We must convert the given data into the SI system.

Given: $|q| = 1.6 \times 10^{-19}$ C (from Table 15.1)   *Find:* m (mass of a methane molecule)
$\qquad r = d/2 = (5.0 \text{ cm})/2 = 0.025 \text{ m}$
$\qquad B_2 = 6.7 \times 10^{-3}$ T
$\qquad v = 1.0 \times 10^3$ m/s
$\qquad \theta = 90°$ (from Fig. 19.25)

We write that the molecule's centripetal force ($F_c = mv^2/r$) is equal to the magnetic force ($F_m = qvB_2 \sin 90° = qvB_2$):

$$\frac{mv^2}{r} = qvB_2$$

We solve for m and put in the given values:

$$m = \frac{qB_2 r}{v} = \frac{(1.6 \times 10^{-19}\,\text{C})(6.7 \times 10^{-3}\,\text{T})(0.025\,\text{m})}{1.0 \times 10^3\,\text{m/s}} = 2.7 \times 10^{-26}\,\text{kg}$$

Follow-up Exercise. In this Example, if the magnetic field between the parallel plates of the velocity selector, which are 10.0 mm apart, is set at 5.00×10^{-2} T, what voltage must be applied to the plates?

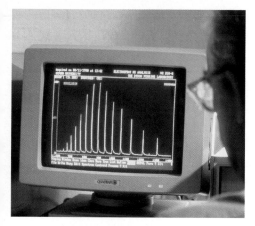

•**FIGURE 19.26 Mass spectrometer** Computer display of a mass spectrometer. The molecule being analyzed is myoglobin, a protein that stores oxygen in muscle tissue. Each of the peaks on the display represents the mass of a particular ionized fragment of the molecule. Such patterns, called *mass spectra,* can help determine the composition and structure of large molecules. The mass spectrometer can also be used to identify tiny amounts of a particular molecule in a complex mixture.

Mass spectrometers have many functions in modern laboratories (•Fig. 19.26). For example, they are used to help establish the age of ancient rocks and more recent human artifacts by measuring the relative abundance of different isotopes that they contain. (This method, called radioactive dating, is discussed in Section 29.3.) Among the other roles of these devices are tracking short-lived intermediates in studies of the biochemistry of living organisms; helping determine the structure of large organic molecules; analyzing the composition of complex mixtures, such as a sample of smog-laden air or the output of a petroleum refinery; and detecting tiny amounts of impurities in metal alloys and semiconductors.

Another interesting application of the principles that underlie the magnetic mass spectrometer can be found in the *momentum selector* (•Fig. 19.27). This device is used extensively in accelerator laboratories throughout the world as a means of selecting particles having a desired momentum (and thus, in effect, of selecting the particle energy). For example, it is often essential to learn how a certain nuclear reaction depends on the energy of the bombarding particles. In future cancer therapy using particle beams, it may be important to control the kinetic energy of the particles very precisely in order to ensure that they penetrate just to the tumor depth and not beyond.

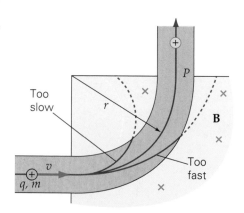

•**FIGURE 19.27 Momentum selector** By selecting only those particles traveling in a path of a particular radius, we are selecting only those particles having a specific linear momentum and kinetic energy. Particles with a higher or a lower momentum follow paths that do not take them through the exit opening of the beam tube. Such precise energy determination is important in basic nuclear physics research and in possible future cancer therapy using charged particle beams.

The Electronic Balance

Traditional laboratory balances measure mass by balancing the weight force of an unknown mass against that of a known mass. The newer, digital electronic balances (•Fig. 19.28a) work on a different principle. There is still a suspended beam with a pan on one end that holds the object to be weighed (or, more correctly, massed), but there is no known mass. The balancing downward force is supplied by current-carrying coils of wire in the field of a permanent magnet (Fig. 19.28b). The coils move up and down in the cylindrical gap of the magnet, and the downward force is proportional to the current in the coils. The mass of the object in the pan is determined from the coil current that produces a force just sufficient to balance the beam.

The current required to produce balance is controlled automatically. This control is done by means of light photosensing and a feedback loop. When the beam is balanced and horizontal, a knife-edge obstruction cuts off part of the light from a source that falls on a photosensitive "electric eye." The resistance of the electric eye depends on the amount of light falling on it. This resistance controls the current that an amplifier sends through the coil. For example, if the beam tilts so that the knife-edge raises and more light strikes the eye, the current in the coil is increased to counterbalance the tilting. In this manner, the beam is electronically maintained in nearly horizontal equilibrium. The current that keeps the beam in the horizontal position is read out on a digital ammeter that is calibrated in grams or milligrams instead of amperes.

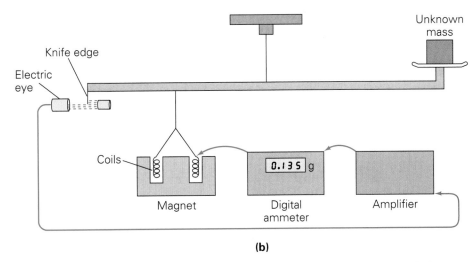

(a) **(b)**

•**FIGURE 19.28 Electronic balance** (a) A digital electronic balance. (b) Diagram of the principle of an electronic balance. The balance force is supplied by electromagnetism.

*19.7 The Earth's Magnetic Field

Objectives: To (a) state some of the general characteristics of the Earth's magnetic field, (b) explain the theory about its possible source, and (c) discuss some of the ways in which it affects the Earth's local environment.

The magnetic field of the Earth was used for centuries before people had any clues about its origin. In ancient times, navigators used lodestones or magnetized needles to locate north. Some other forms of life, including certain bacteria and homing pigeons, also use the Earth's magnetic field for navigation. (See Insight on p. 622.)

An early study of magnetism was done by the English scientist Sir William Gilbert around 1600. In investigating the magnetic field of a specially cut spherical lodestone that simulated the Earth, he came to believe that the Earth as a whole acts as a magnet. Gilbert thought that the field might be produced by a large body of permanently magnetized material within the Earth.

In fact, the Earth's external magnetic field, or the geomagnetic field, does have approximately the configuration that would be produced by a large interior bar magnet with the south pole of the bar magnet pointing north, as •Fig. 19.29 shows. The magnitude of the horizontal component of the Earth's magnetic field at the magnetic equator is about 10^{-5} T (about 0.1 G), and the vertical component at the geomagnetic poles is about 10^{-4} T (about 1 G). It has been calculated that for a ferromagnetic material of maximum magnetization to produce this field, it would have to occupy only about 0.01% of the Earth's volume.

The idea of a ferromagnet of this size within the Earth may not seem unreasonable at first, but this cannot be a correct model. The interior temperature of the Earth is well above the Curie temperatures of iron and nickel, the ferromagnetic materials believed to be the most abundant in the Earth's interior. For iron, this temperature is 770°C, which is attained at a depth of only 100 km. The temperature of the Earth is higher at greater depths. Thus the existence of such a permanent internal magnet is impossible.

The fact that an electric current produces a magnetic field has led scientists to speculate that the Earth's magnetic field is associated with motions in the liquid outer core. (Recall that the magnetic field of a loop of wire or a solenoid is similar to that of a bar magnet—compare Figs. 19.9 and 19.10 to Fig. 19.1.) These motions may be associated in some way with the Earth's rotation. We know that Jupiter and Saturn have magnetic fields much larger than that of Earth; these planets are

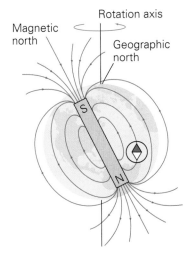

•**FIGURE 19.29 Geomagnetic field** The Earth's magnetic field is similar to that of a bar magnet. However, a permanent solid magnet could not exist within the Earth because of the high temperatures there. The field is believed to be associated with motions in the liquid outer core deep within the Earth.

Magnetism in Nature

For centuries, humans have relied on compasses to provide directional information for navigation according to the Earth's magnetic field (Fig. 19.29). In contrast, certain living creatures have their own built-in directional sensors. For example, some species of bacteria have been determined to be *magnetotactic*— that is, able to sense or feel the presence and direction of the Earth's magnetic field.

The initial experiments* were done in the 1980s on bacteria commonly found in the mud of bogs, swamps, and ponds. When, in a laboratory magnetic field, a droplet of muddy water was viewed under a microscope, one species of bacteria always migrated in the direction of the field (Fig. 1)— just as these bacteria do in their natural environment with the

Earth's field. Furthermore, when these bacteria were killed and therefore could no longer migrate, they maintained their alignment with the magnetic field even when its direction changed. It became apparent that members of this species act like magnetic dipoles or compass needles. Once aligned with the field, they migrate along magnetic field lines simply by moving their *flagella*, whiplike appendages.

What makes these bacteria living compasses? Even among known magnetotactic species, "new" bacteria (formed by cell division) do not have this magnetotactic sense. However, if they live in a solution containing a minimum concentration of iron, they are able to synthesize a chain of small magnetic particles (Fig. 2). Oddly enough, these internal compasses have the same chemical composition as the original slivers of naturally occurring ore used as compasses by

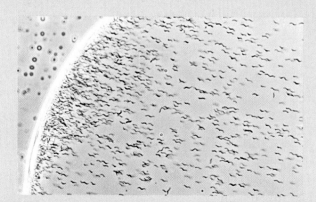

FIGURE 1 Magnetotactic bacteria in migration Bacteria in a drop of muddy water, as viewed under a microscope, align along the direction of the applied magnetic field (north to the left) and accumulate at the edge. When the field is reversed, so is the migration direction.

*See, for example, "Magnetic Navigation in Bacteria," R. P. Blakemore and R. B. Frankel, *Scientific American,* December 1981. We are indepted to Professor Frankel for several interesting discussions on this topic.

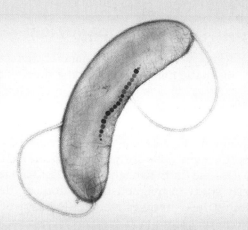

FIGURE 2 An elliptical magnetotactic bacterium A freshwater magnetotactic bacterium, shown in an electron micrograph. Two whiplike appendages, or flagella, are clearly visible along with a chain of magnetite particles.

largely gaseous and rotate very rapidly. Mercury and Venus have very weak magnetic fields; these planets are more like Earth in density and rotate relatively slowly.

Several theoretical models have been proposed to explain the origin of the Earth's magnetic field. For example, it has been suggested that it arises from currents associated with thermal convection cycles in the liquid outer core caused by heat from the inner core. But the details of the mechanism are still not clear.

The axis of the Earth's magnetic field does not lie along its rotational axis, which defines the geographic poles. Hence the north magnetic pole and the geographic North Pole do not coincide (see Fig. 19.29): The magnetic pole is several thousand kilometers south of the geographic North Pole (true north). The south magnetic pole is displaced by even more from its geographic pole, so the magnetic axis is not even a straight line through the center of the Earth!

A compass indicates the direction of *magnetic north,* not "true," or geographic, north. The angular difference in these two directions is called the *magnetic declination* (•Fig. 19.30). As shown on the map, the magnetic declination varies for different locations. Knowing the variations in the magnetic declination is particularly important in accurate navigation, as you can imagine.

ancient sailors: magnetite (Fe_3O_4). The individual particles in the chain are approximately 50 nm across, and the chain of a mature bacterium typically contains about twenty of them. Each particle is a single magnetic domain.

Thus the bacteria are passively steered by their internal compasses. But why is it biologically important for these bacteria to follow the Earth's magnetic field direction? An important piece of the puzzle was found while investigators were studying the same species from Southern Hemisphere waters. These bacteria migrate *opposite* the direction of the Earth's field. Recall that in the Northern Hemisphere the Earth's field inclines downward and that the reverse is true in the Southern Hemisphere. These facts lead scientists to be-lieve that the bacteria are using the field direction for survival (Fig. 3). Since oxygen is toxic to these bacteria, they are most likely to survive in the muddy, nutrient-rich depths, and the Earth's field direction points them that way. Their directional sense also aids them near the equator; there it does not direct them downward but instead keeps them at a constant depth, thus avoiding an upward migration to the deadly oxygen-rich surface waters.

In addition to bacteria, evidence of magnetic field navigation has been found in such diverse places as bees, butterflies, homing pigeons, and dolphins. It is hoped that continued research in this area might provide insight into possible magnetic sensing by other organisms.

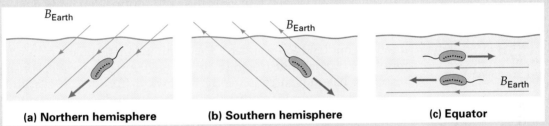

(a) Northern hemisphere   **(b) Southern hemisphere**   **(c) Equator**

FIGURE 3   Survival of the fittest?   (a) In the Northern Hemisphere, where the Earth's magnetic field inclines downward, magnetotactic bacteria follow the field to the nutrient-rich depths. **(b)** In the Southern Hemisphere, the Earth's field is inclined upward, but the bacteria migrate opposite the field and so are able to stay in deep waters. **(c)** Around the equator, the bacteria move parallel to the water surface and thus are kept away from the shallow, oxygen-rich (and hence poisonous) waters.

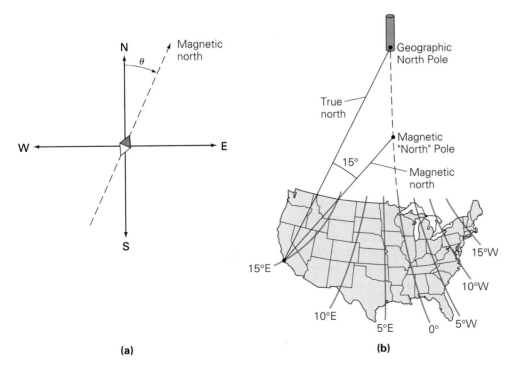

•FIGURE 19.30   Magnetic declination   (a) The angular difference between magnetic north and "true," or geographic, north is called the magnetic declination. **(b)** The magnetic declination varies with location and time. The map shows *isogonic* (same magnetic declination) lines for the continental United States. For locations on the 0° line, magnetic north is in the same direction as true (geographic) north. On either side of this line, a compass has an easterly or westerly variation. For example, on a 15°E line, a compass has an easterly declination of 15° (magnetic north is 15° east of true north).

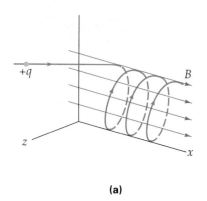

(a)

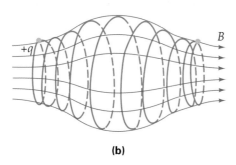

(b)

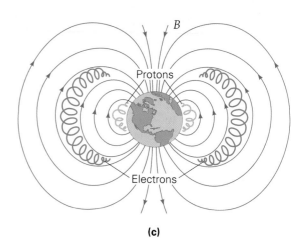

Protons

Electrons

(c)

•**FIGURE 19.31 Magnetic confinement** (a) A charged particle entering a uniform magnetic field at an angle other than 90° moves in a spiraling path. (b) In a nonuniform, bulging magnetic field, particles spiral back and forth as though confined in a magnetic bottle. (c) Particles are trapped in the Earth's magnetic field, and the regions where they are concentrated are called Van Allen belts.

•**FIGURE 19.32 Aurora borealis—the northern lights** This spectacular aurora display is believed to be caused by energetic solar particles trapped in the Earth's magnetic field. The particles excite or ionize air atoms; on de-excitation (or recombination), light is emitted. Aurorae are most commonly observed in the polar regions, where the solar particles are concentrated by the Earth's magnetic field.

The Earth's magnetic field also exhibits a variety of fluctuations. On a long time scale, the Earth's magnetic poles have switched polarity at various times in the past, most recently about 700 000 years ago. During a period of reversed polarity, the south magnetic pole is near the south geographic pole—the opposite of today's polarity. We are not certain why such pole reversals have occurred.

On a shorter time scale, the magnetic poles do not always remain in the same locations but tend to "wander." The north magnetic pole has recently been moving about 1° of latitude (roughly 110 km or 70 mi) per decade. For some unknown reason, it has moved consistently northward from its 1904 latitude position of 69°N, and westward, crossing the 100°W longitudinal meridian. This long-term polar drift means that the magnetic declination map (Fig. 19.30b) varies with time and must be updated periodically.

On a still shorter time scale, there are sometimes dramatic daily shifts of as much as 80 km (50 mi), followed by a return to the starting positions. These shifts are thought to be caused by charged particles from the Sun that reach the Earth's upper atmosphere and set up currents that change the overall magnetic field.

Charged particles from the Sun and cosmic rays entering the Earth's magnetic field give rise to other phenomena. A charged particle, entering a uniform magnetic field *not* perpendicular to the field, spirals in a helix (•Fig. 19.31a). This is because the component of the particle's velocity parallel to the field does not change. The motions of charged particles in a nonuniform field are quite complex. However, for a bulging field such as that depicted in Fig. 19.31b, the particles spiral back and forth as though in a magnetic "container," a "magnetic bottle." This is an important consideration in confining a plasma (a gas of charged particles) in fusion research (Chapter 30).

An analogous phenomenon occurs in the Earth's magnetic field, giving rise to regions where there are concentrations of charged particles. Two large donut-shaped regions at altitudes of several thousand kilometers are called the *Van Allen belts* (Fig. 19.31c). It is in the lower Van Allen belt that light emissions called *aurorae* occur—the aurora borealis, or northern lights, in the Northern Hemisphere and the aurora australis, or southern lights, in the Southern Hemisphere. These eerie, flickering lights are most commonly observed in the Earth's polar regions but have been seen at all latitudes (•Fig. 19.32).

It is believed that an aurora is created when charged solar particles are trapped in the Earth's magnetic field. Maximum aurora activity occurs after a solar disturbance, such as a solar flare. Solar flares are violent magnetic storms on the Sun that spew out enormous quantities of charged particles and radiation. Trapped in the Earth's magnetic field, the charged particles are guided toward the polar regions. There they excite or ionize oxygen and nitrogen atoms in the atmosphere. When the excited atoms return to their normal state and ions regain their normal number of electrons, light is emitted (Chapter 27), producing the beautiful glow of the aurora.

Chapter Review

Important Terms

pole–force law *or* law of poles   *600*
magnetic field   *601*
electromagnetism   *602*
tesla (T)   *602*
right-hand force rule   *603*
right-hand source rule   *605*

ferromagnetic materials   *609*
magnetic domains   *609*
magnetic permeability   *610*
Curie temperature   *610*
right-hand force rule for a
   current-carrying wire   *611*

magnetic moment   *614*
dc motor   *616*
cathode ray tube (CRT)   *617*
mass spectrometer   *618*

Important Concepts

- Opposite magnetic poles attract, and like poles repel.
- The magnetic field is defined by the force it exerts on moving charges. The SI unit of the magnetic field is the tesla (T).
- The direction of the magnetic force exerted on charged particles moving in a magnetic field is determined by the right-hand force rule.
- The direction of the magnetic field produced by a current or moving charge is determined by the right-hand source rule.
- In ferromagnetic materials, the electron spins align, creating domains where the magnetic fields of indi-

vidual electrons add constructively and thus produce very large net fields. The magnetic permeability of ferromagnetic materials is many times that of a vacuum, reflecting the fact that these materials greatly enhance an external magnetic field.
- Above the Curie temperature, the magnetic domains in a ferromagnet become thermally disordered and the material loses its "permanent" magnetic field.
- The direction of the magnetic moment of a current loop is specified by a right-hand rule. An external magnetic field will exert a torque on a current loop, and the magnetic moment vector tends to align itself with the field direction.

Important Equations

Magnitude of the Magnetic Force on a Moving Charged Particle (q):

$$F = qvB \sin \theta \qquad (19.3)$$

Directional right-hand force rule for moving charges: When the fingers of the right hand are pointed in the direction of **v** and then curled toward the vector **B**, the extended thumb points in the direction of the force **F** on a *positive* charge. (**F** is in the opposite direction for a negative charge.)

Magnitude of the Magnetic Field near a Long, Straight, Current-Carrying Wire:

$$B = \frac{\mu_o I}{2\pi d} \qquad (19.4)$$

(where $\mu_o = 4\pi \times 10^{-7}$ T·m/A, called the magnetic permeability of free space)

Directional right-hand source rule: When a current-carrying wire is grasped with the right hand, the extended thumb pointing in the direction of the current, the curled fingers indicate the directional sense of the magnetic field.

Magnitude of the Magnetic Field at the Center of a Circular Loop of Current-Carrying Wire:

$$B = \frac{\mu_o I}{2r} \qquad (19.5)$$

Magnitude of the Magnetic Field at the Center of a Solenoid (along the axis):

$$B = \frac{\mu_o NI}{L} \qquad (19.6)$$

or

$$B = \mu_o nI \qquad \text{(where } n = N/L) \qquad (19.7)$$

Magnitude of Force on a Straight, Current-Carrying Wire:

$$F = ILB \sin \theta \qquad (19.11)$$

Directional right-hand force rule for a current-carrying wire: When the fingers of the right hand are pointed in the direction of the conventional current I and then curled toward the **B** vector, the extended thumb points in the direction of the force on the wire.

Magnitude of Torque on a Single Current-Carrying Loop:

$$\tau = IAB \sin \theta \qquad (19.12)$$

(where IA is called the magnetic moment, m, of the loop: $m = IA$)

Magnitude of Torque on a Current-Carrying Coil (of N loops):

$$\tau = NIAB \sin \theta \qquad (19.14)$$

Exercises

19.1 Magnets, Magnetic Poles, and Magnetic Field Direction

1. When the ends of two bar magnets are near each other, they attract one another. The ends must be (a) one north, the other south, (b) one south, the other north, (c) both north, (d) both south, (e) either (a) or (b).

2. A compass placed in a magnetic field directed east to west will point (a) perpendicular to the field lines, (b) with the south pole to the east, (c) with the south pole to the west, (d) at a nonzero angle ($\theta < 90°$) to the field lines.

3. If you look directly down on the S pole of a bar magnet, the magnetic field points (a) to the right, (b) to the left, (c) away from you, (d) toward you.

4. Given two identical iron bars, one of which is a permanent magnet and the other unmagnetized, how could you tell the difference by using only the two bars?

5. List some of the similarities and differences between electric and magnetic forces.

19.2 Magnetic Field Strength and Magnetic Force

6. Can a magnetic field set a stationary electron in motion? If so, how?

7. ⊙ A proton moves vertically upward in a uniform magnetic field and deflects to the right as you watch it. What is the magnetic field direction? (a) directly away from you, (b) directly toward you, (c) to the right, (d) to the left.

8. ⊙ If a negatively charged particle were moving downward along the right edge of this page, which way should a magnetic field be oriented so that the particle would initially be deflected to the left?

9. An electron passes through a magnetic field without being deflected. What do you conclude about the orientation between the magnetic field and the velocity of the electron?

10. ■ ⊙ Three particles enter a uniform magnetic field as shown in •Fig. 19.33a. Particles 1 and 3 have equal speeds and charges of the same magnitude. What can you say about (a) the charges of the particles and (b) their masses?

11. ■ ⊙ You want to deflect a positively charged particle in an "S" path as shown in Fig. 19.33b, using only magnetic fields. (a) Explain how this could be done. (b) How does the magnitude of the emerging velocity compare to that of the initial velocity?

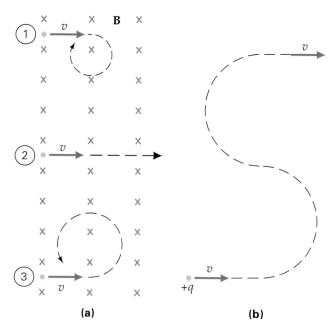

•FIGURE 19.33 **Charges in motion** See Exercises 10 and 11.

12. ■ A positive charge of 0.25 C moves horizontally to the right at a speed of 2.0×10^2 m/s and enters a magnetic field directed vertically downward. If it experiences a force of 20 N, what is the magnetic field strength?

13. ■ A charge of 0.050 C moves vertically in a field of 0.080 T that is oriented horizontally. What speed must the charge have to experience a force of 10 N?

14. ■ A magnetic field can be used to determine the sign of charge carriers. Consider a wide conducting strip in a magnetic field oriented as shown in •Fig. 19.34. The charge carriers are deflected by the magnetic force and accumulate on one side of the strip, giving rise to the measurable voltage across it. (This is known as the *Hall effect*.) If the sign of the charge carriers is unknown (they are either positive charges moving as indicated by the arrows in the figure, or negative charges moving in the opposite direction), how does the measured voltage polarity or sign allow the sign of the charge to be determined?

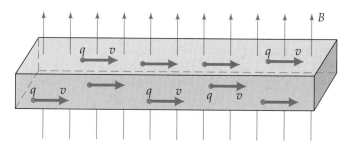

•FIGURE 19.34 **The Hall effect** See Exercise 14.

15. ■■ A beam of protons is accelerated to a speed of 5.0×10^6 m/s in a particle accelerator and emerges horizontally from the accelerator into a uniform magnetic field. What **B** field would cancel the force of gravity and keep the beam moving exactly horizontally?

16. ■■ An electron travels at a uniform speed of 3.0×10^6 m/s in the $+x$ direction. It then enters a uniform magnetic field and experiences a maximum force of 5.0×10^{-19} N in the $-y$ direction. What are the magnitude and direction of the magnetic field?

17. ■■ An electron travels at a speed of 2.0×10^4 m/s through a uniform magnetic field whose magnitude is 1.2×10^{-3} T. What is the magnitude of the magnetic force on the electron if its velocity vector and the magnetic field vector (a) are perpendicular, (b) make an angle of 45°, or (c) are parallel?

18. ■■ What angle(s) does a particle's velocity vector have to make with the magnetic field direction for the particle to be subjected to half the maximum possible magnetic force?

19. ■■■ A horizontal proton beam in a particle accelerator is accelerated to a speed of 3.0×10^5 m/s. The beam enters a uniform magnetic field of 0.50 T oriented at an upward angle of 37° relative to the direction of the beam. (a) What is the initial acceleration of a proton in the beam? (b) If the beam were made up of electrons, what would be the difference in the force on the particles as the beam entered the magnetic field?

19.3 Electromagnetism—The Source of Magnetic Fields

20. A long, straight wire is parallel to the ground and carries a steady current to the east. At a point directly below the wire, what is the direction of the magnetic field the wire produces? (a) north, (b) east, (c) south, (d) west.

21. You are looking directly into one end of a long solenoid. The magnetic field at its center points at you. What is the direction of the current in the solenoid? (a) clockwise, (b) counterclockwise, (c) directly toward you, (d) directly away from you.

22. A long, straight current-carrying wire is oriented vertically. On its east side, the field it creates points south. What is the current direction? (a) up, (b) down.

23. A circular current-carrying loop is laying flat on a table and creates a field, at the loop's center, that points vertically upward. If you look straight down on the loop, what is the current direction? (a) clockwise, (b) counterclockwise.

24. ■ A long, straight wire carries a current of 2.5 A. Find the magnitude of the magnetic field 25 cm from the wire.

25. ■ In a physics lab, a student discovers that the magnitude of the magnetic field in a specific location near a long wire is 4.0 μT. If the wire carries a current of 5.0 A, what is the distance from the wire to that location?

26. ■ The magnetic field at the center of a 50-turn coil of radius 15 cm is 0.80 mT. Find the current in the coil.

27. ■ At the center of a solenoid 0.20 m long with 100 turns is a magnetic field of 1.5 mT. Find the current in the coil.

28. ■■ Two long, parallel wires separated by 50 cm carry currents of 4.0 A each in a horizontal direction. Find the magnetic field midway between the wires if the currents are (a) in the same direction and (b) in opposite directions.

29. ■■ Two long, parallel wires separated by 0.20 m carry equal currents of 1.5 A in the same direction. Find the magnitude of the net magnetic field 0.15 m away from each wire on the side opposite the other wire (•Fig. 19.35).

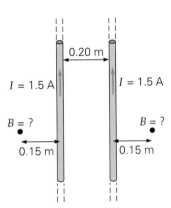

•**FIGURE 19.35 Magnetic field summation** See Exercise 29.

30. ■■ Two long, parallel wires carry currents of 8.0 A and 2.0 A (•Fig. 19.36). (a) What is the magnitude of the magnetic field at the point midway between the wires? (b) Where on a line perpendicular to and joining the wires is the magnetic field zero?

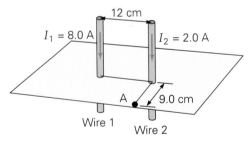

•**FIGURE 19.36 Parallel current-carrying wires** See Exercises 30, 31, 32, 64, and 65.

31. ■■ Find the magnetic field (magnitude and direction) at point A in Fig. 19.36, which is located 9.0 cm away from wire 2 on a line perpendicular to the line joining the wires.

32. ■■ Suppose that the current in wire 1 in Fig. 19.36 were in the opposite direction. (a) What would be the magnetic field at a point midway between the wires? (b) Where on a line perpendicular to and joining the wires would the magnetic field be zero?

33. ■■ A circular loop of wire with a diameter of 12 cm is in the horizontal plane and carries a current of 1.8 A counterclockwise, as viewed from above. What is the magnetic field (magnitude and direction) at the center of the loop?

34. ■■ How much current must flow in a circular loop with a radius of 10 cm to produce a magnetic field at its center that is approximately the same magnitude as the horizontal component of the Earth's magnetic field at the equator?

35. ■■ A coil of four circular loops of radius 5.0 cm carries a current of 2.0 A clockwise, as viewed along a line perpendicular to the circles formed by the loops. What is the magnetic field at the center of the coil?

36. ■■ A circular loop of wire with a radius of 5.0 cm carries a current of 1.0 A. Another circular loop of wire is concentric with (has a common center with) the first and carries a current of 2.0 A. The magnetic field at the center of the loops is zero. What is the radius of the second loop?

37. ■■ A solenoid 10 cm long is wound with 1000 turns of wire. How much current must exist in the windings to produce a magnetic field of 4.0×10^{-4} T at the solenoid's center?

38. ■■ A solenoid is wound with 200 turns per centimeter. An outer layer of insulated wire with 180 turns per centimeter is wound over the first layer of wire. In operation, the inner coil carries a current of 10 A and the outer coil carries a current of 15 A in the direction opposite to that of the current in the inner coil (•Fig. 19.37). What is the magnitude of the magnetic field at the central axis of this doubly wound solenoid?

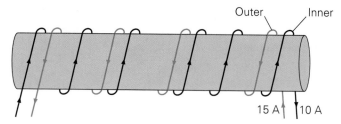

•FIGURE 19.37   **Double it up?**   See Exercise 38.

39. ■■■ Two long, perpendicular wires carry currents of 15 A as illustrated in •Fig. 19.38. What is the magnitude of the magnetic field at the midpoint of the line joining the wires?

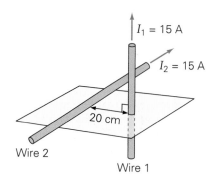

•FIGURE 19.38   **Perpendicular current-carrying wires**
See Exercises 39 and 69.

40. ■■■ Four wires running through the corners of a square with sides of length a, as shown in •Fig. 19.39, carry equal currents I. Calculate the magnetic field at the center of the square in terms of these parameters.

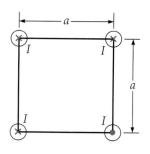

•FIGURE 19.39   **Current-carrying wires in a square array**   See Exercise 40.

41. ■■■ A charged particle (charge q, mass m) moves in a horizontal plane at right angles to a uniform vertical magnetic field B. (a) What is the frequency (f) of the particle's circular motion in terms of q, B, and m? (This frequency is called the *cyclotron frequency*.) (b) Show that the time required for any charged particle to make one complete revolution is independent of its speed and radius. (c) Compute the path radius and frequency if the particle is an electron with a speed $v = 10^5$ m/s and if the field strength is $B = 10^{-4}$ T.

19.4 Magnetic Materials

42. The main source of magnetism in magnetic materials is from (a) electron orbits, (b) electron spin, (c) magnetic poles, (d) nuclear properties.

43. When a ferromagnetic material is placed in an external magnetic field, (a) the domain orientation may change slightly, (b) the domain boundaries may change, (c) new domains are created, (d) both (a) and (b).

44. ■ If you are looking down on a circular current loop with the current circulating counterclockwise, what is the direction of the magnetic field at the center of the loop?

45. ■ If you are looking down on the electron orbit in a hydrogen atom and the electron orbits counterclockwise, what is the direction of the magnetic field the electron produces at the site of the proton?

46. ■■ A solenoid with 100 turns per centimeter has an iron core with a relative permeability of 2000. The solenoid carries a current of 1.2 A. (a) What is the magnetic field at the central axis of the solenoid? (b) How much greater is the magnetic field with the iron core than it would be without it?

47. ■■■ What is the magnetic field at the center of the circular orbit of the electron in a hydrogen atom, which has an orbital radius of 0.0529 nm? [Hint: Find the electron's period by considering the centripetal force.]

19.5 Magnetic Forces on Current-Carrying Wires

48. A long, straight, horizontal wire is located on the equator and carries a current directed toward the east. What is the direction of the force on it due to the Earth's magnetic field? (a) east, (b) west, (c) south, (d) upward?

49. Two straight wires are parallel to each other. If the currents in the wires are in the same direction, will the wires attract or repel each other? What if the currents are opposite?

50. What happens to the length of a spring when a large current passes through it?

51. Is it possible to orient a current loop in a uniform magnetic field such that there is no torque on the loop? Describe.

52. ■ A straight, horizontal segment of wire is 1.0 m long and carries a current of 5.0 A in the $+x$ direction in a magnetic field of 0.30 T that is directed vertically downward (the $-z$ direction). What is the force on the wire (magnitude and direction)?

53. ■ A 2.0-meter length of straight wire carries a current of 20 A in a uniform magnetic field of 50 mT whose field lines make an angle of $37°$ with the current direction. Find the force on the wire.

54. ■ Use a right-hand rule to find the direction of the current in a wire in a magnetic field that results in the force on the wire shown for each case in •Fig. 19.40.

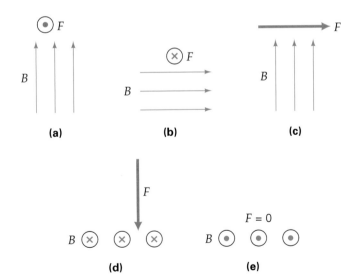

•FIGURE 19.40 **The right-hand rule** See Exercise 54.

55. ■■ A straight wire 50 cm long conducts a current of 4.0 A directed vertically upward. If the wire experiences a force of 1.0×10^{-2} N in the eastward direction due to a magnetic field at right angles to its length, what are the magnitude and direction of the magnetic field?

56. ■■ A horizontal magnetic field of 1.0×10^{-4} T is at an angle of $30°$ to the direction of the current in a straight, horizontal wire 75 cm long. If the wire carries a current of 15 A, what is the magnitude of the force on the wire?

57. ■■ A wire carries a current of 10 A in the $+x$ direction in a uniform magnetic field of 0.40 T. Find the magnitude of the force per unit length and direction of the force on the wire if the magnetic field is directed in (a) the $+x$ direction, (b) the $+y$ direction, (c) the $+z$ direction, (d) the $-y$ direction, and (e) the $-z$ direction.

58. ■■ A straight wire 25 cm long is oriented vertically in a uniform horizontal magnetic field of 0.30 T in the $-x$ direction. What current (including direction) would cause the wire to experience a force of 0.050 N in the $+y$ direction?

59. ■■ A wire carries a current of 10 A in the $+x$ direction. Find the force per unit length on the wire if the magnetic field has components $B_x = 0.020$ T and $B_y = 0.040$ T.

60. ■■ A set of jumper cables used to start a car from another car's battery is connected to the terminals of both batteries. If 15 A of current exists in the cables when the second car is started and the cables are parallel and 15 cm apart, what is the force per unit length on the cables?

61. ■■ Find the force per unit length on each of two long, straight, parallel wires that are 24 cm apart when one carries a current of 2.0 A and the other a current of 4.0 A in the same direction.

62. ■■ Two long, straight, parallel wires 10 cm apart carry equal currents of 3.0 A in opposite directions. What is the force per unit length on the wires?

63. ■■ A nearly horizontal dc power line on the equator carries a current of 1000 A directly eastward. If the Earth's magnetic field at the location is 5.0×10^{-5} T due north, what are the magnitude and direction of the magnetic force on a 15-meter section of this line?

64. ■■ What is the force per unit length on wire 1 in Fig. 19.36?

65. ■■ What is the force per unit length on wire 2 in Fig. 19.36?

66. ■■ A long wire is vertically below a rigidly mounted second wire (•Fig. 19.41). Each has the same current, and

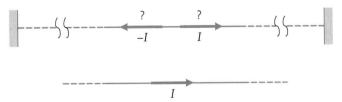

•FIGURE 19.41 **Magnetic suspension** The bottom wire is magnetically attracted to the top (rigidly fixed) wire. See Exercise 66.

the lower wire has a linear mass density of 1.5×10^{-3} kg/m. (a) How should the current directions in the wires be arranged so that the lower wire is in equilibrium ("floats")? (b) What is the current magnitude?

67. ■■ A wire is bent as shown in •Fig. 19.42 and placed in a magnetic field with a magnitude of 1.0 T in the indicated direction. Find the magnitude of the force on each segment of the wire if $x = 50$ cm and the wire carries a current of 5.0 A.

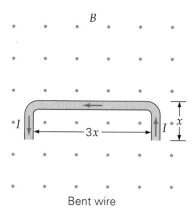

•FIGURE 19.42 **Current-carrying wire in a magnetic field** See Exercise 67.

68. ■■ A rectangular loop of wire whose dimensions are 20 cm by 30 cm carries a current of 1.5 A. (a) What is the magnitude of the magnetic moment of the loop? (b) How should the loop be oriented in a uniform magnetic field to obtain the maximum torque on it?

69. ■■■ Two straight wires are positioned at right angles to each other as in Fig. 19.38. What is the net force on each wire? Is there a net torque on each wire?

70. ■■■ A rectangular wire loop with a cross-sectional area of 0.20 m² carries a current of 0.25 A. The loop is free to rotate about an axis that is perpendicular to a uniform magnetic field strength of 0.30 T. The plane of the loop is initially at an angle of 30° to the direction of the magnetic field. What is the magnitude of the torque on the loop?

19.6 Applications of Electromagnetism

71. 💿 A mass spectrometer (a) can be used to determine the masses of atoms and molecules, (b) requires charged particles, (c) can be used to determine relative abundances of isotopes, (d) all of these.

72. In a galvanometer, the deflection of the needle is directly proportional to (a) the current in the wire coil, (b) the number of turns in the coil, (c) the magnetic field, (d) all of these.

73. Explain the operation of the doorbell and door chimes illustrated in •Fig. 19.43.

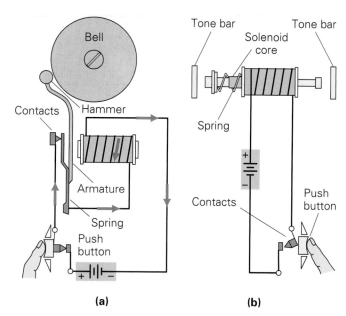

(a)   (b)

•**FIGURE 19.43 Electromagnetic applications** Both (a) a doorbell and (b) door chimes have electromagnets. See Exercise 73.

74. Why can a nearby magnet distort the display of a computer monitor or television picture tube (•Fig. 19.44)?

•**FIGURE 19.44 Magnetic disturbance** See Exercise 74.

75. ■ An ionized deuteron (a particle with a +1 charge) passes through a velocity selector whose magnetic and electric fields have magnitudes of 40 mT and 8.0 kV/m, respectively. Find the speed of the ion.

76. ■ In a velocity selector, a uniform magnetic field of 1.5 T from a large magnet and two parallel plates with a separation distance of 1.5 cm produce a perpendicular electric field. What voltage should be applied across the plates so that (a) a singly charged ion traveling at a speed of 8.0×10^4 m/s will pass through undeflected or (b) a doubly charged ion traveling at the same speed will pass through undeflected?

77. ■ A charged particle travels undeflected through perpendicularly crossed electric and magnetic fields whose magnitudes are 3000 N/C and 30 mT, respectively. Find the speed of the particle if it is (a) a proton or (b) an alpha particle.

(An alpha particle is a helium nucleus, a positive ion with a double positive charge.)

78. ■■ 💿 In a mass spectrometer, a singly charged ion having a particular velocity is selected by using a magnetic field of 0.10 T perpendicular to an electric field of 1.0×10^3 V/m. The same magnetic field is used to deflect the ion, which moves in a circular path with a radius of 1.2 cm. What is the mass of the ion?

79. ■■ 💿 In a mass spectrometer, a doubly charged ion having a particular velocity is selected by using a magnetic field of 100 mT perpendicular to an electric field of 1.0 kV/m. The same magnetic field is used to deflect the ion in a circular path with a radius of 15 mm. Find (a) the mass of the ion and (b) the kinetic energy of the ion. (c) Does the kinetic energy of the ion increase in the circular path? Explain.

80. ■■ In an experimental technique for treating deep tumors, unstable positively charged pions (π^+, elementary particles with a mass of 2.25×10^{-28} kg) are aimed to penetrate the flesh and to disintegrate at the tumor site, releasing energy to kill the cancer cells. If pions with a kinetic energy of 10 keV are required and if a velocity selector with an electric field strength of 2.0×10^3 V/m is used, what must be the magnetic field strength?

81. ■■■ In a momentum selector, a beam of protons enters a magnetic field. Some of the protons make exactly a one-quarter circular arc with a radius of 0.50 m. If the magnetic field is oriented at right angles to the proton's velocity direction, what is the field's magnitude if the exiting protons have a kinetic energy of 10 keV?

*19.7 The Earth's Magnetic Field

82. The Earth's magnetic field (a) has poles that coincide with the geographic poles, (b) is produced by internal ferromagnetic material, (c) reverses polarity every few hundred years, (d) none of these.

83. Aurorae (see Fig. 19.32) occur (a) only in the Northern Hemisphere, (b) in the lower Van Allen belt, (c) because of pole reversals, (d) predominantly when there are no solar disturbances.

84. What will be the direction of the force experienced by an electron near the equator due to the Earth's magnetic field if the electron's velocity is directed (a) due south, (b) northwest, (c) upward?

85. What is the polarity of the magnetic pole near the Earth's geographic South Pole?

Additional Exercises

86. A proton is accelerated through a potential difference of 3.0 kV. It then enters a region between two parallel plates that are separated by 10 cm and have a potential difference of 250 V. Find the magnitude of the magnetic field

(perpendicular to **E**) needed to allow the proton to pass undeflected between the plates.

87. A solenoid 10 cm long has 3000 turns of wire and carries a current of 5.0 A. A 2000-turn coil of wire of the same length as the solenoid surrounds the solenoid and is concentric with it (shares a common center). The outer coil carries a current of 10 A in the same direction as that of the current in the solenoid. Find the magnetic field at their common center.

88. A horizontal beam of electrons travels at a speed of 1.0×10^3 m/s along a north-to-south line in a discharge tube. What is the force on the electrons due to the downward vertical component of the Earth's magnetic field if the field has a magnitude of 5.0×10^{-5} T at that location?

89. A proton enters a uniform magnetic field of 0.80 T such that the proton follows a circular path with a radius of 4.6 cm. What is the kinetic energy of the proton in circular motion?

90. A current of 10 A is maintained in a square loop. When placed in a magnetic field of 500 mT at an angle of 50° relative to the loop's plane, the loop experiences a torque of 0.15 m·N about an axis through the loop's center and parallel to one of the sides. Find the length of each side of the square.

91. A proton is accelerated from rest through a potential difference of 1.0 kV. It enters a uniform magnetic field of 4.5 mT that is perpendicular to the direction of its motion. (a) Find the radius of the circular path of the proton. (b) Calculate the period of revolution of the proton.

92. 💿 A beam of protons travels north at 2.0×10^2 m/s. The beam passes through the space between two horizontal parallel plates, where a constant electric field and a constant magnetic field are at right angles to one another. If the electric field has a magnitude of 100 V/m and is directed upward from the bottom to the top plate, what must the magnitude and direction of the magnetic field be to allow the beam to pass undeflected? [Hint: Make a sketch of the situation.]

93. How fast should an electron travel at right angles to a magnetic field of 1.0×10^{-4} T so that the magnetic force just balances the gravitational force on it?

94. A particle with a charge of 4.0×10^{-8} C moves at a speed of 3.0×10^2 m/s through a magnetic field in the direction at which the magnetic force on the particle is maximum. If the force on the particle is 1.8×10^{-6} N, what is the magnitude of the magnetic field?

95. A current-carrying loop is rectangular and has dimensions of 20 cm by 30 cm. The loop carries a current of 10 A and is in a uniform magnetic field of 50 mT directed parallel to the plane of the loop. Find the torque on the loop.

CHAPTER
20

Electromagnetic Induction

Insights

- Electromagnetic Induction in Action
- CFCs and Ozone Layer Depletion

As we saw in Chapter 19, an electric current produces a magnetic field. But the mutual relationship of electricity and magnetism does not stop there. In this chapter we shall see that under certain conditions a magnetic field can be used to produce an electric current.

How is this done? Chapter 19 considered only constant magnetic fields. No current is induced in a loop of wire that is stationary in a constant magnetic field. However, if the magnetic field changes with time, or if the wire loop moves across or is rotated in the field, a current *is* induced in the wire.

The uses of this interrelationship of electricity and magnetism are legion. One example is the playing of a cassette tape. The music you hear was encoded as tiny variations in a magnetic field. These variations produce electrical impulses, which are amplified and drive the speakers. The speakers, in turn, utilize electromagnetic interactions to translate the electrical impulses back into audible sound. Similar processes are involved when information is stored on or retrieved from a disk in your computer.

On a larger scale, consider the generation of the electricity that we use to power our cassette players, computers, and so many other devices that we use every day. Think how dependent we are on this form of energy: Electric power is the basis of our modern civilization. A great deal of the energy that we extract from the environment is converted to electricity before being put to use. At hydroelectric power plants such as that in the photo (on the Nantahala River in North Carolina), one of the oldest and simplest power sources on Earth—falling water—is used to generate the electricity. In terms of basic physical principles, the gravitational potential energy of water above the wheel is converted into kinetic energy (as the water is allowed to fall), and some of this kinetic energy is transformed into electric energy. But regardless of the ultimate source of the energy—the burning of oil, coal, or gas, a nuclear reactor, or falling water—the actual conversion to electricity is accomplished by means of magnetic fields. This chapter examines the underlying electromagnetic principles that make such conversion possible.

20.1 Induced Emf's: Faraday's Law and Lenz's Law

Objectives: **To (a) define magnetic flux and explain how induced emfs are created by changing magnetic flux, and (b) calculate the magnitude and predict the polarity of an induced emf.**

A magnet held stationary near a conducting wire loop does not induce a current in that loop (•Fig. 20.1a). If the magnet is moved toward the loop, however, as shown in Fig. 20.1b, the deflection of the galvanometer needle indicates that there is a current in the loop during the motion. If the magnet is moved away from the loop, as shown in Fig. 20.1c, the galvanometer needle is deflected in the opposite direction, which indicates a reversal of the current's direction.

Deflections of the galvanometer needle, indicating the presence of *induced currents*, also occur if the loop is moved toward or away from a stationary magnet.

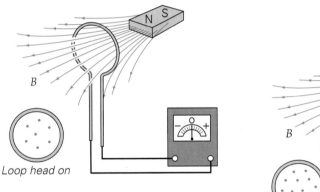

(a) No motion between magnet and loop

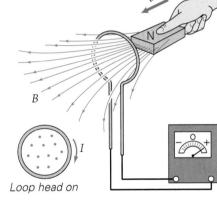

(b) Magnet is moved toward loop

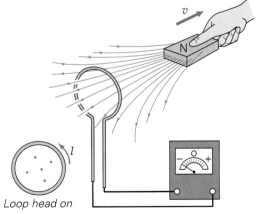

(c) Magnet is moved away from loop

(d)

•**FIGURE 20.1 Electromagnetic induction** (a) When there is no relative motion between the magnet and the wire loop, the number of field lines through the loop (in this case, 7) is constant, and the galvanometer (measuring current) shows no deflection. (b) Moving the magnet toward the loop increases the number of field lines passing through the loop (now 12), and an induced current is detected. (c) Moving the magnet away from the loop decreases the number of field lines passing through the loop (5). The induced current is in the opposite direction, as indicated by the opposite needle deflection. (d) In an actual experiment of this type, a solenoid, a coil with a large number of loops, is used to increase this effect.

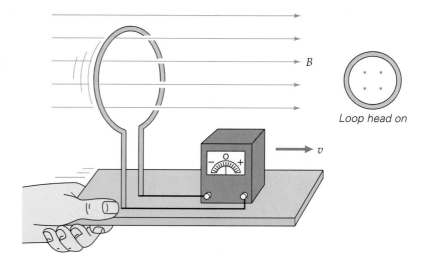

Loop head on

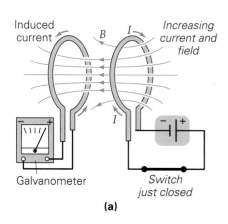

(a)

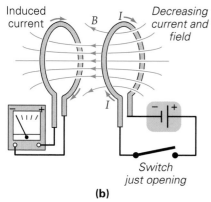

(b)

•FIGURE 20.3 **Mutual induction** **(a)** When the switch is closing in the right-loop circuit, the current buildup (typically over a few milliseconds) produces a changing magnetic field that passes through the other loop, inducing a current in it. **(b)** When the switch is opened, the magnetic field collapses and the number of field lines through the left-hand loop decreases. The induced current in this loop is then in the opposite direction.

The effect depends on *relative* motion of the loop and magnet. The magnitude of the induced current depends on the speed of that relative motion. However, experimentally there is a noteworthy exception. If a loop is moved with a velocity *parallel* to a *uniform* magnetic field, as shown in •Fig. 20.2, no current is induced in the loop.

Yet another way to induce a current in a stationary wire loop is to vary the *current* in another loop close to it. When the switch in the battery-powered circuit in •Fig. 20.3a is closed, the current in its loop goes from zero to a constant value in a short time. During this time only, the magnetic field caused by this current and passing through the other loop increases, and the galvanometer needle deflects. When the current in the loop in the battery circuit is at its maximum (constant) value, the resulting magnetic field is also constant, and the galvanometer reads zero. Similarly, when the switch is opened (Fig. 20.3b), the current and the field decrease to zero, and the galvanometer deflects in the opposite direction. In both cases, the deflection and induced current occur only when the current and the magnetic field are *changing*.

The current induced in a loop is caused by an induced electromotive force (emf) due to a process called **electromagnetic induction**. Recall from Chapter 17 that an emf represents energy capable of moving charges around a circuit. For example, a battery is a chemical source of emf. In the case of a moving magnet and a stationary loop (Fig. 20.1), we say that an emf is *induced* in the loop, thereby causing the current. For the case of *two* stationary loops (Fig. 20.3), where a changing current in one circuit induces an emf in the other circuit, we speak of *mutual induction*.

Experiments on electromagnetic induction were done independently by Michael Faraday in England and Joseph Henry in the United States about 1830. Faraday realized that the important factor in electromagnetic induction was the time rate of change of the number of magnetic field lines passing through the loop area. That is,

an induced emf is produced in a loop by changing the number of magnetic field lines passing through the plane of the loop.

(Note that this is the case for the situations in Figs. 20.1 and 20.3. Think about the situation in Fig. 20.2.)

Magnetic Flux

Since the induced emf or induced current in a loop depends on the change in the number of field lines through it, the ability to quantify the number of field lines through the loop at any time is useful. Consider a loop of wire in a uniform mag-

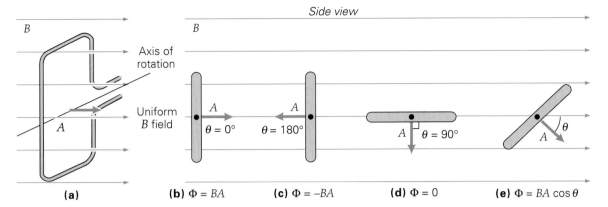

(a)　　　**(b)** $\Phi = BA$　　**(c)** $\Phi = -BA$　　**(d)** $\Phi = 0$　　**(e)** $\Phi = BA \cos \theta$

•**FIGURE 20.4   Magnetic flux** **(a)** Magnetic flux (Φ) is a measure of the number of field lines passing through an area (A). The area can be represented by a vector **A** perpendicular to the plane of the area. **(b)** When the plane of a rotating loop is perpendicular to the field and $\theta = 0°$, then $\Phi = \Phi_{max} = BA$. **(c)** When $\theta = 180°$, the magnetic flux has the same magnitude but is opposite in direction: $\Phi = -\Phi_{max} = -BA$. **(d)** When $\theta = 90°$, then $\Phi = 0$. **(e)** As the loop is rotated from an orientation perpendicular to the field to one more nearly parallel to the field, less area is open to the field lines and the flux decreases. In general, $\Phi = BA \cos \theta$.

netic field **B**, as illustrated in •Fig. 20.4a. The number of field lines through the loop depends on its orientation relative to the **B** field. To describe this orientation, we use a vector **A** normal to the plane of the loop to represent the loop's area with a directional sense. **A** is called the *area vector*, and its magnitude is equal to the area of the loop. The orientation of the loop can then be described by the angle θ, which is the angle between **B** and **A**. In Fig. 20.4a, $\theta = 0°$.

In general, a relative measure of the number of field lines passing through a particular area (the area within a loop, in our case) is given by the **magnetic flux** (Φ), which is defined as

$$\Phi = BA \cos \theta \qquad (20.1)$$

SI unit of magnetic flux: tesla–meter squared (T·m^2), or weber (Wb)

A unit of magnetic flux is the weber (Wb). Recall from Section 19.2 that B has the units Wb/m^2 (or tesla, T). Since the cosine is unitless, the units of flux, Φ (or BA) are (Wb/m^2)(m^2) = Wb, which is T·m^2 in the SI. Thus 1 T·m^2 = 1 Wb.

The orientation of the loop to the magnetic field affects the number of field lines passing through it, and this is accounted for by the cosine term in Eq. 20.1. Let us consider several possible orientations:

- If **B** and **A** are parallel ($\theta = 0°$), then the magnetic flux is positive and a maximum: $\Phi_{max} = BA \cos 0° = +BA$, and the maximum number of field lines pass through the loop (Fig. 20.4b).
- If **B** and **A** are opposite ($\theta = 180°$), then the magnitude of the magnetic flux is a maximum again, but of opposite sign: $\Phi_{180°} = BA \cos 180° = -BA = -\Phi_{max}$ (Fig. 20.4c).
- **B** and **A** are perpendicular, there are no field lines through the loop, and the flux is zero: $\Phi_{90°} = BA \cos 90° = 0$ (Fig. 20.4d).
- For situations at intermediate angles, the flux is less than the maximum value (Fig. 20.4e). We can interpret $A \cos \theta$ as the *effective* area of the loop perpendicular to the field lines (•Fig. 20.5a). Alternatively, we can think of $B \cos \theta$ as the perpendicular component of the *field* through the full area of the loop, A, as shown in Fig. 20.5b. Thus $\Phi = (B \cos \theta)A$, and the result is the same regardless of the interpretation.

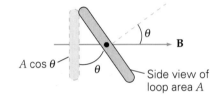

$\Phi = (A \cos \theta)B$

(a)

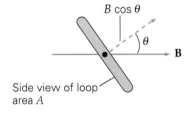

$\Phi = A(B \cos \theta)$

(b)

•**FIGURE 20.5   Magnetic flux through a loop: an alternative interpretation** Instead of defining the flux **(a)** in terms of the magnetic field (B) passing through a reduced area ($A \cos \theta$), we can define it **(b)** in terms of the perpendicular component of the magnetic field ($B \cos \theta$) passing through A. Either way, Φ is a measure of the number of field lines passing through A and is given by $\Phi = BA \cos \theta$.

Faraday's Law of Induction and Lenz's Law

From his experiments, Faraday concluded that the emf induced in a coil of N loops depends on the time rate of change of the number of field lines through all the loops, or the *time rate of change of the magnetic flux through all the loops*. This dependence, known as **Faraday's law of induction**, is expressed mathematically as

$$\mathscr{E} = -N \frac{\Delta \Phi}{\Delta t} \qquad \text{induced emf} \qquad (20.2)$$

where $\Delta \Phi$ is the change in flux through *one loop*. In a coil of N loops of wire, the total change of flux will be $N\Delta\Phi$. Note that $\mathscr{E}$ in Eq. 20.2 is an *average* value over the time interval Δt.

The minus sign is included in Eq. 20.2 to give an indication of the *polarity* or *direction* of the induced emf, which is found by considering the resulting induced current and its effect, according to **Lenz's law**:

> An induced emf in a metal loop or coil gives rise to a current whose magnetic field opposes the *change* in magnetic flux that produced it.

This means that the magnetic field *due to the induced current* is in a direction that tends to keep the flux through the loop from changing. For example, if the flux increases in the $+x$ direction, the magnetic field due to the induced current will be in the $-x$ direction. This tends to cancel the increase in the flux, or *oppose the change*, since magnetic fields add vectorially. Essentially, the magnetic field due to the induced current tries to maintain the status quo value of the magnetic flux. This effect is sometimes called "electromagnetic inertia," by analogy with the tendency of objects to resist changes in their velocity. In the long run, the induced current cannot maintain the magnetic flux as unchanged. However, during the time the flux through the coil is changing, the induced magnetic field will tend to cancel the change.

The induced current direction is given by a right-hand rule: With the fingers of the right hand pointing in the direction of the induced field, the extended thumb points in the direction of the induced current. This is the right-hand rule used to find the direction of a magnetic field produced by a current (Chapter 19), applied in reverse, so to speak. Try applying Lenz's law to the loops in Fig. 20.3.

Lenz's law incorporates the conservation of energy. To see this, suppose that the magnetic field due to an induced current *added* to the original field—that is, it increased the flux. This increased flux would give a greater induced emf and current, which would produce a greater magnetic field, which in turn would give a greater induced current, and so on. Such a something-for-nothing situation would be incompatible with the law of the conservation of energy—a situation that has never been observed.

For the case of an emf induced in a loop by a moving magnet (Fig. 20.1), recall that a current-carrying loop has a magnetic field similar to that of a bar magnet. The induced current sets up a magnetic field whose effect is such that the loop acts like a bar magnet with a polarity that opposes the motion of the real bar magnet. Thus, there is opposition to the motion; work must be done to move the magnet.

Substituting the expression for the magnetic flux (Φ) given by Eq. 20.1 into Eq. 20.2 shows that a change in flux can result from changes other than just a change in magnetic field strength. In general,

$$\mathscr{E} = -\frac{N\Delta\Phi}{\Delta t} = -\frac{N\Delta(BA\cos\theta)}{\Delta t}$$

Hence there are three quantities that can change with time to produce an induced emf: (1) the strength of the magnetic field B, (2) the area of the loop A, and (3) the angle θ. Expanding the equation to see (and label) these terms explicitly, we have

$$\mathcal{E} = -N\left[(A\cos\theta)\left(\frac{\Delta B}{\Delta t}\right) + (B\cos\theta)\left(\frac{\Delta A}{\Delta t}\right) + BA\left(\frac{\Delta(\cos\theta)}{\Delta t}\right)\right] \quad (20.3)$$

$$\qquad\qquad (1) \qquad\qquad\qquad (2) \qquad\qquad\qquad (3)$$

1. Term *(1)* represents a flux change due to a *time-varying magnetic field*, with the area (A) and orientation (θ) constant. A time-varying magnetic field is easily obtained by using a time-varying current or by moving a magnet in the space near a coil, as in Fig. 20.1 (or by moving the coil in the space near the magnet).

2. Term *(2)* represents a flux change due to a *time-varying loop area*, with the magnetic field (**B**) and orientation (θ) constant. This could occur if a loop were being stretched or flattened in a plane, as in Fig. 20.6, or if it had an adjustable circumference (imagine an adjustable loop around a balloon being blown up).

3. Term *(3)* represents a change in flux resulting from a *change in orientation of the loop with time*, with the magnetic field (**B**) and loop area (A) constant. This occurs when a coil (made up of N loops) is rotated in a magnetic field. The change in the number of field lines through a single loop in this case is evident in the sequential views of the rotating loop in Fig. 20.4.

Perhaps the most common and useful way of inducing an emf by rotating a coil will be considered separately in Section 20.2. The emfs resulting from the first two terms in Eq. 20.3 are analyzed in the following three Examples. (Also, see the Insight on p. 638 for some applications of electromagnetic induction.)

Conceptual Example 20.1

Fields in the Fields: Electromagnetic Induction

In rural areas where electric power lines carry current to big cities, it is possible to generate very small amounts of electricity from the lines by means of induction in a single conducting loop. The overhead power lines carry relatively large currents that periodically reverse in direction 60 times per second. How would you orient the plane of the loop to produce maximum induced current if the power lines run in a north-to-south direction? (a) Parallel to the Earth's surface, (b) perpendicular to the surface in the north–south direction, (c) perpendicular to the surface in the east–west direction, (d) the orientation of the loop would not affect the size of the induced current.

Reasoning and Answer. Making a sketch would help here. Magnetic field lines from long wires are circular, centered on the wire (see Fig. 19.8). The direction of the magnetic field at ground level would be parallel to the surface but would alternate in direction, from east to west and back to east again, every $\frac{1}{60}$ s. We can immediately eliminate answer (d), because we know that the flux *does* depend on the orientation of the loop relative to the field (see Eq. 20.1 and Fig. 20.4). Answer (a) cannot be correct, for in this orientation the loop would never have magnetic flux passing through it, and thus the flux would be constant (zero), so there would be no induced emf. If the loop is oriented perpendicular to the surface but in the east–west direction, there would also never be any magnetic flux through it, so answer (c) can also be ruled out. Hence the answer is (b). If the loop is oriented perpendicular to the surface with its plane in the north–south direction, the flux through it would vary from zero to its maximum value and back 60 times per second, and the emf induced in the loop would be a maximum.

Follow-up Exercise. (a) Suggest possible ways of increasing the amount of current produced by the arrangement discussed in this Example by changing only properties of the loop and not the overhead wire. (b) Explain clearly why this method would not work if the overhead wires carried a constant dc current.

Electromagnetic Induction in Action

You have probably used applications of electromagnetic induction many times without knowing it. One of these is magnetic tape—either audio or video. In an audio tape recorder, a plastic tape coated with a film of iron oxide or chromium oxide runs past a recording head, which consists of a coil wound around an iron core with a gap (Fig. 1). Current in the coil produces a magnetic field in the gap, magnetizing the tape's film. The strength and direction of the magnetization in the film are determined by the gap field, which is determined by the current pulses that are generated by sound from a microphone.

To reproduce the sound, the tape is run by the same head or one similar to it. As the tape passes through the head area, the changing flux due to the magnetization of the film on the tape induces an emf in the coil, matching the original voltage or current fluctuations. These are amplified and converted to sound in a speaker (which also commonly uses electromagnetic induction; see Exercise 7 and Fig. 20.25). A tape can be erased by passing it over the gap in a head that has a high-frequency alternating current (ac) input. The alternating magnetic field produced by this ac current "randomizes" the magnetic domains in the magnetized areas on the tape. The film is left in a demagnetized condition as it leaves the gap.

Electromagnetic induction is also used in an electrical safety device called a *ground fault circuit interrupter* (GFCI). A GFCI can be plugged into a wall outlet or installed as part of a home circuit. It senses any abrupt change in current in an electrical appliance plugged into the outlet or circuit and turns off the appliance, protecting you from electrical shock.

The principle of the GFCI is illustrated in Fig. 2. It consists of a sensing coil and a circuit breaker. The coil is wrapped on an iron ring, through which pass the wires carrying current to and from the protected outlet or circuit. This arrangement is sometimes called a *differential transformer* because it can sense a difference in the currents carried by the two wires.

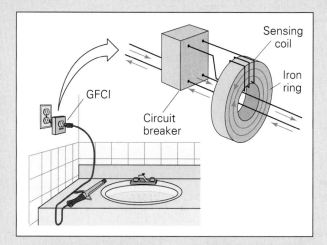

FIGURE 2 Ground fault circuit interrupter (GFCI) A safety device that quickly detects currents to ground and opens the circuit.

The opposite currents in the two wires produce opposite magnetic fields, which are concentrated in the iron ring. With alternating current, the directions of the fields are constantly changing. However, since the fields are always equal and opposite, the total field is normally zero. Hence, the net flux through the coil is zero, and no emf is induced in the sensing coil.

However, suppose that a wire breaks inside an appliance and touches its metal casing (Section 18.5) or that a plugged-in hair dryer falls into a sink full of water. If you then touch the appliance or put a hand in the water, some of the current will pass through your body to ground (a fault to ground). The currents in the wires passing through the ring are then not equal, giving rise to a nonzero magnetic flux. A changing flux causes an induced emf in the sensing coil; the resulting current trips the circuit breaker, opening the circuit.

All of this happens in about 30 ms (30/1000 of a second) in response to an induced current of 4 to 6 mA. Notice that this is much less than the current that would normally trip a protective circuit breaker, which is usually set at 20 or 30 A. The GFCI does not protect you from receiving a shock, but it does limit the time the hazard exists and the potential for serious injury.

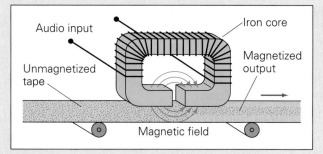

FIGURE 1 Tape recorder head Current pulses in the coil on the iron core produce magnetic fields that magnetize the metallic coating on the tape as it passes by.

Example 20.2 ■ A Potential Hazard to Computer Equipment: Induced Currents

(a)

Electrical instruments can be damaged or destroyed if they are in the vicinity of a rapidly changing magnetic field. Because a changing field can give rise to a changing magnetic flux within the instuments, the resulting induced emfs can create induced currents large enough to do harm. Consider a laptop computer's speaker that is near a single wire carrying a household current (•Fig. 20.6a). At a distance of 30 cm, a maximum ac current of 10 A would produce a maximum magnetic field of about 7.0×10^{-6} T. Both the current and the field would reverse direction every half-cycle, or $\Delta t = 1/120$ s.

The speaker's coil consists of 1000 circular wire loops and has a total resistance of 1.0 Ω and a radius of 3.0 cm. Since the coil is small, assume for simplicity that the magnetic field strength at any instant is constant over the coil's area, which is perpendicular to the field direction (Fig. 20.6b). According to the manufacturer, the coil can carry only 25 mA before it is damaged. What are (a) the magnitude of the average induced emf in the coil during one half-cycle and (b) the magnitude of the average induced current in the coil? (Will the speaker's coil survive?)

Thinking It Through. (a) The flux goes from a (maximum) positive to a (maximum) negative value in 1/120 s, so we can use the definition of magnetic flux, Eq. 20.1 with $\theta = 0°$ and $\theta = 180°$ (why?), to determine the magnetic fluxes and the flux change. We can then calculate the average induced emf from Eq. 20.2. (b) Once we know emf, we can calculate the induced current from Ohm's law, $I = V/R$, and compare that result to 25 mA.

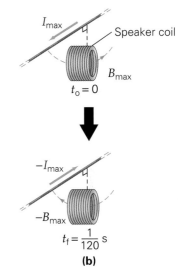

(b)

Solution. Listing the data, we have

Given: $B_o = +B_{max} = 7.0 \times 10^{-6}$ T pointing one way
$\quad\quad\quad B_f = -B_{max} = 7.0 \times 10^{-6}$ T pointing the opposite way
$\quad\quad\quad \Delta t = 1/120$ s $= 8.3 \times 10^{-3}$ s
$\quad\quad\quad N = 1000$ loops
$\quad\quad\quad R = 1.0$ Ω
$\quad\quad\quad r = 3.0$ cm $= 3.0 \times 10^{-2}$ m
$\quad\quad\quad I_{max} = 25$ mA

Find: (a) $\mathscr{E}$ (average induced emf in coil)
(b) I (average induced current in coil)

(a) First we must compute the circular loop area: $A = \pi r^2 = \pi(3.0 \times 10^{-2} \text{ m})^2 = 2.8 \times 10^{-3}$ m². Then the initial flux through *one* loop is given by Eq. 20.1:

$$\Phi_o = B_o A \cos \theta = (7.0 \times 10^{-6} \text{ T})(2.8 \times 10^{-3} \text{ m}^2)(\cos 0°) = 1.96 \times 10^{-8} \text{ T·m}^2/\text{loop}$$

Since the final flux is the negative of this result ($\theta = 180°$), the change in flux through one loop is

$$\Delta\Phi = \Phi_f - \Phi_o = -\Phi_o - \Phi_o = -2\Phi_o = -3.9 \times 10^{-8} \text{ T·m}^2/\text{loop}$$

For the magnitude of the average induced emf, we can ignore the minus sign. From Eq. 10.2, the result is

$$\mathscr{E} = N\frac{|\Delta\Phi|}{\Delta t} = (1000 \text{ loops})\frac{|-3.9 \times 10^{-8} \text{ T·m}^2/\text{loop}|}{8.3 \times 10^{-3} \text{ s}} = 4.7 \times 10^{-3} \text{ V}$$

(b) To get the magnitude of the average induced current, we substituted $\mathscr{E}$ for V into Ohm's law, because here the emf is *induced*, not a steady dc emf from a battery:

$$I = \frac{\mathscr{E}}{R} = \frac{4.7 \times 10^{-3} \text{ V}}{1.0 \text{ Ω}} = 4.7 \times 10^{-3} \text{ A} = 4.7 \text{ mA}$$

This value does not exceed the allowed speaker current of 25 mA. The coil of this speaker will likely survive.

Follow-up Exercise. During the half-cycle oscillation in this Example, what is the average electric power, and how much total joule heat energy is dissipated in the speaker coil?

•**FIGURE 20.6 Computer hazard?** **(a)** An emf can be induced in an electrical component, such as the internal speaker coil of a laptop computer, **(b)** near a wire carrying an ac current. If the resulting induced current exceeds the manufacturer's limit, the component can be damaged or destroyed. See Example 20.2.

Conceptual Example 20.3

A Basic Electric Generator: Turning Mechanical Work into Electrical Current

Rarely does term (2) of Eq. 20.3 apply in everyday applications because it is difficult to design practical ways to change the size of a loop or coil area. However, we can use the situation in •Fig. 20.7 to study the essence of electrical energy generation. An external force does work on the bar as the bar is pulled to the right; the circuit area is immersed in a magnetic field. Thus the flux through that area changes with time, inducing a current. How does the induced emf depend on the speed v at which the bar is pulled? (a) It is independent of v. (b) It increases with v. (c) It decreases with v.

Reasoning and Answer. The induced emf depends on the *rate* at which the flux through the loop area is changing. Since the speed of the bar determines the rate that the loop area is increasing and since the flux depends directly on the loop area, (a) cannot be correct.

Let us calculate the flux symbolically to help us choose between (b) and (c). The magnetic field is perpendicular to the loop area, so the flux is $\Phi = BA$. Since B is constant, the change in flux must be due to the area change: $\Delta\Phi = B(\Delta A)$. Figure 20.7 shows that $\Delta A = L\,\Delta x = Lv\,\Delta t$. Because the change in area is directly proportional to the speed of the bar, so is the flux change and therefore so is the induced emf. The answer is (b): A faster pull makes for a larger emf. The expression for the magnitude of the induced emf is $|\mathcal{E}| = |\Delta\Phi|/\Delta t = BLv$.

This Example shows what is at the heart of electrical energy generation. Any magnetic flux change can be used to produce electrical energy. Consider a weight dropping over a pulley attached to the bar by a string. The work done by gravity would be converted into electrical energy in the loop.

Follow-up Exercise. In this Example (and looking from above in Fig. 20.7), what is the direction of the induced current?

The emf produced by the movement of the rod in Example 20.3 is known as *motional emf*. What would happen if the bar moved through the field on an insulated frame? With no conducting loop, and thus no complete circuit, there could be no sustained current. Yet the electrons in the bar would be moving through a magnetic field and so would experience a force in a direction given by the right-hand force rule. So there would still be an induced emf, electrons would move in response to it, and the bar would quickly become polarized, with one end positive (from the deficiency of electrons) and the other negative (from the surplus of electrons). Once the bar was polarized, there would be no further movement of charge, but an induced emf would still exist. It is through this mechanism that a very small emf (but no long-term currents!) develops across the wings of an airplane in flight as they "cut through" the Earth's magnetic field lines.

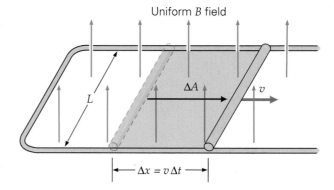

•**FIGURE 20.7   Motional emf**   As the metal rod is pulled on the metal frame, the area of the rectangular loop varies with time. A current is induced in the loop as a result of the changing flux. See Conceptual Example 20.3.

20.2 Generators and Back Emf

Objectives: To (a) understand the operation of electrical generators and calculate the emf produced by an ac generator, and (b) explain the origin of back emf and its effect on the behavior of motors.

As was pointed out in Section 20.1, one method to induce an emf or current in a loop is through a change in the loop's orientation, or a change in its effective area. Note again in Fig. 20.4 that the number of field lines through the loop changes as it *rotates*; that is, the *effective* area of the loop (the area perpendicular to the field) is $A \cos \theta$. This way of producing a flux change is the principle of operation of electrical generators.

Generators

A *generator* is a device that converts mechanical energy into electrical energy. Basically, the function of a generator is the reverse of that of a motor. In unusual instances, a generator can be run "backward" as a motor, and vice versa.

Note: Motors are discussed in Section 19.6.

A battery supplies direct current (dc). That is, the polarity of the voltage and the current's direction do not change. Generators can produce either direct current or **alternating current (ac)**, for which the polarity of the voltage and the direction of the current periodically change. Here we will analyze only **ac generators**, since the electricity used in homes and industry is primarily ac.

An ac generator is sometimes called an *alternator*. The basic elements of a simple ac generator are shown in •Fig. 20.8. A wire loop (called an armature) is mechanically rotated in a magnetic field by some external means. The rotation of the loop causes the magnetic flux through the loop to change, and a current is induced in the wire loop. The ends of the loop are connected to an external circuit by means of slip rings and brushes. (In practice, generators have many loops, or windings, on their armatures.)

Note: Although redundant, the expression ac current is often used, as is ac voltage (usually abbreviated VAC, for volts ac).

When the loop is rotated with a constant angular speed (ω), the angle (θ) between the magnetic field vector and the area vector of the loop (which is perpendicular to the plane of the loop) changes with time: $\theta = \omega t$ (assuming $\theta = 0°$ at $t = 0$). As a result, the cross-sectional area of the loop perpendicular to the magnetic field lines changes with time, and from Eq. 20.1 the flux will vary with time according to

$$\Phi = BA \cos \theta = BA \cos \omega t$$

By Faraday's law, the induced emf (which also now varies with time) for a rotating coil of N loops is then

$$\mathscr{E} = -N \frac{\Delta \Phi}{\Delta t} = -NBA \left(\frac{\Delta (\cos \omega t)}{\Delta t} \right)$$

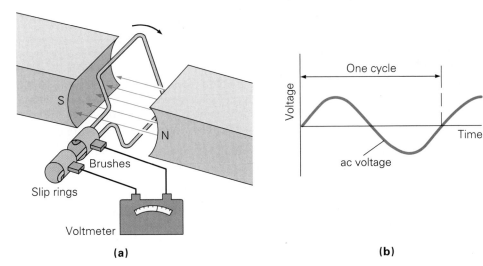

One cycle

Voltage

Time

ac voltage

(a)

(b)

•**FIGURE 20.8 A simple ac generator (a)** The rotation of a wire loop in a magnetic field produces **(b)** a voltage output whose polarity reverses with each half-cycle. This alternating voltage gives rise to an alternating current.

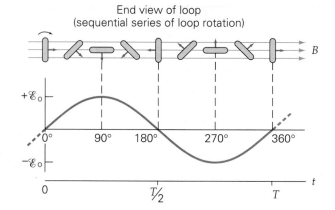

•FIGURE 20.9 $\mathscr{E} = \mathscr{E}_o \sin \omega t$ A graph of the sinusoidal generator output, together with an end view of the corresponding orientations of the loop during a cycle.

It can be shown by mathematical methods beyond the scope of this book that $\Delta(\cos \omega t)/\Delta t = -\omega \sin \omega t$. Thus, the instantaneous value of the emf at any time t is

$$\mathscr{E} = (NBA\omega) \sin \omega t$$

where $NBA\omega$ is the *maximum* magnitude of the emf, which occurs whenever $\sin \omega t = \pm 1$. If we designate $NBA\omega$ as $\mathscr{E}_o$, the maximum value of the emf, then

Sinusoidal ac emf

$$\mathscr{E} = \mathscr{E}_o \sin \omega t \qquad (20.4)$$

Since the value of the sine function varies between $+1$ and -1, the sign, or polarity, of the emf changes with time (•Fig. 20.9). Note from the figure that the emf has its maximum value $\mathscr{E}_o$ when $\theta = 90°$ or $\theta = 270°$, just when the plane of the rotating loop is parallel to field lines. This is because the *change* in flux is greatest at these points, where the flux's sign changes.

The direction of the current also periodically changes, which is why it is called *alternating* current. Since $\omega = 2\pi f$, Eq. 20.4 can be written as

$$\mathscr{E} = \mathscr{E}_o \sin 2\pi f t \qquad \text{alternator emf} \qquad (20.5)$$

where f is the rotational frequency (in hertz) of the generator's armature. The ac frequency in the United States is 60 Hz (or 60 cycles/s). Some other nations commonly use 50 Hz.

Keep in mind that Eqs. 20.4 and 20.5 give the instantaneous value of the emf and that $\mathscr{E}$ varies between $+\mathscr{E}_o$ and $-\mathscr{E}_o$ over one period. You will learn how to determine more practical time-averaged values for ac voltage and current in Chapter 21.

Example 20.4 ■ An ac Generator: Free and Renewable Electric Energy

A farmer decides to use a waterfall on his property to create a small hydroelectric power plant. He builds a coil consisting of 1500 closely packed circular loops of wire with a radius of 20 cm and constructs his generator to spin at 60 Hz in a magnetic field. To generate a maximum emf of 120 V, what must be the magnitude of the magnetic field?

Thinking It Through.   We can apply our result for the maximum emf ($\mathscr{E}_o$) from our previous discussion.

Solution.

Given: $\mathscr{E}_o = 120$ V   *Find:*   magnetic field magnitude (B)
$\quad\quad\quad N = 1500$ loops
$\quad\quad\quad r = 20$ cm $= 0.20$ m
$\quad\quad\quad f = 60$ Hz

The generator's maximum (or peak) emf is given by $\mathcal{E}_o = NBA\omega$. Because $\omega = 2\pi f$ and, for a circle, $A = \pi r^2$, we can rewrite this expression as $\mathcal{E}_o = 2\pi^2 NBr^2 f$. Solving for B, we get

$$B = \frac{\mathcal{E}_o}{2\pi^2 Nr^2 f} = \frac{120 \text{ V}}{2\pi^2(1500)(0.20 \text{ m})^2(60 \text{ Hz})} = 1.7 \times 10^{-3} \text{ T}$$

which is on the order of thirty times that of the Earth's field.

Follow-up Exercise. By what percentage would the maximum emf increase if the frequency regulator on the farmer's generator failed and the coils began turning at 70 Hz instead of 60 Hz?

In most large-scale ac generators (power plants), the armature is stationary and magnets are revolved about it. The revolving magnetic field produces a time-varying flux through the coils of the armature and thus an ac output. The mechanical energy required to spin the magnets in the generator is supplied by a turbine (•Fig. 20.10). Turbines are typically powered by steam generated from the heat of combustion of fossil fuels or by nuclear reactions, but they can also be rotated by falling water (hydroelectricity). Thus the basic difference among the various types of power plants is the *source* of the energy that turns the turbines.

Back Emf

Although their main job is to convert electric energy into mechanical energy, motors also generate emfs at the same time. Like a generator, a motor has a rotating armature in a magnetic field. The induced emf in this case is called a **back emf**, $\mathcal{E}_b$ (or sometimes a "counter" emf), because its polarity is opposite to that of the line voltage driving the motor and tends to reduce the current in the armature coils. (What would happen if this were not the case?)

If V is the line voltage, then the net voltage driving the motor is $V_{net} = V - \mathcal{E}_b$. Thus for a motor with a coil of internal resistance R, the current the motor draws when in operation can be found from Ohm's law:

$$I = \frac{V_{net}}{R} = \frac{V - \mathcal{E}_b}{R}$$

or

$$\mathcal{E}_b = V - IR \qquad \begin{array}{l} \textit{back emf} \\ \textit{of a motor} \end{array} \qquad (20.6)$$

Remember that V is the *line* voltage.

(a)

(b)

•**FIGURE 20.10 Electrical generation** (a) The gravitational potential energy of water trapped behind the Glen Canyon dam on the Colorado River in Arizona is used (b) to turn turbines such as these, which generate electric energy in much larger quantities than can the hydroelectric plant in the opening photo.

The back emf of a motor depends on the rotational speed of the armature and builds up from zero to some maximum value as the armature goes from rest to its normal operating speed. On startup, the back emf is zero (why?), so the starting current is a maximum (Eq. 20.6). Ordinarily, a motor turns something; that is, it has a mechanical load. Without a load, the armature speed will increase until the back emf has built up until it almost equals the line voltage, with just enough current in the coils to overcome friction and joule heat loss. Under normal load conditions, the back emf will be less than the line voltage. The larger the load, the slower the motor will rotate and the smaller the back emf will be. If a motor is overloaded and turns very slowly, the back emf may be reduced so much that the current becomes large enough (since V_{net} increases as $\mathscr{E}_b$ decreases) to burn out the coils. Thus, the back emf is involved in the regulation of a motor's operation.

A back emf in a dc motor circuit can be represented in a circuit diagram as a battery with polarity opposite that of the driving voltage (•Fig. 20.11).

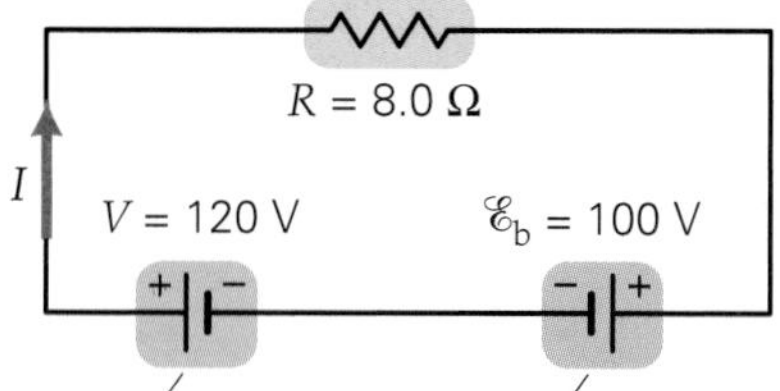

•**FIGURE 20.11 Back emf** The back emf in the armature of a dc motor can be represented as a battery with polarity opposite that of the driving source.

Example 20.5 ■ Getting up to Speed: Back Emf in a dc Motor

A dc motor with a resistance of 8.0 Ω in its windings operates on 120 V. With a normal load, there is a back emf of 100 V when the motor reaches full speed (see Fig. 20.11). What are (a) the starting current drawn by the motor and (b) the armature current at operating speed under a normal load?

Thinking It Through. We assume the resistance of the winding is constant (ohmic). (a) The only difference between startup and full speed is that there is essentially no back emf at startup. The net voltage and resistance determine the current, so we can apply Eq. 20.6 (rearranged). (b) At operating speed, the back emf opposes the driving voltage. The net or effective voltage is the difference in these two voltages (again Eq. 20.6).

Solution.

Given: $R = 8.0\ \Omega$ *Find:* (a) I_s (starting current)
        $V = 120\ V$ (b) I (operational current)
        $\mathscr{E}_b = 100\ V$

(a) From Eq. 20.6, the current in the windings is given by

$$I_s = \frac{V}{R} = \frac{120\ V}{8.0\ \Omega} = 15\ A$$

(b) Here

$$I = \frac{V - \mathscr{E}_b}{R} = \frac{120\ V - 100\ V}{8.0\ \Omega} = 2.5\ A$$

Note that, with no back emf, the starting current of the motor is relatively large. (When a big motor, such as that of a central air conditioning unit for a whole building, starts up, you may notice the lights momentarily dim because of the large starting current that it draws.) There is danger of coil burnout at low operating speeds. In some instances, resistors are temporarily connected in series with a motor's coil to protect the windings from burning out as a result of large starting currents.

Follow-up Exercise. (a) In this Example, how much energy is required to bring the motor to operating speed if it takes 10 s and the back emf averages 50 V during that time? (b) Compare this amount to the amount of energy required to keep the motor running for 10 s once it has reached its operating conditions.

Since motors and generators are opposites, so to speak, and a back emf develops in a motor, you may be wondering whether a back force develops in a generator. The answer is yes. When an operating generator is not connected to an external circuit, no current flows and there is no force on the coils of the armature due to the magnetic field. However, when the generator delivers power to a circuit and current flows in the coils (current-carrying wires in a magnetic field), there is a force that produces a *counter torque*, which opposes the rotation of the armature. As more current is drawn, the counter torque increases and a greater driving force is needed to turn the armature. Therefore, the higher the current output of a generator, the greater the energy expended (fuel consumed) in overcoming the counter torque.

20.3 Transformers and Power Transmission

Objectives: To (a) explain transformer action in terms of Faraday's law, (b) calculate the output of step-up and step-down transformers, and (c) understand the importance of transformers in electric energy delivery systems.

Electricity is transmitted by power lines over long distances. It is desirable to minimize the I^2R losses (joule heat) that can occur in these transmission lines. The resistance of a line is fixed, so reducing the I^2R losses means reducing the current. However, the power output of a generator is determined by its current and voltage outputs ($P = IV$), and for a fixed voltage (such as 120 V), a reduction in current would mean a reduced power output. It might appear that there is no way to reduce the current while maintaining the power supplied. Fortunately, however, electromagnetic induction can be applied to reduce power transmission losses by increasing the voltage and simultaneously reducing the current in just such a way that the delivered *power* is essentially unchanged. This is done with a device called a transformer.

A simple **transformer** consists of two coils of insulated wire wound on the same (closed) iron core. When ac voltage is applied to the input coil, or **primary coil**, the alternating current gives rise to an alternating magnetic flux that is concentrated in the iron core. The changing flux also passes through the output coil, or **secondary coil**, inducing an alternating voltage and current in it.

The induced voltage in the secondary coil differs from the voltage in the primary coil depending on the ratio of the numbers of turns in the two coils. By Faraday's law, the secondary voltage is given by

Note: For transformers, it is customary to use the term *voltage* rather than *emf*.

$$V_s = -N_s \frac{\Delta \Phi}{\Delta t}$$

where N_s is the number of turns in the secondary coil. The changing flux in the primary coil produces a back emf equal to

$$V_p = -N_p \frac{\Delta \Phi}{\Delta t}$$

where N_p is the number of turns in the primary coil. If the resistance of the primary coil is neglected, this emf must be equal in magnitude to the external voltage applied to it (why?). Then, forming a ratio gives

$$\frac{V_s}{V_p} = \frac{-N_s(\Delta \Phi / \Delta t)}{-N_p(\Delta \Phi / \Delta t)}$$

or

$$\frac{V_s}{V_p} = \frac{N_s}{N_p} \tag{20.7}$$

Here it is assumed that the core concentrates the field so the flux through each coil is the same—that is, no magnetic field lines stray outside the iron core.

If the transformer is assumed to be 100% efficient (no energy losses), the power input is equal to the power output. Since $P = IV$,

$$I_p V_p = I_s V_s \tag{20.8}$$

Although some energy is always lost, this is a good approximation, since a well-designed transformer can have an efficiency of more than 95%. (The sources of

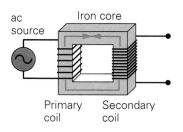
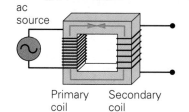

(a) Step-up transformer: high-voltage (low-current) output

(b) Step-down transformer: low-voltage (high-current) output

(c) Transformer circuit symbol

(d)

•FIGURE 20.12 **Transformers** (a) A step-up transformer has more turns in the secondary coil than in the primary. (b) A step-down transformer has more turns in the primary coil than in the secondary. (c) The circuit symbol for a transformer reflects its structure to some extent. (d) Large transformers used in electrical power transmission systems (to be discussed shortly).

energy losses will be discussed shortly.) Then, with Eq. 20.7, the transformer currents and voltages are related to the turn ratio by the relationship

$$\frac{I_p}{I_s} = \frac{V_s}{V_p} = \frac{N_s}{N_p} \tag{20.9}$$

With these equations, it is easy to see how a transformer affects the voltage and current. In terms of the output,

$$V_s = \left(\frac{N_s}{N_p}\right)V_p \quad \text{and} \quad I_s = \left(\frac{N_p}{N_s}\right)I_p \tag{20.10}$$

That is, if the secondary coil has more windings than the primary coil does ($N_s > N_p$, or $N_s/N_p > 1$) as in •Fig. 20.12a, the voltage is stepped up, or ($V_s > V_p$). This type of arrangement is called a **step-up transformer**. Notice, however, in this arrangement there is less current in the secondary than in the primary ($N_p/N_s < 1$ and $I_s < I_p$). For example, if the primary coil of a transformer has 50 turns and the secondary has 100 turns, $N_s/N_p = 2$ and $N_p/N_s = \frac{1}{2}$. Thus, a 220-volt input at 10 A will be stepped up to a 440-volt output at 5.0 A.

The opposite situation, in which the secondary coil has fewer turns than the primary, characterizes a **step-down transformer** (Fig. 20.12b). In this case, the voltage is stepped down, or reduced, and the current is increased. A step-up transformer may be used as a step-down transformer simply by reversing the output and input connections.

Step-up transformer: $N_s > N_p$, or $N_s/N_p > 1$ (more turns on secondary than on primary)

Step-down transformer: $N_p < N_s$, or $N_s/N_p < 1$ (more turns on primary than on secondary)

Note: The terms "step-up" and "step-down" refer to what happens to the primary or input *voltage*, not to the current. The effect on the current is the reverse of the effect on the voltage.

Example 20.6 ■ Transformer Orientation: Step-Up or Step-Down?

A transformer has 50 turns on its primary coil and 100 turns on its secondary coil. (a) If the primary is connected to a 120-volt source, what is the voltage output of the secondary? (b) If the transformer is operated in reverse and the 120-volt input is applied to the 100-turn coil, what would be the voltage output?

Thinking It Through. (a) Since the number of turns is greater in the secondary than the primary coil, this is a step-up transformer. In (b) the situation is reversed—this is a step-down transformer. In both cases, Eq. 20.10 applies for the secondary voltage.

Solution.

Given: (a) $N_p = 50$ **Find:** (a) V_s (secondary voltage output)
$N_s = 100$ (b) V_s
$V_p = 120$ V
(b) $N_p = 100$
$N_s = 50$
$V_p = 120$ V

(a) For this step-up transformer with $N_s > N_p$ and a turn ratio of $N_s/N_p = 100/50 = 2$,

$$V_s = \left(\frac{N_s}{N_p}\right)V_p = (2)(120 \text{ V}) = 240 \text{ V}$$

Since the voltage is stepped up by a factor of 2, the current must be stepped down and only half as large (show this).

(b) For this step-down transformer with $N_p > N_s$ and a turn ratio of $N_s/N_p = 50/100 = \frac{1}{2}$,

$$V_s = \left(\frac{N_s}{N_p}\right)V_p = (\tfrac{1}{2})(120 \text{ V}) = 60 \text{ V}$$

As you might expect, in this case, the voltage is stepped down and is half as large, while the current is stepped up by a factor of 2.

Follow-up Exercise. (a) In this Example, which connection would you use for European operation (at 240 V) of American appliances? Explain. (b) What would happen if you used the other connection while in Europe? Show a calculation for a 1200-watt hair dryer assumed to be ohmic and designed for use at 120 V in the United States.

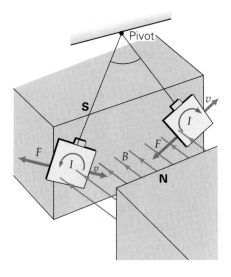

(a)

The preceding relationships apply only to ideal transformers, but actual transformers do have some energy losses. First, there is always some flux leakage; that is, not all of the flux passes through the secondary coil. In some transformers, one of the insulated coils is wound on top of the other rather than having the two on separate "legs" of a core to help avoid flux leakage and reduce the size of the transformer.

Second, when ac current flows in the primary coil, the changing magnetic flux through the loops gives rise to an induced emf in that coil. This is called *self-induction*. By Lenz's law, the self-induced emf will oppose the change in current and will thus limit the current (similar to a back emf in a motor). Think of self-induction as a kind of electromagnetic inertia—like the inertia of material bodies, it opposes change. Third, some energy is also lost due to the resistances of the coil wires (I^2R losses), but this loss is generally small.

A fourth cause of energy loss is **eddy currents** in the transformer core. To increase the density of the magnetic flux, the core is made of a highly permeable material, but such materials are usually good conductors. The changing magnetic flux sets up swirling movements of charge, or eddy currents, in the core material, and these dissipate energy.

To reduce this effect, transformer cores are made of thin sheets of material (usually iron) laminated with an insulating glue between them. The insulating layers between the sheets break up the eddy currents or confine them to the sheets, greatly reducing energy loss due to them. Well-designed transformers generally have less than 5% internal energy loss.

An effect of eddy currents can be demonstrated by swinging a plate made of a conductive but nonmagnetic metal, such as aluminum, through a magnetic field, as illustrated in •Fig. 20.13a. Eddy currents are set up in the plate as a result of its motion in the field and thus a changing magnetic flux through the plate's area. By Lenz's law, eddy currents are induced, opposing the flux change.

On its swing into the field (the left-hand plate position in Fig. 20.13a), a counterclockwise current is induced (apply Lenz's law to show this). This is equivalent

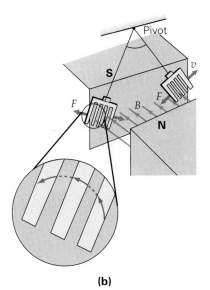

(b)

•**FIGURE 20.13 Eddy currents**
(a) Eddy currents are induced in a nonmagnetic conductive plate moving in a magnetic field. The induced currents oppose the change in flux, and a retarding force opposes the motion. Note that the currents reverse direction as the plate swings through the field. **(b)** If a plate with slits is used, the plate swings more freely.

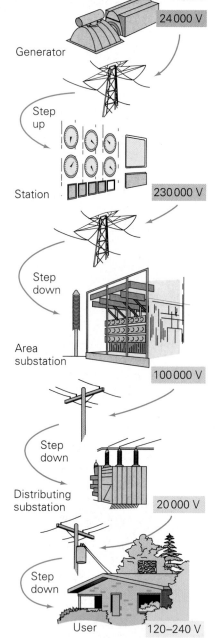

24 000 V — Generator

Step up

Station — **230 000 V**

Step down

Area substation — **100 000 V**

Step down

Distributing substation — **20 000 V**

Step down

User — **120–240 V**

•**FIGURE 20.14 Power transmission** A diagram of a typical electrical power distribution system.

to the plate having a north magnetic pole near the permanent magnet's north pole (likewise for the south pole). Hence there is a repulsive magnetic force on the plate as it enters the field region. The current in the plate is reversed as it leaves the field, giving rise to an attractive magnetic force. Both forces slow the plate's motion.

The breaking up of the eddy currents can be demonstrated by using a plate with slits (Fig. 20.13b). When this plate swings between the magnet's poles, it swings relatively freely because the eddy currents are much reduced by the gaps (slits). Thus the magnetic force on the plate is also reduced.

The damping effect of eddy currents is applied in the braking systems of rapid-transit rail cars. When an electromagnet on the car is turned on, it applies a magnetic field to a metal wheel or the rail. The repulsive force due to the induced eddy currents acts as a braking force. As the car slows, the eddy currents decrease, allowing a smooth braking action.

Power Transmission

For power transmission over long distances, transformers provide a means to increase (step up) the voltage and reduce the current of a generator's output to cut down the resistive I^2R losses. A schematic diagram of an ac power distribution system is shown in •Fig. 20.14. The voltage output of the generator is stepped up and transmitted over long distances to an area substation near the consumers. There the voltage is stepped down. There are further voltage step-downs at distributing substations and utility poles before 120-volt and 240-volt electricity are supplied to homes and businesses.

The following Example illustrates the benefits of being able to step up the voltage (and step down the current) for electrical power transmission.

Example 20.7 ■ **Cutting Your Losses: Power Transmission at High Voltage**

A small hydroelectric power plant produces energy in the form of electric current at 10 A and a voltage of 440 V. The voltage is stepped up to 4400 V (by an ideal transformer) for transmission over 40 km of power line, which has a resistance of 0.50 Ω/km. (A power transmission line has two wires for a complete circuit; however, for simplicity, take the length given to be the total wire length rather than doubling it.) (a) What percentage of the original energy would have been lost in transmission if the voltage had not been stepped up? (b) What percentage of the original energy is lost with the voltage stepped up?

Thinking It Through. (a) We can calculate the power output from $P = IV$ and compare that to the power loss in the wire, $P = I^2R$. We know the wire length and can therefore calculate its resistance. (b) We must use Eq. 20.10 to account for the stepped-up voltage and stepped-down current. We can then repeat the previous calculation.

Solution.

Given: $I_p = 10$ A
$V_p = 440$ V
$V_s = 4400$ V
$R_o/L = 0.50$ Ω/km
$L' = 40$ km

Find: (a) Percentage energy loss without voltage step-up
(b) Percentage energy loss with voltage step-up

(a) The power output of the generator is

$$P = I_pV_p = (10 \text{ A})(440 \text{ V}) = 4400 \text{ W}$$

The resistance of 40 km of power line is

$$R = \left(\frac{R_o}{L}\right)L' = \left(\frac{0.50 \text{ Ω}}{\text{km}}\right)(40 \text{ km}) = 20 \text{ Ω}$$

The wire energy loss rate (joules per second or watts) in transmitting a current of 10 A is

$$P_{loss} = I_p{}^2R = (10 \text{ A})^2(20 \ \Omega) = 2000 \text{ W}$$

Thus, the percentage of the produced energy lost to joule heat in the wires is

$$\% \text{ loss} = \frac{P_{loss}}{P} \times 100\% = \frac{2000 \text{ W}}{4400 \text{ W}} \times 100\% = 45\%$$

(b) When the voltage is stepped up to 4400 V, the transmitted current is

$$I_s = \left(\frac{V_p}{V_s}\right)I_p = \left(\frac{440 \text{ V}}{4400 \text{ V}}\right)(10 \text{ A}) = 1.0 \text{ A}$$

(The voltage was stepped up by a factor of 10, so the current is stepped down by the same factor.) The power loss in this case is

$$P_{loss} = I_s{}^2R = (1.0 \text{ A})^2(20 \ \Omega) = 20 \text{ W}$$

The percentage loss is reduced from 2000 W by a factor of 100 (since the current is reduced by a factor of 10 and the power depends on the *square* of the current):

$$\% \text{ loss} = \frac{P_{loss}}{P} \times 100\% = \frac{20 \text{ W}}{4400 \text{ W}} \times 100\% = 0.45\%$$

Follow-up Exercise. Some heavy-duty electrical appliances, such as water pumps, can be wired to sockets of 240 V or 120 V. Their power rating is the same regardless of the voltage at which they run. (a) Explain the efficiency advantage of operating such appliances at the higher voltage. (b) For a 1-horsepower pump (746 W), estimate the ratio of the power lost in the wires at 240 V to the loss at 120 V (assuming all resistances are ohmic and the connection wires are the same).

Example 20.7 shows the advantage of using high-voltage, or high-tension, transmission lines (•Fig. 20.15a). However, there is a practical limit to the degree of voltage step-up. At very high voltages, the molecules in the air surrounding a power line may be ionized by the induced electric fields, forming a conducting path to nearby trees, buildings, or the ground surface. This is referred to as a leakage loss. Crews from electric companies are continually clearing foliage from near power lines, and long insulators are used to hold the high-voltage wires away from the metal towers (Fig. 20.15b). Leakage losses are generally greater during wet weather because moist air is more easily ionized than dry air. Under certain conditions such as fog, you may see the arcing or corona discharge from a high-voltage line and/or hear the accompanying crackling noise.

20.4 Electromagnetic Waves

Objectives: To (a) explain the physical nature, origin, and means of propagation of electromagnetic waves, and (b) describe the properties and uses of various types of electromagnetic waves.

Electromagnetic waves (or *radiation*) were considered as a means of heat transfer in Section 11.4. Now you are ready to understand more fully the production and characteristics of electromagnetic radiation. As the name implies, these waves have both electric and magnetic properties, which can be described by quantities you have studied.

James Clerk Maxwell showed that four fundamental relationships could completely describe all observed electromagnetic phenomena. Maxwell also used this set of equations to predict the existence of waves of an electromagnetic nature. Because of his contributions, the set of equations is known as **Maxwell's equations**, although they were for the most part developed individually by other scientists (for example, Faraday discovered the law of induction).

Essentially, Maxwell's equations combine the electric field and the magnetic field into a single electromagnetic field. The apparently separate fields are symmetrically related in the sense that either one can create the other. This symmetry

(a)

(b)

•**FIGURE 20.15   Long lines**
(a) Long-distance power lines carry electric power at high voltage. The use of high voltages reduces power losses in transmission. **(b)** Workers maintaining insulators at a power substation. Large insulators are needed to reduce leakage from high-voltage lines.

Note: The Scottish physicist James Clerk Maxwell (1831–1879) unified the electric and magnetic fields.

is evident in the equations as presented in their advanced mathematical form (not shown here). For our purposes, a qualitative description is sufficient:

A time-varying magnetic field produces a time-varying electric field.

A time-varying electric field produces a time-varying magnetic field.

Changing B produces changing E. Changing E produces changing B.

The first statement is simply the observation that, as we saw in Section 20.1, a changing magnetic flux gives rise to an induced emf and current in a wire. The second statement is crucial to understanding the self-propagating characteristic of electromagnetic waves. Together, these two characteristics enable these waves to travel through a vacuum, whereas all other waves require a supporting medium.

According to Maxwell's theory, electromagnetic waves are produced by *accelerating* electric charges, such as an electron oscillating in simple harmonic motion. For example, the electron could be one of the many nearly free electrons in the metal antenna of a radio transmitter, driven by an electrical oscillator with a frequency of about 10^6 Hz. As such an electron oscillates, it continually accelerates and decelerates and thus radiates an electromagnetic wave (•Fig. 20.16a). The continual oscillations of many such charges due to the alternating voltage in the transmitter produce time-varying electric and magnetic fields in the immediate vicinity of the antenna. The electric field, shown in red in Fig. 20.16a, is in the plane of the paper and continually changes direction, as does the magnetic field (shown in blue and into and out of the paper by a right-hand rule).

Both the electric and the magnetic fields carry energy and propagate outward with the speed of light (c in a vacuum, 3.00×10^8 m/s). Maxwell's equations show that, except in the immediate vicinity of the source, the electromagnetic waves at a fixed instant of time are plane waves (Fig. 20.16b). The electric field (**E**) is perpendicular to the magnetic field (**B**), and each varies sinusoidally with time. Both **E** and **B** are in phase and perpendicular to the direction of wave propagation. Thus,

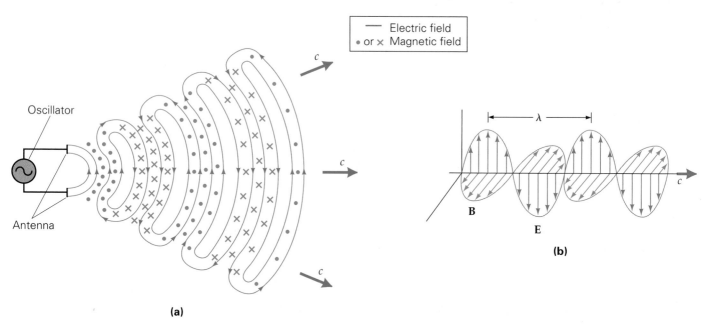

(a)

(b)

•**FIGURE 20.16   Source of electromagnetic waves**   Electromagnetic waves are produced, fundamentally, by accelerating electric charges. **(a)** The charges are moved in simple harmonic motion by an oscillating voltage source connected to a metal antenna. As the polarity of the antenna and the direction of the current in the wire periodically change, alternating electric and magnetic fields propagate outward. The electric field **E** and the magnetic field **B** are in phase and are perpendicular to one another and to the direction of propagation of the wave. Thus electromagnetic waves are transverse waves. **(b)** At large distances from the source, these curved wavefronts become almost planar.

electromagnetic waves are transverse waves, but the *fields*, not any material or medium, oscillate perpendicularly to the direction of propagation. As each wave builds up and collapses over time, it creates the other; the process is repeated again and again, giving rise to a traveling wave.

Radiation Pressure

An electromagnetic wave carries energy. Consequently, it can do work and can exert a force on a material it strikes. Consider light striking an electron on a surface (•Fig. 20.17). The electric field of the electromagnetic wave does work on the electron; assume that this work gives the electron a downward velocity (**v**) as shown in the figure. Recall from Chapter 19 that a moving charged particle in a magnetic field experiences a force. As a result, there is a magnetic force on the electron due to the magnetic field component of the light wave. By the right-hand rule for the magnetic force on a moving charged particle, the force on the electron is initially in the direction shown in Fig. 20.17. That is, the electromagnetic wave produces a force on the electron in the direction in which the wave is propagating and therefore exerts a force on the surface to which the electron is bound.

The radiation force per area is called the **radiation pressure**. Radiation pressure is negligible for most common situations, but it can be of importance in some atmospheric and astronomical phenomena. For example, radiation pressure plays a key role in determining the direction in which the tail of a comet points. The sunlight delivers energy to the comet "head," which consists of frozen ices and dust. Some of this material is evaporated as the comet nears the Sun, and the evaporated gases are pushed away from the Sun by radiation pressure. Thus the tail points generally away from the Sun, whether the comet is approaching or leaving the Sun's vicinity.

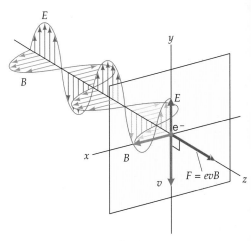

•**FIGURE 20.17 Radiation pressure** The electric field of an electromagnetic wave that strikes a surface acts on an electron, giving it a velocity (**v**). The magnetic field then exerts a force on the moving charge in the direction of propagation of the incident light. (Check this, using the magnetic right-hand force rule.)

Note: Even the light from your desk lamp exerts a very, very small force on your desk. However, all such effects are negligible in everyday settings.

Example 20.8 ■ The Speed of Light (and Other EM Waves)

Viking space probes landed on Mars in 1976 and sent radio and TV signals back to Earth. How much longer did it take for a signal to reach us when Mars was farthest from Earth than when it was closest to us? The average distances of Mars and Earth from the Sun are 229 million km (142 million mi) and 150 million km (93 million mi), respectively. Assume circular orbits with these average distances as radii.

Thinking It Through.   This situation calls for a time–distance calculation. To find the *difference* in the travel times of the radio signals, we first find the *difference* in the distances. The planets are farthest apart when they are on opposite sides of the Sun, when they are separated by a distance ($d_M + d_E$). The planets are closest when they are aligned on the same side of the Sun; their separation distance is then ($d_M - d_E$). (Draw a diagram to help you visualize these relationships.)

Solution.   Listing the data, we have

Given: $d_M = 229 \times 10^6$ km   *Find:* Δt (time difference)
                 (average distance of Mars from Sun)
             $d_E = 150 \times 10^6$ km
                 (average distance of Earth from Sun)

The difference in the separation distances is

$$\Delta d = (d_M + d_E) - (d_M - d_E) = 2d_E$$

$$= 2(150 \times 10^6 \text{ km}) = 3.0 \times 10^{11} \text{ m}$$

(Notice that the difference is twice the Earth's distance from the Sun, so the distance of Mars from the Sun was not needed. This situation illustrates the advantage of doing operations algebraically before putting in numerical values.)

Radio waves are electromagnetic waves and travel at "the speed of light," c. Since $\Delta d = c\,\Delta t$,

$$\Delta t = \frac{\Delta d}{c} = \frac{3.0 \times 10^{11}\,\text{m}}{3.0 \times 10^{8}\,\text{m/s}} = 1.0 \times 10^{3}\,\text{s} \quad (\text{or } 16.7\,\text{min})$$

Follow-up Exercise. The *Sojourner* roving Martian vehicle (see Example 2.1, p. 34) is moving at 0.050 m/min and heading for a disastrous collision with a rock 2.0 m ahead. If Mars is closest to the Earth, then how long does it take a video signal from *Sojourner* to reach Earth and an immediate return "stop" signal to reach *Sojourner*? Will the rover be spared the collision?

Conceptual Example 20.9

Sailing the Sea of Space: Radiation Pressure in Action

Some scientists have proposed that a relatively light spacecraft with a huge sail could be built and launched from a space station in Earth orbit. With little or no power of its own, it would use the pressure of sunlight to propel it to the outer planets. To get the maximum propulsive force, what kind of surface should the sail have? (a) Shiny and reflective. (b) Dark and absorptive. (c) Surface characteristics would not matter.

Reasoning and Answer. At first glance, you might think that the answer is (c). However, as we have seen, radiation possesses energy and is capable of exerting force, and it transfers momentum to whatever it strikes. Thus, we must think of the interaction between the radiation and the sail in terms of conservation of momentum, as discussed in Section 6.3. If the radiation is absorbed, the situation is analogous to a completely inelastic collision, and the sail would acquire all of the momentum (call it p) originally possessed by the radiation.

However, if the radiation is *reflected*, the situation is analogous to a completely elastic collision. Since the momentum of the radiation after the collision would be equal to its original momentum in magnitude but opposite in direction, its momentum would be reversed (from $+p$ to $-p$). To conserve momentum, the momentum transferred to the shiny sail would be twice as great ($2p$) as that for the dark sail. Because force is the rate of change of momentum, reflective sails would experience, on average, twice as much force as absorptive ones. The answer is (a).

Follow-up Exercise. (a) Would the sail in this Example provide less or more acceleration as the interplanetary sailing ship moved farther from the Sun? (b) How could this change in acceleration be counteracted?

Types of Electromagnetic Waves

Electromagnetic waves are classified by ranges of frequencies or wavelengths in a spectrum. Recall from Chapter 13 that frequency and wavelength are inversely related by the traveling-wave relationship $\lambda = c/f$. The greater the frequency, the shorter the wavelength, and vice versa. The electromagnetic spectrum is continuous, so the limits of the various ranges are approximate (•Fig. 20.18). Table 20.1 lists these frequency and wavelength ranges for the general types of electromagnetic waves.

Power Waves. Electromagnetic waves of frequency 60 Hz result from currents moving back and forth (alternating) in electrical circuits. As Table 20.1 indicates, these power waves have a wavelength of 5.0×10^{6} m, or 5000 km (more than 3000 mi). Waves of this low frequency are of little practical use. They may occasionally produce a so-called 60-hertz hum on your stereo or introduce unwanted electrical noise in delicate instruments. More seriously, concerns have been expressed about possible health effects of these waves. Some early research tended to suggest that very low-frequency fields may have potentially harmful biologi-

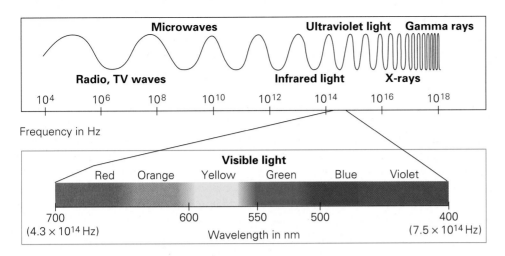

•FIGURE 20.18 The electromagnetic spectrum The spectrum of frequencies or wavelengths is divided into regions, or ranges. Note that the visible region is a very small part of the total electromagnetic spectrum. The wavelengths are given in nanometers: 1 nm = 10^{-9} m. (Top illustrative wavelengths are not drawn to scale.)

TABLE 20.1 Classification of Electromagnetic Waves

Type of Wave	Approximate Frequency Range (Hz)	Approximate Wavelength Range (m)	Source
Power waves	60	5×10^6	Electric currents
Radio waves			Electric circuits
AM	(0.53×10^6)–(1.7×10^6)	570–186	
FM	(88×10^6)–(108×10^6)	3.4–2.8	
TV	(54×10^6)–(890×10^6)	5.6–0.34	
Microwaves	10^9–10^{11}	10^{-1}–10^{-3}	Special vacuum tubes
Infrared radiation	10^{11}–10^{14}	10^{-3}–10^{-7}	Warm and hot bodies
Visible light	(4.0×10^{14})–(7.0×10^{14})	10^{-7}	Sun and lamps
Ultraviolet radiation	10^{14}–10^{17}	10^{-7}–10^{-10}	Very hot bodies and special lamps
X-rays	10^{17}–10^{19}	10^{-10}–10^{-12}	High-speed electron collisions and atomic processes
Gamma rays	above 10^{19}	below 10^{-12}	Nuclear reactions, processes in particle accelerators, and natural radioactivity

cal effects on cells and tissues. However, recent surveys indicate that this is not the case in daily life.

Radio and TV Waves. Radio and TV waves are generally in the frequency range from 500 kHz to about 1000 MHz. The AM (amplitude-modulated) band runs from 530 to 1710 kHz (1.71 MHz). Higher frequencies, up to 54 MHz, are used for "short-wave" bands. TV bands range from 54 MHz to 890 MHz. The FM (frequency-modulated) radio band runs from 88 to 108 MHz, which lies in a gap between channels 6 and 7 of the TV band. Cellular phones use radio waves to transmit voice communication in the ultrahigh frequency (UHF) band, with frequencies similar to those of radio waves used for television channels 13 and higher.

Early global communications used the "short-wave" bands, as do amateur (ham) radio operators today. But how are the normally straight-line radio waves transmitted around the curvature of the Earth? This is accomplished by reflection off ionic layers in the upper atmosphere. Energetic particles from the Sun ionize gas molecules, giving rise to several ion layers. Certain of these layers reflect radio waves below a specific frequency. By "bouncing" radio waves off these layers, we can send radio transmissions beyond the horizon, to any region of the Earth.

Such reflection of radio waves requires the ionic layers to have uniform density. When, from time to time, a solar disturbance produces a shower of energetic particles that upsets this uniformity, a communications "blackout" can occur as the radio waves are scattered in many directions rather than reflected in straight

•FIGURE 20.19 Global village
Radio and TV transmissions from around the world, relayed by orbiting satellites, can be picked up by backyard and rooftop antennas.

•FIGURE 20.20 Slow down
Radar speed guns based on the Doppler effect employ radiation in the microwave region of the spectrum. Here the speed of a car is being monitored.

lines. To avoid such disruptions, global communications have, in the past, relied largely on transoceanic cables. Now we also have communications satellites, which can provide line-of-sight transmission to any point on the globe (•Fig. 20.19).

Microwaves.   Microwaves, with frequencies in the gigahertz (GHz) range, are produced by special vacuum tubes (called klystrons and magnetrons). Microwaves are used in communications and radar applications. In addition to its many roles in navigation and guidance, radar provides the basis for the speed guns used to time such things as baseball pitches, tennis serves, and motorists (•Fig. 20.20). When radar waves are reflected from a moving object, their wavelength is shifted by the Doppler effect (Section 14.5). The amount of the shift indicates the velocity of the object toward or away from the observer. Another very common use of microwaves today is in microwave ovens (see the Insight on p. 380).

Infrared Radiation.   The infrared region of the electromagnetic spectrum lies adjacent to the low-frequency or long-wavelength end of the visible spectrum. The frequency at which a warm body emits radiation depends on the body's temperature. Such a body emits electromagnetic waves of many frequencies, but the frequency of the maximum intensity characterizes the radiation. A body at about room temperature emits radiation in the far infrared region (farthest from the visible region).

Recall from Section 11.4 that infrared radiation is sometimes referred to as heat rays. This is because water molecules, which are present in most materials, readily absorb infrared wavelengths. When they do, their random thermal motion is increased—they heat up and heat their surroundings. Infrared lamps are used in therapeutic applications and to keep food warm in cafeterias. Infrared radiation is also associated with maintaining the Earth's warmth or average temperature through the greenhouse effect. Incoming visible light (which passes relatively easily through the atmosphere) is absorbed by the Earth's surface and re-radiated as infrared (longer-wavelength) radiation, which is trapped by greenhouse gases such as carbon dioxide and water vapor.

Visible Light.   The visible region occupies only a very small portion of the total electromagnetic spectrum. It runs from about 4×10^{14} Hz to about 7×10^{14} Hz, or a wavelength range of about 700 to 400 nm, respectively (Fig. 20.18). Only the radiation in this region can activate the receptors in our eyes. Visible light emitted or reflected from the objects around us provides us with much information about our world. Visible light and optics will be discussed in Chapters 22 to 25.

It is interesting to note that not all animals are sensitive to the same range of wavelengths. For example, snakes can detect infrared radiation, and the visible range of many insects extends well into the ultraviolet. It is also interesting to note that the sensitivity range of our eyes conforms closely to the spectrum of wavelengths emitted by the Sun, both having maxima in the yellow–green region. If the Sun emitted mostly infrared radiation, for example, the surface of the Earth would appear very dark to us.

Ultraviolet Radiation.   Beyond the violet end of the visible region lies the ultraviolet frequency range. Ultraviolet (or UV) radiation is produced by special lamps and very hot bodies. The Sun emits large amounts of ultraviolet radiation, but fortunately most of it received by the Earth is absorbed in the ozone (O_3) layer in the atmosphere at an altitude of about 40 to 50 km. Because the ozone layer plays a protective role, there is concern about its depletion by chlorofluorocarbon gases (such as Freon, once commonly used in refrigerators) that drift upward and react with the ozone. (See the Insight on p. 656.)

Most ultraviolet radiation is absorbed by certain molecules in ordinary glass. Therefore, you cannot get a tan or a sunburn through glass windows. Sunglasses are now labeled to indicate the UV protection standards they meet in shielding the eyes from this potentially harmful radiation. Welders wear special glass goggles or face masks to protect their eyes from the large amounts of ultraviolet radiation produced by the arcs of welding torches. Similarly, it is important to shield your eyes from a sunlamp or from snow-covered surfaces. The ultraviolet component of sunlight reflected from snow-covered surfaces can produce snowblindness in unprotected eyes.

X-rays. Beyond the ultraviolet region of the electromagnetic spectrum is the important X-ray region. We are familiar with X-rays primarily through medical applications. X-rays were discovered accidentally in 1895 by the German physicist Wilhelm Roentgen (1845–1923) when he noted the glow of a piece of fluorescent paper caused by some mysterious radiation coming from a cathode ray tube. Because of the apparent mystery involved, these were named x-radiation or X-rays.

The basic elements of an X-ray tube are shown in •Fig. 20.21. An accelerating voltage, typically of several thousand volts, is applied across the electrodes in a sealed, evacuated tube. Electrons emitted from the heated negative electrode are thus accelerated toward the positive anode (target). When electrons strike the anode material, they are slowed down by the repulsive interaction with the atomic electrons of the material. They can also transfer some of their energy to the atoms themselves by exciting the atomic electrons of those atoms. During the deceleration of the electrons, some of their kinetic energy loss is converted to electromagnetic energy in the form of X-rays. When the atomic electrons de-excite, their loss of energy also appears in the form of X-rays.

Similar processes take place in today's color television picture tubes, which use high voltages and electron beams. When the high-speed electrons are stopped as they hit the screen, they can emit X-rays into the environment. Fortunately, all modern televisions have the shielding necessary to protect viewers from exposure to this radiation. In the early days of color television, this was not always the case (and hence we were warned not to sit too close to the screen).

As you will learn in Chapter 27, the energy of electromagnetic radiation depends on its frequency. High-frequency X-rays have very high energies and can cause cancer, skin burns, and other harmful effects. However, at low intensities, X-rays can be used with relative safety to view the internal structure of the human body and other opaque objects.* X-rays can pass through materials that are opaque to other types of radiation. The denser the material, the greater its absorption of X-rays and the less intense the transmitted radiation will be. For example, as X-rays pass through the human body, many more X-rays are absorbed by bone than by tissue. If the transmitted radiation is directed onto a photographic plate or film, the exposed areas show variations in intensity that form a picture of internal structures.

The combination of the computer with modern X-ray machines permits the formation of three-dimensional images by means of a technique called *computerized tomography*, or CT (•Fig. 20.22).

Gamma Rays. The electromagnetic waves of the uppermost frequency range of the known electromagnetic spectrum are called gamma rays (γ-rays). This high-frequency radiation is produced in nuclear reactions, in particle accelerators, and also in certain types of nuclear radioactivity. Gamma rays will be discussed in Chapter 29.

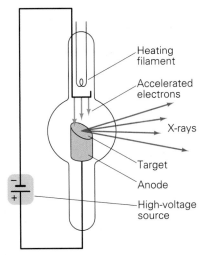

•**FIGURE 20.21 The X-ray tube** Electrons accelerated through a large voltage strike a target electrode. There they slow down and interact with (excite) the atomic electrons of the target material. Energy is emitted in the form of X-rays during both the "braking" process and the de-excitation of the atoms in the material.

---

*Most health scientists believe that there is no safe "threshold" level for X-rays or other energetic radiation—that is, no level of exposure that is completely risk-free—and that some of the dangerous effects are cumulative over a lifetime. People should therefore avoid unnecessary medical X-rays or any other unwarranted exposure to "hard" radiation (Chapter 29). However, properly used, X-rays can be an extremely useful diagnostic tool capable of saving lives.

CFCs and Ozone Layer Depletion

We hear a lot about the ozone layer these days. What is it, and why should it concern us? At an altitude of 30 km (about 20 mi) in the atmosphere exists a concentration of ozone called the *ozone layer*. The ozone molecule has a molecular formula of O_3 and consists of three oxygen atoms bound together. The formation of ozone starts when oxygen molecules (O_2) dissociate into oxygen atoms (O) by absorbing solar UV radiation ($O_2 + UV \rightarrow O + O$). The resulting atomic oxygen can combine with an oxygen molecule to form ozone ($O_2 + O \rightarrow O_3$).

Ozone is very unstable in the presence of sunlight, however. When ozone absorbs UV radiation, it dissociates into atomic and molecular oxygen ($O_3 + UV \rightarrow O + O_2$). Moreover, an oxygen atom may react with an ozone molecule to form two ordinary oxygen molecules ($O + O_3 \rightarrow O_2 + O_2$). Both of these reactions destroy ozone. All of these processes occur simultaneously in the ozone layer, so a natural balance between ozone production and ozone destruction is maintained.

The ozone layer is vitally important to the Earth's living creatures—ourselves included—because it absorbs most of the UV radiation that reaches our planet from the Sun. It thus shields us from these energetic rays, which are potentially harmful. For example, most cases of *melanoma*, an extremely dangerous form of skin cancer, are thought to be caused by excessive UV exposure. Overexposure to UV light is also responsible for the burning of skin. All people (except albinos) have a pigment in their skin called *melanin*. Exposure to UV light induces the production of melanin, causing the skin to tan, but overexposure can cause the skin to burn (sunburn). Creams and lotions containing UV-absorbing chemicals (sunscreens) are recommended by most physicians for people who must be exposed to sunlight.

However, the protective blanket of ozone is threatened by human activity. The cause has been the release into the atmosphere of a class of chemicals called *chloroflourocarbons* (CFCs). Until the 1990s, CFCs were widely used as refrigerants (Freon is one commercial CFC). We depend on refrigerants in our home and automobile air conditioners as well as in our refrigerators. CFCs were also widely used as propellants in spray cans, plastic foam blowing agents, and industrial solvents.

As gases, these CFCs escaped into the atmosphere and became a cause for worldwide concern. In the lower atmosphere, there is no known mechanism for destroying CFC molecules, and they rise slowly (over 10 to 30 years) to the altitude of the ozone layer. There the CFC molecules absorb UV sunlight, dissociate, and release highly reactive single chlorine (Cl) atoms:

$$CFC \xrightarrow{(UV)} Cl + \text{other molecules}$$

These atoms in turn react with and destroy the ozone molecules in a repeating cycle:

$$Cl + O_3 \rightarrow ClO + O_2$$

and finally

$$ClO + O \rightarrow Cl + O_2$$

Notice that the chlorine atom reappears and is available for another reaction. It is estimated that these chlorine atoms can persist in the atmosphere for a year or more. During this time, a single chlorine atom can destroy as many as 100 000 ozone molecules.

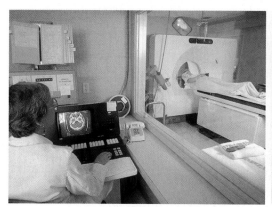
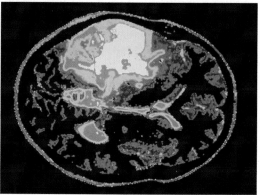

(a)   (b)

•**FIGURE 20.22 CT scan** In an ordinary X-ray image, the entire thickness of the body is projected onto the film. Internal structures often overlap, making details hard to distinguish. In CT—computerized tomography (from the Greek words *tomo*, "slice," and *graph*, "picture")—X-ray beams scan across a slice of the body. **(a)** The transmitted radiation is recorded by a series of detectors and processed by a computer. Using information from multiple slices, the computer can construct a three-dimensional image. Any single slice can also be displayed for study, as on the monitor. CT scans typically provide physicians with much more information than can be obtained from a conventional X-ray. **(b)** CT image of a brain with a benign tumor.

Attention was drawn to the fate of the ozone layer in 1985 when a hole in that layer was discovered over Antarctica (Fig. 1). Current observations show a similar but smaller hole over the North Pole regions. These discoveries prompted companies to seek safer substitutes for CFCs. For example, in the 1990s, new air conditioners in automobiles use a different refrigerant—one that is much safer for the ozone layer than are CFCs. Finding a refrigerant that produces *no* harm to the ozone layer is a primary goal of scientists today.

The use of these new refrigerants involves tradeoffs. They are less efficient and therefore could cause up to a 3% increase in electrical energy usage. This could add to consumer cost and cause an increase in carbon dioxide emissions, since most of our electricity is still generated by the burning of fossil fuels. More carbon dioxide (CO_2) could in turn enhance the greenhouse effect and contribute to global warming (Chapter 11).

It is estimated that for every 1% reduction in the ozone layer, an additional 2% of the Sun's UV radiation reaches the Earth's surface. The direct effect of increased UV exposure on humans is likely to be a higher incidence of melanoma. In fact, a rise in the number of deadly melanoma cases has already been reported in Australia and other southern regions where the ozone hole is the largest. Eye damage, such as cataracts, is also likely to rise. Globally, higher UV exposure could kill many of the oceanic microorganisms that play a key role in the food chain. An increase in UV radiation could also warm the Earth's surface, thus combining with the greenhouse effect, possibly causing coastline flooding as a result of polar cap melting.

The problem has alarmed many political leaders as well as scientists. In 1987, 24 nations signed the *Montreal Protocol on Substances that Deplete the Ozone Layer*. In 1990, more than 90 nations agreed to phase out CFCs by 2000. Because measurements indicated worse depletion than expected, industrialized protocol nations ceased CFC production in 1996. Developing countries have a 10-year grace period. Over 160 nations have now signed the treaty.

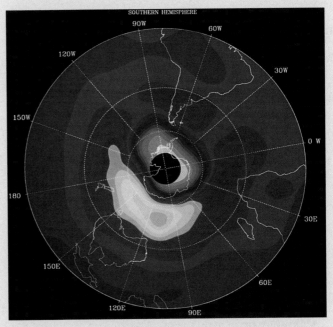

FIGURE 1 Ozone hole Satellite photo of the depletion of ozone over the South Pole (late September 1998). Pink shades indicate the regions of lowest ozone concentrations.

Chapter Review

Important Terms

electromagnetic induction *634*
magnetic flux *635*
Faraday's law of induction *636*
Lenz's law *636*
alternating current (ac) *641*
ac generators *641*

back emf *643*
transformer *645*
primary coil *645*
secondary coil *645*
step-up transformer *646*
step-down transformer *646*

eddy currents *647*
electromagnetic waves
  (radiation) *649*
Maxwell's equations *649*
radiation pressure *651*

Important Concepts

- Electromagnetic induction refers to the creation of induced emfs whenever the magnetic flux through a coil is changed. Faraday's law of induction states that the magnitude of the induced emf is equal to the time rate of change of the magnetic flux.

- Lenz's law states that when a change in magnetic flux induces an emf, the resulting current is in such a direction as to create a magnetic field that tends to oppose the change in flux.

- An ac current alternates in direction periodically.

- Generators (ac or dc) are the reverse of electric motors. When coils are spun in a magnetic field (or the field is spun while the coils are kept fixed), an emf is induced in the coils and can be used to create an electric current.

- Back emf is a reverse emf created by induction in motors when their armature is rotated in a magnetic field. The back emf partially cancels the voltage that drives the motor.

- A transformer is a device that changes the voltage supplied to it by means of induction. If the voltage is increased, the output current is reduced, and vice versa.

- An electromagnetic wave consists of mutually perpendicular, time-varying electric and magnetic fields that propagate at a constant speed in vacuum ($c = 3.00 \times 10^8$ m/s). The different types of electromagnetic radiation differ in frequency and thus in wavelength ($c = \lambda f$).

- Maxwell's equations are a set of four equations that describe all magnetic and electric field phenomena.

- Radiation carries energy and momentum and thus can exert force.

Important Equations

Magnetic Flux:

$$\Phi = BA \cos \theta \qquad (20.1)$$

Faraday's Law of Induction:

$$\mathscr{E} = -N \frac{\Delta \Phi}{\Delta t} \qquad (20.2)$$

$$= -N \left[(A \cos \theta)\left(\frac{\Delta B}{\Delta t}\right) + (B \cos \theta)\left(\frac{\Delta A}{\Delta t}\right) + BA\left(\frac{\Delta (\cos \theta)}{\Delta t}\right) \right] \qquad (20.3)$$

Generator emf (where $\mathscr{E}_o = NBA\omega$):

$$\mathscr{E} = \mathscr{E}_o \sin \omega t = \mathscr{E}_o \sin 2\pi f t \qquad (20.4, 20.5)$$

Back emf of a motor:

$$\mathscr{E}_b = V - IR \qquad (20.6)$$

Currents, voltages, and turn ratio for a transformer:

$$\frac{I_P}{I_s} = \frac{V_s}{V_P} = \frac{N_s}{N_P} \qquad (20.9)$$

Exercises

20.1 Induced Emf's: Faraday's Law and Lenz's Law

1. A unit of magnetic flux is (a) Wb, (b) T·m², (c) T·m/A, (d) both (a) and (b).

2. Magnetic flux through a loop can change due to a change in (a) the area of the coil, (b) the magnetic field strength, (c) the orientation of the loop, (d) all of the above.

3. A bar magnet is dropped through a coil of wire as shown in •Fig. 20.23. (a) Describe what is observed on the galvanometer by sketching a graph of $\mathscr{E}$ versus t. (b) Does the magnet fall freely? Explain.

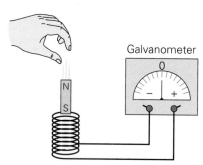

•FIGURE 20.23 **A time-varying magnetic field** What will the galvanometer measure? See Exercise 3.

4. 🔘 For an induced current to appear in a loop of wire, (a) there must be a large magnetic flux through the loop, (b) the loop must be parallel to the magnetic field, (c) the loop must be perpendicular to the magnetic field, (d) the magnetic flux through the loop must vary with time.

5. In Fig. 20.1b, what is the direction of the induced current in the loop if the approaching pole were a south pole?

6. Two identical strong magnets are dropped simultaneously into two vertical tubes of the same dimensions by two students (•Fig. 20.24). One tube is made of copper, and the other is made of plastic. From which tube will the magnet emerge first? Why?

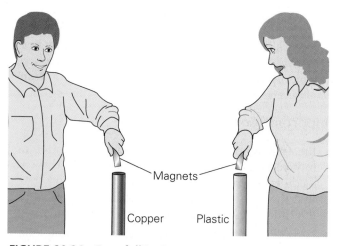

•FIGURE 20.24 **Free fall?** See Exercise 6.

7. ■ A basic telephone has both a speaker-transmitter and a receiver (•Fig. 20.25). Until the advent of digital phones in the 1990s, the transmitter had a diaphragm coupled to a carbon chamber (called the button), which contained loosely packed granules of carbon. As the diaphragm vibrated because of incident sound waves, the pressure on the granules varied, causing them to be more or less closely packed. As a result, the resistance of the button changed. The receiver converted electrical impulses to sound. Applying the principles of electricity and magnetism you have learned, explain the basic operation of this type of telephone.

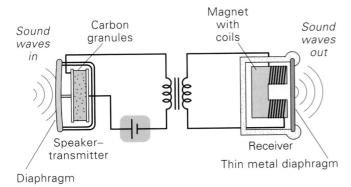

•FIGURE 20.25 Telephone operation See Exercise 7.

8. ■ A circular loop of wire that encloses an area of 0.015 m² is in a uniform magnetic field of 0.30 T. What is the flux through the loop if its plane is (a) parallel to the field, (b) at an angle of 37° to the field, and (c) perpendicular to the field?

9. ■ A circular loop with a radius of 20 cm is positioned in various orientations in a uniform magnetic field of 0.15 T. Find the magnetic flux if the normal to the plane of the loop is (a) perpendicular to the magnetic field, (b) parallel to the magnetic field, and (c) at an angle of 40° to the magnetic field.

10. ■ ◯ A conductive loop enclosing an area of 0.020 m² is perpendicular to a uniform magnetic field of 0.30 T. If the field goes to zero in 0.0045 s, what is the magnitude of the average emf induced in the loop?

11. ■ A loop in the form of a right triangle with one side of 40.0 cm and a hypotenuse of 50.0 cm lies in a plane perpendicular to a uniform magnetic field of 550 mT. What is the flux through the loop?

12. ■ A square coil of wire with 10 turns is in a magnetic field of 0.25 T. The total flux through the coil is 0.50 T·m². Find the effective area of one turn if the field (a) is perpendicular to the plane of the coil and (b) makes an angle of 60° with the plane of the coil.

13. ■■ What is the magnetic flux (due to its own magnetic field) through the cross section of an ideal solenoid whose windings have a radius of 3.0 cm if the turn density is 250 turns/m and a current of 1.5 A flows through the wire?

14. ■■ A magnetic field perpendicular to the plane of a wire loop with an area of 0.40 m² decreases by 0.20 T in 10⁻³ s. What is the magnitude of the average value of the emf induced in the loop?

15. ■■ ◯ A square loop of wire with sides of length 40 cm experiences a uniform, perpendicular magnetic field of 100 mT. If the field goes to zero in 0.010 s, what is the magnitude of the average emf induced in the loop?

16. ■■ The magnetic flux through one turn of a 60-turn coil of wire is reduced from 35 Wb to 5.0 Wb in 0.10 s. The average induced current in the coil is 3.6 × 10⁻³ A. Find the total resistance of the wire.

17. ■■ If the magnetic flux through a single loop of wire increases by 30 T·m² and an average current of 40 A was induced in the wire (of resistance 2.5 Ω), over what period of time was the flux increased?

18. ■■ Over 0.20 s, a coil of wire with 50 loops has an average induced emf of 9.0 V due to a changing magnetic field perpendicular to the plane of the coil. The radius of the coil is 10 cm and the initial value of the magnetic field is 1.5 T. Assuming the field decreased with time, what is the final value of the field?

19. ■■ A single strand of wire of adjustable length is wound around the circumference of a round balloon that has a diameter of 20 cm. A uniform magnetic field with a magnitude of 0.15 T is perpendicular to the plane of the loop. If the balloon is blown up such that its diameter d and the diameter of the wire loop increase to 40 cm in 0.040 s (•Fig. 20.26), what is the magnitude of the average value of the emf induced in the loop?

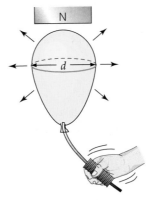

•FIGURE 20.26 Pumping energy
See Exercise 19.

20. ■■ The magnetic field perpendicular to the plane of a wire loop with an area of 0.10 m² changes with time as shown in •Fig. 20.27. What is the average emf induced in the loop for each segment of the graph (for example, from 0 to 2.0 ms)?

21. ■■ A boy carries a metal rod, with its length oriented in the east–west direction, at a uniform speed due north; he holds the rod parallel to the ground. (a) Will there be an induced emf when the rod is at the Equator? (b) Where would the largest emf be generated in the rod?

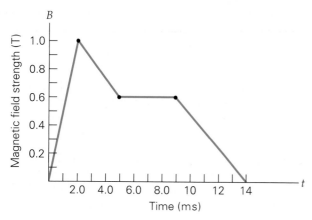

•FIGURE 20.27 **Magnetic field versus time** See Exercise 20.

22. ■■ A metal airplane with a wing span of 30 m flies horizontally at a constant speed of 320 km/h in a region where the vertical component of the Earth's magnetic field is 5.0×10^{-5} T. What is the induced motional emf across its wing tips?

23. ■■ Suppose that the metal rod in Fig. 20.7 is 20 cm long and is moving at a speed of 10 m/s in a magnetic field of 0.30 T, but the metal frame is covered with an insulating material. Find (a) the magnitude of the induced emf across the rod and (b) the current in the rod.

24. ■■ The flux through a fixed loop of wire changes uniformly from +40 Wb to −20 Wb in 1.5 ms. What are (a) the significance of the negative flux and (b) the average induced emf in the loop?

25. ■■■ A coil of wire with 10 turns and a cross-sectional area of 0.055 m² is placed in a magnetic field of 1.8 T and oriented such that the area is perpendicular to the field. The coil is then flipped by 90° in 0.25 s and ends up with the area parallel to the field (•Fig. 20.28). What is the magnitude of the average emf induced in the coil?

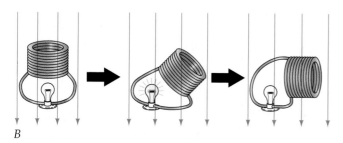

•FIGURE 20.28 **Flipping the coil** See Exercise 25.

26. ■■■ A uniform magnetic field of 0.50 T penetrates a double-incline block as shown in •Fig. 20.29. Determine the magnetic flux through each surface of the block.

27. ■■■ A length of 20-gauge copper wire (of diameter 0.8118 mm) is formed into a circular loop with a radius of 20 cm. A magnetic field perpendicular to the plane of the loop increases from zero to 14 mT in 0.25 s. Find the average electrical energy dissipated in the process.

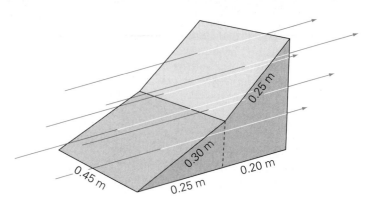

•FIGURE 20.29 **Magnetic flux** See Exercise 26. (Not drawn to scale.)

20.2 Generators and Back Emf

28. Increasing the rotating area in an ac generator (a) increases the frequency of rotation, (b) decreases the maximum induced emf, or (c) increases the maximum induced emf.

29. The back emf of a motor depends on (a) the input voltage, (b) the input current, (c) the armature's rotational speed, (d) none of these.

30. What is the orientation of the armature loop in a simple ac generator when the value of (a) the emf is a minimum, (b) the magnetic flux is a minimum? Explain why for each and also why they do not occur at the same orientation.

31. ■ A student has a bright idea for a generator: For the arrangement shown in •Fig. 20.30, the magnet is pulled down and released. With a highly elastic spring, the student thinks there should be a relatively continuous electrical output. What is wrong with this idea?

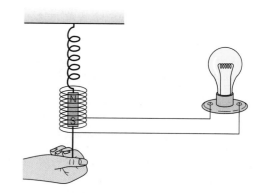

•FIGURE 20.30 **Inventive genius?** See Exercise 31.

32. ■ An emergency ac generator operates at a rotation frequency of 60 Hz. If the output voltage is a maximum (in magnitude) at $t = 0$, when is it next (a) a maximum (magnitude), (b) zero, and (c) its initial value?

33. ■ A student makes a simple generator by using a square loop 10 cm on each side. The loop is then rotated at a frequency of 60 Hz in a magnetic field of 0.015 T. (a) What

is the maximum value of the emf output? (b) What would be the maximum value if 10 such loops were used?

34. ■ A simple ac generator consists of a coil with 10 turns of wire, each loop having an area of 50 cm². The coil rotates in a uniform magnetic field of 350 mT with a frequency of 60 Hz. (a) Write an equation showing how the emf of the generator varies as a function of time. (b) Compute the maximum emf.

35. ■■ A 60-hertz sinusoidal ac voltage has a maximum value of 120 V. What is the voltage 1/180 s after that voltage has a value of zero? (Is there more than one such answer? If so, list all possible answers. Explain why there might be more than one.)

36. ■■ A simple ac generator with a maximum emf of 40 V is to be constructed from loops of wire with a radius of 0.15 m, a rotational frequency of 60 Hz, and a magnetic field of 200 mT. How many loops of wire will be needed?

37. ■■ A simple ac generator has a rotational frequency of 60 Hz and a maximum emf of 100 V. Assume zero emf at start-up. What is the instantaneous emf (a) 1/240 s after start-up and (b) 1/120 s after the emf passes through zero in changing to a negative polarity?

38. ■■ The armature of a simple ac generator has 20 circular loops of wire with a radius of 10 cm. It is rotated with a frequency of 60 Hz in a uniform magnetic field of 800 mT. What is the maximum emf induced in the loops, and how often is this value attained?

39. ■■ The armature of an ac generator has 100 turns, which are rectangular loops measuring 8.0 cm by 12 cm. The generator has a sinusoidal voltage output with an amplitude of 24 V. If the magnetic field of the generator is 250 mT, with what frequency does the armature turn?

40. ■■ Two students displayed their ac generators in a science fair. The generator made by student A has a loop area of 100 cm² rotating in a magnetic field of 20 mT at 60 Hz; the one made by student B has a loop area of 75 cm² rotating in a magnetic field of 200 mT at 120 Hz. Which one generates the greater maximum emf? Justify your answer mathematically.

41. ■■ A motor has a resistance of 2.50 Ω and draws a current of 4.00 A when operating at its normal speed on a 110-volt line. What is the back emf of the motor?

42. ■■ The starter motor in an automobile has a resistance of 0.40 Ω in its armature windings. The motor operates on 12 V and has a back emf of 10 V when running at normal operating speed. How much current does the motor draw (a) when running at its operating speed and (b) when starting up?

43. ■■ When running at its operating speed, a 240-volt dc motor with an armature whose resistance is 1.50 Ω draws a current of 16.0 A. (a) What is the back emf of the motor when it is operating normally? (b) What is the starting current? (Assume there is no additional resistance.) (c) What series resistance would be required to limit the starting current to 25 A?

20.3 Transformers and Power Transmission

44. A step-up transformer has (a) more windings in the primary coil, (b) more windings in the secondary coil, (c) the same number of windings in the primary and secondary coils.

45. The power transmitted by an ideal step-down transformer is (a) greater in the primary coil, (b) greater in the secondary coil, (c) the same in primary and secondary coils.

46. To reduce resistance losses, electric power is transmitted (a) at low voltage, (b) at high current, (c) by using step-up transformers at the generating plant, (d) none of these.

47. ■ The voltage for ignition by a spark plug in an automobile is supplied by what is referred to as the *coil*, which is actually a pair of induction coils (•Fig. 20.31). Explain how the voltage of the car's battery (12 V) is raised to as high as 25 kV by this device. Explain the function of the distributor.

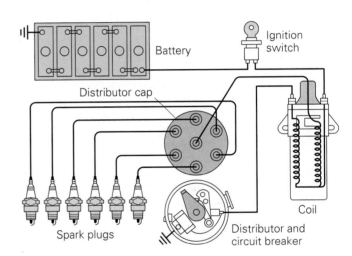

•FIGURE 20.31   Auto ignition with coil   See Exercise 47.

48. ■ The secondary coil of an ideal transformer has 450 turns, and the primary coil has 75. (a) Which type of transformer is this? (b) What is the ratio of the current in the primary coil to the current in the secondary? (c) What is the ratio of the voltage in the primary coil to the voltage in the secondary?

49. ■ An ideal transformer steps 8.0 V up to 2000 V, and the 4000-turn secondary coil carries 2.0 A. (a) Find the number of turns in the primary coil. (b) Find the current in the primary.

50. ■ The primary coil of an ideal transformer has 720 turns, and the secondary coil has 180 turns. If the primary carries 15 A at a voltage of 120 V, what are (a) the voltage and (b) the current output of the secondary?

51. ■ The transformer in the power supply for a computer 100MB ZIP drive changes a 120-volt input to a 5.0-volt output (•Fig. 20.32). Find the ratio of the number of turns in the primary coil to the number of turns in the secondary.

•**FIGURE 20.32   Step down before use**   See Exercise 51.

52. ■ The primary coil of an ideal transformer is connected to a 120-volt source and draws 10 A. The secondary coil has 800 turns and a current of 4.0 A. (a) What is the voltage across the secondary? (b) How many turns are in the primary?

53. ■ An ideal transformer has 840 turns in its primary coil and 120 turns in its secondary. If the primary draws 2.50 A at 110 V, what are (a) the current and (b) the voltage output of the secondary?

54. ■■ The efficiency ϵ of a transformer is defined as the ratio of the power output to the power input: $\epsilon = I_s V_s / I_p V_p$. Show that in terms of the ratios of currents and voltages given in Eq. 20.9 (for an ideal transformer), an efficiency of 100% is obtained. What does this imply?

55. ■■ The specifications of a transformer for a small appliance read: Input, 120 V, 6.0 W; Output, 9.0 V, 300 mA. (a) Is this an ideal transformer? (b) If not, what is its efficiency?

56. ■■ A circuit component operates at 20 V and 0.50 A. A transformer with 300 turns in its primary coil is used to convert 120-volt household electricity to the proper voltage. (a) How many turns must the secondary coil have? (b) How much current is in the primary?

57. ■■ A transformer in a door chime steps down the voltage from 120 V to 6.0 V and supplies a current of 0.50 A to the chime mechanism. (a) What is the turn ratio of the transformer? (b) What is the current input to the transformer?

58. ■■ An ac generator supplies 20 A at 440 V to a 10 000-volt power line. If the step-up transformer has 150 turns in its primary coil, how many turns are in the secondary?

59. ■■ The electricity supplied in Exercise 58 is transmitted over a line 80.0 km long with a resistance of 0.80 Ω/km. (a) How many kilowatt-hours are saved in 5.00 h by stepping up the voltage? (b) At $0.10/kWh, how much of a saving (to the nearest $10) is this to the consumer in a 30-day month, assuming the energy is supplied continuously?

60. ■■ At an area substation, the power-line voltage is stepped down from 100 000 V to 20 000 V. If 10 MW of power is delivered to the 20 000-volt circuit, what are the primary and secondary currents in the transformer?

61. ■■■ A voltage of 200 000 V in a transmission line is reduced to 100 000 V at an area substation, to 7200 V at a distributing substation, and finally to 240 V at a utility pole outside a house. (a) What turn ratio N_s/N_p is required for each reduction step? (b) By what factor is the transmission-line current stepped up in each voltage step-down? (c) What is the overall factor by which the current is stepped up from transmission line to utility pole?

62. ■■■ An electric-generating plant produces electric energy at 50 A and 20 kV. The electricity is transmitted 25 km over transmission lines whose resistance is 1.2 Ω/km. (a) What would the power loss be if the energy is transmitted at 20 kV? (b) To what value should the output voltage of the generator be stepped up to decrease the energy loss by a factor of 15?

63. ■■■ Electrical power is transmitted through a power line 175 km long of resistance 1.2 Ω/km. The generator output is 50 A at its operating voltage of 440 V. This voltage has a single step up for transmission to 44 kV. (a) How much power is lost as joule heat? (b) What turn ratio of a transformer at the power delivery point would provide an output voltage of 220 V? (Neglect the voltage drop along the line.)

20.4 Electromagnetic Waves

64. An antenna is connected to a battery. Will the antenna emit an electromagnetic wave?

65. Relative to the blue end of the visible spectrum, the yellow and green regions have (a) higher frequencies, (b) longer wavelengths, (c) shorter wavelengths, (d) both (a) and (c).

66. Which of these electromagnetic waves has the lowest frequency: (a) ultraviolet, (b) infrared, (c) X-ray, (d) microwave?

67. On a cloudy summer day, you work outside and feel cool. Yet that evening you find that you have a sunburn. Why? Explain in terms of infrared and ultraviolet radiations.

68. ■ Find the frequencies of electromagnetic waves with wavelengths of (a) 2.0 m, (b) 25 m, (c) 75 m.

69. ■ The frequency ranges of the radio AM band, radio FM band, and the TV band are 530 kHz to 1710 kHz, 88 MHz to 108 MHz, and 54 MHz to 890 MHz, respectively. What are the corresponding wavelength ranges for these bands?

70. ■ In determining the distance to a cloud, a meteorologist in a TV station uses radar and notes that a time of 0.24 ms elapses between the sending and return of a radar pulse. How far away is the cloud?

71. ■ How long does it take for a laser beam to travel from the Earth to a reflector on the Moon and back? Take the distance of the Moon to be 2.4×10^5 mi. (This experiment was done when the *Apollo* flights of the early 1970s left laser reflectors on the lunar surface to determine the Earth–Moon distance accurately.)

72. ■■ Orange light has a wavelength of 600 nm, and that of green light is 510 nm. What is their frequency difference?

73. ■■ A wire radio antenna is called a quarter-wavelength antenna because its length is equal to one-quarter of the wavelength. If you were going to make such antennae for the AM and FM radio bands by using the mid frequency of each band, what lengths of wire would you use?

74. ■■ Microwave ovens can have cold spots and hot spots due to standing electromagnetic waves analogous to standing wave nodes and antinodes in strings (•Fig. 20.33). If your microwave has cold spots (nodes) approximately every 5.0 cm, what frequency waves does it use to cook?

•**FIGURE 20.33 Cold spots?** See Exercise 74.

Additional Exercises

75. A 120-volt ac motor has a resistance of 4.0 Ω. When operating at normal speed, the motor develops a back emf of 110 V. What are (a) the start-up current of the motor and (b) the current when it is operating at normal speed?

76. An electric door bell operates at 4.5 V. If this voltage is obtained from standard 120-volt household voltage, which of the windings in the bell's transformer has more turns, and how many times more than the other?

77. A circular wire loop of 10 turns is in a uniform magnetic field of 150 mT. The flux through the loop is 0.12 T·m². What is the radius of the loop if the normal to the plane of the loop makes an angle of 45° with the field?

78. A power line transmits electric energy at a voltage of 1.50 kV to a residential area. There a transformer with 500 turns in its primary coil steps down the voltage to 240 V. How many turns are in its secondary coil?

79. A pivoted coil of wire is rotated 50 times per second in a uniform magnetic field. How often does the induced emf in the coil have a value of zero?

80. A 120-volt dc motor draws a current of 6.0 A and has a back emf of 96 V at its operating speed. (a) What starting current is required by the motor? (Assume that there is no additional resistance.) (b) What series resistance would be required to limit the starting current to 15 A?

81. In •Figure 20.34, a metal bar of length L moves in a region with a constant magnetic field. That field is directed into the page and has a magnitude of 250 mT. What is (a) the direction of the current through the resistor? (b) the current?

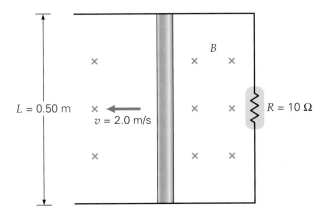

•**FIGURE 20.34 Motional emf** See Exercise 81.

82. An ideal transformer has 120 turns in its primary coil and 840 turns in its secondary. If 14 A is in the primary with a voltage of 120 V, what are (a) the voltage and (b) the current output of the secondary?

83. A loop of wire in the form of a square 1.0 m on each side is perpendicular to a uniform magnetic field of 0.75 T. What is the flux through the loop?

84. An ideal transformer has 80 turns in its primary coil and 360 turns in its secondary. If 12 V applied to the primary coil gives a 3.0-amp current, what are the (a) current, (b) voltage output, and (c) power output of the secondary?

85. A square conductive loop 50 cm on each side has a resistance of 0.200 Ω. Find the rate at which a magnetic field perpendicular to the plane of the loop must change with time to cause an average of 5.0 J/s to be expended in the loop.

86. A coil with 150 concentric loops of wire and a radius of 12 cm is used as the armature of a generator with a magnetic field of 0.80 T. With what frequency should the armature be rotated to make the polarity change every 0.010 s?

87. The transformer on a utility pole steps the voltage down from 20 000 V to 220 V for a college science building. If the building uses 6.6 kW of power, what are the primary and secondary currents in the transformer?

88. Gamma rays have typical wavelengths on the order of 3×10^{-15} m (nuclear diameter size). What is the frequency of typical gamma ray radiation?

89. In some countries, the voltage is delivered at 50 Hz instead of the 60 Hz as in the United States. (a) What is the rotation rate of the turbines that drive the electric generators for such voltage delivery? (b) Would different transformers be needed if the step-up and step-down requirements of the delivery systems were to be the same? Explain carefully, including ideal and nonideal transformers.

AC Circuits

Insight

- Oscillator Circuits: Broadcasters of Electromagnetic Radiation

Direct current circuits have many uses. When you start your car in the morning or search for a lost object with a pocket flashlight, you are making use of direct current. But the neon lights in the photo operate on alternating current (ac). The electric power delivered to our homes and offices is also ac, and most of the devices we use every day require alternating current.

There are several reasons for this. For one thing, almost all electric power is produced by generators using electromagnetic induction, and such generators (as you saw in Chapter 20) produce ac. Moreover, the voltage and current in ac circuits can be changed by transformers in a way that allows economical transmission of electric power over very long distances. But perhaps the most important reason that ac is used so universally is that ac circuits are extremely versatile. The alternation of the current produces electromagnetic effects that can be exploited in a wide variety of devices. Every time you tune a radio to a favorite station, for example, you are taking advantage of a special property of ac circuits.

In common dc circuits, we are concerned only with ohmic resistances. There is ohmic resistance in ac circuits, too, but there are other factors that affect the flow of charge. Recall that once a capacitor in a dc circuit has been fully charged, it offers infinite resistance (an open circuit). However, in an ac circuit, this is not the case. Alternating voltage continually charges and discharges a capacitor, so there is both current in the circuit and opposition to that current. Coils of wire also oppose an ac current through induction. An example, the back emf of a motor, was described in Chapter 20.

In this chapter, we will look at some of the basic principles of ac circuits. Forms of Ohm's law and expressions for power will be developed specifically for ac circuits. Finally, we will explore the phenomenon of resonance—the condition for maximum energy transfer in an ac circuit—and some of its important practical applications.

21.1 Resistance in an AC Circuit

Objectives: To (a) specify how voltage, current, and power vary with time in an ac circuit, (b) understand the concepts of rms and peak values, and (c) learn how resistors respond under ac conditions.

An ac circuit contains an ac voltage source and one or more other circuit elements. The circuit diagram for an ac circuit with a single resistive element is shown in •Fig. 21.1. If the source output varies sinusoidally, as is the case for a simple generator (Section 20.2), the voltage across the resistor varies with time in accordance with the equation

$$V = V_o \sin \omega t = V_o \sin 2\pi ft \qquad (21.1)$$

where ω is the angular frequency ($\omega = 2\pi f$). The voltage oscillates between values of $+V_o$ and $-V_o$, where V_o is the **peak voltage**.

AC Current and Power

Under ac conditions, the current through the resistor also oscillates—that is, its direction and magnitude change. From the relationship between voltage, resistance, and current in a resistor, the current in the resistor can be expressed as a function of time:

$$I = \frac{V}{R} = \frac{V_o}{R} \sin 2\pi ft$$

The right-hand side can be conveniently rewritten as

$$I = I_o \sin 2\pi ft \qquad (21.2)$$

where the amplitude of the current is $I_o = V_o/R$ and is called the **peak current**.

Both current and voltage for an ac circuit are plotted versus time in the graph in •Fig. 21.2. Note that they are in step, or *in phase*, with each other, both reaching zero values and maxima at the same times. The current oscillates and has corresponding positive and negative values during each cycle. Thus, the sum of all the instantaneous current values (positive and negative) over one complete cycle is zero, so the *average current is zero*. Mathematically, this reflects the fact that the time-averaged value of the sine function over one or more *complete* (360°) cycles is zero. Using overbars to denote a time-averaged value, we can express this fact as $\overline{\sin \theta} = \overline{\sin 2\pi ft} = 0$. Similarly, $\overline{\cos \theta} = 0$.

The fact that the *average* current is zero, however, does not mean that there is no joule heating (I^2R losses). The dissipation of electrical energy depends on the collisions of electrons with the atoms that make up the wire, regardless of the direction of the current in that wire.

The instantaneous power is obtained from the instantaneous current (Eq. 21.2). The power is a function of time because the current varies with time:

$$P = I^2R = I_o^2R \sin^2 2\pi ft \qquad (21.3)$$

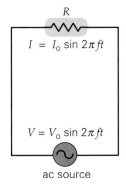

•**FIGURE 21.1   A purely resistive circuit**   The ac source delivers a sinusoidal voltage to the circuit, and the voltage across and current through the resistor are sinusoidal.

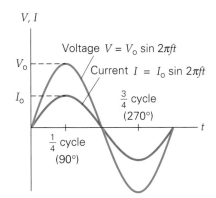

•**FIGURE 21.2   Voltage and current in phase**   In a purely resistive ac circuit, the voltage and current are in step, or in phase, with zero values and maxima occurring at the same times.

Even though the current changes sign each half-cycle, the square of the current, I^2, is always positive. Thus the average value of I^2R is *not* zero, even though the average value of I is zero. The average, or mean, value of I^2 is

$$\overline{I^2} = \overline{I_o^2 \sin^2 2\pi ft} = I_o^2 \, \overline{\sin^2 2\pi ft}$$

Using the trigonometric identity $\sin^2 \theta = \frac{1}{2}(1 - \cos 2\theta)$, we can write $\overline{\sin^2 \theta} = \frac{1}{2}(1 - \overline{\cos 2\theta})$, and since $\overline{\cos 2\theta} = 0$ (just as $\overline{\cos \theta} = 0$), we have $\overline{\sin^2 \theta} = \frac{1}{2}$. Thus,

$$\overline{I^2} = I_o^2 \, \overline{\sin^2 2\pi ft} = \tfrac{1}{2}I_o^2 \tag{21.4}$$

The average power is therefore

$$\overline{P} = \overline{I^2}R = \tfrac{1}{2}I_o^2 R \tag{21.5}$$

It is customary to write the ac power in the same form as the dc power ($P = I^2R$). To do this, a special value of current is used:

$$I_{rms} = \sqrt{\overline{I^2}} = \sqrt{\tfrac{1}{2}I_o^2} = \frac{I_o}{\sqrt{2}} = \frac{\sqrt{2}}{2}I_o = 0.707 I_o \tag{21.6}$$

where I_{rms} is called the **rms current**, or **effective current**. (Here *rms* stands for root-mean-square, indicating the square *root* of the *mean* value of the *square* of the current.) Then with $I_{rms}^2 = (I_o/\sqrt{2})^2 = \tfrac{1}{2}I_o^2$, the effective or average power is

$$\overline{P} = \tfrac{1}{2}I_o^2 R = I_{rms}^2 R \tag{21.7}$$

This power is the average power that is equivalent to the oscillating power averaged over time (•Fig. 21.3).

AC Voltage

The peak values of voltage and current for a resistor are related by $V_o = I_o R$. Hence we can use a development similar to that for rms current to show that the **rms voltage**, or **effective voltage**, is given by

$$V_{rms} = \frac{V_o}{\sqrt{2}} = \frac{\sqrt{2}}{2}V_o = 0.707 V_o \tag{21.8}$$

In previous chapters we have, in fact, calculated currents and voltages for ac cases. What we have been finding all along are the rms values of these quantities. For resistors under ac conditions, then, we can use dc ideas as long as we realize we are solving for rms values. For alternating current, the relationship between rms values of current and rms values of voltage across a resistor is

$$V_{rms} = I_{rms}R \qquad \begin{array}{l}\textit{voltage across}\\\textit{a resistor}\end{array} \tag{21.9}$$

Combining Eqs. 21.9 and 21.7, we have equivalent expressions for power under ac conditions:

$$\begin{aligned}\overline{P} &= I_{rms}^2 R\\ &= I_{rms} V_{rms}\\ &= \frac{V_{rms}^2}{R}\end{aligned} \tag{21.10}$$

It is customary to measure and specify rms values for ac quantities. For example, the household line voltage of 120 V, as you may have guessed by now, is an rms value with a peak voltage of

$$V_o = \sqrt{2}\, V_{rms} = 1.414(120 \text{ V}) = 170 \text{ V}$$

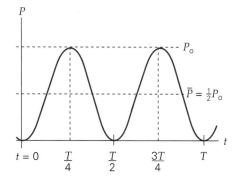

•**FIGURE 21.3 Power variation with time in a resistor** Although both current and voltage oscillate in direction (sign), their product (power) is a positive oscillating quantity. Because of the way in which the rms values of current and voltage are defined, the average power is one-half the peak power.

The rms and peak values of ac current and voltage are shown in the graphs in •Fig. 21.4. Another voltage designation sometimes used is the *peak-to-peak* value, V_{p-p}, which is the total voltage range from the most negative voltage polarity in the cycle to the most positive (see Fig. 21.4b):

$$V_{p-p} = 2V_o$$

For the purpose of significant figures in calculations, we will assume the frequency of our voltages (for example, 60 Hz) to be exact.

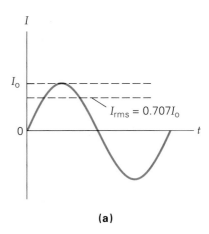

(a)

Example 21.1 ■ A Light Bulb: RMS Values versus Peak Values

A lamp with a 60-watt bulb is plugged into a 120-volt outlet. (a) What are the rms and peak currents through the lamp? (b) What is the resistance of the bulb? (Neglect the resistance of the wiring.)

Thinking It Through. (a) Since we know the average power and rms voltage, we can find the rms current from Eq. 21.10. From the rms current, we can use Eq. 21.6 to calculate the peak current. (b) The resistance is then found from Eq. 21.9.

Solution. We list the average power and the rms voltage of the source.

Given: $\overline{P} = 60$ W   *Find:* (a) I_{rms} and I_o (rms and peak currents)
        $V_{rms} = 120$ V   (b) R (resistance)

(a) The rms current is

$$I_{rms} = \frac{\overline{P}}{V_{rms}} = \frac{60 \text{ W}}{120 \text{ V}} = 0.50 \text{ A}$$

and the peak current is gotten by rearranging Eq. 21.6:

$$I_o = \sqrt{2}\, I_{rms} = \sqrt{2}\,(0.50 \text{ A}) = 0.71 \text{ A}$$

(b) The resistance of the bulb is

$$R = \frac{V_{rms}}{I_{rms}} = \frac{120 \text{ V}}{0.50 \text{ A}} = 240 \text{ } \Omega$$

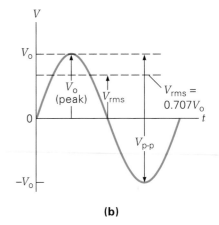

(b)

•**FIGURE 21.4 Root-mean-square (rms) current and voltage** The rms values of **(a)** current and **(b)** voltage are 0.707, or $1/\sqrt{2}$, times the peak (maximum) values. The voltage is sometimes described by the peak-to-peak (V_{p-p}) value.

Follow-up Exercise. What would be the (a) rms current and (b) peak current in a 60-watt light bulb in Great Britain, where the house rms voltage is 240 V at 50 Hz? (c) What would the resistance of a 60-watt light bulb be in Great Britain compared to one designed for operation in the United States? Why are they different?

Conceptual Example 21.2

Across the Pond: British versus American Electrical Systems

In European countries, the normal line voltage is 240 V. If a British tourist in the United States plugged in a hair dryer brought from home, you might expect the dryer (a) not to operate at all, (b) to operate normally, (c) to operate poorly, (d) to burn out.

Reasoning and Answer. British small appliances are designed to operate on 240 V. With half that voltage, an appliance would operate, but not as normally designed. With decreased voltage, there would be decreased current ($I = V/R$) and greatly reduced joule heating ($P = V^2/R$). Assuming the same appliance resistance R, with half the voltage, there would be only one-fourth the power output. Thus, the heating element of the hair dryer might get warm, but it would certainly not work to the user's expectations, so the answer is (c). Also, the decreased current might cause the blower motor to run slower than normal.

Fortunately, most people do not make this mistake, because the plugs and sockets are different (as are 240-volt plug and sockets different from the 120-volt plug and sockets in the United States). Several different types of plugs are used in various countries. When on foreign travel with appliances from home, you should carry a converter/adapter kit (•Fig. 21.5). It contains a selection of different plugs into which a voltage converter

•**FIGURE 21.5 Converter and adapters** In countries that have 240-volt line voltages, a conversion to 120 V is needed to operate normal U.S. appliances properly. Note the different types of plugs for different countries. The small plugs go into the foreign sockets, and the converter prongs fit into the back of a socket. A U.S. standard two-prong plug fits into the converter, which has a 120-volt output. See Conceptual Example 21.2.

Note: For the remainder of this chapter, the rms subscript is deleted for convenience.

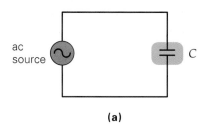

(a)

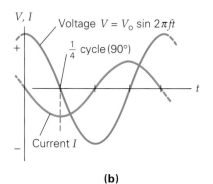

(b)

•**FIGURE 21.6 A purely capacitive circuit** (a) In a circuit with only capacitance, (b) the current leads the voltage by 90°, or one-quarter cycle. Half of a full cycle of voltage and current, shown, corresponds to Fig. 21.7.

fits. The converter is a solid-state device that converts 240 V to 120 V for U.S. travelers and vice versa for tourists visiting the United States. (Note that some travel appliances are designed for dual voltage and will operate on either 120 V or 240 V.)

Follow-up Exercise. Foreign electrical systems generally operate at a frequency of 50 Hz—that is, 240 V, 50 Hz as compared to 120 V, 60 Hz as in the United States. Assuming the correct use of a voltage converter/adapter, what effect do you think this *frequency* difference might have on an electric razor brought by the British tourist in this Exercise? On his plug-in electric clock?

For ac circuits including only resistors, the formulas derived for dc circuits can be used if the rms, or effective, values of the quantities are inserted. *For convenience and simplicity, the rms subscript will be omitted from now on, as is normally done—but keep in mind that rms values are given for ac quantities unless otherwise specified.* Thus it is important to realize that we are working with only rms or peak quantities, not their values at specific times (instantaneous values).

21.2 Capacitive Reactance

Objectives: To (a) explain the behavior of capacitors in ac circuits and (b) calculate their effect on ac current (capacitive reactance).

As we saw in Chapter 16, when a capacitor is connected to a voltage source in a dc circuit, current will exist only for the short time required to charge the capacitor. As charge accumulates on the capacitor's plates, the voltage across them increases, opposing the current. That is, a capacitor in a dc circuit will limit or oppose the current as the capacitor charges. When the capacitor is fully charged, the current in the circuit falls to zero.

When a capacitor is in a circuit with an ac source, as shown in •Fig. 21.6a, the capacitor limits, or regulates, the current but does not completely prevent the flow of charge. This is because the capacitor is alternately charged and discharged as the current and voltage reverse each half-cycle.

Plots of ac current and voltage versus time for a circuit with a capacitor are shown in Fig. 21.6b. Let's look at the changing conditions of the capacitor to explain the variation of those parameters (•Fig. 21.7).

- We arbitrarily choose our starting point in the cycle (called $t = 0$) as the point of maximum voltage* (Fig. 21.7a). Thus the capacitor is fully charged ($Q_o = CV_o$) in the polarity shown. Since the plates cannot accommodate more charge, there is no current in the circuit.

- As the voltage decreases, the capacitor begins to discharge, giving rise to a counterclockwise current* (Fig. 21.7b).

- The current reaches its maximum value when the voltage drops to zero and the capacitor plates are completely discharged (Fig. 21.7c). This occurs at exactly one-quarter of the way through the cycle ($t = T/4$).

- The voltage source now reverses polarity and starts to increase in magnitude. The capacitor begins to charge, this time with the opposite polarity (Fig. 21.7d). With the plates initially uncharged, there is no opposition to the current (other than wire resistance), so the current is initially at its maximum value. However, as the plates accumulate charge, they also inhibit the current; thus the current decreases in magnitude.

- Halfway through the cycle ($t = T/2$), the capacitor is fully charged but opposite in polarity to its starting situation (Fig. 21.7e). The current has dropped to zero, and the voltage again is at its maximum magnitude but opposite the polarity of the initial voltage.

*We have arbitrarily chosen the starting voltage polarity as positive (+ on the graph in Fig. 21.6) and the counterclockwise current flow to be negative (− on the graph).

During the next half-cycle (not shown in Fig. 21.7), the process is reversed. As the voltage starts to decline, the current returns, but this time in the clockwise (positive) direction. It attains its maximum value three-quarters of the way through the cycle, when the voltage has once again momentarily dropped to zero. As the voltage rises again to its exact starting value and polarity, the current begins to fall as the capacitor charges. Eventually the current drops to zero, and the voltage source and the capacitor are at their maximum voltage and charge, respectively, with the same polarity as the starting conditions. The circuit is back to its $t = 0$ condition.

Note that the current and voltage are *not* in step, or in phase, in this situation. The current reaches its maximum a quarter-cycle *ahead* of the voltage. The relationship between the current and the voltage is commonly stated in this way:

In a purely capacitive ac circuit, the *current* leads the *voltage* by 90°, or by one-quarter ( ) cycle.

As in a dc circuit, a capacitor provides opposition to the charging process in an ac circuit, but it is not totally limiting (as it is in a dc open-circuit condition). The quantitative measure of the opposition of a capacitor to current is referred to as its **capacitive reactance** (X_C). In an ac circuit, capacitive reactance is given by

$$X_C = \frac{1}{2\pi f C} = \frac{1}{\omega C} \qquad \begin{array}{l}\textit{capacitive}\\\textit{reactance}\end{array} \qquad (21.11)$$

SI unit of capacitive reactance: ohm (Ω), or second per farad (s/F)

where $\omega = 2\pi f$, C is the capacitance (in farads), and f is the frequency (in Hz). Like resistance, reactance is measured in ohms (Ω). Can you show that the ohm is equivalent to the second per farad?

Equation 21.11 shows that the reactance is inversely proportional to both the capacitance (C) and the voltage frequency (f). Recall from Chapter 16 that capacitance is the charge per voltage ($C = Q/V$). For a particular voltage, therefore, the greater the capacitance, the more charge the capacitor can accommodate for a chosen frequency. Putting more charge on the plates requires a larger flow of charge (a greater current), which means the opposition to the current, or the reactance, must be smaller.

The greater the frequency of the ac driving source, the shorter the charging time in each cycle. A shorter charging time means less charge accumulation on the plates and therefore less opposition to the current. Thus, the capacitive reactance is inversely proportional to the frequency as well as to the capacitance. (Note in Eq. 21.11 that if $f = 0$, as is the case for dc, the capacitive reactance is infinite and there is no current flow, which corresponds to an open circuit.)

The capacitive reactance is related to both the voltage across the capacitor and the current by an equation that has the same form as $V = IR$ for pure resistances:

$$V = IX_C \qquad \begin{array}{l}\textit{voltage across}\\\textit{a capacitor}\end{array} \qquad (21.12)$$

(Remember that V and I are rms values.) Consider the following Example about a capacitor connected to an ac voltage source.

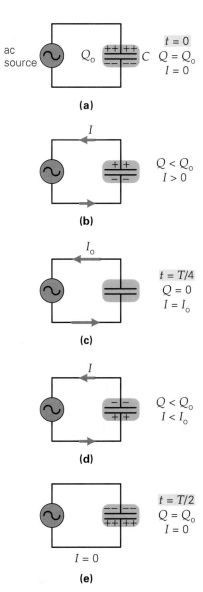

•**FIGURE 21.7   A capacitor under ac conditions**   This sequence shows the voltage, charge, and current in a circuit containing only a capacitor and an ac voltage source. All five circuit diagrams taken together represent physically what is plotted in the first half of the cycle (from $t = 0$ to $t = T/2$) in the graph shown in Fig. 21.6b.

Example 21.3 ■ Current Under AC Conditions: Capacitive Reactance

A 15.0-microfarad capacitor is connected to a 120-volt, 60-hertz source. What are (a) the capacitive reactance and (b) the current (rms and peak) in the circuit?

Thinking It Through.   We can determine the capacitive reactance from the given values of the capacitance and the driving frequency by using Eq. 21.11. (b) We can then calculate the (rms) current from the reactance and (rms) voltage via Eq. 21.12. The relationship in Eq. 21.6 then gives the peak current.

Solution. As previously stated, we assume the 60 Hz frequency to be exact, thus our answers are to three significant figures.

Given: $C = 15.0 \ \mu\text{F} = 15.0 \times 10^{-6} \text{ F}$ *Find:* (a) X_C (capacitive reactance)
$V = 120 \text{ V}$ (b) I (current—both rms
$f = 60 \text{ Hz (exactly)}$ and peak)

(a) The capacitive reactance is

$$X_C = \frac{1}{2\pi f C} = \frac{1}{2\pi(60 \text{ Hz})(15.0 \times 10^{-6}\text{F})} = 177 \ \Omega$$

(b) Then, the (rms) current is

$$I = \frac{V}{X_C} = \frac{120 \text{ V}}{177 \ \Omega} = 0.678 \text{ A}$$

The peak current is thus

$$I_o = \sqrt{2}I = \sqrt{2}(0.678 \text{ A}) = 0.959 \text{ A}$$

Thus the current oscillates between $+0.959$ A and -0.959 A and is ahead of the voltage by 90°.

Follow-up Exercise. In this Example, (a) what is the peak voltage, and (b) what frequency would give the same current if the capacitance were twice as large?

21.3 Inductive Reactance

Objectives: **To (a) explain what an inductor is, (b) explain the behavior of inductors in ac circuits, and (c) calculate the effect of inductors on ac current (inductive reactance).**

Inductance is a measure of the opposition a circuit element presents to a time-varying current (by Lenz's law). In principle, all circuit elements (even resistors) have some inductance. A coil of wire with negligible resistance has only inductance. When placed in a circuit with a time-varying current, such a coil, called an **inductor**, will exhibit a reverse voltage, or back emf, in opposition to the changing current. The changing current through the coil produces a changing magnetic field and flux. The back emf is the induced emf (by Faraday's law) in opposition to this changing flux. Since this emf is induced in the inductor as a result of its own changing magnetic field, we call this phenomenon *self-induction*. (Nearby coils, such as the primary and secondary sides of a transformer, can induce emfs in one another—a phenomenon known as *mutual inductance*—which we will not cover.)

The self-induced emf is given by Faraday's law (Eq. 20.2): $\mathscr{E} = -N \ \Delta\Phi/\Delta t$. Here, since the geometry of the coil is fixed, N (the number of loops) is constant. The time rate of change of the flux, $\Delta\Phi/\Delta t$, is proportional to $\Delta I/\Delta t$, the time rate of change of the current in the coil (since it is the coil current that produces the magnetic field responsible for the changing flux in the coil). Thus, for the case of the fixed-geometry inductor, the back emf is proportional to the time rate of change of the current and is oppositely directed.

We can express this relationship in equation form with a constant of proportionality:

$$\mathscr{E} = -L\left(\frac{\Delta I}{\Delta t}\right) \tag{21.13}$$

where L is the inductance of the coil (more properly, its *self*-inductance). As the equation suggests, inductance has units of volt-seconds per ampere (V·s/A). This combination is called a **henry** (H) after Joseph Henry (1797–1878), an American physicist and early investigator of electromagnetic induction. The smaller unit the millihenry (mH) is commonly used (1 mH = 10^{-3} H).

The opposition that an inductor presents to the current in an ac circuit depends on the value of the inductance and on the frequency of the voltage. This relationship is expressed quantitatively by the circuit's **inductive reactance** (X_L),

$$X_L = 2\pi fL = \omega L \qquad (21.14)$$

Inductive reactance: opposition to current

SI unit of inductive reactance: ohm (Ω), or henry per second (H/s)

where f is the frequency of the driving source, $\omega = 2\pi f$, and L is the inductance. Like capacitive reactance, inductive reactance is measured in ohms (Ω), which can be shown to be equivalent to the henry per second.

Note that the inductive reactance is directly proportional to both the inductance (L) of the coil and the frequency (f) of the voltage source. The inductance of a coil is a constant that depends on the number of turns, the coil's diameter, its length, and the material of the core (if any). The greater the inductance, the greater the opposition to current. The frequency of the ac driving source plays a role because the more rapidly the current in the coil changes, the greater the rate of change in its magnetic flux and thus the greater the self-induced emf that opposes the current.

In terms of X_L, the equation for the voltage across an inductor has a mathematical form similar to that for resistors:

$$V = IX_L \qquad \begin{array}{l}\textit{voltage across} \\ \textit{an inductor}\end{array} \qquad (21.15)$$

The circuit symbol for an inductor and the graphs of voltage across the inductor and the current in the circuit are shown in •Fig. 21.8. When an inductor is connected to an ac voltage source, maximum voltage corresponds to zero current. When the voltage drops to zero, the current is a maximum. This happens because as the voltage changes polarity [causing the magnetic (self) flux through the inductor to drop to zero], the inductor acts to prevent this change in accordance with Lenz's law, so the induced emf creates a current. Thus the current *lags* a one-quarter cycle behind the voltage, a relationship commonly expressed as follows:

In a purely inductive ac circuit, the *voltage* leads the *current* by 90°, or one-quarter ( ) cycle.

The phase relationships of current and voltage for purely inductive and purely capacitive circuits are opposite. A phrase that may help you remember this is

ELI the ICE man

With E representing voltage (for *emf*) and I representing current, *ELI* indicates that with an inductance (L) the voltage leads the current (I). Similarly, *ICE* tells you that with a capacitance (C) the current leads the voltage.

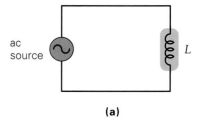

(a)

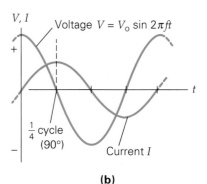

(b)

•**FIGURE 21.8 A purely inductive circuit (a)** In a circuit with only inductance, **(b)** the voltage leads the current by 90°, or one-quarter cycle.

Example 21.4 ■ Current Opposition Without Resistance: Inductive Reactance

A 125-millihenry inductor is connected to a 120-volt, 60-hertz source. What are (a) the inductive reactance and (b) the rms current in the circuit?

Thinking It Through. Since we know the inductance and the frequency, we can get the inductive reactance from Eq. 21.14 and the current from Eq. 21.15.

Solution. We list the given data and again assume the frequency is exact.

Given: $L = 125$ mH $= 0.125$ H   *Find:* (a) X_L (inductive reactance)
        $V = 120$ V   (b) I (rms current)
        $f = 60$ Hz (exact)

(a) The inductive reactance is

$$X_L = 2\pi fL = 2\pi(60 \text{ Hz})(0.125 \text{ H}) = 47.1 \ \Omega$$

(b) Then, the rms current is

$$I = \frac{V}{X_L} = \frac{120 \text{ V}}{47.1 \ \Omega} = 2.55 \text{ A}$$

Follow-up Exercise. In this Example, (a) what is the peak current, and (b) what voltage-source frequency would give the same current for an inductor with triple the inductance?

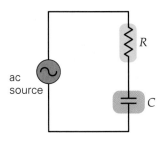

(a) Circuit diagram

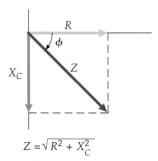

$$Z = \sqrt{R^2 + X_C^2}$$

(b) Phase diagram

•**FIGURE 21.9 A series RC circuit**
(a) In a series RC circuit, **(b)** the impedance Z is the phasor sum of the resistance R and the capacitive reactance X_C.

Impedance: effective opposition to current

Note: The word *impedance* refers to the overall circuit opposition to current (Z). We reserve the words *reactance* and *resistance* for individual element oppositions.

21.4 Impedance: RLC Circuits

Objectives: To **(a)** calculate currents and voltages when various reactive circuit elements are present in ac circuits, **(b)** use phase diagrams to calculate overall impedance and rms currents, and **(c)** understand and use the concept of the power factor in ac circuits.

The previous sections considered purely capacitive or purely inductive circuits, in which only *nonresistive* opposition to current was found. However, it is impossible to have purely reactive circuits, since there is always some resistance—at minimum, that from the connecting wires or the wire in an inductor. Therefore, real ac circuits usually have some resistance, and the current is impeded by *both* resistances and reactances (capacitive and/or inductive). Analysis of some simple circuits will illustrate these effects.

Series RC Circuit

Suppose that an ac circuit consists of a voltage source, a resistor, and a capacitive element connected in series, as illustrated in •Fig. 21.9a. The phase difference between the current and the voltage is different for each of these circuit elements. As a result, a special graphical method is used to find the effective opposition to the current in the circuit. This method employs a *phase diagram*.

In a **phase diagram**, such as the one shown in Fig. 21.9b, the resistance and reactance of the circuit are given vectorlike properties and their magnitudes are represented as arrows called **phasors**. On a set of x–y coordinate axes, the resistance is plotted on the positive x axis, since the voltage–current phase difference for a resistor is zero. The capacitive reactance is plotted along the negative y axis, to reflect a phase difference ϕ of $-90°$. (A negative phase angle implies that the voltage lags behind the current, as is the case for a capacitor.)

The phasor sum is the effective, or net, opposition to the current, which we call the **impedance** (Z). Phasors are added in the same way vectors are (as they must be because the effects are not in phase). For the series RC circuit,

$$Z = \sqrt{R^2 + X_C^2} \quad \begin{array}{l} \textit{series RC circuit} \\ \textit{impedance} \end{array} \quad (21.16)$$

The unit of impedance is the ohm.

The generalization of Ohm's law to circuits containing capacitors and inductors along with resistors is

$$V = IZ \quad \begin{array}{l} \textit{generalization} \\ \textit{of Ohm's law} \\ \textit{to ac circuits} \end{array} \quad (21.17)$$

(Once again, remember that current and voltage symbols here mean rms values.)

To see how phasors are used with an RC circuit, consider the following Example. Take particular note of part (b), where there is an *apparent* violation of Kirchhoff's loop theorem—explained by phase differences between the voltages across the two elements of the RC circuit.

Example 21.5 ■ Is Kirchhoff's Loop Theorem Valid for ac Circuits? RC Impedance

A series RC circuit has a resistance of 100 Ω and a capacitance of 15.0 μF. (a) What is the (rms) current in the circuit when driven by a 120-volt, 60-hertz source? (b) Compute the (rms) voltage across each circuit element and the two elements combined. Compare it to that of the voltage source. Is Kirchhoff's loop theorem satisfied?

Thinking It Through. (a) Note that we have added a resistor to the circuit in Example 21.3. By using phasors, we will combine the capacitive reactance and the resistance to get the impedance (Eq. 21.16). From Eq. 21.17, the impedance and voltage will then give the current. (b) The current is the same at all points at any given time in a series circuit, so we can use the result of part (a) to calculate the desired voltages. The total rms voltage across both elements is found, not by simply adding them, but by recalling that the individual voltages are out of phase by 90°.

Solution. We are working with three significant figures again because the frequency of 60 Hz is assumed to be exactly known.

Given: $R = 100\ \Omega$
$C = 15.0\ \mu\text{F} = 15.0 \times 10^{-6}\ \text{F}$
$V = 120\ \text{V}$
$f = 60\ \text{Hz (exact)}$

Find: (a) I (rms current)
(b) V_C (rms voltage across capacitor)
V_R (rms voltage across resistor)
$V_R + V_C$ (total rms voltage)

(a) In Example 21.3, we calculated the reactance for the capacitor at this frequency to be $X_C = 177\ \Omega$. We use this directly in Eq. 21.16 to get the circuit impedance:

$$Z = \sqrt{R^2 + X_C^2} = \sqrt{(100\ \Omega)^2 + (177\ \Omega)^2} = 203\ \Omega$$

Since $V = IZ$, the current is

$$I = \frac{V}{Z} = \frac{120\ \text{V}}{203\ \Omega} = 0.591\ \text{A}$$

(b) Using Eq. 21.17 first for the resistor alone ($Z = R$), we have

$$V_R = IR = (0.591\ \text{A})(100\ \Omega) = 59.1\ \text{V}$$

For the capacitor alone ($Z = X_C$),

$$V_C = IX_C = (0.591\ \text{A})(177\ \Omega) = 105\ \text{V}$$

The algebraic sum of these two rms voltages adds up to 164 V, which is *not* the same as the rms value of the voltage source. This does not, however, mean violation of Kirchhoff's loop theorem, which is based on conservation of energy per unit charge. At *any instant*, the source voltage does, in fact, equal the voltages across the capacitor and resistor *if you account for phase differences*. Thus we expect the rms values to agree if the combination of the voltages is calculated properly. Since they are 90° out of phase, the rms values can be combined, using the Pythagorean theorem to get the total voltage:

$$V_{R+C} = \sqrt{V_R^2 + V_C^2} = \sqrt{(59.1\ \text{V})^2 + (105\ \text{V})^2} = 120\ \text{V}$$

When the individual voltage differences are combined, taking into account that they are out of phase and so never peak at the same time, we see that Kirchhoff's law *is* valid. The total voltage across both elements is equal (rms) to the voltage supplied by the source.

Follow-up Exercise. (a) How would the result in part (a) of this Example change if the circuit were driven by a voltage source with the same rms voltage but oscillating at 120 Hz? (b) Is the resistor or the capacitor responsible for the change?

Series RL Circuit

The analysis of a series RL circuit (•Fig. 21.10) is similar to that of a series RC circuit. However, the inductive reactance is plotted along the positive y axis in the phase diagram to reflect a phase difference of +90°. (A positive phase angle implies that the voltage *leads* the current, as is the case for an inductor.)

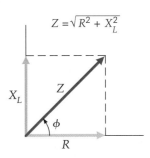

(a) Circuit diagram

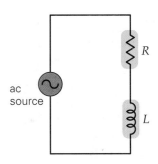

(b) Phase diagram

•**FIGURE 21.10 A series RL circuit** (a) In a series RL circuit, (b) the impedance Z is the phasor sum of the resistance R and the inductive reactance X_L.

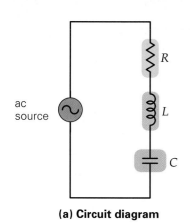

(a) Circuit diagram

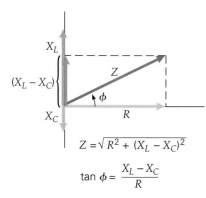

$$Z = \sqrt{R^2 + (X_L - X_C)^2}$$

$$\tan \phi = \frac{X_L - X_C}{R}$$

(b) Phase diagram

•**FIGURE 21.11   A series RLC circuit   (a)** In a series RLC circuit, **(b)** the impedance Z is the phasor sum of the resistance R and the total reactance $(X_L - X_C)$.

Note: $V = IZ$ can be applied to *any* circuit as long as the impedance Z is properly calculated, using phasors.

The impedance is then

$$Z = \sqrt{R^2 + X_L{}^2} \qquad \begin{array}{l}\textit{series RL circuit}\\ \textit{impedance}\end{array} \qquad (21.18)$$

(For any ac circuit, $V = IZ$ always applies.)

Series RLC Circuit

An ac circuit may contain all three circuit elements—resistor, inductor, and capacitor—shown in series in •Fig. 21.11. Again, phasor addition is used to determine the impedance by a component method. Summing the vertical components (the inductive and capacitive reactances) gives the total reactance ($X_L - X_C$, since X_C is defined as a negative phasor). The impedance is then the phasor sum of the resistance and the total reactance. The phase diagram shows that the impedance is

$$Z = \sqrt{R^2 + (X_L - X_C)^2} \qquad \begin{array}{l}\textit{series RLC circuit}\\ \textit{impedance}\end{array} \qquad (21.19)$$

Once again $V = IZ$.

The **phase angle** (ϕ) between the voltage from the source and the current in the circuit is given by

$$\tan \phi = \frac{X_L - X_C}{R} \qquad \begin{array}{l}\textit{phase angle in series}\\ \textit{RLC circuit}\end{array} \qquad (21.20)$$

Note that if X_L is greater than X_C (as shown in Fig. 21.11b), the phase angle is positive ($+\phi$) and the circuit is said to be an **inductive circuit,** since the nonresistive part of the impedance is dominated by the inductor. If X_C is greater than X_L, the phase angle is negative ($-\phi$) and the circuit is said to be a **capacitive circuit,** because the capacitor dominates the reactance.

A summary of impedances and phase angles for the three circuit elements and various series combinations considered is given in Table 21.1. Consider the RLC circuit in the following Example.

TABLE 21.1   Impedances and Phase Angles for Series Circuits

Circuit element(s)	Impedance, Z (in Ω)	Phase angle, ϕ
R	R	0°
C	X_C	−90°
L	X_L	+90°
RC	$\sqrt{R^2 + X_C{}^2}$	negative
RL	$\sqrt{R^2 + X_L{}^2}$	positive
RLC	$\sqrt{R^2 + (X_L - X_C)^2}$	positive if $X_L > X_C$; negative if $X_C > X_L$

Example 21.6 ■ All Together Now: Impedance in an RLC Circuit

A series RLC circuit has a resistance of 25.0 Ω, a capacitance of 50.0 μF, and an inductance of 0.300 H. If the circuit is driven by a 120-volt, 60-hertz source, what are (a) the impedance of the circuit, (b) the current in the circuit, and (c) the phase angle between the current and the voltage?

Thinking It Through.   (a) To calculate the overall impedance from Eq. 21.19, we must first calculate the individual reactances. (b) The current is computed from the generalization of Ohm's law, $V = IZ$ (Eq. 21.17) and (c) the phase angle from Eq. 21.20.

Solution. We are given all the necessary data, which we list.

Given: $R = 25.0 \, \Omega$ *Find:* (a) Z (impedance)
$C = 50.0 \, \mu\text{F} = 5.00 \times 10^{-5} \, \text{F}$ (b) I (current)
$L = 0.300 \, \text{H}$ (c) ϕ (phase angle)
$V = 120 \, \text{V}$
$f = 60 \, \text{Hz}$ (exact)

(a) The individual reactances are

$$X_C = \frac{1}{2\pi f C} = \frac{1}{2\pi(60 \, \text{Hz})(5.00 \times 10^{-5} \, \text{F})} = 53.1 \, \Omega$$

and

$$X_L = 2\pi f L = 2\pi(60 \, \text{Hz})(0.300 \, \text{H}) = 113 \, \Omega$$

Then,

$$Z = \sqrt{R^2 + (X_L - X_C)^2} = \sqrt{(25.0 \, \Omega)^2 + (113 \, \Omega - 53.1 \, \Omega)^2} = 64.9 \, \Omega$$

(b) Since $V = IZ$, we have

$$I = \frac{V}{Z} = \frac{120 \, \text{V}}{64.9 \, \Omega} = 1.85 \, \text{A}$$

(c) Solving $\tan \phi = (X_L - X_C)/R$ for the phase angle gives

$$\phi = \tan^{-1}\left(\frac{X_L - X_C}{R}\right) = \tan^{-1}\left(\frac{113 \, \Omega - 53.1 \, \Omega}{25.0 \, \Omega}\right) = +67.3°$$

We should have expected a positive phase angle. How can you tell from the results of part (a)?

Follow-up Exercise. In a series RLC circuit as in this Example, how would the inductive and capacitive reactances compare if we wanted the current to be a maximum?

You should now appreciate the importance of phasor diagrams in finding the impedance and the current in ac circuits. However, you might be wondering about the usefulness of the phase angle ϕ. To illustrate its use, we must look at the *power loss* for our RLC circuit, which illustrates another important benefit derived from using phasor diagrams.

Power Factor for a Series RLC Circuit

The power, or energy dissipated per unit time, is an interesting feature of an RLC circuit. *There are no power losses associated with pure capacitances and inductances in an ac circuit.* Capacitors and inductors simply store energy and then give it back. Ideally, they have no resistance to produce joule heating. For example, during a half-cycle, a capacitor in an ac circuit is charged and energy is stored in the electric field between the plates. During the next half-cycle, the capacitor discharges and returns the charge to the voltage source. Similarly, in an inductor (with negligible resistance in its coil wires), the energy is stored in the magnetic field associated with the current through it; the current builds up, drops to zero, and then reverses during each cycle.*

Thus, *the only element that dissipates energy in an RLC series circuit is the resistor.* As you learned earlier, the average power dissipated by a resistor is $P = I^2 R$ (where I is the rms value). The power can also be expressed in terms of the current and voltage, but the voltage in this case *must be that across the resistor* (V_R), since this is the only dissipative element. That is, the power dissipated in a series RLC circuit is

$$P = P_R = IV_R$$

---

*In Chapter 16, you saw that the energy stored in the electric field of a capacitor is given by $U_C = \frac{1}{2}CV^2$. Similarly, it can be shown that the energy stored in the magnetic field of an inductor is given by $U_L = \frac{1}{2}LI^2$, where L is the inductance and I is the current in the inductor.

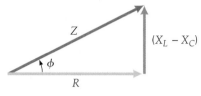

(a) Phasor triangle

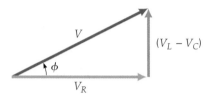

(b) Equivalent voltage triangle

•**FIGURE 21.12 Phasor and voltage triangles** The voltages across the components of a series RLC circuit are given by $V_R = IR$, $V_L = IX_L$, and $V_C = IX_C$. Since the current is the same through each, **(a)** the phasor triangle can be converted to **(b)** a voltage triangle. Note that $V_R = V \cos \phi$.

The voltage across the resistor is conveniently found from a voltage triangle that corresponds to the phasor triangle (•Fig. 21.12). The voltages across the components of an RLC circuit are given by $V_R = IR$, $V_L = IX_L$, and $V_C = IX_C$. Combining the last two voltages, we can write $(V_L - V_C) = I(X_L - X_C)$. If we multiply each of the phasor triangle legs in Fig. 21.12a by the current I, we obtain an equivalent voltage triangle (Fig. 21.12b). As this triangle shows, the voltage across the resistor (V_R) depends on the phase angle:

$$V_R = V \cos \phi \qquad (21.21)$$

The term $\cos \phi$ is called the **power factor**. From Fig. 21.11,

$$\cos \phi = \frac{R}{Z} \qquad \text{series RLC power factor} \qquad (21.22)$$

The power in terms of the power factor is

$$P = IV \cos \phi \qquad \text{series RLC power} \qquad (21.23)$$

With power dissipated only in the resistance ($P = I^2R$), we can use Eq. 21.22 to express the average power as

$$P = I^2 Z \cos \phi \qquad \text{series RLC power} \qquad (21.24)$$

Note that $\cos \phi$ varies from a maximum of $+1$ (when $\phi = 0$) to a minimum of 0 (when $\phi = \pm\pi/2$). When $\phi = 0$, the circuit is said to be *completely resistive*. That is, there is maximum power dissipation (as though the circuit contained only a resistor). The power factor decreases as the phase angle increases in either direction [$+\phi$ or $-\phi$, since $\cos(-\phi) = \cos \phi$], and the circuit becomes more inductive or capacitive. At $\phi = +90°$, the circuit is *completely inductive*; at $\phi = -90°$, it is *completely capacitive*. For these cases, the circuit contains only an inductor or a capacitor, respectively, so no power is dissipated ($P = IV \cos 90° = 0$). In practice, since there is always some resistance; a circuit cannot be completely inductive or capacitive. It is possible, however, for an RLC circuit to appear to be completely resistive even if it contains both a capacitor and an inductor, as we shall see when we study resonance in Section 21.5. Let's look at our previous RLC Example with an emphasis on power.

Example 21.7 ■ Example 21.6 Revisited: Power Factor

How much power is dissipated in the circuit described in Example 21.6?

Thinking It Through. We can compute the power factor from the resistance (R) and impedance (Z) used in Eq. 21.22. The power can then be calculated from Eq. 21.23.

Solution.

Given: See Example 21.6   *Find:* P (power)

In Example 21.6, the circuit was found to have an impedance of $Z = 64.9 \ \Omega$, and its resistance was $R = 25.0 \ \Omega$. The power factor of the circuit is therefore

$$\cos \phi = \frac{R}{Z} = \frac{25.0 \ \Omega}{64.9 \ \Omega} = 0.385$$

Using other data from Example 21.6 gives

$$P = IV \cos \phi = (1.85 \text{ A})(120 \text{ V})(0.385) = 85.5 \text{ W}$$

This is less than the power that would be dissipated without a capacitor and inductor.

Follow-up Exercise. In this Example, (a) what frequency *would* make the circuit completely resistive, and (b) what would be the power dissipated in the resistor then?

21.5 Circuit Resonance

Objectives: To (a) understand the concept of resonance in ac circuits and
(b) calculate the resonance frequency of an RLC circuit.

When the power factor cos ϕ of an RLC series circuit is equal to 1, there is maximum power dissipation. For a particular voltage, the current in the circuit is then a maximum, since the impedance is a minimum. This occurs when the inductive and capacitive reactances effectively cancel each other—when they are 180° out of phase and equal in magnitude. This can be made to happen in any RLC circuit by choosing the appropriate driving (source) frequency.

The key to finding the correct frequency is to realize that both the inductive and capacitive reactances are frequency-dependent, so the impedance also depends on the frequency. From the expression for the impedance of an RLC circuit, $Z = \sqrt{R^2 + (X_L - X_C)^2}$, we can see that the impedance is a minimum when the inductive and capacitive reactances are equal in magnitude. This will occur at a particular value of the frequency, which we will designate f_o. We can find f_o by setting $X_L - X_C = 0$, or, from the expressions for the two reactances,

$$2\pi f_o L = \frac{1}{2\pi f_o C}$$

We can then solve this equation for the frequency:

$$f_o = \frac{1}{2\pi\sqrt{LC}} \qquad (21.25)$$

Series RLC resonance frequency

Equation 21.25 gives the frequency for the condition of minimum impedance, which is called the **resonance frequency**. A graph of reactance versus frequency is shown in •Fig. 21.13a. The X_C and X_L curves intersect at f_o, where their values are equal ($X_L = X_C$).

A Physical Explanation of Resonance

The physical explanation of resonance in a series RLC circuit is worth exploring. We know that the resistor voltage is in phase with the current, so let us reference our phase angles to the voltage across the resistor. The voltage across the

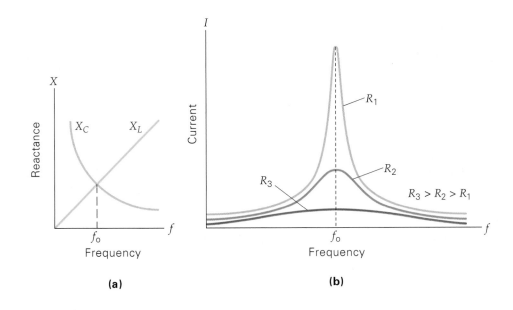

(a)

(b)

•FIGURE 21.13 Resonance frequency for series RLC circuit (a) At the resonance frequency (f_o), the capacitive and inductive reactances are equal ($X_L = X_C$). On a graph of X versus f, this is the frequency at which the X_C and X_L curves intersect. (b) On a graph of I versus f, the current is a maximum at f_o. The curve becomes sharper and narrower as the resistance in the circuit decreases.

capacitor lags the current, and thus the voltage across the resistor, by 90°. The voltage across the inductor leads the current, and thus the voltage across the resistor, by 90°. Hence *the capacitor and inductor voltages are always 180° out of phase, or of opposite polarity.* (This is embodied graphically in the phase diagram, Fig. 21.11b.)

Normally, however, the capacitive and inductive voltages are not of the same magnitude and so do *not* cancel completely. Thus, the voltage across the resistor is generally less than the voltage of the source, and the power dissipated in the resistor (joule heat) is less than the maximum. However, if the capacitive and inductive voltages do cancel exactly, then the full source voltage appears across the resistor, the power factor is 1, and we have maximum power output, or resonance conditions.

Conceptual Example 21.8

Is Resonance Possible? RL and RC Circuits

Is it possible for resonance to occur in an RL or RC circuit?

Reasoning and Answer. With only one reactive element (capacitor *or* inductor) in each circuit, there is no possibility of two reactance voltages exactly cancelling, as can happen in an RLC circuit. Hence no resonance is possible.

Follow-up Exercise. As the frequency of the voltage source increases, what happens to the power dissipated in (a) an RL circuit and (b) an RC circuit? Explain.

Resonance Applications

When a series RLC circuit is driven at its resonance frequency, the impedance is a minimum. Thus both the power transfer from the source to the circuit and the current in the circuit are a maximum. Since $X_L = X_C$, then $Z_{min} = R$, and the impedance is completely resistive. In terms of the power factor, $\cos \phi = R/Z = R/R = 1$; thus the power factor is a maximum and there is maximum power dissipation in the circuit.

A graph of current versus frequency is shown in Fig. 21.13b for several different resistances in an RLC circuit with fixed capacitance and inductance. Notice that the currents have maximum values at f_0. Also, notice how the curve becomes sharper and narrower as the resistance decreases.

Resonant circuits have a wide variety of applications, for example, in the tuning mechanism of a radio. Each radio station has an assigned broadcast frequency at which radio waves are transmitted. The oscillating electric and magnetic fields of the radio waves set electrons in the receiving antenna into regular back-and-forth motion—that is, they produce an alternating current—as a voltage source in the circuit would do. (This reception process is the reverse of the *production* of electromagnetic waves by oscillating electrons in a transmitting antenna.)

Signals broadcast by many different stations generally reach a radio, but the receiver circuit selectively picks up only the one with a frequency at or near its resonance frequency. Most radios allow you to alter this resonance frequency to "tune in" different stations. Variable air capacitors in the tuning circuit have long been used for this purpose (•Fig. 21.14). More compact variable capacitors in some smaller radios have a polymer dielectric between thin plates. The polymer sheets help maintain the plate separation and increase the capacitance, allowing makers to use plates of smaller area. (Recall that $C = K\epsilon_0 A/d$.) In other radios, variable capacitors are replaced by solid-state devices.

Note: See Section 20.4 and the Insight on p. 680.

Note: See Eq. 16.15, Section 16.4.

AM versus FM: Resonance in Radio Reception

When you switch from the AM band to the FM band in some radios, you are effectively changing the capacitance of the receiving circuit, assuming constant inductance. Is the capacitance (a) increased or (b) decreased when you make this change?

Reasoning and Answer. Since FM stations broadcast at significantly higher frequencies than do AM stations (see Table 20.1), we must increase the resonance frequency of the radio receiver to receive a significant amount of power from them. To increase the resonance frequency of the receiver requires reducing either the inductance or the capacitance. Since the capacitance is changed, we must be switching to a lower capacitance than that used to tune in AM stations. Thus the answer is (b).

Follow-up Exercise. Based on the resonance curves shown in Fig. 21.13b, can you explain how it is possible to pick up two radio stations *simultaneously* on your radio? (You may have encountered this phenomenon, particularly between two large cities. Two stations are sometimes granted licenses for broadcasting on closely spaced frequencies under the assumption that they are far enough apart that they are never received by the same receiver. However, under certain atmospheric conditions, this may not be true.)

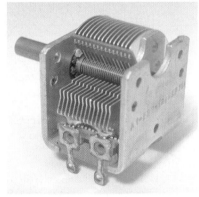

•**FIGURE 21.14 Variable air capacitor** Rotating the movable plates between the fixed plates changes the overlap area and thus the capacitance. Such capacitors were common in tuning circuits in older radios.

Example 21.10 ■ That Certain Frequency: Resonance in an RLC Circuit

A series RLC circuit has a resistance of 50.0 Ω, a capacitance of 6.00 nF, and an inductance of 28.0 mH. The circuit is connected to a wide-range, adjustable-frequency voltage source with an output of 25.0 V. (a) What is the resonance frequency of the circuit? (b) How much current is in the circuit when it is in resonance? (c) What is the voltage across each circuit element for this condition?

Thinking It Through. (a) Since we know the circuit's capacitance and inductance, we can calculate the resonance frequency, f_o, directly from Eq. 21.25. (b,c) At the resonance frequency, $X_L = X_C$ and $Z = Z_{min} = R = 50.0$ Ω. The current and voltages are obtained from the generalized version of Ohm's law ($V = IZ$).

Solution. We list the values given:

Given: $R = 50.0$ Ω
  $C = 6.00$ nF $= 6.00 \times 10^{-9}$ F
  $L = 28.0$ mH $= 28.0 \times 10^{-3}$ H
  $V = 25.0$ V

Find: (a) f_o (resonance frequency)
  (b) I (current)
  (c) V_R, V_C, V_L (voltages)

(a) The resonance frequency is

$$f_o = \frac{1}{2\pi\sqrt{LC}} = \frac{1}{2\pi\sqrt{(28.0 \times 10^{-3}\,\text{H})(6.00 \times 10^{-9}\,\text{F})}}$$

$$= 12.3 \times 10^3\,\text{Hz} = 12.3\,\text{kHz}$$

(b) The current in the circuit is

$$I = \frac{V}{Z} = \frac{V}{R} = \frac{25.0\,\text{V}}{50.0\,\Omega} = 0.500\,\text{A}$$

(c) The inductive reactance at f_o is

$$X_L = 2\pi f_o L = 2\pi(12.3 \times 10^3\,\text{Hz})(28.0 \times 10^{-3}\,\text{H})$$

$$= 2.16 \times 10^3\,\Omega = 2.16\,\text{k}\Omega$$

Oscillator Circuits: Broadcasters of Electromagnetic Radiation

To generate the high-frequency electromagnetic waves used in radio communications and television (Fig. 1), electrons must be made to oscillate with high frequencies in electronic circuits. This can be accomplished with resonant RLC circuits. Such circuits are called *oscillator circuits*, because they "oscillate" at resonance frequencies determined by their inductive and capacitive elements.

When the resistance R in an RLC circuit is very small, the circuit is essentially an LC circuit. Energy in such a circuit oscillates back and forth between the inductor and capacitor at a frequency f. This is the natural frequency of the circuit and is equal to its resonance frequency (Eq. 21.25). Some energy is dissipated by the small resistance. However, in an ideal LC circuit with no ohmic resistance, the oscillation would continue indefinitely.

To understand this energy oscillation better, consider an ideal LC parallel circuit as shown in Fig. 2a. Let's assume that the capacitor is initially charged and the switch is then closed. The following sequence of events then occurs:

1. The capacitor would discharge instantaneously (since RC = 0) were it not for the necessity of the current having to pass through the coil. When the switch is closed (at t = 0), the current through the coil is zero (Fig. 2a). As the current builds up, so does the magnetic field in the coil. By Lenz's law, the increasing magnetic field and change of flux through the coil induces an emf in such a direction as to oppose the current increase. Because of this opposition, the capacitor takes some time to discharge.

2. When the capacitor is (instantaneously) in a fully discharged state (Fig. 2b), the energy initially stored in it has been transferred to the inductor, where it creates a magnetic field. Recall that since R = 0, no energy has been lost because of joule heating. At this time (one-quarter of a cycle or period), the magnetic field and the current in the coil are at their maximum values and thus all the energy is stored in the inductor.

3. As the magnetic field collapses from its maximum value, an emf that opposes the collapse is induced in the coil. This emf is in a direction that would tend to continue the current in the coil even as the current is decreasing (Lenz's law again). The polarity of the emf is now opposite what it was in step 1. Thus, current continues to transport charge to the capacitor plates and charges the capacitor with a polarity the reverse of its initial value.

4. When the capacitor is fully charged (with reverse polarity), it has the same energy that it had initially (Fig. 2c). This occurs halfway through the cycle, or half a period from the start. The magnetic field in the coil is now zero, as is the current in the circuit.

5. The capacitor then begins to discharge and the previous four steps are repeated, setting up an "oscillation" in the circuit.

For an ideal case, these oscillations would continue indefinitely. The oscillation of energy between a capacitor and an inductor is analogous to the exchange of potential and kinetic energies for a mass oscillating on an ideal spring, which is shown next to the circuit at corresponding times during a typical half-cycle. (See the bar graphs next to the circuits in Fig. 2.)

FIGURE 1 Broadcast antennas These antennas are located atop the World Trade Center in New York City.

At resonance, we know that $X_C = X_L = 2.16$ kΩ. Then, using the relationship between voltage and current, we get the voltage *magnitudes*:

$$V_R = IR = (0.500 \text{ A})(50.0 \ \Omega) = 25.0 \text{ V}$$

$$V_L = IX_L = (0.500 \text{ A})(2.16 \times 10^3 \ \Omega) = 1.08 \times 10^3 \text{ V}$$

$$V_C = IX_C = (0.500 \text{ A})(2.16 \times 10^3 \ \Omega) = 1.08 \times 10^3 \text{ V}$$

Follow-up Exercise.   In this Example, note that the magnitudes of the voltages across the capacitor and inductor are much higher than the voltage source. Kirchhoff's loop theorem appears to be violated. Explain why this is not really the case.

In operation, a small ac voltage of the proper frequency applied to the circuit is sufficient to maintain strong oscillations between the electric field of the capacitor and the magnetic field of the inductor. An LC circuit is sometimes called a "tank circuit," which suggests its energy storage capability. Without voltage input, however, the oscillator would run down as a result of I^2R losses and radiation losses (as its mechanical counterpart would be slowed down by frictional losses).

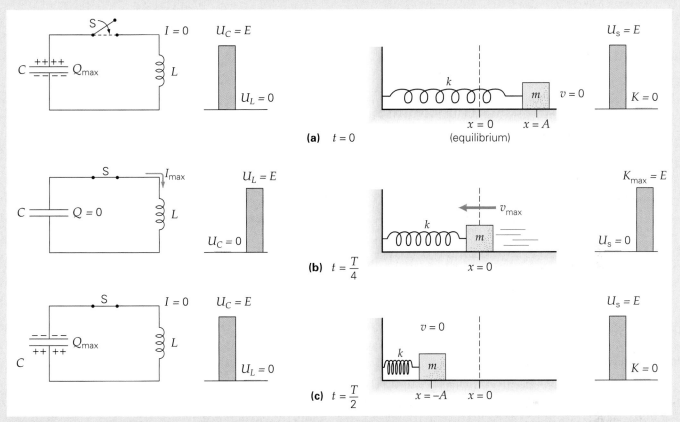

FIGURE 2   An oscillating LC circuit If the resistance is negligible, this circuit will oscillate indefinitely. Half a cycle is shown. Essentially, energy is transferred back and forth between magnetic and electric potential energy. The oscillating electrons in the wire will give off electromagnetic radiation at the circuit's oscillating frequency, $f_o = 1/(2\pi\sqrt{LC})$. A frictionless spring–mass oscillator is shown at the same points in the half-cycle. The spring is analogous to the capacitor (both store potential energy), and the kinetic energy (due to mass motion) is analogous to the inductor's stored energy that depends on charge motion (current).

Chapter Review

Important Terms

peak voltage *665*
peak current *665*
rms (or effective) current *666*
rms (or effective) voltage *666*
capacitive reactance *669*
inductance *670*

inductor *670*
henry (H) *670*
inductive reactance *671*
phase diagram *672*
phasors *672*
impedance *672*

phase angle *674*
inductive circuit *674*
capacitive circuit *674*
power factor *676*
resonance frequency *677*

Important Concepts

- In an ac circuit, voltage and current vary sinusoidally with time. The rms (root mean square), or effective, voltage and current are less than the peak, or maximum, value that these quantities attain during a cycle of oscillation.

- Capacitors and inductors do not eliminate the current in an ac circuit but offer some opposition to it, termed capacitive reactance and inductive reactance, respectively. Both quantities vary with the frequency of the circuit, but in opposite ways.

- In a purely capacitive ac circuit, the current leads the voltage by 90°. In a purely inductive circuit, the voltage leads the current by 90°.

- Impedance is the generalization of resistance to include not only resistance but also inductive and capacitive reactances.

- Phasors are vector-like quantities that allow the resistance of resistors and the reactances of capacitors and inductors to be graphically represented, and impedance to be calculated, under ac conditions.

- The power factor of an RLC circuit is a measure of how close the circuit is to expending the maximum power.

- The resonance frequency of an RLC circuit is the frequency at which the circuit dissipates maximum power.

Important Equations

Instantaneous Voltage in an ac Circuit:

$$V = V_0 \sin \omega t = V_0 \sin 2\pi f t \qquad (21.1)$$

Instantaneous Current in an ac Circuit (containing resistance only): (where $I_0 = V_0/R$)

$$I = I_0 \sin 2\pi f t \qquad (21.2)$$

Rms (or Effective) Current:

$$I_{rms} = \frac{I_0}{\sqrt{2}} = 0.707 I_0 \qquad (21.6)$$

Effective Power in an ac Circuit

$$\overline{P} = I_{rms}^2 R \qquad (21.7)$$

Rms (or Effective) Voltage:

$$V_{rms} = \frac{V_0}{\sqrt{2}} = 0.707 V_0 \qquad (21.8)$$

Relationship between Voltage and Current in an ac Circuit (resistor only):

$$V_{rms} = I_{rms} R \qquad (21.9)$$

Capacitive Reactance:

$$X_C = \frac{1}{2\pi f C} = \frac{1}{\omega C} \qquad (21.11)$$

Relationship between Voltage and Current (capacitor only):

$$V = IX_C \qquad (21.12)$$

Inductive Reactance:

$$X_L = 2\pi f L = \omega L \qquad (21.14)$$

Relationship between Voltage and Current (inductor only):

$$V = IX_L \qquad (21.15)$$

Ohm's Law Generalized to ac Circuits:

$$V = IZ \qquad (21.17)$$

Impedance for a Series RLC Circuit:

$$Z = \sqrt{R^2 + (X_L - X_C)^2} \qquad (21.19)$$

Phase Angle Between Voltage and Current in a Series RLC Circuit:

$$\tan \phi = \frac{X_L - X_C}{R} \qquad (21.20)$$

Power Factor of a Series RLC Circuit:

$$\cos \phi = \frac{R}{Z} \qquad (21.22)$$

Power in Terms of Power Factor:

$$P = IV \cos \phi \qquad (21.23)$$

or

$$P = I^2 Z \cos \phi \qquad (21.24)$$

Resonance Frequency of a Series RLC Circuit:

$$f_0 = \frac{1}{2\pi\sqrt{LC}} \qquad (21.25)$$

Exercises

21.1 Resistance in an AC Circuit

1. Which of the following voltages has the greatest magnitude for a sinusoidal ac voltage: (a) V_o, (b) V_{rms}, or (c) V_{p-p}?

2. If the average current in an ac circuit is zero, why isn't the average power zero?

3. The voltage and current of a resistor in an AC circuit are *in phase*. What does that mean?

4. ■ What are the peak voltages of a 120-volt ac line and a 240-volt ac line?

5. ■ An ac circuit has an rms current of 5.0 A. What is the peak current?

6. ■ The maximum (magnitude) potential difference across a resistor in an ac circuit is 156 V. Find the effective voltage.

7. ■ How much ac current must flow through a 10-ohm resistor to produce a power of 15 W?

8. ■ An ac circuit with a resistance of 5.0 Ω has an effective current of 0.75 A. (a) Find the rms voltage and peak voltage. (b) Find the average power dissipated by the resistance.

9. ■ A hair dryer rated at 1200 W is plugged into a 120-volt outlet. Find (a) the effective current; (b) the peak current; (c) the resistance of the dryer.

10. ■■ The voltage across a resistor varies according to the following equation: $V = (170\ \text{V})(\sin 120\,t)$. What are (a) the effective voltage across the resistor and (b) the period of the voltage cycle?

11. ■■ An ac voltage is applied to a 25-ohm resistor, and it dissipates 500 W of power. Find (a) the effective and peak currents and (b) the effective and peak voltages.

12. ■■ An ac voltage source has an amplitude of 85 V and a frequency of 60 Hz. (a) If $V = 0$ at $t = 0$, what is the instantaneous voltage at $t = 2.0$ s? (b) What is the effective or rms voltage?

13. ■■ An ac voltage source has an rms value of 120 V. The voltage starts from zero and rises to its maximum value in 4.2 ms. Write an expression for the voltage as a function of time.

14. ■■ What are the resistance and rms current of a 100-watt, 120-volt computer monitor?

15. ■■ Find the effective and peak currents through a 40-watt, 120-volt light bulb.

16. ■■ A 50-kilowatt heater is connected to a 240-volt ac source. Find (a) the peak current and (b) the peak voltage.

17. ■■ A circuit has a current given by $I = (8.0\ \text{A}) \sin (40\pi t)$ with an applied voltage $V = (60\ \text{V}) \sin (40\pi t)$. What is the average power delivered to the circuit?

18. ■■ The current and voltage outputs of an ac generator have peak values of 2.5 A and 16 V, respectively. What is the effective power output of the generator?

19. ■■■ The current through a 60-ohm resistor is given by $I = (2.0\ \text{A}) \sin (380t)$. (a) What is the frequency of the current? (b) the effective current? (c) How much power is dissipated by the resistor? (d) Write an equation expressing the voltage as a function of time. (e) Write an equation expressing the power as a function of time. (f) Show that the rms power obtained from (e) is the same as your answer to (c).

20. ■■■ Rotating coils are often used to measure magnetic fields. A coil of 40 circular loops with a radius of 7.5 cm rotates at 60 Hz about an axis along a diameter perpendicular to a uniform magnetic field. An effective voltage of 50 V, as read from an ac voltmeter, is induced in the coil. What is the magnitude of the magnetic field?

21.2 Capacitive Reactance
and
21.3 Inductive Reactance

21. In a purely capacitive ac circuit, (a) the current and voltage are in phase, (b) the current leads the voltage, (c) the current lags the voltage, (d) none of the above.

22. The unit of capacitive reactance is (a) F, (b) $(F \cdot s)^{-1}$, (c) Ω, (d) Ω·m.

23. Explain why at low frequencies a capacitor acts as an ac open circuit while an inductor acts as a short circuit.

24. The unit of inductance is (a) H/s, (b) H, (c) Ω, (d) Ω·s.

25. In an ac circuit, what can oppose current other than ohmic resistance, and why?

26. Explain these statements: "The voltage across an inductor leads the current by 90°" and "The voltage across a capacitor lags the current by 90°."

27. ■ A capacitor of what capacitance would have a reactance of 100 Ω in a 60-hertz ac circuit?

28. ■ Find the frequency at which a 25-microfarad capacitor has a reactance of 25 Ω.

29. ■ A 2.0-microfarad capacitor is connected across a 60-hertz voltage source, and a current of 2.0 mA is measured on an ammeter. What is the capacitive reactance of the circuit?

30. ■ A 50-millihenry inductor is connected in a circuit to a source that delivers 120 V at 60 Hz. (a) What is the inductive reactance of the circuit? (b) How much current is in the circuit? (c) What is the phase angle between the current and the supplied voltage? (Assume negligible resistance.)

31. ■ How much current is delivered to a circuit with a 50-microfarad capacitor if the circuit includes an ac generator with an output of 120 V and 60 Hz?

32. ■ In Exercise 31, how much power is delivered to the circuit?

33. ■■ A variable capacitor in a circuit with a 120-volt, 60-hertz source is initially at a value of 0.25 μF and is then increased to 0.40 μF. What is the percentage change in the current in the circuit?

34. ■■ An inductor has a reactance of 90 Ω in a 60-hertz ac circuit. What is its inductance?

35. ■■ Find the frequency at which a 250-millihenry inductor has a reactance of 400 Ω.

36. ■■ A 150-millihenry inductor is in a circuit with a 60-hertz voltage source, and a current of 1.6 A is measured on an ammeter. (a) What is the voltage of the source? (b) What is the phase angle between the current and that voltage?

37. ■■ What inductance has the same reactance in a 120-volt, 60-hertz circuit as a capacitance of 10 μF?

38. ■■ A circuit with a single capacitor is connected to a 120-volt, 60-hertz source. What is the capacitance if there is a current of 0.20 A in the circuit?

21.4 Impedance: RLC Circuits
and
21.5 Circuit Resonance

39. The impedance of an RLC circuit depends on (a) frequency, (b) inductance, (c) capacitance, (d) all of these.

40. If the capacitance of a series RLC circuit is decreased, (a) the capacitive reactance increases, (b) the inductive reactance increases, (c) the current remains constant, (d) the power factor remains constant.

41. When a series RLC circuit is driven at its resonance frequency, (a) energy is dissipated by only the resistive element, (b) the power factor is one, (c) there is maximum energy transfer to the circuit, (d) all of these.

42. What is the impedance of an RLC circuit at resonance?

43. ■ A coil in a 60-hertz circuit has a resistance of 100 Ω and an inductance of 0.45 H. Calculate (a) the coil's reactance and (b) the circuit's impedance.

44. ■ A series RC circuit has a resistance of 200 Ω and a capacitance of 25 μF and is driven by a 120-volt, 60-hertz source. (a) Find the capacitive reactance and impedance of the circuit. (b) How much current is drawn from the source?

45. ■ A series RL circuit has a resistance of 100 Ω and an inductance of 100 mH and is driven by a 120-volt, 60-hertz source. (a) Find the inductive reactance and the impedance of the circuit. (b) How much current is drawn from the source?

46. ■ An RC circuit has a resistance of 250 Ω and a capacitance of 6.0 μF. If the circuit is driven by a 60-hertz source, find (a) the capacitive reactance and (b) the impedance of the circuit.

47. ■ An RL circuit has a resistance of 100 Ω and an inductive reactance of 50 Ω. What is the phase angle of the circuit?

48. ■■ A series RLC circuit has a resistance of 25 Ω, an inductance of 0.30 H, and a capacitance of 8.0 μF. At what frequency should the circuit be driven for the maximum power to be transferred from the driving source?

49. ■■ In a series RLC circuit, $R = X_C = X_L = 40$ Ω for a particular driving frequency. If the driving frequency is doubled, what will be the impedance of the circuit?

50. ■■ A resistor, an inductor, and a capacitor have values of 500 Ω, 500 mH, and 3.5 μF, respectively, and are connected in series to a power supply of 240 V, 60 Hz. Is the circuit driven in resonance? If not, what values of resistance and inductance would be required for this to occur?

51. ■■ How much power is dissipated in the circuit described in Exercise 50?

52. What is the resonance frequency of an RLC circuit with a resistance of 100 Ω, an inductance of 100 mH, and a capacitance of 5.00 μF?

53. ■■ A tuning circuit in an antique radio receiver has a fixed inductance of 0.50 mH and a variable capacitor

(•Fig. 21.15). If the circuit is tuned to a radio station broadcasting at 980 kHz on the AM dial, what is the capacitance of the capacitor?

•FIGURE 21.15   **Stay tuned**   See Exercises 53, 54 and 67.

54. ■■ What would be the range of the variable capacitor in Exercise 53 for tuning over the complete AM band? [Hint: See Table 20.1.]

55. ■■ Find the currents supplied by the ac source for all possible connections in •Fig. 21.16.

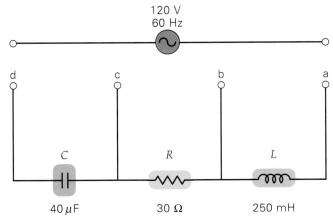

•FIGURE 21.16   **A series RLC circuit**   See Exercise 55.

56. ■■ A coil with a resistance of 30 Ω and an inductance of 0.15 H is connected to a 120-volt, 60-hertz source. (a) How much current is in the circuit? (b) What is the phase angle between the current and the source voltage?

57. ■■ A small welder operates on 120 V at 60 Hz. In operation, it uses 1200 W of power, and the power factor is 0.75. Find the effective current in the circuit containing the machine.

58. ■■ A circuit connected to a 220-volt, 60-hertz power supply has the following components connected in series: a 10-ohm resistor, a coil with an inductive reactance of 120 Ω, and a capacitor with a reactance of 120 Ω. Compute the magnitude of the voltage across (a) the resistor, (b) the inductor, and (c) the capacitor.

59. ■■ A series RLC circuit has a resistance of 25 Ω, a capacitance of 0.80 μF, and an inductance of 250 mH. The circuit is connected to a variable-frequency source with a fixed rms output of 12 V. If the source frequency is set at the circuit's resonance frequency, what is the magnitude of the rms voltage across each of the circuit elements?

60. ■■ In Exercises 58 and 59, the sum of the magnitudes of the three rms voltages is much greater than the source voltage. How do you explain the discrepancy?

61. ■■ The signal generator in •Fig. 21.17 can supply ac voltage of variable frequency. (a) What is the impedance with a driving frequency of 60 Hz? (b) Is the circuit driven in resonance? If not, find the resonance frequency of the circuit.

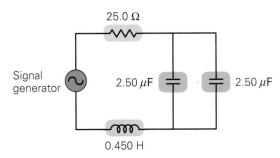

•FIGURE 21.17   **Tune to resonance**   See Exercise 61.

62. ■■■ A series RLC circuit with a resistance of 400 Ω has capacitive and inductive reactances of 300 Ω and 500 Ω, respectively. (a) What is the power factor of the circuit? (b) If the circuit operates at 60 Hz, an additional capacitor of what capacitance should be connected with the original capacitor to give a power factor of one, and how should the capacitors be connected?

63. ■■■ A series RLC circuit has components with these values: R = 50 Ω, L = 0.15 H, and C = 20 μF. The circuit is driven by a 120-volt, 60-hertz source. What is the power loss of the circuit as a percentage of its power loss when in resonance?

Additional Exercises

64. A 60.0-watt light bulb operates on 120-volt household voltage. What are (a) the peak voltage across the bulb and (b) the peak current in the circuit?

65. The rms current in an ac circuit is 0.25 A. What is the peak-to-peak current?

66. A 120-volt, 60-hertz source is connected across a 0.40-henry inductor. (a) Find the current through the inductor. (b) Find the phase angle between the current and the supplied voltage.

67. A radio receiver circuit with an inductance of 1.50 μH is tuned to an FM station at 98.9 MHz by adjusting a variable

capacitor (see Fig. 21.15). What is the capacitance of the capacitor when the circuit is tuned to the station?

68. A circuit connected to a 110-volt, 60-hertz source contains a 50-ohm resistor and a coil with an inductance of 100 mH. Find (a) the reactance of the coil, (b) the impedance of the circuit, (c) the current in the circuit, and (d) the power dissipated by the coil.

69. Calculate the phase angle between the current and the supplied voltage in Exercise 68.

70. A 1.0-microfarad capacitor is connected to a 120-volt, 60-hertz source. (a) What is the capacitive reactance of the circuit? (b) How much current is in the circuit? (c) What is the phase angle between the current and the supplied voltage?

71. The circuit in •Fig. 21.18a is called a *low-pass filter* because a large current is delivered to the load resistor (resistance R_L) only by a low-frequency source. The circuit in Fig. 21.18b

is called a *high-pass filter* because a large current is delivered to the load resistor only by a high-frequency source. Show mathematically why the circuits have these characteristics.

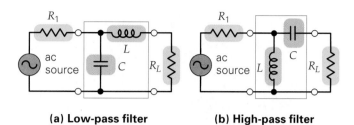

(a) Low-pass filter   **(b) High-pass filter**

•**FIGURE 21.18   Low-pass and high-pass filters**   See Exercise 71.

72. In a series RLC circuit with an inductance of 750 mH, what values of resistance and capacitance would give the circuit a resonance frequency of 60 Hz?

Geometrical Optics: Reflection and Refraction of Light

We live in a visual world, surrounded by images. How these images are created is something we take largely for granted—until we see something that we can't easily account for, as in the photo above. Optics is the study of light and vision. Human vision requires light, specifically what we call visible light. The broad definition of visible light is any radiation in or near the visible region of the electromagnetic spectrum. This definition includes infrared and ultraviolet radiations. Similar optical properties are shared by all electromagnetic waves.

In this chapter we will investigate the basic optical phenomena of reflection and refraction. The principles that govern reflection explain the behavior of mirrors, while those that govern refraction explain the properties of lenses. With the aid of these and other optical principles, we can understand many optical phenomena that we experience every day: why a glass prism spreads out light into a spectrum of colors, the cause of mirages, how rainbows are created, why the legs of a person standing in a lake or swimming pool seem to shorten—and why the underwater swimmer in the photo is being shadowed by a phantom double. We will also explore some less familiar territory, including the fascinating field of fiber optics.

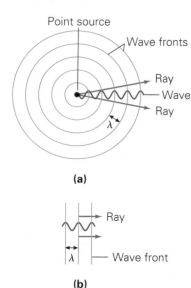

(a)

(b)

FIGURE 22.1 Wave fronts and rays A wave front is defined by adjacent points on a wave that are in phase, such as those along wave crests or troughs. **(a)** Near a point source, the wave fronts are circular in two dimensions and spherical in three dimensions. **(b)** Far from a point source, the wave fronts are approximately linear or planar. A line perpendicular to a wave front in the direction of the wave's propagation is called a ray.

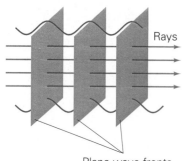

•FIGURE 22.2 Light rays A plane wave front travels in a straight line in the direction of its rays. A beam of light can be represented by a group of parallel rays (or by a single ray).

Note that the title of the chapter is geometrical optics. A simple geometrical approach, using straight lines and angles, can be used to investigate many aspects of reflection and refraction. For these purposes we need not be concerned with the physical (wave) nature of electromagnetic waves. The principles of geometrical optics will be introduced here and applied in greater detail in the study of mirrors and lenses in Chapter 23.

22.1 Wave Fronts and Rays

Objective: **To define and explain the concepts of wave fronts and rays.**

Waves, electromagnetic or other, are conveniently described in terms of wave fronts. A **wave front** is the line or surface defined by adjacent portions of a wave that are in phase. If an arc is drawn along one of the crests of a circular water wave moving out from a point source, all the particles on the line will be in phase (•Fig. 22.1a). A line along a wave trough would work equally well (and would also form a circle). For a three-dimensional spherical wave, such as sound or light emitted from a point source, the wave front is a spherical surface rather than a circle.

Far from the source, the curvature of a short segment of a circular or spherical wave front is small. Such a segment may be approximated as a linear wave front (in two dimensions) or a **plane wave front** (in three dimensions), just as we take the surface of the Earth to be locally flat (Fig. 22.1b). A plane wave front can also be produced directly by a linear, elongated source such as a long bulb filament. In a uniform medium, wave fronts propagate outward from the source at a wave speed characteristic of the medium. We saw this for sound waves in Chapter 14, and the same occurs for light, although at a much faster speed. The speed of light is fastest in a vacuum: $c = 3.00 \times 10^8$ m/s.

The geometrical description of a wave in terms of wave fronts tends to neglect the fact that the wave is actually sinusoidal, like those studied in Chapter 13. This simplification is carried a step further in the concept of a ray. As illustrated in Fig. 22.1, a line drawn perpendicular to a series of wave fronts and pointing in the direction of propagation is called a **ray**. Note that a ray points in the direction of the energy flow of a wave. A plane wave front is assumed to travel in a straight line in a medium in the direction of its rays. A beam of light can be represented by a group of parallel rays or simply as a single ray (•Fig. 22.2). The representation of light as rays is adequate and convenient for describing many optical phenomena.

The use of the geometrical representations of wave fronts and rays to explain phenomena such as the reflection and refraction of light is called **geometrical optics**. However, certain other phenomena, such as the interference of light, cannot be treated in this manner and must be explained in terms of actual wave characteristics. These phenomena will be considered in Chapter 24.

22.2 Reflection

Objectives: **To (a) explain the law of reflection and (b) distinguish between regular (specular) and irregular (diffuse) reflections.**

The reflection of light is an optical phenomenon of enormous importance—if light were not reflected to our eyes by objects around us, we wouldn't see them at all. **Reflection** involves the absorption and re-emission of light by means of complex electromagnetic vibrations in the atoms of the reflecting medium. However, the phenomenon is easily described using rays.

A light ray incident on a surface is described by an **angle of incidence** (θ_i). This angle is measured relative to a line perpendicular, or normal, to the reflecting surface, a line commonly referred to as a *normal* (•Fig. 22.3). Similarly, the reflected ray is described by an **angle of reflection** (θ_r). The relationship between these angles is given by the **law of reflection**:

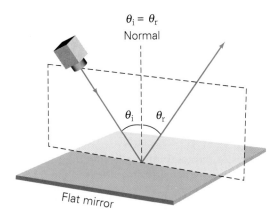

$\theta_i = \theta_r$
Normal

θ_i θ_r

Flat mirror

•**FIGURE 22.3 The law of reflection**
According to the law of reflection, the angle of incidence (θ_i) is equal to the angle of reflection (θ_r). Note that the angles are measured relative to a normal (a line perpendicular to the reflecting surface). The normal and the incident and reflected rays lie in the same plane.

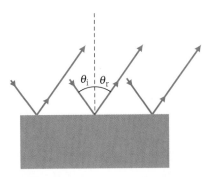

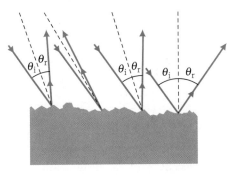

(a) Regular, or specular, reflection

(b) Irregular, or diffuse, reflection

•**FIGURE 22.4 Regular (specular) reflection and irregular (diffuse) reflection (a)** When a light beam is reflected from a smooth surface and the reflected rays are parallel, the reflection is said to be regular or specular. **(b)** Reflected rays from a relatively rough surface, such as this page, are not parallel; the reflection is said to be irregular or diffuse. (Note that the law of reflection still applies locally to each individual ray.)

The angle of incidence is equal to the angle of reflection:

$$\theta_i = \theta_r \qquad (22.1)$$

Another attribute of reflection is that the incident ray, the reflected ray, and the normal all lie in the same plane, which is sometimes called the plane of incidence.

When the reflecting surface is smooth, the reflected rays of a beam of light are parallel (•Fig. 22.4a). This is called **regular**, or **specular**, **reflection**. Reflection from a flat mirror is regular. If the reflecting surface is rough, however, the light rays are reflected in nonparallel directions because of the irregular nature of the surface (Fig. 22.4b). This is termed **irregular**, or **diffuse**, **reflection**. The reflection of light from this page is an example of diffuse reflection.

Note in Fig. 22.4 that the law of reflection still applies locally to both regular and diffuse reflection. However, the type of reflection involved determines whether we see images from a reflecting surface. In regular reflection, the reflected, parallel rays produce an image (•Fig. 22.5). Diffuse reflection does not produce an image because the light is reflected in various directions.

Experience with friction and direct investigations show that all surfaces are rough on a microscopic scale. What then determines whether reflection is specular or diffuse? In general, if the dimensions of the surface irregularities are greater than the wavelength of the light, the reflection is diffuse. Therefore, to make a good mirror, glass (with a metal coating) or metal must be polished until the surface irregularities are about the same size as the wavelength of light. Recall from Chapter 20 that the wavelength of visible light is on the order of 10^{-7} m.

It is because of diffuse reflection that we are able to see illuminated objects, such as the Moon. If the Moon's spherical surface were smooth, only the reflected sunlight from a small region would come to an observer on Earth, and only that small

•**FIGURE 22.5 Regular reflection** Regular, or specular, reflection from a smooth water surface produces an almost perfect mirror image of salt mounds at this Australian salt mine.

illuminated area would be seen. Also, you can see the beam of light from a flash-light or spotlight because of diffuse reflection from dust and particles in the air.

All That Glitters: Reflecting on Reflection

When you see the Sun or Moon over a lake or the ocean, you often observe a long swath of light that is sometimes called a "glitter path" (•Fig. 22.6). This reflection pattern is seen when (a) the surface of the water is disturbed by waves or ripples, (b) the reflection from the water surface is diffuse reflection, (c) the water surface is particularly smooth, as in Fig. 22.5.

Reasoning and Answer. The fact that you see patches of light in the glitter path indicates that there is regular, or specular, reflection in these areas, so (b) is not correct. However, if the water had a smooth mirror surface, we would see the Sun's or Moon's reflection exactly as we observe it directly in the sky, so (c) is also ruled out. The answer must be (a).

However, we should be able to justify this answer by explaining why it is correct. We realize that because the water surface has waves, only certain segments of the surface are oriented so as to reflect the Sun's or Moon's image toward us at any instant. In effect, we have specular reflection, not from one big mirror surface, but from a series of little mirrors, like the facets of a disco ball, with the reflecting facets changing almost randomly from moment to moment as the water surface undulates.

Follow-up Exercise. When following an 18-wheel truck, you may see a sign on the back stating, "If you can't see my mirror, I can't see you." What does this mean?

•**FIGURE 22.6 A glitter path**
See Conceptual Example 22.1.

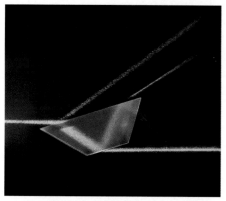

•**FIGURE 22.7 Reflection and refraction** A beam of light is incident on a trapezoidal prism from the left. Part of the beam is reflected, and part is refracted. The refracted beam is partially reflected and partially refracted at the bottom glass–air surface. (What happens to the reflected portion?)

22.3 Refraction

Objectives: To (a) explain refraction in terms of Snell's law and the index of refraction, and (b) give examples of refractive phenomena.

Refraction refers to the change in direction of a wave at a boundary where the wave passes from one medium into another. In general, when a wave is incident on a boundary between media, some of its energy is reflected and some is transmitted. For example, when light traveling in air is incident on a transparent material such as glass (•Fig. 22.7), it is partially reflected and partially transmitted. But the direction in which the transmitted light is propagated is different from the direction of the incident light, so the light is said to have been refracted, or bent.

We can analyze this phenomenon conveniently by using a geometrical method developed by the Dutch physicist Christian Huygens (1629–1696). **Huygens' principle** for wave propagation states the following:

> Every point on an advancing wave front can be considered to be a source of secondary waves, or *wavelets*, and the line or surface tangent to all these wavelets defines a new position of the wave front.

Huygens' principle is applied to incident and transmitted wave fronts at a media boundary as shown in •Fig. 22.8a. The wave speeds are different in the two media. In this case, $v_1 > v_2$. (The speed of light varies in different media and, in general, is less in denser media. Intuitively, you might expect the passage of light to take longer through a medium with more atoms per volume. And, in fact, the speed of light in water is about 75% of that in air or a vacuum.)

The distances the wave fronts travel in a time t are $v_1 t$ in medium 1 and $v_2 t$ in medium 2 (Fig. 22.8b). As a result of the smaller wave speed in the second medium, the direction of the transmitted wave front is different from that of the incident wave front. The particles of the second medium are driven or set in motion by the incident wave disturbance, and thus the waves' frequency is the same in both media. However, the wavelengths are different because of the different wave speeds ($v = \lambda f$).

The change in the direction of wave propagation is described by the **angle of refraction**. In Fig. 22.8b, the angle of incidence is θ_1 and the angle of refraction is

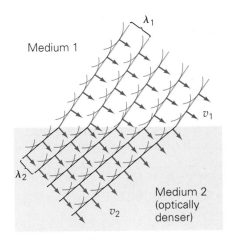
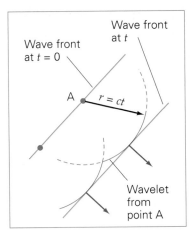
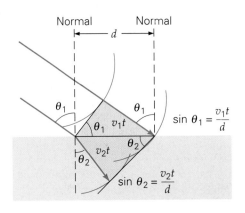

(a)

(b)

•**FIGURE 22.8 Huygens' principle**
(a) Each point on a wave front is considered to be a wavelet source. The line or surface tangent to these wavelets defines a new position of the wave front. When a wave enters an optically denser medium, its speed is less and the wave fronts (or rays) are bent, or refracted. **(b)** The geometry for derivation of Snell's law.

θ_2. From the geometry of two parallel rays, where d is the distance between the normals to the boundary at the points where the rays are incident,

$$\sin \theta_1 = \frac{v_1 t}{d} \quad \text{and} \quad \sin \theta_2 = \frac{v_2 t}{d}$$

Combining these two equations in ratio form gives

$$\frac{\sin \theta_1}{\sin \theta_2} = \frac{v_1}{v_2} \qquad \textit{Snell's law} \qquad (22.2)$$

This expression is known as **Snell's law**.

This law was named for the Dutch physicist Willebord Snell (1591–1626), who discovered it.

Thus, light is refracted when passing from one medium into another because the speed of light is different in the two media (see the Insight on p. 692). The speed of light is greatest in a vacuum, and it is convenient to compare the speed of light in other media to this constant value (c). This is done by defining a ratio called the **index of refraction** (n):

$$n = \frac{c}{v} = \frac{\text{(speed of light in a vacuum)}}{\text{(speed of light in a medium)}} \qquad (22.3)$$

Index of refraction: a ratio of speeds

As a ratio of speeds, the index of refraction is a unitless quantity. The indices of refraction of several substances are given in Table 22.1. Note that these values are for a specific wavelength of light. This is specified because v, and consequently n, are slightly different for different wavelengths ($n = c/v = c/\lambda_m f$, where λ_m is the wavelength of light in a particular medium). The values of n given in Table 22.1 will be used in examples and problems in this chapter for all wavelengths of light in the visible region, unless noted otherwise.

Remember that the frequency of light does not change when it enters another medium, but the wavelength of light in a material differs from the wavelength of that light in a vacuum, as can be easily shown:

$$n = \frac{c}{v} = \frac{\lambda f}{\lambda_m f}$$

or

$$n = \frac{\lambda}{\lambda_m} \qquad (22.4)$$

Note: When light is refracted,
• its speed and wavelength are changed;
• its frequency remains unchanged.

The wavelength of light in the medium is then $\lambda_m = \lambda/n$. Note that n is always greater than 1 because the speed of light in a vacuum is greater than the speed of light in any material ($c > v$). Therefore, $\lambda_m < \lambda$.

Understanding Refraction: A Marching Analogy

The refraction of light is not as easy to understand or visualize as reflection. We say that light is bent because waves have different speeds in different media. Intuitively, we might expect light to slow down when it enters a denser medium. The transmission of light through a transparent medium involves complex atomic absorption and emission processes, but it makes sense to suppose that these processes would take longer in a denser medium. This is indeed the case. The speed of light in water is 75% of that in air or in a vacuum, and in glass it is about 67% or less (depending on the type of glass). What is difficult to visualize is why light is bent or changes direction because of a change in speed.

To give some insight into this phenomenon, let's consider an analogy of a band marching across a field (Fig. 1). Part of the field is wet and muddy, and the marching column enters this region obliquely (at an angle of incidence). As the marchers in a row enter the wet, slippery region, they keep marching with the same cadence (frequency). However, slipping in the mud, the stride (wavelength) of the marchers is shorter, so they are slowed down.

The band members in the far end of the same row are still on dry ground and continue on with their original stride. The effect of the change in speed is a change in direction when the band enters the second medium. (A similar change in direction produced by changes in marching speeds is seen when a marching column turns a corner. The marchers nearest the corner deliberately shorten their stride and slow down, al-

lowing those farther from the corner to swing around and complete their wider turn.) We might think of the marching rows as wave fronts. As in refraction, the frequency (cadence) remains the same, but the wavelength, speed, and direction change (to keep a row aligned) on entering another medium.

FIGURE 1   Marching analogy for refraction   On obliquely entering a muddy field, a marching row changes direction slightly, analogous to the refraction of a wave front.

TABLE 22.1   Indices of Refraction (at $\lambda = 590$ nm)*

Substance	n
Air	1.00029
Water	1.33
Ethyl alcohol	1.36
Fused quartz	1.46
Glycerine	1.47
Polystyrene	1.49
Oil (typical value)	1.50
Glass (by type)†	1.45–1.70
crown	1.52
flint	1.66
Zircon	1.92
Diamond	2.42

*One nanometer (nm) is 10^{-9} m.

†Crown glass is a soda–lime silicate glass; flint glass is a lead–alkali silicate glass. Flint glass is more dispersive than crown glass (Section 22.5).

Example 22.2 ■ The Speed of Light in Water: Index of Refraction

What is the speed of light in water?

Thinking It Through.   If we know the index of refraction of a medium, then we can obtain the speed of light in the medium from Eq. 22.3.

Solution.

Given: $n = 1.33$ (from Table 22.1)   *Find:* v (speed of light in H_2O)
        $c = 3.00 \times 10^8$ m/s (known)

Since $n = c/v$,

$$v = \frac{c}{n} = \frac{3.00 \times 10^8 \text{ m/s}}{1.33} = 2.26 \times 10^8 \text{ m/s}$$

Note that $1/n = v/c = 1/1.33 = 0.75$, so v is 75% of the speed of light in a vacuum.

Follow-up Exercise.   The speed of light in a particular liquid is 2.40×10^8 m/s. What is the index of refraction of the liquid?

Hence the index of refraction n is a measure of the speed of light in a transparent material, or technically, a measure of the *optical density* of a material. For example, the speed of light in water is less than that in air, so water is said to be optically denser than air. (Optical density in general correlates with mass density. However, in some instances, a material with a greater optical density than another has a lower mass density.) So, the greater the index of refraction of a material, the greater its optical density and the smaller the speed of light in the material.

For practical purposes, the index of refraction is measured in air rather than in a vacuum, since the speed of light in air is very close to c, and

$$n_{air} \approx \frac{c}{c} = 1$$

(From Table 22.1, $n_{air} \approx 1.00029$.)

The index of refraction of a material can be determined experimentally by using Snell's law:

$$\frac{\sin \theta_1}{\sin \theta_2} = \frac{v_1}{v_2} = \frac{c/n_1}{c/n_2} = \frac{n_2}{n_1}$$

or

$$n_1 \sin \theta_1 = n_2 \sin \theta_2 \qquad \begin{array}{l} \textit{Snell's law} \\ \textit{(another form)} \end{array} \qquad (22.5)$$

where n_1 and n_2 are the indices of refraction for the first and second media, respectively.

Example 22.3 ■ Angle of Refraction: Snell's Law

Light in air is incident on a piece of crown glass at an angle of 37.0° (relative to the normal). What is the angle of refraction in the glass?

Thinking It Through. The alternate form of Snell's law (Eq. 22.5) is most practical in this case. (Why?)

Solution. Listing the given quantities, we have

Given: $\theta_1 = 37.0°$ *Find:* θ_2 (angle of refraction)
        $n_1 = 1.00$ (air)
        $n_2 = 1.52$ (crown glass, Table 22.1)

To find the angle of refraction, we use

$$\sin \theta_2 = \frac{n_1 \sin \theta_1}{n_2} = \frac{(1.00)(\sin 37.0°)}{1.52} = 0.396$$

and

$$\theta_2 = \sin^{-1}(0.396) = 23.3°$$

Follow-up Exercise. The Example 22.2 Follow-up Exercise states that the speed of light in a liquid is 2.40×10^8 m/s. This value would be difficult to measure directly. Instead, it was found experimentally that a beam of light entering the liquid from air at an angle of incidence of 37.0° had an angle of refraction of 28.8° in the liquid. What is the speed of light in the liquid?

Note that Eq. 22.5 is a very practical form of Snell's law. If the first medium is air, then $n_1 \approx 1$ and $n_2 \approx \sin \theta_1 / \sin \theta_2$. Thus, only the angles of incidence and refraction need to be measured to determine the index of refraction of a material experimentally. If the index of refraction of a material is known, that value can be used in the practical form of Snell's law to find the angle of refraction for a given angle of incidence. Also note that the sine of the angle of refraction is inversely proportional to the index of refraction, $\sin \theta_2 \approx \sin \theta_1 / n_2$. Hence, for a given angle of incidence, the greater the index of refraction, the smaller $\sin \theta_2$ and the smaller the angle of refraction θ_2.

More generally, the following relationships can be easily deduced from Eq. 22.5:

- If the second medium is more optically dense than the first medium ($n_2 > n_1$), the refracted ray is bent *toward* the normal ($\theta_2 < \theta_1$), as illustrated in ●Fig. 22.9a.
- If the second medium is less optically dense than the first medium ($n_2 < n_1$), the refracted ray is bent *away from* the normal ($\theta_2 > \theta_2$), as illustrated in Fig. 22.9b.

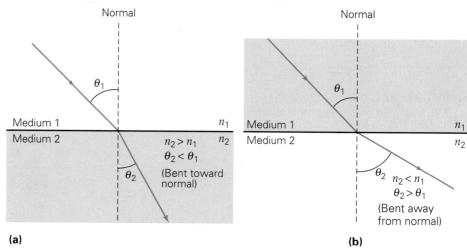

Normal

θ_1

Medium 1 n_1
Medium 2 $n_2 > n_1$ n_2
            $\theta_2 < \theta_1$
θ_2 (Bent toward
             normal)

(a)

Normal

θ_1

Medium 1 n_1
Medium 2 n_2

θ_2 $n_2 < n_1$
             $\theta_2 > \theta_1$
             (Bent away
             from normal)

(b)

•**FIGURE 22.9   Index of refraction and ray deviation   (a)** When the second medium is more optically dense than the first ($n_2 > n_1$), the refracted ray is bent toward the normal, as in the case of light entering water from air. **(b)** When the second medium is less optically dense than the first ($n_2 < n_1$), the refracted ray is bent away from the normal. [This is the case if the ray in part (a) is traced in reverse, going from medium 2 to medium 1.]

Example 22.4 ■ A Glass Table-top: More about Refraction

A beam of light traveling in air strikes the plate-glass top of a coffee table at an angle of incidence of 45° (•Fig. 22.10). The glass has an index of refraction of 1.5. (a) What is the angle of refraction for the light transmitted into the glass? (b) If the glass plate is 2.0 cm thick, what is the lateral distance between the point where the ray enters the glass and the point where it emerges (horizontal distance the light ray travels through the glass—x in the figure)? (c) Prove that the emergent beam is parallel to the incident beam, that is, that $\theta_4 = \theta_1$.

Thinking It Through: This Example involves (a) an application of Snell's law, (b) a bit of geometry, and (c) a double dose of Snell's law.

Solution.   Listing the data, we have

Given: $\theta_1 = 45°$ *Find:* (a) θ_2 (angle of refraction)
          $n_1 = 1.0$ (air)   (b) x (lateral distance)
          $n_2 = 1.5$ (c) Show that $\theta_4 = \theta_1$
          $y = 2.0$ cm

(a)   Using the alternate form of Snell's law, Eq. 22.5, with $n_1 = 1.0$ gives

$$\sin \theta_2 = \frac{\sin \theta_1}{n_2} = \frac{\sin 45°}{1.5} = \frac{0.707}{1.5} = 0.47$$

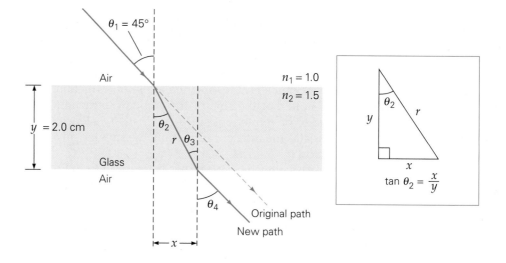

•**FIGURE 22.10   Double refraction**   The refracted ray travels laterally (sideways) a distance x in the glass, and the emergent ray is parallel to the original ray. See Example 22.4.

Thus,

$$\theta_2 = \sin^{-1}(0.47) = 28°$$

Note that the beam is bent toward the normal.

(b) You can see from the inset in Fig. 22.10 that the lateral distance (x) is related to θ_2 by $\tan \theta_2 = x/y$, so

$$x = y \tan \theta_2 = (2.0 \text{ cm})(\tan 28°) = (2.0 \text{ cm})(0.53) = 1.1 \text{ cm}$$

(c) If $\theta_1 = \theta_4$, then the emergent ray is parallel to the incident ray. Applying Snell's law to the beam at both surfaces gives

$$n_1 \sin \theta_1 = n_2 \sin \theta_2$$

and

$$n_2 \sin \theta_3 = n_1 \sin \theta_4$$

From the figure, we see that $\theta_2 = \theta_3$. Therefore,

$$n_1 \sin \theta_1 = n_1 \sin \theta_4$$

or

$$\theta_1 = \theta_4$$

Thus, the emergent beam is parallel to the incident beam but displaced laterally a distance x from the normal at the point of entry.

Follow-up Exercise. If the glass in this Example had $n = 1.7$, would the lateral displacement be the same, larger, or smaller?

Conceptual Example 22.5

Lighting the Fish Tank: Refraction and Wavelength

A beam of monochromatic (single-frequency) light is directed at the side of a fish tank, so the light passes from air to glass and from glass to water. The wavelength of the light in the water (a) is the same as that in air, (b) is independent of the wavelength in the glass, (c) does not change at the glass–water interface, (d) is shorter than that in the glass.

Reasoning and Answer. First, we need to have an idea of the relative magnitudes of the optical densities of the materials or their indices of refraction. As can be seen from Table 22.1, $n_{\text{glass}} > n_{\text{water}} > n_{\text{air}}$.

Then, looking at the possible answers, we can eliminate (a) since the wavelength of light in a solid or liquid is less than that in vacuum or in air, as shown previously (from Eq. 22.4, $\lambda_m = \lambda/n$). Similarly, since the indices of refraction of glass and water are different, the wavelength of the light changes at this interface, so (c) is not the answer. Noting that $n_{\text{glass}} > n_{\text{water}}$, we know that the wavelength is shorter in the more optically dense medium (glass), so (d) is eliminated.

This leaves only (b), which is the correct answer, but it is important to understand why the wavelength of light in water is independent of that in glass. Let's look at the interface of two general material media with indices n_2 and n_1. Forming a ratio,

$$\frac{n_2}{n_1} = \frac{c/v_2}{c/v_1} = \frac{v_1}{v_2} = \frac{\lambda_1 f}{\lambda_2 f} = \frac{\lambda_1}{\lambda_2} \quad \text{or} \quad \lambda_2 = \left(\frac{n_1}{n_2}\right)\lambda_1$$

Then, applying the wavelength-index conditions at the interfaces,

$$\text{(air–glass)} \quad \lambda_{\text{glass}} = \frac{\lambda_{\text{air}}}{n_{\text{glass}}}$$

$$\text{(glass–water)} \quad \lambda_{\text{water}} = \left(\frac{n_{\text{glass}}}{n_{\text{water}}}\right)\lambda_{\text{glass}}$$

Combining, we get

$$\lambda_{\text{water}} = \left(\frac{n_{\text{glass}}}{n_{\text{water}}}\right)\frac{\lambda_{\text{air}}}{n_{\text{glass}}} = \frac{\lambda_{\text{air}}}{n_{\text{water}}}$$

We can see that the wavelength of the light in water is independent of that in glass, since the latter does not appear in the final equation. This is to be expected since the speed of light in a transparent medium, and thus the wavelength for light of a constant frequency, is characteristic of the medium itself and not of effects at interfaces.

Follow-up Exercise. A light source with a particular frequency in air is submersed in water in a fish tank. The beam travels in the water, through double plate glass at the side of the tank (each glass plate having a different n), and into air. In general, what are (a) the frequency and (b) the wavelength of the light when it emerges in the outside air?

Refraction is very common in everyday life and explains many things we observe. Let's look at refraction in action.

An apparently "wet" road and hot air rising: These are common observations on a highway on a hot summer day. The refraction of light by the warm air near the road surface gives rise to the observed "wet" spot, which is really a mirage (•Fig. 22.11a). The term *mirage* generally brings to mind a thirsty person in the desert "seeing" a pool of water that really isn't there. This is not the mind playing tricks but an optical illusion. Such water images are images of the sky produced by light that is refracted toward the observer (Fig. 22.11b).

We also "see" hot air rising from a hot road surface. Although we cannot really see air, gases have a property that makes it possible for us to sense them visually when they are heated. The index of refraction of a gas is proportional to its density, which in turn is inversely proportional to its temperature. Thus, you can perceive regions of air that have different densities (temperatures) because as the light passes through the rising air, the light rays are refracted. This property can give rise to a distorted image, which seems to shimmer as the air density fluctuates from moment to moment (Fig. 22.11c). Ordinarily, the air near the road's surface has a relatively uniform density and you do not see these effects.

(a)

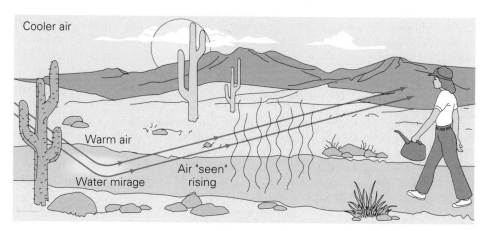

(b)

•**FIGURE 22.11 Refraction in action** (a) A "wet spot" mirage. The road is actually dry; what looks like water is a refracted image of part of the sky. (b) The mirage is produced when light from the sky is refracted by warm air near the road surface. (c) Such refraction also enables us to perceive heated air as it rises. Although we do not see the air itself, objects viewed through the turbulent updrafts are distorted by refraction and seem to shimmer.

(a)

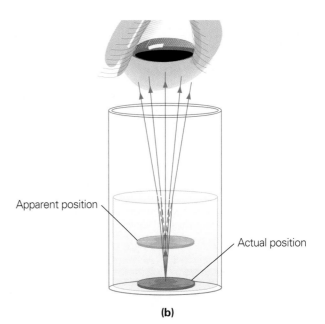

(b)

•**FIGURE 22.12   Refractive effects**
(a) The light is bent, and because we tend to think of light as traveling in straight lines, the fish is not where we think it is.
(b) Because of refraction, the coin appears to be closer than it actually is (normal incidence).

Twinkle, twinkle, little star: Refraction is also responsible for the twinkling of stars. At night, the air through which starlight travels to our eyes continually varies in density because of temperature variations and turbulence. The resulting refraction causes the stars' images to shimmer and fluctuate in brightness—the familiar "twinkling." (A similar, more local, effect is observed in the shimmering distortions of writing or pictures on the bottom of a swimming pool.)

Not where it should be: You may have experienced a refractive effect while trying to reach for something under water, such as a fish (•Fig. 22.12a). We are used to light traveling in straight lines from objects to our eyes, but the light reaching our eyes from a submerged object is bent at the air–water interface. (Note in the figure that the ray is bent away from the normal.) As a result, the object appears to be closer to the surface than it actually is, and therefore we tend to miss the object when reaching for it. For the same reason, a coin in a glass of water will appear closer than it really is (Fig. 22.12b) and the legs of a person standing in water seem shorter than their actual length.

Some extra sunlight: Atmospheric refraction accounts for the fact that the Sun (or Moon) can be seen for a short time before it actually rises above, or after it actually sets below, the horizon. The denser air near the Earth refracts the light over the horizon. A couple of other refractive solar effects are discussed in the Insight on p. 698.

22.4   Total Internal Reflection and Fiber Optics

Objectives: To (a) describe internal reflection and (b) give examples of fiber-optic applications.

An interesting phenomenon occurs when light traveling in one medium is incident on the boundary with another medium that is less optically dense, such as when light goes *from* water *into* air. As you know, in such a case a transmitted beam will be bent away from the normal, and Snell's law tells you that the greater the angle of incidence, the greater the angle of refraction. That is, the more the angle of incidence increases, the farther the refracted ray diverges from the normal.

However, there is a limit. For a certain angle of incidence called the **critical angle** (θ_c), the angle of refraction is 90°, and the refracted ray is directed along the boundary. But what happens if the angle of incidence is even larger? If the angle of incidence is greater than the critical angle ($\theta_1 > \theta_c$), the light isn't transmitted

A Flattened Sun? An Anti-Sun??

Atmospheric refraction lengthens the day, so to speak, by allowing us to see the Sun just before it actually rises above the horizon and just after it actually sets below the horizon. Note a couple of other refraction effects when the Sun sets (or rises).

You may have noticed that the Sun on the horizon sometimes appears to be flattened, with its horizontal dimension greater than its vertical dimension (Fig. 1a). This effect is the result of temperature and density variations in the denser air along the horizon. These variations occur predominantly vertically, so light from the top and bottom portions of the Sun is refracted differently as that light passes through different atmospheric densities with different indices of refraction.

But what is that emerging red image below the Sun in Fig. 1b? It continues to grow and meet the Sun (Fig. 1c–e). It is known as an "anti-Sun." This rarely observed phenomenon also results from refraction. The different refractive layers of air give rise to a lensing effect. (As we shall see in Chapter 23, refraction is the principle of lenses, as used in eyeglasses.) Essentially, what is seen on the horizon is an image of the lower portion of the Sun. As the Sun descends (sets), this refracted image rises to meet it, giving the appearance of two Suns merging.

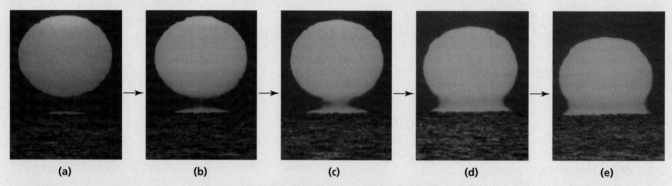

(a)   (b)   (c)   (d)   (e)

FIGURE 1   What's going on?   (a) The Sun on the horizon commonly appears flattened as a result of atmospheric refraction. **(b)** But what is emerging from below the horizon—another Sun? Many stars are binary (double) stars orbiting around each other; a planet in such a system would have two suns. That's not the case here; these photos were taken at Greenport, New York. (© Edward Pascuzzi) **(c–e)** This so-called anti-Sun and the Sun appear to merge.

but is internally reflected (•Fig. 22.13). This condition is called **total internal reflection.** The reflection process can be almost 100% efficient. Because of internal reflection, prisms can be used as mirrors (•Fig. 22.14).

The critical angle for total internal reflection can be obtained from Snell's law. If $\theta_1 = \theta_c$ in the optically denser medium, $\theta_2 = 90°$ and

(a)   (b)

•FIGURE 22.13   Internal reflection   (a) When light enters a less optically dense medium, it is refracted away from the normal. At a critical angle (θ_c), the light is refracted along the interface (common boundary) of the media. At an angle greater than the critical angle ($\theta_1 > \theta_c$), there is total internal reflection. Can you estimate the critical angle in the photograph? **(b)** Total internal reflection occurs for incident angles outside a cone with an apex angle of $2\theta_c$.

$$\frac{\sin \theta_1}{\sin \theta_2} = \frac{\sin \theta_c}{\sin 90°} = \frac{n_2}{n_1}$$

Since $\sin 90° = 1$,

$$\sin \theta_c = \frac{n_2}{n_1} \qquad where \; n_1 > n_2 \qquad (22.6)$$

If the second medium is air, $n_2 \approx 1$ and the critical angle for total internal reflection at the boundary from a medium into air is given by

$$\sin \theta_c = \frac{1}{n} \qquad \begin{array}{l} where \; n \; is \; the \; index \; of \\ refraction \; of \; the \; medium \\ (medium–air \; interface \; only) \end{array} \qquad (22.7)$$

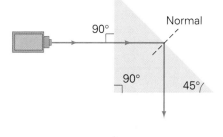

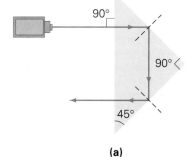

(a)

Example 22.6 ■ A View from the Pool: Critical Angle

(a) What is the critical angle for light traveling in water and incident on a water–air boundary? (b) If a diver submerged in a pool looked up at the water surface at an angle of $\theta < \theta_c$, what would she see? (Neglect any thermal or motional effects.)

Thinking It Through. (a) The critical angle is given by Eq. 22.7. (b) As shown in Fig. 22.13b, θ_c forms a cone of vision for viewing from above the water.

Solution.

Given: $n = 1.33$ (from Table 22.1)   *Find:* (a) θ_c (critical angle)
                                             (b) View for $\theta < \theta_c$

(a) The critical angle is

$$\theta_c = \sin^{-1}\left(\frac{1}{n}\right) = \sin^{-1}\left(\frac{1}{1.33}\right) = 48.8°$$

(b) Using Fig. 22.13, mentally trace the rays in reverse for light coming from all angles in three dimensions. Light coming from the above-water 180° panorama could be viewed in a cone with a half-angle of 48.8°. As a result, objects would appear distorted. An underwater panoramic view is seen in ●Fig. 22.15.

Follow-up Exercise. What would the diver see when looking up at the water surface at an angle of $\theta > \theta_c$?

Internal reflections enhance the brilliance of cut diamonds. The critical angle for a diamond–air surface is

$$\theta_c = \sin^{-1}\left(\frac{1}{n}\right) = \sin^{-1}\left(\frac{1}{2.42}\right) = 24.4°$$

A so-called brilliant-cut diamond has many facets, or faces (58 in all—33 on the upper face and 25 on the lower). Light entering the main and upper facets above

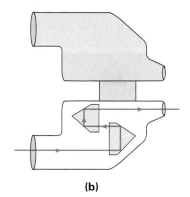

(b)

●**FIGURE 22.14 Internal reflection in a prism** (a) Because the critical angle of glass is less than 45°, prisms with 45° and 90° angles can be used to reflect light through 90° and 180°. (b) Internal reflection of light by prisms in binoculars make this instrument much shorter than a telescope.

●**FIGURE 22.15 Panoramic and distorted** An underwater view of the surface of a swimming pool in Hawaii.

•**FIGURE 22.16 Diamond brilliance** (a) Internal reflection gives rise to a diamond's brilliance. (b) The brilliant-cut diamond has 58 facets, or faces. Light entering the upper facets is internally reflected and reemerges through these facets, giving the diamond a sparkling brilliance. The depth proportions are critical. If a stone is too shallow or too deep, light will be lost through the lower facets.

(a)

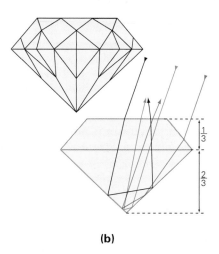

(b)

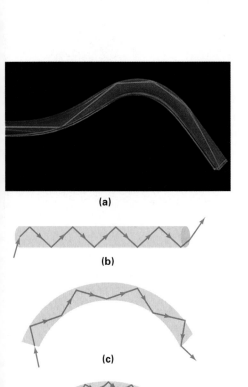

(a)

(b)

(c)

(d)

•**FIGURE 22.17 Light pipes** (a) Total internal reflection in an optical fiber. (b) When light is incident on the end of a cylindrical form of transparent material such that the internal angle of incidence is greater than the critical angle of the material, the light is reflected down the length of the light pipe. (c) Light is also transmitted along curved light pipes by internal reflection. (d) As the diameter of the rod or fiber becomes smaller, the number of reflections per unit length increases.

the critical angle is internally reflected in the diamond. The light then emerges from the upper facets, giving rise to the diamond's brilliance (•Fig. 22.16).

Fiber Optics

When a fountain is illuminated from below, the light is transmitted along the curved streams of water. This phenomenon was first demonstrated in 1870 by the British scientist John Tyndall (1820–1893), who showed light was "conducted" along the curved path of a stream of water flowing from a hole in the side of a container. This phenomenon is observed because light is internally reflected along the water stream.

Internal reflection forms the basis of **fiber optics**, a fascinating field centered on the use of transparent fibers to transmit light. Multiple internal reflections make it possible to "pipe" light along a transparent rod, even if the rod is curved (•Fig. 22.17). Note from the figure that the smaller the diameter of the light pipe, the more internal reflections it has. In a small fiber, there can be as many as several hundred internal reflections per centimeter.

Internal reflection is an exceptionally efficient process. Optical fibers can be used to transmit light over very long distances with losses of only about 25% per kilometer. These losses are primarily due to fiber impurities, which scatter the light. Transparent materials have different degrees of transmission. Fibers are made of certain plastics and special glass for maximum transmission, and the greatest efficiency is achieved with infrared radiation, for which there is less scattering.

The greater efficiency of multiple internal reflections over multiple external reflections can be illustrated by considering a good reflecting surface, such as a plane mirror, which at best has a reflectivity of about 95%. Suppose two *plane mirrors* (common flat mirrors) are placed facing and nearly parallel to each other. In this manner a beam from an object between the mirrors and near one end will be reflected back and forth, in a manner similar to the reflections in the light pipe in Fig. 22.17b.

After each reflection, the beam intensity would be 95% of the incident beam from the preceding reflection. The intensity I of the reflected beam after n reflections is given by

$$I = 0.95^n I_o$$

where I_o is the initial beam intensity before the first reflection. After 14 reflections,

$$I = 0.95^{14} I_o = 0.49I_o$$

Thus, after only 14 reflections, the intensity is reduced to less than half (49%). For 100 reflections, $I = 0.006I_o$, and there is only 0.6% of the initial intensity. Compare this to about 75% in optical fibers over a kilometer in length with *thousands and thousands* of reflections. An illustration of multiple reflections in nearly parallel mirrors is shown in •Fig. 22.18.

•**FIGURE 22.18   Multiple reflections**   Multiple reflections from nearly parallel mirrors.

(a)

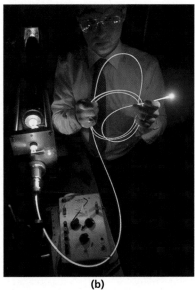

(b)

•**FIGURE 22.19   Fiber-optic bundle   (a)** Hundreds or even thousands of extremely thin fibers **(b)** are grouped together to make an optical fiber colored blue by a laser.

Fibers whose diameters are about 10 μm (10^{-5} m) are grouped together in flexible bundles that are 4 to 10 mm in diameter and up to several meters in length, depending on the application (•Fig. 22.19). A fiber bundle with a cross-sectional area of 1 cm^2 can contain as many as 50 000 individual fibers. To prevent light from being transmitted between fibers in contact with each other, they are coated with a thin film. (See the Insight on p. 703.)

22.5   Dispersion

Objective: **To explain dispersion and some of its effects.**

Light of a single frequency is called *monochromatic light* (from the Greek *mono*, meaning "one," and *chroma*, meaning "color"). Visible light that contains all the component frequencies, or colors, is termed *white light*. Sunlight is white light. When a beam of white light passes through a glass prism as shown in •Fig. 22.20a, it is spread out, or dispersed, into a spectrum of colors. This phenomenon led Newton to believe

(a)

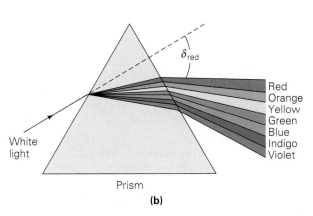

(b)

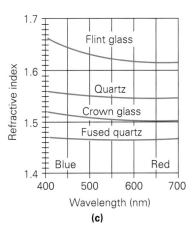

(c)

•**FIGURE 22.20   Dispersion   (a)** White light is dispersed into a spectrum of colors by glass prisms. **(b)** In a dispersive medium, the index of refraction varies slightly with wavelength. Red light, longest in wavelength, has the smallest index of refraction and is refracted least. The angle between the incident beam and an emergent ray is the angle of deviation (δ) for that ray. (The angles are exaggerated for clarity.) **(c)** Variation of the index of refraction with wavelength for some common transparent media.

that sunlight is a mixture of component colors. When the beam enters the prism, the component colors, corresponding to different wavelengths of light, must be refracted at slightly different angles so they spread out into a spectrum (Fig. 22.20b).

The formation of a spectrum indicates that the index of refraction of glass must be slightly different for different wavelengths, and this is true of many transparent media (Fig. 22.20c). The reason has to do with the speed of light. In a vacuum, the speed of light c is the same for all wavelengths, but in a dispersive medium the speed of light is slightly different for different wavelengths. Since the index of refraction n of a medium is a function of the speed of light in that medium ($n = c/v = c/\lambda f$), the index of refraction is then different for different wavelengths. It follows from Snell's law that light of different wavelengths is refracted at different angles.

We can summarize by saying that in a transparent material with different indices of refraction for different wavelengths of light, in refracting there is a separation of the wavelengths, and the material is said to exhibit **dispersion**. Dispersion varies with different media and can be neglected in many instances. Also, because the differences in the indices of refraction for different wavelengths are small, a representative value at some specified wavelength can be used for general purposes (see Table 22.1).

Example 22.7 ■ Forming a Spectrum: Dispersion

The index of refraction of a particular transparent material is 1.4503 for the red end ($\lambda = 700$ nm) of the visible spectrum and 1.4698 for the blue end ($\lambda = 400$ nm). If white light is incident on a prism of this material as in Fig. 22.20b at an angle of 45°, what is the angular dispersion of the visible spectrum inside the prism?

Thinking It Through. The angle of refraction is given by Snell's law, and we can compute the angle of refraction for each end component of the visible spectrum. The angular dispersion of the light inside the prism is the difference in these angles of refraction.

Solution.

Given: (red) $n_r = 1.4503$ for $\lambda_r = 700$ nm   *Find:* $\Delta\theta$ (angular dispersion)
          (blue) $n_b = 1.4698$ for $\lambda_b = 400$ nm
          $\theta_1 = 45°$

Using Eq. 22.1 with $n_1 = 1.00$ (air), we get

$$\sin\theta_{2_r} = \frac{\sin\theta_1}{n_{2_r}} = \frac{\sin 45°}{1.4503} = 0.487\,56 \quad \text{and} \quad \theta_{2_r} = 29.180°$$

Similarly,

$$\sin\theta_{2_b} = \frac{\sin\theta_1}{n_{2_b}} = \frac{\sin 45°}{1.4698} = 0.481\,09 \quad \text{and} \quad \theta_{2_b} = 28.757°$$

So

$$\Delta\theta = \theta_{2_r} - \theta_{2_b} = 29.180° - 28.757° = 0.423°$$

This is not much deviation, but as the light travels, it spreads out and emerges from the prism, and the dispersion becomes evident (Fig. 22.20a).

Follow-up Exercise. If green light exhibits an angular dispersion of 0.156° from the red light, what is the index of refraction for green light in the Example material?

A good example of a dispersive material is diamond, which is about five times more dispersive than glass. In addition to the brilliance resulting from internal reflections off many facets (Fig. 22.16), a cut diamond shows a play of colors, or "fire," resulting from the dispersion of the internally reflected light.

Another dramatic example of dispersion is the production of a rainbow, as discussed in the Insight on p. 704.

Fiber Optics

Fiber optics can be used for purely decorative purposes, such as lamps, but a much more important application involves the piping of light to and images from hard-to-reach places. To do this, the ends of a fiber bundle are cut and polished to form a flexible *fiberscope*. A beam of light can be transmitted along the bundle to illuminate an area, even if the bundle is bent and twisted (Fig. 1). Equally important, an image or picture can be transmitted back by a fiberscope. Light travels through one set of fibers to illuminate an object and, after reflection, travels back in another set. The overall image thus has a mosaic pattern, like a newspaper picture. The smaller the elements in the mosaic, the finer the detail. Thus, a fiberscope has a very large number of extremely fine fibers. A transmitted image can be magnified by a lens for viewing.

There are a number of interesting applications of fiber optics. You're probably aware that many telephone transmissions are accomplished by means of fiber optics. Light signals, rather than electrical signals, are transmitted through optical telephone lines. Optical fibers have lower energy losses than current-carrying wires do, particularly at higher frequencies, and can carry much more data. Also, these fibers are lighter than metal wires, have greater flexibility, and are not affected by electromagnetic disturbances (electric and magnetic fields) since they are made of materials that are electrical insulators.

Fiber optics has been widely applied in medicine. Endoscopes, instruments used to view internal portions of the human body, previously consisted of lens systems in long narrow tubes. Some contained a dozen or more lenses and produced relatively poor images. Also, because the lenses had to be aligned in certain ways, the tubes had to have rigid sections, which limited the endoscope's maneuverability. Such an endoscope could be inserted down the throat into the stomach to observe the stomach lining. However, there would be blind spots due to the curvature of the stomach and the inflexibility of the instrument.

The flexibility of fiber bundles has eliminated this problem. Lenses are used at the end of the fiber bundles to focus the light, and a prism is used to change the direction for its return. The incident light is usually transmitted by an outer layer of fiber bundles, and the image is returned through a central core of fibers. Mechanical linkages allow maneuverability. The end of a fiber endoscope can be equipped with devices to obtain specimens of the viewed tissues for biopsy (diagnostic examination) or even to perform surgical procedures (Fig. 2). For example, you may have heard of arthroscopic surgery being performed on the knees of injured athletes. The arthroscope that is now routinely used for inspecting and repairing damaged joints is simply a fiber endoscope fitted with appropriate surgical implements.

A fiber-optic cardioscope used for direct observation of heart valves typically has a fiber bundle about 4 mm in diameter and 30 cm long. Such a cardioscope passes easily to the heart through the jugular vein, which is about 15 mm in diameter. To displace the blood and provide a clear field of view for observations and photographing, a transparent balloon at the tip of the cardioscope is inflated with saline (salt water) solution.

Another application of fiber optics is the coding and decoding of information. To make a "coded" image of a classified picture, for example, the component fibers of a bundle are deliberately misaligned and randomly interwoven. As a result, a transmitted image is jumbled and unrecognizable unless it is viewed through an identically interwoven bundle that "decodes" it.

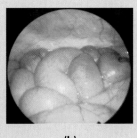

(b)

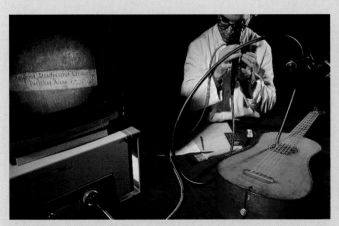

FIGURE 1   Fibroscopy   A technician uses a fiberscope to inspect the interior of a guitar. The label on the inside (image displayed on screen) reveals that it was made by Stradivarius in 1711.

(a)

FIGURE 2   Endoscopy   (a) Surgeons use a fiber-optic endoscope to perform surgery. **(b)** An endoscopic view of the intestines.

The Rainbow

FIGURE 1 Double rainbow Notice that the colors of the rainbows are reversed: In the lower, primary rainbow, the colors run vertically from blue to red, whereas in the upper, secondary rainbow, they run from red to blue.

We have all been fascinated by the beautiful array of colors of a rainbow (Fig. 1). With the optical principles learned in this chapter, we are now in a position to understand the formation of this spectacular display.

A rainbow is produced by refraction, dispersion, and internal reflection of light within water droplets. When millions of water droplets remain suspended in the air after a rainstorm, we see a multicolored arc whose colors run from violet along the lower part of the spectrum (in order of wavelength) to red along the upper. Below the arc, the light from the droplets combines to form a region of brightness. Occasionally, more than one rainbow is seen; the main, or primary, rainbow is sometimes accompanied by a fainter and higher secondary rainbow.

The light that forms the primary rainbow is reflected once inside each water droplet (Fig. 2). Being also refracted and dispersed, the light is spread out into a spectrum of col-

ors. However, because of the conditions for refraction and internal reflection in water, the angles of deviation (between incoming and outgoing rays) for violet to red light lie within a narrow range of 40°–42° (Fig. 2b). This means that you can see a rainbow only when the Sun is behind you, so the dispersed light is reflected to you through these angles.

The less frequently seen secondary rainbow is caused by a double internal reflection (Fig. 2c). This results in an arc whose vertical color sequence is the inverse of the primary rainbow's. The angles of deviation in this case lie between 50.5° for red light and 54° for violet light.

We generally see only rainbow arcs, because the formation by water droplets is cut off at the ground. If you were on a cliff, you might see a complete circular rainbow (Fig. 2c). Also, the higher the Sun is in the sky, the less of a rainbow you will be able to see from the ground. In fact, you won't see a primary rainbow if the Sun's angle above the horizon is greater than 42°. The primary rainbow can still be seen from a height, however. As an observer's elevation increases, more of the arc becomes visible. It is common to see a completely circular rainbow from an airplane. You may also have seen a circular rainbow in the spray from a garden hose.

QUESTION: Can there be a third-order, or tertiary, rainbow?

ANSWER: Yes, a third-order rainbow resulting from three internal reflections in a water droplet is possible.

In a book entitled *Opticks*, Isaac Newton wrote, "The Light which passes through a drop of Rain after two Refractions, and three or more Reflexions, is scarcely strong enough to cause a sensible rainbow." (By "sensible," he meant intense enough to be seen.) However, his friend Edmund Halley (for whom the famous comet was named) showed that the reason the tertiary rainbow is not seen is not because the reflected light lacks intensity, but rather because the tertiary arc is formed in the general direction of the Sun and cannot be seen against the brightness of the sky.

Chapter Review

Important Terms

wave front   *688*

plane wave front   *688*

ray   *688*

geometrical optics   *688*

reflection   *688*

angle of incidence   *688*

angle of reflection   *688*

law of reflection   *688*

regular (specular) reflection   *689*

irregular (diffuse) reflection   *689*

refraction   *690*

Huygen's principle   *690*

angle of refraction   *690*

Snell's law   *691*

index of refraction   *691*

critical angle   *697*

total internal reflection   *698*

fiber optics   *700*

dispersion   *702*

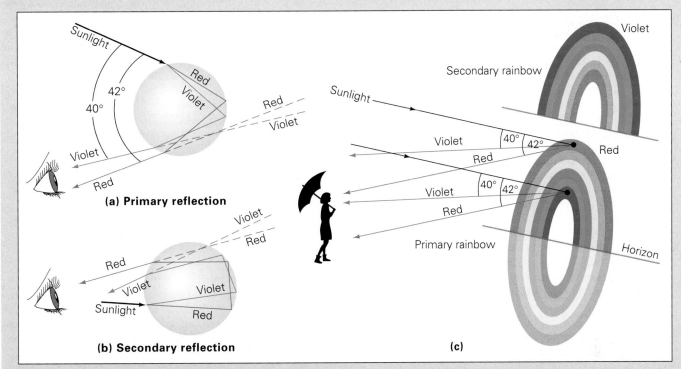

FIGURE 2   The rainbow   Rainbows are created by the refraction, dispersion, and internal reflection of sunlight.
(a) A single internal reflection gives rise to the primary rainbow. **(b)** A double internal reflection produces a secondary rainbow. **(c)** In the primary rainbow, an observer sees red light at the top of the bow—that is, from the higher droplets—because red light is deviated most. (The other color components from these droplets pass above the observer's eyes.) Similarly, violet or blue is seen from the lower droplets.

Important Concepts

- Law of reflection: The angle of incidence equals the angle of reflection (as measured from the normal to the reflecting surface).
- The index of refraction of any medium is the ratio of the speed of light in a vacuum to its speed in that medium.
- The angle of refraction is given by Snell's law. If the second medium is more optically dense, the refracted

ray is bent toward the normal; if less dense, the ray is bent away from the normal.
- If the second medium is less dense that the first, light will be totally internally reflected if the angle of incidence exceeds the critical angle.
- Light is dispersed in some media because different wavelengths have slightly different indices of refraction.

Important Equations

Law of Reflection:

$$\theta_i = \theta_r \tag{22.1}$$

Snell's Law:

$$\frac{\sin \theta_1}{\sin \theta_2} = \frac{v_1}{v_2} \quad \text{or} \quad n_1 \sin \theta_1 = n_2 \sin \theta_2 \tag{22.2, 22.5}$$

Index of Refraction:

$$n = \frac{c}{v} = \frac{\lambda}{\lambda_m} \tag{22.3, 22.4}$$

Critical Angle at Boundary between Two Materials (where $n_1 > n_2$):

$$\sin \theta_c = \frac{n_2}{n_1} \tag{22.6}$$

Critical Angle at Medium–Air Boundary: (where n is the index of refraction of the medium)

$$\sin \theta_c = \frac{1}{n} \tag{22.7}$$

Exercises*

22.1 Wave Fronts and Rays

1. A wave front is (a) always circular, (b) parallel to a ray, (c) described by a surface of equal phase, (d) none of these.

2. A ray (a) is perpendicular to the direction of energy flow, (b) is always parallel to other rays, (c) is perpendicular to a series of wave fronts, (d) illustrates the wave nature of light.

22.2 Reflection

3. For regular, or specular, reflection, (a) the angle of incidence equals the angle of reflection, (b) the rays of a reflected beam are not parallel, (c) the incident ray, the reflected ray, and the normal lie in the same plane, (d) both (a) and (c).

4. For irregular, or diffuse, reflection, (a) the angle of incidence equals the angle of reflection, (b) the rays of a reflected beam are not parallel, (c) the incident ray, the reflected ray, and the local normal lie in the same plane, (d) all of these.

5. Explain why the sunbeams are visible in •Fig. 22.21.

•**FIGURE 22.21** **Sunbeams** See Exercise 5.

6. Is the angle of reflection *always* equal to the angle of incidence? Is the angle of reflection *always* less than the angle of incidence?

7. ■ The angle of incidence of a light ray on a mirrored surface is 35°. What is the angle between the incident and reflected rays?

8. ■ A beam of light is incident on a plane mirror at an angle of 32° relative to the normal. What is the angle between the reflected rays and the surface of the mirror?

*Assume angles to be exact.

9. ■ A beam of light is incident on a plane mirror at an angle of 43° relative to the mirror surface. What is the angle between the reflected ray and the normal?

10. ■ A light ray incident on a plane mirror is at an angle of 55° relative to the mirror surface. At what angle is the reflected ray relative to the surface?

11. ■ Two people stand, 3.0 m in front of a large plane mirror and spaced 5.0 m apart, in a dark room. At what angle of incidence should one of them shine a flashlight on the mirror so that the reflected beam directly strikes the other person?

12. ■■ Two upright plane mirrors touch along one edge, where their planes make an angle of 60°. If a beam of light is directed onto one of the mirrors at an angle of incidence of 40° and is reflected onto the other mirror, what will be the angle of reflection of the beam from the second mirror?

13. ■■ Two plane mirrors are placed 50 cm apart with their mirrored surfaces parallel and facing each other. If the mirrors are 25 cm wide, at what angle should a beam of light be incident at one end of one mirror so that it just strikes the far end of the other?

14. ■■ It is difficult to drive during the night after rain because of the glare. Instead of seeing the road, you see the images of buildings, trees, and so on. What causes this glare? Why is there no such glare when the road is dry? [Hint: Think of specular and diffuse reflections.]

15. ■■ If you hold a 30-centimeter square plane mirror 45 cm from your eyes and can just see the full length of a 8.5-meter flag pole behind you, how far are you from the pole? [Hint: A diagram is helpful here.]

16. ■■■ Two plane mirrors, M_1 and M_2, are placed together as illustrated in •Fig. 22.22. (a) If the angle α between the mirrors is 70° and the angle of incidence θ_{i_1} of a light ray incident on M_1 is 35°, what is the angle of reflection θ_{r_2} from M_2? (b) If $\alpha = 115°$ and $\theta_{i_1} = 60°$, what is θ_{r_2}?

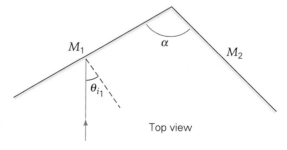

•**FIGURE 22.22** **Plane mirrors together** See Exercises 16 and 17.

17. ■■■ For the plane mirrors in Fig. 22.22, what angles α and θ_{i_1} would allow a ray to be reflected back in the direction from which it came (parallel to the incident ray)?

22.3 Refraction
and
22.4 Total Internal Reflection and Fiber Optics

18. 💿 Light refracted at the boundary of two different media (a) is bent toward the normal when $n_1 > n_2$, (b) is bent away from the normal when $n_1 > n_2$, (c) has the same angle of refraction as the angle of incidence, (d) always experiences a decrease in speed.

19. The index of refraction (a) is always greater than or equal to one, (b) is inversely proportional to the speed of light in a medium, (c) is inversely proportional to the wavelength of light in the medium, (d) all of these.

20. 💿 As light travels from one medium to another, does its wavelength change? its frequency? its speed?

21. The critical angle for total internal reflection at a medium–air boundary (a) is independent of the wavelength of the light in the medium, (b) is greater for a material with a smaller index of refraction, (c) may be greater than 90°, (d) none of these.

22. Under what conditions will total internal reflection occur?

23. Explain why the pencil in •Fig. 22.23 appears almost severed.

•**FIGURE 22.23   Refraction effect** See Exercise 23.

24. The photos in •Fig. 22.24 were taken with a camera on a tripod at a fixed angle. There is a penny in the container, but only its tip is seen initially. However, when water is added, more of the coin is seen. Why?

•**FIGURE 22.24   You barely see it, but then you do** See Exercise 24.

25. Two hunters, one with bow and arrow and the other with a laser gun, see a fish under water. They both aim directly where they see it. Which one, the arrow or the laser beam, has a better chance to hit the fish? Explain.

26. ■ The speed of light in one material is 2.07×10^8 m/s. What is the index of refraction of the material?

27. ■ Is the speed of light greater in diamond or in zircon? Express the difference as a percentage.

28. ■ 💿 A beam of light enters water at an angle of 60° relative to the normal of the water surface. Find the angle of refraction.

29. ■ 💿 Light passes from air into water. If the angle of refraction is 20°, what is the angle of incidence?

30. ■ 💿 A beam of light traveling in air is incident on a transparent plastic material at an angle of incidence of 50°. The angle of refraction is 35°. What is the index of refraction of the plastic?

31. ■ 💿 (a) What is the critical angle of diamond in air? (b) For total internal reflection to occur, should the light be directed from air to diamond or from diamond to air?

32. ■ 💿 The critical angle for a certain type of glass in air is 41.8°. What is the index of refraction of the glass?

33. ■■ A beam of light in air is incident on the surface of a slab of fused quartz. Part of the beam is transmitted into the quartz at an angle of refraction of 30° relative to a normal to the surface, and part is reflected. What is the angle of reflection?

34. ■■ 💿 A beam of light is incident on a flat piece of polystyrene at an angle of 55° relative to a surface normal. What angle does the refracted ray make with the plane of the surface?

35. ■■ Monochromatic blue light that has a frequency of 6.5×10^{14} Hz enters a piece of flint glass. What are the frequency and wavelength of the light in the glass?

36. ■■ 💿 Light passes from material A, which has an index of refraction of $\frac{4}{3}$, into material B, which has an index of refraction of $\frac{5}{4}$. Find the ratio of the speed of light in material B to the speed of light in material A.

37. ■■ In Exercise 36, what is the ratio of the light's wavelength in material B to that in material A?

38. ■■ A He–Ne (helium–neon) laser beam ($\lambda = 632.8$ nm in air) is directed through ethyl alcohol. What are the wavelength and frequency of the light in the alcohol?

39. ■■ An object immersed in water appears closer to the surface than it actually is. Using •Fig. 22.25, show that the apparent depth d' for small angles of refraction is

$d' = d/n$, where n is the index of refraction of the water. [Hint: Recall that for small angles, $\tan\theta \approx \sin\theta$.]

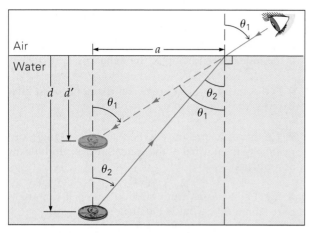

•FIGURE 22.25 **Apparent depth?** See Exercise 39. (For small angles only; enlarged for clarity.)

40. ■■ A fish tank is made of glass with an index of refraction of 1.50. A person shines a light beam on the glass at an incident angle of 40° to see a fish inside. Is the fish illuminated? Justify your answer.

41. ■■ A beam of light passes through a 45°−90°−45° prism (•Fig. 22.26). (a) What can be said about the index of refraction of the prism? (b) What if the prism were under water?

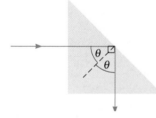

•FIGURE 22.26 **Internal reflection in a prism** See Exercise 41.

42. ■■ A 45°−90°−45° prism is made of a material with an index of refraction of 1.85. Can the prism be used to deflect a beam of light by 90° (a) in air or (b) in water?

43. ■■ A light ray in air is incident on a glass plate 10.0 cm thick at an angle of 40°. The glass has an index of refraction of 1.65. The emerging ray on the other side of the plate is parallel to the incident ray but laterally displaced. (a) How far is the emerging ray displaced relative to the normal? (b) What is the perpendicular distance between the original direction of the ray and the direction of the emerging ray?

44. ■■ 🌀 A person lying at poolside looks over the edge of the pool and sees a bottle cap on the bottom directly below, where the pool depth is 3.2 m. How far below the water surface does the bottle cap appear to be? (See Exercise 39.)

45. ■■ 🌀 What percentage of the actual depth is the apparent depth of an object submerged in water if the observer is looking almost straight downward. (See Exercise 39.)

46. ■■ At what angle to the surface must a diver submerged in a lake look toward the surface to see the setting Sun?

47. ■■ A submerged diver shines a light toward the surface of the water at angles of incidence of 40° and 50°. Can a person on the shore see a beam of light emerging from the surface in either case? Justify your answer mathematically.

48. ■■ To a submerged diver looking upward through the water, the altitude of the Sun (the angle between the Sun and the horizon) appears to be 45°. What angle is the Sun's actual altitude?

49. ■■ A coin lies on the bottom of a pool under 1.5 m of water and 0.90 m from the side wall (•Fig. 22.27). If a light beam is incident on the water surface at the wall, at what angle (θ) relative to the wall must the beam be directed so it will illuminate the coin?

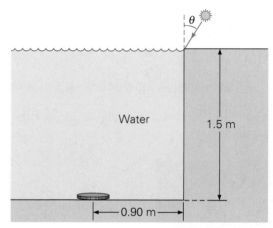

•FIGURE 22.27 **Find the coin** See Exercise 49. (Not drawn to scale.)

50. ■■ Describe a method for measuring the index of refraction of the fluids in Fig. 22.9. Determine the index of refraction of the liquid.

51. ■■ A light beam traveling upward in a plastic material with an index of refraction of 1.60 is incident on an upper horizontal air interface at an angle of 45°. (a) Is the beam transmitted? (b) Suppose the upper surface of the plastic material is covered with a layer of liquid with an index of refraction of 1.20. What happens in this case?

52. ■■ A crown-glass plate 2.5 cm thick is placed over a newspaper. How far beneath the top surface of the plate would the print appear to be if you were looking almost vertically downward through the plate? (See Exercise 39.)

53. ■■■ An outdoor circular fish pond has a diameter of 4.0 m and a uniform full depth of 1.50 m. A fish halfway down in the pond and 0.50 m from the near side can just see the full height of a 1.8-meter-tall person. How far away from the edge of the pond is the person?

54. ■■■ A cube of flint glass sits on a newspaper on a table. By looking into one of the vertical sides of the cube, is it possible to see the portion of the newspaper covered by the glass?

55. ■■■ Two glass prisms are placed together (•Fig. 22.28). (a) If a beam of light strikes the face of one of the prisms at normal incidence as shown, at what angle θ does the beam emerge from the other prism? (b) At what angle of incidence would the beam be refracted along the prism interface?

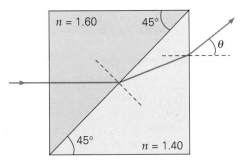

•FIGURE 22.28 Joined prisms   See Exercise 55.

22.5 Dispersion

56. Dispersion occurs for (a) monochromatic light in reflection, (b) monochromatic light in refraction, (c) polychromatic light in reflection, (d) polychromatic light in refraction.

57. A transparent material (a) shows dispersion for $\theta_1 = 0°$, (b) has different indices of refraction for different wavelengths, (c) changes the frequency of a particular light wave, (d) none of these.

58. What causes light of different frequencies to separate upon being refracted?

59. A glass prism disperses white light into a spectrum. Can a second glass prism be used to recombine the spectral components? Explain.

60. ◐ A light beam consisting of two colors, A and B, is sent through a prism. Color A is refracted more than color B. Which color has a longer wavelength? Explain.

61. ■■ (a) If glass is dispersive, why don't we see a spectrum of colors when sunlight passes through a window pane? (b) Does dispersion occur for polychromatic light incident on a dispersive medium at an angle of 0°? Explain. (Are the speeds of each color of light the same in the medium?)

62. ■ The index of refraction of crown glass is 1.515 for red light and 1.523 for blue light. Find the angle separating rays of the two colors in a piece of crown glass if their angle of incidence is 37°.

63. ■■ ◐ White light passes through a prism made of crown glass and strikes an air interface at an angle of 41.15°. Using the indices of refraction given in Exercise 62, describe what happens.

64. ■■ ◐ A beam of light with red and blue components of wavelengths 670 nm and 425 nm, respectively, strikes a slab of fused quartz at an incident angle of 30°. On refraction, the different components are separated by an angle of

0.001 31 rad. If the index of refraction of the red light is 1.4925, what is the index of refraction of the blue light?

65. ■■■ ◐ A beam of red light is incident on an equilateral prism as shown in •Fig. 22.29. (a) If the index of refraction of red light is 1.400, at what angle θ does the beam emerge from the other face of the prism? (b) Suppose the incident beam were white light. What would be the angular separation of the red and blue components in the emergent beam if the index of refraction of blue light is 1.403? (c) if the index of refraction of blue light were 1.405?

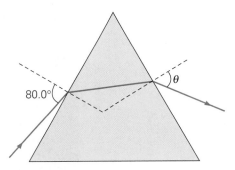

•FIGURE 22.29 Prism revisited   See Exercise 65.

Additional Exercises

66. An opaque container that is empty except for a single coin is 15 cm deep. When looking into the container at a viewing angle of 50° relative to the vertical side of the container, you see nothing on the bottom. When the container is filled with water, you see the coin (from the same viewing angle) on the bottom of and just beyond the side of the container. (See Fig. 22.24.) How far is the coin from the side of the container?

67. Light travels from a material whose index of refraction is $n_1 = 2.0$ into another material, whose index of refraction is $n_2 = 1.7$. The opposite parallel surface of the second material is exposed to the air. (a) Find the critical angle at which light will be reflected at the interface of the materials. (b) Find the angle at which the light will pass through the interface and then be reflected at the opposite surface of the second material.

68. Light strikes a surface at an angle of 50° relative to the surface. What is the angle of reflection?

69. A beam of light traveling in water strikes a surface of a transparent material at an angle of incidence of 45°. If the angle of refraction in the material is 35°, what is the index of refraction of the material?

70. Yellow-green light of wavelength 550 nm is incident on the surface of a flat piece of crown glass at an angle of 40°. What is (a) the angle of refraction of the light? (b) the speed of the light in the glass? (c) the wavelength of the light in the glass?

71. Light strikes water perpendicular to the surface. What is the angle of refraction?

Insights

- Right–Left or Front–Back Reversal?
- Fresnel Lenses

Think of what life would be like if there were no mirrors in bathrooms or on cars and if no one could get glasses. Imagine a world without optical images of any kind—no photographs, no movies, no TV. Think about how little we'd know about the universe if there were no telescopes with which to observe distant planets and stars; how little we'd know about biology and medicine if there were no microscopes with which to see bacteria and our own cells. We sometimes forget how dependent we are on mirrors and lenses.

The first mirror was probably the reflecting surface of a pool of water. Later, people discovered that polished metals and glass have reflective properties. They must also have noticed that when they looked at things through glass, the objects looked different, depending on the shape of the glass. In some cases, the objects appeared to be enlarged (magnified) or inverted, as in the photo above. In time, people learned to shape glass purposefully into lenses, opening the way for eventual development of the many optical devices we now take so much for granted.

The optical properties of mirrors and lenses are based on the principles of reflection and refraction of light, introduced in Chapter 22. In this chapter you'll learn how mirrors and lenses work. Among other things, you'll discover why the image in the photo is upside down, whereas your image in an ordinary flat mirror is right-side up—but doesn't seem to comb its hair with the same hand you use!

23.1 Plane Mirrors

Objectives: To (a) describe the characteristics of plane mirrors and (b) explain apparent right–left reversals.

Mirrors are smooth reflecting surfaces, usually made of polished metal or glass that has been coated with some metallic substance. As you know, even an uncoated piece of glass such as a window pane can act as a mirror (see Demonstration 12 on p. 712). However, when one side of a piece of glass is coated with a compound of tin, mercury, or silver, the reflectivity of the glass is increased, and light is not transmitted through the coating. A mirror may be front-coated or back-coated, depending on the application.

When you look directly into a mirror, you see the reflected images of yourself and objects around you (apparently on the other side of the surface). The geometry of a mirror's surface affects the size, orientation, and type of image. An *image* is the visual counterpart of an object produced by reflection (mirrors), refraction (lenses), or the passage of rays through a small hole.

A mirror with a flat surface is called a **plane mirror**. How images are formed by a plane mirror is illustrated by the ray diagram in •Fig. 23.1. An image appears to be behind or "inside" the mirror. This is because when the mirror reflects a ray from the object to the eye (Fig. 23.1a), the ray appears to us to originate from behind the mirror. Reflected rays from the top and bottom of an object are shown in Fig. 23.1b. In actuality, light rays coming from all points on the side of the object facing the mirror are reflected, and a complete image is observed.

The image formed in this way *appears* to be behind the mirror. Such an image is called a **virtual image**. Light rays appear to diverge from virtual images but do not actually do so. However, spherical mirrors (discussed in Section 23.2) can produce images through which light actually passes. This type of image is called a **real image**. An example of a real image is one that you see on a movie screen.

Notice in Fig. 23.1b the positions or distances of the object and image from the mirror. Quite logically, the distance of an object from a mirror is called the *object distance* (d_o), and the distance its image appears to be behind the mirror is called the *image distance* (d_i). By geometry of similar triangles, it can be shown that $d_o = d_i$, which means that *the image formed by a plane mirror appears to be at a distance behind the mirror that is equal to the distance of the object in front of the mirror* (see Exercise 15).

We are interested in various characteristics of images. One of these is the size of an image compared to that of its object. This comparison is expressed in terms of the **lateral magnification** (M)*, which is defined as a ratio:

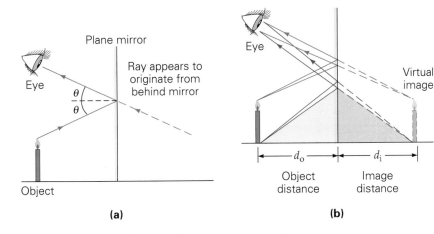

(a)　　　　　**(b)**

•**FIGURE 23.1 Image formed by a plane mirror** (a) A ray from a point on the object is reflected in the mirror according to the law of reflection. (b) Rays from various points on the object produce an extended image. Because the two shaded triangles are similar, the image distance d_i (the distance of the image from the mirror) is equal to the object distance d_o. That is, the image appears to be the same distance behind the mirror as the object is in front of the mirror. The rays appear to diverge from the image position. In this case, the image is said to be virtual.

---

*Lateral, or height distance, magnification (M) is distinguished from angular, or angular distance, magnification (m) in Chapter 25. In this chapter we will refer to M simply as the magnification, or magnification factor, for convenience.

Demonstration 12 ■ A Candle Burning Underwater?

It would appear so, but you know this is not possible. It's a reflection of an image.

(a) The black frame holds a pane of glass, which acts both as a window and as a plane mirror. The burning candle seen in the water is the image of the candle on the front stand. There is a container of water on a similar stand behind the glass, but no burning candle.

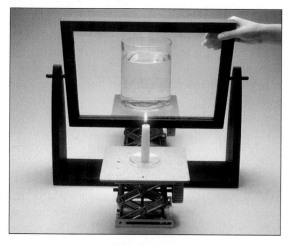

(b) The effect can be removed by tilting the glass—the image can no longer be seen from this viewing point. (Something to do with the law of reflection. What?)

$$M = \frac{\text{image height}}{\text{object height}} = \frac{h_i}{h_o} \tag{23.1}$$

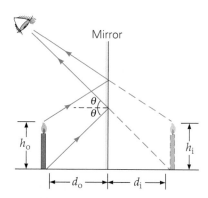

•FIGURE 23.2 Magnification
The lateral, or height distance, magnification is given by $M = h_i/h_o$. If we write this relationship as $h_i = Mh_o$, then the M is a magnification factor. For a plane mirror, $M = 1$, which means that $h_i = h_o$ and the image is the same height as the object.

Referring to •Fig. 23.2, you should be able to show by similar triangles that $h_i = h_o$, so $M = 1$ for a plane mirror and there is no magnification. That is, you and your image in a plane mirror are the same size. Note that M is a dimensionless magnification *factor*, since $h_i = Mh_o$.

We use a lighted candle as an object to allow us to address another image characteristic, its orientation—that is, whether the image is upright or inverted with respect to the orientation of the object. (In sketching ray diagrams, an arrow is a convenient object for this purpose.) For a plane mirror, the image is always upright (or erect). This means that the image is oriented in the same vertical direction as the object. Note that in Fig. 23.2, both the object and the image are oriented upward. (If both the object and image were oriented downward, the image would still be described as being upright because it has the same orientation as the object.)

With other types of mirrors, such as spherical mirrors (which we will consider shortly), it is possible to have inverted images. In this case, the vertical orientation of the image is opposite that of the object.

The main characteristics of an image formed by a plane mirror are summarized in Table 23.1. For another interesting plane mirror characteristic, see the Insight on p. 713.

TABLE 23.1   Characteristics of Images Formed by Plane Mirrors

The image distance is equal to the object distance ($d_i = d_o$). That is, the image appears to be as far behind the mirror as the object is in front.

The image is virtual, upright, and unmagnified ($M = 1$).

Right–Left or Front–Back Reversal?

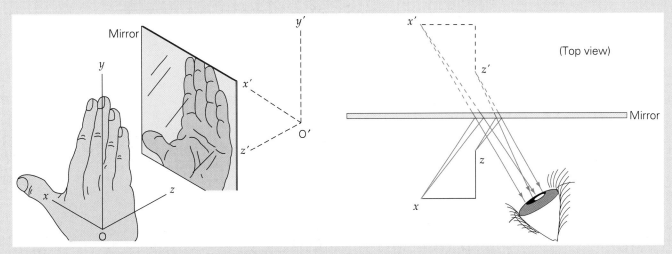

FIGURE 1   Symmetry reversal in a plane mirror   **(a)** An image in a plane mirror is said to show right–left reversal relative to the object. However, the thumb in both object and image is at the bottom of the mirror and points in the +*x* direction. There is a front-to-back reversal of the *z*–*z'* axes inasmuch as they point toward each other. This doesn't occur for the *x*–*x'* and *y*–*y'* axes. **(b)** The front–back reversal of the *z* axis is consistent with the law of reflection. It is this reversal that gives rise to the apparent right–left reversal.

A characteristic of a plane mirror that you have probably noticed is the *apparent* right–left reversal of images. If you stand in front of a mirror and raise your right hand, your mirror image will raise its left hand (Fig. 1a). Similarly, if you part your hair on one side, the part will appear on the other side of your image's head.

As shown in the ray diagram in Fig. 1b, this apparent reversal is consistent with the law of reflection. However, note the *z* axis. Here there is a front–back reversal. That is, the *z'* axis (the reflected image of the *z* axis) has been reversed, such that it points in the direction opposite that of the *z* axis, back toward the plane of the mirror. (Note that this is not the case for the *x* and *y* axes.) So, we see that a plane mirror image reproduces exactly all object points in the dimensions *parallel* to the mirror surface but reverses the sequential order of things in the direction *perpendicular* to the mirror surface, giving rise to a front–back reversal.

It is this front–back reversal that produces the apparent right–left reversal. Note in Fig. 1a that with the right-hand palm facing the mirror, the image appears to be that of a left-hand palm. This is because the sequential order of the parts of the right hand have been reversed in the perpendicular, hand-thickness direction (that is, along the *z* axis).

But it is also because we are symmetrical creatures, with left hands that are "mirror images" of our right hands! It's easier to understand this point if you imagine what you would see in the mirror if your left and right sides were as different as your head and your feet—for example, if you had a right hand but a left wing. Then it would be clear that even if you could step into the mirror and turn around, you would not match your mirror image. In fact, no movements

or rotations can ever get you and your mirror image to coincide. (Even our highly symmetrical body plans can't entirely conceal this fact; for example, your image's heart isn't in the right place!)

When working with mirrors, it is good to keep in mind that right and left are directional *senses* (like clockwise and counterclockwise) rather than fixed directions referenced to a coordinate system. For example, consider the images shown in Fig. 2. Notice that the mirror does not reverse the sequence of the objects (letters) in the direction parallel to its surface, but does give a front–back reversal in the perpendicular direction, which is noticeable only for the nonsymmetric Ls. Also note that, with the letters positioned in this way, the front–back reversal of the Ls appears not as a right–left reversal, but as a top-to-bottom reversal.

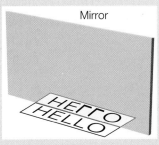

FIGURE 2   Hello!   The letters of HELLO on a piece of paper placed on a surface next to a vertical mirror form an image that shows no right–left reversal in the sequence of letters. However, each nonsymmetric L shows a front–back reversal, which in this instance appears as a top–bottom reversal.

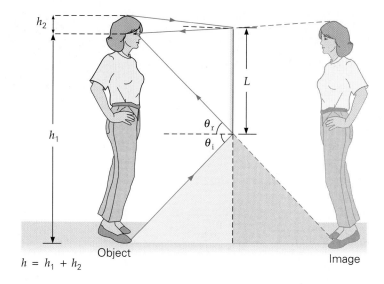

•**FIGURE 23.3 Seeing it all** The minimum height, or vertical length, of a plane mirror needed for a person to see his or her complete (head-to-toe) image turns out to be half the person's height. See Example 23.1.

$h = h_1 + h_2$

Object · Image

Example 23.1 ■ All of Me: Minimum Mirror Length

What is the minimum vertical length of a plane mirror needed for a person to be able to see a complete (head-to-toe) image (•Fig. 23.3)?

Thinking It Through. Applying the law of reflection, we see in the figure that two triangles are formed by the rays needed for the image to be complete. These triangles relate the person's height with the minimum mirror length.

Solution. To determine this length, consider the situation shown in Fig. 23.3. With a mirror of minimum length, a ray from the top of the head would be reflected at the top of the mirror, and the ray from the feet would be reflected at the bottom of the mirror. The length L of the mirror is then the distance between the dashed lines perpendicular to the mirror at its top and bottom.

However, these lines are also the normals for the ray reflections. By the law of reflection, they bisect the angles between incident and reflected rays; that is, $\theta_i = \theta_r$. Then, because their respective triangles on each side of the dashed normal are similar, the length of the mirror from its bottom to a point even with the person's eyes is $h_1/2$, where h_1 is the person's height from the feet to the eyes. Similarly, the small upper length of the mirror is $h_2/2$ (the vertical distance between the eyes and the top of mirror). Thus,

$$L = \frac{h_1}{2} + \frac{h_2}{2} = \frac{h_1 + h_2}{2} = \frac{h}{2}$$

where h is the person's total height.

Hence, for a person to see his or her complete image in a plane mirror, the minimum height or vertical length of the mirror must be half the height of the person.

Follow-up Exercise. What effect does a person's distance from the mirror have on the minimum mirror length required to produce his or her complete image?

23.2 Spherical Mirrors

Objectives: To (a) distinguish between converging and diverging spherical mirrors, (b) describe images and their characteristics, and (c) determine these image characteristics from ray diagrams and the spherical mirror equation.

As the name implies, a **spherical mirror** is a reflecting surface with spherical geometry. •Figure 23.4 shows that if a portion of a sphere of radius R is sliced off along a plane, the severed section has the shape of a spherical mirror. Either the inside or outside of such a section can be reflective. For inside reflections, the mirror is

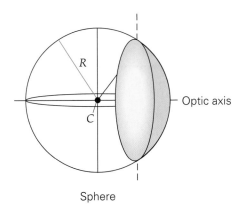 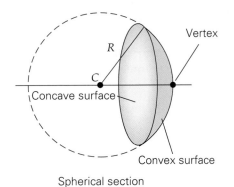

Sphere

Spherical section

a **concave mirror** (think of looking into a cave to remember the indented surface of a concave mirror). For outside reflections, the mirror is a **convex mirror**.

The radial line through the center of the spherical mirror is called the *optic axis*, and it intersects the mirror surface at the *vertex* of the spherical section (Fig. 23.4). The point on the optic axis that corresponds to the center of the sphere of which the mirror forms a section is called the **center of curvature** (*C*). The distance between the vertex and the center of curvature is equal to the radius of the sphere and is called the **radius of curvature** (*R*).

When rays parallel to the optic axis are incident on a concave mirror, the reflected rays intersect, or converge, at a common point called the **focal point** (*F*)*. As a result, a concave mirror is called a **converging mirror** (•Fig. 23.5a).

Similarly, a beam parallel to the optic axis of a convex mirror diverges on reflection, as though the reflected rays came from a focal point behind the mirror's surface (Fig. 23.5b). Thus, a convex mirror is called a **diverging mirror**. A diverging mirror is shown in •Fig. 23.6. When you see diverging rays, your brain interprets or assumes there is an object from which the rays *appear* to diverge, even though there is none there. The true object is somewhere else.

The distance from the vertex to the focal point of a spherical mirror is called the **focal length**, *f* (see Fig. 23.5). The focal length is related to the radius of curvature by this simple equation:

$$f = \frac{R}{2} \qquad \begin{array}{l}\textit{focal length,}\\ \textit{spherical mirror}\end{array} \qquad (23.2)$$

Note:
Concave mirror = converging mirror
Convex mirror = diverging mirror

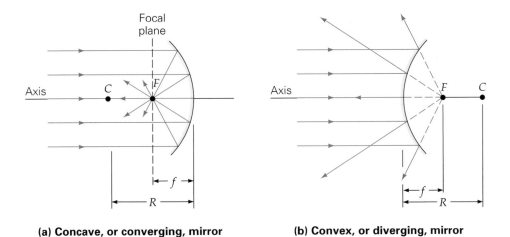

(a) Concave, or converging, mirror

(b) Convex, or diverging, mirror

•FIGURE 23.5 **Focal point**
(a) Rays parallel and close to the optic axis of a concave spherical mirror converge at the focal point, *F*. **(b)** Rays parallel to the optic axis of a convex spherical mirror are reflected along paths as though they came from a focal point behind the mirror.

*This is the case when the width of the mirror or the illuminated area is small compared to the radius of curvature—that is, for small angles of reflection.

•FIGURE 23.6 Diverging mirror
Note by reverse-ray tracing in
Fig. 23.5b that a diverging (convex)
spherical mirror gives an expanded
field of view, as can be seen in this
store-monitoring mirror.

Ray Diagrams

The characteristics of images formed by spherical mirrors can be determined from geometrical optics (introduced in Chapter 22). The method involves drawing rays emanating from one or more points on an object. The law of reflection ($\theta_i = \theta_r$) applies, and three key rays are defined with respect to the mirror's geometry as follows:

1. A **parallel ray** is a ray that is incident along a path parallel to the optic axis and is reflected through the focal point (as are all rays near and parallel to the axis).

2. A **chief ray**, or **radial ray**, is a ray that is incident through the center of curvature (C) of the mirror. Since the chief ray is incident normal to the mirror's surface, this ray is reflected back along its incident path, through point C.

3. A **focal ray** is a ray that passes through (or appears to go through) the focal point and is reflected parallel to the optic axis. (It is a mirror image, so to speak, of a parallel ray.)

These rays are illustrated in the ray diagrams in •Fig. 23.7 for concave and convex mirrors according to the numbers in the list above. It is customary to use the tip of the object (for example, the head of an arrow or the flame of a candle) as the origin point of the rays. This makes it easy to see whether the image is upright or inverted. The corresponding point of the image is at the point of intersection of the rays. Also, the candle is arbitrarily taken to be upright with its base on the optic axis.

Keep in mind, however, that *properly traced rays from any point on the object can be used to find the image.* Every point on a visible object acts as an emitter of light. For example, for a candle, the flame emits its own light, but every point on the candle reflects light.

Note in Fig. 23.7a that for a concave (converging) mirror with an object at a distance greater than the radius of curvature ($d_o > R$), the reflected rays converge and define the location of the flame tip of the image, which is formed in front of the mirror surface. The image is said to be *real.* The reflected rays converge and pass through the image (point). As a result, the real image could be seen on a screen (for example, a piece of white paper) that is positioned at a distance d_i from the concave mirror. The image is real, inverted, and smaller than the object.

The rays reflected from a convex (diverging) mirror diverge (Fig. 23.7b). Projecting the rays behind the convex mirror to determine where they appear to intersect indicates that the image is *virtual,* analogous to the virtual image formed behind a plane mirror. The reflected rays diverge and only appear to pass through the image. (The apparent intersection point of the rays indicates the position of the tip of the candle flame.) The image is virtual, upright, and smaller than the object.

Since the reflected rays of an object at any distance from a convex mirror diverge, *a diverging mirror always forms a virtual image.* The diverging rays would not form an image on a screen. A real image can be formed on a screen, but a virtual image cannot.

However, a converging mirror does *not* always form a real image. For a converging spherical mirror, the characteristics of the image change with the distance of the object from the mirror. Dramatic changes take place at two points: C (the center of curvature) and F (the focal point). These points divide the optic axis into

Converging (concave) mirror   **Diverging (convex) mirror**

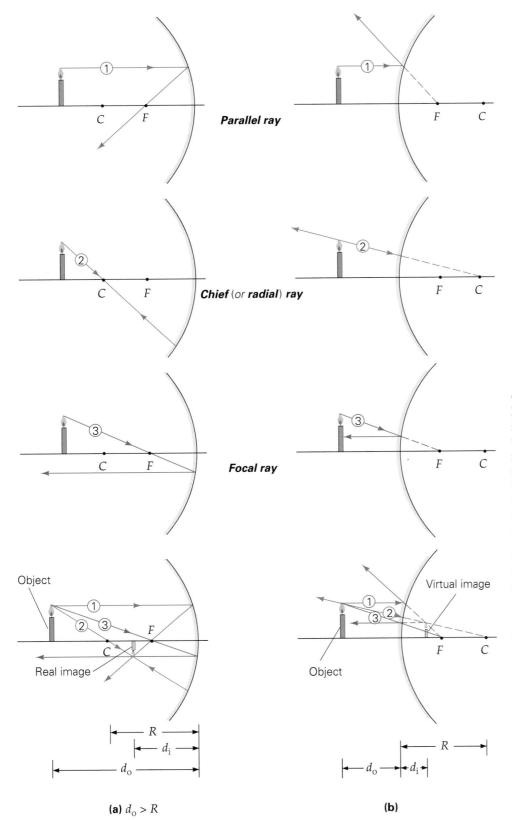

Parallel ray

Chief (or *radial*) *ray*

Focal ray

Object

Real image

Virtual image

Object

R

d_i

d_o

(a) $d_o > R$

R

d_o d_i

(b)

•**FIGURE 23.7   Ray diagrams**
Ray diagrams to find the images for spherical mirrors can be drawn with three rays: (1) A parallel ray is reflected through the focal point F for a converging mirror; it appears to come from the internal focal point for a diverging mirror. (2) A chief (or radial) ray is incident through the center of curvature C and is reflected back along its path of incidence for a converging mirror; it appears to be reflected from the internal center of curvature for a diverging mirror. (3) A focal ray passing through the focal point F is reflected parallel to the optic axis for a converging mirror; it appears to pass through the internal focal point for a diverging mirror. **(a)** Ray diagram of a converging mirror for $d_o > R$. With the rays coming from the tip of the object arrow, their intersection defines the location of the tip of the (real) image relative to the optic axis. **(b)** Ray diagram of a diverging mirror. Here the image is virtual and behind, or inside, the mirror.

•FIGURE 23.8 Concave mirrors

(a) For a concave, or converging, mirror, the object is located within one of three regions defined by the center of curvature (C) and the focal point (F) or at one of these two points. For $d_o > R$, the image is real, inverted, and smaller, as shown by the ray diagram in Fig. 23.7a. (See Demonstration 13 for $d_o = R$.) (b) For $R > d_o > f$, the image will also be real and inverted but enlarged, or magnified. Only two rays are needed to locate the image, and the focal ray is omitted for clarity. (c) For an object at the focal point F, or $d_o = f$, the image is said to be formed at infinity. (d) For $d_o < f$, the image will be virtual, upright, and enlarged.

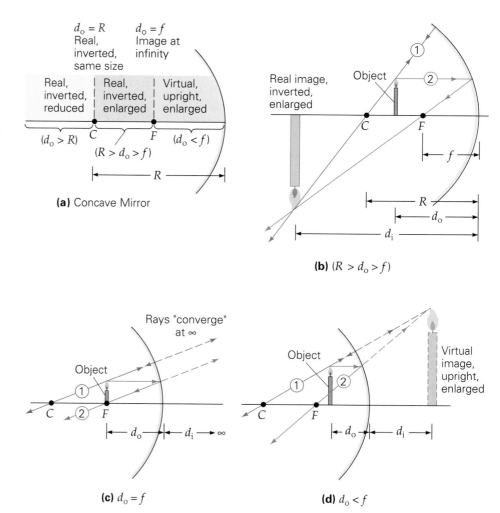

(a) Concave Mirror

(b) $(R > d_o > f)$

(c) $d_o = f$

(d) $d_o < f$

three regions (•Fig. 23.8a)*: $d_o > R$, $R > d_o > f$, and $d_o < f$. Let's start with an object in the region farthest from the mirror $(d_o > R)$ and move toward the mirror.

- The case of $d_o > R$ has already been dealt with in Fig. 23.7a.
- The case of $d_o = R$ is shown in Demonstration 13. The image is real, inverted, and the same size as the object.
- When $R > d_o > f$, an enlarged, inverted, real image is formed (Fig. 23.8b). The image is magnified when the object is inside the center of curvature C.
- When $d_o = f$, the object is at the focal point (Fig. 23.8c). The reflected rays are parallel, and the image is said to be formed at infinity. This expresses the idea that parallel lines converge at infinity (like railroad tracks that appear to converge at a great distance). The focal point F is a special "cross-over" point.
- When $d_o < f$, the object is inside the focal point (between the focal point and the mirror's surface). A virtual image is formed (Fig. 23.8d).
- When $d_o > f$, the image is real when $d_o < f$, the image is virtual. We can't see an image formed at infinity, but we say the image is formed there because of symmetry with the case in Fig. 23.8c. When an object is at "infinity"—it is so far away that rays emanating from it are essentially parallel—its image is formed at F. By reverse ray tracing, rays from an object at a great ("infinite") distance from the mirror are shown to be essentially parallel when near the mirror and to form an image on a screen aligned in the focal plane. This fact provides a method for determining the focal length of a concave mirror.

*Only two rays are needed to determine the image, and we will use only two, the parallel and chief rays, for illustration clarity. However, draw the third type of ray, a focal ray, as a check.

Demonstration 13 ■ A Candle Burning at Both Ends

Or is it? Notice that one flame is burning downward, which is rather strange. It's an illusion done with a spherical concave mirror.

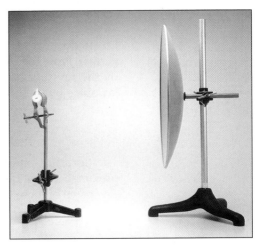

(a) When an object is at the center of curvature of a spherical concave mirror, a real image is formed that is inverted and the same size as the object, and the image distance is the same as the object distance. What is seen here is a horizontal candle burning at one end (flame up) and its overlapping image (flame down). Viewed at the same level, the inverted flame image appears to be at the opposite end of the candle.

(b) A side view showing the burning end of the horizontal candle in front of the spherical mirror.

The position and size of the image can be determined graphically from ray diagrams drawn to scale. However, these can be determined more quickly by analytical methods. The distances and focal length can be shown to be related by what is known as the **spherical mirror equation:**

$$\frac{1}{d_o} + \frac{1}{d_i} = \frac{1}{f} = \frac{2}{R} \qquad \textit{spherical mirror equation} \qquad (23.3)$$

Note that this equation can be written in terms of either the radius of curvature R or the focal distance f, since by Eq. 23.2, $f = R/2$.

The image distance d_i is often the quantity to be found. An alternative form of this equation that avoids the need for the common denominator of fractions is

$$d_i = \frac{d_o f}{d_o - f} \qquad \begin{array}{l} \textit{spherical mirror equation} \\ \textit{in different form} \end{array} \qquad (23.4)$$

The **magnification factor** M for a spherical mirror can also be found analytically. This quantity is expressed in terms of the image and object distances:

$$M = -\frac{d_i}{d_o} \qquad \begin{array}{l} \textit{magnification equation} \\ \textit{for a spherical mirror} \end{array} \qquad (23.5)$$

Note: A helpful hint for remembering that the magnification is d_i over d_o is that the ratio is in alphabetical order (i over o).

Note: $|M|$ is the absolute value of M: its magnitude without regard to sign. For example, $|+2| = |-2| = 2$.

The minus sign is added by convention to indicate the orientation of the image as given below. Hence, if $|M| > 1$, the image is magnified, or larger than the object. If $|M| < 1$, the image is reduced, or smaller than the object. [Note that, for mirrors, the lateral magnification M, also called the *magnification factor* or simply *magnification*, is conveniently expressed in terms of the image distance d_i and the object distance d_o rather than the image and object heights (Eq. 23.1). The origin of Eq. 23.5 follows as optional coverage.]

Derivation of the Spherical Mirror Equation   Students often wonder where various equations originate. The spherical mirror equation can be derived with the aid of a little trigonometry. Consider the ray diagram in ●Fig. 23.9. The object and image distances (d_o and d_i) and the heights of the object and image (h_o and h_i) are shown. Note that these lengths make up the bases and heights of triangles formed by the ray reflected at the vertex (V). These triangles ($O'VO$ and $I'VI$) are similar, since by the law of reflection their angles at V are equal. Hence, we can write

$$\frac{h_i}{h_o} = \frac{d_i}{d_o} \tag{1}$$

The (focal) ray through F also forms similar triangles, $O'FO$ and VFA (in the approximation that the mirror is small compared to its radius). (Why are they similar?) The bases of these triangles are $VF = f$ and $OF = d_o - f$. Then, if VA is taken to be h_i,

$$\frac{h_i}{h_o} = \frac{VF}{OF} = \frac{f}{d_o - f} \tag{2}$$

Equating Eqs. 1 and 2, we have

$$\frac{d_i}{d_o} = \frac{f}{d_o - f} \tag{3}$$

Multiplying both sides of this equation by d_o gives

$$d_i = \frac{d_o f}{d_o - f}$$

which is the spherical mirror equation in the form of Eq. 23.4.

We can go one step further and get the magnification factor M. Recall that the lateral magnification is the ratio of the height of the image (h_i) to the height of the object (h_o)—that is, $M = h_i/h_o$ (Eq. 23.1). And, by Eq. 1 above,

$$M = \frac{h_i}{h_o} = -\frac{d_i}{d_o}$$

where the minus sign is inserted as a convention.

The signs on the various quantities are very important in the application of Eqs. 23.3 through 23.5. We will use the sign convention summarized in Table 23.2.

Examples 23.2 through 23.4 show how these equations and sign convention are used for spherical mirrors. In general, this usually involves finding the image

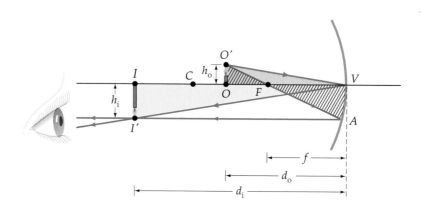

●FIGURE 23.9 Spherical mirror equation
The rays provide the geometry, through similar triangles, for the derivation of the spherical mirror equation.

TABLE 23.2 Sign Convention for Spherical Mirrors (assignment of + and − signs)

Focal length
Concave (converging) mirror: $+f$ (or $+R$)
Convex (diverging) mirror: $-f$ (or $-R$)

Object distance d_o (same for both concave and convex mirrors)
$+d_o$ when the object is in front of the mirror (real object)
$-d_o$ when the object is behind the mirror (virtual object)*

Image distance d_i (same for both mirrors)
$+d_i$ when the image is formed in front of the mirror (real image)
$-d_i$ when the image is formed behind the mirror (virtual image)

Image orientation M (same for both mirrors)
$+M$ when the image is upright with respect to the object
$-M$ when the image is inverted with respect to the object

---

*In a combination of two (or more) mirrors, the image formed by the first mirror is the object of the second mirrror (and so on). If this image–object falls behind the second mirror, it is referred to as a *virtual* object, and the object distance is taken to be negative (−). This is more important for lens combinations, as we will see in Section 23.3, and is mentioned here only for completeness.

of an object; you will be asked where the image is formed (d_i) and what the image characteristics are. These characteristics tell whether the image is real or virtual, upright or inverted, and larger or smaller (magnified or reduced).

Example 23.2 ■ What Kind of Image? Behavior of a Concave Mirror

A concave mirror has a radius of curvature of 30 cm. If an object is placed (a) 45 cm, (b) 20 cm, and (c) 10 cm from the mirror, where is the image formed and what are its characteristics? (Specify real or virtual, upright or inverted, and larger or smaller for each image.)

Thinking It Through. Here we are given R, from which we can compute $f = R/2$. We are also given three different object distances, which we can apply in Eq. 23.3, along with Eq. 23.4.

Solution.

Given: $R = 30$ cm
  so $f = R/2 = 15$ cm
  (a) $d_o = 45$ cm
  (b) $d_o = 20$ cm
  (c) $d_o = 10$ cm

Find: d_i and image characteristics
for given object distances

Note that these object distances correspond to the regions shown in Fig. 23.8a. There is no need to convert the distances to meters, but all distances must be in the same unit (centimeters here).

(a) In this case, the object distance is greater than the radius of curvature ($d_o > R$), and

$$\frac{1}{d_o} + \frac{1}{d_i} = \frac{1}{f}$$

or

$$\frac{1}{45} + \frac{1}{d_i} = \frac{1}{15}$$

where the units have been omitted for simplicity. Then

$$\frac{1}{d_i} = \frac{2}{45}$$

or

$$d_i = \frac{45}{2} = +22.5 \text{ cm}$$

and

$$M = -\frac{d_i}{d_o} = -\frac{22.5 \text{ cm}}{45 \text{ cm}} = -\frac{1}{2}$$

Thus, the image is real (positive d_i), inverted (negative M), and half as large as the object ($|M| = \frac{1}{2}$).

(b) Here $R > d_o > f$, and the object is between the focal point and center of curvature. Using Eq. 23.4, we have

$$d_i = \frac{d_o f}{d_o - f} = \frac{(20 \text{ cm})(15 \text{ cm})}{20 \text{ cm} - 15 \text{ cm}} = 60 \text{ cm}$$

and

$$M = -\frac{d_i}{d_o} = -\frac{60 \text{ cm}}{20 \text{ cm}} = -3.0$$

In this case, the image is real, inverted, and three times the size of the object.

(c) For this case, $d_o < f$, and the object is inside the focal point. Then,

$$d_i = \frac{d_o f}{d_o - f} = \frac{(10 \text{ cm})(15 \text{ cm})}{10 \text{ cm} - 15 \text{ cm}} = -30 \text{ cm}$$

Then

$$M = -\frac{d_i}{d_o} = -\frac{(-30 \text{ cm})}{10 \text{ cm}} = +3.0$$

In this case, the image is virtual, upright, and three times the size of the object.

From the denominator of the right-hand side of the equation for d_i (Eq. 23.4), you can see that d_i will always be negative when d_o is less than f. Therefore, a virtual image is always formed for an object inside the focal point of a converging mirror.

Draw representative ray diagrams for each of these cases to see that the image characteristics are correct.

Follow-up Exercise.   For the converging mirror in this Example, where is the image formed and what are its characteristics if the object is at 30 cm, or $d_o = R$?

Problem-Solving Hint

When using the spherical mirror equations to find image characteristics, it is helpful to make a quick sketch of the ray diagram for the situation first. Doing this shows you the image characteristics and helps you avoid making mistakes when applying the sign convention. *The ray diagram and the mathematical solution must agree.*

Example 23.3 ■ Similarities and Differences: Behavior of a Convex Mirror

An object is 30 cm in front of a diverging mirror that has a focal length of 10 cm. Where is the image, and what are its characteristics?

Thinking It Through.   Keeping in mind the sign convention (Table 23.2) for convex (diverging) mirrors, we use the same equations as in Example 23.2. However, we know that the image distance will be negative and that the image will be virtual. (Why?)

Solution.

Given: $d_o = 30$ cm   *Find:* d_i and image characteristics
            $f = -10$ cm

Note that the focal length is negative for a convex mirror (see Table 23.2). Using Eq. 23.4, we have

$$d_i = \frac{d_o f}{d_o - f} = \frac{(30 \text{ cm})(-10 \text{ cm})}{30 \text{ cm} - (-10 \text{ cm})} = -7.5 \text{ cm}$$

Then

$$M = -\frac{d_i}{d_o} = -\frac{(-7.5 \text{ cm})}{30 \text{ cm}} = +0.25$$

Thus, the image is virtual (negative d_o) and upright (positive M) and is 0.25 times (one quarter) the size (height) of the object.

Follow-up Exercise. As we pointed out previously, a diverging mirror always forms a virtual image. What about the other characteristics of the image: its orientation and magnification? Can any general statements be made about them?

Example 23.4 ■ In the Distance: Finding the Focal Length of a Concave Mirror

Where is the image formed by a concave mirror if the object is at infinity? (An object at a great distance from a mirror, relative to the mirror's dimensions, may be considered to be at infinity.)

Thinking It Through. This provides an interesting application of the spherical mirror equation. Think back to ray tracing: An object at the focal point of a concave mirror forms an image at infinity (Fig. 23.8c). (Might the reverse be true?)

Solution. With $d_o = \infty$, we have

$$\frac{1}{d_o} + \frac{1}{d_i} = \frac{1}{\infty} + \frac{1}{d_i} = \frac{1}{f} \quad \text{or} \quad d_i = f$$

Thus, the image is real ($+d_i$) and is formed at the focal point (or in the focal plane for an extended object—that is, the plane that both passes through the focal point and is perpendicular to the optic axis).

This result provides an experimental means of determining the focal length of such a mirror. A screen (a piece of white paper) is adjusted until an image of a distant object ($d_o \gg f$) is formed on it. The screen is then in the focal plane of the mirror. (Does the magnification equation hold for this special case?)

Follow-up Exercise. In the limit, a plane mirror can be thought of as a section of a sphere having an infinite radius of curvature ($R = \infty$). What do the spherical mirror equation and magnification factor tell you about the image characteristics of a plane mirror?

Spherical Aberration of Mirrors

Technically, our descriptions of image characteristics for spherical mirrors are true only for objects near the optic axis—that is, only for small angles of reflection. If these conditions do not hold, the images will be blurred (out of focus) or distorted, because not all of the rays will converge in the same plane. As illustrated in •Fig. 23.10, incident parallel rays far from the optic axis do not converge at the focal point. The farther the incident ray is from the axis, the more distant is its reflected ray from the focal point. This effect is called **spherical aberration.**

Spherical aberration does not occur with a parabolic mirror. (As the name implies, a section through the center of a parabolic mirror has the form of a parabola.) All of the incident rays parallel to the optic axis of such a mirror have a common focal point. For this reason, parabolic mirrors are used in most astronomical telescopes, as we will see in Chapter 24. However, these mirrors are more difficult to make than spherical mirrors (and therefore more expensive).

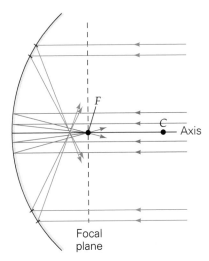

•**FIGURE 23.10 Spherical aberration for a mirror**
According to the small-angle approximation, rays parallel to and near the mirror's axis converge at the focal point. However, when parallel rays not near the axis are reflected, they converge in front of the focal point. This effect, called spherical aberration, gives rise to blurred images.

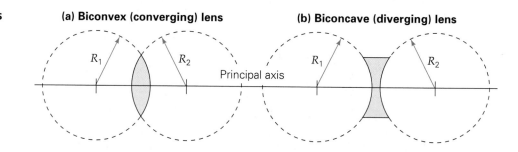

•FIGURE 23.11 Spherical lenses
Spherical lenses have surfaces defined by two spheres, and the surfaces are either convex or concave. **(a)** Biconvex and **(b)** biconcave lenses are shown here. If $R_1 = R_2$, a lens is spherically symmetric.

(a) Biconvex (converging) lens   **(b) Biconcave (diverging) lens**

23.3 Lenses

Objectives: To **(a)** distinguish between converging and diverging lenses, **(b)** describe images and their characteristics, and **(c)** find image characteristics by using ray diagrams and the thin-lens equation.

The word *lens* comes from the Latin word for lentil, a seed whose shape is similar to that of a common lens. An optical **lens** is made from some transparent material (most commonly glass but sometimes plastics or crystals). One or both surfaces usually have a spherical contour. *Biconvex* spherical lenses (both surfaces convex) and *biconcave* spherical lenses (both surfaces concave) are illustrated in •Fig. 23.11.

The properties of lenses are due to the refraction of light passing through them. When light rays pass through a lens, they are bent, or deviated from their original paths, according to the law of refraction. You studied refraction by prisms in Chapter 22. To analyze lens refraction, we can approximate a biconvex lens by two prisms placed base to base, as shown in •Fig. 23.12a.

A biconvex lens is a **converging lens**: Incident light rays parallel to the lens axis converge at a focal point on the opposite side of the lens. You may have focused the Sun's rays with a magnifying glass (a biconvex, or converging, lens) and have witnessed the concentration of radiant energy that results (•Fig. 23.13). The parallel rays coming from the Sun or some other distant object (at infinity) converge at the focal point. This fact provides a way for experimentally determining the focal length of a converging lens.

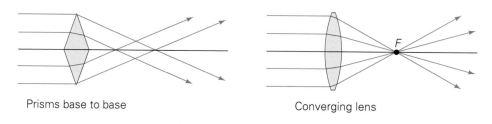

Prisms base to base   Converging lens

(a) Biconvex (converging) lens

•FIGURE 23.12 Converging and diverging lenses **(a)** A biconvex, or converging, lens can be approximated by two prisms placed base to base. For a thin biconvex lens, rays parallel to the axis converge at the focal point F. **(b)** A biconcave, or diverging, lens can be approximated by two prisms placed tip to tip. Rays parallel to the axis of a biconcave lens appear to diverge from a focal point on the incident side of the lens.

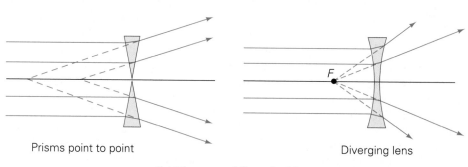

Prisms point to point   Diverging lens

(b) Biconcave (diverging) lens

Conversely, a biconcave lens can be approximated by two prisms placed point to point (Fig. 23.12b). A biconcave lens is a **diverging lens**: Incident parallel rays emerge from the lens as though they emanated from a focal point on the incident side of the lens.

There are several types of converging and diverging lenses (•Fig. 23.14). Meniscus lenses are the type most commonly used for corrective eyeglasses. In general, a converging lens is thicker at its center than at its periphery, and a diverging lens is thinner at its center than at its periphery. This discussion will be limited to spherically symmetric biconvex and biconcave lenses, for which both surfaces have the same radius of curvature. (A special type of convex lens, along with an inversion oddity, is shown in Demonstration 14.)

When light passes through a lens, it is refracted and displaced laterally, as shown in Example 22.4 for a piece of glass (Fig. 22.10). If a lens is thick, this displacement may be fairly large and can complicate the analysis of the lens's characteristics. This problem does not arise with thin lenses, for which the refractive displacement of transmitted light is negligible. Our discussion will be limited to thin lenses.

Like a spherical mirror, a lens with spherical geometry has *for each lens surface* a center of curvature, a radius of curvature, a focal point, and a focal length. The focal points are at equal distances on either side of a thin lens. However, for a spherical lens, $f \neq R/2$, as it is for a spherical mirror (Eq. 23.2). Usually, only the focal length of a lens is specified rather than the radius of curvature.

The general rules for drawing ray diagrams for lenses are similar to those for spherical mirrors, but some modifications are necessary since light passes through a lens in either direction. Opposite sides of a lens are generally distinguished as the object side and image side. The object side is the side on which an object is positioned, and the image side is the opposite side of the lens (where a real image would be formed). The three rays from a point on an object are drawn as follows:

•**FIGURE 23.13 Burning glass** A magnifying glass (converging lens) can be used to focus the Sun's rays to a spot—with incendiary results. Do not try this at home!

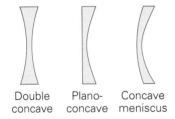
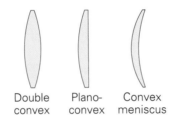

| Double concave | Plano-concave | Concave meniscus | Double convex | Plano-convex | Convex meniscus |

Diverging lenses   **Converging lenses**

•**FIGURE 23.14 Lens shapes** Lens shapes vary widely and are normally categorized as converging or diverging. In general, a converging lens is thicker at its center than at the periphery, and a diverging lens is thinner at its center than at the periphery.

Demonstration 14 ■ Inverted Image

A demonstration of how a cylindrical convex lens produces an inverted image.

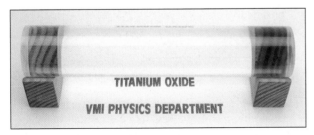

(a) Observe the words TITANIUM OXIDE as placed alongside a cylindrical plastic rod. Note: A similar inversion occurs for spherical convex lenses.

(b) Viewed through the rod, the letters in TITANIUM are inverted, but the letters in OXIDE appear not to be. What is going on?

1. A **parallel ray** is a ray that is parallel to the lens axis on incidence and that after refraction either (a) passes through the focal point on the image side of a converging lens *or* (b) appears to diverge from the focal point on the object side of a diverging lens.

2. A **chief ray**, or **central ray**, is a ray that passes through the center of the lens and is undeviated.*

3. A **focal ray** is a ray that (a) passes through the focal point on the object side of a converging lens *or* (b) appears to pass through the focal point on the image side of a diverging lens, and after refraction is parallel to the lens axis.

These numbered rays are shown in •Fig. 23.15 for converging and diverging lenses. Focal rays are drawn only in the initial diagrams. As with spherical mirrors, only

*This definition incorporates the thin-lens approximation (negligible lateral deviation).

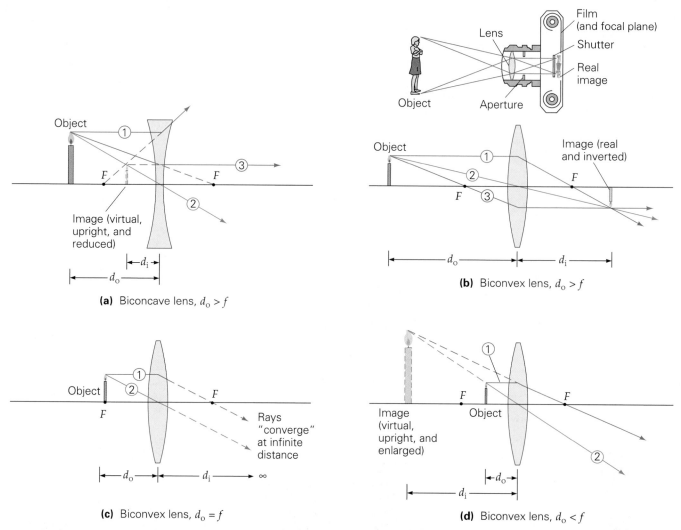

•**FIGURE 23.15 Ray diagrams for lenses** As in ray diagrams for spherical mirrors, (1) parallel rays, (2) chief rays, and (3) focal rays can be drawn to analyze lenses. **(a)** The basic rays for a diverging biconcave lens. Note a virtual image is formed (on the object side of the lens). **(b)** A converging biconvex lens forms a real object when $d_o > f$, as shown in the ray diagram, along with a practical example. **(c)** and **(d)** Ray diagrams for a converging lens with $d_o = f$ and $d_o < f$, respectively. The focal rays have been omitted in these diagrams for clarity. For $d_o = f$, the image is formed at infinity, and for $d_o < f$, the image is virtual. (Note that for simplicity, the refraction of the rays is depicted as if it occurred at the center of each lens. In reality, it would occur at the air–glass and glass–air surfaces of each lens.)

two rays are needed to determine the image, and we will use the parallel and chief rays. (As in the case of mirrors, however, it is generally a good idea to include a third ray in your diagrams as a check.)

For a lens, the image is real when formed on the side of the lens opposite the object (on the image side, •Fig. 23.16a) and virtual when formed on the same side of the lens as the object (on the object side, Fig. 23.16b). Another way of looking at this is that rays converge at a real image and appear to diverge from a virtual image. Note that for a diverging lens (Fig. 23.15a), the rays appear to diverge from the virtual image on the object side of the lens after being refracted. Like diverging mirrors, diverging lenses can form only virtual images. Also, as with spherical mirrors, a real image from a lens can be formed on a screen but a virtual image cannot.

Regions could be similarly defined for the object distance with a converging lens, as was done for a converging mirror in Fig. 23.8a. Here, an object distance of $d_o = 2f$ for a converging lens has a significance similar to that of $d_o = R = 2f$ for a converging mirror. However, the center of curvature for a converging lens does not have the same distinction as it does for a mirror, since $R \neq 2f$ for a lens.

The image distances and characteristics for a lens can also be found analytically. The equations for thin symmetrical biconvex and biconcave lenses are identical to those for spherical mirrors. The **thin-lens equation** is

$$\frac{1}{d_o} + \frac{1}{d_i} = \frac{1}{f} \quad \text{or} \quad d_i = \frac{d_o f}{d_o - f} \qquad \text{thin-lens equation} \quad (23.6)$$

The **magnification factor** is given by

$$M = -\frac{d_i}{d_o} \qquad \begin{array}{l} \text{magnification equation} \\ \text{for a thin lens} \end{array} \quad (23.7)$$

(A thin lens is one for which the thickness of the lens is assumed negligible compared to its focal length.) The sign conventions for these thin-lens equations (Table 23.3) are similar to those for spherical mirrors.

Just as when we work with mirrors, it is helpful to sketch a ray diagram before working a lens problem analytically.

(a)

(b)

•**FIGURE 23.16 Images formed by a converging lens** (a) When the distance of the object from the lens is greater than the focal length ($d_o > f$), an inverted, reduced, real image is formed on the opposite (image) side of the lens. **(b)** When the distance of the object from the lens is less than the focal length ($d_o < f$), an upright, virtual image is formed on the object side of the lens. It appears to be larger and more distant than the actual object but cannot be projected on a screen.

TABLE 23.3 Sign Convention for Thin Lenses (assignment of + and − signs)

Focal length
Converging lens (sometimes called a *positive* lens): $+f$
Diverging lens (sometimes called a *negative* lens): $-f$

Object distance (d_o)
$+d_o$ when the object is in front of the lens (real object—the usual case)
$-d_o$ when the object is behind the lens (virtual object)*

Image distance d:
$+d_i$ when the image is formed on the opposite (image) side of the lens from the object (real image)
$-d_i$ when the image is formed on the same (object) side of the lens as the object (virtual image)

Image orientation M
$+M$ when the image is upright with respect to the object
$-M$ when the image is inverted with respect to the object

*In a combination of two (or more) lenses, the image formed by the first lens is taken as the object of the second lens (and so on). If this image–object falls behind the second lens, it is referred to as a virtual object, and the object distance is taken to be negative (−).

Example 23.5 ■ Two Images: Behavior of a Converging Lens

A biconvex lens has a focal length of 12 cm. Where is the image formed, and what are its characteristics, for an object (a) 18 cm and (b) 4 cm from the lens?

Thinking It Through. With the focal length (f) and the object distances (d_o) given, we can directly apply Eq. 23.6 to find the image distances (d_i). Sketch ray diagrams for both of these cases to give yourself an idea of the image characteristics.

Solution.

Given: $f = 12$ cm
(a) $d_o = 18$ cm
(b) $d_o = 4$ cm (taken to be exact for simplicity)

Find: d_i and image characteristics for both parts

(a) Using the fractional form of Eq. 23.6, we have

$$\frac{1}{d_o} + \frac{1}{d_i} = \frac{1}{f}$$

or

$$\frac{1}{18} + \frac{1}{d_i} = \frac{1}{12}$$

With a common denominator,

$$\frac{2}{36} + \frac{1}{d_i} = \frac{3}{36}$$

Thus,

$$\frac{1}{d_i} = \frac{1}{36} \quad \text{or} \quad d_i = 36 \text{ cm}$$

Then

$$M = -\frac{d_i}{d_o} = -\frac{36}{18} = -2$$

The image is real (positive d_i), inverted ($-M$), and twice as tall as the object ($|M| = 2$).

(b) Using the alternative form of Eq. 23.6 for d_i gives

$$d_i = \frac{d_o f}{d_o - f} = \frac{(4 \text{ cm})(12 \text{ cm})}{4 \text{ cm} - 12 \text{ cm}} = -6 \text{ cm}$$

Then

$$M = -\frac{d_i}{d_o} = -\frac{(-6 \text{ cm})}{4 \text{ cm}} = +1.5$$

In this case, the image is virtual (on the object side of the lens), upright, and magnified by a factor of 1.5.

Follow-up Exercise. Suppose the lens in this Example were biconcave, or diverging, with a focal length of 12 cm. What is the magnification of an object at $d_o = 4$ cm? (As you might expect, diverging lenses, like diverging spherical mirrors, form only *upright, reduced, virtual* images. See Exercise 56.)

Example 23.6 ■ Two Images Revisited: Finding the Transition Point

As the object distance of a biconvex lens is varied, at what point does the real image go from being reduced to being magnified?

Thinking It Through. This is similar to the case of a converging mirror (Section 23.2), in which the image was reduced on one side of a particular location ($d_o = R = 2f$) and magnified on the other. However, for a thin lens, $R \neq 2f$, but, as we shall see, the results are similar.

Solution. For a biconvex (converging) lens, $R \neq 2f$. However, an object distance $d_o = 2f$, or a distance $2f$ from the lens, does have this significance for the size of the real image. Substituting $2f$ for d_o in Eq. 23.6 gives

$$\frac{1}{d_o} + \frac{1}{d_i} = \frac{1}{2f} + \frac{1}{d_i} = \frac{1}{f}$$

This equation can be solved for d_i, but by inspection $1/d_i = 1/2f$, and

$$d_i = 2f$$

Thus,

$$M = \frac{d_i}{d_o} = -\frac{2f}{2f} = -1$$

That is, when the object is at a distance of twice the focal length from the lens, a real, inverted image the same size as the object is formed at an equal distance on the opposite side of the lens.

It can be easily shown that for $d_o > 2f$, the image is smaller than the object, and for $2f > d_o > f$, the image is larger. (You can do this analytically or by quickly sketching two ray diagrams.)

Follow-up Exercise. What is the image magnification for an object placed at $d_o = 2f$ for a biconcave (diverging) lens?

Conceptual Example 23.7

Half an Image?

A converging lens forms an image on a screen, as shown in •Fig. 23.17a. Then the lower half of the lens is blocked, as shown in Fig. 23.17b. As a result, (a) only the top half of the original image will be visible on the screen; (b) only the bottom half of the original image will be visible on the screen; (c) the entire image will be visible.

Reasoning and Answer. At first thought, you might imagine that blocking off half of the lens would eliminate half of the image. However, rays from *every* point on the object pass through *all parts* of the lens. Thus, the upper half of the lens can form a total image (as could the lower half), so the answer is (c).

You might confirm this conclusion by drawing a chief ray in Fig. 23.17b. Or, you might use the scientific method and experiment—particularly if you wear eyeglasses. Block off the bottom part of your glasses, and you will find that you can still read through the top part (unless you wear bifocals).

Follow-up Exercise. Can you think of any property of the image that *would* be affected by blocking off half of the lens? Explain.

A special type of lens that you may have encountered is discussed in the Insight on p. 730.

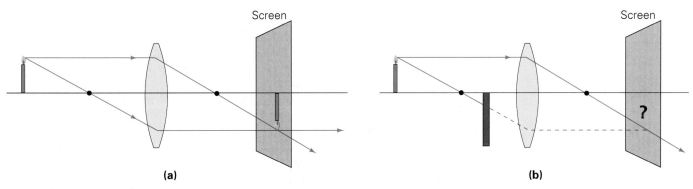

(a)   **(b)**

•**FIGURE 23.17   Half a lens, half an image?   (a)** A converging lens forms an image on a screen. **(b)** The lower half of the lens is blocked. What happens to the image? See Conceptual Example 23.7.

Fresnel Lenses

To focus or to produce a large beam of parallel light rays, a sizable converging lens is necessary. The large mass of glass necessary to form such a lens is bulky and heavy; moreover, the thick lens absorbs some of the light and is likely to show distortions. A French physicist named Augustin Fresnel (Fre-nel'; 1788–1827) developed a solution to this problem for lenses used in lighthouses.

Fresnel recognized that the refraction of light takes place at the surfaces of a lens. Hence, a lens could be made thinner—even flat—by removing glass from the interior, as long as the refracting properties of the surfaces were not changed.

This can be accomplished by cutting a series of concentric grooves in the surface of the lens (Fig. 1a). Note that the surface of each remaining curved segment is nearly parallel to the corresponding surface of the original lens. Together,

the concentric segments refract light like the original biconvex lens (Fig. 1b). In effect, the lens has simply been slimmed down by the removal of unnecessary glass between the refracting surfaces.

A lens with such a series of concentric curved surfaces is called a Fresnel lens. Such lenses are widely used in overhead projectors and in beacons (Fig. 1c). A Fresnel lens is very thin and therefore much lighter than a conventional biconvex lens with the same optical properties. Also, Fresnel lenses are easily molded from plastic—often with one flat side (plano-convex) so the lens can be attached to a glass surface.

One disadvantage of Fresnel lenses is that concentric circles are visible when an observer is looking through such a lens and when an image produced by one is projected on a screen, as when we use an overhead projector.

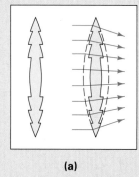

(a)

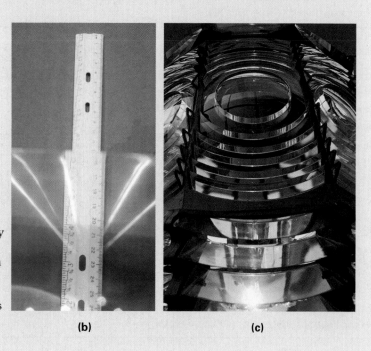

(b)   **(c)**

FIGURE 1   Fresnel lens   (a) The focusing action of a lens comes from refraction at its surfaces. It is therefore possible to reduce the thickness of a lens by cutting away glass in concentric grooves, leaving a set of curved surfaces with the same refractive properties as the lens from which they were derived. **(b)** A flat Fresnel lens with concentric curved surfaces magnifies like a biconvex converging lens. **(c)** An array of Fresnel lenses produces focused beams in this Boston harbor light. (Fresnel lenses were in fact developed for use in lighthouses.)

Combinations of Lenses

Many optical instruments such as microscopes and telescopes (Chapter 25) use a combination of lenses, or a compound lens system. When two or more lenses are used in combination, we can determine the overall image produced by considering the lenses individually in sequence. That is, the image formed by the first lens is the object for the second lens, and so on. For this reason, we present the principles of lens combinations before considering the specifics of the real-life applications.

If the first lens produces an image in front of the second lens, that image is treated as a real object for the second lens (•Fig. 23.18a). If, however, the lenses are close enough together that the image from the first lens is not formed before the rays pass through the second lens (Fig. 23.18b), then a modification must be made in the sign convention. In this case, the image from the first lens is treated as a *virtual* object for the second lens, and the object distance for it is taken to be *negative* in the lens equation (Table 23.3).

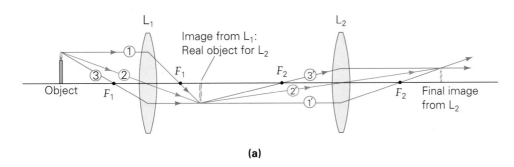

(a)

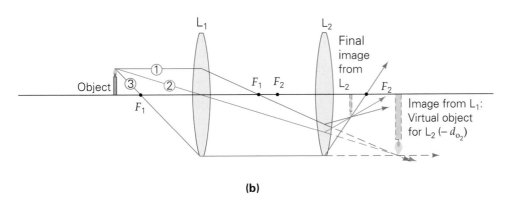

(b)

•FIGURE 23.18 Lens combinations The final image produced by a compound lens system can be found by treating the image of one lens as the object for the adjacent lens. (a) If the image of the first lens (L_1) is formed in front of the second lens (L_2), the object for the second lens is said to be real. (Note that rays 1', 2', and 3' are the parallel, chief, and focal rays, respectively, for L_2. They are *not* continuations of rays 1, 2, and 3, the parallel, chief, and focal ray for L_1.) (b) If the rays pass through the second lens before the image is formed, the object for the second lens is said to be virtual, and the object distance for the second lens is taken to be negative.

The total magnification (M_{total}) of a compound lens system is the product of the individual magnification factors of the component lenses. For example, for a two-lens system as in Fig. 23.18,

$$M_{total} = M_1 M_2 \qquad (23.8)$$

The conventional signs for M_1 and M_2 carry through the product to indicate from the sign of M_{total} whether the final image is upright or inverted. (See Exercise 70.)

Example 23.8 ■ A Special Offer: A Lens Combo and a Virtual Object

Consider two lenses similar to those illustrated in Fig. 23.18b. Suppose the object is 20 cm in front of lens L_1, which has a focal length of 15 cm. Lens L_2, with a focal length of 12 cm, is 26 cm from L_1. What is the location of the final image, and what are its characteristics?

Thinking It Through. This is a double application of the thin-lens equation. The lenses are treated successively. The image of lens L_1 becomes the object of lens L_2. We must keep the quantities distinctly labeled and the distances appropriately referenced.

Solution. We have

Given: $d_{o_1} = 20$ cm   *Find:* d_{i_2} and image characteristics
$f_1 = 15$ cm
$f_2 = 12$ cm
$D = 26$ cm (distance between lenses)

The first step is to apply the thin-lens equation (Eq. 23.6) and the magnification factor (Eq. 23.7) to L_1:

$$d_{i_1} = \frac{d_{o_1} f_1}{d_{o_1} - f_1} = \frac{(20 \text{ cm})(15 \text{ cm})}{20 \text{ cm} - 15 \text{ cm}} = 60 \text{ cm (real)}$$

$$M_1 = -\frac{d_{i_1}}{d_{o_1}} = -\frac{60 \text{ cm}}{20 \text{ cm}} = -3.0 \text{ (inverted and magnified)}$$

The image from L_1 becomes the object for L_2. This image is then $d_{i_1} - D = 60 \text{ cm} - 26 \text{ cm} = 34$ cm on the right, or image, side of L_2. Therefore, it is a virtual object (see Table 23.3), and $d_{o_2} = -34$ cm. (The d_o for virtual objects is taken to be negative.)

Then we apply the equations to the second lens L_2:

$$d_{i_2} = \frac{d_{o_2} f_2}{d_{o_2} - f_2} = \frac{(-34 \text{ cm})(12 \text{ cm})}{-34 \text{ cm} - 12 \text{ cm}} = 8.9 \text{ cm (real)}$$

and

$$M_2 = -\frac{d_{i_2}}{d_{o_2}} = \frac{-8.9 \text{ cm}}{(-34 \text{ cm})} = 0.26 \text{ (upright and reduced)}$$

(Note: The virtual object for L_2 was inverted, and "upright" means that the final image is also inverted.)

The total magnification M_{total} is then

$$M_{\text{total}} = M_1 M_2 = (-3.0)(0.26) = -0.78$$

The sign is carried through with the magnifications. We determine that the final real image is located 8.9 cm on the right (image) side of L_2 and that it is inverted (minus) relative to the initial object and reduced.

Follow-up Exercise. Suppose the object in Fig. 23.18b were located 30 cm in front of L_1. Where would the final image be formed in this case, and what would be its characteristics?

23.4 Lens Aberrations

Objectives: To (a) describe some common lens aberrations and (b) explain how they be can be reduced or corrected.

Lenses, like mirrors, can also have aberrations. Here are some common ones.

Spherical Aberration

The discussion of lenses thus far has focused on thin lenses, for which the lateral deviation of light due to refraction is negligible. In general, with spherical mirrors and lenses, light rays parallel to and near the axis will be reflected or refracted to converge at the focal point. Like spherical mirrors, however, converging lenses may show **spherical aberration**, an effect that occurs when parallel rays passing through different regions of a lens do not come together on a common focal plane.

In general, rays close to the axis of a converging lens are refracted less and come together at a point farther from the lens than do rays passing through the periphery (•Fig. 23.19a). The place where the transmitted light beam has the smallest cross section is called the *circle of least confusion*, and the best (least distorted) image is formed at this location.

Spherical aberration can be minimized by using an aperture to reduce the effective area of the lens, so that only light rays near the axis are transmitted. Also, combinations of converging and diverging lenses can be used. The aberration of one lens can be compensated for (nullified) by the optical properties of another lens.

Chromatic Aberration

Chromatic aberration is an effect that occurs because the index of refraction of the material making up a lens is not the same for all wavelengths (that is, the material is dispersive). When white light is incident on a lens, the transmitted rays of different wavelengths (colors) do not have a common focal point, and images of different colors are produced at different locations (Fig. 23.19b).

This dispersive aberration can be minimized, but not eliminated, by using a compound lens system consisting of lenses of different materials, such as crown glass and flint glass. The lenses are chosen so that the dispersion produced by one is compensated for by opposite dispersion produced by the other. With a properly constructed two-component lens system, called an *achromatic doublet* (achromatic means without color), the images of any two selected colors can be made to coincide.

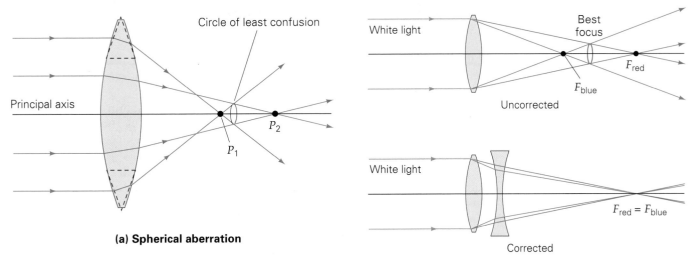

(a) Spherical aberration

(b) Chromatic aberration

•**FIGURE 23.19 Lens aberrations** **(a)** Spherical aberration. In general, rays closer to the axis of a lens are refracted less and come together at a point farther from the lens than do rays passing through the periphery of the lens. The smallest cross section of the transmitted beam is called the circle of least confusion, and the best (least distorted) image will be formed in its plane. **(b)** Chromatic aberration. Because of dispersion, different wavelengths (colors) of light are focused in different planes, which results in distortion of the overall image (top). This aberration can be corrected by using a converging lens and a diverging lens of different materials that compensate for each other's dispersion. The lens pair is called an achromatic doublet (bottom).

Astigmatism

A circular beam of light along the lens axis forms a circular illuminated area on the lens. When incident on a converging lens, the parallel beam converges at the focal point. However, when a circular cone of light from an off-axis source falls on the convex spherical surface of a lens some distance away, the light forms an elliptical illuminated area on the lens. The rays entering along the major and minor axes of the ellipse then focus at different points after passing through the lens. This condition is called **astigmatism**.

With different focal points in different planes, the images in both planes are blurred. For example, the image of a point is no longer a point but two separated short line images (blurred points). As with spherical aberration, the best image is formed somewhere between the images at the location of the circle of least confusion. Astigmatism can be reduced by reducing the effective area of the lens with an aperature or by adding a cylindrical lens to compensate.

*23.5   The Lens Maker's Equation

Objectives: To **(a)** describe the lens maker's equation and **(b)** explain how its application differs from that of the thin-lens equation.

The biconvex and biconcave thin lenses considered so far in this chapter have been relatively easy to analyze. However, there are a variety of other shapes of lenses, as illustrated in Fig. 23.14. For them, the analysis becomes more involved, but it is important to know the focal lengths of such lenses for optical considerations. (Lenses are ground for specific purposes or applications.)

Lens refraction depends on the shapes of the surfaces and also on the index of refraction of the lens. These properties are related to the focal length of a thin lens by the **lens maker's equation,** which gives the focal length of a thin lens in air ($n_{air} = 1$):

$$\frac{1}{f} = (n - 1)\left(\frac{1}{R_1} - \frac{1}{R_2}\right) \qquad \text{(for thin lens in air)} \qquad (23.9)$$

$+R$	when C is on the side of the lens from which light emerges, or on the back side of lens
$-R$	when C is on side of the lens on which light is incident, or on the front side of the lens
$R = \infty$	for a plane (flat) surface
$+f$	converging (positive) lens
$-f$	diverging (negative) lens

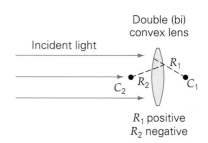

R_1 positive
R_2 negative

•FIGURE 23.20 Centers of curvature Lenses, such as this biconvex lens, have two centers of curvature, which define the signs of the radii of curvature. See Table 23.3.

where n is the index of refraction of the lens material and R_1 and R_2 are the radii of curvature of the first (front side) and second (back side) lens surfaces, respectively. (The first surface is the one on which light from an object is first incident.) The equation locates the focal point at which light passing through a lens will converge for a converging lens or the focal point from which transmitted light appears to diverge for a diverging lens.

A sign convention is required for the lens maker's equation, and a common one is summarized in Table 23.4. The signs depend on the location of a surface's center of curvature (C) relative to the side of the lens on which light is incident or to the side from which light emerges.

If the lens is in a fluid other than air, then the first term in parentheses in Eq. 23.9 becomes $(n/n_m) - 1$, where n and n_m are the indices of refraction of the lens material and the fluid medium, respectively. Now we can see why some converging lenses in air become diverging when submerged in water: If $n_m > n$, then f is negative, and the lens is diverging.

The lens maker's equation allows us to see why the focal length of a biconvex lens is not equal to half the radius of curvature ($f \neq R/2$), as it was for a spherical mirror (Section 23.2). The sign convention for the the lens maker's equation is applied to a biconvex lens in •Fig. 23.20. Note the centers of curvature C_1 and C_2. The radii of curvature extend from these, and by our convention, R_1 is positive and R_2 is negative. (Do you agree?) Because the two sides of the lens have equal curvature, R_1 and R_2 have the same length, so let's say that $R_1 = +R$ and $R_2 = -R$. Substituting into Eq. 23.9, we have

$$\frac{1}{f} = (n - 1)\left[\frac{1}{R} - \left(\frac{-1}{R}\right)\right] = (n - 1)\left(\frac{2}{R}\right)$$

or

$$f = \frac{R}{2(n - 1)}$$

To have $f = R/2$ requires that the index of refraction for the glass lens be $n = 2$. Known glasses do not have an index of refraction this large (see Table 22.1).

Lens Power: Diopters

Notice that the lens maker's equation (Eq. 23.9) gives the inverse focal length $1/f$. Optometrists use this inverse relationship to express the *lens power* (P) of a lens in units called **diopters** (abbreviated D). The lens power is the reciprocal of the focal length of the lens expressed in *meters*:

$$P \text{ (expressed in diopters)} = \frac{1}{f \text{ (expressed in meters)}} \qquad 23.10$$

So, $1\,D = 1\,m^{-1}$. The lens maker's equation gives a lens's power ($1/f$) in diopters if the radii of curvature are expressed in meters.

If you wear glasses, you may have noticed that the prescription the optometrist gave you for your eyeglass lenses was written in terms of diopters. Converging and diverging lenses are referred to as positive ($+$) and negative ($-$) lenses, respectively. Thus, if an optometrist prescribes a corrective lens with a power of $+2$ diopters, it is a converging lens with a focal length of

$$f = \frac{1}{P} = \frac{1}{+2D} = \frac{1}{2\,m^{-1}} = 0.50\,m = 50\,cm$$

The greater the power of the lens in diopters, the shorter its focal length and the more converging or diverging it is. A "stronger" prescription lens (greater lens power) has a smaller f.

Chapter Review

Important Terms

plane mirror   711
virtual image   711
real image   711
lateral magnification   711
spherical mirror 714
concave (converging) mirror   715
convex (diverging) mirror   715
center of curvature   715
radius of curvature   715
focal point   715
focal length   715

parallel ray (for mirror)   716
chief (radial) ray (for mirror)   716
focal ray (for mirror)   716
spherical mirror equation   719
magnification factor (for spherical
   mirror)   719
spherical aberration (for mirror)   723
lens   724
converging (biconvex) lens   724
diverging (biconcave) lens   725
parallel ray (for lens)   726

chief (central) ray (for lens)   726
focal ray (for lens)   726
thin-lens equation   727
magnification factor
   (for lens)   727
spherical aberration
   (for lens)   732
chromatic aberration   732
astigmatism   733
*lens maker's equation   733
*diopters   734

Important Concepts

- Plane mirrors form virtual, upright, and unmagni-fied images.
- Spherical mirrors are either concave (converging) or convex (diverging). Diverging spherical mirrors al-ways form upright, reduced, virtual images.
- Bispherical lenses are either convex (converging) or concave (diverging). Diverging spherical lenses al-ways form upright, reduced, virtual images.

- In finding the location and determining the charac-teristics of the image formed by a mirror or lens an-alytically, an initial quick ray diagram sketch is helpful and recommended.
- The thin-lens equation relates focal length, object distance, and image distance.
- *The lens maker's equation is used to compute the grinding radii for the desired focal length of a lens.

Important Equations

Focal Length of Spherical Mirror:

$$f = \frac{R}{2} \tag{23.2}$$

Spherical Mirror Equation[†]:

$$\frac{1}{d_o} + \frac{1}{d_i} = \frac{1}{f} = \frac{2}{R} \quad \text{or} \quad d_i = \frac{d_o f}{d_o - f} \tag{23.3, 23.4}$$

Thin-Lens Equation[†] (where $f \neq R/2$):

$$\frac{1}{d_o} + \frac{1}{d_i} = \frac{1}{f} \quad \text{or} \quad d_i = \frac{d_o f}{d_o - f} \tag{23.6}$$

Magnification Factor[†] (spherical mirror and thin lens):

$$M = -\frac{d_i}{d_o} \tag{23.5, 23.7}$$

Total Magnification with a Two-Lens System:

$$M_{total} = M_1 M_2 \tag{23.8}$$

Lens Maker's Equation (thin lens in air only):

$$\frac{1}{f} = (n - 1)\left(\frac{1}{R_1} - \frac{1}{R_2}\right) \tag{23.9}$$

Lens Power in Diopters (where f is in meters):

$$P = \frac{1}{f} \tag{23.10}$$

Exercises

23.1 Plane Mirrors

1. 💿 A plane mirror (a) has a greater image distance than it has an object distance, (b) produces a virtual, upright, unmagnified image, (c) changes the vertical orientation of an object, (d) reverses top and bottom.

[†]See Tables 23.2 through 23.4 for sign conventions.

2. 💿 A plane mirror (a) can be used to magnify, (b) produces both real and virtual images, (c) always produces a virtual image, (d) forms images by diffuse reflection.

3. (a) A transparent window pane can serve as a mirror, but usually only when it is dusk or dark outside. Why? (b) When looking at a reflecting window pane at night, you may see two similar images. Why? (c) One-way, half-silvered mirrors reflect on one side and can be seen through

on the other. (This effect is sometimes used on sunglasses.) What is the principle used in making a one-way mirror? [Hint: At night a window pane may be a one-way mirror.]

4. Day–night rearview mirrors are common in cars. You tilt the mirror backward, and the intensity and glare of headlights behind you at night are reduced (•Fig. 23.21). The mirror is wedge-shaped and is silvered on the back. The effect has to do with front-surface and back-surface reflections. The unsilvered front surface reflects about 5% of incident light; the silvered back surface reflects about 90% of the incident light. Explain how the day–night mirror works.

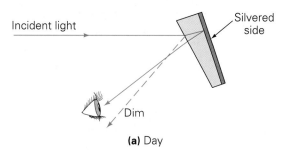

(a) Day

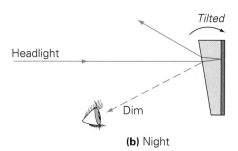

(b) Night

•**FIGURE 23.21   Automobile day–night mirror**   See Exercise 4.

5. When you stand in front of a plane mirror, there is a right–left reversal. (a) Why is there not a top–bottom reversal of your body? (b) Could you effect an apparent top–bottom reversal by positioning your body differently?

6. What is the magnification of a plane mirror?

7. Why do some emergency vehicles have ƎƆИA⅃UᙠMA (•Fig. 23.22) written on the front?

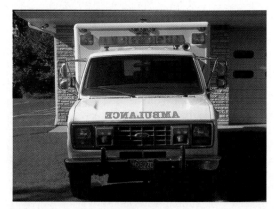

•**FIGURE 23.22   Backward and reversed**   See Exercise 7.

8. ■ 🌀 A person stands 2.0 m away from the reflecting surface of a plane mirror. (a) What is the apparent distance between the person and his or her image? (b) What are the image characteristics?

9. ■ An object 5.0 cm tall is placed 40 cm from a plane mirror. Find (a) the distance from the object to the image, (b) the height of the image, and (c) its magnification.

10. ■ Standing 2.5 m in front of a plane mirror with your camera, you decide to take a picture of yourself. To what distance should the camera be manually focused so that you will get a complete image?

11. ■■ A small dog sits at a distance of 1.5 m in front of a plane mirror. (a) Where is the dog's image? (b) If the dog jumps at the mirror with a speed of 0.50 m/s, how fast does the dog approach its image?

12. ■■ A woman fixing the hair on the back of her head holds a hand (plane) mirror 30 cm in front of her face so as to look into a plane mirror on the bathroom wall behind her. If she is 90 cm from the wall mirror, approximately how far does the image of the back of her head appear in front of her?

13. ■■ You observe multiple images when you stand between plane mirrors on opposite walls in a dance studio. If you stand 3.0 m from the mirror on the north wall and 5.0 m from the mirror on the south wall, what are the image distances for the first two images in both mirrors?

14. ■■ A woman 1.70 m tall stands 3.0 m in front of a plane mirror. (a) What is the minimum height the mirror must be to allow the woman to view her complete image from head to foot? Assume that her eyes are 10 cm below the top of her head. (b) What would be the minimum height of the mirror if she stood 5.0 m away?

15. ■■ Prove that $d_o = d_i$ for plane mirror. [Hint: Refer to Fig. 23.2, and use similar triangles.]

16. ■■■ Draw ray diagrams showing how three images of an object are formed in two plane mirrors at right angles as shown in •Fig. 23.23a. [Hint: Consider rays from both ends of the object in the drawing for each image.] Figure 23.23b

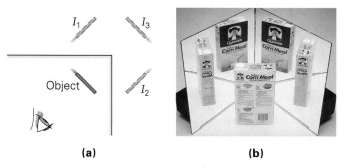

(a)   **(b)**

•**FIGURE 23.23   Two mirrors—multiple images**   See Exercise 16.

shows a similar situation from a different point of view that gives four images. Explain the extra image in this case.

23.2 Spherical Mirrors

17. ⚫ Which one of the following statements concerning spherical mirrors is correct? (a) A converging mirror alone can produce an inverted virtual image. (b) A diverging mirror alone can produce an inverted virtual image. (c) A diverging mirror can produce an inverted real image. (d) A converging mirror can produce an upright real image.

18. ⚫ The image produced by a convex mirror is always (a) virtual and upright, (b) real and upright, (c) virtual and inverted, (d) real and inverted.

19. (a) What is the purpose of using a dual mirror on a car or truck, such as the one shown in ●Fig. 23.24? (b) Some rearview mirrors on the passenger side of automobiles have the warning "OBJECTS IN MIRROR ARE CLOSER THAN THEY APPEAR." Explain why this is so. (c) Could a TV satellite dish as shown in Fig. 20.19 be considered a converging mirror? Explain.

●**FIGURE 23.24   Mirror applications**   See Exercise 19.

20. (a) If you look into a shiny spoon, you see an inverted image on one side and an upright image on the other. (Try it.) Why? (b) Could you see upright images on both sides? Explain.

21. (a) A 10-centimeter-tall mirror is advertised as a "Full-view mini mirror. See your full body in 10 cm." How can this be? (b) A popular novelty item consists of a concave mirror with a ball suspended at or slightly inside the center of curvature (●Fig. 23.25). When the ball swings

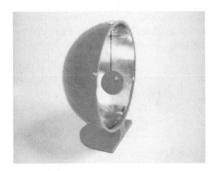

●**FIGURE 23.25   Spherical mirror toy**   See Exercise 21.

toward the mirror, its image grows larger and suddenly fills the whole mirror. The image appears to be jumping out of the mirror. Explain what is happening.

22. How can the focal length be determined experimentally for a concave mirror?

23. ■ A spherical mirror has a radius of curvature of 10 cm. What is the focal length of this mirror?

24. ■ ⚫ An object 3.0 cm tall is placed 20 cm from the front of a concave mirror with a radius of curvature of 30 cm. Where is the image formed, and how tall is it?

25. ■ ⚫ If the object in Exercise 24 is moved to a position 10 cm from the front of the mirror, what will the characteristics of the image be?

26. ■ ⚫ If an object is 30 cm in front of a convex mirror that has a focal length of 60 cm, how far behind the mirror will the image appear to an observer? How tall will the image be?

27. ■ A candle with a flame 1.5 cm tall is placed 5.0 cm from the front of a concave mirror. A virtual image is produced that is 10 cm from the vertex of the mirror. (a) Find the focal length and radius of curvature of the mirror. (b) How tall is the image of the flame?

28. ■■ ⚫ An object 3.0 cm tall is placed at different locations in front of a concave mirror whose radius of curvature is 30 cm. Determine the location of the image and its characteristics when the object distance is (a) 40 cm, (b) 30 cm, (c) 15 cm, and (d) 5.0 cm.

29. ■■ ⚫ Draw a ray diagram for each case in Exercise 28.

30. ■■ ⚫ Using the spherical mirror equation and the magnification factor, show that for a concave mirror with $d_o < f$, the image of an object is always virtual, upright, and magnified.

31. ■■ ⚫ Using the spherical mirror equation and the magnification factor, show that for a convex mirror, the image of an object is always virtual, upright, and reduced.

32. ■■ The erect image of an object 18 cm in front of a convex mirror is half the size of the object. What is the focal length of the mirror?

33. ■■ A concave shaving mirror is constructed so that a man at a distance of 20 cm from the mirror sees his image magnified 1.5 times. What is the radius of curvature of the mirror?

34. ■■ A concave mirror has a magnification of +3.0 for an object placed 50 cm in front of it. (a) What type of image is produced? (b) Find the radius of curvature of the mirror.

35. ■■ ⚫ A child looks at a reflecting Christmas tree ball that has a diameter of 9.0 cm and sees an image of her face that is half the real size. How far is the child's face from the ball?

36. ■■ A dentist uses a spherical mirror that produces an upright image that is magnified four times. What kind of mirror is it? What is its focal length in terms of the object distance?

37. ■■ A 15-centimeter-long pencil is placed with its eraser on the optic axis of a concave mirror and its point directed upward at a distance of 20 cm in front of the mirror. The radius of curvature of the mirror is 30 cm. (a) Where is the image of the pencil formed, and what are its characteristics? (b) Draw a ray diagram for the situation in which the pencil point is directed downward from the optic axis.

38. ■■ ◐ A pill bottle 3.0 cm tall is placed 12 cm in front of a mirror. An upright image 9.0 cm tall is formed. What kind of mirror is it, and what is its radius of curvature?

39. ■■ ◐ An amusement park spherical mirror shows anyone who stands 2.5 m in front of it an upright image two times the person's height. What is its radius of curvature?

40. ■■ ◐ For values of d_o from 0 to ∞, (a) sketch graphs of (1) d_i versus d_o and (2) M versus d_o for a converging mirror. (b) Sketch similar graphs for a diverging mirror.

41. ■■■ The front surface of a wooden cube 5.0 cm on each side is placed a distance of 30 cm in front of a converging mirror having a focal length of 20 cm. Where is the image of the front surface of the cube located, and what are its characteristics?

42. ■■■ A section of a sphere is mirrored on both sides. If the magnification of an object is 11.8 when the section is used as a concave mirror, what is the magnification of an object at the same distance in front of the convex side?

43. ■■■ ◐ A concave mirror has a radius of curvature of 20 cm. For what two object distances will the image have twice the height of the object?

44. ■■■ ◐ A convex mirror is on the exterior of the passenger side of many cars. If the focal length of such a mirror is −40.0 cm, what will be the location and height of the image of a car that is 2.0 m tall and (a) 100 m behind the mirror and (b) 10.0 m behind the mirror? (See Exercise 19b.)

45. ■■■ ◐ Two students in a physics laboratory each have a concave mirror with the same radius of curvature, 40 cm. Each student places an object in front of a mirror. The image in both mirrors is three times the size of the object. However, when the students compare notes, they find that the object distances are not the same. Is this possible? If so, what are the object distances?

23.3 Lenses
and
23.4 Lens Aberrations

46. A diverging lens (a) must have at least one concave surface, (b) always produces a virtual image, (c) is thinner at its center than at the periphery, (d) all of these.

47. ◐ The image produced by a diverging lens is always (a) virtual and magnified, (b) real and magnified, (c) virtual and reduced, (d) real and reduced.

48. A lens aberration that is caused by dispersion is called (a) spherical aberration, (b) chromatic aberration, (c) refractive aberration, (d) none of these.

49. Explain why a fish in a spherical fish bowl appears larger than it really is.

50. ◐ Does a converging lens ever form a virtual image of a real object?

51. ■ ◐ An object is placed 50.0 cm in front of a converging lens of focal length 10.0 cm. What are the image distance and the lateral magnification?

52. ■ ◐ An object placed 30 cm in front of a converging lens forms an image 15 cm behind the lens. What is the focal length of the lens?

53. ■ ◐ A converging lens with a focal length of 20 cm is used to produce an image on a screen that is 2.0 m from the lens. What is the object distance?

54. ■■ ◐ An object 4.0 cm tall is in front of a converging lens of focal length 22 cm. Where is the image formed, and what are its characteristics, if the object distance is (a) 15 cm and (b) 36 cm? Sketch ray diagrams.

55. ■■ ◐ An object is placed in front of a biconcave lens whose focal length is 18 cm. Where is the image located, and what are its characteristics, if the object distance is (a) 10 cm and (b) 25 cm? Sketch ray diagrams for each.

56. ■■ Using the thin-lens equation and the magnification factor, show that for a spherical diverging lens, the image of a real object is always virtual, upright, and reduced.

57. ■■ A biconvex lens has a focal length of 0.12 m. Where on the lens axis should an object be placed in order to get (a) a real, enlarged image with a magnification of 2.0 and (b) a virtual, enlarged image with a magnification of 2.0?

58. ■■ (a) Design a single-lens projector that will form a sharp image on a screen 4.0 m away with the transparent slides 6.0 cm from the lens. (b) If the object on a slide is 1.0 cm tall, how tall will the image on the screen be, and how should the slide be placed in the projector?

59. ■■ A simple single-lens camera (biconvex lens) is used to photograph a man 1.7 m tall who is standing 4.0 m from the camera. If the man's image fills the length of a frame of film (35 mm), what is the focal length of the lens?

60. ■■ ◐ An object 5.0 cm tall is 10 cm from a concave lens. The resulting image is one-fifth as large as the object. What is the focal length of the lens?

61. ■■ An object is placed 40 cm from a screen. (a) At what point between the object and the screen should a converging lens with a focal length of 10 cm be placed so it will produce a sharp image on the screen? (b) What is the lens's magnification?

62. ■■ Using •Fig. 23.26, derive (a) the thin-lens equation and (b) the magnification equation for a thin lens. [Hint: Use similar triangles.]

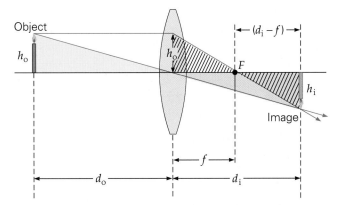

•**FIGURE 23.26   The thin-lens equation**   The geometry for deriving the thin-lens equation (and magnification factor). Note the two sets of similar triangles. See Exercise 62.

63. ■■ (a) For a biconvex lens, what is the *minimum* distance between an object and its image if the image is real? (b) What is the distance if the image is virtual? [Hint: Use ray diagrams and the thin lens equation.]

64. ■■ With a magnifying glass, a biology student on a field trip views a small insect. If she sees the insect magnified by a factor of 3.5 when the glass is held 3.0 cm from it, what is the focal length of the lens?

65. ■■ (a) If a book is held 30 cm from an eyeglass lens with a focal length of −45 cm, where is the image of the print formed? (b) If an eyeglass lens with a focal length of +57 cm is used, where is the image formed?

66. ■■ For the arrangement shown in •Fig. 23.27, an object is placed 0.40 m in front of the converging lens, which has a focal length of 0.15 m. If the concave mirror has a focal length of 0.13 m, where is the final image formed, and what are its characteristics?

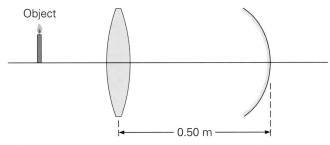

•**FIGURE 23.27   Lens–mirror combination**   See Exercise 66.

67. ■■ A simple camera uses a single converging lens with a focal length of 4.5 cm. A picture is taken of a 26-centimeter-tall physics book standing on a table at a distance of 1.5 m from the camera. If a sharp image is formed on the film, how far is the film from the lens? (b) What is the height of the image of the book on the film?

68. ■■■ 🌐 The geometry of a compound microscope, which consists of two converging lenses, is shown in •Fig. 23.28. (More detail on microscopes is given in Chapter 25.) The objective lens and the eyepiece lens have focal lengths of 2.8 mm and 3.3 cm, respectively. If an object is located 3.0 mm from the objective lens, where is the final image located, and what type of image is it?

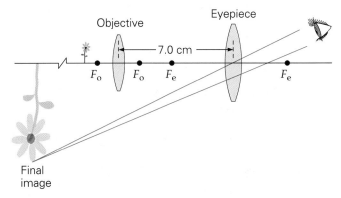

•**FIGURE 23.28   Compound microscope**   See Exercise 68.

69. ■■■ Two converging lenses, L_1 and L_2, have focal lengths of 30 cm and 20 cm, respectively. The lenses are placed 60 cm apart along the same axis, and an object is placed 50 cm from L_1 on the side opposite L_2. Where is the image formed relative to L_2, and what are its characteristics?

70. ■■■ For a lens combination, why is the total magnification $M_{total} = M_1 M_2$? [Hint: Think about the definition of magnification.]

71. ■■■ Show that for thin lenses that have focal lengths f_1 and f_2 and are in contact, the effective focal length (f) is given by

$$\frac{1}{f} = \frac{1}{f_1} + \frac{1}{f_2}$$

*23.5 The Lens Maker's Equation

72. The power of a lens is expressed in units of (a) watts, (b) diopters, (c) meters, (d) both (b) and (c).

73. When a lens with $n = 1.60$ is immersed in water, (a) is there a change in the focal length of the lens? (b) What would be the case for a submerged lens with $n = 1.30$?

74. A converging lens in air is submerged in a fluid whose index of refraction is greater than that of the lens. Is the lens still converging?

75. What is the focal length of a rectangular glass block?

76. ■■ An optometrist prescribes a corrective lens with a power of +1.5 diopters. The lens maker will start with glass blank with an index of refraction of 1.6 and a convex front surface whose radius of curvature is 20 cm. To what radius of curvature should the other surface be ground?

77. ■■ A plastic plano-concave lens has a radius of curvature of 50 cm for its concave surface. If the index of refraction of the plastic is 1.35, what is the power of the lens?

78. ■■■ A diver wants to use a biconvex lens that is made of glass whose index of refraction is 1.6 and that has radii of curvature of 30 cm for one surface and 40 cm for the other. (a) What is the focal length of the lens if it is used in the air? (b) What is the focal length if she takes it under water?

Additional Exercises

79. To photograph a full Moon, a photographer uses a single-lens camera having a focal length of 60 mm. What will be the diameter of the Moon's image on the film? [Note: Data about the Moon are inside the back cover.]

80. 🌐 A virtual image of magnification 0.50 is produced when an object is placed in front of a spherical mirror. (a) What type of mirror is it? (b) Find the radius of curvature of the mirror if the object is 7.0 cm in front of it.

81. A bottle 6.0 cm tall is located 75 cm from the concave surface of a mirror with a radius of curvature of 50 cm. Where is the image located, and what are its characteristics?

82. A boy runs with a speed of 4.5 km/h toward a plane mirror. At what speed, in meters per second, does his image approach him?

83. A method of determining the focal length of a diverging lens is called *autocollimation*. As • Fig. 23.29 shows, first a sharp image of a light source is projected on a screen by a converging lens. Second, the screen is replaced with a plane mirror. Third, a diverging lens is placed between the converging lens and the mirror. Light will then be reflected by the mirror back through the compound lens system, and an image will be formed on a screen near the light source. This image is made sharp by adjusting the distance between the diverging lens and the mirror. The distance at which the image is clearest is equal to the focal length of the lens. Explain why this is true.

84. Show that the magnification for objects near the optic axis of a convex mirror is given by $|M| = d_i/d_o$. [Hint: Use a ray diagram with rays reflected at the mirror's vertex.]

85. A biconvex lens produces a real, inverted image of an object that is magnified 2.5 times when the object is 20 cm from the lens. What is the focal length of the lens?

86. Two lenses, each having a power of +10 diopters, are placed 20 cm apart along the same axis. If an object is 60 cm from the first lens on the side opposite the second

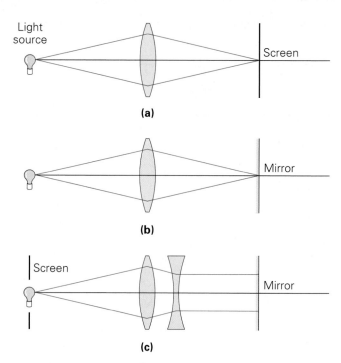

(a)

(b)

(c)

•**FIGURE 23.29 Autocollimation** See Exercise 83.

lens, where is the final image relative to the first lens, and what are its characteristics?

87. (a) For values of d_o from 0 to ∞, sketch graphs of (1) d_i versus d_o and (2) M versus d_o for a converging lens. (b) Sketch similar graphs for a diverging lens.

88. 🌐 An object is 15 cm from a converging lens whose focal length is 10 cm. On the opposite side of that lens, at a distance of 60 cm, is a converging lens with a focal length of 20 cm. Where is the final image formed, and what are its characteristics?

89. (a) Show that a ray parallel to the optic axis of a biconvex lens is refracted toward the axis at the incident surface and again at the exit surface. (b) Show that the refraction effect also holds for a biconcave lens, but with deflections away from the axis.

90. An object 3.0 cm tall is placed 15 cm from the front of a convex mirror whose focal length is 25 cm. Find the location of the image and its height.

91. A converging glass lens with an index of refraction of 1.62 has a focal length of 30 cm in air. What is the focal length when the lens is submerged in water?

92. The image of an object located 30 cm from a concave mirror is formed on a screen located 20 cm from the mirror. What is the mirror's radius of curvature?

93. A concave cosmetic mirror produces a virtual image 1.5 times the size of a person whose face is 20 cm from the mirror. (a) Draw a ray diagram of this situation. (b) What is the focal length of the mirror?

Physical Optics: The Wave Nature of Light

Insights

- Nonreflecting Lenses
- LCDs and Polarized Light

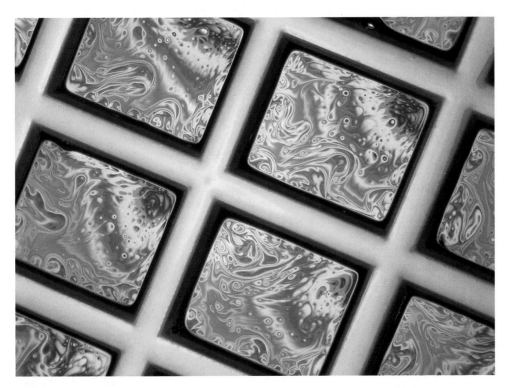

It's always intriguing to see brilliant colors produced by objects that we know don't ordinarily have any colors of their own. The glass of a prism, for example, clear and transparent by itself, nevertheless gives rise to a whole array of colors when white light passes through it. Prisms, like the water droplets that produce the rainbow, don't create anything that isn't already there. They merely separate the different wavelengths that make up white light and spread them out.

The phenomena of reflection and refraction are conveniently analyzed by using geometrical optics. Ray diagrams show what happens when light is reflected from a mirror or passed through a lens. However, some other phenomena involving light, such as the interference patterns in the chapter opening photo, cannot be adequately explained or described with rays, since this technique ignores the wave nature of light. Other such phenomena include diffraction and polarization.

Physical optics, or **wave optics**, takes into account those wave properties that geometrical optics ignores. The wave theory of light leads to satisfactory explanations of those phenomena that cannot be analyzed with rays. Thus, this chapter again considers waveforms.

24.1 Young's Double-Slit Experiment

Objectives: To (a) explain how Young's experiment demonstrated the wave nature of light and (b) compute the wavelength of light from experimental results.

It has been stated that light behaves like a wave, but no proof of this has been given. How would you go about demonstrating the wave nature of light? One method that involves the use of interference was first devised in 1801 by the English scientist Thomas Young (1773–1829). **Young's double-slit experiment** not only demonstrates the wave nature of light but also allows the measurement of its wavelengths.

Recall from the discussion of wave interference in Sections 13.4 and 14.4 that superimposed waves may interfere constructively or destructively. Total constructive interference occurs when two crests are superimposed, and total destructive interference occurs when a crest and a trough of two identical waves are superimposed. Interference can be observed with water waves (•Fig. 24.1), for which constructive and destructive interference produce obvious interference patterns.

The interference of light waves is not as easily observed because of their relatively short wavelengths ($\approx 10^{-7}$ m). Also, stationary interference patterns are produced only with *coherent sources*—sources that produce light waves having a constant phase relationship to one another. For example, for constructive interference to occur at some point, the waves meeting at that point must be in phase. As the waves meet, a crest must always overlap a crest and a trough must always overlap a trough. If a phase difference develops between the waves over time, the interference pattern changes, and a stable or stationary pattern will not be set up.

In an ordinary light source, the atoms are excited randomly, and the emitted light waves fluctuate in amplitude and frequency. Thus, light from two such sources is incoherent and does not produce a stationary interference pattern. Interference does occur, but the phase difference between the interfering waves changes so fast that the interference effects are not discernible. To obtain two coherent sources, a barrier with one narrow slit is placed in front of a light source, and a barrier with two slits are positioned symmetrically in front of the first (•Fig. 24.2a). Waves prop-

Note: Compare Fig. 24.1 with Fig. 14.7a.

•**FIGURE 24.1 Water wave interference** The constructive and destructive interference of water waves from two coherent sources in a ripple tank produce interference patterns.

•**FIGURE 24.2 Double-slit interference** (a) The coherent waves from two slits are shown in blue (top slit) and red (bottom slit). The waves spread out as a result of diffraction from narrow slits ($\lambda = d$). The waves interfere, producing alternating maxima and minima, or bright and dark fringes, on the screen. (b) An interference pattern. Note the symmetry of the pattern about the central maximum ($n = 0$).

Light source

S_1

S_2

Single slit

Double slit

Screen

Max ($n = 2$)
Min
Max ($n = 1$)
Min
Max ($n = 0$)
Min
Max ($n = 1$)
Min
Max ($n = 2$)

$n = 4$
$n = 3$
$n = 2$
$n = 1$
$n = 0$
$n = 1$
$n = 2$
$n = 3$
$n = 4$

(a)

(b)

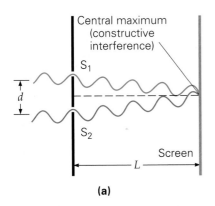 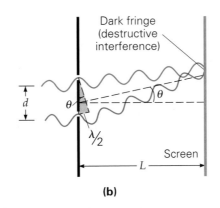 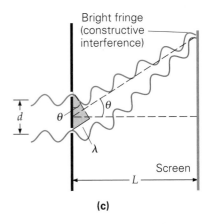

(a)   **(b)**   **(c)**

•**FIGURE 24.3 Interference** The interference that produces bright or dark fringes depends on the difference in the path lengths of the light from the two slits. **(a)** The path-length difference at the position of the central maximum is zero, so the waves arrive in phase and interfere constructively. **(b)** At the position of the first dark fringe, the path-length difference is $\lambda/2$, and the waves interfere destructively. **(c)** At the position of the first bright fringe, the path-length difference is λ, and the interference is constructive.

agating out from the single slit are in phase, and the double slits then act as two coherent sources by separating each wave into two parts. Any random changes in the light from the original source will thus occur for the light passing through both slits, and the phase difference will be constant.

A series of bright lines is then observed on a screen placed relatively far from the slits (Fig. 24.2b). This pattern results from *wave interference.*

To help analyze Young's experiment, let's imagine that light with a single wavelength (monochromatic light) is used. Because of diffraction, or the bending of the light around the corners of the slits, the waves spread out and interfere as illustrated in Fig. 24.2a. Coming from two coherent "sources," the interfering waves produce a stable interference pattern on the screen. The pattern consists of a bright central maximum (•Fig. 24.3a) and a series of symmetrical dark (Fig. 24.3b) and bright (Fig. 24.3c) side fringes, which mark the positions at which destructive and constructive interference occurs. The bright fringes decrease in intensity on either side of this central maximum. Hence, the interference pattern demonstrates the wave nature of light.

Measuring the wavelength of light requires us to look at the geometry of Young's experiment, as shown in •Fig. 24.4. Let the screen be a distance L from the slits and P be a point at the center of an arbitrary maximum, or bright side fringe. P is located a distance y from the center of the central maximum and at an angle θ relative to a normal line between the slits. The slits S_1 and S_2 are separated by a distance d. Note that the light path from slit S_1 to P is longer than the path from slit S_2 to P. As the figure shows, the path-length difference (ΔL) is

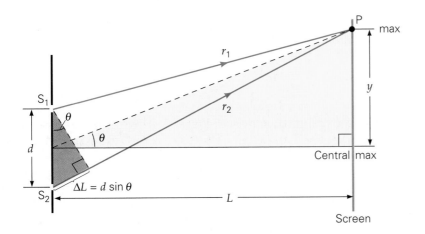

•**FIGURE 24.4 Geometry of Young's double-slit experiment** The difference in the path lengths for light traveling from the two slits to a point P of an arbitrary maximum, or bright fringe, is $r_2 - r_1 = \Delta L$, which forms a side of the small shaded triangle. Because the barrier with the slits is parallel to the screen, the angle between r_2 and the barrier (at S_2, in the small shaded triangle) is equal to the angle between r_2 and the screen. When L is much greater than y, that angle is almost identical to the angle between the screen and the dashed line, which is an angle in the large shaded triangle. The two shaded triangles are then almost exactly similar, and the angle at S_1 in the small triangle is almost exactly equal to θ. Thus, $\Delta L = d \sin \theta$. By the condition for constructive interference, $d \sin \theta = n\lambda$, where $n = 0, 1, 2, 3, \ldots$ is the order of interference. (Not drawn to scale. Assume $d \ll L$.)

$$\Delta L = d \sin \theta$$

The fact that the angle in the small shaded triangle is equal to θ can be shown by a simple geometrical argument involving similar triangles, as described in the caption of Figure 24.4.

The relationship of the phase difference of two waves to their path-length difference was discussed in Chapter 14 for the interference of sound waves. The conditions for interference hold for any sinusoidal waves, including light waves. Constructive interference occurs at any point where the path-length difference between two coherent, in-phase waves is an integral number of wavelengths:

$$\Delta L = n\lambda \quad \text{for } n = 0, 1, 2, 3, \ldots \qquad \begin{array}{l} \textit{condition for} \\ \textit{constructive interference} \end{array} \quad (24.1)$$

Similarly, for destructive interference, the path-length difference is an odd number of half-wavelengths:

$$\Delta L = \frac{m\lambda}{2} \quad \text{for } m = 1, 3, 5, \ldots \qquad \begin{array}{l} \textit{condition for} \\ \textit{destructive interference} \end{array} \quad (24.2)$$

Thus, in Fig. 24.4, point P will be the location of a bright fringe (constructive interference) if

$$d \sin \theta = n\lambda \quad \text{for } n = 0, 1, 2, 3, \ldots \qquad \begin{array}{l} \textit{condition for} \\ \textit{bright fringes} \end{array} \quad (24.3)$$

where n is called the order number. The zeroth-order fringe ($n = 0$) corresponds to the central maximum, the first-order fringe ($n = 1$) is the first bright fringe on either side of the central maximum, and so on. As the path-length difference varies from point to point, so do the phase difference and the resulting interference.

The wavelength can therefore be determined by measuring d and θ for a particular order bright fringe (other than the central maximum). The angle θ is related to the displacement of a fringe (y), which is conveniently measured from a photograph of the interference pattern, such as Fig. 24.2b. If P is a point at the center of a bright fringe, the distance from the center of the central maximum to that fringe is (see Fig. 24.4)

$$y = L \tan \theta = L \left(\frac{\sin \theta}{\cos \theta} \right) \qquad (24.4)$$

If θ is small ($y \ll L$), $\cos \theta \approx 1$ and $\tan \theta$ can be approximated by $\sin \theta$. Hence,

$$y \approx L \sin \theta \quad \text{or} \quad \sin \theta \approx \frac{y}{L} \qquad (24.5)$$

Substituting this expression for $\sin \theta$ into Eq. 24.3 and then solving for y gives a good approximation of the distance of the nth bright fringe (y_n) from the central maximum on either side:

$$y_n \approx \frac{nL\lambda}{d} \qquad \begin{array}{l} \textit{lateral distance to bright fringe} \\ \textit{(for small } \theta \textit{ only)} \end{array} \quad (24.6)$$

Thus, the wavelength of the light is

Wavelength of light in Young's experiment from geometrical analyses

$$\lambda \approx \frac{y_n d}{nL} \qquad \begin{array}{l} \text{for } n = 1, 2, 3, \ldots \\ \textit{(for small } \theta \textit{ only)} \end{array} \quad (24.7)$$

A similar analysis gives the distances to the dark fringes (see Exercise 10a).

From Eq. 24.3, we see that, except for the zeroth order fringe, $n = 0$ (the central maximum), the positions of the fringes depend on wavelength—different λ values give different values of $\sin \theta$ and therefore of θ. Hence, when the experimenter uses white light, as Young did, the central fringe is white, but the other orders contain a spectrum of colors. By measuring the positions of the color fringes within a particular order, Young was able to determine the wavelengths of the colors of visible light.

Example 24.1 ■ Measuring the Wavelength of Light: Young's Double-Slit Experiment

In a lab experiment, monochromatic light passes through two narrow slits that are 0.050 mm apart. The interference pattern is observed on a white wall 1.0 m from the slits, and the second-order bright fringe is 2.4 cm from the center of the central maximum. (a) What is the wavelength of the light? (b) What is the distance between the second-order and third-order bright fringes?

Thinking It Through. (a) We can use Eq. 24.7 to find the wavelength. (b) We could compute y_3 from Eq. 24.6, and then get the distance between the second-order and third-order fringes ($y_3 - y_2$). However, the bright fringes for a given wavelength of light are evenly spaced; in general, the distance between adjacent bright fringes is constant (for small θ).

Solution. From the problem we have

Given: $d = 0.050$ mm
$L = 1.0$ m $= 10^3$ mm
$y_2 = 2.4$ cm $= 24$ mm
$n = 2$

Find: (a) λ (wavelength)
(b) $y_3 - y_2$ (distance between $n = 2$ and $n = 3$)

(a) Note: We use all distances in millimeters for convenience; we don't have to convert to meters. Using Eq. 24.7 gives

$$\lambda = \frac{y_n d}{nL} = \frac{(24 \text{ mm})(0.050 \text{ mm})}{2(10^3 \text{ mm})} = 6.0 \times 10^{-4} \text{ mm} = 6.0 \times 10^{-7} \text{ m}$$

This value is 600 nm, which is the wavelength of yellow-orange light (see Fig. 20.18).

(b) Using a general approach for n and $n + 1$, we get

$$y_{n+1} - y_n = \frac{(n + 1)L\lambda}{d} - \frac{nL\lambda}{d} = \frac{L\lambda}{d}$$

In this case, the distance between successive fringes is

$$\frac{L\lambda}{d} = \frac{(1.0 \text{ m})(6.0 \times 10^{-7} \text{ m})}{0.050 \times 10^{-3} \text{ m}} = 1.2 \times 10^{-2} \text{ m} = 1.2 \text{ cm}$$

Follow-up Exercise. Suppose white light were used instead of monochromatic light in this Example. What would be the separation distance of the red ($\lambda = 700$ nm) and blue ($\lambda = 400$ nm) components in the second-order fringe?

24.2 Thin-Film Interference

Objectives: To (a) describe how thin films produce colorful displays and (b) give some examples of practical applications of thin-film interference.

Have you ever wondered what causes the rainbowlike colors when light is reflected from a thin film of oil or a soap bubble? This effect—**thin-film interference**—is a result of the interference of light reflected from opposite surfaces of the film and may be readily understood in terms of wave interference.

First, however, you need to see how the phase of a light wave is affected by reflection. Recall from Chapter 13 that a wave pulse on a rope undergoes a 180° phase shift (change) when reflected from a rigid support and no phase shift when reflected from a free support (●Fig. 24.5). Similarly, as the figure shows, the phase change for the reflection of light waves at a boundary depends on the optical densities, or the indices of refraction, of the two materials:

- A light wave traveling in one medium and reflected from the boundary of a second medium whose index of refraction is greater than that of the first medium ($n_2 > n_1$) undergoes a 180° phase change.
- If the reflecting medium has the smaller index of refraction ($n_2 < n_1$), there is no phase change.

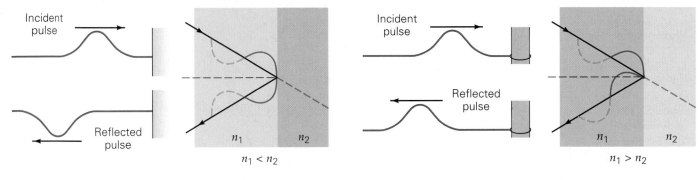

(a) Fixed end: 180° phase shift   **(b)** Free end: zero phase shift

•**FIGURE 24.5 Reflection and phase shifts** The phase changes that light waves undergo on reflection are analogous to those for pulses in strings. **(a)** The phase of a pulse in a string is shifted by 180° on reflection from a fixed end, and so is the phase of a light wave when it is reflected from a more optically dense medium. **(b)** A pulse in a string has a phase shift of zero (is not shifted) when reflected from a free end. Analogously, a light wave is not phase-shifted when reflected from a less optically dense medium.

To understand why you see colors in an oil film (on water or on a wet road), consider the reflection of monochromatic light from a thin film, as illustrated in •Fig. 24.6. The path length of the wave in the film depends on the angle of incidence (why?), but for simplicity we will assume normal (perpendicular) incidence for the light, even though the rays are drawn at an angle in the figure for clarity.

The oil film has a greater index of refraction than air, and the light reflected from the air–oil interface undergoes a 180° phase shift. The transmitted waves pass through the oil film and are reflected at the oil–water interface. In general, the index of refraction of oil is greater than that of water (see Table 22.1); that is, $n_2 < n_1$, so a reflected wave in this instance is *not* phase shifted.

You might think that if the path length of the wave in the oil film ($2t$, twice the thickness, down and back) were an integral number of wavelengths [$2t = 2(\lambda'/2) = \lambda'$ in Fig. 24.6], then the waves reflected from the two surfaces would interfere constructively. But keep in mind that the wave reflected from the top surface undergoes a 180° phase shift. The reflected waves from the two sur-

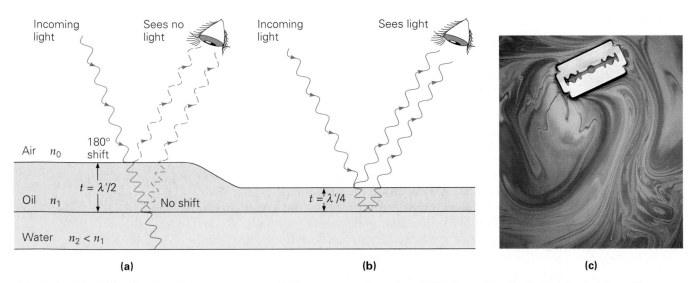

•**FIGURE 24.6 Thin-film interference** For an oil film on water, there is a 180° phase shift for light reflected from the air–oil interface and a zero phase shift at the oil–water interface. **(a)** Destructive interference occurs if the oil film has a thickness of $\lambda'/2$ for normal incidence. (Waves are displaced and angled for clarity.) **(b)** Constructive interference occurs with a minimum film thickness of $\lambda'/4$. **(c)** Thin-film interference in an oil slick. Different film thicknesses give rise to the reflections of different colors.

faces are therefore *out of phase* for this film thickness and interfere destructively. This means that light of this wavelength is not reflected, only transmitted.

Similarly, if the path length of the wave in the film were an odd number of half-wavelengths [$2t = 2(\lambda'/4) = \lambda'/2$ in Fig. 24.6], the reflected waves would be in phase (as a result of a 180° phase shift of the incident wave) and would interfere constructively. Light of this wavelength would be reflected from the oil film.

Keep in mind that the path length is expressed in terms of the wavelength of light *in the film*, λ'. Recall from Section 22.3 that the wavelength is different in different media. Here, $\lambda' = \lambda/n$, where λ' is the wavelength in the oil, λ the wavelength in air, and n the index of refraction of the oil. This can be seen from $n = c/v = f\lambda/f\lambda' = \lambda/\lambda'$.

Because oil and soap films generally have different thicknesses in different regions, particular wavelengths (colors) of white light interfere constructively in different regions and are reflected. As a result, a vivid display of various colors appears, which may change if the film thickness changes with time (see Demonstration 15). Thin-film interference may be seen if two glass slides are stuck together with an air film between them (•Fig. 24.7a). The bright colors of a peacock, an example of colorful interference in nature, are a result of layers of fibers in its feathers. Light reflected from successive layers interferes constructively, giving bright colors, even though the feather has no pigment of its own. Since the condition for constructive interference depends on the angle of incidence, the color pattern changes somewhat with the viewing angle and motion of the bird.

From this analysis of thin-film interference, you can see that the word "destructive" in this context does not imply that energy is destroyed. Destructive interference is simply a description of a physical fact—that a light wave is not present at a particular location but is somewhere else. When destructive interference occurs for light incident on the surfaces of a thin film, you know that the light is not reflected but that the incident beam (energy) is transmitted. Similarly, the mathematical description of Young's double-slit experiment in Section 24.1 tells you that you *should not expect* to find light at the locations of the dark fringes. The light is at the locations of the bright fringes, and the interference analysis shows that the light is merely redistributed.

A practical application of thin-film interference is described in the Insight on p. 749 dealing with nonreflective coatings for lenses. The situation discussed there differs from that of the previous oil–water example, because glass has a greater index of refraction than the nonreflecting film. Consequently, phase shifts of incident light take place at both the film and glass surfaces. In such a case, the condition for destructive interference is that the path-length difference for normally (perpendicularly) incident light is $\lambda'/2$, where λ' is the wavelength of the light in the film. Thus, $2t = \lambda'/2$, and the minimum film thickness that will produce destructive interference between the reflected waves is $t = \lambda'/4$. The wavelength of the light in the film (λ') is related to that in air (λ) by $\lambda' = \lambda/n$. Substituting this expression into the previous one gives

$$t = \frac{\lambda}{4n_1} \qquad \begin{array}{l}\textit{minimum film thickness} \\ \textit{(for } n_2 > n_1 > n_0)\end{array} \qquad (24.8)$$

where n_2, n_1, and n_0 are the indices of refraction of the lens, film, and air, respectively.

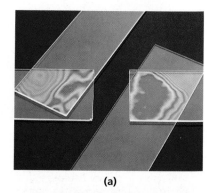

(a)

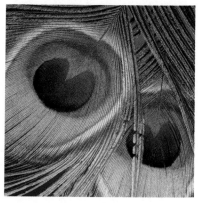

(b)

•**FIGURE 24.7 Thin-film interference** (a) A thin air film between microscope slides gives colorful patterns. (b) Multilayer interference in a peacock's feathers gives rise to bright colors. The brilliant throat colors of hummingbirds are produced the same way.

Destructive interference with a nonreflective coating

Example 24.2 ■ Nonreflective Coatings: Thin-film Interference

A glass lens ($n = 1.60$) is coated with a thin, transparent film of magnesium fluoride (MgF_2, $n = 1.38$) to make the lens nonreflecting. What is the minimum film thickness for the lens to be nonreflecting for incident light of wavelength 550 nm?

Thinking It Through. We can use Eq. 24.8 directly to get an idea of the minimum film thickness for a nonreflective coating.

A vivid display of colors from soap bubbles illuminated with white light from below.

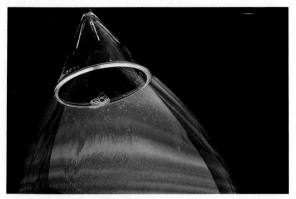

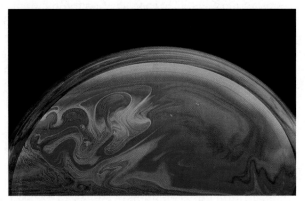

(a) Interference of light reflected from the inner and outer soap-bubble surfaces produces the colors. As the film thickness increases from top to bottom, the spectrum of colors is repeated several times.

(b) Swirling colors can be produced by blowing gently tangent to the film.

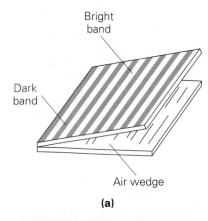

(a)

(b)

•FIGURE 24.8 Optical flatness
(a) An optical flat is used to check the smoothness of a reflecting surface. The flat is placed so that there is an air wedge between it and the surface. **(b)** If the surface is smooth, a regular, or symmetric, interference pattern is seen.

Solution.

Given: $n_1 = 1.38$ (for film)   ***Find:*** t (film thickness)
         $n_2 = 1.60$ (for lens)
         $\lambda = 550$ nm

Since $n_2 > n_1$,

$$t = \frac{\lambda}{4n_1} = \frac{550 \text{ nm}}{4(1.38)} = 99.6 \text{ nm}$$

which is quite thin ($\approx 10^{-5}$ cm). Relative to atoms, which have diameters on the order of 10^{-10} m, or 10^{-8} cm, the film is 10^3 atoms thick.

Follow-up Exercise.   For the glass lens in this Example to reflect rather than transmit the incident light through the lens, what would be the minimum film thickness? (Such a coating specific for infrared radiation on windows could be useful in hot climates.)

Optical Flats and Newton's Rings

The phenomenon of thin-film interference is used to check the smoothness and uniformity of optical components such as mirrors and lenses. **Optical flats** are made by grinding and polishing glass plates until they are as flat and smooth as possible. The degree of flatness can be checked by putting two such plates together at a slight angle so that a thin air wedge is between them (•Fig. 24.8a). If the plates are smooth and flat, a regular interference pattern of bright and dark fringes, or bands, appears (Fig. 24.8b). This pattern is a result of the uniformly varying differences in path lengths between the plates. Any irregularity in the pattern indicates an irregularity in at least one plate. Once a good optical flat is verified, it can be used to check the flatness of a reflecting surface, such as that of a precision mirror.

A similar technique is used to check the smoothness and symmetry of lenses. When a curved lens is placed on an optical flat, the air wedge is a ring below the periphery of the circular lens (•Fig. 24.9a). The regular interference pattern in this case is a set of concentric bright and dark circular fringes (Fig. 24.9b). They are called **Newton's rings**, after Isaac Newton, who first described this interference effect. Lens irregularities give rise to a distorted fringe pattern.

Nonreflecting Lenses

You may have noticed the blue–purple tint of the coated optical lenses used in cameras and binoculars. The coating makes the lenses "nonreflecting." If a lens is nonreflecting, the incident light is almost totally transmitted. Complete transmission of light is desirable for the exposing of photographic film and for viewing objects with binoculars.

A lens is made nonreflecting by coating it with a thin film of a material that has an index of refraction between the indices of refraction of air and glass (Fig. 1). If the coating is a quarter-wavelength thick, the difference in path length between the reflected rays is $\lambda'/2$, where λ' is the wavelength of light in the coating. In this case, both reflected waves undergo a 180° phase shift, and they are out of phase for a path-length difference of $\lambda'/2$ and interfere destructively. That is, the incident light is transmitted, and the coated lens is nonreflecting.

Note that a quarter-wavelength thickness of film is specific for a particular wavelength of light. The thickness is usually chosen to be a quarter-wavelength of yellow–green light ($\lambda \approx 550$ nm), to which the human eye is most sensitive. The wavelengths at the red and blue ends of the visible region are reflected, giving the coated lens its bluish-purple tint (Fig. 2). Sometimes other quarter-wavelength thicknesses are chosen, giving rise to other hues, such as amber or reddish purple.

The coating on a nonreflecting lens actually serves a double purpose. It promotes nonreflection from the front of the lens and reduces back reflection. Some of the light transmitted through a lens is reflected from the back surface. This could be reflected again from the front surface of an uncoated lens, giving rise to poor images. However, the proper film thickness allows back reflections to be transmitted through the coating.

Nonreflective coatings are also applied to the surfaces of solar cells, which convert light into electrical energy (Chapter 27). Since the thickness of such a coating is wavelength-dependent, like that on a nonreflecting lens, not all of the light is transmitted. However, the reflective losses can be decreased from around 30% to 10%, making the cell more efficient.

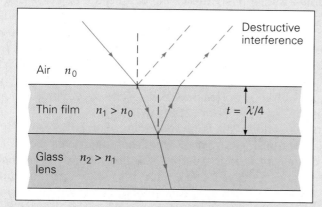

FIGURE 1   Nonreflective coating   For a thin film on a glass lens, there is a 180° phase shift at each interface when the index of refraction of the film is less than that of the glass. As a result, destructive interference occurs for a minimum film thickness of $\lambda'/4$, and the waves are transmitted rather than reflected, making the lens surface nonreflecting.

FIGURE 2   Coated lenses   The nonreflective coating on binocular and camera lenses generally produces a characteristic bluish-purple hue. (Why?)

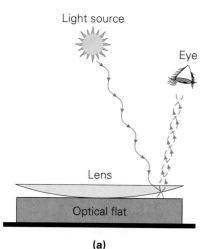

(a)

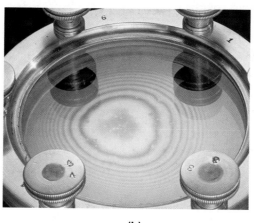

(b)

•FIGURE 24.9   Newton's rings
(a) A lens placed on an optical flat forms a ring-shaped air wedge, which gives rise to interference.
(b) The resulting interference pattern is a set of concentric rings called Newton's rings. Lens irregularities produce a distorted pattern.

24.3 Diffraction

Objectives: To (a) define diffraction and (b) give examples of diffractive effects.

In geometrical optics, light is represented by rays and pictured as traveling in straight lines. If this model represented the real nature of light, however, there would be no interference effects in Young's double-slit experiment. Instead, there would be only two bright slit images on the screen with a well-defined shadow area where no light enters. But we do see interference patterns, which means that the light must deviate from a straight-line path and enter the regions that would otherwise be in shadow. According to Huygens' principle (p. 690), the waves spread out from the slits; this deviation of light is called **diffraction**. Diffraction generally occurs when waves pass through small openings or around sharp edges or corners. The diffraction of water waves is shown in •Fig. 24.10. (See also Fig. 13.18.)

As Fig. 13.18 shows, there are different degrees of bending or diffraction. The amount of diffraction depends on the wavelength of the wave and, the size of the opening or object. In general, *the longer the wavelength compared to the width of the opening or object, the greater the diffraction.* This generality is illustrated in •Fig. 24.11.

In Fig. 24.11a, the wavelength is shorter than the width of the object ($\lambda < d$), and there is little diffraction—the large shadow zone behind the object is free of waves. (This corresponds to Fig. 13.18a for an opening. The waves pass through with little diffraction, and the shadow zones are around the corners of the opening.) In Fig. 24.11b, with the wavelength and width of the object about equal ($\lambda \approx d$); there is some diffraction but still a noticeable shadow zone. Finally, in Fig. 24.11c, with the wavelength much longer than the object's width ($\lambda > d$), there is much diffraction and little or no shadow zone. The wave bends around and passes on almost as if the object were not there.

The diffraction of sound (Chapter 14) is quite evident. Someone can talk to you from another room or around the corner of a building, and even in the absence of reflections, you can easily hear the person. Audible sound waves have wavelengths on the order of centimeters to meters. Thus, the wavelengths of sound are longer than or about the same width as ordinary objects and openings, and diffraction readily occurs for sound.

Visible light, however, has wavelengths on the order of 10^{-7} m (Chapter 20), and diffraction phenomena for light waves often go unnoticed. Careful observation will reveal that a shadow boundary is blurred or fuzzy. On close inspection, you can see a pattern of bright and dark fringes (•Fig. 24.12). These interference patterns are evidence of the diffraction of the light around the edge of the object.

(a)

(b)

•**FIGURE 24.10 Water wave diffraction** **(a)** Diffraction of water waves can be observed at the seashore when conditions are suitable and water comes through an opening. See Fig. 13.18 for a laboratory demonstration of water wave diffraction. **(b)** Satellite image of water waves at a depth of about 213 m (700 ft) in the Strait of Gibraltar.

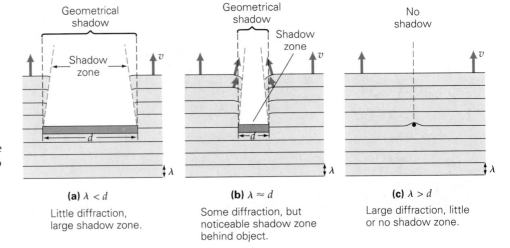

•**FIGURE 24.11 Wavelength and object dimensions** In general, the longer the wavelength compared to the width of an object or opening, the greater the diffraction. Without diffraction, there would be a shadow zone without waves behind the object.

(a) $\lambda < d$
Little diffraction, large shadow zone.

(b) $\lambda \approx d$
Some diffraction, but noticeable shadow zone behind object.

(c) $\lambda > d$
Large diffraction, little or no shadow zone.

To illustrate diffraction, consider a single slit in a barrier (•Fig. 24.13). Suppose that a single slit of width w is illuminated with monochromatic light whose wavelength (λ) is much shorter than the slit width. An interference pattern consisting of a bright central maximum and a symmetrical array of bright fringes (regions of constructive interference) on both sides is observed on a screen at a distance L from the slit (where $L \gg w$).

Because the width of the slit is very much greater than the wavelength of the light, the slit with light passing through cannot be treated as a point source of Huygens' wavelets. However, various points on the wave front passing through the slit can be considered to be such point sources. Then, the interference of those wavelets can be analyzed much as we analyzed double-slit interference earlier.

The analysis will not be done here, but you will notice that the stated result is very similar in form to that for Young's double slit. However, for the single slit, dark fringes (regions of destructive interference) rather than bright fringes are analyzed. From the geometry, the condition for *destructive interference*, or interference minima (dark fringes), can be shown to be

$$w \sin \theta = m\lambda \quad \text{for } m = 1, 2, 3, \ldots \qquad \begin{array}{l}\textit{condition for}\\ \textit{dark fringes}\end{array} \quad (24.9)$$

where θ is the angle of a particular minimum designated by $m = 1, 2, 3, \ldots$ on either side of the central bright fringe. (There is no $m = 0$. Why?)

Also, a small-angle approximation, $\sin \theta \approx y/L$, can be made in some instances in obtaining a formula for the lateral displacement of a particular interference minimum on the screen. This relationship gives a good approximation of the distances of the dark fringes on either side of the center of the central maximum:

$$y_m = \frac{mL\lambda}{w} \quad \text{for } m = 1, 2, 3, \ldots \qquad (24.10)$$

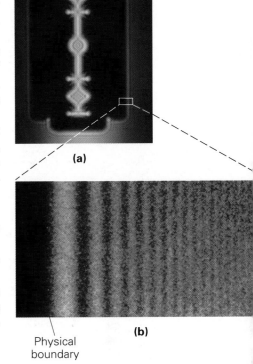

(a)

Physical boundary

(b)

•**FIGURE 24.12 Diffraction in action** (a) Diffraction patterns produced by a razor blade. (b) A close-up view of the fringes formed at the edge of the blade.

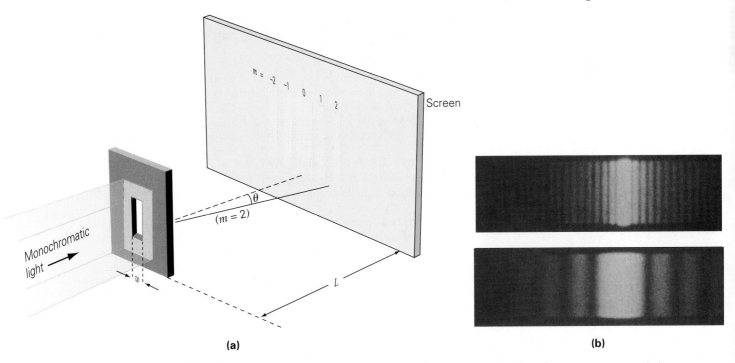

(a)

(b)

•**FIGURE 24.13 Single-slit diffraction** (a) The diffraction of light by a single slit gives rise to an interference pattern consisting of a large bright central maximum and a symmetric array of side fringes. (b) The widths of the fringes depend on the width of the slit—the narrower the slit, the wider the fringes.

Note that Eq. 24.10 has the same form as Eq. 24.6, the equation for the displacement of bright fringes in double-slit interference. Here w is the width of the slit, and there d is the distance between the slits.

The qualitative predictions from Eq. 24.10 are quite interesting and instructive:

- For a given slit width (w), the longer the wavelength (λ), the wider the diffraction pattern.
- For a given wavelength (λ), the narrower the slit width (w), the wider the diffraction pattern.
- The width of the central maximum is twice the width of the side maxima (each maximum other than the central one).

Let's consider each of these predictions.

As the slit is made narrower, the central maximum and the side fringes spread out and become larger (Fig. 24.13b). Equation 24.10 is not applicable to very small slit widths (because of the small-angle approximation). If the slit is decreased until it is the same order of magnitude as the wavelength of the light, the central maximum spreads out over the whole screen. That is, diffraction becomes dramatically evident when the width of the slit (or object) is about the same as the wavelength. Diffraction effects are most easily observed when $\lambda/w \geq 1$, or $\lambda \geq w$.

Conversely, if the slit is made wider for a given wavelength of light, the diffraction pattern becomes narrower. The fringes move closer together and eventually become difficult to distinguish when λ is much smaller than w ($\lambda \ll w$). The pattern then appears as a fuzzy shadow around the central maximum, which is the illuminated image of the slit. This type of pattern is observed for the image produced by sunlight entering a dark room through a hole in a curtain. Such an observation led early experimenters to investigate the wave nature of light. The acceptance of this concept was, in large part, due to the explanation of diffraction offered by physical optics.

Conceptual Example 24.3

Diffraction and Radio Reception

Driving with the car radio on in the city or in mountainous areas of the country, you have probably noticed that on certain broadcast bands the quality of your radio reception can vary sharply from place to place, with stations seeming to fade out and reappear. What do you think might cause this variation, and which band would you expect to be least affected by it? (a) Weather (162 MHz); (b) FM (88–108 MHz); (c) AM (525–1610 kHz).

Reasoning and Answer. Radio waves, like light, are electromagnetic waves and so tend to travel in straight lines long distances from their sources. Although they are generally better than visible light at penetrating solid materials, they can still be blocked by objects in their path—especially if the objects are massive (such as hills) or are made largely of metal (such as buildings).

However, because of diffraction, radio waves can also "wrap around" obstacles or "fan out" as they pass through openings, just as sound waves do, provided their wavelength is at least roughly the size of the obstacle or opening. The longer the wavelength, the greater will be the diffraction, and so the less likely the radio waves are to be obstructed.

To determine which band benefits most by such diffraction, we need to find the wavelengths corresponding to the given frequencies, using the relationship $\lambda f = c$. When we do this, we find that AM radio waves, with $\lambda = 186$–571 m, are the longest of the three bands specified by a factor of about 100. We might conclude that weather and FM broadcasts would tend to be blocked by large objects such as hills or buildings, whereas AM broadcasts are likely to be diffracted around such objects or through the openings between them. Thus, the answer is (c).

Follow-up Exercise: During half-time at a football game, when a marching band faces you, you can easily hear the woodwind instruments (clarinet, saxophone, and so on) and the brass instruments (trumpet, trombone, and so on). Yet, when the players march away from you, the brass instruments sound muted but you can hear the woodwinds quite well. Why?

Notice from Fig. 24.13 that the central maximum differs in width from the side maxima. The central maximum turns out to be twice as wide, which can be shown as follows. Taking the width of the central maximum to be the distance between the bounding minima on each side ($m = 1$), or a width of $2y_1$, from Eq. 24.10 with $y_1 = L\lambda/w$,

$$2y_1 = \frac{2L\lambda}{w} \quad \textit{width of central maximum} \quad (24.11)$$

Similarly, the width of the bright side fringes is given by

$$y_2 - y_1 = y_3 - y_2 = \frac{L\lambda}{w}$$

Or, in general,

$$y_{m+1} - y_m = \frac{L\lambda}{w} \quad \textit{width of side fringes} \quad (24.12)$$

Thus, the width of the central maximum is twice that of the side fringes.

Example 24.4 ■ Width of the Central Maximum: Single-slit Diffraction

Monochromatic blue light ($\lambda = 425$ nm) passes through a slit whose width is 0.50 mm. What is the width of the central maximum on a screen located 1.0 m from the slit?

Thinking It Through. This situation is a direct application of Eq. 24.11.

Solution.

Given: $\lambda = 425$ nm $= 4.25 \times 10^{-5}$ cm   *Find:* $2y_1$ (width of
$\quad\quad\quad w = 0.50$ mm $= 0.050$ cm $\quad\quad\quad\quad$ central maximum)
$\quad\quad\quad L = 1.0$ m $= 10^2$ cm

Equation 24.11 gives

$$2y_1 = \frac{2L\lambda}{w} = \frac{2(10^2 \text{ cm})(4.25 \times 10^{-5} \text{ cm})}{0.050 \text{ cm}} = 0.17 \text{ cm}$$

Note that diffraction effects would not be readily observed in this case since $\lambda \ll w$.

Follow-up Exercise. By what factor would the width of the central maximum change if red light ($\lambda = 700$ nm) were used in this Example?

Diffraction Gratings

Bright and dark fringes result from diffraction and interference when light passes through a single slit or double slits. As the number of slits is increased, the bright lines become sharper (narrower) and the dark lines broader. The intensity of each line is less when the light has to pass through many narrow slits, but even so, the sharp lines are useful in optical analysis of light sources and other applications. Arrangements of large numbers of parallel, closely spaced slits are called **diffraction gratings** (•Fig. 24.14a).

Diffraction gratings were first made of fine strands of wire. Their effects were similar to what can be seen by viewing a candle flame through a feather held close to the eye. Better gratings have a large number of fine lines or grooves on glass or metal surfaces. If light is transmitted through a grating, it is called a *transmission grating*. However, *reflection gratings* are more common. The closely spaced grooves of a compact disk act as a reflection grating, giving rise to their familiar iridescent

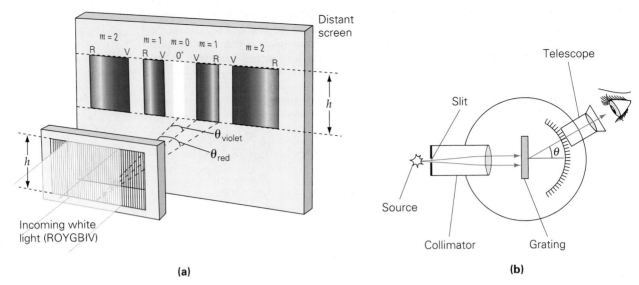

•FIGURE 24.14 Diffraction grating (a) A diffraction grating produces a sharply defined interference pattern. In each side fringe, components of different wavelengths are separated, since the deviation depends on wavelength: $\theta = \sin^{-1}(n\lambda/d)$. **(b)** As a result, gratings are used in spectrometers to determine the wavelengths present in a beam of light by measuring their angles of diffraction and to separate the various wavelengths for further analysis.

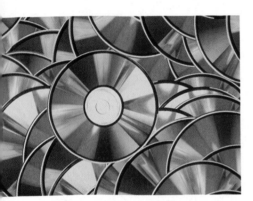

•FIGURE 24.15 Diffraction effects The narrow grooves of compact disks (CDs) act as reflection diffraction gratings, producing colorful displays.

Note: d is the distance between adjacent slits.

sheen (•Fig. 24.15). Commercial gratings are made by depositing a thin film of aluminum on an optically flat surface and then removing some of the reflecting metal by cutting regularly spaced, parallel lines.

Precision diffraction gratings are made using two coherent laser beams, intersecting at an angle. The beams expose a layer of photosensitive material, which is then etched. The spacing of the grating lines is determined by the intersection angle of the beams. Precision gratings may have 30 000 lines per centimeter or more and are therefore expensive and difficult to fabricate. Most gratings used in laboratory instruments are replica gratings, which are plastic castings of high-precision master gratings.

Diffraction gratings are widely used in *spectroscopy* (the study of spectra), where they have almost replaced prisms. The forming of a spectrum and the measurement of wavelengths by means of a grating depend only on geometrical measurements such as lengths and/or angles. Wavelength determination with a prism, in contrast, depends also on the dispersive characteristics of the glass or other material of which the prism is made.

It can be shown that the condition for interference maxima for a grating illuminated with monochromatic light is identical to that for a double slit. The expression is

$$d \sin \theta = n\lambda \quad \text{for } n = 0, 1, 2, 3, \ldots \qquad \begin{array}{l} \textit{interference maxima} \\ \textit{for a grating} \end{array} \quad (24.13)$$

where n is the order of interference and θ is the angle of deviation of a particular wavelength. The zeroth-order maximum corresponds to the central maximum of the diffraction pattern. The spacing between adjacent slits (d) is obtained from the number of lines per unit length of the grating, $d = 1/N$. For example, if $N = 5000$ lines/cm,

$$d = \frac{1}{N} = \frac{1}{5000} = 2.0 \times 10^{-4}\,\text{cm}$$

In contrast to a prism, which deviates red light least and violet light most, a diffraction grating produces the least angle of deviation for violet light (short λ) and the greatest for red light (long λ). Also, a prism disperses white light into a single spectrum. A diffraction grating, however, produces a number of spectra, one for each order other than $n = 0$. There is no deviation of the components of the light for the zeroth order ($\sin \theta = 0$ for all wavelengths), so the central maximum is a white image. However, a spectrum is produced for each higher order.

The sharp spectra produced by diffraction gratings are used in instruments called *spectrometers* (Fig. 24.14b). In a spectrometer, materials are illuminated with light of various wavelengths to find which ones are strongly transmitted or reflected. The diffraction pattern helps identify the material. The grating is rotated so that the sample is illuminated with a succession of different wavelengths. The wavelengths may also be in the infrared and ultraviolet regions.

The number of spectral orders produced by a grating depends on the wavelength of the light (which may be infrared, visible, or ultraviolet) and on the grating's spacing (d). From Eq. 24.13, since $\sin \theta$ cannot exceed 1 ($\sin \theta \le 1$),

$$\sin \theta = \frac{n\lambda}{d} \le 1$$

The order number n is therefore limited as follows (why?):

$$n \le \frac{d}{\lambda} \qquad\qquad (24.14) \quad \textbf{Limit of spectral order}$$

Example 24.5 ■ A Diffraction Grating: Line Spacing and Spectral Orders

A particular diffraction grating produces an $n = 2$ spectral order at a deviation angle of 30° for light with a wavelength of 500 nm. (a) How many lines per centimeter does the grating have? (b) If the grating were illuminated with white light, how many orders of the *complete* visible spectrum would be produced?

Thinking It Through. (a) To find the number of lines per centimeter (N) the grating has, we need to know the grating spacing (d), since $N = 1/d$. With the given data, we can find d from Eq. 24.13. (b) Equation 24.14 describes the order number limit—one complete order for each integer in n. A fractional part of an integer is not a *complete* order.

Solution.

Given: $n = 2$
$\lambda = 500$ nm $= 5.00 \times 10^{-7}$ m
$\theta = 30°$

Find: (a) N (lines/cm)
(b) Number of orders

(a) Using Eq. 24.13, we get the grating spacing:

$$d = \frac{n\lambda}{\sin \theta} = \frac{2(5.00 \times 10^{-7}\,\text{m})}{\sin 30°} = 2.00 \times 10^{-6}\,\text{m} = 2.00 \times 10^{-4}\,\text{cm}$$

Then

$$N = \frac{1}{d} = \frac{1}{2.00 \times 10^{-4}\,\text{cm}} = 5000\,\text{lines/cm}$$

(b) The greatest angle of deviation for any order is for the longest wavelength, which is that of red light in the visible spectrum ($\lambda = 700$ nm $= 7.00 \times 10^{-7}$ m). With a spacing of $d = 2.00 \times 10^{-6}$ m for this grating, by Eq. 24.14, the order numbers for the visible spectrum are limited to

$$n \le \frac{d}{\lambda} = \frac{2.00 \times 10^{-6}\,\text{m}}{7.00 \times 10^{-7}\,\text{m}} = 2.86$$

Thus, for a grating with this spacing, only two complete orders are observed, or four complete visible spectra—two on each side of the central maximum. A large portion of the spectrum is seen in the third order, but this portion is limited to wavelengths of

$$\lambda \le \frac{d}{n} = \frac{2.00 \times 10^{-6}\,\text{m}}{3} = 6.67 \times 10^{-7}\,\text{m} = 667\,\text{nm}$$

Thus, the red end of the visible spectrum is not complete in the third spectral order.

Follow-up Exercise. Suppose the grating in this Example had 10 000 lines/cm. Using this grating, how many orders of the complete visible spectrum would be observed for white light? (Before computing the answer, would you expect more or fewer orders to be seen?)

Note that it is possible for spectra produced by diffraction gratings to overlap at higher orders. That is, the angles of deviation for different orders may be the same for two different wavelengths. The spacing must be greater than or equal to the wavelength ($d \geq \lambda$) to have interference for the first spectral order. However, if the spacing is much greater than the wavelength ($d \gg \lambda$), the difference in the deviation angles for nearby wavelengths is quite small. This difference increases for higher orders, but then the overlapping of spectra becomes a problem. Thus, there is an optimal spacing of the lines on a diffraction grating for each spectral region of interest—infrared, visible, or ultraviolet. The spacing is typically chosen to be between 3λ and 6λ, where λ is the median (middle) wavelength of the spectral region.

X-ray Diffraction

The wavelength of any electromagnetic wave can be determined via a diffraction grating with the appropriate spacing. Diffraction was used to determine the wavelengths of X-rays early in the 20th century. Experimental evidence indicated that the wavelengths of X-rays were probably around 10^{-8} cm, but it is impossible to construct a diffraction grating with this spacing. Around 1913, Max von Laue, a German physicist, suggested that the regular spacing of the atoms in a crystalline solid might make the crystal act as a diffraction grating for X-rays, since the atomic spacing in crystals is on the order of 10^{-8} cm. X-rays were directed at crystals, and diffraction patterns were indeed observed.

• Figure 24.16a illustrates diffraction by the planes of atoms in a crystal such as sodium chloride. The path-length difference is $2d \sin \theta$ (twice $d \sin \theta$), where d

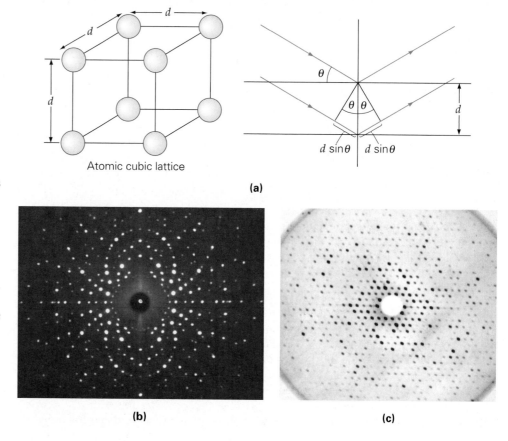

Atomic cubic lattice

(a)

•**FIGURE 24.16 Crystal diffraction** (a) The array of atoms in a crystal lattice structure acts as a diffraction grating, and X-rays are diffracted from the planes of atoms. With a lattice spacing of d, the path-length difference for the X-rays diffracted from adjacent planes is $2d \sin \theta$. (b) X-ray diffraction pattern of a crystal of potassium sulfate. By analyzing the geometry of such patterns, investigators can deduce the structure of the crystal and the position of its various atoms. (c) X-ray diffraction pattern of the protein hemoglobin, which carries oxygen in the blood.

(b)

(c)

is the distance between the crystal's internal planes. Thus, the condition for constructive interference is

$$2d \sin \theta = n\lambda \quad \text{for } n = 1, 2, 3, \ldots \qquad (24.15)$$

This relationship is known as **Bragg's law**, after W. L. Bragg (1890–1971), the British physicist who first derived it.

The wavelengths of X-rays were experimentally determined by this means, and X-ray diffraction is now used to investigate the internal structure, not only of simple crystals, but of large, complex biological molecules such as proteins and DNA. Because of their short wavelengths, X-rays provide a diffraction "probe" for investigating the interatomic spacings of molecules.

24.4 Polarization

Objectives: **To (a) explain light polarization and (b) give examples of polarization, both in the environment and in commercial applications.**

When you think of polarized light, you may visualize polarizing (or Polaroid™) sunglasses, since this is one of the more common applications of polarization. When something is polarized, it has a preferential direction, or orientation (think of a polarized electrical plug, as described in Chapter 18). In terms of light waves, **polarization** refers to the orientation of the transverse wave oscillations.

Note: See Fig. 18.23.

Recall from Chapter 20 that light is an electromagnetic wave with oscillating electric and magnetic field vectors (**E** and **B**, respectively) perpendicular to the direction of propagation. Light from most sources consists of a large number of electromagnetic waves emitted by the atoms of the source. Each atom produces a wave with a particular orientation, corresponding to the direction of the atomic vibration. However, since electromagnetic waves are produced by numerous atoms, all orientations of the **E** and **B** fields are possible in the composite light emitted. When the field vectors are randomly oriented, the light is said to be *unpolarized*. This is commonly represented schematically in terms of the electric field vector as shown in •Fig. 24.17a.

Note: Refer to Figure 20.16.

(a) Unpolarized

If there is some preferential orientation of the field vectors, the light is said to be *partially polarized* (Fig. 24.17b). If the field vectors oscillate in only *one* plane, the light is *plane polarized*, or *linearly polarized* (Fig. 24.17c). Note that polarization is evidence that light is a transverse wave. True longitudinal waves, such as sound waves in air, cannot be polarized because there are no two-dimensional vibrations.

(b) Partially polarized

Light can be polarized in several ways. Polarization by reflection and double refraction will be discussed here. Polarization by scattering will be considered in Section 24.5.

(c) Plane (linearly) polarized

Polarization by Reflection

When a beam of unpolarized light strikes a smooth, transparent medium such as glass, the beam is partially reflected and partially transmitted. The reflected light may be completely polarized, partially polarized, or unpolarized, depending on the angle of incidence. The unpolarized case occurs for $0°$, or normal incidence. As the angle of incidence is varied, the reflected light is partially polarized. The electric field components parallel to the surface are reflected more strongly, producing this partial polarization. However, at one particular angle of incidence, the reflected beam is completely polarized. At this angle, moreover, the refracted beam is partially polarized (•Fig. 24.18).

David Brewster (1781–1868), a Scottish physicist, found that the complete polarization of the reflected beam occurs when the reflected and refracted beams are $90°$ apart. The incident angle at which this polarization occurs is called the **polarizing angle** (θ_p) or the **Brewster angle**, and it is specific for a given material.

•**FIGURE 24.17 Polarization**
Polarization is represented by the orientation of the plane of vibration of the electric field vectors. **(a)** When the vectors are randomly oriented (as viewed along the direction of propagation), the light is unpolarized. **(b)** With preferential orientation of the field vectors, the light is partially polarized. **(c)** When the vectors are in one plane, the light is plane polarized, or linearly polarized.

•FIGURE 24.18 Polarization by reflection When the reflected and refracted components of a beam of light are 90° apart, the reflected component is linearly polarized, and the refracted component is partially polarized. This occurs when $\theta_1 = \theta_p = \tan^{-1} n$. For a stack of glass plates (or layers of thin film), the intensity of the reflected polarized beam increases with each reflection (the sum of the reflected arrows in the diagram). In this manner, the refracted beam becomes more polarized (there are fewer vertical beam components in the diagram).

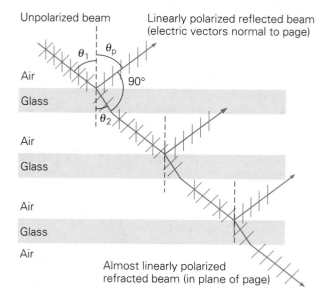

As shown in Fig. 24.18, the reflected and refracted beams are 90° apart. Because $\theta_1 = \theta_p$,

$$\theta_1 + 90° + \theta_2 = 180°$$

Then

$$\theta_1 + \theta_2 = 90° \quad \text{or} \quad \theta_2 = 90° - \theta_1$$

By Snell's law (Chapter 22), for incidence in air,

$$\frac{\sin \theta_1}{\sin \theta_2} = n$$

In this case, $\sin \theta_2 = \sin(90° - \theta_1) = \cos \theta_1$. Thus,

$$\frac{\sin \theta_1}{\sin \theta_2} = \frac{\sin \theta_1}{\cos \theta_1} = \tan \theta_1 = n$$

With $\theta_1 = \theta_p$, we get

$$\tan \theta_p = n \quad \text{and} \quad \theta_p = \tan^{-1} n \qquad (24.16)$$

(Since glass is dispersive, the Brewster angle also depends to some degree on the wavelength of the incident light.)

As shown in Fig. 24.18, in a stack of glass plates, the reflections from successive surfaces increase the intensity of the reflected polarized beam. Note that the refracted beam becomes more linearly polarized with successive refractions. This effect is sometimes referred to as *polarization by refraction*. In practical applications, several thin films of a transparent material are used instead of glass plates.

θ_p, **polarizing or Brewster angle: incident angle for complete polarization of reflected light**

Example 24.6 ■ Sunlight on a Pond: Polarization by Reflection

Sunlight is reflected from the smooth surface of a pond. What is the Sun's altitude when the polarization of the reflected light is greatest?

Thinking It Through. Incident light at the Brewster angle has the greatest polarization on reflection, so we must find the altitude of the Sun for this condition. The altitude is represented by the angle between the Sun and the horizon.

Solution. Here, we obtain the index of refraction of water from Table 22.1:

Given: $n = 1.33$ (Table 22.1)   *Find:* θ (altitude angle for greatest polarization)

The angle of incidence is the angle complementary to the altitude angle (draw a sketch), so

$$\theta = 90° - \theta_p$$

where θ_p is the polarizing angle for maximum or complete polarization.

Using Eq. 24.16, we get

$$\theta_p = \tan^{-1} n = \tan^{-1}(1.33) = 53°$$

So

$$\theta = 90° - \theta_p = 90° - 53° = 37°$$

Follow-up Exercise. Light is incident on a flat, transparent material with an index of refraction of 1.38. At what angle of refraction would the transmitted light have the greatest polarization?

It is interesting to note that a rainbow, or light from a rainbow, is partially polarized. This polarization occurs because the reflection angle inside a water droplet is somewhat close to the Brewster angle. The plane of polarization is tangential to the bow at each point.

Polarization by Double Refraction (Birefringence and Dichroism)

When monochromatic light travels in glass, its speed is the same in all directions and is characterized by a single index of refraction. Any material like this is said to be *isotropic*, meaning that it has the same characteristics in all directions. Some crystalline materials, such as quartz, calcite, and ice, are *anisotropic* with respect to the speed of light; that is, the speed of light is different in different directions within the material. Anisotropy gives rise to some unique optical properties, one of which is that the index of refraction may vary with the direction of propagation. Such materials are said to be doubly refracting, or to exhibit **birefringence**, and polarization is involved.

For example, a beam of unpolarized light incident on a birefringent crystal of calcite ($CaCO_3$, calcium carbonate) is illustrated in •Fig. 24.19. When the beam propagates at an angle to a particular crystal axis, the beam is doubly refracted and separated into two components, or rays. These two rays are linearly polarized in mutually perpendicular directions. One ray, called the *ordinary (o) ray*, passes through the crystal in an undeflected path and is characterized by an index of refraction (n_o). The second ray, called the *extraordinary (e) ray*, is refracted and is characterized by an index of refraction (n_e). The particular axis direction indicated by dashed lines in Fig. 24.19a is called the optic axis. Along this direction, $n_o = n_e$, and nothing extraordinary is noted about the transmitted light.

Some birefringent crystals, such as tourmaline, exhibit the interesting property of absorbing one of the polarized components more than the other. This

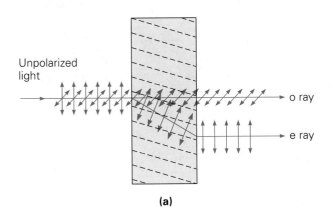

Unpolarized light

o ray

e ray

(a)

(b)

•**FIGURE 24.19 Double refraction, or birefringence** (a) Unpolarized light incident normal to the surface of a birefringent crystal and at an angle to a particular direction in the crystal (dashed lines) is separated into two components. The ordinary (o) ray and the extraordinary (e) ray are plane polarized in mutually perpendicular directions. (b) Birefringence seen through a calcite crystal.

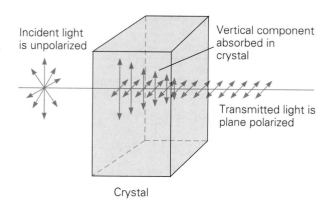

•**FIGURE 24.20 Dichroism** Dichroic birefringent crystals selectively absorb one polarized component more than the other. If the crystal is thick enough, the emerging beam is linearly polarized.

Incident light is unpolarized

Vertical component absorbed in crystal

Transmitted light is plane polarized

Crystal

property is called **dichroism**. If a dichroic crystal is sufficiently thick, the more strongly absorbed component may be completely absorbed. In that case, the emerging beam is plane polarized (•Fig. 24.20).

Another dichroic crystal is quinine sulfide periodide (commonly called herapathite, after W. Herapath, an English physician who discovered its polarizing properties in 1852). This crystal was of great practical importance in the development of modern polarizers. Around 1930, Edwin H. Land, an American scientist, found a way to align tiny, needle-shaped dichroic crystals in sheets of transparent celluloid. The result was a thin sheet of polarizing material that was given the commercial name Polaroid.

Better polarizing films have been developed by using synthetic polymer materials instead of celluloid. During the manufacturing process, this kind of film is stretched to align the long molecular chains of the polymer. With proper treatment, the outer (valence) electrons of the molecules can move along the oriented chains. As a result, light with **E** vectors parallel to the oriented chains is readily absorbed, but light with **E** vectors perpendicular to the chains is transmitted. The direction perpendicular to the orientation of the molecular chains is called the **transmission axis**, or the **polarization direction**. Thus, when unpolarized light falls on a polarizing sheet, the sheet acts as a polarizer and transmits polarized light (•Fig. 24.21).

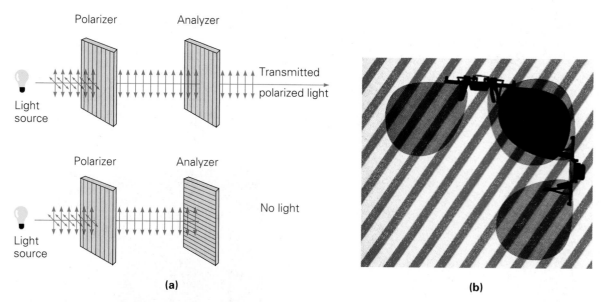

Polarizer   Analyzer

Light source

Transmitted polarized light

Polarizer   Analyzer

Light source

No light

(a)   (b)

•**FIGURE 24.21 Polarizing sheets** (a) When polarizing sheets are oriented so that their polarization directions are the same, the emerging light is polarized. The first sheet acts as a polarizer, and the second as an analyzer. (b) When one of the sheets is rotated 90° and the polarization directions are perpendicular (crossed polarizers), little light (ideally, none) is transmitted.

The human eye cannot distinguish between polarized and unpolarized light. To tell whether light is polarized, we must use an *analyzer,* which may be a sheet of polarizing film. As shown in Fig. 24.21, if the transmission axis of an analyzer is parallel to the plane of polarization of polarized light, there is maximum transmission. If the transmission axis of the analyzer is perpendicular to the plane of polarization, little light (ideally, none) will be transmitted. In general, the intensity of the transmitted light is given by

$$I = I_{\text{o}} \cos^2 \theta$$

where θ is the angle between the transmission axes of the polarizer and analyzer. This expression is known as *Malus's law,* after its discoverer, French physicist E. L. Malus (1775–1812).

Now you can understand the principle of polarizing sunglasses. Light reflected from a smooth surface is partially polarized, as you learned earlier in this section. The direction of polarization is chiefly in the plane of the surface (see Fig. 24.18). Light reflected from water or snow can be so intense that it gives rise to visual glare (•Fig. 24.22). To reduce this, the polarizing lenses of glasses are oriented with their transmission axes vertical so that some of the partially polarized light from reflective surfaces is blocked or absorbed.

Polarizing glasses whose lenses show different polarization directions are used to view some 3-D movies. The pictures are projected on the screen by two projectors that transmit slightly different images, photographed by cameras a short distance apart. The projected light is linearly polarized but in directions that

(a)

•**FIGURE 24.22 Glare reduction** **(a)** Light reflected from a horizontal surface is partially polarized in the horizontal plane. When sunglasses are oriented so that their polarizing direction is vertical, the horizontally polarized component of such light is not transmitted. The intensity of the reflected light reaching the eye is diminished, so glare is reduced. **(b)** Polarizing filters for cameras exploit the same principle. The photo at right was taken with such a filter. Note the reduction in reflections from the store window.

(b)

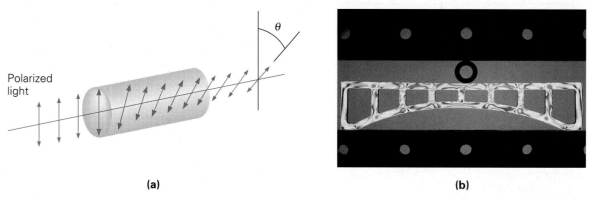

Polarized light

θ

(a)   **(b)**

•**FIGURE 24.23 Optical activity** (a) Some substances have the property of rotating the polarization direction of linearly polarized light. This ability, which depends on the molecular structure, is called optical activity. **(b)** Glasses and plastics become optically active under stress, and the points of greatest stress are apparent when the material is viewed through crossed polarizers. Engineers can thus test plastic models of structural elements to see where the greatest stresses will occur when the models are "loaded" (subjected to the forces that they will experience in actual use). Here a model of a suspension bridge strut is being analyzed.

are mutually perpendicular. The lenses of the 3-D glasses also have polarization directions that are perpendicular. Thus, one eye sees the image from one projector, and the other eye sees the image from the other projector. The brain interprets the slight difference in perspective of the two images as depth, or a third dimension, just as it does in normal vision.

Some transparent materials have the ability to rotate the direction of polarization of linearly polarized light. This property, called **optical activity**, is due to the molecular structure of the material (•Fig. 24.23). The rotation may be clockwise or counterclockwise, depending on the molecular orientation. Optically active molecules include those of certain proteins, amino acids, and sugars.

Glasses and plastics become optically active when under stress. The greatest rotation of the direction of polarization of the transmitted light occurs in the regions where the stress is the greatest. Viewing the stressed piece of material through crossed Polaroids allows the points of greatest stress to be identified. This determination is called optical stress analysis (Fig. 24.23b).

Another use of polarizing films, liquid crystal displays, is described in the Insight on p. 764.

24.5 Atmospheric Scattering of Light

Objectives: To (a) define scattering and (b) explain why the sky is blue and sunsets are red.

When light is incident on a suspension of particles, such as the molecules of air, some of the light may be absorbed and reradiated. This process is called **scattering**. The scattering of sunlight in the atmosphere produces some interesting effects, including the polarization of sky light (that is, sunlight that has been scattered by the atmosphere), the blueness of the sky, and the redness of sunsets and sunrises.

Atmospheric scattering causes the sky light to be polarized. When unpolarized sunlight is incident on air molecules, the electric field of the light wave sets electrons of the molecules into vibration. The vibrations are complex, but the accelerating charges emit radiation, like the vibrating charges in the antenna of a radio broadcast station (see Section 20.4). The intensity of this emitted radiation is strongest along a line perpendicular to the oscillation. And, as illus-

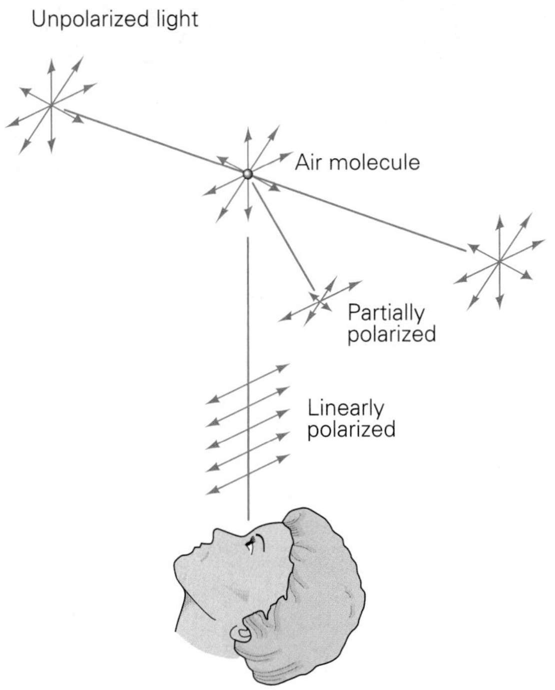

Unpolarized light

Air molecule

Partially polarized

Linearly polarized

•FIGURE 24.24 **Polarization by scattering** When incident unpolarized sunlight is scattered by a gas molecule in the air, the light in the plane perpendicular to the direction of the incident ray is linearly polarized. Light scattered at some arbitrary angle is partially polarized. An observer at a right angle (90°) to the direction of the incident sunlight receives plane-polarized light.

trated in •Fig. 24.24, an observer viewing from an angle of 90° with respect to the direction of the sunlight will receive linearly polarized light because of the horizontal charge oscillations. The light also has a vertically polarized component. At other viewing angles, both components are present, and sky light seen through a polarizing filter appears partially polarized because of the stronger component.

Since the scattering of light with the greatest degree of polarization is at a right angle to the direction of the Sun, at sunrise and sunset the scattered light from directly overhead has the greatest degree of polarization. The polarization of sky light can be observed by viewing the sky through a polarizing filter (or a polarizing sunglass lens) and rotating the filter. Light from different regions of the sky will be transmitted in different degrees, depending on its degree of polarization. It is believed that some insects, such as bees, use polarized sky light to determine navigational directions relative to the Sun.

Why the Sky Is Blue

The scattering of sunlight by air molecules causes the sky to look blue. This effect is not due to polarization but is caused by the selective absorption of light. As oscillators, air molecules have resonant frequencies in the ultraviolet region. Consequently, the sunlight is preferentially scattered, with light in the nearby blue end of the visible region scattered more than that in the red end.

For particles such as air molecules, which are much smaller than the wavelength of light, the scattering S is inversely proportional to the wavelength to the fourth power (that is, $S \propto 1/\lambda^4$). This wavelength relationship is called **Rayleigh scattering** after Lord Rayleigh (1842–1919), a British physicist who derived it. This inverse relationship predicts that light of the shorter-wavelength, or blue, end of the spectrum will be scattered more than light of the longer-wavelength, or red, end. The scattered blue light is rescattered in the atmosphere and eventually is directed toward the ground. This is the sky light we see, so the sky appears blue.

LCDs and Polarized Light

LCDs (liquid crystal displays) are commonplace on watches, calculators, gas pumps, and even some television and computer monitors. The name "liquid crystal" may seem self-contradictory. Normally, when a crystalline solid melts, the resulting liquid no longer has an orderly atomic or molecular arrangement. Some organic compounds, however, pass through an intermediate state in which the molecules may rearrange somewhat but still maintain the overall order that is characteristic of a crystal.

A liquid crystal flows like a liquid, but its optical properties may depend on the order of its molecules. For example, certain liquid crystals with orderly arrangements are transparent. But the crystalline order can easily be disturbed by applied electrical forces. The disordered liquid then scatters light and is opaque.

A common type of LCD, called a twisted-nematic display, makes use of the effect of a liquid crystal on polarized light (Fig. 1). A display such as those commonly used on wristwatches and calculators is made by sandwiching a layer of liquid crystal material between two glass plates that have fine parallel grooves, or channels, in their surfaces. One of the plates is then rotated or twisted 90°. The molecules in contact with the plates remain parallel to the grooves, and the result is a molecular orientation that rotates through 90° between the plates. In this configuration, the liquid crystal is optically active and will rotate the direction of polarization of linearly polarized light.

The plates are then placed between crossed polarizing sheets and backed with a mirrored surface. Light entering and passing through the LCD is polarized, rotated 90°, re-

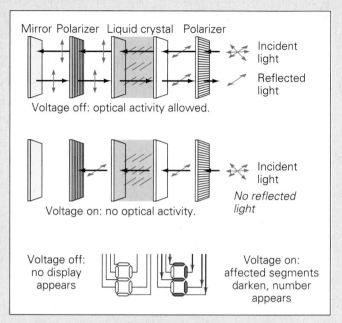

FIGURE 1 Liquid crystal display (LCD) A twisted-nematic display is an application involving the optical activity of a liquid crystal and crossed polarizers. When the crystalline order is disoriented by an electric field from an applied voltage, the liquid crystal loses its optical activity in that region, and light is not transmitted and reflected. Numerals and letters are formed by applying voltages to segments of a block display.

Example 24.7 ■ The Red and the Blue: Rayleigh Scattering

How much more is light at the blue end of the visible spectrum scattered by air molecules than is light at the red end?

Thinking It Through. We know that Rayleigh scattering is proportional to $1/\lambda^4$ and that light from the blue end of the spectrum (shorter wavelength) is scattered more. "How much more" implies a factor or ratio.

Solution. The Rayleigh scattering relationship is $S \propto 1/\lambda^4$, where S is the amount, or intensity, of scattering for a particular wavelength. Thus, you can form the ratio

$$\frac{S_{\text{blue}}}{S_{\text{red}}} = \left(\frac{\lambda_{\text{red}}}{\lambda_{\text{blue}}}\right)^4$$

The blue end of the spectrum (violet light) has a wavelength of about $\lambda_{\text{blue}} = 400$ nm, and red light has a wavelength of about $\lambda_{\text{red}} = 700$ nm. Inserting these values gives

$$\frac{S_{\text{blue}}}{S_{\text{red}}} = \left(\frac{\lambda_{\text{red}}}{\lambda_{\text{blue}}}\right)^4 = \left(\frac{700 \text{ nm}}{400 \text{ nm}}\right)^4 = 9.4 \quad \text{or} \quad S_{\text{blue}} = 9.4 S_{\text{red}}$$

Thus, blue light is scattered almost ten times as much as red light is.

Follow-up Exercise. What wavelength of light is scattered twice as much as red light? What color light is this?

flected, and again rotated 90° by its components. After the return trip through the liquid crystal, the polarization direction of the light is the same as that of the initial polarizer. Thus, the light is transmitted and leaves the display unit. Because of the reflection and transmission, the display appears to be a light color (usually light gray) when illuminated with white, unpolarized light.

By using an analyzer, we can readily show the light from the bright regions of an LCD to be polarized. The whole display appears dark when the analyzer is properly oriented (Fig. 1). You may have noticed this effect if you have ever tried to see the time on the LCD of a wristwatch while you were wearing polarizing sunglasses (Fig. 2).

The dark numbers or letters on an LCD are formed by applying an electric field to parts of the liquid crystal layer. The molecules of the liquid crystal are polar, so an electric

field can disorient them. Transparent, electrically conductive film coatings arranged in a seven-block pattern are put onto the glass plates. Each block, or display segment, has a separate electrical connection. When a voltage is applied across one or more of the segments, the electric field disorients the molecules of the liquid crystal in that area, and their optical activity is lost. (Note that all the numerals 0 through 9 can be formed by the segmented display.) The incident polarized light passing through the disoriented regions of the liquid crystal is absorbed by the second polarizer. Thus, these regions are opaque and appear dark. To produce images on small TV monitors, the display segments of the LCD are coupled so that many of them can be energized with a single lead.

One of the major advantages of LCDs is their low power consumption. Other similar displays, such as those using red light-emitting diodes (LEDs), produce light themselves, using relatively large amounts of power. LCDs produce no light but instead use reflected light.

Some liquid crystals respond to temperature changes, and the orientation of the molecules affects the light-scattering properties of the crystal. Different wavelengths, or colors, of light are selectively scattered. The color of a liquid crystal can thus be an indication of the surrounding temperature, and liquid crystals are used to make thermometers.

Finally, there is a new and emerging display technology called the polymer dispersed liquid crystal (PDLC) display. This type of display consists of a liquid crystal and polymer mixture, which has the property of electrically controlled light scattering. That is, the transparency of the display can be varied from opaque to clear by changing the applied voltage. A major advantage of PDLC displays is that they do not require polarizers, making them more cost effective. One emerging practical application of PDLC technology is the use of "smart" windows and shutters that control interior light levels.

FIGURE 2 Polarized light The light from an LCD is polarized, as can be shown by using polarizing sunglasses as an analyzer.

Conceptual Example 24.8

Scattering: Particle Size Makes a Difference

(a) Why does a polarizing filter used in color photography improve the contrast between clouds and the sky (•Fig. 24.25)? **(b)** "Once in a blue moon" indicates something that doesn't occur very often. Is it possible for the Moon to appear blue?

Reasoning and Answers. **(a)** Scattering effects depend on particle size relative to the wavelength of light. As we learned, if the scattering particles are much smaller than the wavelength, the (Rayleigh) scattering is preferential, depending strongly on the wavelength. The sky is blue as a result of preferential scattering. If the particles are larger than the wavelength, the scattering is not preferential, and visible light of all wavelengths (white light) is scattered. Hence, clouds appear white because of scattering from relatively large water droplets.

Light scattered from small particles (air molecules) is polarized to a much greater degree than is light scattered from larger particles (water droplets). As a result, when

•FIGURE 24.25 Contrast
A cloud photo taken with a polarizing filter. Note the darkness of the blue sky compared to the whiteness of the cloud. See Conceptual Example 24.8.

you use a polarizing filter in the proper orientation, the skylight is blocked more than the light from the clouds. The greater degree of darkening of the sky on the film then gives increased contrast between the sky and the clouds.

(b) In folklore, a second full moon in a calendar month, which doesn't occur too often, is referred to as a blue moon. This has nothing to do with the Moon appearing blue, but scattering does.

We have considered scattering from large and small particles but not from particles with sizes on the same order as the wavelength of light. In this range, red light is preferentially scattered more than blue light for particles of certain size. As a result, the red component of moonlight (reflected sunlight) coming through such particles might be scattered to the extent that the transmitted light has a bluish hue, and a "blue Moon" is seen. This effect does not occur often. The necessary particle size generally comes from large forest fires or volcanic eruptions. When the atmosphere has a large concentration of such particles, a "blue" Sun may also be observed.

Follow-up Exercise.   Clouds usually appear white because of scattered light. Considering the previous discussion of particle size, what does this indicate about clouds?

Keep in mind that all colors are present in sky light but that the dominant color is blue. (The sky doesn't appear violet because our eyes are more sensitive to blue and there is more blue than violet light in sunlight.) You may have noticed that the sky is more blue directly overhead than toward the horizon and appears white just above the horizon. (Look for this effect the next time you are outside on a clear day.) This is because there are relatively fewer air molecules—scatterers—directly overhead than toward the horizon. Multiple scatterings occur in the denser air near the horizon, and the recombination of scattered light gives rise to the white appearance. (Analogously, if you add a drop or two of milk to a glass of water and illuminate the suspension with intense white light, the scattered light has a bluish hue. But a glass of undiluted milk is white because of multiple scatterings.) In the same way, atmospheric pollution may impart a milky white appearance to most or all of the sky.

Why Sunsets and Sunrises Are Red

When the Sun is near the horizon, the sunlight travels a greater distance through the denser air near the Earth's surface. Since the light therefore undergoes a great deal of scattering, you might think that only the least scattered light, the red light, would reach observers on the Earth's surface. This would explain red sunsets. However, it has been shown that the dominant color of white light after only molecular scattering is orange. Thus, there must be other scattering that shifts the light from the setting (or rising) Sun toward the red (•Fig. 24.26).

Red sunsets have been found to result from the scattering of sunlight by atmospheric gases *and* by small foreign particles. These particles are not necessary for the blueness of the sky but are necessary for deep red sunsets and sunrises. (This is why spectacular red sunsets are observed in the months after a large volcanic eruption that puts tons of particulate matter into the atmosphere.) Red sunsets occur most often when there is a high-pressure air mass to the west, since the concentration of particles is generally greater in high-pressure air masses than in low-pressure air masses. Similarly, red sunrises occur most often when there is a high-pressure air mass to the east.

Now you can understand the old saying "Red sky at night, sailors' delight. Red sky in the morning, sailors take warning." (See also the Biblical quote Matthew 16:1–4.) Fair weather generally accompanies high-pressure air masses because they are associated with reduced cloud formation. Most of the United States lies in the Westerlies wind zone, in which air masses generally move from west to east. A red sky at night is thus likely to indicate a fair-weather, high-pressure air mass to the

•**FIGURE 24.26   Red sky at night** A spectacular red sunset over a mountaintop observatory in Chile. The red sky results from the scattering of sunlight by atmospheric gases and small solid particles. A directly observed reddening Sun is due to the scattering of the wavelengths toward the blue end of the spectrum in the direct line of sight.

(a)   (b)

•FIGURE 24.27   **And now for something completely different**   **(a)** During the day, the sky of Mars is red, because carbon dioxide preferentially scatters red wavelengths. **(b)** Can you explain why Martian sunrises and sunsets have a bluish cast?

west that will be coming your way. A red sky in the morning means the high-pressure air mass has passed and poor weather may set in.

As a final note, how would you like a sky that is normally *red*? Then try Mars, the "red planet." The thin Martian atmosphere is about 95% carbon dioxide (CO_2). The CO_2 molecule is more massive than an oxygen (O_2) or a nitrogen (N_2) molecule. As a result, CO_2 molecules have a lower resonant frequency (longer wavelength), and preferentially scatter the red end of the visible spectrum. Hence, the Martian sky is red during the day, adding to the true red of the iron oxide-laden ground (•Fig. 24.27). And what of the color of sunrises and sunsets on Mars? Think about it....

Chapter Review

Important Terms

physical (wave) optics   *741*
Young's double-slit
   experiment   *742*
thin-film interference   *745*
optical flats   *748*
Newton's rings   *748*
diffraction   *750*

diffraction grating   *753*
Bragg's law   *757*
polarization   *757*
polarizing (Brewster) angle   *757*
birefringence   *759*
dichroism   *760*

transmission axis (polarization
   direction)   *760*
optical activity   *762*
scattering   *762*
Rayleigh scattering   *763*
LCD (liquid crystal display)   *764*

Important Concepts

• Young's double-slit experiment provides evidence of the wave nature of light and a way to measure the wavelength of light ($\approx 10^{-7}$ m).

• Light reflected at a media boundary for which $n_2 > n_1$ undergoes a 180° phase change. If $n_2 < n_1$, there is no phase change on reflection.

• Polarization is the preferential orientation of the electromagnetic field vectors of light and is evidence that light is a transverse wave. Light can be polarized by reflection, double refraction (birefringence), and scattering.

• Reflection phase changes at boundaries contribute to thin-film interference, which depends on film thickness.

• In general, the longer the wavelength compared to the width of an opening or object, the greater the diffraction.

• The intensity of Rayleigh scattering is inversely proportional to the fourth power of the wavelength. The blueness of Earth's sky results from the preferential scattering of sunlight by air molecules.

Important Equations

Bright Fringe Condition (double-slit interference):
$$d \sin \theta = n\lambda \quad \text{for } n = 0, 1, 2, 3, \ldots \quad (24.3)$$

Wavelength Measurement (double-slit interference for small θ only):
$$\lambda \approx \frac{y_n d}{nL} \quad \text{for } n = 1, 2, 3, \ldots \quad (24.7)$$

Nonreflecting Film Thickness (minimum):
$$t = \frac{\lambda}{4n} \quad (\text{for } n_2 > n_1) \quad (24.8)$$

Dark Fringe Condition (single-slit diffraction):
$$w \sin \theta = m\lambda \quad \text{for } m = 1, 2, 3, \ldots \quad (24.9)$$

Lateral Displacement of Dark Fringes (single-slit diffraction):
$$y_m = \frac{mL\lambda}{w} \quad \text{for } m = 1, 2, 3, \ldots \quad (24.10)$$

Interference Maxima for a Diffraction Grating:
$$d \sin \theta = n\lambda \quad \text{for } n = 0, 1, 2, 3, \ldots \quad (24.13)$$
where $d = 1/N$ and N is the number of lines per unit length.

Limit of the Order Number:
$$n \leq \frac{d}{\lambda} \quad (24.14)$$

Brewster (polarizing) Angle:
$$\tan \theta_p = n \quad (24.16)$$

Exercises

24.1 Young's Double-Slit Experiment

1. ● In a Young's experiment using monochromatic light, if the slit spacing d decreases, the interference fringe spacing will (a) decrease, (b) increase, (c) remain unchanged, (d) disappear.

2. If the path-length difference between two identical and coherent beams is 2λ when they arrive at a point on a screen, the point will be (a) bright, (b) dark, (c) multicolored, (d) gray.

3. ● When white light is used in Young's experiment, many bright fringes with a spectrum of colors are seen. In a given fringe, is the red end or the blue end closer to the central maximum?

4. What would be observed with the set-up for Young's experiment if the light passing through the slit were not coherent?

5. Television pictures often flutter when an airplane passes by (•Fig. 24.28). Explain a possible cause of this fluttering based on interference effects.

6. Will two flashlights held close to each other produce a stationary interference fringe pattern in a Young's double-slit experiment? Explain.

7. ■ In the development of Young's experiment, a small-angle approximation ($\tan \theta \approx \sin \theta$) was used to find the lateral displacements of the bright and dark fringes. How good is this approximation? For example, what is the percentage error for $\theta = 15°$?

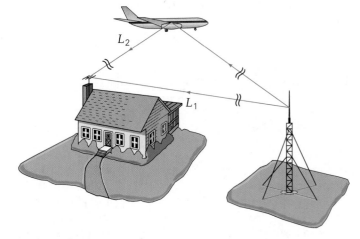

•**FIGURE 24.28 Interference** See Exercise 5.

8. ■ To study wave interference, a student uses two speakers driven by the same sound wave of wavelength 0.50 m. If the distances from a point to the speakers differ by 0.75 m, will the waves interfere constructively or destructively?

9. ■ Two point sources of coherent sound are located at (2.0 m, 5.0 m) and (4.0 m, 3.0 m) in the x-y plane. What is the longest possible wavelength of the sound waves at the point (15 m, 20 m) if (a) they interfere destructively and (b) they have a path-length difference of $\lambda/8$?

10. ■■ (a) Derive a relationship giving the locations of the dark fringes in a Young's double-slit experiment. What is the distance between the dark fringes? (b) For a third-order destructive interference (the third dark side fringe

from the central maximum), what is the path-length difference between that location and the two slits?

11. ■■ An interference pattern is formed on a screen when light of wavelength 550 nm is incident on two parallel slits 50 μm apart. The second-order bright fringe is 4.5 cm from the center of the central maximum. How far from the slits is the screen?

12. ■■ Monochromatic light passes through two narrow slits 0.25 mm apart and forms an interference pattern on a screen 1.5 m away. If light of wavelength 680 nm is used, what is the distance from the center of the central maximum to the center of the third-order bright fringe?

13. ■■ In a double-slit experiment using monochromatic light, a screen is placed 1.25 m from the slits, which have a separation distance of 0.0250 mm. The third-order bright fringe is 6.60 cm from the center of the central maximum. Find (a) the wavelength of the light and (b) the lateral displacement of the second-order bright fringe.

14. ■■ 💿 Monochromatic light illuminates two parallel slits that are 0.20 mm apart. The adjacent bright lines of the interference pattern on a screen 1.5 m away from the slits are 0.45 cm apart. What is the wavelength and color of the light?

15. ■■ 💿 Two parallel slits 1.0 mm apart are illuminated with monochromatic light of wavelength 640 nm, and an interference pattern is observed on a screen 3.00 m away. (a) What is the separation between adjacent interference maxima? (b) If the distance between the slits were increased to 1.5 mm, how would this change affect the separation of the maxima?

16. ■■ Yellow-green light ($\lambda = 550$ nm) is used in a double-slit experiment in which the slit separation distance is 1.75×10^{-4} m. With the screen located 2.00 m from the slits, determine (a) the angular separation in radians between the central maximum and the second-order bright fringe and (b) the lateral displacement of the fringe.

17. ■■ In a double-slit experiment with monochromatic light and a screen at a distance of 1.50 m, the angle between the second-order bright fringe and the central maximum is 0.0230 rad. If the separation distance of the slits is 0.0350 mm, what are (a) the wavelength and color of the light and (b) the lateral displacement of the fringe?

18. ■■■ What would be the qualitative effect on the spacing of the interference fringes if the apparatus for a Young's double-slit experiment were completely immersed in still water? (b) What would the lateral displacement in Exercise 12 be if the entire system were immersed in still water?

19. ■■■ 💿 Light of two different wavelengths is used in a double-slit experiment. The location of the third-order bright fringe for yellow-orange light ($\lambda = 600$ nm) coin-

cides with the location of the fourth-order bright fringe for the other color's light. What is the wavelength of the other light?

24.2 Thin-Film Interference

20. For a thin film with $n_1 > n_o$ and $n_1 > n_2$, where n_1 is the index of refraction of the film, a film thickness for constructive interference of the reflected light is (a) $\lambda'/4$, (b) $\lambda'/2$, (c) λ', (d) both (a) and (b).

21. For a thin film with $n_o < n_1 < n_2$, where n_1 is the index of refraction of the film, the minimum film thickness for destructive interference of the reflected light is (a) $\lambda'/4$, (b) $\lambda'/2$, (c) λ'.

22. What would be the effect on thin-film interference if the incident light is not perpendicular to the film?

23. Most lenses used in cameras are coated with thin films and appear bluish purple when viewed with reflected light. What wavelengths are not visible in the reflected light?

24. When destructive interference of two waves occurs at a certain location, there is no energy at that location. Is this situation a violation of the conservation of energy? Explain.

25. ■ Light of wavelength 550 nm in air is normally incident on a glass plate ($n = 1.5$) whose thickness is 1.1×10^{-5} m. (a) What is the thickness of the glass in terms of the wavelength in glass? (b) Will they interfere constructively or destructively?

26. ■ A lens with an index of refraction of 1.6 is to be coated with a material ($n = 1.4$) that will make the lens non-reflecting for red light ($\lambda = 700$ nm) normally incident on the lens. What is the minimum required thickness of the coating?

27. ■ Magnesium fluoride ($n = 1.38$) is frequently used as a lens coating to make nonreflecting lenses. What is the difference in the minimum film thickness required for maximum transmission of blue light ($\lambda = 400$ nm) and red light ($\lambda = 700$ nm)?

28. ■■ A solar cell is designed to have a nonreflective coating of a transparent material whose index of refraction is 1.22. (a) What is the minimum thickness of the film for light with a wavelength of 550 nm? (b) Does the index of refraction of the underlying material in the solar cell make a difference?

29. ■■ A thin layer of oil ($n = 1.5$) floats on water. Destructive interference is observed for light of wavelengths 480 nm and 600 nm, each at a different location. Find the two minimum thicknesses of the oil film assuming normal incidence.

30. ■■ It is desired to make a nonreflecting lens for an infrared radiation detector. If the coating material has an index of refraction of 1.20, what should the film thickness be for radiation with a frequency of 3.75×10^{14} Hz?

31. ■■ Two parallel plates are separated by a small distance as illustrated in •Fig. 24.29. If the top plate is illuminated with light from a He–Ne laser ($\lambda = 632.8$ nm), for what minimum separation distances will the light be (a) reflected back to the observer from air layer surfaces and (b) transmitted through the plates? [Note: $t = 0$ is *not* an answer for (b).]

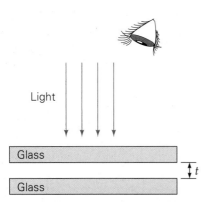

•**FIGURE 24.29   Reflection or transmission?**   See Exercise 31.

32. ■■■ An *air wedge* as shown in •Fig. 24.30 can be used to measure small dimensions such as the diameter of a thin wire. If the top glass plate is illuminated with monochromatic light, describe the observed interference pattern: (a) Express the locations of the bright interference fringes in terms of wedge thicknesses measured from the apex of the wedge. (b) Show that the number of dark fringes is given by $m' = 2t/\lambda_{air}$, where t is the maximum thickness of the wedge.

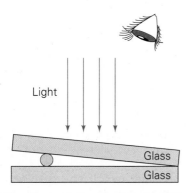

•**FIGURE 24.30   Air wedge**   See Exercises 32 and 33.

33. ■■■ The glass plates in Fig. 24.30 are separated by a thin, round filament. When the top plate is illuminated normally with light of wavelength 550 nm, the filament lies directly below the sixth bright fringe. What is the diameter of the filament?

24.3 Diffraction

34. ◐ In a single-slit diffraction pattern, (a) all maxima have the same width, (b) the central maximum is twice as wide as the side maxima, (c) the side maxima are twice as wide as the central maximum, (d) none of the above.

35. ◐ As the number of lines per unit length of a diffraction grating increases, the number of spectral orders that can be seen for a given wavelength (a) increases, (b) decreases, (c) remains unchanged.

36. (a) What does Eq. 24.11, for the width of the central maximum for the case of single-slit diffraction, predict when $w = \lambda$? Is the prediction correct? Explain. (b) Why is there no zeroth order ($n = 0$) in Bragg's law for crystal diffraction?

37. In the discussion of single-slit diffraction, the length of the slit is assumed to be much greater than the width. What changes would be observed in the diffraction pattern if the length were comparable to the width of the slit?

38. ■ A slit of width 0.20 mm is illuminated with monochromatic light of wavelength 480 nm, and a diffraction pattern is formed on a screen 1.0 m away from the slit. (a) What is the width of the central maximum? (b) What are the widths of the second-order and third-order bright fringes?

39. ■ A slit 0.025 mm wide is illuminated with red light ($\lambda = 680$ nm). How wide are (a) the central maximum and (b) the side maxima of the diffraction pattern formed on a screen 1.0 m from the slit?

40. ■ At what angle will the second-order diffraction maximum be seen from a diffraction grating of spacing 1.25 μm when illuminated by light of wavelength 550 nm?

41. ■ A Venetian blind is essentially a diffraction grating—not for visible light, but for waves with longer wavelengths. If the spacing between the slats of a blind is 2.5 cm, (a) for what wavelength is there a first-order maximum at an angle of 10°? (b) What type of radiation is this?

42. ■■ A single slit of width 0.50 mm is illuminated with monochromatic light ($\lambda = 680$ nm). A screen is placed 1.80 m from the slit to observe the fringe pattern. (a) What is the angle between the second dark fringe ($m = 2$) and the central maximum? (b) What is the lateral displacement of this dark fringe?

43. ■■ What is the width of the central maximum for the single-slit arrangement in Exercise 42?

44. ■■ A certain crystal gives a deflection angle of 25° for the first-order diffraction of monochromatic X-rays with a frequency of 5.0×10^{17} Hz. What is the lattice spacing of the crystal?

45. ■■ 🔵 Find the locations of the blue (λ = 420 nm) and red (λ = 680 nm) components of the first- and second-order spectra in a pattern formed on a screen 1.5 m from a diffraction grating with 7500 lines/cm.

46. ■■ How many fringes of interference maxima occur when monochromatic light of wavelength 560 nm illuminates a diffraction grating that has 10 000 lines/cm?

47. ■■ In a particular diffraction pattern, the red component (700 nm) in the second-order spectrum is deviated at an angle of 20°. (a) How many lines per centimeter does the grating have? (b) If the grating is illuminated with white light, how many bright fringes of the complete visible spectrum are produced?

48. ■■ 🔵 White light whose components have wavelengths from 400 nm to 700 nm illuminates a diffraction grating with 4000 lines/cm. Do the first and second orders overlap? Justify your answer.

49. ■■ 🔵 Visible white light traveling in air contains components with wavelengths from about 400 nm to 700 nm. If the white light illuminates a diffraction grating with 8000 lines/cm, what is the angular width of the first-order spectrum produced?

50. ■■ A diffraction grating with 8000 lines/cm is illuminated with a beam of monochromatic red light from a He–Ne laser (λ = 632.8 nm). How many side maxima would be formed in the diffraction pattern, and at what angles would they be observed?

51. ■■■ 🔵 Show that for a diffraction grating the violet (λ = 400 nm) portion of the third-order spectrum overlaps the yellow-orange (λ = 600 nm) portion of the second-order spectrum regardless of the grating's spacing.

52. ■■■ A teacher standing in a doorway 1.0 m wide blows a whistle with a frequency of 1000 Hz to summon children from the playground (•Fig. 24.31). Two boys are playing on the swings 100 m away from the school building. The boy on the left tells the other they had better go, but the boy on the right, at an angle of 19.6° from a line normal to the doorway, asks why?—he has not heard the whistle. Is this possible? (Take the speed of sound in air to be 335 m/s.)? Justify your answer mathematically.

24.4 Polarization

53. Light can be polarized by (a) reflection, (b) refraction, (c) scattering, (d) all of these.

54. The Brewster angle depends on (a) the index of refraction of a material, (b) Bragg's law, (c) internal reflection, (d) interference.

55. Given two pairs of sunglasses, could you tell if one or both were polarizing?

56. How would you hold an analyzer to detect the polarization of a rainbow? Could you block out the polarized light completely? Explain.

57. Suppose that you held two polarizing sheets in front of you and looked through both of them. How many times would you see the sheets lighten and darken (a) if one were rotated through one complete rotation, (b) if both were rotated through one complete rotation at the same rate in opposite directions, (c) if both were rotated through one complete rotation at the same rate in the same direction, and (d) if both were rotated through one complete rotation in opposite directions, but one twice as fast as the other?

58. How can you use the Brewster angle to determine the index of refraction of a material?

59. ■ Some types of glass have a range of indices of refraction of about 1.4 to 1.7. What is the range of the polarizing angle for these glasses?

60. ■ If the polarizing angle for a certain material in air is 58°, what is the index of refraction of the material?

61. ■■ A beam of light is incident on a glass plate (n = 1.62), and the reflected ray is completely polarized. What is the angle of refraction for the beam?

62. ■■ The critical angle for internal reflection in a certain medium is 45°. What is the Brewster angle for light externally incident on the medium?

63. ■■ The angle of incidence is adjusted so there is maximum linear polarization for the reflection of light from a transparent piece of plastic with n = 1.22. However, some of the light is transmitted. What is the angle of refraction?

64. ■■ Sunlight is reflected off a plate glass window (n = 1.55). What would be the Sun's altitude (angle above the horizon) for the reflected light to be completely polarized?

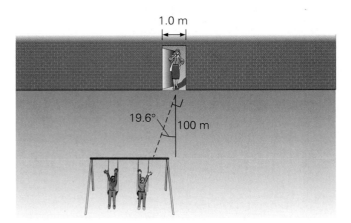

•**FIGURE 24.31 Moment of truth** See Exercise 52. (Not drawn to scale)

65. ■■ The Brewster angle in Eq. 24.16 was derived for incident light in air. Show that in general, $\theta_p = \tan^{-1}(n_2/n_1)$.

66. ■■ Find the Brewster angle for a piece of glass ($n = 1.60$) that is submerged in water.

67. ■■■ A plate of crown glass is covered with a layer of water. A beam of light traveling in air is incident on the water and partially transmitted. Is there any angle of incidence for which the light reflected from the water–glass interface will have maximum linear polarization? Justify your answer mathematically.

24.5 Atmospheric Scattering of Light

68. Which of the following colors is scattered the most in the atmosphere: (a) blue, (b) yellow, (c) red, (d) color makes no difference?

69. Explain why the sky is red in the morning and evening and blue during the day.

70. (a) Why does the sky not have a uniform blueness? (b) What color would an astronaut on the Moon see when looking at the sky or into space?

Additional Exercises

71. If the Brewster angle for a glass plate is 1.05 rad, what is the index of refraction of the glass?

72. If the slit width in a single-slit experiment were doubled, the distance to the screen reduced by one-third, and the wavelength of the light changed from 600 nm to 450 nm, how would the width of the bright fringes be affected?

73. A glass plate has an index of refraction of 1.5. At what angle of incidence will light be reflected from the plate with the maximum linear polarization?

74. When illuminated with monochromatic light, a diffraction grating with 1000 lines/cm produces a first-order maximum 4.50 cm from the central maximum on a screen 1.00 m away. What is the wavelength of the light?

75. A thin air wedge between two flat glass plates forms bright and dark interference bands when illuminated with normally incident monochromatic light (see Fig. 24.8.) (a) Show that the thickness of the air wedge changes by $\lambda/2$ from one bright fringe to the next, where λ is the wavelength of the light. (b) What would be the change in the wedge thickness between bright fringes if the space were filled with a liquid with an index of refraction n?

76. A diffraction grating is designed to have the second-order maxima at $\pm 10°$ from the central maximum for the red end ($\lambda = 700$ nm) of the visible spectrum. How many lines per centimeter does the grating have?

77. Two parallel slits 0.75 mm apart are illuminated with monochromatic light of wavelength 480 nm. Find the angle between the center of the central maximum and the center of an adjacent bright fringe formed on a screen 1.2 m away from the slits.

78. Monochromatic light of wavelength 500 nm falls on two slits separated by a distance of 40 μm. What is the distance between the first-order and third-order bright fringes formed on a screen 1.0 m away from the slits?

79. A film on a lens is 1.0×10^{-7} m thick and is illuminated with white light. The index of refraction of the film is 1.4. For what wavelength of light will the lens be nonreflecting?

80. What is the spectral order limit of a diffraction grating with 9000 lines/cm when the grating is illuminated by white light?

81. In a double-slit experiment that uses monochromatic light, the angular separation between the central maximum and the second-order bright fringe is 0.16°. What is the wavelength of the light if the distance between slits is 0.35 mm?

82. A camera lens is coated with a thin layer of a material that has an index of refraction of 1.35. This coating makes the lens nonreflecting for light of wavelength 450 nm (in air) that is normally incident on the lens. What is the thickness of the thinnest film that will make the lens nonreflecting?

Optical Instruments

Insights

- Biometrics: Your Body Is Unique
- Telescopes Using Nonvisible Radiation

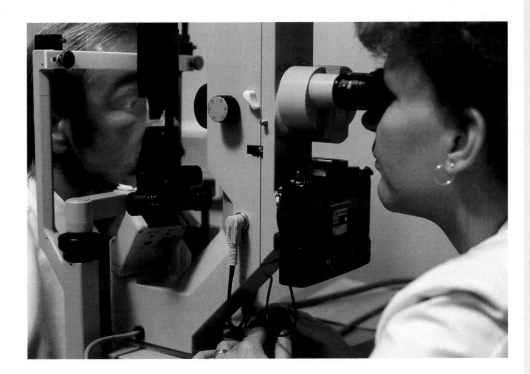

Around 1609, Galileo constructed an astronomical telescope and turned it skyward. He saw spots on the Sun, craters on the Moon, and four satellites circling Jupiter. About half a century later, a Dutch amateur lens grinder made a simple microscope and viewed microbes for the first time.

What would our knowledge of the world around us be like if these events had never happened? Perhaps some of our modern sciences would have survived. There would be chemistry, for example, but it would likely be severely limited in extent. But planets, stars, and galaxies would have remained nothing but mysterious points of light, and bacteria would still be unknown!

The basic function of optical instruments is to improve and extend our powers of visual observation. Indeed, vision is our chief means of acquiring information about the world. Mirrors and lenses are used in a variety of optical instruments, including those in medicine. The ophthalmoscope in the photo is used to examine the retina of the eye. The device contains a light source and a series of lenses on a rotatable selector that allows the retina to be brought into focus.

Mirrors and lenses were discussed in Chapter 23 and other optical phenomena in Chapter 24. The principles developed in those chapters can be easily applied to the study of optical instruments. In this chapter, you will learn more about microscopes and telescopes, and about the factors that limit these devices. You'll also discover more about the essential optical instrument without which all the others would be of little use: the human eye.

25.1 The Human Eye

Objectives: To (a) describe the optical workings of the eye and (b) explain some common vision defects and how they are corrected.

The human eye is the most fundamental optical instrument, since without it the field of optics would not exist. The eye is analogous to a simple camera in several respects (•Fig. 25.1). A simple camera consists of a converging lens, which is used to focus images on light-sensitive film at the back of the camera's interior chamber. (Recall from Chapter 23 that for relatively distant objects, a converging lens produces a smaller, inverted, real image.) An adjustable diaphragm, or opening, and a shutter control the amount of light entering the camera.

The eye too has a converging lens that focuses images on the light-sensitive lining on the rear surface inside the eyeball. (Other transparent media in the eye also help create the image by refracting incoming light.) The eyelid might be thought of as a shutter; however, the shutter of the camera, which controls the exposure time, is generally opened only for a fraction of a second, and the eyelid is normally open for continuous exposure. The human nervous system performs a function analogous to that of a shutter in analyzing image signals from the eye at a rate of 20 to 30 times per second. The eye might therefore be likened to a movie or video camera, which exposes a similar number of frames (images) per second.

Although the optical functions of the eye are relatively simple, its physiological functions are quite complex. As Fig. 25.1b shows, the eyeball is a nearly spherical chamber. It has an internal diameter of about 1.5 cm and is filled with a jellylike substance called the *vitreous humor*. The eyeball has a white outer covering called the *sclera*, part of which is visible as the white of the eye. Light enters the eye through a curved, transparent tissue called the *cornea* and passes into

> **Note:** Image formation by a converging lens is discussed in Section 23.3; see Figure 23.15b.

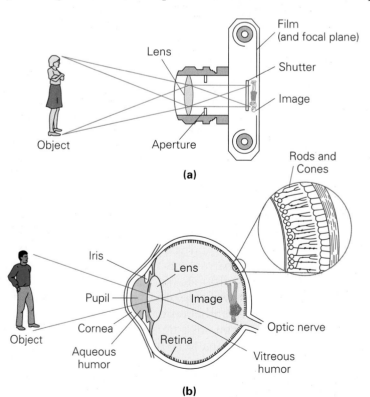

•**FIGURE 25.1 Camera and eye analogy**
In some respects, a simple camera is similar to the human eye. An image is formed on the film in a camera and on the retina of the eye. (The complex refractive properties of the eye are not shown because multiple refractive media are involved.)

a clear fluid known as the *aqueous humor*. Behind the cornea is a circular diaphragm, the *iris*, whose central hole is called the *pupil.* The iris contains the pigment that determines eye color. Through muscle action, the iris can change the area of the pupil (2 to 8 mm in diameter), thereby controlling the amount of light entering the eye.

Behind the iris is a *crystalline lens,* a converging lens composed of microscopic glassy fibers. When tension is exerted on the lens by attached muscles, the glassy fibers slide over each other, causing the shape and focal length of the lens to change, thus focusing the image properly.

On the back interior wall of the eyeball is a light-sensitive surface called the **retina.** From the retina, the optic nerve relays signals to the brain. The retina is composed of nerves and two types of light receptors, or photosensitive cells, called **rods** and **cones** because of their shapes. The rods are more sensitive to light than the cones are and distinguish light from dark in low light intensities (twilight vision). The cones, conversely, can distinguish frequency ranges of sufficiently intense light, which the brain interprets as colors. Most of the cones are clustered together around a central region of the retina. The more numerous rods are outside this region and distributed nonuniformly over the retina.

Rods—twilight vision
Cones—color vision

The focusing adjustment of the eye is unlike that of a simple camera. A camera lens has a constant focal length, and the image distance is varied by moving the lens to produce sharp images on the film for different object distances. In the eye, the image distance is constant, and the focal length of the lens is varied to produce sharp images on the retina for different object distances.

When the eye is focused on distant objects, the muscles are relaxed, and the crystalline lens is thinnest and has a power of about 20 D (diopters). Recall that the power (*P*) of a lens in diopters (D) is the reciprocal of its focal length *in meters.* When the eye is focused on closer objects, the lens becomes thicker, and the radius of curvature and focal length are decreased. The lens power may increase to 30 D, or even more in young children. The adjustment of the focal length of the crystalline lens is called *accommodation.* (Look at a nearby object and then at one in the distance, and notice how fast accommodation takes place. It's practically instantaneous.)

Note: This relationship is presented in Eq. 23.10, Section 23.5.

The extremes of the range over which distinct vision (sharp focus) is possible are known as the far point and the near point. The *far point* is the greatest distance at which the normal eye can see objects clearly; it is taken to be infinity. The *near point* is the position closest to the eye at which objects can be seen clearly; it depends on the extent the lens can be deformed (thickened) by accommodation. The range of accommodation gradually diminishes with age as the crystalline lens loses its elasticity. That is, the near point gradually recedes with age. The approximate positions of the near point at various ages are listed in Table 25.1.

Note: The eye sees clearly between its far point and near point.

Children can see sharp images of objects that are within 10 cm (4 in.) of their eyes, and the crystalline lens of a normal young adult eye can be deformed to produce sharp images of objects as close as 12 to 15 cm (5 to 6 in.). However, at about the age of 40, the near point normally moves beyond 25 cm (10 in.). You may have noticed people over the age of 40 holding reading material away from their eyes to bring it within the range of accommodation. When the print gets too small or the arm too short, corrective reading glasses are the solution. The recession of the near point with age is not considered an abnormal defect of vision, since it proceeds at about the same rate in all normal eyes.

Another use of eye characteristics is discussed in the Insight on p. 776.

TABLE 25.1 Approximate Near Points of the Normal Eye at Different Ages

Age (years)	Near Point (centimeters)
10	10
20	12
30	15
40	25
50	40
60	100

Vision Defects

Speaking of the "normal" eye implies the existence of eyes with defective vision. That this is the case is apparent from the number of people who wear glasses or contact lenses. The eyes of many people cannot accommodate within the normal

Biometrics: Your Body Is Unique

Certain physical characteristics differentiate you from everyone else on the Earth (as best we know). The name *biometrics* refers to various methods that have been devised to analyze or measure these characteristics for identification purposes. The most common of these methods is fingerprinting, and we mentioned voice prints in Chapter 14. However, there are more, and they are coming into increasing use.

Here are some of the anatomical features and their characteristics that are believed to make you unique:

- Your hand—shape; size; thickness, width, and length of fingers; and size of knuckles.

- Your face—shape, size, and distances among facial features.

- Your finger—distances and points of reference in a finger scan rather than a fingerprint with ink.

- Your retina—the vascular pattern on the retina of your eye.

- Your iris—the pattern of light versus dark areas (more detailed than a fingerprint).

Let's focus on iris identification, which is used in national security applications and even at some bank automatic teller machines (ATMs). Iris recognition-based systems use simple black-and-white video cameras, but ATMs use infrared light, which works even under poor lighting conditions at night.

Figure 1 shows an iris recognition at an ATM. The customer stands in front of the machine, and the identification device images the face and determines the position of an eye. The system then lays a circular grid over the video picture of the eye and generates a human "bar code" from the light and dark areas of the iris. This code is checked against one in the customer's identification file. The whole process takes about 2 to 3 s. If there's no match, there's no money.

Iris images are more popular than retina scans because of ease of use. For the latter, a person has to come eyeball-to-lens with the machine.

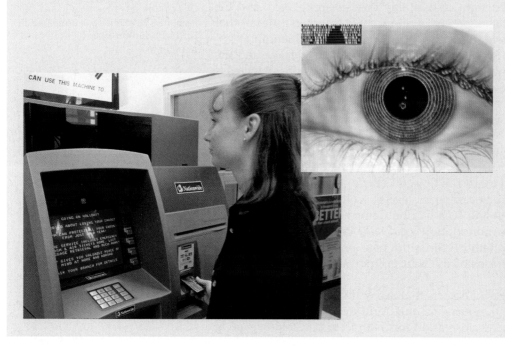

FIGURE 1 Biometrics in action Computerized iris recognition systems are used at some ATM machines. Each person's iris is unique, and the computer divides the iris into eight concentric circles to recognize the unique features—a human "bar code" that is checked against the customer's identification file. No match, no money.

range of 25 cm to infinity. These people have one of the two most common visual defects: nearsightedness (myopia) or farsightedness (hyperopia). Both of these conditions can usually be corrected (•Fig. 25.2).

Nearsightedness: difficulty in seeing distant objects

Nearsightedness (or *myopia*) is the ability to see nearby objects clearly but not distant objects. That is, the far point is not infinity but some nearer point. When an object beyond the far point is viewed, the rays come to focus in front of the retina (Fig. 25.2b). As a result, the image on the retina is blurred, or out of focus. As the object is moved closer to the eye, its image moves back toward the retina. If the object is moved within the far point, a sharp image is seen.

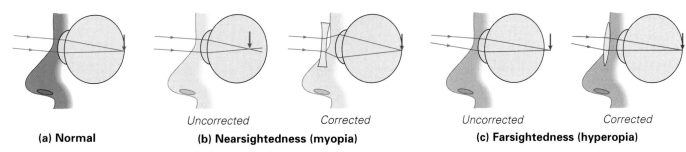

	Uncorrected	Corrected	Uncorrected	Corrected
(a) Normal	**(b) Nearsightedness (myopia)**		**(c) Farsightedness (hyperopia)**	

•**FIGURE 25.2   Nearsightedness and farsightedness**   (a) The normal eye produces sharp images on the retina for objects located between its near point and its far point. The image is real, inverted, and always smaller than the object (why?). Here, the object is a distant, upward-pointing arrow (not shown), and the light rays come from its tip. **(b)** In a nearsighted eye, the image of a *distant* object is focused *in front of* the retina. This defect is corrected with a diverging lens. **(c)** In a farsighted eye, the image of a *nearby* object is focused *behind* the retina. This defect is corrected with a converging lens. (Not drawn to scale.)

Nearsightedness arises because the eyeball is too long or perhaps because the curvature of the cornea is too great. Whatever the reason, the images of distant objects are focused in front of the retina. Appropriate diverging lenses correct this condition. Such a lens causes the rays to diverge, and the eye focuses the image farther back so that it falls on the retina.

Farsightedness (or *hyperopia*) is the ability to see distant objects clearly but not nearby objects. That is, the near point is not at the normal position but at some point farther from the eye. The image of an object that is closer to the eye than the near point is formed behind the retina (Fig. 25.2c). Farsightedness arises because the eyeball is too short or perhaps because of insufficient curvature of the cornea. A similar farsighted condition (*presbyopia*) occurs when the near point recedes with age, as discussed previously.

Farsightedness is usually corrected with appropriate converging lenses. Such a lens causes the rays to converge, and the eye is then able to focus the image on the retina. Converging lenses are also used to correct the farsightedness associated with the natural recession of the near point with age. Older people often must wear reading glasses, which have converging lenses.

Farsightedness: difficulty in seeing nearby objects

Example 25.1 ■ Correcting Nearsightedness: Use of a Diverging Lens

A certain nearsighted person cannot see objects clearly when they are more than 78 cm from either eye. What power must corrective lenses have if this person is to see distant objects clearly? Assume that the lenses are in eyeglasses and are 3.0 cm in front of the eye (see •Fig. 25.3).

Thinking It Through.   For nearsightedness, the corrective lens must be diverging and must form the image of a distant object at the eye's far point.

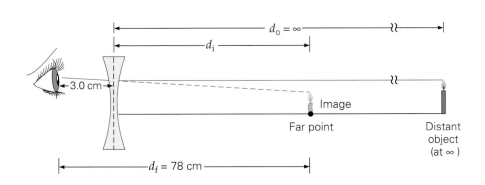

•**FIGURE 25.3   Correcting nearsightedness**   A diverging lens is used. See Example 25.1. (Not drawn to scale.)

Note: Review Examples 23.5 and 23.6.

Note: Image formation by a diverging lens is discussed in Section 23.3; see Figure 23.15a.

Solution. Listing the given distances, we have

Given: $d_f = 78$ cm (far point) *Find:* P (in diopters)
$d = 3.0$ cm

This is a lens problem similar to those in Chapter 23. Recall that the power of a lens is $P = 1/f$ (Eq. 23.10). We can use the thin-lens equation (Eq. 23.6) to find f or $1/f$ if we can determine the object and image distances, d_o and d_i.

The lens must effectively put the image of a distant object ($d_o = \infty$) at the far point d_f, which is 78 cm from the eye. The image, which acts as an object for the eye, is then within the range of accommodation. The image distance is *measured from the lens*, which is 3.0 cm from the eye, so $d_i = -75$ cm: $d_i = d_f - d = 78$ cm $- 3.0$ cm $= 75$ cm (see the figure, which is not drawn to scale). A minus sign is used because the image is virtual, being on the object side of the lens. (You may recall from Chapter 23 that diverging lenses can form only virtual images.)

Then, using the thin-lens equation, we get

$$\frac{1}{f} = \frac{1}{d_o} + \frac{1}{d_i} = \frac{1}{\infty} - \frac{1}{75 \text{ cm}} = -\frac{1}{75 \text{ cm}} \quad \text{or} \quad f = -75 \text{ cm}$$

The minus sign indicates a diverging lens, as we expected. From Eq. 23.10, we get

$$P = \frac{1}{f} = \frac{1}{-0.75 \text{ m}} = -1.3 \text{ D}$$

A negative, or diverging, lens with a power of 1.3 D is needed.

Follow-up Exercise. Suppose a mistake was made in this Example and a "corrective" lens of $+1.3$ D were used. What effect would this have?

If the near point is changed by corrective diverging lenses, as in Example 25.1, and this causes another viewing problem, bifocal lenses can be used. Bifocals were invented by Ben Franklin, who glued two lenses together. They are now made by grinding lenses with different curvatures in two different regions. Both nearsightedness and farsightedness can be treated at the same time with bifocals. (Trifocals are also available, with lenses having three different curvatures.)

A more modern technique to correct nearsightedness is shown in •Fig. 25.4. It involves the use of a laser that vaporizes body tissue. A flap of material on the cornea surface is cut and pulled up. The laser is then used to shape the exposed surface of the cornea, which changes its refractive characteristics so as to have the image of a distant object fall on the retina. The corneal flap is then replaced. This procedure eliminates the need for corrective lenses. It is relatively painless, and recovery time for vision is on the order of 24 hours.

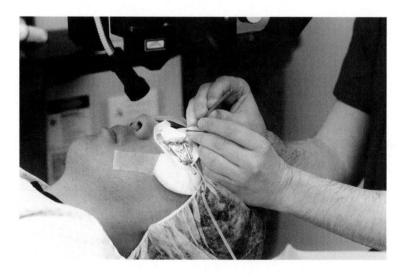

•FIGURE 25.4 Lasers and nearsightedness The surgeon cuts and retracts a flap of the cornea in preparation for shaping the cornea with a laser; the flap will be replaced. Some patients regain their sight within a day after this procedure. Notice that the surgeon is wearing no latex gloves. The fine chalk dust used on the gloves as a lubricant could contaminate the eye.

Example 25.2 ■ Correcting Farsightedness: Use of a Converging Lens

A farsighted person has a near point of 75 cm for one eye and a near point of 100 cm for the other. What powers should contact lenses have to allow the person to see an object clearly at a distance of 25 cm?

Thinking It Through. For farsightedness, the corrective lens must be converging and must form the image of an object at the normal eye's near point.

Solution. The eyes are distinguished as 1 and 2, and the image distances are negative (why?).

Given: $d_{i_1} = -75 \text{ cm} = -0.75 \text{ m}$ *Find:* P_1 and P_2 (lens power for each eye)
$d_{i_2} = -100 \text{ cm} = -1.0 \text{ m}$
$d_o = 25 \text{ cm} = 0.25 \text{ m}$

The optics of a person's eyes are usually different, and a different lens may be needed for each eye. In this case, each lens is to form an image at its eye's near point of an object that is at a distance (d_o) of 25 cm. The image will then act as an object within the eye's range of accommodation. This situation corresponds to a person wearing reading glasses. (For the sake of clarity, the lens in •Fig. 25.5a is not in contact with the eye.)

The image distances are negative since the images are virtual. With contact lenses, the distance from the eye and the distance from the lens are taken to be the same. Then

$$P_1 = \frac{1}{f_1} = \frac{1}{d_o} + \frac{1}{d_{i_1}} = \frac{1}{0.25 \text{ m}} - \frac{1}{0.75 \text{ m}} = \frac{2}{0.75 \text{ m}} = +2.7 \text{ D}$$

and

$$P_2 = \frac{1}{f_2} = \frac{1}{d_o} + \frac{1}{d_{i_2}} = \frac{1}{0.25 \text{ m}} - \frac{1}{1.0 \text{ m}} = \frac{3}{1.0 \text{ m}} = +3.0 \text{ D}$$

Note that these are positive, or converging, lenses, as expected.

Follow-up Exercise. A mistake is made in grinding the corrective lenses in this Example—the left lens is ground to the prescription intended for the right eye, and vice versa. What effect would this have?

Another common defect of vision is **astigmatism**, which is usually due to a refractive surface, normally the cornea or crystalline lens, being out of round (nonspherical). As a result, the eye has different focal lengths in different planes (•Fig. 25.6a). Points may appear as lines, and the image of a line may be distinct in one direction and blurred in another or blurred in both directions in the *circle of least confusion* (the location of the least distorted image). A test for astigmatism is given in Fig. 25.6b.

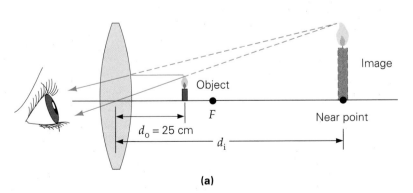

(a)

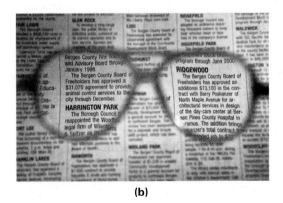

(b)

•**FIGURE 25.5   Reading glasses and correcting farsightedness   (a)** When an object at the normal near point (25 cm) is viewed through reading glasses with converging lenses, the image is formed farther away but within the eye's range of accommodation (beyond the receded near point). **(b)** Small print as viewed through the lens of reading glasses. See Example 25.2.

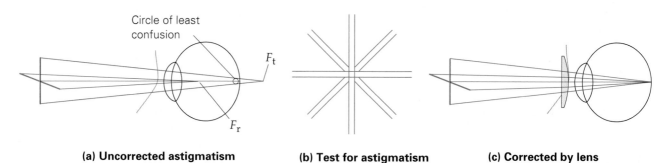

| (a) Uncorrected astigmatism | (b) Test for astigmatism | (c) Corrected by lens |

•**FIGURE 25.6 Astigmatism** When one of the eye's refracting components is not spherical, the eye has different focal lengths in different planes. **(a)** The effect occurs because rays along the transverse and radial axes are focused at different points: F_t and F_r, respectively. **(b)** If the eye is astigmatic, some or all of the lines in this diagram will appear blurred. **(c)** Nonspherical lenses, such as plano-convex cylindrical lenses, are used to correct astigmatism.

Astigmatism can be corrected with lenses that have greater curvature in the plane in which the cornea or crystalline lens has deficient curvature (Fig. 25.5c). Astigmatism is lessened in bright light because the pupil of the eye becomes smaller and the circle of least confusion is reduced.

You have probably heard of 20/20 vision. But what is it? *Visual acuity* is a measure of how vision is affected by object distance. This quantity is commonly determined by using a chart of letters placed at a given distance from the eyes. The result is usually expressed as a fraction: The *numerator* is the distance at which the test eye sees a standard symbol, such as the letter E clearly; the *denominator* is the distance at which the letter is seen clearly by a *normal* eye. A 20/20 (test/normal) rating, which is sometimes called "perfect" vision, means that at a distance of 20 ft, the eye being tested can see standard-sized letters as clearly as can a normal eye.

A person's eyes can have differing visual acuity; for example, one eye may have 20/20 vision and the other 20/30. The latter means that the eye can read standard-sized letters from a chart at a distance of 20 ft and someone with normal vision could do so at 30 ft. That is, the eye in question has to be closer to the chart than a normal eye would to see the letters clearly. Similarly, a person with 20/15 vision would have a particularly sharp eye. The test eye could see an object clearly from a distance of 20 ft, whereas the normal eye would have to be a distance of 15 ft to do so. Thus the greater the ratio or fraction, the better the visual acuity of the eye.

25.2 Microscopes

Objectives: To (a) distinguish between lateral and angular magnification and
(b) describe simple and compound microscopes and their magnifications.

Microscopes are used to magnify objects so that we can see more detail or see features that are normally indiscernible. Two basic types of microscope will be considered here.

The Magnifying Glass (A Simple Microscope)

When we look at an object in the distance, it appears to be very small. As it is brought or comes closer to our eyes, it appears larger. How large an object appears depends on the size of the image on the retina. This size is related to the angle subtended by the object (•Fig. 25.7): The greater the angle, the larger the image.

When we want to examine detail, or look at something closely, we bring it close to our eyes so that it subtends a greater angle. For example, you may examine the detail of a figure in this book by bringing it closer to your eyes. You'll see the greatest amount of detail when the book is at your near point (assum-

Observed size of fly

θ

Actual size of fly

(a) Narrow angle **(b) Wider angle**

•**FIGURE 25.7 Magnification and angle (a)** How large an object appears is related to the angle subtended by the object. **(b)** The angle and the size of an object are increased with a converging lens.

ing you are not wearing glasses). If your eyes were able to accommodate to shorter distances, an object brought very close to them would be further magnified. However, as you can easily prove by bringing this book very close to your eyes, images are blurred when objects are at distances inside the near point.

A **magnifying glass**, which is simply a single convex lens (sometimes called a simple microscope), forms a clear image of an object that is closer than the near point. In such a position, an object subtends a greater angle and therefore appears larger, or magnified. The lens produces a virtual image beyond the near point on which the eye focuses. If a hand-held magnifying glass is used, its position is usually adjusted until this image is seen clearly.

As illustrated in •Fig. 25.8, the angle subtended by an object is much greater when a magnifying glass is used. The magnification of an object *viewed through a magnifying glass* is expressed in terms of this angle. This **angular magnification**, or magnifying power, is symbolized by m. The angular magnification is defined as the ratio of the angular size of the object viewed through the magnifying glass (θ) to the angular size of the object viewed without the magnifying glass (θ_o):

$$m = \frac{\theta}{\theta_o} \qquad \textit{angular magnification} \qquad (25.1)$$

(This m is not the same as M, the lateral magnification, a ratio of heights: $M = h_i/h_o$.)

The maximum angular magnification occurs when the image seen through the glass is at the eye's near point, $d_i = -25$ cm, since this is as close as it can be clearly seen. (A value of 25 cm will be assumed to be typical for near point in this discussion. The minus sign is used because the image is virtual.) The corresponding object distance can be calculated from the thin-lens equation, Eq. 23.6:

$$d_o = \frac{d_i f}{d_i - f} = \frac{(-25 \text{ cm})f}{-25 \text{ cm} - f}$$

or

$$d_o = \frac{25 f}{25 + f} \qquad (25.2)$$

where f must be in centimeters.

Note: Angular magnification is not the same as lateral magnification, which is discussed in Section 23.1 (see Eq. 23.1).

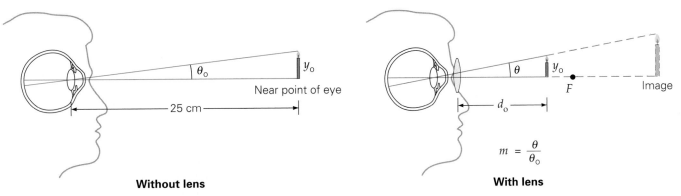

Without lens **With lens**

Near point of eye

25 cm

θ_o y_o

θ y_o F Image

d_o

$$m = \frac{\theta}{\theta_o}$$

•**FIGURE 25.8 Angular magnification** The angular magnification (m) of a lens is defined as the ratio of the angular size of an object viewed through the lens to the angular size of the object viewed without the lens, $m = \theta/\theta_o$.

The angular sizes of the object are related to its height by

$$\tan \theta_o = \frac{y_o}{25} \quad \text{and} \quad \tan \theta = \frac{y_o}{d_o}$$

(see Fig. 25.8). Assuming that a small-angle approximation ($\tan \theta \approx \theta$) is valid gives

$$\theta_o \approx \frac{y_o}{25} \quad \text{and} \quad \theta \approx \frac{y_o}{d_o}$$

Then the maximum angular magnification can be expressed as

$$m = \frac{\theta}{\theta_o} = \frac{y_o/d_o}{y_o/25} = \frac{25}{d_o}$$

Substituting for d_o from Eq. 25.2 gives

$$m = \frac{25}{25f/(25 + f)}$$

which simplifies to

Magnifying glass magnification

$$m = 1 + \frac{25 \text{ cm}}{f} \qquad \begin{array}{l} \textit{angular magnification} \\ \textit{for image at near point (25 cm)} \end{array} \qquad (25.3)$$

Thus, lenses with shorter focal lengths give greater angular magnifications.

In the derivation of Eq. 25.3, the object for the unaided eye was taken to be at the near point, as was the image viewed through the lens. Actually, the normal eye can focus on an image located anywhere between the near point and infinity. At the extreme where the image is at infinity, the eye is more relaxed (the muscles attached to the crystalline lens are relaxed, and the lens is thin). For the image to be at infinity, the object must be at the focal point of the lens. In this case,

$$\theta \approx \frac{y_o}{f}$$

and the angular magnification is

$$m = \frac{25 \text{ cm}}{f} \qquad \begin{array}{l} \textit{angular magnification} \\ \textit{for image at infinity} \end{array} \qquad (25.4)$$

Mathematically, it seems that the magnifying power can be increased to any desired value by using lenses with short enough focal lengths. Physically, however, lens aberrations limit the practical range of a single magnifying glass to about 3× or 4× (read as "three ex" and "four ex"), or a sharp image magnification of three or four times the size of the object when used normally.

Example 25.3 ■ It's Elementary: Angular Magnification of a Magnifying Glass

Sherlock Holmes uses a converging lens with a focal length of 12 cm to examine the fine detail of some cloth fibers found at the scene of a crime. (a) What is the maximum magnification given by the lens? (b) What is the magnification for relaxed eye viewing?

Thinking It Through. Equations 25.3 and 25.4 apply here. Part (a) asks for the maximum magnification, which is discussed in the derivation of Eq. 25.3 and occurs when the image formed by the lens is at the near point of the eye. For part (b), note that the eye is most relaxed when viewing distant objects.

Solution.

Given: $f = 12$ cm   *Find:* (a) m (d_i = near point)
                                    (b) m ($d_i = \infty$)

(a) For Equation 25.3 the near point was taken to be 25 cm:

$$m = 1 + \frac{25 \text{ cm}}{f} = 1 + \frac{25 \text{ cm}}{12 \text{ cm}} = 3.1$$

(b) Equation 25.4 gives the magnification for the image formed by the lens at infinity:

$$m = \frac{25 \text{ cm}}{f} = \frac{25 \text{ cm}}{12 \text{ cm}} = 2.1$$

Follow-up Exercise. Taking the maximum practical magnification of a magnifying glass to be 4×, which would have the longer focal length, a glass for near-point viewing or one for distant viewing, and how much longer?

The Compound Microscope

A compound microscope provides greater magnification than is attained with a single lens, or simple microscope. A basic **compound microscope** consists of a pair of converging lenses, each of which contributes to the magnification (•Fig. 25.9a). The converging lens with a relatively short focal length ($f_o < 1$ cm) is known as the **objective**. It produces a real, inverted, and enlarged image of an object positioned slightly beyond its focal point. The other lens, called the **eyepiece**, or **ocular**, has a longer focal length (f_e is a few centimeters) and is positioned so that the image formed by the objective falls just *inside* its focal point. This lens forms a magnified virtual image that is viewed by the observer. In essence, the objective acts as a projector, and the eyepiece is a simple magnifying glass used to view the projected image.

> **Note:** It might be useful to review Section 23.3 and Fig. 23.16.

The total magnification (m_{total}) of a lens combination is the product of the magnifications produced by the two lenses. The image formed by the objective is larger than its object by a factor M_o equal to the lateral magnification ($M_o = d_i/d_o$, with the minus sign omitted). In Fig. 25.9a, note that the image distance for the objective lens is approximately equal to L, the distance between the lenses. That is,

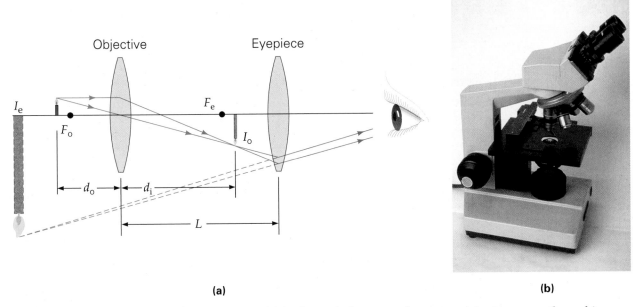

(a)   **(b)**

•**FIGURE 25.9   The compound microscope**   (a) In the optical system of a compound microscope, the real image formed by the objective falls just within the focal point of the eyepiece (F_e) and acts as an object for this lens. An observer looking through the eyepiece sees an enlarged image. **(b)** A compound microscope.

$d_i \approx L$. (The image I_o is formed by the objective just inside the focal point of the eyepiece, which has a very short focal length.) Also, since the object is very close to the focal point of the objective, $d_o \approx f_o$. With these approximations,

$$M_o \approx \frac{L}{f_o}$$

The angular magnification of the eyepiece for an object at its focal point is given by Eq. 25.4:

$$m_e = \frac{25 \text{ cm}}{f_e}$$

Since the object for the eyepiece (the image formed by the objective) is very near the focal point of the eyepiece, a good approximation is

$$m_{total} = M_o m_e = \left(\frac{L}{f_o}\right)\left(\frac{25 \text{ cm}}{f_e}\right)$$

or

Compound microscope magnification

$$m_{total} = \frac{(25 \text{ cm})L}{f_o f_e} \qquad \begin{array}{l} \textit{angular magnification of} \\ \textit{compound microscope} \end{array} \qquad (25.5)$$

where f_o, f_e, and L are in centimeters.

Example 25.4 ■ A Compound Microscope: Finding Magnification

A microscope has an objective with focal length of 10 mm and an eyepiece with a focal length of 4.0 cm. The lenses are positioned 20 cm apart in the barrel. Determine the approximate total magnification of the microscope.

Thinking It Through. This is a direct application of Eq. 25.5.

Solution.

Given: $f_o = 10 \text{ mm} = 1.0 \text{ cm}$ *Find:* m_{total} (total magnification)
        $f_e = 4.0 \text{ cm}$
        $L = 20 \text{ cm}$

Using Eq. 25.5, we get

$$m_{total} = \frac{(25 \text{ cm})L}{f_o f_e} = \frac{(25 \text{ cm})(20 \text{ cm})}{(1.0 \text{ cm})(4.0 \text{ cm})} = 125\times$$

(Note the relatively short focal length of the objective.)

Follow-up Exercise. If the focal length of the eyepiece in this Example were doubled, how would this affect the length of the microscope for the same magnification? (Express the change as a percentage.)

A modern compound microscope is shown in Fig. 25.9b. Interchangeable eyepieces with magnifications from about 5× to over 100× are available. For standard microscopic work in biology or medical laboratories, 5× and 10× eyepieces are normally used. Microscopes are often equipped with rotating turrets, which usually contain three objectives for different magnifications, such as 10×, 43×, and 97×. These objectives and the 5× and 10× eyepieces can be used in various combinations to provide magnifying powers from 50× to 970×. The maximum magnification with a compound microscope is about 2000×.

    Opaque objects are usually illuminated with a light source placed above them. Specimens that are transparent, such as cells or thin sections of tissues on glass slides, are illuminated with a light source beneath the microscope stage so that light passes through the specimen. A modern microscope is usually equipped with a light condenser (converging lens) and diaphragm below the stage, which are used to concentrate the light and control its intensity. A microscope may have an internal light source. The light is reflected into the condenser from a mirror. Older microscopes have two mirrors with reflecting surfaces: one a plane mirror for reflecting light from a high-intensity external source, and the other a concave mirror for converging low-intensity light such as sky light.

25.3 Telescopes

Objectives: **To (a) distinguish between refractive and reflective telescopes and (b) describe the advantages of each.**

Telescopes apply the optical principles of mirrors and lenses to improve our ability to see distant objects. Used for both terrestrial and astronomical observations, telescopes allow some objects to be viewed in greater detail and other fainter or more distant objects simply to be seen. Basically, there are two types of telescopes: refracting and reflecting. These types depend on the gathering and converging of light by lenses and mirrors, respectively.

Refracting Telescope

The principle underlying one type of **refracting telescope** is similar to that of a compound microscope. The major components of this type of telescope are objective and eyepiece lenses, as illustrated in •Fig. 25.10. The objective is a large converging lens with a long focal length, and the movable eyepiece has a relatively short focal length. Rays from a distant object are essentially parallel and form an image (I_o) at the focal point (F_o) of the objective. This image acts as an object for the eyepiece, which is moved until the image lies just inside its focal point (F_e). A large, inverted, virtual image (I_e) is seen by an observer.

For relaxed viewing, the eyepiece is adjusted so that its image (I_e) is at infinity, which means that the objective image (I_o) is at the focal point of the eyepiece (f_e). As Fig. 25.10 shows, the distance between the lenses is then the sum of the focal lengths ($f_o + f_e$), which is the length of the telescope tube. The magnifying power of a telescope focused for the final image at infinity can be shown to be

$$m = -\frac{f_o}{f_e} \qquad \begin{array}{l}\textit{angular magnification}\\ \textit{of refracting telescope}\end{array} \qquad (25.6)$$

(see Exercise 53), where the minus is inserted to indicate the image is inverted as in our lens sign convention in Section 23.3. Thus, to achieve the greatest magnification,

Refracting telescope magnification

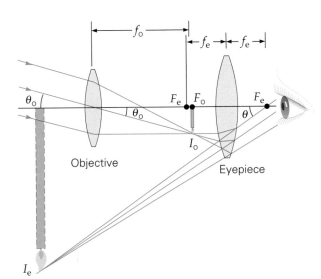

Objective

Eyepiece

•**FIGURE 25.10 The refracting astronomical telescope**   In an astronomical telescope, rays from a distant object form an intermediate image at the focal point of the objective (F_o). The eyepiece is moved so that the image is at or slightly inside its focal point (F_e). An observer sees an enlarged image at infinity (shown at a finite distance here for illustration).

Galileo did not invent the telescope, but he built one after hearing about a Dutch instrument that could be used for distant viewing.

the focal length of the objective should be made as great as possible and the focal length of the eyepiece as short as possible.

The telescope illustrated in Fig. 25.10 is called an **astronomical telescope**. The final image produced by an astronomical telescope is inverted, but this poses little problem to astronomers. (Why?) However, someone viewing an object on Earth through a telescope finds it more convenient to have an upright image. A telescope in which the final image is upright is called a **terrestrial telescope**. An upright final image can be obtained in several ways; two are illustrated in •Fig. 25.11.

In the telescope diagrammed in Fig. 25.11a, a diverging lens is used as an eyepiece. This type of terrestrial telescope is referred to as a Galilean telescope, because Galileo built one in 1609. The light rays pass through the eyepiece before the image is formed, and the image acts as a "virtual" object for the eyepiece (see Section 23.3). An observer sees a magnified, upright, virtual image. (Note that with a diverging lens and negative focal length $(-f_e)$, Eq. 25.6 gives a $+m$, indicating an upright image.)

Galilean telescopes have several disadvantages, most notably very narrow fields of view and limited magnification. A better type of terrestrial telescope, illustrated in Fig. 25.11b, uses a third lens, called the *erecting lens,* between converging objective and eyepiece lenses. If the image is formed by the objective at a distance that is twice the focal length of the intermediate erecting lens $(2f_i)$, then the lens merely inverts the image without magnification, and the telescope magnification is still given by Eq. 25.6.

However, to achieve the upright image in this way requires a greater telescope length. Using the intermediate erecting lens to invert the image increases the length of the telescope by four times the focal length of the erecting lens $(2f_i$ on each side). An erecting lens is used in a spyglass, which is telescoped to achieve the necessary length. The inconvenient length can be avoided by using internally reflecting prisms. This is the principle of prism binoculars, which are double telescopes—one for each eye (•Fig. 25.12).

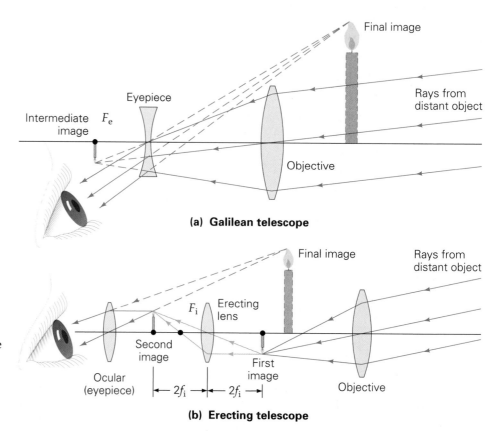

(a) **Galilean telescope**

(b) **Erecting telescope**

•**FIGURE 25.11 Terrestrial telescopes** **(a)** A Galilean telescope uses a diverging lens as an eyepiece, producing upright, virtual images. **(b)** Another way to produce upright images is to use a converging "erecting" lens between the objective and eyepiece in an astronomical telescope. This addition elongates the telescope, but the length can be shortened by using internally reflecting prisms.

Example 25.5 ■ Astronomical Telescope—and Longer Terrestrial Telescope

An astronomical telescope has an objective lens with a focal length of 30 cm and an eyepiece with a focal length of 9.0 cm. (a) What is the magnification of the telescope? (b) If an erecting lens with a focal length of 7.5 cm is used to convert the telescope to a terrestrial type, what is the overall length of the telescope tube?

Thinking It Through. Equation 25.6 applies directly in (a). In (b), the erecting lens elongates the telescope by four times the focal length of the lens ($4f_i$) to the length of the scope (Fig. 25.11b).

Solution. Listing the data, we have

Given: $f_o = 30$ cm
$f_e = 9.0$ cm
$f_i = 7.5$ cm (intermediate erecting lens)

Find: (a) m (magnification)
(b) L (length of telescope tube)

(a) The magnification is given by Eq. 25.6:

$$m = -\frac{f_o}{f_e} = -\frac{30 \text{ cm}}{9.0 \text{ cm}} = -3.3\times$$

where the minus sign indicates that the final image is inverted.

(b) Taking the length of the astronomical tube to be the distance between the lenses, this is the sum of their focal lengths.

$$L_1 = f_o + f_e = 30 \text{ cm} + 9.0 \text{ cm} = 39 \text{ cm}$$

The overall length is then

$$L = L_1 + L_2 = 39 \text{ cm} + 4f_i = 39 \text{ cm} + 4(7.5 \text{ cm}) = 69 \text{ cm}$$

Hence the telescope length is over two-thirds of a meter, with an upright image but the same magnification. (Why?)

Follow-up Exercise. A terrestrial telescope 66 cm in length has an intermediate erecting lens with a focal length of 12 cm. It is desired to reduce the length to a more manageable 50 cm. What is the focal length of an erecting lens that would do this?

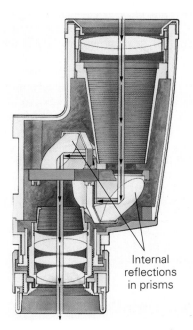

•**FIGURE 25.12 Prism binoculars** A schematic cutaway view of one ocular, showing the internal reflections in the prisms, which reduce the overall length.

Conceptual Example 25.6

Constructing a Telescope

A student is given two converging spherical lenses, one with a focal length of 5.0 cm and the other with a focal length of 20 cm. To construct a telescope to best view distant objects with these lenses, the student should hold the lenses (a) more than 25 cm apart, (b) less than 25 cm but more than 20 cm apart, (c) less than 20 cm but more than 5.0 cm apart, or (d) less than 5.0 cm apart. Specify which lens should be used as the eyepiece.

Reasoning and Answer. The only type of telescope that can be constructed with two converging lenses is an astronomical telescope. In this type of telescope, the lens with the longer focal length is used as an objective lens to produce a real image of a distant object. That image is then viewed with the lens with the shorter focal length (the eyepiece), used as a simple magnifier.

If the object is at a great distance, a real image is formed by the objective lens in the focal plane of the lens. This image acts as the object for the eyepiece, which is positioned so that the image/object lies just inside its focal point so as to produce a large, inverted second image (Fig. 25.10).

This means that the two lenses must be *slightly* less than 25 cm apart. Thus, answer (a) is not correct. Answers (c) and (d) are also not correct because the eyepiece would be too close to the objective to get the large secondary image needed for optimal viewing of a distant object. In these cases, the rays would pass through the second lens before the image was formed, and a *reduced* image might be produced (see Section 23.3). Thus, (b), with the objective image just inside the eyepiece focal point, is the correct answer.

Follow-up Exercise.   A third converging lens with a focal length of 4.0 cm is used with the above two lenses to produce a terrestrial telescope in which the third lens does nothing more than invert the image. How should the lenses be positioned, and how far apart should they be, for the final image to be of maximum size and upright?

Reflecting Telescope

For viewing the Sun, Moon, and nearby planets, magnification is necessary to see details. However, even with the highest feasible magnification, stars appear only as faint points of light. For distant stars and galaxies, it is more important to gather enough light so that the object can be seen at all. The intensity of light from a distant source is very weak. In many instances, such a source can be detected only when the light is gathered and focused on a photographic plate over a long period of time.

Recall from Section 14.3 that intensity is energy per unit time per *area*. Thus, more light can be gathered if the size of the objective is increased. This increases the distance at which the telescope can detect faint objects such as distant galaxies. (Recall that light intensity of a point source is inversely proportional to the *square* of the distance between the source and the observer.) However, producing a large lens involves difficulties associated with glass quality, grinding, and polishing. Compound lens systems are required to reduce aberrations, and a very large lens may sag under its own weight, producing further aberrations. The largest objective lens in use has a diameter of 40 in. (102 cm) and is part of the refracting telescope of the Yerkes Observatory at Williams Bay, Wisconsin.

These problems are reduced with a **reflecting telescope** utilizing a large, concave, front-surface parabolic mirror (•Fig. 25.13). A parabolic mirror does not exhibit spherical aberration, and a mirror has no inherent chromatic aberration. High-quality glass is not needed, since the light is reflected by a mirrored front surface. And only one surface has to be ground, polished, and silvered.

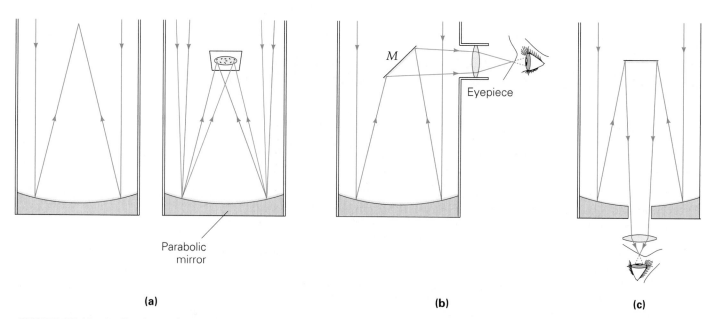

(a)   (b)   (c)

•**FIGURE 25.13   Reflecting telescopes**   A concave mirror can be used in a telescope to converge light to form an image of a distant object. **(a)** The image may be at the prime focus, or **(b)** a small mirror and lens can be used to focus the image outside the telescope, called a Newtonian focus. **(c)** Another arrangement, called a Cassegrain focus, uses a mirror to form the image below the main mirror.

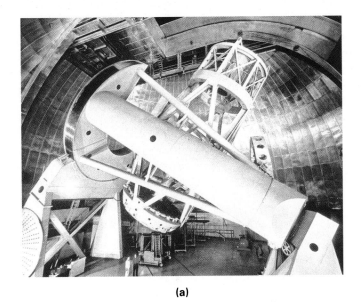

(a)

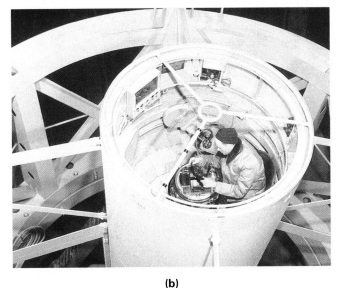

(b)

•**FIGURE 25.14   Reflecting telescopes**   **(a)** The Hale telescope on Palomar Mountain in California, with a mirror 5.1 m (200 in.) in diameter, is the largest single-mirror reflecting telescope in the United States. The mirror is at the bottom center, just above the heads of the people. **(b)** An astronomer in the observing capsule near the top of the telescope tube, at the prime focus. (Compare Figure 25.13a.)

Until recently, the largest single-mirror telescope was the Hale Observatories reflecting telescope, on Palomar Mountain in California, with a mirror 5.1 m (200 in.) in diameter (•Fig 25.14a). In the Hale design, the observer actually sits in a cage at the prime focus of the telescope (Fig. 25.14b).

Even though reflecting telescopes have advantages over refracting telescopes, they also have their own problems. Like a large lens, a large mirror may sag under its own weight, and the weight necessarily increases with the size of the mirror. The weight factor is also a cost factor in construction of a telescope, since the supporting elements for a heavier mirror must be more massive.

These problems are being addressed by new technologies. One approach is to use an array of small mirrors, coordinated so as to function as a single large mirror. Examples include the twin Keck telescopes at Mauna Kea in Hawaii. Each has a mirror consisting of 36 hexagonal segments that are computer-positioned to give the effect of a 10-meter mirror.

Another way of extending our view into space is to put telescopes into orbit about the Earth. Above the atmosphere, the view is unaffected by the twinkling effect of atmospheric turbulence and refraction, and there is no background problem from city lights. In 1990, the optical Hubble Space Telescope (HST) was launched into orbit (•Fig. 25.15). Even with a mirror diameter of only 2.4 m, its privileged position has allowed the HST to produce images seven times clearer than those formed by Earthbound telescopes.

However, not all telescopes are optical, as the Insight on p. 792 points out.

25.4   Diffraction and Resolution

Objectives: To **(a)** describe the relationship of diffraction and resolution and **(b)** state and explain Rayleigh's criterion.

The diffraction of light places a limitation on our ability to distinguish objects that are close together when we use microscopes or telescopes. This effect can be understood by considering two point sources located far from a narrow slit of width d (•Fig. 25.16). The sources could be distant stars, for example. In the absence of

•**FIGURE 25.15   Hubble Space Telescope (HST)**   Late in 1993, astronauts from the Space Shuttle *Endeavor* visited the HST in orbit. They installed corrective equipment that compensated for many of the telescope's optical flaws and repaired or replaced other malfunctioning systems. Since then the HST has produced a wealth of spectacular images and helped astronomers make many important discoveries.

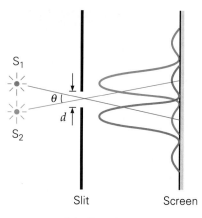

(a) Resolved

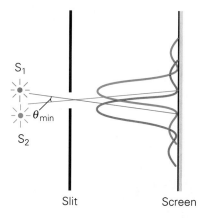

(b) Just resolved

•FIGURE 25.16 Resolution Two light sources in front of a slit produce diffraction patterns. **(a)** When the angle subtended by the sources at the slit is large enough for the diffraction patterns to be distinguishable, the images are said to be resolved. **(b)** At smaller angles, the central maxima are closer together. At θ_{min} the central maximum of one image's diffraction pattern falls on the first dark fringe of the other image's pattern, and the images are said to be just resolved. For smaller angles, the patterns are unresolved.

diffraction, two bright spots, or images, would be observed on a screen. As you know from Section 24.3, however, the slit diffracts the light, and each image consists of a central maximum with a pattern of weaker bright and dark fringes on either side. If the sources are close together, the two central maxima may overlap. Then the images cannot be distinguished, or are *unresolved*. For the images to be *resolved*, the central maxima must not overlap appreciably.

In general, images of two sources can be resolved if the central maximum of one falls at or beyond the first minimum (dark fringes) of the other. This generally accepted limiting condition for the **resolution** of two diffracted images—that is, the ability to distinguish both images as separate—was first proposed by Lord Rayleigh (1842–1919), a British physicist. The condition is known as the **Rayleigh criterion:**

Two images are said to be just resolved when the central maximum of one image falls on the first minimum of the diffraction pattern of the other image.

The Rayleigh criterion can be expressed in terms of the angular separation (θ) of the sources (see Fig. 25.16). The first minimum ($m = 1$) for a single-slit diffraction pattern satisfies this relationship:

$$d \sin \theta = \lambda$$

or

$$\sin \theta = \frac{\lambda}{d}$$

This equation gives the minimum angular separation for two images to be just resolved according to the Rayleigh criterion. In general, for visible light, the wavelength is much smaller than the slit width ($\lambda \ll d$), so a good approximation is $\sin \theta \approx \theta$. The limiting, or minimum, angle of resolution (θ_{min}) for a slit of width d is then

$$\theta_{min} = \frac{\lambda}{d} \qquad \begin{array}{l} \textit{minimum angle of resolution} \\ \textit{(for a slit)} \end{array} \qquad (25.7)$$

(Note that θ_{min} is a pure number and therefore must be expressed in radians.) Thus, the images of two sources will be *distinctly* resolved if the angular separation of the sources is greater than λ/d.

The apertures (openings) of cameras, microscopes, and telescopes are generally circular. Thus, there is a circular diffraction pattern around the central maximum, in the form of a bright circular disk (•Fig. 25.17). Detailed analysis for a circular aperture shows that the minimum angular separation for the images of two objects to be just resolved is

$$\theta_{min} = \frac{1.22\lambda}{D} \qquad \begin{array}{l} \textit{minimum angle of resolution} \\ \textit{(for a circular aperture)} \end{array} \qquad (25.8)$$

where D is the diameter of the aperture and θ_{min} is in radians.

Equation 25.8 applies to the objective lens of a microscope or telescope, which may be considered to be a circular aperture for light. According to the equation, to make θ_{min} small so objects close together can be resolved, the aperture should be as large as possible. This is another reason for using large lenses (and mirrors) in telescopes, in addition to their greater light-gathering power.

Example 25.7 ■ Eye and Telescope: Evaluating Resolution with the Rayleigh Criterion

Determine the minimum angle of resolution by the Rayleigh criterion for (a) the pupil of the eye (daytime diameter of about 4.0 mm), (b) the Yerkes Observatory refracting telescope (diameter of 102 cm), both (a) and (b) for visible light with a wavelength of 660 nm; and (c) a radio telescope 25 m in diameter for radiation with a wavelength of 21 cm.

Thinking It Through. This is a comparison of θ_{min} for different diameter apertures—a direct application of Eq. 25.8.

Solution.

Given: (a) $D = 4.0$ mm $= 4.0 \times 10^{-3}$ m
$\lambda = 660$ nm $= 6.60 \times 10^{-7}$ m
(b) $D = 102$ cm $= 1.02$ m
$\lambda = 660$ nm $= 6.60 \times 10^{-7}$ m
(c) $D = 25$ m
$\lambda = 21$ cm $= 0.21$ m

Find: (a) θ_{min} (minimum angles
(b) θ_{min} of resolution)
(c) θ_{min}

(a) For the eye,

$$\theta_{min} = \frac{1.22\lambda}{D} = \frac{1.22(6.60 \times 10^{-7}\,\text{m})}{4.0 \times 10^{-3}\,\text{m}} = 2.0 \times 10^{-4}\,\text{rad}$$

(b) For the light telescope,

$$\theta_{min} = \frac{1.22(6.60 \times 10^{-7}\,\text{m})}{1.02\,\text{m}} = 7.9 \times 10^{-7}\,\text{rad}$$

(*Note:* The resolution of Earth-bound telescopes with large-diameter objectives is usually not limited by diffraction but by other effects, such as atmospheric turbulence, and have a θ_{min} on the order of 10^{-6} rad.)

(c) For the radio telescope,

$$\theta_{min} = \frac{1.22(0.21\,\text{m})}{25\,\text{m}} = 0.010\,\text{rad}$$

The smaller the angular separation, the better the resolution. What do the results tell you?

Follow-up Exercise. As noted in Section 25.3, the Hubble Space Telescope has a mirror of 2.4 m. How does its resolution compare to Earthbound telescopes? [See the note in (b) above. Atmospheric effects are not a problem for the HST. Why?]

For a microscope, it is convenient to specify the actual separation (s) between two point sources. Since the objects are usually near the focal point of the objective, to a good approximation

$$\theta_{min} = \frac{s}{f} \quad \text{or} \quad s = f\theta_{min}$$

where f is the focal length of the lens. (Here s is taken as the arc length subtended by θ_{min}, and $s = r\theta_{min} = f\theta_{min}$.) Then, using Eq. 25.8, we get

$$s = f\theta_{min} = \frac{1.22\lambda f}{D} \qquad \begin{matrix}\textit{resolving power} \\ \textit{of a microscope}\end{matrix} \qquad (25.9)$$

This minimum distance between two points whose images can be just resolved is called the **resolving power** of the lens. Practically, the resolving power of a microscope indicates the ability of the objective to distinguish fine detail in specimens' structures. For another real-life example of resolution, see •Fig. 25.18.

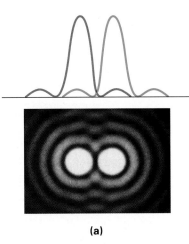

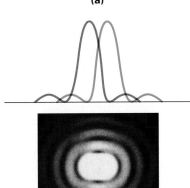

(a)

(b)

•**FIGURE 25.17 Circular aperture resolution** (a) When the angular separation of two objects is large enough, the images are well resolved. (Compare Fig. 25.16a.) (b) The minimum angular separation for two images to be just resolved is given by the Rayleigh criterion: The central maximum of the diffraction pattern of one image falls on the first minimum of the diffraction pattern of the other image. (Compare Fig. 25.16b.) Objects with smaller angular separations produce images that overlap so much that they cannot be clearly distinguished as individual images.

Conceptual Example 25.8

Viewing from Space: the Great Wall of China

The Great Wall of China was originally about 2400 km (1500 mi) long, with a base width of about 6.0 m and a top width of about 3.7 m. Several hundred kilometers of the Wall remain intact (•Fig. 25.19, p. 793). The Wall is said to be the only human construction that can be seen with the unaided eye by an astronaut orbiting the Earth. Using the result from part (a) of Example 25.7, see if this is true. (Neglect any atmospheric effects).

Telescopes Using Nonvisible Radiation

The word *telescope* usually brings to mind visual observations. However, the visible region is a very small part of the electromagnetic spectrum, and celestial objects emit radiation of many other types, including radio waves. This fact was discovered accidentally in 1931 by an electrical engineer named Carl Jansky while he was working on the problem of static interference with intercontinental radio communications. Jansky found an annoying static hiss that came from a fixed direction in space, apparently from a celestial source. It was soon apparent that radio waves are another source of astronomical information, and radio telescopes were built.

A radio telescope operates similarly to a reflecting light telescope. A reflector with a large area collects and focuses the radio waves at a point where a detector picks up the signal (Fig. 1). The parabolic collector, called a *dish*, is covered with metal wire mesh or metal plates. Since the wavelengths of radio waves range from a few millimeters to several meters, wire mesh is a good reflecting surface for such waves.

Radio telescopes supplement optical telescopes and provide some definite advantages. Radio waves pass freely through the huge clouds of dust that hide a large part of our galaxy from visual observation. Also, radio waves easily penetrate the Earth's atmosphere, which reflects and scatters a large percentage of the incoming visible light.

The Earth's atmosphere absorbs many wavelengths of electromagnetic radiation and completely blocks some from reaching the ground. Water vapor, a strong absorber of infrared radiation, is concentrated in the lower part of the atmosphere near the surface of the Earth. Thus, observations with infrared telescopes are made from high-flying aircraft or from orbiting spacecraft.

The first orbiting infrared observatory was launched in 1983. Not only are atmospheric interferences eliminated in space, but a telescope may be cooled to a very low temperature without becoming coated with condensed water vapor. Cooling the telescope helps eliminate interfering infrared radiation generated by the telescope itself. The orbiting tele-

FIGURE 1   Radio telescopes   Several of the dish antennae that make up the Very Large Array (VLA) radio telescope near Socorro, New Mexico. There are 27 movable dishes, each 25 m in diameter, forming the array along a Y-shaped railway network. The data from all the antennae are combined to produce a single radio image. In this way it is possible to attain a resolution equivalent to that of one giant radio dish.

scope launched in 1983 was cooled with liquid helium to about 10 K; it carried out an infrared survey of the entire sky.

The atmosphere is virtually opaque to ultraviolet radiation, X-rays, and gamma rays from distant sources. Orbiting satellites with telescopes sensitive to these types of radiation have mapped out portions of the sky, and other surveys are planned.

At the altitudes reached by orbiting satellites, even observations in the visible region are not affected by air motion and temperature refraction. Perhaps in the not too distant future, a permanently staffed orbiting observatory carrying a variety of telescopes will help expand our knowledge of the universe.

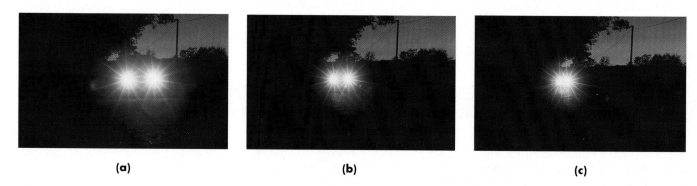

(a)   (b)   (c)

•**FIGURE 25.18   Real-life resolution**   (a), (b), (c) A sequence of automobile headlights at greater and greater distances. In (c), the headlights are almost unresolved through the circular aperture of the camera (or your eye).

Reasoning and Answer. Despite the length of the Wall, it would not be visible from space unless its *width* subtends the minimum angle of resolution for the eye of an observing astronaut ($\theta_{min} = 2.0 \times 10^{-4}$ rad from Example 25.7). Guard towers with roofs wider than 6.0 m were located every 180 m, so let's take the maximum observable width dimension to be 7.0 m. [Actually, it is the circular arc length that subtends the angle, but at such a radius, the chord length (width) is very nearly equal. Refer to Example 7.2, and make yourself a sketch.]

Let's assume that the astronaut is just able to distinguish the Wall. Recall that $s = r\theta$ (Eq. 7.3), where s is an approximation of the maximum observable width of the Wall and r is the radial (height) distance. Then, the astronaut would have to be at a radial distance of

$$r = \frac{s}{\theta} = \frac{7.0 \text{ m}}{2.0 \times 10^{-4} \text{(rad)}} = 3.5 \times 10^4 \text{ m} = 35 \text{ km} \ (= 22 \text{ mi})$$

So, above 35 km, the Wall would not be able to be seen with the unaided eye.

Orbiting satellites are 300 km (190 mi) or more above the Earth so as to be above the denser part of the atmosphere (why?). Hence, for an astronaut to see the Great Wall, he or she would have to come back very near to Earth. The statement about the ability to see the Wall from space is false.

Follow-up Exercise. What would be the minimum objective diameter of a telescope that would allow an astronaut orbiting the Earth at an altitude of 300 km to see the Great Wall? (Take all conditions to be the same.)

•**FIGURE 25.19 The Great Wall** The walkway of the Great Wall of China, which was built as a fortification along China's northern border.

Note from Eq. 25.8 that higher resolution can be gained by using radiation of a shorter wavelength. Thus, a telescope with an objective of a given size will have greater resolution with violet light than with red light. For microscopes, it is possible to increase resolving power by shortening the wavelengths of the light used to create the image. This can be done with a specialized objective called an *oil immersion lens*. When such a lens is used, a drop of transparent oil fills the space between the objective and the specimen. Recall that the wavelength of light in oil is $\lambda_m = \lambda/n$, where n is the index of refraction of the oil and λ the wavelength of light in air. With values of n of about 1.50 or higher, the wavelength is significantly reduced, and the resolving power increased proportionally.

Note: The relationship between wavelength and index of refraction is given in Section 22.3; see Eq. 22.4.

*25.5 Color

Objective: **To relate color vision and light.**

In general, physical properties are fixed or absolute. For example, a particular radiation has a certain frequency or wavelength. However, visual perception of this radiation may vary from person to person. How we see radiation gives rise to color vision.

Color Vision

Color is perceived because of a physiological response to excitation by light of the cone receptors in the retina of the human eye. (Many animals have no cone cells and live in a black-and-white world.) The cones are sensitive to light with frequencies between 7.5×10^{14} Hz and 4.3×10^{14} Hz (wavelengths between 400 nm and 700 nm). Different frequencies of light are perceived by the brain as different colors. The association of a color with a particular frequency is subjective and may vary from person to person. As pitch is to sound and hearing, color is to light and vision.

Color vision is not well understood. One of the most widely accepted theories is that there are three types of cones in the retina, responding to different parts of the visible spectrum, particularly in the red, green, and blue regions (•Fig. 25.20). Presumably, the types of cones absorb light of specific ranges of frequencies and functionally overlap one another to form combinations that are interpreted by the brain as various colors of the spectrum. For example, when

Note: The relationship between pitch and frequency for sound is discussed in Section 14.6.

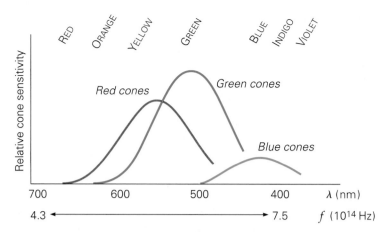

•FIGURE 25.20 Sensitivity of cones
Different types of cones in the retina of the human eye may respond to different frequencies of light to give three general color responses: red, green, and blue.

red and green cones are stimulated equally by light of a particular frequency, the brain interprets the color as yellow. But when the red cones are stimulated more strongly than the green cones, the brain senses orange. *Color blindness* is believed to result when one type of cone is missing.

As Fig 25.20 shows, the human eye is not equally sensitive to all colors. Some colors evoke a greater brightness response than others do. The wavelength of maximum visual sensitivity is about 550 nm, in the yellow-green region.

Conceptual Example 25.9

The Color of Light: Frequency or Wavelength?

What determines the color of light? (a) intensity, (b) frequency, (c) wavelength, (d) both frequency and wavelength.

Reasoning and Answer.   Clearly, (a) is not the answer; if it were, objects around you would change color every time the light was dimmed or brightened in a room. The other possible choices, however, may seem perplexing. You probably noticed in Figs. 20.18 and 25.20 that both wavelengths and frequencies were specified for the various colors. And in Chapter 24, we commonly talked about the color of light in terms of wavelength—for example, blue light ($\lambda = 425$ nm). This practice is based on the relationship $\lambda f = c$, where f and λ have an exact relationship. Thus, you may think that you don't have sufficient basis for deciding among the other possible answers. However, you may possess more evidence than you realize, as the following thought experiment shows.

Suppose you have an object that appears red when observed in air; light coming from the object has a wavelength of $\lambda = 700$ nm. Then you submerge the object in a swimming pool and jump in to observe it from under water. According to Eq. 22.4, with $n = 1.33$ for water, the wavelength of the light from the submerged object would be

$$\lambda' = \frac{\lambda}{n} = \frac{700 \text{ nm}}{1.33} = 526 \text{ nm}$$

If color depended on wavelength, the red object would appear green under water! But this is not what you observe in the real world. Thus (c) is not correct, nor is (d). Therefore, the answer must be (b). But why frequency? Remember from Section 22.3 that frequency is the wave parameter that doesn't change when the same light travels in different media—wavelength and wave speed change, but not frequency (or color).

So, our color perception depends on frequency. When we refer to the wavelength of a color, this is for the fixed speed of light in vacuum ($\lambda f = c$), which is commonly assumed in most situations involving the wavelength of light.

Follow-up Exercise.   Suppose a light source that appears blue in air ($\lambda = 450$ nm) is viewed underwater. (a) What will be the observed speed, wavelength, and frequency of the light from this object? (b) What color will it appear to the observer? (c) If color perception depended on wavelength rather than on frequency, what color would it appear?

The theory of color vision described above is based on the experimental fact that beams of varying intensities of red, green, and blue light will produce most colors. The red, blue, and green from which we interpret a full spectrum of colors are called the **additive primary colors.** When light beams of the additive primaries are projected on a white screen so they overlap, other colors will be produced, as illustrated in •Fig. 25.21. This technique is called the **additive method of color production.** Triad dots consisting of three phosphors that emit the additive primary colors when excited are used in television picture tubes to produce colored images.

Note in Fig. 25.21 that a certain combination of the primary colors appears white to the eye. Also, many *pairs* of colors appear white to the eye when combined. The colors of such pairs are said to be **complementary colors.** The complement of blue is yellow, that of red is cyan, and that of green is magenta. As the figure shows, the complementary color of a particular primary is the combination or sum of the other two primaries. Hence, the primary and its complement together appear white.

Edwin H. Land (the developer of Polaroid™ film) has shown that when the proper mixtures of only two wavelengths (colors) of light are passed through black and white transparencies (no color), the wavelengths produce images of various colors. Land wrote, *"In this experiment we are forced to the astonishing conclusion that the rays are not in themselves color-making. Rather they are bearers of information that the eye uses to assign appropriate colors to various objects in an image."** From Land's experiments, it seems that information about colors other than for the two wavelengths used is developed in the brain via the cone cells. However, our knowledge about color vision is far from complete.

Objects have a color when they are illuminated with white light because they reflect (scatter) or transmit the light of the frequency of that color. The other frequencies of the white light are absorbed. For example, when white light strikes a ripe red apple, only the waves in the red portion of the spectrum are reflected—all others (and thus all other colors) are absorbed. Similarly, when white light passes through a piece of transparent red glass, or a filter, only the red rays are transmitted. This occurs because the color pigments in the glass are selective absorbers.

Pigments are mixed to form various colors, such as in the production of paints and dyes. You are probably aware that mixing yellow and blue paints produces green. This is because the yellow pigment absorbs most of the wavelengths except those in the yellow and nearby regions of the visible spectrum, and the blue pigment absorbs most of the wavelengths except those in the blue and nearby regions. The wavelengths in the intermediate green region, adjacent to both the yellow and blue wavelengths, are not strongly absorbed by either pigment, and therefore the mixture appears green. The same effect can be accomplished by passing white light through stacked yellow and blue filters. The light coming through both filters appears green.

Mixing pigments results in the subtraction of colors. The color that is perceived is created by whatever is *not* absorbed by the pigment—that is, not subtracted from the original beam. This is the principle of what is called the **subtractive method of color production.** Three particular pigments—cyan, magenta, and yellow—are called the **subtractive primary pigments.** Various combinations of two of the three subtractive primaries produce the three additive primary colors (red, blue, and green), as illustrated in •Fig. 25.22. When the subtractive primaries are mixed in the proper proportions, the mixture appears black (all wavelengths are absorbed). Painters often refer to the subtractive primaries as red, yellow, and blue. They are loosely referring to magenta (purplish-red), yellow, and cyan ("true" blue). Mixing these paints in the proper proportions produces a broad spectrum of colors.

Note in Fig. 25.22 that the magenta pigment essentially subtracts the color green where it overlaps with cyan and yellow. As a result, magenta is sometimes

*From "Experiments in Color Vision," by Edwin H. Land, in *Scientific American*, May 1959, pp. 84–99.

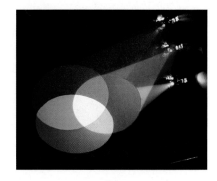

(a)

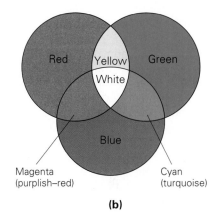

(b)

•**FIGURE 25.21 Additive method of color production** When light beams of the primary colors (red, blue, and green) are projected onto a white screen, mixtures of them produce other colors. Varying the intensities of the beams allow most colors to be produced.

Note: Recall from Chapter 22 that the colors of the visible spectrum may be remembered by ROY G. BIV.

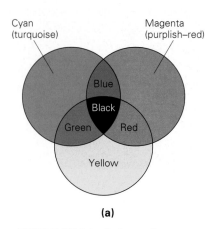

(a)

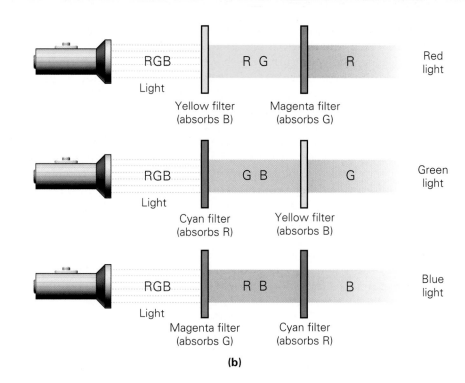

(b)

•FIGURE 25.22 **Subtractive method of color production**
(a) When the primary pigments (cyan, magenta, and yellow) are mixed, different colors are produced by subtractive absorption; for example, the mixing of yellow and magenta produces red. When all three pigments are mixed and all the wavelengths of visible light are absorbed, the mixture appears black. **(b)** Subtractive color mixing, using filters. The principle is the same as in part (a). Each pigment selectively absorbs certain colors, removing them from the white light. The colors that remain are what we see.

referred to as "minus green." If a magenta filter were placed in front of a green light, no light would be transmitted. Similarly, cyan is called "minus red," and yellow is called "minus blue."

An example of subtractive color mixing is a photographer's use of a yellow filter to bring out white clouds on black and white film. This filter absorbs blue from the sky, darkening it relative to the clouds, which reflect white light. Hence, the contrast between the two is enhanced. What type of filter would you use to darken green vegetation on black and white film? To lighten it?

Chapter Review

Important Terms

retina *775*
rods *775*
cones *775*
nearsightedness *776*
farsightedness *777*
astigmatism *779*
magnifying glass (simple microscope) *781*
angular magnification *781*

compound microscope *783*
objective *783*
eyepiece (ocular) *783*
refracting telescope *785*
astronomical telescope *786*
terrestrial telescope *786*
reflecting telescope *788*
resolution *790*
Rayleigh criterion *790*

resolving power *791*
additive primary colors *795*
additive method of color production *795*
complementary colors *795*
subtractive method of color production *795*
subtractive primary pigments *795*

Important Concepts

• Nearsighted people cannot see distant objects clearly. Farsighted people cannot see nearby objects clearly. These conditions may be corrected by diverging and converging lenses, respectively.

• The magnification of a magnifying glass (or simple microscope) is expressed in terms of angular magnification (*m*), as distinguished from the lateral magnification (*M*, Chapter 23).

- The objective of a compound microscope has a relatively short focal length, and the eyepiece, or ocular, has a longer focal length. Both contribute to the total magnification.
- A refracting telescope uses a converging lens to gather light, and a reflecting telescope uses a converging mirror. The image created by either one is magnified by the eyepiece.

- Diffraction places a limit on the ability to resolve, or distinguish, objects that are close together. Two images are said to be just resolved when the central maximum of one image falls on the first minimum of the diffraction pattern of the other image (Rayleigh criterion).
- Color is our brain's interpretation of the frequency of light.

Important Equations

Angular Magnification:

$$m = \frac{\theta}{\theta_o} \qquad (25.1)$$

Magnification of a Magnifying Glass With Image at Near Point (25 cm):

$$m = 1 + \frac{25 \text{ cm}}{f} \qquad (25.3)$$

Magnification of a Magnifying Glass With Image at Infinity:

$$m = \frac{25 \text{ cm}}{f} \qquad (25.4)$$

Angular Magnification of a Compound Microscope (with L, f_o, and f_e in centimeters):

$$m_{total} = M_o m_e = \frac{(25 \text{ cm})L}{f_o f_e} \qquad (25.5)$$

Magnification of a Refracting Telescope:

$$m = -\frac{f_o}{f_e} \qquad (25.6)$$

Minimum Angle of Resolution for a Slit:

$$\theta_{min} = \frac{\lambda}{d} \qquad (25.7)$$

Minimum Angle of Resolution for a Circular Aperture: (diameter D)

$$\theta_{min} = \frac{1.22\lambda}{D} \qquad (25.8)$$

Resolving Power:

$$s = f\theta_{min} = \frac{1.22\lambda f}{D} \qquad (25.9)$$

Exercises

25.1 The Human Eye*

1. The rods of the retina (a) are responsible for 20/20 vision, (b) are responsible for black and white vision at twilight, (c) are responsible for color vision, (d) focus light.

2. A vision defect that commonly occurs due to normal aging is (a) astigmatism, (b) nearsightedness, (c) farsightedness, (d) accommodation.

3. The focal length of the crystalline lens of the human eye varies with muscle action. Discuss the shape (curvature) of the lens when looking at distant and close objects.

4. Is the image formed on the retina inverted or upright?

5. People and other animals often exhibit "red eye" when photographed with a flash camera. Light reflected from the retina is red because of blood vessels near the surface. Some cameras have an anti-red eye option, which when activated gives a quick flash before the longer picture-taking flash. Explain how this option reduces red eye.

*Assume corrective lenses are in contact with the eye (contact lenses) unless otherwise stated.

6. Which parts of the camera correspond to the iris, crystalline lens, and retina of the eye?

7. ■ What are the powers of (a) a converging lens of focal length 20 cm, (b) a diverging lens of focal length −50 cm?

8. ■ The far point of a certain nearsighted person is 300 cm. Which type of lens should be prescribed to enable the person to see more distant objects clearly?

9. ■ A certain farsighted person has a near point of 50 cm. Which type of lens should an optometrist prescribe to enable the person to see objects clearly as close as 25 cm?

10. ■■ A woman cannot see objects clearly when they are farther than 12.5 m away. (a) Is she nearsighted or farsighted? (b) Which type of lens of what power (in diopters) will allow her to see distant objects clearly?

11. ■■ A nearsighted woman has an uncorrected far point of 200 cm. Which type of contact lens of what power would correct this condition?

12. ■■ A farsighted professor can just see the print in a book clearly when she holds the book at arm's length (0.80 m

from the eyes). (a) Which type of lens will allow her to read the text at the normal near point? (b) What is its focal length?

13. ■■ To correct a case of hyperopia, an optometrist prescribes positive contact lenses that effectively move the patient's near point from 100 cm to 25 cm. (a) What is the power of the lenses? (b) When wearing the lenses, will the patient be able to see distant objects clearly, or will she have to take them out?

14. ■■ A farsighted person with a near point of 0.95 m gets contact lenses and can then read a newspaper held at a distance of 25 cm. What is the power of the lenses? (Assume that the lenses are the same for both eyes.)

15. ■■ A farsighted man is unable to focus on objects nearer than 1.5 m. (a) Which type of lens will allow him to focus on the print of a book held 25 cm from his eyes? (b) What power should it be?

16. ■■ A certain myopic man has a far point of 150 cm. (a) What power must a contact lens have to allow him to see distant objects clearly? (b) If he is able to read print at 25 cm while wearing his contacts, is his near point less than 25 cm? (c) Give an indication of his age with normal recession of near point.

17. ■■ A middle-aged man starts to wear eyeglasses with lenses of +2.0 D that allow him to read a book held as closely as 25 cm. Several years later, he finds that he must hold a book no closer than 33 cm to read it clearly with the same glasses, so he gets new glasses. What is the power of the new lenses? (Assume both lenses are the same.)

18. ■■ A college professor can see objects clearly only if they are between 70 and 500 cm from her eyes. Her optometrist prescribes bifocals (•Fig. 25.23) that enable her to see distant objects through the top half of the lenses and read students' papers at a distance of 25 cm through the lower half. What are the powers of the top and bottom lenses? (Assume both lenses are the same.)

Nearsightedness correction

Farsightedness correction

•FIGURE 25.23 **Bifocals** See Exercises 18 and 21.

19. ■■ A nearsighted student wears contact lenses to correct for a far point that is 4.00 m from her eyes. When she is not wearing her contact lenses, her near point is 20 cm. What is her near point when she is wearing her contacts?

20. ■■■ A nearsighted boy has a far point located 7.5 m from one eye. (a) If a corrective lens is worn 2.0 cm from the eye, what would be the power of the lens necessary for the boy to see distant objects? (b) What would be the power necessary if a contact lens were used?

21. ■■■ Bifocal glasses are used to correct nearsightedness and farsightedness (Fig. 25.23). If the near points in the right and left eyes are 35.0 cm and 45.0 cm, respectively, and the far point is 220 cm for both eyes, what are the powers of the lenses prescribed for the glasses? (Assume the glasses are worn 3.00 cm from the eyes.)

25.2 Microscopes*

22. A magnifying glass (a) is a concave lens, (b) forms a clear image of an object that is closer than the near point, (c) magnifies by effectively increasing the angle the object subtends, (d) both (b) and (c).

23. A compound microscope has (a) unlimited magnification, (b) two lenses of the same focal length, (c) a diverging objective lens, (d) an eyepiece of relatively long focal length.

24. With an object at the focal point of a magnifying glass, the magnification is given by $m = (25 \text{ cm})/f$ (Eq. 25.4). According to this equation, the magnification could be increased indefinitely by using lenses with shorter focal lengths. Why, then, do we need compound microscopes?

25. When you use a simple convex lens such as a magnifying glass, where should you put the object: farther away than the focal length or inside the focal length? Explain.

26. ■ Using the small-angle approximation, compare the angular sizes of a car 1.0 m in height when at distances of (a) 500 m and (b) 1025 m.

27. ■ An object is placed 10 cm in front of a converging lens with a focal length of 18 cm. What are (a) the lateral magnification and (b) the angular magnification?

28. ■ A biology student uses a converging lens to examine the details of a small insect. If the focal length of the lens is 12 cm, what is the maximum angular magnification?

29. ■ A physics student uses a converging lens with a focal length of 15 cm to read a small measurement scale. (a) What is the maximum magnification that can be obtained? (b) What is the magnification for viewing of the relaxed eye?

30. ■ A student uses a magnifying glass to examine the details of a microcircuit in the lab. If the lens has a focal length of 8.0 cm and a virtual image is formed at the student's near point (25 cm), (a) how far from the circuit is the lens held, and (b) what is the magnification?

*The normal near point should be taken as 25 cm unless otherwise specified.

31. ■■ A detective looks at a fingerprint with a magnifying glass whose power is +3.5 D. What is the maximum magnification of the print?

32. ■■ A lens with a power of +10 D is used as a simple microscope. (a) To be examined through the lens, how close can an object be brought to the eye? (b) What is the angular magnification at this point?

33. ■■ What is the maximum magnification of a magnifying glass with a power of +3.0 D for (a) a person with a near point of 25 cm and (b) a person with a near point of 10 cm?

34. ■■ A compound microscope has a distance of 15 cm between lenses and an ocular with a focal length of 8.0 mm. What power should the objective have to give a total magnification of 360×?

35. ■■ The distance between lenses of a compound microscope is 15 cm. The focal length of the eyepiece is 3.0 cm, and the focal length of the objective is 0.35 cm. Find the total magnification of the microscope.

36. ■■ The focal length of the objective lens of a compound microscope is 4.5 mm. The eyepiece has a focal length of 3.0 cm. If the distance between the lenses is 18 cm, what is the magnification of a viewed image?

37. ■■ A compound microscope has an objective lens with a focal length of 0.50 cm and an ocular with a focal length of 3.25 cm. The separation distance between the lenses is 22 cm. A student with a normal near point uses the microscope. (a) What is the total magnification? (b) Compare the total magnification (as a percentage) with the magnification of the eyepiece alone as a simple magnifying glass.

38. ■■ A 150× microscope has an objective whose focal length is 0.75 cm. If the distance between lenses is 20 cm, find the focal length of the eyepiece.

39. ■■ A specimen is 5.0 mm from the objective of a compound microscope, which has a power of +250 D. What must the magnifying power of the eyepiece be if the total magnification of the specimen is 100×?

40. ■■■ Referring to •Fig. 25.24, show that the magnifying power for a magnifying glass held at a distance d from the eye is given by

$$m = \left(\frac{25}{f}\right)\left(1 - \frac{d}{D}\right) + \frac{25}{D}$$

when the actual object is located at the near point (25 cm). [Hint: Use a small-angle approximation, and note that $y_i/y_o = d_i/d_o$ by similar triangles.]

41. ■■■ A magnifying glass with a focal length of 10 cm is held 4.0 cm from the eyes to view the small print of a book. What is the magnification if the magnifying glass is 5.0 cm from the book? [Hint: See Exercise 40.]

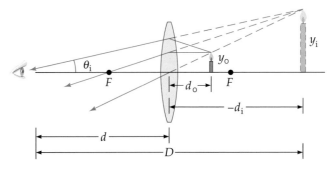

•**FIGURE 25.24 Power of a magnifying glass** See Exercise 40.

42. ■■■ A modern microscope is equipped with a turret having three objectives with focal lengths of 16 mm, 4.0 mm, and 1.6 mm and interchangeable eyepieces of 5.0× and 10×. A specimen is positioned such that each objective produces an image 150 mm from the objective. What are the least and greatest magnifications possible?

25.3 Telescopes

43. An inverted image is produced by (a) a terrestrial telescope, (b) an astronomical telescope, (c) a Galilean telescope, (d) all of these.

44. Compared to large refracting telescopes, large reflecting telescopes have the advantage of (a) greater light-gathering capability, (b) freedom from chromatic aberration, (c) lower cost, (d) all of these.

45. In Fig. 25.13b, part of the light entering the concave mirror is obstructed by a small plane mirror used to redirect the rays to a viewer. Does this mean that only a portion of a star can be seen? How does the size of the obstruction affect the image?

46. ■ Find the magnification of a telescope whose objective has a focal length of 50 cm and whose eyepiece has a focal length of 2.5 cm.

47. ■ An astronomical telescope has an objective and an eyepiece whose focal lengths are 60 cm and 15 cm, respectively. What are (a) the magnifying power and (b) the length of the telescope?

48. ■■ A telescope has an ocular with a focal length of 10 mm. If the length of the tube is 1.5 m, what is the angular magnification of the telescope when it is focused for an object at infinity?

49. ■■ An astronomical telescope has objective and eyepiece lenses with focal lengths of 87.5 cm and 8.00 mm, respectively. (a) What is the magnification of the telescope, and (b) what is its approximate length?

50. ■■ The three lenses of a terrestrial telescope have focal lengths of 40 cm, 20 cm, and 15 cm for the objective, erecting lens, and eyepiece, respectively. (a) What is the magnification of the telescope for an object at infinity? (b) What is the length of the telescope barrel? (c) Does the erecting lens affect the magnification of the telescope? Explain.

51. ■■ Two astronomical telescopes have the following characteristics:

| | Objective | Eyepiece | |
Telescope	Focal length (cm)	Focal length (cm)	Objective diameter
1	90.0	0.84	75
2	85.0	0.77	60

Which telescope would you choose (a) for magnification? (b) For resolution?

52. ■■ A telescope has an angular magnification of 50× and a barrel 1.02 m long. What are the focal lengths of the objective and the eyepiece?

53. ■■■ Referring to •Fig. 25.25, show that the angular magnification of a refracting telescope focused for the final image at infinity is $m = f_o/f_e$. (Since telescopes are designed for viewing distant objects, the angular size of an object viewed with the unaided eye is the angular size of the object at its actual location rather than at the near point, as is true for a microscope.)

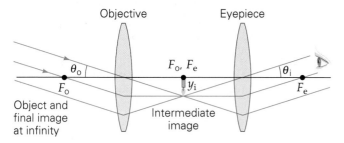

•FIGURE 25.25   Angular modification of a refracting telescope   See Exercise 53.

25.4 Diffraction and Resolution*

54. The images of two sources are said to be resolved when (a) the central maxima of the diffraction patterns fall on each other, (b) the first bright fringes of the diffraction patterns fall on each other, (c) the central maximum of one diffraction pattern falls on the first dark fringe of the other, (d) none of these.

55. For a telescope with a circular aperture, the minimum angle of resolution (a) is greater for red light than for blue light, (b) is independent of the frequency of the light, (c) is directly proportional to the radius of the aperture, (d) is independent of the area of the aperture.

56. The purpose of using oil immersion lenses on microscopes is to (a) reduce the size of microscopes, (b) increase the magnification, (c) increase the wavelength of light so as to in-

*Ignore atmospheric blurring unless otherwise stated.

crease the resolving power, (d) reduce the wavelength of light so as to increase the resolving power.

57. When an optical instrument is designed, high resolution is often desired to observe fine details. Does higher resolution mean smaller or larger minimum angle of resolution? Explain.

58. ■ According to the Rayleigh criterion, what is the minimum angle of resolution for two point sources of red light ($\lambda = 680$ nm) in the diffraction pattern produced by a single slit with a width of 0.55 mm?

59. ■ The minimum angular separation of the images of two identical monochromatic point sources in a single-slit diffraction pattern is 0.0055 rad. If a slit width of 0.10 mm is used, what is the wavelength of the sources?

60. ■ What is the resolution limit due to diffraction for the Yerkes Observatory refracting telescope (102 cm, or 40 in., in diameter) for light with a wavelength of 550 nm?

61. ■ What is the resolution due to diffraction for the Hale telescope at Mount Palomar with its 200-inch-diameter mirror for light with a wavelength of 550 nm? Compare this value to the resolution limit for the Yerkes Observatory telescope in Exercise 60.

62. ■■ The maximum diameter of the eye's pupil at night is about 7.0 mm. What is the minimum angle of separation for two 550-nanometer sources from which light passes through the pupil of the eye?

63. ■■ Some African tribesmen claim to be able to see the moons of Jupiter with the unaided eye. If two moons of Jupiter are at a minimum distance of 3.1×10^8 km away from Earth and at a maximum separation distance of 3.0×10^6 km, is it possible? Assume that the moons reflect sufficient light and that their observation is not restricted by Jupiter. [Hint: See Exercise 62.]

64. ■■ Assuming that the headlights of a car are point sources 1.7 m apart, what is the maximum distance from an observer to the car at which the headlights are distinguishable from each other? [Hint: See Exercise 62.]

65. ■■ A refracting telescope with a lens whose diameter is 30.0 cm is used to view a binary star system that emits light in the visible region. (a) What is the minimum angular separation of the pair of stars for them to be barely resolved? (b) If the binary star is a distance of 6.00×10^{20} km from the Earth, what is the lateral separation between the stars of the pair? (Assume that a line joining the stars is perpendicular to our line of sight.)

66. ■■ The objective of a microscope is 2.50 cm in diameter and has a focal length of 30.0 mm. (a) If yellow light with a wavelength of 570 nm is used to illuminate a specimen, what is the minimum angular separation of two fine details of the specimen for them to be just resolved? (b) What is the resolving power of the lens?

67. ■■■ A microscope with an objective 1.20 cm in diameter is used to view a specimen via light from a mercury source with a wavelength of 546.1 nm. (a) What is the limiting angle of resolution? (b) If details finer than those achieved in (a) are to be observed, what color of light in the visible spectrum would have to be used? (c) If an oil immersion lens were used ($n_{oil} = 1.50$), what would be the change (expressed as a percentage) in the resolving power?

*25.5 Color

68. An additive primary color is (a) blue, (b) green, (c) red, (d) all of these.

69. A subtractive primary color is (a) cyan, (b) yellow, (c) magenta, (d) all of these.

70. Describe how the American flag would appear if it were illuminated with light of each of the primary colors.

71. Can white be obtained by the subtractive method of color production? Explain. It is sometimes said that black is the absence of all color or that a black object absorbs all incident light. If so, why do we see black objects?

72. Several beverages, such as root beer, when poured into a glass develops a "head" of foam. Why is the foam generally white or light, while the liquid is dark?

73. White light is incident on two filters as shown in •Fig. 25.26. Complete the color rays to show which color of light emerges from the yellow filter.

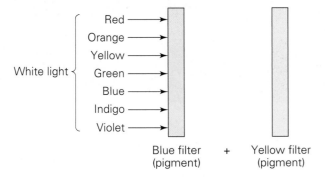

•FIGURE 25.26   Color absorption   See Exercise 73.

Additional Exercises

74. A myopic woman has eyeglasses with a lens that corrects for a far point of 130 cm. What is the power of the lens?

75. A radio telescope has a diameter of 300 m and can be tuned to wavelengths as short as 4.0 m. What is the minimum angular separation of two stars that can be distinguished by the telescope?

76. A terrestrial telescope uses an objective and eyepiece with focal lengths of 45 cm and 15 cm, respectively. What should the focal length of the erecting lens be if the overall length of the telescope is to be 0.80 m?

77. A nearsighted man wears eyeglasses whose lenses have a power of −0.15 D. How far away is his far point?

78. When viewing an object with a magnifying glass whose focal length is 15 cm, Karen positions the lens so that there is minimum eyestrain. What is the observed magnification?

79. Light from two monochromatic point sources, both of wavelength 550 nm, is incident on a slit that is 0.050 mm wide. What is the minimum angle of resolution in the diffraction pattern formed by the slit?

80. A compound microscope has an objective with a focal length of 4.00 mm and an eyepiece with a magnification of 10.0×. If the objective and eyepiece are 15.0 cm apart, what is the total magnification of the microscope?

81. A person using a magnifying glass with a focal length of 12 cm views an object at the focal point of the lens. What is the magnification?

82. A refracting telescope has an objective with a focal length of 50 cm and an eyepiece with a focal length of 15 mm. The telescope is used to view an object that is 10 cm high and located 50 m away. What is the apparent angular height of the object as viewed through the telescope?

83. An eyeglass lens with a power of +2.8 D allows a far-sighted child to read a book held at a distance of 25 cm from her eyes. At what distance must she hold the book to read it without glasses?

84. A student views the details of a dollar bill with a magnifying glass, achieving its maximum magnification of 3×. What is the focal length of the magnifying glass?

85. A microscope has objective and eyepiece lenses that are 15 cm apart and have focal lengths of 7.5 mm and 10 mm, respectively. What is the total magnification of the microscope?

86. The amount of light reaching the film in a camera depends on the lens aperture (the effective area) as controlled by the diaphragm. The f number is the ratio of the focal length of the lens to its effective diameter. For example, an f/8 setting means that the diameter of the aperture is one-eighth of the focal length of the lens. The lens setting is commonly referred to as the *f-stop*. (a) Determine how much light each of the following lens settings admits to the camera compared to f/8: (1) f/3.2 and (2) f/16. (b) The exposure time of a camera is controlled by the shutter speed. If a photographer correctly uses a lens setting of f/8 with a film exposure time of 1/60 s, what exposure time should he use to get the same amount of light exposure if he sets the f-stop at f/5.6?

87. From a spacecraft in orbit 150 km above the Earth's surface, an astronaut wishes to observe her home town as she passes over it. What size features will she be able to identify with the unaided eye, neglecting atmospheric effects? [Hint: Estimate the diameter of the human iris.]

Relativity

You might not think so at first glance, but this photograph tells us something very remarkable about the universe we inhabit. The bright shapes are galaxies, each consisting of many billions of stars—trillions, probably, for the larger ones. They are very far from us—billions of light years away. The faint arcs that make the photograph resemble a spider's web represent light from galaxies far more distant still!

What is remarkable about these fine wisps of light, however, is not the billions of years that they have been traveling to reach us, but the path they have taken. Light usually travels in straight lines, yet the light from these distant galaxies has been warped by the gravitational field of the galaxies in the foreground, creating the arcs in the photo.

The fact that light can be affected by gravity in this way was predicted by Albert Einstein as a consequence of his theory of relativity. The theory of **relativity** originated from the analysis of physical phenomena when speeds approach that of light. Indeed, relativity created a revolution in physics by causing us to rethink our basic understanding of space, time, and gravitation. It successfully challenged Newtonian concepts that had dominated scientific thinking for 250 years.

The impact of relativity has been especially great in those branches of science concerned with the extremes of physical reality—the subatomic realm of nuclear and particle physics (discussed in Chapters 29 and 30), where time intervals and distances are almost inconceivably small, and the cosmic realm, where time intervals and distances are almost unimaginably large. All modern theories about the birth, evolution, and ultimate fate of our universe are inextricably linked to our understanding of relativity.

In this chapter, you will learn how Einstein's theory of relativity explains the changes in length and time that we observe in rapidly moving objects, the equivalence of energy and mass, and the bending of light rays by gravitational fields—phenomena that appear very strange from the classical Newtonian viewpoint.

26.1 Classical Relativity

Objectives: To (a) summarize the concepts of classical relativity and relative velocities, (b) define inertial and noninertial reference frames, and (c) explain the reasoning behind the ether hypothesis.

Physics is concerned with the description of the world around us, and this endeavor depends on observations and measurements. We expect some aspects of nature to be consistent and unvarying. That is, the ground rules by which nature plays should be consistent, and our descriptions of nature or physical principles should not change from observation to observation. We emphasize this fact by referring to such principles as "laws"—for example, the laws of motion. Not only have physical laws proved valid over time, but they are the same for all observers.

In other words, a physical principle or law should be the same regardless of the observer's frame of reference. When we make a measurement or perform an experiment, we do so with reference to a particular frame or coordinate system, usually the laboratory, which is considered to be at rest. The experiment may also be observed by a passer-by (in motion relative to the laboratory). On comparing notes, the experimenter and the observer should find the results of the experiment or the physical principles involved to be the same. Physicists believe that the laws of nature are the same regardless of the way you observe them. Measured quantities may vary and descriptions may be different, but the laws that these quantities obey must be the same for all observers.

For example, suppose that you are standing at rest and observe two cars traveling in the same direction on a straight road at speeds of 60 km/h and 90 km/h. Even though we rarely say it, it is clear that these speeds are measured relative to your reference frame—the ground. However, a person in the car traveling at 60 km/h observes the other car traveling at 30 km/h *relative to her reference frame*— the car in which she is riding. (What does someone in the car traveling at 90 km/h observe?) In general, what each person observes is *relative velocity*—that is, the velocity relative to his or her reference frame.

Note: Review the discussion of relative velocities in Section 3.3.

In measuring relative velocity, there seems to be no "true" rest frame. Any reference frame can be considered to be at rest if the observer moves along with it. We can, however, make a distinction between what are called inertial and noninertial reference frames. An **inertial reference frame** is defined as one in which Newton's first law of motion holds. That is, in an inertial system, an object on which there is no net force remains stationary or moves with a constant velocity. Since Newton's first law holds in this reference frame, his second law ($\mathbf{F} = m\mathbf{a}$) applies to all objects within this system.

Note: The relationship between Newton's first and second laws is discussed in Section 4.3.

Conversely, in a noninertial (or accelerating) reference frame, an object with no net force on it would appear to accelerate. Note, however, that it is actually the frame that is accelerating, not the object. If such an object is observed from an accelerated reference frame, then Newton's second law will not correctly describe the motion of any object. One example of a noninertial reference frame is

Inertial versus noninertial reference frames

an automobile. When you accelerate from a stop sign, a coffee cup perched on the dashboard may *appear,* when viewed from the car's noninertial reference frame, to accelerate backward without being acted on by any force. In fact, the noninertial observer would have to invoke a "fictitious," or fake, backward force to explain the cup's apparent acceleration! From the inertial frame of a sidewalk observer, however, the cup simply tends to stay put (in accordance with the first law) as the car accelerates out from under it.

Any reference frame moving with a constant velocity relative to an inertial reference frame is itself also an inertial frame. Given a constant relative velocity, no acceleration effects are introduced in comparing one frame to another. In such cases, $\mathbf{F} = m\mathbf{a}$ can be used by observers in *either* frame (with their own coordinate values) to analyze a dynamic situation, and both observers will come to similar conclusions. That is, the law will hold in both frames. Thus, for the laws of mechanics, there is no one inertial frame of reference that is preferred over another. This is sometimes called the **principle of classical,** or **Newtonian, relativity:**

The laws of mechanics are the same in all inertial reference frames.

The Absolute Reference Frame: The Ether

In the late 1800s, with the development of the theories of electricity and magnetism, some serious questions arose. Maxwell's equations predicted light to be an electromagnetic wave that travels with a speed of 3.00×10^8 m/s in vacuum. But *relative to what reference frame* does light have this speed? Classically, the speed of light, c, measured from different frames of reference would be expected to differ— possibly to be even greater than 3.00×10^8 m/s if you were approaching the light wave—by vector addition of velocities.

Consider •Fig. 26.1: A person on the truck is in a reference frame moving in relation to another with a constant velocity $\mathbf{v}'$ and throws a ball with a velocity $\mathbf{v}_b$. Then the "stationary" observer would say that the ball had a velocity of $\mathbf{v} = \mathbf{v}' + \mathbf{v}_b$ relative to the stationary reference frame. For example, suppose a truck were moving with a speed of 20 m/s relative to the ground and a ball were thrown with a speed of 10 m/s relative to the truck in the direction of the truck's motion. The ball would have a speed of 20 m/s + 10 m/s = 30 m/s relative to the ground.

Now suppose that a person on the truck turned on a flashlight, projecting a beam of light. In this case, we have $\mathbf{v} = \mathbf{v}' + \mathbf{c}$, and the speed of light measured by an observer on the ground should be greater than 3.00×10^8 m/s. Classically then, the speed of light could have any value, depending on the observer's frame of reference.

It was reasoned, therefore, that the particular light speed of 3.00×10^8 m/s must be referenced to a unique frame, just as the speed of sound at room temperature, about 345 m/s, is referenced to the frame in which the air is at rest, or still. Moreover, as a wave, light was thought to require some medium of transport, just as sound or water waves do. This was quite natural, since experience showed that *all* other traveling waves must be transmitted through the stress or distortion of some medium. Since the Earth receives light from the Sun and from distant

•**FIGURE 26.1 Relative velocity**
(a) According to a stationary observer on the ground, $\mathbf{v} = \mathbf{v}' + \mathbf{v}_b$. **(b)** Similarly, the velocity of light would classically be measured to be $\mathbf{v} = \mathbf{v}' + \mathbf{c}$, with a magnitude greater than c. (Velocity vectors are not drawn to scale—why?)

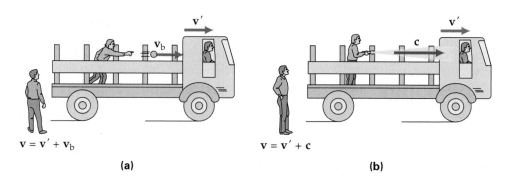

$\mathbf{v} = \mathbf{v}' + \mathbf{v}_b$ $\mathbf{v} = \mathbf{v}' + \mathbf{c}$

(a)   **(b)**

stars and galaxies, it was thought that the light-transporting medium in which the speed of light is 3.00×10^8 m/s must permeate all space. This medium was called *luminiferous ether*, or simply **ether**.

The idea of an undetected ether became popular in the latter part of the nineteenth century. Maxwell, whose work laid the foundations for our understanding of electromagnetic waves, believed in the existence of an etherlike substance:

> Whatever difficulties we may have in forming a consistent idea of the constitution of the ether, there can be no doubt that the interplanetary and interstellar spaces are not empty, but are occupied by a material substance or body which is certainly the largest, and probably the most uniform body of which we have any knowledge.

It would seem then that Maxwell's equations, which describe the propagation of light, did *not* satisfy the Newtonian relativity principle as did other physical laws. On the basis of the preceding discussion, a preferential or special inertial reference frame would seem to exist, one that could be considered absolutely at rest—the ether frame.

This was the state of affairs toward the end of the nineteenth century when scientists set out to investigate whether they had come upon a new dimension of physics or perhaps a flaw in what were considered to be established principles. One of the first attempts was to test the ether theory. If it could be proved that the ether existed, then presumably a true absolute rest frame could finally be identified. This was the purpose of the Michelson–Morley experiment.

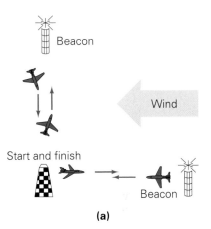

(a)

26.2 The Michelson–Morley Experiment

Objectives: **To explain (a) the general concept and operation of the Michelson–Morley experiment, (b) its result, and (c) the effect on the ether concept.**

If the ether permeated all space, as was thought, the Earth itself would be surrounded by this medium. In its orbit around the Sun, the Earth presumably moved through the ether, like an airplane moving through still air. Relative to the airplane, there is air motion or wind, and similarly the orbiting Earth would experience an "ether wind."

If the ether wind existed, this would provide a means to detect the ether experimentally. Observers could carefully measure the speed of light on Earth in several directions—for example, parallel and perpendicular to the orbital velocity of the Earth—and compare the results. Because of the Earth's motion relative to the ether, the speeds should be different in different directions.

As an analogy, consider two airplanes flying at the same air speed and the same distances back and forth as illustrated in •Fig. 26.2a. The plane flying perpendicular to the wind must be directed slightly into the wind on both legs to stay on a straight-line course, an action that slows the plane somewhat. The other plane flies into the wind on the first leg and with the wind on the return leg. The velocities at which the planes fly relative to the ground are given by vector addition. It is a simple matter to show that the plane flying perpendicular to the wind takes less time to complete the two-leg trip than does the other plane.

It seemed that a similar effect *should* be seen if the airplanes are replaced by light beams and the atmospheric wind by the ether wind (Fig. 26.2b). However, this is more easily said than done. The average speed of the Earth in its orbit is about 30 km/s, compared to 300 000 km/s for the speed of light. In the late nineteenth century, the American physicist Albert A. Michelson (1852–1931) devised a technique for detecting the extremely small travel-time differences by using the interference of light by designing an extremely sensitive **interferometer**.

The basic principle of Michelson's interferometer is shown in •Fig. 26.3. A monochromatic light beam is incident on a partially silvered glass plate (P), which

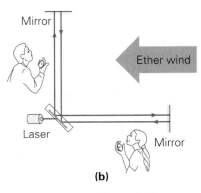

(b)

•**FIGURE 26.2 Ether wind detection** (a) When two airplanes fly with equal air speeds over equal distances, the plane flying perpendicular to the wind direction takes less time to fly a round trip. (b) Classically, the ether wind should give rise to a similar difference in the times for light to be reflected back and forth.

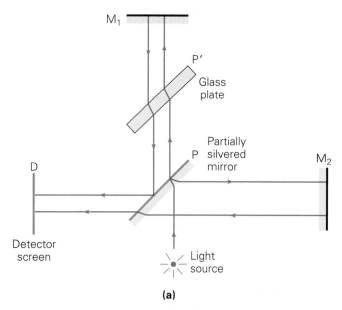

(b)

•**FIGURE 26.3 Michelson interferometer** **(a)** If one of the arms of the interferometer were in the direction of the ether wind, the beams should differ in phase when they arrive at the detector, and an interference pattern should be produced. (The light rays have been offset from one another for clarity.) **(b)** A laboratory model of the Michelson interferometer.

acts as a beam splitter, dividing the incident beam into transmitted and reflected components. The beams travel to the plane mirrors (M_1 and M_2) and are reflected back toward P. There, part of each beam is reflected and transmitted to the detector screen (D). Another glass plate (P′) is inserted in one beam, so both beams pass through an equal thickness of glass.

Since both beams are from the same initial beam, they are coherent and, on combining according to the superposition principle, interfere according to their relative phase relationship. The combination leads to constructive or destructive interference. This is determined by the difference in the path lengths of the beams.

Consider now the interferometer (and the Earth) that is traveling through the stationary ether with a velocity **v** and is arranged so that one of its perpendicular arms is parallel to **v** (•Fig. 26.4). Since c is the speed of the beam relative to the ether, to us the beam travels to M_2 at $c - v$. On the return trip, the relative velocity has a magnitude of $c + v$. The total time for the round trip is

$$t_2 = \frac{L_2}{c - v} + \frac{L_2}{c + v} = \frac{2L_2c}{c^2 - v^2} \tag{26.1}$$

Note: Interference is discussed in Sections 13.4, 14.4, and 24.1.

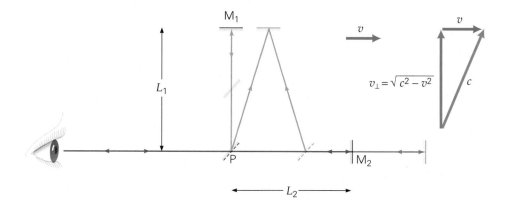

•**FIGURE 26.4 Interferometer light paths** An interferometer moving through the stationary ether would be observed to have the light paths shown.

For the light beam traveling between P and M_1, the speed of light in the direction perpendicular to **v** is $v_\perp = \sqrt{c^2 - v^2}$ (as shown in the vector diagram in Fig. 26.4). The speed is the same for the return path, so the time for the round trip on this arm is $t_1 = 2L_1/v_\perp$, or

$$t_1 = \frac{2L_1}{\sqrt{c^2 - v^2}} \qquad (26.2)$$

Then, if $L_1 = L_2 = L$, we have, after rearranging terms,

$$\Delta t = t_2 - t_1 = \frac{2L}{c}\left[\frac{1}{1 - (v^2/c^2)} - \frac{1}{\sqrt{1 - (v^2/c^2)}}\right] \qquad (26.3)$$

If there were a time difference, the beams would differ in phase when they arrive at the detector, and thus an interference pattern should be observed. From the interference fringes, we could in principle determine the speed of the light source relative to the ether. Notice, however, that the results depend on the *square* of the ratio of the speed of the Earth (relative to the ether) to the speed of light, a number almost assuredly very small, and much, much smaller when squared. Thus this experimental setup needed to be, and was, very sensitive.

It is difficult to ensure experimentally that $L_1 = L_2$ to the required degree of accuracy. This problem can be resolved by rotating the apparatus by 90°, which interchanges the arms. (With arms of equal length, the time difference, Δt, when the arms are interchanged becomes the negative of the value obtained from Eq. 26.3.) What would be observed then is a *shift* in the interference pattern or fringes.

Michelson performed this experiment with his colleague E. W. Morley in 1887. The interferometer was sensitive enough to detect the predicted fringe shift. But much to their surprise, Michelson and Morley found *no fringe shift*. Perhaps, they thought, the experiment had been done just at a time when the Earth was nearly at rest relative to the ether. To check this possibility, many measurements were made at different times (day and night, and at different seasons), always with the same null result. Since then, essentially these same experiments have been repeated with better and better accuracy, and the null result has never changed.

Where then was the ether? Several hypotheses were suggested to explain the null result of the **Michelson–Morley experiment.** One suggested that the ether in the vicinity of the Earth was dragged along with the Earth, so the interferometer was at rest with respect to the ether (that is, $v = 0$). But because of the Earth's orbital motion, the viewing of a star perpendicular to the plane of the orbital plane throughout the year requires the telescope to be consistently tilted in the direction of the Earth's motion. This effect, called the *aberration of starlight,* was explained in 1725. The aberration requires that the Earth move with respect to the ether and *not* be stationary relative to it (as would be the case if the ether were dragged along).

An Irish physicist, George F. FitzGerald, proposed that as a result of the motion through the ether, perhaps the linear dimensions of objects in the direction of motion contracted or became smaller. If this occurred by a fractional amount of $\sqrt{1 - (v^2/c^2)}$, then the null result would be explained. This hypothesis cannot be experimentally investigated, however, since the measurement instrument would also contract by the same fractional amount.

Another scientist, H. A. Lorentz, who also suggested this possibility, considered it in terms of possible changes in the electromagnetic forces between the atoms of a material due to the motion. This suggested length contraction, which is known as the *Lorentz–FitzGerald contraction*, might be termed the right idea but for the wrong reason. In fact, a length contraction was predicted by a theory proposed by Albert Einstein in 1905. Einstein's idea, however, had to do *not* with a mechanical shrinkage but rather with the fundamentals of the way we measure lengths and, indeed, the very nature of space and time. Einstein's special theory of relativity and some of its predictions, which forever changed the way physicists view space and time, are described in the following sections.

26.3 The Postulates of Special Relativity and the Relativity of Simultaneity

Objectives: To explain (a) how the two postulates of relativity imply the relativity of simultaneity and (b) how the relativity of simultaneity leads to length contraction.

The failure of the Michelson–Morley experiment to detect the ether left the scientific community in a quandary. The inconsistencies between Newtonian mechanics and electromagnetic theory remained unexplained. Many physicists were convinced that the experiment needed to be more accurate; they could not believe that light did not need a medium through which to propogate. These problems were resolved, however, by a theory introduced in 1905 by Albert Einstein (•Fig. 26.5). Interestingly, the Michelson–Morley experiment was apparently *not* a motivation for the development of his theory of relativity. Einstein later could not recall whether he had even known about the experiment at the time he formulated his ideas (while working as a clerk in the Swiss Patent Office!).

Einstein's key insight was based on his idea that the laws of mechanics should not be the only ones to obey the relativity principle. He reasoned that nature should be more symmetrical than that—*all* laws should be included under the relativity principle. In Einstein's view, the inconsistencies in electromagnetic theory were due to the assumption that an absolute rest frame (the ether) existed. His theory did away with the need for an ether by placing all laws of physics on an equal footing, thus eliminating any way of measuring the absolute speed of an inertial reference frame.

The first of the two postulates on which relativity is based is thus an extension of the Newtonian principle of relativity, which applies only to the laws of mechanics. In Einstein's theory, the **principle of relativity** applies to *all* the laws of physics, including those of electricity and magnetism:

> **Postulate I (principle of relativity):** All the laws of physics are the same in all inertial reference frames.

The first postulate implies that all inertial reference frames are equivalent, with physical laws being the same in all of them. Thus there exists no type of experiment, performed entirely within a reference frame, that would enable an observer in that frame to detect its absolute motion. This means that *there is no such entity as an absolute reference frame* having unique physical properties that would distinguish it from all other inertial reference frames. The first postulate seems quite reasonable, since we might expect the basic laws of nature to be the same for all inertial observers. We have no reason to think that nature would play favorites.

Einstein's second postulate involves the speed of light. As he saw it, a problem came from applying the vector addition of velocities to the velocity of light. With the appropriate relative velocity between inertial systems, the observed speed of light could have any value. With $(c + v)$, it could be greater than c; with $(c - v)$, it could be less than c; and with $(c - c)$, it could even reduce to zero! The last possibility was the source of Einstein's question, "What would I see if I rode a beam of light?" In the same frame as the electromagnetic wave, the electric and magnetic field vectors would be seen to be static—that is, not vary with time. Time variation of these two fields is fundamental to the propagation of light. Such static fields were not consistent with electromagnetic theory.

To avoid such inconsistencies, Einstein formulated his second postulate, which can be stated as follows:

> **Postulate II (constancy of the speed of light):** The speed of light in vacuum has the same value in all inertial systems.

These two postulates form the basis of Einstein's **special theory of relativity**. The "special" designation indicates that it deals only with the special case of inertial ref-

•FIGURE 26.5 Einstein and Michelson A 1931 photo shows Einstein (right) with Michelson during a meeting in Pasadena, California.

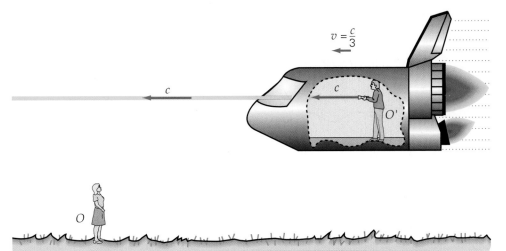

•FIGURE 26.6 Constancy of the
speed of light Two observers in
different inertial frames measure
the speed of the same beam of
light. The observer in frame O'
measures a speed of c inside the
ship. Classically, we might expect
the observer in frame O to measure
a speed of $c + c/3 = 4c/3$ as the
beam passes her, but she measures
c relative to her reference frame,
the ground.

erence frames. The *general* theory of relativity, discussed later in this chapter, deals
with the more general case of noninertial or accelerating reference frames.

The second postulate is perhaps more difficult to accept than the first. It states
that two observers in different inertial reference frames measure the speed of light
to be c independent of the speed of the source and/or the speed of the observer.
For example, if a person moving toward you with a constant velocity turned on
a light, both you and that person would measure the speed of light to be c, regardless of the magnitude of your relative velocity (•Fig. 26.6).

The second postulate is essential to the validity of the first and consistent with
the null result of the Michelson–Morley experiment. By doing away with the ideas
of the ether and an absolute reference frame, Einstein was able to reconcile the apparently fundamental differences between mechanics and electromagnetism. Even
so, the second postulate seemed to go against common sense. But keep in mind
that we have very little everyday experience in dealing with speeds near the speed
of light! The ultimate test of any theory is provided by the scientific method. What
does Einstein's theory predict, and has it been experimentally verified? As we
shall see, the answer is a resounding "yes" for every experiment yet performed.

The postulates of special relativity can be better understood by imagining simple situations to see what these postulates predict. Einstein did this himself in
what he called *gedanken,* or "thought," experiments. Let us begin with a series of
famous Einstein gedanken experiments related to simultaneity and the fundamental idea of how we measure length and time. We will also look at some experimental evidence that supports the nonclassical predictions of the theory.

The Relativity of Simultaneity

In everyday life we think of two events that are simultaneous to one person as
also being simultaneous to all other observers. What could be more obvious? Simultaneous events are those that occur "at the same time," and doesn't that mean
the same thing to all observers? No—at least, not at high speeds!

To see this, let's think of an inertial reference frame (called O) in which two
events are designed to be simultaneous. For example, we could arrange for two
small firecrackers (located at points A and B on the x axis) to explode when a switch
controlling a voltage source, placed midway between them, is flipped to the "on"
position (•Fig. 26.7a). Let us also equip the observer in this frame with a light receptor at point R, midway between the two firecrackers, capable of detecting
whether two light flashes do, in fact, arrive at the same time. (We could place the
light receptor at *any* location, but then the reception times would have to be corrected for time differences due to the unequal distances light would have to travel.
To avoid these complications, we place our simultaneity detector at the midpoint.)

•**FIGURE 26.7   The relativity of simultaneity** (a) An observer in reference frame O triggers two explosions (at points A and B) simultaneously. A light receptor R located midway between them records the two light signals as arriving at the same time. **(b)** An observer midway between the two explosions, but in frame O', moving with respect to O, sees the burned marks made by the two explosions on the x' axis but sees the explosion at A happen before that at B. **(c)** The situation as viewed from O'. The observer in O' sees the explosion at A before that at B. To him, O is moving to the left.

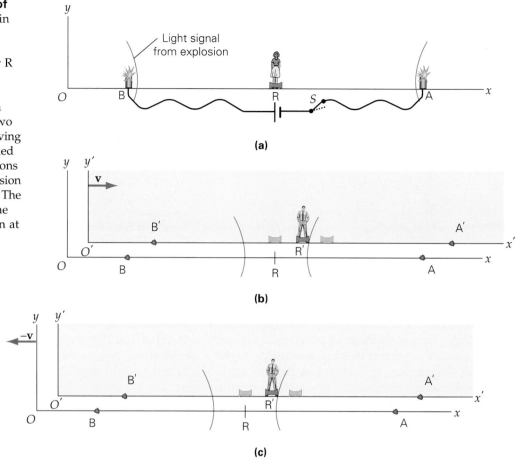

After detonation, the light receptor records that the two explosions went off simultaneously *in the O frame*. But consider the same two explosions as seen by another observer in a different inertial frame, O'. Viewed from O, the O' frame is moving to the right at a speed v. The observer in O' has equipped himself with a *series* of light receptors on his x' axis, because he is not sure which one will end up midway between the explosion points.

After the explosions, there are burn marks on both the x axis (at A and B) and x' axis (at A' and B'), as shown in Fig. 26.7b. He can use these marks to identify the particular O' light receptor (call it R') that was located midway between A' and B'. But when he reviews the data from this receptor, he finds that it did *not* record the explosions simultaneously. (The speed of light, c, is the same in both reference frames; why?) This fact does not cause the O observer to doubt his conclusion, however—he has a simple explanation of what happened. As he sees the situation, during the time it took for the light to get to R', that receptor had moved some distance toward A and away from B. Consequently, the O' receptor received the flash from A before that from B.

Which observer is correct? It's hard to find any objection to the conclusion reached by the observer in O—but isn't the observer in O' making a mistake? It should be obvious to him that he is moving with respect to the firecrackers. Why doesn't he realize this and take his motion into account? Since his light receptor was moving toward A and away from B, it should not surprise him that it recorded the flash from A before the flash from B. All he has to do is allow for this motion in his calculations, and he will conclude that the flashes "really" were simultaneous.

But if we reason this way, we are ignoring the postulates of relativity! We are unconsiously assuming that when we look at this situation from O (as we are doing in Fig. 26.7a,b), we are looking at what "really" happened from the vantage point of the frame that is "really" at rest. But, according to the first postulate, no inertial reference frame is more valid than any other, and none can be considered absolutely at rest. The observer in O' doesn't think of himself as moving—to him, the O frame is moving and O' is the rest frame. He observes the firecrackers moving past him with a speed v (Fig. 26.7c), but this motion would not affect his conclusions. To him the explosions occurred at A' and B', equally distant from R'. They were recorded at *different* times, so they were not simultaneous.

You might wonder whether we could somehow arrange it for the O' observer to agree that the two explosions did occur simultaneously. To do this, the observer in O would have to delay firecracker A relative to B so that R' would receive the two signals at the same time. However, the two observers would still disagree as to whether the events were simultaneous—because now they would no longer be simultaneous in O!

What are we to make of this curious situation? In a nonrelativistic world, one of the observers would have to be wrong. But as we have seen, both observers performed the measurements correctly. Neither one did anything wrong procedurally, or used faulty instruments, or made any errors in logic. So we must conclude that both are correct. The results may seem paradoxical, but this is the outcome when we apply the two postulates of special relativity.

Finally, we note that there is nothing special about a firecracker explosion. Any "happening" at a particular point in space at a particular time—a karate kick, a clock striking the hour, a soap bubble bursting—would have done just as well. Such a happening is called an **event** in the language of relativity. Thus we can summarize our conclusions by saying that

> Events that are simultaneous in a particular inertial reference frame may *not* be simultaneous as measured in a different inertial frame.

So we must give up the notion that simultaneity is an absolute concept. *Length contraction* and *time dilation*, discussed in the following section, are just two of the many relativistic results that follow directly from the two postulates and the relativity of simultaneity.

Notice that if the relative speed is slow, then the lack of simultaneity is completely undetectable. Thus, since we *are* accustomed to everyday low speeds, we come to the erroneous conclusion that simultaneity is absolute. Most relativistic effects have this property. That is, the departure from familiar experience is not apparent when the speeds are much less than the speed of light. Since we never travel at speeds approaching that of light, it is hardly surprising that these conclusions seem strange to us.

Conceptual Example 26.1

Agreeing to Disagree: The Relativity of Simultaneity

From Fig. 26.7, estimate the relative speed of the two observers. If the relative speed were only 10 m/s, would there be better agreement on simultaneity? Why?

Reasoning and Answer. The best way to estimate their relative speed is to compare the distance between B and B' in Fig. 26.7b to the distance the light has traveled from B. Using this method, we can see that the O' frame has moved about 30% as far as the light wave has in the same time. Therefore, the relative speed is approximately 30% of the speed of light or $v \approx 0.3c$.

At 10 m/s, the two frames would not have moved a noticeable distance during the whole sequence; thus both observers would agree on simultaneity.

Follow-up Exercise. Show that two events that occur simultaneously on the y axis of O *are* perceived as simultaneous by an observer in O', regardless of the relative speed, as long as the relative motion is along their common $x–x'$ axes.

It should be clear that to measure the length of a moving object correctly, we must mark the positions of both ends *simultaneously*. However, we have just seen that two inertial observers in relative motion will *disagree* on the concept of simultaneity. Thus, we expect that they will also disagree on the length of the object. It also turns out to be true that they disagree on time intervals. These two effects, relativistic length contraction and time dilation, are the subjects of the next section.

26.4 The Relativity of Length and Time: Length Contraction and Time Dilation

Objectives: To (a) understand the concepts of time dilation and length contraction, and (b) calculate the relationship between time intervals and lengths observed in different inertial frames.

Let us try yet another thought experiment related to the measurement of time intervals in different inertial frames. To compare time intervals in moving inertial reference frames, let's use a special "relativistic" light pulse clock as illustrated in •Fig. 26.8a. A tick or unit time interval on the clock corresponds to the time for a light pulse to make a round trip between the source and the mirror. It is assumed that two observers in two reference frames, O and O', have identical light clocks and that the clocks run at the same rate when both are at rest relative to one another. *For an observer at rest with respect to one of these clocks*, the time interval for a round trip of a light pulse is

$$\Delta t_o = \frac{2L}{c} \qquad (26.4)$$

Now consider O' to be in motion with a constant velocity of magnitude v relative to O. With its own identical clock (at rest in O') O' also measures the same time interval—Δt_o. But, as seen by O, the clock in the O' system is moving, and the path of the light pulse forms the sides of a triangle, as shown in Fig. 26.8b. Thus, O sees a longer light path for the clock in the moving system. From the O

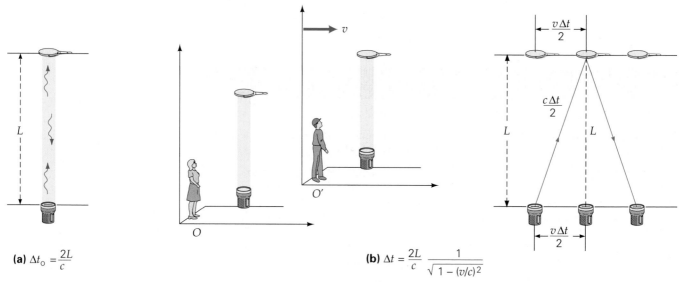

(a) $\Delta t_o = \dfrac{2L}{c}$

(b) $\Delta t = \dfrac{2L}{c} \dfrac{1}{\sqrt{1 - (v/c)^2}}$

•**FIGURE 26.8 Time dilation**   **(a)** A light clock that measures time in units of round-trip reflections of light pulses. The time for light to travel up and back is $\Delta t_o = 2L/c$. **(b)** An observer in O measures a time interval of $\Delta t = (2L/c)[1/\sqrt{1 - (v/c)^2}]$ on the clock in the O' frame. Thus the moving clock appears to run slowly to the O observer.

reference frame, the geometry of the path of the light pulse is a triangle in the figure. We can apply the Pythagorean theorem:

$$\frac{(c\,\Delta t)^2}{4} = \frac{(v\,\Delta t)^2}{4} + L^2$$

where Δt is the time interval of the O' clock as measured by the observer in O.

It should now be clear why O and O' measure different time intervals. Since the light path in the clock moving relative to frame O is longer, and the speed of light is the same for all observers, then the light in the moving clock must take a longer time to cover the path. Thus *from the O reference frame,* the moving clock runs slower—the ticks occur at a slower rate. We can calculate how much slower by solving the preceding equation for Δt:

$$\Delta t = \frac{2L}{c}\left[\frac{1}{\sqrt{1-(v/c)^2}}\right] \qquad (26.5)$$

But the time interval (Δt_o) is measured by either observer on a clock at rest in his or her *own* frame and is given by Eq. 26.4. So the measured time interval on the clock moving with respect to an inertial frame is longer than the interval on a clock at rest in that frame. Combining the equations, we have

$$\Delta t = \frac{\Delta t_o}{\sqrt{1-(v/c)^2}} \qquad \text{\textit{time dilation}} \qquad (26.6)$$

Since $\sqrt{1-(v/c)^2}$ is less than 1 (why?), we have $\Delta t > \Delta t_o$, and O measures a longer time interval on the O' clock than does observer O' on the O' clock. This effect is called **time dilation**. With longer time between ticks, the O' clock appears to O to run more slowly than the O clock. The situation is symmetric and relative. The observer in O' would say that the clock in the O frame ran slowly relative to the O' clock. Thus, because of the constancy of the speed of light, we have the qualitative statement of time dilation:

Definition of:
Time dilation

> Moving clocks are observed to run more slowly than clocks at rest in the observer's own frame of reference.

This effect, like all relativistic effects, is significant only if the relative speeds are close to that of light.

To distinguish between the two time intervals, we use the term **proper time**. As with most measurements, it is usually "proper" or normal to be at rest with respect to a clock when a time interval is measured. In our development, the proper time can be seen to be Δt_o, or the time interval in the reference frame in which the clock is at rest. Stated another way, the proper time interval between two events is that interval measured by an observer who is at rest relative to the events and sees those events occur *at the same location in space.* (What are the two events for the light-pulse clock? Are they at the same location in O' for the O' clock?) In Fig. 26.8, the observer in O sees the events by which the time interval of the O' clock is measured at different locations: Because the clock is moving, the starting event (the light pulse leaving) occurs at a different location in O than the ending event (the light pulse returning). Thus Δt is *not* the proper time.

Definition of:
Proper time

Many of the equations of special relativity can be written more simply if we represent the expression $1/\sqrt{1-(v/c)^2}$ as a dimensionless factor symbolized by γ (gamma):

$$\gamma = \frac{1}{\sqrt{1-(v/c)^2}} \qquad (26.7)$$

The factor γ is always greater than or equal to 1. (When is it equal to 1?) Notice that as v approaches c, γ approaches infinity. Since an infinite time interval is not physically possible, we expect that *relative speeds equal to or greater than that of light are not possible.*

The values of γ for several values of v (expressed as fractions of c) are given in Table 26.1. The values illustrate how the speed of an object must be an appreciable fraction of the speed of light for relativistic effects to be observed. Notice that at a speed of 10% that of light, γ differs from 1 by only 1%.

The time dilation equation (Eq. 26.6) can then be written more compactly as

$$t = \frac{t_o}{\sqrt{1 - (v/c)^2}} = \gamma t_o \tag{26.8}$$

where the deltas have been eliminated and it is understood that the t's represent time *intervals.* In words, the time interval measured on a moving clock is γ times the proper time interval. Thus the proper time interval is always the shortest time interval between two events.

For example, suppose you observed a clock at rest in a system moving at a constant velocity (magnitude 0.600c) relative to you. The factor γ would be 1.25 (calculate or refer to Table 26.1). Because of the time dilation effect, when 20 min had elapsed as recorded on that clock, you would observe a time of $t = \gamma t_o = (1.25)(20\ \text{min}) = 25$ min to have elapsed on *your* clock. The 20 min is the proper time, since the events defining the 20-minute interval took place at the same location (that of the "moving" clock—for example, for readings at 8:00 and 8:20). Thus the "moving" clock runs more slowly (20 min elapsed as opposed to 25 min on your clock) when viewed by an observer moving relative to it.

Finally, note that the time dilation effect cannot apply solely to our artificial light-pulse clock. It must be true for *all* moving clocks (that is, anything that keeps to a rhythm or frequency, including the heart). If this were not the case—if mechanical watches, for example, did not exhibit time dilation—then a watch and a light-pulse clock would run at different rates in a given inertial frame, and an observer in that frame would have a way of telling whether he or she was moving by making a comparison *solely within that frame.* Since that violates Postulate I, all moving clocks, regardless of their nature, must exhibit the same time dilation effect.

Let's take a look at an actual time dilation effect in nature.

Note: The proper time interval t_o is always less than the dilated time interval t.

TABLE 26.1 Some Values of $\gamma = \dfrac{1}{\sqrt{1 - (v/c)^2}}$

v	γ*
0	1.00
0.100c	1.01
0.200c	1.02
0.300c	1.05
0.400c	1.09
0.500c	1.15
0.600c	1.25
0.700c	1.40
0.800c	1.67
0.900c	2.29
0.950c	3.20
0.990c	7.09
0.995c	10.0
0.999c	22.4
c	∞

*To three significant figures

Example 26.2 ■ Muon Decay Viewed From the Ground: Time Dilation Verified by Experiment

There are many subatomic particles in nature. One of these, called a muon, is commonly created in the Earth's atmosphere when cosmic rays (mostly protons) collide with the nuclei of the gas molecules of the air. Once created, muons then approach the Earth with a speed near that of light (typically about 0.998c).

However, muons are unstable and quickly decay into other particles. The average lifetime of a muon at rest is 2.20×10^{-6} s. In this time, a muon would travel a distance of

$$d = v_o t = 0.998c(2.20 \times 10^{-6}\ \text{s}) = [0.998(3.00 \times 10^8\ \text{m/s})](2.20 \times 10^{-6}\ \text{s}) = 659\ \text{m}$$

This is only 0.659 km, or less than half a mile. Since muons are created at altitudes of 10 to 15 km, we would expect very few of them to reach the Earth's surface. However, an appreciable number of muons *do* reach the Earth. Using time dilation ideas, explain this apparent paradox.

Thinking It Through.   The paradox arises because the preceding calculation does *not* take time dilation into account. That is, a muon decays by its own "internal clock," as measured in its own reference frame (this is a proper time interval, since in the rest frame of the muon, "birth" and "death" events take place at the same location). However, to an observer on Earth, any clock in the muon reference frame would appear to run more slowly than a clock on Earth (•Fig. 26.9).

Solution. In a time interval $t_o = 2.20 \times 10^{-6}$ s on the muon "internal clock," the Earth clock reads a time interval that is greater by a factor of γ, which we can compute from Eq. 26.7:

$$\gamma = \frac{1}{\sqrt{1 - (v/c)^2}} = \frac{1}{\sqrt{1 - (0.998c/c)^2}} = 15.8$$

Thus from Eq. 26.8 the muon lifetime, according to an Earth observer, is

$$t = \gamma t_o = (15.8)(2.20 \times 10^{-6}\,\text{s}) = 3.48 \times 10^{-5}\,\text{s}$$

The muon travel distance is then

$$d = vt = 0.998(3.0 \times 10^8\,\text{m/s})(3.48 \times 10^{-5}\,\text{s})$$

$$= 1.04 \times 10^4\,\text{m} = 10.4\,\text{km}$$

Therefore, the theory of relativity predicts that the muons would travel a much greater distance in our frame than is predicted by classical physics. Since t_o is an *average* value, some muons travel distances actually greater than 10.4 km and are thus observed at the Earth's surface.

Follow-up Exercise. In this Example, what speed would enable the muons to travel twice as far (20.8 km) relative to the Earth? Would they have to travel twice as fast?

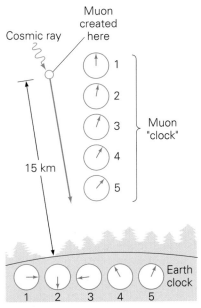

•**FIGURE 26.9 Experimental evidence of time dilation** Muons are observed at the surface of the Earth, as predicted by the theory of relativity. See Example 26.2.

Problem-Solving Hint

When working time dilation problems, you must identify the proper time interval t_o. To do this, first identify (1) the events that define the time interval and (2) a clock that measures it. Then, if the events occur in the same location in that reference frame, an observer at rest in the same frame (at rest relative to the clock) measures the proper time interval t_o. Since $t = \gamma t_o$ and $\gamma > 1$, the proper time interval t_o is always less than the dilated time interval t.

Length Contraction

As was mentioned in Section 26.3, the null result of the Michelson–Morley experiment could be artificially explained if it were assumed that the length of the interferometer arm in the direction of motion somehow contracted by a factor of $\sqrt{1 - (v/c)^2}$, the Lorentz–FitzGerald contraction. Length contraction is also a consequence of special relativity. However, its origin is not some artificial ether interaction with interatomic forces, but instead is explained by the application of the two postulates of special relativity.

To measure the length of an object that is not at rest in our reference frame, we must mark both ends simultaneously. Consider again the two inertial reference frames in relative motion that we used in our discussion of simultaneity, and imagine a measuring stick lying on the x axis at rest in O (•Fig. 26.10a). If the observer in O marks the ends simultaneously, the observer in O' will observe end A marked before B (why?). Thus for O' to make a correct length measurement, we must ask O to delay the marking of end A relative to B (Fig. 26.10b). We can picture O setting off explosions that would make burn marks in both reference frames (on both x and x' axes), with the information about these events traveling out at the speed c in all directions as visible light. When the ends are marked correctly for O', all that needs to be done is subtract the two positions to get the length of the stick. Notice that it will be *less* than that measured by O.

•FIGURE 26.10 Measuring
lengths correctly To measure the
length of a moving object correctly,
the ends must be marked
simultaneously. **(a)** When the
observer in frame O marks the
ends simultaneously, the observer
in frame O' does not agree and
observes A marked before B,
resulting in too long a length from
the O' view. **(b)** When O delays
A relative to B by just the correct
amount (how do we know from
the sketch?), then O' measures the
correct length from his point of
view. The length measured by O'
is less than the rest (or proper)
length measured by O.

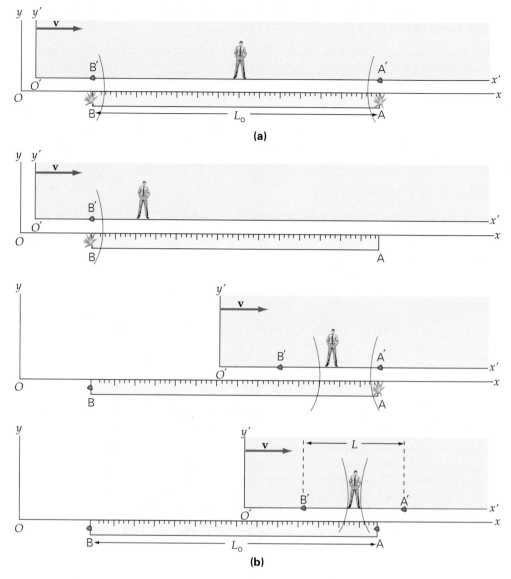

Again we ask, "Which observer is correct?" Both are, since *both have measured the length correctly* in their own frames. Neither thinks that the other observer has done things correctly, but each is satisfied with his or her own measurement. The observer in O' would have measured the same length as O *if he were willing to overlook the fact that the burn marks were not made simultaneously*, as in Fig. 26.10a, but this is *not* the correct way to measure the length of the moving stick from the O' point of view. The lack of agreement on simultaneity leads to the following qualitative statement of length contraction:

> The length of an object is largest for the observer at rest with respect to the object and smaller for observers in motion with respect to it (in the direction of the relative motion of two inertial observers).

Once again this effect is entirely negligible at speeds that are low compared to c.

The distance between two points *as measured by the observer at rest with respect to the two points* is designated by L_0 and called the **proper length**. The proper length, sometimes known as the *rest length*, is the largest possible length measured between these two points. The term *proper* has nothing to do with the "correctness" of the measurement, because as we have seen, each observer is measuring correctly in his or her own frame.

Definition of:
Proper length

Using a gedanken "experiment," let us develop a mathematical expression for length contraction. Consider a rod that is at rest in frame O. Consistent with our understanding of the term *proper*, we see that the observer in frame O is the proper length measurer for this rod. Thus we call the length of the rod that he measures L_o. The length of this same rod is measured by an observer in O', who is traveling with a constant speed v in the direction parallel to the stick (•Fig. 26.11). She does this by using the clock she is holding to measure the time interval required for the two ends of the rod to pass her. Since she measures the proper time interval (how do we know this?), we call this time t_o. In her reference frame, the rod is moving to the left with speed v, so the length that she measures, L, is given by $L = vt_o$ (speed × time).

Observer O could also measure the length of the rod by the same means. To him, the observer in O' is moving to the right past the rod. If he notes the times on his clock when O' passes the ends of the rod, he will measure a time interval t, and for him the length of the stick would be $L_o = vt$. (The v's are the same because the two observers agree on their relative speed.) Then, dividing one length by the other, we have

$$\frac{L}{L_o} = \frac{vt_o}{vt} = \frac{t_o}{t} \qquad (26.9)$$

But we know from our discussion of time dilation that $t = \gamma t_o$ (Eq. 26.8), or $t_o/t = 1/\gamma$. Equation 26.9 then becomes

$$L = \frac{L_o}{\gamma} = L_o\sqrt{1 - (v/c)^2} \qquad \begin{array}{l} \textit{length} \\ \textit{contraction} \end{array} \qquad (26.10)$$

Since γ is always greater than 1, we have $L < L_o$, or a **length contraction** by a factor of $1/\gamma = \sqrt{1 - (v/c)^2}$, as expected from our previous qualititative discussion of simultaneity.

Thus the measurements of both time intervals and space intervals (lengths and distances) are affected by relative motion. Remember that the length contraction occurs only along the direction of motion. As a result, in a relativistic world, objects would be contracted only in the dimensions that are parallel to the direction of their relative motion. To see the effects of length contraction, consider the following high-speed Example.

Definition of:
Length contraction

Note: The proper length L_o is always greater than the contracted length L.

•FIGURE 26.11 Derivation of length contraction The observer O measures the time it takes for the observer in O' to move past the ends of the rod. Similarly, the observer in O' measures the time it takes for the ends of the rod to pass her. The observer in O' is the *proper time measurer*; she measures the shortest possible time between these two events. The measured lengths of the rod are not the same. The observer in O is the *proper length measurer*; he measures the longest possible length.

Example 26.3 ■ Warp Speed: Length Contraction and Time Dilation

An observer sees a spaceship, measured as 100 m long when at rest, pass by in uniform motion with a speed of $0.500c$ (●Fig. 26.12). While the observer is watching the spaceship, a time of 2.00 s elapses on a clock on board the ship. (a) What would the observer measure the length of the moving spaceship to be? (b) What time interval elapses on the observer's clock for the 2.00-second interval measured on the ship's clock?

Thinking It Through. The 100 m is the proper length, because it is measured when the spaceship is at rest with respect to the observer. The time interval of 2.00 s is the proper time interval, because the beginning and end of the interval are measured by the same clock in the same location. We can then use (a) Eq. 26.10 to calculate the contracted length and (b) Eq. 26.8 to determine the dilated time interval.

Solution. We list the given quantities:

Given: $L_o = 100$ m (proper length)
$v = 0.500c$
$t_o = 2.00$ s (proper time interval)

Find: (a) L (contracted length)
(b) t (dilated time interval)

(a) By calculation or from Table 26.1, $\gamma = 1.15$ for $v = 0.500c$, and the length contraction is given by Eq. 26.10:

$$L = \frac{L_o}{\gamma} = \frac{100 \text{ m}}{1.15} = 87.0 \text{ m}$$

As expected, the observer measures the length of the spaceship to be considerably shorter than its proper length.

(b) The time interval (t) measured by the observer is longer than the proper time interval (t_o) and is given by Eq. 26.8:

$$t = \gamma t_o = 1.15(2.00 \text{ s}) = 2.30 \text{ s}$$

Follow-up Exercise. In this Example, calculate the time it takes the spaceship to pass a given point in the observer's reference frame, as seen by (a) a person in the spaceship and (b) the observer watching the ship move by. (c) Explain why these time intervals are not the same.

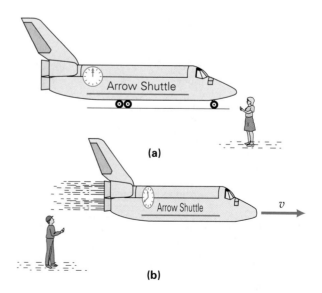

(a)

(b)

●**FIGURE 26.12 Length contraction and time dilation** As a result of length contraction, moving objects are observed to be shorter, or contracted in the direction of motion, and moving clocks are observed to run more slowly because of time dilation. See Example 26.3.

Example 26.4 ■ Muon Decay Revisited: Egos Rescued

We saw in Example 26.2 that many more muons reach the Earth's surface than can be accounted for in nonrelativistic terms. Experimenters on Earth who observe the muons moving at high speed explain this phenomenon in terms of relativistic time dilation. Since a hypothetical observer *on the muon* could not invoke time dilation (why not?), how would she explain the fact that the average muon makes it to the surface? Which explanation is correct?

Thinking It Through. In Example 26.2, we discussed a way of verifying time dilation by using muon decay. A muon traveling at $v = 0.998c$ decays by its own clock (proper time) in $t_o = 2.20 \times 10^{-6}$ s. A muon would travel a distance of only 659 m in this time. Yet muons are observed on Earth and so must travel more than 10 km, on the average, before decaying. This paradox was explained by a time dilation effect for an Earth observer, who sees the muon's clock running slow, thus enabling it to travel farther (before decay) than you would expect.

How does this situation look to a hypothetical observer on the muon? For such an observer, the muon clock is correct—and it registers a time interval that is *not* sufficient for the muon to reach the Earth's surface! How can this be reconciled with the observation of the Earth-bound observer? After all, two observers cannot disagree on the end result: We know that a significant fraction of the muons do, indeed, make it to sea level. To the muon observer, it is the distance to the surface that moves by quickly—so we suspect her reasoning involves length contraction.

Solution. The apparent paradox disappears when length contraction is taken into account. Then both observations are consistent with the experimental results but for different reasons. For our imaginary observer on the muon, the muon clock reads correctly ($t_o = 2.20 \times 10^{-6}$ s), but the observed travel distance—a length—is shorter because of length contraction. With $\gamma = 15.8$ for $v = 0.998c$, a travel length of 10.0 km in the Earth frame (proper length L_o) is measured by the observer on the muon to be L:

$$L = \frac{L_o}{\gamma} = \frac{10.0 \text{ km}}{15.8} = 0.633 \text{ km} = 633 \text{ m}$$

To travel this distance would take a time of

$$t = \frac{L}{v} = \frac{L}{0.998c} = \frac{633 \text{ m}}{0.998(3.00 \times 10^8 \text{ m/s})} = 2.11 \times 10^{-6} \text{ s}$$

This is roughly the average muon lifetime *in the muon's frame!* Thus, through relativistic considerations *both observers agree that many muons reach the Earth* (the experimental result). The Earth observer explains this result by saying that the muon clock is running slow (time dilation). The observer on the muon says, "No, the clock is fine—but the distance we have to travel is considerably less than you claim" (length contraction). Who is correct? Both are. *The reasoning is different for different observers, but the result is the same.*

Follow-up Exercise. Muons are created with a range of speeds in the upper atmosphere. What is the speed of a muon if it decays 5.00 km from its creation point as measured by the Earth observer? (Assume its proper lifetime is 2.20 μs.)

The Twin Paradox

Time dilation gave rise to another popular relativistic topic—the **twin paradox**, or **clock paradox**. According to special relativity, a clock moving relative to an observer runs more slowly than one in that observer's frame. For example, with $\gamma = 4$, for a proper time interval of 15 min, an hour would elapse on the observer's clock. Similarly, 1 y of proper time in the system moving relative to the observer takes 4 y in the observer's frame. Since our heartbeat and age are measured by proper time, the question arises: Do you age more quickly than a person moving relative to you?

One way to explore this question is through a gendanken experiment: Consider identical twins, one of whom goes on a high-speed space journey. Will the space traveler come back younger than his Earth-bound twin? Or will the space twin see the Earth twin age more slowly? However, *both* can't be right.

The solution to this paradox lies in the fact that in leaving and returning to Earth, the space twin must experience *accelerations* and so is *not* always in an inertial reference frame. The stay-at-home twin does not feel the forces associated with speed and directional changes that the traveling twin experiences. Thus the two twins are individually "marked," and their experiences are *not* symmetrical. However, if the acceleration periods (at the start, at turnaround, and at return) occupy only a small part of the total trip time, special relativity can be applied. The result: The space-traveling twin does indeed return younger than the Earth-bound twin, and both agree. (Einstein's general theory, which considers accelerating systems, confirms this result.)

While traveling at a constant velocity on the outward and return trips, which we assume occupies the total trip, both twins measure the same relative speed. However, the proper *length* of the trip is measured in the Earth twin's reference frame with fixed beginning and end points. The traveling twin thus measures a length contraction, or shorter trip distance. Traveling at the same relative speed, the space twin travels a shorter time and returns home younger than his Earth-bound twin *according to the traveling twin. According to the Earth twin*, the traveling twin's heartbeat (internal clock) runs slower so the traveling twin returns home younger than the Earth-bound twin! Thus there is no disagreement, the traveler returns having aged less (fewer heartbeats) than the nontraveler. At extremely high speeds, it would even be possible for the traveler to come back and find many generations of Earthlings long since born and died.

The twin paradox has been experimentally verified with atomic clocks. One of these clocks was flown around the world in jet planes. It was necessary to use these extremely accurate clocks to detect any effects because of the small γ factors, or low relative speeds. The time elapsed as recorded by the traveling clock was compared to that measured by an identical clock that stayed at home. These clocks were accurate enough to measure the predicted effect (with corrections made for accelerations), and time dilation was indeed experimentally verified.* To see numerical results for a hypothetical twin paradox, consider the next Example.

Example 26.5 ■ To the Stars: Relativity and Space Travel

Assuming special relativity, ignoring the accelerations at the start, end, and turnaround points, and assuming that no time is spent at the destination, (a) find the speed at which a space explorer would need to travel to make the round trip to a nearby star 100 ly (light-years) away in only 20.0 y of traveler time. [The light-year is defined as the distance light travels (in vacuum) in 1 y.] (b) How much time elapses on Earth during this trip?

Thinking It Through. Let us calculate a one-way trip. The explorer measures the one-way proper time (half of 20.0 y, or 10.0 y), because his clock is present for the start and end of the trip. People on Earth measure the (one-way) proper length, 100 ly—proper because its end markers (the Earth and the star) are at rest with respect to the Earth. (a) To find the speed, we can use either the Earth-bound explanation (time dilation) or that of the traveler (length contraction). (b) To get the travel time from the Earth's viewpoint, we can divide the (proper) distance by the explorer's speed, from (a).

Solution. We list the given quantities:

Given: $L_0 = 100$ ly (proper length)   *Find:* (a) v (speed of the traveler)
$t_0 = 10.0$ y (one-way   (b) t (time elapsed on Earth)
proper time)

---

*Hafele and Keating, "Around the World Atomic Clocks: Relativistic Time Gains Observed," *Science*, vol. 117, no. 4044 (July 14, 1972), pp. 166–170.

(a) For the traveler, the length to use is the contracted version of the 100 ly. According to the traveler, a one-way trip takes

$$t_o = \frac{\text{distance traveled}}{\text{speed}} = \frac{L}{v} = \frac{L_o\sqrt{1 - \left(\dfrac{v}{c}\right)^2}}{v}$$

This can be solved for v, giving

$$v = \frac{c}{\sqrt{1 + \left(\dfrac{ct_o}{L_o}\right)^2}} = \frac{c}{\sqrt{1 + \left(\dfrac{10.0 \text{ ly}}{100 \text{ ly}}\right)^2}} = 0.995c$$

so v is 99.5% the speed of light. Notice that we have used the fact that the distance light travels in 10.0 y (ct_o) is, by definition, 10.0 ly. We did not have to convert to meters because we expressed both distances (ct_o and L_o) in light-years.

(b) The people on Earth would observe this traveler covering a total of 200 ly at 0.995c, so the round-trip time would be 200 ly/0.995c, or about 201 y, compared to 20.0 y for the traveler. The traveler would come back to find that everyone who was alive on Earth when he left had long since died.

Follow-up Exercise. Using time dilation considerations, show that the Earth observers in this Example calculate that 201 of their years elapse during the trip. (Carry intermediate results to 5 decimal places, and round to 3 significant figures at the end.)

26.5 Relativistic Kinetic Energy, Momentum, Total Energy, and Mass–Energy Equivalence

Objectives: To (a) understand the relativistically correct expressions for kinetic energy, momentum, and total energy when objects move near the speed of light, (b) understand the equivalence of mass and energy, and (c) use the relativistically correct expressions to calculate energy and momentum in particle interactions.

The ramifications of special relativity are particularly important in particle physics, in which the speeds of the particles can be appreciable fractions of the speed of light. Such speeds are possible, for example, in modern particle accelerators. Many of our results from classical mechanics are incorrect when applied to high-speed particles. One of the most important quantities is kinetic energy. Does it have a form different from the expression that we used when we studied Newton's laws ($K = \frac{1}{2}mv^2$)? Yes, as we will now see.

Relativistic Kinetic Energy

Einstein showed that kinetic energy at high speeds is indeed given by an expression different from the one we are used to. As in classical mechanics, it increases with speed, but in a different way. The **relativistic kinetic energy** of a particle of mass m moving with speed v is

$$K = \left[\frac{1}{\sqrt{1 - (v/c)^2}} - 1\right]mc^2 = (\gamma - 1)mc^2 \qquad \begin{array}{l}\textit{relativistic}\\\textit{kinetic energy}\end{array} \qquad (26.11)$$

It can be shown that this expression becomes the more familiar $K = \frac{1}{2}mv^2$ when $v \ll c$.

According to Eq. 26.11, as v approaches c, the kinetic energy of an object approaches infinity. In other words, to accelerate an object to $v = c$ would require an infinite amount of energy, which is not possible. Thus special relativity predicts that no object can travel as fast as or faster than the speed of light. Hence, c is the maximum and unattainable speed limit in the universe. Time dilation and

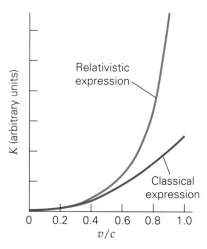

•FIGURE 26.13 Relativistic versus classical kinetic energy The kinetic energy variation with particle speed, expressed as a fraction of c, is shown for the relativistically correct expression and for the classical expression. The classical expression becomes negligibly different from the relativistic one for speeds less than about $0.2c$.

length contraction (Eqs. 26.6 and 26.10) also imply this. If a reference frame could travel with a speed c, then time intervals would become infinite (time would, in effect, stand still) and the lengths of objects would contract to nothing (in effect, they would vanish).

Particle accelerators can accelerate charged particles to very high speeds by using electric and magnetic fields. There is complete agreement between the experimentally measured kinetic energies of these charged particles and Eq. 26.11. A comparison of the relativistically correct expression for kinetic energy and the classical (low-speed) expression is shown in •Fig. 26.13.

Relativistic Momentum

In special relativity, the definition of momentum must also be changed. **Relativistic momentum** is

$$\mathbf{p} = \frac{m\mathbf{v}}{\sqrt{1 - \left(\dfrac{v}{c}\right)^2}} = \gamma m\mathbf{v} \qquad \textit{relativistic momentum} \quad (26.12)$$

Momentum still retains its vector character—that is, the momentum vector ($\mathbf{p}$) is still in the direction of the particle's velocity vector. Linear momentum, from Eq. 26.12, is still conserved.

Relativistic Total Energy and Rest Energy; the Equivalence of Mass and Energy

In classical mechanics, the total mechanical energy of an object is the sum of its kinetic and potential energies ($E = K + U$). When there is no potential energy, this reduces to $E = K$; that is, all the total energy is in the form of kinetic energy, *classically*. However, Einstein was able to show that the *relativistic total energy* of such an object is given by

Relativistic total energy

$$E = \frac{mc^2}{\sqrt{1 - \left(\dfrac{v}{c}\right)^2}} = \gamma mc^2 \qquad \begin{array}{l}\textit{(total energy}\\\textit{of an object,}\\\textit{with } U = 0)\end{array} \quad (26.13)$$

This means that even when a body is at rest and thus has no kinetic energy ($K = 0$), it still has an energy of mc^2. This minimum energy that a body always possesses is called its **rest energy**, E_o:

Rest energy

$$E_o = mc^2 \qquad (26.14)$$

According to Eq. 26.13, the total energy of an object decreases as v decreases—which also happens according to the classical equation ($E = K = \frac{1}{2}mv^2$). But when $v = 0$, the total energy of the particle, according to special relativity, does *not* drop to zero, as is predicted by the classical equation. Since $K = (\gamma - 1)mc^2$, we can express the total energy of a particle as the sum of its kinetic and rest energies. To see this, note that the relativistic kinetic energy expression (Eq. 26.11) can be rewritten as

$$K = \gamma mc^2 - mc^2 = E - E_o$$

Thus an alternative to Eq. 26.13 is

$$E = K + E_o = K + mc^2 \qquad (26.15)$$

(When a particle has potential energy U, the total energy is $E = K + U + mc^2$.)

A direct relationship between the total energy and the rest energy can be obtained by rewriting Eq. 26.13 (replacing mc^2 with E_o):

$$E = \gamma E_o \qquad\qquad (26.16)$$

For $v = 0$ (or $K = 0$) and thus $\gamma = 1$, we have $E = E_o$, as noted above.

Equation 26.14 expresses Einstein's famous **mass–energy equivalence**, which points out that *mass is a form of energy*. An object has energy even at rest—its rest energy. Thus we think of mass as another form of energy. It is impossible in nuclear and particle physics to keep the law of conservation of energy if we do not adopt this idea. If the equivalence of mass and energy is adopted, however, and the expressions we have developed here are used, then total system energy is conserved.

Mass–energy equivalence

The mass–energy equivalence does not mean that we can convert mass into useful energy at will. If we could, our energy problems would be solved, since we have a lot of mass on the Earth! Significant practical conversion of mass into other forms of energy, such as heat, does take place on a limited scale in nuclear reactors used for generating electric energy (discussed in Chapter 30). For now, the important point is that an object's mass must be considered to be one form of energy.

The (rest) energy equivalent of an object depends on its mass, which varies from object to object ($E_o = mc^2$). For example, the mass of an electron in SI units is 9.109×10^{-31} kg, and its rest energy is

$$E_o = mc^2 = (9.109 \times 10^{-31} \text{ kg})(2.998 \times 10^8 \text{ m/s})^2 = 8.187 \times 10^{-14} \text{ J}$$

or

$$E_o = (8.187 \times 10^{-14} \text{ J})\left(\frac{1 \text{ eV}}{1.602 \times 10^{-19} \text{ J}}\right) = 5.110 \times 10^5 \text{ eV} = 0.5110 \text{ MeV}$$

In modern physics, particle energies are commonly expressed in energy units of electron volts because charged particles are often given kinetic energy by accelerating them through potential differences measured in volts.

How do you know whether you need to use the relativistically correct formulas or can "get away" with the classical ones? The general rule of thumb is that if the object's speed is 10% of the speed of light or less, then the error in using the nonrelativistic kinetic energy expression is less than 1% (the classical expression is low). At $v < 0.1c$, the object's kinetic energy is less than 0.5% of its rest energy. So a commonly accepted practice is to use this speed and kinetic energy region as a dividing line:

> For speeds below 10% of the speed of light or kinetic energies less than 0.5% of the rest energy, the error in using the nonrelativistic formulas is less than 1% and it is then usually acceptable to use the nonrelativistic expressions.

Thus, an electron with a kinetic energy of 50 eV would qualify as a nonrelativistic electron because 50 eV is 10 000 times (only 0.01%) smaller than its rest energy. A proton with a kinetic energy of 0.511 MeV would also qualify as nonrelativistic because its rest energy is about 939 MeV, and therefore its kinetic energy is only about 0.5% of its rest energy. An electron with a kinetic energy of 0.511 MeV would, however, be *highly* relativistic because its kinetic energy is the same as its rest energy. Thus, what counts is the kinetic energy relative to the rest energy of the particle, or the speed of the particle relative to that of light—even the dividing line is relative! The following Example shows how particle energies are calculated by using relativistically correct expressions.

Example 26.6 ■ A Speedy Electron: Accelerating Energy

(a) How much work is required to give an electron (initially at rest) a speed of $0.900c$? (b) How much error would you have made if you had used the nonrelativistic expression?

Thinking It Through. (a) By the work–energy theorem, the work needed is equal to the electron's gain in kinetic energy. The gain in kinetic energy is then the final kinetic energy, since the initial kinetic energy is zero. From the text discussion, we know the rest energy of the electron—$E_o = mc^2 = 0.5110$ MeV. We are given its speed; its kinetic energy follows directly from Eq. 26.11. (b) The nonrelativistic kinetic energy is simply $\frac{1}{2}mv^2$.

Solution.   We list the data:

Given: $v = 0.900c$ *Find:* (a) W (work required)
           $E_o = 0.5110$ MeV (from text)   (b) K (nonrelativistic kinetic energy)

(a)   The relativistic kinetic energy is conveniently given by Eq. 26.11:

$$K = mc^2(\gamma - 1) = E_o(\gamma - 1)$$

With $v = 0.900c$, by calculation or from Table 26.1, $\gamma = 2.29$, and

$$K = E_o(\gamma - 1) = (0.5110 \text{ MeV})(2.29 - 1) = 0.659 \text{ MeV}$$

Thus, by the work–energy theorem, we know that 0.659 MeV of work is required to accelerate an electron to a speed of 0.900c.

(b)   Let's see what happens if we use the nonrelativistic expression. A shortcut is to multiply and divide by c^2, which enables us to use E_o:

$$K_{\text{nonrel}} = \tfrac{1}{2}mv^2 = \tfrac{1}{2}(mc^2)\left(\frac{v}{c}\right)^2 = \tfrac{1}{2}(0.5110 \text{ MeV})(0.900)^2 = 0.207 \text{ MeV}$$

As we suspected, the nonrelativistic kinetic energy expression produces results too low when the particle's speed is near that of light.

Follow-up Exercise.   In this Example, (a) what is the relativistically correct total energy of the electron? (b) What would the electron's total energy be had we used the nonrelativistic formula?

Example 26.7 ■ When One Plus One Doesn't Equal Two: Conservation of Relativistic Momentum and Energy

An elementary particle of mass m, moving to the right with speed $v = 0.800c$ after leaving an accelerator, collides with and sticks to an identical particle initially at rest. (a) What is the speed of the combined particles? (b) What is the mass of the combined particles? (c) What is the kinetic energy of the combined particles? [Parts (b) and (c) are to be answered in terms of m and c.]

Thinking It Through.   We can analyze collisions by using conservation of momentum and of total energy. Here we must use the relativistic expressions. The collision is completely inelastic (why?). In (a), the combined particle must be moving to the right after the collision to conserve the direction of the momentum. The magnitude of the single particle's momentum beforehand must equal the magnitude of the combined particle's momentum afterward. For (b) and (c), we will use their combined speed.

Solution.

Given: $v = 0.800c$ *Find:* (a) v' (speed after collision)
           $m = $ mass of one particle   (b) m' (mass of combined particles)
                                                    (c) K' (kinetic energy of combined particles)

(a)   We know that the gamma factor for the incoming particle is

$$\gamma = 1/\sqrt{1 - (v/c)^2} = 1/\sqrt{1 - (0.800)^2} = 1.67$$

Thus we can write momentum conservation as

$$p_i = p_f$$

Using the expression for relativistic momentum (Eq. 26.12), we have

$$\gamma mv = \gamma' m' v'$$

or

$$1.67m(0.800c) = \gamma'm'v'$$

where γ' is the gamma factor of the combined particles after the collision.

We use energy conservation and equate the total initial energy before the collision (which includes the total energy of the moving particle and the rest energy of the "target") to that after the collision (which is the total energy of the combined particle):

$$E_i = E_f$$

or, because $E = \gamma E_o$,

$$1.67mc^2 + mc^2 = \gamma'm'c^2$$

If we cancel the speed of light and divide it into the momentum result, we obtain

$$\frac{1.67m(0.800c)}{2.67m} = \frac{\gamma'm'v'}{\gamma'm'} = v'$$

Thus $v' = 0.500c$.

(b) From the speed afterward, v', we have $\gamma' = 1/\sqrt{1 - (0.500)^2} = 1.15$. Then we can use the energy equation above to solve for the mass of the combined particles:

$$2.67mc^2 = 1.15m'c^2$$

or

$$m' = 2.32m$$

The mass of the combined particles is greater than $2m$ because some of the initial kinetic energy is converted into mass (energy).

(c) For the kinetic energy afterward, we have

$$K' = (\gamma' - 1)m'c^2 = 0.15(2.32mc^2) = 0.348mc^2$$

As a check on energy, the initial total energy is $2.67mc^2$. The final total energy is $E' = \gamma'm'c^2 = 1.15(2.32mc^2) = 2.67mc^2$ as expected.

Follow-up Exercise. In terms of m, repeat the calculations in this Example for two objects, each of mass m and traveling at a speed of $0.800c$, that collide head-on and stick together. Compare your answers to those obtained from a nonrelativistic analysis.

26.6 The General Theory of Relativity

Objectives: To (a) explain the principle of equivalence and (b) specify some of the predictions of general relativity.

Special relativity applies to inertial systems but not to accelerating systems. Accelerating systems require an extremely complex theory, which was described by Einstein in several papers published about 1915. Called the **general theory of relativity**, this is essentially a gravitational theory.

The Equivalence Principle

An important aspect of general relativity is the **principle of equivalence:**

> An inertial reference frame in a uniform gravitational field is physically equivalent to a reference frame, in the absence of a gravitational field, that is experiencing a constant acceleration.

What this basically means is that

> No experiment performed in a closed system can distinguish between the effects of a gravitational field and the effects of an acceleration.

According to the principle of equivalence, an observer in an accelerating system would find the effects of a gravitational field and those of her own acceleration to be equivalent and indistinguishable. Rotating frames are excluded, since rotational acceleration can be distinguished from gravitational acceleration by releasing an

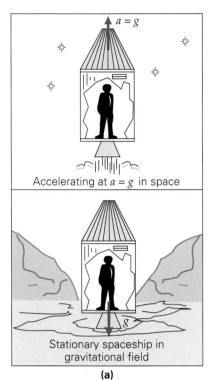

Accelerating at a = g in space

Stationary spaceship in gravitational field

(a)

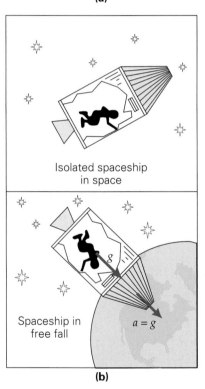

Isolated spaceship in space

Spaceship in free fall

(b)

•**FIGURE 26.14 The principle of equivalence** (a) The astronaut can perform no experiment in a closed spaceship that would determine whether he was in a gravitational field or an accelerating system. (b) Similarly, an inertial frame without gravity cannot be distinguished from free fall in a gravitational field.

object on a smooth, almost frictionless table. (Why?) Thus we stick with only linearly accelerating systems.

To understand the equivalence principle, consider the situations illustrated in •Fig. 26.14. Imagine yourself as an astronaut in a closed spaceship. You drop a pencil, and it accelerates to the floor. What does this mean? According to the equivalence principle, it could mean (a) that you are in a gravitational field or (b) that you are in an accelerating system (Fig. 26.14a). Any experiment performed inside the ship can't determine which. Whether the spaceship is in free space and accelerating with an acceleration a or is in a gravitational field with $g = -a$, the pencil would accelerate to the floor in either case. In your closed system (you are *not* allowed to look outside!), there is nothing you can do to determine whether the pencil's fall is a gravitational or an acceleration effect.

Suppose that the pencil does *not* begin to fall but instead remains suspended in midair next to you. This could mean that (a) you are in an inertial frame with no gravity or (b) you are in free fall in a gravitational field. Once again, in the closed spaceship (Fig. 26.14b), you have no way of knowing whether you are in free space (negligible gravity) or accelerating toward the Earth in free fall. (Think of how a released pencil would appear to float to an observer in a closed, freely falling elevator.) Because the effects of gravitation and acceleration are equivalent, you have no way of telling whether both are present, or neither is.

Light and Gravitation

The principle of equivalence of general relativity leads to an important prediction—that light is bent in a gravitational field. To see how this prediction arises, let's use a gedanken experiment. Suppose a beam of light traverses a spaceship that is accelerating rapidly. If the spaceship were stationary, light emitted at point A would arrive at point B (•Fig. 26.15a). However, because the spaceship is accelerating during the finite time it takes the light to traverse the ship, the light arrives at point C.

Now consider the situation as observed by a person on board the spaceship (Fig. 26.15b). From her point of view, she would observe the same effect of light emitted at A and arriving at C—*as though the light path were bent.* The acceleration of the rocket produces the effect, so by the equivalence principle we conclude that light path should also be bent in a gravitational field.

Under everyday conditions this effect must be very small, since we have observed no evidence of gravitational light bending in the Earth's gravitational field. However, this prediction of Einstein's theory was experimentally verified in 1919 during a solar eclipse. Distant stars appear to be motionless and are measured to have a constant angular distance separating them at night, when they can be clearly seen (•Fig. 26.16a). Light from one of the stars may pass near the Sun, which has a relatively strong gravitational field. However, any bending would not be observed on Earth because most of the starlight would be masked by the glare of the Sun itself and by the sunlight scattered in the atmosphere (which is why we can't see stars during the day).

But during a total solar eclipse, when the Moon comes between the Earth and the Sun, an observer in the Moon's shadow (the umbra) can see stars (Fig. 26.16b). If the light from a star passing near the Sun were bent, then the star would have an apparent location different from its actual position. As a result, the angular distance between stars would be measured to be slightly larger. Einstein's theory predicted the angular difference between the apparent and actual positions of the stars to be 1.75 seconds of arc, or an angle of about 0.000 05°. The experimental angular distance was found to be 1.61 ± 0.30 seconds of arc, in agreement with the calculation.

General relativity describes a gravitational field as a "warping" of space and time, as illustrated in Fig. 26.16c. A light beam follows the curvature of space–time like a ball rolling on a surface. The bending of light was verified during subsequent solar eclipses and also by signals from space probes. For example, signals from a *Viking* lander on the surface of Mars passed through the gravitational field

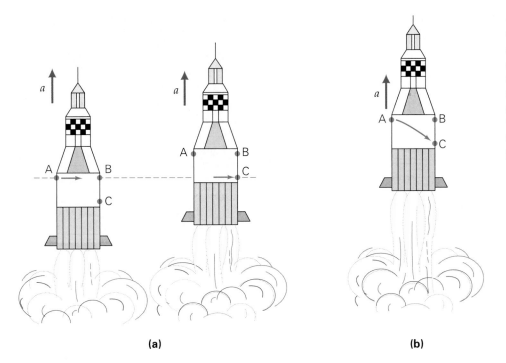

•FIGURE 26.15 Light bending
(a) Light traversing an accelerating, closed rocket from point A arrives at point C. (b) In the accelerating system, the light path would appear to be bent. Since an acceleration produces this effect, by the principle of equivalence, light should also be bent in a gravitational field.

of the Sun when Mars was on the far side of the Sun. The signals were observed to be delayed by about 100 μs. The delay was caused by the signals passing through the space–time warp, or "sink," of the Sun.

Gravitational Lensing.   Another effect of the gravitational bending of light is known as *gravitational lensing*. In the late 1970s, a double quasar was discovered. (A quasar is a powerful astronomical radio source.) The fact that it was a double quasar was not unusual, but everything about the two quasars seemed to be exactly the same, except that one was fainter than the other. It was suggested that perhaps there was

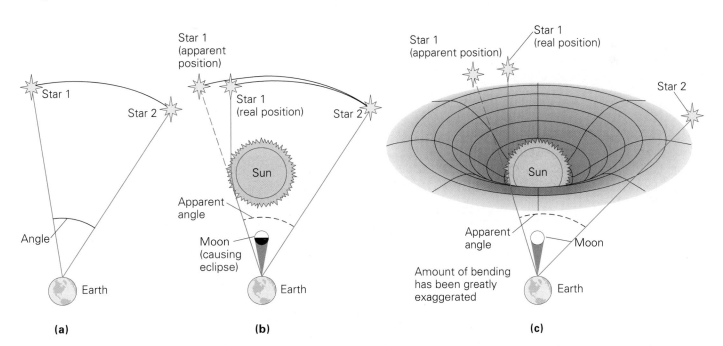

•FIGURE 26.16 **Gravitational attraction of light**   (a) Normally, two distant stars are observed to be a certain angular distance apart. (b) During a solar eclipse, the star behind the Sun can still be seen because of the effect of solar gravity on the starlight, which is evidenced by a larger measured angular separation. (c) General relativity views a gravitational field as a warping of space and time. A simplified analogy is the surface of a warped rubber diaphragm or sheet.

•**FIGURE 26.17   Gravitational lensing**   (a) The bending of light by a massive object such as a galaxy or cluster of galaxies can give rise to multiple images of a more distant object. (b) The discovery of what appeared to be four images of the same quasar (the "Einstein Cross") suggested the possibility of gravitational lensing. On investigation, a faint intervening galaxy was found.

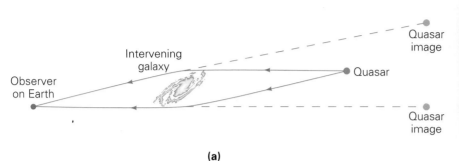

(a)

(b)

•**FIGURE 26.18   Gravitational lensing and dark matter**   (a) Two similar images of a remote galaxy on either side of an intervening cluster of galaxies were observed by the Hubble Space Telescope. (b) The effect was thought to result from gravitational lensing by galaxies containing large amounts of dark matter—matter that emits no electromagnetic radiation and so cannot be detected by ordinary observations. Gravitational lensing may offer a method to detect dark matter.

only one quasar and that, somewhere between it and the Earth, a massive but optically faint object had bent its electromagnetic radiation, producing multiple images. The subsequent detection of a faint galaxy between the two quasar images confirmed this hypothesis, and other examples have since been discovered (•Fig. 26.17).

Gravitational lenses can be considered to be a test of general relativity, but this concept has also given general relativity a new role in astronomy. By examining multiple images of a distant galaxy or quasar, their relative intensities, and so on, astronomers can gain information about an intervening galaxy or cluster of galaxies whose gravitational field causes the bending of light.

Such an effect has been observed by the Hubble Space Telescope. On a photo of a remote galaxy, two mirror images of the structure were observed on opposite sides of the picture (•Fig. 26.18a). These were believed to be caused by the gravitational lensing of an intervening cluster of galaxies containing much *dark matter:* matter that does not emit electromagnetic radiation and so cannot be detected by regular observations (Fig. 26.18b). How much matter the universe contains is an important question for scientists. Is there enough matter for gravitational attraction to slow and stop the Big Bang expansion of the universe? It is estimated that there might be 10 times more matter in the universe than is actually observed, but we do not know what this dark matter consists of, although several theories have been proposed. Gravitational lensing may provide a method to detect and measure such dark matter indirectly.

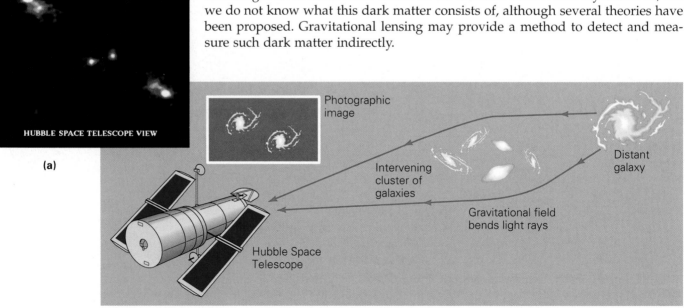

(a)

(b)

Black Holes. The idea that gravity can affect light finds its most extreme application in the concept of a black hole. A **black hole** is thought to form from the gravitationally collapsed remnant of a massive star, typically many times the mass of our Sun.* Such an object has a density so great and a gravitational field so intense that nothing can escape it. In terms of our space–time warp analogy, a black hole would be a bottomless pit in the fabric of space–time. Even light can't escape the intense gravitational field of a black hole—hence the blackness.

An estimate of the size of a black hole can be obtained by using the concept of escape speed. Recall from Section 7.6 that the escape speed from the surface of a spherical body of mass M and radius R is given by

$$v_{esc} = \sqrt{\frac{2GM}{R}} \tag{26.17}$$

If light does not escape from a black hole, the hole's escape speed must exceed the speed of light. The critical radius of a sphere around a black hole from which light could not escape is obtained by substituting $v_{esc} = c$ into Eq. 26.17. We solve for R:

$$R = \frac{2GM}{c^2} \tag{26.18}$$

Schwarzschild radius

This value is called the **Schwarzschild radius**, after Karl Schwarzschild (1873–1916), a German astronomer who developed the concept. The boundary of a sphere of radius R defines the **event horizon**. Any event occurring within this horizon is invisible to an observer outside, since light cannot escape.

Note: A collapsing star becomes a black hole if it collapses beyond its Schwarzschild radius or event horizon.

The event horizon is not necessarily the radius of a black hole. It gives only the limiting distance within which light cannot escape from a black hole. The mass inside would continue to collapse, making the black hole smaller than its event horizon. But information about what goes on inside the event horizon can never reach us, so questions such as "What does it look like inside the event horizon?" are meaningless.

Example 26.8 ■ If the Sun Were a Black Hole: Schwarzschild Radius

What would be the Schwarzschild radius if our Sun collapsed to a black hole? ($M_S = 2.0 \times 10^{30}$ kg)

Thinking It Through. This is a straightforward calculation using Eq. 26.18.

Solution.

Given: $M_S = 2.0 \times 10^{30}$ kg   *Find:* R (Schwarzschild
$\quad\quad G = 6.67 \times 10^{-11}$ N·m²/kg² (known)   radius)
$\quad\quad c = 3.00 \times 10^8$ m/s (known)

Keeping all the quantities in SI units, we have

$$R = \frac{2GM_S}{c^2} = \frac{2(6.67 \times 10^{-11}\ \text{N·m}^2/\text{kg}^2)(2.0 \times 10^{30}\ \text{kg})}{(3.00 \times 10^8\ \text{m/s})^2} = 3.0 \times 10^3\ \text{m} = 3.0\ \text{km}$$

Less than 2 mi! (The Sun's radius is now about 7.0×10^5 km.) However, our Sun will not become a black hole. This fate befalls only stars much more massive than the Sun.

Follow-up Exercise. Once black holes form, they continuously draw in matter, expanding and increasing their Schwarzschild radius. How many times more massive would our Sun have to be for its Schwarzschild radius to extend to, and swallow up, Mercury, the innermost planet? (The average distance from the Sun to Mercury is 57.9×10^9 m.)

---

*It is speculated that black holes may originate in other ways, such as the collapse of entire star clusters in the center of a galaxy or at the beginning of the Universe during the Big Bang. Some current theories (combining quantum mechanics and general relativity) hint that black holes might emit material particles and radiation and thus "evaporate," but none have been discovered in this process. We will limit our discussion to stellar collapse.

If nothing, including radiation, escapes a black hole from inside its event horizon, how then might we observe or locate one? This question is addressed in the Insight on p. 831.

Gravitational Red Shift. Another prediction of general relativity is that gravity affects time by causing it to slow down—the greater the gravitational field, the greater the slowing of time. There is evidence of this effect in what is called the *gravitational red shift* of light from the Sun. The gravitational field of the Sun is quite large, and if the slowing of time in a gravitational field is correct, the electronic vibrations in the Sun's atoms should be slower. As a result, light emitted by these atoms would have a lower frequency and be shifted toward the red end of the visible spectrum (hence the name *red* shift).

Relativistic effects apply to all systems, no matter how complicated. Although these effects are most readily observed for atomic particles and light, they apply as well to real clocks, human beings, and so on. Although we don't commonly observe them, relativistic effects form the bases of some good science fiction and even some (not so good?) poetry, such as the following variation on a limerick that appeared in the British magazine *Punch* in 1923:

> A precocious student named Bright
> Could travel much faster than light;
> He departed one day
> In a relative way
> And arrived home the previous night.

*26.7 Relativistic Velocity Addition

Objectives: To (a) understand the necessity for a relativistic velocity addition equation and (b) apply it to simple relative velocity calculations.

As we have seen, the postulates of special relativity affect our classical concepts of distance and time. Since velocity involves these two quantities, we might expect that relative velocities would also be affected, and this is the case.

Recall from Section 26.1 that there is a problem with vector addition when light is involved. Ordinarily, if someone on a truck moving at 20 m/s throws a ball at 10 m/s in the direction of motion, an observer on the ground would say that the velocity of the ball is 20 m/s + 10 m/s = 30 m/s in the direction of the motion. However, if the ball is replaced by a beam of light (Fig. 26.1), such vector addition gives a speed greater than c, which violates the second postulate of relativity.

Note: Review Section 6.6 and Fig. 6.24.

The same problem occurs with objects moving at appreciable fractions of the speed of light. Consider the rocket separation shown in ●Fig. 26.19. After separation, the jettisoned stage has a velocity $\mathbf{v}$ with respect to the Earth and the rocket has a velocity $\mathbf{u}'$ with respect to the jettisoned stage. Then, by addition of velocity vectors, the velocity $\mathbf{u}$ of the rocket payload with respect to Earth should be given by

$$\mathbf{u} = \mathbf{v} + \mathbf{u}'$$

However, assuming relativistic speeds, suppose $v = 0.50c$ and $u' = 0.60c$. Then we would have $u = v + u' = 0.50c + 0.60c = 1.10c$. Thus, classical velocity addition says that an Earth observer would see the rocket traveling at a speed greater than

●**FIGURE 26.19 Relativistic velocity addition** After jettisoning, the rocket has a velocity $\mathbf{u}'$ with respect to the jettisoned stage and the jettisoned stage has a velocity $\mathbf{v}$ with respect to the Earth. For relativistic velocities, the velocity $\mathbf{u}$ of the rocket with respect to the Earth is not given by ordinary vector velocity addition; it must be obtained from a relativistic equation.

Black Holes

The idea of black holes has caught the public's fancy. Try to imagine something so dense that nothing—not even light—can escape from it. It is so dense that a tablespoon full of its matter would contain more mass than Mt. Everest. According to the theory of stellar evolution, black holes could result from the collapse of stars having much greater mass than the Sun. But gathering experimental proof of the existence of a black hole is another matter.

The boundary of the "blackness" of the hole is the surface of the event horizon at its Schwarzschild radius. The boundary of a black hole is not a sphere of matter but the distance at which the gravitational force is sufficiently strong to keep even light from escaping. The actual size of the black hole's matter may be much less. What form or size matter takes inside a black hole is not known and in principle can never be known. Even if we could send a probe "inside" a black hole (that is, closer than its Schwarzschild radius), the probe could never send data back to us. (Why?)

How then could we observe a black hole? Light passing nearby would be bent, but it is unlikely that we would ever observe this, considering the vastness of space. The most likely possibility comes from observing binary star systems, which are common. Cygnus X-1, the first X-ray source discovered in the constellation Cygnus, provides the best evidence so far for a black hole in our galaxy. (Many X-ray sources have been discovered through observations from rockets and satellites. Recall that X-rays do not penetrate the atmosphere. Most of these X-ray sources, however, are not massive enough to qualify as black holes.)

Looking at the location of the Cygnus X-1 X-ray source in visible light (Fig. 1), we observe a giant star whose spectrum shows periodic Doppler red and blue shifts, indicating an orbital motion. (Binary stars orbit about each other—more precisely, about their common center of mass.) Measuring the precise orbital data could in principle tell us the mass of both stars. In Cygnus X-1 there is a companion to the giant star that *seems* to have more than enough mass to qualify as a black hole.

It is speculated that in binary star systems such as Cygnus X-1, one member has become a black hole. Matter drawn from the other, more normal, star would create an *accretion disk* of spiraling matter around the hole (Fig. 2). The matter falling into the disk would be accelerated and heated. Collisions and deceleration would produce a characteristic spectrum of X-rays, and the black hole would appear as an X-ray source.

A similar mechanism may explain the enormous energy output of *active galaxies* and *quasars*. It has been proposed that the centers of these brilliant objects, which produce vast amounts of radiation, might contain black holes with masses millions or even billions of times that of the Sun. Indeed, recent observations suggest that even many normal galaxies, including our own Milky Way galaxy, may harbor black holes in their cores.

Today, most physicists and astronomers believe that black holes do exist. As more definitive studies of binary systems are conducted, solid proof of the existence of black holes is expected to emerge.

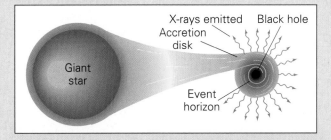

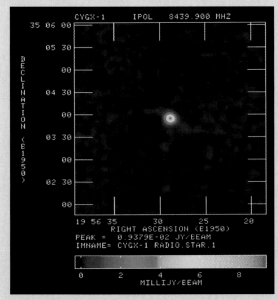

FIGURE 1   Cygnus X-1   The overexposed large dark spot at the center is a giant star believed to be a companion of the X-ray source and possible black hole, Cygnus X-1.

FIGURE 2   X-rays and black holes   Matter drawn from the "normal" member of a binary star system forms a spiraling accretion disk around the hole. Matter falling into the disk is accelerated, and collisions give rise to the emission of X-rays.

the speed of light in vacuum. However, as we saw in Section 26.4, no object can travel faster than the speed of light.

Einstein recognized that since length and time are different with respect to different reference frames, classical velocity vector addition is not applicable in relativistic cases. He showed the correct equation for motion in a straight line to be

Relativistic velocity addition

$$u = \frac{v + u'}{1 + \dfrac{vu'}{c^2}} \qquad \text{(straight line only)} \qquad (26.19)$$

where the velocities have the same meanings as above and sign notation is used to indicate velocity directions. Notice that the observed velocity u is *reduced* by a factor of $1/(1 + vu'/c^2)$ from that of the classical result.

It is important to identify clearly the velocities in working problems, so we list them again generally:

v = velocity of object 1 with respect to an inertial observer (such as on Earth)

u' = velocity of object 2 with respect to an object 1

u = velocity of object 2 with respect to an inertial observer (such as on Earth)

Signs are used to indicate velocity directions. For example, if object 2 were approaching object 1, then u' would have a negative value in Eq. 26.19. Let's see how Eq. 26.19 avoids the pitfall of classical velocity addition in the following Examples.

Example 26.9 ■ Faster Than the Speed of Light: Relativistic Velocity Addition for Objects

For the rocket separation in Fig. 26.19, let the velocities be $v = 0.50c$ and $u' = 0.60c$. According to relativistic velocity addition, what is the velocity of the rocket nose cone as seen by an observer on Earth?

Thinking It Through. We can apply Eq. 26.19. We expect u to be less than c, unlike the classical result of $1.10c$.

Solution. The velocities are taken to be in the positive direction.

Given: $v = +0.50c$ (velocity of jettisoned stage with respect to Earth)
$u' = +0.60c$ (velocity of rocket with respect to jettisoned stage)

Find: u (velocity of rocket with respect to Earth)

We have

$$u = \frac{v + u'}{1 + \dfrac{vu'}{c^2}} = \frac{+0.50c + 0.60c}{1 + \dfrac{(+0.50c)(+0.60c)}{c^2}} = \frac{+1.1c}{1.3} = +0.85c$$

Note: It is usually convenient to express velocities as fractions of c.

As expected, the result is less than c.

Follow-up Exercise. Repeat this Example with both objects moving at 40% of the speed of light. Classically, the answer is $0.80c$, which does *not* violate the second postulate. Is the relativistically correct result lower than the classical result, as expected?

Example 26.10 ■ Can Light Travel Faster Than Light? Relativistic Velocity Addition for Light

Suppose a spacecraft is moving in free space with a constant velocity of magnitude $0.25c$ relative to an observer. An astronaut on board the spacecraft sends out two light beams, one (a) in the direction of the craft's motion and the other (b) back toward the observer. What are the speeds of the light beams relative to the observer?

Thinking It Through. Again we can apply Eq. 26.19. However, here one of the "objects" is a beam of light. If we take care of our velocity signs, we should get c for both answers, according to the second postulate of special relativity.

Solution.

Given: $v = 0.25c$

    (a) $u' = c$ (in the direction of motion)

    (b) $u' = -c$ (opposite the direction of motion)

Find: u (velocity of the light beam for both cases)

(a) We have

$$u = \frac{v + u'}{1 + \dfrac{vu'}{c^2}} = \frac{+0.25c + c}{1 + \dfrac{(+0.25c)(c)}{c^2}} = \frac{1.25c}{1.25} = +c$$

(b) Similarly, with $u' = -c$,

$$u = \frac{v + u'}{1 + \dfrac{vu'}{c^2}} = \frac{+0.25c - c}{1 + \dfrac{(+0.25c)(-c)}{c^2}} = \frac{-0.75c}{0.75} = -c$$

Hence our relativistic velocity addition formula gives answers consistent with special relativity's second postulate—the constancy of the speed of light, as we expect, because the two postulates are what gave rise to all the relativistic formulas in this chapter!

Follow-up Exercise. Two rocket travelers approach one another, each traveling at $0.90c$ relative to a third inertial frame. Each shoots a light beam at the other. Calculate the speed that each measures for the other's light beam as it passes.

Chapter Review

Important Terms

relativity *802*
inertial reference frame *803*
principle of classical (Newtonian)
 relativity *804*
ether *805*
interferometer *805*
Michelson–Morley
 experiment *807*
principle of relativity *808*

constancy of the speed
 of light *808*
special theory of relativity *808*
event *811*
time dilation *813*
proper time *813*
proper length *816*
length contraction *817*
twin (clock) paradox *819*

relativistic kinetic energy *821*
relativistic momentum *822*
rest energy *822*
mass–energy equivalence *823*
general theory of relativity *825*
principle of equivalence *825*
black hole *829*
Schwarzschild radius *829*
event horizon *829*

Important Concepts

- An inertial (nonaccelerating) reference frame is one in which Newton's first law holds.

- The principle of relativity (special relativity postulate I) states that all the laws of physics are the same in all inertial reference frames.

- The constancy of the speed of light (special relativity postulate II) states that the speed of light in a vacuum is the same for all inertial reference frames.

- Clocks moving relative to an observer in an inertial reference frame run more slowly than clocks at rest in the observer's frame (relativistic time dilation).

- Lengths of objects moving relative to an observer in an inertial reference frame are less in the dimension of relative motion than if the object were at rest in the observer's frame (relativistic length contraction).

- The relativistic mass–energy equivalence, $E_o = mc^2$, tells us that the rest energy is the energy equivalent of an object's mass. An object has rest energy even if it has no kinetic or potential energy.

- The principle of equivalence, the guiding postulate of general relativity, states that no experiment performed in a closed system can distinguish between the effects of a gravitational field and the effects of an acceleration.

Important Equations

γ (dimensionless relativistic factor):

$$\gamma = \frac{1}{\sqrt{1 - (v/c)^2}} \qquad (26.7)$$

Time Dilation:

$$\Delta t = \frac{\Delta t_o}{\sqrt{1 - (v/c)^2}} \qquad (26.6)$$

and

$$t = \frac{t_o}{\sqrt{1 - (v/c)^2}} = \gamma t_o \qquad (26.8)$$

Length Contraction:

$$L = \frac{L_o}{\gamma} = L_o \sqrt{1 - (v/c)^2} \qquad (26.10)$$

Relativistic Kinetic Energy:

$$K = (\gamma - 1)mc^2 \qquad (26.11)$$

Relativistic Momentum:

$$\mathbf{p} = \gamma m \mathbf{v} \qquad (26.12)$$

Rest Energy:

$$E_o = mc^2 \qquad (26.14)$$

Relativistic Total Energy:

$$E = K + E_o = K + mc^2 = \gamma mc^2 \quad (26.13, 26.15)$$

Total Energy and Rest Energy:

$$E = \gamma E_o \qquad (26.16)$$

Schwarzschild Radius:

$$R = \frac{2GM}{c^2} \qquad (26.18)$$

*Relativistic Velocity Addition (straight line):

$$u = \frac{v + u'}{1 + \dfrac{vu'}{c^2}} \qquad (26.19)$$

Exercises

Note: Assume c to be exact at 3.00×10^8 m/s. Consider speeds given in terms of c as having the same number of significant figures as the accompanying numerical coefficient. (For example, $0.85c$ has two significant figures.)

26.1 Classical Relativity
and
26.2 The Michelson–Morley Experiment

1. An object free of all forces exhibits a changing velocity in a certain reference frame. Then (a) the frame is inertial, (b) $\mathbf{F} = m\mathbf{a}$ applies in this frame, (c) the laws of mechanics are the same in this reference frame as in all inertial frames, (d) none of these.

2. The principle of classical relativity (a) did not seem to apply to electricity and magnetism, (b) required a constant speed of light different from that predicted by Maxwell's equations, (c) implied that nothing can travel as fast as light.

3. The existence of the ether (a) would provide a special or absolute reference frame, (b) is necessary for light propagation, (c) made Maxwell's equations relativistically correct, (d) both (a) and (b).

4. The Michelson–Morley experiment (a) showed the expected interference fringes, (b) proved the existence of the ether, (c) gave a null result, (d) was based on the Lorentz–FitzGerald contraction.

5. The Michelson–Morley experiment was based on the (a) interference, (b) refraction, (c) diffraction, (d) polarization property of light.

6. We live on a rotating Earth and therefore are in an accelerating, noninertial system. How then can we apply Newton's laws of motion on the Earth?

7. If the speed of light is a constant (that is, the speeds of light in the two arms of the Michelson interferometer are the same regardless of the motion of the Earth), can you explain the null result of the Michelson–Morley experiment?

8. ■ Car A is traveling eastward at 85 km/h. Car B is traveling with a speed of 65 km/h. Find the relative velocity of car B with respect to car A if car B is traveling (a) westward and (b) eastward.

9. ■ A person 1.20 km away from you fires a gun, and you observe the muzzle flash. A wind of 10.0 m/s is blowing. How long will it take the sound to reach you if the wind is (a) toward you and (b) toward the person who fired the gun? (Take the speed of sound to be 345 m/s.)

10. ■ A small airplane has an air speed (speed with respect to air) of 200 km/h. Find the airplane's ground speed if there is (a) a headwind of 35 km/h and (b) a tailwind of 25 km/h.

11. ■ A speedboat can travel with a speed of 50 m/s in still water. If the boat is in a river that has a flow speed of 5.0 m/s, find the maximum and minimum values of the boat's speed relative to an observer on the riverbank.

12. ■■ A small boat can travel with a speed of 0.50 m/s in still water. The boat heads directly across a river 50 m wide with a flow speed of 0.15 m/s. (a) How long will it take the boat to reach the opposite shore? (b) How far

downstream will the boat travel? (c) How would the boat have to be steered to arrive at a point across the river directly opposite the departure point?

13. ■■ The swimmer analogy of the Michelson–Morley experiment is illustrated in •Fig. 26.20. Two swimmers start out from point A at the same time and swim with a constant speed c relative to the water. One swimmer swims directly across the river to point C and back to A; the other swimmer swims to point B and back to A. Do the swimmers reach A at the same time? Justify your answer mathematically.

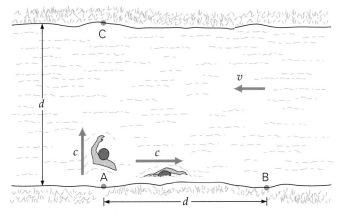

•**FIGURE 26.20 The swimmer analogy** See Exercise 13.

14. ■■■ The apparatus used by the French scientist Armand Fizeau in 1849 for measuring the speed of light is illustrated in •Fig. 26.21. Teeth on a rotating wheel periodically interrupt a beam of light. The light flashes travel to a plane mirror and are reflected back to an observer. Show that if the wheel is rotated so that light passing through one tooth gap reaches the mirror and is reflected to the observer through the adjacent gap, the speed of light is given by $c = 2fNL$, where N is the number of gaps in the wheel; f is the frequency of, or number of revolutions per second made by, the rotating wheel; and L is the distance between the wheel and the mirror.

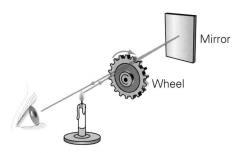

•**FIGURE 26.21 Fizeau's apparatus** See Exercise 14.

15. ■■■ One proposed explanation of the null result of the Michelson–Morley experiment suggested that the interferometer arm in the direction of motion is shortened by a factor of $\sqrt{1 - (v/c)^2}$ (the Lorentz–FitzGerald contraction). (This hypothesis was disproved by making the lengths of the arms of the interferometer unequal.) Show how this contraction would give a null result.

26.3 The Postulates of Special Relativity and the Relativity of Simultaneity

16. The special theory of relativity (a) applies to inertial systems, (b) applies to noninertial systems, (c) applies to both inertial and noninertial systems, (d) disproves the constancy of the speed of light.

17. Events that are simultaneous in one inertial reference frame are (a) always simultaneous in other inertial reference frames, (b) never simultaneous in other inertial reference frames, (c) sometimes simultaneous in other inertial reference frames, (d) none of these.

18. In a space war, a warrior on a spaceship heads directly toward a laser; he is traveling at a speed of $0.85c$. At what speed does the warrior see the light from the laser approach him?

19. In a gedanken experiment, you mark the ends of your spaceship simultaneously (according to you) while also making marks on your friend's ship. He finds himself, coincidentally, in the exact middle of the marks in his reference frame. If he is traveling parallel to the length of your ship (from rear to front) at a speed of $0.5c$, he will observe (a) the marks made simultaneously, (b) the front mark before the rear mark, (c) the rear mark before the front mark.

20. In the gedanken experiment in Exercise 19, for your friend to agree that the marks are made simultaneously, you must (a) make the marks simultaneously in your ship, (b) make the front mark before the rear mark, (c) make the rear mark before the front mark.

21. In the gedanken experiment shown in •Fig. 26.22, two events in the same inertial reference frame (O) are related by cause and effect: (1) A gun at the origin fires a bullet along the x axis with a speed of 300 m/s. (Assume that there are no gravitational or frictional forces.) (2) The bullet hits a target at $x = +300$ m. (a) What is the velocity of the inertial reference frame that would determine the proper time between these two events? (b) Use qualitative arguments to show that these two events could not be viewed simultaneously by any inertial observer. (Note: This shows that special relativity preserves the time *sequence* of two events if they are related as cause and effect. To all observers, the gun fires before the bullet hits the target.)

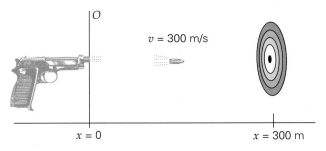

•**FIGURE 26.22 A thought experiment** See Exercise 21.

22. In a gedanken experiment (•Fig. 26.23), two events that cannot be related by cause and effect occur in the same inertial reference frame (O): (1) Strobe light A, at the origin of the x axis, flashes. (2) Strobe light B, located at $x = +600$ m, flashes 1.0 μs later. B's flash cannot be caused by A's flash (light travels only 300 m in the time between the two events). (a) Show that there exists another inertial reference frame, traveling at less than c, in which these two events would be observed to occur simultaneously. (b) What is the direction of the velocity of the reference frame in (a)? Relativity does *not* preserve the time sequence of two events if they are not related as cause and effect. (Note: Since one event doesn't cause the other, no physical principles are violated if observers see them in reversed order.)

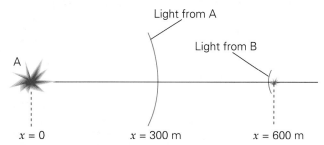

•**FIGURE 26.23 Another thought experiment** See Exercise 22.

26.4 The Relativity of Length and Time: Length Contraction and Time Dilation

23. An observer sees a friend passing by her in a rocketship that has a uniform velocity and a magnitude of an appreciable fraction of the speed of light. The observer knows her friend to be 5 ft 8 in. tall; he is standing such that his length is perpendicular to their relative velocity. To the observer, he will appear (a) taller than 5 ft 8 in., (b) shorter than 5 ft 8 in., (c) exactly 5 ft 8 in. tall.

24. A farm boy (who knows some physics) wants to store a pole 5 m long in a storage shed 4 m long. He claims that if he runs sufficiently fast, the pole will fit in the shed as a result of length contraction. Is this possible from his viewpoint? Explain.

25. On returning from a round trip through space at high speeds (close to c), would you find yourself younger than if you had stayed home?

26. ■ A student observes 10 min to elapse on her clock. How much time does she observe to elapse during this interval on a clock in the window of her friend's spacecraft, moving by the student with a relative speed of 0.90c?

27. ■ You have a pulse rate of 80 beats/min. What would your pulse rate be according to your physics professor, who is moving with a speed of 0.85c relative to you?

28. ■ You fly your 15.0-meter-long spaceship, parallel to the ship's length, with a speed of c/3 relative to your friend. How long is your spaceship, as observed by your friend?

29. ■ What length will an observer measure a field 100 m long to be when she is moving by it, parallel to the field's length, with a speed of 0.75c? Which length is the proper length?

30. ■■ The proper lifetime of a muon is 2.20 μs. If the muon has a lifetime of 34.8 μs according to an observer on Earth, what is the muon's speed, as a fraction of c, relative to the observer?

31. ■■ One 25-year-old twin takes a round trip through space while the other twin remains on Earth. The traveling twin moves at a speed of 0.95c for 39 y, according to Earth time. Assuming that special relativity applies, what are their ages when the traveling twin returns to Earth?

32. ■■ Alpha Centauri, the binary star closest to our solar system, is about 4.3 ly away. If a spaceship traveled this distance with a constant speed of 0.60c relative to Earth, how much time would elapse (a) on a clock on the spaceship? (b) on an Earth-based clock?

33. ■■ A cylindrical spaceship of length 35.0 m and diameter 8.25 m, traveling in the direction of its cylindrical axis, passes by the Earth with a relative speed of 2.44×10^8 m/s. What are the dimensions of the ship, as measured by an Earth observer?

34. ■■ A "flying wedge" spaceship is a right triangle in side view. At rest, the base and altitude (which form the 90° angle) measure 40.0 m and 15.0 m, respectively. As the ship moves with a speed of 0.900c past an observer on Earth, what does she (a) measure the area of the side of the ship to be? (b) calculate the angle of the triangular "nose" to be?

35. ■■ How fast must a meterstick be moving, parallel to its length, relative to an observer so that he measures its length to be 50 cm?

36. ■■ The distance to Planet X is 1.00 ly. How long does it take a spaceship to reach X, according to the pilot of the space-ship, if the speed of the ship is 0.700c relative to X?

37. ■■ A pole vaulter at the Relativistic Olympics sprints past you to do a vault with a speed of 0.65c. When he is at rest, his pole is 7.0 m long. What length do you perceive the pole to be as he passes you? (What assumption do you need to make to arrive at your answer?)

38. ■■ What is the length contraction (ΔL) of an automobile 5.00 m long when it is traveling at 100 km/h? [Hint: For $x \ll 1$, $\sqrt{1 - x^2} \approx 1 - (x^2/2)$.]

39. ■■ An observer in a certain reference frame cannot understand why there is a problem in converting to the metric system because she notes the length of a meterstick in another moving reference frame to be the same length as her yardstick (parallel to the meterstick). Explain this situation quantitatively.

40. ■■■ Sirius is about 9.0 ly from Earth. To reach the star by spaceship in 12 y (ship time), how fast must you travel?

26.5 Relativistic Kinetic Energy, Momentum, Total Energy, and Mass–Energy Equivalence

41. The special theory of relativity shows relativistic effect(s) for the fundamental properties of (a) length, (b) mass, (c) time, (d) all of these.

42. The total energy of a relativistic moving particle is (a) $E = mv^2$, (b) $E = \gamma E_o$, (c) $E = K + \gamma mc^2$, (d) $E = mc^2$.

43. The special theory of relativity places an upper limit on the speed an object can have. Are there similar limits on energy and momentum?

44. Is it physically possible to accelerate a mass to the speed of light if a continuous force is acting on the mass?

45. ■ If an electron has a kinetic energy of 2 keV, could the classical expression for kinetic energy be used to compute its speed accurately? What if the kinetic energy of the electron were 2 MeV?

46. ■ An electron travels at a speed of $0.600c$. What is its total energy?

47. ■ An electron is accelerated from rest through a potential difference of 2.50 MV. Find the electron's (a) speed, (b) kinetic energy, and (c) momentum.

48. ■ How fast must an object travel for its total energy to be (a) 1% more than its rest energy, and (b) 99% more than its rest energy?

49. ■ A home uses about 1.5×10^4 kWh of electricity per year. How much matter would have to be converted to energy (assuming 100% efficiency) to supply energy to a city with 250 000 homes? (Are you surprised by the answer?)

50. ■ The United States uses 3.0 trillion kWh of electricity annually. If this electrical energy were supplied by nuclear generating plants, how much nuclear mass would have to be converted to energy, assuming a production efficiency of 25%?

51. ■ To travel to a nearby star, a spaceship is accelerated to $0.99c$ to take advantage of time dilation. If the ship has a mass of 3.0×10^6 kg, how much work must be done to get it up to speed from rest? Compare this value to the annual electricity usage of the United States (see Exercise 50).

52. ■■ An electron has a total energy of 2.8 MeV. What is its momentum?

53. ■■ How much energy, in kiloelectron volts, is required to accelerate an electron from rest to $0.50c$?

54. ■■ The kinetic energy of an electron is 60% of its total energy. Find the (a) speed and (b) momentum of the electron.

55. ■■ Sketch graphs of the (a) momentum and (b) energy of an object as a function of speed. (Use relativistically correct expressions.)

56. ■■ A proton moves with a speed of $0.35c$. What are its (a) total energy (b) kinetic energy, and (c) relativistic momentum?

57. ■■ A proton moving with a constant velocity has a total energy 2.5 times its rest energy. What are (a) the proton's speed and (b) kinetic energy?

58. ■■ The Sun has a mass of 1.989×10^{30} kg and radiates at a rate of 3.827×10^{23} kW. Assuming it loses mass at a constant rate, (a) how much mass does the Sun lose in 1 h? (b) How long will it take the Sun to lose 1% of its mass?

59. ■■ A beam of electrons is accelerated from rest to a speed of $0.950c$ in a particle accelerator. In megaelectron volts, what are the (a) kinetic energy and (b) total energy of an electron in the beam?

60. ■■ Phase changes require latent heat (Chapter 11). If 1 kg of ice at 0°C is converted to liquid at 0°C, how much more mass would the liquid have as a result of energy absorbed?

61. ■■ A nickel has a mass of 5.00 g. If this mass could be completely converted to electric energy, how long would it keep a 100-watt light bulb lit?

62. ■■■ Using the relationship $E = \gamma mc^2$, show that the total energy E and relativistic momentum magnitude p are related by $E^2 = p^2c^2 + (mc^2)^2$.

63. ■■■ In a proton linear accelerator, protons are accelerated to a kinetic energy of 600 MeV. (a) How do we know that the 600 MeV must be kinetic energy and not total energy? (b) What is the speed of these protons? (c) What is their momentum?

26.6 The General Theory of Relativity

64. The general theory of relativity (a) provides a theoretical basis for explaining a gravitational force, (b) applies to rotating systems, (c) applies only to inertial systems, (d) refutes the principle of equivalence.

65. One of the predictions of the general theory is (a) mass–energy equivalence, (b) the constancy of the speed of light, (c) the twin paradox, (d) the bending of light in a gravitational field.

66. The Schwarzschild radius is the radius of (a) a black hole, (b) a star, (c) an event horizon, (d) none of the above.

67. Suppose a meterstick were dropped toward the event horizon of a black hole. Describe what the effects might be on the meterstick.

68. ■ Compare the radii of Earth's and Jupiter's event horizons if these two planets were black holes. ($M_E = 6.0 \times 10^{24}$ kg, $M_J = 318M_E$)

69. ■■ If the Sun gravitationally collapsed and became a black hole, what would be its average density at that moment? (See Example 26.8.)

70. ■■ A black hole has an event horizon of 5.00×10^3 m. (a) What is the mass of the "hole"? (b) Set a lower limit on the density of the black hole.

71. ■■ An apparatus like that in •Fig. 26.24 was given to Albert Einstein on his 76th birthday by Eric M. Rogers, a physics professor at Princeton University. The goal is to get the ball into the cup without touching the ball. (Jiggling the pole up and down will not do it.) Einstein solved the puzzle immediately and then confirmed his answer with an experiment. How did Einstein get the ball into the cup? [Hint: He used a fundamental concept of general relativity.]†

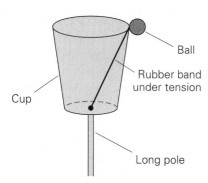

•**FIGURE 26.24 How to get the ball in the cup** See Exercise 71.

*26.7 Relativistic Velocity Addition

72. ■ After jettisoning a stage, a rocket has a velocity of $+0.20c$ relative to the jettisoned stage. An observer on Earth sees the jettisoned stage moving with a velocity of $+7.5 \times 10^7$ m/s, relative to her, in the same direction as the rocket. What is the velocity of the rocket relative to the Earth observer?

73. ■ In moving away from Planet Z, a spacecraft fires a probe with speed $0.15c$, relative to the spacecraft, back toward Z. If the speed of the spacecraft is $0.40c$ relative to Z, what is the velocity of the probe as seen by an observer on Z?

74. ■■ A rocket launched outward from Earth has a speed of $0.100c$ relative to the Earth. The rocket is directed toward an incoming meteor that may hit the planet. If the meteor moves with a speed of $0.250c$ relative to the rocket and directly toward it, what is the velocity of the meteor as observed from Earth?

75. ■■■ In a colliding beam apparatus, two beams of protons are aimed at each other. The first proton moves with a speed of $0.800c$ to the right; the second moves with a speed of $0.900c$ to the left. Both speeds are measured relative to the laboratory frame of reference. What are (a) the velocity

†This problem is adapted from R. T. Weidner, *Physics*, Allyn and Bacon, 1985, p. 333.

of the second proton relative to that of the first, and (b) the velocity of the first proton relative to that of the second?

Additional Exercises

76. What is the rest energy of a proton (in megaelecton volts)?

77. A spaceship travels with a speed of $0.60c$ relative to an observer in an inertial frame. How much time does a clock on board the spaceship appear to lose in 1 day, according to that observer?

78. An electron travels at a speed of 9.5×10^7 m/s. What is its total energy?

79. A relativistic rocket is measured to be 50 m long, 2.5 m high, and 2.0 m wide. To an inertial observer, what are its dimensions when the rocket travels at $0.65c$ in the direction of its length?

80. If the mass of 1.0 kg of coal (or any substance) could be completely converted into energy, how many kilowatt-hours of energy would be produced?

81. Imagine that you are moving with a speed of $0.80c$ past a person who is reading this chapter in his textbook. If he takes 30 min to read the chapter by his clock, how much time would you observe to elapse on your clock?

82. In electron–positron annihilation, an electron and a positron (with the same mass as the electron but carrying a positive charge) collide, and both masses are converted to electromagnetic radiation. Assuming both particles are at rest when they annihilate, what is the total energy of the radiation?

83. A spaceship has a length of 150 m in its rest frame. An observer in another, inertial frame measures the length of the spaceship as 110 m, with a relative velocity parallel to that length. What is the relative speed of the reference frames?

84. In the Michelson–Morley experiment, a difference in phase for the beams could be detected if the apparatus were rotated by 90°, since the resulting interference pattern would change. Also, this would take care of the problem of assuming that the interferometer arm lengths are exactly equal, which should not be done arbitrarily. Show that the difference in the time differences of beam travel for the nonrotated and rotated conditions is

$$\Delta t - \Delta t' = \frac{2}{c}(L_1 + L_2)\left[\frac{1}{1 - (v^2/c^2)} - \frac{1}{\sqrt{1 - (v^2/c^2)}}\right]$$

where $\Delta t'$ is the time difference of the rotated condition and L_1 and L_2 are the lengths of the interferometer arms.

85. At typical nuclear power plants, refueling occurs every 18 months. Assuming that a plant has operated continuously since the last refueling and produces 1.2 GW of electrical power at an efficiency of 33%, how much less massive are the fuel rods at the end of the 18 months than at the start? (Assume 30-day months.)

Quantum Physics

Lasers are used in a variety of everyday contexts—in bar-code scanners at store checkout counters, in computer printers, in surgery, and in laser light shows, such as this one in Paris. You own a laser yourself if you have a CD player.

The laser is a practical application of principles that revolutionized physics when they were first developed in the period from about 1900 to 1930, one of the most productive in the history of physics. Einstein's theory of special relativity (Chapter 26) helped resolve the problems that classical physics had in describing particles moving at speeds comparable to that of light. However, there were other troublesome areas in which classical theory did not agree with experimental results. Scientists devised new hypotheses based on nonclassical approaches, ushering in a profound revolution in our understanding of the physical world. Chief among these new theories was the idea that light is *quantized*—made up of discrete amounts of energy. This hypothesis led to the formulation of a new set of principles and a new branch of physics, *quantum mechanics.*

Quantum theory demonstrated that particles often exhibit wave properties and waves frequently behave like particles. It also showed that in the realm of atomic (and smaller) dimensions, explanations have to deal in *probabilities* rather than in the precisely determined values of classical theory.

A detailed treatment of quantum mechanics is quite complex mathematically. However, a general overview of the important results is essential to an understanding of physics as we know it today. Thus, the important developments of "quantum" physics are presented in this chapter, and an introduction to quantum mechanics is provided in Chapter 28.

27.1 Quantization: Planck's Hypothesis

Objectives: To (a) define blackbody radiation and use Wien's law and (b) understand how Planck's hypothesis paved the way for quantum ideas.

One of the problems scientists were having at the end of the nineteenth century was how to explain the spectra of electromagnetic radiation (light) emitted by hot objects. This is sometimes called **thermal radiation**. You learned in Chapter 11 that the total intensity of the emitted radiation is proportional to the fourth power of the absolute Kelvin temperature (T^4). Since the temperatures of all objects are above absolute zero, all objects emit thermal radiation.

At everyday temperatures, this radiation is in the infrared region and so is not visible. At temperatures of about 1000 K, an object emits radiation in the long-wavelength end of the visible spectrum and has a reddish glow. A hot electric stove burner is a good example. Increased temperature causes the radiation to shift to shorter wavelengths and the color to change to yellow-orange. Above a temperature of about 2000 K, an object glows yellowish-white, like the glowing filament of a light bulb, and gives off all the visible colors.

Although we see only a particular (dominant) color, there is a continuous spectrum (all wavelengths) at all temperatures, as illustrated in •Fig. 27.1a. The curves shown do not correspond to any real radiating body; they are for an idealized blackbody. A **blackbody** is an ideal system that absorbs (and emits) all radiation that is incident on it.

An absorbing blackbody can be approximated by a small hole leading to an internal cavity inside a block of material (Fig. 27.1b). Radiation falling on the hole enters the cavity and is reflected back and forth by the cavity walls. If the hole is very small in comparison to the surface area of the cavity, only a small amount of radiation will make its way out of the hole. Since nearly all radiation incident

Note: Review Eq. 11.5.

Blackbody: an ideal absorber and emitter

Note: The concept of a blackbody was introduced in Section 11.4.

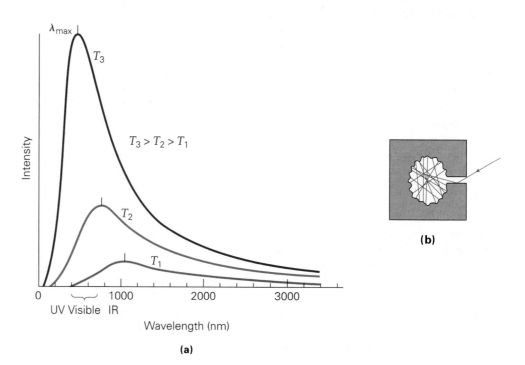

•FIGURE 27.1 **Thermal radiation** **(a)** Intensity versus wavelength curves for the thermal radiation from an idealized blackbody at different temperatures. Notice how the location of maximum intensity shifts to shorter wavelengths with increasing temperature. **(b)** A blackbody can be approximated by a small hole leading to an interior cavity in a block of material.

on the hole is absorbed, the hole, viewed from the outside, is an excellent approximation of the surface of a blackbody.

Two things happen as the temperature increases in Fig. 27.1a: More radiation is emitted at every wavelength, and the wavelength of the maximum-intensity component (λ_{max}) becomes shorter. This wavelength shift obeys a relationship named **Wien's displacement law:**

$$\lambda_{max}T = 2.90 \times 10^{-3} \text{ m·K} \qquad (27.1)$$

Wien's displacement law—change of λ_{max} with temperature

where λ_{max} is the wavelength of the radiation (in meters) at which maximum intensity occurs and T is the temperature of the body (in kelvins).

Wien's law can be used to determine the wavelength of the maximum spectral component if the temperature is known, or the temperature of the emitter if the wavelength of the strongest emission is known. Thus, it can be used to estimate the temperatures of stars from their radiation spectrum, as the following Example shows.

Example 27.1 ■ Solar Colors: Using Wien's Law

The visible surface of the Sun is the gaseous photosphere from which radiation escapes. At the top of the photosphere, the temperature is 4500 K; at a depth of about 260 km, the temperature is 6800 K. If the Sun behaves as a blackbody, what are the wavelengths of the radiation of maximum intensity for these temperatures?

Thinking It Through. Wien's displacement law (Eq. 27.1) is directly applicable for determining the wavelengths.

Solution.

Given: $T_1 = 4500$ K   *Find:* λ_{max} (for different temperatures)
$T_2 = 6800$ K

We have

$$\text{At top: } \lambda_{max} = \frac{2.90 \times 10^{-3} \text{ m·K}}{4500 \text{ K}} = 6.44 \times 10^{-7} \text{ m} = 644 \text{ nm}$$

$$\text{At depth of 260 km: } \lambda_{max} = \frac{2.90 \times 10^{-3} \text{ m·K}}{6800 \text{ K}} = 4.26 \times 10^{-7} \text{ m} = 426 \text{ nm}$$

Hence, at the Sun's surface, the emitted radiation of maximum intensity is in the orange-red region of the visible spectrum (Section 20.4). As the temperature increases with photosphere depth, the wavelength of the radiation of maximum intensity shifts toward the blue end of the spectrum; at 260 km, λ_{max} is near the violet end. Some of the emitted radiation will be in the ultraviolet region, most of which is ordinarily absorbed by the ozone layer in the Earth's atmosphere (see the Insight on p. 656).

Note: Review Fig. 20.18.

Follow-up Exercise. What would you expect to be the dominant color from the following stars? (a) Betelgeuse, average surface temperature = 3.00×10^3 K; (b) Rigel, average surface temperature = 1.00×10^4 K

The Ultraviolet Catastrophe and the Planck Hypothesis

Classically, thermal radiation results from the thermal agitation and acceleration of electric charges near the surface of an object. Electric charges in thermal agitation would undergo many different accelerations, so a continuous spectrum of emitted radiation would be expected. Classical theoretical calculations to predict the radiation emitted by a blackbody result in an intensity *inversely* related to the emitted wavelength.

At long wavelengths, this classical prediction agrees fairly well with experimental data. However, at shorter and shorter wavelengths, the agreement worsens. Contrary to experimental observations (including Wien's law), classical theory predicts that the radiation intensity should increase as the wavelength goes to zero (•Fig. 27.2). This classical prediction is sometimes called the *ultraviolet catastrophe*—"ultraviolet" because the difficulty occurs for short wavelengths beyond the violet end of the visible spectrum and "catastrophe" because it predicts that the emitted intensity or energy will become infinitely large.

The failure of classical electromagnetic theory to explain the characteristics of thermal radiation led Max Planck (1858–1947), a German physicist, to reexamine the phenomenon. In 1900, Planck formulated a theory that correctly predicted the observed distribution of the blackbody radiation spectrum, but only by introducing a radical new idea. Planck showed that if it were assumed that the thermal oscillators (the atoms emitting the radiation) have only *discrete*, or particular, amounts of energy rather than a continuous distribution of energies, or any arbitrary energy, then theory would agree with experiment.

These discrete amounts of energy of thermal oscillators were related to the frequency *f* of oscillation by

$$E_n = n(hf) \quad \text{for } n = 1, 2, 3, \dots \tag{27.2}$$

That is, the energy occurs only in integral multiples of hf. The symbol h represents a constant known as **Planck's constant** and has a value (to three significant figures) of

$$h = 6.63 \times 10^{-34} \, \text{J·s}$$

Note: As with other fundamental constants, for calculations we will take *h* to be *exact* at 6.63×10^{-34} J·s.

Planck's hypothesis—the birth of quantum physics

Equation 27.2 is called **Planck's hypothesis.** Rather than considering oscillator energy to be a continuous quantity as it is classically, according to Planck's hypothesis the energy was *quantized*, or occurred only in discrete amounts. The smallest possible amount of oscillator energy, according to Eq. 27.2 with $n = 1$, is

$$E_1 = hf \tag{27.3}$$

All other permitted values of the energy are integral multiples of hf. The quantity hf is called a **quantum** of energy (from the Latin word *quantus*, meaning "how much"). Thus, Planck's hypothesis implies that the energy of each oscillator can change only by the absorption or emission of energy *in quantum amounts*.

Although the theoretical predictions agreed with experiment, Planck himself was not convinced of the validity of his quantum hypothesis. However, the concept of quantization was then extended to explain other phenomena that could not be explained classically. Despite Planck's hesitation, the quantum hypothesis earned him a Nobel Prize in 1918.

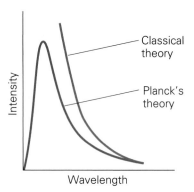

•**FIGURE 27.2 The ultraviolet catastrophe** Classical theory predicts the intensity of thermal radiation is inversely related to the emitted wavelength. Thus the intensity of the radiation should become infinitely large as the wavelength approaches zero. In contrast, Planck's quantum theory predicts the observed radiation distribution.

27.2 Quanta of Light: Photons and the Photoelectric Effect

Objectives: To (a) describe the photoelectric effect, (b) explain how it can be understood by assuming that light energy is carried by particles, and (c) summarize the properties of photons.

The concept of the quantization of light was introduced in 1905 by Albert Einstein in a paper about light absorption and emission, at the same time he published his famous paper on special relativity! In contrast to Planck's hypothesis, which was made from a practical need to have a "theory" that fit the blackbody data, Einstein reasoned more fundamentally that energy quantization is a fundamental property of electromagnetic waves. He suggested that if the energy of the thermal oscillators in a hot substance is quantized, then it necessarily followed that, to conserve energy, *the emitted radiation should also be quantized.* He therefore boldly proposed that

> …the radiant energy from a point source is not distributed continuously throughout an increasingly larger region, but, instead, this energy consists of a finite number of spatially localized energy quanta which, moving without subdividing, can only be absorbed and created in whole units.

A quantum or packet of light is referred to as a **photon.** Each photon has a definite amount of energy; that amount depends on the frequency f of the light:

$$E = hf \qquad (27.4)$$

Energy of a photon of light with frequency f

This idea suggests that light can behave as discrete quanta (plural of quantum), or "particles" of energy, rather than as waves. Equation 27.4 is thus the mathematical "connection" between the *wave nature* of light (a wave of frequency f) and the *particle nature* (photons each with an energy E). Given light of a certain frequency or wavelength, we can use this equation to calculate the amount of energy in each photon, or vice versa.

Einstein used this quantum concept to explain the **photoelectric effect,** another phenomenon for which the classical physical description was inadequate. Certain metallic materials are *photosensitive.* That is, when light strikes their surface, electrons can be emitted. The radiant energy supplies the work necessary to free the electrons from the material's surface. Such materials are used to make photocells, as illustrated in •Fig. 27.3a. A voltage is maintained between the anode and the cathode. When light strikes the cathode, which is photosensitive, electrons are emitted. These emitted *photoelectrons* complete the circuit, and a current is registered on the ammeter.

When a photocell is illuminated with monochromatic (single-frequency) light of different intensities, characteristic curves are obtained as shown in Fig. 27.3b. For positive voltages, the anode is positive and attracts the electrons. Under these conditions $(+V)$, the *photocurrent* (the flow of electrons—current—released by photons of light) essentially does not vary with applied voltage. This is because *all* of the emitted electrons reach the anode, and any increase in voltage has no effect on the current.

As would be expected classically, the current is proportional to the intensity of the incident light—the greater the intensity $(I_2 > I_1$ in Fig. 27.3b), the more energy there is to free additional electrons.

A measure of the kinetic energy of the electrons being emitted from the photoelectric material can be gained by reversing the voltage across the electrodes. As the voltage is decreased and then made negative (so the electrons are now repelled from the anode), the photocurrent decreases. Because of the retarding voltage, only the electrons with kinetic energy greater than $e|V|$ can make it to the negative plate and contribute to the current. (The kinetic energy is converted into electric potential energy as the electrons approach the now negatively charged plate.)

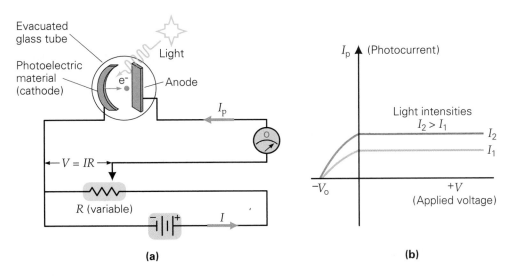

(a)   (b)

•**FIGURE 27.3   The photoelectric effect and characteristic curves** (a) Incident light on the photoelectric material in a photocell (or phototube) causes the emission of electrons, and a current flows in the circuit. The voltage applied to the tube can be changed by means of the variable resistor. (b) As the plots of current versus voltage for two intensities of monochromatic light show, the current is saturated, or constant, as the voltage is increased. However, for negative voltages (by a reversal of the battery polarity), the current goes to zero at a particular stopping potential $-V_o$, which is independent of intensity.

At some voltage $-V_o$, called the **stopping potential**, no current will flow between the electrodes. The kinetic energy of even the fastest electron is completely converted into electric potential energy at this voltage. Thus the maximum kinetic energy (K_{max}) of the photoelectrons—electrons emitted from the photomaterial—is related to the stopping potential by

Note: Recall from Eq. 16.2 that $\Delta U_e = qV$.

$$K_{max} = \Delta U_e$$

or

$$K_{max} = eV_o \qquad (27.5)$$

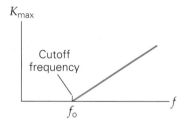

•**FIGURE 27.4   Maximum kinetic energy and frequency**
The maximum kinetic energy (K_{max}) of the photoelectrons is a linear function of the frequency of the incident light. Below a certain cutoff frequency f_o, no photoemission occurs regardless of light intensity.

When the frequency of the incident light is varied, the maximum kinetic energy of the electrons depends linearly on the frequency (•Fig. 27.4). No *photoemission* (the emission of electrons from a material after it has absorbed light energy) is observed to occur below a certain *cutoff frequency f_o*. Moreover, the electron emission begins instantaneously, with no observable time delay, when a photomaterial is illuminated by light with $f > f_o$, even if the light intensity is very low.

These four key characteristics of photoemission are summarized in Table 27.1 along with their correspondence with classical theoretical predictions. Only the first observation is predicted correctly by classical wave theory. In this theory, the electric field of a light wave interacts with the electrons at the surface of the material and sets them into oscillation. The intensity (energy) of an oscillating electric field is proportional to the amplitude squared of the field vector, so the average kinetic energy of the emitted electrons is proportional to the intensity of the light. Therefore, it is difficult to understand why the *maximum* kinetic energy of the photoelectrons is independent of the intensity, as is observed. Also, according to wave theory, photoemission should occur for *any* frequency of light, not just above a certain frequency.

The immediate emission of electrons raises an even more serious problem for wave theory. According to the wave model, the time required for an electron to acquire enough energy to be freed from a material can be on the order of *minutes* for extremely low light intensities.

An explanation of the photoelectric effect was put forth in 1905 by Einstein, using the quantum idea suggested by Planck in his theory of thermal oscillators. As was stated previously, Einstein proposed that the energy of light occurs in quantum bundles, or photons, each with energy $E = hf$. Thus, a light *wave* of frequency f could transfer energy to the electrons in a photomaterial only in discrete amounts equal to hf. In other words, electromagnetic energy (light) must be treated

TABLE 27.1 Photoelectric Effect Characteristics	
Characteristic	*Predicted by wave theory?*
1. The photocurrent is proportional to the intensity of the light.	Yes
2. The maximum kinetic energy of the emitted electrons is dependent on the frequency of the light but not on its intensity.	No
3. No photoemission occurs for light with a frequency below a certain cutoff frequency f_o regardless of its intensity.	No
4. A photocurrent is observed immediately when the light frequency is greater than f_o even if the light intensity is extremely low.	No

not as a wave, but as a particle in this case! An electron thus absorbs either the whole photon energy or none at all.

The electrons of a photomaterial are bound to that material by attractive forces. Therefore, work must be done to free an electron, and some of the absorbed photon energy goes into doing just that. Thus, when a photon of energy hf is absorbed, by the conservation of energy we have

$$E = K + \phi$$

or

$$(27.6)$$

$$hf = K + \phi$$

where ϕ is the work or energy needed to free an electron from the material.

Part of the energy of the absorbed photon goes into the work of freeing the electron, and the rest is carried off by the emitted electron as kinetic energy.

In the photoelectric effect, the common unit of energy for E, K, and ϕ is the electron volt (eV), introduced in Chapter 16. Recall that, to three significant figures, $1.00 \text{ eV} = 1.60 \times 10^{-19} \text{ J}$.

The least tightly bound electron will have the maximum kinetic energy (K_{max}) (why?). The energy needed to free the least bound electron is called the **work function ϕ_0** of the material. The energy conservation equation is then

Work function: the *minimum* energy needed to dislodge an electron

$$\underset{\substack{incident \\ photon \\ energy}}{hf} = \underset{\substack{maximum \\ kinetic\ energy \\ of\ dislodged \\ electron}}{K_{max}} + \underset{\substack{minimum\ work \\ needed\ to \\ dislodge\ electron}}{\phi_0} \qquad (27.7)$$

Other electrons require more energy than the minimum, thus their kinetic energy will be less than K_{max}. Some typical numerical values are shown in the next Example.

Example 27.2 ■ The Photoelectric Effect: Electron Speed and Stopping Potential

The work function of a particular metal is 2.00 eV. (a) If the metal is illuminated with monochromatic light of wavelength 550 nm, what will be the maximum kinetic energy of the emitted electrons and (b) their maximum speed? (c) What is the required stopping potential?

Thinking It Through. (a) By energy conservation (Eq. 27.7), the maximum kinetic energy is the difference between the incoming photon energy and the work function. (b) We can get the speed from the kinetic energy, since we know the mass of an electron ($m = 9.11 \times 10^{-31}$ kg). (c) The stopping potential is determined by requiring that all of the kinetic energy be converted to electric potential energy.

Solution. We first convert the data into SI units.

Given: $\phi_0 = (2.00 \text{ eV})(1.60 \times 10^{-19} \text{ J/eV})$ *Find:* (a) K_{max} (maximum kinetic energy)
$\qquad = 3.20 \times 10^{-19} \text{ J}$ (b) v_{max} (maximum speed)
$\qquad \lambda = 550 \text{ nm} = 5.50 \times 10^{-7} \text{ m}$ (c) V_0 (stopping potential)

(a) The photon energy of light with the given wavelength is, from $\lambda f = c$,

$$E = hf = \frac{hc}{\lambda} = \frac{(6.63 \times 10^{-34} \text{ J·s})(3.00 \times 10^8 \text{ m/s})}{5.50 \times 10^{-7} \text{ m}} = 3.62 \times 10^{-19} \text{ J}$$

Then

$$K_{max} = E - \phi_0 = 3.62 \times 10^{-19} \text{ J} - 3.20 \times 10^{-19} \text{ J}$$

$$= 0.42 \times 10^{-19} \text{ J} = 0.26 \text{ eV}$$

(b) We solve for v_{max} from $K_{max} = \frac{1}{2}mv_{max}^2$:

$$v_{max} = \sqrt{\frac{2K_{max}}{m}} = \sqrt{\frac{2(0.42 \times 10^{-19} \text{ J})}{9.11 \times 10^{-31} \text{ kg}}} = 3.04 \times 10^5 \text{ m/s}$$

(c)   The stopping potential is related to K_{max} by $K_{max} = eV_o$:

$$V_o = \frac{K_{max}}{e} = \frac{0.42 \times 10^{-19}\,\text{J}}{1.60 \times 10^{-19}\,\text{C}} = 0.26\,\text{V}$$

Follow-up Exercise.   We use light of a different wavelength in this Example and get a stopping voltage of 0.50 V. What is the wavelength of this new light? Explain why this wavelength requires a larger stopping voltage.

Einstein's quantum theory was consistent with *all* the experimental characteristics of the photoelectric effect. Increasing the light intensity increases the number of photons and thus also the number of electrons dislodged (the photocurrent). An increase in intensity, however, does *not* change the energy of individual photons, which depends solely on the frequency of the light: $E = hf$. Therefore K_{max} is predicted to be independent of intensity but linearly dependent on the frequency of the incident light—both results have been observed experimentally.

Einstein's theory also explained the existence of a limiting frequency. In the Einstein interpretation, the photon energy depends on frequency. Below a certain (cutoff) frequency, the photons do not have enough energy to dislodge electrons, so no current is observed. When $K_{max} = 0$, we can find the cutoff frequency f_o from Eq. 27.7:

$$hf_o = K_{max} + \phi_o = 0 + \phi_o$$

or

Threshold frequency, or cutoff frequency

$$f_o = \frac{\phi_o}{h} \qquad (27.8)$$

In this case, the photon has just enough energy to free an electron from the material but no extra energy to give it kinetic energy. (Notice that the graph in Fig. 27.4 provides a means to determine the value of h, which is the slope of the line.)

Note: A photon of energy hf_o will barely dislodge an electron, but the electron will have essentially no kinetic energy.

The frequency f_o is sometimes called the **threshold frequency.** Only light with a frequency above the threshold f_o will dislodge electrons. If the incident light has a frequency less than f_o, no matter how many photons are available (that is, no matter how intense the light), the photons will not have enough energy to cause photoemission. No time delay in the emission of electrons would be expected in the quantum theory, since the energy is absorbed by an electron in a concentrated bundle instead of being continuously supplied and distributed over an area as in wave theory. A typical threshold frequency is calculated in the next Example.

Example 27.3 ■ The Photoelectric Effect: Threshold Frequency and Wavelength

What are the threshold frequency and corresponding wavelength for the metal described in Example 27.2?

Thinking It Through.   Example 27.2 gives the work function, so Eq. 27.8 allows us to determine the threshold frequency. The threshold wavelength is then computed from $\lambda = c/f$.

Solution.   Listing the data, we have

Given: $\phi_o = 2.00\,\text{eV}$ *Find:* f_o (threshold frequency)
         $= 3.20 \times 10^{-19}\,\text{J}$ λ_o (wavelength at threshold frequency)
        (from Example 27.2)

Solving for f_o from $\phi_o = hf_o$ (Eq. 27.8), we get

$$f_o = \frac{\phi_o}{h} = \frac{3.20 \times 10^{-19}\,\text{J}}{6.63 \times 10^{-34}\,\text{J·s}} = 4.83 \times 10^{14}\,\text{Hz}$$

The wavelength corresponding to the threshold frequency is

$$\lambda_\text{o} = \frac{c}{f_\text{o}} = \frac{3.00 \times 10^8 \text{ m/s}}{4.83 \times 10^{14} \text{ Hz}} = 6.21 \times 10^{-7} \text{ m} = 621 \text{ nm}$$

Any frequency lower than 4.83×10^{14} Hz, or any wavelength longer than 621 nm, would not yield photoelectrons. Notice that this wavelength lies in the red end of the electromagnetic spectrum, so yellow light, for example, would dislodge electrons, but deeper red or infrared would not.

Follow-up Exercise. In this Example, what would the stopping voltage be if the frequency of the light were twice the cutoff frequency?

Problem-Solving Hint

In photon calculations, we are often given the wavelength of the light rather than the frequency and need to find the corresponding photon energy. Instead of first calculating the frequency ($f = c/\lambda$), then the energy in joules ($E = hf$), and finally converting to electron volts, we can do all this in one step. We simply use the relationship $E = hf = hc/\lambda$ with hc expressed in electron-volt-nanometers (eV·nm). The value of this useful constant is

$$hc = (6.63 \times 10^{-34} \text{ J·s})(3.00 \times 10^8 \text{ m/s}) = 1.99 \times 10^{-25} \text{ J·m}$$

$$= \frac{(1.99 \times 10^{-25} \text{ J·m})(10^9 \text{ nm/m})}{1.60 \times 10^{-19} \text{ J/eV}}$$

$$= 1.24 \times 10^3 \text{ eV·nm (to 3 significant figures)}$$

This shortcut can save time and effort in working problems and simplifies the estimation of the photon energy associated with light of a given wavelength (or vice versa). Thus, if you are given orange light, for example, of wavelength 600 nm, you need only divide to find that each photon carries approximately 2 eV of energy:

$$E = \frac{hc}{\lambda} = \frac{1.24 \times 10^3 \text{ eV·nm}}{600 \text{ nm}} = 2.07 \text{ eV}$$

There are many applications of the photoelectric effect. The fact that the current produced by photocells is proportional to the light intensity makes them ideal for use in photographers' light meters. Photocells are also used in solar energy applications to convert sunlight to electricity. They are not yet efficient enough (about 20–25%) to produce commercial electricity at a price competitive with non-renewable power sources. But this pollution-free method has great potential as nonrenewable fuels such as oil are depleted.

Another common application of the photocell is in the electric eye (•Fig. 27.5a). As long as light strikes the photocell, current flows in the circuit. Interrupting or blocking the light opens the circuit in the relay (magnetic switch), which controls some device. A common application of the electric eye is to turn on street lights automatically at night. A safety application is shown in Fig. 27.5b. The threshold frequencies of some metals lie outside the visible region. Hence, non-visible infrared light can be used in such applications as burglar alarms and home protection systems.

Einstein's theory of the photoelectric effect represented a great success for the idea of quantization. It is interesting to note that Einstein won the Nobel Prize in physics in 1921 for his theory of the photoelectric effect rather than for his more famous theory of special relativity (Chapter 26). Two years later, the quantum idea scored another triumph, which we will see in Section 27.3.

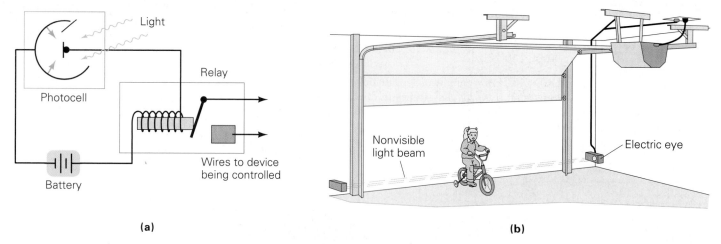

•**FIGURE 27.5 Photoelectric applications** **(a)** A diagram of an electric eye circuit. When light strikes a photocell, there is a current in the circuit. Interrupting the light beam opens the circuit in the relay or magnetic switch, which controls some device. **(b)** An electric eye can be used in an automatic garage door. When the door starts downward, any interruption of the electric eye beam will cause the door to stop, protecting children or pets that pass under the descending door.

27.3 Quantum "Particles": The Compton Effect

Objectives: To (a) understand how the photon model of light explains scattering of light from electrons (the Compton effect) and (b) calculate the wavelength of the scattered light in the Compton effect.

In 1923, the American physicist Arthur H. Compton (1892–1962) explained a phenomenon he observed in the scattering of X-rays from a graphite (carbon) block by considering the radiation to be composed of quanta. His explanation of the observed effect provided additional convincing evidence that, at least in certain types of experiments, light, and electromagnetic radiation in general, is composed of quanta, or "particles" of energy called photons.

Compton had observed that when a beam of monochromatic X-rays was scattered by a material, the scattered radiation had a wavelength slightly longer than the wavelength of the incident beam. Also, the change in the wavelength was dependent on the angle through which the X-rays were scattered but not on the nature of the scattering material (•Fig. 27.6). This phenomenon came to be known as the **Compton effect**.

According to the wave model, the scattered radiation should have the same frequency or wavelength as the incident radiation. The electrons in the atoms of the scattering material absorb the radiation and oscillate at the same frequency as the incident wave. On being emitted by these electrons, the scattered radiation should have the same frequency in all directions.

By Einstein's photon theory, the energy of one quantum is directly proportional to the frequency of the light ($E = hf$), so a change in frequency or a shift in wavelength would indicate a change in photon energy. The wavelength increased, so the scattered photons had *less* energy than the incident ones. The effect of the scattering angle on energy reminded Compton of scattering in the elastic collision of particles. Could the same principles apply in the scattering of quantum "particles"?

Pursuing this idea, Compton assumed that a photon behaves like a particle or billiard ball in collision with other particles. He also considered only metals, where there are lots of essentially free electrons (the loosely bound ones responsible for the high electrical conductivity of metals). Compton reasoned that if an incident photon collides with an electron initially at rest, the photon transfers some energy and momentum to the electron. Hence, the energy and frequency of the scattered photon should be decreased ($E = hf$) and its wavelength increased ($\lambda = c/f$). If we assume the collision is elastic, energy and momentum must be

Note: Elastic collisions are discussed in Section 6.4.

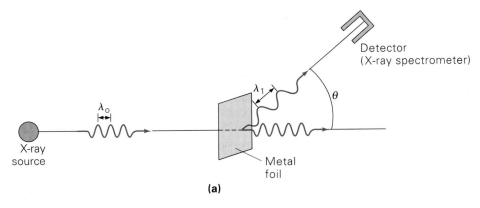

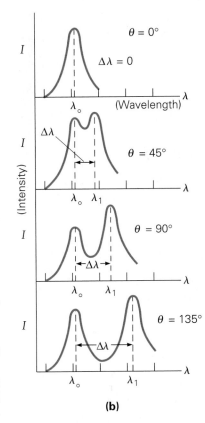

(a)

•FIGURE 27.6 X-ray scattering (a) When X-rays of a single wavelength are scattered by the almost-free electrons in metal foil, that incident wavelength is increased in the scattered X-rays. (b) The wavelength shift grows as the scattering angle increases.

(b)

conserved. Applying these principles, Compton showed that the shift in the wavelength of the photon scattered at an angle θ is given by

$$\Delta\lambda = \lambda_1 - \lambda_o = \lambda_C(1 - \cos\theta) \qquad \begin{array}{c} \textit{Compton} \\ \textit{effect} \end{array} \qquad (27.9)$$

where λ_o is the wavelength of the incident photon and λ_1 that of the scattered photon. The constant $\lambda_C = h/m_ec = 2.43 \times 10^{-12}$ m $= 2.43 \times 10^{-3}$ ¡nm is called the *Compton wavelength* of the electron. The equation correctly predicted the observed wavelength shift, and Compton was awarded a Nobel Prize in 1927.

Note that the maximum wavelength increase is $\Delta\lambda_{max} = 2\lambda_C = 4.86 \times 10^{-3}$ nm. It occurs when the photon is scattered directly backward—that is, for $\theta = 180°$ and $\cos\theta = -1$. (Why?) Since this is the maximum wavelength *change*, it will be negligible for incident wavelengths several thousand times this value or for wavelengths larger than several nanometers (strong UV). In other words, the Compton effect is experimentally negligible for ultraviolet light and longer wavelengths. It is significant only for X-ray and gamma ray scattering from electrons.

Example 27.4 ■ X-Ray Scattering: The Compton Effect

A monochromatic beam of X-rays of wavelength 1.35×10^{-10} m is scattered by a metal foil. By what percentage is the wavelength shifted if the scattered X-rays are observed at an angle of 90°?

Thinking It Through. The change or shift in the wavelength, $\Delta\lambda$, is given by Eq. 27.9 with $\theta = 90°$, and the fractional change is $\Delta\lambda/\lambda_o$. The percentage change should be positive, because the scattered light has a longer wavelength than the incident light has.

Solution. Listing the data, we have

Given: $\lambda_o = 1.35 \times 10^{-10}$ m   *Find:* % change in wavelength
          $\theta = 90°$

$$\frac{\Delta\lambda}{\lambda_o} = \frac{\lambda_C}{\lambda_o}(1 - \cos\theta) = \frac{2.43 \times 10^{-12} \text{ m}}{1.35 \times 10^{-10} \text{ m}}(1 - \cos 90°) = 1.80 \times 10^{-2}$$

so

$$\frac{\Delta\lambda}{\lambda_o} \times 100\% = 1.80\%$$

Follow-up Exercise. In this Example, (a) what would the fractional change be for backscattering (180°)? (b) Is it larger or smaller than the Example answer? Why?

(a)

(b)

•**FIGURE 27.7 Gas discharge tubes** These luminous glass tubes are gas discharge tubes, in which atoms of various gases emit light when electrically excited. Each gas radiates its own characteristic wavelengths. **(a)** Only some "neon lights" actually contain neon, which glows with a red hue; **(b)** other gases produce other colors.

Einstein's and Compton's successes in explaining electromagnetic phenomena in terms of quanta left scientists with two apparently competing theories of electromagnetic radiation. Classically, the radiation is pictured as a continuous wave, and this theory satisfactorily explains such wave-related phenomena as interference and diffraction. Conversely, quantum theory was necessary to explain the photoelectric and Compton effects correctly.

The two theories gave rise to a description that is called the **dual nature of light.** That is, light sometimes behaves as a wave and at other times as photons or "particles." The relationship between the wave and particle nature of matter in general will be considered in Chapter 28.

27.4 The Bohr Theory of the Hydrogen Atom

Objectives: To (a) understand how the Bohr model of the hydrogen atom explains atomic emission and absorption spectra, (b) calculate the energies and wavelengths of emitted photons for transitions in atomic hydrogen, and (c) understand how the generalized concept of atomic energy levels can explain other atomic phenomena.

In the 1800s, much experimental work was done with gas discharge tubes—for example, those containing hydrogen, neon, and mercury vapor (•Fig. 27.7). (The common neon "lights" are gas discharge tubes.) Normally, light from an incandescent source (say, a light bulb) exhibits a *continuous spectrum.* However, when light emissions from these tubes were analyzed, discrete spectra, or line spectra, were observed (•Fig. 27.8). Light coming directly from such a tube gives a *bright-line spectrum,* or **emission spectrum,** indicating that only certain wavelengths are emitted. In general, a discrete line spectrum is characteristic of the individual atoms or molecules of a particular material and can identify that material spectroscopically (Fig. 27.8a,b).

Atoms absorb light as well as emitting it. If white light is passed through a relatively cool gas, certain frequencies or wavelengths are missing, or absorbed. The result is an **absorption spectrum**—a series of dark lines superimposed on a continuous spectrum (Fig. 27.8c). When the absorption and emission lines of a particular gas were compared, they were found to occur at the same frequencies. The reason for line spectra was not understood at the time, but they provided a clue to the electron structure of the atoms.

Hydrogen, with its relatively simple visible spectrum, received much attention. It is also the simplest atom, consisting of only one electron and one proton. In the

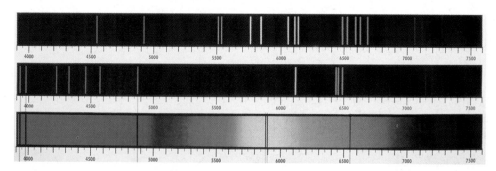

•**FIGURE 27.8 Spectra** When a gas is excited by heat or electricity and the light it emits is separated into its component wavelengths by a prism or diffraction grating, the result is a bright-line, or emission, spectrum, such as these emission spectra of **(a)** barium and **(b)** calcium. Each atom or molecule emits a characteristic pattern of discrete wavelengths. **(c)** When a continuous spectrum consisting of all wavelengths is passed through a cool gas, a series of dark lines is observed. Each line represents a "missing" wavelength—a particular wavelength the gas has absorbed. The wavelengths absorbed by any substance are the same ones it emits when excited. This absorption spectrum of the Sun shows several prominent absorption lines; the gases of the solar atmosphere produce it.

late nineteenth century, the Swiss physicist J. J. Balmer found an empirical formula that gives the wavelengths of the spectral lines of hydrogen in the visible region:

$$\frac{1}{\lambda} = R\left(\frac{1}{2^2} - \frac{1}{n^2}\right) \quad \text{for } n = 3, 4, 5, \ldots \tag{27.10}$$

where R is called the *Rydberg constant* and has a value of 1.097×10^{-2} nm^{-1}. The spectral lines of hydrogen in the visible region, called the **Balmer series**, were found to fit the formula, but it was not understood why. Similar formulas were found to fit the spectral line series in the ultraviolet and infrared regions.

An explanation of the spectral lines was given in a theory of the hydrogen atom put forth in 1913 by the Danish physicist Niels Bohr (1885–1962). Bohr assumed that the electron of the hydrogen atom orbits the proton in a circular orbit (analogous to a planet orbiting the Sun). The electrical Coulomb force (Chapter 15) supplies the necessary centripetal force for the circular motion, so

Note: Review Eq. 7.11 and Eq. 15.2.

$$F = \frac{mv^2}{r} = \frac{ke^2}{r^2} \tag{27.11}$$

where e is the magnitude of the charge of the proton and the electron, v the electron's orbital speed, and r the radius of its orbit (•Fig. 27.9).

The total energy of the atom is the sum of its kinetic and potential energies:

Note: Recall from Chapter 16 that the electric potential energy of two point charges is $U_e = kq_1q_2/r$.

$$E = K + U_e = \tfrac{1}{2}mv^2 - \frac{ke^2}{r}$$

From Eq. 27.11, the kinetic energy can be written $\tfrac{1}{2}mv^2 = ke^2/2r$. With this, the total energy becomes

$$E = \frac{ke^2}{2r} - \frac{ke^2}{r} = -\frac{ke^2}{2r} \tag{27.12}$$

Notice that E is negative, indicating that the system is bound, but that as the radius approaches infinity, E approaches zero. With $E = 0$, the electron would no longer be bound to the proton, and the atom, having lost its electron, would be ionized.

Thus far, only classical principles have been applied. At this point in the theory, Bohr made a radical assumption—radical in the sense that he introduced a quantum concept to attempt to explain atomic line spectra.

Bohr assumed that *the angular momentum of the electron was quantized* and could have only discrete values that were integral multiples of $h/2\pi$, where h is Planck's constant.

Recall that in a circular orbit of radius r, the angular momentum L is given by mvr (Eq. 8.14). Thus, Bohr's assumption translates into the equation

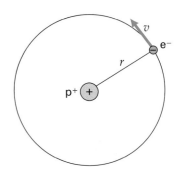

•**FIGURE 27.9 The Bohr model of the hydrogen atom** The electron of the hydrogen atom is pictured as revolving around the nuclear proton in a circular orbit, similar to the way a planet orbits the Sun.

$$mvr = n\left(\frac{h}{2\pi}\right) \quad \text{for } n = 1, 2, 3, 4, \ldots \tag{27.13}$$

The integer n is called a *quantum number*, specifically the **principal quantum number**. From this assumption, the orbital speed of the electron (v) can be found. Its quantized values are given by

$$v = \frac{nh}{2\pi mr}$$

Putting this expression for v into Eq. 27.11 and solving for r, we find that

$$r_n = \left(\frac{h^2}{4\pi^2 ke^2 m}\right)n^2 \quad \text{for } n = 1, 2, 3, 4, \ldots \tag{27.14}$$

Here we use the subscript on r to indicate that, for the electron orbits, only certain radii are possible as determined by the principal quantum number n. The energy for a particular orbit can be found by substituting this expression for r into Eq. 27.12:

$$E_n = -\left(\frac{2\pi^2 k^2 e^4 m}{h^2}\right)\frac{1}{n^2} \quad \text{for} \quad n = 1, 2, 3, 4, \dots \tag{27.15}$$

where the energy is also written as E_n to show its dependence on n. The quantities in the parentheses on the right-hand sides of Eqs. 27.14 and 27.15 are constants and can be evaluated numerically. The important results are

Principal quantum number, n:
specifies orbit size and energy

$$r_n = 0.0529n^2 \text{ nm} \tag{27.16}$$

$$E_n = \frac{-13.6}{n^2} \text{ eV} \quad \text{for} \quad n = 1, 2, 3, 4, \dots$$

orbital radii and energies (for hydrogen atom) (27.17)

where radius and energy are expressed in atomic-scale units—nanometers and electron volts, respectively. The use of these expressions is shown in Example 27.5.

Example 27.5 ■ A Bohr Orbit: Radius and Energy

Find the orbital radius and energy of an electron in a hydrogen atom characterized by the principal quantum number $n = 2$.

Thinking It Through. We can use Eqs. 27.16 and 27.17 with $n = 2$.

Solution. For $n = 2$, we have

$$r_2 = 0.0529n^2 \text{ nm} = 0.0529(2)^2 \text{ nm} = 0.212 \text{ nm}$$

and

$$E_2 = \frac{-13.6}{n^2} \text{ eV} = \frac{-13.6}{2^2} \text{ eV} = -3.40 \text{ eV}$$

Follow-up Exercise. In this Example, what is the speed of the orbiting electron?

However, there was still a classical problem with Bohr's theory. Classically, an accelerating electron radiates electromagnetic energy, and even in discrete circular orbits the electron is accelerating centripetally. Thus, an orbiting electron should lose energy and spiral into the nucleus, as the orbit of an Earth satellite might be expected to decay because of frictional losses. This doesn't happen in the hydrogen atom, so Bohr had to make another nonclassical assumption:

> Bohr postulated that the hydrogen electron does not radiate energy when it is in a bound, discrete orbit but does so only when it makes a transition to another orbit.

Energy Levels

Note: Review the concept of potential energy wells presented in Sections 5.4 and 7.5 (see Figs. 5.14 and 7.20).

Ground state: $n = 1$
Excited states: $n \geq 2$

The possible allowed orbits of the hydrogen electron are commonly expressed in terms of their energy as illustrated in •Fig. 27.10. In this context, we simply refer to the electron as being in a particular "energy level" or state. The electron, being bound to the nuclear proton, is in an electric potential energy well, much like the Moon is in a gravitational potential well. The principal quantum number designates the energy level. The lowest energy level ($n = 1$) is called the **ground state**. The energy levels above the ground state are called **excited states**. For example, $n = 2$ is the first excited state (as is the orbit in Ex. 27.5), and so on.

The electron is normally in the ground state and must be given energy to excite it, or to raise it to an excited state. The hydrogen electron can be excited only by discrete amounts. The energy levels are somewhat analogous to the rungs of a ladder. Just as a person goes up and down a ladder, an electron goes up and down the energy ladder in discrete steps. Notice, however, that the energy rungs of the hydrogen atom are not evenly spaced.

If enough energy is absorbed to raise the electron to the top of the energy well so the electron is no longer bound, the atom is ionized. For example, to raise a hydrogen electron from the ground state to $n = \infty$ requires 13.6 eV of energy. How-

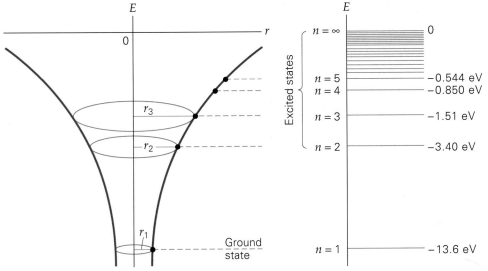

•FIGURE 27.10 Orbits and energy levels of the hydrogen electron The Bohr theory predicts that the hydrogen electron can occupy only certain orbits having discrete radii. Each allowed orbit has a corresponding energy, conveniently displayed as an energy level. The lowest energy level ($n = 1$) is the ground state; those above ($n > 1$) are excited states. The orbits are shown at left, with orbital radius plotted against the $1/r$ electrical potential of the proton. The electron in the ground state is deepest in the potential well, analogous to the gravitational potential well of Fig. 7.20. (Neither r nor the energy levels are drawn to scale—why?)

ever, if the electron is already part way up the ladder (or higher up in the well)—that is, in an excited state—less energy is needed to get it to the top of the well. The energy of the electron in any state is E_n, and the energy needed to raise it to the top of the well is $-E_n$, which is called the **binding energy** of the electron. (Since E_n is negative—the electron is in a negative potential well—then $-E_n$ is positive and equal to the energy needed to be *given* to the electron to raise it to the top of the well.)

An electron generally does not remain in an excited state for long; it decays, or makes a transition, to a lower energy level, in a very short time. The time an electron spends in an excited state is called the **lifetime** of the excited state. For many states, the lifetime is about 10^{-8} s. In making a transition to a lower state, energy is emitted as a photon of light (•Fig. 27.11). The energy ΔE of the photon is equal in magnitude to the energy *difference* of the levels:

$$\Delta E = E_{n_i} - E_{n_f} = \frac{-13.6}{n_i^2}\,\text{eV} - \left(\frac{-13.6}{n_f^2}\,\text{eV}\right)$$

or

$$\Delta E = 13.6\left(\frac{1}{n_f^2} - \frac{1}{n_i^2}\right)\text{eV} \qquad \begin{array}{l}\textit{H-atom emitted}\\ \textit{photon energy}\end{array} \quad (27.18)$$

where the subscripts i and f refer to the initial and final states, respectively. Since $\Delta E = hf = hc/\lambda$, only photons of particular frequencies and wavelengths are emitted. These correspond to the various transitions between energy levels.

The different values of the final principal quantum number refer to the energy level in which the electron starts or ends in the various emission or absorption series. The original Balmer series in the visible region corresponds to $n_f = 2$. There is only one series entirely in the ultraviolet range, the *Lyman* series, in which all the transitions end (in the case of emission) or start (in the case of absorption) with $n = 1$ (the ground state). There are many series entirely in the infrared region, most notably the *Paschen* series, which ends or starts with the electron in the second excited state, $n = 3$.

Usually we are interested in the wavelength of the light, since experimentally, that is what is commonly measured. The wavelength of the emitted light, λ, is obtained from the relationship between photon energy and light frequency (Eq. 27.4):

$$\lambda = \frac{hc}{\Delta E} = \frac{1.24 \times 10^3}{\Delta E \text{ (in eV)}}\,\text{nm} \qquad (27.19)$$

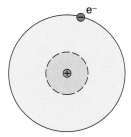

Excited atom

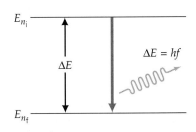

De-excitation

•FIGURE 27.11 Electron transitions and photon emission When an electron in the excited state of an atom decays (makes a transition) to a lower orbit or energy level, a photon is emitted. The energy of the photon is equal to the energy difference between the two energy levels.

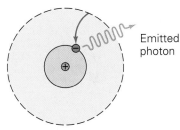

where ΔE is expressed in electron volts and the constant hc has been replaced by its numerical value of 1.24×10^3 eV·nm (see the Problem-Solving Hint on p. 847). The relationship between the emitted light wavelength and principle quantum numbers is shown in the following Example.

Example 27.6 ■ Investigating the Balmer Series: Visible Light from Hydrogen

What is the wavelength of the emitted light when excited electrons in hydrogen atoms undergo transitions from the $n = 3$ energy level to the $n = 2$ energy level?

Thinking It Through. The emitted photon has an energy equal to the energy difference between the two energy levels (Eq. 27.18). We can then convert to wavelength by using Eq. 27.19. The light will be visible because it is from the Balmer series ($n_f = 2$).

Solution.

Given: $n_i = 3$ *Find:* λ (wavelength of emitted light)
$\quad\quad\quad n_f = 2$

The energy of the emitted photon is equal to the magnitude of the atom's change in energy (which itself is negative; why?). Thus we have

$$\Delta E = 13.6\left(\frac{1}{n_f{}^2} - \frac{1}{n_i{}^2}\right) \text{eV} = 13.6\left(\frac{1}{4} - \frac{1}{9}\right) \text{eV} = 1.89 \text{ eV}$$

Using Eq. 27.19 with ΔE expressed in electron volts, we get

$$\lambda = \frac{1.24 \times 10^3 \text{ eV·nm}}{\Delta E} = \frac{1.24 \times 10^3 \text{ eV·nm}}{1.89 \text{ eV}} = 656 \text{ nm}$$

which is in the red portion of the visible spectrum. Check for agreement with Balmer's original empirical results (Eq. 27.10).

Follow-up Exercise. Light of what wavelength would be just sufficient to ionize a hydrogen atom if it started in its first excited state?

We see that in the Bohr model of hydrogen, the electron can make transitions only between discrete energy levels, thus emitting photons of discrete energies, or light of discrete wavelengths, thus producing spectral lines (•Fig. 27.12). The wavelengths of the various transitions, computed from Eq. 27.19, correspond to the experimentally observed spectral lines for all the series.

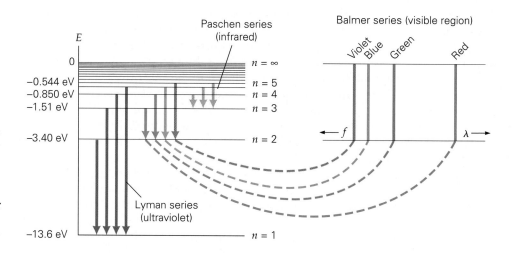

•**FIGURE 27.12 Hydrogen spectrum** Transitions may occur between two or more energy levels as the excited electron returns to the ground state. Transitions to the $n = 2$ state give spectral lines that lie in the visible region of the spectrum and are called the Balmer series. Transitions to other levels give rise to other series as shown.

The Balmer Series: Entirely Visible?

We know that four wavelengths of the Balmer series are in the visible range. Are there more in the series, and if so, what type of light are they likely to be: (a) infrared, (b) visible, (c) ultraviolet?

Reasoning and Answer. The Balmer emission series ends with the electron in the first excited state ($n = 2$). We know that red light is emitted when the electron drops from the $n = 3$ level. The other three visible lines are emitted during the downward transitions from $n = 4$, 5, and 6. (Check that all the wavelengths are between about 400 and 700 nm.) However, there is no reason a transition cannot occur from, say, $n = 7$ to $n = 2$. This is still in the Balmer series, but it involves a larger energy difference than the visible emissions. Hence the emitted photons will have larger energies, corresponding to smaller wavelengths. Wavelengths smaller than visible are likely to lie in the ultraviolet range, so the answer is (c). (This can be confirmed by calculation.)

Follow-up Exercise. Which type of light (visible, ultraviolet, or infrared) provides the minimum energy to ionize a hydrogen atom initially in its second excited state? Explain.

Although Bohr's results gave excellent agreement with the experiment for hydrogen as well as singly ionized helium and doubly ionized lithium, it could not successfully handle multielectron atoms. Bohr's theory was incomplete in the sense that it patched new quantum ideas into a basically classical framework. The theory contains many correct concepts, but a complete and correct description of the atom did not come until the full theory of quantum mechanics was developed (Chapter 28). Nevertheless, the idea of discrete energy levels enables us to understand qualitatively such phenomena as *fluorescence.*

Fluorescence. Note in Fig. 27.12 that an electron excited to a higher energy level does not necessarily return to the ground state in a single jump. Often, several routes involving different sequences of transitions may be possible. For example, an electron in the $n = 4$ excited level of hydrogen could drop to either of the intermediate levels ($n = 3$ or $n = 2$)—or even to both in succession—before falling back to the ground state. This fact underlies the phenomenon of fluorescence, exhibited by a number of natural and synthetic substances.

In **fluorescence**, an electron in an excited state returns to the ground state in two or more steps, like a ball bouncing down a flight of stairs. At each step a photon is emitted. Since each drop represents a smaller energy transition than the original upward boost, each emitted photon must have a lower energy, and thus a longer wavelength, than the original exciting photon. For example, the atoms of many minerals can be excited by absorbing ultraviolet (UV) light and fluoresce, or glow, in the visible region when they de-excite (•Fig. 27.13). A variety of living organisms, from corals to butterflies, manufacture fluorescent pigments that emit visible light.

27.5 A Quantum Success: The Laser

Objective: **To understand some of the practical applications of the quantum hypothesis—in particular, the laser.**

The development of the laser was a major technological success derived from theoretical atomic physics. Although many scientific discoveries have been made accidentally and many applications or inventions have come about by trial and error—including Roentgen's discovery of X-rays and Edison's electric lamp—the laser was developed on purpose. Via quantum physics, the principle of the **laser** was first predicted theoretically, then designed, built, and finally applied. (The word *laser* is an acronym that stands for *light amplification by stimulated emission of radiation.* Stimulated emission will be discussed shortly.)

(a) Visible illumination

(b) UV illumination

•**FIGURE 27.13 Fluorescence** Many minerals glow at visible wavelengths when illuminated by invisible ultraviolet light ("black light"). The visible light is produced when atoms excited by the UV fall back to lower energy levels in two or more smaller steps.

This relatively new invention has found widespread application. For example, laser light beams are used in medicine to control bleeding, to weld torn retinas, and to treat skin cancer. In industry, lasers are used to drill holes, to weld, and to cut. They are extremely accurate in surveying, and laser light is used to carry telephone and television signals over fiber optic cables. There are laser printers, laser pickups in video and compact disc players, and lasers in supermarket checkouts.

The concept of atomic energy levels allows us to understand this important technological instrument. Bohr's model of the atom explains spectral absorption and emission lines in terms of quantum jumps between energy levels. However, other questions arise. Since there are usually several lower energy levels, what determines the energy level to which an excited electron will decay? Also, how long does an electron stay in an excited state before making a downward transition? These questions were answered by the use of quantum mechanics, which treats electrons as waves (Chapter 28). The general results are that different transitions have different probabilities of occurring and, crucial to laser application, the time an electron stays in an excited state varies with the atom and the state.

Usually, an excited electron makes a transition to a lower energy level almost immediately, remaining in an excited state for only about 10^{-8} s. However, the lifetimes of an excited electron in some energy levels are appreciable. An energy level in which an excited electron remains for some time is called a **metastable state. Phosphorescent** materials are examples of substances made up of atoms with metastable states. These materials are used on luminous watch dials, toys, and items that "glow in the dark." When a phosphorescent material is exposed to light, atomic electrons are excited to higher energy levels. Many of the excited electrons return to their normal state very quickly. However, there also exist metastable states in which electrons may remain for seconds, minutes, even more than an hour. Consequently, the material, made of many such atoms, emits light and can glow for some time (•Fig. 27.14).

A major consideration in laser operation is the emission process. As •Fig. 27.15 shows, absorption and spontaneous emission of radiation can occur between two energy levels. That is, a photon is absorbed and a photon is emitted almost immediately. However, when the higher energy state is metastable, there is another possible emission process called **stimulated emission.** Einstein proposed this process in 1919. If a photon with an energy equal to an allowed transition strikes an atom already in an excited metastable state, it may stimulate an electron to make a transition to a lower energy level, with the emission of another photon. Thus two photons with the same frequency and phase then go off in the same direction.

Notice that stimulated emission is an amplification process—one photon in, two out. But, this is not a case of getting something for nothing, since the atom must be initially excited and energy is needed to boost it to a higher state. This initial excitation and stimulated emission process is somewhat analogous to pumping water to a roof-top reservoir for later use. Stimulated emission thus provides a way to amplify the incoming light.

•FIGURE 27.14 Phosphorescence When atoms in a phosphorescent material are excited, some of them do not immediately return to the ground state but remain in metastable states for periods of time. In this exhibit at the San Francisco Exploratorium, phosphorescent walls and floor continue to glow for about 30 s after being illuminated, retaining the shadows of children present when the phosphors were exposed to light.

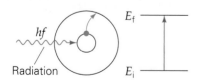

(a) Absorption

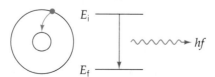

(b) Spontaneous emission

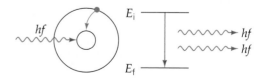

(c) Stimulated emission

•FIGURE 27.15 Photo absorption and emission **(a)** In absorption, a photon is absorbed, and the atom is excited to a higher energy level. **(b)** After a short time, the atom spontaneously decays to a lower energy level with the emission of a photon. **(c)** If another photon with an energy equal to an allowed transition strikes an excited atom, stimulated emission occurs, and two photons with the same frequency and phase go off in the same direction as that of the incident photon.

Ordinarily, when light passes through a material, photons are more likely to be absorbed than to give rise to stimulated emission. This is because there are normally many more atoms in their ground state than in their excited states. However, under specially arranged circumstances, it is possible to prepare a material so that more atoms are in an excited metastable state than in the ground state. This condition is known as a **population inversion**. In this case, there may be more stimulated emission than absorption, and the net result is an amplification. *Population inversion and stimulated emission are the two basic conditions necessary for laser action.*

There are many types of lasers capable of producing light of different wavelengths. The helium–neon (He–Ne) gas laser is probably the most familiar, since it is used for classroom demonstrations and laboratory experiments. The characteristic reddish-pink light produced by the He–Ne laser ($\lambda = 632.8$ nm) can also be observed in an optical scanning system at supermarket checkouts. The gas mixture is about 85% helium and 15% neon. Essentially, the helium is used for energizing and the neon for amplification. The gas mixture is subjected to a high voltage discharge of electrons produced by a radio-frequency power supply (converted to dc). The helium atoms are excited by collision with these electrons (•Fig. 27.16a). This process is referred to as *pumping*. Energy is pumped into the system, and the helium atoms are pumped into an excited state (20.61 eV).

The excited state in the He atom has a relatively long lifetime (about 10^{-4} s) and has almost the same energy (20.61 eV) as an excited energy level in the Ne atom (20.66 eV). Because this lifetime is so long, there is a good chance that before an excited He atom can spontaneously emit a photon, that atom will collide with a Ne atom. When a collision occurs, energy can be transferred to the Ne atom. Light is produced when the Ne electron drops down to a specific lower level. However, the lifetime of the 20.66-electron-volt neon state is also relatively long. The delay causes a piling up of neon atoms in this state, leading to a population inversion.

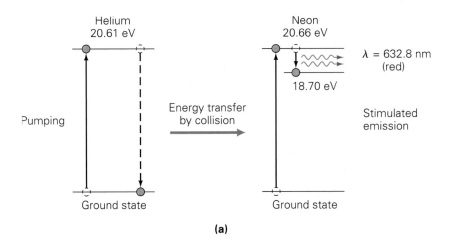

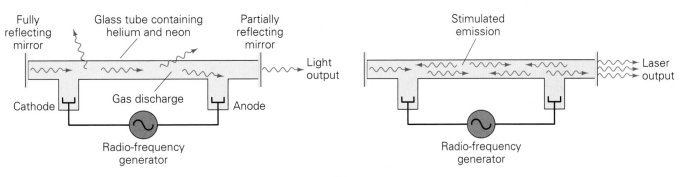

(b)

•FIGURE 27.16 **The helium–neon laser** (a) Helium atoms are excited, or pumped into an excited state, by collision with electrons. In the process, energy is transferred from helium atoms to neon atoms, and stimulated emission occurs with the emission of red light. (b) Reflections from the end mirrors of the laser tube set up an intense, coherent light beam (due to the stimulated emission) parallel to the axis of the tube.

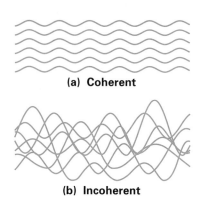

(a) Coherent

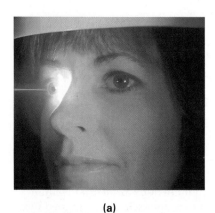

(b) Incoherent

•**FIGURE 27.17 Coherent light**
(a) Laser light is monochromatic (single frequency or color) and coherent, or in phase. **(b)** Light waves from sources in which atoms emit randomly are incoherent, or out of phase.

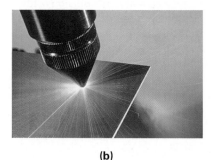

(a)

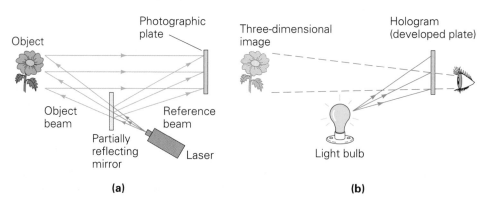

(b)

•**FIGURE 27.18 Laser applications**
(a) Laser eye surgery. Such procedures, which minimize bleeding and scarring, are used to repair detached retinas, treat glaucoma, correct retinal abnormalities produced by diabetes, and alleviate various other problems. **(b)** An industrial laser cuts through a steel plate.

The stimulated emission and amplification of the light emitted by the neon atoms are enhanced by reflections from mirrors placed at the end of the laser tube (Fig. 27.16b). Some excited Ne atoms spontaneously emit photons in all directions, and these photons induce stimulated emissions. In stimulated emission, the two photons leave the atom in the same direction as that of the incident photon. Photons traveling in the direction of the tube axis are reflected back through the tube by the end mirrors. The photons, in reflecting back and forth, cause even more stimulated emissions, and an intense, highly directional, coherent, monochromatic beam develops along the tube axis. Part of the beam emerges through one of the end mirrors, which is only partially silvered.

The monochromatic (single-frequency), coherent (in-phase), and directional properties of laser light give it unique properties (•Fig. 27.17). Light from sources such as incandescent lamps is incoherent. The atoms emit randomly and at different frequencies (many different transitions). As a result, the light is out of phase, or incoherent. Such beams spread out and become less intense. The properties of laser light allow the formation of a very narrow beam, which with amplification can be very intense.

Some laser applications are shown in •Fig. 27.18, and another is discussed in the Insight on p. 859. Although laser beams are used to weld detached retinas in the eye, viewing a laser beam directly can be quite hazardous, because the beam is focused on the retina in a very small area. If the beam intensity and the viewing time are sufficient, the photosensitive cells of the retina can be burned and destroyed by this concentrated energy.

Another interesting application of laser light is the production of three-dimensional images in a process called **holography** (from the Greek word *holos*, meaning "whole"). The process does not use lenses as ordinary image-forming processes do, yet it recreates the original scene in three dimensions. The key to holography is the coherent property of laser light, which gives the light waves a definite spatial relationship to each other.

In the photographic process of making a *hologram*, an arrangement such as that illustrated in •Fig. 27.19 is used. Part of the light from the laser (the object beam) passes through a partially reflecting mirror to the object. The other part, or reference beam, is reflected to the film. The light incident on the object is also reflected to the film, and it interferes with the reference beam. The film records the interference pattern of the two light beams, which essentially imprints on the film the information carried from the object by the light wavefronts.

When the film is developed, the interference pattern bears no resemblance to the object and appears as a meaningless pattern of light and dark areas. However, when the wavefront information is reconstructed by passing light through the

•**FIGURE 27.19 Holography** **(a)** The coherent light from a laser is split into reference and object beams. The interference patterns of these beams are recorded on a photographic plate. **(b)** When the developed plate, or hologram, is illuminated, the viewer sees a three-dimensional image.

The Compact Disc

The compact disc (CD) is a relatively new method of data storage; most of us are familiar with it in terms of musical recordings. The CD system was introduced in 1980. Now music CD's outsell tapes and have nearly supplanted long-playing records. The disc itself is only 12 cm (5 in.) in diameter, but it can store more than 6 billion bits of information. This is the equivalent of more than 1000 floppy disks or over 275 000 pages of text that could be stored on a CD for display on a television monitor. For audio use, a CD can store 74 min of music, and the sound reproduction is virtually unaffected by dust, scratches, or fingerprints on the disc.

Information on the disc is in the form of raised areas called "pits," which are separated by flat areas called "land." The surface is coated with a thin layer of aluminum to reflect the laser beam that "reads" the information (Fig. 1). The pits are arranged in a spiral track like the grooves in a phonograph record, but the track is much narrower. These tracks are about 60 times closer together than the grooves on a long-playing album.

In the read-out system, a laser beam, which comes from a small semiconductor (solid state) laser, is applied from be-low the disc and focused on the aluminum coating of the track. The disc rotates at about 3.5 to 8 revolutions per second as the laser beam follows the spiral track. When the beam strikes a land area between two pits, it is reflected. When the beam spot overlaps a land area and a pit, the light reflected from the different areas interferes, causing fluctuations in the reflected beam. To make the fluctuations more distinct, the raised pit thickness (t) is chosen to be one-quarter wavelength of the laser beam. Then, reflected light from land areas travels an additional path length of one-half wavelength, and destructive interference occurs when the two parts of the reflected beam combine (Section 24.1). As a result, there is less reflected intensity when the beam passes over the edge of a pit than when passing over a land area alone.

The reflected beam of varying intensity strikes a photodiode (solid state photocell), which reads the information. The fluctuations of the reflected light convey the coded information as a series of binary numbers (zeros and ones). A pit represents a binary 1, and a land area is read as a binary 0. The signals are electronically converted back into sound.

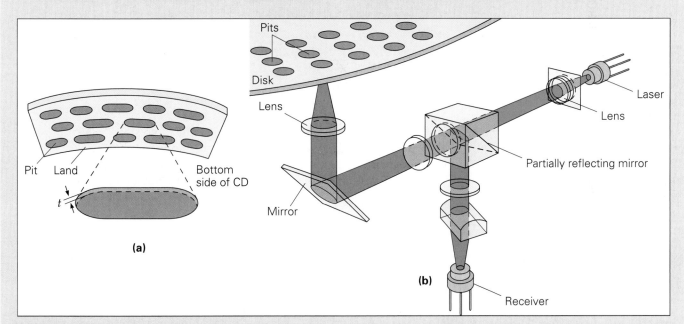

FIGURE 1 The compact disc (a) The information on the disc is recorded in the form of raised areas called pits, which are separated by flat areas called land. The pits or imprints are on the bottom of the disc. (b) The surface is coated with a thin layer of aluminum to reflect the laser beam, which "reads" the information.

film, a three-dimensional image is perceived (•Fig. 27.20). If part of the three-dimensional image is hidden from view, you can see it by moving your head to one side, just as you would to see a hidden part of a real object.

Holography has many potential applications. Using coherent ultrasonic sound waves to form a hologram could provide a three-dimensional view of internal structures of the human body. Also, holograms made with X-ray lasers (yet to be developed) could provide better resolution for an in-depth view of microscopic structures. Holographic three-dimensional television may some day be commonplace.

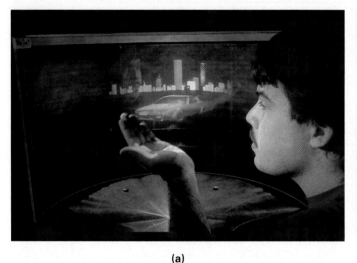

(a)

(b)

•**FIGURE 27.20 Holographic images** (a) A three-dimensional hologram can be viewed from any angle, just as if it were a real object. (b) Holographic projection is used to display flight information at the pilot's eye level in an airplane cockpit.

Chapter Review

Important Terms

thermal radiation *840*
blackbody *840*
Wien's displacement law *841*
Planck's constant *842*
Planck's hypothesis *842*
quantum *842*
photon *843*
photoelectric effect *843*
stopping potential *844*
work function *845*

threshold frequency *846*
Compton effect *848*
dual nature of light *850*
Bohr theory of the hydrogen
  atom *850*
emission spectrum *850*
absorption spectrum *850*
Balmer series *851*
principal quantum number *851*
ground state *852*

excited states *852*
binding energy *853*
lifetime *853*
fluorescence *855*
laser *855*
metastable state *856*
phosphorescent *856*
stimulated emission *856*
population inversion *857*
holography *858*

Important Concepts

- Thermal radiation, typically produced by hot objects, has a continuous spectrum.
- A blackbody is an ideal system that absorbs and emits all the radiation that falls on it.
- Wien's displacement law states that the wavelength of maximum intensity for radiation from a blackbody is inversely related to its temperature.
- Planck's constant (h) is the fundamental proportionality constant between energy and frequency of thermal oscillators as well as frequency of a light wave and energy of the corresponding photons.
- The Compton effect is the wavelength increase of light scattered by electrons or other charged particles.
- The dual nature of light means that light must be thought of as having both particle and wave natures.
- The Bohr theory of the hydrogen atom treats the electron as a classical particle in circular orbit around the

proton, held in orbit by the electric attraction force. The angular momentum of the electron is quantized in integral multiples of Planck's constant, h.

- The ground state of a hydrogen atom is the state of lowest energy for the atomic electron (in the smallest orbit, $n = 1$). Excited states ($n > 1$) have greater energies.
- A metastable state is a relatively long-lived excited atomic state.
- Stimulated emission can occur when a photon prematurely causes a downward atomic transition yielding a photon identical to itself.
- A laser uses population inversion, produced by metastable excited atomic states, to produce light amplification through stimulated emission.

Important Equations

Wien's Displacement Law:
$$\lambda_{max}T = 2.90 \times 10^{-3} \text{ m·K} \qquad (27.1)$$

Photon Energy:
$$E = hf \qquad (27.4)$$
(where Planck's constant is $h = 6.63 \times 10^{-34}$ J·s)

K_{max} and Stopping Potential in the Photoelectric Effect:
$$K_{max} = eV_o \qquad (27.5)$$

Energy Conservation in the Photoelectric Effect:
$$E = hf = K + \phi \qquad (27.6)$$

Energy Conservation for Least Bound Electron in the Photoelectric Effect:
$$hf = K_{max} + \phi_o \qquad (27.7)$$

Work Function and Threshold Frequency in the Photoelectric Effect:
$$f_o = \frac{\phi_o}{h} \qquad (27.8)$$

Compton Equation (scattering from free electrons):
$$\Delta\lambda = \lambda_1 - \lambda_o = \lambda_C(1 - \cos\theta) \qquad (27.9)$$
(where $\lambda_C = h/m_e c = 2.43 \times 10^{-12}$ m $= 2.43 \times 10^{-3}$ nm)

Bohr Theory Orbit Radius (for hydrogen atom):
$$r_n = 0.0529n^2 \text{ nm} \quad n = 1, 2, 3, 4, \ldots \qquad (27.16)$$

Bohr Theory Electron Energy (for hydrogen atom):
$$E_n = \frac{-13.6}{n^2} \text{ eV} \quad n = 1, 2, 3, 4, \ldots \qquad (27.17)$$

Emitted Photon Energy (from Bohr theory for hydrogen atom):
$$\Delta E = 13.6\left(\frac{1}{n_f^2} - \frac{1}{n_i^2}\right) \text{ eV} \qquad (27.18)$$

Bohr Theory Photon Wavelength:
$$\lambda = \frac{1.24 \times 10^3}{\Delta E \text{ (in eV)}} \text{ nm} \qquad (27.19)$$

Exercises

Note: Take h to be exactly 6.63×10^{-34} J·s in terms of significant figures, and use $hc = 1.24 \times 10^3$ eV·nm (3 significant figures).

27.1 Quantization: Planck's Hypothesis

1. A blackbody (a) absorbs all radiation incident on it, (b) re-emits all radiation incident on it, (c) has a wavelength at which maximum radiation intensity is inversely proportional to the blackbody's temperature, (d) all of these.

2. Planck's hypothesis (a) was the basis of Wien's law, (b) called for the intensity of blackbody radiation to be proportional to $1/\lambda^4$, (c) quantized the energy of thermal atomic oscillators, (d) predicted the ultraviolet catastrophe.

3. Does a blackbody at 600 K emit twice as much total radiation as it does when its temperature is 300 K? Explain.

4. Some stars appear reddish, and some others are blue. Which of the two types has the lower surface temperature?

5. ■ Sketch a graph of temperature versus (a) frequency and (b) wavelength of the most intense radiation component of blackbody radiation.

6. ■ Find the approximate temperature of a red star that emits light with a wavelength of maximum emission of 700 nm (deep red).

7. ■ What are the wavelength and frequency of the most intense radiation component from a blackbody with a temperature of 0°C?

8. ■ Assuming that our skin is at our normal body temperature (37°C), what is the wavelength of the radiation compo-

nent of maximum intensity emitted by our bodies? In what region of the electromagnetic spectrum is this wavelength?

9. ■ The walls of a blackbody cavity are at a temperature of 27°C. What is the frequency of the radiation of maximum intensity?

10. ■■ What would be the change in frequency of the most intense spectral component of a blackbody that was initially at 200°C and was heated to 400°C?

11. ■■ What is the minimum quantized energy of a thermal oscillator producing radiation at λ_{max} in a blackbody with a temperature of 100°C?

12. ■■ The temperature of a blackbody is 1000 K. If the intensity of the emitted radiation of 2.0 W/m² is due entirely to the most intense frequency component, how many quanta are emitted per second per square meter? (Assume that each quantum has the minimum allowed energy.)

27.2 Quanta of Light: Photons and the Photoelectric Effect

13. In the photoelectric effect, classical theory predicted that (a) no photoemission can occur below a threshold frequency, (b) the photocurrent is proportional to the light intensity, (c) the kinetic energy of the emitted electron is dependent on frequency but not on intensity, (d) a photocurrent is observed immediately.

14. A photocurrent is observed when (a) the light frequency is above the threshold frequency, (b) the energy of the photons is greater than the work function, (c) the light frequency is below the threshold frequency, (d) both (a) and (b).

15. If electromagnetic radiation is made up of quanta, why don't we detect the discrete packets of energy, for example, when we listen to a radio?

16. In terms of biological damage, ionization does more damage when you stand in front of a very weak (low power) beam of X-ray radiation than in front of a stronger beam of red light. How does the photon concept explain this paradoxical situation?

17. In the photoelectric effect, if the frequency of the radiation is below a certain cutoff value, no photoelectrons will be observed *no matter how intense* the radiation is. Why does this fact favor a particle theory of light over a wave theory?

18. ■ What is the energy of a quantum of light with a frequency of 5.0×10^{14} Hz?

19. ■ A photon has an energy of 3.3×10^{-15} J. What wavelength and type of radiation is this photon?

20. ■ Which has more energy, and how many times more: a quantum of violet light or a quantum of red light from the opposite ends of the visible spectrum?

21. ■ A beam of ultraviolet light has a wavelength of 150 nm. How much energy does a photon of UV light have, in (a) joules and (b) electron volts?

22. ■ If the intensity of the light producing a photocurrent is doubled, how is that current affected?

23. ■ The photoelectrons emitted by an illuminated surface have a maximum kinetic energy of 3.0 eV. If the intensity of the light is tripled, what is the maximum kinetic energy of photoelectrons now?

24. ■■ If a 100-watt light bulb gives off 2.50% of its energy as visible light, how many photons of visible light are given off in 1.00 min? (Use an average visible wavelength of 550 nm.)

25. ■■ •Figure 27.21 shows a graph of stopping potential versus frequency for a photoelectric material. Determine

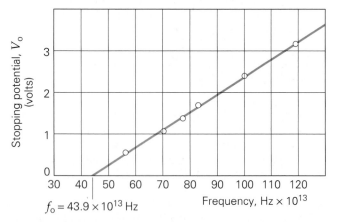

•FIGURE 27.21 **Stopping potential versus frequency**
See Exercise 25.

(a) Planck's constant and (b) the work function of the material.

26. ■■ A metal with a work function of 2.40 eV is illuminated with a beam of monochromatic light. The stopping potential for the emitted electrons is 2.50 V. What is the wavelength of the light, in nanometers?

27. ■■ What is the lowest frequency of light that can cause the release of electrons from a metal that has a work function of 2.8 eV?

28. ■■ The threshold wavelength for emission from a metallic surface is 500 nm. Calculate the maximum speed of emitted photoelectrons for light of wavelength (a) 400 nm, (b) 500 nm, and (c) 600 nm.

29. ■■ In Exercise 28, what is the work function of the metal, in electron volts?

30. ■■ The work function of a photoelectric material is 3.5 eV. If the material is illuminated with monochromatic light ($\lambda = 300$ nm), (a) find the stopping potential of emitted photoelectrons. (b) What is the cutoff frequency?

31. ■■ In studying a solid for possible use as a solar cell for the direct conversion of light into electricity, materials engineers used a beam of monochromatic blue light ($\lambda = 420$ nm) to observe the emission of photoelectrons. The maximum kinetic energy of these electrons is 1.00×10^{-19} J. If this solid is to be used as a solar electricity generator, can we expect that the red end of the solar spectrum ($\lambda = 700$ nm) will dislodge electrons from it? Justify your answer mathematically.

32. ■■ Gold has a work function of 4.82 eV. If a block of gold is illuminated with ultraviolet light ($\lambda = 160$ nm), what are (a) the maximum kinetic energy (in electron volts) of the emitted photoelectrons and (b) the threshold frequency?

33. ■■ Sodium and silver have work functions of 2.46 eV and 4.73 eV, respectively. (a) If the surfaces of both metals are illuminated with the same monochromatic light, photoelectrons from which metal will have the greater speed on emission? (b) What wavelengths (in nanometers) are required for photoelectrons to be emitted from each metal surface?

34. ■■■ When a certain photoelectric material is illuminated with red light ($\lambda = 700$ nm) and then with blue light ($\lambda = 400$ m), the maximum kinetic energy of the photoelectrons emitted by the blue light is twice that of photoelectrons emitted by the red light. What is the work function of the material?

35. ■■■ When the surface of a particular metal is illuminated with monochromatic light of various frequencies, the stopping potentials for the photoelectrons are:

Frequency (in hertz)

9.9×10^{14}	7.6×10^{14}	6.2×10^{14}	5.0×10^{14}

Stopping potential (in volts)

2.6	1.6	1.0	0.60

Plot these data, and determine the experimental value of Planck's constant and the work function of the metal.

27.3 Quantum "Particles": The Compton Effect

36. The Compton effect was first observed by using (a) visible light, (b) infrared radiation, (c) ultraviolet light, (d) X-rays.

37. The wavelength shift for Compton scattering is a maximum when (a) the photon scattering angle is $90°$, (b) the electron scattering angle is $90°$, (c) the shift is equal to the Compton wavelength, (d) the incident photon is backscattered.

38. A photon can undergo Compton scattering from a molecule as well as from a free electron. How does the maximum wavelength shift for Compton scattering from a molecule as a unit compare to that from a free electron? Explain.

39. In the center of the Sun, most of the electromagnetic energy is in the form of X-rays. By the time it reaches the surface, however, it is mostly in the visible range. Explain how a large number of successive Compton scatterings can contribute to this wavelength change.

40. In Compton scattering, why does the scattered photon always have a longer wavelength than the incident photon?

41. ■ Calculate the maximum wavelength shift for Compton scattering from a free electron.

42. ■ What is the change in wavelength when monochromatic X-rays are scattered by free electrons through an angle of $30°$?

43. ■ A monochromatic beam of X-rays of wavelength 2.80×10^{-10} m is scattered by a metal foil. What is the wavelength of the scattered X-rays at an angle of $45°$ from the direction of the incident beam?

44. ■ X-rays of wavelength 0.0045 nm are used in a scattering experiment. If the X-rays are scattered through an angle of $53°$, what is the wavelength of the scattered radiation?

45. ■■ A photon with an energy of 10.0 keV is scattered by a free electron. If the electron recoils with a kinetic energy of 0.200 keV, what is the wavelength of the scattered photon?

46. ■■ X-rays scattered from the electrons in a carbon target show a wavelength shift of 0.000 326 nm. What is the scattering angle?

47. ■■ If the Compton shift of a photon scattered by an electron is 1.25×10^{-4} nm, what is the scattering angle?

48. ■■■ A photon can undergo the Compton effect by scattering from any particle, such as a proton. What is (a) the Compton wavelength of a proton? (b) the ratio of maximum wavelength shifts for scattering by an electron to that by a proton?

27.4 The Bohr Theory of the Hydrogen Atom

49. In his theory of the hydrogen atom, Bohr postulated the quantization of (a) energy, (b) centripetal acceleration, (c) light, (d) angular momentum.

50. An excited hydrogen atom emits light when its electron (a) makes a transition to a lower energy level, (b) is excited to a higher energy level, (c) is in the ground state.

51. Explain why the Bohr theory is applicable only to the hydrogen atom and to hydrogen-like atoms, such as singly ionized helium, doubly ionized lithium, and other one-electron systems.

52. Does it take more energy to ionize (free) the electron of a hydrogen atom that is in an excited state than one in the ground state? Explain.

53. Very accurate measurements of the wavelengths of light emitted by a hydrogen atom indicate that all wavelengths are slightly longer than expected from the Bohr theory. How might the conservation of momentum help explain this? [Hint: Photons carry momentum and energy, both of which must be conserved.]

54. ■ Find the energy required to excite a hydrogen electron from (a) the ground state to $n = 2$ and (b) $n = 2$ to $n = 3$.

55. ■ In Exercise 54, classify the type of light needed to create the transitions.

56. ■ Find the energy needed to ionize a hydrogen atom whose electron is in the (a) $n = 2$ state and (b) $n = 3$ state.

57. ■ What frequency of light would excite the electron of a hydrogen atom from state (a) $n = 2$ to $n = 5$ and (b) $n = 2$ to $n = \infty$?

58. ■ Find the radius of the electron orbit in a hydrogen atom for each of the following states: (a) $n = 2$, (b) $n = 4$, (c) $n = 5$.

59. ■ Scientists are now studying "large" atoms—atoms with orbits approaching everyday sizes. For the diameter of the orbit of hydrogen to be 10^{-6} m—close to the diameter of a dust particle ($1 \, \mu$)—what excited state is required? (Give an approximate principal quantum number.)

60. ■ Use the Bohr theory to find the value of the Rydberg constant in the empirical formula for hydrogen (Eq. 27.10).

61. ■ Find the energy of the hydrogen electron in each of the following states: (a) $n = 3$, (b) $n = 6$, (c) $n = 10$.

62. ■■ Find the binding energy of the hydrogen electron in each of the following states: (a) $n = 3$, (b) $n = 5$, (c) $n = 10$.

63. ■■ The ionization energy of a hydrogen atom is 13.6 eV. Would the absorption of light of frequency 7.00×10^{15} Hz cause a hydrogen atom to be ionized? If so, what would be the kinetic energy of the emitted electron?

64. ■■ A hydrogen electron in the ground state is excited to the $n = 5$ energy level. The electron makes a transition directly to the $n = 2$ level before returning to the ground state. (a) What are the wavelengths of the emitted photons? (b) Would any of the photons emitted be in the visible part of the spectrum?

65. ■■ For which one of the following transitions of a hydrogen electron is the photon of greatest energy emitted: (a) $n = 5$ to $n = 3$, (b) $n = 6$ to $n = 2$, (c) $n = 2$ to $n = 1$? Justify your answer mathematically.

66. ■■ In the hydrogen spectrum, the series of lines called the Lyman series results from transitions to the $n = 1$ energy level. What is the longest wavelength in this series, and in which region of the electromagnetic spectrum does it lie?

67. ■■ What is the binding energy for the ground-state electron in the following hydrogen-like ions: (a) He^+; (b) Li^{2+}?

68. ■■ A hydrogen atom absorbs a photon of wavelength 486 nm. (a) How much energy did the atom absorb? (b) What were the initial and final states of the hydrogen electron?

69. ■■■ Show that the speeds of an electron in the Bohr orbits of the hydrogen atom are given (to two significant figures) by $v_n = (2.2 \times 10^6 \text{ m/s})/n$.

70. ■■■ Show that the total energy of an electron in a hydrogen atom in an orbit of radius 0.0529 nm is -13.6 eV.

71. ■■■ In Excercise 70, (a) how much of the energy is potential, and how much is kinetic? (b) How do the magnitudes of these energies compare?

27.5 A Quantum Success: The Laser

72. Which of the following is not an essential for laser action? (a) population inversion, (b) phosphorescence, (c) pumping, (d) stimulated emission.

73. Why does a laser emit only one or a few colors, whereas a light bulb emits a continuous spectrum?

74. In what sense is a laser an "amplifier" of energy? Explain why this does not violate conservation of energy.

Additional Exercises

75. An FM radio station broadcasts at a frequency of 98.9 MHz and radiates 750 kW of power. How many photons are radiated from the station's antenna each second?

76. A photon can be treated mathematically as a "massless" particle. Show that the momentum of a photon is given by $p = h/\lambda$. [Hint: Use the relativistic relationship $E^2 = p^2c^2 + (mc^2)^2$, which relates the total energy and momentum (Exercise 62, Chapter 26)].

77. Referring to •Fig. 27.22, derive the equation for the wavelength shift in Compton scattering by applying the

conservation of momentum and energy to an elastic collision. [Hint: Consider total relativistic energy. Also, analyze momentum classically with components as in Chapter 6, but remember that $p = h/\lambda$ for a photon; see Exercise 76.]

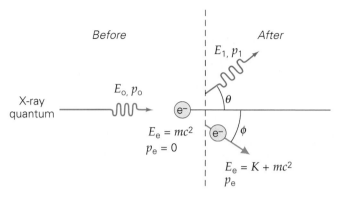

•**FIGURE 27.22 Compton scattering** See Exercise 77.

78. Light of wavelength 340 nm is incident on a metal surface and ejects electrons that have a maximum speed of 3.5×10^5 m/s. What is the work function of the metal?

79. What are the energies of the photons required to excite a hydrogen electron in the ground state to states (a) $n = 3$ and (b) $n = 5$?

80. How many photons of red light ($\lambda = 700$ nm) would it take to have 1.0 J of energy?

81. What is the incident energy of each photon in a beam of monochromatic X-rays that have a wavelength of 4.5×10^{-10} m at a Compton scattering angle of 37°?

82. What is the frequency of the most intense spectral component from a blackbody at room temperature (20°C)?

83. The work function of a particular metal is 5.0 eV. (a) What is the frequency of light that causes electrons to be emitted with a maximum kinetic energy of 2.0×10^{-19} J? (b) What is the stopping potential of the electrons?

84. For what scattering angle would the wavelength shift for Compton scattering be 25% of the Compton wavelength?

85. How many orbital transitions of the electron in a hydrogen atom result in the emission of light in the visible region of the spectrum (400–700 nm)?

86. Explain why, in the photoelectric effect, the photocurrent reduces gradually to zero as we turn up the retarding voltage. In other words, why does the current not stay at a large constant value until we reach the stopping potential and then suddenly drop to zero? [Hint: See Fig. 27.3b, and recall that all electrons in a solid are not bound equally.]

87. What is the frequency of revolution of the electron about a hydrogen nucleus for the $n = 1$ orbit?

Quantum Mechanics and Atomic Physics

Just a few decades ago, if someone had claimed to have a photograph of an atom, they would have been laughed at. Today, a device called the scanning tunneling microscope routinely produces images such as the photograph above. The blue shapes represent iron atoms, neatly arranged on a copper surface.

The basis of this type of microscope is a quantum-mechanical phenomenon called tunneling. Tunneling reflects two fundamental features of the subatomic realm: the probabilistic character of quantum processes and the wave nature of particles. These features explain how particles may turn up in places where, according to classical notions, they have no right to be.

Around 1925, a new kind of physics based on the synthesis of wave and quantum ideas was introduced. This new theory, called **quantum mechanics**, attempts to combine the wave–particle duality into a single consistent description. It revolutionized scientific thought and provided the basis of our present understanding of microscopic (and submicroscopic) phenomena.

In this chapter we will present some of the basic ideas of quantum mechanics to show how they describe waves and particles. We will also discuss some of the practical applications made possible by the quantum-mechanical view of nature, such as the electron microscope and magnetic resonance imaging (MRI).

28.1 Matter Waves: The De Broglie Hypothesis

Objectives: To (a) explain de Broglie's hypothesis, (b) calculate the "wavelength" of a matter wave, and (c) specify under what circumstances the wave nature of matter will be observable.

Note: These relationships are discussed in Section 26.4.

Since a photon travels at the speed of light, we treat it relativistically as a "massless" particle. If this were not the case, it would have infinite energy. (At $v = c$, the dimensionless relativistic factor γ becomes infinite and its total energy, $E = \gamma mc^2$, would also become infinite.) It can be shown that the total energy E and relativistic momentum p are related by the equation $E^2 = p^2c^2 + (mc^2)^2$. (See Exercise 62, Chapter 26.) A photon ($m = 0$) then has a total energy of $E = pc$, or a relativistic momentum of $p = E/c$. Photon energy can also be written $E = hf = hc/\lambda$, so the momentum a photon carries is related to its wavelength by

Note: See Eq. 27.4 in Section 27.2.

Photon momentum

$$p = \frac{E}{c} = \frac{hf}{c} = \frac{h}{\lambda} \qquad \textit{(photons only)} \qquad (28.1)$$

Thus, the energy in electromagnetic waves of wavelength λ can be thought of as being carried by photon "particles," each having a momentum of h/λ.

Since nature exhibits a great deal of symmetry, the French physicist Louis de Broglie (1892–1987) thought that there might be a wave–particle symmetry. That is, if light sometimes behaves like a particle, perhaps material particles, such as electrons, also have wave properties. In 1924, de Broglie hypothesized that a moving particle has a wave associated with it. He proposed that the wavelength of a material particle is related to the particle's momentum ($p = mv$) by an equation similar to that of a photon (Eq. 28.1). The **de Broglie hypothesis** states that

A particle with momentum of magnitude p has a wave associated with it. This wave has a wavelength of

De Broglie wavelength of a moving particle

$$\lambda = \frac{h}{p} = \frac{h}{mv} \qquad (28.2)$$

These waves associated with moving particles were called **matter waves** or, more commonly, **de Broglie waves**, and were thought to somehow influence or "guide" particle motion. We might think of an electromagnetic wave as the associated wave for a photon. However, de Broglie waves associated with particles such as electrons and protons are *not* electromagnetic waves. The de Broglie equation for the wavelength gives no clue as to the nature of the wave associated with a particle that has mass.

Needless to say, de Broglie's hypothesis met with a great deal of skepticism. The idea that the motions of photons were somehow governed by the electromagnetic wave properties of light did not seem unreasonable. But to extend this idea to say that the motion of a particle *with mass* is somehow governed by the properties of an associated wave was difficult to accept. Moreover, there was no evidence at the time that particles exhibited any wave properties such as interference and diffraction.

Note: The Bohr model is outlined in Section 27.4.

In support of his hypothesis, de Broglie showed how it could give an interpretation of the quantization of the angular momentum in Bohr's theory of the hydrogen atom. Recall that to get agreement with the hydrogen atomic spectrum, Bohr had to make the ad hoc hypothesis that the angular momentum of the orbiting electron was quantized in integer multiples of $h/2\pi$. De Broglie argued that for a *free* particle, the associated wave would have to be a *traveling* wave. However, the *bound* electron of a hydrogen atom travels repeatedly in discrete circular orbits according to the Bohr theory. The associated matter wave would then be expected to be a *standing* wave. For a standing wave to be produced, an integral number of wavelengths had to fit into a given orbital circumference. The

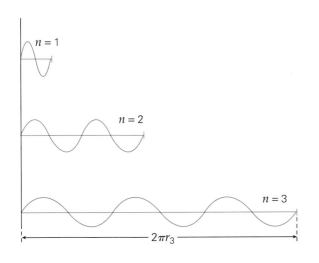

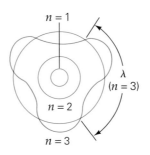

•FIGURE 28.1 **De Broglie waves and Bohr orbits** Similar to standing waves in a stretched string, de Broglie waves form circular standing waves on the circumferences of the Bohr orbits. The number of wavelengths in a particular orbit of radius r, shown here for $n = 3$, is equal to the principal quantum number of that orbit. (Not drawn to scale.)

wave must meet itself, so to speak, with the same phase. That is, on returning to the same spot, it must reinforce itself constructively, much as a standing wave in a circular string might do. This idea is similar to the linear case for a standing wave in a stretched string fixed at both ends (•Fig. 28.1).

Note: Compare the discussion of standing waves in Section 13.5.

The circumference of a Bohr orbit of radius r_n is $2\pi r_n$, where n is the quantum number. De Broglie equated this length to an integral number of electron "wavelengths":

$$2\pi r_n = n\lambda \quad \text{for} \quad n = 1, 2, 3, \ldots$$

Using the de Broglie equation, $\lambda = h/mv$, for the electron wavelength, and recalling that the angular momentum for a circular orbit is $L = mvr$, we have

Note: See Eq. 8.14 for a reminder about angular momentum.

$$2\pi r_n = \frac{nh}{mv} \quad \text{or} \quad L_n = mvr_n = n\left(\frac{h}{2\pi}\right) \quad \text{for} \quad n = 1, 2, 3, \ldots$$

Thus, according to de Broglie, the angular momentum is quantized just as Bohr proposed. It was clear that de Broglie's hypothesis was consistent with Bohr's postulate for the hydrogen atom. For orbits other than those allowed by the Bohr theory, the de Broglie wave for the orbiting electron would *not* close on itself. That is, the circumference of a disallowed orbit would *not* be equal to an integral number of wavelengths. This would give rise to destructive interference, and the resulting wave would have zero amplitude. This concept seems to imply that *the amplitude of the de Broglie wave must then be related, in some way, to the location of the electron.* This idea became a fundamental cornerstone of modern quantum mechanics.

If particles really are associated with a wave, why then have we never observed wave effects for them? As the next Example points out, effects such as diffraction are significant only when the wavelength λ is roughly the same size as the object or opening it meets (Section 24.3). If λ is much smaller than these dimensions, diffraction is negligible.

Example 28.1 ■ Should Ballplayers Worry About Diffraction? De Broglie Wavelength

A pitcher throws a fastball to his catcher at 40 m/s through a square strike-zone practice target—a sheet of canvas with an opening 50 cm on each side. If the ball's mass is 0.15 kg, (a) what is the wavelength of the de Broglie wave associated with the ball? (b) Should the catcher expect diffraction as the ball passes through this opening?

Thinking It Through. (a) We can use the de Broglie relationship (Eq. 28.2) to calculate the wavelength of the matter wave associated with the ball. (b) Is the wavelength much larger than, much smaller than, or approximately the same size as the opening?

Solution.

Given: $v = 40$ m/s 　　　　　　　　　　*Find:* (a) λ (de Broglie wavelength)
　　　　$d = 50$ cm $= 0.50$ m 　　　　　　　　　　　(b) Is diffraction likely?
　　　　$m = 0.15$ kg

(a) We have

$$\lambda = \frac{h}{mv} = \frac{6.63 \times 10^{-34}\,\text{J·s}}{(0.15\,\text{kg})(40\,\text{m/s})} = 1.1 \times 10^{-34}\,\text{m}$$

(b) Recall from Section 24.3 that for a wave to exhibit significant diffraction when passing through an opening, its wavelength must be similar in size to the opening. Since $\lambda = 1.1 \times 10^{-34}$ m $\ll 0.50$ m, this is *not* the case here. With a wavelength so small compared with the opening, the wave travels straight through with no noticeable diffraction.

Follow-up Exercise. In this Example, (a) how fast would the ball have to be thrown for diffractive effects to become important? (b) At that speed, how long would the ball take to travel the 20 m to the plate? (Compare your answer with the age of the universe, about 15 billion years. Do you think this movement would be noticeable?)

From Example 28.1, it is little wonder that we don't directly observe any effects due to the wave nature of matter. We cannot then hope to detect the wave properties of ordinary objects. However, particles with very small masses traveling at low speeds are another story, as the following Example shows.

Example 28.2 ■ A Whole Different Ballgame: de Broglie Wavelength of an Electron

(a) What is the de Broglie wavelength of the wave associated with an electron that has been accelerated from rest through a potential of 50.0 V? (b) Compare your answer to the typical distance between atoms in a solid crystal, about 10^{-10} m. Would you expect diffraction to occur as these electrons pass between such atoms?

Thinking It Through. (a) We use the de Broglie relationship (Eq. 28.2), but first we must calculate the electron's speed by using energy conservation—the conversion of electric potential energy into kinetic energy. (b) For diffractive effects to be important, we must have $\lambda \approx 10^{-10}$ m.

Solution.

Given: $V = 50.0$ V 　　　　　　　　　　*Find:* (a) λ (de Broglie wavelength)
　　　　$m = 9.11 \times 10^{-31}$ kg (known) 　　　　　　　(b) Is diffraction likely?

(a) The magnitude of the potential energy lost by the electron is $|\Delta U_e| = eV$ and is equal to its kinetic energy gain ($\Delta K = \frac{1}{2}mv^2$ since $K_o = 0$). Equating the two magnitudes enables us to calculate the speed:

$$\tfrac{1}{2}mv^2 = eV \quad \text{or} \quad v = \sqrt{\frac{2eV}{m}}$$

so

$$v = \sqrt{\frac{2(1.60 \times 10^{-19}\,\text{C})(50.0\,\text{V})}{9.11 \times 10^{-31}\,\text{kg}}} = 4.19 \times 10^6\,\text{m/s}$$

Thus, the electron's de Broglie wavelength is

$$\lambda = \frac{h}{mv} = \frac{6.63 \times 10^{-34}\,\text{J·s}}{(9.11 \times 10^{-31}\,\text{kg})(4.19 \times 10^6\,\text{m/s})} = 1.74 \times 10^{-10}\,\text{m}$$

Note: Diffraction from crystal lattices is discussed in Section 24.3.

(b) Since this result is the same order of magnitude as the opening between atoms, diffraction might be observed. Thus, using electrons and a crystal lattice might help us prove de Broglie's hypothesis.

Follow-up Exercise. In this Example, what if the particle were a proton? (Give a numerical answer.) In other words, would the de Broglie wave of a proton be more or less likely to diffract than that of an electron under the same conditions? Explain.

In Example 28.2, if we changed the accelerating voltage, we would have to go through the entire calculation of v and λ again, using the electron mass, electron charge, Planck's constant, and so on. Instead, it can be convenient to use the known values of m, e, and h to derive a nonrelativistic formula for the wavelength of an electron accelerated through a potential difference V. We begin by expressing the kinetic energy in terms of momentum. Since $p = mv$, the kinetic energy, $\frac{1}{2}mv^2$, can be written as $p^2/2m$. Then, by energy conservation,

$$\frac{p^2}{2m} = eV \quad \text{or} \quad p = \sqrt{2meV}$$

Thus for these conditions the de Broglie wavelength is

$$\lambda = \frac{h}{p} = \frac{h}{\sqrt{2meV}} = \sqrt{\frac{h^2}{2meV}}$$

Inserting the values of h, e, and m and rounding the result to three significant figures gives (can you show this?)

$$\lambda = \sqrt{\frac{1.50}{V}} \times 10^{-9} \text{ m} = \sqrt{\frac{1.50}{V}} \text{ nm} \quad \begin{array}{l}\textit{(nonrelativistic}\\ \textit{electrons only)}\end{array} \quad (28.3)$$

where V is in volts. Thus, if $V = 50.0$ V,

$$\lambda = \sqrt{\frac{1.50}{50.0}} \times 10^{-9} \text{ m} = 0.173 \text{ nm}$$

(Note that this answer differs in the last digit from that found in Example 28.2, because there the values for h, e, and m were rounded to three significant figures *before* we performed the calculations.)

You may wish to derive a comparable formula for a proton, to use in solving similar problems. (What quantities would you have to change?)

Note: This alternative de Broglie wavelength expression for electrons accelerated through a potential difference, in which V must be in volts, applies *only to nonrelativistic electrons.*

In 1927, two physicists in the United States, C. J. Davisson and L. H. Germer, used a crystal to diffract a beam of electrons, thereby demonstrating a wavelike property of particles. A single crystal of nickel was used in the experiment. The crystal was cut to expose a spacing of $d = 0.215$ nm between the lattice planes. When a beam of electrons was directed normally onto the crystal face, a maximum in the intensity of the scattered electrons was observed at an angle of 50° relative to the surface normal (•Fig. 28.2). The scattering was most intense for an accelerating potential of 54.0 V.

According to wave theory, constructive interference due to waves reflected from two lattice planes (a distance d apart) should occur at certain angles of scattering θ. The theory predicts that the first-order maximum should be observed at an angle given by

$$d \sin \theta = \lambda$$

This condition in Fig. 28.2 requires a wavelength of

$$\lambda = d \sin \theta = (0.215 \text{ nm}) \sin 50° = 0.165 \text{ nm}$$

We use Eq. 28.3 to determine the de Broglie wavelength of the electrons:

$$\lambda = \sqrt{\frac{1.50}{V}} \text{ nm} = \sqrt{\frac{1.50}{54.0}} \text{ nm} = 0.167 \text{ nm}$$

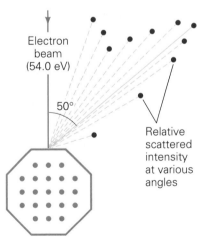

•**FIGURE 28.2 The Davisson–Germer experiment**
When a beam of electrons of kinetic energy 54.0 eV is incident on the face of a nickel crystal, a maximum in the scattering intensity is observed at an angle of 50°.

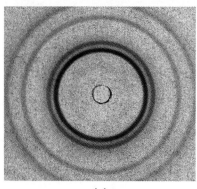

(a)

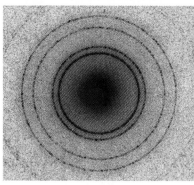

(b)

•**FIGURE 28.3 Diffraction patterns** Comparison of **(a)** an X-ray diffraction pattern and **(b)** an electron diffraction pattern leaves little doubt that electrons exhibit wavelike properties.

The agreement of the wavelengths was excellent within the limits of experimental accuracy, and the Davisson–Germer experiment gave convincing proof of the validity of de Broglie's matter wave hypothesis.

Another experiment, carried out by G. P. Thomson in Great Britain in the same year, added further proof. Thomson passed a beam of energetic electrons through a thin metal foil. The diffraction pattern of the electrons was the same as that of X-rays. A comparison of such patterns, as shown in •Fig. 28.3, leaves little doubt that particles exhibit wavelike properties. A practical application of this fact is given in the Insight on p. 872.

28.2 The Schrödinger Wave Equation

Objectives: To understand qualitatively (a) the reasoning that underlies the Schrödinger wave equation and (b) this equation's use in finding particle wave functions.

De Broglie's hypothesis predicts that moving particles have associated waves that somehow govern or describe their behavior. However, it does not tell us the *form* of these waves, only their wavelengths. To have a useful theory, we need an equation that will give us the form of the matter waves. In particular, we need the wave form of a particle moving under the influence of a force—for example, an electron orbiting a proton under the influence of the electric force. Also, we must know the rules under which these waves govern particle motion. In 1926, Erwin Schrödinger, an Austrian physicist, presented his ideas on this subject. They included a general equation that describes the de Broglie matter waves, how they travel, and their interpretation.

The de Broglie wave is described by some mathematical function. (Recall how wave motion was described by equations in Section 13.2.) In general, this function is commonly denoted by the symbol ψ (the Greek letter psi, pronounced "sigh"). Called the **wave function,** ψ describes the wave as a function of time and space. For example, the wave function of a (classical) traveling wave of amplitude A, location x, wavelength λ, and angular frequency ω is $\psi = A \sin [(2\pi x/\lambda) - \omega t]$.

Since the de Broglie wave governs a particle's motion, it would seem reasonable that the wave function is associated with the particle's energy, both kinetic and potential. Recall from the discussion of conservation of energy in Section 5.5 that, for a conservative mechanical system, the sum of the kinetic and potential energies is a constant:

$$K + U = E = \text{a constant}$$

where the sum of the kinetic and potential energies equals the total mechanical energy, E. Schrödinger proposed a similar equation for the de Broglie matter waves involving the wave function ψ. **Schrödinger's wave equation** has the general form

$$(K + U)\psi = E\psi \qquad (28.4)$$

That is, the wave function is associated with the energy of a system. When Eq. 28.4 is applied to various situations with known expressions for K and U, complex mathematical equations result that are beyond the scope of this text.* The idea is to solve these equations for ψ (for a given force represented by a potential energy function U) and thus determine its mathematical expression. But once we solve for ψ, what is its physical significance?

In fact, during the early development of quantum mechanics (sometimes called wave mechanics), it was not clear how ψ should be interpreted. After much

---

*Although Eq. 28.4 looks like a multiplication of the previous equation by ψ, it is much more complex. For example, K is no longer $\frac{1}{2}mv^2$ but is instead replaced by a quantity that enables us to *extract* the kinetic energy from ψ.

thought and investigation, it was concluded that ψ^2 (the wave function squared) represents the probability of finding a particle at a certain position and time.

This interpretation involves the amplitude of ψ. Recall from Chapter 13 that the energy or intensity of a classical wave is proportional to the square of its amplitude. Similarly, the intensity of a light wave is proportional to E^2, where E is the electric field amplitude. In terms of "particle" photons, the *intensity* of a light beam is proportional to the number (n) of photons in the beam, so $n \propto E^2$. That is, the number of photons is proportional to the square of the electric field amplitude of the wave.

Note: Review Eq. 13.5, Section 13.1.

The wave function ψ is interpreted in an analogous manner. The wave function generally varies in magnitude in space and time. If ψ describes a beam of electrons, then ψ^2 will be proportional to the number of electrons that should be found in a small volume centered on a given location at a particular time. However, when there is a *single* particle, ψ^2 represents the probability of finding the electron at various locations. *Thus, in quantum mechanics the square of a particle's wave function is proportional to the probability of finding the particle at a given location in space and time, or the* **probability density**.

Note: More precisely, the *square* of the absolute value of the wave function solution to Schrödinger's equation, $|\psi|^2$, gives the probability of finding a particle at a location.

The interpretation of ψ^2 as the probability of finding a particle at a particular place altered the idea that the electron of the hydrogen atom could be found only in orbits at discrete distances from the nucleus, as described in the Bohr theory. When the Schrödinger equation was solved for the hydrogen atom, the radial wave functions for each energy level were found to have nonzero values at almost any distance from the nucleus. Thus, quantum mechanically, there is some nonzero probability (ψ^2) of finding the electron in a given energy level at almost *any* distance from the nucleus. For example, the relative probability that an electron in the ground state ($n = 1$) is at a given radial distance from the proton is shown in •Fig. 28.4a. The *maximum* probability coincides with the Bohr radius of 0.0529 nm, but it is possible (that is, there is a finite probability) that the electron could be at almost any distance from the nucleus, including inside the nucleus! The wave function exists for distances beyond 0.20 nm, but there is little chance of finding an electron in the ground state beyond this distance. The probability

Note: The Bohr model of the hydrogen atom is discussed in Section 27.4.

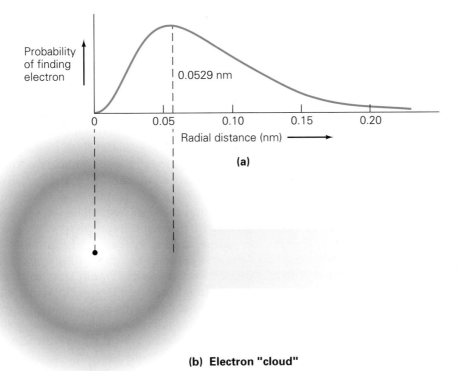

(a)

(b) Electron "cloud"

•**FIGURE 28.4 Electron probability: the hydrogen atom** **(a)** The square of the wave function is the probability of finding the hydrogen electron here, assumed to be in the ground state ($n = 1$), as a function of radial distance from the proton. The electron has the greatest probability of being at distance of 0.0529 nm, which matches the radius of the first Bohr orbit. **(b)** The probability distribution gives rise to the idea of an electron cloud around the nucleus. The cloud density reflects the probability density.

The Electron Microscope

The de Broglie hypothesis helped set the stage for the development of an important practical application of electron beams—the *electron microscope*. As we have seen from the discussion of the Davisson–Germer and Thomson experiments, the wave properties of the electron are not just conceptual or mathematical abstractions. Electrons can undergo diffraction, as do light waves. Does this imply that electrons can also be focused like light waves? Yes, and in fact we can focus electron "waves" to form images, though the means employed are somewhat different from those used with light waves in a light microscope. As Example 28.2 shows, accelerated electrons have very short wavelengths. With such short wavelengths, it is possible to obtain greater magnification and finer resolution than any light microscope can provide. (Recall the relationship of resolving power to wavelength in Section 25.4.) The resolving power of standard electron microscopes is on the order of a few nanometers.

Technological developments made in the 1920s, in particular the focusing of electron beams by magnetic coils, permitted the construction of the first electron microscope in Germany in 1931. In a *transmission electron microscope*, an electron beam is directed onto a very thin specimen. Different numbers of electrons pass through different parts of the specimen depending on its structure. The transmitted beam is then brought to focus by a magnetic objective coil. The general components of electron and light microscopes are analogous (Fig. 1), but an electron microscope is housed in a high-vacuum chamber so that the electrons do not interact with air molecules. As a result, an electron microscope

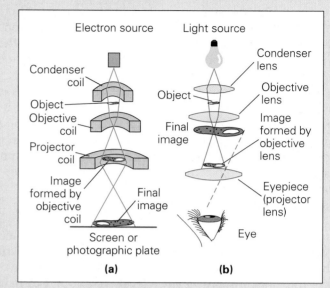

FIGURE 1   Electron and light microscopes
A comparison of the elements of **(a)** an electron microscope and **(b)** a compound light microscope. The light microscope is drawn upside down for a better comparison.

looks nothing like a light microscope (Fig. 2). A light microscope image (Fig. 3a) is limited to a magnification of about 2000×, whereas magnifications up to 100 000× can be achieved with an electron microscope (Fig. 3b).

density distribution gives rise to the idea of an *electron cloud* around the nucleus (Fig. 28.4b). The cloud density reflects the probability density that the electron is in a particular region.

Thus, quantum mechanics uses the wave functions of particles to predict the probability of particle phenomena. This also applies to light and photon "particles." For example, the square of a classical electromagnetic wave function, which is the de Broglie wave for its photons, gives the probability of finding the photons at a location in space. In a single-slit diffraction experiment, we know that many photons will strike the region of the central maximum on a screen, but fewer photons will arrive in the regions of the first bright fringes, even fewer in the regions of the second bright fringes, none in the dark regions between them, and so on. The quantum mechanical probability distribution or density on the screen has the same relative distribution as the classical intensity or brightness.

If this experiment could be done one photon at a time, *we could not predict exactly* to which region it would go. Its wave function would pass through both slits, and the two diffracted waves would interfere at the screen. We could, therefore, give only the *probability* that it would land in a particular spot on the screen. Thus, each photon lands at a spot, with a probability proportional to the square of its wave function at that point, which is the sum of the two diffracted waves from the two slits. Finally, after many such photons land, the classical two-slit interference pattern (fringes) of light waves would eventually build up.

Note: Single-slit diffraction is discussed in Section 24.3.

FIGURE 2 An electron microscope The microscope is housed in a cylindrical vacuum chamber, at left.

Another difference is that the final lens in an electron microscope, called the projector coil, has to project a real image onto a fluorescent screen or photographic film, since the eye cannot perceive an electron image directly. Specimens for transmission electron microscopy must be very thin. Special techniques allow the preparation of specimen sections as thin as 10 nm to 20 nm (only about 100 atoms thick).

The surfaces of thicker objects can be examined by the *reflection* of the electron beam from the surface. This is done with the more recently developed *scanning electron microscope*. A beam spot is scanned across the specimen by means of deflecting coils, much as is done in a television tube. Surface irregularities cause directional variations in the intensity of the reflected electrons, which gives contrast to the image. The specimens must be coated with a thin layer of metal (such as gold or aluminum) to make them conducting. Otherwise, they would be charged up nonuniformly from the electron beam and distort the image. Through such techniques, an electron microscope gives pictures with a remarkable three-dimensional quality, such as those in Fig. 3c.

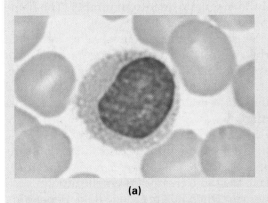

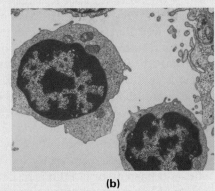

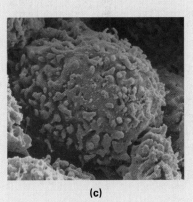

(a)   (b)   (c)

FIGURE 3 Lymphocytes (white blood cells) Images produced by **(a)** a light microscope, **(b)** a transmission electron microscope (TEM), and **(c)** a scanning electron microscope (SEM).

Another quantum-mechanical result that runs counter to our everyday experiences with large objects is called *tunneling.* As we have seen, a basic theme of quantum mechanics is that a particle's location is governed by a probability wave, and we know that the behavior of waves is different from that of particles. In classical physics, there are regions where particles are forbidden because of energy considerations. These are regions where a particle's potential energy would be greater than its total energy. The particle is not allowed in such regions because it would have a negative kinetic energy ($E - U = K < 0$), which is impossible. In such situations we say that the particle's location is limited by a *potential energy barrier.*

In certain instances, however, quantum mechanics predicts a small, but finite, probability of the particle's wave function penetrating the barrier and thus of the particle itself being found on the other side of the barrier. We say that there is a certain probability of the particle "tunneling" through the barrier. Such tunneling is observed to occur and forms the basis of the scanning tunneling microscope (STM), which creates images with resolution on the order of the size of a single

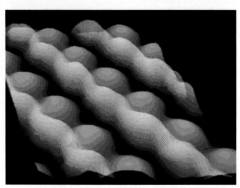

atom (•Fig. 28.5). Barrier penetration is also used to explain certain nuclear decay processes that classically should not occur, as we shall see in Chapter 29.

28.3   Atomic Quantum Numbers and the Periodic Table

Objective: To understand the structure of the periodic table in terms of quantum mechanical electron orbits and the Pauli exclusion principle.

The Hydrogen Atom

When the Schrödinger equation was solved for the hydrogen atom, the results predicted the allowed energy levels to be the same as those in the Bohr theory (Section 27.4). Their energy values depended only on the principal quantum number n. However, in addition, the solution gave two other quantum numbers, which are designated as ℓ and m_ℓ. Three quantum numbers are needed because the electron is described in three dimensions.

The quantum number ℓ is called the **orbital quantum number**. It is associated with the orbital angular momentum of the electron. For each value of n, the ℓ quantum number has integer values from zero up to $\ell = (n - 1)$. For example, if $n = 3$, the three possible values of ℓ are 0, 1, and 2. •Figure 28.6a shows three orbits with different angular momentum but the same energy. The number of different ℓ values for a given n value is equal to n. In the hydrogen atom, the energy of the electron in orbit depends only on n. Thus orbits with the same n value but different ℓ values have the same energy and are said to be *degenerate*.

The quantum number m_ℓ is called the **magnetic quantum number**. The name originated from experiments that applied an external *magnetic* field to a sample. They showed that a particular energy level (one with a given value of n and ℓ) of a hydrogen atom consists of several orbits with slightly different energies *only when in the magnetic field*. Thus, in the absence of a field, there was additional energy degeneracy. Clearly, there must be more to the description of the orbit than just n and ℓ. The quantum number m_ℓ was introduced to enumerate the number of levels that existed for a given orbital quantum number ℓ. Under normal conditions, the energy of the atom does not depend on either of these quantum numbers.

The magnetic quantum number m_ℓ is associated with the orientation of the orbital angular momentum vector **L** in space (Fig. 28.6b). If there is no external magnetic field, then all orientations of **L** have the same energy. (Magnetic dipoles such as current loops have different energies *only* when in a magnetic field, depending on their orientation.) For each value of ℓ, m_ℓ possesses values from zero to $\pm\ell$. That is, $m_\ell = 0, \pm 1, \pm 2, \ldots, \pm\ell$. For example, an orbit described by $n = 3$ and $\ell = 2$ can have m_ℓ values of $-2, -1, 0, +1, +2$. In other words, this orbit's angular momentum vector **L** has five orientations, each with identical energies if no magnetic field is present. In general, for a given value of ℓ, there will be $2\ell + 1$ possible values of m_ℓ. With $\ell = 2$, there are five values: $2\ell + 1 = (2 \times 2) + 1 = 5$.

This was not the end of the story. The use of high-resolution optical spectrometers indicated that each emission line of hydrogen is in fact two very closely spaced lines. Thus each emitted wavelength is actually two. This splitting is called spectral *fine structure*. So, for a given orbit, a fourth quantum number was needed! To describe the electron orbit correctly and completely required the concept of the electron **spin quantum number**, m_s. It is associated with the *intrinsic* angular momentum of the electron. This property, called *electron spin*, is sometimes described in analogy with the angular momentum associated with a spinning object (Fig. 28.6). If an electron were a spinning particle, its intrinsic angular momentum (spin) could possess only two orientations: "spin up" and "spin down." The fine structure of an atom was eventually found to be the result of the electron spin having two orientations with respect to the atom's internal magnetic field, produced by the electron's orbital motion. In analogy to a magnetic moment in a magnetic field, the atom possesses a slightly different energy when "lined up" with the field than

•**FIGURE 28.5   Scanning tunneling microscopy** Atoms of the semiconductor gallium arsenide. The scanning tunneling microscope, developed in the 1970s and 1980s, uses the quantum phenomenon called *vacuum tunneling*. The tip of a metal needle probe is moved across the contours of a metallized specimen surface. Applying a small voltage between the probe and the surface causes electrons to tunnel through the vacuum. The tunneling current is extremely sensitive to the separation of the needle tip and the surface. As a result, when the probe is scanned across the sample, surface features as small as atoms show as variations in the tunneling current that can be processed to produce three-dimensional images. (The closer the feature is to the probe, the more electrons can tunnel and thus the more current results.)

ℓ = orbital quantum number;
m_ℓ = magnetic quantum number;
m_s = spin quantum number

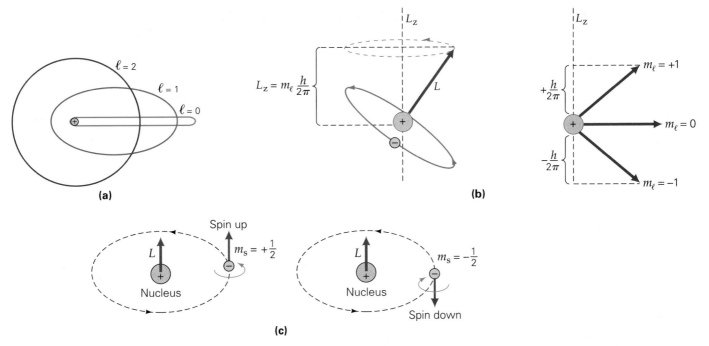

•FIGURE 28.6   **Classical interpretation of orbital quantum numbers**   (a) For planetary orbits, the energy depends only on the size (long axis) of the orbit. This provides a classical model for the subshell energy degeneracy. Here we plot the classical electron orbits around a proton for $n = 3$, showing the three subshells with different angular momentum but the same energy. The circular orbit has the maximum angular momentum; the narrowest orbit would actually pass through the proton classically and thus has zero angular momentum. **(b)** Here **L** is the vector angular momentum about the nucleus associated with the orbit. Because the energy of an orbit is independent of its spatial orientation, all orbits with the same ℓ but different values m_ℓ have the same energy. There are $2\ell + 1$ possible orientations for a given ℓ. The value of m_ℓ tells the *component* of the angular momentum vector in a given direction, as shown. **(c)** The spin can be either up or down, depending on the rotational sense of the electron. However, electron spin is strictly a quantum mechanical property and should not be identified with the physical spin of a macroscopic body except for conceptualization.

when "upside down" in the field. Thus the words "up" and "down" were coined and are relative to the field direction.

An electron's spin is fundamentally a purely quantum-mechanical property of the particle; it is not analogous to a spinning top. It is characterized by the spin quantum number m_s, which can have only two values, taken to be $m_s = \pm\frac{1}{2}$. For our example of an orbit with $n = 3$ and $\ell = 2$, any value of m_ℓ would offer two possible spin orientations and thus two possible sets of the four quantum numbers: $n = 3$, $\ell = 2$, $m_\ell = +1$, and $m_s = +\frac{1}{2}$; and $n = 3$, $\ell = 2$, $m_\ell = +1$, and $m_s = -\frac{1}{2}$. Both sets would have nearly the same energy. Thus the orbit's energy is almost independent of the electron spin direction. This results in yet another energy degeneracy: The energy levels of a hydrogen atom are, to a very high degree, determined solely by the primary quantum number, n.

The four quantum numbers for the hydrogen atom are summarized in Table 28.1. A particularly useful property of the spin of a proton and of atomic nuclei in general is discussed in the Insight on p. 876.

TABLE 28.1   Hydrogen Atom Quantum Numbers

Quantum Number	Symbol	Allowed Values	Number of Allowed Values
Principal	n	$1, 2, 3, \ldots, \infty$	No limit
Orbital angular momentum	ℓ	$0, 1, 2, 3, \ldots, (n-1)$	n (for each n)
Orbital magnetic	m_ℓ	$0, \pm1, \pm2, \pm3, \ldots, \pm\ell$	$2\ell + 1$ (for each ℓ)
Spin	m_s	$\pm\frac{1}{2}$	2 (for each m_ℓ)

Magnetic Resonance Imaging (MRI)

Magnetic resonance imaging, or MRI, also known as NMR (for nuclear magnetic resonance), has become an increasingly common and important medical technique for visualizing parts of the human body (Fig. 1). It is based on a quantum mechanical spin phenomenon.

As we saw in Chapter 20, a current-carrying loop in a magnetic field experiences a torque that tends to orient the loop in the field. Specifically, the torque tends to orient the magnetic moment of the loop parallel to the field much like a compass. In this chapter we saw that electrons possess a quantum mechanical property described as intrinsic electron angular momentum, or spin. This gives rise to a *spin magnetic moment*, and when atoms are placed in a magnetic field, the energy levels are split into two closely spaced levels that give rise to fine structure in their spectra.

Nuclei exhibit similar properties when placed in magnetic fields. These properties can be understood in terms of nuclear magnetic moments. Various nuclei have different magnetic moments, but to get a general understanding of MRI, we will consider the simplest, the hydrogen atom, with a nucleus consisting of a single proton. Hydrogen is also the most common nucleus used in medical applications, since it is the most abundant element in the human body. The spin angular momentum (or magnetic moment orientation) of the proton in the hydrogen nucleus can take on only two values, similar to that of electron spin. These values are commonly referred to as "spin up" and "spin down," in reference to the orientation of the magnetic moment in a magnetic field (Fig. 2a). With spin up, the magnetic moment is parallel to the field (like a compass); with spin down, it is antiparallel to the field (with more energy—in a less stable configuration).

In terms of energy levels, in the presence of a magnetic field, there are two energy levels for the nucleus (Fig. 2b). The spin down level has the higher energy, and the energy difference of the levels is proportional to the magnitude of the magnetic field: $\Delta E \propto B$. The transition to the higher level can be made through the absorption of a photon of energy ΔE.

In an MRI imaging apparatus, the sample to be studied (usually, a region of the human body) is placed in a uniform

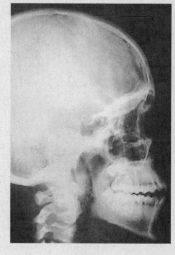

(a)

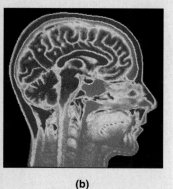

(b)

FIGURE 1   Diagnostic images (a) An X-ray of a human head. (b) A magnetic resonance image (MRI) of a human head. The amount of detail captured, especially in the soft tissues of the brain, makes such images very useful for medical diagnosis.

magnetic field **B**, the magnitude of which determines the energy (or frequency) of the photon needed to cause a transition: $hf = \Delta E \propto B$. The photons to trigger this transition are supplied by a pulsed beam of radio frequency (rf) radiation applied to the sample. If the frequency of the radiation corresponds to photons with energy exactly equal to the energy difference between the levels (ΔE), then many nuclei will absorb photons from the rf beam and be excited into the higher, spin-down energy level. This frequency that drives such excitation is known as a "resonance" frequency, $f = \Delta E/h$; hence the name magnetic resonance imaging.

A diagram and photo of a typical MRI device are shown in Fig. 3. Large coils produce the magnetic field. (Notice the solenoid arrangement in Fig. 3a.) Other coils produce the rf signals ("rf photons") that cause the nuclei to "flip their spin"

Multielectron Atoms

The Schrödinger equation for atoms with more than one electron (multielectron atoms) cannot be solved exactly. However, a solution can be found, to a workable approximation, in which each electron occupies a state characterized by a set of hydrogen-like quantum numbers. As might be expected with repulsive forces between electrons, the description of electron energy in a multielectron atom is much more complicated. For one thing, the energy depends not only on the principal quantum number n, but also on the orbital quantum number ℓ. This gives rise to a subdivision of energy levels. Thus, in multielectron atoms, the degeneracies of the hydrogen atom energy levels are removed, and the energy of an orbit generally depends on all four quantum numbers.

It is common to refer to the energy levels with the same n value as a **shell** and to ℓ levels of that shell as **subshells**. That is, atomic electrons with the same

Note: n quantum numbers designate energy shells; ℓ quantum numbers designate energy subshells.

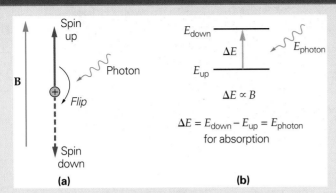

FIGURE 2 Nuclear spin **(a)** In a uniform magnetic field, the spin angular momentum, or spin magnetic moment, of a hydrogen nucleus (a proton) can have only two values—called "spin up" and "spin down" in reference to the direction of the external magnetic field. **(b)** This gives rise to two energy levels for the nucleus. Energy must be absorbed to "flip" the proton spin.

or be excited from the lower to the upper energy level. The resulting absorption of energy is detected, as is emitted radiation coming from a return transition to the lower state. The regions that produce the greatest absorption (or reemission) are those with the greatest concentration of the particular nucleus to which the apparatus is "tuned" by the choice of B and f. Images are produced by means of computerized tomography, similar to that used in X-ray CT scans (see Fig. 20.21). The result is a two- or three-dimensional image (Fig. 1b), which provides a great deal of diagnostic medical information.

Although a variety of atoms or nuclei exhibit nuclear magnetic resonance, most medical work is done with hydrogen atoms because of the varying water content of tissue. For example, muscle tissue has more water than does fatty tissue, so there is a distinct contrast in radiation intensity. Fatty deposits in blood vessels are distinct from the tissue of the vessel walls. (The water-rich blood is not seen because it has moved to a new tomography plane by the time it reradiates.) A tumor with a water content different from that of the surrounding tissue would show up in an MRI image.

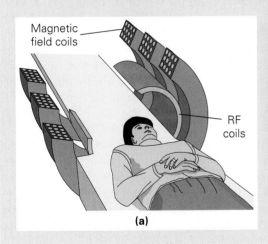

FIGURE 3 MRI **(a)** A diagram and **(b)** a photograph of the apparatus used for magnetic resonance imaging.

n value are said to be in the same electron shell. Electrons with the same n and ℓ values are said to be in the same electron subshell.

The ℓ subshells can be designated by numbers, as discussed previously. However, it is common to use letters instead. The letters s, p, d, f, g, ... correspond to the values of $\ell = 0$, 1, 2, 3, 4, ..., respectively. (This designation derives from historical spectroscopic notation.) After f, the letters go alphabetically (Table 28.2). This letter designation helps avoid confusion with the principal quantum number n, as we shall see.

Since an electron's energy depends on both n and ℓ in multielectron atoms, both quantum numbers are used to label atomic energy levels (a shell and subshell). For distinction, we write n as a number, followed by the letter that stands for the value of ℓ. For example, $1s$ denotes an energy level with $n = 1$ and $\ell = 0$; $2p$ is for $n = 2$ and $\ell = 1$; $3d$ for $n = 3$ and $\ell = 2$, and so on.

TABLE 28.2 Subshell Designations

ℓ Value	Letter Designation
$\ell = 0$	s
$\ell = 1$	p
$\ell = 2$	d
$\ell = 3$	f
$\ell = 4$	g
$\ell = 5$	h
...	...

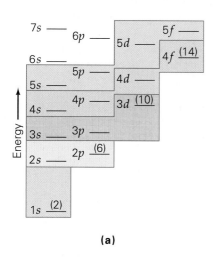

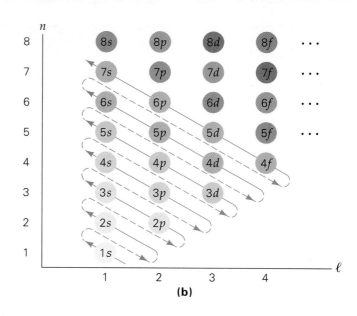

•FIGURE 28.7 Energy levels of a multielectron atom **(a)** The shell–subshell (n–ℓ) sequence shows that the energy levels are not evenly spaced and that the sequence of energy levels has numbers out of order. For example, the $4s$ level lies below the $3d$ level. The maximum number of electrons for a subshell, $2(2\ell + 1)$, is shown in parentheses on representative levels. (The vertical energy differences may not be drawn to scale.) **(b)** A convenient way to remember the energy level order of a multielectron atom is to list the n versus ℓ values as shown here. The diagonal lines then give the energy levels in ascending order.

Also, it is common to refer to the m_ℓ values as representing *orbitals*. For example, a $2p$ energy level has three orbitals corresponding to the m_ℓ values of -1, 0, and $+1$ (for the p subshell or $\ell = 1$). The spin quantum number also applies to electrons in multielectron atoms, but no special name is given to it.

As we saw in the Bohr theory, the energy levels of the hydrogen atoms are not evenly spaced but increase sequentially. However, in multielectron atoms, the numerical sequence of the energy levels has numbers out of order. The shell—subshell (n–ℓ notation) energy level sequence of a multielectron atom is shown in •Fig. 28.7a. Notice, for example, how the $4s$ level is below the $3d$ level. Such variations result in part from electrical forces between electrons in multielectron atoms. Also, the electrons in the outer orbits are shielded from the attractive force of the nucleus by the electrons in the inner orbits. Due to its highly elliptical orbit (Fig. 28.6a), an electron in the $4s$ ($\ell = 0$) orbit spends more time near the nucleus, hence is more tightly bound, than if it were in the $3d$ ($\ell = 2$) orbit. A convenient way to remember the level order is given in Fig. 28.7b.

The ground state of a multielectron atom has some similarities to that of the hydrogen atom, with the electron in the $1s$ or lowest energy level. For example, in a multielectron atom, the ground state is the combination of energy levels with the lowest total energy. That is, the electrons are in the lowest possible energy levels. But to identify those levels, we must know how many electrons can occupy a particular energy level. For example, the lithium (Li) atom has three electrons. Can they all be in the $1s$ level? As we shall see, the answer is no.

The Exclusion Principle

Exactly how the electrons of a multielectron atom distribute themselves in the ground-state energy levels is governed by a principle set forth in 1928 by the Austrian physicist Wolfgang Pauli. The **Pauli exclusion principle** states that

> No two electrons in a multielectron atom can have the same set of quantum numbers (n, ℓ, m_ℓ, m_s). That is, no two electrons can be in the same quantum state.

This means that each set of quantum numbers (n, ℓ, m_ℓ, m_s) corresponds to a unique quantum state that can be occupied by one and only one electron.

The limits on the quantum numbers set the limits on the number of electrons that can occupy a given energy level. For example, the $1s$ ($n = 1$, $\ell = 0$) level can have only one m_ℓ value, $m_\ell = 0$, along with only two m_s values, $m_s = \pm\frac{1}{2}$. Thus, there are only two unique sets of quantum numbers (n, ℓ, m_ℓ, m_s) for the $1s$ level: $(1, 0, 0, +\frac{1}{2})$ and $(1, 0, 0, -\frac{1}{2})$, so only two electrons can be in the $1s$ level.

We say such a *shell is full*; all other electrons are excluded from it by Pauli's principle. Thus, for a Li atom, with three electrons, the third electron must occupy the next higher level ($2s$) when the atom is in the ground state. This is illustrated in •Fig. 28.8, along with the ground-state energy levels for some other atoms.

Example 28.3 ■ The Quantum Shell Game: How Many States?

How many possible sets of quantum numbers or electron states are there in (a) the $3p$ subshell and (b) the $4d$ subshell?

Thinking It Through. This is a matter of following the quantum mechanical counting rules we have just learned.

Solution. For each subshell, we replace the letter designation for ℓ by its number:

Given: (a) $3p$ level ($n = 3$, $\ell = 1$)   *Find:*   number of sets of quantum numbers
        (b) $4d$ level ($n = 4$, $\ell = 2$)   of electron states

(a) For a particular subshell, it is the ℓ value that determines the number of states. Recall that there are $(2\ell + 1)$ possible m_ℓ values for a given ℓ. Thus, for $\ell = 1$, there are $[(2 \times 1) + 1] = 3$ values for m_ℓ. Each of these can have two m_s values ($\pm\frac{1}{2}$), making six different combinations of (n, ℓ, m_ℓ, m_s), or six states.

     Thus, in general, the number of possible states for a given ℓ value is $2(2\ell + 1)$, since there are two possible "spin states" for each orbital state:

Number of electron states for the given ℓ value $= 2(2\ell + 1) = 2[(2 \times 1) + 1] = 6$

(b) We could do long-hand counting as in part (a), but instead let's use the general result. The $4d$ level with $\ell = 2$ has

$$\text{Number of states} = 2(2\ell + 1) = 2[(2 \times 2) + 1] = 10$$

These results are summarized in Table 28.3. Notice that the total number of states in a given shell (n) is $2n^2$. For example, for the $n = 2$ shell, the total number of states for its combined s and p subshells ($\ell = 0, 1$) is $2n^2 = 2(2)^2 = 8$. This means that up to eight electrons can be accommodated in the $n = 2$ shell: two in the $2s$ subshell and six in the $2p$ subshell.

Follow-up Exercise. How many electrons could be accommodated in the $1s$ subshell if there were no spin quantum number?

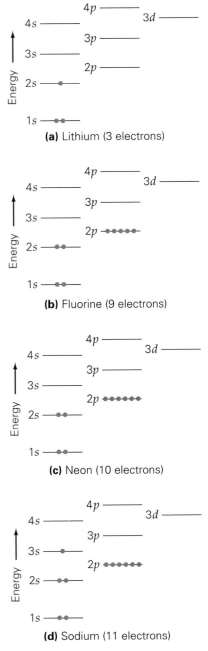

(a) Lithium (3 electrons)

(b) Fluorine (9 electrons)

(c) Neon (10 electrons)

(d) Sodium (11 electrons)

•**FIGURE 28.8 Filling subshells** The electron subshell distributions for several unexcited atoms (in their ground state) according to the Pauli exclusion principle. The s subshell can have a maximum of two electrons, and the p subshell a maximum of six electrons.

TABLE 28.3 Possible Sets of Quantum Numbers and States

Electron Shell n	Subshell ℓ	Subshell Notation	Orbitals m_ℓ	Number of Orbitals (m_ℓ) in Subshell ($2\ell + 1$)	Number of States (m_s) in Subshell $2(2\ell + 1)$	Total Electron States for Shell $2n^2$
1	0	$1s$	0	1	2	2
2	0	$2s$	0	1	2	8
	1	$2p$	1, 0, −1	3	6	
3	0	$3s$	0	1	2	
	1	$3p$	1, 0, −1	3	6	18
	2	$3d$	2, 1, 0, −1, −2	5	10	
4	0	$4s$	0	1	2	
	1	$4p$	1, 0, −1	3	6	32
	2	$4d$	2, 1, 0, −1, −2	5	10	
	3	$4f$	3, 2, 1, 0, −1, −2, −3	7	14	

Electron configuration: a shorthand notation for the electron quantum states occupied in a ground-state atom

Electron Configurations. We can build up the electron structure of atoms, so to speak, by putting an increasing number of electrons in the lower energy subshells [hydrogen (H), 1 electron; helium (He), 2 electrons; lithium (Li), 3 electrons, and so on], as was done for four elements in Fig. 28.8. However, rather than draw diagrams each time we want to represent the electron arrangement in a ground-state atom, we can use a shorthand notation called the **electron configuration**.

In this notation, we write the subshells in order of increasing energy and designate the number of electrons in each subshell with a superscript. For example, $3p^5$ means that a $3p$ subshell is occupied by five electrons:

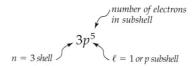

The electron configurations for the atoms shown in Fig. 28.8 can thus be written as

Li	(3 electrons)	$1s^2 2s^1$
F	(9 electrons)	$1s^2 2s^2 2p^5$
Ne	(10 electrons)	$1s^2 2s^2 2p^6$
Na	(11 electrons)	$1s^2 2s^2 2p^6 3s^1$

In writing an electron configuration, when one subshell is filled, you go on to the next higher one. Notice that the superscripts in any configuration must add up to the total number of electrons in the atom.

The energy spacing between adjacent subshells is not uniform, as Figs. 28.7a and 28.8 show. In general, there are relatively large energy gaps between the s subshells and the subshells below them. (Compare the $4s$ with the $3p$ in Fig. 28.7a.) Those just below the s subshells are usually p subshells, with the exception of the lowest—the $1s$ is below the $2s$ (and there are none below the $1s$). The gaps between other subshells—for example, between the $3s$ and the $3p$ above it, or between the $4d$ and $5p$—are considerably smaller.

This unevenness in energy differences gives rise to periodic large energy gaps represented by vertical lines between certain subshells in the electron configuration:

$$1s^2 | 2s^2 2p^6 | 3s^2 3p^6 | 4s^2 3d^{10} 4p^6 | 5s^2 4d^{10} 5p^6 | 6s^2 4f^{14} 5d^{10} 6p^6 | \ldots$$

(number of states) (2)   (8)   (8)   (18)   (18)   (32)

Electron periods: sets of energy levels with about the same energy

The subshells *between* the lines have only slightly different energies. We refer to a grouping of subshells (for example, $2s^2 2p^6$) that have about the same energy as an **electron period**.

These electron periods are the basis of the periodic table of elements. With your present knowledge of electron configurations, you are now in a position to understand the periodic table better than the person who originally developed it.

The Periodic Table of Elements

By 1860, over 60 chemical elements had been discovered. Several attempts had been made to classify the elements into some orderly arrangement, but none were very satisfactory. It had been noted in the early 1800s that the elements could be listed in such a way that similar chemical properties recurred periodically throughout the list. With this idea, in 1869 a Russian chemist, Dmitri Mendeleev (pronounced men-duh-*lay*-eff), formulated an arrangement of the elements based on this periodic property. The modern version of his **periodic table of elements** is used today and can be seen on the walls of just about every science building (•Fig. 28.9).

Periods: horizontal rows

Mendeleev arranged the known elements in rows, which are called **periods**, in order of increasing atomic masses. When he came to an element that had chemical properties similar to those of one of the previous elements, he went back and put this element below the previous similar one. In this manner, he formed both

The periodic table of elements

GROUP I	GROUP II					Transition elements						GROUP III	GROUP IV	GROUP V	GROUP VI	GROUP VII	GROUP VIII
1 **H** 1.01 $1s^1$																	2 **He** 4.00 $1s^2$
3 **Li** 6.94 $2s^1$	4 **Be** 9.01 $2s^2$											5 **B** 10.81 $2p^1$	6 **C** 12.01 $2p^2$	7 **N** 14.01 $2p^3$	8 **O** 16.00 $2p^4$	9 **F** 19.00 $2p^5$	10 **Ne** 20.18 $2p^6$
11 **Na** 22.99 $3s^1$	12 **Mg** 24.31 $3s^2$											13 **Al** 26.98 $3p^1$	14 **Si** 28.09 $3p^2$	15 **P** 30.97 $3p^3$	16 **S** 32.07 $3p^4$	17 **Cl** 35.45 $3p^5$	18 **Ar** 39.95 $3p^6$
19 **K** 39.10 $4s^1$	20 **Ca** 40.08 $4s^2$	21 **Sc** 44.96 $3d^14s^2$	22 **Ti** 47.88 $3d^24s^2$	23 **V** 50.94 $3d^34s^2$	24 **Cr** 52.00 $3d^54s^1$	25 **Mn** 54.94 $3d^54s^2$	26 **Fe** 55.85 $3d^64s^2$	27 **Co** 58.93 $3d^74s^2$	28 **Ni** 58.69 $3d^84s^2$	29 **Cu** 63.55 $3d^{10}4s^1$	30 **Zn** 65.39 $3d^{10}4s^2$	31 **Ga** 69.72 $4p^1$	32 **Ge** 72.61 $4p^2$	33 **As** 74.92 $4p^3$	34 **Se** 78.96 $4p^4$	35 **Br** 79.90 $4p^5$	36 **Kr** 83.80 $4p^6$
37 **Rb** 85.47 $5s^1$	38 **Sr** 87.62 $5s^2$	39 **Y** 88.96 $4d^15s^2$	40 **Zr** 91.22 $4d^25s^2$	41 **Nb** 92.91 $4d^45s^1$	42 **Mo** 95.94 $4d^55s^1$	43 **Tc** (98) $4d^55s^2$	44 **Ru** 101.07 $4d^75s^1$	45 **Rh** 102.91 $4d^85s^1$	46 **Pd** 106.42 $4d^{10}5s^6$	47 **Ag** 107.87 $4d^{10}5s^1$	48 **Cd** 112.41 $4d^{10}5s^2$	49 **In** 114.82 $5p^1$	50 **Sn** 118.71 $5p^2$	51 **Sb** 121.76 $5p^3$	52 **Te** 127.60 $5p^4$	53 **I** 126.90 $5p^5$	54 **Xe** 131.29 $5p^6$
55 **Cs** 132.91 $6s^1$	56 **Ba** 137.33 $6s^2$	57 **La** 138.91 $5d^16s^2$ *	72 **Hf** 178.49 $5d^26s^2$	73 **Ta** 180.95 $5d^36s^2$	74 **W** 183.85 $5d^46s^2$	75 **Re** 186.21 $5d^56s^2$	76 **Os** 190.2 $5d^66s^2$	77 **Ir** 192.22 $5d^76s^2$	78 **Pt** 195.08 $5d^96s^1$	79 **Au** 196.97 $5d^{10}6s^1$	80 **Hg** 200.59 $5d^{10}6s^2$	81 **Tl** 204.36 $6p^1$	82 **Pb** 207.2 $6p^2$	83 **Bi** 208.98 $6p^3$	84 **Po** (209) $6p^4$	85 **At** (210) $6p^5$	86 **Rn** (222) $6p^6$
87 **Fr** (223) $7s^1$	88 **Ra** 226.03 $7s^2$	89 **Ac** 227.03 $6d^17s^2$ †	104 **Rf** (261) $6d^27s^2$	105 **Db** (262) $6d^37s^2$	106 **Sg** (263) $6d^47s^2$	107 **Bh** (264) $6d^57s^2$	108 **Hs** (265) $6d^67s^2$	109 **Mt** (268) $6d^77s^2$	110 (269)	111 (272)	112 (277)						

Key (example entry):
Atomic number — 26 **Fe** — Symbol
58.85 — Atomic mass
$3d^64s^2$ — Outer electron configuration

(Lanthanides) *

58 **Ce** 140.12 $5d^14f^16s^2$	59 **Pr** 140.91 $4f^36s^2$	60 **Nd** 144.24 $4f^46s^2$	61 **Pm** (145) $4f^56s^2$	62 **Sm** 150.36 $4f^66s^2$	63 **Eu** 151.96 $4f^76s^2$	64 **Gd** 157.25 $5d^14f^76s^2$	65 **Tb** 158.93 $5d^14f^86s$	66 **Dy** 162.50 $4f^{10}6s^2$	67 **Ho** 164.93 $4f^{11}6s^2$	68 **Er** 167.26 $4f^{12}6s^2$	69 **Tm** 168.93 $4f^{13}6s^2$	70 **Yb** 173.04 $4f^{14}6s^2$	71 **Lu** 174.97 $5d^14f^{14}6s^2$

(Actinides) †

90 **Th** 232.04 $6d^27s^2$	91 **Pa** 231.04 $5f^26d^17s^2$	92 **U** 238.03 $5f^36d^17s^2$	93 **Np** 237.05 $5f^46d^17s^2$	94 **Pu** (244) $5f^66d^07s^2$	95 **Am** (243) $5f^76d^07s^2$	96 **Cm** (247) $5f^76d^17s^2$	97 **Bk** (247) $5f^86d^17s^2$	98 **Cf** (251) $5f^{10}6d^07s^2$	99 **Es** (252) $5s^{11}6d^07s^2$	100 **Fm** (257) $5f^{12}6d^07s^2$	101 **Md** (258) $5f^{13}6d^07s^2$	102 **No** (259) $5f^{14}6d^07s^2$	103 **Lr** (260) $5f^{14}6d^17s^2$

•**FIGURE 28.9 The periodic table of elements**   The elements are arranged in order of increasing atomic or proton number. Horizontal rows are called periods, and vertical columns are called groups. The elements in a group have similar chemical properties. Each atomic mass represents an average of that element's isotopes, weighted to reflect their relative abundance in our immediate environment. The masses have been rounded to two decimal places; more precise values are given in Appendices IV and V. (A value in parentheses represents the mass number of the best-known or longest-lived isotope of an unstable element. Elements 110–112 have not yet been named.) See Appendix IV for an alphabetical listing of elements.

horizontal rows of elements and vertical columns called **groups**, or families of elements with similar properties. The table was later rearranged in order of increasing atomic or proton number (the number of protons in the nucleus of an atom is given by the number at the top left of the element boxes in Fig. 28.9) to resolve some inconsistencies. Notice that if atomic masses were used, cobalt and nickel, atomic numbers 27 and 28, respectively, would fall in different columns.

With only 65 elements, there were vacant spaces in Mendeleev's table. The elements for these spaces were yet to be discovered. Because the missing elements were part of a sequence and had properties similar to those of other elements in a group, Mendeleev could predict their masses and chemical properties. Less than 20 y after Mendeleev devised his table, showing chemists what to look for to find the undiscovered elements, three of the missing elements were in fact discovered.

Notice that the periodic table puts the elements into seven horizontal rows or periods, in order of increasing atomic or proton number. The first period has only two elements. Periods 2 and 3 have eight elements, and periods 4 and 5 have 18 elements. Recall that the s, p, d, and f subshells can contain a maximum of 2, 6, 10, and 14 electrons [$2(2\ell + 1)$], respectively. You should begin to see a correlation between these numbers and the arrangements of elements in the periodic table.

The periodicity of the periodic table can be understood in terms of the electron configurations of the atoms. For $n = 1$, the electrons are in one of two s states ($1s$); for $n = 2$, electrons go into the $2s$ and $2p$ states, which gives a total of 10 electrons; and so on. Thus, for a given element, its period number is equal to the highest n shell containing electrons in the atom. Notice the electron configurations for the elements in Fig. 28.9. Also, compare the electron periods given earlier, as defined by energy gaps, and the periods in the periodic table (•Fig. 28.10). There is a one-to-one correlation, so the periodicity comes from energy level considerations in atoms.

Chemists refer to *representative elements*, which we see from Fig. 28.9 are those in which the last electron enters an s or p subshell. In *transition elements*, the last electron enters a d subshell; and in *inner transition elements*, the last electron en-

Shell (last to be filled)	Subshells	Number of electrons in subshell, $2(2\ell + 1)$	Corresponding period in periodic table
$n = 7$	$7p$	6	Period 7 (32 elements)
	$6d$	10	
	$5f$	14	
	$7s$	2	
$n = 6$	$6p$	6	Period 6 (32 elements)
	$5d$	10	
	$4f$	14	
	$6s$	2	
$n = 5$	$5p$	6	Period 5 (18 elements)
	$4d$	10	
	$5s$	2	
$n = 4$	$4p$	6	Period 4 (18 elements)
	$3d$	10	
	$4s$	2	
$n = 3$	$3p$	6	Period 3 (8 elements)
	$3s$	2	
$n = 2$	$2p$	6	Period 2 (8 elements)
	$2s$	2	
$n = 1$	$1s$	2	Period 1 (2 elements)

•FIGURE 28.10 **Electron periods** The periods of the periodic table are related to electron configurations. The last n shell to be filled is equal to the period number. The electron periods and the corresponding periods of the table are defined by relatively large energy gaps between successive subshells (such as between $4s$ and $3p$) of the atoms.

ters an *f* subshell. So that the periodic table is not unmanageably wide, the *f* subshells are usually placed in two rows at the bottom of the table. Each row is given a name—the *lanthanide series* and the *actinide series*—based on its position within the period.

Finally, you can also understand why elements in vertical columns or groups have similar chemical properties. As you probably know from chemistry classes, the chemical properties of an atom, such as its ability to react and form compounds, depends almost entirely on the outermost electrons in the atom—that is, the electrons in the outermost *unfilled* shell. It is these electrons, called *valence electrons*, that form chemical bonds with other atoms. Because of the way the elements are arranged in the table, the outermost electron configurations of all the atoms in any one group are very similar. The atoms in such a group would be expected to have similar chemical properties, and they do. For example, notice the first two groups at the left of the table. They have one and two outermost electrons in an *s* subshell, respectively. These elements are all highly reactive metals that form compounds having many similarities. The group at the far right, the noble gases, has completely filled subshells (and thus a full shell) and is at the ends of electron periods, or just before a large energy gap. These gases are chemically very nonreactive and can form compounds (via chemical bonds) only under very special conditions.

Conceptual Example 28.4

Combining Atoms: Performing Chemistry on the Periodic Table

Simple combinations of atoms, called molecules, can form if two atoms come together and share outer electrons. This sharing process is called *covalent bonding*. In this "shared custody" scheme, both atoms find it energetically beneficial (that is, they lower their combined total energies) to have the equivalent of a filled outer shell of electrons, if only on a part-time basis. To which atom would an oxygen atom most likely bond covalently: (a) neon, Ne; (b) fluorine, F; or (c) hydrogen, H?

Reasoning and Answer. Looking at the periodic chart, we see that oxygen is two electrons away from having a full shell. Thus it could share two electrons by bonding to a partner (or partners) with the opposite problem—having just one or two electrons beyond a full shell. We can eliminate (a), neon, because it has a full outer shell. Choice (b), fluorine, is one electron shy of having a full shell and is in a similar situation to oxygen. Hence fluorine is not likely to combine covalently with oxygen. The remaining candidate, hydrogen, has one electron less than a full shell. That is, if a hydrogen atom adds one electron, it attains an electron configuration like that of stable helium. Each of two hydrogen atoms can do this part of the time by sharing with an oxygen atom. (The rest of the time both hydrogen atoms must give their electron to the oxygen atom for oxygen to have a full $n = 2$ shell.) So our answer is (c). This combination of two hydrogen atoms, covalently bonded to a single oxygen atom, has the molecular formula H_2O—it is water.

Follow-up Exercise. In this Example, (a) what would be the electron configuration of the oxygen in the water molecule at some instant when it had custody of the electrons from both hydrogen atoms? (b) What would be the net charge on the oxygen in this case?

28.4 The Heisenberg Uncertainty Principle

Objective: **To understand the inherent quantum mechanical limits on the accuracy of physical observations.**

An important aspect of quantum mechanics has to do with measurement and accuracy. In classical mechanics, there is no limit to the accuracy of a measurement. Theoretically, by continual refinement of a measurement instrument and procedure, the accuracy could be improved to any degree so as to give *exact* values. This resulted in a *deterministic* view of nature. For example, if you know both the

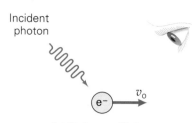

Incident
photon

v_0

(a) Before collision

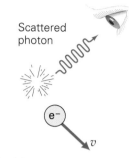

Scattered
photon

v

(b) After collision

•**FIGURE 28.11 Measurement-induced uncertainty** **(a)** To measure the position and momentum (or velocity) of an electron, at least one photon must collide with the electron and be scattered toward the eye or detector. **(b)** In the collision process, energy and momentum are transferred to the electron, which induces uncertainty in the velocity.

Minimum uncertainties on momentum and position

position and the velocity of an object *exactly* at a particular time, you can determine exactly where it will be in the future and where it was in the past (assuming you knew all the future and past forces on it).

However, quantum theory predicts otherwise and sets limits on the accuracy of measurements. This idea was introduced in 1927 by the German physicist Werner Heisenberg, who had developed another approach to quantum mechanics that complemented Schrödinger's wave theory. The **Heisenberg uncertainty principle** as applied to position and momentum (or velocity) can be stated as follows:

It is impossible to know simultaneously an object's exact position and momentum.

This concept is often illustrated with a simple thought experiment. Suppose that you wanted to measure the position and momentum (actually the velocity) of an electron. In order for you to "see," or locate, the electron, at least one photon must bounce off the electron and come to your eye, as illustrated in •Fig. 28.11. However, in the collision process, some of the photon's energy and momentum are transferred to the electron (as in the Compton effect, described in Section 27.3).

After the collision, the electron recoils. Thus, in the very process of trying to locate the position very accurately (trying to make the uncertainty of position Δx very small), you induce uncertainty into your knowledge of the electron's velocity or momentum ($\Delta \mathbf{p} = m \, \Delta \mathbf{v}$), since the process of determining its position sent the electron flying off in a different direction. We do not observe this in the everyday macroscopic world because the uncertainty produced by viewing an object is negligible: Light cannot appreciably alter the motion or position of an ordinary-sized object.

For our subatomic case, the position of an electron could be measured *at best* to an accuracy of about the wavelength λ of the incident light—that is, $\Delta x \approx \lambda$ (recall the resolving power of light). The photon "particle" we use for this has a momentum of $p = h/\lambda$. Since we cannot tell how much of this momentum might be transferred during collision, the final momentum of the electron would have an uncertainty on the order of $\Delta p \approx h/\lambda$.

The product of these two individual uncertainties is at least as large as h, since

$$(\Delta p)(\Delta x) \approx \left(\frac{h}{\lambda}\right)(\lambda) = h$$

This equation relates the *minimum* uncertainties or maximum accuracies of *simultaneous* measurements of the momentum and position. In actuality, the uncertainties could be much worse depending on the amount of light (number of photons) used, the apparatus, and the technique. Through detailed theoretical calculations, Heisenberg found that, for the one-dimensional case, *at very best*,

$$(\Delta p)(\Delta x) \geq \frac{h}{2\pi} \tag{28.5}$$

That is, one version of Heisenberg's uncertainty principle states that the product of the *minimum* uncertainties of simultaneous position and momentum measurements is on the order of Planck's constant divided by 2π (that is, about 10^{-34} J·s).

In trying to locate the position of the particle accurately (that is, to make Δx small), we make the uncertainty in the momentum larger ($\Delta p \approx h/2\pi \, \Delta x$) and vice versa. To take the extreme case, if we could measure the exact location of a particle ($\Delta x \rightarrow 0$), we would have no idea about its momentum ($\Delta p \rightarrow \infty$). Thus, the measurement process itself limits the accuracy to which we can simultaneously measure position and momentum. In Heisenberg's words, "Since the measuring device has been constructed by the observer . . . we have to remember that what we observe is not nature in itself but nature exposed to our method of questioning." To see how the Heisenberg uncertainty principle affects the microscopic and macroscopic worlds, consider the following Example.

Example 28.5 ■ Electron versus Bullet: The Uncertainty Principle

An electron and a 20-gram bullet, both moving linearly, are measured to have equal speeds of 300 m/s to an accuracy of $\pm 0.010\%$. What is the minimum uncertainty in the position of each?

Thinking It Through. We can solve the Heisenberg uncertainty principle (Eq. 28.5) for Δx since we know Δp. We expect the electron to be affected much more by the uncertainty principle because of its very small mass.

Solution. Listing the quantities, we have

Given: $m_e = 9.11 \times 10^{-31}$ kg (known)   *Find:* Δx_e and Δx_b (minimum uncertainties
$m_b = 20$ g $= 0.020$ kg   in position)
$v = 300$ m/s $\pm 0.010\%$

The magnitude of the uncertainty is 0.010% of the speed for both the electron and the bullet. Thus the uncertainty is

$$(300 \text{ m/s})(0.00010) = 0.030 \text{ m/s}$$

so

$$v = 300 \text{ m/s} \pm 0.010\% = 300 \text{ m/s} \pm 0.030 \text{ m/s}$$

The *total* uncertainty in speed is *twice* this amount because the measurements can be off by 0.010% above or below the measured values, so

$$\Delta v = 0.060 \text{ m/s}$$

Then, for the electron, the minimum position uncertainty is

$$\Delta x_e = \frac{h}{2\pi \, \Delta p} = \frac{h}{2\pi m_e \, \Delta v} = \frac{6.63 \times 10^{-34} \text{ J·s}}{2\pi(9.11 \times 10^{-31} \text{ kg})(0.060 \text{ m/s})} = 0.0019 \text{ m} = 1.9 \text{ mm}$$

Similarly, we can calculate the bullet's minimum position uncertainty as

$$\Delta x_b = \frac{h}{2\pi m_b \, \Delta v} = \frac{6.63 \times 10^{-34} \text{ J·s}}{2\pi(0.020 \text{ kg})(0.060 \text{ m/s})} = 8.8 \times 10^{-32} \text{ m}$$

Notice that the uncertainty in the bullet's position (much smaller than nuclear diameters!) is many orders of magnitude less than that of the electron. The uncertainty for relatively massive objects traveling at ordinary speeds is negligible. However, for electrons, the answer of 1.9 mm is very significant, as this is many times larger than an atom.

Follow-up Exercise. In this Example, what would be the minimum uncertainty in the electron's speed for the minimum uncertainty in its position to be on the order of atomic dimensions, or 0.10 nm?

Another form of the uncertainty principle relates uncertainties in energy and time. Consider the position of the electron in the previous thought experiment to be known with an uncertainty of $\Delta x \approx \lambda$. The photon used to detect the particle travels with a speed c, and it takes a time of $\Delta x/c \approx \lambda/c$ for this photon to traverse a distance equal to the uncertainty in the particle's position. Thus, the time when the particle is at the measured position is uncertain by about

$$\Delta t \approx \lambda/c$$

Since we can't tell whether the photon transfers some or all of its energy $(E = hf = hc/\lambda)$ to the particle, the uncertainty in the energy of the electron is

$$\Delta E \approx \frac{hc}{\lambda}$$

Then the product of these uncertainties is at least

$$(\Delta E)(\Delta t) \approx \left(\frac{hc}{\lambda}\right)\left(\frac{\lambda}{c}\right) = h$$

As with the position–momentum relationship, a more exact and detailed analysis shows that at very best we have

Minimum uncertainties in energy and time

$$(\Delta E)(\Delta t) \geq \frac{h}{2\pi} \qquad (28.6)$$

This form of the Heisenberg uncertainty principle indicates that the energy of an object may be uncertain by an amount ΔE for a time $\Delta t \approx h/(2\pi \Delta E)$. During this time, the energy is uncertain and might not even be conserved. This is an important consideration in particle interactions, as you will learn in Chapter 30.

Notice that we cannot measure the energy of a particle exactly unless we take an infinite amount of time to do so. If a measurement of energy is carried out in a time Δt, then it must be uncertain by an amount ΔE. For example, the measurement of the frequency of a photon emitted by an atomic electron is a measurement of the energy associated with the transition from an excited state to the ground state. The measurement must be carried out in a time comparable to the time the electron is in the excited state—that is, the lifetime of the excited state. As a result, the observed emission line in the frequency spectrum has a finite energy width, since $\Delta E = h \, \Delta f$ (•Fig. 28.12). This *natural broadening* was ignored in Chapter 27, where spectral lines were considered to have exact frequencies (that is, to have no uncertainty).

Note: Spectral lines are discussed in Section 27.4.

28.5 Particles and Antiparticles

Objectives: To understand (a) the relationship between particles and antiparticles and (b) the energy requirements for pair production.

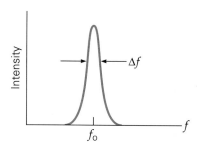

•**FIGURE 28.12 Natural line broadening** Because a measurement must be carried out in a time comparable to the lifetime (Δt) of an electron in an excited state, the energy of that state is uncertain by an amount $\Delta E = h \, \Delta f$. The observed emission line has a width of Δf, rather than being a line of single frequency f_0 with no width.

The quantum mechanics of Schrödinger and Heisenberg was successful in explaining observations and in predicting new atomic phenomena. When the quantum theory was extended to include relativistic considerations by the British physicist Paul A. M. Dirac in 1928, something new and very different was predicted—a particle called the **positron**. The positron should have the same mass as the electron but should carry a *positive* charge. The oppositely charged positron is said to be the **antiparticle** of the electron.

The positron was first observed experimentally in 1932 by the American physicist C. D. Anderson in cloud-chamber experiments with cosmic rays. The curvature of the particle tracks in a magnetic field showed two types of particles (•Fig. 28.13). The tracks indicated that both particles had the same mass, but their spiral curvatures were opposite. The magnetic force acts in opposite directions on particles of opposite charge. Thus the two types of particles must be oppositely charged. Anderson had in fact discovered the positron.

Because charge is conserved, a positron can be created only with the simultaneous creation of an electron; this process is called **pair production**. In Anderson's experiment, positrons were observed to be emitted from a thin lead plate exposed to cosmic rays from outer space, which contain highly energetic X-rays. Pair production occurs when an X-ray photon comes near a nucleus—the nuclei of the lead atoms of the plate in the Anderson experiment. In this process, the photon goes out of existence, and an "electron pair" (an electron and a positron) is created, as illustrated in •Fig. 28.14, in a conversion of electromagnetic (photon) energy into mass. By the conservation of energy,

$$hf = 2m_e c^2 + K_{e^-} + K_{e^+} + K_{nuc}$$

•FIGURE 28.13 **Cloud-chamber photograph of pair production** In this false-color cloud-chamber photograph, a gamma-ray photon (not visible but about to enter the region in a downward direction) interacts with a nearby atomic nucleus (point P) to produce an electron and a positron (green and red spiral tracks at top). In the process, the photon dislodges an orbital electron (the nearly vertical green track). An external magnetic field causes the electron and positron to be deflected in paths of opposite curvature. A similar event is recorded in the bottom half of the photo. (Why might the paths of the particles created in this case show less deflection?)

where hf is energy of the photon, $2m_ec^2$ is the total mass–energy equivalent of the electron–positron pair, and the K's are the kinetic energies of the particles and the recoil nucleus. Because the recoil nucleus has relatively large mass, its kinetic energy is usually negligible, so to a good approximation

$$hf = 2m_ec^2 + K_{e^-} + K_{e^+}$$

From the energy equation, the minimum energy occurs when the pair is produced at rest—when K_{e^-} and K_{e^+} are zero. Thus, the *minimum* photon energy to produce an electron–positron pair is given by

$$E_{min} = hf = 2m_ec^2 = 1.022 \text{ MeV} \qquad (28.7)$$

(Recall from Section 26.5 that the rest energy of an electron is $m_ec^2 = 0.511$ MeV.) This minimum photon energy is called the *threshold energy for pair production.*

But if they are created by cosmic rays, why are positrons not commonly found in nature? For example, why are they not evident in ordinary chemical processes? Positrons are taken out of existence by a process called **pair annihilation**. When an energetic positron appears in pair production, it loses kinetic energy in collisions

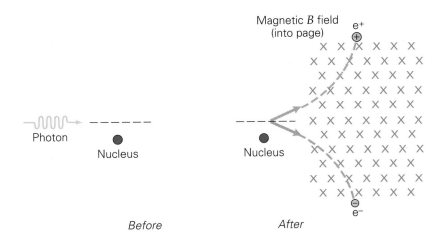

•FIGURE 28.14 **Pair production** An electron (e^-) and a positron (e^+) can be created when an energetic photon comes near a heavy nucleus.

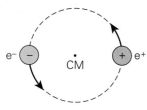

Before annihilation
(positronium atom)

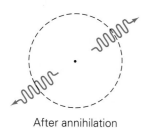

After annihilation
(photons)

•**FIGURE 28.15 Pair annihilation**
The disappearance of a positronium atom is signaled by the appearance of two photons, each with an energy of 0.511 MeV. Why do we not expect one photon with an energy of 1.022 MeV? (CM is center of mass.)

as it passes through matter. Finally, moving at a low speed, it combines with an electron of that material and forms a hydrogen-like atom, called a *positronium atom*, in which a positron substitutes for a proton. The positronium atom is unstable and quickly decays ($\approx 10^{-10}$ s) into two photons, each with an energy of 0.511 MeV (•Fig. 28.15). Pair annihilation is then a direct conversion of mass into electromagnetic energy—the inverse of pair production, so to speak. These processes are striking examples of the mass–energy equivalence predicted by Einstein ($E = mc^2$)—that is, mass is a form of energy.

All subatomic particles have been found to have antiparticles, which can be produced by cosmic rays entering our atmosphere from outer space and/or by nuclear processes. For example, there is an antiproton with the same mass as a proton but with a negative charge. Even a neutral particle such as the neutron has an antiparticle—the antineutron. In our environment, there is a preponderance of electrons, protons, and neutrons. When antiparticles are created, they quickly combine with their respective particles in annihilation processes.

It is conceivable that antiparticles predominate in some parts of the universe. If so, the atoms of the **antimatter** in this region would consist of negatively charged nuclei composed of antiprotons and antineutrons, surrounded by positively charged positrons (antielectrons). It would be difficult to distinguish a region of antimatter visibly, since the physical behavior of antimatter atoms would presumably be the same as those of ordinary matter. However, if antimatter and ordinary matter came into contact, they would annihilate each other in an explosive release of energy.

Chapter Review

Important Terms

Important Concepts

- The de Broglie hypothesis assigns a wavelength to material particles, by analogy with the assignment of momentum to light particles (photons).

- A quantum-mechanical wave function is the probability wave associated with a particle. The Schrödinger wave equation lets us calculate the wave function in different situations. The probability density, the square of the wave function, gives the relative probability of finding a particle at a particular location.

- The orbital quantum number determines the allowed value of the angular momentum for an electron orbit. The magnetic quantum number determines the orientation of the electron orbit with respect to a given axis. The spin quantum number is a purely quantum-mechanical property of a particle. In a classical analogy, it indicates whether the spin angular momentum of an electron is oriented up or down.

- The Pauli exclusion principle states that in a given atom, no two electrons can have exactly the same set of quantum numbers.
- The Heisenberg uncertainty principle states that you cannot simultaneously measure both the position and the momentum (or velocity) of a particle exactly. The same holds true for the particle's energy and the period of time during which that energy is measured.
- Pair production refers to the creation of a particle and its antiparticle by high-energy photons. The reverse process is pair annihilation, in which a particle and antiparticle annihilate and their energy is converted into photon energy.

Important Equations

Momentum of a Photon:

$$p = \frac{E}{c} = \frac{hf}{c} = \frac{h}{\lambda} \tag{28.1}$$

De Broglie Wavelength of a Moving Particle:

$$\lambda = \frac{h}{p} = \frac{h}{mv} \tag{28.2}$$

Electron Wavelength When Accelerated Through Potential V (nonrelativistic):

$$\lambda = \sqrt{\frac{1.50}{V}} \times 10^{-9}\,\text{m} = \sqrt{\frac{1.50}{V}}\,\text{nm} \tag{28.3}$$

Heisenberg Uncertainty Principle (two forms):

$$(\Delta p)(\Delta x) \geq \frac{h}{2\pi} \tag{28.5}$$

$$(\Delta E)(\Delta t) \geq \frac{h}{2\pi} \tag{28.6}$$

Condition for Electron–Positron Pair Production:

$$E_{\min} = hf \geq 2m_e c^2 = 1.022\,\text{MeV} \tag{28.7}$$

Exercises

28.1 Matter Waves: The De Broglie Hypothesis

1. The momentum of a photon is (a) zero, (b) equal to c, (c) proportional to its frequency, (d) proportional to its wavelength, (e) given by the de Broglie hypothesis.

2. The Davisson–Germer experiment (a) dealt with X-ray spectra, (b) verified the Heisenberg uncertainty principle, (c) supported the Pauli exclusion principle, (d) demonstrated the wavelike properties of electrons.

3. The de Broglie hypothesis predicts that a wave associated with any object has momentum. Why do we not observe the wave nature of a moving car?

4. ■ What is the de Broglie wavelength associated with a 1000-kilogram car moving at 25 m/s?

5. ■ What is the de Broglie wavelength of (a) an electron and (b) a proton, both moving with a speed of 100 m/s?

6. ■ Calculate the de Broglie wavelength of a 70-kilogram person running with a speed of 2.0 m/s.

7. ■■ A proton and an electron are accelerated from rest through a potential difference V. What is the ratio of the de Broglie wavelength of an electron to that of a proton (to two significant figures)?

8. ■■ Electrons are accelerated from rest through a potential difference of 250 kV. If the potential difference is increased to 600 kV, what is the ratio of the new de Broglie wavelength to the original?

9. ■■ An electron is accelerated from rest through a potential difference that gives it a matter wave with a wavelength of 0.010 nm. What is the potential difference?

10. ■■ Charged particles are accelerated through a potential difference V. By what factor would the de Broglie wavelength of the particles change if the voltage were doubled?

11. ■■ A proton traveling with a speed of 4.5×10^4 m/s is accelerated to a faster speed by a potential difference of 37 V. By what percentage does the de Broglie wavelength of the proton change?

12. ■■ What is the energy of a beam of electrons that exhibits a first-order maximum at an angle of 25° when diffracted by a crystal grating with a spacing between the lattice planes of 0.15 nm?

13. ■■ What is the de Broglie wavelength of the Earth in its orbit about the Sun? (Assume a circular orbit.)

14. ■■ According to the Bohr theory of the hydrogen atom, the speed of the electron in the first Bohr orbit is 2.19×10^6 m/s. (a) What is the wavelength of the matter wave associated with the electron? (b) How does this wavelength compare with the circumference of the first Bohr orbit?

15. ■■ You want to observe with an electron microscope details of a molecule on the order of 0.25 nm wide. Through what potential difference must the electrons be accelerated so that they have a de Broglie wavelength of 0.25 nm?

16. The wave function solution to the Schrödinger equation (a) can never be found, (b) is the probability of finding a particle, (c) functionally describes the de Broglie wave of a particle, (d) none of these.

17. The square of a particle's wave function is interpreted as being (a) the energy of the particle, (b) the probability of locating the particle, (c) the quantum number of a state, (d) the basis of the Pauli exclusion principle.

18. ■■■ A particle in a box is constrained to move in one dimension, like the bead on a wire illustrated in ●Fig. 28.16. Assuming that no forces act on the particle in the interval $0 < x < L$ and that it hits a perfectly rigid wall, we can think of it as a particle at the bottom of an infinite well with $U = 0$. (a) Show that the spatial wave function for the particle is $\psi_n = A \sin(n\pi x/L)$ for $n = 1, 2, 3, \ldots$. (b) Show that the kinetic energy of the particle is given by $K_n = n^2[h^2/(8mL^2)]$, where m is the particle's mass. [Hint: (a) Consider boundary conditions such as those of a standing string wave. (b) Recall that the nonrelativistic expression of kinetic energy is $K = p^2/(2m)$ and that the wave function involves the de Broglie wavelength.]

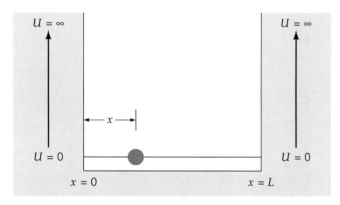

●FIGURE 28.16   **Particle in a box**   See Exercises 18 and 19.

19. ■■■ (a) Sketch the wave functions of the particle for the first three states in Exercise 18. (b) Where is the particle most likely to be found? What is the probability of finding it there? Sketch the probability densities for these states.

28.3 Atomic Quantum Numbers and the Periodic Table

20. The ℓ quantum number of a hydrogen atom state (a) determines the total energy of that state, (b) is associated with the angular momentum of the electron in that state, (c) is associated with the orientation of the angular momentum, (d) is associated with the electron spin.

21. The quantum number m_ℓ (a) determines the energy of the electron, (b) tells whether the electron is spinning up or down, (c) gives the orientation of the angular momentum of the electron, (d) all of these.

22. What information does the quantum number n give about an orbit in a hydrogen atom?

23. The quantum number m_s (a) is purely a quantum mechanical concept, (b) arises from the orbital motion of an electron, (c) is due to actual electron spinning, (d) all of these.

24. Niels Bohr set forth a correspondence principle stating that quantum mechanics and classical physics are in general agreement when the quantum numbers are very large. Discuss this principle in terms of the hydrogen atom.

25. In terms of quantum theory, what is the basis of the periodic table of elements? What do the elements in a particular group have in common?

26. ■ (a) How many possible sets of quantum numbers are there for $n = 2$ and $n = 3$ shells? (b) Explicitly write the $(n, \ell, m_\ell,$ and $m_s)$ sets for these levels.

27. ■ How many possible sets of quantum numbers are there for the subshells (a) $\ell = 0$ and (b) $\ell = 3$?

28. ■ Which has more possible sets of quantum numbers associated with it, $n = 2$ or $\ell = 2$?

29. ■ An electron in a multielectron atom has a magnetic quantum number of $m_\ell = 2$. What are the minimum values of (a) ℓ and (b) n for that electron?

30. ■■ Draw the ground-state energy level diagrams like those in Fig. 28.8 for (a) nitrogen, N, and (b) potassium, K.

31. ■■ Identify the atoms of each of the following ground-state electron configurations: (a) $1s^2 2s^2$, (b) $1s^2 2s^2 2p^3$, (c) $1s^2 2s^2 2p^6$, (d) $1s^2 2s^2 2p^6 3s^2 3p^4$.

32. ■■ Draw schematic diagrams for the electrons in the subshells of (a) sodium, Na, and (b) argon, Ar, atoms in the ground state.

33. ■■ Write the ground-state electron configurations for atoms of: (a) boron, B, (b) calcium, Ca, (c) zinc, Zn, (d) tin, Sn.

34. ■■■ What would be the first two inert or noble gases if there were no electron spin?

35. ■■■ How would the electronic structure of lithium differ if electron spin has three possible orientations instead of just two?

28.4 The Heisenberg Uncertainty Principle

36. If the uncertainty in the position of a moving particle increases, (a) the particle may be located more exactly, (b) the uncertainty in its momentum decreases, (c) the uncertainty in its velocity increases, (d) none of these.

37. By the uncertainty principle, to measure the exact energy of a particle requires (a) special equipment, (b) infinite time, (c) uncertainty in the momentum, (d) none of these.

38. Why is it impossible to measure accurately the position and velocity of a particle simultaneously?

39. ■ To what minimum uncertainty can the position of an electron of momentum $(3.284\,70 \pm 0.000\,25) \times 10^{-30}$ kg·m/s be measured?

40. ■ What is the minimum uncertainty in the speed of an electron that is known to be somewhere between 0.050 nm and 0.10 nm from a proton?

41. ■ What is the minimum uncertainty in the speed of a 0.50-kilogram ball that is known to be 1.0000 ± 0.0005 cm from the edge of a table?

42. ■■ The energy of an electron of 2.00 keV is known to $\pm 3.00\%$. How accurately can the electron's position be measured?

43. ■■ If an excited state of an atom is known to have a lifetime of 1.0×10^{-7} s, what is the minimum error within which the energy of the state can be measured?

44. ■■ The energy of the first excited state of a hydrogen atom is -0.34 eV $\pm 0.000\,30$ eV. What is the average lifetime for this state?

45. ■■ How many times greater is the width of a spectral line due to natural broadening for a transition from an excited state with a lifetime of 10^{-12} s than for one from a state with a lifetime of 10^{-8} s?

28.5 Particles and Antiparticles

46. Pair production involves (a) the production of two electrons, (b) the production of two positrons, (c) a positronium atom, (d) a certain threshold energy.

47. Due to momentum considerations, pair annihilation cannot result in the emission of how many photons? (a) one, (b) two, (c) several

48. Science fiction writers have suggested that matter and antimatter could be combined as a source of energy. (The starship *Enterprise* on "Star Trek" had antimatter engines with the antimatter being stored in antimatter "pods.") Speculate on the difficulties of storing antimatter similar to the way we currently store fuel in containers.

49. ■ A photon with a frequency of 2.5×10^{20} Hz passes near a nucleus. Will pair production occur? Justify your answer mathematically.

50. ■ What is the energy of each photon produced in electron–positron pair annihilation?

51. ■ What would be the required energy for a photon that could cause the production of a proton–antiproton pair?

52. ■■ A muon has a negative charge like that of an electron, but its mass is 207 times greater. What would be the required energy for a photon that could cause the pair production of a muon and an antimuon?

Additional Exercises

53. If the spacing between the lattice planes of a particular crystal is 0.190 nm, what voltage is needed to accelerate electrons (from rest) into a beam that exhibits a first-order diffraction maximum at an angle of 45°?

54. With what accuracy would you have to measure the position of a moving electron so that its speed is uncertain by 0.500 cm/s?

55. A gamma-ray photon has an energy of 7.5 MeV. What are (a) its momentum and (b) the wavelength of the associated light wave?

56. An electron is accelerated from rest through a potential difference of 5.00 kV $\pm 3.0\%$. What is the minimum uncertainty in the position of the electron after the acceleration?

57. Show that for a particle moving freely in one dimension between perfectly rigid boundaries $(-L < x < L)$, the wave functions are $\psi_n = A \cos(n\pi x/2L)$, where $n = 1, 3, 5\ldots$ and $\psi_n = A \sin(n\pi x/2L)$, where $n = 2, 4, 6 \ldots$.

58. In Exercise 57, show that the allowed kinetic energies are four times less than those of Exercise 18. Why?

59. Where is the particle of Exercise 57 most likely to be found in its ground state?

60. What is the de Broglie wavelength for the matter wave associated with (a) a 150-gram ball thrown at 20 m/s and (b) a 1200-kilogram automobile traveling at 90 km/h?

61. Is it possible for the number of quantum states for a given n quantum number to be equal to the number of states for a given ℓ quantum number? Justify your answer.

62. Assuming that both particles in a proton–antiproton pair are initially at rest, what are the (a) energy and (b) frequency of each of the emitted photons (to two significant figures)?

63. Scientists can create "muonic" atoms by bombarding targets with beams of negative muons (see Exercise 52). These muons would be captured into similar orbitals (different energies and sizes, however) as electrons. In principle, would the Pauli exclusion principle prohibit them from sharing the lower orbitals that are filled with electrons? Explain.

64. Suppose a positronium atom were moving along the x axis when an annihilation happened. Sketch the emission paths of the two photons. Would they go off exactly back to back? Explain.

CHAPTER

29

The Nucleus

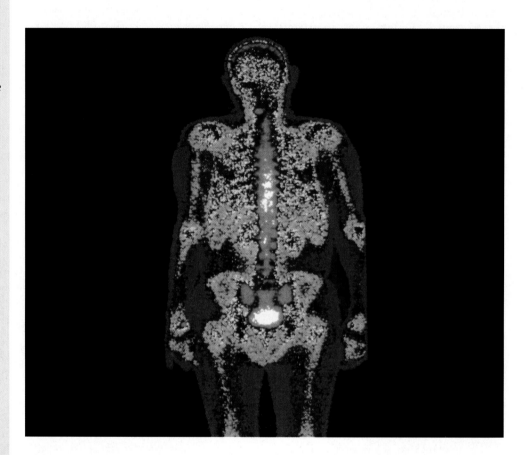

This high-tech skeletal image (a bone scan) was created by radiation from a radioactive source. *Radiation* and *radioactivity* are words that produce anxiety, but we often overlook the many beneficial uses to which radiation can be put. For instance, exposure to high-energy radiation can cause cancer—yet precisely the same sort of radiation can be useful in the diagnosis of cancer and even in the treatment of cancer.

The bone scan shown above was created by radiation released in the body when unstable nuclei spontaneously break apart—a process we call radioactive decay. But why are some nuclei so much more likely to decay than others? What determines which nuclei are likely to be radioactive, the rate at which they break down, and what particles they emit when they do so? These are some of the questions you'll explore in this chapter. You'll also learn how radiation is detected and measured, as well as more about its dangers and uses.

In addition, the study of radioactivity and nuclear stability gives us some general ideas about the nature of the nucleus, the energy it possesses, and how this energy can be released. The release of nuclear energy has become one of our major energy sources. This will be considered in Chapter 30. First, let's take a look at the nucleus itself.

29.1 Nuclear Structure and the Nuclear Force

Objectives: To (a) distinguish between the Thompson and Rutherford–Bohr models of the atom, (b) specify some of the basic properties of the strong nuclear force, and (c) understand nuclear notation.

It is evident from the emission of electrons from heated filaments (*thermionic emission*) and the photoelectric effect that atoms contain electrons. Since an atom is normally electrically neutral, it must also contain positive charge equal in magnitude to the total charge on all the electrons in the atom. Also, since the mass of an electron is small in comparison to the mass of even the lightest atoms, most of the atomic mass appears to be associated with that positive charge.

Based on these observations, J. J. Thomson, a British physicist who had experimentally proven the existence of the electron in 1897, proposed a model of the atom. In the Thomson model, the negatively charged electrons were uniformly distributed within a continuous sphere of positive charge. It was called the "plum pudding" model because the electrons in the positive charge were analogous to the raisins in a plum pudding. The region of positive charge was assumed to have a radius on the order of 10^{-10} m, or 0.1 nm, based on calculations of the bulk properties of matter.

Our modern model of atomic structure is quite different. This model pictures all of an atom's positive charges, and practically all of its mass, as concentrated in a central "nucleus," which is surrounded by the orbiting negatively charged electrons. The concept of an atomic nucleus was proposed by the British physicist Ernest Rutherford (1871–1937). Combined with Bohr's theory of orbiting electrons (Section 27.4), this idea led to the simplistic "solar system" model or **Rutherford–Bohr model** of the atom.

Rutherford's insight came from the results of alpha particle scattering experiments performed in his laboratory around 1911. An alpha (α) particle is a doubly positively charged particle that is naturally emitted from some radioactive materials (to be discussed further in Section 29.2). A beam of alpha particles from a radioactive source was directed at a thin gold foil "target," and the deflection angles of the scattered particles were observed (•Fig. 29.1). Such experiments were designed to investigate the distribution of mass and electric charge within atoms.

An alpha particle is over 7000 times more massive than an electron. Thus, the Thomson model would predict only small deflections, the result of collisions with the very light electrons as an alpha particle passed through a gold atom (•Fig. 29.2). Surprisingly, however, Rutherford and his colleagues observed alpha particles scattered at appreciable angles. In about 1 in 8000 instances, the alpha particles were actually *back-scattered*, or scattered through angles greater than 90° (•Fig. 29.3).

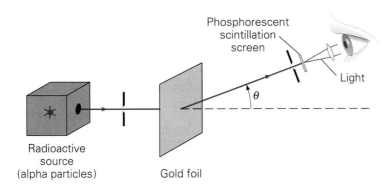

•**FIGURE 29.1 Rutherford's scattering experiment**   A beam of alpha particles from a radioactive source was scattered by gold nuclei in a thin foil, and the scattering was observed as a function of the scattering angle θ. The observer detects the light (viewed through a lens) given off by a scintillating phosphorescent screen.

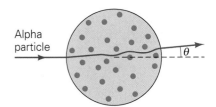

•**FIGURE 29.2 The plum pudding model**   With Thomson's plum pudding model of the atom, massive alpha particles were expected to be only slightly deflected by collisions with the electrons (blue dots) in the atom. The experimental results were quite different.

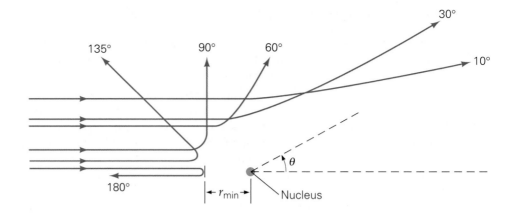

•FIGURE 29.3 **Rutherford scattering** A compact, dense atomic nucleus with a positive charge accounts for the observed scattering. An alpha particle in a head-on collision with the nucleus would be scattered directly backward ($\theta = 180°$) after coming within a distance r_{min} of the nucleus. At this scale, the electron orbits (about the nucleus) are too large to show.

Calculations showed that the probability of this back-scattering taking place with a Thomson model and a thin foil (10^{-4} cm) was miniscule—certainly much less than 1 in 8000. As Rutherford described the back-scattering, "It was almost as incredible as if you had fired a 15-inch shell at a piece of tissue paper and it came back and hit you."

The experimental results led Rutherford to the concept of an atomic nucleus:

> On consideration, I realized that this scattering backward must be the result of a single collision, and when I made calculations I saw that it was impossible to get anything of that order of magnitude unless you took a system in which the greater part of the mass of the atom was concentrated in a minute nucleus. It was then that I had the idea of an atom with a minute massive center carrying a charge.

If all of the positive charge of a target atom were concentrated in a very small region in the atom, then an alpha particle coming very close to this region would experience a large deflecting force. The mass of this nucleus of charge would be larger than that of the alpha particle, since most of the atomic mass is associated with the positive charge. Thus, back-scattering would be more likely.

A simple calculation can give an idea of the size of the atomic nucleus. For a head-on collision, an alpha particle would attain its distance of closest approach to the nucleus (r_{min} in Fig. 29.3). That is, the alpha particle approaching the nucleus would be stopped at a distance r_{min} by the repulsive Coulomb force and accelerated back along its original path. Assuming a spherical or point-charge distribution, the electric potential is $V = kZe/r$, where Z is the **atomic number,** or the number of protons in the nucleus, and Ze is the nuclear total charge. The work done by the Coulomb force in stopping the alpha particle is $W = qV = (2e)V$, since the alpha particle has a double positive electronic charge. By the work–energy theorem, this is equal to the initial kinetic energy of the alpha particle:

Z, the atomic number: the number of protons in the nucleus

$$\tfrac{1}{2} mv^2 = \frac{k(2e)Ze}{r_{min}}$$

so

$$r_{min} = \frac{4kZe^2}{mv^2} \qquad (29.1)$$

From the known value of the energy of the alpha particles from the particular source and other known values, r_{min} is found to be about 10^{-14} m. This value is an *upper limit* for the nuclear radius, since the alpha particle does not reach the nucleus itself.

Although the nuclear model of the atom is useful, the nucleus is much more than simply a region of positive charge. We now know that the atomic nucleus is composed of two types of particles—protons and neutrons—which are collectively referred to as **nucleons**. The nucleus of the common hydrogen atom is a single proton. Rutherford suggested that the hydrogen nucleus be named *proton*

Nucleon: proton or neutron—a particle in the nucleus

(from a Greek word meaning "first") after he became convinced that there is no positively charged particle in an atom less massive than the hydrogen nucleus. A neutron is an electrically neutral particle with a mass slightly greater than that of a proton. The existence of the neutron was not experimentally verified until 1932.

The Nuclear Force

Of the forces in the nucleus, we know there is an attractive gravitational force between nucleons (protons and neutrons). But in Chapter 15, we saw that the gravitational force is negligible in comparison with the repulsive electrical force between protons, which are positively charged. Taking only these repulsive forces into account, we might expect the nucleus to fly apart. Yet the nuclei of most atoms are stable, so there must be an additional force that holds the nucleus together. This strongly attractive force is called the **strong nuclear force,** or simply the *nuclear force.*

Note: Review Example 15.4, Section 15.3.

The exact expression for the nuclear force is extremely complex. However, some general features of this force are as follows:

- The nuclear force is strongly attractive and much larger in magnitude than either the electrostatic force or gravitational force.
- The nuclear force is very short-ranged; that is, a nucleon interacts only with its nearest neighbors, over distances on the order of 10^{-15} m.
- The nuclear force is independent of electric charge; that is, it acts between *any* two nucleons—between two protons, a proton and a neutron, or two neutrons.

Thus, nuclear protons in close proximity repel each other by the electric force but attract each other (and nearby neutrons) by the strong nuclear force. Having no electric charge, neutrons are attracted to nearby protons and neutrons only by the nuclear force (gravitational force being negligible).

Nuclear Notation

To describe the nuclei of different atoms, it is convenient to use the notation illustrated in •Fig. 29.4a. The chemical symbol of the element is used with subscripts and a superscript. The left subscript is the atomic number (Z), which indicates the number of protons in the nucleus. An alternative (and more descriptive) term, which we will generally use in this book, is **proton number**. For atoms, which are electrically neutral, this is equal to the number of orbital electrons.

The number of protons in the nucleus of an atom determines the species of the atom—the element to which the atom belongs. In the example in Fig. 29.4b, the proton number $Z = 6$ indicates that this is a carbon nucleus. The proton number defines which chemical symbol is used. Electrons can be removed (or added) to an atom to form an ion, *but this does not change its species.* For example, a nitrogen atom with an electron removed, N^+, is still nitrogen—a nitrogen *ion.* The number of electrons can vary by ionization, but the proton number cannot vary without producing an atom of a different element. In that sense, adding the label Z is redundant because the chemical symbol implies the proton number.

The left superscript on the chemical symbol is called the **mass number** (A)—the total number of protons and neutrons in the nucleus. Since protons and neutrons have roughly equal masses, the mass numbers of nuclei give a relative comparison of nuclear masses. For the example in Fig. 29.4b, the carbon nucleus, the mass number is $A = 12$ because there are 6 protons and 6 neutrons. The number of neutrons, called the **neutron number** (N), is sometimes indicated by a subscript on the right side, but this is usually omitted because it can be determined from A and Z; that is, $N = A - Z$. Similarly, the proton number is routinely omitted because the chemical symbol uniquely specifies Z.

The atoms of an element, even though they must have the same number of protons in their nuclei, may have different numbers of neutrons. For example, nuclei

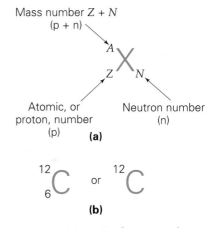

•**FIGURE 29.4   Nuclear notation** **(a)** A complete representation of the contents of a nucleus is given by the chemical symbol of the element with the mass number A (number of protons and neutrons) as a left superscript and the proton (atomic) number Z as a left subscript. The neutron number N may be shown as a right subscript, but both Z and N are routinely omitted because the letter symbol tells you Z, and $N = A - Z$. **(b)** The two most common nuclear notations for a nucleus of one of the isotopes of carbon.

of different carbon atoms ($Z = 6$) may contain 6, 7, or 8 neutrons. In nuclear notation, these atoms would be

$$^{12}_{6}C_6 \quad ^{13}_{6}C_7 \quad ^{14}_{6}C_8$$

Atoms whose nuclei have the same number of protons but different numbers of neutrons are called **isotopes.** This example shows three isotopes of carbon.

Isotopes are like members of a family: They all have the same Z number and the same surname, but they are distinguishable by the number of neutrons in their nuclei and therefore their mass. Isotopes are referred to by their mass numbers; for example, the previous isotopes of carbon are called carbon-12, carbon-13, and carbon-14. There are other isotopes of carbon: ^{11}C, ^{15}C, and ^{16}C. A particular nuclear species or isotope of any element is called a **nuclide.** Thus, we have displayed six nuclides of carbon. Generally, only a few isotopes of a given species are stable. In this case, although it is relatively long-lived, ^{14}C is unstable. Only ^{12}C and ^{13}C, in fact, are stable isotopes of carbon.

Another important family of isotopes is that of hydrogen, which has three isotopes or nuclides: ^{1}H, ^{2}H, and ^{3}H. These isotopes are not generally referred to as hydrogen-1 and so on but are given special names. The isotope ^{1}H is called ordinary hydrogen or simply hydrogen; ^{2}H is called *deuterium*. Deuterium, sometimes known as heavy hydrogen, can combine with oxygen to form heavy water (written D_2O). The third isotope of hydrogen, ^{3}H, is called *tritium*, which is unstable.

Isotopes of an element: same number of protons, different numbers of neutrons

Note: Isotopes in one family have almost the same orbital electron structure and thus very similar chemical properties.

Nuclide: a particular nuclear species or isotope

29.2 Radioactivity

Objectives: To (a) define the term *radioactivity*, (b) distinguish among alpha, beta, and gamma decay, and (c) write nuclear decay equations.

Most elements have stable isotopes. It is the atoms of these stable nuclides with which we are most familiar in the environment. However, the nuclei of some isotopes are unstable and disintegrate spontaneously (decay), emitting energetic particles and photons. Such isotopes are said to be *radioactive* or to exhibit **radioactivity.** For example, the tritium isotope of hydrogen has a radioactive nucleus. The prefix *radio-* refers to the emitted nuclear radiation, which in the modern context may be a particle or a wave. A radioactive isotope is "active" when it is emitting radiation. Of the nearly 1200 known unstable nuclides, only a small number occur naturally. The others are produced artificially (Chapter 30).

We know that radioactivity is unaffected by normal physical or chemical processes, such as heat, pressure, and chemical reactions. Since chemical processes involve only outer electrons, the source of radioactivity must lie deeper in the atom—that is, in the nucleus. This instability cannot be explained directly by a simple imbalance of attractive and repulsive forces in the nucleus, because the nuclear disintegrations of a given isotope occur at a fixed rate. Classically, we would expect identical nuclei to do the same thing at the same time. Therefore, radioactive decay suggests quantum mechanical probability effects.

The discovery of radioactivity is credited to the French scientist Henri Becquerel. In 1896, while studying the fluorescence of a uranium compound, Becquerel discovered that a photographic plate in the vicinity of a sample had been darkened even though the compound had not been activated by exposure to light and was not fluorescing. Apparently the darkening was caused by some new type of radiation emitted from the compound. In 1898, Pierre and Marie Curie announced the discovery of two radioactive elements, radium and polonium, which they had isolated from uranium pitchblende ore (•Fig. 29.5).

Experiment shows that the radiation emitted by radioactive isotopes is of three different kinds. When a radioisotope is placed in a chamber so that the emitted radiation passes through a magnetic field to a photographic plate (•Fig. 29.6), the radiations expose the plate, producing identifying spots. The positions of the spots show that some isotopes emit radiation that is deflected to the left; some, radiation

•**FIGURE 29.5 The Curies** Marie Sklodowska Curie (1867–1934) was born in Poland and studied in France. There she met and married Pierre Curie (1859–1906), who was a physicist well known for his work on crystals and magnetism. In 1903, Madame Curie (as she is commonly known) and Pierre shared the Nobel Prize in physics with Henri Becquerel (1852–1908) for their work on radioactivity. She was also awarded the Nobel prize in chemistry in 1911 for the discovery of radium and the study of its properties. The Curies are shown on the cover of a 1904 magazine.

that is deflected to the right; and some, radiation that is undeflected. These spots are characteristic of what came to be known as *alpha, beta,* and *gamma* radiations.

From the opposite deflections of two of the types of radiation in the magnetic field, it is evident that positively charged particles are emitted from nuclei undergoing alpha decay and that negatively charged particles are emitted in beta decay. The degree of deflection shows that alpha particles must be more massive than beta particles. The undeflected gamma radiation (gamma rays) is electrically neutral.

Detailed investigations of the three different radiations reveal that

- **Alpha particles** are doubly charged ($+2e$) particles containing two protons and two neutrons. They are identical to the nucleus of the helium atom ($^{4}_{2}\text{He}$).
- **Beta particles** are electrons.*
- **Gamma rays** are particles, or quanta, or photons, of electromagnetic energy.

For a few radioactive elements, two spots are found on the film, indicating that some elements decay by two different modes.

Let's now look at what happens to a decaying nucleus in each of these three radioactive decay modes.

Alpha Decay

When an alpha particle is ejected from a radioactive nucleus, the nucleus loses two protons and two neutrons, so the mass number (A) is decreased by four, $\Delta A = -4$; and the proton number (Z) is decreased by two, $\Delta Z = -2$. Since the parent nucleus loses two protons, the resulting *daughter nucleus* must be the nucleus of another element as defined by the new proton number. (The original and resulting nuclei are commonly referred to as the parent and daughter nuclei, respectively.) Thus, the **alpha decay** process is one of nuclear *transmutation,* in which the nuclei of one element change into the nuclei of a lighter element.

An example of an isotope or nuclide that undergoes alpha decay is polonium-214. The decay process is represented as a nuclear equation (usually written without neutron numbers), similar to a chemical equation except that it refers to nuclei:

$$\underset{polonium}{^{214}_{84}\text{Po}} \rightarrow \underset{lead}{^{210}_{82}\text{Pb}} + \underset{\substack{alpha\ particle\\(helium\ nucleus)}}{^{4}_{2}\text{He}}$$

The totals of the mass numbers and the proton numbers must be equal on each side of the equation: $(214 = 210 + 4)$ and $(84 = 82 + 2)$, respectively. This reflects the fact that *two conservation laws apply to all nuclear processes.* The first is the **conservation of nucleons:** The total number of nucleons (A) remains constant in any process. The second is the familiar **conservation of charge,** when applied to nuclear reactions. These conservation laws allow us to predict the daughter nucleus, as the following Example illustrates.

Example 29.1 ■ Uranium's Daughter: Alpha Decay

A $^{238}_{92}\text{U}$ nucleus undergoes alpha decay. What is the resulting daughter nucleus?

Thinking It Through. We apply nucleon conservation to predict the daughter's proton number. From that, we can get the element name from the periodic table.

Solution. Since $\Delta Z = -2$ for alpha decay, the uranium-238 (^{238}U) nucleus loses two protons, and the daughter nucleus has a proton number $Z = 92 - 2 = 90$, the proton number of thorium (see the periodic table, Fig. 28.9). The equation for this decay is

---

*Electrons are emitted in the most common type of beta decay; but there are other types, as will be discussed shortly.

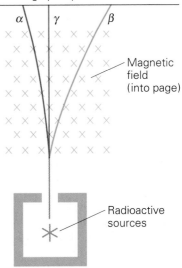

Photographic plate

•**FIGURE 29.6 Nuclear radiation** Radiations from radioactive sources can be distinguished by passing them through a magnetic field. Alpha and beta particles are deflected. Applying the right-hand magnetic force rule, we find that alpha particles are positively charged and beta particles are negatively charged. The radii of curvature (not drawn to scale) allow the particles to be distinguished by mass. Gamma rays are not deflected and so are uncharged; they are quanta of electromagnetic energy.

Alpha particle: a helium nucleus

Note: In any nuclear process, the number of nucleons and total charge is conserved.

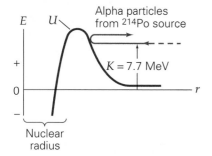

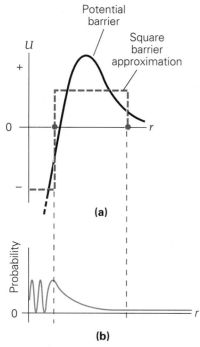

•**FIGURE 29.7 Potential barrier for alpha particles** Alpha particles from radioactive polonium with energies of 7.7 MeV do not have enough energy to overcome the electrostatic potential barrier of the ^{238}U nucleus and are scattered.

•**FIGURE 29.8 Tunneling effect, or barrier penetration** (a) The potential barrier presented by a nucleus to an alpha particle can be approximated by a barrier. (b) The square of the alpha particle wave function (the probability of finding it at a given location) is shown. Inside the nucleus it has a high likelihood of being found (large amplitude), but it can "tunnel" through the barrier and appear outside the nucleus with a much smaller, but nonzero, probability.

$$^{238}_{92}\text{U} \rightarrow {}^{234}_{90}\text{Th} + {}^{4}_{2}\text{He}$$

(Sometimes the helium nucleus is written as α or ${}^{4}_{2}\alpha$.)

Follow-up Exercise. Using high-energy accelerators, it is possible to add an alpha particle to a nucleus, essentially the reverse of the reaction in this Example. Write this nuclear reaction and predict the resulting nucleus if an alpha particle is added to a ^{12}C nucleus in this fashion.

The energies of the alpha particles from radioactive sources are typically a few megaelectron volts (MeV, introduced in Section 16.2). For example, the energy of the alpha particle emitted during the decay of ^{214}Po is about 7.7 MeV, and that from ^{238}U decay is about 4.14 MeV. Alpha particles from such radioactive sources were used in the scattering experiments that led to the Rutherford nuclear model. Scattering experiments give some idea of the size of the nucleus, as you learned in Section 29.1. In addition, they can shed light on the internal structure of the nucleus and the forces within it.

Outside the nucleus, the repulsive electric force increases as an alpha particle nears the nucleus. Inside the nucleus, however, the strongly attractive nuclear force dominates. These conditions are depicted in •Fig. 29.7 in a graph of the potential energy, U, as a function of r, the distance from the center of the nucleus. Outside the nucleus, U varies as $1/r$ since it represents the electrical potential. Within the nucleus, U is opposite in sign and has the form of a negative potential well because of the strongly attractive nuclear force. This diagram could represent the potential energy of an alpha particle approaching a ^{238}U nucleus.

Consider alpha particles from a ^{214}Po source incident on a thin foil of ^{238}U. Rutherford scattering takes place as a result of a high potential energy barrier that the incident alpha particles do not have enough energy to overcome. Instead, they are scattered. If an incident particle did have enough energy to overcome, or cross, this Coulomb barrier, it could enter the nucleus. In this case, a nuclear reaction might result (to be discussed in Chapter 30).

However, the ^{238}U nucleus itself undergoes alpha decay, *emitting an alpha particle* with an energy of 4.4 MeV. This fact appears to contradict the scattering experiment. How can these lower-energy alpha particles cross a potential barrier from the inside that higher-energy incident alpha particles cannot cross from the outside? Classically, this is impossible, since it violates the conservation of energy, and thus the outside of the potential barrier is referred to as a classically forbidden region. However, quantum mechanics offers an explanation.

Quantum mechanics predicts a finite probability of finding a particle in a classically forbidden region if the wave function exists in the region. If the alpha particle exists as an entity in the nucleus, its wave function will tail off (in nonoscillatory behavior) in the barrier region and reach the outside. There it appears as a wave of much smaller amplitude (•Fig. 29.8). There would then be a finite probability (ψ^2, Section 28.2) of finding the alpha particle *outside* the nucleus. This is called **tunneling**, or **barrier penetration**, since the alpha particle has a small but nonzero chance or probability of tunneling through the barrier.

The probability of finding an alpha particle outside the ^{238}U nucleus is extremely small, as reflected in the extremely slow decay rate of ^{238}U. Some other nuclei decay very rapidly by alpha decay. The height and width of the potential barrier determine the probability that an alpha particle can escape from the nucleus. Thus, the barrier controls the decay rate (as we will see in Section 29.3).

Beta Decay

The emission of an electron (a beta particle) in a nuclear decay process might seem contradictory to the proton–neutron model of the nucleus. Note that the electron emitted in beta decay is not an orbital electron but one emitted *from* the nucleus, although it is *not* part of the original nucleus. In fact, the process of **beta decay** indicates that *the electron is created in the nucleus itself*. For example, the equation for ^{14}C beta decay is

$$\underset{\text{carbon}}{^{14}_{6}\text{C}} \quad \rightarrow \quad \underset{\text{nitrogen}}{^{14}_{7}\text{N}} \quad + \quad \underset{\substack{\text{beta particle} \\ \text{(electron)}}}{^{0}_{-1}\text{e}}$$

The parent carbon nucleus has six protons and eight neutrons, whereas the daughter nitrogen nucleus has seven protons and seven neutrons. Notice that the notation for the electron specifies a mass (or nucleon) number of zero (why?) and a charge number of -1. Thus both nucleon number and electric charge are conserved.

In this type of beta decay, the neutron number of the parent nucleus decreases by one ($\Delta N = -1$), and the proton number of the daughter nucleus increases by one ($\Delta Z = +1$). The nucleon number is unchanged. This indicates that, in essence, *a neutron within the nucleus decays into a proton and an electron*:

$$\underset{\text{neutron}}{^{1}_{0}\text{n}} \quad \rightarrow \quad \underset{\text{proton}}{^{1}_{1}\text{p}} \quad + \quad \underset{\text{electron}}{^{0}_{-1}\text{e}} \qquad \textit{(basic } \beta^{-} \textit{ decay)}$$

(This equation also describes the behavior of a *free* neutron, which is unstable when outside a nucleus.) In nuclear notation, the neutron has a mass number of 1 and a proton number of zero (why?).

This process enables an unstable nucleus to move closer to stability in *two* ways. The ^{14}C, for example, has one too many neutrons to be stable (the most massive stable isotope is ^{13}C). The net result of beta decay is to decrease the neutron number and increase the proton number, as we shall see later both of which work in the direction of greater stability. In the ^{14}C example, ^{14}N is stable. We should also note that there is another elementary particle, called a *neutrino*, emitted in beta decay. For simplicity, it is not shown in the nuclear decay equations. Its place in beta decay will be discussed in Chapter 30.

There are two modes of beta decay, β^{-} and β^{+}, as well as a competing process called *electron capture*. The **β^{-} decay** involves the emission of an electron as above. Isotopes that decay by this means do so because they have too many neutrons (compared to protons) to be stable. However, unstable isotopes that undergo **β^{+} decay,** or *positron decay*, which involves the emission of a positron ($^{0}_{+1}\text{e}$), have too many protons compared to neutrons. (We introduced positrons in Section 28.5.) As a result, the net effect of β^{+} decay is to convert a proton into a neutron. An example of β^{+} decay is

β^{-}: **an** electron

β^{+}: **a** positron

$$\underset{\text{oxygen}}{^{15}_{8}\text{O}_{7}} \quad \rightarrow \quad \underset{\text{nitrogen}}{^{15}_{7}\text{N}_{8}} \quad + \quad \underset{\text{positron}}{^{0}_{+1}\text{e}}$$

Positron emission is also accompanied by a neutrino (of a different type than that in β^{-} decay), which we again omit until Chapter 30. As in β^{-} decay, the mass numbers of the parent and daughter nuclei do not change, but the proton number of the daughter nucleus in this case is one less than that of the parent nucleus.

The third related process, **electron capture (EC),** involves the absorption of *orbital* electrons by a nucleus. The net result is a daughter nucleus the same as would have been produced by positron decay. Usually these two types of decay compete, since they produce the same, more stable daughter nucleus. This makes physical sense because the inward absorption of an electron is electrically equivalent to the outward emission of a positron. In the language of quantum mechanics, there can be a nonzero probability for *both* processes to happen to a given nucleus. A specific example of electron capture is

$$\underset{\substack{\text{orbital} \\ \text{electron}}}{^{0}_{-1}\text{e}} \quad + \quad \underset{\text{beryllium}}{^{7}_{4}\text{Be}} \quad \rightarrow \quad \underset{\text{lithium}}{^{7}_{3}\text{Li}}$$

Note: Electron shells are introduced in Section 28.3.

As in β^+ decay, a proton changes into a neutron, but no beta particle is emitted in electron capture. An electron in the innermost shell, called the K shell from spectroscopic notation, is most likely to be captured, an event referred to as *K-capture*. Electrons in more distant orbits (for example, *L-capture*) occur with much less probability. Since there is no charged particle emission, the process is detected by observing the emission of characteristic X-rays produced when an orbiting electron from an outer shell makes a downward transition into a K-shell vacancy. Because X-ray emission must take place *after* the K-capture, the X-rays are characteristic of the daughter nucleus, not the parent.

Gamma Decay

Gamma ray: a quantum or photon of electromagnetic energy

In **gamma decay**, the nucleus emits a gamma (γ) ray, or a photon of electromagnetic energy. The emission of a gamma ray by a nucleus in an excited state is analogous to the emission of a photon by an excited atom. This may come about because of an energetic collision with another particle or, more commonly, because the nucleus is a daughter nucleus left in an excited state after a radioactive decay.

Nuclei possess energy levels analogous to those of atomic electrons. However, the nuclear energy levels are much farther apart and more complicated than those of an atom. The nuclear energy levels are on the order of kiloelectron volts (keV) and megaelectron volts (MeV) apart, compared to a few electron volts between atomic levels. As a result, gamma rays are very energetic, having frequencies greater than those of X-rays and thus much shorter wavelengths. Consider the decay of ^{61}Ni from an excited nuclear state to one of lesser energy:

$$\underset{\substack{nickel \\ (excited)}}{^{61}_{28}\text{Ni}^*} \quad \rightarrow \quad \underset{nickel}{^{61}_{28}\text{Ni}} \quad + \quad \underset{gamma\ ray}{\gamma}$$

The asterisk indicates that the ^{61}Ni nucleus is in an excited state. De-excitation results in the emission of a gamma ray. Note that *in gamma decay, the mass and proton numbers do not change*. The daughter nucleus in this case is simply the parent nucleus with less energy. As an example of gamma emission following beta decay, consider the following Example.

Example 29.2 ■ Two for One: Beta Decay and Gamma Decay

Natural cesium consists of a single stable isotope $^{133}_{55}$Cs. An unstable isotope $^{137}_{55}$Cs is a common fragment of the fission process (for example, when the $^{235}_{92}$U nucleus splits, or "fissions"). When $^{137}_{55}$Cs decays, its daughter nucleus is sometimes left in an excited state, giving off a gamma ray with an energy of 0.662 MeV to produce a stable nucleus. (a) Predict the beta decay mode (β^+ or β^-) for the $^{137}_{55}$Cs nucleus, and find the daughter product. (b) Write the gamma decay equation showing the daughter in its final stable configuration.

Thinking It Through. (a) The $^{137}_{55}$Cs has too many neutrons to be stable. (Compare to ^{133}Cs.) This excess implies that β^- decay will convert a neutron into a proton. (b) Gamma decay is a de-excitation of the daughter nucleus in (a).

Solution. The data are as follows:

Given: Initial nucleus $^{137}_{55}$Cs   *Find:* (a) Decay particle and daughter, first decay reaction
                                            (b) Final decay showing gamma decay to final stable state

(a) The proton number will increase by one; thus the daughter will be barium (see the periodic table, Fig. 28.9). We write down the decay, indicating that barium is left in an excited state ready to decay via gamma emission, and ignore the emitted neutrino:

$$\underset{cesium}{^{137}_{55}\text{Cs}} \quad \rightarrow \quad \underset{\substack{barium \\ (excited)}}{^{137}_{56}\text{Ba}^*} \quad + \quad \underset{electron}{^{0}_{-1}\text{e}}$$

(b) The gamma emission of the barium is

$$^{137}_{55}Ba^* \rightarrow \ ^{137}_{55}Ba \ + \ \gamma$$

barium   barium   0.662 MeV
(excited)   gamma ray

Sometimes this total process is written in two stages to show the sequential behavior:

$$^{137}_{55}Cs \rightarrow \ ^{137}_{56}Ba^* \ + \ ^{0}_{-1}e$$

cesium   barium   electron
          (excited)

$$\rightarrow \ ^{137}_{56}Ba \ + \ \gamma$$

barium   0.662 MeV
          gamma ray

Follow-up Exercise. An unstable isotope of sodium, ^{22}Na, can be produced in nuclear reactors. Write its β^+ decay equation, and predict the daughter nucleus.

Radiation Penetration

The absorption or degree of penetration of nuclear radiations is an important consideration in applications such as radioisotope treatment of cancer and nuclear shielding (for example, around a nuclear reactor). Also, the absorption of nuclear radiation is used to monitor and automatically control the thickness of metal and plastic sheets and films in fabrication processes.

The three types of radiation (alpha, beta, and gamma) are absorbed quite differently. Along their paths, the electrically charged alpha and beta particles interact with the atoms of a material and may produce ionizations of these atoms. The greater the charge and the slower the particle, the greater the energy transfer and ionization along the path, which determine the degree of penetration. Also, the penetration depends on the density of the material.

- Alpha particles are doubly charged, have a larger mass, and generally move more slowly. A few centimeters of air or a sheet of paper will usually completely stop them.
- Beta particles are much less massive and singly charged. They can travel a few meters in air or a few millimeters in aluminum before being stopped.
- Gamma rays are uncharged and are therefore more penetrating. A significant portion of a beam of high-energy gamma rays can penetrate a centimeter or more of a dense material such as lead. (Lead is commonly used as shielding for harmful X-rays, which are less energetic than gamma rays.)

Radiation passing through matter can do considerable damage. Metals and other structural materials can become brittle and lose their strength when exposed to strong radiation, such as that in nuclear reactors (Chapter 30) and the intense cosmic radiation to which space vehicles are exposed. In biological tissue, the radiation damage is chiefly due to ionizations in living cells (Section 29.5). We are exposed to normal background radiation from radioisotopes in the environment and to cosmic radiation from outer space. The intensity of most such radiation is usually too low to be harmful. However, concern has been expressed that flight crews who spend many hours aboard high-flying jet aircraft may receive excessive exposure to such cosmic radiations.

As pointed out earlier, of the nearly 1200 known unstable nuclides, only a small number occur naturally. Most of the radioactive nuclides found in nature occur as products of the decay series of heavy nuclei: a series of continual radioactive decay into lighter elements. The ^{238}U decay series is given in •Fig. 29.9. It stops when the stable isotope ^{206}Pb is reached. Note that some nuclides in the series decay by two modes. Radon (^{222}Rn) is part of this decay series; this radioactive gas has received a great deal of attention because it can accumulate in significant amounts in poorly ventilated buildings (see the Insight on p. 902).

Note: See also Figs. 29.23 (^{237}Np) and 29.24 (^{239}Pu).

•FIGURE 29.9 Decay series of uranium-238
On this plot of N versus Z, a diagonal transition from right to left is an alpha decay process; a diagonal transition from left to right is a β^- decay process (how can you tell?). The decay series continues until the stable nucleus ^{206}Pb is reached.

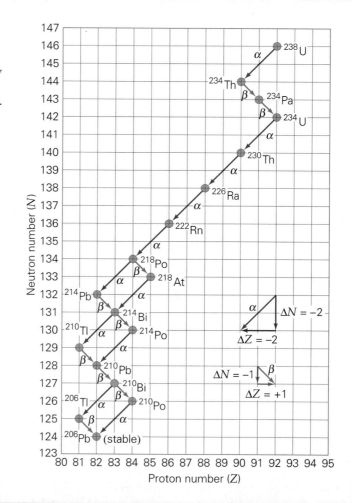

Radon in the Home

The widespread concern over radon gas accumulations in homes and other buildings is a relatively recent worry; two decades ago most people had never heard of radon. Radon is a chemically inert gas belonging to the noble gas group, along with helium, neon, and others. However, it is radioactive and is part of the ^{238}U decay series (see Fig. 29.9). As such, it occurs naturally, the product of uranium deposits that exist in many areas. The amount of radon released from the ground depends on the local geology.

Because radon gas is chemically inert, it doesn't react with anything in the soil and so is not immobilized in the ground. Also, the half-life of radon is such that significant concentrations can persist for days. Thus there is adequate time for it to seep up into homes. It enters through holes and cracks in the basement floor or walls or around pipes. Radon gas can also be dissolved in water, brought into the home with the water supply and released on aeration. It can also be released from some common building materials, such as bricks and concrete blocks.

What is the problem with this inert gas that is a natural part of our environment? When inhaled by humans, it can decay inside the lungs. Radon alpha decays into polonium-218. Polonium-218 is not a gas, nor is it chemically inert, but it is

radioactive. It can lodge in the lungs, where it too undergoes alpha decay, into ^{214}Pb. (Polonium-218 also beta decays—see Fig. 29.9—but more than 99% of its decay is by the alpha mode.)

As was discussed in the text, the penetration range of alpha particles is quite small. However, from inside the lungs these particles can penetrate the cells of the lung lining, causing damage and possibly inducing lung cancer (Section 29.5).

Since radon occurs naturally in the environment, the next immediate questions are: How do I know how much radon I have, and how much radon in my home is too much? You can purchase sampling kits to test the level of radon in your home. There is some controversy over the level of radon considered to be safe. The U.S. Environmental Protection Agency has set an activity standard of 4 picocuries (pCi)* per liter of air as an "action level." If your home shows a level of radon gas greater than this, you should consult the appropriate state agency for retesting and possible corrective actions, such as sealing cracks in the basement and around pipes coming through walls. Water sources may also need checking. Testing is particularly important in colder climates, where homes are insulated and sealed and closed for long periods of time.

*A curie (Ci) is a unit of radioactivity discussed later in this chapter.

29.3 Decay Rate and Half-Life

Objectives: To (a) explain the concepts of activity, decay constant, and half-life of a radioactive sample, and (b) use radioactive decay to find the age of objects.

The nuclei of a sample of a radioactive nuclide do not decay all at once but randomly at a characteristic rate that is unaffected by any external stimulus. No one can tell exactly when a particular nucleus will decay. What can be determined is how many nuclei in a sample will decay over a period of time.

The **activity** of a sample of radioactive nuclide is the number of nuclear disintegrations, or decays, per second. For a given amount of material, the activity decreases with time, as fewer and fewer radioactive nuclei remain. Each nuclide has its own characteristic rate of decrease. The rate at which the number of parent nuclei (N) decreases is proportional to the number present:

$$\frac{\Delta N}{\Delta t} \propto N$$

This rate can be written in equation form with a constant of proportionality:

$$\frac{\Delta N}{\Delta t} = -\lambda N \qquad (29.2)$$

Decay rate of a radioisotope

where the constant λ is called the **decay constant;** λ has SI units of s^{-1} and is unique to each nuclide. The larger the decay constant, the greater the rate of decay. The minus sign in Eq. 29.2 indicates that N is decreasing. The sample's *activity* is the magnitude of $\Delta N/\Delta t$, in decays per second, without a minus sign. We will use $\Delta N/\Delta t$ for activity, but keep in mind that we mean a *positive* number. (See the usage in Example 29.3.)

Note: Activity = number of decays per second = $|\Delta N/\Delta t|$. We will omit the absolute value signs and write the activity as $\Delta N/\Delta t$, but keep in mind that we mean a *positive* number.

The way in which the number of parent nuclei N decreases with time (compared to the initial number N_o) is illustrated in •Fig. 29.10. The number N is said to decay *exponentially* since the graph follows an exponential function $e^{-\lambda t}$ with time, where λ is the decay constant and $e \approx 2.718$ is the base of natural logarithms. Equation 29.2 can be solved for the number N of the remaining undecayed parent nuclei at a time t:

$$N = N_o e^{-\lambda t} \qquad (29.3)$$

where N_o is the initial number of nuclei (at $t = 0$).

The decay rate of a nuclide is commonly expressed in terms of its half-life rather than the decay constant. The **half-life** ($t_{1/2}$) is the time it takes for half of the radioactive nuclei in a sample to decay. This is the time corresponding to $N_o/2$ parent nuclei in Fig. 29.10—that is, $N/N_o = 1/2$. In the same time, activity also

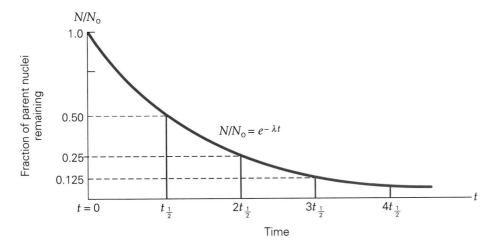

•**FIGURE 29.10 Radioactive decay versus time** The fraction of the remaining parent nuclei (N/N_o) in a radioactive sample plotted as a function of time gives an exponential decay curve, the shape or steepness of which depends on the decay constant λ or the half-life $t_{1/2}$.

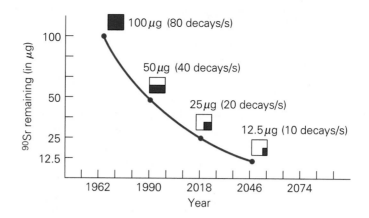

•FIGURE 29.11 Radioactive decay and half-life As shown here for strontium-90, after each half-life ($t_{1/2}$ = 28 y), only half of the amount of ^{90}Sr present at the start of that time period remains, and the activity (decays per second) has decreased by half. The other half of the sample has decayed to ^{90}Y via beta decay.

decreases to the same fractional amount, since it is proportional to the number of nuclei present. That is, the number of decays per second (the activity) itself decreases by half in this instance. Rather than counting nuclei, the decay *rate* is what is usually measured to determine the half-life. In other words, we usually monitor the rate at which the decay particles are emitted until this rate decays by a certain fraction (not always the half-life, especially if the half-life is long).

For example, the graph in •Fig. 29.11 shows that the half-life of strontium-90 (^{90}Sr) is 28 y. Another way of looking at what occurs in a half-life is to consider the amount, or mass, of the parent material. As illustrated in Fig. 29.11, if there were initially 100 micrograms (μg) of ^{90}Sr, only 50 μg would remain at the end of 28 y. The other 50 μg would have decayed by beta emission,

$$\underset{strontium}{^{90}_{38}\text{Sr}} \longrightarrow \underset{yttrium}{^{90}_{39}\text{Y}} + \underset{electron}{^{0}_{-1}\text{e}}$$

and the sample would contain both strontium and yttrium nuclei. After another 28 y, another half of the strontium nuclei would decay, leaving 25 μg, and so on.

The half-lives of radioactive nuclides vary greatly, as Table 29.1 shows. Nuclides with very short half-lives are generally created in nuclear reactions, which

TABLE 29.1 The Half-Lives of Some Radioactive Nuclides

Nuclide	Primary Decay Mode	Half-Life
Beryllium-8 ($^{8}_{4}$Be)	α	1×10^{-16} s
Polonium-213 ($^{213}_{84}$Po)	α	4×10^{-16} s
Oxygen-19 ($^{19}_{8}$O)	β^-	27 s
Fluorine-17 ($^{17}_{9}$F)	β^+, EC	66 s
Polonium-218 ($^{218}_{84}$Po)	α, β^-	3.05 min
Technetium-104 ($^{104}_{33}$Te)	β^-	18 min
Krypton-76 ($^{76}_{36}$Kr)	EC	14.8 h
Magnesium-28 ($^{28}_{12}$Mg)	β^-	21 h
Iodine-123 ($^{123}_{53}$I)	EC	13.3 h
Radon-222 ($^{222}_{86}$Rn)	α	3.82 d
Cobalt-60 ($^{60}_{27}$Co)	β^-	5.3 y
Strontium-90 ($^{90}_{38}$Sr)	β^-	28 y
Radium-226 ($^{226}_{88}$Ra)	α	1600 y
Carbon-14 ($^{14}_{6}$C)	β^-	5730 y
Plutonium-239 ($^{239}_{94}$Pu)	α	2.4×10^4 y
Uranium-238 ($^{238}_{92}$U)	α	4.5×10^9 y
Rubidium-87 ($^{87}_{37}$Rb)	β^-	4.7×10^{10} y

will be considered in Chapter 30. If these nuclides did exist when the Earth was formed, they would have long since decayed. In fact, this is the case for the elements technetium (Tc) and promethium (Pm), but they can be produced in laboratories. Conversely, the half-life of the naturally occurring ^{238}U isotope is 4.5×10^9 y, which is close to the estimated age of the Earth. Thus, about half of the original ^{238}U present when the Earth was formed exists today.

The longer the half-life of a nuclide, the more slowly it decays and the smaller the decay constant, λ. The half-life and the decay constant have an inverse relationship, or $t_{1/2} \propto 1/\lambda$. This can be seen from Eq. 29.3. The half-life is the time it takes for the number of nuclei present, or the activity, to decrease by half: $N = N_o/2$. Thus,

$$\frac{N}{N_o} = e^{-\lambda t_{1/2}} = \tfrac{1}{2}$$

But since

$$e^{-0.693} \approx \tfrac{1}{2}$$

by comparison of the exponents, to three significant figures,

$$t_{1/2} = \frac{0.693}{\lambda} \qquad (29.4) \qquad \text{Relationship between decay constant and half-life}$$

The concept of half-life is very important in medical applications, as is shown in Example 29.3.

Example 29.3 ■ Before You Send the Patient Home: Half-Life and Activity

The half-life of iodine-131, used in thyroid treatments, is 8.0 d. At a certain time, an amount of ^{131}I containing 4.0×10^{14} nuclei is known to be in a patient's thyroid gland. (a) What will be the ^{131}I activity in the thyroid? (b) How many ^{131}I nuclei remain after 1.0 d?

Thinking It Through. (a) Equation 29.4 enables us to determine the decay constant λ from the half-life. The initial activity is then given by λN_o (Eq. 29.2 at $t = 0$). (b) To get N, we must use Eq. 29.3 and the e^x button on a calculator.

Solution. We are given the elapsed time and the half life. Listing the data and converting the half-life into seconds, we have

Given: $t_{1/2} = 8.0$ d $= 6.9 \times 10^5$ s   *Find:* (a) $\Delta N/\Delta t$ (activity)
$N_o = 4.0 \times 10^{14}$ nuclei   (b) N (number of undecayed nuclei)
$t = 1.0$ d

(a) The decay constant with $t_{1/2}$ in seconds is

$$\lambda = \frac{0.693}{t_{1/2}} = \frac{0.693}{6.9 \times 10^5 \text{ s}} = 1.0 \times 10^{-6} \text{ s}^{-1}$$

Then, based on the initial number of undecayed nuclei, N_o, the initial activity is

$$\Delta N/\Delta t = \lambda N_o = (1.0 \times 10^{-6} \text{ s}^{-1})(4.0 \times 10^{14}) = 4.0 \times 10^8 \text{ decay/s}$$

(b) With $t = 1.0$ d and $\lambda = 0.693/t_{1/2} = 0.693/8.0$ d $= 0.087$ d^{-1}, we have

$$N = N_o e^{-\lambda t} = (4.0 \times 10^{14} \text{ nuclei}) e^{-(0.087 \text{ d}^{-1})(1.0 \text{ d})}$$

$$= (4.0 \times 10^{14} \text{ nuclei}) e^{-0.087} = (4.0 \times 10^{14} \text{ nuclei})(0.917) = 3.7 \times 10^{14} \text{ nuclei}$$

The e^x function calculator button is sometimes the inverse of the ln x function. Become familiar with it. Here we calculated $e^{-0.087} \approx 0.917$ to three significant figures.

Follow-up Exercise. Suppose that in this Example the attending physician would not allow the patient to be released from the hospital until the activity was one-sixty-fourth of its original level. (a) How long would the patient have to remain in observation? (b) In practice, the time is much shorter. Can you think of a possible biological reason?

The "strength" of a radioactive sample or source at a given time is specified by its activity. A common unit of radioactivity is named in honor of Pierre and Marie Curie. One **curie (Ci)** is defined as

$$1 \text{ Ci} \equiv 3.70 \times 10^{10} \text{ decays/s}$$

This definition is based on the activity of 1.00 g of pure radium. The curie is the traditional unit for expressing radioactivity. However, the proper SI unit is the **becquerel (Bq),** which is defined as

$$1 \text{ Bq} \equiv 1 \text{ decay/s}$$

Therefore,

$$1 \text{ Ci} = 3.70 \times 10^{10} \text{ Bq}$$

Even with the present-day emphasis on SI units, radioactive sources are commonly rated in curies. The curie is a relatively large unit, however, so the millicurie (mCi), the microcurie (μCi), and sometimes even smaller multiples are used. Teaching laboratories, for example, typically use samples with activities on the order of microcuries. A source strength is calculated in the following Example.

Example 29.4 ■ Declining Source Strength: Get a Half-Life!

A ^{90}Sr beta source has an initial activity of 10.0 mCi. How many decays per second will be taking place after 84.0 y (approximately one human lifetime)?

Thinking It Through. We look up the half-life in Table 29.1 and use the fact that in each successive half-life, the activity decreases by half from what it was at the start of that interval.

Solution.

Given: Initial activity = 10.0 mCi   *Find:* $\Delta N / \Delta t$ (activity, in
　　　　　　$t = 84.0$ y   　　　　　　decays per second)
　　　　　　$t_{1/2} = 28.0$ y (from Table 29.1)

After three half-lives, the activity will be only $\frac{1}{8}$ as great ($\frac{1}{2} \times \frac{1}{2} \times \frac{1}{2} = \frac{1}{8}$), and the strength of the source will then be

$$\frac{\Delta N}{\Delta t} = 10.0 \text{ mCi} \times \frac{1}{8} = 1.25 \text{ mCi} = 1.25 \times 10^{-3} \text{ Ci}$$

In terms of decays per second or becquerels, we have

$$\frac{\Delta N}{\Delta t} = (1.25 \times 10^{-3} \text{ Ci})\left(3.70 \times 10^{10} \frac{\text{decays/s}}{\text{Ci}}\right)$$

$$= 4.63 \times 10^7 \text{ decays/s} = 4.63 \times 10^7 \text{ Bq}$$

Follow-up Exercise. For the material in this Example, the radiation safety officer tells you this sample can go into a low-level waste disposal canister only when its activity drops to one-millionth of its current activity (in other words, when it drops from 10.0 mCi to 10.0 pCi, a factor of a million). Using your calculator for trial and error, estimate, to two significant figures, how long you will have to keep it under its present cover. [*Hint:* On your calculator, 2 raised to what power produces about a million?]

Radioactive Dating

Because their decay rates are constant, radioactive nuclides can be used as nuclear clocks. As we have seen, the half-life of a radioactive nuclide can be used to project how much of a given amount of material will exist in the future. Similarly, by using the half-life to project backward in time, scientists can determine the age of objects containing radioactive nuclides. As you might surmise, some idea of the initial composition or initial amount of a nuclide must be known.

To illustrate the principle of radioactive dating, let's look at how it is done with ^{14}C. **Carbon-14 dating** is used on materials that were once part of living things, or the remnants of objects made from or containing such materials (such as wood, bone, leather, or parchment). The process depends on the fact that living things (including yourself) contain a known amount of radioactive ^{14}C. The concentration is very small, about one ^{14}C atom for every 7.2×10^{11} atoms of ordinary ^{12}C. Even so, this concentration cannot be due to ^{14}C present when the Earth was formed, since a ^{14}C half-life of $t_{1/2} = 5730$ y is brief in comparison to the estimated age of the Earth (over 4.0×10^9 y).

The observed concentration of ^{14}C is accounted for by its continuous production in the upper atmosphere. Cosmic rays from outer space cause reactions that produce neutrons (•Fig. 29.12). The neutrons are absorbed by the nuclei of the nitrogen atoms of the air, which in turn decay by emitting a proton to produce ^{14}C by the reaction

$$^{14}_{7}N + ^{1}_{0}n \rightarrow ^{14}_{6}C + ^{1}_{1}H$$

Note: The cosmic ray production of ^{14}C is an example of a nuclear reaction that induces a nuclear transmutation. Such reactions will be studied in more detail in Chapter 30.

Recall that the ^{14}C then decays by beta decay ($^{14}_{6}C \rightarrow ^{14}_{7}N + ^{0}_{-1}e$). Although the intensity of incident cosmic rays may not be constant over time, the concentration of ^{14}C in the atmosphere is relatively constant because of atmospheric mixing and the fixed decay rate.

The ^{14}C is oxidized into carbon dioxide (CO_2), so a small fraction of the CO_2 molecules in the air are radioactive. Plants take in this radioactive CO_2 by photosynthesis, and animals ingest the plant material. As a result, the concentration of ^{14}C in living organic matter is the same as the concentration in the atmosphere, one part in 7.2×10^{11}. However, once an organism dies, the ^{14}C is *not* replenished, and the concentration decreases with radioactive decay (with $t_{1/2} = 5730$ y). *The concentration of ^{14}C in dead matter relative to that in living things can then be used to establish when the organism died.* Since radioactivity is generally measured in terms of activity, we must first know the ^{14}C activity in a living organism. This is calculated in the following Example.

Example 29.5 ■ We the Living: Natural Carbon-14 Activity

Determine the average ^{14}C activity, in decays per minute per gram of natural carbon, found in living organisms if the concentration of ^{14}C is the same as that in the atmosphere.

Thinking It Through. From the text we know the concentration of ^{14}C relative to that of ^{12}C. To calculate the ^{14}C activity, we need the decay constant (λ), which we can get from the half-life ($t_{1/2} = 5730$ y, from Table 29.1), and the number of ^{14}C atoms (N) per gram. Carbon has an atomic mass of 12.0, so N can be found from Avogadro's number N_A (6.02×10^{23}) and the number of moles, $n = N/N_A$ (see Section 10.3).

Solution.

Given: $\quad \dfrac{^{14}C}{^{12}C} = \dfrac{1}{7.2 \times 10^{11}} = 1.4 \times 10^{-12}$ $\qquad$ *Find:* $\quad$ Average $\dfrac{\Delta N}{\Delta t}$ per gram

For a half-life of $t_{1/2} = (5730 \text{ y})(5.26 \times 10^5 \text{ min/y}) = 3.01 \times 10^9$ min, the decay constant is

$$\lambda = \frac{0.693}{t_{1/2}} = \frac{0.693}{3.01 \times 10^9 \text{ min}} = 2.30 \times 10^{-10} \text{ min}^{-1}$$

For 1.0 g of carbon, the number of moles is $n = 1.0 \text{ g}/(12 \text{ g/mol}) = \frac{1}{12}$ mol, so

$$N = nN_A = (\tfrac{1}{12} \text{ mol})(6.02 \times 10^{23} \text{ C nuclei/mol}) = 5.0 \times 10^{22} \text{ C nuclei (per gram)}$$

The number of ^{14}C nuclei per gram is given by the concentration factor

$$N\left(\frac{^{14}C}{^{12}C}\right) = (5.0 \times 10^{22} \text{ nuclei/g})(1.4 \times 10^{-12}) = 7.0 \times 10^{10} \text{ }^{14}C \text{ nuclei/g}$$

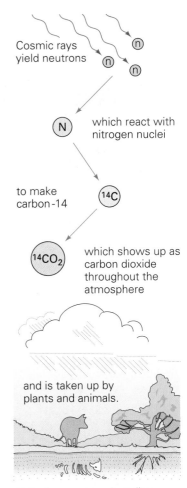

Cosmic rays yield neutrons

which react with nitrogen nuclei

to make carbon-14

which shows up as carbon dioxide throughout the atmosphere

and is taken up by plants and animals.

But when an organism dies, no fresh carbon-14 replaces the carbon-14 decaying in its tissues, and the carbon-14 radioactivity decreases by half every 5730 y.

•**FIGURE 29.12 Carbon-14 radioactive dating** The formation of carbon-14 in the atmosphere and its entry into the biosphere.

The activity, or number of decays per gram of carbon per minute (to two significant figures), is

$$\frac{\Delta N}{\Delta t} = \lambda N = (2.30 \times 10^{-10}\,\text{min}^{-1})(7.0 \times 10^{10}\,{}^{14}\text{C/g}) = 16 \quad \frac{{}^{14}\text{C decays}}{\text{g}\cdot\text{min}}$$

Thus, if we determined that an artifact such as a bone or a piece of cloth had an activity of 8.0 counts per gram of carbon per minute, the original living organism would have died about one half-life, or about 5700 y, ago. This would put the date of the artifact near 3700 B.C.

Follow-up Exercise. Suppose your instruments could measure ${}^{14}\text{C}$ beta emissions only down to 1.0 decay/g·min. How far back (to two significant figures) could you estimate the ages of dead organisms?

Now consider how the activity calculated in Example 29.5 can be used to date ancient organic finds.

Example 29.6 ■ Old Bones: Carbon-14 Dating

•FIGURE 29.13 **Dating old bones** See Example 29.6.

A bone is unearthed in an archeological dig (•Fig. 29.13). Laboratory analysis determines that there are 20 beta emissions per minute from 10 g of carbon in the bone. What is the approximate age of the bone?

Thinking It Through. We know the current activity, and the initial activity of the living sample in Example 29.5; we can work backward to determine the time elapsed.

Solution.

Given: Activity = 20 decays/min in 10 g of carbon.   *Find:* Age of bone
            = 2.0 decays/g·min

Assuming that the bone had the normal concentration of ${}^{14}\text{C}$ when the organism died, at the time of death the ${}^{14}\text{C}$ activity would be 16 decays/g·min (Example 29.5). Afterward, the decay rate would decrease by half for each half-life:

$$16 \xrightarrow{t_{1/2}} 8 \xrightarrow{t_{1/2}} 4 \xrightarrow{t_{1/2}} 2 \text{ decays/g·min}$$

So with the observed activity, the ${}^{14}\text{C}$ in the bone would have gone through approximately three half-lives, or the bone would be three half-lives old. Thus with $t_{1/2} = 5730$ y (Table 29.1), to two significant figures, we have

$$\text{Age} = 3.0 t_{1/2} = 3.0(5730\text{ y}) = 1.7 \times 10^4\text{ y}$$

Follow-up Exercise. Let us try to date radioactively the time of production (to two significant figures) of the potassium currently present on Earth. Isotopic data indicate that the stable nuclide ${}^{39}\text{K}$ and the long-lived nuclide ${}^{40}\text{K}$ are the only isotopes of potassium that currently exist. The ${}^{40}\text{K}$ nuclide decays with a half-life of 1.28×10^9 y (from Appendix V). The present abundance ratio of the two potassium isotopes is ${}^{39}\text{K}/{}^{40}\text{K} \approx 14$. Taking this ratio to be 16 for simplicity, at what time in the past would the abundances of the two isotopes have been equal? (According to present theories dealing with the formation of the elements and of our solar system, this date should provide an estimate of the age of the Earth. Compare your answer to the age of the Earth, which is estimated at 4.6×10^9 y.)

The limit of radioactive carbon dating depends on the ability to measure the very low activity in an old organic sample. Current techniques give an age dating limit of about 40 000–50 000 years, depending on the size of the sample. After about nine half-lives, the radioactivity of a carbon sample has decreased so much that it is barely measurable (less than 2 decays per gram per *hour*). Another radioactive dating process uses lead-206 (${}^{206}\text{Pb}$) and uranium-238 (${}^{238}\text{U}$). This dating method is used extensively in geology because of the long half-life of ^{238}U. Lead-206 is the stable end isotope of the ^{238}U decay series (see Fig. 29.9). If a rock sample contains both of these isotopes, the lead is assumed to be a decay product of the uranium that was there when the rock first formed. Thus, the ratio of ^{206}Pb/^{238}U is a measure of the geologic age of the rock.

29.4 Nuclear Stability and Binding Energy

Objectives: To (a) state which proton and neutron number combinations result in stable nuclei, (b) explain the pairing effect and magic numbers in relation to nuclear stability, and (c) calculate nuclear binding energy.

Now that we have considered some of the properties of unstable isotopes, we turn to the stable isotopes. Stable isotopes exist naturally for all elements having proton numbers from 1 to 83, except for those with $Z = 43$ (technetium) and $Z = 61$ (promethium). The nuclear interactions that give rise to nuclear stability are extremely complicated. However, by looking at some of the general properties of stable nuclei, it is possible to obtain general criteria for nuclear stability. Thus we will be able to determine whether a particular nuclide should be stable or unstable.

Nucleon Populations

One of the first considerations is the relative number of protons and neutrons in stable nuclei. Nuclear stability must be related to the dominance of either the repulsive Coulomb force between protons or the attractive nuclear force between nucleons. This force dominance depends on the ratio of protons and neutrons.

For stable nuclei of low mass numbers (about $A < 40$), the ratio of neutrons to protons (N/Z) is approximately 1. That is, the number of protons and the number of neutrons are equal or nearly equal. For example, $^4_2$He, $^{12}_6$C, $^{23}_{11}$Na, and $^{27}_{13}$Al have the same or about the same number of protons as neutrons. For stable nuclei of higher mass numbers ($A > 40$), the number of neutrons exceeds the number of protons ($N/Z > 1$). The heavier the nuclei, the more the neutrons outnumber the protons.

This trend is illustrated in •Fig. 29.14, a plot of the neutron number (N) versus proton number (Z) for stable nuclei. The heavier stable nuclei lie above the

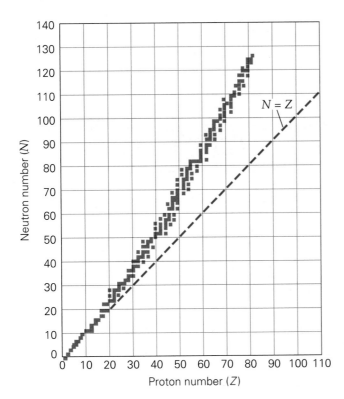

•**FIGURE 29.14 A plot of *N* versus *Z* for stable nuclei** For nuclei with mass numbers $A < 40$ ($Z < 20$ and $N < 20$), the number of protons and the number of neutrons are equal or nearly equal. For nuclei with $A > 40$, the number of neutrons exceeds the number of protons and the nuclei lie above the $N = Z$ line.

$N = Z$ line, so $N > Z$ for them. Examples of heavy stable nuclei include $^{62}_{28}\text{Ni}$, $^{114}_{50}\text{Sn}$, $^{208}_{82}\text{Pb}$, and $^{209}_{83}\text{Bi}$. (Bismuth is the heaviest element that has a stable isotope.*) The extra neutrons in such heavy stable nuclei act as "spacers" between the protons, reducing the increased electrical repulsion and increasing stability.

Radioactive decay "adjusts" the proton and neutron numbers of an unstable nuclide until a stable nuclide on the stability curve in Fig. 29.14 is reached. Since alpha decay decreases the numbers of protons and neutrons by equal amounts, alpha decay alone would give nuclei with neutron populations that are *larger* than that of the stable nuclides on the curve. However, β^- decay *following* alpha decay can lead to a stable combination, since the effect of β^- decay is the loss of a neutron and the gain of a proton. Thus very heavy nuclei undergo a chain, or sequence, of alpha and beta decays until a stable nucleus is reached, as was illustrated in Fig. 29.9 for ^{238}U.

Pairing Effect

Many stable nuclei have even numbers of both protons and neutrons and very few have odd numbers of *both* protons and neutrons. A survey of the stable isotopes (Table 29.2) shows that 168 stable nuclei have this even–even combination, 107 are even–odd or odd–even, and only four contain odd numbers of both protons and neutrons. These four are isotopes of the elements with the four lowest odd proton numbers: ^2_1H, ^6_3Li, $^{10}_5\text{B}$, and $^{14}_7\text{N}$.

The dominance of even–even combinations indicates that the protons and neutrons in nuclei tend to "pair up." That is, two protons pair up and, separately, two neutrons pair up. Aside from the four exceptions, all odd–odd nuclei are unstable. If there is only one odd nucleon (odd–even or even–odd combination), the stability is less than for even–even nuclei.

This **pairing effect** provides a qualitative criterion for stability. For example, you might expect the aluminum isotope $^{27}_{13}\text{Al}$ to be stable (even–odd), but not $^{26}_{13}\text{Al}$ (odd–odd). This is actually the case.

The *general criteria for nuclear stability* can be summarized as follows:

1. All isotopes with a proton number greater than 83 ($Z > 83$) are unstable.
2. (a) Most even–even nuclei are stable.
   (b) Many odd–even or even–odd nuclei are stable.
   (c) Only four odd–odd nuclei are stable (^2_1H, ^6_3Li, $^{10}_5\text{B}$, and $^{14}_7\text{N}$).
3. (a) Stable nuclei with mass numbers less than 40 ($A < 40$) have approximately the same number of protons and neutrons.
   (b) Stable nuclei with mass numbers greater than 40 ($A > 40$) have more neutrons than protons.

TABLE 29.2 Pairing Effect of Stable Nuclei

Proton Number	Neutron Number	Number of Stable Nuclei
Even	Even	168
Even	Odd ⎫	107
Odd	Even ⎭	
Odd	Odd	4

Criteria for nuclear stability

Conceptual Example 29.7

Running Down the Checklist: Nuclear Stability

Is the sulfur isotope $^{38}_{16}\text{S}$ likely to be stable?

Reasoning and Answer. We apply the general criteria for nuclear stability.

1. *Satisfied.* Isotopes with $Z > 83$ are unstable. With $Z = 16$, this criterion is satisfied.
2. *Satisfied.* The isotope $^{38}_{16}\text{S}_{22}$ has an even–even nucleus, so it could be stable.
3. *Not satisfied.* Here $A < 40$, but $Z = 16$ and $N = 22$ are not approximately equal.

Therefore, the ^{38}S isotope is likely to be unstable. (The nucleus is unstable and decays by β^- emission, since it is neutron-rich.)

Follow-up Exercise. (a) List likely isotopes of copper ($Z = 29$). (b) Apply the criteria to see which of those isotopes should be stable. Use Appendix V to check your conclusions.

---

*Bismuth-209 alpha decays, but with a half-life of 2×10^{18} y; for practical purposes, it is stable.

Binding Energy

An important quantitative aspect of nuclear stability is the binding energy of the nucleons. Binding energy can be calculated by considering the mass–energy equivalence of the nuclear masses.

Note: We introduced binding energy in Section 27.4.

Since nuclear masses are so small in relation to the SI standard kilogram, another standard, the **atomic mass unit (u)**, is used to measure them. A *neutral atom* of ^{12}C is defined as having a mass of exactly 12 u. Thus, since a neutral ^{12}C atom has a mass of 1.9927×10^{-26} kg,

$$1 \, u = \frac{1.9927 \times 10^{-26} \, kg}{12} = 1.6606 \times 10^{-27} \, kg$$

We can then express the masses of the various particles in atomic mass units (Table 29.3). The listed energy equivalents reflect Einstein's $E = mc^2$ mass–energy equivalence relationship from Eq. 26.14. Thus, a mass of 1 u has an energy equivalent of

$$mc^2 = (1.6606 \times 10^{-27} \, kg)(2.9977 \times 10^8 \, m/s)^2 = 1.4922 \times 10^{-10} \, J$$

$$= \frac{1.4922 \times 10^{-10} \, J}{1.602 \times 10^{-13} \, J/MeV} = 931.5 \, MeV$$

as given in the first entry of the table. We will use 931.5 MeV/u as a handy conversion factor to avoid having to multiply by c^2.

Note that in Table 29.3, the proton and the hydrogen atom ($^1_1$H) are listed separately, having slightly different masses. This difference is due to the mass of the atomic electron. We deal almost exclusively with the masses of neutral atoms (nucleons plus Z electrons) rather than strictly with nuclei, since the atomic masses are what are measured. Keep this in mind. We will soon be interested in very small mass differences, and the electron mass can be significant. We must balance out the masses of the electrons when we compare masses in nuclear processes.

We can now look at nuclear stability in terms of energy. For example, if you compare the mass of a helium-4 nucleus to the total mass of nucleons that make it up, a significant inequity emerges: A neutral helium atom (including its *two* electrons) has a mass of 4.002 603 u. (Atomic masses of various isotopes are given in Appendix V.) The total mass of two protons (with *two* electrons, or actually two ^{1}H atoms) and two neutrons is, by addition,

$$2m(^1H) = 2.015\,650 \, u$$

$$2m_n = \underline{2.017\,330 \, u}$$

$$\text{total} \quad 4.032\,980 \, u$$

This total is greater than the mass of the helium atom (4.002 603 u). The helium nucleus is less massive than the sum of its parts by an amount

TABLE 29.3 Particle Masses and Energy Equivalents

Particle	Mass u	Mass kg	Equivalent Energy (MeV)
	1 (exact)	1.6606×10^{-27}	931.5
Electron	0.000 548	9.1095×10^{-31}	0.511
Proton	1.007 27	$1.672\,65 \times 10^{-27}$	938.28
$^1_1$H atom	1.007 825	$1.673\,56 \times 10^{-27}$	938.79
Neutron	1.008 665	$1.675\,00 \times 10^{-27}$	939.57

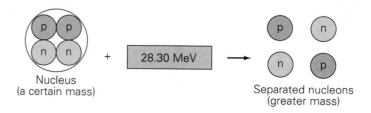

•**FIGURE 29.15 Binding energy** Here 28.30 MeV of energy is required to separate a helium nucleus into free protons and neutrons. Conversely, if two protons and two neutrons combine to form a helium nucleus, 28.30 MeV of energy would be released. This is the binding energy of the nucleus.

$$\Delta m = [2m(^1\text{H}) + 2m_\text{n}] - m(^4\text{He})$$

$$= 4.032\,980 \text{ u} - 4.002\,603 \text{ u} = 0.030\,377 \text{ u}$$

Mass defect Δm: The difference between the mass of a nucleus and the sum of its nucleon masses

where the two electronic masses subtract out, since the mass of two hydrogen atoms was used in place of two protons. This mass difference, called the **mass defect,** has an energy equivalent of

$$(0.030\,377 \text{ u})(931.5 \text{ MeV/u}) = 28.30 \text{ MeV}$$

Binding energy: total energy needed to separate the nucleus into its individual nucleons

This energy is the **total binding energy** (E_b) of the ^{4}He nucleus—that is, the amount of energy released when the separate nucleons combined to form a nucleus.

In general, for any nucleus, the total binding energy is given by

$$E_\text{b} = (\Delta m)c^2 \qquad \begin{array}{l}\textit{total binding}\\\textit{energy}\end{array} \qquad (29.5)$$

Binding energy (alternate): the energy released when the nucleus is formed from its individual nucleons

where Δm is the mass defect. An alternate interpretation of the binding energy is that it represents the energy required to separate the constituent nucleons completely into free particles. This is illustrated in •Fig. 29.15 for the helium nucleus. Exactly 28.30 MeV of energy is necessary to separate the four nucleons.

We can gain some insight into the nature of the nuclear force by considering the *average binding energy per nucleon* for stable nuclei. This quantity is the total binding energy of a nucleus divided by the total number of nucleons, or E_b/A, where A is the mass number (number of protons plus number of neutrons). For example, for the helium nucleus (^{4}He) in Fig. 29.15, the average binding energy per nucleon is

$$\frac{E_\text{b}}{A} = \frac{28.30 \text{ MeV}}{4} = 7.075 \text{ MeV/nucleon}$$

Compared to the binding energy of atomic electrons (13.6 eV for a hydrogen electron in the ground state), nuclear binding energies are millions of times larger, indicative of a very strong binding force.

Example 29.8 ■ The Stablest of the Stable: Binding Energy per Nucleon

Compute the average binding energy per nucleon of the iron-56 nucleus ($^{56}_{26}$Fe).

Thinking It Through. We can find the atomic mass of the iron isotope in Appendix V and the other needed masses in Table 29.3. We can then compute the mass defect, Δm; the total binding energy, E_b; and lastly the average binding energy per nucleon, E_b/A.

Solution.

Given: $^{56}_{26}$F mass = 55.934 939 u   *Find:* E_b/A (average binding energy
        $^1_1$H mass = 1.007 825 u   per nucleon)
        $^1_0$n mass = 1.008 665 u

(Notice that we use the masses of the iron *atom* and the hydrogen *atom* rather than the nuclear masses; why?)

The mass defect is the difference between the mass of the iron atom and the mass of its individual nucleons. The nucleons have a total mass of

$$26m(^1\text{H}) = 26(1.007\,825\text{ u}) = 26.203\,450\text{ u}$$

$$30m_n = 30(1.008\,665\text{ u}) = 30.259\,950\text{ u}$$

$$\text{total} \quad 56.463\,400\text{ u}$$

and

$$\Delta m = [26m(^1\text{H}) + 30m_n] - m(^{56}\text{Fe}) = 56.463\,400\text{ u} - 55.934\,939\text{ u} = 0.528\,461\text{ u}$$

The total binding energy is then gotten from the energy equivalence of 1 u:

$$E_b = (\Delta m)c^2 = (0.528\,461\text{ u})(931.5\text{ MeV/u}) = 492.3\text{ MeV}$$

Our iron nuclide has 56 nucleons, so the average binding energy per nucleon is

$$\frac{E_b}{A} = \frac{492.3\text{ MeV}}{56} = 8.791\text{ MeV/nucleon}$$

Follow-up Exercise. (a) To illustrate the pairing effect, compare the average binding energy per nucleon for ^{4}He (calculated previously to be 7.075 MeV) to that of ^{3}He. (Find atomic masses in Appendix V.) (b) How does your answer reflect the pairing idea?

If E_b/A is calculated for various nuclei and plotted versus mass number, the values generally lie along a curve as shown in •Fig. 29.16. The value of E_b/A rises rapidly with increasing A for light nuclei and starts to level off (around $A = 15$) at about 8.0 MeV/per nucleon, with a maximum value of about 8.8 MeV/per nucleon. The maximum of the curve occurs in the vicinity of iron, which has the most stable nucleus (see Example 29.8). For $A > 60$, the E_b/A values decrease slowly, indicating that the nucleons are on average less tightly bound.

The maximum in the curve gives an important indication that will be considered in Chapter 30. If a large nucleus could be split, or fissioned, into two lighter nuclei, the nucleons would be more tightly bound and energy would be released in the process. Similarly, on the low-mass side of the maximum, if we could fuse together two light nuclei into a heavier nucleus, this would create a more tightly bound nucleus, and energy would also be released.

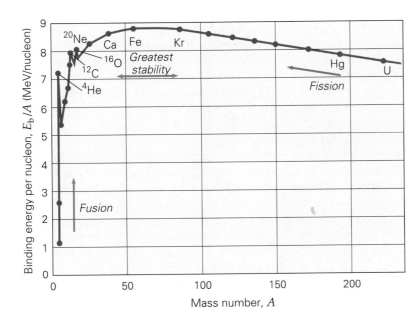

•**FIGURE 29.16 A plot of binding energy per nucleon versus mass number** If the binding energy per nucleon (E_b/A) is plotted versus mass number (A), the curve has a maximum, which indicates that the nuclei in this region are on average the most tightly bound together and have the greatest stability.

In general, the E_b/A curve shows that, except for very light nuclei, the binding energy per nucleon does not change a great deal and has a value of

$$E_b/A \approx 8 \text{ MeV/nucleon}$$

Since E_b/A is relatively constant for most nuclei, an approximation is that

$$E_b \propto A \qquad \textit{(experimental approximation)}$$

In other words, the *total* binding energy is (approximately) proportional to the mass number or total number of nucleons.

This proportionality indicates a characteristic of the nuclear force that is quite different from the electrical force. Suppose that the attractive nuclear force acts between *all* the pairs of nucleons in a nucleus. Each pair of nucleons would then contribute to the total binding energy. In a nucleus containing A nucleons, there are $A(A-1)/2$ pairs, so there would be $A(A-1)/2$ contributions to the total binding energy. For nuclei for which $A \gg 1$ (heavier nuclei), $A(A-1) \approx A^2$, and we would expect that

$$E_b \propto A^2 \qquad \textit{(if nucleon–nucleon force is long range)}$$

But in actuality $E_b \propto A$, which indicates that a given nucleon is *not* bound to all the other nucleons. This phenomenon, called *saturation*, implies that the nuclear forces act over a very short range and that a particular nucleon interacts only with its nearest neighbors.

Magic Numbers

We are familiar with the concept of filled shells in atoms. At the end of every row of the periodic chart, there exists an inert gas (such as neon, Ne). An inert element consists of single atoms that tend not to form bonds with other atoms. Such atoms are relatively unreactive because their outer shells of electrons are full. Gaining or losing an electron would be very costly in terms of energy and is therefore unlikely to occur.

There is an analogous effect in the nucleus. Although the concepts of individual nucleon "orbits" and "filled shells" inside the nucleus are hard to visualize, experimental evidence does indicate the existence of "closed nuclear shells" when the number of protons *or* neutrons is 2, 8, 20, 28, 50, 82, or 126. Important work on the nuclear shell model was done by Maria Goeppert–Mayer (1906–1972), a German-born physicist (•Fig. 29.17).

Solid evidence of the existence of such **magic numbers** is provided by the number of stable isotopes of various elements. If an element has a magic number of protons, it also has an unusually high number of stable isotopes. Far from a magic number, there may be only one or two (or even no) stable isotopes per element. Aluminum, for example, with 13 protons, has just one stable isotope, ^{27}Al. But tin, with $Z = 50$ (a magic number), has *ten* stable isotopes ranging from $N = 62$ to $N = 74$. Neighboring indium, with $Z = 49$, has only two stable isotopes, and antimony, with $Z = 51$, also has only two.

The second piece of evidence of magic numbers is related to binding energies. Experiments can be done, for instance, with high-energy gamma ray photons, which knock out single nucleons from a nucleus (called the *photonuclear effect*) in a manner analogous to the photoelectric effect in metals. It is found experimentally that a nuclide such as tin (with a proton magic number) requires about 2 MeV *more* photon energy to eject the least bound proton from its nucleus than a nuclide that is not at a magic number. *Thus, magic numbers are associated with extra large binding energies, another sure sign of higher stability.*

Consider the following calculations for the removal of single neutrons from two different isotopes—one at a magic number, and the other not.

•**FIGURE 29.17** **Maria Goeppert-Mayer** A pioneer in the field of nuclear physics.

■ Exploring Magic Numbers: The Photonuclear Effect

Compare the minimum photon energy needed to remove the least bound neutron from $^{40}_{20}Ca$ to that required to remove the least bound neutron from $^{40}_{18}Ar$. The necessary atomic masses (in u) are supplied below. [Here we are comparing neutron removal from a nucleus with a magic number of neutrons (20 for calcium) to neutron removal from a nucleus with a nonmagic number of neutrons (22 for argon). In both cases, all the neutrons are paired—ensuring a magic number effect comparison.]

Thinking It Through. It requires energy input to break loose a neutron in either case. Thus we expect the total mass (neutron plus product nucleus) after the removal to be more than the initial nucleus mass. This mass difference, expressed in energy units (MeV), is the minimum energy input needed. Essentially, we are asked to compute the binding energy of the least bound neutron.

Solution. Using the masses from Appendix V, we have

Given: ^{40}Ca mass = 39.962 591 u   *Find:* Minimum energy needed to
^{39}Ca mass = 38.970 719 u   remove a neutron from
^{40}Ar mass = 39.962 383 u ^{40}Ca and ^{40}Ar
^{39}Ar mass = 38.964 314 u
$^{1}_{0}n$ mass = 1.008 665 u

For ^{40}Ca, the mass after breakup (at minimum energy, the products have no kinetic energy) can be found as follows:

$$^{39}_{20}Ca = 38.970 719 \text{ u}$$

$$^{1}_{0}n = \underline{1.008 665 \text{ u}}$$

$$\text{total} \quad 39.979 384 \text{ u} \quad \textit{(mass afterward)}$$

and the mass difference is

$$\Delta m = (^{39}Ca + {}^{1}_{0}n) - {}^{40}Ca = 39.979 384 \text{ u} - 39.962 591 \text{ u} = 0.016 793 \text{ u}$$

The binding energy for the neutron (expressed in MeV) is then

$$E = (0.016 793 \text{ u})(931.5 \text{ MeV/u}) = 15.64 \text{ MeV}$$

For ^{40}Ar, the mass after breakup is

$$^{39}_{20}Ar = 38.964 314 \text{ u}$$

$$^{1}_{0}n = \underline{1.008 665 \text{ u}}$$

$$\text{total} \quad 39.972 979 \text{ u}$$

and

$$\Delta m = (^{39}Ar + {}^{1}_{0}n) - {}^{40}Ar = 39.972 979 \text{ u} - 39.962 383 \text{ u} = 0.010 596 \text{ u}$$

The binding energy for the neutron is then

$$E = (0.010 596 \text{ u})(931.5 \text{ MeV/u}) = 9.870 \text{ MeV}$$

Thus in this instance, it takes about 6 MeV *more* energy to remove a paired neutron from a pair that completes a closed nuclear shell (a magic number) than from a pair that does not complete a shell.

Follow-up Exercise. (a) Calculate the minimum gamma ray energy required for the removal of a proton from ^{12}C and a neutron from ^{12}C. (The necessary masses are given in Appendix V.) (b) In ^{12}C all the protons are paired, as are the neutrons. There is also no magic number effect. Can you explain why removal of the proton takes less energy than removal of the neutron?

29.5 Radiation Detection and Applications

Objectives: To (a) gain insight into the operating principles of various nuclear radiation detectors, (b) investigate the medical and biological effects of radiation exposure, and (c) study some of the practical uses and applications of radiation.

Detecting Radiation

Since our senses cannot detect the products of radioactive decay directly, detection must be done by indirect means. For example, people who work with radioactive materials and X-rays usually wear film badges, which indicate cumulative exposure to radiation by the degree of darkening of the film when developed. However, more immediate and quantitative methods of detection are desirable, and a variety of instruments have been developed for this purpose.

These **radiation detectors** are based on the ionization or excitation of atoms by the passage of energetic particles through matter. Alpha and beta particles are electrically charged particles, and they transfer energy to atoms by electrical interactions, removing electrons and creating ions. Gamma rays can also produce ionization by the photoelectric effect on an atom or by Compton scattering (Section 27.3). Alternatively, they may *produce* electrons by pair production (Section 28.5), if the energy of the photon is large enough. The particles produced by all these interactions are what is "detected" by a radiation detector.

One of the most common radiation detectors is the *Geiger counter*, which was developed chiefly by Hans Geiger, a student and then colleague of Ernest Rutherford. The principle of the Geiger counter is illustrated in •Fig. 29.18. A voltage of 800–1000 V is applied across the wire electrode and outer electrode (a metallic tube) of the Geiger tube. The tube contains a gas (such as argon) at low pressure. When an ionizing particle enters the tube through a thin window at one end, the particle ionizes some atoms of the gas. The freed electrons are accelerated toward the positive wire anode. On their way, they strike and ionize other gas atoms. This process multiplies, and the resulting "avalanche" discharge produces a current pulse. The pulse is amplified and sent to an electronic counter that counts the pulses or the number of particles detected. The pulses can also be used to drive a loudspeaker so that particle detection is heard as an audible click.

Another method of detection, one of the oldest, is used in the *scintillation counter* (•Fig. 29.19). In this counter, atoms of a phosphor material (such as sodium iodide, NaI) are excited by an incident particle, and visible light is emitted when the atoms return to their ground state. The light pulse is converted to an electrical pulse by a photoelectric material. This pulse is amplified in a *photomultiplier tube*, which consists of a series of electrodes of successively higher potential. The photoelectrons are accelerated toward the first electrode and acquire sufficient energy to cause several secondary electrons from ionization to be emitted when they

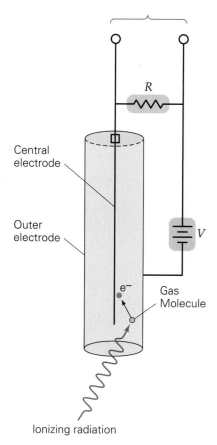

•FIGURE 29.18 The Geiger counter The radiation ionizes a gas atom, freeing an electron that is accelerated toward the central (positive) electrode. On the way, this electron produces additional electrons through ionization, resulting in a current pulse that is detected as a voltage across the external resistor *R*.

•FIGURE 29.19 The scintillation counter A photon emitted by a phosphor atom excited by an incoming particle causes the emission of a photoelectron from the photocathode. Accelerated through a potential difference in a photomultiplier tube, the photoelectrons free secondary electrons when they collide with successive electrodes at higher potentials. After several steps, a relatively weak scintillation is converted into a measurable electric pulse.

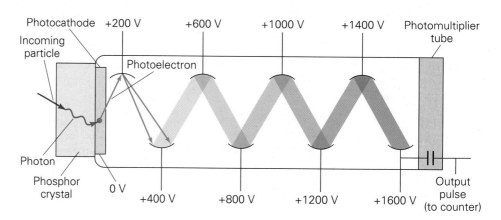

strike the electrode. This process continues, and relatively weak scintillations are converted into sizable electrical pulses, which are then counted electronically.

In a *solid state* or *semiconductor detector,* charged particles passing through a semiconductor material produce electron-hole pairs (solid state ionization, so to speak). With a voltage across the material, the collected electron-hole pairs give rise to electric signals, which can be amplified and counted.

Counting methods are used to determine the number of particles produced by decaying nuclei of a radioactive isotope. Other methods of detection allow the tracks of charged particles to be seen or recorded visually. These methods may include the cloud chamber, the bubble chamber, or the spark chamber. In the first two, vapors and liquids are supercooled and superheated, respectively, by suddenly varying the volume and pressure. The *cloud chamber* was developed early in the 1900s by C. T. R. Wilson, a British atmospheric physicist. In the chamber, supercooled vapor condenses into droplets on ionized molecules created along the path of an energetic particle. When the chamber is illuminated, the droplets scatter the light, making the path visible (•Fig. 29.20).

The *bubble chamber,* which was invented by the American physicist D. A. Glazer in 1952, uses a similar principle. A reduction in pressure causes a liquid to be superheated and able to boil. Ions produced along the path of an energetic particle become sites for bubble formation, and a trail of bubbles is created. When a magnetic field is applied across any of these chambers, charged particles are deflected, and the energy of a particle can be calculated from the radius of curvature of the particle's path. Since the bubble chamber uses a liquid, commonly liquid hydrogen, the density of atoms in it is much greater than in the vapor of a cloud chamber. Thus tracks are more readily observable, and bubble chambers have displaced cloud chambers.

Gamma rays do not leave visible tracks in a bubble chamber. However, their presence can be detected indirectly, for example, by electrons that have been Compton scattered and by the creation of electron–positron pairs.

The path of a charged particle is registered by a series of sparks in a *spark chamber*. Basically, a charged particle passes between a pair of electrodes that have a high potential difference and are immersed in an inert (noble) gas. The charged particle causes the ionization of gas molecules, giving rise to a visible spark or flash between the electrodes as the released electrons travel to the positive electrode. A spark chamber is merely an array of such electrode pairs in the form of parallel plates or wires. A series of sparks, which can be photographed, marks the particle's path.

•**FIGURE 29.20 Cloud chamber tracks** The circular track in this cloud chamber photograph was made by a positron in a strong magnetic field. (Can you explain the approximately circular path of the particle in terms of the orientation of the magnetic field relative to the positron's velocity?)

Biological Effects and Medical Applications of Radiation

In medicine, nuclear radiation has both advantages and disadvantages. It can be used beneficially in the diagnosis and treatment of some diseases but also is potentially harmful if it is not properly handled and administered. Nuclear radiation and X-rays can penetrate human tissue without pain or any other sensation. However, early investigators quickly learned that large doses or repeated small doses led to red skin, lesions, and other conditions.

The chief hazard of radiation is damage to living cells, primarily due to ionization. Ions, particularly complex ions or radicals produced by ionizing radiation, may be highly reactive (for example, a hydroxyl ion OH^- from water). Such reactive ions interfere with the normal chemical operations of the cell. If enough cells are damaged or killed by radiation, cell reproduction might not be fast enough, and the irradiated tissue could eventually die.

In other instances, genetic damage, or mutation, may occur in a chromosome in the cell nucleus. If the affected cells are sperm or egg cells (or their precursors), any children that they produce may suffer from various birth defects. If the damaged cells are ordinary body cells, they may become cancerous, losing their normal form and reproducing in a rapid and uncontrolled manner, producing a

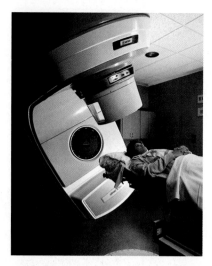

•**FIGURE 29.21 Nuclear medicine** A particle accelerator used in cancer treatment.

The gray unit was named in honor of Louis Harold Gray, a British radiobiologist whose studies laid the foundation for measuring absorbed dose.

TABLE 29.4 Typical Relative Biological Effectiveness (RBE) of Radiations

Type	RBE (or QF)
X-rays and gamma rays	1
Beta particles	1.2
Slow neutrons	4
Fast neutrons and protons	10
Alpha particles	20

malignant tumor. The human cells most susceptible to radiation damage are those of the reproductive organs, bone marrow, and lymph nodes.

Although radiation can cause cancer, it can also be used to treat cancer. Localized doses of radiation are used to destroy cancer cells. The radiation may be gamma rays from a radioactive ^{60}Co source or X-rays. Other particles produced in accelerators are also used to treat cancer (•Fig. 29.21). The electrical charge and energy of the particles determine the penetrating power of the radiation. Also, radioactive compounds that are preferentially absorbed by certain tissues or organs can be administered into the body.

Radiation Dosage. An important consideration in radiation therapy and radiation safety is the amount, or *dose*, of radiation. Several quantities are used to describe this amount in terms of *exposure, absorbed dose,* and *equivalent dose.* The earliest unit of dosage, the **roentgen (R)**, was based on exposure and defined in terms of the ionization produced in air. (One roentgen is the quantity of X-rays or gamma rays required to produce an ionization charge of 2.58×10^{-4} C/kg in air.)

The roentgen has been largely replaced by the rad (*r*adiation *a*bsorbed *d*ose), which is an absorbed dose unit. One **rad** is an absorbed dose of radiation energy of 10^{-2} J per kilogram of absorbing *material.* The rad is not an SI unit. The SI unit for absorbed dose is the **gray (Gy):**

$$1\,\text{Gy} \equiv 1\,\text{J/kg} \equiv 100\,\text{rad}$$

More meaningful biologically than the roentgen, these *physical* units give the energy absorbed per mass. But it is helpful to have some means of measuring the *biological damage* produced by radiation, because equal doses of different types of radiation produce different effects. For example, a relatively massive alpha particle with a charge of $+2e$ moves through tissue rather slowly with a great deal of electrical interaction. The ionizing collisions thus occur close together along a short penetration path, and more localized damage is done than by a fast-moving electron or gamma ray.

This effect, or effective dose, is measured in terms of the **rem** unit (*r*oentgen or *r*ad *e*quivalent *m*an). The different degrees of effectiveness of different particles are characterized by the **relative biological effectiveness (RBE)**, or *quality factor* (QF), which has been tabulated for the various particles in Table 29.4. The RBE is defined in terms of the number of rads of X-rays or gamma rays that produce the same biological damage as 1 rad of a given radiation. (Note in Table 29.4 that gamma rays have an RBE of 1.)

The effective dose is given by the product of the dose in rads and the appropriate RBE:

$$\text{effective dose (in rems)} = \text{dose (in rads)} \times \text{RBE} \qquad (29.6)$$

Thus, 1 rem of any type of radiation does approximately the same amount of biological damage. For example, a 20-rem effective dose of alpha particles does the same damage as a 20-rem dose of X-rays. But to administer this dose, 20 rad of X-rays is needed compared to only 1 rad of alpha particles.

The SI unit of absorbed dose is the gray, and the SI unit of effective dose is called the **sievert (Sv):**

$$\text{effective dose (in sieverts)} = \text{dose (in grays)} \times \text{RBE} \qquad (29.7)$$

Since 1 Gy $\equiv$ 100 rad, it follows that 1 Sv $\equiv$ 100 rem. A summary of the radiation units is given in Table 29.5.

It is difficult to set a maximum permissible radiation dosage, but the general standard for humans is an average dose of 5 rem/y after the age of 18 with no more than 3 rem in any 3-month period. In the United States, the normal average annual dose per capita is about 200 mrem (millirem). About 125 mrem comes from the natural background of cosmic rays and naturally occurring radioactive iso-

TABLE 29.5   Radiation Units

Unit	Basis
roentgen (R)	1 R—the quantity of X-rays or gamma rays that produces an ionization charge of 2.58×10^{-4} C/kg in air
rad (*radiation absorbed dose*)	1 rad—an absorbed dose of radiation of 10^{-2} J/kg
gray (Gy)	SI absorbed dose unit 1 Gy = 1 J/kg = 100 rad
rem (*rad equivalent man*) effective dose (in rem) = dose (in rad) $\times$ RBE	Effective dose. Relative effectiveness depends on type of radiation and is characterized by RBE (relative biological effectiveness). See Table 29.4.
sievert (Sv) effective dose (in Sv) = dose (in Gy) $\times$ RBE	SI unit of effective dose 1 Sv = 100 rem

topes in the soil, building materials, and so on. The remainder is chiefly from diagnostic medical applications, mostly X-rays (•Fig. 29.22), and from miscellaneous sources such as television tubes.

Diagnostic Applications.   Radioactive isotopes offer an important technique for diagnostic procedures. For example, a radioactive isotope, or *radioisotope,* behaves chemically like a stable isotope of the element and can participate in its normal chemical reactions. If we add radioisotopes to molecules, the molecules can then be used as tracers.

Many body functions can be studied by monitoring the location and activity of the tracer molecules as they are absorbed during body processes. For example, the activity of the thyroid gland in hormone production can be determined by monitoring its iodine uptake with radioactive iodine-123. This isotope emits gamma rays and has a half-life of 13.3 h. Iodine is required by the body, and when ingested it collects in the thyroid gland, near the throat. The uptake of radioactive iodine can be monitored by a gamma detector to see if there is an abnormality.

Similarly, radioactive solutions of iodine and gold are quickly absorbed by the liver. This can be done safely because the isotopes have short half-lives. One of the most commonly used diagnostic tracers is technetium-99 (^{99}Tc). It has a

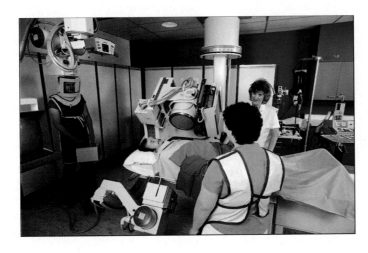

•**FIGURE 29.22   X-ray precautions**   Although the radiation dosage represented by an individual diagnostic X-ray is generally relatively small, there is no threshold level for radiation exposure. That is, no exposure, no matter how low, can be guaranteed to be totally safe. Moreover, radiation effects can accumulate over time. For this reason, lead-lined aprons are used to protect parts of the patient's body that are not being photographed, as well as the technicians operating the X-ray equipment. It is especially important to shield the reproductive organs, since damage to reproductive cells could result in the production of children with birth defects or cause sterility.

(a)

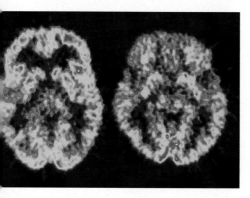

(b)

•**FIGURE 29.23 PET (a)** A PET scanner monitors brain activity after the administration of glucose-containing radioactive isotopes. Note the array of detectors for the gamma radiation produced when an emitted positron is annihilated. **(b)** PET scans of a normal brain (left) and the brain of a schizophrenic patient (right).

convenient half-life of 6 h and combines with a large variety of compounds. Detectors outside the body can scan a region and record the activity so that a complete activity image can be constructed.

It is possible to image gamma ray activity in a single plane, or "slice," through the body. A gamma detector is moved around the patient to measure the emission intensity from many angles. A complete image can then be constructed by using computer-assisted tomography, as in X-ray CT. This process is referred to as single-photon emission tomography (SPET). Another technique, positron emission tomography (PET), uses tracers that are positron emitters such as ^{11}C and ^{15}O. When a positron is emitted, it is quickly annihilated, and two gamma rays are produced. Recall that the photons have equal energies and fly off in nearly opposite directions, conserving momentum (Section 28.5). The gamma rays are recorded simultaneously by a ring of detectors surrounding the patient (•Fig. 29.23).

Domestic and Industrial Applications of Radiation

Common application of radioactivity in the home is the smoke detector. In the smoke detector, a weak radioactive source ionizes the air molecules, setting up a small current in the detector circuit. If smoke enters the detector, the ions there become attached to the smoke particles. This causes a reduction in the current. The current drop is electronically sensed, and an alarm is triggered (•Fig. 29.24).

Industry also makes good use of radioactive isotopes. Radioactive tracers are used to determine flow rates in pipes, to detect leaks, and to study corrosion and wear. Also, it is possible to radioactivate certain compounds at a particular stage in a process by irradiating them with particles, generally neutrons. This technique is called **neutron activation analysis** and is also an important method of identifying elements in a sample. Until its development, the chief methods of identification were chemical and spectral analyses. In both of these methods, a fairly large amount of a sample has to be destroyed in the analysis procedure. As a result, a sample may not be large enough for analysis, or small traces of elements in a sample may go undetected. Neutron activation analysis has the advantage over these methods on both scores: Only minute samples are needed, and the method can detect even tiny amounts of an element.

A typical neutron activation process starts with californium-252, an unstable neutron emitter:

$$\text{(source)} \qquad ^{252}_{98}\text{Cf} \rightarrow \, ^{251}_{98}\text{Cf} + \, ^{1}_{0}\text{n}$$

The neutrons are used to bombard, for example, ^{14}N and are absorbed, leaving the nitrogen nucleus in an excited state:

$$^{1}_{0}\text{n} + \, ^{14}_{7}\text{N} \;\rightarrow\; ^{15}_{7}\text{N*} \;\rightarrow\; ^{15}_{7}\text{N} + \gamma$$
$$\qquad\qquad\quad \textit{excited} \qquad \textit{gamma}$$
$$\qquad\qquad\quad \textit{nucleus} \qquad\; \textit{ray}$$

•**FIGURE 29.24 Smoke detector** In **(a)** the common smoke detector, **(b)** a weak radioactive source ionizes the air and sets up a small current. Smoke particles in the detector reduce the current, causing an alarm to sound.

(a)

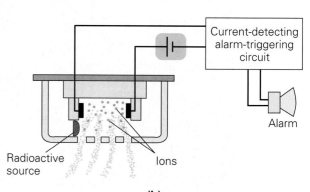

Current-detecting alarm-triggering circuit

Alarm

Radioactive source

Ions

(b)

The excited nitrogen-15 nucleus decays with the emission of a gamma ray with a distinctive energy. Other neutron-activated nuclei may decay by other modes, such as beta decay.

Nitrogen activation is commonly used to screen for bombs in airport luggage since virtually all explosives contain nitrogen. By using neutron activation and analyzing the energy of any gamma ray emission coming from a suitcase, we can check for an explosive device in the suitcase. Other materials in the suitcase may contain nitrogen too, so manual checks are made to confirm any suspicious findings.

As a final example, note that the U.S. government has given permission for the use of gamma radiation in the processing of poultry. The radiation kills bacteria and helps preserve the food and in no way makes the nuclei in the food radioactive. There are, however, continuing concerns with this process from food health professionals. Even though the gamma emission cannot make the meat radioactive, it could change some of the *chemical* bonding through ionization effects. This possibility has prompted concerns about whether these radiations can affect the chemical structure of the meat—making it unsafe to eat—enough to warrant further study.

Chapter Review

Important Terms

Rutherford–Bohr model   *893*
atomic number   *894*
nucleon   *894*
strong nuclear force   *895*
proton number   *895*
mass number   *895*
neutron number   *895*
isotope   *896*
nuclide   *896*
radioactivity   *896*
alpha particle   *897*
beta particle   *897*
gamma ray   *897*
alpha decay   *897*
conservation of nucleons   *897*

conservation of charge   *897*
tunneling   *898*
barrier penetration   *898*
beta decay   *899*
β^- decay   *899*
β^+ decay   *899*
electron capture   *899*
gamma decay   *900*
activity   *903*
decay constant   *903*
half-life   *903*
curie (Ci)   *906*
becquerel (Bq)   *906*
carbon-14 dating   *907*
pairing effect   *910*

atomic mass unit (u)   *911*
mass defect   *912*
total binding energy   *912*
magic number   *914*
radiation detectors   *916*
roentgen (R)   *918*
rad   *918*
gray (Gy)   *918*
rem   *918*
relative biological effectiveness
 (RBE)   *918*
sievert (Sv)   *918*
neutron activation analysis   *920*

Important Concepts

- The Rutherford–Bohr model of the atom is a planetary model, with negative electrons orbiting the positively charged nucleus, in which almost all the mass of the atom is located.
- The strong nuclear force is the short-range attractive force between nucleons that is responsible for holding the nucleus together against the repulsive electric force of its protons.
- Isotopes of an element are nuclides that differ only in the number of neutrons they contain.
- Radioactivity refers to any type of nuclear instability that results in the emission of particles and photons.
- Nuclei may undergo radioactive decay by the emission of an alpha particle, or helium nucleus (α de-

cay); a beta particle, either an electron (β^- decay) or a positron (β^+ decay); or a gamma ray, a high-energy photon of electromagnetic radiation (γ decay). Nuclei can also change by capturing one of the atom's orbital electrons (electron capture).
- The half-life of a nuclide is the time required for the number of undecayed nuclei in a sample (and hence the radioactive decay rate or activity of the sample) to fall to half of its initial value.
- Total binding energy is the minimum amount of energy needed to separate a nucleus into its constituent nucleons.
- The effective dose of radiation is determined by the energy deposited per kilogram of material and the type of particle depositing the energy (RBE).

Important Equations

Decay Rate of a Radioisotope:

$$\frac{\Delta N}{\Delta t} = -\lambda N \qquad (29.2)$$

$$(1 \text{ Ci} = 3.70 \times 10^{10} \text{ decays/s} = 3.70 \times 10^{10} \text{ Bq})$$
$$(\text{activity} = |\Delta N/\Delta t| = \lambda N)$$

Number of Undecayed Nuclei:

$$N = N_0 e^{-\lambda t} \qquad (29.3)$$

Half-Life and Decay Constant:

$$t_{1/2} = \frac{0.693}{\lambda} \qquad (29.4)$$

Total Binding Energy:

$$E_b = (\Delta m)c^2 \qquad (29.5)$$

[direct conversion of mass–energy equivalent is most easily done if Δm is expressed in u, then $E_b = (\Delta m)(931.5 \text{ MeV/u})$]

Effective Dose:

Dose (in rem) = dose (in rad) × RBE $\qquad (29.6)$

Dose (in Sv) = dose (in Gy) × RBE $\qquad (29.7)$

Exercises

29.1 Nuclear Structure and the Nuclear Force

1. The concept of the atomic nucleus resulted from the work of (a) Bohr, (b) Thompson, (c) Rutherford, (d) Geiger.

2. Carbon-12 (a) has the same number of protons and neutrons, (b) is a nuclide, (c) is an isotope of carbon, (d) all of these.

3. The strong nuclear force (a) binds the orbital electrons to the atomic nucleus, (b) has a longer range than the electrostatic force, (c) acts only between identical particles, (d) overcomes the force of repulsion between protons in the nucleus.

4. (a) Is the nuclear force attractive or repulsive? (b) What particles can it act between? (c) Is it a long-ranged or short-ranged force?

5. What do the isotopes of an element have in common? How do they differ?

6. ■ Two isotopes of magnesium are ^{24}Mg and ^{25}Mg. For each isotope, how many protons, neutrons, and electrons are in an atom with that isotope as its nucleus if that atom (a) is electrically neutral, (b) is actually an ion with a $-2e$ charge, (c) is actually an ion with a $+1e$ charge?

7. ■ (a) Using symbols H, D, and T, write the nuclear notations for the isotopes of hydrogen and (b) write the chemical symbols for six different forms of water. (c) Identify the radioactive forms of water.

8. ■ Oxygen has three stable isotopes with 8, 9, and 10 neutrons, respectively. Write the nuclear notations for these isotopes.

9. ■ An isotope of potassium has the same number of neutrons as the nuclide ^{40}Ar. Write the nuclear notation for the potassium isotope.

10. ■ For an isotope of uranium that has a mass number of 238, what are the number of protons, neutrons, and electrons in a neutral atom?

11. ■■ A theoretical result for the approximate nuclear radius (R) is $R = R_0 A^{1/3}$, where $R_0 = 1.2 \times 10^{-15}$ m and A is the mass number of the nucleus. (a) Find the nuclear radii of these noble gas atoms: He, Ne, Ar, Kr, Xe, and Rn. (b) How does the order of magnitude of your answers compare to the approximate size of an atom, roughly 0.1 nm?

29.2 Radioactivity

12. The conservation of nucleons and the conservation of charge apply to (a) alpha decay only, (b) beta decay only, (c) gamma decay only, (d) all nuclear processes.

13. The neutron number is not conserved in a beta decay. Is this a violation of the conservation of nucleons? Explain.

14. ■ Write the nuclear equations expressing (a) the beta decay of $^{60}_{27}$Co, and (b) the alpha decay of $^{226}_{88}$Ra.

15. ■ Write the nuclear equations for (a) the alpha decay of ^{237}Np, (b) the β^- decay of ^{32}P, (c) the β^+ decay of ^{56}Co, (d) electron capture in ^{56}Co, and (e) the gamma decay of ^{42}K.

16. ■ Tritium is radioactive. (a) Would you expect it to β^+ or β^- decay? (b) What is the daughter nucleus? Is it stable? [Hint: Write the nuclear equation for each.]

17. ■ Polonium-214 can decay by alpha emission followed by beta decay, ending as ^{210}Bi. Write the nuclear equations for these sequential decay processes.

18. ■ What are the main differences between alpha, beta, and gamma rays?

19. ■ A ^{209}Pb nucleus results from both alpha–beta sequential decays and beta–alpha sequential decays. What was the grandparent nucleus? (Show this for both decay routes by writing the nuclear equations for these decay processes.)

20. ■■ Complete the following nuclear decay equations:

(a) $^{8}_{4}Be \rightarrow ^{4}_{2}He +$ ____

(b) $^{240}_{94}Po \rightarrow ^{97}_{38}Sr + ^{139}_{56}Ba +$ ____

(c) $^{47}_{21}Sc^{*} \rightarrow ^{47}_{21}Sc +$ ____

(d) $^{29}_{11}Na \rightarrow ^{0}_{-1}e +$ ____

21. ■■ Complete the following nuclear equations:

(a) $^{238}_{92}U \rightarrow ^{234}_{90}Th +$ ____

(b) $^{40}_{19}K \rightarrow ^{40}_{20}Ca +$ ____

(c) $^{236}_{92}U \rightarrow ^{131}_{53}I + 3(^{1}_{0}n) +$ ____

(d) $^{23}_{11}Na + \gamma \rightarrow$ ____

(e) $^{22}_{11}Na + ^{0}_{-1}e \rightarrow$ ____

22. ■■■ Actinium-227 ($^{227}_{89}Ac$) decays by alpha decay or by beta decay and is part of a decay sequence like that in Fig. 29.9. Each daughter nucleus then decays to $^{223}_{88}Ra$, which subsequently decays to $^{215}_{84}Po$. Write the nuclear equations for the decay process in the decay series from ^{227}Ac to ^{215}Po.

23. ■■■ The decay series for ^{237}Np is shown in •Fig. 29.25. (a) Identify the decay modes and each of the nuclei in the sequence. (b) Tell why each nucleus is likely to decay.

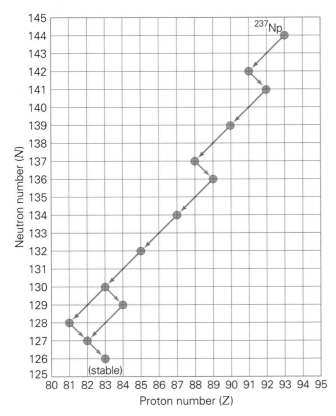

•FIGURE 29.25 Neptunium-237 decay series See Exercise 23.

29.3 Decay Rate and Half-Life

24. After one half-life, a sample of a particular radioactive material (a) is half as massive, (b) has its half-life reduced by half, (c) is no longer radioactive, (d) has its activity reduced by half.

25. In two half-lives, the activity of a radioactive sample will have decreased by what percentage? (a) 25%, (b) 50%, (c) 75%, (d) 87.5%.

26. Which physical or chemical properties affect the decay rate or half-life of a radioactive isotope?

27. One radioactive nuclide has a decay constant that is half that of another. If the two nuclides start with the same number of undecayed nuclei, will twice as many of its nuclei decay in a given time? Explain.

28. What are the (a) half-life and (b) decay constant for a stable isotope?

29. ■ A radioactive sample undergoes 2.50×10^{6} decays/s. What is the activity of the sample (a) in curies and (b) in becquerels?

30. ■ At present, a laboratory radioactive beta source has an activity of 20 mCi. (a) What is the present decay rate in decays per second? (b) Assuming that one beta particle is emitted per decay, how many are emitted per minute?

31. ■■ The half-life of a radioactive isotope is 1.0 h. What fraction of a sample of this isotope would be left after (a) 3.0 h and (b) 1.0 d?

32. ■■ A sample of ^{104}Tc has an activity of 10 mCi. Estimate the activity of the sample after 1 h ($t_{1/2} = 18$ min).

33. ■■ What period of time is required for a sample of radioactive tritium (^{3}H) to lose 80.0% of its activity? Tritium has a half-life of 12.3 y.

34. ■■ Which has the greater decay constant, $^{28}_{12}Mg$ or $^{104}_{32}Tc$? By what factor is it greater?

35. ■■ Iodine-131 is introduced into a patient in a medical diagnostic procedure (see Example 29.3). To three significant figures, what percentage of the sample remains after exactly one day if all of the ^{131}I is retained in the patient's thyroid gland?

36. ■■ A sample of bone has 4.0 beta decays/min for each gram of carbon. Approximately how old is the bone?

37. ■■ Show that the number N of radioactive nuclei remaining in a sample after n half-lives is given by

$$N = \frac{N_{o}}{2^{n}} = \left(\frac{1}{2}\right)^{n} N_{o}$$

where N_{o} is the initial number of nuclei.

38. ■■ Some ancient writings on parchment are found sealed in pottery in a cave. If ^{14}C dating shows the parchment to be 28 650 y old, what percentage of the ^{14}C atoms still remain in the sample, compared to the number when the parchment was made?

39. ■■ The activity of a beta source decreases by 87.5% in 54 min. What is the radioactive nuclide? (See Table 29.1.)

40. ■■ For a sample of ^{60}Co, how long would it take (to the nearest whole year) for its activity to reduce to 20% of the original activity?

41. ■■ A soil sample contains 40 μg of ^{90}Sr. Approximately how much ^{90}Sr will be in the sample 150 y from now?

42. ■■ (a) What is the decay constant of ^{17}F? (b) How long will it take for the activity of a sample of ^{17}F to decrease to 10% of its initial value ($t_{1/2} = 66.0$ s)?

43. ■■ Francium-223 ($^{223}_{87}$Fr) has a half-life of 21.8 min. (a) If there is initially a 25.0-milligram sample of this isotope, how many nuclei are present? (b) How many nuclei will be present 1 h and 49 min later?

44. ■■ A basement room containing radon gas ($t_{1/2} = 3.82$ d) is sealed air-tight. If 7.50×10^{10} radon atoms are trapped in the room at the time, (a) estimate how many radon atoms would be in the room after 1 week. (b) Radon undergoes alpha decay. Will there be the same number of daughter nuclei as decayed parent nuclei at the end of 30 d? Explain.

45. ■■ In 1898, Pierre and Marie Curie isolated about 10 mg of ^{226}Ra from 8 tons of uranium ore. If this sample had been placed in a museum, how much of the radium would remain by the year 2100?

46. ■■■ An old artifact contains 250 g of carbon and has an activity of 475 decays/min. What is the approximate age of the artifact, to the nearest thousand years?

47. ■■■ The recoverable reserves of high-grade ^{238}U ore in the United States are estimated to be about 500 000 tons. (High-grade ore contains about 10 kg of ^{238}U$_3$O$_8$ per ton.) Neglecting any geological changes, what mass of ^{238}U existed in this high-grade ore when the Earth formed about 4.6 billion y ago? [Hint: See Appendix V.]

48. ■■■ Nitrogen-13, with a half-life of 10 min, decays with the emission of positrons. (a) Write the nuclear equation for this decay process. (b) If a pure sample of ^{13}N has a mass of 0.0015 kg, what is the activity in 35 min? (c) What percentage of the sample is ^{13}N at this time?

29.4 Nuclear Stability and Binding Energy

49. For nuclei with a mass number greater than 40, which of the following is correct? (a) The number of protons is approximately equal to the number of neutrons, (b) the number of protons exceeds the number of neutrons, (c) all are stable up to $Z = 92$, (d) none of these.

50. The average binding energy per nucleon of the daughter nucleus in a decay process is (a) greater than, (b) less than, (c) equal to that of the parent nucleus.

51. ■ Determine whether each of the three isotopes of hydrogen is likely to be stable or unstable.

52. ■ Using Appendix V, list those stable nuclei that are "doubly magic." That is, list those that have magic numbers for *both* neutrons and protons.

53. ■ From which of the following pairs of nuclei would you expect it to be easiest to remove a neutron? State your reasoning in each case: (a) $^{16}_{8}$O or $^{17}_{8}$O, (b) $^{40}_{20}$Ca or $^{42}_{20}$Ca, (c) $^{10}_{5}$B or $^{11}_{5}$B, (d) $^{208}_{82}$Pb or $^{209}_{83}$Bi.

54. ■ Write in nuclear notation the most abundant stable isotope of the nuclide with a proton magic number of 28. (See Appendix V.)

55. ■ Only two isotopes of Sb (antimony, $Z = 51$) are stable. Pick the two stable isotopes from among the following: (a) ^{120}Sb, (b) ^{121}Sb, (c) ^{122}Sb, (d) ^{123}Sb, (e) ^{124}Sb.

56. ■ Determine which of the following nuclides are likely to be unstable: (a) ^{23}Na, (b) ^{50}V, (c) ^{209}Bi, (d) ^{209}Po, (e) ^{44}Ca.

57. ■ The total binding energy of $^{2}_{1}$H is 2.224 MeV. Compute (in atomic mass units) the mass of a ^{2}H nucleus from the known mass of a proton and a neutron.

58. ■■ Use Avogadro's number, $N_A = 6.02 \times 10^{23}$ atoms/mol, to show that 1 u $= 1.66 \times 10^{-27}$ kg. (Recall that a ^{12}C atom has a mass of exactly 12 u.)

59. ■■ The mass of $^{12}_{6}$C is 12.000 000 u. (a) What is the total binding energy of this nucleus? (b) What is the average binding energy per nucleon?

60. ■■ The mass of $^{16}_{8}$O is 15.994 915 u. What is E_b/A for this nucleus?

61. ■■ Which isotope of hydrogen has the smaller average binding energy per nucleon, deuterium or tritium? (Justify your answer mathematically.)

62. ■■ Near high neutron areas, such as a nuclear reactor, neutrons will be absorbed by protons (hydrogen nuclei in water molecules) and will give off a characteristic energy gamma ray in the process. What is the energy of the gamma ray (to three significant figures)?

63. ■■ How much energy (to four significant figures) would be required to separate completely all the nucleons of a ^{14}N nucleus, the atom of which has a mass of 14.003 074 u?

64. ■■ Calculate the binding energy of the last neutron in the $^{40}_{19}$K nucleus. [Hint: Compare the mass of $^{40}_{19}$K with the mass of $^{39}_{19}$K plus the mass of a neutron.]

65. ■■ If an alpha particle could be removed intact from an ^{27}Al nucleus ($m = 26.981 541$ u), then a ^{23}Na nucleus ($m = 22.989 770$ u) would remain. How much energy would be required to do this?

66. ■■ On average, are nucleons more tightly bound in an ^{27}Al nucleus or in a ^{23}Na nucleus?

67. ■■ The atomic mass of $^{235}_{92}$U is 235.043 925 u. Find the average binding energy per nucleon of this nuclide.

68. ■■■ The total binding energy of $^6_3$Li is 32.0 MeV. What is the mass (in atomic mass units) of the ^{6}Li nucleus?

69. ■■■ The mass of $^8_4$Be is 8.005 305 u. (a) Which is less, the total mass of two alpha particles or the mass of the ^{8}Be nucleus? (b) Which is greater, the total binding energy of the ^{8}Be nucleus or the total binding energy of two alpha particles? (c) On this basis alone, would you expect the ^{8}Be nucleus to decay spontaneously into two alpha particles?

29.5 Radiation Detection and Applications

70. The detector that uses light emitted from excited atoms to detect radiation is the (a) Geiger counter, (b) scintillation counter, (c) solid state detector.

71. A unit of radiation dosage is the (a) rad, (b) gray, (c) sievert, (d) all of these.

72. A basic assumption of radiocarbon dating is that the cosmic ray intensity has been generally constant for the last 40 000 y or so. Suppose it were found that the intensity was much less 10 000 y ago than it is today. How would this affect ^{14}C dating?

73. A patient receives a dose of radiation of 1.0 rad from X-rays and a dose of 1.0 rad from an alpha source. (a) Which has the greater biological (damage) effectiveness? (b) What are the doses in rems?

74. ■ A technician working at a nuclear reactor facility is exposed to a slow neutron radiation and receives a dose of 1.25 rad. (a) How much energy is absorbed by 200 g of the worker's tissue? (b) Was the maximum permissible radiation dosage exceeded?

75. ■ A person working with nuclear isotopes for 2 months receives doses of 0.5 rad from a gamma source, 0.3 rad from a slow-neutron source, and 0.1 rad from an alpha source. Was the maximum permissible radiation dosage exceeded?

76. ■ In a diagnostic procedure, a hospital patient ingests 80 mCi of ^{198}Au ($t_{1/2}$ = 2.7 d). What is the activity at the end of 1 month if none of the gold is eliminated from the body by biological functions?

77. ■■ Neutron activation analysis was recently performed on pieces of hair taken from the exiled Napoleon after he died in 1821. The samples were found to contain abnormally high levels of arsenic, which supported a theory that his death was not due to natural causes. If β^- emissions from ^{76}As were detected to come from the samples, which arsenic isotope was actually present in the hair, and what nucleus remained after this beta decay?

Additional Exercises

78. A fossil specimen is believed to be about 18 000 y old. How could this be confirmed?

79. The binding energy per nucleon of $^4_2$He is 7.075 MeV. (a) What is the total binding energy? (b) What is the mass of a ^{4}He nucleus?

80. Starting with $^{234}_{92}$U, a decay sequence of four alpha decays and two beta minus decays occurs. What is the resulting nucleus of the sequence?

81. A sample of ^{215}Bi that beta decays ($t_{1/2}$ = 2.4 min) contains Avogadro's number of nuclei. How many bismuth nuclei are present after (a) 10 min and (b) 1.0 h? (c) What are the activities, in curies and becquerels, at these times, and is this a realistic laboratory radioactive source?

82. A bone yields, on average, 1.00 beta emission per gram of carbon per minute. About how old is the bone?

83. Complete the following nuclear decay equations:
    (a) $^{47}_{21}$Sc → $^{47}_{22}$Ti + ____
    (b) $^{226}_{88}$Ra → $^4_2$He + ____
    (c) $^{237}_{93}$Np → $^0_{-1}$e + ____
    (d) $^{210}_{84}$Po* → $^{210}_{84}$Po + ____
    (e) $^{11}_6$C → $^{11}_5$B + ____

84. •Figure 29.26 shows the decay series for ^{239}Pu. (a) Identify each of the nuclei in this decay process. (b) Compare the totals of the half-lives before and after the ^{227}Ac nuclei.

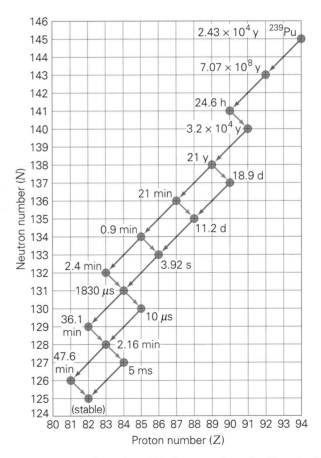

•FIGURE 29.26 **Plutonium-239 decay series** See Exercise 84.

85. The value of E_b/A for $^{238}_{92}U$ is 7.58 MeV/nucleon. (a) What is the total binding energy of this nucleus? (b) What is its mass?

86. What are the mass equivalent energies of (a) a ^{12}C atom and (b) an alpha particle?

87. Determine which of the following isotopes are likely to be stable: (a) ^{20}C, (b) ^{100}Sn, (c) ^{6}Li, (d) ^{9}Be, (e) ^{22}Na.

88. Thorium-229 ($^{229}_{90}Th$) undergoes alpha decay. (a) What is the daughter nucleus of this process? Write the decay equation. (b) This daughter nucleus undergoes beta decay. What is the resulting granddaughter nucleus? Write the decay equation.

89. When a $^{6}_{3}Li$ nucleus is struck by a proton, an alpha particle and a product nucleus are released. What is the product nucleus?

90. Gold-198 beta decays with a half-life of 2.7 d. (a) What is the activity of a sample of 1.0 g of pure $^{198}_{79}Au$? (b) What is the activity of the sample at the end of 1 month?

91. Assuming that nuclei are spherical (they are approximately so in most cases), use the nuclear radius $R = R_o A^{1/3}$, where $R_o = 1.2 \times 10^{-15}$ m and A is the mass number of the nucleus, to (a) show that the average nucleon density in a nucleus is 1.4×10^{44} nucleons/m^3 and (b) estimate the nuclear density in kilograms per cubic meter.

92. What percentage of a sample containing $^{3}_{1}H$, which has a half-life of 12.33 y, will remain after 6.00 y?

93. (a) Compare the average nuclear density results of Exercise 91 to the average atomic density of a hydrogen atom with a radius of 0.0529 nm. Assume the atom to be spherical. (b) Why are they so different if the nuclear mass and atomic mass are essentially the same? (c) Does your number for hydrogen agree with the density of hydrogen gas at normal temperatures and pressures?

94. Calculate the kinetic energy of an alpha particle emitted by $^{232}_{92}U$. Neglect the recoil velocity of the daughter nucleus, $^{228}_{90}Th$.

Nuclear Reactions and Elementary Particles

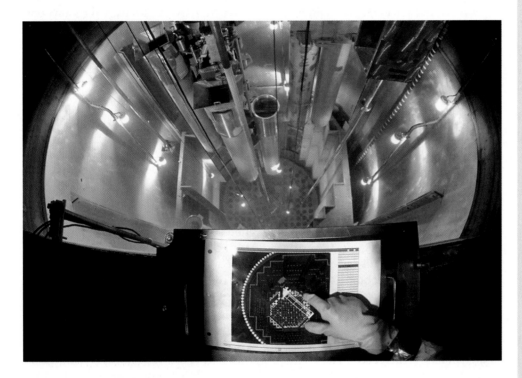

Today, more than half a century after the first nuclear reactor was built, these devices remain the center of intense controversy. Some people see them as essential to solving the world's energy problems. To others, they embody all that is wrong and dangerous about modern technology. The reactor shown above is a perfect symbol of the ongoing debate. It is located at the Three Mile Island (TMI) plant in Pennsylvania—the site of one of the largest and most widely publicized nuclear accidents to date. In this photo, you are looking into the heart of one of the two remaining TMI plants, both of them running safely.

In this chapter, you'll learn about the nuclear fission reactions that power such reactors. You'll also learn how scientists are trying to harness even more powerful fusion reactions, which occur within the Sun and other stars, and might one day provide a safe and virtually inexhaustible supply of clean energy for our planet.

We will also look at elementary particles and their interactions. Investigations have shown that a variety of particles other than the proton and neutron are associated with the nucleus. The discovery of these particles has given us insight into the puzzling world of the nucleus and what holds it together, and has provided a clearer understanding of the evolution of the universe.

30.1 Nuclear Reactions

Objectives: To (a) use charge and nucleon conservation to write nuclear reaction equations, and (b) understand and use the concepts of Q value and threshold energy to analyze nuclear reactions.

Ordinary chemical reactions between atoms and molecules involve only the orbital electrons. Thus the atoms of the reaction do not lose their nuclear identities in the process. Conversely, in **nuclear reactions**, the original nuclei are converted into the nuclei of other elements. Scientists first became aware of this type of reaction during the study of the nucleus by bombarding various nuclei with energetic particles.

The first induced nuclear reaction was produced by Ernest Rutherford in 1919. Nitrogen was bombarded with alpha particles from a natural source (bismuth-214). The occasional particles that came from the reactions were identified as protons. Rutherford reasoned that an alpha particle colliding with a nitrogen nucleus must sometimes induce a reaction that produces a proton. As a result, the nitrogen nucleus is *artificially transmuted* into an oxygen nucleus:

The first nuclear reaction

$$^{14}_{7}\text{N} \quad + \quad ^{4}_{2}\text{He} \quad \rightarrow \quad ^{17}_{8}\text{O} \quad + \quad ^{1}_{1}\text{H}$$

nitrogen	alpha particle	oxygen	proton
(14.003 074 u)	(4.002 603 u)	(16.999 133 u)	(1.007 825 u)

(The atomic masses are given for later use.)

This reaction and many others like it actually form a short-lived intermediate *compound nucleus* in a highly excited state. For example, the preceding reaction can be written more precisely as

$$^{14}_{7}\text{N} + ^{4}_{2}\text{He} \rightarrow (^{18}_{9}\text{F}^{*}) \rightarrow ^{17}_{8}\text{O} + ^{1}_{1}\text{H}$$

The fluorine nucleus, $^{18}_{9}\text{F}$, formed in an excited state, is indicated by the usual asterisk. A compound nucleus typically rids itself of excess energy by ejecting a particle or particles, in this case the proton. Since a compound nucleus lasts only a very short time, it is commonly omitted from the nuclear reaction equation.

Think of the implication of Rutherford's discovery. One element had been converted into another! This was the age-old dream of the alchemists, although their main goal was to change common metals, such as mercury and lead, into gold. This and many other transmutations can be initiated today with machines called *particle accelerators*, which accelerate charged particles to very high energies (●Fig. 30.1). When these particles strike target nuclei, they can initiate nuclear reactions. One nuclear reaction that is initiated when protons strike nuclei of mercury is

$$^{200}_{80}\text{Hg} \quad + \quad ^{1}_{1}\text{H} \quad \rightarrow \quad ^{197}_{79}\text{Au} \quad + \quad ^{4}_{2}\text{He}$$

mercury	proton	gold	alpha particle
(199.968 321 u)	(1.007 825 u)	(196.966 56 u)	(4.002 603 u)

In this reaction, mercury is converted into gold, so it would seem that modern physics has fulfilled the alchemists' dream. However, making such small amounts of gold in an accelerator costs far more than the gold is worth.

Reactions such as those written above have the general form

$$A + a \rightarrow B + b$$

where the uppercase letters represent the nuclei and the lowercase letters represent the particles. Such reactions are often written in a shorthand notation:

$$A(a, b)B$$

For example, in this form, the two previous reactions are written as

$$^{14}\text{N}(\alpha, \text{p})^{17}\text{O} \quad \text{and} \quad ^{200}\text{Hg}(\text{p}, \alpha)^{197}\text{Au}$$

The periodic table (Fig. 28.9) lists over 100 elements, but only 90 elements occur naturally on Earth. Elements with proton numbers greater than uranium

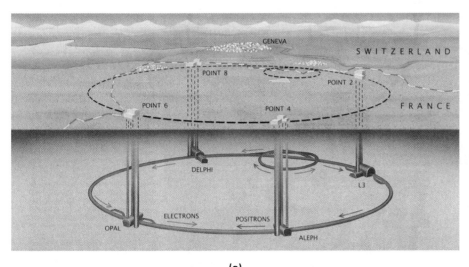

(a) **(b)**

•**FIGURE 30.1 Particle accelerators** **(a)** The Large Electron–Positron (LEP) Collider on the border between France and Switzerland, near Geneva, has a circumference of 26.7 km. The electrons and positrons that it accelerates in opposite directions to speeds approaching that of light collide with an energy of over 100 billion eV (100 GeV). **(b)** A 4500-ton particle detector at the Fermi National Accelerator Laboratory in Batavia, Illinois. Such devices are used to measure the trajectories and energies of subatomic particles, created in the nuclear reactions that occur when a beam of accelerated particles smashes into a target.

($Z = 92$), as well as technetium ($Z = 43$) and promethium ($Z = 61$), are created artificially by nuclear reactions. The name *technetium* comes from the Greek word *technetos*, meaning "artificial," and technetium was the first unknown element to be created by artificial means. Elements up to $Z = 112$ have been confirmed.

Conservation of Mass–Energy and the *Q* Value

In every nuclear reaction, the total relativistic energy ($E = K + mc^2$) is conserved as shown in Chapter 26. Consider the reaction by which nitrogen is converted into oxygen, $^{14}\text{N}(\alpha, \text{p})^{17}\text{O}$. By the conservation of total relativistic energy,

$$(K_\text{N} + m_\text{N}c^2) + (K_\alpha + m_\alpha c^2) = (K_\text{O} + m_\text{O}c^2) + (K_\text{p} + m_\text{p}c^2)$$

where the subscripts refer to the particular particle or nucleus. Rearranging the equation, we have

$$K_\text{O} + K_\text{p} - (K_\text{N} + K_\alpha) = (m_\text{N} + m_\alpha - m_\text{O} - m_\text{p})c^2$$

The **Q value** of the reaction is defined as

$$Q = (K_\text{O} + K_\text{p}) - (K_\text{N} + K_\alpha) = \Delta K \qquad (30.1)$$

Q value: energy released or absorbed in a reaction

The change in kinetic energy, Q, can be positive or negative depending on whether kinetic energy increases or decreases. Thus the Q value is a measure of the energy released or absorbed in a reaction. Equation 30.1 can alternately be expressed in terms of the masses:

$$Q = (m_\text{N} + m_\alpha - m_\text{O} - m_\text{p})c^2 \qquad (30.2)$$

In terms of a general reaction of the form $\text{A} + \text{a} \rightarrow \text{B} + \text{b}$,

$$Q = (m_\text{A} + m_\text{a} - m_\text{B} - m_\text{b})c^2 = (\Delta m)c^2 \qquad (30.3)$$

The Q value is the difference in the mass-equivalent energies of the reactants and the products of a reaction, mass can be converted into kinetic energy and vice versa.

The mass difference, Δm, can be positive or negative. From the equation for Q (Eq. 30.3), we see that if mass is gained during the reaction, Q is negative; if mass is lost, Q is positive. When mass is gained, there is a conversion of kinetic energy into mass; when mass is lost, there is a conversion of mass into kinetic energy. Moreover, if Q is negative, the reaction *requires* a certain amount of kinetic energy before the mass of the products can be attained. That is, there is a minimum amount of kinetic energy required for the reaction to proceed. To see this more clearly, let us look at Rutherford's original reaction in some detail.

Using the masses given under the reactants and products in the equation on p. 928 for the $^{14}\text{N}(\alpha, \text{p})^{17}\text{O}$ reaction, we obtain

$$Q = (m_\text{N} + m_\alpha - m_\text{O} - m_\text{p})c^2$$
$$= [(14.003\,074 \text{ u} + 4.002\,603 \text{ u}) - (16.999\,133 \text{ u} + 1.007\,825 \text{ u})]c^2$$
$$= (-0.001\,281 \text{ u})c^2$$

or, using the mass–energy equivalence factor from Section 29.4,

$$Q = (-0.001\,281 \text{ u})(931.5 \text{ MeV/u}) = -1.193 \text{ MeV}$$

Endoergic: energy input required for the reaction to proceed

The negative Q value indicates that energy was absorbed in the reaction. When Q is negative, the reaction is said to be **endoergic** (or endothermic). That is, energy (*ergic*) must be put into (*endo*) the reaction for it to proceed. In endoergic reactions, the kinetic energy of the reacting particles is at least partially converted into mass.

Exoergic: energy released during the reaction

When the Q value of a reaction is positive, energy is released, and the reaction is said to be **exoergic** (or exothermic). That is, energy is produced by (*exo*) the reaction. In this case, some mass is converted into energy in the form of increased kinetic energy of the reaction products.

Example 30.1 ■ A Possible Energy Source? Q Value of a Reaction

Is the following reaction endoergic or exoergic?

$$\underset{\substack{\text{deuteron} \\ (2.014\,102 \text{ u})}}{^{2}_{1}\text{H}} + \underset{\substack{\text{deuteron} \\ (2.014\,102 \text{ u})}}{^{2}_{1}\text{H}} \rightarrow \underset{\substack{\text{helium} \\ (3.016\,029 \text{ u})}}{^{3}_{2}\text{He}} + \underset{\substack{\text{neutron} \\ (1.008\,665 \text{ u})}}{^{1}_{0}\text{n}}$$

Thinking It Through. The nature of the reaction is determined by the Q value: endoergic, $Q < 0$; exoergic, $Q > 0$. The Q value is given by Eq. 30.3, so we first calculate the mass difference (Δm) of the reaction from the given masses.

Solution. Subtracting the final masses from the initial masses, we have

$$\Delta m = 2m_\text{D} - m_\text{He} - m_\text{n}$$
$$= 2(2.014\,102 \text{ u}) - 3.016\,029 \text{ u} - 1.008\,665 \text{ u} = 0.003\,51 \text{ u}$$

Since the difference is positive, mass has been lost and kinetic energy has been gained. Thus the reaction is exoergic. The Q value is calculated by converting the mass difference (in atomic mass units) to its energy equivalent:

$$Q = (0.003\,51 \text{ u})(931.5 \text{ MeV/u}) = +3.27 \text{ MeV}$$

Follow-up Exercise. Show that the following reaction is endoergic, and calculate its Q value:

$$\underset{\substack{\text{carbon} \\ (12.000\,000 \text{ u})}}{^{12}_{6}\text{C}} + \underset{\substack{\text{helium} \\ (4.002\,603 \text{ u})}}{^{4}_{2}\text{He}} \rightarrow \underset{\substack{\text{carbon} \\ (13.003\,355 \text{ u})}}{^{13}_{6}\text{C}} + \underset{\substack{\text{helium} \\ (3.016\,029 \text{ u})}}{^{3}_{2}\text{He}}$$

Note that Q values are most easily computed from the mass difference, expressed in atomic mass units, by using the mass–energy conversion factor derived in Chapter 29 (see Table 29.3). This method eliminates the need for c^2 numerically and gives the Q value directly in megaelectron volts (MeV).

Radioactive decay is considered a special type of nuclear reaction with one reactant nucleus and two (or more) product nuclei. The Q value of radioactive decay is always positive $(Q > 0)$, because kinetic energy is released via the emitted energetic particles. This is sometimes called the *disintegration energy*. The meanings of the positive and negative Q values are summarized in Table 30.1.

When the Q value of a reaction is negative, the mass of the products is greater than the mass of the reactants. Such reactions will not occur unless a minimum amount of kinetic energy is initially available. You might therefore think that if a particle incident on a stationary nucleus had at least a kinetic energy equal to the Q value of the reaction $(K = |Q|)$*, then the reaction would be able to occur. However, if all the kinetic energy were converted to mass, there would be none left over, and the particles would be at rest after the reaction. But this means that momentum would *not* be conserved, since the incoming particle had momentum and the products had none.

Hence, in an endoergic reaction, for the products of the reaction to conserve linear momentum, the kinetic energy of the incident particle must be *greater* than $|Q|$. Just how much greater depends on the Q value and on the particle masses. The minimum kinetic energy (K_{min}) that an incident particle needs to initiate an endoergic reaction is called the **threshold energy**. For nonrelativistic energies, the threshold energy is

$$K_{min} = \left(1 + \frac{m_a}{M_A}\right)|Q| \qquad \begin{array}{l}\textit{(stationary}\\\textit{target only)}\end{array} \qquad (30.4)$$

TABLE 30.1 Interpretation of Q Values

Q value	Effect
Positive $(Q > 0)$	Exoergic, some mass converted into energy (mass of reactants greater than mass of products)
Negative $(Q < 0)$	Endoergic, some kinetic energy converted into mass (mass of products greater than mass of reactants)

Threshold energy of an endoergic reaction

where m_a and M_A are the masses of the incident particle and the stationary target nucleus, respectively. In Eq. 30.4 the factor multiplying $|Q|$ is greater than one (why?), so as expected $K_{min} > |Q|$. The use of the threshold energy is shown in the next Example.

Example 30.2 ■ Nitrogen into Oxygen: Threshold Energy

What is the threshold energy for the reaction $^{14}N(\alpha, p)^{17}O$?

Thinking It Through. The Q value for this reaction was calculated previously in the text. To get the threshold energy, we use Eq. 30.4, for which we need the masses of the alpha particle and the ^{14}N nucleus.

Solution. We have the following data:

Given: $m_a = m_\alpha = 4.002\,603$ u   *Find:* K_{min} (threshold energy)
       $M_A = m_N = 14.003\,074$ u
       $Q = -1.193$ MeV

Using Eq. 30.4, we can write

$$K_{min} = \left(1 + \frac{m_\alpha}{M_N}\right)|Q| = \left(1 + \frac{4.002\,603\text{ u}}{14.003\,074\text{ u}}\right)(1.193\text{ MeV}) = 1.534\text{ MeV}$$

Note that K_{min} is significantly larger than $|Q|$.

---

*Kinetic energy is written in terms of $|Q|$, the absolute value of Q, because kinetic energy cannot be negative. The sign of Q arises from the mass difference and indicates the gain or loss of mass during the reaction. If Q is negative, we want K_{min} to be positive; hence the use of absolute value.

Follow-up Exercise. In this Example, how much of the threshold kinetic energy would go into increasing the mass of the system, and how much remained, in the final state, as kinetic energy? Explain your reasoning.

Reaction Cross Sections

Due to the probabilistic nature of quantum mechanics, more than one reaction is usually possible when a particle collides with a nucleus. As we have seen for an endoergic reaction, the incident particle must have at least a minimum kinetic energy. When a particle has more kinetic energy than the threshold energies of the *several* possible reactions, any of the reactions may occur. A measure of the probability that a particular reaction will occur is called the **cross section** of that reaction.

The cross section, or probability, of a particular reaction depends on many factors. Usually the cross section depends on the kinetic energy of the initiating particle, sometimes very dramatically. For positively charged incident particles, such as protons and alpha particles, the Coulomb barrier of the nucleus influences the reaction cross section. Thus for charged particles, the probability of a given reaction occurring generally increases with the kinetic energy of the incident particle.

Note: Review Figs. 29.7 and 29.8.

Being electrically neutral, neutrons are unaffected by the Coulomb barrier. As a result, the cross section of a given reaction can be quite large, even for low-energy neutrons, in particular for reactions such as $^{27}\text{Al}(n, \gamma)^{28}\text{Al}$. Such *neutron capture reactions* can be identified by the characteristic gamma ray energies emitted by the final nucleus. For low energies, neutron reaction cross sections are often proportional to $1/v$, where v is the neutron's speed. That is, the probability of a reaction appears to be proportional to the time a neutron spends near a nucleus. (Classically, the time to transit the nucleus is $t \approx d/v$, where d is the nuclear diameter, so $t \propto 1/v$.) As the neutron energy increases, the cross section varies a great deal, as •Fig. 30.2 shows. The peaks in the curve, called *resonances*, are associated with nuclear energy levels in the nucleus being formed. If the neutron's energy is "just right" to bring the final nucleus to one of its preferred energy levels, or states, there is a higher than average probability that neutron absorption will occur.

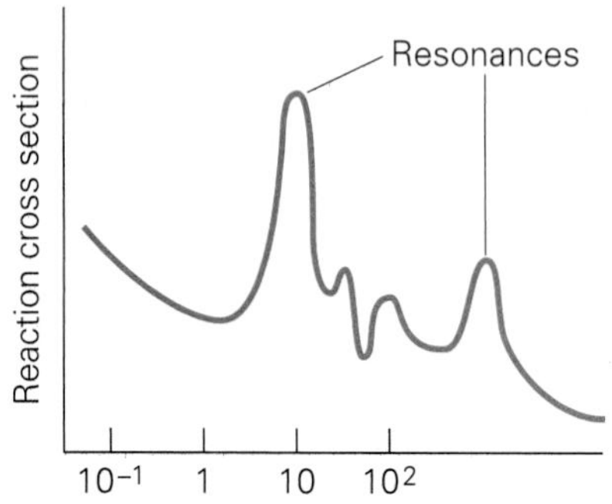

•**FIGURE 30.2 Reaction cross section** A typical graph of neutron reaction cross section versus energy. The peaks where the probabilities of reactions are greatest are called resonances.

Nuclear fission: "splitting" of the nucleus

30.2 Nuclear Fission

Objectives: To understand (a) the process of nuclear fission, (b) the nature and cause of a nuclear chain reaction, and (c) the basic principles involved in the operation of nuclear reactors.

In early attempts to make heavier elements artificially, uranium, the heaviest element known at the time, was bombarded with neutrons. An unexpected result was that the uranium nuclei sometimes broke apart, or "split" into fragments. In the later 1930s, these fragments were identified as the nuclei of lighter elements. The process was dubbed *fission* after the biological fission process of cell division.

In a **fission reaction**, a heavy nucleus divides into two lighter nuclei with the emission of neutrons. Mass is converted into kinetic energy during this process; that released energy is carried off primarily by the neutrons and fission fragments. Some heavy nuclei undergo *spontaneous fission*, but at very slow rates. However, fission can be *induced*, and this is the important process in energy production. For example, when a ^{235}U nucleus absorbs an incident neutron, it may fission by the reaction

$$^{235}_{92}\text{U} + {}^{1}_{0}\text{n} \rightarrow ({}^{236}_{92}\text{U}^*) \rightarrow {}^{140}_{54}\text{Xe} + {}^{94}_{38}\text{Sr} + 2({}^{1}_{0}\text{n})$$

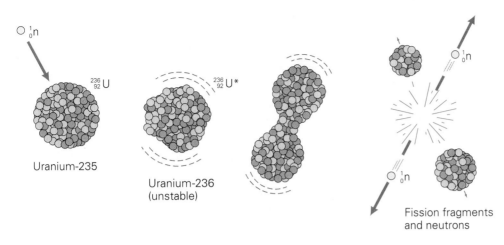

•FIGURE 30.3 **Liquid drop model of fission** When an incident neutron is absorbed by a fissionable nucleus, such as ^{235}U, the unstable compound nucleus (^{236}U) undergoes violent oscillations and breaks apart like a liquid drop, typically emitting two or more neutrons and yielding two radioactive fragments.

$^{1}_{0}$n

$^{236}_{92}$U

Uranium-235

$^{236}_{92}$U*

Uranium-236
(unstable)

$^{1}_{0}$n

$^{1}_{0}$n

Fission fragments
and neutrons

The capture of a neutron results in the formation of an excited uranium-236 nucleus. According to the *liquid drop model,* the ^{236}U nucleus undergoes violent oscillations and becomes distorted like a liquid drop (•Fig. 30.3). The separation of the nucleons into different parts of the "drop" weakens the nuclear force. The repulsive electrical force between the parts causes the nucleus to split, or fission.

The preceding reaction is only one way in which ^{235}U, the fissionable isotope of uranium, can fission. Other ways include (compound nuclei omitted)

$$^{1}_{0}\text{n} + {}^{235}_{92}\text{U} \rightarrow {}^{141}_{56}\text{Ba} + {}^{92}_{36}\text{Kr} + 3({}^{1}_{0}\text{n})$$

$$^{1}_{0}\text{n} + {}^{235}_{92}\text{U} \rightarrow {}^{150}_{60}\text{Nd} + {}^{81}_{32}\text{Ge} + 5({}^{1}_{0}\text{n})$$

Only certain nuclei undergo fission, and for them the probability of a fission reaction depends on the energy of the incident neutrons. For example, the largest cross sections of fission reactions for ^{235}U and ^{239}Pu are for "slow" neutrons with energies less than 1 eV. However, for ^{232}Th, "fast" neutrons with energies of 1 MeV or more are most likely to trigger a fission reaction.

We can obtain an estimate of the energy released in an exoergic fission reaction from the E_b/A curve for stable nuclei (Fig. 29.15). When a nucleus with a large mass number A, such as uranium, splits into two nuclei, it is in effect moving up along the downward-sloping tail of this curve to a more stable state. As a result, the average binding energy per nucleon increases from about 7.8 MeV to approximately 8.8 MeV. The energy liberated is about 1 MeV per nucleon in the fission products. In the fission reaction on p. 932, 140 + 94 = 234 nucleons are bound in the products. Thus the energy release is approximately

$$\frac{1 \text{ MeV}}{\text{nucleon}} \times 234 \text{ nucleons} \approx 234 \text{ MeV}$$

At first glance, this might not seem like much energy, since 200 MeV is only about 3×10^{-11} J. By way of comparison, when you pick up your textbook from the desk, you expend about 5 J of energy. In fact, 200 MeV is less than 0.1 percent of the energy equivalent of the mass of the ^{235}U nucleus, which is approximately (235 nucleons)(939 MeV/nucleon) = 2.2×10^5 MeV. Nevertheless, on a percentage basis, it is many times larger than the energy release from chemical reactions, such as the burning of oil or coal.

Practical amounts of energy can be obtained when huge numbers of these decays occur per second. This is accomplished by means of a **chain reaction**. For example, suppose a ^{235}U nucleus fissions with the release of two neutrons (•Fig. 30.4). Ideally, the neutrons can then initiate two more fission reactions, a process that releases four neutrons. These neutrons may initiate more reactions, and the process multiplies, the number of neutrons doubling with each generation. When this occurs uniformly with time, the neutron production rate grows exponentially.

•**FIGURE 30.4 Fission chain reaction** The neutrons that result from one fission event can initiate other fission reactions, and so on. When enough fissionable material is present, the sequence of reactions can be adjusted to be self-sustaining (a chain reaction).

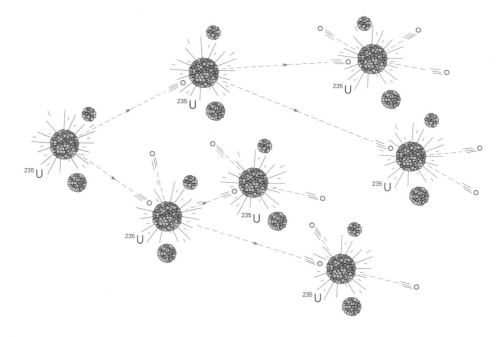

Note: A chain reaction requires a critical mass.

To maintain a sustained chain reaction, there must be an adequate quantity of fissionable material. The minimum mass required to produce a sustained chain reaction is called the **critical mass**. This means that there is enough fissionable material such that at least one neutron from each fission event, on average, goes on to fission another nucleus.

Several factors determine the critical mass. Most evident is the amount of fissionable material. If the quantity of material is small, many neutrons will escape from the sample before inducing a fission event, and the chain reaction will die out. Also, nuclides other than ^{235}U in the sample may absorb neutrons, thereby limiting the chain reaction. As a result, the purity of the fissionable isotope affects the critical mass.

Natural uranium is made up of the isotopes ^{238}U and ^{235}U. The natural concentration of the fissionable ^{235}U isotope is only about 0.7%. The remaining 99.3% is ^{238}U, which may absorb neutrons and trigger a nuclear reaction other than fission, thereby preventing those neutrons from contributing to a chain reaction of ^{235}U fission. To have more fissionable ^{235}U nuclei in a sample and reduce the critical mass, the ^{235}U can be concentrated. This enrichment varies from 3–5% ^{235}U for reactor-grade material used in electrical generation to over 99% for weapons-grade material. This is an important difference, since it is desirable that an energy-generating nuclear reactor *not* explode like an atomic bomb nor utilize fuel capable of being easily made into a bomb. Reactor-grade fuel still requires significant refining to bring it up to weapons-grade material.

Chain reactions take place almost instantly. If such a reaction proceeds uncontrolled, the quick and enormous release of energy from the fissioning nuclei causes an explosion. (See Demonstration 16.) This is the principle of the atomic bomb. In such a bomb, several subcritical pieces of fuel are suddenly imploded to form a critical mass. The chain reaction goes out of control, releasing an enormous amount of energy—a bomb. For the practical production of nuclear energy, the chain reaction process must somehow be controlled. We shall now see how this is done in **nuclear reactors**.

Nuclear Reactors

The Power Reactor.   Currently, the only practical nuclear reactor design for generating electrical power is based on the fission chain reaction. The first nuclear reactor for the purpose of electrical generation began operation in 1957 in Shippingport, Pennsylvania. A typical design for the components of a nuclear reactor

Demonstration 16 ■ A Simulated Chain Reaction

A simulation of how neutrons from fission reactions induce reactions in other nuclei such that the process grows in a chain reaction.

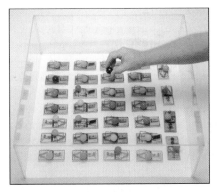

(a) Here, the "nuclei" are mousetraps loaded with a single Super Ball (neutron) that will be emitted when a "fission" occurs. A Plexiglas enclosure ensures a "critical mass," since the balls will be reflected rather than escaping.

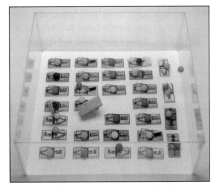

(b) The chain reaction begins when a ball is tossed in.

(c) The chain reaction grows, with an increasing number of "fissions" and release of energy.

is shown in •Fig. 30.5. There are five key elements to a reactor: fuel rods, core, coolant, control rods, and moderator.

Tubes packed with pellets of uranium oxide form the **fuel rods**, which are located in the reactor **core**, the central portion of the reactor. A typical commercial reactor contains fuel rods bundled in fuel assemblies of approximately 200 rods each. Coolant flows around the rods to remove the heat energy from the chain reaction. The reactors used in the United States are light water reactors, which means that ordinary water is used as a **coolant** to remove heat. However, the hydrogen

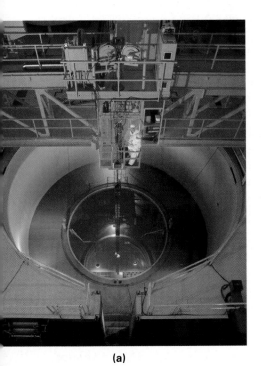

(a)

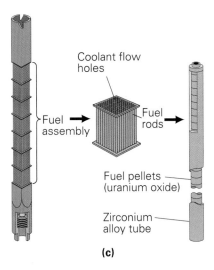

Control rod operating mechanism

Control rod tubes

Coolant ports

Control rod extension

Outlets for pressurized water to steam generator

Control rod

Insulation to prevent heat loss

Fuel assembly

Carbon steel reactor vessel

Inlets for pressurized water from steam generator

(b)

Coolant flow holes

Fuel assembly

Fuel rods

Fuel pellets (uranium oxide)

Zirconium alloy tube

(c)

•**FIGURE 30.5 Nuclear reactor** (a) The interior of a reactor during refueling. (b) A schematic diagram of a reactor vessel. (c) A fuel rod and its assembly.

nuclei of ordinary water have a tendency to capture neutrons and form deuterium, removing neutrons from the chain reaction. Hence enriched uranium with 3–5% ^{235}U must be used to achieve a critical mass if we use light water as a coolant.

There are reactors other than light water types. Many Canadian reactors use heavy water, D_2O. The advantage is that the deuterium does not readily absorb neutrons, and therefore the fuel can be of lower uranium enrichment. However, D_2O must be separated from normal water, H_2O, requiring energy. (Deuterium, although stable, constitutes less than 0.02% of normally occurring hydrogen.)

The chain reaction rate and therefore the energy output of a reactor are controlled by boron or cadmium **control rods**, which can be inserted into or withdrawn from the reactor core. Cadmium and boron have a very high cross section for absorbing neutrons. By inserting the control rods between the fuel rod assemblies, neutrons are removed from the chain reaction. The control rods are adjusted so that chain reactions release energy at a steady rate. The object is to create a sustainable fission chain reaction in which the average fission produces only one more fission. If even slightly more than one is produced, the reaction can get out of control. If the ratio is less than one, the chain reaction quickly dies out.

In practice, the energy production is monitored electronically, and the control rods are operated by computers that push them in if the power level rises and pull them out if it begins to fall. The operation of the reactor is stabilized by this *feedback* technique. For refueling, or in an emergency, the control rods are fully inserted, and enough neutrons are removed to curtail the chain reaction and shut down the reactor. However, even with the chain reaction shut down, water must circulate to prevent heat build-up due to the continuing decay of radioactive fission products in the fuel rods. If not, damage to the fuel rods can result.

The water flowing through the fuel rod assemblies acts not only as a coolant, but also as a **moderator**. The fission cross section of ^{235}U is largest for slow neutrons (kinetic energies less than 1 eV). The neutrons emitted from a fission reaction, however, are fast neutrons with average energies of 2 MeV. These are slowed down, or their speed is moderated, by collisions. The hydrogen atoms in water are very effective in slowing down neutrons because their masses are nearly equal to the neutron mass. Recall from Chapter 6 that there is maximum energy transfer in a head-on collision of particles of equal mass. Not all collisions are head-on, but it takes only about 20 collisions to moderate fast neutrons down to energies of less than 1 eV.

Note: This result was obtained in Section 6.4.

Other materials, such as graphite (carbon), can be used as moderators. Because of their relatively heavy nuclei, not as much energy is lost per collision. The first experimental proof that a chain reaction was feasible was accomplished in a graphite reactor in 1942 by a team of scientists working with Enrico Fermi on the World War II Manhattan Project. The reactor was called a "pile" because it essentially consisted of a pile of graphite blocks. (A carbon reactor was involved in the 1986 accident at Chernobyl in the former Soviet Union.)

The Breeder Reactor. In general, in a commercial power reactor fuel assembly, the ^{238}U goes along for the ride. However, ^{238}U has an appreciable fission cross section for *fast* neutrons. Since not all the neutrons are slowed by enough collisions with water molecules, ^{238}U reactions can occur, including this conversion via successive beta decays:

$$^{1}_{0}n + {}^{238}_{92}U \rightarrow {}^{239}_{92}U^* \rightarrow {}^{239}_{93}Np + {}^{0}_{-1}e$$

(*fast*)
$$\hookrightarrow {}^{239}_{94}Pu + {}^{0}_{-1}e$$

Note: A breeder reactor doesn't produce something from nothing. It uses the kinetic energy of the neutrons, wasted in a nonbreeder reactor, to create fissionable ^{239}Pu from the nonfissionable ^{238}U component of uranium.

Plutonium-239, with a half-life of 24 000 y, is *fissionable* and serves as additional fuel that prolongs the time to reactor refueling. It is possible to promote actively this conversion of ^{238}U in a reactor by reduced moderation. When, on average, one or more neutrons from ^{235}U fission are absorbed by ^{238}U nuclei to produce ^{239}Pu, the same amount or more of fissionable fuel is produced (^{239}Pu) as is consumed (^{235}U). This is the principle of the **breeder reactor**.

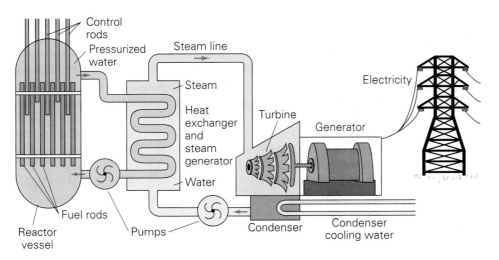

•FIGURE 30.6 Pressurized water reactor The components of a pressurized water reactor. The heat energy from the reactor core is carried away by the circulating water. The water in the reactor is pressurized so that it can be heated to high temperatures for more efficient heat removal. The energy is used to generate steam, which drives the turbine that turns the generator to produce electric energy.

To utilize this fuel in other reactors or in nuclear weapons, we must chemically separate the plutonium from the uranium and fission fragments. This process is not trivial, involving the application of chemical methods to radioactive material. In addition, plutonium is chemically very reactive, so it is likely to contaminate any living tissue with which it comes in contact. It is a natural alpha emitter, and minute amounts of plutonium breathed into lung tissue can cause lung cancer. In the United States, the spent fuel rods from (nonbreeder) power reactors contain significant quantities of fissionable plutonium. Spent fuel rods are primarily stored on-site in borated water pools. (Why boron, and why underwater?)

Developmental work on the breeder reactor in the United States was essentially stopped in the 1970s. France went on to develop operational breeder reactors that provide nuclear fuel for power reactors. France and other nations are highly dependent on nuclear energy, for their electrical energy needs.

Electrical Generation.   The components of a typical pressurized water reactor used in the United States are illustrated in •Fig. 30.6. The heat generated by the controlled chain reaction is carried away by the water surrounding the rods in the fuel assembly. The water in the reactor is pressurized to between 100 and 200 atm so that it can be heated to temperatures over 300°C for more efficient heat removal. The hot water is pumped to a heat exchanger, where the energy is transferred to the water of a steam generator. Notice that the reactor coolant and the exchanger water are in separate, closed systems. (Why?)

High-pressure steam then turns a turbine that operates an electrical energy generator, as would be the case for any nonnuclear power plant. The steam is then cooled and condensed after turning the turbine.

Note: Recall from Section 11.3 that the boiling point of water increases with pressure.

Nuclear Reactor Safety

Nuclear energy is used to generate a substantial amount of the electricity in the world. More than twenty-five countries now produce nuclear-generated electricity. Several hundred nuclear reactor units are in operation throughout the world, with more than 100 units in the United States. With the increasing number of nuclear-generating facilities comes the fear of nuclear accidents and the subsequent release of radioactive materials into the environment.

If the coolant of a light water reactor is lost, the chain reaction would stop because the coolant is also the moderator. Without it, there are very few slow neutrons. However, decay reactions among the fission fragments, some of which have half-lives of hundreds of years, would continue. In such a **LOCA** (loss-of-coolant accident), the fuel rods might then become hot enough (several thousand degrees Celsius) for the cladding (the outside covering) to melt and fracture. Once this oc-

(a)

(b)

•**FIGURE 30.7 Nuclear safety**
(a) An aerial photo showing damage to the reactor at Chernobyl. This accident released large amounts of radioactive materials into the environment, with very dire consequences.
(b) Experimental equipment being installed in an underground cavern to help test its suitability for the storage of radioactive nuclear wastes. The safety of such storage in the long term is still being studied and debated.

curs, the hot, fissioning mass could fall into the water on the floor of the containment vessel and cause a steam explosion, a hydrogen explosion, or both. This might rupture the walls of the containment vessel and allow radioactive fragments into the environment. Even if the walls were not breached, the hot "melt" could burn through the floor of the containment building into the underground environment, eventually reaching ground level and the atmosphere (a situation called the *China syndrome*).

A partial **meltdown** occurred at the Three Mile Island (TMI) generating plant near Middletown, Pennyslvania, in 1979. This was a LOCA in which a small amount of radioactive steam was vented into the atmosphere. Inside one reactor vessel, now sealed, electronic robots have discovered heavy damage to the fuel rods.

The April 1986 nuclear accident at Chernobyl, in the Ukraine, was a meltdown caused by human error but magnified by the inherent instability that results from the use of carbon as a moderator. The reactor had 1660 fuel assemblies encased in 1700 tons of graphite blocks. When the cooling was inadvertently removed, the chain reaction went out of control—something that could not happen in a light water reactor—producing a tremendous rise in temperature. The resulting explosions blew the top off the building (•Fig. 30.7a). When the fuel rods melted, the graphite blocks burned like a massive charcoal barbeque, spewing radioactive smoke into the air. Winds carried this radioactive smoke over much of Europe and over the North Pole into Canada and the United States, where significant amounts of core fission fragments (such as ^{131}I, ^{90}Sr, and ^{137}Cs) were detected.

In western Russian and eastern Europe, there was (and still is) concern over possible health effects. Although small quantities of radioactive iodine are safely used for thyroid diagnostic purposes, large amounts can cause thyroid cancer. Only a few dozen people died as a direct result of the Chernobyl accident, but it is feared that many more will eventually die of cancer related to the radiation.

Even if nuclear reactors operate safely (their safety record, particularly in the United States, is a very good one), there remains the problem of radioactive waste. The fission fragments are radioactive and have long half-lives. As a result, nuclear waste will be a continuing problem for years. The United States recently opened its first waste site in New Mexico. Where and how to seal, safely transport, bury, and guard nuclear waste will be important decisions for many generations to come (Fig. 30.7b).

30.3 Nuclear Fusion

Objectives: To **(a)** explain the fundamental difference between fusion and fission, **(b)** calculate energy releases in fusion reactions, and **(c)** understand how fusion might eventually provide a source of electric energy.

Another type of nuclear reaction that can produce energy is *fusion*. In a **fusion reaction**, light nuclei fuse to form a heavier nucleus, releasing energy in the process ($Q > 0$). A simple fusion reaction—the fusion of two deuterium nuclei ($^{2}_{1}$H), sometimes called a D–D reaction—was examined in Example 30.1. There we showed that this reaction releases 3.27 MeV of energy per fusion.

Another example is the fusion of deuterium and tritium (a D–T reaction):

$$\underset{(2.014\,102\ u)}{^{2}_{1}\text{H}} \quad + \quad \underset{(3.016\,049\ u)}{^{3}_{1}\text{H}} \quad \rightarrow \quad \underset{(4.002\,603\ u)}{^{4}_{2}\text{He}} \quad + \quad \underset{(1.008\,665\ u)}{^{1}_{0}\text{n}}$$

Using the given masses, you should be able to show there is a release of 17.6 MeV per fusion.

Fusion reactions release very little energy in comparison to the more than 200 MeV released from a fission reaction. However, equal-mass samples of hydrogen and uranium have many, many more hydrogen nuclei than uranium nuclei. Thus, per kilogram, the fusion of hydrogen gives almost three times the energy released from the fission of uranium.

In a sense, our lives depend on nuclear fusion because it is the source of energy for stars, including the Sun. In the initial stages of its life, a star's light energy output results from the fusion of hydrogen into helium. One such sequence of fusion reactions that is believed to be responsible for the Sun's energy output is:

$$_1^1H + _1^1H \rightarrow _1^2H + _{+1}^0e + \nu$$

(where ν represents a neutrino, a particle to be discussed in the next section). Then protons and these deuterons fuse:

$$_1^1H + _1^2H \rightarrow _2^3He + \gamma$$

and finally two of these ^3He nuclei fuse:

$$_2^3He + _2^3He \rightarrow _2^4He + _1^1H + _1^1H$$

The net effect of this sequence, called the *proton–proton cycle*, is that four protons ($_1^1H$) combine to form one helium nucleus ($_2^4He$) plus two positrons ($_{+1}^0e$), two gamma rays (γ), and two neutrinos (ν) and energy release:

$$4(_1^1H) \rightarrow _2^4He + 2(_{+1}^0e) + 2\gamma + 2\nu + \text{energy} \qquad (Q = 24.7 \text{ MeV})$$

In a star such as our Sun, the gamma ray photons (Compton) scatter off nuclei on their way to the surface. Each scattering results in a reduction in energy until each photon has only a few electron volts of energy. Thus, on reaching the surface, they have become visible light. In our Sun, fusion involves only the central 10% of the Sun's mass. Even so, it has been going on for about 5 billion years and should continue, approximately as it is, for another 5 billion years. After that, further fusion processes in the Sun's central core will create heavier elements such as lithium and carbon from the helium. In the next Example, consider the enormous number of fusion reactions required to power the Sun.

Example 30.3 ■ Still Going? The Fusion Power of the Sun

Incoming sunlight falls on the Earth's upper atmosphere at about $1.40 \times 10^3 \text{ W/m}^2$. Assume that the Sun's energy is produced by the proton–proton cycle. Assume also that only 10% of its total hydrogen is available for fusion. Calculate (a) the mass loss of the Sun per second, (b) the mass of hydrogen lost per second, and (c) the expected lifetime of the Sun.

Thinking It Through. (a) To find the mass loss rate, we first need to calculate the total power output of the Sun. This is best done by imagining the power flow through an imaginary sphere—centered on the Sun and with a radius the size of the Earth's orbit—and then calculating its area. From that the solar power per square meter, we can get the Sun's total power output. The conversion to the Sun's mass loss rate is then based on mass–energy equivalence (Table 29.3). (b) We can convert the mass loss rate to the hydrogen loss rate because each fusion reaction is a net loss (conversion) of four protons into one helium nucleus. (c) We must find the time it takes for total conversion, assuming the rate is constant, from the total useable mass and the rate of conversion.

Solution. The data we need are:

Given: $R_{E-S} = 1.50 \times 10^8 \text{ km}$
$\qquad = 1.50 \times 10^{11} \text{ m}$
$\qquad M_{Sun} = 2.00 \times 10^{30} \text{ kg}$
$\qquad P_{Sun}/A = 1.40 \times 10^3 \text{ W/m}^2$

Find: (a) $\Delta m/\Delta t$ (overall mass loss rate)
(b) $\Delta m_H/\Delta t$ (hydrogen mass loss rate)
(c) Δt (Sun's lifetime)

(a) The surface area of our imaginary sphere (which will intercept all of the Sun's energy) is

$$A = 4\pi R_{E-S}^2 = 4\pi (1.5 \times 10^{11} \text{ m})^2 = 2.83 \times 10^{23} \text{ m}^2$$

Thus the total power output of the Sun is

$$P_{Sun} = (1.40 \times 10^3 \text{ W/m}^2)(2.83 \times 10^{23} \text{ m}^2) = 3.96 \times 10^{26} \text{ W} = 3.96 \times 10^{26} \text{ J/s}$$

To find the equivalent mass loss rate, we first convert this power to MeV per second:

$$\frac{3.96 \times 10^{26} \, \text{J/s}}{1.60 \times 10^{-13} \, \text{J/MeV}} = 2.48 \times 10^{39} \, \text{MeV/s}$$

Then, using the mass–energy equivalence, we have the mass loss rate

$$\frac{\Delta m}{\Delta t} = \frac{(2.48 \times 10^{39} \, \text{MeV/s})(1.66 \times 10^{-27} \, \text{kg/u})}{931.5 \, \text{MeV/u}} = 4.42 \times 10^{9} \, \text{kg/s}$$

(b) This mass loss is a result of many fusions per second, each losing a miniscule amount of mass:

$$\Delta m \text{ per fusion} = 4m_{\text{H}} - m_{\text{He}} = 4(1.007\,825 \, \text{u}) - 4.002\,603 \, \text{u} = 0.028\,697 \, \text{u}$$

or

$$\Delta m \text{ per fusion} = (0.028\,697 \, \text{u})(1.66 \times 10^{-27} \, \text{kg/u}) = 4.76 \times 10^{-29} \, \text{kg/fusion}$$

Thus the number of fusions occurring per second is

$$\text{fusions per second} = \frac{4.42 \times 10^{9} \, \text{kg/s}}{4.76 \times 10^{-29} \, \text{kg/fusion}} = 9.29 \times 10^{37} \, \text{fusions/s}$$

Each fusion eliminates four protons, so the amount of *hydrogen* (H) lost per second is

$$\Delta m_{\text{H}}/\Delta t = (9.29 \times 10^{37} \, \text{fusions/s})[4(1.67 \times 10^{-27}) \, \text{kg of H lost/fusion}]$$

$$= 6.21 \times 10^{11} \, \text{kg of H lost/s}$$

(c) At this rate, 10% of the Sun's hydrogen supply (10% of its total mass) would last

$$\Delta t = \frac{(0.10)(2.00 \times 10^{30} \, \text{kg})}{6.21 \times 10^{11} \, \text{kg/s}} = 3.22 \times 10^{17} \, \text{s} \approx 10 \text{ billion y}$$

Since we believe the Sun formed about 5 billion years ago, it is middle-aged.

Follow-up Exercise. In this Example, how many kilograms of helium are produced in the Sun per second (to three significant figures)?

Fusion as a Source of Energy Produced on the Earth

In several ways, fusion appears to be the ideal energy source of the future. Enough deuterium exists in the oceans, in the form of heavy water, to supply our needs for centuries. Controlled fusion does not depend on a chain reaction, so there is less danger of the release of radioactive material. Also, fusion products tend to have relatively short half-lives. For example, tritium has a half-life of only 12.3 y, compared to hundreds or thousands of years for fission products.

However, there are many unresolved technical problems in the use of fusion to produce electric energy. A primary problem is that very high temperatures are needed to *initiate* fusion reactions, because energy is needed to overcome the Coulomb repulsion between the fusing nuclei. Temperatures on the order of millions of degrees are needed to initiate *thermonuclear fusion reactions,* as they are called. The problem is in confining sufficient energy in a reaction region to maintain these high temperatures. Because of this, practical fusion reactors have not been achieved. Uncontrolled fusion has been demonstrated in the form of the hydrogen (H) bomb. In this case, the fusion reaction is initiated by an implosion created by a small atomic (fission) bomb. This provides the necessary density and temperatures to begin the fusion process.

At such high temperatures, electrons are stripped from their nuclei and a gas of positively charged ions and free, negatively charged electrons is created. Such a gas of charged particles is called a **plasma**. Plasmas have a number of special physical properties. They are sometimes referred to as the fourth phase of matter, a term used in 1879 by William Crookes, an English chemist, who generated plasmas in gas-discharge tubes, or Crookes tubes.

Plasma: a fourth phase of matter

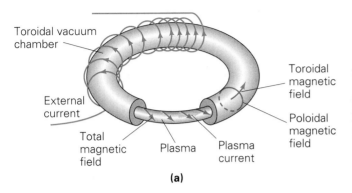

•**FIGURE 30.8 Magnetic confinement** Magnetic confinement is one method by which controlled nuclear fusion might be achieved. **(a)** Tokamak configuration, showing the *B* field generated by external currents. The magnetic field confines the plasma in the ring. **(b)** The Princeton Tokamak Fusion Test Reactor (TFTR).

(a)

(b)

The technological problem involved in the confinement of a plasma is being approached in two ways: magnetic confinement and inertial confinement. Since a plasma is a gas of charged particles, it can be controlled and manipulated by using electric and magnetic fields. In **magnetic confinement**, magnetic fields are used to hold the plasma in a confined space, a so-called magnetic bottle (see Fig. 19.31b).

Once a plasma is confined, electric fields are used to produce in it electric currents that raise its temperature. Temperatures of 100 million kelvins have been achieved in a design that uses a donut-shape called a *tokamak* (•Fig. 30.8).

To initiate fusion, in addition to a very high temperature, there are also minimum requirements on the plasma density and confinement time. The trick is to put all these together in a fusion reactor. The generation of several megawatts of power for less than a second in a magnetically confined plasma is typical of the best results so far. But for commercial applications, much higher power levels must be delivered continuously.

Inertial confinement depends on implosion techniques. Hydrogen fuel pellets are either dropped or positioned in a reactor chamber (•Fig. 30.9). Pulses of laser, electron, or ion beams are then used to implode the pellet, producing compression and high densities and temperatures. Fusion would occur if the pellet stayed together for a sufficient time, which depends on its inertia (hence the name, inertial confinement). At this time, lasers and particle beams are not powerful enough to produce sustainable fusion by this means. Practical energy production from fusion is not expected until well into the twenty-first century, if at all.

30.4 Beta Decay and the Neutrino

Objectives: To (a) explain why the neutrino is necessary to account for observed beta decay data, (b) specify some of the physical properties of neutrinos, and (c) write complete beta decay equations.

As described in Section 29.2, beta decay *appears* to be a two-body decay process in which certain unstable nuclei emit an electron or a positron. Examples of β^- and β^+ decays are

$$\underset{(14.003242\ u)}{^{14}_{6}\text{C}} \quad \rightarrow \quad \underset{(14.003074\ u)}{^{14}_{7}\text{N}} \quad + \quad ^{0}_{-1}\text{e} \qquad \beta^- \ decay$$

$$\underset{(13.005739\ u)}{^{13}_{7}\text{N}} \quad \rightarrow \quad \underset{(13.003355\ u)}{^{13}_{6}\text{C}} \quad + \quad ^{0}_{+1}\text{e} \qquad \beta^+ \ decay$$

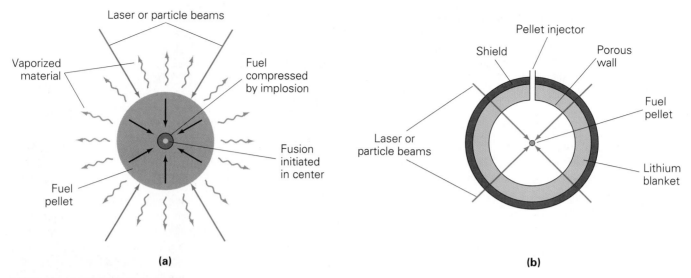

Laser or particle beams

Vaporized material

Fuel compressed by implosion

Fuel pellet

Fusion initiated in center

(a)

Pellet injector

Shield

Porous wall

Fuel pellet

Laser or particle beams

Lithium blanket

(b)

(c)

•**FIGURE 30.9 Inertial confinement** Controlled nuclear fusion might be achieved by inertial confinement. **(a)** A fuel pellet implosion. The compression is enhanced by the vaporization of the outer shell material of the pellet. **(b)** A possible fusion reactor cavity. The lithium blanket captures neutrons from the fusion reactions, producing tritium, which can then be used as a fuel. **(c)** The world's most powerful laser, Nova, created a brief burst of fusion in a tiny fuel capsule. In the 50 trillionths of a second it lasted, the reaction gave off ten trillion neutrons.

However, more than this is apparently going on in these processes: When analyzed in detail, they appear to violate several conservation laws.

One of the difficulties arises with energy. The energy released in the above β^- process can be calculated from the mass difference of the nuclei:*

$$\Delta m = 14.003\,242 \text{ u} - 14.003\,074 \text{ u} = 0.000\,168 \text{ u}$$

and

$$Q = (0.000\,168 \text{ u})(931.5 \text{ MeV/u}) = 0.156 \text{ MeV}$$

Thus, the decay reaction has a Q value or disintegration energy of 0.156 MeV. The electron, being a light particle, would be expected to be emitted always with a kinetic energy of 0.156 MeV. However, this is not what happens.

When the kinetic energies of the electrons from beta decay are measured, a *continuous spectrum* of energies is observed up to $K_{max} \approx Q$ as shown in •Fig. 30.10. (See Conceptual Example 30.4.) That is, most electrons are emitted with less kinetic energy than expected. This would appear to be a violation of the conservation of energy, since not all the energy is accounted for.

Nor is this the only difficulty. Observations of individual beta decay processes show that the emitted electron and the daughter nucleus from a stationary parent nucleus do not always leave the disintegration site in opposite directions. Thus, they appear to violate the conservation of linear momentum.

There is also an apparent violation of the conservation of angular momentum. Nucleons and electrons have spin quantum numbers of $\frac{1}{2}$ (Section 28.3). For a nucleus with an even number of nucleons (such as ^{14}C), the quantum number of the total angular momentum will be an integer, since an even number of spin $\frac{1}{2}$ particles is present. When an electron is created in the decay process, an odd number of spin $\frac{1}{2}$ particles is present, and the quantum number of the total angular momentum after the decay will be a half-integer. Therefore, the total angular momentum would not be the same before and after spontaneous decay.

What, then, is the problem? We would hope that our conservation laws are not invalid. An alternative is that these apparent violations are telling us something about nature that we do not yet recognize. The apparent difficulties can be resolved if we assume that, in addition to the electron, another unobserved particle of intrinsic spin $\frac{1}{2}$ were emitted in beta decay.

*Using the atomic mass of the daughter ^{14}N (with seven electrons) is necessary to take into account the emitted electron since the ^{14}N resulting from beta decay would have only the six electrons that orbit the parent ^{14}C nucleus.

This explanation and the existence of such a particle were first proposed in 1930 by Wolfgang Pauli. The particle was christened the **neutrino** (meaning "little neutral one"). For charge to be conserved in beta decay, the new particle had to be electrically neutral. The fact that the neutrino had not been observed suggested that it must interact very weakly with matter. Later, scientists discovered that the neutrino interacts with matter through a second nuclear force, much weaker than the strong force, called the *weak interaction,* or the weak nuclear force. (This force is described in Section 30.5.)

Observations of beta decay suggest that the neutrino has zero mass and therefore travels with the speed of light. It also has linear momentum p with a total energy $E = pc$ and has a spin quantum number of $\frac{1}{2}$. In 1956, a particle with these properties was detected experimentally, and the existence of the neutrino was established. The nuclear equations for the previous beta decays can now be written correctly as

$$^{14}_{6}\text{C} \rightarrow {}^{14}_{7}\text{N} + {}^{0}_{-1}\text{e} + \bar{\nu}_{e}$$

and

$$^{13}_{7}\text{N} \rightarrow {}^{13}_{6}\text{C} + {}^{0}_{+1}\text{e} + \nu_{e}$$

where the Greek letter ν (nu) is the symbol for the neutrino. The symbol with a bar over it represents an *antineutrino.* The "overbar" notation is a common way of indicating an antiparticle. Two different neutrinos are associated with common beta decay. In general, a neutrino is emitted in β^+ decay, and an antineutrino is emitted in β^- decay. The subscript "e" identifies the neutrinos as associated with electron–positron beta decay. As we will see in Section 30.5, there are other types of neutrinos.

The *anti*neutrino is fundamentally associated with the decay of a free neutron:

$$\text{n} \rightarrow \text{p} + \text{e}^{-} + \bar{\nu}_{e}$$

The neutrino ν_{e} is associated with the decay of a proton into a neutron and a positron. However, a free proton cannot decay by positron emission because that would violate the conservation of energy—the products would be more massive than the proton. But because of binding energy effects, this can, and does, occur during β^+ decay of a *nucleus* with the emission of a neutrino.

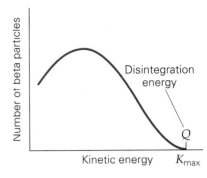

•**FIGURE 30.10 Beta ray spectrum** For a typical beta decay process, most beta particles are emitted with $K < Q$, leaving energy unaccounted for.

Conceptual Example 30.4

Having It All? Maximum Kinetic Energy in Beta Decay

Consider the beta decay of an unstable nucleus initially at rest. As a specific example, take the classic decay $^{14}_{6}\text{C} \rightarrow {}^{14}_{7}\text{N} + {}^{0}_{-1}\text{e} + \bar{\nu}_{e}$. We know that Q is the amount of energy released during the decay. Is it possible for the maximum kinetic energy of the emitted beta particle to be *exactly* equal to Q?

Reasoning and Answer. In beta decay, as in all interactions, we expect total energy and momentum to be conserved. Here energy is conserved, and the loss of mass shows up as the gain of kinetic energy of the product particles. Since there is initially no momentum, there should be no total momentum after the decay. Even in the simplest case, if the daughter nucleus does not recoil, the neutrino must go off in a direction exactly opposite that of the beta particle to preserve zero momentum. Thus, even in the most favorable case, the beta particle cannot get all of Q as its kinetic energy. Some amount (however small) must go to either the neutrino or the daughter nucleus (or, in general, to both) to cancel the momentum of the beta particle. Thus for the beta particle, $K_{max} < Q$.

Follow-up Exercise. A typical energy spectrum of emitted beta particles is shown in Fig. 30.10. What this graph indicates is that the most likely energy of the beta particles is somewhere between the maximum allowable and zero. What would the decay product trajectories look like after the decay if the emitted beta particle had almost no kinetic energy? (There is a small probability that this will happen.)

30.5 Fundamental Forces and Exchange Particles

Objectives: To (a) understand the quantum mechanical description of forces and (b) classify the various forces according to their strengths, properties, ranges, and virtual particles.

The forces involved in everyday activities are complicated because of the large numbers of atoms that make up ordinary objects. Contact forces between two hard objects, for example, are due to the electromagnetic forces between atoms. Looking at the fundamental interactions between particles makes things simpler. On this level, there are only four known **fundamental forces**: the *gravitational force*, the *electromagnetic force*, the *strong nuclear force*, and the *weak nuclear force*.

The most familiar of these are the gravitational and electromagnetic forces. Gravity acts between all particles, while the electromagnetic force is restricted to charged particles.* Both forces obey an inverse-square relationship: They decrease with increasing particle separation distance and therefore have an infinite range.

Classically, the action-at-a-distance model of these forces uses the concept of a field. For example, a charged particle is considered to be surrounded by an electric field, through which it *interacts* with other charged particles. Quantum mechanics, however, provides an alternative description of how forces are transmitted. This process is pictured as an exchange of particles, analogous to you and another person interacting by tossing a ball back and forth.

The creation of such particles would seem to violate the conservation of energy. However, within the limitations of the uncertainty principle (Section 28.4), a particle can be created for a *short time* with no outside energy input. Thus, over *long* time intervals, energy is conserved. For *extremely* short time intervals, the uncertainty principle *permits* a large *uncertainty* in energy ($\Delta E \propto 1/\Delta t$), creating a particle is allowed, and energy conservation can be briefly violated. The created particle is absorbed before it is ever detected, so we never observe energy nonconservation. A particle created and absorbed in such a manner is called a **virtual particle.** In this sense, *virtual* means "undetected."

The fundamental forces are considered to be carried by virtual **exchange particles.** The exchange particle for the four forces differ in mass. The greater the mass of the particle, the greater the energy ΔE required to create it and the shorter time Δt it exists. Since a massive particle can exist for only a short time, the distance it can travel and hence the range of interaction for the associated force would be small. That is, *the range of a force associated with an exchange particle is inversely proportional to the mass of that exchange particle.*

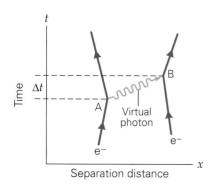

•**FIGURE 30.11 Feynman diagram of an electron–electron interaction** The interacting electrons undergo a change in energy and momentum due to the exchange of a virtual photon, which is created at A and absorbed at B in a time (Δt) that is consistent with the uncertainty principle.

Photon: the exchange particle of the electromagnetic force

The Electomagnetic Force and the Photon

The exchange particle of the electromagnetic force is a (virtual) **photon.** As a "massless" particle, it has an infinite range, as is required for the electromagnetic force. A particle exchange is visualized by using a *Feynman diagram* as in •Fig. 30.11. This graph shows how the exchange idea gives rise to the repulsion between two colliding electrons. Such space-time diagrams are named after American scientist and Nobel Prize winner Richard Feynman (1918–1988), who used them to analyze electrodynamic interactions in his theory named *quantum electrodynamics.* The important points are the vertices of the diagram. One electron is considered to create a virtual photon at point A and the other electron to absorb it at point B. Each of the electrons undergoes a change in energy and in momentum (including their direction) by virtue of the photon exchange and resulting force.

*Relativity shows that both electric and magnetic forces are components of a single force—the electromagnetic force.

The Strong Nuclear Force and Mesons

Using the concept of exchange particles, Japanese physicist Hideki Yukawa (1907–1981) proposed in 1935 that the short-range strong nuclear force between two nucleons is associated with an exchange particle called the **meson**. An estimate of the meson's mass can be made from the uncertainty principle. If a nucleon were to create a meson, the conservation of energy would be (undetectably) violated by an amount of energy at least as great as the meson mass, or

Meson: the exchange particle of the strong nuclear force

$$\Delta E = (\Delta m)c^2 = m_m c^2$$

where m_m is the mass of the meson.

By the uncertainty principle, the meson would have to be absorbed in the exchange process in a time on the order of

$$\Delta t = \frac{h}{2\pi \Delta E} = \frac{h}{2\pi m_m c^2}$$

In this time, the meson could travel a distance R (for range) not greater than

$$R = c\,\Delta t = \frac{h}{2\pi m_m c} \qquad (30.5)$$

Taking this distance to be the experimental range of the nuclear force ($R \approx 1.4 \times 10^{-15}$ m) and solving for m_m gives

$$m_m \approx 270 m_e$$

where m_e is the electron's mass. Thus, if Yukawa's meson exists, it should have a mass about 270 times that of an electron.

Virtual mesons of an exchange process cannot be directly observed. But if sufficient energy were involved in the collision of nucleons, real mesons might be created through the energy available in the collision. These real mesons could then be detected. At the time of Yukawa's prediction, there were no known particles with masses between that of the electron (m_e) and that of the proton ($m_p = 1836 m_e$).

In 1936, Yukawa's prediction seemed to come true when a new particle with a mass of about $200 m_e$ was discovered in cosmic rays. Originally called the μ (mu) meson, and now just **muon**, the particle was shown to have two charged varieties, $\pm e$, with a mass of

$$m_{\mu^\pm} = 207 m_e$$

However, further investigations showed that the muon did not behave like the strongly interacting particle of Yukawa's theory. In particular, the interaction of muons with matter was very weak. The muons from cosmic radiation could penetrate a large mass of material, as evidenced by their detection in deep mines.

This situation was a source of controversy and confusion for years. But in 1947, more particles in the appropriate mass range were discovered in cosmic radiation. These particles (one positive, one negative, and one with no charge) were called π (pi) mesons (primary mesons), and now more commonly **pions**. Measurement showed the masses of the pions to be

$$m_{\pi^\pm} = 273 m_e \qquad \text{and} \qquad m_{\pi^\circ} = 264 m_e$$

Moreover, pions interact strongly with matter. The pion fulfilled the requirements of Yukawa's theory, and this meson was generally accepted as the particle primarily responsible for the transmission of the strong nuclear force. Feynman diagrams for some nucleon–nucleon interactions are shown in •Fig. 30.12.

Free mesons are unstable. For example, the π^+ particle decays in about 10^{-8} s into a muon and another type of neutrino:

$$\pi^+ \rightarrow \mu^+ + \nu_\mu$$

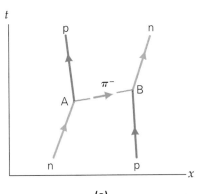

(a)

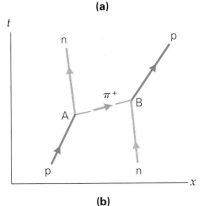

(b)

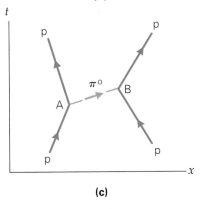

(c)

•**FIGURE 30.12 Feynman diagrams of nucleon–nucleon (strong nuclear force) interactions** Some possible nucleon–nucleon interactions through the exchange of virtual pions. **(a)** An n–p reaction, **(b)** a p–n reaction, and **(c)** a p–p scattering. (Diagrams (a) and (b) are called "charge exchange" interactions. Why?)

The ν_μ is a *muon neutrino,* which differs from the electron neutrino produced in beta decay. The muons can also decay into positrons and electrons with the emission of both types of neutrinos. For example, the positive muon decay scheme is

$$\mu^+ \rightarrow e^+ + \nu_e + \bar{\nu}_\mu$$

The Weak Nuclear Force and the W particle

The discrepancies in beta decay discussed in the preceding section led to another discovery. Electrons and neutrinos are emitted from unstable nuclei, but there was evidence that they did not exist *inside* the nucleus before the decay. Enrico Fermi proposed that these particles did in fact *not* exist before being emitted but were created in the decaying radioactive nucleus. For β^- decay, this would mean that a neutron was in some way *transmuted* into three particles:

$$n \rightarrow p^+ + e^- + \bar{\nu}_e$$

Experiments confirmed that free neutrons disintegrate by this decay scheme, with a half-life of 10.4 min. But which force could cause a neutron to disintegrate in this manner? Since the neutron was observed outside the nucleus (free), none of the known forces, including the strong nuclear force, was applicable. It was reasoned that some other force must be acting in beta decay. Decay rate measurements indicated that the force was extraordinarily weak—weaker than the electromagnetic force but still much stronger than the gravitational force. Thus, the **weak nuclear force** was discovered.

It was originally thought that the weak interaction was localized, without any measurable range. We now know that the weak force has a range of about 10^{-17} m. In terms of an exchange particle, the virtual carriers must be much more massive than the pions of the strong nuclear force. The virtual exchange particles of the weak nuclear force were called **W particles.** *W (weak)* particles have masses about 100 times that of a proton, which explains the extremely short range of the force. The existence of the W particle was confirmed in the 1980s when accelerators were built with enough energy to create the first real W particles.

The weak force is the only force that acts on neutrinos, which explains why they are so difficult to detect. Research has shown that the weak force is involved in the transmutation of other subatomic particles. In general, the weak force is limited to transmuting the identities of particles within the nucleus. The only way it manifests its existence in the outside world is through the emitted neutrinos. For example, the Sun's fusion reactions create neutrinos that continuously pass through the Earth.

One highly noticeable but infrequent announcement of the weak force at work occurs during the explosion of a supernova. In a *supernova,* the collapse of the core of an aging star gives rise to a huge release of neutrinos. In a relatively "nearby" supernova (a mere 1.5×10^{18} km away), observed in 1987, a burst of neutrinos was detected at essentially the same time as the light flash. Since both signals arrived after traveling extremely long distances, this result can be used to set an upper limit on the mass of the neutrino. If it had mass, its speed would be less than that of light and it would arrive later.

The Gravitational Force and Gravitons

The exchange particles of the gravitational force are called **gravitons**. There is still no firm evidence of the existence of this massless particle. (Why must gravitons be massless?) Ongoing experiments to detect the graviton have as yet proven unsuccessful due to the relative weakness of its interaction. A comparison of the relative strengths of the four fundamental forces is given in Table 30.2.

W particle: an exchange particle of the weak nuclear force

Graviton: the exchange particle of the gravitational force

---

*The weak force is actually carried by *three* exchange particles: W^+, W^-, and Z^0 (neutral).

TABLE 30.2 Fundamental Forces

Force	Relative Strength	Action Distance	Exchange Particle	Particles with Interaction
Strong nuclear	1	Short range ($\approx 10^{-15}$ m)	Pion (π meson)	Hadrons*
Electromagnetic	10^{-3}	Inverse square (infinite)	Photon	Electrically charged
Weak nuclear	10^{-8}	Extremely short range ($\approx 10^{-17}$ m)	W particle†	All
Gravitational	10^{-45}	Inverse square (infinite)	Graviton	All

*Hadrons are particles discussed in Section 30.6.
†Three particles are involved, as described in Section 30.8.

30.6 Elementary Particles

Objectives: To (a) classify the various elementary particles into families and
(b) understand the different properties of the various families of elementary particles.

The fundamental building blocks of matter are referred to as **elementary particles**. Simplicity reigned when it was thought that an atom was an indivisible particle and therefore was *the* elementary particle. Early in the twentieth century, the proton, neutron, and electron were discovered to be constituents of atoms. It was then hoped that these three are Nature's elementary particles. However, scientists now know of a huge variety of subatomic particles and are working to simplify and reduce this list to a smaller set of truly elementary particles—building blocks for all the other "composite" particles.

Several systems classify elementary particles on the basis of their various properties. One classification uses the distinction of nuclear force interactions. Particles that interact via the weak nuclear force but not the strong force are called **leptons** ("light ones"). The lepton family includes the electron, muon, and their neutrinos. Other particles, called **hadrons**, are the only ones to experience or interact by the strong nuclear force. These include the proton, neutron, and pion. Let's look briefly at the lepton and hadron families.

Leptons: weak force interactions but no strong interactions

Hadrons: strong force interactions

Leptons

The most familiar lepton is the electron. It is the only lepton that exists naturally in atoms. There is no evidence of any internal structure, at least down to the measurement limit of 10^{-17} m. Thus, at present the electron is considered to be a point particle.

Muons were first observed in cosmic rays. They are electrically charged and about 200 times more massive than an electron. Since they also appear not to have any internal structure, they are sometimes called heavy electrons. Muons are unstable and decay in about 2×10^{-6} s, according to the following scheme:

Note: Muon decay was treated in Examples 26.2 and 26.4 as evidence of time dilation.

$$\mu^- \rightarrow e^- + \bar{\nu}_e + \nu_\mu$$

A third charged lepton is known as a tau (τ^-) particle, or **tauon.** It has a mass about twice that of a proton. The electron, muon, and tauon are negatively charged, have no apparent internal structure, and have positively charged antiparticles.

The remaining leptons are neutrinos, which are electrically neutral. Neutrinos are present in cosmic rays and are emitted by the Sun and by some radioactive decays. Neutrinos appear to have no mass and thus travel at the speed of light. They feel neither the electromagnetic force nor the strong force and so pass through matter almost as if it weren't there.

There are three types of neutrinos, each associated with a different charged lepton ($e^\pm$, $\mu^\pm$, $\tau^\pm$). They are named, not surprisingly, the electron neutrino (ν_e), the muon neutrino (ν_μ), and the tau neutrino (ν_τ). There is an antineutrino for each of these, for a total of six different neutrinos.

This completes the list of leptons. With a total of six leptons plus their antiparticles, there are twelve different leptons in all. Current theories predict that there should be no others.

Hadrons

Another family of elementary particles is the hadrons. All hadrons interact by the strong force, the weak force, and gravity. The electrically charged members can also interact by the electromagnetic force.

The hadrons are subdivided into *baryons* and *mesons*. **Baryons** include the familiar nucleons—the proton and neutron. They are distinguished from mesons in that they possess half-integer intrinsic spin ($\frac{1}{2}$ or $\frac{3}{2}$). Except for the stable proton, baryons decay into products that eventually include a proton (recall, for example, the beta decay of a neutron into a proton, from Section 29.2).

Mesons, which include pions, have integer spin values (0 or 1) and eventually decay into leptons and photons. For example, the neutral pion decays into two gamma rays (why must there be two?).

The large number of hadrons suggests that they may be composites of other truly elementary particles. Some help came to sorting out the hadron "zoo" in 1963 when Murray Gell-Mann and George Zweig of Caltech put forth the *quark theory*, which we discuss in Section 30.7.

Leptons and hadrons and their properties are summarized in Table 30.3.

TABLE 30.3 Some Elementary Particles and Their Properties

Family Name	Particle Name	Particle Symbol	Antiparticle Symbol	Rest Energy (MeV)	Lifetime* (s)
Lepton	Electron	e^-	e^+	0.511	stable
	Muon	μ^-	μ^+	105.7	2.2×10^{-6}
	Tauon	τ^-	τ^+	1784	$\approx 3 \times 10^{-13}$
	Electron neutrino	ν_e	$\overline{\nu}_e$	0†	stable
	Muon neutrino	ν_μ	$\overline{\nu}_\mu$	0†	stable
	Tauon neutrino	ν_τ	$\overline{\nu}_\tau$	0†	stable
Hadron					
Mesons	Pions	π^+	π^-	139.6	2.6×10^{-8}
		π^0	same	135.0	8.4×10^{-17}
	Kaons	K^+	K^-	493.7	1.2×10^{-8}
		K^0	$\overline{K}^0$	497.7	8.9×10^{-11}
Baryons	Proton	p	$\overline{p}$	938.3	stable (?)§
	Neutron	n	$\overline{n}$	939.6	9×10^2
	Lamba	Λ^0	$\overline{\Lambda}^0$	1116	2.6×10^{-10}
	Sigma	Σ^+	$\overline{\Sigma}^-$	1189	8.0×10^{-10}
		Σ^0	$\overline{\Sigma}^0$	1192	6×10^{-20}
		Σ^-	$\overline{\Sigma}^+$	1197	1.5×10^{-10}
	Xi	Ξ^0	$\overline{\Xi}^0$	1315	2.9×10^{-10}
		Ξ^-	$\overline{\Xi}^+$	1321	1.6×10^{-10}
	Omega	Ω^-	Ω^+	1672	8.2×10^{-11}

*Lifetimes are expressed to two significant figures or less.

†Neutrinos have either zero mass or very small masses. Experiments yield only upper limits; for example, it is known that the mass of the electron neutrino is less than 7×10^{-6} MeV.

§Electroweak theory predicts that the proton is unstable with a half-life of 1000 trillion times the age of the universe.

30.7 The Quark Model

Objectives: To (a) become familiar with the quark model and quark properties, and (b) understand how the quark model accounts for the properties of baryons and mesons.

Gell-Mann and Zweig proposed that hadrons are *not* elementary particles in the sense that they are not fundamental building blocks. Hadrons, they theorized, are composite particles composed of truly elementary (fundamental) particles. They named these particles **quarks** (taken from James Joyce's *Finnegan's Wake*). However, Gell-Mann and Zweig noted that since leptons and photons are essentially point particles, they *were* likely to be truly elementary particles with no internal structure.

Noting that some hadrons are electrically charged, Gell-Mann and Zweig reasoned that quarks must also possess charge. Their initial **quark model** consisted of three different quarks (with fractional charges) and their antiparticles (antiquarks). Table 30.4 shows that to account for hadrons found later, the list of quarks had to be expanded to include six types.

However, in 1963 three quarks were all that was needed to construct the hadrons known at that time. These quarks were named the *up* quark (*u*), the *down* quark (*d*), and the *strange* quark (*s*). By combining three quarks, the relatively heavy hadrons, the baryons (whose name means "heavy ones") could be built. Quark–antiquark pairs could account for the lighter hadrons, the mesons. Thus Gell-Mann and Zweig had proposed a radical new idea:

> Quarks are the fundamental particles of the hadron family.

Since several quarks have to combine to give the charge on the hadron, it was clear that they have fractions of an electron charge, *e*. Thus the theory proposed that *u*, *d*, and *s* quarks have charges of $+\frac{2}{3}e$, $-\frac{1}{3}e$, and $-\frac{1}{3}e$, respectively. The antiquarks, designated by overbars, such as $\bar{u}$, have opposite charges (for example, the charge on the $\bar{u}$ is $-\frac{2}{3}e$). Thus three quark combinations could produce any baryons. For instance, the quark composition of the proton and neutron would be *uud* and *udd*, respectively. Mesons could be constructed from various combinations of a quark and an antiquark, such as $u\bar{d}$ for the positive pion, π^+ (•Fig. 30.13).

TABLE 30.4 Types of Quarks*

Name	Symbol	Charge
Up	*u*	$+\frac{2}{3}e$
Down	*d*	$-\frac{1}{3}e$
Strange	*s*	$-\frac{1}{3}e$
Charm	*c*	$+\frac{2}{3}e$
Top (Truth)	*t*	$+\frac{2}{3}e$
Bottom (Beauty)	*b*	$-\frac{1}{3}e$

*Antiquarks are designated by an overbar and have the opposite charge.

Note: The three original quark names (up, down, and strange) are arbitrary and therefore should not be taken literally. They serve simply to identify the quarks.

Note: Three quarks in proper combination account for all known baryons. Quark–antiquark pairs in proper combination account for all known mesons.

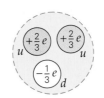 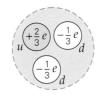

Proton (*uud*)
$q = +e$

Neutron (*udd*)
$q = 0$

(a)

Conceptual Example 30.5

Building Mesons: Quark Engineering, Inc.

Using Table 30.4, explain why it is not possible to build a meson from two quarks.

Reasoning and Answer. In the Table, notice that all the positively charged quarks have a charge of $+\frac{2}{3}e$. Thus any two of them would add up to a meson charge of $+\frac{4}{3}e$, which is not an integral number of electron charges *e*, contrary to observations. Similar reasoning holds if we use two negatively charged quarks: Since they all have the same charge $(-\frac{1}{3}e)$, any two of them add up to a total charge of $-\frac{2}{3}e$. But again, particles are never observed with fractional charge. Finally, consider combining one positively charged quark and one negatively charged one. The net charge of this combination is $+\frac{1}{3}e$, also not in agreement with observation. Thus no combination of two quarks (or two antiquarks) can be used to produce a meson.

Follow-up Exercise. The antiparticle of the positive pion (π^+) is the negative pion (π^-). What is the quark structure of the negative pion? Show that it is composed of the antiquarks of the quarks that make up the positive pion.

The discovery of new elementary particles in the 1970s led to the addition of the last three quark types: *charm* (*c*); *top*, or *truth* (*t*); and *bottom*, or *beauty* (*b*). Today there is firm experimental evidence of the existence of all six and their antiparticles. How many more such particles will be needed to keep up with our growing "zoo" of elementary particles? The present picture includes the following

Positive pion, π^+ ($u\bar{d}$)
$q = +e$

(b)

•**FIGURE 30.13 Hadronic quark structure** (a) Three combinations of quarks can be used to construct all the baryons, such as the proton and neutron. (b) Quark–antiquark combinations can be used to construct all the mesons, such as the positive pion.

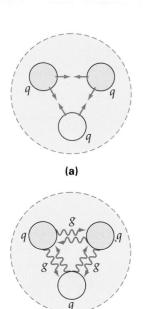

•FIGURE 30.14 Quarks, color charge, confinement, and gluons (a) Quarks of different color charge attract each other via the color force, which keeps them confined (here inside a baryon—how do you know this?). (b) Gluons (wiggly arrows) are exchanged between quarks of different color charge, creating the color force—analogous to virtual photons and the electric force.

truly elementary particles: leptons, quarks (and antiparticles), and exchange particles such as the photon. The hope is that this list will not need to be expanded.

Quark Confinement, Color Charge, and Gluons

In our quark discussion, one thing has been missing—direct experimental observation of a quark. Unfortunately, even in the most energetic particle collisions, *a free quark has never been observed.* We now believe that they are permanently confined within their particles by a springlike force. That is, a force exists between quarks that grows as they separate from one another. This force grows very rapidly with distance and prevents the ejection of a quark from its particle. This phenomenon is called **quark confinement.**

To understand the force between quarks and to clear up some problems with apparent violation of the Pauli exclusion principle (see atomic structure, Section 28.3), quarks were endowed with another characteristic called **color charge,** or simply *color.* There are three types of color charge: red, green, and blue. (These names have nothing to do with visual color.) In analogy to electric charge, the quark confinement force exists because different color charges attract each other. Recall from Section 30.5 that the electromagnetic force is due to virtual photon exchanges between charged particles. Similarly, the force between quarks of different color is due to exchanges of virtual particles called **gluons** (•Fig. 30.14). This force is sometimes called the **color force.** This theory is named *quantum chromodynamics* (QCD) (*chromo* for color) in analogy to Feynman's quantum *electro*dynamics (QED), which successfully explains the electromagnetic force.

The force between quarks can be extended to explain the force between hadrons—the strong nuclear force. Consider our previous explanation of the strong force between a neutron and proton as shown in Fig. 30.12a: the exchange of a virtual negative pion. More fundamentally, we can qualitatively depict this force in terms of an exchange of quarks between the hadrons (•Fig. 30.15).

30.8 Force Unification Theories and the Early Universe

Objectives: To (a) become familiar with current attempts to unify the four fundamental forces and (b) understand why elementary particle interactions might hold the key to the very early evolution of the universe.

Unification Theories

Early in the twentieth century, Einstein was one of the first to conjecture that it might be possible to unify the fundamental forces of nature—that these four apparently very different forces might be different manifestations of one force. Each manifestation would appear under a different set of physical conditions; for example, an electrically neutral particle would not exhibit the electromagnetic portion of the unified force. Since then, it has been the dream of physicists to understand the fundamental interactions in the universe through just one force. Attempts to unify the various forces are called **unification theories.**

Actually, the first step toward unification was taken in the nineteenth century by Maxwell when he combined the electric and magnetic forces into a single electromagnetic force. Einstein later showed that the two are connected by relativity. The next major step occurred in the 1960s when Sheldon Glashow, Abdus Salam, and Steven Weinberg successfully combined the electromagnetic force with the weak force into a single **electroweak force.** For their efforts, they were awarded a Nobel Prize in 1979.

How can such apparently very different forces be unified? Their exchange particles are so different—the massless photon (γ) for the electromagnetic force and the massive W particle for the weak force. However, it was thought that at extremely high energies, the mass–energy of the W particle is negligible compared to its total energy, in effect making it massless like the photon. To see this, recall from Section 26.5 that the total energy of a particle is the sum of its kinetic and rest energies ($E = K + mc^2$). When $K \gg mc^2$ (that is, high energy), the particle's rest energy is negligible compared to its kinetic energy—perhaps, Glashow, Salam, and Weinberg reasoned, making the W particle more "photon-like" than was previously thought.

In this theory, weakly interacting particles such as electrons and neutrinos carry a *weak charge*. This weak charge is responsible for the exchange of particles, thus creating the combined electroweak force. The electroweak unification theory predicted the existence of three electroweak exchange particles: W^+ and W^- when there is charge exchange, and the neutral Z^o when there is no charge transfer (•Fig. 30.16). The eventual discovery of these particles led to a Nobel Prize for Carlo Rubbia and Simon van der Meer in 1984.

Scientists are now attempting to unify the electroweak force with the strong nuclear force. If successful, this **grand unified theory (GUT)** would reduce the fundamental forces to two. In most GUT attempts, several dozen exchange particles are required. In addition, leptons and quarks are combined into one family and can change into each other through the exchange of these particles. Scientists are mildly optimistic that experimental verification of one of the GUT candidates might occur early in the twenty-first century. For example, most of the current GUT theories predict that the proton should be unstable and decay with a half-life of about 10^{32} years, which is 10^{22} longer than the age of the universe! Experiments to look for this decay are currently under way, so far with no success. The experiments require huge amounts of water (protons) to make up for the expected very slow decay rate.

The ultimate unification would be to fold the gravitational force into the GUT, creating a single **superforce.** How this would be done, or even *if* it can ever be done, is not clear at this point. One problem is that, although the three components of the GUT can be represented as force fields in space and time, in our current view, gravity *is* space and time!

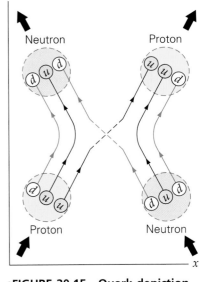

•**FIGURE 30.15 Quark depiction of strong nuclear force** Instead of envisioning the n–p interaction as an exchange of a virtual π^- particle, we can use a model of quark exchange. In this exchange, a pair of quarks (equivalent to a π^-) is transferred. A proton becomes a neutron and vice versa.

Evolution of the Universe and the Superforce

An interesting connection now exists between elementary particle physicists and astrophysicists interested in the evolution of the universe—in particular, its very early evolution. This connection occurs because, according to present ideas, the universe began ten to fifteen billion years ago from a huge explosion named the *Big Bang.* It is theorized that temperatures during the first 10^{-45} seconds of the universe were on the order of 10^{32} K. This corresponds to particle kinetic energies of about 10^{22} MeV, high enough that the rest energy of even the most massive elementary particles would be negligible. In effect, they would be massless, perhaps placing the four forces on an equal footing.

As the universe expanded and cooled, these early elementary particles condensed into what we see today. First protons, neutrons, and electrons formed; in turn they combined into atoms and eventually molecules. Over billions of years, the average

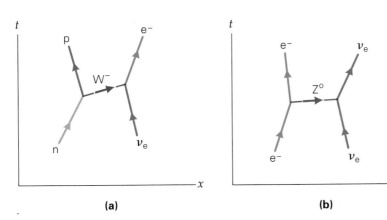

(a)　　　　　　(b)

•**FIGURE 30.16 Weak force interactions** The Feynman diagrams for **(a)** the neutrino-induced conversion of a neutron into a proton and an electron through the exchange of a W^- particle and **(b)** the scattering of a neutrino by an electron through the exchange of a neutral Z^o particle.

temperature of the universe has cooled off to its present value of about 3 K. In the process, scientists think that the superforce symmetry (equal footing of all four components) has been lost, leaving us with four very different-looking forces that are really components of a superforce.

It is hoped that future experiments might tell us more about the early moments of the universe and thus about the superforce. The ultimate goal of physics might even be within reach—to understand the basic interactions that govern the universe. While great strides are being made, it is likely to take well into the twenty-first century, if not further, to achieve this goal.

Chapter Review

Important Terms

nuclear reactions *928*	LOCA *937*	graviton *946*
Q value *929*	meltdown *938*	elementary particles *947*
endoergic *930*	fusion reaction *938*	leptons *947*
exoergic *930*	plasma *940*	hadrons *947*
threshold energy *931*	magnetic confinement *941*	tauon *947*
cross section *932*	inertial confinement *941*	baryons *948*
fission reaction *932*	neutrino *943*	quarks *949*
chain reaction *933*	fundamental forces *944*	quark model *949*
critical mass *934*	virtual particle *944*	quark confinement *950*
nuclear reactor *934*	exchange particle *944*	color charge *950*
fuel rods *935*	photon *944*	gluon *950*
core *935*	meson *945*	color force *950*
coolant *936*	muon *945*	unification theories *950*
control rods *936*	pion *945*	electroweak force *950*
moderator *936*	weak nuclear force *946*	grand unified theory *951*
breeder reactor *936*	W particle *946*	superforce *951*

Important Concepts

- The Q value of a reaction or decay is the energy released or absorbed in the process; it appears as a change in the total mass of the system.

- In fission, an unstable heavy nucleus usually decays by splitting into two fragments and several neutrons, which together have less total mass than the original nucleus; kinetic energy is thus released.

- A chain reaction occurs when the neutrons released from one fission trigger other fissions, which trigger further fissions, and so on.

- In fusion, two light nuclei fuse, producing a nucleus with less total mass than the original nuclei; kinetic energy is thus released.

- A plasma, produced when matter is completely ionized, is a mixture of positive ions and free electrons.

- Beta decay produces a daughter nucleus, an electron or positron, and an antineutrino or neutrino.

- Exchange particles are virtual particles believed to be associated with various forces. The pion (a meson) is the exchange particle primarily responsible for the strong nuclear force.

- The weak nuclear force, transmitted by the W particle, is primarily responsible for beta decay and the instability of the neutron.

- Hadrons are the family of elementary particles involved with the strong force. Nucleons, along with mesons, are part of the hadron family.

- Leptons are the family of elementary particles involved with the weak force. Electrons, muons, and neutrinos comprise the lepton family.

- Quarks are elementary, fractionally charged particles that make up hadrons.

- Charge color describes the color force between quarks. Quark exchange is the fundamental explanation for the strong nuclear force.

- The electroweak force is the name given to the unified electromagnetic and weak forces.

- The grand unified theory (GUT) is an attempt to unify the electroweak force with the strong nuclear force.

- Superforce is the sought-after unification of the four fundamental forces.

Important Equations

Q Value (for reactions of the type A + a → B + b):
$$Q = (m_A + m_a - m_b - m_B)c^2 = (\Delta m)c^2 \qquad (30.3)$$

Threshold Energy:
$$K_{min} = \left(1 + \frac{m_a}{M_A}\right)|Q| \qquad (30.4)$$

Range of Exchange Particle:
$$R = c\,\Delta t = \frac{h}{2\pi m_m c} \qquad (30.5)$$

Exercises

30.1 Nuclear Reactions

1. The Q value of a reaction (a) is given by the difference in the kinetic energies of the products and reactants, (b) is equal to the mass–energy difference between the initial and final systems, (c) can be positive or negative, (d) all of these.

2. To initiate an endoergic reaction, a particle incident on a stationary nucleus must have (a) a minimum disintegration energy, (b) a kinetic energy equal to the Q value, (c) a kinetic energy less than a certain threshold energy, (d) a kinetic energy larger than the Q value, which depends on m_a/M_A.

3. How does the threshold energy of an incident particle in a nuclear reaction vary with the mass of the target nuclei? (Sketch a graph if you can.) What is the interpretation of $Q = 0$?

4. ■ Complete the following nuclear reactions:
   (a) $^{1}_{0}n + ^{40}_{18}Ar \rightarrow \underline{\hspace{1cm}} + ^{0}_{-1}e$
   (b) $^{1}_{0}n + ^{235}_{92}U \rightarrow ^{98}_{40}Zr + \underline{\hspace{1cm}} + 3(^{1}_{0}n)$
   (c) $^{1}_{0}n + ^{235}_{92}U \rightarrow ^{133}_{51}Sb + ^{99}_{41}Nb + \underline{\hspace{1cm}}$
   (d) $\underline{\hspace{1cm}}(\alpha, p)^{17}_{8}O$
   (e) $^{137}_{56}Ba\,(n,\underline{\hspace{1cm}})^{137}_{55}Cs$

5. ■ Complete the following nuclear reactions:
   (a) $^{13}_{6}C + ^{1}_{1}H \rightarrow \gamma + \underline{\hspace{1cm}}$
   (b) $^{10}_{5}B + ^{4}_{2}He \rightarrow ^{12}_{6}C + \underline{\hspace{1cm}}$
   (c) $^{27}_{13}Al\,(\alpha, n)\underline{\hspace{1cm}}$
   (d) $^{14}_{7}N\,(\alpha, p)\underline{\hspace{1cm}}$
   (e) $^{13}_{6}C\,(p, \alpha)\underline{\hspace{1cm}}$

6. ■ Write the compound nuclei for the reactions in Exercise 4.

7. ■ Give the compound nuclei for the reactions in Exercise 5.

8. ■ Find the energy required to initiate the following reaction: $^{13}_{6}C + ^{1}_{1}H \rightarrow ^{4}_{2}He + ^{10}_{5}B$.

9. ■ What would the daughter nuclei in the following decay equations be? Will the reactions occur spontaneously?
   (a) $^{22}_{10}Ne \rightarrow \underline{\hspace{1cm}} + ^{0}_{-1}e$
   (b) $^{226}_{88}Ra \rightarrow \underline{\hspace{1cm}} + ^{4}_{2}He$
   (c) $^{16}_{8}O \rightarrow \underline{\hspace{1cm}} + ^{4}_{2}He$

10. ■ Show that the Q value of the reaction $^{1}_{1}H + ^{2}_{1}H \rightarrow ^{3}_{2}He + \gamma$ is 5.49 MeV.

11. ■ Find the Q value for the alpha decay of ^{238}U:

$$^{238}_{92}U \rightarrow ^{234}_{90}Th + ^{4}_{2}He$$
$$(238.050\,786\ u) \quad (234.043\,583\ u) \quad (4.002\,603\ u)$$

Would you expect Q to be positive or negative?

12. ■■ Find the threshold energy of the following reaction:

$$^{16}_{8}O + ^{1}_{0}n \rightarrow ^{13}_{6}C + ^{4}_{2}He$$
$$(15.994\,915\ u) \quad (1.008\,665\ u) \quad (13.003\,355\ u) \quad (4.002\,603\ u)$$

13. ■■ Find the threshold energy of the following reaction:

$$^{3}_{2}He + ^{1}_{0}n \rightarrow ^{2}_{1}H + ^{2}_{1}H$$
$$(3.016\,029\ u) \quad (1.008\,665\ u) \quad (2.014\,102\ u) \quad (2.014\,102\ u)$$

14. ■■ Find the threshold energy of the following reaction:

$$^{13}_{6}C + ^{1}_{1}H \rightarrow ^{1}_{0}n + ^{13}_{7}N$$
$$(13.003\,355\ u) \quad (1.007\,825\ u) \quad (1.008\,665\ u) \quad (13.005\,739\ u)$$

15. ■■ Determine the minimum kinetic energy of an incident alpha particle that will initiate the following reaction:

$$^{14}_{7}N + ^{4}_{2}He \rightarrow ^{17}_{8}O + ^{1}_{1}H$$
$$(14.003\,074\ u) \quad (4.002\,603\ u) \quad (16.999\,131\ u) \quad (1.007\,825\ u)$$

16. ■■ Is the following reaction endoergic or exoergic? Prove your answer.

$$^{7}_{3}Li + ^{1}_{1}H \rightarrow ^{4}_{2}He + ^{4}_{2}He$$
$$(7.016\,005\ u) \quad (1.007\,825\ u) \quad (4.002\,603\ u) \quad (4.002\,603\ u)$$

17. ■■ Is the reaction $^{200}Hg\,(p, \alpha)^{197}Au$ endoergic or exoergic? Prove your answer. (The reaction on p. 928 has mass values.)

18. ■■ Determine the Q value of the following reaction:

$$^{9}_{4}\text{Be} \quad + \quad ^{4}_{2}\text{He} \quad \rightarrow \quad ^{12}_{6}\text{C} \quad + \quad ^{1}_{0}\text{n}$$

$(9.012\,183\ u) \qquad (4.002\,603\ u) \qquad (12.000\,000\ u) \qquad (1.008\,665\ u)$

19. ■■ What is the minimum kinetic energy of a proton that will initiate the reaction $^{3}_{1}\text{H}(p, d)^{2}_{1}\text{H}$? (Here d stands for a deuterium nucleus called the *deuteron*.)

20. ■■ A stationary ^{226}Ra nucleus decays and emits an alpha particle of kinetic energy 4.706 MeV. Find the kinetic energy of the recoiling daughter nucleus.

21. ■■■ The same type of incident particle is used for two endoergic reactions. In one reaction, the mass of the target nucleus is 15 times greater than the particle, and in the other reaction 20 times greater. If the Q value of the first reaction is three times that of the second, which has the greater threshold energy, and how many times greater (to three significant figures; assume data are exact)?

22. ■■■ To simulate neutrons hitting the nuclei of a thin foil, n (where n is an integer ≥ 1) pie plates of radius R ($\ll L$ and W) are randomly fixed on a rectangular wall with dimensions L and W. If you throw a very small (point) object at the wall, what would be the probability of hitting a plate in terms of n, R, L, and W?

23. ■■■ The units of cross section in SI are meters squared. When neutron resonances were first discovered, they were much, much larger than any cross section previously measured; someone said they were "as big as a barn." Hence a more common unit of cross section is the *barn* (b) defined as $1\ \text{b} \equiv 10^{-28}\ \text{m}^2$. Using the empirical equation for the radius of a nucleus, $R = R_0 A^{1/3}$, where $R_0 \approx 1.2 \times 10^{-15}$ m $= 1.2$ fm, estimate the geometrical cross section (in barns) of (a) ^{12}C, (b) ^{56}Fe, (c) ^{208}Pb, and (d) ^{238}U.

24. ■■■ Assume that the average kinetic energy of ions in a plasma is given by the equation for the kinetic energy of the atoms in an ideal gas ($\frac{1}{2}mv^2 = \frac{3}{2}k_BT$). Assume also that fusion occurs when the ions approach each other within a distance of the upper limit for the nuclear diameter ($R = 10^{-12}$ cm). Calculate the temperature required for the fusion of two deuterium ions. (Boltzmann's constant is $k_B = 1.38 \times 10^{-23}$ J/K.)

30.2 Nuclear Fission
and
30.3 Nuclear Fusion

25. Nuclear fission (a) is endoergic, (b) occurs only for ^{235}U, (c) releases about 500 MeV of energy per fission, (d) requires a critical mass for a sustained reaction.

26. A nuclear reactor (a) can operate on natural (unenriched) uranium, (b) has its chain reaction controlled by neutron-absorbing materials, (c) can be partially controlled by the amount of moderator, (d) all of these.

27. A nuclear fusion reaction (a) has a negative Q value, (b) can occur spontaneously, (c) is an example of "splitting" the atom, (d) releases less than 50 MeV of energy per fusion.

28. Controlled fusion requires (a) no critical mass, (b) confinement, (c) the formation of a plasma, (d) all of these.

29. The energy produced in fission reactions is carried off as kinetic energies of the products. How is this energy converted to heat in a nuclear reactor?

30. ■ Find the approximate energy released in these fission reactions: (a) $^{235}_{92}\text{U} + ^{1}_{0}\text{n} \rightarrow 2$ fission fragments $+ 5$ neutrons and (b) $^{239}_{94}\text{Pu} + ^{1}_{0}\text{n} \rightarrow 2$ fission fragments $+ 2$ neutrons.

31. ■ Calculate the amounts of energy released in the following fusion reactions: (a) $^{1}_{1}\text{H} + ^{1}_{0}\text{n} \rightarrow ^{2}_{1}\text{H} + \gamma$ and (b) $^{3}_{2}\text{He} + ^{3}_{2}\text{He} \rightarrow ^{4}_{2}\text{He} + 2(^{1}_{1}\text{H})$.

32. ■ Calculate the amounts of energy released in the following fusion reactions: (a) $^{2}_{1}\text{H} + ^{2}_{1}\text{H} \rightarrow ^{3}_{2}\text{He} + ^{1}_{0}\text{n}$ and (b) $^{2}_{1}\text{H} + ^{3}_{1}\text{H} \rightarrow ^{4}_{2}\text{He} + ^{1}_{0}\text{n}$.

33. ■■ Using water as a moderator works well because the proton and neutron have nearly the same mass. For a head-on elastic collision, we might expect a neutron to lose all of its kinetic energy in one collision, whereas for an "almost miss" we might expect it to lose essentially none. Assume that on average the neutron loses 60% of its kinetic energy in each collision. Estimate how many collisions are needed to reduce a neutron of kinetic energy 2.0 MeV to one of only 0.02 eV (approximately "thermal").

30.4 Beta Decay and the Neutrino

34. In the absence of a neutrino, what is not conserved in beta decay: (a) energy, (b) linear momentum, (c) angular momentum, (d) all of these?

35. A neutrino interacts with matter by (a) an electrical interaction, (b) a strong interaction, (c) a weak interaction, (d) both (b) and (c).

36. Why is it so difficult to detect neutrinos experimentally?

37. ■ A neutrino created in a beta decay has 2.65 MeV of energy. What is the de Broglie wavelength of the neutrino?

38. ■■ In Exercise 37, what is the momentum of the beta particle plus the daughter nucleus? Include the direction and your reasoning.

39. ■■ In Exercise 37, if the disintegration energy was 3.51 MeV, what would be (a) the maximum possible kinetic energy of the beta particle? (b) its momentum? (c) the kinetic energy and momentum of the daughter nucleus?

40. ■■ Show that the disintegration energy for β^- decay is $Q = (m_P - m_D - m_e)c^2 = (M_P - M_D)c^2$, where m_P and m_D are the masses of the parent and daughter *nuclei*, respectively, m_e is the electron mass, and M_P and M_D are the masses of the neutral *atoms*.

41. ■■ What is the maximum kinetic energy of the electron emitted when a ^{12}B nucleus beta decays into a ^{12}C nucleus? (See Exercise 40.)

42. ■■ The kinetic energy of an electron emitted from a ^{32}P nucleus that beta decays into a ^{32}S nucleus is 1.00 MeV. What is the energy of the accompanying neutrino? Neglect the recoil energy of the daughter. (See Exercise 40.)

43. ■■ Show that the disintegration energy for β^+ decay is $Q = (m_P - m_D - m_e)c^2 = (M_P - M_D - 2m_e)c^2$, where m_P and m_D are the masses of the parent and daughter *nuclei*, respectively, m_e is the electron mass, and M_P and M_D are the masses of the neutral *atoms*.

44. ■■■ The kinetic energy of a positron emitted from the β^+ decay of a ^{13}N nucleus into a ^{13}C nucleus is 0.250 MeV. What is the energy of the accompanying neutrino in the process? (Neglect the recoil energy of the daughter nucleus, and see Exercise 43.)

45. ■■■ On the basis of the Q values given in Exercises 40 and 43, what are the mass requirements of the parent *atoms* if β^- and β^+ processes are to be energetically possible?

30.5 Fundamental Forces and Exchange Particles

46. Virtual particles (a) form virtual images, (b) exist only for the short time permitted by the uncertainty principle, (c) make up positrons, (d) can be observed in exchange processes.

47. The exchange particle of the weak nuclear force is the (a) pion, (b) W particle, (c) muon, (d) positron.

48. If virtual exchange particles are unobservable, how is their existence verified?

49. ■ Assuming the range of the nuclear force to be on the order of 10^{-15} m, predict the mass of the exchange particle.

50. ■■ A high-speed proton collides with a nucleus and travels a distance of 5.0×10^{-16} m on average before a reaction takes place. Which type of interaction is this, and during what time period does the interaction take place?

51. ■■ By what minimum energy is the conservation of energy "violated" during a neutral pi meson exchange process?

52. ■■ How long is the conservation of energy "violated" in a neutral pi meson exchange process?

53. ■■ A W particle in a weak interaction has a mass-equivalent energy of 1.00 GeV. What is the approximate range for the weak interaction with this particle?

30.6 Elementary Particles
and
30.7 The Quark Model

54. Particles that interact by the strong nuclear force are called (a) muons, (b) hadrons, (c) W particles, (d) leptons.

55. Quarks make up which of the following particles? (a) hadrons, (b) muons, (c) Z particles, (d) all of these.

56. What is meant by quark flavor and color? Can these be changed? Explain.

57. With so many types of hadrons, why aren't fractional electronic charges observed?

58. Distinguish between baryons and mesons.

59. Which particle has more mass, (a) π^+ or π°, (b) K^+ or K°, (c) Σ^+ or Σ°, (d) Ξ° or Ξ^-? [Hints: See Table 30.3.]

60. Show that the quark combination of a proton is *uud*.

61. What is the quark combination of a neutron?

30.8 Force Unification Theories and the Early Universe

62. The grand unified theory would reduce the fundamental forces to (a) one, (b) two, (c) three, (d) four.

63. The magnetic force is part of the (a) electroweak force, (b) weak force, (c) strong force.

Additional Exercises

64. Complete the following nuclear reactions:
(a) $^6_3Li + ^1_1H \rightarrow ^3_2He + \underline{\quad}$
(b) $^{58}_{28}Ni + ^2_1H \rightarrow ^{59}_{28}Ni + \underline{\quad}$
(c) $^{235}_{92}U + ^1_0n \rightarrow ^{138}_{54}Xe + 5^1_0n + \underline{\quad}$
(d) $^9_4Be\ (\alpha, n)\underline{\quad}$
(e) $^{16}_8O\ (n, p)\underline{\quad}$

65. Give the compound nuclei for the reactions in Exercise 64.

66. Compute the Q values (to three significant figures) of the following fusion reactions: (a) $^2_1H + ^3_1H \rightarrow ^4_2He + ^1_0n$ and (b) $^2_1H + ^2_1H \rightarrow ^3_1H + ^1_1H$.

67. Find the energy released in the beta decay of ^{14}C.

68. Show that the Q value for electron capture is given by $Q = (m_P + m_e - m_D)c^2 = (M_P - M_D)c^2$, where m_P and m_D are the masses of the parent and daughter *nuclei*, respectively, m_e is the electron mass, and M_P and M_D are the masses of the neutral *atoms*.

69. (a) In the electron capture conversion of a ^7Be nucleus into a ^7Li nucleus, what is the energy of the emitted neutrino? (Neglect daughter recoil, and see Exercise 68.) (b) Is it energetically possible for ^7Be to β^+ decay into ^7Li?

70. Complete the following nuclear reactions:
(a) $^4_2He + ^{14}_7N \rightarrow ^{17}_8O + \underline{\quad}$
(b) $^{13}_7N + ^1_0n \rightarrow ^4_2He + \underline{\quad}$
(c) $^{92}_{40}Zr\ (p, \alpha)\underline{\quad}$
(d) $^3_2He\ (\gamma, p)\underline{\quad}$

Appendices

Appendix I Mathematical Relationships

Algebraic Relationships

$(a + b)^2 = a^2 + 2ab + b^2$
$(a - b)^2 = a^2 - 2ab + b^2$
$(a^2 - b^2) = (a + b)(a - b)$

Quadratic Formula

If $ax^2 + bx + c = 0$, then $x = \dfrac{-b \pm \sqrt{b^2 - 4ac}}{2a}$

Powers and Exponents

$x^0 = 1$

$x^1 = x$ $\qquad x^{-1} = \dfrac{1}{x}$

$x^2 = x \cdot x$ $\qquad x^{-2} = \dfrac{1}{x^2}$ $\qquad x^{\frac{1}{2}} = \sqrt{x}$

$x^3 = x \cdot x \cdot x$ $\qquad x^{-3} = \dfrac{1}{x^3}$ $\qquad x^{\frac{1}{3}} = \sqrt[3]{x}$

etc. $\qquad$ etc. $\qquad$ etc.

$x^a \cdot x^b = x^{(a+b)}$
$x^a / x^b = x^{(a-b)}$
$(x^a)^b = x^{ab}$

Logarithms

If $x = a^n$, then $n = log_a x$.

common logarithms: base 10
(assumed when abbreviation "log" is used, unless another base is specified)

$log\ 10^x = x$

$log\ xy = log\ x + log\ y$
$log\ x/y = log\ x - log\ y$
$log\ x^y = y\ log\ x$

natural logarithms: base $e = 2.718\,28\ldots$ (abbreviated "ln")

$ln\ e^x = x$
$log\ x = 0.434\,29\ ln\ x$
$ln\ x = 2.3026\ log\ x$

Geometric and Trigonometric Relationships

Areas and Volumes of Some Common Shapes

Circle: $\quad A = \pi r^2 = \dfrac{\pi d^2}{4}$ (area)

$\qquad\qquad c = 2\pi r = \pi d$ (circumference)

Triangle: $\quad A = \frac{1}{2}ab$

Sphere: $\quad A = 4\pi r^2$
$\qquad\qquad V = \frac{4}{3}\pi r^3$

Cylinder: $\quad A = \pi r^2$ (end)
$\qquad\qquad A = 2\pi rh$ (body)
$\qquad\qquad A = 2(\pi r^2) + 2\pi rh$ (total)
$\qquad\qquad V = \pi r^2 h$

Definitions of Trigonometric Functions

$sin\ \theta = \dfrac{y}{r}$ $\qquad cos\ \theta = \dfrac{x}{r}$ $\qquad tan\ \theta = \dfrac{sin\ \theta}{cos\ \theta} = \dfrac{y}{x}$

$\theta°$ (rad)	$sin\ \theta$	$cos\ \theta$	$tan\ \theta$
0° (0)	0	1	0
30° ($\pi/6$)	0.500	0.866	0.577
45° ($\pi/4$)	0.707	0.707	1.00
60° ($\pi/3$)	0.866	0.500	1.73
90° ($\pi/2$)	1	0	$\to \infty$

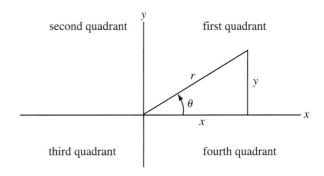

For very small angles:

$\cos \theta \approx 1 \qquad \sin \theta \approx \theta \text{ (radians)}$

$\tan \theta = \dfrac{\sin \theta}{\cos \theta} \approx \theta \text{ (radians)}$

The sign of a trigonometric function depends on the quadrant, or the signs of x and y; for example, in the second quadrant $(-x, y)$, $-x/r = \cos \theta$, and $y/r = \sin \theta$. The sign can also be assigned by using the reduction formulas.

Reduction Formulas

	(θ in second quadrant)	(θ in third quadrant)	(θ in fourth quadrant)
$\sin \theta =$	$\cos(\theta - 90°) =$	$-\sin(\theta - 180°) =$	$-\cos(\theta - 270°)$
$\cos \theta =$	$-\sin(\theta - 90°) =$	$-\cos(\theta - 180°) =$	$\sin(\theta - 270°)$

Fundamental Identities

$\sin^2 \theta + \cos^2 \theta = 1$

$\sin 2\theta = 2 \sin \theta \cos \theta$

$\cos 2\theta = \cos^2 \theta - \sin^2 \theta = 2 \cos^2 \theta - 1 = 1 - 2 \sin^2 \theta$

$\sin^2 \theta = \frac{1}{2}(1 - \cos 2\theta)$

$\cos^2 \theta = \frac{1}{2}(1 + \cos 2\theta)$

For half-angle ($\theta/2$) identities, replace θ with $\theta/2$; for example:

$\sin^2 \theta/2 = \frac{1}{2}(1 - \cos \theta) \qquad \cos^2 \theta/2 = \frac{1}{2}(1 + \cos \theta)$

$\sin(\alpha \pm \beta) = \sin \alpha \cos \beta \pm \cos \alpha \sin \beta$

$\cos(\alpha \pm \beta) = \cos \alpha \cos \beta \mp \sin \alpha \sin \beta$

$\tan(\alpha \pm \beta) = \dfrac{\tan \alpha \pm \tan \beta}{1 \mp \tan \alpha \tan \beta}$

Law of Cosines

For a triangle with angles A, B, and C with opposite sides a, b, and c, respectively:

$a^2 = b^2 + c^2 - 2bc \cos A$

(with similar results for $b^2 =$ and for $c^2 =$).

If $A = 90°$, this equation reduces to the Pythagorean theorem:

$a^2 = b^2 + c^2 \quad$ (of the form $r^2 = x^2 + y^2$)

Law of Sines

For a triangle with angles A, B, and C with opposite sides a, b, and c, respectively:

$\dfrac{a}{\sin A} = \dfrac{b}{\sin B} = \dfrac{c}{\sin C}$

Appendix II Kinetic Theory of Gases

The basic assumptions are as follows:

1. All the molecules of a pure gas have the same mass (m) and are in continuous and completely random motion. (The mass of each molecule is so small that the effect of gravity on it is negligible.)

2. The gas molecules are separated by large distances and occupy a volume that is negligible compared to these distances.

3. The molecules exert no forces on each other except when they collide.

4. Collisions of the molecules with one another and with the walls of the container are perfectly elastic.

The magnitude of the force exerted on the wall of the container by a gas molecule colliding with it is $F = \Delta p/\Delta t$. Assuming that the direction of the velocity (v_x) is normal to the wall, the magnitude of the average force is

$$F = \frac{\Delta(mv)}{\Delta t} = \frac{mv_x - (-mv_x)}{\Delta t} = \frac{2mv_x}{\Delta t} \qquad (1)$$

After striking one wall of the container, which for convenience is assumed to be a cube with sides of dimensions L, the molecule recoils in a straight line. Suppose that the molecule reaches the opposite wall without colliding with any other molecules along the way. The molecule then travels the distance L in a time equal to L/v_x. After the collision with that wall, again assuming no collisions on the return trip, the round trip will take $\Delta t = 2L/v_x$. Thus, the number of collisions per unit time a molecule makes with a particular wall is $v_x/2L$, and the average force of the wall from successive collisions is

$$F = \frac{2mv_x}{\Delta t} = \frac{2mv_x}{2L/v_x} = \frac{mv_x^2}{L} \qquad (2)$$

The random motions of the many molecules produce a relatively constant force on the walls, and the pressure (p) is the total force on a wall divided by its area:

$$p = \frac{\Sigma F_i}{L^2} = \frac{m(v_{x_1}^2 + v_{x_2}^2 + v_{x_3}^2 + \cdots)}{L^3} \qquad (3)$$

The subscripts refer to individual molecules.

The average of the squares of the speeds is given by

$$\overline{v_x^2} = \frac{v_{x_1}^2 + v_{x_2}^2 + v_{x_3}^2 + \cdots}{N}$$

where N is the number of molecules in the container. In terms of this average, Eq. 3 can be written as

$$p = \frac{Nm\overline{v_x^2}}{L^3} \qquad (4)$$

However, the molecules' motions occur with equal frequency along any one of the three axes, so $\overline{v_x^2} = \overline{v_y^2} = \overline{v_z^2}$, and $\overline{v^2} = \overline{v_x^2} + \overline{v_y^2} + \overline{v_z^2} = 3\overline{v_x^2}$. Then,

$$\sqrt{\overline{v^2}} = v_{rms}$$

where v_{rms} is called the root-mean-square (rms) speed. Substituting this result into Eq. 4 along with V for L^3 (since L^3 is the volume of the cubical container) gives

$$pV = \tfrac{1}{3}Nmv_{rms}^2 \qquad (5)$$

This result is true even though collisions between molecules were ignored. Statistically, these collisions average out, so the number of collisions with each wall is as described. This result is also independent of the shape of the container. A cube merely simplifies the derivation.

We combine this result with the empirical perfect gas law:

$$pV = Nk_BT = \tfrac{1}{3}Nmv_{rms}^2$$

The average kinetic energy per gas molecule is thus proportional to the absolute temperature of the gas:

$$\overline{K} = \tfrac{1}{2}mv_{rms}^2 = \tfrac{3}{2}k_BT \qquad (6)$$

The collision time is negligible compared with the time between collisions. Some kinetic energy will be momentarily converted to potential energy during a collision; however, this potential energy can be ignored because each molecule spends a negligible amount of time in collisions. Therefore, by this approximation, the total kinetic energy is the internal energy of the gas, and the internal energy of a perfect gas is directly proportional to its absolute temperature.

Appendix III Planetary Data

Name	Equatorial Radius (km)	Mass (Compared to Earth's)*	Mean Density (× 10³ kg/m³)	Surface Gravity (Compared to Earth's)	Semimajor Axis × 10⁶ km	Semimajor Axis AU†	Period Years	Period Days	Eccentricity	Inclination to Ecliptic
Mercury	2439	0.0553	5.43	0.378	57.9	0.3871	0.24084	87.96	0.2056	7°00′26″
Venus	6052	0.8150	5.24	0.894	108.2	0.7233	0.61515	224.68	0.0068	3°23′40″
Earth	6378.140	1	5.515	1	149.6	1	1.00004	365.25	0.0167	0°00′14″
Mars	3397.2	0.1074	3.93	0.379	227.9	1.5237	1.8808	686.95	0.0934	1°51′09″
Jupiter	71398	317.89	1.36	2.54	778.3	5.2028	11.862	4337	0.0483	1°18′29″
Saturn	60000	95.17	0.71	1.07	1427.0	9.5388	29.456	10760	0.0560	2°29′17″
Uranus	26145	14.56	1.30	0.8	2871.0	19.1914	84.07	30700	0.0461	0°48′26″
Neptune	24300	17.24	1.8	1.2	4497.1	30.0611	164.81	60200	0.0100	1°46′27″
Pluto	1500–1800	0.02	0.5–0.8	~0.03	5913.5	39.5294	248.53	90780	0.2484	17°09′03″

*Planet's mass/Earth's mass, where $M_E = 6.0 \times 10^{24}$ kg
†Astronomical unit, 1 AU = 1.5×10^8 km, the average distance between the Earth and the Sun.

Appendix IV Alphabetical Listing of the Chemical Elements (perodic table on p. 881.)

Element	Symbol	Atomic Number (Proton Number)	Atomic Mass	Element	Symbol	Atomic Number (Proton Number)	Atomic Mass	Element	Symbol	Atomic Number (Proton Number)	Atomic Mass
Actinium	Ac	89	227.0278	Chromium	Cr	24	51.996	Hydrogen	H	1	1.00794
Aluminum	Al	13	26.98154	Cobalt	Co	27	58.9332	Indium	In	49	114.82
Americium	Am	95	(243)	Copper	Cu	29	63.546	Iodine	I	53	126.9045
Antimony	Sb	51	121.757	Curium	Cm	96	(247)	Iridium	Ir	77	192.22
Argon	Ar	18	39.948	Dubnium	Db	105	(262)	Iron	Fe	26	55.847
Arsenic	As	33	74.9216	Dysprosium	Dy	66	162.50	Krypton	Kr	36	83.80
Astatine	At	85	(210)	Einsteinium	Es	99	(252)	Lanthanum	La	57	138.9055
Barium	Ba	56	137.33	Erbium	Er	68	167.26	Lawrencium	Lr	103	(260)
Berkelium	Bk	97	(247)	Europium	Eu	63	151.96	Lead	Pb	82	207.2
Beryllium	Be	4	9.01218	Fermium	Fm	100	(257)	Lithium	Li	3	6.941
Bismuth	Bi	83	208.9804	Fluorine	F	9	18.998403	Lutetium	Lu	71	174.967
Bohrium	Bh	107	(264)	Francium	Fr	87	(223)	Magnesium	Mg	12	24.305
Boron	B	5	10.81	Gadolinium	Gd	64	157.25	Manganese	Mn	25	54.9380
Bromine	Br	35	79.904	Gallium	Ga	31	69.72	Meitnerium	Mt	109	(268)
Cadmium	Cd	48	112.41	Germanium	Ge	32	72.561	Mendelevium	Md	101	(258)
Calcium	Ca	20	40.078	Gold	Au	79	196.9665	Mercury	Hg	80	200.59
Californium	Cf	98	(251)	Hafnium	Hf	72	178.49	Molybdenum	Mo	42	95.94
Carbon	C	6	12.011	Hahnium	Ha	105	(262)	Neodymium	Nd	60	144.24
Cerium	Ce	58	140.12	Hassium	Hs	108	(265)	Neon	Ne	10	20.1797
Cesium	Cs	55	132.9054	Helium	He	2	4.00260	Neptunium	Np	93	237.048
Chlorine	Cl	17	35.453	Holmium	Ho	67	164.9304	Nickel	Ni	28	58.69

Element	Symbol	Atomic Number (Proton Number)	Atomic Mass	Element	Symbol	Atomic Number (Proton Number)	Atomic Mass	Element	Symbol	Atomic Number (Proton Number)	Atomic Mass
Niobium	Nb	41	92.9064	Rhodium	Rh	45	102.9055	Thallium	Tl	81	204.383
Nitrogen	N	7	14.0067	Rubidium	Rb	37	85.4678	Thorium	Th	90	232.0381
Nobelium	No	102	(259)	Ruthenium	Ru	44	101.07	Thulium	Tm	69	168.9342
Osmium	Os	76	190.2	Rutherfordium	Rf	104	(261)	Tin	Sn	50	118.710
Oxygen	O	8	15.9994	Samarium	Sm	62	150.36	Titanium	Ti	22	47.88
Palladium	Pd	46	106.42	Scandium	Sc	21	44.9559	Tungsten	W	74	183.85
Phosphorus	P	15	30.973 76	Seaborgium	Sg	106	(263)	Uranium	U	92	238.0289
Platinum	Pt	78	195.08	Selenium	Se	34	78.96	Vanadium	V	23	50.9415
Plutonium	Pu	94	(244)	Silicon	Si	14	28.0855	Xenon	Xe	54	131.29
Polonium	Po	84	(209)	Silver	Ag	47	107.8682	Ytterbium	Yb	70	173.04
Potassium	K	19	39.0983	Sodium	Na	11	22.989 77	Yttrium	Y	39	88.9059
Praseodymium	Pr	59	140.9077	Strontium	Sr	38	87.62	Zinc	Zn	30	65.39
Promethium	Pm	61	(145)	Sulfur	S	16	32.066	Zirconium	Zr	40	91.22
Protactinium	Pa	91	231.0359	Tantalum	Ta	73	180.9479			110	(269)
Radium	Ra	88	226.0254	Technetium	Tc	43	(98)			111	(272)
Radon	Rn	86	(222)	Tellurium	Te	52	127.60			112	(277)
Rhenium	Re	75	186.207	Terbium	Tb	65	158.9254				

Appendix V Properties of Selected Isotopes

Atomic Number (Z)	Element	Symbol	Mass Number (A)	Atomic Mass*	Abundance (%) or Decay Mode† (if radioactive)	Half-Life (if radioactive)
0	(Neutron)	n	1	1.008 665	β^-	10.6 min
1	Hydrogen	H	1	1.007 825	99.985	
	Deuterium	D	2	2.014 102	0.015	
	Tritium	T	3	3.016 049	β^-	12.33 y
2	Helium	He	3	3.016 029	0.000 14	
			4	4.002 603	≈ 100	
3	Lithium	Li	6	6.015 123	7.5	
			7	7.016 005	92.5	
4	Beryllium	Be	7	7.016 930	EC, γ	53.3 d
			8	8.005 305	2α	6.7×10^{-17} s
			9	9.012 183	100	
5	Boron	B	10	10.012 938	19.8	
			11	11.009 305	80.2	
			12	12.014 353	β^-	20.4 ms
6	Carbon	C	11	11.011 433	β^+, EC	20.4 ms
			12	12.000 000	98.89	
			13	13.003 355	1.11	
			14	14.003 242	β^-	5730 y
7	Nitrogen	N	13	13.005 739	β^-	9.96 min
			14	14.003 074	99.63	
			15	15.000 109	0.37	
8	Oxygen	O	15	15.003 065	β^+, EC	122 s
			16	15.994 915	99.76	
			18	17.999 159	0.204	
9	Fluorine	F	19	18.998 403	100	
10	Neon	Ne	20	19.992 439	90.51	
			22	21.991 384	9.22	
11	Sodium	Na	22	21.994 435	β^+, EC, γ	2.602 y
			23	22.989 770	100	

Atomic Number (Z)	Element	Symbol	Mass Number (A)	Atomic Mass*	Abundance (%) or Decay Mode (if radioactive)	Half-Life (if radioactive)
			24	23.990 964	β^-, γ	15.0 h
12	Magnesium	Mg	24	23.985 045	78.99	
13	Aluminum	Al	27	26.981 541	100	
14	Silicon	Si	28	27.976 928	92.23	
			31	30.975 364	β^-, γ	2.62 h
15	Phosphorus	P	31	30.973 763	100	
			32	31.973 908	β^-	14.28 d
16	Sulfur	S	32	31.972 072	95.0	
			35	34.969 033	β^-	87.4 d
17	Chlorine	Cl	35	34.968 853	75.77	
			37	36.965 903	24.23	
18	Argon	Ar	40	39.962 383	99.60	
19	Potassium	K	39	38.963 708	93.26	
			40	39.964 000	β^-, EC, γ, β^+	1.28×10^9 y
20	Calcium	Ca	30	39.962 591	96.94	
24	Chromium	Cr	52	51.940 510	83.79	
25	Manganese	Mn	55	54.938 046	100	
26	Iron	Fe	56	55.934 939	91.8	
27	Cobalt	Co	59	58.933 198	100	
			60	59.933 820	β^-, γ	5.271 y
28	Nickel	Ni	58	57.935 347	68.3	
			60	59.930 789	26.1	
			64	63.927 968	0.91	
29	Copper	Cu	63	62.929 599	69.2	
			64	63.929 766	β^-, β^+	12.7 h
			65	64.927 792	30.8	
30	Zinc	Zn	64	63.929 145	48.6	
			66	65.926 035	27.9	
33	Arsenic	As	75	74.921 596	100	
35	Bromine	Br	79	78.918 336	50.69	
36	Krypton	Kr	84	83.911 506	57.0	
			89	88.917 563	β^-	3.2 min
38	Strontium	Sr	86	85.909 273	9.8	
			88	87.905 625	82.6	
			90	89.907 746	β^-	28.8 y
39	Yttrium	Y	89	89.905 856	100	
43	Technetium	Tc	98	97.907 210	β^-, γ	4.2×10^6 y
47	Silver	Ag	107	106.905 095	51.83	
			109	108.904 754	48.17	
48	Cadmium	Cd	114	113.903 361	28.7	
49	Indium	In	115	114.903 88	95.7; β^-	5.1×10^{14} y
50	Tin	Sn	120	119.902 199	32.4	
53	Iodine	I	127	126.904 477	100	
			131	130.906 118	β^-, γ	8.04 d
54	Xenon	Xe	132	131.904 15	26.9	
			136	135.907 22	8.9	
55	Cesium	Cs	133	132.905 43	100	
56	Barium	Ba	137	136.905 82	11.2	
			138	137.905 24	71.7	
			144	143.922 73	β^-	11.9 s

Atomic Number (Z)	Element	Symbol	Mass Number (A)	Atomic Mass*	Abundance (%) or Decay Mode (if radioactive)	Half-Life (if radioactive)
61	Promethium	Pm	145	144.912 75	EC, α, γ	17.7 y
74	Tungsten (Wolfram)	W	184	183.950 95	30.7	
76	Osmium	Os	191	190.960 94	β^-, γ	15.4 d
			192	191.961 49	41.0	
78	Platinum	Pt	195	194.964 79	33.8	
79	Gold	Au	197	196.966 56	100	
80	Mercury	Hg	202	201.970 63	29.8	
81	Thallium	Tl	205	204.974 41	70.5	
			210	209.990 069	β^-	1.3 min
82	Lead	Pb	204	203.973 044	β^-, 1.48	1.4×10^{17} y
			206	205.974 46	24.1	
			207	206.975 89	22.1	
			208	207.976 64	52.3	
			210	209.984 18	α, β^-, γ	22.3 y
			211	210.988 74	β^-, γ	36.1 min
			212	211.991 88	β^-, γ	10.64 h
			214	213.999 80	β^-, γ	26.8 min
83	Bismuth	Bi	209	208.980 39	100	
			211	210.987 26	α, β^-, γ	2.15 min
84	Polonium	Po	210	209.982 86	α, γ	138.38 d
			214	213.995 19	α, γ	164 μs
86	Radon	Rn	222	222.017 574	α, β	3.8235 d
87	Francium	Fr	223	223.019 734	α, β^-, γ	21.8 min
88	Radium	Ra	226	226.025 406	α, γ	1.60×10^3 y
			228	228.031 069	β^-	5.76 y
89	Actinium	Ac	227	227.027 751	α, β^-, γ	21.773 y
90	Thorium	Th	228	228.028 73	α, γ	1.9131 y
			232	232.038 054	100; α, γ	1.41×10^{10} y
92	Uranium	U	232	232.037 14	α, γ	72 y
			233	233.039 629	α, γ	1.592×10^5 y
			235	235.043 925	0.72; α, γ	7.038×10^8 y
			236	236.045 563	α, γ	2.342×10^7 y
			238	238.050 786	99.275; α, γ	4.468×10^9 y
			239	239.054 291	β^-, γ	23.5 min
93	Neptunium	Np	239	239.052 932	β^-, γ	2.35 d
94	Plutonium	Pu	239	239.052 158	α, γ	2.41×10^4 y
95	Americium	Am	243	243.061 374	α, γ	7.37×10^3 y
96	Curium	Cm	245	245.065 487	α, γ	8.5×10^3 y
97	Berkelium	Bk	247	247.070 03	α, γ	1.4×10^3 y
98	Californium	Cf	249	249.074 849	α, γ	351 y
99	Einsteinium	Es	254	254.088 02	α, γ, β^-	276 d
100	Fermium	Fm	253	253.085 18	EC, α, γ	3.0 d
101	Mendelevium	Md	255	255.0911	EC, α	27 min
102	Nobelium	No	255	255.0933	EC, α	3.1 min
103	Lawrencium	Lr	257	257.0998	α	$\approx$35 s
104	Rutherfordium	Rf	261	261.1087	α	1.1 min
105	Hahnium	Ha	262	262.1138	α	0.7 min

*The masses given throughout this table are those for the neutral atom, including the Z electrons.
†EC stands for electron capture.

Answers to Follow-up Exercises

Chapter 1

1.1 A mass of 1000 kg has a weight of 2200 lb. The weight of a metric ton is equivalent to the British *long* ton (2200 lb), or 200 lb greater than the British *short* ton (2000 lb).

1.2 Yes, m = m.

1.3 (a) 50 mi/h [(0.447 m/s)/(mi/h)] = 22 m/s. (b) Since (1609 m/mi) = 1 and (1 h/3600 s) = 1, then multiplying the two factors gives 1. Then, trading denominators, we get (1609 m/3600 s)(h/mi) = 1. Thus, 0.447 m/s = 1 mi/h.

1.4 13.3 times

1.5 $1 \text{ m}^3 = 10^6 \text{ cm}^3$

1.6 (a) To get the smallest numerical value, we must make the number in the numerator as small as possible and the number in the denominator as large as possible. It is clear from Example 1.6 that, since 1 mi = 1.61 km, the distance in the numerator should be in miles. Similarly, since there are about 4 L in 1 gal, the volume in the denominator should be in liters. Expressing fuel economy in miles per liter will give the smallest numerical value. (b) To demonstrate the comparative numerical values, we convert one set of units to the other three. Arbitrarily selecting 1 mi/gal, on converting, we obtain 1 mi/gal = 1.61 km/gal = 0.264 mi/L = 0.425 km/L.

1.7 (a) $7.0 \times 10^5 \text{ kg}^2$ (b) 3.02×10^2 (no units)

1.8 (a) 23.70 (b) 22.09

1.9 $2.3 \times 10^{-3} \text{ m}^3$, or $2.3 \times 10^3 \text{ cm}^3$

1.10 16.9 m

1.11 $1.03 \times 10^3 \text{ cm}^2$

1.12 1.8×10^{11} metric tons (180 000 million metric tons, or 3000 times the annual U.S. output)

Chapter 2

2.1 (a) No. Using meters per minute, the answer would be in minutes. (b) 0.200 m/min (or 0.003 33 m/s)

2.2 (a) Not if there is motion; $d \neq 0$, so *avg.sp.* $= d/t \neq 0$. A zero speed would simply imply that $d \neq 0$, or the object is at rest. (b) 4.4×10^4 mi/h (44 000 mi/h—pretty fast)

2.3 33 km/h

2.4 No. The + and − signs indicate the vector directions with respect to the reference axis. If the velocity and acceleration are in *opposite* directions, a moving object will slow down. For example, suppose that a car traveling in the negative x direction ($-v_o$) experiences an acceleration in the positive x direction ($+a$). The car would slow down, so a positive acceleration ($+a$) can produce a deceleration. Similarly, if both the velocity and acceleration are in the *same* direction, the car will speed up. For example, if the car is initially traveling in the negative x direction ($-v_o$), a negative acceleration ($-a$) will speed it up in that direction.

2.5 9.0 m/s in the direction of the original motion

2.6 96 m (A lot quicker, isn't it?)

2.7 0.904 s (The positive is root taken—why?)

2.8 The speeds of the cars as a function of time are given by Eq. 2.7, $v = v_o + at$. With $v_o = 0$, $v = at$. Hence, speed is proportional to t (that is, $v \propto t$); if car B accelerates twice as long, it will have twice the speed. (No t^2 dependence here.)

2.9 20 ft (Speed in a school zone *can* make a difference.)

2.10 1.16 s longer

2.11 Time for bill to fall its length = 0.179 s. This is less than the average reaction time (0.192 s) computed in the Example, so most people cannot catch bill.

2.12 $y_u = y_d = 5.12$ m, as measured from reference $y = 0$ at release point.

Chapter 3

3.1 $v_x = -0.40$ m/s (in $-x$ direction), $v_y = +0.30$ m/s

3.2 $x = 9.00$ m, $y = 12.6$ m (same)

3.3 Relative to the ground, or "fixed" reference frame

3.4 $\mathbf{v} = 0\,\hat{\mathbf{x}} + (3.7 \text{ m/s})\,\hat{\mathbf{y}}$

3.5 No, the boat would curve back and forth with both components acting.

3.6 $d = 524.976$ m $= 525$ m (rounding differences.)

3.7 14.5° W of N

3.8 $\mathbf{v} = (8.25 \text{ m/s})\,\hat{\mathbf{x}} + (-22.1 \text{ m/s})\,\hat{\mathbf{y}}$

3.9 $(x, y) = (73.8 \text{ m}, 7.50 \text{ m})$

3.10 Yes, the ball is in the air 1.77 s before reaching the hoop.

3.11 Consider the velocity components of the player's motion. Note that for the vertical motion near the maximum height (similar to Fig. 3.17b), the upward speed (v_y) is quite small, going to zero at the maximum height and then increasing from zero in the descent. During this time, the combination of the slow vertical motions and the constant horizontal motion gives the illusion that the player hangs in the air.

Chapter 4

4.1 6.0 m/s in the direction of the net force

4.2 (a) 11 lb (b) weight in pounds ÷ 2.2 lb/kg

4.3 8.3 N

4.4 (a) 50° above the $+x$ axis
(b) $\mathbf{v} = (9.8 \text{ m/s})\,\hat{\mathbf{x}} + (4.5 \text{ m/s})\,\hat{\mathbf{y}}$

4.5 2000 N

4.6 $a = \dfrac{(m_2 - m_1) - f}{m_1 + m_2}$

4.7 (a) 7.35 N (b) neglecting air resistance, 7.35 N, downward

4.8 (a) $m_2 > 1.7$ kg (b) $\theta < 17°$

4.9 Move supports closer together so angle (and $\sin\theta$) are greater, and T ($\propto 1/\sin\theta$) is reduced.

4.10 $\mu_s = 1.41\mu_k$ (for three cases in Table 4.1)

4.11 F varies with angle, with the angle for minimum applied force being around 33° in this case (greater forces are required for 20° and 50°). In general, the optimum angle depends on the coefficient of friction.

4.12 $mg\sin\theta - f_k = ma$, block accelerates down the plane. No, would have to measure acceleration.

4.13 Air resistance depends not only on speed, but also on size and shape. If the heavier ball were larger, it would have more exposed area to collide with air molecules, and the retarding force would build up faster. Depending on the size difference, the heavier ball might reach terminal velocity first and the lighter ball strike the ground first. Or the balls might reach terminal velocity together.

Chapter 5

5.1 −20 J

5.2 Work is done initially in lifting the handles and load. But when you move the wheelbarrow forward at a constant height, no work is done by the vertical component because that component is perpendicular to the motion.

5.3 No, speed would decrease (and if it stopped, the block would then move up the plane).

5.4 $W_{x_1} = 0.034$ J, $W_x = 0.64$ J (measured from x_o)

5.5 No, $W_2/W_1 = 4$, or 4 times as much

5.6 Here we have $m_s = m_g/2$ as before. However, $v_s/v_g = (6.0 \text{ m/s})/(4.0 \text{ m/s}) = \frac{3}{2}$. Using a ratio, $K_s/K_g = \frac{9}{8}$, and the safety still has more kinetic energy than the guard. (Answer could also be obtained from direct calculations of kinetic energies, but for a relative comparison, a ratio is usually quicker.)

5.7 $W_3/W_2 = 1.4$, or 40% larger. More work, but a smaller percentage increase.

5.8 $\Delta K_{total} = 0$, $\Delta U_{total} = 0$

5.9 9.9 m/s

5.10 To determine whether the speed depends on the mass of a ball *in principle*, look at the physical relationships that apply to the situation. Here, energy is the major consideration, and as always, we should keep in mind the versatility of the conservation of energy in analyzing phenomena. The mechanical energy is conserved while the balls are in flight, so let's consider the initial energy (E_o) of a ball and its energy (E) just before striking the ground. By the conservation of energy,

$$E_o = E$$
$$\tfrac{1}{2}mv_o^2 + mgh = \tfrac{1}{2}mv^2$$

and

$$v = \sqrt{v_o^2 + 2gh}$$

The mass cancels and doesn't appear in the equation. Thus, speed is independent of mass. (Recall that in free fall, all objects or projectiles fall with the same vertical acceleration g—see Section 2.5.)

5.11 0.025 m

5.12 (a) 59% (b) $E_{loss}/t = mg(y/t) = mgv = (60mg)$ J/s

5.13 52%

5.14 (a) The replacement motor does the same work in twice the time, or half the work in the same time. (b) This motor does the same work in half the time, or twice the work in the same time.

5.15 (a) No. This would mean that all the energy input could be converted into useful work with no losses. Practically, friction and losses are always present. (b) The work output would be greater than the energy input—you'd get more out than you put in. That is, energy would be created. (That would be nice, but it would violate the conservation of energy, one of the cornerstones of physics.)

Chapter 6

6.1 5.0 m/s. Yes, this is 18 km/h or 11 mi/h, a speed at which humans can run.

6.3 $(-3.0 \text{ kg·m/s})\hat{x} + (4.0 \text{ kg·m/s})\hat{y}$

6.4 It would increase to 60 m/s: greater speed, longer drive, ideally. (There is also a directional consideration.)

6.5 (a) Yes, for the m_1/m_2/Earth system. But with m_2 attached to the Earth, the mass of this part of the system would be vastly greater than that of m_2, so its change in velocity would be negligible. (b) Assuming the ball is tossed in the + direction: for the tosser, $v_t = -0.50$ m/s; for the catcher, $v_c = 0.48$ m/s. For the ball: $p = 0$, $+25$ kg·m/s, $+1.2$ kg·m/s.

6.6 yes: $x_1 = \dfrac{m_1 v_{1_o}^2}{f}$, and $x_2 = \dfrac{m_2 v_{2_o}^2}{2f}$, then $\dfrac{x_1}{x_2} = 1$,

and $x_1 = x_2$.

6.7 (a) Yes, $K_o = K_1 + K_2$. (b) No, the y component of momentum is zero before the collision, so it must be zero afterward; thus one puck must be below the $+x$ axis and one above.

6.8 No; all of the kinetic energy cannot be lost to make the dent. The momentum after the collision cannot be zero, since it was not zero initially. Thus, the balls must be moving and have kinetic energy. This can also be seen from Eq. 6.11: $K_f/K_i = m_1/(m_1 + m_2)$, and K_f cannot be zero (unless m_1 is zero, which is not possible).

6.9 5.0 m

6.10 No. If $v_1 = v_2$, then $m_1 - m_2 = 2m_1$, which requires that $m_2 = -m_1$. Negative mass is not possible, so $m_1 - m_2$ is always less than $2m_1$ and the velocities cannot be equal.

6.11 All of the balls swing out, but to different degrees. With $m_1 > m_2$, the stationary ball (m_2) moves off with a greater speed after collision than the incoming, heavier ball (m_1), and the heavier ball's speed is reduced after collision, in accordance with Eq. 6.16 (see Fig. 6.14b). Hence, a "shot" of momentum is passed along the row of balls with equal mass (see Fig. 6.14a), and the end ball swings out with the same speed as was imparted to m_2. Then, the process is repeated: m_1, *now moving more slowly*, collides again with the initial ball in the row (m_2), and another, but smaller, shot of momentum is passed down the row. The new end ball in the row receives less kinetic energy than the one that swung out just a moment previously, and so doesn't swing as high. This process repeats itself instantaneously for each ball, with the observed result that all of the balls swing out to different degrees.

6.12 $M = 11$ kg, initially at origin

6.13 $(X_{CM}, Y_{CM}) = (0.47 \text{ m}, 0.10 \text{ m})$; same location as in Example, two-thirds of the length of the bar from m_1. Note: The location of the CM does not depend on the frame of reference.

6.14 Yes, the CM does not move.

Chapter 7

7.1 1.61×10^3 m $= 1.61$ km (about a mile)

7.2 1% (still a pretty good approximation)

7.3 (a) 4.7 rad/s, 0.38 m/s; 4.7 rad/s, 0.24 m/s (b) To equalize the running distances, because the curved sections of the track have different radii and thus different lengths.

7.4 120 rpm

7.5 (a) 86% (b) No; the astronaut still has mass and thus, with gravity acting, has weight by definition. (See the end of Section 7.6.)

7.6 106 rpm

7.7 The string cannot be exactly horizontal; it must make some small angle to the horizontal so that there will be an upward component of the tension force to balance the ball's weight.

7.8 No; it depends on mass: $F_c = \mu_s mg$.

7.9 No. Both masses have the same angular frequency or speed ω, and $a_c = r\omega^2$, so actually $a_c \propto r$. Remember, $v = 2\pi r/T$, and note that $v_2 > v_1$, with $a_c = v^2/r$.

7.10 The direction of both $\boldsymbol{\omega}$ and $\boldsymbol{\alpha}$ for the CD would be downward, perpendicular to the plane of the CD.

7.11 -0.031 rad/s^2

7.12 0.0028 m/s^2 (a large force, but a small acceleration)

7.13 (a) No, they do not vary linearly; $\Delta U = 2.4 \times 10^9$ J, only a 9.1% increase. (b) A smaller negative value means a greater energy in the Earth's negative potential well (higher up in the well).

7.14 This is the amount of *negative* work done by an external force or agent when the masses are brought together. The masses attract each other gravitationally, so without an external force they would accelerate toward each other and collide. An external agent must apply forces in opposite directions to prevent this and get the masses in the configuration, hence the work is negative. To separate the masses by infinite distances, an equal amount of positive work (against gravity) would have to be done.

7.15 4.7 times

7.16 Because of air resistance, the spacecraft would lose more energy and continue its downward spiral. With this frictional effect, the spacecraft would heat up. (Recall that meteors burn up in the atmosphere.) To keep a spacecraft from overheating, a material is used that evaporates from the outer surface tiles. This requires work, which is supplied by heat energy. Enough heat is carried off for the spacecraft to remain relatively cool. (See latent heat, Chapter 11.)

7.17 You wouldn't want to apply a forward thrust as in this Example, since the radius of the orbit of your spacecraft would increase and the spacecraft traveled more slowly, falling farther behind. By using a reverse thrust (Example 7.16), you would move to a lower orbit and speed up. Once you had overtaken the station, you could apply an appropriate forward thrust to put the spacecraft back into the proper orbit for docking.

Chapter 8

8.1 $s = r\theta = (0.12 \text{ m})(1.7) = 0.20$ m; $s = v_{CM}t = (0.10 \text{ m/s}) (2.00 \text{ s}) = 0.20$ m

8.2 The weights of the balls and the forearm produce torques that tend to cause rotation in the direction opposite that of the applied torque.

8.3 $T \propto 1/\sin \theta$, and as θ gets smaller, so does $\sin \theta$ and T increases. In the limit, $\sin \theta \to 0$ and $T \to \infty$.

8.4 $\Sigma \tau$: $Nx - m_1gx_1 - m_2gx_2 - m_3gx_3 = (200 \text{ g})g(50 \text{ cm}) - (25 \text{ g})g(0 \text{ cm}) - (75 \text{ g})g(20 \text{ cm}) - (100 \text{ g})g(85 \text{ cm}) = 0$, where $N = Mg$.

8.5 No. With f_{s_2}, the reaction force N would not generally be the same (f_{s_2} and N are perpendicular components of the force exerted on the ladder by the wall). In this case, we still have $N = f_{s_1}$, but $Ny - (m_1g)x_1 - (m_mg)x_m - f_{s_2}x_3 = 0$.

8.6 5 bricks

8.7 (d) no (equal masses) (e) Yes; with larger mass farther from axis of rotation, $I = 360$ kg·m^2.

8.8 The long pole (or your extended arms) increases the moment of inertia by placing more mass farther from the axis of rotation (the tightrope or rail). When the walker leans to the side, a gravitational torque tends to produce a rotation about the axis of rotation, causing a fall. However, with a greater rotational inertia (greater I), the walker has time to shift his or her body so that the center of gravity is again over the rope or rail and thus again in (unstable) equilibrium. With very flexible poles, the CG may be below the wire, thus ensuring stability.

8.9 $t = 0.63$ s

8.10 $\alpha = \dfrac{2mg - (2\tau_f/R)}{(2m + M)R}$; $\dfrac{N}{\text{kg·m}}$, and $\dfrac{N}{\text{kg·m}} = \dfrac{\text{kg·m/s}^2}{\text{kg·m}} = \dfrac{1}{\text{s}^2}$

8.11 No; in general, they have different moments of inertia.

8.12 The yo-yo would roll back and forth, oscillating about the critical angle.

8.13 (a) 0.24 m (b) The force of *static* friction, f_s, acts at the point of contact, which is always instantaneously at rest and so does no work. Some frictional work may be done due to rolling friction, but this is considered negligible for hard objects and surfaces.

8.14 $v_{CM} = 2.2$ m/s; using a ratio, 1.4 times greater; no rotational energy

8.15 You already know the answer: 5.6 m/s. (It doesn't depend on the mass of the ball.)

Chapter 9

9.1 (a) +0.10% (b) 39 kg

9.2 2.3×10^{-4} L, or 2.3×10^{-7} m^3

9.3 3.03×10^4 N (or 6.82×10^3 lb—about 3.4 tons!) This is roughly the force on your back right now. Our bodies don't collapse under atmospheric pressure because cells are filled with incompressible fluids (mostly water!), bone, and muscle, which react with an equal outward pressure (equal and opposite forces). As with forces, it is a pressure *difference* that gives rise to dynamic effects.

9.4 $d_o = \sqrt{\dfrac{F_o}{F_i}} d_i = \sqrt{\dfrac{1}{10}}(8.0 \text{ cm}) = 2.5$ cm

9.5 10.3 m (about 34 ft). You can see why we don't use water barometers.

9.6 Pressure in veins is lower than that in arteries (120/80).

9.7 greater, by 7.75×10^2

9.8 The object would sink, so the buoyant force is less than the object's weight. Hence, the scale would have a reading greater than 40 N. Note that with a greater density, the object would not be as large and less water would be displaced.

9.9 11%

9.10 The water level falls. When in the boat, the block is supported by buoyant forces; that is, it is essentially floating and displaces its own weight in water. When the block is in the water, however, it displaces only its own *volume* of water. Since the block is denser than water, less water is displaced in the second case than in the first.

9.11 −18%

9.12 69%

9.13 As the water falls, speed (v) increases and area (A) must decrease to have Av = a constant.

9.14 0.38 m

Chapter 10

10.1 (a) −40°C (b) You should immediately know the answer—this is the temperature at which the Fahrenheit and Celsius temperatures are numerically equal.

10.2 (a) $T_R = T_F + 460$ (b) $T_R = \frac{9}{5}T_C + 492$ (c) $T_R = \frac{9}{5}T_K$

10.3 96°C

10.4 273°C; no, not on Earth

10.5 50 C°

10.6 It depends on the metal of the bar. If the thermal expansion coefficient (α) of the bar is less than that of iron, it will not expand as much and not be as long as the diameter of the circular ring after heating. However, if the bar's α is greater than that of iron, the bar will expand more than the ring and the ring will be distorted.

10.7 Basically, the situations would be reversed. Faster cooling would be achieved by submerging the ice in Example 10.7—the cooler water would be less dense and would rise, promoting mixing. For a lake with cooling at the surface, cooler, less-dense water would remain at the surface until minimum density was achieved. With further cooling, the denser water would sink and freezing would occur from the bottom up.

10.8 v_{rms}, 1.69%; K, 3.41%

10.9 The rotational kinetic energy for oxygen is the difference between the total energies, 2.44×10^3 J. The oxygen is less massive and so has the higher v_{rms}.

Chapter 11

11.1 21.4 m (about the height of a five-story building)

11.2 -2.5×10^4 J

11.3 40°C

11.4 Mercury has the greater temperature change—29.9 times greater than that of water.

11.5 The final temperature (T_f) would be less, so the computed value of c would be lower.

11.6 2.28×10^6 J

11.7 24°C

11.8 42 J/s

11.9 No; the air spaces provide good insulation because air is a poor conductor. The many small pockets of air between the body and the outer garment form an insulating layer that minimizes conduction and so retards the loss of body heat. (There is little convection in the small spaces.) Similarly, for a diver in a wet suit, a thin film of water acts as an insulating layer.

11.10 2.1×10^4 s (350 min, or 5.8 h)

11.11 Drapes reduce heat loss by limiting radiation through the window and by keeping convection currents away from the glass.

Chapter 12

12.1 Pumping quickly amounts to an adiabatic compression of the gas in the pump cylinder (why?). Work done *on* a system is negative ($-W$), and since $Q = 0$, the first law of thermodynamics tell us that the change in internal energy is positive since $\Delta U = Q - W = 0 - (-W) = +W$. Thus all the work the bicyclist does goes into increasing the internal energy U. Since the internal energy of a gas is proportional to its temperature, there will be a corresponding increase in temperature: $\Delta T \propto \Delta U > 0$.

12.2 122°C

12.3 -1.22×10^3 J/K

12.4 Overall zero entropy change requires $|\Delta S_w| = |\Delta S_m|$ or $|Q_w/T_w| = |Q_m/T_m|$. Because the system is isolated, the magnitudes of the two heat flows *must* be the same $|Q_w| = |Q_m|$. Thus no overall entropy change requires the water and the metal to have the same average temperature $\overline{T}_w = \overline{T}_m$. This is not possible, unless they are initially at the *same* temperature. This represents a very uninteresting situation, involving no heat flow! So if there is any heat flow (due to initial temperature differences), zero overall entropy change is impossible.

12.5 As the blades of grass break down, they become more disordered. Thus they experience an entropy increase. The overall universe entropy must also increase, since this degradation is a naturally occurring process.

12.6 An increase in humidity would inhibit evaporation from the bird's head, thus decreasing the dunking rate. A humidity decrease would increase the dunking rate. A change in atmospheric pressure would have no significant effect on the liquid in the bird's body and therefore no noticeable effect on its operation.

12.7 150 J/cycle

12.8 (a) $COP_{ref} = 3.3$, $COP_{hp} = 4.3$ (b) COP_{ref}

12.9 an increase of 7.5%

Chapter 13

13.1 No, its maximum speed is $(\sqrt{k/m})A = 4.0$ m/s. Thus it is traveling at 75% of the maximum speed.

13.2 0.49 J

13.3 (a) 0.63 m/s (b) $x = (0.20\,\text{mm})\cos(1000\pi t)$, $x = -0.19$ mm at $t = 1.1 \times 10^{-3}$ s

13.4 9.76 m/s²; no. Since this is less than the accepted value at sea level, the park is probably at an altitude above sea level.

13.5 (a) 0.50 m (b) 0.10 Hz

13.6 440 Hz

13.7 increase the tension

Chapter 14

14.1 (a) 2.3 (b) 10.2

14.2 (a) 858 m (b) $\Delta t = 0.05$ s

14.3 It would be greatest in He, because it has the smallest molecular mass. (It would be lowest in oxygen, which has the largest molecular mass.)

14.4 (a) The dB scale is logarithmic, not linear. (b) 3.16×10^{-6} W/m²

14.5 No, $I_2 = 316I_1$.

14.6 65 dB

14.7 destructive interference: $\Delta L = 2.5\lambda = 5(\lambda/2)$, and $m = 5$ No sound would be heard if the waves from the speakers had equal amplitudes. Of course, during a concert the sound would not be single-frequency tones but would have a variety of frequencies and amplitudes. Listeners at certain locations might not hear certain parts of the audible spectrum, but this probably wouldn't be noticed.

14.8 toward, 431 Hz; past, 369 Hz

14.9 With the source and the observer traveling in the same direction at the same speed, their relative velocity would be zero. That is, the observer would consider the source to be stationary. Since the speed of source and observer is subsonic, the sound from the source would overtake the observer without a shift in frequency. Generally, for motions involved in a Doppler shift, the word *toward* is associated with an *increase* in frequency and *away* with a *decrease* in frequency. Here, the source and observer remain a constant distance apart. (What would be the case if the speeds were supersonic?)

14.10 768 Hz; yes

Chapter 15

15.1 (a) positive charge (b) $+2.0$ μC (c) deficiency (d) It loses 1.3×10^{13} electrons.

15.2 No; if the comb were positive, it would polarize the paper in the reverse direction and still attract it.

15.3 $(0.17\,\mathrm{N})\hat{\mathbf{x}}$ As the x distance increases, the angle θ decreases. Thus the x components approach a value equal to the magnitude of each force. Their total approaches a value equal to twice each force's magnitude.

15.4 0.12 m

15.5 $F_e/F_g = 4.16 \times 10^{42}$; F_g is much *less* than F_e because the electron's mass is considerably smaller than a proton's, and F_e (magnitude) doesn't change.

15.6 The field is zero to the left of q_1 at $x = -0.60$ m. The fields *oppose* one another both to the left of q_1 and the right of q_2. Both point to the right between the charges and cannot cancel there. Since q_1 is the smaller of the two in magnitude, the location of zero net field must be closer to it than to q_2.

15.7 $\mathbf{E} = (-65.8\,\mathrm{N/C})\hat{\mathbf{x}} + (360\,\mathrm{N/C})\hat{\mathbf{y}}$, or 366 N/C at 79.6° above the $-x$ axis.

15.8 By symmetry, the y components cancel and the net electric field is in the $+x$ direction and given by $E_{net} = 2E \cos\theta$. For large distances, $\theta \to 0°$ and $\cos\theta \to 1$. Similarly, $x \to r$ and E from each charge becomes $E = kq/r^2 \to kq/x^2$. Thus $E_{net} = 2E\cos\theta \to 2(kq/x^2)(1) = k(2q)/x^2$. At large values of x, the separation distance d between the charges appears to be zero, so it is not in the answer. The charges appear to be superimposed and located at the origin. Thus the field approaches that of a single point charge with a charge of $Q = 2q$.

15.9 (a) E points upward, from ground to cloud.
(b) 5.6×10^{11} electrons/m^2

15.10 Positive charge would move to the outside surface. The leaves of the electroscope attached to the outside would deflect. The leaves of the electroscope attached to the inside surface would not deflect.

15.11 Their sign is negative since electric field lines points toward negative charges—inward at our Gaussian surface. In general, the net number of lines passing through a Gaussian surface is proportional to the net charge enclosed by it. In this case, the net charge is negative and so is the net count on the field lines.

Chapter 16

16.1 (a) $\Delta U_e = +7.20 \times 10^{-19}$ J (b) ΔV is unchanged (why?). (c) ΔV_{mid} is unchanged. (d) $v = 4.65 \times 10^4$ m/s

16.2 E is in the $-x$ direction, so the proton gains electric potential energy and lose kinetic energy as it travels in the $+x$ direction. By equating the electric potential energy gain to the loss in kinetic energy (the proton stops, so $\Delta K = K - K_o = -K_o$ here), the electric field strength is $E = 9.39 \times 10^4$ V/m.

16.3 (a) $\Delta U_e = +4.35 \times 10^{-18}$ J (b) The electron moves to a region of lower electric potential since it moves farther from the positively charged proton.

16.4 $U_{12} = -3.27 \times 10^{-19}$ J, $W = +3.27 \times 10^{-19}$ J

16.5 (a) No, you know only the potential difference, not the absolute value of the potentials. There is no mention of where the value of zero electric potential has been assigned. (b) Under storm conditions, E is upward. Thus the potential decreases upward and the higher potential surface is closer to the Earth's surface.

16.6 $d = 2.7 \times 10^{-11}$ m; this value is much smaller than an atomic diameter, so such a capacitor is not practical.

16.7 $V = 7.91 \times 10^3$ V

16.8 If the spacing increased, C would decrease. Since $Q = CV$, then $Q \propto C$ when the voltage across the capacitor is kept constant. Hence, Q would decrease and charge would flow away from the capacitor: $\Delta Q = -3.30 \times 10^{-12}$ C.

16.9 $V_1 = 8.00$ V; $V_2 = 4.00$ V; $U_{C_1} = 8.00 \times 10^{-5}$ J; $U_{C_2} = 4.00 \times 10^{-5}$ J

16.10 (a) $Q_1 = 8.0 \times 10^{-7}$ C; $Q_2 = 1.6 \times 10^{-6}$ C; $Q_3 = 2.4 \times 10^{-6}$ C
(b) $U_{C_1} = 3.2 \times 10^{-6}$ J; $U_{C_2} = 6.4 \times 10^{-6}$ J; $U_{C_3} = 4.8 \times 10^{-6}$ J

Chapter 17

17.1 1.0×10^9 s, or about 32 y!

17.2 (a) 12 Ω (b) 4.8×10^2 Ω (c) 1.2×10^3 Ω

17.3 0.82 m

17.4 $R = 0.67$ Ω; for a sensitive electrical thermometer, use material with a high temperature coefficient of resistivity to get a large resistance change for a given temperature change.

17.5 print mode: $R = 82.3$ Ω; standby mode: $R = 1.8$ kΩ

17.6 First form: $P_o = I_o V_o$. Since the resistor is ohmic, when $V_o \to 2V_o$, then $I_o \to 2I_o$ and the power becomes $P \to (2I_o)(2V_o) = 4I_o V_o = 4P_o$.
Second form: $P_o = V_o^2/R_o$. Since the resistor is ohmic, when $V_o \to 2V_o$, the power becomes $P_o \to (2V_o)^2/R_o = 4V_o^2/R_o = 4P_o$.
Third form: $P_o = I_o^2 R_o$. Since the resistor is ohmic, when $V_o \to 2V_o$, then $I_o \to 2I_o$ and the power becomes $P \to (2I_o)^2 R_o = 4I_o^2 R_o = 4P_o$.

17.7 The dryer's power would increase, because the wire's cross-sectional area is larger, thus reducing its resistance ($P = V^2/R$). The new power rating would increase by 0.25 kW to 1.45×10^3 W. Since the power goes up, this could potentially be dangerous, since the filament could overheat and start a fire.

17.8 8.3 h

17.9 At best, power plants produce electric energy with efficiencies of 35% (ignoring joule heat losses during transmission). Thus, in terms of primary fuels, 35% is the maximum efficiency of *any electrical appliance*. The electric hot water heater would then convert 95% of the delivered electric energy to heat. The net result is that 95% of 35%, or only 33%, of the original fuel energy content is actually used to heat water. For a natural gas system, the gas is delivered at essentially no energy cost to the water heater, where it can be burned. Thus approximately 95% of the gas's heat content goes into heating the water, and gas heating systems are overall almost three times more efficient than the electric type.

Chapter 18

18.1 (a) series: $P_1 = 4.0$ W, $P_2 = 8.0$ W, $P_3 = 12.0$ W; parallel: $P_1 = 144$ W, $P_2 = 72$ W, $P_3 = 48$ W (b) In series, the most power is dissipated in the largest resistance. In parallel, the most power is dissipated in the smallest resistance. (c) The series arrangement requires less power: It dissipates less total power, because it has a larger equivalent resistance and carries a much smaller current at the same voltage.

18.2 The voltage of each light bulb increases by a factor of 18/17. Since the power dissipated in each is proportional to the *square* of the voltage (resistance stays the same), the power increases by a factor of $(18/17)^2 = 1.12$, or a 12% increase.

18.3 $P_1 = I_1^2 R_1 = 54.0$ W, $P_2 = I_2^2 R_2 = 9.0$ W, $P_3 = I_3^2 R_3 = 0.87$ W, $P_4 = I_4^2 R_4 = 2.55$ W, $P_5 = I_5^2 R_5 = 5.63$ W; the sum of the power dissipated in each resistor is 72.1 W, rounded to three significant figures. For the battery, we have agreement to within rounding error since $P_{batt} = I_{batt}V_{batt} = 72.0$ W.

18.4 (a) Since there are no junctions, $I_1 = I_2$, which we call I. Using the loop theorem, we have $12 - 6I - 3I = 0$, so $I = 1.33$ A. (b) $V_1 = IR_1 = 8.0$ V, $V_2 = IR_2 = 4.0$ V

18.5 At the junction, we still have $I_1 = I_2 + I_3$ (Eq. 1). Using the loop theorem around loop 3 in the clockwise direction gives $6 - 6I_1 - 9I_2 = 0$ (Eq. 2). For loop 1, we still get $6 - 6I_1 - 12 - 2I_3 = 0$ (Eq. 3). Solve Eq. 1 for I_2 and substitute into Eq. 2. Then solve Eq. 2 and Eq. 3 simultaneously for I_1 and I_3. All answers are the same as those in the Example.

18.6 4.2 s

18.7 10 A

18.8 1.99×10^{-4} A, or 199 μA

Chapter 19

19.1 south

19.2 (a) Since the proton has a positive charge, it would initially be deflected in the opposite direction, or +z. Use the right-hand force rule. (b) 0.10 T

19.3 1.5×10^3 A

19.4 5.0×10^3 turns/m, or 1.5×10^3 turns in the solenoid length of 0.30 m

19.5 (a) 3.3×10^{-3} N/m (b) The force is repulsive, since the currents are oppositely directed. You should be able to show this by using the right-hand source and force rules.

19.6 0.13 A·m^2, about 1.4×10^{22} times larger than that due to the orbital motion of the electron

19.7 0.50 V

Chapter 20

20.1 (a) any way that will increase the flux, such as increasing the loop area or the number of loops, changing to a lower resistance wire would also help (b) A dc current creates a steady magnetic field, thus a steady flux and no emf.

20.2 $\overline{P} = 2.2 \times 10^{-4}$ W, $E = \overline{P}t = 1.8 \times 10^{-6}$ J

20.3 clockwise

20.4 17%

20.5 (a) 6.1×10^3 J (b) 5.0×10^2 J, 12 times more energy

20.6 (a) You would use it as a step-down transformer—the primary coil would have 100 windings and the secondary only 50, because U.S. appliances are designed to work at half of the European voltage of 240 V. (b) The voltage would double to 480 V. The appliance, designed to run at 120 V, would have four times the required voltage. Since $P = V^2/R$, with ohmic resistance, the power would go up by a factor of 16 to about 19 kW. It would burn up, melt, and perhaps start a fire.

20.7 (a) Higher voltages allow lower current usage. This in turn reduces joule heat losses in the delivery wires and in the motor windings, making more energy available for doing mechanical work and therefore a more efficient use of the electric energy. (b) Since the voltage is doubled, the current is halved. The heat loss in the wire is proportional to the *square* of the current. Thus losses will be cut by a factor of 4 or be reduced to 25% of their value at 120 V.

20.8 The round-trip time for electromagnetic radiation is about 8.8 min. In that time, *Sojourner* will travel about 0.44 m and will not collide with the rock.

20.9 (a) With increasing distance, the Sun's light intensity drops, as would the force of light pressure on the sail. In turn, the ship's acceleration would be reduced. (b) by enlarging the sail area to catch more light

Chapter 21

21.1 (a) 0.25 A (b) 0.35 A (c) 9.6×10^2 Ω, larger than the 240 Ω required for a bulb of the same power in the United States. The voltage in Great Britain is larger than that in the United States; to keep the power constant, the current must be reduced by using a larger resistance.

21.2 If the motor of the razor runs off the voltage frequency, then it, and the electric clock as well, might run slower.

21.3 (a) $\sqrt{2}(120$ V$) = 170$ V (b) 30 Hz

21.4 (a) $\sqrt{2}(2.55$ A$) = 3.61$ A (b) 20 Hz

21.5 (a) would increase to 0.896 A (b) The capacitor; with a frequency increase, X_C decreases. Since resistance is independent of frequency, it remains constant and overall Z decreases.

21.6 They would be equal.

21.7 (a) 41.1 Hz (b) 576 W

21.8 (a) As the frequency driving an RL circuit increases, inductive reactance increases ($X_L = 2\pi fL$). Thus Z *increases* since $Z = \sqrt{R^2 + X_L^2}$, and current I decreases. Since the power loss occurs entirely across the resistor (why?) ($P = I^2R$), power decreases. (b) As the frequency in an RC circuit increases, capacitive reactance *decreases* ($X_C = 1/2\pi fC$). Thus Z also decreases. This results in an increase in both current and power.

21.9 If you have a receiver tuned to a frequency *between* the two station frequencies, you would not receive the maximum strength signal from either station but there might be enough power from each to hear them simultaneously.

21.10 The voltage across the inductor opposes the voltage across the capacitor. Since they have equal magnitudes, when we add the voltages, including their polarity, they cancel: 25.0 V + $(1.08 \times 10^3$ V$) - (1.08 \times 10^3$ V$) = 25.0$ V. We get the source voltage, in agreement with Kirchhoff's loop theorem.

Chapter 22

22.1 Light travels in straight lines; by the law of reflection, with $\theta_i = \theta_r$, if you can see someone in a mirror, that person can see you. Conversely, if you can't see the trucker's mirror, then he or she can't see your image in that mirror and won't know that your car is behind the truck.

22.2 $n = 1.25$

22.3 By Snell's law, $n_2 = 1.25$, so $v = c/n_2 = 2.40 \times 10^8$ m/s.

22.4 It would be smaller.

22.5 (a) The frequency of the light is unchanged in the different media, so the emerging light has the same frequency as that of the source. (b) The wavelength in air is independent of the water and glass media, as can be shown by adding another step (medium) to the Example solution. By reverse analysis, $\lambda_{air} = n_{water}\lambda_{water} = (c/v_{water})\lambda_{water} = c/f$. Thus, the wavelength in air is c/f.

22.6 Because of total internal reflections, the diver would see the reflection of something on the sides and/or bottom of the pool. (Use reverse ray tracing.)

22.7 $n = 1.4574$

Chapter 23

23.1 No effect. Note that the solution to the Example does not include the distance. The geometry of the situation is the same regardless of the distance from the mirror.

23.2 $d_i = 30$ cm, real, inverted, and same size. Note from the Example results (a) and (b) that when the object is closer to the mirror than is the center of curvature, the image becomes magnified.

23.3 The image always upright and reduced (and virtual).

23.4 $d_i = -d_o$ and $M = +1$; virtual (behind mirror, same distance as object is in front), upright, and same size

23.5 $|M| = 0.75$

23.6 $|M| = \frac{1}{3}$ (Recall that the image is always reduced with a diverging lens.)

23.7 Blocking off half of the lens would halve the *amount* of light focused at the image plane, so the resulting image would be less bright but still full size.

23.8 3.0 cm to right of L_2; real, inverted, and $|M| = \frac{3}{4}$

Chapter 24

24.1 $\Delta y = y_r - y_b = 1.20 \times 10^{-2}$ m

24.2 twice as thick, $t = 199$ nm

24.3 In brass instruments, the sound comes from a relatively large, flared opening. There is thus little diffraction, so most of the energy is radiated in the forward direction. In woodwind instruments, much of the sound comes from tone holes along the column of the instrument. These holes are small compared to the wavelength of the sound, so there is appreciable diffraction. As a result, the sound is radiated in nearly all directions, even backward.

24.4 The width would increase by a factor of 1.65.

24.5 fewer; $n \leq 1.43$, so $n = 0, 1$

24.6 $\theta_2 = 35.9°$

24.7 $\lambda = 588$ nm; yellow

24.8 It follows that the particles making up clouds scatter all wavelengths of incident (sun) light. Clouds are made up of small water droplets and ice crystals (high clouds). But, though small, they are still much larger than the wavelengths of light, so there is no preferential scattering. (Storm clouds, which have more, and larger, droplets appear dark because they block out the sunlight that is strongly scattered on the other side of the cloud.)

Chapter 25

25.1 It wouldn't work, a real image would form on person's side of lens ($d_i = +75$ cm).

25.2 For an object at $d_o = 25$ cm, the image for eye 1 would be formed at 1.0 m; this is beyond the near point for that eye, so the object could be seen clearly. The image for eye 2 would be formed at 0.77 m; this is inside the near point for that eye, so the object would not be seen clearly.

25.3 glass for near-point viewing, 2.0 cm longer

25.4 length doubles

25.5 $f_i = 8.0$ cm

25.6 The erecting lens (of focal length f_e) should go between the objective and the eyepiece, positioned a distance of $2f_e$ from the image formed by the objective, which acts as an object. The erecting lens then produces an inverted image of the same size at $2f_e$ on the opposite side of the lens, which acts as an object for the eyepiece. The use of the erecting lens lengthens the telescope by $4f_e$.

25.7 3.4×10^{-7} rad, an order of magnitude better

25.8 3.46×10^{-2} m

25.9 (a) $v = 2.26 \times 10^8$ m/s, $\lambda' = 338$ nm, $f' = 6.67 \times 10^{14}$ Hz (b) blue (unchanged) (c) Wavelength of 338 nm lies in the nonvisible ultraviolet region, so the object would not be seen or would appear black.

Chapter 26

26.1 Light waves from two simultaneous events on the y axis meet at some midpoint receptor on the y axis. Since there is no relative motion along that axis, a simultaneous recording of the two events will also be recorded along the y' axis. Hence the two observers agree on simultaneity for this situation.

26.2 $v = 0.9995c$; no, not twice as fast. Travel is limited to less than c, so this is only about a 0.15% increase.

26.3 (a) 0.667 μs (b) 0.580 μs (c) The observer watching the ship measures the proper time interval. To the person on the ship, that time interval is dilated.

26.4 $v = 0.991c$

26.5 The traveler measures the proper time interval of 20.0 y, but Earth inhabitants measure the dilated version of this (why?). The gamma factor is based on a recalculated value for traveler speed $v = 0.995\,04c$. Keeping five places after the decimal and then rounding to three significant figures we get

$$\gamma = 1/\sqrt{1 - (0.995\,04c/c)^2} = 10.049\,88$$

and

$$t = \gamma t_o = (10.049\,88)(20.0\ \text{y}) = 200.99\ \text{y} \approx 201\ \text{y}$$

26.6 (a) 1.17 MeV (b) 0.207 MeV

26.7 (a) Total momentum is zero both nonrelativistically and relativistically (both particles have the same mass but opposite velocities, so their equal and opposite momenta cancel to zero). Hence the final speed is the same using both approaches, $v' = 0$ (b) Nonrelativistically, the combined mass afterward is $m' = 2m$, exactly. Relativistically, the combined mass is $m' = 3.34m$. (c) Both approaches give a final kinetic energy of zero, since the combined particle must be at rest due to momentum conservation.

26.8 1.95×10^7 more massive

26.9 $u = +0.69c$, which, as expected, is lower in magnitude than the nonrelativistic (and wrong) result of $+0.80c$.

26.10 The velocity addition equation always yields the result c for light. Here the results are $-c$ for the observer heading in the $+x$ direction and $+c$ for the observer headed in the $-x$ direction, as expected.

Chapter 27

27.1 (a) The wavelength is 967 nm, which is infrared. With some emissions in the red region, it would appear reddish. (b) The wavelength is 290 nm, which is ultraviolet. With significant visible emissions, it would appear bluish-white.

27.2 496 nm, shorter than in the Example, indicating a larger photon energy. Thus the maximum kinetic energy of the photoelectrons is larger, requiring a higher stopping voltage.

27.3 2.00 V

27.4 (a) 3.60×10^{-2} (wavelength will increase by 3.60%) (b) This is a larger increase than that in the Example, because in a head-on collision the photon transfers the maximum possible amount of energy to the electron. The photon will lose more energy than in the "glancing" collision of the Example.

27.5 1.09×10^6 m/s

27.6 365 nm

27.7 The minimum energy required for ionization from the $n = 3$ (second excited state) in hydrogen is 1.51 eV (see Fig. 27.12). If an incoming photon carried exactly 1.51 eV, it could be absorbed and the electron would escape the atom, but essentially with no kinetic energy. The atom would "barely" be ionized. To carry 1.51 eV, the light needs a wavelength of $\lambda = hc/E = 1.24 \times 10^3$ eV·nm/1.51 eV = 821 nm, which is infrared.

Chapter 28

28.1 (a) 8.8×10^{-33} m/s (b) 2.3×10^{33} s, or 7.2×10^{25} y—about 4.8×10^{15} times longer than the age of the universe. This movement would not be noticeable.

28.2 The proton's de Broglie wavelength under the same conditions is 4.1×10^{-12} m, about twenty times smaller than atomic spacing distances. Thus protons would *not* be expected to exhibit significant diffraction effects under conditions in which electrons might.

28.3 Only one electron; a second electron would have all three quantum numbers (n, ℓ, and m_ℓ) the same as the first, which is not allowed by the Pauli exclusion principle.

28.4 (a) $1s^2\, 2s^2\, 2p^6$ (b) $-2e$, or -3.2×10^{-19} C

28.5 1.2×10^6 m/s

Chapter 29

29.1 $^{12}_{6}\text{C} + ^{4}_{2}\text{He} \rightarrow ^{16}_{8}\text{O}$; the resulting nucleus is therefore ^{16}O.

29.2 Neglecting the emitted neutrino, the decay is $^{22}_{11}\text{Na} \rightarrow ^{22}_{10}\text{Ne} + ^{0}_{+1}\text{e}$. The daughter nucleus is ^{22}Ne.

29.3 (a) 48 d (reducing the activity by a factor of 64 requires six half lives; $1/2^6 = 1/64$) (b) The biological process of excretion in the body also remove the ^{131}I from the body. Radioactive decay is not the only way to eliminate it.

29.4 The closest integer is 20, since $2^{20} \approx 1.1 \times 10^6$. (Neither 19 nor 21 works; $2^{19} \approx 0.52 \times 10^6$ and $2^{21} \approx 2.1 \times 10^6$.) Thus it takes 20 half-lives, or about 560 y.

29.5 to four ^{14}C half lives, or 2.3×10^4 y

29.6 About four half lives, or 5.1×10^9 y; this is in decent agreement with the estimate given (using other methods) but is an *overestimate* (why?).

29.7 (a) Starting with 29 protons and 29 neutrons: $^{58}_{29}\text{Cu}$, $^{59}_{29}\text{Cu}$, $^{60}_{29}\text{Cu}$, $^{61}_{29}\text{Cu}$, $^{62}_{29}\text{Cu}$, $^{63}_{29}\text{Cu}$, $^{64}_{29}\text{Cu}$, and so on. We delete the odd–odd isotopes (why?) and get the most likely (stable) isotopes: $^{59}_{29}\text{Cu}$, $^{61}_{29}\text{Cu}$, $^{63}_{29}\text{Cu}$, $^{65}_{29}\text{Cu}$, and so on. (b) We can further trim the list by deleting those with $N \approx Z$ and those with N significantly larger than Z (why?). From Fig. 29.14, for this proton number, we expect neutron numbers in the mid-30s. Hence we would guess just $^{63}_{29}\text{Cu}$ and $^{65}_{29}\text{Cu}$. According to Appendix V, these are in fact the only two stable isotopes of copper.

29.8 (a) The average binding energy per nucleon for ^{3}He is 2.573 MeV/nucleon, which is considerably smaller than the value of 7.075 MeV/nucle for ^{4}He. (b) Unlike ^{3}He, all protons and neutrons in ^{4}He *are* paired, making it the more tightly bound of the two isotopes.

29.9 (a) Removal of a single proton from ^{12}C takes only 15.96 MeV; removal of a single neutron takes a minimum of 18.72 MeV. (b) One factor is the proton repulsion due to other protons. We expect protons, in general, to be less bound to a nucleus than are neutrons.

Chapter 30

30.1 $Q = -15.63$ MeV, so it is endoergic (takes energy to happen).

30.2 The increase in mass has an energy equivalent of 1.193 MeV. The rest of the incident kinetic energy (1.534 MeV − 1.193 MeV, or 0.341 MeV) must be distributed between the kinetic energies of the proton and of the oxygen.

30.3 6.17×10^{11} kg of He produced per fusion

30.4 The neutrino and the daughter nucleus would recoil in opposite directions, conserving linear momentum (assuming the original nucleus was at rest and therefore had zero linear momentum).

30.5 For the negative pion (π^-) to have a charge of $-e$, its quark structure must be $\bar{u}d$. For the same reason, the quark structure of the positive pion (π^+) is $u\bar{d}$, and the antiparticle of the π^+ would have the composition of $\bar{u}\bar{\bar{d}}$. However, the antiquark of an antiquark is the original quark—for example, $\overline{(\bar{u})} = u$, and so on. Hence the antiparticle of the π^+ has the quark structure $u\bar{\bar{d}} = \bar{u}d$, which is the π^-, as expected.

Answers to Odd-numbered Exercises

Chapter 1

1. (b)
3. (c)
5. A duodecimal system might have a dime worth 12¢ and a dollar worth 12 "dimes," or $1.44.
7. 1 nautical mile = 6076 ft
9. (d)
11. (d)
13. (d)
15. $m^2 = m^2$
17. no, $V = \pi d^3/6$
19. yes; m/s = m/s − m/s
21. yes; $[L]^2 = [L]^2 + [L]^2$
23. 1/s, or s^{-1}
25. (a) kg·m^2/s^2 (b) yes
27. (a)
29. (a) yd (b) dm (c) cm
31. 37 000 000 times
33. 63.7 m, 60.9 m, 6.20 m, and 19.5 m
35. 68 L
37. metric; $9.4 \times 10^2\ m^2$
39. (a) 341 m/s (b) 0.268 s
41. (a) 16 km/h for each 10 mi/h (b) 105 km/h
43. (a) 59 mL (b) 3.5 oz
45. 6.1 cm
47. (a) 26 m^3 (b) $9.2 \times 10^2\ ft^3$
49. (a) $2.4 \times 10^3\ kg/m^3$ (b) 80 kg
51. (a)
53. 5.05 cm; 5.05×10^{-1} dm; 5.05×10^{-2} m
55. no; only one doubtful digit; best 25.48 cm
57. (a) 1.0 m (b) 8.0 cm (c) 16 kg (d) $1.5 \times 10^{-2}\ \mu s$
59. (a) 10.1 m (b) 775 km (c) 2.55×10^{-3} kg (d) 9.30×10^7 mi
61. $1.27 \times 10^{14}\ m^2$
63. 470 cm^2
65. (a) 2.0 kg·m/s (b) 2.1 kg·m/s, rounding difference
67. (c)
69. 1.03×10^3 g = 1.03 kg
71. (a) 52% (b) 64 g, 20 g
73. 9.47×10^{15} m
75. 0.87 m
77. same area for both, 1.3 cm^2
79. 25 min
81. 17 m
83. yes, $1.21/gal
85. 42 cm^2
87. $4.9 \times 10^2\ cm^3$
89. 0.60 L
91. not reasonable (56 mi/h)
93. $4.48 \times 10^3\ cm^3$

Chapter 2

1. (c)
3. (a)
5. no final position can be given; may be 0 to 500 m from start
7. no
9. 1.65 m down
11. 30 km, yes
13. (a) 0.50 m/s (b) 8.3 min
15. 45 m at 27° west of north
17. (a) 2.7 cm/s (b) 1.9 cm/s
19. (a) 7.35 s (b) no
21. (a) 1.0 m/s; 0; 1.3 m/s; 2.8 m/s; 0; 1.0 m/s (b) 1.0 m/s; 0; 1.3 m/s; −2.8 m/s; 0; 1.0 m/s (c) 1.0 m/s; 0; 0; −3.8 m/s (d) −0.89 m/s
23. 8.7×10^2 km
25. 60.0 mi/h; no
27. impossible
29. yes, velocity is changing due to direction change
31. v_o
33. 1.85 m/s^2
35. 3.7 s
37. 15.5 m/s
39. $a_{0\text{-}4} = 2.0\ m/s^2$; $a_{4\text{-}10} = 0$; $a_{10\text{-}18} = -1.0\ m/s^2$
41. 150 s
43. (a)
45. no, 9.9 m/s^2
47. 3.5 m/s^2
49. (a) 81.4 km/h (b) 0.794 s
51. 3.09 s
53. 23 s
55. 1.43×10^{-4} s
57. 75 m/s
59. no, x = 13.3 m
61. 96 m
63. $x - x_o = \dfrac{v_o + v}{2}(t - t_o)$;
$v = v_o + a(t - t_o)$;
$x - x_o = v_o(t - t_o) + \frac{1}{2}a(t - t_o)^2$;
$v^2 = v_o^2 + 2a(x - x_o)$
The x in Eq. 2.9 stands for displacement; the x here stands for position.
65. (d)
67. 9.80 m/s^2 downward
69. 78.4 m
71. (a) straight line, negative slope (b) parabola
73. 15.9 m
75. 67 m
77. slightly less than 8.0 m/s
79. (a) 48 m (b) 38 m/s downward
81. (a) same, 1.14 s (b) 11.2 m/s, −11.2 m/s (equal but opposite)
83. 3.61 m (no)
85. hits 14 cm in front of the prof.
87. (a) 2.45 (b) Moon: 99.2 m; 22.0 s, Earth: 16.5 m; 3.67 s
89. (a) −20.9 m/s (b) 2.87 s
91. (a) 21 m/s (b) 19 m
93. 45 mi/h
95. 71 m
97. 16.6 m/s
99. 1.2×10^2 m
101. (a) 27 m/s (b) 4.3 s
103. 51.5 m
105. (a) 5.0 m/s^2 (b) 13 m/s (c) and (d) see ISM

Chapter 3

1. (a)
3. (a) velocity increases or decreases
(b) parabolic path (c) circle
5. 5.0 m/s^2, 53° above +x axis
7. (a) 6.0 m/s (b) 3.6 m/s
9. (a) 70 m (b) 0.57 min, 0.43 min
11. x = 1.75 m; y = −1.75 m
13. (a) x = 6.56 m; y = 3.25 m (b) 5.41 m/s at 13.9° above +x axis
15. 1.0 m/s
17. (a) 4.3 m/s (b) 2.1×10^2 m
19. opposite direction, same direction
21. v_x = 8.66 m/s; v_y = 5.00 m/s
23. yes
25.

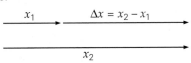

27. 109 mi/h
29. (a)

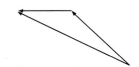

(b) 372 mi at 28.4° north of west
31. (a) (−3.4 cm) $\hat{x}$ − (2.9 cm) $\hat{y}$ (b) 4.5 cm, 63° above −x axis (c) (4.0 cm) $\hat{x}$ − (6.9 cm) $\hat{y}$
33. (a) (14.4 N) $\hat{y}$ (b) 12.7 N at 85.0° above +x axis
35. 16 m/s at 79° above the −x axis
37. 23 m/s at 3.0° above the +x axis
39. 8.5 N at 21° below −x axis
41. parallel 30 N, perpendicular 40 N
43. 27 m at 72° above −x axis
45. 102 mi/h at 61.1° north of west
47. 242 N at 48° below −x axis
49. (a) face (b) back and face (c) back
51. 13 s
53. (a) +85 km/h (b) −5.0 km/h
55. 4.0 min
57. 146 s = 2.43 min
59. (a) 68 m (b) 0.76 m/s, 11° relative to shore
61. (a)

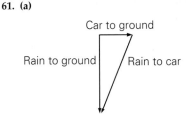

(b) 4.7 m/s
63. (a) v_{BA} = 39 km/h, 40° north of east (b) v_{AB} = 39 km/h, 40° south of west
65. see Fig. 3.17
67. the vertical motion does not affect the horizontal motion
69. 2.7×10^{-13} m
71. 40 m
73. 26 m/s and 15 m/s
75. yes, 0.11 m

A-15

77. (a) 0.77 m **(b)** would not fall back in
79. (a) 1.15 km **(b)** 30.6 s **(c)** 6.13 km
81. 35° or 55°
83. (a) 26 m **(b)** 23 m/s at 68° below horizontal
85. 40.9 m/s, 11.9° above horizonal
87. the pass is short
89. (a) 66.0 m/s **(b)** too long
91. $(4.0 \text{ m/s}) \hat{x} + (2.0 \text{ m/s}) \hat{y}$; 4.5 m/s at 27° above the $+x$ axis
93. 1.3 m at 43° south of east
95. 6.02 m/s
97. (a) $y_{max} = 10.2$ m (at $x = 20.4$ m) **(b)** yes, barely
99. (a) 30.3° or 59.7° **(b)** no
101. (a) 21.7 m/s **(b)** 33.3 m/s, 49° below horizontal
103. (a) 53° above or below $+x$ axis **(b)** ±8.0 m/s
105. 6.1×10^2 km

Chapter 4

1. no, no net force
3. (c)
5. no, same mass, same inertia
7. balloon moves **(a)** forward **(b)** backward
9. the dishes at rest tend to remain at rest
11. $m_{Al} = 1.35 m_{water}$
13. $\mathbf{F}_3 = (-7.6 \text{ N}) \hat{x}$
15. (b)
17. longer contact time, smaller force
19. 1.7 kg
21. 7.0×10^6 N
23. 75.5 kg
25. 78 N, 18 lb
27. (a) yes **(b)** no, 2.7 kg (1 lb)
29. (a) 1.2 m/s² **(b)** same
31. 2.40 m/s²
33. (a) 0.133 m/s² **(b)** 300 N, constant v
35. 8.9×10^4 N
37. (c)
39. yes, forces on different objects (one on ball, one on bat) cannot cancel
41. brick exerts a force of 800 N on the fist, no
43. 1.5 m/s², opposite to hers
45. yes, if it accelerates downward at 1.8 m/s²
47. (a) 735 N **(b)** 735 N **(c)** 585 N
49. 2.0 m/s², 19° north of east **(b)** 1.3 m/s² 30° south of east
51. (a) 0.96 m/s² **(b)** 2.6×10^2 N
53. 64 m
55. (a) 3.0 m/s² **(b)** 2.7 m/s²
57. (a) 0.711 m/s² **(b)** 1067 N
59. (a) 2.5 m/s² right **(b)** 2.0 m/s² left
61. 10 m/s²
63. 3.9×10^2 N
65. 1.2 m/s², m_1 up and m_2 down **(b)** 21 N
67. (a) no friction **(b)** opposite its direction of velocity **(c)** sideways **(d)** forward, in direction of velocity
69. kinetic friction (sliding) is less than static friction (rolling)
71. increasing normal force, therefore friction; yes, more difficult to accelerate
73. treads displace water, wide and smooth tires increase friction

75. 2.7×10^2 N
77. (a) zero **(b)** 3.1 m/s²
79. 0.064
81. 0.28 m
83. (a) 30° **(b)** 22°
85. 0.73
87. (a) 26 N **(b)** 21 N
89. 0.33
91. (a) between 0.72 kg and 1.7 kg **(b)** between 0.88 kg and 1.5 kg
93. 1.5×10^3 N
95. no
97. (a) 75 N **(b)** 7.3 N
99. 0.32 N
101. 5.2 m
103. no

Chapter 5

1. (d)
3. (a) no **(b)** yes **(c)** no **(d)** yes
5. positive on the way down, negative on the way up; not constant; maximum at points B and D ($\theta = 0°$ or 180°); minimum at points A and C ($\theta = 90°$)

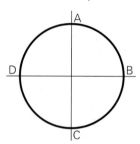

7. −98 J
9. 6.65 m
11. 3.7 J
13. 1.47×10^5 J
15. 2.4×10^3 J
17. (a) 1.48×10^5 J **(b)** -1.23×10^5 J **(c)** 2.50×10^4 J
19. (c)
21. no, more work because the force increases as the spring stretches
23. 2.5 cm
25. 1.25×10^5 N/m
27. 0.21 J
29. (a) 15 kg **(b)** 5.1 kg more
31. −3.0 J
33. Because the heel can lift from the blade, the blade will be in contact with the ice longer. This will make the displacement a bit greater (longer stride), so more work is done. More work translates to faster speed according to the work–energy theorem.
35. 19 J
37. small car is four times as long
39. (a) 45 J **(b)** 21 m/s
41. 4.2 m/s in +x direction
43. 8.3×10^3 J
45. (d)
47. 4.9×10^2 J
49. 5.9 J
51. (a) 8.0 J **(b)** 5.6 J

53. (a) $U_b = -44$ J, $U_a = 66$ J **(b)** to the attic: -1.1×10^2 J; to the ground: -1.1×10^2 J; to the basement: -1.1×10^2 J
55. (d)
57. Due to the conservation of energy, the initial potential energy is equal to the final potential energy ($U_i = U_f$), so the initial height is equal to the final height ($h_i = h_f$).
59. Each time you land on the trampoline, you coil your legs on the way down, then push down on the trampoline as you land, thereby compressing it more. Storing more energy, you rebound and go higher. The limit is determined by the "spring constant" of the trampoline and how much you can compress the trampoline.
61. (a) $K = 15.0$ J, $U = 0$, $E = 15.0$ J **(b)** $K = 7.65$ J, $U = 7.35$ J, $E = 15.0$ J **(c)** $K = 0$, $U = 15.0$ J, $E = 15.0$ J
63. (a) $K = 20$ J, $U = 60$ J **(b)** 8.9 m/s **(c)** 80 J
65. (a) 1.03 m **(b)** 0.841 m **(c)** 2.32 m/s
67. (a) 11 m/s **(b)** no **(c)** 7.7 m/s
69. (a) 2.7 m/s **(b)** 0.38 m **(c)** 29°
71. 12 m/s
73. no, $E = Pt$, so you are paying for energy
75. the one that arrives first, due to a shorter time interval
77. 97 W
79. 5.7×10^{-5} W
81. 6.7×10^2 J
83. 1.8 hp
85. (a) 2.3 hp; the horse is working hard **(b)** 10 hp
87. $0.36
89. 13 J
91. (a) $+2.2 \times 10^5$ J **(b)** -2.2×10^5 J
93. (a) at least 2.9 m/s **(b)** yes, if starting with $v_o < 2.9$ m/s; insufficient energy
95. (a) 52 km/h = 14 m/s **(b)** 37 km/h = 10 m/s
97. no, because $E_{top} > E_{bottom}$
99. (a) 5.5×10^2 W **(b)** 0.74 hp

Chapter 6

1. (b)
3. no, mass is also a factor
5. 75 kg
7. the lineman
9. 24 J
11. 4.05 kg·m/s in the direction opposite the initial velocity v_o
13. (a) 3.7 kg·m/s down **(b)** 7.0 kg·m/s down
15. 3.5 m/s in the same direction or 8.2 m/s in the opposite direction
17. $\Delta p = (-1.1 \text{ kg·m/s}) \hat{x} + (-2.8 \text{ kg·m/s}) \hat{y}$
19. 6.5×10^3 N
21. (a) 491 kg·m/s downward **(b)** yes; 519 kg·m/s downward
23. According to the impulse–momentum theorem, a shorter contact time will result in a greater force if all other factors (m, v_o, v) remain the same.
25. (a), (b) Increasing the contact time Δt reduces the average force F.
27. 6.0×10^3 N

29. (a) 1.2×10^3 N **(b)** 1.2×10^4 N
31. 1.2×10^2 N in the direction opposite initial velocity v_o
33. 1.1×10^3 N; 4.7×10^3 N
35. 8.2 N upward
37. 0.057 s
39. Air moves backward and the boat moves forward according to momentum conservation. If a sail were installed behind the fan on the boat, the boat would not go forward, because the forces between the fan and the sail are internal forces of the system.
41. Throw something or even blow a strong breath of air out of your mouth.
43. 1.0 m/s westward
45. 1.67 m/s in the original direction of the bullet
47. 7.6 m/s, 12° above the +x axis
49. (a) 11 m/s to the right **(b)** 22 m/s to the right **(c)** at rest (0)
51. 0.78 m/s
55. (a)
57. Momentum is a vector, and kinetic energy is a scalar.
59. $v_1 = -0.48$ m/s; $v_2 = +0.020$ m/s
61. $v_p = -1.8 \times 10^6$ m/s; $v_\alpha = +1.2 \times 10^6$ m/s
63. 2.1×10^5 J
65. 13.9 m/s at 53.1° south of east
67. no, 1.1 kg·m/s at 291°
69. (a) $m_1/m_2 = \frac{1}{2}$ **(b)** $v_2 = \frac{2}{3}v_{1o}$
71. (a) $v = v_o/90$ **(b)** 2.5×10^2 m/s **(c)** no, 99%
73. (b) $e_{elastic} = 1.0$; $e_{inelastic} = 0$
75. It lowers the center of mass and increases the moment of inertia.
77. (d)
79. (a) $(0, -0.45$ m$)$ **(b)** no, only that they are equidistant from CM
81. (a) 4.6×10^6 m from the center of the Earth **(b)** 1.8×10^6 m below the surface of the Earth
83. $(8.0$ m, $0)$
85. remains at the center of the sheet
89. (a) The 65-kilogram skater travels 3.3 m, and the 45-kilogram skater travels 4.7 m.
(b) same distances as in part (a)
91. (a) no, only if the masses are the same **(b)** in the same direction
93. (a) $(4.6d, 1.3d)$ **(b)** $(3.9d, 6.4d)$ **(c)** $(4.3d, 3.8d)$
95. (a) 2.3 kg·m/s at 45° below the $-x$ axis **(b)** Not necessarily, a collision does not have to occur; the momentum would be the same as in part (a).
97. (a) 6.0×10^4 kg·m/s **(b)** 7.5×10^3 N
99. (a) 4.0 m/s **(b)** 4.0 m/s
101. (a) 1.7 m/s **(b)** 32 J
103. (a) 28% **(b)** 1.3×10^7 m/s
105. $(8.7$ cm, 15 cm$)$

Chapter 7

1. (c)
3. $(2.5$ m, 53°$)$
5. (a) 0.26 rad **(b)** 0.79 rad **(c)** 1.6 rad **(d)** 2.1 rad

7. 4.7 cm
9. (a) 4.00 rad **(b)** 229°
11. (a) 1 rad = 57.3° **(b)** 2π rad = 360°
13. 10.7 rad
15. (a) 0.53° = 9.2×10^{-3} rad **(b)** 1.9° = 3.4×10^{-2} rad
17. No, we can cut six such pieces and one 0.28-radian piece.
19. (b)
21. yes; no
23. The point farthest from the center has the greatest tangential speed, and the point closest to the center has the smallest tangential speed.
25. 2.4 rev/day
27. 1.8 s
29. B
31. (a) 0.84 rad/s **(b)** 3.4 m/s, 4.2 m/s
33. (a) 7.27×10^{-5} rad/s **(b)** 1.99×10^{-7} rad/s
35. (a) 60 rad **(b)** 3.6×10^3 m
37. (d)
39. There is insufficient centripetal force (provided by friction and adhesive forces) on the water droplets, so they fly out along a tangent and the clothes get dry.
41. The inertia of our body tends to keep us moving forward along a straight line (Newton's first law), and the car makes a turn by the centripetal force between the tires and the road. So we feel as if we were being thrown outward.
43. 0.049 rad/s, or about 680 rev/day
45. 1.3 m/s
47. no; 1.33 m/s² > 1.25 m/s²
49. 13.4°
51. 9.9 m/s
53. (a) $v = \sqrt{rg}$ **(b)** $(5/2)r$
55. (b) The angle is independent of the mass. **(c)** $\tan\theta = \dfrac{v^2}{gr} - \mu_s$
57. (d)
59. No, a car in circular motion always has centripetal acceleration.
61. 1.1×10^{-3} rad/s²
63. 7.5 rev
65. (a) 1.82 rad/s² **(b)** 28.7 rev
67. (a) 53 s **(b)** $\mathbf{a} = (8.5$ m/s²$)\,\hat{\mathbf{r}} + (1.4$ m/s²$)\,\hat{\mathbf{t}}$
69. (b)
71. No. Gravity acts on the astronauts and the spacecraft, providing the necessary centripetal force for the orbit, so g is not zero and, by definition, there is weight $(w = mg)$. The "floating" occurs because the spacecraft and astronauts are "falling" (accelerating toward Earth at the same rate).
73. yes, if we also know the radius of the Earth
75. 9.80 m/s²
77. (a) 2.4×10^{20} N toward the Sun **(b)** 6.4×10^{20} N toward the Sun
79. 8.0×10^{-10} N, toward the opposite corner
81. 3.4×10^5 m
83. The gravitational force by the Sun is greater by 1.7×10^2 times.
85. (a) -2.5×10^{-10} J **(b)** 0
87. (c)

89. (a) to get more velocity relative to space **(b)** The tangential speed of the Earth is higher in Florida.
91. (a) 3.0×10^{-19} s²/m³ **(b)** 3.0×10^{-19} s²/m³
93. 1.53×10^9 m
95. 2.0 min
97. (a) 0 **(b)** 1.2×10^3 N **(c)** 1.8×10^3 N
99. (a) 1.10×10^{10} m/s² **(b)** $(1.12 \times 10^9)g$
101. 1.1×10^4 m/s
103. 1 y²/AU³

Chapter 8

1. (a)
3. Yes; rolling motion is a good example.
5. (b)
7. 0.10 m
9. (a) 0.50 m/s **(b)** 1.0 m/s
11. yes
13. 0.58 rotations
15. (a)
17. Yes, the toy clown is in stable equilibrium. Its center of gravity is directly below the tightrope. If the clown leans to one side, his own weight will restore its equilibrium position. If the weights are removed, the clown will be in a unstable equilibrium and he will fall.
19. 3.3×10^2 N
21. 6 stable (faces) and 20 unstable (12 edges and 8 corners)
23. 2.3 m
27. 1.6×10^2 N
29. 3.6 N
31. $m_2 = 0.20$ kg, $m_3 = 0.50$ kg, $m_4 = 0.40$ kg
33. Yes; the center of gravity of each stick is at or to the left of the edge of the table.
35. 1.2 m from left end of board
37. (a) to satisfy$\sum\tau = 0$ **(b)** 0.19 **(c)** 3.5 m/s
39. (a)
41. The moment of inertia depends on how mass is distributed about an axis. Physically, it means that under a constant torque, the angular acceleration depends on the location of the axis of rotation.
43. When the paper is pulled quickly (a large force is required to accelerate the roll), the force the paper can provide is not great enough to accelerate the paper roll. However, if the paper is pulled slowly, the paper is strong enough to accelerate the roll because the force required is smaller. The amount of paper on the roll affects the results: The more paper on the roll, the easier to tear.
45. (a) 22.5 kg·m² **(b)** 62.5 kg·m² **(c)** 85.0 kg·m²
47. 0.64 m·N
49. (a) 8.0 kg·m² **(b)** 7.5 kg·m² **(c)** yes
51. 1.1 rad
53. 3.5 times
55. 1.2 m/s²
57. 2.4 N
59. (a) 1.5g **(b)** 67-cm position
61. 6.5 m/s²

63. $\theta = \tan^{-1} \dfrac{7\mu_s}{2}$

65. (c)

67. (a) 28 J (b) 14 W

69. 0.47 m·N

71. 0.16 m

73. cylinder goes higher by 7.1%

75. (a) 1.31×10^8 J (b) 1.46×10^6 W

77. (a) 29% (b) 40% (c) 50%

79. (a) $v = \sqrt{gR}$ (b) $h = 2.7R$
(c) weightlessness

81. Walking toward the center decreases the moment of inertia and so increases the rotational speed.

83. The arms and legs are put into these positions to decrease the moment of inertia, thereby increasing the rotational speed.

85. The cat manipulates its body to change the moment of inertia to rotate or flip over. It twists part of the body one way and then rotates the feet the other way.

87. 1.4 rad/s

89. $L_{\text{rot}} = 2.4 \times 10^{29}$ kg·m^2/s;
$L_{\text{rev}} = 2.8 \times 10^{34}$ kg·m^2/s

91. $T_2 = 3.3T_1$

93. (a) 4.3 rad/s (b) $K = 1.1K_o$ (c) the work done by the skater in tucking her arms

95. (a) The lazy Susan will rotate in the opposite direction. (b) 0.56 rad/s (c) no, 2.1 rad; the Earth would have to rotate slower in the eastward direction. This would result in a longer day.

97. 52.4 J

99. (a) 4.9 m/s^2 (b) 9.8 rad/s^2

101. zero

103. (a) 2.5 cm (b) 36 cm

105. 1.3 rad/s

107. $T_1 = 21$ N; $T_2 = 15$ N

109. smaller side; 1.3 times

Chapter 9

1. (c)

3. It will have less strain for a given stress or greater stress for a given strain.

5. N/m

7. 0.020

9. 1.1×10^{11} N/m^2

11. 47 N

13. 3.92×10^{-5} m

15. 2.0×10^{-7} m

17. 6.7×10^3 N/m^2

19. (a) Ethyl alcohol has the smallest B (greatest compressibility).
(b) $p_w/p_{ea} = 2.2$

21. 5.4×10^{-5}

23. (c)

25. When the bowl is full, the atmospheric pressure on the water does not allow water out of the bottle. When the water level in the bowl drops below the neck of the bottle, air bubbles in and water flows out until the pressures are equalized again. No.

27. Bicycle tires have a much smaller contact area with the ground, so they need a higher pressure to balance the weight of the bicycle and the rider.

29. 10 m; no

31. 22 cm

33. 6.39×10^{-4} m^2

35. 1.6×10^4 N = 3600 lb

37. The air density is not a constant but decreases rapidly with altitude.

39. 1.07×10^5 Pa

41. (a) 1.1×10^8 Pa (b) 1.9×10^6 N

43. 470 N

45. 2.6 mm

47. (a)

49. There is no change. As the ice melts, the volume of the newly converted water decreases; however, the ice that was initially above the water surface is now under the water. This compensates for the decrease in volume. It does not matter whether the ice is hollow.

51. 10 000 metric tons

53. (a) water displacement (b) no

55. 33 N; same

57. no

59. 0.33 m

61. 17.7 m

63. no

65. (a)

67. The water on the inside of the curtain increases the speed of the air inside. This in turn decreases the pressure inside, according to Bernoulli's principle. The pressure difference moves the curtain inward.

69. (a) The air flow above the paper decreases the pressure there. This creates a pressure difference, and a lift force results. (b) The egg is kept aloft by the pressure of the air coming out of the end of the tube. As the eggs moves to one side, a change in the flow speed around the egg creates an inward pressure that makes the egg move back to midstream.

71. 4

73. (a) 3.5 cm^3/s (b) 0.031% (c) The slow speed allows time for the exchange of substances such as oxygen between the blood and the tissues.

75. (a) 0.99 m/s; 1.7 m/s; 2.2 m/s; 2.6 m/s
(b) 0.44 m, from $y = 20$ cm

77. 2.2 Pa

79. (b)

81. (a)

83. 2.0×10^2 Pa

85. 99.99 cm

87. (a) 0.09 m (b) 8.1 kg

89. 1.32×10^5 N = 2.96×10^4 lb

91. 8.5×10^2 kg/m^3

93. 0.45 m

95. 2.2 m/s

97. (a) $v_s = 11$ cm/s; $v_n = 66$ cm/s
(b) 1.2×10^3 Pa

99. (a) 3.3×10^4 N (b) 3.8×10^4 N

Chapter 10

1. (d)

3. Not necessarily; internal energy U does not depend solely on temperature. It also depends on mass.

5. (a) 538°C (b) −18°C (c) −29°C
(d) −40°C

7. 77.8°C

9. 56.7°C; −62°C

11. 136°F; −128°F

13. (a) −101 F° (b) 27 C°

15. (c)

17. (a) increases (b) constant

19. The pressure of the gas is held constant. If the temperature increases, so does the volume and vice versa, according to the ideal gas law. Therefore temperature is determined by volume.

21. The balloons collapsed.

23. (a) −273°C (b) −23°C (c) 0°C (d) 52°C

25. 300 K is lower

27. (a) 2.2 mol (b) 2.5 mol (c) 3.0 mol
(d) 2.5 mol

29. 1.2×10^5 Pa

31. 0.0247 m^3

33. increases by 4 times

35. 33.4 lb/in^2

37. (a) expand (b) 10.6%

39. (d)

41. (a) The ice moves upward. (b) The ice moves downward. (c) copper

43. Water expands when cooled from 4°C to 2°C as it exhibits abnormal expansion between 0°C and 4°C.

45. no; yes

47. 0.06%

49. 0.0027 cm

51. (a) 60.07 cm (b) 3.91×10^{-3} cm^2; yes, the flow speed will be affected.

53. ring; 353°C

55. 0.48 gal; morning

57. 13.4×10^3 kg/m^3

59. (a) 1.05×10^3 °C (b) no

61. (c)

63. The gases diffuse through the porous membrane, but the helium gas diffuses faster because it has a smaller mass. Eventually there will be equal concentrations of the two gases on both sides of the container.

65. (a) 6.1×10^{-21} J (b) 7.7×10^{-21} J

67. (a) 6.21×10^{-21} J (b) 1.37×10^3 m/s

69. O$_2$ by 1.22 times

71. 273°C

73. (b)

75. 6.1×10^3 J

77. 0.46 gal

79. 6.8×10^{-4} m

81. 7.8 cm

83. 33 C°

85. (a) 3.3×10^{-2} mol (b) 2.0×10^{22} molecules (c) 0.94 g

87. (a) 18 F° (b) 5.6 C°

89. 9.9×10^{20}

93. 3.3 cm

95. 2.026×10^5 Pa

Chapter 11

1. (d)

3. 2.54×10^5 J

5. 5.86×10^3 W

7. (a) 60 000 times (b) 83 h

9. Water has a high specific heat, so it takes longer to cool off than does the air.

11. Yes, a negative heat corresponds to removal of heat.

13. only on the temperature difference

15. (a) copper **(b)** 2.1×10^4 J

17. 1.7×10^6 J

19. 0.13 kg

21. 0.18 C°

23. 55 L

25. (a) 3.1×10^2 J/(kg·C°) **(b)** The final temperature will be higher, and the measured specific heat value will be in error and higher than the accepted value.

27. 21.5°C

29. (d)

31. Different substances have different internal energies, different molecular structures, and different intrinsic heat values. These quantities affect the temperature at which phase changes take place due to the different effects of the addition or removal of heat on different substances. Latent heats will also be different for different substances because of different molecular structures, or bonds. The latent heat energy goes into breaking these bonds.

33. The latent heat of vaporization of water is high; the condensing of steam on the skin releases 2.26×10^6 J of heat.

35. 1.84×10^6 J more

37. 8.0×10^4 J

39. 4.1×10^3 J

41. 0.81 kg

43. 11°C

45. 0.94 L or 0.94 kg

47. 4.7×10^{14} J; yes

49. 0.17 L

51. convection

53. (a) It convects the heat from the hot soup to the cooler air. **(b)** No; the ice blocks air flow and cooling of the air. Also, ice is a poor conductor.

55. They increase the surface area for better conduction and radiation.

57. The double-walled and partially evacuated container counteracts conduction and convection because both processes depend on a medium to transfer the heat. (The double walls are more for holding the partially evacuated region than for reducing conduction and convection.) The mirrored interior minimizes radiation.

59. 4.5 times

61. 4.54×10^6 L

63. (a) 5.5×10^5 W **(b)** 73 kg; not reasonable

65. 7.8 h

67. (a) 4.9 in. **(b)** 6.9 in.

69. 1.7

71. 2.3 cm

73. 7.8×10^5 J

75. 1.4×10^8 J

77. 65°C

79. 88.7°C

81. 1.9 g; heat loss to the skates and to the environment

83. Al by 1.8×10^4 J

85. 2.3×10^6 J

Chapter 12

1. (c)

3. (a)

5.

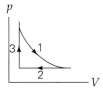

7. (b)

9. (d)

11. This is an adiabatic compression. When the plunger is pushed in, the work done goes into the increase of the internal energy of the air. The increase in internal energy increases the temperature of the air and causes the paper to catch fire.

13. (a) 400 J added **(b)** $\Delta U = 0$, so T constant

15. (a) decrease **(b)** zero **(c)** -500 J

17. (a) 3.3×10^3 J **(b)** yes, -2.3×10^3 J

19. (a) zero **(b)** $-p_1V_1$ (on the system) **(c)** $-p_1V_1$ (out of the system)

21. 2.09×10^3 J

23. 3.6×10^4 J

25. (a) AB: -1.66×10^3 J; BC: zero; CD: 3.31×10^3 J; DA: zero **(b)** $\Delta U = 0$, $Q = W = 1.65 \times 10^3$ J **(c)** 800 K

27. (b)

29. Energy must be created for the change in entropy to be negative.

31. 1.2×10^3 J/K

33. -2.1×10^2 J/K

35. -26 J/K

37. 126°C

39. 1.33 J/K

41. (a) 2.73×10^4 J **(b)** adiabatic or isentropic

43. (a) 61.0 J/K **(b)** -57.8 J/K **(c)** 3.2 J/K

45. (b)

47. Heat could be completely converted to work for a single process (not a cycle), such as an isothermal process of an ideal gas.

49. No, the warm air rises to the higher altitude, and gravity and buoyancy do work. Since it is a natural process with work input, the entropy increases and the second law is not violated.

51. 25%

53. 1.47×10^5 J

55. 40%

57. (a) 6.6×10^8 J **(b)** 27%

59. (a) 6.1×10^5 J **(b)** 1.9×10^6 J

61. 3.0 kW

63. (a) 2.4×10^5 J **(b)** 10 C°

65. (a) 5.6×10^5 W **(b)** 4.5×10^5 W

67. (a)

69. (a) No, this change does not make the cycle a Carnot cycle because a Carnot cycle consists of two adiabatic and two isothermal processes. **(b)** There are heat transfers for all four legs. There are heat inputs for Q_3 and Q_4 and heat outputs for Q_1 and Q_2. The transfers at legs 1 and 3 occur at constant temperatures as the processes are isothermal.

71. Diesel engines run hotter because diesel fuel burns at a higher temperature. According to Carnot efficiency, the higher the hot reservoir temperature, the higher the efficiency.

73. 0°C

75. (a) 6.7% **(b)** Probably not, due to the poor return; fossil fuels are much better.

77. 9.1×10^3 J

79. 100 C°

81. raising the high-temperature reservoir's temperature

83. (a) 50% **(b)** 39 kW

85. (a) 25% **(b)** $T_{\text{hot}} = (4/3)T_{\text{cold}}$

87. (a) 64% **(b)** ϵ_c is the upper limit of the efficiency. In reality, much more energy is lost than is ideally predicted.

89. (a) 13 **(b)** no, COP = 11

95. (a) isobaric expansion **(b)** 146 J

97. 3.2×10^3 J/K

99. 0.10 J/K

101. 3500 J

103. The process is an adiabatic free expansion. There is no transfer of heat, and the gas does no work. So the internal energy of the gas remains unchanged according to the first law of thermodynamics.

Chapter 13

1. (b)

3. (a) four times as large **(b)** twice as large

5. (b)

7. $4A$

9. 0.025 s

11. 41 N/m

13. (a) 10^{-12} s **(b)** 63 m/s

15. (a) $x = \pm A$ **(b)** $x = 0$ **(c)** $x = \pm A$

17. (a) 1.0 m/s **(b)** at the equilibrium position **(c)** 0.90 m/s

19. decreases by a factor of $1/\sqrt{2}$

21. (b) 1.8×10^2 times more

23. (a) 2.5 m/s **(b)** 2.5 m/s **(c)** 2.7 m/s, equilibrium position

25. (a) 0.14 m **(b)** 10.0 cm (original position)

27. $\sqrt{2}$ times as large

29. No; the tangent goes to ∞.

31. 1.0 kg

33. (a) 2.0 s **(b)** 0.50 Hz

35. (a) $x = A \sin \omega t$ **(b)** $x = A \cos \omega t$

37. (a) 5.0 cm **(b)** 10 Hz **(c)** 0.10 s

41. 9.787 m/s²

43. (a) (8.0 cm) cos (6.3 rad/s)t **(b)** 90°, or $\pi/2$ rad **(c)** -8.0 cm, or at the other amplitude

45. increase by 2.4 times

47. (5.0 cm) cos ($\pi t/4$)

51. (d)

53. (a) transverse and longitudinal **(b)** longitudinal **(c)** longitudinal

55. 0.34 m

57. 1.5 m/s

59. 4.7×10^{14} Hz

61. (a) 188 m to 545 m **(b)** 2.78 m to 3.41 m

63. 3.0 m/s

65. (a) 3.8×10^2 s **(b)** yes **(c)** P waves take 1.6×10^3 s; S waves do not go through the liquid core.

67. (a) 0.20 s **(b)** 2.0 s

69. (b)

71. reflection

73. (d)

75. At a frequency of $f_o/2$ (or a period of $2T_o$), the swing is pushed every other oscillation. So only half the energy is going into the swing, but it is pushed smoothly.

77. (a) 200 Hz **(b)** 300 Hz

79. (a) 6.0 m **(b)** 2.0 m

81. $n = 5$

83. 10 Hz, 20 Hz, 30 Hz, and 40 Hz

85. 503 Hz

87. 0.22 kg, 0.055 kg, 0.024 kg, 0.014 kg

89. (a) 12 cm **(b)** 3.0 cm, 9.0 cm, 15 cm

91. 5.0 m/s; 2.5×10^2 m/s^2

93. 210 Hz

95. (a) Since $v = \sqrt{\dfrac{F_T}{\mu}}$, doubling the tension

will make the speed $\sqrt{2}$ times as large. The wavelength is unchanged, as it is determined by the length of the string and the mode of vibration. **(b)** The frequency is increased by a factor of $\sqrt{2}$ due to the speed increase.

97. 1.1 cm/s

99. 3.0 s

Chapter 14

1. (b)

3. (a)

5. Some insects produce sounds that are not in our audible range.

7. They arrive at the same time.

11. 32°C

13. 1.5 km

15. (a) 1.29 m **(b)** 1.34 m

17. (a) 0.81 s **(b)** 0.78 s

19. −1.75%

21. 583 m

23. −4.5%

25. (b)

27. yes, an intensity below the threshold intensity

29. four times as large

31. (a) 0 dB **(b)** 120 dB

33. 20 dB

35. (a) 10^{-7} W/m^2 **(b)** 10^{-3} W/m^2

37. (a) 63 dB **(b)** 83 dB **(c)** 113 dB

39. (a) 10^{-3} W/m^2, 10^{-8} W/m^2
(b) $I_M/I_L = 10^5$

41. 10 bands

43. $I_B = 0.56 I_A$; $I_C = 0.25 I_A$; $I_D = 0.17 I_A$

45. (a) 3.2×10^{-3} W/m^2 **(b)** 16

47. 10^5 bees

49. (b)

51. (a) no **(b)** increasing sound frequency

53. The varying sound intensity is caused by the interference effect. At certain locations, interference is constructive; at other locations, it is destructive.

55. 0.172 m

57. 267 Hz and 261 Hz

59. 834 Hz

61. 3.3 Hz

63. 90°

65. (a) 1.74 **(b)** 555 m/s

67. 28 m/s

69. (a)

71. The spacing (change in length) depends not only on the frequency difference Δf, but also on the values of the frequencies themselves. For different frequencies, ΔL is therefore different.

73. As the level of water increases in the bottle, the length of the air column above the water decreases. This decrease in length of the air column decreases the wavelength and increases the frequency.

75. 510 Hz

77. (a) f_2 does not exist, only odd harmonics in this case **(b)** 322 Hz

81. 0.194 m; 0.583 m; 0.972 m

83. destructively

85. increases by 1.44 times

87. speed of sound

89. 7.5×10^{-3} J

91. 38 vibrations

93. approaching: 755 Hz; moving away: 652 Hz

95. 1.1 m

Chapter 15

1. (c)

3. (a) We know there are two types of charges because attractive and repulsive forces can be produced by different combinations of just two types of charges. **(b)** no effect

5. decrease; increase

7. 3.1×10^{14} electrons

9. (a) $+4.8 \times 10^{-9}$ C **(b)** 2.7×10^{-20} kg

11. negative

13. No, the wall is neutral but polarized.

15. The spheres can be charged through polarization by induction. If you bring a positively charged object near one of the two spheres, the sphere near that object will have a net negative charge, and the other sphere will have a net positive charge (polarization by induction). Then separate the two spheres (with the positively charged object nearby), and they will have opposite charges according to charge conservation.

17. (d)

19. Both the Sun and the planets are electrically neutral.

21. (a) same **(b)** 1/4 as large **(c)** 1/2 as large

23. (a) 5.8×10^{-11} N **(b)** zero

25. 2.24 m

27. (a) 50 cm **(b)** 50 cm

29. (a) $x = 0.25$ m **(b)** nowhere
(c) $x = -0.94$ m for $\pm q_3$

31. 2.3×10^{39}

33. (a) 96 N at 39° below the $+x$ axis
(b) 61 N at 84° above the $-x$ axis

35. (b)

37. (b)

39. It is determined by the relative density or spacing of the field lines. The closer the lines, the greater the magnitude.

41. Since the charge is positive, the electric field inside the sphere is not zero. Field lines run radially outward to the sphere's inside surface and stop at the induced negative charges on that surface. The lines reappear on the outside surface (positively charged) and continue radially outward as if emanating from the central point charge. If the charge were negative, the field lines would reverse direction.

43. 2.0×10^5 N/C

45. 1.2×10^{-7} m

47. 1.0×10^{-7} N/C upward; 5.6×10^{-11} N/C downward

49. $\mathbf{E} = (2.2 \times 10^5$ N/C$)\hat{\mathbf{x}}$ − $(4.1 \times 10^5$ N/C$)\hat{\mathbf{y}}$, or 4.6×10^5 N/C at 61° below the $+x$ axis

51. 5.4×10^6 N/C toward the charge of -4.0 μC

53. 3.8×10^7 N/C in the $+y$ direction

55. 15 μC/m^2

57. $\mathbf{E} = (-4.4 \times 10^6$ N/C$)\hat{\mathbf{x}}$ + $(7.3 \times 10^7$ N/C$)\hat{\mathbf{y}}$

59. (b)

61. no

63. (a) zero **(b)** $+Q$ **(c)** $-Q$ **(d)** $+Q$

65. (a) zero **(b)** $\dfrac{kQ}{r^2}$ **(c)** zero **(d)** $\dfrac{kQ}{r^2}$

67.

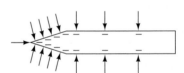

69. (a)

71. The net charges are equal and opposite in sign.

73. 10 field lines entering (negative)

75. no, just more positives than negatives: net positive

77. 3.1×10^{12} electrons

79. (a) 1.6×10^{-7} s **(b)** $(-0.25$ m, 0)

81. $+3.0$ μC on the $-y$ axis or -3.0 μC on the $+y$ axis

83.

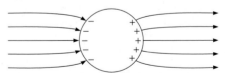

85. If d is zero, the positive and negative charges overlap, so the electric field is zero.

87. It will still rotate, but it will be drawn into the region where the field is highest, negative end first.

89. 4.9 mm

Chapter 16

1. (d)

3. (a)

5. Approaching a negative charge means moving toward a region of larger negative potential values—that is, losing potential. Positive charges tend to move toward regions of lower potential, thereby losing potential energy and gaining kinetic energy (speeding up).

7. The electron will lose potential energy.

9. (b)

11. (a) 2.7 μC **(b)** positive to negative

13. 1.4 $\times$ 10^6 m/s

15. (a) 1.3 $\times$ 10^6 m/s down **(b)** It loses potential energy.

17. (a) 0.90 m **(b)** -6.7×10^3 V (-6.7 kV), a decrease

19. (a) $+6.2 \times 10^{-19}$ J **(b)** -6.2×10^{-19} J **(c)** $+4.8 \times 10^{-19}$ J

21. 1.1 J

23. (a) $+0.27$ J **(b)** no

25. -0.72 J

27. 3.1 $\times$ 10^5 V

29. -1.3×10^6 V

31. (a) 5.9 $\times$ 10^7 m/s **(b)** 5.9 $\times$ 10^{-9} s

33. (a)

35.

37. yes

39. (a)

41. It increases to a larger positive value.

43. An electron volt is $e \, \Delta V = q \, \Delta V = \Delta U$. It is also the kinetic energy gained by an electron that goes through a potential difference of 1 V. So, it is a unit of energy. And 1 GeV is larger than 1 MeV by 1000 times.

45. Equipotential surfaces cannot cross. Since electric field lines are perpendicular to the equipotential surfaces, electric field lines cannot cross either.

47. 1.0 cm

49. $+6.0$ V

51. 12.6 m

53. $+300$ eV $= +4.8 \times 10^{-17}$ J

55. (a) 2.0 $\times$ 10^7 eV **(b)** 2.0 $\times$ 10^4 keV **(c)** 20 MeV **(d)** 2.0 $\times$ 10^{-2} GeV **(e)** 3.2 $\times$ 10^{-12} J

57. 6.2 $\times$ 10^7 m/s (proton); 4.4 $\times$ 10^7 m/s (alpha particle)

59. (a) 3.5 V; 1.1 $\times$ 10^6 m/s **(b)** 41 kV; 3.8 $\times$ 10^7 m/s **(c)** 5.0 kV; 4.2 $\times$ 10^7 m/s

61. $+2.8$ V

63. (b)

65. 2.4 $\times$ 10^{-5} C

67. 0.71 mm

69. (a) 4.2 $\times$ 10^{-9} C **(b)** 2.5 $\times$ 10^{-8} J

71. 2.2 V

73. (d)

75. The charged rubber rod polarizes the stream of water, and the stream is attracted to it.

77. 3.1 $\times$ 10^{-9} C; 3.7 $\times$ 10^{-8} J

79. 2.4

81. (b)

83. equal capacitance

85. 6.0 μF

87. (a) 3.0 μC **(b)** 9.0 μC

89. max. 6.5 μF; min. 0.67 μF

91. (a) 2.4 μC for C_1, 2.4 μC for C_2, 1.2 μC for C_3, 3.6 μC for C_4 **(b)** 6.0 V for all the capacitors

93. $-\dfrac{kq^2}{d}$

95. -1.1×10^2 eV

97. (a) 8.4 $\times$ 10^{-13} V **(b)** 1.5 $\times$ 10^{-9} V **(c)** In (a), the top is $+$; in (b) top is $-$.

99. 2.6 $\times$ 10^{-16} J

Chapter 17

1. (b)

3. (a)

5. As the internal resistance increases, the voltage across it also increases. This decreases the terminal voltage.

7. (a) 3.0 V **(b)** 1.5 V

9. (a) 24 V **(b)** the two 6.0-volt batteries in series, combined in parallel with the 12-volt battery

11. (d)

13. 8.3 min

15. 4.7 $\times$ 10^{20} electrons

17. 3.6 $\times$ 10^5 C

19. (a) 5.9 $\times$ 10^{16} protons **(b)** 1.9 $\times$ 10^5 J/s

21. (d)

23. (a) 11.4 V **(b)** 0.32 Ω

25. (a) same **(b)** 1/4 the current

27. Aluminum wire is thicker than copper by 29%

29. 1.0 V

31. 1.3 $\times$ 10^{-2} Ω

33. 3.0 $\times$ 10^{19} electrons/s

35. The shorter wire carries 4 times the current.

37. 0.13 Ω

39. 4.6 mΩ

41. not ohmic

43. It increases by 1.6 times.

45. 8.2°C to 32°C

47. 0.77 A

49. (d)

51. (d)

53. The wire in a 60-watt bulb would be thicker.

55. 2.0 $\times$ 10^3 W

59. 18 min

61. (a) 4.3 $\times$ 10^3 J **(b)** 13 Ω

63. 58 Ω

65. (a) 0.60 kWh **(b)** \$0.09

67. (a) 0.15 A **(b)** 1.4 $\times$ 10^{-4} $\Omega \cdot$m **(c)** 2.3 W

69. (a) 1.1 $\times$ 10^2 J **(b)** 6.8 J

71. 21 Ω

73. $R_{120}/R_{60} = 4/3$

75. \$151

77. 64 C°

79. (a) 0.33 A **(b)** 3.6 $\times$ 10^2 Ω

81. It increases to 1.78 kW.

83. copper at 117 °C or aluminum at -73°C

85. 3.5 $\times$ 10^6 J

87. (a) 5.0 $\times$ 10^{-2} A **(b)** 5.0 V

89. (a) 4.0 $\times$ 10^2 Ω **(b)** 4.0 W **(c)** 4.8 $\times$ 10^2 J

91. 1.1 $\times$ 10^{22} electrons

93. Since $P = VI$, a higher voltage lowers the current if P is kept constant. The resistance of the transmission lines is a fixed number. With lower current, the power loss (I^2R loss) on the lines is reduced: $P_{\text{loss}} = I^2 R = \dfrac{P^2}{V^2} R \propto \dfrac{1}{V^2}$. So if V is raised by a factor of 10, P_{loss} will be reduced by 1/100.

95. 6.6 $\times$ 10^{-6} m/s

Chapter 18

1. (b)

3. No, not generally; however, if all resistances are equal, the voltages across them are also the same.

5. The 5-ohm resistor gets the most power, because it gets the most voltage and all resistors have the same current.

7. (a) series: 60 Ω **(b)** parallel: 5.5 Ω

9. 30 Ω

11. (a) 30 Ω **(b)** 0.30 A **(c)** 1.4 W

13. (a) 0.57 Ω **(b)** 6.0 V **(c)** 9.0 W

15. 3.0 $\mu\Omega$

17. (a) 1.0 A **(b)** 1.0 A **(c)** 2.0 W; 4.0 W; 6.0 W **(d)** $P_{\text{sum}} = P_{\text{total}} = 12$ W

19. $I = 1.0$ A for all; $V_{8.0} = 8.0$ V; $V_{4.0} = 4.0$ V

21. 2.7 Ω

23. (a)

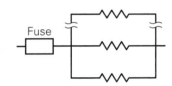

(b) 31

25. 0.25 A, 2.5 V

27. (a) 1.0 A; 0.50 A; 0.50 A **(b)** 20 V; 10 V **(c)** 30 W

29. no

31. 100 s = 1.7 min

33. (a) 0.085 A **(b)** 7.0 W; 2.6 W; 0.24 W; 0.41 W

35. 35 W

37. $I_1 = 2.7$ A; $I_2 = 3.3$ A; $I_3 = I_4 = 0.55$ A

39. (d)

41. positive

45. $I_1 = 1.0$ A; $I_2 = I_3 = 0.50$ A

47. $I_1 = 0.33$ A left; $I_2 = 0.33$ A right

49. $I_1 = 3.75$ A up; $I_2 = 1.25$ A left; $I_3 = 1.25$ A right

51. 0.664 A left; 0.786 A right; 1.450 A up; 0.770 A down; 0.016 A down; 0.664 A down

53. (c)

55. (a) $V_R = V_o$, $V_C = 0$ **(b)** $V_R = 0.37V_o$, $V_C = 0.63V_o$ **(c)** $V_R = 0$, $V_C = V_o$

57. 2.0 MΩ

59. (a) 1.50 MΩ **(b)** 11.4 V

61. (a) 9.4 $\times$ 10^{-4} C **(b)** $V_C = 24$ V; $V_R = 0$

63. (a) 0 V; 2.0 μA **(b)** 1.7 $\times$ 10^{-6} C

65. (c)

67. (a) An ammeter has very low resistance, so if it were connected in parallel in a circuit, the circuit current would be very high and the galvanometer could burn out. **(b)** A voltmeter has very high resistance, so if it were connected in series in a circuit, it would read the voltage of the source, because it has the highest resistance (most probably) and therefore the largest voltage drop among the circuit elements.

69. 7.4 kΩ

71. 50 kΩ

73. 0.20 mA

75. **(b)** If $R_a = 0$, $R = \dfrac{V}{I}$ (that is, the measurement is "perfect").

77. **(c)**

79. Since current is caused by voltage, a high voltage can produce high "harmful" current even though the resistance of a body is high.

81. It is safer to jump. If you step off the car one foot at a time, there will be a high voltage between your feet. If you jump, the voltage between your feet is zero, because both feet will be at the same potential all the time.

83. 75 Ω

85. 6.0 Ω

87. 8.1 J/s

89. **(a)** 1.0 A; 0.40 A; 0.20 A; 0.40 A **(b)** 100 W; 4.0 W; 2.0 W; 4.0 W

91. 0.440 A; 0.323 A; 0.117 A; 0.0878 A; 0.0293 A

95. 11.8 A

97. **(a)** 4.47 V **(b)** 0.447 A

99. 10 mΩ, 2.0 mΩ, and 1.0 mΩ

Chapter 19

1. **(e)**

3. **(c)**

5. Similarities: Two kinds of poles, north and south (charges, positive and negative); like poles (charges) repel, and unlike poles (charges) attract. Differences: Poles come in pairs (single charge can exist).

7. **(b)**

9. They are either parallel or opposite.

11. **(a)** The bottom half would have a magnetic field directed into the page, and the top half would have a magnetic field directed out of the page. **(b)** the same

13. 2.5×10^3 m/s

15. 2.0×10^{-14} T to the left, looking in the direction of the velocity

17. **(a)** 3.8×10^{-18} N **(b)** 2.7×10^{-18} N **(c)** zero

19. **(a)** 8.6×10^{12} m/s^2; horizontal toward right, looking in direction of velocity. **(b)** same magnitude but in opposite direction

21. **(b)**

23. **(b)**

25. 0.25 m

27. 2.4 A

29. both 2.9×10^{-6} T

31. 1.4×10^{-5} T at 38° below a horizontal line to the left

33. 1.9×10^{-5} T toward observer (up)

35. 1.0×10^{-4} T away from observer

37. 3.2×10^{-2} A

39. 4.2×10^{-5} T

41. **(a)** $f = \dfrac{qB}{2\pi m}$ **(b)** $T = \dfrac{2\pi m}{qB}$, independent of m and v **(c)** 5.7×10^{-3} m; $f = 2.8 \times 10^6$ Hz

43. **(d)**

45. away from you

47. 12 T

49. attract for currents in the same direction, repel for opposite currents

51. yes, with the plane of the loop perpendicular to the magnetic field

53. 1.2 N perpendicular to the plane of current and field

55. 5.0×10^{-3} T north to south

57. **(a)** zero **(b)** 4.0 N/m in $+z$ direction **(c)** 4.0 N/m in $-y$ direction **(d)** 4.0 N/m in $-z$ direction **(e)** 4.0 N/m in $+y$ direction

59. 0.40 N/m in $+z$ direction

61. 6.7×10^{-6} N/m, attractive

63. 0.75 N upward

65. 2.7×10^{-5} N/m toward wire 1

67. Left segment: 2.5 N to the left; right segment: 2.5 N to the right; top segment: 7.5 N upward

69. zero, yes

71. **(d)**

73. Pushing the button in both cases completes the circuit. The current through the wires activates the electromagnet. Doorbell: This causes the clapper to be attracted and ring the bell. However, this breaks the armature contact and opens the circuit. Holding the button causes this to repeat, so the bell rings continuously. Door chimes: When the circuit is completed, the electromagnet attracts the core and compresses the spring. Inertia causes it to hit one tone bar, and the spring force then sends the core in the opposite direction to strike the other bar.

75. 2.0×10^5 m/s

77. **(a)** 1.0×10^5 m/s **(b)** the same v, independent of charge

79. **(a)** 4.8×10^{-26} kg **(b)** 2.4×10^{-18} J **(c)** no, work equals zero

81. 2.9×10^{-2} T

83. **(b)**

85. north magnetic pole

87. 0.44 T

89. 1.0×10^{-14} J

91. **(a)** 1.0 m **(b)** 1.5×10^{-5} s

93. 5.6×10^{-7} m/s

95. 3.0×10^{-2} m·N

Chapter 20

1. **(d)**

3. **(a)** When the bar magnet enters the coil, the needle deflects to one side. When the magnet leaves the coil, the needle reverses direction.

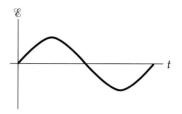

(b) no

5. counterclockwise (in head-on view).

7. Sound waves cause the resistance of the button to change as described. This changes the current, so the sound wave produces electrical pulses that travel through the phone lines and to a receiver. The receiver has a coil wrapped around a magnet. The pulses create a varying magnetic field as they pass through the coil, causing the diaphragm in the receiver to vibrate in air and producing sound waves.

9. **(a)** zero **(b)** 1.9×10^{-2} T·m^2 **(c)** 1.4×10^{-2} T·m^2

11. 3.3×10^{-2} T·m^2

13. 1.3×10^{-6} T·m^2

15. 1.6 V

17. 0.30 s

19. 0.35 V

21. **(a)** No, the emf is zero. **(b)** near the Earth's magnetic poles

23. **(a)** 0.60 V **(b)** zero

25. 4.0 V

27. 3.0×10^{-4} J

29. **(c)**

31. As it moves up and down in the coil, the magnet will induce a current in the coil, which will light the bulb. However, the magnet produces the current at the expense of the kinetic energy and potential energy. The magnet's motion will damp out rather rapidly (Lenz's law).

33. **(a)** 0.057 V **(b)** 0.57 V

35. ± 104 V; the initial voltage direction was not specified, so there are two possible directions.

37. **(a)** 100 V **(b)** zero

39. 16 Hz

41. 100 V

43. **(a)** 216 V **(b)** 160 A **(c)** 8.1 Ω

45. **(c)**

47. The dc voltage must first be converted to changing voltage. This is achieved by the points in older cars and electronic devices in newer cars. The coil functions as a step-up transformer to give high-voltage pulses. The distributor distributes these pulses to different spark plugs to ignite the gas–air mixture.

49. **(a)** 16 **(b)** 5.0×10^2 A

51. 24:1

53. **(a)** 17.5 A **(b)** 15.7 V

55. **(a)** no **(b)** 45%

57. **(a)** $N_s/N_p = 1:20$ **(b)** 2.5×10^{-2} A

59. **(a)** 128 kWh **(b)** $1840

61. **(a)** $N_s/N_p = 1:2$; 1:14; 1:30 **(b)** 2.0; 14; 30 **(c)** 833

63. **(a)** 53 W **(b)** $N_p/N_s = 200$

65. (b)

67. Infrared (heat) radiation is absorbed by clouds (water molecules), but the sunburning ultraviolet radiation is not.

69. AM: 1.8×10^2 m to 5.7×10^2 m; FM: 2.8 m to 3.4 m; TV: 0.34 m to 5.6 m

71. 2.6 s

73. AM: 67 m; FM: 0.77 m

75. (a) 30 A **(b)** 2.5 A

77. 0.60 m

79. once every 0.01 s

81. (a) up **(b)** 25 mA

83. 0.75 T·m^2

85. 4.0 T/s

87. 0.33 A and 30 A

89. (a) 50 rotations/s **(b)** No, if they are ideal (without losses), the same turn ratios would work. In reality, the 50-hertz delivery would have fewer losses due to eddy currents because these currents would be generated less frequently.

Chapter 21

1. (c)

3. The voltage and current reach maximum at the same time, reach minimum at the same time, are zero at the same, and so on.

5. 7.1 A

7. 1.2 A

9. (a) 10 A **(b)** 14 A **(c)** 12 Ω

11. (a) 4.5 A; 6.3 A **(b)** 112 V; 158 V

13. $V = (170 \text{ V}) \sin 119\pi t$

15. 0.33 A; 0.47 A

17. 2.4×10^2 W

19. (a) 60 Hz **(b)** 1.4 A **(c)** 1.2×10^2 W
(d) $V = (120 \text{ V}) \sin 380t$
(e) $P = (240 \text{ W}) \sin^2 380t$

21. (b)

23. For a capacitor, the *lower the frequency,* the longer the charging time in each cycle. A longer charging time means more charge accumulation on the plates and therefore *more* opposition to the current. For an inductor, the *lower the frequency,* the more slowly the current in the inductor changes, the smaller the rate of change in its magnetic flux, and thus the *smaller* the self-induced emf that opposes the current.

25. In an ac circuit, a capacitor can oppose current, because as the capacitor charges, the voltage across its plates increases. Also, an inductor can oppose current, because the induced emf opposes the change in flux.

27. 2.65×10^{-5} F

29. 1.3×10^3 Ω

31. 2.3 A

33. 60%, an increase

35. 255 Hz

37. 0.70 H

39. (d)

41. (d)

43. (a) 1.7×10^2 Ω **(b)** 2.0×10^2 Ω

45. (a) 38 Ω; 1.1×10^2 Ω **(b)** 1.1 A

47. $+27°$

49. 72 Ω

51. 50 W

53. 5.3×10^{-11} F

55. ab: 1.3 A; ac: 1.2 A; bc: 4.0 A; cd: 1.8 A; bd: 1.6 A; ad: 2.9 A

57. 13 A

59. $(V_{\text{rms}})_R = 12$ V; $(V_{\text{rms}})_L = 2.7 \times 10^2$ V; $(V_{\text{rms}})_C = 2.7 \times 10^2$ V

61. (a) 362 Ω **(b)** no; 1.1×10^2 Hz

63. 30%

65. 0.71 A

67. 1.7×10^{-12} F

69. 37°

71. The capacitor opposes current more strongly at lower frequencies, and the inductor opposes current more strongly at high frequencies. In Fig. 21.18a, the inductor in series with R_L filters out the high-frequency current, so only the low-frequency current reaches R_L. In Fig. 21.18b, the capacitor in series with R_L filters out the low-frequency current, so only the high-frequency current reaches R_L.

Chapter 22

1. (c)

3. (d)

5. They are visible because of the diffuse reflections off the particulate matter in the air.

7. 70°

9. 47°

11. 40°

13. 27°

15. 12 m

17. $\alpha = 90°$, any θ_{i_1}

19. (d)

21. (b)

23. The angles of refractions are different for air–glass and for water–glass interfaces.

25. The laser beam has a better chance to hit the fish. Due to refraction, the fish appears to the first hunter at a location different from its true location. The laser beam obeys the same law of refraction and retraces the light by which the second hunter sees the fish.

27. 26% greater in zircon

29. 27°

31. (a) 24° **(b)** diamond to air

33. 47°

35. 6.5×10^{14} Hz; 2.8×10^{-7} m

37. 16/15

41. (a) 1.41 **(b)** 1.88

43. (a) 4.2 cm **(b)** 3.2 cm

45. 75%

47. seen for 40° but not for 50°

49. 43°

51. (a) no, $\theta_c = 38.7°$ **(b)** transmitted, $\theta_c = 48.6°$

53. 2.0 m

55. (a) 12.5° **(b)** 26.2°

57. (b)

59. No, the light will be further dispersed by the second prism.

61. (a) The angle of incidence is approximately zero. **(b)** No (no); the speeds are different.

63. Red is transmitted, and blue is internally reflected.

65. (a) 21.7° **(b)** 0.22° **(c)** 0.37°

67. (a) 58° **(b)** 30°

69. 1.64

71. 0°; not refracted

Chapter 23

1. (b)

3. (a) During the day, light passes both ways through the pane, so it is difficult to see the reflection due to the light coming through. At night, there is little light coming through the pane, so the reflections are much more clear. **(b)** The two images are due to reflections on both sides of the pane of glass, producing two similar images. **(c)** It is a combination of half-silvering and bright light on one side and dark on the other.

5. (a) The left–right reversal is an apparent one, caused by the front–back reversal. Right and left are directional senses (like clockwise and counterclockwise) rather than fixed directions referenced to a coordinate system. **(b)** no

7. When viewed by a driver through a rearview mirror, the right–left reversal property of the image formed by a plane mirror makes it read "AMBULANCE" to the driver in front.

9. (a) 0.80 m **(b)** 5.0 cm **(c)** $+1.0$

11. (a) 1.5 m behind the mirror
(b) 1.0 m/s

13. 3.0 m and 13 m behind north mirror; 5.0 m and 11 m behind south mirror

17. (d)

19. (a) The plane mirror gives a large view of the area immediately around that side of the truck. The small convex mirror gives a wide-angle perspective of the road in back of both sides of the truck. **(b)** These are convex mirrors that give a better field of view, but the images are smaller than the objects and hence appear closer than they actually are. **(c)** Yes, it can be considered a converging mirror, because it collects a large amount of radio waves and focuses them onto a small area.

21. (a) The image is smaller than the object, so it is possible to "see your full body in 10 cm" in a diverging mirror. **(b)** As the ball swings toward the mirror and approaches the focal point, the image enlarges. An enlarged image appears to be closer to the observer's eyes, so it appears to move toward the observer and therefore to jump out of the mirror.

23. 5.0 cm

25. $d_i = -30$ cm; $h_i = 9.0$ cm; image is virtual, upright, and magnified.

27. (a) $f = 10$ cm, $R = 20$ cm **(b)** 3.0 cm

29. (a)

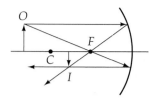

(b)

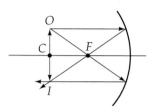

(c)

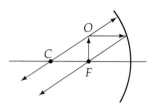

(d)

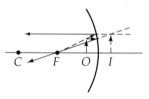

33. 120 cm

35. 2.3 cm

37. (a) $d_i = 60$ cm; $M = -3.0$, real and inverted

(b)

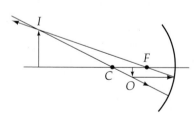

39. 10 m

41. $d_i = 60$ cm; image is real, inverted, and magnified.

43. 5.0 cm, 15 cm

45. yes: 13 cm, 27 cm

47. (c)

49. When the fish is inside the focal point, the image is upright, virtual, and magnified.

51. $d_i = 12.5$ cm; $M = -0.25$

53. 22 cm

55. (a) $d_i = -6.4$ cm; $M = 0.64$, virtual and upright

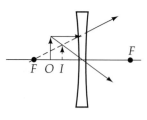

(b) $d_i = -10.5$ cm; $M = 0.42$, virtual and upright

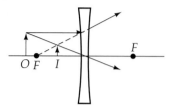

57. (a) 18 cm **(b)** 6.0 cm

59. 8.1 cm

61. (a) 20 cm **(b)** $M = -1$

63. (a) $d = 4f$ **(b)** approaches zero

65. (a) -18 cm **(b)** -90 cm

67. (a) 4.6 cm **(b)** 0.80 cm, inverted

69. 8.6 cm; $M = 0.86$, real and inverted

73. (a) focal length decreases by a factor of 0.34 **(b)** The diverging lens becomes the converging lens and vice versa, and f decreases by a factor of 0.075.

75. ∞; it has no focusing power.

77. -0.70 D

79. 0.55 mm (inverted)

81. $d_i = 37.5$ cm; $h_i = 3.0$ cm, real and inverted

83. The image formed by the converging lens is at the mirror. This image is the object for the diverging lens. If the mirror is at the focal point of the diverging lens, the rays refracted after the diverging lens will be parallel to the axis. These rays will be reflected back parallel to the axis by the mirror and will form another image at the mirror. This second image is now the object for the converging lens. By reversing the rays, a sharp image is formed at the location on the screen where the original object is. Therefore the distance from the diverging lens to the mirror is the focal length of the diverging lens.

85. 14 cm

87. (a) $d_i = \dfrac{d_o f}{d_o - f} = \dfrac{f}{1 - f/d_o}$

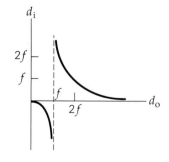

$|M| = \dfrac{d_i}{d_o} = \dfrac{f}{d_o - f}$

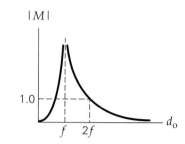

(b) $d_i = \dfrac{d_o(-f)}{d_o + f} = \dfrac{-f}{1 + f/d_o}$

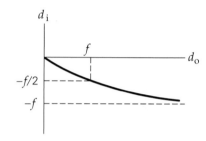

$|M| = \dfrac{d_i}{d_o} = \dfrac{-f}{d_o + f}$

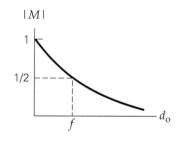

89. (a)

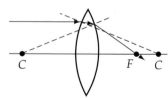

(b)

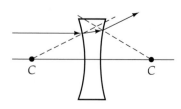

91. 85 cm

93. (a)

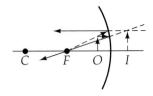

(b) 60 cm

Chapter 24

1. (b)

3. blue

5. The path-length difference will change because the airplane is moving. This change results in a change in the condition of interference—constructive is no longer constructive, and so on. Therefore the pictures flutter.

7. 3.4%

9. (a) 0.80 m **(b)** 3.2 m

11. 2.0 m

13. (a) 4.40×10^{-7} m **(b)** 4.40 cm

15. (a) 1.9 mm **(b)** decreases to 1.2 mm

17. (a) $\lambda = 4.00 \times 10^{-7}$ m, violet **(b)** 3.45 cm

19. 450 nm

21. (a)

23. all wavelengths except bluish purple

25. (a) 30λ **(b)** destructively

27. 5.4×10^{-8} m

29. 160 nm; 200 nm

31. (a) 158.2 nm **(b)** 316.4 nm

33. 1.51×10^{-6} m

35. (b)

37. a second diffraction pattern perpendicular to the first

39. (a) 5.4 cm **(b)** 2.7 cm

41. (a) 4.3 mm **(b)** microwave

43. 4.9 mm

45. blue: $x_1 = 0.50$ m, 18.4°, $x_2 = 1.2$ m, 39.2°; red: $x_1 = 0.89$ m, 30.7°, x_2 not possible

47. (a) 2.44×10^3 lines/cm **(b)** 11

49. 15.4°

53. (d)

55. Looking through a lens of each pair, rotate one of the lenses. If the intensity changes as the glasses are rotated, both pairs are polarized.

57. (a) twice **(b)** four times **(c)** none **(d)** six times

59. 54° to 60°

61. 31.7°

63. 39.3°

67. no

69. Blue light scatters more efficiently than red light. In the morning and evening, the blue component of the light from the Sun is scattered more in the denser atmos- phere near the Earth, so we see red when we look in the direction of the rising or setting Sun. During the day, we see mainly the blue component from over- head scattering.

71. 1.74

73. 56°

75. (b) $\lambda/2n$

77. 6.4×10^{-4} rad

79. 560 nm

81. 4.9×10^{-7} m

Chapter 25

1. (b)

3. According to the lens maker's equa- tion, the eye focuses by changing the shape of its lens to adjust the focal length to form a sharp image. The image distance is fairly constant, as is the distance from the lens to the retina. From the thin-lens equation, the eye must have a short focal length for looking at close objects, so the radius is small; the eye must have a long focal length for looking at distant objects, so the radius is large.

5. The pre-flash occurs before the aperture is open and the film exposed. The bright light causes the iris to reduce the area of the pupil so that when the second flash comes momentarily, you don't have a wide open- ing through which you get the red-eye re- flection from the retina.

7. (a) +5.0 D **(b)** −2.0 D

9. converging

11. diverging, −5.0 D

13. (a) +3.0 D **(b)** take them out

15. (a) converging **(b)** +3.3 D

17. +3.0 D

19. 21 cm

21. right: +1.42 D, −0.46 D; left: +2.16 D, −0.46 D

23. (d)

25. inside the focal length

27. (a) 2.3× **(b)** 2.5×

29. (a) 2.7× **(b)** 1.7×

31. 1.9×

33. (a) 1.8× **(b)** 1.3×

35. 360×

37. (a) 340× **(b)** 3900%

39. 25×

41. 3.6×

43. (b)

45. No, the whole star can still be seen. The obstruction will reduce the intensity or brightness of the image.

47. (a) 4.0× **(b)** 75 cm

49. (a) 110× **(b)** 88.3 cm

51. (a) 2 **(b)** 1

55. (a)

57. smaller

59. 550 nm

61. 1.32×10^{-7} rad; Yerkes/Hale = 4.98

63. yes

65. (a) 1.63×10^{-6} rad **(b)** 9.76×10^{17} m

67. (a) 5.55×10^{-5} rad **(b)** blue **(c)** 33.3%

69. (d)

71. Since white is obtained by adding col- ors, it cannot be obtained by the subtrac- tive method. That method subtracts colors, and the one we see is the one that is not absorbed. Black objects do not absorb all wavelengths of light. We see the objects because we perceive the extremely faint light as black. (Think of twilight vision.)

73. White light enters the blue filter, which allows the green, blue, and violet to pass through; when it passes through the yellow filter, only the green emerges.

75. 1.6×10^{-2} rad

77. 6.7 m

79. 1.1×10^{-2} rad

81. 2.1×

83. 83 cm

85. 500×

87. objects as large as typical houses

Chapter 26

1. (d)

3. (a)

5. (a)

7. Since the speed of light is constant in all directions through all frames of refer- ences and no ether (wind) would be pre- sent, there is no extra time difference be- tween the two beams and no fringe shift is observed.

9. (a) 3.38 s **(b)** 3.58 s

11. 55 m/s; 45 m/s

13. no

17. (c)

19. (b), since your friend is traveling to the front, he sees the front mark first.

21. (a) 300 m/s

23. (c), her friend does appear the same height because the height is perpendicular to the velocity.

25. Yes, you will find yourself younger af- ter a high-speed space trip (close to c) due to time dilation.

27. 42 beats/min

29. 66 m; 100 m

31. Earth twin is 64 y; traveling twin is 37 y.

33. length of 20.4 m and diameter 8.25 m

35. 0.87c

37. 5.3 m; assume relative velocity parallel to length of stick

39. She is traveling at 0.40c in the direction of the lengths of the two sticks.

41. (d)

43. No, there are no such limits on mo- mentum and energy as $p = \gamma mv$ and $E = \gamma mc^2$.

45. yes; no

47. (a) 0.985c **(b)** 2.50 MeV **(c)** 1.56×10^{-21} kg·m/s

49. 0.15 kg

51. 1.6×10^{24} J, or 150 000 times more than 1.08×10^{19} J = 3 trillion kWh

53. 79 keV

55. (a)

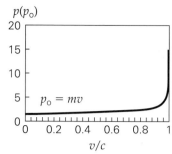

(b)

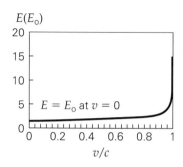

57. (a) $0.92c$ **(b)** 1.4×10^3 MeV
59. (a) 1.13 MeV **(b)** 1.64 MeV
61. 1.43×10^5 y
63. (a) $E_o = 939$ MeV > 600 MeV
(b) $0.792c$ **(c)** 6.50×10^{-19} kg·m/s
65. (d)
67. The stick would be elongated or "stretched" by the gravity gradient (difference).
69. 1.8×10^{19} kg/m³
71. Drop the cup with the pole vertical.
73. $0.27c$, same direction as spacecraft
75. (a) $0.988c$ to the left **(b)** $0.988c$ to the right
77. 4.8 h
79. length 38 m; height 2.5 m; width 2.0 m
81. 50 min
83. $0.68c$
85. 1.9 kg

Chapter 27

1. (d)
3. No, it is actually $2^4 = 16$ times since it depends on T^4.
5. (a)

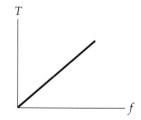

(b)

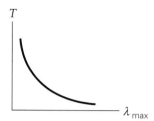

7. 1.06×10^{-5} m; 2.83×10^{13} Hz
9. 3.1×10^{13} Hz
11. 2.56×10^{-20} J
13. (b)
15. The packets of radio signals arrive in such large quantity and so quickly that our ear cannot distinguish between discrete arrivals. It is similar to a movie, in which the picture is actually single frames passing in front of us rapidly, but we see it as a smooth and continuous moving picture. Likewise, although the radio signals are single frames, we hear them continuously.
17. The energy of radiation is proportional to its intensity, according to the wave theory, and its frequency, according to the particle theory. It takes a certain amount of energy to eject a photoelectron. Since only the frequency, not the intensity, matters in this case, it favors the particle theory.
19. $\lambda = 6.0 \times 10^{-11}$ m, X-ray
21. (a) 1.32×10^{-18} J **(b)** 8.27 eV
23. 3.0 eV
25. (a) 6.7×10^{-34} J·s **(b)** 2.9×10^{-19} J
27. 6.8×10^{14} Hz
29. 2.48 eV
31. no
33. (a) sodium **(b)** $\lambda_{silver} < 262$ nm; $\lambda_{sodium} < 50^4$ nm
35. 6.6×10^{-34} J·s; 1.5 eV

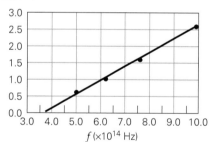

37. (d)
39. 200 000 scatterings can shift the wavelength from X-ray to visible.
41. 4.86×10^{-3} nm
43. 0.281 nm
45. 0.127 nm
47. 18.5°
49. (d)
51. The theory applies only to atoms with a single electron, because it does not include electron–electron interactions.

53. According to momentum conservation, the atom recoils, so it carries some kinetic energy and therefore the energy of the photon is smaller than expected. Thus the wavelength of the photon is longer.
55. (a) ultraviolet **(b)** visible (red)
57. (a) 6.89×10^{14} Hz **(b)** 8.21×10^{14} Hz
59. $n \approx 100$
61. (a) -1.51 eV **(b)** -0.378 eV **(c)** -0.136 eV
63. yes; 15.4 eV
65. (a) 0.967 eV **(b)** 2.97 eV **(c)** 10.2 eV; (c)
67. (a) 54.4 eV **(b)** 122 eV
69. $(2.2 \times 10^6$ m/s$)/n$
71. (a) potential $= -27.2$ eV; kinetic $= +13.6$ eV **(b)** $|U| = 2K$, potential energy is twice as large in magnitude.
73. Through stimulated emission, the photon from a laser is caused by electron transitions between two discrete energy levels, so there are only a few colors (frequencies). A light bulb emits thermal radiation at many frequencies.
75. 1.14×10^{31} photons
79. (a) 12.1 eV **(b)** 13.1 eV
81. 4.4×10^{-16} J
83. (a) 1.5×10^{15} Hz **(b)** 1.3 V
85. four
87. 6.59×10^{15} Hz

Chapter 28

1. (c)
3. The wavelength associated with a moving car is too short for us to observe.
5. (a) 7.28×10^{-6} m **(b)** 3.97×10^{-9} m
7. $\lambda_e/\lambda_p = 43$
9. 1.5×10^4 V
11. -53% (a decrease)
13. 3.7×10^{-63} m
15. 24 V
17. (b)
19. (a)

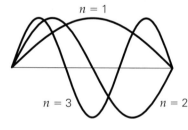

(b)

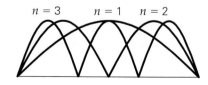

21. (c)
23. (a)
25. The periodic table groups elements according to the values of quantum numbers n and ℓ. The elements within a group have the

same or very similar electronic configurations for the outermost electrons.

27. (a) 2 (b) 14
29. (a) 2 (b) 3
31. (a) Be (b) N (c) Ne (d) S
33. (a) $1s^2 2s^2 2p^1$ (b) $1s^2 2s^2 2p^6 3s^2 3p^6 4s^2$
(c) $1s^2 2s^2 2p^6 3s^2 3p^6 3d^{10} 4s^2$
(d) $1s^2 2s^2 2p^6 3s^2 3p^6 3d^{10} 4s^2 4p^6 4d^{10} 5s^2 5p^2$
35. It would be $1s^3$ since the first state can now populate three electrons.
37. (b)
39. 0.21 m
41. 2.1×10^{-29} m/s
43. 1.1×10^{-27} J
45. 10^4 times
47. (a)
49. yes
51. 1.9 GeV
53. 83.5 V
55. (a) 4.0×10^{-21} kg·m/s (b) 1.7×10^{-13} m
59. $x = 0$
61. Yes, when $n = 1$ and $\ell = 0$.
63. no

Chapter 29

1. (c)
3. (d)
5. The isotopes of an element have the same number of protons but a different number of neutrons.
7. (a) ^{1}H; ^{2}D; ^{3}T (b) H_2O; D_2O; T_2O; HDO; HTO; DTO (c) T_2O, HTO, DTO
9. $^{41}_{19}$K
11. (a) $R_{He} = 1.9 \times 10^{-15}$ m;
$R_{Ne} = 3.3 \times 10^{-15}$ m; $R_{Ar} = 4.1 \times 10^{-15}$ m;
$R_{Kr} = 5.3 \times 10^{-15}$ m; $R_{Xe} = 6.1 \times 10^{-15}$ m;
$R_{Rn} = 7.3 \times 10^{-15}$ m (b) They are roughly ten thousand to fifty thousand times smaller.
13. No; in beta decay, a proton, which is a nucleon, is created.
15. (a) $^{237}_{93}$Np $\rightarrow$ $^{233}_{91}$Pa $+$ $^4_2$He
(b) $^{32}_{15}$P $\rightarrow$ $^{32}_{16}$S $+$ $^{0}_{-1}$e (c) $^{56}_{27}$Co $\rightarrow$ $^{56}_{26}$Fe $+$ $^{0}_{+1}$e
(d) $^{56}_{27}$Co $+$ $^{0}_{-1}$e $\rightarrow$ $^{56}_{26}$Fe (e) $^{42}_{19}$K* $\rightarrow$ $^{42}_{19}$K $+$ γ
17. α-decay: $^{214}_{84}$Po $\rightarrow$ $^{210}_{82}$Pb $+$ $^4_2$He;
β-decay: $^{210}_{82}$Pb $\rightarrow$ $^{210}_{83}$Bi $+$ $^{0}_{-1}$e
19. α–β: $^{213}_{83}$Bi; β–α: $^{213}_{83}$Bi
21. (a) $^4_2$He (b) $^{0}_{-1}$e (c) $^{102}_{39}Y$ (d) $^{23}_{11}$Na*
(e) $^{22}_{10}$Ne
23. (a) α to ^{233}Pa; β to ^{233}U; α to ^{229}Th; α to ^{225}Ra; β to ^{225}Ac; α to ^{221}Fr; α to ^{217}At; α to ^{213}Bi; α to ^{209}Tl or β to ^{213}Po; β to ^{209}Pb or α to ^{209}Pb; β to ^{209}Bi (b) They decay because there are too many neutrons in the nucleus.
25. (c)
27. No, decay is exponential, not linear.
29. (a) 6.76×10^{-5} Ci (b) 2.50×10^6 Bq
31. (a) 1/8 (b) $1/2^{24}$, or about 6×10^{-6}% of original
33. 28.6 y
35. 91.7%
39. ^{104}Tc
41. 0.98 μg
43. (a) 6.75×10^{19} nuclei (b) 2.11×10^{18} nuclei
45. 9.2 mg
47. 7.6×10^6 kg

49. (d)
51. ^{1}H and ^{2}H are stable, and ^{3}H is unstable because its nucleus has too many neutrons.
53. (a) $^{17}_8$O, because it has an unpaired neutron beyond a magic number. (b) $^{42}_{20}$Ca, because the magic number difference makes it more stable (both are paired). (c) $^{10}_5$B, because it lacks pairing. (d) Approximately the same, as both have 126 neutrons (paired and magic) and the different number of protons does not affect neutrons.
55. (b) and (d); others are odd–odd
57. 2.013 553 u
59. (a) 92.2 MeV (b) 7.68 MeV/nucleon
61. deuterium
63. 104.7 MeV
65. 10.1 MeV
67. 7.59 MeV/nucleon
69. (a) two alphas (b) two alphas (c) Yes, since two alphas gives a more stable arrangement.
71. (d)
73. (a) α particles (b) X-rays: 1.0 rem; α particles: 20 rem
75. yes
77. $^{75}_{33}$As, $^{76}_{34}$Se ($^1_0$n $+$ $^{75}_{33}$As $\rightarrow$ $^{76}_{33}$As $\rightarrow$ $^{76}_{34}$Se $+$ $^{0}_{-1}$e)
79. (a) 28.3 MeV (b) 4.001 501 u
81. (a) 3.4×10^{22} nuclei (b) 1.8×10^{16} nuclei (c) 9.7×10^{19} decays/min;
5.2×10^{15} decays/min; no
83. (a) $^{0}_{-1}$e (b) $^{222}_{86}$Rn (c) $^{237}_{94}$Pu (d) γ (e) $^{0}_{+1}$e
85. (a) 1.80×10^3 MeV (b) 237.997 821 u
87. (c) and (d)
89. $^3_2$He
91. (b) 2.3×10^{17} kg/m^3
93. (a) Atomic density is about 10^{14} times less than the nuclear density. (b) The masses are the same but are averaged over a much larger volume. (c) No, the average atomic density is much higher than hydrogen gas (most gases are typically $\approx$ 2 kg/m^3) because the gas density includes mostly empty space between gas molecules.

Chapter 30

1. (d)
3. As the target mass M_A becomes very large, $K_{min} \rightarrow |Q|$. If $Q = 0$, the incident particle would not need kinetic energy to start the reaction.

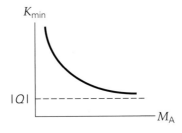

5. (a) $^{14}_7$N (b) $^2_1$H (c) $^{30}_{15}$P (d) $^{17}_8$O (e) $^{10}_5$B
7. (a) $^{14}_7$N* (b) $^{14}_7$N* (c) $^{31}_{15}$P* (d) $^{18}_9$F* (e) $^{14}_7$N*
9. (a) $^{22}_{11}$Na; no (b) (b) $^{222}_{86}$Rn; yes (c) $^{12}_6$C; no

11. +4.28 MeV
13. 4.36 MeV
15. 1.53 MeV
17. exoergic
19. 5.38 MeV
21. the first reaction, by 3.05 times
23. (a) 0.24 b (b) 0.66 b (c) 1.6 b (d) 1.7 b
25. (d)
27. (d)
29. The reaction products collide with the core materials and coolant and in the process produce heat, which converts water to steam that powers a turbine.
31. (a) 2.22 MeV (b) 12.9 MeV
33. 36 collisions
35. (c)
37. 4.68×10^{-13} m
39. (a) 0.86 MeV (b) 1.41×10^{-21} kg·m/s opposite that of the neutrino (c) both zero
41. 13.37 MeV
45. β^-: $M_P > M_D$; β^+: $M_P > M_D + 2m_e$
47. (b)
49. 3.5×10^{-28} kg
51. 135 MeV
53. 1.98×10^{-16} m
55. (a)
57. All hadrons contain quarks and/or antiquarks. Quarks are not believed to exist freely outside the nucleus.
59. (a) π^+ (b) K° (c) Σ° (d) Ξ^-
61. udd
63. (a)
65. (a) $^7_4$Be* (b) $^{60}_{29}$Cu* (c) $^{236}_{92}$U* (d) $^{13}_6$C* (e) $^{17}_8$O*
67. 0.156 MeV
69. (a) 0.86 MeV (b) no

Photo Credits

All Demonstration photos: Michael Freeman

Chapter 1—CO.1 Photo Researchers, Inc. **Fig. 1.2b** U.S. Department of Commerce **Fig. 1.3c** U.S. Department of Commerce **Fig. 1.5** Pearson Education/PH College **Fig. 1.7a** Jerry Wilson/Lander College **Fig. 1.7b** Fundamental Photographs **Fig. 1.8** Visuals Unlimited **Fig. 1.9** John Smith **Fig. 1.16** Liaison Agency, Inc. **Fig. 1.18** Pearson Education/PH College **Fig. 1.19** Pearson Education/PH College

Chapter 2—CO.2 Corbis Corporation Media **Fig. 2.2** John Smith **Fig. 2.3** Liaison Agency, Inc. **Fig. 2.7** Focus on Sports Inc. **Fig. 2.15b** James Sugar/Black Star **IN, p. 52 Fig. 1** North Wind Picture Archives **IN, p. 52 Fig. 2** J. M. Charles Rapho/Photo Researchers, Inc. **Fig. 2.16a-b** Pearson Education/PH College

Chapter 3—CO.3 FPG International LLC **Fig. 3.12** Tony Stone Images **Fig. 3.13c** Photri-Microstock **Fig. 3.16a** Richard Megna/Educational Development Center/Fundamental Photographs **Fig. 3.18a** Photo Researchers, Inc. **Fig. 3.18b** Rainbow **Fig. 3.22** Focus on Sports Inc.

Chapter 4—CO.4 Rainbow **Fig. 4.1** The Granger Collection **Fig. 4.4** John Smith **IN, p. 112 Fig. 1** AP/Wide World Photos **Fig. 4.17a-b** Arnold & Brown **Fig. 4.20a** Photo Researchers, Inc. **Fig 4.20b** Gabe Palmer/Mug Shots **Fig. 4.25** PhotoEdit **Fig. 4.27** Photo Researchers, Inc. **Fig. 4.29** Arnold & Brown **Fig. 4.31** Allsport Photography USA Inc. **Fig. 4.32** Arnold & Brown **Fig. 4.39** The Stock Market **Fig. 4.40L-R** The Goodyear Tire and Rubber Company **Fig. 4.41** Focus on Sports Inc.

Chapter 5—CO.5 Asarco, Inc. **Fig. 5.9a** Guntram Gerst/Peter Arnold, Inc. **Fig. 5.9b** Michael Collier **Fig. 5.11a** Ken Straiton/Photo Researchers, Inc. **Fig. 5.11b** Vince Streano/The Image Works **Fig. 5.19** The Image Works **Fig. 5.21** SuperStock, Inc. **IN, p. 163 Fig. 1** Michael Probst/AP/Wide World Photos **Fig. 5.22** Daemmrich/Stock Boston **Fig. 5.25** Bont Skates Pty Ltd **Fig. 5.27** David Lissy/Focus on Sports, Inc.

Chapter 6—CO.6 John Cordes/Focus on Sports, Inc. **Fig. 6.1a** Gary S. Settles/Photo Researchers, Inc. **Fig. 6.1b** David G. Curran/Rainbow **Fig. 6.1c** Stephen J. Krasemann/Photo Researchers, Inc. **Fig. 6.5a** Omikron/Science Source/Photo Researchers, Inc. **IN, p. 180 Fig. 1** Llewellyn/Uniphoto Picture Agency **Fig. 6.8a** M. Hans/Vandystadt/Photo Researchers, Inc. **Fig. 6.8b** Globus Brothers/The Stock Market **Fig. 6.11a** Ann Purcell/Photo Researchers, Inc. **Fig. 6.11b** H.P. Merten/The Stock Market **Fig. 6.15** Richard Megna/Fundamental Photographs **Fig. 6.16a** Richard Megna/Fundamental Photographs **Fig. 6.16b** R.D. Rubic/Precision Chromes, Inc. **Fig. 6.20** Paul Silverman/Fundamental Photographs **Fig. 6.22** Focus on Sports, Inc. **Fig. 6.23b-c** NASA Headquarters **Fig. 6.27** Jonathan Watts/Science Photo Library/Photo Researchers, Inc. **Fig. 6.29** Runk/Schoenberger/Grant Heilman Photography, Inc. **Fig. 6.33** Addison Geary/Stock Boston

Chapter 7—CO.7 Tony Savino/The Image Works **Fig. 7.6b** Tom Tracy/The Stock Market **Fig. 7.7** SuperStock, Inc. **Fig. 7.11** Chris Priest/Science Photo Library/Photo Researchers, Inc. **Fig. 7.13c** Michael Livenston/The Stock Market **Fig. 7.14** Focus on Sports, Inc. **Fig. 7.24a** NASA/Science Source/Photo Researchers, Inc. **Fig. 7.24c** Johan Elbers **Fig. 7.24d** C. Seghers/Photo Researchers, Inc. **IN, p. 242 Fig. 1** Photo Researchers, Inc. **Fig. 7.33** NASA Headquarters **Fig. 7.34** Frank Labua/Pearson Education/PH College **Fig. 7.36** Alexander Lowry/Photo Reseachers, Inc.

Chapter 8—CO.8 Jerry Wachter/Focus on Sports, Inc. **Fig. 8.11a** Richard Hutchings/Photo Researchers, Inc. **Fig. 8.11b** Ed. Degginger/Color-Pic, Inc. **Fig. 8.11c** Jean-Marc Loubat/Agence Vandystadt/Photo Researchers, Inc. **IN, p. 264 Fig. 1** (*FPG* International LLC) **Fig. 8.12b** Beaura Katherine Ringrose **IN, p. 266 Fig. 1** Albert Einstein (TM) represented by The Roger Richman Agency, Inc. Beverly Hills, CA 90012 www.hollywoodlegends.com. Photo courtesy of the Archives California Institute of Technology **Fig. 8.25a** Frank Labua/Pearson Education/PH College **Fig. 8.25b** Frank Labua/Pearson Education/PH College **Fig. 8.25c** NASA Headquarters **Fig. 8.28a** Vince Streano/The Image Works **Fig. 8.28b** NASA Headquarters **Fig. 8.29** Jerry Wilson **Fig. 8.43b** Tony Savino/The Image Works **Fig. 8.46** Gerard Lacz/Natural History Photographic Agency

Chapter 9—CO.9 Focus on Sports, Inc. **IN, p. 300 Fig. 1** Jonathan Watts/Science Photo Library/Photo Researchers, Inc. **Fig. 9.11** James Holmes/Reed Nurse/Science Photo Library/Photo Researchers, Inc. **IN, p. 309 Fig. 2** Blair Seitz/Photo Researchers, Inc. **Fig. 9.12** Jan Halaska/Photo Researchers, Inc. **Fig. 9.14** Compliments of Clearly Canadian Beverage Corporation **Fig. 9.15** Bill Curtsinger/Photo Researchers, Inc. **Fig. 9.17b** Diane Schiumo/ Fundamental Photographs **Fig. 9.23b** Hermann Eisenbeiss/Photo Researchers, Inc. **Fig. 9.24a** David Spears/Science Photo Library/Photo Researchers, Inc. **Fig. 9.24b** Richard Steedman/The Stock Market **Fig. 9.26** Patrick Watson/Medichrome/The Photo Shop **Fig. 9.28** Underwood & Underwood/Corbis **Fig. 9.29** Frank Labua/Pearson Education/PH College **Fig. 9.30a-b** Courtesy of CENCO **Fig. 9.34** Michael J. Howell/Stock Boston **Fig. 9.35a-b** Stephen T. Thornton **Fig. 9.35c** John Smith **Fig. 9.36** Stephen T. Thornton **Fig. 9.39** Tom Cogill

Chapter 10—CO.10 Jim Corwin/Photo Researchers, Inc. **Fig. 10.2c** Frank Labua/Pearson Education/PH College **Fig. 10.3a** Richard Megna/Fundamental Photographs **Fig. 10.3b** Leonard Lessin/Peter Arnold, Inc. **Fig. 10.5b** John Smith **IN, p. 337 Fig. 1** Maximilian Stock/Science Photo Library/Photo Researchers, Inc. **Fig. 10.6** Sinclair Stammers/Science Photo Library/Photo Researchers, Inc. **Fig. 10.11a** Richard Choy/Peter Arnold, Inc. **Fig. 10.11b** Joe Sohm/The Image Works **Fig. 10.15** Paul Silverman/Fundamental Photographs **Fig. 10.20** Stephen T. Thornton

Chapter 11—CO.11 Tom Cogill **Fig. 11.2** Jerry Wilson **Fig. 11.4** John Smith **Fig. 11.10** Frank Labau/Pearson Education/PH College **Fig. 11.11c** Richard Lowenberg/Science Source/Photo Researchers, Inc. **Fig. 11.15** Dr. Ray Clark and Mervyn Goff/Science Photo Library/Photo Researchers, Inc. **Fig. 11.16** John Smith **Fig. 11.17** Carl Glassman/The Image Works **Fig. 11.19** John Smith **Fig. 11.20** Leonard Lessin/Peter Arnold, Inc. **Fig. 11.23** Joe Sohm/Photo Researchers, Inc.

Chapter 12—CO.12 Helen Marcus/Photo Researchers, Inc. **IN, p. 401 Fig. 2** Garry Ladd/Photo Researchers Inc. **Fig. 12.8** Lowell Georgia/Photo Researchers, Inc. **Fig. 12.10** Leonard Lessin/Peter Arnold, Inc.

Chapter 13—CO.13 J.R. Berintenstein/Photo Researchers, Inc. **Fig. 13.9** John Matchett/Picture Perfect USA, Inc. **IN, p. 436 Fig. 1** Peter Menzel/Peter Arnold, Inc. **IN, p. 436 Fig. 2** H. Yamaguchi/Sygma **IN, p. 437 Fig. 4b** Russell D. Curtis/Photo Researchers, Inc. **IN, p. 437 Fig. 4c** Russell D. Curtis/Photo Researchers, Inc. **Fig. 13.14** Education Development Center, Inc. **Fig. 13.18** Fundamental Photographs **Fig. 13.19a** Richard Megna/Fundamental Photographs **Fig. 13.20** Richard Megna/Fundamental Photographs **Fig. 13.21** SuperStock, Inc. **Fig. 13.22** Lawrence Migdale/Stock Boston **IN, p. 445 Fig. 1** University of Washington Libraries, Special Collections Farquharson 12 **Fig. 13.26** Courtesy of CENCO **Fig. 13.28** Courtesy of Central Scientific Company

Chapter 14—CO.14 Michael A. Keller/The Stock Market **Fig. 14.1b** Leonard Lessin/Peter Arnold, Inc. **Fig. 14.3a** NASA Headquarters **Fig. 14.3b** Merlin D. Tuttle/Photo Researchers, Inc. **Fig. 14.3c** Howard Sochurek/Woodfin Camp & Associates **Fig. 14.12b** Philippe Plaily/Science Photo Library/Photo Researchers, Inc. **Fig. 14.12c** Jim Kahnweiler/Positive Images **IN, p. 473 Fig. 1** University of Illinois, Department of Atmospheric Sciences **IN, p. 473 Fig. 2** David R. Frazier/Photo Researchers, Inc. **Fig. 14.14** Jeff Greenberg/Visuals Unlimited **Fig. 14.16** Michael Furman Photographer Ltd./The Stock Market **Fig. 14.18b** Richard Megna/Fundamental Photographs

Chapter 15—CO.15 Zefa/London/The Stock Market **Fig. 15.7a** Jerry Wilson **Fig. 15.7c** Charles D. Winters/Photo Researchers, Inc. **IN, p. 500, Fig. 1b** Keith Kent/Peter Arnold, Inc. **IN, p. 495, Fig. 1c** Philippe Wojazer/Reuters/Corbis-Bettmann

Chapter 16—CO.16 Richard Megna/Fundamental Photographs **Fig. 16.13** Spencer Grant/Photo Researchers, Inc. **Fig. 16.16a** Larry Mulvehill/Photo Researchers, Inc.

Chapter 17—CO.17 Phil Jude/Photo Researchers, Inc. **IN, p. 543 Fig. 1** Courtesy of International Business Machines Corporation. Unauthorized use not permitted **Fig. 17.7** Alfred Pasieka/Science Photo Library/Photo Researchers, Inc. **Fig. 17.11a** Ronald Brown/Arnold & Brown **Fig. 17.11b** Frank Labua/Pearson Education/PH College **IN, p. 555, Fig. 1 and Fig. 2** Frank Labua/Pearson Education/PH College **Fig. 17.12** NASA/Mark Marten/Photo Researchers, Inc. **Fig. 17.13** Frank Labua/Pearson Education/PH College

Chapter 18—CO.18 Richard Megna/Fundamental Photographs Fig. 18.14a Courtesy of Central Scientific Company IN, p. 584, Fig. 2 Zircon Fig. 18.17c Richard Megna/Fundamental Photographs Fig. 18.19b Frank Labua/Pearson Education/PH College Fig 18.19c Paul Silverman/ Fundamental Photographs Fig. 18.20b Frank Labua/Pearson Education/PH College Fig. 18.20c U.S. Department of Energy/Science Photo Library/Photo Researchers, Inc. Fig. 18.23a-b Frank Labua/Pearson Education/PH College Fig. 18.41 M. Antman/ The Image Works

Chapter 19—CO.19 Peter Menzel/Stock Boston Fig. 19.1 Courtesy of Central Scientific Company Fig. 19.3 Richard Megna/Fundamental Photographs Fig. 19.4b Richard Megna/Fundamental Photographs Fig. 19.8a Richard Megna/Fundamental Photographs Fig. 19.9a Richard Megna/Fundamental Photographs Fig. 19.10a Richard Megna/ Fundamental Photographs Fig. 19.13c Grant Heilman Photography Fig. 19.23b Jerry Wilson Fig. 19.26 James Holmes/Oxford Centre for Molecular Sciences/Science Photo/Photo Researchers, Inc. Fig. 19.28 Courtesy of Central Scientific Company IN, p. 622 Fig 1 and Fig. 2 Richard Blakemore, University of New Hampshire Fig. 19.32 Pekka Parviainen/Science Photo Library/Photo Researchers, Inc.

Chapter 20—CO.20 Tim Barnwell/Stock Boston Fig. 20.1d Richard Megna/Fundamental Photographs IN, p. 638 Fig. 2b Richard Megna/Fundamental Photographs Fig. 20.6a Beaura Katherine Ringrose Fig. 20.10a U.S. Department of the Interior Fig. 20.10b Jim Steinberg/Photo Researchers, Inc. Fig. 20.12d Westinghouse Electric Corp. Fig. 20.15a Malcolm Fife PhotoDisc, Inc. Fig. 20.15b U.S. Department of Energy/Mark Marten/Photo Researchers, Inc. Fig. 20.19 Munshi Ahmed Photography Fig. 20.20 AP/Wide World Photos IN, p. 648 Richard Megna/Fundamental Photographs Fig. 20.22a Larry Mulvehill/Science Source/Photo Researchers, Inc. Fig. 20.22b GCA/CNRI/Phototake NYC IN, p. 657 Fig. 1 National Oceanic and Atmospheric Administration Fig. 20.32 Beaura Katherine Ringrose

Chapter 21—CO.21 Jeffry Myers/The Stock Market Fig. 21.5 Frank Labua/Pearson Education/PH College Fig. 21.14 Frank Labua/ Pearson Education/PH College IN, p. 670, Fig. 1 Rip Griffith/Photo Researchers, Inc. Fig. 21.15 Tom Lyle/Medichrome/The Stock Shop, Inc.

Chapter 22—CO.22 Alise and Mort Pechter/The Stock Market Fig. 22.5 Peter M. Fisher/The Stock Market Fig. 22.6 Photri/The Stock Market Fig. 22.7 Richard Megna/Fundamental Photographs Fig. 22.9 Richard Megna/Fundamental Photographs Fig. 22.11a Kent Wood/Photo Researchers, Inc. Fig. 22.11c DiMaggio/Kalish/Peter Arnold, Inc. IN, p. 698 Fig 1a-e © Edward Pascuzzi Fig. 22.13 Ken Kay/Fundamental Photographs Fig. 22.15 Stuart Westmoreland/Corbis Fig. 22.16a © Gemological Institute of America Fig. 22.17a Courtesy of Central Scientific Company Fig. 22.18 Richard Megna/ Fundamental Photographs Fig. 22.19a Nick Koudis/PhotoDisc, Inc. Fig. 22.19b Hank Morgan/Photo Researchers, Inc. Fig. 22.20a David Parker/Photo Researchers, Inc. IN, p. 703 Fig. 1 P. Gontier/The Image Works IN, p. 703 Fig. 2 SIU/Photo Researchers Inc. IN, p. 704 Fig. 1 Doug Johnson/Photo Researchers, Inc. Fig. 22.21 Martin Pond/Science Photo Library/Photo Researchers, Inc. Fig. 22.23 SIU/Photo Researchers, Inc. Fig. 22.24 Frank Labua/Pearson Education/PH College

Chapter 23—CO.23 Peticolas/Megna/Photo Researchers, Inc. Fig. 23.6 Paul Silverman/Fundamental Photographs Fig. 23.13 Richard Hutchings/Photo Researchers, Inc. Fig. 23.16a John Smith Fig. 23.16b Richard Megna/Fundamental Photographs IN, p. 730 Fig. 1b John Smith IN, p.730 Fig. 1c Bohdan Hrynewych/Stock Boston Fig. 23.22 Frank Labua/Pearson Education/PH College Fig. 23.23b Michael Freeman Fig. 23.24 Tom Tracy/The Stock Market Fig. 23.25 John Smith

Chapter 24—CO.24 Adrienne Hart Davis/Science Photo Library/Photo Researchers, Inc. Fig. 24.1 Richard Megna/Fundamental Photographs Fig. 24.2b From the "Atlas of Optical Phenomena," Michel Cagnet, Maurice Francon, Jean Claude Thrierr. (c) by Springer-Verlag OHG, Berlin, 1962. Fig. 24.6c Peter Aprahamian/Photo Researchers, Inc. Fig. 24.7a David Parker/Science Photo Library/Photo Researchers Fig. 24.7b Gregory G. Dimijian, M.D./Photo Researchers, Inc. Fig. 24.8b H.R. Bramaz/Peter Arnold, Inc. IN, p. 749, Fig. 2 Kristen Brochmann/Fundamental Photographs Fig. 24.9b Ken Kay/ Fundamental Photographs Fig. 24.10a Courtesy of Jay Boleman Fig. 24.10b NASA/Science Photo Library/Photo Researchers, Inc. Fig. 24.12 Ken Kay/Fundamental Photographs Fig. 24.13b Eduquip-Malacalaster Corporation Fig. 24.15 Dan McCoy/Rainbow Fig. 24.16b A.J. Stosick Fig. 24.16c Courtesy of Dr. M.F. Perutzle/Medical Research Council Fig. 24.19b Ed Degginger Fig. 24.21b Diane Schiumo/ Fundamental Photographs Fig. 24.22b Nina Barnett Photography Fig. 24.23b Peter Aprahamian/Sharples Stress Engineers Ltd./Science Photo Library/Photo Researchers, Inc. IN, p. 765, Fig. 2 Leonard Lessin/Peter Arnold, Inc. Fig. 24.25 Jim Baron/The Image Finders Fig. 24.26 Roger Ressmeyer/© 1994 CORBIS Fig. 24.27 Photri-Microstock

Chapter 25—CO.25 Patrick Bennett/Corbis IN, p. 776 Fig. 1 James King-Holmes/Science Photo Library/Photo Researchers, Inc. Fig. 25.5b Frank Labua/Pearson Education/PH College Fig. 25.9b Rocher/Jerrican/Photo Researchers, Inc. Fig. 25.14a-b California Institute of Technology/Palomar Observatory Fig. 25.15 NASA Headquarters IN, p. 792 Fig. 1 Tony Craddock/Science Photo Library/Photo Researchers, Inc. Fig. 25.18 The Image Finders Fig. 25.19 Bill Bachmann/Photo Researchers, Inc. Fig. 25.21a Fritz Goro/Life Magazine, Time Warner, Inc.

Chapter 26—CO.26 Space Design International Fig. 26.3b PASCO Scientific Fig. 26.5 Science Photo Library/Photo Researchers, Inc. Fig. 26.17b Space Telescope Science Institute IN, p. 831, Fig. 1 National Radio Astronomy Laboratory Fig. 2 David Hardy/Science Photo Library/Photo Researchers, Inc.

Chapter 27—CO.27 Stephane Husain/Liaison Agency, Inc. Fig. 27.7a Larry Albright Fig. 27.7b Ann Purcell/Photo Researchers, Inc. Fig. 27.8 Wabash Instrument Corp./Fundamental Photographs Fig. 27.13 Gary Retherford/Photo Researchers, Inc. Fig. 27.14 Dan McCoy/Rainbow Fig. 27.18a Will & Deni McIntyre/Photo Researchers, Inc. Fig. 27.18b Michal Rizza/Stock Boston Fig. 27.20a Hank Morgan/Rainbow Fig. 27.20b Chuck O'Rear/Westlight/Corbis

Chapter 28—CO.28 Courtesy of International Business Machines Corporation. Unauthorized use not permitted. Fig. 28.3a Hank Morgan/Rainbow Fig. 28.3b Will & Deni McIntyre/Photo Researchers, Inc. IN, p. 873, Fig. 2 Lawrence Migdale/Stock Boston; Fig. 3a Manfred Kage/Peter Arnold, Inc.; Fig. 3b Dr. R. Kessel/Peter Arnold, Inc.; Fig. 3c David Scharf/Peter Arnold, Inc. Fig. 28.5 Courtesy of International Business Machines Corporation. Unauthorized use not permitted. Almaden Research Center IN, p. 876, Fig. 1a Omikron/Science Source/Photo Researchers, Inc.; Fig. 1b Mehau Kulyk/Science Photo Library/Photo Researchers, Inc.; Fig. 3b Will & Deni McIntyre/Photo Researchers, Inc. Fig. 28.13 Lawrence Berkeley/Science Photo Library/Photo Researchers, Inc.

Chapter 29—CO.29 Roger Tully/Tony Stone Images Fig. 29.5 Mary Evans Picture Library/Photo Researchers, Inc. Fig. 29.13 Tom McHugh/Photo Researchers, Inc. Fig. 29.17 AP/Wide World Photos Fig. 29.20 Lawrence Berkeley Lab/Photo Researchers, Inc. Fig. 29.21 Larry Mulvehill/Photo Researchers, Inc. Fig. 29.22 Larry Mulvehill/Photo Researchers, Inc. Fig. 29.23a Dan McCoy/Rainbow Fig. 29.23b Monte S. Buchsbaum, M.D. Mount Sinai School of Medicine, New York, NY Fig. 29.24a S. Wanke/Photo Disk, Inc.

Chapter 30—CO.30 Alexander Tsiaras/Stock Boston Fig. 30.1b Fermilab Visual Media Services Fig. 30.5a Tom Tracy/The Stock Market Fig. 30.7a Igor Kostin/Imago/Sygma Fig. 30.7b U.S. Dept. of Energy/Science Photo Library/Photo Researchers, Inc. Fig. 30.8b Plasma Physics Laboratory, Princeton University Fig. 30.9c Gary Stone/Lawrence Livermore National Laboratory Jerry Wilson author photo, Jeremey Kwasney

Index

Interactive Journey through Physics Correlation Key

Text Exercise #	IJTP Unit	Topic	Subtopic	Simulation #
2.20	Mechanics	1D-2D Motion	Horizontal Motion	Simulation 2
2.21	Mechanics	1D-2D Motion	Horizontal Motion	Simulation 2
2.30	Mechanics	1D-2D Motion	Horizontal Motion	Simulation 1
2.44	Mechanics	1D-2D Motion	Horizontal Motion	Simulation 1
2.52	Mechanics	1D-2D Motion	Horizontal Motion	Simulation 3
2.65	Mechanics	1D-2D Motion	Free Fall	Simulation 2
2.70	Mechanics	1D-2D Motion	Free Fall	Simulation 1
2.71	Mechanics	1D-2D Motion	Free Fall	Simulation 3
2.82	Mechanics	1D-2D Motion	Free Fall	Simulation 1
3.52	Mechanics	1D-2D Motion	Frames of Reference	Simulation
3.59	Mechanics	1D-2D Motion	Boats	Simulation
3.78	Mechanics	1D-2D Motion	Projectile Motion	Simulation 1
3.79	Mechanics	1D-2D Motion	Projectile Motion	Simulation 1
3.92	Mechanics	1D-2D Motion	Boats	Simulation
3.96	Mechanics	1D-2D Motion	Projectile Motion	Simulation 2
3.101	Mechanics	1D-2D Motion	Projectile Motion	Simulation 3
3.105	Mechanics	1D-2D Motion	Airplanes	Simulation 2
4.12	Mechanics	Forces	Vector Components	Simulation 1
4.13	Mechanics	Forces	Vector Components	Simulation 1
4.14	Mechanics	Forces	Vector Components	Simulation 1
4.46	Mechanics	Forces	Elevator	Simulation 1
4.47	Mechanics	Forces	Elevator	Simulation 1
4.52	Mechanics	Forces	Inclined Planes	Simulation 1
4.58	Mechanics	Forces	Vector Components	Simulation 3
4.78	Mechanics	Forces	Friction	Simulation 1
4.79	Mechanics	Forces	Friction	Simulation 1
4.80	Mechanics	Forces	Friction	Simulation 2
4.82	Mechanics	Forces	Friction	Simulation 2
4.83	Mechanics	Forces	Friction	Simulation 2
4.85	Mechanics	Forces	Friction	Simulation 2
4.87	Mechanics	Forces	Vector Components	Simulation 3
4.88	Mechanics	Forces	Vector Components	Simulation 2
4.98	Mechanics	Forces	Inclined Planes	Simulation 3
5.6	Mechanics	Energy	Work	Simulation 1
5.9	Mechanics	Energy	Work	Simulation 1
5.19	Mechanics	Energy	Work	Simulation 2
5.34	Mechanics	Energy	Kinetic Energy	Simulation
5.38	Mechanics	Energy	Friction	Simulation
5.40	Mechanics	Energy	Friction	Simulation
5.41	Mechanics	Energy	Friction	Simulation
5.52	Mechanics	Energy	Potential Energy	Simulation 1
5.55	Mechanics	Energy	Work and Energy	Simulation 1
5.60	Mechanics	Energy	Potential Energy	Simulation 1

5.67	Mechanics	Energy	Work and Energy	Simulation 2
5.70	Mechanics	Energy	Potential Energy	Simulation 2
5.71	Mechanics	Energy	Work and Energy	Simulation 1
5.88	Mechanics	Energy	Work and Energy	Simulation 1
5.95	Mechanics	Energy	Kinetic Energy	Simulation
5.97	Mechanics	Energy	Work and Energy	Simulation 1
6.41	Mechanics	Forces	Collisions	Simulation 1
6.42	Mechanics	Forces	Collisions	Simulation 1
6.49	Mechanics	Forces	Collisions	Simulation 2
6.55	Mechanics	Forces	Collisions	Simulation 2
6.58	Mechanics	Forces	Collisions	Simulation 2
6.62	Mechanics	Forces	Collisions	Simulation 2
6.91	Mechanics	Forces	Collisions	Simulation 2
6.99	Mechanics	Forces	Collisions	Simulation 1
7.21	Mechanics	Energy	Circular Motion I	Simulation 1
7.23	Mechanics	Energy	Circular Motion I	Simulation 1
7.30	Mechanics	Energy	Circular Motion I	Simulation 1
7.31	Mechanics	Energy	Circular Motion I	Simulation 1
7.45	Mechanics	Energy	Circular Motion I	Simulation 2
7.48	Mechanics	Energy	Circular Motion I	Simulation 2
7.60	Mechanics	Energy	Circular Motion II	Simulation 1
7.61	Mechanics	Energy	Circular Motion II	Simulation 1
7.63	Mechanics	Energy	Circular Motion II	Simulation 3
7.65	Mechanics	Energy	Circular Motion II	Simulation 3
7.66	Mechanics	Energy	Circular Motion II	Simulation 3
7.87	Mechanics	Energy	Planets	Simulation 1
7.90	Mechanics	Energy	Planets	Simulation 2
8.46	Mechanics	Energy	Circular Motion II	Simulation 2
8.47	Mechanics	Energy	Circular Motion II	Simulation 2
8.74	Mechanics	Rotation	Rotation & Translation	Simulation
8.76	Mechanics	Rotation	Rotation & Translation	Simulation
8.77	Mechanics	Rotation	Rotation & Translation	Simulation
8.91	Mechanics	Rotation	Angular momentum	Simulation
8.92	Mechanics	Rotation	Angular momentum	Simulation
8.93	Mechanics	Rotation	Angular momentum	Simulation
8.98	Mechanics	Rotation	Rotation & Transfer	Simulation
10.1	Thermodynamics	Temp Changes	Temperature Scales	Simulation
10.2	Thermodynamics	Temp Changes	Temperature Scales	Simulation
10.15	Thermodynamics	Temp Changes	Temperature Scales	Simulation
10.16	Thermodynamics	Gas Laws	PV plots	Simulation 2
10.22	Thermodynamics	Temp Changes	Temperature Scales	Simulation
10.28	Thermodynamics	Gas Laws	PT plots	Simulation 1
10.30	Thermodynamics	Gas Laws	PV = nRT	Simulation
10.31	Thermodynamics	Gas Laws	PV = nRT	Simulation
10.37	Thermodynamics	Gas Laws	TV plots	Simulation 1
10.46	Thermodynamics	Temp Changes	Temperature Scales	Simulation
10.62	Thermodynamics	Temp Changes	Kinetic Theory	Simulation
10.65	Thermodynamics	Temp Changes	Kinetic Theory	Simulation
10.66	Thermodynamics	Temp Changes	Kinetic Theory	Simulation
10.67	Thermodynamics	Temp Changes	Kinetic Theory	Simulation
10.68	Thermodynamics	Temp Changes	Kinetic Theory	Simulation
10.69	Thermodynamics	Temp Changes	Kinetic Theory	Simulation
10.95	Thermodynamics	Temp Changes	Kinetic Theory	Simulation
11.11	Thermodynamics	Temp Changes	Calorimetry	Simulation
11.13	Thermodynamics	Temp Changes	Calorimetry	Simulation
11.15	Thermodynamics	Temp Changes	Calorimetry	Simulation

11.16	Thermodynamics	Temp Changes	Calorimetry	Simulation
11.20	Thermodynamics	Temp Changes	Calorimetry	Simulation
11.83	Thermodynamics	Temp Changes	Calorimetry	Simulation
12.4	Thermodynamics	Heat and Work	First Law	Simulation
12.5	Thermodynamics	Heat and Work	Special Processes	Simulation 1, 2, 3, 4
12.6	Thermodynamics	Heat and Work	Special Processes	Simulation 1, 2, 3, 4
12.8	Thermodynamics	Heat and Work	First Law	Simulation
12.9	Thermodynamics	Heat and Work	First Law	Simulation
12.12	Thermodynamics	Heat and Work	Special Processes	Simulation 3
12.14	Thermodynamics	Heat and Work	First Law	Simulation
12.15	Thermodynamics	Heat and Work	Special Processes	Simulation 4
12.17	Thermodynamics	Heat and Work	Special Processes	Simulation 2
12.22	Thermodynamics	Heat and Work	Cycles	Simulation 1
12.23	Thermodynamics	Heat and Work	Cycles	Simulation 1
12.25	Thermodynamics	Heat and Work	Cycles	Simulation 1
12.66	Thermodynamics	Heat and Work	Carnot Cycle	Simulation
12.92	Thermodynamics	Heat and Work	Special Processes	Simulation 2
12.95	Thermodynamics	Heat and Work	Special Processes	Simulation 2
13.2	Mechanics	Vibrations	SHM	Simulation 2
13.3	Mechanics	Vibrations	SHM	Simulation 2
13.6	Mechanics	Vibrations	SHM	Simulation 2
13.12	Mechanics	Vibrations	SHM	Simulation 2
13.18	Mechanics	Vibrations	SHM	Simulation 2
13.19	Mechanics	Vibrations	SHM	Simulation 2
13.22	Mechanics	Vibrations	SHM	Simulation 2
13.27	Mechanics	Vibrations	Pendulums	Simulation
13.30	Mechanics	Vibrations	Pendulums	Simulation
13.33	Mechanics	Vibrations	Pendulums	Simulation
13.35	Mechanics	Vibrations	SHM	Simulation 1
13.36	Mechanics	Vibrations	SHM	Simulation 1
13.37	Mechanics	Vibrations	SHM	Simulation 1
13.38	Mechanics	Vibrations	SHM	Simulation 1
13.41	Mechanics	Vibrations	Pendulums	Simulation
13.42	Mechanics	Vibrations	SHM	Simulation 1
13.45	Mechanics	Vibrations	Pendulums	Simulation
13.79	Mechanics	Vibrations	Standing Waves	Simulation 2
13.80	Mechanics	Vibrations	Standing Waves	Simulation 2
13.89	Mechanics	Vibrations	Standing Waves	Simulation 2
13.91	Mechanics	Vibrations	SHM	Simulation 1
13.93	Mechanics	Vibrations	Standing Waves	Simulation 2
13.97	Mechanics	Vibrations	SHM	Simulation 1
14.58	Mechanics	Vibrations	Doppler Effect	Simulation
14.59	Mechanics	Vibrations	Doppler Effect	Simulation
14.60	Mechanics	Vibrations	Doppler Effect	Simulation
14.62	Mechanics	Vibrations	Doppler Effect	Simulation
14.67	Mechanics	Vibrations	Doppler Effect	Simulation
15.20	E&M	Fixed Charges	Coulomb's Law	Simulation
15.21	E&M	Fixed Charges	Coulomb's Law	Simulation
15.24	E&M	Fixed Charges	Coulomb's Law	Simulation
15.25	E&M	Fixed Charges	Coulomb's Law	Simulation
15.26	E&M	Fixed Charges	Coulomb's Law	Simulation
15.37	E&M	Fixed Charges	Electric Field	Simulation 3
15.46	E&M	Fixed Charges	Electric Field	Simulation 1
15.47	E&M	Fixed Charges	Electric Field	Simulation 3
15.48	E&M	Fixed Charges	Electric Field	Simulation 1
15.50	E&M	Fixed Charges	Electric Field	Simulation 1
15.52	E&M	Fixed Charges	Electric Field	Simulation 1

15.53	E&M	Fixed Charges	Electric Field	Simulation 2
15.54	E&M	Moving Charges	Constant E Field	Simulation
15.57	E&M	Fixed Charges	Electric Field	Simulation 2
16.13	E&M	Moving Charges	Through a Potential	Simulation
16.16	E&M	Fixed Charges	Electric Potential	Simulation 1
16.84	E&M	Circuits	Capacitors	Simulation 1, 2
16.85	E&M	Circuits	Capacitors	Simulation 2
17.22	E&M	Circuits	Ohm's Law	Simulation 2
17.30	E&M	Circuits	Ohm's Law	Simulation 2
17.46	E&M	Circuits	Ohm's Law	Simulation 2
18.1	E&M	Circuits	Resistors	Simulation 1
18.2	E&M	Circuits	Resistors	Simulation 2
18.3	E&M	Circuits	Resistors	Simulation 1
18.4	E&M	Circuits	Resistors	Simulation 2
18.5	E&M	Circuits	Resistors	Simulation 1
18.6	E&M	Circuits	Resistors	Simulation 2
18.7	E&M	Circuits	Resistors	Simulation 1, 2
18.52	E&M	Circuits	RC Circuits	Simulation 2
18.53	E&M	Circuits	RC Circuits	Simulation 1
18.55	E&M	Circuits	RC Circuits	Simulation 1
18.56	E&M	Circuits	RC Circuits	Simulation 1
18.58	E&M	Circuits	RC Circuits	Simulation 2
18.59	E&M	Circuits	RC Circuits	Simulation 2
18.60	E&M	Circuits	RC Circuits	Simulation 1
18.61	E&M	Circuits	RC Circuits	Simulation 1
18.62	E&M	Circuits	RC Circuits	Simulation 1
19.7	E&M	Moving Charges	Constant B Field	Simulation 1
19.8	E&M	Moving Charges	Constant B Field	Simulation 1
19.10	E&M	Moving Charges	Constant B Field	Simulation 1
19.11	E&M	Moving Charges	Constant B Field	Simulation 1
19.71	E&M	Moving Charges	Constant B Field	Simulation 2
19.78	E&M	Moving Charges	Constant B Field	Simulation 2
19.79	E&M	Moving Charges	Constant B Field	Simulation 2
19.92	E&M	Moving Charges	E and B Fields	Simulation
20.4	E&M	Moving Charges	Induced EMF	Simulation 1
20.10	E&M	Moving Charges	Induced EMF	Simulation 1
20.15	E&M	Moving Charges	Induced EMF	Simulation 1
22.18	L&O	Refraction	Index of Refraction	Simulation
22.20	L&O	Refraction	Index of Refraction	Simulation
22.28	L&O	Refraction	Snell's Law	Simulation 1
22.29	L&O	Refraction	Snell's Law	Simulation 1
22.30	L&O	Refraction	Snell's Law	Simulation 1
22.31	L&O	Refraction	Critical Angle	Simulation
22.32	L&O	Refraction	Critical Angle	Simulation
22.34	L&O	Refraction	Snell's Law	Simulation 1
22.36	L&O	Refraction	Index of Refraction	Simulation
22.44	L&O	Refraction	Snell's Law	Simulation 2
22.45	L&O	Refraction	Snell's Law	Simulation 2
22.60	L&O	Refraction	Prisms	Simulation 1
22.63	L&O	Refraction	Prisms	Simulation 1
22.64	L&O	Refraction	Prisms	Simulation 1
22.65	L&O	Refraction	Prisms	Simulation 1
23.1	L&O	Reflection	Law of Reflection	Simulation
23.2	L&O	Reflection	Law of Reflection	Simulation
23.8	L&O	Reflection	Law of Reflection	Simulation

23.17	L&O	Reflection	Mystery Mirr	Simulation
23.18	L&O	Reflection	Convex Mirro	Simulation 2
23.24	L&O	Reflection	Concave Mirr	Simulation 2
23.25	L&O	Reflection	Concave Mirr	Simulation 2
23.26	L&O	Reflection	Convex Mirro	Simulation 2
23.28	L&O	Reflection	Concave Mirro	Simulation 2
23.29	L&O	Reflection	Concave Mirro	Simulation 1
23.30	L&O	Reflection	Concave Mirro	Simulation 2
23.31	L&O	Reflection	Convex Mirror	Simulation 2
23.35	L&O	Reflection	Convex Mirror	Simulation 2
23.38	L&O	Reflection	Mystery Mirror	Simulation
23.39	L&O	Reflection	Mystery Mirror	Simulation
23.40	L&O	Reflection	Mirror Eq/Magication	Simulation 1, 2
23.43	L&O	Reflection	Concave Mirror	Simulation 2
23.44	L&O	Reflection	Convex Mirror	Simulation 2
23.45	L&O	Reflection	Concave Mirror	Simulation 2
23.47	L&O	Refraction	Diverging Lens	Simulation 2
23.50	L&O	Refraction	Converging Lens	Simulation 1
23.51	L&O	Refraction	Converging Lens	Simulation 2
23.52	L&O	Refraction	Lens Equation	Simulation
23.53	L&O	Refraction	Converging Lens	Simulation 2
23.54	L&O	Refraction	Converging Lens	Simulation 2
23.55	L&O	Refraction	Diverging Lens	Simulation 2
23.60	L&O	Refraction	Lens Equation	Simulation
23.68	L&O	Refraction	Converging Lens	Simulation 1
23.80	L&O	Reflection	Mystery Mirror	Simulation
23.88	L&O	Refraction	Converging Lens	Simulation 1
24.1	L&O	Interference	Double Slit	Simulation 1
24.3	L&O	Interference	Double Slit	Simulation 2
24.14	L&O	Interference	Double Slit	Simulation 3
24.15	L&O	Interference	Double Slit	Simulation 1
24.19	L&O	Interference	Double Slit	Simulation 4
24.35	L&O	Diffraction	Grating	Simulation 1
24.34	L&O	Diffraction	Single Slit	Simulation
24.45	L&O	Diffraction	Grating	Simulation 2
24.48	L&O	Diffraction	Grating	Simulation 2
24.49	L&O	Diffraction	Grating	Simulation 2
24.51	L&O	Diffraction	Grating	Simulation 2

Recommended System Requirements for CD-ROM:

MACINTOSH, 25MHz, 68030 processor or PowerMac, System 7.x, 16 MB RAM, 256 color VGA or Super VGA Monitor, 2x CD-ROM drive

WINDOWS, 486/25MHz processor, Windows 3.x or Windows 95, 16 MB RAM, 256 color VGA monitor, 2x CD-ROM drive

Mathematical Symbols

$=$	is equal to
$\neq$	is not equal to
$\approx$	is approximately equal to
$\sim$	about
$\propto$	is proportional to
$>$	is greater than
$\geq$	is greater than or equal to
$\gg$	is much greater than
$<$	is less than
$\leq$	is less than or equal to
$\ll$	is much less than
$\pm$	plus or minus
$\mp$	minus or plus
$\bar{x}$	average value of x
Δx	change in x
$|x|$	absolute value of x
Σ	sum of
∞	infinity

The Greek Alphabet

Alpha	A	α	Nu	N	ν
Beta	B	β	Xi	Ξ	ξ
Gamma	Γ	γ	Omicron	O	o
Delta	Δ	δ	Pi	Π	π
Epsilon	E	ε	Rho	P	ρ
Zeta	Z	ζ	Sigma	Σ	σ
Eta	H	η	Tau	T	τ
Theta	Θ	θ	Upsilon	Y	υ
Iota	I	ι	Phi	Φ	ϕ
Kappa	K	κ	Chi	X	χ
Lambda	Λ	λ	Psi	Ψ	ψ
Mu	M	μ	Omega	Ω	ω

Quadratic Formula

If $ax^2 + bx + c = 0$, then

$$x = \frac{-b \pm \sqrt{b^2 - 4ac}}{2a}$$

Values of Some Useful Numbers

$\pi = 3.141\,59\ldots$ $\sqrt{2} = 1.414\,21$

$e = 2.718\,28\ldots$ $\sqrt{3} = 1.732\,05$

Trigonometric Relationships

Definitions of Trigonometric Functions

$$\sin\theta = \frac{y}{r} \qquad \cos\theta = \frac{x}{r} \qquad \tan\theta = \frac{\sin\theta}{\cos\theta} = \frac{y}{x}$$

$\theta°$ (rad)	$\sin\theta$	$\cos\theta$	$\tan\theta$
0° (0)	0	1	0
30° ($\pi/6$)	0.500	$\sqrt{3}/2 \approx 0.866$	$\sqrt{3}/3 \approx 0.577$
45° ($\pi/4$)	$\sqrt{2}/2 \approx 0.707$	$\sqrt{2}/2 \approx 0.707$	1.00
60° ($\pi/3$)	$\sqrt{3}/2 \approx 0.866$	0.500	$\sqrt{3} \approx 1.73$
90° ($\pi/2$)	1	0	∞

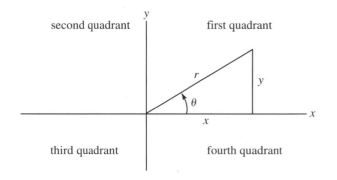